29. $\int x(ax^2 + c)^n \, dx = \frac{1}{2a} \frac{(ax^2 + c)^{n+1}}{n+1} \qquad n \neq -1$

30. $\int \frac{x}{ax^2 + c} \, dx = \frac{1}{2a} \ln |ax^2 + c|$

31. $\int \sqrt{x^2 \pm p^2} \, dx = \frac{1}{2}[x\sqrt{x^2 \pm p^2} \pm p^2 \ln |x + \sqrt{x^2 \pm p^2}|]$

32. $\int \sqrt{p^2 - x^2} \, dx = \frac{1}{2}\left(x\sqrt{p^2 - x^2} + p^2 \sin^{-1}\frac{x}{p}\right)$

33. $\int \frac{dx}{\sqrt{x^2 \pm p^2}} = \ln |x + \sqrt{x^2 \pm p^2}|$

34. $\int \sqrt{(p^2 - x^2)^3} \, dx = \frac{1}{4}\left[x\sqrt{(p^2 - x^2)^3} + \frac{3p^2 x}{2}\sqrt{p^2 - x^2} + \frac{3p^4}{2} \sin^{-1}\frac{x}{p}\right]$

Expressions Containing $ax^2 + bx + c$

35. $\int \frac{dx}{ax^2 + bx + c} = \frac{1}{\sqrt{b^2 - 4ac}} \ln \left|\frac{2ax + b - \sqrt{b^2 - 4ac}}{2ax + b + \sqrt{b^2 - 4ac}}\right| \qquad b^2 > 4ac$

36. $\int \frac{dx}{ax^2 + bx + c} = \frac{2}{\sqrt{4ac - b^2}} \tan^{-1} \frac{2ax + b}{\sqrt{4ac - b^2}} \qquad b^2 < 4ac$

37. $\int \frac{dx}{ax^2 + bx + c} = -\frac{2}{2ax + b} \qquad b^2 = 4ac$

38. $\int \frac{dx}{(ax^2 + bx + c)^{n+1}} = \frac{2ax + b}{n(4ac - b^2)(ax^2 + bx + c)^n} + \frac{2(2n - 1)a}{n(4ac - b^2)} \int \frac{dx}{(ax^2 + bx + c)^n}$

39. $\int \frac{x \, dx}{ax^2 + bx + c} = \frac{1}{2a} \ln |ax^2 + bx + c| - \frac{b}{2a} \int \frac{dx}{ax^2 + bx + c}$

40. $\int \frac{dx}{\sqrt{ax^2 + bx + c}} = \frac{1}{\sqrt{a}} \ln |2ax + b + 2\sqrt{a}\sqrt{ax^2 + bx + c}| \qquad a > 0$

41. $\int \frac{dx}{\sqrt{ax^2 + bx + c}} = \frac{1}{\sqrt{-a}} \sin^{-1} \frac{-2ax - b}{\sqrt{b^2 - 4ac}} \qquad a < 0$

42. $\int \frac{x \, dx}{\sqrt{ax^2 + bx + c}} = \frac{\sqrt{ax^2 + bx + c}}{a} - \frac{b}{2a} \int \frac{dx}{\sqrt{ax^2 + bx + c}}$

43. $\int \sqrt{ax^2 + bx + c} \, dx = \frac{2ax + b}{4a} \sqrt{ax^2 + bx + c} + \frac{4ac - b^2}{8a} \int \frac{dx}{\sqrt{ax^2 + bx + c}}$

Expressions Containing $\sin ax$

44. $\int \sin^2 ax \, dx = \frac{x}{2} - \frac{\sin 2ax}{4a}$

45. $\int \sin^3 ax \, dx = -\frac{1}{a} \cos ax + \frac{1}{3a} \cos^3 ax$

46. $\int \sin^n ax \, dx = -\frac{\sin^{n-1} ax \cos ax}{na} + \frac{n-1}{n} \int \sin^{n-2} ax \, dx \quad n \text{ positive integer}$

47. $\int \frac{dx}{1 \pm \sin ax} = \mp \frac{1}{a} \tan \left(\frac{\pi}{4} \mp \frac{ax}{2}\right)$

Expressions Containing cos *ax*

48. $\int \cos^2 ax \, dx = \dfrac{x}{2} + \dfrac{\sin 2ax}{4a}$

49. $\int \cos^3 ax \, dx = \dfrac{1}{a} \sin ax - \dfrac{1}{3a} \sin^3 ax$

50. $\int \cos^n ax \, dx = \dfrac{\cos^{n-1} ax \sin ax}{na} + \dfrac{n-1}{n} \int \cos^{n-2} ax \, dx$

Expressions Containing Algebraic and Trigonometric Functions

51. $\int x \sin ax \, dx = \dfrac{1}{a^2} \sin ax - \dfrac{1}{a} x \cos ax$

52. $\int x \cos ax \, dx = \dfrac{1}{a^2} \cos ax + \dfrac{1}{a} x \sin ax$

53. $\int x^n \sin ax \, dx = -\dfrac{1}{a} x^n \cos ax + \dfrac{n}{a} \int x^{n-1} \cos ax \, dx$

54. $\int x^n \cos ax \, dx = \dfrac{1}{a} x^n \sin ax - \dfrac{n}{a} \int x^{n-1} \sin ax \, dx$ *n positive*

55. $\int \sin ax \cos bx \, dx = -\dfrac{\cos(a-b)x}{2(a-b)} - \dfrac{\cos(a+b)x}{2(a+b)}$ $a^2 \neq b^2$

Expressions Containing Exponential and Logarithmic Functions

56. $\int xe^{ax} \, dx = \dfrac{e^{ax}}{a^2}(ax - 1), \quad \int xb^{ax} \, dx = \dfrac{xb^{ax}}{a \ln b} - \dfrac{b^{ax}}{a^2 (\ln b)^2}$ $b > 0$

57. $\int x^n e^{ax} \, dx = \dfrac{1}{a} x^n e^{ax} - \dfrac{n}{a} \int x^{n-1} e^{ax} \, dx$ *n positive*

58. $\int e^{ax} \sin bx \, dx = \dfrac{e^{ax}}{a^2 + b^2}(a \sin bx - b \cos bx)$

59. $\int e^{ax} \cos bx \, dx = \dfrac{e^{ax}}{a^2 + b^2}(a \cos bx + b \sin bx)$

60. $\int x^n \ln ax \, dx = x^{n+1} \left[\dfrac{\ln ax}{n+1} - \dfrac{1}{(n+1)^2} \right]$ $n \neq -1$

Expressions Containing Inverse Trigonometric Functions

61. $\int \sin^{-1} ax \, dx = x \sin^{-1} ax + \dfrac{1}{a} \sqrt{1 - a^2 x^2}$

62. $\int \cos^{-1} ax \, dx = x \cos^{-1} ax - \dfrac{1}{a} \sqrt{1 - a^2 x^2}$

63. $\int \csc^{-1} ax \, dx = x \csc^{-1} ax + \dfrac{1}{a} \ln |ax + \sqrt{a^2 x^2 - 1}|$

64. $\int \sec^{-1} ax \, dx = x \sec^{-1} ax - \dfrac{1}{a} \ln |ax + \sqrt{a^2 x^2 - 1}|$

65. $\int \tan^{-1} ax \, dx = x \tan^{-1} ax - \dfrac{1}{2a} \ln(1 + a^2 x^2)$

66. $\int \cot^{-1} ax \, dx = x \cot^{-1} ax + \dfrac{1}{2a} \ln(1 + a^2 x^2)$

ALGEBRA

Roots of $ax^2 + bx + c = 0$: $x = \dfrac{-b \pm \sqrt{b^2 - 4ac}}{2a}$ (quadratic formula)

$d^n - c^n = (d - c)(d^{n-1} + d^{n-2}c + d^{n-3}c^2 + \cdots + c^{n-1})$

$a + ar + ar^2 + \cdots + ar^{n-1} = \dfrac{a(1 - r^n)}{1 - r}, \, r \neq 1$ (sum of finite geometric series)

$k! = 1 \cdot 2 \cdot 3 \cdots k$

$(1 + x)^n = 1 + nx + \binom{n}{2} x^2 + \cdots + \binom{n}{k} x^k + \cdots + x^n$ (binomial theorem)

$\binom{n}{k} = \dfrac{n!}{k!(n-k)!} = \dfrac{n}{1} \dfrac{(n-1)}{2} \dfrac{(n-2)}{3} \cdots \dfrac{(n-k+1)}{k}$

CALCULUS AND ANALYTIC GEOMETRY

Third Edition

Sherman K. Stein
Professor of Mathematics
University of California, Davis

McGraw-Hill Book Company

New York
St. Louis
San Francisco
Auckland
Bogotá
Hamburg
Johannesburg
London
Madrid
Mexico
Montreal
New Delhi
Panama
Paris
São Paulo
Singapore
Sydney
Tokyo
Toronto

CALCULUS AND ANALYTIC GEOMETRY

Copyright © 1982, 1977, 1973 by McGraw-Hill, Inc. All rights reserved. Printed in the United States of America. Except as permitted under the United States Copyright Act of 1976, no part of this publication may be reproduced or distributed in any form or by any means, or stored in a data base or retrieval system, without the prior written permission of the publisher.

1234567890 RMRM 898765432

ISBN 0-07-061153-X

This book was set in Times New Roman. The editors were Shelly Levine Langman, Peter R. Devine, and Stephen Wagley; the designer was Joseph Gillians; the production supervisor was Charles Hess.
New drawings were done by J & R Services, Inc.
The cover art was done by Joseph Gillians;
the cover photograph was taken by Al Green.
Rand McNally & Company was printer and binder.

Library of Congress Cataloging in Publication Data

Stein, Sherman K.
    Calculus and analytic geometry.

    Includes index.
    1. Calculus.   2. Geometry, Analytic.   I. Title.
QA303.S857   1982   515'.16   81-4058
ISBN 0-07-061153-X   AACR2

*To
Joshua,
Rebecca,
and
Susanna*

Arnold Toynbee, *Experiences*, Oxford University Press, pp. 12–13, 1969.

... at about the age of sixteen, I was offered a choice which, in retrospect, I can see that I was not mature enough, at the time, to make wisely. This choice was between starting on the calculus and, alternatively, giving up mathematics altogether and spending the time saved from it on reading Latin and Greek literature more widely. I chose to give up mathematics, and I have lived to regret this keenly after it has become too late to repair my mistake. The calculus, even a taste of it, would have given me an important and illuminating additional outlook on the Universe, whereas, by the time at which the choice was presented to me, I had already got far enough in Latin and Greek to have been able to go farther with them unaided. So the choice that I made was the wrong one, yet it was natural that I should choose as I did. I was not good at mathematics; I did not like the stuff.... Looking back, I feel sure that I ought not to have been offered the choice; the rudiments, at least, of the calculus ought to have been compulsory for me. One ought, after all, to be initiated into the life of the world in which one is going to have to live. I was going to have to live in the Western World... and the calculus, like the full-rigged sailing ship, is... one of the characteristic expressions of the modern Western genius.

CONTENTS

Preface   xi

0   An Overview of Calculus and This Book   1
- 0.1   The Derivative   1
- 0.2   The Integral   5
- 0.3   Survey   9

1   Preliminaries   11
- 1.1   Coordinate Systems and Graphs (Review)   11
- 1.2   Lines and Their Slopes   22
- 1.3   Functions   31
- 1.4   Composite Functions   39
- 1.S   Summary   44

2   Limits and Continuous Functions   48
- 2.1   The Limit of a Function   48
- 2.2   Computation of Limits   56
- 2.3   Asymptotes and Their Use in Graphing   67
- 2.4   Precise Definitions of Limits (Optional)   73
- 2.5   The Limit of $(\sin \theta)/\theta$ as θ Approaches 0   89
- 2.6   Continuous Functions   90
- 2.7   The Maximum-Value Theorem and the Intermediate-Value Theorem   97
- 2.S   Summary   103

3   The Derivative   110
- 3.1   Four Problems with One Theme   110
- 3.2   The Derivative   118
- 3.3   The Derivative and Continuity; Antiderivatives   126
- 3.4   The Derivatives of the Sum, Difference, Product and Quotient   132
- 3.5   The Derivatives of the Trigonometric Functions   143
- 3.6   Composite Functions and the Chain Rule   149
- 3.S   Summary   158

4 Applications of the Derivative — 166

4.1	Rolle's Theorem and the Mean-Value Theorem	166
4.2	Using the Derivative and Limits When Graphing a Function	177
4.3	Concavity and the Second Derivative	185
4.4	Motion and the Second Derivative	196
4.5	Applied Maximum and Minimum Problems	204
4.6	Implicit Differentiation	215
4.7	The Differential	219
4.S	Summary	226

5 The Definite Integral — 235

5.1	Estimates in Four Problems	235
5.2	Summation Notation and Approximating Sums	241
5.3	The Definite Integral	247
5.4	The Fundamental Theorems of Calculus	260
5.5	Properties of the Antiderivative and the Definite Integral	270
5.6	Proofs of the Two Fundamental Theorems of Calculus (Optional)	277
5.S	Summary	281

6 Topics in Differential Calculus — 287

6.1	Review of Logarithms, the Number e	287
6.2	The Derivatives of the Logarithmic Functions	297
6.3	The Natural Logarithm Defined as a Definite Integral (Optional)	304
6.4	Inverse Functions and the Derivative of b^x	309
6.5	The Derivatives of the Inverse Trigonometric Functions	319
6.6	Related Rates	329
6.7	Separable Differential Equations and Growth	335
6.8	L'Hôpital's Rule	349
6.9	The Hyperbolic Functions and Their Inverses	359
6.10	Exponential Functions Defined in Terms of Logarithms (Optional)	365
6.S	Summary	370

7 Computing Antiderivatives — 380

7.1	Substitution in an Antiderivative and in a Definite Integral	381
7.2	Integration by Parts	392
7.3	Using a Table of Integrals (Optional)	399
7.4	How to Integrate Certain Rational Functions	404
7.5	The Integration of Rational Functions by Partial Fractions	410
7.6	How to Integrate Powers of Trigonometric Functions	418
7.7	How to Integrate Rational Functions of $\sin \theta$ and $\cos \theta$	427
7.8	How to Integrate Rational Functions Involving $\sqrt{a^2 - x^2}$, $\sqrt{a^2 + x^2}$, $\sqrt{x^2 - a^2}$, or $\sqrt[n]{ax + b}$	431
7.9	What to Do in the Face of an Integral	438
7.S	Summary	446

8 Applications of the Definite Integral — 454

8.1	Computing Area by Parallel Cross Sections	454
8.2	Computing Volume by Parallel Cross Sections	461
8.3	How to Set Up a Definite Integral	468
8.4	Computing the Volume of a Solid of Revolution	475

		8.5	The Centroid of a Plane Region	484
		8.6	Improper Integrals	494
		8.7	Work	503
		8.8	The Average of a Function over an Interval	509
		8.9	Estimates of Definite Integrals	514
		8.S	Summary	520
9	**Plane Curves and**		9.1 Polar Coordinates	529
	Polar Coordinates	**529**	9.2 Area in Polar Coordinates	534
		9.3	Parametric Equations	540
		9.4	Arc Length and Speed on a Curve	548
		9.5	Area of a Surface of Revolution	556
		9.6	The Angle between a Line and a Tangent Line	562
		9.7	The Second Derivative and the Curvature of a Curve	569
		9.S	Summary	575
10	**Series and**		10.1 Sequences	581
	Related Topics	**580**	10.2 Series	588
		10.3	The Integral Test	594
		10.4	The Comparison Test and the Ratio Test	600
		10.5	The Alternating-Series and Absolute-Convergence Tests	608
		10.6	Power Series	618
		10.7	Manipulating Power Series	626
		10.8	Taylor's Formula	635
		10.9	Complex Numbers (Optional)	646
		10.10	The Relation between the Exponential Function and the Trigonometric Functions (Optional)	659
		10.11	Linear Differential Equations with Constant Coefficients (Optional)	664
		10.12	The Binomial Theorem for Any Exponent (Optional)	673
		10.13	Newton's Method for Solving an Equation	678
		10.S	Summary	682
11	**Algebraic Operations**		11.1 The Algebra of Vectors	692
	on Vectors	**692**	11.2 Vectors in Space	698
		11.3	The Dot Product of Two Vectors	707
		11.4	Lines and Planes	717
		11.5	Determinants of Orders Two and Three	725
		11.6	The Cross Product of Two Vectors	733
		11.S	Summary	740
12	**Partial Derivatives**	**747**	12.1 Graphs of Equations	747
		12.2	Functions and Level Curves	756
		12.3	Partial Derivatives	762
		12.4	The Change Δf and the Differential df	768
		12.5	The Chain Rules	773
		12.6	Directional Derivatives and the Gradient	782
		12.7	Critical Points	789

		12.8	Second-Order Partial Derivatives and Relative Extrema	795
		12.9	The Taylor Series for $f(x, y)$ (Optional)	800
		12.S	Summary	807

13	**Definite Integrals over Plane and Solid Regions**	**814**	13.1	The Definite Integral of a Function over a Region in the Plane	814
			13.2	How to Describe a Plane Region by Coordinates	822
			13.3	Computing $\int_R f(P)\,dA$ by Introducing Rectangular Coordinates in R	827
			13.4	Computing $\int_R f(P)\,dA$ by Introducing Polar Coordinates in R	839
			13.5	Mappings from a Plane to a Plane (Optional)	846
			13.6	Magnification, Change of Coordinates, and the Jacobian (Optional)	853
			13.7	The Definite Integral of a Function over a Region in Space	861
			13.8	Describing Solid Regions with Cylindrical or Spherical Coordinates	871
			13.9	Computing $\int_R f(P)\,dV$ with Cylindrical or Spherical Coordinates	878
			13.S	Summary	885

14	**The Derivative of a Vector Function**	**892**	14.1	The Derivative of a Vector Function	893
			14.2	Properties of the Derivative of a Vector Function	900
			14.3	Vectors Perpendicular to a Surface; the Tangent Plane	904
			14.4	Lagrange Multipliers (Optional)	909
			14.5	The Acceleration Vector	917
			14.6	The Unit Vectors **T** and **N**	922
			14.7	The Scalar Components of the Acceleration Vector Along **T** and **N**	928
			14.8	Newton's Law Implies Kepler's Three Laws (Optional)	933
			14.S	Summary	939

15	**Green's Theorem, the Divergence Theorem, and Stokes' Theorem**	**945**	15.1	Vector and Scalar Fields	945
			15.2	Line Integrals	953
			15.3	Conservative Vector Fields	964
			15.4	Green's Theorem	976
			15.5	Surface Integrals	988
			15.6	The Divergence Theorem	997
			15.7	Stokes' Theorem	1005
			15.8	Maxwell's Equations (Optional)	1014
			15.S	Summary	1020

Appendixes		**S1**	A.	Real Numbers	S1
			B.	Some Topics in Algebra	S8
			C.	Exponent	S18
			D.	Trigonometry	S26

E.	Conic Sections		S38
	E.1 Conic Sections		S38
	E.2 Translation of Axes and the Graph of $Ax^2 + Cy^2 + Dx + Ey + F = 0$		S45
	E.3 Rotation of Axes and the Graph of $Ax^2 + Bxy + Cy^2 + Dx + Ey + F = 0$		S48
	E.4 Conic Sections in Polar Coordinates		S53
F.	Theory of Limits		S57
	F.1 Proofs of Some Theorems about Limits		S57
	F.2 Definitions of Other Limits		S61
G.	The Interchange of Limits		S64
	G.1 The Equality of f_{xy} and f_{yx}		S64
	G.2 The Derivative of $\int_a^b f(x, y)\, dx$ with Respect to y		S66
	G.3 The Interchange of Limits		S68
H.	Tables		S72
	H.1 Exponential Function		S72
	H.2 Natural Logarithms (Base e)		S73
	H.3 Common Logarithms (Base 10)		S74
	H.4 Trigonometric Functions (Degrees)		S75
	H.5 Trigonometric Functions (Radians)		S76

Answers to Selected Odd-Numbered Problems and Guide Quizzes   S78

List of Symbols   S171

Index   S175

PREFACE

The goal of this edition remains the same as that of the first two editions: To provide the student and the instructor with a readable, flexible text that covers the important topics of single and multivariable calculus as simply and clearly as possible.

Organizational changes

When both users and nonusers of the second edition made the same suggestions in the survey conducted by the publisher, I accepted their advice. As a result, the antiderivative and the definite integral are treated much earlier. Limits precede the derivative. An optional section on ε, δ has been added. The treatment of the number e has been drastically revised.

Plane curves, applications of the definite integral, series, and multiple integrals are now covered in single chapters. The discussion of line integrals and Green's theorem and its generalizations has been reorganized and expanded. Vectors are treated before partial derivatives so that directional derivatives and the gradient can be treated with partial derivatives.

There have been some deletions and many additions, such as optional sections on complex numbers, the relation between the exponential function and the trigonometric functions, separable and linear differential equations, and the role of the Jacobian in change of variables. Two new overview sections, "What to do in the face of an integral" (Sec. 7.9) and "How to set up a definite integral" (Sec. 8.3), should prove helpful. Optional sections on Kepler's laws and Maxwell's equations have been added, with the former developed through a sequence of exercises. The appendixes now include a review of algebra and a treatment of change of coordinates.

The derivatives of the trigonometric, exponential, and logarithmic functions are still done quite early because of their importance in applications and the students' need for extensive practice with them. Furthermore, the arguments that obtain the derivatives of these functions reveal more clearly the idea of a limit than does the algebra that produces the derivative of a polynomial (in which Δx can be set equal to 0 with impunity). Also, L'Hôpital's rule is presented early in order to make it available throughout the course.

Pedagogical changes

Once again I have strengthened the chapter summaries, which students find very helpful; they provide an emphasis and perspective that individual sections cannot. Also I have added many more asides in the margin to guide the student. Figures have been revised and many new ones added; they are now numbered.

Applications

The number and variety of applications have been increased. This has been done primarily through exercises since I wanted to keep the main exposition uncluttered. Applications vary in length from a brief mention in an exercise to one- or two-page presentations in the text. They are listed in the index under "applications."

Exercises

I have not counted the exercises, but there are more than enough of all degrees of difficulty. Exercises before the single box (■) are routine. These generally now come in pairs, with each odd-numbered exercise comparable to the following even-numbered exercise. They focus principally on definitions and drill, and so should not constitute a full homework assignment. Exercises between the single box and double box may involve more steps or computations. Exercises after the double box may be more challenging or offer alternative perspectives or further applications. Often the most interesting (but not necessarily the most difficult) problems are to be found here. The back of the book contains answers to the odd-numbered exercises and guide quizzes. Calculator exercises are included when appropriate.

Epsilon, delta

Section 2.4, which is new, is devoted to the ε, δ definition of limits. It begins with the definition of $\lim_{x \to \infty} f(x) = \infty$, for which the concept, the diagrams, and the details are easiest. Then, after dealing with $\lim_{x \to \infty} f(x) = L$, it turns to $\lim_{x \to a} f(x) = L$. This section may be omitted (it is marked "optional") or it may be covered in one to three lectures, depending on the depth of treatment.

Level of difficulty

The level of difficulty is controlled by the choice of sections and exercises and by the pace. The exposition has been kept as simple as possible, with a strong emphasis on motivation. The text can serve students of widely varying abilities and interests, such as those in engineering, the physical sciences, mathematics, economics, and biology.

Differential equations

The text now contains solutions to two types of differential equations: separable and linear with constant coefficients. The first are included because of their use in the differential equations of natural growth and inhibited growth, the second, because they suffice for almost all elementary physical applications.

Complex numbers

Students who do not meet complex numbers in calculus could easily bypass them completely. In subsequent courses that do make use of complex numbers it is often assumed that the student has "surely" met them somewhere—in high school or in calculus. Therefore Sec. 10.9 is devoted to the complex numbers. The following section obtains the equation $e^{i\theta} = \cos \theta + i \sin \theta$, thus giving a major application of series and demonstrating the connection between the exponential and the trigonometric functions. I encourage the instructor to include these sections, though they are marked optional, even at the sacrifice of some traditional material.

Duration

There is enough material for a three-semester course. Since the number of class meetings per week ranges from three to five, it is impossible to give a uniform guide to what should be covered each quarter or semester. As a rule of thumb one section corresponds to one class meeting, though several are longer and some shorter. There are 121 non-summary sections in Chapters 0 to 15. Of these, sixteen are marked optional.

The following table describes a maximum (complete) and a minimum (core only) treatment, with remarks on certain sections. Most instructors will steer a course somewhere between the two listed. The instructor's manual has a more detailed commentary as well as answers to the even-numbered exercises, including sketches of solutions to the non-routine exercises.

	MAXIMUM		MINIMUM	
	Lectures	Comment	Lectures	Comment
Chapter				
0	2	Survey of calculus and text	0	Left to student to read
1	4		1	Precalculus material, but mention Secs. 1.3, 1.4
2	10	Two days on Sec. 2.2; three days on Sec. 2.4, perhaps a bit of Appendix F	6	Omit Sec. 2.4
3	6		6	
4	8	Two days on Sec. 4.5	7	
5	6		5	Omit Sec. 5.6
6	13	Two days on each of Secs. 6.7, 6.8, 6.9	8	Omit Secs. 6.3 and 6.10
7	9		6	Omit Secs. 7.3 and 7.9; coalesce Secs. 7.6, 7.7, 7.8
8	9		7	Omit Sec. 8.7; touch Sec. 8.8 lightly
9	7		6	Omit Sec. 9.6
10	13		8	Omit Secs. 10.9 to 10.13
11	6		5	Assume Sec. 11.5
12	9		7	Omit Secs. 12.8, 12.9
13	10	Two days on Sec. 13.3	7	Omit Secs. 13.5, 13.6
14	8		5	Omit Secs. 14.7, 14.8
15	9	Two days on Sec. 15.3	7	Omit Sec. 15.8
Total	129		91	
Appendix				
A	1		0	
B	2	Precalculus material, some to be treated early in course	0	Precalculus material, used as reference by students
C	1		0	
D	2		0	
E	4		0	
F	3	ε, δ continued	0	
G	3	A sample of advanced calculus	0	
Total	16			

Options

Chapter 0 may be left to the student to read. If the class is adequately prepared, Chap. 1 may be omitted or the last two sections emphasized. In Chap. 2 there is the choice of omitting Sec. 2.4 on ε, δ.

Chapter 6 offers a choice in the way logarithms are treated. The approach in Secs. 6.1, 6.2, and 6.4 assumes the exponential functions as given and that $\lim_{h \to 0} (1 + h)^{1/h}$ exists. It grows naturally out of the student's precalculus experience and provides an opportunity to review the manipulations of exponents and logarithms. However, instructors who wish to define the logarithm as an integral are free to follow Secs. 6.3 and 6.10 and de-emphasize Secs. 6.1 and 6.2. (If the informal approach in Secs. 6.1, 6.2, and 6.4 is followed, most of Chap. 6 can be done before Chap. 5, that is, Secs. 6.1, 6.2, 6.4 to 6.6, 6.8 and 6.9.)

The next choice is how much attention will be given the special integration techniques in Secs. 7.6 to 7.9.

In Chap. 10, after completing the standard topics in series, there are several optional sections. Sections 10.9 and 10.10, taken together, introduce complex numbers and exhibit a major application of series. Section 10.11, on linear differential equations with constant coefficients, depends at one point on Sec. 10.10.

Sections 13.5 and 13.6 present an optional unit, an intuitive treatment of the Jacobian, its significance as a measure of local magnification, and its use in the change of variables in an integral.

Only Secs. 14.1, 14.2, 14.3, and 14.6 in Chap. 14 are needed in Chap. 15.

Solutions manual

Complete solutions to all odd-numbered exercises and guide quizzes are available to the student in a manual prepared to accompany this text.

Acknowledgments

At each stage of this revision two former graduate students at Davis, Anthony Barcellos and Dean Hickerson, scrutinized every sentence, every formula, every diagram, every marginal note (adding, incidentally, many of their own), and every line of type. Harsh taskmasters, both conscientiously represented instructor and student; their dedication and thoroughness significantly improved much of the exposition.

The revision also benefits from suggestions I received from colleagues at Davis, in particular Henry Alder, Carl Carlson, G. Donald Chakerian, David Mead, Washek Pfeffer, and Evelyn Silvia.

Daniel Drucker of Wayne State University was the main outside critic during the revision. He combined a meticulous attention to detail with a sense of broad organization.

Valuable contributions were made also by Larry Curnett, Bellevue Community College; Mark Bridger, Northeastern University; Augustus J. Garver, University of Missouri; David Finkel, Bucknell University; John C. Higgins, Brigham Young University; James Hurley, University of Connecticut; Melvin D. Lax, California State University, Long Beach; Peter A. Lindstrom, Genesee Community College; Jeffrey McLean, Ohio Northern University; Joel Stemple, CUNY, Queens College; and Lawrence A. Trivieri, Mohawk Valley Community College.

Several students worked on parts of the text and the accompanying solutions manual, reviewing manuscript, doing exercises, and checking answers. For their labors I wish to thank Dana Reneau, Judy Clarke, Colin Missel, Mark O'Donnell, Karen Thomason, Kevin Zumbrun, and Ed Bazo.

My appreciation also goes to Shelly Langman, Carol Napier, and Stephen Wagley at McGraw-Hill for their enthusiastic and skillful support in this revision. In spite of the little leeway in choice and order of topics granted any author of a basic text, the publisher has encouraged me to offer fresh options if they are needed by users of calculus.

One final remark. Special care has been taken to keep errors to a minimum. Galleys and page proofs received four independent readings. Each exercise was worked by at least three people; answers in the back of the book were checked against page proofs. Though it is every author's dream to produce the error-free book, no one, to my knowledge, has ever achieved that aspiration. I would therefore appreciate your calling to my attention any errors that may still remain.

Sherman K. Stein

AN OVERVIEW OF CALCULUS AND THIS BOOK

There are two main concepts in calculus: the derivative and the integral. Underlying both is the theme of limits. This chapter introduces these ideas informally, tells where they appear in the text, and offers a glimpse into their history. The reader may wish to turn back to these pages from time to time to maintain a broad perspective, which is otherwise too easily lost in the day-by-day details of definitions, theorems, and applications.

0.1 The derivative

Figure 0.1

The tangent line to a circle at a given point P can be found as follows. First draw the radius from the center of the circle to P and then construct the line through P perpendicular to that radius. That line is tangent to the circle. (See Fig. 0.1.)

But how would we construct the tangent line at a point P on a curve that is not a circle? For instance, how would we find the tangent line at the point P on the curve in Fig. 0.2 which is described by the equation $y = x^2$?

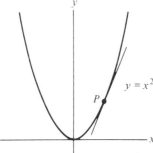

Figure 0.2

The mathematician Pierre Fermat (pronounced "Fair-MAH") published a general method for solving this type of problem in 1637 in his *Methodus ad Disquirendam Maximam et Minimam* (A method for finding maxima and minima). This method amounts to an algorithm for calculating the steepness or "slope" of a curve at any point on that curve. Once we know the slope, we can easily draw the tangent line.

As the title of Fermat's work indicates, information about tangent lines can be of aid in finding high and low points on a curve. A glance at Fig. 0.3 suggests that at such points the tangent is horizontal. Thus the search for high and low points on a curve can be simplified: one can find where the tangent lines are horizontal.

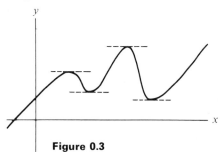

Figure 0.3

Finding tangent lines is only one of the many applications of the *derivative*, one of the two central concepts of calculus. For instance, the derivative also serves to measure how rapidly a quantity changes. To be specific, consider an object that falls under the influence of gravity. As it falls, its speed increases. How can we find that varying speed if we know for any positive number t the distance the object falls during the first t seconds? The derivative provides a formula for that speed.

Chapter 3 introduces the derivative, and Chaps. 4 and 6 present some of its main applications. Chapter 2, on limits, provides the foundation on which both the derivative and the integral rest. It turns out that the derivative is just a special case of a limit.

The following computations will convey some idea of limits and derivatives. They were done on a hand-held calculator that has keys for logarithms, exponents, and trigonometric functions. Such a calculator is a boon to the calculus student, for computations made with it can bring abstract ideas down to earth. However, a calculator may give misleading answers. If it displays numbers to, say, only eight decimal places, it would be dangerous to work with numbers as small as 0.000000001.

The first example illustrates the notion of a limit; the second explores a particular limit that arises in the study of the derivative.

EXAMPLE 1   Examine what happens to $\sqrt{x^2 + 3x} - x$ as x gets very large.

SOLUTION   Make a table with the aid of a calculator. For instance, for $x = 10$, we find that

$\approx$ *means*
"*approximately equals.*"

$$\sqrt{x^2 + 3x} - x = \sqrt{10^2 + 3 \cdot 10} - 10 = \sqrt{130} - 10 \approx 11.402 - 10 = 1.402.$$

0.1 THE DERIVATIVE

x	$\sqrt{x^2 + 3x}$	$\sqrt{x^2 + 3x} - x$
10	11.402	1.402
20	21.448	1.448
50	51.478	1.478
100	101.489	1.489
1000	1001.499	1.499

It appears that, as x gets larger, $\sqrt{x^2 + 3x} - x$ gets nearer and nearer to 1.5. Though we are not sure, this seems a likely guess. We suspect that the "limit" of $\sqrt{x^2 + 3x} - x$ as x gets arbitrarily large is 1.5. ∎

EXAMPLE 2 An object moves t^3 feet during the first t minutes of its journey. Estimate how fast it is moving after 2 minutes.

SOLUTION To estimate the speed, compute how far the object moves during a small interval of time, say from time $t = 2$ minutes to time $t = 2.1$ minutes.

At the beginning of this time interval, the object has already traveled $2^3 = 8$ feet. By the end of the time interval it has traveled $2.1^3 = 9.261$ feet. So, during the interval of 0.1 minute it moves 1.261 feet. Its average speed during the short interval of time is therefore

$$\text{Average speed} = \frac{\text{Distance}}{\text{Time}} = \frac{1.261}{0.1} = 12.61 \text{ feet per minute.}$$

This provides an estimate of the speed at time $t = 2$ minutes.

To get better estimates, use smaller intervals of time.

For the time interval from $t = 2$ to $t = 2.01$ minutes the average speed is

This computation uses the y^x key of the calculator.

$$\frac{2.01^3 - 2^3}{0.01} = \frac{8.120601 - 8}{0.01} = \frac{0.120601}{0.01} = 12.0601 \text{ feet per minute.}$$

But we could consider even shorter intervals of time. Moreover, we could consider short intervals of time that *end* at $t = 2$ rather than start at $t = 2$. For instance, the average speed from time $t = 1.9$ to $t = 2$ minutes is

$$\frac{2^3 - 1.9^3}{0.1} = \frac{8 - 6.859}{0.1} = \frac{1.141}{0.1} = 11.41 \text{ feet per minute.}$$

The estimates, 12.61, 12.0601, and 11.41 feet per minute, are just approximations of the speed at time $t = 2$ minutes.

What we really want to find is what happens to the quotient

Chapter 2 will show how to find this limit exactly.

$$\frac{t^3 - 8}{t - 2},$$

as t gets nearer and nearer the number 2. Just as in Example 1, we are studying a "limit," in this case the limit of $(t^3 - 8)/(t - 2)$ as t approaches 2.

This particular limit is called "the derivative of t^3 at $t = 2$." ∎

Exercises

A calculator would come in handy in these exercises.

1 (a) Complete this table.

x	$\sqrt{x^2 + 2x}$	$\sqrt{x^2 + 2x} - x$
1		
5		
10		
100		
1000		

(b) On the basis of (a), what number do you think $\sqrt{x^2 + 2x} - x$ approaches as x gets very large?

2 (a) Complete this table.

x	$\sqrt[3]{x^3 + x}$	$\sqrt[3]{x^3 + x} - x$
1		
10		
100		
1000		

(b) On the basis of (a), what number do you think $\sqrt[3]{x^3 + x} - x$ approaches as x gets very large?

3 (a) Complete this table.

x	$x^3 - 1$	$x - 1$	$(x^3 - 1)/(x - 1)$
0.5			
0.9			
0.99			
0.999			

(b) On the basis of (a), what number do you think $(x^3 - 1)/(x - 1)$ approaches as x gets nearer and nearer to 1?

4 (a) Complete this table.

x	$x^3 - 1$	$x^2 - 1$	$(x^3 - 1)/(x^2 - 1)$
2			
1.5			
1.1			
1.01			
1.001			
0.9			
0.99			

(b) On the basis of (a), what number do you think $(x^3 - 1)/(x^2 - 1)$ approaches as x gets nearer and nearer to 1?

5 An object travels t^3 feet in the first t minutes of its journey. Estimate its speed when $t = 1$ as follows.
(a) How far does it move during the time interval from $t = 1$ to $t = 1.01$ minutes?
(b) What is its average speed during the time interval given in (a)?
(c) What is its average speed during the time interval from $t = 1$ to $t = 1.001$ minutes?
(d) What is its average speed during the time interval from $t = 0.999$ minutes to $t = 1$ minute?
(e) On the basis of (b), (c), and (d), what do you think its speed at time $t = 1$ minute is?

6 An object falls $16t^2$ feet in t seconds.
(a) How far does it fall during the interval of time from $t = 2$ seconds to $t = 2.01$ seconds?
(b) What is its average speed during the time interval from $t = 2$ seconds to $t = 2.01$ seconds?
(c) What is its average speed during the time interval from $t = 2$ seconds to $t = 2.001$ seconds?
(d) What is its average speed during the time interval from $t = 1.999$ seconds to $t = 2$ seconds?
(e) On the basis of (b), (c), and (d), what do you think its speed at time $t = 2$ seconds is?

7 (a) Complete this table.

x	2^x	$2^x - 1$	$\dfrac{2^x - 1}{x}$
1			
0.5			
0.1			
0.01			
0.001			
-0.001			

(b) On the basis of (a), what do you think happens to $(2^x - 1)/x$ as x gets closer and closer to 0?

8 In this exercise angles are measured in **degrees**. For instance $\sin 30 = 0.5$.
(a) Complete this table.

x	$\sin x$	$\dfrac{\sin x}{x}$
30		
10		
5		
1		
0.1		

(b) On the basis of (a), what do you think happens to $(\sin x)/x$ as x gets closer and closer to 0? (Express your guess to five decimal places.)

0.2 The integral

The derivative provides information at a point or at a particular instant, so-called *local* information. The *integral*, the other major concept of calculus, does just the opposite. It is a tool for obtaining the numerical value of some overall quantity from local information. For instance, we may know that the length of the vertical line segment in Fig. 0.4 is x^2 for each x

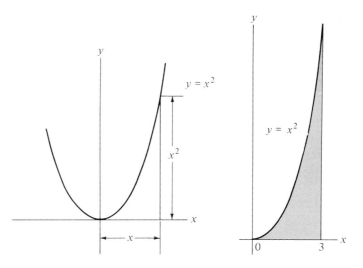

Figure 0.4　　　　　　　**Figure 0.5**

The integral can be used to find area.

and want to find the total shaded area in Fig. 0.5. What makes the problem both difficult and interesting is that the length x^2 is not constant but varies with x. This should be contrasted with the problem of finding the area of a rectangle, which is simple to calculate because the parallel line segments all have the same length, as shown in Fig. 0.6. With the aid of the

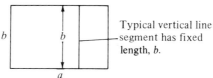

Figure 0.6

integral, developed in Chap. 5, we will be able to find the shaded area in Fig. 0.5 almost as easily as we calculate the area of the rectangle in Fig. 0.6.

The integral can be used to find the distance an object moves if we know its speed at any time.

The integral is also used in the theory of motion. Assume that we know the speed of a moving object at any time and wish to calculate the total distance the object travels during a certain interval of time. If the speed is constant, the problem is easy: The distance is then obtained by multiplying the fixed speed by the length of time. But if the speed is not constant, if it changes from moment to moment, the problem is not so simple. However, with the aid of the integral, we will be able to find the total distance if we have a formula for the speed at each instant.

The derivative takes us from the global to the local, for instance, from total distance to the speed at any time. The integral takes us from the local to the global, for instance, from the speed at any time to total dis-

The fundamental theorem of calculus links the derivative and the integral.

tance. This close connection is exploited in Chap. 5 to show that derivatives can be used to evaluate integrals, a result known as the Fundamental Theorem of Calculus. This intimate connection between the derivative and the integral was observed by Isaac Newton in 1669 (and published in 1711) and, independently, by Gottfried Leibniz (pronounced "LIBE-nits") in 1673 (and published in 1714) during their development of calculus into a structure with techniques applicable to a broad class of problems.

Before the invention of calculus the area under each curve was found by some special technique. For instance, Fermat in 1636 and Cavalieri in 1639 managed to find the area under a curve of the form $y = x^n$ and James Gregory in 1668 found the area under the curve $y = 1/\cos x$. Calculus develops a uniform approach for computing such areas with the aid of integrals.

In Chap. 7 techniques for computing integrals are presented. In Chaps. 8 and 9 the integral is applied in geometric problems, such as calculating the volume and surface area of a sphere and the length of a curve, as well as in other disciplines, such as physics, economics, and biology.

To give some idea of the integral—and why it, like the derivative, involves limits—let us try to estimate the area of the shaded region in Fig. 0.5.

Since we can compute the area of a rectangle precisely, let us estimate the area in Fig. 0.5 by approximating it with a staircase of narrow rectangles, as in Figs. 0.7 and 0.8. The rectangles in Fig. 0.7 *underestimate* the area under the curve; the rectangles in Fig. 0.8 *overestimate* the area. In Examples 1 and 2 the computations for specific choices of rectangles are carried out in detail.

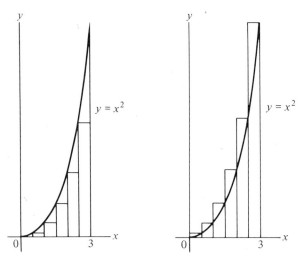

Figure 0.7   Figure 0.8

EXAMPLE 1 Figure 0.9 shows three rectangles, all of width 1. Find their total area.

SOLUTION Each rectangle has width 1. The heights of the rectangles can be found by using the formula for the curve, $y = x^2$. The height of the lowest rectangle is therefore 1^2, the next has height 2^2, and the tallest has height 3^2. Their total area is

We use only three rectangles to keep the arithmetic simple.

0.2 THE INTEGRAL 7

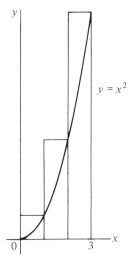

Figure 0.9

$$\underbrace{1^2}_{\text{height}} \cdot \underbrace{1}_{\text{width}} + \underbrace{2^2}_{\text{height}} \cdot \underbrace{1}_{\text{width}} + \underbrace{3^2}_{\text{height}} \cdot \underbrace{1}_{\text{width}} = 1 + 4 + 9 = 14.$$

The area under the curve is therefore less than 14. ∎

EXAMPLE 2   Figure 0.10 shows three rectangles, all of width 1. Find their total area.

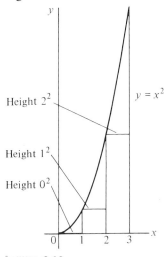

Figure 0.10

SOLUTION   In this case the sum of the areas of the rectangles is

$$0^2 \cdot 1 + 1^2 \cdot 1 + 2^2 \cdot 1 = 0 + 1 + 4 = 5.$$

The first rectangle, of height 0, is just a line segment.

The area under the curve is therefore larger than 5. ∎

The computations in Examples 1 and 2 show that the area under the curve is somewhere between 5 and 14, which is quite a wide margin. To get closer bounds, use narrower rectangles. Figures 0.11 and 0.12 each show 10 rectangles, all of width $\frac{3}{10} = 0.3$. The total area of the rectangles in Fig. 0.11 is 10.395, and the total area of those in Fig. 0.12 is 7.695. So the area under the curve is trapped between the numbers 10.395 and 7.695, which represents a much smaller margin of error than 5 and 14, which were found in Examples 1 and 2.

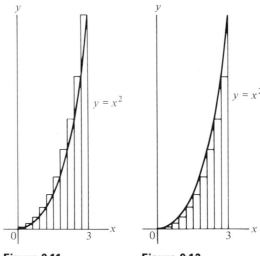

Figure 0.11   Figure 0.12

Chapter 5 shows how to compute such a limit.

It seems likely that the more rectangles we use and the narrower they are, the better their total area will approximate the area under the curve. This suggests that, to find the area under the curve, we should discover what happens to the total area of the approximating staircase of rectangles as the widths of the rectangles are chosen smaller and smaller. Just as in the preceding section we again are confronting the notion of a *limit*. The limit of the total area of the staircase of rectangles as their widths approach 0 will give us the area under the curve. This limit is called "the integral of x^2 from 0 to 3."

Exercises

1. (a) Draw the four rectangles of width $\frac{3}{4}$ which, like the rectangles of Example 1, *overestimate* the area under $y = x^2$ from 0 to 3.
   (b) Complete this table. The rectangles are labeled from left to right.

Rectangle	Height	Width	Area
First			
Second			
Third			
Fourth			

   (c) Find the total area of the rectangles in (b).

2. (a) Draw the four rectangles of width $\frac{3}{4}$ which, like the rectangles in Example 2, *underestimate* the area under $y = x^2$ from 0 to 3.
   (b) Find their total area. (The first one has area 0.)

3. Like Exercise 1, with five rectangles of width $\frac{3}{5}$ instead of four.

4. Like Exercise 2, with five rectangles of width $\frac{3}{5}$ instead of four.

5. Using 10 rectangles all of width $\frac{3}{10}$, verify that the area under the curve $y = x^2$ from 0 to 3 is (a) less than 10.395 and (b) more than 7.695.

6. Consider the area under the curve $y = x^2$ from 1 to 2.
   (a) Using five rectangles of width $\frac{1}{5}$, find an estimate of the area which is too large.

(b) Again using five rectangles of width $\frac{1}{5}$, find an estimate of the area which is too small.

7 Consider the area of the region under the curve $y = x^3$ from $x = 0$ to $x = 1$.
  (a) Compute the total area of the four rectangles in Fig. 0.13. Each has width $\frac{1}{4}$.
  (b) Like part (a), with eight rectangles of width $\frac{1}{8}$.
  (c) Using eight rectangles of width $\frac{1}{8}$ that lie *below* the curve $y = x^3$, show that the area under $y = x^3$ from 0 to 1 is greater than $\frac{49}{256} = 0.19140625$.

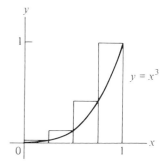

Figure 0.13

0.3 Survey

The theory and applications of the derivative and the integral occupy Chaps. 1 to 9. Chapter 10, on series, has a different flavor, though it makes significant use of the derivative and integral. It turns out that certain expressions, though not polynomials, can be represented as "polynomials of infinite degree." For instance, for x between -1 and 1,

$$\frac{1}{1-x} = 1 + x + x^2 + x^3 + \cdots.$$

(The more summands you add up, the closer you get to $1/(1-x)$.)
  As another instance,

$$\sin x = x - \frac{x^3}{1 \cdot 2 \cdot 3} + \frac{x^5}{1 \cdot 2 \cdot 3 \cdot 4 \cdot 5} - \frac{x^7}{1 \cdot 2 \cdot 3 \cdot 4 \cdot 5 \cdot 6 \cdot 7} + \cdots.$$

(The angle is in radians, not degrees.) Chapter 10 examines what happens when you add more and more numbers that are getting smaller and smaller. Do the sums get arbitrarily large? Do they approach some specific number?

Most of the ideas and techniques in the first 10 chapters were known by the middle of the eighteenth century, the contributions of Newton and Leibniz and their successors, such as the Bernoulli brothers, Brook Taylor, and Leonhard Euler (pronounced "oiler").

Chapters 12 and 13 extend the notions of the derivative and the integral to higher dimensions.

Chapters 11, 14, and 15, further generalizing the same two concepts, introduce vector analysis, a development of the nineteenth century, which was put in its present form about a hundred years ago. Vector analysis is the language of electricity and magnetism as well as of fluid flow.

Appendixes F and G return to the subject of limits and introduce some of the topics treated in advanced calculus.

The reader who wishes to learn the history of calculus, the roots of which reach back to Archimedes and earlier, may wish to consult some of the references listed at the end of this section.

Since calculus may be the only mathematics some students meet in their studies, it should be emphasized that it is by no means all of mathematics. Algebra, combinatorics, logic, computations and algorithms, topology, analysis, and geometry are a few of the other branches of mathematics that are rich in applications and where new discoveries are still being made throughout the world.

References

History of Calculus

E. T. Bell, *Men of Mathematics*, Simon and Schuster, New York, 1937. (In particular Chap. 6 on Newton and Chap. 7 on Leibniz.)

Carl B. Boyer, *The History of the Calculus and its Conceptual Development*, Dover, New York, 1959.

Carl B. Boyer, *A History of Mathematics*, Wiley, New York, 1968.

M. J. Crowe, *A History of Vector Analysis*, Notre Dame, 1967 (In particular Chap. 5).

C. H. Edwards, Jr., *The Historical Development of the Calculus*, Springer-Verlag, New York, 1979 (In particular, Chap. 8, "The Calculus According to Newton," and Chap. 9, "The Calculus According to Leibniz.")

Morris Kline, *Mathematical Thought from Ancient to Modern Times*, Oxford, New York, 1972. (In particular, Chap. 17.)

Introductions to other areas of mathematics

Richard Courant and Herbert Robbins, *What Is Mathematics?*, Oxford, New York, 1960.

Ross Honsberger, ed., *Mathematical Plums*, Dolciani Mathematical Expositions, Number 4, Mathematical Association of America, 1979.

Morris Kline, *Mathematics in the Modern World, Readings from Scientific American*, W. H. Freeman, San Francisco, 1968.

Lynn Arthur Steen, *Mathematics Today, Twelve Informal Essays*, Springer-Verlag, New York, 1979.

Sherman K. Stein, *Mathematics, the Man-made Universe*, W. H. Freeman, San Francisco, 1975.

PRELIMINARIES

This chapter reviews coordinate systems, graphs, lines and their slopes, and functions—concepts that play a key role throughout calculus.

Other precalculus material of a more specialized nature is to be found in Appendixes A to D.

Included are such topics as inequalities, absolute value, rationalizing an expression, completing the square, the quadratic formula, the binomial theorem, finite geometric series, exponents, and trigonometry. They will be referred to when needed.

REFERENCE TO APPENDIX

In Sec 1.3: Closed interval, etc. Appendix A

1.1 Coordinate systems and graphs (review)

Just as each point on a line can be described by a number, each point in the plane can be described by a pair of numbers. To do this, choose two perpendicular lines furnished with identical scales, as in Fig. 1.1. One is called the x axis and the other, the y axis. Usually the x axis is horizontal, as in Fig. 1.1. Any point P in the plane can then be described by a pair of numbers. The line through P parallel to the y axis meets the x axis at a number x, called the x *coordinate* or *abscissa* of P. The line through P parallel to the x axis meets the y axis at a number y, called the y *coordinate* or *ordinate* of P. P is then denoted (x, y), as in Fig. 1.2. The point $(0, 0)$, where the two axes cross, is called the *origin*.

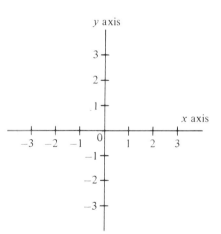

Figure 1.1

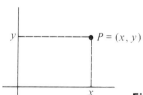

Figure 1.2

The two axes cut the plane into four parts, called *quadrants*, numbered as in Fig. 1.3. In the first quadrant both the x and y coordinates are positive; in the second, x is negative and y is positive; in the third, both x and y are negative; in the fourth, x is positive and y is negative.

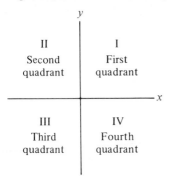

Figure 1.3

EXAMPLE 1 Plot the points $(1, 2)$, $(-3, 1)$, $(-\frac{1}{2}, -2)$, and $(3, -1)$ and specify the quadrant in which each lies.

SOLUTION As shown in Fig. 1.4, the point $(1, 2)$ is in the first quadrant, $(-3, 1)$ is in the second, $(-\frac{1}{2}, -2)$ is in the third, and $(3, -1)$ is in the fourth.

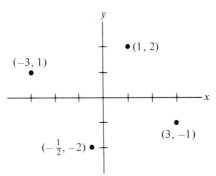

Figure 1.4

THE DISTANCE BETWEEN TWO POINTS

The distance d between two points $P_1 = (x_1, y_1)$ and $P_2 = (x_2, y_2)$ can be found with the aid of the Pythagorean theorem. Form a right triangle whose hypotenuse is the line segment joining P_1 to P_2 and whose legs are parallel to the axes, as in Fig. 1.5.

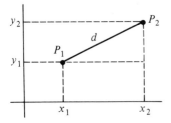

Figure 1.5

1.1 COORDINATE SYSTEMS AND GRAPHS (REVIEW)

The length of the horizontal leg is $x_2 - x_1$ if $x_2 \geq x_1$, as in Fig. 1.5. However, if $x_1 > x_2$, the length is $-(x_2 - x_1)$. Similarly, the length of the vertical leg is either $y_2 - y_1$ or its negative. In either case, since the negative of a number has the same square as the number, we have, by the Pythagorean theorem,

Recall that $(-a)^2 = a^2$.

$$d^2 = (x_2 - x_1)^2 + (y_2 - y_1)^2 \tag{1}$$

$$\text{or}\quad d = \sqrt{(x_2 - x_1)^2 + (y_2 - y_1)^2}. \tag{2}$$

Keep both (1) and (2) in mind. While (2) gives d explicitly, (1) is often preferable in computations since it does not involve square roots.

EXAMPLE 2 Find the distance between the points $(-1, 3)$ and $(2, -5)$.

SOLUTION Let $(x_1, y_1) = (-1, 3)$ and $(x_2, y_2) = (2, -5)$.

Then $\quad d^2 = (2 - (-1))^2 + ((-5) - 3)^2 = 3^2 + (-8)^2 = 73.$

Thus $\quad d = \sqrt{73}.$

We could just as well have labeled the points in the opposite order, $(x_1, y_1) = (2, -5)$ and $(x_2, y_2) = (-1, 3)$. The arithmetic is slightly different, but the result is the same:

$$d^2 = ((-1) - 2)^2 + (3 - (-5))^2 = (-3)^2 + 8^2 = 73$$

and $\quad d = \sqrt{73}.$ ∎

CIRCLES The distance formula gives us a way of dealing algebraically with the geometric notion of a circle of radius r and center $(0, 0)$. A point (x, y) lies on this circle if its distance from $(0, 0)$ is r, that is, if

The circle $x^2 + y^2 = r^2$

$$\sqrt{(x - 0)^2 + (y - 0)^2} = r,$$

or, more simply, if

$$x^2 + y^2 = r^2.$$

For the sake of brevity we may speak of "the circle $x^2 + y^2 = r^2$," which is the circle of radius r centered at the origin.

EXAMPLE 3 Determine which of these points lie on the circle of radius 13 and center at the origin: $(5, 12), (-5, 12), (10, 7)$.

SOLUTION To test whether the point (x, y) lies on the circle of radius 13 and center at the origin, check whether

$$x^2 + y^2 = 13^2,$$

that is, whether $\quad x^2 + y^2 = 169.$

14 PRELIMINARIES

Since $\qquad 5^2 + 12^2 = 25 + 144 = 169,$

the point (5, 12) lies on the circle. So does the point (−5, 12), since $(-5)^2 + 12^2 = 25 + 144 = 169.$

Does (10, 7) also lie on this circle? We find that $10^2 + 7^2 = 100 + 49 = 149.$ Thus (10, 7) does not lie on the circle. (See Fig. 1.6.)

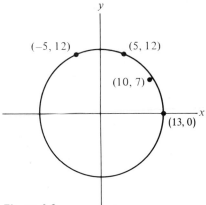

Figure 1.6

GRAPHS

The graph of an equation consists of the points for which the equation holds.

The circle shown in Fig. 1.6 is called "the graph of the equation $x^2 + y^2 = 169$." The *graph* of any equation involving one or both of the letters x and y consists of those points (x, y) whose coordinates satisfy the equation. Some examples will illustrate a few of the graphs referred to often in later chapters.

THE PARABOLA

DEFINITION *Parabola.* The graph of $y = ax^2$, for any constant $a \neq 0$, is called a *parabola* in standard position.

EXAMPLE 4 Graph the parabola $y = x^2$.

SOLUTION To begin, find a few points on the graph by choosing some specific values of x and calculating the corresponding values of y. Let us use $x = 0, 1, 2, 3, -1, -2,$ and -3 and fill in this table:

Other equations also give rise to parabolas. See Example 7 and Appendix E.

x	0	1	2	3	−1	−2	−3
$y = x^2$	0	1	4	9	1	4	9

The parabola is infinite; only a small part can be shown.

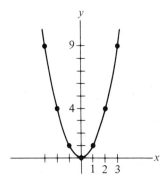

Figure 1.7

1.1 COORDINATE SYSTEMS AND GRAPHS (REVIEW)

The table provides seven points on the graph of $y = x^2$, as shown in Fig. 1.7. Note that, as $|x|$ increases, so does $y = x^2$. The graph goes arbitrarily high. ∎

EXAMPLE 5 Graph the cubic $y = x^3$.

SOLUTION Again begin with a table:

x	0	1	2	−1	−2
$y = x^3$	0	1	8	−1	−8

Note that, as x increases, so does x^3. The graph of $y = x^3$ is shown in Fig. 1.8.

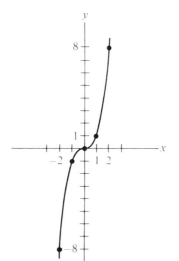

The graph of $y = x^3$ crosses the x axis "horizontally," not at an angle.

Figure 1.8 ∎

LINES

DEFINITION The graph of $y = mx + b$, for any constants m and b, is called a *line*.

EXAMPLE 6 Graph the line $y = 2x + 1$.

SOLUTION First make a short table of values:

x	0	1	−1
$y = 2x + 1$	1	3	−1

Thus the points $(0, 1)$, $(1, 3)$, and $(-1, -1)$ are on the line. Any two determine the line, which is shown in Fig. 1.9; the third point serves as a check. ∎

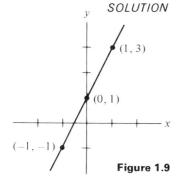

Figure 1.9

SYMMETRY

The graphs of $x^2 + y^2 = r^2$, $y = x^2$, and $y = x^3$ possess some symmetry. For instance, the part of the parabola $y = x^2$ left of the y axis is a mirror image of the part right of the y axis. Had we known this in advance, we could have sketched the graph in the first quadrant and then copied this graph in the second quadrant. Since many graphs have some symmetry, the following tests for symmetry can be quite useful.

DEFINITION

Symmetry with respect to the y axis

The graph of an equation is *symmetric with respect to the y axis* if the equation is unchanged when x is replaced by $-x$.

For example, when x is replaced by $-x$ in the equation $y = x^2$, we obtain the equation $y = (-x)^2$, which simplifies to the original equation $y = x^2$. If x appears only to *even* powers in an equation, the graph of that equation will be symmetric with respect to the y axis.

DEFINITION

Symmetry with respect to the x axis

The graph of an equation is *symmetric with respect to the x axis* if the equation is unchanged when y is replaced by $-y$.

For example, the graph of $x^2 + y^2 = 169$ is symmetric with respect to the x axis, since the equation $x^2 + (-y)^2 = 169$ reduces to $x^2 + y^2 = 169$, the original equation. The graph is also symmetric with respect to the y axis, since $(-x)^2 + y^2 = 169$ also reduces to $x^2 + y^2 = 169$.

The graph of $y = x^3$ is not symmetric with respect to the x axis or the y axis. But it does have the symmetry described in the next definition.

DEFINITION

Symmetry with respect to the origin

The graph of an equation is *symmetric with respect to the origin* if the equation is unchanged when y is replaced by $-y$ and x by $-x$ at the same time.

If in the equation $y = x^3$, we replace y by $-y$ and x by $-x$, we obtain the equation $-y = (-x)^3$, which a little algebra reduces to $-y = -x^3$, and thus $y = x^3$, the original equation. Thus the cubic $y = x^3$ is symmetric with respect to the origin.

Figure 1.10 is a memory device showing the three types of symmetry.

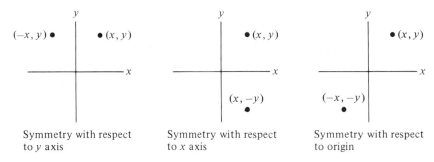

Figure 1.10

INTERCEPTS

The points where a graph meets the axes are frequently of aid in sketching the graph.

DEFINITION *Intercepts.* The x coordinates of the points where a graph meets the x axis are the *x intercepts* of the graph. The y coordinates of the points where a graph meets the y axis are the *y intercepts* of the graph.

How to find intercepts

To find the x intercepts of a graph, set $y = 0$ in its equation. To find the y intercepts, set $x = 0$.

For instance, to find the x intercepts of the circle $x^2 + y^2 = 169$, set $y = 0$, obtaining the equation

$$x^2 = 169.$$

The x intercepts are therefore 13 and -13, which can be checked by inspecting Fig. 1.6.

EXAMPLE 7 Find the intercepts of the graph of $y = x^2 - 4x - 5$.

SOLUTION To find the x intercepts, set $y = 0$, obtaining

$$0 = x^2 - 4x - 5.$$

By the quadratic formula,
$$x = \frac{4 \pm \sqrt{16 + 20}}{2}$$
$$= \frac{4 \pm \sqrt{36}}{2}$$
$$= \frac{4 \pm 6}{2}$$
$$= 5 \text{ and } -1.$$

To find y intercepts, set $x = 0$, obtaining

$$y = 0^2 - 4 \cdot 0 - 5$$
$$= -5.$$

There is only one y intercept, namely -5. The graph and intercepts are shown in Fig. 1.11. ∎

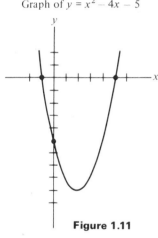

Graph of $y = x^2 - 4x - 5$

Figure 1.11

This is a parabola.

THE ELLIPSE

The equation of a circle, $x^2 + y^2 = r^2$, can be written in the form

$$\frac{x^2}{r^2} + \frac{y^2}{r^2} = 1.$$

A generalization of this equation provides the equation of the oval or egg-shaped curve known as an ellipse.

18 PRELIMINARIES

Ellipses not in standard position are discussed in Appendix E.

DEFINITION *Ellipse* (in standard position). Let a and b be positive numbers. The graph of

$$\frac{x^2}{a^2} + \frac{y^2}{b^2} = 1 \qquad (3)$$

is called an *ellipse* in standard position.

Note that an ellipse is symmetric with respect to both axes. Moreover, its x intercepts are a and $-a$; its y intercepts are b and $-b$. The graph of (3) is shown in Fig. 1.12. For $a > b$, the ellipse is wider than high; for $b > a$, it is higher than wide; and if $a = b$, the ellipse is a circle of radius a.

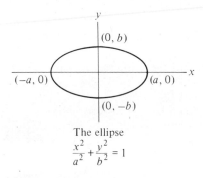

The ellipse $\frac{x^2}{a^2} + \frac{y^2}{b^2} = 1$

Figure 1.12

EXAMPLE 8 Sketch the ellipse

$$\frac{x^2}{4} + \frac{y^2}{8} = 1.$$

SOLUTION In this case $a^2 = 4$ and $b^2 = 8$; so $a = 2$ and $b = \sqrt{8}$, which is about 2.8. First plot the four points where the ellipse crosses the axes, then fill in the curve smoothly freehand, as shown in Fig. 1.13.

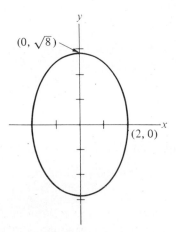

Figure 1.13

THE HYPERBOLA

DEFINITION

Other hyperbolas are discussed in Appendix E.

Hyperbola (in standard position). Let a and b be positive numbers. The graph of

$$\frac{x^2}{a^2} - \frac{y^2}{b^2} = 1 \qquad (4)$$

is called a *hyperbola* in standard position.

Though Eq. (4) looks a lot like (3), its graph is quite different. First of all, the hyperbola given by (4) has no y intercept. To see why, consider the equation,

$$\frac{0^2}{a^2} - \frac{y^2}{b^2} = 1,$$

which reduces to $\quad y^2 = -b^2.$

Since b^2 is positive, $-b^2$ is negative. But there is no real number y whose square is negative. Thus the hyperbola (4) has no y intercept.

Second, the hyperbola extends arbitrarily far from the origin. To show this, solve (4) for y:

$$\frac{x^2}{a^2} - \frac{y^2}{b^2} = 1$$

$$\frac{x^2}{a^2} = 1 + \frac{y^2}{b^2}$$

$$\frac{y^2}{b^2} = \frac{x^2}{a^2} - 1$$

$$y^2 = b^2\left(\frac{x^2}{a^2} - 1\right)$$

$$y^2 = \frac{b^2}{a^2}(x^2 - a^2)$$

$$y = \pm\frac{b}{a}\sqrt{x^2 - a^2}.$$

For $|x| \geq a$, the expression in the radical is not negative. So, for $|x| \geq a$, hence for large x, there will be a corresponding value of y (namely $y = + (b/a)\sqrt{x^2 - a^2}$).

How to graph a hyperbola

To graph the hyperbola $x^2/a^2 - y^2/b^2 = 1$, follow these steps:

1. Plot the points $(a, 0)$ and $(-a, 0)$, where the hyperbola meets the x axis.

2. Plot the point (a, b) and the line through it and the origin. Do the same for the point $(a, -b)$.

3. Sketch the hyperbola freehand, using the two lines in step (2) as guides. (As shown in Exercise 40, the hyperbola approaches these lines arbitrarily closely.) The two lines are called *asymptotes* of the hyperbola.

Asymptotes of a hyperbola

EXAMPLE 9 Graph the hyperbola

$$\frac{x^2}{9} - \frac{y^2}{4} = 1.$$

SOLUTION In this case, $a^2 = 9$ and $b^2 = 4$; so $a = 3$ and $b = 2$. The x intercepts are 3 and -3. One asymptote passes through $(0, 0)$ and $(3, 2)$, and the other through $(0, 0)$ and $(3, -2)$. With the aid of the intercepts and the asymptotes, the graph is easy to sketch freehand. (See Fig. 1.14.) Note that since x and y appear only to even powers, the graph is symmetric with respect to both axes.

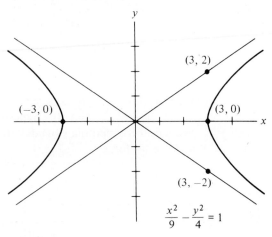

Figure 1.14 ∎

Another equation whose graph is a hyperbola

As mentioned in Appendix E, when $B^2 - 4AC$ is positive, the graph of $Ax^2 + Bxy + Cy^2 + Dx + Ey + F = 0$ is a hyperbola (or, in special cases, a pair of intersecting lines). In particular, the graph of $xy = a$, where a is a non-zero constant, is a hyperbola.

EXAMPLE 10 Graph the hyperbola $xy = 1$.

SOLUTION Since the product of x and y is 1, neither x nor y can be 0. Thus the hyperbola $xy = 1$ has no intercepts. Moreover, since $x \neq 0$, we can solve for y in terms of x:

$$y = 1/x$$

and make a table of values:

x	$\frac{1}{10}$	10	1	2	$-\frac{1}{10}$	-10	-1	-2
$y = 1/x$	10	$\frac{1}{10}$	1	$\frac{1}{2}$	-10	$-\frac{1}{10}$	-1	$-\frac{1}{2}$

With the aid of these eight points, sketch the graph, which consists of two identical pieces. Far to the right and to the left the graph approaches the x axis. Also, as x approaches 0, the graph approaches the y axis. In this case, the x and y axes are the asymptotes of the hyperbola. (See Fig. 1.15.)

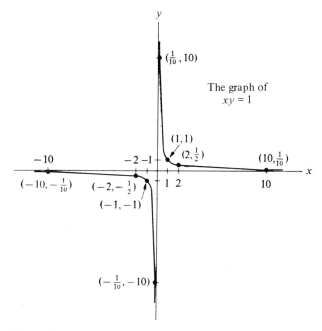

Figure 1.15

The discussions of the parabola, ellipse, and hyperbola in this section are algebraic. For their geometric definitions, see Appendix E.

SUMMARY

Geometry	Algebra
point	(x, y)
distance between two points	$\sqrt{(x_2 - x_1)^2 + (y_2 - y_1)^2}$
circle (radius r, center at origin)	$x^2 + y^2 = r^2$
line	$y = mx + b$
parabola	$y = ax^2, a \neq 0$
ellipse	$\dfrac{x^2}{a^2} + \dfrac{y^2}{b^2} = 1, a > 0, b > 0$
hyperbola	$\dfrac{x^2}{a^2} - \dfrac{y^2}{b^2} = 1, a > 0, b > 0$
	and $xy = a, a \neq 0$

The table lists the equations of parabolas, ellipses, and hyperbolas, only when they are conveniently located in the xy plane. Appendix E defines and examines these notions more generally.

In addition, symmetry and intercepts were discussed, as well as the asymptotes of hyperbolas.

Exercises

In Exercises 1 and 2 plot the points and state in which quadrant each lies.

1. (a) $(-3, 5)$ (b) $(4, -3)$
   (c) $(2, 7)$ (d) $(-2, -7)$
2. (a) $(1, 3)$ (b) $(-3, -\frac{1}{2})$
   (c) $(5, -3)$ (d) $(-3, 4)$

In Exercises 3 and 4 find the distances between the given points.

3. (a) $(4, 1)$ and $(-2, 9)$ (b) $(6, 3)$ and $(1, 15)$
   (c) $(4, 0)$ and $(7, 0)$
4. (a) $(3, 4)$ and $(5, 6)$ (b) $(-3, -2)$ and $(4, -2)$
   (c) $(-3, -4)$ and $(3, 4)$

In Exercises 5 to 16 graph the equations.

5. $x^2 + y^2 = 49$ 6. $x^2 + y^2 = 1$
7. $y = 2x^2$ 8. $y = -2x^2$
9. $y = -x^2$ 10. $y = -x^2/2$
11. $y = -x^3$ 12. $y = 2x^3$
13. $y = x^4$ (Include the points for which $x = 0, \frac{1}{2}$, and 1.)
14. $y = x^5$ (Include the points for which $x = 0, \frac{1}{2}$, and 1.)
15. $y = x - 1$ 16. $y = -2x + 3$

In Exercises 17 to 22 find the x and y intercepts, if there are any.

17. $y = 2x + 6$ 18. $y = 3x - 6$
19. $xy = 6$ 20. $x^2 - y^2 = 1$
21. $y = 2x^2 + 5x - 3$ 22. $y = 4 - x^2$

In Exercises 23 to 28 graph the equations.

23. $\dfrac{x^2}{25} + \dfrac{y^2}{16} = 1$ 24. $\dfrac{x^2}{16} + \dfrac{y^2}{25} = 1$

25. $\dfrac{x^2}{16} - \dfrac{y^2}{25} = 1$ 26. $\dfrac{x^2}{25} - \dfrac{y^2}{16} = 1$

27. $xy = 8$ 28. $xy = -1$

29. (a) What symmetry does the graph of $x = y^2$ have?
    (b) Graph $x = y^2$.
30. (a) What symmetry does the graph of $y^2 = x^3$ have?
    (b) Graph $y^2 = x^3$.

31. Graph $y = x - x^2$.
32. Graph $y = 1 + x^2$.
33. Graph $y = (x - 1)^2$.
34. Graph $y = \sqrt{x}$.
35. Graph $y = |x|$. ($|x|$ is the absolute value of x.)
36. Graph $y = \sqrt[3]{x}$.

37. Find an equation of the circle whose center is $(2, 1)$ and whose radius is 7.
38. (a) Graph $y = x^2 + 4x + 9$.
    (b) What point on the graph in (a) has the smallest y coordinate? (*Hint:* First rewrite $x^2 + 4x + 9$ by completing the square.)
39. Let $y = ax^2 + bx + c$, where a is positive and the discriminant, $b^2 - 4ac$, is positive. (a) Find the value of x for which y is as small as possible. (*Hint:* First rewrite $ax^2 + bx + c$ by completing the square.) (b) Show that the number x in (a) is the average of the two x intercepts of the graph.
40. This exercise shows why the part of the hyperbola $x^2/a^2 - y^2/b^2 = 1$ in the first quadrant approaches the line $y = bx/a$ when x is large.

    Consider the point $P = (x, y)$ on the hyperbola, $x \geq a$, $y \geq 0$.
    (a) Show that $y = (b/a)\sqrt{x^2 - a^2}$.
    (b) Let $Q = (x, bx/a)$ be the point on the line $y = bx/a$ with the same x coordinate as P. Show that Q is near P for large x, by showing that

    $$\frac{bx}{a} - \frac{b}{a}\sqrt{x^2 - a^2}$$

    approaches 0 when x is large. (*Hint:* Rationalize by multiplying by $(x + \sqrt{x^2 - a^2})/(x + \sqrt{x^2 - a^2})$.)

1.2 Lines and their slopes

The vertical change, $y_2 - y_1$, is the "rise"; the horizontal change, $x_2 - x_1$, is the "run."

Consider a line that is not parallel to the y axis. Select distinct points $P_1 = (x_1, y_1)$ and $P_2 = (x_2, y_2)$ on it as in Fig. 1.16. As we move from P_1 to P_2, the vertical change is $y_2 - y_1$ and the horizontal change is $x_2 - x_1$, as shown in Fig. 1.17. The quotient, $(y_2 - y_1)/(x_2 - x_1)$ is a measure of the steepness of the line and is called its *slope*.

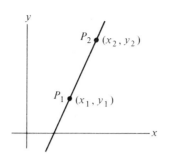
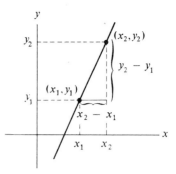

Figure 1.16   **Figure 1.17**

SLOPE

DEFINITION   *Slope of a line.* Consider a line that is not parallel to the y axis. Let $P_1 = (x_1, y_1)$ and $P_2 = (x_2, y_2)$ be two distinct points on the line. The *slope m* of the line is

Slope is rise over run.
$$m = \frac{y_2 - y_1}{x_2 - x_1}.$$

The slope of a line parallel to the y axis is not defined. Accordingly, a vertical line is said to have "no slope."

EXAMPLE 1   Find the slope of the line through the points $(1, 4)$ and $(3, 5)$.

SOLUTION   Let P_1 be $(1, 4)$ and P_2 be $(3, 5)$. Then

$$m = \frac{5 - 4}{3 - 1} = \frac{1}{2}.$$ ∎

In finding the slope in Example 1 we could have chosen P_1 to be $(3, 5)$ and P_2 to be $(1, 4)$. The arithmetic would be slightly different, but the result would be the same, since in this case

$$m = \frac{4 - 5}{1 - 3} = \frac{-1}{-2} = \frac{1}{2}.$$

The order of the two points does not affect the slope, since

$$\frac{y_2 - y_1}{x_2 - x_1} = \frac{y_1 - y_2}{x_1 - x_2}.$$

Moreover the slope of a line does not depend on the particular pair of points selected on it. If one pair is $P_1 = (x_1, y_1)$ and $P_2 = (x_2, y_2)$ and the other pair is $P'_1 = (x'_1, y'_1)$ and $P'_2 = (x'_2, y'_2)$, then

24 PRELIMINARIES

$$\frac{y_2 - y_1}{x_2 - x_1} = \frac{y'_2 - y'_1}{x'_2 - x'_1}.$$

This equation follows from the fact that the ratios between the lengths of corresponding sides of similar triangles are equal. (See Fig. 1.18.)

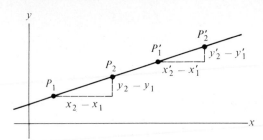

Figure 1.18

EXAMPLE 2 Find the slopes of the lines through the given points. Then draw the lines.
(a) (0, 2) and (1, 8)   (b) (1, 1) and (5, 2)
(c) (2, 6) and (4, 3)   (d) (5, 1) and (−2, 1)

SOLUTION (a) $m = \dfrac{8 - 2}{1 - 0} = 6$ (b) $m = \dfrac{2 - 1}{5 - 1} = \dfrac{1}{4}$

(c) $m = \dfrac{3 - 6}{4 - 2} = -\dfrac{3}{2}$ (d) $m = \dfrac{1 - 1}{(-2) - 5} = 0$

The lines are graphed in Fig. 1.19.

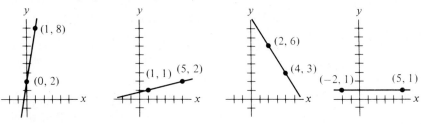

Figure 1.19

The meaning of positive and negative slopes

As suggested by the lines in Example 2, a *positive slope* tells us that as we move from left to right on the line, we go up. (As the x coordinate increases, so does the y coordinate.) A *negative slope* tells us that as we move from left to right on the line, we go down. (As the x coordinate increases, the y coordinate decreases.) A slope of 0 corresponds to a *horizontal* line, that is, a line parallel to the x axis.

TESTS FOR PARALLEL OR PERPENDICULAR LINES

Given the slopes of two lines, we can determine whether they are parallel or perpendicular.

TEST FOR PARALLEL LINES A line of slope m_1 is parallel to a line of slope m_2 if and only if $m_1 = m_2$.

The test for perpendicularity is a little fancier.

TEST FOR PERPENDICULARITY A line of slope m_1 is perpendicular to a line of slope m_2 if and only if the product of the two slopes is -1, $m_1 m_2 = -1$. (In other words,

$$m_2 = -1/m_1.$$

The slope m_2 is the negative of the reciprocal of m_1.) It is assumed that neither line is parallel to the y axis.

Proofs of these two tests are outlined in Exercises 35 and 36.

EXAMPLE 3 Determine whether the line through $(-1, -2)$ and $(2, 2)$ is parallel to the line through $(6, 5)$ and $(1, 1)$.

SOLUTION The slope of the first line is

$$\frac{2-(-2)}{2-(-1)} = \frac{4}{3}.$$

The slope of the second line is

$$\frac{1-5}{1-6} = \frac{-4}{-5} = \frac{4}{5}.$$

Since $\frac{4}{3} \neq \frac{4}{5}$, the lines are not parallel. ∎

EXAMPLE 4 Determine whether the line through $(1, 5)$ and $(4, 4)$ is perpendicular to the line through $(2, 1)$ and $(3, 4)$.

SOLUTION The first line has slope

$$\frac{4-5}{4-1} = \frac{-1}{3}.$$

The second line has slope

$$\frac{4-1}{3-2} = 3.$$

The product of the two slopes is $3(-\frac{1}{3}) = -1$, so the two lines are perpendicular. ∎

THE SLOPE OF THE LINE $y = mx + b$

In Sec. 1.1 it was mentioned that the graph of $y = mx + b$ (where m and b are fixed numbers) is a line. To find the slope of this line select any two points (x_1, y_1) and (x_2, y_2), $x_1 \neq x_2$, on the line. Thus

$$y_1 = mx_1 + b \quad \text{and} \quad y_2 = mx_2 + b.$$

The slope determined by these two points is

$$\frac{y_2 - y_1}{x_2 - x_1} = \frac{(mx_2 + b) - (mx_1 + b)}{x_2 - x_1}$$

$$= \frac{mx_2 - mx_1}{x_2 - x_1} \qquad \text{canceling the } b\text{'s}$$

$$= \frac{m(x_2 - x_1)}{x_2 - x_1} \qquad \text{factoring out } m$$

$$= m.$$

The slope of $y = mx + b$ is m, the coefficient of x.

Since the result m does not depend on which pair of points is selected on the line, we say simply that the *slope of the line* is m. In short, the slope of the line $y = mx + b$ is the coefficient of x; the number b does not influence the slope. (The y intercept of the line is equal to b.)

The slope-intercept equation of a line

The equation $y = mx + b$ is known as the *slope-intercept* equation of a straight line. (See Fig. 1.20.) If $m = 0$, the line is parallel to the x axis and has the equation $y = 0x + b$, which is just $y = b$. In this case the symbol x does not appear.

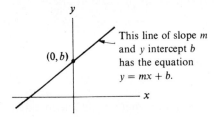

Figure 1.20

The equation of a line parallel to the y axis

No line of the form $y = mx + b$ is parallel to the y axis. The equation of a line parallel to the y axis has the form $x = a$, where a is a constant; the symbol y does not appear. Figure 1.21 shows the graphs of the equations $y = -2$ and $x = 3$.

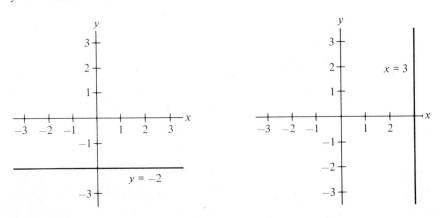

Figure 1.21

How to graph $y = mx + b$

With the knowledge that m is the slope and b is the y intercept, it is easy to graph the line $y = mx + b$. The procedure is illustrated in Example 5.

1.2 LINES AND THEIR SLOPES

EXAMPLE 5 Graph the lines (a) $y = 2x + 3$ and (b) $y = 4x/3 + 1$.

SOLUTION (a) $y = 2x + 3$. Here the y intercept is 3, so the point $P = (0, 3)$ is on the line. Plot this point. To obtain a second point on the line, write the slope 2 in the form $\frac{2}{1}$. Plot a second point, Q, 1 unit to the right of P and 2 units above it, as shown in Fig. 1.22. The line through P and Q has slope 2 and y intercept 3.

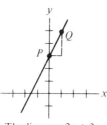

The line $y = 2x + 3$ The line $y = -4x/3 + 1$

Figure 1.22   **Figure 1.23**

A way to sketch a line if it has rational slope

(b) $y = -4x/3 + 1$. The y intercept is 1 and the slope is $-\frac{4}{3}$. The point $P = (0, 1)$ is on the line. Next draw the point Q that is 3 units to the right of P and 4 units lower, as shown in Fig. 1.23. The line through P and Q has slope $-\frac{4}{3}$ and y intercept 1. ∎

Sometimes we may wish to find an equation of the line that passes through a certain point $P_1 = (x_1, y_1)$ and has a certain slope m. To find such an equation, consider the typical point $P = (x, y)$ on the line, $x \neq x_1$. The slope determined by P and P_1 must be the same as the prescribed slope m, that is,

$$\frac{y - y_1}{x - x_1} = m.$$

How to find the equation of a line through a given point with a given slope

Clearing the denominator provides an equation in x and y, which is called the *point-slope* equation of the line.

THE POINT-SLOPE EQUATION The line through (x_1, y_1) of slope m is described by the equation

$$y - y_1 = m(x - x_1).$$

EXAMPLE 6 Find the point-slope equation of the line through $(1, 3)$ of slope -2.

SOLUTION The point-slope formula gives

$$y - 3 = -2(x - 1).$$

This equation could be rewritten in several ways. For instance, $y - 3 = -2x + 2$ or $y = -2x + 5$ or $2x + y = 5$. ∎

The point-slope formula gives a quick way to find an equation of the line through two given points $P_1 = (x_1, y_1)$ and $P_2 = (x_2, y_2)$. (Assume that $x_1 \neq x_2$. If $x_1 = x_2$, the line is parallel to the y axis and has the equation $x = x_1$.)

The slope of the line through P_1 and P_2 is

$$\frac{y_2 - y_1}{x_2 - x_1}.$$

How to find an equation of the line through two given points

Since the line passes through (x_1, y_1) and has slope $(y_2 - y_1)/(x_2 - x_1)$, its point-slope equation is

$$y - y_1 = \frac{(y_2 - y_1)}{(x_2 - x_1)}(x - x_1).$$

EXAMPLE 7 Find an equation of the line through $(-1, 4)$ and $(2, 3)$.

SOLUTION The slope of the line is

$$\frac{3 - 4}{2 - (-1)} = -\frac{1}{3}.$$

Since the point $(-1, 4)$ is on the line, the line has the point-slope equation

$$y - 4 = -\frac{1}{3}(x - (-1)).$$

Thus
$$y - 4 = -\frac{x}{3} - \frac{1}{3}$$

or
$$y = -\frac{x}{3} + \frac{11}{3},$$

which is the slope-intercept equation of the line. ∎

THE GRAPH OF THE EQUATION $Ax + By + C = 0$

Let A, B, and C be fixed numbers such that at least one of A and B is not zero. With the aid of the slope-intercept formula it will be shown that the graph of the equation $Ax + By + C = 0$ is a line. For this reason the equation $Ax + By + C = 0$ is said to be "a linear equation in x and y."

The graph of $Ax + By + C = 0$ is a line.

To show that the graph of $Ax + By + C = 0$ is a line, consider two cases: (1) $B = 0$ and (2) $B \neq 0$.

Case (1): $B = 0$. In this instance the equation is just $Ax + C = 0$, which is equivalent to $x = -C/A$. So the graph is a line parallel to the y axis.

Case (2): $B \neq 0$. In this case rewrite the equation $Ax + By + C = 0$ as follows:

$$By = -Ax - C$$

or
$$y = -\frac{A}{B}x - \frac{C}{B}.$$

The graph is therefore the line of slope $-A/B$ and y intercept $-C/B$, by the slope-intercept formula.

EXAMPLE 8 Find the intercepts and slope of the line $2x + 3y - 6 = 0$.

SOLUTION To find the x intercept, set $y = 0$, obtaining
$$2x + 3 \cdot 0 - 6 = 0$$
$$2x = 6$$
$$x = 3.$$

The x intercept is 3.

To find the y intercept, set $x = 0$, obtaining
$$2 \cdot 0 + 3y - 6 = 0$$
$$3y = 6$$
$$y = 2.$$

The y intercept is 2.

To find the slope, rewrite the equation $2x + 3y - 6 = 0$ in the slope-intercept form by solving for y:
$$2x + 3y - 6 = 0$$
$$3y = -2x + 6$$
$$y = \left(-\tfrac{2}{3}\right)x + 2.$$

The slope is $-\tfrac{2}{3}$. ∎

Exercises

In Exercises 1 to 6 plot the points and find the slopes of the lines through them.

1 $(-1, 1)$ and $(4, 2)$
2 $(3, 1)$ and $(-2, -1)$
3 $(-1, 4)$ and $(3, -1)$
4 $(1, 5)$ and $(5, 1)$
5 $(1, 7)$ and $(11, 7)$
6 $(-2, -5)$ and $(0, -5)$
7 Describe the slope of each line in Fig. 1.24 as positive, negative, or 0.

8 Describe the slope of each line in Fig. 1.25 as positive, negative, or 0.

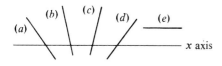

Figure 1.24

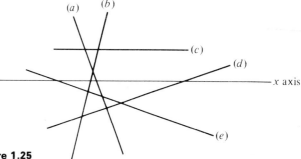

Figure 1.25

30 PRELIMINARIES

9 A line L has slope 4. (a) What is the slope of a line parallel to L? (b) What is the slope of a line perpendicular to L?

10 A line L has slope $-\frac{1}{2}$. (a) What is the slope of a line parallel to L? (b) What is the slope of a line perpendicular to L?

11 Is the line through $(2, 4)$ and $(3, 7)$ parallel to the line through $(1, 7)$ and $(-1, 1)$?

12 Is the line through $(1, 2)$ and $(-3, 5)$ parallel to the line through $(3, -1)$ and $(-1, 4)$?

13 Is the line through $(0, 0)$ and $(4, 6)$ perpendicular to the line through $(5, -1)$ and $(3, 2)$?

14 Is the line through $(-1, -2)$ and $(6, 2)$ perpendicular to the line through $(2, 3)$ and $(8, -7)$?

In Exercises 15 and 16 find the slope-intercept equations for the lines:

15 (a) with y intercept 2 and slope 3, (b) with y intercept -2 and slope $\frac{2}{3}$, (c) with y intercept 0 and slope -3.

16 (a) with y intercept $\frac{3}{4}$ and slope $-\frac{1}{2}$, (b) with y intercept -2 and slope 2, (c) with y intercept 3 and slope $\frac{3}{5}$.

In Exercises 17 and 18 find the slopes and y intercepts of the given lines and graph them.

17 (a) $y = 3x - 1$ (b) $y = -2x + 1$ (c) $y = 3x/5$

18 (a) $y = -x/2 + 3$ (b) $y = -3x + 4$ (c) $y = -2x/3 + 1$

In Exercises 19 and 20 find point-slope equations of the given lines:

19 (a) slope 3, passing through $(1, 2)$,
    (b) slope -2, passing through $(3, -1)$.

20 (a) slope $\frac{4}{3}$, passing through the origin.
    (b) slope 0, passing through $(2, 5)$.

In Exercises 21 and 22 find point-slope equations of the lines through the given points.

21 (a) $(1, 2)$ and $(5, 3)$ (b) $(-1, 2)$ and $(3, 1)$
    (c) $(4, 5)$ and $(2, 3)$.

22 (a) $(4, 4)$ and $(6, 6)$ (b) $(-2, -5)$ and $(3, 4)$
    (c) $(0, 0)$ and $(3, 5)$.

In Exercises 23 and 24 find the slope-intercept equations of the given lines and graph the lines.

23 (a) $x + y + 1 = 0$ (b) $-2x + y = 0$
    (c) $2x + 3y - 12 = 0$

24 (a) $x - y = 0$ (b) $-2x - 5y + 10 = 0$
    (c) $x - 2y = 4$

25 Does the line through $(4, 1)$ and $(7, 2)$ pass through the point (a) $(10, 3)$? (b) $(14, 5)$?

26 Find an equation of the line with y intercept 2 and perpendicular to the line $y = (-2x/3) + 7$.

27 Find an equation of the line through $(1, 3)$ and perpendicular to the line through $(4, 1)$ and $(2, 5)$.

28 Find an equation of the line parallel to the line $y = x + 6$ and passing through the origin.

29 Find an equation of the line parallel to $y = 3x + 2$ with y intercept 5.

Exercises 30 to 33 concern another form of an equation of a line, the "intercept form," which is convenient for working with the intercepts.

30 Let a and b be nonzero numbers. Show that the line $x/a + y/b = 1$ has x intercept a and y intercept b. (See Fig. 1.26.) The equation $x/a + y/b = 1$ is called the *intercept* equation of the line through $(a, 0)$ and $(0, b)$.

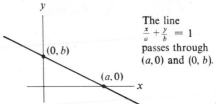

Figure 1.26

31 Find the slope of the line whose intercept equation is $x/a + y/b = 1$.

32 Find the intercept equation of the line with (a) x intercept 3 and y intercept 5, (b) x intercept -1 and y intercept 2, (c) x intercept $-\frac{1}{2}$ and y intercept -3.

33 Find the x intercept of the line (a) with slope 4 and y intercept -7, (b) through $(1, 3)$ with slope $-\frac{2}{3}$, (c) $x/3 - y/4 = 1$, (d) $-2x + 3y = 12$.

■ ■

34 Find the intersection of the line through $(0, 1)$ and $(4, 7)$ with the line through $(3, 3)$ and $(5, 1)$.

35 This exercise outlines a proof that parallel lines have equal slopes. Let L and L' be parallel lines. Let the fixed distance between them, measured parallel to the y axis, be d. That is, if (x, y) is on L, then $(x, y + d)$ is on L'. Let $P_1 = (x_1, y_1)$ and $P_2 = (x_2, y_2)$ be points on L. Then $P'_1 = (x_1, y_1 + d)$ and $P'_2 = (x_2, y_2 + d)$ are points on L'.
(a) Find the slope of L using P_1 and P_2.
(b) Find the slope of L' using P'_1 and P'_2.
Putting (a) and (b) together shows that the slopes of parallel lines are equal.

36 This exercise outlines a proof that the product of the slopes of perpendicular lines (neither parallel to the y axis) is -1. Assume that L_1 and L_2 are perpendicular lines. Let L_1 have slope m_1 and L_2 have slope m_2. For convenience, assume that both lines pass through the origin. A sketch shows that one line has positive slope and one line has negative slope. Say that m_1 is positive and m_2 is negative.
(a) Show that the point $(1, m_1)$ lies on L_1.
(b) Show that the point $(-m_1, 1)$ lies on L_2. (Recall that L_2 is perpendicular to L_1.)
(c) Deduce that $m_2 = -1/m_1$. (*Hint:* First draw the right triangle with vertices $(0, 0)$, $(1, 0)$, and $(1, m_1)$ and the right triangle with vertices $(0, 0)$, $(-m_1, 0)$, and $(-m_1, 1)$.)
(d) If the two perpendicular lines do not pass through the origin, show that the product of their slopes is -1. (*Hint:* Use (c) and Exercise 35.)

1.3 Functions

The area A of a square depends on the length of its side x and is given by the formula

$$A = x^2.$$

Similarly, the distance s (in feet) that a freely falling object drops in the first t seconds is described by the formula

$$s = 16t^2.$$

Each choice of t determines a unique value for s. For instance, when $t = 3$, $s = 16 \cdot 3^2 = 144$.

Both of these formulas illustrate the mathematical notion of a function.

DEFINITION *Function.* Let X and Y be sets. A *function* (from X to Y) is a rule or method for assigning to each element in X a unique element in Y.

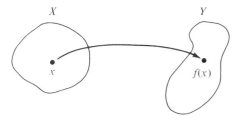

Figure 1.27

The notion of a function is schematized in Fig. 1.27.

A function may be given by a formula, as are the functions $A = x^2$ and $s = 16t^2$. In daily life a function is often indicated by a table. For instance, a table of populations of the cities in the United States can be thought of as a function that assigns to each city the size of its population. In this case X is the set of cities and Y is the set of positive integers.

A function is often denoted by the symbol f. The element that the function assigns to the element x is denoted $f(x)$ (read "f of x"). In practice, though, almost everyone speaks interchangeably of the function f or the function $f(x)$.

EXAMPLE 1 Let $f(x) = x^2$ for each real number x. Compute
(a) $f(3)$, (b) $f(2)$, (c) $f(-2)$.

SOLUTION (a) $f(3) = 3^2 = 9$.
(b) $f(2) = 2^2 = 4$.
(c) $f(-2) = (-2)^2 = 4$. ■

DEFINITIONS *The domain and range of a function* Let X and Y be sets and let f be a function from X to Y. The set X is called the *domain* of the function. If $f(x) = y$, y is called the *value* of f at x. The set of all values of the function is called the *range* of the function.

When the function is given by a formula, the domain is usually understood to consist of all the numbers for which the formula is defined.

EXAMPLE 2 Find the domain and range of the function given by the formula

$$f(x) = x^2.$$

SOLUTION Since x^2 is defined for every real number x, the domain of the squaring function x^2 consists of the entire x axis.

The range of f consists of all real numbers that are of the form x^2 for some real number x. The number 9 is in the range since $9 = 3^2$; the number 5 is in the range since $5 = (\sqrt{5})^2$. Zero is also in the range since $0 = 0^2$. In fact, every nonnegative real number r is in the range since r can be expressed as the square of a real number: $r = (\sqrt{r})^2$.

However, no negative number is in the range, since no negative number is the square of a real number.

Thus the range consists precisely of the nonnegative real numbers. ∎

EXAMPLE 3 Let $f(x) = 1/x$. Find the domain and range of f.

SOLUTION Since division by 0 is meaningless, the domain consists of all real numbers except 0.

What is the range of $1/x$? For instance, $\tfrac{1}{2}$ is in the range since

$$f(2) = \tfrac{1}{2}.$$

Also, -3 is in the range since

$$f(-\tfrac{1}{3}) = \frac{1}{-\tfrac{1}{3}} = -3.$$

More generally, any real number other than 0 is in the range. To show this, consider a real number b, $b \neq 0$. We must show that there is a real number x such that

$$f(x) = b,$$

that is,

$$\frac{1}{x} = b.$$

This last equation is equivalent to $1 = xb$ or

$$x = \frac{1}{b}.$$

So, if $x = 1/b$, $f(x) = b$. (To check this, compute $f(1/b)$: $f(1/b) = 1/(1/b) = b$.)

The number 0 is not in the range, since there is no number x such that $1/x = 0$.

By coincidence the range and domain of this function $1/x$ coincide; both consist of all real numbers except 0. ∎

EXAMPLE 4 Find the domain and range of $f(x) = 2 + \sqrt{x - 1}$.

SOLUTION For $2 + \sqrt{x-1}$ to be meaningful, the square root of $x - 1$ must make sense, that is, $x - 1$ must not be negative. Thus the domain consists of all numbers x such that

$$x - 1 \geq 0,$$
or, equivalently,
$$x \geq 1.$$

That is, the domain is the interval $[1, \infty)$. (See Appendix A for this notation.) As x varies from 1 to larger numbers, $f(x)$ increases from $f(1) = 2 + \sqrt{1-1}$ to arbitrarily large values. Thus the range of f consists of all numbers greater than or equal to 2, that is, the interval $[2, \infty)$. ∎

Input and output

Independent and dependent variables

The value $f(x)$ of a function f at x is also called the *output*; x is called the *input* or *argument*. If $y = f(x)$, the symbol x is called the *independent variable* and the symbol y is called the *dependent variable*.

If both the inputs and outputs of a function are numbers, we shall call the function *numerical*. In some more advanced courses such a function is also called a real function of a real variable.

Several of the keys on a hand-held calculator correspond to functions. One such function is the $\sqrt{x}$-key. The input can be any nonnegative number. If you punch in a negative number, the calculator may flash a warning that you have given it a number not in the domain of the function. The reciprocal or $1/x$-key represents another function. The sum or +-key represents a function of a different type. In this case the input is a pair of numbers x and y, and the output is their sum, a single number $x + y$.

THE GRAPH OF A NUMERICAL FUNCTION

If both the domain and range of a function consist of real numbers it is possible to draw a picture that displays the behavior of the function.

DEFINITION *Graph* of a numerical function. Let f be a numerical function. The *graph* of f consists of those points (x, y) such that $y = f(x)$.

For instance, the graph of the squaring function $f(x) = x^2$ consists of the points (x, y) such that $y = x^2$. It is the parabola sketched in Example 4 of Sec. 1.1. The graph of the cubing function $y = x^3$ is shown in Example 5 of Sec. 1.1. The next example shows how to use a table to graph a function.

EXAMPLE 5 Graph the function $f(x) = 1/(1 + x^2)$.

SOLUTION Since $1 + x^2$ is never 0, the domain of the function consists of the entire x axis. Pick a few convenient inputs x and calculate the corresponding outputs, as shown in this table:

x	0	1	2	3
$f(x) = \dfrac{1}{1 + x^2}$	1	$\tfrac{1}{2}$	$\tfrac{1}{5}$	$\tfrac{1}{10}$

For any x, $x^2 \geq 0$, so $1 + x^2 \geq 1$ and $\dfrac{1}{1 + x^2} \leq 1$.

Since x appears only to an even power, the graph is symmetric with respect to the y axis; there is no need to evaluate the function for negative x. Plotting the four calculated points suggests the general shape of the graph. (See

Fig. 1.28.) When $|x|$ is large, $y = 1/(1 + x^2)$ is small. This means that for large $|x|$ the graph approaches the x axis.

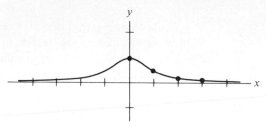

Figure 1.28

Note that the range of the function consists of all positive numbers less than or equal to 1. ∎

Not every curve is the graph of a function. For instance, the circle $x^2 + y^2 = 169$ shown in Example 3 of Sec. 1.1 is not the graph of a function. The reason is that a function assigns to a given input a *single* number as the output. A line parallel to the y axis therefore meets the graph of a function in at most one point. A glance at the graph of $x^2 + y^2 = 169$ will show that there are many lines parallel to the y axis that meet the graph in more than one point.

There is a visual test for deciding whether a curve in the plane is the graph of a function $y = f(x)$:

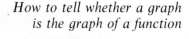

If each line parallel to the y axis either misses the curve or else meets the curve in only one point, then the curve is the graph of a function $y = f(x)$. If some line parallel to the y axis meets the curve at least twice, then the curve is not the graph of a function $y = f(x)$. (See Fig. 1.29.)

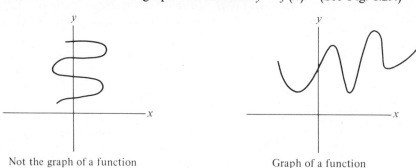

Not the graph of a function   Graph of a function

Figure 1.29

The graph of $x^2 + y^2 = 169$ can be cut into two pieces, each of which is the graph of a function. To see this, solve the equation $x^2 + y^2 = 169$ for y:

$$y^2 = 169 - x^2$$

$$y = \sqrt{169 - x^2} \quad \text{or} \quad y = -\sqrt{169 - x^2}.$$

In Fig. 1.30 the graph of the function $y = \sqrt{169 - x^2}$ is the top semicircle, while the graph of $y = -\sqrt{169 - x^2}$ is the bottom semicircle. Note that the domain of the function $f(x) = \sqrt{169 - x^2}$ is the closed interval $[-13, 13]$ and its range is $[0, 13]$.

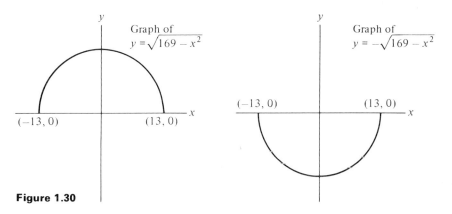

Figure 1.30

VARIOUS VIEWS OF A NUMERICAL FUNCTION

As already mentioned, we may think of a numerical function $y = f(x)$ as a set in the plane that meets each line parallel to the y axis at most once. The point (x, y) records the information that when the input is x, the output is y, or $f(x)$, as in Fig. 1.31.

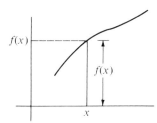

Figure 1.31

A function as a table with two rows

A function can also be thought of as a table consisting of two rows, one row for inputs and one row for outputs. There may be repetitions in the outputs but not in the inputs. Many functions are given this way. For instance, the following table records the rate in millions of barrels per day at which the United States imported petroleum from 1948 to 1980. The corresponding graph is shown in Fig. 1.32.

Year	1948	1949	1950	1951	1952	1953	1954	1955	1956	1957	1958
Imports	0.51	0.65	0.85	0.84	0.95	1.03	1.05	1.25	1.44	1.57	1.70
Year	1959	1960	1961	1962	1963	1964	1965	1966	1967	1968	1969
Imports	1.78	1.82	1.92	2.08	2.12	2.26	2.47	2.57	2.54	2.84	3.17
Year	1970	1971	1972	1973	1974	1975	1976	1977	1978	1979	1980
Imports	3.42	3.93	4.74	6.26	6.11	6.06	7.31	8.81	5.57	5.84	5.79

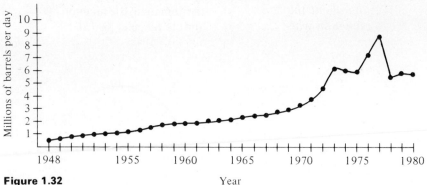

Figure 1.32

A function as an input-output machine

It is sometimes helpful to think of a function as a machine. When you insert the number x into the machine, the number $f(x)$ falls out, as illustrated in Fig. 1.33.

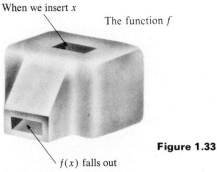

Figure 1.33

A function as a projector from line to line

On several occasions it will be illuminating to picture a function f as a projection from a slide to a screen. Both the slide and screen are lines. The function, through some ingenious lens, projects the point on the slide given by the number x to the point on the screen given by the number $f(x)$, as shown in Fig. 1.34.

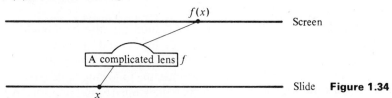

Figure 1.34

EVALUATING A FUNCTION

Beginning in Chap. 2 we shall examine how the output of a function changes as we change the input. To prepare for the algebraic manipulations that will be needed, we pause now to illustrate some of the techniques.

EXAMPLE 6 Let f be the squaring function $f(x) = x^2$. Compute $f(2+3)$.

SOLUTION For any number x, $f(x)$ is the square of that number. Thus
$$f(2+3) = (2+3)^2$$
$$= 5^2 = 25. \blacksquare$$

Warning about the value of a function when the input is a sum

Warning: A common error is to assume that $f(2 + 3)$ is somehow related to $f(2) + f(3)$. For most functions there is no relation between the two numbers. In the case of the function x^2, $f(2) + f(3) = 2^2 + 3^2 = 4 + 9 = 13$, but, as was shown in Example 6, $f(2 + 3) = 25$.

EXAMPLE 7 Let f be the cubing function $f(x) = x^3$. Evaluate $f(2 + 0.1) - f(2)$.

SOLUTION
$$f(2 + 0.1) - f(2) = f(2.1) - f(2)$$
$$= (2.1)^3 - 2^3$$
$$= 9.261 - 8$$
$$= 1.261. \quad \blacksquare$$

In Examples 6 and 7 the inputs are specific numbers. Often the inputs will be indicated by algebraic expressions instead. The next three examples show how to deal with such inputs.

EXAMPLE 8 Let f be the squaring function $f(x) = x^2$. Simplify the expression
$$f(a + b) - f(a) - f(b)$$
as far as possible.

SOLUTION
$$f(a + b) - f(a) - f(b) = (a + b)^2 - a^2 - b^2$$
$$= a^2 + 2ab + b^2 - a^2 - b^2$$
$$= 2ab. \quad \blacksquare$$

EXAMPLE 9 Let $f(x) = x^3$. Let d and c be two distinct numbers. Express the following quotient as simply as possible:
$$\frac{f(d) - f(c)}{d - c}.$$

SOLUTION

This uses the factorization of $d^n - c^n$ in Appendix B.

$$\frac{f(d) - f(c)}{d - c} = \frac{d^3 - c^3}{d - c} \qquad \text{definition of } f$$
$$= \frac{(d - c)(d^2 + dc + c^2)}{d - c} \qquad \text{factoring}$$
$$= d^2 + dc + c^2 \qquad \text{canceling.} \quad \blacksquare$$

EXAMPLE 10 Let $f(x) = 1/(x + 1)$. Express the quantity
$$\frac{f(x + h) - f(x)}{h}$$
as simply as possible. (Assume that $h \neq 0$, $x \neq -1$, and $x + h \neq -1$ to avoid division by 0.)

SOLUTION

$$\frac{f(x+h)-f(x)}{h} = \frac{\dfrac{1}{x+h+1} - \dfrac{1}{x+1}}{h} \quad \text{definition of } f$$

$$= \frac{\dfrac{(x+1)-(x+h+1)}{(x+1)(x+h+1)}}{h} \quad \text{algebra}$$

$$= \frac{(x+1)-(x+h+1)}{(x+1)(x+h+1)h} \quad \text{algebra: } \frac{\left(\dfrac{a}{b}\right)}{c} = \frac{a}{bc}$$

$$= \frac{x+1-x-h-1}{(x+1)(x+h+1)h} \quad \text{algebra}$$

$$= \frac{-h}{(x+1)(x+h+1)h} \quad \text{algebra}$$

$$= \frac{-1}{(x+1)(x+h+1)} \quad \text{canceling } h. \quad \blacksquare$$

Exercises

In Exercises 1 to 10 graph the functions.
1 $f(x) = 3x$
2 $f(x) = -2x$
3 $f(x) = 3x^2$
4 $f(x) = 1 + x^2$
5 $f(x) = -x^2 + 1$
6 $f(x) = -3x^2 + 2$
7 $f(x) = x^2 - x$
8 $f(x) = x^2 + 2x + 1$
9 $f(x) = 2/(1 + x^2)$
10 $f(x) = 1/(1 + 2x^2)$

In Exercises 11 to 20 describe the domain and range of each function.
11 $f(x) = \sqrt{x}$
12 $f(x) = \sqrt{x+1}$
13 $f(x) = \sqrt{4-x^2}$
14 $f(x) = \sqrt{4+x^2}$
15 $f(x) = 3/x^2$
16 $f(x) = 1/(x+1)$
17 $f(x) = 1/x^3$
18 $f(x) = 1/x^4$
19 $f(x) = 1/\sqrt{x}$
20 $f(x) = 1/(1-x^2)$

In Exercises 21 and 22 which sets are the graphs of functions and which are not?

21

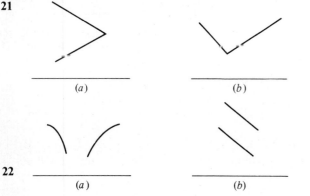

22

In each of Exercises 23 to 28 compute as decimals the outputs of the given function for the given inputs.
23 $f(x) = 3x$ (a) 2 (b) -1 (c) 0 (d) $\tfrac{1}{3}$
24 $f(x) = x^3 - x$ (a) 1 (b) 0 (c) -1 (d) 10
25 $f(x) = x + 1$ (a) -1 (b) 3 (c) 1.25 (d) 0
26 $f(x) = 1/(1 + x)$ (a) -3 (b) 3 (c) 9 (d) 99
27 $f(x) = x^3$ (a) $1 + 2$ (b) $4 - 1$
28 $f(x) = 1/x^2$ (a) $5 - 3$ (b) $4 - 6$

In Exercises 29 to 36 for the given functions evaluate and simplify the given expressions. (Assume that no denominator is 0.)
29 $f(x) = x^3; f(a+1) - f(a)$

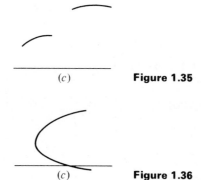

Figure 1.35

Figure 1.36

30 $f(x) = 1/x$; $f(a+h) - f(a)$

31 $f(x) = 1/x^2$; $\dfrac{f(d) - f(c)}{d - c}$

32 $f(x) = 1/(2x+1)$; $\dfrac{f(x+h) - f(x)}{h}$

33 $f(x) = 3x + 2$; $\dfrac{f(x+h) - f(x-h)}{2h}$

34 $f(x) = 5x^2 + 3x + 2$; $\dfrac{f(a+b) - f(a)}{b}$

35 $f(x) = x + 1/x$; $\dfrac{f(u) - f(v)}{u - v}$

36 $f(x) = 3 - 1/x$; $\dfrac{f(x+h) - f(x)}{h}$

■

37 Graph $y = x^2 + x$ after filling in this table.

x	-2	-1	0	$\tfrac{1}{2}$	1	2	3
$x^2 + x$							

38 Graph $f(x) = x(x+1)(x-1)$.
(a) For which values of x is $f(x) = 0$?
(b) Where does the graph cross the x axis?
(c) Where does the graph cross the y axis?

39 A complicated lens projects a linear slide to a linear screen as shown in Fig. 1.37, which indicates the paths of four of the light rays. Let $f(x)$ be the image on the screen of x on the slide.
(a) What are $f(0), f(1), f(2), f(3)$?
(b) Fill in this table.

x	0	1	2	3
$f(x)$				

(c) Plot the four points in (b).
(d) Which is larger, $f(3) - f(2)$ or $f(2) - f(1)$?

40 The graph in Fig. 1.38 shows the average price of a gallon of gasoline (including taxes) in the United States as a function of time. Thus $f(x)$ is the price of a gallon of gasoline in year x. Estimate $f(x+5)/f(x)$ for (a) $x = 1960$, (b) 1970, (c) 1975.

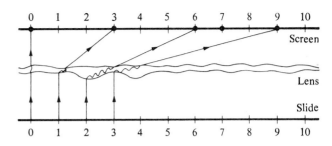

Figure 1.37

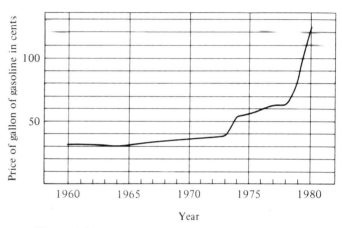

Figure 1.38

41 (Calculator) Let $f(x) = x^2$. Evaluate
$$[f(3+h) - f(3)]/h$$
to four decimal places for h equal to (a) 1, (b) 0.01, (c) -0.01, (d) 0.0001.

42 (Calculator) Let $f(x) = x^3$. Evaluate
$$[f(2+h) - f(2)]/h$$
to four decimal places for h equal to (a) 1, (b) 0.05, (c) 0.0001, (d) -0.001.

43 For which of the following functions is $f(a+b)$ equal to $f(a) + f(b)$ for all positive numbers a and b?
(a) $f(x) = x^2$ (b) $f(x) = 3x$ (c) $f(x) = -4x$
(d) $f(x) = \sqrt{x}$ (e) $f(x) = 2x + 1$

44 For which of the following functions is $f(ab)$ equal to $f(a)f(b)$ for all positive numbers a and b?
(a) $f(x) = 3x$ (b) $f(x) = x^3$ (c) $f(x) = 1/x$
(d) $f(x) = \sqrt{x}$ (e) $f(x) = x + 1$

1.4 Composite functions

This section describes a way of building up functions by applying one function to the output of another. For instance, the function
$$y = (1 + x^2)^{100}$$

is built up by raising $1 + x^2$ to the one-hundredth power. That is,

$$y = u^{100}, \quad \text{where } u = 1 + x^2.$$

Similarly, the function

$$y = \sqrt{1 + 2x^2}$$

is built up by taking the square root of $1 + 2x^2$,

$$y = \sqrt{u}, \quad \text{where } u = 1 + 2x^2.$$

The theme common to these two examples is spelled out in the following definition.

COMPOSITE FUNCTIONS

DEFINITION *Composition of functions.* Let f and g be functions. The function that assigns to each x the value

$$f(g(x))$$

is called the *composition of f and g* and is denoted $f \circ g$. It is defined for each x such that $g(x)$ is in the domain of f. Thus, if

$$y = f(u) \quad \text{and} \quad u = g(x),$$

then

$$y = h(x),$$

where

$$h(x) = f(g(x)),$$

or

$$y = (f \circ g)(x).$$

($f \circ g$ is read as "f circle g.")

In ordinary English the definition says, "To compute $f \circ g$, first apply g, and then apply f to the result."

Thinking of functions as input-output machines, we may consider $f \circ g$ as the machine built by hooking the machine for f onto the machine for g, as shown in Fig. 1.39.

The output of the g machine $g(x)$, becomes the input for the f machine.

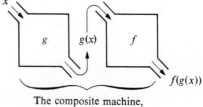

The composite machine, denoted h or $f \circ g$

Figure 1.39

For our purposes, it is also instructive to think of composite functions in terms of projections from slides to screens. If g is interpreted as some complicated projection from a slide to a screen (see Sec. 1.3) and f as a projection from that screen to a second screen, then $h = f \circ g$ is the projection from the slide to the second screen, as in Fig. 1.40.

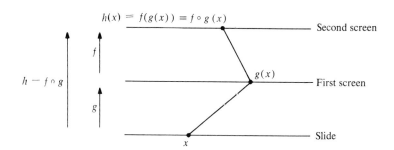

Figure 1.40

A function can be the composition of more than two functions. For example, $\sqrt{(1 + x^2)^5}$ is the composition of three functions. First $1 + x^2$ is formed; then the fifth power; then a square root. More formally, the assertion that $y = \sqrt{(1 + x^2)^5}$ is the same as saying that

$$y = \sqrt{u}, \quad u = v^5, \quad \text{and} \quad v = 1 + x^2.$$

EXAMPLE 1 Write $y = 2^{(x^2)}$ as a composition of functions.

SOLUTION $y = 2^u$, where $u = x^2$. ∎

EXAMPLE 2 Let $f(x) = 1 + 2x$ and $g(x) = x^2$. Compute $(f \circ g)(x)$ and $(g \circ f)(x)$. Are they equal?

SOLUTION
$$(f \circ g)(x) = f(g(x))$$
$$= f(x^2)$$
$$= 1 + 2x^2. \qquad (f \text{ of } any \text{ input is } 1 + \text{twice that input.})$$

$$(g \circ f)(x) = g(f(x))$$
$$= g(1 + 2x)$$
$$= (1 + 2x)^2. \qquad (g \text{ of } any \text{ input is the square of that input.})$$

Since the function $(1 + 2x)^2$ is not equal to $1 + 2x^2$, $f \circ g$ is not equal to $g \circ f$. ∎

EXAMPLE 3 Let $f(x) = -x$. Compute $(f \circ f)(x)$.

SOLUTION

$$(f \circ f)(x) = f(f(x))$$
$$= f(-x)$$
$$= -(-x)$$
$$= x.$$

Thus $(f \circ f)(x) = x$. ∎

EXAMPLE 4   Let f be the cubing function, $f(x) = x^3$, and g the cube-root function, $g(x) = \sqrt[3]{x}$. Compute $(f \circ g)(x)$ and $(g \circ f)(x)$.

SOLUTION

$$(f \circ g)(x) = f(g(x))$$
$$= f(\sqrt[3]{x})$$
$$= (\sqrt[3]{x})^3$$
$$= x.$$

$$(g \circ f)(x) = g(f(x))$$
$$= g(x^3)$$
$$= \sqrt[3]{x^3}$$
$$= x. \quad \blacksquare$$

Inverse functions   In Example 4, f and g "reverse" the effect of each other. For instance, $f(2) = 2^3 = 8$ and $g(8) = \sqrt[3]{8} = 2$. Two functions in this relation are said to be "inverses" of each other. The concept of inverse functions will be developed more fully, when needed, in Chap. 6.

Even and odd functions   Certain functions behave nicely when composed with the function $-x$. That is, their values at $-x$ are closely related to their values at x. The following definitions make this precise.

DEFINITION   *Even function.* A function f such that $f(-x) = f(x)$ is called an even function.

Consider, for instance, $f(x) = x^4$. We have

Recall that $(-1)^4 = 1$.
$$f(-x) = (-x)^4 = x^4 = f(x).$$

Thus $f(x) = x^4$ is an even function. In fact, for any *even* integer n, $f(x) = x^n$ is an even function (hence the name).

DEFINITION   *Odd function.* A function f such that $f(-x) = -f(x)$ is called an odd function.

The function $f(x) = x^3$ is odd since

$$f(-x) = (-x)^3 = -(x^3) = -f(x).$$

For any odd integer n, $f(x) = x^n$ is an odd function.

Most functions are neither even nor odd. For instance, $x^3 + x^4$ is neither even nor odd since $(-x)^3 + (-x)^4 = -x^3 + x^4$, which is neither $x^3 + x^4$ nor $-(x^3 + x^4)$. However, many functions used in calculus happen to be even or odd. The graph of such a function is symmetric with respect to the y axis or with respect to the origin, as will now be shown.

Symmetry of an even function Consider an even function f. Assume that the point (a, b) is on the graph of f. That means that $f(a) = b$. Since f is even, $f(-a) = b$. Consequently the point $(-a, b)$ is also on the graph of f. In other words, the graph of an even function is symmetric with respect to the y axis. (See Fig. 1.41.)

Symmetry of an odd function Let (a, b) be on the graph of an odd function f. That is, $f(a) = b$. Since f is odd, $f(-a) = -b$, which tells us that the point $(-a, -b)$ is also on the graph of f. Consequently the graph of an odd function is symmetric with respect to the origin. (See Fig 1.42.)

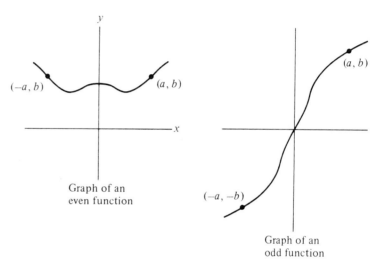

Figure 1.41   **Figure 1.42**

Exercises

In each of Exercises 1 to 4 find the function $y = f(x)$ defined by the composition of the given functions.
1. $y = u^3$, $u = x^2$
2. $y = 1 + u^2$, $u = 1 + x$
3. $y = \sqrt{u}$, $u = 1 + 2v$, $v = x^3$
4. $y = 1/u$, $u = 3 + v$, $v = x^2$

In Exercises 5 to 8 express the given functions as compositions of two or more simpler functions.
5. $y = (1 + x^3)^2$
6. $y = \dfrac{1}{1 + x^2}$
7. $y = \sqrt[3]{\dfrac{1}{1 + x^2}}$
8. $y = [(1 + x^2)^5]^3$

In Exercises 9 to 12 compute the compositions of the given functions.
9. $w = y^3$, $y = 3t$
10. $w = v^{10}$, $v = u^4$, $u = 1 + x$
11. $w = \sqrt[5]{y}$, $y = 1 + u^2$, $u = 1/x$
12. $y = w^3$, $w = 1/u$, $u = 1 + 3x$

■

13. Let $f(x) = 1 + 2x$. For which constants a and b does the function $g(x) = a + bx$ commute with f, that is, $(g \circ f)(x) = (f \circ g)(x)$?

44 PRELIMINARIES

14 Which of the following functions are even? odd? neither?
(a) $3x^2 + 5x^4$, (b) $|x|$, (c) $|x+1|$, (d) $x^3 - 4x^5$,
(e) $x^2/(1+x)$, (f) $x^3/(1+x^5)$, (g) $1 + x^2 + x^5$,
(h) $4 + x^2$, (i) $3 + x^5$, (j) $x^3/(1+x^4)$,
(k) $x^2/(1+x^4)$

15 Show that the constant function $f(x) = 0$ is both odd and even.

■ ■

16 Let $f(x) = 2x^2 - 1$ and $g(x) = 4x^3 - 3x$. Show that $(f \circ g)(x) = (g \circ f)(x)$. [Rare indeed are pairs of polynomials that commute with each other, as you may convince yourself by trying to find more. Of course, any two powers, such as x^3 and x^4, commute. (The composition of x^3 and x^4 in either order is x^{12}, as may be checked.)]

17 Assuming that 0 is in the domain of the odd function f, find $f(0)$.

18 Let $f(x) = 1/(1-x)$. What is the domain of f? of $f \circ f$? of $f \circ f \circ f$? Show that $(f \circ f \circ f)(x) = x$ for all x in the domain of $f \circ f \circ f$.

1.S Summary

This chapter discussed graphs and functions.

Section 1.1, on graphs, described algebraically some of the more common curves: the circle, parabola, ellipse, and hyperbola.

Lines were the subject of Sec. 1.2. The slope of a line was defined, and the line $y = mx + b$ was shown to have slope m.

Section 1.3 defined functions and suggested various ways of thinking of them when they are numerical: as a set in the xy plane that meets each line parallel to the y axis at most once; as a table of inputs and outputs; as a screen-to-slide projection.

Section 1.4 described how complicated functions can be built up from simpler functions by composition.

VOCABULARY AND SYMBOLS

coordinate system, x axis, y axis	slope
abscissa or x coordinate	function
ordinate or y coordinate	domain and range
origin	input, argument
quadrant	output, value
distance formula	composite function
graph	even function
symmetry	odd function
intercepts	

KEY FACTS The distance d between points (x_1, y_1) and (x_2, y_2) is given by the distance formula

$$d = \sqrt{(x_2 - x_1)^2 + (y_2 - y_1)^2}.$$

It is often preferable to rewrite this as

$$d^2 = (x_2 - x_1)^2 + (y_2 - y_1)^2$$

to avoid square roots.

AIDS IN GRAPHING

Find any x or y intercepts.
Determine whether there are any symmetries.

SOME COMMON GRAPHS

$y = mx + b$	line
$y = 0$	x axis
$x = 0$	y axis
$y = ax^2, a \neq 0$	parabola ($\cup$-shaped if $a > 0$, $\cap$-shaped if $a < 0$)
$y = x^3$	cubic ($\sim$-shaped)
$y = 1/x$	hyperbola
$xy = a, a \neq 0$	hyperbola
$x^2 + y^2 = r^2$	circle of radius r, center $(0, 0)$
$\dfrac{x^2}{a^2} + \dfrac{y^2}{b^2} = 1$	ellipse
$\dfrac{x^2}{a^2} - \dfrac{y^2}{b^2} = 1$	hyperbola (To draw it, plot the intercepts $(a, 0)$ and $(-a, 0)$, the line through the origin and (a, b), and the line through the origin and $(a, -b)$. They are asymptotes.)

EQUATIONS OF A LINE

Given	Equation
Slope m and y intercept b	$y = mx + b$
Slope m and point (x_1, y_1)	$y - y_1 = m(x - x_1)$
Points (x_1, y_1) and (x_2, y_2)	Use preceding formula with $m = \dfrac{y_2 - y_1}{x_2 - x_1}.$

Lines with the same slope are parallel. Lines with slopes m_1 and m_2 are perpendicular if $m_1 m_2 = -1$.

FUNCTIONS A function f assigns to each input from one set an output in a second set. If both the inputs and outputs are real numbers, the function is called *numerical* or a *real function of a real variable*. The graph of a numerical function is usually a curve or line.

Care should be taken when evaluating a function, especially if the input is given by a formula. There is usually no relation between $f(a+b)$ and the outputs $f(a)$ and $f(b)$.

Guide quiz on chap. 1

1. Show that the points $(2, 2+\sqrt{3})$, $(5, 2)$, and $(2, 2-\sqrt{3})$ are vertices of an equilateral triangle by computing the distances between them.
2. Graph (a) $x^2 + y^2 = 3$ (b) $x^2 - y^2 = 1$ (Show asymptotes.)
   (c) $\dfrac{x^2}{1} + \dfrac{y^2}{4} = 1$ (d) $xy = 4$
   (e) $y = 3x - 2$ (f) $y = -3x^2$
3. (a) Graph the lines $y = 5x + 1$ and $y = -x/4 + 2$.
   (b) Are they perpendicular?
4. Let $f(x) = 2x^2 - 1/x$. Express the fraction
$$\frac{f(x+h) - f(x)}{h}$$
   as simply as possible. (Assume that none of h, x, $x+h$ are 0.)
5. Express the function $y = \sqrt{(1+2x)^3} + \sqrt{1+2x}$ as the composition of three functions.

Review exercises for chap. 1

In each of Exercises 1 to 6 give the domain and range of the function.
1. $x^{1/5}$
2. $x^{1/6}$
3. $x^{2/5}$
4. x^2
5. $1/\sqrt{x+1}$
6. $\sqrt{4-x^2}$

In Exercises 7 to 10 evaluate and express as decimals.

7. $f(2+0.1) - f(2)$ if $f(x) = x^2$
8. $f(2-0.1) - f(2)$ if $f(x) = x^2 + 3$
9. $f(4) - f(3)$ if $f(x) = x^3 + 2x$
10. $f(1.25) - f(1)$ if $f(x) = 1/x$

In Exercises 11 to 16 evaluate and simplify using algebra.

11. $\dfrac{f(a+h) - f(a)}{h}$ if $f(x) = 1/(x+1)$
12. $\dfrac{f(2+h) - f(2)}{h}$ if $f(x) = 3 + 2x + x^2$
13. $\dfrac{f(1+h) - f(1)}{h}$ if $f(x) = x^3$
14. $f(3b) - f(4b)$ if $f(x) = -x$
15. $\dfrac{f(u) - f(v)}{u - v}$ if $f(x) = x^3 - 3x - 2$
16. $f(x) + f(-x)$ if $f(x) = x$

In Exercises 17 to 30 graph the functions.

17. $\sqrt{4 - x^2}$
18. $-\sqrt{9 - x^2}$
19. $\sqrt{x^2 - 1}$
20. $-\sqrt{x^2 - 5}$
21. $x + |x|$
22. $|x| - \sqrt{x^2}$
23. $1/(x-1)^2$
24. $1/2^x$
25. $x^{5/2}$
26. $x^{2/5}$
    (Suggestion: Use a calculator.)
27. 3^{-x}
28. 1^x
29. $(x-1)^2$
30. $-x^2 + x + 4$
31. For which of the following functions is $f(b+1)$ equal to $3f(b)$?
    (a) $f(x) = 2^x$ (b) $f(x) = 3^x$
    (c) $f(x) = 3^x/2$ (d) $f(x) = x/3$
32. For which of the following functions is $f(a+b) = f(a)f(b)$?
    (a) $f(x) = -1$ for all x (b) $f(x) = 2^x$
    (c) $f(x) = 5x$ (d) $f(x) = 3^{-x}$
    (e) $f(x) = 1$ for all x (f) $f(x) = 3^x/7$

33. For which x is $x^3(x-1)(x+2)$ positive?
34. (Calculator) Graph $f(x) = x/2^x$.
35. (Calculator) Let $f(x) = (1+x)^{1/x}$ for $x > 0$.
    (a) Fill in this table:

x	0.001	0.01	0.1	1	2	10	100
$f(x)$							

    (b) With the aid of (a) graph f.
36. (Calculator) Let $f(x) = x^x$ for $x > 0$.
    (a) Evaluate the function for the inputs 1, 0.5, 0.4, 0.3, 0.1, and 0.001.
    (b) With the aid of (a) graph f.
37. Graph $f(x) = x^2 - 2x$.
38. Find the slope of the line through (a) (1.3, 2) and (1.4, 2.2), (b) (a, b) and (c, d), (c) (x, x^2) and (a, a^2).

39 (a) Sketch the lines $\frac{x}{3} + \frac{y}{4} = 1$ and $\frac{x}{4} - \frac{y}{3} = 1$.

  (b) Find their slopes.
  (c) Are they perpendicular?

40 Is the line through $(-2, 1)$ and $(3, 8)$ parallel to the line through $(3, -2)$ and $(10, 8)$?

41 A line has the equation $y = mx + b$. What can be said about m if the line
  (a) is nearly vertical (almost parallel to the y axis)?
  (b) is nearly horizontal (almost parallel to the x axis)?
  (c) slopes downward as you move to the right?
  (d) slopes upward as you move to the right?

42 Compute $f(2 + 0.5)$ $f(2)$ if (a) $f(x) = x^2 - 3x + 4$;
  (b) $f(x) = x/(x + 1)$.

43 Write in the form $y = mx + b$ the equation of the line through $(1, 2)$ and $(0, 5)$.

44 Show that $A = (0, 0)$, $B = (4, 2)$, $C = (6, -2)$, $D = (2, -4)$ are the vertices of a rectangle by demonstrating that
  (a) opposite sides AB and CD are parallel,
  (b) opposite sides BC and DA are parallel,
  (c) adjacent sides AB and BC are perpendicular.

■ ■

45 Find a point P on the x axis such that the line through P and $(1, 1)$ is perpendicular to the line through P and $(-3, 4)$.

LIMITS AND CONTINUOUS FUNCTIONS

2

This chapter develops the concept of a limit, which lies at the heart of calculus and provides the foundation for both the derivative and the integral. The first two sections present the basic definitions and properties of limits. Section 2.3 treats limits in graphing. Section 2.4, which is optional, offers a more precise formulation of limits. Section 2.5 evaluates an important and instructive trigonometric limit. The final two sections define the continuous functions and describe their fundamental properties.

REFERENCES TO APPENDIXES

In Sec. 2.1: Absolute value, open interval, closed interval, etc., Appendix A

In Sec. 2.2: Rationalizing an expression, the binomial theorem, Appendix B

In Sec. 2.4: Absolute value and inequalities, Appendix A

In Sec. 2.5: Radian measure, sin x, cos x, tan x, and $\sin^2 x + \cos^2 x = 1$, Appendix D

2.1 The limit of a function

Three examples will introduce the notion of the limit of a numerical function. After them, the concept of a limit will be defined.

EXAMPLE 1   Let $f(x) = 2x^2 + 1$. What happens to $f(x)$ as x is chosen closer and closer to 3?

SOLUTION   Let us make a table of the values of $f(x)$ for some choices of x near 3.

x	3.1	3.01	3.001	2.999	2.99	2.9
$f(x)$	20.22	19.1202	19.012002	18.988002	18.8802	17.82

When x is close to 3, $2x^2 + 1$ is close to $2 \cdot 3^2 + 1 = 19$. We say that "the limit of $2x^2 + 1$ as x approaches 3 is 19" and write

The "limit" notation

$$\lim_{x \to 3} (2x^2 + 1) = 19. \quad \blacksquare$$

Example 1 presented no obstacle. The next example offers a slight challenge.

EXAMPLE 2   Let $f(x) = (x^3 - 1)/(x^2 - 1)$. Note that this function is not defined when $x = 1$, for, when x is 1, both numerator and denominator are 0. But we have every right to ask: How does $f(x)$ behave when x is *near* 1 but is *not* 1 itself?

SOLUTION   First make a brief table of values of $f(x)$, to four decimal places, for x near 1. Choose some x larger than 1 and some x smaller than 1. For instance,

$$f(1.01) = \frac{1.01^3 - 1}{1.01^2 - 1} = \frac{1.030301 - 1}{1.0201 - 1} = \frac{0.030301}{0.0201} \approx 1.5075.$$

x	1.1	1.01	0.9	0.99
$\dfrac{x^3 - 1}{x^2 - 1}$	1.5762	1.5075	1.4263	1.4925

(If you have a calculator handy, evaluate $(x^3 - 1)/(x^2 - 1)$ at 1.001 and 0.999 as well.)

Two influences operate on
$\dfrac{x^3 - 1}{x^2 - 1}.$

There are two influences acting on the fraction $(x^3 - 1)/(x^2 - 1)$ when x is near 1. On the one hand, the numerator $x^3 - 1$ approaches 0; thus there is an influence pushing the fraction toward 0. On the other hand, the denominator $x^2 - 1$ also approaches 0; division by a small number tends to make a fraction large. How do these two opposing influences balance out?

The algebraic identities

$$x^3 - 1 = (x^2 + x + 1)(x - 1)$$

and

$$x^2 - 1 = (x + 1)(x - 1)$$

enable us to answer the question.

Rewrite the quotient $(x^3 - 1)/(x^2 - 1)$ as follows:

$$\frac{x^3 - 1}{x^2 - 1} = \frac{(x^2 + x + 1)(x - 1)}{(x + 1)(x - 1)}, \qquad x \neq 1$$

$$= \frac{x^2 + x + 1}{x + 1}, \qquad x \neq 1.$$

So the behavior of $(x^3 - 1)/(x^2 - 1)$ for x near 1, but not equal to 1, is the same as the behavior of $(x^2 + x + 1)/(x + 1)$ for x near 1, but not equal to 1. Thus

$$\lim_{x \to 1} \frac{x^3 - 1}{x^2 - 1} = \lim_{x \to 1} \frac{x^2 + x + 1}{x + 1}.$$

Now, as x approaches 1, $x^2 + x + 1$ approaches 3 and $x + 1$ approaches 2. Thus

$$\lim_{x \to 1} \frac{x^2 + x + 1}{x + 1} = \frac{3}{2},$$

from which it follows that

$$\lim_{x \to 1} \frac{x^3 - 1}{x^2 - 1} = \frac{3}{2}.$$

Note that $\frac{3}{2} = 1.5$, which is closely approximated by $f(1.01)$ and $f(0.99)$. ■

The notation $\to$ for "approaches"

The arrow $\to$ will stand for "approaches." According to Example 2,

$$\text{as } x \to 1, \quad \frac{x^3 - 1}{x^2 - 1} \to \frac{3}{2}.$$

This notation will be used in the next example and often later.

EXAMPLE 3   Consider the function f defined by

$$f(x) = \frac{x}{|x|}.$$

The domain of this function consists of every number except 0. For instance,

$$f(3) = \frac{3}{|3|} = \frac{3}{3} = 1,$$

and

$$f(-2) = \frac{-2}{|-2|} = \frac{-2}{2} = -1.$$

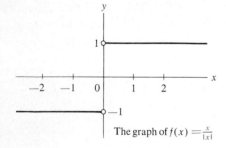

The graph of $f(x) = \frac{x}{|x|}$

Figure 2.1

The "hollow dot" notation for a missing point

When x is positive, $f(x) = 1$. When x is negative, $f(x) = -1$. This is shown in Fig. 2.1. The graph does not intersect the y axis, since f is not defined for $x = 0$. The hollow circles at $(0, 1)$ and $(0, -1)$ indicate that those points are not on the graph. What happens to $f(x)$ as $x \to 0$?

SOLUTION   As $x \to 0$ through positive numbers, $f(x) \to 1$, since $f(x) = 1$ for any positive number. As $x \to 0$ through negative numbers, $f(x) \to -1$, since $f(x) = -1$ for any negative number.

When x is near 0, it is *not* the case that $f(x)$ is near one specific number. Thus

$$\lim_{x \to 0} f(x)$$

does *not* exist, that is,
$$\lim_{x \to 0} \frac{x}{|x|}$$

does *not* exist. However, if $a \neq 0$,

$$\lim_{x \to a} f(x)$$

does exist, being 1 when a is positive and -1 when a is negative. Thus $\lim_{x \to a} f(x)$ exists for all a other than 0. ∎

Whether a function f has a limit at a has nothing to do with $f(a)$ itself. In fact a might not even be in the domain of f. See, for instance, Examples 2 and 3. In Example 1, $a\,(=3)$ happened to be in the domain of f, but that fact did not influence the reasoning. It is only the behavior of $f(x)$ for x near a that concerns us.

These three examples provide a background for describing the limit concept which will be used throughout the text.

A more formal definition of limit is to be found in Sec. 2.4.

Consider a function f and a number a which may or may not be in the domain of f. In order to discuss the behavior of $f(x)$ for x near a, we must know that the domain of f contains numbers arbitrarily close to a. Note how this assumption is built into each of the following definitions.

DEFINITION   *Limit of $f(x)$ at a.* Let f be a function and a some fixed number. Assume that the domain of f contains open intervals (c, a) and (a, b) for some number $c < a$ and some number $b > a$, as shown in Fig. 2.2. If, as x approaches a, both from the left and from the right, $f(x)$ approaches a specific number L, then L is called the *limit of $f(x)$ as x approaches a.*

Figure 2.2

This is written

$$\lim_{x \to a} f(x) = L,$$

or $\qquad$ as $\quad x \to a, \quad f(x) \to L.$

"$\lim_{x \to 3}$" *is text version of* "$\lim_{x \to 3}$"

In Example 1, we found $\lim_{x \to 3} (2x^2 + 1) = 19$, which illustrates this definition for $a = 3$ and $f(x) = 2x^2 + 1$. The fact that 3 happens to be in the domain of f is irrelevant. Example 2 showed that

$$\lim_{x \to 1} \frac{x^3 - 1}{x^2 - 1} = \frac{3}{2},$$

which illustrates this definition for $a = 1$ and $f(x) = (x^3 - 1)/(x^2 - 1)$. The fact that $f(x)$ is not defined for $x = 1$ did not affect the reasoning.

52 LIMITS AND CONTINUOUS FUNCTIONS

Example 3 concerns the behavior of $f(x) = x/|x|$ when x is near 0. As $x \to 0$, $f(x)$ does not approach a specific number. However, as x approaches 0 through positive numbers, $f(x) \to 1$. Also, as x approaches 0 through negative numbers, $f(x) \to -1$. This behavior illustrates the idea of a one-sided limit, which will now be defined.

DEFINITION

Right-hand limit

Right-hand limit of $f(x)$ at a. Let f be a function and a some fixed number. Assume that the domain of f contains an open interval (a, b) for some number $b > a$. If, as x approaches a from the right, $f(x)$ approaches a specific number L, then L is called the *right-hand limit of $f(x)$ as x approaches a*. This is written

$$\lim_{x \to a^+} f(x) = L$$

or   as $x \to a^+$, $f(x) \to L$.

The assertion that

$$\lim_{x \to a^+} f(x) = L$$

is read "the limit of f of x as x approaches a from the right is L" or "as x approaches a from the right, $f(x)$ approaches L."

Left-hand limit

The left-hand limit is defined similarly. The only difference is the demand that the domain of f contain an open interval of the form (c, a), for some number $c < a$; $f(x)$ is examined as x approaches a from the left. The notations for the left-hand limit are

$$\lim_{x \to a^-} f(x) = L$$

or   as $x \to a^-$, $f(x) \to L$.

As Example 3 showed,

$$\lim_{x \to 0^+} \frac{x}{|x|} = 1 \quad \text{and} \quad \lim_{x \to 0^-} \frac{x}{|x|} = -1.$$

We could also write, for instance,

as $x \to 0^+$, $\dfrac{x}{|x|} \to 1$.

Note that if both the right-hand and the left-hand limits of f exist at a and are equal, then $\lim_{x \to a} f(x)$ exists. But if the right-hand and left-hand limits are not equal, then $\lim_{x \to a} (x)$ does not exist.

The next example reviews the three limit concepts.

EXAMPLE 4   Figure 2.3 shows the graph of a function f whose domain is the closed interval $[0, 5]$.

2.1 THE LIMIT OF A FUNCTION

(a) Does $\lim_{x \to 1} f(x)$ exist?
(b) Does $\lim_{x \to 2} f(x)$ exist?
(c) Does $\lim_{x \to 3} f(x)$ exist?

SOLUTION (a) Inspection of the graph shows that

$$\lim_{x \to 1^-} f(x) = 1 \quad \text{and} \quad \lim_{x \to 1^+} f(x) = 2.$$

Though the two one-sided limits exist, they are not equal. Thus $\lim_{x \to 1} f(x)$ does not exist. In short, "f does not have a limit as $x \to 1$."
(b) Inspection of the graph shows that

$$\lim_{x \to 2^-} f(x) = 3 \quad \text{and} \quad \lim_{x \to 2^+} f(x) = 3.$$

Thus $\lim_{x \to 2} f(x)$ exists and is 3. Incidentally, the solid dot at $(2, 2)$ shows that $f(2) = 2$. This information, however, plays no role in our examination of the limit of $f(x)$ as $x \to 2$.
(c) Inspection shows that

$$\lim_{x \to 3^-} f(x) = 2 \quad \text{and} \quad \lim_{x \to 3^+} f(x) = 2.$$

Thus $\lim_{x \to 3} f(x)$ exists and is 2. Incidentally, the fact that $f(3) = 2$ is irrelevant in determining $\lim f(x)$. ∎

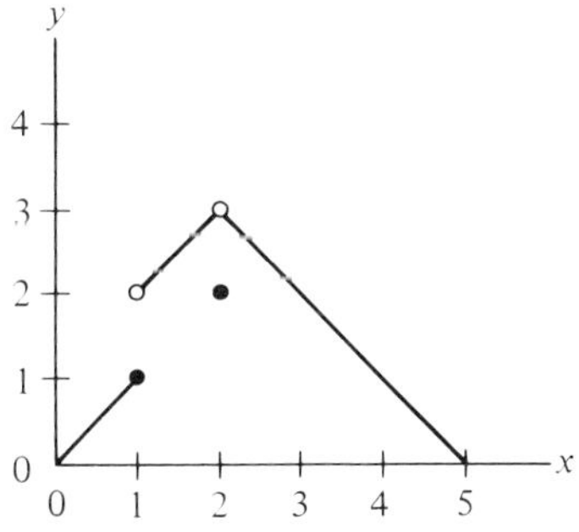

Figure 2.3

● means the point is present;
○ means that it is not.

EXAMPLE 5 Figure 2.4 shows the graph of a function whose domain is the closed interval $[0, 6]$.

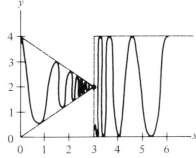

For $x < 3$, the graph goes up and down infinitely often between the dashed lines.

For $x > 3$, the graph goes up and down infinitely often between $y = 0$ and $y = 4$.

Figure 2.4

(a) Does $\lim_{x \to 3^-} f(x)$ exist?
(b) Does $\lim_{x \to 3^+} f(x)$ exist?

SOLUTION (a) As $x \to 3$ from the left, $f(x) \to 2$. Thus $\lim_{x \to 3^-} f(x)$ exists and is 2.

(b) As $x \to 3$ from the right, $f(x)$ does not approach one single number. Thus $\lim_{x \to 3+} f(x)$ does not exist. ∎

A function as wild as the one described in Example 5 will not be of major concern in calculus. However, it does serve to clarify the definitions of right-hand limit and left-hand limit, just as the notion of sickness illuminates our understanding of health.

One final point should be made. Consider

$$\lim_{x \to 3} \frac{1+x^2}{1+x^2}.$$

For all x, $(1+x^2)/(1+x^2) = 1$. Thus

$$\lim_{x \to 3} \frac{1+x^2}{1+x^2} = \lim_{x \to 3} 1$$
$$= 1.$$

More generally, if $f(x)$ is equal to a fixed number L for all x, then

$$\lim_{x \to a} f(x) = L.$$

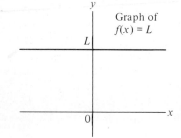

Figure 2.5

Constant functions

In this case, it may seem a little strange to say that "the limit of L is L" or "L approaches L." In practice, though, this offers no difficulty.

A function that assigns the same output to all inputs is called *constant*. Its graph is a line parallel to the x axis, as in Fig. 2.5. The function

$$f(x) = 7,$$

which assigns the output 7 to all inputs, is a constant function. Another instance of a constant function is

$$f(x) = 1^x.$$

Exercises

In Exercises 1 to 14 find the limits, all of which exist. Use intuition and, if needed, algebra.

1. $\lim_{x \to 5} (x + 7)$
2. $\lim_{x \to 1} (4x - 2)$
3. $\lim_{x \to 2} \frac{x^2 - 4}{x - 2}$
4. $\lim_{x \to 3} \frac{x^2 - 9}{x - 3}$
5. $\lim_{x \to 1} \frac{x^4 - 1}{x^3 - 1}$
6. $\lim_{x \to 1} \frac{x^6 - 1}{x^3 - 1}$
7. $\lim_{x \to 3} \frac{1}{x + 2}$
8. $\lim_{x \to 5} \frac{3x + 5}{4x}$
9. $\lim_{x \to 3} 25$
10. $\lim_{x \to 3} \pi^2$
11. $\lim_{x \to 0+} \sqrt{x}$
12. $\lim_{x \to 1+} \sqrt{4x - 4}$
13. $\lim_{x \to 1+} \frac{x - 1}{|x - 1|}$
14. $\lim_{x \to 1-} \frac{x - 1}{|x - 1|}$

In Exercises 15 to 24 decide whether the limits exist and, if they do, evaluate them.

15. $\lim_{h \to 1} \frac{(1 + h)^2 - 1}{h}$
16. $\lim_{h \to 0} \frac{(1 + h)^2 - 1}{h}$

17 $\lim_{x \to 2} \dfrac{\frac{1}{x} - \frac{1}{2}}{x - 2}$ 18 $\lim_{x \to 3} \dfrac{\frac{1}{x} - \frac{1}{2}}{x - 2}$

19 $\lim_{x \to 0} \dfrac{\sqrt{x + 4} - 2}{x}$ (*Hint*: Rationalize the numerator.)

20 $\lim_{x \to 0} \dfrac{\sqrt{x + 4} + 2}{x}$

21 $\lim_{x \to 4^+} (\sqrt{x - 4} + 2)$ 22 $\lim_{x \to 9} (\sqrt{x - 4} + \sqrt{x + 4})$

23 $\lim_{x \to 0} 64^x$ 24 $\lim_{x \to 1} \dfrac{3^x - 3}{2^x}$

In each of Exercises 25 and 26 there is a graph of a function. Decide which of the given limits exist, and evaluate those that do.

25 (See Fig. 2.6.) (*a*) $\lim_{x \to 0^+} f(x)$ (*b*) $\lim_{x \to 1} f(x)$

(*c*) $\lim_{x \to 2^-} f(x)$ (*d*) $\lim_{x \to 2^+} f(x)$

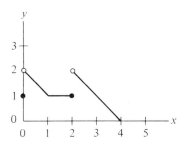

Figure 2.6

26 (See Fig. 2.7.) (*a*) $\lim_{x \to 1} f(x)$ (*b*) $\lim_{x \to 2} f(x)$

(*c*) $\lim_{x \to 3} f(x)$ (*d*) $\lim_{x \to 4^-} f(x)$

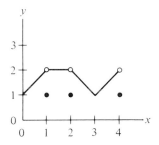

Figure 2.7

27 Find (*a*) $\lim_{x \to 4} (x - 4)$, (*b*) $\lim_{x \to 4} (\sqrt{x} - 2)$,

(*c*) $\lim_{x \to 4} \dfrac{x - 4}{\sqrt{x} - 2}$.

28 Find (*a*) $\lim_{x \to -2} (x + 2)$, (*b*) $\lim_{x \to -2} (x^3 + 8)$,

(*c*) $\lim_{x \to -2} \dfrac{x^3 + 8}{x + 2}$.

29 (Calculator) Let $f(x) = (3^x - 1)/(2^x - 1)$.
(*a*) Fill in this table:

x	1	0.1	0.01	0.001	-1	-0.01	-0.001
$f(x)$							

(*b*) On the basis of (*a*) do you think $\lim_{x \to 0} f(x)$ exists? If so, estimate this limit. (In Chap. 6 it will be shown to exist.)

30 (Calculator) Let $f(x) = (1 + x)^{1/x}$ for $x > -1$, $x \neq 0$.
(*a*) Compute $f(x)$ for $x = 1, 0.1, 0.01,$ and 0.001.
(*b*) Compute $f(x)$ for $x = -0.1, -0.01, -0.001$.
(*c*) It will be proved in Chap. 6 that $\lim_{x \to 0} f(x)$ exists. On the basis of (*a*) and (*b*) estimate its value.

31 Figure 2.8 shows a graph of a function that goes up and down infinitely often between the dashed lines, both to the right and to the left of 3. Does $\lim_{x \to 3} f(x)$ exist? If so, what is it?

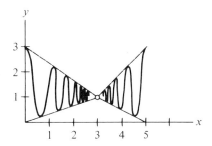

Figure 2.8

32 Define a certain function f as follows:

$$f(x) = \begin{cases} 1 & \text{if } x \text{ is an integer} \\ 0 & \text{if } x \text{ is not an integer} \end{cases}$$

(*a*) Graph f.
(*b*) Does $\lim_{x \to 3} f(x)$ exist?
(*c*) Does $\lim_{x \to 3.5} f(x)$ exist?
(*d*) For which numbers a does $\lim_{x \to a} f(x)$ exist?

33 (*a*) Graph the function f given by the formula

$$f(x) = |x| - x.$$

(*b*) For which numbers a does $\lim_{x \to a} f(x)$ exist?

56 LIMITS AND CONTINUOUS FUNCTIONS

34 Define f as follows:

$$f(x) = \begin{cases} x & \text{if } x \text{ is rational} \\ -x & \text{if } x \text{ is not rational} \end{cases}$$

(a) What does the graph of f look like? (A dotted curve may be used to indicate that points are missing.)
(b) Does $\lim_{x \to 1} f(x)$ exist?
(c) Does $\lim_{x \to \sqrt{2}} f(x)$ exist?
(d) Does $\lim_{x \to 0} f(x)$ exist?
(e) For which numbers a does $\lim_{x \to a} f(x)$ exist?

35 Define $f(x) = \begin{cases} x^2 & \text{if } x \text{ is rational} \\ x^3 & \text{if } x \text{ is irrational} \end{cases}$

(a) What does the graph of f look like? (See the advice for Exercise 34(a).)
(b) Does $\lim_{x \to 2} f(x)$ exist?
(c) Does $\lim_{x \to 1} f(x)$ exist?
(d) Does $\lim_{x \to 0} f(x)$ exist?
(e) For which numbers a does $\lim_{x \to a} f(x)$ exist?

36 (Calculator) Let $f(x) = x^x$ for $x > 0$.
(a) Fill in this table:

x	0.1	0.2	0.3	0.4	0.5	1.0
x^x						

(b) What do you think is the smallest value of x^x for x in $(0, 1)$?
(c) Do you think that $\lim_{x \to 0^+} x^x$ exists? If so, what do you think it is?

2.2 Computations of limits

Certain frequently used properties of limits should be put on the record. Let f and g be two functions and assume that

$$\lim_{x \to a} f(x) \quad \text{and} \quad \lim_{x \to a} g(x)$$

both exist. Then

Properties of limits

(1) $\lim_{x \to a} (f(x) + g(x)) = \lim_{x \to a} f(x) + \lim_{x \to a} g(x)$.

(2) $\lim_{x \to a} (f(x) - g(x)) = \lim_{x \to a} f(x) - \lim_{x \to a} g(x)$.

(3) $\lim_{x \to a} kf(x) = k \lim_{x \to a} f(x)$, for any constant k.

(4) $\lim_{x \to a} f(x)g(x) = \lim_{x \to a} f(x) \lim_{x \to a} g(x)$.

(5) $\lim_{x \to a} \dfrac{f(x)}{g(x)} = \dfrac{\lim_{x \to a} f(x)}{\lim_{x \to a} g(x)}$, if $\lim_{x \to a} g(x) \neq 0$.

(6) $\lim_{x \to a} f(x)^{g(x)} = \left(\lim_{x \to a} f(x) \right)^{\lim_{x \to a} g(x)}$ if $\lim_{x \to a} f(x) > 0$.

These properties have been tacitly assumed in Sec. 2.1. Properties (1) to (5) are treated in Appendix F, which employs the precise definitions of limits given in the optional Sec. 2.4. Property (6) depends on results in Sec. 6.10.

Property (1), for instance, asserts that $\lim_{x \to a} (f(x) + g(x))$ exists and equals the sum of the two given limits. This property extends to any finite sum of functions. For example, if $\lim_{x \to a} f(x)$, $\lim_{x \to a} g(x)$, and $\lim_{x \to a} h(x)$ exist, then

$$\lim_{x \to a} (f(x) + g(x) + h(x)) \text{ exists,}$$

and $\quad \lim_{x \to a} (f(x) + g(x) + h(x)) = \lim_{x \to a} f(x) + \lim_{x \to a} g(x) + \lim_{x \to a} h(x).$

Similarly, property (4) extends to the product of any finite number of functions.

EXAMPLE 1 If $\lim_{x \to 3} f(x) = 4$ and $\lim_{x \to 3} g(x) = 5$, discuss $\lim_{x \to 3} f(x)/g(x)$.

SOLUTION By property (5), $\lim_{x \to 3} f(x)/g(x)$ exists and

$$\lim_{x \to 3} \frac{f(x)}{g(x)} = \frac{4}{5}.$$

No further information about f and g is needed to determine the limit of $f(x)/g(x)$ as $x \to 3$. ∎

EXAMPLE 2 If $\lim_{x \to 3} f(x) = 0$ and $\lim_{x \to 3} g(x) = 0$, discuss $\lim_{x \to 3} f(x)/g(x)$.

SOLUTION In contrast to Example 1, in this case property (5) gives no information, since $\lim_{x \to 3} g(x) = 0$. It is necessary to have more information about f and g.

For instance, if

$$f(x) = x^2 - 9 \quad \text{and} \quad g(x) = x - 3,$$

then $\quad \lim_{x \to 3} f(x) = 0 \quad$ and $\quad \lim_{x \to 3} g(x) = 0$

and the limit of the quotient is:

$$\lim_{x \to 3} \frac{x^2 - 9}{x - 3} = \lim_{x \to 3} \frac{(x + 3)(x - 3)}{x - 3} \quad (x \neq 3)$$

$$= \lim_{x \to 3} (x + 3)$$

$$= 6.$$

Loosely put, "when x is near 3, $x^2 - 9$ is about 6 times as large as $x - 3$."

A different choice of f and g could produce a different limit for the quotient $f(x)/g(x)$. To be specific, let

$$f(x) = (x - 3)^2 \quad \text{and} \quad g(x) = x - 3.$$

Then $\quad \lim_{x \to 3} f(x) = 0 \quad$ and $\quad \lim_{x \to 3} g(x) = 0,$

and the limit of the quotient is

$$\lim_{x \to 3} \frac{(x - 3)^2}{x - 3} = \lim_{x \to 3} (x - 3) \quad (x \neq 3)$$

$$= 0.$$

If we know only that $f(x) \to 0$ and $g(x) \to 0$ as $x \to a$, we do not know how $f(x)/g(x)$ behaves as $x \to a$.

In this case we could say, "$(x-3)^2$ approaches 0 much faster than $x-3$ does, when $x \to 3$."

In short, the information that $\lim_{x \to 3} f(x) = 0$ and $\lim_{x \to 3} g(x) = 0$ is not enough to tell us how $f(x)/g(x)$ behaves as $x \to 3$. ∎

LIMITS AS $x \to \infty$

Sometimes it is useful to know how $f(x)$ behaves when x is a very large positive number (or a negative number of large absolute value). Example 3 serves as an illustration and introduces a variation on the theme of limits.

EXAMPLE 3 Determine how $f(x) = 1/x$ behaves for (a) large positive inputs, (b) negative inputs of large absolute value.

SOLUTION (a) First make a table of values:

x	10	100	1000
$1/x$	0.1	0.01	0.001

As x gets arbitrarily large, $1/x$ approaches 0.
(b) This is similar to (a). For instance

$$f(-1000) = -0.001.$$

As negative numbers x are chosen of arbitrarily large absolute value, $1/x$ approaches 0. ∎

The notation $\lim_{x \to \infty} f(x) = L$

Rather than writing, "as x gets arbitrarily large through positive values, $f(x)$ approaches the number L," it is customary to use the shorthand

$$\lim_{x \to \infty} f(x) = L.$$

This is read, "as x approaches infinity, $f(x)$ approaches L," or "the limit of $f(x)$ as x approaches infinity is L." For instance

Since ∞ is not a number, the case $x \to \infty$ is distinct from the case $x \to a$.

$$\lim_{x \to \infty} \frac{1}{x} = 0.$$

More generally, for any fixed positive exponent a,

$$\lim_{x \to \infty} \frac{1}{x^a} = 0.$$

The notation $\lim_{x \to -\infty} f(x) = L$

Similarly, the assertion that "as negative numbers x are chosen of arbitrarily large absolute value, $f(x)$ approaches the number L" is abbreviated to

$$\lim_{x \to -\infty} f(x) = L.$$

For instance

$$\lim_{x \to -\infty} \frac{1}{x} = 0.$$

2.2 COMPUTATIONS OF LIMITS

The six properties of limits stated at the beginning of the section hold when "$x \to a$" is replaced by "$x \to \infty$" or by "$x \to -\infty$."

It could happen that as $x \to \infty$, a function $f(x)$ becomes and remains arbitrarily large and positive. For instance, as $x \to \infty$, x^3 gets arbitrarily large. The shorthand for this is

The notation $\lim_{x \to \infty} f(x) = \infty$

$$\lim_{x \to \infty} f(x) = \infty.$$

For instance,
$$\lim_{x \to \infty} x^3 = \infty.$$

It is important, when reading the shorthand,

$$\lim_{x \to \infty} f(x) = \infty,$$

to keep in mind that "∞" is not a number. The limit does not exist. Properties (1) to (6) cannot, in general, be applied in such cases.

Other notations, such as $\lim_{x \to \infty} f(x) = -\infty$ or $\lim_{x \to -\infty} f(x) = \infty$ are defined similarly. For instance,

$$\lim_{x \to -\infty} x^3 = -\infty.$$

It can be shown that if, as $x \to \infty$, $f(x) \to \infty$ and $g(x) \to L > 0$, then $\lim_{x \to \infty} f(x)g(x) = \infty$. This fact is used in the next example.

EXAMPLE 4 Discuss the behavior of $2x^3 - 11x^2 + 12x$ when x is large.

SOLUTION First consider x positive and large. The three terms, $2x^3$, $-11x^2$, and $12x$, all become of large absolute value. To see how $2x^3 - 11x^2 + 12x$ behaves for large positive x, factor out x^3:

This factoring shows the importance of the highest power.

$$2x^3 - 11x^2 + 12x = x^3\left(2 - \frac{11}{x} + \frac{12}{x^2}\right). \tag{1}$$

Now, since $11/x$ and $12/x^2 \to 0$ as $x \to \infty$,

$$\lim_{x \to \infty}\left(2 - \frac{11}{x} + \frac{12}{x^2}\right) = 2.$$

Moreover, as $x \to \infty$, $x^3 \to \infty$. Thus

$$\lim_{x \to \infty} x^3\left(2 - \frac{11}{x} + \frac{12}{x^2}\right) = \infty;$$

hence
$$\lim_{x \to \infty} (2x^3 - 11x^2 + 12x) = \infty.$$

60 LIMITS AND CONTINUOUS FUNCTIONS

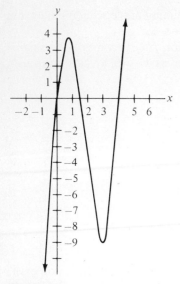

Figure 2.9

Now consider x negative and of large absolute value. The argument is similar. Use (1), and notice that $\lim_{x \to -\infty} x^3 = -\infty$ and

$$\lim_{x \to -\infty} \left(2 - \frac{11}{x} + \frac{12}{x^2}\right) = 2.$$

It follows that

$$\lim_{x \to -\infty} (2x^3 - 11x^2 + 12x) = -\infty.$$

This completes the discussion. It is of interest, however, to graph $f(x) = 2x^3 - 11x^2 + 12x$ and see what is happening for $|x|$ large. This is done in Fig. 2.9. Since $\lim_{x \to \infty} (2x^3 - 11x^2 + 12x) = \infty$ the graph rises arbitrarily high as $x \to \infty$. Since $\lim_{x \to -\infty} (2x^3 - 11x^2 + 12x) = -\infty$ the graph goes arbitrarily far down as $x \to -\infty$. ∎

General form of a polynomial

Example 4 generalizes to any polynomial. (A *polynomial* is a function of the form $a_n x^n + a_{n-1} x^{n-1} + \cdots + a_0$, where $a_0, a_1, \ldots, a_n$ are fixed real numbers and n is a nonnegative integer. If a_n is not 0, n is the *degree* of the polynomial. The numbers $a_0, a_1, \ldots, a_n$ are the *coefficients*.)

Limits of a polynomial as $x \to \infty$ or as $x \to -\infty$

Let $f(x)$ be a polynomial of degree at least 1 and with the lead coefficient a_n positive. Then

$$\lim_{x \to \infty} f(x) = \infty.$$

If the degree of f is even, then

$$\lim_{x \to -\infty} f(x) = \infty.$$

But if the degree of f is odd, then

$$\lim_{x \to -\infty} f(x) = -\infty.$$

EXAMPLE 5 Determine how

$$f(x) = \frac{x^3 + 6x^2 + 10x + 2}{2x^3 + x^2 + 5}$$

behaves for arbitrarily large positive numbers x.

SOLUTION As x gets large, the numerator $x^3 + 6x^2 + 10x + 2$ grows large, influencing the quotient to become large. On the other hand, the denominator also grows large, influencing the quotient to become small. An algebraic device will help reveal what happens to the quotient. We have

$$f(x) = \frac{x^3 + 6x^2 + 10x + 2}{2x^3 + x^2 + 5} = \frac{x^3\left(1 + \frac{6}{x} + \frac{10}{x^2} + \frac{2}{x^3}\right)}{x^3\left(2 + \frac{1}{x} + \frac{5}{x^3}\right)}$$

$$= \frac{1 + \frac{6}{x} + \frac{10}{x^2} + \frac{2}{x^3}}{2 + \frac{1}{x} + \frac{5}{x^3}}.$$

Now we can see what happens to $f(x)$ when x is large.

As x increases, $6/x \to 0$, $10/x^2 \to 0$, $2/x^3 \to 0$, $1/x \to 0$, and $5/x^3 \to 0$. Thus

$$f(x) \to \frac{1 + 0 + 0 + 0}{2 + 0 + 0} = \frac{1}{2}.$$

So, as x gets arbitrarily large through positive values, the quotient $(x^3 + 6x^2 + 10x + 2)/(2x^3 + x^2 + 5)$ approaches $\frac{1}{2}$.

In short $$\lim_{x \to \infty} \frac{x^3 + 6x^2 + 10x + 2}{2x^3 + x^2 + 5} = \frac{1}{2}. \blacksquare$$

The technique used in Example 5 applies to any function that can be written as the quotient of two polynomials. Such a function is called a *rational function*.

How to find the limit of a rational function as $x \to \infty$ or as $x \to -\infty$

Let $f(x)$ be a polynomial and let ax^n be its term of highest degree. Let $g(x)$ be another polynomial and let bx^m be its term of highest degree. Then

$$\lim_{x \to \infty} \frac{f(x)}{g(x)} = \lim_{x \to \infty} \frac{ax^n}{bx^m} \quad \text{and} \quad \lim_{x \to -\infty} \frac{f(x)}{g(x)} = \lim_{x \to -\infty} \frac{ax^n}{bx^m}.$$

(The proofs of these facts are similar to the argument used in Example 5.) In short, when working with the limit of a quotient of two polynomials as $x \to \infty$ or as $x \to -\infty$, disregard all terms except the one of highest degree in each of the polynomials. The next chapter illustrates this technique.

EXAMPLE 6 Examine the following limits:

(a) $\displaystyle\lim_{x \to \infty} \frac{3x^4 + 5x^2}{-x^4 + 10x + 5}$ (b) $\displaystyle\lim_{x \to \infty} \frac{x^3 - 16x}{5x^4 + x^3 - 5x}$

(c) $\displaystyle\lim_{x \to -\infty} \frac{x^4 + x}{6x^3 - x^2}$

SOLUTION By the preceding observations,

(a) $\displaystyle\lim_{x \to \infty} \frac{3x^4 + 5x^2}{-x^4 + 10x + 5} = \lim_{x \to \infty} \frac{3x^4}{-x^4} = \lim_{x \to \infty} (-3) = -3.$

LIMITS AND CONTINUOUS FUNCTIONS

(b) $\lim\limits_{x\to\infty} \dfrac{x^3 - 16x}{5x^4 + x^3 - 5x} = \lim\limits_{x\to\infty} \dfrac{x^3}{5x^4} = \lim\limits_{x\to\infty} \dfrac{1}{5x} = 0.$

(c) $\lim\limits_{x\to-\infty} \dfrac{x^4 + x}{6x^3 - x^2} = \lim\limits_{x\to-\infty} \dfrac{x^4}{6x^3} = \lim\limits_{x\to-\infty} \dfrac{x}{6} = -\infty.$ ∎

The technique of factoring out a power of x applies more generally than just to polynomials, as the next example illustrates.

EXAMPLE 7 Examine (a) $\lim\limits_{x\to\infty} \dfrac{\sqrt{3x^2 + x}}{x}$ and (b) $\lim\limits_{x\to-\infty} \dfrac{\sqrt{3x^2 + x}}{x}$.

SOLUTION Before beginning the solution, note that if x is positive, $\sqrt{x^2} = x$, but if x is negative, $\sqrt{x^2} = -x$.

Recall that $\sqrt{a^2} = |a|$.

(a) $\lim\limits_{x\to\infty} \dfrac{\sqrt{3x^2 + x}}{x} = \lim\limits_{x\to\infty} \dfrac{\sqrt{x^2(3 + x/x^2)}}{x}$

$= \lim\limits_{x\to\infty} \dfrac{x\sqrt{3 + 1/x}}{x}$

$= \lim\limits_{x\to\infty} \sqrt{3 + 1/x}$

$= \sqrt{3}.$

(b) $\lim\limits_{x\to-\infty} \dfrac{\sqrt{3x^2 + x}}{x} = \lim\limits_{x\to-\infty} \dfrac{\sqrt{x^2(3 + x/x^2)}}{x}$

$= \lim\limits_{x\to-\infty} \dfrac{-x\sqrt{3 + 1/x}}{x}$

$= \lim\limits_{x\to-\infty} -\sqrt{3 + 1/x}$

$= -\sqrt{3}.$ ∎

The final step in Example 7 deserves some comment, for it is a big leap that otherwise may pass unnoticed. It is assumed that

$$\lim\limits_{x\to-\infty} \sqrt{3 + \dfrac{1}{x}} = \sqrt{3}$$

because

$$\lim\limits_{x\to-\infty} \left(3 + \dfrac{1}{x}\right) = 3.$$

This conclusion depends on a property of the square-root function, which will now be described.

Let $f(x) = \sqrt{x}$ and let $g(x) = 3 + 1/x$. Then, in Example 7 it was assumed that

Taking the limit "inside"

$$\lim\limits_{x\to-\infty} f(g(x)) = f\left(\lim\limits_{x\to-\infty} g(x)\right).$$

For the functions f commonly met in calculus this switch of the order of "lim" and "f" is justified. Section 2.6 investigates this type of function f in detail.

EXAMPLE 8 Examine $\lim_{x \to \infty} (\sqrt{x^2 + x} - x)$.

SOLUTION As $x \to \infty$, both $\sqrt{x^2 + x}$ and x approach ∞. It is not immediately clear how their difference $\sqrt{x^2 + x} - x$ behaves. It is necessary to use a little algebra and rationalize the expression:

$$\lim_{x \to \infty} (\sqrt{x^2 + x} - x) = \lim_{x \to \infty} (\sqrt{x^2 + x} - x) \frac{(\sqrt{x^2 + x} + x)}{(\sqrt{x^2 + x} + x)}$$

$$= \lim_{x \to \infty} \frac{x^2 + x - x^2}{\sqrt{x^2 + x} + x}$$

$$= \lim_{x \to \infty} \frac{x}{\sqrt{x^2 + x} + x}$$

$$= \lim_{x \to \infty} \frac{x}{\sqrt{x^2(1 + 1/x)} + x}$$

$$= \lim_{x \to \infty} \frac{x}{x(\sqrt{1 + 1/x} + 1)}$$

$$= \lim_{x \to \infty} \frac{1}{\sqrt{1 + 1/x} + 1}$$

$$= \tfrac{1}{2}. \quad \blacksquare$$

The result in Example 8 may be surprising, but a little calculation would have suggested that the limit is $\tfrac{1}{2}$. For instance $\sqrt{100^2 + 100} - 100 \approx 100.499 - 100 = 0.499$.

INFINITE LIMITS AS $x \to a$

The next example concerns a case in which $f(x)$ becomes arbitrarily large as x approaches a fixed real number.

EXAMPLE 9 How does $f(x) = 1/x$ behave when x is near 0?

SOLUTION The reciprocal of a small number x has a large absolute value. For instance, when $x = 0.01$, $1/x = 100$; when $x = -0.01$, $1/x = -100$. Thus, as x approaches 0 from the right, $1/x$, which is positive, becomes arbitrarily large. The notation for this is

$$\lim_{x \to 0^+} \frac{1}{x} = \infty.$$

64 LIMITS AND CONTINUOUS FUNCTIONS

As x approaches 0 from the left, $1/x$, which is negative, has arbitrarily large absolute values. The notation for this is

$$\lim_{x \to 0^-} \frac{1}{x} = -\infty. \blacksquare$$

The behavior of $1/x$, described in Example 9, is quite different from that of $1/x^2$. Since x^2 is positive whether x is positive or negative, and since $1/x^2$ is large when x is near 0, we have

The notation "$\lim_{x \to 0} \frac{1}{x^2} = \infty$" is useful, though the limit does not exist since ∞ is not a number.

$$\lim_{x \to 0^+} \frac{1}{x^2} = \infty \quad \text{and} \quad \lim_{x \to 0^-} \frac{1}{x^2} = \infty.$$

In this case we may write

$$\lim_{x \to 0} \frac{1}{x^2} = \infty,$$

meaning that "as $x \to 0$, both from the right and from the left, $1/x^2$ becomes arbitrarily large through positive values." We can also write

$$\lim_{x \to 0} \frac{1}{|x|} = \infty,$$

but there is no corresponding statement for $\lim_{x \to 0} 1/x$.

AN EXPONENTIAL LIMIT The final example concerns the limit of an exponential function.

EXAMPLE 10 Examine $\lim_{x \to \infty} 1.01^x$.

SOLUTION Let $f(x) = 1.01^x$. As x increases, so does $f(x)$. This table lists a few of its values to three decimal places.

1.01^x grows slowly at first, then increases rapidly.

x	0	1	2	10	100	1000
$f(x) = 1.01^x$	1	1.01	1.020	1.105	2.705	20,959.156

The function grows slowly at first, but then quite rapidly.

With the aid of the binomial theorem, we can establish that as $x \to \infty$, $1.01^x \to \infty$. Since the function is increasing, consider only $x = n$, a positive integer. Then we have, for $n \geq 2$, that

$$1.01^n = (1 + \tfrac{1}{100})^n = 1 + n(\tfrac{1}{100}) + \text{more positive terms},$$

by the binomial theorem. Thus

2.2 COMPUTATIONS OF LIMITS

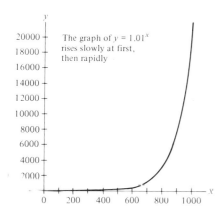

The graph of $y = 1.01^x$ rises slowly at first, then rapidly

Figure 2.10

$$1.01^n > 1 + \frac{n}{100},$$

for $n \geq 2$. As $n \to \infty$, $(1 + n/100) \to \infty$ also. Thus $\lim_{x \to \infty} 1.01^x = \infty$. The graph of $y = 1.01^x$ is shown in Fig. 2.10. ∎

A similar argument establishes that, if $b > 1$, then

$$\lim_{x \to \infty} b^x = \infty.$$

From this it follows that if $0 < b < 1$, then

$$\lim_{x \to \infty} b^x = 0.$$

(See Exercises 47 and 48.)

LIST OF VARIOUS LIMITS

The many different types of limits all have the same flavor. Rather than spell each out in detail, we list some of the typical cases that arise.

Notation	In words	Concept	Example
$\lim_{x \to a} f(x) = L$	As x approaches a, $f(x)$ approaches L	Discussed in Sec. 2.1.	$\lim_{x \to 3} (2x + 1) = 7$
$\lim_{x \to \infty} f(x) = L$	As x approaches (positive) infinity, $f(x)$ approaches L	$f(x)$ is defined for all x beyond some number and, as x gets large through positive values, $f(x)$ approaches L.	$\lim_{x \to \infty} \frac{1}{x} = 0$
$\lim_{x \to -\infty} f(x) = L$	As x approaches negative infinity, $f(x)$ approaches L	$f(x)$ is defined for all x to the left of some number and, as the negative number x takes on large absolute values, $f(x)$ approaches L.	$\lim_{x \to -\infty} \frac{x+1}{x} = 1$
$\lim_{x \to \infty} f(x) = \infty$	As x approaches infinity, $f(x)$ approaches positive infinity.	$f(x)$ is defined for all x beyond some number and, as x gets large through positive values, $f(x)$ becomes and remains arbitrarily large and positive.	$\lim_{x \to \infty} x^3 = \infty$
$\lim_{x \to a^+} f(x) = \infty$	As x approaches a from the right, $f(x)$ approaches (positive) infinity.	$f(x)$ is defined in some open interval (a, b), $b > a$, and, as x approaches a from the right, $f(x)$ becomes and remains arbitrarily large and positive.	$\lim_{x \to 0^+} \frac{1}{x} = \infty$
$\lim_{x \to a^+} f(x) = -\infty$	As x approaches a from the right, $f(x)$ approaches negative infinity.	$f(x)$ is defined in some open interval (a, b), $b > a$, and, as x approaches a from the right, $f(x)$ becomes negative and $|f(x)|$ becomes and remains arbitrarily large.	$\lim_{x \to 1^+} \frac{1}{1-x} = -\infty$
$\lim_{x \to a} f(x) = \infty$	As x approaches a, $f(x)$ approaches (positive) infinity.	$f(x)$ is defined for some open intervals (c, a) and (a, b), $c < a < b$, and, as x approaches a from either side, $f(x)$ becomes and remains arbitrarily large and positive.	$\lim_{x \to 0} \frac{1}{x^2} = \infty$

Other notations, such as

$$\lim_{x \to a} f(x) = -\infty. \qquad \lim_{x \to a^-} f(x) = \infty, \quad \text{and} \quad \lim_{x \to \infty} f(x) = -\infty$$

are defined similarly.

Exercises

In Exercises 1 to 39 examine the given limits and compute those which exist.

1. $\lim_{x \to \infty} (x^5 - 100x^4)$
2. $\lim_{x \to \infty} (-4x^5 + 35x^2)$
3. $\lim_{x \to -\infty} (6x^5 + 21x^3)$
4. $\lim_{x \to -\infty} (19x^6 + 5x - 300)$
5. $\lim_{x \to -\infty} (-x^3)$
6. $\lim_{x \to -\infty} (-x^4)$
7. $\lim_{x \to \infty} \dfrac{6x^3 - x}{2x^{10} + 5x + 8}$
8. $\lim_{x \to \infty} \dfrac{100x^9 + 22}{x^{10} + 21}$
9. $\lim_{x \to \infty} \dfrac{x^5 + 1066x^2 - 1492x}{2x^4 - 1982}$
10. $\lim_{x \to \infty} \dfrac{6x^3 - x^2 + 5}{3x^3 - 100x + 1}$
11. $\lim_{x \to \infty} \dfrac{x^3 + 1}{x^4 + 2}$
12. $\lim_{x \to -\infty} \dfrac{5x^3 + 2x}{x^{10} + x + 7}$
13. $\lim_{x \to 0^+} \dfrac{1}{x^4}$
14. $\lim_{x \to 0^-} \dfrac{1}{x^4}$
15. $\lim_{x \to 0^+} \dfrac{1}{x^3}$
16. $\lim_{x \to 0^-} \dfrac{1}{x^3}$
17. $\lim_{x \to \infty} (\sqrt{x^2 + 100} - x)$
18. $\lim_{x \to 1} (\sqrt{x^2 + 5} - \sqrt{x^2 + 3})$
19. $\lim_{x \to \infty} (\sqrt{x^2 + 100x} - x)$
20. $\lim_{x \to \infty} (\sqrt{x^2 + 100x} - \sqrt{x^2 + 50x})$
21. $\lim_{x \to \infty} \dfrac{\sqrt{4x^2 + 2x + 1}}{3x}$
22. $\lim_{x \to -\infty} \dfrac{\sqrt{9x^2 + x + 3}}{6x}$
23. $\lim_{x \to \infty} \dfrac{\sqrt{4x^2 + x}}{\sqrt{9x^2 - 3x}}$
24. $\lim_{x \to -\infty} \dfrac{\sqrt{x^2 + 3x + 1}}{\sqrt{16x^2 + x + 2}}$
25. $\lim_{x \to \infty} 2^x$
26. $\lim_{x \to \infty} (1.0001)^x$
27. $\lim_{x \to \infty} (2/3)^x$
28. $\lim_{x \to \infty} (0.999)^x$
29. $\lim_{x \to -\infty} 3^x$
30. $\lim_{x \to -\infty} (\sqrt{2})^x$

31. (a) $\lim_{x \to 1^+} 1/(x-1)$ (b) $\lim_{x \to 1^-} 1/(x-1)$
    (c) $\lim_{x \to 1} 1/(x-1)$
32. (a) $\lim_{x \to -1^-} 1/(x+1)^2$ (b) $\lim_{x \to -1^+} 1/(x+1)^2$
    (c) $\lim_{x \to -1} 1/(x+1)^2$
33. (a) $\lim_{x \to \infty} 3^x/4^x$ (b) $\lim_{x \to 0} 3^x/4^x$ (c) $\lim_{x \to -\infty} 3^x/4^x$
34. (a) $\lim_{x \to \infty} (2^x + 2^{-x})$ (b) $\lim_{x \to -\infty} (2^x + 2^{-x})$
    (c) $\lim_{x \to 0} (2^x + 2^{-x})$
35. $\lim_{x \to \infty} \dfrac{1 + 3^{-x}}{2 + 4^{-x}}$
36. $\lim_{x \to \infty} \dfrac{1 + 2^x}{2^x}$
37. $\lim_{x \to \infty} \dfrac{3^x + 4^x}{4^x}$
38. (a) $\lim_{x \to 1^+} \dfrac{1}{2 - 2^x}$ (b) $\lim_{x \to 1^-} \dfrac{1}{2 - 2^x}$
    (c) $\lim_{x \to 1} \dfrac{1}{2 - 2^x}$
39. (a) $\lim_{x \to 3^+} \dfrac{|x-3|}{x-3}$ (b) $\lim_{x \to 3^-} \dfrac{|x-3|}{x-3}$
    (c) $\lim_{x \to 3} \dfrac{|x-3|}{x-3}$

In Exercises 40 to 42 information is given about functions f and g. In each case decide whether the limit asked for can be determined on the basis of that information. If it can, give its value. If it cannot, show by specific choices of f and g that it cannot.

40. Given that $\lim_{x \to \infty} f(x) = 0$ and $\lim_{x \to \infty} g(x) = 1$, discuss
    (a) $\lim_{x \to \infty} (f(x) + g(x))$ (b) $\lim_{x \to \infty} |(f(x)/g(x))|$

(c) $\lim_{x \to \infty} f(x)g(x)$ (d) $\lim_{x \to \infty} (g(x)/f(x))$
(e) $\lim_{x \to \infty} g(x)/|f(x)|$

41. Given that $\lim_{x \to \infty} f(x) = \infty$ and $\lim_{x \to \infty} g(x) = \infty$, discuss
    (a) $\lim_{x \to \infty} (f(x) + g(x))$ (b) $\lim_{x \to \infty} (f(x) - g(x))$
    (c) $\lim_{x \to \infty} f(x)g(x)$ (d) $\lim_{x \to \infty} (g(x)/f(x))$

42. Given that $\lim_{x \to \infty} f(x) = 1$ and $\lim_{x \to \infty} g(x) = \infty$, discuss
    (a) $\lim_{x \to \infty} (f(x)/g(x))$ (b) $\lim_{x \to \infty} f(x)g(x)$
    (c) $\lim_{x \to \infty} (f(x) - 1)g(x)$

43. Let $P(x)$ be a polynomial of degree n, with lead term ax^n, $a > 0$, and let $Q(x)$ be a polynomial of degree m, with lead term bx^m, $b > 0$. Examine $\lim_{x \to \infty} P(x)/Q(x)$ if (a) $m = n$, (b) $m < n$, (c) $m > n$.

44. A function f is defined as follows: $f(x) = x$ if x is an integer, and $f(x) = -x$ if x is not an integer. (a) Graph f. (b) Examine $\lim_{x \to \infty} f(x)$. (c) Examine $\lim_{x \to \infty} |f(x)|$.

45. Examine $\lim_{x \to \infty} \cos x$. (Hint: Graph $y = \cos x$. See Appendix D if necessary.)

46. A function f is defined as follows: $f(x) = 2$ if x is an integer and $f(x) = 3$ if x is not an integer. (a) Graph f. (b) Discuss $\lim_{x \to \infty} f(x)$. (c) Discuss $\lim_{x \to 2} f(x)$.

47. Let $b > 1$. Show that $\lim_{x \to \infty} b^x = \infty$. *Hint:* Write $b = 1 + c$, $c > 0$, and use the binomial theorem to show that $(1 + c)^n > 1 + nc$ for $n \geq 2$.

48. Let $0 < b < 1$. Show that $\lim_{x \to \infty} b^x = 0$. *Hint:* Write b as $1/k$, where $k > 1$, and use Exercise 47.

2.3 Asymptotes and their use in graphing

Horizontal asymptotes

If $\lim_{x \to \infty} f(x) = L$, where L is a real number, the graph of $y = f(x)$ gets arbitrarily close to the horizontal line $y = L$ as x increases. The line $y = L$ is called a *horizontal asymptote* of the graph of f. An asymptote is defined similarly if $f(x) \to L$ as $x \to -\infty$.

Vertical asymptotes

If $\lim_{x \to a^+} f(x) = \infty$ or if $\lim_{x \to a^-} f(x) = \infty$, the graph of $y = f(x)$ resembles the vertical line $x = a$ for x near a. The line $x = a$ is called a *vertical asymptote* of the graph of f. A similar definition holds if $\lim_{x \to a^+} f(x) = -\infty$ or if $\lim_{x \to a^-} f(x) = -\infty$.

Figure 2.11 shows some of these asymptotes. The graph of $f(x) = 1/x$ in Fig. 1.15 of Sec. 1.1 has both horizontal and vertical asymptotes. Tilted asymptotes, such as those in the hyperbola shown in Fig. 1.14 of Sec. 1.1 can also occur.

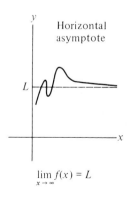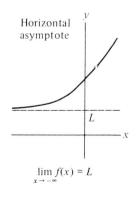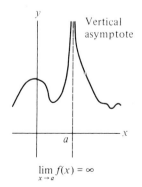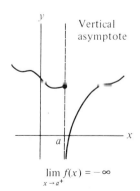

Figure 2.11

HORIZONTAL AND VERTICAL ASYMPTOTES IN GRAPHING

Some examples of graphing rational functions will show the usefulness of asymptotes.

EXAMPLE 1   Graph

$$f(x) = \frac{1}{x(x-1)}.$$

SOLUTION   Note that when $x = 0$ or when $x = 1$ the denominator $x(x-1)$ is 0. Thus 0 and 1 are not in the domain of the function. More important, when x is near 0 or near 1 the quotient $1/[x(x-1)]$ has large absolute value. Thus the lines $x = 0$ and $x = 1$ are vertical asymptotes for the graph. To decide how the graph approaches these asymptotes, it is necessary to examine the sign of $f(x)$ for x near 0 and for x near 1.

Consider x near 0. If x is a very small *positive* number, then

$$f(x) = \frac{1}{x}\frac{1}{x-1}$$

is the product of $1/x$, which is a large positive number, and $1/(x-1)$, which is near $1/(-1) = -1$. So, for x small and positive, $f(x)$ is negative and of large absolute value, that is,

$$\lim_{x \to 0^+} f(x) = -\infty.$$

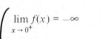

Figure 2.12

This fact is recorded in Fig. 2.12.

If x is near 0, but negative, then $1/x$ is negative and of large absolute value. Again $1/(x-1)$ is near -1. Thus

$$\lim_{x \to 0^-} f(x) = \lim_{x \to 0^-} \frac{1}{x}\frac{1}{x-1} = \infty.$$

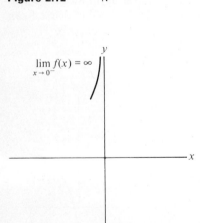

Figure 2.13

This fact is recorded in Fig. 2.13.

Next, how does $f(x)$ behave near $x = 1$? Consider first x near 1 but larger than 1. Then

$$f(x) = \frac{1}{x}\frac{1}{x-1}$$

is a large positive number, since $1/x$ is near 1 and $x - 1$ is a small positive number. Thus

$$\lim_{x \to 1^+} f(x) = \infty.$$

Similarly, $\quad\lim_{x \to 1^-} f(x) = -\infty.$

These two facts are recorded in Fig. 2.14. Piecing together these three figures suggests that the graph of f for x in or near the interval $(0, 1)$ looks

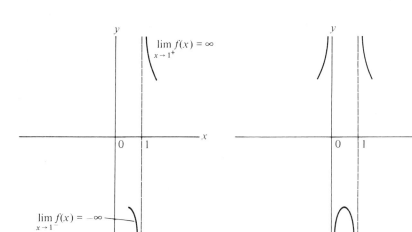

Figure 2.14

Figure 2.15

something like Fig. 2.15. (How high the curve goes for x in $(0, 1)$ is the type of question answered in Chap. 4 with the aid of the derivative. However, Exercise 21 describes a method that works for this particular function.)

How does $f(x)$ behave when $|x|$ is large? Since

$$\lim_{x \to \infty} \frac{1}{x(x-1)} = 0 \quad \text{and} \quad \lim_{x \to -\infty} \frac{1}{x(x-1)} = 0,$$

the x axis is a horizontal asymptote (both for x positive and for x negative.) In both cases the graph approaches the x axis from above, not from below, since the function is positive when $|x|$ is large. The graph of f must look something like Fig. 2.16.

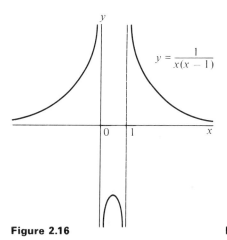

Figure 2.16

EXAMPLE 2 Using asymptotes, graph

$$f(x) = \frac{1}{(x-1)^2}.$$

70 LIMITS AND CONTINUOUS FUNCTIONS

The square of any nonzero real number is positive.

SOLUTION When $x = 1$, the function is undefined. However, when x is near 1, $1/(x-1)^2$ is a large positive number, since $(x-1)^2$ is a small positive number. Thus

$$\lim_{x \to 1} \frac{1}{(x-1)^2} = \infty.$$

This means that the graph of $f(x) = 1/(x-1)^2$ approaches the upper part of the vertical asymptote $x = 1$ both from the right and from the left as shown in Fig. 2.17.

Since

$$\lim_{x \to \infty} \frac{1}{(x-1)^2} = 0 \quad \text{and} \quad \lim_{x \to -\infty} \frac{1}{(x-1)^2} = 0,$$

the x axis is a horizontal asymptote.

All this information is incorporated in Fig. 2.18.

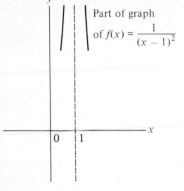

Figure 2.17

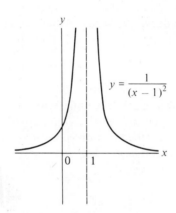

Figure 2.18 ■

OTHER ASYMPTOTES As mentioned, tilted asymptotes can occur, as in the case of the hyperbola in standard position. But much simpler graphs can have tilted asymptotes, as Example 3 illustrates.

EXAMPLE 3 Graph

$$f(x) = \frac{x^2 + 1}{x}.$$

SOLUTION After dividing x into $x^2 + 1$, we can write

$$f(x) = x + \frac{1}{x}.$$

When $|x|$ is large, $f(x)$ differs from x by the small quantity $1/x$. So when $|x|$ is large, the graph of f is close to the line $y = x$. When x is negative, $f(x) = x + 1/x$ is smaller than x, since $1/x$ is negative. So for x negative the graph of f lies below the line $y = x$. Similar reasoning

2.3 ASYMPTOTES AND THEIR USE IN GRAPHING

shows that for x positive the graph of f lies above the line $y = x$. This information is recorded in Fig. 2.19.

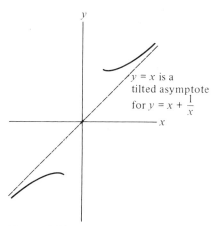

Figure 2.19

Next search for any vertical asymptotes. Near $x = 0$ the function becomes arbitrarily large. In fact,

$$\lim_{x \to 0^+} \left(x + \frac{1}{x} \right) = \infty \quad \text{and} \quad \lim_{x \to 0^-} \left(x + \frac{1}{x} \right) = -\infty.$$

The y axis is a vertical asymptote. The graph in Fig. 2.20 incorporates the information found from considering the tilted and vertical asymptotes.

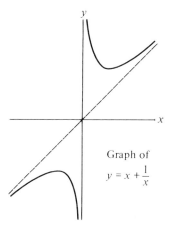

It is shown in Appendix E3 that the graph in Fig. 2.20 is a hyperbola. **Figure 2.20**

In the next example a curve, not a line, plays the role of an asymptote.

EXAMPLE 4 Graph $$f(x) = x^2 + \frac{1}{x}.$$

SOLUTION When x is large, $1/x$ is near 0. The graph of $f(x) = x^2 + 1/x$, therefore,

72 LIMITS AND CONTINUOUS FUNCTIONS

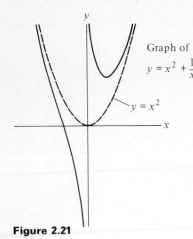

Graph of $y = x^2 + \frac{1}{x}$

$y = x^2$

Figure 2.21

resembles the graph of the parabola $y = x^2$ when x is large. Moreover, for positive x, $f(x)$ is larger than x^2 and, for negative x, $f(x)$ is smaller than x^2.

At $x = 0$ there is a vertical asymptote, and

$$\lim_{x \to 0^+} f(x) = \infty \quad \text{and} \quad \lim_{x \to 0^-} f(x) = -\infty.$$

The graph of f is shown in Fig. 2.21. ∎

Exercises

In Exercises 1 to 14 use asymptotes to sketch the graphs of the functions.

1 $f(x) = \dfrac{1}{x - 2}$

2 $f(x) = \dfrac{1}{x + 3}$

3 $f(x) = \dfrac{1}{(x + 1)^2}$

4 $f(x) = \dfrac{1}{(x + 1)^3}$

5 $y = \dfrac{1}{x^2 - x}$ (*Hint:* Factor the denominator.)

6 $y = \dfrac{1}{x^3 - x}$

7 $y = \dfrac{1}{x^4 - x^2}$

8 $y = \dfrac{1}{x^3 + x^2}$

9 $y = \dfrac{x(x - 1)}{x^2 + 1}$ (Also show the x intercepts.)

10 $y = \dfrac{(x - 1)(x - 2)}{x^2(x - 3)}$ (Also show the x intercepts.)

11 $y = \dfrac{x^3 + 2x^2 + x + 4}{x^2}$

12 $y = \dfrac{x^2 - 4}{x + 4}$ (Also show the x intercepts.)

13 $y = \dfrac{x^3}{x^2 + 1}$

14 $y = \dfrac{x^3}{x^2 - 1}$

15 Graph $y = 2^x + 2^{-x}$.

16 Graph $y = 2^x - 2^{-x}$.

17 Graph $y = \dfrac{x^2}{x^2 + 1}$.

18 Graph $y = x^3 + x^{-1}$.

19 Let a, b, and c be constants, $a \neq 0$. Show that $y = (ax^2 + bx + c)/(x + 1)$ has a tilted asymptote and give its equation.

20 Let $P(x)$ be a polynomial of degree m and $Q(x)$ be a polynomial of degree n. For which values of m and n does the graph of $y = P(x)/Q(x)$ have (a) a horizontal asymptote? (b) a tilted asymptote?

21 In Chap. 4 a general method for finding low and high points on curves will be given. However, elementary algebra can find the high point on the graph of $f(x) = 1/[x(x - 1)]$ in Example 1 for x in $(0, 1)$.
  (a) Show that $x(x - 1) = (x - \tfrac{1}{2})^2 - \tfrac{1}{4}$.
  (b) Using (a), find the minimum value of $x(x - 1)$ for all possible values of x.
  (c) Using (b), find the maximum value of $1/[x(x - 1)]$ for x in $(0, 1)$.
  (d) What are the coordinates of the highest point P on that part of the graph that lies between $x = 0$ and $x = 1$?

2.4 Precise definitions of limits (optional)

In the definitions of the limits considered in Secs. 2.1 and 2.2 appear such phrases as "x approaches a," "$f(x)$ approaches a specific number," "as x gets larger," and "$f(x)$ becomes and remains arbitrarily large." Such phrases, though appealing to the intuition and conveying the sense of a limit, are not precise. The definitions seem to suggest moving objects and call to mind the motion of a pencil point as it traces out the graph of a function.

This informal approach was adequate during the early development of calculus, from Leibniz and Newton in the seventeenth century through the Bernoullis, Euler, and Gauss in the eighteenth and early nineteenth centuries. But by the mid nineteenth century, mathematicians, facing more complicated functions and more difficult theorems, no longer could depend solely on intuition. They realized that glancing at a graph was no longer adequate to understand the behavior of functions—especially if theorems covering a broad class of functions were needed.

It was Weierstrass who developed, in the period 1841–1856, a way to define limits without any hint of motion or of pencils tracing out graphs. His approach, on which he lectured after joining the faculty at the University of Berlin in 1859, has since been followed by pure and applied mathematicians throughout the world. Even an undergraduate advanced calculus course depends on Weierstrass's approach.

In this section we examine how Weierstrass would define each of the three following concepts:

(1) $\lim_{x \to \infty} f(x) = \infty$, (2) $\lim_{x \to \infty} f(x) = L$, and (3) $\lim_{x \to a} f(x) = L$.

Throughout, "f" refers to a numerical function.

THE PRECISE DEFINITION OF $\lim_{x \to \infty} f(x) = \infty$

Recall the definition of "$\lim_{x \to \infty} f(x) = \infty$" given in the table in Sec. 2.2.

Informal *Informal definition of* $\lim_{x \to \infty} f(x) = \infty$: $f(x)$ is defined for all x beyond some number and, as x gets large through positive values, $f(x)$ becomes and remains arbitrarily large and positive.

To take us part way to the precise definition, let us reword the informal definition, paraphrasing it in the following definition, which is still informal.

Reworded *Reworded informal definition of* $\lim_{x \to \infty} f(x) = \infty$:

First,

there is a number c such that $f(x)$ is defined for all $x > c$.

Second,

if x is sufficiently large and positive, then
$f(x)$ is necessarily large and positive.

The precise definition parallels the reworded definition.

74 LIMITS AND CONTINUOUS FUNCTIONS

Precise

> *Precise definition* of $\lim_{x \to \infty} f(x) = \infty$:
>
> First,
>
> > there is a number c such that $f(x)$ is defined for all $x > c$.
>
> Second,
>
> > for each number E there is a number D such that for all $x > D$, it is true that
> >
> > $$f(x) > E.$$

The "challenge and reply" approach to limits

Think of the number E as a challenge (of an Enemy) and D as the reply (of a Defender). The *larger* E is, the *larger* D must usually be. Only if a number D (which depends on E) can be found for *every* number E, can we assert that "$\lim_{x \to \infty} f(x) = \infty$."

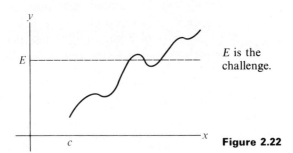

E is the challenge.

Figure 2.22

To picture the idea behind the precise definition, consider the graph in Fig. 2.22 of a function f for which $\lim_{x \to \infty} f(x) = \infty$. For each possible choice of a horizontal line, say, at height E, if you are far enough to the right on the graph of f, you stay above that line. That is, there is a number D such that if $x > D$, then $f(x) > E$, as illustrated in Fig. 2.23.

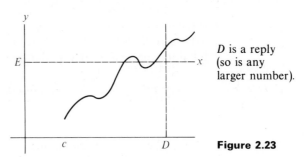

D is a reply (so is any larger number).

Figure 2.23

Examples 1 and 2 illustrate how the precise definition is used.

EXAMPLE 1 Using the precise definition, show that

$$\lim_{x \to \infty} 2x = \infty.$$

2.4 PRECISE DEFINITIONS OF LIMITS (OPTIONAL)

SOLUTION Let E be any number. We must show that there is a number D such that whenever $x > D$, it follows that $2x > E$. (For example, if $E = 100$, then $D = 50$ would do. It is indeed the case that if $x > 50$, then $2x > 100$.) The number D will depend on E.

Now, the inequality $2x > E$ is equivalent to

$$x > \frac{E}{2}.$$

In other words, if $x > E/2$, then $2x > E$. So $D = E/2$ suffices. That is, for $x > D \;(= E/2)$, $2x > E$. We conclude immediately that

$$\lim_{x \to \infty} 2x = \infty. \;\blacksquare$$

The response D is not unique.

In Example 1 a formula was provided for a suitable D in terms of E, namely $D = E/2$. For instance, when $E = 1000$, $D = 500$ suffices. In fact, any larger value of D also is suitable. If $x > 600$, it is still the case that $2x > 1000$ (since, then, $2x > 1200$). If one value of D is a satisfactory response to a given challenge E, then any larger value of D also is a satisfactory response.

EXAMPLE 2 Using the precise definition, show that

See Appendix D for a review of $\sin x$.

$$\lim_{x \to \infty} \left(\frac{x}{2} + \sin x\right) = \infty.$$

SOLUTION Let E be any number. We must exhibit a number D, depending on E, such that $x > D$ forces

$$\frac{x}{2} + \sin x > E. \tag{1}$$

Now, $\sin x \geq -1$ for all x. So, if we can force

$$\frac{x}{2} > E + 1, \tag{2}$$

then it will follow that

$$\frac{x}{2} + \sin x > E.$$

Inequality (2) is equivalent to

$$x > 2(E + 1).$$

Thus $D = 2(E + 1)$ will suffice. That is,

if $x > 2(E + 1)$, then $\frac{x}{2} + \sin x > E.$

76 LIMITS AND CONTINUOUS FUNCTIONS

To verify this assertion, we must check that $D = 2(E + 1)$ is a satisfactory reply to E. Assume that $x > 2(E + 1)$. Then

$$\frac{x}{2} > E + 1$$

and

$$\sin x \geq -1.$$

Adding these last two inequalities gives

If $a > b$ and $c \geq d$ then $a + c > b + d$.

$$\frac{x}{2} + \sin x > (E + 1) - 1,$$

or, simply,

$$\frac{x}{2} + \sin x > E,$$

which is (1).

Thus we are permitted to assert that

$$\lim_{x \to \infty} \left(\frac{x}{2} + \sin x \right) = \infty. \quad \blacksquare$$

The graph of the function examined in Example 2 appears in Fig. 2.24. Note that the function does not always increase as x increases. Nevertheless, as x increases, the function does tend to become and remain large, despite the small dips downward.

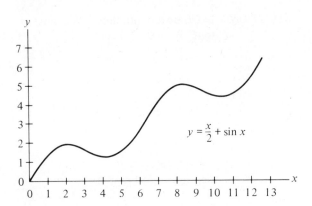

Figure 2.24

THE PRECISE DEFINITION OF $\lim_{x \to \infty} f(x) = L$

Recall the definition of "$\lim_{x \to \infty} f(x) = L$" given in the table of Sec. 2.2. L refers to a fixed number.

Informal Informal definition of $\lim_{x \to \infty} f(x) = L$: $f(x)$ is defined for all x beyond some number, and

as x gets large through positive values, $f(x)$ approaches L.

2.4 PRECISE DEFINITIONS OF LIMITS (OPTIONAL)

Again we reword this definition before offering the precise definition.

Reworded

Reworded informal definition of $\lim_{x \to \infty} f(x) = L$:

First,

there is a number c such that $f(x)$ is defined for all $x > c$.

Second,

if x is sufficiently large and positive, then $f(x)$ is necessarily near L.

Again, the precise definition parallels the reworded informal definition. In order to make precise the phrase "$f(x)$ is necessarily near L," we shall use the absolute value of $f(x) - L$ to measure the distance from $f(x)$ to L. The following definition says that "if x is large enough, then

$$|f(x) - L|$$

is as small as we please."

Precise

Precise definition of $\lim_{x \to \infty} f(x) = L$:

First,

there is a number c such that $f(x)$ is defined for all $x > c$.

Second,

for each positive number ε, there is a number D such that for all $x > D$, it is true that

$$|f(x) - L| < \varepsilon.$$

ε (epsilon) is the Greek letter corresponding to the English letter e.

The positive number ε is the challenge, and D is a response. The smaller ε is, the larger D usually must be chosen. The geometric meaning of the precise definition of $\lim_{x \to \infty} f(x) = L$ is shown in Fig. 2.25.

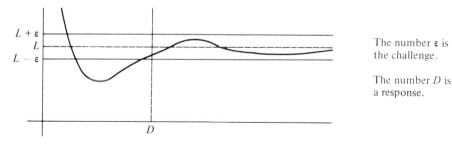

The number ε is the challenge.

The number D is a response.

Figure 2.25

The narrower the band, the larger D must usually be.

Draw two lines parallel to the x axis, one of height $L + \varepsilon$ and one of height $L - \varepsilon$. They are the two edges of an endless band of width 2ε. Assume that for each positive ε, a number D can be found, depending on ε, such that the part of the graph to the right of $x = D$ lies within the band. Then we say that "as x approaches ∞, $f(x)$ approaches L" and write

$$\lim_{x \to \infty} f(x) = L.$$

EXAMPLE 3 Use the precise definition of "$\lim_{x \to \infty} f(x) = L$" to show that

$$\lim_{x \to \infty} \left(1 + \frac{1}{x}\right) = 1.$$

SOLUTION Here $f(x) = 1 + 1/x$, which is defined for all $x > 0$. The number L is 1. We must show that for each positive number ε, however small, there is a number D such that, for all $x > D$,

$$\left|\left(1 + \frac{1}{x}\right) - 1\right| < \varepsilon. \tag{3}$$

The inequality (3) reduces to

$$\left|\frac{1}{x}\right| < \varepsilon.$$

Since we shall consider only $x > 0$, this inequality is equivalent to

$$\frac{1}{x} < \varepsilon. \tag{4}$$

Multiplying (4) by the positive number x yields the equivalent inequality

$$1 < \varepsilon x. \tag{5}$$

Division of (5) by the positive number ε yields

$$\frac{1}{\varepsilon} < x$$

or

$$x > \frac{1}{\varepsilon}.$$

These steps are reversible. This shows that $D = 1/\varepsilon$ is a suitable reply to the challenge ε. If $x > 1/\varepsilon$, then

$$\left|\left(1 + \frac{1}{x}\right) - 1\right| < \varepsilon.$$

According to the precise definition of "$\lim_{x \to \infty} f(x) = L$," we may conclude that

$$\lim_{x \to \infty} \left(1 + \frac{1}{x}\right) = 1. \quad \blacksquare$$

2.4 PRECISE DEFINITIONS OF LIMITS (OPTIONAL)

The graph of $f(x) = 1 + 1/x$, shown in Fig. 2.26, reinforces the argument. It seems plausible that no matter how narrow a band someone may place around the line $y = 1$, it will always be possible to find a number D such that the part of the graph to the right of $x = D$ stays within that band. In Fig. 2.26 the typical band is shown shaded.

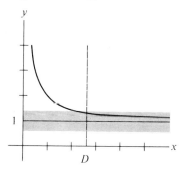

Figure 2.26

THE PRECISE DEFINITION OF $\lim_{x \to a} f(x) = L$

Recall the informal definition given in Sec. 2.2.

Informal   Informal definition of $\lim_{x \to a} f(x) = L$: Let f be a function and a some fixed number. Assume that the domain of f contains open intervals (c, a) and (a, b) for some number $c < a$ and some number $b > a$.

If, as x approaches a, both from the left and from the right, $f(x)$ approaches a specific number L, then L is called the limit of $f(x)$ as x approaches a. This is written

$$\lim_{x \to a} f(x) = L.$$

Keep in mind that a need not be in the domain of f. Even if a happens to be in the domain of f, the value $f(a)$ plays no role in determining whether $\lim_{x \to a} f(x) = L$.

Reworded   Reworded informal definition of $\lim_{x \to a} f(x) = L$:

First,

> there are numbers c and b, $c < a < b$, such that $f(x)$ is defined for all x in (c, a) and all x in (a, b).

Second,

> if x is sufficiently close to a but not equal to a, then $f(x)$ is necessarily near L.

The "ε, δ" definition of "$\lim_{x \to a} f(x) = L$"   The precise definition parallels the reworded informal definition. The letter δ that appears in it is the lower case Greek "delta," equivalent to the English letter d.

80 LIMITS AND CONTINUOUS FUNCTIONS

Precise

> Precise definition of $\lim_{x \to a} f(x) = L$:
>
> First,
>
> > there are numbers c and b, $c < a < b$, such that $f(x)$ is defined for all x in (c, a) and all x in (a, b).
>
> Second,
>
> > for each positive number ε there is a positive number δ such that for all x that satisfy the inequality
> >
> > $$0 < |x - a| < \delta$$
> >
> > it is true that
> >
> > $$|f(x) - L| < \varepsilon.$$

The meaning of $0 < |x - a| < \delta$

The inequality $0 < |x - a|$ that appears in the definition is just a fancy way of saying "x is not a." The inequality $|x - a| < \delta$ asserts that x is within a distance δ of a. The two inequalities may be combined as the single statement $0 < |x - a| < \delta$, which describes the open interval $(a - \delta, a + \delta)$ from which a is deleted. This deletion is made since the value $f(a)$ plays no role in the definition of $\lim_{x \to a} f(x)$.

Once again ε is the challenge. The response is δ. Usually, the smaller ε is, the smaller δ will have to be.

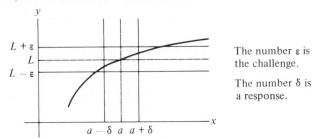

Figure 2.27

The number ε is the challenge.

The number δ is a response.

The geometric significance of the precise definition of "$\lim_{x \to a} f(x) = L$" is shown in Fig. 2.27. The narrow horizontal band of width 2ε is again the challenge. The response is a sufficiently narrow vertical band, of width 2δ, such that the part of the graph within that vertical band (except perhaps at $x = a$) also lies in the challenging horizontal band of width 2ε. In Fig. 2.28 the vertical band shown is not narrow enough to meet the challenge of the horizontal band shown. But the vertical band shown in Fig. 2.29 is sufficiently narrow.

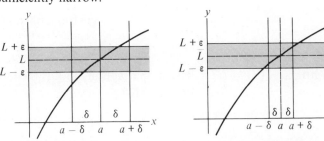

Figure 2.28

δ is not small enough.

Figure 2.29

δ is small enough.

Assume that for each positive number ε, it is possible to find a positive number δ such that the parts of the graph between $x = a - \delta$ and $x = a$ and between $x = a$ and $x = a + \delta$ lie within the given horizontal band. Then we say that "as x approaches a, $f(x)$ approaches L." The narrower the horizontal band around the line $y = L$, the smaller δ usually must be.

EXAMPLE 4 Use the precise definition of "$\lim_{x \to a} f(x) = L$" to show that

$$\lim_{x \to 0} x^2 = 0.$$

SOLUTION In this case $a = 0$ and $L = 0$. Let ε be a positive number. We wish to find a positive number δ such that for $0 < |x - 0| < \delta$, it follows that $|x^2 - 0| < \varepsilon$.

Since $|x|^2 = |x^2|$, we are asking, "for which x is

$$|x|^2 < \varepsilon?"$$

This inequality is satisfied when

$$|x| < \sqrt{\varepsilon}.$$

In other words, when $|x| < \sqrt{\varepsilon}$, it follows that $|x^2 - 0| < \varepsilon$. Thus $\delta = \sqrt{\varepsilon}$ suffices.

(For instance, when $\varepsilon = 1$, $\delta = \sqrt{1} = 1$ is a suitable response. When $\varepsilon = 0.01$, $\delta = 0.1$ suffices.) ∎

EXAMPLE 5 Use the precise definition of "$\lim_{x \to a} f(x) = L$" to show that

$$\lim_{x \to 2} (3x + 5) = 11.$$

SOLUTION Here $a = 2$ and $L = 11$. Let ε be a positive number. We wish to find a number $\delta > 0$ such that for $0 < |x - 2| < \delta$, it follows that $|(3x + 5) - 11| < \varepsilon$.

So let us find out for which x it is true that $|(3x + 5) - 11| < \varepsilon$. This inequality is equivalent to

$$|3x - 6| < \varepsilon$$

or

$$3|x - 2| < \varepsilon$$

or

$$|x - 2| < \frac{\varepsilon}{3}.$$

Thus $\delta = \varepsilon/3$ is an adequate response. If $0 < |x - 2| < \varepsilon/3$, then $|(3x + 5) - 11| < \varepsilon$. ∎

The reasoning in the next example is a little more involved.

EXAMPLE 6 Use the precise definition of "$\lim_{x \to a} f(x) = L$" to show that

$$\lim_{x \to 2} 3x^3 = 24.$$

SOLUTION *Before* trying to meet the challenge of the number ε, let us explore how the size of $|x - 2|$ influences the size of $|3x^3 - 24|$.

Experiment with δ before worrying about ε.

To begin, note that

$$|3x^3 - 24| = |3x^3 - 3 \cdot 2^3|$$
$$= 3|x^3 - 2^3|$$

or $\quad |3x^3 - 24| = 3|x - 2| \, |x^2 + 2x + 4|.$ \hfill (6)

So, as long as $|x - 2|$ is small and $|x^2 + 2x + 4|$ is kept from being large, $|3x^3 - 24|$ will be kept small.

Consider a positive number δ that is also less than 1. If $|x - 2| < \delta$, then we also have $|x - 2| < 1$. Thus x is in the open interval (1, 3). Consequently $|x| < 3$,

Think of $|x - 2| < 1$ in terms of distance: x is within 1 unit of 2.

and $\quad |x^2 + 2x + 4| \le |x^2| + 2|x| + |4| \quad$ (triangle inequality)
$$\le 9 + 6 + 4 = 19.$$

In light of (6), if $|x - 2| < \delta$ and $\delta < 1$, it follows that

$$|3x^3 - 24| < 3 \cdot \delta \cdot 19 = 57\delta. \tag{7}$$

If the challenge number ε is now given, any number δ that is less than 1 and also less than $\varepsilon/57$ will serve as a response. According to (7), for $|x - 2|$ less than such a δ,

$$|3x^3 - 24| < 57 \frac{\varepsilon}{57} = \varepsilon.$$

This shows that $\qquad\qquad \lim_{x \to 2} 3x^3 = 24.$ ■

SUMMARY This section introduced the precise definitions of some of the limits on which calculus is based. It takes time, patience, and practice to be able to apply these definitions. In Appendix F the precise definition of "$\lim_{x \to a} f(x) = L$" is used to justify the basic properties of limits listed at the beginning of Sec. 2.2. For instance, it is proved there that if $\lim_{x \to a} f(x) = A$ and $\lim_{x \to a} g(x) = B$, then $\lim_{x \to a} (f(x) + g(x)) = A + B$.

Exercises

Exercises 1 to 14 concern the precise definition of "$\lim_{x \to \infty} f(x) = \infty$."

1 Let $f(x) = 3x$. (a) Find a number D such that, for $x > D$, it follows that $f(x) > 600$.

(b) Find another number D such that, for $x > D$, it follows that $f(x) > 600$.
(c) What is the smallest number D such that, for $x > D$, it follows that $f(x) > 600$?

2 Let $f(x) = 4x$. (a) Find a number D such that, for $x > D$, it follows that $f(x) > 1000$.
   (b) Find another number D such that, for $x > D$, it follows that $f(x) > 1000$.
   (c) What is the smallest number D such that, for $x > D$, it follows that $f(x) > 1000$?
3 Let $f(x) = 5x$. Find a number D such that, for all $x > D$, (a) $f(x) > 2000$, (b) $f(x) > 10{,}000$.
4 Let $f(x) = 6x$. Find a number D such that, for all $x > D$, (a) $f(x) > 1200$, (b) $f(x) > 1800$.

In Exercises 5 to 12 use the precise definition of "$\lim_{x \to \infty} f(x) = \infty$" to establish the given limits.

5 $\lim_{x \to \infty} 3x = \infty$ 6 $\lim_{x \to \infty} 4x = \infty$
7 $\lim_{x \to \infty} (x + 5) = \infty$ 8 $\lim_{x \to \infty} (x - 600) = \infty$
9 $\lim_{x \to \infty} (2x + 4) = \infty$ 10 $\lim_{x \to \infty} (3x - 1200) = \infty$
11 $\lim_{x \to \infty} (4x + 100 \cos x) = \infty$
12 $\lim_{x \to \infty} (2x - 300 \cos x) = \infty$

13 Let $f(x) = x^2$. (a) Find a number D such that, for $x > D$, $f(x) > 100$.
   (b) Let E be any nonnegative number. Find a number D such that, for $x > D$, it follows that $f(x) > E$.
   (c) Let E be any negative number. Find a number D such that, for $x > D$, it follows that $f(x) > E$.
   (d) Using the precise definition of "$\lim_{x \to \infty} f(x) = \infty$," show that $\lim_{x \to \infty} x^2 = \infty$.
14 Using the precise definition of "$\lim_{x \to \infty} f(x) = \infty$," show that $\lim_{x \to \infty} x^3 = \infty$.

Exercises 15 to 22 concern the precise definition of "$\lim_{x \to \infty} f(x) = L$."

15 Let $f(x) = 3 + 1/x$ if $x \neq 0$. (a) Find a number D such that, for $x > D$, it follows that $|f(x) - 3| < \frac{1}{10}$.
   (b) Find another number D such that, for $x > D$, it follows that $|f(x) - 3| < \frac{1}{10}$.
   (c) What is the smallest number D such that, for $x > D$, it follows that $|f(x) - 3| < \frac{1}{10}$?
   (d) Using the precise definition of "$\lim_{x \to \infty} f(x) = L$," show that $\lim_{x \to \infty} (3 + 1/x) = 3$.
16 Let $f(x) = 2/x$ if $x \neq 0$. (a) Find a number D such that, for $x > D$, it follows that $|f(x) - 0| < \frac{1}{100}$.
   (b) Find another number D such that, for $x > D$, it follows that $|f(x) - 0| < \frac{1}{100}$.
   (c) What is the smallest number D such that, for $x > D$, it follows that $|f(x) - 0| < \frac{1}{100}$?
   (d) Using the precise definition of "$\lim_{x \to \infty} f(x) = L$," show that $\lim_{x \to \infty} 2/x = 0$.

In Exercises 17 to 22 use the precise definition of "$\lim_{x \to \infty} f(x) = L$" to establish the given limits.

17 $\lim_{x \to \infty} [(\sin x)/x] = 0$ (*Hint:* $|\sin x| \leq 1$ for all x.)
18 $\lim_{x \to \infty} [(x + \cos x)/x] = 1$ 19 $\lim_{x \to \infty} 4/x^2 = 0$
20 $\lim_{x \to \infty} [(2x + 3)/x] = 2$ 21 $\lim_{x \to \infty} [1/(x - 100)] = 0$
22 $\lim_{x \to \infty} [(2x + 10)/(3x - 5)] = \frac{2}{3}$

Exercises 23 to 26 concern the precise definition of "$\lim_{x \to a} f(x) = L$."

23 (See Example 4.) Let $f(x) = 9x^2$. (a) Find $\delta > 0$ such that, for $0 < |x - 0| < \delta$, it follows that $|9x^2 - 0| < \frac{1}{100}$.
   (b) Let ε be any positive number. Find a positive number δ such that, for $0 < |x - 0| < \delta$, it follows that $|9x^2 - 0| < \varepsilon$.
   (c) Show that $\lim_{x \to 0} 9x^2 = 0$.
24 Let $f(x) = x^3$ (a) Find $\delta > 0$ such that, for $0 < |x - 0| < \delta$, it follows that $|x^3 - 0| < \frac{1}{1000}$.
   (b) Show that $\lim_{x \to 0} x^3 = 0$.
25 (See Example 5.) Let $f(x) = 2x + 5$. (a) Find $\delta > 0$ such that, for $0 < |x - 3| < \delta$, it follows that $|f(x) - 11| < 30$.
   (b) Find $\delta > 0$ such that, for $0 < |x - 3| < \delta$, it follows that $|f(x) - 11| < 1$.
   (c) Find $\delta > 0$ such that, for $0 < |x - 3| < \delta$, it follows that $|f(x) - 11| < \frac{1}{10}$.
   (d) Show that $\lim_{x \to 3} (2x + 5) = 11$.
26 Let $f(x) = (3x + 2)/2$. (a) Find $\delta > 0$ such that, for $0 < |x - 1| < \delta$, it follows that $|f(x) - \frac{5}{2}| < 0.001$.
   (b) Show that $\lim_{x \to 1} (3x + 2)/2 = \frac{5}{2}$.

■

In Exercises 27 to 35 develop precise definitions of the given limits. Phrase your definitions in terms of a challenge number, E or ε, and a reply, D or δ. In each case show on a graph the geometric meaning of your definition.

27 $\lim_{x \to \infty} f(x) = -\infty$ 28 $\lim_{x \to -\infty} f(x) = \infty$
29 $\lim_{x \to -\infty} f(x) = L$ 30 $\lim_{x \to -\infty} f(x) = -\infty$
31 $\lim_{x \to a^+} f(x) = L$ 32 $\lim_{x \to a^+} f(x) = \infty$
33 $\lim_{x \to a^+} f(x) = -\infty$ 34 $\lim_{x \to a} f(x) = \infty$
35 $\lim_{x \to a} f(x) = -\infty$

36 Let $f(x) = 5$ for all x. (a) Using the precise definition of "$\lim_{x \to a} f(x) = L$," show that $\lim_{x \to 3} f(x) = 5$. (b) Using the precise definition of "$\lim_{x \to \infty} f(x) = L$," show that $\lim_{x \to \infty} f(x) = 5$.

■ ■

37 (a) Show that, if $0 < \delta < 1$ and $|x - 3| < \delta$, then $|x^2 - 9| < 7\delta$.
   (b) From (a) deduce that $\lim_{x \to 3} x^2 = 9$.
38 (a) Show that, if $0 < \delta < 1$ and $|x - 4| < \delta$, then

84 LIMITS AND CONTINUOUS FUNCTIONS

$$|\sqrt{x} - 2| < \frac{\delta}{\sqrt{3} + 2}.$$

(*Hint:* Rationalize $\sqrt{x} - 2$.)
(b) From (a) deduce that $\lim_{x \to 4} \sqrt{x} = 2$.

39 (a) Show that, if $0 < \delta < 1$ and $|x - 3| < \delta$, then $|x^2 + 5x - 24| < 12\delta$. (*Hint:* Factor $x^2 + 5x - 24$.)
(b) From (a) deduce that $\lim_{x \to 3} (x^2 + 5x) = 24$.

40 (a) Show that, if $0 < \delta < 1$ and $|x - 2| < \delta$, then

$$\left|\frac{1}{x} - \frac{1}{2}\right| < \frac{\delta}{2}.$$

(b) From (a) deduce that $\lim_{x \to 2} 1/x = \frac{1}{2}$.

41 Show that $\lim_{x \to 1} (x^2 + x + 1) = 3$, using the precise definition of the limit.

42 (a) (Calculator) Using some large inputs x, make a conjecture concerning the value of L in

$$\lim_{x \to \infty} (\sqrt{x^2 + 4x} - x) = L.$$

(b) Verify your conjecture, using the precise definition of the limit.

43 Show that the assertion "$\lim_{x \to 2} 3x = 5$" is false. To do this, it is necessary to exhibit a positive number ε such that there is no response number $\delta > 0$. (*Hint:* Draw a picture.)

44 Show that each of the following assertions is false. In each exhibit an appropriate challenge number, E or ε, such that there is no reply number, D.
(a) $\lim_{x \to \infty} \sin x = 10$. (b) $\lim_{x \to \infty} \sin x = \infty$.
(c) $\lim_{x \to \infty} \sin x = 0$. (d) $\lim_{x \to \infty} \sin x = 1$.

2.5 The limit of $(\sin \theta)/\theta$ as θ approaches 0

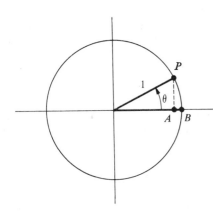

Figure 2.30

The limits evaluated in Secs. 2.1 and 2.2 were found by algebraic means, such as factoring, rationalizing, or canceling. But some of the most important limits in calculus cannot be found so easily. To reinforce the concept of a limit and also to prepare for the calculus of trigonometric functions, we shall determine

$$\lim_{\theta \to 0} \frac{\sin \theta}{\theta}.$$

To begin, a sketch (shown in Fig. 2.30) of a small angle in the unit circle suggests what the limit is. Since an angle is measured in radians, and the circle has radius 1, the length of the arc PB is θ. By the definition of $\sin \theta$, the length of $PA = \sin \theta$. Thus

$$\frac{\sin \theta}{\theta} = \frac{\text{Length of side } PA}{\text{Length of arc } PB}.$$

When θ is small, so are the lengths of PA and PB. However, for small θ, PA looks so much like the arc PB that it seems likely that the quotient

$$\frac{\text{Length of side } PA}{\text{Length of arc } PB}$$

is near 1. This suggests that

$$\lim_{\theta \to 0} \frac{\sin \theta}{\theta} = 1.$$

A few computations easily performed on a calculator also suggest that $\lim_{\theta \to 0} (\sin \theta)/\theta = 1$. This table shows $\sin \theta$ and $(\sin \theta)/\theta$ to five significant figures.

θ	1	0.1	0.01
$\sin \theta$	0.84147	0.099833	0.0099998
$\dfrac{\sin \theta}{\theta}$	0.84147	0.99833	0.99998

The reader is invited to try $\theta = 0.001$ and 0.0001. Be sure to set the calculator for radians, not degrees. (If your calculator cannot be set for radians, multiply θ by $180/\pi$ to convert to degrees. Then calculate the sine.)

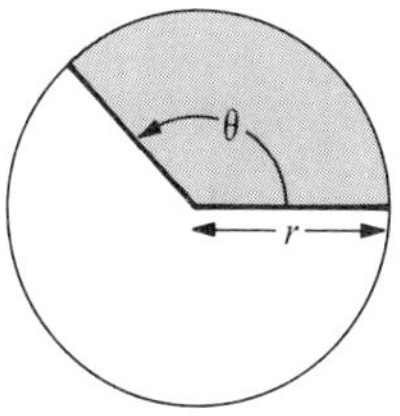

The geometric argument that $\lim_{\theta \to 0} (\sin \theta)/\theta = 1$ depends on a comparison of three areas. For this reason it will be necessary to develop a formula for the area of a sector of a circle subtended by an angle of θ radians, as in Fig. 2.31. When the angle is 2π, the sector is the entire circle of radius r; hence it has an area of πr^2. Since the area of a sector is proportional to θ, it follows that

$$\frac{\text{Area of sector}}{\pi r^2} = \frac{\theta}{2\pi}.$$

The shaded sector corresponds to the angle θ.

Figure 2.31

From this equation it follows that

Formula for area of a sector

$$\text{Area of sector} = \frac{\theta}{2\pi} \pi r^2 = \frac{\theta r^2}{2}.$$

(This formula will be used in later chapters. It is safer to memorize the proportion that led to it than the formula itself. It is easy to forget the denominator 2 and also that the number π does *not* appear.)

The next theorem describes the behavior of $(\sin \theta)/\theta$ when θ is near 0.

THEOREM 1 Let $\sin \theta$ denote the sine of an angle of θ radians. Then

$$\lim_{\theta \to 0} \frac{\sin \theta}{\theta} = 1.$$

PROOF It will be enough to consider only $\theta > 0$, since

$$\frac{\sin(-\theta)}{-\theta} = \frac{-\sin \theta}{-\theta} = \frac{\sin \theta}{\theta}.$$

Moreover, it will be convenient to restrict θ to be less than $\pi/2$.

We shall compare the areas of three regions in Fig. 2.32, two sectors and a triangle. One sector, OAC, has angle θ and radius equal to $\cos \theta$. Sector OBP has angle θ and radius 1. Triangle OAP has base $\cos \theta$ and altitude $\sin \theta$.

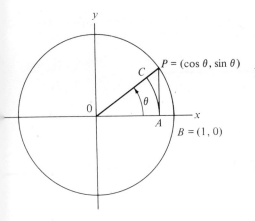

Figure 2.32

As inspection of Fig. 2.32 shows,

Area of sector OAC < Area of triangle OAP < Area of sector OBP. (1)

Now,
$$\text{Area of sector } OAC = \frac{\theta(\cos\theta)^2}{2}$$

$$\text{Area of triangle } OAP = \tfrac{1}{2} \cdot \text{Base} \cdot \text{Altitude} = \frac{\cos\theta \cdot \sin\theta}{2},$$

and
$$\text{Area of sector } OBP = \frac{\theta \cdot 1^2}{2} = \frac{\theta}{2}.$$

So inequalities (1) take the form

$$\frac{\theta \cos^2 \theta}{2} < \frac{\cos\theta \sin\theta}{2} < \frac{\theta}{2}. \qquad (2)$$

Multiplying by the positive number 2 and dividing by the positive number $\theta \cos\theta$ yields the inequalities

$$\cos\theta < \frac{\sin\theta}{\theta} < \frac{1}{\cos\theta}. \qquad (3)$$

A glance at the unit circle suggests that

$$\lim_{\theta \to 0} \cos\theta = 1$$

and hence that

$$\lim_{\theta \to 0} \frac{1}{\cos\theta} = \frac{1}{1} = 1.$$

Thus, as $\theta \to 0$, both $\cos\theta$ and $1/\cos\theta$ approach 1. Hence, $(\sin\theta)/\theta$, trapped between $\cos\theta$ and $1/\cos\theta$, must also approach 1. Thus

$$\lim_{\theta \to 0} \frac{\sin\theta}{\theta} = 1,$$

as had been anticipated. ∎

We shall also need the limit

$$\lim_{\theta \to 0} \frac{1 - \cos\theta}{\theta}$$

in the next chapter. It is not obvious what this limit is, if indeed it exists. As $\theta \to 0$, the numerator $1 - \cos\theta$ approaches 0; so does the denominator. The numerator influences the quotient to become small, while the denominator influences the quotient to become large. The following theorem shows that the numerator is the stronger influence, causing the quotient to approach 0 as θ approaches 0.

THEOREM 2 Let $\cos \theta$ denote the cosine of an angle of θ radians. Then

$$\lim_{\theta \to 0} \frac{1 - \cos \theta}{\theta} = 0.$$

PROOF

$$\frac{1 - \cos \theta}{\theta} = \frac{1 - \cos \theta}{\theta} \frac{1 + \cos \theta}{1 + \cos \theta} \qquad (\theta \neq 0, \cos \theta \neq -1)$$

$$= \frac{1 - \cos^2 \theta}{\theta(1 + \cos \theta)}$$

$$= \frac{\sin^2 \theta}{\theta(1 + \cos \theta)}$$

$$= \frac{\sin \theta}{\theta} \frac{\sin \theta}{1 + \cos \theta}.$$

Thus

$$\lim_{\theta \to 0} \frac{1 - \cos \theta}{\theta} = \lim_{\theta \to 0} \left(\frac{\sin \theta}{\theta} \frac{\sin \theta}{1 + \cos \theta} \right)$$

$$= 1 \frac{0}{1 + 1}$$

$$= 0.$$

Consequently, $\lim_{\theta \to 0} \dfrac{1 - \cos \theta}{\theta} = 0.$

This implies that, when θ is small, $1 - \cos \theta$ is much smaller than θ. ∎

A sketch of the unit circle will make Theorem 2 plausible. Figure 2.33 shows a small angle θ in the unit circle. The arc length along the circle from P to B is θ. Since $\overline{OA} = \cos \theta$, $\overline{AB} = 1 - \cos \theta$. It appears from Fig. 2.33 that the ratio $\overline{AB}/(\text{arc length } PB)$ is small when θ is small. This suggests that $\lim_{\theta \to 0} (1 - \cos \theta)/\theta = 0$, as Theorem 2 asserts.

Some other limits involving sine and cosine can be found with the aid of $\lim_{\theta \to 0} (\sin \theta)/\theta$ and $\lim_{\theta \to 0} (1 - \cos \theta)/\theta$, as shown in the following examples.

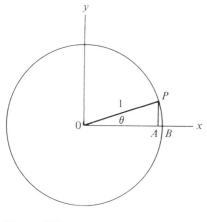

Figure 2.33

EXAMPLE 1 Find $\lim_{x \to 0} (\sin 5x)/5x$.

SOLUTION Observe that as $x \to 0$, $5x \to 0$. Let $\theta = 5x$. Thus

$$\lim_{x \to 0} \frac{\sin 5x}{5x} = \lim_{\theta \to 0} \frac{\sin \theta}{\theta} = 1. \ \blacksquare$$

EXAMPLE 2 Find $\lim_{x \to 0} (\sin 5x)/2x$.

88 LIMITS AND CONTINUOUS FUNCTIONS

SOLUTION A little algebra permits one to exploit the result found in Example 1.

$$\lim_{x \to 0} \frac{\sin 5x}{2x} = \lim_{x \to 0} \frac{\sin 5x}{5x} \cdot \frac{5x}{2x}$$

$$= \lim_{x \to 0} \frac{\sin 5x}{5x} \cdot \frac{5}{2}$$

$$= 1 \cdot \frac{5}{2}$$

$$= \frac{5}{2}. \quad \blacksquare$$

EXAMPLE 3 Find $\lim_{x \to 0} \sin 3x / \sin 2x$.

SOLUTION First rewrite the quotient as follows:

$$\frac{\sin 3x}{\sin 2x} = \frac{\sin 3x}{3x} \cdot 3x \cdot \frac{2x}{\sin 2x} \cdot \frac{1}{2x}$$

$$= \frac{3}{2} \cdot \frac{\sin 3x}{3x} \cdot \frac{2x}{\sin 2x}.$$

Now, $\lim_{\theta \to 0} (\sin \theta)/\theta = 1$ and $\lim_{\theta \to 0} \theta/\sin \theta = 1$. Thus

$$\lim_{x \to 0} \frac{3}{2} \cdot \frac{\sin 3x}{3x} \cdot \frac{2x}{\sin 2x} = \frac{3}{2} \cdot \lim_{x \to 0} \frac{\sin 3x}{3x} \cdot \lim_{x \to 0} \frac{2x}{\sin 2x}$$

$$= \frac{3}{2} \cdot 1 \cdot 1$$

$$= \frac{3}{2}.$$

Consequently $\lim_{x \to 0} \frac{\sin 3x}{\sin 2x} = \frac{3}{2}. \quad \blacksquare$

From a practical point of view this section showed that, if angles are measured in radians, then the sine of a small angle is "roughly" the angle itself, that is

$$\sin x \approx x.$$

A useful fact: for small x,
$\sin x \approx x$
$\tan x \approx x$.

This is another way of saying that when x is small the quotient $(\sin x)/x$ is close to 1. In engineering and physics $\sin x$ is often replaced by x when x is small. Moreover, $\tan x$ may also be replaced by x for small x. That this is a reasonable estimate is justified by the fact that

$$\lim_{x \to 0} \frac{\tan x}{x} = \lim_{x \to 0} \frac{\frac{\sin x}{\cos x}}{x}$$

$$= \left(\lim_{x \to 0} \frac{\sin x}{x}\right)\left(\lim_{x \to 0} \frac{1}{\cos x}\right) = 1 \cdot 1$$
$$= 1.$$

So $\tan x \approx x$ for small x.

Exercises

1. What is the area of a sector of a circle of (a) radius 3 and angle $\pi/2$? (b) radius 1 and angle θ? (c) radius 2 and angle θ?
2. What is the area of the sector of a circle of radius 6 inches subtended by an angle of (a) $\pi/4$ radians? (b) 3 radians? (c) $45°$?

In Exercises 3 to 16 examine the limits.

3. $\lim_{x \to 0} \dfrac{\sin x}{2x}$
4. $\lim_{x \to 0} \dfrac{\sin 2x}{x}$
5. $\lim_{x \to 0} \dfrac{\sin 3x}{5x}$
6. $\lim_{x \to 0} \dfrac{2x}{\sin 3x}$
7. $\lim_{\theta \to 0} \dfrac{\sin^2 \theta}{\theta}$
8. $\lim_{h \to 0} \dfrac{\sin h^2}{h^2}$
9. $\lim_{\theta \to 0} \dfrac{\tan^2 \theta}{\theta}$
10. $\lim_{\theta \to 0} \theta \cot \theta$
11. $\lim_{\theta \to 0} \dfrac{1 - \cos \theta}{\theta^2}$
12. $\lim_{\theta \to 0^-} \dfrac{1 - \cos \theta}{\theta^3}$
13. $\lim_{\theta \to 0^+} \dfrac{1 - \cos \theta}{\theta^3}$
14. $\lim_{x \to 0} \dfrac{\sin^2 x}{x^2}$
15. $\lim_{\theta \to 0^+} \dfrac{1}{\sin \theta}$
16. $\lim_{\theta \to 0^-} \dfrac{1}{\sin \theta}$

■

17. (a) What is the domain of the function $f(x) = (\sin x)/x$?
    (b) Show that f is an even function, that is, $f(-x) = f(x)$.
    (c) Find $\lim_{x \to \infty} f(x)$. Hint: For all x, $|\sin x| \leq 1$.
    (d) For which x is $f(x) = 0$?
    (e) Fill in the following table to two decimal places, using $\pi \approx 3.14$:

x	0.1	$\pi/2$	$3\pi/2$	2π	$\dfrac{5\pi}{2}$	3π	$7\pi/2$
$\sin x$							
$\dfrac{\sin x}{x}$							

(f) Graph f for $x > 0$.
(g) Graph f for $x < 0$.
(h) What is $\lim_{x \to 0} f(x)$?

18. (a) What is the domain of the function $g(x) = (1 - \cos x)/x$?
    (b) Show that g is an odd function, that is, $g(-x) = -g(x)$.
    (c) Find $\lim_{x \to \infty} g(x)$. Hint: $|1 - \cos x| \leq 2$ for all x.
    (d) For which x is $g(x) = 0$?
    (e) Fill in the following table to two decimal places, using $\pi \approx 3.14$:

x	0.1	$\pi/2$	$3\pi/2$	2π	3π
$1 - \cos x$					
$g(x) = \dfrac{1 - \cos x}{x}$					

(f) Graph g for $x > 0$.
(g) Graph g for $x < 0$.
(h) What is $\lim_{x \to 0} g(x)$?

19. Examine $\lim_{x \to \pi/2} \dfrac{1 - \sin x}{x - \dfrac{\pi}{2}}$. Hint: Let $\theta = x - \dfrac{\pi}{2}$.

20. Examine $\lim_{x \to 0^-} \dfrac{\sin^2 x}{x^3}$.

■ ■

Exercises 21 to 25 concern the function $f(x) = \sin(1/x)$ and its graph. Use the same scale on both axes, with the distance from 0 to 1 at least 10 centimeters.

21. (a) Show that for any nonzero integer n, $f(1/(n\pi)) = 0$.
    (b) Plot the points on the graph of f with x coordinate $1/\pi$, $1/(2\pi)$, $1/(3\pi)$, $1/(4\pi)$, and $1/(5\pi)$.
22. (a) Show that, for any integer n, $f(1/[(2n + \tfrac{3}{2})\pi]) = -1$.
    (b) Plot the points on the graph of f with $x = 1/(\tfrac{3}{2}\pi)$, $1/(\tfrac{7}{2}\pi)$, and $1/(\tfrac{11}{2}\pi)$.
23. (a) Show that, for any integer n, $f(1/[(2n + \tfrac{1}{2})\pi]) = 1$.
    (b) Plot the points on the graph of f with $x = 1/(\pi/2)$, $1/(5\pi/2)$, and $1/(9\pi/2)$.
24. What is $\lim_{x \to \infty} \sin(1/x)$?
25. (a) With the aid of Exercises 21 to 24, graph f for $x > 0$.
    (b) Does $\lim_{x \to 0^+} f(x)$ exist?
26. This exercise concerns the graph of $f(x) = x \sin x$.
    (a) For which x is $f(x) = 0$?

(b) For which x is $f(x) = x$?
(c) For which x is $f(x) = -x$?
(d) For $x \geq 0$, plot the points given in (a), (b), and (c). (There are an infinite number of them; just sketch a few.)
(e) With the aid of (d) graph f for $x \geq 0$.
(f) Does $\lim_{x \to 0^+} f(x)$ exist?
(g) Does $\lim_{x \to \infty} f(x)$ exist?

27 (Calculator) Examine the behavior of $(\theta - \sin \theta)/\theta^3$ for θ near 0.

28 (Calculator) Examine the behavior of $(\cos \theta - 1 + \theta^2/2)/\theta^4$ for θ near 0.

29 (See Exercise 11.) (a) Show that for small θ, $\cos \theta \approx 1 - \theta^2/2$.
(b) Use (a) to estimate cos 0.1 and cos 0.01.
(c) Compare the values in (b) to the values for $\cos \theta$ found in a table or by a calculator.

30 (A test of intuition) Intuition suggested that $\lim_{\theta \to 0} (\sin \theta)/\theta = 1$, which turned out to be correct. Try your intuition on another limit associated with the unit circle shown in Fig. 2.34.
(a) What do you think happens to the quotient

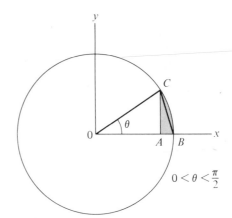

Figure 2.34

$$\frac{\text{Area of triangle } ABC}{\text{Area of shaded region}}$$

as $\theta \to 0$? More precisely what does your intuition suggest is the limit of that quotient as $\theta \to 0$?
(b) Estimate the limit in (a) using $\theta = 0.01$. *Note:* The limit is determined in Chap. 6. This question arose during some research in geometry. I guessed wrong, as has everyone I have asked.

2.6 Continuous functions

Imagine that you had the following information about some function f:

x	0.9	0.99	0.999
$f(x)$	2.93	2.9954	2.9999997

What would you expect the output $f(1)$ to be?

It would be quite a shock to be told that $f(1)$ is, say, 625. A reasonable function should present no such surprise. The expectation is that $f(1) = 3$. More generally, we expect the output of a function at the input a to be closely connected with the outputs of the function at inputs that are near a. The functions of interest in calculus usually behave in the expected way; they offer no spectacular gaps or jumps. Their graphs consist of curves or lines, not wildly scattered points. The technical term for these functions is "continuous," which will be defined in this section.

CONTINUITY AT A POINT

The following three definitions express in terms of limits our expectation that $f(a)$ is determined by the values of $f(x)$ for x near a.

DEFINITION *Continuity from the right at a number a.* Assume that $f(x)$ is defined at a and in some open interval (a, b), $b > a$. Then the function f is *continuous at a from the right* if

1 $\lim_{x \to a^+} f(x)$ exists and

2 that limit is $f(a)$.

Figure 2.35 illustrates right-sided continuity at a.

DEFINITION *Continuity from the left at a number a.* Assume that $f(x)$ is defined at a and in some open interval (c, a), $c < a$. Then the function f is *continuous at a from the left* if

1 $\lim_{x \to a^-} f(x)$ exists, and

2 that limit is $f(a)$.

Figure 2.36 illustrates this type of continuity.

The next definition applies if the function is defined in some open interval that includes the number a. It essentially combines the first two definitions.

DEFINITION *Continuity at a number a.* Assume that $f(x)$ is defined at a and in some open interval (c, b) that contains the number a. Then the function f is *continuous at a* if

1 $\lim_{x \to a} f(x)$ exists and

2 that limit is $f(a)$.

This third definition amounts to asking that the function be continuous both from the right and from the left at a. It is illustrated by Fig. 2.37.

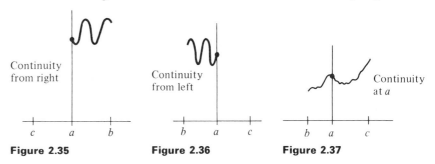

Figure 2.35   **Figure 2.36**   **Figure 2.37**

Some examples will show how the definitions are applied and why all three definitions have to be made.

EXAMPLE 1 Let $f(x)$ be the largest integer that is less than or equal to x, denoted $[x]$. Graph f and determine at which numbers it is continuous.

SOLUTION Begin with a table.

The function $[x]$ is also denoted $\lfloor x \rfloor$, and called the "floor" of x.

x	0	0.5	0.9	1.0	1.5	1.9
$f(x) = [x]$	0	0	0	1	1	1

92 LIMITS AND CONTINUOUS FUNCTIONS

An application of the greatest integer function: How old are you?

As the table suggests,

$$f(x) = 0 \text{ for } 0 \leq x < 1 \quad \text{and} \quad f(x) = 1 \text{ for } 1 \leq x < 2.$$

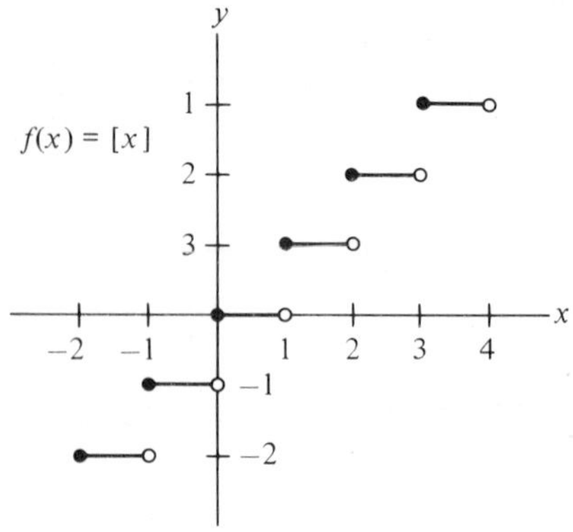

The graph of f, which consists of horizontal line segments, is shown in Fig. 2.38.

Is f continuous at $a = 1$? Inspection of the graph shows that

$$\lim_{x \to 1^+} f(x) = 1 \quad \text{and} \quad \lim_{x \to 1^-} f(x) = 0.$$

Since these limits are not equal, $\lim_{x \to 1} f(x)$ does not exist. No matter what $f(1)$ is, there is no chance for the function to be continuous at $a = 1$.

Similar reasoning shows that f is not continuous at any integer a.

Is f continuous at $a = \frac{1}{2}$? Inspection of the graph shows that as $x \to \frac{1}{2}$, $f(x) \to 0$. That is,

$$\lim_{x \to 1/2} f(x) = 0.$$

Moreover, $\quad f(\tfrac{1}{2}) = 0.$

Since $\lim_{x \to 1/2} f(x)$ exists and equals $f(\tfrac{1}{2})$, f is continuous at $a = \tfrac{1}{2}$.

Similar reasoning shows that f is continuous at any number a that is not an integer.

Incidentally, at $a = 1$ (or at any integer) f is continuous from the right but not continuous from the left. ∎

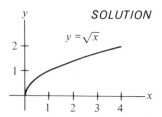

Figure 2.38

EXAMPLE 2 Let $f(x) = x^2$ for all x. Show that f is continuous at $a = 3$.

SOLUTION As $x \to 3$, $f(x) = x^2$ approaches 9, that is,

$$\lim_{x \to 3} x^2 = 9.$$

The approach is informal and intuitive. See Sec. 2.4 or Appendix F for a more rigorous approach.

Next, compute $f(3)$, which is 3^2 or 9. Since $\lim_{x \to 3} f(x)$ exists and equals $f(3)$, f is continuous at 3. (In fact f is continuous at each real number.) ∎

EXAMPLE 3 Let $f(x) = \sqrt{x}$ for $x \geq 0$. Show that f is continuous from the right at $a = 0$.

SOLUTION As the graph of $f(x) = \sqrt{x}$ in Fig. 2.39 reminds us, the domain of f does not contain an open interval around 0. It is meaningful to speak of "continuity from the right" at 0 but not of "continuity from the left."

Since $\sqrt{x}$ approaches 0 as x approaches 0, $\lim_{x \to 0^+} f(x) = 0$. Is this limit the same as $f(0)$? Since $f(0) = \sqrt{0} = 0$, the answer is "yes." In short, f is continuous from the right at 0. ∎

Figure 2.39

All three forms of continuity must be called upon in the case of $f(x) = \sqrt{1 - x^2}$, whose domain is $[-1, 1]$ and whose graph is a semicircle, as

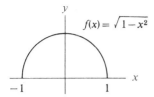

Figure 2.40

shown in Fig. 2.40. At $a = -1$, $f(x) = \sqrt{1 - x^2}$ is continuous from the right. At $a = 1$ it is continuous from the left. At any number a in $(-1, 1)$ it is continuous.

CONTINUOUS FUNCTIONS

We will want to call the functions x^2, $\sqrt{x}$, and $\sqrt{1 - x^2}$ "continuous." The greatest integer function $[x]$ will not be continuous. The following definitions define the notion of "continuous function"; they depend on the type of domain of the function.

DEFINITION *Continuous function if the domain is the x axis or is made up of open intervals.* Let f be a function whose domain is the x axis or is made up of open intervals. Then f is a *continuous function* if it is continuous at each number a in its domain.

Thus x^2 is a continuous function. So is $1/x$, whose domain consists of the intervals $(-\infty, 0)$ and $(0, \infty)$. Though this function explodes at 0, that does not prevent it from being a continuous function. The key to being continuous is that the function is continuous at each number in its domain. The number 0 is not in the domain of $1/x$.

Continuous functions on closed intervals

Only a slight modification of the definition is necessary to cover functions whose domains involve closed intervals. We will say that a function whose domain is the closed interval $[a, b]$ is *continuous* if it is continuous at each point in the open interval (a, b), continuous from the right at a, and continuous from the left at b. Thus $\sqrt{1 - x^2}$ is continuous on the interval $[-1, 1]$.

In a similar spirit, we say that a function with domain $[a, \infty)$ is continuous if it is continuous at each point in (a, ∞) and continuous from the right at a. Thus $\sqrt{x}$ is a continuous function. A similar definition covers functions whose domains are of the form $(-\infty, b]$.

WHICH FUNCTIONS ARE CONTINUOUS?

Examples of continuous functions

Many of the functions met in algebra and trigonometry are continuous. For instance, 2^x, $\sqrt{x}$, $\sin x$, $\tan x$, and any polynomial are continuous. So is any rational function (the quotient of two polynomials). (This is proved in Appendix F.) Moreover, algebraic combinations of continuous functions are continuous. For example, since x^3 and $\sin x$ are continuous, so are $x^3 + \sin x$, $x^3 - \sin x$, and $x^3 \sin x$. The function $x^3/\sin x$, which is not defined when $\sin x = 0$, is continuous on its domain. The following definitions are needed to make these statements general.

DEFINITION *Sum, difference, product, and quotient of functions.* Let f and g be two functions. The functions, $f + g$, $f - g$, fg, and f/g are defined as follows.

$$(f + g)(x) = f(x) + g(x) \quad \text{for } x \text{ in the domains of both } f \text{ and } g.$$
$$(f - g)(x) = f(x) - g(x) \quad \text{for } x \text{ in the domains of both } f \text{ and } g.$$

94 LIMITS AND CONTINUOUS FUNCTIONS

$$(fg)(x) = f(x)g(x) \qquad \text{for } x \text{ in the domains of both } f \text{ and } g.$$

$$\left(\frac{f}{g}\right)(x) = \frac{f(x)}{g(x)} \qquad \text{for } x \text{ in the domains of both } f \text{ and } g$$
$$\text{and } g(x) \neq 0.$$

If f and g are defined at least in an open interval that includes the number a and if f and g are continuous at a, then so are $f + g, f - g$, and fg. Moreover, if $g(a) \neq 0, f/g$ is also continuous at a. (Proofs of these statements are to be found in Appendix F.)

A function obtained by the composition of continuous functions is also continuous. That is, if the function g is continuous at a and the function f is continuous at $g(a)$, then the composition, $f \circ g$, is continuous at a. For instance, the function $\sqrt[3]{1 + x^2}$ is continuous since both the polynomial $1 + x^2$ and the cube-root function are continuous.

A useful property of a continuous function

Note carefully the role of continuity in the following example.

EXAMPLE 4 Find $\displaystyle\lim_{x \to 0} \sqrt[3]{\frac{\sin x}{x}}$.

SOLUTION As $x \to 0$, $(\sin x)/x \to 1$. Moreover the cube-root function is continuous. As $x \to 0$, therefore,

The limit can go inside.

$$\sqrt[3]{\frac{\sin x}{x}} \to \sqrt[3]{1} = 1. \quad \blacksquare$$

Example 4 generalizes, as follows. Let f be a continuous function. If g is some other function for which $\lim_{x \to a} g(x)$ exists and is in the domain of f, and $g(x)$ is in the domain of f for x near a, then

For continuous f, "f" and "lim" can be switched.

$$\lim_{x \to a} f(g(x)) = f\left(\lim_{x \to a} g(x)\right).$$

In Example 4, $f(x) = \sqrt[3]{x}$ and $g(x) = (\sin x)/x$. See also the discussion after Example 7 in Sec. 2.2.

SOME MORE ILLUSTRATIONS

The next example shows that the graph of a function can have a sharp corner even though the function is continuous.

EXAMPLE 5 Determine at which numbers a the absolute-value function $f(x) = |x|$ is continuous.

SOLUTION The graph of $f(x) = |x|$ is shown in Fig. 2.41. For each number a, inspection of the graph shows that $\lim_{x \to a} f(x)$ exists and equals $f(a)$. So f is continuous at each number a.

Since f is continuous at each number a, it is a continuous function. $\blacksquare$

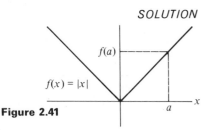
Figure 2.41

The next example explores continuity and limits in terms of the graph of a function.

EXAMPLE 6 Figure 2.42 is the graph of a certain function whose domain is the x axis. (Obviously, the entire graph cannot be shown.)
(a) At which a does $\lim_{x \to a} f(x)$ not exist?
(b) At which a is f not continuous?

SOLUTION (a) Consider $a = 2$. We have

$$\lim_{x \to 2^-} f(x) = 2 \quad \text{and} \quad \lim_{x \to 2^+} f(x) = 1.$$

Since these limits are not equal, $\lim_{x \to 2} f(x)$ does not exist.
Consider $a = 4$. Since $\lim_{x \to 4^+} f(x) = \infty$, $\lim_{x \to 4^+} f(x)$ does not exist. Hence $\lim_{x \to 4} f(x)$ does not exist.
At $a = 3$, $\lim_{x \to 3^-} f(x) = 2$ and $\lim_{x \to 3^+} f(x) = 2$. Thus $\lim_{x \to 3} f(x)$ exists and equals 2. The fact that $f(3) = 3$ has no influence on whether $\lim_{x \to 3} f(x)$ exists. At all a other than 2 and 4, $\lim_{x \to a} f(x)$ exists.
(b) Since $\lim_{x \to 2} f(x)$ and $\lim_{x \to 4} f(x)$ do not exist, f cannot be continuous at 2 and 4. Furthermore, since $\lim_{x \to 3} f(x)$ does not equal $f(3)$, f is not continuous at 3. However, f is continuous at all numbers a other than 2, 3, and 4. ∎

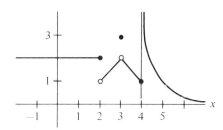

Figure 2.42

As Example 6 shows, a function whose domain is the x axis can fail to be continuous at a given number a for either of two reasons: First, $\lim_{x \to a} f(x)$ might not exist. Second, when $\lim_{x \to a} f(x)$ does exist, $f(a)$ might not equal that limit.

Exercises

In each of Exercises 1 to 8 there is a graph of a function f defined on $[0, 1]$. In each case answer the question on the basis of the graph. *Note:* Only if $\lim_{x \to a} f(x) = L$, L a real number, is the limit said to exist.

1 (Fig. 2.43) (a) Does $\lim_{x \to 1/2} f(x)$ exist? If so, evaluate it. (b) Is f continuous at $\frac{1}{2}$?

2 (Fig. 2.44) (a) Does $\lim_{x \to 1/2} f(x)$ exist? If so, evaluate it. (b) Is f continuous at $\frac{1}{2}$?

3 (Fig. 2.45) (a) Does $\lim_{x \to 1/2} f(x)$ exist? If so, evaluate it. (b) Is f continuous at $\frac{1}{2}$?

4 (Fig. 2.46) (a) Does $\lim_{x \to 1^-} f(x)$ exist? If so, evaluate it. (b) Is f continuous at 1?

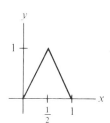

Figure 2.43

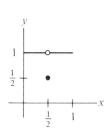

Figure 2.44

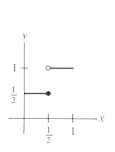

Figure 2.45

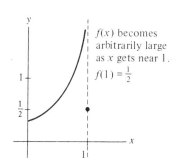

Figure 2.46

5 (Fig. 2.47) (a) Does $\lim_{x\to 0^+} f(x)$ exist? If so, evaluate it. (b) Is f continuous at 0?

6 (Fig. 2.48) (a) Does $\lim_{x\to 0^+} f(x)$ exist? If so, evaluate it. (b) Is f continuous at 0?

7 (Fig. 2.49) (a) Does $\lim_{x\to 1/4} f(x)$ exist?
(b) Is f continuous at $\tfrac{1}{4}$?
(c) Does $\lim_{x\to 3/4} f(x)$ exist?
(d) Is f continuous at $\tfrac{3}{4}$?
(e) Is f continuous at $\tfrac{1}{2}$?
(f) Is f continuous at 0?

8 (Fig. 2.50) (a) Does $\lim_{x\to 1/2} f(x)$ exist?
(b) Is f continuous at $\tfrac{1}{2}$?

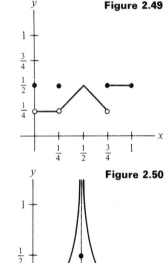

Figure 2.49

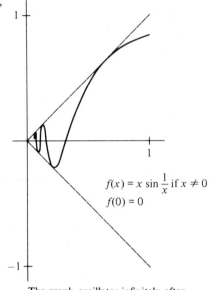

$f(x) = x \sin \dfrac{1}{x}$ if $x \neq 0$
$f(0) = 0$

The graph oscillates infinitely often between the dotted lines.

Figure 2.48

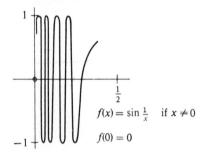

$f(x) = \sin \dfrac{1}{x}$ if $x \neq 0$
$f(0) = 0$

The graph oscillates infinitely often between the lines $y = 1$ and $y = -1$.

Figure 2.47

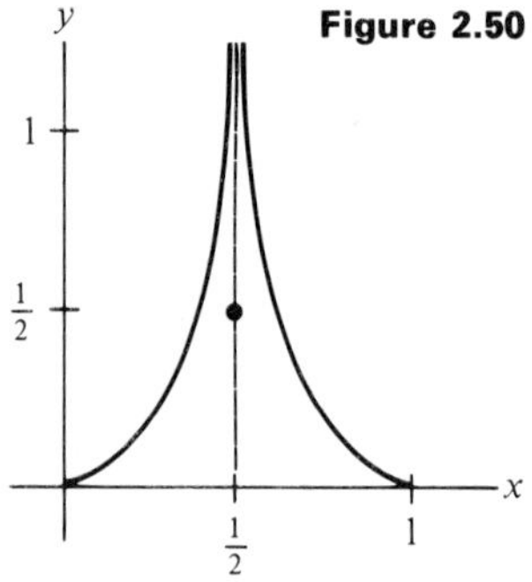

Figure 2.50

9 Let $f(x)$ equal the least integer that is greater than or equal to x. For instance, $f(3) = 3$, $f(3.4) = 4$, $f(3.9) = 4$. (This function is sometimes denoted $\lceil x \rceil$ and called the "ceiling" of x.)
(a) Graph f. (b) Does $\lim_{x\to 4^-} f(x)$ exist? If so, what is it?
(c) Does $\lim_{x\to 4^+} f(x)$ exist? If so, what is it?
(d) Does $\lim_{x\to 4} f(x)$ exist? If so, what is it?
(e) Is f continuous at 4? (f) Where is f continuous?
(g) Where is f not continuous?

10 Let f be the "nearest integer, with rounding down" function. That is,

$$f(x) = \begin{cases} \text{the integer nearest to } x \text{ if } x \text{ is not midway between two consecutive integers,} \\ x - \tfrac{1}{2} \text{ if } x \text{ is midway between two consecutive integers.} \end{cases}$$

For instance, $f(1.4) = 1$, $f(1.5) = 1$, and $f(1.6) = 2$.
(a) Graph f.
(b) Does $\lim_{x\to 3.5^-} f(x)$ exist? If so, evaluate it.
(c) Does $\lim_{x\to 3.5^+} f(x)$ exist? If so, evaluate it.
(d) Does $\lim_{x\to 3.5} f(x)$ exist? If so, evaluate it.
(e) Is f continuous at 3.5?
(f) Where is f continuous?
(g) Where is f not continuous?

11 Let $f(x) = (1 - \cos x)/x$ for $x \neq 0$. Is it possible to define $f(0)$ in such a way that f is continuous throughout the x axis?

12 Let $f(x) = (x^3 - 1)/(x - 1)$ if $x \neq 1$. Is it possible to define $f(1)$ in such a way that f is continuous throughout the x axis?

13 Let $f(x) = 2 - x$ if $x < 1$ and let $f(x) = x^2$ if $x > 1$.
(a) Graph f.
(b) Can $f(1)$ be defined in such a way that f is continuous through the x axis?

14 Let $f(x) = x^2$ for $x < 1$ and let $f(x) = 2x$ for $x > 1$.
(a) Graph f.
(b) Can $f(1)$ be defined in such a way that f is continuous throughout the x axis?

15 Let $f(x) = 0$ for $x < 1$ and let $f(x) = (x - 1)^2$ for $x > 1$.
(a) Graph f.
(b) Can $f(1)$ be defined in such a way that f is continuous throughout the x axis?

16 Let $f(x) = x + |x|$.
(a) Graph f.
(b) Is f continuous at 0?

17 Let $f(x) = 2^{1/x}$ for $x \neq 0$.
(a) Find $\lim_{x\to \infty} f(x)$. (b) Find $\lim_{x\to -\infty} f(x)$.
(c) Does $\lim_{x\to 0^+} f(x)$ exist?
(d) Does $\lim_{x\to 0^-} f(x)$ exist?

(e) Graph f, incorporating the information in parts (a) to (d).
(f) Is it possible to define $f(0)$ in such a way that f is continuous throughout the x axis?

18. Let $f(x) = 2^{-1/x^2}$ for $x \neq 0$.
    (a) Find $\lim_{x \to \infty} f(x)$.   (b) Find $\lim_{x \to 0^+} f(x)$.
    (c) Graph f.
    (d) Is it possible to define $f(0)$ in such a way that f is continuous throughout the x axis?

19. (a) Graph $f(x) = \sin x$ for x in $[-2\pi, 2\pi]$.
    (b) Is f continuous?

20. Let $f(x) = x$ for rational x and let $f(x) = x^2$ for irrational x.
    (a) Sketch the graph of f. (Use a dotted curve to indicate a curve from which points are missing.)
    (b) At which inputs, if any, is f continuous?

21. Let $f(x) = x^2$ for rational x and let $f(x) = x^3$ for irrational x.
    (a) Sketch the graph of f. (Use a dotted curve to indicate a curve from which points are missing.)
    (b) At which inputs, if any, is f continuous?

■ ■

22. Let f be a continuous function defined for all x. Assume that $f(x) = 0$ when x is rational. Explain why $f(x) = 0$ when x is irrational as well.

23. Let f and g be continuous functions defined for all x. Assume that $f(x) = g(x)$ for all rational x. Deduce that $f(x) = g(x)$ for all real numbers x. (*Hint:* Exercise 22 may be of use.)

24. This exercise determines all continuous functions f such that $f(x + y) = f(x)f(y)$ for all real numbers x and y and such that the values of f are always positive.
    (a) Let b be a fixed positive number. Let $f(x) = b^x$. Check that $f(x + y) = f(x)f(y)$. (Part (b) will show that there are no other functions that satisfy the stated conditions.)
    (b) Assume that f is a continuous function such that $f(x) > 0$ for all x and $f(x + y) = f(x)f(y)$ for all x and y. Let $f(1) = c$.
    (1) Show that $f(n) = c^n$ for any positive integer n.
    (2) Show that $f(0) = 1$.
    (3) Show that $f(n) = c^n$ for any negative integer n.
    (4) Show that $f(1/n) = \sqrt[n]{c}$ for any positive integer n.
    (5) Show that $f(m/n) = (\sqrt[n]{c})^m$ for any integer m and positive integer n.
    (6) By (5), $f(x) = c^x$ for any rational number x. Assuming that f is continuous and that the exponential function c^x is continuous, deduce that $f(x) = c^x$ for all real numbers x. (*Hint:* See Exercise 23.)

25. Let f be a continuous function whose domain is the x axis and which has the property that
$$f(x + y) = f(x) + f(y)$$
for all numbers x and y. This exercise shows that f must be of the form $f(x) = cx$ for some constant c. (The function cx does satisfy the equation, $f(x + y) = f(x) + f(y)$, since $c(x + y) = cx + cy$.)
    (a) Let $f(1) = c$. Show that $f(2) = 2c$.
    (b) Show that $f(0) = 0$.
    (c) Show that $f(-1) = -c$.
    (d) Show that for any positive integer n, $f(n) = cn$.
    (e) Show that for any negative integer n, $f(n) = cn$.
    (f) Show that $f(\tfrac{1}{2}) = c/2$.
    (g) Show that for any nonzero integer n, $f(1/n) = c/n$.
    (h) Show that for any integer m and positive integer n, $f(m/n) = c(m/n)$.
    (i) Show that for any irrational number x, $f(x) = cx$. (This is where the continuity of f enters. See Exercise 23.) Parts (h) and (i) together complete the solution.

26. (Calculator) The reason 0^0 is not defined. It might be hoped that if the positive number b and the number x are both close to 0, then b^x might be close to some fixed number. If that were so, it would suggest a definition of 0^0. The following computations using the y^x key show that, when b and x are both near 0, b^x is not near any specific number. Compute (a) $(0.001)^{0.001}$, (b) $(0.0000001)^{0.1}$. Consider also $(0.001)^0$ and $0^{0.001}$.

2.7 The maximum-value theorem and the intermediate-value theorem

Continuous functions have two properties of particular importance in calculus: the "maximum-value" property and the "intermediate-value" property. Both are quite plausible, and a glance at the graph of a "typical" continuous function may persuade us that they are true of all continuous functions. No proofs will be offered; they depend on the precise definitions of limits given in Sec. 2.4 and are part of an advanced calculus course.

98 LIMITS AND CONTINUOUS FUNCTIONS

THE MAXIMUM-VALUE THEOREM

The first theorem asserts that a function that is continuous throughout the closed interval $[a, b]$ takes on a largest value somewhere in the interval. (It also take on a smallest value.)

MAXIMUM-VALUE THEOREM   Let f be continuous throughout the closed interval $[a, b]$. Then there is at least one number in $[a, b]$ at which f takes on a maximum value. That is, for some number c in $[a, b]$

$$f(c) \geq f(x) \quad \text{for all } x \text{ in } [a, b]. \quad \blacksquare$$

(Similarly, f takes on a minimum value somewhere in the interval.)

To persuade yourself that this theorem is plausible, imagine sketching the graph of a continuous function. As your pencil moves along the graph from some point on the graph to some other point on the graph, it passes through a highest point and also through a lowest point. (See Fig. 2.51.)

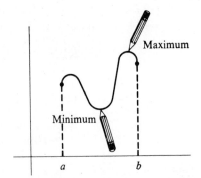

As a pencil runs along the graph of a continuous function from one point to another, it passes through at least one maximum point and at least one minimum point.

Figure 2.51

The maximum-value theorem guarantees that a maximum value exists, but it does *not* tell *how* to find it. The problem of finding the maximum value (and minimum value) is discussed in Chap. 4.

EXAMPLE 1   Let $f(x) = \cos x$ and $[a, b] = [0, 3\pi]$. Find all numbers in $[0, 3\pi]$ at which f takes on a maximum value. Also find all numbers in $[0, 3\pi]$ at which f takes on a minimum value.

SOLUTION   Figure 2.52 is a graph of $\cos x$ for x in $[0, 3\pi]$. Inspection of the graph shows that the maximum value of $\cos x$ for $0 \leq x \leq 3\pi$ is 1, and it is

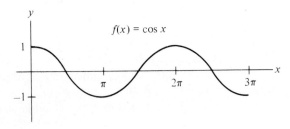

Figure 2.52

attained when $x = 0$ and when $x = 2\pi$. The minimum value is -1, which is attained when $x = \pi$ and when $x = 3\pi$. ∎

Extreme values

"*Extrema*" *is the plural of* "*extremum.*"

The maximum and minimum values of a function are frequently called its *extreme values* or *extrema*. Thus the extreme values of cos x for x in $[0, 3\pi]$ are 1 and -1.

To apply the maximum-value theorem, we must know that the function is continuous and the interval is closed (that is, contains its endpoints). The next three examples show that if either of these assumptions is deleted, the conclusion no longer need hold. In Examples 2 and 3 the interval is not closed; in Example 4 the function is not continuous.

EXAMPLE 2   Let $f(x) = 1/(1 - x^2)$ and (a, b) be the open interval $(-1, 1)$. Show that f does not have a maximum value for x in (a, b).

SOLUTION   For x near 1, $f(x)$ gets arbitrarily large since the denominator $1 - x^2$ is close to 0. The graph of f, for x in $(-1, 1)$, is shown in Fig. 2.53. This function is continuous throughout the open interval $(-1, 1)$, but there is no number c in $(-1, 1)$ at which f takes on a maximum value. However, f does take on a minimum value, $f(0) = 1$.

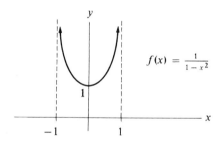

Figure 2.53   ∎

EXAMPLE 3   Let $f(x) = 1/x$ and let (a, b) be the open interval $(0, 1)$. Show that f does not have a maximum value in (a, b).

SOLUTION   Figure 2.54 shows the pertinent part of the graph of $1/x$. Since $\lim_{x \to 0^+} 1/x = \infty$, the function has no maximum value.

Moreover, the function has no minimum value for x in $(0, 1)$. It does take on values arbitrarily close to 1 for inputs that are close to 1, but there is no number in the open interval $(0, 1)$ at which $f(x)$ is equal to 1. ∎

In the next example the interval is closed but the function is not continuous.

EXAMPLE 4   Let $f(x) = x$ if x is not an integer and let $f(x) = 0$ if x is an integer. Show that f does not assume a maximum value on the interval $[0, 1]$.

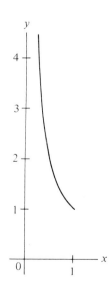

Figure 2.54

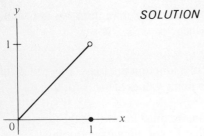

Figure 2.55

SOLUTION First graph f for x in $[0, 1]$, as in Fig. 2.55. Inspection of the graph shows that at no number in $[0, 1]$ does f take on a maximum value. As $x \to 1$ from the left, $f(x) \to 1$, but $f(x)$ does not attain the value 1. (Incidentally f takes on its minimum value 0 for the interval $[0, 1]$ at two places, at $x = 0$ and at $x = 1$.) ∎

THE INTERMEDIATE-VALUE THEOREM

The next theorem says that a function which is continuous throughout an interval takes on all values between any two of its values.

INTERMEDIATE-VALUE THEOREM Let f be continuous throughout the closed interval $[a, b]$. Let m be any number between $f(a)$ and $f(b)$. (That is, $f(a) \leq m \leq f(b)$ if $f(a) \leq f(b)$, or $f(a) \geq m \geq f(b)$ if $f(a) \geq f(b)$.) Then there is at least one number c in $[a, b]$ such that $f(c) = m$. ∎

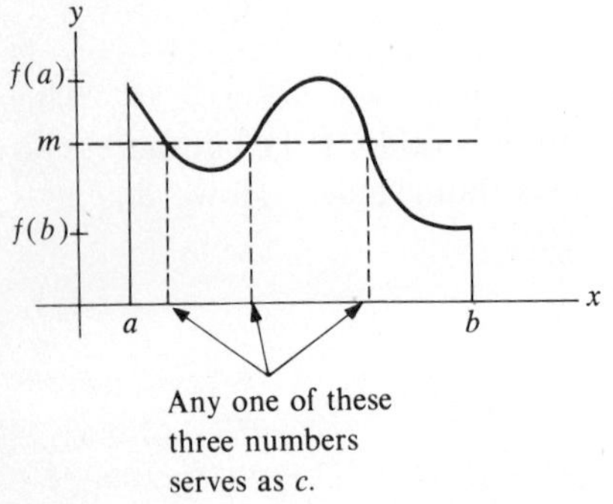

In ordinary English, the intermediate-value theorem reads: A continuous function defined on $[a, b]$ takes on all values between $f(a)$ and $f(b)$. Pictorially, it asserts that a horizontal line of height m must meet the graph of f at least once if m is between $f(a)$ and $f(b)$, as shown in Fig. 2.56. In other words, when you move a pencil along the graph of a continuous function from one height to another, the pencil passes through all intermediate heights.

Even though the theorem guarantees the existence of c, it does *not tell how* to find it. To find c, we must solve an equation, namely $f(c) = m$.

Any one of these three numbers serves as c.

Figure 2.56

EXAMPLE 5 Use the intermediate-value theorem to show that the equation $2x^3 + x^2 - x + 1 = 5$ has a solution in the interval $[1, 2]$.

SOLUTION Let $P(x) = 2x^3 + x^2 - x + 1$. Then

$$P(1) = 2 \cdot 1^3 + 1^2 - 1 + 1 = 3$$

and

$$P(2) = 2 \cdot 2^3 + 2^2 - 2 + 1 = 19.$$

The intermediate-value theorem can help guarantee a solution.

Since P is continuous and 5 is between $P(1) = 3$ and $P(2) = 19$, we may apply the intermediate-value theorem to P in the case $a = 1$, $b = 2$, and $m = 5$. Thus there is at least one number c between 1 and 2 such that $P(c) = 5$. This completes the answer.

(To get a more accurate estimate for a number c such that $P(c) = 5$, find a shorter interval for which the intermediate-value theorem can be applied. For instance, $P(1.2) \approx 4.7$ and $P(1.3) \approx 5.8$. By the intermediate-value theorem, there is a number c in $[1.2, 1.3]$ such that $P(c) = 5$.) ∎

EXAMPLE 6 Show that the equation $x^5 - 2x^2 + x + 11 = 0$ has at least one real root.

SOLUTION For x large and positive the polynomial $P(x) = x^5 - 2x^2 + x + 11$ is positive (since $\lim_{x \to \infty} P(x) = \infty$). Thus there is a number b such that $P(b) > 0$. Similarly, for x negative and of large absolute value, $P(x)$

is negative (since $\lim_{x \to -\infty} P(x) = -\infty$). Select a number a such that $P(a) < 0$.

The number 0 is between $P(a)$ and $P(b)$. Since P is continuous on the interval $[a, b]$, there is a number c in $[a, b]$ such that $P(c) = 0$. This number c is a real solution to the equation $x^5 - 2x^2 + x + 11 = 0$. ∎

Any polynomial of odd degree has a real root.

Note that the argument in Example 6 applies to any polynomial of *odd* degree. The argument does not hold for polynomials of even degree, since the equations $x^2 + 1 = 0$, $x^4 + 1 = 0$, $x^6 + 1 = 0$, and so on, have no real solutions.

EXAMPLE 7 Use the intermediate-value theorem to show that there is a number c in $[1, 2]$ such that $4 - c = 2^c$.

SOLUTION Introduce the function $f(x) = (4 - x) - 2^x$, the difference of $4 - x$ and 2^x, and show that there is a number c in $[1, 2]$ such that $f(c) = 0$. Observe that

$$f(1) = 4 - 1 - 2^1 = 1,$$
$$f(2) = 4 - 2 - 2^2 = -2,$$

and that f is continuous.

Since 0 is between $f(1)$ and $f(2)$, we may apply the intermediate-value theorem with $a = 1$, $b = 2$, and $m = 0$. The theorem assures us that there is at least one number c in $[1, 2]$ such that $f(c) = 0$, that is, $4 - c - 2^c = 0$. From this it follows that $4 - c = 2^c$. ∎

In Example 7 the intermediate-value theorem does not tell what c is. The graphs of $4 - x$ and 2^x in Fig. 2.57 suggest that c is unique and about 1.4. Further arithmetic, done on a calculator, shows that $c \approx 1.39$.

From Examples 6 and 7 we make two observations:

1. If a continuous function defined on an interval is positive somewhere in the interval and negative somewhere in the interval, then it must be 0 at some number in that interval.
2. To show that two functions are equal at some number in an interval, show that their difference is 0 at some number in the interval.

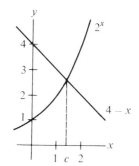

Figure 2.57

Exercises

1. Does the function $(x^3 + x^4)/(1 + 5x^2 + x^6)$ have (a) a maximum value for x in $[1, 4]$? (b) a minimum value for x in $[1, 4]$?
2. Does the function $2^x - x^3 + x^5$ have (a) a maximum value for x in $[-3, 10]$? (b) a minimum value for x in $[-3, 10]$?
3. Does the function x^3 have a maximum value for x in (a) $[2, 4]$? (b) $[-3, 5]$? (c) $(1, 6)$?
4. Does the function x^4 have a minimum value for x in (a) $[-5, 6]$? (b) $(-2, 4)$? (c) $(3, 7)$? (d) $(-4, 4)$?
5. Does the function $2 - x^2$ have (a) a maximum value for x in $(-1, 1)$? (b) a minimum value for x in $(-1, 1)$?
6. Does the function $2 + x^2$ have (a) a maximum value for x in $(-1, 1)$? (b) a minimum value for x in $(-1, 1)$?

102 LIMITS AND CONTINUOUS FUNCTIONS

7 Show that the equation $x^5 + 3x^4 + x - 2 = 0$ has at least one root in the interval $[0, 1]$.
8 Show that the equation $x^5 - 2x^3 + x^2 - 3x + 1 = 0$ has at least one root in the interval $[1, 2]$.

In Exercises 9 to 15 verify the intermediate-value theorem for the specified function f, the interval $[a, b]$, and the indicated value m. Find all c's in each case.

9 Function $3x + 5$; interval $[1, 2]$; $m = 10$.
10 Function $x^2 - 2x$; interval $[-1, 4]$; $m = 5$.
11 Function $\sin x$; interval $[\pi/2, 11\pi/2]$; $m = 0$.
12 Function $\cos x$; interval $[0, 5\pi]$; $m = 0$.
13 Function $\cos x$; interval $[0, 5\pi]$; $m = \frac{1}{2}$.
14 Function 2^x; interval $[0, 3]$; $m = 4$.
15 Function $x^3 - x$; interval $[-2, 2]$; $m = 0$.
16 Show that the equation $2^x - 3x = 0$ has a solution in the interval $[0, 1]$.
17 Show that the equation $x + \sin x = 1$ has a solution in the interval $[0, \pi/2]$.
18 Show that the equation $x^3 = 2^x$ has a solution in the interval $[1, 2]$.
19 Use the intermediate-value theorem to show that the equation $3x^3 + 11x^2 - 5x = 2$ has a solution.

■

20 Let $f(x) = 1/x$, $a = -1$, $b = 1$, $m = 0$. Note that $f(a) \leq 0 \leq f(b)$. Is there at least one c in $[a, b]$ such that $f(c) = 0$? If so, find c; if not, does this imply that the intermediate-value theorem is sometimes false?
21 Let $f(x) = \sin x$, $a = 0$, $b = 5\pi/2$.
    (a) Draw the graph of f from $x = 0$ to $x = 5\pi/2$.
    (b) For $m = 0$, find all the c's guaranteed by the intermediate-value theorem.
    (c) How many c's are there for various choices of m, $f(a) \leq m \leq f(b)$?
22 Let $P(x) = a_n x^n + a_{n-1} x^{n-1} + \cdots + a_0$ be a polynomial of odd degree n and with lead coefficient a_n positive. Show that there is at least one real number r such that $P(r) = 0$.
23 (This continues Exercise 22.) The *factor theorem* from algebra asserts that the number r is a root of the polynomial $P(x)$ if and only if $x - r$ is a factor of $P(x)$. For instance, 2 is a root of the polynomial $x^2 - 3x + 2$ and $x - 2$ is a factor of the polynomial: $x^2 - 3x + 2 = (x - 2)(x - 1)$.
    (a) Use the factor theorem and Exercise 22 to show that every polynomial of odd degree has a factor of degree one.
    (b) Show that none of the polynomials $x^2 + 1$, $x^4 + 1$, or $x^{100} + 1$ has a first degree factor.
    (c) Check that $x^4 + 1 = (x^2 + \sqrt{2}x + 1)(x^2 - \sqrt{2}x + 1)$. (It can be shown using complex numbers that every polynomial is the product of polynomials of degrees at most two.)

24 Let f and g be two continuous functions defined at least on the interval $[a, b]$. Assume that $f(a) < g(a)$ and that $f(b) > g(b)$. Prove that there is a number c in (a, b) such that $f(c) = g(c)$. *Hint:* Apply the intermediate-value theorem to the function h defined by $h(x) = f(x) - g(x)$.

■ ■

A set in the plane bounded by a curve is *convex* if for any two points P and Q in the set the line segment joining them also lies in the set. (See Fig. 2.58.)

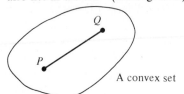

Figure 2.58

Circles, triangles, and parallelograms are convex sets. The quadrilateral shown in Fig. 2.59 is not convex. Convex sets will be referred to in the following exercises and occasionally in the exercises of later chapters.

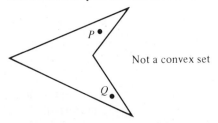

Figure 2.59

In Exercises 25 to 31 you will need to define various functions geometrically. *You may assume that they are continuous.*

25 Let L be a line in the plane and let K be a convex set. Show that there is a line parallel to L that cuts K into two pieces of equal area. *Hint:* Consider all lines parallel to L that meet K and notice how they divide K. Apply the intermediate-value theorem to an appropriate function.
26 Let P be a point in the plane and let K be a convex set. Show that there is a line through P that cuts K into two pieces of equal area.
27 Let K_1 and K_2 be two convex sets in the plane. Show that there is a line that simultaneously cuts K_1 into two pieces of equal area and cuts K_2 into two pieces of equal area.
28 Let K be a convex set in the plane. Show that there is a line that simultaneously cuts K into two pieces of equal area and cuts the boundary of K into two pieces of equal length.

29 Let K be a convex set. Show that there are two perpendicular lines that cut K into four pieces of equal area. (It is not known whether it is always possible to find two perpendicular lines that divide K into four pieces whose areas are $\frac{1}{8}, \frac{1}{8}, \frac{3}{8}$, and $\frac{3}{8}$ of the area of K, with the parts of equal area sharing an edge, as in Fig. 2.60.)

30 Let K be a convex set. Show that there is at least one square circumscribed about K (that is, a square that contains K and such that each edge of the square touches K).

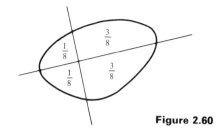

Figure 2.60

2.S Summary

The two central ideas of this chapter are limits and continuous functions. The first two sections examined limits in an informal and computational way. In Sec. 2.3 limits were used to find asymptotes. The optional Sec. 2.4 gave a precise (rigorous) definition of the limit concept. The next section determined two limits needed in the next chapter:

$$\lim_{\theta \to 0} \frac{\sin \theta}{\theta} = 1 \quad \text{and} \quad \lim_{\theta \to 0} \frac{1 - \cos \theta}{\theta} = 0.$$

The final two sections, Secs. 2.6 and 2.7, concerned continuous functions. The definition of a continuous function is phrased in terms of behavior of the function at and near each number in an interval. Though the definition of continuity depends on the definition of limit, the easiest way to think of a function that is continuous throughout an interval is that its graph is a curve that can be drawn without lifting pencil from paper. Though this graph may have sharp corners it does not have holes or points stranded from the rest of the graph.

Two properties of continuous functions, known as the maximum-value property and the intermediate-value property, were discussed. They will be used often in later chapters.

VOCABULARY AND SYMBOLS

limit $\lim_{x \to a} f(x)$, $\lim_{x \to \infty} f(x)$, etc.
right-hand limit $\lim_{x \to a^+} f(x)$
left-hand limit $\lim_{x \to a^-} f(x)$
constant function
polynomial function
rational function
asymptote (vertical, horizontal, tilted)
greatest integer less than or equal to x, $[x]$
continuous at a
continuous on an interval
continuous
sum, difference, product, quotient of functions $f + g, f - g, fg, f/g$
maximum value, minimum value, extreme value
maximum-value theorem
intermediate-value theorem

104 LIMITS AND CONTINUOUS FUNCTIONS

KEY FACTS For the definition of the various limits, such as

$$\lim_{x \to a} f(x) = L, \qquad \lim_{x \to \infty} f(x) = L, \quad \text{and} \quad \lim_{x \to -\infty} f(x) = L,$$

see the table in Sec. 2.2. (The precise definitions of limits in Sec. 2.4 are not covered in this summary.)

PROPERTIES OF LIMITS

If $\lim_{x \to a} f(x)$ and $\lim_{x \to a} g(x)$ both exist, then

$$\lim_{x \to a} (f(x) + g(x)) = \lim_{x \to a} f(x) + \lim_{x \to a} g(x)$$

$$\lim_{x \to a} (f(x) - g(x)) = \lim_{x \to a} f(x) - \lim_{x \to a} g(x)$$

$$\lim_{x \to a} f(x)g(x) = \lim_{x \to a} f(x) \lim_{x \to a} g(x)$$

$$\lim_{x \to a} f(x)/g(x) = \lim_{x \to a} f(x)/\lim_{x \to a} g(x), \quad \text{if } \lim_{x \to a} g(x) \neq 0$$

$$\lim_{x \to a} f(x)^{g(x)} = \left(\lim_{x \to a} f(x)\right)^{\lim_{x \to a} g(x)}, \quad \text{if } \lim_{x \to a} f(x) > 0.$$

LIMITS OF RATIONAL FUNCTIONS

$$\lim_{x \to \infty} \frac{ax^n + \cdots}{bx^m + \cdots} = \lim_{x \to \infty} \frac{ax^n}{bx^m}$$

(The degree of the numerator is n and the degree of the denominator is m.)

Consequently,

$$\lim_{x \to \infty} \frac{ax^n + \cdots}{bx^m + \cdots} = \begin{cases} a/b & \text{if } m = n \\ 0 & \text{if } n < m \\ \infty \text{ or } -\infty & \text{if } n > m \text{ (depending on the signs of } a \text{ and } b\text{).} \end{cases}$$

Similar assertions hold for $x \to -\infty$.

CONTINUITY

Let $f(x)$ be defined for all x. Then f is *continuous* at $x = a$ if

  1 $\lim_{x \to a} f(x)$ exists

and

  2 $\lim_{x \to a} f(x)$ equals $f(a)$.

A similar definition holds if f, though not defined for all x, is defined at least on some open interval that includes the number a.

For the definitions of "continuous from the right" and "continuous from the left" see Sec. 2.6.

A function that is continuous at each point of an open interval is said to be *continuous* on that interval. Similar definitions cover functions whose domains are closed intervals.

If f and g are continuous on the same interval so are $f+g$, $f-g$, and fg. Moreover, f/g is continuous wherever $g(x)$ is not 0. The composition of continuous functions is continuous.

If f is continuous and $\lim_{x \to a} g(x)$ exists and is in the domain of f, then

$$\lim_{x \to a} f(g(x)) = f\left(\lim_{x \to a} g(x)\right).$$

("It is legal to move 'lim' past 'f' if f is continuous.")

MAXIMUM-VALUE THEOREM

Let f be continuous throughout the closed interval $[a, b]$. Then there is at least one number in $[a, b]$ at which f takes on a maximum value. That is, for some number c in $[a, b]$,

$$f(c) \geq f(x)$$

for all x in $[a, b]$.

(A corresponding minimum-value theorem also holds.)

INTERMEDIATE-VALUE THEOREM

Let f be continuous throughout the closed interval $[a, b]$. Let m be any number between $f(a)$ and $f(b)$. Then there is at least one number c in $[a, b]$ such that

$$f(c) = m.$$

In particular, if f is continuous throughout $[a, b]$ and if one of $f(a)$ and $f(b)$ is negative and the other is positive, then there is a number c in $[a, b]$ such that $f(c) = 0$.

With the aid of the preceding fact, it was shown that a polynomial of odd degree has at least one real root. In other words, the graph of a polynomial of odd degree always crosses the x axis at least once.

Guide quiz on limits and continuous functions

1. Define "$\lim_{x \to a} f(x) = L$." (The informal definition, not the one in Sec. 2.4.)
2. Define "$\lim_{x \to \infty} f(x) = L$." (The informal definition, not the one in Sec. 2.4.)
3. Figure 2.61 is the graph of a function f whose domain is $[1, 5]$.

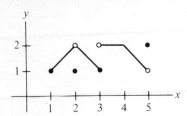

Figure 2.61

(a) Does $\lim_{x \to 2} f(x)$ exist?
(b) Is f continuous at 2?
(c) Does $\lim_{x \to 3} f(x)$ exist?
(d) Does $\lim_{x \to 5} f(x)$ exist?
(e) Is f continuous on $[1, 5]$?
(f) Is f continuous on $(3, 5)$?

4. Examine the following limits:

(a) $\lim_{x \to 1} (x^2 + 5x)$
(b) $\lim_{x \to \infty} \dfrac{3x^4 - 100x + 3}{5x^4 + 7x - 1}$
(c) $\lim_{x \to 0} \dfrac{3x^4 - 100x + 3}{5x^4 + 7x - 1}$
(d) $\lim_{x \to -\infty} \dfrac{500x^3 - x^2 - 5}{x^4 + x}$
(e) $\lim_{t \to 0} \dfrac{\sin 3t}{6t}$
(f) $\lim_{x \to -\infty} \dfrac{-6x^5 + 4x}{x^2 + x + 5}$
(g) $\lim_{x \to \infty} 2^{-x}$
(h) $\lim_{x \to 0} \dfrac{x^3 + 8}{x + 2}$
(i) $\lim_{x \to -2} \dfrac{x^3 + 8}{x + 2}$
(j) $\lim_{x \to 0} \sin \dfrac{1}{x}$
(k) $\lim_{x \to \infty} \sin x$
(l) $\lim_{x \to \infty} \dfrac{1 + 3 \cos x}{x^2}$
(m) $\lim_{x \to \infty} (\sqrt{4x^2 + 5x} - \sqrt{4x^2 + x})$
(n) $\lim_{x \to 16} \dfrac{\sqrt{x} - 4}{x - 16}$

5. If $\lim_{x \to a} f(x) = 3$ and $\lim_{x \to a} g(x) = 4$, what, if anything, can be said about
(a) $\lim_{x \to a} f(x)g(x)$?
(b) $\lim_{x \to a} f(x)/g(x)$?
(c) $\lim_{x \to a} (f(x) + g(x))$?
(d) $\lim_{x \to a} (f(x) - 3)/(g(x) - 4)$?
(e) $\lim_{x \to a} (f(x) - 3)^{g(x)}$?

6. If $\lim_{x \to a} f(x) = 0$ and $\lim_{x \to a} g(x) = \infty$, what, if anything, can be said about
(a) $\lim_{x \to a} (f(x) + g(x))$?
(b) $\lim_{x \to a} f(x)/g(x)$?
(c) $\lim_{x \to a} f(x)^{g(x)}$?
(d) $\lim_{x \to a} (2 + f(x))^{g(x)}$?
(e) $\lim_{x \to a} f(x)g(x)$?

7. (a) State the assumptions in the maximum-value theorem.
   (b) State the conclusion of the maximum-value theorem.
8. (a) State the assumptions in the intermediate-value theorem.
   (b) State the conclusion of the intermediate-value theorem.

Review exercises for chap. 2

In Exercises 1 to 36 examine the limits. Evaluate those that exist. Determine those that do not exist and, among these, the ones that are infinite.

1. $\lim_{x \to 1} \dfrac{x^3 + 1}{x^2 + 1}$
2. $\lim_{x \to 1} \dfrac{x^3 - 1}{x^2 - 1}$
3. $\lim_{x \to 2} \dfrac{x^4 - 16}{x^3 - 8}$
4. $\lim_{x \to 0} \dfrac{x^4 - 16}{x^3 - 8}$
5. $\lim_{x \to \infty} \dfrac{x^7 - x^2 + 1}{2x^7 + x^3 + 300}$
6. $\lim_{x \to -\infty} \dfrac{x^9 + 6x + 3}{x^{10} - x - 1}$
7. $\lim_{x \to -\infty} \dfrac{x^3 + 1}{x^2 + 1}$
8. $\lim_{x \to -\infty} \dfrac{x^4 + x^2 + 1}{3x^2 + 4}$
9. $\lim_{x \to 4} \dfrac{\sqrt{x} - 2}{x - 4}$
10. $\lim_{x \to 81} \dfrac{x - 81}{\sqrt{x} - 9}$
11. $\lim_{x \to \infty} (\sqrt{x^2 + 2x + 3} - \sqrt{x^2 - 2x + 3})$
12. $\lim_{x \to \infty} (\sqrt{2x^2} - \sqrt{2x^2 - 6x})$
13. $\lim_{x \to 1^+} \dfrac{1}{x - 1}$
14. $\lim_{x \to 1^-} \dfrac{1}{x - 1}$
15. $\lim_{x \to 3^-} [2x]$ (greatest-integer function)
16. $\lim_{x \to 3^+} [2x]$ (greatest-integer function)
17. $\lim_{x \to 0^+} 2^{1/x}$
18. $\lim_{x \to 0^-} 2^{1/x}$

19. $\lim_{x \to \infty} 2^{1/x}$

20. $\lim_{x \to -\infty} 2^{1/x}$

21. $\lim_{x \to \infty} \frac{(x+1)(x+2)}{(x+3)(x+4)}$

22. $\lim_{x \to -\infty} \frac{(x+1)^{100}}{(2x+50)^{100}}$

23. $\lim_{x \to \pi/2} \frac{\cos x}{1 + \sin x}$

24. $\lim_{x \to \pi/2} \frac{\cos x}{1 - \sin x}$

25. $\lim_{x \to 0} \frac{\sin x}{3x}$

26. $\lim_{x \to \infty} \frac{\sin x}{3x}$

27. $\lim_{x \to \pi/2^+} \cos x$

28. $\lim_{x \to \pi/2^+} \sec x$

29. $\lim_{x \to 0^-} \sin x$

30. $\lim_{x \to 0^-} \csc x$

31. $\lim_{x \to \infty} \sin \frac{1}{x}$

32. $\lim_{x \to \infty} x \sin \frac{1}{x}$

33. $\lim_{x \to \pi/4} x^2 \cos x$

34. $\lim_{x \to \infty} x^2 \cos x$

35. $\lim_{\theta \to \infty} (\cos^2 \theta + \sin^2 \theta)$

36. $\lim_{\theta \to \infty} (\cos^2 \theta - \sin^2 \theta)$

In Exercises 37 to 42 exhibit specific functions f and g that meet all three conditions. (The answers are not unique.)

37. $\lim_{x \to 0} f(x) = 0$, $\lim_{x \to 0} g(x) = 0$,
    and $\lim_{x \to 0} f(x)/g(x) = 5$

38. $\lim_{x \to \infty} f(x) = 0$, $\lim_{x \to \infty} g(x) = \infty$,
    and $\lim_{x \to \infty} f(x)g(x) = 20$

39. $\lim_{x \to \infty} f(x) = 0$, $\lim_{x \to \infty} g(x) = \infty$,
    and $\lim_{x \to \infty} f(x)g(x) = \infty$

40. $\lim_{x \to \infty} f(x) = \infty$, $\lim_{x \to \infty} g(x) = \infty$,
    and $\lim_{x \to \infty} (f(x) - g(x)) = 3$

41. $\lim_{x \to \infty} f(x) = \infty$, $\lim_{x \to \infty} g(x) = \infty$,
    and $\lim_{x \to \infty} (f(x) - g(x)) = \infty$

42. $\lim_{x \to \infty} f(x) = \infty$, $\lim_{x \to \infty} g(x) = \infty$,
    and $\lim_{x \to \infty} f(x)/g(x) = \infty$

43. Does $x + \sin x$ have a maximum value for x in (a) $[0, 100]$? (b) $[0, \infty)$?

44. Does $x^3 + x + 1$ have a minimum value for x in $[-100, 5]$? (b) $(-\infty, 5]$?

45. Does $1/(1 + x^2)$ have (a) a maximum value for x in $(-1, 1)$? (b) a minimum value for x in $(-1, 1)$?

46. Does $1/x^3$ have a maximum value for x in (a) $[2, 100]$? (b) $[2, \infty)$? A minimum value for x in (c) $[2, 100]$? (d) $[2, \infty)$?

47. Show that the equation $x^5 = 2^x$ has a solution (a) less than 2, (b) greater than 2.

48. Show that the equation $x^3 - 2x^2 - 3x + 1 = 0$ has a solution (a) less than 0, (b) in $[0, 2]$, (c) larger than 2.

■

49. As $\theta \to 0$, what happens to the quotients

$$\frac{\text{Length of } AP}{\text{Length of arc } BP}$$

and

$$\frac{\text{Length of } AB}{\text{Length of arc } BP},$$

where A, B, and P are shown in Fig. 2.62?

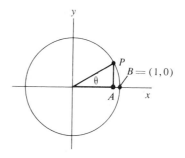

Figure 2.62

50. Assume that $\lim_{x \to 3} f(x) = 0$ and $\lim_{x \to 3} g(x) = 0$. What, if anything, can be said about
    (a) $\lim_{x \to 3} (f(x) - g(x))$? (b) $\lim_{x \to 3} \sin f(x)$?
    (c) $\lim_{x \to 3} \cos f(x)$? (d) $\lim_{x \to 3} f(x)g(x)$?
    (e) $\lim_{x \to 3} [f(x)]^3/g(x)$?

In each of Exercises 51 to 53 verify the intermediate-value theorem for the indicated function, closed interval, and value.

51. 2^x, $[1, 8]$, $m = 4$.
52. $\sin x$, $[0, 9\pi/2]$, $m = \frac{1}{2}$.
53. $\tan x$, $[0, \pi/3]$, $m = 1$.

54. Find $\lim_{x \to 0} \frac{\tan x - \sin x}{x}$.

55. Find $\lim_{x \to 0} \frac{\tan x - \sin x}{x^2}$.

56. Let $f(x) = \sin(1/x)$ if $x \neq 0$. Is it possible to define $f(0)$ in such a way that f is continuous throughout the x axis?

57. Let $f(x) = x \sin(1/x)$ if $x \neq 0$. Is it possible to define $f(0)$ in such a way that f is continuous throughout the x axis?

108 LIMITS AND CONTINUOUS FUNCTIONS

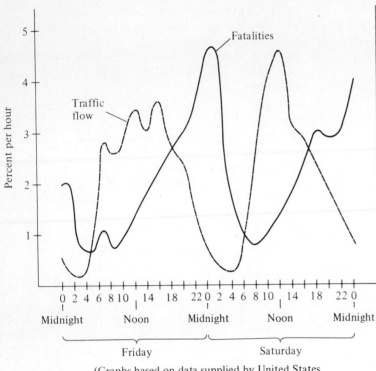

Figure 2.63

(Graphs based on data supplied by United States Department of Transportation and by Caltrans.)

58 The amount of automobile traffic and the rate of fatal automobile accidents vary with the day of the week and the time of the day. Let $f(x)$ = number of automobiles on the roads at time x. Let $g(x)$ = number of fatal accidents per hour at time x.
(a) What function defined in terms of f and g would best represent the danger or risk of driving at time x?
(b) Figure 2.63 is the graph of f and g on a typical 48-hour period, Friday and Saturday. (The vertical scale is proportional to the numbers of cars and fatal accidents.)

The units on the vertical scale depend on the particular road or freeway. On the basis of Fig. 2.63 estimate what is the most dangerous time to drive and what is the safest. How many times as risky is the first in comparison to the second? ■ ■

59 Two citizens are arguing about
$$\lim_{x \to \infty} \left(\frac{3x^2 + 2x}{x + 5} - 3x \right).$$
The first claims, "For large x, $2x$ is small in comparison to $3x^2$, and 5 is small in comparison to x. So $(3x^2 + 2x)/(x + 5)$ behaves like $3x^2/x = 3x$. Hence the limit in question is 0." Her companion replies, "Nonsense. After all,
$$\frac{3x^2 + 2x}{x + 5} = \frac{3x + 2}{1 + (5/x)},$$
which clearly behaves like $3x + 2$ for large x. Thus the limit in question is 2, not 0."
Settle the argument.

60 Let f be a function defined on the x axis, with the property that
$$f(x + y) = f(x) + f(y)$$
for all x and y. Assume that $x > y$ implies that $f(x) \geq f(y)$. Prove that $f(x) = cx$ for some constant c.

61 (See Exercise 60.) Let f be a function defined on the x axis with the property that
$$f(x + y) = f(x) + f(y) \quad \text{and} \quad f(xy) = f(x)f(y)$$
for all x and y.
(a) Prove that $x > y$ implies that $f(x) \geq f(y)$. Hint: If $x > y$, then there is a number u such that $x = y + u^2$.
(b) Prove that $f(x) = x$ for all x, or else that $f(x) = 0$ for all x.

62 Let f be a continuous function defined on the x axis. Assume that $f(f(f(x))) = x$ for all x. Prove that $f(x) = x$ for all x. Hint: First show that, if $f(0) > 0$, there is a number b such that $f(b) < b$.

63 Let f be a continuous function such that $f(x)$ is in $[0, 1]$ when x is in $[0, 1]$. Prove that there is at least one number c in $[0, 1]$ such that $f(c) = c$. *Hint:* Consider the function g given by $g(x) = f(x) - x$.

64 Examine (a) $\lim_{x \to 0^+} (4^x + 3^x)^{1/x}$
(b) $\lim_{x \to 0^-} (4^x + 3^x)^{1/x}$ (c) $\lim_{x \to \infty} (4^x + 3^x)^{1/x}$.

65 Let f be a continuous function such that $f(f(x)) = x$ for all x. Prove that there is at least one number c such that $f(c) = c$.

66 If f is a function, then by a *chord of* f we shall mean a line segment whose ends are on the graph of f. Now let f be continuous throughout $[0, 1]$, and let $f(0) = f(1) = 0$.
(a) Explain why there is a horizontal chord of f of length $\frac{1}{2}$.
(b) Explain why there is a horizontal chord of f of length $1/n$, where $n = 1, 2, 3, 4, \ldots$.
(c) Must there exist a horizontal chord of f of length $\frac{2}{3}$?
(d) What is the answer to (c) if we also demand that $f(x) \geq 0$ for all x in $[0, 1]$?

THE DERIVATIVE

This chapter introduces one of the most important concepts of calculus, the derivative. The approach uses geometric and physical illustrations, but a few examples and exercises show that the derivative has far more varied applications.

> **REFERENCES TO APPENDIXES**
> In Sec. 3.2: factoring of $d^n - c^n$, Appendix B
> In Sec. 3.5: identities for $\sin(A+B)$ and $\cos(A+B)$, $\cot x$, $\sec x$, $\csc x$, Appendix D

3.1 Four problems with one theme

This section discusses four problems which at first glance may seem unrelated. The first one concerns the slope of a tangent line to a curve. The second involves velocity. The final two concern magnification and density. A little arithmetic will quickly show that they are all just different versions of one mathematical idea.

SLOPE

PROBLEM 1 *Slope.* What is the slope of the tangent line to the graph of $y = x^2$ at the point $P = (2, 4)$, as shown in Fig. 3.1?

For the present, by the *tangent line* to a curve at a point P on the curve shall be meant the line through P that has the "same direction" as the curve at P. This will be made precise in the next section.

3.1 FOUR PROBLEMS WITH ONE THEME 111

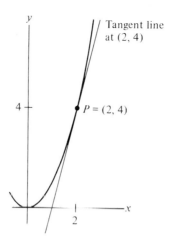

Figure 3.1

SOLUTION One approach consists of carefully drawing the parabola $y = x^2$ and trying to draw the line tangent to the curve at $(2, 4)$. Though this is a reasonable method, its precision is limited (as some of the exercises will show). In this particular example, where the tangent line is so nearly vertical, a slight error in guessing the angle that the tangent line makes with the x axis can cause a large error in estimating the slope.

We shall choose a different approach, one which is perfectly accurate. As a start, compute the slope of a line that approximates the tangent line at $P = (2, 4)$. To do this, take a point Q near P on the curve $y = x^2$, say the point $Q = (2.1, 2.1^2)$, and compute the slope of the line passing through P and Q, as shown in Figs. 3.2 and 3.3.

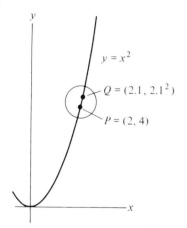

Figure 3.2

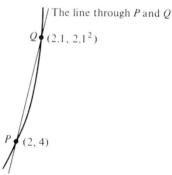

Figure 3.3

The slope of the line through P and Q is

$$\frac{2.1^2 - 2^2}{2.1 - 2},$$

which is

$$\frac{4.41 - 4}{0.1} = \frac{0.41}{0.1}$$

$$= 4.1.$$

112 THE DERIVATIVE

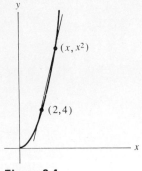

Figure 3.4

x could be > 2 or < 2.

Observe the use of limits.

Thus an estimate of the slope of the tangent line is 4.1. Note that in making this estimate there was no need to draw the curve.

To obtain a better estimate, we could repeat the process, using the line through $P = (2, 4)$ and $Q = (2.01, 2.01^2)$. Rather than do this, it is simpler to consider a *typical* point Q. That is, consider the line through $P = (2, 4)$ and $Q = (x, x^2)$ when x is near 2. (See Fig. 3.4.)

This line has slope
$$\frac{x^2 - 2^2}{x - 2}.$$

To find out what happens to this quotient as x gets closer to 2, apply the techniques of limits. Replace $x^2 - 2^2$ by $(x + 2)(x - 2)$. Thus for $x \neq 2$ the slope of the line through P and Q is

$$\frac{(x + 2)(x - 2)}{x - 2},$$

which equals $x + 2$. As x gets closer to 2, $x + 2$ approaches 4. Thus the tangent line at $(2, 4)$ has slope 4. ∎

SPEED

PROBLEM 2 *Speed.* A rock falls $16t^2$ feet in t seconds. What is its speed after 2 seconds?

SOLUTION For practice, make an estimate by finding the average speed of the rock during a short time interval, say from 2 to 2.01 seconds. At the start of this interval the rock had fallen $16(2^2) = 64$ feet. By the end it had fallen $16(2.01)^2 = 16(4.0401) = 64.6416$ feet. So, during 0.01 seconds it fell 0.6416 feet. Its average speed during this time interval is

$$\frac{0.6416}{0.01} = 64.16 \text{ feet per second},$$

an estimate of the speed at time $t = 2$ seconds.

Though we will keep $t > 2$, estimates could just as well be made with $t < 2$.

Rather than making another estimate with the aid of a still shorter interval of time, let us consider the typical time interval, from 2 to t seconds, $t > 2$. During this short time of $t - 2$ seconds the rock travels $16t^2 - 16 \cdot 2^2 = 16(t^2 - 2^2)$ feet, as shown in Fig. 3.5. The average speed of the rock during this period is

$$\frac{16(t^2 - 2^2)}{t - 2} \text{ feet per second}.$$

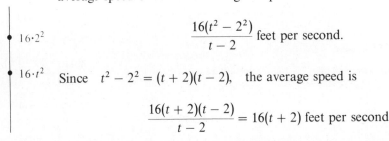

Figure 3.5

Since $t^2 - 2^2 = (t + 2)(t - 2)$, the average speed is

$$\frac{16(t + 2)(t - 2)}{t - 2} = 16(t + 2) \text{ feet per second}$$

during the time interval from 2 to t seconds, $t > 2$.

As t gets closer to 2 this average speed approaches

Observe the use of limits.
$$16(2 + 2) = 64 \text{ feet per second.}$$

Thus the speed of the rock at time 2 seconds (the so-called "instantaneous speed") is 64 feet per second. ■

Even though Problems 1 and 2 seem unrelated at first, their solutions turn out to be practically identical: The slope in Problem 1 is approximated by the quotient

$$\frac{x^2 - 2^2}{x - 2},$$

and the speed in Problem 2 is approximated by the quotient

$$\frac{16t^2 - 16 \cdot 2^2}{t - 2} = 16 \frac{t^2 - 2^2}{t - 2}.$$

The only difference between the solutions is that the second quotient had a 16, and a t instead of an x.

The answers to Problems 1 and 2 involve almost the same limits.
The similarity between the two problems can also be expressed in terms of limits. In Problem 1,

$$\text{Slope} = \lim_{x \to 2} \frac{x^2 - 2^2}{x - 2};$$

in Problem 2,
$$\text{Speed} = \lim_{t \to 2} \frac{16t^2 - 16 \cdot 2^2}{t - 2}.$$

MAGNIFICATION

The third problem concerns magnification, a concept that occurs in everyday life. For instance, photographs can be blown up or reduced in size, with each part magnified (or shrunk) by the same factor. However, magnification may vary from point to point, as with a curved mirror. If the projection of an interval of length L is an interval of length L^*, we say that the interval on the photograph is magnified by the factor L^*/L.

PROBLEM 3 *Magnification.* A light, two lines (a slide and a screen), and a complicated lens are placed as in Fig. 3.6. This arrangement projects the point on the bottom line, whose coordinate is x, to the point on the top line, whose coordinate is x^2. For example, 2 is projected onto 4, and 3 onto 9. The projection of the interval [2, 3] is [4, 9], which is five times as long. The magnification of the interval [2, 3] is said to be 5. The projection of the interval $[0, \frac{1}{3}]$ is $[0, \frac{1}{9}]$, which is only one-third as long. In this case the interval is magnified by a factor of $\frac{1}{3}$. For large x the lens magnifies to a great extent; for x near 0 the lens markedly reduces. What is the "magnification at $x = 2$"?

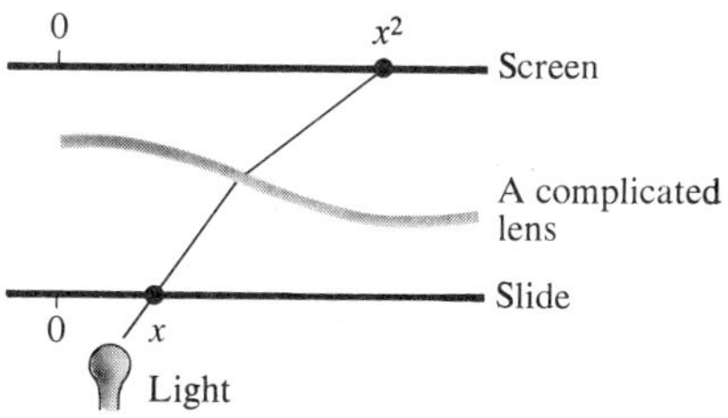

Figure 3.6

114   THE DERIVATIVE

SOLUTION   The lens projects the point having the coordinate 2 onto the point having the coordinate 2^2. More concisely, the image of 2 is $2^2 = 4$. The image of 3 is $3^2 = 9$; the image of 5 is $5^2 = 25$; and so on. Let us join some sample points to their images by straight lines, as in Fig. 3.7. This diagram shows that the interval [2, 3] on the slide is magnified to become the interval [4, 9] on the screen, a fivefold magnification. Similarly, [3, 4] on the slide has as its image on the screen [9, 16], a sevenfold magnification. The magnifying power of the lens increases from left to right.

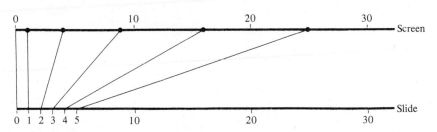

Figure 3.7

To estimate the magnification at 2 on the slide, examine the projection of a small interval in the vicinity of 2. Let us see what the image of [2, 2.1] is on the screen. Since the image of 2 is 2^2 and the image of 2.1 is 2.1^2, the image of the interval [2, 2.1] of length 0.1 is the interval $[2^2, 2.1^2]$ of length

$$2.1^2 - 2^2 = 0.41.$$

The magnifying factor over the interval [2, 2.1] is

$$\frac{0.41}{0.1} = 4.1$$

This number, 4.1, is an estimate of the magnification at $x = 2$.

You have probably guessed the next step. Rather than go on and consider the magnification of another specific interval, such as the interval [2, 2.01], we go directly to a typical interval with 2 as its left end.

For convenience we use $x > 2$. Estimates could be made with $x < 2$ also.   The image of the interval [2, x], where x is greater than 2, is the interval $[2^2, x^2]$. Since [2, x] has length $x - 2$ and $[2^2, x^2]$ has length $x^2 - 2^2$, the magnification of the interval [2, x] is

$$\frac{x^2 - 2^2}{x - 2} = x + 2, \qquad x > 2.$$

As already observed, when x approaches 2, this quotient approaches 4. Thus the magnification at 2 is 4. In terms of limits

$$\text{Magnification at 2} = \lim_{x \to 2} \frac{x^2 - 2^2}{x - 2}. \quad \blacksquare$$

DENSITY

The next problem is concerned with density, which is a measure of the heaviness of a material. Density is defined as the quotient

Density is mass divided by volume.

$$\frac{\text{Total mass}}{\text{Total volume}}.$$

Water has a density of 1 gram per cubic centimeter, while the density of lead is 11.3 grams per cubic centimeter. Air has a density of only 0.0013 grams per cubic centimeter. The density of an object may vary from point to point. For instance, the density of matter near the center of the earth is much greater than that near the surface. In fact, the *average* density of the earth is 5.5 grams per cubic centimeter, more than five times that of water.

The idea of density provides a concrete analog of several mathematical ideas and will be referred to frequently in later chapters. The next problem concerns a string of varying density. This density will be considered in terms of grams per linear centimeter, rather than grams per cubic centimeter. The matter is imagined as a continuous distribution, not composed of isolated molecules.

PROBLEM 4 *Density.* The mass of the left-hand x centimeters of a nonhomogeneous string 10 centimeters long is x^2 grams, as shown in Fig. 3.8. For instance, the left half has a mass of 25 grams, while the whole string has a mass of 100 grams. Clearly the right half is denser than the left half. What is the density, in grams per centimeter, of the material at $x = 2$?

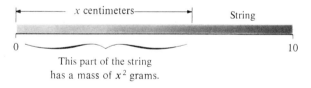

Figure 3.8

SOLUTION To estimate the density of the string 2 centimeters from its left end, examine the mass of the material in the interval $[2, 2.1]$. (See Fig. 3.9.)

Figure 3.9

The material in the interval $[2, 2.1]$ has a mass of $2.1^2 - 2^2$ grams, which equals $4.41 - 4 = 0.41$ gram. Thus, the average density for this interval is $0.41/0.1 = 4.1$ gram per centimeter.

Rather than make another estimate, consider the density in the typical small interval $[2, x]$. The mass in this interval is

$$x^2 - 2^2 \quad \text{grams.}$$

The interval has length $x - 2$ centimeters.

Thus the density of matter in this interval is

$$\frac{x^2 - 2^2}{x - 2} = x + 2 \quad \text{grams per centimeter.}$$

As x approaches 2, this quotient approaches 4, and we say that the density 2 centimeters from the left end of the string is 4 grams per centimeter.

In terms of limits,

$$\text{Density at 2} = \lim_{x \to 2} \frac{x^2 - 4}{x - 2}. \quad \blacksquare$$

From a mathematical point of view, the problems of finding the slope of the tangent line, the speed of the rock, the magnification of the lens, and the density along the string are the same. Each leads to the same type of quotient as an estimate. In each case the behavior of this quotient is studied as x (or t) approaches 2. In each case the answer is a limit.

The underlying mathematical theme is explored in the next section, which introduces the derivative.

Exercises

Exercises 1 to 8 concern *slope*.

1 Using the method of Problem 1, find the slope of the tangent line to $y = x^2$ at the point $(3, 3^2) = (3, 9)$.

2 Using the method of Problem 1, find the slope of the tangent line to $y = x^2$ at the point $(\frac{1}{2}, (\frac{1}{2})^2) = (\frac{1}{2}, \frac{1}{4})$.

3 Using the method of Problem 1, find the slope of the tangent line to $y = x^2$ at the point $(-2, (-2)^2) = (-2, 4)$.

4 Using the method of Problem 1, find the slope of the tangent line to $y = x^2$ at the point $(1, 1^2) = (1, 1)$.

5 Using the method of Problem 1, find the slope of the tangent line to $y = x^3$ at $(2, 2^3) = (2, 8)$. *Hint:* The identity $d^3 - c^3 = (d^2 + dc + c^2)(d - c)$ will be helpful.

6 Using the method of Problem 1, find the slope of the tangent line to $y = x^3$ at $(1, 1^3) = (1, 1)$. *Hint:* The identity $d^3 - c^3 = (d^2 + dc + c^2)(d - c)$ will be helpful.

7 (a) Using the method of Problem 1, find the slope of the tangent line to $y = x^2$ at $(0, 0)$.
 (b) Sketch the graph of $y = x^2$ and the tangent line at $(0, 0)$.

8 (a) Using the method of Problem 1, find the slope of the tangent line to $y = x^3$ at $(0, 0)$.
 (b) Sketch the graph of $y = x^3$ and the tangent line at $(0, 0)$. (Be especially careful when sketching the graph near $(0, 0)$.)

Exercises 9 to 14 concern *speed*.

9 Use the method of Problem 2 to find the speed of the rock after 3 seconds.

10 Use the method of Problem 2 to find the speed of the rock after $\frac{1}{2}$ second.

11 Use the method of Problem 2 to find the speed of the rock after 1 second.

12 Use the method of Problem 2 to find the speed of the rock after $\frac{1}{4}$ second.

13 A certain object travels t^3 feet in the first t seconds.
 (a) How far does it travel during the time interval from 2 to 2.1 seconds?
 (b) What is its average speed during that time interval?
 (c) Let t be any number larger than 2. Find the average speed of the object from time 2 to time t seconds.

(d) Find the speed of the object at time 2 seconds by letting t approach 2 in part (c).

14 A certain object travels t^3 feet in the first t seconds.
(a) Find its average speed during the time interval from 3 to 3.01 seconds.
(b) Find its average speed during the time interval from 3 to t seconds, $t > 3$.
(c) By letting t approach 3, in part (b), find the speed of the object at time 3 seconds.

In the slope problem the nearby point Q was always pictured as being to the right of P. The point Q can just as well have been chosen to the left of P. Exercises 15 and 16 illustrate this case.

15 Consider the parabola $y = x^2$.
(a) Find the slope of the line through $P = (2, 4)$ and $Q = (1.9, 1.9^2)$.
(b) Find the slope of the line through $P = (2, 4)$ and $Q = (1.99, 1.99^2)$.
(c) Find the slope of the line through $P = (2, 4)$ and $Q = (x, x^2)$, where $x < 2$.
(d) Show that as x approaches 2, the slope in (c) approaches 4.

16 Consider the curve $y = x^3$.
(a) Find the slope of the line through $P = (2, 2^3)$ and $Q = (1.9, 1.9^3)$.
(b) Find the slope of the line through $P = (2, 2^3)$ and $Q = (x, x^3)$, where $x < 2$.
(c) Show that as x approaches 2 the slope in (b) approaches 12.

The next two exercises are intended to emphasize the limitation of graphs in finding the slope of a tangent line.

17 (a) Draw the curve $y = x^2$ as carefully as you can.
(b) Draw as carefully as you can the tangent line at (4, 16).
(c) Using a ruler or the scale on your graph, estimate the slope of the line you drew in (b).
(d) Find the slope of the line through (4, 16) and the nearby point (4.01, (4.01)^2).
(e) Find the slope of the line through (4, 16) and the nearby point (3.99, (3.99)^2).
(f) How does your result in (c) compare with those in (d) and (e)?

18 (a) Draw the curve $y = x^2$ as carefully as you can.
(b) Draw by eye the tangent line at (−1, 1).
(c) What is the slope of the line you drew in (b)?
(d) Examining the appropriate quotients, show that the tangent line at (−1, 1) has slope −2.

Exercises 19 to 22 concern *magnification*.

19 By what factor does the lens in Problem 3 magnify the interval (a) [1, 1.1]? (b) [1, 1.01]? (c) [1, 1.001]?
(d) Find the magnification at 1.

20 By what factor does the lens in Problem 3 magnify the interval (a) [3, 3.1]? (b) [3, 3.01]? (c) [3, 3.001]?
(d) Find the magnification at 3.

21 By what factor does the lens in Problem 3 magnify the interval (a) [0.49, 0.5]? (b) [0.499, 0.5]? (c) Find the magnification at 0.5 by examining the magnification of intervals of the form $[x, 0.5]$, where $x < 0.5$.

22 By what factor does the lens in Problem 3 magnify the interval (a) [1.49, 1.5]? (b) [1.499, 1.5]? (c) Find the magnification at 1.5 by examining the typical interval of the form $[x, 1.5]$, where $x < 1.5$.

Exercises 23 and 24 concern *density*.

23 The left x centimeters of a string have a mass of x^2 grams.
(a) What is the mass in the interval [3, 3.01]?
(b) Using the interval [3, 3.01], estimate the density at 3.
(c) Using the interval [2.99, 3], estimate the density at 3.
(d) By considering intervals of the form $[3, x]$, $x > 3$, find the density at the point 3 centimeters from the left end.
(e) By considering intervals of the form $[x, 3]$, $x < 3$, find the density at the point 3 centimeters from the left end.

24 The left x centimeters of a string have a mass of x^2 grams.
(a) What is the mass in the interval [2, 2.01]?
(b) Using the interval [2, 2.01], estimate the density at 2.
(c) Using the interval [1.99, 2], estimate the density at 2.
(d) By considering intervals of the form $[2, x]$, $x > 2$, find the density at the point 2 centimeters from the left end.
(e) By considering intervals of the form $[x, 2]$, $x < 2$, find the density at the point 2 centimeters from the left end.

■

The next two exercises show that the idea common to the four problems in this section also appears in biology and economics.

25 A certain bacterial culture has a mass of t^2 grams after t minutes of growth.
(a) How much does it grow during the time interval [2, 2.01]?
(b) What is its rate of growth during the time interval [2, 2.01]?
(c) What is its rate of growth when $t = 2$?

26 A thriving business has a profit of t^2 million dollars in its first t years. Thus, from time $t = 3$ to time $t = 3.5$ (the first half of its fourth year) it has a profit of $(3.5)^2 − 3^2$ million dollars. This gives an annual rate of

$$\frac{(3.5)^2 - 3^2}{0.5} = 6.5 \text{ million dollars per year.}$$

(a) What is its annual rate of profit during the time interval [3, 3.1]?
(b) What is its annual rate of profit during the time interval [3, 3.01]?
(c) What is its annual rate of profit after 3 years?

27. (a) Graph the curve $y = 2x^2 + x$.
   (b) By eye, draw the tangent line to the curve at the point (1, 3). Using a ruler, estimate the slope of this tangent line.
   (c) Sketch the line that passes through the point (1, 3) and the point $(x, 2x^2 + x)$.
   (d) Find the slope of the line in (c).
   (e) Letting x get closer and closer to 1, find the slope of the tangent line at (1, 3). How close was your estimate in (b)?

28. An object travels $2t^2 + t$ feet in t seconds.
   (a) Find its average speed during the interval of time $[1, t]$, where t is greater than 1.
   (b) Letting t get closer and closer to 1, find the speed at time 1.

29. Find the slope of the tangent line to the curve $y = x^2$ of Problem 1 at the typical point $P = (x, x^2)$. To do this, consider the slope of the line through P and the nearby point $Q = (w, w^2)$ and let w approach x.

30. Find the speed of the falling rock of Problem 2 at any time t. To do this, consider the average speed during the time interval $[t, w]$ and then let w approach t.

31. Find the magnification of the lens in Problem 3 at the typical point x by considering the magnification of the short interval $[x, w]$, where $w > x$, and then let w approach x.

32. Find the density of the string in Problem 4 at a typical point x centimeters from the left end. To do this, consider the mass in a short interval $[x, w]$, where $w > x$, and let w approach x.

33. Does the tangent line to the curve $y = x^2$ at the point (1, 1) pass through the point (6, 12)?

34. An astronaut is traveling from left to right along the curve $y = x^2$. When she shuts off the engine, she will fly off along the line tangent to the curve at the point where she is at that moment. At what point should she shut off the engine in order to reach the point (a) (4, 9)? (b) (4, −9)? *Hint:* See Exercise 29.

35. (a) Sketch the curve $y = x^3 - x^2$.
   (b) Using the method of the nearby point, find the slope of the tangent line to the curve at the typical point $(x, x^3 - x^2)$. *Hint:* Let the nearby point be $(w, w^3 - w^2)$.
   (c) Find all points on the curve where the tangent line is horizontal.
   (d) Find all points where the tangent line has slope 1.

36. Answer the same questions as in Exercise 35 for the curve $y = x^3 - x$.

37. See Exercises 34 and 36. Where can an astronaut who is traveling from left to right along the curve $y = x^3 - x$ shut off the engine and pass through the point (2, 2)?

3.2 The derivative

The solution of the slope problem in Sec. 3.1 (as well as those of the magnification problem and the density problem) led to the limit

$$\lim_{x \to 2} \frac{x^2 - 2^2}{x - 2}.$$

The speed problem involved a similar limit,

$$\lim_{t \to 2} 16 \frac{t^2 - 2^2}{t - 2}.$$

These limits arose from the particular formulas x^2 and $16t^2$ that had been picked. In each case we formed a *difference quotient*,

$$\frac{\text{Difference in outputs}}{\text{Difference in inputs}},$$

and examined its limit as the change in the inputs was made smaller and smaller.

3.2 THE DERIVATIVE

The whole procedure can be carried out for functions other than x^2 and $16t^2$ and at numbers other than 2.

The four problems in Sec. 3.1 had one theme in common. The underlying common theme of the four problems in Sec. 3.1 is the important mathematical concept, the *derivative* of a numerical function, which will now be defined.

In the following definition x is fixed and w approaches x.

DEFINITION *The derivative of a function at the number x.* Let f be a function that is defined at least in some open interval that contains the number x. If

$$\lim_{w \to x} \frac{f(w) - f(x)}{w - x}$$

The f' notation exists, it is called the *derivative* of f at x and is denoted $f'(x)$.

If the function f is defined only to the right of x, in an interval of the form $[x, b)$, then in the preceding definition $w \to x$ would be replaced by $w \to x^+$. The function is then said to be "differentiable on the right." A similar stipulation is made if f is defined only in an interval of the form $(a, x]$, *The derivative at an endpoint* and the function is said to be "differentiable on the left."

The numerator, $f(w) - f(x)$, is the change, or difference, in the outputs; the denominator, $w - x$, is the change in the inputs. (See Fig. 3.10.) Keep in mind that w can be either to the right or left of x. Similarly, $f(w)$ can be either larger or smaller than $f(x)$.

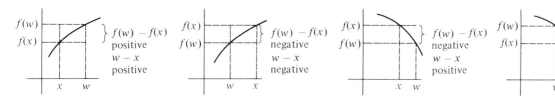

Figure 3.10

A few examples will illustrate the concept of the derivative.

EXAMPLE 1 Find the derivative of the squaring function at the number 2.

SOLUTION In this case, $f(x) = x^2$ for any input x. By definition, the derivative of this function at 2 is

$$\lim_{w \to 2} \frac{w^2 - 2^2}{w - 2}.$$

In Sec. 3.1, t and x played the role of w in this type of computation. To find this limit, proceed as in Sec. 3.1:

$$\lim_{w \to 2} \frac{w^2 - 2^2}{w - 2} = \lim_{w \to 2} \frac{(w + 2)(w - 2)}{w - 2}$$

$$= \lim_{w \to 2} (w + 2) \qquad w \neq 2$$

$$= 4.$$

We say that "the derivative of the function x^2 at 2 is 4." ∎

120 THE DERIVATIVE

The next example determines the derivative of the squaring function at any input, not just at 2.

EXAMPLE 2 Find the derivative of the function x^2 at any number x.

SOLUTION By definition, the derivative at x is

$$\lim_{w \to x} \frac{w^2 - x^2}{w - x} = \lim_{w \to x} \frac{(w + x)(w - x)}{w - x}$$

$$= \lim_{w \to x} (w + x) \qquad w \neq x$$

$$= x + x$$

$$= 2x.$$

The derivative of the squaring function at x is $2x$. ∎

That the derivative of the function x^2 is the function $2x$ is denoted

$$(x^2)' = 2x.$$

A warning on notation

This notation is convenient when dealing with a specific function. (*Warning*: Don't replace x by a specific number in this notation. For instance, do not write that $(3^2)'$ equals $2 \cdot 3$. This is not correct.)

The result in Example 2 can be interpreted in terms of each of the four problems in Sec. 3.1. For example, we now know from Example 2 that the slope of the tangent line to the parabola $y = x^2$ at the point (x, x^2) is $2x$. In particular, the slope of the tangent line at $(2, 2^2)$ is $2 \cdot 2 = 4$, a result found in Sec. 3.1. Also, according to the formula for the derivative, $(x^2)' = 2x$, the slope of the tangent line to $y = x^2$ at $(-2, (-2)^2)$ is $2 \cdot (-2) = -4$ and at $(0, 0)$ is $2 \cdot 0 = 0$. A glance at Fig. 3.11 shows that these are reasonable results. The derivative of x^2 is a function. It assigns to the number x the slope of the tangent line to the parabola $y = x^2$ at the point (x, x^2).

The next two examples illustrate the idea of the derivative with functions other than x^2.

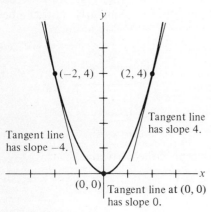

Figure 3.11

EXAMPLE 3 Find $f'(x)$ if $f(x) = x^3$.

SOLUTION In this case, $f(w) = w^3$ and $f(x) = x^3$. The derivative of the function at x is therefore

Keep in mind that x is fixed and that $w \to x$.

$$\lim_{w \to x} \frac{w^3 - x^3}{w - x} = \lim_{w \to x} \frac{(w^2 + wx + x^2)(w - x)}{w - x}$$

$$= \lim_{w \to x} (w^2 + wx + x^2) \qquad w \neq x$$

$$= x^2 + x^2 + x^2$$

$$= 3x^2.$$

The derivative of x^3 at x is $3x^2$. ∎

3.2 THE DERIVATIVE

In view of Example 3 we may say that "the derivative of the function x^3 is the function $3x^2$" and write $(x^3)' = 3x^2$.

Example 3 tells us, for instance, that the slope of the graph of $y = x^3$ is never negative (since x^2 is never negative). Moreover, at $x = 0$ the slope is $3 \cdot 0^2 = 0$. Thus the tangent line to the curve $y = x^3$ at the origin is horizontal, as is shown in Fig. 3.12.

It may seem strange that a tangent line can *cross the curve*, as this tangent line does. However, the basic property of a tangent line is that it indicates the direction of a curve at a point. In high school geometry, where only tangent lines to circles are considered, the tangent line never crosses the curve.

Example 3 can also be interpreted in terms of the speed of a moving object. If an object moves x^3 feet in the first x seconds, its speed after x seconds is $3x^2$ feet per second.

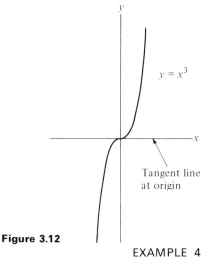

Figure 3.12

EXAMPLE 4 Find the derivative of the square root function $f(x) = \sqrt{x}$.

SOLUTION Since the domain of $\sqrt{x}$ contains no negative numbers, assume that $x \geq 0$. Then, by definition of the derivative,

When finding a derivative, be sure to write "lim" at each step.

$$f'(x) = \lim_{w \to x} \frac{\sqrt{w} - \sqrt{x}}{w - x} \qquad w \neq x, \quad x \geq 0, \quad w \geq 0$$

$$= \lim_{w \to x} \frac{\sqrt{w} - \sqrt{x}}{w - x} \frac{\sqrt{w} + \sqrt{x}}{\sqrt{w} + \sqrt{x}} \qquad \text{rationalize the numerator}$$

$$= \lim_{w \to x} \frac{w - x}{(w - x)(\sqrt{w} + \sqrt{x})}$$

$$= \lim_{w \to x} \frac{1}{\sqrt{w} + \sqrt{x}}.$$

There are two cases to consider, $x > 0$ and $x = 0$.
Case 1: If $x > 0$,

$$f'(x) = \lim_{w \to x} \frac{1}{\sqrt{w} + \sqrt{x}}$$

$$= \frac{1}{\sqrt{x} + \sqrt{x}}$$

$$= \frac{1}{2\sqrt{x}}.$$

The derivative of the square root function at x is

$$\frac{1}{2\sqrt{x}}, \qquad \text{if } x > 0.$$

Case 2: If $x = 0$, then necessarily $w > 0$, and we are considering in fact the right-hand limit

122 THE DERIVATIVE

$$\lim_{w \to 0^+} \frac{1}{\sqrt{w} + \sqrt{0}},$$

which is infinite. Thus the derivative of $\sqrt{x}$ does not exist at 0. ∎

Memorize $(\sqrt{x})' = \dfrac{1}{2\sqrt{x}}$.

x	0	1	4	9	16	25
$\sqrt{x}$	0	1	2	3	4	5

According to Example 4, $(\sqrt{x})' = 1/(2\sqrt{x})$. Is this result reasonable? It says that, when x is large, the slope of the tangent line at $(x, \sqrt{x})$ is near 0 (since $1/(2\sqrt{x})$ is near 0). Let us draw the graph and see. First we make a brief table, as shown in the margin. With the aid of these six points, the graph can be sketched. (See Fig. 3.13.) For points far to the right on the graph the tangent line is indeed almost horizontal, as the formula $1/(2\sqrt{x})$ suggests. When x is near 0, the derivative $1/(2\sqrt{x})$ is large. The graph gets steeper and steeper near $x = 0$.

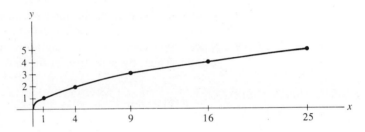

Figure 3.13

THE DERIVATIVES OF x^n AND $x^{1/n} = \sqrt[n]{x}$

In Examples 2 and 3 it was shown that

$$(x^2)' = 2x \quad \text{and} \quad (x^3)' = 3x^2.$$

Both are special cases of the following theorem.

THEOREM 1 For each positive integer n,

$$(x^n)' = nx^{n-1}.$$

PROOF It is necessary to find what happens to the quotient

$$\frac{w^n - x^n}{w - x}$$

Use the factoring of $d^n - c^n$ in Appendix B.

when $w \to x$. The identity

$$w^n - x^n = (w^{n-1} + w^{n-2}x + w^{n-3}x^2 + \cdots + x^{n-1})(w - x)$$

shows that the quotient equals

$$\frac{(w^{n-1} + w^{n-2}x + w^{n-3}x^2 + \cdots + x^{n-1})(w - x)}{w - x}$$

which equals

$$w^{n-1} + w^{n-2}x + w^{n-3}x^2 + \cdots + x^{n-1}, \quad w \neq x.$$

As $w \to x$, this approaches

$$x^{n-1} + x^{n-2}x + x^{n-3}x^2 + \cdots + x^{n-1}$$
$$= x^{n-1} + x^{n-1} + x^{n-1} + \cdots + x^{n-1}$$
$$= nx^{n-1} \quad \text{since there are } n \text{ summands.}$$

In short, $$\lim_{w \to x} \frac{w^n - x^n}{w - x} = nx^{n-1},$$

for any positive integer n. This proves the theorem. ∎

Direct application of this theorem yields, for instance:

The derivative of x^4 is $4x^{4-1} = 4x^3$.

The derivative of x^3 is $3x^{3-1} = 3x^2$.

The derivative of x^2 is $2x^{2-1} = 2x$.

The derivative of x^1 is $1x^0 = 1$ (in agreement with the fact that the line given by the formula $y = x$ has slope 1).

The next theorem generalizes the fact that $(\sqrt{x})' = 1/(2\sqrt{x})$, which can be written

$$(x^{1/2})' = \tfrac{1}{2}x^{1/2 - 1}.$$

THEOREM 2 For each positive integer n,

$$(x^{1/n})' = \frac{1}{n}x^{1/n - 1}$$

(for those x at which both $x^{1/n}$ and $x^{1/n - 1}$ are defined).

PROOF

$$(x^{1/n})' = \lim_{w \to x} \frac{w^{1/n} - x^{1/n}}{w - x}.$$

The denominator will be rewritten with the aid of the identity

$$d^n - c^n = (d - c)\underbrace{(d^{n-1} + d^{n-2}c + \cdots + c^{n-1})}_{n \text{ summands}}$$

as follows:

$$w - x = (w^{1/n})^n - (x^{1/n})^n$$
$$= (w^{1/n} - x^{1/n})((w^{1/n})^{n-1} + (w^{1/n})^{n-2}(x^{1/n}) + \cdots + (x^{1/n})^{n-1}).$$

Use $d = w^{1/n}$ and $c = x^{1/n}$.

Thus

$$(x^{1/n})' = \lim_{w \to x} \frac{w^{1/n} - x^{1/n}}{(w^{1/n} - x^{1/n})((w^{1/n})^{n-1} + (w^{1/n})^{n-2}x^{1/n} + \cdots + (x^{1/n})^{n-1})}$$

$$= \lim_{w \to x} \frac{1}{(w^{1/n})^{n-1} + (w^{1/n})^{n-2}x^{1/n} + \cdots + (x^{1/n})^{n-1}}$$

$$= \frac{1}{(x^{1/n})^{n-1} + (x^{1/n})^{n-2}x^{1/n} + \cdots + (x^{1/n})^{n-1}}$$

$$= \frac{1}{(x^{1/n})^{n-1} + (x^{1/n})^{n-1} + \cdots + (x^{1/n})^{n-1}}$$

$$= \frac{1}{n(x^{1/n})^{n-1}}$$

$$= \frac{1}{nx^{(n-1)/n}} \qquad \text{power of a power}$$

$$= \frac{1}{n} \frac{1}{x^{1-1/n}}$$

$$= \frac{1}{n} x^{1/n - 1}. \quad \blacksquare$$

$D(x^a) = ax^{a-1}$ *for fixed rational exponent a.*

Theorems 1 and 2 both fit into the same pattern: to find the derivative of x^a, for a fixed exponent a, lower the exponent by 1 and multiply by the original exponent. In Sec. 3.6 this rule will be extended to all rational exponents a and in Chap. 6 to all irrational exponents as well. At this point the rule is established only when the exponent is a positive integer or the reciprocal of a positive integer.

EXAMPLE 5   Use Theorem 2 to find $(\sqrt[3]{x})'$ at $x = 8$.

SOLUTION $\sqrt[3]{x} = x^{1/3}$. Thus,

$$(\sqrt[3]{x})' = (x^{1/3})'$$

$$= \frac{1}{3} x^{1/3 - 1}$$

$$= \frac{1}{3} x^{-2/3}$$

$$= \frac{1}{3} \frac{1}{x^{2/3}}.$$

In particular, at $x = 8$, the derivative is

$$\frac{1}{3} \cdot \frac{1}{8^{2/3}} = \frac{1}{3} \cdot \frac{1}{4}$$

$$= \frac{1}{12}. \quad \blacksquare$$

FOUR INTERPRETATIONS OF THE DERIVATIVE

Now that we have the concept of the derivative, we are in a position to define *tangent line, speed, magnification,* and *density,* terms used only intuitively until now. These definitions are suggested by the similarity of the computations made in the four problems in Sec. 3.1.

The slope of a nonvertical line was defined as the quotient $(y_2 - y_1)/(x_2 - x_1)$, where $P_1 = (x_1, y_1)$ and $P_2 = (x_2, y_2)$ are any distinct points on the line. Now it is possible to define the slope of a curve at a point on the curve.

(In all five definitions it is assumed that the derivative exists.)

DEFINITION *Slope of a curve.* The *slope* of the graph of the function f at $(x, f(x))$ is the derivative of f at x.

DEFINITION *Tangent line to a curve.* The *tangent line* to the graph of the function f at the point $P = (x, y)$ is the line through P that has a slope equal to the derivative of f at x.

DEFINITION *Velocity and speed of a particle moving on a line.* The *velocity* at time t of an object, whose position on a line at time t is given by $f(t)$, is the derivative of f at time t. The *speed* of the particle is the absolute value of the velocity.

Note the distinction between velocity and speed. Velocity can be negative; speed is either positive or 0.

DEFINITION *Magnification of a linear projector.* The *magnification* at x of a lens that projects the point x of one line onto the point $f(x)$ of another line is the derivative of f at x.

DEFINITION *Density of material.* The *density* at x of material distributed along a line in such a way that the left-hand x centimeters have a mass of $f(x)$ grams is equal to the derivative of f at x.

Slope, velocity, magnification, and density are just interpretations of the derivative. The derivative itself is a purely mathematical concept, a special limit formed in a certain way from a function:

$$\lim_{w \to x} \frac{f(w) - f(x)}{w - x}.$$

Exercises

In Exercises 1 to 16 use the definition of the derivative to find the derivatives of the given functions.

1. x^4
2. x^5
3. $2x$
4. $5x$
5. $x^2 + 3$
6. $4x^2 + 5$
7. $-5x^2 + 4x$
8. $2x^3 - 4x^2 + 5$
9. $7\sqrt{x}$
10. $x^2 + 3\sqrt{x}$
11. $1/x$
12. $1/(x + 1)$

126 THE DERIVATIVE

13 $1/x^2$ 14 $\sqrt{x} + 5/x$
15 $3 - 6/x$ 16 $1/x^3$

In Exercises 17 to 24, use Theorems 1 and 2 to find the derivatives of the given functions at the given numbers.

17 x^4 at -1 18 x^4 at $\frac{1}{2}$
19 x^5 at a 20 x^5 at $\sqrt{2}$
21 $\sqrt[3]{t}$ at -8 22 $\sqrt[3]{t}$ at -1
23 $\sqrt[4]{x}$ at 16   24 $\sqrt[4]{x}$ at 81

25 Let $f(x) = x^4$. (a) What is the slope of the line joining $(1, 1)$ to $(1.1, 1.1^4)$? (b) What is the slope of the tangent line to the curve at the point $(1, 1)$?

26 An object travels t^3 feet in the first t seconds.
(a) What is its average velocity from time $t = 2$ to time $t = 2.01$?
(b) What is its average velocity from time $t = 1.99$ to time $t = 2$?
(c) What is its velocity at time $t = 2$?

27 A lens projects x on the slide to x^4 on the screen.
(a) How much does it magnify the interval $[1, 1.01]$?
(b) What is its magnification at $x = 1$?

28 The left x centimeters of a string have a mass of x^3 grams. (a) What is the average density of the interval $[2, 2.01]$? (b) What is the average density of the interval $[1.99, 2.01]$? (c) What is the density at $x = 2$?

29 (a) Show that the tangent line to the curve $y = x^3$ at $(1, 1)$ passes through $(2, 4)$.
(b) Use (a) to draw the tangent to the curve $y = x^3$ at $(1, 1)$. (Note that it is not necessary to draw the curve.)

30 (a) Using the definition of the derivative, find the derivative of $x + x^2$.
(b) Graph $y = x + x^2$.
(c) For which x is the derivative in (a) equal to 0?
(d) Using (c), find at which point on the graph of $y = x + x^2$ the slope is 0.
(e) In view of (d), what do you think is the smallest possible value of $x + x^2$?

31 Figure 3.14 shows the graph of a function f whose domain is the interval $[0, 4]$. [Solid dots indicate $f(1)$ and $f(2)$.]

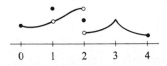

Figure 3.14

(a) For which numbers a does $\lim_{x \to a} f(x)$ *not* exist?
(b) At which numbers a is f not continuous?
(c) At which numbers a is f not differentiable?

3.3 The derivative and continuity; antiderivatives

After presenting another notation for the difference quotient

$$\frac{f(w) - f(x)}{w - x}$$

and the derivative $f'(x)$, this section shows the relation between "having a derivative" and "being continuous." It concludes by introducing the notion of an "antiderivative."

THE Δ NOTATION FOR THE DIFFERENCE QUOTIENT

There is another notation for the difference quotient,

$$\frac{\text{Change in output}}{\text{Change in input}},$$

that is especially useful for obtaining properties of derivatives.

We used

$$\frac{f(w) - f(x)}{w - x},$$

where x is fixed and $w \to x$.

3.3 THE DERIVATIVE AND CONTINUITY; ANTIDERIVATIVES

Δ is the capital Greek letter corresponding to the English D.

It is also common to give the difference or change $w - x$ the name Δx ("delta x"),

$$w - x = \Delta x$$

or

$$w = x + \Delta x.$$

The difference quotient then takes the form

$$\frac{f(x + \Delta x) - f(x)}{\Delta x},$$

Δx is not a product.

and the derivative is defined as

$$f'(x) = \lim_{\Delta x \to 0} \frac{f(x + \Delta x) - f(x)}{\Delta x}.$$

Furthermore, the difference in the outputs is often named Δf or Δy:

$$f(x + \Delta x) - f(x) = \Delta f$$

and so

$$f(x + \Delta x) = f(x) + \Delta f.$$

The latter equation says that "the value of the function at $x + \Delta x$ is equal to the value of the function at x plus the change in the function." With Δx denoting the change in the inputs and Δf denoting the change in the outputs, we have

$$f'(x) = \lim_{\Delta x \to 0} \frac{\Delta f}{\Delta x}.$$

Sometimes the change Δx is denoted h. Then the derivative is defined as

$$\lim_{h \to 0} \frac{f(x + h) - f(x)}{h}.$$

Figure 3.15 illustrates the Δ notation for the difference quotient.

Despite their variety all these notations describe the same idea and yield the same definition of the derivative. However, a change of notation may lead to different algebraic steps, as illustrated in Example 1.

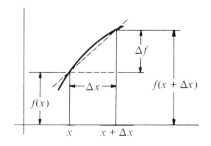

Figure 3.15

EXAMPLE 1 Find $(x^2)'$ using the Δ notation.

SOLUTION By the definition of the derivative, the derivative of the squaring function at x is

$$\lim_{\Delta x \to 0} \frac{(x + \Delta x)^2 - x^2}{\Delta x}.$$

Since

$$(x + \Delta x)^2 = x^2 + 2x \cdot \Delta x + (\Delta x)^2,$$

128 THE DERIVATIVE

the limit equals $\quad \lim\limits_{\Delta x \to 0} \dfrac{x^2 + 2x \cdot \Delta x + (\Delta x)^2 - x^2}{\Delta x}.$

Then we have

$$\begin{aligned}
(x^2)' &= \lim_{\Delta x \to 0} \frac{2x\,\Delta x + (\Delta x)^2}{\Delta x} & &\text{algebra} \\
&= \lim_{\Delta x \to 0} \frac{\Delta x(2x + \Delta x)}{\Delta x} & &\text{factoring} \\
&= \lim_{\Delta x \to 0} (2x + \Delta x) & &\text{canceling } \Delta x \neq 0 \\
&= 2x.
\end{aligned}$$

So the derivative of x^2 is $2x$, in agreement with the result in Example 1 of the preceding section. ∎

It would be of value to compare the two different calculations of $(x^2)'$.

NOTATIONS FOR THE DERIVATIVE   The derivative $f'(x)$ is also commonly denoted

$$\frac{df}{dx} \quad \text{or} \quad D(f).$$

If $f(x)$ is denoted y, the derivative is also denoted

$\dfrac{dy}{dx}$ *is the notation used by Leibniz.*

$$\frac{dy}{dx} \quad \text{or} \quad D(y).$$

For instance, $\quad \dfrac{d(x^3)}{dx} = 3x^2 \quad \text{or} \quad D(x^3) = 3x^2.$

If the formula for the function is long, it is customary to write

$$\frac{d}{dx}(f(x)) \quad \text{instead of} \quad \frac{d(f(x))}{dx}.$$

For instance, the notation

$$\frac{d}{dx}(x^3 - x^2 + 6x + \sqrt{x})$$

is clearer than $\quad \dfrac{d(x^3 - x^2 + 6x + \sqrt{x})}{dx}.$

If the variable is denoted by some letter other than x, such as t (for time), we write, for instance,

$$\frac{d(t^3)}{dt} = 3t^2, \quad \frac{d}{dt}(t^3) = 3t^2, \quad \text{and} \quad D(t^3) = 3t^2.$$

dy and dx are without meaning at this point.

Keep in mind that in the notations *df/dx* and *dy/dx*, the symbols *df, dy,* and *dx* have no meaning by themselves. The symbol *dy/dx* should be thought of as a single entity, just like the numeral 8, which we do not think of as formed of two 0's.

The dot notation

In the study of motion, Newton's dot notation is often used. If x is a function of time t, then $\dot{x}$ denotes the derivative dx/dt.

DIFFERENTIABLE FUNCTIONS

A function f is said to be *differentiable* at the number x if the derivative f' exists at x. If f is differentiable at each number x in some interval it is said to be differentiable throughout that interval.

A very small piece of the graph of a differentiable function looks almost like a straight line, as shown in Fig. 3.16. In this sense the differentiable functions are even "better" than the continuous functions. The following theorem shows that a differentiable function is necessarily continuous. However, a function can be continuous without being differentiable, as Example 2 will show.

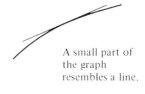

A small part of the graph resembles a line.

Figure 3.16

THEOREM If f is differentiable at a, then it is continuous at a.

PROOF We must show that $\lim_{x \to a} f(x)$ exists and equals $f(a)$. To accomplish both goals at one time, we will show that

$$\lim_{x \to a} (f(x) - f(a)) = 0.$$

For then
$$\lim_{x \to a} f(x) = \lim_{x \to a} [(f(x) - f(a)) + f(a)]$$
$$= 0 + f(a) = f(a).$$

To begin, note that for $x \neq a$,

$$f(x) - f(a) = \frac{f(x) - f(a)}{x - a}(x - a).$$

Thus

$$\lim_{x \to a} (f(x) - f(a)) = \lim_{x \to a} \frac{f(x) - f(a)}{x - a}(x - a)$$
$$= \lim_{x \to a} \frac{f(x) - f(a)}{x - a} \lim_{x \to a} (x - a) \quad \text{since both limits exist}$$
$$= f'(a) \cdot 0$$
$$= 0.$$

Thus f is continuous at a. ∎

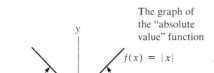

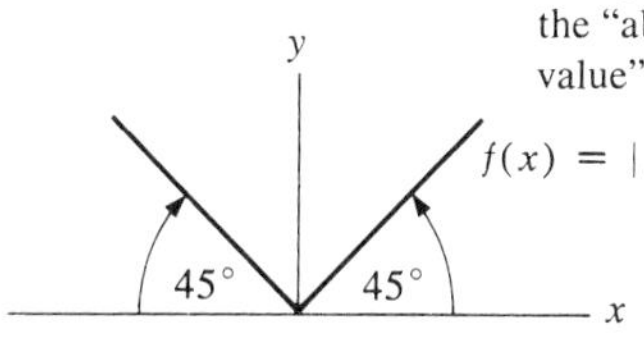

The graph of the "absolute value" function $f(x) = |x|$

Figure 3.17

The converse of this theorem is not true. The function $|x|$ is continuous at 0, but, as the graph in Fig. 3.17 suggests, it is not differentiable at 0. This is demonstrated in Example 2.

EXAMPLE 2   Show that the absolute value function $f(x) = |x|$ is not differentiable at 0.

SOLUTION   Since $\lim_{x \to 0} |x| = 0 = f(0)$, the function $f(x) = |x|$ is continuous at $x = 0$. To show that f is not differentiable at $x = 0$, we must show that

$$\lim_{x \to 0} \frac{|x| - |0|}{x - 0}$$

does not exist. To accomplish this, we will examine the right- and left-hand limits at 0.

As $x \to 0$ from the right, x is positive. Thus

$$\lim_{x \to 0^+} \frac{|x| - |0|}{x - 0} = \lim_{x \to 0^+} \frac{x - 0}{x - 0}$$

$$= \lim_{x \to 0^+} \frac{x}{x}$$

$$= \lim_{x \to 0^+} 1$$

$$= 1.$$

As $x \to 0$ from the left, x is negative. Thus

$$\lim_{x \to 0^-} \frac{|x| - |0|}{x - 0} = \lim_{x \to 0^-} \frac{-x - 0}{x - 0}$$

$$= \lim_{x \to 0^-} \frac{-x}{x}$$

$$= \lim_{x \to 0^-} (-1)$$

$$= -1.$$

Since the left-hand limit at 0 does not equal the right-hand limit at 0,

$$\lim_{x \to 0} \frac{|x| - |0|}{x - 0}$$

does not exist. Thus $|x|$ is not differentiable at 0. ∎

ANTIDERIVATIVES

If f and F are two functions and f is the derivative of F, then F is called an *antiderivative* of f. For instance, since $(x^3)' = 3x^2$, the function x^3 is an antiderivative of $3x^2$. This concept will play an important role beginning with Chap. 5. As more derivatives are computed later in this chapter, prac-

$x^3 + 2001$ is also an antiderivative of $3x^2$, as can be checked.

Exercises

In Exercises 1 to 10 use the Δ notation to find the given derivatives.

1. $\dfrac{d(x^3)}{dx}$
2. $\dfrac{d(x^4)}{dx}$

   (*Suggestion:* In Exercises 1 and 2 use the binomial theorem in Appendix B to expand the power of $x + \Delta x$.)

3. $\dfrac{d(\sqrt{x})}{dx}$
4. $\dfrac{d(3\sqrt{x})}{dx}$
5. $\dfrac{d(5x^2)}{dx}$
6. $\dfrac{d}{dx}(5x^2 + 3x + 2)$
7. $D\left(\dfrac{3}{x}\right)$
8. $\dfrac{d}{dx}\left(\dfrac{5}{x^2}\right)$
9. $\dfrac{d}{dx}\left(\dfrac{3}{x} - 4x + 2\right)$
10. $\dfrac{d}{dx}(x^3 - 5x + 1982)$

11. Let $f(x) = x^2$. Find Δf if (a) $x = 1$ and $\Delta x = 0.1$, (b) $x = 3$ and $\Delta x = -0.1$.
12. Let $f(x) = 1/x$. Find Δf if (a) $x = 2$ and $\Delta x = \frac{1}{5}$, (b) $x = 2$ and $\Delta x = -\frac{1}{8}$.
13. Using the Δ notation, find the derivatives of
    (a) $6x + 3$,
    (b) $6x + 7$,
    (c) $6x - 273$,
    (d) $6x + C$ for any constant C.
    (e) Give five different antiderivatives of the constant function $f(x) = 6$.
14. (a) Using the Δ notation, find the derivative of $x^3/3 + C$ for any constant C.
    (b) How many different antiderivatives does the function x^2 have?

15. Using the notation $(f(w) - f(x))/(w - x)$ for the difference quotient, find the derivative of
    (a) x^4,
    (b) $17x^4$,
    (c) kx^4 for any constant k.
    (d) Give an example of an antiderivative for the function x^3.

16. (a) Using the notation $[f(w) - f(x)]/(w - x)$ for the difference quotient, find the derivative of kx^5 for any constant k.
    (b) Give an example of an antiderivative for the function x^4.
17. (a) Let k and C be constants. Find the derivative of $kx^6 + C$ both by the "w and x" notation and by the Δ notation.
    (b) Give three different antiderivatives of x^5.
18. Figure 3.18 is the graph of a hypothetical function f.

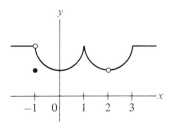

Figure 3.18

(a) For which numbers a does $\lim_{x \to a} f(x)$ exist, but f is not continuous at a?
(b) For which numbers a is f continuous at a, but not differentiable at a?

19. Let $f(x) = x^3$. (a) Graph f.
    (b) On the graph indicate x, $x + \Delta x$, Δx, $f(x)$, $f(x + \Delta x)$, and Δf for $x = 2$ and $\Delta x = 0.3$.
20. Using the definition of the derivative, find

$$\dfrac{d}{dx}(2x^3 - 3x^2 + 4x - 5).$$

21. Using the definition of the derivative, find

$$D\left(\dfrac{1}{3x + 5}\right).$$

22. Using the Δ notation, find $D(\sin x)$. *Suggestion:* Use the identity for $\sin(A + B)$.
23. Using the Δ notation, find $D(\cos x)$.

132 THE DERIVATIVE

3.4 The derivatives of the sum, difference, product, and quotient

This section develops methods for finding derivatives of functions, or what is called "differentiating" functions. With these methods it will be a routine matter to find, for instance, the derivative of

$$\frac{(1+\sqrt{x})(x^3+1)}{x^2+5x+3},$$

without going back to the definition of the derivative and (at great effort) finding the limit of a difference quotient.

Before developing the methods, it will be useful to find the derivative of any constant function.

THEOREM 1 The derivative of a constant function is 0; in symbols

$$(c)' = 0 \quad \text{or} \quad \frac{dc}{dx} = 0 \quad \text{or} \quad D(c) = 0.$$

PROOF Let c be a fixed number and let f be the constant function, $f(x) = c$ for all inputs x. Then

$$f'(x) = \lim_{\Delta x \to 0} \frac{f(x + \Delta x) - f(x)}{\Delta x}.$$

Since the function f has the same output c for all inputs,

$$f(x + \Delta x) = c \quad \text{and} \quad f(x) = c.$$

Thus
$$f'(x) = \lim_{\Delta x \to 0} \frac{c - c}{\Delta x}$$

$$= \lim_{\Delta x \to 0} \frac{0}{\Delta x}$$

$$= \lim_{\Delta x \to 0} 0 \quad (\text{since } \Delta x \neq 0)$$

$$= 0.$$

The horizontal line is the graph of $f(x) = c$.

This shows that the derivative of any constant function is 0 for all x. ∎

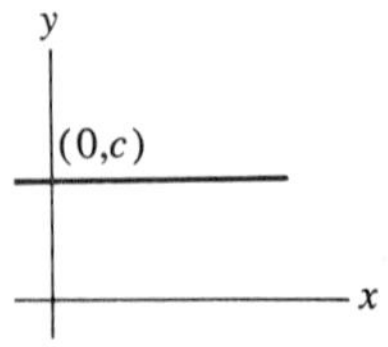

Figure 3.19

From two points of view, Theorem 1 is no surprise: Since the graph of $f(x) = c$ is a horizontal line, it coincides with each of its tangent lines, as can be seen in Fig. 3.19. Also, if we think of x as time and $f(x)$ as the position of a particle, Theorem 1 implies that a stationary particle has zero velocity.

EXAMPLE 1 Differentiate the function π^3.

SOLUTION The number π^3, in this case, is short for the constant function

3.4 THE DERIVATIVES OF THE SUM, DIFFERENCE, PRODUCT, AND QUOTIENT

$$f(x) = \pi^3$$

for all x. Since f is constant,

$$f'(x) = 0.$$

$c' = 0$ no matter how fancy the constant c may be.

Thus
$$(\pi^3)' = 0. \blacksquare$$

THE DERIVATIVES OF $f + g$ AND $f - g$

The next theorem asserts that if the functions f and g have derivatives at a certain number x, then so does their sum $f + g$, and

$$\frac{d}{dx}(f + g) = \frac{df}{dx} + \frac{dg}{dx}.$$

In other words, "The derivative of the sum is the sum of the derivatives." A similar formula holds for the derivative of $f - g$.

THEOREM 2 If f and g are differentiable functions, then so is $f + g$. Its derivative is given by the formula

$$(f + g)' = f' + g'.$$

Similarly,
$$(f - g)' = f' - g'.$$

PROOF Give the function $f + g$ the name u. That is,

$$u(x) = f(x) + g(x).$$

Then $\quad u(x + \Delta x) = f(x + \Delta x) + g(x + \Delta x).$

So, $\quad \Delta u = u(x + \Delta x) - u(x)$
$$= [f(x + \Delta x) + g(x + \Delta x)] - [f(x) + g(x)]$$
$$= [f(x + \Delta x) - f(x)] + [g(x + \Delta x) - g(x)]$$
$$= \Delta f + \Delta g.$$

Thus $\quad u'(x) = \lim_{\Delta x \to 0} \frac{\Delta u}{\Delta x}$

$$= \lim_{\Delta x \to 0} \frac{\Delta f + \Delta g}{\Delta x}$$

$$= \lim_{\Delta x \to 0} \left(\frac{\Delta f}{\Delta x} + \frac{\Delta g}{\Delta x} \right)$$

$$= \lim_{\Delta x \to 0} \frac{\Delta f}{\Delta x} + \lim_{\Delta x \to 0} \frac{\Delta g}{\Delta x}$$

$$= f'(x) + g'(x).$$

Hence $f + g$ is differentiable, and

$$(f + g)' = f' + g'.$$

A similar argument applies to $f - g$. ∎

Theorem 2 extends to any finite number of differentiable functions. For example,

$$(f + g + h)' = f' + g' + h'$$

and

$$(f - g + h)' = f' - g' + h'.$$

EXAMPLE 2 Using Theorem 2, differentiate $x^2 + x^3$.

SOLUTION

$$\frac{d}{dx}(x^2 + x^3) = \frac{d}{dx}(x^2) + \frac{d}{dx}(x^3) \qquad \text{Theorem 2}$$

$$= 2x + 3x^2 \qquad D(x^n) = nx^{n-1}. \quad \blacksquare$$

EXAMPLE 3 Differentiate $x^4 - \sqrt{x} + 5$.

SOLUTION

$$\frac{d}{dx}(x^4 - \sqrt{x} + 5) = \frac{d}{dx}(x^4) - \frac{d}{dx}(\sqrt{x}) + \frac{d}{dx}(5) \qquad \text{Theorem 2}$$

$$= 4x^3 - \frac{1}{2\sqrt{x}} + 0$$

$$= 4x^3 - \frac{1}{2\sqrt{x}}. \quad \blacksquare$$

THE DERIVATIVES OF fg AND cf

The following theorem concerning the derivative of the product of two functions may be surprising for it turns out that the derivative of the product is *not* the product of the derivatives. The formula is more complicated than that for the derivative of the sum, hence a little harder to apply. (It asserts that the derivative of the product is "the first function times the derivative of the second plus the second function times the derivative of the first.")

THEOREM 3 If f and g are differentiable functions, then so is fg. Its derivative is given by the formula

The product rule

$$(fg)' = fg' + gf'.$$

PROOF Call the function fg simply u, that is,

$$u(x) = f(x)g(x).$$

Then

$$u(x + \Delta x) = f(x + \Delta x)g(x + \Delta x).$$

3.4 THE DERIVATIVES OF THE SUM, DIFFERENCE, PRODUCT, AND QUOTIENT

Rather than subtract directly, first write

$$f(x + \Delta x) = f(x) + \Delta f \quad \text{and} \quad g(x + \Delta x) = g(x) + \Delta g.$$

Then
$$\begin{aligned} u(x + \Delta x) &= [f(x) + \Delta f][g(x) + \Delta g] \\ &= f(x)g(x) + f(x)\,\Delta g + g(x)\,\Delta f + \Delta f\,\Delta g. \end{aligned}$$

Hence
$$\begin{aligned} \Delta u &= u(x + \Delta x) - u(x) \\ &= f(x)g(x) + f(x)\,\Delta g + g(x)\,\Delta f + \Delta f\,\Delta g - f(x)g(x) \\ &= f(x)\,\Delta g + g(x)\,\Delta f + \Delta f\,\Delta g, \end{aligned}$$

and
$$\frac{\Delta u}{\Delta x} = f(x)\frac{\Delta g}{\Delta x} + g(x)\frac{\Delta f}{\Delta x} + \Delta f\frac{\Delta g}{\Delta x}.$$

As $\Delta x \to 0$, $\Delta g/\Delta x \to g'(x)$, $\Delta f/\Delta x \to f'(x)$, and, because f is differentiable (hence continuous), $\Delta f \to 0$. It follows that

$$\lim_{\Delta x \to 0} \frac{\Delta u}{\Delta x} = f(x)g'(x) + g(x)f'(x) + 0 \cdot g'(x).$$

The formula for $(fg)'$ was discovered by Leibniz in 1676. (His first guess was wrong.)

Therefore u is differentiable and

$$u' = fg' + gf'. \quad \blacksquare$$

EXAMPLE 4 Find $\dfrac{d}{dx}((x^2 + x^3)(x^4 - \sqrt{x} + 5))$.

SOLUTION (Note that in Examples 2 and 3 the derivatives of both $x^2 + x^3$ and $x^4 - \sqrt{x} + 5$ were found.)
By Theorem 3,

$$\frac{d}{dx}((x^2 + x^3)(x^4 - \sqrt{x} + 5))$$

$$= (x^2 + x^3)\frac{d}{dx}(x^4 - \sqrt{x} + 5) + (x^4 - \sqrt{x} + 5)\frac{d}{dx}(x^2 + x^3)$$

$$= (x^2 + x^3)\left(4x^3 - \frac{1}{2\sqrt{x}}\right) + (x^4 - \sqrt{x} + 5)(2x + 3x^2). \quad \blacksquare$$

A special case of the formula $(fg)' = fg' + gf'$ occurs so frequently that it is singled out in Theorem 4.

THEOREM 4 If c is a constant function and f is a differentiable function, then cf is differentiable and its derivative is given by the formula,

$$(cf)' = cf'.$$

136 THE DERIVATIVE

PROOF By Theorem 3,

$$(cf)' = cf' + c'f \qquad \text{derivative of a product}$$
$$= cf' + 0 \cdot f \qquad \text{derivative of constant is 0}$$
$$= cf'.$$

Thus $(cf)' = cf'$. ∎

In other notations for the derivative, Theorem 4 is expressed as

$$\frac{d(cf)}{dx} = c\frac{df}{dx} \quad \text{and} \quad D(cf) = cD(f).$$

A constant factor can go past the derivative symbol.

Theorem 4 asserts that "it is legal to move a constant factor outside the derivative symbol."

EXAMPLE 5 Find $D(6x^3)$.

SOLUTION

$$D(6x^3) = 6D(x^3) \qquad \text{6 is constant}$$
$$= 6 \cdot 3x^2 \qquad D(x^n) = nx^{n-1}$$
$$= 18x^2. \quad \blacksquare$$

EXAMPLE 6 Find $D(x^5/11)$.

SOLUTION

$$D\left(\frac{x^5}{11}\right) = D\left(\frac{1}{11}x^5\right)$$
$$= \frac{1}{11}D(x^5)$$
$$= \frac{1}{11}5x^4$$
$$= \frac{5}{11}x^4. \quad \blacksquare$$

$$\frac{d}{dx}\left(\frac{f}{c}\right) = \frac{1}{c}\frac{d}{dx}(f)$$

Example 6 generalizes to the fact that, for a nonzero constant c,

$$\left(\frac{f}{c}\right)' = \frac{f'}{c}.$$

The formula for the derivative of the product extends to the product of several differentiable functions. For instance,

$$(fgh)' = f'gh + fg'h + fgh'.$$

In each summand only one derivative appears. (See Exercise 49.)

3.4 THE DERIVATIVES OF THE SUM, DIFFERENCE, PRODUCT, AND QUOTIENT

EXAMPLE 7 Differentiate $\sqrt{x}(x^2 - 2)(2x^3 + 1)$.

SOLUTION By the preceding remark,

$$[\sqrt{x}(x^2 - 2)(2x^3 + 1)]' = (\sqrt{x})'(x^2 - 2)(2x^3 + 1) + \sqrt{x}(x^2 - 2)'(2x^3 + 1) + \sqrt{x}(x^2 - 2)(2x^3 + 1)'$$

$$= \frac{1}{2\sqrt{x}}(x^2 - 2)(2x^3 + 1) + \sqrt{x}(2x)(2x^3 + 1) + \sqrt{x}(x^2 - 2)(6x^2). \blacksquare$$

Any polynomial can be differentiated by the methods already developed, as Example 8 illustrates.

EXAMPLE 8 Differentiate $6x^8 - x^3 + 5x^2 + \pi^3$.

Differentiate a polynomial "term by term."

SOLUTION
$$(6x^8 - x^3 + 5x^2 + \pi^3)' = (6x^8)' - (x^3)' + (5x^2)' + (\pi^3)'$$
$$= 6 \cdot 8x^7 - 3x^2 + 5 \cdot 2x + 0$$
$$= 48x^7 - 3x^2 + 10x. \blacksquare$$

THE DERIVATIVES OF f/g AND $1/g$

It will next be shown that if the functions f and g are differentiable at a number x, and if $g(x) \neq 0$, then f/g is differentiable at x. The formula for $(f/g)'$ is a bit messy; a suggestion for remembering it is given after the proof.

THEOREM 5 If f and g are differentiable functions, then so is

$$\frac{f}{g},$$

The quotient rule and $\left(\dfrac{f}{g}\right)' = \dfrac{gf' - fg'}{g^2}$ (where $g(x)$ is not 0).

PROOF Denote the quotient function f/g by u, that is,

$$u(x) = \frac{f(x)}{g(x)}.$$

Then
$$u(x + \Delta x) = \frac{f(x + \Delta x)}{g(x + \Delta x)}$$

(Since we consider only values of x such that $g(x) \neq 0$ and g is continuous, for Δx sufficiently small, $g(x + \Delta x) \neq 0$.)

Before computing $\lim_{\Delta x \to 0} (\Delta u/\Delta x)$, express the numerator Δu as simply as possible in terms of $f(x)$, Δf, $g(x)$, and Δg.

$$\Delta u = u(x + \Delta x) - u(x) \quad \text{by definition of } \Delta u$$

138 THE DERIVATIVE

$$= \frac{f(x + \Delta x)}{g(x + \Delta x)} - \frac{f(x)}{g(x)} \qquad \text{by definition of the function } u$$

$$= \frac{f(x) + \Delta f}{g(x) + \Delta g} - \frac{f(x)}{g(x)} \qquad \text{by definition of } \Delta f \text{ and } \Delta g$$

$$= \frac{g(x)(f(x) + \Delta f) - f(x)(g(x) + \Delta g)}{(g(x) + \Delta g)g(x)} \qquad \text{placing over common denominator}$$

$$= \frac{g(x)f(x) + g(x)\,\Delta f - f(x)g(x) - f(x)\,\Delta g}{(g(x) + \Delta g)g(x)} \qquad \text{multiplying out}$$

$$= \frac{g(x)\,\Delta f - f(x)\,\Delta g}{(g(x) + \Delta g)g(x)} \qquad \text{canceling}$$

After this simplification, $u'(x)$ can be found as follows:

$$u'(x) = \lim_{\Delta x \to 0} \frac{\Delta u}{\Delta x} = \lim_{\Delta x \to 0} \frac{\dfrac{g(x)\,\Delta f - f(x)\,\Delta g}{(g(x) + \Delta g)g(x)}}{\Delta x}$$

$$= \lim_{\Delta x \to 0} \frac{g(x)\,\Delta f - f(x)\,\Delta g}{(g(x) + \Delta g)g(x)\,\Delta x} \qquad \text{algebra: } \frac{\left(\dfrac{a}{b}\right)}{c} = \frac{a}{bc}$$

$$= \lim_{\Delta x \to 0} \frac{g(x)\dfrac{\Delta f}{\Delta x} - f(x)\dfrac{\Delta g}{\Delta x}}{(g(x) + \Delta g)g(x)} \qquad \text{algebra: } \frac{ab - cd}{ef} = \frac{a\dfrac{b}{f} - c\dfrac{d}{f}}{e}$$

$$= \frac{g(x)f'(x) - f(x)g'(x)}{g(x)g(x)} \qquad \text{taking limits.}$$

This establishes the formula for $(f/g)'$. ∎

A suggestion for using the formula for the derivative of a quotient.

Word of advice: When using the formula for $(f/g)'$, first write down the parts where g^2 and g appear:

$$\frac{g}{g^2}.$$

In that way you will get the denominator correct and have a good start on the numerator. You may then go on to complete the numerator, *remembering that it has a minus sign*:

$$\left(\frac{f}{g}\right)' = \frac{gf' - fg'}{g^2}.$$

EXAMPLE 9 Compute $(x^2/(x^3 + 1))'$, showing each step in detail.

3.4 THE DERIVATIVES OF THE SUM, DIFFERENCE, PRODUCT, AND QUOTIENT

SOLUTION Step 1

$$\left(\frac{x^2}{x^3+1}\right)' = \frac{(x^3+1)\cdots}{(x^3+1)^2} \qquad \text{write denominator and start numerator}$$

Step 2

$$= \frac{(x^3+1)(x^2)' - (x^2)(x^3+1)'}{(x^3+1)^2} \qquad \text{complete numerator, remembering minus sign}$$

Step 3

$$= \frac{(x^3+1)(2x) - x^2(3x^2)}{(x^3+1)^2} \qquad \text{computing derivatives}$$

Step 4

$$= \frac{2x^4 + 2x - 3x^4}{(x^3+1)^2} \qquad \text{algebra}$$

Step 5

$$= \frac{2x - x^4}{(x^3+1)^2} \qquad \text{algebra: collecting} \qquad \blacksquare$$

As Example 9 illustrates, the techniques for differentiating polynomials and quotients suffice to differentiate any rational function.

The next example uses the formulas for the derivatives of the product and the quotient.

EXAMPLE 10 Differentiate

$$\frac{(x^3+1)\sqrt{x}}{x^2}, \qquad \text{where } x > 0.$$

SOLUTION

$$D\left(\frac{(x^3+1)\sqrt{x}}{x^2}\right)$$

$$= \frac{x^2 D((x^3+1)\sqrt{x}) - (x^3+1)\sqrt{x}\, D(x^2)}{(x^2)^2} \qquad D\left(\frac{f}{g}\right)$$

$$= \frac{x^2[(x^3+1)D(\sqrt{x}) + \sqrt{x}\, D(x^3+1)] - (x^3+1)\sqrt{x}(2x)}{x^4} \qquad D(fg)$$

$$= \frac{x^2\left[(x^3+1)\frac{1}{2\sqrt{x}} + \sqrt{x}(3x^2)\right] - (x^3+1)\sqrt{x}(2x)}{x^4} \qquad D(x^n)$$

This can be simplified further, but there is no need to here.

The result can be simplified a little by algebra. A factor x can be cancelled in numerator and denominator. Also, coefficients can be placed at the front of the terms. This gives

$$D\left(\frac{(x^3+1)\sqrt{x}}{x^2}\right) = \frac{x[(x^3+1)/(2\sqrt{x}) + 3\sqrt{x}\, x^2] - 2\sqrt{x}(x^3+1)}{x^3}. \qquad \blacksquare$$

The following corollary is just a special case of the formula for $(f/g)'$. Since it is needed often, it is worth memorizing.

COROLLARY 1 $\left(\dfrac{1}{g}\right)' = -\dfrac{g'}{g^2}$ (where $g(x)$ is not 0).

PROOF
$\left(\dfrac{1}{g}\right)' = \dfrac{g \cdot (1)' - 1 \cdot g'}{g^2}$ derivative of a quotient

$= \dfrac{g \cdot 0 - g'}{g^2}$ derivative of a constant

The special case $\left(\dfrac{1}{g}\right)'$ $= -\dfrac{g'}{g^2}.$ ∎

EXAMPLE 11 Find $D(1/(2x^3 + x + 5))$.

SOLUTION By the formula for $(1/g)'$,

$$D\left(\dfrac{1}{2x^3 + x + 5}\right) = \dfrac{-D(2x^3 + x + 5)}{(2x^3 + x + 5)^2}$$

$$= \dfrac{-(6x^2 + 1)}{(2x^3 + x + 5)^2}.$$ ∎

In Sec. 3.2 it was shown that $D(x^n) = nx^{n-1}$ when n is a positive integer. The next corollary shows that the same formula holds when n is a negative integer.

COROLLARY 2 If n is a negative integer, $n = -1, -2, -3, \ldots$, then

$$(x^n)' = nx^{n-1}.$$

PROOF Let $n = -m$, where m is a positive integer. Then

$(x^n)' = (x^{-m})'$

$= \left(\dfrac{1}{x^m}\right)'$

$= \dfrac{-(x^m)'}{(x^m)^2}$ $\quad D\left(\dfrac{1}{f}\right) = \dfrac{-f'}{f^2}$

$= \dfrac{-mx^{m-1}}{x^{2m}}$ $\quad D(x^m) = mx^{m-1}$ for m a positive integer

$= -mx^{-m-1}$ algebra

$= nx^{n-1}$ $n = -m.$ ∎

EXAMPLE 12 Use Corollary 2 to find the derivative of $1/x$.

SOLUTION $\dfrac{d(1/x)}{dx} = \dfrac{d(x^{-1})}{dx}$

$= (-1)x^{-1-1}$ $\quad D(x^n) = nx^{n-1}$ for any integer n

$= -x^{-2}.$

Memorize $\left(\dfrac{1}{x}\right)' = -\dfrac{1}{x^2}$. Thus $\dfrac{d(1/x)}{dx} = \dfrac{-1}{x^2}$. ∎

EXAMPLE 13 Find $D(1/x^3)$.

SOLUTION
$$D\left(\frac{1}{x^3}\right) = D(x^{-3})$$
$$= -3x^{-3-1}$$
$$= -3x^{-4}$$
$$= \frac{-3}{x^4}. \blacksquare$$

EXAMPLE 14 Differentiate $(1/2t)^5$.

SOLUTION
$$D\left(\left(\frac{1}{2t}\right)^5\right) = D\left(\frac{1}{32t^5}\right)$$
$$= D\left(\frac{1}{32} \cdot \frac{1}{t^5}\right)$$
$$= \frac{1}{32} D(t^{-5}) \qquad D(cf) = cD(f)$$
$$= \frac{1}{32}(-5)t^{-6} \qquad D(x^n) = nx^{n-1}$$
$$= -\frac{5}{32}\frac{1}{t^6}. \blacksquare$$

The derivative of x^n in case $n = 0$ is 0 since $x^0 = 1$ for $x \neq 0$. This agrees with the formula nx^{n-1}. Thus $(x^n)' = nx^{n-1}$ for any integer.

SUMMARY OF THIS SECTION

$(f + g)' = f' + g' \qquad (f - g)' = f' - g'$

$(fg)' = fg' + gf' \qquad \left(\dfrac{f}{g}\right)' = \dfrac{gf' - fg'}{g^2}$

$\left(\dfrac{1}{f}\right)' = \dfrac{-f'}{f^2}$

c denotes a constant function.

$c' = 0 \qquad (cf)' = cf' \qquad \left(\dfrac{f}{c}\right)' = \dfrac{f'}{c}$

$(x^n)' = nx^{n-1} \qquad$ for any integer n

The derivative of a polynomial is the sum of the derivatives of its terms.

Exercises

In Exercises 1 to 34 differentiate with the aid of formulas, *not* by using the definition of the derivative.

1. $x^5 - 2x^2 + 3$
2. $x^3 + 5x^2 + 2$
3. $2x^4 - 6x^2 + 5x + 2$
4. $x^4 - x - \sqrt{2}$
5. $(x^2 + 3x + 1)(x^3 - 2x)$
6. $(5x^7 - x^2 + 4x + 2)(3x^2 - 7)$
7. $\dfrac{3x^4 - x^2 + 5x + 2}{7}$
8. $\dfrac{2x^3 + 3x + \pi^2}{10}$
9. $5\sqrt{x}$
10. $7\sqrt[3]{x}$
11. $12/x$
12. $-5/x^3$
13. $\dfrac{3+x}{3+x^2}$
14. $\dfrac{1+x}{2-x}$
15. $\dfrac{t^2 - 3t + 1}{t^3 + 1}$
16. $\dfrac{s^3 + 2s}{5s^2 + s + 2}$
17. $(1 + \sqrt{x})(x^3)$
18. $\left(\dfrac{2}{x} + \sqrt[3]{x}\right)(x^3 + 1)$
19. $(2x)^3$
20. $\left(\dfrac{3}{x}\right)^5$
21. $1 - \dfrac{1}{x} + \dfrac{1}{x^2}$
22. $x^2 + x + \dfrac{1}{x} + \dfrac{1}{x^2}$
23. $\dfrac{1}{x^3 + 2x + 1}$
24. $\dfrac{1}{x + \sqrt{x}}$
25. $\dfrac{(x^3 + x + 1)(x^2 - 1)}{5x^2 + 3}$
26. $\dfrac{(x^3 + 1)(5 + \sqrt[3]{x})}{x^5}$
27. $(2x + 1)^2$ *Hint:* Write it as $(2x + 1)(2x + 1)$.
28. $(3w^2 - 2w + 5)^2$ *Hint:* Write it as the product $(3w^2 - 2w + 5)(3w^2 - 2w + 5)$.
29. $\dfrac{1 + (1/x)}{1 - (1/x)}$
30. $\left(x + \dfrac{2}{x}\right)(x^3 + 6x + 1)$
31. $\dfrac{1}{\sqrt{x}}$
32. $\dfrac{1}{\sqrt[3]{x}}$
33. $\dfrac{(x^2 + 3x + 1)\sqrt{x}}{(x^3 - 5x + 2)}$
34. $\dfrac{\sqrt[3]{x}\sqrt{x}}{x^3 + 1}$

In each of Exercises 35 to 37 find the slope of the given curve at the point with the given x coordinate.

35. $y = x^3 - x^2 + 2x$, at $x = 1$
36. $y = 1/(2x + 1)$, at $x = 2$
37. $y = \sqrt{x}(x^2 + 2)$, at $x = 4$

In each of Exercises 38 to 40 the distance an object travels in the first t seconds is given by the formula. Find the velocity and the speed (= absolute value of velocity) at the given time t.

38. $2t^4 + t^3 + 2t$, $t = 1$
39. $5\sqrt{t}$, $t = 9$
40. $\sqrt[4]{t/9}$, $t = 16$

In each of Exercises 41 to 43 a lens projects x on the linear slide to a point on the linear screen given by the formula. Find the magnification at the given point x.

41. $\sqrt[3]{x}$, $x = 8$
42. $4x^2 + x + 2$, $x = 2$
43. $x^3 + 5x$, $x = \tfrac{1}{2}$

In each of Exercises 44 and 45 the mass of the left x centimeters of a string is given by the indicated formula. Find the density at the point x.

44. $x\sqrt[3]{x}$, $x = 8$
45. $x/(x + 1)$, $x = 2$
46. Give two antiderivatives for each of these functions:
    (a) x^3 (b) x (c) $1/x^2$ (d) $1/x^3$.

■ ■

47. Show that if f, g, and h are differentiable functions, then $(f + g + h)' = f' + g' + h'$. *Hint:* Use Theorem 2 twice after writing $f + g + h$ as $(f + g) + h$.
48. Using the definition of the derivative, prove that $(f - g)' = f' - g'$.
49. Show that if f, g, and h are differentiable functions, then $(fgh)' = f'gh + fg'h + fgh'$. *Hint:* First write fgh as $(fg)h$. Then use the formula for the derivative of the product of two functions.
50. Let f be a differentiable function. (a) Show that $D(f^2) = 2ff'$. (b) Using Exercise 49, show that $D(f^3) = 3f^2f'$.
51. Obtain the formula $(1/g)' = -g'/g^2$, in Corollary 2, by introducing the function $h(x) = 1/g(x)$ and using the definition of the derivative to find h'.
52. (Economics; elasticity) Economists try to forecast the impact a change in price will have on demand. For instance, if the price of a gallon of gasoline is raised 10 cents, either by the oil company or by the government (in the form of a tax), how much petroleum will be conserved? To deal with such problems economists use the concept of the *elasticity of demand*, which will now be defined. Let $y = f(x)$ be the demand for a product as a function of the price x, that is, the amount that will be bought at the price x in a given time. Assume that y is a differentiable function of x.

(a) Is the derivative y' in general positive or is it negative?
(b) What are the dimensions of y' if y is measured in gallons and x is measured in cents?
(c) Why is the ratio $(\Delta y/y)/(\Delta x/x)$ called a "dimensionless quantity"? (*Note:* If gasoline is measured in liters instead of gallons, this ratio does not change.)
(d) The *elasticity* of demand at the price x is defined as

$$\varepsilon = \lim_{\Delta x \to 0} \frac{\Delta y/y}{\Delta x/x}$$

Show that

$$\varepsilon = (x/y)y'.$$

(e) Find ε if a 1 percent increase in the price causes a 2 percent decrease in demand.
(f) Find ε if a 2 percent increase in price causes a 1 percent decrease in demand.
(g) If $|\varepsilon| > 1$, the demand is called *elastic*; if $|\varepsilon| < 1$, it is called *inelastic*. Why?
(h) Show that $y = x^{-3}$ has a constant elasticity (that is, the elasticity is independent of x).

3.5 The derivatives of the trigonometric functions

The derivatives of the six trigonometric functions—sin x, cos x, tan x, sec x, csc x, and cot x—will be found in this section.

THE DERIVATIVES OF sin x AND cos x

In order to find $\dfrac{d}{dx}(\sin x)$ and $\dfrac{d}{dx}(\cos x)$, it will be necessary to make use of the limits

$$\lim_{\Delta x \to 0} \frac{\sin \Delta x}{\Delta x} = 1 \quad \text{and} \quad \lim_{\Delta x \to 0} \frac{1 - \cos \Delta x}{\Delta x} = 0,$$

which were found in Sec. 2.5.

THEOREM 1 The derivative of the sine function is the cosine function; symbolically,

$$(\sin x)' = \cos x \quad \text{or} \quad \frac{d}{dx}(\sin x) = \cos x.$$

PROOF The derivative at x of a function f is defined as

$$\lim_{\Delta x \to 0} \frac{f(x + \Delta x) - f(x)}{\Delta x}.$$

In this case f is the function "sine," and the limit under consideration is

$$\lim_{\Delta x \to 0} \frac{\sin(x + \Delta x) - \sin x}{\Delta x}.$$

Keep in mind that x is fixed while $\Delta x \to 0$. As $\Delta x \to 0$, the numerator approaches

$$\sin x - \sin x = 0,$$

144 THE DERIVATIVE

When finding a derivative, we run into "zero-over-zero."

while the denominator Δx also approaches 0. Since the expression 0/0 is meaningless, it is necessary to change the form of the quotient

$$\frac{\sin(x + \Delta x) - \sin x}{\Delta x}$$

before letting Δx approach 0.

Let us use the formula

$$\sin(A + B) = \sin A \cos B + \cos A \sin B$$

in the case $A = x$ and $B = \Delta x$, obtaining

$$\sin(x + \Delta x) = \sin x \cos \Delta x + \cos x \sin \Delta x.$$

Then the numerator, $\sin(x + \Delta x) - \sin x$, takes the form

$$\sin x \cos \Delta x + \cos x \sin \Delta x - \sin x = \sin x (\cos \Delta x - 1) + \cos x \sin \Delta x$$
$$= -\sin x (1 - \cos \Delta x) + \cos x \sin \Delta x.$$

Therefore

$$\lim_{\Delta x \to 0} \frac{\sin(x + \Delta x) - \sin x}{\Delta x} = \lim_{\Delta x \to 0} \frac{-\sin x (1 - \cos \Delta x) + \cos x \sin \Delta x}{\Delta x}$$
$$= \lim_{\Delta x \to 0} \left(-\sin x \, \frac{1 - \cos \Delta x}{\Delta x} + \cos x \, \frac{\sin \Delta x}{\Delta x} \right)$$
$$= (-\sin x)(0) + (\cos x)(1)$$
$$= \cos x.$$

In short, the derivative of the sine function is the cosine function. This concludes the proof. ∎

The formula obtained in Theorem 1 provides interesting information about the graph of $y = \sin x$. Since

$$\frac{d}{dx}(\sin x) = \cos x,$$

the derivative of the sine function when $x = 0$ is $\cos 0$, which is 1. This implies that the slope of the curve $y = \sin x$, when $x = 0$, is 1. Consequently, the graph of $y = \sin x$ passes through the origin at an angle of $\pi/4$ radians (45°). See Fig. 3.20.

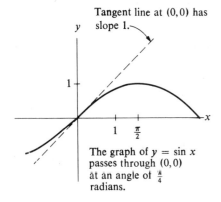

Tangent line at $(0,0)$ has slope 1.

The graph of $y = \sin x$ passes through $(0,0)$ at an angle of $\frac{\pi}{4}$ radians.

Figure 3.20

THEOREM 2 The derivative of the cosine function is the negative of the sine function; symbolically,

$$(\cos x)' = -\sin x. \quad \blacksquare$$

3.5 THE DERIVATIVES OF THE TRIGONOMETRIC FUNCTIONS

We omit the proof, which is similar to that of Theorem 1. It makes use of the trigonometric identity

$$\cos(A + B) = \cos A \cos B - \sin A \sin B,$$

and is left as Exercise 19.

THE DERIVATIVES OF sec x AND csc x

Recall that

$$\sec x = \frac{1}{\cos x}, \qquad \tan x = \frac{\sin x}{\cos x},$$

$$\csc x = \frac{1}{\sin x}, \quad \text{and} \quad \cot x = \frac{\cos x}{\sin x}.$$

In the next two theorems the derivatives of these four functions are obtained.

THEOREM 3
$$(\sec x)' = \sec x \tan x$$
and
$$(\csc x)' = -\csc x \cot x.$$

PROOF
$$(\sec x)' = \left(\frac{1}{\cos x}\right)' = \frac{-(\cos x)'}{\cos^2 x} \qquad \text{using } \left(\frac{1}{g}\right)' = \frac{-g'}{g^2}$$

$$= \frac{-(-\sin x)}{\cos^2 x}$$

$$= \frac{\sin x}{\cos^2 x}$$

$$= \frac{1}{\cos x} \frac{\sin x}{\cos x}$$

$$= \sec x \tan x.$$

Thus $(\sec x)' = \sec x \tan x$.

Next,
$$(\csc x)' = \left(\frac{1}{\sin x}\right)'$$

$$= \frac{-(\sin x)'}{\sin^2 x}$$

$$= \frac{-\cos x}{\sin^2 x}$$

$$= -\frac{1}{\sin x} \frac{\cos x}{\sin x}$$

$$= -\csc x \cot x. \blacksquare$$

THE DERIVATIVES OF tan x AND cot x

THEOREM 4 $(\tan x)' = \sec^2 x$ and $(\cot x)' = -\csc^2 x$.

PROOF
$$(\tan x)' = \left(\frac{\sin x}{\cos x}\right)'$$
$$= \frac{\cos x \, (\sin x)' - \sin x \, (\cos x)'}{\cos^2 x}$$
$$= \frac{\cos x \cos x - \sin x \, (-\sin x)}{\cos^2 x}$$
$$= \frac{\cos^2 x + \sin^2 x}{\cos^2 x}$$
$$= \frac{1}{\cos^2 x}$$
$$= \sec^2 x.$$

Next,
$$(\cot x)' = \left(\frac{\cos x}{\sin x}\right)'$$
$$= \frac{\sin x \, (\cos x)' - \cos x \, (\sin x)'}{\sin^2 x}$$
$$= \frac{\sin x \, (-\sin x) - \cos x \, (\cos x)}{\sin^2 x}$$
$$= \frac{-\sin^2 x - \cos^2 x}{\sin^2 x}$$
$$= \frac{-1}{\sin^2 x}$$
$$= -\csc^2 x. \quad \blacksquare$$

This table summarizes the six formulas. Note that the derivatives of the cosine, cosecant, and cotangent have minus signs.

The derivatives of the cofunctions (cosine, cotangent, cosecant) have minus signs.

TABLE OF TRIGONOMETRIC DERIVATIVES

$(\sin x)' = \cos x$	$(\cos x)' = -\sin x$
$(\tan x)' = \sec^2 x$	$(\cot x)' = -\csc^2 x$
$(\sec x)' = \sec x \tan x$	$(\csc x)' = -\csc x \cot x$

EXAMPLE 1 Differentiate $x - \sin x \cos x$.

SOLUTION $D(x - \sin x \cos x) = D(x) - D(\sin x \cos x)$

3.5 THE DERIVATIVES OF THE TRIGONOMETRIC FUNCTIONS

$$= 1 - [\sin x \, D(\cos x) + \cos x \, D(\sin x)]$$
$$= 1 - [\sin x \, (-\sin x) + \cos x \, (\cos x)]$$
$$= 1 + \sin^2 x - \cos^2 x.$$

Since $1 - \cos^2 x = \sin^2 x$, this expression can be simplified to $\sin^2 x + \sin^2 x$, or simply $2 \sin^2 x$. ∎

The next section will present a shortcut for differentiating sin 2x. The derivative is *not* cos 2x. The next example uses a trigonometric identity to find $(\sin 2x)'$.

EXAMPLE 2 Differentiate sin 2x.

SOLUTION $\sin 2x = 2 \sin x \cos x$.

Thus,
$$D(\sin 2x) = D(2 \sin x \cos x)$$
$$= 2 D(\sin x \cos x)$$
$$= 2 [\sin x \, D(\cos x) + \cos x \, D(\sin x)]$$
$$= 2 [\sin x \, (-\sin x) + \cos x \, (\cos x)]$$
$$= 2 (\cos^2 x - \sin^2 x)$$
$$= 2 \cos 2x.$$

In short, the derivative of sin 2x is 2 cos 2x. ∎

EXAMPLE 3 Find $D(x^3 \sec x)$.

SOLUTION
$$D(x^3 \sec x) = x^3 D(\sec x) + \sec x \, D(x^3)$$
$$= x^3 \sec x \tan x + \sec x \, (3x^2),$$

which is usually written $x^3 \sec x \tan x + 3x^2 \sec x$ for clarity. ∎

EXAMPLE 4 Find $\dfrac{d}{dx}\left(\dfrac{\csc x}{\sqrt{x}}\right)$.

SOLUTION
$$\frac{d}{dx}\left(\frac{\csc x}{\sqrt{x}}\right) = \frac{\sqrt{x}\left[\dfrac{d}{dx}(\csc x)\right] - \csc x \left[\dfrac{d}{dx}(\sqrt{x})\right]}{(\sqrt{x})^2}$$

$$= \frac{\sqrt{x}(-\csc x \cot x) - \csc x \, \dfrac{1}{2\sqrt{x}}}{x}.$$

Since $\sqrt{x}/x = 1/\sqrt{x}$ and $x\sqrt{x} = x^{3/2}$, this can be simplified a little to

$$\frac{-\csc x \cot x}{\sqrt{x}} - \frac{\csc x}{2x^{3/2}}. \quad ∎$$

148 THE DERIVATIVE

WHY RADIAN MEASURE IS USED IN CALCULUS

Figure 3.21

Throughout this section angles are measured in radians, as is generally done in calculus. What would the derivatives of sine and cosine be if angles were measured in degrees instead? Let Sin x denote the sine of an angle of x degrees. The graph of $y = \text{Sin } x$, where angle is measured in degrees, has a slope near 0 for all x, as shown in Fig. 3.21. Indeed the graph of $y = \text{Sin } x$ is now practically a horizontal line. We are tempted to change the vertical scale to stretch the graph vertically, but to do so would change the slopes, the very object of our inquiry.

If you look over the proof of Theorem 1, you will see that the key step is finding out that

$$\lim_{\Delta x \to 0} \frac{\sin \Delta x}{\Delta x} = 1.$$

If angle is measured in degrees, this limit is no longer 1. To see what the limit is, observe that an angle of x degrees is the same as an angle of $\pi x/180$ radians. Thus

$$\frac{\text{Sin } x}{x} = \frac{\sin (\pi x/180)}{x},$$

where, on the right side of the equation, the angle is measured in radians. Now

$$\frac{\sin (\pi x/180)}{x} = \frac{\sin (\pi x/180)}{\pi x/180} \cdot \frac{\pi}{180}.$$

Thus $\displaystyle\lim_{x \to 0} \frac{\text{Sin } x}{x} = \lim_{x \to 0} \frac{\sin (\pi x/180)}{\pi x/180} \cdot \frac{\pi}{180} = 1 \cdot \frac{\pi}{180} = \frac{\pi}{180}.$

In short, when angles are measured in degrees,

$$\lim_{x \to 0} \frac{\text{Sin } x}{x} = \frac{\pi}{180} \approx 0.017.$$

Even if angles are measured in degrees, and Cos x means the cosine of an angle of x degrees, it is still true that

$$\lim_{x \to 0} \frac{1 - \text{Cos } x}{x} = 0,$$

as the reader may check by examining the proof in Sec. 2.5 that this limit is 0 when angles are measured in radians.

Imitation of the steps in the proof of Theorem 1 shows that if the angle is measured in degrees, then

$$(\text{Sin } x)' = \frac{\pi}{180} \text{Cos } x \approx 0.017 \text{ Cos } x.$$

The great advantage of radian measure for angles is now evident: It makes the formula for the derivative of the sine function (indeed, of any trigonometric function) easier. There is no extra constant, such as $\pi/180$, to memorize. This goes back basically to the simplicity of

$$\lim_{\Delta x \to 0} \frac{\sin \Delta x}{\Delta x}$$

when angles are measured in radians; it is just 1.

Exercises

In each of Exercises 1 to 8 differentiate the function and simplify your answer.

1 $\sin x - x \cos x$
2 $\cos x + x \sin x$
3 $2x \sin x + 2 \cos x - x^2 \cos x$
4 $3x^2 \sin x - 6 \sin x - x^3 \cos x + 6x \cos x$
5 $\tan x - x$
6 $-\cot x - x$
7 $2x \cos x - 2 \sin x + x^2 \sin x$
8 $(3x^2 - 6) \cos x + (x^2 - 6) \sin x$

In Exercises 9 to 18 differentiate the functions.

9 $\dfrac{1 + \sin x}{\cos x}$
10 $\dfrac{1 - \sin x}{\cos x}$
11 $\dfrac{1 + 3 \sec x}{\tan x}$
12 $x^3 \sec x$
13 $\dfrac{\csc x}{\sqrt[3]{x}}$
14 $3 \csc x + 2 \tan x$
15 $\sin x \tan x$
16 $x^2 \cos x \cot x$
17 $\dfrac{\cot x}{1 + x^2}$
18 $\dfrac{x}{\sec x}$

∎

19 Using the identity for $\cos (A + B)$, show that $D(\cos x) = -\sin x$.
20 Differentiate $\cos 2x$.
21 Find the slope of the curve $y = \cos x$ at the point for which x is
   (a) 0 (b) $\pi/6$ (c) $\pi/4$ (d) $\pi/3$ (e) $\pi/2$.

22 (a) Find the slope of the curve $y = \tan x$ when $x = \pi/4$. (b) Using (a), estimate the angle that the tangent line to the curve $y = \tan x$ at $(\pi/4, 1)$ makes with the x axis. Recall that this angle is called the "angle of inclination" of the line.
23 A mass bobbing up and down on the end of a spring has the y coordinate $y = 3 \sin t$ centimeters at time t seconds.
   (a) How high does it go? (b) How low? (c) What is its velocity when $t = 0$ and when $t = \pi$? (d) What is its speed when $t = 0$ and when $t = \pi$?
24 The height of the ocean surface above (or below) mean sea level is, say, $y = 2 \sin t$ feet at t hours.
   (a) Find the rate at which the tide is rising or falling at time t.
   (b) Is the surface rising most rapidly at low tide or when it is at mean sea level?
25 At what angle does the graph of $y = \tan x$ cross the x axis?
26 What is the angle of inclination of the tangent line to the curve $y = \sin x$ at the point $(\pi, 0)$?
27 Give two antiderivatives for each of these functions:
   (a) $\cos x$ (b) $5 \cos x$ (c) $\sin x$ (d) $-3 \sin x$.

■ ■

28 Using the definition of the derivative and the identity for $\sin (A + B)$, show that $(\sin 7x)' = 7 \cos 7x$.
29 Using the definition of the derivative and the identity for $\cos (A + B)$, show that $D(\cos 11x) = -11 \sin 11x$.

3.6 Composite functions and the chain rule

Composite functions were discussed in Sec. 1.4.

The differentiation techniques obtained so far do not enable us to differentiate such functions as

$$\sqrt{1+x^2} \quad \text{or} \quad \sin x^3.$$

We could differentiate $(1+2x)^{100}$, but only with great effort, by first expanding $(1+2x)^{100}$ to form a polynomial of degree 100 and then differentiating that polynomial. This section develops a shortcut for differentiating composite functions, such as $\sqrt{1+x^2}$, $\sin x^3$, and $(1+2x)^{100}$, which are built up from simpler functions.

THE CHAIN RULE

If f and g are differentiable functions, is the composite function $h = f \circ g$ also differentiable? If so, what is its derivative? More concretely: If $y = f(u)$ and $u = g(x)$, then y is a function of x. How can we find dy/dx?

Our experience with projectors suggests the answer. Let $h = f \circ g$. The effect of the first projector, which takes x on the slide to $g(x)$ on the first screen, is to magnify by a factor $g'(x)$. Then what is the effect of the second projector f, which takes $g(x)$ on the first screen to $f(g(x))$ on the second screen? (See Fig. 3.22.) It magnifies by the factor f' evaluated at $g(x)$, that

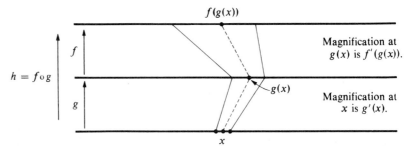

Figure 3.22

is, $f'(g(x))$. But when one magnification is followed by another, the total effect is the *product* of the two magnifications. For instance, a twofold magnification, followed by a threefold, is a sixfold magnification. We suspect, therefore, that $h'(x)$, the total magnification, is equal to $f'(g(x)) \cdot g'(x)$.

In the differential notation for derivatives, this formula becomes much shorter. Let

$$y = f(u), \quad \text{where} \quad u = g(x),$$

and let

$$y = h(x)$$

be the composite function

$$h = f \circ g.$$

Then our suspicion is that the derivative of h is given by the formula

$$\frac{dy}{dx} = \frac{dy}{du}\frac{du}{dx}. \tag{1}$$

(This, written in Leibniz notation, is a much more suggestive formula. If the symbol du had meaning by itself, it would look as if the du's cancel.)

3.6 COMPOSITE FUNCTIONS AND THE CHAIN RULE **151**

Formula (1), known as the chain rule, will be proved at the end of the section after several examples illustrate how it is applied.

THE CHAIN RULE (informal statement)

An easily remembered form of the chain rule

If y is a differentiable function of u and u is a differentiable function of x, then y is a differentiable function of x and

$$\frac{dy}{dx} = \frac{dy}{du}\frac{du}{dx}.$$

The equation

$$\frac{dy}{dx} = \frac{dy}{du}\frac{du}{dx}$$

is read as "derivative of y with respect to x equals derivative of y with respect to u times derivative of u with respect to x." This says "total magnification of the two projections equals magnification of second projection times magnification of first projection."

The chain rule in D notation

In the D notation the chain rule reads

$$D_x(y) = D_u(y)D_x(u).$$

The subscripts are needed to indicate the appropriate variable. $D_x(y)$ is the derivative of y with respect to x, that is, $\lim_{\Delta x \to 0} (\Delta y / \Delta x)$. $D_u(y)$ is the derivative of y with respect to u, that is, $\lim_{\Delta u \to 0} (\Delta y / \Delta u)$.

HOW TO USE THE CHAIN RULE

Some examples will show the power of the chain rule, which is the technique most often used when computing derivatives. At first, each calculation will be displayed in full detail. However, Example 6 indicates how the calculations are done after sufficient practice.

EXAMPLE 1 Differentiate $\sqrt{1 + x^2}$.

SOLUTION $y = \sqrt{1 + x^2}$ is a composite function,

$$y = \sqrt{u} \quad \text{where} \quad u = 1 + x^2.$$

By the chain rule,

$$\frac{dy}{dx} = \frac{dy}{du} \cdot \frac{du}{dx},$$

or

$$\frac{d}{dx}(\sqrt{1 + x^2}) = \frac{d}{du}(\sqrt{u}) \cdot \frac{d}{dx}(1 + x^2)$$

$$= \frac{1}{2\sqrt{u}} \cdot 2x$$

152 THE DERIVATIVE

$$= \frac{x}{\sqrt{u}}$$

$$= \frac{x}{\sqrt{1+x^2}}.$$

In short, $\dfrac{d}{dx}(\sqrt{1+x^2}) = \dfrac{x}{\sqrt{1+x^2}}.$ ∎

EXAMPLE 2 Differentiate $\sin x^3$.

SOLUTION $y = \sin x^3$ can be expressed as

$$y = \sin u \quad \text{where} \quad u = x^3.$$

By the chain rule,

$$\frac{d}{dx}(\sin x^3) = \frac{d}{du}(\sin u) \cdot \frac{d}{dx}(x^3)$$

$$= \cos u \cdot 3x^2$$

$$= 3x^2 \cos u$$

$$= 3x^2 \cos x^3. \quad \blacksquare$$

Note that the answer is not $\cos x^3$.

EXAMPLE 3 Differentiate $(1 + 2x)^{100}$.

SOLUTION $y = (1 + 2x)^{100}$ is the composition of

$$y = u^{100} \quad \text{and} \quad u = 1 + 2x.$$

By the chain rule,

$$\frac{d}{dx}((1+2x)^{100}) = \frac{d}{du}(u^{100}) \cdot \frac{d}{dx}(1+2x)$$

$$= 100u^{99} \cdot 2$$

$$= 200(1+2x)^{99}. \quad \blacksquare$$

Note that the answer is not $100(1 + 2x)^{99}$.

The chain rule extends to a function built up as the composition of three or more functions. For instance, if

$$y = f(u), \quad u = g(v), \quad \text{and} \quad v = h(x),$$

then y is a function of x and it can be shown that

An extended form of the chain rule

$$\frac{dy}{dx} = \frac{dy}{du} \frac{du}{dv} \frac{dv}{dx}.$$

The next example applies this fact.

3.6 COMPOSITE FUNCTIONS AND THE CHAIN RULE

EXAMPLE 4 Differentiate $\sqrt{(1+x^2)^5}$.

SOLUTION $y = \sqrt{(1+x^2)^5}$ can be expressed as

$$y = \sqrt{u}, \quad u = v^5, \quad \text{and} \quad v = 1 + x^2.$$

Then
$$\frac{dy}{dx} = \frac{dy}{du}\frac{du}{dv}\frac{dv}{dx}$$

or
$$\frac{d}{dx}(\sqrt{(1+x^2)^5}) = \frac{d}{du}(\sqrt{u})\frac{d}{dv}(v^5)\frac{d}{dx}(1+x^2)$$

$$= \frac{1}{2\sqrt{u}} \cdot 5v^4 \cdot 2x$$

$$= \frac{5v^4 x}{\sqrt{u}}$$

$$= \frac{5(1+x^2)^4 x}{\sqrt{v^5}}$$

$$= \frac{5(1+x^2)^4 x}{\sqrt{(1+x^2)^5}}. \quad \blacksquare$$

EXAMPLE 5 Differentiate $\sin^2 3x$.

SOLUTION $y = \sin^2 3x$ can be written as

$$y = u^2, \quad u = \sin v, \quad v = 3x.$$

Thus
$$\frac{d}{dx}(\sin^2 3x) = \frac{d}{du}(u^2)\frac{d}{dv}(\sin v)\frac{d}{dx}(3x)$$

$$= (2u)(\cos v)(3)$$

$$= 6u \cos v$$

$$= 6 \sin v \cos v$$

$$= 6 \sin 3x \cos 3x. \quad \blacksquare$$

EXAMPLE 6 Compute $\dfrac{d}{dx}(x^2 \sin^5 2x)$.

SOLUTION First of all, by the formula for the derivative of the product,

$$\frac{d}{dx}(x^2 \sin^5 2x) = x^2 \frac{d}{dx}(\sin^5 2x) + \sin^5 2x \frac{d}{dx}(x^2).$$

The chain rule is needed for computing

$$\frac{d}{dx}(\sin^5 2x).$$

Without all the details (that is, introduction of the letters u and v and exhibition of the various functions in detail), the computation looks like this:

$$\frac{d}{dx}(\sin^5 2x) = (5\sin^4 2x)(\cos 2x)(2)$$

$$= 10\sin^4 2x \cos 2x.$$

Thus $\quad \dfrac{d}{dx}(x^2 \sin^5 2x) = x^2(10\sin^4 2x \cos 2x) + (\sin^5 2x)(2x)$

$$= 10x^2 \sin^4 2x \cos 2x + 2x \sin^5 2x. \quad \blacksquare$$

y	$\dfrac{dy}{dx}$
u^n	$nu^{n-1}\dfrac{du}{dx}$
$\sin u$	$\cos u \dfrac{du}{dx}$
$\cos u$	$-\sin u \dfrac{du}{dx}$

As these examples suggest, the chain rule is the most important tool in the computation of derivatives.

The table on the left records a few special cases of the chain rule. They are used so often that they are worth memorizing. In each case u is a differentiable function of x.

When first working with the chain rule it is safest to write down every step, showing the various functions with the aid of the letters u, v, and so on. However, with practice, it will not be necessary to record every detail.

The next example develops an idea for use later.

EXAMPLE 7 Let y be a differentiable function of x. Then y^2 is also a differentiable function of x. Express its derivative with respect to x, $D_x(y^2)$, in terms of the derivative of y with respect to x, $D_x(y)$.

SOLUTION Denote y^2 by w. Then

$$w = y^2 \quad \text{where } y \text{ is a function of } x.$$

By the chain rule, $\quad D_x(w) = D_y(w)D_x(y)$

$$= 2yD_x(y)$$

In short, $\quad D_x(y^2) = 2yD_x(y). \quad \blacksquare$

THE DERIVATIVE OF $x^{m/n}$ In Secs. 3.2 and 3.3 it was shown that $D(x^n) = nx^{n-1}$ for any integer and $D(x^{1/n}) = (1/n)x^{1/n-1}$ for any positive integer. These are special cases of the formula for the derivative of $x^{m/n}$ for any rational exponent m/n. This formula will now be obtained with the aid of the chain rule.

THEOREM 1 Let m be an integer and n a positive integer.

Then $\quad D(x^{m/n}) = \dfrac{m}{n} x^{(m/n)-1}.$

PROOF Let $y = x^{m/n} = (x^{1/n})^m$. Then y is a composite function:

$$y = u^m \quad \text{where} \quad u = x^{1/n}.$$

By the chain rule,

$$\frac{dy}{dx} = \frac{d(u^m)}{du} \cdot \frac{d(x^{1/n})}{dx}$$

$$= mu^{m-1} \cdot \frac{1}{n} x^{(1/n)-1} \qquad \text{known formulas}$$

$$= m(x^{1/n})^{m-1} \cdot \frac{1}{n} x^{(1/n)-1}$$

$$= mx^{(m-1)/n} \cdot \frac{1}{n} x^{(1/n)-1} \qquad \text{power of a power}$$

$$= \frac{m}{n} x^{(m-1)/n + (1/n) - 1} \qquad \text{basic law of exponents}$$

$$= \frac{m}{n} x^{(m/n)-1} \quad \blacksquare$$

EXAMPLE 8 Use the formula for $D(x^{m/n})$ to differentiate

$$\sqrt[3]{x}\sqrt{x}.$$

SOLUTION
$$\sqrt[3]{x}\sqrt{x} = x^{1/3} x^{1/2}$$
$$= x^{1/3 + 1/2}$$
$$= x^{5/6}.$$

Then
$$D(x^{5/6}) = \frac{5}{6} x^{(5/6)-1}$$
$$= \frac{5}{6} x^{-1/6}$$
$$= \frac{5}{6} \frac{1}{\sqrt[6]{x}}. \quad \blacksquare$$

PROOF OF THE CHAIN RULE

THEOREM 2 *The chain rule.* If f and g are differentiable functions and $g(x)$ is in the domain of f for all x in the domain of g, then $h = f \circ g$ is also differentiable, and

$$h'(x) = f'(g(x)) \cdot g'(x).$$

More briefly, if $y = f(u)$ and $u = g(x)$, then $y = h(x)$ and

$$\frac{dy}{dx} = \frac{dy}{du} \frac{du}{dx}.$$

PROOF To examine $h'(x)$, it is necessary to go back to the definition of the derivative,

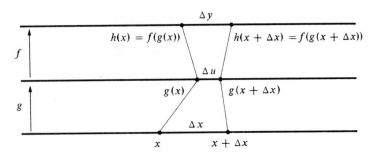

Figure 3.23

$$h'(x) = \lim_{\Delta x \to 0} \frac{\Delta y}{\Delta x}.$$

The computation will involve Δx, $\Delta u = \Delta g$, and Δy, shown in Fig. 3.23. That is, Δx, which is not 0, determines a number Δu, the change in u,

$$\Delta u = g(x + \Delta x) - g(x),$$

and a number Δy, the change in y,

$$\Delta y = h(x + \Delta x) - h(x).$$

It is important to note that, since g is differentiable, $\Delta u \to 0$ as $\Delta x \to 0$. However, it could happen that Δu is 0, even though Δx is not 0. Since in a moment we shall wish to divide by Δu, we make an extra assumption, namely that

$$g'(x) \text{ is not } 0.$$

Since
$$g'(x) = \lim_{\Delta x \to 0} \frac{\Delta u}{\Delta x},$$

this implies that, when Δx is sufficiently small, Δu is not 0.

With this assumption, which most functions satisfy at most values of x, the proof is short:

$$h'(x) = \lim_{\Delta x \to 0} \frac{\Delta y}{\Delta x}$$

$$= \lim_{\Delta x \to 0} \frac{\Delta y}{\Delta u} \frac{\Delta u}{\Delta x} \qquad \text{algebra}$$

$$= \lim_{\Delta x \to 0} \frac{\Delta y}{\Delta u} \lim_{\Delta x \to 0} \frac{\Delta u}{\Delta x}$$

$$= \lim_{\Delta u \to 0} \frac{\Delta y}{\Delta u} \lim_{\Delta x \to 0} \frac{\Delta u}{\Delta x} \qquad \text{since } \Delta u \to 0 \text{ as } \Delta x \to 0$$

$$= \frac{dy}{du} \frac{du}{dx}.$$

[The special case, where $g'(x) = 0$, is discussed in Exercise 58.] This concludes the proof. ∎

Exercises

Exercises 1 to 8 concern the notion of a composite function. In each case show that the function is a composition of other functions by introducing the necessary symbols, such as u and v, and describing the functions.

1. $(x^3 + x^2 - 2)^{50}$
2. $(\sqrt{x} + 1)^{10}$
3. $\sqrt{x + 3}$
4. $\sqrt[3]{1 + x^2}$
5. $\sin 2x$
6. $\sin^2 x$
7. $\cos^3 2x$
8. $\sqrt[3]{(1 + 2x)^{50}}$

In Exercises 9 to 38 differentiate the functions.

9. $(2x^3 - x)^{40}$
10. $(x^4 - 6)^{100}$
11. $\left(1 + \dfrac{3}{x}\right)^4$
12. $\left(1 - \dfrac{2}{x^2}\right)^5$
13. $\sqrt{x^3 + x + 2}$
14. $\sqrt[3]{x^3 + 8}$
15. $\sin 3x$
16. $(1 + 4x)^{10}$
17. $\dfrac{x^3 \tan (1/x)}{1 + x^2}$
18. $\sqrt{1 + x^2} \cot^3 5x$
19. $(2x + 1)^5 (3x + 1)^7$
20. $x^5 (x^2 + 2)^3 \cos^3 5x$
21. $x^2 \sin^5 3x$
22. $x^3 \cos^2 3x \sin^2 2x$
23. $\dfrac{1}{(2x + 3)^5}$ *Shortcut:* First write it as $(2x + 3)^{-5}$.
24. $\dfrac{1}{(3x - 5)^7}$
25. $\dfrac{(2x + 1)^3}{(3x + 1)^4}$
26. $\dfrac{x^2}{(x^2 + 1)^3}$
27. $\dfrac{(x^3 - 1)^5 \cot^3 5x}{x}$
28. $\dfrac{\sqrt{1 - x^2} \sec^2 \sqrt{x}}{1 + x}$
29. $\left(\dfrac{1 + 2x}{1 + 3x}\right)^4$
30. $\left(\dfrac{x^2 + 3x + 5}{2x - 1}\right)^5$
31. $\tan^2 \sqrt{x}$
32. $(2x + 1)^3 \cot^5 x^3$
33. $\dfrac{1}{\sqrt{1 - x^2}}$
34. $\dfrac{x}{\sqrt{1 - x^2}}$
35. $\sqrt{5(3x - 2)^4 + 1}$
36. $\sqrt[3]{(2x + 1)^2 + x}$
37. $\cos 3x \sin 4x$
38. $\sec 2x \tan 5x$

Exercises 39 to 44 concern the formula $D(x^{m/n}) = (m/n)x^{m/n - 1}$. In each case use it when finding the derivative of the indicated function.

39. $x^{7/5}$
40. $\sqrt[3]{x^4}$
41. $(x^3 + 2)^{7/9}$
42. $\sqrt[3]{(x^2 + x)^2}$
43. $\sqrt[3]{x} \cdot x^3$
44. $\sqrt[3]{(2x + 1)^2}$

In each of Exercises 45 to 48 y is a differentiable function of x; express the indicated derivative in terms of y and dy/dx.

45. $\dfrac{d}{dx}(y^5)$
46. $\dfrac{d}{dx}(1/y)$
47. $\dfrac{d}{dx}(\sin y)$
48. $\dfrac{d}{dx}(\cos y^2)$

∎

49. Let $y = 3u$ and $u = 5x$. Then y is a composite function of x. (a) Compute dy/dx by the chain rule. (b) Compute dy/dx by first expressing y in terms of x.

In each of Exercises 50 to 53 give an example of an antiderivative of the given function.

50. $(2x + 1)^4$
51. $\cos 3x$
52. $x^{4/3}$
53. $\sin 2x$

In Exercises 54 to 56 differentiate the given functions and simplify your answers as far as possible.

54. $\dfrac{2(9x - 2)}{135} \sqrt{(3x + 1)^3}$
55. $-\tfrac{1}{3} \cos 3x + \tfrac{1}{9} \cos^3 3x$
56. $\dfrac{3x}{8} - \dfrac{3 \sin 10x}{80} - \dfrac{\sin^3 5x \cos 5x}{20}$

∎ ∎

57. We used the "magnification" interpretation of the derivative to guess the chain rule. The "slope" interpretation is not as illuminating. Let us see what the chain rule asserts about slopes.
    (a) Let f and g be differentiable functions, such that $g(1) = 2$ and $f(2) = 3$. Show that $(1, 2)$ is on the graph of g, $(2, 3)$ is on the graph of f, and $(1, 3)$ is on the graph of $f \circ g$.
    (b) If the slope of the graph of g at the point $(1, 2)$ is 5, and the slope of the graph of f at $(2, 3)$ is 7, what is the slope of the graph of $f \circ g$ at $(1, 3)$?

58. This exercise outlines the proof of the chain rule when $g'(x) = 0$.
    (a) Show that in this case it is sufficient to prove that
    $$\lim_{\Delta x \to 0} \dfrac{\Delta y}{\Delta x} = 0.$$
    (b) There are two types of Δx, those for which $\Delta u \neq 0$ and those for which $\Delta u = 0$. Show that, as $\Delta x \to 0$ through values of the first type, then $\Delta y/\Delta x \to 0$. *Hint:* Write $\Delta y/\Delta x = (\Delta y/\Delta u)(\Delta u/\Delta x)$.

158 THE DERIVATIVE

(c) Show that, when Δx is of the second type, Δy is 0; hence

$$\frac{\Delta y}{\Delta x} = 0.$$

Thus, as $\Delta x \to 0$ through values of the second type, $\Delta y/\Delta x \to 0$.

(d) Combine (b) and (c) to show that, as $\Delta x \to 0$, $\Delta y/\Delta x \to 0$.

59 Combine the formula for $D(1/x)$ with the chain rule to show that $D(1/g) = -g'/g^2$ for any differentiable function g.

3.S Summary

This chapter defined the derivative of a function as

$$\lim_{w \to x} \frac{f(w) - f(x)}{w - x} \quad \text{or} \quad \lim_{\Delta x \to 0} \frac{f(x + \Delta x) - f(x)}{\Delta x}$$

if this limit exists. Informally, the derivative is

$$\text{The limit of } \frac{\text{difference in outputs}}{\text{difference in inputs}},$$

as the difference in inputs approaches 0.

The derivative measures how quickly the value of a function changes. If a slight change in the input causes a relatively large change in the output, the derivative will be large. Most of the chapter was spent developing techniques for computing derivatives without having to return to the definition of the derivative and calculate a (perhaps horrendous) limit each time you want to know the derivative of some function. These labor-saving methods are listed later in this summary.

The derivative was motivated by slope, velocity, magnification, and density—concepts drawn from geometry or the physical world. The derivative has many more applications, and we will cite four more. But whenever the rate at which some quantity changes is studied, a derivative will surely enter the picture.

Biology Let $P(t)$ be a differentiable function that estimates the size of a population at time t. Then the derivative $P'(t)$ tells how fast the population is increasing (if $P'(t) > 0$) or decreasing (if $P'(t) < 0$) at time t.

Physiology Let $Q(t)$ be the amount of blood, in cubic centimeters, that flows through an artery during the first t seconds of an experiment. Then the derivative $Q'(t)$ is the rate, in cubic centimeters per second, at which blood flows through the artery at time t.

Economics Let $C(x)$ be the cost in dollars of producing x refrigerators. (In reality x is an integer; in economic theory and practice it is convenient to assume that $C(x)$ is defined for all real numbers in some interval and that $C(x)$ is a differentiable function.) The derivative $C'(x)$ is called the *marginal cost*. This marginal cost, as we will now show, is roughly the cost of producing the $(x + 1)$st refrigerator. The actual cost of producing refrigerator number $x + 1$ is the cost of producing the first $x + 1$ refrigerators, less the cost of producing the first x refrigerators. So the cost of producing the $(x + 1)$st refrigerator is $C(x + 1) - C(x)$, which equals

$$\frac{C(x + 1) - C(x)}{1},$$

which, by the definition of the derivative, is an approximation of $C'(x)$. Or, looked at the opposite way, $C'(x)$ is an approximation of the ratio $[C(x + 1) - C(x)]/1$, the cost of the $(x + 1)$st refrigerator.

Similarly, if $R(x)$ is the total revenue received for x refrigerators, then the derivative $R'(x)$ is called the *marginal revenue*, which can be thought of as the extra revenue obtained by selling the $(x + 1)$st refrigerator.

Energy   Let $Q(t)$ be the total amount of crude oil in the earth at time t, measured in barrels. (One barrel holds 42 gallons.) The derivative $Q'(t)$ tells how fast $Q(t)$ changes. If no new reserves are being formed, then $Q'(t)$ is negative, approximately $-50{,}000{,}000$ barrels per day. Estimates of $Q(t)$ for $t = 1980$ vary but are on the order of $2 \cdot 10^{12}$ barrels (two trillion barrels). If $Q'(t)$ remains constant, all known and conjectured reserves would be used up in about a century.

Predictions of the rate at which petroleum—or any other nonrenewable natural resource—will be used depend on estimates of derivatives.

The following table is worth careful study. The bottom row describes the derivative, which is the underlying mathematical concept, free of any particular interpretation. Each of the other lines describes one of its many *applications* or *interpretations*.

If we interpret x as	and $f(x)$ as	then $\dfrac{f(x + \Delta x) - f(x)}{\Delta x}$ is	and, as Δx approaches 0, the quotient approaches
The abscissa of a point in the plane	The ordinate of that point	The slope of a certain line	The slope of a tangent line
Time	The location of a particle moving on a line	An average velocity over a time interval	The velocity at time x
A point on a linear slide	Its projection on a linear screen	An average magnification	The magnification at x
A location on a non-uniform string	The mass from 0 to x	An average density	The density at x
Time	Mass of a bacterial culture at time x	An average growth rate over a time interval	The growth rate at time x
Time	Total profit up to time x	An average rate of profit over a time interval	The rate of profit at time x
Just a number: the input	A number depending on x: the output	A quotient: the change in the output divided by the change in the input	The derivative evaluated at x (the rate of change of the function with respect to x)

VOCABULARY AND SYMBOLS

difference quotient $\dfrac{f(w)-f(x)}{w-x}$ or $\dfrac{f(x+\Delta x)-f(x)}{\Delta x}$

derivative f', $\dfrac{d}{dx}(f)$, $\dfrac{df}{dx}$, $\dfrac{dy}{dx}$, $D(f)$, and the dot notation $\dot{x} = \dfrac{dx}{dt}$

differentiable
Δx (change in input)   velocity and speed   tangent to a curve
Δf (change in output)   magnification   antiderivative
slope of a curve   density   chain rule

KEY FACTS

The derivative is defined as a limit:

$$\lim_{w \to x} \frac{f(w)-f(x)}{w-x} \quad \text{or} \quad \lim_{\Delta x \to 0} \frac{f(x+\Delta x)-f(x)}{\Delta x},$$

if the limit exists.

If $f'(x)$ exists at a particular number x, then f is said to be differentiable at that number. A function that is differentiable throughout an interval is said to be differentiable on that interval. Most functions met in applications are differentiable throughout their domains with perhaps the exception of a few isolated points. For instance, $\sqrt{x}$ is differentiable throughout $(0, \infty)$ but not at 0.

Wherever a function is differentiable it is necessarily continuous. However a function can be continuous at a number yet not be differentiable there. For example $|x|$ is continuous at 0 but not differentiable there. (In 1831 Bolzano constructed a function that is continuous everywhere but differentiable nowhere!)

The computational formulas and techniques obtained in the chapter are recorded in the following two tables.

FORMULAS		
f	f'	Remark
c	0	Constant function.
x	1	
x^a	ax^{a-1}	Rational a, fixed. This includes the case when a is an integer or reciprocal of a positive integer.
$\sqrt{x}$	$\dfrac{1}{2\sqrt{x}}$	Though a special case of x^a, worth memorizing.
$\dfrac{1}{x}$	$\dfrac{-1}{x^2}$	Though a special case of x^a, worth memorizing.
$a_n x^n + \cdots + a_1 x + a_0$	$na_n x^{n-1} + \cdots + a_1$	Differentiate polynomials term by term.
$\sin x$	$\cos x$	
$\cos x$	$-\sin x$	
$\tan x$	$\sec^2 x$	Remember that the derivatives of the "co" functions have the minus sign.
$\cot x$	$-\csc^2 x$	
$\sec x$	$\sec x \tan x$	
$\csc x$	$-\csc x \cot x$	

Combining these formulas with the chain rule shows, for instance, that,

$$\frac{d}{dx}((u(x))^a) = a(u(x))^{a-1}u'(x)$$

and
$$\frac{d}{dx}(\sin u(x)) = (\cos u(x))u'(x),$$

where $u(x)$ is a differentiable function of x.

TECHNIQUES

$(cf)' = cf'$ $\left(\dfrac{f}{c}\right)' = \dfrac{f'}{c}$ (c constant)

$(f + g)' = f' + g'$ $(f - g)' = f' - g'$

$(fg)' = fg' + gf'$ $\left(\dfrac{f}{g}\right)' = \dfrac{gf' - fg'}{g^2}$

$\left(\text{memory aid: begin by writing } \dfrac{g \cdots}{g^2}\right)$

$\left(\dfrac{1}{g}\right)' = -\dfrac{g'}{g^2}$

And, most important, most often used, the chain rule:
If $y = f(u)$ and $u = g(x)$, then

$$\frac{dy}{dx} = \frac{dy}{du}\frac{du}{dx}.$$

Guide quiz on chap. 3

1. Define the derivative.
2. *Use the definition* of the derivative to compute
   (a) $\dfrac{d}{dx}(5x^3 - 2x + 2)$ (b) $\dfrac{d}{dx}\left(\dfrac{5}{3x + 2} + 6x\right)$
   (c) $\dfrac{d}{dx}(3 \sin 2x)$ (d) $\dfrac{d}{dx}(x^{-2})$

3. Using formulas developed in the chapter, differentiate:
   (a) $5\sqrt{x}$ (b) $x^2\sqrt{3 - 2x^2}$ (c) $\cos 5x$
   (d) $(1 + x^2)^{3/4}$ (e) $\sqrt[3]{\tan 6x}$ (f) $x^3 \sin 5x$
   (g) $\dfrac{1}{\sqrt{2x + 1}}$ (h) $(2x^5 - x^3)^{-4}$ (i) $\sqrt[3]{x^3 - 3}$
   (j) $\dfrac{2x^3 + 2}{3x + 1}$ (k) $\dfrac{1}{5x^2 + 1}$
   (l) $\dfrac{1}{(3x + 2)^{10}}$ (m) $(1 + 2x)^5 x^3 \sec 3x$
   (n) $\csc \sqrt{x}$ (o) $(1 + 3 \cot 4x)^{-2}$

4. On a sketch of the graph of a typical function f,
   (a) show the line whose slope is $[f(w) - f(x)]/(w - x)$;
   (b) show the tangent line at the point $(x, f(x))$.

5. (a) Sketch the graph of $y = 3x^2 + 5x + 6$.
   (b) By inspection of your graph, estimate the x coordinate of the point where the tangent line is horizontal.
   (c) Using the derivative, solve (b) precisely.

6. Without sketching the graph of $y = x^4$, draw the line that is tangent to it at the point $(\tfrac{1}{2}, \tfrac{1}{16})$.

7. (a) Graph $y = x^3 - 12x$ with the aid of this table.

x	-2	-1	0	1	2	3
$x^3 - 12x$						

(b) Evaluate $(x^3 - 12x)'$.
(c) Find all x such that the derivative in (b) has the value 0.
(d) At what points on the graph in (a) is the tangent line horizontal? Specify both the x and y coordinates.

8 Let f be a differentiable function and let w and x be numbers such that $w > x$. Interpret $f(w) - f(x)$, $w - x$, and their quotient $[f(w) - f(x)]/(w - x)$ if
(a) $f(x)$ is the height of a rocket x seconds after lift-off;
(b) $f(x)$ is the number of bacteria in a bacterial culture at time x;
(c) $f(x)$ is the mass of the left-hand x centimeters of a rod;
(d) $f(x)$ is the position of the image on the linear screen of the point x on the linear slide.

9 A bug is wandering on the x axis. At time t seconds it is at the point $x = t^2 - 2t$. Assume that distance is measured in meters.
(a) What is the bug's velocity at time t?
(b) What is the bug's velocity at $t = \frac{1}{4}$?
(c) What is the bug's speed when $t = \frac{1}{4}$?
(d) Is the bug moving to the right or to the left when $t = \frac{1}{4}$?

10 Give an antiderivative for
(a) x^2 (b) $1/x^2$ (c) $\sin 3x$
(d) $3x^2 + 4x + 5$ (e) $\sec^2 x$ (f) $\sin x \cos x$.

Review exercises for chap. 3

In Exercises 1 to 6 use the definition of the derivative to differentiate the given functions.
1 $5x^3$
2 $\sqrt{3x}$
3 $1/(x + 3)$
4 $(2x + 1)^2$
5 $\cos 3x$
6 $\sin 5x$

In Exercises 7 to 36 find the derivatives of the given functions.
7 $2x^5 + x^3 - x$
8 $t^4 - 5t^2 + 2$
9 $\dfrac{x^2}{4x + 1}$
10 $\dfrac{(3x + 1)^4}{(2x - 1)^2}$
11 $\sqrt{3x^2 + 2x + 4}$
12 $\sqrt{5x^2 - x}$
13 $\sqrt[3]{(2t - 1)^2}$
14 $(t^2 + 1)^{3/4}$
15 $\sin^2 5x$
16 $\cos^3 7x$
17 $\dfrac{(5x + 1)^4}{7}$
18 $\dfrac{(3x - 2)^{-5}}{11}$
19 $x \sin 3x$
20 $x^2 \cos 4x$
21 $\tan^2 \sqrt[3]{1 + 2x}$
22 $\left(\dfrac{\sin 2x}{1 + \tan 3x}\right)^3$
23 $\dfrac{x^3 \cos 2x}{1 + x^2}$
24 $\dfrac{x^2 \sin 5x}{(2x + 1)^3}$
25 $\sqrt[3]{1 + \sqrt{x}}$
26 $\sqrt[4]{x + \sqrt[3]{x}}$
27 $\sqrt[3]{(\cot 5x)^7}$
28 $\sqrt[5]{(\csc 3x)^{11}}$
29 $\sqrt{8x + 3}$
30 $\sqrt{5x - 1}$
31 $\dfrac{x^2}{x^3 + 1}$
32 $\dfrac{x^3 + 1}{x^2}$
33 $((x^2 + 3x)^4 + x)^{-5/7}$
34 $\left(\dfrac{3x + 1}{2x + 1}\right)^4$
35 $\dfrac{x\sqrt{2x + 1} \cos^2 6x}{5}$
36 $\dfrac{5(1 + x^2)^3}{x\sqrt{2x + 1}}$

In Exercises 37 to 40 the number a is a constant. In each case differentiate the given function and simplify your answer.

37 $\dfrac{1}{a^2} \sin ax - \dfrac{1}{a} x \cos ax$

38 $\dfrac{x^2}{4} - \dfrac{x \sin 2ax}{4a} - \dfrac{\cos 2ax}{8a^2}$

39 $\dfrac{x}{2} - \dfrac{\sin 2ax}{4a}$

40 $-\dfrac{1}{a} \cos ax + \dfrac{1}{3a} \cos^3 ax$

41 The height of a ball thrown straight up is $64t - 16t^2$ feet after t seconds.
(a) Show that its velocity after t seconds is $64 - 32t$ feet per second.
(b) What is its velocity when $t = 0$? $t = 1$? $t = 2$? $t = 3$?
(c) What is its speed when $t = 0$? $t = 1$? $t = 2$? $t = 3$?
(d) For what values of t is the ball rising? falling?

42 In the study of the seepage of irrigation water into soil, equations such as $y = \sqrt{t}$ are sometimes used. The equation says that the water penetrates $\sqrt{t}$ feet in t hours.
(a) What is the physical significance of the derivative $1/(2\sqrt{t})$?
(b) What does (a) say about the rate at which water penetrates the soil when t is large?

43 Find an equation of the tangent line to the curve $y = x^3 - 2x^2$ at $(1, -1)$.

44 Find an equation of the tangent line to the curve $y = 2x^4 - 6x^2 + 8$ at $(2, 16)$.

45 (a) The left-hand x centimeters of a rod have a mass of $3x^4$ grams. What is its density at $x = 1$?
(b) Devise a magnification problem mathematically equivalent to (a).
(c) Devise a velocity problem mathematically equivalent to (a).

46 The left-hand x centimeters of a string have a mass of $\sqrt{x}$ grams. What is its density when x is (a) $\frac{1}{4}$? (b) 1? (c) Is its density defined at $x = 0$?

47 A snail crawls $\sqrt{t}$ feet in t seconds. What is its speed when t is (a) $\frac{1}{9}$? (b) 1? (c) 4? (d) 9?

48 Sketch a graph of $y = x^3$.
(a) Why can the tangent line to this graph at $(0, 0)$ *not* be defined as "the line through $(0, 0)$ that meets the graph just once"?
(b) Why can the tangent line to this graph at $(1, 1)$ *not* be defined as "the line through $(1, 1)$ that meets the graph just once"?
(c) How is the tangent line at any point on the graph defined?

49 It costs a certain firm $C(x) = 1000 + 5x + x^2/200$ dollars to produce x calculators, for $x \leq 400$.
(a) How much does it cost to produce 0 calculators? (This represents start-up costs which are independent of the number produced.)
(b) What is the marginal cost $C'(x)$?
(c) What is the marginal cost when $x = 10$?
(d) Compute $C(11) - C(10)$, the cost of producing the eleventh calculator.

50 After t hours a certain bacterial population has a mass of $500 + t^3$ grams. Find the rate at which it grows (in grams per hour) when
(a) $t = 0$ (b) $t = 1$ (c) $t = 2$.

51 A lens projects the point x on the x axis onto the point x^3 on a linear screen.
(a) How much does it magnify the interval $[2, 2.1]$?
(b) How much does it magnify the interval $[1.9, 2]$?
(c) What is its magnification at 2?

52 Let f be the function whose value at x is $4x^2$.
(a) Compute $[f(2.1) - f(2)]/0.1$.
(b) What is the interpretation of the quotient in (a) if $f(x)$ denotes the total profit of a firm (in millions of dollars) in its first x years?
(c) What is the interpretation of the quotient in (a) if $f(x)$ denotes the height of the ordinate in the graph of $y = 4x^2$?
(d) What is the interpretation of the quotient in (a) if $f(x)$ is the distance a particle moves in the first x seconds?

53 If the function f records the weight of a person (dependent on age), then we may think of the derivative f' as _____.

54 (a) If the function f records the trade-in value of a car (dependent on its age), then we may think of the derivative f' as _____.
(b) When is the derivative in (a) negative? positive? Which is the more usual case?

55 (Economics) A certain growing business firm makes a profit of t^2 million dollars in its first t years.
(a) How much profit does it make during its third year, that is, from time $t = 2$ to time $t = 3$?
(b) How much profit does it make from time $t = 2$ to time $t = 2.5$ (a duration of half a year)?
(c) Using (b), show that its average rate of profit from time $t = 2$ to time $t = 2.5$ is 4.5 million dollars per year.
(d) Find its "rate of profit at time $t = 2$" by considering short intervals of time from 2 to t, $t > 2$, and letting t approach 2.

56 (Biology) A certain increasing bacterial population has a mass of t^2 grams after t hours.
(a) By how many grams does the population increase from time $t = 3$ hours to time $t = 4$ hours?
(b) By how many grams does it increase from time $t = 3$ hours to $t = 3.01$ hours?
(c) By how many grams does it increase from time $t = 3$ hours to time t hours, where t is larger than 3?
(d) Using (c), show that the average rate of growth from time 3 to time t, $t > 3$, is $3 + t$ grams per hour. As t approaches 3, the average growth rate approaches 6 grams per hour, which is called "the growth rate at time 3 hours."

57 In each of these functions, y denotes a differentiable function of x. Express the derivative of each with respect to x in terms of y and dy/dx.
(a) y^3 (b) $\cos y$ (c) $1/y$.

58 Let $f(t)$ be the height in miles of the cloud top above burst height t minutes after the explosion of a 1-megaton nuclear bomb. Figure 3.24 is a graph of this function.

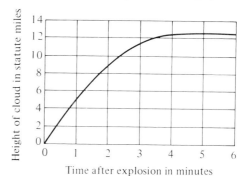

Height of cloud top above burst height at various times after a 1-megaton explosion for a moderately low air burst

Figure 3.24

(Note that the vertical and horizontal scales are different.) Assume that the cloud is not dispersed.
(a) What is the physical meaning of $f'(t)$?
(b) As t increases, what happens to $f'(t)$?
(c) As t increases, what happens to $f(t)$?
(d) Esimate how rapidly the cloud is rising at the time of explosion.
(e) Estimate how rapidly the cloud is rising 1 minute after the explosion.
In (d) and (e) make the estimate by drawing a tangent line. The estimate will be in miles per minute.

59 Give two antiderivatives for each of these functions:
(a) $4x^3$ (b) x^3 (c) $x^4 + x^3 + \cos x$
(d) $x^3 + \sin x$ (e) $(x^2 + 1)^2$
Hint: For (e), first expand $(x^2 + 1)^2$.

60 (a) Is $D(x^2 + x^3) = D(x^2) + D(x^3)$?
(b) Is $D(x^2 x^3) = D(x^2)D(x^3)$?
(c) Is $D(x^2 - x^3) = D(x^2) - D(x^3)$?
(d) Is $D\left(\dfrac{x^2}{x^3}\right) = \dfrac{D(x^2)}{D(x^3)}$?

61 Figure 3.25 shows the graph of a function f. (a) At which numbers a does $\lim_{x \to a} f(x)$ not exist? (b) At which numbers a does $\lim_{x \to a} f(x)$ exist yet f is not continuous at a? (c) Where is f continuous but not differentiable?

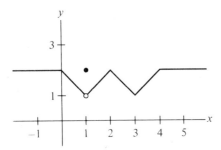

Figure 3.25

The next exercise briefly introduces a concept (that will later be treated in detail) which might be met early in an elementary physics or chemistry class.

62 In the case of a function of more than one variable we may differentiate with respect to any one of the variables, treating the others as constants. For instance, the derivative of x^3y^2z with respect to x is $3x^2y^2z$; the derivative of x^3y^2z with respect to y is $2x^3yz$; the derivative with respect to z is x^3y^2. Such derivatives are called *partial derivatives*, and are denoted with the symbol ∂ (called "del") as follows:

$$\dfrac{\partial}{\partial x}(x^3y^2z) = 3x^2y^2z,$$

$$\dfrac{\partial}{\partial y}(x^3y^2z) = 2x^3yz,$$

$$\dfrac{\partial}{\partial z}(x^3y^2z) = x^3y^2.$$

Compute the following partial derivatives:

(a) $\dfrac{\partial}{\partial x}\left(\dfrac{x^2z}{y}\right)$ (b) $\dfrac{\partial}{\partial y}\left(\dfrac{x^2z}{y}\right)$ (c) $\dfrac{\partial}{\partial z}\left(\dfrac{x^2z}{y}\right)$

(d) $\dfrac{\partial}{\partial x}(\cos 3x \sin 4y)$ (e) $\dfrac{\partial}{\partial y}(\cos 3x \sin 4y)$

63 Tell what is wrong with this alleged proof that $2 = 1$: Observe that $x^2 = x \cdot x = x + x + \cdots + x$ (x times). Differentiation with respect to x yields the equation $2x = 1 + 1 + \cdots + 1$ (x 1s). Thus $2x = x$. Setting $x = 1$ shows that $2 = 1$.

64 Define $f(x)$ to be $x^2 \sin(1/x)$ if $x \neq 0$, and $f(0)$ to be 0.
(a) Show that f has a derivative at 0, namely 0.
Hint: Investigate $\lim\limits_{\Delta x \to 0} \dfrac{f(\Delta x) - f(0)}{\Delta x}$.
(b) Show that f has a derivative at $x \neq 0$.
(c) Show that the derivative of f is not continuous at $x = 0$.

65 Let $V(r) = 4\pi r^3/3$, the volume of a sphere of radius r. Let $S(r) = 4\pi r^2$, the surface area of a sphere of radius r.
(a) Show that $V'(r) = S(r)$.
(b) With the aid of a picture showing concentric spheres of radii r and $r + \Delta r$, explain why the result in (a) is plausible.

Express the limits in Exercises 66 and 67 as derivatives, and evaluate them.

66 $\lim\limits_{w \to 2} \dfrac{(1 + w^2)^3 - 125}{w - 2}$

67 $\lim\limits_{\Delta x \to 0} \dfrac{\sin\sqrt{3 + \Delta x} - \sin\sqrt{3}}{\Delta x}$

Exercises 68 and 69 concern the functions b^x.

68 None of the formulas or techniques of this chapter enable us to differentiate an exponential function b^x. Using the definition of the derivative in the Δ notation show that

$$D(b^x) = c \cdot b^x,$$

where c is equal to

$$\lim\limits_{\Delta x \to 0} \dfrac{b^{\Delta x} - 1}{\Delta x}.$$

At this point we are not able to show that this limit exists.

69 (Calculator) This continues Exercise 68. Using $\Delta x = 0.0001$, estimate

(a) $\lim\limits_{\Delta x \to 0} \dfrac{2^{\Delta x} - 1}{\Delta x}$ (b) $\lim\limits_{\Delta x \to 0} \dfrac{3^{\Delta x} - 1}{\Delta x}$

(c) $\lim\limits_{\Delta x \to 0} \dfrac{2.71828^{\Delta x} - 1}{\Delta x}$

The base b that is most convenient for calculus is the one for which $(b^x)' = b^x$. As (c) suggests, such a base is approximately 2.71828. The actual value is an endless decimal and is denoted e. This will be discussed in Chap. 6.

70 Find all continuous functions f, whose domains are the x axis, such that $f(0) = 5$ and, for each x, $f(x)$ is an integer.

71 (a) Draw a freehand curve indicating a typical function f.
(b) Label on it the three points $P_0 = (x, f(x))$, $P_1 = (x + h, f(x + h))$, and $P_2 = (x - h, f(x - h))$.

(c) Show that the slope of the line through P_1 and P_2 is

$$\dfrac{f(x + h) - f(x - h)}{2h}.$$

(d) For a differentiable function, what do you think is the value of

$$\lim\limits_{h \to 0} \dfrac{f(x + h) - f(x - h)}{2h}?$$

(e) Compute the limit in (d) if $f(x) = x^3$.

72 (Contributed by David G. Mead) (a) Sketch the curves $y = x^2 + 1$ and $y = -x^2$. (b) Find equations of the lines L that are tangent to both curves simultaneously.

APPLICATIONS OF THE DERIVATIVE

Chapter 1, besides reviewing some precalculus concepts, developed the notion of a function. Chapter 2 introduced the limit concept and, through it, the notion of a continuous function. The derivative, a particular limit, was defined in Chap. 3, which also offered some of the more common interpretations of the derivative and ways to compute it. The present chapter applies the derivative to graphing, to the study of motion, and to finding the maximum and minimum values of a function.

4.1 Rolle's theorem and the mean-value theorem

This section presents three closely related theorems that are the basis for many applications of the derivative. It also argues for their plausibility and illustrates them with specific functions. The proofs, deferred to the end of the section, show how each theorem follows from its predecessor.

THE DERIVATIVE AT AN EXTREMUM

Let f be a differentiable function defined at least on the closed interval $[a, b]$. Because it is differentiable it is necessarily continuous. As mentioned in Sec. 2.7, the function f must therefore take on a maximum value for some number c in $[a, b]$. That is, for some number c in $[a, b]$,

$$f(c) \geq f(x)$$

for all x in $[a, b]$. What can be said about $f'(c)$, the derivative at c?

First, if c is neither a nor b, that is, c is in the open interval (a, b), the maximum would appear as in Fig. 4.1. It seems likely that a tangent to the graph at $(c, f(c))$ would be parallel to the x axis, in which case

$$f'(c) = 0.$$

4.1 ROLLE'S THEOREM AND THE MEAN-VALUE THEOREM

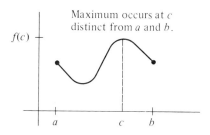
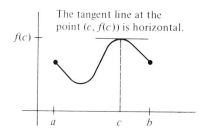

Figure 4.1

If, instead, the maximum occurs at an endpoint of the interval, at a or b, as the graph in Fig. 4.2 illustrates, the derivative at such a point need not be 0. In this graph the maximum occurs at b, where the derivative is not 0.

The case in which the maximum (or minimum) occurs away from the ends of the interval, that is, in the interior of the interval, is so important that we state it as a theorem.

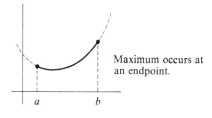

Figure 4.2

THEOREM OF THE INTERIOR EXTREMUM

Let f be a function defined at least on the open interval (a, b). If f takes on an extreme value at a number c in this interval and if $f'(c)$ exists, then

$$f'(c) = 0.$$

Michel Rolle was a 17th-century mathematician.

This theorem will be exploited beginning in Sec. 4.2 to find maximum and minimum values of a function. If an extreme value occurs within an open interval and the derivative exists there, the derivative must be 0 there. In the present section it will be used to establish Rolle's theorem. To motivate Rolle's theorem, we introduce the notion of a chord of a graph.

DEFINITION *Chord of f.* A line segment joining two points on the graph of a function f is called a *chord of f*.

Assume that a certain differentiable function f has a chord parallel to the x axis, as in Fig. 4.3. It seems reasonable that the graph will then have at least one horizontal tangent line. (In the case shown, there are three such lines tangent to the graph, as is indicated in Fig. 4.4.) This is the substance of the next theorem.

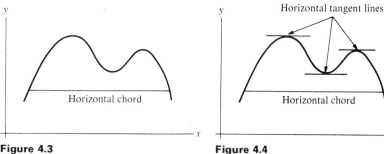

Figure 4.3   **Figure 4.4**

168 APPLICATIONS OF THE DERIVATIVE

ROLLE'S THEOREM Let f be a continuous function on the closed interval $[a, b]$ and have a derivative at all x in the open interval (a, b). If $f(a) = f(b)$, then there is at least one number c in (a, b) such that $f'(c) = 0$. ∎

EXAMPLE 1 Verify Rolle's theorem for the case $f(x) = x^2 - 2x + 5$ and $[a, b] = [0, 2]$.

SOLUTION Note that $f(0) = 5 = f(2)$. Also, f is continuous and f' exists (even at a and b though this is not necessary to apply Rolle's theorem). According to Rolle's theorem, there is a number c in $(0, 2)$ for which $f'(c) = 0$. It is easy to find such a c for this function, since $f'(x) = 2x - 2$. Setting the derivative $2x - 2 = 0$, we see that $c = 1$ (in this case c is unique). ∎

EXAMPLE 2 Verify Rolle's theorem for the case $f(x) = \cos x$ and $[a, b] = [\pi, 5\pi]$.

SOLUTION Note that $f(\pi) = -1 = f(5\pi)$. Since $\cos x$ is differentiable for all x it is continuous on $[\pi, 5\pi]$ and differentiable on $(\pi, 5\pi)$. According to Rolle's theorem, there must be at least one number c in $(\pi, 5\pi)$ for which $(\cos x)'$ is 0. Now, $(\cos x)' = -\sin x$. Thus there should be at least one solution of the equation

$$-\sin x = 0$$

in the open interval $(\pi, 5\pi)$. As can be checked, the equation has three such solutions, namely 2π, 3π, and 4π. ∎

The next example has a slight twist since the function is not differentiable at the ends of the interval $[a, b]$.

EXAMPLE 3 Verify Rolle's theorem for the case $f(x) = \sqrt{1 - x^2}$ and $[a, b] = [-1, 1]$.

SOLUTION Observe that $f(-1) = 0 = f(1)$, that f is continuous, and that $f'(x) = -x/\sqrt{1 - x^2}$, which is defined for all x in $(-1, 1)$. Rolle's theorem then guarantees that there is at least one number c in $(-1, 1)$ such that $f'(c) = 0$. We can find c by setting $f'(c) = 0$:

$$\frac{-c}{\sqrt{1 - c^2}} = 0.$$

Thus $c = 0$ (and this happens to be unique). ∎

The next example shows that it is necessary to assume in Rolle's theorem that f is differentiable throughout (a, b).

EXAMPLE 4 Can Rolle's theorem be applied to $f(x) = |x|$ in the interval $[a, b] = [-2, 2]$?

SOLUTION First, $f(-2) = 2 = f(2)$; second, f is continuous. The function, however, fails to be differentiable at 0. (See Example 2 in Sec. 3.3.) Not all

4.1 ROLLE'S THEOREM AND THE MEAN-VALUE THEOREM

of the hypotheses of Rolle's theorem hold; thus there is no need for the conclusion to hold. Indeed, there is no number c such that $f'(c) = 0$, as a glance at the graph of f in Fig. 4.5 shows. ∎

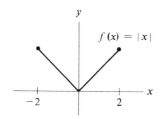

Figure 4.5

THE MEAN-VALUE THEOREM

Rolle's theorem asserts that, if the graph of a function has a horizontal chord, then it has a tangent line parallel to that chord. The mean-value theorem is a generalization of Rolle's theorem, since it concerns any chord of f, not just horizontal chords.

In geometric terms the theorem asserts that, if you draw a chord for the graph of a well-behaved function (as in Fig. 4.6), then somewhere above or below that chord the graph has at least one tangent line parallel to the chord.

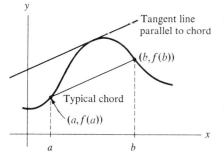

Figure 4.6

Let us translate this geometric statement into the language of functions. Call the ends of the chord $(a, f(a))$ and $(b, f(b))$. The slope of the chord is then

$$\frac{f(b) - f(a)}{b - a},$$

while the slope of the tangent line at a typical point $(x, f(x))$ on the graph is

$$f'(x).$$

The mean-value theorem then asserts there is at least one number c in the open interval (a, b) such that

$$f'(c) = \frac{f(b) - f(a)}{b - a}. \quad \blacksquare$$

MEAN-VALUE THEOREM Let f be a continuous function on the closed interval $[a, b]$ and have a derivative at every x in the open interval (a, b). Then there is at least one number c in the open interval (a, b) such that

$$f'(c) = \frac{f(b) - f(a)}{b - a}. \quad \blacksquare$$

EXAMPLE 5 Verify the mean-value theorem for $f(x) = 2x^3 - 8x + 1$, $a = 1$, and $b = 3$.

SOLUTION
$$f(a) = f(1) = 2 \cdot 1^3 - 8(1) + 1 = -5$$
and
$$f(b) = f(3) = 2 \cdot 3^3 - 8(3) + 1 = 31.$$

According to the mean-value theorem there is at least one number c between $a = 1$ and $b = 3$ such that

$$f'(c) = \frac{31 - (-5)}{3 - 1} = \frac{36}{2} = 18.$$

Let us find c explicitly. Since $f'(x) = 6x^2 - 8$, we need to solve the equation
$$6x^2 - 8 = 18,$$
that is,
$$6x^2 = 26$$
$$x^2 = \frac{26}{6}.$$

The solutions are $\sqrt{\frac{13}{3}}$ and $-\sqrt{\frac{13}{3}}$. But only $\sqrt{\frac{13}{3}}$ is in $(1, 3)$. Hence there is only one number, namely $\sqrt{\frac{13}{3}}$, that serves as the c whose existence is guaranteed by the mean-value theorem. ∎

The interpretation of the derivative as slope suggested the mean-value theorem. What does the mean-value theorem say when the derivative is interpreted, say, as velocity? This question is considered in Example 6.

EXAMPLE 6 A car moving on the x axis has the x coordinate $f(t)$ at time t. At time a its position is $f(a)$. At some later time b its position is $f(b)$. What does the mean-value theorem assert for this car?

SOLUTION The quotient
$$\frac{f(b) - f(a)}{b - a}$$
equals
$$\frac{\text{Change in position}}{\text{Change in time}},$$
or "average velocity" for the interval of time $[a, b]$. The mean-value theorem asserts that at some time during this period the velocity of the car must equal its average velocity. To be specific, if a car travels 210 miles in 3 hours, then at some time its speedometer must read 70 miles per hour. ∎

There are several ways of writing the mean-value theorem. For example, the equation
$$f'(c) = \frac{f(b) - f(a)}{b - a}$$
is equivalent to
$$f(b) - f(a) = (b - a)f'(c),$$
hence to
$$f(b) = f(a) + (b - a)f'(c).$$

In this form, the mean-value theorem asserts that $f(b)$ is equal to $f(a)$ plus a quantity that involves the derivative f'. The following important corollaries exploit this alternative view of the mean-value theorem.

COROLLARY 1 If the derivative of a function is 0 throughout an interval, then the function is constant throughout that interval.

PROOF Let s and t be any two numbers in the interval and let the function be denoted by f. To prove the corollary, it suffices to prove that

$$f(t) = f(s).$$

By the mean-value theorem there is a number c between s and t such that

$$f(t) = f(s) + (t - s)f'(c).$$

But $f'(c) = 0$, since $f'(x)$ is 0 for all x in the given interval. Hence

$$f(t) = f(s) + (t - s)(0),$$

which proves that $\quad f(t) = f(s)$. ∎

When Corollary 1 is interpreted in terms of motion, it is quite plausible. It asserts that, if a particle has zero velocity for a period of time, then it does not move during that time.

EXAMPLE 7 Use Corollary 1 to show that $f(x) = \cos^2 3x + \sin^2 3x$ is a constant. Find the constant.

SOLUTION $f'(x) = -6 \cos 3x \sin 3x + 6 \sin 3x \cos 3x = 0$. Corollary 1 says that f is constant. To find the constant, just evaluate f at some specific number, say at 0. We have $f(0) = \cos^2(3 \cdot 0) + \sin^2(3 \cdot 0) = \cos^2 0 + \sin^2 0 = 1$. Thus

$$\cos^2 3x + \sin^2 3x = 1$$

for all x. This should be no surprise since, by the Pythagorean theorem, $\cos^2 \theta + \sin^2 \theta = 1$. ∎

COROLLARY 2 If two functions have the same derivatives throughout an interval, then they differ by a constant. That is, if $f'(x) = g'(x)$ for all x in an interval, then there is a constant C such that $f(x) = g(x) + C$.

PROOF Define a third function h by the equation

$$h(x) = f(x) - g(x).$$

Then
$$h'(x) = f'(x) - g'(x) = 0.$$

Since the derivative of h is 0, Corollary 1 implies that h is constant, that is,

$$h(x) = C$$

for some fixed number C. Thus

$$f(x) - g(x) = C$$

or
$$f(x) = g(x) + C,$$

and the corollary is proved. ∎

172 APPLICATIONS OF THE DERIVATIVE

Is Corollary 2 plausible when the derivative is interpreted as slope? In this case, the corollary asserts that, if the graphs of two functions have the property that their tangent lines at points with the same x coordinate are parallel, then one graph can be obtained from the other by raising (or lowering) it by an amount C. If you sketch two such graphs (as in Fig. 4.7), you will see that the corollary is reasonable.

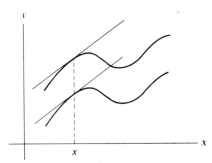

If two graphs have parallel tangent lines at all pairs of points with the same x coordinate, then one graph is obtainable from the other by raising or lowering it.

Figure 4.7

EXAMPLE 8 What functions have a derivative equal to $2x$ everywhere?

SOLUTION One such function is x^2; another is $x^2 + 25$. For any constant C, $D(x^2 + C) = 2x$. Are there any other possibilities? Corollary 2 tells us there are not. For if f is a function such that $f'(x) = 2x$, then $f'(x) = (x^2)'$ for all x. Thus the functions f and x^2 differ by a constant, say C, that is,

$$f(x) = x^2 + C.$$

The only functions whose derivatives are $2x$ are of the form $x^2 + C$. ∎

Antiderivatives are unique up to a constant difference C.

Example 8 shows that every antiderivative of the function $2x$ must be of the form $x^2 + C$ for some constant C. More generally, if $F(x)$ is a particular antiderivative of the function $f(x)$ on an interval, then any other antiderivative of $f(x)$ there must be of the form $F(x) + C$ for some constant C.

Corollary 1 asserts that if $f'(x) = 0$ for all x, then f is constant. What can be said about f if $f'(x)$ is *positive* for all x? In terms of the graph of f, this assumption implies that all the tangent lines slope upward. It is reasonable to expect that, as we move from left to right on the graph in Fig. 4.8, the y coordinate increases. In Corollary 3 this is proved.

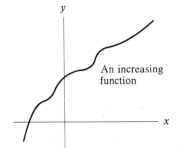

An increasing function

Figure 4.8

DEFINITION *Increasing (decreasing) function.* If $f(x_1) > f(x_2)$, whenever $x_1 > x_2$, then f is an *increasing* function. If $f(x_1) < f(x_2)$ whenever $x_1 > x_2$, then f is a *decreasing* function.

COROLLARY 3 If f is continuous on $[a, b]$ and has a positive derivative on the open interval (a, b), then f is increasing on the interval $[a, b]$. If f is continuous on $[a, b]$ and has a negative derivative on the open interval (a, b), then f is decreasing on the interval $[a, b]$.

PROOF We prove the "increasing" case. Take two numbers x_1 and x_2 such that

$$a \leq x_2 < x_1 \leq b.$$

By the mean-value theorem,

$$f(x_1) = f(x_2) + (x_1 - x_2)f'(c)$$

for some c between x_1 and x_2. Since $x_1 - x_2$ is positive, and since $f'(c)$ is assumed to be positive, it follows that

$$(x_1 - x_2)f'(c) > 0.$$

Thus $f(x_1) > f(x_2)$, and the corollary is proved. (The "decreasing" case is proved similarly.) ∎

EXAMPLE 9 Show that the function $f(x) = x^5 + x$ is increasing on any interval.

SOLUTION In this case

$$f'(x) = 5x^4 + 1.$$

Since $x^4 \geq 0$ for all real numbers x, $f'(x) \geq 1$ for all real numbers. Since $f'(x)$ is positive, f is an increasing function. ∎

As Example 9 illustrates, the derivative can be quite useful in showing that a function is increasing. In Sec. 4.2 this technique will be used in graphing functions.

PROOFS OF THE THREE THEOREMS

PROOF OF THE INTERIOR EXTREMUM THEOREM

If a number is ≥ 0 and ≤ 0, then it must be 0.

We will prove this for the case of a maximum, the other case being similar. Assume that $f(c)$ is the maximum value of f on $[a, b]$, that c is in (a, b), and that $f'(c)$ exists. It is to be shown that $f'(c) = 0$. We shall prove that $f'(c) \leq 0$ and that $f'(c) \geq 0$. From this it will follow that $f'(c)$ must be 0.

Consider the quotient

$$\frac{f(c + \Delta x) - f(c)}{\Delta x}$$

used in defining $f'(c)$. Take Δx so small that $c + \Delta x$ is in the interval $[a, b]$. Since $f(c)$ is the maximum value of $f(x)$ for x in $[a, b]$,

$$f(c + \Delta x) \leq f(c).$$

Hence
$$f(c + \Delta x) - f(c) \leq 0.$$

Therefore, when Δx is positive,

$$\frac{f(c + \Delta x) - f(c)}{\Delta x} \quad \text{is negative or } 0.$$

Consequently, as $\Delta x \to 0$ through positive values,

$$\frac{f(c + \Delta x) - f(c)}{\Delta x},$$

being negative or 0, cannot approach a positive number. Thus

$$f'(c) = \lim_{\Delta x \to 0} \frac{f(c + \Delta x) - f(c)}{\Delta x} \leq 0.$$

If, on the other hand, Δx is negative, then the denominator of

$$\frac{f(c + \Delta x) - f(c)}{\Delta x}$$

is negative, and the numerator is still ≤ 0. Hence, for negative Δx,

$$\frac{f(c + \Delta x) - f(c)}{\Delta x} \geq 0$$

(the quotient of two negative numbers being positive). Thus as $\Delta x \to 0$ through negative values, the quotient approaches a number ≥ 0. Hence $f'(c) \geq 0$.

Since $0 \leq f'(c) \leq 0$, $f'(c)$ must be 0, and the theorem is proved. ∎

PROOF OF ROLLE'S THEOREM

Since f is continuous, it has a maximum value M and a minimum value m for x in $[a, b]$. Certainly $m \leq M$.

If $m = M$, f is constant, and $f'(x) = 0$ for all x in $[a, b]$. Then, any number x in (a, b) will serve as the desired number c.

If $m < M$, then the minimum and maximum cannot both occur at the ends of the interval a and b, since $f(a) = f(b)$. One of them, at least, occurs at a number c, $a < c < b$. And, by the interior-extremum theorem, at that c, $f'(c)$ is 0. This proves Rolle's theorem. ∎

PROOF OF THE MEAN-VALUE THEOREM

We shall prove the theorem by introducing a function to which Rolle's theorem can be applied. The chord through $(a, f(a))$ and $(b, f(b))$ is part of a line L, whose equation is, let us say, $y = g(x)$. Let $h(x) = f(x) - g(x)$, which represents the difference between the graph of f and the line L for a given x. It is clear from Fig. 4.9 that $h(a) = h(b)$, since both equal 0.

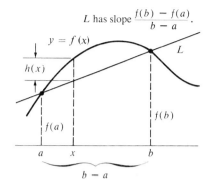

Figure 4.9

By Rolle's theorem, there is at least one number c in the open interval (a, b) such that

$$h'(c) = 0.$$

But $$h'(c) = f'(c) - g'(c).$$

Since $y = g(x)$ is the equation of the line through $(a, f(a))$ and $(b, f(b))$, $g'(x)$ is the slope of the line, that is,

$$g'(x) = \frac{f(b) - f(a)}{b - a}$$

for all x.

Hence $$0 = h'(c) = f'(c) - \frac{f(b) - f(a)}{b - a}.$$

In short $$f'(c) = \frac{f(b) - f(a)}{b - a},$$

and the mean-value theorem is proved. ∎

Exercises

Exercises 1 and 2 concern the theorem of the interior extremum.

1 Consider the function $f(x) = x^2$ only for x in $[-1, 2]$.
  (a) Graph the function $f(x)$ for x in $[-1, 2]$.
  (b) What is the maximum value of $f(x)$ for x in $[-1, 2]$?
  (c) Does $f'(x)$ exist at the maximum?
  (d) Does $f'(x)$ equal 0 at the maximum?
  (e) Does $f'(x)$ equal 0 at the minimum?

2 Consider the function $f(x) = \sin x$ for x in $[0, \pi]$.
  (a) Graph the function $f(x)$ for x in $[0, \pi]$.
  (b) Does $f'(x)$ equal 0 at the maximum value of $f(x)$ for x in $[0, \pi]$?
  (c) Does $f'(x)$ equal 0 at the minimum value of $f(x)$ for x in $[0, \pi]$?

Exercises 3 to 8 concern Rolle's theorem.

3 (a) Graph $f(x) = x^{2/3}$ for x in $[-1, 1]$.
  (b) Show that $f(-1) = f(1)$.
  (c) Is there a number c in $(-1, 1)$ such that $f'(c) = 0$?
  (d) Why does this function not contradict Rolle's theorem?

4 (a) Graph $f(x) = 1/x^2$ for x in $[-1, 1]$.
  (b) Show that $f(-1) = f(1)$.
  (c) Is there a number c in $(-1, 1)$ such that $f'(c) = 0$?
  (d) Why does this function not contradict Rolle's theorem?

In each of Exercises 5 to 8 verify that the given function satisfies the hypotheses of Rolle's theorem. Find all numbers c that satisfy the conclusion of the theorem.

5 $\sin x$ and $[0, 2\pi]$
6 $x^3 - x + 2$ and $[-1, 1]$
7 $x^4 - 2x^2 + 1$ and $[-2, 2]$
8 $\sin x + \cos x$ and $[0, 4\pi]$

In Exercises 9 to 16 find explicitly all values of c which satisfy the mean-value theorem for the given functions and intervals.

9 $x^3 - x^2$ and $[-1, 2]$
10 $x^3 - x$ and $[-1, 2]$
11 $5x$ and $[1, 3]$
12 x^3 and $[-2, 1]$
13 $\cos x$ and $[0, 6\pi]$
14 $\sin x$ and $[0, 5\pi]$
15 $x^3 + 3x^2 + 3x$ and $[1, 2]$
16 $x^3 + x^2 - x$ and $[0, 1]$

17 (a) Differentiate $\sec^2 x$ and $\tan^2 x$.
  (b) The derivatives in (a) are equal. Corollary 2 then asserts that there is a constant C such that $\sec^2 x = \tan^2 x + C$. Find that constant.

18 (a) Differentiate $\csc^2 x$ and $\cot^2 x$.
  (b) The derivatives in (a) are equal. Find the constant C (promised by Corollary 2) such that $\csc^2 x = \cot^2 x + C$.

In Exercises 19 and 20 find all antiderivatives of the given functions.

19 (a) $8x^3$ (b) $\sin 2x$ (c) $1/x^2$ (d) $1/(x+1)^3$

20 (a) $10x^4$ (b) $\sec^2 3x$ (c) $x^3 + 5x^2 + 1$ (d) $(2x+1)^{10}$

■

21 Show that the function $3x^3 - 6x^2 + 4x$ is increasing over all intervals.

22 Show that the function $7x + 3\sin 2x$ is increasing over all intervals.

The answers to Exercises 23 and 24 should be phrased in colloquial English. (Section 3.1 introduced the concepts of density and magnification.)

23 State the mean-value theorem in terms of density and mass. Let x be the distance from the left end of a string and $f(x)$ the mass of the string from 0 to x. When stated in these terms, does the mean-value theorem seem reasonable?

24 State the mean-value theorem in terms of a slide and a screen. Let x denote the position on the (linear) slide and $f(x)$ denote the position of the image on the screen. In optical terms, what does the mean-value theorem say?

25 Differentiate for practice:

(a) $\sqrt{1-x^2} \sin 3x$ (b) $\dfrac{\sqrt[3]{x}}{x^2+1}$ (c) $\tan \dfrac{1}{(2x+1)^2}$

26 Assume that f has a derivative for all x, that $f(3) = 7$, and that $f(8) = 17$. What can we conclude about f' at some number?

27 Which of the corollaries to the mean-value theorem implies that (a) if two cars on a straight road have the same velocity at every instant, they remain a fixed distance apart? (b) if all the tangents to a curve are horizontal, the curve is a horizontal straight line? Explain in each case.

28 At time t seconds a thrown ball has the height $f(t) = -16t^2 + 32t + 40$ feet.
(a) Show that after 2 seconds it returns to its initial height $f(0)$.
(b) What does Rolle's theorem imply about the velocity of the ball?
(c) Verify Rolle's theorem in this case by computing the numbers c which it asserts exist.

29 Consider the function f given by the formula $f(x) = x^3 - 3x$.
(a) At which numbers x is $f'(x) = 0$?
(b) Use the theorem of the interior extremum to show that the maximum value of $x^3 - 3x$ for x in $[1, 5]$ occurs either at 1 or at 5.
(c) What is the maximum value of $x^3 - 3x$ for x in $[1, 5]$?

30 Let $f(x) = (x^2 - 4)/(x + \frac{1}{2})$. Observe that $f(2) = f(-2)$.
(a) What does Rolle's theorem say about f?
(b) For which values of x is $f'(x) = 0$?

■ ■

31 Let f and g be two functions differentiable on (a, b) and continuous on $[a, b]$. Assume that $f(a) = g(a)$ and that $f'(x) < g'(x)$ for all x in (a, b). Prove that $f(b) < g(b)$.

32 (See Exercise 31.)
(a) Show that $\tan x > x$ if $0 < x < \pi/2$.
(b) What does (a) say about certain lengths related to the unit circle?

33 Show that between any two roots of a polynomial there is at least one root of its derivative.

34 (a) Assume that every polynomial of degree 5 has at most five real roots. Use Rolle's theorem to prove that every polynomial of degree 6 has at most six real roots. *Hint:* Use Exercise 33.
(b) Use the technique of (a) to prove that a polynomial of degree n has at most n real roots.

35 (a) Recall the definition of $g(x)$ in the proof of the mean-value theorem, and show that

$$g(x) = f(a) + \frac{x-a}{b-a}[f(b) - f(a)].$$

(b) Using (a), show that

$$g'(x) = \frac{f(b) - f(a)}{b - a}.$$

36 Let $f(x) = 2x^5 - 10x + 5$.
(a) Show that $f'(x)$ is positive for x in $(1, \infty)$ and in $(-\infty, -1)$ but negative for x in $(-1, 1)$.
(b) Compute $f(-1)$ and $f(1)$ and sketch a graph of f. (Don't try to find its x intercepts.)
(c) Using (a) and (b), show that the equation $2x^5 - 10x + 5 = 0$ has exactly three real roots.

Remark: It is proved in advanced algebra that none of the roots of the equation in (c) can be expressed in terms of combinations of square roots, cube roots, fourth roots, fifth roots, ... However, the roots of polynomials of degree 2, 3, or 4 can be expressed in terms of such roots. (The quadratic formula takes care of the case of degree 2. There is no corresponding formula for degree 5; no such formula can *ever* be found.)

4.2 Using the derivative and limits when graphing a function

The x and y intercepts

The primitive and inefficient way to graph a function is to make a table of values, plot many points, and draw a curve through the points (hoping that the chosen points adequately represent the function). Chapters 1 and 2 refined the technique somewhat. It was also pointed out that the x and y intercepts are of aid for they tell where the graph meets the x and y axes. Furthermore, horizontal and vertical *asymptotes* were discussed; they can be of use in sketching the graph for large $|x|$ and also near a number where the function becomes infinite (usually because a denominator is 0). For instance, the line $x = 1$ is a vertical asymptote of $1/(x - 1)$; the line $y = 0$ is a horizontal asymptote of the same curve. The line $x = \pi/2$ is a vertical asymptote of the curve $y = \tan x$.

This section shows how to use the derivative and limits to help graph a function. Of particular interest will be these questions:

Where is the derivative equal to 0?

Where is the derivative positive? Negative?

How does the function behave for $|x|$ large?

The answers will tell a good deal about the general shape of a particular graph; it will then not be necessary to plot so many specific points on the graph.

THE DERIVATIVE AND GRAPHING

First a few helpful definitions.

DEFINITION *Critical number* and *critical point.* A value of x at which $f'(x) = 0$ is called a *critical number* for the function f. The corresponding point $(x, f(x))$ on the graph of f is a *critical point* on that graph.

DEFINITION *Relative maximum (local maximum).* The function f has a *relative maximum* (or *local maximum*) at the number c if there is an open interval (a, b) around c such that $f(c) \geq f(x)$ for all x in (a, b) that lie in the domain of f. A *local* or *relative minimum* is defined analogously.

DEFINITION *Global maximum.* The function f has a *global maximum* (or *absolute maximum*) at the number c if $f(c) \geq f(x)$ for all x in the domain of f. A *global minimum* is defined analogously.

Note that a global maximum is necessarily a local maximum as well. A local maximum is like the summit of a single mountain; a global maximum corresponds to Mount Everest.

Figure 4.10 illustrates the notions of critical point, local maximum, global maximum, local minimum, and global minimum in the graph of a hypo-

178 APPLICATIONS OF THE DERIVATIVE

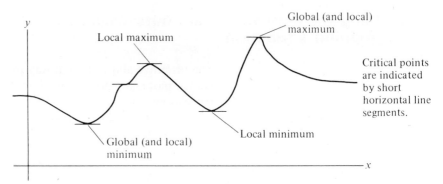

Figure 4.10

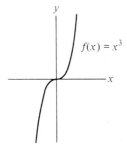

Figure 4.11

thetical function. Any given function may have none of these, or some, or all.

By the theorem of the interior extremum in Sec. 4.1, there is a close relation between a local maximum (or minimum) and critical points. If a local maximum occurs at a number c that lies within some open interval within the domain of f, then $f'(c) = 0$. This means that c is a critical number. However, a critical point need not be a local maximum or local minimum. This is illustrated by the function x^3, whose derivative $3x^2$ is 0 at 0. Thus $c = 0$ is a critical number of the function x^3. A glance at Fig. 4.11 shows that x^3 has neither a local maximum nor a local minimum at 0.

To determine whether a function has a local maximum or minimum (or neither) at c, it is not enough to know that $f'(c) = 0$. It is also important to know how the derivative behaves for inputs near c.

THE FIRST-DERIVATIVE TEST

The following test for local maximum or local minimum is an immediate consequence of the fact that, when the derivative is positive, the function increases and, when it is negative, it decreases.

First-derivative test for local maximum (or local minimum) at $x = c$

Let f be a function and let c be a number in its domain. Assume that numbers a and b exist such that $a < c < b$ and

1 f is continuous on the open interval (a, b);

2 f is differentiable on the open interval (a, b), except possibly at c;

3 $f'(x)$ is positive for all $x < c$ in the interval and is negative for all $x > c$ in the interval.

Then f has a *local maximum* at c. A similar test, with "positive" and "negative" interchanged, holds for a *local minimum*. (See Figs. 4.12 and 4.13.)

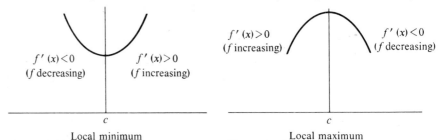

Figure 4.12 Local minimum **Figure 4.13** Local maximum

Informally, the derivative test says, "If the derivative changes sign at c, then the function has either a local minimum or a local maximum." To decide which it is, just make a crude sketch of the graph near $(c, f(c))$ to show on which side of c the function is increasing and on which side it is decreasing.

Note that in the derivative test there is no assumption that the derivative exists at $x = c$. Thus the test applies to $f(x) = |x|$, which has a local minimum at $x = 0$, where the derivative changes from -1 to 1. However, $f(x) = |x|$ is not differentiable at 0, as shown in Example 2 of Sec. 3.3.

EXAMPLES OF GRAPHING

EXAMPLE 1 Graph $$f(x) = \frac{2x - 5}{x - 1}.$$

SOLUTION Note that at $x = 1$ the function is not defined.

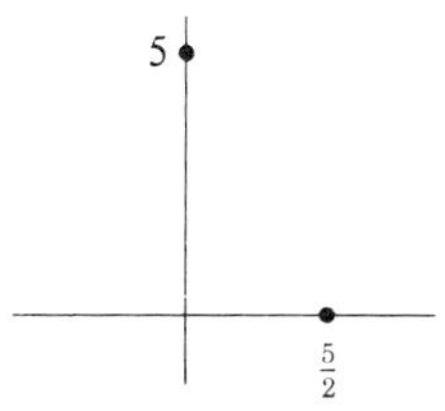

Next observe that $f(0) = 5$ and that $f(x) = 0$ only when the numerator, $2x - 5$, is 0. Thus the y intercept is 5 and the only x intercept is $x = \frac{5}{2}$, the solution of the equation $2x - 5 = 0$. This information is recorded in Fig. 4.14.

Next, determine the critical numbers of f. To do this, compute $f'(x)$:

$$f'(x) = D\left(\frac{2x - 5}{x - 1}\right) = \frac{(x - 1) \cdot 2 - (2x - 5) \cdot 1}{(x - 1)^2}$$

$$= \frac{3}{(x - 1)^2}.$$

Figure 4.14

No critical numbers Since the numerator is never 0, there are no critical numbers.

Where is the function increasing? Decreasing? Since $f'(x) = 3/(x - 1)^2$, the derivative is positive throughout the domain of the function. The function is always increasing.

Always increasing

How does the function behave when $|x|$ is large? We have

$$\lim_{x \to \infty} \frac{2x - 5}{x - 1} = \lim_{x \to \infty} \frac{2 - 5/x}{1 - 1/x} = \frac{2}{1} = 2.$$

Similarly, $$\lim_{x \to -\infty} \frac{2x - 5}{x - 1} = 2.$$

Horizontal asymptote Thus the line $y = 2$ is an asymptote of the graph both far to the right and far to the left. Since the function is *increasing*, the graph, for $|x|$ large, resembles the sketch in Fig. 4.15.

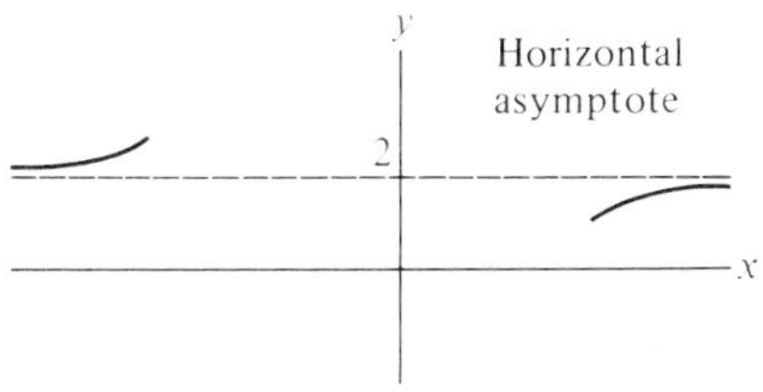

Are there any vertical asymptotes? In other words, are there any numbers a near which the function becomes arbitrarily large? Since

$$f(x) = \frac{2x - 5}{x - 1},$$

Figure 4.15

Vertical asymptote and $x - 1$ is small when x is near 1, near $a = 1$ the function "blows up." If x is near 1 and right of 1, the numerator is near $2 \cdot 1 - 5 = -3$, hence negative, and the denominator is a small positive number. Thus

$$\lim_{x \to 1^+} \frac{2x-5}{x-1} = -\infty.$$

Similarly, $\quad\lim_{x \to 1^-} \frac{2x-5}{x-1} = \infty.$

The line $x = 1$ is a vertical asymptote.

With this information, the graph can be completed. It is shown in Fig. 4.16.

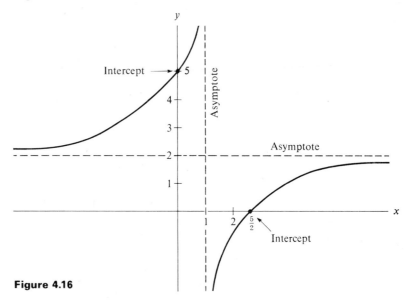

Figure 4.16

The next example illustrates the graphing of a polynomial.

EXAMPLE 2 Sketch the graph of $f(x) = 2x^3 - 3x^2 - 12x$.

SOLUTION Note that $f(x)$, being a polynomial, is defined for all x.
Intercepts Since $f(0) = 2 \cdot 0^3 - 3 \cdot 0^2 - 12 \cdot 0 = 0$, the y intercept is 0. To find the x intercepts it is necessary to solve the equation

$$f(x) = 0.$$

In the case of this function the equation can be solved easily:

$$2x^3 - 3x^2 - 12x = 0,$$

or $\quad x(2x^2 - 3x - 12) = 0.$

Either $x = 0$ or $\quad 2x^2 - 3x - 12 = 0.$

The latter equation can be solved by the quadratic formula:

$$x = \frac{-(-3) \pm \sqrt{(-3)^2 - 4(2)(-12)}}{2 \cdot 2}$$

4.2 USING THE DERIVATIVE AND LIMITS WHEN GRAPHING A FUNCTION

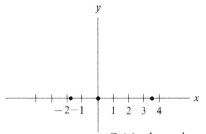

Figure 4.17   *Critical numbers*

$$= \frac{3 \pm \sqrt{9 + 96}}{4}$$

$$= \frac{3 \pm \sqrt{105}}{4}.$$

These two solutions are approximately -1.8 and 3.3.
The intercepts are recorded in Fig. 4.17.
When is $f'(x) = 0$? We have

$$\begin{aligned} f'(x) &= 6x^2 - 6x - 12 \\ &= 6(x^2 - x - 2) \\ &= 6(x - 2)(x + 1). \end{aligned}$$

Thus $f'(x) = 0$ when $\quad 6(x - 2)(x - 1) = 0,$

that is, when $\quad\quad\quad x = 2 \quad \text{or} \quad x = -1.$

At these critical numbers the function has the values

$$f(2) = 2 \cdot 2^3 - 3 \cdot 2^2 - 12 \cdot 2 = -20$$

and $\quad f(-1) = 2(-1)^3 - 3(-1)^2 - 12(-1) = 7.$

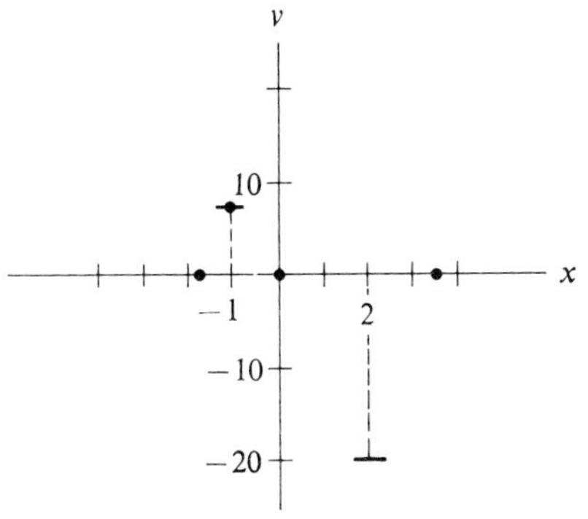

Figure 4.18 records the data gathered so far.

Next, examine the sign of $f'(x)$ to determine where the function is increasing and where it is decreasing. Recall that $f'(x) = 6(x - 2)(x + 1)$ and use the accompanying chart as an aid.

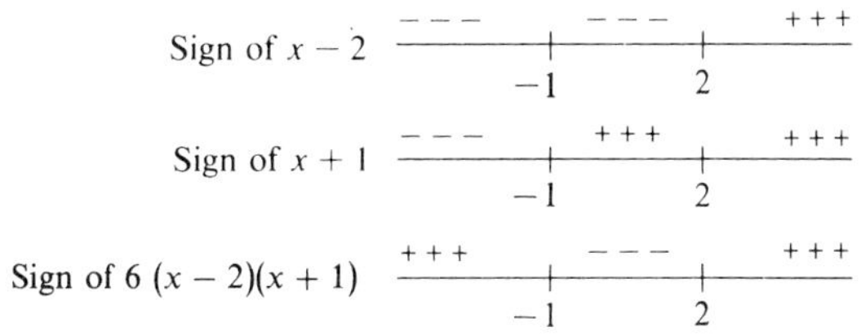

Figure 4.18 (Vertical scale foreshortened)

```
                          - - -      - - -      + + +
         Sign of x - 2   ─────────┼─────────┼─────────
                                 -1         2

                          - - -      + + +      + + +
         Sign of x + 1   ─────────┼─────────┼─────────
                                 -1         2

                          + + +      - - -      + + +
   Sign of 6(x - 2)(x + 1) ─────────┼─────────┼─────────
                                 -1         2
```

Where increasing or decreasing

Thus the function is increasing for $x < -1$ and for $x > 2$; it is decreasing for $-1 < x < 2$. The information gathered so far is recorded in Fig. 4.19.

Finally, consider the behavior of $f(x) = 2x^3 - 3x^2 - 12x$ when $|x|$ is large. Since

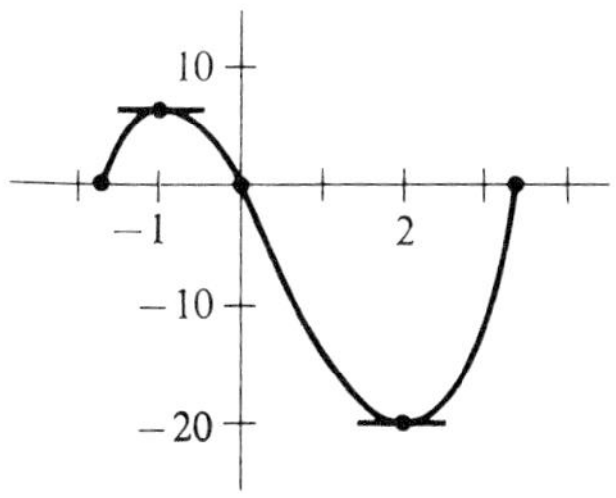

$$\lim_{x \to \infty} f(x) = \lim_{x \to \infty} (2x^3 - 3x^2 - 12x)$$

$$= \lim_{x \to \infty} x^3 \left(2 - \frac{3}{x} - \frac{12}{x^2} \right)$$

$$= \infty,$$

Figure 4.19

No asymptotes   the graph does not have a horizontal asymptote as $x \to \infty$. Similar reasoning shows that it has no horizontal asymptote as $x \to -\infty$. In fact,

$$\lim_{x \to -\infty} f(x) = -\infty.$$

With this last information the curve can be sketched. The graph (with the y axis compressed) appears in Fig. 4.20.

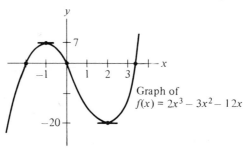

Graph of $f(x) = 2x^3 - 3x^2 - 12x$

The graph crosses the x axis at $x = 0$,
$\frac{3 - \sqrt{105}}{4} \approx -1.8$, and $\frac{3 + \sqrt{105}}{4} \approx 3.3$.   **Figure 4.20**

There is a local maximum at $x = -1$, a local minimum at $x = 2$, but no global maximum or minimum. ∎

EXAMPLE 3   Graph $f(x) = 3x^4 - 4x^3$. Discuss relative maxima and minima.

SOLUTION   To find the intercepts, note that $f(0) = 0$ and that $3x^4 - 4x^3 = 0$ when $x^3(3x - 4) = 0$, that is, when $x = 0$ or $x = \frac{4}{3}$. The derivative is

$$f'(x) = 12x^3 - 12x^2$$
$$= 12x^2(x - 1).$$

The critical numbers are the solutions of the equation

$$12x^2(x - 1) = 0,$$

namely 0 and 1.

How does the sign of $f'(x) = 12x^2(x - 1)$ behave when x is near 0? For $x < 0$, $12x^2$ is positive and $x - 1$ is negative; hence $12x^2(x - 1)$ is negative. For $0 < x < 1$, $12x^2$ is positive and $x - 1$ is still negative. Thus the sign of $f'(x)$ does *not* change as x passes through 0. In fact, since $f'(x)$ remains negative (except at 0), the function f is decreasing for $x \leq 1$. Thus there is no relative maximum or minimum at $x = 0$.

How does the sign of $f'(x) = 12x^2(x - 1)$ behave when x is near 1? The factor $12x^2$ remains positive, but $x - 1$ changes sign from negative to positive. Hence at $x = 1$ the function has a local minimum.

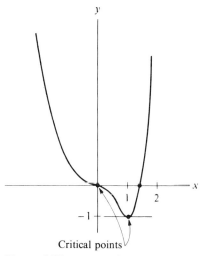

Figure 4.21

Writing
$$f(x) = 3x^4 - 4x^3$$
$$= x^4\left(3 - \frac{4}{x}\right)$$

shows that, when $|x|$ is large, $f(x)$ behaves like $3x^4$ (since $4/x$ is then near 0). Since $3x^4$ becomes arbitrarily large when x is large, the function has no global maximum. The graph in Fig. 4.21 shows the x intercepts and the critical points. Note that at $x = 1$ a global minimum occurs. ∎

THE MAXIMUM OVER A CLOSED INTERVAL

In many applied problems we are interested in the behavior of a differentiable function just over some closed interval $[a, b]$. Such a function will have a global maximum for that interval by the maximum-value theorem of Sec. 2.7. That maximum can occur either at an endpoint—a or b—or else at some number c in the open interval (a, b). In the latter case c must be a critical number, for $f'(c) = 0$ by the interior-maximum theorem of Sec. 4.1.

Figures 4.22 and 4.23 show some of the ways in which a relative or global maximum or minimum can occur for a function considered only on a closed interval $[a, b]$.

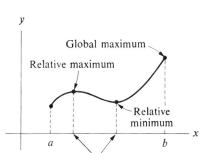

The derivative is 0 at these two numbers.

The global maximum occurs at an end. (The derivative need not be 0.)

Figure 4.22

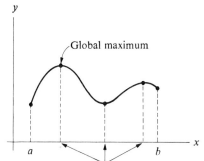

The derivative is 0 at these three numbers.

The global maximum occurs at a number other than a or b. (The derivative is 0.)

Figure 4.23

The major point to keep in mind is that the maximum value of a differentiable function f on a closed interval occurs:

1 at an endpoint of the interval, or

2 at a critical number (where $f'(x) = 0$).

EXAMPLE 4 Find the maximum value of $f(x) = x^3 - 3x^2 + 3x$ for x in $[0, 2]$.

SOLUTION First compute f at the ends of the interval, 0 and 2:

$$f(0) = 0 \quad \text{and} \quad f(2) = 2.$$

Next, compute $f'(x)$, which is $3x^2 - 6x + 3$. When is $f'(x) = 0$? When

$$3x^2 - 6x + 3 = 0,$$
or
$$3(x^2 - 2x + 1) = 0,$$
or
$$3(x - 1)^2 = 0.$$

Thus 1 is the only critical number, and it lies in the interval $[0, 2]$.

The maximum of f must therefore occur either at an endpoint of the interval (at 0 or 2) or at the only critical number, 1. It is necessary to calculate $f(1)$ to determine where the maximum occurs:

$$f(1) = 1^3 - 3 \cdot 1^2 + 3 \cdot 1$$
$$= 1.$$

Since $f(0) = 0$, $f(2) = 2$, and $f(1) = 1$, the maximum value is 2, occurring at the endpoint 2.

Now that the problem is solved, it may be instructive to sketch the graph of the function. Since

$$f'(x) = 3(x - 1)^2$$

is positive for all x other than 1, the function is increasing. Figure 4.24 shows how the graph looks. Observe that the minimum occurs at 0. ∎

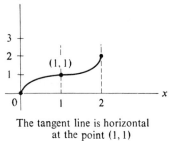

The tangent line is horizontal at the point $(1, 1)$

Figure 4.24

The accompanying flowchart summarizes the method for maximizing a differentiable function on a closed interval.

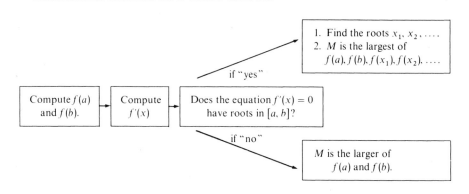

Exercises

In each of Exercises 1 to 8 find all critical numbers of the given function and use the first derivative test to determine whether a local maximum, a local minimum, or neither, occurs there.

1. x^5
2. x^6
3. $(x - 1)^3$
4. $(x - 1)^4$
5. $3x^4 + x^3$
6. $2x^3 + 3x^2$
7. $x \sin x + \cos x$
8. $x \cos x - \sin x$

In Exercises 9 to 24 graph the given functions, showing any intercepts, asymptotes, critical points, or local or global extrema.

9 $x^3 - 3x^2 + 3x$
10 $x^4 - 4x^3 + 4x^2$
11 $x^4 - 4x + 3$
12 $x^5 + 5x$
13 $x^2 - 6x + 5$
14 $2x^2 + 3x + 5$
15 $x^4 + 2x^3 - 3x^2$
16 $2x^3 + 3x^2 - 6x$
17 $\dfrac{3x + 1}{3x - 1}$
18 $\dfrac{x}{x + 1}$
19 $\dfrac{x}{x^2 + 1}$
20 $\dfrac{x}{x^2 - 1}$
21 $\dfrac{1}{2x^2 - x}$
22 $\dfrac{1}{x^2 - 3x + 2}$
23 $\dfrac{x^2 + 3}{x^2 - 4}$
24 $\dfrac{\sqrt{x^2 + 1}}{x}$

Exercises 25 to 32 concern functions whose domains are restricted to closed intervals. In each find the maximum and the minimum value for the given function over the given interval.

25 $x^2 - x^4$; [0, 1].
26 $4x - x^2$; [0, 5].
27 $4x - x^2$; [0, 1].
28 $2x^2 - 5x$; [-1, 1].
29 $x^3 - 2x^2 + 5x$; [-1, 3].
30 $x/(x^2 + 1)$; [0, 3].
31 $x^2 + x^4$; [0, 1].
32 $(x + 1)/\sqrt{x^2 + 1}$; [0, 3].

■

In Exercises 33 to 40 graph the functions.

33 $\dfrac{x - 1}{x^2 + 1}$
34 $x + \dfrac{1}{x}$
35 $\dfrac{\sin x}{1 + 2\cos x}$
36 $\dfrac{\sqrt{x^2 - 1}}{x}$
37 $\dfrac{1}{(x - 1)^2(x - 2)}$
38 $\dfrac{3x^2 + 5}{x^2 - 1}$
39 $2x^{1/3} + x^{4/3}$
40 $\dfrac{3x^2 + 5}{x^2 + 1}$

■ ■

41 Graph $3 \sin x + \cos x$.
42 Let f and g be polynomials without a common root.
   (a) Show that if the degree of g is odd, the graph of f/g has a vertical asymptote.
   (b) Show that if f and g have the same degree, the graph of f/g has a horizontal asymptote.
   (c) Show that if the degree of f is less than the degree of g, the graph of f/g has a horizontal asymptote.

4.3 Concavity and the second derivative

Velocity is the rate at which distance changes. The rate at which velocity changes is called *acceleration*. Thus if $y = f(t)$ denotes position at time t on a line, then the derivative $dy/dt = \dot{y}$ equals the velocity, and the derivative of the derivative, that is,

$$\frac{d}{dt}\left(\frac{dy}{dt}\right)$$

equals the acceleration.

THE HIGHER DERIVATIVES

The second derivative The derivative of the derivative of a function $y = f(x)$ is called the *second derivative* of the function. It is denoted by any of the following notations:

$$\frac{d^2y}{dx^2}, \quad D^2y, \quad y'', \quad f'', \quad D^2f, \quad f^{(2)}, \quad f^{(2)}(x), \quad \frac{d^2f}{dx^2}$$

If $y = f(t)$, where t denotes time, the second derivative d^2y/dt^2 is also denoted $\ddot{y}$.

For instance, if $y = x^3$,

$$\frac{dy}{dx} = 3x^2$$

and

$$\frac{d^2y}{dx^2} = 6x.$$

Other ways of denoting the second derivative of this function are

$$D^2(x^3) = 6x,$$

$$\frac{d^2(x^3)}{dx^2} = 6x,$$

and

$$(x^3)'' = 6x.$$

y	$\dfrac{dy}{dx}$	$\dfrac{d^2y}{dx^2}$
x^3	$3x^2$	$6x$
$\dfrac{1}{x}$	$-\dfrac{1}{x^2}$	$\dfrac{2}{x^3}$
$\sin 5x$	$5\cos 5x$	$-25\sin 5x$

The table in the margin lists dy/dx, the first derivative, and d^2y/dx^2, the second derivative, for a few functions. The first derivative of f, $f'(x)$, is also denoted $f^{(1)}(x)$.

Most functions f met in applications of calculus can be differentiated repeatedly in the sense that Df exists, the derivative of Df, namely D^2f, exists, the derivative of D^2f exists, and so on. The derivative of the second derivative,

$$\frac{d}{dx}\left(\frac{d^2y}{dx^2}\right)$$

is called the *third derivative* and is denoted many ways, such as

$$\frac{d^3y}{dx^3}, \quad D^3y, \quad y''', \quad f''', \quad f^{(3)}, \quad f^{(3)}(x), \quad \frac{d^3f}{dx^3}.$$

The *fourth derivative* $f^{(4)}(x)$ is defined as the derivative of the third derivative and is represented by similar notations. Similarly, $f^{(n)}(x)$ is defined for $n = 5, 6, \ldots$. The derivatives $f^{(n)}(x)$ for $n \geq 2$ are called the *higher derivatives* of f.

EXAMPLE 1 Compute $f^{(n)}(x)$ if $f(x) = x^3 - 2x^2 + x + 5$ and n is a positive integer.

SOLUTION

$$f^{(1)}(x) = \frac{df}{dx} = 3x^2 - 4x + 1$$

$$f^{(2)}(x) = \frac{d}{dx}(f^{(1)}(x)) = 6x - 4$$

$$f^{(3)}(x) = \frac{d}{dx}(f^{(2)}(x)) = 6$$

$$f^{(4)}(x) = 0, \quad \text{since } f^{(3)}(x) \text{ is constant.}$$

Since $f^{(4)}(x)$ is constant, its derivative $f^{(5)}(x)$ is 0 for all x. Similarly, $f^{(6)}(x) = 0$, $f^{(7)}(x) = 0$, and so on. ∎

As Example 1 may suggest, for any polynomial $f(x)$ of degree at most 3, $f^{(n)}(x) = 0$ for all integers $n \geq 4$. The next example is quite different.

EXAMPLE 2 Compute $f^{(n)}(x)$ if $f(x) = \cos x$.

SOLUTION

$$f^{(1)}(x) = \frac{d}{dx}(\cos x) = -\sin x$$

$$f^{(2)}(x) = \frac{d}{dx}(f^{(1)}(x)) = \frac{d}{dx}(-\sin x) = -\cos x$$

$$f^{(3)}(x) = \frac{d}{dx}(-\cos x) = \sin x$$

$$f^{(4)}(x) = \frac{d}{dx}(\sin x) = \cos x$$

$$f^{(5)}(x) = \frac{d}{dx}(\cos x) = -\sin x$$

$$f^{(6)}(x) = \frac{d}{dx}(-\sin x) = -\cos x$$

Note that $f^{(4)}(x) = f(x)$, $f^{(5)}(x) = f^{(1)}(x)$, and so on. The higher derivatives repeat every four steps. ∎

Of the higher derivatives the second derivative is the one most commonly used. Derivatives higher than the second are of aid in the study of the error in approximating a quantity by a particular formula, for instance, a function by a polynomial. (These will be discussed in Sec. 10.8.) However, Exercises 49 to 51 of this section illustrate the utility of the higher derivatives by using them to obtain the binomial theorem.

Whether the first derivative is positive, negative, or 0 tells a good deal about a function and its graph. This section will explore the geometric significance of the second derivative being positive, negative, or 0. The following section will show how the second derivative is used in the study of motion.

CONCAVE UPWARD AND CONCAVE DOWNWARD

Assume that $f''(x)$ is positive for all x in the open interval (a, b). Since f'' is the derivative of f', it follows that f' is an increasing function throughout the interval (a, b). In other words, as x increases, the slope of the graph of $y = f(x)$ increases as we move from left to right on that part of the graph corresponding to the interval (a, b). The slope may increase from negative to positive values as in Fig. 4.25. Or the slope may be positive throughout (a, b) and increasing, as in Fig. 4.26. Or the slope may be negative throughout (a, b) and increasing as in Fig. 4.27.

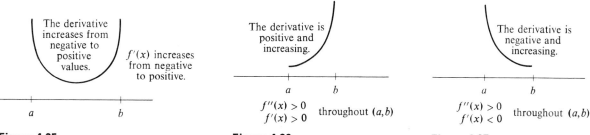

Figure 4.25   Figure 4.26   Figure 4.27

As you drive along such a graph from left to right, your car keeps turning to the left.

DEFINITION *Concave upward.* A function f whose first derivative is increasing throughout the open interval (a, b) is called *concave upward* in that interval.

A concave upward curve lies above its tangent lines and below its chords.

Figure 4.28

Note that, when a function is concave upward, it is shaped like part of a cup (Concave *UP*ward).

It can be proved that where a curve is concave upward it lies *above its tangent lines* and *below its chords*, as shown in Fig. 4.28. (See Exercises 44 to 46.)

As was observed, in an interval where $f''(x)$ is positive, the function $f'(x)$ is increasing, and so the function f is concave upward. However, if a function is concave upward, $f''(x)$ is not necessarily positive. For instance, $y = x^4$ is concave upward over any interval, since the derivative $4x^3$ is increasing. The second derivative $12x^2$ is not always positive; at $x = 0$ it is 0.

If, on the other hand, $f''(x)$ is negative throughout (a, b), then f' is a decreasing function and the graph of f looks like part of the curve in Fig. 4.29.

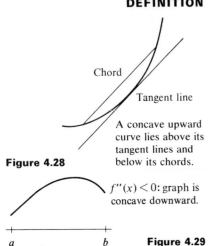

$f''(x) < 0$: graph is concave downward.

Figure 4.29

DEFINITION *Concave downward.* A function whose first derivative is decreasing throughout an open interval (a, b) is called *concave downward* in that interval.

Where a function is concave downward, it lies *below its tangent lines* and *above its chords.*

EXAMPLE 3 Where is the graph of $f(x) = x^3$ concave upward? Concave downward?

SOLUTION Compute the second derivative. Since $D(x^3) = 3x^2$, $D^2(x^3) = 6x$.

Clearly $6x$ is positive for all positive x and negative for all negative x. The graph, shown in Fig. 4.30, is concave upward if $x > 0$ and concave downward if $x < 0$. Note that the sense of concavity changes at $x = 0$. When you drive along this curve from left to right, your car turns to the right until you pass through (0, 0). Then it starts turning to the left.

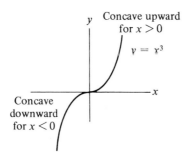

Figure 4.30

■

4.3 CONCAVITY AND THE SECOND DERIVATIVE

EXAMPLE 4 Consider the function $f(x) = \sin x$ for x in $[0, 2\pi]$. Where is the graph concave upward? Concave downward?

SOLUTION

$$f(x) = \sin x$$
$$f'(x) = \cos x$$
$$f''(x) = -\sin x$$

The second derivative, $-\sin x$, is negative for $0 < x < \pi$. It is positive for $\pi < x < 2\pi$. Therefore the graph is concave downward for x in $(0, \pi)$ and concave upward for x in $(\pi, 2\pi)$, as shown in Fig. 4.31. ∎

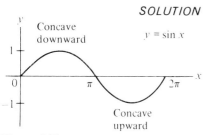

Figure 4.31

INFLECTION POINTS The sense of concavity is a useful tool in sketching the graph of a function. Of special interest in Examples 3 and 4 is the presence of a point on the graph where the sense of concavity changes. Such a point is called an *inflection point*.

DEFINITION *Inflection point* and *inflection number*. Let f be a function and let a be a number. Assume that there are numbers b and c such that $b < a < c$ and

1. f is continuous on the open interval (b, c);

2. f is concave upward in the interval (b, a) and concave downward in the interval (a, c), or vice versa.

Then the point $(a, f(a))$ is called an *inflection point* or *point of inflection*. The number a is called an *inflection number*.

Observe that if the second derivative changes sign at the number a, then a is an inflection number.

If the second derivative exists at an inflection point, it must be 0. But there can be an inflection point even if f'' is not defined there, as shown by the next example, which is closely related to Example 3.

EXAMPLE 5 Examine the concavity of $y = x^{1/3}$.

SOLUTION Here

$$y' = \tfrac{1}{3} x^{-2/3},$$

and

$$y'' = \frac{1}{3} \cdot \frac{-2}{3} x^{-5/3}.$$

Neither y' nor y'' is defined at 0; however, the sign of y'' changes at 0. When x is negative, y'' is positive; when x is positive, y'' is negative. The concavity switches from upward to downward at $x = 0$. The graph is shown in Fig. 4.32. ∎

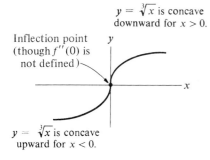

Figure 4.32

When graphing a function f, find where $f(x) = 0$, where $f'(x) = 0$, and where $f''(x) = 0$, if the solutions are easy. Determine where $f'(x)$ is positive and where it is negative. Determine also where $f''(x)$ is positive and where it is negative. The following table contrasts the interpretations of the signs of f, f', and f''. (It is assumed that f, f', and f'' are continuous.)

	Is Positive	Is Negative	Changes Sign
Where the ordinate $f(x)$	The graph is above the x axis.	The graph is below the x axis.	The graph crosses the x axis.
Where the slope $f'(x)$	The graph slopes upward.	The graph slopes downward.	The graph has a horizontal tangent and a relative maximum or minimum.
Where $f''(x)$	The graph is concave upward (like a cup).	The graph is concave downward.	The graph has an inflection point.

Keep in mind that the graph can have an inflection point at a, even though the second derivative is not defined at a (Example 5). Similarly, a graph can have a maximum or minimum at a, even though the first derivative is not defined at a. (Consider $f(x) = |x|$ at $a = 0$.)

EXAMPLE 6 Graph $f(x) = 2x^3 - 3x^2$, showing intercepts, critical points, and inflection points.

SOLUTION To find x intercepts, set $f(x) = 0$:

$$2x^3 - 3x^2 = 0$$
$$x^2(2x - 3) = 0$$
$$x^2 = 0 \quad \text{or} \quad 2x - 3 = 0.$$

Thus the x intercepts are

$$x = 0 \quad \text{and} \quad x = \tfrac{3}{2}.$$

The y intercept is simply $f(0)$, which is 0.
To find critical numbers, set $f'(x) = 0$:

$$f'(x) = (2x^3 - 3x^2)' = 6x^2 - 6x = 0$$
$$6x(x - 1) = 0$$
$$x = 0 \quad \text{or} \quad x - 1 = 0.$$

There are two critical numbers, $x = 0$ and $x = 1$.
To find inflection numbers, determine where the sign of $f''(x)$ changes. Since the second derivative exists everywhere, $f''(x)$ will exist and be 0 at an inflection number. So set $f''(x) = 0$ and check whether the second derivative changes sign at any of the solutions. We have

$$f''(x) = (6x^2 - 6x)' = 12x - 6.$$

The equation $$12x - 6 = 0$$
has only one solution: $$x = \tfrac{1}{2}.$$

But is $x = \tfrac{1}{2}$ an inflection number? To check, we must see whether the second derivative changes sign at $\tfrac{1}{2}$. The second derivative is $12x - 6 = 6(2x - 1)$. For $x > \tfrac{1}{2}$, the second derivative is positive; for $x < \tfrac{1}{2}$, the second derivative is negative. Thus $x = \tfrac{1}{2}$ is an inflection number.

The inputs of interest are 0 (which is an x intercept and a critical number), $\tfrac{3}{2}$ (which is an x intercept), 1 (which is a critical number), and $\tfrac{1}{2}$ (which is an inflection number). Compute the outputs for these inputs in order to plot the key points on the graph.

x	0	$\tfrac{1}{2}$	1	$\tfrac{3}{2}$
$f(x) = 2x^3 - 3x^2$	0	$-\tfrac{1}{2}$	-1	0

Figure 4.33 displays the information gathered so far. Noting that $\lim_{x \to \infty} (2x^3 - 3x^2) = \infty$ and that $\lim_{x \to -\infty} (2x^3 - 3x^2) = -\infty$, we complete the graph freehand, as shown in Fig. 4.34.

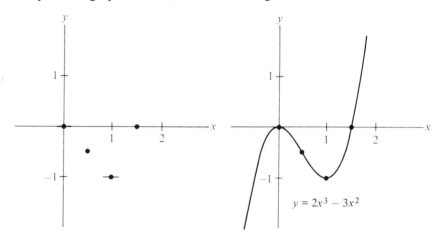

Figure 4.33   **Figure 4.34**

(Try completing it yourself; if you don't have local extrema at $(0, 0)$ and $(1, -1)$, then you will force there to be more critical numbers.) ∎

THE SECOND DERIVATIVE AND LOCAL EXTREMA

The second derivative is also useful in testing whether at a critical number there is a relative maximum or relative minimum. For instance, let a be a critical number for the function f and assume that $f''(a)$ happens to be negative. If f'' is continuous in some open interval that contains a, then $f''(a)$ remains negative for a suitably small open interval that contains a. This means that the graph of f is concave downward near $(a, f(a))$, hence lies below its tangent lines. In particular, it lies below the horizontal tangent line at the critical point $(a, f(a))$, as illustrated in Fig. 4.35. Thus the function has a *relative maximum* at the critical number a. This observation suggests the following test for a relative maximum or minimum.

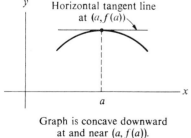

Graph is concave downward at and near $(a, f(a))$.

Figure 4.35

APPLICATIONS OF THE DERIVATIVE

THEOREM

The second derivative test for extrema

Second-derivative test for relative maximum or minimum. Let f be a function such that $f'(x)$ is defined at least on some open interval containing the number a. Assume that $f''(a)$ is defined. If

$$f'(a) = 0$$
and
$$f''(a) < 0,$$

then f has a local maximum at a.
 Similarly, if

$$f'(a) = 0$$
and
$$f''(a) > 0,$$

then f has a local minimum at a.

PROOF Assume that $f'(a) = 0$ and that $f''(a)$ is negative. Then for x sufficiently close to a the difference quotient

$$\frac{f'(x) - f'(a)}{x - a}$$

must be negative. Since $f'(a) = 0$,

$$\frac{f'(x)}{x - a}$$

is negative.

When x is greater than a, the denominator $x - a$ is positive; thus $f'(x)$ is negative. When x is less than a, $x - a$ is negative; thus $f'(x)$ is positive.

Since the sign of $f'(x)$ changes from positive to negative at a as x increases, there is a local maximum at $x = a$.

A similar argument shows that if $f''(a)$ is positive at a critical number, then the function f has a local minimum at a. ∎

EXAMPLE 7 Find all local extrema of the function $f(x) = x^4 - 2x^3$.

SOLUTION Since f is differentiable throughout its domain, any local extremum can occur only at a critical number. So begin by finding the critical numbers, as follows:

$$f'(x) = (x^4 - 2x^3)' = 4x^3 - 6x^2 = 0$$
$$x^2(4x - 6) = 0.$$

Thus $x^2 = 0$ or $4x - 6 = 0$.

The critical numbers are therefore

$$x = 0 \quad \text{and} \quad x = \tfrac{3}{2}.$$

Now use the second derivative to determine whether either of these corresponds to a local extremum.

The second derivative is

$$f''(x) = (4x^3 - 6x^2)' = 12x^2 - 12x.$$

At $x = \frac{3}{2}$ we have

$$f''(\tfrac{3}{2}) = 12(\tfrac{3}{2})^2 - 12(\tfrac{3}{2}) = 27 - 18 = 9,$$

which is positive. Since $f'(\tfrac{3}{2}) = 0$ and $f''(\tfrac{3}{2}) > 0$, f has a local minimum at $x = \tfrac{3}{2}$.

How about the other critical number, $x = 0$? In this case

$$f''(0) = 12 \cdot 0^2 - 12 \cdot 0 = 0.$$

Since $f''(0) = 0$, the second-derivative test tells us nothing about the critical number 0. Instead, we must resort to the first-derivative test and examine the sign of $f'(x) = 4x^3 - 6x^2 = x^2(4x - 6)$ for x near 0. For x sufficiently near 0, whether to the right of 0 or to the left, x^2 is positive and $4x - 6$ is negative. Thus $f'(x)$ is negative for x near 0. Since f is a decreasing function near 0, it has neither a local maximum nor a local minimum at 0. (As may be checked, there happens to be an inflection point at $(0, 0)$.) The function is graphed in Fig. 4.36.

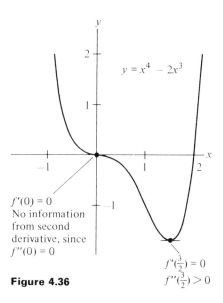

Figure 4.36

SUMMARY OF THIS SECTION This section, which examined the geometric significance of the second derivative, introduced the concepts of *concave upward*, *concave downward*, *inflection point*, and *inflection number*. It also developed a second-derivative test for a local maximum ($f'(x) = 0$ and $f''(x)$ negative) and for a local minimum ($f'(x) = 0$ and $f''(x)$ positive). If $f'(x) = 0$ and $f''(x) = 0$, no

194 APPLICATIONS OF THE DERIVATIVE

conclusion can be drawn without further analysis. Perhaps there is an inflection point (as occurred in Example 7); perhaps there is a local minimum (this is the case with the function $y = x^4$ at 0, as may be checked by graphing); perhaps there is a local maximum (as is the case with the function $y = -x^4$).

The following table shows the steps in graphing a function. Of course, not all these steps will apply for each function. For instance, there may not be any intercepts or asymptotes. A critical number can be examined by either the first-derivative test or the second-derivative test to determine whether it gives a relative maximum or relative minimum; which method is preferable depends on the function.

General Procedure for Graphing a Function

	Calculations	Geometric Meaning
Domain	1. Find where $f(x)$ is defined.	Find horizontal extent of graph.
Intercepts	2. Find $f(0)$ and the values of x for which $f(x) = 0$.	Find where graph crosses the axes.
Critical numbers	3. Find where $f'(x) = 0$.	Find where tangent line is horizontal.
	4. Compute $f(x)$ at all critical numbers.	Data needed for critical points.
Increasing, decreasing	5. Find the values of x for which $f'(x)$ is positive and those for which $f'(x)$ is negative.	Find where graph goes up and where it goes down as pencil moves to the right.
Horizontal asymptotes	6. Find $\lim_{x \to \infty} f(x)$ and $\lim_{x \to -\infty} f(x)$.	Find horizontal asymptotes or general behavior when $|x|$ is large.
Vertical asymptotes	7. Find the values of a where $\lim_{x \to a^+} f(x)$ or $\lim_{x \to a^-} f(x)$ is infinite.	Find vertical asymptotes (where the graph "blows up").
Concavity and inflection points	8. Find the values of x for which $f''(x)$ is positive and those for which $f''(x)$ is negative. Note where it changes sign.	Find where the graph is concave upward and where it is concave downward. Note inflection points.
	9. Sketch the graph, showing intercepts, critical points, asymptotes, local and global maxima and minima, and inflection points.	

Exercises

In Exercises 1 to 8 compute $f^{(1)}$, $f^{(2)}$, $f^{(3)}$, and $f^{(4)}$ for the given functions f.

1. $x^2 - x + 5$
2. $2x^2 - 5x$
3. x^3
4. $x^4 + 5x^2$
5. $\sin 2x$
6. $\cos 3x$
7. $1/x^2$
8. $1/(x+1)$

In Exercises 9 and 10 find d^2f/dx^2 for the given functions f.

9. (a) $\tan 2x$ (b) $\sqrt{1 + 2x}$ (c) $\cot \sqrt{x}$
10. (a) $\csc 3x$ (b) $(1 + 3x)^{10}$ (c) $\sec x^2$

In Exercises 11 to 28 graph the functions, showing any relative maxima, relative minima, and inflection points.

11. $x^3 - x^2$
12. $x^3 + x^2$
13. $x^4 + 2x^3$
14. $x^3 - 3x^2$
15. $x^4 - 4x^3$
16. $3x^5 - 5x^4$
17. $\dfrac{1}{1 + x^2}$
18. $\dfrac{1}{1 + x^4}$
19. $x^3 - 6x^2 - 15x$
20. $\dfrac{x^2}{2} + \dfrac{1}{x}$
21. $\dfrac{1}{1 + x^3}$
22. $\dfrac{1}{1 + x^5}$
23. $\tan x$
24. $\sin x + \sqrt{3} \cos x$
25. $\dfrac{1}{1 + 3x^2}$
26. $\dfrac{x}{1 + 3x^2}$
27. $(x - 1)^4$
28. $(x - 1)^5$

In each of Exercises 29 to 36 sketch the general appearance of the graph of the given function near (1, 1) on the basis of the information given. Assume that f, f', and f'' are continuous.

29 $f(1) = 1, f'(1) = 0, f''(1) = 1$
30 $f(1) = 1, f'(1) = 0, f''(1) = -1$
31 $f(1) = 1, f'(1) = 0, f''(1) = 0$ (Sketch four possibilities.)
32 $f(1) = 1, f'(1) = 0, f''(1) = 0, f''(x) < 0$ for $x < 1$ and $f''(x) > 0$ for $x > 1$
33 $f(1) = 1, f'(1) = 0, f''(1) = 0$ and $f''(x) < 0$ for x near 1
34 $f(1) = 1, f'(1) = 1, f''(1) = -1$
35 $f(1) = 1, f'(1) = 1, f''(1) = 0$ and $f''(x) < 0$ for $x < 1$ and $f''(x) > 0$ for $x > 1$
36 $f(1) = 1, f'(1) = 1, f''(1) = 0$ and $f''(x) > 0$ for x near 1

■

37 Let $f(x) = ax^2 + bx + c$, where a, b, and c are constants, $a \neq 0$. Show that f has no inflection points.
38 Let $f(x) = ax^3 + bx^2 + cx + d$, where a, b, c, and d are constants, $a \neq 0$. Show that f has exactly one inflection point.
39 Figure 4.37 appeared in "Energy Use in the United States Food System," by John S. Steinhart and Carol E. Steinhart, in *Perspectives on Energy*, edited by Lon C. Ruedisili and Morris W. Firebaugh, Oxford, New York, 1975. The graph shows farm output as a function of energy input.

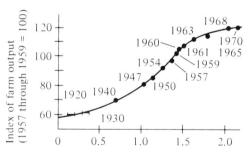

Figure 4.37 Farm output as a function of energy input to the United States food system, 1920 through 1970.

(a) What is the practical significance of the fact that the function has a positive derivative?
(b) What is the practical significance of the inflection point?

40 Let f be a function such that $f''(x) = (x - 1)(x - 2)$.
(a) For which x is f concave upward?
(b) For which x is f concave downward?
(c) List its inflection numbers.
(d) Find a specific function f whose second derivative is $(x - 1)(x - 2)$.

41 Let f be a function such that $f''(x) = (x - 1)^3(x - 2)^2$.
(a) Where is $f''(x) = 0$?
(b) What are the inflection numbers of f?

■ ■

42 A certain function $y = f(x)$ has the property that
$$y' = \sin y + 2y + x.$$
Show that at a critical number the function has a local minimum.
43 Sketch the graph of a function f such that for all x:
(a) $f(x) > 0, f'(x) > 0, f''(x) > 0$;
(b) $f'(x) < 0, f''(x) < 0$.
(c) Can there be a function such that $f(x) > 0, f'(x) < 0, f''(x) < 0$ for all x? Explain.
44 Let f be a function such that $f(0) = 0 = f(1)$ and $f''(x) \geq 0$ for all x in [0, 1].
(a) Using a sketch, explain why $f(x) \leq 0$ for all x in [0, 1].
(b) Without a sketch prove that $f(x) \leq 0$ for all x in [0, 1].
45 (See Exercise 44.) Prove that if f is a function such that $f''(x) > 0$ for all x, then the graph of $y = f(x)$ lies below its chords; i.e.,
$$f(ax_1 + (1 - a)x_2) < af(x_1) + (1 - a)f(x_2)$$
for any a in (0, 1), and for any x_1 and x_2, $x_1 \neq x_2$.
46 Prove, without using a picture, that where the graph of f is concave upward it lies above its tangent.
47 Prove (without referring to a picture) that if the graph of f lies above its tangent lines for all x in $[a, b]$, then $f''(x) \geq 0$ for all x in $[a, b]$. [The case $y = x^4$ shows that we should not try to prove that $f''(x) > 0$.]
48 (Contributed by David Hayes) Let f be a function that is continuous for all x and differentiable for all x other than 0. Figure 4.38 is the graph of its derivative $f'(x)$, as a function of x.

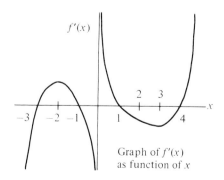

Graph of $f'(x)$ as function of x

Figure 4.38

(a) Answer the following questions about f (not about f'). Where is f increasing? decreasing? concave upward? concave downward? What are the critical numbers? Where do any relative maxima or relative minima occur?

(b) Assuming that $f(0) = 1$, graph a hypothetical function f that satisfies the conditions given.

Exercises 49 to 51 obtain the binomial theorem with the aid of the higher derivatives. The binomial theorem asserts that for any positive integer n the coefficient of x^k in the expansion of $(1 + x)^n$ is

$$\frac{n(n-1)(n-2)\cdots(n-k+1)}{1 \cdot 2 \cdot 3 \cdots k} = \frac{n!}{k!(n-k)!} = \binom{n}{k}.$$

(This is proved in a precalculus course by counting permutations and combinations.) Exercises 49 and 50 are for practice.

49 Let $f(x) = (1 + x)^2 = a_0 + a_1 x + a_2 x^2$, where the constants a_0, a_1, and a_2 are to be determined with the aid of the higher derivatives of f.
(a) Compute $f(0)$, $f'(0)$, and $f''(0)$, using the formula $f(x) = (1 + x)^2$.

(b) Compute $f(0)$, $f'(0)$, and $f''(0)$, using the formula $f(x) = a_0 + a_1 x + a_2 x^2$.
(c) Comparing the results in (a) and (b), show that

$$a_0 = 1, \quad a_1 = 2, \quad \text{and} \quad a_2 = 1.$$

This shows that $(1 + x)^2 = 1 + 2x + x^2$, which establishes the binomial theorem for $n = 2$.

50 Let $f(x) = (1 + x)^3 = a_0 + a_1 x + a_2 x^2 + a_3 x^3$, where the constants a_0, a_1, a_2, and a_3 are to be determined.
(a) Compute $f(0)$, $f^{(1)}(0)$, $f^{(2)}(0)$, and $f^{(3)}(0)$, using the formula $f(x) = (1 + x)^3$.
(b) Compute the quantities in (a), using the formula $f(x) = a_0 + a_1 x + a_2 x^2 + a_3 x^3$.
(c) Comparing the results in (a) and (b), show that $(1 + x)^3 = 1 + 3x + 3x^2 + x^3$.

51 Let $f(x) = (1 + x)^n = a_0 + a_1 x + a_2 x^2 + \cdots + a_n x^n$, the polynomial obtained when $(1 + x)^n$ is multiplied out. Compute $f^{(k)}(0)$ in two different ways, as in the preceding two exercises, to obtain a formula for a_k and thus establish the binomial theorem in general.

4.4 Motion and the second derivative

In the preceding section the second derivative was interpreted in terms of the graph of a function. In the present section the second derivative will be used in the study of motion, where it represents acceleration. The following example illustrates how the second derivative provides information about an object moving under the influence of gravity.

EXAMPLE 1 A falling rock drops $16t^2$ feet in the first t seconds. Find its velocity and acceleration.

SOLUTION Place the y axis in the usual position, with 0 at the beginning of the fall and the part with positive values above 0, as in Fig. 4.39. At time t the object has the y coordinate

$$y = -16t^2.$$

The velocity is $\quad (-16t^2)' = -32t \quad$ feet per second,

and the acceleration is

$$(-32t)' = -32 \text{ feet per second per second.}$$

The velocity changes at a constant rate. That is, the acceleration is constant. ∎

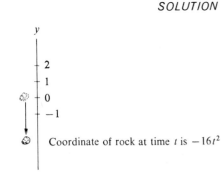

Figure 4.39

The second derivative represents acceleration in other contexts, as the next example shows.

4.4 MOTION AND THE SECOND DERIVATIVE

EXAMPLE 2 Translate this news item into calculus: "The latest unemployment figures can be read as bearing out the forecast that the recession is nearing its peak. Though unemployment continues to increase, it is doing so at a slower rate than before."

SOLUTION Let y be a differentiable function of t that approximates the number of people unemployed at time t. As time changes, so does y:

$$y = f(t).$$

The rate of change in unemployment is the derivative

$$\frac{dy}{dt}.$$

The news that "unemployment continues to increase" is recorded by the inequality

$$\frac{dy}{dt} > 0.$$

There is optimism in the article. The rate of increase, dy/dt, is itself declining ("unemployment continues to increase ... at a *slower rate* than before"). The function dy/dt is decreasing. Thus its derivative

$$\frac{d}{dt}\left(\frac{dy}{dt}\right)$$

is negative:

$$\frac{d^2y}{dt^2} < 0.$$

Economics need not be a dismal science. If the first derivative is bleak, maybe the second derivative is promising.

In short, the bad news is that dy/dt is positive. But there is good news: d^2y/dt^2 is negative.

The promise that "the recession is nearing its peak" amounts to the prediction that soon dy/dt will be 0, and then switch sign to become negative. In short, a local maximum in the graph of $y = f(t)$ appears in the economists' crystal ball. ∎

FINDING THE POSITION FROM THE ACCELERATION

The remaining examples focus on the use of the second derivative in the study of motion. Knowing the initial position and initial velocity and the acceleration at all times, it is possible to calculate the position at all times.

EXAMPLE 3 In the simplest motion, no forces act on a moving particle. Assume that a particle is moving on the x axis and no forces act on it. Let its location at time t seconds be

$$x = f(t) \quad \text{feet.}$$

If at time $t = 0$, $x = 3$ feet and the velocity is 5 feet per second, determine $f(t)$.

198 APPLICATIONS OF THE DERIVATIVE

SOLUTION The assumption that no force operates on the particle means that $d^2x/dt^2 = 0$. Call the velocity v. Then

$$\frac{dv}{dt} = \frac{d^2x}{dt^2} = 0.$$

Now, v is itself a function of time. Since its derivative is 0, v must be constant:

$$v(t) = C$$

for some constant C. Since $v(0) = 5$, the constant C must be 5.

To find the position x as a function of time, note that

$$\frac{dx}{dt} = 5.$$

This equation implies that x must be of the form

$$x = 5t + K$$

for some constant K. Now, when $t = 0$, $x = 3$. Thus $K = 3$. In short, at any time t seconds, the particle is at

$$x = 5t + 3 \quad \text{feet.} \quad \blacksquare$$

The next example concerns the case in which the acceleration is constant, but not zero.

EXAMPLE 4 A ball is thrown straight up, with a speed of 64 feet per second, from a cliff 96 feet above the ground. Where is the ball t seconds later? When does it reach its maximum height? How high above the ground does the ball rise? When does the ball hit the ground? Assume that there is no air resistance and that the acceleration due to gravity is constant.

SOLUTION Introduce a vertical coordinate axis to describe the position of the ball. It is more natural to call it the y axis, and so velocity is dy/dt, and acceleration is d^2y/dt^2. Place the origin at ground level and let the positive part of the y axis be above the ground, as in Fig. 4.40.

At time $t = 0$, the velocity $dy/dt = 64$, since the ball is thrown up at a speed of 64 feet per second. (If it had been thrown down, dy/dt would be -64.) As time increases, dy/dt decreases from 64 to 0 (when the ball reaches the top of its path and begins its descent) and continues to decrease through negative values as the ball falls down to the ground. Since v is decreasing, the acceleration dv/dt is negative. The (constant) value of dv/dt, obtained from experiments, is approximately -32 feet per second per second.

From the equation

$$\frac{dv}{dt} = -32$$

Velocity is an antiderivative of acceleration.   it follows that

$$v = -32t + C,$$

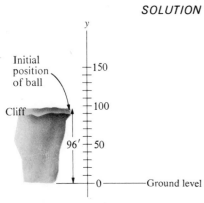

Figure 4.40

where C is some constant. To find C, recall that $v = 64$ when $t = 0$. Thus

$$64 = -32 \cdot 0 + C,$$

and $C = 64$. Hence $v = -32t + 64$ for any time t until the ball hits the ground. Now $v = dy/dt$, so

$$\frac{dy}{dt} = -32t + 64.$$

Position is an antiderivative of velocity.

This equation implies that

$$y = -16t^2 + 64t + K,$$

where K is a constant. To find K, make use of the fact that $y = 96$ when $t = 0$. Thus

$$96 = -16 \cdot 0^2 + 64 \cdot 0 + K,$$

and $K = 96$.

We have obtained a complete description of the position of the ball at any time t while it is in the air:

$$y = -16t^2 + 64t + 96.$$

This, together with $v = -32t + 64$, provides answers to many questions about the ball's flight.

Maximum height

When does it reach its maximum height? When $v = 0$; that is, when $-32t + 64 = 0$, or when $t = 2$ seconds.

How high above the ground does the ball rise? Simply compute y when $t = 2$. This gives $-16 \cdot 2^2 + 64 \cdot 2 + 96 = 160$ feet.

Hitting the ground

When does the ball hit the ground? When $y = 0$. Find y such that

$$y = -16t^2 + 64t + 96 = 0.$$

Division by -16 yields the simpler equation

$$t^2 - 4t - 6 = 0,$$

which has the solution

$$t = \frac{4 \pm \sqrt{16 + 24}}{2},$$

or simply

$$t = 2 \pm \sqrt{10}.$$

Since $2 - \sqrt{10}$ is negative, and the ball cannot hit the ground before it is thrown, the physically meaningful solution is $2 + \sqrt{10}$. The ball lands $2 + \sqrt{10}$ seconds after it is thrown; it is in the air for about 5.2 seconds.

The graphs of y, v, and speed, as functions of time, provide another perspective of the motion of the ball, as shown in Fig. 4.41. Of course, the actual path of the ball is restricted to a vertical line and looks somewhat as pictured in Fig. 4.42.

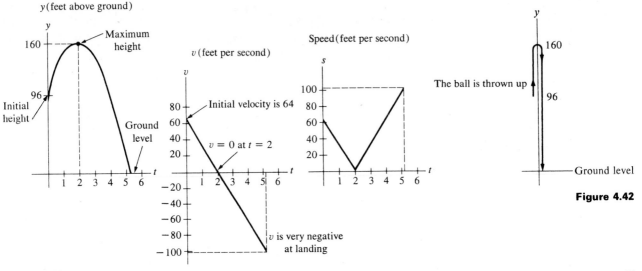

Figure 4.41

Figure 4.42

Reasoning like that in Example 4 establishes the following description of motion in which the acceleration is constant.

MOTION UNDER CONSTANT ACCELERATION

Assume that a particle moving on the y axis has a constant acceleration a at any time. Assume that at time $t = 0$ it has an initial velocity v_0 and has the initial y coordinate y_0. Then at any time $t \geq 0$ its y coordinate is

$$y = \frac{a}{2} t^2 + v_0 t + y_0.$$

Example 4 is the special case, $a = -32$, $v_0 = 64$, and $y_0 = 96$.

ESCAPE VELOCITY (optional)

In fact, the acceleration due to gravity is not constant. It varies inversely as the square of the distance from the center of the earth. However, it is almost constant if the particle moves only small distances—such as a few miles up; so, treating it as constant in practical engineering is justified. The next example concerns motion subject to an acceleration that is not constant. More complicated, it is included as a little detour for the interested reader, perhaps to be read later when reviewing. The mathematical basis for determining the escape velocity was developed in Newton's *Principia*, published in 1687. There he investigated not only the inverse square law but other laws of attraction, such as the *inverse cube*, which are not known to occur in nature.

EXAMPLE 5 Find the initial velocity that a payload must have in order that it will "coast to infinity" rather than fall back to earth. Assume that it is launched straight up.

SOLUTION Begin by studying the motion of a projectile fired with an initial velocity of v_0 miles per second from the surface of the earth. For this purpose introduce a coordinate system whose origin is at the center of the earth. Let r denote the distance of the projectile from the center of the earth. (See Fig. 4.43.)

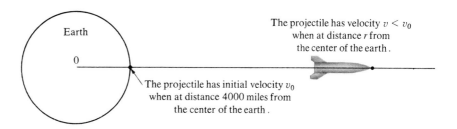

Figure 4.43

The velocity of the projectile is defined as

$$v = \frac{dr}{dt},$$

and its acceleration as $d^2r/dt^2 = dv/dt$. Assume that the acceleration due to gravity is proportional to the force of gravity on the particle; that is,

$$\text{Acceleration} = \frac{dv}{dt} = \frac{-K}{r^2}, \tag{1}$$

where K is some positive constant. (The negative sign in (1) reminds us that gravity slows the projectile down.)

Before analyzing (1) further, we determine K. At the surface of the earth, where $r = 4000$ miles, the acceleration due to gravity is -32 feet per second per second, which is approximately -0.006 miles per second per second. Thus K satisfies the equation

$$-0.006 = \frac{-K}{4000^2},$$

and

$$K = (4000)^2(0.006). \tag{2}$$

Now return to (1), which links velocity, time, and distance. It is possible to eliminate time from (1) by using the chain rule:

$$\frac{dv}{dt} = \frac{dv}{dr}\frac{dr}{dt} = \frac{dv}{dr}v.$$

Thus (1) now reads
$$v\frac{dv}{dr} = \frac{-K}{r^2}, \qquad (3)$$

an equation linking velocity and distance. Observe that (3) is equivalent to the equation

$$\frac{d}{dr}\left(\frac{v^2}{2}\right) = \frac{d}{dr}\left(\frac{K}{r}\right).$$

Since the functions $v^2/2$ and K/r have the same derivative with respect to r, they differ by a constant. Thus

$$\frac{v^2}{2} = \frac{K}{r} + C, \qquad (4)$$

where C is constant. To determine C, again use information available at the surface of the earth, namely, $v = v_0$ when $r = 4000$. From (4) it follows that

$$\frac{v_0^2}{2} = \frac{K}{4000} + C,$$

and
$$C = \frac{v_0^2}{2} - \frac{K}{4000}. \qquad (5)$$

Combining (2), (4), and (5) yields

$$\frac{v^2}{2} = \frac{K}{r} + \left(\frac{v_0^2}{2} - \frac{K}{4000}\right)$$

$$= \frac{v_0^2}{2} + K\left(\frac{1}{r} - \frac{1}{4000}\right)$$

$$= \frac{v_0^2}{2} + (4000)^2(0.006)\left(\frac{1}{r} - \frac{1}{4000}\right).$$

Hence
$$v^2 = v_0^2 + (4000)^2(0.012)\left(\frac{1}{r} - \frac{1}{4000}\right). \qquad (6)$$

Equation (6) describes v as a function of r. If v in (6) is never 0, that is, if the payload never reaches a maximum distance from the earth, then the payload will not fall back to the earth. Thus, by (6), if v_0 is such that the equation

$$0 = v_0^2 + (4000)^2(0.012)\left(\frac{1}{r} - \frac{1}{4000}\right) \qquad (7)$$

has no solution r, then v_0 is large enough to send the payload on an endless journey. To find such v_0, rewrite (7) as

$$(4000)(0.012) - v_0^2 = (4000)^2(0.012)\left(\frac{1}{r}\right). \qquad (8)$$

If the left side of (8) is greater than 0, then there is a solution for r, and the payload will reach a maximum distance; if the left side of (8) is less than or equal to 0, then there is no solution for r. The smallest v_0 that satisfies the inequality $(4000)(0.012) - v_0^2 \leq 0$ is

$$v_0 = \sqrt{(4000)(0.012)} = \sqrt{48} \approx 6.93 \text{ miles per second.}$$

The escape velocity is 6.93 miles per second, which is about 25,000 miles per hour. ∎

Escape velocity is $\sqrt{2}$ times orbit velocity. In Sec. 14.5, it will be shown that the velocity necessary to maintain an object in a circular orbit at the surface of the earth is about 4.90 miles per second, which is approximately 18,000 miles per hour. This orbit velocity equals the escape velocity divided by $\sqrt{2}$.

Exercises

1. Translate into calculus the following news report about the leaning tower of Pisa. "The tower's angle from the vertical was increasing more rapidly." *Suggestion:* Let $\theta = f(t)$ be the angle of deviation from the vertical at time t.

    Incidentally, the tower, begun in 1174 and completed in 1350, is 179 feet tall and leans about 14 feet from the vertical. Each day it leans, on the average, another $\frac{1}{5000}$ inch.

2. Translate this news headline into calculus: "Gasoline prices increase more slowly." *Suggestion:* Let $G(t)$ be the price of a gallon of gasoline at time t.

Exercises 3 to 6 concern Example 4.

3. In Example 4 the origin of the y axis is at ground level. If the origin is located on top of the cliff, what would be the formulas for y and v as functions of t?
    (a) How long after the ball in Example 4 is thrown does it pass by the top of the cliff?
    (b) What are its speed and velocity then?

4. If the ball in Example 4 had simply been dropped from the cliff, what would y be as a function of time? How long would the ball fall?

5. In view of the result of Exercise 4 interpret physically each of the three terms on the right side of the formula $y = -16t^2 + 64t + 96$.

6. What is a possible physical interpretation of the solution $2 - \sqrt{10}$ in Example 4?

7. Show that, if $y = a \sin kt + b \cos kt$, where a and b and k are constants, then $D^2(y) = -k^2 y$. Incidentally, this describes the motion of a weight at the end of a spring, bobbing up and down, or any motion whose acceleration is proportional (with a negative constant of proportionality) to the displacement from a certain point. Such motion is called *harmonic*.

8. Let $y = (t - 1)^{2/3}$. Show that y satisfies the "differential equation"

$$\frac{d^2y}{dt^2} = -\frac{2}{9}\frac{1}{y^2}.$$

This differential equation says that the acceleration of y is inversely proportional to the square of y. It describes the motion of an object "coasting to infinity" away from the earth. Note that its velocity approaches 0.

9. At time $t = 0$ a particle is at $y = 3$ feet and has a velocity of -3 feet per second; it has a constant acceleration of 6 feet per second per second. Find its position at any time t.

10. At time $t = 0$ a particle is at $y = 10$ feet and has a velocity of 8 feet per second; it has a constant acceleration of -8 feet per second per second. (a) Find its position at any time t. (b) What is its maximum y coordinate?

11. At time $t = 0$ a particle is at $y = 0$ and has a velocity of 0 feet per second. Find its position at any time t if its acceleration is always -32 feet per second per second.

12. At time $t = 0$ a particle is at $y = -4$ feet and has a velocity of 6 feet per second; it has a constant acceleration of -32 feet per second per second. (a) Find its position at any time t. (b) What is its largest y coordinate?

∎

13. (a) Find all functions f whose second derivatives are 0 for all x.
    (b) Find all functions f such that $f(0) = 4$, $f'(0) = 5$, and $f^{(2)}(x) = 0$ for all x.

14 Find all functions whose second derivatives are $5x$ for all x.

15 Find all functions whose second derivatives are $\cos 3x$ for all x.

16 Find all functions f such that $d^3f/dx^3 = 6$ for all x.

17 Let $y = f(t)$ describe the motion on the y axis of an object whose acceleration has the constant value a. Show that

$$y = \frac{a}{2}t^2 + v_0 t + y_0,$$

where v_0 is the velocity when $t = 0$, and y_0 is the position when $t = 0$.

18 A car accelerates with constant acceleration from 0 (rest) to 60 miles per hour in 15 seconds. How far does it travel in this period? Be sure to do your computations either all in seconds or all in hours; for instance, 60 miles per hour is 88 feet per second.

19 The reaction time of a driver is about 0.6 second. If a car can decelerate at 16 feet per second per second, find the total distance covered if the car is braked at (*a*) 60 miles per hour, (*b*) 30 miles per hour, (*c*) 20 miles per hour.

20 Show that a ball thrown straight up from the ground takes as long to rise as to fall back to its initial position. How does the velocity with which it strikes the ground compare with its initial velocity? Consider the same question for its speed.

21 A mass at the end of a spring oscillates. At time t seconds its position (relative to its position at rest) is $y = 6 \sin t$ inches.
(*a*) Graph y as a function of t.

(*b*) What is the maximum displacement of the mass from its rest position?
(*c*) Show that its acceleration is proportional to its displacement y.
(*d*) Where is it when its speed is maximum?
(*e*) Where is it when the absolute value of its acceleration is maximum?

■ ■

Exercises 22 to 27 relate to Example 5, the launch of the payload.

22 If a payload is launched with the velocity of 7 miles per second, which is greater than its escape velocity, what happens to its velocity far out in its journey? In other words, determine $\lim_{r \to \infty} v$.

23 If we launch a payload with a speed of 6 miles per second, how far will it go from the center of the earth?

24 At what speed must we launch a payload if it is to reach the moon, 240,000 miles from the center of the earth? (Disregard the gravitational field of the moon.)

25 When a payload is launched with precisely the escape velocity, what happens to its velocity far out in its journey? In other words, determine $\lim_{r \to \infty} v$.

26 (Disregard air resistance.) In order to propel an object 100 miles straight up, what must the launching velocity be if it is assumed that (*a*) the force of gravity varies as in Example 5? (*b*) the force of gravity is constant?

27 Could it happen that a projectile shot straight out from the earth neither returns nor travels to "infinity" but approaches a certain finite limiting position? (Disregard other gravitational fields than the earth's.)

4.5 Applied maximum and minimum problems

One of the most important applications of calculus is obtaining the most efficient design of a product. Frequently the problem of minimizing cost or maximizing the volume of a certain object reduces to minimizing or maximizing some function $f(x)$. In that case, the methods developed in Secs. 4.2 and 4.3 may be called on. They are the use of critical points, the first-derivative test, and the second-derivative test. Recall that, when maximizing or minimizing a function over a closed interval, it is essential to consider also the values of the function at the endpoints.

The five examples that follow are typical. The only novelty is the challenge of how to translate each problem into the terminology of functions.

EXAMPLE 1 If we cut four congruent squares out of the corners of a square piece of cardboard 12 inches on each side, we can fold up the four remaining flaps to obtain a tray without a top. What size squares should be cut in order to maximize the volume of the tray?

4.5 APPLIED MAXIMUM AND MINIMUM PROBLEMS

SOLUTION Let us remove squares of side x, as shown in Figs. 4.44 and 4.45. Folding on the dotted lines, we obtain a tray of volume

$$V(x) = (12 - 2x)^2(x) = 4x^3 - 48x^2 + 144x.$$

Since each side of the cardboard square has length 12 inches, the only values of x which make sense are those in the closed interval $[0, 6]$. Thus, we wish to find the number x in $[0, 6]$ that maximizes $V(x)$.

Notice that $V(x) = (12 - 2x)^2(x)$ is small when x is near 0 (that is, when we try to economize by making the height of the tray small) and small when x is near 6 (that is, when we try to economize by making the base small). We have a "two-influence" problem; to find the best balance between them, we use calculus.

The maximum value of $V(x)$ for x in $[0, 6]$ occurs either at 0, 6, or at a critical number (where $V'(x) = 0$). Now, $V(0) = 0$ (the tray has height 0), and $V(6) = 0$ (the tray has a base of area 0). These are minimum values for the volume, certainly not the maximum volume, so the maximum must occur at some critical number in $(0, 6)$.

Next compute $V'(x)$:

$$\begin{aligned} V'(x) &= (4x^3 - 48x^2 + 144x)' \\ &= 12x^2 - 96x + 144 \\ &= 12(x^2 - 8x + 12) \\ &= 12(x - 6)(x - 2). \end{aligned}$$

The equation $\qquad 12(x - 6)(x - 2) = 0$

has two roots in $[0, 6]$, namely 2 and 6. The critical numbers are 2 and 6. As already remarked, the maximum does not occur at 0 or 6. Hence it occurs at $x = 2$. When $x = 2$, the volume is

$$\begin{aligned} V(2) &= [12 - 2 \cdot 2]^2(2) \\ &= 8^2 \cdot 2 \\ &= 128 \text{ cubic inches.} \end{aligned}$$

This is the largest possible volume and is obtained when the length of the cut is 2 inches.

As a matter of interest, let us graph the function V, showing its behavior for all x, not just for values of x significant to the problem. Note in Fig. 4.46 that at $x = 6$ the tangent is horizontal. ∎

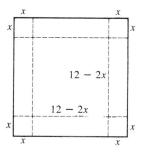

A piece of cardboard

Figure 4.44

Tray

Figure 4.45

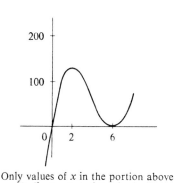

Only values of x in the portion above $[0, 6]$ correspond to physically realizable trays

Figure 4.46

EXAMPLE 2 A couple have enough wire to construct 100 feet of fence. They wish to use it to form three sides of a rectangular garden, one side of which is along a building, as shown in Fig. 4.47.

What shape garden should they choose in order to enclose the largest possible area?

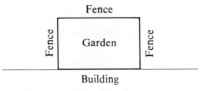

Fence | Garden | Fence
Building
The three sides of the garden not along the building total 100 feet.

Figure 4.47

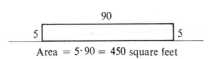

Figure 4.48

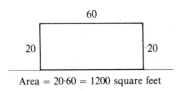

Figure 4.49

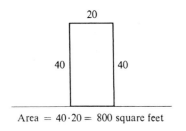

Figure 4.50

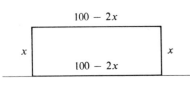

Figure 4.51

SOLUTION Figs. 4.48–4.50 show some possible ways of laying out the 100 feet of fence. For convenience, let x denote the length of the side of the garden that is perpendicular to the building, and y the length of the side parallel to the building. Since 100 feet of fencing is available, $2x + y = 100$. When $x = 5$, the area is 450 square feet. When x has increased to 20, y has decreased to 60 feet, and the area is 1200 square feet. When $x = 40$, $y = 20$, and the area is only 800 square feet. It is not immediately clear how to choose x to maximize the area of the garden. What makes the problem interesting is that, when you increase one dimension of the rectangle, the other automatically decreases. The area, which is the product of the two dimensions, is subject to two opposing forces, one causing it to increase, the other to decrease. This type of problem is easily solved with the aid of the derivative.

First of all, express the area A of the garden in terms of x and y:

$$A = xy.$$

Then use the equation $100 = 2x + y$ to express y in terms of x:

$$y = 100 - 2x.$$

Thus the area A is

$$A = x(100 - 2x) \quad \text{square feet.}$$

(See Fig. 4.51.)

Clearly $0 \leq x \leq 50$. Thus the problem now has become: Maximize $f(x) = x(100 - 2x)$ for x in $[0, 50]$.

In this case $f(x) = 100x - 2x^2$; hence

$$f'(x) = 100 - 4x.$$

Set the derivative equal to 0:

$$0 = 100 - 4x,$$

or

$$4x = 100;$$

hence

$$x = 25.$$

Thus 25 is the only critical number for the function. The maximum of f occurs either at 25 or at one of the ends of the interval, 0 or 50. Now,

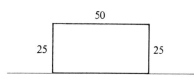

Figure 4.52

$$f(0) = 0[100 - 2 \cdot 0] = 0,$$
$$f(50) = 50[100 - 2 \cdot 50] = 0,$$
$$f(25) = 25[100 - 2 \cdot 25] = 1250$$

Thus the maximum possible area is 1250 square feet, and the fence should be laid out as shown in Fig. 4.52. ∎

Examples 1 and 2 illustrate the general procedure for solving applied maximum (or minimum) problems.

PROCEDURE FOR FINDING A MAXIMUM

1. Name the various quantities in the problem by letters, such as x, y, A, V.
2. Express the quantity to be maximized in terms of one or more other letters.
3. By eliminating variables, express the quantity to be maximized as a function of one variable.
4. Maximize the function obtained in step 3 by examining its derivative.

Example 3 also illustrates this procedure, as applied instead to a minimization problem.

EXAMPLE 3 Of all the tin cans that enclose a volume of 100 cubic inches, which requires the least metal?

SOLUTION Denote the radius of a can of volume 100 cubic inches by r, and its height by h. The can may be flat or tall. If the can is flat, the side uses little metal, but then the top and bottom bases are large. If the can is shaped like a mailing tube, then the two bases require little metal, but the curved side requires a great deal of metal. (See Fig. 4.53.) What is the ideal compromise between these two extremes?

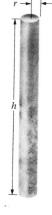
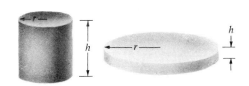

A tin can of volume 100 cubic inches; $\pi r^2 h = 100$. The can may be flat or long

Figure 4.53

The surface area S of the can is given by

$$S = 2\pi r^2 + 2\pi r h, \qquad (1)$$

which accounts for the two circular bases and the side. Since the amount of metal in the can is proportional to S, it suffices to minimize S.

In the tin can under consideration the radius and height are related by the constraint

$$\pi r^2 h = 100. \qquad (2)$$

In order to express S as a function of one variable, use Eq. (2) to eliminate either r or h. Choosing to eliminate h, we solve Eq. (2) for h:

$$h = \frac{100}{\pi r^2}.$$

Substitution into Eq. (1) yields

$$S = 2\pi r^2 + 2\pi r \frac{100}{\pi r^2},$$

or

$$S = 2\pi r^2 + \frac{200}{r}. \qquad (3)$$

Equation (3) expresses S as a function of just one variable r. The domain of this function for our purposes is $(0, \infty)$, since the tin can has a positive radius.

Compute dS/dr:

$$\frac{dS}{dr} = 4\pi r - \frac{200}{r^2} \qquad (4)$$

$$= \frac{4\pi r^3 - 200}{r^2}.$$

This derivative is 0 only when

$$4\pi r^3 = 200, \qquad (5)$$

that is, when

$$r = \sqrt[3]{50/\pi}.$$

Thus $r = \sqrt[3]{50/\pi}$ is the only critical number. Does it in fact provide a minimum?

First let us check by the second-derivative test. Differentiation of Eq. (1) yields

$$\frac{d^2S}{dr^2} = 4\pi + \frac{400}{r^3}.$$

When $r = \sqrt[3]{50/\pi}$, it follows that $r^3 = 50/\pi$. Thus, when $r = \sqrt[3]{50/\pi}$,

$$\frac{d^2S}{dr^2} = 4\pi + \frac{400\pi}{50}$$

$$= 12\pi.$$

Since d^2S/dr^2 is positive, and of course $dS/dr = 0$, we have a local minimum. But is it a global minimum?

Using the first derivative to test for global minimum

The first derivative will enable us to answer this question. Recall that

$$\frac{dS}{dr} = \frac{4\pi r^3 - 200}{r^2}.$$

At the critical number the numerator is 0. If r is less than the critical number, the numerator, hence the quotient, is *negative*. If r is larger than the critical number, the quotient is positive. Thus the function decreases for $0 < r < \sqrt[3]{50/\pi}$ and increases for $r > \sqrt[3]{50/\pi}$. Thus the critical number indeed provides an absolute or global minimum.

Using the second derivative to test for global minimum

The same conclusion could have been reached with the aid of the second derivative,

$$\frac{d^2S}{dr^2} = 4\pi + \frac{400}{r^3},$$

which is clearly positive for all positive r. A critical point on a curve that is concave upward everywhere is a global minimum for that curve. ∎

EXAMPLE 4 Find the dimensions of the rectangle of largest area that can be inscribed in a circle of radius a.

SOLUTION A circle of radius a and a typical inscribed rectangle are shown in Fig. 4.54. Label the lengths of the sides w and h, as shown in Fig. 4.55. The area of the typical rectangle is

$$A = hw. \tag{6}$$

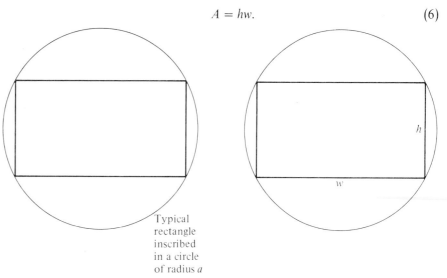

Figure 4.54   **Figure 4.55**

But h and w are linked to each other. (If h is small, w is large; if h is large, w is small.)

In order to express the formula $A = hw$ as a function of one variable,

To find the "missing equation" it may be necessary to draw the "missing line."

either h or w, it is necessary to find the relation between them. The relation must express the fact that the rectangle is inscribed in a circle of radius a.

The distance from any point P on the circle to its center C is a, as shown in Fig. 4.56. The radius CP is the hypotenuse of a right triangle whose two legs have lengths $w/2$ and $h/2$, as in Fig. 4.57. The Pythagorean theorem then provides an equation linking w and h:

$$\left(\frac{w}{2}\right)^2 + \left(\frac{h}{2}\right)^2 = a^2,$$

or, equivalently,
$$w^2 + h^2 = 4a^2. \tag{7}$$

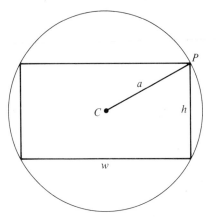

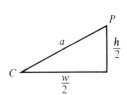

Figure 4.57

Figure 4.56

Solving (7) for h, say, gives
$$h = \sqrt{4a^2 - w^2}. \tag{8}$$

Taken together, (6) and (8) show how the area of the inscribed rectangle varies with w:
$$A(w) = hw = w\sqrt{4a^2 - w^2}. \tag{9}$$

The geometry is done; the calculus begins. $A(w)$ is defined and continuous for w in $[0, 2a]$. Thus $A(w)$ has a maximum either at 0, at $2a$, or at a critical number.

Maximizing $A(w)$ is equivalent to maximizing $(A(w))^2$. The latter problem avoids square roots, as you can check.

To find any critical numbers, differentiate $A(w)$:

$$A'(w) = w \cdot \frac{1}{2} \cdot \frac{(-2w)}{\sqrt{4a^2 - w^2}} + \sqrt{4a^2 - w^2}$$

$$= \frac{-w^2}{\sqrt{4a^2 - w^2}} + \sqrt{4a^2 - w^2}$$

$$= \frac{-w^2 + 4a^2 - w^2}{\sqrt{4a^2 - w^2}}$$

$$= \frac{4a^2 - 2w^2}{\sqrt{4a^2 - w^2}}. \tag{10}$$

For $A'(w)$ to be 0, the numerator of (10) must be 0:

$$4a^2 - 2w^2 = 0$$
$$2w^2 = 4a^2$$
$$w^2 = 2a^2$$
$$w = \sqrt{2}\,a.$$

(The negative square root of $2a^2$ lies outside the interval $[0, 2a]$.)

Thus the maximum value of $A(w)$ is the maximum of the three numbers $A(0)$, $A(2a)$, and $A(\sqrt{2}\,a)$. A quick computation using (9) or a glance at Fig. 4.56 shows that $A(0) = 0$ and $A(2a) = 0$. Thus the maximum must occur at $w = \sqrt{2}\,a$.

By (8), the corresponding h is

$$h = \sqrt{4a^2 - (\sqrt{2}a)^2}$$
$$= \sqrt{4a^2 - 2a^2}$$
$$= \sqrt{2a^2}$$
$$= \sqrt{2}\,a.$$

The rectangle of largest area inscribed in the circle is a square. ∎

Exercises

(Formulas for various volumes and areas are listed inside the cover.)

Exercises 1 to 4 are related to Example 1. In each case find the length of the cut that maximizes the volume of the tray. The dimensions of the cardboard are given.

1. 5 inches by 5 inches
2. 5 inches by 7 inches
3. 4 inches by 8 inches
4. 6 inches by 10 inches
5. Solve Example 2, expressing A in terms of y instead of x.
6. Solve Example 2 if there is instead enough wire to construct 160 feet of fence.
7. Solve Example 3, expressing S in terms of h instead of r.
8. Of all cylindrical tin cans without a top that contain 100 cubic inches, which requires the least material?
9. Of all enclosed rectangular boxes with square bases that have a volume of 1000 cubic inches, which uses the least material?
10. Of all topless rectangular boxes with square bases that have a volume of 1000 cubic inches, which uses the least material?
11. Solve Example 4, but express the area of the rectangle not in terms of w and h, but in terms of the angle θ shown in Fig. 4.58.

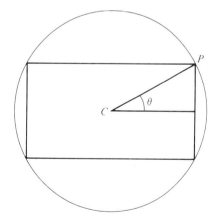

Figure 4.58

12. Find the dimensions of the rectangle of largest perimeter that can be inscribed in a circle of radius a.
13. Show that of all rectangles of a given perimeter, the square has the largest area. *Suggestion:* Call the fixed perimeter p and keep in mind that it is constant.

14 Show that of all rectangles of a given area, the square has the shortest perimeter. *Suggestion:* Call the fixed area A and keep in mind that it is constant.

15 A rancher wants to construct a rectangular corral. He also wants to divide the corral by a fence parallel to one of the sides. He has 240 feet of fence. What are the dimensions of the corral of largest area he can enclose?

16 A river has a 45° turn, as indicated in Fig. 4.59. A rancher wants to construct a corral bounded on two sides by the river and on two sides by 1 mile of fence ABC, as shown. Find the dimensions of the corral of largest area.

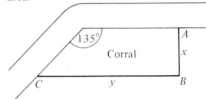

Figure 4.59

17 (a) How should one choose two nonnegative numbers whose sum is 1 in order to maximize the sum of their squares?
(b) To minimize the sum of their squares?

18 How should one choose two nonnegative numbers whose sum is 1 in order to maximize the product of the square of one of them and the cube of the other?

19 An irrigation channel made of concrete is to have a cross section in the form of an isosceles trapezoid, three of whose sides are 4 feet long. See Fig. 4.60. How should the trapezoid be shaped if it is to have the maximum possible area? Consider the area as a function of x and solve.

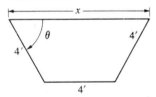

Figure 4.60

20 (a) Solve Exercise 19, expressing the area as a function of θ instead of x.
(b) Do the answers in (a) and Exercise 19 agree?

In Exercises 21 to 24 use the fact that the combined length and girth (distance around) of a package to be sent through the mails cannot exceed 100 inches.

21 Find the dimensions of the right circular cylinder of largest volume that can be sent through the mail.

22 Find the dimensions of the right circular cylinder of largest surface area that can be sent through the mail.

23 Find the dimensions of the rectangular box with square base of largest volume that can be sent through the mail.

24 Find the dimensions of the rectangular box with square base of largest surface area that can be sent through the mail.

Exercises 25 to 30 concern "minimal cost" problems.

25 A cylindrical can is to be made to hold 100 cubic inches. The material for its top and bottom costs twice as much per square inch as the material for its side. Find the radius and height of the most economical can. *Warning:* This is not the same as Example 3.
(a) Would you expect the most economical can in this problem to be taller or shorter than the solution to Example 3? (Use common sense, not calculus.)
(b) For convenience, call the cost of 1 square inch of the material for the side k cents. Thus the cost of 1 square inch of the material for the top and bottom is $2k$ cents. (The precise value of k will not affect the answer.) Show that a can of radius r and height h costs

$$C = 4k\pi r^2 + 2k\pi rh \quad \text{cents.}$$

(c) Find r that minimizes the function C in (b). Keep in mind during any differentiation that k is constant.
(d) Find the corresponding h.

26 A rectangular box with a square base is to hold 100 cubic inches. Material for the sides costs twice as much per square inch as the material for the top and bottom.
(a) If the base has side x and the height is y, what does the box cost?
(b) Find the dimensions of the most economical box.

27 A rectangular box with a square base is to hold 100 cubic inches. Material for the top costs 2 cents per square inch; material for the sides costs 3 cents per square inch; material for the bottom costs 5 cents per square inch. Find the dimensions of the most economical box.

28 The cost of operating a certain truck (for gasoline, oil, and depreciation) is $(15 + s/9)$ cents per mile when it travels at a speed of s miles per hour. A truck driver earns $9 per hour. What is the most economical speed at which to operate the truck during a 600-mile trip?
(a) If you considered only the truck, would you want s to be small or large?
(b) If you considered only the expense of the driver's wages, would you want s to be small or large?
(c) Express cost as a function of s and solve. (Be sure to put the costs all in terms of cents or all in terms of dollars.)
(d) Would the answer be different for a 1000-mile trip?

29 A government contractor who is removing earth from a large excavation can route trucks over either of two roads. There are 10,000 cubic yards of earth to move. Each truck holds 10 cubic yards. On one road the cost per truck load is $1 + 2x^2$ cents, where x trucks use that road; the function records the cost of congestion. On the other road the cost is $2 + x^2$ cents per truckload when x trucks use that road. How many trucks should be dispatched to each of the two roads?

30 On one side of a river 1 mile wide is an electric power station; on the other side, s miles upstream, is a factory. (See Fig. 4.61.) It costs 3 dollars per foot to run cable over land, and 5 dollars per foot under water. What is the most economical way to run cable from the station to the factory?

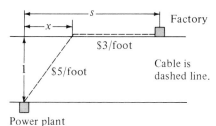

Figure 4.61

(a) Using no calculus, what do you think would be (approximately) the best route if s were very small? if s were very large?

(b) Solve with the aid of calculus, and draw the routes for $s = \frac{1}{2}, \frac{3}{4}, 1$, and 2.

(c) Solve for arbitrary s.

Warning: Minimizing the length of cable is not the same as minimizing its cost.

31 What are the dimensions of the right circular cylinder of largest volume that can be inscribed in a sphere of radius a?

32 The stiffness of a rectangular beam is proportional to the product of the width and the cube of the height of its cross section. What shape beam should be cut from a log in the form of a right circular cylinder of radius r in order to maximize its stiffness?

33 A rectangular box-shaped house is to have a square floor. Three times as much heat per square foot enters through the roof as through the walls. What shape should the house be if it is to enclose a given volume and minimize heat entry? (Assume no heat enters through the floor.)

34 (See Fig. 4.62.) Find the coordinates of the points $P = (x, y)$, with $y \leq 1$, on the parabola $y = x^2$, that (a) minimize $\overline{PA}^2 + \overline{PB}^2$, (b) maximize $\overline{PA}^2 + \overline{PB}^2$.

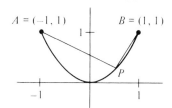

Figure 4.62

35 The speed of traffic through the Lincoln Tunnel in New York City depends on the density of the traffic. Let S be the speed in miles per hour and let D be the density in vehicles per mile. The relation between S and D was seen to be approximated closely by the formula

$$S = 42 - \frac{D}{3},$$

for $D \leq 100$.

(a) Express in terms of S and D the total number of vehicles that pass through the tunnel in an hour.

(b) What value of D will maximize the flow in (a)?

36 When a tract of timber is to be logged, a main logging road is built from which small roads branch off as feeders. The question of how many feeders to build arises in practice. If too many are built, the cost of construction would be prohibitive. If too few are built, the time spent moving the logs to the roads would be prohibitive. The formula for total cost,

$$y = \frac{CS}{4} + \frac{R}{VS},$$

is used in a logger's manual to find how many feeder roads are to be built. R, C, and V are known constants: R is the cost of road at "unit spacing"; C is the cost of moving a log a unit distance; V is the value of timber per acre. S denotes the distance between the regularly spaced feeder roads. (See Fig. 4.63.) Thus the cost y

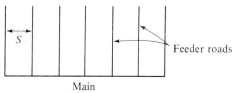

Figure 4.63

is a function of S, and the object is to find that value of S that minimizes y. The manual says, "To find the desired S set the two summands equal to each other and solve:

$$\frac{CS}{4} = \frac{R}{VS}.$$"

Show that the method is valid.

37 A delivery service is deciding how many warehouses to set up in a large city. The warehouses will serve similarly shaped regions of equal area A and, let us assume, an equal number of people.

(a) Why would transportation costs per item presumably be proportional to $\sqrt{A}$?

(b) Assuming that the warehouse cost per item is inversely proportional to A, show that C, the cost of transportation and storage per item, is of the form $t\sqrt{A} + w/A$, where t and w are appropriate constants.

(c) Show that C is a minimum when $A = (2w/t)^{2/3}$.

■ ■

38 A pipe of length b is carried down a long corridor of width $a < b$ and then around corner C. (See Fig. 4.64.) During the turn y starts out at 0, reaches a maximum, and then returns to 0. (Try this with a short stick.) Find that maximum in terms of a and b. *Suggestion:* Express y in terms of a, b, and θ; θ is a variable, while a and b are constants.

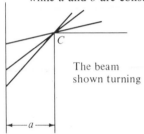
The beam shown turning
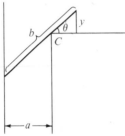

Figure 4.64

39 Fig. 4.65 shows two corridors meeting at a right angle. One has width 8; the other, width 27. Find the length of the longest pipe that can be carried horizontally from one hall, around the corner and into the other hall. *Suggestion:* Do Exercise 38 first.

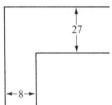

Figure 4.65

40 Two houses, A and B, are a distance p apart. They are distances q and r, respectively, from a straight road, and on the same side of the road. Find the length of the shortest path that goes from A to the road, and then on to the other house B.
(a) Use calculus.
(b) Use only elementary geometry. *Hint:* Introduce an imaginary house C such that the midpoint of B and C is on the road and the segment BC is perpendicular to the road; that is, "reflect" B across the road to become C.

41 The base of a painting on a wall is a feet above the eye of an observer, as shown in Fig. 4.66. The vertical side of the painting is b feet long. How far from the wall should the observer stand to maximize the angle that the painting subtends? *Hint:* It is more convenient to maximize $\tan \theta$ than θ itself. Recall that

$$\tan(A - B) = \frac{\tan A - \tan B}{1 + \tan A \tan B}.$$

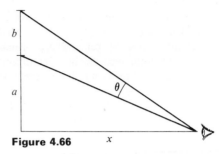

Figure 4.66

42 Find the point P on the x axis such that the angle APB in Fig. 4.67 is maximal. *Suggestion:* Note hint in Exercise 41.

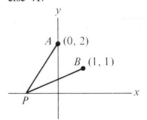

Figure 4.67

43 (*Economics*) Let p denote the price of some commodity and y the number sold at that price. To be concrete, assume that $y = 250 - p$ for $0 \leq p \leq 250$. Assume that it costs the producer $100 + 10y$ dollars to manufacture y units. What price p should the producer choose in order to maximize total profit, that is, "revenue minus cost"?

44 (*Leibniz on light*) A ray of light travels from point A to point B in Fig. 4.68 in minimal time. The point A is in

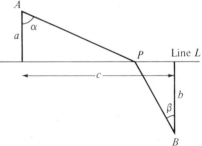

Figure 4.68

one medium, such as air or a vacuum. The point B is in another medium, such as water or glass. In the first medium light travels at velocity v_1 and in the second at velocity v_2. The media are separated by line L. Show that, for the path APB of minimal time,

$$\frac{\sin \alpha}{v_1} = \frac{\sin \beta}{v_2}.$$

Leibniz solved this problem with calculus in a paper published in 1684. (The result is called Snell's law of refraction.)

Leibniz then wrote, "other very learned men have

sought in many devious ways what someone versed in this calculus can accomplish in these lines as by magic." (See C. H. Edwards Jr., *The Historical Development of the Calculus*, p. 259, Springer-Verlag, New York.)

45 The following calculation occurs in an article by Manfred Kochen, "On Determining Optimum Size of New Cities.": The net utility to the total client-centered system is

$$U = \frac{RLv}{A} n^{1/2} - nK - \frac{ALc}{v} n^{-1/2}.$$

All symbols except U and n are constant; n is a measure of decentralization. Regarding U as a differentiable function of n, we can determine when $dU/dn = 0$. This occurs when

$$\frac{RLv}{2A} n^{-1/2} - K + \frac{ALc}{2v} n^{-3/2} = 0.$$

This is a cubic equation for $n^{-1/2}$.
(a) Check that the differentiation is correct.
(b) Of what cubic polynomial is $n^{-1/2}$ a root?

4.6 Implicit differentiation

Sometimes a function $y = f(x)$ is given indirectly by an equation that relates x and y. For instance, consider the equation

$$x^2 + y^2 = 25. \tag{1}$$

This equation can be solved for y:

$$y^2 = 25 - x^2,$$
$$y = \sqrt{25 - x^2},$$
or
$$y = -\sqrt{25 - x^2}.$$

There are thus two continuous functions that satisfy (1).
   The equation

$$x^2 + y^2 = 25$$

is said to describe the function $y = f(x)$ *implicitly*. The equations

$$y = \sqrt{25 - x^2} \quad \text{and} \quad y = -\sqrt{25 - x^2}$$

are said to describe the function $y = f(x)$ *explicitly*.

IMPLICIT DIFFERENTIATION

It is possible to differentiate a function given implicitly without having to solve for the function and express it explicitly. An example will illustrate the method, which is simply to differentiate both sides of the equation that defines the function implicitly. This procedure is called *implicit differentiation*.

EXAMPLE 1   Let $y = f(x)$ be the continuous function that satisfies the equation

$$x^2 + y^2 = 25$$

such that $y = 4$ when $x = 3$. Find dy/dx when $x = 3$ and $y = 4$.

APPLICATIONS OF THE DERIVATIVE

SOLUTION Differentiating both sides of the equation

$$x^2 + y^2 = 25$$

Note that the chain rule is used to find $\frac{d(y^2)}{dx}$. See Example 7 in Sec. 3.6.

with respect to x yields $\frac{d}{dx}(x^2 + y^2) = \frac{d}{dx}(25)$,

$$2x + 2y\frac{dy}{dx} = 0.$$

Hence

$$x + y\frac{dy}{dx} = 0.$$

In particular, when $x = 3$ and $y = 4$,

$$3 + 4\frac{dy}{dx} = 0,$$

and therefore

$$\frac{dy}{dx} = -\frac{3}{4}.$$

The problem could also be solved by differentiating $\sqrt{25 - x^2}$. But the algebra involved is more complicated, since it is necessary to differentiate a square root. ∎

In the next example implicit differentiation is the only way to find the derivative, for in this case there is no formula expressible in terms of trigonometric and algebraic functions giving y explicitly in terms of x.

EXAMPLE 2 Assume that the equation

$$2xy + \pi \sin y = 2\pi$$

defines a function $y = f(x)$. Find dy/dx when $x = 1$ and $y = \pi/2$. (Note that $x = 1$ and $y = \pi/2$ satisfy the equation.)

SOLUTION Implicit differentiation yields

$$\frac{d}{dx}(2xy + \pi \sin y) = \frac{d(2\pi)}{dx},$$

$$2\left(x\frac{dy}{dx} + y\frac{dx}{dx}\right) + \pi (\cos y)\frac{dy}{dx} = 0,$$

or

$$2x\frac{dy}{dx} + 2y + \pi (\cos y)\frac{dy}{dx} = 0.$$

For $x = 1$ and $y = \pi/2$ this last equation becomes

$$2 \cdot 1\frac{dy}{dx} + 2\frac{\pi}{2} + \pi \left(\cos \frac{\pi}{2}\right)\frac{dy}{dx} = 0$$

or

$$2\frac{dy}{dx} + 2\frac{\pi}{2} = 0.$$

Hence

$$\frac{dy}{dx} = -\frac{\pi}{2}. \quad \blacksquare$$

IMPLICIT DIFFERENTIATION AND EXTREMA

Example 3 of Sec. 4.5 answered the question, "Of all the tin cans that enclose a volume of 100 cubic inches, which requires the least metal?" The radius of the most economical can is $\sqrt[3]{50/\pi}$. From this and the fact that its volume is 100 cubic inches, its height could be calculated. In the next example implicit differentiation is used to answer the same question. Not only will the algebra be simpler than before, but the answer will provide more information, since also the general shape—the proportion between height and radius—is revealed. Before reading the next example, it would be instructive to read over the solution in Sec. 4.5.

EXAMPLE 3   Of all the tin cans that enclose a volume of 100 cubic inches, which requires the least metal?

SOLUTION   The height h and radius r of any can of volume 100 cubic inches are related by the equation

$$\pi r^2 h = 100. \tag{2}$$

The surface area S of the can is

$$S = 2\pi r^2 + 2\pi r h. \tag{3}$$

Consider h, and hence S, as functions of r. However, *it is not necessary to find these functions explicitly.*

Differentiation of (2) and (3) with respect to r yields

All we use about 100 is that it is a constant.

$$\pi\left(r^2 \frac{dh}{dr} + 2rh\right) = \frac{d(100)}{dr} = 0 \tag{4}$$

and

$$\frac{dS}{dr} = 4\pi r + 2\pi\left(r\frac{dh}{dr} + h\right). \tag{5}$$

Since when S is a minimum, $dS/dr = 0$, we have

$$0 = 4\pi r + 2\pi\left(r\frac{dh}{dr} + h\right). \tag{6}$$

Equations (4) and (6) yield, with a little algebra, a relation between h and r, as follows.

Factoring πr out of (4) and 2π out of (6) shows that

$$r\frac{dh}{dr} + 2h = 0$$

and

$$2r + r\frac{dh}{dr} + h = 0. \tag{7}$$

Elimination of dh/dr from (7) yields

$$2r + r\left(\frac{-2h}{r}\right) + h = 0,$$

which simplifies to

$$2r = h. \tag{8}$$

On your next visit to a supermarket, note the ratio of height to diameter of the cans.

Equation (8) asserts that the height of the most economical can is the same as its diameter. Moreover, this is the ideal shape, no matter what the prescribed volume happens to be. (Equation (4) follows from (2) merely because 100 is constant.)

The specific dimensions of the most economical can are found by combining the equations

$$2r = h \tag{8}$$

and
$$\pi r^2 h = 100. \tag{2}$$

Elimination of h from these two equations shows that

$$\pi r^2 (2r) = 100,$$

or
$$r^3 = \frac{50}{\pi};$$

hence
$$r = \sqrt[3]{\frac{50}{\pi}},$$

and
$$h = 2r = 2\sqrt[3]{50/\pi}. \quad \blacksquare$$

As in the case of Example 3, implicit differentiation finds the proportions of a general solution before finding the exact values of the variables. Often it is the proportion, rather than the (perhaps messier) explicit values, that gives more insight into the answer. For instance, Eq. (8) tells that the diameter equals the height for the most economical can.

The procedure illustrated in Example 3 is quite general. It may be of use when maximizing (or minimizing) a quantity that at first is expressed as a function of two variables which are linked by an equation. The equation that links them is called the *constraint*. In Example 3 the constraint is $\pi r^2 h = 100$.

The constraint

General procedure for using implicit differentiation in an applied maximum problem

---

HOW TO USE IMPLICIT DIFFERENTIATION IN AN EXTREMUM PROBLEM

1. Name the various quantities in the problem by letters, such as x, y, A, V.
2. Express the quantity to be maximized (or minimized) in terms of other letters, such as x and y.
3. Obtain an equation relating x and y. (This equation is called a constraint.)
4. Differentiate implicitly both the constraint and the expression to be maximized (or minimized), interpreting all the various quantities to be functions of x (or, perhaps, of y).
5. Set the derivative of the expression to be maximized (or minimized) equal to 0 and combine with the derivative of the constraint to obtain an equation relating x and y at a maximum (or minimum).
6. Step 5 gives only a relation or proportion between x and y at an extremum. If the explicit values of x and y are desired, find them by using the fact that x and y also satisfy the constraint.

Exercises

In Exercises 1 to 4 find dy/dx at the indicated values of x and y in two ways: explicitly (solving for y first) and implicitly.

1. $xy = 4$ at $(1, 4)$.
2. $x^2 - y^2 = 3$ at $(2, 1)$.
3. $x^2 y + xy^2 = 12$ at $(3, 1)$.
4. $x^2 + y^2 = 100$ at $(6, -8)$.

In Exercises 5 to 8 find dy/dx at the given points by implicit differentiation.

5. $\dfrac{2xy}{\pi} + \sin y = 2$ at $(1, \pi/2)$.
6. $2y^3 + 4xy + x^2 - 7$ at $(1, 1)$.
7. $x^5 + y^3 x + yx^2 + y^5 - 4$ at $(1, 1)$.
8. $x + \tan xy = 2$ at $(1, \pi/4)$.
9. Solve Example 3 by implicit differentiation, but differentiate (2) and (3) with respect to h instead of r.
10. What is the shape of the cylindrical can of largest volume that can be constructed with a given surface area? Do not find the radius and height of the largest can; find the ratio between them. *Suggestion:* Call the surface area S and keep in mind that it is constant.

In Exercises 11 to 16 solve by implicit differentiation:

11. Example 2 of Sec. 4.5.
12. Example 4 of Sec. 4.5.
13. Exercise 13 of Sec. 4.5.
14. Exercise 14 of Sec. 4.5.
15. Exercise 23 of Sec. 4.5.
16. Exercise 24 of Sec. 4.5.

Exercise 17 shows how to find y'' if y is given implicitly.

17. Assume that $y(x)$ is a differentiable function of x and that $x^3 y + y^4 = 2$. Assume that $y(1) = 1$. Find $y''(1)$, following these steps.
    (a) Show that $x^3 y' + 3x^2 y + 4y^3 y' = 0$.
    (b) Use (a) to find $y'(1)$.
    (c) Differentiate the equation in (a) to show that
    $$x^3 y'' + 6x^2 y' + 6xy + 4y^3 y'' + 12 y^2 (y')^2 = 0.$$
    (d) Use the equation in (c) to find $y''(1)$. *Hint:* $y(1)$ and $y'(1)$ are known.
18. Find $y''(1)$ if $y(1) = 2$ and $x^5 + xy + y^5 = 35$.
19. Find $y'(1)$ and $y''(1)$ if $y(1) = 0$ and $\sin y = x - x^3$.
20. Find $y''(2)$ if $y(2) = 1$ and $x^3 + x^2 y - xy^3 = 10$.

■ ■

Exercises 21 and 22 obtain by implicit differentiation the formulas for differentiating $x^{1/n}$ and $x^{m/n}$ with the assumption that they are differentiable functions.

21. Let n be a positive integer. Assume that $y = x^{1/n}$ is a differentiable function of x. From the equation $y^n = x$ deduce by implicit differentiation that $y' = (1/n)x^{1/n - 1}$.
22. Let m be a nonzero integer and n a positive integer. Assume that $y = x^{m/n}$ is a differentiable function of x. From the equation $y^n = x^m$ deduce by implicit differentiation that $y' = (m/n)x^{m/n - 1}$.

4.7 The differential

Let $f(x)$ denote the outdoor temperature measured in degrees at time x, measured in hours. At a certain time, say $x = 3$ P.M., the temperature is 70°. What is a reasonable estimate of the temperature a quarter of an hour later? If nothing more is known, the safest estimate is 70°. This estimate is essentially based on the assumption that the function is continuous: Its values do not fluctuate wildly. However, if in addition it is known that the temperature is increasing at 3 P.M. at the rate of 4° per hour, then a more accurate estimate can be made: A quarter of an hour later the temperature has probably increased about 1° and has become about 71°. This more refined estimate is essentially based on the assumption that the temperature function is differentiable and that its derivative remains constant (or almost constant) during the 15-minute interval.

The present section discusses this use of the derivative to estimate the change in a function, and thereby the resulting value of a function. The reasoning will be geometric, utilizing the interpretation of the derivative as slope. *The idea is that a very short piece of the graph around a point P, of a differentiable function, looks straight and closely resembles a short segment of the tangent line to the graph at P.* (See Fig. 4.69.) This suggests that the tangent line can be used to estimate the change in the functional value caused by a small change in x.

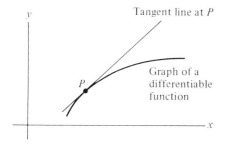

Figure 4.69

220 APPLICATIONS OF THE DERIVATIVE

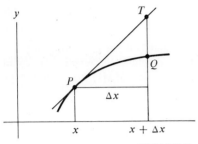

Figure 4.70

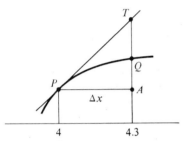

Figure 4.71

CHANGE ALONG THE TANGENT LINE

Let us look at the details. As the x coordinate changes from x to $x + \Delta x$, a point on the graph moves from P to Q, and a point on the tangent line to the graph at P from P to T. (See Fig. 4.70.) When Δx is small, the points T and Q are close together. Hence it seems reasonable that the y coordinate of T is a good estimate of $f(x + \Delta x)$, the y coordinate of Q, if Δx is small. Let us see if this is so in an example.

EXAMPLE 1 Let f be the square root function $f(x) = \sqrt{x}$. Let $x = 4$ and $\Delta x = 0.3$. In this case $P = (4, \sqrt{4}) = (4, 2)$. Compute the difference between the y coordinates of Q and T.

SOLUTION To aid the calculations, introduce the point A, at which the horizontal line through P meets the vertical line through T. Note in Fig. 4.71 that the y coordinate of Q is equal to $f(4) + \overline{AQ}$, where $\overline{AQ}$ is the change caused by moving from $x = 4$ to $x = 4.3$ on the graph; that is,

$$f(4.3) = f(4) + \overline{AQ} = f(4) + \Delta f.$$

(In this example $\overline{AQ}$ is positive; for other functions it may be negative.)
To compute $\overline{AT}$, observe that

$$\frac{\overline{AT}}{\overline{PA}} = \text{slope of tangent line at } P = f'(4).$$

Hence
$$\overline{AT} = f'(4) \cdot \overline{PA}$$
$$= f'(4) \cdot \Delta x$$
$$= f'(4) \cdot (0.3).$$

Since $f(x) = \sqrt{x}$, $\quad f'(x) = \dfrac{1}{2\sqrt{x}}.$

Thus $\quad f'(4) = \dfrac{1}{2\sqrt{4}} = \dfrac{1}{4} = 0.25.$

Consequently, $\overline{AT}$, the vertical change along the tangent line, is

$$(0.25)(0.3) = 0.075.$$

The y coordinate of T is therefore

$$2 + 0.075 = 2.075.$$

On the other hand, the y coordinate of Q is $\sqrt{4.3} \approx 2.0736$. Thus, the y coordinates of Q and T differ by very little, approximately

$$2.0736 - 2.075 = -0.0014.$$

This example shows that the y coordinate of T is an excellent approximation to the y coordinate of Q. ∎

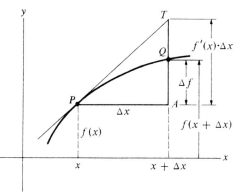

Figure 4.72

The method used in Example 1 applies to any differentiable function. The idea is to use vertical change along the tangent line to estimate vertical change along the graph.

The graph in Fig. 4.72 shows the situation for any differentiable function: The change along the tangent line is $\overline{AT}$. Since

$$\frac{\overline{AT}}{\Delta x} = \text{slope of tangent line at } P$$

$$= f'(x),$$

it follows that $\overline{AT} = f'(x)\,\Delta x.$

Hence $f'(x)\,\Delta x$ is a good estimate of Δf, the change along the graph, when Δx is small.

THE DIFFERENTIAL

The estimate $f'(x)\,\Delta x$ is of both practical and theoretical interest. For this reason it is given a name.

DEFINITION *Differential.* If f is a differentiable function, and x and Δx are numbers, the product $f'(x)\,\Delta x$ is called the *differential* of f at x. It is denoted by df. (If the notation $y = f(x)$ is used, the differential is also denoted dy.)

Thus $\qquad f(x + \Delta x) = f(x) + \Delta f \approx f(x) + df.$

In short $\qquad f(x + \Delta x) \approx f(x) + df$

or, more explicitly, $\quad f(x + \Delta x) \approx f(x) + f'(x)\,\Delta x.$

The differential df is a function of two variables x and Δx. For instance if $y = f(x) = \sqrt{x}$, then

$$df = \frac{1}{2\sqrt{x}}\,\Delta x,$$

or, in the dy notation, $\qquad dy = \dfrac{1}{2\sqrt{x}}\,\Delta x.$

Figure 4.73 shows Δx and $df = dy$. If the tangent lies below the graph, then the diagram appears as shown in Fig. 4.74 (and df underestimates Δf).

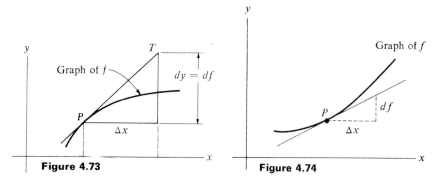

Figure 4.73   **Figure 4.74**

222 APPLICATIONS OF THE DERIVATIVE

In Example 1, the value of the differential was computed when $x = 4$ and $\Delta x = 0.3$. Let us see how close df is to Δf in another example.

EXAMPLE 2 Let $y = f(x) = x^3$, $x = 5$, $\Delta x = 0.1$. Compute df and Δf and compare.

SOLUTION In this case $f'(x) = 3x^2$ and $df = 3x^2 \, \Delta x$. When $x = 5$ and $\Delta x = 0.1$,

$$df = 3 \cdot 5^2 \cdot (0.1) = 7.5.$$

Now, Δf is defined as $f(5 + \Delta x) - f(5)$, that is,

$$(5.1)^3 - 5^3 = 132.651 - 125 = 7.651.$$

Again, df, which is 7.5, is a good estimate of Δf, which is 7.651. ∎

If the derivative of a function is known, so is its differential. For example,

$$d(\tan x) = \sec^2 x \, \Delta x$$
$$d(x^5) = 5x^4 \, \Delta x$$

and

$$d(x) = 1 \, \Delta x = \Delta x.$$

$dx = \Delta x$ Notice that $d(x) = \Delta x$. For this reason it is customary to write Δx also as dx. The differential of f, then, is also written as

$$df = f'(x) \, dx,$$

or

$$dy = f'(x) \, dx.$$

Thus we can also write that

$$d(\tan x) = \sec^2 x \, dx$$

and

$$d(x^5) = 5x^4 \, dx.$$

The origin of the symbol $\dfrac{dy}{dx}$ The symbols dy and dx now have meaning individually. It is meaningful to divide both sides of the equation

$$dy = f'(x) \, dx$$

by dx, obtaining $dy \div dx = f'(x).$

This is the origin of the symbol dy/dx for the derivative. It goes back to Leibniz at the end of the seventeenth century when dx denoted a number "vanishingly small," blasted by Bishop Berkeley in 1734 as "a ghost of a departed quantity." Now, however, dx denotes a number and dy is defined as $f'(x) \, dx$.

If f and g are two differentiable functions, then

$$d(f + g) = df + dg.$$

To show this, use the definition of differentials:

$$d(f + g) = (f + g)' \, dx$$
$$= (f' + g') \, dx$$
$$= f' \, dx + g' \, dx$$
$$= df + dg.$$

Similarly, $d(f - g) = df - dg$, $d(fg) = f \, dg + g \, df$, and

$$d\left(\frac{f}{g}\right) = \frac{g \, df - f \, dg}{g^2} \qquad g \neq 0.$$

(See Exercise 41.)

It will be instructive to compare dy and $\Delta y \, (= \Delta f)$ for various values of dx $(= \Delta x)$ both large and small. Let us use the function x^3 and $x = 5$. Then $f'(x) = 3x^2 = 75$. Hence

$$dy = 75 \, dx,$$

while
$$\Delta y = (5 + dx)^3 - 5^3$$
$$= 125 + 75 \, dx + 15(dx)^2 + (dx)^3 - 125$$
$$= 75 \, dx + 15(dx)^2 + (dx)^3.$$

Note that for small dx, the terms $15(dx)^2$ and $(dx)^3$ become extremely small in comparison with $75 \, dx$.

This table contains information about the cubing function $y = x^3$ for various choices of dx, when $x = 5$.

dx	dy	Δy
3	225	387
2	150	218
1	75	91
0.1	7.5	7.651
0.01	0.75	0.751501
0	0	0
-1	-75	-61

$dy = 75 \, dx$;

$\Delta y = (5 + dx)^3 - 5^3$.

Note that the smaller dx is, the better dy approximates Δy. Keep in mind that it is usually easier to compute dy than Δy.

APPLYING THE DIFFERENTIAL IN ESTIMATES

Examples 3 and 4 show how the differential can be used to estimate the output of a function or a change in output.

EXAMPLE 3 Use a differential to estimate $\sqrt{67}$.

SOLUTION The object is to estimate the value of the square root function $f(x) = \sqrt{x}$ at the input $x = 67$.

224 APPLICATIONS OF THE DERIVATIVE

In this case, $f(64)$ is known and so is $f'(64)$:

$$f(64) = \sqrt{64} = 8 \quad \text{and} \quad f'(64) = \frac{1}{2\sqrt{64}} = \frac{1}{16}.$$

Now, $\qquad 67 = 64 + 3,$

so Δx (or dx) is 3. Therefore,

$$\begin{aligned}\sqrt{67} = f(64 + 3) &\approx f(64) + df \\ &= f(64) + f'(64)(3) \\ &= 8 + \tfrac{1}{16}(3) \\ &= 8 + \tfrac{3}{16} \\ &= 8.1875.\end{aligned}$$

Thus $\qquad \sqrt{67} \approx 8.1875.$

(A calculator or square root table shows that to three decimal places, $\sqrt{67} \approx 8.185$. So the estimate obtained by the differential is not far off.) ∎

The method used in Example 3 amounts to the following general procedure.

How to use a differential to estimate an output of a function

---
TO ESTIMATE $f(b)$

1 Find a number a near b at which $f(a)$ and $f'(a)$ are easy to calculate.
2 Find $\Delta x = b - a$. (Δx may be positive or negative.)
3 Compute $f(a) + f'(a)\,\Delta x$. This is an estimate of $f(b)$.

EXAMPLE 4 The side of a cube is measured with an error of at most 1 percent. What percent error may this cause in calculating the volume of the cube?

SOLUTION Let x be the length of a side of the cube and V its volume. Let dx denote the possible error in measuring x. The relative error

$$\frac{dx}{x}$$

is at most 0.01 in absolute value. That is, $|dx|/x \leq 0.01$.

Estimating relative error

The differential dV is an estimate of the actual error in calculating the volume. Thus

$$\frac{dV}{V}$$

is an estimate of the relative error in the volume.

Since
$$dV = d(x^3)$$
$$= 3x^2\,dx,$$

it follows that
$$\frac{dV}{V} = \frac{3x^2\,dx}{x^3}$$
$$= 3\frac{dx}{x}.$$

Therefore the relative error in the volume is about three times the relative error in measuring the side, hence at most about 3 percent. ∎

THE UNDERLYING THEOREM

In what sense is the differential df a good estimate of the change Δf? After all, when Δx is small, both df and Δf are small. To claim that df differs very little from Δf when both are near 0, therefore, sheds no light. What is important is that their ratio is near 1 when Δx is small. This is proved in the following theorem.

THEOREM Let f be differentiable at a number x and assume that $f'(x) \neq 0$. Then

$$\lim_{\Delta x \to 0} \frac{\Delta f}{df} = 1.$$

PROOF Consider the quotient

$$\frac{\Delta f}{df} = \frac{\Delta f}{f'(x)\,\Delta x} = \frac{\Delta f}{\Delta x}\frac{1}{f'(x)}.$$

(Since $f'(x) \neq 0$, division is permissible.) Thus

$$\lim_{\Delta x \to 0} \frac{\Delta f}{df} = \lim_{\Delta x \to 0} \frac{\Delta f}{\Delta x}\frac{1}{f'(x)} = f'(x)\frac{1}{f'(x)} = 1.$$

This concludes the proof. ∎

Exercises

In Exercises 1 to 6, compare dy and Δy for the given functions and values of x and dx, and represent them on graphs of the functions.

1. x^2 at $x = 1$ and $dx = 0.3$.
2. x^3 at $x = \frac{1}{2}$ and $dx = 0.1$.
3. x^3 at $x = 1$ and $dx = -0.1$.
4. $\sqrt{x}$ at $x = 9$ and $dx = 0.6$.
5. $\sqrt[3]{x}$ at $x = 27$ and $dx = -3$.
6. $\sin x$ at $x = \pi/4$ and $dx = \pi/12$.
7. (a) Compute the differential of the function $1/x$ when $x = 1$ and $dx = 0.02$.
   (b) Use (a) to show that $1/1.02$ is approximately 0.98.
8. (a) Compute the differential of the function $1/x$ when $x = 1$ and $dx = h$.
   (b) Use (a) to show that, when h is small, $1 - h$ is a good estimate of $1/(1 + h)$.
9. (a) Compute the differential of the function $\sqrt{x}$ when $x = 1$ and $dx = h$.
   (b) Use (a) to show that, when h is small, $1 + h/2$ is a good estimate of $\sqrt{1 + h}$.
   (c) Compute $1 + h/2$ and $\sqrt{1 + h}$ when $h = 0.21$. By how much do they differ?
10. (a) Compute the differential of the function $(1 + x)^4$ when $x = 0$ and $dx = h$.

(b) Use (a) to show that, when h is small, $1 + 4h$ is a good estimate of $(1 + h)^4$.
(c) Compute $1 + 4h$ and $(1 + h)^4$ when $h = 0.01$. By how much do they differ?

11 Fill in this table for the function $y = x^3$ and the indicated values of x and dx.

x	dx	dy	Δy	$\Delta y/dy$
3	1			
3	-0.5			
1	0.1			
2	-0.1			

In Exercises 12 to 20 calculate the differentials, expressing them in terms of x and dx.

12 $d(\sqrt[3]{1 + x})$
13 $d(1/x^3)$
14 $d(\sqrt{1 + x^2})$
15 $d(\sin 2x)$
16 $d(\cos 5x)$
17 $d(\csc x)$
18 $d(\tan x^3)$
19 $d\left(\dfrac{\cot 5x}{x}\right)$
20 $d(x^3 \sec^2 5x)$

21 The side of a square is measured with an error of at most 5 percent. Estimate the largest percent error this may induce in the measurement of the area.

■

Exercises 22 to 39 concern the use of differentials in estimating functional values. Example 3 illustrates the technique. In each case estimate the given quantity.

22 $\sqrt{65}$
23 $\sqrt{61}$
24 $\sqrt{103}$
25 $\sqrt{98}$
26 $\sqrt[3]{28}$
27 $\sqrt[3]{25}$
28 $\tan\left(\dfrac{\pi}{4} + 0.01\right)$
29 $\tan\left(\dfrac{\pi}{4} - 0.01\right)$
30 $\sin\left(\dfrac{\pi}{3} + 0.02\right)$
31 $\sin\left(\dfrac{\pi}{3} - 0.02\right)$
32 $(1.03)^5$
33 $\sin 0.13$
34 $\sqrt{15.7}$
35 $1/4.03$

36 $\sin 2°$ (*Warning*: First translate into radians. Use $\pi \approx 3.14$.)
37 $\sin 32°$
38 $\tan 3°$
39 $\cos 28°$

40 Let $f(x) = x^2$, the area of a square of side x, shown in Fig. 4.75.

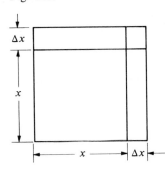

Figure 4.75

(a) Compute df and Δf in terms of x and Δx.
(b) In the square in the diagram, shade the part whose area is Δf.
(c) Shade the part of the square in (b) whose area is df.

41 Prove that, if f and g are two differentiable functions, then
(a) $d(f - g) = df - dg$;
(b) $d(fg) = f\,dg + g\,df$;
(c) $d\left(\dfrac{f}{g}\right) = \dfrac{g\,df - f\,dg}{g^2}$.

■ ■

42 Let y be a differentiable function of u, and u a differentiable function of x. Then $dy = D_u(y)\,du$ and $du = D_x(u)\,dx$. But y is a composite function of x, and one writes: $dy = D_x(y)\,dx$. Show that the two values of dy are equal.

43 (See Exercise 40.) Let $f(x) = x^3$, the volume of a cube of side x.
(a) Compute df and Δf in terms of x and Δx.
(b) Draw a diagram analogous to Fig. 4.75 showing cubes of sides x and $x + \Delta x$.
(c) Indicate in the diagram in (b) the part whose volume is df.

4.S Summary

This chapter applied the derivative in four areas: graphs, motion, extrema, and estimates.

Section 4.1 provided the foundation for the chapter. The theorem of the interior extremum showed that there is a close tie between a maximum or minimum value and a derivative being equal to 0, though the two conditions are not equivalent. The mean-value theorem, which followed from Rolle's theorem, showed that a function whose derivative is positive is increasing and that two functions that have the same derivative differ only by a constant.

Section 4.2 described a procedure for graphing functions which involves intercepts, asymptotes, and the first derivative. It is useful to know where the first derivative is 0, where it is positive, where it is negative, and where it changes sign.

The higher derivatives were introduced in Sec. 4.3, which also showed how the second derivative can be used in graphing.

The notions of concave upward, concave downward, inflection point, and inflection number were introduced. (In advanced texts "concave upward" is often called "convex" and "concave downward" is called "concave.")

Section 4.4 discussed the use of the second derivative to represent acceleration. In particular, motion under constant acceleration was described completely. (The position turns out to be a second degree polynomial in time t.)

Section 4.5, which contained no new mathematical ideas, applied the techniques for finding extrema to such problems as finding a maximum volume or a minimal cost.

Implicit differentiation and its use in certain extremum problems was the subject of Sec. 4.6.

The final section, Sec. 4.7, introduced the differential $dy = f'(x)\,dx$. It may be thought of as a function of x and dx that records change along a tangent line. It provides an estimate, dy, of the change, Δy, along the curve.

VOCABULARY AND SYMBOLS

theorem of the interior extremum
chord of f
Rolle's theorem
mean-value theorem
increasing (decreasing) function
critical number
critical point
relative (local) maximum or minimum
global maximum or minimum
first-derivative test for local extremum
second derivative d^2y/dx^2, y'', f'', $f^{(2)}$, $D^2(y)$

higher derivatives $d^n y/dx^n$, $f^{(n)}$, $D^n(f)$
acceleration
concave upward, concave downward
inflection point
inflection number
second-derivative test for local extremum
implicit function
implicit differentiation
constraint
differential dy, df

KEY FACTS

THEOREM OF THE INTERIOR EXTREMUM

("Maximum" case) Let f be defined at least on (a, b). Let c be a number in (a, b) such that $f(c) \geq f(x)$ for all x in (a, b). If f is differentiable at c, then $f'(c) = 0$.

The "minimum" case is similar.

ROLLE'S THEOREM

Let f be continuous on $[a, b]$ and differentiable on (a, b). If $f(a) = f(b)$, then there is at least one number c in (a, b) such that $f'(c) = 0$.

Informally, Rolle's theorem asserts that, if a graph of a differentiable function has a horizontal chord, then it has a horizontal tangent line.

MEAN-VALUE THEOREM

Let f be continuous on $[a, b]$ and differentiable on (a, b). Then there is at least one number c in (a, b) such that

$$f'(c) = \frac{f(b) - f(a)}{b - a}.$$

Informally, the mean-value theorem asserts that, for any chord on the graph of a differentiable function, there is a tangent line parallel to it.

The conclusion of the mean-value theorem may also be written as

$$f(b) = f(a) + f'(c)(b - a),$$

for some number c in (a, b).

INFORMATION PROVIDED BY f' AND f''

Where f' is positive, f is increasing.
Where f'' is positive, f is concave upward.
Where f' is negative, f is decreasing.
Where f'' is negative, f is concave downward.
Where $f' = 0$, f may have an extremum.
Where $f'' = 0$, f may have an inflection point.
First-derivative test for local maximum at c: $f'(c) = 0$ and f' changes from positive to negative at c.
Second-derivative test for local maximum at c: $f'(c) = 0$ and $f''(c)$ is negative.
First-derivative test for local minimum at c: $f'(c) = 0$ and f' changes from negative to positive at c.
Second-derivative test for local minimum at c: $f'(c) = 0$ and $f''(c)$ is positive.
(If $f'(c)$ or $f''(c)$ does not exist, it is best to study the behavior of $f(x)$ and $f'(x)$ for x near c.)

Two functions, defined over the same interval, with equal derivatives, differ by a constant. (From this it follows, for instance, that if $F'(x) = 2x$, then $F(x)$ must be of the form $x^2 + C$ for some constant C.) This fact will be of use in the next chapter where antiderivatives will be needed.

The second derivative records acceleration, the rate of change of velocity. If the acceleration is constant, the function is given by the formula,

$$y = \frac{a}{2} t^2 + v_0 t + y_0,$$

where a = acceleration, v_0 = initial velocity, and y_0 = initial positon.

In Sec. 4.3 a table summarized the general procedure for graphing a function, using information about f, f', and f''. That table is a useful review. The following table summarizes the procedure for graphing f in terms of questions about f, f', and f'':

How to graph using f, f', f''

What are the intercepts?
What are the critical numbers?
Where is the function increasing? decreasing?
Are there any local maxima or minima?
Where is the second derivative positive? negative? zero?
Where is the curve concave upward? concave downward?
Are there any inflection points?
Are there any vertical or horizontal asymptotes?
What happens when $|x| \to \infty$?

To find the derivative of a function given implicitly, differentiate the defining equation, remembering that the chain rule may be needed. Differentiation of the resulting equation will then give an equation for the second derivative.

The quantity, $df = f'(x)\Delta x$ (or $f'(x)dx$), the change along a tangent line, is an estimate of Δf, the change along the curve. When Δx is small, Δf and df will be small and their ratio will be near one.

If you know $f(a)$ and $f'(a)$ and if b is near a, then you can estimate $f(b)$ by

$$f(b) \approx f(a) + f'(a)(b - a).$$

This amounts to the same thing as the approximations

$$f(x + \Delta x) \approx f(x) + f'(x)\Delta x$$

or

$$f(x + \Delta x) \approx f(x) + df.$$

(There is another way of looking at the differential. Instead of thinking of $\Delta f/\Delta x$ as an approximation of $f'(x)$, think of $f'(x)$ as an approximation of $\Delta f/\Delta x$:

$$\frac{\Delta f}{\Delta x} \approx f'(x)$$

or $\Delta f \approx f'(x)\Delta x$.)

Guide quiz on chap. 4

1. (a) State all the assumptions in Rolle's theorem.
   (b) State the conclusion of Rolle's theorem.
2. (a) State all the assumptions in the mean-value theorem.
   (b) State the conclusion of the mean-value theorem.
3. What does each of the following imply about the graph of a function?
   (a) As you move from left to right, $f(x)$ changes sign at a from positive to negative.

(b) As you move from left to right, $f'(x)$ changes sign at a from positive to negative.
(c) As you move from left to right, $f''(x)$ changes sign at a from positive to negative.

4 Use the mean-value theorem to show that a function whose derivative is zero throughout an interval is constant.

5 (a) Prove that $\tan x - x$ is an increasing function of x, when $0 \leq x < \pi/2$.
(b) Deduce that $\tan x > x$ for x in $(0, \pi/2)$.
(c) From (b) obtain the inequality $x \cos x - \sin x < 0$, if x is in $(0, \pi/2)$.
(d) Prove that $\sin x/x$ is a decreasing function for x in $(0, \pi/2)$.

6 (a) Describe all functions whose derivatives equal $\sin 3x$.
(b) How are you sure that you have found all possibilities in (a)?

7 How should one choose two nonnegative numbers whose sum is 1 in order to minimize the sum of the square of one and the cube of the other?

8 A track of a certain length L is to be laid out in the shape of two semicircles at the ends of a rectangle, as shown in Fig. 4.76. Find the relative proportion of the radius of the circle r and the length of the straight section x if the track is to enclose a maximum area. Discuss also the case of minimum area.

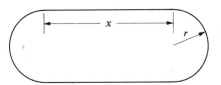

Figure 4.76

9 Using differentials, show that a good estimate (a) of $\sqrt{25 + dx}$ is $5 + dx/10$; (b) of $\sqrt{a^2 + dx}$ is $a + dx/2a$ (if a is positive). (c) Use the formula in (b) to estimate $\sqrt{65.6}$.

10 Graph $1/(x^2 - 3x + 2)$.

11 Find $y'(0)$ and $y''(0)$ if $y(0) = 2$ and $y^3 + x^2 y + x^3 = 8$.

12 Give all functions whose derivatives are
(a) $x^3 + 2x^2$ (b) $1/x^3$
(c) $5 \sin 2x$ (d) $\dfrac{x}{\sqrt{1 + x^2}}$

13 Use differentials to show that for small x, $\tan x \approx x$ (angle in radians).

14 Graph $f(x) = 3x^4 - 16x^3 + 24x^2$.

15 Find the fourth derivative of (a) $2x^5 - 1/x$ (b) $\cos 2x$ (c) $17x^3 - 5x + 2$ (d) $\sqrt{x}$

16 (a) How do you go about finding the global maximum of a differentiable function on a closed interval?
(b) Describe two different tests for a relative (local) minimum at a critical point.

Review exercises for chap. 4

1 Show that the equation $x^5 + 2x^3 - 2 = 0$ has exactly one solution in the interval $[0, 1]$. (Why does it have at least one? Use Rolle's theorem to show that there is at most one.)

2 Show that the equation $3 \tan x + x^3 = 2$ has exactly one solution in the interval $[0, \pi/4]$.

3 Let $f(x) = 1/x$.
(a) Show that $f'(x)$ is negative for all x in the domain of f.
(b) If $x_1 > x_2$, is $f(x_1) < f(x_2)$?

4 A rancher wishes to fence in a rectangular pasture 1 square mile in area, one side of which is along a road. The cost of fencing along the road is higher and equals 5 dollars a foot. The fencing for the other three sides costs 3 dollars a foot. What is the shape of the most economical pasture?

5 Graph $y = \dfrac{1}{x^2} + \dfrac{1}{x - 1}$.

6 Translate the following excerpt from a news article into the terminology of calculus.

With all the downward pressure on the economy, the first signs of a slowing of inflation seem to be appearing. Some sensitive commodity price indexes are down; the overall wholesale price index is rising at a slightly slower rate.

7 (a) Graph $y = \sqrt{x}$ for $0 \leq x \leq 5$.
(b) Compute dy for $x = 4$ and $dx = 1$.
(c) Compute Δy for $x = 4$ and $\Delta x = 1$.
(d) Using the graph in (a), show dy and Δy.

8 Fill in this table.

Interpretation of $f(x)$	Interpretation of $f'(x)$
The y coordinate in a graph of $y = f(x)$	
Total distance traveled up to time x	
Projection by lens of point x	
Size of population at time x	
Total mass of left x centimeters of string	
Velocity at time x	

9 Explain why each of these two proposed definitions of a tangent line is inadequate:
(1) A line L is tangent to a curve at a point P if L meets the curve only at P.

(2) A line L is tangent to a curve at a point P if L meets the curve at P and does not cross the curve at P.

10. A window is made of a rectangle and an equilateral triangle, as shown in Fig. 4.77. What should the dimensions be to maximize the area of the window if its perimeter is prescribed?

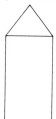

Figure 4.77

11. A wire of length L is to be cut into two pieces. One piece will be shaped into an equilateral triangle and the other into a square. How should the wire be cut in order to
    (a) minimize the sum of the areas of the triangle and square?
    (b) maximize the sum of the areas?

12. A square foot of glass is to be melted into two shapes. Some of it will be the thin surface of a cube. The rest will be the thin surface of a sphere. How much of the glass should be used for the cube and how much for the sphere if
    (a) their total volume is to be a minimum?
    (b) their total volume is to be a maximum?

13. If f is defined for all x, $f(0) = 0$, and $f'(x) \geq 1$ for all x, what is the most that can be said about $f(3)$? Explain.

14. (a) Using differentials, show that $\sqrt[3]{8 + h} \approx 2 + h/12$ when h is small.
    (b) What is the percent error when $h = 1$? $h = -1$?

15. For what value of the exponent a is the function $y = x^a$ a solution to the differential equation
$$\frac{dy}{dx} = -y^2?$$

16. Find all functions f such that
$$\frac{d^2 f}{dx^2} = x.$$

17. Graph $y = \sqrt{x}/(1 + x)$.
18. Graph $y = x^4 - 12x^3 + 54x^2$.
19. Show that the equation $2x^7 + 3x^5 + 6x + 10 = 0$ has exactly one real solution.
20. For each of the following give an example of a function whose domain is the x axis, such that
    (a) f has a global minimum at $x = 1$, but 1 is not a critical number.
    (b) f has an inflection point at $(0, 0)$, but $f''(0)$ is not defined.
    (c) $f'(2) = 0$, but f does not have a local extremum at $x = 2$.
    (d) $f''(2) = 0$, but 2 is not an inflection number.

21. Differentiate for practice:
    (a) $\dfrac{2x^3 - x}{x + 2}$ (b) $x^5 \sqrt{1 + 3x}$ (c) $\dfrac{(2x - 1)^5}{7}$
    (d) $\sin^4 \sqrt{x}$ (e) $\cos(1/x^3)$ (f) $\tan \sqrt{1 - x^2}$

22. A rectangular box with a square base is to be constructed. Material for the top and bottom costs a cents per square inch and material for the sides costs b cents per square inch.
    (a) For a given cost, what shape has the largest volume? (Express your answer in terms of the ratio between height and dimension of base.)
    (b) For a given volume, what shape is most economical?

23. A person can walk 3 miles per hour on grass and 5 miles per hour on sidewalk. She wishes to walk from point A to point B, shown in Fig. 4.78, in the least time.

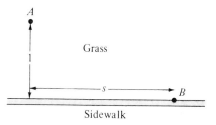

Figure 4.78

What route should she follow if s is (a) 1? (b) 2? (c) $\frac{4}{3}$?

24. Use a differential to estimate each of the following for small h.
    (a) $\sec\left(\dfrac{\pi}{3} + h\right)$ (b) $\sqrt[3]{1 + h^2}$ (c) $1/(1 - h)^2$

25. Use a differential to estimate each of the following.
    (a) $\sin 5°$ (First translate to radians.)
    (b) $(0.996)^3$

26. Find $y'(1)$ if $y(1) = 1$ and $\tan\left(\dfrac{\pi}{4} xy\right) + y^3 + x = 3$.

27. Let $f(x) = x \sin x$.
    (a) Show that $f(0) = f(\pi)$.
    (b) Use Rolle's theorem to show that there is a number c in $(0, \pi)$ such that $\tan c = -c$.
    (c) Graph the equations $y = -x$ and $y = \tan x$ to determine how many such numbers c there are. (Consider points where the two graphs intersect.)

28. The derivative of a certain function f is 5 when x is 2.
    (a) If $f(x)$ is the distance in feet that a rocket travels in x seconds, about how far does it travel from $x = 2$ to $x = 2.1$ seconds?
    (b) If $f(x)$ is the projection of x on a slide, about how long is the projection of the interval $[2, 2.1]$?
    (c) If $f(x)$ is the depth in feet that water penetrates the soil in the first x hours, how much does the water penetrate in the 6 minutes from 2 to 2.1 hours?

29. A certain function $y = f(x)$ has the property that
$$\frac{dy}{dx} = 3y^2.$$
Show that
$$\frac{d^2y}{dx^2} = 18y^3.$$

30. If dy/dx is proportional to x^2, show that d^2y/dx^2 is proportional to x.

31. If dy/dx is proportional to y^2, show that d^2y/dx^2 is proportional to y^3.

32. Using differentials, show that
$$\sin\left(\frac{\pi}{6} + h\right) \approx \frac{1}{2} + \frac{\sqrt{3}}{2} h$$
for small h.

33. Show that, if
$$\frac{dy}{dx} = 3y^4,$$
then
$$\frac{d^2y}{dx^2} = 36y^7.$$

34. Find the maximum value of $\sin^2 \theta \cos \theta$.

∎

35. What are the dimensions of the rectangle of largest area that can be inscribed in the ellipse $x^2/a^2 + y^2/b^2 = 1$? (Assume that the sides of the rectangle are parallel to the axes.)

36. Show that the equation $x^5 - 6x + 3 = 0$ has exactly three real roots.

37. Of all squares that can be inscribed in a square of side a what is the side of the one of smallest area?

38. What point on the parabola $y = x^2$ is closest to the point $(3, 0)$?

39. Let f be a differentiable function and let A be a point not on the graph of f. Show that if B is the point on the graph closest to A, then the segment AB is perpendicular to the tangent line to the curve at B.

40. Find the volume of the largest right circular cone that can be inscribed in a sphere of radius a.

41. In the theory of *inhibited growth* it is assumed that the growing quantity y approaches some limiting size M. Specifically one assumes that the rate of growth is proportional both to the amount present and to the amount left to grow:
$$\frac{dy}{dt} = ky(M - y).$$
Prove that the graph of y as a function of time has an inflection point when the amount y is exactly half the amount M.

42. (a) The area of a circle of radius r is πr^2. Use a differential to estimate the change in the area when the radius changes from r to $r + dr$.
    (b) The circumference of a circle of radius r is $2\pi r$. Explain why $2\pi r$ appears in the answer to (a).

43. The graph of a certain function is shown in Fig. 4.79. List the x coordinates of (a) relative maxima, (b) relative minima, (c) critical points, (d) global maximum, (e) global minimum.

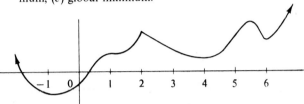

Figure 4.79

44. How should one choose two positive numbers whose product is 2 in order to minimize the sum of their squares?

45. The left-hand x centimeters of a string 12 centimeters long have a mass of $18x^2 - x^3$ grams.
    (a) What is its density x centimeters from the left-hand end?
    (b) Where is its density greatest?

46. Of all right circular cones with fixed volume V, which shape has the least surface area, including the area of the base? (The area of the curved part of a cone of slant height l and radius r is πrl.)

47. Of all right circular cones with fixed surface area A (including the area of the base), which shape has the largest volume?

48. What point on the line $y = 3x + 7$ is closest to the origin? (Instead of minimizing the distance, it is much more convenient to minimize the square of the distance. Doing so avoids square roots.)

49. (a) What is the maximum value of the function $y = 3 \sin t + 4 \cos t$?
    (b) What is the maximum value of the function $y = A \sin kt + B \cos kt$, where A, B, and k are nonzero constants?

50. Let $f(x) = (x - 1)^n(x - 2)$, where n is an integer, $n \geq 2$.
    (a) Show that $x = 1$ is a critical number.
    (b) For which values of n will $x = 1$ provide a relative maximum? a relative minimum? neither?

51. Let p and q be constants.
    (a) Show that, if $p > 0$, the equation $x^3 + px + q = 0$ has exactly one real root.
    (b) Show that, if $4p^3 + 27q^2 < 0$, the cubic equation $x^3 + px + q = 0$ has three distinct real roots.

52. Let m be a number ≥ 1. Prove that $(1 + x)^m > 1 + mx$ for any $x > 0$. This is known as Bernoulli's inequality. (*Note:* The binomial theorem in Appendix B implies this inequality for integer values of m.)

53 Consider the function $f(x) = x^3 + ax^2 + c$. Show that if $a < 0$ and $c > 0$, then f has exactly one negative root.

54 Let f have a derivative for all x.
  (a) Is every chord of the graph of f parallel to some tangent to the graph of f?
  (b) Is every tangent to the graph of f parallel to some chord of the graph of f?

55 Given an example of a function f such that $f(x)$ is defined for all x, f is differentiable and increasing, and yet the derivative of f is *not* positive for all x.

56 (a) Sketch the graph of $y = 1/x$.
  (b) Estimate by eye the point (x_0, y_0) on the graph closest to $(3, 1)$.
  (c) Show that x_0 is a solution of the equation $x^4 - 3x^3 + x - 1 = 0$.
  (d) Show that the equation in (c) has a root between 2.9 and 3.

■ ■

57 Show that, if $f'(x) \neq 0$, then dy is a good approximation to Δy when dx is small, in the sense that

$$\lim_{dx \to 0} \frac{\Delta y - dy}{dx} = 0.$$

58 A swimmer stands at a point A on the bank of a circular pond of diameter 200 feet. He wishes to reach the diametrically opposite point B by swimming to some point P on the bank and walking the arc PB along the bank. If he swims 100 feet per minute and walks 200 feet per minute, to what point P should he swim in order to reach B in the shortest possible time?

59 Let f be differentiable everywhere. Assume that $f'(a) = 0$ and $f'(b) = 1$. Prove that there is a number c, with $a < c < b$, such that $f'(c) = \tfrac{1}{2}$. *Warning:* f' need not be continuous.

60 In each case below decide if there is a function that meets all the conditions. If there is, sketch the graph of such a function. If there is none, indicate why not.
  (a) $f'(x) > 0$ and $f''(x) < 0$ for all x.
  (b) $f(x) > 0$ and $f''(x) < 0$ for all x.
  (c) $f(x) > 0$ and $f'(x) > 0$ for all x.

61 Is this proposed proof of the mean-value theorem correct? *Proof:* Tilt the x and y axes until the x axis is parallel to the given chord. The chord is now "horizontal," and we may apply Rolle's theorem.

62 An express subway train starts with a constant acceleration, then travels at its maximum speed, and finally slows down at a constant (negative) deceleration. It requires 120 seconds to go nonstop from 42d Street to 72d Street and 96 seconds to go from 72d Street to 96th Street.
  (a) Sketch the graph of speed as a function of time in each case.
  (b) What is the train's maximum speed? Assume that there are 20 blocks in a mile.

Exercises 63 to 67 concern tilted asymptotes, asymptotes that are neither horizontal nor vertical.

63 The line $y = ax + b$ is called an asymptote to the graph of the function f if

$$\lim_{x \to \infty} [f(x) - (ax + b)] = 0.$$

or if

$$\lim_{x \to -\infty} [f(x) - (ax + b)] = 0.$$

  (a) Show that

$$\lim_{x \to \infty} \left[\frac{2x^2 + 3x + 5}{x + 1} - (2x + 1) \right] = 0.$$

  (b) Show that

$$\lim_{x \to -\infty} \left[\frac{2x^2 + 3x + 5}{x + 1} - (2x + 1) \right] = 0.$$

  (c) Graph $f(x) = (2x^2 + 3x + 5)/(x + 1)$.

64 (a) Find a tilted asymptote to the graph of

$$f(x) = \frac{3x^2 + 2x + 3}{2x - 5}.$$

  (b) Graph f.

65 (a) Find a tilted asymptote to the graph of

$$f(x) = \sqrt{x^2 + 2x}.$$

  (b) Graph f.

66 Let a and b be positive constants. Let

$$f(x) = \frac{b}{a}\sqrt{x^2 - a^2}.$$

Show that the graph of f has a tilted asymptote.

67 Let $P(x)$ be a polynomial of degree $n \geq 1$ and let $Q(x)$ be a polynomial of degree $m \geq 1$. What relation must hold between m and n if the graph of $f(x) = P(x)/Q(x)$ has
  (a) the x axis as asymptote?
  (b) a horizontal asymptote other than the x axis?
  (c) a tilted asymptote?

The next three exercises are related.

68 Let f and g be differentiable functions such that $f(0) = g(0)$. Assume that for $x > 0$, $f'(x) > g'(x)$. Deduce that for $x > 0$, $f(x) > g(x)$. *Hint:* Corollary 3 in Sec. 4.1 may help.

69 (a) Starting with the inequalities

$$0 < \cos x < 1,$$

which are valid for all x in $(0, \pi/2)$, obtain the inequalities

$$0 < \sin x < x$$

for the same interval.

(b) Deduce from the inequalities obtained in (a) that
$$-1 < -\cos x < -1 + \frac{x^2}{2}$$
for x in $(0, \pi/2)$.

(c) Obtain, in order, the following inequalities for x in $(0, \pi/2)$:
$$-x < -\sin x < -x + \frac{x^3}{2 \cdot 3}$$
$$1 - \frac{x^2}{2} < \cos x < 1 - \frac{x^2}{2} + \frac{x^4}{4 \cdot 3 \cdot 2}$$
$$x - \frac{x^3}{3 \cdot 2} < \sin x < x - \frac{x^3}{3 \cdot 2} + \frac{x^5}{5 \cdot 4 \cdot 3 \cdot 2}.$$

70 Continuing Exercise 69, show that, for x in $(0, \pi/2)$,

(a) $1 - \dfrac{x^2}{2} + \dfrac{x^4}{4 \cdot 3 \cdot 2} - \dfrac{x^6}{6 \cdot 5 \cdot 4 \cdot 3 \cdot 2}$
$$< \cos x < 1 - \frac{x^2}{2} + \frac{x^4}{4 \cdot 3 \cdot 2}$$

(b) $x - \dfrac{x^3}{3 \cdot 2} + \dfrac{x^5}{5 \cdot 4 \cdot 3 \cdot 2} - \dfrac{x^7}{7 \cdot 6 \cdot 5 \cdot 4 \cdot 3 \cdot 2}$
$$< \sin x < x - \frac{x^3}{3 \cdot 2} + \frac{x^5}{5 \cdot 4 \cdot 3 \cdot 2}.$$

We see from (b) that if x is not too large, then
$$\sin x \approx x - \frac{x^3}{3!} + \frac{x^5}{5!}.$$
(We use the factorial notation.) In fact, if $|x| \le 1$, the error is at most $1/7! = \frac{1}{5040}$.

71 Does every polynomial of even degree $n \ge 2$ have at least one critical point? at least one inflection point? a global maximum or minimum?

72 Does every polynomial of odd degree $n \ge 3$ have at least one critical point? A global maximum or minimum?

Exercises 73 and 74 are related.

73 Let f be a function whose second derivative is of the form $f^{(2)}(x) = (x - a)^k g(x)$, where k is a positive integer, a is a fixed number, and g is a continuous function such that $g(a) \ne 0$. (a) Show that if k is odd, then a is an inflection number. (b) Show that if k is even, then a is not an inflection number.

74 Explain why a polynomial of odd degree at least 3 always has at least one inflection point. *Suggestion:* Let f be the polynomial and let $a_1, a_2, \ldots, a_j$ be the roots of its second derivative. Then $f^{(2)}(x)$ can be written as $(x - a_1)^{k_1}(x - a_2)^{k_2} \cdots (x - a_j)^{k_j} g(x)$, where g is a polynomial with no real roots.

75 Find the shape (the proportions) of the right circular cylinder of maximal surface area (including the top and bottom) inscribed in a sphere of radius a.

76 A ladder of length b leans against a wall of height a, $a < b$. What is the maximal horizontal distance that the ladder can extend beyond the wall if its base rests on the horizontal ground?

77 (a) Use the mean-value theorem to show that the function $f(x) = \sin x$ satisfies the inequality
$$|f(x_1) - f(x_2)| \le |x_1 - x_2|$$
for all numbers x_1 and x_2.

(b) Find all functions that satisfy the inequality
$$|f(x_1) - f(x_2)| \le |x_1 - x_2|^{1.01}$$
for all x_1 and x_2.

78 The potential energy in a diatomic molecule is given by the formula
$$U(r) = U_0 \left(\left(\frac{r_0}{r}\right)^{12} - 2 \left(\frac{r_0}{r}\right)^6 \right),$$
where U_0 and r_0 are constants and r is the distance between the atoms. For which value of r is $U(r)$ a minimum?

THE DEFINITE INTEGRAL

Chapters 3 and 4 were concerned with the derivative, which gives local information, such as the slope at a particular point on a curve or the velocity at a particular time. The present chapter introduces the second major concept of calculus, the definite integral.

In contrast to the derivative, the definite integral gives overall global information. For instance, if the velocity of a moving object is known at every instant, the definite integral of the velocity gives the total distance that the object moves. Also, the definite integral measures the area of the region bounded by a curve if a formula for the curve is given.

Surprisingly, the derivative turns out to be the main tool for evaluating definite integrals.

REFERENCE TO APPENDIX

In Sec. 5.3: the inequality

$$c^2(d-c) < \frac{d^3}{3} - \frac{c^3}{3} < d^2(d-c), \qquad 0 \le c < d,$$

Appendix B.

5.1 Estimates in four problems

Just as Chap. 3 introduced the derivative by four problems, this chapter introduces the definite integral by four problems. At first glance these problems may seem unrelated, but by the end of the section it will be clear that they represent one basic problem in various guises.

236 THE DEFINITE INTEGRAL

AN AREA PROBLEM

PROBLEM 1   Find the area of the region bounded by the curve $y = x^2$, the x axis, and the vertical line $x = 3$, as shown in Fig. 5.1.

An estimate for the area under $y = x^2$

An estimate of the area can be made using a staircase of six rectangles, as in Fig. 5.2.

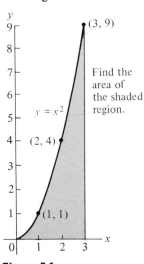

Find the area of the shaded region.

Figure 5.1

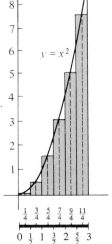

The area of the six rectangles approximates the area under the curve.

Figure 5.2

First break the interval from 0 to 3 into six smaller intervals, each of length $\frac{1}{2}$. Then above each small interval draw the rectangle whose height is that of the curve $y = x^2$ above the midpoint of that interval. The total area of the six rectangles is easily computed. This is equal to

$$\left(\tfrac{1}{4}\right)^2\left(\tfrac{1}{2}\right) + \left(\tfrac{3}{4}\right)^2\left(\tfrac{1}{2}\right) + \left(\tfrac{5}{4}\right)^2\left(\tfrac{1}{2}\right) + \left(\tfrac{7}{4}\right)^2\left(\tfrac{1}{2}\right) + \left(\tfrac{9}{4}\right)^2\left(\tfrac{1}{2}\right) + \left(\tfrac{11}{4}\right)^2\left(\tfrac{1}{2}\right),$$

which reduces to $\frac{286}{32}$ or 8.9375.

We have *not* computed the area. The preceding computation provides only an *estimate*, 8.9375, of the area.

A TOTAL MASS PROBLEM

PROBLEM 2   A thin nonuniform string 3 centimeters long is made of a material that is very light near one end and very heavy near the other end. In fact, at a distance of x centimeters from the left end it has a density of $5x^2$ grams per centimeter. Find the mass of the string, shown in Fig. 5.3.

An estimate for the mass of the nonuniform string

Let us cut the string into six sections of equal length, as in Fig. 5.4.

The density of the string in each of the six pieces varies less than it does over the whole length of the string. If the density were constant in a section, we would have, since

$$\text{Density} = \frac{\text{mass}}{\text{length}},$$

$$\text{Mass} = \text{density} \cdot \text{length}.$$

Figure 5.3

Figure 5.4

We shall treat each of the six sections as having throughout a constant density equal to that at its midpoint. To obtain an estimate of the mass of each of the six sections, let us multiply the density at the midpoint of the section by the length of the section.

The left section has a density of $5(\frac{1}{4})^2$ grams per centimeter at its midpoint, $\frac{1}{4}$, and thus has a mass of about $5(\frac{1}{4})^2(\frac{1}{2})$ gram. The next section, from $\frac{1}{2}$ to 1, has a density of $5(\frac{3}{4})^2$ at its midpoint, and thus has a mass of about $5(\frac{3}{4})^2(\frac{1}{2})$ grams. An estimate of the mass of each of the four other sections can be made similarly. An estimate of the total mass of the nonuniform string is then the sum

$$5(\tfrac{1}{4})^2(\tfrac{1}{2}) + 5(\tfrac{3}{4})^2(\tfrac{1}{2}) + 5(\tfrac{5}{4})^2(\tfrac{1}{2}) + 5(\tfrac{7}{4})^2(\tfrac{1}{2}) + 5(\tfrac{9}{4})^2(\tfrac{1}{2}) + 5(\tfrac{11}{4})^2(\tfrac{1}{2}).$$

This sum is five times the sum in Prob. 1 and hence equals 5(8.9375), a little less than 45 grams. More important is the similarity in form between this sum and the sum used in the first problem.

A TOTAL DISTANCE PROBLEM

PROBLEM 3 An engineer drives a car whose clock and speedometer work, but whose odometer (mileage recorder) is broken. On a 3-hour trip out of a congested city into the countryside she begins at a snail's pace and, as the traffic thins, she gradually speeds up. Indeed, she notices that after traveling t hours her speed is $8t^2$ miles per hour. Thus after the first $\frac{1}{2}$ hour she is crawling along at 2 miles per hour, but after 3 hours she is traveling at 72 miles per hour. How far does the engineer travel in 3 hours?

An estimate for the total distance

The speed during the 3-hour trip varies from 0 to 72 miles per hour. During shorter time intervals such a wide fluctuation will not occur.

As in the first two problems, cut the 3 hours of the trip into six equal intervals, each $\frac{1}{2}$ hour long, and use them to make an estimate of the total distance covered. Represent time by a line segment, cut into six parts of equal length, as in Fig. 5.5.

Figure 5.5

To estimate the distance the engineer travels in the first $\frac{1}{2}$ hour, multiply her speed at $\frac{1}{4}$ hour by the duration of the first interval of time, $\frac{1}{2}$ hour. Since her speed at time t is $8t^2$, after $\frac{1}{4}$ hour her speed is $8(\frac{1}{4})^2$ miles per hour. Thus during the first $\frac{1}{2}$ hour the engineer travels about $8(\frac{1}{4})^2(\frac{1}{2})$ mile. During the second $\frac{1}{2}$ hour she travels about $8(\frac{3}{4})^2(\frac{1}{2})$ miles.

Making similar estimates for each of the other $\frac{1}{2}$-hour periods, we obtain this estimate for the length of the trip:

$$8(\tfrac{1}{4})^2(\tfrac{1}{2}) + 8(\tfrac{3}{4})^2(\tfrac{1}{2}) + 8(\tfrac{5}{4})^2(\tfrac{1}{2}) + 8(\tfrac{7}{4})^2(\tfrac{1}{2}) + 8(\tfrac{9}{4})^2(\tfrac{1}{2}) + 8(\tfrac{11}{4})^2(\tfrac{1}{2}).$$

This sum is eight times the sum in Prob. 1, hence equals

$$8 \cdot (8.9375) = 71.5 \text{ miles}.$$

Keep in mind that this is only an estimate of the length of the trip.

A VOLUME PROBLEM

PROBLEM 4 Find the volume of a tent with a square floor of side 3 feet, whose pole, 3 feet long, rises above a corner of the floor. The tent is shown in Fig. 5.6. It can also be thought of as the surface obtained when the piece of paper, shown in Fig. 5.7, is folded along the dotted lines and the free edges taped in such a way that A, B, C, and D come together (to become P).

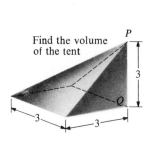

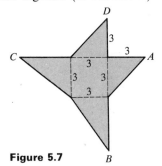

Figure 5.6   **Figure 5.7**

An estimate for the volume

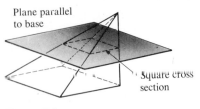

Figure 5.8

Observe that the cross section of the tent made by any plane parallel to the base is a square, as shown in Fig. 5.8.

This time we cut a vertical line, representing the pole, into six sections of equal length. Then we approximate each slab by a flat rectangular box, $\frac{1}{2}$ foot high. The cross section of the smallest box is obtained by passing a horizontal plane through the midpoint of the highest of the six sections. The remaining five boxes are determined in a similar manner, as shown in Figs. 5.9 and 5.10. As the side view of the six boxes shows, the square cross section of the top box has a side equal to $\frac{1}{4}$ foot, the box below it has a side equal to $\frac{3}{4}$ foot, and so on until we reach the bottom box, whose side is $\frac{11}{4}$ feet.

Since the volume of a box is just the area of its base times its height, the total volume of the six boxes is

$$(\tfrac{1}{4})^2(\tfrac{1}{2}) + (\tfrac{3}{4})^2(\tfrac{1}{2}) + (\tfrac{5}{4})^2(\tfrac{1}{2}) + (\tfrac{7}{4})^2(\tfrac{1}{2}) + (\tfrac{9}{4})^2(\tfrac{1}{2}) + (\tfrac{11}{4})^2(\tfrac{1}{2}) \text{ cubic feet.}$$

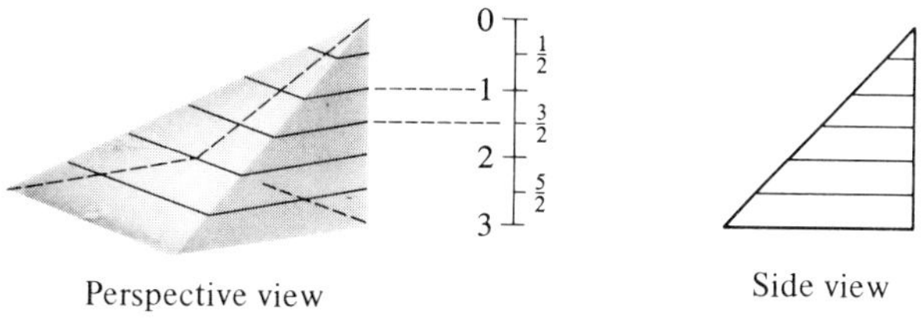

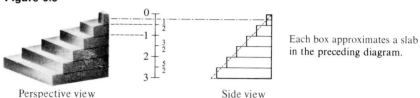

Perspective view   Side view

Figure 5.9

Perspective view   Side view

Each box approximates a slab in the preceding diagram.

Figure 5.10

This is the same sum that was met in estimating the area under the curve $y = x^2$. Thus the volume of the tent is estimated as 8.9375 cubic feet.

None of the four problems is yet solved; in each case all we have is an estimate. In the next section the precise answers will be found.

Exercises

To show the similarity of the four problems, the interval [0, 3] has been cut into six sections each time, and the midpoint of each section used to determine the cross section, density, speed, or area. Of course, we are free to cut the interval into more or fewer sections and to use a point other than the midpoint in each section. Furthermore, there is no need to restrict the sections to be of equal length. Exercises 1 to 5 concern other estimates for the same problems.

1. (a) Estimate the area in Prob. 1 by using three sections, each of length 1, and the midpoint each time.
   (b) Estimate the same area by using the same three sections, but now use the y coordinate of the point on $y = x^2$ above the right end of each section to determine the rectangle.
   (c) Draw the three rectangles used in (b). Is their total area more or less than the area under the curve?
   (d) Proceed as in (b) and (c), but use the point on $y = x^2$ above the left endpoint of each section.
   (e) Using information gathered in (b) and (d), complete this sentence: The area in Prob. 1 is certainly less than _____ but larger than _____.

2. Cutting the interval from 0 to 3 into five sections of equal length, estimate the area in Prob. 1 by finding the sum of the areas of five rectangles whose heights are determined by
   (a) midpoints,
   (b) right endpoints,
   (c) left endpoints.
   (d) Using information gathered in (b) and (c), complete this sentence: The area in Prob. 1 is certainly less than _____ but larger than _____.

3. Estimate the mass of the string in Prob. 2 by cutting it into five sections of equal length. For an estimate of the mass of each of these sections use the density at (a) the midpoint of each section, (b) the right endpoint, (c) the left endpoint. (d) On the basis of (b) and (c), the mass in Prob. 2 is less than _____ but larger than _____.

4. Cutting the interval of 3 hours into five periods of $\frac{3}{5}$ hour each, estimate the length of the engineer's trip in Prob. 3. For the approximate velocity in each period use the speedometer reading at (a) the middle of the period, (b) the end of the period, (c) the beginning of the period. (d) In view of (b) and (c) the length of the trip is less than _____ but larger than _____.

5. Make an estimate for each of the four problems, using in each case the accompanying division into four sections. As the points where the cross section, density, or velocity is computed, use $\frac{1}{2}$, $\frac{3}{2}$, 2, and $\frac{14}{5}$ (one of these is in each of the four sections). See Fig. 5.11.

Figure 5.11

6 Estimate the area between the curve $y = x^3$, the x axis, and the vertical line $x = 6$ using a division into (a) three sections of equal length and midpoints; (b) six sections of equal length and midpoints; (c) six sections of equal length and left endpoints; (d) same as (c) but right endpoints.

7 (Calculator) Estimate the volume of the tent in Prob. 4 by cutting [0, 3] into ten sections of equal length and taking cross sections at (a) left endpoints, (b) right endpoints.

8 (Calculator) Estimate the area under $y = x^2$ and above [0, 1] by cutting [0, 1] into ten sections of equal length and using (a) left endpoints, (b) right endpoints.

∎

9 A business which now shows no profit is to increase its profit flow gradually in the next 3 years until it reaches a rate of 9 million dollars per year. At the end of the first half year the rate is to be $\frac{1}{4}$ million dollars per year; at the end of 2 years, 4 million dollars per year. In general, at the end of t years, where t is any number between 0 and 3, the rate of profit is to be t^2 million dollars per year. Estimate the total profit during the next 3 years if the plan is successful. Use six intervals of equal length and midpoints.

10 Estimate the area under $y = x^2$ and directly above the interval [1, 5] by the midpoint method with the aid of a partition of [1, 5] into (a) four sections of equal length, (b) eight sections of equal length.

11 A right circular cone has a height of 3 feet and a radius of 3 feet, as shown in Fig. 5.12. Estimate its volume by the sum of the volumes of six cylindrical slabs, just as we estimated the volume of the tent with the aid of six rectangular slabs. In particular, (a) show with the aid of a diagram how the same sections and midpoints we used determine six cylinders, and (b) compute their total volume.

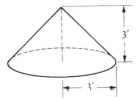

Right circular cone of height 3 feet and radius 3 feet

Figure 5.12

12 (Calculator or table in Appendix H) Estimate the area of the region under the curve $y = \sin x$ and above the interval $[0, \pi/2]$, cutting the interval as shown in Fig. 5.13 and using (a) left endpoints, (b) right endpoints. All but the last section are of the same size.

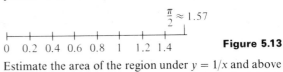

Figure 5.13

13 Estimate the area of the region under $y = 1/x$ and above [1, 2], using five sections of equal length and (a) left endpoints, (b) right endpoints.

14 Differentiate for practice:

(a) $(1 + x^2)^{4/3}$

(b) $\dfrac{(1 + x^3) \sin 3x}{\sqrt[3]{5x}}$

(c) $\dfrac{3x}{8} + \dfrac{3x \sin 4x}{32} + \dfrac{\cos^3 2x \sin 2x}{8}$

(d) $\dfrac{3}{8(2x + 3)^2} - \dfrac{1}{4(2x + 3)}$

(e) $\dfrac{\cos^3 2x}{6} - \dfrac{\cos 2x}{2}$

(f) $x^3 \sqrt{x^2 - 1} \tan 5x$

15 Give an example of a function F whose derivative is

(a) $(x + 2)^3$, (b) $(x^2 + 1)^2$, (c) $x \sin x^2$,

(d) $x^3 + \dfrac{1}{x^3}$, (e) $\dfrac{1}{\sqrt{x}}$.

∎ ∎

16 The kinetic energy of an object, for example, a bullet or car, of mass m and speed v is defined as $mv^2/2$ ergs. (Here mass is measured in grams and speed in centimeters per second.) Now, in a certain machine a uniform rod, 3 centimeters long and weighing 32 grams, rotates once per second around one of its ends. Estimate the kinetic energy of this rod by cutting it into six sections, each $\frac{1}{2}$ centimeter long, and taking as the "speed of a section" the speed of its midpoint.

17 Draw an accurate graph of $y = x^2$ and the six rectangles with heights equal to the ordinates of the curve at the midpoints that we used to estimate the area. Does each of these rectangles underestimate or overestimate the area under $y = x^2$ and above the base of the rectangle? (Form your opinion on the basis of your drawing.)

18 This exercise concerns the area of the region under the curve $1/x$.

(a) Estimate the area under $y = 1/x$ and above [1, 2], using five sections of equal length and left endpoints.

(b) Estimate the area under $y = 1/x$ and above [3, 6], using five sections of equal length and left endpoints.

(c) The answers to (a) and (b) are the same. Would

they be the same if you used 100 sections of equal length (instead of 5)? Explain.
(d) Write a short paragraph explaining why the area under $y = 1/x$ above $[1, 2]$ equals the area under $y = 1/x$ above $[3, 6]$.
(e) Let a and b be numbers greater than 1. Explain why the area under $1/x$ and above $[1, a]$ equals the area under $1/x$ and above $[b, ab]$.
(f) For $t > 1$, let $G(t)$ equal the area under the curve $1/x$ and above $[1, t]$. Show that, for a and b greater than 1, $G(ab) = G(a) + G(b)$.
(g) What function f studied in precalculus resembles the function G in that $f(ab) = f(a) + f(b)$?

5.2 Summation notation and approximating sums

In Sec. 5.1 sums of a particular form were formed and computed. Such sums play an essential role in the theory of the definite integral, developed in this chapter, and in applications in several later chapters. Since such sums will be needed often, let us introduce a convenient notation for them, the so-called *sigma notation*, named after the Greek letter Σ, which corresponds to the S of Sum.

$\sum$ *notation or summation notation*

SIGMA NOTATION

The sigma notation is useful in dealing with sums in which the summands all have the same general form.

DEFINITION *Sigma notation.* Let $a_1, a_2, \ldots, a_n$ be n numbers. The sum $a_1 + a_2 + \cdots + a_n$ will be denoted by the symbol $\sum_{i=1}^{n} a_i$ or $\sum_{i=1}^{n} a_i$, which is read as "the sum of a sub i as i goes from 1 to n."

In the sigma notation the formula for the typical summand is given, as is a description of where the summation starts and ends.

EXAMPLE 1 Write the sum $1^2 + 2^2 + 3^2 + 4^2$ in the sigma notation.

SOLUTION Since the ith summand is the square of i and the summation extends from $i = 1$ to $i = 4$, we have

$$1^2 + 2^2 + 3^2 + 4^2 = \sum_{i=1}^{4} i^2.$$

Simple arithmetic shows that the sum is equal to 30. ∎

EXAMPLE 2 Compute $\sum_{i=1}^{3} 2^i$.

SOLUTION This is short for the sum $2^1 + 2^2 + 2^3$, which is $2 + 4 + 8$, or 14. ∎

In the definition of the sigma notation the letter i (for "index") was used. Any letter, such as j or k, would do just as well. Such an index is sometimes called a *summation index* or *dummy index*.

EXAMPLE 3 Compute $\sum_{j=1}^{4} \frac{1}{j}$.

SOLUTION This is short for $\frac{1}{1} + \frac{1}{2} + \frac{1}{3} + \frac{1}{4}$, which is approximately 2.083. ∎

The summation notation has two properties which will be of use in coming chapters. First of all, if c is a fixed number, then

$$\sum_{i=1}^{n} ca_i = ca_1 + ca_2 + \cdots + ca_n = c(a_1 + a_2 + \cdots + a_n) = c\sum_{i=1}^{n} a_i.$$

Two useful facts:

$$\sum_{i=1}^{n} ca_i = c \sum_{i=1}^{n} a_i$$

and

$$\sum_{i=1}^{n} (a_i + b_i) = \sum_{i=1}^{n} a_i + \sum_{i=1}^{n} b_i$$

Thus

$$\sum_{i=1}^{n} ca_i = c \sum_{i=1}^{n} a_i.$$

This distributive rule is read as "a constant factor can be moved past $\sum$." Second,

$$\sum_{i=1}^{n} (a_i + b_i) = (a_1 + b_1) + (a_2 + b_2) + \cdots + (a_n + b_n)$$
$$= (a_1 + a_2 + \cdots + a_n) + (b_1 + b_2 + \cdots + b_n)$$
$$= \sum_{i=1}^{n} a_i + \sum_{i=1}^{n} b_i.$$

This is a direct consequence of the rules of algebra.

EXAMPLE 4 Compute $\sum_{i=1}^{4} \left(i^2 + \frac{1}{i} \right)$.

SOLUTION This may be rewritten as $\sum_{i=1}^{4} i^2 + \sum_{i=1}^{4} \frac{1}{i}$.

By Examples 1 and 3, the sum, to three decimals, is $30 + 2.083 = 32.083$. ∎

EXAMPLE 5 What is the value of $\sum_{i=1}^{5} 3$?

SOLUTION In this case $a_i = 3$ for each index i. Each summand has the value 3. Thus

$$\sum_{i=1}^{5} 3 = 3 + 3 + 3 + 3 + 3 = 15.$$

$$\sum_{i=1}^{n} c = cn$$

More generally, if c is a fixed number not depending on i, then $\sum_{i=1}^{n} c = cn$. ∎

The next example shows how to interpret the sigma notation when the index does not start at 1.

EXAMPLE 6 Compute $\sum_{i=2}^{6} 5i$ (read as "the sum of $5i$ as i goes from 2 to 6").

SOLUTION This is short for $5 \cdot 2 + 5 \cdot 3 + 5 \cdot 4 + 5 \cdot 5 + 5 \cdot 6,$

which equals $5(2 + 3 + 4 + 5 + 6),$

or 100. ∎

A useful fact to note is that

$$\sum_{i=1}^{m} a_i + \sum_{i=m+1}^{n} a_i = \sum_{i=1}^{n} a_i, \qquad 1 < m < n.$$

For instance, $\sum_{i=1}^{4} i^2 + \sum_{i=5}^{10} i^2 = \sum_{i=1}^{10} i^2.$

EXAMPLE 7 Let b_0, b_1, b_2, b_3 be four numbers. Form the three differences

$$a_1 = b_1 - b_0, \qquad a_2 = b_2 - b_1, \qquad a_3 = b_3 - b_2.$$

Show that $\sum_{i=1}^{3} a_i = b_3 - b_0.$

SOLUTION $\sum_{i=1}^{3} a_i = a_1 + a_2 + a_3 = (b_1 - b_0) + (b_2 - b_1) + (b_3 - b_2).$

Cancellations of b_1 and $-b_1$ and of b_2 and $-b_2$ show that

$$a_1 + a_2 + a_3 = b_3 - b_0. \qquad ∎$$

Example 7 can easily be generalized from four to any finite list of numbers. If $b_0, b_1, \ldots, b_n$ are $n + 1$ numbers, and $a_i = b_i - b_{i-1}$, $i = 1, \ldots, n$, then

$$\sum_{i=1}^{n} a_i = \sum_{i=1}^{n} (b_i - b_{i-1}) = b_n - b_0.$$

Telescoping sums In short, the sum $\sum_{i=1}^{n} (b_i - b_{i-1})$ "telescopes" to $b_n - b_0$.

SIGMA NOTATION FOR THE APPROXIMATING SUMS

The sums used in Sec. 5.1 to approximate the area, mass, distance, or volume were all made the same way. Consider, for instance, how an approximating sum for the area under x^2 and above the interval $[a, b]$ is formed.

First the interval $[a, b]$ is partitioned into smaller sections, perhaps all of equal length, perhaps not. There could be any finite number of sections. Say that there are n sections. These sections are determined by choosing $n - 1$ numbers in (a, b), $x_1, x_2, \ldots, x_{n-1}$,

$$a < x_1 < x_2 < \cdots < x_{n-1} < b.$$

For convenience, introduce

$$x_0 = a \qquad \text{and} \qquad x_n = b.$$

244 THE DEFINITE INTEGRAL

The ith section, $i = 1, 2, \ldots, n$ has the left endpoint x_{i-1} and the right endpoint x_i, as shown in Fig. 5.14. The typical section is shown in Fig. 5.15. For instance, the first section is $[x_0, x_1]$, which is $[a, x_1]$, the second section is $[x_1, x_2], \ldots,$ the nth section is $[x_{n-1}, x_n]$, which is $[x_{n-1}, b]$. The length of the ith section is $x_i - x_{i-1}$, which is often denoted Δx_i.

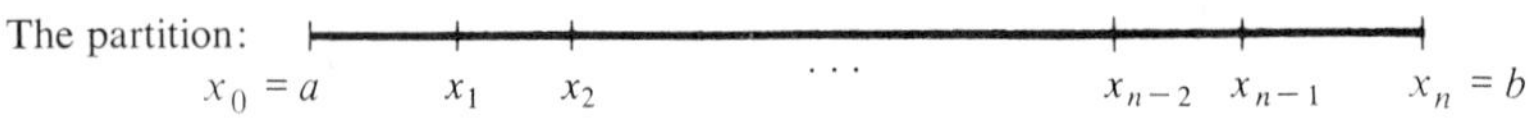

The ith section is $[x_{i-1}, x_i]$.

$\Delta x_i = x_i - x_{i-1}$

The partition:

$x_0 = a \quad x_1 \quad x_2 \quad \cdots \quad x_{n-2} \quad x_{n-1} \quad x_n = b$

Figure 5.14

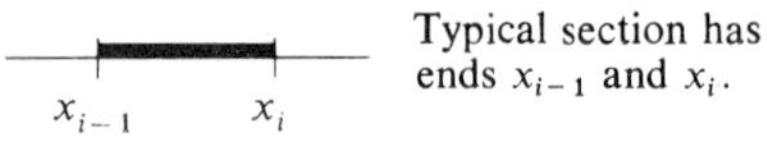

Typical section has ends x_{i-1} and x_i.

$x_{i-1} \quad x_i$

Figure 5.15

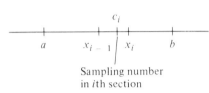

Figure 5.16

c_i is sampling number.

After the partition or division into n sections is formed, a number is selected in each section at which to evaluate x^2. The number chosen in $[x_{i-1}, x_i]$ could be its left endpoint x_{i-1}, its right endpoint x_i, its midpoint $(x_{i-1} + x_i)/2$, or any point whatsoever in the section. To allow for the most general possible choice, denote the number chosen in the ith section by c_i. The number c_i is called a *sampling number*. It is shown in Fig. 5.16.

The next step is to evaluate the function x^2 at each c_i and form the sum with n summands,

$$c_1^2(x_1 - x_0) + c_2^2(x_2 - x_1) + \cdots + c_i^2(x_i - x_{i-1}) + \cdots + c_n^2(x_n - x_{n-1}). \quad (1)$$

It takes a long time to write down the sum (1) or to read it aloud. The $\sum$ notation compresses it to the simple expression,

$$\sum_{i=1}^{n} c_i^2(x_i - x_{i-1}), \quad (2)$$

which is read, "the sum from 1 to n of $c_i^2(x_i - x_{i-1})$." If the length $x_i - x_{i-1}$ is denoted Δx_i, (2) reduces to

$$\sum_{i=1}^{n} c_i^2 \, \Delta x_i.$$

Note that the numbers a and b, which are essential in forming these sums, do not even appear in the formulas (1) and (2). It must be kept in mind that $x_0 = a$ and $x_n = b$.

EXAMPLE 8 Write in sigma notation the typical approximating sum for the area of the region under $f(x) = x^2$ and above $[2, 7]$, using right endpoints as sampling numbers.

SOLUTION In this case $a = 2$, $b = 7$, and $c_i = x_i$. The typical sum is

$$\sum_{i=1}^{n} x_i^2(x_i - x_{i-1})$$

$$\text{or} \qquad \sum_{i=1}^{n} x_i^2 \, \Delta x_i.$$

It is to be understood that

$$2 = x_0 < x_1 < \cdots < x_{n-1} < x_n = 7. \blacksquare$$

But x^2 is just one possible function. Similar approximating sums can be formed for any other function, such as x^3, $1/x$, or $\sin x$. The typical approximating sum for any function $f(x)$ and any interval $[a, b]$ in the domain of the function is formed just like (1). First a partition of $[a, b]$ is determined,

$$a = x_0 < x_1 < \cdots < x_{n-1} < x_n = b.$$

Then a sampling number c_i is picked in the ith section, $i = 1, 2, \ldots, n$. The function $f(x)$ is evaluated at each c_i and finally the approximating sum is formed:

$$f(c_1)(x_1 - x_0) + \cdots + f(c_i)(x_i - x_{i-1}) + \cdots + f(c_n)(x_n - x_{n-1}).$$

In $\sum$ notation this reduces to

$$\sum_{i=1}^{n} f(c_i)(x_i - x_{i-1})$$

$$\text{or} \qquad \sum_{i=1}^{n} f(c_i) \, \Delta x_i.$$

An approximating sum is also called a Riemann sum. Such an approximating sum is also called a Riemann sum in honor of the nineteenth-century mathematician, Georg Riemann, who made many fundamental contributions to the theory of integral calculus.

Exercises

In Exercises 1 to 4 evaluate the sums.

1 (a) $\sum_{i=1}^{3} i$ (b) $\sum_{i=1}^{4} 2i$ (c) $\sum_{d=1}^{3} d^2$.

2 (a) $\sum_{i=2}^{4} i^2$ (b) $\sum_{j=2}^{4} j^2$ (c) $\sum_{i=1}^{3} (i^2 + i)$.

3 (a) $\sum_{i=1}^{4} 1^i$ (b) $\sum_{k=2}^{6} (-1)^k$ (c) $\sum_{j=1}^{150} 3$.

4 (a) $\sum_{i=3}^{5} \frac{1}{i}$ (b) $\sum_{i=0}^{4} \cos 2\pi i$ (c) $\sum_{i=1}^{3} 2^{-i}$.

In Exercises 5 to 8 write in the sigma notation. (Do not evaluate.)

5 (a) $1 + 2 + 2^2 + 2^3 + \cdots + 2^{100}$;
   (b) $x^3 + x^4 + x^5 + x^6 + x^7$;
   (c) $\frac{1}{3} + \frac{1}{4} + \cdots + \frac{1}{102}$.

6 (a) $\frac{1}{2} + \frac{1}{3} + \cdots + \frac{1}{100}$;
   (b) $\frac{1}{3} + \frac{1}{5} + \frac{1}{7} + \frac{1}{9} + \frac{1}{11}$;
   (c) $\frac{1}{1^2} + \frac{1}{3^2} + \frac{1}{5^2} + \cdots + \frac{1}{101^2}$.

7 (a) $x_0^2(x_1 - x_0) + x_1^2(x_2 - x_1) + x_2^2(x_3 - x_2)$;
   (b) $x_1^2(x_1 - x_0) + x_2^2(x_2 - x_1) + x_3^2(x_3 - x_2)$.

8 (a) $8t_0^2(t_1 - t_0) + 8t_1^2(t_2 - t_1) + \cdots + 8t_{99}^2(t_{100} - t_{99})$;
   (b) $8t_1^2(t_1 - t_0) + 8t_2^2(t_2 - t_1) + \cdots + 8t_n^2(t_n - t_{n-1})$.

In Exercises 9 and 10 evaluate the telescoping sums.

9 (a) $\sum_{i=1}^{100} (2^i - 2^{i-1})$;

(b) $\sum_{i=2}^{100} \left(\frac{1}{i} - \frac{1}{i-1}\right)$;

(c) $\sum_{i=1}^{50} \left(\frac{1}{2i+1} - \frac{1}{2(i-1)+1}\right)$.

10 (a) $\sum_{i=1}^{100} \left(\frac{x_i^3}{3} - \frac{x_{i-1}^3}{3}\right)$;

(b) $\sum_{i=5}^{70} \left(\frac{1}{x_i} - \frac{1}{x_{i-1}}\right)$.

11 Writing out each sum in longhand, show that

(a) $\sum_{i=1}^{3} a_i = \sum_{j=1}^{3} a_j = \sum_{k=1}^{3} a_k$;

(b) $\sum_{i=1}^{3} (a_i + 4) = 12 + \sum_{j=2}^{4} a_{j-1}$.

12 Writing out each sum in longhand, show that

(a) $\sum_{i=1}^{3} (a_i - b_i) = \sum_{i=1}^{3} a_i - \sum_{i=1}^{3} b_i$;

(b) $\sum_{i=1}^{2} a_i b_i$ is *not* always equal to $\sum_{i=1}^{2} a_i \cdot \sum_{i=1}^{2} b_i$;

(c) $\sum_{i=1}^{3} \left(\sum_{j=1}^{3} b_j\right) a_i = \sum_{j=1}^{3} \left(\sum_{i=1}^{3} a_i\right) b_j$.

In Exercises 13 to 18 evaluate the approximating sum $\sum_{i=1}^{n} f(c_i)(x_i - x_{i-1})$. In each case the interval is $[1, 3]$ and the partition consists of four sections of equal length. Express answers to two decimal places.

13 $f(x) = 3x$, $c_i = x_i$
14 $f(x) = 3x$, $c_i = x_{i-1}$
15 $f(x) = 5$, $c_i = x_i$
16 $f(x) = x^3 + x$, $c_i = x_i$
17 $f(x) = 1/x$, $c_1 = 1.25$, $c_2 = 1.8$, $c_3 = 2.2$, $c_4 = 3$
18 $f(x) = 2^x$, $c_i = x_{i-1}$

19 Let $f(x) = 11$ for all x. Consider the partition $x_0 = a < x_1 < \cdots < x_n = b$ and let c_i be in $[x_{i-1}, x_i]$. Evaluate $\sum_{i=1}^{n} f(c_i)(x_i - x_{i-1})$.

20 Write in $\sum$ notation the typical approximating sum for the area of the region under x^3 and above $[1, 7]$ in which the sampling numbers c_i are (a) left endpoints, (b) right endpoints, (c) midpoints.

In Exercises 21 to 23 evaluate $\sum_{i=1}^{n} f(c_i)(x_i - x_{i-1})$ for the given data.

21 $f(x) = \sqrt{x}$, $x_0 = 1$, $x_1 = 3$, $x_2 = 5$, $c_1 = 1$, $c_2 = 4$ $(n = 2)$

22 $f(x) = \sqrt[3]{x}$, $x_0 = 0$, $x_1 = 1$, $x_2 = 4$, $x_3 = 10$, $c_1 = 0$, $c_2 = 1$, $c_3 = 8$ $(n = 3)$

23 $f(x) = 1/x$, $x_0 = 1$, $x_1 = 1.25$, $x_2 = 1.5$, $x_3 = 1.75$, $x_4 = 2$, $c_1 = 1$, $c_2 = 1.25$, $c_3 = 1.6$, $c_4 = 2$ $(n = 4)$

■ ■

24 Let $S_n = 1 + 2 + 3 + \cdots + n = \sum_{i=1}^{n} i$. We outline a shortcut for computing this sum, illustrating the method in the case $n = 20$.

(a) $S_{20} = 1 + 2 + 3 + \cdots + 19 + 20$.
$S_{20} = 20 + 19 + 18 + \cdots + 2 + 1$.
Adding the two lines, show that
$2S_{20} = 21 + 21 + 21 + \cdots + 21 + 21$
(20 summands).

(b) From (a) deduce that $S_{20} = 21 \cdot 20/2$ (= 210).
(c) Show that $S_n = [(n + 1)n]/2$.
(d) Use (c) to compute $\sum_{i=1}^{100} i$.

25 Write the expression

$$c^{n-1} + c^{n-2}d + c^{n-3}d^2 + \cdots + d^{n-1}$$

in the sigma notation.

26 (a) By inspection of Fig. 5.17 show, without using any arithmetic, that $1 + 3 + 5 + 7 = 4^2$.

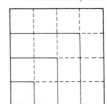

Figure 5.17

(b) Use the idea in (a) to find a short formula for the sum of the first n odd positive integers.
(c) Find a formula for the ith positive odd integer.
(d) Write the sum of the first n positive odd integers in sigma notation.

27 (a) Show that $2i + 1 = (i + 1)^2 - i^2$.
(b) Deduce from (a) a formula for the sum of the first n positive odd integers.

28 (a) Show that $i = [(i + 1)^2 - i^2 - 1]/2$.
(b) Use (a) to show that $\sum_{i=1}^{n} i = (n + 1)^2/2 - 1/2 - n/2$.
(c) From (b) obtain the formula in Exercise 24(c).

29 (a) Show that $(i + 1)^3 - i^3 = 3i^2 + 3i + 1$.
(b) Use (a) to show that

$$3 \sum_{i=1}^{n} i^2 + 3 \sum_{i=1}^{n} i + \sum_{i=1}^{n} 1 = (n + 1)^3 - 1.$$

(c) Evaluate $\sum_{i=1}^{n} 1$.
(d) Using Exercise 24 or 28, evaluate $\sum_{i=1}^{n} i$.
(e) Solving the equation in (b) for $\sum_{i=1}^{n} i^2$, obtain the formula

$$\sum_{i=1}^{n} i^2 = \frac{n(n+1)(2n+1)}{6}.$$

(*f*) Check the formula in (*e*) for the case $n = 5$.

30 (Calculator) Let $S_n = \sum_{i=1}^{n} 1/[i(i+1)]$.
  (*a*) Compute S_2, S_3, S_4, and S_5 to at least four decimal places.
  (*b*) What do you think happens to S_n as $n \to \infty$?

31 (Calculator) Let $S_n = \sum_{i=n}^{2n} 1/i$. For example,
$$S_3 = \tfrac{1}{3} + \tfrac{1}{4} + \tfrac{1}{5} + \tfrac{1}{6}.$$
  (*a*) Compute S_1, S_2, S_3, and S_4 to at least three decimal places.
  (*b*) What do you think happens to S_n as $n \to \infty$?
  (*c*) Show that $S_n > \tfrac{1}{2}$ for all positive integers n.

5.3 The definite integral

There are two main concepts in calculus: the derivative and the definite integral. This section defines the definite integral and uses it to solve the four problems in Sec. 5.1. In Sec. 5.1 there were only *estimates* of the four quantities; now we will obtain their exact values.

THE DEFINITE INTEGRAL

In Sec. 5.1 sums of the form

$$\sum_{i=1}^{n} f(c_i)(x_i - x_{i-1})$$

were used to estimate certain quantities such as area, mass, distance, and volume. The larger *n* is and the shorter the sections $[x_{i-1}, x_i]$ are, the closer we would expect these approximating sums to be to the quantity we are trying to find. We are really interested in *what happens to these approximating sums as all the sections in the partition are chosen smaller and smaller.* This leads to the notion of the definite integral of a function over an interval, which will be defined after we introduce a measure of the "fineness" of a partition.

DEFINITION *Mesh.* The mesh of a partition is the length of the longest section (or sections) in the partition.

For instance, the partition used in Sec. 5.1 has mesh equal to $\tfrac{1}{2}$.

DEFINITION *The definite integral of a function f over an interval $[a, b]$.* If *f* is a function defined on $[a, b]$ and the sums $\sum_{i=1}^{n} f(c_i)(x_i - x_{i-1})$ approach a certain number as the mesh of partitions of $[a, b]$ shrinks toward 0 (no matter how the sampling number c_i is chosen in $[x_{i-1}, x_i]$), that certain number is called the *definite integral of f over $[a, b]$.*

Notation for the definite integral: $\int_a^b f(x)\,dx$

The definite integral is also called the *definite integral of f from a to b* and the *integral of f from a to b*. The symbol for this number is $\int_a^b f(x)\,dx$. The symbol $\int$ comes from the letter S of Sum; the dx traditionally suggests a small section of the *x* axis and will be more meaningful and useful later. It is important to realize that area, mass, distance traveled, and volume are merely applications of the definite integral. (It is a mistake to link the definite integral too closely with one of its applications, just as it narrows our understanding of the number 2 to link it always with the idea of two fingers.)

Slope, velocity, magnification, and density are particular interpretations or applications of the derivative, which is a purely mathematical concept defined as a limit:

$$\text{Derivative of } f \text{ at } x = \lim_{\Delta x \to 0} \frac{f(x + \Delta x) - f(x)}{\Delta x}.$$

Similarly, area, total distance, mass, and volume are just particular interpretations of the definite integral, which is also defined as a limit:

Recall that $x_0 = a$ and $x_n = b$.

$$\text{Definite integral of } f \text{ over } [a, b] = \lim_{\text{mesh} \to 0} \sum_{i=1}^{n} f(c_i)(x_i - x_{i-1}).$$

In advanced calculus it is proved that, if f is continuous, then

$$\lim_{\text{mesh} \to 0} \sum_{i=1}^{n} f(c_i)(x_i - x_{i-1})$$

exists; that is, a continuous function always has a definite integral. For emphasis we record this fact, an important result in advanced calculus, as a theorem.

THEOREM *Existence of the definite integral.* Let f be a continuous function defined on $[a, b]$. Then the approximating sums

$$\sum_{i=1}^{n} f(c_i)(x_i - x_{i-1})$$

approach a single number as the mesh of the partition of $[a, b]$ approaches 0. Hence $\int_a^b f(x)\, dx$ exists. ∎

To bring the definition down to earth, let us use it to evaluate the definite integral of a constant function.

EXAMPLE Let f be the function whose value at any number x is 4; that is, f is the constant function given by the formula $f(x) = 4$. Use only the definition of the definite integral to compute

$$\int_1^3 f(x)\, dx.$$

SOLUTION In this case a typical partition has $x_0 = 1$ and $x_n = 3$. The approximating sum

$$\sum_{i=1}^{n} f(c_i)(x_i - x_{i-1})$$

becomes
$$\sum_{i=1}^{n} 4(x_i - x_{i-1})$$

since, no matter how the sampling number c_i is chosen, $f(c_i) = 4$. Now

$$\sum_{i=1}^{n} 4(x_i - x_{i-1}) = 4 \sum_{i=1}^{n} (x_i - x_{i-1}) = 4(x_n - x_0),$$

since the sum is telescoping. Since $x_n = 3$ and $x_0 = 1$, it follows that all approximating sums have the same value, namely

$$4(3 - 1) = 8.$$

It does not matter whether the mesh is small or where the c_i are picked in each section. Thus, as the mesh approaches 0,

$$\sum_{i=1}^{n} f(c_i)(x_i - x_{i-1})$$

approaches 8. Indeed, the sums are always 8. Thus

$$\int_{1}^{3} 4 \, dx = 8. \quad \blacksquare$$

The definite integral $\int_{1}^{3} 4 \, dx$ was found using only its definition. However, the result could have been anticipated by considering a string that occupies the interval $[1, 3]$ on the x axis and has the constant density 4 grams per centimeter throughout its length. Since the string is of uniform density, its mass is just

Density · length = $4 \cdot 2 = 8$ grams.

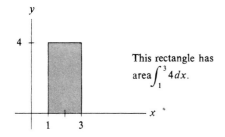

This rectangle has area $\int_{1}^{3} 4 \, dx$.

Figure 5.18

We could also have guessed the value of $\int_{1}^{3} 4 \, dx$ by interpreting the definite integral as an area. To do so, draw a region in the plane whose cross section has length 4 for all lines perpendicular to the x axis, as in Fig. 5.18. This region may be taken to be a rectangle of height 4 (and base coinciding with the interval $[1, 3]$). Since the area of a rectangle is its base times its height, it follows again that $\int_{1}^{3} 4 \, dx = 8$.

Similar reasoning shows that, for any constant function that has the fixed value c,

$$\int_{a}^{b} c \, dx = c(b - a).$$

Though area is the most intuitive of the interpretations, physical scientists, if they want to think of the definite integral concretely, should think of it as giving total mass if we know the density everywhere. This interpretation carries through easily to higher dimensions; the area interpretation does not.

It is the concept of the definite integral that links the four problems of Sec. 5.1, which are summarized below.

PROBLEM 1 The area under the curve $y = x^2$ and above $[0, 3]$ equals the definite integral $\int_0^3 x^2\, dx$.

PROBLEM 2 The mass of the string whose density is $5x^2$ g/cm equals the definite integral $\int_0^3 5x^2\, dx$.

PROBLEM 3 The distance traveled by the engineer whose speed is $8t^2$ miles per hour at time t equals the definite integral $\int_0^3 8t^2\, dt$ (the t reminding us of time).

PROBLEM 4 The volume of the tent equals the definite integral $\int_0^3 x^2\, dx$.

It is somewhat satisfying to have reduced all four problems to one problem, that of evaluating the definite integral of x^2 from 0 to 3, $\int_0^3 x^2\, dx$. Once it is found, all four problems are solved.

But it is one thing to define a certain limit; it is quite a different matter to evaluate it. Recall that the derivative was defined in Sec. 3.2, but it took several sections to develop techniques for computing it. Just as we resorted to algebraic identities to differentiate x^2, x^3, and $\sqrt{x}$ in Sec. 3.2, we will use an algebraic inequality to compute $\int_0^3 x^2\, dx$.

EVALUATING $\int_a^b x^2\, dx$

It takes no more work to evaluate $\int_a^b x^2\, dx$ than $\int_0^3 x^2\, dx$, so let us evaluate $\int_a^b x^2\, dx$, $0 \leq a < b$. Though the reasoning will be free of any particular interpretation, the definite integral $\int_a^b x^2\, dx$ can be thought of as the area of the region under the curve $y = x^2$ and above the interval $[a, b]$, as shown in Fig. 5.19.

The key to the evaluation will be the inequality

$$c^2(d - c) < \frac{d^3}{3} - \frac{c^3}{3} < d^2(d - c), \tag{1}$$

which is valid if $0 \leq c < d$. (See Appendix B.)

To begin, consider partitions of the interval $[a, b]$ and choose $x_0, x_1, \ldots, x_n$ such that

$$a = x_0 < x_1 < \cdots < x_n = b.$$

Then examine the approximating sums

$$\sum_{i=1}^n c_i^2 (x_i - x_{i-1}), \tag{2}$$

when the mesh is small.

Since the function x^2 is increasing for $x \geq 0$,

$$x_{i-1}^2 \leq c_i^2 \leq x_i^2$$

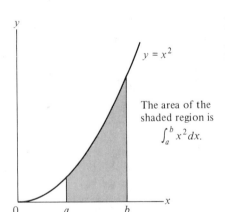

The area of the shaded region is $\int_a^b x^2\, dx$.

Figure 5.19

for $i = 1, 2, \ldots, n$. Consequently the sum $\sum_{i=1}^{n} c_i^2(x_i - x_{i-1})$ is squeezed between the overestimating sum $\sum_{i=1}^{n} x_i^2(x_i - x_{i-1})$ and the corresponding underestimating sum $\sum_{i=1}^{n} x_{i-1}^2(x_i - x_{i-1})$.

$$\sum_{i=1}^{n} x_{i-1}^2(x_i - x_{i-1}) \leq \sum_{i=1}^{n} c_i^2(x_i - x_{i-1}) \leq \sum_{i=1}^{n} x_i^2(x_i - x_{i-1}). \quad (3)$$

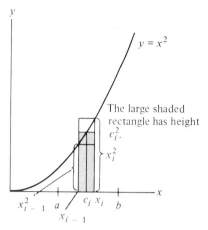

Figure 5.20

The rectangles corresponding to the typical summands from these three sums are shown in Fig. 5.20.

The problem is reduced to finding out what happens to the underestimating sum

$$\sum_{i=1}^{n} x_{i-1}^2(x_i - x_{i-1}) \quad (4)$$

and the overestimating sum

$$\sum_{i=1}^{n} x_i^2(x_i - x_{i-1}) \quad (5)$$

as the mesh $\to 0$. If we can show that they both approach the same number, and we can find that number, then

$$\sum_{i=1}^{n} c_i^2(x_i - x_{i-1}),$$

trapped between them, must approach that same number.

Consider inequality (1), with c replaced by x_{i-1} and d replaced by x_i. It becomes

$$x_{i-1}^2(x_i - x_{i-1}) < \frac{x_i^3}{3} - \frac{x_{i-1}^3}{3} < x_i^2(x_i - x_{i-1}).$$

Thus $\sum_{i=1}^{n} x_{i-1}^2(x_i - x_{i-1}) < \sum_{i=1}^{n} \left(\frac{x_i^3}{3} - \frac{x_{i-1}^3}{3}\right) < \sum_{i=1}^{n} x_i^2(x_i - x_{i-1}).$

But the sum $\sum_{i=1}^{n} \left(\frac{x_i^3}{3} - \frac{x_{i-1}^3}{3}\right)$

is a telescoping sum with many cancellations:

$$\left(\frac{x_1^3}{3} - \frac{x_0^3}{3}\right) + \left(\frac{x_2^3}{3} - \frac{x_1^3}{3}\right) + \left(\frac{x_3^3}{3} - \frac{x_2^3}{3}\right) + \cdots + \left(\frac{x_n^3}{3} - \frac{x_{n-1}^3}{3}\right). \quad (6)$$

Everything in (6) cancels except $-x_0^3/3$ and $x_n^3/3$, leaving only

$$\frac{x_n^3}{3} - \frac{x_0^3}{3}.$$

252 THE DEFINITE INTEGRAL

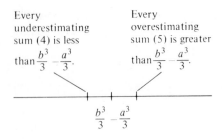

Every underestimating sum (4) is less than $\frac{b^3}{3} - \frac{a^3}{3}$.

Every overestimating sum (5) is greater than $\frac{b^3}{3} - \frac{a^3}{3}$.

Figure 5.21

We will not make use of the theorem of this section, which guarantees that x^2, being continuous, has a definite integral.

Every underestimating sum (4) is less than $\frac{b^3}{3} - \frac{a^3}{3}$.

Every overestimating sum (5) is greater than $\frac{b^3}{3} - \frac{a^3}{3}$.

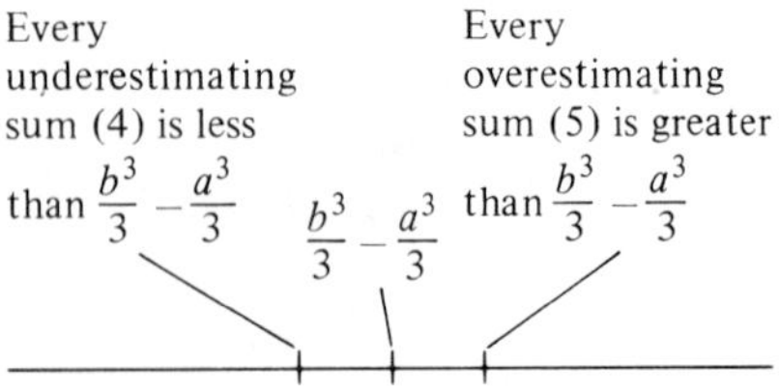

But when all the sections $[x_{i-1}, x_i]$ are chosen shorter and shorter, the difference between the overestimating and underestimating sums approaches 0.

Figure 5.22

Precise answers to the four problems

But $x_0 = a$ and $x_n = b$. Thus

$$\sum_{i=1}^{n}\left(\frac{x_i^3}{3} - \frac{x_{i-1}^3}{3}\right) = \frac{b^3}{3} - \frac{a^3}{3}.$$

Consequently, any underestimating sum is less than $b^3/3 - a^3/3$ and any overestimating sum is greater than $b^3/3 - a^3/3$. (See Fig. 5.21.) There is good reason to suspect that these sums both approach the number $b^3/3 - a^3/3$ as the mesh $\to 0$. All that remains is to show that the overestimating sums get arbitrarily close to the underestimating sums as the mesh $\to 0$. So let us look at the difference between them.

The difference between the two sums is

$$\sum_{i=1}^{n} x_i^2(x_i - x_{i-1}) - \sum_{i=1}^{n} x_{i-1}^2(x_i - x_{i-1}) = \sum_{i=1}^{n} (x_i^2 - x_{i-1}^2)(x_i - x_{i-1}).$$

If the mesh is $\leq p$, then

$$\sum_{i=1}^{n} (x_i^2 - x_{i-1}^2)(x_i - x_{i-1}) \leq \sum_{i=1}^{n} (x_i^2 - x_{i-1}^2)p.$$

But
$$\sum_{i=1}^{n} (x_i^2 - x_{i-1}^2)p = p\sum_{i=1}^{n} (x_i^2 - x_{i-1}^2) \quad \text{a telescoping sum}$$
$$= p(x_n^2 - x_0^2)$$
$$= p(b^2 - a^2).$$

When p is small, the product $p(b^2 - a^2)$ is also small. The overestimating and underestimating sums must therefore approach the same number when the mesh $\to 0$, and the number is $b^3/3 - a^3/3$. (See Fig. 5.22.)

Consequently, $\quad \int_a^b x^2\, dx = \frac{b^3}{3} - \frac{a^3}{3} \qquad 0 \leq a < b.$ \hfill (7)

Note that the computations were carried out without any thought of area, mass, distance, or volume.

In particular, when $a = 0$ and $b = 3$, (7) becomes

$$\int_0^3 x^2\, dx = \frac{3^3}{3} - \frac{0^3}{3} = 9.$$

From this result we immediately conclude that:

The area under $y = x^2$ and above $[0, 3]$ is 9 square units.
The engineer in Prob. 2 of Sec. 5.1 travels $8 \cdot 9 = 72$ miles.
The mass of the string in Prob. 3 of Sec. 5.1 is $5 \cdot 9 = 45$ grams.
The volume of the tent in Prob. 4 of Sec. 5.1 is 9 cubic feet.
It took a lot of work and the special inequality (1) to find

$\int_a^b x^2\,dx$. Fortunately, the next section in this chapter develops a powerful technique for evaluating many, *but not all*, definite integrals without having to refer to their definition in terms of approximating sums.

FOUR GENERAL APPLICATIONS OF THE DEFINITE INTEGRAL

It would be appropriate at this time to summarize the four main applications of the definite integral in full generality. Each has already been illustrated by one of the four problems in Sec. 5.1.

AREA OF A PLANE REGION AS A DEFINITE INTEGRAL

Let S be some region in the plane whose area is to be found. Let L be a line in the plane which will be considered to be the x axis. Each line in the plane and perpendicular to L meets S in what shall be called a cross section. (If the line misses S, the cross section is empty.) See Figs. 5.23 to 5.25. Let the coordinate on L where the typical line meets L be x. The length of the typical cross section is denoted by $f(x)$. Assume that the lines that are perpendicular to L and that meet S intersect L in an interval whose ends are a and b. (In Prob. 1 of Sec. 5.1, L is the x axis, $f(x) = x^2$, $a = 0$, and $b = 3$.)

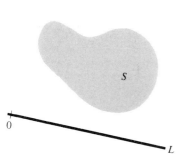
Figure 5.23

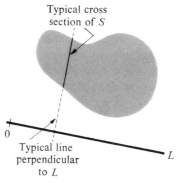
Figure 5.24

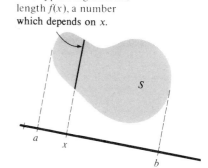
Figure 5.25

To estimate the area of S, proceed just as for the region under $y = x^2$. First cut $[a, b]$ into n sections by means of the numbers $x_0 = a$, x_1, x_2, ..., $x_n = b$, as in Fig. 5.26.

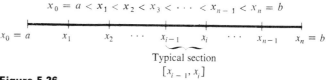
Figure 5.26

In each of these sections select a number at random. In the section $[x_0, x_1]$ select c_1, in $[x_1, x_2]$ select c_2, and so on. Simply stated, in the ith interval, $[x_{i-1}, x_i]$, select c_i.

With this choice of x_i's and c_i's, form a set of rectangles, whose typical

254 THE DEFINITE INTEGRAL

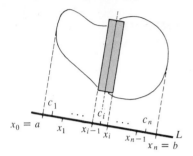

A typical rectangle has base $x_i - x_{i-1}$, height $f(c_i)$ and area $f(c_i)(x_i - x_{i-1})$.

Figure 5.27

member is shown in Fig. 5.27. Thus $\sum_{i=1}^{n} f(c_i)(x_i - x_{i-1})$ is an estimate of the area of S. As the lengths $x_i - x_{i-1}$ are chosen to be smaller and smaller, we would expect that these sums tend toward the area of S.

But, by the definition of the definite integral, the sums

$$\sum_{i=1}^{n} f(c_i)(x_i - x_{i-1})$$

approach

$$\int_a^b f(x)\,dx$$

as the mesh $\to 0$. Thus

Area of $S = \int_a^b f(x)\,dx$, where $f(x)$ is the length of a cross section of S.

In short "area is the definite integral of cross-sectional length." In practical terms, this tells us that if we can compute definite integrals, then we can compute areas.

THE MASS OF A STRING AS A DEFINITE INTEGRAL

A string is made of a material whose density may vary from point to point. (Such a string is called *nonuniform* or *nonhomogeneous*.) How would its total mass be computed if its density at each point is known?

First, place the string somewhere on the x axis and denote by $f(x)$ its density at x. The string occupies an interval $[a, b]$ on the axis, as in Fig. 5.28.

Nonuniform string

Figure 5.28

Then cut the string into n sections $[x_{i-1}, x_i]$, $i = 1, 2, \ldots, n$, as in Fig. 5.29.

Figure 5.29

In each section the density is almost constant. So the mass of the ith section is approximately

$$f(c_i)(x_i - x_{i-1}),$$

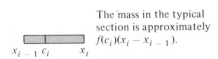

The mass in the typical section is approximately $f(c_i)(x_i - x_{i-1})$.

Figure 5.30

where c_i is some point in $[x_{i-1}, x_i]$. (See Fig. 5.30.) Thus $\sum_{i=1}^{n} f(c_i)(x_i - x_{i-1})$ is an estimate of the mass of the string. And, what is more important, it seems plausible that, as the mesh of the partition approaches 0, the sum $\sum_{i=1}^{n} f(c_i)(x_i - x_{i-1})$ approaches the mass of the string. (The case when $a = 0$, $b = 3$, and $f(x) = 5x^2$ is Prob. 2 in Sec. 5.1.)

But, by the definition of the definite integral, the sums

$$\sum_{i=1}^{n} f(c_i)(x_i - x_{i-1})$$

approach

$$\int_{a}^{b} f(x)\, dx$$

as the mesh $\to 0$. Thus

$$\text{Mass} = \int_{a}^{b} f(x)\, dx, \text{ where } f(x) \text{ is the density at } x.$$

In short, "mass is the definite integral of density."

THE DISTANCE TRAVELED AS A DEFINITE INTEGRAL

An engineer takes a trip that begins at time a and ends at time b. Imagine that at any time t during the trip her velocity is $f(t)$, depending on the time t. How far does she travel? (The case in which $a = 0$, $b = 3$, $f(t) = 8t^2$ is Prob. 3 in Sec. 5.1.)

First, cut the time interval $[a, b]$ into smaller intervals by a partition and estimate the trip's length by summing the estimates of the distance the engineer travels during each of the time intervals. (See Fig. 5.31.)

Figure 5.31

During a small interval of time, the velocity changes little. We thus expect to obtain a reasonable estimate of the distance covered during the ith time interval $[t_{i-1}, t_i]$ by observing the speedometer reading at some instant T_i in that interval, $f(T_i)$, and computing the product $f(T_i)(t_i - t_{i-1})$. Thus $\sum_{i=1}^{n} f(T_i)(t_i - t_{i-1})$ is an estimate of the length of the trip. Moreover, as the mesh of the partition approaches zero, the sum $\sum_{i=1}^{n} f(T_i)(t_i - t_{i-1})$ approaches the length of the trip.

Since these sums also approach the definite integral $\int_{a}^{b} f(t)\, dt$ we have

$$\text{Total distance} = \int_{a}^{b} f(t)\, dt, \text{ where } f(t) \text{ is the velocity at time } t.$$

In short "distance is the definite integral of velocity."

THE VOLUME OF A SOLID REGION AS A DEFINITE INTEGRAL

Suppose that we wish to compute the volume of a solid S, and we happen to know the area $A(x)$ of each cross section made by planes in a fixed direction (see Fig. 5.32). In the case of the tent in Prob. 4 of Sec. 5.1, $a = 0$, $b = 3$, and $A(x) = x^2$.

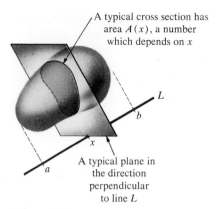

Figure 5.32

Every partition of $[a, b]$ and selection of c's provides an estimate of the volume of S, the sum of the volumes of slabs. A typical slab is shown in perspective and side views in Figs. 5.33 and 5.34.

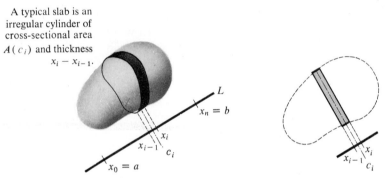

Figure 5.33   **Figure 5.34**

Thus $\sum_{i=1}^{n} A(c_i)(x_i - x_{i-1})$ is an estimate of the volume of the solid S. As the mesh of the partition shrinks, the slabs become thin, and the sum of their volumes becomes a more and more accurate estimate of the volume of S. As the mesh of the partition approaches 0, we see that the sum $\sum_{i=1}^{n} A(c_i)(x_i - x_{i-1})$ approaches the volume of S.

From this it follows that

$$\text{Volume of } S = \int_a^b A(x)\, dx, \text{ where } A(x) \text{ is the cross-sectional area at } x.$$

In short, "volume is the definite integral of cross-sectional area."

The following table shows the similarities of these four general types of problems. To emphasize these similarities, all the functions, whether cross-sectional length, density, velocity, or cross-sectional area, are denoted by the same symbol $f(x)$.

Spend some time examining this table. The concepts it describes will be used often.

$f(x)$	$\sum_{i=1}^{n} f(c_i)(x_i - x_{i-1})$	$\int_a^b f(x)\, dx$
Variable length of cross section of set in plane	Approximation to area of the set in the plane	The area of the set in the plane
Variable density of string	Approximation to mass of the string	The mass of the string
Variable velocity	Approximation to the distance traveled	The distance traveled
Variable area of cross section of a solid	Approximation to the volume of the solid	The volume of the solid

FORMAL DEFINITION OF $\int_a^b f(x)\, dx$

The definition of the definite integral $\int_a^b f(x)\, dx$ given earlier in this section is rather informal in the sense that it uses terms like "approaches a certain number" and "shrinks toward 0." The following definition, in the spirit of Sec. 2.4, is more precise.

DEFINITION *The definite integral of f over $[a, b]$.* The number L is the definite integral of f over $[a, b]$ if the following condition is met. For any positive number ε (however small) there must exist a positive number δ, depending on ε, such that any sum

$$\sum_{i=1}^{n} f(c_i)(x_i - x_{i-1})$$

formed with any partition of $[a, b]$ of mesh less than δ (no matter how c_i is chosen in $[x_{i-1}, x_i]$) differs from L by less than ε, that is, satisfies the inequality

$$\left| \sum_{i=1}^{n} f(c_i)(x_i - x_{i-1}) - L \right| < \varepsilon.$$

Exercises

1. Find the mesh of each of these partitions of $[1, 6]$:
   (a) $x_0 = 1, x_1 = 2, x_2 = 3, x_3 = 4, x_4 = 5, x_5 = 6$;
   (b) $x_0 = 1, x_1 = 3, x_2 = 5, x_3 = 6$;
   (c) $x_0 = 1, x_1 = 4, x_2 = 4.5, x_3 = 5, x_4 = 5.5, x_5 = 6$.

2. A partition of $[a, b]$ is formed by cutting it into n sections of equal length. What is the mesh of this partition?

3. A string occupying the interval $[a, b]$ has the density $f(x)$ grams per centimeter at x. Let $x_0, \ldots, x_n$ be a partition of $[a, b]$ and let $c_1, \ldots, c_n$ be sampling numbers. What is the physical interpretation of

    (a) $x_i - x_{i-1}$?
    (b) $f(c_i)$?
    (c) $f(c_i)(x_i - x_{i-1})$?
    (d) $\sum_{i=1}^{n} f(c_i)(x_i - x_{i-1})$?
    (e) $\int_a^b f(x)\, dx$?

4. A rocket moving with a varying speed travels at $f(t)$ miles per second at time t. Let $t_0, \ldots, t_n$ be a partition

of $[a, b]$, and let $T_1, \ldots, T_n$ be sampling numbers. What is the physical interpretation of
(a) $t_i - t_{i-1}$?
(b) $f(T_i)$?
(c) $f(T_i)(t_i - t_{i-1})$?
(d) $\sum_{i=1}^{n} f(T_i)(t_i - t_{i-1})$?
(e) $\int_a^b f(t)\,dt$?

5 Water is flowing into a lake at the rate of $f(t)$ gallons per second at time t. Answer the five questions in Exercise 4 for this interpretation of the function f, using the same partition and sampling numbers as in Exercise 4.

6 A business firm has a flow of income at the rate of $f(t)$ dollars per year at time t. Consider f to be a continuous function. (Thus in a short period of time of Δt years near time t it earns approximately $f(t)\Delta t$ dollars.) Let $t_0, t_1, \ldots, t_n$ be a partition of the time interval $[a, b]$ and let $T_1, T_2, \ldots, T_n$ be sampling numbers. Answer the five questions in Exercise 4 for this interpretation of the function.

Exercises 7 to 18 concern the definite integral, free of any particular interpretation.

7 Estimate $\int_1^5 x^2\,dx$, using a partition into four sections of equal length and, as sampling points, (a) left endpoints, (b) right endpoints.

8 Estimate $\int_1^4 5x\,dx$, using a partition into three sections of equal length and midpoints as sampling points.

9 Let $f(x) = 3$ for all x. Show that any approximating sum for $\int_2^6 f(x)\,dx$ equals 12.

10 Show that if $f(x) = c$, constant, then any approximating sum for $\int_a^b f(x)\,dx$ equals $c(b - a)$.

11 Estimate $\int_1^3 (1/x)\,dx$, using a partition into four sections of equal length and, for sampling points, (a) left endpoints, (b) right endpoints.

12 Estimate $\int_0^3 2^x\,dx$, using a partition into three sections of equal length and, as sampling points, (a) left endpoints, (b) right endpoints.

13 Estimate $\int_0^1 x^3\,dx$, using a partition into five sections of equal length and, as sampling points, (a) left endpoints, (b) right endpoints.

14 (Calculator) Estimate $\int_0^1 x^3\,dx$, using a partition into 10 sections of equal length and left endpoints as sampling points.

15 Estimate $\int_0^{\pi/2} \sin x\,dx$, using the partition $x_0 = 0$, $x_1 = \pi/6$, $x_2 = \pi/4$, $x_3 = \pi/3$, and $x_4 = \pi/2$ and, as sampling points, (a) left endpoints, (b) right endpoints. Express the answers to two decimal places.

16 (Calculator) Estimate $\int_0^1 \sqrt{x}\,dx$, using a partition into 10 sections of equal length and, as sampling points, (a) left endpoints, (b) right endpoints, (c) midpoints.

17 Estimate $\int_1^3 (1/x^2)\,dx$, using a partition into four sections of equal length and left endpoints as sampling points.

18 Estimate $\int_1^4 (2x + 1)\,dx$, using a partition into three sections of equal length and right endpoints as sampling points.

In the text it was shown that $\int_a^b x^2\,dx = b^3/3 - a^3/3$ for $0 \leq a < b$. Use this formula to solve Exercises 19 to 22.

19 Find the area of the region under $y = x^2$ and above $[2, 5]$.

20 Find the area of the region under $y = x^2$ and above $[1, 4]$.

21 The density of a string is x^2 grams per centimeter at the point x. Find the mass in the interval $[1, 3]$.

22 An object is moving with a velocity of t^2 feet per second at time t seconds. How far does it travel during the time interval $[2, 5]$?

■

23 Figure 5.35 shows the graph of $y = x$ and the region below it and above the interval $[a, b]$, $0 \leq a < b$. Using the formula for the area of a trapezoid (see inside front cover), show that $\int_a^b x\,dx = b^2/2 - a^2/2$.

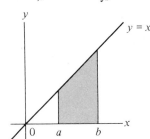

Figure 5.35

24 (See Exercise 23.)
(a) Set up an appropriate definite integral $\int_a^b f(x)\,dx$ which equals the volume of the headlight in Fig. 5.36 whose cross section by a typical plane perpendicular to the x axis at x is a circle of radius $\sqrt{x/\pi}$.

Figure 5.36

(b) Evaluate the definite integral in (a) with the aid of Exercise 23.

The next two exercises concern the curve $y = 1/x^2$.
25 (a) Show that if $0 \le c < d$, then $c^2(d-c) < cd(d-c) < d^2(d-c)$.
   (b) Use (a) and approximating sums to show that the area of the region under the curve $y = 1/x^2$ and above the interval $[a, b]$, where $b > a > 0$, is $1/a - 1/b$.
26 (See Exercise 25.) A tiring turtle travels at the velocity $1/t^2$ feet per second at time $t > 0$. How far does it travel (a) from time $t = 1$ to time $t = 2$ seconds? (b) from time $t = 2$ to time $t = 3$ seconds?
27 Review the argument that shows that if the mesh is $\le p$, then any two approximating sums formed for $\int_a^b x^2 \, dx$ from the same partition differ by at most $p(b^2 - a^2)$.
   (a) Show that any approximating sum for $\int_1^3 x^2 \, dx$ formed with a partition of mesh less than 0.01 differs from $\int_0^3 x^2 \, dx$ by at most 0.09.
   (b) Assume that you do not know the value of $\int_1^3 x^2 \, dx$. If you wanted to estimate $\int_1^3 x^2 \, dx$ by an approximating sum with an error at most 0.001, how fine a mesh should you use?

Exercises 28 and 29 depend on the factorization of $d^n - c^n$ given in Appendix B.
28 (a) Show that if $0 \le c < d$, then
$$c^3(d-c) < \frac{d^4}{4} - \frac{c^4}{4} < d^3(d-c).$$
   (b) In the text it was shown that if $0 \le a < b$, then $\int_a^b x^2 \, dx = b^3/3 - a^3/3$. Use a similar argument to show that $\int_a^b x^3 \, dx = b^4/4 - a^4/4$.
29 Use the formula in Exercise 28(b) to answer these questions:
   (a) What is the area of the region under $y = x^3$ and above $[1, 3]$?
   (b) What is the distance an object moves from time $t = 1$ to time $t = 3$ if its velocity at time t is t^3 feet per second?
   (c) A solid is formed by revolving the region below the curve $y = x^{3/2}$ and above $[1, 3]$ around the x axis. What is its volume?

■ ■

30 Show that the volume of a right circular cone of radius a and height h is $\pi a^2 h/3$. (*Suggestion:* First show that a cross section by a plane perpendicular to the axis of the cone and a distance x from the vertex is a circle of radius ax/h.)
31 If you make three tents out of paper with the pattern given in Sec. 5.1, you will be able to fit them together easily to form a cube. If you have more geometric intuition than time, you might prefer to see that this is so by examining Fig. 5.37. Use this information to find

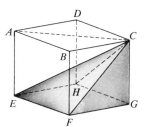

Figure 5.37

The three tents have vertex C and base $GFEH$, vertex C and base $ABFE$, vertex C and base $DHEA$.

the volume of the tent directly. (This trick solves all four problems in Sec. 5.1 since the answer to one determines the answers to the other three.)

The next two exercises concern a property of the function $1/x$ first noticed in the seventeenth century. (See Exercise 18 in Sec. 5.1.)
32 (a) Write out the typical approximating sum for the area under the curve $y = 1/x$ and above the interval $[1, 2]$, using left endpoints as the c_i and the typical partition $x_0 = 1, x_1, \ldots, x_n = 2$.
   (b) Show that $3x_0, 3x_1, \ldots, 3x_n$ is a partition of the interval $[3, 6]$. Show that if the left endpoints are used to form the approximating sum for the area under $y = 1/x$ and above $[3, 6]$, then this sum has the same value as the one obtained in (a).
   (c) Show that the area under the curve $y = 1/x$ and above $[1, 2]$ equals the area under the curve $y = 1/x$ and above $[3, 6]$.
33 (See Exercise 32.) (a) Show that, if $A(t)$ is the area under the curve $y = 1/x$ and above $[1, t]$, then $A(2) = A(6) - A(3)$. (b) Show that, if $x > 1$ and $y > 1$, then $A(x) = A(xy) - A(y)$. (c) By (b), $A(xy) = A(x) + A(y)$ for x and y greater than 1. What famous functions f have the property that $f(xy) = f(x) + f(y)$ for all positive x and y?
34 (See Exercise 28.) Show that if $0 \le a < b$ and m is a positive integer, then
$$\int_a^b x^m \, dx = \frac{b^{m+1}}{m+1} - \frac{a^{m+1}}{m+1}.$$
35 (a) Sketch a graph of $y = 1/(1 + x^2)$ for $0 \le x \le 1$.
   (b) Let A be the area under the graph in (a) and above $[0, 1]$.
   (c) Use elementary geometry to show $\frac{3}{4} < A < 1$.
   (d) Use a partition of $[0, 1]$ into five sections to obtain lower and upper estimates of A. It will be shown that $\int_0^1 1/(1 + x^2) \, dx = \pi/4$, hence that $A \approx 0.7854$.
36 Let f be a decreasing function for x in $[a, b]$. Consider approximating sums for $\int_a^b f(x) \, dx$ with mesh $\le p$.
   (a) Show that any approximating sum formed with left endpoints as sampling numbers differs from the corresponding sum formed with right endpoints as sam-

pling numbers (and the same partition) by at most $p(f(a) - f(b))$.

(b) If you want to use an approximating sum to estimate $\int_a^b f(x)\, dx$ with an error at most 0.001, how fine a mesh should you use?

37 For each positive integer n form an approximating sum for $\int_0^3 x\, dx$ as follows: Partition $[0, 3]$ into n sections, each of length $3/n$. As sampling numbers choose c_i to be the right endpoint of the ith section.

(a) Draw the partition.
(b) What is the numerical value of x_0? x_1? x_2? x_i?
(c) Show that $A_n = \sum_{i=1}^{n} (3i/n)(3/n)$ is an estimate of $\int_0^3 x\, dx$.
(d) Exercise 24(c) of Sec. 5.2 contains the formula

$$\sum_{i=1}^{n} i = \frac{n(n+1)}{2}.$$

Use that formula to show that $A_n = \tfrac{9}{2}(1 + 1/n)$.
(e) When n is large, what number does A_n approach?
(f) With the aid of (e), show that $\int_0^3 x\, dx = \tfrac{9}{2}$.

38 In Exercise 29 of Sec. 5.2 it was shown that

$$\sum_{i=1}^{n} i^2 = \frac{n(n+1)(2n+1)}{6}.$$

This formula can be used to find $\int_0^b x^2\, dx$, as follows. Partition $[0, b]$ into n sections of equal length. Form an approximating sum using right endpoints.

(a) Show that $x_i - x_{i-1} = b/n$.
(b) Show that $x_i = ib/n$.
(c) Show that

$$\sum_{i=1}^{n} f(x_i)(x_i - x_{i-1}) = \frac{(n+1)(2n+1)b^3}{6n^2}.$$

(d) Find

$$\lim_{n \to \infty} \frac{(n+1)(2n+1)b^3}{6n^2}.$$

(This limit is $\int_0^b x^2\, dx$.)

5.4 The fundamental theorems of calculus

This section shows that there is an intimate connection between the definite integral and the derivative. This relationship, expressed in the fundamental theorems of calculus, provides a tool for computing many, but not all, definite integrals without having to form a single approximating sum. The argument will be intuitive. Section 5.6 approaches the fundamental theorems of calculus from a purely mathematical point of view.

THE FIRST FUNDAMENTAL THEOREM OF CALCULUS

In Chap. 3 it was shown that *velocity is the derivative of the distance*. In Sec. 5.3 it was shown that *the definite integral of velocity is the change in distance*. These two facts suggest that there is a close relation between derivatives and definite integrals. In order to express this relation mathematically, let us introduce some mathematical symbols to describe these observations about velocity and change in distance.

For convenience, assume velocity positive (so particle moves to the right).

Let x denote time, and let $F(x)$ denote the coordinate of a particle moving on a line. Then the velocity at time x is the derivative $F'(x)$. The change in distance of the moving particle from time a to time b is

Final coordinate − initial coordinate = $F(b) - F(a)$.

The assertion, "The definite integral of velocity is the change in distance," now reads mathematically

$$\int_a^b F'(x)\, dx = F(b) - F(a).$$

5.4 THE FUNDAMENTAL THEOREMS OF CALCULUS

This equation generalizes the statement, "Rate · time = distance," which is valid for a particle moving at a constant speed.

Recall also from Chap. 3 that *density is the derivative of mass.* In Sec. 5.3 it was shown that *mass is the definite integral of density.* Specifically, let x denote the distance of a point from the left end of a string whose density may not be constant. Let $F(x)$ be the mass of the string situated to the left of x. Then $F'(x)$ is the density of the string at x. The assertion, "The definite integral of density is mass," now reads

$$\int_a^b F'(x)\,dx = F(b) - F(a),$$

where a and b are the coordinates of the endpoints of a typical section of string. See Fig. 5.38. The equation relating F', $F(a)$, and $F(b)$ generalizes the simple statement, "Density · length = mass," which is valid for a homogeneous string.

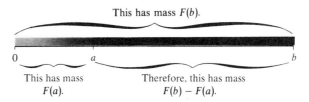

Figure 5.38

These two physical arguments suggest that there is a single purely mathematical theorem in the background. Moreover, computational evidence from Sec. 5.3 also points in the same direction for the particular function $F(x) = x^3/3$. It was proved in Sec. 5.3 that

$$\int_a^b x^2\,dx = \frac{b^3}{3} - \frac{a^3}{3}, \qquad 0 \le a < b.$$

Define the function F by the formula

$$F(x) = \frac{x^3}{3}.$$

Then

$$F'(x) = x^2,$$

and the equation

$$\int_a^b x^2\,dx = \frac{b^3}{3} - \frac{a^3}{3}$$

now reads

$$\int_a^b F'(x)\,dx = F(b) - F(a).$$

Thus three separate lines of reasoning all suggest the general and purely mathematical result:

This equation links the derivative and the definite integral.

$$\int_a^b F'(x)\, dx = F(b) - F(a),$$

that is, "The definite integral of the derivative of a function over an interval is simply the difference in the values of the function at the ends of the interval."

As it stands, the conjecture is not quite correct. A function F may have such a wild derivative F' that $\int_a^b F'(x)\, dx$ does not exist; that is, the approximating sums do not approach a single number as the mesh of the partition approaches 0. But, as mentioned in Sec. 5.3, it is proved in advanced calculus that if a function f is continuous on the interval $[a, b]$, then $\int_a^b f(x)\, dx$ *does* exist. We shall accept this as an assumption, for it takes a long time to prove.

The observations about velocity, density, and $\int_a^b x^2\, dx$ all suggest the following theorem, whose proof will be presented in Sec. 5.6.

FIRST FUNDAMENTAL THEOREM OF CALCULUS

If f is continuous on $[a, b]$ and if F is an antiderivative of f,

then
$$\int_a^b f(x)\, dx = F(b) - F(a). \quad \blacksquare$$

The symbol f is introduced for F' with a view toward applying this theorem to the computation of definite integrals. It says, "If you want to compute $\int_a^b f(x)\, dx$, search for a function F whose derivative is f, that is,

$$F' = f.$$

Then $\int_a^b f(x)\, dx$ is just $F(b) - F(a)$."

The following examples will exhibit the power and the limitations of the fundamental theorem of calculus. (Generally the adjective "first" is omitted.)

EXAMPLE 1 Find the area of the region under the curve $y = \cos x$, above the x axis, and between $x = 0$ and $x = \pi/2$. (See Fig. 5.39.)

SOLUTION As was shown in Sec. 5.3, area is the definite integral of the cross-sectional length. In this case

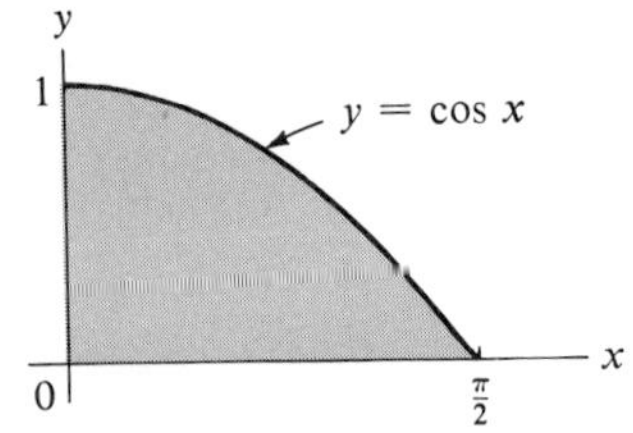

$$\text{Area} = \int_0^{\pi/2} \cos x\, dx.$$

The fundamental theorem of calculus asserts that if we can find a function F such that

$$F'(x) = \cos x,$$

then the definite integral can be evaluated easily, as $F(\pi/2) - F(0)$. Now,

Figure 5.39

5.4 THE FUNDAMENTAL THEOREMS OF CALCULUS

in Chap. 3 it was shown that the derivative of the sine function is the cosine function. So let

$$F(x) = \sin x.$$

The fundamental theorem of calculus then says

$$\int_0^{\pi/2} \cos x \, dx = F\left(\frac{\pi}{2}\right) - F(0)$$

$$= \sin \frac{\pi}{2} - \sin 0$$

$$= 1 - 0$$

$$= 1. \ \blacksquare$$

EXAMPLE 2 In Sec. 5.3 it was shown that

$$\int_a^b x^2 \, dx = \frac{b^3}{3} - \frac{a^3}{3},$$

when both a and b are nonnegative and $a < b$. What does the fundamental theorem of calculus say about $\int_a^b x^2 \, dx$ for any a and b, $a < b$?

SOLUTION The function $F(x) = x^3/3$ is an antiderivative of x^2. According to the fundamental theorem of calculus,

$$\int_a^b x^2 \, dx = F(b) - F(a) = \frac{b^3}{3} - \frac{a^3}{3}.$$

This holds even if a or b is negative provided that a is less than b. $\blacksquare$

The next two examples form a fable whose moral should be remembered.

EXAMPLE 3 Compute $\int_0^{\pi/4} x \cos x \, dx$.

SOLUTION To apply the fundamental theorem of calculus, it is necessary to find an antiderivative of $x \cos x$. It happens that the derivative of the function $x \sin x + \cos x$ equals $x \cos x$. (Chapter 7 presents some methods for finding an antiderivative, when it is possible to do so.) The fundamental theorem of calculus asserts that if

$$F(x) = x \sin x + \cos x,$$

The letters FTC under the equality sign record the use of the fundamental theorem of calculus.

then
$$\int_0^{\pi/4} x \cos x \, dx \underset{\text{FTC}}{=} F\left(\frac{\pi}{4}\right) - F(0)$$

$$= \left(\frac{\pi}{4} \sin \frac{\pi}{4} + \cos \frac{\pi}{4}\right) - (0 \cdot \sin 0 + \cos 0)$$

$$= \left(\frac{\pi}{4} \cdot \frac{\sqrt{2}}{2} + \frac{\sqrt{2}}{2}\right) - (0 \cdot 0 + 1)$$

$$= \frac{\pi\sqrt{2}}{8} + \frac{\sqrt{2}}{2} - 1. \quad \blacksquare$$

EXAMPLE 4   Compute $\int_0^{\pi/4} x \tan x \, dx$.

ATTEMPT AT SOLUTION   To apply the fundamental theorem of calculus, it is necessary to find a function F such that

$$F'(x) = x \tan x.$$

Elementary functions

As will be shown, there is such a function F. However, mathematicians have proved that F is *not an elementary function*. This is, F is not expressible in terms of polynomials, logarithms, exponentials, trigonometric functions, or any composition of these functions. We are therefore blocked, for the fundamental theorem of calculus is of use in computing $\int_a^b f(x) \, dx$ only if f is "nice" enough to be the derivative of an elementary function. $\blacksquare$

The moral of these last two examples is this: It is not easy to tell by glancing at f whether the desired F is elementary. After all, $x \tan x$ looks no more complicated than $x \cos x$, yet it is not the derivative of an elementary function, while $x \cos x$ is.

As another example, $\cos \sqrt{x}$ looks more complicated than $\cos x^2$. Yet it turns out that $\cos \sqrt{x}$ is the derivative of an elementary function, while $\cos x^2$ is not. (It is not hard to check that $(2\sqrt{x} \sin \sqrt{x} + 2 \cos \sqrt{x})'$ is $\cos \sqrt{x}$.)

Every time that we differentiate an elementary function we get an elementary function. For instance, $(\sqrt{x})' = 1/(2\sqrt{x}), (x^3 - 3x^2)' = 3x^2 - 6x$, and $(\tan x)' = \sec^2 x$. In Sec. 6.2 it will be shown that the derivative of a logarithm function is elementary, and in Sec. 6.4 that the derivative of an exponential function, such as 10^x, is elementary. In Sec. 6.5 it will be shown that the derivatives of the inverse trigonometric functions, such as arcsine of x, are also elementary. But if you start with an elementary function f and search for an elementary function F whose derivative is to be f, you may be frustrated—not because it may be hard to find F—but because it may be that no such F exists.

THE SECOND FUNDAMENTAL THEOREM OF CALCULUS

There is a second theorem, closely related to the first fundamental theorem of calculus, which is also called the fundamental theorem of calculus. It describes the connection between the derivative and the definite integral in a different way from the first fundamental theorem of calculus.

Let f denote the density, in grams per centimeter, of a string. That is, at a distance of t centimeters from the left end, the string has density $f(t)$ grams per centimeter. (The letter t rather than x is used since, in a moment, the letter x will be needed to describe the endpoint of an interval.)

5.4 THE FUNDAMENTAL THEOREMS OF CALCULUS

The total mass of the string from $t = a$ to $t = b$ is

$$\int_a^b f(t)\, dt \quad \text{grams}.$$

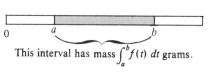

Figure 5.40

(See Fig. 5.40.) Now, keep a fixed and consider intervals of the form $[a, x]$, where $a < x \leq b$.

Define a function G as follows. For $x > a$, let

$$G(x) = \int_a^x f(t)\, dt.$$

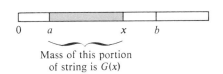

Figure 5.41

Then $G(x)$ is the mass in the part of the string situated in the interval $[a, x]$, shown in Fig. 5.41.

What is the derivative of G? By the definition of the derivative,

$$G'(x) = \lim_{\Delta x \to 0} \frac{G(x + \Delta x) - G(x)}{\Delta x}$$

$$= \lim_{\Delta x \to 0} \frac{\Delta G}{\Delta x}.$$

The numerator, $\qquad G(x + \Delta x) - G(x) = \Delta G$

is the mass of the string in the interval $[x, x + \Delta x]$, shown in Fig. 5.42.

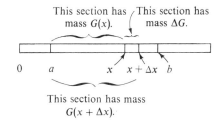

Figure 5.42

Thus the quotient $\qquad \dfrac{\Delta G}{\Delta x}$

is an estimate of the density at x. Consequently,

$$\lim_{\Delta x \to 0} \frac{\Delta G}{\Delta x} = \text{density at } x.$$

But the density at x is given as $f(x)$. Therefore we conclude that

$$G'(x) = f(x).$$

In short, $\qquad \dfrac{d}{dx}\left(\int_a^x f(t)\, dt \right) = f(x).$

Putting the above equation in words we would say, "The derivative of the definite integral of f from a to x is $f(x)$."

Let us check whether this is true in a particular case, where the formula for G is known, say when $f(t) = t^2$.

EXAMPLE 5 Let $f(t) = t^2$ and let $\qquad G(x) = \int_a^x t^2\, dt.$

Is $G'(x) = x^2$?

266 THE DEFINITE INTEGRAL

SOLUTION In this case we may compute $G(x)$ explicitly:

$$G(x) = \int_a^x t^2 \, dt \underset{\text{FTC}}{=} \frac{x^3}{3} - \frac{a^3}{3}.$$

Now
$$G'(x) = \frac{d}{dx}\left(\frac{x^3}{3} - \frac{a^3}{3}\right) = \frac{3x^2}{3} - 0 = x^2.$$

(Remember that a is constant.) As expected, $G'(x)$ does equal x^2. ∎

The assertion
$$\frac{d}{dx}\left(\int_a^x f(t) \, dt\right) = f(x)$$

is the substance of the second fundamental theorem of calculus. It says, "The derivative of the definite integral of f with respect to the right end coordinate of the interval is simply f evaluated at that coordinate."

SECOND FUNDAMENTAL THEOREM OF CALCULUS

Let f be continuous on an open interval containing the interval $[a, b]$. Let

$$G(x) = \int_a^x f(t) \, dt$$

for $a \leq x \leq b$. Then G is differentiable on $[a, b]$ and its derivative is f; that is,

$$G'(x) = f(x). \quad \blacksquare$$

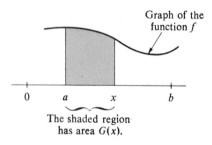

The shaded region has area $G(x)$.

Figure 5.43

Considerations of density and mass suggested the second fundamental theorem. In view of its importance it is worthwhile to pause and see how considerations of area would also suggest the same conclusion.

Let f be a continuous function such that $f(x)$ is positive for x in $[a, b]$. For x in $[a, b]$ let $G(x)$ be the area of the region under the graph of f and above the interval $[a, x]$, shown in Fig. 5.43. Let Δx be a small positive number. Then $G(x + \Delta x)$ is the area under the graph of f and above the interval $[a, x + \Delta x]$. Consequently,

$$\Delta G = G(x + \Delta x) - G(x)$$

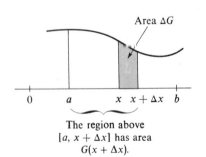

The region above $[a, x + \Delta x]$ has area $G(x + \Delta x)$.

Figure 5.44

is the area of the narrow strip shaded in Fig. 5.44. When Δx is small, the narrow shaded strip above $[x, x + \Delta x]$ resembles a rectangle of base Δx and height $f(x)$, with area $f(x)\Delta x$. Therefore it seems reasonable that, when Δx is small,

$$\frac{\Delta G}{\Delta x} \text{ is approximately } f(x).$$

In short, it seems plausible that

$$\lim_{\Delta x \to 0} \frac{\Delta G}{\Delta x} = f(x).$$

Briefly, $$G'(x) = f(x).$$

Now $$G(x) = \int_a^x f(t)\,dt,$$

since area is the definite integral of the cross-sectional length. Thus considerations of area lead to the same formula as did considerations of mass, namely

$$\frac{d}{dx}\left(\int_a^x f(t)\,dt\right) = f(x).$$

This is again the second fundamental theorem.

As a consequence of this theorem, every continuous function is the derivative of some function. This is stated as a corollary for emphasis.

COROLLARY Let f be continuous on an interval $[a, b]$. Then f is the derivative of some function.

PROOF Let $F(x) = \int_a^x f(t)\,dt$. Then, by the second fundamental theorem,

$$F'(x) = f(x),$$

that is, $$\frac{d}{dx}\left(\int_a^x f(t)\,dt\right) = f(x).$$

This proves the corollary. ∎

EXAMPLE 6 Find a function whose derivative is $\sin x^2$.

SOLUTION The function F defined by

$$F(x) = \int_0^x \sin t^2\,dt$$

is such a function. ∎

The information obtained in Example 6 is of no use in trying to exploit the fundamental theorem of calculus to compute, say, $\int_0^{\pi/4} \sin x^2\,dx$. Though there is a function whose derivative is $\sin x^2$, there is no *elementary* function whose derivative is $\sin x^2$; hence other means must be used to compute $\int_0^{\pi/4} \sin x^2\,dx$. (The proof that there is no such elementary function is complicated and is given in graduate-level courses.)

EXAMPLE 7 Differentiate the function

$$f(x) = \int_0^{x^2} \sqrt{1 + t^2}\,dt.$$

SOLUTION The upper limit of integration is x^2, not x. The chain rule will be needed. Let

How to differentiate a definite integral if upper limit is a function of x

Then
$$y = \int_0^{x^2} \sqrt{1+t^2}\, dt.$$

$$y = \int_0^u \sqrt{1+t^2}\, dt, \quad \text{where} \quad u = x^2.$$

By the second fundamental theorem of calculus,

$$\frac{dy}{du} = \sqrt{1+u^2}.$$

The chain rule then says that

$$\frac{dy}{dx} = \frac{dy}{du} \cdot \frac{du}{dx} = \sqrt{1+u^2} \cdot 2x$$
$$= \sqrt{1+(x^2)^2} \cdot 2x$$
$$= 2x\sqrt{1+x^4}. \quad \blacksquare$$

Exercises

1. Let k be a constant and let a be a fixed rational number, $a \neq -1$. Show by differentiation that $kx^{a+1}/(a+1)$ is an antiderivative of kx^a.
2. Use the formula in Exercise 1 to find an antiderivative of
   (a) $5x^3$, (b) $-4x^2$, (c) x^{-4}, (d) $5/x^3$,
   (e) $\sqrt{x}$, (f) $1/\sqrt{x}$, (g) $\sqrt[3]{x}$, (h) $1/x^2$.

In Exercises 3 to 12 use the fundamental theorem of calculus to evaluate the given quantities.

3. The area of the region under the curve $3x^2$ and above $[1, 4]$.
4. The area of the region under the curve $1/x^2$ and above $[2, 3]$.
5. The area of the region under the curve $6x^4$ and above $[-1, 1]$.
6. The area of the region under the curve $\sqrt{x}$ and above $[25, 36]$.
7. The distance an object travels from time $t = 1$ second to time $t = 2$ seconds, if its speed at time t seconds is t^5 feet per second.
8. The distance an object travels from time $t = 1$ second to time $t = 8$ seconds if its speed at time t seconds is $7\sqrt[3]{t}$ feet per second.
9. The total mass of a string in the section $[1, 2]$ if its density at x is $5x^3$ grams per centimeter.
10. The total mass of a string in the section $[\frac{1}{4}, 1]$ if its density at x is $4\sqrt{x}$ grams per centimeter.
11. The volume of a solid located between a plane at $x = 1$ and a plane located at $x = 5$ if the cross-sectional area of the solid by a plane corresponding to x is $6x^3$ square centimeters. (Assume that the planes are all perpendicular to the x axis.)
12. Like Exercise 11, except that the typical cross-sectional area is $1/x^3$ instead of $6x^3$.

In Exercises 13 to 20 use the fundamental theorem of calculus to evaluate the definite integrals.

13. $\int_1^2 x^3\, dx$
14. $\int_{-1}^1 x^3\, dx$
15. $\int_0^3 6x\, dx$
16. $\int_1^4 (1/x^2)\, dx$
17. $\int_4^9 5\sqrt{x}\, dx$
18. $\int_1^9 (1/\sqrt{x})\, dx$
19. $\int_1^8 \sqrt[3]{x^2}\, dx$
20. $\int_2^3 (4/x^3)\, dx$

Exercises 21 to 30 concern the second fundamental theorem of calculus. In each of Exercises 21 to 24 find the derivative of the given function f of x, $x > 1$, in two ways. First evaluate the function with the aid of the first fundamental theorem of calculus, and then differentiate the result. In the second approach, use the second fundamental theorem of calculus.

21. $\int_1^x t^4\, dt$
22. $\int_1^x (1/t^2)\, dt$
23. $\int_1^{x^2} 10\sqrt{t}\, dt$
24. $\int_1^{x^3} (t^2 + 5t)\, dt$

In Exercises 25 to 30 use the second fundamental theorem of calculus to differentiate the given functions of x.

25. $\int_1^x t^5\, dt$
26. $\int_1^{\sqrt{x}} (1/t^3)\, dt$

27 $\int_0^{\sin x} \sqrt{1+t^3}\, dt$ 28 $\int_{-1}^{\sqrt{x}} 3^t\, dt$

29 $\int_0^x \sin t^2\, dt$ 30 $\int_1^{x^3} \sqrt{t}\sqrt{t+1}\sqrt{t+2}\, dt$

■

Exercises 31 to 36 concern the volume of the solid obtained by revolving a region in the xy plane around the x axis.

31 Let $y = f(x)$ be nonnegative for x in $[a, b]$. The region below $y = f(x)$ and above $[a, b]$ is revolved around the x axis to form a "solid of revolution," as shown in Fig. 5.45.

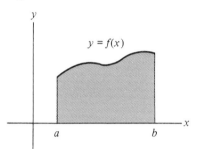

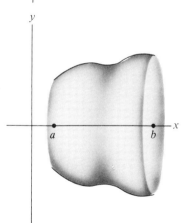

Figure 5.45

(a) What is the area of the cross section of the solid made by a plane perpendicular to the x axis and passing through the point on the x axis with coordinate x?

(b) Using (a), obtain a definite integral that equals the volume of the solid.

In Exercises 32 to 36 a solid of revolution is formed by revolving around the x axis the region under the graph of the given function and above the given interval. Find its volume using the formula developed in Exercise 31.

32 x^3, $[1, 2]$
33 $x^{3/2}$, $[0, 1]$
34 $1/x$, $[1, 2]$
35 $x + 1$, $[2, 3]$
36 $5x$, $[0, 1]$
37 In Example 1, $F(x) = \sin x$ was used to compute $\int_0^{\pi/2} \cos x\, dx$.
   (a) Show that the function $5 + \sin x$ also has a derivative equal to $\cos x$.
   (b) Use $F(x) = 5 + \sin x$ to evaluate $\int_0^{\pi/2} \cos x\, dx$.

38 If $f(x)$ is positive, then $\int_a^b f(x)\, dx$ represents the area of the region below the graph of f and above the interval $[a, b]$. If $f(x)$ is negative, then $\int_a^b f(x)\, dx$ is *negative*. It is the *negative* of the area of the region above the graph and below $[a, b]$, as shown in Fig. 5.46.

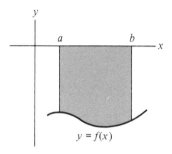

Figure 5.46

Use this observation to find the area of the region (a) above $y = \sin x$ and below $[\pi, 2\pi]$; (b) above $y = x^3$ and below $[-2, -1]$.

■ ■

In each of Exercises 39 and 40 there are three definite integrals. Two are easy to evaluate with the aid of the fundamental theorem of calculus, while the other, like $\int_0^{\pi/4} x \tan x\, dx$, cannot be evaluated by it. Decide which integrals can be evaluated by the fundamental theorem and evaluate them.

39 (a) $\int_0^1 \sqrt[3]{x + x^2}\, dx$ (b) $\int_0^1 \sqrt[3]{1 + x}\, dx$

    (c) $\int_0^1 \dfrac{1}{\sqrt[3]{1 + x}}\, dx$.

40 (a) $\int_0^1 x\sqrt{x^2 + 1}\, dx$ (b) $\int_0^1 x^2 \sqrt[3]{x^3 + 1}\, dx$

    (c) $\int_0^1 \sqrt[3]{x^3 + 1}\, dx$

41 A plane at a distance x from the center of a sphere of radius r, $0 \leq x \leq r$, meets the sphere in a circle.
   (a) Show that the radius of the circle is $\sqrt{r^2 - x^2}$.
   (b) Show that the area of the circle is $\pi r^2 - \pi x^2$.
   (c) Using the fundamental theorem, find the volume of the sphere.

42 There is no elementary function whose derivative is $\sqrt{\sin x}$. Consider the curve $y = \sqrt{\sin x}$ for x in $[0, \pi]$.
(a) Set up a definite integral for the area of the region under the curve and above $[0, \pi]$.
(b) Set up a definite integral for the volume of the solid obtained by revolving the region in (a) around the x axis. (*Hint*: See Exercise 31.)
(c) Show that the fundamental theorem of calculus is useful in finding the volume in (b) but not in finding the area in (a).

5.5 Properties of the antiderivative and the definite integral

This section describes some of the important properties of the antiderivative and the definite integral and gives a notation for the antiderivative. Frequently an antiderivative is called an *integral* or *indefinite integral*. There is a danger that "integral" will become confused with "definite integral." It should be kept in mind that the definite integral $\int_a^b f(x)\,dx$ is defined as a number, a limit of certain sums, while an integral or antiderivative is a function. Sometimes, in applications, both the definite integral and the indefinite integral are called integrals; it takes a clear mind and mastery of the definitions to keep the ideas separate. The table of integrals in a mathematical handbook is primarily a table of antiderivatives (functions); it is usually followed by a short section that lists the values of a few common definite integrals (numbers).

Inside the cover of this book is a short table of antiderivatives or "indefinite integrals."

NOTATION An antiderivative of f is denoted $\int f(x)\,dx$.

Thus if $F' = f$, we write $F = \int f(x)\,dx$ and say "F is an antiderivative of f" or, for convenience, "$F(x)$ is an antiderivative of $f(x)$." For instance, we say "x^3 is an antiderivative of $3x^2$." Note that $x^3 = \int 3x^2\,dx$ and $x^3 + 1 = \int 3x^2\,dx$. That does *not* imply that $x^3 = x^3 + 1$.

PROPERTIES OF ANTIDERIVATIVES

For any constant C, $(x^3 + C)' = 3x^2$. Thus

$$x^3 + C = \int 3x^2\,dx.$$

There are no other antiderivatives of $3x^2$. If you have found one antiderivative F for a function f, then any other antiderivative of f is of the form $F(x) + C$ for some constant C. This is established in the following theorem.

THEOREM 1 If F and G are both antiderivatives of f on an interval $[a, b]$, then there is a constant C such that

$$F(x) = G(x) + C.$$

PROOF The functions F and G have the same derivative f. By Corollary 2 in Sec. 4.1 they must differ by a constant. This proves the theorem. ∎

5.5 PROPERTIES OF THE ANTIDERIVATIVE AND THE DEFINITE INTEGRAL

In tables C is usually omitted.

When writing down an antiderivative it is best to add the constant C. (It will be needed in the study of differential equations.) For example

$$\int 5\, dx = 5x + C,$$

$$\int x^3\, dx = \frac{x^4}{4} + C,$$

and

$$\int \sin 2x\, dx = -\frac{\cos 2x}{2} + C.$$

Any property of derivatives implies a corresponding property of antiderivatives. The next theorem records two of the most important properties of antiderivatives.

THEOREM 2 Assume that f and g are functions with antiderivatives $\int f(x)\, dx$ and $\int g(x)\, dx$. Then the following hold:

Properties of antiderivatives

(1) $\int cf(x)\, dx = c \int f(x)\, dx$ for any constant c

(2) $\int (f(x) + g(x))\, dx = \int f(x)\, dx + \int g(x)\, dx$

(3) $\int (f(x) - g(x))\, dx = \int f(x)\, dx - \int g(x)\, dx$

PROOF

(1) It is necessary to show that the derivative of $c \int f(x)\, dx$ is $cf(x)$. The differentiation follows:

$$\frac{d}{dx}\left(c \int f(x)\, dx\right) = c\frac{d}{dx}\left(\int f(x)\, dx\right) \qquad \text{a constant moves past the derivative symbol}$$

$$= cf(x) \qquad \text{definition of } \int f(x)\, dx.$$

(2) It is necessary to show that the derivative of $\int f(x)\, dx + \int g(x)\, dx$ is $f(x) + g(x)$:

$$\frac{d}{dx}\left(\int f(x)\, dx + \int g(x)\, dx\right) = \frac{d}{dx}\left(\int f(x)\, dx\right) + \frac{d}{dx}\left(\int g(x)\, dx\right)$$

$$= f(x) + g(x).$$

(3) The proof of (3) is similar to that of (2). ∎

The last two parts of Theorem 2 extend to any finite number of functions. For instance,

$$\int (f(x) - g(x) + h(x))\, dx = \int f(x)\, dx - \int g(x)\, dx + \int h(x)\, dx.$$

THEOREM 3 Let a be a rational number other than -1. Then

$$\int x^a\, dx = \frac{x^{a+1}}{a+1} + C.$$

272 THE DEFINITE INTEGRAL

PROOF $\left(\dfrac{x^{a+1}}{a+1} + C\right)' = \dfrac{(a+1)x^{a+1-1}}{a+1} = x^a.$ ∎

EXAMPLE 1 Find $\int (2x^5 - 3x^2 + 4)\, dx.$

SOLUTION

$\int (2x^5 - 3x^2 + 4)\, dx = \int 2x^5\, dx - \int 3x^2\, dx + \int 4\, dx$ Theorem 2

$\qquad = 2\int x^5\, dx - 3\int x^2\, dx + \int 4\, dx$ Theorem 2

A single constant is enough.

$\qquad = 2\dfrac{x^6}{6} - 3\dfrac{x^3}{3} + 4x + C$ Theorem 3

$\qquad = \dfrac{x^6}{3} - x^3 + 4x + C.$ ∎

As Example 1 illustrates, an antiderivative of any polynomial is again a polynomial.

NOTATION $F(b) - F(a)$ is abbreviated to $F(x)\Big|_a^b$.

EXAMPLE 2 Evaluate $\displaystyle\int_1^2 \dfrac{1}{x^2}\, dx$ by the fundamental theorem of calculus.

SOLUTION $\displaystyle\int_1^2 \dfrac{1}{x^2}\, dx = \dfrac{-1}{x}\Big|_1^2 = \dfrac{-1}{2} - \dfrac{-1}{1} = \dfrac{-1}{2} + 1 = \dfrac{1}{2}.$ ∎

The fundamental theorem of calculus asserts that

$\displaystyle\underbrace{\int_1^2 \dfrac{1}{x^2}\, dx}_{\text{The definite integral; a limit of sums.}} \quad = \quad \underbrace{\int \dfrac{1}{x^2}\, dx\,\Big|_1^2}_{\substack{\text{An antiderivative: a function; the value of this}\\ \text{function at 1 is subtracted from its value at 2.}}}$

The symbols on the right and left of the equal sign are so similar that it is tempting to think that the equation is obvious or says nothing whatever. *Beware:* That compact equation is in fact a terse statement of the (first) fundamental theorem of calculus.

DEFINITION *Integrand.* In the definite integral $\int_a^b f(x)\, dx$ and in the antiderivative $\int f(x)\, dx$, $f(x)$ is called the *integrand*.

The related processes of computing $\int_a^b f(x)\, dx$ and of finding an antiderivative $\int f(x)\, dx$ are both called *integrating* $f(x)$. Thus integration refers to two separate but related problems: computing a number $\int_a^b f(x)\, dx$ or finding a function $\int f(x)\, dx$. The fundamental theorem of calculus states that the second process may be of use in computing $\int_a^b f(x)\, dx$.

PROPERTIES OF THE DEFINITE INTEGRAL

In the notation for the definite integral $\int_a^b f(x)\,dx$, b is larger than a. It will be useful to be able to speak of "the definite integral from a to b" even if b is less than or equal to a. The following definitions meet this need and will be used in the next section in the proofs of the two fundamental theorems of calculus.

DEFINITION *The integral from a to b, where b is less than a.* If b is less than a, then

$$\int_a^b f(x)\,dx = -\int_b^a f(x)\,dx.$$

EXAMPLE 3 Compute $\int_3^0 x^2\,dx$, the integral from 3 to 0 of x^2.

SOLUTION The symbol $\int_3^0 x^2\,dx$ is defined as $-\int_0^3 x^2\,dx$. As was shown in Sec. 5.3, $\int_0^3 x^2\,dx = 9$. Thus

$$\int_3^0 x^2\,dx = -9. \quad\blacksquare$$

DEFINITION *The integral from a to a.* $\int_a^a f(x)\,dx = 0$.

Remark: The definite integral is defined with the aid of partitions. Rather than permit partitions to have sections of length 0, it is simpler just to make the above definition.

The point of making these two definitions is that now the symbol $\int_a^b f(x)\,dx$ is defined for any numbers a and b and any continuous function f. It is no longer necessary that a be less than b.

The definite integral has several properties, some of which will be used in this section and some later in the text.

PROPERTIES OF THE DEFINITE INTEGRAL

Let f and g be continuous functions and let c be a constant. Then

(1) $\int_a^b cf(x)\,dx = c\int_a^b f(x)\,dx.$

(2) $\int_a^b (f(x) + g(x))\,dx = \int_a^b f(x)\,dx + \int_a^b g(x)\,dx.$

(3) $\int_a^b (f(x) - g(x))\,dx = \int_a^b f(x)\,dx - \int_a^b g(x)\,dx.$

(4) If $f(x) \geq 0$ for all x in $[a, b]$, $a < b$, then

$$\int_a^b f(x)\,dx \geq 0.$$

(5) If $f(x) \geq g(x)$ for all x in $[a, b]$, $a < b$, then

$$\int_a^b f(x)\,dx \geq \int_a^b g(x)\,dx.$$

274   THE DEFINITE INTEGRAL

(6) If a, b, and c are numbers, then

$$\int_a^c f(x)\,dx + \int_c^b f(x)\,dx = \int_a^b f(x)\,dx.$$

(7) If m and M are numbers and $m \leq f(x) \leq M$ for all x between a and b,

$$m(b-a) \leq \int_a^b f(x)\,dx \leq M(b-a) \quad \text{if} \quad a < b,$$

and

$$m(b-a) \geq \int_a^b f(x)\,dx \geq M(b-a) \quad \text{if} \quad b < a.$$

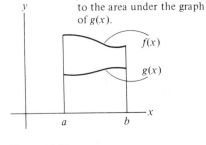

The area under the graph of $f(x)$ is greater than or equal to the area under the graph of $g(x)$.

Figure 5.47

Interpretation of the definite integral as an area (for positive integrands) makes these seven assertions plausible. For instance, (5) amounts then to the assertion that if one plane region contains another, then it has at least as large an area as the region it contains. (See Fig. 5.47.)

In the case that $a < c < b$ and $f(x)$ assumes only positive values, (6) asserts that the area of the region below the graph of f and above the interval $[a, b]$ is the sum of the areas of the regions below the graph and above the smaller intervals $[a, c]$ and $[c, b]$. This is certainly plausible.

Figure 5.48 expresses (6) geometrically.

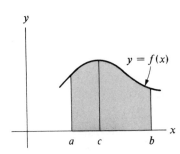

The area from a to c plus the area from c to b is equal to the area from a to b.

Figure 5.48

The inequalities in (7) compare the area under $f(x)$ with the areas of two rectangles, one of height M and one of height m. (See Fig. 5.49.)

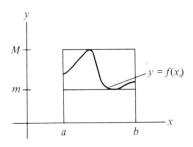

Figure 5.49

Properties (1) to (7) can be proved by examining approximating sums and seeing how they behave as the mesh approaches 0. (See Exercises 43 and 44 for instance.)

A mean-value theorem for definite integrals follows from these properties.

MEAN VALUE THEOREM FOR DEFINITE INTEGRALS Let a and b be numbers, and let f be a continuous function defined for x between a and b. Then there is a number c between a and b such that

$$\int_a^b f(x)\, dx = f(c)(b - a).$$

PROOF Consider the case $a < b$. Let M be the maximum and m the minimum of $f(x)$ for x in $[a, b]$. [Recall the maximum- (and minimum-) value theorem of Sec. 2.7.] By (7),

$$m \leq \frac{\int_a^b f(x)\, dx}{b - a} \leq M.$$

By the intermediate-value theorem of Sec. 2.7 there is a number c in $[a, b]$ such that

$$f(c) = \frac{\int_a^b f(x)\, dx}{b - a},$$

and the theorem is proved. (The case $b < a$ can be obtained from the case $a < b$.) ∎

If $f(x)$ is positive, $\int_a^b f(x)\, dx$ can be interpreted as the area of the region below $y = f(x)$ and above $[a, b]$. The mean-value theorem then asserts that there is a rectangle whose base is $[a, b]$ and whose height is $f(c)$ that has the same area as the region for some c in $[a, b]$. This rectangle is shown in Fig. 5.50.

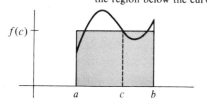

The area of the rectangle is the same as the area of the region below the curve.

Figure 5.50

Exercises

In Exercises 1 to 10 evaluate the antiderivatives, adding in each case the constant C. Check each answer by differentiating it.

1 $\int (2x - x^3 + x^5)\, dx$

2 $\int \left(6x^2 + \frac{1}{\sqrt{x}}\right) dx$

3 $\int \left(3x + 4 \sin 2x - \frac{1}{4x^2}\right) dx$

4 $\int (\sin 2x + \cos 4x)\, dx$

5 $\int \sqrt{2x + 5}\, dx$

6 $\int \frac{1}{(2x + 1)^3}\, dx$

7 $\int \tan 2x \sec 2x\, dx$

8 $\int \csc^2 3x\, dx$

9 $\int (1 - \cos 2x)\, dx$

10 $\int \sin^2 x\, dx$ (*Hint:* Use a trigonometric identity.)

11 Compute: (a) $(\int x^5 \, dx)'$, (b) $(\int \sin x^5 \, dx)'$.

12 What guarantees that

$$\int_0^{\pi/2} \sin x \, dx = \int \sin x \, dx \Big|_0^{\pi/2} ?$$

13 Compute: (a) $\int x^2 \, dx \Big|_1^2$ (b) $\int \frac{1}{(x+1)^2} \, dx \Big|_1^2$

(c) $\int \sin 3x \, dx \Big|_1^2$.

14 Compute: (a) $\int \cos x \, dx \Big|_0^{\pi/4}$ (b) $\int \sec^2 x \, dx \Big|_0^1$

(c) $\int \frac{1}{\sqrt[3]{x}} \, dx \Big|_1^2$.

15 Compute: (a) $x^2 \Big|_1^2$ (b) $x^2 \Big|_{-1}^1$

(c) $\sin x \Big|_0^\pi$ (d) $\cos x \Big|_0^\pi$.

16 Find $\int 1 \, dx$. (This is usually written $\int dx$.)

17 (a) Is $\int x^2 \, dx$ a function or is it a number?
(b) Is $\int x^2 \, dx \Big|_1^3$ a function or is it a number?
(c) Is $\int_1^3 x^2 \, dx$ a function or is it a number?

18 (a) Which of these two numbers is defined as a limit of sums:

$$\int x^2 \, dx \Big|_1^3 \quad \text{or} \quad \int_1^3 x^2 \, dx ?$$

(b) Why are the two numbers in (a) equal?

19 True or false: (a) Every elementary function has an elementary derivative. (b) Every elementary function has an elementary antiderivative.

20 True or false: (a) $\sin x^2$ has an elementary antiderivative, (b) $\sin x^2$ has an antiderivative.

In Exercises 21 and 22 verify the equations quoted from a table of antiderivatives (integrals). Just differentiate each of the alleged antiderivatives and see whether you obtain the integrand. The number a is constant in each case.

21 $\int x^2 \sin ax \, dx = \frac{2x}{a^2} \sin ax + \frac{2}{a^3} \cos ax - \frac{x^2}{a} \cos ax + C.$

22 $\int x \sin^2 ax \, dx = \frac{x^2}{4} - \frac{x \sin 2ax}{4a} - \frac{\cos 2ax}{8a^2} + C.$

In Exercises 23 to 26 evaluate the expressions.

23 $\int_0^0 2^{x^2} \, dx$ **24** $\int_1^0 x^2 \, dx$

25 $\int_2^1 (12x^3 - 2x) \, dx$ **26** $\int_3^2 (x+1)^{-2} \, dx$

In Exercises 27 to 30 differentiate $f(x)$.

27 $f(x) = \int_x^{17} \sin^3 t \, dt$ (*Hint*: First rewrite it as $-\int_{17}^x \sin^3 t \, dt$.)

28 $f(x) = \int_{x^2}^{17} \sin^3 2t \, dt$

29 $f(x) = \int_{2x}^{3x} t \tan t \, dt$ (*Hint*: First rewrite it as $\int_{2x}^0 t \tan t \, dt + \int_0^{3x} t \tan t \, dt$.)

30 $f(x) = \int_{x^2}^{x^3} \sqrt[3]{1+t^2} \, dt$

In Exercises 31 to 34 find d^2y/dx^2.

31 $y = \int_{1984}^x \sin t^2 \, dt$

32 $y = \int_{1776}^x t^2 \sqrt{1+2t} \, dt$

33 $y = \int_x^{1492} \frac{t^3 \sin 2t}{\sqrt{1+3t}} \, dt$

34 $y = \int_{x^2}^{1865} t^2 \sqrt[3]{1+2t} \cos 3t \, dt$

35 (a) If f is continuous, and $3 \leq f(x) \leq 5$ for all x in $[2, 6]$, what can be said about $\int_2^6 f(x) \, dx$?
(b) Translate (a) into a statement about a string whose density at x is $f(x)$ grams per centimeter.

36 If f is continuous and $x^2 \leq f(x) \leq x^3$ for all x in $[1, 2]$, what can be said about $\int_1^2 f(x) \, dx$?

In Exercises 37 to 40 find at least one number c whose existence is guaranteed by the mean-value theorem for integrals.

37 $\int_0^{4\pi} \sin x \, dx$ **38** $\int_1^3 x^2 \, dx$

39 $\int_1^3 x^{-3} \, dx$ **40** $\int_0^1 (x - x^2) \, dx$

■

41 Verify that $\int_a^c f(x) \, dx + \int_c^b f(x) \, dx = \int_a^b f(x) \, dx$ when (a) $c = a$; (b) $c = b$; (c) $a = b$.

42 Let $v(t)$ be the velocity of an object at time t moving on a straight path. The object's speed is $|v(t)|$.
(a) What is the physical interpretation of the area under the graph of $|v(t)|$ from t_1 to t_2?
(b) What is the physical significance of the slope of the graph of $v(t)$?

43 Justify property (4) of definite integrals by showing that every approximating sum for $\int_a^b f(x) \, dx$ is ≥ 0.

44 Justify property (7) for definite integrals by showing that every approximating sum for $\int_a^b f(x)\,dx$ is $\leq M(b-a)$ and $\geq m(b-a)$, if $a < b$.

■ ■

45 Show that the mean-value theorem for definite integrals of continuous functions is a consequence of the mean-value theorem for differentiable functions stated in Sec. 4.1.

46 (Optimal replacement) The following argument appears in *Optimal Replacement Policy*, by D. W. Jorgenson, J. J. McCall, and R. Radner, pp. 92–93, Rand McNally, Chicago, 1967:

The average cost per unit good time, $V(N)$, is

$$V(N) = \frac{N + K}{\int_0^N R(t)\,dt}, \qquad (*)$$

where K is a constant, "the imputed down time," and N is the time of replacement. To determine the optimum preparedness maintenance policy, $V(N)$ is minimized with respect to replacement age N. Differentiation of (*) with respect to N yields the condition

$$\frac{dV}{dN} = \frac{\int_0^N R(t)\,dt - (N+K)R(N)}{[\int_0^N R(t)\,dt]^2} = 0.$$

Verify that the differentiation is correct.

5.6 Proofs of the two fundamental theorems of calculus (optional)

Geometric and physical intuition suggested the two fundamental theorems of calculus. It may be of value to give a mathematical proof for these two theorems, a proof which is independent of any particular interpretation of the definite integral. This will offer a chance to think of the definite integral as a purely mathematical concept.

The proof of the second fundamental theorem will be given first.

The second fundamental theorem asserts that the derivative of $G(x) = \int_a^x f(t)\,dt$ is $f(x)$. (See Sec. 5.4 for its full statement.)

PROOF OF THE SECOND FUNDAMENTAL THEOREM

It is necessary to study the behavior of the quotient

$$\frac{\Delta G}{\Delta x} = \frac{G(x + \Delta x) - G(x)}{\Delta x}$$

as $\Delta x \to 0$ (x is fixed). The number Δx may be positive or negative.

We have

$$\Delta G = \int_a^{x+\Delta x} f(t)\,dt - \int_a^x f(t)\,dt.$$

By Property (6) in Sec. 5.5

$$\int_a^{x+\Delta x} f(t)\,dt = \int_a^x f(t)\,dt + \int_x^{x+\Delta x} f(t)\,dt.$$

Hence

$$\Delta G = \int_a^x f(t)\,dt + \int_x^{x+\Delta x} f(t)\,dt - \int_a^x f(t)\,dt$$

$$= \int_x^{x+\Delta x} f(t)\,dt.$$

278 THE DEFINITE INTEGRAL

Now, by the mean-value theorem for definite integrals,

$$\Delta G = \int_{x}^{x+\Delta x} f(t)\,dt = f(c)((x + \Delta x) - x) \quad \text{for some } c \text{ between } x \text{ and } x + \Delta x$$

$$= f(c)\,\Delta x.$$

Thus
$$\frac{\Delta G}{\Delta x} = f(c)$$

for some number c between x and $x + \Delta x$. (Note that c is not constant but depends on the choice of Δx and the fixed number x.)

Now it will be possible to compute

$$\lim_{\Delta x \to 0} \frac{\Delta G}{\Delta x}$$

because, by the above reasoning,

$$\lim_{\Delta x \to 0} \frac{\Delta G}{\Delta x} = \lim_{\Delta x \to 0} f(c).$$

Since c is between x and $x + \Delta x$, as $\Delta x \to 0$, c must approach x. Since f is continuous at x and c is between x and $x + \Delta x$, it follows that

$$\lim_{\Delta x \to 0} f(c) = f(x).$$

This proves that G is differentiable and that $G' = f$. ∎

Remark: It may be illuminating to follow the various steps in this proof if we interpret "definite integral" as "area." If $f(x)$ is positive for all x in $[a, b]$, then $G(x)$ can be thought of as the area under the curve $y = f(t)$ from a to x. Then for $\Delta x > 0$, $G(x + \Delta x) - G(x) = \Delta G$ represents the area of a narrow strip above the interval $[x, x + \Delta x]$. Choose c in such a way that the rectangle with base Δx and height $f(c)$ has an area equal to the area of the shaded strip in Fig. 5.52. Then $\Delta G/\Delta x = f(c)$, the height of the rectangle chosen as described. As Δx approaches 0, $f(c)$ approaches $f(x)$, since f is continuous. See Figs. 5.51 to 5.53.

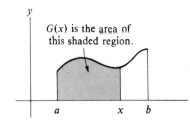

Figure 5.51

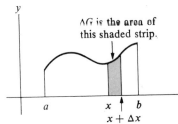

Figure 5.52

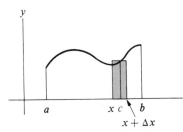

The area of the rectangle of base Δx and height $f(c)$ equals the area of the shaded strip.

Figure 5.53

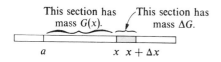

Figure 5.54

Remark: We can also interpret the proof of the second fundamental theorem in terms of density and mass. If we interpret f as density and $G(x) = \int_a^x f(t)\,dt$ as the mass in the interval $[a, x]$, then ΔG is the mass in the short interval $[x, x + \Delta x]$. (See Fig. 5.54.) Thus $\Delta G/\Delta x$ is the average density in the interval $[x, x + \Delta x]$ and equals the density somewhere in that interval,

$$\frac{\Delta G}{\Delta x} = f(c) \qquad \text{for some } c \text{ between } x \text{ and } x + \Delta x.$$

As $\Delta x \to 0$, $f(c) \to f(x)$. Hence

$$G'(x) = f(x).$$

EXAMPLE Let f be continuous. Find $\dfrac{d}{dx}\left(\displaystyle\int_x^b f(t)\,dt\right)$.

In other words, differentiate the definite integral with respect to the *lower* limit of integration.

SOLUTION To do this, write $\displaystyle\int_x^b f(t)\,dt = -\int_b^x f(t)\,dt.$

Thus the derivative equals

$$\frac{d}{dx}\left(-\int_b^x f(t)\,dt\right) = -\frac{d}{dx}\left(\int_b^x f(t)\,dt\right) = -f(x).$$

Note the minus sign. ∎

Remark: The minus sign in the answer in the Example makes sense if $f(x)$ is positive, $x < b$, and $\int_x^b f(t)\,dt$ is thought of as area (or mass). As x increases, the area (or mass) *decreases*, and its derivative should be negative. (See Fig. 5.55.)

Now let us turn to the proof of the first fundamental theorem.

The first fundamental theorem, which asserts that if $f = F'$, then $\int_a^b f(x)\,dx = F(b) - F(a)$, follows from the second, as will now be shown. (See Sec. 5.4 for its complete statement.)

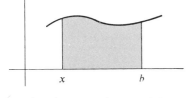

As x increases, the area of the shaded region decreases.

Figure 5.55

PROOF OF THE FIRST FUNDAMENTAL THEOREM Let $G(u) = \int_a^u f(x)\,dx$. (The letter u is introduced to describe the interval of integration $[a, u]$, since x has already been used to describe the inputs of the function f.) By the second fundamental theorem,

$$G'(u) = f(u).$$

Since it is assumed that $\qquad F'(u) = f(u),$

it follows that the functions G and F have the same derivative. Thus they differ by a constant (see Corollary 2, Sec. 4.1):

280 THE DEFINITE INTEGRAL

Then
$$G(u) = F(u) + C, \quad C \text{ constant.}$$
$$G(b) - G(a) = [F(b) + C] - [F(a) + C]$$
$$= F(b) - F(a).$$

But
$$G(a) = \int_a^a f(x)\,dx = 0,$$

and
$$G(b) = \int_a^b f(x)\,dx.$$

Hence
$$G(b) - G(a) = \int_a^b f(x)\,dx - 0$$
$$= \int_a^b f(x)\,dx.$$

Consequently,
$$\int_a^b f(x)\,dx = G(b) - G(a)$$
$$= F(b) - F(a),$$

and the theorem is proved. ∎

Exercises

1 The first fundamental theorem can be proved directly, without referring to the second fundamental theorem. Assume that f is continuous, $f = F'$, and $\int_a^b f(x)\,dx$ exists. The steps are outlined as follows:

(a) Given x_{i-1} and x_i, $x_{i-1} < x_i$, in $[a, b]$, show that there is a number c_i in $[x_{i-1}, x_i]$ such that
$$F(x_i) - F(x_{i-1}) = F'(c_i)(x_i - x_{i-1}).$$

(b) Given x_{i-1} and x_i in $[a, b]$, show that there is a number c_i in $[x_{i-1}, x_i]$ such that
$$f(c_i)(x_i - x_{i-1}) = F(x_i) - F(x_{i-1}).$$

(c) Let $x_0 = a, x_1, \ldots, x_n = b$ determine a partition of $[a, b]$ into n sections. Show that, if the sampling numbers c_i are chosen as in (b), then
$$\sum_{i=1}^{n} f(c_i)(x_i - x_{i-1}) = F(b) - F(a).$$

(d) Use (c) to show that
$$\lim_{\text{mesh} \to 0} \sum_{i=1}^{n} f(c_i)(x_i - x_{i-1}) = F(b) - F(a)$$

[even if the c_i are not chosen as in (b)]. This proves the first fundamental theorem directly.

2 Assume that the function f is defined for all x and has a continuous derivative. Assume that $f(0) = 0$ and that $0 < f'(x) \leq 1$.

(a) Prove that
$$\left[\int_0^1 f(x)\,dx\right]^2 \geq \int_0^1 [f(x)]^3\,dx.$$

Hint: Prove a more general result, namely that the inequality holds when the upper limit of integration 1 is replaced by t, $0 \leq t \leq 1$.

(b) Give an example where equality occurs in (a).

3 Letting $f(x)$ denote the velocity of an object at time x, interpret each step of the proof of the second fundamental theorem in terms of velocity and distance.

4 Let f be a function such that $\int_0^x f(t)\,dt = (f(x))^2$ for $x \geq 0$. Assume that $f(x) > 0$ for $x > 0$.

(a) Find $f(0)$.

(b) Find $f(x)$ for $x > 0$.

5.S Summary

Four problems led to the concept of the definite integral: area of the region under a curve, total mass of a string of varying density, total distance traveled when the speed is varying, and volume of a certain tent.

All four problems required the same procedure: choosing partitions of some interval, sampling numbers, and then forming approximating sums,

$$\sum_{i=1}^{n} f(c_i)(x_i - x_{i-1}).$$

In Sec. 5.1 the interval was $[0, 3]$, and the function f was the squaring function.

The typical sum

$$\sum_{i=1}^{n} f(c_i)(x_i - x_{i-1})$$

is *not* the definite integral, any more than $[f(x + \Delta x) - f(x)]/\Delta x$ is the derivative. A definite integral $\int_a^b f(x)\, dx$ is the limit of the approximating sums as their mesh is chosen smaller and smaller:

$$\int_a^b f(x)\, dx = \lim_{\text{mesh}\to 0} \sum_{i=1}^{n} f(c_i)(x_i - x_{i-1}).$$

The table below shows at a glance why the definite integral is related to mass, area, volume, and distance. In each case consider only functions with nonnegative values. Study this table carefully. It records the core ideas of the chapter.

Function	Interpretation of typical summand	Approximating sum	Definite integral	Meaning of definite integral
Density	$f(c_i)(x_i - x_{i-1})$ is estimate of mass in $[x_i, x_{i-1}]$.	$\sum_{i=1}^{n} f(c_i)(x_i - x_{i-1})$	$\int_a^b f(x)\, dx$	Mass
Length of cross section of a plane region by a line	$f(c_i)(x_i - x_{i-1})$ is area of an approximating rectangle.	$\sum_{i=1}^{n} f(c_i)(x_i - x_{i-1})$	$\int_a^b f(x)\, dx$	Area
Area of cross section of a solid by a plane	$f(c_i)(x_i - x_{i-1})$ is volume of a thin approximating slab.	$\sum_{i=1}^{n} f(c_i)(x_i - x_{i-1})$	$\int_a^b f(x)\, dx$	Volume
Speed	$f(T_i)(t_i - t_{i-1})$ is estimate of distance covered from time t_{i-1} to time t_i.	$\sum_{i=1}^{n} f(T_i)(t_i - t_{i-1})$	$\int_a^b f(t)\, dt$	Distance
Just a function (no application in mind)	$f(c_i)(x_i - x_{i-1})$ is just a product of two numbers.	$\sum_{i=1}^{n} f(c_i)(x_i - x_{i-1})$	$\int_a^b f(x)\, dx$	Just a number (no application in mind)

The close tie between the derivative and the definite integral is expressed in the two fundamental theorems of calculus:

FIRST FUNDAMENTAL THEOREM

If f is continuous and is the derivative of the function F, then

$$\int_a^b f(x)\, dx = F(b) - F(a).$$

SECOND FUNDAMENTAL THEOREM

If f is continuous, then

$$\frac{d}{dx}\left(\int_a^x f(t)\, dt\right) = f(x).$$

The first fundamental theorem is also called the fundamental theorem and is abbreviated by the letters FTC. It provides a tool for computing many definite integrals. If an antiderivative of f is elementary, then FTC is of use. But there are elementary functions, for instance, $\sqrt{1 + x^3}$, which are not derivatives of elementary functions. In these cases it may be necessary to estimate the definite integral, say, by an approximating sum or by methods discussed in Sec. 8.9.

Any function whose derivative is the function f is called an antiderivative of f and is denoted $\int f(x)\, dx$. Any two antiderivatives of a function over an interval differ by a constant.

The second fundamental theorem implies that every continuous function is the derivative of some function. More specifically, it tells the rate of change of a definite integral as the interval over which the integral is computed is changed.

The FTC is *not* a theorem about area or mass. It is a theorem about the limit of the sums $\sum_{i=1}^n f(c_i)(x_i - x_{i-1})$. In many applications, it is first shown that a certain quantity (area, distance, volume, mass, etc.) is estimated by sums of that type, and then the FTC is called on. However, it may or may not be of use.

VOCABULARY AND SYMBOLS

sigma notation $\sum_{i=1}^n a_i$

partition

section of a partition

sampling number c_i

mesh

definite integral $\int_a^b f(x)\, dx$

telescoping sum

$\sum_{i=1}^n (b_i - b_{i-1}) = b_n - b_0$

approximating sum

$\sum_{i=1}^n f(c_i)(x_i - x_{i-1})$

first fundamental theorem of calculus = fundamental theorem of calculus = first fundamental theorem = fundamental theorem = FTC.

$$\int_a^b f(x)\,dx = -\int_b^a f(x)\,dx \quad \text{if } b < a; \qquad \int_a^a f(x)\,dx = 0$$

second fundamental theorem of calculus = second fundamental theorem
elementary function indefinite integral
integral integrand
antiderivative $\int f(x)\,dx$ $F(x)\Big|_a^b$

KEY FACTS   If f is continuous, $\int_a^b f(x)\,dx$ exists. (This is assumed.)

If $f = F'$, then $\int_a^b f(x)\,dx = F(b) - F(a)$ (first fundamental theorem).

$\dfrac{d}{dx}\left(\int_a^x f(t)\,dt\right) = f(x)$ (second fundamental theorem).

$\int x^a\,dx = \dfrac{x^{a+1}}{a+1} + C, \quad a \neq -1.$

PROPERTIES OF ANTIDERIVATIVES

1 $\int cf(x)\,dx = c\int f(x)\,dx$ (c constant).

2 $\int (f(x) + g(x))\,dx = \int f(x)\,dx + \int g(x)\,dx.$

3 $\int (f(x) - g(x))\,dx = \int f(x)\,dx - \int g(x)\,dx.$

PROPERTIES OF DEFINITE INTEGRALS

1 $\int_a^b cf(x)\,dx = c\int_a^b f(x)\,dx$ (c constant).

2 $\int_a^b (f(x) + g(x))\,dx = \int_a^b f(x)\,dx + \int_a^b g(x)\,dx.$

3 $\int_a^b (f(x) - g(x))\,dx = \int_a^b f(x)\,dx - \int_a^b g(x)\,dx.$

4 If $f(x) \geq 0$, then $\int_a^b f(x)\,dx \geq 0$ ($a < b$).

5 If $f(x) \geq g(x)$, then $\int_a^b f(x)\,dx \geq \int_a^b g(x)\,dx$ ($a < b$).

6 $\int_a^b f(x)\,dx = \int_a^c f(x)\,dx + \int_c^b f(x)\,dx.$

7 If m and M are constants, $m \leq f(x) \leq M$, then

$$m(b-a) \leq \int_a^b f(x)\,dx \leq M(b-a) \quad (a < b).$$

> **MEAN-VALUE THEOREM FOR INTEGRALS**
>
> There is a number c in $[a, b]$ such that
>
> $$\int_a^b f(x)\,dx = f(c)(b-a).$$
>
> (The full statement is in Sec. 5.5.)

But just because there are formulas for computing some definite integrals, do not forget that a definite integral is a limit of sums. There are two reasons for keeping this fundamental concept clear:

1. In many applications in science the concept of the definite integral is more important than its use as a computational tool.
2. Many definite integrals cannot be evaluated by a formula. Some of the more important of these have been tabulated to several decimal places and published in handbooks of mathematical tables.

To emphasize that the fundamental theorem of calculus does not dispose of all definite integrals, here is a little assortment of elementary functions whose integrals are not elementary:

$$\sqrt{x}\sqrt[n]{1+x}, \quad \sqrt[n]{1+x^2} \quad \text{for } n = 3, 4, 5, \ldots$$

$$\sqrt{1+x^n} \quad \text{for } n = 3, 4, 5, \ldots$$

$$\frac{\sin x}{x}, \quad x \tan x, \quad \frac{2^x}{x}, \quad 2^{x^2}$$

Even when an elementary antiderivative exists, given the easy access to calculators and computers, we may prefer to get a good estimate of the definite integral by calculating an approximating sum rather than by struggling to find an antiderivative. (Programmable calculators can provide estimates of definite integrals accurate to several decimal places.)

Guide quiz on chap. 5

1. Define (a) $\int_a^b f(x)\,dx$, (b) $\int f(x)\,dx$.
2. Estimate $\int_0^3 dx/(1+x^2)$ by using an approximating sum with three sections of equal length and, as sampling points, (a) left endpoints, (b) right endpoints.
3. (a) Differentiate $1/(2x+3)$.
   (b) Find the area under the curve $y = 1/(2x+3)^2$ and above the interval $[0, 1]$.
4. The curve $y = 1/x$ is rotated around the x axis. Find the volume of the solid enclosed by the resulting surface, between $x = 1$ and $x = 4$ (see Fig. 5.56).

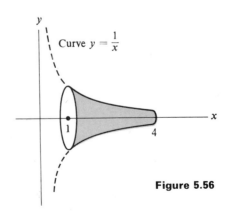

Figure 5.56

5 Find (a) $\lim_{\Delta x \to 0} \dfrac{\int_2^{5+\Delta x} \sin x^2 \, dx - \int_2^5 \sin x^2 \, dx}{\Delta x}$

(b) $\dfrac{d^2}{dx^2}\left[\left(\int_0^{x^2} \dfrac{dt}{\sqrt{1-5t^3}}\right)^2\right]$

6 (a) What is meant by an "elementary function"?
(b) Give an example of an elementary function whose antiderivative is not elementary.
(c) Give an example of a function that is not elementary, yet has an elementary derivative.

Review exercises for chap. 5

In each of Exercises 1 to 6 find the area of the region under the given curve and above the given interval.

1 $y = 2x^3$; [1, 2]   2 $y = 6x^2 + 10x^4$; [−1, 2]
3 $y = \sin 3x$; [0, $\pi/6$]   4 $y = 3\cos 2x$; [$\pi/6$, $\pi/4$]
5 $y = \dfrac{1}{x^3}$; [2, 3]   6 $y = \dfrac{1}{(x+1)^4}$; [0, 1]

In each of Exercises 7 to 16 give a formula for the antiderivatives of the given function (including the constant C).

7 $\sec^2 x$ 8 $\sec^2 3x$
9 $\sec x \tan x$ 10 $5 \sec 3x \tan 3x$
11 $4 \csc x \cot x$ 12 $-\csc 5x \cot 5x$
13 $(x^3 + 1)^2$ (Suggestion: First multiply it out.)
14 $\left(x + \dfrac{1}{x}\right)^2$ 15 $100x^{19}$ 16 $\dfrac{24}{x^3}$

17 (a) Differentiate $(x^3 + 1)^6$.
(b) Find $\int (x^3 + 1)^5 x^2 \, dx$.
18 (a) Differentiate $\sin x^3$.
(b) Find $\int x^2 \cos x^3 \, dx$.
19 Write these sums without using the sigma notation:

(a) $\sum_{j=1}^{3} d^j$ (b) $\sum_{k=1}^{4} x^k$

(c) $\sum_{i=0}^{3} i 2^{-i}$ (d) $\sum_{i=2}^{5} \dfrac{i+1}{i}$

(e) $\sum_{i=2}^{4} \left(\dfrac{1}{i} - \dfrac{1}{i+1}\right)$ (f) $\sum_{i=1}^{4} \sin \dfrac{\pi i}{4}$.

20 Write these sums without sigma notation and then evaluate:

(a) $\sum_{i=1}^{100} (2^i - 2^{i-1})$ (b) $\sum_{i=0}^{100} (2^{i+1} - 2^i)$

(c) $\sum_{i=1}^{100} \left(\dfrac{1}{i} - \dfrac{1}{i+1}\right)$.

21 Let n be a positive integer and f a function.
(a) Show that $\sum_{i=1}^{n} f(i/n)(1/n)$ is an approximating sum for the definite integral $\int_0^1 f(x) \, dx$.

(b) What is the length of the ith section of the partition in (a)?
(c) What is the mesh of the partition?
(d) Where does the sampling number c_i lie in the ith section?

22 Explain why $\dfrac{1}{100} \sum_{i=1}^{100} f(i/100)$ is an estimate of $\int_0^1 f(x) \, dx$.

23 What definite integrals are estimated by the following sums?

(a) $\sum_{i=1}^{200} \left(\dfrac{i}{100}\right)^3 \dfrac{1}{100}$ (b) $\sum_{i=1}^{100} \left(\dfrac{i-1}{100}\right)^4 \dfrac{1}{100}$

(c) $\sum_{i=101}^{300} \left(\dfrac{i}{100}\right)^5 \dfrac{1}{100}$.

24 (See Exercise 23.)
(a) Show that $(1/n^3) \sum_{i=1}^{n} i^2$ is an approximation of $\int_0^1 x^2 \, dx$.
(b) Compute the sum in (a) when $n = 4$.
(c) Compute $\int_0^1 x^2 \, dx$.
(d) Find $\lim_{n \to \infty} (1/n^3) \sum_{i=1}^{n} i^2$.

25 (a) Compute $\dfrac{d}{dx}\left(\int_1^x 3t^2 \sin 2t^3 \, dt\right)$.

(b) Compute $\dfrac{d}{dx}\left(\int_1^{x^3} \sin 2t \, dt\right)$.

(c) Using (a) and (b), show that

$$\int_1^x 3t^2 \sin 2t^3 \, dt = \int_1^{x^3} \sin 2t \, dt.$$

26 How often should a machine be overhauled? This depends on the rate $f(t)$ at which it depreciates and the cost A of overhaul. Denote the time interval between overhauls by T.
(a) Explain why you would like to minimize $g(T) = [A + \int_0^T f(t) \, dt]/T$.
(b) Find dg/dT.
(c) Show that when $dg/dT = 0$, $f(T) = g(T)$.
(d) Is this reasonable?

27. Give an example of a function F such that $F(4) = 0$ and $F'(x) = \sqrt[3]{1 + x^2}$.

28. As a stone is lowered into water, we record the volume of water it displaces. When x inches are submerged, the stone displaces $V(x)$ cubic inches of water. How can we find the area of the cross section of the stone made by the plane of the surface of the water when it is submerged to a depth of x inches? Assume we know $V(x)$ for all x.

29. An unmanned satellite automatically reports its speed every minute. If a graph is drawn showing speed as a function of time during the flight, what is the physical interpretation of the area under the curve and above the time axis? Explain.

■ ■

30. A number is *dyadic* if it can be expressed as the quotient of two integers m/n, where n is a power of 2. (These are the fractions into which an inch is usually divided.) Between any two numbers lies an infinite set of dyadic numbers and also an infinite set of numbers that are not dyadic. With this background we shall define a function f that does *not* have a definite integral over the interval $[0, 1]$, as follows.

$$\text{Let } f(x) = \begin{cases} 0 & \text{if } x \text{ is dyadic}; \\ 3 & \text{if } x \text{ is not dyadic}. \end{cases}$$

(a) Show that for any partition of $[0, 1]$ it is possible to choose sampling numbers c_i such that

$$\sum_{i=1}^{n} f(c_i)(x_i - x_{i-1}) = 3.$$

(b) Show that for any partition of $[0, 1]$ it is possible to choose sampling numbers c_i such that

$$\sum_{i=1}^{n} f(c_i)(x_i - x_{i-1}) = 0.$$

(c) Why does f not have a definite integral over the interval $[0, 1]$?

31. Let f be an increasing function on $[a, b]$. Let p be a positive number. Show that any two approximating sums for $\int_a^b f(x)\,dx$ formed using the same partition of mesh $\leq p$ differ by at most $p(f(b) - f(a))$.

32. Let f be a continuous function that has a derivative nowhere. (There are such functions.) Construct a function that has a derivative everywhere but a second derivative nowhere.

33. A man whose jeep has a vertical windshield drives a mile through a vertical rain consisting of drops that are uniformly distributed and falling at a constant rate. Should he go slow or fast in order to minimize the amount of rain that strikes the windshield?

34. Prove that if f is an increasing function on the interval $[a, b]$ and $f = F'$, then $\int_a^b f(x)\,dx$ exists and equals $F(b) - F(a)$. (Do not assume the fundamental theorem of calculus.)

35. Prove that if g is differentiable and $|g'(x)| < k$ for all x in $[a, b]$, then the function $f(x) = g(x) + kx$ is increasing on $[a, b]$.

36. (See Exercises 34 and 35.) Prove that if f is a differentiable function whose derivative f' is bounded between two fixed numbers over the interval $[a, b]$, and if $f = F'$, then $\int_a^b f(x)\,dx$ exists and equals $F(b) - F(a)$. (Do not assume that a continuous function has a definite integral.)

TOPICS IN DIFFERENTIAL CALCULUS

This chapter obtains the formulas for the derivatives of the logarithmic functions $\log_b x$, the exponential functions b^x, and the inverse trigonometric functions. These formulas, together with the differentiation formulas already obtained, enable us to differentiate any elementary function. Sections 6.6 and 6.7 then apply derivatives to problems concerning rates of change and growth. Section 6.8 presents a method, known as l'Hôpital's rule, for finding many limits. In Sec. 6.9 the hyperbolic functions, certain combinations of exponential functions, are defined. Optional Secs. 6.3 and 6.10 present an alternative approach to logarithms and exponentials: first the logarithm is defined as an integral and then the exponential is obtained indirectly from it.

6.1 Review of logarithms, the number e

If a is rational and *not equal to* -1, then

$$\int x^a \, dx = \frac{x^{a+1}}{a+1} + C. \tag{1}$$

The formula works, for instance, when $a = -1.01$:

$$\int x^{-1.01} \, dx = \frac{x^{-1.01+1}}{-1.01+1} + C$$
$$= \frac{x^{-0.01}}{-0.01} + C$$
$$= -100 x^{-0.01} + C.$$

288 TOPICS IN DIFFERENTIAL CALCULUS

But formula (1) breaks down when $a = -1$.

In view of the fundamental theorem of calculus, it would be helpful to have a formula for $\int x^a \, dx$ when $a = -1$, that is, a formula for

$$\int \frac{dx}{x}.$$

The function $\dfrac{1}{x}$ leads us to study logarithms.

It turns out that the antiderivative of $1/x$ is a logarithm, as will be shown in the next section and, hence, is an elementary function.

Before examining the derivatives of the logarithm functions, it would be prudent to review what they are since they suffer from undeserved neglect in many a precalculus course that overemphasizes polynomials.

REVIEW OF LOGARITHMS

Consider the question $\qquad 3^? = 9;$

read as "3 raised to what power equals 9?" The answer, whatever its numerical value might be, is called "the logarithm of 9 to the base 3." Since

$$3^2 = 9,$$

we say that "the logarithm of 9 to the base 3 is 2." The general definition of logarithm follows:

DEFINITION *Logarithm.* If b and c are positive numbers, $b \neq 1$, and

$$b^x = c,$$

Remember one thing: A logarithm is an exponent!

then the number x is the *logarithm* of c to the base b, and is written

$$\log_b c.$$

Any exponential equation $b^x = c$ may be translated into a logarithmic equation $x = \log_b c$, just as any English statement may be translated into French. This table illustrates some of these translations. Read it over several times, perhaps aloud, until you can, when covering a column, fill in the correct translation of the other column.

Exponential language	Logarithmic language
$3^? = 9$	$\log_3 9 = 2$
$7^0 = 1$	$\log_7 1 = 0$
$10^3 = 1{,}000$	$\log_{10} 1{,}000 = 3$
$10^{-2} = 0.01$	$\log_{10} 0.01 = -2$
$9^{1/2} = 3$	$\log_9 3 = \tfrac{1}{2}$
$8^{2/3} = 4$	$\log_8 4 = \tfrac{2}{3}$
$8^{-1} = \tfrac{1}{8}$	$\log_8 \tfrac{1}{8} = -1$
$5^1 = 5$	$\log_5 5 = 1$
$b^1 = b$	$\log_b b = 1$

Since $b^x = c$ is equivalent to $x = \log_b c$, it follows that

$$b^{\log_b c} = c.$$

The equality $b^{\log_b c} = b$ is not deep; it just restates the definition of a logarithm.

EXAMPLE 1 Find $\log_5 125$.

SOLUTION Look for an answer to the question

"5 to what power equals 125?"

or, equivalently, for a solution of the equation

$$5^x = 125.$$

Since $5^3 = 125$, the answer is 3.

$$\log_5 125 = 3 \quad \blacksquare$$

EXAMPLE 2 Find $\log_{10} \sqrt{10}$.

SOLUTION By the definition of $\log_{10} \sqrt{10}$,

$$10^{\log_{10} \sqrt{10}} = \sqrt{10}.$$

Now, $\qquad\qquad 10^{1/2} = \sqrt{10}.$

Thus $\qquad\qquad \log_{10} \sqrt{10} = \tfrac{1}{2}.$

In words, "The power to which we must raise 10 to get $\sqrt{10}$ is $\tfrac{1}{2}$." As a calculator shows, $\sqrt{10} \approx 3.162$.

Thus $\qquad\qquad \log_{10} 3.162 \approx \tfrac{1}{2} = 0.5.$ $\quad\blacksquare$

In order to get an idea of the logarithm as a function, consider logarithms to the familiar base 10,

$$y = \log_{10} x.$$

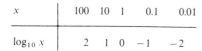

Begin with a table, as shown in the margin. We must restrict ourselves to $x > 0$. A negative number, such as -1, cannot have a logarithm, since there is no power of 10 that equals -1. Similarly 0 does not have a logarithm since the equation $10^x = 0$ has no solution. The domain of the function "log to the base 10" consists of the positive real numbers.

With the aid of the five points in the table, the graph is easy to sketch and is shown in Fig. 6.1. The graph lies to the right of the y axis. Far to the right it rises slowly; not until x reaches 100 does the y coordinate reach 2. When x is a small positive number, observe that $\log_{10} x$ is a negative number of large absolute value. Moreover, as x increases, so does $\log_{10} x$. Though x is restricted to being positive, $\log_{10} x$ takes on all values, positive and negative. Logarithms to the base 10 are called *common*

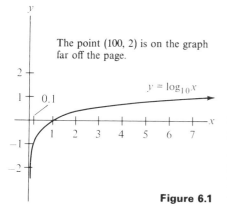

Figure 6.1

TOPICS IN DIFFERENTIAL CALCULUS

logarithms. A brief table of common logarithms is to be found in Appendix H and, in more detail, in any handbook of mathematical tables. Many calculators have a $\log_{10}$-key, usually labeled "log."

PROPERTIES OF LOGARITHMS

Since each exponential equation $b^x = c$ translates into the corresponding logarithmic equation $x = \log_b c$, every property of exponentials must carry over to some property of logarithms. For instance, the information that

$$b^0 = 1$$

translates, in the language of logarithms, to

$$\log_b 1 = 0.$$

The logarithm of 1 in any base b is 0.

The equation $$b^1 = b$$

amounts to saying that the logarithm of b (the base) in the base b is 1,

$$\log_b b = 1.$$

Thus $\log_{10} 10 = 1$ and $\log_{2.718}(2.718) = 1$.

The following table lists the fundamental properties of exponential functions together with the corresponding properties of logarithms.

Exponents	Logarithms
$b^0 = 1$	$\log_b 1 = 0$
$b^{1/2} = \sqrt{b}$	$\log_b \sqrt{b} = \frac{1}{2}$
$b^1 = b$	$\log_b b = 1$
$b^{x+y} = b^x b^y$	$\log_b cd = \log_b c + \log_b d$
$b^{-x} = \dfrac{1}{b^x}$	$\log_b \left(\dfrac{1}{c}\right) = -\log_b c$
$b^{x-y} = \dfrac{b^x}{b^y}$	$\log_b \left(\dfrac{c}{d}\right) = \log_b c - \log_b d$
$(b^x)^y = b^{xy}$	$\log_b c^m = m \log_b c$

Of these identities for logarithms the most fundamental is

$$\log_b cd = \log_b c + \log_b d,$$

which asserts that the log of the product is the sum of the logs of the factors. The proof of this is short and instructive. (Proofs of the others are left as exercises.)

THEOREM 1 For any positive numbers c and d and for any base b,

The log of the product
$$\log_b cd = \log_b c + \log_b d.$$

PROOF By the definition of the logarithm as an exponent,

$$c = b^{\log_b c} \quad \text{and} \quad d = b^{\log_b d}. \tag{2}$$

Thus
$$cd = b^{\log_b c} b^{\log_b d}.$$

By the basic law of exponents ($b^x b^y = b^{x+y}$),

$$b^{\log_b c} b^{\log_b d} = b^{\log_b c + \log_b d} \tag{3}$$

Combining (2) and (3) shows that

$$cd = b^{\log_b c + \log_b d}.$$

So the exponent to which b must be raised to get cd is

$$\log_b c + \log_b d.$$

In other words, the logarithm of cd to the base b is $\log_b c + \log_b d$. ∎

EXAMPLE 3 Use the last line in the preceding table to evaluate

$$\log_9 (3^7) \quad \text{and} \quad \log_5 \sqrt[3]{25^2}.$$

SOLUTION
$$\log_9 (3^7) = 7 \log_9 3 = 7(\tfrac{1}{2}) = \tfrac{7}{2}$$
$$\log_5 \sqrt[3]{25^2} = \log_5 (25)^{2/3} = \tfrac{2}{3} \log_5 25 = (\tfrac{2}{3})2 = \tfrac{4}{3}. \ \blacksquare$$

The next example shows how logarithms can be used to solve equations in which the unknown appears in an exponent.

EXAMPLE 4 Find x if $5 \cdot 3^x \cdot 7^{2x} = 2$.

SOLUTION First rewrite the equation as

$$3^x \cdot 7^{2x} = \tfrac{2}{5} = 0.4,$$

and then take logarithms to the base 10 of both sides:

$$\log_{10} (3^x 7^{2x}) = \log_{10} 0.4$$
$$\log_{10} 3^x + \log_{10} 7^{2x} = \log_{10} 0.4 \quad \text{log of a product}$$
$$x \log_{10} 3 + 2x \log_{10} 7 = \log_{10} 0.4 \quad \text{log of an exponential}$$
$$x = \frac{\log_{10} 0.4}{\log_{10} 3 + 2 \log_{10} 7} \quad \text{solving for } x$$
$$x \approx \frac{-0.3979}{0.4771 + 2(0.8451)} \quad \text{calculator or table}$$
$$x \approx -0.1836 \ \blacksquare$$

TOPICS IN DIFFERENTIAL CALCULUS

THE LIMIT OF $(1 + h)^{1/h}$ AS $h \to 0$

When we find the derivative of $\log_b x$ by examining the limit

$$\lim_{\Delta x \to 0} \frac{\log_b (x + \Delta x) - \log_b x}{\Delta x}$$

we will encounter a limit not met in earlier chapters, namely,

$$\lim_{h \to 0} (1 + h)^{1/h} \tag{4}$$

It is helpful to study this limit (4) in advance.

Two influences on $(1 + h)^{1/h}$

There are two influences operating on $(1 + h)^{1/h}$ as $h \to 0$. (For simplicity, consider only $h > 0$.) First, the base gets near 1, so there is a chance that $(1 + h)^{1/h}$ gets near 1. But the exponent $1/h$ gets arbitrarily large. Thus there is a chance that $(1 + h)^{1/h}$ gets large since the quantity $1 + h$ is raised to a large power.

Some computations on a calculator suggest how these two influences balance. The following table gives the values of $(1 + h)^{1/h}$ to five decimal places for several small values of h, both positive and negative.

When $h = 0.1$, $(1 + h)^{1/h}$
$= (1.1)^{10} \approx 2.59374$

h	0.1	0.001	0.0001	0.00001	-0.0001	-0.00001
$1/h$	10	1000	10,000	100,000	$-10,000$	$-100,000$
$1 + h$	1.1	1.001	1.0001	1.00001	0.9999	0.99999
$(1 + h)^{1/h}$	2.59374	2.71692	2.71815	2.71827	2.71842	2.71830

As $h \to 0$ through positive values, $(1 + h)^{1/h}$ appears to approach a number whose decimal expansion begins 2.718. Similarly, as $h \to 0$ through negative values, $(1 + h)^{1/h}$ seems to approach the same number.

Just as numerical evidence suggested in Sec. 2.5 that $(\sin \theta)/\theta \to 1$ as $\theta \to 0$, numerical evidence suggests that, as $h \to 0$, $(1 + h)^{1/h} \to$ a number approximately equal to 2.718. In Sec. 2.5 it was shown by comparing the areas of a triangle and two sectors of a circle that the quotient $(\sin \theta)/\theta$ does indeed approach 1 as $\theta \to 0$. However, there is no such direct demonstration that $(1 + h)^{1/h}$ has a limit as $h \to 0$. (Section 6.10 contains an indirect proof involving integrals and derivatives. See also Exercise 30 in this section.)

It will be assumed that $\lim_{h \to 0} (1 + h)^{1/h}$ exists.

In order to obtain the derivative of the logarithm functions expeditiously, we will assume that

$$\lim_{h \to 0} (1 + h)^{1/h}$$

exists. This limit is called e in honor of the eighteenth century mathematician Leonhard Euler (pronounced "oiler"). Its value, correct to nine decimal places, is given in the following definition.

6.1 REVIEW OF LOGARITHMS, THE NUMBER e

DEFINITION *The number e.*

e is not a repeating decimal; the next digit is 4.

$$e = \lim_{h \to 0} (1+h)^{1/h} \approx 2.718281828. \tag{5}$$

Since $(1+h)^{1/h} \to e$ as $h \to 0$, we may say that

$$(1 + \text{small number})^{\text{reciprocal of } same \text{ small number}} \approx e.$$

Consequently, for $x \neq 0$ and fixed,

$$\lim_{\Delta x \to 0} \left(1 + \frac{\Delta x}{x}\right)^{x/\Delta x} = e. \tag{6}$$

It is this limit that will be needed in differentiating the logarithm functions. Another consequence of (5) is that

$$\lim_{n \to \infty} \left(1 + \frac{1}{n}\right)^n = e, \tag{7}$$

$n = \dfrac{1}{(1/n)}$ where n refers to integers. The exponent n is the reciprocal of the small number $1/n$. The limit (7) is frequently met in applications of limits. In a moment it appears in a discussion of compound interest.

The number e is as conspicuous in calculus as the number π is in the study of the circle. Its importance will be demonstrated in the next section.

COMPOUND INTEREST AND e (optional)

Say that you put A dollars in an account at the beginning of some interest period. This period could be 1 year or 6 months or 3 months or an even shorter time period. The bank agrees to pay an interest rate r for money left on deposit for that period. In practice r is around 6 percent $= 0.06$ per year or 0.03 for 6 months. During the time period the value of the account would increase by rA dollars and amount to a total of

$$A + rA = (1+r)A \text{ dollars.}$$

So, to find the final amount, just multiply the initial amount by $1 + r$.

$$\text{Amount at end of an interest period} = (1+r)(\text{initial amount}). \tag{8}$$

Now assume that the interest rate is very generous, 100 percent per year, that is, $r = 1$ for the year. One dollar left in the bank a year would grow to

$$(1+1)1 = 2 \text{ dollars,}$$

by formula (8).

If instead the bank compounds interest twice a year, the interest rate for each of the 6-month periods would be 50 percent or $r = 0.50$, or simply $r = \frac{1}{2}$. At the end of the first 6 months an initial 1 dollar would grow to

$$(1 + \tfrac{1}{2})1 = 1 + \tfrac{1}{2} \text{ dollars,}$$

again by formula (8). During the second 6 months this amount, $1 + \tfrac{1}{2}$, would grow to

$$(1 + \tfrac{1}{2})(1 + \tfrac{1}{2}) = (1 + \tfrac{1}{2})^2 = 2.25 \text{ dollars.}$$

again by formula (8).

A competing bank offers to compound interest 3 times a year, at 4-month intervals. Thus $r = \tfrac{1}{3}$. At the end of the first 4 months 1 dollar would grow to

$$(1 + \tfrac{1}{3})1 = 1 + \tfrac{1}{3} \text{ dollars.}$$

During the second period of 4 months, this amount, $1 + \tfrac{1}{3}$ dollars, left on deposit, grows to

$$(1 + \tfrac{1}{3})(1 + \tfrac{1}{3}) = (1 + \tfrac{1}{3})^2 \text{ dollars.}$$

During the final 4 months, the amount $A = (1 + \tfrac{1}{3})^2$ grows to

$$(1 + \tfrac{1}{3})A = (1 + \tfrac{1}{3})(1 + \tfrac{1}{3})^2 = (1 + \tfrac{1}{3})^3 \approx 2.37 \text{ dollars.}$$

Other bankers join in, one compounding 4 times a year (quarterly), another daily, and another hourly. When interest is compounded 4 times a year, a deposit of 1 dollar would grow to

$$(1 + \tfrac{1}{4})^4 \approx 2.44 \text{ dollars}$$

in a year. If interest is deposited n times a year, 1 dollar would grow to

$$\left(1 + \frac{1}{n}\right)^n \text{ dollars}$$

in a year. The following table shows the amount in the account at the end of the year if 1 dollar is compounded n times at an annual interest rate of 100 percent.

	Simple Interest	Semi-annually	Every 4 Months	Quarterly	Monthly	Daily	Hourly
Number of times compounded (n)	1	2	3	4	12	365	8760
Value at end of year (rounded off)	2	$2.25	$2.37037	$2.44141	$2.61304	$2.71457	$2.71812

The more often the interest is computed, the more the account grows. Moreover, since

$$\lim_{n \to \infty} \left(1 + \frac{1}{n}\right)^n = e \approx 2.71828,$$

there is little to be gained by compounding every second. In the limit, at so-called "continuous compounding," an account of 1 dollar grows to e dollars in a year.

In these computations the annual interest rate was taken as an unrealistic 100 percent. Assume instead that the interest rate is 10 percent per year. Similar reasoning shows that a deposit of 1 dollar grows to

$$\left(1 + \frac{0.10}{n}\right)^n \text{ dollars,}$$

if interest is compounded n times a year. In Exercise 27 it is shown that

$$\lim_{n \to \infty} \left(1 + \frac{0.10}{n}\right)^n = e^{0.10} \approx 1.105.$$

So, even if compounded continuously, 1 dollar would grow to only 1.105 dollars in a year. This is equivalent to an annual interest rate of 10.5 percent not compounded.

Exercises

1. Translate these equations into the language of logarithms.
   (a) $2^5 = 32$ (b) $3^4 = 81$
   (c) $10^{-3} = 0.001$ (d) $5^0 = 1$
   (e) $1{,}000^{1/3} = 10$ (f) $49^{1/2} = 7$

2. Translate these equations into the language of logarithms.
   (a) $8^{2/3} = 4$ (b) $10^3 = 1{,}000$ (c) $10^{-4} = 0.0001$
   (d) $3^0 = 1$ (e) $10^{1/2} = \sqrt{10}$ (f) $(\tfrac{1}{2})^{-2} = 4$

3. (a) Fill in this table.

x	$\tfrac{1}{9}$	$\tfrac{1}{3}$	1	3	9
$\log_3 x$					

   (b) Plot the five points in (a) and graph $y = \log_3 x$.

4. (a) Fill in this table.

x	$\tfrac{1}{16}$	$\tfrac{1}{4}$	1	2	4	8	16
$\log_4 x$							

   (b) Plot the seven points in (a) and graph $y = \log_4 x$.

5. Translate these equations into the language of exponents.
   (a) $\log_2 7 = x$ (b) $\log_5 2 = s$
   (c) $\log_3 \tfrac{1}{3} = -1$ (d) $\log_7 49 = 2$

6. Translate these equations into the language of exponents.
   (a) $\log_{10} 1{,}000 = 3$ (b) $\log_5 \tfrac{1}{25} = -2$
   (c) $\log_{1/2}(\tfrac{1}{4}) = 2$ (d) $\log_{64} 128 = \tfrac{7}{6}$

7. Evaluate
   (a) $2^{\log_2 16}$ (b) $2^{\log_2 (1/2)}$
   (c) $2^{\log_2 7}$ (d) $2^{\log_2 g}$

8. Evaluate
   (a) $10^{\log_{10} 100}$ (b) $10^{\log_{10} 0.01}$
   (c) $10^{\log_{10} 7}$ (d) $10^{\log_{10} p}$

9. Evaluate
   (a) $\log_3 \sqrt{3}$ (b) $\log_3 (3^5)$ (c) $\log_3 (\tfrac{1}{27})$

10. If $\log_4 A = 2.1$, evaluate
    (a) $\log_4 A^2$ (b) $\log_4 (1/A)$ (c) $\log_4 16A$

In each of Exercises 11 to 14 solve for x.

11. $2 \cdot 3^x = 7$ 12. $3 \cdot 5^x = 6^x$
13. $3^{5x} = 27^x$ 14. $10^{2x} 3^{2x} = 5$

Exercises 15 to 20 concern the relationships between logarithms to different bases.

15. (a) Evaluate $\log_2 8$, $\log_8 2$, and then their product.
    (b) Show that, for any two bases a and b,
    $$(\log_a b) \times (\log_b a) = 1.$$
    Suggestion: Start with $a^{\log_a b} = b$ and take $\log_b$ of both sides.

16 Let a and b be fixed bases.
   (a) Prove that the quotient
   $$\frac{\log_b x}{\log_a x}$$
   is independent of x. In fact, it equals $\log_b a$. *Suggestion:* Start with $a^{\log_a x} = x$.
   (b) What does (a) imply about the relation between the graphs of $y = \log_b x$ and $y = \log_a x$?

17 This exercise shows how to use a calculator's log-key (log to the base 10) to find logarithms to any base b. Say that you wish to find $\log_b c$. Let $x = \log_b c$.
   (a) Show that $b^x = c$.
   (b) Take $\log_{10}$ of both sides of the equation in (a) to show that
   $$x = \frac{\log_{10} c}{\log_{10} b}.$$
   In short, to find $\log_b c$, first find $\log_{10} c$ and then divide by $\log_{10} b$.

In Exercises 18 to 20 use the method of Exercise 17 to evaluate the logarithms.

18 (Calculator) (a) $\log_3 5$ (b) $\log_2 7$ (c) $\log_2 e$ ($e \approx 2.718$)
19 (Calculator) (a) $\log_3 (\tfrac{1}{2})$ (b) $\log_{3.5} (4.17)$ (c) $\log_{0.5} (6)$
20 (Calculator) (a) $\log_{20} 30$ (b) $\log_5 (0.003)$ (c) $\log_2 1{,}000$

■

21 Find $\lim_{h \to 0} (1 + 2h)^{4/h}$. *Suggestion:* First write
   $$(1 + 2h)^{4/h} = [(1 + 2h)^{1/(2h)}]^8.$$
   What happens to the quantity in brackets as $h \to 0$?

22 (See Exercise 21.) Evaluate the following limits:
   (a) $\lim_{x \to 0} (1 - x)^{1/x}$ (b) $\lim_{x \to 0} (1 + x^2)^{1/x}$
   (c) $\lim_{x \to 0} (1 + x)^{1/x^2}$

23 (a) If $b^3 = c$, then we may write $3 = $ _____.
   (b) If $b^3 = c$, then we may write $b = $ _____.

24 From the fact that $(b^x)^y = b^{xy}$, deduce that $\log_b c^m = m \log_b c$. (*Hint:* Write c as $b^{\log_b c}$ and consider c^m.)

25 From the fact that $b^{x-y} = b^x/b^y$, deduce that $\log_b (c/d) = \log_b c - \log_b d$. (*Hint:* Write c as $b^{\log_b c}$ and d as $b^{\log_b d}$.)

26 In the Richter scale, the intensity M of an earthquake is related to the energy E of the earthquake by the formula $\log_{10} E = 11.4 + 1.5 M$. ($E$ is measured in ergs.)
   (a) If one earthquake has a thousand times the energy of another, how much larger is its Richter rating M?
   (b) What is the ratio of the energy of the San Francisco earthquake of 1906 ($M = 8.3$) with that of the Eureka earthquake of 1980 ($M = 7$)?
   (c) What is the Richter rating of a 10 megaton H-bomb, that is, of an H-bomb whose energy is equivalent to that in 10 million tons of TNT? (One ton of TNT contains 4.2×10^6 ergs.)
   For a fuller discussion of the Richter scale see the *Encyclopaedia Britannica, Macropedia,* vol. 6, pp. 71–72.

27 Let r be fixed. Show that $\lim_{n \to \infty} (1 + r/n)^n = e^r$. *Hint:* See the technique described in Exercise 21.

■ ■

28 Prove that $\log_3 2$ is irrational. (*Hint:* Assume that it is rational, that is, equal to m/n for some integers m and n, and obtain a contradiction.)

29 Using approximating sums and the definition of a definite integral, show that
   (a) $\int_1^2 dx/x < 1$ and   (b) $\int_1^3 dx/x > 1$.

30 Pretend that you have *never heard of e* and $\lim_{h \to 0} (1 + h)^{1/h}$. This exercise outlines an argument using areas that $\lim_{h \to 0} (1 + h)^{1/h}$ exists. For simplicity, consider only $h > 0$.
   (a) Sketch the curves $y = 1/x$ and $y = 1/x^{1-h}$ relative to the same axes. Note that for $x > 1$ the second curve lies above the first curve.
   (b) For a given number h (rational) find the number $A(h)$ such that
   $$\int_1^{A(h)} \frac{dx}{x^{1-h}} = 1.$$
   (c) Using Exercise 29, show that there is a number B such that $\int_1^B dx/x = 1$. Note that $2 < B < 3$.
   (d) Using (a), (b), and (c), show that, for $h > 0$, $(1 + h)^{1/h} < B$.
   (e) Why would you expect, on the basis of geometric intuition, that $\lim_{h \to 0} (1 + h)^{1/h} = B$. This number B is, of course, the number e. (A similar argument works for $h < 0$.)

31 (The slide rule) Logarithms are the basis for the design of the slide rule, a device for multiplying or dividing two numbers, to three significant figures. Slide rules were common from the early part of the seventeenth century to their recent eclipse by the hand-held calculator. A slide rule consists of two sticks (or circular disks), of which one is fixed and the other is free to slide next to it. A scale in introduced on each stick by placing the number N at a distance $\log_e N$ inches from the left end of the stick. (Any base would do as well as e.) Thus 1 is

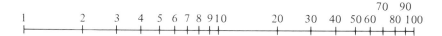

Figure 6.2

placed at the left end of each stick and 10 at $\log_e 10 \approx 2.3$ inches from the left end. The scale of the top stick is at the bottom edge; the scale of the lower stick is at the top edge. Each stick has the scale shown in Fig. 6.2.
(a) Make two such scales on paper or cardboard.
(b) Use them to multiply 4 times 25, as follows. Place the 1 of the lower stick at the 4 of the upper stick. Above the 25 of the lower stick appears the product. Check that this works with your two sticks.
(c) Explain precisely why the slide rule works.
(d) How would you use it to divide two numbers?

6.2 The derivatives of the logarithmic functions

This section obtains the derivative of $\log_b x$, where b is any fixed positive number other than 1. We will see that an antiderivative of $1/x$ is a logarithmic function.

In the proof of Theorem 1 it will be assumed that the function $\log_b x$ is continuous. (The continuity of $\log_b x$ is shown in Sec. 6.3.)

THEOREM 1 The derivative of the function $\log_b x$ is

$$\frac{\log_b e}{x}$$

for all positive numbers x.

PROOF It is necessary to compute

$$\lim_{\Delta x \to 0} \frac{\log_b (x + \Delta x) - \log_b x}{\Delta x}.$$

Recall that

$$f'(x) = \lim_{\Delta x \to 0} \frac{f(x + \Delta x) - f(x)}{\Delta x},$$

by definition of the derivative.

Before letting $\Delta x \to 0$, rewrite the difference quotient,

$$\frac{\log_b (x + \Delta x) - \log_b x}{\Delta x},$$

using algebra and properties of logarithms, as follows:

$$\frac{\log_b (x + \Delta x) - \log_b x}{\Delta x}$$

$$= \frac{\log_b \left(\frac{x + \Delta x}{x}\right)}{\Delta x} \qquad \log_b (c/d) = \log_b c - \log_b d$$

$$= \frac{1}{\Delta x} \log_b \left(1 + \frac{\Delta x}{x}\right) \qquad \text{algebra}$$

$$= \log_b \left(1 + \frac{\Delta x}{x}\right)^{1/\Delta x} \qquad \log_b c^m = m \log_b c$$

$$= \log_b \left(\left(1 + \frac{\Delta x}{x}\right)^{x/\Delta x}\right)^{1/x} \qquad \text{power of a power}$$

$$= \frac{1}{x} \log_b \left(1 + \frac{\Delta x}{x}\right)^{x/\Delta x} \qquad \log_b c^m = m \log_b c$$

After these manipulations, it is easy to take limits:

$$\lim_{\Delta x \to 0} \frac{\log_b (x + \Delta x) - \log_b x}{\Delta x}$$

$$= \lim_{\Delta x \to 0} \frac{1}{x} \log_b \left(1 + \frac{\Delta x}{x}\right)^{x/\Delta x}$$

$$= \frac{1}{x} \lim_{\Delta x \to 0} \log_b \left(1 + \frac{\Delta x}{x}\right)^{x/\Delta x} \qquad 1/x \text{ is fixed}$$

$$= \frac{1}{x} \log_b \left(\lim_{\Delta x \to 0} \left(1 + \frac{\Delta x}{x}\right)^{x/\Delta x}\right) \qquad \log_b \text{ is continuous}$$

$$= \frac{1}{x} \log_b e \qquad \text{definition of } e$$

Thus $\log_b$ has a derivative,

$$(\log_b x)' = \frac{\log_b e}{x}. \quad \blacksquare$$

In particular,

$$(\log_{10} x)' = \frac{\log_{10} e}{x}.$$

As a table of logarithms or a calculator shows, $\log_{10} e \approx 0.434$.

Thus
$$(\log_{10} x)' \approx \frac{0.434}{x}. \tag{1}$$

It is interesting to compare (1) with the slope of the graph of $\log_{10} x$ shown in Fig. 6.1. According to (1), the slope of the graph is always positive. The slope approaches 0 as $x \to \infty$; the slope approaches ∞ as $x \to 0$ through positive values. These conclusions are consistent with the general appearance of the graph.

THE MOST CONVENIENT BASE FOR LOGARITHMS IN CALCULUS

Which is the best of all possible bases b to use? More precisely, for which base b does the formula

$$\frac{\log_b e}{x}$$

take its simplest form? Certainly not $b = 10$. It would be nice to choose the base b in such a way that

$$\log_b e = 1,$$

Why e is used as a base for logarithms

that is, b^1 must equal e. In this case b is e. The best of all bases to use for logarithms is e. The derivative of the $\log_e$ function is given by

$$\frac{d}{dx}(\log_e x) = \frac{\log_e e}{x} = \frac{1}{x}.$$

In this case there is no constant, such as 0.434, to memorize.

The natural logarithm

For this reason, the base e is preferred in calculus. We shall write $\log_e x$ as $\ln x$, the *natural logarithm* of x. (Only for purposes of arithmetic, such as multiplying with the aid of logarithms, is base 10 preferable.) Most handbooks of mathematical tables include tables of $\log_{10} x$ (common logarithm) and $\ln x$ (natural logarithm). It would be helpful at this point to browse through both tables, shortened versions of which appear in Appendix H.

Scientific calculators usually have a ln*-key* ($\log_e$) *and a* log*-key* ($\log_{10}$).

A simple equation summarizes much of this section:

$$\boxed{\frac{d}{dx}(\ln x) = \frac{1}{x} \quad x > 0.}$$

Many texts and tables use log x *to denote the natural logarithm.*

It is well worth memorizing the following statement: The derivative of the natural logarithm function $\ln x$ is the reciprocal function $1/x$.

EXAMPLE 1 Find the area under the curve $y = 1/x$ and above the interval $[1, 100]$.

SOLUTION The area equals the definite integral

$$\int_1^{100} \frac{1}{x} \, dx.$$

By the fundamental theorem of calculus,

$$\int_1^{100} \frac{1}{x} \, dx = \ln x \Big|_1^{100}$$

$$= \ln 100 - \ln 1$$

$$= \ln 100 \approx 4.605. \quad \blacksquare$$

EXAMPLE 2 Find $(\ln(x^2 + 1))'$.

SOLUTION Let $y = \ln(x^2 + 1)$. Then

$$y = \ln u \quad \text{where} \quad u = x^2 + 1.$$

By the chain rule,

$$\frac{dy}{dx} = \frac{dy}{du} \frac{du}{dx}$$

$$= \frac{d}{du}(\ln u)\,\frac{d}{dx}(x^2+1)$$

$$= \frac{1}{u}\cdot 2x$$

$$= \frac{2x}{x^2+1}.$$

Thus the derivative of $\ln(x^2+1)$ is $2x/(x^2+1)$. ∎

EXAMPLE 3 Graph $y = \ln x$. At what angle does it cross the x axis?

SOLUTION The graph is shown in Fig. 6.3. Note that at $x = e \approx 2.718$, the y coordinate is 1. The graph crosses the x axis at $(1, 0)$. Since the derivative of $\ln x$ is $1/x$, the slope of the curve $y = \ln x$ at $x = 1$ is $1/1$, which is 1. Since the slope of the tangent line at $(1, 0)$ is 1, the tangent makes an angle of $\pi/4$ radians (45°) with the x axis.

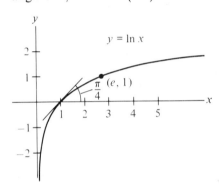

Figure 6.3

THE ANTIDERIVATIVE OF $1/x$

Since $(\ln x)' = 1/x$, $\int \dfrac{1}{x}\,dx = \ln x + C$. \hfill (2)

However (2) makes no sense if x is negative, since only positive numbers have logarithms. The next theorem provides an antiderivative for $1/x$, even where x is negative.

THEOREM 2

$$\int \frac{1}{x}\,dx = \ln|x| + C \quad \text{for } x > 0 \text{ or for } x < 0 \qquad (3)$$

PROOF It is necessary to show that the derivative of $\ln|x|$ is $1/x$. For positive x this has already been done. Now consider x negative.
For $x < 0$, $|x| = -x$. Thus, for negative x,

6.2 THE DERIVATIVES OF THE LOGARITHMIC FUNCTIONS

$$\frac{d}{dx}(\ln|x|) = \frac{d}{dx}(\ln(-x))$$

$$= \frac{1}{(-x)} \frac{d(-x)}{dx} \quad \text{by chain rule, as in Example 2}$$

$$= \frac{1}{-x}(-1)$$

$$= \frac{1}{x}.$$

This completes the proof. ∎

EXAMPLE 4 Compute $\int_{-3}^{-1} dx/x$.

SOLUTION

$$\int_{-3}^{-1} \frac{dx}{x} \underset{\text{FTC}}{=} \ln|x| \Big|_{-3}^{-1} \quad \text{by Theorem 2}$$

$$= \ln|-1| - \ln|-3|$$

$$= \ln 1 - \ln 3$$

$$= 0 - \ln 3$$

$$= -\ln 3. \quad \blacksquare$$

Careful, $\int_a^b \frac{dx}{x}$ does not exist if $a < 0 < b$.

The next theorem shows that the logarithm function enables us to integrate many functions besides $1/x$.

THEOREM 3 Let $f(x)$ be a differentiable function. Then, if $f(x) \neq 0$,

$$\boxed{\int \frac{f'(x)}{f(x)} dx = \ln|f(x)| + C.}$$

PROOF Let $y = \ln|f(x)|$. Then

$$y = \ln|u| \quad \text{where} \quad u = f(x).$$

By the chain rule and Theorem 2,

$$\frac{dy}{dx} = \frac{1}{u} f'(x)$$

$$= \frac{f'(x)}{f(x)}. \quad \blacksquare$$

How to integrate $\dfrac{f'(x)}{f(x)}$

With the aid of the formula in Theorem 3 we can integrate the quotient of two functions if the numerator is exactly the derivative of the denominator. In fact, if the numerator is a constant times the derivative of the denominator, we can still use the formula, as the next example shows.

EXAMPLE 5 Compute

$$\int \frac{x^2 \, dx}{x^3 + 1}.$$

SOLUTION The numerator is not quite the derivative of the denominator. However, the numerator is $\frac{1}{3}$ times the derivative of the denominator, that is,

$$x^2 = \tfrac{1}{3} \cdot 3x^2.$$

With this observation, the integration of $x^2/(x^3 + 1)$ is quick:

$$\int \frac{x^2 \, dx}{x^3 + 1} = \int \frac{(\tfrac{1}{3}) 3x^2 \, dx}{x^3 + 1}$$

$$= \frac{1}{3} \int \frac{3x^2 \, dx}{x^3 + 1} \qquad\qquad \int cf(x) \, dx = c \int f(x) \, dx$$
$$\text{(c constant)}$$

$$= \tfrac{1}{3} \ln |x^3 + 1| + C. \quad \blacksquare$$

LOGARITHMIC DIFFERENTIATION

The next example presents a special case of implicit differentiation called *logarithmic differentiation*. This is a method for differentiating a function whose logarithm is simpler than the function itself.

EXAMPLE 6 Differentiate

$$y = \frac{\sqrt[3]{x} \sqrt{(1 + x^2)^3}}{x^{4/5}}.$$

SOLUTION Rather than compute dy/dx directly, take logarithms of both sides of the equation first, obtaining

$$\ln y = \tfrac{1}{3} \ln x + \tfrac{3}{2} \ln (1 + x^2) - \tfrac{4}{5} \ln x.$$

Then differentiate this equation implicitly:

$$\frac{1}{y} \frac{dy}{dx} = \frac{1}{3x} + \frac{3 \cdot 2x}{2(1 + x^2)} - \frac{4}{5} \frac{1}{x}.$$

Solving for dy/dx yields

$$\frac{dy}{dx} = y \left(\frac{1}{3x} + \frac{3x}{1 + x^2} - \frac{4}{5x} \right)$$

$$= \frac{\sqrt[3]{x} \sqrt{(1 + x^2)^3}}{x^{4/5}} \left(\frac{1}{3x} + \frac{3x}{1 + x^2} - \frac{4}{5x} \right).$$

The reader is invited to find dy/dx directly from the explicit formula for y. Doing so will show the advantage of logarithmic differentiation. ∎

SUMMARY OF THIS SECTION

The derivative of $\log_b x$ was computed and the natural logarithm function $\ln x = \log_e x$ was introduced. The following table records derivatives and antiderivatives involving logarithms.

Derivative	Antiderivative
$(\ln x)' = \dfrac{1}{x}$	$\displaystyle\int \dfrac{dx}{x} = \ln x + C,\ x > 0$
$(\ln |x|)' = \dfrac{1}{x}$	$\displaystyle\int \dfrac{dx}{x} = \ln |x| + C,\ x \neq 0$
$(\ln f(x))' = \dfrac{f'(x)}{f(x)}$	$\displaystyle\int \dfrac{f'(x)\,dx}{f(x)} = \ln f(x) + C,\ f(x) > 0$
$(\ln |f(x)|)' = \dfrac{f'(x)}{f(x)}$	$\displaystyle\int \dfrac{f'(x)\,dx}{f(x)} = \ln |f(x)| + C,\ f(x) \neq 0$
$(\log_b x)' = \dfrac{\log_b e}{x}$	Not needed as antiderivative

To differentiate a function $y = f(x)$ whose logarithm is simple, differentiate the function $\ln y$ implicitly. Then solve for the derivative y'.

Before differentiating $\ln f(x)$, first simplify by using laws of logarithms.

Exercises

In Exercises 1 to 8 differentiate the functions.

1. $\ln(1 + x^2)$
2. $x^3 \ln(1 + x^2)$
3. $\dfrac{\ln x}{x},\ x > 0$
4. $(\ln x)^3,\ x > 0$
5. $\sin 4x \ln 3x$
6. $\sec 5x \ln(1 + x^4)$
7. $(\ln(\sin x))^2,\ 0 < x < \pi$
8. $\sin(\ln(1 + x^6))$

In Exercises 9 to 14 differentiate and simplify your answers.

9. $\ln |2x + 3|$
10. $\dfrac{x}{3} - \dfrac{1}{9}\ln|3x + 1|$
11. $\dfrac{2}{25(5x + 2)} + \dfrac{1}{25}\ln|5x + 2|$
12. $x + 3 - 6\ln|x + 3| - \dfrac{9}{x + 3}$
13. $\ln\left|x + \sqrt{x^2 - 5}\right|$
14. $\ln\left|x + \sqrt{x^2 + 1}\right|$

In Exercises 15 to 20 first simplify by using laws of logarithms; then differentiate.

15. $\dfrac{1}{5}\ln\dfrac{x}{3x + 5},\ x > 0.$
16. $-\dfrac{1}{3x} + \dfrac{5}{9}\ln\left|\dfrac{3x + 5}{x}\right|$
17. $\dfrac{1}{10}\ln\left|\dfrac{5 + x}{5 - x}\right|$
18. $\sqrt{x^2 + 1}\ \ln\dfrac{\sqrt{x^2 + 1} - 1}{x}$
19. $\ln[(x^2 + 1)^3(x^5 + 1)^4]$
20. $\ln\dfrac{\sqrt{2x + 1}\sqrt[3]{3x + 2}}{(x^2 + 1)^5}$

In Exercises 21 to 24 differentiate by logarithmic differentiation.

21 $(1 + 3x)^5(\sin 3x)^6$

22 $\sqrt{1 + x^2}\sqrt[3]{(1 + \cos 3x)^5}$

23 $\dfrac{(\sec 4x)^{5/3} \sin^3 2x}{\sqrt{x}}$

24 $\dfrac{(\cot 5x)^3}{\sqrt[3]{x}(x^3 + 2)^{5/2}}$

■

Exercises 25 and 26 concern the derivative of $\log_b x$.

25 Differentiate $\log_{10} \sqrt[3]{x}$.

26 Differentiate $\log_2 ((x^2 + 1)^3(\sin 3x))$.

27 (a) Evaluate $\ln e^n$ for the positive integer n.
    (b) Show that $\ln x$ is an increasing function.
    (c) Deduce from (a) and (b) that $\lim_{x \to \infty} \ln x = \infty$.

28 Find $\lim_{x \to 0^+} \ln x$. *Suggestion:* Write $x = 1/t$.

In Exercises 29 and 30 find the indicated antiderivatives.

29 (a) $\displaystyle\int \dfrac{5\,dx}{5x + 1}$ (b) $\displaystyle\int \dfrac{x\,dx}{x^2 + 5}$

    (c) $\displaystyle\int \dfrac{\cos x\,dx}{\sin x}$ (d) $\displaystyle\int \dfrac{(1/x)\,dx}{\ln x}$

30 (a) $\displaystyle\int \dfrac{dx}{3x + 2}$ (b) $\displaystyle\int \dfrac{\sin x\,dx}{\cos x}$ (c) $\displaystyle\int \dfrac{(6x + 1)\,dx}{3x^2 + x + 5}$

31 Find the area of the region under the curve $y = 1/x$ and above the interval
    (a) $[1, e]$, (b) $[1, 10]$, (c) $[1/e^{10}, e^3]$.

32 Find the area of the region above the curve $y = 1/x$ and below the interval
    (a) $[-e^3, -e^2]$, (b) $[-10, -1]$.

■ ■

33 (a) Find the area of the region under the curve $y = 1/x$ and above $[1, b]$.
    (b) Using the result in (a), show that the area of the region under $y = 1/x$ and above $[1, \infty)$ is infinite.
    (c) The region below $y = 1/x$ and above $[1, \infty)$ is revolved around the x axis. Find the volume of that part of the resulting solid situated to the left of $x = b$, $b > 1$. (See Exercise 31 of Sec. 5.4.)
    (d) Show that the volume of the unbounded solid in (c) is finite.
    (e) In view of (b) it is impossible to paint the region in (b) with a finite amount of paint. However, in view of (d), we could fill the unbounded solid of revolution with a finite amount of paint, then dip the region in (b) into the paint. That would paint the region in (b) with a finite amount of paint. What is wrong?

6.3 The natural logarithm defined as a definite integral (optional)

Recall how the logarithm and exponential functions were defined. First the exponential function b^x ($b > 0$) was built up in stages: $b^n = b \cdot b \cdots b$ (n times) for $n = 1, 2, 3, \ldots$; $b^0 = 1$; $b^{-n} = 1/b^n$ for $n = 1, 2, 3, \ldots$; $b^{1/n} = \sqrt[n]{b}$, the positive nth root of b for $n = 1, 2, 3, \ldots$; $b^{m/n} = (b^{1/n})^m$ for m an integer and n a positive integer; and finally $b^x = \lim b^{m/n}$ as $m/n \to x$ for irrational x.

A thorough treatment of the exponential functions based on this approach encounters many difficulties, such as: How do we know that b has an nth root? If $b > 1$, is b^x increasing? Does $\lim_{m/n \to x} b^{m/n}$ exist? After answering these questions we would still be left with showing that b^x is continuous and that $b^x b^y = b^{x+y}$.

Assuming that b^x has these desired properties, we then defined $\log_b x$. Then, to obtain the derivative of $\log_b x$ we had to assume that

$$\lim_{h \to 0} (1 + h)^{1/h}$$

exists and that $\log_b x$ is a continuous function.

6.3 THE NATURAL LOGARITHM DEFINED AS A DEFINITE INTEGRAL (OPTIONAL)

The present section takes a completely different approach. It first defines a function $L(x)$ with the aid of the definite integral. (It will turn out that $L(x) = \ln x$.) Then it defines e by setting a definite integral equal to 1. The fact that $L'(x) = 1/x$ will follow from the second fundamental theorem of calculus. In Sec. 6.10 the exponential function will be defined as the inverse of the function $L(x)$ and it will be shown that $\lim_{h \to 0} (1 + h)^{1/h}$ exists and equals e.

Basic to the following argument is the assumption that a continuous function has a definite integral over each interval $[a, b]$ in its domain.

THE FUNCTION $L(x)$

The key step is the introduction of a function $L(x)$, which will turn out to be $\ln x$.

DEFINITION *The function $L(x)$.*

$$L(x) = \int_1^x (1/t)\, dt, \qquad x > 0.$$

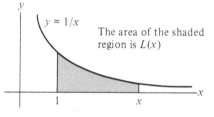

Figure 6.4

Observe that $L(1) = \int_1^1 (1/t)\, dt = 0$; if $x > 1$, $L(x) > 0$ and is the area of the shaded region in Fig. 6.4. If $0 < x < 1$, then $L(x) = \int_1^x (1/t)\, dt = -\int_x^1 (1/t)\, dt$, the negative of the area of the shaded region in Fig. 6.5. Thus, if $0 < x < 1$, we have $L(x) < 0$. $L(x)$ resembles a logarithm function also in that $L(x)$ is defined only for $x > 0$.

By the second fundamental theorem of calculus,

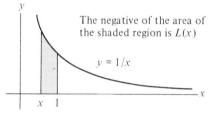

Figure 6.5

$$L'(x) = \frac{1}{x}. \qquad (1)$$

The information that $L(1) = 0$ and that $L'(x) = 1/x$ already shows that $L(x) = \ln x$. After all, since $L(x)$ and $\ln x$ have the same derivative, they differ by a constant,

$$L(x) = \ln x + C. \qquad (2)$$

To find the constant C, set $x = 1$ in (2), obtaining

$$L(1) = \ln (1) + C.$$

Since both $L(1)$ and $\ln (1)$ equal 0, it follows that $C = 0$. Thus $L(x) = \ln x$.

It is reassuring to know that $L(x)$, though defined so differently, is $\ln x$, defined in Sec. 6.2. However, in this section, *no use will be made of this knowledge*, since the present goal is to build up the theory of logarithms from scratch, with minimal assumptions.

Incidentally, the fact that the area under the curve $y = 1/x$ behaves like a logarithm was first noted by Gregory St. Vincent and his friend A. A. de Sarasa in 1647, some twenty years before the work of Newton and Leibniz.

PROPERTIES OF THE FUNCTION $L(x)$

PROPERTY 1 $L(xy) = L(x) + L(y)$, for $x, y > 0$.

PROOF   Hold y fixed at the value k. It will be shown that

$$L(xk) = L(x) + L(k), \quad x > 0.$$

Introduce two functions of x,

$$f(x) = L(xk) \quad \text{and} \quad g(x) = L(x) + L(k).$$

We wish to show that $f(x) = g(x)$.
To begin, compute their derivatives:

$$\begin{aligned} f'(x) &= (L(xk))' \\ &= \frac{1}{xk}(xk)' \quad &&\text{by (1) and the chain rule} \\ &= \frac{k}{xk} \\ &= \frac{1}{x}, \end{aligned}$$

and

$$\begin{aligned} g'(x) &= (L(x) + L(k))' \\ &= L'(x) \quad &&k \text{ is a constant} \\ &= \frac{1}{x}. \end{aligned}$$

Since $f(x)$ and $g(x)$ have the same derivatives, they differ by a constant, that is,

$$L(xk) = L(x) + L(k) + C \quad\quad C \text{ constant.} \tag{3}$$

To find C, set $x = 1$ in (3), obtaining

$$L(k) = L(1) + L(k) + C.$$

Since $L(1) = 0$, this equation shows that $C = 0$, and Property 1 is established. ∎

PROPERTY 2   For any integer n, $L(x^n) = nL(x)$, $x > 0$.

PROOF   Differentiate both $L(x^n)$ and $nL(x)$:

$$(L(x^n))' = \frac{1}{x^n} n x^{n-1} \quad \text{by (1) and the chain rule}$$

6.3 THE NATURAL LOGARITHM DEFINED AS A DEFINITE INTEGRAL (OPTIONAL)

$$= \frac{n}{x}$$

and $\quad (nL(x))' = \dfrac{n}{x} \quad$ by (1).

Thus there is a constant C such that

$$L(x^n) = nL(x) + C. \tag{4}$$

Setting $x = 1$ in (4) shows that

$$L(1) = nL(1) + C$$

or $\quad\quad\quad\quad\quad\quad\quad 0 = 0 + C.$

Hence $C = 0$ and Property 2 is established. ∎

Property 2, with $n = -1$, implies that

$$L(x^{-1}) = (-1)L(x),$$

or, simply

$$L\left(\frac{1}{x}\right) = -L(x),$$

an equation that should not come as a surprise.

PROPERTY 3 $\quad \lim\limits_{x \to \infty} L(x) = \infty \quad$ and $\quad \lim\limits_{x \to 0^+} L(x) = -\infty.$

PROOF Since $L'(x)$ is positive, $L(x)$ is an increasing function. Moreover, $L(2) > 0$ and $L(2^n) = nL(2)$. Thus $L(2^n)$ gets arbitrarily large as $n \to \infty$, so

$$\lim_{x \to \infty} L(x) = \infty.$$

To show that $\quad\quad \lim\limits_{x \to 0^+} L(x) = -\infty,$

replace x by $1/t$, where $t \to \infty$. Then we have

$$\lim_{x \to 0^+} L(x) = \lim_{t \to \infty} L\left(\frac{1}{t}\right)$$

$$= \lim_{t \to \infty} (-L(t)) \quad \text{by Property 2}$$

$$= -\lim_{t \to \infty} L(t)$$

$$= -\infty. \quad ∎$$

THE NUMBER e

Since $L(1) = 0$ and $L(x)$ is an increasing continuous function that takes on arbitrarily large values, there must exist a unique number x such that

$$L(x) = 1.$$

The intermediate-value theorem guarantees that there is at least one x. Why is there only one?

DEFINITION e is the solution to the equation $L(x) = 1$, that is,

$$L(e) = 1.$$

In Sec. 6.10, after the exponential functions are defined, it will be shown that $e = \lim_{h \to 0} (1 + h)^{1/h}$.

Exercises

1 Show that L is continuous.
2 Show that the graph of $L(x)$ is concave downward.
3 (a) Compute the area of the eight rectangles of the same width in Fig. 6.6.

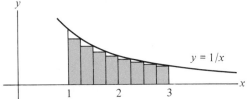

Figure 6.6

(b) From (a) deduce that $L(3) > 1$.
(c) From (b) deduce that $e < 3$.
4 Using an argument like that in Exercise 3, show that $e > 2$.
5 Let n be an integer larger than 1. By considering appropriate rectangles of width 1, show that

$$L(n) > \frac{1}{2} + \frac{1}{3} + \frac{1}{4} + \cdots + \frac{1}{n}.$$

6 (a) Let n be an integer larger than 1. Show that

$$L(n+1) < \frac{1}{1} + \frac{1}{2} + \frac{1}{3} + \cdots + \frac{1}{n}.$$

(b) From (a) deduce that $\lim_{n \to \infty} \sum_{i=1}^{n} 1/i = \infty$. (This was first proved by Oresme around the year 1360.)

Exercises 7 and 8 outline a way to compute $L(2)$ with any degree of accuracy desired.

7 (a) By using a partition of $[1, 2]$ of n sections of equal length and left endpoints as sampling points, show that

$$L(2) < \frac{1}{n} + \frac{1}{n+1} + \cdots + \frac{1}{2n-1}.$$

(b) Deduce that $L(2) < 0.74$.

8 Show that for any positive integer n,

$$L(2) > \frac{1}{n+1} + \frac{1}{n+2} + \cdots + \frac{1}{2n}.$$

Suggestion: Form a suitable partition of $[1, 2]$ and use right endpoints as sampling points.

∎

9 (a) Prove that for $b > 1$, we have $\int_a^{ab} (1/x)\, dx = \int_1^b (1/x)\, dx$ by observing that if the sum $\sum_{i=1}^n (1/c_i)(x_i - x_{i-1})$ is an approximation of $\int_1^b (1/x)\, dx$, then $\sum_{i=1}^n (1/ac_i)(ax_i - ax_{i-1})$ is an approximation of $\int_a^{ab} (1/x)\, dx$.
(b) If $a, b > 1$, deduce that $\int_1^{ab} (1/x)\, dx - \int_1^a (1/x)\, dx = \int_1^b (1/x)\, dx$.
(c) From (b), show that, if $a, b > 1$, then $L(ab) = L(a) + L(b)$.

∎ ∎

10 Let n be an integer larger than 1. The area of the shaded region in Fig. 6.7 is equal to

$$\frac{1}{1} + \frac{1}{2} + \frac{1}{3} + \cdots + \frac{1}{n-1} - L(n).$$

Using geometric intuition (no calculus), show that the area of the shaded region is (*a*) less than 1, (*b*) greater than $\frac{1}{2}$.

As $n \to \infty$, the area of the shaded region approaches a number, denoted γ (gamma), called Euler's constant. To three decimal places, $\gamma \approx 0.577$. It is not known whether γ is rational.

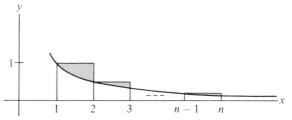

Figure 6.7

6.4 Inverse functions and the derivative of b^x

For a decreasing function, output decreases as input increases.

Consider a continuous increasing function $y = f(x)$ defined on some finite or infinite interval. As the input x increases, the output y increases as well. This implies that different inputs have different outputs. To put it another way, "if you know the output, you can figure out the input," or "the output determines the input." This means that the input can be thought of as a function of the output. (A similar conclusion holds for a decreasing function.)

For example, the cubing function $y = f(x) = x^3$ increases throughout its entire domain, which is the x axis. For a given output y, it is possible to solve for the corresponding input x, as follows:

$$y = x^3,$$

so

$$x = \sqrt[3]{y}.$$

This second function, the "cube-root function," is called the *inverse* of the cubing function.

DEFINITION *The inverse function.* Let $y = f(x)$ be a continuous increasing function on some interval. The function g that assigns to each output the corresponding unique input is called the *inverse* of f. That is, if $y = f(x)$, then $x = g(y)$. The inverse of a continuous decreasing function is defined similarly.

The function $y = f(x) = x^2$ is not increasing. It does in some cases assign the same output to different inputs. For instance

$$f(5) = f(-5),$$

since $5^2 = 25$ and $(-5)^2 = 25$. However, the squaring function is an increasing function if it is considered only for $x \geq 0$. In that case, if $x \geq 0$ and $y = f(x)$, then we have

$$y = x^2,$$

310 TOPICS IN DIFFERENTIAL CALCULUS

The inverse of squaring is taking the square root.

and can solve uniquely for x as a function of y.

$$x = \sqrt{y}.$$

One-to-one functions

So the square-root function is the inverse of the squaring function (if the domain of the squaring function is restricted to the nonnegative numbers).

Any function that assigns distinct outputs to distinct inputs is called *one-to-one*. An increasing function is one-to-one, as is a decreasing function. It is usually possible to restrict the domain of a differentiable function enough so that it is one-to-one on that domain. (See Fig. 6.8.)

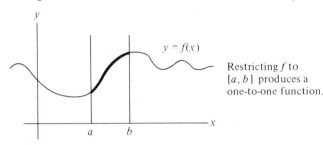

Restricting f to $[a, b]$ produces a one-to-one function.

Figure 6.8

SOME WELL-KNOWN INVERSE FUNCTIONS

As mentioned, the cube-root function is the inverse of the cubing function and the square-root function is the inverse of the squaring function (on an appropriate domain). Some other famous functions come in pairs. For instance, the function 10^x is increasing on the entire x axis and so has an inverse function. What is that inverse function?

The inverse of the exponential function 10^x

If $y = 10^x$, we must express x in terms of y. But the exponent x is the logarithm of y to the base 10,

$$x = \log_{10} y.$$

Thus the inverse of the exponential function 10^x is the logarithmic function $\log_{10} y$.

The function $y = \log_{10} x$ is increasing, so it too has an inverse function. What is it? If $y = \log_{10} x$, how is x expressed in terms of y? Take the case $y = 2$. If

$$2 = \log_{10} x,$$

what is x? What number has a logarithm of 2 in the base 10? The answer is 10^2 or 100. Thus, if

$$2 = \log_{10} x, \quad \text{then} \quad x = 10^2.$$

More generally, if

$$y = \log_{10} x, \quad \text{then} \quad x = 10^y.$$

In other words, the inverse of the logarithm function is the exponential function.

Inverse functions come in pairs, each reversing the effect of the other. This table lists some pairs of inversely related functions.

Function f	Inverse Function g
Cubing, $y = x^3$	Cube root, $x = \sqrt[3]{y}$
Cube root, $y = \sqrt[3]{x}$	Cubing, $x = y^3$
Squaring, $y = x^2$	Square root, $x = \sqrt{y}$
Square root, $y = \sqrt{x}$	Squaring, $x = y^2$
Exponential, base 10, $y = 10^x$	Logarithm, base 10, $x = \log_{10} y$
Logarithm, base 10, $y = \log_{10} x$	Exponential, base 10, $x = 10^y$
Natural logarithm, $y = \ln x$	Exponential, base e, $x = e^y$
Exponential, base e, $y = e^x$	Natural logarithm, $x = \ln y$

(In the next section inverses of the trigonometric functions will be considered.)

INVERSE FUNCTIONS ON A CALCULATOR

Most scientific calculators have a $\sqrt{x}$ key and an x^2 key. Each "undoes" the effect of the other. For instance, if you press the

1 5 key
2 then the $\sqrt{x}$ key
3 then the x^2 key

the calculator will end up displaying 5.

If you reverse the order of the square-root and squaring keys, a similar effect results. If you press

1 the 5 key
2 then the x^2 key
3 then the $\sqrt{x}$ key

the final output, the composition of the two functions, should be 5.

To minimize the number of keys, many calculators have a special "inverse" key, sometimes labeled "inv." The inv key, in combination with the e^x key (also labeled "exp"), provides the natural logarithm. If you press

The inverse key might be labeled "2nd F" or have a special color and no label.

1 the 2 key
2 then the inv key
3 then the e^x key,

the calculator will display a decimal estimate of ln 2. (Sometimes "inv" is combined with ln x to give e^x.)

The inverse trig functions are called arccos, arcsin, and arctan, etc.

In some calculators the inverse key is labeled "arc" and is used only in conjunction with a "sin," "cos," or "tan" key. The inverse trig functions are called arccos, arcsin, and arctan, etc. They will be discussed in the next section.

GRAPH OF AN INVERSE FUNCTION

EXAMPLE 1 Determine the inverse of the "doubling" function f defined by $f(x) = 2x$ and then graph it.

SOLUTION If $y = 2x$, there is only one value of x for each value of y, and it is obtained by solving the equation $y = 2x$ for x: $x = y/2$. Thus f is one-to-one and its inverse function g is the "halving" function: If y is the input in the function g, then the output is $y/2$.

For instance, $f(3) = 6$ and $g(6) = 3$. Thus $(3, 6)$ is on the graph of f, and $(6, 3)$ is on the graph of g. Since it is customary to reserve the x axis for inputs, we should write the formula for g, the "halving" function, as

$$g(x) = \frac{x}{2}.$$

Thus f has the formula $\quad y = 2x, \quad$ doubling,

and g has the formula $\quad y = \dfrac{x}{2}, \quad$ halving.

The graphs of f and g are lines (see Fig. 6.9); one has slope 2, and the other has slope $\tfrac{1}{2}$. ∎

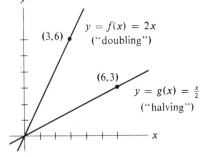

Figure 6.9

EXAMPLE 2 Graph $y = 10^x$ and its inverse function $y = \log_{10} x$ on the same axes.

SOLUTION In order to graph the logarithm and exponential functions, it is advisable to prepare brief tables first.

x	10	1	0.1
$\log_{10} x$	1	0	-1

x	1	0	-1
10^x	10	1	0.1

Graph the functions with the aid of the tables, as in Fig. 6.10. ∎

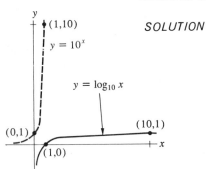

Figure 6.10

Note the relation between the two tables in Example 2. One is obtained from the other by switching inputs and outputs. This is the case for any one-to-one function and its inverse. If a and b are the numerical entries in a column for one function, then b and a are the entries in a column for the inverse function. Also, note the relation between the two graphs in Examples 1 and 2. One graph is obtained from the other by reflecting it across the line $y = x$. This can be done because, if (a, b) is on the graph of one function, then (b, a) is on the graph of the other. If you fold the paper along the line $y = x$, the point (b, a) comes together with the point (a, b), as you will note in Fig. 6.11. This relation between the graphs holds for any one-to-one function and its inverse.

These example are typical of the correspondence between a one-to-one function and its inverse. Perhaps the word *reverse* might be more descriptive than *inverse*. One final matter of notation: We have used the letter g to denote the inverse of f. It is common to use the symbol f^{-1} (read as "f inverse") to denote the inverse function. We preferred to delay its use

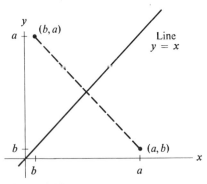

Figure 6.11

The symbol f^{-1} denotes the inverse function.

because its resemblance to the reciprocal notation might cause confusion. It should be clear from the examples that f^{-1} does *not* mean to divide 1 by f. The symbol "inv f" would be unambiguous. However, it is longer than the symbol "f^{-1}" and the weight of tradition is behind f^{-1}.

THE DERIVATIVE OF AN INVERSE FUNCTION

Similar conclusions hold if f' is negative.

Let f have a positive derivative throughout some infinite interval or some finite interval $[a, b]$. It then has an inverse function f^{-1}. The graphs of f and f^{-1} are shown in Fig. 6.12.

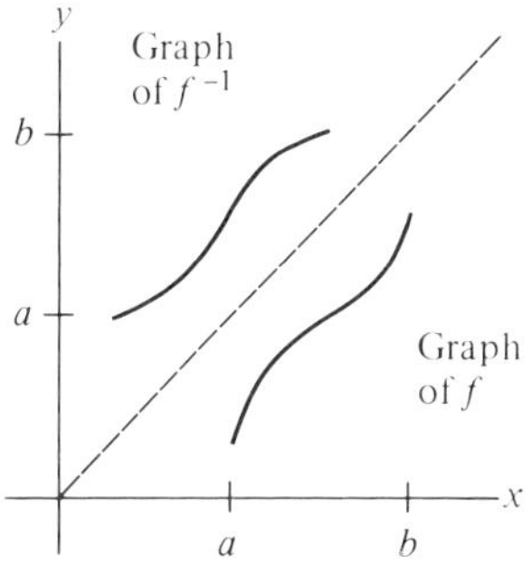

It is proved in advanced calculus that f^{-1} is also differentiable. A glance at Fig. 6.12 should make that result plausible. (Folding the graph of f across the line $y = x$ also carries its tangent lines along.)

Our optical intuition will suggest the relation between the derivatives of f and f^{-1}. Think of f as the projection from a slide to the screen and f^{-1} as "reverse" projection that assigns to each point on the screen the point on the slide that projects onto it. Now compare the magnifications. If f magnifies by a certain ratio at x, say 3, then we expect f^{-1} to shrink by a factor 3 at $f(x)$, that is, "magnify" by a factor $\frac{1}{3}$. Intuitively, if f has a derivative at x, we expect f^{-1} to have a derivative at the number $f(x)$ and, in fact,

$$(f^{-1})'(f(x)) = \frac{1}{f'(x)}, \qquad \text{if} \quad f'(x) \neq 0.$$

Figure 6.12

$$(f^{-1})'(y) = \frac{1}{f'(x)}$$

That is, $(f^{-1})'(y) = 1/f'(x)$, where $y = f(x)$. This fact will be assumed in what follows.

Let $y = f(x)$ denote the given function and $x = f^{-1}(y)$ denote the inverse function. Let us apply the chain rule to the special case

$$x = f^{-1}(y), \qquad \text{where} \quad y = f(x).$$

By the chain rule,
$$\frac{dx}{dx} = \frac{dx}{dy} \cdot \frac{dy}{dx},$$

or
$$\boxed{1 = \frac{dx}{dy} \cdot \frac{dy}{dx}.} \qquad (1)$$

Equation (1) is important. It will be used in this section to differentiate b^x and in the next to differentiate the inverse trigonometric functions.

THE DERIVATIVES OF e^x AND b^x

THEOREM 1 *The derivative of e^x.* Let $y = e^x$ be the exponential function with base e.

Then
$$\frac{d(e^x)}{dx} = e^x.$$

PROOF Let $y = e^x$. This function is the inverse of the natural logarithmic function, that is,

$$y = e^x \quad \text{is equivalent to} \quad x = \ln y.$$

Since the derivative dx/dy exists, the derivative dy/dx exists, wherever dx/dy is not 0.

Since $x = \ln y$,

we have
$$1 = \frac{dx}{dx} = \frac{dx}{dy} \cdot \frac{dy}{dx},$$

or
$$1 = \frac{d}{dy}(\ln y) \cdot \frac{dy}{dx}.$$

Hence
$$1 = \frac{1}{y} \cdot \frac{dy}{dx},$$

Thus
$$\frac{dy}{dx} = y$$
$$= e^x.$$

e^x is equal to its own derivative!

In short,
$$\frac{dy}{dx} = e^x,$$

and the theorem is proved. ∎

EXAMPLE 3 Find the derivative of e^{3x}.

SOLUTION Let $y = e^{3x}$. Then $y = e^u$, where $u = 3x$.

Thus
$$\frac{dy}{dx} = \frac{dy}{du}\frac{du}{dx} \quad \text{chain rule}$$
$$= \frac{d(e^u)}{du}\frac{d(3x)}{dx}$$
$$= e^u \cdot 3 \quad \text{Theorem 1}$$
$$= e^{3x} \cdot 3$$
$$= 3e^{3x}. \quad ∎$$

The formula for the derivative of e^x is quite simple. Let us next compute the derivative of 10^x.

EXAMPLE 4 Find the derivative of 10^x.

SOLUTION Write 10 as a power of e: $\quad 10 = e^{\ln 10}$.

Then $\quad 10^x = (e^{\ln 10})^x$.

By the "power-of-a-power" rule,

$$(e^{\ln 10})^x = e^{(\ln 10)x}.$$

6.4 INVERSE FUNCTIONS AND THE DERIVATIVE OF b^x

Thus
$$10^x = e^{(\ln 10)x}.$$

Since $\ln 10$ is a constant, this problem is similar to Example 3. Let $y = e^{(\ln 10)x}$.

This can be written as $y = e^u$, where $u = (\ln 10)x$.

Then
$$\frac{dy}{dx} = \frac{d(e^u)}{du} \frac{d}{dx}[(\ln 10)x] \quad \text{chain rule}$$
$$= e^u \cdot \ln 10 \quad \text{Theorem 1}$$
$$= e^{(\ln 10)x} \cdot \ln 10$$
$$= 10^x \cdot \ln 10$$
$$= (\ln 10) \cdot 10^x.$$

Note that $\dfrac{d(10^x)}{dx} \approx (2.3)10^x$. Thus
$$\frac{d(10^x)}{dx} = (\ln 10) \cdot 10^x. \quad \blacksquare$$

Similarly, for any base b,
$$\frac{d(b^x)}{dx} = (\ln b)b^x.$$

Why e is the ideal base for an exponential function

(Since it is easy to forget the coefficient $\ln b$, it may be safer to write b^x as $e^{x \ln b}$ and differentiate than to memorize this formula.) The coefficient $\ln b$ equals 1 only if the base b is chosen to be e; then there is nothing to remember, except $(e^x)' = e^x$. That is why e is the most convenient base for an exponential function in calculus.

EXAMPLE 5 Differentiate
$$f(x) = \frac{e^{ax}}{a^3}(a^2x^2 - 2ax + 2),$$

where a is a nonzero constant.

SOLUTION First bring a^3, which is constant, to the front of the formula by writing the function as
$$f(x) = \frac{1}{a^3}e^{ax}(a^2x^2 - 2ax + 2).$$

Now differentiate:
$$f'(x) = \frac{1}{a^3}[e^{ax}(a^2x^2 - 2ax + 2)]' \quad \text{derivative of constant times a function}$$
$$= \frac{1}{a^3}[e^{ax}(a^2 \cdot 2x - 2a) + (a^2x^2 - 2ax + 2)ae^{ax}] \quad \text{derivative of product, chain rule, Theorem 1}$$
$$= \frac{1}{a^3}[(2a^2x - 2a + a^3x^2 - 2a^2x + 2a)e^{ax}]$$
$$= x^2e^{ax} \quad \text{after canceling.} \quad \blacksquare$$

THE DERIVATIVE OF x^a FOR ANY a

The next theorem generalizes the formula $(x^a)' = ax^{a-1}$ from rational a to any real number a.

THEOREM 2 Let a be a fixed real number. Then for $x > 0$

$$\frac{d(x^a)}{dx} = ax^{a-1}.$$

PROOF Let $y = x^a$. Since
$$x = e^{\ln x},$$
$$y = (e^{\ln x})^a.$$

Hence, by the power of a power rule,

$$y = e^{a \ln x}.$$

This can be written as $y = e^u$, where $u = a \ln x$.

Hence $\qquad \dfrac{dy}{dx} = \dfrac{d(e^u)}{du} \dfrac{d}{dx}(a \ln x) \qquad$ chain rule

$$= e^u \frac{a}{x}$$

$$= e^{a \ln x} \frac{a}{x}$$

$$= x^a \frac{a}{x}$$

$$= ax^{a-1}.$$

This proves the theorem. ∎

Incidentally, this provides a second proof that $(x^a)' = ax^{a-1}$ for positive x and rational a.

For example, $d(x^\pi)/dx = \pi x^{\pi-1}$ and $(x^{\sqrt{2}})' = \sqrt{2}(x^{\sqrt{2}-1})$.

DIFFERENTIATION OF $f(x)^{g(x)}$

At this point we can differentiate x^a (the exponent is fixed and the base varies) and b^x (the base is fixed and the exponent varies). What if both the base and the exponent vary? For instance, how would we differentiate

$$x^x?$$

One way is to rewrite the base as $e^{\ln x}$. Then

$$x^x = (e^{\ln x})^x = e^{x \ln x};$$

since $e^{x \ln x}$ has a fixed base, it can be differentiated by the chain rule, as e^{3x}

was differentiated in Example 3. However, a simpler way is to use logarithmic differentiation, as shown in Example 6.

EXAMPLE 6 Use logarithmic differentiation to find $(x^x)'$.

SOLUTION Let
$$y = x^x.$$

We wish to find y'.
Begin by taking logarithms,

$$\ln y = \ln (x^x)$$

or
$$\ln y = x \ln x \qquad (\ln c^m = m \ln c).$$

Then differentiate this last equation with respect to x, obtaining:

$$\frac{1}{y} y' = (x \ln x)'$$

$$= x \cdot \frac{1}{x} + (\ln x)1$$

or
$$\frac{y'}{y} = 1 + \ln x$$

Solve for y':

$$y' = y(1 + \ln x).$$

Thus
$$y' = x^x(1 + \ln x). \blacksquare$$

How to differentiate $(f(x))^{g(x)}$ To differentiate $y = f(x)^{g(x)}$ differentiate $\ln y = \ln (f(x)^{g(x)}) = g(x) \ln f(x)$ implicitly and solve for y'.

Exercises

Exercises 1 to 18 concern inverse functions.
The functions in Exercises 1 to 6 have inverse functions. In each case graph the given function and also the inverse function relative to the same axes.

1   $y = 3x$ 2 $y = 3^x$
3   $y = \log_4 x$ 4 $y = \sqrt[3]{x}$
5   $y = 2x + 1$ 6 $y = x^5$

In Exercises 7 to 10 determine whether the given functions are one-to-one on the specified intervals.

7   x^2;   (a) $[-1, 1]$ (b) $[2, 3]$
            (c) $[-4, -3]$ (d) $(-\infty, 0]$
8   x^5;   (a) $[1, 2]$ (b) $[-1, 3]$
            (c) $[0, \infty)$ (d) the x axis
9   $\sin x$;   (a) $[0, \pi]$ (b) $[\pi/2, 3\pi/2]$
             (c) $[\pi/4, 3\pi/4]$ (d) $[-\pi/2, \pi/2]$

10   $\cos x$;   (a) $[0, \pi]$ (b) $[\pi/2, 3\pi/2]$
             (c) $[\pi/4, 3\pi/4]$ (d) $[0, 2\pi]$

In each of Exercises 11 to 18 show that the given function is one-to-one by examining its derivative. Then find its inverse.

11   $1/x^2$, $x > 0$ 12 e^{-3x}
13   10^{x^2}, $x \geq 0$ 14 $1/(x - 1)$, $x > 1$
15   $3x + 5$ 16 $x^{3/5}$
17   $\sqrt[3]{1 + x^5}$, $x > -1$ 18 $x + x^2$, $x > -\frac{1}{2}$

In Exercises 19 to 34 differentiate the given functions.

19   e^{x^2} 20 xe^{-4x}
21   $x^2 e^{2x}$ 22 $(\sin 2x)(\sin e^{-x})$
23   2^{-x^2} 24 $3^{\sqrt{x}}$

25 $x^{(x^2)}$
26 $(2 + \cos x)^{\sin x}$
27 $e^{\ln 3x}$
28 $\ln e^{x^2}$
29 $x^{\tan 3x}$
30 $(\tan \sqrt{x})^{e^{-x}}$
31 $\dfrac{e^{-4x} \cos 5x}{1 + e^x}$
32 $\dfrac{10^{x^2} \cot 5x}{\ln(1 + x^2)}$
33 $x^{\sqrt{3}} (\sin 3x) e^{x^2}$
34 $\dfrac{x^\pi \tan e^x}{2^x}$

In Exercises 35 to 38 differentiate and simplify. (The numbers a, b, and c are positive constants.)

35 $\dfrac{e^{ax}(ax - 1)}{a^2}$

36 $\dfrac{xb^{ax}}{a \ln b} - \dfrac{b^{ax}}{a^2 (\ln b)^2}$

37 $\dfrac{e^{ax}}{a^2 + b^2}(a \sin bx - b \cos bx)$

38 $\dfrac{1}{ac} \ln(b + ce^{ax})$

In each of Exercises 39 to 42 a region in the plane is described. (a) Find its area. (b) Find the volume of the solid formed by revolving the region around the x axis. (See Exercise 31 in Sec. 5.4.)

39 Under e^{3x}, above $[1, 5]$.
40 Under $5e^{-2x}$, above $[0, \ln 2]$.
41 Under 10^x, above $[0, 3]$.
42 Under 2^{-x}, above $[-4, 1]$.

■

43 Graph $y = e^{-x^2}$, showing (a) critical points, (b) inflection points, (c) asymptotes.

The result in Exercise 44 is used in Exercises 45 to 50.

44 (Calculator) (a) Compute x/e^x for $x = 1, 5,$ and 10.
    (b) On the basis of (a), what do you think $\lim_{x \to \infty} x/e^x$ is?
    (c) Compute $x^2 e^{-x}$ for $x = 1, 10,$ and 20.
    (d) On the basis of (c), what do you think $\lim_{x \to \infty} x^2 e^{-x}$ is? The methods of Sec. 6.8 show that $\lim_{x \to \infty} x^n/e^x$ ($= \lim_{x \to \infty} x^n e^{-x}) = 0$, for any positive integer n. This result may be assumed when doing Exercises 45 to 50. (See Exercises 58 and 59 of this section for a short proof using the binomial theorem.)

In each of Exercises 45 to 50 find (a) intercepts, (b) critical points, (c) local maxima or minima, (d) inflection points, and (e) asymptotes. (f) Graph the function.

45 $f(x) = xe^{-x}$
46 $f(x) = x^2 e^{-x}$
47 $f(x) = x^3 e^{-x}$
48 $f(x) = x2^{-x}$
49 $f(x) = (x - x^2)e^{-x}$
50 $f(x) = xe^x$

51 Using results of this section and the definition of the derivative, evaluate:

(a) $\lim_{h \to 0} \dfrac{e^h - 1}{h}$ (b) $\lim_{x \to 1} \dfrac{2^x - 2}{x - 1}$

(c) $\lim_{h \to 0} \dfrac{10^h - 1}{h}$

52 Let $f(x) = 5 + (x - x^2)e^x$.
    (a) Show that $f(0) = f(1)$.
    (b) Find all numbers c in $(0, 1)$ whose existence is guaranteed by Rolle's theorem, stated in Sec. 4.1.

53 Let A and k be constants. Show that the derivative of Ae^{kx} is proportional to Ae^{kx}.

54 (a) Graph $y = e^x$ and $y = -x$ on the same axes.
    (b) Using the graphs in (a), graph $y = e^x - x$.
    (c) Find all points on the graph in (b) where the tangent line is horizontal.

55 (Calculator) The formula $y = e^{-t} \sin t$ describes a decaying alternating current. Consider t in $[0, 4\pi]$.
    (a) Fill in this table and plot the resulting points.

t	0	$\dfrac{\pi}{2}$	π	$\dfrac{3\pi}{2}$	2π	$\dfrac{5\pi}{2}$	3π	$\dfrac{7\pi}{2}$	4π
$e^{-t} \sin t$									

(b) Graph $y = e^{-t} \sin t$ for t in $[0, 2\pi]$.
(c) Find all points on the graph in (b) where the tangent line is horizontal.

56 When doing this exercise disregard all work done in this section. Pretend that you are back in Chap. 3, where the derivative was defined. Let f be the exponential function given by the formula $f(x) = 10^x$.
(a) Copy and complete this table.

x	-2	-1	0	1
10^x	0.01			10

(b) Graph the function for x in $[-2, 1]$. (The same scale should be used for both axes.)
(c) Using a ruler, draw what you think would be the tangent line at $(0, 1)$.
(d) Using a ruler, estimate the slope of the line you drew in (c).
(e) Show that the derivative of 10^x at $x = 0$ is

$$\lim_{h \to 0} \dfrac{10^h - 1}{h}.$$

(f) The limit in (e) is far from obvious. What does (d) suggest as an estimate of this limit?
(g) Let the limit in (e) be denoted c. Show that $(10^x)' = c10^x$.

57 (a) Show that $y = x^5 + 3x$ is a one-to-one function.
(b) Does it have an inverse function?
(c) Can you find a formula for the inverse function?

58 (Review the binomial theorem in Appendix B.) This exercise shows that if $b > 1$, then $\lim_{x \to \infty} x/b^x = 0$.
(a) Show that for x sufficiently large, x/b^x is a decreasing function.
(b) Write $b = 1 + c$, $c > 0$. Using the binomial theorem for $n > 2$, show that

$$b^n > 1 + nc + \frac{n(n-1)}{2}c^2 \quad \text{if} \quad n > 2.$$

(c) From (b) deduce that

$$\lim_{n \to \infty} \frac{n}{b^n} = 0.$$

(d) From (a) and (c) deduce that $\lim_{x \to \infty} x/b^x = 0$.

59 (This continues Exercise 58.) Show that if $n > 3$ and $b > 1$, then

$$b^n > 1 + nc + \frac{n(n-1)}{2}c^2 + \frac{n(n-1)(n-2)}{3!}c^3.$$

Then, modeling your argument on Exercise 58, show that $\lim_{x \to \infty} x^2/b^x = 0$.

A similar argument shows that for any $b > 1$ and any positive a, $\lim_{x \to \infty} x^a/b^x = 0$. (The case $a \leq 0$ is trivial. Why?)

6.5 The derivatives of the inverse trigonometric functions

The derivative of $\frac{1}{2} \ln(1 + x^2)$ is $x/(1 + x^2)$. However, up to this point we have no function whose derivative is simply $1/(1 + x^2)$. Nor do we have any function whose derivative is $1/\sqrt{1 - x^2}$ or $\sqrt{1 - x^2}$. Such functions, which are needed in integral calculus, will be obtained in this section. Surprisingly, they turn out to involve trigonometric functions.

ARCTAN x AND ITS DERIVATIVE

First we will provide a function whose derivative is $1/(1 + x^2)$.

Consider the function $y = \tan x$ in the open interval $-\pi/2 < x < \pi/2$. As x increases in this interval, $\tan x$ increases. (See Fig. 6.13.) Thus the function $\tan x$ is one-to-one if the domain is restricted to be between $-\pi/2$ and $\pi/2$. Note that as $x \to \pi/2$ or $x \to -\pi/2$, $|\tan x|$ gets very large.

The graph of the inverse function g is obtained by folding the graph of $y = \tan x$ across the line $y = x$. As x gets large, note in Fig. 6.14 that $g(x) \to \pi/2$.

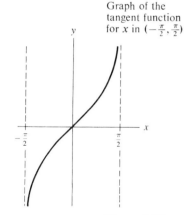

Graph of the tangent function for x in $(-\frac{\pi}{2}, \frac{\pi}{2})$

Figure 6.13

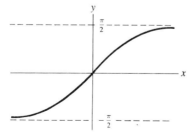

Graph of the inverse of the tangent function

Figure 6.14

Remember: arctan $x = \tan^{-1} x$ is an angle measured in radians, a dimensionless number.

The inverse of the tangent function is called the *arctangent* function, and is written arctan x or $\tan^{-1} x$. (This is not the reciprocal of tan x, which is written cot x, $1/\tan x$, or $(\tan x)^{-1}$ to avoid confusion.) As an example, since $\tan \pi/4 = 1$,

$$\arctan 1 = \pi/4 \quad \text{or} \quad \tan^{-1} 1 = \pi/4.$$

Observe that the domain of the arctan function is the entire x axis, and that, when $|x|$ is large, arctan x is near $\pi/2$ or $-\pi/2$.

It is frequently useful to picture the tangent and arctangent functions in terms of the unit circle, as shown in Figs. 6.15 and 6.16.

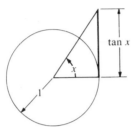
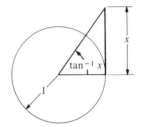

Figure 6.15 **Figure 6.16**

x	$\tan^{-1} x$
-1000	
-100	
-10	
-1	
0	
1	
10	
100	
1000	

The inverse tangent function is easily evaluated on most scientific calculators. To get a feel for this function fill in the table in the margin, remembering to set the calculator for radians (not degrees).

Theorem 1 obtains the derivative of $\tan^{-1} x$. The proof makes use of the identity $\sec^2 \theta = 1 + \tan^2 \theta$. (To obtain this identity divide both sides of $\cos^2 \theta + \sin^2 \theta = 1$ by $\cos^2 \theta$.)

THEOREM 1 $\quad \dfrac{d}{dx}(\tan^{-1} x) = \dfrac{1}{1 + x^2}.$

PROOF Let $y = \tan^{-1} x$. The problem is to find dy/dx. By the definition of the inverse tangent function, $x = \tan y$. Note that $dx/dy = \sec^2 y$.

As in the preceding section,

$$\frac{dx}{dx} = \frac{dx}{dy} \cdot \frac{dy}{dx}$$

$$1 = \sec^2 y \, \frac{dy}{dx}.$$

Hence
$$\frac{dy}{dx} = \frac{1}{\sec^2 y}$$

$$= \frac{1}{1 + \tan^2 y}$$

$$= \frac{1}{1 + x^2}.$$

Memorize the formula
$$(\tan^{-1} x)' = \frac{1}{1 + x^2}.$$

This completes the proof. ∎

6.5 THE DERIVATIVES OF THE INVERSE TRIGONOMETRIC FUNCTIONS

EXAMPLE 1 Find $\frac{d}{dx}(\tan^{-1}\sqrt{x})$.

SOLUTION Theorem 1 and the chain rule are needed. Let

$$y = \tan^{-1}\sqrt{x};$$

then

$$y = \tan^{-1} u, \quad \text{where} \quad u = \sqrt{x}.$$

thus

$$\frac{dy}{dx} = \frac{dy}{du} \cdot \frac{du}{dx}$$

$$= \frac{d}{du}(\tan^{-1} u) \frac{d}{dx}(\sqrt{x})$$

$$= \frac{1}{1+u^2} \cdot \frac{1}{2\sqrt{x}}$$

$$= \frac{1}{1+x} \cdot \frac{1}{2\sqrt{x}}$$

$$= \frac{1}{2\sqrt{x}(1+x)}. \quad \blacksquare$$

ARCSIN x AND ITS DERIVATIVE

We turn next to the inverse of the sine function.

The sine function is not one-to-one. For instance, $\sin(\pi/4) = \sqrt{2}/2 = \sin(3\pi/4)$. However, if the domain is restricted to $[-\pi/2, \pi/2]$ a one-to-one function results. The function $y = \sin x$ increases from -1 to 1 as x goes from $-\pi/2$ to $\pi/2$.

The inverse function is called the *arcsine* and is written arcsin x or $\sin^{-1} x$. Since $\sin \pi/2 = 1$, we have, for instance,

$$\arcsin 1 = \frac{\pi}{2} \quad \text{or} \quad \sin^{-1} 1 = \frac{\pi}{2}.$$

x	$\sin^{-1} x$
-1	
-0.6	
-0.2	
0	
0.2	
0.6	
1	

Both these latter equations say, "the angle whose sine is 1 is $\pi/2$." For more practice with $\sin^{-1} x$, use a calculator to fill in the table in the margin.

We graph the sine and arcsine functions in Figs. 6.17 and 6.18.

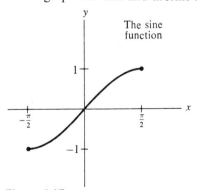

Figure 6.17

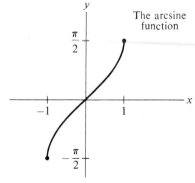

Figure 6.18

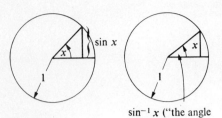

Figure 6.19

sin⁻¹ x ("the angle whose sine is x")

It is also useful to visualize these two functions in terms of the unit circle; they are depicted in Fig. 6.19. Note that

$$-\pi/2 \leq \arcsin x \leq \pi/2.$$

The proof of the next theorem is similar to that of Theorem 1.

THEOREM 2 $\dfrac{d}{dx}(\sin^{-1} x) = \dfrac{1}{\sqrt{1-x^2}}.$

PROOF Let $y = \sin^{-1} x$, hence $x = \sin y$. Thus

$$\frac{dx}{dx} = \frac{d}{dx}(\sin y)$$

$$= \frac{d}{dy}(\sin y)\frac{dy}{dx}$$

or $\qquad 1 = \cos y \, \dfrac{dy}{dx}.$

Hence $\qquad \dfrac{dy}{dx} = \dfrac{1}{\cos y}.$

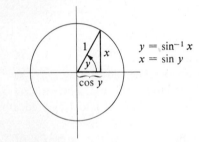

Figure 6.20

To express $\cos y$ in terms of x, draw the unit circle and indicate on it that $x = \sin y$. Inspection of Fig. 6.20 shows that $\cos y$ is positive (since $-\pi/2 \leq y \leq \pi/2$) and

$$\cos^2 y + x^2 = 1.$$

Thus $\cos y = \sqrt{1-x^2},$ the positive square root,

and dy/dx takes the form $\qquad \dfrac{dy}{dx} = \dfrac{1}{\sqrt{1-x^2}}.$

This proves that $\qquad \dfrac{d}{dx}(\sin^{-1} x) = \dfrac{1}{\sqrt{1-x^2}}.$ ∎

EXAMPLE 2 Find the derivative of $\sin^{-1} \dfrac{3x}{4}.$

SOLUTION Let $y = \sin^{-1} \dfrac{3x}{4}.$ This is a composite function, with

$$y = \sin^{-1} u, \qquad \text{where} \qquad u = \frac{3x}{4}.$$

Now $\qquad \dfrac{dy}{du} = \dfrac{1}{\sqrt{1-u^2}} \qquad$ and $\qquad \dfrac{du}{dx} = \dfrac{3}{4}.$

Thus $\qquad \dfrac{dy}{dx} = \dfrac{dy}{du} \cdot \dfrac{du}{dx}$

6.5 THE DERIVATIVES OF THE INVERSE TRIGONOMETRIC FUNCTIONS

$$= \frac{1}{\sqrt{1-u^2}} \cdot \frac{3}{4} = \frac{1}{\sqrt{1-(3x/4)^2}} \cdot \frac{3}{4}$$

$$= \frac{1}{\sqrt{1-9x^2/16}} \cdot \frac{3}{4}$$

$$= \frac{\sqrt{16}}{\sqrt{16-9x^2}} \cdot \frac{3}{4}$$

$$= \frac{3}{\sqrt{16-9x^2}} \cdot \blacksquare$$

EXAMPLE 3 Differentiate $x\sqrt{1-x^2} + \sin^{-1} x$.

SOLUTION
$$\frac{d}{dx}(x\sqrt{1-x^2} + \sin^{-1} x) = \frac{d}{dx}(x\sqrt{1-x^2}) + \frac{d}{dx}(\sin^{-1} x)$$

$$= x \frac{d}{dx}(\sqrt{1-x^2}) + \sqrt{1-x^2}\frac{dx}{dx} + \frac{1}{\sqrt{1-x^2}}$$

$$= x \cdot \frac{1}{2} \cdot \frac{-2x}{\sqrt{1-x^2}} + \sqrt{1-x^2} + \frac{1}{\sqrt{1-x^2}}$$

$$= \frac{-x^2}{\sqrt{1-x^2}} + \sqrt{1-x^2} + \frac{1}{\sqrt{1-x^2}}$$

$$= \frac{1-x^2}{\sqrt{1-x^2}} + \sqrt{1-x^2}$$

$$= \sqrt{1-x^2} + \sqrt{1-x^2}$$

$$= 2\sqrt{1-x^2}. \blacksquare$$

Note that Example 3 provides an antiderivative for $2\sqrt{1-x^2}$,

$$\int 2\sqrt{1-x^2}\, dx = x\sqrt{1-x^2} + \sin^{-1} x + C.$$

Why inverse trig functions are important

The integrals of many algebraic functions involve inverse trigonometric functions. It is this fact that makes the inverse trigonometric functions important in calculus.

ARCCOS x AND ITS DERIVATIVE

x	$\cos^{-1} x$
-1	
-0.5	
-0.2	
0	
0.2	
0.5	
1	

The function $\cos x$ is decreasing for x in $[0, \pi]$. Thus it has an inverse function, which assigns to each number in $[-1, 1]$ a number in $[0, \pi]$. The inverse function is called the *arccosine* function, denoted arccos x or $\cos^{-1} x$.

For instance, what is

$$\cos^{-1} \tfrac{1}{2}?$$

It is the angle between 0 and π whose cosine is $\tfrac{1}{2}$. That angle is $\pi/3$. Thus

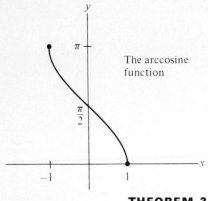

Figure 6.21

$$\cos^{-1} \frac{1}{2} = \frac{\pi}{3} \approx 1.05.$$

For practice with $\cos^{-1} x$ you might fill in the table in the margin on p. 323, using a calculator. The graph of $y = \cos^{-1} x$ is shown in Fig. 6.21.

Reasoning like that in the proof of Theorem 2 shows that

$$(\cos^{-1} x)' = \frac{-1}{\sqrt{1-x^2}}.$$

(The negative sign reflects the fact that $\cos^{-1} x$ is a decreasing function.)

THEOREM 3 $\quad \dfrac{d}{dx}(\cos^{-1} x) = \dfrac{-1}{\sqrt{1-x^2}}$ ∎

Since the derivative of $\cos^{-1} x$ differs from the derivative of $\sin^{-1} x$ only by the constant factor -1, it is not needed for finding antiderivatives. In most integral tables $\sin^{-1} x$ is used in preference to $\cos^{-1} x$.

ARCSEC x AND ITS DERIVATIVE

x	$\sec^{-1} x$
-10	
-2	
-1	
1	
2	
10	

After $\tan^{-1} x$ and $\sin^{-1} x$ the next most important function for the computation of antiderivatives is the inverse of the secant function. Recall that $\sec x = 1/\cos x$; it is graphed in Appendix D. Since $|\sec x| \geq 1$, the inverse of the secant function is defined only for inputs of absolute value ≥ 1. For $|x| \geq 1$, define $\sec^{-1} x$ to be that angle in $[0, \pi]$ whose secant is x. If the secant of an angle is x, then the cosine of that angle is $1/x$. Thus

$$\sec^{-1} x = \cos^{-1} \frac{1}{x}.$$

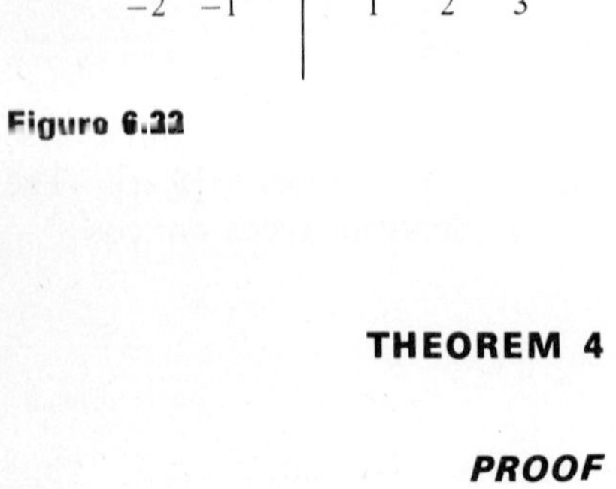

This formula, which is sometimes taken as a definition of the inverse secant function, makes it possible to compute $\sec^{-1} x$ on a calculator. For practice, you might fill in the table in the margin. The graph of $\sec^{-1} x$ is shown in Fig. 6.22.

To compute arcsecants, express their inverse nature in words. For instance, to evaluate $y = \sec^{-1} 2$, reason as follows:

$$y = \text{angle whose secant is } 2$$
$$= \text{angle whose cosine is } \tfrac{1}{2}$$
$$= \pi/3.$$

Figure 6.22

Thus $\quad \sec^{-1} 2 = \dfrac{\pi}{3}.$

THEOREM 4 $\quad \dfrac{d}{dx}(\sec^{-1} x) = \dfrac{1}{|x|\sqrt{x^2-1}},\quad$ where $\quad |x| > 1.$

PROOF Let $\quad y = \sec^{-1} x.$

6.5 THE DERIVATIVES OF THE INVERSE TRIGONOMETRIC FUNCTIONS

Then
$$x = \sec y,$$

and
$$\frac{dx}{dx} = \frac{dx}{dy} \cdot \frac{dy}{dx},$$

or
$$1 = \sec y \tan y \, \frac{dy}{dx}.$$

Hence
$$\frac{dy}{dx} = \frac{1}{\sec y \tan y}$$
$$= \frac{1}{x \tan y}.$$

All that remains is to express $\tan y$ in terms of x.

Since
$$\sec^2 y = 1 + \tan^2 y,$$
or
$$x^2 = 1 + \tan^2 y,$$
it follows that
$$\tan y = \pm\sqrt{x^2 - 1}, \quad \text{where} \quad |x| > 1.$$

Which sign is to be chosen? If $x > 1$, $y = \sec^{-1} x$ is in the range $(0, \pi/2)$; thus $\tan y$ is positive. If $x < -1$, $y = \sec^{-1} x$ is in the range $(\pi/2, \pi)$; thus $\tan y$ is negative. But, in both cases $x \tan y$ is positive. Thus

$$\frac{dy}{dx} = \frac{1}{|x|\sqrt{x^2 - 1}}.$$

(Note that this derivative is positive, in agreement with the graph of $y = \sec^{-1} x$; its tangent lines slope upward.) ∎

EXAMPLE 4 Differentiate $y = \sec^{-1} 5x$.

SOLUTION The chain rule is required.

Here
$$y = \sec^{-1} u, \quad \text{where} \quad u = 5x.$$

Thus
$$\frac{dy}{dx} = \frac{1}{|u|\sqrt{u^2 - 1}} \cdot 5$$
$$= \frac{1}{|5x|\sqrt{25x^2 - 1}} \cdot 5$$
$$= \frac{1}{|5||x|\sqrt{25x^2 - 1}} \cdot 5$$
$$= \frac{1}{|x|\sqrt{25x^2 - 1}}. \quad \blacksquare$$

The inverses of the remaining two trigonometric functions, cot x and csc x, will not be needed. For the record, they may be defined as follows:

$$\cot^{-1} x = \frac{\pi}{2} - \tan^{-1} x \quad \text{and} \quad \csc^{-1} x = \sin^{-1} \frac{1}{x}.$$

Their derivatives are given by the formulas:

$$(\cot^{-1} x)' = \frac{-1}{1 + x^2} \quad \text{and} \quad (\csc^{-1} x)' = \frac{-1}{|x|\sqrt{x^2 - 1}},$$

which add nothing of value to our tools for integration.

SUMMARY OF THIS SECTION

With this section, the roster of elementary functions is completed. Any function that can be obtained by algebraic means and composition from polynomials, nth roots, logarithms, exponentials, the six trigonometric functions, and their inverses is called elementary. For instance, the function

$$\frac{\sqrt{1 + x^3} - \sin^{-1}(e^{-x^2})}{x^2 \tan(\ln 3x)}$$

is elementary. The formulas for differentiation assure us that its derivative is also an elementary function. However, as we saw in Chap. 5, an antiderivative of an elementary function need not be elementary. Chapter 7 develops techniques for finding the antiderivatives of many elementary functions whose antiderivatives are elementary.

The key points in this section are the definitions of $\tan^{-1} x$, $\sin^{-1} x$, and $\sec^{-1} x$ and the calculations of their derivatives.

Function	Derivative	
$\tan^{-1} x$	$\dfrac{1}{1 + x^2}$	
$\sin^{-1} x$	$\dfrac{1}{\sqrt{1 - x^2}}$	
$\sec^{-1} x$	$\dfrac{1}{|x|\sqrt{x^2 - 1}}$	(If x is positive, the absolute value sign can be omitted.)

With each derivative comes a corresponding antiderivative:

$$\int \frac{dx}{1 + x^2} = \tan^{-1} x + C$$

$$\int \frac{dx}{\sqrt{1 - x^2}} = \sin^{-1} x + C, \quad |x| < 1$$

$$\int \frac{dx}{x\sqrt{x^2-1}} = \sec^{-1} x + C \qquad x > 1$$

The chain rule shows that for a differentiable function $u(x)$,

$$(\tan^{-1} u(x))' = \frac{u'(x)}{1 + (u(x))^2}$$

$$(\sin^{-1} u(x))' = \frac{u'(x)}{\sqrt{1 - (u(x))^2}}$$

$$(\sec^{-1} u(x))' = \frac{u'(x)}{|u(x)|\sqrt{(u(x))^2 - 1}}.$$

Exercises

Exercises 1 to 16 concern the three inverse functions, $\tan^{-1} x$, $\sin^{-1} x$, and $\sec^{-1} x$.

1. Draw a circle of radius 10 centimeters and use it, a centimeter ruler, and a protractor to estimate
   (a) $\tan^{-1} 1.5$ (b) $\tan^{-1} 0.7$ (c) $\tan^{-1}(-1.2)$
   (d) $\sin^{-1} 0.4$ (e) $\sin^{-1}(-0.5)$ (f) $\sin^{-1} 0.8$
   If your protractor reads degrees, turn your answer into radians by dividing by 57 (an approximation of $180/\pi$).

2. Evaluate
   (a) $\sin^{-1} 1$ (b) $\tan^{-1} 1$
   (c) $\sin^{-1}(-\sqrt{3}/2)$ (d) $\tan^{-1}(-\sqrt{3})$
   (e) $\sec^{-1} \sqrt{2}$.

3. Use a calculator to find
   (a) $\tan^{-1} 3$ (b) $\tan^{-1}(-2)$
   (c) $\sin^{-1} 0.4$ (d) $\sec^{-1} 3$

4. What happens when you try to find arcsin 2 on your calculator? Why?

5. Which of these are *meaningless*?
   (a) $\cos^{-1} 1.5$ (b) $\sec^{-1} 1.5$
   (c) $\tan^{-1} 1.5$ (d) $\sec^{-1} 0.3$

In Exercises 6 to 8 use a calculator to fill in the tables (angles in radians). Then use the data to graph the functions.

6.

x	-20	-10	-1	0	1	10	20
$\tan^{-1} x$							

7.

x	-1	-0.8	-0.6	-0.4	-0.2	0	0.2	0.4	0.6	0.8	1
$\sin^{-1} x$											

8.

x	-20	-10	-1	1	10	20
$\sec^{-1} x$						

(Use the fact that $\sec^{-1} x = \cos^{-1}(1/x)$.)

Exercises 9 and 10 obtain relations between inverse trigonometric functions. They depend on the right triangle shown in Fig. 6.23 and the fact that

$$\sin \theta = b/c, \qquad \cos \theta = a/c, \qquad \text{and} \qquad \tan \theta = b/a.$$

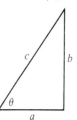

Figure 6.23

9. (See Fig. 6.24.) Show that, for $x \geq 0$,

$$\sin^{-1} \frac{x}{\sqrt{x^2 + 1}} = \tan^{-1} x.$$

Figure 6.24

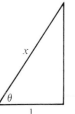

Figure 6.25

10. (See Fig. 6.25.) (a) Express $\cos \theta$ in terms of x.
    (b) Express $\sec \theta$ in terms of x.
    (c) Verify that for $x \geq 1$, $\sec^{-1} x = \cos^{-1}(1/x)$.

In Exercises 11 to 16 evaluate the expressions without recourse to a calculator. A sketch of the unit circle or of an appropriate right triangle may help.

11. $\sin(\tan^{-1} 1)$ 12. $\tan(\sec^{-1} 2)$

13. $\tan(\sin^{-1}(-\sqrt{2}/2))$
14. $\tan(\sin^{-1}(\sqrt{3}/2))$
15. $\sin(\sin^{-1} 0.3)$
16. $\sin(\tan^{-1} 0)$

In Exercises 17 to 46 differentiate the given functions.

17. $\sin^{-1} 5x$
18. $\tan^{-1} 3x$
19. $\sec^{-1} 3x$
20. $\sin^{-1} e^{-x}$
21. $\tan^{-1} \sqrt[3]{x}$
22. $-\dfrac{1}{3}\sin^{-1}\dfrac{3}{x}$
23. $x^2 \sec^{-1}\sqrt{x}$
24. $\dfrac{1}{\sin^{-1} 2x}$
25. $\sin 3x \sin^{-1} 3x$
26. $x^3 \tan^{-1} 2x$
27. $\dfrac{x \sec^{-1} 3x}{e^{2x}}$
28. $\arcsin(2x - 3)$
29. $\arctan \sqrt{x}$
30. $\operatorname{arcsec} \sqrt{x}$
31. $\ln \sec^{-1}\sqrt{x}$
32. $\ln((\sin^{-1} 5x)^2)$
33. $\dfrac{x}{\tan^{-1} 10^x}$
34. $10^{\sec^{-1} 2x}$
35. $\sin^{-1} x - \sqrt{1 - x^2}$
36. $2^x \cdot \log_3 x \cdot \sec 3x$
37. $(\tan^{-1} 2x)^3$
38. $(\sin^{-1}\sqrt{x - 1})^4$
39. $\dfrac{x}{2}\sqrt{2 - x^2} + \sin^{-1}(x/\sqrt{2})$
40. $\sqrt{3x^2 - 1} - \tan^{-1}\sqrt{3x^2 - 1}$
41. $\tfrac{2}{3}\sec^{-1}\sqrt{3x^5}$
42. $\dfrac{1}{2}\left[(x - 3)\sqrt{6x - x^2} + 9\sin^{-1}\dfrac{x - 3}{3}\right]$
43. $\sqrt{1 + x}\sqrt{2 - x} - 3\sin^{-1}\sqrt{\dfrac{2 - x}{3}}$
44. $x \sin^{-1} 3x + \tfrac{1}{3}\sqrt{1 - 9x^2}$
45. $x(\sin^{-1} 2x)^2 - 2x + \sqrt{1 - 4x^2}\sin^{-1} 2x$
46. $x \tan^{-1} 5x - \tfrac{1}{10}\ln(1 + 25x^2)$

■

In Exercises 47 to 49 differentiate the given functions. Note that quite different functions may have very similar derivatives.

47. (a) $\ln(x + \sqrt{x^2 - 9})$
    (b) $\sin^{-1}\dfrac{x}{3}$

48. (a) $-\dfrac{1}{5}\ln\dfrac{5 + \sqrt{25 - x^2}}{x}$
    (b) $-\dfrac{1}{5}\sin^{-1}\dfrac{5}{x}$

49. (a) $\ln\dfrac{\sqrt{2x^2 + 1} - 1}{x}$
    (b) $\sec^{-1} x\sqrt{2}$

50. (a) Show that
$$\int \dfrac{dx}{\sqrt{1 - a^2 x^2}} = \dfrac{1}{a}\sin^{-1} ax + C \quad (a > 0).$$
Use (a) to find
(b) $\displaystyle\int \dfrac{dx}{\sqrt{1 - 25x^2}},$ (c) $\displaystyle\int \dfrac{dx}{\sqrt{1 - 3x^2}}.$

51. (a) Show that
$$\int \dfrac{dx}{\sqrt{a^2 - x^2}} = \sin^{-1}\dfrac{x}{a} + C \quad (a > 0).$$
Use (a) to find
(b) $\displaystyle\int \dfrac{dx}{\sqrt{25 - x^2}}$ (c) $\displaystyle\int \dfrac{dx}{\sqrt{5 - x^2}}.$

The next two exercises offer an interesting contrast.

52. (a) Sketch $y = 1/(1 + x^2)$.
    (b) Find the area under the curve in (a) and above $[0, b]$, $b > 0$.
    (c) Would you say that the area under the curve and above $[0, \infty)$ is finite or infinite? If finite, what is it?

53. (a) Sketch $y = 1/(1 + x)$ for $x \geq 0$.
    (b) Find the area under the curve in (a) and above $[0, b]$, $b > 0$.
    (c) Would you say that the area under the curve and above $[0, \infty)$ is finite or infinite? If finite, what is it?

54. Consider the curve $y = \sqrt{1 - x^2}$ for x in $[0, 1]$.
    (a) Sketch the curve. *Suggestion:* Show it is part of a circle.
    (b) Show that
$$\int \sqrt{1 - x^2}\, dx = \dfrac{x\sqrt{1 - x^2}}{2} + \dfrac{\sin^{-1} x}{2} + C.$$
    (c) Find the area below $y = \sqrt{1 - x^2}$ and above $[0, \tfrac{1}{2}]$.
    (d) Find the area below $y = \sqrt{1 - x^2}$ and above $[0, 1]$.

Exercises 55 to 57 are related.

55. (a) Compute the differential $d(\tan^{-1} x)$.
    (b) Use (a) to estimate $\tan^{-1} 1.1$. *Suggestion:* $\tan^{-1} 1$ is known. Use $\pi \approx 3.14$.

56. (a) Compute the differential $d(\sin^{-1} x)$.
    (b) Use (a) to estimate $\sin^{-1} 0.47$.

57 (a) Compute the differential $d(\sec^{-1} x)$.
 (b) Use (a) to estimate $\sec^{-1} 2.08$.

58 Find the area below $y = 1/(x\sqrt{x^2 - 1})$ and above $[\sqrt{2}, 2]$.

59 (a) Differentiate $\tan^{-1} ax$, where a is a constant.
 (b) Find the area under $y = 1/(1 + 3x^2)$ and above $[0, 1]$.

■ ■

60 Show that $\tan^{-1}\tfrac{1}{2} + \tan^{-1}\tfrac{1}{3} = \pi/4$. *Hint:* Use a trigonometric identity.

6.6 Related rates

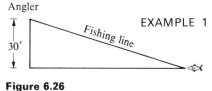

Figure 6.26

Figure 6.27

Sometimes the rate at which one quantity is changing is known, and we wish to find the rate at which some related quantity is changing. Example 1 is typical of such problems and indicates a general method of attacking them.

EXAMPLE 1 An angler has a fish at the end of his line, which is reeled in at 2 feet per second from a bridge 30 feet above the water. At what speed is the fish moving through the water when the amount of line out is 50 feet? 31 feet? Assume the fish is at the surface of the water. (See Fig. 6.26.)

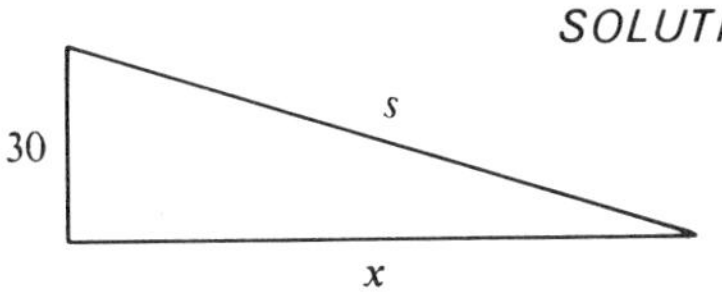

SOLUTION Let s be the length of the line and x the horizontal distance of the fish from the bridge. (See Fig. 6.27.)

Since the line is reeled in at the rate of 2 feet per second,

$$\frac{ds}{dt} = -2.$$

The rate at which the fish moves through the water is given by the derivative

$$\frac{dx}{dt}.$$

The problem is to find dx/dt when $s = 50$ and also when $s = 31$.

The quantities x and s are related by the equation given by the Pythagorean theorem:

$$x^2 + 30^2 = s^2.$$

Both x and s are functions of time t. Thus both sides of the equation may be differentiated with respect to t, yielding

$$\frac{d(x^2)}{dt} + \frac{d(30^2)}{dt} = \frac{d(s^2)}{dt},$$

or

$$2x \frac{dx}{dt} + 0 = 2s \frac{ds}{dt},$$

Hence

$$x \frac{dx}{dt} = s \frac{ds}{dt}.$$

This last equation provides the tool for answering the questions.

Since
$$\frac{ds}{dt} = -2,$$

$$x\frac{dx}{dt} = s(-2).$$

Hence
$$\frac{dx}{dt} = \frac{-2s}{x},$$

When $s = 50$,
$$x^2 + 30^2 = 50^2,$$
from which it follows that $x = 40$.

Thus, when 50 feet of line are out, the speed is

$$\frac{2s}{x} = \frac{2 \cdot 50}{40} = 2.5 \text{ feet per second.}$$

When $s = 31$, $\quad x^2 + 30^2 = 31^2;$

hence
$$x = \sqrt{31^2 - 30^2}$$
$$= \sqrt{961 - 900}$$
$$= \sqrt{61}.$$

Thus, when 31 feet of line are out, the fish is moving at the speed of

$$\frac{2s}{x} = \frac{2 \cdot 31}{\sqrt{61}} = \frac{62}{\sqrt{61}}$$

$$\approx 7.9 \text{ feet per second.} \blacksquare$$

General procedure for finding related rates

The method used in Example 1 applies to many related rate problems. This is the general procedure, broken into three steps:

1 Find an equation relating the varying quantities.

WARNING: Differentiate, then substitute the specific numbers for the variables. If you reversed the order, you would just be differentiating constants.

2 Differentiate both sides of the equation implicitly with respect to time or some other appropriate variable.

3 Use the equation obtained in step 2 to determine the unknown rate from the given rates. It is at this point that you substitute numbers for the variables and known rates of change.

EXAMPLE 2   A woman on the ground is watching a jet through a telescope as it approaches at a speed of 10 miles per minute at an altitude of 7 miles. At what rate is the angle of the telescope changing when the horizontal distance of the jet from the woman is 24 miles? When the jet is directly above the woman?

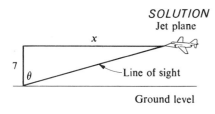

Figure 6.28

SOLUTION To begin, sketch a diagram and label the parts that are of interest, as has been done in Fig. 6.28. Observe that

$$\frac{dx}{dt} = -10 \text{ miles per minute.}$$

The rate at which θ changes, $\dfrac{d\theta}{dt}$, is to be found.

Step 1 consists of finding an equation relating θ and x. One such equation is

$$\tan \theta = \frac{x}{7}.$$

In step 2 this equation is differentiated with respect to time:

$$\frac{d}{dt}(\tan \theta) = \frac{d}{dt}(x/7);$$

hence, by the chain rule,

$$\sec^2 \theta \, \frac{d\theta}{dt} = \frac{1}{7} \frac{dx}{dt}.$$

Since

$$\frac{dx}{dt} = -10,$$

it follows that

$$\frac{d\theta}{dt} = \frac{1}{7} \frac{(-10)}{\sec^2 \theta}$$

$$= \frac{-10}{7 \sec^2 \theta}$$

$$= -\frac{10}{7} \cos^2 \theta \text{ radians per minute.}$$

The negative sign in this formula shows that θ is decreasing.

When $x = 24$, let us find $d\theta/dt$. First it is necessary to compute $\cos \theta$ in the triangle in Fig. 6.29. Since

$$s^2 = 7^2 + 24^2$$
$$= 49 + 576$$
$$= 625,$$

it follows that $s = 25$.

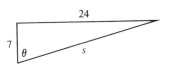

Figure 6.29

Thus

$$\cos \theta = \frac{7}{25},$$

and

$$\frac{d\theta}{dt} = -\frac{10}{7} \left(\frac{7}{25}\right)^2$$

$$= -\tfrac{70}{625}$$
$$= -0.112 \text{ radian per minute}$$
$$\approx -6° \text{ per minute.}$$

When the plane is directly above the woman, $x = 0$, $\theta = 0$, $\cos \theta = 1$, and the formula

$$\frac{d\theta}{dt} = -\frac{10}{7} \cos^2 \theta$$

shows that
$$\frac{d\theta}{dt} = -\frac{10}{7} \cdot 1^2$$
$$= -\tfrac{10}{7} \text{ radians per minute}$$
$$\approx -82° \text{ per minute.}$$

The telescope moves much more rapidly when the plane is directly overhead. ■

The method described in Example 1 for determining unknown rates from known ones extends to finding an unknown acceleration. Just differentiate another time. Example 3 illustrates the procedure.

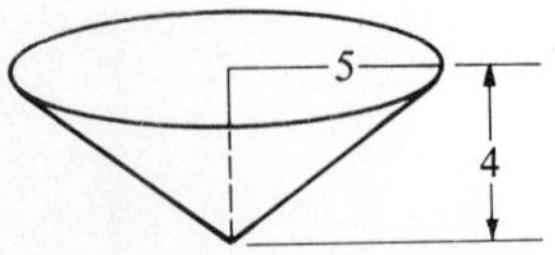

EXAMPLE 3 Water flows into a conical tank at the constant rate of 3 cubic meters per second. The radius of the cone is 5 meters and its height is 4 meters. Let $h(t)$ represent the height of the water above the bottom of the cone at time t. Find dh/dt (the rate at which the water is rising in the tank) and d^2h/dt^2, when the tank is filled to a height of 2 meters. (See Fig. 6.30.)

Figure 6.30

SOLUTION Let $V(t)$ be the volume of water in the tank at time t. The data imply that

$$\frac{dV}{dt} = 3,$$

and hence
$$\frac{d^2V}{dt^2} = 0.$$

To find dh/dt and d^2h/dt^2, first obtain an equation relating V and h.

When the tank is filled to the height h, the water forms a cone of height h and radius r. (See Fig. 6.31.) By similar triangles,

$$\frac{r}{h} = \frac{5}{4},$$

or $r = \tfrac{5}{4}h.$

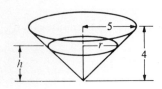

Figure 6.31

Thus $V = \tfrac{1}{3}\pi r^2 h = \tfrac{1}{3}\pi(\tfrac{5}{4}h)^2 h = \tfrac{25}{48}\pi h^3.$

The equation relating V and h is

$$V = \frac{25\pi}{48} h^3. \tag{1}$$

From here on, the procedure is automatic: Just differentiate as often as needed.

Differentiating once (using the chain rule) yields

$$\frac{dV}{dt} = \frac{25\pi}{48} \frac{d(h^3)}{dh} \frac{dh}{dt},$$

or

$$\frac{dV}{dt} = \frac{25\pi}{16} h^2 \frac{dh}{dt}. \quad (2)$$

Since $dV/dt = 3$ all the time, and $h = 2$ at the moment of interest, substitute these values into (2), obtaining

$$3 = \frac{25\pi}{16} 2^2 \frac{dh}{dt}.$$

Hence

$$\frac{dh}{dt} = \frac{12}{25\pi} \text{ meters per second}$$

Do not differentiate $\frac{dh}{dt} = \frac{12}{25\pi}$.

when $h = 2$.

To find d^2h/dt^2, differentiate (2), obtaining

$$\frac{d^2V}{dt^2} = \frac{25\pi}{16} \left(h^2 \frac{d^2h}{dt^2} + 2h \frac{dh}{dt} \frac{dh}{dt} \right). \quad (3)$$

Since $d^2V/dt^2 = 0$ all the time, and, when $h = 2$, $dh/dt = 12/(25\pi)$, (3) implies that

$$0 = \frac{25\pi}{16} \left[2^2 \frac{d^2h}{dt^2} + 2 \cdot 2 \left(\frac{12}{25\pi} \right)^2 \right]. \quad (4)$$

Solving Eq. (4) for d^2h/dt^2 shows that

$$\frac{d^2h}{dt^2} = \frac{-144}{625\pi^2} \text{ meters per second per second.}$$

Since d^2h/dt^2 is negative, the rate at which the water rises in the tank is slowing down. In general, the higher the water, the slower it rises. Even though V changes at a constant rate, h does not. ■

Exercises

Exercises 1 and 2 are related to Example 1.
1. How fast is the fish moving through the water when it is 1 foot horizontally from the bridge?
2. The angler in Example 1 decides to let the line out as the fish swims away. The fish swims away at a constant speed of 5 feet per second relative to the water. How fast is the angler paying out his line when the horizontal distance from the bridge to the fish is (a) 1 foot? (b) 100 feet?
3. A 10-foot ladder is leaning against a wall. If a person pulls the base of the ladder away from the wall at the rate of 1 foot per second, how fast is the top going down the wall when the base of the ladder is (a) 6 feet from the wall? (b) 8 feet from the wall? (c) 9 feet from the wall?
4. A kite is flying at a height of 300 feet in a horizontal wind. When 500 feet of string are out, the kite is pulling the string out at a rate of 20 feet per second. What is the wind velocity?

5 A beachcomber walks 2 miles per hour along the shore as the beam from a rotating light 3 miles offshore follows him. (See Fig. 6.32.)

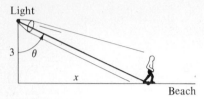

Figure 6.32

(a) Intuitively, what do you think happens to the rate at which the light rotates as the beachcomber walks further and further along the shore away from the lighthouse?
(b) Letting x describe the distance of the beachcomber from the point on the shore nearest the light, and θ the angle of the light, obtain an equation relating θ and x.
(c) With the aid of (b) show that $d\theta/dt = 6/(9 + x^2)$ (radians per hour).
(d) Does the formula in (c) agree with your guess in (a)?

6 A man 6 feet tall walks at the rate of 5 feet per second away from a lamp that is 20 feet high. At what rate is his shadow lengthening when he is (a) 10 feet from the lamp? (b) 100 feet from the lamp?

7 The length of a rectangle is increasing at the rate of 7 feet per second, and the width is decreasing at the rate of 3 feet per second. When the length is 12 feet and the width is 5 feet, find the rate of change of (a) the area, (b) the perimeter, (c) the length of the diagonal.

8 A shrinking spherical balloon loses air at the rate of 1 cubic inch per second. At what rate is its radius changing when the radius is (a) 2 inches? (b) 1 inch? (The volume V of a sphere of radius r is $4\pi r^3/3$.)

9 Bulldozers are moving earth at the rate of 1000 cubic yards per hour onto a conically shaped hill whose height remains equal to its radius. At what rate is the height of the hill increasing when the hill is (a) 20 yards high? (b) 100 yards high? (The volume of a cone of radius r and height h is $\pi r^2 h/3$.)

10 The lengths of the two legs of a right triangle depend on time. One leg, whose length is x, increases at the rate of 5 feet per second, while the other, of length y, decreases at the rate of 6 feet per second. At what rate is the hypotenuse changing when $x = 3$ feet and $y = 4$ feet? Is the hypotenuse increasing or decreasing then?

11 Two sides of a triangle and their included angle are changing with respect to time. The angle increases at the rate of 1 radian per second, one side increases at the rate of 3 feet per second, and the other side decreases at the rate of 2 feet per second. Find the rate at which the area is changing when the angle is $\pi/4$, the first side is 4 feet long, and the second side is 5 feet long. Is the area decreasing or increasing then?

12 A large spherical balloon is being inflated at the rate of 100 cubic feet per minute. At what rate is the radius increasing when the radius is (a) 10 feet? (b) 20 feet?

Exercises 13 to 16 concern acceleration. The notation $\dot{x}$ for dx/dt, $\dot\theta$ for $d\theta/dt$, $\ddot{x}$ for d^2x/dt^2, and $\ddot\theta$ for $d^2\theta/dt^2$ is common in physics.

13 What is the acceleration of the fish described in Example 1 when the length of line is (a) 300 feet? (b) 31 feet?

14 Find $\ddot\theta$ in Example 2 when the horizontal distance from the jet is (a) 7 miles, (b) 1 mile.

15 A particle moves on the parabola $y = x^2$ in such a way that $\dot{x} = 3$ throughout the journey. Find formulas for (a) $\dot{y}$ and (b) $\ddot{y}$.

16 Call one acute angle of a right triangle θ. The adjacent leg has length x and the opposite leg has length y.
(a) Obtain an equation relating x, y, and θ.
(b) Obtain an equation involving $\dot{x}$, $\dot{y}$, and $\dot\theta$ (and other variables).
(c) Obtain an equation involving $\ddot{x}$, $\ddot{y}$, and $\ddot\theta$ (and other variables).

■

17 A woman is walking on a bridge that is 20 feet above a river as a boat passes directly under the center of the bridge (at a right angle to the bridge) at 10 feet per second. At that moment the woman is 50 feet from the center and approaching it at the rate of 5 feet per second. (a) At what rate is the distance between the boat and woman changing at that moment? (b) Is the rate at which they are approaching or separating increasing or is it decreasing?

18 A spherical raindrop evaporates at a rate proportional to its surface area. Show that the radius shrinks at a constant rate.

19 A couple is on a Ferris wheel when the sun is directly overhead. The diameter of the wheel is 50 feet, and its speed is 0.1 revolution per second. (a) What is the speed of their shadows on the ground when they are at a two-o'clock position? (b) A one-o'clock position? (c) Show that the shadow is moving its fastest when they are at the top, and its slowest when they are at the three-o'clock position.

20 Water is flowing into a hemispherical kettle of radius 5 feet at the constant rate of 1 cubic foot per minute.
(a) At what rate is the top surface of the water rising when its height above the bottom of the kettle is 3 feet? 4 feet? 5 feet?
(b) If $h(t)$ is the depth in feet at time t, find $\ddot{h}$ when $h = 3$, 4, and 5.

■ ■

21 The atmospheric pressure at an altitude of x kilometers is approximately $1000(0.88)^x$ millibars. A rocket is rising at the rate of 5 kilometers per second vertically. At what rate is the atmospheric pressure changing (in millibars per second) when the altitude of the rocket is (a) 1 kilometer? (b) 50 kilometers?

6.7 Separable differential equations and growth

The abbreviation for "differential equation" is D.E. or O.D.E. (for ordinary differential equation).

An equation that involves one or more of the derivatives of a function is called a *differential equation*. For example, Sec. 4.4 examined the differential equation for motion under constant acceleration,

$$\frac{d^2y}{dt^2} = a \quad (a \text{ constant}). \tag{1}$$

Finding an antiderivative $F(x)$ for a function $f(x)$ amounts to solving the differential equation

$$\frac{dF}{dx} = f(x) \tag{2}$$

for the unknown function $F(x)$.

Solution of a D.E. A *solution* of a differential equation is any function that satisfies the equation. To *solve* a differential equation means to find all its solutions. In Sec. 4.4 it was shown that the most general solution of (1) is

$$y = \frac{at^2}{2} + v_0 t + y_0,$$

where v_0 and y_0 are constants. The *most general* solution of

$$\frac{dF}{dx} = x^2 \tag{3}$$

is

$$F(x) = \frac{x^3}{3} + C,$$

where C represents an arbitrary constant.

A differential equation can be much more complicated than (1) or (3), as the differential equation

$$\left(\frac{d^2y}{dx^2}\right)^2 + 3x\frac{dy}{dx} + \sin x = \frac{d^3y}{dx^3} \tag{4}$$

Order of a D.E. illustrates. The *order* of a differential equation is the highest order of the derivatives that appear in it. Thus (1) is of order two, (2) and (3) are of order one, and (4) is of order three.

This section examines a special and important type of first-order differential equation, called *separable*. After showing how to solve it, we will apply it to the study of natural growth and decay and also to inhibited growth.

SEPARABLE DIFFERENTIAL EQUATIONS

A *separable* differential equation is one that can be written in the form

Separable D.E.
$$\frac{dy}{dx} = \frac{f(x)}{g(y)}, \tag{5}$$

where $f(x)$ and $g(y)$ are differentiable functions. Such an equation can be solved by *separating the variables*, that is, bringing all the x's to one side and all the y's to the other side to obtain the following equation in differentials:

$$g(y)\,dy = f(x)\,dx. \qquad (6)$$

This is solved by integrating both sides:

$$\int g(y)\,dy = \int f(x)\,dx + C. \qquad (7)$$

Some examples will illustrate the technique.

EXAMPLE 1 Solve

$$\frac{dy}{dx} = \frac{2x}{3y}, \qquad (y > 0).$$

SOLUTION Separating the variables, we obtain

$$3y\,dy = 2x\,dx.$$

Thus

$$\int 3y\,dy = \int 2x\,dx + C$$

or

$$\frac{3y^2}{2} = x^2 + C. \qquad (8)$$

Equation (8) determines y as a function of x implicitly. Each choice of C produces a solution. ∎

EXAMPLE 2 Solve the differential equation

$$\frac{dy}{dx} = \frac{2y}{x} \qquad (x, y > 0). \qquad (9)$$

SOLUTION At first glance the equation does not appear to be of the form (5). However, it can be rewritten in the form

$$\frac{dy}{dx} = \frac{(1/x)}{(1/2y)},$$

so it has the form of a separable differential equation. Separation of the variables is not hard:

$$\frac{dy}{dx} = \frac{2y}{x}$$

$$\frac{dy}{2y} = \frac{dx}{x}.$$

6.7 SEPARABLE DIFFERENTIAL EQUATIONS AND GROWTH

Hence
$$\int \frac{dy}{2y} = \int \frac{dx}{x} + C$$

Since x, y assumed > 0,   or
$\ln|x| = \ln x$, $\ln|y| = \ln y$.

$$\tfrac{1}{2} \ln y = \ln x + C. \tag{10}$$

In this case let us solve for y explicitly:

$$\ln y = 2 \ln x + 2C$$

$$\begin{aligned} y &= e^{2 \ln x + 2C} & &\text{definition of natural logarithm} \\ &= e^{2 \ln x} e^{2C} & &\text{basic law of exponents} \\ &= (e^{\ln x})^2 e^{2C} & &\text{power of a power} \\ &= x^2 e^{2C}. \end{aligned}$$

Since e^{2C} is an arbitrary positive constant, call it k. Thus the most general solution of (9) is

$$y = kx^2. \tag{11}$$

As a check on this solution, see if $y = kx^2$ satisfies (9):

$$\frac{d(kx^2)}{dx} \stackrel{?}{=} \frac{2(kx^2)}{x},$$

$$2kx \stackrel{?}{=} \frac{2kx^2}{x}.$$

Yes, it checks. ∎

The solution of a separable differential equation (in fact, any first order differential equation) will generally involve one arbitrary constant. Each choice of that constant determines a specific function that satisfies the differential equation.

THE DIFFERENTIAL EQUATIONS OF NATURAL GROWTH AND DECAY

The next example treats a differential equation that is important in the study of growth and decay. It arises in such diverse areas as biology, ecology, physics, chemistry, and economic forecasting. Its application will be illustrated later in this section.

EXAMPLE 3 Solve the differential equation

$$\frac{dy}{dx} = ky \quad (y > 0), \tag{12}$$

where k is a nonzero constant.

SOLUTION Separation of the variables yields

$$\frac{dy}{y} = k \, dx$$

338 TOPICS IN DIFFERENTIAL CALCULUS

$$\int \frac{dy}{y} = \int k\, dx + C$$

$$\ln y = kx + C$$

$$y = e^{kx+C}$$

$$y = e^C e^{kx}.$$

Denote the arbitrary positive constant e^C by the letter A. Then

$$y = Ae^{kx}. \tag{13}$$

The most general solution of $dy/dx = ky$ is $y = Ae^{kx}$. ∎

THE DIFFERENTIAL EQUATION OF INHIBITED GROWTH

Example 4 solves a differential equation that arises in the study of bounded growth. The solution will be applied later in the section. At one point in the solution the algebraic identity

$$\frac{M}{y(M-y)} = \frac{1}{y} + \frac{1}{M-y}$$

will be needed. (Check this identity before reading Example 4.)

EXAMPLE 4 Solve the differential equation

$$\frac{dy}{dx} = ky\left(1 - \frac{y}{M}\right) \tag{14}$$

where k and M are positive constants and $0 < y < M$.

SOLUTION Separate the variables and integrate:

$$\frac{dy}{y[1-(y/M)]} = k\, dx$$

$$\int \frac{dy}{y[1-(y/M)]} = \int k\, dx + C$$

$$\int \frac{M\, dy}{y(M-y)} = \int k\, dx + C \qquad \text{algebra}$$

$$\int \left(\frac{1}{y} + \frac{1}{M-y}\right) dy = \int k\, dx + C \qquad \text{algebraic identity}$$

$$\int \frac{dy}{y} + \int \frac{dy}{M-y} = \int k\, dx + C$$

$$\ln y - \ln(M-y) = kx + C \qquad \begin{array}{l} 0 < y < M, \text{ so } \ln|y| = \ln y \\ \ln|M-y| = \ln(M-y) \end{array}$$

$$\ln\left(\frac{y}{M-y}\right) = kx + C$$

6.7 SEPARABLE DIFFERENTIAL EQUATIONS AND GROWTH

$$\frac{y}{M-y} = e^{kx+C} \qquad \text{definition of ln}$$

$$\frac{M-y}{y} = e^{-kx-C} \qquad \text{taking reciprocals}$$

$$\frac{M-y}{y} = e^{-C}e^{-kx} \qquad \text{basic law of exponents}$$

$$= ae^{-kx} \qquad \text{setting } e^{-C} = a$$

$$M - y = ae^{-kx}y$$

$$M = (1 + ae^{-kx})y$$

Finally, $$y = \frac{M}{1 + ae^{-kx}}. \qquad \blacksquare \qquad (15)$$

NATURAL GROWTH AND DECAY

The change in the size of the world's population is determined by two basic variables: the birth rate and the death rate. If they are equal, the size of the population remains constant. But if one is larger than the other, the population either grows or shrinks.

The birth rate and death rate are determined by different variables, and therefore it would be sheer coincidence if they were equal. The birth rate is influenced, for instance, by age at marriage, infant mortality rate, age at weaning, attitudes toward birth control and abortion, woman's self-image, and the role of children in supporting aged parents. The death rate is influenced by such factors as public health, wars, nutrition, pollution, and stress. Since man has in general brought famines and epidemics under control, he has brought the death rate below the birth rate.

Consequently, for the past 2 centuries the world population has increased dramatically, as shown in Fig. 6.33. The size of the world population in 1650 is estimated to have been about 0.5 billion; in 1750 about 0.7 billion; in 1850 about 1.1 billion; in 1980 about 4.5 billion. The population has been growing at an accelerating pace. If back in 1800 you had moved every human being to China, then that total number would just match the present population of China.

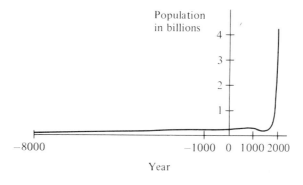

Figure 6.33

What will the population be in the year 2000 if the present rate of growth continues? To answer this question we must describe the size of the population mathematically.

Let $P(t)$ denote the size of the population at time t. Actually, $P(t)$ is an integer, and the graph of P has "jumps" whenever someone is born or dies. However, assume that P is a "smooth" (differentiable) function that approximates the size of the population.

The derivative P' then records the rate of change of the population. If social, medical, and technological factors remain constant, then it is reasonable to expect the rate of growth $P'(t)$ to be proportional to the size of the population $P(t)$: A large population will produce more babies in a year than a small population. More precisely, there is a fixed number k, independent of time, such that

The D.E. of natural growth and decay

$$P'(t) = kP(t). \tag{16}$$

This is the differential equation of *natural growth* (or *decay*, if k is negative). The constant k is called the *growth constant*.

To forecast the population it is necessary to solve (16). But (16) is just (12) expressed in different letters. By Example 3,

$$P(t) = Ae^{kt}. \tag{17}$$

What is the meaning of the constant A? To find out, set $t = 0$ in (17), obtaining

$$P(0) = Ae^{k \cdot 0}$$

or

$$P(0) = A.$$

The meaning of A Thus A is the amount or size of the population at the "initial time" $t = 0$. To find the meaning of k, first let us interpret e^k.

Let

$$b = e^k.$$

Then (17) takes the simpler form

$$P(t) = Ab^t. \tag{18}$$

Thus the function P at time $t + 1$, one unit after t, has the value,

$$P(t + 1) = Ab^{t+1},$$

and

$$\frac{P(t + 1)}{P(t)} = \frac{Ab^{t+1}}{Ab^t} = b.$$

Thus

$$P(t + 1) = bP(t).$$

The meaning of b The constant b is the ratio between the population at time $t + 1$ and the population at time t.

6.7 SEPARABLE DIFFERENTIAL EQUATIONS AND GROWTH

In 1980 the world population was growing at the relative rate of 1.8 percent ($= 0.018$) per year. Thus $b = 1 + 0.018 = 1.018$. But $b = e^k$. So

$$k = \ln b = \ln 1.018$$
$$\approx 0.0178.$$

Note that k, the growth constant, is fairly close to the growth rate per unit time, 1.8 percent, or 0.018, per year.

The relative growth rate per unit time, r

In describing growing populations newspapers report the relative growth per year, which is usually in the 1 to 3 percent range. If this number is denoted r, then

$$b = 1 + r.$$

Use a differential to show that for small r, $\ln(1 + r) \approx r$

Since r is small, $k = \ln(1 + r) \approx r$. (See Exercise 21.) Thus for small r, the growth constant k is approximately the same as the observed relative growth rate per unit time, r,

For small r, $k \approx r$ and $b \approx 1 + k$.

$$k \approx r \qquad \text{(if } r \text{ is small)}.$$

Or to put it another way,

$$b = e^k \approx 1 + k.$$

EXAMPLE 5 The population of the United States in 1980 was estimated to be about 225 million and to be growing at 0.7 percent per year. (a) Estimate the population in the year 2000. (b) When will the population double?

SOLUTION In this case the relative growth rate per unit time r is 0.007. Thus

$$b = 1.007$$

Introduce a time scale with $t = 0$ corresponding to the year 1980. Let $P(t)$ be the population of the United States at time t, that is, in the year $1980 + t$. Since $P(0) = 225$ million,

$$P(t) = 225(1.007)^t \text{ million}$$

at time t.

(a) To find the population in the year 2000, which is 20 years after the time $t = 0$, compute $P(20)$:

$$P(20) = 225(1.007)^{20} \text{ million}$$
$$\approx 225(1.150) \text{ million} \quad \text{(use } y^x \text{ key on calculator)}$$
$$\approx 259 \text{ million}.$$

So, if the growth rate continues at 0.7 percent per year, the population will be 259 million in the year 2000.

(b) To find when the population will double, solve the equation

$$2 \cdot 225 = 225(1.007)^t$$

or
$$2 = 1.007^t. \tag{19}$$

To solve (19), take ln of both sides,

$$\ln 2 = t \ln 1.007.$$

Since
$$\ln 1.007 \approx 0.007,$$
$$\ln 2 \approx (0.007)t$$

Since population figures are not precise, this estimate is good enough.

or
$$t \approx \frac{\ln 2}{0.007}$$
$$\approx \frac{0.69}{0.007}$$
$$\approx 99 \text{ years.}$$

Thus in about a century the population would double. ∎

The doubling time, t_2

The time it takes for a quantity growing in accord with the formula $P(t) = Ab^t$ to double is called its *doubling time*. As in Example 4, the doubling time, denoted t_2, is given by the formula

$$t_2 = \frac{\ln 2}{\ln b}.$$

But if the growth rate per unit time, r, is small, then $\ln b \approx r$, and

An estimate for t_2 when r is small

$$t_2 \approx \frac{\ln 2}{r} \approx \frac{0.69}{r}. \tag{20}$$

Equation (20) gives a quick way to estimate doubling time. For instance, in 1980 the less developed countries had an annual growth rate $r \approx 2.2$ percent $= 0.022$. Therefore their population may double in

$$t_2 \approx \frac{0.69}{0.022} \approx 31 \text{ years.}$$

Natural decay

A substance can *decrease* (decay) at a rate proportional to the amount present. In this case the same differential equation still holds,

$$P'(t) = kP(t),$$

but $P'(t)$ is negative, and therefore k is negative. The next example illustrates this type of decay.

6.7 SEPARABLE DIFFERENTIAL EQUATIONS AND GROWTH

EXAMPLE 6

Half life $t_{1/2}$

Carbon 14 (chemical symbol ^{14}C), one of the three isotopes of carbon, is radioactive and decays at a rate proportional to the amount present. In about 5730 years only half of the material is left. This is expressed by saying that its *half-life* is 5730 years. The half-life is denoted by $t_{1/2}$. Find the constant k in the formula Ae^{kt} in this case.

SOLUTION Let $P(t) = Ae^{kt}$ be the amount present after t years. If A is the initial amount, then

$$P(5730) = \frac{A}{2}.$$

Thus
$$Ae^{k \cdot 5730} = \frac{A}{2},$$

or
$$e^{5730k} = \tfrac{1}{2}.$$

Thus
$$5730k = \ln \tfrac{1}{2} = -\ln 2 \approx -0.69.$$

Solving for k, we have
$$k \approx \frac{-0.69}{5730} \approx -0.00012.$$

Thus
$$P(t) \approx Ae^{-0.00012t}.$$

Exercise 38 shows how this formula may be used to determine the age of fossils and other once-living things. ∎

The graphs of natural growth and natural decay look quite different. The first gets arbitrarily steep as time goes on; the second approaches the t axis and becomes almost horizontal. (See Fig. 6.34.)

If a quantity is increasing subject to natural growth, its graph may quickly shoot off an ordinary size piece of paper. The next example describes a special way of graphing an equation of the form $y = Ae^{kt}$.

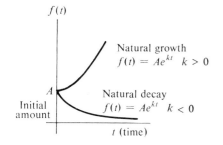

Figure 6.34

EXAMPLE 7 Let $y = Ae^{kt}$. Let $Y = \ln y$. Show that the graph of Y, considered as a function of t, is a straight line.

SOLUTION
$$Y = \ln y$$
$$= \ln Ae^{kt}$$
$$= \ln A + \ln e^{kt}$$
$$= \ln A + kt.$$

The graph is thus a straight line with Y intercept $\ln A$ and slope k. The diagram in Fig. 6.35 contrasts the graphs of y and $Y = \ln y$ as functions of t. ∎

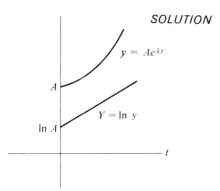

Figure 6.35

Special graph paper called *semilog*, based on the observation in Example 7, is available. It enables one to graph $Y = \ln y$ directly without having to compute logarithms. On the vertical axis the number y is placed a distance from the x axis proportional to $\ln y$, just as on a slide rule for multiplication. (See Exercise 47.)

INHIBITED GROWTH In many cases of growth there is obviously a finite upper bound M, which the population size must approach. It is reasonable to assume (or to take as a model) that

The D.E. of inhibited growth

$$\frac{dP}{dt} = kP(t)\left(1 - \frac{P(t)}{M}\right). \tag{21}$$

This equation asserts that the population $P(t)$ grows at a rate proportional both to itself and to the fraction left to grow. Thus as $P(t)$ approaches its limiting size M, the fraction

$$1 - \frac{P(t)}{M}$$

approaches 0 and acts as a damper in (21), decreasing the rate of change dP/dt.

Growth satisfying (21) is called *logistic*, *inhibited*, or *sigmoidal* (the last because its graph, shown in Fig. 6.36, is S-shaped). It was first proposed in the 1840s by P. F. Verhulst.

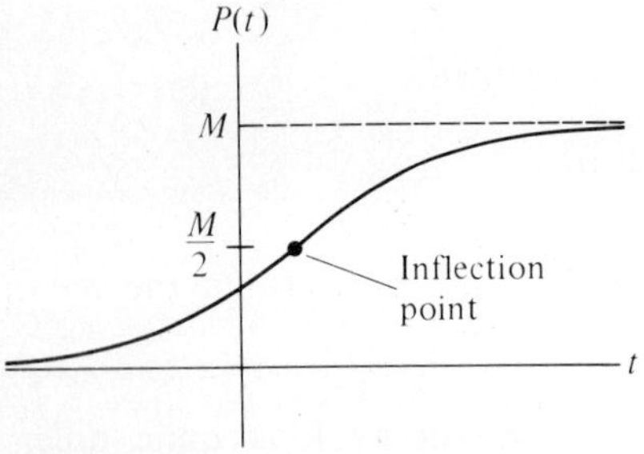

By Example 4,

$$P(t) = \frac{M}{1 + ae^{-kt}}. \tag{22}$$

Figure 6.36

The logistic curve is symmetric with respect to its inflection point.

Note that as $t \to \infty$, $e^{-kt} \to 0$, so $P(t) \to M$, as desired. The logistic curve has an inflection point at height $M/2$ (see Exercise 39). Therefore if experimental data indicated an inflection point at a height h, then the ultimate population limit could be conjectured as $2h$.

EXAMPLE 8 What is the significance of the constant a in (22)?

SOLUTION Setting $t = 0$ in (22) yields

$$P(0) = \frac{M}{1 + a}.$$

Thus

$$(1 + a)P(0) = M$$

$$P(0) + aP(0) = M$$

and

$$a = \frac{M - P(0)}{P(0)}$$

$$= \frac{M}{P(0)} - 1.$$

The logistic growth function expressed in terms of initial and ultimate populations

The constant a is one less than the ratio between the limiting population M and the initial population $P(0)$.

Thus
$$P(t) = \frac{M}{1 + \{[M/P(0)] - 1\}e^{-kt}}. \quad \blacksquare$$

SUMMARY OF THIS SECTION

The separable differential equation

$$\frac{dy}{dx} = \frac{g(x)}{h(y)}$$

is solved by separating the variables and integrating,

$$\int h(y)\, dy = \int g(x)\, dx + C.$$

The differential equation of natural growth or decay,

$$\frac{dP}{dt} = kP(t)$$

has the general solution

$$P(t) = Ae^{kt},$$

or
$$P(t) = Ab^t,$$

where $b = e^k$.

In the case of natural growth, $k > 0$ and $b > 1$. Let $b = 1 + r$. Then r is the relative growth rate per unit time. When r is small,

$$k \approx r \quad \text{and thus} \quad b \approx 1 + k.$$

The doubling time t_2 is equal to $\ln 2/\ln b$. When r is small,

$$t_2 \approx \frac{\ln 2}{r} \approx \frac{0.69}{r}.$$

In the case of natural decay, $k < 0$ and $0 < b < 1$. The half-life, the time required to decay to half the initial quantity, is

$$t_{1/2} = \frac{\ln\left(\frac{1}{2}\right)}{\ln b} = -\frac{\ln 2}{\ln b}.$$

If $b = 1 + r$ (r negative) and r is small, then

$$t_{1/2} \approx \frac{\ln 2}{|r|} \approx \frac{0.69}{|r|}.$$

The differential equation of inhibited growth is

$$\frac{dP}{dt} = kP\left(1 - \frac{P}{M}\right).$$

The general solution is

$$P(t) = \frac{M}{1 + \{[M/P(0)] - 1\}e^{-kt}}.$$

where $P(0)$ is the initial population. As $t \to \infty$, $P(t) \to M$. The graph of P has one inflection point; at that point the y coordinate is $M/2$.

Exercises

Exercises 1 to 20 concern differential equations.
In Exercises 1 to 6 check by substitution that the given functions satisfy the given differential equations.

1. $y = Ae^{kx}$; $dy/dx = ky$.
2. $y = ke^{x^2/2}$; $dy/dx = xy$.
3. $y = e^{-3x}$; $d^2y/dx^2 = 9y$.
4. $y = Ae^{-3x} + Be^{3x}$; $d^2y/dx^2 = 9y$ (A and B constants).
5. $y = A\cos 3x + B\sin 3x$; $d^2y/dx^2 = -9y$.
6. $y = M/(1 + ae^{-kx})$; $dy/dx = ky(1 - y/M)$. (This checks the result in Example 4.)

In Exercises 7 to 20 solve the differential equations.

7. $dy/dx = x^3/y^4$
8. $dy/dx = y^4/x^3$
9. $dy/dx = y/x$
10. $dy/dx = e^y/x$
11. $dy/dx = \sqrt{1 - y^2}/(1 + x^2)$
12. $dy/dx = (y + 4)/(x\sqrt{x^2 - 1})$, $x > 1$, $y > 0$
13. $x + y^2 \dfrac{dy}{dx} = 0$
14. $x + y \dfrac{dy}{dx} = 3$
15. $y \dfrac{dy}{dx} - (1 + y^2)x^2 = 0$
16. $(3 + x)e^{2y} \dfrac{dy}{dx} = 1$
17. $\dfrac{1}{\sqrt{y^2 - 1}} \dfrac{dy}{dx} = yx^2$
18. $\dfrac{(1 + \sqrt[3]{1 + 2y})}{\cos 3x} \dfrac{dy}{dx} = 5$
19. $\dfrac{dy}{dx} = \dfrac{\sqrt{1 - y^2}}{\sqrt{3 - x}}$, $x < 3$, $|y| < 1$
20. $(1 + 2y)^3 \dfrac{dy}{dx} = \sec^2 3x$

Exercises 21 to 38 concern natural growth and decay.

21. Use differentials to show that, for k small,
    (a) $\ln(1 + k) \approx k$ and (b) $e^k \approx 1 + k$.
22. A population is growing at the rate of 2 percent per year.
    (a) What is b? (b) Find k to four decimal places.
23. The amount of a certain growing substance increases at the rate of 10 percent per hour. Find
    (a) r (b) b (c) k (d) t_2.
24. A quantity is increasing according to the law of natural growth. The amount present at time $t = 0$ is A. It will double when $t = 10$.
    (a) Express the amount in the form Ae^{kt} for suitable k.
    (b) Express the amount in the form Ab^t for suitable b.
25. The mass of a certain bacterial culture after t hours is $10 \cdot 3^t$ grams.
    (a) What is the initial amount?
    (b) What is the growth constant k?
    (c) What is the percent increase in any period of 1 hour?
26. Let $f(t) = 3 \cdot 2^t$.
    (a) Solve the equation $f(t) = 12$.
    (b) Solve the equation $f(t) = 5$.
    (c) Find k such that $f(t) = 3e^{kt}$.
27. In 1980 the world population was about 4.5 billion and was increasing at the rate of 1.8 percent per year. If it continues to grow at that rate, when will it (a) double? (b) quadruple? (c) reach 100 billion?
28. The population of Latin America has a doubling time of 27 years. By what percent does it grow per year?

29 In 1980 the United States population was 225 million and increasing at the rate of 0.7 percent per year and the population of Mexico was 62 million and increasing at the rate of 2.4 percent per year. When will the two populations be the same size if they continue to grow at the same rates?

30 A bacterial culture grows from 100 to 400 grams in 10 hours according to the law of natural growth.
   (a) How much was present after 3 hours?
   (b) How long will it take the mass to double? quadruple? triple?

31 At 1 P.M. a bacterial culture weighed 100 grams. At 4:30 P.M. it weighed 250 grams. Assuming that it grows at a rate proportional to the amount present find (a) at what time it will grow to 400 grams, (b) its growth constant, and (c) the growth rate per hour.

32 The population of Russia in 1980 was 255 million and growing at the relative rate of 1.2 percent per year and the population of the United States was 225 million and growing at the relative rate of 0.7 percent per year. When will the population of Russia be twice as large as that of the United States?

33 See Example 6. How much carbon 14 remains after (a) 11,460 years? (b) 2000 years?

34 The half-life of radium is about 1600 years.
   (a) From this, find k in the expression of Ae^{kt}.
   (b) How long does it take 75 percent of the radium to disintegrate?
   (c) Solve (b) without using calculus.
   (d) How long will it take for 90 percent of the radium to disintegrate?
   (e) Without calculus, show that the answer to (d) is between 4800 and 6400 years.

35 A bacterial culture grows at a rate proportional to the amount present. From 9 to 11 A.M. it increases from 100 to 200 grams. What will be its weight at noon?

36 A disintegrating radioactive substance decreases from 12 grams to 11 grams in 1 day. Find its half-life.

37 A radioactive substance disintegrates at the instantaneous rate of 0.05 gram per day when its mass is 10 grams.
   (a) How much of the substance will remain after t days if the initial amount is A?
   (b) What is its half-life?

38 If the carbon 14 concentration in the carbon from a plant or piece of wood of unknown age is half that of the carbon 14 concentration in a present-day live specimen, then it is about 5730 years old. Show that, if A_c and A_u are the radioactivities of samples prepared from contemporary and from undated materials, respectively, then the age of the undated material is about $t = 8300 \ln (A_c/A_u)$. (This method is dependable up to an age of about 70,000 years.) See Radiocarbon Dating in *Encyclopedia of Science and Technology*, McGraw-Hill, New York, 1971.

39 Show that at the inflection point of the logistic curve the y coordinate is $M/2$. Use (21) and differentiate both sides of the equation.

40 Solve Exercise 39, but this time use the explicit formula for $P(t)$ given by (22).

■

41 The differential equation

$$L \frac{di}{dt} + Ri = E$$

occurs in the study of electrical circuits. L, R, and E are constants that describe the inductance, resistance, and voltage; i is the current; and t is time. Assume that di/dt is positive. Then $E - Ri$ is positive as well.
   (a) Solve for di/dt in terms of i.
   (b) The equation in (a) is separable. Solve it and express the answer in terms of R, L, E, and i_0, the initial current.

42 Find all functions $y = f(t)$ such that

$$\frac{dy}{dt} = k(y - A)$$

where k and A are constants. For negative k this is Newton's law of cooling; y is the temperature of some heated object at time t. The room temperature is A. The differential equation $dy/dt = k(y - A)$ says, "The object cools at a rate proportional to the difference between its temperature and the room temperature."

43 This is an excerpt from a news article published in 1975. There are 57,000 babies born daily in India, or 21 million a year. With 8 million deaths, the annual population increase is 13 million. The population, nearly 570 million, is expected to reach a billion by the end of the century.
   (a) Assuming natural growth, estimate the population of India in the year 2000.
   (b) What is the relative growth rate per year?
   (c) What is the growth constant k?

44 (a) Assume that k_1 and k_2 are constants and that y_1 and y_2 are functions such that $dy_1/dx = k_1 y_1$ and $dy_2/dx = k_2 y_2$. Without solving for y_1 and y_2 explicitly, show that $d(y_1 y_2)/dx = (k_1 + k_2) y_1 y_2$.
   (b) If energy consumption per person in the United States increases at the rate of 1 percent per year and the population grows at the rate of 0.7 percent per year, at what rate does the energy consumption in the nation increase? *Hint:* The answer is not 1.7 percent.

45 A bank offers to pay 8 percent annual interest compounded over a very small time interval Δt. Let $A(t)$ be the amount in the account at time t.
  (a) Assume that Δt is measured in years. Explain why $\Delta A \approx 0.08 A(t)\, \Delta t$.
  (b) From (a) deduce that $dA/dt = 0.08 A(t)$ in the limiting case as $\Delta t \to 0$.
  (c) Under "continuous compounding" of interest, $dA/dt = 0.08 A(t)$. Deduce that in one year of such compounding an initial amount earns about 8.33 percent interest.

46 (Doomsday equation) A differential equation of the form $dP/dt = kP^{1.01}$ is called a *doomsday equation*. The rate of growth is just slightly higher than that for natural growth.
  (a) Solve the equation.
  (b) Show that there is a finite number t_1 such that $\lim_{t \to t_1} P(t) = \infty$. Naturally, t_1 is Doomsday.

47 Figure 6.37 has a logarithmic vertical axis. Since only one axis is logarithmic, the axes are called "semilogarithmic."

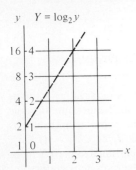

Figure 6.37

  (a) By inspection of the graph, find Y as a function of x.
  (b) From (a) find y as a function of x.

48 Figure 6.38 shows semilogarithmic graph paper. Graph $y = 3 \cdot 2^x$ on it.

Figure 6.38

49 Semilogarithmic graph paper is available in bookstores where engineering forms are sold. Figure 6.39 is part of one such sheet, which is $8\frac{1}{2}$ by 11 inches. For this exercise either copy Fig. 6.39 or buy a piece of semilog paper.

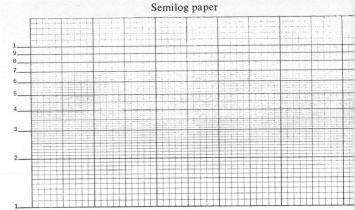

Figure 6.39

The population of the world in 1950 was 2.5 billion; in 1970, 3.7 billion; in 1980, 4.5 billion.
  (a) Plot these data on semilog paper like that in Fig. 6.39. (They should lie on a straight line.)
  (b) Draw a line through the three points in (a) and use it to estimate the population in the year 2000.
  (c) Using the line drawn in (b), estimate when the world population will reach 10 billion.

■ ■

50 Find all functions $f(x)$ such that $f(x) = 3 \int_0^x f(t)\, dt$.

51 Let $I(x)$ be the intensity of sunlight at a depth of x meters in the ocean. As x increases, $I(x)$ decreases.
  (a) Why is it reasonable to assume that there is a constant k (negative) such that $\Delta I \approx kI(x)\, \Delta x$ for small Δx?
  (b) Deduce that $I(x) = I(0)e^{kx}$, where $I(0)$ is the intensity of sunlight at the surface. Incidentally, sunlight at a depth of 1 meter is only $\frac{1}{4}$ as intense as at the surface.

52 A particle moving through a liquid meets a "drag" proportional to the velocity; that is, its acceleration is proportional to its velocity. Let x denote its position and v its velocity at time t.

  (a) Show that there is a positive constant k such that
  $$\frac{dv}{dt} = -kv.$$
  (b) Show that there is a constant A such that
  $$v = Ae^{-kt}.$$
  (c) Show that there is a constant B such that
  $$x = -\frac{1}{k} Ae^{-kt} + B.$$

(d) How far does the particle travel as t goes from 0 to ∞?

53 A company is founded with a capital investment A. The plan is to have its rate of investment proportional to its total investment at any time. Let $f(t)$ denote the rate of investment at time t.
  (a) Show that there is a constant k such that $f(t) = k[A + \int_0^t f(x)\,dx]$ for any $t \geq 0$.
  (b) Differentiate the relation in (a), and, with the aid of the equation you obtain, find a formula for f.

6.8 L'Hôpital's rule

Pronounced "Low-pee-tal's rule"

The problem of finding a limit has arisen in graphing a curve and will appear often in later chapters. Fortunately, there are some general techniques for computing a wide variety of limits. This section discusses one of the most important of these methods, l'Hôpital's rule, which concerns the limit of a quotient of two functions.

If f and g are functions and a is a number such that

$$\lim_{x \to a} f(x) = 2 \quad \text{and} \quad \lim_{x \to a} g(x) = 3,$$

then

$$\lim_{x \to a} \frac{f(x)}{g(x)} = \frac{2}{3}.$$

This problem presents no difficulty; no more information is needed about the functions f and g. But if

$$\lim_{x \to a} f(x) = 0 \quad \text{and} \quad \lim_{x \to a} g(x) = 0,$$

then finding

$$\lim_{x \to a} \frac{f(x)}{g(x)}$$

may present a serious problem. For instance, Sec. 2.5 was dedicated to showing that

$$\lim_{\theta \to 0} \frac{\sin \theta}{\theta} = 1.$$

Here $f(\theta) = \sin \theta \to 0$ and $g(\theta) = \theta \to 0$ as $\theta \to 0$. The quotient $f(\theta)/g(\theta)$ approaches 1. In that same section it was proved that

$$\lim_{\theta \to 0} \frac{1 - \cos \theta}{\theta} = 0.$$

In this second limit, the numerator rushes toward 0 so much faster than the denominator that the quotient approaches 0. These two examples serve to point out that, if you know *only* that

$$\lim_{x \to a} f(x) = 0 \quad \text{and} \quad \lim_{x \to a} g(x) = 0,$$

350 TOPICS IN DIFFERENTIAL CALCULUS

you do not have enough information to determine

$$\lim_{x \to a} \frac{f(x)}{g(x)}.$$

L'HÔPITAL'S RULE: ZERO-OVER-ZERO CASE

Theorem 1 describes a general technique for dealing with the troublesome quotient

$$\frac{f(x)}{g(x)}$$

when $\qquad f(x) \to 0 \qquad$ and $\qquad g(x) \to 0.$

It is known as the *zero-over-zero case* of l'Hôpital's rule.

THEOREM 1 *L'Hôpital's rule (zero-over-zero case).* Let a be a number and let f and g be differentiable over some open interval (a, b). Assume also that $g'(x)$ is not 0 for any x in that interval. If

$$\lim_{x \to a^+} f(x) = 0, \qquad \lim_{x \to a^+} g(x) = 0,$$

and

$$\lim_{x \to a^+} \frac{f'(x)}{g'(x)} = L,$$

then

$$\lim_{x \to a^+} \frac{f(x)}{g(x)} = L. \quad \blacksquare$$

Similar rules hold for $x \to a^-$, $x \to a$, $x \to \infty$, and $x \to -\infty$. Of course, corresponding changes must be made in the hypotheses.

Before worrying about *why* this theorem is true, we illustrate its use by an example.

EXAMPLE 1 Find

$$\lim_{x \to 1^+} \frac{x^5 - 1}{x^3 - 1}.$$

SOLUTION In this case

$$a = 1, \qquad f(x) = x^5 - 1, \qquad \text{and} \qquad g(x) = x^3 - 1.$$

All the assumptions of l'Hôpital's rule are satisfied. In particular,

$$\lim_{x \to 1^+} (x^5 - 1) = 0 \qquad \text{and} \qquad \lim_{x \to 1^+} (x^3 - 1) = 0.$$

According to l'Hôpital's rule,

$$\lim_{x \to 1^+} \frac{x^5 - 1}{x^3 - 1} = \lim_{x \to 1^+} \frac{(x^5 - 1)'}{(x^3 - 1)'},$$

if the latter limit exists.

Now, $$\lim_{x \to 1^+} \frac{(x^5 - 1)'}{(x^3 - 1)'} = \lim_{x \to 1^+} \frac{5x^4}{3x^2} \quad \text{differentiation of numerator and denominator}$$
$$= \lim_{x \to 1^+} \tfrac{5}{3}x^2 \quad \text{algebra}$$
$$= \tfrac{5}{3}.$$

This can also be solved by factoring $x^5 - 1$ and $x^3 - 1$.

Thus $$\lim_{x \to 1^+} \frac{x^5 - 1}{x^3 - 1} = \frac{5}{3}. \quad \blacksquare$$

A complete proof of Theorem 1 may be found in Exercises 189 and 190 of Sec. 6.S. Let us pause long enough here to make the theorem plausible.

Argument for a special case of Theorem 1

To do so, consider the *special case* where f, f', g, and g' are all continuous throughout an open interval containing a. Assume that $g'(x) \neq 0$ throughout the interval. Since $\lim_{x \to a^+} f(x) = 0$ and $\lim_{x \to a^+} g(x) = 0$, it follows by continuity that $f(a) = 0$ and $g(a) = 0$. Then

$$\lim_{x \to a^+} \frac{f(x)}{g(x)} = \lim_{x \to a^+} \frac{f(x) - f(a)}{g(x) - g(a)} \quad \text{since } f(a) = 0 \text{ and } g(a) = 0$$

$$= \lim_{x \to a^+} \frac{\dfrac{f(x) - f(a)}{x - a}}{\dfrac{g(x) - g(a)}{x - a}} \quad \text{algebra}$$

$$= \frac{\lim_{x \to a^+} \dfrac{f(x) - f(a)}{x - a}}{\lim_{x \to a^+} \dfrac{g(x) - g(a)}{x - a}} \quad \text{limit of quotients}$$

$$= \frac{f'(a)}{g'(a)} \quad \text{by definition of } f'(a) \text{ and } g'(a)$$

$$= \frac{\lim_{x \to a^+} f'(x)}{\lim_{x \to a^+} g'(x)} \quad f' \text{ and } g' \text{ are continuous}$$

$$= \lim_{x \to a^+} \frac{f'(x)}{g'(x)} \quad \text{``limit of quotient'' property}$$

$$= L \quad \text{by assumption.}$$

Consequently, $$\lim_{x \to a^+} \frac{f(x)}{g(x)} = L.$$

Sometimes it may be necessary to apply l'Hôpital's rule more than once, as in the next example.

EXAMPLE 2   Find $$\lim_{x \to 0} \frac{\sin x - x}{x^3}.$$

SOLUTION   As $x \to 0$, both numerator and denominator approach 0. By l'Hôpital's rule,

$$\lim_{x \to 0} \frac{\sin x - x}{x^3} = \lim_{x \to 0} \frac{\cos x - 1}{3x^2}.$$

Repeated application of l'Hôpital's rule   But as $x \to 0$, both $\cos x - 1 \to 0$ and $3x^2 \to 0$. So use l'Hôpital's rule again:

$$\lim_{x \to 0} \frac{\cos x - 1}{3x^2} = \lim_{x \to 0} \frac{-\sin x}{6x}.$$

Or recall from Sec. 2.5 that
$$\lim_{x \to 0} \frac{\sin x}{x} = 1.$$

Both $\sin x$ and $6x$ approach 0 as $x \to 0$. Use l'Hôpital's rule yet another time:

$$\lim_{x \to 0} \frac{-\sin x}{6x} = \lim_{x \to 0} \frac{-\cos x}{6}$$
$$= -\tfrac{1}{6}.$$

So, after three applications of l'Hôpital's rule, we find that

$$\lim_{x \to 0} \frac{\sin x - x}{x^3} = -\frac{1}{6}. \quad \blacksquare$$

Sometimes a limit may be simplified before l'Hôpital's rule is applied. For instance, consider

$$\lim_{x \to 0} \frac{(\sin x - x)\cos^5 x}{x^3}.$$

Since $\lim_{x \to 0} \cos^5 x = 1$, we have

$$\lim_{x \to 0} \frac{(\sin x - x)\cos^5 x}{x^3} = \left(\lim_{x \to 0} \frac{\sin x - x}{x^3}\right)(1),$$

which, by Example 2, is $-\tfrac{1}{6}$. This shortcut saves a lot of work, as may be checked by finding the limit using l'Hôpital's rule without separating $\lim_{x \to 0} \cos^5 x$.

L'HÔPITAL'S RULE: INFINITY-OVER-INFINITY CASE

Theorem 1 concerns the problem of finding the limit of $f(x)/g(x)$ when both $f(x)$ and $g(x)$ approach 0, the zero-over-zero case of l'Hôpital's rule. But a similar problem arises when both $f(x)$ and $g(x)$ get arbitrarily large as $x \to a$ or as $x \to \infty$. The behavior of the quotient $f(x)/g(x)$ will be influenced by how rapidly $f(x)$ and $g(x)$ become large.

For example,

$$\begin{aligned}\lim_{x\to\infty}\frac{x^2+5x+2}{x^3+x-1} &= \lim_{x\to\infty}\frac{x^2[1+(5/x)+(2/x^2)]}{x^3[1+(1/x^2)-(1/x^3)]}\\ &= \lim_{x\to\infty}\frac{1}{x}\left[\frac{1+(5/x)+(2/x^2)}{1+(1/x^2)-(1/x^3)}\right]\\ &= 0\cdot 1 = 0.\end{aligned}$$

On the other hand,

$$\begin{aligned}\lim_{x\to\infty}\frac{4x+1}{2x} &= \lim_{x\to\infty}\left(2+\frac{1}{2x}\right)\\ &= 2.\end{aligned}$$

In this case, the numerator is increasing about twice as rapidly as the denominator.

The next theorem presents a form of l'Hôpital's rule that covers the case in which $f(x)\to\infty$ and $g(x)\to\infty$. It is called the *infinity-over-infinity* case of l'Hôpital's rule.

THEOREM 2 *L'Hôpital's rule (infinity-over-infinity case).* Let f and g be defined and differentiable for all x larger than some fixed number. Then, if

$$\lim_{x\to\infty} f(x)=\infty, \qquad \lim_{x\to\infty} g(x)=\infty,$$

L'Hôpital's rule for the infinity-over-infinity case

and

$$\lim_{x\to\infty}\frac{f'(x)}{g'(x)}=L,$$

it follows that

$$\lim_{x\to\infty}\frac{f(x)}{g(x)}=L.$$

A similar result holds for $x\to a$, $x\to a^-$, $x\to a^+$, or $x\to-\infty$. Moreover, $\lim_{x\to\infty} f(x)$ and $\lim_{x\to\infty} g(x)$ could both be $-\infty$, or one could be ∞ and the other $-\infty$. ■

The proof of this is left to an advanced calculus course. However, it is easy to see why it is plausible. Imagine that $f(t)$ and $g(t)$ describe the location on the x axis of two cars at time t. Call the cars the f-car and the g-car. See Fig. 6.40. Their velocities are therefore $f'(t)$ and $g'(t)$. These two cars are on endless journeys. But let us assume that as time $t\to\infty$ the f-car tends to travel at a speed closer and closer to L times the speed of the g-car. That is, assume that

$$\lim_{t\to\infty}\frac{f'(t)}{g'(t)}=L.$$

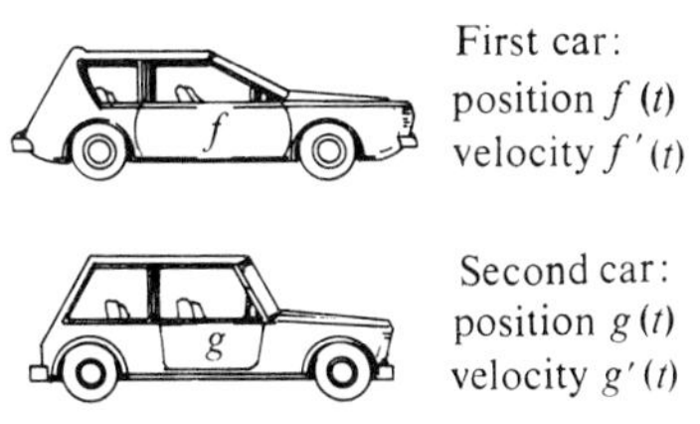
f
First car:
position f (t)
velocity f ′ (t)
g
Second car:
position g (t)
velocity g′ (t)

Figure 6.40

No matter how the two cars move in the short run, it seems reasonable that in the *long run* the f-car will tend to travel about L times as far as the g-car; that is,

$$\lim_{t \to \infty} \frac{f(t)}{g(t)} = L.$$

Though not a proof, this argument does provide a perspective for a rigorous proof.

EXAMPLE 3 Use l'Hôpital's rule to find $\lim_{x \to \infty} \dfrac{x}{e^x}$.

SOLUTION By Theorem 2,

$$\lim_{x \to \infty} \frac{x}{e^x} = \lim_{x \to \infty} \frac{x'}{(e^x)'}$$

$$= \lim_{x \to \infty} \frac{1}{e^x} = 0. \blacksquare$$

EXAMPLE 4 Find $\lim_{x \to \infty} \dfrac{x^3}{2^x}$.

SOLUTION Since both numerator and denominator approach ∞ as $x \to \infty$, l'Hôpital's rule may be applied. It asserts that

$$\lim_{x \to \infty} \frac{x^3}{2^x} = \lim_{x \to \infty} \frac{3x^2}{2^x \ln 2} \qquad \text{(if the latter limit exists)}$$

$$= \lim_{x \to \infty} \frac{6x}{2^x (\ln 2)^2} \qquad \text{using the rule again}$$

$$= \lim_{x \to \infty} \frac{6}{2^x (\ln 2)^3} \qquad \text{again l'Hôpital's rule}$$

$$= 0.$$

Thus as $x \to \infty$, 2^x grows much faster than x^3. $\blacksquare$

This table shows that for small x, 2^x is smaller than x^3, but it catches up around $x = 10$. Example 4 shows that for large x, 2^x is much larger than x^3.

x	1	2	3	4	5	6	7	8	9	10	11	12
2^x	2	4	8	16	32	64	128	256	512	1,024	2,048	4,096
x^3	1	8	27	64	125	216	343	512	729	1,000	1,331	1,728

The next example conveys a warning.

EXAMPLE 5 Find $\lim_{x \to \infty} \dfrac{x - \cos x}{x}$. (1)

SOLUTION   Both numerator and denominator approach ∞ as $x \to \infty$. Trying l'Hôpital's rule, we obtain

$$\lim_{x\to\infty} \frac{x - \cos x}{x} = \lim_{x\to\infty} \frac{1 + \sin x}{1}.$$

L'Hôpital's rule may fail to provide an answer.

But $\lim_{x\to\infty} (1 + \sin x)$ does not exist.

What can we conclude about (1)? Nothing at all. L'Hôpital's rule says that if $\lim_{x\to\infty} f'/g'$ exists, then $\lim_{x\to\infty} f/g$ exists. It says nothing about the case when $\lim_{x\to\infty} f'/g'$ does not exist.

It is not difficult to evaluate (1) directly, as follows:

$$\lim_{x\to\infty} \frac{x - \cos x}{x} = \lim_{x\to\infty} \left(1 - \frac{\cos x}{x}\right) \quad \text{algebra}$$

$$= 1 - 0 \quad (\text{since } |\cos x| \leq 1)$$

$$= 1. \quad \blacksquare$$

VARIATIONS OF L'HÔPITAL'S RULE

EXAMPLE 6   Find
$$\lim_{x\to 0^+} \frac{\ln x}{1/x}.$$

SOLUTION   In this case $f(x) = \ln x$ and $g(x) = 1/x$. Note that

$$\lim_{x\to 0^+} \ln x = -\infty \quad \text{and} \quad \lim_{x\to 0^+} \frac{1}{x} = \infty.$$

An analog of Theorem 2, with $x \to 0^+$, asserts that

$$\lim_{x\to 0^+} \frac{\ln x}{1/x} = \lim_{x\to 0^+} \frac{1/x}{-1/x^2}$$

$$= \lim_{x\to 0} (-x)$$

$$= 0.$$

Thus
$$\lim_{x\to 0^+} \frac{\ln x}{1/x} = 0.$$

In short, $1/x$ gets large (in absolute value) much more quickly than does $\ln x$ as $x \to 0^+$. $\blacksquare$

TRANSFORMING SOME LIMITS SO L'HÔPITAL'S RULE APPLIES

The zero-times-infinity case

Many limit problems can be transformed to limits to which l'Hôpital's rule applies. For instance, the problem of finding

$$\lim_{x\to 0^+} x \ln x$$

does not seem to be related to l'Hôpital's rule, since it does not involve the quotient of two functions. As $x \to 0^+$, one factor, x, approaches 0 and the other factor, $\ln x$, approaches $-\infty$. It is not obvious how their product, $x \ln x$, behaves as $x \to 0^+$. But a little algebraic manipulation transforms it into a problem to which l'Hôpital's rule applies:

$$x \ln x = \frac{\ln x}{1/x}.$$

In the latter form the limit was determined in Example 6. Thus

$$\lim_{x \to 0^+} x \ln x = 0.$$

The remaining examples illustrate other limits that can be found by first relating them to limit problems to which l'Hôpital's rule applies.

EXAMPLE 7   Find
$$\lim_{x \to 0^+} x^x.$$

SOLUTION

The zero-to-the-zero case

Since this limit involves an exponential, not a quotient, it does not fit directly into l'Hôpital's rule. But a little algebraic manipulation will change the problem to one covered by l'Hôpital's rule.

Let $\qquad y = x^x.$

Then $\qquad \ln y = \ln x^x$

$\qquad\qquad\qquad = x \ln x.$

By the remarks preceding this example,

$$\lim_{x \to 0^+} \ln y = 0.$$

If $\ln y \to 0$, then y must approach 1.

Thus $\qquad \lim_{x \to 0^+} y = \lim_{x \to 0} e^{\ln y} = e^0 = 1.$

In short, $\qquad \lim_{x \to 0^+} x^x = 1.$ ∎

SUMMARY OF THIS SECTION

L'Hôpital's rule (zero-over-zero case) asserts that if $f(x) \to 0$ and $g(x) \to 0$ as $x \to a$ and if $\lim_{x \to a} f'(x)/g'(x)$ exists, then $\lim_{x \to a} f(x)/g(x)$ exists and

$$\lim_{x \to a} \frac{f(x)}{g(x)} = \lim_{x \to a} \frac{f'(x)}{g'(x)}.$$

($x \to a$ can be replaced by $x \to a^+$ or by $x \to \infty$, etc.)

A similar rule holds in the infinity-over-infinity case.

Some other limits can be brought into a form to which l'Hôpital's rule can be applied:

Form	Method
$f(x)g(x); f(x) \to 0, g(x) \to \infty$	Rewrite as $\dfrac{g(x)}{1/f(x)}$.
$f(x)^{g(x)}; f(x) \to 1, g(x) \to \infty$ $f(x)^{g(x)}; f(x) \to 0, g(x) \to 0$	Write $y = f(x)^{g(x)}$, take $\ln y$, find limit of $\ln y$ and then the limit of $y = e^{\ln y}$.

Exercises

In Exercises 1 to 12 check that l'Hôpital's rule applies and use it to find the limits.

1. $\lim\limits_{x \to 2} \dfrac{x^3 - 8}{x^2 - 4}$

2. $\lim\limits_{x \to 1} \dfrac{x^7 - 1}{x^3 - 1}$

3. $\lim\limits_{x \to 0} \dfrac{\sin 3x}{\sin 2x}$

4. $\lim\limits_{x \to 0} \dfrac{\sin x^2}{(\sin x)^2}$

5. $\lim\limits_{x \to \infty} \dfrac{x^3}{e^x}$

6. $\lim\limits_{x \to \infty} \dfrac{x^5}{3^x}$

7. $\lim\limits_{x \to 0} \dfrac{1 - \cos x}{x^2}$

8. $\lim\limits_{x \to 0} \dfrac{\sin x - x}{(\sin x)^3}$

9. $\lim\limits_{x \to 0} \dfrac{\tan 3x}{\ln (1 + x)}$

10. $\lim\limits_{x \to 1} \dfrac{\cos (\pi x/2)}{\ln x}$

11. $\lim\limits_{x \to \infty} \dfrac{(\ln x)^2}{x}$

12. $\lim\limits_{x \to 0} \dfrac{\sin^{-1} x}{e^{2x} - 1}$

In each of Exercises 13 to 18 transform the problem into one to which l'Hôpital's rule applies; then find the limit.

13. $\lim\limits_{x \to 0} (1 - 2x)^{1/x}$

14. $\lim\limits_{x \to 0} (1 + \sin 2x)^{\csc x}$

15. $\lim\limits_{x \to 0^+} (\sin x)^{(e^x - 1)}$

16. $\lim\limits_{x \to 0^+} x^2 \ln x$

17. $\lim\limits_{x \to 0^+} (\tan x)^{\tan 2x}$

18. $\lim\limits_{x \to 1} x^{3/\ln x}$

In Exercises 19 to 46 find the limits. Use any method. *Warning*: l'Hôpital's rule, carelessly applied, may give a wrong answer.

19. $\lim\limits_{x \to \infty} \dfrac{2^x}{3^x}$

20. $\lim\limits_{x \to \infty} \dfrac{2^x + x}{3^x}$

21. $\lim\limits_{x \to \infty} \dfrac{\log_2 x}{\log_3 x}$

22. $\lim\limits_{x \to 1} \dfrac{\log_2 x}{\log_3 x}$

23. $\lim\limits_{x \to \infty} \left(\dfrac{1}{x} - \dfrac{1}{\sin x} \right)$

24. $\lim\limits_{x \to \infty} (\sqrt{x^2 + 3} - \sqrt{x^2 + 4x})$

25. $\lim\limits_{x \to \infty} \dfrac{x^2 + 3 \cos 5x}{x^2 - 2 \sin 4x}$

26. $\lim\limits_{x \to \infty} \dfrac{e^x - 1/x}{e^x + 1/x}$

27. $\lim\limits_{x \to 0} \dfrac{3x^3 + x^2 - x}{5x^3 + x^2 + x}$

28. $\lim\limits_{x \to \infty} \dfrac{3x^3 + x^2 - x}{5x^3 + x^2 + x}$

29. $\lim\limits_{x \to \infty} \dfrac{\sin x}{4 + \sin x}$

30. $\lim\limits_{x \to \infty} 5 \sin 3x$

31. $\lim\limits_{x \to 1^+} (x - 1) \ln (x - 1)$

32. $\lim\limits_{x \to \pi/2} \dfrac{\tan x}{x - (\pi/2)}$

33. $\lim\limits_{x \to 0} (\cos x)^{1/x}$

34. $\lim\limits_{x \to 0^+} x^{1/x}$

35. $\lim\limits_{x \to \infty} \dfrac{\sin 2x}{\sin 3x}$

36. $\lim\limits_{x \to 1} \dfrac{x^2 - 1}{x^3 - 1}$

37. $\lim\limits_{x \to 0} \dfrac{xe^x(1 + x)^3}{e^x - 1}$

38. $\lim\limits_{x \to 0} \dfrac{xe^x \cos^2 6x}{e^{2x} - 1}$

39. $\lim\limits_{x \to 0} (\csc x - \cot x)$

40. $\lim\limits_{x \to 0} \dfrac{\csc x - \cot x}{\sin x}$

41. $\lim\limits_{x \to 0} \dfrac{5^x - 3^x}{\sin x}$

42. $\lim\limits_{x \to 0} \dfrac{\tan^5 x - \tan^3 x}{1 - \cos x}$

43. $\lim\limits_{x \to 2} \dfrac{x^3 + 8}{x^2 + 5}$

44. $\lim\limits_{x \to \pi/4} \dfrac{\sin 5x}{\sin 3x}$

45. $\lim\limits_{x \to 0} \left(\dfrac{1}{1 - \cos x} - \dfrac{2}{x^2} \right)$

46. $\lim\limits_{x \to 0} \dfrac{\sin^{-1} x}{\tan^{-1} 2x}$

In Exercises 47 to 49 find the limits. Try l'Hôpital's rule first, but it may not help.

47. $\lim\limits_{x \to 0} \dfrac{\sqrt{x^2 - 5x + 2}}{x}$

48 $\lim_{x \to \infty} \dfrac{x + \sqrt{x}}{x - \sqrt{x}}$

49 $\lim_{x \to \infty} \dfrac{e^{\sqrt{x}}}{e^x}$

50 Find $\lim_{x \to 0} \dfrac{\int_1^{1+x} e^{t^2}\, dt}{\int_2^{2+x} e^{t^2}\, dt}$

In Exercises 51 to 55 graph the functions, showing intercepts, critical points, inflection points, and asymptotes.

51 $(\ln x)/x$ **52** $(\ln x)/x^2$
53 xe^{-x} **54** $x^2 e^{-x}$ **55** e^x/x

■ ■

56 Graph $y = (1 + x)^{1/x}$ for $x > -1$, $x \neq 0$, showing (a) where y is decreasing, (b) asymptotes, and (c) behavior of y for x near 0.

57 Show that for any polynomial $P(x)$ of degree at least one, $\lim_{x \to \infty} (\ln x)/P(x) = 0$.

58 In R. P. Feynman, *Lectures on Physics*, Addison-Wesley, Reading, Mass., this remark appears. "Here is the quantitative answer of what is right instead of kT. This expression,

$$\dfrac{\hbar\omega}{e^{\hbar\omega/kT} - 1}$$

should, of course, approach kT as $\omega \to 0$ or as $T \to \infty$. See if you can prove that it does—learn how to do the mathematics." Do the mathematics.

59 Find $\lim_{x \to 0} \left(\dfrac{1 + 2^x}{2} \right)^{1/x}$.

60 In Fig. 6.41, the unit circle is centered at the origin; BQ is a vertical tangent line; the length of BQ is the same as the arc length BP. Prove that the x coordinate of R approaches -2 as $P \to B$.

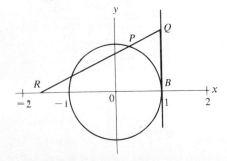

Figure 6.41

61 *Warning:* As Albert Einstein observed, "Common sense is the deposit of prejudice laid down in the mind before the age of 18." Exercise 30 of Sec. 2.5 asked the reader to guess a certain limit. Now that limit will be computed. In Fig. 6.42, which shows a circle, let $f(\theta) =$ area of triangle ABC and let $g(\theta) =$ area of the shaded region formed by deleting triangle OAC from the sector OBC. Clearly, $0 < f(\theta) < g(\theta)$.

(a) What would you guess is the value of $\lim_{\theta \to 0} f(\theta)/g(\theta)$?

(b) Find $\lim_{\theta \to 0} f(\theta)/g(\theta)$.

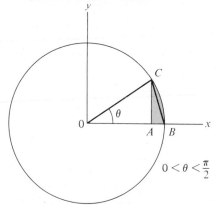

$0 < \theta < \dfrac{\pi}{2}$

Figure 6.42

62 Linus proposes this proof for Theorem 1. Since

$$\lim_{x \to a^+} f(x) = 0 \quad \text{and} \quad \lim_{x \to a^+} g(x) = 0,$$

I will define $f(a) = 0$ and $g(a) = 0$. Next I consider $x > a$ but near a. I now have continuous functions f and g defined on the closed interval $[a, x]$ and differentiable on the open interval (a, x). So, using the mean-value theorem, I conclude that there is a number c, $a < c < x$, such that

$$\dfrac{f(x) - f(a)}{x - a} = f'(c) \quad \text{and} \quad \dfrac{g(x) - g(a)}{x - a} = g'(c).$$

Since $f(a) = 0$ and $g(a) = 0$, these equations tell me that

$$f(x) = (x - a)f'(c) \quad \text{and} \quad g(x) = (x - a)g'(c).$$

Thus $\quad \dfrac{f(x)}{g(x)} = \dfrac{f'(c)}{g'(c)}.$

Hence $\quad \lim_{x \to a^+} \dfrac{f(x)}{g(x)} = \lim_{n \to a^+} \dfrac{f'(c)}{g'(c)}$
$\qquad\qquad = L.$

Alas, Linus made one error. What is it?

63 (Economics) In Eugene Silberberg, *The Structure of Economics*, McGraw-Hill, New York, this argument appears:

Consider the production function

$$y = k(\alpha x_1^{-\rho} + (1 - \alpha)x_2^{-\rho})^{-1/\rho},$$

where k, α, x_1, x_2 are positive constants and $\alpha < 1$. Taking limits as $\rho \to 0^+$, we find that

$$\lim_{\rho \to 0^+} y = k x_1^{\alpha} x_2^{1-\alpha},$$

which is the Cobb-Douglas function, as expected.
Fill in the details.

6.9 The hyperbolic functions and their inverses

Certain combinations of the exponential functions e^x and e^{-x} occur often enough in differential equations and engineering to be given names. This section defines these so-called *hyperbolic functions* and obtains their basic properties. Since the letter x will be needed later for another purpose, we will use the letter t when writing the two preceding exponentials, namely e^t and e^{-t}.

THE HYPERBOLIC COSINE

DEFINITION *The hyperbolic cosine.* Let t be a real number. The hyperbolic cosine of t, denoted cosh t, is given by the formula

Pronounced as written, "cosh", rhyming with "gosh"

$$\cosh t = \frac{e^t + e^{-t}}{2}.$$

To graph cosh t, note first that

$$\cosh(-t) = \frac{e^{-t} + e^{-(-t)}}{2}$$

$$= \frac{e^{-t} + e^t}{2}$$

$$= \cosh t.$$

Since $\cosh(-t) = \cosh t$, the cosh function is even, and so its graph is symmetric with respect to the vertical axis. Furthermore, cosh t is the sum of two terms,

$$\cosh t = \frac{e^t}{2} + \frac{e^{-t}}{2}.$$

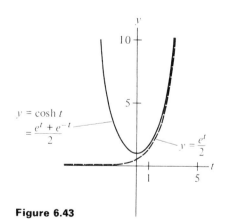

Figure 6.43

For $|t| \to \infty$, the graph of $y = \cosh t$ is asymptotic to the graph of $y = \dfrac{e^t}{2}$ or $y = \dfrac{e^{-t}}{2}$.

As $t \to \infty$, the second term $e^{-t}/2$ approaches 0. Thus for $t > 0$ and large, the graph of cosh t is just a little above that of $e^t/2$. This information, together with the fact that $\cosh 0 = (e^0 + e^{-0})/2 = 1$, is the basis for Fig. 6.43.

The curve $y = \cosh t$ in Fig. 6.43 is called a *catenary* (from the Latin *catena*, meaning "chain"). A chain or rope, suspended from its ends, forms a curve that is part of a catenary.

THE HYPERBOLIC SINE

The other hyperbolic function used in practice is defined as follows.

DEFINITION   *The hyperbolic sine.* Let t be a real number. The hyperbolic sine of t, denoted sinh t, is given by the formula

"sinh" is pronounced "sinch," rhyming with "pinch."

$$\sinh t = \frac{e^t - e^{-t}}{2}.$$

sinh t is an odd function.

It is a straightforward matter to check that $\sinh 0 = 0$ and $\sinh(-t) = -\sinh t$, so that the graph of sinh t is symmetric with respect to the origin. Moreover it lies below the graph of $e^t/2$. However, as $t \to \infty$, the two graphs approach each other since $e^{-t}/2 \to 0$ as $t \to \infty$. Figure 6.44 shows the graph of sinh t.

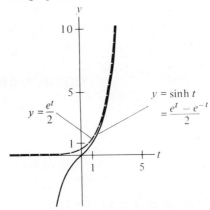

Figure 6.44

Note the contrast between sinh t and sin t. As t becomes large, the hyperbolic sine becomes large, $\lim_{t \to \infty} \sinh t = \infty$ and $\lim_{t \to -\infty} \sinh t = -\infty$. There is a similar contrast between cosh t and cos t. While the trigonometric functions are periodic, the hyperbolic functions are not.

Why they are called hyperbolic functions

Example 1 shows why the functions $(e^t + e^{-t})/2$ and $(e^t - e^{-t})/2$ are called hyperbolic.

EXAMPLE 1   Show that for any real number t the point

$$x = \cosh t, \quad y = \sinh t$$

lies on the hyperbola $x^2 - y^2 = 1$.

SOLUTION   Compute $\cosh^2 t - \sinh^2 t$ and see whether it equals 1. We have

$$\cosh^2 t - \sinh^2 t = \left(\frac{e^t + e^{-t}}{2}\right)^2 - \left(\frac{e^t - e^{-t}}{2}\right)^2$$

$$= \frac{e^{2t} + 2e^t e^{-t} + e^{-2t}}{4} - \frac{e^{2t} - 2e^t e^{-t} + e^{-2t}}{4}$$

$$= \frac{2 + 2}{4} \quad \text{cancellation}$$

$$= 1.$$

6.9 THE HYPERBOLIC FUNCTIONS AND THEIR INVERSES

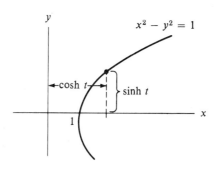

Figure 6.45

$\cosh^2 t - \sinh^2 t = 1$ Observe that, since $\cosh t \geq 1$, the point $(\cosh t, \sinh t)$ is on the right half of the hyperbola $x^2 - y^2 = 1$, as shown in Fig. 6.45. ∎

DERIVATIVES OF THE HYPERBOLIC FUNCTIONS

The four other hyperbolic functions, namely, the hyperbolic tangent, the hyperbolic secant, the hyperbolic cotangent, and the hyperbolic cosecant, are defined as follows:

These are defined like the circular functions, tan, sec, cot, csc.

$$\tanh t = \frac{\sinh t}{\cosh t} \qquad \text{sech } t = \frac{1}{\cosh t}$$

$$\coth t = \frac{\cosh t}{\sinh t} \qquad \text{csch } t = \frac{1}{\sinh t}$$

Each can be expressed in terms of exponentials. For instance,

$$\tanh t = \frac{(e^t - e^{-t})/2}{(e^t + e^{-t})/2} = \frac{e^t - e^{-t}}{e^t + e^{-t}}.$$

The derivatives of the six hyperbolic functions can be computed directly. For instance,

$$(\cosh t)' = \frac{(e^t + e^{-t})'}{2} = \frac{e^t - e^{-t}}{2} = \sinh t.$$

The following table lists the six derivatives.

Function	Derivative
$\cosh t$	$\sinh t$
$\sinh t$	$\cosh t$
$\tanh t$	$\text{sech}^2 t$
$\coth t$	$-\text{csch}^2 t$
$\text{sech } t$	$-\text{sech } t \tanh t$
$\text{csch } t$	$-\text{csch } t \coth t$

Notice that the formulas, except for the minus signs, are like those for the derivatives of the trigonometric functions.

THE INVERSE HYPERBOLIC FUNCTIONS

Inverse hyperbolic functions appear on some calculators and in integral tables. Just as the hyperbolic functions are expressed in terms of the exponential function, each inverse hyperbolic function can be expressed in terms of a logarithm. They provide useful antiderivatives as well as solutions to some differential equations.

Consider the inverse of sinh t first. Since sinh t is increasing, it is one-to-one; there is no need to restrict its domain. To find its inverse, it is necessary to solve the equation

$$x = \text{sinh } t$$

Finding the inverse of the hyperbolic sine

for t as a function of x. The steps are straightforward:

$$x = \frac{e^t - e^{-t}}{2}$$

$$2x = e^t - \frac{1}{e^t}$$

$$2xe^t = (e^t)^2 - 1$$

or
$$(e^t)^2 - 2xe^t - 1 = 0. \tag{1}$$

Equation (1) is quadratic in the unknown e^t. By the quadratic formula,

$$e^t = \frac{2x \pm \sqrt{(2x)^2 + 4}}{2}$$

$$= x \pm \sqrt{x^2 + 1}.$$

Since $e^t > 0$ and $\sqrt{x^2 + 1} > x$, the $+$ sign is kept and the $-$ sign rejected. Thus

$$e^t = x + \sqrt{x^2 + 1}$$

or
$$t = \ln(x + \sqrt{x^2 + 1}).$$

Consequently, the inverse of the function sinh t is given by the formula

Formula for $\sinh^{-1} x$
$$\sinh^{-1} x = \ln(x + \sqrt{x^2 + 1}).$$

Computation of $\tanh^{-1} x$ is a little different. Since the derivative of tanh t is $\text{sech}^2 t$, the function tanh t is increasing and has an inverse. However, $|\tanh t| < 1$, as will be shown in a moment, and so the inverse function will be defined only for $|x| < 1$. To see that $|\tanh t| < 1$, observe that

$$\tanh t = \frac{e^t - e^{-t}}{e^t + e^{-t}}. \tag{2}$$

6.9 THE HYPERBOLIC FUNCTIONS AND THEIR INVERSES

$-1 < \tanh t < 1$ so tanh is bounded, and cosh and sinh are unbounded (the exact opposite of their trig analogs).

The numerator in (2) is always less than the denominator. However, as $t \to \infty$, $e^{-t} \to 0$; thus $\lim_{t \to \infty} \tanh t = 1$. It is not hard to show that $\tanh(-t) = -\tanh t$. Thus $\lim_{t \to -\infty} \tanh t = -1$.

We find the inverse of $x = \tanh t$ as follows:

$$x = \tanh t = \frac{e^t - e^{-t}}{e^t + e^{-t}}$$

$$= \frac{e^t - (1/e^t)}{e^t + (1/e^t)}.$$

So $\quad x = \dfrac{e^{2t} - 1}{e^{2t} + 1} \quad$ multiplying both numerator and denominator by e^t.

Thus $\quad xe^{2t} + x = e^{2t} - 1$

$$1 + x = e^{2t}(1 - x)$$

$$e^{2t} = \frac{1 + x}{1 - x}$$

$$2t = \ln\left(\frac{1 + x}{1 - x}\right)$$

$$t = \frac{1}{2} \ln\left(\frac{1 + x}{1 - x}\right).$$

Formula for $\tanh^{-1} x$ Thus $\quad \tanh^{-1} x = \dfrac{1}{2} \ln\left(\dfrac{1 + x}{1 - x}\right) \quad |x| < 1.$

Inverses of the other four hyperbolic functions are computed similarly. Their formulas are included in the following table.

The derivatives are found as in Example 2.

cosh x and sech x are one-to-one for $x \geq 0$

Function	Formula	Derivative	Domain
$\cosh^{-1} x$	$\ln(x + \sqrt{x^2 - 1})$	$\dfrac{1}{\sqrt{x^2 - 1}}$	$x \geq 1$
$\sinh^{-1} x$	$\ln(x + \sqrt{x^2 + 1})$	$\dfrac{1}{\sqrt{x^2 + 1}}$	x axis
$\tanh^{-1} x$	$\dfrac{1}{2} \ln\left(\dfrac{1 + x}{1 - x}\right)$	$\dfrac{1}{1 - x^2}$	$|x| < 1$
$\coth^{-1} x$	$\dfrac{1}{2} \ln\left(\dfrac{x + 1}{x - 1}\right)$	$\dfrac{1}{1 - x^2}$	$|x| > 1$
$\operatorname{sech}^{-1} x$	$\ln\left(\dfrac{1 + \sqrt{1 - x^2}}{x}\right)$	$\dfrac{-1}{x\sqrt{1 - x^2}}$	$0 < x \leq 1$
$\operatorname{csch}^{-1} x$	$\ln\left(\dfrac{1}{x} + \sqrt{1 + \dfrac{1}{x^2}}\right)$	$\dfrac{-1}{|x|\sqrt{1 + x^2}}$	$x \neq 0$

EXAMPLE 2 Verify the formula

$$\int \frac{dx}{a^2 - x^2} = \frac{1}{a} \tanh^{-1} \frac{x}{a} + C, \qquad a > 0, |x| < a,$$

included in many integral tables.

SOLUTION Let

$$y = \frac{1}{a} \tanh^{-1} \frac{x}{a} + C$$

$$= \frac{1}{a} \frac{1}{2} \ln \left(\frac{1 + x/a}{1 - x/a} \right) + C \qquad \text{from preceding table}$$

$$= \frac{1}{2a} \ln \left(\frac{a + x}{a - x} \right) + C \qquad \text{algebra}$$

$$= \frac{1}{2a} [\ln (a + x) - \ln (a - x)] + C. \qquad \ln (A/B) = \ln A - \ln B$$

Differentiation then yields

$$y' = \frac{1}{2a} \left(\frac{1}{a + x} + \frac{1}{a - x} \right)$$

$$= \frac{1}{2a} \cdot \frac{2a}{a^2 - x^2}$$

$$= \frac{1}{a^2 - x^2}. \quad \blacksquare$$

Exercises

Exercises 1 to 14 concern the hyperbolic functions.
1. Show that $\cosh t + \sinh t = e^t$.
2. Show that $\cosh t - \sinh t = e^{-t}$.

In Exercises 3 to 6 differentiate the given functions and express the derivatives in terms of hyperbolic functions.
3. $\sinh t$
4. $\tanh t$
5. $\operatorname{sech} t$
6. $\coth t$

In Exercises 7 to 14 differentiate the functions. Be sure to use the chain rule.
7. $\cosh 3x$
8. $\sinh 5x$
9. $\tanh \sqrt{x}$
10. $\operatorname{sech} (\ln x)$
11. $e^{3x} \sinh x$
12. $\dfrac{(\tanh 2x)(\operatorname{sech} 3x)}{\sqrt{1 + 2x}}$
13. $(\cosh 4x)(\coth 5x)(\operatorname{csch} x^2)$
14. $\dfrac{(\coth 5x)^{5/2}(\tanh 3x)^{1/3}}{3x + 4 \cos x}$ (Call it y and use logarithmic differentiation.)

Exercises 15 to 22 concern the inverse hyperbolic functions.

In Exercises 15 to 18 obtain the given formulas, using the technique illustrated in the text.
15. $\cosh^{-1} x = \ln (x + \sqrt{x^2 - 1}), x \geq 1$
16. $\operatorname{sech}^{-1} x = \ln \left(\dfrac{1 + \sqrt{1 - x^2}}{x} \right), 0 < x \leq 1$
17. $\coth^{-1} x = \dfrac{1}{2} \ln \left(\dfrac{x + 1}{x - 1} \right), |x| > 1$
18. $\operatorname{csch}^{-1} x = \ln \left(\dfrac{1}{x} + \sqrt{1 + \dfrac{1}{x^2}} \right), x \neq 0$

In Exercises 19 to 22 verify the integration formulas by differentiation (as in Example 2).

19. $\int \dfrac{dx}{\sqrt{x^2 - 1}} = \cosh^{-1} x + C, x > 1$
20. $\int \dfrac{dx}{\sqrt{x^2 + 1}} = \sinh^{-1} x + C$
21. $\int \dfrac{dx}{x\sqrt{1 - x^2}} = -\operatorname{sech}^{-1} x + C, 0 < x < 1$

22 $\int \frac{dx}{x\sqrt{1+x^2}} = -\operatorname{csch}^{-1} x + C, \quad x > 0$

∎

In Exercises 23 to 29 use the definitions of the hyperbolic functions to verify the given identities.

23 $\tanh(-x) = -\tanh x$
24 $\operatorname{sech}^2 x + \tanh^2 x = 1$
25 (a) $\cosh(x+y) = \cosh x \cosh y + \sinh x \sinh y$;
   (b) $\sinh(x+y) = \sinh x \cosh y + \cosh x \sinh y$.
26 $\tanh(x+y) = \dfrac{\tanh x + \tanh y}{1 + \tanh x \tanh y}$.
27 (a) $\cosh(x-y) = \cosh x \cosh y - \sinh x \sinh y$;
   (b) $\sinh(x-y) = \sinh x \cosh y - \cosh x \sinh y$.
28 (a) $\cosh 2x = \cosh^2 x + \sinh^2 x$;
   (b) $\sinh 2x = 2 \sinh x \cosh x$.
29 (a) $2 \sinh^2(x/2) = \cosh x - 1$;
   (b) $2 \cosh^2(x/2) = \cosh x + 1$.
30 At what angle does the graph of $y = \sinh x$ cross the x axis?
31 (Calculator) (a) Compute $\cosh t$ and $\sinh t$ for $t = -3, -2, -1, 0, 1, 2,$ and 3. (b) Plot the seven points $(\cosh t, \sinh t)$ given by (a). They should lie on the hyperbola $x^2 - y^2 = 1$.
32 (Calculator) (a) Compute $\cosh t$ and $e^t/2$ for $t = 0, 1, 2, 3,$ and 4. (b) Using the data in (a) graph $\cosh t$ and $e^t/2$ relative to the same axes.
33 (Calculator) (a) Compute $\tanh x$ for $x = 0, 1, 2,$ and 3. (b) Using the data in (a) and the fact that $\tanh(-x) = -\tanh x$, graph $y = \tanh x$.

∎ ∎

34 Some integral tables contain the formulas

$$\int \frac{dx}{\sqrt{ax+b}\sqrt{cx+d}} = \begin{cases} \dfrac{2}{\sqrt{-ac}} \tan^{-1} \sqrt{\dfrac{-c(ax+b)}{a(cx+d)}}, & ac < 0 \\ \dfrac{2}{\sqrt{ac}} \tanh^{-1} \sqrt{\dfrac{c(ax+b)}{a(cx+d)}}, & ac > 0 \end{cases}$$

The first formula is used if a and c have opposite signs, the second if they have the same signs. Check each of the formulas by differentiation.

35 One of the applications of hyperbolic functions is to the study of motion in which the resistance of the medium is proportional to the square of the velocity. Suppose that a body starts from rest and falls x meters in t seconds. Let g (a constant) be the acceleration due to gravity. It can be shown that there is a constant V such that

$$x = \frac{V^2}{g} \ln \cosh \frac{gt}{V}.$$

(a) Find the velocity $v(t) = dx/dt$ as a function of t.
(b) Show that $\lim_{t \to \infty} v(t) = V$.
(c) Compute the acceleration dv/dt as a function of t.
(d) Show that the acceleration equals $g - g(v/V)^2$.
(e) What is the limit of the acceleration as $t \to \infty$?

36 Two particles repel each other with a force proportional to the distance x between them. There is thus a constant k such that

$$\frac{d^2x}{dt^2} = k^2 x.$$

(a) Show that, for any constants A and B, the function $x = A \cosh kt + B \sinh kt$ satisfies the given differential equation.
(b) If at time $t = 0$ the particles are a distance a apart and motionless, show that $A = a$ and $B = 0$. (Thus $x = a \cosh kt$.)

6.10 Exponential functions defined in terms of logarithms (optional)

In Sec. 6.3 the function $L(x) = \int_1^x dt/t$, $x > 0$, was introduced and shown to have the familiar properties of logarithms. It was pointed out that *if* all the gaps in the development of $\ln x$ were filled in, then $L(x)$ could be shown to coincide with $\ln x$. The number e was defined in Sec. 6.3 by the condition, $L(e) = 1$, that is

$$\int_1^e \frac{1}{x} dx = 1.$$

The present section introduces a function E, which will turn out to be the exponential function to the base e. However, it is important in this section to make no use at all of b^x, as defined in Appendix C, since the object of this section is to construct these functions with the fewest assumptions.

THE FUNCTION E

Recall that the function L is increasing for $x > 0$. It thus has an inverse.

DEFINITION *The function E.* The inverse of the function L will be denoted E.

Since $L(1) = 0$ and E is the inverse of L, $E(0) = 1$, just like an exponential function. Moreover, since $L(e) = 1$, it follows that $E(1) = e$. Keep in mind that $E(L(x)) = x$ and $L(E(x)) = x$. Theorem 1 shows that the function E has the basic property of an exponential function.

THEOREM 1 $E(x + y) = E(x)E(y)$.

PROOF Write x as $L(u)$ and y as $L(v)$. Then

Every number x is the log of some number.

$$E(x + y) = E(L(u) + L(v))$$

$$ = E(L(uv)) \qquad L(uv) = L(u) + L(v)$$

$$ = uv \qquad E \text{ and } L \text{ are inverses}$$

$$ = E(x)E(y). \blacksquare$$

Next the exponential function b^x will be defined for $b > 0$. In the traditional approach, there is the identity

$$b^x = e^{x \ln b}.$$

This suggests how to define the function b^x in the present approach.

DEFINITION *The exponential function b^x.* If $b > 0$, define b^x to be

$$E(xL(b)).$$

THEOREM 2 If b is positive, then

$$b^0 = 1, \tag{1}$$

$$b^1 = b, \tag{2}$$

$$b^{x+y} = b^x b^y. \tag{3}$$

PROOF (1) $b^0 = E(0 \cdot L(b)) = E(0) = 1$.

(2) $b^1 = E(1 \cdot L(b)) = E(L(b)) = b$.

(3) $b^{x+y} = E((x + y)L(b)) \qquad$ definition of b^{x+y}

6.10 EXPONENTIAL FUNCTIONS DEFINED IN TERMS OF LOGARITHMS (OPTIONAL)

$$= E(xL(b) + yL(b)) \quad \text{algebra}$$
$$= E(xL(b))E(yL(b)) \quad \text{Theorem 1}$$
$$= b^x b^y. \quad \text{definition of } b^x \text{ and } b^y \quad \blacksquare$$

According to the definition of b^x, when the base b is chosen to be e, we have

$$e^x = E(xL(e)) = E(x \cdot 1) = E(x).$$

So e^x is another name for $E(x)$. But do not think of it as the function e^x that was built up laboriously in Appendix C first for integer x, then for rational x, and finally, by limits, for irrational x.

THEOREM 3 $L(b^x) = xL(b)$.

PROOF
$$b^x = E(xL(b)) \quad \text{definition of } b^x$$
$$L(b^x) = L(E(xL(b)))$$
$$= xL(b) \quad L \text{ and } E \text{ are inverses of each other.} \quad \blacksquare$$

THEOREM 4 $(b^x)^y = b^{xy}$.

PROOF Since L is one-to-one it suffices to show that $L(b^{xy}) = L((b^x)^y)$. This will be established by three applications of Theorem 3:

$$L(b^{xy}) = xy\, L(b) \quad \text{Theorem 3}$$
while
$$L((b^x)^y) = y\, L(b^x) \quad \text{Theorem 3}$$
$$= y(x\, L(b)) \quad \text{Theorem 3}$$
$$= xy\, L(b) \quad \text{algebra}$$

Thus $L(b^{xy}) = L((b^x)^y)$ and the proof is complete. $\blacksquare$

THE FUNCTION $\log_b x$

If $b \neq 1$, the function b^x is one-to-one. To show this, assume that

$$b^{x_1} = b^{x_2}, \tag{4}$$

and show that $x_1 = x_2$. By the definition of b^x, (4) can be written

$$E(x_1 L(b)) = E(x_2 L(b)).$$

Since E is one-to-one,

$$x_1 L(b) = x_2 L(b).$$

Since $b \neq 1$, $L(b) \neq 0$, and cancellation of $L(b)$ yields $x_1 = x_2$.
Since b^x is one-to-one, it has an inverse.

DEFINITION $\log_b x$ is the inverse of the function b^x.

Theorem 5 shows that $\log_b x$ has the traditional properties of a logarithm.

THEOREM 5 If $b > 0$, $b \neq 1$, then

$$\log_b 1 = 0, \tag{5}$$
$$\log_b b = 1, \tag{6}$$
$$\log_b xy = \log_b x + \log_b y, \quad x, y > 0, \tag{7}$$
$$\log_b x^y = y \log_b x, \quad x > 0. \quad \blacksquare \tag{8}$$

The proof is left to the reader.

THE DERIVATIVE OF THE FUNCTION E

Since E is the inverse of the increasing differentiable function L, it is also differentiable. To find its derivative, let

$$y = E(x); \tag{9}$$

hence
$$x = L(y). \tag{10}$$

Implicit differentiation of (10) with respect to x yields

$$1 = \frac{1}{y}\frac{dy}{dx};$$

thus
$$\frac{dy}{dx} = y.$$

In short,
$$\frac{d}{dx}(E(x)) = E(x). \tag{11}$$

This proves the next theorem.

THEOREM 6 The derivative of E is E:

$$\frac{d}{dx}(E(x)) = E(x). \quad \blacksquare$$

With the help of Theorem 6 it is not hard to prove the following generalization.

THEOREM 7 $\dfrac{d}{dx}(b^x) = b^x L(b). \quad \blacksquare$

The proofs of Theorem 7 and of the following theorem are left to the reader.

THEOREM 8 $\dfrac{d}{dx}(\log_b x) = \dfrac{\log_b e}{x}.$ ∎

THE LIMIT OF $(1+x)^{1/x}$ AS $x \to 0$

It is now possible to show that

$$\lim_{x \to 0} (1+x)^{1/x} = e.$$

(Keep in mind that e is defined in Sec. 6.3 by the demand that $L(e) = 1$, that is, $\int_1^e dt/t = 1$.)

THEOREM 9 $\lim_{x \to 0} (1+x)^{1/x} = e.$

PROOF The derivative of $L(x)$ at $x = 1$ is $1/1$ or 1. On the other hand, by the definition of the derivative,

$$L'(1) = \lim_{x \to 0} \frac{L(1+x) - L(1)}{x}$$

$$= \lim_{x \to 0} \frac{L(1+x)}{x} \qquad \text{since } L(1) = 0$$

$$= \lim_{x \to 0} \frac{1}{x} L(1+x) \qquad \text{algebra}$$

$$= \lim_{x \to 0} L((1+x)^{1/x}). \qquad \text{Theorem 3}$$

Since

$$1 = \lim_{x \to 0} L((1+x)^{1/x}),$$

$$E(1) = E\left(\lim_{x \to 0} L((1+x)^{1/x})\right).$$

See Example 4 of Sec. 2.6. Since E is continuous, E and lim can be switched; thus

$$E(1) = \lim_{x \to 0} E(L(1+x)^{1/x}). \tag{12}$$

Since $E(1) = e$, and E and L are inverse functions, (12) implies that

$$e = \lim_{x \to 0} (1+x)^{1/x},$$

as required. ∎

Theorem 9 shows that the definition of e in Sec. 6.1 as a limit is consistent with the definition in Sec. 6.3, where it is defined with the aid of a definite integral.

Exercises

In the exercises prove the given statements, using the definitions in Sec. 6.3 and this section.

1 Theorem 5. **2** Theorem 7. **3** Theorem 8.

6.S Summary

This chapter completed the task of computing the derivatives of all elementary functions. The notions of one-to-one functions and inverse functions were the keys to differentiating b^x and the inverse trigonometric functions.

In Sec. 6.6, a discussion of related rates, it was shown how to find the rate at which one quantity changes if the rates of change of the other quantities to which it is related by an equation are known.

Differential equations were introduced in Sec. 6.7. Separable differential equations were solved and applied to natural growth (and decay) and inhibited growth.

L'Hôpital's rule in Sec. 6.8 provides a tool for dealing with many limits that are of the form $\lim_{x \to a} f(x)/g(x)$, where $f(x)$ and $g(x)$ both approach 0 or both approach ∞. Some other limits can be transformed to the type covered by l'Hôpital's rule.

Section 6.9, which considered hyperbolic functions and their inverses, is specialized and intended as a reference. It will not be reviewed in this summary.

Sections 6.3 and 6.10 offered an alternative approach to logarithms and exponentials, one that begins with the definition of the natural logarithm as an integral, $\int_1^x dt/t$. It has several advantages over the approach in Appendix C. For instance, b^x for $b > 0$ is defined for all x at one blow, the differentiability of logarithms and exponentials is easily established, and the existence of $\lim_{x \to 0} (1 + x)^{1/x}$ is easily demonstrated. These sections will not be reviewed further in the summary.

VOCABULARY AND SYMBOLS

$e = \lim_{x \to 0} (1 + x)^{1/x} \approx 2.718$
$\ln x \,(= \log_e x)$
one-to-one
inverse function f^{-1} ("INV" on some calculators)
inverse trigonometric function   arcsin x, $\sin^{-1} x$, etc.
differential equation   D.E.
solution of a differential equation
order of a differential equation
separable differential equation
natural growth or decay $dP/dt = kP$
relative rate of growth per unit time r
growth constant k (negative k for decay)
doubling time t_2
half-life $t_{1/2}$
inhibited growth $dP/dt = kP(1 - P/M)$
l'Hôpital's rule
hyperbolic functions

KEY FACTS The following table lists the derivatives found in Chap. 3 and in Secs. 6.2, 6.4, and 6.5.

Formulas

Function	Derivative
x^a	ax^{a-1}
$(u(x))^a$	$a(u(x))^{a-1}u'(x)$
$\sqrt{x}$	$\dfrac{1}{2\sqrt{x}}$
$\sin x$	$\cos x$
$\cos x$	$-\sin x$ (remember the $-$)
$\tan x$	$\sec^2 x$
$\cot x$	$-\csc^2 x$ (remember the $-$)
$\sec x$	$\sec x \tan x$
$\csc x$	$-\csc x \cot x$ (remember the $-$)
$\ln x$	$\dfrac{1}{x}, x > 0$
$\ln|x|$	$\dfrac{1}{x}, x \neq 0$
e^x	e^x
b^x	$(\ln b)\, b^x$
$\sin^{-1} x$	$\dfrac{1}{\sqrt{1-x^2}}$
$\tan^{-1} x$	$\dfrac{1}{1+x^2}$
$\sec^{-1} x$	$\dfrac{1}{|x|\sqrt{x^2-1}}$

It is not necessary to memorize the formulas for the derivatives of $\cos^{-1}$, $\cot^{-1}$, and $\csc^{-1}$. (They are obtained by putting minus signs in front of the last three formulas in the list.)

In a related-rate problem find an equation linking the variables and then differentiate implicitly with respect to time. To find acceleration, differentiate again.

A separable D.E. has the form

$$\frac{dy}{dx} = \frac{g(x)}{h(y)}.$$

(It may also have the form $dy/dx = g(x)h(y)$ or $dy/dx = h(y)/g(x)$.) The equation is solved by bringing all x's to one side and all y's to the other (separating the variables), then integrating:

$$\int h(y)\, dy = \int g(x)\, dx + C.$$

> ## TECHNIQUES
>
> $$(f+g)' = f' + g' \qquad (f-g)' = f' - g'$$
>
> $$(cf)' = cf' \qquad \left(\frac{f}{c}\right)' = \frac{f'}{c} \quad c \text{ constant}$$
>
> $$(fg)' = fg' + gf' \qquad \left(\frac{f}{g}\right)' = \frac{gf' - fg'}{g^2}$$
>
> chain rule (the most important technique):
>
> If $y = f(x)$ and $u = g(x)$,
>
> then $\dfrac{dy}{dx} = \dfrac{dy}{du} \cdot \dfrac{du}{dx}$.
>
> logarithmic differentiation   implicit differentiation

L'Hôpital's rule concerns the behavior of $f(x)/g(x)$ in case numerator and denominator both approach 0 or both approach ∞. Then, if

$$\lim_{x \to a} \frac{f'(x)}{g'(x)} \text{ exists,}$$

$$\lim_{x \to a} \frac{f(x)}{g(x)} = \lim_{x \to a} \frac{f'(x)}{g'(x)}.$$

The forms "zero times infinity," "one to the infinity," and "zero to the zero" can be reduced to l'Hôpital's form. In the last two cases: take logs, find the limit of the logs, and then be sure to complete the problem by finding the original limits.

Guide quiz on chap. 6 (except Secs. 6.3, 6.9, and 6.10)

1 Differentiate: (a) $e^{\sin^{-1} 3x}$
   (b) $\sin((\tan^{-1} 4x)^2)$ (c) $(x^3 + 5x)x^{\sqrt{x}}$
   (d) $\dfrac{\tan 5x}{\sec^{-1} 2x}$ (e) $\left(\dfrac{x^{-3}(\ln(x^2+1))^8}{\sqrt[3]{\cos^2 2x}}\right)^5$

2 Differentiate and simplify your answer.
   (a) $\dfrac{e^{ax}}{a^2 + b^2}(a \cos bx + b \sin bx)$
   (b) $x \sin^{-1} ax + \dfrac{1}{a}\sqrt{a^2 - x^2}$
   (c) $x \tan^{-1} ax - \dfrac{1}{2a} \ln(1 + a^2 x^2)$
   (d) $x \sec^{-1} ax - \dfrac{1}{a} \ln(ax + \sqrt{a^2 x^2 - 1})$

3 The quantities x and y, which are differentiable functions of t, are related by the equation

$$xy + e^y = e + 1$$

When $t = 0$, $x = 1$, $y = 1$, $\dot{x} = 2$, and $\ddot{x} = 3$. Find $\dot{y}$ and $\ddot{y}$ when $t = 0$.

4 A load of concrete M hangs from a rope which passes over a pulley B. (See Fig. 6.46.) A construction worker at C pulls the rope as she walks away at 5 feet per second. The level of the pulley is 10 feet higher than her hand. At what rate is the load rising when the length $\overline{BC}$ is (a) 15 feet? (b) 100 feet?

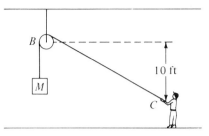

Figure 6.46

5. Solve: (a) $\dfrac{dy}{dx} = e^{-2y} x^3$ (b) $\dfrac{dy}{dx} = \dfrac{4y^2 + 1}{y}$

6. Radon has a half-life of 3.825 days. How long does it take for radon to diminish to only 10 percent of its original amount?

7. Examine these limits: (a) $\lim\limits_{x \to 0} \dfrac{\sin 2x}{\tan^{-1} 3x}$

    (b) $\lim\limits_{x \to \infty} \dfrac{\sin 2x}{\tan^{-1} 3x}$ (c) $\lim\limits_{x \to (\pi/2)^-} (\sec x - \tan x)$

    (d) $\lim\limits_{x \to 0} (1 - \cos 2x)^x$ (e) $\lim\limits_{x \to (\pi/2)^-} \dfrac{\int_0^x \tan^2 \theta \, d\theta}{\tan \theta}$

8. Arrange these functions in order of increasing size for large x: x^3, $\ln x$, 1.001^x, 2^x, $\log_{10} x$.

9. Let $f(x) = (\ln x)/x$.
    (a) What is the domain of f?
    (b) Find any x intercepts of f.
    (c) Find any critical numbers of f.
    (d) Use the second-derivative test to find whether there is a local maximum or minimum at any critical number.
    (e) For which x is f increasing? Decreasing?
    (f) Does f have a global minimum? A global maximum?
    (g) Find $\lim_{x \to 0^+} f(x)$.
    (h) Find $\lim_{x \to \infty} f(x)$.
    (i) Graph f.

10. (a) What is meant by "e"?
    (b) What is e to three decimal places?

11. Assume that a calculator has a $\log$ key ($\log_{10}$) but no $\ln$ key. How could you use the $\log$ key to find $\ln 7$?

Review exercises for chap. 6

The next chapter, which treats the problem of finding an antiderivative, assumes a mastery of the formal computation of derivatives. Do as many of the first 86 exercises as you can find time for.
In Exercises 1 to 86 differentiate and simplify. (Assume that the input for any logarithm is positive.)

1. $\sqrt{1 + x^3}$
2. 5^{x^2}
3. $\sqrt{x}$
4. $\dfrac{1}{\sqrt{x}}$
5. $\cos^2 3x$
6. $\sin^{-1} 3x$
7. $\sqrt{x^3}$
8. $\dfrac{1}{3} \tan^{-1} \dfrac{x}{3}$
9. $\sqrt{\sin x}$
10. $\dfrac{\cos 5x}{x^2}$
11. $\cot x^2$
12. $\dfrac{1}{x^3}$
13. $\dfrac{e^{x^2}}{2x}$
14. $x^2 \sin 3x$
15. $x^{5/6} \sin^{-1} x$
16. $e^{\sqrt{x}}$
17. $x^2 e^{3x}$
18. $\sin^2 2x$
19. $\ln (\sec 3x + \tan 3x)$
20. $\dfrac{1}{5} \tan^{-1} \dfrac{x}{5}$
21. $\cos \sqrt{x}$
22. $e^{-x} \tan x^2$
23. $\ln (\sec x + \tan x)$
24. $3 \cot 5x + 5 \csc 3x$
25. $\dfrac{1}{\sqrt{6 + 3x^2}}$
26. $\ln (\sin 2x)$
27. $\sqrt{\dfrac{2}{15}(5x + 7)^3}$
28. $x \cos 5x + \sin 5x$
29. $\dfrac{x}{3} - \dfrac{4}{9} \ln (3x + 4)$
30. $\dfrac{\sqrt{4 - 9x^2}}{x}$
31. $(1 + x^2)^5 \sin 3x$
32. $\cos [\log_{10} (3x + 1)]$
33. $(2x^3 - 2x + 5)^4$
34. $(\sin 3x)^5$
35. $(1 + 2x)^5 \cos 3x$
36. $\dfrac{\sin 3x}{(x^2 + 1)^5}$
37. $\cos 3x \sin 4x$
38. $\tan \sqrt{x}$
39. $\csc 3x^2$
40. $\dfrac{1}{\sqrt{1 - x^2}}$
41. $x \left(\dfrac{x^2}{1 + x} \right)^3$
42. $\left(\dfrac{1 + 2x}{1 + 3x} \right)^4$
43. $x \sec 3x$
44. $[\ln (x^2 + 1)]^3$
45. $\dfrac{(x^3 - 1)^3 (x^{10} + 1)^4}{(2x + 1)^5}$
46. $\dfrac{x}{\sqrt{1 - x^2}}$
47. $\ln (x + \sqrt{x^2 + 1})$
48. $\dfrac{\sin^3 2x}{x^2 + x}$
49. $e^{-x} \tan^{-1} x^2$

50. $\log_{10}(x^2+1)$
51. $2e^{\sqrt{x}}(\sqrt{x}-1)$
52. $\sin^{-1}\dfrac{x}{5}$
53. $\dfrac{\ln x}{x^2}$
54. $(x^2+1)\sin 2x$
55. $\dfrac{\sin^2 x}{\cos x}$
56. $x^5 - 2x + \ln(2x+3)$
57. $(\sec^{-1} 3x)x^2 \ln(1+x^2)$
58. $(\ln(1+x^2))^3$
59. $x - 2\ln(x-1) + \dfrac{1}{x+1}$
60. $\ln\sqrt{\dfrac{1+x^2}{1+x^3}}$
61. $\ln\left(\dfrac{1}{6x^2+3x+1}\right)$
62. $\ln(\sqrt{4+x}\sqrt[3]{x^2+1})$
63. $\ln\left[\dfrac{(5x+1)^3(6x+1)^2}{(2x+1)^4}\right]$
64. $\dfrac{1}{8}\left[\ln(2x+1) + \dfrac{2}{2x+1} - \dfrac{1}{2(2x+1)^2}\right]$
65. $\dfrac{1}{2}\left(x\sqrt{9-x^2} + 9\sin^{-1}\dfrac{x}{3}\right)$
66. $\dfrac{1}{8}\left[2x+3 - 6\ln(2x+3) - \dfrac{9}{2x+3}\right]$
67. $\tan 3x - 3x$
68. $\dfrac{1}{\sqrt{6}}\tan^{-1}\left(x\sqrt{\dfrac{2}{3}}\right)$
69. $\ln(x+\sqrt{x^2+25})$
70. $\dfrac{e^x(\sin 2x - 2\cos 2x)}{5}$
71. $x\sin^{-1} x + \sqrt{1-x^2}$
72. $x\tan^{-1} x - \tfrac{1}{2}\ln(1+x^2)$
73. $\tfrac{1}{3}\ln(\tan 3x + \sec 3x)$
74. $\tfrac{1}{6}\tan^2 3x + \tfrac{1}{3}\ln\cos 3x$
75. $-\tfrac{1}{3}\cos 3x + \tfrac{1}{9}\cos^3 3x$
76. $\tfrac{1}{8}e^{-2x}(4x^2+4x+2)$
77. $e^{3x}\sin^2 2x \tan x$
78. $\dfrac{1}{3}\ln\left(\dfrac{\sqrt{2x+3}-3}{\sqrt{2x+3}+3}\right)$
79. $\dfrac{x}{2}\sqrt{4x^2+3} + \dfrac{3}{4}\ln(2x+\sqrt{4x^2+3})$
80. $\sqrt{1+\sqrt[3]{x}}$
81. $\dfrac{x^3(x^4-x+3)}{(x+1)^2}$
82. $\dfrac{1}{1+\csc 5x}$
83. $(1+3x)^{x^2}$
84. $2^x 5^{x^2} 7^{x^3}$
85. $\dfrac{x\ln x}{(1+x^2)^5}$
86. $x^3 \cot^3\sqrt{x^3}$

87. (a) Fill in this table for the function $\log_2 x$:

x	$\tfrac{1}{8}$	$\tfrac{1}{4}$	$\tfrac{1}{2}$	1	2	4	8
$\log_2 x$							

(b) Plot the seven points in (a) and graph $\log_2 x$.
(c) What is $\lim_{x\to\infty} \log_2 x$?
(d) What is $\lim_{x\to 0^+} \log_2 x$?

88. (a) Use a differential to show that for small x, $\ln(1+x) \approx x$.
(b) Use (a) to estimate $\ln 1.1$.

89. (a) Use a differential to show that for small x, $\log_{10}(1+x) \approx 0.434 x$.
(b) Use (a) to estimate $\log_{10} 1.05$.

In Exercises 90 to 100 check the equations by differentiation. The letters a and b refer to constants.

90. $\displaystyle\int \ln ax\, dx = x\ln ax - x + C$

91. $\displaystyle\int x\ln ax\, dx = \dfrac{x^2}{2}\ln ax - \dfrac{x^2}{4} + C$

92. $\displaystyle\int x^2 \ln ax\, dx = \dfrac{x^3}{3}\ln ax - \dfrac{x^3}{9} + C$

93. $\displaystyle\int \dfrac{dx}{x\ln ax} = \ln(\ln ax) + C$

94. $\displaystyle\int \dfrac{dx}{\sin x} = \ln|\csc x - \cot x| + C$

95. $\displaystyle\int \tan ax\, dx = -\dfrac{1}{a}\ln|\cos ax| + C$

96. $\displaystyle\int \tan^3 ax\, dx = \dfrac{1}{2a}\tan^2 ax + \dfrac{1}{a}\ln|\cos ax| + C$

97. $\displaystyle\int \dfrac{dx}{\sin ax \cos ax} = \dfrac{1}{a}\ln|\tan ax| + C$

98. $\displaystyle\int \dfrac{dx}{\sqrt{ax^2+b}} = \dfrac{1}{\sqrt{a}}\ln|x\sqrt{a}+\sqrt{ax^2+b}| + C \ (a>0)$

99. $\displaystyle\int \dfrac{x\,dx}{(ax+b)^2} = \dfrac{b}{a^2(ax+b)} + \dfrac{1}{a^2}\ln|ax+b| + C$

100. $\displaystyle\int (\ln ax)^2\, dx = x(\ln ax)^2 - 2x\ln ax + 2x + C$

101 (a) Differentiate $\ln\left|\dfrac{1+x}{1-x}\right|$.

   (b) Find the area of the region under the curve $y = 1/(1-x^2)$ and above $[0, \tfrac{1}{2}]$.

102 Find the area of the region under the curve $y = x^3/(x^4 + 5)$ and above $[1, 2]$.

103 Differentiate and then simplify your answer:
$$\ln\left|\dfrac{\sqrt{ax+b}-b}{\sqrt{ax+b}+b}\right|.$$

104 Assume that $\log_{10} 2 \approx 0.30$ and $\log_{10} 3 \approx 0.48$. From this information estimate
   (a) $\log_{10} 4$
   (b) $\log_{10} 5$
   (c) $\log_{10} 6$
   (d) $\log_{10} 8$
   (e) $\log_{10} 9$
   (f) $\log_{10} 1.5$
   (g) $\log_{10} 1.2$
   (h) $\log_{10} 1.33$
   (i) $\log_{10} 20$
   (j) $\log_{10} 200$
   (k) $\log_{10} 0.006$

105 Find these integrals:
   (a) $\displaystyle\int \dfrac{3x^2 + 1}{x^3 + x - 6}\,dx$
   (b) $\displaystyle\int \dfrac{\cos 2x}{\sin 2x}\,dx$
   (c) $\displaystyle\int \dfrac{dx}{5x + 3}$
   (d) $\displaystyle\int \dfrac{dx}{(5x + 3)^2}$

106 (Calculator) (a) Evaluate $(1 - 2h)^{1/h}$ for $h = 0.01$.
   (b) Find $\lim_{h\to 0} (1 - 2h)^{1/h}$.

107 Differentiate:
   (a) $\ln\left(\dfrac{\sqrt[3]{\tan 4x}\,(1-2x)^5}{\sqrt[3]{(1+3x)^2}}\right)$
   (b) $\dfrac{(1+x^2)^3\sqrt{1+x}}{\sin 3x}$

108 The graph in Fig. 6.47 shows the tangent line to the curve $y = \ln x$ at a point (x_0, y_0). Prove that AB has length 1, independent of the choice of (x_0, y_0).

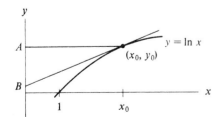

Figure 6.47

109 Find (a) $\lim_{x\to 1} \log_x 2$, (b) $\lim_{x\to\infty} \log_x 2$. *Hint:* What is the relation between $\log_a b$ and $\log_b a$?

110 The *information content* or *entropy* of a binary source (such as a telegraph that transmits dots and dashes) whose two values occur with probabilities p and $1-p$ is defined as $H(p) = -p \ln p - (1-p)\ln(1-p)$, where $0 < p < 1$. Show that H has a maximum at $p = \tfrac{1}{2}$. The practical significance of this result is that, for maximum flow of information per unit time, the two values should, in the long-run, appear in equal proportions.

111 (See Exercise 110.) Let p be fixed so that $0 < p < 1$. Define $M(q) = -p\ln q - (1-p)\ln(1-q)$. Show that $H(p) \le M(q)$ for $0 < q < 1$ and that equality holds if and only if $p = q$.

In each of Exercises 112 to 114 evaluate the limit by first showing that it is the derivative of a certain function at a specific number.

112 $\displaystyle\lim_{x\to 3} \dfrac{\ln(1+2x) - \ln 7}{x - 3}$

113 $\displaystyle\lim_{x\to 0} \dfrac{\ln(2+x) - \ln 2}{x}$

114 $\displaystyle\lim_{x\to 0} \dfrac{\ln(1+x)}{x}$

115 Graph $y = \ln|x|$.

116 Compute: (a) $\displaystyle\int \dfrac{dx}{(3x+2)^2}$ (b) $\displaystyle\int \dfrac{dx}{\sqrt{3x+2}}$
   (c) $\displaystyle\int \dfrac{dx}{3x+2}$

117 Find the coordinates of the point P on the curve $y = \ln x$ such that the line through $(0, 0)$ and P is tangent to the curve.

118 Use logarithmic differentiation, as in Example 6 of Sec. 6.2, to find the derivative of
   (a) $\dfrac{x^{3/5}(1+2x)^4 \sin^3 2x}{\tan^2 5x}$
   (b) $\dfrac{x^3}{\sqrt[3]{x^3 + x^2}\cos 4x}$

119 Find $y'(0)$ if $y = f(x)$ is a function satisfying the equation
$$\ln(1+y) + xy = \ln 2$$
and $y(0) = 1$.

120 Does the graph of $y = x^4 - 4\ln x$ have (a) any local maxima? (b) local minima? (c) inflection points?

In Exercises 121 to 159 examine the limits.

121 $\displaystyle\lim_{x\to\infty} x^{1/\log_2 x}$

122 $\displaystyle\lim_{x\to\pi/4} \dfrac{\sin x - \sqrt{2}/2}{x - \pi/4}$

123 $\displaystyle\lim_{h\to 0} \dfrac{e^{3+h} - e^3}{h}$

124 $\displaystyle\lim_{x\to 1} \dfrac{\sin \pi x}{x - 1}$

125 $\displaystyle\lim_{x\to 0} \dfrac{\cos\sqrt{x} - 1}{\tan x}$

126 $\displaystyle\lim_{x\to\infty} 2^x e^{-x}$

127 $\lim_{x \to 0} (1 + 2x^2)^{1/x^2}$

128 $\lim_{x \to 0} (1 + 3x)^{1/x}$

129 $\lim_{x \to -\infty} \dfrac{e^x - e^{-x}}{e^x + e^{-x}}$

130 $\lim_{x \to \infty} \dfrac{x^2 + 5}{2x^2 + 6x}$

131 $\lim_{x \to 0} \dfrac{1 - \cos x}{x + \tan x}$

132 $\lim_{x \to 0^+} (\sin x)^{\sin x}$

133 $\lim_{x \to 0} \dfrac{\sin 2x}{e^{3x} - 1}$

134 $\lim_{x \to \infty} \dfrac{(x^2 + 1)^5}{e^x}$

135 $\lim_{x \to \infty} \dfrac{3x - \sin x}{x + \sqrt{x}}$

136 $\lim_{x \to 0} \dfrac{xe^x}{e^x - 1}$

137 $\lim_{x \to 2} \dfrac{x^3 - 8}{x^2 - 4}$

138 $\lim_{x \to 0} \dfrac{\sin x^2}{x \sin x}$

139 $\lim_{x \to \pi/2} \dfrac{\sin x}{1 + \cos x}$

140 $\lim_{x \to 1} \dfrac{e^x + 1}{e^x - 1}$

141 $\lim_{x \to 2} \dfrac{x^3 + 8}{x^2 + 2}$

142 $\lim_{x \to \pi/4} \dfrac{\cos x^2}{\cos x}$

143 $\lim_{x \to 3^+} \dfrac{\ln (x - 3)}{x - 3}$

144 $\lim_{x \to 0^+} \dfrac{1 - \cos x}{x - \tan x}$

145 $\lim_{x \to \infty} \dfrac{\cos x}{x}$

146 $\lim_{x \to 2} \dfrac{5^x + 3^x}{x}$

147 $\lim_{x \to 0} \dfrac{5^x - 3^x}{x}$

148 $\lim_{x \to \infty} [(x^2 - 2x)^{1/2} - x]$

149 $\lim_{x \to 0} (1 + 3x)^{2/x}$

150 $\lim_{x \to \infty} \dfrac{e^{-x}}{x^2}$

151 $\lim_{x \to \infty} \dfrac{\ln (x^2 + 1)}{\ln (x^2 + 8)}$

152 $\lim_{x \to \infty} e^{-x} \ln x$

153 $\lim_{x \to 0^+} \sin x \ln x$

154 $\lim_{x \to \pi/2^-} (1 - \sin x)^{\tan x}$

155 $\lim_{x \to \infty} (2^x - x^{10})^{1/x}$

156 $\lim_{x \to 0^+} \left(\dfrac{\sin x}{x} \right)^{1/x}$

157 $\lim_{x \to \infty} \dfrac{(x^3 - x^2 - 4x)^{1/3}}{x}$

158 $\lim_{x \to \pi/4} \dfrac{\sin x}{x}$

159 $\lim_{h \to 0} \dfrac{e^{3+h} - e^3}{1 - h}$

160 Graph $y = xe^{-x}/(x + 1)$, showing intercepts, critical points, relative maxima and minima, and asymptotes.

161 (a) Why does calculus use radian measure?
(b) Why does calculus use the base e for logarithms?
(c) Why does calculus use the base e for exponentials?

162 Using the e^x key, the ln key and the × (multiplication) but not the y^x key, how would you calculate 3^{80} on a calculator?

163 In which cases below is it possible to determine $\lim_{x \to a} f(x)^{g(x)}$ without further information about the functions?
(a) $\lim_{x \to a} f(x) = 0; \lim_{x \to a} g(x) = 7.$
(b) $\lim_{x \to a} f(x) = 2; \lim_{x \to a} g(x) = 0.$
(c) $\lim_{x \to a} f(x) = 0; \lim_{x \to a} g(x) = 0.$
(d) $\lim_{x \to a} f(x) = 0; \lim_{x \to a} g(x) = \infty.$
(e) $\lim_{x \to a} f(x) = \infty; \lim_{x \to a} g(x) = 0.$
(f) $\lim_{x \to a} f(x) = \infty; \lim_{x \to a} g(x) = -\infty.$

164 In which cases below is it possible to determine $\lim_{x \to a} f(x)/g(x)$ without further information about the functions?
(a) $\lim_{x \to a} f(x) = 0; \lim_{x \to a} g(x) = \infty.$
(b) $\lim_{x \to a} f(x) = 0; \lim_{x \to a} g(x) = 1.$
(c) $\lim_{x \to a} f(x) = 0; \lim_{x \to a} g(x) = 0.$
(d) $\lim_{x \to a} f(x) = \infty; \lim_{x \to a} g(x) = -\infty.$

165 (a) State the assumptions in the zero-over-zero case of l'Hôpital's rule.
(b) State the conclusion.

166 Prove that (a) $(\sin^{-1} x)' = 1/\sqrt{1 - x^2}$;
(b) $(e^x)' = e^x$; (c) $(\tan^{-1} x)' = \dfrac{1}{1 + x^2}$.

167 What is the inverse of each of these functions?
(a) ln x   (b) e^x
(c) x^3 (d) 3x
(e) $\sqrt[3]{x}$ (f) $\sin^{-1} x$

168 (a) Let $f(x) = 5^{7x} 6^{8x+3}$. Show that $f'(x)$ is proportional to $f(x)$.
(b) Does (a) contradict the theorem that asserts that the only functions whose derivatives are proportional to the functions are of the form Ae^{kx}?

169 The graph in Fig. 6.48 shows the tangent line to the curve $y = e^x$ at a point (x_0, y_0). Find the length of the segment AB.

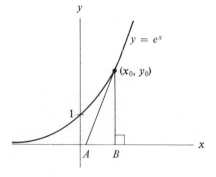

Figure 6.48

170 A bank pays 5 percent interest per year.
(a) Show that, if it compounds interest n times per year, a deposit of 1 dollar grows to the value $(1 + 0.05/n)^{nt}$ dollars in t years.
(b) Show that $\lim_{n \to \infty} (1 + 0.05/n)^{nt} = e^{0.05t}$.
(c) Show that, if n in (a) is large, 1 dollar grows to 2 dollars in about 14 years. Notice that this is about 70 percent of the time it would take to double if interest were not compounded, that is, if earned interest were not left in the account.

Exercises 171 to 175 are based on D. L. Meadows, D. H. Meadows, J. Randers, and W. W. Behrens III, *The Limits to Growth*, Universe Books, New York, 1972. All the time intervals are approximations.

171 Let $Y(t)$ be the amount of some natural resource that is consumed from time $t = 0$ to time $t > 0$. For our purposes $t = 0$ corresponds to the year 1970. Let $c(t)$ be the rate at which that resource is being consumed. Since population and industry are growing, $c(t)$ is an increasing function. Assume that $c(t)$ is increasing exponentially; that is, there are constants A and k such that $c(t) = Ae^{kt}$. Show that $Y(t) = A(e^{kt} - 1)/k$.

172 The amounts of such natural resources as aluminium, natural gas, and petroleum are finite. The formula obtained in Exercise 171 enables one to estimate how long a given resource will last subject to various assumptions.

Let R be the amount of a given resource remaining at time $t = 0$.
(a) Show that if the rate of consumption remains constant, then the resource will last R/A years. This is called the *static index* and is denoted by s.
(b) Show that, if the rate of consumption continues to grow exponentially, the resource will last

$$\frac{\ln(ks + 1)}{k} \quad \text{years.}$$

This is called the *exponential index*. The letter s denotes the static index, defined in (a).

173 The static index of aluminum is 100 years (known global reserves are $1.17 \cdot 10^9$ tons). The projected growth rate for consumption is 6.4 percent a year. Thus $c(t) = A(1.064)^t$. However, $e^{0.064}$ is a very good approximation to 1.064 and is used in the cited reference to obtain the formula $c(t) = Ae^{0.064t}$.
(a) Show that the exponential index is 31 years (approximately). Show that, if the global reserves are five times as large as quoted, the static index will be 500 years and the exponential index will be 55 years.

(b) Show that, if $k = 0.032$ instead of 0.064, the static index doubles but the exponential index will be 45 years. (Assume known global reserves.)

174 The known global reserves of natural gas are $1.14 \cdot 10^{15}$ cubic feet. The static index is 38 years. If consumption continues to increase at 4.7 percent per year, show that the exponential index (a) is 22 years, (b) is 49 years if global reserves are 5 times as large.

175 The known global reserves of petroleum are $455 \cdot 10^8$ barrels. The static index is 31 years. If the rate of consumption continues to increase at 3.9 percent per year, show that the exponential index is 20 years and, if reserves are five times as large, 50 years.

176 It was once conjectured that the speed of a ball falling from rest is proportional to the distance s that it drops.
(a) Show that, if this conjecture were correct, s would grow exponentially as a function of time.
(b) With the aid of (a) show that the speed would also grow exponentially.
(c) Recalling that the initial speed is 0, show that (b) leads to an absurd conclusion.

In fact, ds/dt is proportional to time t rather than to distance s.

177 A large lake is polluted. As a result of a court order, no new pollutants are added, and gradually the pollutants are removed from the lake by the streams. Assume that the volume of the lake remains constant and that the pollutants are thoroughly mixed. Let $P(t)$ be the amount of pollutants t years after the court order.
(a) Why is $\Delta P \approx kP(t)\,\Delta t$ for small Δt? (b) Is k positive or negative? (c) Show that $P(t) = P(0)e^{kt}$, where $P(0)$ is the initial pollution.

178 Find the limit of $(1^x + 2^x + 3^x)^{1/x}$ as (a) $x \to 0$, (b) $x \to \infty$, (c) $x \to -\infty$.

■ ■

179 According to a Chinese riddle, the lilies in a pond doubled in number each day. At the end of a month of 30 days the lilies finally covered the entire pond. On what day was exactly half the pond covered?

180 Assume that $\lim_{x \to 0} f(x) = 1$ and $\lim_{x \to 0} g(x) = 1$. Which of the following limits can be determined without any more information? Give their values. Which can not? In those cases, give examples that show the limits are not determined.

(a) $\lim_{x \to 0} f(x)g(x)$
(b) $\lim_{x \to 0} (f(x) + g(x))$
(c) $\lim_{x \to 0} \frac{f(x)}{g(x)}$
(d) $\lim_{x \to 0} \frac{f(x) - 1}{g(x)}$

(e) $\lim_{x \to 0} \dfrac{f(x) - 1}{g(x) - 1}$ (f) $\lim_{x \to 0} (1 - f(x))^{g(x)}$

(g) $\lim_{x \to 0} (1 - f(x))^{1 - g(x)}$

181 Figure 6.49 shows a triangle ABC and a shaded region cut from the parabola $y = x^2$ by a horizontal line. Find the limit, as $x \to 0$, of the ratio between the area of the triangle and the area of the shaded region.

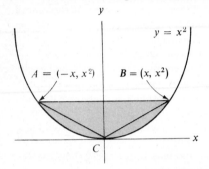

Figure 6.49

■ ■

182 If
$$\lim_{t \to \infty} f(t) = \infty = \lim_{t \to \infty} g(t)$$
and
$$\lim_{t \to \infty} \frac{\ln f(t)}{\ln g(t)} = 1,$$
must
$$\lim_{t \to \infty} \frac{f(t)}{g(t)} = 1?$$

183 If
$$\lim_{t \to \infty} f(t) = \infty = \lim_{t \to \infty} g(t)$$
and
$$\lim_{t \to \infty} \frac{f(t)}{g(t)} = 3,$$
what can be said about
$$\lim_{t \to \infty} \frac{\ln f(t)}{\ln g(t)}?$$

(Do not assume that f and g are differentiable.)

184 Give an example of functions f and g such that $\lim_{x \to 0} f(x) = 1$, $\lim_{x \to 0} g(x) = \infty$, and
$$\lim_{x \to 0} f(x)^{g(x)} = 2.$$

185 Let f be a function such that f, f', and f'' are continuous, $f(x) \geq 0, f(0) = 0, f'(0) = 0$, and $f''(0) > 0$. The graph of f is indicated in Fig. 6.50. Find the limit as $x \to 0^+$ of the quotient

$$\frac{\text{Area under curve and above } [0, x]}{\text{Area of triangle } OAP}.$$

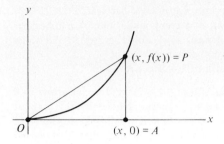

Figure 6.50

186 A certain fish increases in number at a rate proportional to the size of the population. In addition, it is being harvested at a constant rate. Let $P(t)$ be the size of the fish population at time t.
(a) Show that there are positive constants h and k such that for small Δt, $\Delta P \approx kP\,\Delta t - h\,\Delta t$.
(b) Find a formula for $P(t)$ in terms of $P(0)$, h, and k.

187 Let $f(x) = (1 + x)^{1/x}$ for $x > -1$, $x \neq 0$. Let $f(0) = e$.
(a) Show that f is continuous at $x = 0$.
(b) Show that f is differentiable at $x = 0$. What is $f'(0)$?

188 A salesman, trying to persuade a tycoon to invest in Standard Coagulated Mutual Fund, shows him the accompanying graph which records the value of a similar investment made in the fund in 1950. "Look! In the first 5 years the investment increased $1000," the salesman observed, "but in the past 5 years it increased by $2000. It's really improving. Look at the slope from 1970 to 1975, which you can see clearly in Fig. 6.51."

The tycoon replied, "Hogwash; in order to present an unbiased graph, you should use semilog paper. Though your graph is steeper from 1970 to 1975, in fact, the rate of return is less than from 1950 to 1955. Indeed, that was your best period."

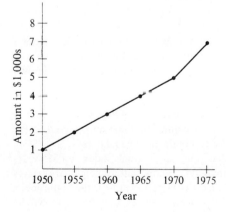

Figure 6.51

(a) If the percentage return on the accumulated investment remains the same over each 5-year period as the first 5-year period, sketch the graph.
(b) Explain the tycoon's reasoning.

189 The proof of Theorem 1 in Sec. 6.8, to be outlined in Exercise 190, depends on the following *generalized mean-value theorem*.

Generalized mean-value theorem. Let f and g be two functions that are continuous on $[a, b]$ and differentiable on (a, b). Furthermore, assume that $g'(x)$ is never 0 for x in (a, b). Then there is a number c in (a, b) such that

$$\frac{f(b) - f(a)}{g(b) - g(a)} = \frac{f'(c)}{g'(c)}.$$

(a) During a given time interval one car travels twice as far as another car. Use the generalized mean-value theorem to show that there is at least one instant when the first car is traveling exactly twice as fast as the second car.
(b) To prove the generalized mean-value theorem introduce a function h:

$$h(x) = f(x) - f(a) - \frac{f(b) - f(a)}{g(b) - g(a)} [g(x) - g(a)].$$

Show that $h(b) = 0$ and $h(a) = 0$. Then apply Rolle's theorem to h.

Remark: The function h is geometrically quite similar to the function h used in the proof of the mean-value theorem in Sec. 4.1. It is easy to check that $h(x)$ is the vertical distance between the point $(f(x), g(x))$ and the line through $(f(a), g(a))$ and $(f(b), g(b))$.

190 This exercise proves Theorem 1 of Sec. 6.8, l'Hôpital's rule in the zero-over-zero case. Assume the hypotheses of that theorem. Define $f(a) = 0$ and $g(a) = 0$, so that f and g are continuous at a.

Note that $\quad \dfrac{f(x)}{g(x)} = \dfrac{f(x) - f(a)}{g(x) - g(a)}$

and apply the generalized mean-value theorem from Exercise 189.

191 Let f and g be functions defined on some open interval. Assume that they have only positive values and that they are differentiable and possess second derivatives which are not 0 in the interval considered. Let $F(x) = \ln f(x)$ and let $G(x) = \ln g(x)$.
(a) If F is concave upward, must f be concave upward?
(b) If f is concave upward, must F be concave upward?
(c) If f and g are concave upward, must $f + g$ be concave upward?
(d) If f and g are concave upward, must fg be concave upward?
(e) If F and G are concave upward, must $\ln fg$ be concave upward?
(f) If F and G are concave upward, must $\ln (f + g)$ be concave upward?
In each case explain your answer.

COMPUTING ANTIDERIVATIVES

In Chaps. 3 and 6 it was shown how to compute the derivative of any elementary function. In Chap. 5 the computation of many definite integrals was reduced to the computation of antiderivatives. In the present chapter some techniques for finding antiderivatives will be developed.

The problem of computing antiderivatives differs from that for computing derivatives in two important aspects. First of all, some elementary functions, such as e^{x^2}, do not have elementary antiderivatives. Second, a slight change in the form of an integrand can cause a great change in the form of its antiderivative; for instance,

$$\int \frac{1}{x^2 + 1} \, dx = \tan^{-1} x + C \qquad \text{and} \qquad \int \frac{x}{x^2 + 1} \, dx = \frac{1}{2} \ln (x^2 + 1) + C.$$

A few moments of browsing through a table of antiderivatives (usually called a *table of integrals*) will yield many such examples. A short table of integrals may be found on the inside covers of this book.

To be convenient, a table of antiderivatives should be short; it should not try to anticipate every antiderivative that may arise in practice. Sometimes it is necessary to transform a problem into one listed in the table or else to solve it without the aid of the table. It will often be quicker to use these techniques than to thumb through the pages of the table, even when the table lists the answer.

It is strongly recommended that the reader purchase a standard handbook of mathematical tables, such as Burington's *Handbook of Mathematical Tables and Formulas*. Such a handbook also contains extensive lists of key facts from algebra, geometry, trigonometry, analytic geometry, statistics, etc.

The substitution method in Sec. 7.1 and integration by parts in Sec. 7.2 are the most general techniques in the chapter. Section 7.3 shows how to use an integral table. The methods in Secs. 7.4 to 7.8 are more specialized, applying to integrands of particular types. The concluding Sec. 7.9 addresses the question, "When I'm faced with an integral, how do I decide which method to use?"

7.1 Substitution in an antiderivative and in a definite integral

This section describes the substitution technique that changes the form of an integral, preferably to that of an easier integral. Before describing this technique, we collect some basic facts about integrals in order to have a supply of integrands that can be "integrated at a glance."

SOME BASIC INTEGRALS

Every formula for a derivative provides a corresponding formula for an antiderivative or integral. For instance, since

$$\left(\frac{x^3}{3}\right)' = x^2,$$

it follows that

$$\int x^2 \, dx = \frac{x^3}{3} + C.$$

The following miniature integral table lists a few formulas that should be memorized. Each can be checked by differentiation.

$$\int x^a \, dx = \frac{x^{a+1}}{a+1} + C \quad \text{for } a \neq -1$$

$$\int \frac{1}{x} \, dx = \ln|x| + C$$

$$\int e^x \, dx = e^x + C$$

$$\int \sin x \, dx = -\cos x + C \quad \text{(Remember the minus sign.)}$$

$$\int \cos x \, dx = \sin x + C$$

$$\int \frac{1}{\sqrt{1-x^2}} \, dx = \sin^{-1} x + C$$

$$\int \frac{1}{1+x^2} \, dx = \tan^{-1} x + C$$

$$\int \frac{1}{|x|\sqrt{x^2-1}} \, dx = \sec^{-1} x + C$$

The following general formula is needed often enough to be included in this brief list:

$$\int \frac{f'}{f} \, dx = \ln|f| + C.$$

---

382 COMPUTING ANTIDERIVATIVES

The last entry in the table states that if the numerator is precisely the derivative of the denominator, then the natural logarithm of the absolute value of the denominator is an antiderivative of the quotient.

A constant can be moved past the integral sign.

If c is constant, then

$$\int cf(x)\, dx = c \int f(x)\, dx \quad \text{and} \quad \int \frac{f(x)\, dx}{c} = \frac{1}{c} \int f(x)\, dx.$$

An antiderivative of the sum of functions can be found by adding antiderivatives of the functions.

For two functions f and g,

$$\int [f(x) + g(x)]\, dx = \int f(x)\, dx + \int g(x)\, dx.$$

Similarly, $\quad \int [f(x) - g(x)]\, dx = \int f(x)\, dx - \int g(x)\, dx.$

A few examples will show how these formulas serve to integrate many functions.

EXAMPLE 1 Find

Antiderivative of a polynomial

$$\int (2x^4 - 3x + 2)\, dx.$$

SOLUTION

$$\int (2x^4 - 3x + 2)\, dx = \int 2x^4\, dx - \int 3x\, dx + \int 2\, dx$$

$$= 2 \int x^4\, dx - 3 \int x\, dx + 2 \int 1\, dx$$

One constant of integration is enough.

$$= 2 \frac{x^5}{5} - 3 \frac{x^2}{2} + 2x + C. \quad \blacksquare$$

EXAMPLE 2 Find

Antiderivative of f'/f

$$\int \frac{4x^3}{x^4 + 1}\, dx.$$

SOLUTION The numerator is precisely the derivative of the denominator. Hence

$$\int \frac{4x^3}{x^4 + 1}\, dx = \ln |x^4 + 1| + C.$$

Since $x^4 + 1$ is always positive, the absolute-value sign is not needed, and

$$\int \frac{4x^3}{x^4 + 1}\, dx = \ln (x^4 + 1) + C. \quad \blacksquare$$

EXAMPLE 3 Find

Antiderivative of x^a

$$\int \sqrt{x}\, dx.$$

SOLUTION

$$\int \sqrt{x}\, dx = \int x^{1/2}\, dx$$

$$= \frac{x^{1+1/2}}{1+\frac{1}{2}} + C$$

$$= \tfrac{2}{3} x^{3/2} + C$$

$$= \tfrac{2}{3} (\sqrt{x})^3 + C. \blacksquare$$

EXAMPLE 4   Find
$$\int \frac{1}{x^3} \, dx.$$

SOLUTION
$$\int \frac{1}{x^3} \, dx = \int x^{-3} \, dx$$

$$= \frac{x^{-3+1}}{-3+1} + C$$

$$= -\tfrac{1}{2} x^{-2} + C$$

$$= -\frac{1}{2x^2} + C. \blacksquare$$

EXAMPLE 5   Find
$$\int \left(3 \cos x - 4 \sin x + \frac{1}{x^2} \right) dx.$$

SOLUTION
$$\int \left(3 \cos x - 4 \sin x + \frac{1}{x^2} \right) dx = 3 \int \cos x \, dx - 4 \int \sin x \, dx + \int \frac{1}{x^2} \, dx$$

$$= 3 \sin x + 4 \cos x - \frac{1}{x} + C. \blacksquare$$

EXAMPLE 6   Find
$$\int \frac{x}{1+x^2} \, dx.$$

SOLUTION

Multiplying the integrand by a constant

If the numerator were $2x$, then the numerator would be the derivative of the denominator, and the antiderivative would be $\ln(1+x^2)$. But the numerator can be multiplied by 2, if we simultaneously divide by 2:

$$\int \frac{x}{1+x^2} \, dx = \frac{1}{2} \int \frac{2x}{1+x^2} \, dx.$$

This step depends on the fact that a constant can be moved past the integral sign:

$$\frac{1}{2} \int \frac{2x}{1+x^2} \, dx = \frac{1}{2} 2 \int \frac{x}{1+x^2} \, dx$$

$$= \int \frac{x}{1+x^2} \, dx.$$

Since $1 + x^2 > 0$, the absolute value is not needed in $\ln(1 + x^2)$.

Thus
$$\int \frac{x}{1+x^2}\,dx = \frac{1}{2}\int \frac{2x}{1+x^2}\,dx$$
$$= \tfrac{1}{2}\ln(1+x^2) + C. \quad \blacksquare$$

NOTATION In Examples 2, 4, and 6 the integrand was separated from the dx. This was done for emphasis. Usually the dx is combined with the integrand:

$$\int \frac{4x^3}{x^4+1}\,dx \quad \text{is written} \quad \int \frac{4x^3\,dx}{x^4+1}$$

and
$$\int 1\,dx \quad \text{is written} \quad \int dx.$$

THE SUBSTITUTION TECHNIQUE

This method involves the introduction of a function that changes the form of the integrand, hopefully to that of a simpler integrand. Several examples will illustrate the mechanics of the method, known as *substitution*. The proof that it works will be given after these examples. Substitution is the most important tool in computing antiderivatives.

EXAMPLE 7 Find $\int \sin(x^2)\,2x\,dx.$

SOLUTION Note that $2x$ is the derivative of x^2. Introduce
$$u = x^2.$$
Then
$$du = 2x\,dx,$$
and
$$\int \sin(x^2)\,2x\,dx = \int \sin u\,du.$$

Now it is easy to find $\int \sin u\,du$:
$$\int \sin u\,du = -\cos u + C.$$

Replacing u by x^2 in $-\cos u$ yields $-\cos(x^2)$. Thus
$$\int \sin(x^2)\,2x\,dx = -\cos x^2 + C.$$

This answer can be checked by differentiation (using the chain rule):
$$\frac{d}{dx}(-\cos x^2 + C) = \sin x^2 \,\frac{d}{dx}(x^2) + 0$$
$$= \sin(x^2)\,2x. \quad \blacksquare$$

7.1 SUBSTITUTION IN AN ANTIDERIVATIVE AND IN A DEFINITE INTEGRAL

EXAMPLE 8 Find $\int e^{x^5} 5x^4 \, dx.$

SOLUTION Introduce $u = x^5$. Then $du = 5x^4 \, dx$,

and
$$\int e^{x^5} 5x^4 \, dx = \int e^u \, du$$
$$= e^u + C$$
$$= e^{x^5} + C. \quad \blacksquare$$

EXAMPLE 9 Find $\int \sin^2 \theta \cos \theta \, d\theta.$

SOLUTION Note that $\cos \theta$ is the derivative of $\sin \theta$ and introduce

$$u = \sin \theta;$$

hence $\quad du = \cos \theta \, d\theta.$

Then
$$\int \underbrace{\sin^2 \theta}_{u^2} \underbrace{\cos \theta \, d\theta}_{du} = \int u^2 \, du$$
$$= \frac{u^3}{3} + C$$
$$= \frac{\sin^3 \theta}{3} + C. \quad \blacksquare$$

EXAMPLE 10 Find $\int (1 + x^3)^5 x^2 \, dx.$

SOLUTION The derivative of $1 + x^3$ is $3x^2$, which differs from x^2 in the integrand only by the constant factor 3. So let

$$u = 1 + x^3;$$

hence $\quad du = 3x^2 \, dx \quad$ and $\quad \dfrac{du}{3} = x^2 \, dx.$

Then
$$\int (1 + x^3)^5 x^2 \, dx = \int u^5 \, \frac{du}{3}$$
$$= \frac{1}{3} \int u^5 \, du$$
$$= \frac{1}{3} \frac{u^6}{6} + C$$
$$= \frac{(1 + x^3)^6}{18} + C.$$

It would be instructive to check this answer by differentiation. $\quad \blacksquare$

In Example 10 note that, if the x^2 were not present in the integrand, the substitution method would not work. To find $\int (1 + x^3)^5 \, dx$, it would be necessary to multiply out $(1 + x^3)^5$ first, a most unpleasant chore.

EXAMPLE 11 Compare the problems of finding these antiderivatives:

$$\int \frac{dx}{\sqrt{1 + x^3}} \quad \text{and} \quad \int \frac{x^2 \, dx}{\sqrt{1 + x^3}}.$$

SOLUTION It turns out that the first antiderivative is *not* an elementary function, while the second is easy, because the x^2 is present.

Since x^2 differs from the derivative of $1 + x^3$ only by a constant factor 3, use the substitution

$$u = 1 + x^3;$$

hence
$$du = 3x^2 \, dx \quad \text{and} \quad \frac{du}{3} = x^2 \, dx.$$

Thus
$$\int \frac{x^2 \, dx}{\sqrt{1 + x^3}} = \int \frac{1}{\sqrt{u}} \frac{du}{3} = \frac{1}{3} \int \frac{du}{\sqrt{u}} = \frac{1}{3} \int u^{-1/2} \, du$$

$$= \frac{1}{3} \frac{u^{1/2}}{\frac{1}{2}} + C$$

$$= \tfrac{2}{3} u^{1/2} + C$$

$$= \tfrac{2}{3} (1 + x^3)^{1/2} + C$$

$$= \tfrac{2}{3} \sqrt{1 + x^3} + C. \quad \blacksquare$$

In the next example the choice of substitution is suggested not by something that can serve as du but rather by the desire to simplify a denominator.

EXAMPLE 12 Find $\int_2^4 \frac{x^2 + 1}{2x - 3} \, dx$.

SOLUTION Since the denominator complicates the problem, try the substitution $u = 2x - 3$.

We have

$$u = 2x - 3;$$

To evaluate $\int \frac{P(x)}{(ax + b)^n} \, dx$,

hence $du = 2 \, dx, \quad dx = \dfrac{du}{2}, \quad \text{and} \quad x = \dfrac{u + 3}{2}.$

where $P(x)$ is a polynomial, let $u = ax + b$.

Thus $\int \dfrac{x^2 + 1}{2x - 3} \, dx = \int \dfrac{[(u + 3)/2]^2 + 1}{u} \dfrac{du}{2}$

$$= \int \frac{u^2 + 6u + 13}{8u} \, du$$

$$= \int \left(\frac{u}{8} + \frac{3}{4} + \frac{13}{8u}\right) du$$

$$= \frac{u^2}{16} + \frac{3u}{4} + \frac{13}{8} \ln |u| + C$$

Express the antiderivative in terms of x rather than u.

$$= \frac{(2x-3)^2}{16} + \frac{3}{4}(2x-3) + \frac{13}{8} \ln |2x-3| + C.$$

Consequently

$[F(x)]_a^b = F(b) - F(a)$, a shorthand for $[F(x)]|_a^b$

$$\int_2^4 \frac{x^2 + 1}{2x - 3} \, dx = \left[\frac{(2x-3)^2}{16} + \frac{3}{4}(2x-3) + \frac{13}{8} \ln |2x-3| + C\right]_2^4$$

$$= \left(\frac{5^2}{16} + \frac{3 \cdot 5}{4} + \frac{13}{8} \ln 5 + C\right) - \left(\frac{1^2}{16} + \frac{3 \cdot 1}{4} + \frac{13}{8} \ln 1 + C\right)$$

The C's cancel.

$$= \tfrac{9}{2} + \tfrac{13}{8} \ln 5. \blacksquare$$

As mentioned in the marginal note, the substitution in Example 12, of the type $u = ax + b$, is often useful. Example 13 is another application of this type of substitution.

EXAMPLE 13 Integral tables include a formula for

$$\int \frac{dx}{(ax+b)^n}, \quad n \neq 1.$$

Use a substitution to find the formula.

SOLUTION Let $u = ax + b$. Then

$$du = a \, dx \quad \text{and} \quad dx = \frac{du}{a}.$$

Thus

$$\int \frac{dx}{(ax+b)^n} = \int \frac{du/a}{u^n}$$

$$= \frac{1}{a} \int u^{-n} \, du$$

$$= \frac{1}{a} \frac{u^{-n+1}}{(-n+1)} + C$$

$$= \frac{(ax+b)^{-n+1}}{a(-n+1)} + C$$

$$= \frac{1}{a(-n+1)(ax+b)^{n-1}} + C. \blacksquare$$

A substitution is worth trying in two cases:

When to substitute **1** The integrand can be written in the form of a product of a special type:

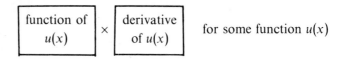

2 The integrand becomes simpler when a part of it is denoted $u(x)$.

SUBSTITUTION IN A DEFINITE INTEGRAL

The substitution technique, or "change of variables," extends to definite integrals, $\int_a^b f(x)\,dx$, with one important proviso. When making the substitution from x to u, be sure to replace the interval $[a, b]$ by the interval whose endpoints are $u(a)$ and $u(b)$. An example will illustrate the necessary change in the limits of integration. (It should be compared with Example 12.)

EXAMPLE 14 Transform the definite integral

$$\int_2^4 \frac{x^2 + 1}{2x - 3}\,dx$$

using the substitution $u = 2x - 3$. Then evaluate the new definite integral.

SOLUTION Let $u = 2x - 3$. Then $du = 2\,dx$, $dx = du/2$, $x = (u + 3)/2$, and, as x goes from 2 to 4, u goes from

$$2 \cdot 2 - 3 = 1 \quad \text{to} \quad 2 \cdot 4 - 3 = 5.$$

This is the last you see of x in the problem. Thus $\int_2^4 \frac{x^2 + 1}{2x - 3}\,dx = \int_1^5 \frac{[(u + 3)/2]^2 + 1}{u}\,\frac{du}{2}$ here $\int_2^4$ is replaced by $\int_1^5$

$$= \int_1^5 \left(\frac{u}{8} + \frac{3}{4} + \frac{13}{8u}\right) du$$

u is positive so $\ln |u| = \ln u.$ $= \left(\frac{u^2}{16} + \frac{3u}{4} + \frac{13}{8}\ln u\right)\Bigg|_1^5$ omit C, since it will cancel

$$= \left(\frac{5^2}{16} + \frac{3 \cdot 5}{4} + \frac{13}{8}\ln 5\right) - \left(\frac{1^2}{16} + \frac{3 \cdot 1}{4} + \frac{13}{8}\ln 1\right)$$

Compare with Example 12. $= \frac{9}{2} + \frac{13}{8}\ln 5.$ ∎

GENERAL DESCRIPTION OF THE SUBSTITUTION TECHNIQUE

In each example the method is basically the same. In order to apply the substitution technique to find

7.1 SUBSTITUTION IN AN ANTIDERIVATIVE AND IN A DEFINITE INTEGRAL

$$\int f(x)\,dx,$$

look for a function $u = h(x)$ such that

$$f(x) = g(h(x))h'(x),$$

for some function g, or more simply

$$f(x)\,dx = g(u)\,du.$$

Then find an antiderivative of g,

$$\int g(u)\,du$$

and replace u by $h(x)$ in this antiderivative.

It is to be hoped that the problem of finding $\int g(u)\,du$ is easier than that of finding $\int f(x)\,dx$. If it is not, try another substitution or one of the methods presented in the rest of the chapter. It is important to keep in mind that there is no simple routine method for antidifferentiation of elementary functions. This is in contrast with the routine that exists for differentiation. Practice in integration pays off in the quick recognition of which technique is most promising.

THE TWO SUBSTITUTION THEOREMS

The proofs of Theorems 1 and 2 show that the substitution technique is based on the chain rule.

THEOREM 1 *Substitution in the indefinite integral.* If

$$f(x) = g(h(x))h'(x)$$

and

$$G(u) = \int g(u)\,du,$$

then

$$G(h(x)) = \int f(x)\,dx.$$

PROOF It is necessary to show that

$$\frac{d}{dx}[G(h(x))] = f(x).$$

The chain rule is the basis of the whole section.

To do so, let $u = h(x)$ and use the chain rule:

$$\frac{d}{dx}[G(h(x))] = \frac{d}{du}[G(u)]\frac{du}{dx}$$

$$= g(u)h'(x)$$

$$= g(h(x))h'(x)$$
$$= f(x).$$

That is,
$$G(h(x)) = \int f(x)\, dx,$$

which proves the theorem. ∎

THEOREM 2 *Substitution in the definite integral.* Let f be a continuous function on the interval $[a, b]$, $u = h(x)$ be a differentiable function on the same interval, and g be a continuous function such that

$$f(x)\, dx = g(u)\, du;$$

that is,
$$f(x) = g(h(x))h'(x).$$

Then
$$\int_a^b f(x)\, dx = \int_{h(a)}^{h(b)} g(u)\, du.$$

PROOF Let $G(u)$ be an antiderivative of $g(u)$. By Theorem 1, $G(h(x))$ is an antiderivative of $f(x)$. Thus

$$\int_a^b f(x)\, dx \underset{\text{FTC}}{=} G(h(x))\Big|_a^b$$
$$= G(h(b)) - G(h(a)).$$

But
$$\int_{h(a)}^{h(b)} g(u)\, du \underset{\text{FTC}}{=} G(h(b)) - G(h(a)).$$

This proves that substitution in the definite integral is valid. ∎

Exercises

In Exercises 1 to 24 compute the antiderivatives. Do *not* use the substitution technique.

1. $\int 5x^3\, dx$

2. $\int \dfrac{dx}{6x^3}$

3. $\int x^{1/3}\, dx$

4. $\int \sqrt[3]{x^2}\, dx$

5. $\int \dfrac{3}{\sqrt{x}}\, dx$

6. $\int \dfrac{1}{\sqrt[4]{x}}\, dx$

7. $\int 5e^{-2x}\, dx$

8. $\int \dfrac{5}{1+x^2}\, dx$

9. $\int \dfrac{6\, dx}{|x|\sqrt{x^2-1}}$

10. $\int \dfrac{5\, dx}{\sqrt{1-x^2}}$

11. $\int \dfrac{x^3}{1+x^4}\, dx$

12. $\int \dfrac{e^x}{1+e^x}\, dx$

13. $\int \dfrac{\sin x}{1+\cos x}\, dx$

14. $\int \dfrac{dx}{1+3x}$

15. $\int \dfrac{1+2x}{x+x^2}\, dx$

16. $\int \dfrac{1+2x}{1+x^2}\, dx$

17. $\int (x^2+3)^2\, dx$

18. $\int (1+e^x)^2\, dx$

19. $\int (1+3x)x^2\, dx$

20. $\int \left(\sqrt{x} - \dfrac{2}{x} + \dfrac{x}{2}\right) dx$

21. $\int x^2\sqrt{x}\, dx$

22. $\int 6\sin x\, dx$

23. $\int \dfrac{1+\sqrt{x}}{x}\, dx$

24. $\int \dfrac{(\sqrt{x}+1)^2}{\sqrt[3]{x}}\, dx$

In Exercises 25 to 38 *use the given substitutions* to find the antiderivatives.

25. $\int (1+3x)^5 3\, dx$; $u = 1 + 3x$.

26. $\int e^{\sin\theta} \cos\theta\, d\theta$; $u = \sin\theta$.

27. $\int \dfrac{x}{\sqrt{1+x^2}}\, dx$; $u = 1 + x^2$.

28. $\int \sqrt{1+x^2}\, x\, dx$; $u = 1 + x^2$.

29. $\int \sin 2x\, dx$; $u = 2x$.

30. $\int \dfrac{e^{2x}}{(1+e^{2x})^2}\, dx$; $u = 1 + e^{2x}$.

31. $\int e^{3x}\, dx$; $u = 3x$.

32. $\int \dfrac{e^{1/x}}{x^2}\, dx$; $u = 1/x$.

33. $\int \dfrac{1}{\sqrt{1-9x^2}}\, dx$; $u = 3x$.

34. $\int \dfrac{t\, dt}{\sqrt{2-5t^2}}$; $u = 2 - 5t^2$.

35. $\int \tan\theta \sec^2\theta\, d\theta$; $u = \tan\theta$.

36. $\int \dfrac{\sin\sqrt{x}}{\sqrt{x}}\, dx$; $u = \sqrt{x}$.

37. $\int \dfrac{(\ln x)^4}{x}\, dx$; $u = \ln x$.

38. $\int \dfrac{\sin(\ln x)}{x}\, dx$; $u = \ln x$.

In Exercises 39 to 60 use appropriate substitutions to find the antiderivatives.

39. $\int (1-x^2)^5 x\, dx$

40. $\int \dfrac{x\, dx}{(x^2+1)^3}$

41. $\int \sqrt[3]{1+x^2}\, x\, dx$

42. $\int \dfrac{\sin\theta}{\cos^2\theta}\, d\theta$

43. $\int \dfrac{e^{\sqrt{t}}}{\sqrt{t}}\, dt$

44. $\int e^x \sin e^x\, dx$

45. $\int \sin 3\theta\, d\theta$

46. $\int \dfrac{dx}{\sqrt{2x+5}}$

47. $\int (x-3)^{5/2}\, dx$

48. $\int \dfrac{dx}{(4x+3)^3}$

49. $\int \dfrac{2x+3}{x^2+3x+2}\, dx$

50. $\int \dfrac{2x+3}{(x^2+3x+5)^4}\, dx$

51. $\int e^{2x}\, dx$

52. $\int \dfrac{dx}{\sqrt{x}(1+\sqrt{x})^3}$

53. $\int x^4 \sin x^5\, dx$

54. $\int \dfrac{\cos(\ln x)\, dx}{x}$

55. $\int \dfrac{x}{1+x^4}\, dx$

56. $\int \dfrac{x^3}{1+x^4}\, dx$

57. $\int \dfrac{x\, dx}{1+x}$

58. $\int \dfrac{x}{\sqrt{1-x^4}}\, dx$

59. $\int \dfrac{\ln 3x\, dx}{x}$

60. $\int \dfrac{\ln x^2\, dx}{x}$

In each of Exercises 61 to 66 use a substitution to transform the definite integral to a more convenient definite integral. Be sure to change the limits of integration. Do *not* evaluate the definite integral.

61. $\int_1^2 e^{x^3} x^2\, dx$

62. $\int_{\pi/2}^{\pi} \sin^3\theta \cos\theta\, d\theta$

63. $\int_0^1 \dfrac{x^2-3}{(x+1)^4}\, dx$

64. $\int_1^2 \dfrac{x^3-x}{(3x+1)^3}\, dx$

65. $\int_1^e \dfrac{(\ln x)^3}{x}\, dx$

66. $\int_0^{\pi/2} \cos^5\theta \sin\theta\, d\theta$

■

In each of Exercises 67 to 70 use a substitution to evaluate the integral.

67. $\int \dfrac{x^2\, dx}{ax+b}$, $a \neq 0$

68. $\int \dfrac{x\, dx}{(ax+b)^2}$, $a \neq 0$

69. $\int \dfrac{x^2\, dx}{(ax+b)^2}$, $a \neq 0$

70. $\int x(ax+b)^n\, dx$, $n \neq -1, -2$

71. Jack claims that $\int 2\cos\theta \sin\theta\, d\theta = -\cos^2\theta$, while Jill claims that the answer is $\sin^2\theta$. Who is right?

72. Jill says, "$\int_0^\pi \cos^2\theta\, d\theta$ is obviously positive." Jack claims, "No, it's zero. Just make the substitution $u = \sin\theta$; hence $du = \cos\theta\, d\theta$. Then I get

$$\int_0^\pi \cos^2\theta\, d\theta = \int_0^\pi \cos\theta \cos\theta\, d\theta$$

$$= \int_0^0 \sqrt{1-u^2}\, du = 0.$$

Simple." (a) Who is right? What is the mistake?

(b) Use the identity $\cos^2 \theta = (1 + \cos 2\theta)/2$ to evaluate the integral without substitution.

■ ■

73. Jill asserts that $\int_{-2}^{1} 2x^2 \, dx$ is obviously positive. "After all, the integrand is never negative and $-2 < 1$." "You're wrong again," Jack replies, "It's negative. Here are my computations. Let $u = x^2$; hence $du = 2x \, dx$. Then

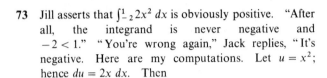

which is obviously negative." Who is right?

7.2 Integration by parts

Integration by parts is used not only to obtain antiderivatives, but also to establish properties of functions. For example, in Sec. 10.8 it provides a way to measure the difference between a function and a certain polynomial that approximates the function. Students going on to a course in differential equations will see integration by parts again when the basic properties of the Laplace transform are proved.

Just as the chain rule is the basis for integration by substitution, the formula for the derivative of a product is the basis for integration by parts.

THEOREM *Integration by parts.* If u and v are differentiable functions and $\int vu' \, dx$ is an antiderivative of vu', then

$$uv - \int vu' \, dx$$

is an antiderivative of uv'. In symbols,

$$\int uv' \, dx = uv - \int vu' \, dx$$

or, in the notation of differentials,

This differential form is the most useful.

$$\int u \, dv = uv - \int v \, du.$$

PROOF Differentiate

$$uv - \int vu' \, dx$$

to see if the result is uv'. We have

$$\left(uv - \int vu' \, dx \right)' = (uv)' - \left(\int vu' \, dx \right)' \qquad \text{derivative of difference}$$

$$= (uv)' - vu' \qquad \text{definition of } \int vu' \, dx$$

$$= uv' + vu' - vu' \quad \text{derivative of product}$$
$$= uv'.$$

Thus $uv - \int vu'\,dx$ is an antiderivative of uv', as was to be shown. ∎

EXAMPLE 1 Find
$$\int xe^x\,dx.$$

SOLUTION To use the formula
$$\int u\,dv = uv - \int v\,du,$$

it is necessary to write
$$\int xe^x\,dx$$

in the form
$$\int u\,dv.$$

(The resulting $\int v\,du$, it is hoped, is easier to find than the original integral $\int u\,dv$.)

The integrand is so simple that there is not much choice. Try
$$u = x \quad \text{and} \quad dv = e^x\,dx;$$

that is, break up the integrand this way:
$$\int \underbrace{x}_{u}\,\underbrace{e^x\,dx}_{dv}.$$

Then find du and v. Since $u = x$, it follows that $du = dx$. Since $dv = e^x\,dx$, we choose $v = e^x$. (Of course v could be $e^x + C$ for any constant C, but choose the simplest v whose derivative is e^x.) Applying integration by parts yields
$$\int \underbrace{x}_{u}\,\underbrace{e^x\,dx}_{dv} = \underbrace{x}_{u}\,\underbrace{e^x}_{v} - \int \underbrace{e^x}_{v}\,\underbrace{dx}_{du}.$$

Is $\int v\,du$ easier than the original integral, $\int u\,dv$? Yes;
$$\int e^x\,dx = e^x.$$

Save the constant of integration till the end. Hence
$$\int xe^x\,dx = xe^x - e^x + C.$$

The reader may check this by differentiation. ∎

EXAMPLE 2 Use integration by parts to find $\int (\sin x)x\,dx$.

SOLUTION Two ways to break up $(\sin x)x\, dx$ come to mind:

(A) $\underbrace{\sin x}_{u}\ \underbrace{x\, dx}_{dv}$

and

(B) $\underbrace{x}_{u}\ \underbrace{\sin x\, dx}_{dv}.$

Let us see what each method gives.

Method A.
$$\int \underbrace{\sin x}_{u}\ \underbrace{x\, dx}_{dv} = \underbrace{\sin x}_{u}\ \underbrace{\frac{x^2}{2}}_{v} - \int \underbrace{\frac{x^2}{2}}_{v}\ \underbrace{\cos x\, dx}_{du}$$

$u = \sin x \qquad du = \cos x\, dx$

$dv = x\, dx \qquad v = x^2/2$

The new integrand is harder than the one we started with. Though $\cos x$ is not harder than $\sin x$, $(x^2/2)\cos x$ is definitely harder than $x \sin x$, for the exponent has increased from 1 to 2. This route being fruitless, we try method B.

Method B. For convenience, first rewrite the integrand, $(\sin x)x$ as $x \sin x$. Then we have

$$\int \underbrace{x}_{u}\ \underbrace{\sin x\, dx}_{dv} = \underbrace{x}_{u}\ \underbrace{(-\cos x)}_{v} - \int \underbrace{(-\cos x)}_{v}\ \underbrace{dx}_{du}$$

$u = x \qquad du = dx$

$dv = \sin x\, dx \qquad v = -\cos x$

This time $\int v\, du$ is easier than $\int u\, dv$; the exponent of x went from 1 down to 0. To finish method B we have to evaluate $\int \cos x\, dx$. All told,

$$\int x \sin x\, dx = -x \cos x + \int \cos x\, dx$$

$$= -x \cos x + \sin x + C. \quad \blacksquare$$

EXAMPLE 3 Find $\int x \ln x\, dx.$

SOLUTION Setting $dv = \ln x\, dx$ is not a wise move, since $v = \int \ln x\, dx$ is not immediately apparent. But setting $u = \ln x$ is promising since $du = d(\ln x) = dx/x$ is much easier than $\ln x$. This second approach goes through smoothly:

That v is messier than dv is no threat in this case.

$$\int x \ln x \, dx = \int \underbrace{\ln x}_{u} \underbrace{x \, dx}_{dv} = \underbrace{\ln x}_{u} \underbrace{\frac{x^2}{2}}_{v} - \int \underbrace{\frac{x^2}{2}}_{v} \underbrace{\frac{dx}{x}}_{du}$$

$$= \frac{x^2 \ln x}{2} - \int \frac{x \, dx}{2}$$

$$= \frac{x^2 \ln x}{2} - \frac{x^2}{4} + C.$$

The result may be checked by differentiation. ∎

Some words of advice

The key to applying integration by parts is the labeling of u and dv. Usually three conditions should be met:

First: v can be found by integrating and should not be too messy.

Second: du should not be messier than u.

Third: $\int v \, du$ should be easier than the original $\int u \, dv$.

For instance, $\int x^2 e^x \, dx$, with $u = x^2$ and $dv = e^x \, dx$, meets these criteria. In this case, $v = e^x$ can be found and is not too messy; $du = d(x^2) = 2x \, dx$ is easier than u; as Example 4 shows, $\int v \, du$ is indeed easier than $\int u \, dv$.

EXAMPLE 4 Find $\int x^2 e^x \, dx.$

SOLUTION

$$\int \underbrace{x^2}_{u} \underbrace{e^x \, dx}_{dv} = \underbrace{x^2}_{u} \underbrace{e^x}_{v} - \int \underbrace{e^x}_{v} \underbrace{2x \, dx}_{du} = x^2 e^x - 2 \int x e^x \, dx$$

$$u = x^2 \qquad du = 2x \, dx$$
$$dv = e^x \, dx \qquad v = e^x.$$

By Example 1, $\int xe^x \, dx = xe^x - e^x.$

Check the answer by differentiation.

Thus $\int x^2 e^x \, dx = x^2 e^x - 2[xe^x - e^x] + C$

$$= x^2 e^x - 2xe^x + 2e^x + C. \quad \blacksquare$$

Integration by parts, with $u = x^3$, could be used to express $\int x^3 e^x \, dx$ in terms of $\int x^2 e^x \, dx$. Another integration by parts, with $u = x^2$, then expresses $\int x^2 e^x \, dx$ in terms of $\int xe^x \, dx$, as was done in Example 4. Each time integration by parts lowers the exponent by 1.

The next example shows how to integrate any inverse trigonometric function. The resulting formula is in the integral tables.

EXAMPLE 5 Find $\int \tan^{-1} x \, dx$.

SOLUTION Recall that the derivative of $\tan^{-1} x$ is $1/(1 + x^2)$, a much simpler function than $\tan^{-1} x$. This suggests the following approach:

Integrating an inverse function by parts

$$\underbrace{\int \tan^{-1} x}_{u} \underbrace{dx}_{dv} = \underbrace{(\tan^{-1} x)}_{u}\underbrace{x}_{v} - \int \underbrace{x}_{v} \underbrace{\frac{dx}{1+x^2}}_{du} \quad \left(du = \frac{dx}{1+x^2}, \quad v = x \right)$$

$$= x \tan^{-1} x - \int \frac{x}{1 + x^2} \, dx.$$

It is easy to compute
$$\int \frac{x}{1 + x^2} \, dx,$$

since the numerator is a constant times the derivative of the denominator:

$$\int \frac{x \, dx}{1 + x^2} = \frac{1}{2} \int \frac{2x}{1 + x^2} \, dx = \frac{1}{2} \ln (1 + x^2).$$

Check by differentiation. Hence $\int \tan^{-1} x \, dx = x \tan^{-1} x - \frac{1}{2} \ln (1 + x^2) + C.$ ∎

Evaluation by integration by parts of a definite integral $\int_a^b f(x) \, dx$, where $f(x) = u(x)v'(x)$, takes the form

$$\int_a^b f(x) \, dx = \int_a^b u \, dv = uv \Big|_a^b - \int_a^b v \, du$$

$$= u(b)v(b) - u(a)v(a) - \int_a^b v(x)u'(x) \, dx.$$

For instance, by Example 5,

$$\int_0^1 \tan^{-1} x \, dx = x \tan^{-1} x \Big|_0^1 - \int_0^1 \frac{x}{1 + x^2} \, dx$$

$$= x \tan^{-1} x \Big|_0^1 - \frac{1}{2} \ln (1 + x^2) \Big|_0^1$$

$$= 1 \tan^{-1} 1 - 0 \tan^{-1} 0 - \frac{1}{2} \ln (1 + 1^2) + \frac{1}{2} \ln (1 + 0^2)$$

$$= \frac{\pi}{4} - \frac{1}{2} \ln 2.$$

EXAMPLE 6 Find $\int e^x \cos x \, dx$.

SOLUTION Proceed as follows:

$$\int \underbrace{e^x}_{u} \underbrace{\cos x \, dx}_{dv} = \underbrace{e^x}_{u} \underbrace{\sin x}_{v} - \int \underbrace{\sin x}_{v} \underbrace{e^x \, dx}_{du},$$

$$du = e^x \, dx, \quad v = \sin x.$$

Repeated integration by parts   It may seem that nothing useful has been accomplished; cos x is replaced by sin x. But watch closely, as the new integral is treated to an integration by parts. Capital letters U and V, instead of u and v, are used to distinguish this computation from the preceding one.

$$\int \underbrace{e^x}_{U} \underbrace{\sin x \, dx}_{dV} = \underbrace{e^x}_{U} \underbrace{(-\cos x)}_{V} - \int \underbrace{(-\cos x)}_{V} \underbrace{e^x \, dx}_{dU} \quad (dU = e^x \, dx, \ V = -\cos x)$$

$$= -e^x \cos x + \int e^x \cos x \, dx.$$

Combining the two yields

$$\int e^x \cos x \, dx = e^x \sin x - \left(-e^x \cos x + \int e^x \cos x \, dx\right)$$

$$= e^x(\sin x + \cos x) - \int e^x \cos x \, dx.$$

This provides an equation for the unknown integral:

$$2\int e^x \cos x \, dx = e^x(\sin x + \cos x).$$

Hence

$$\int e^x \cos x \, dx = \tfrac{1}{2} e^x(\sin x + \cos x) + C. \quad \blacksquare$$

EXAMPLE 7   Many formulas in a table of integrals express the integral of a function that involves the nth power of some expression in terms of the integral of a function that involves the $(n-1)$st or lower power of the same expression. *Reduction or recursion formulas*   These are reduction formulas. Usually they are obtained by an integration by parts. For instance, derive the formula

$$\int \sin^n x \, dx = -\frac{\sin^{n-1} x \cos x}{n} + \frac{n-1}{n}\int \sin^{n-2} x \, dx, \qquad n \geq 2.$$

SOLUTION   First write $\int \sin^n x \, dx$ as $\int \sin^{n-1} x \sin x \, dx$. Then let $u = \sin^{n-1} x$ and $dv = \sin x \, dx$. Thus

$$du = (n-1)\sin^{n-2} x \cos x \, dx \quad \text{and} \quad v = -\cos x.$$

Integration by parts yields

$$\int \underbrace{\sin^{n-1} x}_{u} \underbrace{\sin x \, dx}_{dv} = \underbrace{(\sin^{n-1} x)}_{u}\underbrace{(-\cos x)}_{v} - \int \underbrace{(-\cos x)}_{v}\underbrace{(n-1)\sin^{n-2} x \cos x \, dx}_{du}.$$

But the integral on the right of the above equation is equal to

$$\int (1-n)\cos^2 x \sin^{n-2} x \, dx = (1-n)\int (1-\sin^2 x)\sin^{n-2} x \, dx$$

$$= (1-n)\int \sin^{n-2} x \, dx - (1-n)\int \sin^n x \, dx.$$

Thus

$$\int \sin^n x \, dx = -\sin^{n-1} x \cos x + (n-1) \int \sin^{n-2} x \, dx - (n-1) \int \sin^n x \, dx.$$

Rather than being dismayed by the reappearance of $\int \sin^n x \, dx$, collect like terms:

$$n \int \sin^n x \, dx = -\sin^{n-1} x \cos x + (n-1) \int \sin^{n-2} x \, dx,$$

from which the quoted formula follows. ∎

Exercises

In Exercises 1 to 22, evaluate the integrals by integration by parts.

1. $\int x e^{2x} \, dx$

2. $\int (x+3) e^{-x} \, dx$

3. $\int x \sin 2x \, dx$

4. $\int (x+3) \cos 2x \, dx$

5. $\int x \ln 3x \, dx$

6. $\int (2x+1) \ln x \, dx$

7. $\int_1^2 x^2 e^{-x} \, dx$

8. $\int_0^1 x^2 e^{2x} \, dx$

9. $\int_0^1 \sin^{-1} x \, dx$

10. $\int_0^{1/2} \tan^{-1} 2x \, dx$

11. $\int x^2 \ln x \, dx$

12. $\int x^3 \ln x \, dx$

13. $\int x(3x+5)^{10} \, dx$

14. $\int (x+2)(2x-1)^{50} \, dx$

15. $\int_2^3 (\ln x)^2 \, dx$

16. $\int_2^3 (\ln x)^3 \, dx$

17. $\int_1^e \frac{\ln x \, dx}{x^2}$

18. $\int_e^{e^2} \frac{\ln x \, dx}{x^3}$

19. $\int e^x \sin x \, dx$

20. $\int e^{-2x} \sin 3x \, dx$

21. $\int \frac{\ln(1+x^2) \, dx}{x^2}$

22. $\int x \ln(x^2) \, dx$

23. Obtain this recursion formula, which is usually to be found in a table of integrals:

$$\int \sin^n ax \, dx$$
$$= -\frac{\sin^{n-1} ax \cos ax}{na} + \frac{n-1}{n} \int \sin^{n-2} ax \, dx.$$

24. Obtain this recursion formula ($m, n \geq 0$):

$$\int \sin^m x \cos^n x \, dx$$
$$= -\frac{\sin^{m-1} x \cos^{n+1} x}{m+n} + \frac{m-1}{m+n} \int \sin^{m-2} x \cos^n x \, dx.$$

25. Find $\int \ln(x+1) \, dx$ using
    (a) $u = \ln(x+1)$, $dv = dx$, $v = x$;
    (b) $u = \ln(x+1)$, $dv = dx$, $v = x+1$.
    (c) Which is easier?

In Exercises 26 to 28 obtain recursion formulas for the integrals.

26. $\int x^n e^{ax} \, dx$, n an integer > 0, a a nonzero constant.

27. $\int (\ln x)^n \, dx$, n an integer.

28. $\int x^n \sin x \, dx$, n an integer > 0.

In Exercises 29 to 32 find the integrals. In each case a substitution is required before integration by parts can be used. In Exercises 31 and 32 the notation $\exp(u)$ is used for e^u. This notation is often used for clarity.

29. $\int \sin \sqrt{x} \, dx$

30. $\int \sin \sqrt[3]{x} \, dx$

31. $\int \exp(\sqrt{x}) \, dx$

32. $\int \exp(\sqrt[3]{x}) \, dx$

∎

33 Use the recursion in Example 7 to find

(a) $\int \sin^2 x \, dx$ (b) $\int \sin^4 x \, dx$ (c) $\int \sin^6 x \, dx$.

34 Use the recurion in Example 7 to find

(a) $\int \sin^3 x \, dx$ (b) $\int \sin^5 x \, dx$.

35 $\int [(\sin x)/x] \, dx$ is not elementary. Deduce that $\int \cos x \ln x \, dx$ is not elementary.

36 $\int x \tan x \, dx$ is not elementary. Deduce that $\int (x/\cos x)^2 \, dx$ is not elementary.

In Exercises 37 to 40 find the integrals two ways: (a) by substitution, (b) by integration by parts.

37 $\int x\sqrt{3x+7} \, dx$ **38** $\int \dfrac{x \, dx}{\sqrt{2x+7}}$

39 $\int x(ax+b)^3 \, dx$ **40** $\int \dfrac{x \, dx}{ax+b}, \; a \neq 0$

41 Let I_n denote $\int_0^{\pi/2} \sin^n \theta \, d\theta$, where n is a nonnegative integer.
(a) Evaluate I_0 and I_1.
(b) Using the recursion in Example 7, show that

$$I_n = \frac{n-1}{n} I_{n-2}, \quad \text{for } n \geq 2.$$

(c) Use (b) to evaluate I_2 and I_3.
(d) Use (b) to evaluate I_4 and I_5.
(e) Find a formula for I_n, n odd.
(f) Find a formula for I_n, n even.
(g) Explain why $\int_0^{\pi/2} \cos^n \theta \, d\theta = \int_0^{\pi/2} \sin^n \theta \, d\theta$.

42 Show that

$$\int_0^1 x f''(x) \, dx = 3$$

for every function $f(x)$ that satisfies the following conditions: (i) $f(x)$ is defined for all x, (ii) $f''(x)$ is continuous, (iii) $f(0) = f(1)$, (iv) $f'(1) = 3$.

Exercises 43 and 44 are related.

43 In a certain race, a car starts from rest and ends at rest, having traveled 1 mile in 1 minute. Let $v(t)$ be its velocity at time t, and $a(t)$ be its acceleration at time t. Show that

(a) $\int_0^1 v(t) \, dt = 1$; (b) $\int_0^1 a(t) \, dt = 0$;

(c) $\int_0^1 t a(t) \, dt = -1$.

44 (Continuation of Exercise 43.)
(a) Show that at some time t we have $|a(t)| > 4$.
(b) Show graphically [drawing $v(t)$ as a function of time] that a race can be driven as in Exercise 43, but with $|a(t)| \leq 4.1$ for all t.

Exercise 45 shows how integration by parts can be used to study the approximation of a function by a polynomial.

45 Let f have derivatives of all orders.
(a) Explain why $f(b) = f(0) + \int_0^b f'(x) \, dx$.
(b) Using an integration by parts on the definite integral in (a), with $u = f'(x)$ and $dv = dx$, show that

$$f(b) = f(0) + f'(0)b + \int_0^b f^{(2)}(x)(b-x) \, dx.$$

Hint: Use $v = x - b$.
(c) Similarly, show that

$$f(b) = f(0) + f'(0)b + \frac{f^{(2)}(0)}{2}b^2 + \frac{1}{2}\int_0^b f^{(3)}(x)(b-x)^2 \, dx.$$

(d) Check that (c) is correct for any quadratic polynomial $f(x) = Ax^2 + Bx + C$.
(e) Use another integration by parts on the formula in (c) to obtain the "next formula."

7.3 Using a table of integrals (optional)

There are many methods for finding antiderivatives. The three most frequently used are substitution (described in Sec. 7.1), integration by parts (described in Sec. 7.2), and looking through the three or four hundred formulas in a table of integrals. This section shows how to use an integral table.

First browse through your table of integrals. Notice how the formulas are grouped. They may be arranged as follows:

Group

1. Expressions containing $ax + b$
2. Expressions containing $ax^2 + c$, $ax^n + c$, $x^2 \pm p^2$, and $p^2 - x^2$
3. Expressions containing $ax^2 + bx + c$
4. Miscellaneous algebraic expressions
5. Expressions containing $\sin ax$
6. Expressions containing $\cos ax$
7. Expressions containing $\sin ax$ and $\cos ax$
8. Expressions containing $\tan ax$ or $\cot ax$
9. Expressions containing $\sec ax$ or $\csc ax$
10. Expressions containing $\tan ax$ and $\sec ax$ or $\cot ax$ and $\csc ax$
11. Expressions containing algebraic and trigonometric functions
12. Expressions containing exponential and logarithmic functions
13. Expressions containing inverse trigonometric functions

Some tables, such as the CRC Standard Math Tables and Burington's *Handbook of Mathematical Tables and Formulas*, (McGraw-Hill), use log instead of ln. It is understood that log is taken with respect to base e, *not* base 10.

EXAMPLE 1 Find $\int \sin^6 2x \, dx$ with the aid of the integral tables.

SOLUTION The integrand involves $\sin 2x$, so turn to "Expressions containing $\sin ax$." Here $a = 2$. There are formulas for

$$\int \sin ax \, dx, \quad \int \sin^2 ax \, dx, \quad \int \sin^3 ax \, dx, \quad \int \sin^4 ax \, dx,$$

but for no higher power of $\sin ax$. Instead, this formula is given:

$$\int \sin^n ax \, dx = \frac{-\sin^{n-1} ax \cos ax}{na} + \frac{n-1}{n} \int \sin^{n-2} ax \, dx.$$

This is a *recursion* formula. It reduces $\int \sin^n ax \, dx$ to $\int \sin^{n-2} ax \, dx$ recursively, knocking down the exponent by 2 at each stage. (Exercise 23, Sec. 7.2, shows how this formula is found.) We need the case $n = 6$, $a = 2$, which reads

$$\int \sin^6 2x \, dx = \frac{-\sin^5 2x \cos 2x}{6 \cdot 2} + \frac{5}{6} \int \sin^4 2x \, dx.$$

Now $\int \sin^4 ax \, dx$ is listed explicitly; there is no need to use the recursion for $n = 4$. The table includes a formula for $\int \sin^4 ax \, dx$, giving

The constant of integration is omitted in integral tables.

$$\int \sin^4 2x \, dx = \frac{3x}{8} - \frac{3 \sin 4x}{32} - \frac{\sin^3 2x \cos 2x}{8}.$$

Combination of this with the equation relating $\int \sin^6 2x \, dx$ to $\int \sin^4 2x \, dx$ yields a formula for $\int \sin^6 2x \, dx$. The reader may complete the computation. ∎

EXAMPLE 2 Find these three integrals with the aid of the integral tables:

$$\int \frac{dx}{x^2 - 5x + 3}, \tag{1}$$

$$\int \frac{dx}{x^2 - 2x + 3}, \tag{2}$$

$$\int \frac{dx}{4x^2 - 12x + 9}. \tag{3}$$

SOLUTION Under "Expressions containing $ax^2 + bx + c$" are listed three different formulas for

$$\int \frac{dx}{ax^2 + bx + c}.$$

The three separate cases are $b^2 > 4ac$, $b^2 < 4ac$, and $b^2 = 4ac$. The formulas listed are quite different:

$$\int \frac{dx}{ax^2 + bx + c} = \frac{1}{\sqrt{b^2 - 4ac}} \ln \frac{2ax + b - \sqrt{b^2 - 4ac}}{2ax + b + \sqrt{b^2 - 4ac}}, \quad b^2 > 4ac;$$

$$\int \frac{dx}{ax^2 + bx + c} = \frac{2}{\sqrt{4ac - b^2}} \tan^{-1} \frac{2ax + b}{\sqrt{4ac - b^2}}, \quad b^2 < 4ac;$$

$$\int \frac{dx}{ax^2 + bx + c} = -\frac{2}{2ax + b}, \quad b^2 = 4ac.$$

Many tables omit the absolute value sign in $\ln |u|$. It is assumed that the user will supply it when needed.

In (1), where the denominator is $x^2 - 5x + 3$, $a = 1$, $b = -5$, $c = 3$; hence $b^2 - 4ac = (-5)^2 - 4 \cdot 1 \cdot 3 = 13 > 0$. Thus the first formula applies:

$$\int \frac{dx}{x^2 - 5x + 3} = \frac{1}{\sqrt{13}} \ln \frac{2x - 5 - \sqrt{13}}{2x - 5 + \sqrt{13}}.$$

In (2), where the denominator is $x^2 - 2x + 3$, $a = 1$, $b = -2$, $c = 3$; hence $b^2 - 4ac = (-2)^2 - 4 \cdot 1 \cdot 3 = -8 < 0$. Thus the second formula applies:

$$\int \frac{dx}{x^2 - 2x + 3} = \frac{2}{\sqrt{8}} \tan^{-1} \frac{2x - 2}{\sqrt{8}}.$$

In (3), where the denominator is $4x^2 - 12x + 9$, $a = 4$, $b = -12$, $c = 9$; hence $b^2 - 4ac = (-12)^2 - 4 \cdot 4 \cdot 9 = 144 - 144 = 0$. Thus the third formula applies:

$$\int \frac{dx}{4x^2 - 12x + 9} = \frac{-2}{8x - 12}.$$

The reader may check these three answers simply by differentiating them. ∎

EXAMPLE 3 Find $\int \sqrt{\sin^2 \theta + 5} \cos \theta \, d\theta.$

SOLUTION A search through the expressions containing $\sin \theta$ and $\cos \theta$ will yield no formula covering this case. However, note that $\cos \theta$ is the derivative of $\sin \theta$. This suggests the substitution

Substitution may be needed first.

$$u = \sin \theta;$$

hence
$$du = \cos \theta \, d\theta.$$

Then
$$\int \underbrace{\sqrt{\sin^2 \theta + 5}}_{u^2 + 5} \underbrace{\cos \theta \, d\theta}_{du} = \int \sqrt{u^2 + 5} \, du$$

Now look in the integral tables for $\int \sqrt{u^2 + 5} \, du$. If 5 is written as $(\sqrt{5})^2$, the integrand becomes "an expression in $u^2 + (\sqrt{5})^2$." So look among the "Expressions containing $x^2 \pm p^2$" and find (replacing x by u)

$$\int \sqrt{u^2 + p^2} \, du = \tfrac{1}{2}[u\sqrt{u^2 + p^2} + p^2 \ln(u + \sqrt{u^2 + p^2})].$$

In the problem considered, $p = \sqrt{5}$ and $p^2 = 5$. Thus

$$\int \sqrt{u^2 + 5} \, du = \int \sqrt{u^2 + (\sqrt{5})^2} \, du = \tfrac{1}{2}[u\sqrt{u^2 + 5} + 5 \ln(u + \sqrt{u^2 + 5})].$$

To find $\int \sqrt{\sin^2 \theta + 5} \cos \theta \, d\theta$, replace u by $\sin \theta$ in the formula just obtained. Thus

$$\int \sqrt{\sin^2 \theta + 5} \cos \theta \, d\theta$$
$$= \tfrac{1}{2}[\sin \theta \sqrt{\sin^2 \theta + 5} + 5 \ln(\sin \theta + \sqrt{\sin^2 \theta + 5})]. \quad \blacksquare$$

EXAMPLE 4 Examine $\int \dfrac{x^5 \, dx}{\sqrt{x^3 + 1}}.$

SOLUTION This integral is usually not listed in an integral table. Note that $x^5 \, dx = x^3 x^2 \, dx$ and that $x^2 \, dx$ is almost $d(x^3)$. So let $u = x^3$; hence

$$du = 3x^2 \, dx \quad \text{and} \quad x^2 \, dx = \frac{du}{3}.$$

Thus
$$\int \frac{x^5 \, dx}{\sqrt{x^3 + 1}} = \int \frac{x^3 x^2 \, dx}{\sqrt{x^3 + 1}}$$
$$= \int \frac{u}{\sqrt{u + 1}} \frac{du}{3}$$

$$= \frac{1}{3} \int \frac{u \, du}{\sqrt{u+1}},$$

which is listed in the integral tables under "Expressions containing $ax + b$."

Alternatively, one could have substituted $u = x^3 + 1$. This leads to an easier integral, as the reader may check. Integral tables would not be needed. ■

There are many classes of functions not listed in the integral tables. For instance, there is no formula in the integral tables allowing us to find

$$\int \frac{5x^3 + x - 2}{x^3 - x^2 - x - 2} \, dx.$$

Sections 7.4 and 7.5 show how to integrate quotients of polynomials, the so-called *rational functions*.

Exercises

Integrate with the aid of a table of integrals. For Exercises 1 to 45 the table inside the covers suffices.

1. $\int (3x + 5)^5 \, dx$
2. $\int \frac{dx}{(3x + 7)^3}$
3. $\int \frac{x \, dx}{(2x - 5)^2}$
4. $\int \frac{dx}{x^2(3x - 2)}$
5. $\int \sqrt{5x - 7} \, dx$
6. $\int x\sqrt{2x + 3} \, dx$
7. $\int \frac{dx}{\sqrt{5x + 4}}$
8. $\int \frac{\sqrt{5x + 4}}{x} \, dx$
9. $\int \frac{dx}{x\sqrt{3x - 2}}$
10. $\int \frac{dx}{x\sqrt{3x + 2}}$
11. $\int \frac{dx}{9 + x^2}$
12. $\int \frac{dx}{x^2 + 5}$
13. $\int \frac{dx}{4 - x^2}$
14. $\int \frac{dx}{3 - x^2}$
15. $\int \frac{dx}{5x^2 + 3}$
16. $\int \frac{dx}{5x^2 - 3}$
17. $\int \frac{x \, dx}{3x^2 + 1}$
18. $\int \frac{dx}{x^2(3x + 1)}$
19. $\int \sqrt{x^2 - 1} \, dx$
20. $\int \sqrt{x^2 - 4} \, dx$
21. $\int \sqrt{x^2 - 3} \, dx$
22. $\int \frac{dx}{2x^2 + x + 3}$
23. $\int \frac{x \, dx}{2x^2 + x + 3}$
24. $\int \frac{dx}{\sqrt{x^2 + 3x + 5}}$
25. $\int \frac{dx}{\sqrt{-x^2 + 3x + 5}}$
26. $\int \sqrt{4x - x^2} \, dx$
27. $\int \sqrt{\frac{x + 3}{x - 2}} \, dx$
28. $\int \sin^2 3x \, dx$
29. $\int \sin^7 2x \, dx$
30. $\int \frac{dx}{1 - \sin 3x}$
31. $\int \cos^3 5x \, dx$
32. $\int \sin 3x \cos 2x \, dx$
33. $\int \sec 5x \, dx$
34. $\int x \sin 3x \, dx$
35. $\int x^2 \cos 3x \, dx$
36. $\int 2^x \, dx$
37. $\int x^2 e^{-x} \, dx$
38. $\int x^3 e^{2x} \, dx$
39. $\int e^{-3x} \sin 5x \, dx$
40. $\int x^2 \ln 3x \, dx$
41. $\int \sin^{-1} 3x \, dx$
42. $\int \tan^{-1} 5x \, dx$
43. $\int \sec^{-1} 4x \, dx$
44. $\int \cos 2x \tan 2x \, dx$
45. $\int \frac{dx}{1 + \sin 3x}$

■

404 COMPUTING ANTIDERIVATIVES

After a substitution, evaluate with the aid of a table of integrals.

46 $\int \dfrac{x^5 \, dx}{(5x^3 + 3)^2}$; let $u = x^3$

47 $\int x^2 \sqrt{5x^3 + 2} \, dx$

48 $\int \dfrac{x \, dx}{\sqrt{2x^2 + 3}}$

49 $\int \dfrac{\cos x \, dx}{4 + \sin^2 x}$

50 $\int \dfrac{\sin 2x \, dx}{9 - \cos^2 2x}$

51 $\int \sqrt{(1 + 3x)^2 - 5} \, dx$

52 $\int \dfrac{x^2}{\sqrt{1 - x^6}} \, dx$

53 $\int x \sqrt{x^4 + 9} \, dx$

54 $\int x \sqrt{9 - x^4} \, dx$

7.4 How to integrate certain rational functions

This section shows how to compute

$$\int \dfrac{dx}{(ax + b)^n}, \quad \int \dfrac{dx}{(ax^2 + bx + c)^n}, \quad \text{and} \quad \int \dfrac{x \, dx}{(ax^2 + bx + c)^n}.$$

(The polynomial $ax^2 + bx + c$ is assumed to be irreducible, that is, not the product of two polynomials of degree one.) These three types of integrals will play a basic role in Sec. 7.5.

HOW TO COMPUTE $\int \dfrac{dx}{(ax + b)^n}$

As was shown in Example 13 of Sec. 7.1, the computation of $\int dx/(ax + b)^n$ can be accomplished by the substitution

$$u = ax + b.$$

For instance, $\int \dfrac{dx}{(3x + 2)^5} = \int u^{-5} \dfrac{du}{3}$

$\qquad\qquad\qquad = \dfrac{1}{3} \dfrac{u^{-4}}{-4} + C$

$\qquad\qquad\qquad = \dfrac{-1}{12} \dfrac{1}{(3x + 2)^4} + C.$

$u = 3x + 2$
$du = 3 \, dx$
$\dfrac{du}{3} = dx$

Also, with the same u,

$$\int \dfrac{dx}{3x + 2} = \int \dfrac{du/3}{u} = \dfrac{1}{3} \int \dfrac{du}{u} = \dfrac{1}{3} \ln |u| + C$$

$$= \dfrac{1}{3} \ln |3x + 2| + C.$$

The integral $\int dx/(3x + 2)$ could also be evaluated by noticing that the numerator is almost the derivative of the denominator.

7.4 HOW TO INTEGRATE CERTAIN RATIONAL FUNCTIONS

For any polynomial $P(x)$ the integral

$$\int \frac{P(x)\, dx}{(ax + b)^n} \tag{1}$$

can be computed by making the substitution $u = ax + b$. (Example 12 of Sec. 7.1 illustrates the computations.) Thus a rational function of the form (1) is not hard to integrate. The integral will be the sum of at most three types of functions: a polynomial, a logarithm, and $k/(ax + b)^m$ where $1 \leq m < n$, where k is a constant.

$$\textbf{HOW TO COMPUTE } \int \frac{dx}{ax^2 + bx + c}$$

$ax^2 + bx + c$ is irreducible exactly when $b^2 - 4ac < 0$.

Integration of a rational function becomes more complicated if an irreducible quadratic, $ax^2 + bx + c$, appears in the denominator. (As shown in Exercise 42, $ax^2 + bx + c$ is irreducible exactly when $b^2 - 4ac$ is negative. The case when $ax^2 + bx + c$ is not irreducible turns out to be easier and is treated in Sec. 7.5.)

Some examples will show how to compute $\int dx/(ax^2 + bx + c)$. (Throughout this section it is assumed that the polynomial $ax^2 + bx + c$ is irreducible.)

EXAMPLE 1 Find $\displaystyle\int \frac{dx}{4x^2 + 1}$.

SOLUTION This resembles $\displaystyle\int \frac{dx}{x^2 + 1} = \tan^{-1} x + C$.

For this reason, make the substitution

$u = -2x$ would work too.

$$u^2 = 4x^2 \quad \text{or} \quad u = 2x;$$

hence

$$du = 2\, dx \quad \text{and} \quad \frac{du}{2} = dx.$$

Then $\displaystyle\int \frac{dx}{4x^2 + 1} = \int \frac{1}{u^2 + 1} \frac{du}{2} = \frac{1}{2} \int \frac{du}{u^2 + 1} = \frac{1}{2} \tan^{-1} u + C$

$$= \frac{1}{2} \tan^{-1} 2x + C. \quad \blacksquare$$

EXAMPLE 2 Find $\displaystyle\int \frac{dx}{4x^2 + 9}$.

SOLUTION Again the motivation is provided by the fact that

$$\int \frac{dx}{x^2 + 1} = \tan^{-1} x + C.$$

This time choose u such that $9u^2 = 4x^2$. This substitution is suggested by the equations

$$\frac{1}{4x^2 + 9} = \frac{1}{9u^2 + 9} = \frac{1}{9} \frac{1}{u^2 + 1}.$$

So choose u such that

$$3u = 2x;$$

hence $\quad\quad 3\,du = 2\,dx \quad$ and $\quad \tfrac{3}{2}\,du = dx.$

Thus

$$\int \frac{dx}{4x^2+9} = \int \frac{1}{9u^2+9} \frac{3}{2}\,du = \frac{3}{18} \int \frac{du}{u^2+1} = \frac{1}{6} \tan^{-1} u + C$$

$$= \frac{1}{6} \tan^{-1} \frac{2x}{3} + C.$$

(Note that only at the end is it necessary to solve for u; $u = 2x/3$.) ∎

The next example uses "completing the square," an algebraic technique described in Appendix B.

EXAMPLE 3 Find $\displaystyle\int \frac{dx}{x^2 + 4x + 13}.$

SOLUTION (Since $4^2 - 4 \cdot 1 \cdot 13$ is negative, $x^2 + 4x + 13$ is irreducible.) Begin by completing the square in the denominator:

Completing the square

$$x^2 + 4x + 13 = x^2 + 4x + 2^2 + 13 - 2^2$$
$$= (x+2)^2 + 9.$$

Thus $\displaystyle\int \frac{dx}{x^2+4x+13} = \int \frac{dx}{(x+2)^2+9},$

an integral reminiscent of those in Examples 1 and 2.

To complete the integration, introduce a function u such that

$$9u^2 = (x+2)^2$$

or $\quad\quad 3u = x+2 \quad$ (hence $u = (x+2)/3$)

$3\,du = dx.$

Thus $\displaystyle\int \frac{dx}{(x+2)^2+9} = \int \frac{3\,du}{9u^2+9} = \frac{3}{9} \int \frac{du}{u^2+1}$

$$= \frac{1}{3} \tan^{-1} u + C.$$

Consequently $\int \dfrac{dx}{x^2 + 4x + 13} = \dfrac{1}{3} \tan^{-1} \dfrac{x+2}{3} + C.$ ∎

In the next example the coefficient of x^2 is not 1; completing the square involves a little more algebra.

EXAMPLE 4 Find $\int \dfrac{dx}{4x^2 + 8x + 13}.$

SOLUTION First complete the square in the denominator $4x^2 + 8x + 13$, as follows:

$$4x^2 + 8x + 13 = 4(x^2 + 2x) + 13$$
$$= 4[x^2 + 2x + (\tfrac{2}{2})^2] + 13 - 4(\tfrac{2}{2})^2$$
$$= 4(x+1)^2 + 9.$$

The integral now reads $\int \dfrac{dx}{4(x+1)^2 + 9},$

which resembles Example 2.

Choose the substitution such that $9u^2 = 4(x+1)^2$. Thus choose u so that

$$3u = 2(x+1) \quad \text{(hence } u = 2(x+1)/3\text{).}$$

Consequently $3\,du = 2\,dx$ and $dx = \dfrac{3\,du}{2}.$

The substitution yields

$$\int \dfrac{dx}{4(x+1)^2 + 9} = \int \dfrac{\tfrac{3}{2}\,du}{9u^2 + 9}$$
$$= \dfrac{3}{2} \cdot \dfrac{1}{9} \int \dfrac{du}{u^2 + 1}$$
$$= \dfrac{1}{6} \tan^{-1} u + C.$$

Thus $\int \dfrac{dx}{4x^2 + 8x + 13} = \dfrac{1}{6} \tan^{-1} \dfrac{2(x+1)}{3} + C,$

a result that the skeptical may check by differentiation. ∎

As these examples show, to compute

$$\int \dfrac{dx}{ax^2 + bx + c} \quad (b^2 - 4ac < 0)$$

complete the square and then make a substitution. The integral will be an arctangent.

The integral $\int dx/(ax^2 + bx + c)^n$, $n > 1$, is computed by a recursive formula to be found in Exercise 44.

HOW TO COMPUTE $\int \dfrac{x\,dx}{ax^2 + bx + c}$

The computation of $\int x\,dx/(ax^2 + bx + c)$ can be reduced to that of $\int dx/(ax^2 + bx + c)$, as is shown in Example 5.

EXAMPLE 5 Find
$$\int \frac{x\,dx}{4x^2 + 8x + 13}.$$

SOLUTION If the numerator were $8x + 8$, it would be the derivative of the denominator. The problem would then be covered by the formula

$$\int \frac{f'}{f}\,dx = \ln|f| + C.$$

This prompts the following maneuver:

Now recall Example 4.

$4x^2 + 8x + 13$ *is always positive. In fact, it equals* $4(x + 1)^2 + 9$, *which is never less than 9.*

$$\int \frac{x\,dx}{4x^2 + 8x + 13} = \frac{1}{8}\int \frac{8x\,dx}{4x^2 + 8x + 13}$$

$$= \frac{1}{8}\left(\int \frac{8x + 8}{4x^2 + 8x + 13}\,dx - \int \frac{8}{4x^2 + 8x + 13}\,dx\right)$$

$$= \frac{1}{8}\left[\ln(4x^2 + 8x + 13) - \frac{8}{6}\tan^{-1}\frac{2(x + 1)}{3}\right] + C.$$

Note that the result of Example 4 was used. ∎

SUMMARY OF THIS SECTION

Integrand		Method of Integration
$\dfrac{1}{(ax + b)^n}$		Substitute $u = ax + b$.
$\dfrac{1}{ax^2 + c}$	$a, c > 0$	Substitute so $cu^2 = ax^2$: $u = \sqrt{\dfrac{a}{c}}\,x$.
$\dfrac{1}{ax^2 + bx + c}$		Complete the square, then substitute.
$\dfrac{x}{ax^2 + bx + c}$		First write as $\dfrac{1}{2a}\cdot\dfrac{(2ax + b) - b}{ax^2 + bx + c}$, then break into two parts.

It should be kept in mind that

$$\int \frac{2ax + b}{ax^2 + bx + c}\,dx = \ln|ax^2 + bx + c| + C.$$

Note that, for any constants c_1 and d_1,

$$\int \frac{c_1 x + d_1}{ax^2 + bx + c}\,dx = \int \frac{c_1 x}{ax^2 + bx + c}\,dx + \int \frac{d_1}{ax^2 + bx + c}\,dx.$$

This will be needed in the next section.

Exercises

Compute the integrals in Exercises 1 to 36.

1. $\int \dfrac{dx}{3x - 4}$
2. $\int \dfrac{2\,dx}{3x + 6}$
3. $\int \dfrac{5\,dx}{(2x + 7)^2}$
4. $\int \dfrac{dx}{(4x + 1)^3}$
5. $\int \dfrac{dx}{x^2 + 9}$
6. $\int \dfrac{dx}{9x^2 + 1}$
7. $\int \dfrac{x\,dx}{x^2 + 9}$
8. $\int \dfrac{x\,dx}{x^2 + 2}$
9. $\int \dfrac{2x + 3}{x^2 + 9}\,dx$
10. $\int \dfrac{3x - 5}{x^2 + 9}\,dx$
11. $\int \dfrac{dx}{16x^2 + 25}$
12. $\int \dfrac{dx}{9x^2 + 4}$
13. $\int \dfrac{x\,dx}{16x^2 + 25}$
14. $\int \dfrac{x\,dx}{9x^2 + 4}$
15. $\int \dfrac{x + 2}{9x^2 + 4}\,dx$
16. $\int \dfrac{2x - 1}{9x^2 + 4}\,dx$
17. $\int \dfrac{dx}{2x^2 + 3}$
18. $\int \dfrac{x\,dx}{2x^2 + 3}$
19. $\int \dfrac{dx}{x^2 + 2x + 3}$
20. $\int \dfrac{dx}{x^2 + 2x + 5}$
21. $\int \dfrac{dx}{x^2 - 2x + 3}$
22. $\int \dfrac{x\,dx}{x^2 - 2x + 3}$
23. $\int \dfrac{dx}{2x^2 + x + 3}$
24. $\int \dfrac{dx}{3x^2 - 12x + 13}$
25. $\int \dfrac{dx}{x^2 + 4x + 7}$
26. $\int \dfrac{dx}{x^2 + 4x + 9}$
27. $\int \dfrac{dx}{2x^2 + 4x + 7}$
28. $\int \dfrac{dx}{2x^2 + 6x + 5}$
29. $\int \dfrac{2x\,dx}{x^2 + 2x + 3}$
30. $\int \dfrac{2x\,dx}{x^2 + 2x + 5}$
31. $\int \dfrac{3x\,dx}{5x^2 + 3x + 2}$
32. $\int \dfrac{x\,dx}{5x^2 - 3x + 2}$
33. $\int \dfrac{x + 1}{x^2 + x + 1}\,dx$
34. $\int \dfrac{x + 3}{x^2 + x + 1}\,dx$
35. $\int \dfrac{3x + 5}{3x^2 + 2x + 1}\,dx$
36. $\int \dfrac{x + 5}{2x^2 + 3x + 5}\,dx$

■

37. Compute $\int dx/(ax^2 + c)$, $a, c > 0$.
38. Compute $\int x\,dx/(ax^2 + c)$, $a, c > 0$.
39. (a) Verify that
$$\int \frac{dx}{ax^2 + bx + c}$$
$$= \frac{1}{\sqrt{b^2 - 4ac}} \ln\left|\frac{2ax + b - \sqrt{b^2 - 4ac}}{2ax + b + \sqrt{b^2 - 4ac}}\right| + C,$$
$$b^2 > 4ac.$$

(b) Obtain the formula in (a) by the methods of this section.

40. (a) Verify that $\displaystyle\int \frac{dx}{ax^2 + bx + c}$
$$= \frac{2}{\sqrt{4ac - b^2}} \tan^{-1} \frac{2ax + b}{\sqrt{4ac - b^2}} + C, \quad b^2 < 4ac.$$

(b) Obtain the formula in (a) by the methods of this section.

41. (a) Verify that
$$\int \frac{dx}{ax^2 + bx + c} = -\frac{2}{2ax + b} + C, \quad b^2 = 4ac.$$

(b) Obtain the formula in (a) by the methods of this section.

■ ■

42. (a) Show that if r_1 and r_2 are the roots of the equation $ax^2 + bx + c = 0$, then $a(x - r_1)(x - r_2) = ax^2 + bx + c$. (*Suggestion:* Use the quadratic formula.)

(b) Show that if $ax^2 + bx + c = a(x - r_1)(x - r_2)$, then the numbers r_1 and r_2 are roots of the equation $ax^2 + bx + c = 0$.

(c) Deduce from (a) and (b) that $ax^2 + bx + c$ is irreducible exactly when $b^2 - 4ac < 0$.

43 Verify the following reduction formula by differentiating the right side of the equation. This formula, and the one in the next exercise, can be obtained by integration by parts.

$$\int \frac{x\,dx}{(ax^2+bx+c)^{n+1}} = \frac{-(2c+bx)}{n(4ac-b^2)(ax^2+bx+c)^n}$$
$$- \frac{b(2n-1)}{n(4ac-b^2)}\int \frac{dx}{(ax^2+bx+c)^n}.$$

44 Verify the following recursion formula by differentiating the right side of the equation.

$$\int \frac{dx}{(ax^2+bx+c)^{n+1}} = \frac{2ax+b}{n(4ac-b^2)(ax^2+bx+c)^n} +$$
$$\frac{2(2n-1)a}{n(4ac-b^2)}\int \frac{dx}{(ax^2+bx+c)^n}.$$

45 Use the identity in Exercise 44 and the result of Example 1 to find

$$\int \frac{dx}{(4x^2+1)^2}.$$

46 Use the identity in Exercise 43 and the result of Example 4 to find

$$\int \frac{x\,dx}{(4x^2+8x+13)^2}.$$

7.5 The integration of rational functions by partial fractions

The algebraic technique known as *partial fractions* makes it possible to integrate any rational function. For instance, later in this section it will be shown how to compute the integral

$$\int \frac{x^4+x^3-3x+5}{x^3+2x^2+2x+1}\,dx. \qquad (1)$$

(No integral table lists a form that covers (1).) The technique of partial fractions is also used in differential equations.

This section, which is purely algebraic, depends on this result from advanced algebra: Every rational function can be expressed as a sum of a polynomial (which may be 0) and constant multiples of the three types of functions met in Sec. 7.4,

$$\frac{1}{(ax+b)^n}, \quad \frac{1}{(ax^2+bx+c)^n}, \quad \text{and} \quad \frac{x}{(ax^2+bx+c)^n}. \qquad (2)$$

$ax^2 + bx + c$ is assumed to be irreducible. If not, the second and third types can be expressed in terms of the first type,

$$\left(e.g., \frac{1}{x^2-1} = \frac{\frac{1}{2}}{x-1} - \frac{\frac{1}{2}}{x+1}\right).$$

(Moreover, the representation is unique.) The technique of partial fractions was employed in Example 4 of Sec. 6.7 without proper credit when use was made of the identity

$$\frac{M}{y(M-y)} = \frac{1}{y} + \frac{1}{M-y}.$$

Since any polynomial and each of the three types of rational functions in (2) can be integrated, any rational function can be integrated. The only new question of interest is, "What is the method for expressing a rational function as a sum of these four types of simpler functions?" A general method, which is practical in simpler cases and which has even been programmed for a computer, is presented in this section. The resulting expression is called the *partial-fraction* representation of the rational function.

HOW TO FIND THE PARTIAL-FRACTION REPRESENTATION

To express A/B, where A and B are polynomials, as the sum of partial fractions, follow these steps:

1. Make degree of numerator less than degree of denominator.

Step 1 If the degree of A is *not less* than the degree of B, divide B into A to obtain a quotient and a remainder: $A = QB + R$, where the degree of R is less than the degree of B or else $R = 0$. Then

$$\frac{A}{B} = Q + \frac{R}{B}.$$

Apply the remaining steps to R/B.

EXAMPLE 1 If

$$\frac{A}{B} = \frac{3x^3 + x}{x^2 + 3x + 5},$$

carry out step 1.

SOLUTION Since the degree of the numerator is *not* less than the degree of the denominator, carry out a long division:

$$\begin{array}{r} 3x - 9 \\ x^2 + 3x + 5 \overline{\smash{\big)}\ 3x^3 + 0x^2 + x + 0} \\ \underline{3x^3 + 9x^2 + 15x } \\ -9x^2 - 14x + 0 \\ \underline{-9x^2 - 27x - 45} \\ 13x + 45 \end{array}$$

quotient

remainder

Thus, $\qquad \dfrac{3x^3 + x}{x^2 + 3x + 5} = 3x - 9 + \dfrac{13x + 45}{x^2 + 3x + 5}.$

(To check, just multiply both sides by $x^2 + 3x + 5$.) ∎

Similarly, in the case $\qquad \dfrac{3x^2 + x}{x^2 + 3x + 5},$

a division would be carried out first.

2. Factor denominator.

Step 2 If the degree of A is *less* than the degree of B, then express B as the product of polynomials of degree 1 or 2, where the second-degree factors are *irreducible*. (It can be proved that this is possible. A second-degree polynomial with real coefficients is *irreducible* if it is not the product of polynomials of degree 1 with real coefficients.) To find these factors of degree 1 and 2, except in simple cases, may be quite difficult.

3. List summands of form

$$\frac{k_i}{(ax+b)^i}.$$

Step 3 If $ax + b$ appears exactly n times in the factorization of B, form the sum

$$\frac{k_1}{ax+b} + \frac{k_2}{(ax+b)^2} + \cdots + \frac{k_n}{(ax+b)^n},$$

where the constants $k_1, k_2, \ldots, k_n$ are to be determined later.

4. List summands of form
$$\frac{c_j x + d_j}{(ax^2 + bx + c)^j}.$$

Step 4 If $ax^2 + bx + c$ appears exactly m times in the factorization of B, then form the sum

$$\frac{c_1 x + d_1}{ax^2 + bx + c} + \frac{c_2 x + d_2}{(ax^2 + bx + c)^2} + \cdots + \frac{c_m x + d_m}{(ax^2 + bx + c)^m},$$

where the constants $c_1, c_2, \ldots, c_m$ and $d_1, d_2, \ldots, d_m$ are to be determined later.

5. Find constants k_i, c_j, d_j.

Step 5 Determine the appropriate k's, c's, and d's defined in steps 3 and 4, such that A/B is equal to the sum of all the terms formed in steps 3 and 4 for all factors of B defined in step 2.

Remark: The rational function

$$\frac{c_j x + d_j}{(ax^2 + bx + c)^j}$$

equals

$$\frac{c_j x}{(ax^2 + bx + c)^j} + \frac{d_j}{(ax^2 + bx + c)^j},$$

the sum of two functions that can be integrated by the methods of the preceding section. Combining the two latter quotients into one, as in step 4, saves space.

Remark: If the polynomials $x + 2$ and $2x + 4 \,[= 2(x + 2)]$ both appear in the factorization, express one of them as a constant times the other. In step 2 no irreducible factor should simply be a constant times another irreducible factor.

EXAMPLE 2 Indicate the form of the partial-fraction representation of

$$\frac{A}{B} = \frac{2x^2 + 3x + 3}{(x + 1)^3}.$$

SOLUTION The denominator, if multiplied out, is a polynomial of degree larger than the degree of the numerator. In this case the denominator has only one first-degree factor, $x + 1$ (repeated three times), and no quadratic factors. Only step 3 applies, so that we write

$$\frac{2x^2 + 3x + 3}{(x + 1)^3} = \frac{k_1}{x + 1} + \frac{k_2}{(x + 1)^2} + \frac{k_3}{(x + 1)^3}.$$

The constants $k_1, k_2,$ and k_3 can be found by the method illustrated in Examples 4 to 6, which follow. ∎

7.5 THE INTEGRATION OF RATIONAL FUNCTIONS BY PARTIAL FRACTIONS

EXAMPLE 3 What does the partial-fraction decomposition of

$$\frac{A}{B} = \frac{2x^3 - 6x^2 + 2}{(x + 1)(x^2 + x + 1)^2}$$

look like?

SOLUTION The denominator has degree 5 (if multiplied out), while the numerator has degree 3. Thus step 1 does not apply. The irreducible factors of B are $x + 1$, which appears only once, and $x^2 + x + 1$, which appears to the second power. Both steps 3 and 4 apply, so that we write

$$\frac{2x^3 - 6x^2 + 2}{(x + 1)(x^2 + x + 1)^2} = \frac{k_1}{x + 1} + \frac{c_1 x + d_1}{x^2 + x + 1} + \frac{c_2 x + d_2}{(x^2 + x + 1)^2}. \blacksquare$$

The next example shows how to find the promised constants in a simpler case, where there are fewer constants.

EXAMPLE 4 Express

$$\frac{4x - 7}{x^2 - 3x + 2}$$

as the sum of partial fractions.

SOLUTION Beginning at step 2, factor $x^2 - 3x + 2$ as $(x - 1)(x - 2)$. The denominator B in this case has only first-degree factors; hence step 4 will not apply. Since both $x - 1$ and $x - 2$ appear only once in the factorization, we have $n = 1$ in each case for step 3.

According to step 3 constants k_1 and k_2 exist such that

$$\frac{4x - 7}{x^2 - 3x + 2} = \frac{k_1}{x - 1} + \frac{k_2}{x - 2}. \tag{3}$$

To find k_1 and k_2, multiply both sides of (3) by $x^2 - 3x + 2$, obtaining

$$4x - 7 = k_1(x - 2) + k_2(x - 1). \tag{4}$$

Equation (4) holds for all x, since it is an algebraic identity. Thus it holds when x is replaced by any specific number. To find two equations for the two unknowns k_1 and k_2, we replace the x in (4) by two numbers. Since $x - 1$ vanishes for x equal to 1, and since $x - 2$ vanishes for x equal to 2, for convenience replace x by 1 and by 2. This gives

$$4 \cdot 1 - 7 = k_1(1 - 2) + k_2 \cdot 0 \quad \text{setting } x = 1 \text{ in (4);}$$
$$4 \cdot 2 - 7 = k_1 \cdot 0 + k_2(2 - 1) \quad \text{setting } x = 2 \text{ in (4).}$$

These equations simplify to $\quad -3 = -k_1$

and $\quad\quad\quad\quad\quad\quad\quad\quad\quad 1 = k_2.$

Hence $k_1 = 3$ and $k_2 = 1$, and

$$\frac{4x - 7}{x^2 - 3x + 2} = \frac{3}{x - 1} + \frac{1}{x - 2}. \quad\blacksquare$$

In Example 4 the unknown constants were found by substituting specific values for x. There is another way, called *comparing coefficients*. It depends on the fact that if two polynomials are equal for all x, then their coefficients must be equal. To illustrate this technique, return to Eq. (4).

According to (4), $\quad 4x - 7 = k_1(x - 2) + k_2(x - 1)$

or $\quad\quad\quad\quad\quad 4x - 7 = (k_1 + k_2)x - (2k_1 + k_2).$

Comparing corresponding coefficients, we conclude that

$$\begin{cases} 4 = k_1 + k_2 & \text{comparing coefficients of } x \\ -7 = -2k_1 - k_2 & \text{comparing constant terms.} \end{cases}$$

These simultaneous equations may be solved for k_1 and k_2. Adding them gives

$$-3 = -k_1;$$

hence $k_1 = 3$. Substituting $k_1 = 3$ into $4 = k_1 + k_2$ gives $k_2 = 1$.

EXAMPLE 5 Express

$$\frac{x^2 + 7x + 1}{(x + 2)^2(2x + 1)}$$

as the sum of partial fractions.

SOLUTION The degree of the numerator is less than the degree of the denominator (which is 3). Hence step 1 does not apply. Since the denominator is already factored (a common occurrence in practice), step 2 is done. We shall do steps 3 and 5 simultaneously; step 4 does not apply, since the denominator B has no second-degree irreducible factors.

Since $x + 2$ appears twice in the factorization and $2x + 1$ once,

$$\frac{x^2 + 7x + 1}{(x + 2)^2(2x + 1)} = \frac{k_1}{x + 2} + \frac{k_2}{(x + 2)^2} + \frac{l}{2x + 1}. \quad\quad (5)$$

To find the constants k_1, k_2, l, remove the denominators in (5) by multiplying both sides by $(x + 2)^2(2x + 1)$:

$$x^2 + 7x + 1 = k_1(x + 2)(2x + 1) + k_2(2x + 1) + l(x + 2)^2. \quad\quad (6)$$

7.5 THE INTEGRATION OF RATIONAL FUNCTIONS BY PARTIAL FRACTIONS

Three equations are needed to find the three unknowns k_1, k_2, and l. To obtain them, replace x in (6) by three different numbers in turn. Since $x + 2 = 0$ when $x = -2$, and $2x + 1 = 0$ when $x = -\frac{1}{2}$, replace x by -2 and then by $-\frac{1}{2}$. To obtain a third equation, use $x = 0$. Thus

$$-9 = -3k_2 \qquad \text{setting } x = -2 \text{ in (6);}$$
$$-\tfrac{9}{4} = \left(\tfrac{9}{4}\right)l \qquad \text{setting } x = -\tfrac{1}{2} \text{ in (6);}$$
$$1 = 2k_1 + k_2 + 4l \qquad \text{setting } x = 0 \text{ in (6).}$$

Thus $k_2 = 3$, $l = -1$, and finally $k_1 = 1$. Replacing k_1, k_2, and l in (5) yields

$$\frac{x^2 + 7x + 1}{(x + 2)^2(2x + 1)} = \frac{1}{x + 2} + \frac{3}{(x + 2)^2} - \frac{1}{2x + 1}. \quad \blacksquare$$

EXAMPLE 6 Express

$$\frac{x^4 + x^3 - 3x + 5}{(x + 1)(x^2 + x + 1)}$$

This is (1) in the opening paragraph, as can be checked.

as the sum of partial fractions.

SOLUTION Since the degree of the numerator, 4, is at least as large as the degree of the denominator, 3, step 1 is applicable. Divide by the denominator, $(x + 1)(x^2 + x + 1) = x^3 + 2x^2 + 2x + 1$, as follows:

$$\begin{array}{r}
x - 1 \qquad \text{quotient} \\
x^3 + 2x^2 + 2x + 1 \overline{\smash{\big)}\, x^4 + x^3 + 0x^2 - 3x + 5} \\
\underline{x^4 + 2x^3 + 2x^2 + x} \\
-x^3 - 2x^2 - 4x + 5 \\
\underline{-x^3 - 2x^2 - 2x - 1} \\
-2x + 6 \qquad \text{remainder}
\end{array}$$

Hence

$$\frac{x^4 + x^3 - 3x + 5}{(x + 1)(x^2 + x + 1)} = x - 1 + \frac{-2x + 6}{(x + 1)(x^2 + x + 1)}. \quad (7)$$

Next represent

$$\frac{-2x + 6}{(x + 1)(x^2 + x + 1)}$$

as a sum of partial quotients in accordance with steps 3 and 4. Since $x + 1$ and $x^2 + x + 1$ are irreducible, we seek constants k_1, c_1, and d_1 such that

$$\frac{-2x + 6}{(x + 1)(x^2 + x + 1)} = \frac{k_1}{x + 1} + \frac{c_1 x + d_1}{x^2 + x + 1}. \quad (8)$$

To find k_1, c_1, and d_1, multiply (8) by $(x + 1)(x^2 + x + 1)$, obtaining

$$-2x + 6 = k_1(x^2 + x + 1) + (c_1 x + d_1)(x + 1). \quad (9)$$

Let $x = -1$ (the root of $x + 1 = 0$); then let $x = 0$ and $x = 1$, which are easy numbers to work with, arriving at

$$8 = k_1 \qquad \text{setting } x = -1 \text{ in (9);}$$
$$6 = k_1 + d_1 \qquad \text{setting } x = 0 \text{ in (9);}$$
$$4 = 3k_1 + 2c_1 + 2d_1 \qquad \text{setting } x = 1 \text{ in (9).}$$

The first equation yields $k_1 = 8$, the second $d_1 = -2$, and the third $c_1 = -8$.

Thus (8) takes the form

$$\frac{-2x + 6}{(x + 1)(x^2 + x + 1)} = \frac{8}{x + 1} - \frac{8x + 2}{x^2 + x + 1}. \tag{10}$$

Combining (7) and (10) shows that

$$\frac{x^4 + x^3 - 3x + 5}{(x + 1)(x^2 + x + 1)} = x - 1 + \frac{8}{x + 1} - \frac{8x + 2}{x^2 + x + 1}. \quad \blacksquare \tag{11}$$

No integrations have been carried out in this section, which is purely algebraic. This section, together with the preceding one, shows how to integrate any rational function.

SUMMARY OF THIS SECTION

This flow chart describes the procedure for representing a rational function as a sum of partial fractions.

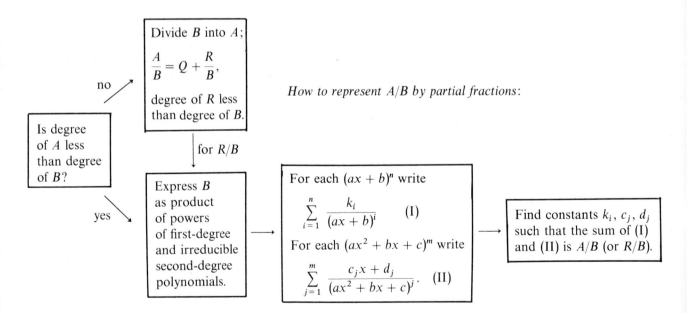

USEFUL FACTS The number of unknown constants equals the degree of B. (Check that you have the right number.)

If $b^2 - 4ac < 0$, $ax^2 + bx + c$ is irreducible. If $b^2 - 4ac \geq 0$, $ax^2 + bx + c$ is reducible.

Exercises

In each of Exercises 1 to 12 indicate the form of the partial-fraction representation of the rational function listed, but do *not* find the constants k_i, c_j, and d_j.

1 $\dfrac{x+3}{(x+1)(x+2)}$

2 $\dfrac{5}{(x-1)(x+3)}$

3 $\dfrac{1}{x^2-4}$

4 $\dfrac{x+3}{x^2-4}$

5 $\dfrac{x^2+3x+1}{(x+1)^3}$

6 $\dfrac{5x+6x^2}{(x+1)^3}$

7 $\dfrac{x^2+x+2}{x^2-1}$

8 $\dfrac{x^3}{x+1}$

9 $\dfrac{x^3}{(x-1)(x+2)}$

10 $\dfrac{x+3}{(x-1)^2(x+2)^2}$

11 $\dfrac{x^4+3x^2}{(x^2+x+1)^3}$

12 $\dfrac{x^7-1}{(x^2+x+1)^3(x+1)^2}$

In Exercises 13 to 28 express the rational functions in terms of partial fractions.

13 $\dfrac{x^2+x+3}{x+1}$

14 $\dfrac{x}{x-3}$

15 $\dfrac{2x^2+2x+3}{2x+1}$

16 $\dfrac{3x-3}{x^2-9}$

17 $\dfrac{-x}{(x+1)(x+2)}$

18 $\dfrac{4x+1}{x(x+1)}$

19 $\dfrac{x^2-3x-1}{x(x+1)^2}$

20 $\dfrac{4x^3-4x^2+x+1}{x^2(x-1)^2}$

21 $\dfrac{2x^2+3x+3}{(x+1)^3}$

22 $\dfrac{2x^3-6x^2+2}{(x+1)(x^2+x+1)^2}$

23 $\dfrac{x+4}{(x+1)^2}$

24 $\dfrac{2x^2+9x+12}{(x+2)(x+3)^2}$

25 $\dfrac{x^5}{x^2+3x+5}$

26 $\dfrac{x^3}{x^2-x-6}$

27 $\dfrac{x^4+x^3+4x^2+1}{x(x^2+x+1)}$

28 $\dfrac{3x^3+x^2-3}{x^2(x^2+3x+3)}$

∎

In each of Exercises 29 to 36 compute the integral of the function in the cited exercise.

29 Exercise 17.   **30** Exercise 18.
31 Exercise 19.   **32** Exercise 20.
33 Exercise 25.   **34** Exercise 26.
35 Exercise 27.   **36** Exercise 28.

37 Compute $\displaystyle\int \dfrac{3x^5+2x^3-x}{(x^2-1)(x^2+1)}\,dx$.

38 (a) If a is a constant other than 0, what is the partial-fraction decomposition of $1/(a^2-x^2)$?
  (b) Using (a), find $\int dx/(a^2-x^2)$.

39 Compute $\displaystyle\int_2^3 \dfrac{dx}{x^3-x}$.

40 Compute $\displaystyle\int_2^3 \dfrac{dx}{x^3+x}$.

41 Solve the differential equation
$$\dfrac{dy}{dx}=\dfrac{y^2-1}{x^2(x+1)}.$$

■ ■

42 Since the only polynomials that are irreducible over the real numbers are the first-degree polynomials and the irreducible second-degree polynomials, the polynomial x^4+1 must be reducible. Factor it. (*Suggestion:* Find constants a and b such that $x^4+1=(x^2+ax+1)(x^2+bx+1)$.)

43 Integrate as easily as possible

(a) $\displaystyle\int \dfrac{x^3\,dx}{x^4+1}$ (b) $\displaystyle\int \dfrac{x\,dx}{x^4+1}$ (c) $\displaystyle\int \dfrac{dx}{x^4+1}$.

44 (a) Write x^4+x^2+1 as the product of irreducible polynomials of second degree.

(b) Compute $\displaystyle\int \dfrac{dx}{x^4+x^2+1}$.

45 There is a partial-fraction representation of rational numbers. The denominators, instead of being irreducible polynomials or their powers, are prime numbers or their powers. (An integer ≥ 2 is prime if it is not the product of positive integers smaller than itself.) A partial fraction has the form k/p^n, where p is prime, n is an integer ≥ 1, and k is an integer, $|k|<p$. Each rational number, A/B, is the sum of an integer (perhaps 0) and fractions of the form k/p^n, where p^n divides B. For instance,

$$\dfrac{19}{24}=\dfrac{0}{2}+\dfrac{1}{2^2}+\dfrac{-1}{2^3}+\dfrac{2}{3} \quad\text{(note: }24=2^3\cdot 3\text{)}$$

$$\dfrac{37}{10}=3+\dfrac{1}{2}+\dfrac{1}{5} \quad\text{(note: }10=2\cdot 5\text{)}.$$

Find a partial-fraction representation of (a) $\tfrac{5}{6}$, (b) $\tfrac{4}{15}$, (c) $\tfrac{19}{15}$, (d) $\tfrac{7}{27}$.

46 (This continues Exercise 45.) (*a*) Find integers x and y such that

$$\frac{1}{77} = \frac{x}{11} + \frac{y}{7},$$

and $|x| < 11$, $|y| < 7$,
(*b*) Find another such pair.

As (*a*) and (*b*) show, the partial fraction representation of rational numbers is not unique. That it is unique for rational functions is far from obvious. It is proved to be unique in an upper-division or graduate algebra course.

47 Find the constants k_1, k_2, and k_3 in Example 2.

7.6 How to integrate powers of trigonometric functions

This section shows how to integrate certain products of powers of the six trigonometric functions, $\cos\theta$, $\sin\theta$, $\tan\theta$, $\sec\theta$, $\csc\theta$, and $\cot\theta$. Since $\tan\theta = \sin\theta/\cos\theta$, $\sec\theta = 1/\cos\theta$, $\csc\theta = 1/\sin\theta$, and $\cot\theta = \cos\theta/\sin\theta$, any such product can be expressed in the form $\cos^m\theta \sin^n\theta$ for some integers m and n, positive, zero, or negative. The methods in this section thus show how to compute $\int \cos^m\theta \sin^n\theta \, d\theta$ for certain convenient combinations of m and n. (The method in the next section shows that $\int \cos^m\theta \sin^n\theta \, d\theta$ is always elementary.) Typical of the integrals that can be evaluated by these methods are

$$\int \sin^2\theta \, d\theta, \quad \int \sin^5\theta \, d\theta, \quad \int \cos^3\theta \sin^4\theta \, d\theta,$$

$$\int \tan^5\theta \sec^3\theta \, d\theta \quad \text{and} \quad \int \sec\theta \, d\theta.$$

The methods are summarized in the table at the end of this section.

HOW TO INTEGRATE $\cos^m\theta \sin^n\theta$

Consider the integral

$$\int \cos^m\theta \sin^n\theta \, d\theta, \tag{1}$$

where m and n are nonnegative integers. The technique for integrating (1) varies with m and n.

If $m = 1$, say, and $n \geq 1$, then (1) becomes

$$\int \cos\theta \sin^n\theta \, d\theta. \tag{2}$$

The substitution $u = \sin\theta$ turns (2) into the easy integral

$$\int u^n \, du.$$

7.6 HOW TO INTEGRATE POWERS OF TRIGONOMETRIC FUNCTIONS

If $m = 0$, then (1) takes the form $\int \sin^n \theta \, d\theta$. A recursive formula was developed in Example 7 of Sec. 7.2, which expressed this integral in terms of $\int \sin^{n-2} \theta \, d\theta$. However, there are other approaches available for the cases $n = 2$ and n odd, as illustrated by Examples 1 and 2.

EXAMPLE 1 Find $\int \sin^2 \theta \, d\theta$ by using the identity

$$\sin^2 \theta = \frac{1 - \cos 2\theta}{2}.$$

SOLUTION

How to find $\int \sin^2 \theta \, d\theta$

$$\int \sin^2 \theta \, d\theta = \int \frac{1 - \cos 2\theta}{2} \, d\theta = \int \tfrac{1}{2} \, d\theta - \int \frac{\cos 2\theta}{2} \, d\theta = \frac{\theta}{2} - \frac{\sin 2\theta}{4} + C$$

$$= \frac{\theta}{2} - \frac{2 \sin \theta \cos \theta}{4} + C = \frac{\theta}{2} - \frac{\sin \theta \cos \theta}{2} + C. \quad \blacksquare$$

The particular definite integrals

$$\int_0^{\pi/2} \sin^2 \theta \, d\theta \quad \text{and} \quad \int_0^{\pi/2} \cos^2 \theta \, d\theta$$

occur so frequently that the following memory device is worth pointing out. A quick sketch of the graphs of $\sin^2 \theta$ and $\cos^2 \theta$ shows that the two definite integrals are equal. Also, since

$$\sin^2 \theta + \cos^2 \theta = 1,$$

$$\int_0^{\pi/2} \sin^2 \theta \, d\theta + \int_0^{\pi/2} \cos^2 \theta \, d\theta = \int_0^{\pi/2} 1 \, d\theta = \frac{\pi}{2}.$$

How to remember $\int_0^{\pi/2} \sin^2 \theta \, d\theta$ and $\int_0^{\pi/2} \cos^2 \theta \, d\theta$

Hence

$$\int_0^{\pi/2} \sin^2 \theta \, d\theta = \frac{\pi}{4} = \int_0^{\pi/2} \cos^2 \theta \, d\theta.$$

In the next example a quick way is shown to find $\int \sin^n \theta \, d\theta$ if n is an *odd* positive integer.

EXAMPLE 2 Find

$$\int \sin^5 \theta \, d\theta.$$

SOLUTION Recall that

$$d(\cos \theta) = -\sin \theta \, d\theta.$$

Thus

$$\int \sin^5 \theta \, d\theta = \int \sin^4 \theta \sin \theta \, d\theta$$

$$= -\int (1 - \cos^2 \theta)^2 \, d(\cos \theta).$$

Letting $u = \cos \theta$, we obtain then

$$\int \sin^5 \theta \, d\theta = -\int (1 - u^2)^2 \, du$$

$$= -\int (1 - 2u^2 + u^4)\, du$$

$$= -\left(u - \frac{2u^3}{3} + \frac{u^5}{5}\right) + C$$

$$= -\cos\theta + \frac{2\cos^3\theta}{3} - \frac{\cos^5\theta}{5} + C. \blacksquare$$

$\cos^m \sin^{odd}$
More generally, to find $\int \cos^m\theta \sin^n\theta\, d\theta$, where m and n are nonnegative integers and n is odd, pair one $\sin\theta$ with $d\theta$ to form $\sin\theta\, d\theta = -d(\cos\theta)$ and use the identity $\sin^2\theta = 1 - \cos^2\theta$ together with the substitution $u = \cos\theta$. The new integrand will be a polynomial in u. A similar approach works on $\int \cos^m\theta \sin^n\theta\, d\theta$ if m is odd, as is illustrated by Example 3.

$\cos^{odd} \sin^m$

EXAMPLE 3 Find $\int \cos^3\theta \sin^4\theta\, d\theta$.

SOLUTION Pair one $\cos\theta$ with $d\theta$ to form

$$\cos\theta\, d\theta = d(\sin\theta),$$

which suggests the substitution $u = \sin\theta$:

$$\int \cos^3\theta \sin^4\theta\, d\theta = \int \cos^2\theta \sin^4\theta\, (\cos\theta\, d\theta)$$

$$= \int (1 - \sin^2\theta)\sin^4\theta\, d(\sin\theta)$$

$$= \int (1 - u^2)u^4\, du$$

$$= \int (u^4 - u^6)\, du$$

$$= \frac{u^5}{5} - \frac{u^7}{7} + C$$

$$= \frac{\sin^5\theta}{5} - \frac{\sin^7\theta}{7} + C. \blacksquare$$

If both m and n are even in $\int \cos^m\theta \sin^n\theta\, d\theta$, the method of Example 3 does not apply. It then helps to use the identities

$$\cos^2\theta = \frac{1 + \cos 2\theta}{2} \quad \text{and} \quad \sin^2\theta = \frac{1 - \cos 2\theta}{2}.$$

Example 1 is an instance of this approach. Example 4 is a more involved case.

7.6 HOW TO INTEGRATE POWERS OF TRIGONOMETRIC FUNCTIONS

EXAMPLE 4 Find $\int_0^{\pi/4} \cos^2 \theta \sin^4 \theta \, d\theta.$

SOLUTION First find an antiderivative of $\cos^2 \theta \sin^4 \theta$:

$$\int \cos^2 \theta \sin^4 \theta \, d\theta = \int \cos^2 \theta (\sin^2 \theta)^2 \, d\theta$$

$$= \int \frac{1 + \cos 2\theta}{2} \left(\frac{1 - \cos 2\theta}{2}\right)^2 d\theta$$

$$= \frac{1}{8} \int (1 + \cos 2\theta)(1 - 2\cos 2\theta + \cos^2 2\theta) \, d\theta$$

$$= \frac{1}{8} \int (1 - \cos 2\theta - \cos^2 2\theta + \cos^3 2\theta) \, d\theta. \quad (3)$$

The four summands in (3) can be integrated separately:

C will be added at the end.

$$\int 1 \, d\theta = \theta, \qquad \int -\cos 2\theta \, d\theta = \frac{-\sin 2\theta}{2}$$

$$\int -\cos^2 2\theta \, d\theta = \int -\frac{1 + \cos 4\theta}{2} \, d\theta = -\frac{\theta}{2} - \frac{\sin 4\theta}{8}$$

$$\int \cos^3 2\theta \, d\theta = \int \cos^2 2\theta \cos 2\theta \, d\theta = \int (1 - \sin^2 2\theta) \cos 2\theta \, d\theta$$

$$= \int (\cos 2\theta - \sin^2 2\theta \cos 2\theta) \, d\theta$$

$$= \frac{\sin 2\theta}{2} - \frac{\sin^3 2\theta}{6}.$$

Thus (3) equals

$$\frac{1}{8}\left(\theta - \frac{\sin 2\theta}{2} - \frac{\theta}{2} - \frac{\sin 4\theta}{8} + \frac{\sin 2\theta}{2} - \frac{\sin^3 2\theta}{6}\right) + C,$$

or

$$\frac{1}{8}\left(\frac{\theta}{2} - \frac{\sin 4\theta}{8} - \frac{\sin^3 2\theta}{6}\right) + C.$$

Consequently

$$\int_0^{\pi/4} \cos^2 \theta \sin^4 \theta \, d\theta = \frac{1}{8}\left(\frac{\theta}{2} - \frac{\sin 4\theta}{8} - \frac{\sin^3 2\theta}{6}\right)\Bigg|_0^{\pi/4}$$

$$= \frac{1}{8}\left[\left(\frac{\pi}{8} - \frac{\sin \pi}{8} - \frac{\sin^3 (\pi/2)}{6}\right) - \left(0 - \frac{\sin 0}{8} - \frac{\sin^3 0}{6}\right)\right]$$

$$= \frac{1}{8}\left(\frac{\pi}{8} - \frac{1}{6}\right). \quad \blacksquare$$

HOW TO INTEGRATE $\tan^m \theta \sec^n \theta$

Recall that $d(\tan \theta) = \sec^2 \theta \, d\theta$ and $d(\sec \theta) = \sec \theta \tan \theta \, d\theta$. These formulas facilitate the computation of $\int \tan^m \theta \sec^n \theta \, d\theta$, m and n nonnegative integers, when m is odd or n is even. When m is odd, form $\sec \theta \tan \theta \, d\theta$; when n is even, form $\sec^2 \theta \, d\theta$.

EXAMPLE 5 Find $\int \tan^5 \theta \sec^3 \theta \, d\theta.$

SOLUTION Recall that $d(\sec \theta) = \tan \theta \sec \theta \, d\theta.$
and that $\tan^2 \theta = \sec^2 \theta - 1.$

Let $u = \sec \theta$ and $du = \sec \theta \tan \theta \, d\theta.$

How to deal with $\tan^{odd} \sec^n$

Then
$$\int \tan^5 \theta \sec^3 \theta \, d\theta = \int \tan^4 \theta \sec^2 \theta (\sec \theta \tan \theta \, d\theta)$$
$$= \int (\sec^2 \theta - 1)^2 \sec^2 \theta (\sec \theta \tan \theta \, d\theta)$$
$$= \int (u^2 - 1)^2 u^2 \, du$$
$$= \int (u^6 - 2u^4 + u^2) \, du$$
$$= \frac{u^7}{7} - \frac{2u^5}{5} + \frac{u^3}{3} + C$$
$$= \frac{\sec^7 \theta}{7} - \frac{2 \sec^5 \theta}{5} + \frac{\sec^3 \theta}{3} + C. \blacksquare$$

EXAMPLE 6 Find $\int \tan^6 \theta \sec^4 \theta \, d\theta.$

SOLUTION Recall that $d(\tan \theta) = \sec^2 \theta \, d\theta.$

So pair $\sec^2 \theta$ with $d\theta$ to form $\sec^2 \theta \, d\theta$. This suggests the substitution

How to deal with $\tan^m \sec^{even}$

$u = \tan \theta \qquad du = \sec^2 \theta \, d\theta.$

Recall also that $\sec^2 \theta = \tan^2 \theta + 1.$

Then
$$\int \tan^6 \theta \sec^4 \theta \, d\theta = \int \tan^6 \theta \sec^2 \theta \sec^2 \theta \, d\theta$$
$$= \int u^6 (u^2 + 1) \, du$$
$$= \int (u^8 + u^6) \, du$$
$$= \frac{u^9}{9} + \frac{u^7}{7} + C$$
$$= \frac{\tan^9 \theta}{9} + \frac{\tan^7 \theta}{7} + C. \blacksquare$$

7.6 HOW TO INTEGRATE POWERS OF TRIGONOMETRIC FUNCTIONS

A slightly different trick disposes of $\int \tan^n \theta \, d\theta$, as the next example illustrates.

EXAMPLE 7 Obtain a recursion formula for $\int \tan^n \theta \, d\theta$.

SOLUTION Keep in mind that $(\tan \theta)' = \sec^2 \theta$ and that $\tan^2 \theta = \sec^2 \theta - 1$. The steps are few:

A recursion for $\int \tan^n \theta \, d\theta$

$$\int \tan^n \theta \, d\theta = \int \tan^{n-2} \theta \tan^2 \theta \, d\theta$$

$$= \int \tan^{n-2} \theta \, (\sec^2 \theta - 1) \, d\theta$$

$$= \int \tan^{n-2} \theta \sec^2 \theta \, d\theta - \int \tan^{n-2} \theta \, d\theta$$

$$= \int u^{n-2} \, du - \int \tan^{n-2} \theta \, d\theta \qquad (u = \tan \theta)$$

$$= \frac{\tan^{n-1} \theta}{n-1} - \int \tan^{n-2} \theta \, d\theta.$$

Repeated application of this recursion eventually produces $\int \tan^1 \theta \, d\theta$ or $\int \tan^0 \theta \, d\theta$. Both are easily computed:

$\int \tan \theta \, d\theta = -\ln |\cos \theta| + C$

$$\int \tan \theta \, d\theta = \int \frac{\sin \theta \, d\theta}{\cos \theta} = -\ln |\cos \theta| + C$$

and

$$\int \tan^0 \theta \, d\theta = \int 1 \, d\theta = \theta + C. \quad \blacksquare$$

EXAMPLE 8 Obtain a recursion formula for $\int \sec^n \theta \, d\theta$.

SOLUTION Write $\sec^n \theta$ as $\sec^{n-2} \theta \sec^2 \theta$ and use integration by parts:

$$\int \underbrace{\sec^{n-2} \theta}_{u} \underbrace{\sec^2 \theta \, d\theta}_{dv} = \underbrace{\sec^{n-2} \theta}_{u} \underbrace{\tan \theta}_{v} - \int \underbrace{\tan \theta}_{v} \underbrace{(n-2) \sec^{n-2} \theta \tan \theta \, d\theta}_{du}$$

$$= \sec^{n-2} \theta \tan \theta - (n-2) \int \sec^{n-2} \theta \tan^2 \theta \, d\theta$$

$$= \sec^{n-2} \theta \tan \theta - (n-2) \int \sec^{n-2} \theta (\sec^2 \theta - 1) \, d\theta$$

$$= \sec^{n-2} \theta \tan \theta - (n-2) \int \sec^n \theta \, d\theta + (n-2) \int \sec^{n-2} \theta \, d\theta.$$

Collecting $\int \sec^n \theta \, d\theta$, we obtain

$$(n-1) \int \sec^n \theta \, d\theta = \sec^{n-2} \theta \tan \theta + (n-2) \int \sec^{n-2} \theta \, d\theta$$

and therefore

$$\int \sec^n \theta \, d\theta = \frac{\sec^{n-2} \theta \tan \theta}{n-1} + \frac{n-2}{n-1} \int \sec^{n-2} \theta \, d\theta. \quad \blacksquare$$

HOW TO INTEGRATE $\sec \theta$ A century before the invention of calculus, the cartographer Gerhardus Mercator had to estimate $\int_0^x \sec \theta \, d\theta$ in order to determine where to place the lines of latitude on his maps. On a Mercator map, a straight line corresponds to a voyage with a constant compass heading, a property of great use to navigators. (See Exercise 44 for the geometry.) Henry Bond in 1645, while scrutinizing a table of $\ln \tan (\alpha/2 + \pi/4)$ conjectured that $\int_0^x \sec \theta \, d\theta = \ln \tan (\alpha/2 + \pi/4)$, but offered no proof. In 1666 Nicolaus Mercator (no relation to Gerhardus) offered the royalties on one of his inventions to the mathematician who could prove that Bond's conjecture was right. Within two years James Gregory provided the missing proof, well before the tools of calculus were available.

This important integral will be evaluated three ways: twice in the next example and once in Exercise 23 of the next section.

EXAMPLE 9 Find $\int \sec \theta \, d\theta$.

SOLUTION *Method I*

$$\int \sec \theta \, d\theta = \int \frac{\sec \theta (\sec \theta + \tan \theta)}{\sec \theta + \tan \theta} \, d\theta$$

To see the motivation, differentiate the answer.

$$= \int \frac{\sec^2 \theta + \sec \theta \tan \theta}{\sec \theta + \tan \theta} \, d\theta$$

$$= \ln |\sec \theta + \tan \theta| + C.$$

While this is the shortest method, it does seem artificial. The next method may seem less contrived.

Method II

$$\int \sec \theta \, d\theta = \int \frac{1}{\cos \theta} \, d\theta$$

$$= \int \frac{\cos \theta}{\cos^2 \theta} \, d\theta$$

$$= \int \frac{\cos \theta}{1 - \sin^2 \theta} \, d\theta.$$

The substitution $\quad u = \sin \theta$
and $\quad du = \cos \theta \, d\theta$

transforms this last integral into the integral of a rational function:

$$\int \frac{du}{1 - u^2} = \frac{1}{2} \int \left(\frac{1}{1 + u} + \frac{1}{1 - u} \right) du$$

Both $1 + u$ and $1 - u$ are positive since $u = \sin \theta$.

$$= \tfrac{1}{2} [\ln (1 + u) - \ln (1 - u)] + C$$

$$= \frac{1}{2} \ln \frac{1 + u}{1 - u} + C.$$

7.6 HOW TO INTEGRATE POWERS OF TRIGONOMETRIC FUNCTIONS

Since $u = \sin \theta$, $\dfrac{1}{2} \ln \dfrac{1+u}{1-u} = \dfrac{1}{2} \ln \dfrac{1+\sin\theta}{1-\sin\theta}$

Note that $\sec\theta + \tan\theta$ *can be negative, but* $(1+\sin\theta)/(1-\sin\theta)$ *cannot be.*

The reader may check that this equals $\ln |\sec\theta + \tan\theta|$ by showing that

$$\frac{1+\sin\theta}{1-\sin\theta} = (\sec\theta + \tan\theta)^2.$$

Neither method gives Bond's conjecture, $\ln \tan(\theta/2 + \pi/4)$. That formula will be obtained directly and in a fairly straightforward way in the next section. However, it is an amusing exercise to show that $\tan(\theta/2 + \pi/4) = \sec\theta + \tan\theta$. From this Bond's conjecture follows immediately. ∎

SUMMARY OF THIS SECTION The following table summarizes the techniques discussed and similar ones for other powers of trigonometric functions.

Integrand	Technique
$\sin^2\theta$	Write $\sin^2\theta$ as $\dfrac{1-\cos 2\theta}{2}$.
$\cos^2\theta$	Write $\cos^2\theta$ as $\dfrac{1+\cos 2\theta}{2}$.
$\sin^n\theta$ (n odd)	Write $\sin^n\theta\,d\theta = \sin^{n-1}\theta(\sin\theta\,d\theta)$ and use $u = \cos\theta$; hence $1 - u^2 = \sin^2\theta$.
$\cos^m\theta \sin^n\theta$ (n odd)	Write $\cos^m\theta \sin^n\theta\,d\theta = \cos^m\theta \sin^{n-1}\theta(\sin\theta\,d\theta)$ and use $u = \cos\theta$; hence $1 - u^2 = \sin^2\theta$.
$\cos^m\theta \sin^n\theta$ (m odd)	Write $\cos^m\theta \sin^n\theta\,d\theta = \cos^{m-1}\theta \sin^n\theta(\cos\theta\,d\theta)$ and use $u = \sin\theta$; hence $1 - u^2 = \cos^2\theta$.
$\cos^m\theta \sin^n\theta$ (m and n positive even integers)	Replace $\cos^2\theta$ by $\dfrac{1+\cos 2\theta}{2}$ and $\sin^2\theta$ by $\dfrac{1-\cos 2\theta}{2}$.
$\tan^m\theta \sec^n\theta$ ($n \geq 2$ even)	Write $\tan^m\theta \sec^n\theta\,d\theta$ as $\tan^m\theta \sec^{n-2}\theta(\sec^2\theta\,d\theta)$ and use $u = \tan\theta$; hence $1 + u^2 = \sec^2\theta$.
$\tan^m\theta \sec^n\theta$ (m odd)	Write $\tan^m\theta \sec^n\theta\,d\theta$ as $\tan^{m-1}\theta \sec^{n-1}\theta(\tan\theta \sec\theta\,d\theta)$ and use $u = \sec\theta$; hence $u^2 - 1 = \tan^2\theta$.
$\tan^n\theta$ ($n \geq 2$)	Write $\tan^n\theta = \tan^{n-2}\theta \tan^2\theta$ $= \tan^{n-2}\theta \sec^2\theta - \tan^{n-2}\theta$ and repeat.
$\tan\theta$	$\int \tan\theta\,d\theta = \ln|\sec\theta| + C$.
$\sec\theta$	$\int \sec\theta\,d\theta = \ln|\sec\theta + \tan\theta| + C$.
$\cot^m\theta \csc^n\theta$ ($n \geq 2$ even)	Write $\cot^m\theta \csc^n\theta$ as $\cot^m\theta \csc^{n-2}\theta(\csc^2\theta\,d\theta)$ and use $u = \cot\theta$; hence $1 + u^2 = \csc^2\theta$.
$\cot^m\theta \csc^n\theta$ (m odd)	Write $\cot^m\theta \csc^n\theta\,d\theta$ as $\cot^{m-1}\theta \csc^{n-1}\theta(\cot\theta \csc\theta\,d\theta)$ and use $u = \csc\theta$; hence $u^2 - 1 = \cot^2\theta$.
$\tan^m\theta \sec^n\theta$ (m even, n even)	Replace $\sec^2\theta$ by $\tan^2\theta + 1$.
$\sec^n\theta$ ($n \geq 2$)	Recursion in Example 8.

Note: For $\int \cos mx \cos nx\,dx$ and $\int \sin mx \cos nx\,dx$ see Exercises 47 and 48. For $\int \sin mx \sin nx\,dx$ see Exercise 200 in Sec. 7.S.

Exercises

In Exercises 1 to 36 find the integrals.

1. $\int \sin^2 2\theta \, d\theta$
2. $\int \cos^2 3\theta \, d\theta$
3. $\int_{\pi/4}^{\pi/2} \sin^2 4\theta \, d\theta$
4. $\int_0^{\pi/10} \cos^2 5\theta \, d\theta$
5. $\int \sin^5 2\theta \, d\theta$
6. $\int \sin^7 \theta \, d\theta$
7. $\int_0^{\pi/2} \cos^3 x \, dx$
8. $\int_0^{\pi/2} \cos^5 x \, dx$
9. $\int \sin^3 \theta \cos^4 \theta \, d\theta$
10. $\int \cos^5 \theta \sin^2 \theta \, d\theta$
11. $\int \cos^3 \theta \sin^3 \theta \, d\theta$
12. $\int \cos^2 \theta \sin^5 \theta \, d\theta$
13. $\int_{\pi/6}^{\pi/4} \cos^4 x \, dx$
14. $\int_0^{\pi} \sin^4 x \, dx$
15. $\int \tan^4 \theta \sec^2 \theta \, d\theta$
16. $\int \tan^4 \theta \sec^4 \theta \, d\theta$
17. $\int \tan^3 \theta \sec \theta \, d\theta$
18. $\int \tan \theta \sec^3 \theta \, d\theta$
19. $\int \tan^3 \theta \, d\theta$
20. $\int \tan^4 \theta \, d\theta$
21. $\int_0^{\pi/4} \sec^4 \theta \, d\theta$
22. $\int_0^{\pi/4} \sec^5 \theta \, d\theta$
23. $\int \sin^2 \theta \cos^3 \theta \, d\theta$
24. $\int \tan^5 \theta \sec^3 \theta \, d\theta$
25. $\int \cot \theta \, d\theta$
26. $\int \cos^2 3\theta \, d\theta$
27. $\int \sec 3\theta \, d\theta$
28. $\int \cot^3 \theta \csc^5 \theta \, d\theta$
29. $\int \cos^5 \theta \sin^3 \theta \, d\theta$
30. $\int \cos^4 \theta \sin^4 \theta \, d\theta$
31. $\int \cot^3 \theta \csc^4 \theta \, d\theta$
32. $\int (\sin \theta + 2 \cos \theta)^2 \, d\theta$
33. $\int \cot^4 \theta \csc^4 \theta \, d\theta$
34. $\int (\tan \theta + 2 \cot \theta)^2 \, d\theta$
35. $\int \sin^3 \theta \tan^2 \theta \, d\theta$
36. $\int \frac{\sin^3 \theta}{\cos^5 \theta} \, d\theta$

In Exercise 41 of Sec. 7.2 the following formula was obtained:

$$\int_0^{\pi/2} \sin^n \theta \, d\theta = \begin{cases} \dfrac{1 \cdot 3 \cdot 5 \cdots (n-1)}{2 \cdot 4 \cdot 6 \cdots n} \dfrac{\pi}{2} & n \text{ even} \\[2mm] \dfrac{2 \cdot 4 \cdot 6 \cdots (n-1)}{3 \cdot 5 \cdot 7 \cdots n} & n \text{ odd} \end{cases}$$

In Exercises 37 to 40 evaluate the definite integrals. The preceding formula may be of aid.

37. $\int_0^{\pi/2} \cos^5 \theta \, d\theta$

   $\left(\textit{Hint}: \text{Why is } \int_0^{\pi/2} \cos^n \theta \, d\theta = \int_0^{\pi/2} \sin^n \theta \, d\theta?\right)$

38. $\int_0^{\pi/2} \sin^6 \theta \, d\theta$
39. $\int_0^{2\pi} \sin^6 \theta \, d\theta$
40. $\int_0^{2\pi} \cos^7 \theta \, d\theta$

41. Find $\int \sin^2 \theta \cos^2 \theta \, d\theta$ by use of the identity $\sin 2\theta = 2 \sin \theta \cos \theta$.

42. Compute $\int \dfrac{d\theta}{1 + \cos \theta}$ with the aid of the identity

$$\cos^2 \frac{\theta}{2} = \frac{1 + \cos \theta}{2}.$$

■ ■

43. Show that $\tan\left(\dfrac{\theta}{2} + \dfrac{\pi}{4}\right) = \sec \theta + \tan \theta$ for $0 \le \theta < \pi/2$.

44. In a Mercator map the meridians (vertical lines of longitude) are spaced at equal distances but the lines of latitude (horizontal lines) are not. Instead they are placed so that "locally the map shrinks horizontal and vertical distances by the same factor." Figure 7.1 shows the

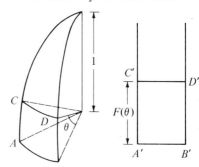

A', B', C', and D' are the images on the map of the points A, B, C, and D on the sphere of radius 1. The arc $\overset{\frown}{CD}$ has latitude θ

Figure 7.1

spherical earth, with radius 1 for convenience and a corresponding section of the map.
(a) Show that if $\widehat{A'B'} = \widehat{AB}$, then $\widehat{C'D'} = \sec\theta\,\widehat{CD}$ where $\widehat{AB}$ and $\widehat{CD}$ are the lengths of the arcs AB and CD respectively.
(b) Let $F(\theta)$ be the distance on the map from $A'B'$ to $C'D'$. In order that the map preserve locally the ratios between vertical and horizontal distances, why should $F'(\theta) = \sec\theta$? *Hint:* Review magnification in Secs. 3.1 and 3.2.
(c) Deduce that $F(\alpha) = \int_0^\alpha \sec\theta\,d\theta$.
(d) In a certain Mercator map the distance from the equator to the 30° latitude line is 3 inches. How far is it from the 30° latitude line to the 60° latitude line?
(e) Can a Mercator map be drawn on a finite piece of paper?
For an extensive discussion of Mercator's map see Philip M. Tuchinsky, *Mercator's World Map and the Calculus,* UMAP Module 206, EDC, UMAP, 55 Chapel St., Newton, Mass. 02160.

45 Find $\int \csc\theta\,d\theta$.
46 Show that
$$\frac{1}{2}\ln\frac{1+\sin\theta}{1-\sin\theta} = \ln|\sec\theta + \tan\theta|.$$

47 (a) Show that
$$\cos mx \cos nx = \tfrac{1}{2}[\cos(m-n)x + \cos(m+n)x].$$
(b) Use (a) to compute $\int \cos mx \cos nx\,dx$.

48 (a) Show that
$$\sin mx \cos nx = \tfrac{1}{2}[\sin(m+n)x + \sin(m-n)x].$$
(b) Use (a) to compute $\int \sin mx \cos nx\,dx$. The computation of $\int \sin mx \sin nx\,dx$ is similar; see Exercise 200 in Sec. 7.S.

7.7 How to integrate any rational function of $\sin\theta$ and $\cos\theta$

This section shows how to integrate a function such as
$$\frac{\sin^3\theta + 5\cos\theta\sin\theta}{3 + 4\cos\theta - 7\sin^2\theta}.$$

Some definitions will be needed before describing the method.

A *polynomial in x and y* is a sum of terms of the form ax^iy^j, where i and j are nonnegative integers and a is a real number. For instance, $1 + x - 2xy + y^2$ and $x^3y - 3xy + x^2y^3$ are polynomials in x and y. The quotient of two such polynomials is called a *rational function of x and y*, and denoted $R(x, y)$. Thus
$$\frac{1 + x - 2xy + y^2}{x^3y - 3xy + x^2y^3}$$
is a rational function of x and y.

If, in $R(x, y)$, x is replaced by $\cos\theta$ and y by $\sin\theta$, we obtain a *rational function of $\cos\theta$ and $\sin\theta$*. Thus
$$R(\cos\theta, \sin\theta) \qquad \frac{1 + \cos\theta - 2\cos\theta\sin\theta + \sin^2\theta}{\cos^3\theta\sin\theta - 3\cos\theta\sin\theta + \cos^2\theta\sin^3\theta}$$
is a rational function of $\cos\theta$ and $\sin\theta$.

The technique described in this section—a particular substitution—reduces the integration of any rational function of $\cos\theta$ and $\sin\theta$ to the integration of a rational function of u. The latter can be accomplished by partial fractions.

DESCRIPTION OF THE METHOD

The method depends on the fact that $\cos \theta$ and $\sin \theta$ can both be expressed as rational functions of $\tan (\theta/2)$.

Consider $-\pi < \theta < \pi$ and let $u = \tan (\theta/2)$. The right triangle in Fig. 7.2 shows the geometry of this substitution for $u \geq 0$, and Fig. 7.3 shows the case when u is negative.

As Figs. 7.2 and 7.3 show,

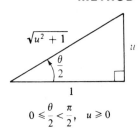

Figure 7.2

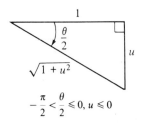

Figure 7.3

$$\cos \frac{\theta}{2} = \frac{1}{\sqrt{1 + u^2}}$$

and

$$\sin \frac{\theta}{2} = \frac{u}{\sqrt{1 + u^2}}.$$

Thus $\cos \theta = \cos^2 \frac{\theta}{2} - \sin^2 \frac{\theta}{2} = \left(\frac{1}{\sqrt{1+u^2}}\right)^2 - \left(\frac{u}{\sqrt{1+u^2}}\right)^2 = \frac{1-u^2}{1+u^2}$

and $\sin \theta = 2 \sin \frac{\theta}{2} \cos \frac{\theta}{2} = 2\left(\frac{u}{\sqrt{1+u^2}}\right)\left(\frac{1}{\sqrt{1+u^2}}\right) = \frac{2u}{1+u^2}.$

In addition, $d\theta$ can be expressed easily in terms of u and du, as will now be shown. Since $u = \tan (\theta/2)$,

$$du = \frac{1}{2} \sec^2 \frac{\theta}{2} \, d\theta$$

$$= \frac{1}{2}\left(1 + \tan^2 \frac{\theta}{2}\right) d\theta$$

$$= \tfrac{1}{2}(1 + u^2) \, d\theta.$$

Thus

$$d\theta = \frac{2 \, du}{1 + u^2}.$$

The substitution $u = \tan \frac{\theta}{2}$

thus leads to the formulas

$$\sin \theta = \frac{2u}{1 + u^2}, \quad \cos \theta = \frac{1 - u^2}{1 + u^2}, \quad \text{and} \quad d\theta = \frac{2 \, du}{1 + u^2}. \tag{1}$$

Though always applicable, this method is not always the most convenient one to use.

This substitution transforms any integral of a rational function of $\cos \theta$ and $\sin \theta$ into an integral of a rational function of u. The resulting rational function can then be integrated by the method of partial fractions.

EXAMPLE 1 Find

$$\int \frac{d\theta}{1 - \sin \theta}.$$

SOLUTION The substitution (1) transforms the integral to

7.7 HOW TO INTEGRATE ANY RATIONAL FUNCTION OF $\sin\theta$ AND $\cos\theta$

$$\int \frac{\frac{2\,du}{1+u^2}}{1 - \frac{2u}{1+u^2}} = \int \frac{2\,du}{1 - 2u + u^2}$$

$$= \int \frac{2\,du}{(1-u)^2}$$

$$= \frac{2}{1-u} + C$$

$$= \frac{2}{1 - \tan(\theta/2)} + C.$$

The identity, $\tan(\theta/2) = \sin\theta/(1 + \cos\theta)$, could be used to express the answer in terms of $\sin\theta$ and $\cos\theta$. ∎

EXAMPLE 2 Compute

$$\int_0^{\pi/2} \frac{d\theta}{4\sin\theta + 3\cos\theta}. \qquad (2)$$

SOLUTION As θ goes from 0 to $\pi/2$, $u = \tan(\theta/2)$ goes from 0 to 1. The substitution (1) transforms (2) into

$$\int_0^1 \frac{\frac{2\,du}{1+u^2}}{4\left(\frac{2u}{1+u^2}\right) + 3\left(\frac{1-u^2}{1+u^2}\right)} = \int_0^1 \frac{2\,du}{8u + 3(1-u^2)}$$

$$= \int_0^1 \frac{2\,du}{8u + 3 - 3u^2}$$

$$= \int_0^1 \frac{2\,du}{(3u+1)(3-u)}$$

$$= \int_0^1 \left(\frac{\frac{3}{5}}{3u+1} + \frac{\frac{1}{5}}{3-u}\right) du \quad \text{partial fractions}$$

$$= \left[\tfrac{1}{5}\ln(3u+1) - \tfrac{1}{5}\ln(3-u)\right]_0^1$$

$$= \left(\tfrac{1}{5}\ln 4 - \tfrac{1}{5}\ln 2\right) - \left(\tfrac{1}{5}\ln 1 - \tfrac{1}{5}\ln 3\right)$$

$$= \frac{\ln 4 - \ln 2 + \ln 3}{5}$$

$$= \frac{\ln 6}{5}. \quad \blacksquare$$

REMARKS ON THE METHOD The substitution (1) transforms any rational function of the six trigonometric functions into a rational function of u. To apply (1) first express $\tan\theta$, $\sec\theta$, $\cot\theta$, or $\csc\theta$ in terms of $\cos\theta$ and $\sin\theta$.

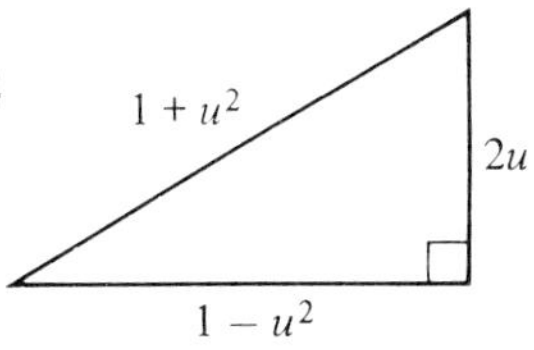

Figure 7.4

Figure 7.4 may be of use in remembering the substitution. Just remember that $(1-u^2)^2 + (2u)^2 = (1+u^2)^2$, an identity that can be checked easily. Inspection of Fig. 7.4 shows that

$$\cos\theta = \frac{1-u^2}{1+u^2} \quad \text{and} \quad \sin\theta = \frac{2u}{1+u^2}.$$

The formula for $d\theta$ resembles that for $\sin\theta$,

$$d\theta = \frac{2\,du}{1+u^2}.$$

Keep in mind that the substitution $u = \tan(\theta/2)$ is called upon only when easier ways, such as those in the preceding section, don't work.

Exercises

In Exercises 1 to 4 use the substitution (1) to transform the integrands to rational functions of u. Do *not* evaluate the resulting integrals.

1. $\displaystyle\int \frac{\sin\theta + \cos\theta}{1 + 2\cos\theta}\,d\theta$

2. $\displaystyle\int \frac{\sin^2\theta\,d\theta}{\sin\theta + \cos\theta}$

3. $\displaystyle\int \frac{3 + \tan\theta}{2 + \tan\theta}\,d\theta$

4. $\displaystyle\int \frac{\sec^2\theta\,d\theta}{1 + 4\sin\theta}$

In Exercises 5 to 20 evaluate the integrals.

5. $\displaystyle\int \frac{d\theta}{4\cos\theta + 3\sin\theta}$

6. $\displaystyle\int \frac{d\theta}{4\cos\theta - 3\sin\theta}$

7. $\displaystyle\int_0^{\pi/4} \frac{d\theta}{1 + 2\cos\theta - \sin\theta}$

8. $\displaystyle\int_0^{\pi/2} \frac{d\theta}{1 + 3\cos\theta + \sin\theta}$

9. $\displaystyle\int \frac{d\theta}{5\cos\theta + 5\sin\theta + 1}$

10. $\displaystyle\int \frac{d\theta}{5 + 3\cos\theta + 4\sin\theta}$

11. $\displaystyle\int \frac{d\theta}{5\sin\theta + 4\sin^2\theta}$

12. $\displaystyle\int \frac{d\theta}{13\sin\theta - 12\sin^2\theta}$

13. $\displaystyle\int \frac{\sin\theta + 2\cos\theta}{1 + \cos\theta}\,d\theta$

14. $\displaystyle\int \frac{3\sin\theta - \cos\theta}{1 + \cos\theta}\,d\theta$

15. $\displaystyle\int \frac{\sin\theta\,d\theta}{4 - 3\tan\theta}$

16. $\displaystyle\int \frac{\sin\theta\,d\theta}{3 + 4\tan\theta}$

17. $\displaystyle\int \frac{d\theta}{4 + 5\cos\theta}$

18. $\displaystyle\int \frac{d\theta}{8 + 17\cos\theta}$

19. $\displaystyle\int \frac{(1 + \cos\theta)\,d\theta}{13 + 18\cos\theta + 5\cos^2\theta}$

20. $\displaystyle\int \frac{(1 + \cos\theta)\,d\theta}{34 + 50\cos\theta + 16\cos^2\theta}$

21. Find
$$\int \frac{\sin\theta\,d\theta}{26\cos\theta + 4\sin\theta + 4\sin\theta\cos\theta + 12\sin^2\theta + 26}.$$

22. Find $\displaystyle\int \frac{3 + \cos\theta}{2 - \cos\theta}\,d\theta.$

23. In Sec. 7.6 $\int \sec\theta\,d\theta$ was computed in two ways.
    (a) Use substitution (1) to show that
    $$\int \sec\theta\,d\theta = \ln\left(\frac{1 + \tan\theta/2}{1 - \tan\theta/2}\right) + C.$$
    (b) Use the identity
    $$\tan(A+B) = \frac{\tan A + \tan B}{1 - \tan A \tan B}$$
    to show that the answer in (a) is the same as $\ln \tan(\theta/2 + \pi/4)$, which was Bond's conjecture.

24 Let $p(u)$, $q(u)$, and $r(u)$ be three polynomials such that $(p(u))^2 + (q(u))^2 = (r(u))^2$. Make a change of variable by

$$\cos \theta = \frac{p(u)}{r(u)} \quad \text{and} \quad \sin \theta = \frac{q(u)}{r(u)}.$$

Show that

$$d\theta = \frac{rq' - qr'}{prr'} dr.$$

This section was based on the fact that

$$(1 - u^2)^2 + (2u)^2 = (1 + u^2)^2.$$

Substitution (1) may therefore be considered independently of the angle $\theta/2$. See Alan H. Schoenfeld, "The Curious Substitution $z = \tan(\theta/2)$ and the Pythagorean Theorem," *Amer. Math. Monthly*, vol. 84, pp. 370–372, May 1977.

7.8 How to integrate rational functions involving $\sqrt{a^2 - x^2}$, $\sqrt{a^2 + x^2}$, $\sqrt{x^2 - a^2}$, or $\sqrt[n]{ax + b}$

The first part of this section describes trigonometric substitutions that turn certain rational functions of quantities that involve square roots into rational functions of $\sin \theta$ and $\cos \theta$; these can be integrated by the method of Sec. 7.7. The second part describes an algebraic substitution that reduces a rational function of x and $\sqrt[n]{ax + b}$ to a rational function of a single variable, which can then be treated by the technique of partial fractions.

THREE TRIGONOMETRIC SUBSTITUTIONS

A rational function of x and $\sqrt{a^2 - x^2}$, $\sqrt{a^2 + x^2}$, or $\sqrt{x^2 - a^2}$ can be integrated by using a trigonometric substitution. If the integrand is a rational function of x and

How to integrate
$R(x, \sqrt{a^2 - x^2})$
$R(x, \sqrt{a^2 + x^2})$
$R(x, \sqrt{x^2 - a^2})$

Case 1 $\sqrt{a^2 - x^2}$; let $x = a \sin \theta$ $\left(a > 0, \quad -\frac{\pi}{2} \leq \theta \leq \frac{\pi}{2}\right)$.

Case 2 $\sqrt{a^2 + x^2}$; let $x = a \tan \theta$ $\left(a > 0, \quad -\frac{\pi}{2} < \theta < \frac{\pi}{2}\right)$.

Case 3 $\sqrt{x^2 - a^2}$; let $x = a \sec \theta$ $\left(a > 0, \quad 0 \leq \theta \leq \pi, \theta \neq \frac{\pi}{2}\right)$.

The motivation behind this general procedure is quite simple. Consider case 1, for instance. If you replace x in $\sqrt{a^2 - x^2}$ by $a \sin \theta$, you obtain

How to make the square root sign in $\sqrt{a^2 - x^2}$ disappear

$$\sqrt{a^2 - x^2} = \sqrt{a^2 - (a \sin \theta)^2} = \sqrt{a^2(1 - \sin^2 \theta)}$$
$$= \sqrt{a^2 \cos^2 \theta}$$
$$= a \cos \theta.$$

(Keep in mind that a and $\cos \theta$ are positive.) The important thing is that *the square root sign disappears.*

Case 3 raises a fine point. We have $a > 0$. However, whenever x is negative, θ is a second quadrant angle, so $\tan \theta$ is *negative*. In that case,

432 COMPUTING ANTIDERIVATIVES

$$\sqrt{x^2 - a^2} = \sqrt{(a \sec \theta)^2 - a^2}$$
$$= a\sqrt{\sec^2 \theta - 1}$$
$$= a\sqrt{\tan^2 \theta}$$

If $c < 0$, $\sqrt{c^2} = -c$.
$$= a(-\tan \theta) \quad \text{since } -\tan \theta \text{ is positive.}$$

In the examples and exercises involving case 3 it will be assumed that x varies through nonnegative values, so that θ remains in the first quadrant and $\sqrt{\sec^2 \theta - 1} = \tan \theta$.

Note that for $\sqrt{a^2 - x^2}$ to be meaningful, $|x|$ must be no larger than a. On the other hand, for $\sqrt{x^2 - a^2}$ to be meaningful, $|x|$ must be at least as large as a. The quantity $\sqrt{a^2 + x^2}$ is meaningful for all values of x.

EXAMPLE 1 Make an appropriate substitution to remove the square root sign in each of these integrals. Do not evaluate.

(a) $\displaystyle\int \frac{x^3 \, dx}{3 + \sqrt{16 - x^2}}$ (b) $\displaystyle\int \frac{x^5 \, dx}{1 + \sqrt{3 + x^2}}$ (c) $\displaystyle\int \frac{(x^2 + 2) \, dx}{x + \sqrt{x^2 - 9}}$

SOLUTION (a) In this case $a = 4$ and the substitution $x = 4 \sin \theta$ is appropriate:

$$\int \frac{x^3 \, dx}{3 + \sqrt{16 - x^2}} = \int \frac{(4 \sin \theta)^3 4 \cos \theta \, d\theta}{3 + \sqrt{16 - (4 \sin \theta)^2}} \qquad dx = d(4 \sin \theta)$$
$$= \int \frac{256 \sin^3 \theta \cos \theta \, d\theta}{3 + \sqrt{16 - 16 \sin^2 \theta}}$$

Since $-\dfrac{\pi}{2} \leq \theta \leq \dfrac{\pi}{2}$,
$\cos \theta$ is not negative
so $\cos \theta = \sqrt{1 - \sin^2 \theta}$.
$$= \int \frac{256 \sin^3 \theta \cos \theta \, d\theta}{3 + 4\sqrt{1 - \sin^2 \theta}}$$
$$= \int \frac{256 \sin^3 \theta \cos \theta \, d\theta}{3 + 4 \cos \theta}$$

(b) In this case $a^2 = 3$, so $a = \sqrt{3}$. The substitution $x = \sqrt{3} \tan \theta$ is appropriate. We then have

$$\int \frac{x^5 \, dx}{1 + \sqrt{3 + x^2}} = \int \frac{(\sqrt{3} \tan \theta)^5 \, d(\sqrt{3} \tan \theta)}{1 + \sqrt{3 + (\sqrt{3} \tan \theta)^2}}$$
$$= \int \frac{27 \tan^5 \theta \sec^2 \theta \, d\theta}{1 + \sqrt{3 + 3 \tan^2 \theta}}$$
$$= \int \frac{27 \tan^5 \theta \sec^2 \theta \, d\theta}{1 + \sqrt{3}\sqrt{1 + \tan^2 \theta}}$$
$$= \int \frac{27 \tan^5 \theta \sec^2 \theta \, d\theta}{1 + \sqrt{3} \sec \theta}.$$

(c) In this case $a = 3$ and the substitution $x = 3 \sec \theta$ is appropriate. Then

7.8 HOW TO INTEGRATE RATIONAL FUNCTIONS INVOLVING ROOTS

$$\int \frac{(x^2 + 2)\,dx}{x + \sqrt{x^2 - 9}} = \int \frac{((3\sec\theta)^2 + 2)\,d(3\sec\theta)}{3\sec\theta + \sqrt{(3\sec\theta)^2 - 9}}$$

$$= \int \frac{(9\sec^2\theta + 2)\,3\sec\theta\tan\theta\,d\theta}{3\sec\theta + 3\sqrt{\sec^2\theta - 1}}$$

$$= \int \frac{(9\sec^2\theta + 2)\sec\theta\tan\theta\,d\theta}{\sec\theta + \tan\theta}. \quad ■$$

EXAMPLE 2 Find $\int x^3\sqrt{16 - x^2}\,dx$.

SOLUTION Let $x = 4\sin\theta$; hence $dx = 4\cos\theta\,d\theta$. Then

$$\int x^3\sqrt{16 - x^2}\,dx = \int (4\sin\theta)^3\sqrt{16 - (4\sin\theta)^2}\,4\cos\theta\,d\theta$$

$$= 4^4 \int \sin^3\theta\sqrt{16 - 16\sin^2\theta}\cos\theta\,d\theta$$

$$= 4^5 \int \sqrt{1 - \sin^2\theta}\,\sin^3\theta\cos\theta\,d\theta$$

$$= 4^5 \int \sin^3\theta\cos^2\theta\,d\theta$$

$$= 4^5 \int \sin^2\theta\cos^2\theta\sin\theta\,d\theta$$

$$= 4^5 \int (1 - \cos^2\theta)\cos^2\theta\sin\theta\,d\theta$$

$$= 4^5 \int (1 - u^2)u^2(-du) \qquad u = \cos\theta$$

$$= 4^5 \int (u^4 - u^2)\,du$$

$$= 4^5 \left(\frac{u^5}{5} - \frac{u^3}{3}\right) + C$$

$$= 4^5 \left(\frac{\cos^5\theta}{5} - \frac{\cos^3\theta}{3}\right) + C. \qquad (1)$$

To express the answer in terms of x, express $\cos\theta$ in terms of x. To do this, use Fig. 7.5, which records the relation $x = 4\sin\theta$ (or $\sin\theta = x/4$). The third side has length $\sqrt{16 - x^2}$, by the Pythagorean theorem.

Fig. 7.5 shows the case $x > 0$; draw the case $x < 0$.

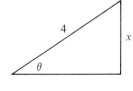

Figure 7.5

Thus
$$\cos\theta = \frac{\sqrt{16-x^2}}{4},$$

and (1) becomes
$$4^5\left[\frac{1}{5}\left(\frac{\sqrt{16-x^2}}{4}\right)^5 - \frac{1}{3}\left(\frac{\sqrt{16-x^2}}{4}\right)^3\right] + C$$

$$= \frac{(\sqrt{16-x^2})^5}{5} - 16\frac{(\sqrt{16-x^2})^3}{3} + C. \quad \blacksquare \quad (2)$$

Had Example 2 asked instead for the value of $\int_{-2}^{2} x^3\sqrt{16-x^2}\,dx$, the solution would have been much simpler. Since the integrand is an odd function, $\int_{-2}^{2} x^3\sqrt{16-x^2}\,dx = 0$. Had Example 2 asked for $\int_{-4}^{2} x^3\sqrt{16-x^2}\,dx$, a convenient computation might run like this:

$$\int_{-4}^{2} x^3\sqrt{16-x^2}\,dx = \int_{-4}^{-2} x^3\sqrt{16-x^2}\,dx + \int_{-2}^{2} x^3\sqrt{16-x^2}\,dx$$

$$= \int_{-4}^{-2} x^3\sqrt{16-x^2}\,dx.$$

To evaluate $\int_{-4}^{-2} x^3\sqrt{16-x^2}\,dx$, use the substitution in Example 2, $x = 4\sin\theta$. As x goes from -4 to -2, $\sin\theta = x/4$ goes from -1 to $-1/2$. Thus θ goes from $-\pi/2$ to $-\pi/6$ and $u = \cos\theta$ goes from 0 to $\sqrt{3}/2$. We would have

$$\int_{-4}^{-2} x^3\sqrt{16-x^2}\,dx = 4^5\int_{0}^{\sqrt{3}/2} (u^4 - u^2)\,du,$$

which can be computed in terms of u by formula (1); formula (2) would not be needed.

EXAMPLE 3 Compute $\int \sqrt{1+x^2}\,dx$.

SOLUTION The identity $\sec\theta = \sqrt{1+\tan^2\theta}$ suggests the substitution described in case 2:

$$x = \tan\theta;$$
hence
$$dx = \sec^2\theta\,d\theta.$$

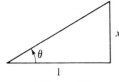

Figure 7.6

(See Fig. 7.6 for the geometry of this substitution.)

Thus
$$\int \sqrt{1+x^2}\,dx = \int \sec\theta\,\sec^2\theta\,d\theta = \int \sec^3\theta\,d\theta.$$

By Examples 8 and 9 in Sec. 7.6,

$$\int \sec^3\theta\,d\theta = \frac{\sec\theta\tan\theta}{2} + \frac{1}{2}\ln|\sec\theta + \tan\theta| + C.$$

7.8 HOW TO INTEGRATE RATIONAL FUNCTIONS INVOLVING ROOTS

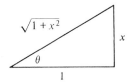

Figure 7.7

To express the antiderivative just obtained in terms of x rather than θ, it is necessary to express $\tan \theta$ and $\sec \theta$ in terms of x. Starting with the definition $x = \tan \theta$, find $\sec \theta$ by means of the relation $\sec \theta = \sqrt{1 + \tan^2 \theta} = \sqrt{1 + x^2}$ as in Fig. 7.7. Thus,

$$\int \sqrt{1 + x^2} \, dx = \frac{x\sqrt{1 + x^2}}{2} + \frac{1}{2} \ln\left(\sqrt{1 + x^2} + x\right) + C. \blacksquare$$

EXAMPLE 4 Compute $\displaystyle\int_4^5 \frac{dx}{\sqrt{x^2 - 9}}$.

SOLUTION Let $x = 3 \sec \theta$; hence $dx = 3 \sec \theta \tan \theta \, d\theta$. (See Fig. 7.8.) Thus

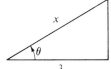

Figure 7.8

$$\int_4^5 \frac{dx}{\sqrt{x^2 - 9}} = \int_{\sec^{-1}(4/3)}^{\sec^{-1}(5/3)} \frac{3 \sec \theta \tan \theta \, d\theta}{\sqrt{9 \sec^2 \theta - 9}}$$

$$= \int_{\sec^{-1}(4/3)}^{\sec^{-1}(5/3)} \frac{\sec \theta \tan \theta \, d\theta}{\tan \theta}$$

$$= \int_{\sec^{-1}(4/3)}^{\sec^{-1}(5/3)} \sec \theta \, d\theta$$

$$= \ln|\sec \theta + \tan \theta| \Big|_{\sec^{-1}(4/3)}^{\sec^{-1}(5/3)} \quad \text{by Example 9 of Sec. 7.6}$$

$$= \ln\left(\frac{5}{3} + \frac{4}{3}\right) - \ln\left(\frac{4}{3} + \frac{\sqrt{7}}{3}\right) \quad \text{using Fig. 7.9 to find } \tan \theta \text{ at the limits of integration.}$$

$$= \ln 3 - \ln\left(\frac{4 + \sqrt{7}}{3}\right)$$

$$= \ln\left(\frac{9}{4 + \sqrt{7}}\right) \blacksquare$$

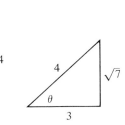

Figure 7.9

A rational function of x and $\sqrt{a^2 - b^2 x^2}$, $\sqrt{a^2 x^2 - b^2}$, or $\sqrt{a^2 + b^2 x^2}$ can be treated by a similar procedure. For instance, to deal with a rational function of x and $\sqrt{a^2 - b^2 x^2}$ make the substitution suggested by

$$b^2 x^2 = a^2 \sin^2 \theta,$$

that is, $$bx = a \sin \theta,$$

or $$x = \frac{a}{b} \sin \theta \quad \text{and} \quad dx = \frac{a}{b} \cos \theta \, d\theta.$$

Then $$\sqrt{a^2 - b^2 x^2} = \sqrt{a^2 - a^2 \sin^2 \theta}$$

$$= a \cos \theta.$$

A rational function of x and $a^2 - b^2 x^2$ is a rational function of x and therefore can be integrated by partial fractions. But in some special cases a trigonometric substitution may provide another solution, as Example 5 illustrates.

436 COMPUTING ANTIDERIVATIVES

EXAMPLE 5 Transform
$$\int \frac{dx}{(4 - 9x^2)^2}$$
into a trigonometric integral.

SOLUTION Assume that $9x^2 < 4$. Let $9x^2 = 4 \sin^2 \theta$, choosing $3x = 2 \sin \theta$; hence

Since $9x^2 < 4$, $|\sin \theta| = \dfrac{|3x|}{2}$ is indeed ≤ 1.

$$dx = \tfrac{2}{3} \cos \theta \, d\theta.$$

Thus
$$\int \frac{dx}{(4 - 9x^2)^2} = \int \frac{\tfrac{2}{3} \cos \theta \, d\theta}{(4 - 4\sin^2 \theta)^2}$$
$$= \frac{2}{3} \int \frac{\cos \theta \, d\theta}{16(1 - \sin^2 \theta)^2}$$
$$= \frac{1}{24} \int \frac{\cos \theta \, d\theta}{\cos^4 \theta}$$
$$= \frac{1}{24} \int \frac{1}{\cos^3 \theta} \, d\theta.$$

Consequently,
$$\int \frac{dx}{(4 - 9x^2)^2} = \frac{1}{24} \int \sec^3 \theta \, d\theta.$$

Recall that $|\sec \theta| \geq 1$.

If $9x^2 > 4$, then the substitution $9x^2 = 4 \sec^2 \theta$ is suggested, with the choice $3x = 2 \sec \theta$. ∎

HOW TO INTEGRATE $R(x, \sqrt[n]{ax + b})$

Let n be a positive integer. Any rational function of x and $\sqrt[n]{ax + b}$ can be transformed into a rational function of u by the substitution

$$u = \sqrt[n]{ax + b}$$

and thus can be integrated by partial fractions.

If $u = \sqrt[n]{ax + b}$, then $u^n = ax + b$. From this it follows that

$$x = \frac{u^n - b}{a} \quad \text{and} \quad dx = \frac{nu^{n-1} \, du}{a}.$$

Two examples illustrate the method.

EXAMPLE 6 Find $\int x^2 \sqrt[3]{4x + 5} \, dx$.

SOLUTION Let $u = \sqrt[3]{4x + 5}$. Then

$$u^3 = 4x + 5, \quad x = \frac{u^3 - 5}{4}, \quad dx = \frac{3u^2 \, du}{4}.$$

Consequently,

$$\int x^2 \sqrt[3]{4x+5}\, dx$$

$$= \int \left(\frac{u^3-5}{4}\right)^2 \cdot u \cdot \frac{3u^2\, du}{4}$$

$$= \frac{3}{64}\int u^3(u^3-5)^2\, du$$

$$= \frac{3}{64}\int u^3(u^6 - 10u^3 + 25)\, du$$

$$= \frac{3}{64}\int (u^9 - 10u^6 + 25u^3)\, du$$

$$= \frac{3}{64}\left(\frac{u^{10}}{10} - \frac{10u^7}{7} + \frac{25u^4}{4}\right) + C$$

$$= \frac{3}{64}\left[\frac{(\sqrt[3]{4x+5})^{10}}{10} - \frac{10}{7}(\sqrt[3]{4x+5})^7 + \frac{25}{4}(\sqrt[3]{4x+5})^4\right] + C. \quad\blacksquare$$

EXAMPLE 7 Express

$$\int \frac{\sqrt{2x+1}}{1+\sqrt[3]{2x+1}}\, dx$$

as the integral of a rational function of a single variable.

SOLUTION At first glance, this problem does not seem to fit into the form "integral of a rational function of x and $\sqrt[n]{ax+b}$." However, both $\sqrt{2x+1}$ and $\sqrt[3]{2x+1}$ are powers of $\sqrt[6]{2x+1}$:

$$\sqrt{2x+1} = (2x+1)^{1/2} = (2x+1)^{3/6} = (\sqrt[6]{2x+1})^3$$

and

$$\sqrt[3]{2x+1} = (2x+1)^{1/3} = (2x+1)^{2/6} = (\sqrt[6]{2x+1})^2.$$

This suggests the substitution $u = \sqrt[6]{2x+1}$; hence

$$u^6 = 2x+1, \qquad x = \frac{u^6-1}{2}, \qquad dx = \frac{6u^5\, du}{2} = 3u^5\, du.$$

Consequently,
$$\int \frac{\sqrt{2x+1}}{1+\sqrt[3]{2x+1}}\, dx = \int \frac{u^3(3u^5\, du)}{1+u^2},$$

an integral that can be computed by partial fractions. $\quad\blacksquare$

Exercises

In Exercises 1 to 30 find the integrals.

1. $\displaystyle\int \frac{\sqrt{1-x^2}\, dx}{x}$

2. $\displaystyle\int \frac{\sqrt{1-x^2}\, dx}{x^2}$

3. $\displaystyle\int_1^{\sqrt{3}} \frac{x^3\, dx}{\sqrt{x^2+1}}$

4. $\displaystyle\int_0^2 \frac{dx}{\sqrt{x^2+4}}$

5. $\displaystyle\int_1^{\sqrt{3}} \frac{dx}{x\sqrt{1+x^2}}$

6. $\displaystyle\int_0^{\sqrt{2}} \frac{dx}{x\sqrt{2+x^2}}$

7. $\int x^3 \sqrt{1 - x^2} \, dx$

8. $\int x^2 \sqrt{1 - x^2} \, dx$

9. $\int_1^2 \sqrt{4 - x^2} \, dx$

10. $\int_{-2/\sqrt{3}}^{2/\sqrt{3}} \sqrt{3 - x^2} \, dx$

11. $\int_4^5 \frac{x^2 \, dx}{\sqrt{x^2 - 9}}$

12. $\int_{3/2}^{3} \frac{x^2 \, dx}{\sqrt{9 - x^2}}$

13. $\int \frac{dx}{\sqrt{9 + x^2}}$

14. $\int \frac{dx}{\sqrt{2 + x^2}}$

15. $\int_3^6 \frac{\sqrt{x^2 - 9} \, dx}{x}$

16. $\int_{2\sqrt{2}}^{4} \frac{\sqrt{x^2 - 4} \, dx}{x^2}$

17. $\int \frac{dx}{\sqrt{x} + 3}$

18. $\int \frac{x \, dx}{\sqrt{x} + 3}$

19. $\int \frac{dx}{x(\sqrt{x} + 1)}$

20. $\int_0^1 \frac{\sqrt{x} \, dx}{1 + \sqrt{x}}$

21. $\int_1^{16} \frac{dx}{\sqrt[4]{x} + \sqrt{x}}$

22. $\int \frac{dx}{\sqrt[3]{x} + \sqrt{x}}$

23. $\int \frac{\sqrt{2x + 1} \, dx}{x}$

24. $\int x^2 \sqrt{2x + 1} \, dx$

25. $\int x \sqrt[3]{3x + 2} \, dx$

26. $\int x(3x + 2)^{5/3} \, dx$

27. $\int \frac{\sqrt{x + 3}}{\sqrt{x - 2}} \, dx$

28. $\int \frac{\sqrt{x + 1} + 3}{\sqrt[3]{x + 1} - 1} \, dx$

29. $\int \frac{dx}{(x - 2)\sqrt{x + 2}}$

30. $\int \frac{x - 1}{\sqrt[3]{x + 1}} \, dx$

■

In Exercises 31 to 40 find the integrals. (a is a positive constant.)

31. $\int \sqrt{a^2 - x^2} \, dx$

32. $\int \sqrt{x^2 - a^2} \, dx$

33. $\int \sqrt{a^2 + x^2} \, dx$

34. $\int \frac{dx}{\sqrt{4 - 9x^2}}$

35. $\int \frac{dx}{\sqrt{25x^2 - 16}}$

36. $\int_{\sqrt{2}}^{2} \sqrt{x^2 - 1} \, dx$

37. $\int_0^{1/2} x^3 \sqrt{1 - x^2} \, dx$

38. $\int \frac{x^3 \, dx}{(1 - x^2)^4}$

39. $\int \frac{\sqrt{x^2 - 4}}{x} \, dx$

40. $\int \frac{\sqrt{4 + x^2}}{x} \, dx$

■ ■

41. Show that any rational function of $\sqrt{x}$ and $\sqrt{x + 1}$ can be integrated by using the substitution $x = \tan^2 \theta$.

42. Show that any rational function of x, $\sqrt{x + a}$, and $\sqrt{x + b}$ can be integrated by introducing the substitution defined by

$$x + a = \frac{1}{4}(a - b)\left(t + \frac{1}{t}\right)^2;$$

hence

$$x + b = \frac{1}{4}(a - b)\left(t - \frac{1}{t}\right)^2.$$

However, it is not the case that every rational function of $\sqrt{x + a}$, $\sqrt{x + b}$, and $\sqrt{x + c}$ has an elementary integral. For instance,

$$\int \frac{1}{\sqrt{x}\sqrt{x - 1}\sqrt{x + 1}} \, dx = \int \frac{1}{\sqrt{x^3 - x}} \, dx$$

is not an elementary function.

43. Show that every rational function of x and $\sqrt[n]{\frac{ax + b}{cx + d}}$ has an elementary antiderivative.

7.9 What to do in the face of an integral

Since the exercises in each section of this chapter were focused on the techniques of that section, it was usually clear what technique to use on a given integral. But what if an integral is met "in the wild," where there is no clue how to evaluate it? This section suggests what to do in this typical situation.

The more integrals you compute, the more quickly you will be able to choose an appropriate technique. There is no substitute for practice. (This section and the summary section offer many further exercises.)

There are only a few techniques on which to draw:

7.9 WHAT TO DO IN THE FACE OF AN INTEGRAL

Substitution	Sec. 7.1
Parts	Sec. 7.2
Partial fractions	Secs. 7.4 and 7.5
Powers of trigonometric functions	Sec. 7.6
$R(\sin\theta, \cos\theta)$	Sec. 7.7
$R(x, \sqrt{a^2-x^2})$, $R(x, \sqrt{a^2+x^2})$, $R(x, \sqrt{x^2-a^2})$	Sec. 7.8
$R(x, \sqrt[n]{ax+b})$	Sec. 7.8
$\int \dfrac{f'(x)\,dx}{f(x)} = \ln\|f(x)\| + C$ (not to be forgotten)	Sec. 7.1

The following recommendations for a strategy of integration are based on my own experience and *Integration, Getting It All Together* by Alan H. Schoenfeld, U 203–205, EDC, UMAP, 55 Chapel St., Newton, Mass. 02160. One possible general strategy consists of answering these questions.

Questions	Examples
Can the integrand be simplified by some algebraic or trigonometric identity?	$\int \dfrac{3+x^2}{x}\,dx = \int \dfrac{3\,dx}{x} + \int \dfrac{x^2\,dx}{x}$ $\int \sin^2\theta\,d\theta = \int\left(\dfrac{1-\cos 2\theta}{2}\right)d\theta$
Will a substitution simplify the integrand? In particular, can the integrand be written as the product $\boxed{\text{function of } u(x)} \times \boxed{u'(x)}$?	$\int (1+x^4)^5 x^3\,dx$ $= \tfrac{1}{4}\int (1+x^4)^5 4x^3\,dx$ $= \tfrac{1}{4}\int u^5\,du \quad (u = 1+x^4)$
Is the integrand of the form f'/f?	$\int \dfrac{dx}{x\ln x} = \int \dfrac{(1/x)\,dx}{\ln x} = \ln\|\ln x\| + C$
Would integration by parts help? Can I write the problem as $\int u\,dv$ in such a way that $\int v\,du$ is easier?	$\int x^4 e^x\,dx = x^4 e^x - \int e^x 4x^3\,dx$ (which reduces exponent of x)
Is the integrand a rational function of x?	$\int \dfrac{dx}{x^3-1} = \int\left(\dfrac{Ax+B}{x^2+x+1} + \dfrac{C}{x-1}\right)dx$ $\left(\text{but do } \int \dfrac{x^2\,dx}{x^3-1} \text{ by the "}f'/f\text{ method"}\right)$
Is the integrand of a special form?	$\int \sin^5\theta\cos^2\theta\,d\theta$ $\int \dfrac{3+2\cos\theta}{5+22\sin\theta}\,d\theta$ $\int x^3\sqrt{9+x^2}\,dx$ $\int \dfrac{\sqrt[3]{x}\,dx}{1+(\sqrt[3]{x})^5}$

The following examples illustrate the strategy.

EXAMPLE 1 $\displaystyle\int \dfrac{x\,dx}{x^2-9}.$

440 COMPUTING ANTIDERIVATIVES

DISCUSSION The substitutions $x = 3 \sec \theta$ or $x = 3 \sin \theta$ would work, depending on whether $|x| > 3$ or $|x| < 3$. However, notice that $x\,dx$ is almost the differential of $x^2 - 9$. This suggests the substitution $u = x^2 - 9$, $du = 2x\,dx$:

$$\int \frac{x\,dx}{x^2 - 9} = \int \frac{du/2}{u} = \frac{1}{2} \int u^{-1}\,du$$

$$= \tfrac{1}{2} \ln |u| + C = \tfrac{1}{2} \ln |x^2 - 9| + C. \quad\blacksquare$$

EXAMPLE 2
$$\int \frac{x\,dx}{1 + x^4}.$$

DISCUSSION Since the integrand is a rational function of x, partial fractions would work. This requires factoring $x^4 + 1$ and then representing $x/(1 + x^4)$ as a sum of partial fractions. With some struggle it can be shown that

$$x^4 + 1 = (x^2 + \sqrt{2}x + 1)(x^2 - \sqrt{2}x + 1).$$

Then constants A, B, C, and D will have to be found such that

$$\frac{x}{1 + x^4} = \frac{Ax + B}{x^2 + \sqrt{2}x + 1} + \frac{Cx + D}{x^2 - \sqrt{2}x + 1}.$$

The method would work but would certainly be tedious.

Try another attack. The numerator x is almost the derivative of x^2. The substitution $u = x^2$ is at least worth testing:

$$u = x^2, \qquad du = 2x\,dx,$$

$$\int \frac{x\,dx}{1 + x^4} = \int \frac{du/2}{1 + u^2},$$

which is easy. The answer is $\tfrac{1}{2} \tan^{-1} u + C = \tfrac{1}{2} \tan^{-1} x^2 + C.$ $\quad\blacksquare$

EXAMPLE 3
$$\int \frac{\sin^3 \theta\,d\theta}{\cos^4 \theta}.$$

DISCUSSION Since this is a rational function of $\sin \theta$ and $\cos \theta$, the substitution $u = \tan (\theta/2)$ would work, but maybe there is an easier way.

The fact that $\sin \theta$ appears to an odd power suggests using $\sin \theta\,d\theta$ to form a differential. That is, introduce $u = \cos \theta$, $du = -\sin \theta\,d\theta$. Then

$$\int \frac{\sin^3 \theta\,d\theta}{\cos^4 \theta} = \int \frac{\sin^2 \theta \sin \theta\,d\theta}{\cos^4 \theta} = \int \frac{(1 - \cos^2 \theta) \sin \theta\,d\theta}{\cos^4 \theta}$$

$$= \int \frac{(1 - u^2)(-du)}{u^4}$$

$$= \int \frac{(u^2 - 1)\,du}{u^4}$$

7.9 WHAT TO DO IN THE FACE OF AN INTEGRAL

$$= \int (u^{-2} - u^{-4})\, du = \frac{u^{-1}}{-1} + \frac{u^{-3}}{3} + C$$

$$= -\sec\theta + \tfrac{1}{3}\sec^3\theta + C. \ \blacksquare$$

EXAMPLE 4
$$\int \frac{1+x}{1+x^2}\, dx.$$

DISCUSSION This is a rational function of x, but partial fractions will not help for the integrand is already in its partial-fraction representation.

The numerator is not the derivative of the denominator, but comes close enough to persuade us to break the integrand into two summands:

$$\int \frac{1+x}{1+x^2}\, dx = \int \frac{dx}{1+x^2} + \int \frac{x\, dx}{1+x^2}.$$

Both of the latter integrals can be done by sight. The first is $\tan^{-1} x + C$ and the second is $\tfrac{1}{2}\ln(1+x^2) + C$. $\blacksquare$

EXAMPLE 5
$$\int \frac{e^{2x}}{1+e^x}\, dx.$$

DISCUSSION At first glance, this integral looks so peculiar that it may not belong in this chapter. However, e^x is a fairly simple function, with $d(e^x) = e^x\, dx$. This suggests trying the substitution $u = e^x$ and seeing what happens:

$$u = e^x, \quad du = e^x\, dx; \quad \text{thus} \quad dx = \frac{du}{e^x} = \frac{du}{u},$$

But what will be done to e^{2x}? Recalling that $e^{2x} = (e^x)^2$, we anticipate no problem:

$$\int \frac{e^{2x}}{1+e^x}\, dx = \int \frac{u^2}{1+u}\frac{du}{u} = \int \frac{u\, du}{1+u},$$

which can be integrated quickly:

Or use long division of $1 + u$ into u.

$$\int \frac{u\, du}{1+u} = \int \frac{u+1-1}{1+u}\, du = \int \left(1 - \frac{1}{1+u}\right) du$$

$$= u - \ln|1+u| + C.$$

$$= e^x - \ln(1+e^x) + C.$$

The same substitution could have been done more elegantly:

$$\int \frac{e^{2x}}{1+e^x}\, dx = \int \frac{e^x(e^x\, dx)}{1+e^x} = \int \frac{u\, du}{1+u}. \ \blacksquare$$

EXAMPLE 6
$$\int \frac{x^3\, dx}{(1-x^2)^5}.$$

DISCUSSION Partial fractions would work, but the denominator, when factored, would be $(1+x)^5(1-x)^5$. There would be 10 unknown constants to find. Look for an easier approach.

Since the denominator is the obstacle, try $u = x^2$ or $u = 1 - x^2$, to see if the integrand gets simpler. Let us examine what happens in each case. Try $u = x^2$ first. Assume that we are interested only in getting an antiderivative for positive x, $x = \sqrt{u}$.

If x were negative, we would use $x = -\sqrt{u}$.

$$u = x^2, \quad du = 2x\, dx, \quad dx = du/2x = du/(2\sqrt{u}).$$

Then
$$\int \frac{x^3\, dx}{(1-x^2)^5} = \int \frac{u^{3/2}}{(1-u)^5} \frac{du}{2\sqrt{u}} = \frac{1}{2}\int \frac{u\, du}{(1-u)^5}.$$

The same substitution could be carried out as follows:

$$\int \frac{x^3\, dx}{(1-x^2)^5} = \int \frac{x^2 x\, dx}{(1-x^2)^5} = \int \frac{u(du/2)}{(1-u)^5} = \frac{1}{2}\int \frac{u\, du}{(1-u)^5}.$$

The substitution $v = 1 - u$ then results in an easy integral, as the reader may check.

Let us try $u = 1 - x^2$, $du = -2x\, dx$:

$$\int \frac{x^3\, dx}{(1-x^2)^5} = \int \frac{x^2(x\, dx)}{(1-x^2)^5} = \int \frac{(1-u)(-du/2)}{u^5} = \int \frac{1}{2}(u^{-4} - u^{-5})\, du,$$

an integral that can be computed without further substitution. So $u = 1 - x^2$ is quicker than $u = x^2$.

Assume $|x| \leq 1$.

Since $1 - x^2 = (\sqrt{1-x^2})^2$, it might be amusing to try the substitution $x = \sin\theta$, $dx = \cos\theta\, d\theta$:

$$\int \frac{x^3\, dx}{(1-x^2)^5} = \int \frac{\sin^3\theta \cos\theta\, d\theta}{(1-\sin^2\theta)^5} = \int \frac{\sin^3\theta \cos\theta\, d\theta}{\cos^{10}\theta}$$

$$= \int \frac{\sin^3\theta\, d\theta}{\cos^9\theta}$$

$$= \int \frac{\sin^2\theta \sin\theta\, d\theta}{\cos^9\theta}$$

$$= \int \frac{(1-\cos^2\theta)\sin\theta\, d\theta}{\cos^9\theta}$$

$$= \int \frac{(1-u^2)(-du)}{u^9} \quad (u = \cos\theta),$$

which is not a difficult integral.

Of the three methods, the substitution $u = 1 - x^2$ is the most efficient. ∎

7.9 WHAT TO DO IN THE FACE OF AN INTEGRAL

EXAMPLE 7
$$\int \frac{\ln x \, dx}{x}.$$

DISCUSSION Integration by parts, with $u = \ln x$ and $dv = dx/x$, may come to mind. In that case $du = dx/x$ and $v = \ln x$:

$$\int \underbrace{\ln x}_{u} \underbrace{\frac{dx}{x}}_{dv} = \underbrace{(\ln x)}_{u}\underbrace{(\ln x)}_{v} - \int \underbrace{\ln x}_{v} \underbrace{\frac{dx}{x}}_{du}.$$

Bringing the $\int \frac{\ln x \, dx}{x}$ all to one side produces the equation

$$2 \int \ln x \, \frac{dx}{x} = (\ln x)^2,$$

from which it follows that

$$\int \ln x \, \frac{dx}{x} = \frac{(\ln x)^2}{2} + C.$$

The method worked, but it is not the easiest one to use. Since $1/x$ is the derivative of $\ln x$, we could have used the substitution $u = \ln x$, $du = dx/x$:

$$\int \frac{\ln x \, dx}{x} = \int u \, du = \frac{u^2}{2} + C = \frac{(\ln x)^2}{2} + C. \blacksquare$$

EXAMPLE 8
$$\int x^3 e^{x^2} \, dx.$$

DISCUSSION Integration by parts may come to mind, since, if $u = x^3$, then $du = 3x^2 \, dx$ is simpler. However, dv must then be $e^{x^2} \, dx$ and force v to be nonelementary. This is a dead end.

So try integration by parts with $u = e^{x^2}$ and $dv = x^3 \, dx$. What will $v \, du$ be? We have $v = x^4/4$ and $du = 2x \, e^{x^2} \, dx$. Thus $v \, du = \frac{1}{2} x^5 e^{x^2} \, dx$, which is worse than the original $u \, dv$. The exponent of x has been raised by 2, from 3 to 5.

But if integration by parts can raise the exponent of x it should also be able to lower it.

This time try $u = x^2$ and $dv = x e^{x^2} \, dx$; thus $du = 2x \, dx$ and $v = e^{x^2}/2$. Integration by parts yields

$$\int x^3 \, e^{x^2} \, dx = \int \underbrace{x^2}_{u} \underbrace{x \, e^{x^2} \, dx}_{dv} = \underbrace{x^2}_{u} \underbrace{\frac{e^{x^2}}{2}}_{v} - \int \underbrace{\frac{e^{x^2}}{2}}_{v} \underbrace{2x \, dx}_{du}$$

$$= \frac{x^2 e^{x^2}}{2} - \frac{e^{x^2}}{2} + C. \blacksquare$$

444 COMPUTING ANTIDERIVATIVES

EXAMPLE 9
$$\int \frac{1-\sin\theta}{\theta+\cos\theta}\,d\theta.$$

DISCUSSION Because θ appears by itself, the integrand is not a rational function of $\sin\theta$ and $\cos\theta$. However, the numerator is the derivative of the denominator, so the integral is $\ln|\theta+\cos\theta|+C$. ∎

EXAMPLE 10
$$\int \frac{1-\sin\theta}{\cos\theta}\,d\theta.$$

DISCUSSION The substitution $u=\tan(\theta/2)$ would work, but there is an easier approach. Break the integrand into two summands:

$$\int \frac{1-\sin\theta}{\cos\theta}\,d\theta = \int \left(\frac{1}{\cos\theta} - \frac{\sin\theta}{\cos\theta}\right)d\theta$$

$$= \int (\sec\theta - \tan\theta)\,d\theta$$

$$= \ln|\sec\theta+\tan\theta| + \ln|\cos\theta| + C.$$

Since $\ln A + \ln B = \ln AB$, the answer can be simplified to

$$\ln(|\sec\theta+\tan\theta|\,|\cos\theta|) + C.$$

But $\sec\theta\cos\theta=1$ and $\tan\theta\cos\theta=\sin\theta$. The answer becomes even simpler:

$$\int \frac{1-\sin\theta}{\cos\theta}\,d\theta = \ln(1+\sin\theta) + C. \quad \blacksquare$$

Exercises

All the integrals in Exercises 1 to 78 are elementary. In each case list the technique or techniques that would be of use. If there is a preferred technique, state what it is. Do *not* evaluate the integrals.

1. $\int \dfrac{1+x}{x^2}\,dx$

2. $\int \dfrac{x^2}{1+x}\,dx$

3. $\int \dfrac{dx}{x^2+x^3}\,dx$

4. $\int \dfrac{x+1}{x^2+x^3}\,dx$

5. $\int \tan^{-1} 2x\,dx$

6. $\int \sin^{-1} 2x\,dx$

7. $\int x^{10} e^x\,dx$

8. $\int \dfrac{\ln x}{x^2}\,dx$

9. $\int \dfrac{\sec^2\theta\,d\theta}{\tan\theta}$

10. $\int \dfrac{\tan\theta\,d\theta}{\sin^2\theta}$

11. $\int \dfrac{x^3}{\sqrt[3]{x+2}}\,dx$

12. $\int \dfrac{x^2}{\sqrt[3]{x^3+2}}\,dx$

13. $\int \dfrac{\cos^3\theta\,d\theta}{\sqrt[3]{\sin\theta}}$

14. $\int \sqrt{\cos\theta}\,\sin\theta\,d\theta$

15. $\int \tan^2\theta\,d\theta$

16. $\int \dfrac{d\theta}{\sec^2\theta}$

17. $\int \dfrac{x}{(\sqrt{9-x^2})^5}\,dx$

18. $\int \dfrac{dx}{(\sqrt{9-x^2})^5}$

19. $\int e^{\sqrt{x}}\,dx$

20. $\int \sin\sqrt{x}\,dx$

21. $\int \dfrac{dx}{(x^2+x+1)^5}$

22. $\int \dfrac{2x+1}{(x^2+x+1)^5}\,dx$

23. $\int \dfrac{dx}{(x^2-4x+3)^2}$

24. $\int \dfrac{(x-2)\,dx}{(x^2-4x+3)^2}$

25. $\int \dfrac{x^5}{x+1}\,dx$

26. $\int \dfrac{x+1}{x^5}\,dx$

27. $\int \dfrac{e^{3x}\,dx}{1+e^x+e^{2x}}$

28. $\int \dfrac{\ln x}{x(1+\ln x)}\,dx$

29. $\int \dfrac{dx}{x\sqrt{x^2+9}}$

30. $\int \dfrac{x\,dx}{\sqrt{x^2+9}}$

31. $\int \dfrac{dx}{(3+\sin x)^2}$

32. $\int \dfrac{\cos x\,dx}{(3+\sin x)^2}$

33. $\int \ln(e^x)\,dx$

34. $\int \ln(\sqrt[3]{x})\,dx$

35. $\int \tan^6 x \sec x\,dx$

36. $\int \sec^6 x \tan x\,dx$

37. $\int \dfrac{x^4-1}{x+2}\,dx$

38. $\int \dfrac{x+2}{x^4-1}\,dx$

39. $\int \dfrac{dx}{\sqrt{x}(3+\sqrt{x})^2}$

40. $\int \dfrac{dx}{(3+\sqrt{x})^3}$

41. $\int (1+\tan\theta)^3 \sec^2\theta\,d\theta$

42. $\int (1+\tan\theta)^3 \sec\theta\,d\theta$

43. $\int \dfrac{e^x+e^{-x}}{e^x-e^{-x}}\,dx$

44. $\int \dfrac{e^{2x}+1}{e^x-e^{-x}}\,dx$

45. $\int \dfrac{(x+3)(\sqrt{x+2}+1)}{\sqrt{x+2}-1}\,dx$

46. $\int \dfrac{(\sqrt[3]{x+2}-1)\,dx}{\sqrt{x+2}+1}$

47. $\int \dfrac{dx}{x^2-9}$

48. $\int (x^2-9)^{10}\,dx$

49. $\int \dfrac{x+7}{(3x+2)^{10}}\,dx$

50. $\int \dfrac{x^3\,dx}{(3x+2)^7}$

51. $\int \dfrac{2^x+3^x}{4^x}\,dx$

52. $\int \dfrac{2^x}{1+2^x}\,dx$

53. $\int \dfrac{(x+\sin^{-1}x)\,dx}{\sqrt{1-x^2}}$

54. $\int \dfrac{x+\tan^{-1}x}{1+x^2}\,dx$

55. $\int x^3\sqrt{1+x^2}\,dx$

56. $\int x(1+x^2)^{3/2}\,dx$

57. $\int \dfrac{x\,dx}{\sqrt{x^2-1}}$

58. $\int \dfrac{r^3}{\sqrt{x^2-1}}\,dx$

59. $\int \sec^2 x \tan^2 x\,dx$

60. $\int \sec^3 x \tan^3 x\,dx$

61. $\int \cos^4\theta \sin^6\theta\,d\theta$

62. $\int \cos^3\theta \sin^6\theta\,d\theta$

63. $\int \sin^6\theta\,d\theta$

64. $\int \sin^5\theta\,d\theta$

65. $\int \dfrac{dx}{(x^2-9)^{3/2}}$

66. $\int \dfrac{x\,dx}{(x^2-9)^{3/2}}$

67. $\int \dfrac{\tan^{-1}x}{1+x^2}\,dx$

68. $\int \dfrac{\tan^{-1}x}{x^2}\,dx$

69. $\int \dfrac{\sin(\ln x)}{x}\,dx$

70. $\int \cos x \ln(\sin x)\,dx$

71. $\int \dfrac{x\,dx}{\sqrt{x^2+4}}$

72. $\int \dfrac{dx}{\sqrt{x^2+4}}$

73. $\int \dfrac{dx}{x^2+x+5}$

74. $\int \dfrac{x\,dx}{x^2+x+5}$

75. $\int \dfrac{x+3}{(x+1)^5}\,dx$

76. $\int \dfrac{x^5+x+\sqrt{x}}{x^3}\,dx$

77. $\int (x^2+9)^{10}x\,dx$

78. $\int (x^2+9)^{10}x^3\,dx$

■

In each of Exercises 79 to 81, (a) state for which integers $n \geq 1$ you can evaluate the integral, (b) evaluate it. (*Warning:* For most n the integral is not elementary. Base your answer on your experience with integrals.)

79. $\int \sqrt{1+x^n}\,dx$ (two answers)

80. $\int (1+x^2)^{1/n}\,dx$ (two answers)

81. $\int (1+x)^{1/n}\sqrt{1-x}\,dx$ (two answers)

■ ■

82. Find $\int \dfrac{dx}{\sqrt{x+2}-\sqrt{x-2}}$.

83. Find $\int \sqrt{1-\cos x}\,dx$.

84. Find $\int \dfrac{\sec^2 x\,dx}{\sqrt{10-\sec^2 x}}$.

85. Find $\int \dfrac{dx}{x(x^{20}+1)}$.

7.S Summary

Method	Description
Substitution	Introduce $u = h(x)$. If $f(x)\, dx = g(u)\, du$, then $\int f(x)\, dx = \int g(u)\, du$.
Substitution in the definite integral	If, in the above substitution, $u = A$ when $x = a$, and $u = B$ when $x = b$, then $$\int_a^b f(x)\, dx = \int_A^B g(u)\, du.$$
Table of integrals	Obtain and become familiar with a table of integrals. Substitution, together with integral tables, will usually be adequate.
Integration by parts	$\int u\, dv = uv - \int v\, du$. Choose u and v so $u\, dv = f(x)\, dx$ and $\int v\, du$ is easier than $\int u\, dv$.
Partial fractions (applies to any rational function of x)	This is an algebraic method. Write the integrand as a sum of a polynomial (if the degree of the numerator is greater than or equal to the degree of the denominator) plus terms of the type $$\frac{k_i}{(ax+b)^i} \quad \text{and} \quad \frac{c_j x + d_j}{(ax^2 + bx + c)^j}.$$ The number of unknown constants is the same as the degree of the denominator. A table of integrals treats the integrals of these two types. (For the first type, use the substitution $u = ax + b$; for the second, complete the square.)
To integrate certain powers of trigonometric functions	There are many special cases. For instance rewrite $$\int \cos^k \theta \sin^{2n+1} \theta\, d\theta$$ as $$\int \cos^k \theta (1 - \cos^2 \theta)^n \sin \theta\, d\theta$$ and make the substitution $u = \cos \theta$. Keep in mind that $d(\sin \theta) = \cos \theta\, d\theta$; $d(\cos \theta) = -\sin \theta\, d\theta$; $d(\sec \theta) = \sec \theta \tan \theta\, d\theta$; $d(\tan \theta) = \sec^2 \theta\, d\theta$. Also, use trigonometric identities, such as $\sin^2 \theta = (1 - \cos 2\theta)/2$ and $\cos^2 \theta = (1 + \cos 2\theta)/2$.
To integrate any rational function of $\cos \theta$ and $\sin \theta$	Let $u = \tan(\theta/2)$. Then $$\cos \theta = \frac{1-u^2}{1+u^2}, \quad \sin \theta = \frac{2u}{1+u^2}, \quad d\theta = \frac{2\, du}{1+u^2},$$ and the new integrand is a rational function of u.
To integrate rational functions of x and one of $\sqrt{a^2 - x^2}, \sqrt{a^2 + x^2}, \sqrt{x^2 - a^2}$, $a^2 - x^2$, $a^2 + x^2$	For $\sqrt{a^2 - x^2}$ or $a^2 - x^2$, let $x = a \sin \theta$. For $\sqrt{a^2 + x^2}$ or $a^2 + x^2$, let $x = a \tan \theta$. For $\sqrt{x^2 - a^2}$, let $x = a \sec \theta$. (Recall the two right triangles shown below.)
To integrate rational functions of x and $\sqrt[n]{ax + b}$	Let $u = \sqrt[n]{ax + b}$, hence $u^n = ax + b$, $nu^{n-1}\, du = a\, dx$, and $x = (u^n - b)/a$. The new integrand is a rational function of u.

The fundamental theorem of calculus, proved in Chap. 5, raised the problem of finding antiderivatives. Now, some very simple and important functions do not have elementary antiderivatives; for instance,

$$\int \frac{\sin x\, dx}{x}, \quad \int e^{x^2}\, dx, \quad \int \frac{dx}{\ln x}, \quad \int x \tan x\, dx,$$

$$\int \frac{\ln x}{x+1}\, dx, \quad \int \sqrt{1 - \frac{\sin^2 x}{4}}\, dx, \quad \text{and} \quad \int \sqrt[3]{x - x^2}\, dx$$

are not elementary. If the definite integral

$$\int_0^1 e^{x^2}\, dx$$

is needed, an estimate must be made, for example, by using an approximating sum. (After all, $\int_0^1 e^{x^2}\, dx$ is defined as a limit of approximating sums; just choose a sum with small mesh.) The "elliptic integral"

$$\int_0^{\pi/2} \sqrt{1 - k^2 \sin^2 x}\, dx,$$

frequently used in engineering, is tabulated for various values of k to four decimal places in most handbooks.

Fortunately many commonly used functions do have elementary antiderivatives, and this chapter has presented a few methods for finding antiderivatives in these cases.

Also, some definite integrals over certain intervals can be simplified before evaluation. For instance, if $f(x)$ is an *even* function, then $\int_{-a}^{a} f(x)\, dx = 2\int_0^a f(x)\, dx$. In the case of an *odd* function, $\int_{-a}^{a} f(x)\, dx = 0$. (For instance, $\int_{-1}^{1} x\, e^{x^2}\, dx = 0$.)

Guide quiz on chap. 7

In Problems 1 to 19 evaluate the integrals.

1. $\int_1^2 \dfrac{x^3\, dx}{1 + x^4}$

2. $\int \sqrt{4 - 9x^2}\, dx$

3. $\int \dfrac{dx}{x^4 - 1}$

4. $\int \tan^5 2x \sec^2 2x\, dx$

5. $\int \dfrac{x^4\, dx}{x^4 - 1}$

6. $\int_0^{4/3} \dfrac{dx}{\sqrt{9x^2 + 16}}$

7. $\int \dfrac{dx}{2\sqrt{x} - \sqrt[4]{x}}$

8. $\int \dfrac{dx}{(x - 3)\sqrt{x + 3}}$

9. $\int \dfrac{dx}{3 - x^2}$

10. $\int \sin^5 2x\, dx$

11. $\int e^x \cos 2x\, dx$

12. $\int \dfrac{dx}{\sin^5 3x}$

13. $\int \dfrac{x^3\, dx}{1 + x^8}$

14. $\int \dfrac{dx}{x\sqrt{4 + x^2}}$

15. $\int \dfrac{x^2\, dx}{\sqrt{x^2 - 9}}$

16. $\int (9 - x^2)^{3/2}\, dx$

17. $\int \dfrac{dx}{3 + \cos x}$

18. $\int \dfrac{dx}{x^2 \sqrt{x^2 + 25}}$

19. $\int \dfrac{x^4 - \sqrt{x}}{x^3}\, dx$

Review exercises for chap. 7

1. (a) By an appropriate substitution, transform this definite integral into a simpler definite integral.

$$\int_0^{\pi/2} \sqrt{(1 + \cos \theta)^3} \sin \theta\, d\theta.$$

(b) Evaluate the new definite integral in (a).

2. Two of these antiderivatives are elementary functions; evaluate them.

(a) $\int \ln x\, dx$ (b) $\int \dfrac{\ln x\, dx}{x}$ (c) $\int \dfrac{dx}{\ln x}$

3 Evaluate:

(a) $\int_1^2 (1 + x^3)^2 \, dx$ (b) $\int_1^2 (1 + x^3)^2 x^2 \, dx$

4 Compute with the aid of a table of integrals:

(a) $\int \dfrac{e^x \, dx}{5e^{2x} - 3}$ (b) $\int \dfrac{dx}{\sqrt{x^2 - 3}}$

5 Compute:

(a) $\int \dfrac{dx}{x^3}$ (b) $\int \dfrac{dx}{\sqrt{x + 1}}$ (c) $\int \dfrac{e^x \, dx}{1 + 5e^x}$

6 Compute:

$$\int \dfrac{5x^4 - 5x^3 + 10x^2 - 8x + 4}{(x^2 + 1)(x - 1)} \, dx$$

7 Compute:

$$\int \dfrac{x^3 \, dx}{(1 + x^2)^4}$$

in two different ways:
(a) by the substitution $u = 1 + x^2$;
(b) by the substitution $x = \tan \theta$.

8 Transform the definite integral

$$\int_0^3 \dfrac{x^3}{\sqrt{x + 1}} \, dx$$

to another definite integral in two different ways (and evaluate):
(a) by the substitution $u = x + 1$;
(b) by the substitution $u = \sqrt{x + 1}$.

9 Compute $\int x^2 \ln (1 + x) \, dx$

(a) without an integral table;
(b) with an integral table.

10 Find $\int \dfrac{x \, dx}{\sqrt{9x^4 + 16}}$

(a) without an integral table;
(b) with an integral table.

11 Compute $\int \dfrac{\sin \theta \, d\theta}{1 + \sin^2 \theta}$

(a) by using the substitution that applies to any rational function of $\cos \theta$ and $\sin \theta$;
(b) by writing $\sin^2 \theta$ as $1 - \cos^2 \theta$ and using the substitution $u = \cos \theta$.

12 (a) Without an integral table, evaluate

$$\int \sin^5 \theta \, d\theta \quad \text{and} \quad \int \tan^6 \theta \, d\theta.$$

(b) Evaluate them with an integral table.

13 Two of these three antiderivatives are elementary. Find them.

(a) $\int \sqrt{1 - 4 \sin^2 \theta} \, d\theta$ (b) $\int \sqrt{4 - 4 \sin^2 \theta} \, d\theta$

(c) $\int \sqrt{1 + \cos \theta} \, d\theta$

14 (a) Transform the definite integral

$$\int_{-1}^4 \dfrac{x + 2}{\sqrt{x + 3}} \, dx$$

into another definite integral, using the substitution $u = x + 3$.
(b) Evaluate the integral obtained in (a).

15 The fundamental theorem can be used to evaluate one of these definite integrals, but not the other. Evaluate one of them.

(a) $\int_0^1 \sqrt[3]{x} \sqrt{x} \, dx$ (b) $\int_0^1 \sqrt[3]{1 - x} \sqrt{x} \, dx$

16 Verify that the following factorizations into irreducible polynomials are correct.
(a) $x^3 - 1 = (x - 1)(x^2 + x + 1)$;
(b) $x^4 - 1 = (x - 1)(x + 1)(x^2 + 1)$;
(c) $x^3 + 1 = (x + 1)(x^2 - x + 1)$;
(d) $x^4 + 1 = (x^2 + \sqrt{2}x + 1)(x^2 - \sqrt{2}x + 1)$.

In Exercises 17 to 24 express as a sum of partial fractions. (Do not integrate.)

17 $\dfrac{2x^2 + 3x + 1}{x^3 - 1}$ **18** $\dfrac{x^4 + 2x^2 - 2x + 2}{x^3 - 1}$

19 $\dfrac{2x - 1}{x^3 + 1}$ **20** $\dfrac{x^4 + 3x^3 - 2x^2 + 3x - 1}{x^4 - 1}$

21 $\dfrac{x^3 - (1 + \sqrt{2})x^2 + (1 - \sqrt{2})x - 1}{x^4 + 1}$

22 $\dfrac{2x + 5}{x^2 + 3x + 2}$

23 $\dfrac{5x^3 + 11x^2 + 6x + 1}{x^2 + x}$ **24** $\dfrac{5x^3 + 6x^2 + 8x + 5}{(x^2 + 1)(x + 1)}$

25 For which values of the nonnegative integers m and n are the following integrations comparatively simple?

(a) $\int \sin^m x \, dx$ (b) $\int \sec^n x \, dx$

(c) $\int \sin^m x \cos^n x \, dx$ (d) $\int \sec^m x \tan^n x \, dx$

(e) $\int \cot^m x \csc^n x \, dx$

26 (a) Develop the reduction formula relating

$$\int \sin^n x \, dx \quad \text{to} \quad \int \sin^{n-2} x \, dx.$$

(b) If n is odd, what technique may be used for finding $\int \sin^n x \, dx$?

In each of Exercises 27 to 32 use an appropriate substitution to obtain an integrand that is a rational function. Do not evaluate.

27 $\int \dfrac{(\sqrt{4-x^2})^3 + 1}{[(4-x^2)^3 + 5]\sqrt{4-x^2}} \, dx$

28 $\int \dfrac{x + \sqrt[3]{x-2}}{x^2 - \sqrt[3]{x-2}} \, dx$

29 $\int \dfrac{(x^2-5)^7}{x^2 + 3 + \sqrt{x^2-5}} \, dx$

30 $\int \dfrac{\cos^2 \theta + \sin \theta}{1 - \sin \theta \cos \theta} \, d\theta$

31 $\int \dfrac{3 \tan^2 \theta + \sec \theta + 1}{2 + \tan \theta + \cos \theta} \, d\theta$

32 $\int \dfrac{(4+x^2)^{1/2}}{5 + (4+x^2)^{3/2}} \, dx$

In Exercises 33 to 174 find the integrals.

33 $\int \dfrac{\cos x \, dx}{\sin^3 x - 8}$

34 $\int \dfrac{dx}{\sqrt{2 + \sqrt{x}}}$

35 $\int \dfrac{\sqrt{x^2+1}}{x^4} \, dx$

36 $\int \dfrac{\sin x \, dx}{1 + 3 \cos^2 x}$

37 $\int \dfrac{\sin x \, dx}{3 + \cos x}$

38 $\int x \sqrt{x^4 - 1} \, dx$

39 $\int x^2 \sqrt{x^3 - 1} \, dx$

40 $\int \sin \sqrt{x} \, dx$

41 $\int \dfrac{dx}{(4+x^2)^2}$

42 $\int (\sqrt[3]{x} + \sqrt[3]{x+1}) \, dx$

43 $\int \sin^2 3x \cos^2 3x \, dx$

44 $\int \sin^3 3x \cos^2 3x \, dx$

45 $\int \tan^4 3\theta \, d\theta$

46 $\int \dfrac{x^2 \, dx}{x^4 - 1}$

47 $\int \dfrac{x^4 + x^2 + 1}{x^3} \, dx$

48 $\int \dfrac{3 \, dx}{\sqrt{1 - 5x^2}}$

49 $\int 10^x \, dx$

50 $\int \dfrac{x^3}{(x^4+1)^3} \, dx$

51 $\int \dfrac{x \, dx}{(x^4+1)^2}$

52 $\int \cos^3 x \sin^2 x \, dx$

53 $\int \cos^2 x \, dx$

54 $\int x \sqrt{x+4} \, dx$

55 $\int x \sqrt{x^2 + 4} \, dx$

56 $\int \dfrac{x+2}{x^2+1} \, dx$

57 $\int \dfrac{x^2 \, dx}{1+x^6}$

58 $\int \sqrt[3]{4x+7} \, dx$

59 $\int x^2 \sin x^3 \, dx$

60 $\int \dfrac{\ln x^4}{x} \, dx$

61 $\int x^4 \ln x \, dx$

62 $\int \dfrac{\tan^{-1} 3x}{1 + 9x^2} \, dx$

63 $\int \dfrac{e^{\sqrt{x}}}{\sqrt{x}} \, dx$

64 $\int \sin (\ln x) \, dx$

65 $\int \ln (x^3 - 1) \, dx$

66 $\int \tan x \, dx$

67 $\int \dfrac{x \, dx}{\sqrt{(x^2+1)^3}}$

68 $\int \dfrac{2 + \sqrt[3]{x}}{x} \, dx$

69 $\int \dfrac{dx}{\sqrt{(x+1)^3}}$

70 $\int \dfrac{2x+3}{x^2 + 3x + 5} \, dx$

71 $\int \dfrac{3 \, dx}{x^2 + 4x + 5}$

72 $\int \dfrac{3 \, dx}{x^2 + 4x - 5}$

73 $\int \dfrac{x \, dx}{1 + \sqrt[3]{x}}$

74 $\int \ln \sqrt{2x-1} \, dx$

75 $\int \dfrac{x^7 \, dx}{\sqrt{x^2+1}}$

76 $\int x^3 \tan^{-1} x \, dx$

77 $\int \dfrac{\tan^{-1} x}{x^2} \, dx$

78 $\int \dfrac{dx}{x^3 + 4x}$

79 $\int e^x \sin 3x \, dx$

80 $\int \sqrt{1 - \cos x} \, dx$

81 $\int \dfrac{dx}{(4-x^2)^{3/2}}$

82 $\int x^{1/4}(1 + x^{1/5}) \, dx$

83 $\int \dfrac{x \, dx}{x^4 - 2x^2 - 3}$

84 $\int \sin^{-1} \sqrt[3]{x} \, dx$

85 $\int \dfrac{x^2 \, dx}{\sqrt[3]{x-1}}$

86 $\int \ln (4 + x^2) \, dx$

87 $\int \dfrac{\sqrt{x^2+4}}{x} \, dx$

88 $\int \sqrt{\tan \theta} \sec^2 \theta \, d\theta$

89 $\int \sec^5 \theta \tan \theta \, d\theta$

90 $\int \tan^6 \theta \, d\theta$

91 $\int \dfrac{dx}{x \sqrt{x^2+9}}$

92 $\int (e^x + 1)^2 \, dx$

93 $\int \dfrac{(1-x)^2}{\sqrt[3]{x}} \, dx$

94 $\int (1 + \sqrt{x}) x \, dx$

95. $\int \sin^2 2x \cos x \, dx$
96. $\int (e^{2x})^3 e^x \, dx$
97. $\int \left(e^x - \frac{1}{e^x}\right)^2 dx$
98. $\int \frac{dx}{(\sqrt{x}+1)(\sqrt{x})}$
99. $\int x \sin^{-1} x^2 \, dx$
100. $\int x \sin^{-1} x \, dx$
101. $\int \frac{dx}{e^{2x} + 5e^x}$
102. $\int \frac{e^x \, dx}{1 - 6e^x + 9e^{2x}}$
103. $\int (2x+1)\sqrt{3x+2} \, dx$
104. $\int \frac{2x^3 + 1}{x^3 - 4x^2} \, dx$
105. $\int \frac{x^2 \, dx}{(x-1)^3}$
106. $\int \frac{dx}{\sqrt{9+x^2}}$
107. $\int \frac{e^x + 1}{e^x - 1} \, dx$
108. $\int \frac{dx}{4x^2 + 1}$
109. $\int (1 + 3x^2)^2 \, dx$
110. $\int \frac{x \, dx}{x^3 + 1}$
111. $\int \frac{x^3 \, dx}{x^3 + 1}$
112. $\int \frac{x^2 \, dx}{\sqrt{2x+1}}$
113. $\int \frac{dx}{\sqrt{2x+1}}$
114. $\int (x + \sin x)^2 \, dx$
115. $\int \frac{x \, dx}{x^4 - 3x^2 - 2}$
116. $\int \frac{x^3 \, dx}{x^4 - 1}$
117. $\int \frac{e^x \, dx}{1 + e^{2x}}$
118. $\int \frac{dx}{x^2 + 5x - 6}$
119. $\int \frac{dx}{x^2 + 5x + 6}$
120. $\int \frac{x \, dx}{2x^2 + 5x + 6}$
121. $\int \frac{4x + 10}{x^2 + 5x + 6} \, dx$
122. $\int \sqrt{4x^2 + 1} \, dx$
123. $\int \frac{dx}{2x^2 + 5x + 6}$
124. $\int \sqrt{-4x^2 + 1} \, dx$
125. $\int \frac{dx}{2x^2 + 5x - 6}$
126. $\int \frac{dx}{2 + 3\sin x}$
127. $\int \frac{dx}{\sin^2 x}$
128. $\int \frac{dx}{3 + 2\sin x}$
129. $\int \frac{dx}{\sin^4 x}$
130. $\int \ln(x^2 + 5) \, dx$
131. $\int \frac{dx}{\sqrt{x - 2\sqrt{3-x}}}$
132. $\int x^3 e^{-5x} \, dx$
133. $\int \sqrt{(1+2x)(1-2x)} \, dx$
134. $\int x \sin 3x \, dx$
135. $\int \frac{2x \, dx}{\sqrt{x^2 + 1}}$
136. $\int \frac{2 \, dx}{\sqrt{x^2 + 1}}$
137. $\int \frac{x^4 + 4x^3 + 6x^2 + 4x - 3}{x^4 - 1} \, dx$
138. $\int \frac{x^3 + 6x^2 + 11x + 5}{(x+2)^2(x+1)} \, dx$
139. $\int \frac{-3x^2 - 11x - 11}{(x+2)^2(x+1)} \, dx$
140. $\int \frac{x^2 - 3x}{(x+1)(x-1)^2} \, dx$
141. $\int \frac{12x^2 + 2x + 3}{4x^3 + x} \, dx$
142. $\int \frac{6x^3 + 2x + \sqrt{3}}{1 + 3x^2} \, dx$
143. $\int \frac{-6x^3 - 13x - 3\sqrt{3}}{1 - 3x^2} \, dx$
144. $\int \frac{x \, dx}{\sqrt{1 - 9x^2}}$
145. $\int \frac{dx}{\sqrt{1 - 9x^2}}$
146. $\int \frac{dx}{x\sqrt{3x^2 - 5}}$
147. $\int \frac{dx}{(3x^2 + 2)^{3/2}}$
148. $\int \frac{dx}{\sin 5x}$
149. $\int \frac{dx}{\cos 4x}$
150. $\int \frac{x^2 \, dx}{1 + 3x^3 + 2x^6}$
151. $\int e^x \sin^2 x \, dx$
152. $\int x^3 \sqrt{1 - 3x^2} \, dx$
153. $\int \sqrt{1 + \sqrt{1 + \sqrt{x}}} \, dx$
154. $\int \frac{x^3 \, dx}{1 - 4x^2}$
155. $\int e^{\sqrt[4]{x}} \, dx$
156. $\int \sec^4 x \, dx$
157. $\int \frac{dx}{\cos^3 x}$
158. $\int \cos^3 x \, dx$
159. $\int x^2 \ln(x^3 + 1) \, dx$
160. $\int \frac{\ln x + \sqrt{x}}{x} \, dx$
161. $\int \frac{(3 + x^2)^2 \, dx}{x}$
162. $\int \frac{dx}{e^x}$
163. $\int \frac{(1 + 3\cos x)^2 \, dx}{\sin x}$
164. $\int (3 + \cos\theta)/(2 - \cos\theta) \, d\theta$
165. $\int (e^{2x} + 1)e^{-x} \, dx$
166. $\int \sqrt{9x^2 - 4} \, dx$

167 $\int \dfrac{2x^2 + 4x + 3}{x^3 + 2x^2 + 3x} dx$

168 $\int \dfrac{dx}{x^4 + 3x^2 + 1}$

169 $\int \dfrac{\sec^2 \theta \, d\theta}{\sqrt{\sec^2 \theta - 1}}$

170 $\int \ln(2x + x^2) \, dx$

171 $\int x \sin^2 x \, dx$

172 $\int \dfrac{dx}{1 + 2e^{3x}}$

173 $\int x \tan^2 x \, dx$

174 $\int \sqrt{1 + \cos 3\theta} \, d\theta$

175 Compute:
(a) $\int \dfrac{dx}{x^2 + 4x + 3}$
(b) $\int \dfrac{dx}{x^2 + 4x + 4}$
(c) $\int \dfrac{dx}{x^2 + 4x + 5}$
(d) $\int \dfrac{dx}{x^2 + 4x - 2}$

176 Compute $\int \dfrac{x^3 \, dx}{(x-1)^2}$

(a) using partial fractions;
(b) using the substitution $u = x - 1$.
(c) Which method is easier?

177 Compute:
(a) $\int \sec^5 x \, dx$
(b) $\int \sec^5 x \tan x \, dx$
(c) $\int \dfrac{\sin x}{(\cos x)^3} dx$

178 (a) Compute $\int \dfrac{x^{2/3} \, dx}{x + 1}$.
(b) What does a table of integrals say about $\int \dfrac{x^{2/3} \, dx}{x + 1}$?

179 Find $\int (x^3/\sqrt{1 + x^2}) \, dx$ with the aid of a table of integrals.

180 Compute $\int x \sqrt[3]{x + 1} \, dx$ using
(a) the substitution $u = \sqrt[3]{x + 1}$,
(b) the substitution $u = x + 1$.

181 Transform $\int (x^2/\sqrt{1 + x}) \, dx$ by each of the substitutions
(a) $u = \sqrt{1 + x}$;
(b) $u = 1 + x$;
(c) $x = \tan^2 \theta$.
(d) Solve the easiest of the resulting problems.

In Exercises 182 to 185 evaluate the integrals.

182 $\int_0^1 (e^x + 1)^3 e^x \, dx$

183 $\int_0^1 (x^4 + 1)^5 x^3 \, dx$

184 $\int_0^e \dfrac{\sqrt{\ln x}}{x} dx$

185 $\int_0^{\pi/2} \dfrac{\cos \theta \, d\theta}{\sqrt{1 + \sin \theta}}$

■

186 The following is a quote from an article on the management of energy resources:

Let $u(t)$ be the rate at which water flows through the turbines at time t. Then $U(t) = \int_0^t u(s) \, ds$ is the total flow during the time $[0, t]$. Let $z(t)$ represent the rate of demand for electric energy at time t, as measured in equivalent water flow. Then the cost incurred during this period of time is $\int_0^t c(z(s) - u(s)) \, ds$, where c is a cost function for imported energy.

Explain why these interpretations of the integrals make sense.

187 Transform the problem of finding $\int (x^3/\sqrt{1 + x^2}) \, dx$ to a different problem, using
(a) integration by parts with $dv = (x \, dx)/\sqrt{1 + x^2}$;
(b) the substitution $x = \tan \theta$;
(c) the substitution $u = \sqrt{1 + x^2}$.

188 Two of these three integrals are elementary. Evaluate them.
(a) $\int \sin^2 x \, dx$
(b) $\int \sin \sqrt{x} \, dx$
(c) $\int \sin x^2 \, dx$

189 Compute $\int x\sqrt{1 + x} \, dx$ in three ways:
(a) let $u = \sqrt{1 + x}$;
(b) let $x = \tan^2 \theta$;
(c) by parts, with $u = x$, $dv = \sqrt{1 + x} \, dx$.

190 One of the functions $\sqrt{x} \sqrt[3]{1 - x}$ and $\sqrt{1 - x} \sqrt[3]{1 - x}$ has an elementary antiderivative F. Find F in the case for which an elementary antiderivative exists.

191 Consider the problem of finding the area under $y = e^{x^2}$ from $x = 0$ to $x = 1$.
(a) Why is the FTC useless in determining this area?
(b) Estimate $\int_0^1 e^{x^2} \, dx$ by utilizing a partition of $[0, 1]$ into the five sections $[0, 0.2]$, $[0.2, 0.4]$, $[0.4, 0.6]$, $[0.6, 0.8]$, $[0.8, 1]$, and as sampling numbers $c_1 = 0.1$, $c_2 = 0.3$, $c_3 = 0.5$, $c_4 = 0.7$, and $c_5 = 0.9$.

192 Assuming that $\int (e^x/x) \, dx$ is not elementary (a theorem of Liouville), prove that $\int 1/\ln x \, dx$ is not elementary.

193 Evaluate $\int_{-1}^1 x e^{x^4} \, dx$

194 The factor theorem asserts that if r is a root of a polynomial $P(x)$, then $x - r$ is a divisor of $P(x)$. Use this theorem to factor each of the following polynomials into irreducible factors.
(a) $x^3 - 1$ (b) $x^3 + 1$ (c) $x^3 - 5$
(d) $x^3 + 2$ (e) $x^4 - 1$ (f) $x^3 - 3x + 2$

195 (See Exercise 194.) Represent in partial fractions:

(a) $\dfrac{x}{x^3 - 1}$ (b) $\dfrac{1}{x^3 - 3x + 2}$

(c) $\dfrac{x^5}{x^3 + 1}$ (d) $\dfrac{1}{x^3 + 8}$

196 (a) Find $\int x^{1/3}(1 + x)^{5/3}\, dx$. (*Hint:* First use the substitution $x = 1/t$.)

(b) Find $\int \sqrt[4]{x(1 + x)^3}\, dx$.

197 One integral table lists the antiderivative of $\int (\sqrt{x^2 + a^2}/x)\, dx$ as

$$\sqrt{x^2 + a^2} - a \ln\left(\frac{a + \sqrt{x^2 + a^2}}{x}\right)$$

while another lists it as

$$\sqrt{x^2 + a^2} + a \ln\left(\frac{\sqrt{x^2 - a^2} - a}{x}\right)$$

Is there an error in one?

■ ■

198 One of these integrals is elementary, the other is not. Evaluate the one that is elementary.

(a) $\int \ln(\cos x)\, dx$.

(b) $\int \cos(\ln x)\, dx$.

199 (a) Explain why $\int x^m e^x\, dx$ is an elementary function for any positive integer m.

(b) Explain why $\int x^m (\ln x)^n\, dx$ is an elementary function for any positive integers m and n.

200 (a) Prove the trigonometric identity

$$\sin mx \sin nx = \tfrac{1}{2}\{\cos[(m - n)x] - \cos[(m + n)x]\}$$

by expanding the right side.

(b) Use it to compute $\int \sin 2x \sin 3x\, dx$.

201 Liouville proved that, if f and g are rational functions and if $\int e^{f(x)} g(x)\, dx$ is an elementary function, then $\int e^{f(x)} g(x)\, dx$ can be expressed in the form $e^{f(x)} w(x)$, where $w(x)$ is a rational function. With the aid of this result, prove that $\int e^x/x\, dx$ is not an elementary function. *Hint:* Assume $[e^x(p/q)]' = e^x/x$, where p and q are relatively prime polynomials. Write $q = x^i r$, where $i \geq 0$ and x does not divide the polynomial r.

202 Three of these six antiderivatives are elementary. Compute them.

(a) $\int x \cos x\, dx$ (b) $\int \dfrac{\cos x}{x}\, dx$

(c) $\int \dfrac{x\, dx}{\ln x}$ (d) $\int \dfrac{\ln x^2}{x}\, dx$

(e) $\int \sqrt{x - 1}\sqrt{x}\sqrt{x + 1}\, dx$

(f) $\int \sqrt{x - 1}\sqrt{x + 1}\, x\, dx$

203 This is an excerpt from an applied chemistry text: "The time for the depth D of the water to fall from 10 to 5 feet equals

$$\int_5^{10} \frac{dD}{\sqrt{D - 4} + 4\sqrt{D - 2}}.$$

The integral is evaluated by first rationalizing the denominator and then ...."

Evaluate the integral. (Use an integral table.)

Exercise 204 is the basis of Exercises 205 to 214.

204 Let p and q be rational numbers. Prove that $\int x^p(1 - x)^q\, dx$ is an elementary function

(a) if p is an integer (*Hint:* if $q = s/t$, let $1 - x = v^t$);

(b) if q is an integer;

(c) if $p + q$ is an integer.

Chebyshev proved that these are the only cases for which the antiderivative in question is elementary. In particular,

$$\int \sqrt{x}\sqrt[3]{1 - x}\, dx \quad \text{and} \quad \int \sqrt[3]{x - x^2}\, dx$$

are not elementary.

205 Deduce from Exercise 204 that $\int \sqrt{1 - x^3}\, dx$ is not elementary.

206 Deduce from Exercise 204 that $\int (1 - x^n)^{1/m}\, dx$, where m and n are positive integers, is elementary if and only if $m = 1$, $n = 1$, or $m = 2 = n$.

207 Deduce from Exercise 204 that $\int \sqrt{\sin x}\, dx$ is not elementary. *Hint:* Let $u = \sin^2 x$.

208 Deduce from Exercise 204 that $\int \sin^a x\, dx$, where a is rational, is elementary if and only if a is an integer.

209 Deduce from Exercise 204 that $\int \sin^p x \cos^q x\, dx$, where p and q are rational, is elementary if and only if p or q is an odd integer or $p + q$ is an even integer.

210 Deduce from Exercise 209 that $\int \sec^p x \tan^q x\, dx$, where p and q are rational, is elementary only if $p + q$ or q is odd, or if p is even.

211 (a) Deduce from Exercise 204 that $\int (x/\sqrt{1 + x^n})\, dx$, where n is a positive integer, is elementary only when $n = 1$, 2, or 4.

(b) Evaluate the integral for $n = 1$, 2, and 4.

212 (a) Deduce from Exercise 204 that $\int (x^2/\sqrt{1 + x^n})\, dx$, where n is a positive integer, is elementary only when $n = 1$, 2, 3, or 6.

(b) Evaluate the integral for $n = 1$, 2, 3, and 6.

213 (a) Using Exercise 204, determine for which positive integers n the integral $\int (x^n/\sqrt{1 + x^4})\, dx$ is elementary.

(b) Evaluate the integral for $n = 3$ and $n = 5$.

214 The following is an excerpt from an engineering text:

"The last equation may be written $\theta = \int c \, dr/(r\sqrt{r^6 - c^2})$, where c is a constant. The integral is easily evaluated by the substitution...."

(a) What substitution did the text recommend?
(b) Using Exercise 204, determine for which positive integers n, $\int (c/(r^n\sqrt{r^6 - c^2})) \, dr$ is elementary.

215 Let f have a derivative everywhere, such that $f' = f$ and $f(0) = 1$. In answering the following, do not make use of the explicit formula for f, $f(x) = e^x$, obtained in Chap. 6.

(a) Show that for any constant k we have
$$[f(x)f(k-x)]' = 0.$$

(b) In view of (a) what kind of function must the product $f(x)f(k-x)$ be?
(c) Prove that $f(x)f(k-x) = f(k)$ for all x.
(d) From (c), prove that $f(x+y) = f(x)f(y)$ for all x and y.

APPLICATIONS OF THE DEFINITE INTEGRAL

The definite integral was defined in Chap. 5 and shown to be useful in computing such quantities as area, volume, total distance (given the velocity), and total mass (given the density). The fundamental theorem of calculus was the key to evaluating many definite integrals. This fact motivated Chap. 7, which presented techniques for finding antiderivatives of a variety of elementary functions.

The present chapter takes up where Chap. 5 left off. Section 8.1 is devoted to the computation of areas and Sec. 8.2 to volumes. Section 8.3 describes a quick approach to setting up definite integrals for a variety of quantities. Further geometric and physical applications of the definite integral are given in Secs. 8.4, 8.5, and 8.7. Section 8.6 concerns integrals in which the interval of integration or the function is unbounded. Section 8.8 defines the average value of a function. Methods of estimating an integral are presented in Sec. 8.9; these are useful if the antiderivative of the integrand is not elementary or if it is tedious to compute.

8.1 Computing area by parallel cross sections

"c" is short for "cross section"

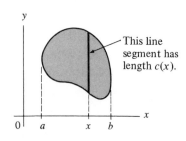

Figure 8.1

In Sec. 5.3 it was shown that the area of a plane region is equal to the integral of its cross-sectional length. Let $c(x)$ denote the length of the intersection with the given region of the vertical line through $(x, 0)$. (See Fig. 8.1.) Then the area of the region is equal to $\int_a^b c(x)\, dx$, where a and b are shown in Fig. 8.1. Note that x need not refer to the x axis of the xy plane; it may refer to any conveniently chosen line in the plane. It may even refer to the y axis; in this case the cross-sectional length would be denoted $c(y)$.

To compute an area:

1 Find a, b, and the cross-sectional length $c(x)$.

2 Evaluate $\int_a^b c(x)\,dx$ by the fundamental theorem of calculus if $\int c(x)\,dx$ is elementary and simple enough to find.

Chapter 7 showed how to accomplish step 2. The present section is concerned primarily with step 1, how to find the cross-sectional length $c(x)$.

HOW TO FIND THE AREA BETWEEN TWO CURVES

Let f and g be two continuous functions such that $f(x) \geq g(x)$ for all x in the interval $[a, b]$. Let R be the region between the curve $y = f(x)$ and the curve $y = g(x)$ for x in $[a, b]$, as shown in Fig. 8.2. In these circumstances the length of the cross section of R made by a line perpendicular to the x axis can be computed in terms of f and g. Inspection of Fig. 8.3 shows that

$$c(x) = f(x) - g(x).$$

In short, to find $c(x)$, subtract the smaller value $g(x)$ from the larger value $f(x)$. The area of R is then given by

Area between two curves

$$\text{Area} = \int_a^b [f(x) - g(x)]\,dx.$$

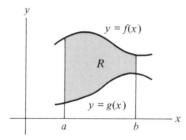

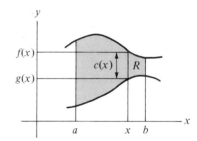

Figure 8.2 **Figure 8.3**

EXAMPLE 1 Find the area of the region shown in Fig. 8.4. The region is bounded by the curve $y = x^2$, the line $y = -\tfrac{3}{2}x$, and the line $x = 2$.

SOLUTION In this case $f(x) = x^2$ and $g(x) = -\tfrac{3}{2}x$. For x in $[0, 2]$, the cross-sectional length is $c(x) = x^2 - (-\tfrac{3}{2}x)$. Thus the area of the region is

$$\int_0^2 [x^2 - (-3x/2)]\,dx.$$

This definite integral can be evaluated by the fundamental theorem of calculus:

$$\int_0^2 [x^2 - (-3x/2)]\,dx = \left(\frac{x^3}{3} + \frac{3x^2}{4}\right)\bigg|_0^2$$

$$= \left[\frac{2^3}{3} + \frac{3 \cdot 2^2}{4}\right] - \left[\frac{0^3}{3} + \frac{3 \cdot 0^2}{4}\right]$$

$$= \frac{17}{3}. \quad \blacksquare$$

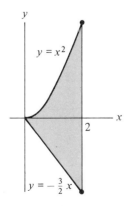

Figure 8.4

APPLICATIONS OF THE DEFINITE INTEGRAL

EXAMPLE 2 Find the area of the region in Fig. 8.4, but this time use cross sections parallel to the x axis.

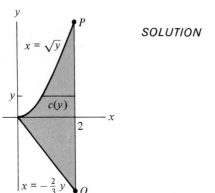

Figure 8.5

y varies from the y coordinate of Q to the y coordinate of P.

SOLUTION Since the cross-sectional length is to be expressed in terms of y, first express the equations of the curves bounding the region in terms of y. The curve $y = x^2$ may be written as $x = \sqrt{y}$, since we are interested only in positive x. The curve $y = -\frac{3}{2}x$ can be expressed as $x = -\frac{2}{3}y$ by solving for x in terms of y. The line $x = 2$ also bounds the region. (See Fig. 8.5.)

For each number y in a certain interval $[a, b]$ that will be determined, the line with y coordinate equal to y meets the region in Fig. 8.5 in a line segment of length $c(y)$. It is necessary to determine the numbers a and b as well as the formula for $c(y)$.

The point P in Fig. 8.5 lies on the parabola $y = x^2$ (or $x = \sqrt{y}$) and has the x coordinate 2. Thus $P = (2, 2^2) = (2, 4)$. The point Q lies on the line $y = -\frac{3}{2}x$ and has x coordinate 2. Thus $Q = (2, -\frac{3}{2} \cdot 2) = (2, -3)$. Consequently a cross section of the region is determined for each number y in the interval $[-3, 4]$. The area of the region is therefore

$$\int_{-3}^{4} c(y) \, dy. \tag{1}$$

Next, a formula for $c(y)$ must be found. For $0 \leq y \leq 4$ the cross section is determined by the line $x = 2$ and the parabola $x = \sqrt{y}$. Thus for $0 \leq y \leq 4$,

$$c(y) = 2 - \sqrt{y},$$

the larger minus the smaller. For $-3 \leq y \leq 0$ the cross section is determined by the line $x = 2$ and the line $x = -\frac{2}{3}y$. Thus for $-3 \leq y \leq 0$,

$$c(y) = 2 - \left(-\frac{2}{3}y\right) = 2 + \frac{2y}{3}.$$

The integral (1) breaks into two separate integrals, each of which is easily evaluated by the fundamental theorem of calculus:

$$\int_{-3}^{0} c(y) \, dy + \int_{0}^{4} c(y) \, dy = \int_{-3}^{0} \left(2 + \frac{2y}{3}\right) dy + \int_{0}^{4} (2 - \sqrt{y}) \, dy$$

$$= \left(2y + \frac{y^2}{3}\right)\bigg|_{-3}^{0} + \left(2y - \frac{2}{3}y^{3/2}\right)\bigg|_{0}^{4}$$

$$= \left(2 \cdot 0 + \frac{0^2}{3}\right) - \left[2(-3) + \frac{(-3)^2}{3}\right]$$

$$+ \left(2 \cdot 4 - \frac{2}{3}4^{3/2}\right) - \left(2 \cdot 0 - \frac{2}{3}0^{3/2}\right)$$

$$= 0 - (-3) + \frac{8}{3} - 0$$

$$= \frac{17}{3}. \quad \blacksquare$$

8.1 COMPUTING AREA BY PARALLEL CROSS SECTIONS

Choose the direction for cross sections to make life easy.

Example 1 needed only one integral, but Example 2 needed two. Moreover, in Example 2 the formula for the cross-sectional length when $0 \leq y \leq 4$ involved $\sqrt{y}$, which is a little harder to integrate than x^2, which appeared in the corresponding formula in Example 1. Though both approaches to finding the area of the region in Fig. 8.4 are valid, the one with cross sections parallel to the y axis is more convenient.

HOW TO FIND CROSS SECTIONS GEOMETRICALLY

Perhaps a region is described not by formulas but geometrically, maybe as a circle, triangle, trapezoid, or some other polygon. Two geometric facts are often of use in finding the cross-sectional length in these circumstances:

THE PYTHAGOREAN THEOREM In a right triangle whose legs have lengths a and b, and whose hypotenuse has length c, $c^2 = a^2 + b^2$.

CORRESPONDING PARTS OF SIMILAR TRIANGLES ARE PROPORTIONAL If a, b, c are the lengths of the sides of one triangle and a', b', c' are the lengths of the corresponding sides of a similar triangle, then $a'/a = b'/b = c'/c$.

In addition, corresponding altitudes of the two triangles are also in the same proportion. (See Fig. 8.6.)

Figure 8.6

EXAMPLE 3   Find the cross-sectional length $c(x)$ if R is a circle of radius 5. Set up a definite integral for the area of R and find the area of R.

SOLUTION   We are free to put the axis anywhere in the plane. Place it in such a way that its origin is below the center of the circle. This provides the simplest formula for $c(x)$, the length of typical chord AB perpendicular to the x axis. This is illustrated in Fig. 8.7.

The axis is tilted as a reminder that not all lines are horizontal or vertical.

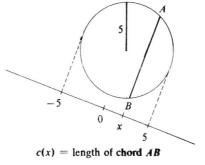

$c(x)$ = length of chord AB

Figure 8.7

Note that $a = -5$ and $b = 5$. To get a feel for $c(x)$ in this case, note by glancing at the diagram that

$$c(-5) = 0, \qquad c(0) = 10 \text{ (the circle's diameter)}, \qquad \text{and} \qquad c(5) = 0.$$

458 APPLICATIONS OF THE DEFINITE INTEGRAL

To find $c(x)$ for any x in the interval $[-5, 5]$, draw the line through the center of the circle and parallel to the x axis. It meets the segment AB at a point M. Call the center of the circle C. Also draw the segment AC, a radius of the circle, as in Fig. 8.8. Then

$$c(x) = \overline{AB} = 2\overline{AM}.$$

To find $\overline{AM}$, use the right triangle ACM. One side CM has length $|x|$, while the hypotenuse has length 5. Hence $5^2 = |x|^2 + \overline{AM}^2$. Since $|x|^2 = x^2$, this equation gives a simple formula for $\overline{AM}^2$,

$$\overline{AM}^2 = 5^2 - x^2 = 25 - x^2.$$

Hence
$$\overline{AM} = \sqrt{25 - x^2},$$

and, since $c(x) = 2\overline{AM}$, $\quad c(x) = 2\sqrt{25 - x^2}.$

Thus the area of the circle is

$$\int_{-5}^{5} 2\sqrt{25 - x^2}\, dx.$$

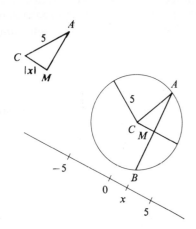

Figure 8.8

To evaluate the integral, make the substitution $x = 5 \sin \theta$, $dx = 5 \cos \theta\, d\theta$. Thus

$$\int_{-5}^{5} 2\sqrt{25 - x^2}\, dx = \int_{-\pi/2}^{\pi/2} 2\sqrt{25 - (5 \sin \theta)^2}\, 5 \cos \theta\, d\theta$$

$$= \int_{-\pi/2}^{\pi/2} 2\sqrt{25 - 25 \sin^2 \theta}\, 5 \cos \theta\, d\theta$$

$$= 50 \int_{-\pi/2}^{\pi/2} \cos \theta \cos \theta\, d\theta$$

$$= 50 \int_{-\pi/2}^{\pi/2} \cos^2 \theta\, d\theta.$$

As θ goes from $-\dfrac{\pi}{2}$ to $\dfrac{\pi}{2}$, x goes from -5 to 5.

Since $\cos(-\theta) = \cos \theta$, $\cos \theta$ is an even function and

$$\int_{-\pi/2}^{0} \cos^2 \theta\, d\theta = \int_{0}^{\pi/2} \cos^2 \theta\, d\theta.$$

Whenever possible, simplify to avoid negative numbers.

Thus
$$\int_{-\pi/2}^{\pi/2} \cos^2 \theta\, d\theta = 2 \int_{0}^{\pi/2} \cos^2 \theta\, d\theta.$$

By the shortcut described in Sec. 7.6,

$$\int_{0}^{\pi/2} \cos^2 \theta\, d\theta = \frac{\pi}{4}.$$

Thus the area of the circle is

Check: Area of circle = $\pi r^2 = \pi \cdot 5^2 = 25\pi.$

$$50 \cdot 2 \cdot \int_{0}^{\pi/2} \cos^2 \theta\, d\theta = 50 \cdot 2 \cdot \frac{\pi}{4} = 25\pi. \quad \blacksquare$$

EXAMPLE 4 Find by integration the area of the triangle whose three vertices are $A = (0, 0)$, $B = (6, 0)$, and $C = (5, 3)$, as shown in Fig. 8.9.

SOLUTION In which direction should the parallel cross sections be taken? If they are parallel to the y axis (vertical cross sections), the formula for $c(x)$ will depend on whether $x \leq 5$ or $x \geq 5$. That will mean that two separate integrals will be required. However, cross sections parallel to the x axis (horizontal cross sections) are much easier and can be found by similar triangles.

To find $c(y)$ for $0 \leq y \leq 3$, note that in Fig. 8.10 triangle CDE is similar to triangle CAB. Since the lengths of corresponding parts of similar triangles are proportional,

$$\frac{c(y)}{6} = \frac{3-y}{3}.$$

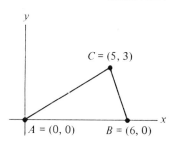

Figure 8.9

(The altitude from C to DE has length $3 - y$; the altitude from C to AB has length 3.) Thus

$$c(y) = \tfrac{6}{3}(3 - y) = 6 - 2y.$$

The area of the triangle is therefore

$$\int_0^3 c(y)\,dy = \int_0^3 (6 - 2y)\,dy$$
$$= (6y - y^2)\Big|_0^3$$
$$= 9. \quad\blacksquare$$

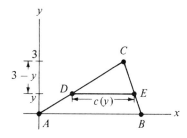

Figure 8.10

As a check,
$$\text{Area} = \tfrac{1}{2} \cdot \text{base} \cdot \text{height}$$
$$= \tfrac{1}{2} \cdot 6 \cdot 3 = 9.$$

Exercises

In each of Exercises 1 to 8 find the area of the region between the two curves and above (or below) the given intervals.
1. $y = x^2$ and $y = x^3$; $[0, 1]$
2. $y = x^2$ and $y = x^3$; $[1, 2]$
3. $y = x^2$ and $y = \sqrt{x}$; $[0, 1]$
4. $y = x^3$ and $y = \sqrt[3]{x}$; $[1, 2]$
5. $y = x^3$ and $y = -x$; $[1, 2]$
6. $y = 1 + x$ and $y = \ln x$; $[1, e]$
7. $y = \sin x$ and $y = \cos x$; $[0, \pi/4]$
8. $y = \sin x$ and $y = \cos x$; $[\pi/2, \pi]$

In Exercises 9 to 16 sketch the finite regions bounded by the given curves. Then find their areas by (a) vertical cross sections, (b) horizontal cross sections.
9. $y = x^2$ and $y = 3x - 2$. (*Suggestion:* To find where the two curves cross, set $x^2 = 3x - 2$ and solve for x.)
10. $y = 2x^2$ and $y = x + 1$
11. $y = 4x$ and $y = 2x^2$
12. $y = x^2$ and $y = 4$
13. $y = 1/x^2$, $y = 0$, $x = 1$, $x = 3$
14. $x = y^2$ and $x = 3y - 2$
15. $y = \sin x$, $y = 0$, $x = \pi/2$ ($x \leq \pi/2$)
16. $y = \tan x$, $y = 0$, $x = \pi/4$ (Consider only $x \geq 0$.)

17. Find the vertical cross-sectional length $c(x)$ if x denotes the x of the x axis and R is the shaded region shown in Fig. 8.11.

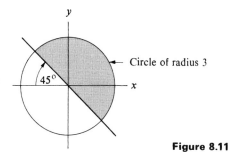

Figure 8.11

18 Find the horizontal cross-sectional length $c(y)$ for the region in Exercise 17.

In Exercises 19 to 21 $c(x)$ and $c(y)$ are distinct functions.

19 (a) Draw the region bordered by $y = \sin x$ and the x axis for x in $[0, \pi]$.
(b) Find $c(x)$, the vertical cross-sectional length.
(c) Find $c(y)$, the horizontal cross-sectional length.

20 (a) Draw the region bordered by $y = x/2$, $y = x - 1$, and the x axis.
(b) Find $c(x)$, the vertical cross-sectional length.
(c) Find $c(y)$, the horizontal cross-sectional length.

21 (a) Draw the region bordered by $y = \ln x$, $y = x/e$, and the x axis.
(b) Find $c(x)$, the vertical cross-sectional length.
(c) Find $c(y)$, the horizontal cross-sectional length.

22 Compute the horizontal cross-sectional length $c(y)$ for the triangle with each of the three choices of the origin and direction of the y axis shown in Fig. 8.12. *Hint:* Use similar triangles in each case. Note that a wise choice of coordinate system can simplify the cross-section function.

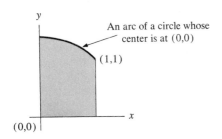

Figure 8.12

23 Find the area of the shaded region in Fig. 8.13 (a) without calculus, (b) using vertical cross sections, (c) using horizontal cross sections. (*Suggestion for (a):* Use the formula for the area of a sector.)

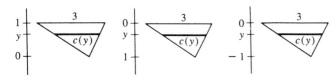

Figure 8.13

24 Find the area between the curves $y^2 = x$ and $y = x - 2$, using (a) horizontal cross sections, (b) vertical cross sections.

25 Sketch the region common to two circles of radius 1 whose centers are a distance 1 apart. Find the area of this region, using (a) vertical cross sections, (b) horizontal cross sections, (c) only elementary geometry, but no calculus.

26 What fraction of the rectangle whose vertices are $(0, 0)$, $(a, 0)$, (a, a^4), $(0, a^4)$, with a positive, is occupied by the region under the curve $y = x^4$ and above $[0, a]$?

27 (a) Guess what the horizontal cross-sectional length $c(y)$ is in the isosceles trapezoid in Fig. 8.14. *Hint:* What is the simplest formula that gives $c(0) = 8$ and $c(2) = 14$?

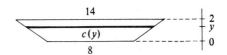

Figure 8.14

(b) Using similar triangles, find $c(y)$.

28 (a) Draw the region R inside the ellipse

$$\frac{x^2}{4} + \frac{y^2}{9} = 1.$$

(b) Find $c(x)$, the vertical cross-sectional length.
(c) Find $c(y)$, the horizontal cross-sectional length.
(d) Find the area of R.

29 Let R be the region cut from a circle of radius 5 by two parallel lines, each a distance 3 from the center, as shown in Fig. 8.15.

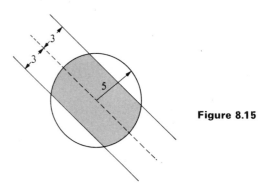

Figure 8.15

(a) Choosing an x axis parallel to the lines, and origin beneath the center, find $c(x)$, the length of cross sections perpendicular to the x axis.
(b) Choosing an x axis perpendicular to the lines, and the origin beneath the center, find $c(x)$, the length of cross sections perpendicular to the x axis.

■ ■

30 Let f be an increasing function with $f(0) = 0$, and assume that it has an elementary antiderivative. Then f^{-1} is an increasing function, and $f^{-1}(0) = 0$. Prove that, if f^{-1} is elementary, then it also has an elementary antiderivative. *Hint:* Observe that the sum of the areas of I and II in Fig. 8.16 is $tf(t)$.

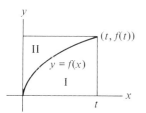

Figure 8.16

31 Show that the shaded area in Fig. 8.17 is $\frac{2}{3}$ the area of the parallelogram $ABCD$. This is an illustration of a theorem of Archimedes concerning sectors of parabolas.

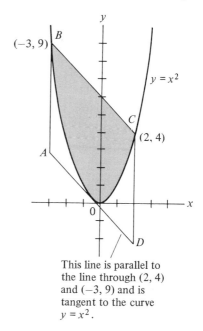

This line is parallel to the line through $(2, 4)$ and $(-3, 9)$ and is tangent to the curve $y = x^2$.

Figure 8.17

8.2 Computing volume by parallel cross sections

In Sec. 5.3 it was shown that the volume V of a spatial region, a "solid," can be expressed as a definite integral of cross-sectional area $A(x)$,

$$V = \int_a^b A(x)\,dx,$$

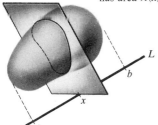

Figure 8.18

as shown in Fig. 8.18. So, to find the volume of some solid, follow these steps:

1 Choose a line L to serve as an x axis.

2 For each plane perpendicular to that axis find the area of the cross section of the solid made by the plane. Call this area $A(x)$.

3 Determine the limits of integration, a and b, for the region.

4 Evaluate the definite integral $\int_a^b A(x)\,dx$.

Most of the effort is usually spent in finding the integrand $A(x)$.

In addition to the Pythagorean theorem and the properties of similar triangles, formulas for the areas of familiar plane figures may be needed:

Figure	Area
Circle of radius r	πr^2
Square of side s	s^2
Equilateral triangle of side s	$\dfrac{\sqrt{3}\,s^2}{4}$
Triangle of height h and base b	$\dfrac{bh}{2}$
Trapezoid of height h and parallel bases b and B	$h\left(\dfrac{b+B}{2}\right)$

Also keep in mind that if corresponding dimensions of similar figures have the ratio k, then their areas have the ratio k^2, that is, the area is proportional to the square of the lengths of corresponding sides.

A few examples will show what may be involved in finding the cross-sectional area $A(x)$.

EXAMPLE 1 Find the volume of the solid whose base is the region bounded by the x axis and the arch of the curve $y = \sin x$ from $x = 0$ to $x = \pi$ and for which each plane section perpendicular to the x axis is a square whose base lies in the region. The solid is shown in Fig. 8.19.

SOLUTION Since the cross section is a square whose side is $\sin x$, its area is $\sin^2 x$. Thus

$$A(x) = \sin^2 x.$$

The definite integral for the volume is therefore

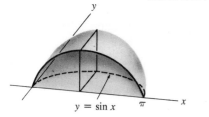

Figure 8.19

Think of the graph of $y = \sin^2 x$ from $x = 0$ to $x = \pi$.

$$\int_0^\pi \sin^2 x \, dx.$$

To evaluate this integral, note that

$$\int_0^\pi \sin^2 x \, dx = 2 \int_0^{\pi/2} \sin^2 x \, dx.$$

By the short cut in Sec. 7.6,

$$\int_0^{\pi/2} \sin^2 x \, dx = \pi/4.$$

Thus the volume of the solid is $2(\pi/4) = \pi/2$. ∎

8.2 COMPUTING VOLUME BY PARALLEL CROSS SECTIONS   463

EXAMPLE 2   Find the volume of a solid triangular pyramid whose base is a right triangle of sides 3, 4, and 5. The altitude of the pyramid is above the vertex of the right angle and has length 2. (See Fig. 8.20.)

SOLUTION   There are three convenient directions in which to define cross sections by planes, namely parallel to each of the three right-triangular faces. Choose, say, planes parallel to the base, as shown in Fig. 8.21.

Introduce an x axis perpendicular to the base triangle and with origin in the plane of that triangle.

The typical cross section is a triangle T. Its area is $A(x) = \frac{1}{2}bh$, where b and h are shown in Fig. 8.21. By similar triangles,

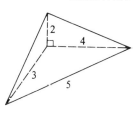

Figure 8.20

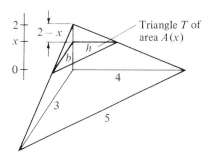

Figure 8.21

$$\frac{b}{3} = \frac{2-x}{2} \quad \text{and} \quad \frac{h}{4} = \frac{2-x}{2}.$$

Thus $b = \frac{3}{2}(2-x)$, $h = 2(2-x)$, and

$$A(x) = \frac{1}{2}bh = (\tfrac{1}{2})(\tfrac{3}{2})(2-x)2(2-x) = \frac{3}{2}(2-x)^2.$$

The same formula for the area of T can also be obtained by comparing T to the base triangle, which has area $\frac{1}{2} \cdot 3 \cdot 4 = 6$. Recall that the areas of similar figures have the ratio k^2, where k is the ratio of their corresponding sides. The corresponding sides of lengths b and 3 have the ratio

$$k = \frac{b}{3} = \frac{2-x}{2},$$

so

$$\frac{\text{Area of } T}{\text{Area of base}} = \left(\frac{2-x}{2}\right)^2$$

or

$$\frac{A(x)}{6} = \left(\frac{2-x}{2}\right)^2.$$

Solving this last equation for $A(x)$ gives

$$A(x) = \frac{3}{2}(2-x)^2.$$

Thus the volume of the pyramid is

$$\int_0^2 \frac{3}{2}(2-x)^2 \, dx.$$

By the fundamental theorem of calculus,

$$\int_0^2 \frac{3}{2}(2-x)^2 \, dx = -\frac{3}{2} \frac{(2-x)^3}{3} \Big|_0^2$$

$$= \left(-\frac{3}{2} \frac{(2-2)^3}{3}\right) - \left(-\frac{3}{2} \frac{(2-0)^3}{3}\right)$$

464 APPLICATIONS OF THE DEFINITE INTEGRAL

$$= 0 + 4$$
$$= 4.$$

The volume of the pyramid is 4 cubic units. ∎

EXAMPLE 3 Find the volume of a sphere of radius a.

SOLUTION Place an x axis in such a way that its origin is beneath the center of the sphere, as in Fig. 8.22. The typical cross section is a circle. Since the area of a circle of radius r is πr^2, all that remains is to find r^2 in terms of x.

To accomplish this, draw a side view of the sphere and the typical cross section, showing r and x clearly, as in Fig. 8.23.

By the Pythagorean theorem,

$$a^2 = |x|^2 + r^2;$$

hence

$$a^2 = x^2 + r^2,$$

and

$$r^2 = a^2 - x^2.$$

Consequently, $\quad A(x) = \pi r^2 = \pi(a^2 - x^2).$

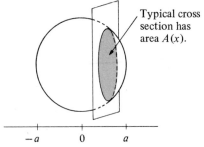

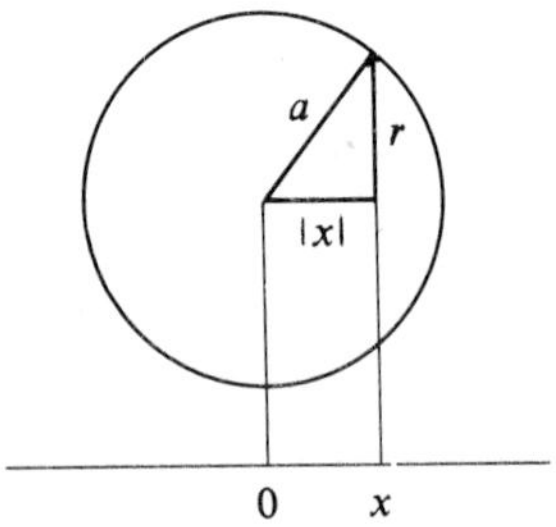

Figure 8.22 Typical cross section has area $A(x)$.

The volume of the sphere of radius a is therefore

$$\int_{-a}^{a} \pi(a^2 - x^2) \, dx.$$

By the fundamental theorem of calculus,

$$\int_{-a}^{a} \pi(a^2 - x^2) \, dx = \pi \left(a^2 x - \frac{x^3}{3} \right) \bigg|_{-a}^{a}$$

$$= \pi \left(a^3 - \frac{a^3}{3} \right) - \pi \left(a^2(-a) - \frac{(-a)^3}{3} \right)$$

$$= \tfrac{4}{3}\pi a^3. \quad \blacksquare$$

Figure 8.23

EXAMPLE 4 A drinking glass has the form of a right circular cylinder of radius a and height h. It is tilted until the water level bisects the base and touches the rim, as shown in Fig. 8.24. The water occupies a region R in space. Find the volume of the water.

Radius of base is a; AB is a diameter.

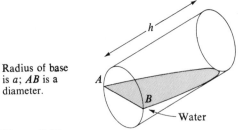

Figure 8.24

8.2 COMPUTING VOLUME BY PARALLEL CROSS SECTIONS

SOLUTION   To develop some spatial intuition, consider four possible directions to use for the parallel plane cross sections:

1. Planes parallel to the surface of the water
2. Planes parallel to the base of the glass
3. Planes perpendicular to the diameter AB
4. Planes parallel to the plane through AB and the axis of the cylinder

As each is considered, the reader should make a sketch in order to develop his or her geometric intuition. (Little that is done in high school or in analytic geometry develops a feeling for space; topics there are usually at most two-dimensional.) Even better, saw a piece out of a cylinder of wood to represent the water. Simpler yet, use a glass of water.

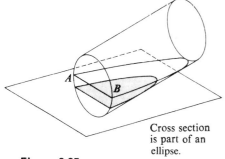

Cross section is part of an ellipse.

Figure 8.25

1. *Planes parallel to the surface of the water.* In this case, each cross section is part of an ellipse. (See Appendix E for a discussion of the ellipse.) The highest cross section, the one through AB, is exactly one half of an ellipse, but the lower ones are smaller portions of ellipses. (See Fig. 8.25.) This is not a good choice, since the area of a portion of an ellipse is hard to compute.

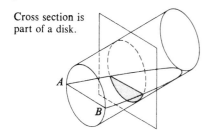

Cross section is part of a disk.

Figure 8.26

2. *Planes parallel to the base of the glass.* (See Fig. 8.26.) In this case each cross section is part of a disk. This is a bit easier than case 1, but still not the most convenient choice.

3. *Planes perpendicular to the diameter AB.* In this case the cross sections are right triangles. (In spite of all the curves in Fig. 8.27, the cross sections have borders made up only of line segments.) Since the area of a right triangle is easy to compute, let us find $A(x)$ in this case.

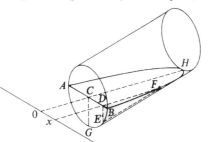

Cross section is a right triangle DEF.

Figure 8.27

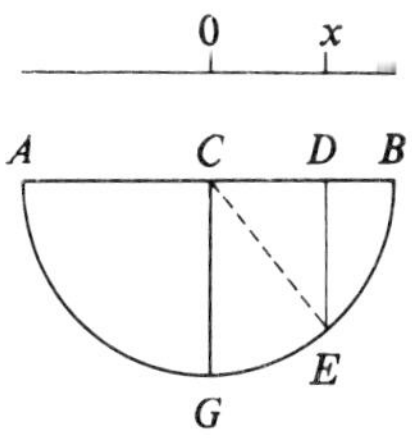

Figure 8.28

View of base

Introduce an x axis parallel to the diameter AB, and locate the origin 0 to correspond with the midpoint of the diameter. Next, label key points, as shown in Figs. 8.27 and 8.28.

The area of the triangle is

$$A(x) = \tfrac{1}{2} \text{ base} \cdot \text{altitude} = \tfrac{1}{2}\overline{DE} \cdot \overline{EF}.$$

First find $\overline{DE}$ in terms of x, using the Pythagorean theorem:

$$a^2 = \overline{CD}^2 + \overline{DE}^2 = x^2 + \overline{DE}^2.$$

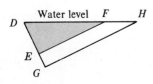

DEF is similar to CGH.

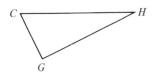

Figure 8.29

(Use right triangle CDE, whose hypotenuse CE is dotted in Fig. 8.28.)

Thus $\overline{DE} = \sqrt{a^2 - x^2}.$

This expresses $\overline{DE}$ in terms of x. Next consider $\overline{EF}$.
The side view in Fig. 8.29 shows two similar triangles,

$$DEF \quad \text{and} \quad CGH.$$

Thus there is this equality relating corresponding parts:

$$\frac{\overline{DE}}{\overline{CG}} = \frac{\overline{EF}}{\overline{GH}}.$$

Since $\overline{DE} = \sqrt{a^2 - x^2},$ $\overline{CG} = a,$ and $\overline{GH} = h,$

the above equality relating corresponding parts becomes

$$\frac{\sqrt{a^2 - x^2}}{a} = \frac{\overline{EF}}{h}.$$

Thus $\overline{EF} = \frac{h}{a}\sqrt{a^2 - x^2}.$

The area of the typical triangle then is

$$A(x) = \tfrac{1}{2}\overline{DE} \cdot \overline{EF}$$

$$= \frac{1}{2}\sqrt{a^2 - x^2} \cdot \frac{h}{a}\sqrt{a^2 - x^2}$$

$$= \frac{h(a^2 - x^2)}{2a}.$$

The volume of the solid is therefore

$$\int_{-a}^{a} \frac{h(a^2 - x^2)}{2a}\,dx.$$

Evaluation of this integral is straightforward:

Since the integrand is an even function, the integral equals twice the integral from 0 to a. This fact could be used to simplify the arithmetic.

$$\int_{-a}^{a} \frac{h(a^2 - x^2)}{2a}\,dx = \frac{h}{2a}\left(a^2 x - \frac{x^3}{3}\right)\bigg|_{-a}^{a}$$

$$= \frac{h}{2a}\left(a^2 \cdot a - \frac{a^3}{3}\right) - \frac{h}{2a}\left(a^2(-a) - \frac{(-a)^3}{3}\right)$$

$$= \frac{ha^2}{3} + \frac{ha^2}{3}$$

$$= \frac{2ha^2}{3}.$$

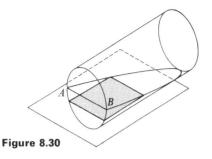

4 *Planes parallel to the plane through AB and the axis of the cylinder.* In this case all four sides are straight and the cross sections are rectangles, as shown in Fig. 8.30. (Even the side at the water surface is straight, for it is the intersection of two planes, the plane of the water surface and the plane that defines the cross section.) This case, like case 3, also leads to a convenient integration. (See Exercise 10.) ∎

Figure 8.30

Exercises

In each of Exercises 1 to 4, (a) draw the solid, (b) set up an integral for its volume, (c) find the volume.

1. The base of the solid is the region in the xy plane bounded by $y = x^2$, the x axis, and the line $x = 1$. A cross section by a plane perpendicular to the x axis is a square one side of which is in the base.

2. The base of the solid is the region in the xy plane bounded by $y = x^2$ and $y = x$. A cross section by a plane perpendicular to the x axis is a square one side of which is in the base.

3. The base of the solid is the region in the xy plane bounded by $y = x^2$ and $y = x$. A cross section by a plane perpendicular to the y axis is a square one side of which is in the base.

4. The base of the solid is the region in the xy plane bounded by $y = e^x$, $x = 1$, $x = 2$, and the x axis. A cross section by a plane perpendicular to the x axis is an equilateral triangle one side of which is in the base.

5. Solve Example 2 if the origin of the x axis, instead of being in the plane of the base triangle, is level with the top vertex, and the positive part of the x axis is downward.

6. Solve Example 2, using cross sections parallel to the triangle with sides of lengths 2 and 4.

7. Solve Example 2, using cross sections parallel to the triangle with sides of lengths 2 and 3.

8. Find the volume of the triangular pyramid shown in Fig. 8.31.

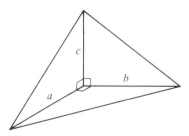

Figure 8.31

9. (a) Make a physical model of the water in Example 4, either by carving soap, modeling clay, or sawing a dowel.
   (b) Mark in the model the outlines of the four types of cross sections discussed in Example 4.

10. Compute $A(x)$ for case 4, the rectangular cross sections of Example 4. Place the x axis perpendicular to the diameter AB and parallel to the base of the glass. Let $x = 0$ correspond to a plane through AB.
    (a) Find the width of the typical rectangle, that is, the length of the side parallel to AB.
    (b) Find the length of the typical rectangle.
    (c) Compute $A(x)$.
    (d) Use (c) to set up a definite integral for the volume of the solid.
    (e) Evaluate the integral in (d).

Exercises 11 to 14 are related.

11. A drinking glass is a right circular cylinder of radius a and height h. It is tilted until the water level just covers the base and touches the rim.
    (a) Draw the water.
    (b) Draw the typical cross sections corresponding to the four cases in Example 4.

12. (See Exercise 11.) Compute $A(x)$, the cross-sectional area, using trapezoidal cross sections. (Place the origin of the x axis on the axis of the cylinder.)

13. (See Exercise 11.) Compute the typical cross-sectional area $A(x)$, using rectangular cross sections. (Place the origin of the x axis on the axis of the cylinder.)

14. Without using any calculus, find the volume of the solid in Exercise 11.

∎

15. A right circular cone has height h and radius a. Consider cross sections parallel to the base. Introduce an x axis such that $x = 0$ corresponds to the vertex of the cone and $x = h$ corresponds to the base.
    (a) Draw the cone, the x axis, and the typical cross section.

(b) Compute $A(x)$, the typical cross-sectional area for planes perpendicular to the x axis.
(c) Set up the definite integral for the volume.
(d) Find the volume.

16 A lumberjack saws a wedge out of a cylindrical tree of radius a. His first cut is parallel to the ground and stops at the axis of the tree. His second cut makes an angle θ with the first cut and meets it along a diameter. Draw the solid.

17 (See Exercise 16.) Place the x axis in such a way that the cross sections of the wedge in Exercise 16 are triangles.
(a) Draw the typical cross section.
(b) Find the area of a typical cross section made by a plane at a distance x from the axis of the tree.
(c) Find the volume of the wedge.

18 (See Exercise 16.) Place the x axis in such a way that the sections of the wedge in Exercise 16 are rectangles.
(a) Draw a typical cross section.
(b) Finds its area if it is made by a plane at a distance x from the axis of the tree.
(c) Find the volume of the wedge.

19 A drill of radius 3 inches bores a hole through a sphere of radius 5 inches, passing symmetrically through the center of the sphere.
(a) Draw the part of the sphere removed by the drill.
(b) Find $A(x)$, the area of a cross section of the region in (a) made by a plane perpendicular to the axis of the drill and at a distance x from the center of the sphere.
(c) Find the volume removed.

■ ■

Exercise 20 provides an easy way to remember the volume of a cone or pyramid.

20 A solid is formed in the following manner. A plane region R and a point P not in that plane are given. The solid consists of all line segments, one of whose ends is P and whose other end is in R. If R has area A and P is a distance h from the plane of R, show that the volume of the solid is $Ah/3$. (See Fig. 8.32.)

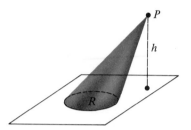

Figure 8.32

21 Find the volume of one octant of the region common to two right circular cylinders of radius 1, whose axes intersect at right angles, as shown in Fig. 8.33.

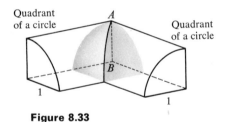

Figure 8.33

8.3 How to set up a definite integral

This section presents an informal shortcut for setting up a definite integral to evaluate some quantity. First the formal and informal approaches are contrasted in the case of setting up the definite integral for area, $A = \int_a^b c(x)\,dx$. Then the informal approach will be illustrated as commonly applied in a variety of fields.

THE FORMAL AND THE INFORMAL APPROACHES CONTRASTED

THE FORMAL APPROACH Recall how the formula $A = \int_a^b c(x)\,dx$ was obtained in Sec. 5.3. The interval $[a, b]$ was partitioned by the numbers $x_0 < x_1 < x_2 < \cdots < x_n$ with $x_0 = a$ and $x_n = b$. A sampling number was chosen in each section $[x_{i-1}, x_i]$. For convenience, let us use x_{i-1} as that sampling number. The sum

$$\sum_{i=1}^{n} c(x_{i-1})(x_i - x_{i-1}) \tag{1}$$

was then formed. As the mesh of the partition approaches 0, the sum (1) approaches the area of the region under consideration. But, by the definition of the definite integral, the sum (1) approaches

$$\int_a^b c(x)\,dx$$

as the mesh of the partition approaches 0. Thus

$$\text{Area} = \int_a^b c(x)\,dx. \qquad (2)$$

That is the "formal" approach to obtain the formula (2). Now consider the "informal" approach, which is just a shorthand for the formal approach.

THE INFORMAL APPROACH The heart of the formal approach is the *local estimate* $c(x_{i-1})(x_i - x_{i-1})$, the area of a rectangle of height $c(x_{i-1})$ and width $x_i - x_{i-1}$, which is shown in Fig. 8.34.

In the informal approach attention is focused on the local approximation. No mention is made of the partition or the sampling numbers or the mesh approaching 0. We illustrate this shorthand approach by obtaining formula (2) informally.

Consider a small positive number dx. What would be a good estimate of the area of the region corresponding to the short interval $[x, x + dx]$ of width dx shown in Fig. 8.35?

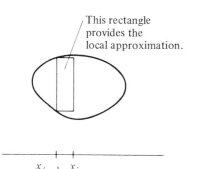

Figure 8.34

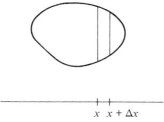

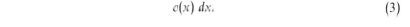

Figure 8.35

The area of the rectangle of width dx and height $c(x)$ shown in Fig. 8.36 would seem to be a plausible estimate. The area of this thin rectangle is

The local approximation

$$c(x)\,dx. \qquad (3)$$

Without further ado, we then write

$$\text{Area} = \int_a^b c(x)\,dx, \qquad (4)$$

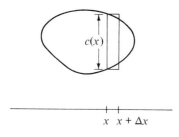

Figure 8.36

which is formula (2). The leap from the local approximation (3) to the definite integral (4) omits many steps of the formal approach. This informal approach is the shorthand commonly used in applications of calculus. It should be emphasized that it is only an abbreviation of the formal approach which deals with approximating sums.

EXAMPLES OF THE INFORMAL APPROACH

EXAMPLE 1 (Volcanic ash settling) After the explosion of a volcano, ash gradually settles from the atmosphere and falls on the ground. Assume that the depth of the ash at a distance x feet from the volcano is Ae^{-kx} feet, where A and k are positive constants. Set up a definite integral for the total volume of ash that falls within a distance b of the volcano.

SOLUTION First estimate the volume of ash that falls on a very narrow ring of width dx and inner radius x centered at the volcano. (See Fig. 8.37.) This estimate can be made since the depth of the ash depends only on the distance from the volcano. On this ring the depth is almost constant.

The area of this ring is approximately that of a rectangle of length $2\pi x$ and width dx. So the area of the ring is approximately

$$2\pi x\, dx.$$

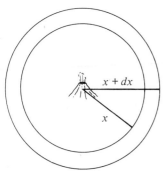

Figure 8.37

(Exercise 1 shows that its area is $2\pi x\, dx + \pi(dx)^2$.)

Though the depth of the ash on this narrow ring is not constant, it does not vary much. A good estimate of the depth throughout the ring is Ae^{-kx}. Thus the volume of ash that falls on the typical ring of inner radius x and outer radius $x + dx$ is approximately

The local approximation

$$Ae^{-kx}2\pi x\, dx \qquad \text{cubic feet.} \tag{5}$$

Once we have the key local estimate (5), we immediately write down the definite integral for the total volume of ash that falls within a distance b of the volcano,

$$\text{Total volume} = \int_0^b Ae^{-kx}2\pi x\, dx.$$

This completes the informal setting up of the definite integral. (It could be evaluated by integration by parts, but this is not our concern at this point.) ∎

Definition of kinetic energy

The second example of the informal approach to setting up definite integrals concerns kinetic energy. The kinetic energy associated with an object of mass m grams and speed v centimeters per second is defined as $mv^2/2$ ergs. If the various parts of the object are not all moving at the same speed, an integral may be needed to express the total kinetic energy.

EXAMPLE 2 A thin rectangular piece of sheet metal is spinning around one of its longer edges 3 times per second, as shown in Fig. 8.38. The length of its shorter edge is 6 centimeters and the length of its longer edge is 10 centimeters. The density of the sheet metal is 4 grams per square centimeter. Find the kinetic energy of the spinning rectangle.

The density is assumed constant.

SOLUTION To find the total kinetic energy of the rotating piece of sheet metal, imagine it divided into narrow rectangles of length 10 centimeters and width dx

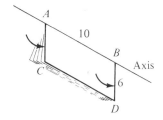

Figure 8.38

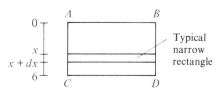

Figure 8.39

centimeters parallel to the edge AB; a typical one is shown in Fig. 8.39. (Introduce an x axis parallel to edge AC with the origin corresponding to A.) Since all points of this typical narrow rectangle move at roughly the same speed, we will be able to estimate its kinetic energy. That estimate will provide the key local approximation of the informal approach to setting up a definite integral.

First of all, the mass of the typical rectangle is

$$4 \cdot 10 \, dx \quad \text{grams},$$

since its area is $10 \, dx$ square centimeters and the density is 4 grams per square centimeter.

Second, we must estimate its speed. The narrow rectangle is spun 3 times per second around a circle of radius x. In 1 second each point in it covers a distance of about

$$3 \cdot 2\pi x = 6\pi x \quad \text{centimeters}.$$

Consequently, the speed of the typical rectangle is

$$6\pi x \quad \text{centimeters per second}.$$

The local estimate of the kinetic energy associated with the typical rectangle is therefore

$$\tfrac{1}{2} \underbrace{40 \, dx}_{\text{mass}} \underbrace{(6\pi x)^2}_{\substack{\text{speed} \\ \text{squared}}} \quad \text{ergs}$$

The local approximation   or simply $720\pi^2 x^2 \, dx$ ergs.   (6)

Having obtained the local estimate (6), we jump directly to the definite integral, and conclude that

$$\begin{array}{c}\text{Total energy of} \\ \text{spinning rectangle}\end{array} = \int_0^6 720\pi^2 x^2 \, dx \quad \text{ergs.} \quad \blacksquare$$

The force exerted by water

The next example concerns water pressure.

The pressure of water at a depth of h feet is $(62.4)h$ pounds per square foot. This pressure exerts a force on a flat horizontal surface submerged at the depth of h feet equal to the product of the pressure and the area of the surface.

Consider, for instance, the horizontal floor of a swimming pool which is 8 feet deep. If the floor has an area of 800 square feet, then the total force of the water on the floor is

$$\underbrace{8(62.4)}_{\text{pressure}} \underbrace{(800)}_{\text{area}} \quad \text{pounds}.$$

472 APPLICATIONS OF THE DEFINITE INTEGRAL

The pressure is exerted in all directions.

The water exerts a pressure on the walls of the pool as well. In fact, at a given depth the pressure is the same in all directions. (A submerged swimmer does not avoid the pressure against her eardrums by turning her head.) But computing the force against the wall is harder than computing the force against the floor, since the pressure is not constant along the wall. A definite integral is required, as shown, informally, in Example 3.

EXAMPLE 3 Let R be part of the submerged vertical wall of a pool. Set up a definite integral for the total force against the surface R.

SOLUTION Begin by introducing a vertical x axis whose origin is at the surface of the water and whose positive part is directed downward. Define a, b, and the cross-section function for R as usual. (See Fig. 8.40.)

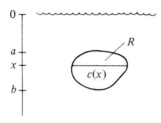

Figure 8.40

Consider the force against a typical narrow strip of R from depth x to depth $x + dx$, where dx is a small positive number, as shown in Fig. 8.41.

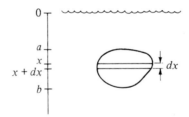

Figure 8.41

The pressure of the water throughout this strip is approximately $62.4\, x$ pounds per square foot. The area of the strip is approximately that of a rectangle of height dx and length $c(x)$. Thus the force against this narrow strip is approximately

The local approximation

$$\underbrace{62.4\, x}_{\text{pressure}} \;\; \underbrace{c(x)\, dx}_{\text{area}} \qquad \text{pounds.}$$

Having obtained a local estimate of the force, we then, using the informal approach, conclude that

This formula will be referred to in later sections.

$$\boxed{\text{Total force against } R = \int_a^b 62.4\, x\, c(x)\, dx \quad \text{pounds.}} \qquad (7)$$

∎

Exercises

Exercises 1 and 2 concern Example 1.

1 In Example 1 the area of the ring with inner radius x and outer radius $x + dx$ was informally estimated to be approximately $2\pi x\, dx$.
(a) Using the formula for the area of a circle, show that the area of the ring is $2\pi x\, dx + \pi(dx)^2$.
(b) Show that the ring has the same area as a trapezoid of height dx and bases of lengths $2\pi x$ and $2\pi(x + dx)$.

2 This exercise presents the formal approach to setting up the definite integral for the quantity of ash discussed in Example 1. First partition the interval $[0, b]$ by $x_0 < x_1 < x_2 < \cdots < x_n$, with $x_0 = 0$ and $x_n = b$. Let c_i be the midpoint of the ith section: $c_i = (x_{i-1} + x_i)/2$.
(a) Show that the area of the ring with inner radius x_{i-1} and outer radius x_i is $2\pi c_i(x_i - x_{i-1})$.
(b) Show that the volume of ash that falls on the ring in (a) is approximately $Ae^{-kc_i}2\pi c_i(x_i - x_{i-1})$.
(c) From (b) deduce that the volume of ash that falls within the circle of radius b is $\int_0^b Ae^{-kx}2\pi x\, dx$.

3 The depth of rain at a distance r feet from the center of a storm is $g(r)$ feet.
(a) Estimate the total volume of rain that falls between a distance r feet and a distance $r + dr$ feet from the center of the storm. (Assume that dr is a small positive number.)
(b) Using (a), set up a definite integral for the total volume of rain that falls between 1000 and 2000 feet from the center of the storm.

4 The following analysis of primitive agriculture is taken from *Is there an Optimum Level of Population?*, edited by S. Fred Singer, McGraw-Hill:

> Consider a circular range of radius a with the home base of production at the center. Let $G(r)$ denote the density of foodstuffs (in calories per square meter) at radius r meters from the home base. Then the total number of calories produced in the range is given by the definite integral _____.

Using the informal approach, work out the definite integral that appeared in the blank.

Exercises 5 and 6 concern Example 2, kinetic energy.

5 The piece of sheet metal in Example 2 is rotated around the line midway between the edges AB and CD at the rate of 5 revolutions per second.
(a) Using the informal approach, obtain a local approximation for the kinetic energy of a narrow strip of the metal.
(b) Using (a), set up a definite integral for the kinetic energy of the piece of sheet metal.
(c) Evaluate the integral in (b).

6 A circular piece of metal of radius 7 centimeters has a density of 3 grams per square centimeter. It rotates 5 times per second around an axis perpendicular to the circle and passing through the center of the circle.
(a) Devise a local approximation for the kinetic energy of a narrow ring in the circle.
(b) With the aid of (a) set up a definite integral for the kinetic energy of the rotating metal.
(c) Evaluate the integral in (b).

Exercises 7 to 10 concern Example 3. Find the total force against the submerged vertical surfaces shown.

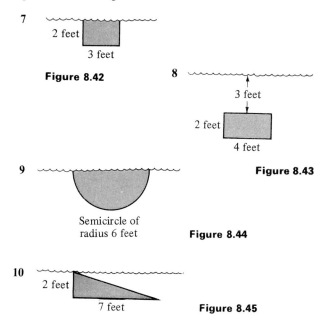

Figure 8.42

Figure 8.43

Semicircle of radius 6 feet

Figure 8.44

Figure 8.45

11 In Example 3 the origin of the x axis was taken at the surface of the water. However, this may be inconvenient for describing the cross sections of the region R. Assume that the origin of the x axis is placed in such a way that the surface of the water has the coordinate $x = k$. Using the informal approach, show that the total force of the water against R is $62.4 \int_b^a (x - k)c(x)\, dx$ pounds.

In Exercises 12 to 16 use the formula developed in Exercise 11 to find the total force against the shaded regions shown in Figs. 8.46 to 8.50. Choose the origin of the x axis conveniently and keep in mind that the positive part of the x axis points down.

12

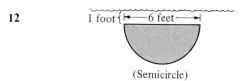

(Semicircle)

Figure 8.46

474 APPLICATIONS OF THE DEFINITE INTEGRAL

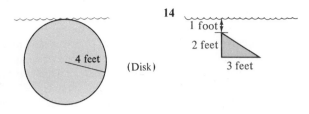

Figure 8.47 **Figure 8.48**

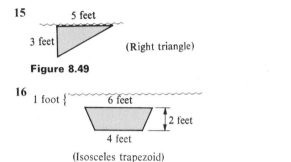

Figure 8.49

Figure 8.50

17. (Poiseuille's law of blood flow) A fluid flowing through a pipe does not all move at the same velocity. The velocity of any part of the fluid depends on its distance from the center of the pipe. The fluid at the center of the pipe moves fastest; the fluid near the wall of the pipe moves slowest. Assume that the velocity of the fluid at a distance x centimeters from the center of the pipe is $g(x)$ centimeters per second.
    (a) Estimate informally the flow of fluid (in cubic centimeters per second) through a thin ring of inner radius r and outer radius $r + dr$ centimeters centered at the axis of the pipe and perpendicular to the axis.
    (b) Using (a), set up a definite integral for the flow (in cubic centimeters per second) of liquid through the pipe. (Let the radius of the pipe be b centimeters.)
    (c) Poiseuille (1797–1869), when studying the flow of blood through arteries, used the function $g(r) = k(b^2 - r^2)$, where k is a constant. Show that in this case the flow of blood through an artery is proportional to the fourth power of the radius of the artery.

18. (Kinetic energy) The density of a rod x centimeters from its left end is $g(x)$ grams per centimeter. The rod has a length of b centimeters. The rod is spun around its left end 7 times per second.
    (a) Estimate the mass of the rod in the section that is between x and $x + dx$ centimeters from the left end. (Assume that dx is small.)
    (b) Estimate the kinetic energy of the mass in (a).
    (c) Set up a definite integral for the kinetic energy of the rotating rod.

19. Obtain the formula $V = \int_a^b A(x)\, dx$ informally.

20. At time t hours, $0 \leq t \leq 24$, a firm uses electricity at the rate of $e(t)$ kilowatts. The rate schedule indicates that the cost per kilowatt-hour at time t is $c(t)$ dollars. Assume that both e and c are continuous functions.
    (a) Estimate the cost of electricity consumed between times t and $t + dt$, where dt is a small positive number.
    (b) Using (a), set up a definite integral for the total cost of electricity for the 24-hour period.

21. (Present value) The *present value* of a promise to pay one dollar t years from now is $g(t)$ dollars.
    (a) What is $g(0)$?
    (b) Why is it reasonable to assume that $g(t) \leq 1$ and that g is a decreasing function of t?
    (c) What is the present value of a promise to pay q dollars t years from now?
    (d) Assume that an investment made now will result in an income flow at the rate of $f(t)$ dollars per year t years from now. (Assume that f is a continuous function.) Estimate informally the present value of the income to be earned between time t and time $t + dt$, where dt is a small positive number.
    (e) On the basis of the local estimate made in (d), set up a definite integral for the present value of all the income to be earned from now to time b years in the future.

22. (Population) Let the number of females in a certain population in the age range from x years to $x + dx$ years, where dx is a small positive number, be approximately $f(x)\, dx$. Assume that on the average, women of age x at the beginning of a calendar year produce $m(x)$ offspring during the year. Assume that both f and m are continuous functions.
    (a) What definite integral represents the number of women between ages a and b years?
    (b) What definite integral represents the total number of offspring during the calendar year produced by women whose age at the beginning of the calendar year was between a and b years?

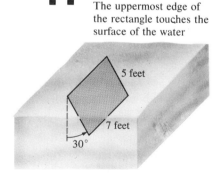

The uppermost edge of the rectangle touches the surface of the water

Figure 8.51

23 Find the force of the water against the rectangle, shown in Fig. 8.51, inclined at an angle of 30° to the vertical and whose top edge lies on the water surface. Use a definite integral (a) in which the interval of integration is vertical; (b) in which the interval of integration is inclined at 30° to the vertical; (c) in which the interval of integration is horizontal. In each case draw a neat picture that shows the interval [a, b] of integration and compute the integrand carefully.

8.4 Computing the volume of a solid of revolution

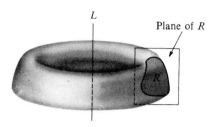

Figure 8.52

Let R be a region in the plane and L a line in the plane. (See Fig. 8.52.) Assume that L does not meet R at all or that L meets R only at points of the boundary. Consider the solid formed by revolving R about L. If L does not meet R, the solid is shaped like a doughnut.

This section presents two ways of computing the volume of such a solid of revolution. One method is by parallel cross sections, as in Sec. 8.2. The other method is by concentric shells.

VOLUME OF A SOLID OF REVOLUTION BY PARALLEL CROSS SECTIONS

Revolving the region below a curve

Let $y = f(x)$ be a continuous function such that $f(x) \geq 0$ for x in the interval $[a, b]$. Let R be the region under $y = f(x)$ and above $[a, b]$. The region R is revolved around the x axis to form a solid of revolution. What is the volume of this solid? (See Figs. 8.53 and 8.54.)

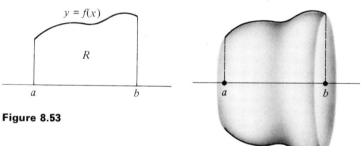

Figure 8.53

Figure 8.54

That volume is the integral of cross-sectional area was shown in Sec. 5.3.

To find the volume, first find the area $A(x)$ of a typical cross section made by a plane perpendicular to the x axis corresponding to the coordinate x. This cross section is a disk of radius $f(x)$, as shown in Fig. 8.55.

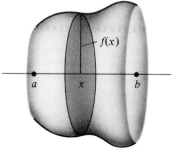

The cross section is a disk of radius $f(x)$.

Figure 8.55

Formula for volume of solid of revolution formed by revolving region below a curve

Thus $A(x) = \pi(f(x))^2$. Since the volume of a solid is the integral of its cross-sectional area, we conclude that

$$\text{Volume of the solid in Fig. 8.54} = \int_a^b \pi(f(x))^2 \, dx. \quad (1)$$

EXAMPLE 1 The region under $y = e^{-x}$ and above $[1, 2]$ is revolved around the x axis. Find the volume of the resulting solid of revolution. (See Fig. 8.56.)

SOLUTION By formula (1), the volume is

$$V = \int_1^2 \pi(e^{-x})^2 \, dx$$

$$= \int_1^2 \pi e^{-2x} \, dx$$

$$\underset{\text{FTC}}{=} \left. \frac{\pi e^{-2x}}{-2} \right|_1^2$$

$$= \frac{\pi}{-2}(e^{-4}) - \left(\frac{\pi}{-2} e^{-2}\right)$$

$$= \frac{\pi}{2}(e^{-2} - e^{-4}). \quad \blacksquare$$

Figure 8.56

Revolving the region between two curves

A similar approach works for finding the volume of the solid of revolution formed when the region between two curves is revolved around the x axis.

Let $y = f(x)$ and $y = g(x)$ be two continuous functions such that $f(x) \geq g(x) \geq 0$ for x in the interval $[a, b]$. Let R be the region bounded by the curves $y = f(x)$ and $y = g(x)$ and above $[a, b]$. The region R is revolved around the x axis to form a solid of revolution. What is the volume of this solid? (See Fig. 8.57.)

The solid may have a hole.

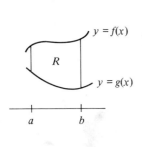

Figure 8.57

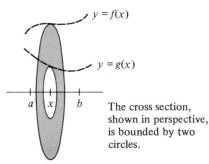

The cross section, shown in perspective, is bounded by two circles.

Figure 8.58

The "washer" method

The cross section of the solid in Fig. 8.57 made by a plane perpendicular to the x axis is a ring, as shown in Fig. 8.58. The ring is bounded by a circle of radius $f(x)$ and a circle of radius $g(x)$. Thus its area $A(x)$ is

This is not $\pi[f(x) - g(x)]^2$.

$$\pi(f(x))^2 - \pi(g(x))^2$$

or $A(x) = \pi[(f(x))^2 - (g(x))^2].$

Consequently,

Formula for volume of solid of revolution formed by revolving region between two curves

$$\text{Volume of solid in Fig. 8.57} = \int_a^b \pi[(f(x))^2 - (g(x))^2]\, dx. \quad (2)$$

EXAMPLE 2 The region between $y = \sqrt{x}$ and $y = x^2$ is revolved around the x axis. Find the volume of the solid of revolution produced.

SOLUTION Figure 8.59 shows the plane region R being revolved. The region R lies above the interval $[0, 1]$ and between the curves $y = \sqrt{x}$ and $y = x^2$. By (2), the volume of the solid of revolution is

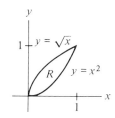

Figure 8.59

$$\int_0^1 \pi[(\sqrt{x})^2 - (x^2)^2]\, dx = \int_0^1 \pi(x - x^4)\, dx$$
$$\underset{\text{FTC}}{=} \pi\left(\frac{x^2}{2} - \frac{x^5}{5}\right)\Big|_0^1$$
$$= \pi(\tfrac{1}{2} - \tfrac{1}{5}) - \pi(\tfrac{0}{2} - \tfrac{0}{5})$$
$$= \frac{3\pi}{10}. \quad\blacksquare$$

VOLUME OF A SOLID OF REVOLUTION BY CONCENTRIC SHELLS

In the method of parallel cross sections the solid of revolution is approximated by very thin disks or washers. But there is a completely different way of viewing a solid of revolution. The solid may also be approximated by concentric hollow pipes (or shells).

Consider a region R and a line L in the plane. Assume, as at the beginning of the section, that L does not meet R or meets R only at points on the boundary of R. In addition, assume that each line in the plane of R which is parallel to L meets R in a line segment or a point, as in Fig. 8.60. The

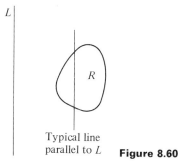

Typical line parallel to L **Figure 8.60**

478 APPLICATIONS OF THE DEFINITE INTEGRAL

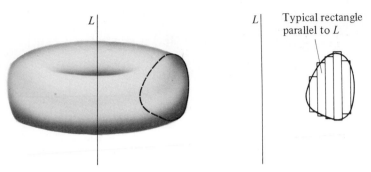

Figure 8.61

Figure 8.62

The line L can be in any direction in the plane.

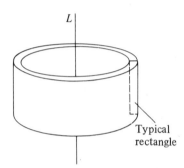

This is the shell formed by revolving the "typical rectangle" in Fig. 8.62 around L.

Figure 8.63

solid of revolution formed by revolving R around L is shown in Fig. 8.61. Approximate R by rectangles, as shown in Fig. 8.62. Then the volume of the solid in Fig. 8.61 is approximated by the volume of the solid of revolution formed by revolving around L the region formed by the rectangles in Fig. 8.62. When the rectangles shown in Fig. 8.62 are revolved around the line L each one sweeps out a cylindrical shell. One of these shells or pipes is shown in Fig. 8.63.

These observations suggest a way of forming a definite integral for the volume of the solid of revolution formed by revolving R around L. The informal approach introduced in Sec. 8.3 will be used. (At the end of the section the formal approach is given.)

Introduce an x axis in the plane of R and perpendicular to L. Assume that L lies to the left of R and cuts the x axis at k and that R lies above the interval $[a, b]$ as in Fig. 8.64.

Estimate the volume of the solid of revolution formed by revolving around L the part of R between those lines parallel to L which meet the x axis at x and $x + dx$. (Assume that dx is a small positive number. See Fig. 8.65.)

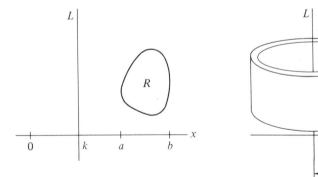

Figure 8.64

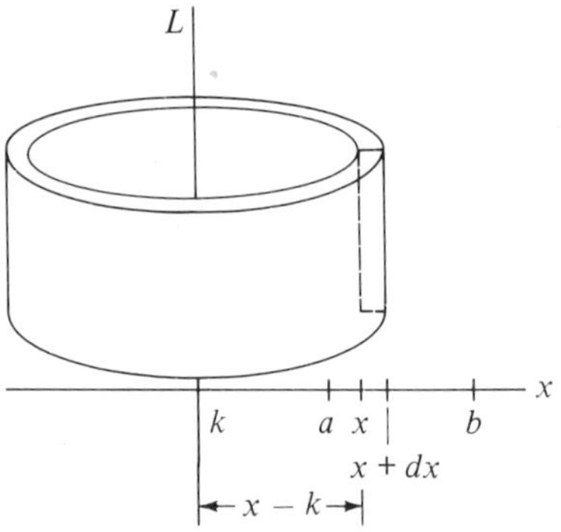

Figure 8.65

To estimate the volume of the shell or tube in Fig. 8.65, begin by letting $c(x)$ be the length of the cross section of R made by a line parallel to L and meeting the coordinate axis at x. Imagine cutting the tube along a direction parallel to L and then laying the tube flat. When laid flat the tube will

The radius is $x - k$, as in Fig. 8.65, so circumference is $2\pi(x - k)$.

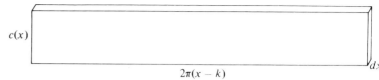

Figure 8.66

resemble a thin slab of thickness dx, width $c(x)$, and length $2\pi(x - k)$, as shown in Fig. 8.66. The volume of the tube therefore is presumably about

This is the key local estimate.

$$2\pi(x - k)c(x)\, dx. \tag{3}$$

With the aid of this local estimate (3), we can then conclude that the volume of the solid of revolution is

Formula for volume of a solid of revolution by shells

$$\boxed{\text{Volume} = \int_a^b 2\pi(x - k)c(x)\, dx.} \tag{4}$$

This is the formula for computing volumes by the "shell technique." If $x - k$ is denoted $R(x)$, the "radius of the shell," then

A shorter version of the formula for volume by shells

$$\boxed{\text{Volume} = \int_a^b 2\pi R(x)c(x)\, dx.} \tag{5}$$

Memory aid: The expression $2\pi R(x)c(x)\, dx$ is the volume of a flat box of height dx, width $c(x)$, and length $2\pi R(x)$, the circumference of a circle of radius $R(x)$. Figure 8.67 gives the shell technique at a glance.

These two diagrams show the essence of the shell technique.

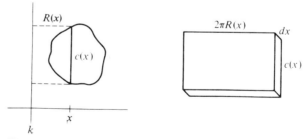

Figure 8.67

EXAMPLE 3 The region between the curves $y = x^2$ and $y = x^3$ is revolved around the y axis. Find the volume of the solid of revolution produced. (It is shaped like a bowl, as shown in Fig. 8.68.)

SOLUTION Use the given x axis to set up the integral $\int_a^b 2\pi R(x)c(x)\, dx$. Since the y axis is the line L about which the region is revolved, $R(x) = x$. Also, $c(x) = x^2 - x^3$. Thus

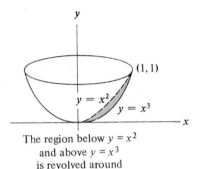

The region below $y = x^2$ and above $y = x^3$ is revolved around the y axis.

Figure 8.68

$$\text{Volume} = \int_0^1 2\pi x(x^2 - x^3)\, dx$$

$$= \int_0^1 2\pi(x^3 - x^4)\, dx$$

$$\underset{\text{FTC}}{=} 2\pi\left[\frac{x^4}{4} - \frac{x^5}{5}\right]_0^1$$

$$= 2\pi\left[\left(\frac{1^4}{4} - \frac{1^5}{5}\right) - \left(\frac{0^4}{4} - \frac{0^5}{5}\right)\right]$$

$$= 2\pi(\tfrac{1}{4} - \tfrac{1}{5})$$

$$= \frac{2\pi}{20} = \frac{\pi}{10}. \quad\blacksquare$$

EXAMPLE 4 The triangle whose vertices are $(0, 0)$, $(2, 0)$, and $(2, 1)$ is revolved around the x axis. Find the volume of the resulting cone. (See Fig. 8.69.)

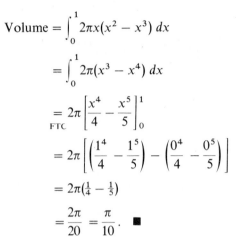

Figure 8.69

Figure 8.70

SOLUTION In this case the shells are formed by partitions of the y axis, not the x axis. (See Fig. 8.70.) The formula for the volume, written with y replacing x, is

$$\int_a^b 2\pi R(y) c(y)\, dy.$$

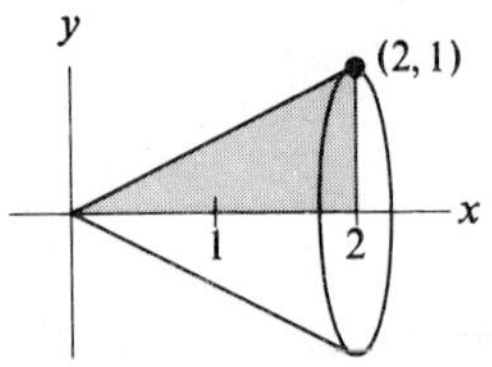

To determine $R(y)$ and $c(y)$, look closely at the given triangle, not at the solid of revolution. Inspection of Fig. 8.71 shows that $R(y) = y$. To find $c(y)$, make use of the equation of the line through $(0, 0)$ and $(2, 1)$, namely $y = x/2$, or $x = 2y$. Inspection of the diagram then shows that

$$c(y) = 2 - 2y.$$

The interval of integration is $[0, 1]$. Thus

$$\text{Volume} = \int_0^1 2\pi y(2 - 2y)\, dy$$

$$= 4\pi \int_0^1 (y - y^2)\, dy$$

$$\underset{\text{FTC}}{=} 4\pi\left(\frac{y^2}{2} - \frac{y^3}{3}\right)\bigg|_0^1$$

$$= 4\pi\left[\left(\frac{1^2}{2} - \frac{1^3}{3}\right) - \left(\frac{0^2}{2} - \frac{0^3}{3}\right)\right]$$

$$= 4\pi\left(\frac{1}{2} - \frac{1}{3}\right)$$

$$= \frac{2\pi}{3}. \quad\blacksquare$$

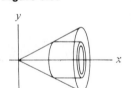

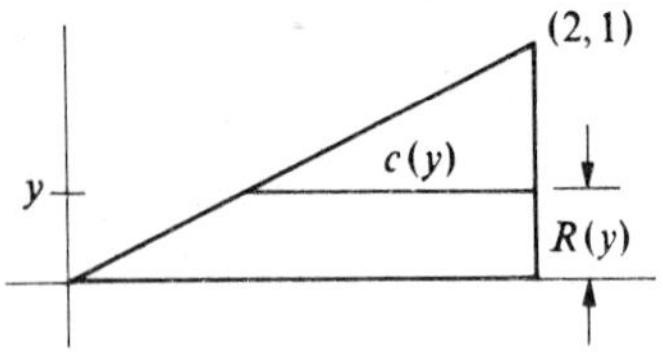

Figure 8.71

8.4 COMPUTING THE VOLUME OF A SOLID OF REVOLUTION

EXAMPLE 5 The region below $y = 1 + \sin x$, above the x axis, and situated between $x = 0$ and $x = 2\pi$ is revolved around the y axis. Find the volume of the resulting solid of revolution, shown in Fig. 8.72.

SOLUTION (Note that cross sections by planes perpendicular to the y axis would be quite messy. For y between 1 and 2, the cross section is one ring, whose radii would be expressed in terms of the function $\sin^{-1}$. For y between 0 and 1, the cross section consists of two pieces.)

The method of concentric shells is easy.

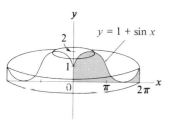

Figure 8.72

$$\text{Volume} = \int_a^b 2\pi R(x) c(x)\, dx = \int_0^{2\pi} 2\pi x(1 + \sin x)\, dx$$
$$= \int_0^{2\pi} 2\pi x\, dx + \int_0^{2\pi} 2\pi x \sin x\, dx.$$

Both integrals can be evaluated with the aid of the fundamental theorem of calculus:

$$\int_0^{2\pi} 2\pi x\, dx = \pi x^2 \Big|_0^{2\pi} = \pi(2\pi)^2 - \pi(0)^2 = 4\pi^3,$$

and
$$\int_0^{2\pi} 2\pi x \sin x\, dx = 2\pi(\sin x - x \cos x)\Big|_0^{2\pi} \quad \text{(integration by parts)}$$
$$= 2\pi[(\sin 2\pi - 2\pi \cos 2\pi) - (\sin 0 - 0 \cos 0)]$$
$$= 2\pi[(0 - 2\pi \cdot 1) - (0 - 0)]$$
$$= -4\pi^2.$$

Thus the volume is $4\pi^3 - 4\pi^2$. ∎

THE SHELL TECHNIQUE (formal approach)

The key tool is the formula for the volume of a cylindrical shell. Assume that its inner radius is R_1, its outer radius R_2, and its height h. (See Fig. 8.73.) The volume of this cylindrical shell equals

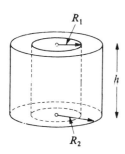

Figure 8.73

$$\text{Area of base} \cdot \text{height} = (\pi R_2^2 - \pi R_1^2)h \quad (\textit{Note: not } \pi(R_2 - R_1)^2 h)$$
$$= \pi(R_2 + R_1)(R_2 - R_1)h$$
$$= 2\pi \frac{R_2 + R_1}{2} h(R_2 - R_1)$$

Thus the volume of the shell is

$$2\pi \cdot \text{average of the inner and outer radii} \cdot \text{height} \cdot \text{thickness}.$$

With the partition

$$a = x_0 < x_1 < \cdots < x_n = b$$

We use the symbol X_i, rather than c_i to avoid confusion with the function $c(x)$. are associated n concentric shells: the ith shell is formed by selecting X_i to be the midpoint $(x_{i-1} + x_i)/2$, of the ith interval; form the rectangle above the interval $[x_{i-1}, x_i]$ whose height is $c(X_i)$, the cross section of R by the line parallel to L and meeting the x axis at X_i, as shown in Fig. 8.74. The ith shell is obtained by revolving this rectangle around the line L.

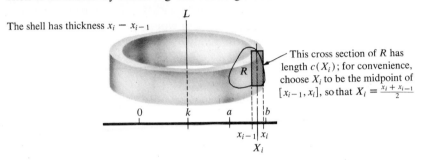

Figure 8.74

The volume of the typical shell is

$$2\pi \cdot \text{average of inner and outer radii} \cdot \text{height} \cdot \text{thickness}$$

$$= 2\pi \left[\frac{(x_{i-1} - k) + (x_i - k)}{2} \right] c(X_i)(x_i - x_{i-1})$$

$$= 2\pi \left(\frac{x_{i-1} + x_i}{2} - k \right) c(X_i)(x_i - x_{i-1})$$

$$= 2\pi(X_i - k)c(X_i)(x_i - x_{i-1}).$$

Now make the assumption that, when the shells are thin, the total volume of these shells is a good estimate of the volume of the solid of revolution. That is, assume that

$$\lim_{\text{mesh} \to 0} \sum_{i=1}^{n} 2\pi(X_i - k)c(X_i)(x_i - x_{i-1}) = \text{volume of solid}.$$

But this limit is the definite integral

$$\int_a^b 2\pi(x - k)c(x)\, dx. \qquad (6)$$

Formula (6) is, as expected, the same as Formula (4).

Exercises

Exercises 1 to 6 concern the parallel cross-section technique. In each of Exercises 1 to 6 a region R in the plane is revolved around the x axis to produce a solid of revolution. In each case (a) draw the region, (b) draw the solid of revolution, (c) draw the typical cross section (a disk or a ring), (d) set up a definite integral for the volume, (e) evaluate the integral.

1. R is bounded by $y = \sqrt{x}$, the x axis, $x = 1$, and $x = 2$.
2. R is bounded by $y = 1/\sqrt{1 + x^2}$, the x axis, $x = 0$, and $x = 1$.

3. R is bounded by $y = x^2$ and $y = x^3$.
4. R is bounded by $y = 1/\sqrt{x}$, $y = 1/x$, $x = 1$, and $x = 2$.
5. R is bounded by $y = \tan x$, $y = \sin x$, $x = 0$, and $x = \pi/4$.
6. R is bounded by $y = \sec x$, $y = \cos x$, $x = \pi/6$, and $x = \pi/3$.

Exercises 7 to 12 concern the concentric-shell technique. In each, a region R in the plane is revolved around a line to produce a solid of revolution. In each case (a) draw the region, (b) draw the solid of revolution, (c) draw the typical approximating shell and label its three dimensions ($c(x)$, $R(x)$, dx) or ($c(y)$, $R(y)$, dy), (d) set up a definite integral for the volume, (e) evaluate the integral.

7. R, bounded by $y = x^2$, $y = 0$, and $x = 1$, is revolved around the y axis.
8. R, given in Exercise 7, is revolved around the x axis.
9. R, the finite region bounded by $y = \sqrt{x}$ and $y = \sqrt[3]{x}$, is revolved around the x axis.
10. R, given in Exercise 9, is revolved around the y axis.
11. R, bounded by $y = \sin x$ and the x axis between $x = 0$ and $x = \pi/2$, is revolved around the y axis.
12. R, given in Exercise 11, is revolved around the x axis.
13. Find the volume of a sphere of radius a by the shell method.
14. Find the volume of a right circular cone of radius a and height h by the shell method.

In Exercises 15 and 16 use the shell technique to compute the volumes of the given solids of revolution.
15. The region bordered by $y = \sqrt{x}$, $x = 1$, $x = 2$, and the x axis is revolved around the line $x = -1$.
16. The region in Exercise 15 is rotated around the line $y = -2$.

17. Find the volume of the solid of revolution formed by revolving the region bounded by $y = 2 + \cos x$, $x = \pi$, $x = 10\pi$, and the x axis around (a) the y axis, (b) the x axis.
18. The disk bounded by the circle $(x - b)^2 + y^2 = a^2$, $0 < a < b$, is revolved around the y axis. Find the volume of the doughnut (torus) produced.
19. The region below $y = \cos x$, above the x axis, and between $x = 0$ and $x = \pi/2$ is revolved around the x axis. Find the volume of the resulting solid of revolution by (a) parallel cross sections, (b) concentric shells.

In Exercises 20 to 22 find the volumes of the solids of revolution given.
20. The region between $y = e^{x^2}$, the x axis, $x = 0$, and $x = 1$ is revolved around the y axis. (It is interesting to note that the fundamental theorem of calculus is of no use in evaluating the area of this region.)
21. The region between $y = \sqrt{1 + x^2}$, $y = 1$, and $x = 1$ is revolved around the line $y = 1$.
22. The region between $y = \ln x$, the x axis, and $x = e$ is revolved around (a) the y axis, (b) the line $y = 1$, (c) the line $y = -1$.

23. Let a and b be positive numbers and $y = f(x)$ be a decreasing differentiable function of x, such that $f(0) = b$ and $f(a) = 0$. Prove that $\int_0^a 2xy\, dx = \int_0^b x^2\, dy$. (a) by considering the volume of a certain solid, (b) by integration by parts.

Exercise 24 shows that from a computational viewpoint the shell technique is not more effective than the cross-section technique.

24. Let f in Exercise 23 be elementary. (a) Show that, if $x^2 f'$ has an elementary integral, so does xf, and conversely. (b) Consider the solid obtained by rotating the region bounded by the curve $y = f(x)$, the x axis, and the y axis around the y axis. Show that its volume expressed by the shell technique involves an elementary integral only when its volume by the cross-section technique involves an elementary integral.

25. Let R be a region in the first quadrant. When it is revolved around the x axis, a solid of revolution is produced. When it is revolved around the y axis, another solid of revolution is produced. Give an example of such a region R with the property that the volume of the first solid cannot be evaluated by the fundamental theorem of calculus, but the volume of the second region can be.

26. (Contributed by Steve Abell) A lead sinker is to be made by revolving a section of a circular disk around the line AB, as shown in Fig. 8.75. The length of the sinker is to be 3 times its maximum radius a, as shown in Fig. 8.75. The volume of the sinker is to be 3 cubic inches. Find the radius of the disk and the maximum radius of the sinker.

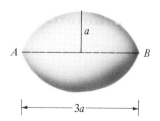

Figure 8.75

484 APPLICATIONS OF THE DEFINITE INTEGRAL

8.5 The centroid of a plane region

This section introduces the centroid of a plane region and shows its relation to the volume of a solid of revolution. In Sec. 8.7 it will also be of use in the study of work.

THE CENTER OF MASS OF n POINT MASSES

A small boy on one side of a seesaw (which we regard as weightless) can balance a bigger boy on the other side. For example, the two boys balance in Fig. 8.76.

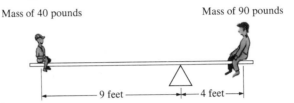

Figure 8.76

The small mass with the long lever arm balances the large mass with the small lever arm. Each contributes the same tendency to turn—but in opposite directions. To be more precise, introduce on the seesaw an x axis with its origin 0 at the fulcrum, the point on which the seesaw rests. Define the moment about 0 of a mass m located at the point x on the x axis to be the product mx. Then the bigger boy has a moment $(90)(4)$, while the smaller boy has a moment $(40)(-9)$. The total moment of the lever-mass system is 0, and the masses balance. (See Fig. 8.77.)

Moment is (mass)·(lever arm), where lever arm can be positive or negative.

Figure 8.77

If a mass m is located on a line at coordinate x, define its moment about the point having coordinate k as the product $m(x - k)$. (See Fig. 8.78.)

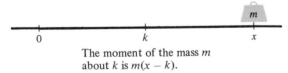

The moment of the mass m about k is $m(x - k)$.

Figure 8.78

Now consider several masses $m_1, m_2, \ldots, m_n$. If mass m_i is located at x_i, with $i = 1, 2, \ldots, n$ then $\sum_{i=1}^{n} m_i(x_i - k)$ is the total moment of all the masses about the point k. If a fulcrum is placed at k then the seesaw rotates clockwise if the total moment is greater than 0, rotates counterclockwise if it is less than 0, and is in equilibrium if the total moment is 0.

EXAMPLE 1 Where should the fulcrum be placed so that the three masses in Fig. 8.79 will be in equilibrium?

8.5 THE CENTROID OF A PLANE REGION   **485**

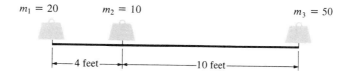

Figure 8.79

SOLUTION   Introduce an x axis with origin at mass m_1 and compute the moments about a typical fulcrum having coordinate k; then select k to make the total moment 0. (See Fig. 8.80.)

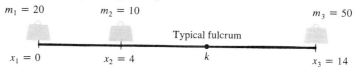

Figure 8.80

The total moment about k is

$$20(0 - k) + 10(4 - k) + 50(14 - k).$$

We seek k such that this expression is equal to 0 or, equivalently,

$$20 \cdot 0 + 10 \cdot 4 + 50 \cdot 14 = k(20 + 10 + 50).$$

Hence $$k = \frac{20 \cdot 0 + 10 \cdot 4 + 50 \cdot 14}{80} = 9.25.$$

This means the fulcrum is to the right of the midpoint, which was to be expected. ∎

The balancing point of masses $m_1, m_2, \ldots, m_n$, located respectively at $x_1, x_2, \ldots, x_n$ on an x axis, is found by solving the equation

$$\sum_{i=1}^{n} m_i(x_i - k) = 0 \qquad (1)$$

for the number k. Expanding (1), we obtain

$$\sum_{i=1}^{n} m_i x_i - \sum_{i=1}^{n} m_i k = 0$$

or

$$\sum_{i=1}^{n} m_i x_i = \sum_{i=1}^{n} m_i k.$$

Thus

$$k \sum_{i=1}^{n} m_i = \sum_{i=1}^{n} m_i x_i$$

and, finally,

$$k = \frac{\sum_{i=1}^{n} m_i x_i}{\sum_{i=1}^{n} m_i}. \qquad (2)$$

The number k given by (2) is called the *center of mass* of the system of

486 APPLICATIONS OF THE DEFINITE INTEGRAL

This idea is the key to the section.

masses. *The center of mass is found by dividing the total moment about 0 by the total mass.*

Finding the center of mass of a finite number of "point masses" involves only arithmetic, no calculus. Now let us turn our attention to finding the center of mass of a continuous distribution of matter in the plane. For this purpose definite integrals will be needed.

THE CENTROID OF A PLANE REGION

Let R be a region in the plane. Imagine that R is occupied by a thin piece of metal that has a density of ρ (rho) grams per square centimeter. Throughout this section ρ will be assumed to be constant (that is, the metal is "homogeneous"). Let L be a line in the plane. (See Fig. 8.81.) Is there a line parallel to L on which R balances? Consider the lines L_1 and L_2 in Fig. 8.82.

The case where ρ is not constant is treated in Chap. 13.

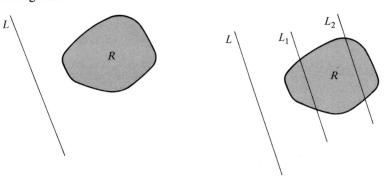

Figure 8.81   **Figure 8.82**

If R were placed on L_1, it would turn one way; if it were placed on L_2, it would turn the other way. It seems reasonable to expect there to be a "balancing" line parallel to L and somewhere between L_1 and L_2. To find that line it is necessary to (a) compute the moment of R about any line L' parallel to L and then (b) find the line L' for which the moment is 0.

Let L' be any line parallel to L. To compute the moment of R about L introduce an x axis perpendicular to L with its origin at its intersection with L. Assume that L' passes through the x axis at the point $x = k$, as in Fig. 8.83. In addition, assume that each line parallel to L meets R either in a line segment or at a point on the boundary of R. The lever arm of the mass distributed throughout R varies from point to point. However, the length of the lever arm is almost constant for the mass located between two lines parallel to L and close to each other.

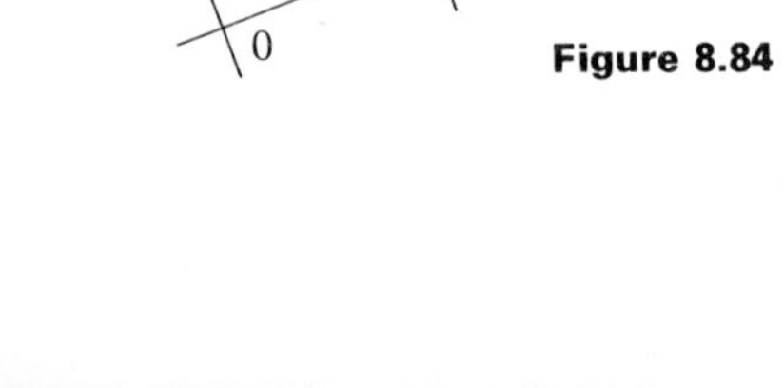

The search for a balancing line begins.

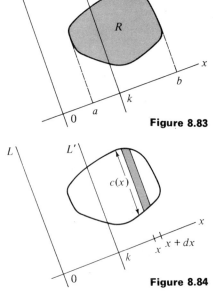

Figure 8.83

Figure 8.84

Consider the moment about L' of the mass in R located between the lines parallel to L and passing through the points on the x axis with coordinates x and $x + dx$, where dx is a small positive number. (See Fig. 8.84.)

Let $c(x)$ be the length of the cross section of R at x. Then the area of that portion of R between the lines passing through x and $x + dx$ is approximately $c(x)\, dx$. The mass of that portion is consequently about $\rho c(x)\, dx$. The lever arm around L' of this mass is about $x - k$, which may be positive or negative. The *local approximation* of the moment of the mass in R about L' is therefore

The all-important local approximation

$$\underbrace{(x-k)}_{\text{lever arm}} \underbrace{\rho\ \underbrace{c(x)\,dx}_{\text{area}}}_{\text{mass}}.$$

Following the informal approach, we conclude that

THE MOMENT OF A PLANE HOMOGENEOUS MASS ABOUT A LINE

$$\boxed{\text{Moment of the mass in } R \text{ about } L' = \int_a^b (x-k)\rho c(x)\,dx.} \qquad (3)$$

Formula (3) is the physicist's definition of the moment of a plane distribution of matter about a line.

The moment around L' may or may not be 0. Let us determine k, hence the line L', such that the moment given by (3) is 0. To do this, solve the equation

$$0 = \int_a^b (x-k)\rho c(x)\,dx$$

for k.

We then have

$$0 = \int_a^b x\rho c(x)\,dx - \int_a^b k\rho c(x)\,dx$$

k and ρ are constants.

$$0 = \rho \int_a^b xc(x)\,dx - k\rho \int_a^b c(x)\,dx$$

$$k\rho \int_a^b c(x)\,dx = \rho \int_a^b xc(x)\,dx$$

ρ cancels.

$$k \int_a^b c(x)\,dx = \int_a^b xc(x)\,dx$$

FORMULA FOR BALANCING LINE $x = k$ (ρ CONSTANT)

$$\boxed{k = \frac{\int_a^b xc(x)\,dx}{\int_a^b c(x)\,dx}.} \qquad (4)$$

The numerator is called "the moment of R about L." The density ρ does not appear in this moment. It is convenient to think of ρ as being 1. Then mass and area have the same numerical value. The denominator $\int_a^b c(x)\,dx$ is the area of R. *Henceforth assume that $\rho = 1$.* Formula (4) shows that there is a unique balancing line $x = k$ parallel to L. Its coordinate is given by

$$k = \frac{\text{Moment of } R \text{ about } L}{\text{Area of } R}.$$

488 APPLICATIONS OF THE DEFINITE INTEGRAL

Assume now that the plane is furnished with an xy coordinate system. There is a unique balancing line parallel to the y axis. Its x coordinate equals

$$\bar{x} = \frac{\int_a^b xc(x)\, dx}{\text{Area of } R}. \tag{5}$$

Similarly, there is a unique balancing line parallel to the x axis. Its y coordinate is given by

$$\bar{y} = \frac{\int_c^d yc(y)\, dy}{\text{Area of } R}, \tag{6}$$

where $c(y)$ is the cross section function of R for lines parallel to the x axis and $[c, d]$ is the interval of integration.

DEFINITION *Centroid of R.* The centroid of R is defined as the point $(\bar{x}, \bar{y})$, where

$$\bar{x} = \frac{\text{Moment of } R \text{ about } y \text{ axis}}{\text{Area of } R}, \qquad \bar{y} = \frac{\text{Moment of } R \text{ about } x \text{ axis}}{\text{Area of } R}$$

or

FORMULA FOR THE CENTROID

$$\bar{x} = \frac{\int_a^b xc(x)\, dx}{\text{Area of } R} \qquad \bar{y} = \frac{\int_c^d yc(y)\, dy}{\text{Area of } R}. \tag{7}$$

It can be shown that the region R balances on any line through its centroid. (See Exercises 32 and 33 in Sec. 13.3.) Moreover, if R is suspended motionless from a string attached at its centroid, it will remain in equilibrium. Note that the centroid is defined purely geometrically.

EXAMPLE 2 Let R be the triangle in the xy plane with vertices at $(0, 0)$, $(6, 0)$, and $(6, 3)$, as shown in Fig. 8.85. Find (a) $\bar{x}$, (b) $\bar{y}$.

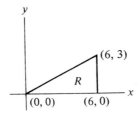

Figure 8.85

SOLUTION (a) To find $\bar{x}$ we first compute the moment of R around the y axis,

$$\int_0^6 xc(x)\, dx,$$

where $c(x)$ is the cross section function for R. To find $c(x)$, use the equation of the line through $(0, 0)$ and $(6, 3)$, namely $y = x/2$. Thus $c(x) = x/2$. Consequently the moment of R around the y axis is

$$\int_0^6 x \cdot \frac{x}{2} dx = \frac{1}{2} \int_0^6 x^2 \, dx$$

$$\underset{\text{FTC}}{=} \frac{1}{2} \left. \frac{x^3}{3} \right|_0^6$$

$$= \frac{1}{2} \cdot \frac{216}{3}$$

$$= 36.$$

To use formula (7), we divide 36 by the area of the triangle, which is $\frac{1}{2} \cdot 6 \cdot 3 = 9$. Thus

$$\bar{x} = \frac{36}{9} = 4.$$

The balancing line parallel to the y axis passes through $x = 4$.
(b) To find $\bar{y}$, first compute the moment $\int_0^3 yc(y) \, dy$. Since the line through $(0, 0)$ and $(6, 3)$ has the equation $y = x/2$, on that line $x = 2y$. Inspection of Fig. 8.85 shows that

$$c(y) = 6 - 2y.$$

Thus
$$\int_0^3 yc(y) \, dy = \int_0^3 y(6 - 2y) \, dy$$

$$= \int_0^3 (6y - 2y^2) \, dy$$

$$\underset{\text{FTC}}{=} \left. \left(3y^2 - \frac{2y^3}{3} \right) \right|_0^3$$

$$= (27 - 18) - 0$$

$$= 9.$$

Hence
$$\bar{y} = \frac{9}{\text{Area of } R} = \frac{9}{9} = 1.$$

Thus
$$(\bar{x}, \bar{y}) = (4, 1),$$

The centroid of a triangle is always two-thirds the way from each vertex to the midpoint of the opposite side.

which is shown in Fig. 8.86. ∎

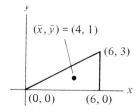

Figure 8.86

Using formula (7), we can show that the centroid of a rectangle is its center. This fact is plausible, since, by symmetry, a rectangle would balance

490 APPLICATIONS OF THE DEFINITE INTEGRAL

on a line through its center and parallel to an edge, as shown in Fig. 8.87.

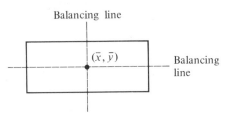

The center of the rectangle is its centroid.

Figure 8.87

ANOTHER FORMULA FOR $\bar{y}$ Consider a rectangle resting on the x axis, as in Fig. 8.88. Let $\bar{y}$ be the y coordinate of its center of mass. Then

$$\bar{y} = \frac{\text{Moment of rectangle about } x \text{ axis}}{\text{Area of rectangle}}$$

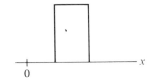

Figure 8.88

or

$$\begin{pmatrix}\text{Moment of rectangle} \\ \text{about } x \text{ axis}\end{pmatrix} = \begin{pmatrix}y \text{ coordinate of} \\ \text{centroid of rectangle}\end{pmatrix} \cdot \begin{pmatrix}\text{Area of} \\ \text{rectangle}\end{pmatrix} \quad (8)$$

Formula (8) is the basis of a shortcut for computing $\bar{y}$ of a region under a curve and above the x axis. We now develop this short cut.

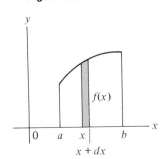

Figure 8.89

Let $y = f(x)$ be a continuous function such that $f(x) \geq 0$ for x in $[a, b]$. Let R be the region below the curve $y = f(x)$ and above $[a, b]$. The moment around the x axis of the portion of R between two lines parallel to the y axis, one with x coordinate x and one with x coordinate $x + dx$, where dx is a small positive number, can be estimated easily. (See Fig. 8.89, where this narrow band is shaded.) The narrow band has area approximately $f(x)\,dx$ and the y coordinate of its center is approximately $f(x)/2$. By (8), its moment around the x axis is approximately

$$\underbrace{\frac{f(x)}{2}}_{y \text{ coordinate of centroid}} \cdot \underbrace{f(x)\,dx}_{\text{area}}. \quad (9)$$

From the local approximation (9), we conclude that

$$\begin{matrix}\text{Moment of } R \\ \text{about } x \text{ axis}\end{matrix} = \int_a^b \frac{(f(x))^2}{2}\,dx. \quad (10)$$

$\bar{y}$ for the region under $y = f(x)$ Consequently, $\bar{y}$, the y coordinate of the centroid of R, equals

$$\boxed{\bar{y} = \frac{\int_a^b \frac{(f(x))^2}{2}\,dx}{\text{Area of } R}.} \quad (11)$$

8.5 THE CENTROID OF A PLANE REGION

This formula is preferable to (7) if R is given as a region under a curve, $y = f(x)$.

EXAMPLE 3 Find the centroid of the semicircular region of radius a shown in Fig. 8.90.

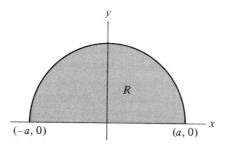

Figure 8.90

SOLUTION By symmetry, $\bar{x} = 0$.

To find $\bar{y}$, use (10) and (11). The function f in this case is given by the formula $f(x) = \sqrt{a^2 - x^2}$. Thus the moment of R about the x axis is

$$\int_{-a}^{a} \frac{(\sqrt{a^2 - x^2})^2}{2} dx = \int_{-a}^{a} \frac{a^2 - x^2}{2} dx$$

The integrand is an even function.

$$= 2 \int_{0}^{a} \frac{a^2 - x^2}{2} dx$$

$$= \int_{0}^{a} (a^2 - x^2) dx$$

$$\underset{\text{FTC}}{=} \left(a^2 x - \frac{x^3}{3} \right) \bigg|_{0}^{a}$$

$$= \left(a^3 - \frac{a^3}{3} \right) - 0$$

$$= \tfrac{2}{3} a^3.$$

Thus
$$\bar{y} = \frac{\tfrac{2}{3} a^3}{\text{Area of } R} = \frac{\tfrac{2}{3} a^3}{\tfrac{1}{2} \pi a^2}$$

$$= \frac{4a}{3\pi}.$$

(Since $4/(3\pi) \approx 0.42$, the center of gravity of R is at a height of about $0.42a$.) ∎

THE CENTROID AND THE VOLUME OF A SOLID OF REVOLUTION

The relation between the centroid of a plane region R and the solid of revolution obtained by revolving R about a line is expressed in the following theorem, due to the fourth-century Greek mathematician Pappus.

THEOREM (*Pappus's theorem*). Let R be a region in the plane and L a line in the plane that either does not meet R or else just meets R at its border. Then the volume of the solid formed by revolving R about L is equal to the product:

(Distance centroid of R is revolved) · (Area of R).

PROOF Introduce an xy coordinate system in such a way that L is the y axis and R lies to the right of it. The volume of the solid of revolution is

$$2\pi \int_a^b xc(x)\, dx \qquad \text{(shell technique)}.$$

Recall that $\bar{x} = \dfrac{\int_a^b xc(x)\, dx}{\text{Area of } R}$. By (7), $\int_a^b xc(x)\, dx = \bar{x} \cdot \text{Area of } R.$

Hence the volume is $2\pi \bar{x} \cdot \text{Area of } R.$

Since $2\pi \bar{x}$ is the distance the centroid of R is revolved, this proves the theorem. ∎

EXAMPLE 4 Use Pappus's theorem to find the volume of the "doughnut" formed by revolving a circle of radius 3 inches about a line 5 inches from the center, as shown in Fig. 8.91.

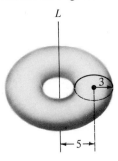

Figure 8.91

SOLUTION In this case the area of R is $\pi \cdot 3^2 = 9\pi$. The centroid of a circle is its center. Hence the distance it is revolved is $2\pi \cdot 5 = 10\pi$. The volume of the doughnut is

$$10\pi \cdot 9\pi = 90\pi^2 \text{ cubic inches}. \quad \blacksquare$$

Pappus's theorem can be used to find the centroid of a region R if the volume of the solid of revolution obtained by revolving R is known. The next example should be contrasted with Example 3.

EXAMPLE 5 Find the centroid of the semicircle R of radius a shown in Fig. 8.92.

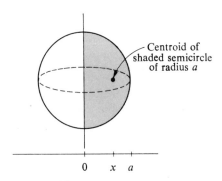

Figure 8.92

SOLUTION By symmetry, the centroid lies somewhere on the radius that is perpendicular to the diameter of the semicircle. Let $\bar{x}$ be its distance from the diameter. When R is revolved about its diameter, it produces a sphere of radius a. The volume of a sphere of radius a is $4\pi a^3/3$.

By Pappus's theorem

$$2\pi \bar{x} \cdot \text{Area of } R = \text{Volume of sphere of radius } a,$$

or

$$2\pi \bar{x} \frac{\pi a^2}{2} = \frac{4}{3}\pi a^3.$$

Consequently

$$\pi^2 a^2 \bar{x} = \frac{4}{3}\pi a^3$$

Compare with Example 3. and thus

$$\bar{x} = \frac{4\pi a^3}{3\pi^2 a^2} = \frac{4a}{3\pi}. \quad \blacksquare$$

TWO KEY FORMULAS IN THIS SECTION

Let R be the region below the curve $y = f(x)$ and above $[a, b]$. Then

$$\bar{x} = \frac{\int_a^b x f(x)\, dx}{\text{Area of } R} \quad \text{and} \quad \bar{y} = \frac{\int_a^b ((f(x))^2/2)\, dx}{\text{Area of } R}.$$

Exercises

In each of Exercises 1 to 6 find the moment of the given region R about the given line L.

1 R is bounded by $y = \sin x$ and the x axis and lies between $x = 0$ and $x = \pi$. L is the x axis.
2 R is bounded by $y = \cos x$ and the x axis and lies between $x = 0$ and $x = \pi/2$. L is the x axis.
3 R is the same as in Exercise 1. L is the y axis.
4 R is the same as in Exercise 2. L is the y axis.
5 R is the rectangle with vertices $(0, 0)$, $(a, 0)$, (a, b), $(0, b)$, where $a > 0$, $b > 0$. L is the x axis.
6 R is the same as in Exercise 5. L is the y axis.

In Exercises 7 to 18 find the centroids of the indicated regions.

7 The triangle bounded by $y = x$, $x = 1$, and the x axis.
8 The triangle bounded by the two axes and the line $x/4 + y/3 = 1$.
9 The triangle bounded by $y = x$, $y = 2x$, and $x = 2$.
10 The region bounded by $y = x^2$, the x axis, and $x = 1$.
11 The region bounded by $y = x^2$ and the line $y = 4$. (R lies in quadrants I and II.)
12 The region bounded by $y = e^x$ and the x axis, between the lines $x = 1$ and $x = 2$.
13 The region bounded by $y = \sin 2x$ and the x axis, between the lines $x = 0$ and $x = \pi/2$.
14 The region bounded by $y = \sqrt{1 + x^2}$ and the x axis, between the lines $x = 1$ and $x = 2$.
15 The region bounded by $y = \ln x$ and the x axis, between the lines $x = 1$ and $x = e$.
16 The region between $y = 1/\sqrt{x^2 - 1}$, between the lines $x = \sqrt{2}$ and $x = 2$.
17 The triangle whose vertices are $(0, 0)$, $(a, 0)$, $(0, b)$, where a and b are positive numbers.
18 The top half of the region within the ellipse $x^2/a^2 + y^2/b^2 = 1$.

Exercise 19 is used in Exercises 20 to 22.

19 Let f and g be continuous functions, such that $f(x) \geq g(x) \geq 0$ for x in $[a, b]$. Let R be the region above $[a, b]$ which is bounded by the curves $y = f(x)$ and $y = g(x)$.
  (a) Set up a definite integral (in terms of f and g) for the moment of R about the y axis.
  (b) Set up a definite integral with respect to x (in terms of f and g) for the moment of R about the x axis.

In Exercises 20 to 22 find (a) the moment of the given region R about the y axis, (b) the moment of R about the x axis, (c) the area of R, (d) $\bar{x}$, (e) $\bar{y}$.

494 APPLICATIONS OF THE DEFINITE INTEGRAL

20 R is bounded by the curves $y = x^2$ and $y = x^3$, between $x = 0$ and $x = 1$.

21 R is bounded by the curves $y = 3^x$ and $y = 2^x$, between $x = 1$ and $x = 2$.

22 R is bounded by the curves $y = x - 1$ and $y = \ln x$, between $x = 1$ and $x = e$.

23 In a letter of 1680 Leibniz wrote:

> Huygens, as soon as he had published his book on the pendulum, gave me a copy of it; and at that time I was quite ignorant of Cartesian algebra and also of the method of indivisibles, indeed I did not know the correct definition of the center of gravity. For, when by chance I spoke of it to Huygens, I let him know that I thought that a straight line drawn through the center of gravity always cut a figure into two equal parts; since that clearly happened in the case of a square, or a circle, an ellipse, and other figures that have a center of magnitude, I imagined that it was the same for all other figures. Huygens laughed when he heard this, and told me that nothing was further from the truth. (Quoted in C. H. Edwards, The Historical Development of the Calculus, Springer-Verlag, New York, 1980, p. 239.)

Give an example showing that "nothing is further from the truth."

■ ■

24 Let a be a constant ≥ 1. Let R be the region below $y = x^a$, above the x axis, between the lines $x = 0$ and $x = 1$.
(a) Sketch R for large a.
(b) Compute the centroid $(\bar{x}, \bar{y})$ of R.
(c) Find $\lim_{a \to \infty} \bar{x}$ and $\lim_{a \to \infty} \bar{y}$.
(d) Show that for large a the centroid of R lies in R.

25 (Contributed by Jeff Lichtman) Let f and g be two continuous functions such that $f(x) \geq g(x) \geq 0$ for x in $[0, 1]$. Let R be the region under $y = f(x)$ and above $[0, 1]$; let R^* be the region under $y = g(x)$ and above $[0, 1]$.
(a) Do you think the center of mass of R is at least as high as the center of mass of R^*? (An opinion only.)
(b) Let $g(x) = x$. Define $f(x)$ to be $\frac{1}{3}$ for $0 \leq x \leq \frac{1}{3}$ and $f(x)$ to be x if $\frac{1}{3} \leq x \leq 1$. (Note that f is continuous.) Find $\bar{y}$ for R and also for R^*. (Which is larger?)

(c) Let a be a constant, $0 \leq a \leq 1$. Let $f(x) = a$ for $0 \leq x \leq a$ and let $f(x) = x$ for $a \leq x \leq 1$. Find $\bar{y}$ for R.
(d) Show that the number a for which $\bar{y}$ defined in part (c) is a minimum is a root of the equation $x^3 + 3x - 1 = 0$.
(e) Show that the equation in (d) has only one real root q. (It is approximately 0.32219.)
(f) For the functions described in (c), how small can the corresponding $\bar{y}$ be? Lichtman showed that if $f(x) \geq x$ for x in $[0, 1]$, then $\bar{y}$ of the corresponding R is at least q.

26 (Using the centroid to compute the force of water against a surface) Let R be a submerged portion of the vertical wall of a pool.
(a) Express the force of the water against R as a definite integral. *Suggestion:* Review Sec. 8.3 and assume that each horizontal line in the plane of R meets R in a line segment or a point.
(b) Using (a), show that the force of the water against R is equal to the pressure at the centroid of R times the area of R.

27 Cut an irregular shape out of cardboard and find three balancing lines for it experimentally. Are they concurrent, that is, do they pass through a common point?

28 This exercise shows that the three medians of a triangle meet at the centroid of the triangle. (A median of a triangle is a line that passes through a vertex and the midpoint of the opposite edge.)

Let R be a triangle with vertices A, B, and C. It suffices to show that the centroid of R lies on the median through C and the midpoint M of the edge AB. Introduce an xy coordinate system such that the origin is at A and B lies on the x axis, as in Fig. 8.93.
(a) Compute $(\bar{x}, \bar{y})$. (b) Find the equation of the median through C and M. (c) Verify that the centroid lies on the median computed in (b).

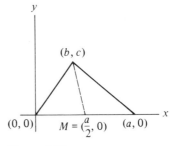

Figure 8.93

8.6 Improper integrals

Consider the volume of the solid obtained by revolving about the x axis the region bordered by $y = 1/x$ and the x axis, to the right of $x = 1$, as shown in

Fig. 8.94. The typical cross section made by a plane perpendicular to the x

What is the volume of this solid?

Figure 8.94

axis is a circle of radius $1/x$. We might therefore be tempted to say that the volume is $\int_1^\infty \pi(1/x)^2 \, dx$. Unfortunately, the symbol $\int_a^\infty f(x) \, dx$ has not been given any meaning so far in this book. The definition of the definite integral involves sums of the form $\sum_{i=1}^n f(c_i)(x_i - x_{i-1})$. If a section in the partition has infinite length, such a sum is meaningless.

It does make sense, however, to examine the volume of that part of the solid from $x = 1$ to $x = b$, where b is some number greater than 1, and then to determine what happens to this volume as $b \to \infty$. In other words, consider $\lim_{b \to \infty} \int_1^b \pi(1/x)^2 \, dx$. Now,

$$\int_1^b \pi\left(\frac{1}{x}\right)^2 dx \underset{\text{FTC}}{=} -\frac{\pi}{x}\bigg|_1^b$$

$$= -\frac{\pi}{b} - \left(-\frac{\pi}{1}\right)$$

$$= \pi - \frac{\pi}{b}.$$

Thus $\lim_{b \to \infty} \int_1^b (\pi/x^2) \, dx = \pi - 0 = \pi$. The volume of the endless solid is finite.

This approach suggests a way to give meaning to the symbol $\int_a^\infty f(x) \, dx$.

IMPROPER INTEGRALS: INTERVAL OF INTEGRATION UNBOUNDED

DEFINITION *Convergent improper integral* $\int_a^\infty f(x) \, dx$. Let f be continuous for $x \geq a$. If $\lim_{b \to \infty} \int_a^b f(x) \, dx$ exists, the function f is said to have a convergent improper integral from a to ∞. The value of the limit is denoted by $\int_a^\infty f(x) \, dx$.

It was shown above that $\int_1^\infty \pi(1/x)^2 \, dx$ is a convergent improper integral, with value π.

DEFINITION *Divergent improper integral* $\int_a^\infty f(x) \, dx$. Let f be a continuous function. If $\lim_{b \to \infty} \int_a^b f(x) \, dx$ does not exist, the function f is said to have a divergent improper integral.

EXAMPLE 1 Determine the area of the region below $y = 1/x$, above the x axis, and to the right of $x = 1$.

496 APPLICATIONS OF THE DEFINITE INTEGRAL

SOLUTION The area in question is given by

$$\int_1^\infty \frac{1}{x}\,dx = \lim_{b\to\infty} \int_1^b \frac{1}{x}\,dx = \lim_{b\to\infty}(\ln b - \ln 1) = \lim_{b\to\infty} \ln b.$$

Divergence due to integral approaching ∞

How does $\ln b$ behave as b increases without bound? First of all, since $D(\ln t) = 1/t$, the function $\ln t$ is increasing. Moreover, since $\ln e = 1$, $\ln e^2 = 2, \ldots, \ln e^n = n$, it follows that $\ln b$ becomes arbitrarily large. Thus $\lim_{b\to\infty} \ln b = \infty$. These observations are summarized in the following statements: The area is infinite, $\int_1^\infty 1/x\,dx = \infty$ or, simply, $\int_1^\infty 1/x\,dx$ is a divergent improper integral. ∎

The improper integral $\int_1^\infty dx/x$ is divergent because $\int_1^b dx/x \to \infty$ as $b \to \infty$. But an improper integral $\int_a^\infty f(x)\,dx$ can be divergent without being infinite. Consider, for instance, $\int_0^\infty \cos x\,dx$. We have

$$\int_0^b \cos x\,dx = \sin x \Big|_0^b$$

$$= \sin b.$$

Divergence due to integral oscillating

As $b \to \infty$, $\sin b$ does not approach a limit nor does it become arbitrarily large. As $b \to \infty$, $\sin b$ just keeps going up and down in the range -1 to 1 infinitely often.

The improper integral $\int_{-\infty}^b f(x)\,dx$ is defined similarly, by considering

$$\int_a^b f(x)\,dx$$

for negative values of a of large absolute value. If

The improper integral

$$\int_{-\infty}^b f(x)\,dx$$

exists, it is denoted

$$\lim_{a\to -\infty} \int_a^b f(x)\,dx$$

$$\int_{-\infty}^b f(x)\,dx.$$

In such a case, the improper integral $\int_{-\infty}^b f(x)\,dx$ is said to be convergent. If

$$\lim_{a\to -\infty} \int_a^b f(x)\,dx$$

does not exist, then the improper integral

$$\int_{-\infty}^b f(x)\,dx$$

is said to be divergent.

The improper integral

$$\int_{-\infty}^\infty f(x)\,dx$$

To deal with improper integrals over the entire x axis, define

$$\int_{-\infty}^\infty f(x)\,dx$$

to be the sum
$$\int_{-\infty}^{0} f(x)\,dx + \int_{0}^{\infty} f(x)\,dx,$$

which will be called convergent if both

$$\int_{-\infty}^{0} f(x)\,dx \quad \text{and} \quad \int_{0}^{\infty} f(x)\,dx$$

are convergent. (If at least one of the two is divergent, $\int_{-\infty}^{\infty} f(x)\,dx$ will be called divergent.)

EXAMPLE 2 Determine the area of the region bounded by the curve $y = 1/(1 + x^2)$ and the x axis, as indicated in Fig. 8.95.

SOLUTION The area in question equals

$$\int_{-\infty}^{\infty} \frac{dx}{1 + x^2}.$$

Now,
$$\int_{0}^{\infty} \frac{dx}{1 + x^2} = \lim_{b \to \infty} \int_{0}^{b} \frac{dx}{1 + x^2}$$

$$\underset{\text{FTC}}{=} \lim_{b \to \infty} (\tan^{-1} b - \tan^{-1} 0)$$

$$= \frac{\pi}{2}.$$

By symmetry,
$$\int_{-\infty}^{0} \frac{dx}{1 + x^2} = \frac{\pi}{2}.$$

Hence
$$\int_{-\infty}^{\infty} \frac{dx}{1 + x^2} = \frac{\pi}{2} + \frac{\pi}{2},$$

and the area in question is π. ∎

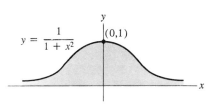

Figure 8.95

Sometimes $\int_{a}^{\infty} f(x)\,dx$ can be shown to be convergent by comparing it to another improper integral $\int_{a}^{\infty} g(x)\,dx$. (See comment before Exercise 38.)

IMPROPER INTEGRALS: INTEGRAND UNBOUNDED

There is a second type of improper integral, in which the function is unbounded in the interval $[a, b]$. If $f(x)$ becomes arbitrarily large in the interval $[a, b]$, then it is possible to have arbitrarily large approximating sums $\sum_{i=1}^{n} f(c_i)(x_i - x_{i-1})$ no matter how fine the partition may be by choosing a c_i that makes $f(c_i)$ large. The next example shows how to get around this difficulty.

EXAMPLE 3 Determine the area of the region bounded by $y = 1/\sqrt{x}$, $x = 1$, and the coordinate axes shown in Fig. 8.96.

SOLUTION Resist for the moment the temptation to write that Area = $\int_{0}^{1} 1/\sqrt{x}\,dx$, for $\int_{0}^{1} 1/\sqrt{x}\,dx$ does not exist according to the definition of the definite integral given in Chap. 5, since its integrand is unbounded in [0, 1]. (Note also that the integrand is not defined at 0.) Instead consider the behavior of $\int_t^1 1/\sqrt{x}\, dx$ as t approaches 0 from the right. Since

$$\int_t^1 \frac{1}{\sqrt{x}}\, dx = 2\sqrt{x}\, \Big|_t^1$$
$$= 2\sqrt{1} - 2\sqrt{t}$$
$$= 2(1 - \sqrt{t}),$$

it follows that

$$\lim_{t \to 0^+} \int_t^1 \frac{dx}{\sqrt{x}} = 2.$$

The area in question is said to be 2.

The reader should check and see that this is the same value for the area that can be obtained by taking horizontal cross sections and evaluating an improper integral from 0 to ∞. ∎

The reasoning in Example 3 motivates the definition of the second type of improper integral, in which the function rather than the interval is unbounded.

DEFINITION *Convergent and divergent improper integrals $\int_a^b f(x)\, dx$.* Let f be continuous at every number in $[a, b]$ except a, and become arbitrarily large for values in (a, b). If $\lim_{t \to a^+} \int_t^b f(x)\, dx$ exists, the function f is said to have a convergent improper integral from a to b. The value of the limit is denoted $\int_a^b f(x)\, dx$. If $\lim_{t \to a^+} \int_t^b f(x)\, dx$ does not exist, the function f is said to have a divergent improper integral from a to b; in brief, $\int_a^b (x)\, dx$ is not defined.

In a similar manner, if f is unbounded only near b, define $\int_a^b f(x)\, dx$ as $\lim_{t \to b^-} \int_a^t f(x)\, dx$.

Example 3 is summarized in the statement, "The improper integral $\int_0^1 1/\sqrt{x}\, dx$ is convergent and has the value 2."

It may happen that a function behaves well everywhere in the interval $[a, b]$ except at the number c, distinct from a and b, where it may be infinite. In that case $\int_a^b f(x)\, dx$ makes no sense. In such a case consider $\int_a^c f(x)\, dx$ and $\int_c^b f(x)\, dx$. If both exist, then the integral $\int_a^b f(x)\, dx$ is said to be convergent and have the value $\int_a^c f(x)\, dx + \int_c^b f(x)\, dx$. More generally, if a function f has an infinite range of values as well as points where it becomes infinite, break the entire integral into the sum of integrals each of which has *only one* of the two basic "troubles," either an infinite range or an endpoint where the function is infinite. For instance, the improper integral $\int_{-\infty}^\infty 1/x^2\, dx$ is troublesome for four reasons: $\lim_{x \to 0^-} 1/x^2 = \infty$, $\lim_{x \to 0^+} 1/x^2 = \infty$, and the range extends infinitely to the left and also to the right. To treat the integral, write it as the sum of four improper integrals of the two basic types:

$$\int_{-\infty}^\infty \frac{1}{x^2}\, dx = \int_{-\infty}^{-1} \frac{1}{x^2}\, dx + \int_{-1}^0 \frac{1}{x^2}\, dx + \int_0^1 \frac{1}{x^2}\, dx + \int_1^\infty \frac{1}{x^2}\, dx.$$

All four of the integrals on the right have to be convergent for $\int_{-\infty}^{\infty} 1/x^2 \, dx$ to be convergent. As a matter of fact, only the first and last are. So $\int_{-\infty}^{\infty} 1/x^2 \, dx$ is divergent.

Just as substitution in a definite integral is valid as long as the same substitution is applied to the limits of integration, substitution in improper integrals is also permissible, as illustrated in Example 4.

EXAMPLE 4 Evaluate $\int_{0}^{\infty} \dfrac{dx}{x^2 + 9}$.

SOLUTION Make the substitution $x = 3u$; hence $dx = 3 \, du$. As x goes from 0 to ∞, $u = x/3$ also goes from 0 to ∞. Thus

$$\int_{0}^{\infty} \frac{dx}{x^2 + 9} = \int_{0}^{\infty} \frac{3 \, du}{9u^2 + 9} = \frac{1}{3} \int_{0}^{\infty} \frac{du}{u^2 + 1}.$$

By Example 2, $\int_{0}^{\infty} \dfrac{du}{u^2 + 1} = \dfrac{\pi}{2}$.

Thus $\int_{0}^{\infty} \dfrac{dx}{x^2 + 9} = \dfrac{1}{3} \left(\dfrac{\pi}{2} \right) = \dfrac{\pi}{6}$. ∎

EXAMPLE 5 Examine the improper integral $\int_{0}^{9} \dfrac{dx}{(x - 1)^{2/3}}$.

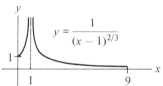

Figure 8.97

SOLUTION At $x = 1$ the integrand is undefined, and near $x = 1$ it is unbounded. It is necessary to examine the two improper integrals

$$\int_{0}^{1} \frac{dx}{(x - 1)^{2/3}} \quad \text{and} \quad \int_{1}^{9} \frac{dx}{(x - 1)^{2/3}}.$$

To treat $\int_{0}^{1} \dfrac{dx}{(x - 1)^{2/3}}$, consider $\lim\limits_{t \to 1^-} \int_{0}^{t} \dfrac{dx}{(x - 1)^{2/3}}$:

$$\int_{0}^{t} \frac{dx}{(x - 1)^{2/3}} = 3(x - 1)^{1/3} \Big|_{0}^{t}$$

$$= 3(t - 1)^{1/3} - 3(0 - 1)^{1/3}.$$

Thus $\lim\limits_{t \to 1^-} \int_{0}^{t} \dfrac{dx}{(x - 1)^{2/3}} = 0 + 3 = 3$.

Hence $\int_{0}^{1} \dfrac{dx}{(x - 1)^{2/3}} = 3$.

Next consider $\lim\limits_{t \to 1^+} \int_{t}^{9} \dfrac{dx}{(x - 1)^{2/3}}$:

500 APPLICATIONS OF THE DEFINITE INTEGRAL

$$\int_t^9 \frac{dx}{(x-1)^{2/3}} = 3(x-1)^{1/3}\Big|_t^9$$
$$= 3(9-1)^{1/3} - 3(t-1)^{1/3}.$$

Thus
$$\lim_{t \to 1^+} \int_t^9 \frac{dx}{(x-1)^{2/3}} = 3 \cdot 8^{1/3} = 6.$$

Hence $\int_0^9 \frac{dx}{(x-1)^{2/3}}$ is convergent and equals $3 + 6 = 9$. ∎

AN IMPROPER INTEGRAL IN ECONOMICS (OPTIONAL)

The final example illustrates the use of improper integrals in economics. Intended for enrichment and perspective, it is not essential for understanding the concept of improper integrals. The reader is invited to read it, if not the first time through the section, then perhaps when reviewing it.

EXAMPLE 6 *Present value of future income.* Both business and government frequently face the question: "What is 1 dollar t years in the future worth today?" Implicit in this question are such considerations as the present value of a business being dependent on its future profit and the cost of a dam being weighed against its future revenue. Determine the present value of a business whose rate of profit t years in the future is $f(t)$ dollars per year.

SOLUTION To begin the analysis, assume that the annual interest rate r remains constant and that 1 dollar deposited today is worth e^{rt} dollars t years from now. This assumption corresponds to continuously compounded interest or to natural growth. Thus A dollars today will be worth Ae^{rt} dollars t years from now. What is the present value of the promise of 1 dollar t years from now? In other words, what amount A invested today will be worth 1 dollar t years from now? To find out, solve the equation

$$Ae^{rt} = 1$$

The present value of $1 t years from now is $$e^{-rt}$.

for A. The solution is
$$A = e^{-rt}. \quad (1)$$

Now consider the present value of the future profit of a business (or future revenue of a dam). Assume that the profit flow t years from now is $f(t)$ dollars per year. This rate may vary within the year; consider f to be a continuous function of time. The profit in the small interval of time dt, from time t to time $t + dt$, would be approximately $f(t)\, dt$. The total future profit $F(T)$ from now, when $t = 0$, to some time T in the future is therefore

$$F(T) = \int_0^T f(t)\, dt. \quad (2)$$

But the *present value* of the future profit is *not* given by (2). It is necessary to consider the present value of the profit earned in a typical short interval of time from t to $t + dt$. According to (1) its present value is approximately

$$e^{-rt} f(t)\, dt$$

Hence the present value of future profit from $t = 0$ to $t = T$ is given by a definite integral:

$$\int_0^T e^{-rt} f(t)\, dt. \qquad (3)$$

The present value of all future profit is therefore the improper integral $\int_0^\infty e^{-rt} f(t)\, dt$.

To see what influence the interest rate r has, denote by $P(r)$ the present value of all future revenue when the interest rate is r; that is,

$$P(r) = \int_0^\infty e^{-rt} f(t)\, dt. \qquad (4)$$

If the interest rate r is raised, then according to (4) the present value of a business declines. An investor choosing between investing in a business or placing his money in a bank account finds the bank account more attractive when r is raised.

Laplace transform   Equation (4) assigns to a profit function f (which is a function of t) a present-value function P, which is a function of r, the interest rate. In the theory of differential equations P is called the *Laplace transform of f*. ∎

Exercises

Exercises 1 to 14 concern improper integrals in which the range of integration is unbounded. In each case determine whether the improper integral is convergent or divergent. Evaluate the convergent ones.

1. $\int_1^\infty \dfrac{dx}{x^3}$ 2. $\int_1^\infty \dfrac{dx}{\sqrt[3]{x}}$

3. $\int_1^\infty \dfrac{\ln x\, dx}{x}$ 4. $\int_0^\infty e^{-x}\, dx$

5. $\int_0^\infty \dfrac{dx}{x^2 + 4}$ 6. $\int_0^\infty \dfrac{dx}{x + 100}$

7. $\int_0^\infty \dfrac{x^3\, dx}{x^4 + 1}$ 8. $\int_{-\infty}^\infty \dfrac{x\, dx}{x^4 + 1}$

9. $\int_1^\infty x^{-1.01}\, dx$ 10. $\int_1^\infty x^{-0.99}\, dx$

11. $\int_0^\infty \dfrac{dx}{(x + 2)^3}$ 12. $\int_2^\infty \dfrac{dx}{x^2 - 1}$

13. $\int_0^\infty \sin 2x\, dx$ 14. $\int_0^\infty \sin^2 x\, dx$

Exercises 15 to 22 concern improper integrals in which the integrand is unbounded. In each case determine whether the improper integral is convergent or divergent. Evaluate the convergent ones.

15. $\int_0^1 x^{-1.01}\, dx$ 16. $\int_0^1 x^{-0.99}\, dx$

17. $\int_0^1 \dfrac{dx}{\sqrt{1 - x}}$ 18. $\int_{-1}^1 \dfrac{dx}{\sqrt[3]{x}}$

19. $\int_0^1 \ln x\, dx$ 20. $\int_0^{\pi/2} \cot x\, dx$

21. $\int_1^2 \dfrac{dx}{x\sqrt{x^2 - 1}}$ 22. $\int_7^{16} \dfrac{dx}{\sqrt[3]{x - 8}}$

∎

In Exercises 23 to 33 determine whether the improper integrals are convergent or divergent. Evaluate the convergent ones.

23. $\int_0^\infty e^{-x} \sin 3x\, dx$

24. $\int_0^\infty x^{-3}\, dx$ (Note that this integral is improper for two reasons. Consider $\int_0^1 dx/x^3$ and $\int_1^\infty dx/x^3$ separately.)

25. $\int_0^\infty \dfrac{dx}{\sqrt[3]{x - 2}}$ 26. $\int_1^\infty \dfrac{dx}{\sqrt{x - 1}}$

27. $\int_0^1 \dfrac{e^x\, dx}{e^x - 1}$ 28. $\int_0^\infty x \ln x\, dx$

29. $\int_0^2 \dfrac{dx}{x^2 - 1}$ 30. $\int_0^\infty \tan^{-1} x\, dx$

31 $\int_0^2 \dfrac{dx}{(x-1)^2}$ **32** $\int_{-\infty}^{\infty} \dfrac{dx}{x^2+2x+1}$

33 $\int_{-\infty}^{\infty} \dfrac{dx}{x^2+2x+2}$

34 Let R be the region between the curves $y = 1/x$ and $y = 1/(x+1)$ to the right of the line $x = 1$. Is the area of R finite or infinite? If it is finite, evaluate it.

35 Let R be the region between the curves $y = 1/x$ and $y = 1/x^2$ to the right of $x = 1$. Is the area of R finite or infinite? If it is finite, evaluate it.

36 Find the area of the region bounded by the curve $y = 1/(x^2 + 6x + 10)$ and the x axis, to the right of the y axis.

37 The following is an excerpt from an article on corporate investment:

> It follows, therefore, that the present value of the incremental stream of benefits resulting from the marginal investment can be described as follows if the maintenance of the stockholder's optimum share value is not to be disturbed at the margin:
>
> $$1 = \int_0^{\infty} (1-b)r'e^{brt}e^{-kt}\, dt, \qquad \text{(i)}$$
>
> or
>
> $$1 = \dfrac{r'(1-b)}{k - rb}. \qquad \text{(ii)}$$

(The constants b, k, r, and r' are positive and $br < k$.) Derive (ii) from (i).

Frequently one can determine whether an improper integral $\int_a^b f(x)\, dx$ is convergent even though f does not have an elementary antiderivative. For instance, if $0 \le f(x) \le g(x)$ and $\int_a^b g(x)\, dx$ is convergent, then it can be proved that $\int_a^b f(x)\, dx$ is convergent. Use this principle in Exercises 38 to 44.

38 Plankton are small football-shaped organisms. The resistance they meet when falling through water is proportional to the integral

$$\int_0^{\infty} \dfrac{dx}{\sqrt{(a^2+x)(b^2+x)(c^2+x)}},$$

where a, b, and c describe the dimensions of the plankton. Is this improper integral convergent or divergent?

39 The function $f(x) = (\sin x)/x$ for $x \ne 0$ and $f(0) = 1$ occurs in communication theory. Show that the energy E of the signal represented by f is finite, where

$$E = \int_{-\infty}^{\infty} [f(x)]^2\, dx.$$

40 Show that $\int_1^{\infty} e^{-x^2}\, dx$ is convergent by showing that it is smaller than $\int_1^{\infty} e^{-x}\, dx$.

41 Show that $\int_1^{\infty} 1/\sqrt{1+x^3}\, dx$ is convergent, but do not try to evaluate it.

42 Show that $\int_0^{\infty} e^{-t} t^{-1/2}\, dt$ is convergent.

43 The following quote is taken from an article on the energy stored in solar ponds:

> The effect of the free surface of the pond on the temperature at the bottom of the pond is given by an expression involving the integral
>
> $$\int_a^{\infty} \left(1 - \dfrac{k}{v^2}\right) e^{-v^2}\, dv.$$

Show that the integral is convergent. (The constants a and k are positive.)

44 In R. P. Feynman, *Lectures on Physics*, Addison-Wesley, Reading, Mass., 1963, appears this remark: "... the expression becomes

$$\dfrac{U}{V} = \dfrac{(kT)^4}{\hbar^3 \pi^2 c^3} \int_0^{\infty} \dfrac{x^3\, dx}{e^x - 1}.$$

This integral is just some number that we can get, approximately, by drawing a curve and taking the area by counting squares. It is roughly 6.5. The mathematicians among us can show that the integral is exactly $\pi^4/15$."

Show at least that the integral is convergent. Consider both $x \to \infty$ and $x \to 0$.

45 (The gamma function) For a real number $n > 0$ define $\Gamma(n)$ to be $\int_0^{\infty} e^{-x} x^{n-1}\, dx$.
(a) Evaluate $\Gamma(1)$.
(b) Show that $\Gamma(n+1) = n\Gamma(n)$.
(c) Using (a) and (b), evaluate $\Gamma(2)$, $\Gamma(3)$, $\Gamma(4)$, and $\Gamma(5)$.
(d) What is the relationship between $n!$ and $\Gamma(n)$?
The gamma function generalizes the factorial, which is defined only at integers, to all positive real numbers. Handbooks of mathematical tables usually include values of the gamma function.

46 Find the error in the following computations. The substitution $x = y^2$, $dx = 2y\, dy$, yields

$$\int_0^1 \dfrac{1}{x}\, dx = \int_0^1 \dfrac{2y}{y^2}\, dy = \int_0^1 \dfrac{2}{y}\, dy = 2\int_0^1 \dfrac{1}{y}\, dy = 2\int_0^1 \dfrac{1}{x}\, dx.$$

Hence

$$\int_0^1 \dfrac{1}{x}\, dx = 2\int_0^1 \dfrac{1}{x}\, dx,$$

from which it follows that $\int_0^1 dx/x = 0$.

47 Find the error in the following computations. Using the substitution $u = 1/x$, $du = -1/x^2\, dx$, we have

$$\int_{-1}^1 \dfrac{1}{1+x^2}\, dx = \int_{-1}^1 \dfrac{1}{1 + 1/u^2}\left(-\dfrac{1}{u^2}\, du\right)$$

$$= -\int_{-1}^1 \dfrac{1}{1+u^2}\, du.$$

Thus $\int_{-1}^1 1/(1+x^2)\, dx$, being equal to its negative, is 0.

48 Show that, if $\int_{-\infty}^{\infty} f(x)\,dx$ is convergent, it equals $\lim_{L\to\infty} \int_{-L}^{L} f(x)\,dx$, but that $\lim_{L\to\infty} \int_{-L}^{L} f(x)\,dx$ may be finite while $\int_{-\infty}^{\infty} f(x)\,dx$ is a divergent improper integral.

49 If the profit flow in Example 6 remains constant, say $f(t) = k > 0$, the total future profit is obviously infinite. Show that the present value is k/r, which is finite.

Let $f(t)$ be a continuous function defined for $t \geq 0$. Assume that, for certain fixed positive numbers r, $\int_0^\infty e^{-rt} f(t)\,dt$ converges and that $e^{-rt} f(t) \to 0$ as $t \to \infty$. Define $P(r)$ to be $\int_0^\infty e^{-rt} f(t)\,dt$. The function P is called the Laplace transform of the function f. (For its economic interpretation see Example 6.) In Exercises 50 to 53 find the Laplace transforms of the given functions.

50 $f(t) = t$ **51** $f(t) = e^t$ (assume $r > 1$)

52 $f(t) = \sin t$ **53** $f(t) = \cos t$

54 Let f and its derivative f' both have Laplace transforms. Let P be the Laplace transform of f and let Q be the Laplace transform of f'. Show that

$$Q(r) = -f(0) + rP(r).$$

55 Let P be the Laplace transform of f. Let a be a positive constant and let $g(t) = f(at)$. Let P be the Laplace transform of f and let Q be the Laplace transform of g. Show that $Q(r) = (1/a)P(r/a)$.

8.7 Work

No work is done if the object is moved horizontally (neglecting friction).

The work required to raise a weight of W pounds a distance D feet is defined to be $W \times D$ foot-pounds. An elevator that lifts a 150-pound person 100 feet thus accomplishes

$$150 \times 100 \text{ foot-pounds,}$$

or 15,000 foot-pounds of work.

When all parts of an object are lifted the same distance, the work is simply the product of two numbers. We now pose a problem in which different parts of an object are lifted different distances.

A tank is filled with water, which weighs 62.4 pounds per cubic foot. The water is pumped out an outlet which is above the level of the water. How much work is accomplished in emptying the tank? See Fig. 8.98. Water at the bottom of the tank must be pumped farther than the water at the top: The lower the water is in the tank, the farther it has to be raised.

To treat this problem mathematically introduce a vertical x axis with positive part below the origin as in Fig. 8.99. The plane perpendicular to

Tank full of water to be emptied

Figure 8.98

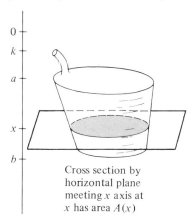

Cross section by horizontal plane meeting x axis at x has area $A(x)$

Figure 8.99

the x axis at level x has area $A(x)$. (The formula for $A(x)$ depends on the shape of the tank and where the origin of the x axis is placed.) Assume that the tank extends from level $x = a$ to level $x = b$, $b > a$, and that the level of the outlet is k. Observe that water at level x is lifted a distance $x - k$ feet.

To find out how much work is accomplished in emptying the tank, consider a thin horizontal slab of water (all the water in the layer is raised about the same distance). The layer consists of all water with x coordinate between x and $x + dx$, where dx is a small positive number. (See Fig. 8.100.)

The side of this slab is usually not vertical.

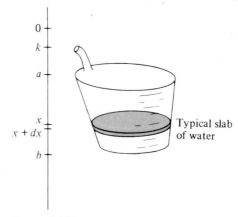

Figure 8.100

Approximate this thin layer by a cylinder whose height is dx and whose base is determined by the cross section at level x and thus has area $A(x)$ square feet.

The side of this slab is vertical.

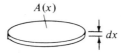

Cylindrical approximation to slab in Fig. 8.100

Figure 8.101

The volume of the cylindrical slab in Fig. 8.101 is $A(x)\, dx$ cubic feet. The work needed to raise the slab is approximately

$$\underbrace{\underbrace{A(x)\, dx}_{\text{volume}} \quad \underbrace{62.4}_{\text{density}}}_{\text{weight}} \quad \underbrace{(x - k)}_{\substack{\text{distance} \\ \text{raised}}} \quad \text{foot-pounds.}$$

Thus the local approximation to the work accomplished is

$$62.4(x - k)A(x)\, dx \quad \text{foot-pounds.}$$

Using the informal approach to setting up definite integrals, we conclude that the total work in emptying the tank is

8.7 WORK

Main formula of this section

$$\text{Total work} = 62.4 \int_a^b (x - k)A(x)\, dx \quad \text{foot-pounds.} \tag{1}$$

EXAMPLE 1 A tank has the shape of a right circular cylinder of radius a and height h, as shown in Fig. 8.102. The outlet is at the top of the tank, which is full of water. (See Fig. 8.102.) How much work is required to empty the tank?

SOLUTION Choose the origin of the x axis to be level with the center of the circular base of the cylinder. The cross section of the tank by a plane at level x is the rectangle shown in Fig. 8.103.

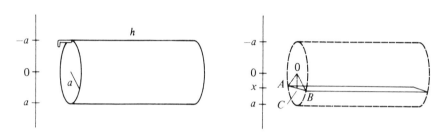

Figure 8.102   **Figure 8.103**

One side of this rectangle has length h. The length of the other side AB is found with the aid of the Pythagorean theorem:

OCB is a right triangle.

$$\left(\frac{\overline{AB}}{2}\right)^2 + x^2 = a^2,$$

so

$$\overline{AB} = 2\sqrt{a^2 - x^2}.$$

Thus $A(x) = 2\sqrt{a^2 - x^2}\, h$. By (1), the total work is

Taking the constant h past the integral sign.

$$62.4h \int_{-a}^{a} (x - (-a)) 2\sqrt{a^2 - x^2}\, dx \quad \text{foot-pounds.} \tag{2}$$

The integral in (2) can be evaluated by the substitution $x = a \sin \theta$, but the following shortcut, which is often of use, is preferable. Note that

$$\int_{-a}^{a} (x - (-a)) 2\sqrt{a^2 - x^2}\, dx = 2 \int_{-a}^{a} x\sqrt{a^2 - x^2}\, dx + 2a \int_{-a}^{a} \sqrt{a^2 - x^2}\, dx. \tag{3}$$

Since $x\sqrt{a^2 - x^2}$ is an odd function, $\int_{-a}^{a} x\sqrt{a^2 - x^2}\, dx = 0$. Since $\int_{-a}^{a} \sqrt{a^2 - x^2}\, dx$ is the area of the top half of a disk of radius a, it equals $\pi a^2/2$. So (3) reduces to πa^3. The work done is $62.4\, \pi a^3$ foot-pounds. ∎

CYLINDRICAL TANKS AND THE CENTROID

A water tank is built in the form of a cylinder of length h feet and base R, a plane region (not necessarily a disk). (See Fig. 8.104.) It is full of water.

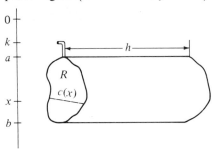

Figure 8.104

R could be a disk, a triangle, a square, etc.

Let the cross section of R at level x have length $c(x)$ feet. By (1), the work accomplished in emptying the tank out the outlet is

$$62.4 \int_a^b (x-k)c(x)h \, dx \qquad \text{foot-pounds.} \qquad (4)$$

In this case, $A(x) = c(x)h$, the area of a rectangle.

If the centroid of R is known, it is not necessary to use (4). It turns out that the work required is the same as if all the water were at the depth of the centroid of R. This fact is the substance of the following theorem.

The centroid serves as the "typical" point.

THEOREM The work required to pump the water out of a full cylindrical tank of the type described is the product

$$W \cdot D \qquad \text{foot-pounds,}$$

where W is the total weight of water in the tank and D is the distance the water at the centroid of R is lifted.

PROOF Place the x axis with its positive part directed downward and with its origin at the outlet. The total work is then

$$62.4h \int_a^b xc(x) \, dx \qquad \text{foot-pounds.}$$

Recall the formula for $\bar{x}$. Since

$$\bar{x} = \frac{\int_a^b xc(x) \, dx}{\text{Area of } R},$$

this equals $\qquad 62.4h\bar{x} \cdot \text{Area of } R \qquad$ foot-pounds.

But $\qquad 62.4h \cdot \text{Area of } R$

is the weight W of water in the tank. Hence

$$\text{Work} = W \cdot \bar{x}.$$

Since $\bar{x}$ is the distance the water at the centroid is lifted, this proves the theorem. ∎

EXAMPLE 2 A tank has a semicircular base of radius a feet and length h feet, as shown in Fig. 8.105. If the tank is full of water, how much work is required to empty it over its rim?

SOLUTION By Example 3 in Sec. 8.5, the depth of the centroid of the disk is $4a/3\pi$ feet. The weight of the water in the tank is $62.4h\pi a^2/2$ pounds. By the theorem, the total work required to empty the tank is

$$\underbrace{\frac{4a}{3\pi}}_{\substack{\text{depth}\\\text{of centroid}}} \underbrace{62.4\pi \frac{a^2 h}{2}}_{\substack{\text{weight of}\\\text{water}}} \quad \text{foot-pounds}$$

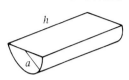

Figure 8.105

or
$$62.4 \frac{2a^3 h}{3} \quad \text{foot-pounds.} \quad \blacksquare$$

WORK IF FORCE OF GRAVITY VARIES

In calculating the work done in emptying a tank we assumed that the force of gravity does not depend on the height of the object. In fact, the farther an object is from the earth, the less it weighs. (Its mass remains constant, but its weight diminishes.) Thus the work required to raise an object 1 foot at sea level is greater than the work required to raise the same object 1 foot at the top of a mountain. However, the difference in altitudes is so small in comparison to the radius of the earth that the difference in work is negligible. On the other hand, when an object is rocketed into space, the fact that the force of gravity diminishes with distance from the center of the earth is critical. Example 3 shows why.

EXAMPLE 3 How much work is required to lift a 1-pound payload from the surface of the earth to "infinity"? If this work should turn out to be infinite, then it would require an infinite amount of fuel to send rockets off on unlimited orbits. Fortunately, it is finite, as will now be shown.

SOLUTION The work W necessary to lift an object a distance x against a constant vertical force F is the product of force times distance,

$$W = F \cdot x.$$

Since the gravitational pull of the earth on the payload *changes* with distance from the earth, an (improper) integral will be needed to express the total work required to lift the load to "infinity."

The payload weighs 1 pound at the surface of the earth. The farther it is from the center of the earth, the less it weighs, for the force of the earth on the mass is inversely proportional to the square of the distance of the mass from the center of the earth. Thus the force on the payload is given by k/r^2 pounds, where k is a constant, which will be determined in a moment, and r is the distance in miles from the payload to the center of the earth. When $r = 4000$ (miles), the force is 1 pound; thus

$$1 = \frac{k}{4000^2}.$$

From this it follows that $k = 4000^2$, and therefore the gravitational force on a 1-pound mass is, in general, $(4000/r)^2$ pounds. As the payload recedes

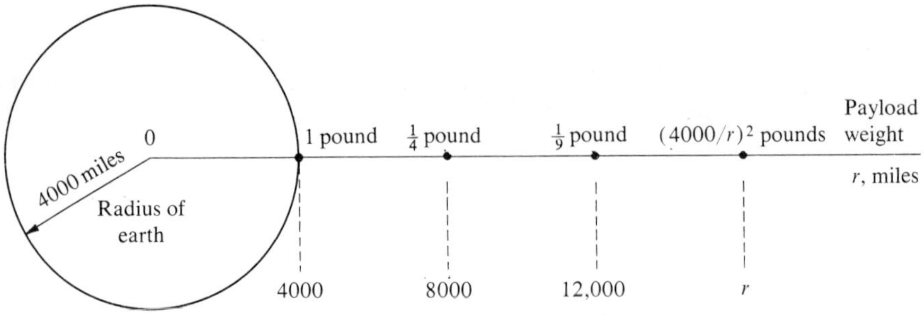

508 APPLICATIONS OF THE DEFINITE INTEGRAL

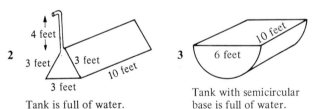

Figure 8.106

from the earth, it loses weight (but not mass), as recorded in Fig. 8.106. The work done in lifting the payload from point r to point $r + dr$ is approximately

$$\underbrace{\left(\frac{4000}{r}\right)^2}_{\text{force}} \underbrace{(dr)}_{\text{distance}} \quad \text{mile-pounds.}$$

Hence the total work required to move the 1-pound mass from the surface of the earth to infinity is given by the improper integral $\int_{4000}^{\infty} (4000/r)^2 \, dr$.

Now, $\int_{4000}^{\infty} \left(\frac{4000}{r}\right)^2 dr = \lim_{b \to \infty} \int_{4000}^{b} \left(\frac{4000}{r}\right)^2 dr = \lim_{b \to \infty} \left.\frac{-4000^2}{r}\right|_{4000}^{b}$

$$= \lim_{b \to \infty} \left(\frac{-4000^2}{b} + \frac{4000^2}{4000}\right) = 4000 \text{ mile-pounds.}$$

The total work is finite because the improper integral is convergent. It is just as if the payload were lifted 4000 miles against a constant gravitational force equal to that at the surface of the earth. ∎

Exercises

In Exercises 1 to 6, find the work required to pump the water out of the tanks shown in Figs. 8.107 to 8.112, either out the spigot, if there is one, or over the rim of the tank.

1

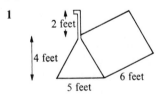

Tank with triangular base is full of water.

Figure 8.107

2 Tank is full of water.

Figure 8.108

3 Tank with semicircular base is full of water.

Figure 8.109

4

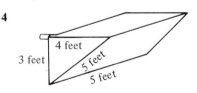

Tank is full of water.

Figure 8.110

5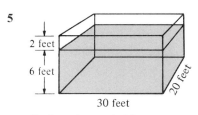

Surface of water is 2 feet below top of tank.

Figure 8.111

6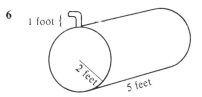

Tank with circular base is full of water.

Figure 8.112

Exercises 7 to 10 refer to Example 3.

7 How much work is done in lifting the 1-pound payload the first 4000 miles of its journey to infinity?

8 (a) If the force of gravity were of the form $k/r^{1.01}$, could a payload be sent to infinity? (b) If the force of gravity were of the form k/r, could a payload be sent to infinity?

9 Assume that the force of gravity obeys an inverse cube law, so that the force on a 1-pound mass a distance r miles from the center of the earth ($r \geq 4000$) is $(4000/r)^3$ pounds. How much work would be required to lift a 1-pound payload from the surface of the earth to "infinity"?

10 If a mass which weighs 1 pound at the surface of the earth were launched from a position 20,000 miles from the center of the earth, how much work would be required to send it to "infinity"?

11 Find the work required to pump the water out of the cylindrical tank with an isosceles trapezoidal base, pictured in Fig. 8.113, through the outlet.

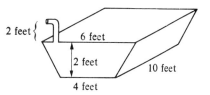

Figure 8.113

12 Use the theorem in this section to solve Example 1.

13 Find the work required to pump the water in a hemispherical tank of radius 3 feet over its side. (The equator of the tank is horizontal and is the top of the tank.)

14 Find the work required to pump the water in a conical tank over its top. The axis of the tank is vertical and its top is a circle of radius 3 feet; its height is 5 feet.

15 The force required to hold one end of a spring stretched x feet from its equilibrium position is proportional to x. The force is of the form kx for some positive constant k.

(a) How much work is required to stretch the string the first foot from its equilibrium position?

(b) How much work is required to stretch it the next foot?

■ ■

16 Geologists, when considering the origin of mountain ranges, estimate the energy required to lift a mountain up from sea level. Assume that two mountains are composed of the same type of matter, which weighs k pounds per cubic foot. Both are right circular cones in which the height is equal to the radius. One mountain is twice as high as the other. The base of each is at sea level. If the work required to lift the matter in the smaller mountain above sea level is W, what is the corresponding work for the larger mountain?

8.8 The average of a function over an interval

The average or, more precisely, the arithmetic mean of n numbers is simply their sum divided by n. Thus the average of 2, 3, and 7 is $(2 + 3 + 7)/3$, or 4. How can we define the average of the function x^2, for x in the interval $[0, 3]$? It makes no sense to find the sum of the squares of all the numbers from 0 to 3 and then divide by the number of those quantities, since the interval $[0, 3]$ contains an infinite set of numbers.

To define the average of a function f over an interval $[a, b]$, proceed as follows. Pick an integer n and then n equally spaced points in the interval

510 APPLICATIONS OF THE DEFINITE INTEGRAL

$[a, b]$. Call these points $x_1, x_2, \ldots, x_n$, with $x_1 = a + (b-a)/n$ and $x_n = b$, as shown in Fig. 8.114.

Figure 8.114

Then
$$\frac{f(x_1) + f(x_2) + \cdots + f(x_n)}{n} \qquad (1)$$

would be a reasonable estimate of the average value of f over $[a, b]$. And, as n increases, this quotient becomes a better estimate of the average.

The sum (1) resembles the approximating sum for a definite integral. A little algebra relates it to such a sum, as follows: first of all,

$$\frac{f(x_1) + f(x_2) + \cdots + f(x_n)}{n} = \frac{1}{b-a}\left[f(x_1)\frac{b-a}{n} + f(x_2)\frac{b-a}{n} + \cdots + f(x_n)\frac{b-a}{n}\right]. \qquad (2)$$

To make the relation between (1) or (2) and $\int_a^b f(x)\, dx$ more evident, note that, for each $i = 1, 2, \ldots, n$ (if x_0 is taken to be a),

$$x_i - x_{i-1} = \frac{b-a}{n}.$$

Hence the right side of (2) is equal to

$$\frac{1}{b-a}[f(x_1)(x_1 - x_0) + f(x_2)(x_2 - x_1) + \cdots + f(x_n)(x_n - x_{n-1})]. \qquad (3)$$

The bracketed expression in (3) is an approximation of $\int_a^b f(x)\, dx$. This observation suggests the following definition of the average of f over $[a, b]$ in terms of the definite integral.

DEFINITION *Average value of a function over an interval.* The average value of f over $[a, b]$ is the quotient

$$\frac{\int_a^b f(x)\, dx}{b-a}.$$

If $f(x)$ is positive for x in $[a, b]$, there is a simple geometric interpretation of the average of the function over the interval. Call the average A; then

$$A = \frac{\int_a^b f(x)\, dx}{b-a} \qquad \text{or} \qquad A(b-a) = \int_a^b f(x)\, dx.$$

Now,
$$\int_a^b f(x)\, dx$$

8.8 THE AVERAGE OF A FUNCTION OVER AN INTERVAL

is the area of the region below the graph of f and above the interval $[a, b]$. The equation

$$A(b - a) = \int_a^b f(x)\, dx$$

asserts that A, the average value of the function, is the height of a rectangle whose base is $(b - a)$ and whose area is equal to the area of the region under the graph of f, as shown in Fig. 8.115.

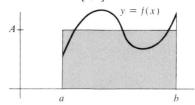

The height of the shaded rectangle is the average value A of f over $[a, b]$, and the area is the same as the area of the region below the curve $y = f(x)$ and above $[a, b]$.

Figure 8.115

EXAMPLE 1 Compute the average value of $\sin x$ for x in the interval $[0, \pi]$.

SOLUTION According to the definition of the average value of a function, this average is

$$\frac{\int_0^\pi \sin x\, dx}{\pi - 0}.$$

Now, $\displaystyle\int_0^\pi \sin x\, dx \underset{\text{FTC}}{=} -\cos x \Big|_0^\pi$

$$= (-\cos \pi) - (-\cos 0)$$
$$= -(-1) - (-1)$$
$$= 1 + 1 = 2.$$

Hence the average value of $\sin x$ for x in $[0, \pi]$ is

$$\frac{2}{\pi},$$

which is about 0.64. ∎

Time average versus distance average

The next two examples show that "average" means "average of a function." It is not enough to ask, "What is the average velocity?" We must ask, "What is the average of velocity with respect to time?" or "What is the average of velocity with respect to distance?" These averages are

512 APPLICATIONS OF THE DEFINITE INTEGRAL

sometimes called the *time average* and the *distance average*, respectively. The answers will usually be different.

EXAMPLE 2 If we travel 30 miles per hour for 1 hour and then 50 miles per hour for another hour, what is the average of our velocity with respect to time?

SOLUTION Denote the velocity at time t by $f(t)$. The average velocity with respect to time is defined as

$$\frac{\int_0^2 f(t)\,dt}{2 - 0}.$$

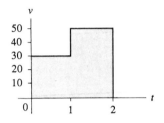

Figure 8.116

The simplest way to compute $\int_0^2 f(t)\,dt$ is to interpret it as the area of the shaded region shown in Fig. 8.116. Since the shaded region has area 80, $\int_0^2 f(t)\,dt = 80$. Thus

$$\text{Average velocity with respect to time} = \frac{\int_0^2 f(t)\,dt}{2 - 0}$$

$$= \frac{80}{2} = 40 \text{ miles per hour.} \quad \blacksquare$$

The notion of average velocity with respect to time coincides with the ordinary notion of average velocity, which is defined as total change in position divided by total time. To see this, consider for simplicity the case in which the velocity $v(t)$ is positive. Then

$$\text{Average velocity with respect to time} = \frac{\int_a^b v(t)\,dt}{b - a} = \frac{\text{Total distance}}{\text{Total time}}.$$

EXAMPLE 3 If we travel 30 miles per hour for 30 miles and then 50 miles per hour for another 50 miles, what is the average of our velocity with respect to distance? (Note that this is the same trip as in Example 2.)

SOLUTION Now the velocity is considered as a function of distance. Denote the velocity after x miles of travel by $g(x)$. The average with respect to distance is therefore

$$\frac{\int_0^{80} g(x)\,dx}{80 - 0}.$$

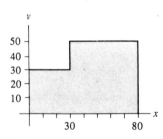

Figure 8.117

To compute the definite integral, interpret it as the area of the shaded region in Fig. 8.117. The area is $30 \cdot 30 + 50 \cdot 50 = 3400$. Therefore,

$$\text{Average velocity with respect to distance} = \frac{\int_0^{80} g(x)\,dx}{80 - 0}$$

$$= \frac{3400}{80} = 42.5 \text{ miles per hour.} \quad \blacksquare$$

Which average is more sensitive to a high velocity maintained during a short period of time?

The reader should pause and explain to his or her own satisfaction why the average with respect to distance ought to be larger than the average with respect to time; Exercises 21 and 22 will answer this question formally, but a practical qualitative explanation is more valuable.

The moral of these two examples is this: When speaking of the average value of a quantity, it is necessary to indicate the variable with respect to which the average is to be computed.

Exercises

In each of Exercises 1 to 12 find the average of the given function over the given interval.

1. x^2, $[-1, 3]$
2. $\sin x$, $[0, 2\pi]$
3. $\ln x$, $[1, e]$
4. $\dfrac{1}{1+x^2}$, $[0, 1]$
5. $\dfrac{1}{1+x}$, $[1, 2]$
6. $\sin^2 x$, $[0, \pi/2]$
7. $\dfrac{1}{x+x^2}$, $[2, 3]$
8. $\dfrac{1}{x^2-x}$, $[2, 3]$
9. $\sqrt{1-x^2}$, $[0, 1]$
10. $\dfrac{1}{x\sqrt{x^2-1}}$, $[\sqrt{2}, 2]$
11. $\tan^{-1} x$, $[0, 1]$
12. xe^{x^2}, $[1, 2]$

13. A person travels 20 miles per hour for $\frac{1}{2}$ hour, 30 miles per hour for 2 hours, and 40 miles per hour for $\frac{1}{2}$ hour.
    (a) Compute the average of velocity with respect to time.
    (b) Compute the average of velocity with respect to distance.

14. A person travels 10 miles per hour for 1 hour, 40 miles per hour for 1 hour, and 70 miles per hour for 1 hour.
    (a) Compute the average of velocity with respect to time.
    (b) Compute the average of velocity with respect to distance.

15. In the first t seconds a falling body drops $16t^2$ feet. Let its position relative to the y axis at time t be $16t^2$. (We aim the positive part of the axis downward.)
    (a) Compute velocity as a function of time. (Note that it is positive.)
    (b) Compute velocity as a function of y.
    (c) Find the average of velocity with respect to time during the first t seconds.
    (d) Find the average of velocity with respect to distance during the first t seconds.

16. Find the average length of the vertical cross section AB of the quadrant of a circle of radius r pictured in Fig. 8.118

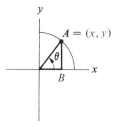

Figure 8.118

    (a) as a function of x,
    (b) as a function of θ.
    (c) Why would you expect the answer to (b) to be less than the answer to (a)?

17. Find the average value of $\sin x^5$ for x in $[-\pi/2, \pi/2]$.
18. Find the average value of $x \cos x^5$ for x in $[-\pi/2, \pi/2]$.
19. Assume that $\lim_{x \to \infty} f(x) = 3$. What happens to the average value of $f(x)$ over $[0, b]$ as $b \to \infty$? Explain.
20. A certain function f is defined throughout $[0, 2]$, but only limited data concerning it are available from experiments. It is known that $f(0) = 3$, $f(\tfrac{1}{4}) = \tfrac{7}{2}$, $f(\tfrac{1}{2}) = 4$, $f(1) = 6$, $f(2) = 8$. What is a sensible estimate of the average value of f over $[0, 2]$?

■ ■

21. This exercise obtains a famous inequality, known as the Schwarz inequality:

$$\int_A^B f(x)g(x)\, dx \le \left\{ \int_A^B [f(x)]^2\, dx \right\}^{1/2} \left\{ \int_A^B [g(x)]^2\, dx \right\}^{1/2}.$$

(a) Prove that if the quadratic polynomial $at^2 + bt + c = 0$ has at most one (real) root, then $b^2 - 4ac \leq 0$. (When does it have exactly one root?)

(b) Let f and g be continuous functions. Define a third function h as follows:

$$h(t) = \int_A^B [tf(x) - g(x)]^2 \, dx.$$

Show that h is a quadratic polynomial in t, $at^2 + bt + c$.

(c) Express the coefficients a, b, and c of h in terms of f and g.

(d) Combining (a), (b), and (c), derive the Schwarz inequality.

(e) When does equality occur in the Schwarz inequality?

In Exercise 22 it is shown that the average of velocity with respect to time is less than or equal to the average of velocity with respect to distance.

22 Assume for convenience that the velocity $v = dx/dt$ is positive, hence that speed equals velocity. Assume that from time a to time b the object moves from x_1 to x_2 on the x axis. Let v as a function of time be given by $v = f(t)$; let v as a function of distance be given by $v = g(x)$.

(a) Show that the average of velocity with respect to distance equals

$$\frac{\int_{x_1}^{x_2} g(x)\, dx}{\int_a^b f(t)\, dt}.$$

(b) Show that $\int_{x_1}^{x_2} g(x)\, dx = \int_a^b [f(t)]^2 \, dt$.

(c) Using the Schwarz inequality from Exercise 21, show that the average of velocity with respect to time is less than or equal to the average of velocity with respect to distance.

23 Let f be a continuous function such that $f(x)$ is always positive. Prove that

$$\left[\int_a^b f(x)\, dx\right] \left[\int_a^b \frac{1}{f(x)}\, dx\right] \geq (b-a)^2.$$

Hint: See Exercise 21.

24 Compute

$$\lim_{b \to a} \frac{\int_a^b f(x)\, dx}{b - a},$$

where f is a continuous function.

25 (a) Show that the average distance between vertices of a regular polygon of n sides inscribed in the unit circle is

$$A_n = \frac{2}{n-1} \cdot \sum_{k=1}^{n-1} \sin \frac{k\pi}{n}.$$

(b) Prove that $\lim_{n \to \infty} A_n = 4/\pi$.

8.9 Estimates of definite integrals

The definite integral $\int_0^1 \sqrt{1-x^3}\, dx$ cannot be evaluated by the fundamental theorem of calculus, since $\sqrt{1-x^3}$ does not have an elementary antiderivative. The method of partial fractions could be used to evaluate $\int_0^1 1/(1+x^5)\, dx$, but the procedure would be long and tedious. In the present section ways of *estimating* a definite integral are presented.

The definite integral $\int_a^b f(x)\, dx$ is, by definition, a limit of sums of the form

$$\sum_{i=1}^n f(c_i)(x_i - x_{i-1}). \tag{1}$$

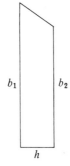

Area = $\frac{(b_1 + b_2) h}{2}$

Figure 8.119

Any such sum consequently provides an estimate of $\int_a^b f(x)\, dx$. However, the two methods described in this section, the trapezoidal method and Simpson's method, generally provide much better estimates for the same amount of arithmetic.

The sum (1) can be thought of as a sum of areas of rectangles. In the trapezoidal method, trapezoids are used instead of rectangles. Recall that the area of a trapezoid of height h and bases b_1 and b_2 is $(b_1 + b_2)h/2$. (See Fig. 8.119.)

THE TRAPEZOIDAL METHOD

Let n be a positive integer. Divide the interval $[a, b]$ into n sections of equal length $h = (b - a)/n$ with

$$x_0 = a, \quad x_1 = a + h, \quad x_2 = a + 2h, \quad \ldots, \quad x_n = b.$$

The sum

$$\frac{f(x_0) + f(x_1)}{2} \cdot h + \frac{f(x_1) + f(x_2)}{2} \cdot h + \cdots + \frac{f(x_{n-1}) + f(x_n)}{2} \cdot h$$

In the trapezoidal method, $h = \dfrac{b-a}{n}$ and f is computed at $n+1$ inputs.

is the *trapezoidal estimate* of $\int_a^b f(x)\,dx$. It is usually written

$$\frac{h}{2}[f(x_0) + 2f(x_1) + 2f(x_2) + \cdots + 2f(x_{n-1}) + f(x_n)]. \qquad (2)$$

Note that $f(x_0)$ and $f(x_n)$ have coefficient 1, while all the other $f(x_i)$'s have coefficient 2. This is due to the double counting of the edges common to two trapezoids.

The diagram in Fig. 8.120 illustrates the trapezoidal approximation for the case $n = 4$. Note that, if f is concave downward, the trapezoidal approximation underestimates $\int_a^b f(x)\,dx$. If f is a linear function, the trapezoidal method, of course, gives the integral exactly.

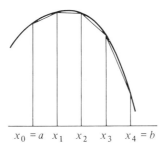

Figure 8.120

EXAMPLE 1 Use the trapezoidal method with $n = 4$ to estimate

$$\int_0^1 \frac{dx}{1 + x^2}.$$

SOLUTION In this case $a = 0$, $b = 1$, and $n = 4$, so $h = (1 - 0)/4 = 1/4$. The successive coefficients in the trapezoidal estimate (2) are 1, 2, 2, 2, and 1.

The trapezoidal estimate is

$$\frac{h}{2}\left[f(0) + 2f\left(\frac{1}{4}\right) + 2f\left(\frac{2}{4}\right) + 2f\left(\frac{3}{4}\right) + f(1)\right].$$

Now $h/2 = \tfrac{1}{4}/2 = \tfrac{1}{8}$. To compute the sum in brackets make a table:

x_i	$f(x_i)$	Coefficient	Summand	Decimal Form
0	$\dfrac{1}{1+0^2}$	1	$1 \cdot \dfrac{1}{1+0}$	1.0000000
$\tfrac{1}{4}$	$\dfrac{1}{1+(\tfrac{1}{4})^2}$	2	$2 \cdot \dfrac{1}{1+\tfrac{1}{16}}$	1.8823529
$\tfrac{2}{4}$	$\dfrac{1}{1+(\tfrac{2}{4})^2}$	2	$2 \cdot \dfrac{1}{1+\tfrac{1}{4}}$	1.6000000
$\tfrac{3}{4}$	$\dfrac{1}{1+(\tfrac{3}{4})^2}$	2	$2 \cdot \dfrac{1}{1+\tfrac{9}{16}}$	1.2800000
$\tfrac{4}{4}$	$\dfrac{1}{1+(\tfrac{4}{4})^2}$	1	$1 \cdot \dfrac{1}{1+1}$	0.5000000

The trapezoidal sum is therefore approximately

$$\tfrac{1}{8}(1 + 1.8823529 + 1.6 + 1.28 + 0.5) = \tfrac{1}{8}(6.2623529)$$
$$\approx 0.782794.$$

Thus $\displaystyle\int_0^1 \frac{dx}{1+x^2} \approx 0.782794.$ ∎

The integral in Example 1 can be evaluated by the fundamental theorem of calculus. It equals $\tan^{-1} 1 - \tan^{-1} 0 = \pi/4 \approx 0.785398$. The trapezoidal estimate is correct to two decimal places.

SIMPSON'S METHOD

In the trapezoidal method a curve is approximated by lines. In Simpson's method a curve is approximated by parabolas. (See Fig. 8.121.) As Exercise 25 shows, Simpson's method is exact if $f(x)$ is a polynomial of degree at most 3.

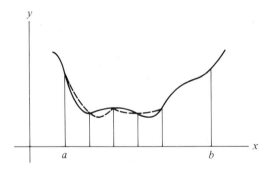

The dashed lines are parts of parabolas.

Figure 8.121

In Simpson's method the interval $[a, b]$ is divided into an *even* number of sections (n is even).

8.9 ESTIMATES OF DEFINITE INTEGRALS

Divide the interval $[a, b]$ into n sections of equal length $h = (b - a)/n$ with

$$x_0 = a, \quad x_1 = a + h, \quad x_2 = a + 2h, \quad \ldots, \quad x_n = b.$$

The sum

Note that $f(x_0)$ and $f(x_n)$ have coefficient 1, while the coefficients of the other $f(x_i)$'s alternate 4, 2, 4, 2, ..., 2, 4.

$$\frac{h}{3}[f(x_0) + 4f(x_1) + 2f(x_2) + 4f(x_3) + \cdots + 2f(x_{n-2}) + 4f(x_{n-1}) + f(x_n)] \quad (3)$$

is the *Simpson estimate* of $\int_a^b f(x)\,dx$.

EXAMPLE 2 Use Simpson's method with $n = 4$ to estimate

$$\int_0^1 \frac{dx}{1 + x^2}.$$

SOLUTION Again $h = \frac{1}{4}$. Simpson's formula (3) takes the form

$$\frac{\frac{1}{4}}{3}\left[f(0) + 4f\left(\tfrac{1}{4}\right) + 2f\left(\tfrac{2}{4}\right) + 4f\left(\tfrac{3}{4}\right) + f(1)\right].$$

x_i	$f(x_i)$	Coefficient	Summand	Decimal Form
0	$\dfrac{1}{1+0^2}$	1	$1 \cdot \dfrac{1}{1+0}$	1.0000000
$\tfrac{1}{4}$	$\dfrac{1}{1+(\frac{1}{4})^2}$	4	$4 \cdot \dfrac{1}{1+\frac{1}{16}}$	3.7647059
$\tfrac{2}{4}$	$\dfrac{1}{1+(\frac{2}{4})^2}$	2	$2 \cdot \dfrac{1}{1+\frac{1}{4}}$	1.6000000
$\tfrac{3}{4}$	$\dfrac{1}{1+(\frac{3}{4})^2}$	4	$4 \cdot \dfrac{1}{1+\frac{9}{16}}$	2.5600000
$\tfrac{4}{4}$	$\dfrac{1}{1+(\frac{4}{4})^2}$	1	$1 \cdot \dfrac{1}{1+1}$	0.5000000

The Simpson approximation of $\int_0^1 dx/(1 + x^2)$ is therefore

$$\tfrac{1}{12}(1 + 3.7647059 + 1.6 + 2.56 + 0.5) = \tfrac{1}{12}(9.4247059)$$
$$\approx 0.785392.$$

Thus
$$\int_0^1 \frac{dx}{1 + x^2} \approx 0.785392. \quad \blacksquare$$

Since $\int_0^1 dx/(1 + x^2) \approx 0.785398$, Simpson's estimate is correct to five decimal places. Simpson's method usually provides a much more accurate estimate than the trapezoidal estimate in Example 1 for the same amount of arithmetic. (As in Examples 1 and 2, each uses the same number of points at which to evaluate the function; the difference is in the weights given the values of the function at those points.)

THE ACCURACY OF THE TRAPEZOIDAL AND SIMPSON'S ESTIMATES

The trapezoidal estimate is exact for polynomials of the form $f(x) = a + bx$. Now, such a polynomial can be thought of as a function whose second derivative is 0 for all x. Therefore, it is reasonable to expect that the accuracy of the trapezoidal method is influenced by the second derivative of f. It can be shown that if M_2 is the maximum value of $|f^{(2)}(x)|$ for x in $[a, b]$, then the error in using the trapezoidal approximation (2) is at most

The error in the trapezoidal method

$$\frac{(b - a)M_2 h^2}{12}.$$

The factor h^2 is the key indicator of the accuracy. It suggests that cutting h in half will cut the error by a factor of four.

As mentioned earlier, Simpson's estimate is exact for any polynomial of the form $f(x) = a + bx + cx^2 + dx^3$, that is, for any function whose fourth derivative is identically 0. The error in using Simpson's method for other functions is measured by the fourth derivative. Let M_4 be the maximum value of $|f^{(4)}(x)|$ for x in $[a, b]$. Then the error in using Simpson's formula (3) in estimating $\int_a^b f(x)\,dx$ is at most

The error in Simpson's method

$$\frac{(b - a)M_4 h^4}{180}.$$

Since $h^4 \to 0$ faster than $h^2 \to 0$ as $h \to 0$, we would expect Simpson's method to be generally more accurate than the trapezoidal. Cutting h in half would tend to cut the error by a factor of 16.

Exercises

Exercises 1 to 6 concern the trapezoidal estimate. In Exercises 1 to 4: (a) Estimate the given integrals by the trapezoidal method, using the given values of n; (b) Evaluate the integrals and calculate the absolute values of the errors in the estimates.

1 $\int_0^1 x^2\,dx$, $n = 3$

2 $\int_0^1 x^2\,dx$, $n = 4$

3 $\int_1^5 \frac{dx}{x}$, $n = 2$

4 $\int_1^7 \frac{dx}{x}$, $n = 6$

5 Estimate $\int_0^1 dx/(1 + x^3)$, using the trapezoidal method with $n = 4$.

6 Estimate $\int_0^1 dx/(1 + x^4)$, using the trapezoidal method with $n = 5$.

In Exercises 7 to 12 estimate the given integrals by Simpson's method, using the given values of n.

7 $\int_1^5 \frac{dx}{x}$, $n = 2$

8 $\int_1^5 \frac{dx}{x}$, $n = 4$

9 $\int_1^7 \frac{dx}{x}$, $n = 2$

10 $\int_1^7 \frac{dx}{x}$, $n = 6$

11 $\int_0^1 \frac{dx}{1 + x^3}$, $n = 4$

12 $\int_0^1 \frac{dx}{1 + x^3}$, $n = 8$

In Exercises 13 to 14, using the given n, estimate the given integrals by (a) the trapezoidal method, (b) Simpson's method.

13 $\int_{1}^{2} \dfrac{e^x \, dx}{x}, \ n = 6$

14 $\int_{1}^{5} e^{-x^2} \, dx, \ n = 6$

■

The "right-point" estimate of $\int_{a}^{b} f(x) \, dx$ is obtained as follows. Select a positive integer n and divide $[a, b]$ into n sections of equal length $h = (b - a)/n$. The points of subdivision are $x_0 < x_1 < \cdots < x_n$, with $x_0 = a$ and $x_n = b$. The right-point estimate is the approximating sum $h(f(x_1) + f(x_2) + \cdots + f(x_n))$. The "left-point" estimate is defined similarly; it is $h(f(x_0) + f(x_1) + \cdots + f(x_{n-1}))$.

15 Show that, if $f(a) = f(b)$, the left-point, right-point and trapezoidal estimates for a given value of h are the same.

16 Show that for a given n the average of the left-point estimate and the right-point estimate equals the trapezoidal estimate.

The next two exercises present cases in which the bounds of maximum error are actually assumed.

17 Show that, if the trapezoidal method with $n = 1$ is used to estimate $\int_0^1 x^2 \, dx$, the error equals $(b - a) M_2 h^2 / 12$, where $a = 0, b = 1, h = 1$, and M_2 is the maximum value of $|d^2(x^2)/dx^2|$ for x in $[0, 1]$.

18 Show that, if Simpson's method with $n = 2$ is used to estimate $\int_0^1 x^4 \, dx$, the error equals $(b - a) M_4 h^4 / 180$, where $a = 0, b = 1, h = \tfrac{1}{2}$, and M_4 is the maximum value of $|d^4(x^4)/dx^4|$ for x in $[0, 1]$.

19 (a) Compute $\int_0^1 dx/(1 + x^3)$ using partial fractions. Hint: $1 + x^3 = (1 + x)(1 - x + x^2)$.
 (b) Estimate $\int_0^1 dx/(1 + x^3)$ using Simpson's method with $n = 6$.
 (c) Compute the absolute value of the error of the estimate made in (b).

■ ■

Exercises 20 to 25 describe the geometric motivation of Simpson's method. Exercises 21, 23, and 24 justify the method.

20 Let $f(x) = Ax^2 + Bx + C$. Show that

$$\int_{-h}^{h} f(x) \, dx = \dfrac{h}{3}(f(-h) + 4f(0) + f(h)).$$

Hint: Just compute both sides.

21 Let f be a function. Show that there is a parabola $y = Ax^2 + Bx + C$ that passes through the three points $(-h, f(-h))$, $(0, f(0))$, and $(h, f(h))$. (See Fig. 8.122.)

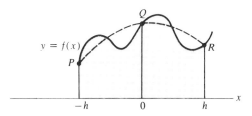

The dashed graph is a parabola, $y = Ax^2 + Bx + C$, through P, Q, and R. The area of the region below the parabola is precisely

$$\dfrac{h}{3}[f(-h) + 4f(0) + f(h)]$$

and is an approximation of the area of the shaded region.

Figure 8.122

22 The equation in Exercise 20 (which was known to the Greeks) is called the prismoidal formula. Use it to compute the volume of
 (a) a sphere of radius a;
 (b) a right circular cone of radius a and height h.

23 Let $f(x) = Ax^2 + Bx + C$. Show that

$$\int_{c-h}^{c+h} f(x) \, dx = \dfrac{h}{3}[f(c - h) + 4f(c) + f(c + h)].$$

Hint: Use the substitution $x = c + t$ to reduce this to Exercise 20.

24 First $[a, b]$ is divided into n sections (n even), which are grouped into $n/2$ pairs of adjacent sections. Over each pair the function is approximated by the parabola that passes through the three points of the graph with x coordinates equal to those that determine the two sections of the pair. (See Fig. 8.121.) The integral of this quadratic function is used as an estimate of the integral of f over each pair of adjacent sections. Show that, when these $n/2$ separate estimates are added, Simpson's formula results.

25 Since Simpson's method was designed to be exact when $f(x) = Ax^2 + Bx + C$, one would expect the error associated with it to involve $f^{(3)}(x)$. By a quirk of good fortune Simpson's method happens to be exact even when $f(x)$ is a cubic, $Ax^3 + Bx^2 + Cx + D$. This suggests that the error involves $f^{(4)}(x)$, not $f^{(3)}(x)$.
 (a) Show that, if $f(x) = x^3$,

$$\int_{-h}^{h} f(x) \, dx = \dfrac{h}{3}(f(-h) + 4f(0) + f(h)).$$

 (b) Show that Simpson's estimate is exact for cubics.

26 There are many other methods for estimating definite integrals. Some old methods, which had been of only theoretical interest because of their messy arithmetic,

have, with the advent of computers, assumed practical importance. This exercise illustrates the simplest of the so-called "Gaussian quadrature formulas." For simplicity consider only integrals over $[-1, 1]$.

(a) Show that

$$\int_{-1}^{1} f(x)\,dx = f\left(\frac{-1}{\sqrt{3}}\right) + f\left(\frac{1}{\sqrt{3}}\right)$$

for $f(x) = 1, x, x^2,$ and x^3.

(b) Let a and b be two numbers, $-1 \leq a < b \leq 1$ such that

$$\int_{-1}^{1} f(x)\,dx = f(a) + f(b)$$

for $f(x) = 1, x, x^2,$ and x^3. Show that $a = -1/\sqrt{3}$ and $b = 1/\sqrt{3}$.

(c) Show that the approximation $\int_{-1}^{1} f(x)\,dx \approx f(-1/\sqrt{3}) + f(1/\sqrt{3})$ has no error when f is a polynomial of degree at most three.

Part (c) suggests that the error in this method involves $f^{(4)}(x)$. It is proved in numerical analysis that the absolute value of the error is at most $M_4/135$, where M_4 is the maximum of $|f^{(4)}(x)|$ for x in $[-1, 1]$. Incidentally, the estimate,

$$\int_{-1}^{1} f(x)\,dx \approx \tfrac{5}{9}f(-\sqrt{\tfrac{3}{5}}) + \tfrac{8}{9}f(0) + \tfrac{5}{9}f(\sqrt{\tfrac{3}{5}}),$$

is exact for polynomials of degree at most five. Gaussian quadrature is discussed in Anthony Ralston and Philip Rabinowitz, *A First Course in Numerical Analysis*, 2d ed., McGraw-Hill, New York, 1978, pp. 98–101.

27 Use the first formula in Exercise 26(c) to estimate $\int_{-1}^{1} dx/(1 + x^2)$.

28 If $f'(a)$ and $f'(b)$ are known, a slight modification of the trapezoidal estimate produces an estimate of $\int_{a}^{b} f(x)\,dx$ about as accurate as that given by Simpson's rule. Let T be the trapezoidal estimate based on n sections of width $h = (b - a)/n$. Let the modified estimate be

$$T^* = T + \frac{(f'(a) - f'(b))h^2}{12}.$$

(a) Show that T^* is exact if $n = 1$ and $f(x) = 1, x, x^2,$ or x^3.

(b) Show that T^* is exact if $n = 1$ and $f(x)$ is any polynomial of degree at most three. *Hint:* Use (a).

(c) Show that T^* is exact if $n = 2$ and $f(x)$ is any polynomial of degree at most three. *Hint:* Use (b) twice, first with $[a, b] = [x_0, x_1]$ and then with $[a, b] = [x_1, x_2]$.

(d) Show that T^* is exact for any positive integer n and any polynomial of degree at most three.

It can be shown that the error in using T^* is at most $M_4(b - a)h^4/720$, which is a quarter of the bound in Simpson's method. See Lynn H. Loomis, *Calculus*, pp. 462–466, Addison-Wesley, Reading, Mass., 1977.

29 (See Exercise 28.) Estimate $\int_{1}^{5} dx/x$ using $n = 4$ and (a) the trapezoidal estimate T, (b) the modified trapezoidal estimate T^*, (c) Simpson's method. (d) Using the fact that $\ln 5 \approx 1.6094379$, find the errors in the three methods and compare them to their theoretical bounds.

8.S Summary

This chapter was concerned primarily with applications of the definite integral. The foundation for the first two sections was laid in Chap. 5, namely, "area is the integral of cross-sectional length" and "volume is the integral of cross-sectional area." Sections 8.1 and 8.2 concentrated on finding those cross sections, $c(x)$ and $A(x)$.

An informal approach to setting up definite integrals was given in Sec. 8.3. Basically, instead of writing the approximating sum

$$\sum_{i=1}^{n} f(c_i)(x_i - x_{i-1})$$

one just determines the individual summand by discovering the "local estimate"

$$f(x)\,dx.$$

This shorthand was used throughout the chapter in setting up integrals for force of a fluid against a submerged surface (Sec. 8.3), volume by shells (Sec. 8.4), moments (Sec. 8.5), and work (Sec. 8.7).

Integrals in which either the interval of integration or the integrand is unbounded ("improper integrals") were discussed in Sec. 8.6. The average of a function $f(x)$ was defined in Sec. 8.8. The concluding Sec. 8.9 described two ways of estimating a definite integral, the trapezoidal method and Simpson's method.

The following table summarizes most of the applications of the definite integral treated in this chapter.

Section	Concept	Memory Aid
8.1	Area $= \int_a^b c(x)\, dx$	Area $= c(x)\, dx$
8.2	Volume $= \int_a^b A(x)\, dx$ (parallel cross sections)	Volume $= A(x)\, dx$
8.3	Force of water $= 62.4 \int_a^b x c(x)\, dx$	Force $= 62.4x \cdot c(x)\, dx$ — Pressure, Area (Surface of water is at $x = 0$.)
8.4	Volume $= \int_a^b 2\pi R(x) c(x)\, dx$ (shells)	(Cut it out and unroll.) Volume $= 2\pi R(x) \cdot c(x) \cdot dx$ — Length, Height, Width

522 APPLICATIONS OF THE DEFINITE INTEGRAL

Section	Concept	Memory Aid
8.5	Moment $= \int_a^b x c(x)\, dx$	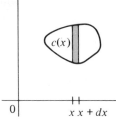 Moment $= \underbrace{x}_{\text{Lever}} \cdot \underbrace{c(x)\, dx}_{\text{Mass}}$ (Density $= 1$)
8.7	Work emptying tank $= 62.4 \int_a^b (x - k) A(x)\, dx$	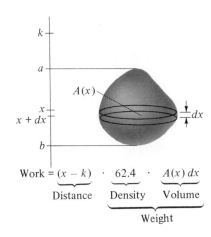 Work $= \underbrace{\underbrace{(x - k)}_{\text{Distance}} \cdot \underbrace{62.4}_{\text{Density}} \cdot \underbrace{A(x)\, dx}_{\text{Volume}}}_{\text{Weight}}$

Keep in mind that volume by shells, force of fluid, and work emptying a tank can easily be computed in terms of centroids.

VOCABULARY AND SYMBOLS

cross-sectional length $c(x)$, $c(y)$
cross-sectional area $A(x)$
solid of revolution
shell technique
lever arm
moment about a line
centroid, center of mass, $(\bar{x}, \bar{y})$
Pappus's theorem
force of water
work
improper integral (convergent and divergent)
average of a continuous function
trapezoidal method
Simpson's method

KEY FACTS Area is equal to $\int_a^b c(x)\,dx$. In particular, if $f(x) \geq g(x)$, the area between $y = f(x)$ and $y = g(x)$ equals $\int_a^b [f(x) - g(x)]\,dx$. Similar formulas hold in any direction; for instance area also equals $\int_c^d c(y)\,dy$.

Volume equals $\int_a^b A(x)\,dx$.

The total force of water against a submerged vertical surface R is $62.4 \int_a^b xc(x)\,dx$ pounds. (The origin of the x axis is at the water surface. The axis is aimed downward.) If the origin is elsewhere and the coordinate of the water surface is k, then the force is $62.4 \int_a^b (x - k)c(x)\,dx$ pounds.

The volume of a solid of revolution may be computed by parallel cross sections,

$$\text{Volume} = \int_a^b \pi[f(x)]^2\,dx, \quad \text{if region revolved around } x \text{ axis lies below}$$

$y = f(x)$ and above $[a, b]$.

$$\text{Volume} = \int_a^b \pi[(f(x))^2 - (g(x))^2]\,dx, \quad \text{if revolved region lies between}$$

$y = f(x)$ and $y = g(x)$ and above $[a, b]$.

By the shell method, if region is revolved around y axis,

$$\text{Volume} = \int_a^b 2\pi R(x)c(x)\,dx.$$

(A similar formula holds if region is revolved around x axis.)

The centroid $(\bar{x}, \bar{y})$ of a plane region R is given by

$$\bar{x} = \frac{\int_a^b xc(x)\,dx}{\text{Area of } R} \quad \text{and} \quad \bar{y} = \frac{\int_c^d yc(y)\,dy}{\text{Area of } R}.$$

If R is the region below $y = f(x)$ and above $[a, b]$, then

$$\bar{x} = \frac{\int_a^b xf(x)\,dx}{\text{Area of } R} \quad \text{and} \quad \bar{y} = \frac{\frac{1}{2}\int_a^b (f(x))^2\,dx}{\text{Area of } R}.$$

The work required to lift the water in a cylindrical tank to a certain level equals

$$62.4h \int_a^b (x - k)c(x)\,dx \text{ foot-pounds.}$$

(The x axis is pointed down and the x coordinate of the level to which the water is lifted is k. The base of the cylinder is some plane region R, whose cross section is given by $c(x)$. The length of the cylinder is h.)

If $f(x)$ is continuous for $x \geq a$, $\int_a^\infty f(x)\,dx$ is an improper integral defined as the limit of $\int_a^b f(x)\,dx$ as $b \to \infty$, if this limit exists. A similar definition holds for $\int_{-\infty}^{b} f(x)\,dx$. The improper integral $\int_{-\infty}^{\infty} f(x)\,dx$ is defined to be the sum of $\int_{-\infty}^{0} f(x)\,dx$ and $\int_{0}^{\infty} f(x)\,dx$, if these exist.

If $f(x)$ is continuous in $[a, b]$ except at b, then the improper integral $\int_{a}^{b} f(x)\,dx$ is defined as the limit of $\int_{a}^{t} f(x)\,dx$ as $t \to b^{-}$. A similar definition holds if f misbehaves near a.

It could happen that $\int_{a}^{\infty} f(x)\,dx$ is improper not only because the interval is unbounded but also because the integrand is unbounded. In that case the integral must be expressed as the sum of improper integrals over subintervals and each examined separately.

THREE USES OF THE CENTROID (tersely stated)

The volume of a solid of revolution formed by revolving R equals

Distance centroid of R moves · Area of R.

The force of water against a submerged flat surface R equals

Pressure at centroid of R · Area of R.

The work required to empty a cylindrical tank of water is the same as if all the water were at the depth of the centroid.

The average of a continuous function f over an interval $[a, b]$ is defined as $\int_{a}^{b} f(x)\,dx / (b - a)$.

TWO WAYS TO ESTIMATE $\int_{a}^{b} f(x)\,dx$

Method	Formula	Weights
Trapezoidal	$\dfrac{h}{2}[f(x_0) + 2f(x_1) + 2f(x_2) + \cdots + 2f(x_{n-1}) + f(x_n)]$	1, 2, 2, 2, ..., 2, 1
Simpson's	$\dfrac{h}{3}[f(x_0) + 4f(x_1) + 2f(x_2) + \cdots + 4f(x_{n-1}) + f(x_n)]$	1, 4, 2, 4, 2, ..., 2, 4, 1

In Simpson's method the number n must be even. In the trapezoidal formula $h/2$ is a factor; in Simpson's method, $h/3$ is a factor. In both methods $h = (b - a)/n$.

In both the inputs x_i are evenly spaced with successive x_i's a distance h apart: $x_i = a + ih$, $i = 0, 1, 2, \ldots, n$.

Guide quiz on chap. 8

In Exercises 1 to 4 R is the region bounded by $y = e^x$ and the x axis, between $x = 1$ and $x = 2$.

1. Find the area of R, using (a) vertical cross sections, (b) horizontal cross sections.
2. (a) Find the moment of R about the x axis.
   (b) Find the moment of R about the y axis.
   (c) Find $(\bar{x}, \bar{y})$, the centroid of R.
3. A solid of revolution is made by revolving R around the line $y = -1$. Find its volume by (a) parallel cross sections, (b) shells, (c) Pappus's theorem.
4. A solid of revolution is made by revolving R around the y axis. Find its volume by the three methods given in Exercise 3.
5. Let R be the region below the curve $y = f(x)$ and above $[a, b]$. Give an intuitive argument that the moment of R around the x axis is $\frac{1}{2}\int_a^b (f(x))^2\, dx$.
6. Find the average value of $f(x) = 1/(9 - x^2)$ over the interval $[1, 2]$.
7. For which exponents a (if any) is
   (a) $\int_1^\infty x^a\, dx$ convergent?
   (b) $\int_0^1 x^a\, dx$ convergent?
   (c) $\int_0^\infty x^a\, dx$ convergent?
8. Show that $\displaystyle\int_0^\infty \frac{dx}{x + e^x}$ is convergent.
9. The integral $\int_1^3 f(x)\, dx$ is to be estimated using seven equally spaced numbers $x_0, x_1, \ldots, x_6$, with $x_0 = 1$ and $x_6 = 3$.
   (a) What is the trapezoidal estimate using these numbers?
   (b) What is Simpson's estimate using these numbers?

Review exercises for chap. 8

In Exercises 1 to 4 set up integrals for the given quantities; do not evaluate them.

1. The area of the region above the parabola $y = x^2$ and below the line $y = 2x$, using (a) vertical cross sections, (b) horizontal cross sections.
2. The volume of the wedge cut from a right circular cylinder of height 5 inches and radius 3 inches by a plane that bisects one base and touches the other base at one point.
3. The volume of the solid obtained by revolving the triangle whose vertices are (2, 0), (2, 1), and (3, 2) about the x axis. (Use the shell technique.)
4. The moment of the region in the first quadrant bounded by $y = x^2$ and $y = x^3$, about the line $y = -2$.

In Exercises 5 to 9
   (a) Find the area of R.
   (b) Find the volume of the solid of revolution formed by revolving R about the x axis.
   (c) The same as (b), but around the y axis.
   (d) The same as (b), but around the line $y = -1$.
5. R is the region below the curve $y = x/(1 + x)$ and above $[1, 2]$.
6. R is the region below $y = 1/(1 + x)^2$ and above $[0, 1]$.
7. R is the region below $y = \sin 2x$ and above $[0, \pi/2]$.
8. R is the region below $y = \sqrt{x^2 - 9}$ and above $[3, 4]$.
9. R is the region below $y = 1/(2x + 1)$ and above $[0, 1]$.

In Exercises 10 to 13 find the moments of the given regions R about the given lines L.

10. R: below $y = \sec x$, above $[\pi/6, \pi/4]$; L: the x axis.
11. R: below $y = (\sin x)/x$, above $[\pi/2, \pi]$; L: the y axis.
12. R: below $y = 1/\sqrt{x^2 + 1}$, above $[0, 1]$; L: the x axis.
13. R: below $y = 1/\sqrt{x^2 + 1}$, above $[0, 1]$; L: the y axis.

In Exercises 14 to 18 find the areas of the given regions R.

14. R is below $y = 1/(x^2 + 3x + 2)$ and above $y = 1/(x^2 + 3x + 4)$, between $x = 0$ and $x = 1$.
15. R is below $y = x$ and above $y = \tan^{-1} x$, between $x = 0$ and $x = 1$.
16. R is below $y = x\sqrt{2x + 1}$ and above $[0, 4]$.
17. R is below $y = \cos^3 x$ and above $y = \sin^3 x$, between $x = 0$ and $x = \pi/4$.
18. R is below $y = 1/\sqrt{4 - x^2}$ and above $y = x/\sqrt{4 - x^2}$, between $x = 0$ and $x = 1$.
19. Find the area of the region between the curves $y = 1/(x^2 - x)$ and $y = 1/(x^3 - x)$, (a) between $x = 2$ and $x = 3$, (b) to the right of $x = 3$.
20. Let R be the region below $y = \tan x$ and above $[0, \pi/4]$.
    (a) Find the area of R.
    (b) Find the moment of R about the x axis.
    (c) Find $\bar{y}$.
    (d) Set up an integral for the moment of R about the y axis. (Don't try to evaluate it; the fundamental theorem of calculus is useless here.)
    (e) Find the volume of the solid of revolution formed by revolving R around the line $y = -1$.
21. Let R be the region below $y = \sin^2 x$ and above $y = \sin^3 x$, between $x = 0$ and $x = \pi/2$.
    (a) Find the area of R.
    (b) Find the moment of R about the x axis. (Recall the formula for $\int_0^{\pi/2} \sin^n x\, dx$.)
    (c) Find $\bar{y}$.

22. Is $\int_0^\infty (dx/\sqrt{x}\sqrt{x+1}\sqrt{x+2})$ convergent or divergent?
23. A drill of radius a inches bores a hole through the center of a sphere of radius b inches, $b > a$, leaving a ring whose height is 2 inches. Find the volume of the ring.
24. A barrel is made by rotating an ellipse around one of its axes and then cutting off equal caps, top and bottom. It is 3 feet high and 3 feet wide at its midsection. Its top and bottom have a diameter of 2 feet. What is its volume?
25. Let R be the region to the right of the y axis, below $y = e^{-x}$ and above the x axis.
    (a) Find the area of R.
    (b) Find the volume of the region obtained by revolving R about the x axis.
    (c) Find the volume of the region obtained by revolving R about the y axis.
26. Let l be a line which intersects the triangle ABC and is parallel to BC. Suppose that l is twice as far from the point A as from the line BC. (See Fig. 8.123.) Show that the centroid of ABC is on l.

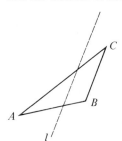

Figure 8.123

27. (a) Develop the shell formula for the volume of a solid of revolution.
    (b) What is the device for remembering the formula?
28. Show that $\int_0^\infty e^{-rx} \sin ax \, dx$ equals $a/(a^2 + r^2)$, where $r > 0$ and a are constant.
29. Show that $\int_0^\infty e^{-rx} \cos ax \, dx$ equals $r/(a^2 + r^2)$, where $r > 0$ and a are constant.
30. Let f be a continuous function such that $f(0) = 2$ and $f(x) \to 3$ as $x \to \infty$. Find the limit of $(1/b) \int_0^b f(x) \, dx$ as (a) $b \to 0$, (b) $b \to \infty$.
31. By interpreting these improper integrals as expressions for the area of a certain region, show that
$$\int_0^\infty \frac{dx}{1+x^2} = \int_0^1 \sqrt{\frac{1-y}{y}} \, dy.$$
32. Define $G(a) = \int_0^\infty a/(1+a^2 x^2) \, dx$. (a) Compute $G(0)$. (b) Compute $G(a)$ if a is negative. (c) Compute $G(a)$ if a is positive. (d) Graph G.
33. Prove that $\int_0^\infty (\sin x^2)/x \, dx = \frac{1}{2} \int_0^\infty (\sin x)/x \, dx$. (It can be shown that these improper integrals are convergent.)
34. Is $\int_0^\infty \frac{dx}{(x-1)^2}$ convergent or divergent?
35. Evaluate $\int_0^\infty e^{-x} \sin(2x+3) \, dx$.
36. (a) Sketch $y = e^{-x}(1 + \sin x)$ for $x \geq 0$.
    (b) The region beneath the curve in (a) and above the positive x axis is revolved around the y axis. Find the volume of the resulting solid.
37. This table shows the temperature $f(t)$ as a function of time.

Time	1	2	3	4	5	6	7
Temperature	81	75	80	83	78	70	60

   (a) Use Simpson's method to estimate $\int_1^7 f(t) \, dt$.
   (b) Use the result in (a) to estimate the average temperature.
38. Estimate the area of the region below $y = \cos x^2$ and above $[0, 1]$, using $n = 6$ and (a) the trapezoidal formula, (b) Simpson's formula.
39. From the fact that $\int (e^x/x) \, dx$ is not elementary, deduce that $\int e^x \ln x \, dx$ is not elementary.
40. Is this computation correct?
$$\int_{-2}^1 \frac{dx}{2x+1} = \frac{1}{2} \ln |2x+1| \Big|_{-2}^1 = \frac{1}{2} \ln 3 - \frac{1}{2} \ln 3 = 0.$$
41. Find the error in the following computations.
$$\int_{-1}^1 \frac{1}{x^2} dx = \frac{-1}{x} \Big|_{-1}^1 = \frac{-1}{1} - \frac{-1}{-1} = -2.$$
(The integrand is positive, yet the integral is negative.)
42. It can be proved that $\int_0^\infty (x^{n-1})/(1+x) \, dx = \pi \csc n\pi$ for $0 < n < 1$. Verify that this equation is correct for $n = \frac{1}{2}$.
43. Compute $\int_0^1 x^4 \ln x \, dx$.
44. Show that $\int_0^\infty \frac{dx}{1+x^4} = \int_0^\infty \frac{x^2 \, dx}{1+x^4}$.

   Hint: Let $x = 1/y$.
45. Show that $\int_0^1 (-\ln x)^3 \, dx = \int_0^\infty x^3 e^{-x} \, dx$.
46. Assume that the density of the earth x miles from its center is $g(x)$ tons per cubic mile. Set up (informally) an integral for the total mass of the earth.
47. Find the centroid of the region bounded by the parabola $y = x^2$ and the line $y = 3x - 2$.
48. Find the centroid of the finite region bounded by $y = 2^x$ and $y = x^2$, to the right of the y axis.
49. Consider the integral $\int_0^4 dx/(1+x^4)$.
    (a) Estimate the integral by the trapezoidal method, $n = 4$.

(b) Estimate the integral by Simpson's method, $n = 4$.
(c) Evaluate the integral, making use of the fact that $x^4 + 1 = (x^2 + \sqrt{2}x + 1)(x^2 - \sqrt{2}x + 1)$. After expressing the integrand in terms of partial fractions, use an integral table.

50 Let f be a function such that $|f^{(2)}(x)| \leq 10$ and $|f^{(4)}(x)| \leq 50$ for all x in $[1, 5]$. If $\int_1^5 f(x)\,dx$ is to be estimated with an error of at most 0.01, how small must h be in (a) the trapezoidal approximation? (b) Simpson's approximation?

■

51 Let T be the trapezoidal estimate of $\int_a^b f(x)\,dx$, using $x_0 = a, x_1, \ldots, x_n = b$. Let M be the "midpoint estimate," $\sum_{i=1}^n f(c_i)(x_i - x_{i-1})$, where $c_i = (x_{i-1} + x_i)/2$. Let S be Simpson's estimate using the $2n + 1$ points $x_0, c_1, x_1, c_2, x_2, \ldots, c_n, x_n$. Show that
$$S = \tfrac{2}{3}M + \tfrac{1}{3}T.$$

Incidentally the bound on the error in the midpoint estimate is half that of the trapezoid estimate. Even so, it is not used much because it involves extra computations.

52 A particle moves on a line in such a way that its time-average velocity over any interval of time $[a, b]$ is the same as its velocity at time $(a + b)/2$. Prove that the velocity $v(t)$ must be of the form $ct + d$ for appropriate constants c and d. Hint: Begin by differentiating the relation $\int_a^b v(t)\,dt = [v((a + b)/2)](b - a)$ with respect to b and with respect to a.

53 A particle moves on a line in such a way that the time-average velocity over any interval of time $[a, b]$ is equal to the average of its velocities at the beginning and the end of the interval of time. Prove that the velocity $v(t)$ must be of the form $ct + d$ for appropriate constants c and d.

54 Is $\int_0^1 \dfrac{\ln x}{1 - x^2}\,dx$ convergent or divergent?

55 Water flows out of a hole in the bottom of a cylindrical tank of radius r and height h at the rate of $\sqrt{y}$ cubic feet per second when the depth of the water is y feet. Initially the tank is full. (See Fig. 8.124.)

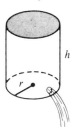

Figure 8.124

(a) How long will it take to become half full?
(b) How long will it take to empty?

56 Is the area under the curve $y = (\ln x)/x^2$, above the x axis and to the right of the line $x = 1$, finite or infinite?

57 (a) Let $G(a) = \int_0^\infty 1/[(1 + x^a)(1 + x^2)]\,dx$. Evaluate $G(0), G(1), G(2)$.
(b) Show, using the substitution $x = 1/y$, that
$$G(a) = \int_0^\infty \frac{x^a\,dx}{(1 + x^a)(1 + x^2)}.$$
(c) From (b), show that $G(a) = \pi/4$, independent of a.

58 There are two values of a for which $\int \sqrt{1 + a \sin^2 \theta}\,d\theta$ is elementary. What are they?

59 From Exercise 58 deduce that there are two values of a for which
$$\int \frac{\sqrt{1 + ax^2}}{\sqrt{1 - x^2}}\,dx$$
is elementary.

60 There are three values of b for which $\int \sqrt{1 + b \cos \theta}\,d\theta$ is elementary. What are they?

61 From Exercise 60 deduce that there are three values of b for which
$$\int \frac{\sqrt{1 + bx}}{\sqrt{1 - x^2}}\,dx$$
is elementary.

■ ■

62 Let f be a function such that $f(x) > 0$. Assume that f has derivatives of all orders and that $\ln f(x) = f(x) \int_0^x f(t)\,dt$. Find (a) $f(0)$, (b) $f^{(1)}(0)$, (c) $f^{(2)}(0)$.

63 Find the number a, $0 \leq a < 2\pi$ that maximizes the function
$$f(a) = \int_0^{2\pi} \sin x \sin(x + a)\,dx.$$

A nonnegative function $f(x)$, such that $\int_{-\infty}^\infty f(x)\,dx = 1$, is called a *probability distribution*. (The probability that a certain variable observed in an experiment is between x and $x + \Delta x$ is approximately $f(x)\,\Delta x$.) The improper integral $\int_{-\infty}^\infty xf(x)\,dx$, if it exists, is called the *mean* of the distribution and is denoted μ. The improper integral $\int_{-\infty}^\infty (x - \mu)^2 f(x)\,dx$, if it exists, is called the *variance* of the distribution and is denoted μ_2. The square root of μ_2 is called the *standard deviation* of the distribution and is denoted σ.

64 Let k be a positive constant. Define $f(x)$ to be $(1/k)e^{-kx}$ if $x > 0$ and 0 if $x \leq 0$.
  (a) Show that $\int_{-\infty}^{\infty} f(x)\, dx = 1$.
  (b) Find μ.
  (c) Find μ_2.
  (d) Find σ.

65 Let k be a positive constant. Define $f(x)$ to be $e^{-x^2/2k^2}/(\sqrt{2\pi}\, k)$. This is a *normal distribution*.
  (a) Show that $\int_{-\infty}^{\infty} f(x)\, dx = 1$. *Suggestion:* Assume that $\int_0^{\infty} e^{-x^2}\, dx = \sqrt{\pi}/2$, a result established in Exercise 24 of Sec. 13.4.
  (b) Find μ.
  (c) Find μ_2.
  (d) Find σ.

PLANE CURVES AND POLAR COORDINATES

Chapters 1 through 8 introduced the derivative, the antiderivative, and the definite integral, as well as a variety of their applications. These concepts form the foundation of calculus. The present chapter uses them in the study of the length of a curve, the area within a curve, the area of a curved surface, and motion along a curve.

9.1 Polar coordinates

Rectangular coordinates are only one of the ways to describe points in the plane by pairs of numbers. In this section another system is described, called *polar coordinates*, which will be used later in this chapter and in subsequent chapters.

The rectangular coordinates x and y describe a point P in the plane as the intersection of a vertical line and a horizontal line. Polar coordinates describe a point P as the intersection of a circle and a ray from the center of that circle. They are defined as follows.

Select a point in the plane and a ray emanating from this point. The point is called the *pole*, and the ray the *polar axis*. (See Fig. 9.1.) Measure positive angles θ counterclockwise from the polar axis and negative angles clockwise. Now let r be a number. To plot the point P that corresponds to the pair of numbers r and θ proceed as follows:

If r is positive, P is the intersection of the circle of radius r whose center is at the pole and the ray of angle θ emanating from the pole.

If r is 0, P is the pole, no matter what θ is.

If r is negative, P is at a distance $|r|$ from the pole on the ray directly opposite the ray of angle θ.

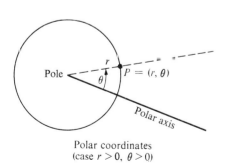

Polar coordinates (case $r > 0$, $\theta > 0$)

Figure 9.1

530 PLANE CURVES AND POLAR COORDINATES

In each case P is denoted (r, θ), and the pair r and θ are called polar coordinates of P. Note that the point (r, θ) is on the circle of radius $|r|$ whose center is the pole. Observe that the pole is the midpoint of the points (r, θ) and $(-r, \theta)$.

EXAMPLE 1   Plot the points $(3, \pi/4)$, $(2, -\pi/6)$, $(-3, \pi/3)$ in polar coordinates.

SOLUTION   To plot $(3, \pi/4)$, go out a distance 3 on the ray of angle $\pi/4$ (shown in Fig. 9.2). To plot $(2, -\pi/6)$, go out a distance 2 on the ray of angle $-\pi/6$. To plot $(-3, \pi/3)$, draw the ray of angle $\pi/3$, and then go a distance 3 in the *opposite* direction from the pole. (See Fig. 9.2)

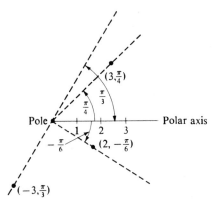

Figure 9.2   ∎

The relation between polar and rectangular coordinates

It is customary to have the polar axis coincide with the positive x axis as in Fig. 9.3. In that case, inspection of the diagram shows the following relation between the rectangular coordinates (x, y), and the polar coordinates of the point P:

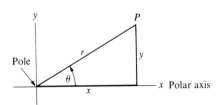

Figure 9.3

and
$$x = r \cos \theta, \qquad y = r \sin \theta,$$
$$r^2 = x^2 + y^2, \qquad \tan \theta = \frac{y}{x}.$$

The polar coordinates of a point are not unique.

Just as we may graph the set of points (x, y), where x and y satisfy a certain equation, so may we graph the set of points (r, θ), where r and θ satisfy a certain equation. It is important, however, to keep in mind that, although each point in the plane is specified by a unique ordered pair (x, y) in rectangular coordinates, there are many ordered pairs (r, θ) in polar coordinates which specify each point. For instance, the point whose rectangular coordinates are $(1, 1)$ has polar coordinates $(\sqrt{2}, \pi/4)$ or $(\sqrt{2}, \pi/4 + 2\pi)$ or $(\sqrt{2}, \pi/4 + 4\pi)$ or $(-\sqrt{2}, \pi/4 + \pi)$ and so on.

EXAMPLE 2   Graph the equation $r = 2 \cos \theta$.

SOLUTION   First make a table, choosing convenient values of θ. Then sketch the points listed in the table, as shown in Fig. 9.4.

θ	0	$\dfrac{\pi}{4}$	$\dfrac{\pi}{3}$	$\dfrac{\pi}{2}$	$\dfrac{3\pi}{4}$	π	$\dfrac{3\pi}{2}$	2π
$r = 2\cos\theta$	2	$\sqrt{2} \approx 1.4$	1	0	$-\sqrt{2} \approx -1.4$	-2	0	2

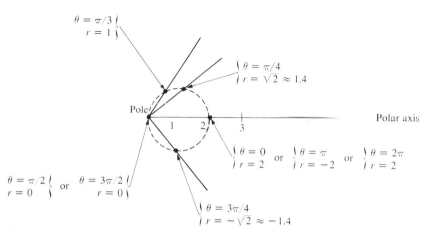

Figure 9.4

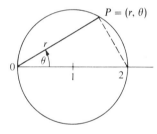

Figure 9.5

The dashed curve in Fig. 9.4 is the graph of $r = 2\cos\theta$.

A little geometry and trigonometry show that the graph of $r = 2\cos\theta$ is a circle. To see this, recall that an angle inscribed in a semicircle is a right angle, and consider a typical point $P = (r, \theta)$ for $r > 0$ on the circle shown in Fig. 9.5. Since $\cos\theta$ equals the adjacent side divided by the hypotenuse in the triangle, $\cos\theta = r/2$ or $r = 2\cos\theta$. ∎

EXAMPLE 3 Transform the equation $r = 2\cos\theta$ into rectangular coordinates.

SOLUTION Since $r^2 = x^2 + y^2$ and $r\cos\theta = x$, first multiply the equation $r = 2\cos\theta$ by r, obtaining

$$r^2 = 2r\cos\theta.$$

Hence $\quad x^2 + y^2 = 2x.$

Example 2 showed that this describes a circle of radius 1 and center $(1, 0)$. This could also be shown by rewriting the equation $x^2 + y^2 = 2x$ in the form

$$(x - 1)^2 + (y - 0)^2 = 1^2.$$

This equation says that the distance from (x, y) to $(1, 0)$ is 1. ∎

EXAMPLE 4 Transform the equation $y = 2$, which describes a horizontal straight line, into polar coordinates.

SOLUTION   Since $y = r \sin \theta$, $r \sin \theta = 2$,

or
$$r = \frac{2}{\sin \theta} = 2 \csc \theta.$$

This is far more complicated than the original equation but is still sometimes useful. ∎

Generally, rectangular coordinates are best for describing straight lines, while polar coordinates are best for describing circles.

EXAMPLE 5   Graph $r = 1 + \cos \theta$.

SOLUTION   Begin by making a table.

θ	0	$\frac{\pi}{4}$	$\frac{\pi}{2}$	$\frac{3\pi}{4}$	π	$\frac{5\pi}{4}$	$\frac{3\pi}{2}$	$\frac{7\pi}{4}$	2π
r	2	$1 + \frac{\sqrt{2}}{2} \approx 1.7$	1	$1 - \frac{\sqrt{2}}{2} \approx 0.3$	0	$1 - \frac{\sqrt{2}}{2} \approx 0.3$	1	$1 + \frac{\sqrt{2}}{2} \approx 1.7$	2

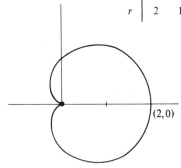

Figure 9.6

The last point is the same as the first. The graph begins to repeat itself. This heart-shaped curve, shown in Fig. 9.6, is called a *cardioid*. ∎

The cardioid in Example 5 is tangent to the x axis at the origin. In general, a polar graph that passes through the origin when $\theta = \theta_0$ is tangent to the ray $\theta = \theta_0$ there.

Spirals turn out to be quite easy to describe in polar coordinates. This will be illustrated by the graph of $r = 2\theta$ in the next example.

EXAMPLE 6   Graph $r = 2\theta$ for $\theta \geq 0$.

SOLUTION   First make a table.

θ	0	$\frac{\pi}{2}$	π	$\frac{3\pi}{2}$	2π	$\frac{5\pi}{2}$	...
r	0	π	2π	3π	4π	5π	...

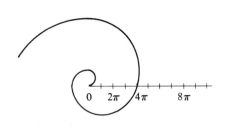

Figure 9.7

Increasing θ by 2π does *not* produce the same value of r. As θ increases, r increases. The graph for $\theta \geq 0$ is an endless spiral, going infinitely often around the pole. It is indicated in Fig. 9.7. ∎

Polar coordinates are also convenient for describing loops arranged like the petals of a flower, as Example 7 shows.

EXAMPLE 7   Graph $r = \sin 3\theta$.

SOLUTION   As θ increases from 0 up to $\pi/3$, 3θ increases from 0 up to π. Thus, r, which is $\sin 3\theta$, goes from 0 up to 1, then back to 0, for θ in $[0, \pi/3]$. This gives one loop of the three loops making up the graph of $r = \sin 3\theta$. For θ in $[\pi/3, 2\pi/3]$, $r = \sin 3\theta$ is negative (or 0). This yields the lower loop in Fig. 9.8. For θ in $[2\pi/3, \pi]$, r is again positive, and we obtain the upper left loop. Further choices of θ lead only to repetition of the loops already shown.

This is called the "three-leaved rose."

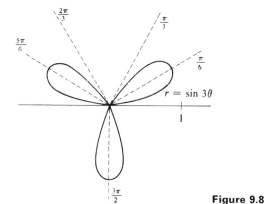

Figure 9.8

The graph of $r = \sin n\theta$ or $r = \cos n\theta$ has n loops when n is an odd integer and $2n$ loops when n is an even integer.

Exercises

1. Plot the points whose polar coordinates are:
   (a) $(1, \pi/6)$ (d) $(-2, \pi/3)$
   (b) $(2, \pi/3)$ (e) $(2, 7\pi/3)$
   (c) $(2, -\pi/3)$ (f) $(0, \pi/4)$
2. Find the rectangular coordinates of the points in Exercise 1.
3. Give at least three pairs of polar coordinates (r, θ) for the point $(3, \pi/4)$,
   (a) with $r > 0$,   (b) with $r < 0$.
4. Find polar coordinates (r, θ), with $0 \leq \theta < 2\pi$ and r positive, for the point whose rectangular coordinates are:
   (a) $(\sqrt{2}, \sqrt{2})$ (d) $(-\sqrt{2}, -\sqrt{2})$
   (b) $(-1, \sqrt{3})$ (e) $(0, -3)$
   (c) $(-5, 0)$ (f) $(1, 1)$

In Exercises 5 to 10 transform the equation into one in rectangular coordinates.

5. $r = \sin \theta$
6. $r = \csc \theta$
7. $r = 3/(4 \cos \theta + 5 \sin \theta)$
8. $r = 4 \cos \theta + 5 \sin \theta$
9. $r = \sin 2\theta$
   Hint: Use the identity for $\sin 2\theta$.
10. $r = 6$

In Exercise 11 to 16 transform the equation into one in polar coordinates.

11. $x + 2y = 3$
12. $x^2 + y^2 = 5$
13. $xy = 1$
14. $x^2 + y^2 = 4x$
15. $x = -2$
16. $y = x^2$

In Exercises 17 to 22 graph the given equations.

17. $r = 1 + \sin \theta$
18. $r = 3 + 2 \cos \theta$
19. $r = 1/\theta, \theta > 0$
20. $r = e^\theta, \theta \geq 0$
21. $r = \cos 3\theta$ (three-leaved rose)
22. $r = \cos 2\theta$ (four-leaved rose)

23. (a) Sketch the curves $r = 1 + \cos \theta$ and $r = 2 \cos \theta$ relative to the same polar axis.
    (b) Where do the curves in (a) intersect? (There are two intersections. Note that they are *not* both obtained by setting $1 + \cos \theta$ equal to $2 \cos \theta$.)
24. (a) Sketch the curves $r = \sin \theta$ and $r = \cos 2\theta$ relative to the same polar axis.
    (b) Where do the curves in (a) intersect? (There are four intersections.)

The curve $r = 1 + a \cos \theta$ (or $r = 1 + a \sin \theta$) is called a limaçon. Its shape depends on the choice of the constant a. For $a = 1$ we have the cardioid of Example 5. Exercises 25 and 26 concern other choices of a.

25. Graph $r = 1 + 2 \cos \theta$. (If $|a| > 1$, the graph of $r = 1 + a \cos \theta$ crosses itself and forms a loop.)

26 Graph

$$r = 1 + \tfrac{1}{2} \cos \theta.$$

27 Obtain the rectangular form of the equation $r^2 = \cos 2\theta$.

28 Graph the curve in Exercise 27, using polar coordinates. Note that, if $\cos 2\theta$ is negative, r is not defined and that, if $\cos 2\theta$ is positive, there are two values of r, $\sqrt{\cos 2\theta}$ and $-\sqrt{\cos 2\theta}$.

■ ■

In Appendix E it is shown that the graph of $r = 1/(1 + e \cos \theta)$ is a parabola if $e = 1$, an ellipse if $0 \leq e < 1$, and a hyperbola if $e > 1$. ("e" here is not related to $e \approx 2.718$.) Exercises 29 to 31 concern such graphs.

29 Find an equation in rectangular coordinates for the curve $r = 1/(1 + \cos \theta)$.

30 (a) Graph

$$r = \frac{1}{1 - \tfrac{1}{2} \cos \theta}.$$

(b) Find an equation in rectangular coordinates for the curve in (a).

31 (a) Graph

$$r = \frac{1}{1 + 2 \cos \theta}.$$

(b) What angles do the asymptotes to the graph in (a) make with the positive x axis?
(c) Find an equation in rectangular coordinates for the curve in (a).

32 (a) Graph

$$r = 3 + \cos \theta.$$

(b) Find the point on the graph in (a) that has the maximum y coordinate.

9.2 Area in polar coordinates

Section 5.3 showed how to compute the area of a region if the lengths of parallel cross sections are known. Sums based on estimating rectangles led to the formula

$$\text{Area} = \int_a^b c(x) \, dx,$$

where $c(x)$ denotes the cross-sectional length. Now consider quite a different situation, in which sectors of a circle, not rectangles, provide an estimate of the area.

Let R be a region in the plane and P a point inside it. Assume that the distance r from P to any point on the boundary of R is known as a function $r = f(\theta)$. (For convenience, assume that any ray from P meets the boundary of R just once, as in Fig. 9.9.)

The cross sections made by the rays from P are *not* parallel. Instead, like spokes in a wheel, they all meet at the point P. It would be unnatural to use rectangles to estimate the area, but it is reasonable to use sectors of circles that have P as a common vertex.

Begin by recalling that in a circle of radius r a sector of central angle θ has area $(\theta/2)r^2$. (See Fig. 9.10.) This formula plays the same role now as the formula for the area of a rectangle did in Sec. 5.3.

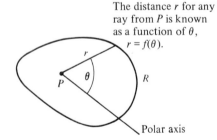

The distance r for any ray from P is known as a function of θ, $r = f(\theta)$.

Figure 9.9

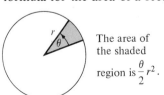

The area of the shaded region is $\dfrac{\theta}{2} r^2$.

Figure 9.10

AREA IN POLAR COORDINATES (informal approach)

Let R be the region bounded by the rays $\theta = \alpha$ and $\theta = \beta$ and by the curve $r = f(\theta)$, as shown in Fig. 9.11.

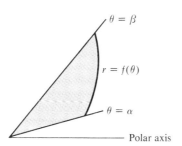

Figure 9.11

Assume $f(\theta) \geq 0$.

To obtain a *local estimate* for the area of R, consider the portion of R between the rays corresponding to the angles θ and $\theta + d\theta$, where $d\theta$ is a small positive number. (See Fig. 9.12.)

Figure 9.12   **Figure 9.13**

This narrow sector of a circle approximates the narrow wedge in Fig. 9.12.

The area of the narrow wedge which is shaded in Fig. 9.12 is approximately that of a sector of a circle of radius $r = f(\theta)$ and angle $d\theta$, shown in Fig. 9.13. The area of the sector in Fig. 9.13 is

This local estimate is the key to computing area in polar coordinates.

$$\frac{[f(\theta)]^2 \, d\theta}{2}. \tag{1}$$

Having found the local estimate of area (1), we conclude that the area of R is

$$\int_\alpha^\beta \frac{[f(\theta)]^2 \, d\theta}{2}.$$

HOW TO FIND AREA IN POLAR COORDINATES

In summary, the area of the region bounded by the rays $\theta = \alpha$ and $\theta = \beta$ and by the curve $r = f(\theta)$ is

$$\int_\alpha^\beta \frac{[f(\theta)]^2}{2} \, d\theta$$

or

$$\int_\alpha^\beta \frac{r^2 \, d\theta}{2}. \tag{2}$$

Formula (2) is applied in Sec. 14.8 to the motion of satellites and planets.

[Assume $f(\theta) \geq 0$.] It must be emphasized that no ray from the origin between α and β can cross the curve twice.

It may seem surprising to find $[f(\theta)]^2$, not just $f(\theta)$, in the integrand. But remember that area has the dimension "length times length." Since θ, given in radians, is dimensionless, being defined as "length of circular arc divided by length of radius," $d\theta$ is also dimensionless. Hence $f(\theta)\, d\theta$, having the dimension of length, not of area, could not be correct. But $\frac{1}{2}[f(\theta)]^2\, d\theta$, having the dimension of area (length times length), is plausible. For rectangular coordinates, in the expression $f(x)\, dx$, both $f(x)$ and dx have the dimension of length, one along the y axis, the other along the x axis; thus $f(x)\, dx$ has the dimension of area.

Memory device

As an aid in remembering the area of the narrow sector in Fig. 9.13, note that it resembles a triangle of height r and base $r\, d\theta$. Its area is in fact

$$\frac{1}{2} \cdot r \cdot r\, d\theta = \frac{r^2\, d\theta}{2}.$$

EXAMPLE 1 Find the area of the region bounded by the curve $r = 3 + \cos\theta$, shown in Fig. 9.14.

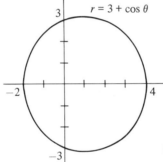

This is not a circle.

Figure 9.14

SOLUTION By the formula just obtained, this area is

$$\int_0^{2\pi} \frac{1}{2}(3 + \cos\theta)^2\, d\theta = \frac{1}{2} \int_0^{2\pi} (9 + 6\cos\theta + \cos^2\theta)\, d\theta$$

$$= \frac{1}{2}\left(9\theta + 6\sin\theta + \frac{\theta}{2} + \frac{\sin 2\theta}{4}\right)\bigg|_0^{2\pi}$$

$$= \frac{1}{2}(19\pi) - \frac{1}{2}(0) = \frac{19\pi}{2}. \quad \blacksquare$$

Observe that any line through the origin intersects the region of Example 1 in a segment of length 6, since $(3 + \cos\theta) + [3 + \cos(\theta + \pi)] = 6$ for any θ. Also, any line through the center of a circle of radius 3 intersects the circle in a segment of length 6. Thus two sets in the plane can have equal corresponding cross-sectional lengths through a fixed point and yet have different areas: the set in Example 1 has area $19\pi/2$, while the circle of radius 3 has area 9π. Knowing the lengths of all the cross sections of a region through a given point is not enough to determine the area of the region!

9.2 AREA IN POLAR COORDINATES

EXAMPLE 2 Find the area of the region inside one of the eight loops of the curve $r = \cos 4\theta$.

SOLUTION To graph one of the loops, start with $\theta = 0$. For that angle, $r = \cos(4 \cdot 0) = \cos 0 = 1$. The point $(r, \theta) = (1, 0)$ is the outer tip of a loop. As θ increases, from 0 to $\pi/8$, $\cos 4\theta$ decreases from $\cos 0 = 1$ to $\cos(\pi/2) = 0$. One of the eight loops is therefore bounded by the rays $\theta = \pi/8$ and $\theta = -\pi/8$. It is shown in Fig. 9.15, which, for good measure, displays all eight loops.

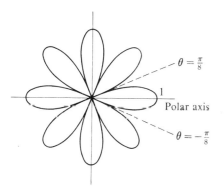

Figure 9.15

The area of the loop which is bisected by the polar axis is

$$\int_{-\pi/8}^{\pi/8} \frac{r^2}{2} d\theta = \int_{-\pi/8}^{\pi/8} \frac{\cos^2 4\theta}{2} d\theta$$

$$= \int_{-\pi/8}^{\pi/8} \frac{1 + \cos 8\theta}{4} d\theta$$

$$= \left(\frac{\theta}{4} + \frac{\sin 8\theta}{32} \right) \Big|_{-\pi/8}^{\pi/8}$$

$$= \left(\frac{\pi}{32} + \frac{\sin \pi}{32} \right) - \left[\frac{-\pi}{32} + \frac{\sin(-\pi)}{32} \right]$$

$$= \frac{\pi}{16}.$$

Incidentally, since the part of the loop above the polar axis has the same area as the part below,

$$\int_{-\pi/8}^{\pi/8} \frac{r^2}{2} d\theta = 2 \int_{0}^{\pi/8} \frac{r^2}{2} d\theta.$$

This shortcut simplifies the arithmetic a little, reducing the chance for error when evaluating the integral. ∎

THE AREA BETWEEN TWO CURVES

Assume that $r = f(\theta)$ and $r = g(\theta)$ describe two curves in polar coordinates and that $f(\theta) \geq g(\theta) \geq 0$ for θ in $[\alpha, \beta]$. Let R be the region between these two curves and the rays $\theta = \alpha$ and $\theta = \beta$, as shown in Fig. 9.16.

The area of R is obtained by subtracting the area within the inner curve $r = g(\theta)$ from the area within the outer curve $r = f(\theta)$. That is,

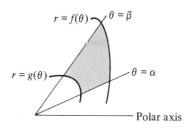

Figure 9.16

$$\text{Area between two curves} = \int_{\alpha}^{\beta} \frac{1}{2}[f(\theta)]^2 \, d\theta - \int_{\alpha}^{\beta} \frac{1}{2}[g(\theta)]^2 \, d\theta$$

$$= \frac{1}{2} \int_{\alpha}^{\beta} \{[f(\theta)]^2 - [g(\theta)]^2\} \, d\theta. \qquad (3)$$

EXAMPLE 3 Find the area of the region between the curves $r = 3 + \sin \theta$ and $r = 2 + \sin \theta$.

SOLUTION Each curve is swept out once as θ goes from 0 to 2π. Moreover, $3 + \sin\theta \geq 2 + \sin\theta \geq 0$. Thus the area between them is

$$\frac{1}{2}\int_0^{2\pi} [(3 + \sin\theta)^2 - (2 + \sin\theta)^2]\, d\theta$$

$$= \frac{1}{2}\int_0^{2\pi} [(9 + 6\sin\theta + \sin^2\theta) - (4 + 4\sin\theta + \sin^2\theta)]\, d\theta$$

$$= \frac{1}{2}\int_0^{2\pi} (5 + 2\sin\theta)\, d\theta$$

$$= \frac{1}{2}(5\theta - 2\cos\theta)\Big|_0^{2\pi}$$

$$= 5\pi. \quad \blacksquare$$

The next example shows that care must be taken when finding the area between two curves given in polar coordinates.

EXAMPLE 4 Find the area of the top half of the region inside the cardioid $r = 1 + \cos\theta$ and outside the circle $r = \cos\theta$.

SOLUTION The region is shown in Fig. 9.17.

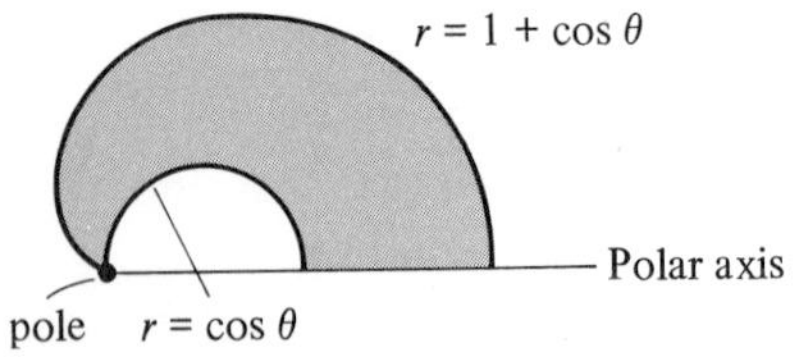

The top half of the circle $r = \cos\theta$ is swept out as θ goes from 0 to $\pi/2$. The top half of the cardioid is swept out as θ goes from 0 to π. Formula (3) for the area between two curves does not cover this situation. The safest thing is to compute the area of the top half of the cardioid and the top half of the circle separately and then subtract the second from the first.

The area of the top half of the cardioid is

$$\frac{1}{2}\int_0^{\pi}(1 + \cos\theta)^2\, d\theta = \frac{1}{2}\int_0^{\pi}(1 + 2\cos\theta + \cos^2\theta)\, d\theta.$$

Now,

$$\int_0^{\pi} \cos\theta\, d\theta = 0$$

since

$$\int_{\pi/2}^{\pi} \cos\theta\, d\theta = -\int_0^{\pi/2} \cos\theta\, d\theta.$$

Also,

$$\int_0^{\pi} \cos^2\theta\, d\theta = 2\int_0^{\pi/2} \cos^2\theta\, d\theta = 2(\pi/4) = \pi/2,$$

by the trick in Sec. 7.6. Thus the area of the top half of the cardioid is

$$\frac{1}{2}\left(\pi + 0 + \frac{\pi}{2}\right) = \frac{3\pi}{4}.$$

The area of the top half of the circle $r = \cos\theta$ is

Or note that it's half the area of a circle of radius $\frac{1}{2}$.

$$\frac{1}{2}\int_0^{\pi/2} \cos^2\theta\, d\theta = \frac{\pi}{8}.$$

Figure 9.17

Thus the area in question is

$$\frac{3\pi}{4} - \frac{\pi}{8} = \frac{5\pi}{8}.$$ ∎

AREA IN POLAR COORDINATES (formal approach)

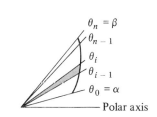

Figure 9.18

We wish to estimate the area of the shaded region in Fig. 9.18. To do so, introduce a partition of $[\alpha, \beta]$:

$$\alpha = \theta_0 < \theta_1 < \cdots < \theta_n = \beta.$$

and estimate the area of R bounded between the rays whose angles are

$$\theta_{i-1} \quad \text{and} \quad \theta_i,$$

as shown in Fig. 9.19.

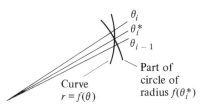

Figure 9.19

Pick any angle θ_i^* in the interval $[\theta_{i-1}, \theta_i]$ and use a sector of a circle whose radius is $f(\theta_i^*)$ to approximate the area of the shaded region. The approximating sector has radius $f(\theta_i^*)$ and angle $\theta_i - \theta_{i-1}$; hence its area is

$$\frac{[f(\theta_i^*)]^2}{2}(\theta_i - \theta_{i-1}).$$

The sum

$$\sum_{i=1}^{n} \frac{[f(\theta_i^*)]^2}{2}(\theta_i - \theta_{i-1})$$

is an estimate of the area of R.

As the mesh of the partition of $[\alpha, \beta]$ approaches 0, these sums become better and better approximations to the area of R, and also approach the definite integral

$$\int_{\alpha}^{\beta} \frac{[f(\theta)]^2}{2} d\theta,$$

which, therefore, is the area of R. This establishes formula (2).

Exoroicoc

In each of Exercises 1 to 6 find the area of the region bounded by the indicated curve and rays.

1 $r = 2\theta$, $\alpha = 0$, $\beta = \dfrac{\pi}{2}$.

2 $r = \sqrt{\theta}$, $\alpha = 0$, $\beta = \pi$.

3 $r = \dfrac{1}{1+\theta}$, $\alpha = \dfrac{\pi}{4}$, $\beta = \dfrac{\pi}{2}$.

4 $r = \sqrt{\sin\theta}$, $\alpha = 0$, $\beta = \dfrac{\pi}{2}$.

5 $r = \tan\theta$, $\alpha = 0$, $\beta = \dfrac{\pi}{4}$.

6 $r = \sec\theta$, $\alpha = \dfrac{\pi}{6}$, $\beta = \dfrac{\pi}{4}$.

7 (a) Graph the curve $r = 2\sin\theta$.
   (b) Compute the area inside it.

8 Find the area within the first turn of the spiral $r = e^\theta$, that is, for $0 \leq \theta \leq 2\pi$.
9 Find the area of the region inside the cardioid $r = 3 + 3\sin\theta$ and outside the circle $r = 3$.
10 (a) Graph the curve $r = \sqrt{\cos 2\theta}$. (Note that r is not defined for all θ.)
   (b) Find the area inside one of its two loops.

In Exercises 11 to 18 find the areas of the regions described.

11 Inside one loop of the curve $r = \sin 3\theta$.
12 Inside one loop of the curve $r = \cos 2\theta$.
13 Inside one loop of the curve $r = \sin 4\theta$.
14 Inside one loop of the curve $r = \cos 5\theta$.
15 Inside one loop of the curve $r = 2\cos 2\theta$, but outside the circle $r = 1$.
16 Inside the cardioid $r = 1 + \cos\theta$, but outside the circle $r = \sin\theta$. (*Suggestion:* Graph both curves first.)
17 Inside the circle $r = \sin\theta$, but outside the circle $r = \cos\theta$.
18 Inside the curve $r = 4 + \sin\theta$, but outside the curve $r = 3 + \sin\theta$.

■

19 (a) Show that the area of the triangle in Fig. 9.20 is $\int_0^\beta \tfrac{1}{2}\sec^2\theta\, d\theta$.

Figure 9.20

(b) From (a) and the fact that the area of a triangle is $\tfrac{1}{2}$(base)(height), show that $\tan\beta = \int_0^\beta \sec^2\theta\, d\theta$.
(c) With the aid of this equation, obtain another proof that $(\tan x)' = \sec^2 x$.

20 Show that the area of the shaded crescent between the two circular arcs is equal to the area of square $ABCD$ (See Fig. 9.21.) This type of result encouraged mathematicians from the time of the Greeks to try to find a method using only straightedge and compass for constructing a square whose area equals that of a given circle. This was proved impossible at the end of the nineteenth century.

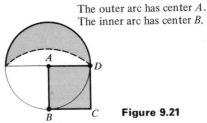

The outer arc has center A.
The inner arc has center B.

Figure 9.21

21 (a) Graph $r = 1/\theta$ for $0 < \theta \leq \pi/2$.
   (b) Is the area of the region bounded by the curve drawn in (a) and the rays $\theta = 0$ and $\theta = \pi/2$ finite or infinite?

■ ■

22 (a) Sketch the curve $r = 1/(1 + \cos\theta)$.
   (b) What is the equation of the curve in (a) in rectangular coordinates?
   (c) Find the area of the region bounded by the curve in (a) and the rays $\theta = 0$ and $\theta = 3\pi/4$, using (2).
   (d) Solve (c) using rectangular coordinates and the equation in (b).

23 A point P in a region R bounded by a closed curve has the property that each chord through P cuts R into two regions of equal area. Must P bisect each chord through P? Explain. (Assume that each chord meets the curve at two points, namely at the ends of the chord.)

24 Let R be a region in the plane and P a point in R such that every line in the plane which passes through P intersects R in an interval of length at least a.
   (a) Make a conjecture about the area of R.
   (b) Prove your conjecture.

25 Prove that, if a region in the plane has the property that any two points in it are within a distance d of each other, then its area is at most $\pi d^2/4$. *Hint:* Use polar coordinates with the pole on the border of the region, and consider $r(\theta)$ and $r(\theta + \pi/2)$.

9.3 Parametric equations

If a ball is thrown horizontally with a speed of 32 feet per second, it falls in a curved path. Air resistance disregarded, its position t seconds later is given by $x = 32t$, $y = -16t^2$ relative to the coordinate system in Fig. 9.22. Here the curve is completely described, not by expressing y as a function of x, but by expressing both x and y as functions of a third variable t. The third variable is called a *parameter* (*para* meaning together, *meter* meaning measure). The equations $x = 32t$, $y = -16t^2$ are called *parametric equations* for the curve.

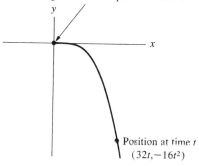

Figure 9.22

In this example it is easy to eliminate t and so find a direct relation between x and y:

$$t = \frac{x}{32},$$

hence
$$y = -16\left(\frac{x}{32}\right)^2 = -\frac{16}{(32)^2}x^2 = -\frac{1}{64}x^2.$$

The path of the falling ball is part of the curve $y = -\frac{1}{64}x^2$.

The next three examples present parametric equations in which the parameter is not time.

EXAMPLE 1 Let
$$\begin{cases} x = \cos 2\theta, \\ y = \sin \theta, \end{cases}$$

describe a curve parametrically. Graph this curve and find an equation linking x and y.

SOLUTION Since x and y are the sine and cosine of angles, $|x|$ and $|y|$ never exceed 1. Moreover when θ increases by 2π, we obtain the same point again. Here is a table showing a few points on the graph.

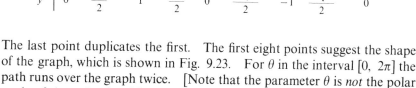

The last point duplicates the first. The first eight points suggest the shape of the graph, which is shown in Fig. 9.23. For θ in the interval $[0, 2\pi]$ the path runs over the graph twice. [Note that the parameter θ is *not* the polar angle of the point (x, y).]

The curve looks like part of a parabola. Fortunately, the equations for x and y are sufficiently simple that θ can be eliminated and a relation between x and y found:

$$x = \cos 2\theta = \cos^2 \theta - \sin^2 \theta$$
$$= 1 - 2 \sin^2 \theta$$
$$= 1 - 2y^2.$$

Thus the path lies on the parabola

$$x = 1 - 2y^2.$$

But note that it is only a small part of the parabola and sweeps out this part infinitely often, like a pendulum. ■

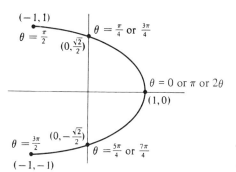

Figure 9.23

542 PLANE CURVES AND POLAR COORDINATES

In Example 1 it was easy to eliminate the parameter θ, thus obtaining a simple equation involving only x and y. In the next example, elimination of θ would lead to a complicated equation involving x and y. One great advantage of parametric equations is that they can provide a simple description of a curve, although it may be difficult or impossible to find an equation in x and y which describes the curve.

EXAMPLE 2 As a bicycle wheel of radius a rolls along, a tack stuck in its circumference traces out a curve called a *cycloid*, which consists of a sequence of arches, one arch for each revolution of the wheel. (See Fig. 9.24.)

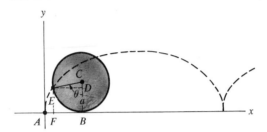

The rolling wheel has radius a.

Note that because the wheel doesn't slip, length of arc $EB = \overline{AB}$, thus $\overline{AB} = a\theta$.

When $\theta = 0$, the tack is on the ground at $x = 0$, $y = 0$.

Figure 9.24

Find the position of the tack as a function of the angle θ through which the wheel turns.

SOLUTION The x coordinate of the tack, corresponding to θ, is

$$\overline{AF} = \overline{AB} - \overline{ED} = a\theta - a \sin \theta$$

and the y coordinate is

$$\overline{EF} = \overline{BC} - \overline{CD} = a - a \cos \theta.$$

Then the position of the tack, as a function of the parameter θ, is

$$\begin{cases} x = a\theta - a \sin \theta, \\ y = a - a \cos \theta. \end{cases}$$

In this case, eliminating θ would lead to a complicated relation between x and y. ∎

Any curve $y = f(x)$ can be given parametrically.

Any curve can be described parametrically. For instance, consider the curve $y = e^x + x$. It is perfectly legal to introduce x itself as a parameter t, and write

$$\begin{cases} x = t \\ y = e^t + t. \end{cases}$$

This device may seem a bit artificial, but it will be useful in the next section in order to apply results for curves expressed by means of parametric equations to curves given in the form $y = f(x)$.

9.3 PARAMETRIC EQUATIONS

SLOPE OF A CURVE GIVEN PARAMETRICALLY

How can we find the slope of a curve which is described parametrically by the equations

$$x = g(t), \quad y = h(t)?$$

The clumsy, perhaps impossible, way is to solve the equation $x = g(t)$ for t as a function of x and plug the result into the equation $y = h(t)$, thus expressing y explicitly in terms of x; then differentiate the result to find dy/dx. Fortunately, there is a very easy way, which we will now describe.

Assume that y is a differentiable function of x. Then, by the chain rule,

$$\frac{dy}{dt} = \frac{dy}{dx} \frac{dx}{dt},$$

from which it follows that

Formula for the slope of a parametrized curve

$$\boxed{\frac{dy}{dx} = \frac{dy/dt}{dx/dt}.} \qquad (1)$$

It is assumed that in formula (1) dx/dt is not 0.

A similar approach gives d^2y/dx^2, as follows:

$$\frac{d^2y}{dx^2} = \frac{d\left(\frac{dy}{dx}\right)}{dx} = \frac{\frac{d}{dt}\left(\frac{dy}{dx}\right)}{\frac{dx}{dt}}.$$

EXAMPLE 3 At what angle does the arch of the cycloid shown in Example 2 meet the x axis at the origin?

SOLUTION The parametric equations of the cycloid are

$$x = a\theta - a\sin\theta, \quad y = a - a\cos\theta.$$

Here θ plays the role of the parameter t. Then

$$\frac{dx}{d\theta} = a - a\cos\theta \quad \text{and} \quad \frac{dy}{d\theta} = a\sin\theta.$$

Consequently,
$$\frac{dy}{dx} = \frac{dy/d\theta}{dx/d\theta} = \frac{a\sin\theta}{a - a\cos\theta}$$

$$= \frac{\sin\theta}{1 - \cos\theta}.$$

When θ is near 0, (x, y) is near the origin. How does the slope, $\sin\theta/(1 - \cos\theta)$, behave as $\theta \to 0^+$? L'Hôpital's rule applies, and we have

$$\lim_{\theta \to 0^+} \frac{\sin \theta}{1 - \cos \theta} = \lim_{\theta \to 0^+} \frac{\cos \theta}{\sin \theta}$$
$$= \infty.$$

Thus the cycloid comes in vertically at the origin. ∎

THE ROTARY ENGINE (optional)

The next two examples use parametric equations to describe the geometric principles of the rotary engine recognized by Felix Wankel in 1954. He found that it is possible for an equilateral triangle to revolve in a certain curve in such a way that its corners maintain contact with the curve and its centroid sweeps out a circle.

EXAMPLE 4 Let e and R be fixed positive numbers and consider the curve given parametrically by

This e is not related to the number $2.718\cdots$.

$$x = e \cos 3\theta + R \cos \theta \quad \text{and} \quad y = e \sin 3\theta + R \sin \theta.$$

Show that an equilateral triangle can revolve in this curve while its centroid describes a circle of radius e.

SOLUTION Figure 9.25 shows the typical point $P = (x, y)$ that corresponds to the parameter value θ. As θ increases by $2\pi/3$ from any given angle, the point Q goes once around the circle of radius e and returns to its initial position. During this revolution of Q the point P moves to a point P_1 whose angle, instead of being θ, is $\theta + 2\pi/3$. Thus, if P is on the curve, so are the points P_1 and P_2 shown in Fig. 9.26; these form an equilateral triangle.

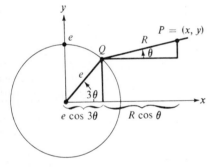

Figure 9.25

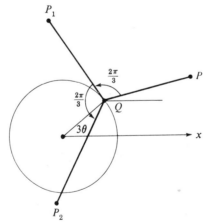

Figure 9.26

Consequently, each vertex of the equilateral triangle sweeps out the curve once, while the centroid Q goes three times around the circle of radius e. ∎

What does the curve described in Example 4 look like? Wankel graphed it without knowing that mathematicians had met it long before in a different setting, described in Example 5, which provides a way of graphing the curve.

EXAMPLE 5 A circle of radius r rolls without slipping around a fixed circle of radius $2r$. Describe the path swept out by a point P located at a distance e from the center of the moving circle, $0 \leq e \leq r$.

SOLUTION Place the rolling circle as shown in Fig. 9.27. Note that the center C of the rolling circle traces out a circle of radius $3r$. Let $R = 3r$.

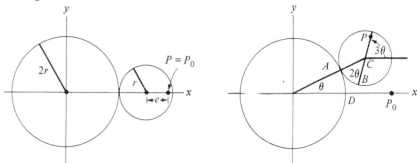

Figure 9.27 **Figure 9.28**

As the little circle rolls counterclockwise around the fixed circle without slipping, the point P traces out a path whose initial point P_0 is shown in Fig. 9.27. The typical point P on the path as the circle rolls around the larger circle is shown in Fig. 9.28. Since the radius of the rolling circle is half that of the fixed circle, angle ACB is 2θ. Thus the angle that CP makes with the x axis is the sum of θ and 2θ, which is 3θ. Consequently, $P = (x, y)$ has coordinates given parametrically as

Recall that in a triangle an exterior angle is the sum of the two opposite interior angles.

$$x = e \cos 3\theta + R \cos \theta \quad \text{and} \quad y = e \sin 3\theta + R \sin \theta.$$

Thus the curve swept out by P is precisely the curve Wankel studied.

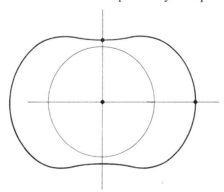

Figure 9.29

Long known to mathematicians as an *epitrochoid*, it appears typically as shown in Fig. 9.29. ∎

In order that the moving rotor in the rotary engine can turn the drive shaft, teeth are placed in it along a circle of radius $2e$ which engage teeth in the drive shaft, which has radius e. (See Fig. 9.30.) For each complete rotation of the rotor the drive shaft completes three rotations.

It was a Stuttgart professor, Othmar Baier, who showed that Wankel's curve was an epitrochoid. This insight was of aid in simplifying the machining of the working surface of the motor.

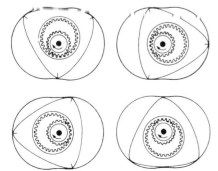

Figure 9.30

Exercises

1 Consider the parametric equations $x = 2t + 1$, $y = t - 1$.

(a) Fill in this table:

t	-2	-1	0	1	2
x					
y					

(b) Plot the five points (x, y) obtained in (a).
(c) Graph the curve given by the parametric equations $x = 2t + 1$, $y = t - 1$.
(d) Eliminate t to find an equation for the graph involving only x and y.

2 Consider the parametric equations $x = t + 1$, $y = t^2$.

(a) Fill in this table:

t	-2	-1	0	1	2
x					
y					

(b) Plot the five points (x, y) obtained in (a).
(c) Graph the curve.
(d) Find an equation in x and y that describes the curve.

3 Consider the parametric equations $x = t^2$, $y = t^2 + t$.

(a) Fill in this table:

t	-3	-2	-1	0	1	2	3
x							
y							

(b) Plot the seven points (x, y) obtained in (a).
(c) Graph the curve given by $x = t^2$, $y = t^2 + t$.
(d) Eliminate t and find an equation for the graph in terms of x and y.

4 Consider the parametric equations $x = 2 \cos t$, $y = 3 \sin t$.

(a) Fill in this table, expressing the entries decimally.

t	0	$\frac{\pi}{4}$	$\frac{\pi}{2}$	$\frac{3\pi}{4}$	π	$\frac{5\pi}{4}$	$\frac{3\pi}{2}$	$\frac{7\pi}{4}$	2π
x									
y									

(b) Plot the eight distinct points in (a).
(c) Graph the curve given by $x = 2 \cos t$, $y = 3 \sin t$.
(d) Using the identity $\cos^2 t + \sin^2 t = 1$, eliminate t.

In Exercises 5 and 6 express the curves parametrically with parameter x.

5 $y = \sqrt{1 + x^3}$
6 $y = \tan^{-1} 3x$

In Exercises 7 and 8 express the curves parametrically with parameter θ.

7 $r = \cos 2\theta$
8 $r = 3 + \cos \theta$

In Exercises 9 to 12 find dy/dx and d^2y/dx^2 for the given curves. (*Remember*: d^2y/dx^2 is not the quotient of d^2y/dt^2 and d^2x/dt^2.)

9 $x = t^3 + t$, $y = t^7 + t + 1$.
10 $x = \sin 3t$, $y = \cos 4t$.
11 $r = \cos 3\theta$.
12 $r = 2 + 3 \sin \theta$.

13 A curve is given parametrically by the equations

$$x = t^5 + \sin 2\pi t \qquad y = t + e^t.$$

Find the slope of the curve at the point corresponding to $t = 1$.
(Try to solve for y explicitly in terms of x. It's impossible.)

14 (a) Letting $t = -1, 0,$ and 1, find three points on the curve

$$\begin{cases} x = t^7 + t^2 + 1, \\ y = 2t^6 + 3t + 1. \end{cases}$$

(b) Can you eliminate t between the two equations?
(c) When t is large, what happens to y/x?
(d) What is the slope of the curve at the point corresponding to $t = 1$?

15 A ball is thrown at an angle α and initial velocity v_0, as sketched in Fig. 9.31. It can be shown that if time is in

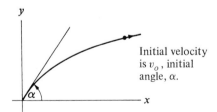

Initial velocity is v_0, initial angle, α.

Figure 9.31

seconds and distance in feet, then t seconds later the ball is at the point

$$\begin{cases} x = (v_0 \cos \alpha)t, \\ y = (v_0 \sin \alpha)t - 16t^2. \end{cases}$$

Eliminate t. The resulting equation shows that the path is a parabola.

16 Eliminate t and plot
$$\begin{cases} x = e^t, \\ y = e^{2t}. \end{cases}$$

17 Let a and b be positive numbers. Consider the curve given parametrically by the equations
$$x = a \cos t, \quad y = b \sin t.$$
(a) Show that the curve is the ellipse
$$\frac{x^2}{a^2} + \frac{y^2}{b^2} = 1.$$
(b) Find the area of the region bounded by the ellipse in (a) by making a substitution that expresses $4 \int_0^a y\, dx$ in terms of an integral in which the variable is t and the range of integration is $[0, \pi/2]$.

18 Consider the curve given parametrically by
$$x = t^2 + e^t, \quad y = t + e^t \quad \text{for } t \text{ in } [0, 1].$$
(a) Plot the points corresponding to $t = 0, \tfrac{1}{2}$, and 1.
(b) Sketch the curve.
(c) Find the slope of the curve at the point $(1, 1)$.
(d) Find the area of the region under the curve and above the interval $[1, e+1]$.

19 In Example 2, what is the value of θ when the point E is (a) at the top of the first arch? (b) at the right end of the first arch? (c) at the top of the second arch? (d) at the right end of the second arch?

■

A curve given in polar coordinates as $r = f(\theta)$ is given in rectangular coordinates parametrically as $x = f(\theta) \cos \theta$, $y = f(\theta) \sin \theta$. This observation is useful in Exercises 20 to 23.

20 Find the slope of the cardioid $r = 1 + \cos \theta$ at the point where it crosses the positive y axis.

21 (a) Find the slope of the cardioid $r = 1 + \cos \theta$ at the point (r, θ).
(b) What happens to the slope in (a) as $\theta \to \pi^-$?

22 Find the slope of the three-leaved rose, $r = \sin 3\theta$, at the point $(r, \theta) = (\sqrt{2}/2, \pi/12)$.

23 The region under one arch of the cycloid
$$\begin{cases} x = a\theta - a \sin \theta, \\ y = a - a \cos \theta, \end{cases}$$
and above the x axis is revolved around the x axis. Find the volume of the solid of revolution produced.

24 The same as the preceding exercise, except the region is revolved around the y axis instead of the x axis. Find the volume (a) by integration, (b) by Pappus's theorem.

■ ■

25 L'Hôpital's rule in Sec. 6.8 asserts that if $\lim_{t \to 0} f(t) = 0$, $\lim_{t \to 0} g(t) = 0$, and $\lim_{t \to 0} [f'(t)/g'(t)]$ exists, then $\lim_{t \to 0} [f(t)/g(t)] = \lim_{t \to 0} [f'(t)/g'(t)]$. Interpret that rule in terms of the parametrized curve $x = g(t)$, $y = h(t)$. (Make a sketch of the curve near $(0, 0)$ and show on it the geometric meaning of the quotients $f(t)/g(t)$ and $f'(t)/g'(t)$.)

26 Let a be a positive constant. Consider the curve given parametrically by the equations $x = a \cos^3 t$, $y = a \sin^3 t$.
(a) Sketch the curve.
(b) Find the slope of the curve at the point corresponding to the parameter value t.

27 Consider a tangent line to the curve in Exercise 26 at a point P in the first quadrant. Show that the length of the segment of that line intercepted by the coordinate axes is a.

This exercise is related to Examples 4 and 5, which concern the rotary engine.

28 In Example 4 a circle of radius r rolled around a circle of radius $2r$. Instead, consider the curve produced by a point P at a distance e from the center of the rolling circle, $0 \leq e \leq r$, if the radius of the fixed circle is $3r$.
(a) Find parametric equations of the curve produced.
(b) Sketch the curve.
(c) The curve in (b) is called a three-lobed epitrochoid. Show that a square rotor can revolve in it, as the triangular rotor did in Example 4. Engines with this design have been tried but are not as efficient as the standard rotary engine.

29 A circle of radius a is situated above the xy plane in a plane that is inclined at an angle to the xy plane. The shadow of the circle cast on the xy plane by light perpendicular to the xy plane is an oval curve. Find parametric equations for this curve and use them to show that the shadow is an ellipse.

30 A *trochoid*, a generalization of a cycloid, is defined as follows. A reflector is attached to a bicycle wheel of radius a at a distance b from the center of the wheel. As the wheel rolls without slipping, the reflector sweeps out a curve, called the trochoid. Find parametric equations for the trochoid. *Suggestion:* Use as parameter the angle that the wheel turns and assume that initially the reflector is directly below the center of the wheel.

31 A circle of radius b rolls without slipping on the inside of a circle of radius a, $a > b$. A point P on the circumference of the moving circle traces out a curve called a *hypocycloid*. Assume that initially the point P is at the

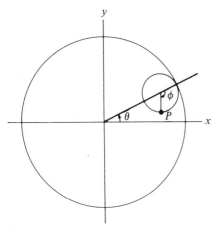

Figure 9.32

point of contact of the two circles and that the coordinates of the center of the moving circle are initially $(a - b, 0)$. Using Fig. 9.32, which shows the angles θ and ϕ, obtain parametric equations for the hypocycloid, as follows:

(a) Show that $b\phi = a\theta$.

(b) Show that $x = (a - b) \cos \theta + b \cos (\phi - \theta)$ and $y = (a - b) \sin \theta - b \sin (\phi - \theta)$.

32 (See Exercise 31.) When $b = a/4$, the equations for the hypocycloid take a very simple form, obtained as follows:
(a) Show that $\sin 3\theta = 3 \sin \theta - 4 \sin^3 \theta$ and $\cos 3\theta = 4 \cos^3 \theta - 3 \cos \theta$.
(b) Using (a) and Exercise 31(b), show that $x = a \cos^3 \theta$ and $y = a \sin^3 \theta$.
(c) From (b) deduce that the hypocycloid in the special case $b = a/4$ has the equation $x^{2/3} + y^{2/3} = a^{2/3}$.
(d) Graph the equation $x^{2/3} + y^{2/3} = 1$.

33 A circle of radius b rolls without slipping on the outside of a circle of radius a. A point P on the circumference of the moving circle traces out a curve called an *epicycloid*. Find parametric equations for an epicycloid. For convenience assume that the center of the fixed circle is $(0, 0)$, the moving circle initially has center $(a + b, 0)$, and initially P is the point of contact of the two circles, $(a, 0)$.

9.4 Arc length and speed on a curve

The path of some particle is given parametrically,

$$\begin{cases} x = g(t), \\ y = h(t). \end{cases}$$

(Think of t as time.) A physicist might ask the following questions: "How far does the particle travel from time $t = a$ to time $t = b$?" "What is the speed of the particle at time t?"

Consider the "distance traveled" question first. The second will then be easy to answer, making use of the derivative. (In our reasoning we shall call the parameter t, and think of it as time, but the results apply to any parameter.)

FORMULA FOR ARC LENGTH (informal approach)

Assume that $x = g(t)$ and $y = h(t)$ have continuous derivatives. Let us make a local estimate of the arc length swept out on the path during the short interval of time from t to $t + dt$.

Recall that $\Delta x \approx dx$ and $\Delta y \approx dy$.

s denotes arc length.

During the time dt the change in the x coordinate, Δx, is approximately $dx = g'(t) \, dt$ and the change in the y coordinate, Δy, is approximately $dy = h'(t) \, dt$. The corresponding change in arc length is approximately ds. (The symbol s will stand for arc length along the path.) Now, a very small piece of the curve resembles a straight line. A relation between dx, dy,

Figure 9.33

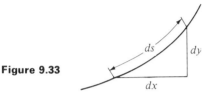

This formula is the key memory device for dealing with arc length.

and *ds* is therefore suggested by Fig. 9.33, which is a "right triangle" whose longest side is almost straight.

In view of the Pythagorean theorem, it is reasonable to suspect that

$$(ds)^2 = (dx)^2 + (dy)^2$$

or
$$ds = \sqrt{(dx)^2 + (dy)^2} \tag{1}$$

Rewriting (1) in the form

$$ds = \sqrt{\left(\frac{dx}{dt}\right)^2 + \left(\frac{dy}{dt}\right)^2}\, dt \tag{2}$$

gives us the local estimate of arc length. From this, we conclude that the arc length of the curve corresponding to *t* in [*a*, *b*] is given by the integral

Formula for arc length of parametrized curve

$$\boxed{\text{Arc length} = \int_a^b \sqrt{\left(\frac{dx}{dt}\right)^2 + \left(\frac{dy}{dt}\right)^2}\, dt.} \tag{3}$$

This formula holds for a curve given parametrically. If the curve is given in the form $y = f(x)$, it may be put in parametric form, with *x* as the parameter:

$$y = f(x), \qquad x = x.$$

Since $dx/dx = 1$, formula (3) for the arc length of the curve $y = f(x)$ for *x* in [*a*, *b*] takes the following form:

Formula for arc length of curve $y = f(x)$

$$\boxed{\text{Arc length} = \int_a^b \sqrt{1 + \left(\frac{dy}{dx}\right)^2}\, dx.} \tag{4}$$

(It is assumed that the derivative dy/dx is continuous.)

Exercise 26 shows that Neil was quite lucky.

Three examples will show how these formulas are applied. The first goes back to the year 1657, when the 20-year old Englishman, William Neil, became the first person to find the length of a curve that was neither a line, a polygon, nor a circle.

EXAMPLE 1 Find the arc length of the curve $y = x^{3/2}$ for *x* in [0, 1].

SOLUTION By (4)

$$\text{Arc length} = \int_0^1 \sqrt{1 + \left(\frac{dy}{dx}\right)^2}\, dx.$$

Since $y = x^{3/2}$, $dy/dx = \tfrac{3}{2}x^{1/2}$. Thus

$$\text{Arc length} = \int_0^1 \sqrt{1 + (\tfrac{3}{2}x^{1/2})^2}\, dx$$

$$= \int_0^1 \sqrt{1 + (9x/4)}\, dx$$

$$= \int_1^{13/4} \sqrt{u} \cdot \tfrac{4}{9}\, du \qquad \text{where } u = 1 + (9x/4),\ du = \tfrac{9}{4}\, dx$$

$$\underset{\text{FTC}}{=} \tfrac{4}{9} \cdot \tfrac{2}{3} u^{3/2} \Big|_1^{13/4}$$

$$= \tfrac{8}{27}[(\tfrac{13}{4})^{3/2} - 1^{3/2}]$$

$$= \frac{8}{27}\left(\frac{13^{3/2}}{8} - 1\right). \quad \blacksquare$$

Incidentally, the length of the curve $y = x^a$, where a is a rational number, usually *cannot* be computed with the aid of the fundamental theorem. The only cases in which it can be computed by the fundamental theorem are $a = 1$ (the graph is the line $y = x$) and $a = 1 + 1/n$, where a is an integer. Exercise 26 treats this question.

EXAMPLE 2 Find the distance s which the ball described at the beginning of Sec. 9.3 travels during the first b seconds.

SOLUTION Here $\qquad x = 32t \qquad$ and $\qquad y = -16t^2$.

Thus $\qquad dx/dt = 32 \qquad$ and $\qquad dy/dt = -32t$.

By formula (3),

$$s = \int_0^b \sqrt{(32)^2 + (-32t)^2}\, dt = 32 \int_0^b \sqrt{1 + t^2}\, dt,$$

a definite integral that can be evaluated with the aid of a table or the substitution $t = \tan \theta$; its value is

$$16b\sqrt{1 + b^2} + 16 \ln(b + \sqrt{1 + b^2}). \quad \blacksquare$$

EXAMPLE 3 Find the length of one arch of the cycloid in Example 2 of Sec. 9.3.

SOLUTION Here the parameter is θ, and we compute $dx/d\theta$ and $dy/d\theta$:

$$\frac{dx}{d\theta} = \frac{d}{d\theta}(a\theta - a \sin \theta) = a - a \cos \theta,$$

and

$$\frac{dy}{d\theta} = \frac{d}{d\theta}(a - a \cos \theta) = a \sin \theta.$$

To complete one arch, θ varies from 0 to 2π. By formula (3), the length of one arch is $\int_0^{2\pi} \sqrt{(a - a \cos \theta)^2 + (a \sin \theta)^2}\, d\theta$. Thus

$$\text{Length of arch} = a\int_0^{2\pi} \sqrt{(1-\cos\theta)^2 + (\sin\theta)^2}\, d\theta$$

$$= a\int_0^{2\pi} \sqrt{1 - 2\cos\theta + (\cos^2\theta + \sin^2\theta)}\, d\theta$$

$$= a\int_0^{2\pi} \sqrt{2 - 2\cos\theta}\, d\theta$$

$$= a\sqrt{2} \int_0^{2\pi} \sqrt{1 - \cos\theta}\, d\theta$$

$$= a\sqrt{2} \int_0^{2\pi} \sqrt{2} \sin\frac{\theta}{2}\, d\theta \qquad \text{trigonometry}$$

$$= 2a \int_0^{2\pi} \sin\frac{\theta}{2}\, d\theta \underset{\text{FTC}}{=} 2a\left(-2\cos\frac{\theta}{2}\right)\Big|_0^{2\pi}$$

$$= 2a[-2(-1) - (-2)(1)] = 8a.$$

While θ varies from 0 to 2π, the bicycle travels a distance $2\pi a \approx 6.28a$, and the tack travels a distance $8a$. ∎

SPEED OF A PARTICLE MOVING ON A CURVE

In practice we are not interested as much in the length of the path as in the speed of the particle as it moves along the path. The work done so far in this section helps us find this speed easily.

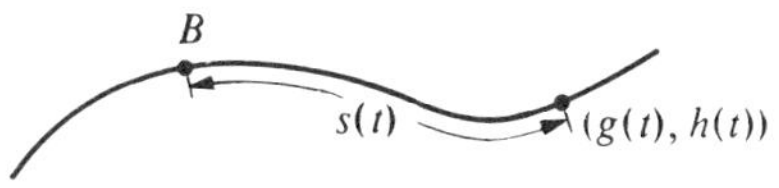

Consider a particle which at time t is at the point $(x, y) = (g(t), h(t))$. Choose a point B on the curve from which to measure distance along the curve, as shown in Fig. 9.34. Let $s(t)$ denote the distance from B to $(g(t), h(t))$. We shall always assume that B has been chosen in such a way that $s(t)$ is an *increasing* function of t.

Figure 9.34

DEFINITION *Speed on a curved path.* If ds/dt exists, it is called the *speed of the particle*.

Since $s(t)$ is assumed to be an increasing function, speed is not negative.

As early as Sec. 3.1 we were able to treat the speed of a particle moving in a straight path. Now it is possible to compute the speed of a particle moving on a curved path.

HOW TO FIND THE SPEED OF A PARTICLE MOVING ON A CURVED PATH

> If a particle at time t is at the point $(x, y) = (g(t), h(t))$, where g and h are functions having continuous derivatives, then its speed at time t is equal to
>
> $$\sqrt{[g'(t)]^2 + [h'(t)]^2}.$$

The argument is short. Let $s(t)$ denote the arc length along the curve from some base point B to the particle at time t.

Now,

$$s(t) = \int_a^t \sqrt{[g'(T)]^2 + [h'(T)]^2} \, dT.$$

(The letter T is introduced, since t is already used to describe the interval of integration.) Differentiation of this relation with respect to t (using the second fundamental theorem of calculus) yields

$$\frac{ds}{dt} = \sqrt{[g'(t)]^2 + [h'(t)]^2}.$$

EXAMPLE 4 Find the speed at time t of the ball described at the beginning of Sec. 9.3.

SOLUTION At time t the ball is at the point

$$(x, y) = (32t, -16t^2).$$

$\dot{x} = \dfrac{dx}{dt}$

$\dot{y} = \dfrac{dy}{dt}$

Thus dx/dt, usually written $\dot{x}$, is 32, and $\dot{y}$ is $-32t$. The speed of the ball is

$$\frac{ds}{dt} = \sqrt{\dot{x}^2 + \dot{y}^2} = \sqrt{32^2 + (-32t)^2}$$

$$= 32\sqrt{1 + t^2} \quad \text{feet per second.} \quad \blacksquare$$

So far in this section curves have been described in rectangular coordinates. Next consider a curve given in polar coordinates by the equation $r = f(\theta)$.

HOW TO FIND THE ARC LENGTH OF $r = f(\theta)$

The length of the curve $r = f(\theta)$ for θ in $[\alpha, \beta]$ is equal to

$$\int_\alpha^\beta \sqrt{[f(\theta)]^2 + [f'(\theta)]^2} \, d\theta$$

or

$$\int_\alpha^\beta \sqrt{r^2 + (r')^2} \, d\theta.$$

(Assume that f has a continuous derivative.)

This formula can be derived from that for the arc length of a parametrized curve in rectangular coordinates, as follows. Find the rectangular coordinates of the point whose polar coordinates are

$$(r, \theta) = (f(\theta), \theta).$$

They are $\begin{cases} x = f(\theta) \cos \theta, \\ y = f(\theta) \sin \theta. \end{cases}$

The curve is now given in rectangular form with parameter θ. Thus its length is

9.4 ARC LENGTH AND SPEED ON A CURVE 553

$$\int_\alpha^\beta \sqrt{\left(\frac{dx}{d\theta}\right)^2 + \left(\frac{dy}{d\theta}\right)^2}\, d\theta.$$

Now, $\dfrac{dx}{d\theta} = f(\theta)(-\sin\theta) + f'(\theta)\cos\theta,$

and $\dfrac{dy}{d\theta} = f(\theta)\cos\theta + f'(\theta)\sin\theta;$

hence

$$\left(\frac{dx}{d\theta}\right)^2 + \left(\frac{dy}{d\theta}\right)^2 = [f(\theta)]^2 \sin^2\theta - 2f(\theta)f'(\theta)\sin\theta\cos\theta + [f'(\theta)]^2 \cos^2\theta +$$
$$[f(\theta)]^2 \cos^2\theta + 2f(\theta)f'(\theta)\sin\theta\cos\theta + [f'(\theta)]^2 \sin^2\theta,$$

which, by the identity $\sin^2\theta + \cos^2\theta = 1$, simplifies to $[f(\theta)]^2 + [f'(\theta)]^2$. This justifies the formula.

EXAMPLE 5 Find the length of the spiral $r = e^{-3\theta}$ for θ in $[0, 2\pi]$.

SOLUTION First compute $r' = \dfrac{dr}{d\theta} = -3e^{-3\theta},$

and then use the formula

$$\text{Arc length} = \int_\alpha^\beta \sqrt{r^2 + (r')^2}\, d\theta$$

$$= \int_0^{2\pi} \sqrt{(e^{-3\theta})^2 + (-3e^{-3\theta})^2}\, d\theta$$

$$= \int_0^{2\pi} \sqrt{e^{-6\theta} + 9e^{-6\theta}}\, d\theta$$

$$= \sqrt{10} \int_0^{2\pi} \sqrt{e^{-6\theta}}\, d\theta$$

$$= \sqrt{10} \int_0^{2\pi} e^{-3\theta}\, d\theta$$

$$\underset{\text{FTC}}{=} \sqrt{10}\, \left.\frac{e^{-3\theta}}{-3}\right|_0^{2\pi}$$

$$= \sqrt{10}\left(\frac{e^{-3\cdot 2\pi}}{-3} - \frac{e^{-3\cdot 0}}{-3}\right)$$

$$= \sqrt{10}\left(\frac{e^{-6\pi}}{-3} + \frac{1}{3}\right)$$

$$= \sqrt{10}\left(\frac{1}{3} - \frac{e^{-6\pi}}{3}\right). \blacksquare$$

554 PLANE CURVES AND POLAR COORDINATES

Memory aid for arc length in polar coordinates

Think of the shaded region in Fig. 9.35 as almost a right triangle. (The "hypotenuse," which is part of the curve $r = f(\theta)$, is not straight. The "leg" of length $r\, d\theta$ is part of a circle of radius r.) It suggests the equation

$$(ds)^2 = (r\, d\theta)^2 + (dr)^2. \tag{5}$$

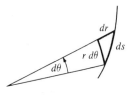

Figure 9.35

Division of (5) by $(d\theta)^2$ gives

$$\left(\frac{ds}{d\theta}\right)^2 = r^2 + \left(\frac{dr}{d\theta}\right)^2;$$

hence

$$\frac{ds}{d\theta} = \sqrt{r^2 + (r')^2}$$

is the integrand needed in computing arc length in polar coordinates.

FORMULA FOR ARC LENGTH (a more formal approach)

Partition the time interval $[a, b]$ and use this partition to inscribe a polygon in the curve of the moving particle, as shown in Fig. 9.36.

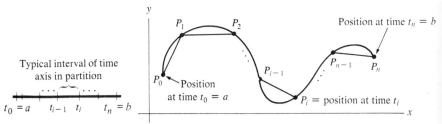

Figure 9.36

We are assuming $x = g(t)$ and $y = h(t)$ have continuous derivatives.

The length of such a polygon should approach the arc length as the mesh of the partition of $[a, b]$ shrinks toward 0 since the points P_i along the curve will get closer and closer together. The length of the typical straight-line segment $P_{i-1}P_i$, where $P_{i-1} = (g(t_{i-1}), h(t_{i-1}))$ and $P_i = (g(t_i), h(t_i))$ is (by the distance formula)

$$\sqrt{[g(t_i) - g(t_{i-1})]^2 + [h(t_i) - h(t_{i-1})]^2}$$

and so the length of the polygon is the sum

$$\sum_{i=1}^{n} \sqrt{[g(t_i) - g(t_{i-1})]^2 + [h(t_i) - h(t_{i-1})]^2}. \tag{6}$$

We shall relate this sum to sums of the type appearing in the definition of a definite integral over $[a, b]$.

By the mean-value theorem there exist numbers T_i^* and T_i^{**}, both in the interval $[t_{i-1}, t_i]$, such that $g(t_i) - g(t_{i-1}) = g'(T_i^*)(t_i - t_{i-1})$ and $h(t_i) - h(t_{i-1}) = h'(T_i^{**})(t_i - t_{i-1})$. Thus the sum (6) can be rewritten as

$$\sum_{i=1}^{n} \sqrt{[g'(T_i^*)]^2 + [h'(T_i^{**})]^2} (t_i - t_{i-1}). \tag{7}$$

If T_i^{**} were equal to T_i^*, then this sum (7) would be an approximating sum used in defining

$$\int_a^b \sqrt{[g'(t)]^2 + [h'(t)]^2}\, dt.$$

To get around this difficulty, notice that, since h' is continuous, $h'(T_i^*)$ is near $h'(T_i^{**})$ when the mesh of the partition of $[a, b]$ is small. If the sum in (7) is a good approximation to the arc length, then presumably so is the sum

$$\sum_{i=1}^{n} \sqrt{[g'(T_i^*)]^2 + [h'(T_i^*)]^2}(t_i - t_{i-1}) \tag{8}$$

In other words, it is reasonable to expect that

$$\text{Arc length} = \lim_{\text{mesh} \to 0} \sum_{i=1}^{n} \sqrt{[g'(T_i^*)]^2 + [h'(T_i^*)]^2}(t_i - t_{i-1})$$

But that limit is precisely the definition of the definite integral

$$\int_a^b \sqrt{[g'(t)]^2 + [h'(t)]^2}\, dt.$$

This shows why this definite integral should yield the arc length.

Exercises

In Exercises 1 to 14 find the arc length of the given curves over the given intervals.

1. $y = x^{3/2}$, x in $[1, 2]$
2. $y = x^{2/3}$, x in $[0, 1]$
3. $y = x^3/3 + 1/(4x)$, x in $[0, 1]$
4. $y = 1/(2x^2) + x^4/16$, x in $[2, 3]$
5. $y = x^{4/3}$, x in $[0, 1]$
6. $y = x^{5/4}$, x in $[0, 1]$
7. $y = (e^x + e^{-x})/2$, x in $[0, 2]$
8. $y = x^2/2 - (\ln x)/4$, x in $[2, 3]$
9. $x = \cos^3 t$, $y = \sin^3 t$, t in $[0, \pi/2]$
10. $x = \cos t + t \sin t$, $y = \sin t - t \cos t$, t in $[\pi/6, \pi/4]$
11. $r = e^\theta$, θ in $[0, 2\pi]$
12. $r = 1 + \cos \theta$, θ in $[0, \pi]$
13. $r = \cos^2(\theta/2)$, θ in $[0, \pi]$
14. $r = \sin^2(\theta/2)$, θ in $[0, \pi]$

In each of Exercises 15 to 18 find the speed of the particle at time t, given the parametric description of its path.

15. $x = 50t$, $y = -16t^2$
16. $x = \sec 3t$, $y = \sin^{-1} 4t$
17. $x = t + \cos t$, $y = 2t - \sin t$
18. $x = \csc 3t$, $y = \tan^{-1} \sqrt{t}$

19. (a) How far does a bug travel from time $t = 1$ to time $t = 2$, if at time t it is at the point (t^2, t^3)?
    (b) How fast is it moving at time t?
    (c) Graph its path relative to an xy coordinate system. Where is it at $t = 1$? At $t = 2$?
    (d) Eliminate t to find y as a function of x.

20. (Calculator) (a) Graph $y = x^3/3$.
    (b) Estimate its arc length from $(0, 0)$ to $(3, 9)$ by an inscribed polygon, whose vertices have x coordinates $0, 1, 2, 3$. A table of square roots or a calculator will be useful.
    (c) Set up a definite integral for the arc length in question.
    (d) Estimate the definite integral in (c) by using a partition of $[0, 3]$ into three sections, each of length 1, and the trapezoid approximation.

21. Assume that a curve is described in rectangular coordinates in the form $x = f(y)$. Show that

$$\text{Arc length} = \int_c^d \sqrt{1 + \left(\frac{dx}{dy}\right)^2}\, dy,$$

where y ranges in the interval $[c, d]$.

556 PLANE CURVES AND POLAR COORDINATES

22 (See Exercise 21.) Consider the arc length of the curve $y = x^{2/3}$ for x in the interval $[1, 8]$.
  (a) Set up a definite integral for this arc length, using x as the parameter.
  (b) Set up a definite integral for this arc length, using y as the parameter.
  (c) Evaluate the easier of the integrals in (a) and (b).

23 (a) At time t a particle has polar coordinates $r = g(t)$, $\theta = h(t)$. How fast is it moving?
  (b) Use the formula in (a) to find the speed of a particle which at time t is at the point $(r, \theta) = (e^t, 5t)$.

24 Let $P = (x, y)$ depend on θ as shown in Fig. 9.37.

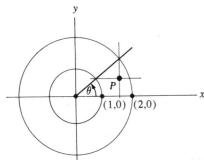

P lies on the horizontal line through $(\cos \theta, \sin \theta)$ and the vertical line through $(2 \cos \theta, 2 \sin \theta)$.

Figure 9.37

  (a) Sketch the curve that P sweeps out.
  (b) Show that $P = (2 \cos \theta, \sin \theta)$.
  (c) Set up a definite integral for the length of the curve described by P. (Do not evaluate it.)
  (d) Eliminate θ and show that P is on the ellipse
  $$\frac{x^2}{4} + \frac{y^2}{1} = 1.$$

■ ■

25 Show that, if $y = \dfrac{x^{m+1}}{m+1} + \dfrac{x^{1-m}}{4(m-1)}$, where m is any number other than 1 or -1, then the definite integral for the arc length of this curve can be computed with the aid of the fundamental theorem of calculus. Consider only arcs corresponding to x in $[a, b]$, $0 < a < b$.

26 Consider the length of the curve $y = x^m$, where m is a rational number. Show that the fundamental theorem of calculus is of aid in computing this length only if $m = 1$ or if m is of the form $1 + 1/n$ for some integer n. *Hint:* The analog of Chebyshev's theorem in Exercise 204 in Sec. 7.S holds for $\int x^p (1 + x)^q \, dx$.

27 Consider the cardioid $r = 1 + \cos \theta$ for θ in $[0, \pi]$. We may consider r as a function of θ or as a function of s, arc length along the curve, measured, say from $(2, 0)$.
  (a) Find the average of r with respect to θ.
  (b) Find the average of r with respect to s. *Hint:* Express all quantities appearing in this average in terms of θ.

28 The function $r = f(\theta)$ describes, for θ in $[0, 2\pi]$, a curve in polar coordinates. Assume r' is continuous and $f(\theta) > 0$. Prove that the average of r as a function of arc length is at least as large as the quotient $2A/s$, where A is the area swept out by the radius and s is the arc length of the curve. When is the average equal to $2A/s$?

9.5 Area of a surface of revolution

Take a cone first.

This section develops a technique for computing the surface area of a solid of revolution, such as a sphere. The approach will be intuitive and will only justify the plausibility of defining the area of a surface of revolution as a certain definite integral.

Begin by considering the area of a rather simple surface of revolution, the curved part of a cone whose base has radius r and whose slant height is l. (See Fig. 9.38.) If this cone is cut along a line through its point and laid flat, it becomes a sector of a circle of radius l, as shown in Fig. 9.39.

Now, the area of a sector of radius l and angle θ (in radians) is $\frac{1}{2}l^2\theta$. Since $\theta = 2\pi r/l$, the area of this sector is $\frac{1}{2}l^2(2\pi r/l)$, which equals $\pi r l$. Thus

$$\text{Area of curved surface of cone} = \pi r l. \qquad (1)$$

Figure 9.38

From (1) we will obtain the key fact needed in this section, namely the formula for the area swept out by revolving a line segment around a line.

9.5 AREA OF A SURFACE OF REVOLUTION 557

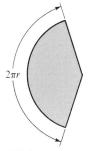

The cone laid flat is a sector of a circle.

Figure 9.39

Consider a line segment of length L in the plane which does not meet a certain line in the plane, called the "axis." (See Fig. 9.40.) When the line segment is revolved around the axis, it sweeps out a curved surface. The area of this surface equals

$$2\pi rL, \qquad (2)$$

where r is the distance from the midpoint of the line segment to the axis. (Note the resemblance to Pappus's theorem.) The surface in Fig. 9.40 is called a "frustum of a cone."

The argument for (2) depends on the fact that the frustum in Fig. 9.40 is the difference of two cones, one of slant height l_1 and radius r_1, and one of slant height l_2 and radius r_2, as shown in Fig. 9.41.

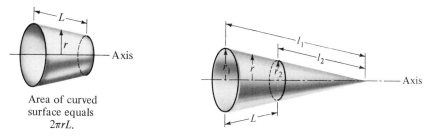

Area of curved
surface equals
$2\pi rL$.

Figure 9.40

Figure 9.41

The area of the frustum in Fig. 9.40 is therefore

$$\pi r_1 l_1 - \pi r_2 l_2. \qquad (3)$$

It will now be shown that (3) reduces to $2\pi rL$, where $r = (r_1 + r_2)/2$, that is, to $\pi(r_1 + r_2)L$.

By similar triangles in Fig. 9.41,

$$\frac{r_1}{r_2} = \frac{l_1}{l_2};$$

hence $r_1 l_2 = r_2 l_1$. This fact will be needed in a moment.

Now,

$$\begin{aligned}
\pi r_1 l_1 - \pi r_2 l_2 &= \pi r_1 (l_2 + L) - \pi r_2 l_2 \\
&= \pi r_1 l_2 + \pi r_1 L - \pi r_2 l_2 \\
&= \pi r_2 l_1 + \pi r_1 L - \pi r_2 l_2 \\
&= \pi r_2 (l_1 - l_2) + \pi r_1 L \\
&= \pi r_2 L + \pi r_1 L \\
&= \pi (r_1 + r_2) L \\
&= 2\pi rL.
\end{aligned}$$

This establishes (2).

Formula (2) is the basis for the following definition for the area of the surface swept out by revolving a curve around an axis. The definition will be justified informally.

DEFINITION *Area of a surface of revolution.* Consider a curve given by the parametric equations $x = g(t)$, $y = h(t)$, where g and h have continuous derivatives and $h(t) \geq 0$. Let C be that portion of the curve corresponding to t in $[a, b]$. Then the area of the surface of revolution formed by revolving C about the x axis is

$$\int_a^b 2\pi h(t) \sqrt{[g'(t)]^2 + [h'(t)]^2} \, dt \tag{4}$$

or, equivalently,

$$\int_a^b 2\pi y \sqrt{\left(\frac{dx}{dt}\right)^2 + \left(\frac{dy}{dt}\right)^2} \, dt.$$

JUSTIFICATION (informal) As the parameter goes from the value t to the value $t + dt$, where dt is a small positive number, the corresponding point on the curves sweeps out a short section of the curve. This short section is almost straight. Call its length ds. Moreover, this section is approximately a fixed distance y from the x axis. The area of the narrow circular band swept out by revolving this section of the curve around the x axis can be estimated with the aid of formula (2). (See Fig. 9.42.) By (2), the area of the band is about

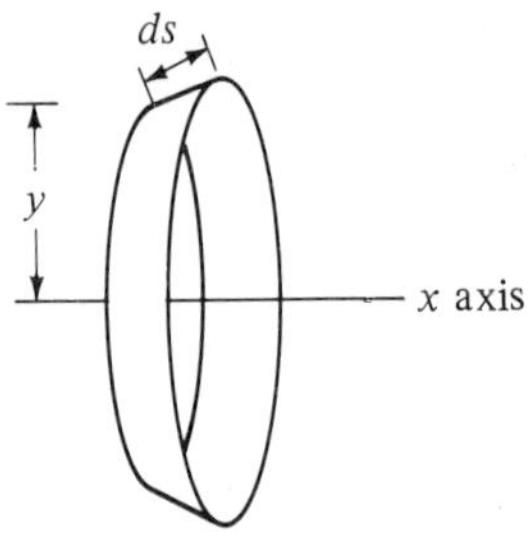

$$2\pi y \, ds. \tag{5}$$

But $ds = \sqrt{(dx/dt)^2 + (dy/dt)^2} \, dt$. So the area of the narrow band in Fig. 9.42 is approximately

$$2\pi y \sqrt{\left(\frac{dx}{dt}\right)^2 + \left(\frac{dy}{dt}\right)^2} \, dt. \tag{6}$$

Figure 9.42

With the aid of this local approximation, we immediately conclude that

$$\text{Surface area} = \int_a^b 2\pi y \sqrt{\left(\frac{dx}{dt}\right)^2 + \left(\frac{dy}{dt}\right)^2} \, dt.$$

This completes the motivation of formula (4).

If a curve is given by $y = f(x)$, where f has a continuous derivative and $f(x) \geq 0$, it may be parameterized by the equations $x = t$ and $y = f(t)$. Then $dx/dt = dx/dx = 1$ and $dy/dt = dy/dx$. Hence the surface area swept out by revolving about the x axis that part of the curve above $[a, b]$ is

A formula for surface area if $y = f(x)$ is revolved around x axis

$$\text{Surface area} = \int_a^b 2\pi y \sqrt{1 + \left(\frac{dy}{dx}\right)^2} \, dx.$$

9.5 AREA OF A SURFACE OF REVOLUTION

As the formulas are stated, they seem to refer only to surfaces obtained by revolving a curve about the x axis. In fact, they refer to revolution about any line. The factor y in the integrand,

$$2\pi y \sqrt{\left(\frac{dx}{dt}\right)^2 + \left(\frac{dy}{dt}\right)^2},$$

is the distance from the typical point on the curve to the axis of revolution. Replace y by R (for *radius*) to free ourselves from coordinate systems. (Use capital R to avoid confusion with polar coordinates.) The expression

$$\sqrt{\left(\frac{dx}{dt}\right)^2 + \left(\frac{dy}{dt}\right)^2} \, dt$$

is simply ds, since $\dfrac{ds}{dt} = \sqrt{\left(\dfrac{dx}{dt}\right)^2 + \left(\dfrac{dy}{dt}\right)^2}$.

The simplest way to write the formula for surface area of revolution is then

GENERAL FORMULA FOR SURFACE AREA

$$\boxed{\text{Surface area} = \int_a^b 2\pi R \, ds,}$$

where the interval $[a, b]$ refers to the parameter s. However, in practice s is seldom used as the parameter. Instead x, y, t', or θ is used and the interval of integration describes the interval through which the parameter varies.

To remember this formula, think of a narrow circular band of width ds and radius R as analogous to the rectangle shown in Fig. 9.43.

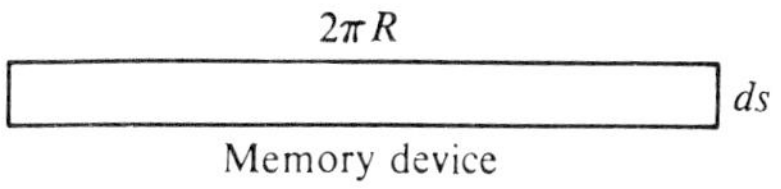

Figure 9.43 Memory device

EXAMPLE 1 Find the area of the surface obtained by revolving around the y axis the part of the parabola $y = x^2$ that lies between $x = 1$ and $x = 2$. (See Fig. 9.44.)

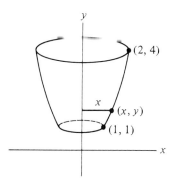

Figure 9.44

560 PLANE CURVES AND POLAR COORDINATES

R is found by inspection of a diagram.

SOLUTION The surface area is $\int_a^b 2\pi R \, ds$. Since the curve is described as a function of x choose x as the parameter. By inspection of Fig. 9.44, $R = x$. Next note that

$$ds = \frac{ds}{dx} dx = \sqrt{1 + \left(\frac{dy}{dx}\right)^2} \, dx$$

The surface area is therefore

$$\int_1^2 2\pi x \sqrt{1 + (2x)^2} \, dx$$

or

$$\int_1^2 2\pi x \sqrt{1 + 4x^2} \, dx.$$

To evaluate the integral use the substitution

$$u = 1 + 4x^2, \quad du = 8x \, dx;$$

hence $x \, dx = du/8$. The new limits of integration are $u = 5$ and $u = 17$. Thus

$$\text{Surface area} = \int_5^{17} 2\pi \sqrt{u} \, \frac{du}{8}$$

$$= \frac{\pi}{4} \int_5^{17} \sqrt{u} \, du$$

$$= \frac{\pi}{4} \cdot \frac{2}{3} u^{3/2} \Big|_5^{17}$$

$$= \frac{\pi}{6}(17^{3/2} - 5^{3/2}). \quad \blacksquare$$

EXAMPLE 2 Find the surface area of a sphere of radius a.

SOLUTION The circle of radius a has the equation $x^2 + y^2 = a^2$. The top half has the equation $y = \sqrt{a^2 - x^2}$. The sphere of radius a is formed by revolving this top half around the x axis. (See Fig. 9.45.) Using the memory device "$2\pi R \, ds$," we have

$$\begin{array}{c}\text{Surface area} \\ \text{of sphere}\end{array} = \int_{-a}^{a} 2\pi R \, \frac{ds}{dx} \, dx.$$

Now, $R = y$ and $ds/dx = \sqrt{1 + (dy/dx)^2}$, where $dy/dx = -x/\sqrt{a^2 - x^2}$. Thus

$$\begin{array}{c}\text{Surface area} \\ \text{of sphere}\end{array} = \int_{-a}^{a} 2\pi y \sqrt{1 + \left(\frac{-x}{\sqrt{a^2 - x^2}}\right)^2} \, dx$$

$$= \int_{-a}^{a} 2\pi \sqrt{a^2 - x^2} \sqrt{1 + \frac{x^2}{a^2 - x^2}} \, dx$$

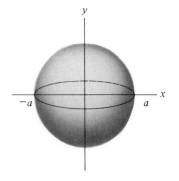

Figure 9.45

All this work for a constant integrand.

$$= \int_{-a}^{a} 2\pi\sqrt{a^2 - x^2}\sqrt{\frac{a^2}{a^2 - x^2}}\, dx$$

$$= \int_{-a}^{a} 2\pi a\, dx$$

$$= 2\pi ax \Big|_{-a}^{a}$$

$$= 4\pi a^2.$$

Exercise 10 has another approach. The surface area of a sphere is 4 times the area of its equatorial cross section. ∎

Exercises

In each of Exercises 1 to 4 set up a definite integral for the area of the indicated surface, using the suggested parameter. Show the radius R on a diagram; do *not* evaluate the definite integrals.

1. The curve $y = x^3$; x in $[1, 2]$; revolved about the x axis; parameter x.
2. The curve $y = x^3$; x in $[1, 2]$; revolved about the line $y = -1$; parameter x.
3. The curve $y = x^3$; x in $[1, 2]$, revolved about the y axis; parameter y.
4. The curve $y = x^3$; x in $[1, 2]$; revolved about the y axis; parameter x.
5. Find the area of the surface obtained by rotating about the x axis that part of the curve $y = e^x$ that lies above $[0, 1]$.
6. Find the area of the surface formed by rotating one arch of the curve $y = \sin x$ about the x axis.
7. One arch of the cycloid given parametrically by the formula $x = \theta - \sin\theta$, $y = 1 - \cos\theta$ is revolved around the x axis. Find the area of the surface produced.
8. The curve given parametrically by $x = e^t \cos t$, $y = e^t \sin t$, t in $[0, \pi/2]$, is revolved around the x axis. Find the area of the surface produced.

∎

The formula in Exercise 9 is used in Exercises 10 to 12.

9. Show that the area of the surface obtained by revolving the curve $r = f(\theta)$, $\alpha \leq \theta \leq \beta$, around the polar axis is

$$\int_{\alpha}^{\beta} 2\pi r \sin\theta \sqrt{r^2 + (r')^2}\, d\theta.$$

10. Use the formula in Exercise 9 to find the surface area of a sphere of radius a.
11. Find the area of the surface formed by revolving the portion of the curve $r = 1 + \cos\theta$ in the first quadrant about (a) the x axis, (b) the y axis. (In (b) the identity $1 + \cos\theta = 2\cos^2(\theta/2)$ may help.)
12. The curve $r = \sin 2\theta$, θ in $[0, \pi/2]$, is revolved around the polar axis. Set up an integral for the surface area.
13. Consider the smallest tin can that contains a given sphere. (The height and diameter of the tin can equal the diameter of the sphere.)

    (a) Compare the volume of the sphere with the volume of the tin can. Archimedes, who obtained the solution about 2200 years ago, considered it his greatest accomplishment. Cicero wrote, about two centuries after Archimedes' death:

    > I shall call up from the dust [the ancient equivalent of a blackboard] and his measuring-rod an obscure, insignificant person belonging to the same city [Syracuse], who lived many years after, Archimedes. When I was quaestor I tracked out his grave, which was unknown to the Syracusans (as they totally denied its existence), and found it enclosed all round and covered with brambles and thickets; for I remembered certain doggerel lines inscribed, as I had heard, upon his tomb, which stated that a sphere along with a cylinder had been set up on the top of his grave. Accordingly, after taking a good look all round (for there are a great quantity of graves at the Agrigentine Gate), I noticed a small column rising a little above the bushes, on which there was the figure of a sphere and a cylinder. And so I at once said to the Syracusans (I had their leading men with me) that I believed it was the very thing of which I was in search. Slaves were sent in with sickles who cleared the ground of obstacles, and when a passage to the place was opened we approached the pedestal fronting us; the epigram was traceable with about half the lines legible, as the latter portion was worn away.

(Cicero, *Tusculan Disputations*, v. 23, translated by J. E. King, Loeb Classical Library, Harvard University, Cambridge, 1950.) Archimedes was killed by a Roman soldier in 212 B.C. Cicero was quaestor in 75 B.C.

(b) Compare the surface area of the sphere with the area of the curved side of the can.

14 (a) Compute the area of the portion of a sphere of radius a that lies between two parallel planes at distances c and $c + h$ from the center of the sphere ($0 \leq c \leq c + h \leq a$).

(b) The result in (a) depends only on h, not on c. What does this mean geometrically?

15 The portion of the curve $x^{2/3} + y^{2/3} = 1$ situated in the first quadrant is revolved around the x axis. Find the area of the surface produced.

In each of Exercises 16 to 23 find the area of the surface formed by revolving the indicated curve about the indicated axis. Leave the answer as a definite integral, but indicate how it could be evaluated by the fundamental theorem of calculus.

16 $y = 2x^3$ for x in $[0, 1]$; about the x axis.
17 $y = 1/x$ for x in $[1, 2]$; about the x axis.
18 $y = x^2$ for x in $[1, 2]$; about the x axis.
19 $y = x^{4/3}$ for x in $[1, 8]$; about the y axis.
20 $y = x^{2/3}$ for x in $[1, 8]$; about the line $y = 1$.
21 $y = x^3/6 + 1/(2x)$ for x in $[1, 3]$; about the y axis.
22 $y = x^3/3 + 1/(4x)$ for x in $[1, 2]$; about the line $y = -1$.
23 The arc in Example 1; about the line $y = -1$.

24 Though the fundamental theorem of calculus is of no use in computing the perimeter of the ellipse $x^2/a^2 + y^2/b^2 = 1$, it is useful in computing the surface area of the "football" formed when the ellipse is rotated about one of its axes. Assuming that $a > b$ and that the ellipse is revolved around the x axis, find that area. Does your answer give the correct formula for the surface area of a sphere of radius a, $4\pi a^2$? (Let $b \to a^-$.)

25 The region bounded by $y = 1/x$ and the x axis and situated to the right of $x = 1$ is revolved around the x axis.
(a) Show that its volume is finite but its surface area is infinite.
(b) Does this mean that an infinite area can be painted by pouring a finite amount of paint into this solid?

26 Check that the formula for surface area agrees with our formula for the area of a cone.

■ ■

27 If the band formed by revolving a line segment is cut along the rotated segment and laid out in the plane, what shape will it be? *Warning*: It is generally *not* a rectangle. With this approach, compute its area.

28 Consider a solid of revolution. Its volume is approximated closely by the sum of the volumes of thin parallel circular slabs. Is the area of the surface of the solid approximated closely by the sum of the areas of the curved surfaces of the slabs?

29 Define the moment of a curve around the x axis to be $\int_a^b y\,ds$, where a and b refer to the range of the parameter, s. The moment of the curve around the y axis is defined as $\int_a^b x\,ds$. The centroid of the curve, $(\bar{x}, \bar{y})$, is defined by setting

$$\bar{x} = \frac{\int_a^b x\,ds}{\text{Length of curve}}, \qquad \bar{y} = \frac{\int_a^b y\,ds}{\text{Length of curve}}.$$

Find the centroid of the top half of the circle $x^2 + y^2 = a^2$.

30 (See Exercise 29.) Show that the area of the surface obtained by revolving a curve around the x axis is equal to the length of the curve times the distance that the centroid of the curve moves. (This is a variant of Pappus's theorem.)

31 Use Exercise 30 to find the surface area of the doughnut formed by revolving a circle of radius a around a line a distance b from its center, $b \geq a$.

32 Use Exercise 30 to find the area of the curved part of a cone of radius a and height h.

33 A disk of radius a is covered by a finite number of strips (perhaps overlapping). Prove that the sum of their widths is at least $2a$. (If the strips are parallel, the assertion is clearly true; do not assume that the strips are parallel.) A strip consists of the points between two parallel lines. *Hint*: Think of the sphere of which the disk is the equatorial cross section.

9.6 The angle between a line and a tangent line

In this section we deal with such questions as: "How do we find the angle between two intersecting curves?" "Why is a reflector parabolic?" "How can we find the angle between the radius arm to a point on a curve in polar coordinates and the tangent line at that point?" All the answers involve the angle between two lines.

THE ANGLE BETWEEN TWO LINES

Consider a line L in the xy plane. It forms an angle of inclination α, $0 \leq \alpha < \pi$, with the positive x axis. The slope m of L is $\tan \alpha$. (If $\alpha = \pi/2$, the slope is not defined.) See Fig. 9.46.

Consider two lines L and L', with angles of inclination α and α', and slopes m and m', respectively, as in Fig. 9.47. There are two (supplementary) angles between the two lines. The following definition serves to distinguish one of these two angles as *the* angle between L and L'.

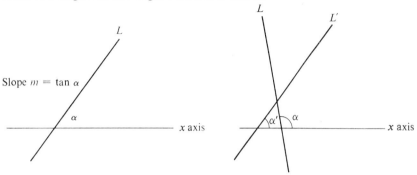

Figure 9.46   **Figure 9.47**

DEFINITION *Angle between two lines.* Let L and L' be two lines in the xy plane, so named that L has the *larger* angle of inclination, $\alpha > \alpha'$. The angle θ between L and L' is defined to be

$$\theta = \alpha - \alpha'.$$

If L and L' are parallel, define θ to be 0.

Figure 9.48 shows θ for some typical L and L'. In each case θ is the counterclockwise angle from L' to L. Note that θ depends on the choice of the x axis, and that $0 \leq \theta < \pi$.

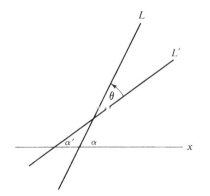

Figure 9.48

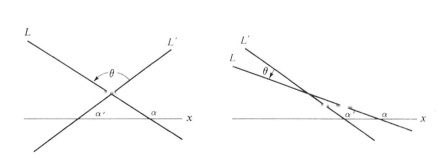

The tangent of θ is easily expressed in terms of the slopes m and m'. We have

$$\tan \theta = \tan (\alpha - \alpha')$$

$$= \frac{\tan\alpha - \tan\alpha'}{1 + \tan\alpha\tan\alpha'} \quad \text{by the identity for } \tan(A - B)$$

$$= \frac{m - m'}{1 + mm'}.$$

Formula for tangent of angle between two lines in terms of their slopes

Thus
$$\tan\theta = \frac{m - m'}{1 + mm'},$$

where m is the slope of the line with larger angle of inclination. If $mm' = -1$, then $\theta = \pi/2$; this corresponds to the fact that, as $mm' \to -1$, $|\tan\theta| \to \infty$.

The angle between two curves

In the first example the formula just developed is applied to find the angle between two intersecting curves. The "angle between two curves" means "the angle between their tangent lines at the intersection."

EXAMPLE 1 The curves $y = x^2$ and $y = \sqrt{x}$ cross at the point $(1, 1)$. Find the angle between them.

SOLUTION The sketch in Fig. 9.49 shows that the tangent line to $y = x^2$ at $(1, 1)$ has a larger inclination than the tangent line to $y = \sqrt{x}$ there. So let L denote the tangent line to $y = x^2$ at $(1, 1)$ and let L' denote the tangent line to $y = \sqrt{x}$ there.

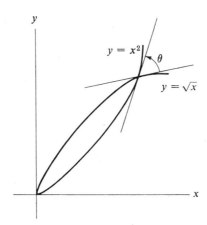

Figure 9.49

Since $(x^2)' = 2x$, L has slope $m = 2 \cdot 1 = 2$; since $(\sqrt{x})' = 1/(2\sqrt{x})$, L' has slope $m' = \frac{1}{2}$. Thus

$$\tan\theta = \frac{2 - \frac{1}{2}}{1 + 2 \cdot \frac{1}{2}} = \frac{\frac{3}{2}}{2}$$

$$= \frac{3}{4}.$$

A trigonometric table or calculator then shows that $\theta \approx 0.64$ radian. ∎

9.6 THE ANGLE BETWEEN A LINE AND A TANGENT LINE

THE REFLECTING PROPERTY OF THE PARABOLA

In the next example it is shown why flashlight reflectors and microwave relay dishes should ideally be parabolic. The argument uses the fact that the angle of reflection of a light ray equals its angle of incidence, as shown in Fig. 9.50.

Path of light reflected from a smooth surface.

Figure 9.50

EXAMPLE 2 Let P be any point on the parabolic curve $y = \sqrt{x}$ and let F be the point $(\frac{1}{4}, 0)$. Show that the angle between the line FP and the tangent line to the parabola at P equals the angle between the x axis and the tangent line at P. (See Fig. 9.51.)

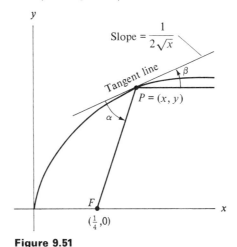

Figure 9.51

SOLUTION The slope of the tangent line at a point on the parabola $y = \sqrt{x}$ is $1/(2\sqrt{x})$. The slope of the line FP is, by the two-point formula,

$$\frac{y - 0}{x - \frac{1}{4}} = \frac{y}{x - \frac{1}{4}}.$$

Let α be the angle between the line FP and the tangent line. Let β be the angle between FP and the line through P parallel to the x axis.

Then $\quad \tan \alpha = \dfrac{\dfrac{y}{x - \frac{1}{4}} - \dfrac{1}{2\sqrt{x}}}{1 + \left(\dfrac{y}{x - \frac{1}{4}}\right)\left(\dfrac{1}{2\sqrt{x}}\right)} \quad$ using formula for tangent of angle between two lines

and $\quad \tan \beta = \dfrac{1}{2\sqrt{x}} \quad$ (slope of tangent line).

To show that $\alpha = \beta$, we show that $\tan \alpha = \tan \beta$, as follows:

$$\tan \alpha = \frac{\dfrac{y}{x - \frac{1}{4}} - \dfrac{1}{2\sqrt{x}}}{1 + \left(\dfrac{y}{x - \frac{1}{4}}\right)\left(\dfrac{1}{2\sqrt{x}}\right)}$$

$$= \frac{2\sqrt{x}\, y - x + \frac{1}{4}}{2(x - \frac{1}{4})\sqrt{x} + y}$$

$$= \frac{2\sqrt{x}\sqrt{x} - x + \frac{1}{4}}{2(x - \frac{1}{4})\sqrt{x} + \sqrt{x}} \qquad (x, y) \text{ lies on the parabola } y = \sqrt{x}.$$

$$= \frac{2x - x + \frac{1}{4}}{2x\sqrt{x} - (\sqrt{x}/2) + \sqrt{x}}$$

$$= \frac{x + \frac{1}{4}}{(2x + \frac{1}{2})\sqrt{x}} = \frac{1}{2\sqrt{x}} = \tan \beta. \quad \blacksquare$$

Similar reasoning shows that rays of light parallel to the x axis reflect off the parabola $y = \sqrt{2cx}$, where c is a positive constant, and pass through the point $(c/2, 0)$, which is called the *focus*. Appendix E offers a geometric definition of the focus.

ANGLE BETWEEN A RAY AND A TANGENT

Consider next a curve described in polar coordinates, $r = f(\theta)$. Let O be the pole and let P be a typical point on the curve. How can we find the angle between the line OP and the tangent to the curve at P? Denote this angle by the letter γ (gamma), as shown in Fig. 9.52.

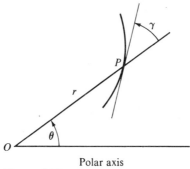

Figure 9.52

The following theorem gives a formula for $\tan \gamma$. Once $\tan \gamma$ is known, γ can be found with the aid of a calculator or tables.

THEOREM Let γ be the angle between the line OP and the tangent line to a point P on the curve $r = f(\theta)$. Then

$$\tan \gamma = \frac{r}{dr/d\theta}. \tag{1}$$

9.6 THE ANGLE BETWEEN A LINE AND A TANGENT LINE 567

PROOF We first find the slope of the tangent line at P, in Fig. 9.52. The tangent line at P has an angle of inclination ϕ and slope dy/dx. The ray from O to P has angle θ and slope y/x, as shown in Fig. 9.53. Thus

$$\tan \gamma = \tan(\phi - \theta)$$
$$= \frac{\tan \phi - \tan \theta}{1 + \tan \phi \tan \theta}$$
$$= \frac{(dy/dx) - (y/x)}{1 + (dy/dx)(y/x)}$$
$$= \frac{x\,dy - y\,dx}{x\,dx + y\,dy} \quad \text{switching to differentials.}$$

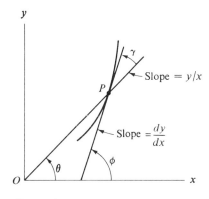

Figure 9.53

To express this last formula in terms of polar coordinates, find the differentials of $x = r\cos\theta$ and $y = r\sin\theta$:

$$dx = -r\sin\theta\,d\theta + \cos\theta\,dr$$

and

$$dy = r\cos\theta\,d\theta + \sin\theta\,dr.$$

Thus

$$\tan \gamma = \frac{r\cos\theta(r\cos\theta\,d\theta + \sin\theta\,dr) - r\sin\theta(-r\sin\theta\,d\theta + \cos\theta\,dr)}{r\cos\theta(-r\sin\theta\,d\theta + \cos\theta\,dr) + r\sin\theta(r\cos\theta\,d\theta + \sin\theta\,dr)}$$
$$= \frac{r^2\,d\theta}{r\,dr} \quad \text{algebra, together with the identity } \cos^2\theta + \sin^2\theta = 1$$
$$= \frac{r}{(dr/d\theta)},$$

which is (1). ∎

Though the justification of formula (1) for $\tan \gamma$ is a bit complicated, a memory aid that appeared in Fig. 9.35, Sec. 9.4 makes the formula quite intuitive, as a glance at Fig. 9.54 will show.

If $d\theta$ is small, the two sides of the sector are almost parallel, the curve locally looks like the tangent line, and the angle δ should be roughly γ. If that is so, it would be reasonable to hope that

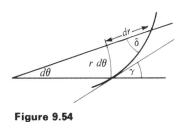

Figure 9.54

$$\tan \gamma \approx \tan \delta \approx \frac{r\,d\theta}{dr} = \frac{r}{dr/d\theta} = \frac{r}{r'}.$$

It is easy to work with γ; if ϕ is needed, express it in terms of γ and θ.

EXAMPLE 3 Let P be a point on the spiral $r = e^\theta$ shown in Fig. 9.55. Find γ and ϕ.

SOLUTION We have
$$\tan \gamma = \frac{r}{r'} = \frac{e^\theta}{(e^\theta)'} = \frac{e^\theta}{e^\theta} = 1.$$

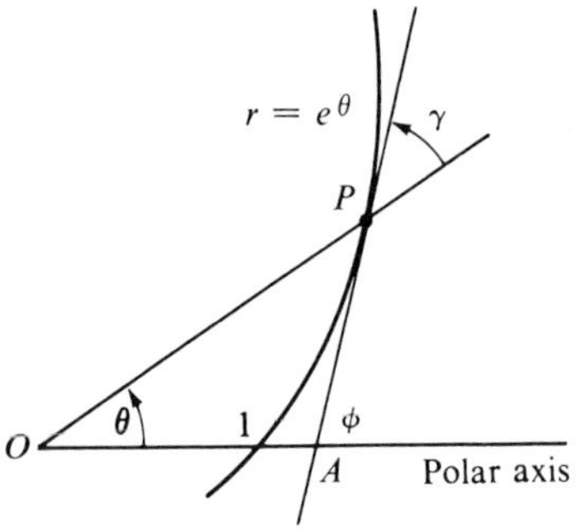

$$r = e^\theta$$

Figure 9.55

The angle is constant, $\gamma = \pi/4$.

Inspection of triangle OAP in Fig. 9.55 shows that

$$\phi = \theta + \gamma = \theta + \pi/4. \quad \blacksquare$$

Exercises

In each of Exercises 1 and 2 find the angle between the two given lines.

1 A line of inclination $\pi/4$ and a line of inclination $3\pi/4$.
2 A line of inclination $5\pi/6$ and a line of inclination $\pi/6$.

In each of Exercises 3 to 6 find the tangent of the angle between two lines with the given slopes and then the angle itself.

3 Slopes 2 and 3.
4 Slopes 2 and $-\tfrac{1}{2}$.
5 Slopes -2 and -3.
6 Slopes $\sqrt{3}$ and $-\sqrt{3}$.

In each of Exercises 7 to 10 find the tangent of the angle between the two curves at the indicated point of intersection.

7 $y = \sin x$ and $y = \cos x$ at $(\pi/4, \sqrt{2}/2)$.
8 $y = x^2$ and $y = x^3$ at $(1, 1)$.
9 $y = e^x$ and $y = e^{-x}$ at $(0, 1)$.
10 $y = \sec x$ and $y = \sqrt{2} \tan x$ at $(\pi/4, \sqrt{2})$.

In Exercises 11 to 14 find γ and ϕ for the given curves and angles θ. Use a table or calculator to estimate γ.

11 $r = e^{\sqrt{-\theta}}$; $\theta = \pi/6$.
12 $r = 1 + \cos \theta$; $\theta = \pi/4$.
13 $r = \sin 2\theta$, $\theta = \pi/6$.
14 $r = \theta$; $\theta = 1$.

15 Show that, if $r = a \sin \theta$ and $0 \le \theta \le \dfrac{\pi}{2}$, then $\gamma = \theta$.
16 Fill in the missing algebra in the proof that $\tan \gamma = r/r'$.
17 Show that, for the cardioid $r = 1 - \cos \theta$, $\gamma = \theta/2$.
18 (a) For the cardioid $r = 1 + \cos \theta$ find $\lim_{\theta \to \pi^-} \gamma$.
  (b) Sketch $r = 1 + \cos \theta$, using the information obtained in (a).
19 If for the curve $r = f(\theta)$, γ always equals θ, what are all the possibilities for f?

20 If for the curve $r = f(\theta)$, γ is independent of θ, what are all the possibilities for f?
21 This exercise explains why "whispering" rooms are elliptical. The ellipse $x^2/a^2 + y^2/b^2 = 1$, $a > b > 0$, has foci at $F = (\sqrt{a^2 - b^2}, 0)$ and $F' = (-\sqrt{a^2 - b^2}, 0)$. Let P be any point on the ellipse, and let T be the tangent line to the ellipse at P as in Fig. 9.56. Show that a sound

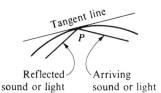

Figure 9.56

wave starting at F, after bouncing off the ellipse at P, will pass through F'. (Assume that sound, like light, is reflected off P in such a way that the angle of reflection equals the angle of incidence.) The whisper of a person standing at one focus of an elliptical room can easily be heard at the other focus, because all sound waves travel the same distance between foci.

22 Consider the curve $r = 1 + a \cos \theta$, where a is fixed, $0 \le a \le 1$.
  (a) Relative to the same polar axis graph the curves corresponding to $a = 0, \tfrac{1}{4}, \tfrac{1}{2}, \tfrac{3}{4}, 1$.
  (b) For $a = \tfrac{1}{4}$ the graph in (a) is convex, but not for $a = 1$. Show that for $\tfrac{1}{2} < a \le 1$ the curve is not convex.
23 Four dogs are chasing each other counterclockwise at the same speed. Initially they are at the four vertices of

a square of side *a*. As they chase each other, they approach the center of the square in spiral paths. How far does each dog travel? (See Fig. 9.7.)

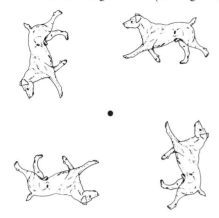

Figure 9.57

(a) First find the equation of the dog's path in polar coordinates; then find its arc length.
(b) Answer the question without calculus.

24 Let $r = f(\theta)$ describe a curve that does not pass through the origin O. Show that, at a point P on the curve closest to O, the line OP is perpendicular to the tangent line to the curve at P. (Assume that f is differentiable.)

25 Show that the parabola $y = \sqrt{2cx}$, where c is positive, has the property that a ray of light parallel to the x axis, reflecting off the parabola, passes through the point $(c/2, 0)$.

26 The formula for $\tan \phi$ (where ϕ is shown in Fig. 9.53) is messier than the formula for $\tan \gamma$. Show that $\tan \phi$ equals

$$\frac{r \cos \theta + r' \sin \theta}{-r \sin \theta + r' \cos \theta}.$$

9.7 The second derivative and the curvature of a curve

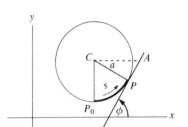

The tangent line turns as P moves counterclockwise and s increases.

Figure 9.58

The rate of change in y with respect to x measures the steepness of a curve. What is a reasonable measure of its "curviness" or curvature? A line, being perfectly straight, has no curvature. It seems reasonable that a line, therefore, should have curvature 0. Moreover, a large circle should have less curvature than a small circle.

When you walk around a small circle, your direction changes much more rapidly than when you walk around a large circle. To make this idea more precise, and to obtain a measure of the curvature of a circle, consider the diagram in Fig. 9.58 of a circle of radius a and a line tangent to it. Start at the bottom of the circle at P_0 and walk counterclockwise. At a distance s along the curve from P_0 the direction is given by the angle ϕ from the positive x axis to the tangent line at P; the angle ϕ depends on s. Define *the curvature of a circle* to be the rate at which ϕ changes with respect to s, that is, $d\phi/ds$.

The next theorem shows that $d\phi/ds$ is small for a large circle and large for a small circle; in fact, it is simply the reciprocal of the radius.

THEOREM 1 For a circle of radius a, swept out counterclockwise, the curvature $d\phi/ds$ is constant and equals $1/a$, the reciprocal of the radius.

PROOF It is necessary to express ϕ as a function of s. To do this, introduce the line CA parallel to the x axis, as in Fig. 9.58. Then ϕ equals angle PAC (alternate interior angles). Thus angle ϕ is the complement of angle PCA (by right triangle CPA). But angle $P_0 CP$ is also the complement of angle PCA. Thus $\phi = $ angle $P_0 CP$, which, by the definition of radian measure, has measure s/a. Thus

$$\phi = \frac{s}{a} \quad \text{and} \quad \frac{d\phi}{ds} = \frac{1}{a},$$

and the theorem is proved. ∎

570 PLANE CURVES AND POLAR COORDINATES

Since $d\phi/ds$, the rate of change of direction with respect to arc length, gives a reasonable measure of curvature for a circle—the larger the circle, the less its curvature—it is common to use it as a measure of curvature for other curves.

Before defining curvature in general, we should discuss the arc length s and the angle ϕ in a little more detail. First of all, let us agree to measure arc length in such a way that it increases as we move along the curve away from the base point.

Second, consider the angle ϕ. There is ambiguity in the choice of ϕ, for if ϕ describes the angle so does $\phi + n\pi$ for any integer n. The particular choice is not of any importance; what does matter is that ϕ should vary continuously as we traverse the curve. Look back at Fig. 9.58 which showed the typical tangent line to a circle of radius a. If we choose $\phi = 0$ for the (horizontal) tangent line at P_0, then our choice of ϕ for all other tangent lines to the circle, as we traverse the circle counterclockwise, is determined. Figure 9.59 shows that, as P goes once around the circle, ϕ increases by 2π.

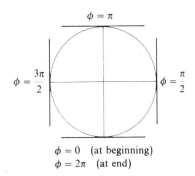

Figure 9.59

If we had chosen ϕ for the tangent line at P_0 initially to be, say, π, then, as we traverse the circle once, ϕ would increase to 3π. Since it is the *rate* at which ϕ changes that concerns us, not its actual value, this ambiguity, which may occur with any curve, does not affect the following definition.

DEFINITION *Curvature.* Assume that a curve is given parametrically, with the parameter of the typical point P being s, the distance along the curve from a fixed point P_0' to P. Let ϕ be the angle between the tangent line at P and the positive part of the x axis. The *curvature* at P is the derivative

$$\frac{d\phi}{ds}$$

(if this derivative exists).

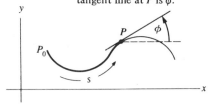

Figure 9.60

Observe that a straight line has zero curvature everywhere, since ϕ is a constant. Note also that in Fig. 9.60 the curvature is negative at the indicated point P, but positive at P_0. If the curve is traversed in the opposite direction, however, the curvature is positive at P and negative at P_0. Inspection of Fig. 9.60 shows that the curvature changes sign at the inflection point. Therefore it is not surprising if the curvature, though defined as a derivative, is intimately connected with the second derivative d^2y/dx^2. But more than the second derivative is involved, as the parabola $y = x^2$ shows (see Fig. 9.61). Far from $(0, 0)$ the parabola $y = x^2$ has little curvature; near $(0, 0)$ it appears to bend more rapidly. Thus the curvature is not constant; but the second derivative d^2y/dx^2 is constant (equaling 2 for all x). As Theorem 2 shows, the curvature is determined by the *pair* of functions d^2y/dx^2 and dy/dx.

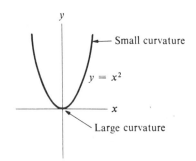

Figure 9.61

HOW TO COMPUTE THE CURVATURE OF $y = f(x)$

THEOREM 2 Assume that in the parameterization of the curve $y = f(x)$, with arc length s from a fixed point P_0 as the parameter, x increases as s increases. Then

9.7 THE SECOND DERIVATIVE AND THE CURVATURE OF A CURVE

$$\text{Curvature} = \frac{d^2y/dx^2}{[1 + (dy/dx)^2]^{3/2}}.$$

PROOF By the chain rule, $\dfrac{d\phi}{ds} = \dfrac{d\phi/dx}{ds/dx}.$

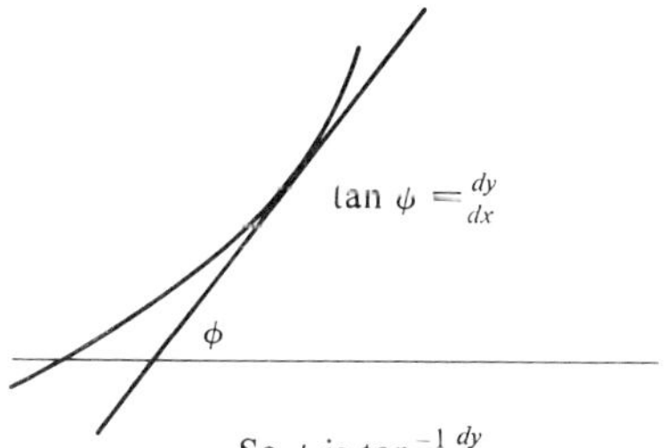

As was shown in Sec. 9.4,

$$\frac{ds}{dx} = \left[1 + \left(\frac{dy}{dx}\right)^2\right]^{1/2}.$$

All that remains is to express $d\phi/dx$ in terms of dy/dx and d^2y/dx^2. Note that, in Fig. 9.62,

$$\tan \phi = \frac{dy}{dx}, \quad \text{(slope of tangent line)}$$

or

$$\phi = \tan^{-1}\frac{dy}{dx} + n\pi,$$

for some fixed integer n. Recall that

$$\frac{d}{du}(\tan^{-1} u) = \frac{1}{1 + u^2}.$$

Thus by the chain rule,

$$\frac{d\phi}{dx} = \frac{1}{1 + (dy/dx)^2} \frac{d}{dx}\left(\frac{dy}{dx}\right)$$

$$= \frac{d^2y/dx^2}{1 + (dy/dx)^2}.$$

Consequently, $\dfrac{d\phi}{ds} = \dfrac{d\phi/dx}{ds/dx} = \dfrac{d^2y/dx^2}{[1 + (dy/dx)^2]\sqrt{1 + (dy/dx)^2}},$

and the theorem is proved. ■

EXAMPLE 1 Find the curvature at a typical point (x, y) on the curve $y = x^2$.

SOLUTION In this case $dy/dx = 2x$ and $d^2y/dx^2 = 2$. Thus the curvature at (x, y) is

$$\text{Curvature} = \frac{d^2y/dx^2}{[1 + (dy/dx)^2]^{3/2}} = \frac{2}{[1 + (2x)^2]^{3/2}}.$$

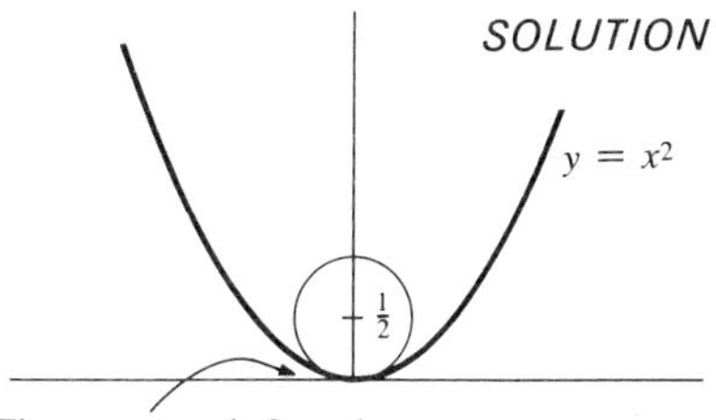

Hence at $(0, 0)$ the curvature is 2, and the curve near the origin resembles a circle of radius $\tfrac{1}{2}$. As $|x|$ increases, the curvature approaches 0, and the curve gets straighter. (See Fig. 9.63.) ■

$\tan \phi = \dfrac{dy}{dx}$

So ϕ is $\tan^{-1}\dfrac{dy}{dx}$ or differs from $\tan^{-1}\dfrac{dy}{dx}$ by a multiple of π.

Figure 9.62

The curvature is 2, and a circle of radius $\tfrac{1}{2}$ fits snugly here.

Figure 9.63

572 PLANE CURVES AND POLAR COORDINATES

Theorem 2 tells how to find the curvature if y is given as a function of x. If a curve is described parametrically, its curvature can be found with the aid of the next theorem. The proof, which rests on Theorem 2, is outlined in Exercise 25.

THEOREM 3 If, as we move along the parametrized curve, $x = g(t)$, $y = f(t)$, to a point P, both x and the arc length s from a point P_0 increase as t increases, then

See Exercise 20 for the formula in polar coordinates.

$$\text{Curvature} = \frac{\dot{x}\ddot{y} - \dot{y}\ddot{x}}{[(\dot{x})^2 + (\dot{y})^2]^{3/2}}.$$

(The dot notation for derivatives shortens the formula; $\dot{x} = dx/dt$, $\ddot{x} = d^2x/dt^2$, etc.) ∎

EXAMPLE 2 The cycloid determined by a wheel of radius 1 has the parametric equations

$$x = \theta - \sin\theta,$$
$$y = 1 - \cos\theta,$$

as shown in Sec. 9.3. (See Fig. 9.64.) Find the curvature at a typical point on this curve.

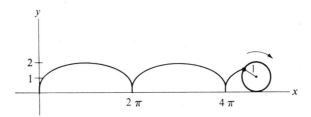

Figure 9.64

SOLUTION To do so, use Theorem 3, noting that θ plays the role of the parameter t. First,

$$\frac{dx}{d\theta} = 1 - \cos\theta, \qquad \frac{dy}{d\theta} = \sin\theta,$$

$$\frac{d^2x}{d\theta^2} = \sin\theta, \qquad \frac{d^2y}{d\theta^2} = \cos\theta.$$

Then, by Theorem 3,

$$\text{Curvature} = \frac{(1-\cos\theta)\cos\theta - (\sin\theta)\sin\theta}{[(1-\cos\theta)^2 + (\sin\theta)^2]^{3/2}}$$

$$= \frac{\cos\theta - (\cos^2\theta + \sin^2\theta)}{[1 - 2\cos\theta + (\cos^2\theta + \sin^2\theta)]^{3/2}}$$

$$= \frac{\cos\theta - 1}{(2 - 2\cos\theta)^{3/2}}$$

$$= \frac{\cos\theta - 1}{2^{3/2}(1 - \cos\theta)^{3/2}}$$

$$= \frac{-1}{\sqrt{8}} \frac{1 - \cos\theta}{(1 - \cos\theta)^{3/2}}$$

$$= \frac{-1}{\sqrt{8}} \frac{1}{(1 - \cos\theta)^{1/2}}.$$

Since $y = 1 - \cos\theta$, the curvature is simply

$$-\frac{1}{\sqrt{8}\sqrt{y}} = -\frac{1}{\sqrt{8y}}. \quad \blacksquare$$

The curvature of the cycloid in Example 2 is negative. This means that $d\phi/ds$ is negative, or ϕ decreases as s increases. This is plausible, for as s increases we move to the right on the arc, and the tangent line turns clockwise: ϕ decreases as a function of arc length s. In general, the sign of the curvature depends on the base point from which arc length is measured and on the direction in which the curve is traversed.

THE RADIUS OF CURVATURE   It was noted in Example 1 that at $(0, 0)$ the parabola $y = x^2$ has curvature 2, the same as a circle of radius $\frac{1}{2}$. In a sense, then, the parabola closely resembles a circle of radius $\frac{1}{2}$ in the vicinity of $(0, 0)$. This observation suggests the following definition.

DEFINITION   *Radius of curvature.* The radius of curvature of a curve at a point is the absolute value of the reciprocal of the curvature:

A large radius of curvature implies a small curvature.

$$\text{Radius of curvature} = \left|\frac{1}{\text{Curvature}}\right|.$$

As can be easily checked, the radius of curvature of a circle of radius a is, fortunately, a.

The cycloid in Example 2 has the radius of curvature at the point (x, y) equal to

$$\left|\frac{1}{(1/\sqrt{8y})}\right| = \sqrt{8y}.$$

In particular, at the top of an arch, the radius of curvature is $\sqrt{8 \cdot 2} = \sqrt{16} = 4$.

The osculating circle   At a given point P on a curve, the osculating circle at P is defined to be that circle which (a) passes through P, (b) has the same slope at P as the curve does, and (c) has the same second derivative there. Its radius is therefore equal to the radius of curvature of the curve at P. Fig. 9.63 shows the osculating circle at the origin for the parabola $y = x^2$. In that case the osculating circle lies on one side of the curve. However, the osculating circle at other points P on the parabola actually *crosses* the parabola at P. In fact, for any curve the osculating circle at a point P where the curvature is neither a maximum nor a minimum generally crosses the curve at P. Though at first this may be surprising, a little reflection will show why it is to be expected.

Think of driving along the curve. If you start at P and move in the direction in which the curve has less curvature than that of the osculating circle at P, then it seems plausible that you will be *outside* the circle. On the other hand, if you start at P and move along the curve in the other direction, in which the curvature of the curve is greater than that of the osculating circle at P, then you will be *inside* the circle. This informal argument suggests that the osculating circle at P crosses the curve at P.

Exercises

In each of Exercises 1 to 6 find the curvature and radius of curvature of the given curve at the given point.

1. $y = x^2$ at $(1, 1)$
2. $y = \cos x$ at $(0, 1)$
3. $y = e^{-x}$ at $(1, 1/e)$
4. $y = \ln x$ at $(e, 1)$
5. $y = \tan x$ at $(\pi/4, 1)$
6. $y = \sec 2x$ at $(\pi/6, 2)$

In Exercises 7 to 10 find the curvature of the given curves for the given values of the parameter.

7. $\begin{cases} x = 2 \cos 3t \\ y = 2 \sin 3t \end{cases}$ at $t = 0$

8. $\begin{cases} x = 1 + t^2 \\ y = t^3 + t^4 \end{cases}$ at $t = 2$

9. $\begin{cases} x = e^{-t} \cos t \\ y = e^{-t} \sin t \end{cases}$ at $t = \pi/6$

10. $\begin{cases} x = \cos^3 \theta \\ y = \sin^3 \theta \end{cases}$ at $\theta = \pi/3$

11. (a) Compute the curvature and radius of curvature for the curve $y = (e^x + e^{-x})/2$.
    (b) Show that the radius of curvature at (x, y) is y^2.

12. Find the radius of curvature along the curve $y = \sqrt{a^2 - x^2}$, where a is a constant. (Since the curve is part of a circle of radius a, the answer should be $1/a$.)

■

13. For what value of x is the radius of curvature of $y = e^x$ smallest?

14. For what value of x is the radius of curvature of $y = x^3$ smallest?

15. (a) Show that where a curve has its tangent parallel to the x axis its curvature is simply the second derivative d^2y/dx^2.
    (b) Show that the absolute value of the curvature is never larger than the absolute value of d^2y/dx^2.

16. An engineer lays out a railroad track as indicated in Fig. 9.65. BC is part of a circle. AB and CD are straight and tangent to the circle. After the first train runs over this track, the engineer was fired because the curvature was not a continuous function. Why should it be?

Figure 9.65

Exercises 17 to 19 are related.

17. Find the radius of curvature at a typical point on the ellipse

$$\begin{cases} x = a \cos \theta, \\ y = b \sin \theta. \end{cases}$$

18. (a) Show, by eliminating θ, that the curve in Exercise 17 is the ellipse

$$\frac{x^2}{a^2} + \frac{y^2}{b^2} = 1.$$

(b) What is the radius of curvature of this ellipse at $(a, 0)$? at $(0, b)$?

19. An ellipse has a major diameter of length 6 and a minor diameter of length 4. Draw the circles that most closely

approximate this ellipse at the four points which lie at the extremities of its diameters. (See Exercises 17 and 18.)

20 Theorem 2 gives a formula for curvature if the curve is given in rectangular form, $y = f(x)$. If the curve is given in polar form, $r = f(\theta)$, show that curvature equals $[r^2 + 2(r')^2 - rr'']/[r^2 + (r')^2]^{3/2}$. *Hint:* Consider the parametric representation of the curve as $x = r \cos \theta$, $y = r \sin \theta$, where $r = f(\theta)$.

21 Use the formula in Exercise 20 to show that the cardioid $r = 1 + \cos \theta$ has curvature $3\sqrt{2}/(4\sqrt{r})$ at (r, θ).

22 Use the formula in Exercise 20 to find the curvature of $r = a \cos \theta$.

23 Use the formula in Exercise 20 to find the curvature of $r = \cos 2\theta$.

24 If, on a curve, $dy/dx = y^3$, express the curvature in terms of y.

25 (a) Show that if x and y are given parametrically as functions of t, then

$$\frac{d^2y}{dx^2} = \frac{\dot{x}\ddot{y} - \dot{y}\ddot{x}}{(\dot{x})^2}.$$

(b) Prove Theorem 3.

■ ■

26 At the top of the cycloid in Example 2 the radius of curvature is twice the diameter of the rolling circle. What would you have guessed the radius of curvature to be at this point? Why is it not simply the diameter of the wheel, since the wheel at each moment is rotating about its point of contact with the ground?

Exercises 27 and 28 are related.

27 Let s denote arc length along a curve. Show that the curvature at a point is equal to $x'y'' - y'x''$ evaluated at that point, where differentiation is with respect to the parameter s, arc length.

28 As in Exercise 27, consider differentiation with respect to arc length. Show that
(a) $(x')^2 + (y')^2 = 1$,
(b) $x'x'' + y'y'' = 0$,
(c) $x'y'' - y'x'' = y''[(x')^2 + (y')^2]/x' = y''/x'$.

9.S Summary

In this chapter we examined various properties of curves. Polar coordinates, useful for describing certain curves, such as cardioids, circles, and n-leaved roses, were defined in Sec. 9.1. The area within a curve given in polar form $r = f(\theta)$ was found in Sec. 9.2. In Sec. 9.3 the parametric form of a curve was described. In physics the parameter is usually time t, but in calculus it is often angle θ or arc length s. Arc length of a curve and speed along a curve were computed in Sec. 9.4 and the area of a surface of revolution in Sec. 9.5. The angle between a line and a tangent line or between two curves was the subject of Sec. 9.6, which was of use in establishing the reflecting property of the parabola. A way to measure the curvature of a curve was defined in Sec. 9.7.

The following table provides memory aids for most of the formulas developed in the chapter.

Section	Concept	Memory Aid
9.2	Area $= \int_{\alpha}^{\beta} \frac{r^2}{2} d\theta$	Area $= \frac{1}{2} r \cdot r\, d\theta$
9.4	Arc length $= \int_a^b \sqrt{\left(\frac{dx}{dt}\right)^2 + \left(\frac{dy}{dt}\right)^2}\, dt$ $= \int_a^b \sqrt{1 + \left(\frac{dy}{dx}\right)^2}\, dx$	$(ds)^2 = (dx)^2 + (dy)^2$

576 PLANE CURVES AND POLAR COORDINATES

Section	Concept	Memory Aid
	Arc length $= \int_{\alpha}^{\beta} \sqrt{r^2 + (r')^2}\, d\theta$	$(ds)^2 = (r\,d\theta)^2 + (dr)^2$
	Speed $= \sqrt{\left(\dfrac{dx}{dt}\right)^2 + \left(\dfrac{dy}{dt}\right)^2}$	
9.5	Area of surface of revolution $= \int_{a}^{b} 2\pi R\, ds$	
9.6	$\tan \gamma = r/r'$	

VOCABULARY AND SYMBOLS polar coordinates (r, θ) curvature
pole, polar axis   radius of curvature
parametric representation   angle between two lines
arc length s angle between tangent line and ray from origin γ
surface of revolution

KEY FACTS Polar and rectangular coordinates are related through the equations $x = r \cos \theta$, $y = r \sin \theta$, $r^2 = x^2 + y^2$, $\tan \theta = y/x$.

The slope of a curve given parametrically as $x = g(t)$, $y = h(t)$ is

$$\frac{dy}{dx} = \frac{dy/dt}{dx/dt} = \frac{\dot{y}}{\dot{x}}.$$

The area $\int_a^b y\, dx$ under a curve given parametrically is obtained by a substitution in which y, dx, and the limits of integration are all expressed in terms of the parameter t. (See Exercise 17(b) of Sec. 9.3.) The parameter may be denoted by a different letter, such as θ, for instance.

For formulas for area, arc length, speed, area of surfaces of revolution, and $\tan \gamma$, see the preceding table.

The angle between two lines in the plane equals the larger inclination minus the smaller inclination. The tangent of this angle is

$$\frac{m - m'}{1 + mm'},$$

where m is the slope of the line of larger inclination and m' is the slope of the other line.

The curvature of a curve is the rate at which the angle of inclination of the tangent line changes with respect to arc length: $d\phi/ds$. The curvature of $y = f(x)$ is

$$\frac{d^2y/dx^2}{[1 + (dy/dx)^2]^{3/2}}.$$

Curvature can also be computed for curves given parametrically, and thus in polar coordinates. (For the latter case $r = f(\theta)$ can be written parametrically in rectangular coordinates as $x = f(\theta)\cos\theta$, $y = f(\theta)\sin\theta$.)

The radius of curvature is the absolute value of the reciprocal of the curvature.

Guide quiz on chap. 9

In Exercises 1 to 7 set up definite integrals for the given quantities, but do not evaluate them.

1. The area of the region within one loop of the curve $r = \cos 5\theta$.
2. The length of the curve $y = x^4$ from $x = 1$ to $x = 2$.
3. The area of the surface obtained by revolving the curve in Exercise 2 around the x axis.
4. The area of the surface obtained by revolving the curve in Exercise 2 around the line $x = 1$.
5. The length of the curve $r = 2 + \sin\theta$, $0 \le \theta \le 2\pi$.
6. The area of the surface obtained by revolving the top half of the curve in Exercise 5 around the polar axis.
7. The length of the curve $x = 5t^2$, $y = \sqrt{t}$, $0 \le t \le 1$.
8. At time t a moving particle is at the point $x = 12t$, $y = -16t^2 + 5t$. What is its speed when $t = 1$?
9. A curve is given parametrically as $x = \tan t$, $y = \sec t$. Eliminate t to find an equation in x and y for the curve.
10. (a) Find the inclination of the tangent line to the curve $r = \cos 2\theta$ at the point for which $\theta = \pi/8$.
    (b) Graph the curve and check that the answer in (a) is reasonable.
11. What is the radius of the circle that best approximates the curve $y = 1/x$ at $(1, 1)$ in the sense that it has the same radius of curvature as the curve does at $(1, 1)$?

Review exercises for chap. 9

1. (a) Develop the formula for area in polar coordinates.
   (b) What is the device for remembering the formula?
2. (a) Develop the formula for arc length with parameter t.
   (b) What is the device for remembering the formula?
3. (a) Develop the formula for the area of a surface of revolution with parameter x.
   (b) What is the device for remembering the formula?
4. (a) Develop the formula for $\tan\gamma$, where γ is the angle between the ray from the pole to the point P on the curve $r = f(\theta)$ and the tangent line to the curve at P.
   (b) Draw the memory device for the formula in (a).
5. (a) Define "curvature."
   (b) From (a), obtain the formula for the curvature of the curve $y = f(x)$.
6. Consider the curve $y = e^x$ for x in $[0, 1]$.
   (a) Set up integrals for its arc length and for the areas of the surfaces obtained by rotating the curve around the x axis and also about the y axis.
   (b) Two of the three integrals are elementary. Evaluate them.
7. Consider the curve $y = \sin x$ for x in $[0, \pi]$. Proceed as in Exercise 6. This time, however, only one of the three integrals is elementary. Evaluate it.
8. Graph $r = 3/(\cos\theta + 2\sin\theta)$ after first finding the rectangular form of the equation.
9. Find the maximum y coordinate of the curve $r = 1 + \cos\theta$.
10. Find the minimum x coordinate of the curve $r = 1 + \cos\theta$.
11. Is the total length of the curve $r = 1/(1 + \theta)$, $\theta \ge 0$, finite or infinite?
12. Is the total length of the curve $r = 1/(1 + \theta^2)$, $\theta \ge 0$, finite or infinite?
13. What is the length of the curve $r = e^{-\theta}$, $\theta \ge 0$?
14. Assume that x and y are functions of t. Obtain the formula for d^2y/dx^2 in terms of the derivatives $\dot{x}, \ddot{x}, \dot{y}, \ddot{y}$.
15. If
$$\begin{cases} x = \cos 2t, \\ y = \sin 3t, \end{cases}$$
express dy/dx and d^2y/dx^2 in terms of t.

16 Find the radius of curvature of the curve $y = \ln x$ at $(e, 1)$.

17 Let $r = e^\theta$, $0 \le \theta \le \pi/4$, describe a curve in polar coordinates. In parts (b) to (f) set up definite integrals for the quantities, and show that they could be evaluated by the fundamental theorem of calculus, but do not evaluate them.
   (a) Sketch the curve.
   (b) The area of the region R below the curve and above the interval $[1, e^{\pi/4}\sqrt{2}/2]$ on the x axis.
   (c) The volume of the solid obtained by revolving R, defined in (b), about the x axis.
   (d) The volume of the solid obtained by revolving R about the y axis.
   (e) The area of the surface obtained by revolving the curve in (a) around the x axis.
   (f) The area of the surface obtained by revolving the curve in (a) around the y axis.
   Rare is the curve for which the corresponding five integrals are all elementary.

18 Find the length of the curve $y = \ln x$ from $x = 1$ to $x = \sqrt{3}$. An integral table will save a lot of work.

∎

19 "If $dy/dx = \dot{y}/\dot{x}$, why is not d^2y/dx^2 equal to $\ddot{y}/\ddot{x}$?" How would you answer this question?

20 A set R in the plane bounded by a curve is convex if, whenever P and Q are points in R, the line segment PQ also lies in R. A curve is convex if it is the boundary of a convex set.
   (a) Show why the average radius of curvature with respect to angle ϕ as you traverse a convex curve equals (length of curve)/2π.
   (b) Deduce from (a) that a convex curve of length L has a radius of curvature equal to $L/2\pi$ somewhere on the curve.

21 Prove that the average value of the curvature as a function of arc length s as you sweep out a convex curve in the counterclockwise direction is 2π/length of curve. (See Exercise 20 for the definition of convex curve.)

22 The flexure formula in the theory of beams asserts that the bending moment M required to bend a beam is proportional to the desired curvature, $M = k/R$, where k is a constant depending on the beam and R is the radius of curvature. A beam is bent to form the parabola $y = x^2$. What is the ratio between the moments required at $(0, 0)$ and at $(2, 4)$?

23 Railroad curves are banked to reduce wear on the rails and flanges. The greater the radius of curvature, the less the curve must be banked. The best bank angle A satisfies the equation $\tan A = v^2/(32R)$, where v is speed in feet per second and R is radius of curvature in feet. A train travels in the elliptical track

$$\frac{x^2}{1000^2} + \frac{y^2}{500^2} = 1$$

(where x and y are measured in feet) at 60 miles per hour (equals 88 feet per second). Find the best angle A at the points $(1000, 0)$ and $(0, 500)$.

24 The larger the radius of curvature of a turn, the faster a given car can travel around that turn. The radius of curvature required is proportional to the square of the maximum speed. Or conversely, the maximum speed around a turn is proportional to the square root of the radius of curvature. If a car moving on the path $y = x^3$ (x and y measured in miles) can go 30 miles per hour at $(1, 1)$ without sliding off, how fast can it go at $(2, 8)$?

25 When calculating the surface-to-volume ratio of the rotary engine, engineers had to determine the arc length of a portion of the epitrochoid.
   (a) Show that the length of a general arc of the epitrochoid given parametrically as in Example 5 of Sec. 9.3 is

$$\int_\alpha^\beta \sqrt{9e^2 + R^2 + 6eR \cos 2\theta}\, d\theta$$

   (b) Show that the integral in (a) equals

$$\int_\alpha^\beta (3e + R)\sqrt{1 - k^2 \sin^2 \theta}\, d\theta.$$

   where $k^2 = 12eR/(3e + R)^2$.
   (c) Show that $k^2 \le 1$. Thus the integral in (b) is an elliptic integral, which is tabulated in many mathematical handbooks.

∎ ∎

26 Let a be a rational number. Consider the curve $y = x^a$ for x in the interval $[1, 2]$. Show that the area of the surface obtained by rotating this curve around the x axis can be evaluated by the fundamental theorem of calculus in the cases $a = 1$ or $1 + 2/n$, where n is any nonzero integer. (These are the only rational a for which the pertinent integral is elementary.) Assume Chebyshev's theorem, which asserts that if p and q are rational numbers, then $\int x^p(1 + x)^q\, dx$ is elementary only when p is an integer, q is an integer, or $p + q$ is an integer.

27 Read the generalized mean value theorem stated in Exercise 189 of Sec. 6.S.
   (a) What does it say about the curve given parametrically by the equations $x = g(t)$, $y = f(t)$? (Here t plays the role of the x in Exercise 189.)
   (b) Let $h(t)$, for t in $[a, b]$, be defined as the vertical distance from $(g(t), f(t))$ to the line that passes through $(g(a), f(a))$ and $(g(b), h(b))$, as in the proof of the mean value theorem in Sec. 4.1. Use the function h to prove the generalized mean value theorem.

28. Let a be a number, and let $x = g(t)$ and $y = h(t)$, $t \geq a$, describe a curve. Let $P = (g(a), h(a))$, and let $Q(t) = (g(t), h(t))$, $t > a$.
    (a) Sketch the chord $PQ(t)$ and the tangent line $T(t)$ at $Q(t)$.
    (b) What is the slope of $PQ(t)$? What is the slope of $T(t)$?
    (c) Under what assumptions will the two slopes in (b) have the same limit as $t \to a^+$?
29. A curve has the property that any of its chords makes equal angles with the tangent lines to the curve at the ends of the chord, as shown in Fig. 9.66. Show that the curve is part of a circle. *Hint:* Keep A fixed and let B vary. Take A as the pole of a polar coordinate system and the tangent line at that point as the polar axis.

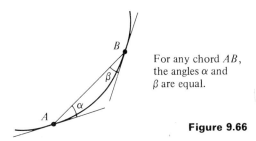

For any chord AB, the angles α and β are equal.

Figure 9.66

SERIES AND RELATED TOPICS

The main theme in this chapter is the approximation of functions by polynomials. For instance, it will be shown that the polynomial

$$1 + x + \frac{x^2}{2!} + \frac{x^3}{3!} + \frac{x^4}{4!}$$

is a good approximation of e^x when x is small. This fact would be useful in estimating

$$\int_1^2 \frac{e^x - 1}{x} dx,$$

an integral that cannot be evaluated by the fundamental theorem of calculus.

Sections 10.1 to 10.5 lay the foundation for the chapter. Sections 10.6 and 10.7 obtain polynomial approximations to such common functions as e^x, $\ln(1 + x)$, $\sin x$, and $\cos x$. Section 10.8 develops a way of showing that a function can be approximated by certain polynomials.

Section 10.9 introduces the complex numbers and Sec. 10.10 uses them to relate trigonometric functions and exponential functions. The relation developed in Sec. 10.10 is used at a key point in Sec. 10.11, which concerns an important type of differential equation. The binomial theorem is generalized from positive integer exponents to any exponent in Sec. 10.12. Section 10.13 presents Newton's method for estimating a root of an equation.

10.1 Sequences

A *sequence* of real numbers,

$$a_1, a_2, a_3, \ldots, a_n, \ldots,$$

is a function that assigns to each positive integer n a number a_n. The number a_n is called the nth *term* of the sequence. For example, the sequence

$$\left(1 + \frac{1}{1}\right)^1, \left(1 + \frac{1}{2}\right)^2, \left(1 + \frac{1}{3}\right)^3, \ldots, \left(1 + \frac{1}{n}\right)^n, \ldots$$

was considered in Sec. 6.1 in the study of the number e. In this case

$$a_n = \left(1 + \frac{1}{n}\right)^n.$$

Sometimes the notation $\{a_n\}$ is used as an abbreviation of the sequence $a_1, a_2, \ldots, a_n, \ldots$. For instance, e involves the sequence $\{(1 + 1/n)^n\}$.

The limit of a sequence   If, as n gets larger, a_n approaches a number L, then L is called the *limit* of the sequence. If the sequence $a_1, a_2, \ldots$ has a limit L, we write

$$\lim_{n \to \infty} a_n = L.$$

For instance, we write

$$\lim_{n \to \infty} \left(1 + \frac{1}{n}\right)^n = e.$$

If a_n becomes and remains arbitrarily large as n gets larger, we write $\lim_{n \to \infty} a_n = \infty$. (The limit does not exist in this case.) For instance, $\lim_{n \to \infty} 2^n = \infty$.

The assertion that "a_n approaches the number L" is equivalent to the assertion that $|a_n - L| \to 0$ as $n \to \infty$. In particular, if $L = 0$, the assertion that $a_n \to 0$ as $n \to \infty$ is equivalent to the assertion that $|a_n| \to 0$ as $n \to \infty$.

To show that $\lim_{n \to \infty} a_n = 0$, show that $\lim_{n \to \infty} |a_n| = 0$.   This means that if the absolute value of a_n approaches 0 as $n \to \infty$, then a_n approaches 0 as $n \to \infty$. This observation will be of use in this section and later.

A sequence need not begin with the term a_1. Later in the chapter sequences of the form $a_0, a_1, a_2, \ldots$ will be considered. In such a case a_0 is called "the zeroth term." Or we may consider a "tail end" of a sequence, a sequence that begins with the term a_k: $a_k, a_{k+1}, a_{k+2}, \ldots$.

The next example introduces a simple but important sequence.

EXAMPLE 1   *The sequence $\{(\tfrac{1}{2})^n\}$.*   A certain radioactive substance decays, losing half its mass in an hour. In the long run how much is left?

SOLUTION   If the initial mass is 1 gram, after an hour only $\tfrac{1}{2}$ gram remains. During the next hour, half of this amount is lost, and half of it remains. Thus after 2 hours

$$(\tfrac{1}{2})^2 = 0.25 \text{ gram}$$

remains. After 3 hours

$$(\tfrac{1}{2})^3 = 0.125 \text{ gram}$$

remains. In general, after n hours

$$(\tfrac{1}{2})^n \text{ gram}$$

remains.

When n is large, the amount remaining is very small. In other words, the sequence

Since $2^n \to \infty$, it follows that $\dfrac{1}{2^n} \to 0$.

$$\{(\tfrac{1}{2})^n\}$$

approaches 0 when n gets large. This is summarized by the equation

$$\lim_{n \to \infty} (\tfrac{1}{2})^n = 0. \quad \blacksquare$$

An important fact about sequences to keep in mind at all times is that the terms of the sequence $\{a_n\}$ may perhaps never equal the value of their limit L but merely approach it arbitrarily closely.

Furthermore, not every sequence has a limit, as the next example illustrates.

EXAMPLE 2 *A sequence that does not have a limit.* Let $a_n = (-1)^n$ for $n = 1, 2, 3, \ldots$. What happens to a_n when n is large?

SOLUTION The first four terms of the sequence are:

$$a_1 = (-1)^1 = -1$$
$$a_2 = (-1)^2 = 1$$
$$a_3 = (-1)^3 = -1$$
$$a_4 = (-1)^4 = 1.$$

The numbers of this sequence continue to alternate $-1, 1, -1, 1, \ldots$. This sequence does not approach a single number. Therefore it does not have a limit. $\blacksquare$

DEFINITION *Convergent and divergent sequences.* A sequence that has a limit is said to *converge* or to be *convergent*. A sequence that does not have a limit is said to *diverge* or to be *divergent*.

The sequence $\{(\tfrac{1}{2})^n\}$ of Example 1 converges to 0. The sequence $\{(-1)^n\}$ of Example 2 is divergent. There is no general procedure for deciding whether a sequence is convergent or divergent. However, there are methods for dealing with most sequences that arise in practice.

THE SEQUENCE r^n

EXAMPLE 3 A certain appliance depreciates in value over the years. In fact, at the end of any year it has only 80 percent of the value it had at the beginning of the year. What happens to its value in the long run if its value when new is $1?

SOLUTION Let a_n be the value of the appliance at the end of the nth year. Thus $a_1 = 0.8$ and $a_2 = (0.8)(0.8) = 0.8^2 = 0.64$. Similarly, $a_3 = 0.8^3$. The question concerns the sequence $\{0.8^n\}$.

This table lists a few values of 0.8^n, rounded off to four decimal places.

n	1	2	3	4	5	10	20
0.8^n	0.8	0.64	0.512	0.4096	0.3277	0.1074	0.0115

The entries in the table suggest that

$$\lim_{n \to \infty} 0.8^n = 0.$$

To verify this assertion, it is necessary to show that the decreasing sequence $\{0.8^n\}$ gets arbitrarily small. To indicate why it does, let us estimate how large n must be so that

$$0.8^n < 0.0001.$$

Taking logarithms to base 10 translates this inequality into the inequality

$$n \log_{10} 0.8 < -4,$$

Division of an inequality by a negative number reverses the direction of the inequality.

or

$$n > \frac{-4}{\log_{10} 0.8}.$$

Note that

$$\frac{-4}{\log_{10} 0.8} \approx \frac{-4}{-0.0969} \approx 41.3.$$

For $n \geq 42$, the number 0.8^n is less than 0.0001. A similar argument shows that 0.8^n approaches 0 as closely as we please when n is large. Consequently,

$$\lim_{n \to \infty} 0.8^n = 0,$$

just as, in Example 1, $\quad \lim_{n \to \infty} (\tfrac{1}{2})^n = 0.$

(The only difference is that $(0.8)^n$ approaches 0 more slowly than $(\tfrac{1}{2})^n$ does.)
In the long run the appliance will be worth less than a nickel, then less than a penny, etc. ∎

The result in Example 3 generalizes to the following theorem.

THEOREM 1 If r is a number in the open interval $(-1, 1)$, then

$$\lim_{n \to \infty} r^n = 0.$$

PROOF If $0 < r < 1$, an argument like that in Example 3 can be given. (See Exercise 31.) If $r = 0$, the sequence $\{r^n\}$ is just the constant sequence $0, 0, 0, \ldots$, which has the limit 0. If $-1 < r < 0$, consider the sequence $\{|r^n|\}$. Since $|r^n| = |r|^n$, we know that $|r^n| \to 0$. Thus $r^n \to 0$ as $n \to \infty$. ∎

THE SEQUENCE $k^n/n!$

In Example 4 a type of sequence is introduced that occurs later in the chapter in the study of $\sin x$, $\cos x$, and e^x.

EXAMPLE 4 Does the sequence defined by $a_n = (3^n/n!)$ converge or diverge?

SOLUTION The first terms of this sequence are computed (to two decimal places) with the aid of this table:

$n!$ is defined in App. B.

n	1	2	3	4	5	6	7	8
3^n	3	9	27	81	243	729	2,187	6,561
$n!$	1	2	6	24	120	720	5,040	40,320
$a_n = \dfrac{3^n}{n!}$	3.00	4.50	4.50	3.38	2.03	1.01	0.43	0.16

Though a_2 is larger than a_1, and a_3 is equal to a_2, from a_4 through a_8, as the table shows, the terms decrease.

The numerator 3^n becomes large as $n \to \infty$, influencing a_n to grow large. But the denominator $n!$ also becomes large as $n \to \infty$, influencing the quotient a_n to shrink toward 0. For $n = 1$ and 2 the first influence dominates, but then, as the table shows, the denominator $n!$ seems to grow faster than the numerator 3^n, forcing a_n toward 0.

To see why
$$\frac{3^n}{n!} \to 0$$

as $n \to \infty$, consider, for instance, a_{10}. Express a_{10} as the product of 10 fractions:

$$a_{10} = \tfrac{3}{1} \tfrac{3}{2} \tfrac{3}{3} \tfrac{3}{4} \tfrac{3}{5} \tfrac{3}{6} \tfrac{3}{7} \tfrac{3}{8} \tfrac{3}{9} \tfrac{3}{10}.$$

The first three fractions are ≥ 1. But all the seven remaining fractions are $\leq \tfrac{3}{4}$. Thus

$$a_{10} < \tfrac{3}{1} \tfrac{3}{2} \tfrac{3}{3} \left(\tfrac{3}{4}\right)^7.$$

Similarly, $\quad a_{100} < \frac{3}{1} \frac{3}{2} \frac{3}{3} \left(\frac{3}{4}\right)^{97}.$

By Theorem 1, $\quad \lim_{n \to \infty} \left(\frac{3}{4}\right)^n = 0.$

Thus $\quad \lim_{n \to \infty} a_n = 0.$

Similar reasoning shows that, for any fixed number k,

This limit will be used often.

$$\lim_{n \to \infty} \frac{k^n}{n!} = 0.$$

This means that the factorial function grows faster than any exponential k^n. ∎

SOME REMARKS ON LIMITS OF SEQUENCES

A definite integral can be expressed as a limit of a sequence, as Example 5 illustrates.

EXAMPLE 5 Express $\int_0^2 x^3 \, dx$ as a limit of a sequence.

SOLUTION For a positive integer n partition $[0, 2]$ into n sections of equal length, $2/n$, using the numbers

$$x_0 = 0, \, x_1 = \frac{2}{n}, \, x_2 = 2 \cdot \frac{2}{n}, \, \ldots, \, x_i = i \cdot \frac{2}{n}, \, \ldots, \, x_n = n \cdot \frac{2}{n} = 2.$$

As the sampling number c_i in the ith section use x_i. Then

$$\sum_{i=1}^{n} (x_i)^3 (x_i - x_{i-1}) = \sum_{i=1}^{n} \left(i \cdot \frac{2}{n}\right)^3 \frac{2}{n} = \frac{16}{n^4} \sum_{i=1}^{n} i^3.$$

is an approximation of $\int_0^2 x^3 \, dx$ and we have

$$\lim_{n \to \infty} \frac{16}{n^4} \sum_{i=1}^{n} i^3 = \int_0^2 x^3 \, dx.$$

Thus the integral is the limit of the sequence $\{a_n\}$, where

$$a_n = \frac{16}{n^4} \sum_{i=1}^{n} i^3. \quad \blacksquare$$

The following theorem will be used several times to show that a sequence converges. It says that if a sequence is nondecreasing ($a_1 \leq a_2 \leq a_3 \leq \cdots$) but does not get arbitrarily large, then it must be convergent.

THEOREM 2 Let $\{a_n\}$ be a nondecreasing sequence with the property that there is a number B such that $a_n \leq B$ for all n. That is, $a_1 \leq a_2 \leq a_3 \leq a_4 \leq \cdots \leq a_n \leq \cdots,$

and $a_n \leq B$ for all n. Then the sequence $\{a_n\}$ is convergent, and a_n approaches a number L less than or equal to B.

Similarly, if $\{a_n\}$ is a nonincreasing sequence and there is a number B such that $a_n \geq B$ for all n, then the sequence $\{a_n\}$ is convergent, and its limit is greater than or equal to B. ∎

The proof, which depends on a fundamental property of the real numbers, called *completeness*, is omitted. (The "completeness" of the real numbers amounts to the fact that there are no holes in the x axis.) Figure 10.1 shows that Theorem 2 is at least plausible. (The a_n's increase but, being less than B, approach some number L, and that number is not larger than B.)

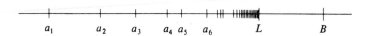

Figure 10.1

Sequences $\{a_n\}$ behave like real functions $f(x)$.

The limits of sequences $\{a_n\}$ behave like the limits of functions $f(x)$, as discussed in Sec. 2.2. The following properties will be used without proof.

Remember that A and B are real numbers (not "infinity").

If $\lim_{n \to \infty} a_n = A$ and $\lim_{n \to \infty} b_n = B$, then

1 $\lim_{n \to \infty} (a_n + b_n) = A + B$.

2 $\lim_{n \to \infty} (a_n - b_n) = A - B$.

3 $\lim_{n \to \infty} a_n b_n = AB$.

4 $\lim_{n \to \infty} \dfrac{a_n}{b_n} = \dfrac{A}{B}$ $(b_n \neq 0, B \neq 0)$.

5 If k is a constant, $\lim_{n \to \infty} ka_n = kA$.

For instance,

$$\lim_{n \to \infty} \left(\frac{3}{n} + \left(\frac{1}{2}\right)^n\right) = 3 \lim_{n \to \infty} \frac{1}{n} + \lim_{n \to \infty} \left(\frac{1}{2}\right)^n$$
$$= 3 \cdot 0 + 0 = 0.$$

Techniques for dealing with $\lim_{x \to \infty} f(x)$ can often be applied to determining $\lim_{n \to \infty} a_n$. The essential point is that

The converse is not true.

if $\lim_{x \to \infty} f(x) = L$, then $\lim_{n \to \infty} f(n) = L$.

Example 6 illustrates this observation.

EXAMPLE 6 Find $\lim_{n \to \infty} (n/2^n)$.

SOLUTION Consider the function $f(x) = (x/2^x)$. By l'Hôpital's rule (infinity-over-infinity case),

$$\lim_{x \to \infty} \frac{x}{2^x} = \lim_{x \to \infty} \frac{1}{2^x \ln 2} = 0.$$

Thus

$$\lim_{n \to \infty} \frac{n}{2^n} = 0. \quad \blacksquare$$

In the optional Sec. 2.4 various limit concepts were given precise, as contrasted to informal, definitions. Those familiar with that section will see that the following definition is in the same spirit.

DEFINITION
Formal definition of limit of a sequence

Limit of a sequence. The number L is the limit of the sequence $\{a_n\}$ if for each $\varepsilon > 0$ there is an integer N such that

$$|a_n - L| < \varepsilon$$

for all integers $n > N$.

Exercises

In each of Exercises 1 to 18 determine whether the sequence with the given value of a_n converges or diverges. If the sequence converges, give its limit.

1 $\{(0.3)^n\}$
2 $\{(0.99)^n\}$
3 $\{1^n\}$
4 $\{(1.01)^n\}$
5 $\{(-0.5)^n\}$
6 $\{(-\frac{9}{10})^n\}$
7 $\{(\frac{1}{3})^n\}$
8 $\{1/n!\}$
9 $\{2^n\}$
10 $\{n!\}$
11 $\left\{3\left(\frac{n+1}{n}\right) + \frac{n}{2^n}\right\}$
12 $\left\{\frac{5n+1}{3n-1} \cdot \frac{6}{n!}\right\}$
13 $\{\cos n\pi\}$
14 $\{\sin 2n\pi\}$
15 $\left\{\left(1 + \frac{2}{n}\right)^n\right\}$
16 $\left\{\left(\frac{n-1}{n}\right)^n\right\}$
17 $\left\{\frac{10^n}{n!}\right\}$
18 $\left\{\frac{(-100)^n}{n!}\right\}$

19 Let $a_n = 100^n/n!$.
   (a) Show that $a_1 < a_2 < \cdots < a_{99}$.
   (b) Show that $a_{99} = a_{100}$.
   (c) Show that $a_{100} > a_{101} > a_{102} > \cdots$.
20 Let $a_n = 200^n/n!$. What is the smallest n such that a_n is larger than the succeeding term a_{n+1}?
21 Examine these sequences for convergence or divergence:
   (a) $\left\{n \sin \frac{1}{n}\right\}$ (b) $\{n \sin \pi n\}$ (c) $\left\{\frac{\sin n}{n}\right\}$
22 How large must n be so that $(0.99)^n$ is less than 0.001?
23 Assume that each year inflation eats away 10 percent of the value of a dollar. Let a_n be the value of a dollar after n years of such inflation.
   (a) Find a_4.   (b) Find $\lim_{n \to \infty} a_n$.
24 (Calculator)
   (a) Compute $4^n/n!$ for $n = 1, 2, 3, 4, 5$, and 15 to three decimal places.
   (b) What is the largest term in the sequence $\{4^n/n!\}$?
   (c) What is $\lim_{n \to \infty} 4^n/n!$?

$\blacksquare \quad \blacksquare$

25 If $\lim_{n\to\infty} x^{2n}/(1+x^{2n})$ exists, call it $f(x)$. (x is fixed, n varies.)
   (a) Compute $f(\frac{1}{2})$, $f(2)$, $f(1)$.
   (b) For which x is $f(x)$ defined? Graph $y = f(x)$.
   (c) Where is f continuous?

In Exercises 26 to 28 determine the given limits by first showing that each limit is a definite integral $\int_a^b f(x)\,dx$ for a suitable interval $[a, b]$ and function f.

26 $\lim_{n\to\infty} \sum_{i=1}^{n} \left(\frac{i}{n}\right)^2 \frac{1}{n}$

27 $\lim_{n\to\infty} \left[\frac{1}{n+1} + \frac{1}{n+2} + \cdots + \frac{1}{2n}\right]$

28 $\lim_{n\to\infty} \sum_{i=1}^{n} \frac{n}{n^2 + i^2}$

29 Let $a_n = \dfrac{1}{1\cdot 2} + \dfrac{1}{2\cdot 3} + \cdots + \dfrac{1}{n(n+1)}$.
   (a) Compute a_n for $n = 1, 2, 3, 4$, in each case expressing a_n as a single fraction.
   (b) Find a short formula for a_n and explain your answer.
   (c) Show that $\lim_{n\to\infty} a_n = 1$.

30 Let
$$a_n = \frac{1}{2^2} + \frac{1}{3^2} + \cdots + \frac{1}{n^2} \text{ for } n \geq 2.$$

Show that $\lim_{n\to\infty} a_n$ exists and is ≤ 1. *Hint*: See Exercise 29 and use Theorem 2.

31 This exercise completes the proof of Theorem 1 in the style of Example 3. Let r be a positive number less than 1. Let p be a positive number. Show that there is an integer n such that $r^n < p$. *Hint*: Use logarithms.

32 This exercise provides a neat proof of Theorem 1 in the case $0 < r < 1$. Let r be a positive number less than 1.
   (a) Using Theorem 2, show that $\lim_{n\to\infty} r^n$ exists.
   (b) Call the limit in (a) L. Why is $rL = L$?
   (c) From (b) deduce that $L = 0$.

33 Let $P(x)$ and $Q(x)$ be polynomials. Assume that $Q(n)$ is not 0 for any positive integer n. What relation must there be between the degree of $P(x)$ and the degree of $Q(x)$ if the sequence $\{P(n)/Q(n)\}$ is convergent?

10.2 Series

Frequently a new sequence is formed by summing terms of a given sequence. Example 1 illustrates this way of constructing a sequence.

EXAMPLE 1 Given the sequence $1, 0.8, 0.8^2, 0.8^3, \ldots$, form a new sequence $\{S_n\}$ as follows:

S is short for "sum."

$$S_1 = 1$$
$$S_2 = 1 + 0.8$$
$$S_3 = 1 + 0.8 + 0.8^2$$

and, in general, $S_n = 1 + 0.8 + 0.8^2 + \cdots + 0.8^{n-1}$.

Each S_n is the sum of the first n terms of the sequence $1, 0.8, 0.8^2, 0.8^3, \ldots$. Examine the behavior of S_n as $n \to \infty$.

SOLUTION
$$S_1 = 1$$
$$S_2 = 1 + 0.8 = 1.8$$
$$S_3 = 1 + 0.8 + 0.8^2 = 1 + 0.8 + 0.64 = 2.44$$
$$S_4 = 1 + 0.8 + 0.8^2 + 0.8^3 = 1 + 0.8 + 0.64 + 0.512 = 2.952$$
$$S_5 = 1 + 0.8 + 0.8^2 + 0.8^3 + 0.8^4$$
$$= 1 + 0.8 + 0.64 + 0.512 + 0.4096 = 3.3616$$

Similar computations show that $S_{50} \approx 4.99993$.

Two influences on $\{S_n\}$

Two influences affect the growth of S_n as n increases. On the one hand, the number of summands increases, causing S_n to get larger. On the other hand, the summands approach 0, so that S_n grows more and more slowly as n increases. Theorem 1 of this section shows that the sequence $\{S_n\}$ converges and that its limit is 5:

$$\lim_{n \to \infty} S_n = 5. \quad \blacksquare$$

The rest of this section extends the ideas introduced in Example 1.

Let $a_1, a_2, a_3, \ldots, a_n, \ldots$ be a sequence. From this sequence a new sequence $S_1, S_2, S_3, \ldots, S_n, \ldots$ can be formed:

$$S_1 = a_1$$
$$S_2 = a_1 + a_2$$
$$S_3 = a_1 + a_2 + a_3$$
$$\ldots\ldots\ldots\ldots\ldots\ldots\ldots$$

Summation notation was introduced in Sec. 5.2.

$$S_n = a_1 + a_2 + a_3 + \cdots + a_n = \sum_{i=1}^{n} a_i.$$

The sequence of sums $S_1, S_2, \ldots$ is called the *series* obtained from the sequence $a_1, a_2, \ldots$. Traditionally, though imprecisely, it is referred to as *the series whose nth term is a_n*. Common notations for the sequence $\{S_n\}$ are $\sum_{n=1}^{\infty} a_n$ and $a_1 + a_2 + a_3 + \cdots + a_n + \cdots$. The sum

$$S_n = a_1 + a_2 + \cdots + a_n = \sum_{i=1}^{n} a_i$$

is called a *partial sum* or the *nth partial sum*. If the sequence of partial sums of a series converges to L, then L is called the *sum* of the series. Frequently one writes $L = a_1 + a_2 + \cdots + a_n + \cdots$. Remember, however, that we do not add an infinite number of numbers; we take the limit of finite sums.

Only a finite number of summands are ever added up.

Example 1 concerns the series whose nth term is 0.8^{n-1}:

$$S_n = 1 + 0.8^1 + 0.8^2 + \cdots + 0.8^{n-1}.$$

It is a special case of a geometric series, which will now be defined.

DEFINITION *Geometric series.* Let a and r be real numbers. The series

App. B treats geometric series with a finite number of terms.

$$a + ar + ar^2 + \cdots + ar^{n-1} + \cdots$$

is called the *geometric series with initial term a and ratio r*.

The series in Example 1 is a geometric series with initial term 1 and ratio 0.8.

THEOREM 1 If $-1 < r < 1$, the geometric series

$$a + ar + \cdots + ar^{n-1} + \cdots$$

converges to $a/(1 - r)$.

PROOF Let S_n be the sum of the first n terms:

$$S_n = a + ar + \cdots + ar^{n-1}.$$

By the formula in Appendix B for the sum of a finite geometric series,

$$S_n = \frac{a(1 - r^n)}{1 - r}.$$

By Sec. 10.1, $\quad \lim_{n \to \infty} r^n = 0.$

Thus $\quad \lim_{n \to \infty} S_n = \frac{a}{1 - r},$

proving the theorem. ∎

In particular, if $a = 1$ and $r = 0.8$, as in Example 1, the geometric series has the sum

$$\frac{1}{1 - 0.8} = \frac{1}{0.2} = 5.$$

Theorem 1 says nothing about geometric series in which the ratio r is ≥ 1 or ≤ -1. The next theorem, which concerns series in general, not just geometric series, will be of use in settling this case.

THEOREM 2 *The nth-term test for divergence.* If $\lim_{n \to \infty} a_n \neq 0$, then the series $a_1 + a_2 + \cdots + a_n + \cdots$ diverges. (The same conclusion holds if $\{a_n\}$ has no limit.)

PROOF Assume that the series $a_1 + a_2 + \cdots$ converges. Since S_n is the sum $a_1 + a_2 + \cdots + a_n$, while S_{n-1} is the sum of the first $n - 1$ terms, it follows that $S_n = S_{n-1} + a_n$, or

$$a_n = S_n - S_{n-1},$$

Let $\quad S = \lim_{n \to \infty} S_n.$

Then we also have $\quad S = \lim_{n \to \infty} S_{n-1},$

since S_{n-1} runs through the same numbers as S_n. Thus

$$\lim_{n \to \infty} a_n = \lim_{n \to \infty} (S_n - S_{n-1})$$
$$= \lim_{n \to \infty} S_n - \lim_{n \to \infty} S_{n-1}$$
$$= S - S$$
$$= 0.$$

So, if a series converges, its nth term must approach 0.

This proves the theorem. ∎

Theorem 2 implies that, if $a \neq 0$ and $r \geq 1$, the geometric series

$$a + ar + \cdots + ar^{n-1} + \cdots$$

diverges. For instance, if $r = 1$,

$$\lim_{n \to \infty} ar^n = \lim_{n \to \infty} a1^n = a,$$

which is not 0. If $r > 1$, then r^n gets arbitrarily large as n increases; hence $\lim_{n \to \infty} ar^n$ does not exist. Similarly, if $r \leq -1$, $\lim_{n \to \infty} ar^n$ does not exist. The above results and Theorem 1 can be summarized by this statement: The geometric series

$$\sum_{n=1}^{\infty} ar^{n-1} = a + ar + ar^2 + \cdots + ar^{n-1} + \cdots,$$

for $a \neq 0$, converges if and only if $|r| < 1$.

WARNING! Even if the nth term approaches 0, the series can diverge!!

Theorem 2 implies that the nth term of a convergent series approaches 0 as n gets large. The next example shows that the converse is not true. If the nth term of a series approaches 0 as n gets large, it does *not* necessarily follow that the series is convergent.

EXAMPLE 2   Show that the series

$$\frac{1}{\sqrt{1}} + \frac{1}{\sqrt{2}} + \frac{1}{\sqrt{3}} + \cdots + \frac{1}{\sqrt{n}} + \cdots$$

diverges.

SOLUTION   Consider $S_n = \dfrac{1}{\sqrt{1}} + \dfrac{1}{\sqrt{2}} + \cdots + \dfrac{1}{\sqrt{n}}.$

Each of the n summands in S_n is $\geq 1/\sqrt{n}$. Hence

$$S_n \geq \underbrace{\frac{1}{\sqrt{n}} + \frac{1}{\sqrt{n}} + \cdots + \frac{1}{\sqrt{n}}}_{n \text{ summands}} = \frac{n}{\sqrt{n}} = \sqrt{n}.$$

As n increases, $\sqrt{n}$ increases without bound. Since $S_n \geq \sqrt{n}$,

$$\lim_{n \to \infty} S_n \text{ does not exist.}$$

$\dfrac{1}{\sqrt{n}} \to 0$ so slowly that $\dfrac{1}{\sqrt{1}} + \dfrac{1}{\sqrt{2}} + \cdots + \dfrac{1}{\sqrt{n}}$ gets arbitrarily large.

In fact, we write $\lim_{n \to \infty} S_n = \infty$.

In short, the series $\dfrac{1}{\sqrt{1}} + \dfrac{1}{\sqrt{2}} + \cdots + \dfrac{1}{\sqrt{n}} + \cdots$

diverges even though its nth term, $1/\sqrt{n}$, approaches 0. ∎

So named by the Greeks because of the role of $\frac{1}{n}$ in musical harmony

In the next example, the nth term approaches 0 much faster than $1/\sqrt{n}$ does. Still the series diverges. The series in this example is called the *harmonic series*. The argument that it diverges is due to the French mathematician Nicolas of Oresme, who presented it about the year 1360.

EXAMPLE 3 Show that the *harmonic series*
$$\frac{1}{1} + \frac{1}{2} + \cdots + \frac{1}{n} + \cdots$$

The harmonic series diverges.

SOLUTION Collect the summands in longer and longer groups in the manner indicated below. (The number of summands is a power of 2, doubling at each step after the first.)

$$\underbrace{\tfrac{1}{1}} + \underbrace{\tfrac{1}{2}} + \underbrace{\tfrac{1}{3} + \tfrac{1}{4}} + \underbrace{\tfrac{1}{5} + \tfrac{1}{6} + \tfrac{1}{7} + \tfrac{1}{8}} + \underbrace{\tfrac{1}{9} + \tfrac{1}{10} + \cdots + \tfrac{1}{16}} + \tfrac{1}{17} + \cdots$$

The sum of the terms in each group is at least $\tfrac{1}{2}$. For instance,

$$\tfrac{1}{5} + \tfrac{1}{6} + \tfrac{1}{7} + \tfrac{1}{8} > \tfrac{1}{8} + \tfrac{1}{8} + \tfrac{1}{8} + \tfrac{1}{8} = \tfrac{4}{8} = \tfrac{1}{2},$$

and
$$\tfrac{1}{9} + \tfrac{1}{10} + \cdots + \tfrac{1}{16} > \tfrac{1}{16} + \tfrac{1}{16} + \cdots + \tfrac{1}{16} = \tfrac{8}{16} = \tfrac{1}{2}.$$

Since the repeated addition of $\tfrac{1}{2}$s produces sums as large as we please, the series diverges. ∎

An important moral: the nth term test is only a test for divergence.

If the series $a_1 + a_2 + \cdots + a_n + \cdots$ converges, it follows that $a_n \to 0$. However, if $a_n \to 0$, it *does not necessarily follow* that $a_1 + a_2 + \cdots + a_n + \cdots$ converges. Indeed, there is no general, practical rule for determining whether a series converges or diverges. Fortunately, a few rules suffice to decide on the convergence or divergence of the most common series; they will be presented in this chapter.

It should be pointed out that, in the case of a convergent series, if $\sum_{n=1}^{\infty} a_n = L$, and c is a number then $\sum_{n=1}^{\infty} c a_n = cL$. If also $\sum_{n=1}^{\infty} b_n = M$, then $\sum_{n=1}^{\infty} (a_n + b_n) = L + M$.

Front ends don't affect convergence.

Keep in mind that you can disregard any finite number of terms when deciding whether a series is convergent or divergent. If you delete a finite number of terms from a series and what is left converges, then the series you started with converges. Another way to look at this is to note that a "front end," $a_1 + a_2 + \cdots + a_n$, does not influence convergence or divergence. It is rather a "tail end," $a_{n+1} + a_{n+2} + \cdots$ that matters. The sum of the series is the sum of any tail end plus the sum of the corresponding front end.

Exercises

In each of Exercises 1 to 8 determine whether the given geometric series converges. If it does, find its sum.

1. $1 + \frac{1}{2} + \frac{1}{4} + \frac{1}{8} + \cdots + (\frac{1}{2})^{n-1} + \cdots$

2. $1 - \frac{1}{3} + \frac{1}{9} - \frac{1}{27} + \cdots + (-\frac{1}{3})^{n-1} + \cdots$

3. $\sum_{n=1}^{\infty} 10^{-n}$

4. $\sum_{n=1}^{\infty} 10^n$

5. $\sum_{n=1}^{\infty} 5(0.99)^n$

6. $\sum_{n=1}^{\infty} 7(-1.01)^n$

7. $\sum_{n=1}^{\infty} 4\left(\frac{2}{3}\right)^n$

8. $-\frac{3}{2} + \frac{3}{4} - \frac{3}{8} + \cdots + \frac{3}{(-2)^n} + \cdots$

In Exercises 9 to 16 determine whether the given series converge or diverge. Find the sums of the convergent series.

9. $-5 + 5 - 5 + 5 - \cdots + (-1)^n 5 + \cdots$

10. $\sum_{n=1}^{\infty} \frac{1}{[1 + (1/n)]^n}$

11. $\sum_{n=1}^{\infty} \frac{2}{n}$

12. $\sum_{n=1}^{\infty} \frac{n}{2n+1}$

13. $\sum_{n=1}^{\infty} 6\left(\frac{4}{5}\right)^n$

14. $\sum_{n=1}^{\infty} 100\left(\frac{-8}{9}\right)^n$

15. $\sum_{n=1}^{\infty} (2^{-n} + 3^{-n})$

16. $\sum_{n=1}^{\infty} (4^{-n} + n^{-1})$

17. This is a quote from an economics text: "The present value of the land, if a new crop is planted at time t, $2t$, $3t$, etc., is

$$P = g(t)e^{-rt} + g(t)e^{-2rt} + g(t)e^{-3rt} + \cdots.$$

Note that each term is the previous term multiplied by e^{-rt}. By the formula for the sum of a geometric series,

$$P = \frac{g(t)e^{-rt}}{1 - e^{-rt}}"$$

Check that the missing step, which simplified the formula for P, was correct.

18. (Calculator) Let

$$S_n = \sum_{i=1}^{n} \frac{1}{i(i+1)}.$$

(a) Compute S_5.
(b) Compute S_{10}.
(c) What do you think happens to S_n as $n \to \infty$?

19. A certain rubber ball, when dropped on concrete, always rebounds 90 percent of the distance it falls.
(a) If the ball is dropped from a height of 6 feet, how far does it travel during the first three descents and ascents?
(b) How far does it travel before coming to rest?

20. A patient takes A grams of a certain medicine every 6 hours. The amount of each dose active in the body t hours later is Ae^{-kt} grams, where k is a positive constant and time is measured in hours.
(a) Show that immediately after taking the medicine for the nth time, the amount active in the body is

$$S_n = A + Ae^{-6k} + Ae^{-12k} + \cdots + Ae^{-6(n-1)k}.$$

(b) If, as $n \to \infty$, $S_n \to \infty$, the patient would be in danger. Does $S_n \to \infty$? If not, what is $\lim_{n \to \infty} S_n$?

■ ■

21. The decimal $0.3333\cdots$ stands for the sum of the geometric series $\sum_{n=1}^{\infty} 3(10^{-n})$. Use this fact to show that $0.3333\cdots$ is equal to $\frac{1}{3}$.

22. Write this decimal

$$0.525252\cdots \quad (52\text{s continuing})$$

as a geometric series and use Theorem 1 to find its value.

23. A gambler tosses a penny until a head appears. On the average, how many times does she toss a penny to get a head? Parts (a) and (b) concern this question.
(a) Experiment with a penny on 10 runs. Each run consists of tossing a penny until heads appears. Average the lengths of the 10 runs.
(b) The probability of a run of length one is $\frac{1}{2}$, since heads must appear on the first toss. The probability of a run of length two is $(\frac{1}{2})^2$. The probability of having heads appear for the first time on the nth toss is $(\frac{1}{2})^n$. It is shown in probability theory that the average number of tosses to get heads is $\sum_{n=1}^{\infty} n/2^n$. (This is a theoretical average approached as the experiment is repeated many times.) Compute $\sum_{n=1}^{8} n/2^n$. (The next exercise sums the infinite series.)

24. This exercise concerns the sum of the series

$$1 + 2x + 3x^2 + 4x^3 + \cdots + nx^{n-1} + \cdots.$$

(a) Use l'Hôpital's rule to show that if $|x| < 1$, then $\lim_{t \to \infty} tx^{t-1} = 0$. ($x$ is fixed and t varies through real numbers.)
(b) Use (a) to show that if $|x| < 1$, $nx^{n-1} \to 0$ as $n \to \infty$.

(c) By differentiating the equation

$$1 + x + x^2 + \cdots + x^n = \frac{1 - x^{n+1}}{1 - x},$$

show that

$$1 + 2x + 3x^2 + \cdots + nx^{n-1}$$
$$= \frac{1 - x^{n+1} + (n+1)x^{n+1} - (n+1)x^n}{(1-x)^2}.$$

(d) With the aid of (c), show that, for $|x| < 1$,

$$\sum_{n=1}^{\infty} nx^{n-1} = \frac{1}{(1-x)^2}.$$

(e) Use (d) to evaluate $\sum_{n=1}^{\infty} n/2^{n-1}$ and then $\sum_{n=1}^{\infty} n/2^n$, the series needed in Exercise 23.

25 Oresme, around the year 1360, summed the series $\sum_{n=1}^{\infty} n/2^n$ by drawing the endless staircase shown in Fig. 10.2, in which each stair has width one and is twice as high as the stair immediately to its right.

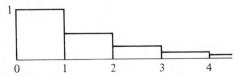

Figure 10.2

(a) By looking at the staircase in two ways, show that

$$1 + \tfrac{1}{2} + \tfrac{1}{4} + \tfrac{1}{8} + \cdots = \tfrac{1}{2} + \tfrac{2}{4} + \tfrac{3}{8} + \cdots.$$

(b) Use (a) to sum $\sum_{n=1}^{\infty} n/2^n$.

26 A plodding snail is at the point $(1/t, (\sin t)/t)$ at time $t \geq 1$. (a) Show that, as $t \to \infty$, the snail approaches the origin. (b) Show that the length of the snail's journey is infinite. *Suggestion:* Sketch the path; don't try to find its arc length by integration.

27 Using the formula for the sum of a geometric series, express the repeating decimal 3.45212121 ... as the quotient of two integers.

28 Deficit spending by the federal government inflates the nation's money supply. However, much of the money paid out by the government is spent in turn by those who receive it, thereby producing additional spending. This produces a chain reaction, called by economists the "multiplier effect." It results in much greater total spending than the government's original expenditure. To be specific, suppose the government spends one billion dollars and that the recipients of that expenditure in turn spend 80 percent while retaining 20 percent. Let S_n be the *total* spending generated after n transactions in the chain, 80 percent of receipts being expended at each step.
(a) Show that $S_n = 1 + 0.8 + 0.8^2 + \cdots 0.8^{n-1}$ billion dollars.
(b) Show that as n increases the total spending approaches 5 billion dollars. (The number 5 is called the "multiplier.")
(c) What would be the total spending if 90 percent of receipts is spent at each step instead of 80 percent?

29 (How banks, with the assistance of the public, create money) If a deposit of A dollars is made at a bank, the bank can lend out most of this amount. However, it cannot lend out all of the amount, for it must keep a reserve to meet the demands of depositors who may withdraw money from their accounts. The government stipulates what this reserve is, usually between 16 percent and 20 percent of the amount deposited. Assume that a bank is allowed to lend 80 percent of the amount deposited. If a person deposits $1000, then the bank can lend another person $800. Assume that this borrower deposits all of the amount; then the bank can lend a third person $640 of that deposit. This process can go on indefinitely, through a fourth person, a fifth, and so on. If this process continues indefinitely, how large will all the deposits total in the long run? Note that this total is much larger than the initial deposit. The banks have created money.

10.3 The integral test

In this section and the next only series with positive terms will be considered. The following test for convergence or divergence applies to many series with positive terms. In particular it will show that the series

$$\frac{1}{1} + \frac{1}{2} + \frac{1}{3} + \frac{1}{4} + \cdots + \frac{1}{n} + \cdots$$

diverges, but that the series

$$\frac{1}{1^{1.01}} + \frac{1}{2^{1.01}} + \frac{1}{3^{1.01}} + \frac{1}{4^{1.01}} + \cdots + \frac{1}{n^{1.01}} + \cdots$$

converges.

THEOREM 1

A function is "positive" if its values are positive for the domain of interest.

Integral test. Let $f(x)$ be a continuous decreasing positive function for $x \geq 1$. Then
(a) if $\int_1^\infty f(x)\,dx$ is convergent, $\sum_{n=1}^\infty f(n)$ is convergent;
(b) if $\int_1^\infty f(x)\,dx$ is divergent, $\sum_{n=1}^\infty f(n)$ is divergent.

PROOF (a) Assume $\int_1^\infty f(x)\,dx$ is convergent.

Consider the total area of the $n-1$ rectangles shown in Fig. 10.3. Each rectangle has width 1. The height of the rectangle above $[1, 2]$ is $f(2)$. Hence the rectangle above $[1, 2]$ has area $f(2) \cdot 1 = f(2)$. Similarly, the rectangle above $[2, 3]$ has area $f(3)$. The total area of the $n-1$ shaded rectangles in Fig. 10.3 is then

$$f(2) + f(3) + \cdots + f(n).$$

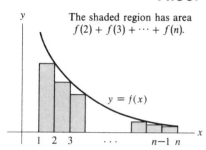

The shaded region has area $f(2) + f(3) + \cdots + f(n)$.

Figure 10.3

Since these rectangles lie below the curve $y = f(x)$,

$$f(2) + f(3) + \cdots + f(n) < \int_1^n f(x)\,dx.$$

Adding $f(1)$ to both sides of this inequality yields the inequality

$$f(1) + f(2) + \cdots + f(n) < f(1) + \int_1^n f(x)\,dx.$$

Since $\int_1^\infty f(x)\,dx$ is a convergent integral, the sums $f(1) + f(2) + \cdots + f(n)$ do not get arbitrarily large. Hence, by Theorem 2 in Sec. 10.1, $\sum_{n=1}^\infty f(n)$ is a convergent series.

(b) Figure 10.4 is the key to showing that, if $\int_1^\infty f(x)\,dx$ is divergent, so is $\sum_{n=1}^\infty f(n)$.

In contrast to Fig. 10.3 each rectangle in Fig. 10.4 has a height equal to the value of f at its *left* abscissa. That is, the rectangle above the interval $[i, i+1]$, $i = 1, 2, \ldots, n$, has height $f(i)$ and thus has area $f(i)$. Since the n shaded rectangles in Fig. 10.4 contain the region under the curve $y = f(x)$ and above the interval $[1, n+1]$, we have

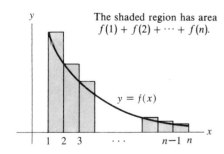

The shaded region has area $f(1) + f(2) + \cdots + f(n)$.

Figure 10.4

$$f(1) + f(2) + \cdots + f(n) > \int_1^{n+1} f(x)\,dx.$$

Since $\int_1^\infty f(x)\,dx = \infty$, it follows that $\sum_{n=1}^\infty f(n)$ is divergent. ∎

Note that the integral test does *not* give the value of the convergent series, $\sum_{n=1}^\infty f(n)$.

The integral test will show that the series

$$\frac{1}{1^{1.01}} + \frac{1}{2^{1.01}} + \frac{1}{3^{1.01}} + \cdots + \frac{1}{n^{1.01}} + \cdots \qquad (1)$$

is convergent. That this series is convergent may be surprising since it resembles the harmonic series

$$\frac{1}{1} + \frac{1}{2} + \frac{1}{3} + \cdots + \frac{1}{n} + \cdots$$

which was shown in Sec. 10.2 to be divergent.

Series (1) is an example of a *p* series, which will now be defined in general.

DEFINITION *p series.* For a fixed real number *p*, the series

$$\frac{1}{1^p} + \frac{1}{2^p} + \frac{1}{3^p} + \cdots$$

is called the *p series* (with exponent *p*).

When $p = 1$, the *p* series is the harmonic series. When $p = 1.01$, the *p* series is the series (1).

THEOREM 2 The *p* series $\sum_{n=1}^{\infty} 1/n^p$ converges if $p > 1$. It diverges if $p \leq 1$.

PROOF Analyze the cases $p < 0$, $p = 0$, and $p > 0$ separately.

If *p* is negative, then $1/n^p = n^{-p} \to \infty$ as $n \to \infty$. By the *n*th term test for divergence, $\sum_{n=1}^{\infty} 1/n^p$ is divergent.

If $p = 0$, then $1/n^p = 1/n^0 = 1$. All the terms of the *p* series in this case are 1. Since the *n*th term does not approach 0 as $n \to \infty$, the *n*th term test for divergence shows that the series diverges.

If *p* is positive, we make use of the integral test. The function $f(x) = 1/x^p = x^{-p}$ is decreasing for $x \geq 1$. Thus the integral test may be applied. Now if $p \neq 1$,

$$\int_1^\infty x^{-p}\,dx = \lim_{b \to \infty} \int_1^b x^{-p}\,dx$$

$$= \lim_{b \to \infty} \frac{x^{1-p}}{1-p}\bigg|_1^b$$

$$= \lim_{b \to \infty} \left(\frac{b^{1-p}}{1-p} - \frac{1}{1-p}\right).$$

For $p > 1$ this limit is $-1/(1-p) = 1/(p-1)$ since

$$\lim_{b \to \infty} b^{1-p} = 0.$$

Thus $\int_1^\infty x^{-p} \, dx$ is convergent. By the integral test the p series converges for $p > 1$.

For $p < 1$, b^{1-p} gets arbitrarily large as $b \to \infty$; hence $\int_1^\infty x^{-p} \, dx$ is divergent and the p series diverges for $p < 1$.

If $p = 1$,
$$\int_1^\infty x^{-p} \, dx = \int_1^\infty \frac{1}{x} \, dx$$
$$= \lim_{b \to \infty} \int_1^b \frac{1}{x} \, dx$$
$$= \lim_{b \to \infty} (\ln b - \ln 1).$$

Since $\ln b$ gets arbitrarily large as $b \to \infty$, $\int_1^\infty 1/x \, dx$ is divergent. Thus the harmonic series diverges. ∎

In particular, series (1) converges, since it is the p series with $p = 1.01$, which is greater than 1. However, the series $\sum_{n=1}^\infty 1/\sqrt{n}$ diverges, since it is the p series with $p = \frac{1}{2}$. (A different argument for the divergence of this series was given in Sec. 10.2.)

ESTIMATING THE ERROR

Let S_n be the sum of the first n terms of a convergent series $\sum_{n=1}^\infty a_n$ whose sum is S. The difference
$$R_n = S - S_n = a_{n+1} + a_{n+2} + a_{n+3} + \cdots$$
is called the *remainder* or *error* in using the sum of the first n terms to approximate the sum of the series. For the series of the special type considered in this section, it is possible to use an improper integral to estimate the error. The reasoning depends once again on comparing a staircase of rectangles with the area under a curve.

Recall that $f(x)$ is a continuous decreasing positive function. The error in using $S_n = f(1) + f(2) + \cdots + f(n) = \sum_{i=1}^n f(i)$ to approximate $\sum_{i=1}^\infty f(i)$ is the sum $\sum_{i=n+1}^\infty f(i)$. This sum is the area of the endless staircase of rectangles shown in Fig. 10.5. Comparing the rectangles with the region under the curve $y = f(x)$, we conclude that

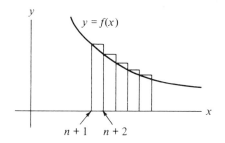

Figure 10.5

$$R_n = a_{n+1} + a_{n+2} + \cdots > \int_{n+1}^\infty f(x) \, dx. \tag{2}$$

Inequality (2) gives a *lower* estimate of the error.

The staircase in Fig. 10.6, which lies below the curve, gives an *upper* estimate of the error. Inspection of Fig. 10.6 shows that

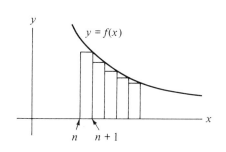

Figure 10.6

$$R_n = a_{n+1} + a_{n+2} + \cdots < \int_n^\infty f(x) \, dx.$$

Putting these observations together yields the following estimate of the error.

THEOREM 3   Let $f(x)$ be a continuous decreasing positive function such that $\int_1^\infty f(x)\,dx$ is convergent. Then the error R_n in using $f(1) + f(2) + \cdots + f(n)$ to estimate $\sum_{i=1}^\infty f(i)$ satisfies the inequality

Estimating the error

$$\int_{n+1}^\infty f(x)\,dx < R_n < \int_n^\infty f(x)\,dx. \quad \blacksquare \qquad (3)$$

EXAMPLE   The first five terms of the p series

$$\frac{1}{1^2} + \frac{1}{2^2} + \frac{1}{3^2} + \cdots + \frac{1}{n^2} + \cdots$$

are used to estimate the sum of the series.
(a) Estimate the error in using just the first five terms.
(b) Estimate $\sum_{n=1}^\infty 1/n^2$.

SOLUTION   (a) By (3), the error R_5 satisfies the inequality

$$\int_6^\infty \frac{dx}{x^2} < R_5 < \int_5^\infty \frac{dx}{x^2}.$$

Now, $\qquad \int_5^\infty \frac{dx}{x^2} = -\frac{1}{x}\Big|_5^\infty = 0 - \left(-\frac{1}{5}\right) = \frac{1}{5}.$

Similarly, $\qquad \int_6^\infty \frac{dx}{x^2} = \frac{1}{6}.$

Thus $\qquad \frac{1}{6} < R_5 < \frac{1}{5}.$

(b) The sum of the first five terms of the series is

$$\frac{1}{1^2} + \frac{1}{2^2} + \frac{1}{3^2} + \frac{1}{4^2} + \frac{1}{5^2} \approx 1.4636.$$

Since the sum of the remaining terms (the "tail-end") is between $\frac{1}{6}$ and $\frac{1}{5}$, the sum of the series is between $1.463 + \frac{1}{6} \approx 1.630$ and $1.464 + \frac{1}{5} = 1.664.$ $\blacksquare$

The integral test works if the function $f(x)$ decreases from some point on.

So far it has been assumed that the function $f(x)$ is decreasing for $x \geq 1$. If, instead, $f(x)$ decreases for $x \geq a$, where a is some constant, similar results hold. For instance $f(x) = x^2/e^x$ increases in the interval $[1, 2]$ and decreases in the interval $[2, \infty)$. The series

$$\frac{1^2}{e^1} + \frac{2^2}{e^2} + \frac{3^2}{e^3} + \cdots + \frac{n^2}{e^n} + \cdots \qquad (4)$$

can be examined with a slight modification of the integral test. Just chop off the first term, that is, $1^2/e^1$. Since $f(x) = x^2/e^x$ is decreasing for $x \geq 2$, the integral test may be applied to the "tail end,"

$$\frac{2^2}{e^2} + \frac{3^2}{e^3} + \cdots + \frac{n^2}{e^n} + \cdots \quad (n \geq 2). \tag{5}$$

The improper integral $\int_0^\infty x^2 e^{-x}\, dx$ can be evaluated and shown to be convergent. Thus (5) is convergent and consequently so is (4).

Exercises

In Exercises 1 to 8 use the integral test to determine whether the series diverge or converge.

1. $\sum_{n=1}^{\infty} \frac{1}{n^{1.1}}$
2. $\sum_{n=1}^{\infty} \frac{1}{n^{0.9}}$
3. $\sum_{n=1}^{\infty} \frac{n}{n^2 + 1}$
4. $\sum_{n=1}^{\infty} \frac{1}{n^2 + 1}$
5. $\sum_{n=2}^{\infty} \frac{1}{n \ln n}$
6. $\sum_{n=1}^{\infty} \frac{1}{n + 1{,}000}$
7. $\sum_{n=1}^{\infty} \frac{\ln n}{n}$ (First check where $(\ln x)/x$ decreases.)
8. $\sum_{n=1}^{\infty} \frac{n^3}{e^n}$ (First check where x^3/e^x decreases.)

In Exercises 9 to 12 use the p test to determine whether the series diverge or converge.

9. $\sum_{n=1}^{\infty} \frac{1}{\sqrt[3]{n}}$
10. $\sum_{n=1}^{\infty} \frac{1}{n^3}$
11. $\sum_{n=1}^{\infty} \frac{1}{n^{1.001}}$
12. $\sum_{n=1}^{\infty} \frac{1}{n^{0.999}}$

In each of Exercises 13 to 16 (a) compute the sum of the first four terms of the series to four decimal places, (b) give upper and lower bounds on the error R_4, (c) combine (a) and (b) to estimate the sum of the series.

13. $\sum_{n=1}^{\infty} 1/n^3$
14. $\sum_{n=1}^{\infty} 1/n^4$
15. $\sum_{n=1}^{\infty} 1/(n^2 + 1)$
16. $\sum_{n=1}^{\infty} 1/(n^2 + n)$

The ideas used in establishing the integral test can also be used to find estimates for the sum of the first n terms of the series $\sum_{n=1}^{\infty} f(n)$ discussed in this section. Exercises 17 to 21 give some examples.

17. Let $f(x)$ be a decreasing continuous positive function for $x \geq 1$. Show, by diagrams, that for $n \geq 2$,

$$\int_1^{n+1} f(x)\, dx < \sum_{i=1}^{n} f(i) < f(1) + \int_1^n f(x)\, dx.$$

18. Use the inequality in Exercise 17 to show that for $n \geq 2$

$$2\sqrt{n+1} - 2 < \sum_{i=1}^{n} \frac{1}{\sqrt{i}} < 2\sqrt{n} - 1.$$

19. Show that for $n \geq 2$,

$$\ln(n+1) < \frac{1}{1} + \frac{1}{2} + \frac{1}{3} + \cdots + \frac{1}{n} < 1 + \ln n.$$

20. (Calculator) Show that the sum of the first million terms of the harmonic series is between 13.8 and 14.9.

21. (a) By comparing the sum with integrals, show that

$$\ln \tfrac{201}{100} < \tfrac{1}{100} + \tfrac{1}{101} + \tfrac{1}{102} + \cdots + \tfrac{1}{200} < \ln \tfrac{200}{99}.$$

   (b) Show that

$$\lim_{n \to \infty} \sum_{i=n}^{2n} \frac{1}{i} = \ln 2.$$

22. (a) Let $f(x)$ be a decreasing continuous positive function for $x \geq 1$, such that $\int_1^\infty f(x)\, dx$ is convergent. Show that

$$\int_1^\infty f(x)\, dx < \sum_{n=1}^{\infty} f(n) < f(1) + \int_1^\infty f(x)\, dx.$$

   (b) Use (a) to estimate $\sum_{n=1}^{\infty} 1/n^2$.

23 Here is an argument that there is an infinite number of primes. Assume that there is only a finite number of primes, $p_1, p_2, \ldots, p_m$.
(a) Show that
$$\frac{1}{1-(1/p_i)} = 1 + \frac{1}{p_i} + \frac{1}{p_i^2} + \frac{1}{p_i^3} + \cdots$$
(b) Show then that
$$\frac{1}{1-1/p_1} \frac{1}{1-1/p_2} \cdots \frac{1}{1-1/p_m} = \sum_{n=1}^{\infty} \frac{1}{n}.$$

(Assume that the series can be multiplied term by term.)
(c) From (b) obtain a contradiction.

24 Cards of unit length from a playing deck are piled on top of each other as shown in Fig. 10.7.

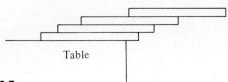

Figure 10.7

(a) If there are n cards, show that it is possible to arrange them so that the total distance that the right edge of the top card extends beyond the table is $\sum_{i=1}^{n} (1/2i)$. (The length of a card is 1.)
(b) Show that it is possible to place three cards in such a pile that the right edge of the top card extends $\frac{11}{12}$ of a card beyond the table.
(c) If you have 52 cards, estimate how far beyond the table the top card can extend.
(d) If you have an unlimited supply of cards, how far beyond the table can you arrange to have the top card extend?

10.4 The comparison test and the ratio test

So far in this chapter three tests for convergence (or divergence) of a series have been presented. The first concerned a special type of series, a geometric series. The second, the nth-term test for divergence, asserts that if the nth term of a series does *not* approach 0, the series diverges. The third, the integral test, applies to certain series of positive terms. In this section, three further tests are developed. They are the ones most frequently applied. This section concerns only series with all terms positive.

THE COMPARISON TEST

THEOREM 1 *The comparison test.*

1 If the series $\quad c_1 + c_2 + \cdots + c_n + \cdots$

converges, where each c_n is positive, and if

$$0 \le p_n \le c_n$$

for each n, then the series

$$p_1 + p_2 + \cdots + p_n + \cdots$$

converges.

2 If the series $\quad c_1 + c_2 + \cdots + c_n + \cdots$

diverges, where each c_n is positive, and if

$$p_n \ge c_n$$

for each n, then the series $p_1 + p_2 + \cdots + p_n + \cdots$ diverges.

PROOF First let us establish case 1. Let the sum of the series $c_1 + c_2 + \cdots$ be C. Let S_n denote the partial sum $p_1 + p_2 + \cdots + p_n$. Then, for each n,

$$S_n = p_1 + p_2 + \cdots + p_n \leq c_1 + c_2 + \cdots + c_n < C.$$

Since the p_n's are nonnegative,

$$S_1 \leq S_2 \leq \cdots \leq S_n \leq \cdots.$$

$S_1 \leq S_2 \leq \cdots < C$ Since each S_n is less than C, Theorem 2 of Sec. 10.1 assures us that the sequence

$$S_1, S_2, \ldots, S_n, \ldots$$

converges to a number L (less than or equal to C). In other words, the series $p_1 + p_2 + \cdots$ converges (and its sum is less than or equal to the sum $c_1 + c_2 + \cdots$).

The second part of the theorem, the divergence test (case 2), follows immediately from the convergence test (case 1). For if the series $p_1 + p_2 + \cdots$ converged, so would the series $c_1 + c_2 + \cdots$, which is assumed to diverge. ∎

To apply the comparison tests, compare a series to one known to converge (or diverge).

EXAMPLE 1   Show that the series $\dfrac{2}{3}\dfrac{1}{1^2} + \dfrac{3}{4}\dfrac{1}{2^2} + \cdots + \dfrac{n+1}{n+2}\dfrac{1}{n^2} + \cdots$

converges.

SOLUTION   The series resembles the series

$$\frac{1}{1^2} + \frac{1}{2^2} + \cdots + \frac{1}{n^2} + \cdots,$$

which was shown by the integral test to be convergent. Since the fraction $(n+1)/(n+2) < 1$,

$$\frac{n+1}{n+2}\frac{1}{n^2} < \frac{1}{n^2}.$$

Thus, by the comparison test for convergence, the series

$$\frac{2}{3}\frac{1}{1^2} + \frac{3}{4}\frac{1}{2^2} + \cdots + \frac{n+1}{n+2}\frac{1}{n^2} + \cdots$$

also converges. ∎

602 SERIES AND RELATED TOPICS

The next test is similar to the comparison test. It is useful when two series of positive terms resemble each other a great deal, even though the terms of one series are not less than the terms of the other.

THEOREM 2 *The limit-comparison test.* Let

$$c_1 + c_2 + \cdots + c_n + \cdots$$

and

$$p_1 + p_2 + \cdots + p_n + \cdots$$

be series of positive terms such that

$$\lim_{n \to \infty} \frac{p_n}{c_n}$$

exists and is not 0. Then

1. If $c_1 + c_2 + \cdots + c_n + \cdots$ converges, so does $p_1 + p_2 + \cdots + p_n + \cdots$.
2. If $c_1 + c_2 + \cdots + c_n + \cdots$ diverges, so does $p_1 + p_2 + \cdots + p_n + \cdots$.

PROOF Let $\lim_{n \to \infty} p_n/c_n = a$. We shall prove case 1. Since as $n \to \infty$, $p_n/c_n \to a$, there must be an integer N such that, for all $n \geq N$, p_n/c_n remains less than, say, $a + 1$. Thus

$$p_N < (a + 1)c_N$$
$$p_{N+1} < (a + 1)c_{N+1}$$
$$\cdots\cdots\cdots\cdots\cdots\cdots\cdots$$
$$p_n < (a + 1)c_n \qquad n \geq N.$$

Now, the series

$$(a + 1)c_N + (a + 1)c_{N+1} + \cdots + (a + 1)c_n + \cdots,$$

being $a + 1$ times the tail end of a convergent series, is itself convergent. By the comparison test,

$$p_N + p_{N+1} + \cdots + p_n + \cdots$$

is convergent. Hence $p_1 + p_2 + \cdots + p_n + \cdots$ is convergent.
Case 2 follows from case 1, as in Theorem 1. ∎

EXAMPLE 2 Show that

$$\sum_{n=1}^{\infty} \frac{(1 + 1/n)^n (1 + (-\tfrac{1}{2})^n)}{2^n}$$

converges.

SOLUTION Note that, as $n \to \infty$, $(1 + 1/n)^n \to e$ and $1 + (-\tfrac{1}{2})^n \to 1$. The major influence is the 2^n in the denominator. So use the limit-comparison test, with the convergent series $c_1 + c_2 + \cdots + c_n + \cdots = 1 + \frac{1}{2} + \frac{1}{4} + \cdots + (1/2^n) + \cdots$, which the given series resembles. Then

$$\lim_{n \to \infty} \frac{\left(1 + \frac{1}{n}\right)^n \left(1 + \left(-\frac{1}{2}\right)^n\right)}{\frac{1}{2^n}} \Big/ \frac{1}{2^n} = \lim_{n \to \infty} \left(1 + \frac{1}{n}\right)^n \left(1 + \left(-\frac{1}{2}\right)^n\right) = e \cdot 1 = e.$$

Since $\sum_{n=1}^{\infty} 2^{-n}$ is convergent, so is the given series. ∎

If $\sum c_n$ converges and $\frac{p_n}{c_n} \to 0$, then $\sum p_n$ converges.

Theorem 2 does not mention the case $\lim_{n \to \infty} p_n/c_n = 0$. That case may be treated separately, as follows. Assume that $\sum_{n=1}^{\infty} c_n$ converges and that $\lim_{n \to \infty} p_n/c_n = 0$. Then for n sufficiently large p_n/c_n is less than, say, 1. That is, there is an integer N such that for all $n \geq N$, $p_n/c_n < 1$. Thus, for $n \geq N$, $p_n < c_n$. By the comparison test, $\sum_{n=N}^{\infty} p_n$ converges. Thus, $\sum_{n=1}^{\infty} p_n$ converges.

If $\sum c_n$ diverges and $\frac{p_n}{c_n} \to 0$, no conclusion.

If $\lim_{n \to \infty} p_n/c_n = 0$ and $\sum_{n=1}^{\infty} c_n$ diverges, no conclusion can be drawn about the convergence or divergence of $\sum_{n=1}^{\infty} p_n$. Consider, for instance, $c_n = 1/\sqrt{n}$. If $p_n = 1/n$, then

$$\lim_{n \to \infty} \frac{p_n}{c_n} = \lim_{n \to \infty} \frac{\frac{1}{n}}{\frac{1}{\sqrt{n}}} = \lim_{n \to \infty} \frac{1}{\sqrt{n}} = 0,$$

and $\sum_{n=1}^{\infty} p_n$ happens to be a divergent series. If c_n is again $1/\sqrt{n}$ but $p_n = 1/n^2$, then $\lim_{n \to \infty} p_n/c_n = \lim_{n \to \infty} (1/n^2)/(1/\sqrt{n}) = 0$, but this time $\sum_{n=1}^{\infty} p_n$ is convergent (since it is the p series for $p = 2$).

THE RATIO TEST

The next test is suggested by the test for the convergence of a geometric series.

THEOREM 3 *The ratio test.* Let $p_1 + p_2 + \cdots + p_n + \cdots$ be a series of positive terms.

1 If $\lim_{n \to \infty} \frac{p_{n+1}}{p_n}$ exists and is less than 1, the series converges.

2 If $\lim_{n \to \infty} \frac{p_{n+1}}{p_n}$ exists and is greater than 1, the series diverges.

PROOF To prove case 1, first let $\lim_{n \to \infty} \frac{p_{n+1}}{p_n} = s < 1.$

Select a number r such that $s < r < 1.$

Then there is an integer N such that, for all $n \geq N$,

$$\frac{p_{n+1}}{p_n} < r;$$

hence
$$p_{n+1} < rp_n.$$

Thus
$$p_{N+1} < rp_N$$
$$p_{N+2} < rp_{N+1} < r(rp_N) = r^2 p_N$$
$$p_{N+3} < rp_{N+2} < r(r^2 p_N) = r^3 p_N$$

and so on.

Thus the terms of the series

$$p_N + p_{N+1} + p_{N+2} + \cdots$$

are less than the corresponding terms of the *geometric series*

$$p_N + rp_N + r^2 p_N + \cdots$$

(except for the first term p_N, which equals the first term of the geometric series). Since $r < 1$, the latter series converges. By the comparison test, $p_N + p_{N+1} + p_{N+2} + \cdots$ converges. Adding in the front end,

$$p_1 + p_2 + \cdots + p_{N-1},$$

still results in a convergent series.

The argument for case 2 is much shorter. If $\lim_{n \to \infty} p_{n+1}/p_n$ is greater than 1, then for all n, from some point on, p_{n+1} is larger than p_n. Thus the nth term of the series $p_1 + p_2 + \cdots$ cannot approach 0. By the nth-term test for divergence the series diverges. This completes the proof. ∎

No information if ratio approaches 1

No mention has been made in Theorem 3 of the case $\lim_{n \to \infty} p_{n+1}/p_n = 1$. The reason for this omission is that anything can happen; the series may diverge or it may converge. (Exercise 27 illustrates these possibilities.) Also, $\lim_{n \to \infty} p_{n+1}/p_n$ may not exist. In that case, one must look to other tests to determine whether the series diverges or converges.

The ratio test is a natural one to try if the nth term of a series involves powers of a fixed number, as the next example shows.

EXAMPLE 3 Show that the series $p + 2p^2 + 3p^3 + \cdots + np^n + \cdots$ converges for any fixed number p for which $0 < p < 1$.

SOLUTION Let a_n denote the nth term of the series. Then

$$a_n = np^n$$

and
$$a_{n+1} = (n+1)p^{n+1}.$$

The ratio between consecutive terms is

$$\frac{a_{n+1}}{a_n} = \frac{(n+1)p^{n+1}}{np^n}$$

$$= \frac{n+1}{n} p.$$

Thus
$$\lim_{n \to \infty} \frac{a_{n+1}}{a_n} = p$$

and the series converges. ∎

EXAMPLE 4 Find for which positive values of x the series

$$\frac{1}{0!} + \frac{x}{1!} + \frac{x^2}{2!} + \frac{x^3}{3!} + \cdots + \frac{x^n}{n!} + \cdots$$

converges and for which it diverges. (Each choice of x determines a specific series with constant terms.)

SOLUTION The nth term, a_n, is $x^n/n!$ Thus

$$a_{n+1} = \frac{x^{n+1}}{(n+1)!},$$

and, therefore,
$$\frac{a_{n+1}}{a_n} = \frac{\frac{x^{n+1}}{(n+1)!}}{\frac{x^n}{n!}}$$

$$= x \frac{n!}{(n+1)!}$$

$$= \frac{x}{n+1}.$$

Since x is fixed,
$$\lim_{n \to \infty} \frac{x}{n+1} = 0.$$

In the next section, it will be shown to converge for negative x too.

By the ratio test, the series converges for all positive x. ∎

The next example uses the ratio test to establish divergence.

EXAMPLE 5 Show that the series $\dfrac{2}{1} + \dfrac{2^2}{2} + \dfrac{2^3}{3} + \cdots + \dfrac{2^n}{n} + \cdots$ diverges.

SOLUTION In this case $a_n = 2^n/n$ and

$$\frac{a_{n+1}}{a_n} = \frac{\frac{2^{n+1}}{n+1}}{\frac{2^n}{n}}$$

$$= \frac{2^{n+1}}{n+1} \frac{n}{2^n}$$

$$= 2\frac{n}{n+1}.$$

Thus
$$\lim_{n \to \infty} \frac{a_{n+1}}{a_n} = 2,$$

which is larger than 1. By the ratio test, the series diverges. ∎

It is not really necessary to call on the powerful ratio test to establish the divergence of the series in Example 5. Since $\lim_{x \to \infty} 2^x/x = \infty$, its nth term gets arbitrarily large; by the nth-term test, the series diverges. Comparison with the harmonic series also demonstrates divergence.

The next test, closely related to the ratio test, is of use when the nth term contains only powers, such as n^n.

THE ROOT TEST

THEOREM 4 *The root test.* Let $\sum_{n=1}^{\infty} p_n$ be a series of positive terms such that $\sqrt[n]{p_n}$ has a limit as $n \to \infty$:

$$\lim_{n \to \infty} \sqrt[n]{p_n} = L.$$

Then

1 if $L < 1$, $\sum_{n=1}^{\infty} p_n$ converges;

2 if $L > 1$, $\sum_{n=1}^{\infty} p_n$ diverges;

3 if $L = 1$, no conclusion can be drawn ($\sum_{n=1}^{\infty} p_n$ may converge or may diverge). ∎

The proof of the root test is outlined in Exercises 25 and 30.

EXAMPLE 6 Use the root test to determine whether

$$\sum_{n=1}^{\infty} \frac{3^n}{n^n}$$

converges or diverges.

SOLUTION
$$\lim_{n \to \infty} \sqrt[n]{\frac{3^n}{n^n}} = \lim_{n \to \infty} \frac{3}{n} = 0.$$

By the root test, the series converges. ∎

Exercises

In Exercises 1 to 4 use the comparison test to determine whether the series converge or diverge.

1. $\sum_{n=1}^{\infty} \dfrac{1}{n^2+3}$

2. $\sum_{n=1}^{\infty} \dfrac{n+2}{(n+1)\sqrt{n}}$

3. $\sum_{n=1}^{\infty} \dfrac{\sin^2 n}{n^2}$

4. $\sum_{n=1}^{\infty} \dfrac{1}{n2^n}$

In Exercises 5 to 8 use the limit comparison test to determine whether the series converge or diverge.

5. $\sum_{n=1}^{\infty} \dfrac{5n+1}{(n+2)n^2}$

6. $\sum_{n=1}^{\infty} \dfrac{2^n+n}{3^n}$

7. $\sum_{n=1}^{\infty} \dfrac{n+1}{(5n+2)\sqrt{n}}$

8. $\sum_{n=1}^{\infty} \dfrac{(1+1/n)^n}{n^2}$

In Exercises 9 to 18 use any test to determine convergence or divergence.

9. $\sum_{n=1}^{\infty} \dfrac{n^3}{2^n}$

10. $\sum_{n=1}^{\infty} \dfrac{(n+1)^2}{n!}$

11. $\sum_{n=1}^{\infty} \dfrac{n!}{n^n}$

12. $\sum_{n=1}^{\infty} \dfrac{4n+1}{(2n+3)n^2}$

13. $\sum_{n=1}^{\infty} \dfrac{1}{n^n}$

14. $\sum_{n=1}^{\infty} \dfrac{(2n+1)(2^n+1)}{3^n+1}$

15. $\sum_{n=1}^{\infty} \dfrac{1+\cos n}{n^2}$

16. $\sum_{n=1}^{\infty} \dfrac{\ln n}{n^2}$

17. $\sum_{n=1}^{\infty} \dfrac{2^n}{(n+1)^n}$

18. $\sum_{n=1}^{\infty} \dfrac{5^n}{n(n^n)}$

In Exercises 19 to 22 determine for which positive numbers x the series (a) converge, (b) diverge.

19. $\sum_{n=1}^{\infty} \dfrac{x^n}{n} = \dfrac{x}{1} + \dfrac{x^2}{2} + \dfrac{x^3}{3} + \cdots + \dfrac{x^n}{n} + \cdots$

20. $\sum_{n=0}^{\infty} \dfrac{x^n}{2^n} = 1 + \dfrac{x}{2} + \dfrac{x^2}{4} + \dfrac{x^3}{8} + \cdots + \dfrac{x^n}{2^n} + \cdots$

21. $\sum_{n=0}^{\infty} \dfrac{2^n x^n}{n!} = 1 + \dfrac{2x}{1} + \dfrac{4x^2}{2} + \dfrac{8x^3}{6} + \cdots + \dfrac{2^n x^n}{n!} + \cdots$

22. $\sum_{n=1}^{\infty} n^5 x^n = x + 2^5 x^2 + 3^5 x^3 + \cdots + n^5 x^n + \cdots$

23. Use the result of Example 4 to show that, for $x > 0$, $\lim_{n \to \infty} x^n/n! = 0$. (This fact was established directly in Sec. 10.1.)

24. Use Exercise 24 of Sec. 10.2 to find the sum of the series in Example 3.

25. This exercise shows that the root test gives no information if $\lim_{n \to \infty} \sqrt[n]{p_n} = 1$.
    (a) Show that for $p_n = 1/n$, $\sum_{n=1}^{\infty} p_n$ diverges and $\lim_{n \to \infty} \sqrt[n]{p_n} = 1$.
    (b) Show that for $p_n = 1/n^2$, $\sum_{n=1}^{\infty} p_n$ converges and $\lim_{n \to \infty} \sqrt[n]{p_n} = 1$.

26. Solve Example 4 using the root test.

27. This exercise shows that the ratio test is useless when $\lim_{n \to \infty} p_{n+1}/p_n = 1$.
    (a) Show that if $p_n = 1/n$, then $\sum_{n=1}^{\infty} p_n$ diverges and $\lim_{n \to \infty} p_{n+1}/p_n = 1$.
    (b) Show that if $p_n = 1/n^2$, then $\sum_{n=1}^{\infty} p_n$ converges and $\lim_{n \to \infty} p_{n+1}/p_n = 1$.

28. (a) Show that $\sum_{n=1}^{\infty} 1/(1+2^n)$ converges.
    (b) Show that the sum of the series in (a) is between 0.64 and 0.77. (Use the first three terms and control the sum of the rest of the series by comparing it to the sum of a geometric series.)

29. (a) Show that $\sum_{n=1}^{\infty} n/[(n+1)2^n]$ converges.
    (b) Show that the sum of the series in (a) is between

    and
    $\tfrac{1}{2} \cdot \tfrac{1}{2} + \tfrac{2}{3} \cdot \tfrac{1}{4} + \tfrac{3}{4} \cdot \tfrac{1}{8}$
    $\tfrac{1}{2} \cdot \tfrac{1}{2} + \tfrac{2}{3} \cdot \tfrac{1}{4} + \tfrac{1}{4}.$

■ ■

30. (Proof of the root test, Theorem 4)
    (a) Assume that $\lim_{n \to \infty} \sqrt[n]{p_n} = L < 1$. Pick any r, $L < r < 1$, and then N such that $\sqrt[n]{p_n} < r$ for $n > N$. Show that $p_n < r^n$ for $n > N$ and compare a tail end of $\sum_{n=1}^{\infty} p_n$ to a geometric series.
    (b) Assume that $\sqrt[n]{p_n} = L > 1$. Pick any number r, $1 < r < L$, and then N such that $\sqrt[n]{p_n} > r$ for $n > N$. Show that $p_n > r^n$ for $n > N$. From this conclude that $\sum_{n=1}^{\infty} p_n$ diverges.

31. Determine whether $\sum_{n=1}^{\infty} (\ln n)^2/n^2$ converges or diverges.

32. Prove the following result, which is used in the statistical theory of stochastic processes: Let $\{a_n\}$ and $\{c_n\}$ be two sequences of nonnegative numbers such that $\sum_{n=1}^{\infty} a_n c_n$ converges and $\lim_{n \to \infty} c_n = 0$. Then $\sum_{n=1}^{\infty} a_n c_n^2$ converges.

10.5 The alternating-series and absolute-convergence tests

The tests for convergence or divergence in Secs. 10.3 and 10.4 concerned series whose terms are positive. This section examines series which may have both positive and negative terms. Two tests for the convergence of such a series are presented. The "alternating-series" test applies to series whose terms alternate in sign, $+, -, +, -, \ldots$ and decrease in absolute value. In the "absolute-convergence" test the signs may vary in any way.

ALTERNATING SERIES

DEFINITION *Alternating series.* If $p_1, p_2, \ldots, p_n, \ldots$ is a sequence of positive numbers, then the series

$$\sum_{n=1}^{\infty} (-1)^{n+1} p_n = p_1 - p_2 + p_3 - p_4 + \cdots + (-1)^{n+1} p_n + \cdots$$

and the series

$$\sum_{n=1}^{\infty} (-1)^n p_n = -p_1 + p_2 - p_3 + p_4 - \cdots + (-1)^n p_n + \cdots$$

are called *alternating series*.

For instance, $1 - \dfrac{1}{3} + \dfrac{1}{5} - \dfrac{1}{7} + \cdots + (-1)^{n+1} \dfrac{1}{2n-1} + \cdots$

and $\qquad -1 + 1 - 1 + 1 - \cdots + (-1)^n + \cdots$

are alternating series.

By the nth-term test, the second series diverges. The following theorem implies that the first series converges.

THEOREM 1 *The alternating-series test.* If $p_1, p_2, \ldots, p_n, \ldots$ is a decreasing sequence of positive numbers such that $\lim_{n \to \infty} p_n = 0$, then the series whose nth term is $(-1)^{n+1} p_n$,

$$\sum_{n=1}^{\infty} (-1)^{n+1} p_n = p_1 - p_2 + p_3 - \cdots + (-1)^{n+1} p_n + \cdots,$$

converges.

PROOF The idea of the proof is easily conveyed by a specific case. For the sake of concreteness and simplicity, consider the series in which $p_n = 1/n$, that is, the series

$$1 - \dfrac{1}{2} + \dfrac{1}{3} - \dfrac{1}{4} + \cdots + (-1)^{n+1} \dfrac{1}{n} + \cdots.$$

Consider first the partial sums of an *even* number of terms, S_2, S_4, S_6, ... For clarity, group the summands in pairs:

$$S_2 = (1 - \tfrac{1}{2})$$
$$S_4 = (1 - \tfrac{1}{2}) + (\tfrac{1}{3} - \tfrac{1}{4}) = S_2 + (\tfrac{1}{3} - \tfrac{1}{4})$$
$$S_6 = (1 - \tfrac{1}{2}) + (\tfrac{1}{3} - \tfrac{1}{4}) + (\tfrac{1}{5} - \tfrac{1}{6}) = S_4 + (\tfrac{1}{5} - \tfrac{1}{6})$$
$$\cdots\cdots\cdots\cdots\cdots\cdots\cdots\cdots\cdots\cdots\cdots\cdots\cdots\cdots$$

Since $\tfrac{1}{3}$ is larger than $\tfrac{1}{4}$, $\tfrac{1}{3} - \tfrac{1}{4}$ is positive. Thus S_4, which equals $S_2 + (\tfrac{1}{3} - \tfrac{1}{4})$, is larger than S_2. Similarly,

$$S_6 > S_4.$$

More generally, then, $\quad S_2 < S_4 < S_6 < S_8 < \cdots.$

The sequence $\{S_{2n}\}$ is increasing.

Next it will be shown that S_{2n} is less than 1, the first term of the given sequence.

First of all $\quad\quad\quad\quad\quad S_2 = 1 - \tfrac{1}{2} < 1.$

Next, consider S_4: $\quad\quad S_4 = 1 - \tfrac{1}{2} + \tfrac{1}{3} - \tfrac{1}{4}$
$$= 1 - (\tfrac{1}{2} - \tfrac{1}{3}) - \tfrac{1}{4}$$
$$< 1 - (\tfrac{1}{2} - \tfrac{1}{3}).$$

Since $\tfrac{1}{2} - \tfrac{1}{3}$ is positive, this shows that

$$S_4 < 1.$$

Similarly, $\quad\quad S_6 = 1 - (\tfrac{1}{2} - \tfrac{1}{3}) - (\tfrac{1}{4} - \tfrac{1}{5}) - \tfrac{1}{6}$
$$< 1 - (\tfrac{1}{2} - \tfrac{1}{3}) - (\tfrac{1}{4} - \tfrac{1}{5})$$
$$< 1.$$

In general then, $\quad\quad S_{2n} < 1$
for all n.

The sequence $\quad\quad\quad S_2, S_4, S_6, \ldots$

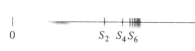

Figure 10.8

is therefore increasing and yet bounded by the number 1, as indicated in Fig. 10.8. By Theorem 2 of Sec. 10.1, $\lim_{n \to \infty} S_{2n}$ exists. Call this limit S. All that remains to be shown is that the numbers

$$S_1, S_3, S_5, \ldots$$

also converge to S.

Note that $\quad\quad\quad S_3 = 1 - \tfrac{1}{2} + \tfrac{1}{3} = S_2 + \tfrac{1}{3}$
$$S_5 = 1 - \tfrac{1}{2} + \tfrac{1}{3} - \tfrac{1}{4} + \tfrac{1}{5} = S_4 + \tfrac{1}{5}.$$

In general, $$S_{2n+1} = S_{2n} + \frac{1}{2n+1}.$$

(The term $1/(2n+1)$ will be p_{2n+1} in the general case.)

Thus
$$\lim_{n \to \infty} S_{2n+1} = \lim_{n \to \infty} \left(S_{2n} + \frac{1}{2n+1} \right)$$
$$= \lim_{n \to \infty} S_{2n} + \lim_{n \to \infty} \frac{1}{2n+1}$$
$$= S + 0$$
$$= S.$$

Since the partial sums $\quad S_2, S_4, S_6, \ldots$

and the partial sums $\quad S_1, S_3, S_5, \ldots$

both have the same limit S, it follows that
$$\lim_{n \to \infty} S_n = S.$$

Thus the sequence $\quad 1 - \tfrac{1}{2} + \tfrac{1}{3} - \tfrac{1}{4} + \tfrac{1}{5} - \cdots \quad$ converges.

A similar argument applies to any alternating series whose nth term approaches 0 and whose terms decrease in absolute value. ∎

Decreasing alternating series   An alternating series whose terms decrease in absolute value as n increases will be called a *decreasing alternating series*. Theorem 1 shows that a decreasing alternating series whose nth term approaches 0 as $n \to \infty$ converges.

EXAMPLE 1   Estimate the sum S of the series
$$1 - \tfrac{1}{2} + \tfrac{1}{3} - \tfrac{1}{4} + \cdots.$$

SOLUTION   These are the first five partial sums:
$$S_1 = 1 = 1.00$$
$$S_2 = 1 - \tfrac{1}{2} = 0.500$$
$$S_3 = 1 - \tfrac{1}{2} + \tfrac{1}{3} \approx 0.500 + 0.3333 = 0.8333$$
$$S_4 = S_3 - \tfrac{1}{4} \approx 0.8333 - 0.250 = 0.5833$$
$$S_5 = S_4 + \tfrac{1}{5} \approx 0.5833 + 0.200 = 0.7833$$

Figure 10.9 is a graph of S_n as a function of n. The sums $S_1, S_3, \ldots$ approach S from above. The sums $S_2, S_4, \ldots$ approach S from below. For instance,
$$S_4 < S < S_5$$
gives the information that $\quad 0.583 < S < 0.784$. (See Fig. 10.10.)

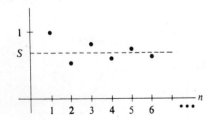

Figure 10.9

10.5 THE ALTERNATING-SERIES AND ABSOLUTE-CONVERGENCE TESTS

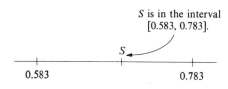

Figure 10.10

The error in estimating the sum of a decreasing alternating series

As Fig. 10.9 suggests, any partial sum of a series satisfying the hypothesis of the alternating-series test differs from the sum of the series by less than the absolute value of the first omitted term. That is, if S_n is the sum of the first n terms of the series and S is the sum of the series, then the error,

$$R_n = S - S_n,$$

has absolute value at most p_{n+1}, which is the absolute value of the first omitted term.

EXAMPLE 2 Does the series

$$\frac{3}{1!} - \frac{3^2}{2!} + \frac{3^3}{3!} - \frac{3^4}{4!} + \frac{3^5}{5!} - \cdots + (-1)^{n+1}\frac{3^n}{n!} + \cdots$$

converge or diverge?

SOLUTION This is an alternating series. By Example 4 of Sec. 10.1, its nth term approaches 0. Let us see whether the absolute values of the terms decrease in size, term by term. These first few absolute values are:

$$\frac{3}{1!} = 3$$

$$\frac{3^2}{2!} = \frac{9}{2} = 4.5$$

$$\frac{3^3}{3!} = \frac{27}{6} = 4.5$$

$$\frac{3^4}{4!} = \frac{81}{24} = 3.375.$$

At first they increase. However, the fourth term is less than the third. Let us show that the rest of the terms decrease in size. For instance,

$$\frac{3^5}{5!} = \frac{3 \cdot 3^4}{5 \cdot 4!} < \frac{3^4}{4!}$$

and, for $n \geq 3$,

$$\frac{3^{n+1}}{(n+1)!} = \frac{3}{n+1}\frac{3^n}{n!} < \frac{3^n}{n!}.$$

By the alternating-series test, the series that begins

$$\frac{3^3}{3!} - \frac{3^4}{4!} + \frac{3^5}{5!} - \frac{3^6}{6!} + \cdots$$

converges. Call its sum S. If the two terms

$$\frac{3}{1!} - \frac{3^2}{2!}$$

are added on, the resulting series still converges and has the sum

$$\frac{3}{1!} - \frac{3^2}{2!} + S. \quad \blacksquare$$

In the alternating-series test the absolute values of the terms must eventually be decreasing.

As Example 2 illustrates, the alternating-series test works as long as the nth term approaches 0 and the terms decrease in size from some point on in the series.

It may seem that any alternating series whose nth term approaches 0 converges. *This is not the case*, as is shown by this series:

$$\frac{2}{1} - \frac{1}{1} + \frac{2}{2} - \frac{1}{2} + \frac{2}{3} - \frac{1}{3} + \frac{2}{4} - \frac{1}{4} + \cdots. \tag{1}$$

whose terms alternate $2/n$ and $-1/n$.

Let S_n be the sum of the first n terms of (1). Then

$$S_2 = \frac{2}{1} - \frac{1}{1} = \frac{1}{1},$$

$$S_4 = \left(\frac{2}{1} - \frac{1}{1}\right) + \left(\frac{2}{2} - \frac{1}{2}\right) = \frac{1}{1} + \frac{1}{2},$$

$$S_6 = \left(\frac{2}{1} - \frac{1}{1}\right) + \left(\frac{2}{2} - \frac{1}{2}\right) + \left(\frac{2}{3} - \frac{1}{3}\right) = \frac{1}{1} + \frac{1}{2} + \frac{1}{3},$$

and, more generally,

$$S_{2n} = \frac{1}{1} + \frac{1}{2} + \frac{1}{3} + \cdots + \frac{1}{n}.$$

Since S_{2n} gets arbitrarily large as $n \to \infty$ (the harmonic series diverges), the sequence (1) diverges.

Also, an alternating series whose terms decrease in size from some point on need not converge. Consider, for instance, the series

$$\frac{2}{1} - \frac{3}{2} + \frac{4}{3} - \frac{5}{4} + \cdots + (-1)^{n+1}\left(\frac{n+1}{n}\right) + \cdots.$$

Since the absolute value of the nth term approaches 1, the nth term does not approach 0. By the nth term test for divergence, the series diverges.

10.5 THE ALTERNATING-SERIES AND ABSOLUTE-CONVERGENCE TESTS

ABSOLUTE CONVERGENCE

Consider a series
$$a_1 + a_2 + \cdots + a_n + \cdots,$$

whose terms may be positive, negative, or zero. It is reasonable to expect it to behave at least as "nicely" as the series

$$|a_1| + |a_2| + \cdots + |a_n| + \cdots,$$

since by making all the terms positive we give the series more chance to diverge. The next theorem confirms this expectation.

THEOREM 2 *Absolute-convergence test.* If the series

$$|a_1| + |a_2| + \cdots + |a_n| + \cdots$$

converges, then so does the series

$$a_1 + a_2 + \cdots + a_n + \cdots.$$

PROOF Since the series $|a_1| + |a_2| + \cdots$ converges, so does the series $2|a_1| + 2|a_2| + \cdots$. (Its sum is $2 \sum_{n=1}^{\infty} |a_n|$.)

Next, introduce the series whose nth term is

$$a_n + |a_n|.$$

Note that if a_n is negative, $a_n + |a_n| = 0$, while if a_n is nonnegative, $a_n + |a_n| = 2|a_n|$. Hence, for all n,

$$0 \leq a_n + |a_n| \leq 2|a_n|.$$

By the comparison test $\sum_{n=1}^{\infty} (a_n + |a_n|)$ converges.

Let
$$\sum_{n=1}^{\infty} (a_n + |a_n|) = A \quad \text{and} \quad \sum_{n=1}^{\infty} |a_n| = B.$$

Now
$$\sum_{n=1}^{k} a_n = \sum_{n=1}^{k} (a_n + |a_n|) - \sum_{n=1}^{k} |a_n|.$$

Thus, as $k \to \infty$, $\sum_{n=1}^{k} a_n \to A - B$, and the theorem is proved. ■

The next example is a typical illustration of how Theorem 2 is applied.

EXAMPLE 3 Show that
$$\frac{1}{1^2} + \frac{1}{2^2} - \frac{1}{3^2} + \frac{1}{4^2} + \frac{1}{5^2} - \frac{1}{6^2} + \cdots.$$

(two positive terms alternating with one negative term) converges.

SOLUTION The series whose nth term is the absolute value of the nth term of the given series is

$$\frac{1}{1^2} + \frac{1}{2^2} + \cdots + \frac{1}{n^2} + \cdots.$$

In Sec. 10.3 this series was shown to converge (by the integral test). By the absolute-convergence test the original series, with $+$'s and $-$'s, converges. ∎

The alternating series $\quad 1 - \tfrac{1}{2} + \tfrac{1}{3} - \tfrac{1}{4} + \cdots$

converges, as shown by Theorem 1. However, when all the terms are replaced by their absolute values, the resulting series, the harmonic series, does not converge; that is,

$$1 + \tfrac{1}{2} + \tfrac{1}{3} + \tfrac{1}{4} + \cdots$$

diverges. Thus the converse of Theorem 2 is false.

The following definitions are frequently used in describing these various cases of convergence or divergence.

DEFINITION *Absolute convergence.* A series $a_1 + a_2 + \cdots$ is said to *converge absolutely* if the series $|a_1| + |a_2| + \cdots$ converges.

Theorem 2 can be stated simply, "If a series converges absolutely, then it converges."

DEFINITION *Conditional convergence.* A series $a_1 + a_2 + \cdots$ is said to *converge conditionally* if it converges but does not converge absolutely.

$1 - \tfrac{1}{2} + \tfrac{1}{3} - \tfrac{1}{4} + \cdots$
converges conditionally.

For instance, the *alternating harmonic series*, $1 - \tfrac{1}{2} + \tfrac{1}{3} - \tfrac{1}{4} + \cdots$, is conditionally convergent.

The next example shows how the absolute convergence test can be combined with other tests to establish convergence of a series.

EXAMPLE 4 Show that

$$\frac{2}{1}\left(\frac{1}{2}\right) - \frac{3}{2}\left(\frac{1}{2}\right)^2 + \frac{4}{3}\left(\frac{1}{2}\right)^3 + \cdots + (-1)^{n+1}\frac{n+1}{n}\left(\frac{1}{2}\right)^n + \cdots \quad (2)$$

converges.

SOLUTION Consider the series of positive terms

$$\frac{2}{1}\left(\frac{1}{2}\right) + \frac{3}{2}\left(\frac{1}{2}\right)^2 + \frac{4}{3}\left(\frac{1}{2}\right)^3 + \cdots + \frac{n+1}{n}\left(\frac{1}{2}\right)^n + \cdots.$$

Its typical term is
$$p_n = \frac{n+1}{n}\left(\frac{1}{2}\right)^n.$$

The presence of the power $(\frac{1}{2})^n$ suggests using the ratio test:

$$\frac{p_{n+1}}{p_n} = \frac{\dfrac{n+2}{n+1}\left(\dfrac{1}{2}\right)^{n+1}}{\dfrac{n+1}{n}\left(\dfrac{1}{2}\right)^n}$$

$$= \frac{n+2}{n+1}\frac{n}{n+1}\cdot\frac{1}{2}.$$

Thus
$$\lim_{n\to\infty}\frac{p_{n+1}}{p_n} = \frac{1}{2},$$

which is less than 1. Consequently, the given series (2), with positive and negative terms, converges absolutely. Thus it converges. ∎

Example 4 suggests the following variant of the ratio test.

THEOREM 3 *The absolute-ratio test.* Let $\sum_{n=1}^{\infty} a_n$ be a series such that

The absolute ratio test

$$\lim_{n\to\infty}\left|\frac{a_{n+1}}{a_n}\right| = L < 1.$$

Then $\sum_{n=1}^{\infty} a_n$ converges. If $L > 1$ or if $\lim_{n\to\infty}|a_{n+1}/a_n| = \infty$, $\sum_{n=1}^{\infty} a_n$ diverges.

PROOF Take the case $L < 1$. By the ratio test, $\sum_{n=1}^{\infty}|a_n|$ converges. Since $\sum_{n=1}^{\infty}|a_n|$ converges, it follows that $\sum_{n=1}^{\infty} a_n$ converges also. The case $L > 1$ is treated in Exercise 36. The case $L = \infty$ can be treated as follows. If $\lim_{n\to\infty}|(a_{n+1}/a_n)| = \infty$, the ratio $|a_{n+1}|/|a_n|$ gets arbitrarily large as $n \to \infty$. So from some point on the numbers $|a_n|$ increase. By the nth term test for divergence, $\sum_{n=1}^{\infty} a_n$ is divergent. ∎

The absolute ratio test simplifies work with minus signs.

Theorem 3 would establish the convergence of the series in Example 4 as follows. Let $a_n = (-1)^{n+1}((n+1)/n)(\frac{1}{2})^n$. Then

$$\left|\frac{a_{n+1}}{a_n}\right| = \left|\frac{(-1)^{n+2}\dfrac{n+2}{n+1}\left(\dfrac{1}{2}\right)^{n+1}}{(-1)^{n+1}\dfrac{n+1}{n}\left(\dfrac{1}{2}\right)^n}\right| = \frac{n+2}{n+1}\cdot\frac{n}{n+1}\cdot\frac{1}{2},$$

which approaches $\frac{1}{2}$ as $n \to \infty$. Thus $\sum_{n=1}^{\infty} a_n$ converges (in fact, absolutely).

EXAMPLE 5 Examine the series

$$\frac{\cos x}{1^2} + \frac{\cos 2x}{2^2} + \frac{\cos 3x}{3^2} + \cdots + \frac{\cos nx}{n^2} + \cdots \qquad (3)$$

for convergence or divergence.

SOLUTION The number x is fixed. The numbers $\cos nx$ may be positive, negative, or zero, in an irregular manner. However, for all n, $|\cos nx| \leq 1$.

Recall that the series

$$\frac{1}{1^2} + \frac{1}{2^2} + \frac{1}{3^2} + \cdots + \frac{1}{n^2} + \cdots$$

converges, as shown in Sec. 10.3. Since $|\cos nx|/n^2 \leq 1/n^2$, the series

$$\frac{|\cos x|}{1^2} + \frac{|\cos 2x|}{2^2} + \frac{|\cos 3x|}{3^2} + \cdots + \frac{|\cos nx|}{n^2} + \cdots$$

Advanced calculus shows that, for $0 \leq x \leq 2\pi$, (3) sums to

$$\frac{3x^2 - 6\pi x + 2\pi^2}{12}.$$

converges by the comparison test. Series (3) thus converges absolutely for all x. Hence it converges. ∎

A series that converges absolutely has the property that no matter how the terms are rearranged the new series converges and has the same sum as the original series. It might be expected that all convergent series have this property, but such is not the case. For instance, the alternating harmonic series

$$\tfrac{1}{1} - \tfrac{1}{2} + \tfrac{1}{3} - \tfrac{1}{4} + \tfrac{1}{5} - \cdots \qquad (4)$$

does not. To show this, rearrange the summands so that two positive summands alternate with one negative summand, as follows:

$$\tfrac{1}{1} + \tfrac{1}{3} - \tfrac{1}{2} + \tfrac{1}{5} + \tfrac{1}{7} - \tfrac{1}{4} + \cdots. \qquad (5)$$

Rearranging a conditionally convergent series is dangerous.

The positive summands in (5) have much more influence than the negative summands. In the battle between the positives and the negatives, the positives will win by a bigger margin in (5) than in (4). In fact, as shown in Exercise 35, the sum of (5) is $\tfrac{3}{2} \ln 2$. (But the sum of (4) is just $\ln 2$, as shown in Exercise 28(e) of Sec. 10.7.)

In advanced calculus it is demonstrated that a conditionally convergent series can be rearranged to converge to any preassigned sum or even to diverge to ∞ or $-\infty$.

Exercises

In Exercises 1 to 8, which concern alternating series, determine which series converge and which diverge. Explain your answers.

1. $\dfrac{1}{2} - \dfrac{2}{3} + \dfrac{3}{4} - \dfrac{4}{5} + \cdots + (-1)^{n+1}\dfrac{n}{n+1} + \cdots$

2. $-\dfrac{1}{1+\frac{1}{2}} + \dfrac{1}{1+\frac{1}{4}} - \dfrac{1}{1+\frac{1}{8}} + \cdots + (-1)^n \dfrac{1}{1+2^{-n}} + \cdots$

3. $\dfrac{1}{\sqrt{1}} - \dfrac{1}{\sqrt{2}} + \dfrac{1}{\sqrt{3}} - \dfrac{1}{\sqrt{4}} + \cdots + (-1)^{n+1}\dfrac{1}{\sqrt{n}} + \cdots$

4. $\dfrac{5}{1!} - \dfrac{5^2}{2!} + \dfrac{5^3}{3!} - \dfrac{5^4}{4!} + \cdots + (-1)^{n+1}\dfrac{5^n}{n!} + \cdots$

5. $\dfrac{3}{\sqrt{1}} - \dfrac{2}{\sqrt{1}} + \dfrac{3}{\sqrt{2}} - \dfrac{2}{\sqrt{2}} + \dfrac{3}{\sqrt{3}} - \dfrac{2}{\sqrt{3}} + \cdots$

6. $\sqrt{1} - \sqrt{2} + \sqrt{3} - \sqrt{4} + \cdots + (-1)^{n+1}\sqrt{n} + \cdots$

7. $\dfrac{1}{3} - \dfrac{2}{5} + \dfrac{3}{7} - \dfrac{4}{9} + \dfrac{5}{11} - \cdots + (-1)^{n+1}\dfrac{n}{2n+1} + \cdots$

8. $\dfrac{1}{1^2} - \dfrac{1}{2^2} + \dfrac{1}{3^2} - \dfrac{1}{4^2} + \cdots + (-1)^{n+1}\dfrac{1}{n^2} + \cdots$

9. Consider the alternating harmonic series
$$\sum_{n=1}^{\infty}(-1)^{n+1}/n.$$
(a) Compute S_5 and S_6 to five decimal places.
(b) Is the estimate S_5 smaller or larger than the sum of the series?
(c) Use (a) and (b) to find two numbers between which the sum of the series must lie.

10. Consider the series $\sum_{n=1}^{\infty}(-1)^{n+1}2^{-n}/n$.
(a) Estimate the sum of the series using S_5.
(b) Estimate the error R_5.

In Exercises 11 to 26 determine which series diverge, converge absolutely, or converge conditionally. Explain your answers.

11. $\sum_{n=1}^{\infty}\dfrac{(-1)^n}{\sqrt[3]{n^2}}$

12. $\sum_{n=1}^{\infty}(-1)^n \ln(1/n)$

13. $\sum_{n=2}^{\infty}\dfrac{(-1)^n}{n \ln n}$

14. $\sum_{n=1}^{\infty}\dfrac{\sin n}{n^{1.01}}$

15. $\sum_{n=1}^{\infty}\left(1 - \cos\dfrac{\pi}{n}\right)$

16. $\sum_{n=1}^{\infty}(-1)^n \cos\left(\dfrac{\pi}{n^2}\right)$

17. $\sum_{n=1}^{\infty}\sin\left(\dfrac{\pi}{n^2}\right)$

18. $\sum_{n=1}^{\infty}\dfrac{(-2)^n}{n!}$

19. $\dfrac{1}{1^2} + \dfrac{1}{2^2} - \dfrac{1}{3^2} + \dfrac{1}{4^2} + \dfrac{1}{5^2} - \dfrac{1}{6^2} + \cdots$

(Two +'s alternating with one −.)

20. $\sum_{n=1}^{\infty}\dfrac{(-3)^n(1+n^2)}{n!}$

21. $\sum_{n=1}^{\infty}\dfrac{\cos n\pi}{2n+1}$

22. $\sum_{n=1}^{\infty}\dfrac{(-1)^n(n+5)}{n^2}$

23. $\sum_{n=1}^{\infty}\dfrac{(-9)^n}{10^n + n}$

24. $\sum_{n=1}^{\infty}\dfrac{(-1)^n}{\sqrt[3]{n}}$

25. $\sum_{n=1}^{\infty}\dfrac{(-1.01)^n}{n!}$

26. $\sum_{n=1}^{\infty}\dfrac{(-\pi)^{2n+1}}{(2n+1)!}$

∎

27. The series $\sum_{n=1}^{\infty}(-1)^{n+1}2^{-n}$ is both a geometric series and a decreasing alternating series whose nth term approaches 0.
(a) Compute S_6 to three decimal places.
(b) Using the fact that the series is a decreasing alternating series, put a bound on R_6.
(c) Using the formula for the sum of a geometric series, compute R_6 exactly.

28. Show that the sum of the series $\sum_{n=1}^{\infty}(-1)^{n+1}/n$ is between 0.62 and 0.77. *Suggestion:* Compute S_6.

29. Show that the sum of the series $\sum_{n=1}^{\infty}(-1)^{n+1}/(2n-1)$ is between 0.72 and 0.84.

30. Show that the sum of the series $\sum_{n=1}^{\infty}(-1)^{n+1}/n!$ is between 0.625 and 0.634.

31. Show that the sum of the series $\sum_{n=1}^{\infty}(-1)^{n+1}/(2n-1)!$ is between 0.833 and 0.842.

■ ■

32 Let $P(x)$ and $Q(x)$ be two polynomials of degree at least one. Assume that for $n \geq 1$, $Q(n) \neq 0$. What relation must there be between the degrees of $P(x)$ and $Q(x)$ if
(a) $P(n)/Q(n) \to 0$ as $n \to \infty$?
(b) $\sum_{n=1}^{\infty} P(n)/Q(n)$ converges absolutely?
(c) $\sum_{n=1}^{\infty} (-1)^n P(n)/Q(n)$ converges conditionally?

33 Prove Theorem 1 as follows:
(a) Show that $S_2 < S_4 < S_6 < \cdots < p_1$.
(b) Show that $\lim_{n \to \infty} S_{2n}$ exists.
(c) Call the limit in (b) "S." Show that $\lim_{n \to \infty} S_{2n+1} = S$.

Exercises 34 and 35 are related.

34 Use elementary geometry to show that the area of that part of the endless staircase shown in Fig. 10.11 above the curve $y = 1/x$, is between $\frac{1}{2}$ and 1.

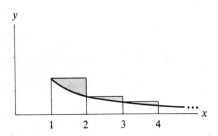

Figure 10.11

The area of the shaded region above the curve is known as Euler's constant γ, whose decimal representation begins 0.577. It is not known whether γ is rational or irrational. Thus, if a_n is defined by $a_n = \sum_{i=1}^{n} 1/i - \ln n$, then $a_n \to \gamma$ as $n \to \infty$.

35 This exercise shows that the sum of the series (5) is $\frac{3}{2} \ln 2$. The argument depends on the formula $\sum_{i=1}^{n} 1/i = \ln n + a_n$, where a_n approaches γ as $n \to \infty$.

(a) Show that $\sum_{i=1}^{n} (1/2i) = (\ln n)/2 + (a_n)/2$.
(b) Show that
$$\tfrac{1}{1} + \tfrac{1}{3} + \tfrac{1}{5} + \cdots + 1/(4n-1)$$
$$= \sum_{i=1}^{2n} 1/(2i-1) = \sum_{i=1}^{4n} 1/i - \tfrac{1}{2} \sum_{i=1}^{2n} 1/i.$$

(c) From (a) and (b) deduce that
$$\sum_{i=1}^{2n} \frac{1}{2i-1} = \ln 4 - \frac{\ln 2}{2} + \frac{\ln n}{2} + a_{4n} - \frac{a_{2n}}{2}.$$

(d) Show that
$$1 + \frac{1}{3} - \frac{1}{2} + \frac{1}{5} + \frac{1}{7} - \frac{1}{4} + \frac{1}{9} + \frac{1}{11} - \frac{1}{6} +$$
$$\cdots + \frac{1}{4n-3} + \frac{1}{4n-1} - \frac{1}{2n}$$

equals $\quad \ln 4 - (\ln 2)/2 + a_{4n} - (a_{2n})/2 - (a_n)/2.$

(e) From (d) deduce that the sum of (5) is $\frac{3}{2} \ln 2$.
In Exercise 28(e) of Sec. 10.7 it is shown that $1 - \frac{1}{2} + \frac{1}{3} - \frac{1}{4} + \cdots = \ln 2$. However, you may use the method of this exercise to obtain this fact as well.

36 This exercise treats the second half of the absolute-ratio test.
(a) Show that if
$$\lim_{n \to \infty} \left| \frac{a_{n+1}}{a_n} \right| = L > 1,$$

then $|a_n| \to \infty$ as $n \to \infty$. *Suggestion:* First show that there is a number r, $r > 1$, such that $|a_{n+1}| > r|a_n|$ for all $n \geq N$ for some integer N.
(b) From (a) deduce that a_n does not approach 0 as $n \to \infty$.

10.6 Power series

Let a be a real number and $\{a_n\}$ be a sequence. The series

$$\sum_{n=0}^{\infty} a_n(x-a)^n = a_0 + a_1(x-a) + a_2(x-a)^2 + \cdots + a_n(x-a)^n + \cdots \qquad (1)$$

Power series is called a *power series* in $x - a$. If $a = 0$, we obtain a power series in x,

$$\sum_{n=0}^{\infty} a_n x^n = a_0 + a_1 x + a_2 x^2 + \cdots + a_n x^n + \cdots. \qquad (2)$$

Maclaurin series A power series in x is also called a Maclaurin series. For instance, the geometric series

$$1 + x + x^2 + \cdots + x^n + \cdots \qquad (3)$$

is a Maclaurin series. If the ratio x has absolute value less than 1, this series converges and has the sum $1/(1 - x)$:

$$\frac{1}{1-x} = \sum_{n=0}^{\infty} x^n \qquad |x| < 1. \tag{4}$$

If $|x| \geq 1$, the geometric series (3) diverges by the nth term test. As (4) illustrates, a power series may converge for some values of x and diverge for other values of x. In addition, (4) shows that some functions that are not polynomials may, however, be represented as power series, which we may think of as "polynomials of infinite degree." This section is concerned with two questions: What can be said about the numbers x for which a power series converges? What common functions can be represented as power series?

Two questions about power series

For the most part we will be concerned with power series in x. At the end of the section we will consider power series in $x - a$.

THE CONVERGENCE OF A MACLAURIN SERIES

For each fixed choice of x, a power series becomes a series with constant terms.

The power series $a_0 + a_1 x + a_2 x^2 + \cdots$ certainly converges when $x = 0$. It may or may not converge for other choices of x. However, as Theorem 1 will show, if the series converges at a certain value c it converges at any number x whose absolute value is less than $|c|$.

THEOREM 1 Let $a_0, a_1, a_2, \ldots$ be a sequence and c a nonzero number. Assume that

$$a_0 + a_1 c + a_2 c^2 + \cdots$$

converges. Then, if $|x| < |c|$,

$$a_0 + a_1 x + a_2 x^2 + \cdots$$

converges. In fact, it converges absolutely.

PROOF Since $\sum_{n=0}^{\infty} a_n c^n$ converges, the nth term $a_n c^n$ approaches 0 as $n \to \infty$. Thus there is an integer N such that, for $n \geq N$, $|a_n c^n| \leq 1$. From now on in the proof, consider only $n \geq N$.

Now,
$$a_n x^n = a_n c^n \left(\frac{x}{c}\right)^n.$$

Since
$$|a_n x^n| = |a_n c^n| \left|\frac{x}{c}\right|^n,$$

it follows that, for $n \geq N$,

$$|a_n x^n| \leq \left|\frac{x}{c}\right|^n \qquad (\text{since } |a_n c^n| \leq 1 \text{ for } n \geq N).$$

The series
$$\sum_{n=N}^{\infty} \left|\frac{x}{c}\right|^n$$

is a geometric series with ratio $|x/c| < 1$. Hence it converges.

Since
$$|a_n x^n| \leq \left|\frac{x}{c}\right|^n$$

for $n \geq N$, the series
$$\sum_{n=N}^{\infty} |a_n x^n|$$

converges (by the comparison test). Thus $\sum_{n=N}^{\infty} a_n x^n$ converges (in fact, absolutely). Putting in the front end $\sum_{n=0}^{N-1} a_n x^n$, we conclude that the series $\sum_{n=0}^{\infty} a_n x^n$ converges absolutely if $|x| < |c|$. ∎

By Theorem 1, the set of numbers x such that $\sum_{n=0}^{\infty} a_n x^n$ converges has no holes. In other words, the set of such x consists of one unbroken piece, which includes the number 0. Moreover if c is in that set, so is the entire open interval $(-|c|, |c|)$. Consequently, either

1. $a_0 + a_1 x + a_2 x^2 + \cdots$ converges for all x, or

2. there is a number R such that $a_0 + a_1 x + a_2 x^2 + \cdots$ converges for all x such that $|x| < R$ but diverges when $|x| > R$.

The radius of convergence   In the second case R is called the *radius of convergence* of the series. In case 1 the radius of convergence is said to be infinite, $R = \infty$. For the geometric series $1 + x + x^2 + \cdots + x^n + \cdots$, $R = 1$, since the series converges when $|x| < 1$ and diverges when $|x| > 1$. (It also diverges when $x = 1$ and $x = -1$.) A power series with radius of convergence R may or may not converge at $x = R$ and at $x = -R$. For convenient reference, these observations are stated as Theorem 2.

THEOREM 2   Associated with the power series $\sum_{n=0}^{\infty} a_n x^n$ is a radius of convergence R. If $R = 0$, the series converges only for $x = 0$. If R is a positive real number, the series converges for $|x| < R$ and diverges for $|x| > R$. If R is ∞, the series converges for all x. ∎

EXAMPLE 1   Use Theorem 1 to find all values of x for which
$$x - \frac{x^2}{2} + \frac{x^3}{3} - \frac{x^4}{4} + \cdots + \frac{(-1)^{n+1} x^n}{n} + \cdots$$

converges.

SOLUTION   Because of the presence of x^n, use the absolute ratio test. The absolute value of the ratio of successive terms is

$$\left| \frac{\frac{(-1)^{n+1}x^{n+1}}{n+1}}{\frac{(-1)^n x^n}{n}} \right| = |x| \frac{n}{n+1}.$$

As $n \to \infty$, $n/(n+1) \to 1$. Thus if $|x| < 1$,

$$\lim_{n \to \infty} |x| \frac{n}{n+1} = |x| < 1.$$

If $|x| > 1$, $$\lim_{n \to \infty} |x| \frac{n}{n+1} = |x| > 1.$$

By the absolute ratio test the series converges for $|x| < 1$ and diverges for $|x| > 1$. All that remains is to see what happens when $x = 1$ or $x = -1$.

Checking the behavior at $x_1 = 1$ For $x = 1$ the alternating harmonic series

$$1 - \tfrac{1}{2} + \tfrac{1}{3} - \tfrac{1}{4} + \cdots$$

is obtained. This series converges, by the alternating-series test. Thus $x - x^2/2 + x^3/3 - x^4/4 + \cdots$ converges when $x = 1$. Hence the series $x - x^2/2 + x^3/3 - x^4/4 + \cdots$ converges for $-1 < x \le 1$.

Checking the behavior at $x = -1$ What about $x = -1$? The series then becomes

$$(-1) - \frac{(-1)^2}{2} + \frac{(-1)^3}{3} - \frac{(-1)^4}{4} + \cdots + \frac{(-1)^{n+1}(-1)^n}{n} + \cdots$$

or

$$-1 - \frac{1}{2} - \frac{1}{3} - \frac{1}{4} - \cdots - \frac{1}{n} - \cdots,$$

which, being the negative of the harmonic series, diverges.

The radius of convergence is $R = 1$. Figure 10.12 records the information obtained about the series.

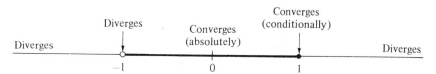

Figure 10.12

EXAMPLE 2 Find the radius of convergence of

$$\sum_{n=0}^{\infty} \frac{x^n}{n!} = 1 + x + \frac{x^2}{2!} + \frac{x^3}{3!} + \cdots + \frac{x^n}{n!} + \cdots.$$

SOLUTION Because of the presence of the power x^n and the factorial $n!$ and the fact that x may be negative, the absolute-ratio test is the logical test to use. The absolute value of the ratio between successive terms is

$$\left| \frac{\frac{x^{n+1}}{(n+1)!}}{\frac{x^n}{n!}} \right| = |x| \frac{n!}{(n+1)!} = \frac{|x|}{n+1}.$$

Since
$$\lim_{n \to \infty} \frac{|x|}{n+1} = 0,$$

the limit of the ratio between successive terms is 0, which is less than 1. Consequently the series converges for all x. That is, the radius of convergence R is infinite. ∎

The next example represents the opposite extreme, $R = 0$.

EXAMPLE 3 Find the radius of convergence of the series

$$\sum_{n=1}^{\infty} n^n x^n = 1x + 2^2 x^2 + 3^3 x^3 + \cdots + n^n x^n + \cdots.$$

SOLUTION The series converges for $x = 0$.

If $x \neq 0$, consider the nth term $n^n x^n$, which can be written as $(nx)^n$. As $n \to \infty$, $|nx| \to \infty$. Thus the nth term does not approach 0 as $n \to \infty$. By the nth term test, the series diverges. In short, the series converges only when $x = 0$. The radius of convergence in this case is $R = 0$. ∎

THE CONVERGENCE OF A POWER SERIES IN $x - a$

Just as a power series in x has an associated radius of convergence, so does a power series in $x - a$. To see this, consider any such power series,

$$\sum_{n=0}^{\infty} a_n (x - a)^n = a_0 + a_1 (x - a) + a_2 (x - a)^2 + \cdots. \tag{5}$$

Let $u = x - a$. Then (5) becomes

$$\sum_{n=0}^{\infty} a_n u^n = a_0 + a_1 u + a_2 u^2 + \cdots. \tag{6}$$

10.6 POWER SERIES **623**

The Maclaurin series (6) has a certain radius of convergence R. That is, (6) converges for $|u| < R$ and diverges for $|u| > R$. Consequently (5) converges for $|x - a| < R$ and diverges for $|x - a| > R$. The number R is called the radius of convergence of the series (5). (R may be infinite.) As Fig. 10.13 suggests, the series $\sum_{n=0}^{\infty} a_n(x - a)^n$ converges in an interval whose midpoint is a. The question marks in Fig. 10.13 indicate that the series may converge or may diverge at the numbers $a - R$ and $a + R$. These cases must be looked at separately.

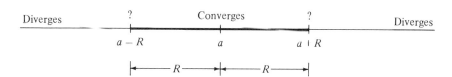

Figure 10.13

These observations are summarized in the following theorem.

THEOREM 3 Associated with the power series $\sum_{n=0}^{\infty} a_n(x - a)^n$ is a radius of convergence R. If $R = 0$, the series converges only for $x = a$. If R is a positive real number, the series converges for $|x - a| < R$ and diverges for $|x - a| > R$. If $R = \infty$, the series converges for all x. ∎

EXAMPLE 4 Find all values of x for which

$$(x - 1) - \frac{(x - 1)^2}{2} + \frac{(x - 1)^3}{3} - \frac{(x - 1)^4}{4} + \cdots \tag{7}$$

converges.

SOLUTION Note that this is Example 1 with x replaced by $x - 1$. Thus $x - 1$ plays the role that x played in Example 1. Consequently, series (7) converges for

$$-1 < x - 1 \leq 1$$

and diverges for all other values of $x - 1$. Its radius of convergence is $R = 1$. The set of values where the series converges is an interval whose midpoint is $x = 1$. The convergence of (7) is recorded in Fig. 10.14.

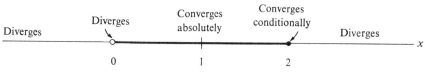

Figure 10.14

EXAMPLE 5 Find the radius of convergence of

$$\sum_{n=0}^{\infty} \left(\frac{x-3}{2}\right)^n. \tag{8}$$

SOLUTION Because the nth term is a power, the absolute-ratio test is the natural test to use. The absolute value of the ratio of successive terms is

$$\left|\frac{\left(\frac{x-3}{2}\right)^{n+1}}{\left(\frac{x-3}{2}\right)^n}\right| = \frac{|x-3|}{2}.$$

By the absolute-ratio test the series (8) converges (absolutely) if

$$\frac{|x-3|}{2} < 1$$

and diverges if

$$\frac{|x-3|}{2} > 1.$$

The inequality $|x-3|/2 < 1$ is equivalent to

$$|x-3| < 2$$

or

$$-2 < x - 3 < 2$$

or (upon the addition of 3 to both sides)

$$1 < x < 5.$$

At $x = 1$, the series (8) becomes the alternating series

$$\sum_{n=0}^{\infty} \left(\frac{1-3}{2}\right)^n = \sum_{n=0}^{\infty} (-1)^n,$$

which, by the nth-term test, diverges. At $x = 5$, the series (8) becomes

$$\sum_{n=0}^{\infty} \left(\frac{5-3}{2}\right)^n = \sum_{n=0}^{\infty} 1^n,$$

which, again by the nth-term test, diverges. Figure 10.15 records this information.

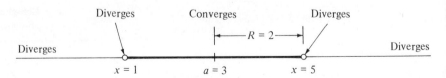

Figure 10.15

Exercises

In Exercises 1 to 18 draw the appropriate diagrams (like Fig. 10.14 or 10.15) showing where the series converge or diverge. Explain your work. Exercises 1 to 10 concern Maclaurin series.

1. $\sum_{n=1}^{\infty} \dfrac{x^n}{n^2}$
2. $\sum_{n=1}^{\infty} \dfrac{x^n}{\sqrt{n}}$
3. $\sum_{n=0}^{\infty} \dfrac{x^n}{3^n}$
4. $\sum_{n=1}^{\infty} n x^n$
5. $\sum_{n=0}^{\infty} \dfrac{x^n}{(2n)!}$
6. $\sum_{n=0}^{\infty} \dfrac{2^n x^n}{n!}$
7. $\sum_{n=0}^{\infty} \dfrac{x^n}{(2n+1)!}$
8. $\sum_{n=0}^{\infty} n! x^n$
9. $\sum_{n=1}^{\infty} \dfrac{(-1)^{n+1} x^n}{n}$
10. $\sum_{n=1}^{\infty} \dfrac{2^n x^n}{n}$

Exercises 11 to 18 concern power series in $x - a$, $a \neq 0$.

11. $\sum_{n=0}^{\infty} \dfrac{(x-2)^n}{n!}$
12. $\sum_{n=1}^{\infty} \dfrac{(x-1)^n}{n 3^n}$
13. $\sum_{n=0}^{\infty} \dfrac{(x-1)^n}{n+3}$
14. $\sum_{n=0}^{\infty} \dfrac{(x-4)^n}{2n+1}$
15. $\sum_{n=1}^{\infty} \dfrac{n(x-2)^n}{2n+3}$
16. $\sum_{n=2}^{\infty} \dfrac{(x-5)^n}{n \ln n}$
17. $\sum_{n=0}^{\infty} \dfrac{(x+3)^n}{5^n}$
18. $\sum_{n=1}^{\infty} n(x+1)^n$

19. Assume that $\sum_{n=0}^{\infty} a_n x^n$ converges when $x = 9$ and diverges when $x = -12$. What, if anything, can be said about
    (a) convergence when $x = 7$?
    (b) absolute convergence when $x = -7$?
    (c) absolute convergence when $x = 9$?
    (d) convergence when $x = -9$?
    (e) divergence when $x = 10$?
    (f) divergence when $x = -15$?
    (g) divergence when $x = 15$?

20. If $\sum_{n=0}^{\infty} a_n 6^n$ converges, what can be said about the convergence of
    (a) $\sum_{n=0}^{\infty} a_n (-6)^n$?
    (b) $\sum_{n=0}^{\infty} a_n 5^n$?
    (c) $\sum_{n=0}^{\infty} a_n (-5)^n$?

21. Prove that, if the series $\sum_{n=0}^{\infty} a_n x^n$ converges whenever x is positive, it converges whenever x is negative.

22. If $\sum_{n=0}^{\infty} a_n (x-3)^n$ converges for $x = 7$, at what other values of x must the series necessarily converge?

23. (a) If the power series $\sum_{n=0}^{\infty} a_n x^n$ diverges when $x = 3$, at which x must it diverge?
    (b) If the power series $\sum_{n=0}^{\infty} a_n (x+5)^n$ diverges when $x = -2$, at which x must it diverge?

24. (a) Letting $a_0 = 1$, $a_1 = -x^2/2!, \ldots$, find the nth term of the series
$$1 - \dfrac{x^2}{2!} + \dfrac{x^4}{4!} - \dfrac{x^6}{6!} + \cdots.$$
   (b) Show that the series in (a) converges for all x.

25. (The binomial series) The binomial theorem in Appendix B asserts that if n is a positive integer,
$$(1 + x)^n = 1 + nx + \dfrac{n(n-1)}{2} x^2 + \cdots + x^n.$$

This fact was proved in Exercises 49 to 51 of Sec. 4.3 by using the higher derivatives of $(1 + x)^n$. Newton generalized the binomial theorem to exponents other than positive integers, as follows:

Let a be a real number that is not 0 or a positive integer and form the series
$$1 + \dfrac{a}{1} x + \dfrac{a(a-1)}{1 \cdot 2} x^2 + \dfrac{a(a-1)(a-2)}{1 \cdot 2 \cdot 3} x^3 + \cdots.$$

(a) Show that, for $x \neq 0$, no term is 0.
(b) Show that, for $a = -1$ and $|x| < 1$, the sum of the series is $(1 + x)^a$.
(c) Show that the series converges absolutely (hence converges) whenever $|x| < 1$.

In Sec. 10.12 it is shown that the series converges to $(1 + x)^a$ whenever $|x| < 1$.

26. If $\sum_{n=0}^{\infty} a_n x^n$ has a radius of convergence 3 and $\sum_{n=0}^{\infty} b_n x^n$ has a radius of convergence 5, what can be said about the radius of convergence of $\sum_{n=0}^{\infty} (a_n + b_n) x^n$?

10.7 Manipulating power series

In advanced calculus it is proved that within its interval of convergence a power series behaves in many ways like a polynomial. In particular it can be differentiated term by term,

$$(a_0 + a_1 x + a_2 x^2 + a_3 x^3 + \cdots)' = a_1 + 2a_2 x + 3a_3 x^2 + \cdots.$$

In Sec. 3.4 it was shown that the sum of a *finite* number of functions can be differentiated term-by-term. The proof of this for power series is far more involved. In Theorem 1 the differentiation rule for power series in x is stated precisely. (A similar theorem holds for power series in $x - a$, but this section concentrates on Maclaurin series.)

DIFFERENTIATING A POWER SERIES

THEOREM 1 *Differentiating a power series.* Assume that $R > 0$ and that $\sum_{n=0}^{\infty} a_n x^n$ converges to $f(x)$ for $|x| < R$. Then for $|x| < R$, f is differentiable, $\sum_{n=1}^{\infty} n a_n x^{n-1}$ converges, and

$$f'(x) = a_1 + 2a_2 x + 3a_3 x^2 + \cdots. \quad \blacksquare$$

The proof of Theorem 1 is left to advanced calculus.

This theorem is *not* covered by the fact that the derivative of the sum of a *finite* number of functions is the sum of their derivatives.

EXAMPLE 1 Obtain a power series for the function $1/(1-x)^2$ from that for $1/(1-x)$.

SOLUTION From the formula for the sum of a geometric series, we know that

$$\frac{1}{1-x} = 1 + x + x^2 + x^3 + \cdots, \qquad |x| < 1.$$

According to Theorem 1, if we differentiate both sides of this equation we obtain a true equation, namely

$$\frac{1}{(1-x)^2} = 0 + 1 + 2x + 3x^2 + \cdots, \qquad |x| < 1.$$

Thus

$$\frac{1}{(1-x)^2} = 1 + 2x + 3x^2 + \cdots = \sum_{n=0}^{\infty} (n+1) x^n, \qquad |x| < 1 \quad \blacksquare$$

Suppose that $f(x)$ has a power-series representation $a_0 + a_1 x + a_2 x^2 + \cdots$; Theorem 1 enables us to find what the coefficients $a_0, a_1, a_2, \ldots$ must be. The formula for a_n appears in Theorem 2. In this theorem $f^{(n)}$ denotes the nth derivative of f; $f^{(1)} = f'$, $f^{(0)}$ stands for f itself, and $0! = 1$.

THEOREM 2 *Formula for a_n.* Assume that $R > 0$ and $f(x) = a_0 + a_1 x + a_2 x^2 + \cdots + a_n x^n + \cdots$ for $|x| < R$. Then

$$a_n = \frac{f^{(n)}(0)}{n!}. \tag{1}$$

PROOF When $x = 0$, $f(0) = a_0 + a_1 0 + a_2 0^2 + \cdots$. Hence

$$f(0) = a_0,$$

Compare this to the argument in Exercises 49–51 in Sec. 4.3, where $f(x) = (1 + x)^n$.

which agrees with (1). To obtain a_1, differentiate $f(x)$ and get

$$f^{(1)}(x) = a_1 + 2a_2 x + 3a_3 x^2 + \cdots + na_n x^{n-1} + \cdots. \tag{2}$$

Set $x = 0$ in (2) and obtain $\quad f^{(1)}(0) = a_1.$

This establishes (1) for $n = 1$.

Getting a_2   To obtain a_2, differentiate (2) and get

$$f^{(2)}(x) = 2a_2 + 3 \cdot 2a_3 x + \cdots + n(n-1)a_n x^{n-2} + \cdots. \tag{3}$$

Letting $x = 0$ gives $\quad f^{(2)}(0) = 2a_2.$

Hence $\quad a_2 = \dfrac{f^{(2)}(0)}{2},$

and (1) is established for $n = 2$.

Getting a_3   To obtain a_3, differentiate (3) as follows:

$$f^{(3)}(x) = 3 \cdot 2a_3 + 4 \cdot 3 \cdot 2a_4 x + \cdots + n(n-1)(n-2)a_n x^{n-3} + \cdots. \tag{4}$$

Set $x = 0$, obtaining $\quad f^{(3)}(0) = 3 \cdot 2a_3,$

or $\quad a_3 = \dfrac{f^{(3)}(0)}{3!}.$

This establishes (1) for $n = 3$ and also shows why the factorial appears in the denominator of (1). The reader should differentiate (4) and verify (1) for $n = 4$. The argument applies for all n. ■

EXAMPLE 2 If e^x can be represented as a Maclaurin series for all x, what must that series be?

SOLUTION Let $\quad e^x = a_0 + a_1 x + a_2 x^2 + \cdots + a_n x^n + \cdots \quad$ for all x. By Theorem 2,

$$a_n = \frac{\text{nth derivative of } e^x \text{ at } x = 0}{n!}.$$

Now $(e^x)' = e^x$, $(e^x)'' = e^x$, and so on. All the higher derivatives of e^x are e^x again. At $x = 0$ they all have the value 1. Thus

$$a_n = \frac{1}{n!}$$

628 SERIES AND RELATED TOPICS

If e^x can be represented as a Maclaurin series, we must have

Maclaurin series for e^x
$$e^x = \frac{1}{0!} + \frac{x}{1!} + \frac{x^2}{2!} + \frac{x^3}{3!} + \frac{x^4}{4!} + \cdots + \frac{x^n}{n!} + \cdots = \sum_{n=0}^{\infty} \frac{x^n}{n!}.$$

(In Sec. 10.8 it will be shown that e^x is indeed represented by this power series.) ■

Calculations similar to those in Example 2 show that, *if* sin x is representable as a Maclaurin series, then

Maclaurin series for sin x
$$\sin x = x - \frac{x^3}{3!} + \frac{x^5}{5!} - \frac{x^7}{7!} + \cdots = \sum_{n=0}^{\infty} \frac{(-1)^n x^{2n+1}}{(2n+1)!}. \tag{5}$$

Differentiation of (5) then yields

Maclaurin series for cos x
$$\cos x = 1 - \frac{x^2}{2!} + \frac{x^4}{4!} - \frac{x^6}{6!} + \cdots = \sum_{n=0}^{\infty} \frac{(-1)^n x^{2n}}{(2n)!}. \tag{6}$$

INTEGRATING A POWER SERIES

The next theorem justifies the term-by-term integration of power series.

THEOREM 3 *Integrating a power series.* Assume that $R > 0$ and

$$f(x) = a_0 + a_1 x + a_2 x^2 + \cdots + a_n x^n + \cdots \quad \text{for} \quad |x| < R.$$

Then

$$a_0 x + \frac{a_1 x^2}{2} + \frac{a_2 x^3}{3} + \cdots + \frac{a_n x^{n+1}}{n+1} + \cdots$$

converges for $|x| < R$, and

$$\int_0^x f(t)\, dt = a_0 x + \frac{a_1 x^2}{2} + \frac{a_2 x^3}{3} + \cdots. \quad ■$$

Note that the t is used to avoid writing $\int_0^x f(x)\, dx$, an expression in which x describes both the interval of integration $[0, x]$ and the independent variable of the function. To avoid use of x with two different meanings, a different letter should be used in the integrand. The next example demonstrates the power of Theorem 3.

EXAMPLE 3 Integrate the Maclaurin series for $1/(1 + x)$ to obtain the Maclaurin series for $\ln(1 + x)$.

SOLUTION   Start with the series

$$\frac{1}{1-x} = 1 + x + x^2 + \cdots \qquad |x| < 1.$$

Replace x by $-x$ and obtain

$$\frac{1}{1+x} = 1 - x + x^2 - x^3 + x^4 - \cdots \qquad |x| < 1.$$

By Theorem 3,

$$\int_0^x \frac{dt}{1+t} = x - \frac{x^2}{2} + \frac{x^3}{3} - \frac{x^4}{4} + \cdots \qquad |x| < 1.$$

Now
$$\int_0^x \frac{dt}{1+t} = \ln(1+t)\Big|_0^x$$
$$= \ln(1+x) - \ln(1+0)$$
$$= \ln(1+x).$$

Thus $\ln(1+x) = x - \dfrac{x^2}{2} + \dfrac{x^3}{3} - \dfrac{x^4}{4} + \cdots \qquad |x| < 1.$ ∎

THE ALGEBRA OF POWER SERIES

In addition to differentiating and integrating power series, we may also add, subtract, multiply, and divide them just like polynomials. The following theorem states the rules for these operations. The first two are easy to establish; proofs of the latter two are reserved for an advanced calculus course.

THEOREM 4   *The algebra of power series.*   Assume that

$$f(x) = a_0 + a_1 x + a_2 x^2 + \cdots \qquad \text{and} \qquad g(x) = b_0 + b_1 x + b_2 x^2 + \cdots$$

for $|x| < R$.   Then for $|x| < R$

1    $f(x) + g(x) = \sum_{n=0}^{\infty} (a_n + b_n) x^n$.
2    $f(x) - g(x) = \sum_{n=0}^{\infty} (a_n - b_n) x^n$.
3    $f(x)g(x) = a_0 b_0 + (a_0 b_1 + a_1 b_0)x + (a_0 b_2 + a_1 b_1 + a_2 b_0)x^2 + \cdots$.
4    $f(x)/g(x)$ is obtainable by long division, if $g(x) \neq 0$ for $|x| < R$. ∎

Two examples will illustrate the usefulness of Theorem 4.

EXAMPLE 4   Find the first few terms of the Maclaurin series for $e^x/(1-x)$.

SOLUTION By Theorem 4, we just multiply the series as we would polynomials:

$$e^x \frac{1}{1-x} = \left(1 + x + \frac{x^2}{2!} + \frac{x^3}{3!} + \cdots\right)(1 + x + x^2 + x^3 + \cdots)$$

$$= 1 \cdot 1 + (1 \cdot 1 + 1 \cdot 1)x + \left(1 \cdot 1 + 1 \cdot 1 + \frac{1}{2!} \cdot 1\right)x^2$$

$$+ \left(1 \cdot 1 + 1 \cdot 1 + \frac{1}{2!} \cdot 1 + \frac{1}{3!} \cdot 1\right)x^3 + \cdots$$

$$= 1 + 2x + \tfrac{5}{2}x^2 + \tfrac{8}{3}x^3 + \cdots \qquad |x| < 1. \blacksquare$$

EXAMPLE 5 Find the first four terms of the Maclaurin series for $e^x/\cos x$.

SOLUTION Write down the Maclaurin series for e^x and $\cos x$ up through the term of degree 3 and arrange the long division as follows.

$$\begin{array}{r}
1 + x + x^2 + \dfrac{2x^3}{3} + \cdots \\
1 + 0x - \dfrac{x^2}{2} + 0x^3 + \cdots \enclose{longdiv}{1 + x + \dfrac{x^2}{2} + \dfrac{x^3}{6} + \cdots} \\
\underline{1 + 0x - \dfrac{x^2}{2} + 0x^3 + \cdots} \\
x + x^2 + \dfrac{x^3}{6} + \cdots \\
\underline{x + 0x^2 - \dfrac{x^3}{2} + \cdots} \\
x^2 + \dfrac{2x^3}{3} + \cdots \\
\underline{x^2 + 0x^3 - \cdots} \\
\dfrac{2x^3}{3} + \cdots \\
\underline{\dfrac{2x^3}{3} + \cdots} \\
\cdots
\end{array}$$

Thus the Maclaurin series for $e^x/\cos x$ begins

$$1 + x + x^2 + \frac{2x^3}{3}. \blacksquare$$

Section 6.8 presented l'Hôpital's rule as a method for dealing with $\lim_{x \to a} f(x)/g(x)$ when both numerator and denominator approach 0 as $x \to a$. The next example shows that to calculate such limits, power series may on occasion be far more efficient.

EXAMPLE 6 Find
$$\lim_{x \to 0} \frac{\sin^2 x - x^2}{(e^{x^2} - 1)^2}.$$

SOLUTION Since
$$\sin x = x - \frac{x^3}{6} + \frac{x^5}{120} - \cdots,$$

the Maclaurin series for the numerator begins

$$\sin^2 x - x^2 = \left(x - \frac{x^3}{6} + \frac{x^5}{120} - \cdots\right)^2 - x^2$$

$$= \left(x^2 - \frac{x^4}{3} + \cdots\right) - x^2$$

$$= -\frac{x^4}{3} + \text{terms of degree more than 4}.$$

To develop a Maclaurin series for the denominator first replace x in the relation

$$e^x = 1 + x + \frac{x^2}{2} + \frac{x^3}{6} + \cdots$$

by x^2 and obtain
$$e^{x^2} = 1 + x^2 + \frac{x^4}{2} + \frac{x^6}{6} + \cdots.$$

Thus
$$(e^{x^2} - 1)^2 = \left(x^2 + \frac{x^4}{2} + \frac{x^6}{6} + \cdots\right)^2$$

$$= x^4 + \text{terms of higher degree}.$$

Hence
$$\frac{\sin^2 x - x^2}{(e^{x^2} - 1)^2} = \frac{-x^4/3 + \text{terms of degree more than 4}}{x^4 + \text{terms of degree more than 4}}$$

$$= \frac{x^4(-\frac{1}{3} + \text{terms of degree at least 1})}{x^4(1 + \text{terms of degree at least 1})}.$$

Cancel the x^4 in numerator and denominator and obtain

$$\frac{\sin^2 x - x^2}{(e^{x^2} - 1)^2} = \frac{-\frac{1}{3} + \text{terms of degree at least 1}}{1 + \text{terms of degree at least 1}}.$$

Thus
$$\lim_{x \to 0} \frac{\sin^2 x - x^2}{(e^{x^2} - 1)^2} = \frac{-\frac{1}{3}}{1} = -\frac{1}{3}.$$

The reader may find the limit by l'Hôpital's rule, but the computations are messier. ∎

POWER SERIES IN $x - a$

The various theorems and methods of this section were stated for power series in x. But analogous theorems hold for power series in $x - a$. Such series may be differentiated and integrated inside the interval in which they converge. For instance, Theorem 2 generalizes to the following assertion.

THEOREM 5 *Formula for a_n.* Assume that $R > 0$ and

$$f(x) = a_0 + a_1(x - a) + a_2(x - a)^2 + \cdots + a_n(x - a)^n + \cdots = \sum_{n=0}^{\infty} a_n(x - a)^n$$

for $|x - a| < R$.

Then $$a_n = \frac{f^{(n)}(a)}{n!}. \quad \blacksquare \qquad (7)$$

The proof is similar to that of Theorem 2: Differentiate n times and replace x by a.

The next example illustrates the use of formula (7).

EXAMPLE 7 If $\sin x$ can be represented as a series in powers of $x - \pi/4$ for all x, find what the series must be.

SOLUTION In this case $f(x) = \sin x$ and $a = \pi/4$. In order to use (7), it is necessary to evaluate all the derivatives of $\sin x$ at $\pi/4$. This table records the computations.

n	$f^{(n)}(x)$	$f^{(n)}(\pi/4)$	$a_n = \dfrac{f^{(n)}(\pi/4)}{n!}$
0	$\sin x$	$\sqrt{2}/2$	$(\sqrt{2}/2)/0!$
1	$\cos x$	$\sqrt{2}/2$	$(\sqrt{2}/2)/1!$
2	$-\sin x$	$-\sqrt{2}/2$	$-(\sqrt{2}/2)/2!$
3	$-\cos x$	$-\sqrt{2}/2$	$-(\sqrt{2}/2)/3!$
4	$\sin x$	$\sqrt{2}/2$	$(\sqrt{2}/2)/4!$
5	$\cos x$	$\sqrt{2}/2$	$(\sqrt{2}/2)/5!$
$\cdots$	$\cdots$	$\cdots$	$\cdots$

The higher derivatives of $\sin x$ repeat in blocks of four:

$$\sin x, \quad \cos x, \quad -\sin x, \quad -\cos x.$$

The series for $\sin x$ in powers of $x - \pi/4$ begins, therefore,

$$\sin x = \frac{\sqrt{2}}{2} + \frac{\sqrt{2}}{2}\left(x - \frac{\pi}{4}\right) - \frac{\sqrt{2}/2}{2!}\left(x - \frac{\pi}{4}\right)^2 - \frac{\sqrt{2}/2}{3!}\left(x - \frac{\pi}{4}\right)^3 + \frac{\sqrt{2}/2}{4!}\left(x - \frac{\pi}{4}\right)^4 + \cdots.$$

Two $+$'s continue to alternate with two $-$'s. $\blacksquare$

Exercises

1 This exercise concerns the Maclaurin series for $\sin x$.
 (a) Copy and fill in this table for $f(x) = \sin x$.

n	$f^{(n)}(x)$	$f^{(n)}(0)$	$f^{(n)}(0)/n!$
0			
1			
2			
3			
4			
5			

 (b) Use (a) to write out the first six terms of the Maclaurin series for $\sin x$ (including terms that are 0).
 (c) Show that the nth nonzero term in the Maclaurin series for $\sin x$ is

$$(-1)^{n+1} \frac{x^{2n-1}}{(2n-1)!}.$$

2 Carry out the analog of Exercise 1 for $f(x) = \cos x$.
3 Find the Maclaurin series for $\cos x$ by differentiating that for $\sin x$. (See Equations (5) and (6).)
4 Differentiate the series $1 + x + x^2/2! + \cdots + x^n/n! + \cdots$ and show that the derivative equals the given series.

In Exercises 5 to 8 determine with the aid of Theorem 2 the first three nonzero terms of the Maclaurin series for the given functions.

5 $\tan x$ **6** $\tan^{-1} x$
7 $\sin^{-1} x$ **8** $\sqrt{1+x}$
9 (a) Show that for $|t| < 1$,

$$\frac{1}{1+t^2} = 1 - t^2 + t^4 - t^6 + \cdots.$$

 (b) Use Theorem 3 to show that for $|x| < 1$,

$$\tan^{-1} x = x - \frac{x^3}{3} + \frac{x^5}{5} - \frac{x^7}{7} + \cdots.$$

 (c) Give the formula for the nth term of the series in (b).

10 Using Theorem 3 show that for $|x| < 1$,
 (a) $\int_0^x \frac{dt}{1+t^3} = x - \frac{x^4}{4} + \frac{x^7}{7} - \frac{x^{10}}{10} + \cdots.$
 (b) Give a formula for the nth term of the series in (a).

In Exercises 11 and 12 obtain the first three nonzero terms in the Maclaurin series for the indicated functions by algebraic operations with known series.

11 $e^x \sin x^{x^2}$ **12** $\dfrac{x}{\cos x}$

In Exercises 13 to 16 use power series to determine the limits.

13 $\lim\limits_{x \to 0} \dfrac{\sin^2 x^3}{(1 - \cos x^2)^3}$ **14** $\lim\limits_{x \to 0} \left(\dfrac{1}{\sin x} - \dfrac{1}{\ln(1+x)} \right)$

15 $\lim\limits_{x \to 0} \dfrac{(e^{x^2} - 1)^3}{\sin^3 x^2}$ **16** $\lim\limits_{x \to 0} \dfrac{\sin x (1 - \cos x)}{e^{x^3} - 1}$

In Exercises 17 to 20 use the formula $a_n = f^{(n)}(a)/n!$ to obtain the indicated series. Write out the first three nonzero terms.
17 Series for $\sin x$ in powers of $x - \pi/6$.
18 Series for $\cos x$ in powers of $x + \pi/4$.
19 Series for e^x in powers of $x - 1$.
20 Series for $\tan x$ in powers of $x - \pi/4$.

■

21 Let $f(x) = a_0 + a_1 x + a_2 x^2 + \cdots$ for $|x| < R$.
 (a) If only even powers appear, that is, $a_n = 0$ for all odd n, show that $f(-x) = f(x)$.
 (b) If only odd powers appear, that is, $a_n = 0$ for all even n, show that $f(-x) = -f(x)$.
22 (a) Use the first five terms of the Maclaurin series for e^x to estimate $e = e^1$.
 (b) Show that the error in (a) is less than the sum of the geometric progression

$$\frac{1}{5!} + \frac{1}{6 \cdot 5!} + \frac{1}{6^2 \cdot 5!} + \cdots.$$

 (c) Deduce from (a) and (b) that

$$2.708 < e < 2.719.$$

23 (a) Noting that $\sqrt{e} = e^{1/2}$, use the first five terms of the Maclaurin series for e^x to estimate $\sqrt{e}$.
 (b) Estimate the error in (a) by comparing it to the sum of a geometric progression.
24 Obtain formula (7) in Theorem 5 for $n = 0, 1, 2, 3$.

In Exercises 25 to 27 use a calculator.
25 (a) Use the first 10 terms of the series $e^x = \sum_{n=0}^{\infty} x^n/n!$ to estimate $e = e^1$.
 (b) Show that the error in the estimate in (a) is less than $11/(10 \cdot 10!) \approx 0.0000003$.
26 (a) Use the first three nonzero terms of the Maclaurin series for $\sin x$ to estimate $\sin \pi/5$ ($= \sin 36°$).
 (b) Show that the error in the estimate in (a) is less than $(\pi/5)^7/7! \approx 0.000008$.

27 The integral $\int_0^1 e^{-x^2}\,dx$ cannot be evaluated by the fundamental theorem of calculus.
(a) Replacing x in the power series $e^x = \sum_{n=0}^\infty x^n/n!$ by $-x^2$, obtain the power series for e^{-x^2}.
(b) Show that
$$\int_0^1 e^{-x^2}\,dx = 1 - \frac{1}{3} + \frac{1}{5\cdot 2!} - \frac{1}{7\cdot 3!} + \cdots.$$
(c) Use (b) to estimate $\int_0^1 e^{-x^2}\,dx$ to three-decimal-place accuracy.

■ ■

28 In this exercise the power series for $\ln(1+x)$ will be obtained without borrowing from advanced calculus the result on integration of power series. It has the further advantage that it takes care of $x = 1$.
(a) Show that, for $t \neq -1$,
$$\frac{1}{1+t} = 1 - t + \cdots + (-1)^{n-1}t^{n-1} + (-1)^n \frac{t^n}{1+t}.$$
(b) Use the identity in (a) to show that, for $x > -1$,
$$\ln(1+x) = x - \frac{x^2}{2} + \frac{x^3}{3} - \cdots + (-1)^{n-1}\frac{x^n}{n}$$
$$+ (-1)^n \int_0^x \frac{t^n}{1+t}\,dt.$$
(c) Show that, if x is in $[0, 1]$, then $\int_0^x t^n/(1+t)\,dt$ approaches 0 as $n \to \infty$. Hint: $1 + t \geq 1$.
(d) Show that, if $-1 < x \leq 0$, then $\int_0^x t^n/(1+t)\,dt$ approaches 0 as $n \to \infty$. Hint: $1 + t \geq 1 + x$.
(e) Conclude that, if $-1 < x \leq 1$, then
$$\ln(1+x) = x - \frac{x^2}{2} + \frac{x^3}{3} - \cdots + (-1)^{n-1}\frac{x^n}{n} + \cdots.$$

(Note that $x = 1$ yields the alternating harmonic series and the equation $\ln 2 = 1 - \frac{1}{2} + \frac{1}{3} - \frac{1}{4} + \cdots$.)

29 In this exercise the power series for $\tan^{-1} x$ will be obtained without using the result from advanced calculus concerning the integration of power series. Moreover it shows that $\tan^{-1} x = x - x^3/3 + x^5/5 - \cdots$, even when $|x| = 1$.
(a) Using the identity in Exercise 28(a), show that
$$\frac{1}{1+t^2} = 1 - t^2 + \cdots + (-1)^{n-1}t^{2n-2} + (-1)^n \frac{t^{2n}}{1+t^2}.$$

(b) From (a) deduce that
$$\tan^{-1} x = x - \frac{x^3}{3} + \frac{x^5}{5} - \cdots + (-1)^{n-1}\frac{x^{2n-1}}{2n-1} +$$
$$(-1)^n \int_0^x \frac{t^{2n}}{1+t^2}\,dt.$$
(c) Show that, if $0 \leq x \leq 1$, $\int_0^x t^{2n}/(1+t^2)\,dt \to 0$ as $n \to \infty$.
(d) From (c) deduce that, for $|x| \leq 1$,
$$\tan^{-1} x = x - \frac{x^3}{3} + \frac{x^5}{5} - \frac{x^7}{7} + \cdots.$$
(e) Show that
$$\sum_{n=0}^\infty \frac{(-1)^n}{2n+1} = 1 - \frac{1}{3} + \frac{1}{5} - \frac{1}{7} + \cdots = \frac{\pi}{4}.$$

30 (a) Let $f(x) = (\sin x)/x$ if $x > 0$ and let $f(0) = 1$. Graph $f(x)$.
(b) Show that, if n is an integer, $n \geq 1$, then
$$\int_{2n\pi}^{(2n+2)\pi} \frac{\sin x}{x}\,dx < \int_{2n\pi}^{(2n+1)\pi} \frac{\pi}{x(x+\pi)}\,dx$$
$$< \int_{2n\pi}^{(2n+1)\pi} \frac{\pi}{x^2}\,dx < \frac{1}{4n^2}.$$
(c) From (a) and (b), deduce that $\int_0^\infty f(x)\,dx$ is convergent.

31 What theorems in the text justify the assertion that
$$\lim_{x \to 0}(a_0 + a_1 x + a_2 x^2 + a_3 x^3 + \cdots) = a_0 ?$$

Assume the series has a nonzero radius of convergence.

32 Show that $\sqrt{x}$ cannot be represented by a Maclaurin series with a nonzero radius of convergence.

33 Show that $\sqrt[3]{x}$ cannot be represented by a Maclaurin series with a nonzero radius of convergence.

34 This exercise presents a function so "flat" at the origin that all its derivatives are 0 there, yet it is not the constant function $f(x) = 0$.
Let $f(x) = e^{-1/x^2}$ if $x \neq 0$, and let $f(0) = 0$.
(a) Show that f is continuous.
(b) Show that f is differentiable.
(c) Show that $f^{(1)}(0) = 0$ and $f^{(2)}(0) = 0$.
(d) Explain why $f^{(n)}(0) = 0$ for all $n \geq 0$.
(e) Show that $f(x)$ is not representable by a Maclaurin series with a nonzero radius of convergence.

10.8 Taylor's formula

In Sec. 10.7 it was pointed out that, if $f(x)$ can be represented as a power series in $x - a$ over some interval around a, then that series must be

$$f(a) + f^{(1)}(a)(x-a) + \frac{f^{(2)}(a)}{2!}(x-a)^2 + \cdots + \frac{f^{(n)}(a)}{n!}(x-a)^n + \cdots. \quad (1)$$

Letting $f^{(0)}(x)$ denote $f(x)$ and recalling that $0! = 1$, we can rewrite series (1) in the form

$$\sum_{n=0}^{\infty} \frac{f^{(n)}(a)}{n!}(x-a)^n. \quad (2)$$

Series (2) may or may not sum to $f(x)$.

The series (2) is called the Taylor series at $x = a$ associated with the function $f(x)$. When $a = 0$, the series (2) is called the Maclaurin series associated with $f(x)$. To show that (2) equals $f(x)$, we must show that

$$\lim_{n \to \infty} \left[f(a) + f^{(1)}(a)(x-a) + \cdots + \frac{f^{(n)}(a)}{n!}(x-a)^n \right] = f(x).$$

In other words, it is necessary to show that

$$\lim_{n \to \infty} \left\{ f(x) - \left[f(a) + f^{(1)}(a)(x-a) + \cdots + \frac{f^{(n)}(a)}{n!}(x-a)^n \right] \right\} = 0. \quad (3)$$

This section obtains short formulas for the expression in the braces in (3). With the aid of these formulas it is then possible to show, for instance, that $e^x = \sum_{n=0}^{\infty} x^n/n!$. (In Sec. 10.7 that equation was obtained with two assumptions: first, that e^x has a power series; second, that it is legal to differentiate a power series term by term.)

THE TAYLOR POLYNOMIALS

DEFINITION *Taylor polynomial of degree n, $P_n(x; a)$.* If the function f has derivatives through order n at a, then the Taylor polynomial of degree n of f at a is

$$f(a) + f'(a)(x-a) + \frac{f^{(2)}(a)}{2!}(x-a)^2 + \cdots + \frac{f^{(n)}(a)}{n!}(x-a)^n.$$

This polynomial of degree n is denoted $P_n(x; a)$. It is just the first $n + 1$ terms, a "front end," of the Taylor series.

EXAMPLE 1 Find the Taylor polynomial of degree 4 associated with e^x at $a = 0$.

SOLUTION In this case $f(x) = e^x$. Repeated differentiation yields

$$f^{(1)}(x) = e^x, \quad f^{(2)}(x) = e^x, \quad f^{(3)}(x) = e^x, \quad \text{and} \quad f^{(4)}(x) = e^x.$$

At $x = 0$ all these derivatives have the value 1. The Taylor polynomial of degree 4 at 0 is therefore

$$P_4(x; 0) = 1 + \frac{1}{1!}(x-0) + \frac{1}{2!}(x-0)^2 + \frac{1}{3!}(x-0)^3 + \frac{1}{4!}(x-0)^4,$$

or simply
$$1 + x + \frac{x^2}{2!} + \frac{x^3}{3!} + \frac{x^4}{4!}. \quad \blacksquare$$

Taking front sections of $P_4(x; 0)$ gives us the Taylor polynomials of lower degrees:

$$P_0(x; 0) = 1$$
$$P_1(x; 0) = 1 + x$$
$$P_2(x; 0) = 1 + x + \frac{x^2}{2!}$$
$$P_3(x; 0) = 1 + x + \frac{x^2}{2!} + \frac{x^3}{3!}$$

In Fig. 10.16 the function $f(x) = e^x$ and the Taylor polynomials up to degree 4 are graphed. Note that the higher the degree, the closer the Taylor polynomial approximates the function e^x near $x = 0$. However, since e^x grows much faster than any polynomial as $x \to \infty$, no fixed polynomial can approximate e^x closely throughout the x axis. Furthermore, as $x \to -\infty$, $e^x \to 0$, but a polynomial of degree at least one approaches ∞ or $-\infty$.

THE REMAINDER: INTEGRAL FORM

DEFINITION *The remainder (error) $R_n(x; a)$.* Let f be a function and let $P_n(x; a)$ be the associated Taylor polynomial of degree n at a. The number $R_n(x; a)$ defined by the equation

$$f(x) = P_n(x; a) + R_n(x; a)$$

is called the *remainder* (or *error*) in using the Taylor polynomial $P_n(x; a)$ to approximate $f(x)$.

In $R_n(x; a)$ think of a and x as fixed and $n \to \infty$.

Our expectation is that, for well-behaved functions f,

$$R_n(x; a) \to 0$$

as $n \to \infty$. If $R_n(x; a) \to 0$ as $n \to \infty$, then $P_n(x; a) \to f(x)$. In other words, the function $f(x)$ is then represented by its Taylor series. Theorem 1 expresses $R_n(x; a)$ as an integral; Theorem 2 expresses $R_n(x; a)$ in terms of a derivative. These formulas enable us to show that for $f(x) = e^x$, $\sin x$, or $\cos x$, $R_n(x; a)$ indeed approaches 0 as $n \to \infty$. They also can be used to establish general properties of functions. (See Exercise 42, for instance.)

10.8 TAYLOR'S FORMULA

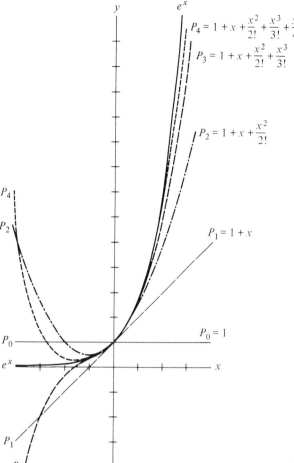

Figure 10.16

The fundamental theorem of calculus asserts that if $f'(x)$ is continuous, then

This equation is the basis of our approach.
$$f(b) - f(a) = \int_a^b f'(x)\,dx. \tag{4}$$

In this form the theorem says that the difference between $f(b)$ and $f(a)$ can be expressed as an integral of the derivative $f'(x)$. To emphasize this fact, we use the letter b instead of x in the following theorem.

THEOREM 1 *Integral form of the remainder.* Assume that a function f has continuous derivatives up through order $n + 1$ in the interval $[a, b]$. Let $P_n(b; a)$ be the Taylor polynomial of degree n associated with f in powers of $b - a$, and let $R_n(b; a)$ be the corresponding remainder:

$$f(b) = P_n(b; a) + R_n(b; a).$$

Then
$$R_n(b; a) = \frac{1}{n!} \int_a^b (b - x)^n f^{(n+1)}(x)\,dx. \tag{5}$$

638 SERIES AND RELATED TOPICS

PROOF First check the case $n = 0$. We have $P_0(b; a) = f(a)$ and

Before reading the proof check the case $n = 0$ yourself.
$$R_0(b; a) = f(b) - P_0(b; a) = f(b) - f(a).$$

By (4),
$$R_0(b; a) = \int_a^b f^{(1)}(x)\, dx. \tag{6}$$

Is (6) a special case of (5)? To see, replace n in (5) by 0 and obtain

$$\frac{1}{0!}\int_a^b (b - x)^0 f^{(0+1)}(x)\, dx,$$

or
$$\int_a^b f^{(1)}(x)\, dx,$$

which is indeed (6). Thus the theorem is true for $n = 0$.

To prove the theorem for $n = 1$, start with

$$f(b) - f(a) = \int_a^b f^{(1)}(x)\, dx \tag{7}$$

Here integration by parts serves as a theoretical tool.

and integrate by parts, as follows:

$$\int_a^b \underbrace{f^{(1)}(x)}_{u}\, \underbrace{dx}_{dv} = \underbrace{f^{(1)}(x)}_{u}\underbrace{[-(b-x)]}_{v}\Big|_a^b - \int_a^b \underbrace{[-(b-x)]}_{v}\underbrace{f^{(2)}(x)\, dx}_{du}$$

Using $v = x$ will not get the theorem, but $v = -(b - x)$ will.

$$du = f^{(2)}(x)\, dx \qquad v = -(b - x)$$

or
$$\int_a^b f^{(1)}(x)\, dx = f^{(1)}(b)[-(b - b)] - f^{(1)}(a)[-(b - a)] +$$

$$\int_a^b (b - x)f^{(2)}(x)\, dx$$

so
$$\int_a^b f^{(1)}(x)\, dx = f^{(1)}(a)(b - a) + \int_a^b (b - x)f^{(2)}(x)\, dx. \tag{8}$$

Combining (7) and (8) shows that

$$f(b) - f(a) = f^{(1)}(a)(b - a) + \int_a^b (b - x)f^{(2)}(x)\, dx$$

so
$$f(b) = f(a) + f^{(1)}(a)(b - a) + \int_a^b (b - x)f^{(2)}(x)\, dx$$

$$= P_1(b; a) + \int_a^b (b - x)f^{(2)}(x)\, dx.$$

10.8 TAYLOR'S FORMULA

Thus
$$R_1(b; a) = \int_a^b (b - x) f^{(2)}(x)\, dx, \tag{9}$$

which establishes (5) for $n = 1$.

Do the integration yourself. It's easier than reading it.

Another integration by parts, starting with the fact that

$$f(b) = f(a) + f^{(1)}(a)(b - a) + \int_a^b (b - x) f^{(2)}(x)\, dx, \tag{10}$$

will prove the theorem for $n = 2$.

We have

Integrate by parts with $u = f^{(2)}(x)$ and $v = \dfrac{-(b-x)^2}{2}$.

$$\int_a^b (b - x) f^{(2)}(x)\, dx = f^{(2)}(x) \frac{-(b-x)^2}{2} \bigg|_a^b - \int_a^b \frac{-(b-x)^2}{2} f^{(3)}(x)\, dx$$

$$= f^{(2)}(b) \frac{-(b-b)^2}{2} - f^{(2)}(a) \frac{-(b-a)^2}{2} +$$

$$\frac{1}{2} \int_a^b (b - x)^2 f^{(3)}(x)\, dx$$

Observe how integration by parts replaces $(b - x) f^{(2)}(x)$ by $(b - x)^2 f^{(3)}(x)$.

or
$$\int_a^b (b - x) f^{(2)}(x)\, dx = f^{(2)}(a) \frac{(b-a)^2}{2} + \frac{1}{2} \int_a^b (b - x)^2 f^{(3)}(x)\, dx. \tag{11}$$

Combining (10) and (11), we have

$$f(b) = f(a) + f^{(1)}(a)(b - a) + f^{(2)}(a) \frac{(b - a)^2}{2} + \frac{1}{2} \int_a^b (b - x)^2 f^{(3)}(x)\, dx$$

or
$$f(b) = P_2(b; a) + \frac{1}{2} \int_a^b (b - x)^2 f^{(3)}(x)\, dx. \tag{12}$$

Equation (12) establishes (5) for $n = 2$.

Do the next integration to reinforce integration by parts.

Another integration by parts, starting with (12), will prove the case $n = 3$. Repeated integration by parts establishes the theorem. ∎

Theorem 1 also holds when $b < a$ and the interval is $[b, a]$.
In the usual notation for the remainder, $R_n(x; a)$, we have

$$R_n(x; a) = \frac{1}{n!} \int_a^x (x - t)^n f^{(n+1)}(t)\, dt. \tag{13}$$

EXAMPLE 2 Use the integral form of the remainder to show that e^x is represented by its Maclaurin series for all x.

SOLUTION In this case $f(x) = e^x$ and $a = 0$. Note that $f^{(n+1)}(t) = e^t$ for all n. Thus

$$R_n(x; 0) = \frac{1}{n!} \int_0^x (x-t)^n e^t \, dt.$$

Consider $x > 0$. In this case, $e^t \leq e^x$ when t is in $[0, x]$. Thus

$$R_n(x; 0) \leq \frac{1}{n!} \int_0^x (x-t)^n e^x \, dt$$

$$= \frac{e^x}{n!} \int_0^x (x-t)^n \, dt$$

$$= -\frac{e^x}{n!} \frac{(x-t)^{n+1}}{n+1} \Big|_{t=0}^{t=x}$$

$$= \frac{e^x x^{n+1}}{(n+1)!}.$$

The notations $t = x$ and $t = 0$ emphasize that t is the variable of integration.

Since $x^{n+1}/(n+1)! \to 0$ as $n \to \infty$,

$$\lim_{n \to \infty} R_n(x; 0) = 0.$$

In other words,

$$e^x = 1 + x + \frac{x^2}{2!} + \frac{x^3}{3!} + \cdots + \frac{x^n}{n!} + \cdots,$$

The case $x \leq 0$ is similar. ∎

In Sec. 10.12 the integral form of the remainder is needed to justify the binomial theorem for any exponent. However, there is a much simpler form of the remainder that is adequate for such functions as $\sin x$ and e^x. It involves a higher derivative but not an integral.

THE REMAINDER: DERIVATIVE FORM

The derivative form for $R_n(x; a)$ follows from the integral form. It is often called Lagrange's formula for the remainder.

THEOREM 2 *Derivative form of the remainder.* Assume that a function f has continuous derivatives of orders through $n + 1$ in an interval that includes the numbers a and b. Let $P_n(b; a)$ be the Taylor polynomial of degree n associated with f in powers of $b - a$. Then

$$R_n(b; a) = f(b) - P_n(b; a) = \frac{f^{(n+1)}(c)}{(n+1)!}(b-a)^{n+1}$$

for some number c between a and b.

PROOF For convenience assume $b > a$. Let M be the maximum value of $f^{(n+1)}(x)$ for x in $[a, b]$ and let m be the minimum value of $f^{(n+1)}(x)$ for x in $[a, b]$. Then,

$$(b - x)^n m \leq (b - x)^n f^{(n+1)}(x) \leq (b - x)^n M$$

for x in $[a, b]$. Thus

$$\int_a^b (b - x)^n m \, dx \leq \int_a^b (b - x)^n f^{(n+1)}(x) \, dx \leq \int_a^b (b - x)^n M \, dx$$

or

$$\frac{m[-(b - x)^{n+1}]}{n + 1}\Big|_{x=a}^{x=b} \leq \int_a^b (b - x)^n f^{(n+1)}(x) \, dx \leq \frac{M[-(b - x)^{n+1}]}{n + 1}\Big|_{x=a}^{x=b}$$

or

$$\frac{m(b - a)^{n+1}}{n + 1} \leq \int_a^b (b - x)^n f^{(n+1)}(x) \, dx \leq \frac{M(b - a)^{n+1}}{n + 1}.$$

Thus

$$\int_a^b (b - x)^n f^{(n+1)}(x) \, dx = \frac{q(b - a)^{n+1}}{n + 1}, \tag{14}$$

where q is some number between m and M. By the intermediate-value theorem, there is a number c in $[a, b]$ such that $f^{(n+1)}(c) = q$. By (5) and (14),

$$R_n(b; a) = \frac{1}{n!}\int_a^b (b - x)^n f^{(n+1)}(x) \, dx$$

$$= \frac{1}{n!} \frac{f^{(n+1)}(c)(b - a)^{n+1}}{n + 1}$$

$$= \frac{f^{(n+1)}(c)(b - a)^{n+1}}{(n + 1)!},$$

which was to be shown.
A similar proof works for $b < a$. ∎

Note that c depends on x and n.

As a special case of Theorem 2, we have the useful formula

$$R_n(x; 0) = \frac{f^{(n+1)}(c) x^{n+1}}{(n + 1)!} \tag{15}$$

for some number c between 0 and x.

EXAMPLE 3 Show that the Maclaurin series associated with $f(x) = \sin x$ represents $f(x)$ for all x.

SOLUTION All that is needed is to show that $R_n(x; 0) \to 0$ as $n \to \infty$. Now, by the derivative form of the remainder,

$$R_n(x; 0) = \frac{f^{(n+1)}(c)x^{n+1}}{(n+1)!},$$

where c is between 0 and x.

If $f(x) = \sin x$, then $f^{(1)}(x) = \cos x$, $f^{(2)}(x) = -\sin x$, $f^{(3)}(x) = -\cos x$, $f^{(4)}(x) = \sin x$, and so on. The higher derivatives are either $\pm \sin x$ or $\pm \cos x$. Thus, for any nonnegative integer n and real number c,

$$|f^{(n+1)}(c)| \leq 1;$$

consequently, $\quad |R_n(x; 0)| = \dfrac{|f^{(n+1)}(c)x^{n+1}|}{(n+1)!} \leq \dfrac{|x|^{n+1}}{(n+1)!},$

which approaches 0 as $n \to \infty$.

Hence the Maclaurin series associated with $\sin x$ represents $\sin x$ for all x. Since that series is $\sum_{n=0}^{\infty} (-1)^n x^{2n+1}/(2n+1)!$,

$$\sin x = x - \frac{x^3}{3!} + \frac{x^5}{5!} - \cdots + (-1)^n \frac{x^{2n+1}}{(2n+1)!} + \cdots.$$

(Terms in the Maclaurin series with value 0 are not shown.) ∎

USING POWER SERIES

The front end of the power series representation of a function $f(x)$ is often taken as an approximation of the function. The next three examples illustrate this use of power series.

EXAMPLE 4 Estimate $\sqrt{e} = e^{1/2}$, using the first four terms of the Maclaurin series for e^x. Discuss the error.

SOLUTION We have

$$f(x) = e^x = 1 + x + \frac{x^2}{2!} + \frac{x^3}{3!} + \cdots.$$

Thus $e^{1/2}$ is approximated by

$$1 + \tfrac{1}{2} + \frac{(\tfrac{1}{2})^2}{2!} + \frac{(\tfrac{1}{2})^3}{3!} = 1 + \tfrac{1}{2} + \tfrac{1}{8} + \tfrac{1}{48} \approx 1.64583.$$

By the derivative form of the remainder, the error is of the form

$$\frac{f^{(4)}(c)(\tfrac{1}{2})^4}{4!}$$

for some number c in $[0, \tfrac{1}{2}]$. Since $f(x) = e^x$, $f^{(4)}(x) = e^x$. Thus the error is of the form

$$\frac{e^c(\tfrac{1}{16})}{24} = \frac{e^c}{384}, \qquad c \text{ in } [0, \tfrac{1}{2}].$$

Since $e^c \le e^{1/2} < 2$ (because $e < 4$), we see that the error is positive but less than $\frac{2}{384} = \frac{1}{192} \approx 0.0052$. Thus, to two decimal places, $e^{1/2} \approx 1.65$. ∎

EXAMPLE 5 In *Optimal Replacement Policy*, a text on the maintenance of equipment, D. W. Jorgenson, J. J. McCall, and R. Radner presented the following argument:

"Suppose that the distribution of time to failure of the equipment is exponential. Then we have

$$\frac{e^{\lambda N}}{\lambda} - N = \frac{1}{\lambda} + K, \qquad (16)$$

an equation for N, the replacement time period. A quadratic approximation to $e^{\lambda N}$ results in the formula

$$N = \sqrt{\frac{2K}{\lambda}}."$$

In this argument λ and K are constants: λ determines the failure distribution and K is the time to effect a repair.

Explain how the formula for N is obtained.

SOLUTION Since

$$e^x = 1 + x + \frac{x^2}{2!} + \frac{x^3}{3!} + \cdots,$$

the Taylor polynomial of degree 2 approximating e^x is

$$1 + x + \frac{x^2}{2}.$$

(For small x this is a good approximation.)

Replacing x by λN gives the approximation

$$e^{\lambda N} \approx 1 + \lambda N + \frac{(\lambda N)^2}{2}. \qquad (17)$$

Combining (16) and (17) yields

$$\frac{1 + \lambda N + \frac{(\lambda N)^2}{2}}{\lambda} - N \approx \frac{1}{\lambda} + K$$

or

$$\frac{1}{\lambda} + N + \frac{\lambda N^2}{2} - N \approx \frac{1}{\lambda} + K$$

or

$$\frac{\lambda N^2}{2} \approx K.$$

Thus

$$N \approx \sqrt{\frac{2K}{\lambda}}. \quad \blacksquare$$

EXAMPLE 6 Use the Maclaurin series for e^x to estimate

$$\int_0^1 e^{-x^2} \, dx.$$

SOLUTION

$$e^x = 1 + x + \frac{x^2}{2!} + \frac{x^3}{3!} + \cdots.$$

Replacing x by $-x^2$ yields

$$e^{-x^2} = 1 - x^2 + \frac{x^4}{2!} - \frac{x^6}{3!} + \cdots. \quad (18)$$

For $0 < |x| \le 1$, series (18) is a decreasing alternating series. Thus

$$1 - x^2 + \frac{x^4}{2!} - \frac{x^6}{3!} < e^{-x^2} < 1 - x^2 + \frac{x^4}{2!} - \frac{x^6}{3!} + \frac{x^8}{4!}.$$

Hence

$$\int_0^1 \left(1 - x^2 + \frac{x^4}{2!} - \frac{x^6}{3!}\right) dx < \int_0^1 e^{-x^2} \, dx < \int_0^1 \left(1 - x^2 + \frac{x^4}{2!} - \frac{x^6}{3!} + \frac{x^8}{4!}\right) dx$$

or

$$1 - \frac{1}{3} + \frac{1}{5 \cdot 2!} - \frac{1}{7 \cdot 3!} < \int_0^1 e^{-x^2} \, dx < 1 - \frac{1}{3} + \frac{1}{5 \cdot 2!} - \frac{1}{7 \cdot 3!} + \frac{1}{9 \cdot 4!}.$$

From this it follows that

$$0.742 < \int_0^1 e^{-x^2} \, dx < 0.748. \quad \blacksquare$$

Exercises

In each of Exercises 1 to 6 find the integral form of the remainder $R_3(x; 0)$ for the given function.

1. $\dfrac{1}{1+x}$
2. $\dfrac{1}{1-x}$
3. $\ln(1+x)$
4. $\ln(1-x)$
5. $\cos x$
6. $\sin x$

In Exercises 7 to 10 find the Taylor polynomials $P_3(x; 0)$ for the given functions.

7. $\sin x$
8. $\cos x$
9. $\tan x$
10. $\sqrt{1+x}$

In Exercises 11 to 14 graph the given functions and the Taylor polynomials $P_0(x; 0)$, $P_1(x; 0)$, $P_2(x; 0)$, and $P_3(x; 0)$. In each exercise graph the functions relative to the same axes.

11. $1/(1+x)$
12. $\ln(1+x)$
13. $\sin x$
14. $\cos x$

In Exercises 15 to 20 show for the given functions that $R_n(x; 0) \to 0$ as $n \to \infty$, using (a) the integral form for $R_n(x; 0)$, (b) the derivative form for $R_n(x; 0)$.

15. e^x
16. e^{-x}
17. $\sin x$
18. $\cos x$
19. $\dfrac{1}{1+x}$, $|x| < 1$
20. $\ln(1+x)$, $|x| < 1$

21. Show that, for $f(x) = \sin x$, $R_n(x; a) \to 0$ as $n \to \infty$ for any x and a.
22. Show that, for $f(x) = e^x$, $R_n(x; a) \to 0$ as $n \to \infty$ for any x and a.

In each of Exercises 23–28 use the first three nonzero terms of a Maclaurin series to estimate the given number. Also use the derivative form of the remainder to put an upper bound on the error.

23. $\sqrt[3]{e}$
24. e^2
25. $\sin 1$
26. $\sin 20°$ Hint: First convert to radians.
27. $\cos 10°$
28. $\cos(1/3)$

29. In Example 4 it was shown that the error is less than $\frac{1}{192}$. Using the derivative form of the remainder, show that the error is greater than $\frac{1}{384}$.
30. Show that $\sin x \approx x - x^3/3!$ with an error at most $\frac{1}{120}$ for $|x| \le 1$.
31. Show that $\cos x \approx 1 - x^2/2 + x^4/24$ with an error at most $\frac{1}{720}$ for $|x| \le 1$.
32. From the Maclaurin series for e^x obtain the Maclaurin series for (a) e^{x^3} (b) e^{-x} (c) $(e^x - 1)/x$.

33. (Calculator) Graph, relative to the same axes, $y = \sin x$ and its Taylor polynomial $x - x^3/6 + x^5/120$.
34. (Calculator) Graph, relative to the same axes, $y = \cos x$ and its Taylor polynomial $y = 1 - x^2/2 + x^4/24$.
35. (a) From the Maclaurin series for $\cos x$ obtain that for $\cos x^2$.
    (b) Use (a) to estimate $\int_0^{1/2} \cos x^2 \, dx$ to three-decimal-place accuracy.
36. Use the Maclaurin series for $\ln(1+x)$ to estimate $\ln(1.1)$ to three-decimal-place accuracy.
37. Estimate $\int_0^1 \sqrt{x} \sin x \, dx$ to two-decimal-place accuracy.
38. Justify this statement, found in a biological monograph: "Expanding the equation

$$a \cdot \ln(x+p) + b \cdot \ln(y+q) = M,$$

we obtain

$$a\left(\ln p + \frac{x}{p} - \frac{x^2}{2p^2} + \frac{x^3}{3p^3} - \cdots\right)$$

$$+ b\left(\ln q + \frac{y}{q} - \frac{y^2}{2q^2} + \frac{y^3}{3q^3} - \cdots\right) = M."$$

39. Justify the second sentence in this statement, quoted from a biological monograph: "Hence the probability of extinction $1 - y$ will be given by $1 - y = e^{-(1+k)y}$. If k is small, y is approximately equal to $2k$."

■ ■

40. Let $f(x)$ be a function such that, for n sufficiently large, $|f^{(n)}(x)| \le n^5$. Deduce that, for all x and a,

$$f(x) = f(a) + f^{(1)}(a)(x-a) + \frac{f^{(2)}(a)}{2!}(x-a)^2 + \cdots.$$

41. Let $g(x)$ be a function which is fairly "flat" at the origin. Specifically, assume that $g(0) = 0$, $g^{(1)}(0) = 0$, and $g^{(2)}(x) = 0$. Assume that $|g^{(3)}(x)| \le M_3$ for all x, where M_3 is a constant. Deduce that, for $x \ge 0$,

$$|g(x)| \le \frac{M_3 x^3}{6}.$$

42. Assume that $f(x)$ has a continuous fourth derivative. Let M_4 be the maximum of $|f^{(4)}(x)|$ for x in $[-1, 1]$. Show that

$$\left|\int_{-1}^{1} f(x)\, dx - f\left(\frac{1}{\sqrt{3}}\right) - f\left(-\frac{1}{\sqrt{3}}\right)\right| \le \frac{7M_4}{270}.$$

Suggestion: Use the representation $f(x) = f(0) + f^{(1)}(0)x + f^{(2)}(0)x^2/2 + f^{(3)}(0)x^3/6 + f^{(4)}(c)x^4/24$, where c depends on x.

43 In R. P. Feynman, *Lectures on Physics*, Addison-Wesley, Reading, Mass., 1963, appears this remark:

Thus the average energy is

$$\langle E \rangle = \frac{\hbar\omega(0 + x + 2x^2 + 3x^3 + \cdots)}{1 + x + x^2 + \cdots}.$$

Now the two sums which appear here we shall leave for the reader to play with and have some fun with. When we are all finished summing and substituting for x in the sum, we should get—if we make no mistakes in the sum—

$$\langle E \rangle = \frac{\hbar\omega}{e^{\hbar\omega/kT} - 1}.$$

This, then, was the first quantum-mechanical formula ever known, or ever discussed, and it was the beautiful culmination of decades of puzzlement.

Have the aforementioned fun, given that $x = e^{-\hbar\omega/kT}$.

10.9 The complex numbers (optional)

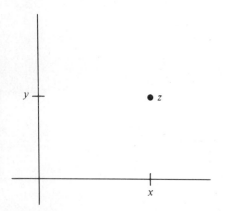

Figure 10.17

Let us think of the number line of real numbers as coinciding with the x axis of an xy coordinate system. This number line, with its addition, subtraction, multiplication, and division, is but a small part of a number system that occupies the plane and which obeys the usual rules of arithmetic. This section describes that system, known as the complex numbers. One of the important properties of the complex numbers is that any nonconstant polynomial has a root; in particular the equation $x^2 = -1$ has two solutions.

THE ALGEBRA OF COMPLEX NUMBERS

By a complex number, z, we shall mean an expression of the form $x + iy$, where x and y are real numbers and i is a symbol such that $i^2 = -1$. This expression will be identified with the point (x, y) in the xy plane, as in Fig. 10.17. Every point in the plane may therefore be thought of as a complex number.

To add or multiply two complex numbers follow the usual rules of arithmetic of real numbers, with one new proviso:

Whenever you see i^2, replace it by -1.

For instance, to add the complex numbers $3 + 2i$ and $-4 + 5i$, just collect like terms:

$$(3 + 2i) + (-4 + 5i) = (3 + (-4)) + (2i + 5i) = -1 + 7i.$$

Addition does not make use of the fact that $i^2 = -1$. However, multiplication does, as Example 1 shows.

EXAMPLE 1 Compute the product $(2 + i)(3 + i)$.

SOLUTION By the distributive law, $a(b + c) = ab + ac$,

$$(2 + i)(3 + i) = (2 + i)3 + (2 + i)i.$$

10.9 THE COMPLEX NUMBERS (OPTIONAL)

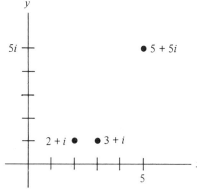

Figure 10.18

-1 *has two square roots.*

Real numbers are on the x axis, imaginary on the y axis.

By the distributive law, $(b + c)a = ba + ca$,

$$(2 + i)3 = 2 \cdot 3 + i3 = 6 + 3i \quad \text{and} \quad (2 + i)i = 2i + i^2 = 2i - 1.$$

Thus

$$(2 + i)(3 + i) = (6 + 3i) + (2i - 1) = 5 + 5i.$$

Figure 10.18 shows the complex numbers $2 + i$, $3 + i$, and their product, $5 + 5i$. ■

Note that $(-i)(-i) = i^2 = -1$. Both i and $-i$ are square roots of -1. It is customary to use the symbol $\sqrt{-1}$ to denote i, not $-i$.

A complex number that lies on the y axis is called *imaginary*. Every complex number z is the sum of a real number and an imaginary number, $z = x + iy$. The number x is called the real part of z, and y is called the imaginary part. One writes Re $z = x$ and Im $z = y$.

We have seen how to add and multiply complex numbers. Subtraction is straightforward. For instance,

$$(3 + 2i) - (4 - i) = (3 - 4) + (2i - (-i))$$
$$= -1 + 3i.$$

Division of complex numbers requires rationalizing the denominator, as Example 2 illustrates.

EXAMPLE 2 Compute

$$\frac{1 + 5i}{3 + 2i}.$$

SOLUTION To divide, "conjugate the denominator":

$$\frac{1 + 5i}{3 + 2i} \cdot \frac{3 - 2i}{3 - 2i} = \frac{3 + 15i - 2i + 10}{9 - 4i^2}$$
$$= \frac{13 + 13i}{13}$$
$$= 1 + i \quad ■$$

Conjugate of z

The solution of Example 2 used the *conjugate* of a complex number. The conjugate of the complex number $z = x + yi$ is the complex number $x - yi$, which is denoted $\bar{z}$. Note that

$$z\bar{z} = (x + yi)(x - yi) = x^2 + y^2$$

and

$$z + \bar{z} = (x + yi) + (x - yi) = 2x.$$

Thus both $z\bar{z}$ and $z + \bar{z}$ are real.

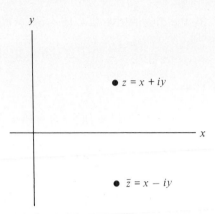

Figure 10.19

Figure 10.19 shows the relation between z and $\bar{z}$, which is the mirror image of z in the x axis.

Every polynomial has a root in the complex numbers.

The complex numbers provide the equation $x^2 + 1 = 0$ with two solutions, i and $-i$. This illustrates an important property of the complex numbers: If $f(x) = a_n x^n + a_{n-1} x^{n-1} + \cdots + a_0$ is any polynomial of degree $n \geq 1$, with real or complex coefficients, then there is a complex number z such that $f(z) = 0$. Though the general proof of this fact requires advanced mathematics, the case $n = 2$ is illustrated in Example 3.

EXAMPLE 3 Solve the quadratic equation $x^2 - 4x + 5 = 0$.

SOLUTION By the quadratic formula, the solutions are

$$z = \frac{-(-4) \pm \sqrt{(-4)^2 - 4 \cdot 1 \cdot 5}}{2}$$

$$= \frac{4 \pm \sqrt{-4}}{2}$$

$$= \frac{4 \pm 2i}{2} = 2 \pm i$$

The solutions are $2 + i$ and $2 - i$.

These solutions can be checked by substitution in the original equation. For instance,

$$(2 + i)^2 - 4(2 + i) + 5 \stackrel{?}{=} 0$$
$$(4 + 4i + i^2) - 8 - 4i + 5 \stackrel{?}{=} 0$$
$$4 + 4i - 1 - 8 - 4i + 5 \stackrel{?}{=} 0$$
$$0 + 0i \stackrel{?}{=} 0$$

Yes, it checks. The solution $2 - i$ can be checked similarly. ∎

THE GEOMETRY OF ADDITION AND MULTIPLICATION

The sum of the complex numbers z_1 and z_2 is the fourth vertex (opposite O) in a parallelogram determined by the origin O and the points z_1 and z_2, as shown in Fig. 10.20.

The origin O is the complex number 0.

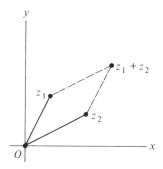

Figure 10.20

The geometric relation between z_1, z_2, and their product, $z_1 z_2$, is described in terms of the magnitude and argument of a complex number.

Each complex number z other than the origin can be written in polar coordinates as $z = (r, \theta)$, where r is positive and θ is determined up to an integer multiple of 2π. If $z = (r, \theta)$, then r is called the *magnitude* of z and is denoted $|z|$. The polar angle θ is called the *argument* of z and is denoted $\arg z$.

The magnitude and argument of z

EXAMPLE 4   (a) Draw all complex numbers P such that $|P| = 3$.
(b) Draw the complex number Q of magnitude 3 and argument $\pi/6$.

SOLUTION   (a) The complex numbers of magnitude 3 form a circle of radius 3 with center at O. (See Fig. 10.21.)
(b) The complex number of magnitude 3 and argument $\pi/6$ is shown in Fig. 10.21.)

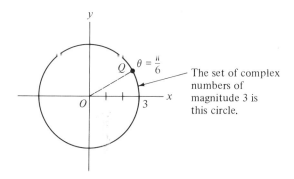

Figure 10.21

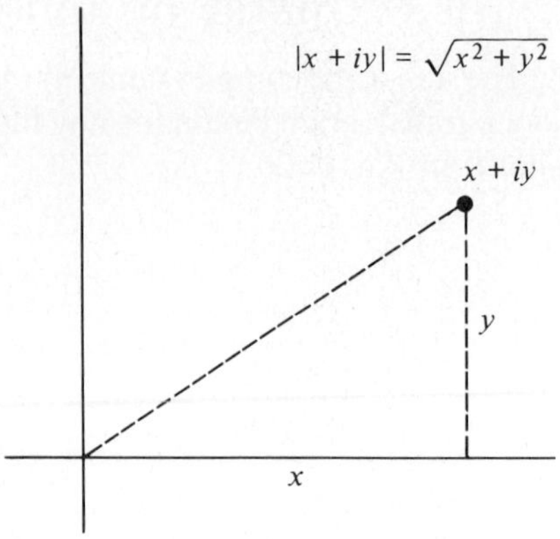

650 SERIES AND RELATED TOPICS

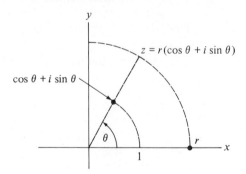

Figure 10.22

Observe that $|x + iy| = \sqrt{x^2 + y^2}$, by the Pythagorean theorem, as shown in Fig. 10.22.

Each complex number $z = x + iy$ other than 0 can be written as the product of a positive real number and a complex number of magnitude 1. To show this first express z in polar form, $z = (r, \theta)$. Then

$$z = r \cos \theta + i\, r \sin \theta$$
$$= r (\cos \theta + i \sin \theta).$$

The number r is a positive real number; the number $\cos \theta + i \sin \theta$ has magnitude 1 since $\sqrt{\cos^2 \theta + \sin^2 \theta} = 1$. Figure 10.23 shows the numbers r and $\cos \theta + i \sin \theta$, whose product is z.

Figure 10.23

The following theorem describes how to multiply two complex numbers if they are given in polar form, that is, in terms of their magnitudes and arguments.

THEOREM Assume that z_1 has magnitude r_1 and argument θ_1 and that z_2 has magnitude r_2 and argument θ_2. Then the product $z_1 z_2$ has magnitude $r_1 r_2$ and argument $\theta_1 + \theta_2$.

PROOF

$$z_1 z_2 = r_1(\cos\theta_1 + i\sin\theta_1)r_2(\cos\theta_2 + i\sin\theta_2)$$
$$= r_1 r_2(\cos\theta_1 + i\sin\theta_1)(\cos\theta_2 + i\sin\theta_2)$$
$$= r_1 r_2[\cos\theta_1\cos\theta_2 - \sin\theta_1\sin\theta_2 + i(\sin\theta_1\cos\theta_2 + \cos\theta_1\sin\theta_2)]$$

(multiplying out)

Recall the identities for $\cos(A+B)$ *and* $\sin(A+B)$.

$$= r_1 r_2(\cos(\theta_1+\theta_2) + i\sin(\theta_1+\theta_2))$$

(by trigonometric identities)

Thus the argument of $z_1 z_2$ is $\theta_1 + \theta_2$ and the magnitude of $z_1 z_2$ is $r_1 r_2$. This proves the theorem. ∎

How to multiply in terms of magnitude and argument

In practical terms, the theorem says, "to multiply two complex numbers just add their arguments and multiply their magnitudes." If $z_1 = (r_1, \theta_1)$ and $z_2 = (r_2, \theta_2)$, then $z_1 z_2 = (r_1 r_2, \theta_1 + \theta_2)$.

EXAMPLE 5 Find $z_1 z_2$ for z_1 and z_2 in Fig. 10.24.

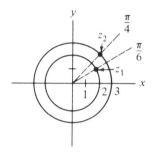

Figure 10.24

SOLUTION z_1 has magnitude 2 and argument $\pi/6$; z_2 has magnitude 3 and argument $\pi/4$. Thus $z_1 z_2$ has magnitude $2 \times 3 = 6$ and argument $\pi/6 + \pi/4 = 5\pi/12$. It is shown in Fig. 10.25.

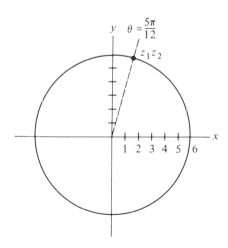

Figure 10.25 ∎

652 SERIES AND RELATED TOPICS

EXAMPLE 6 Using the geometric description of multiplication, find the product of the real numbers -2 and -3.

SOLUTION The number -2 has magnitude 2 and argument π. The number -3 has magnitude 3 and argument π. Thus $(-2) \times (-3)$ has magnitude $2 \times 3 = 6$ and argument $\pi + \pi = 2\pi$. The complex number with magnitude 6 and argument 2π is just our old friend, the real number 6. Thus

As expected, "negative times negative is positive." $(-2) \times (-3) = 6$ in agreement with the customary product of two negative real numbers. (See Fig. 10.26.)

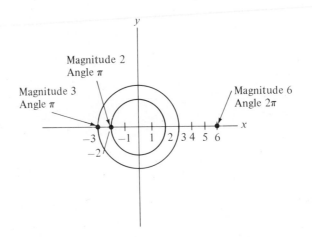

Figure 10.26

See Exercise 17. Division of complex numbers given in polar form is similar, except that the magnitudes are divided and the arguments subtracted:

$$\frac{r_1(\cos \theta_1 + i \sin \theta_1)}{r_2(\cos \theta_2 + i \sin \theta_1)} = \left(\frac{r_1}{r_2}\right)(\cos(\theta_1 - \theta_2) + i \sin(\theta_1 - \theta_2)).$$

EXAMPLE 7 Let $z_1 = 6\left(\cos \frac{\pi}{2} + i \sin \frac{\pi}{2}\right)$ and $z_2 = 3\left(\cos \frac{\pi}{6} + i \sin \frac{\pi}{6}\right)$. Find (a) $z_1 z_2$ and (b) z_1/z_2.

SOLUTION (a)
$$z_1 z_2 = 6 \cdot 3\left(\cos\left(\frac{\pi}{2} + \frac{\pi}{6}\right) + i \sin\left(\frac{\pi}{2} + \frac{\pi}{6}\right)\right)$$
$$= 18\left(\cos \frac{2\pi}{3} + i \sin \frac{2\pi}{3}\right)$$
$$= 18\left(-\frac{1}{2} + i\frac{\sqrt{3}}{2}\right)$$
$$= -9 + 9\sqrt{3}i.$$

(b)
$$\frac{z_1}{z_2} = \frac{6\left(\cos\frac{\pi}{2} + i\sin\frac{\pi}{2}\right)}{3\left(\cos\frac{\pi}{6} + i\sin\frac{\pi}{6}\right)}$$

$$= 2\left(\cos\frac{\pi}{3} + i\sin\frac{\pi}{3}\right)$$

$$= 2\left(\frac{1}{2} + \frac{\sqrt{3}}{2}i\right)$$

$$= 1 + \sqrt{3}i. \blacksquare$$

EXAMPLE 8 Compute $(1+i)(3+2i)$ and check the answer in terms of magnitudes and arguments.

SOLUTION
$$(1+i)(3+2i) = (1+i)(3) + (1+i)(2i)$$
$$= 3 + 3i + 2i + 2i^2$$
$$= 3 + 3i + 2i - 2$$
$$= 1 + 5i$$

As a check, let us see if $|1+5i| = |1+i||3+2i|$. We have

$$|1+5i| = \sqrt{1^2 + 5^2} = \sqrt{26}$$
$$|1+i| = \sqrt{1^2 + 1^2} = \sqrt{2}$$
$$|3+2i| = \sqrt{3^2 + 2^2} = \sqrt{13}.$$

Since $\sqrt{26} = \sqrt{2}\sqrt{13}$, the magnitude of $1+5i$ is the product of the magnitudes of $1+i$ and $3+2i$. Inspection of Fig. 10.27 suggests that the argument of $1+5i$ is indeed the sum of the arguments of $1+i$ and $3+2i$. A calculator shows that $\arg(1+5i) = \tan^{-1} 5 \approx 1.3734$,

$\arg(x+iy) = \tan^{-1}(y/x)$ for $x+iy$ in first or fourth quadrants

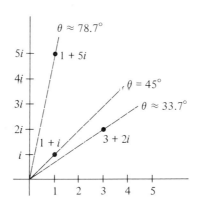

Figure 10.27

$\arg(1+i) = \tan^{-1} 1 \approx 0.7854$, and $\arg(3+2i) = \tan^{-1}(2/3) \approx 0.5880$. Note that $0.7854 + 0.5880 = 1.3734$. $\blacksquare$

654 SERIES AND RELATED TOPICS

The next example involves a quotient whose numerator is in polar form (that is, in terms of magnitude and argument) and whose denominator is in rectangular form (that is, in terms of x and y coordinates). Such a computation frequently arises in the study of alternating currents.

EXAMPLE 9 Let z_1 have magnitude 3 and argument $60°$; let z_2 be $3 + i$. Compute z_1/z_2.

SOLUTION
$$\frac{z_1}{z_2} = \frac{3(\cos 60° + i \sin 60°)}{3 + i}$$
$$= \frac{3\left(\frac{1}{2} + \frac{\sqrt{3}}{2}i\right)}{3 + i}$$
$$= \frac{\left(\frac{3}{2} + \frac{3\sqrt{3}}{2}i\right)}{(3 + i)} \cdot \frac{(3 - i)}{(3 - i)}$$
$$= \frac{\frac{9}{2} + \frac{3\sqrt{3}}{2} + i\left(\frac{9\sqrt{3}}{2} - \frac{3}{2}\right)}{10}$$
$$= \frac{9 + 3\sqrt{3}}{20} + \frac{(9\sqrt{3} - 3)i}{20} \quad\blacksquare$$

RAISING A COMPLEX NUMBER TO A POWER

Let $z = (r, \theta)$. Then $|z^2| = |z||z| = r \cdot r = r^2$ and arg z^2 = arg $(z \cdot z)$ = arg z + arg z = 2 arg z. In short, *to square a complex number, square its magnitude and double its argument.*

How to compute z^n

More generally, for any positive integer n, to compute z^n, find $|z|^n$ and multiply the argument of z by n. In short,

The top equation is known as DeMoivre's law.

$$[r(\cos \theta + i \sin \theta)]^n = r^n(\cos n\theta + i \sin n\theta),$$

or $|z^n| = |z|^n$ and arg z^n = n arg z.

Example 10 illustrates this geometric view of computing powers.

EXAMPLE 10 Let z have magnitude 1 and argument $2\pi/5$. Compute and sketch z, z^2, z^3, z^4, z^5, and z^6.

SOLUTION Since $|z| = 1$, it follows that $|z^2| = |z|^2 = 1^2 = 1$. Similarly, $|z^3| = 1$. For all positive integers n, $|z^n| = 1$; that is, z^n is a point on the unit circle with center 0. All that remains is to examine the arguments of z^2, z^3, etc.:

$$\arg z^2 = 2 \arg z = 2\frac{2\pi}{5} = \frac{4\pi}{5}$$
$$\arg z^3 = 3 \arg z = 3\frac{2\pi}{5} = \frac{6\pi}{5}$$

$$\arg z^4 = 4 \arg z = 4\frac{2\pi}{5} = \frac{8\pi}{5}$$

$$\arg z^5 = 5 \arg z = 5\frac{2\pi}{5} = \frac{10\pi}{5} = 2\pi$$

$$\arg z^6 = 6 \arg z = 6\frac{2\pi}{5} = \frac{12\pi}{5}$$

Since $\arg z^5 = 2\pi$ and $|z^5| = 1$, $z^5 = 1$. Note also that $z^6 = z$, since they have the same magnitude and their arguments differ by 2π. We could also show that $z^6 = z$ as follows: $z^6 = z^5 \cdot z = 1 \cdot z = z$. The powers of z are shown in Fig. 10.28. They form the vertices of a regular pentagon. ∎

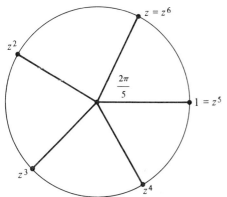

Figure 10.28

The equation $x^5 = 1$ has only one real root, namely 1. However it has five complex roots. For instance, the number z shown in Fig. 10.28 is a solution of $x^5 = 1$, since $z^5 = 1$. Another root is z^2, since $(z^2)^5 = z^{10} = (z^5)^2 = 1^2 = 1$. Similarly, z^3 and z^4 are roots of $x^5 = 1$. The roots are 1, z, z^2, z^3, and z^4.

The powers of i will be needed in the next section. They are $i^2 = -1$, $i^3 = i^2 \cdot i = (-1)i = -i$, $i^4 = i^3 \cdot i = (-i)i = -i^2 = 1$, $i^5 = i^4 \cdot i = -i$, and so on. They repeat in blocks of four: for any integer n, $i^{n+4} = i^n$.

It is often useful to express a complex number $z = x + iy$ in polar form. Note first that $|z| = \sqrt{x^2 + y^2}$. To find $\theta = \arg z$ it is best to sketch z in order to see in which quadrant it lies. Though $\tan \theta = y/x$ we cannot say that $\theta = \tan^{-1}(y/x)$ since $\tan^{-1} u$ lies between $\pi/2$ and $-\pi/2$ for any real number u. However, θ may be a second or third quadrant angle. For instance, to put $z = -2 - 2i$ in polar form, first sketch z, as in Fig. 10.29. We have $|z| = \sqrt{(-2)^2 + (-2)^2} = \sqrt{8}$ and $\arg z = 5\pi/4$. Thus

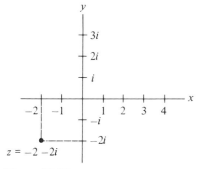

Figure 10.29

$$z = \sqrt{8}\left(\cos\frac{5\pi}{4} + i\sin\frac{5\pi}{4}\right).$$

Note that $\arg z$ is not equal to $\tan^{-1}(-2/(-2))$, which is $\pi/4$.

As early as their sophomore year many students will take a course in electric circuits, where they will find the complex numbers used in the analysis of alternating currents. They will use a different notation, which is described in this table:

Standard	Electrical Engineering
i	j (i is used for current)
$\bar{z}$	z^*
$r(\cos \theta + i \sin \theta)$	$r \angle \theta$

The symbol j is used in the following discussion of alternating current.

ALTERNATING CURRENT AND COMPLEX NUMBERS

The complex numbers, introduced by mathematicians in the course of their pure research, were accepted by them as a legitimate structure early in the nineteenth century. At that time electricity was only an object of laboratory interest; it had little practical significance. Yet before the century was over the discoveries concerning electricity and magnetism were to transform our world, and complex numbers were to serve as a tool in that transformation by simplifying the algebra of alternating currents. The details are worth sketching, not only to show the importance of complex numbers, but to demonstrate how the "pure" knowledge of one generation can become the "practical" technique of a later generation.

In the case of a *direct current*, such as that provided by a battery, the voltage E is constant. This constant voltage, working against a resistance R, produces a constant current I. The three real numbers E, I, and R, are related by the equation

$$E = IR,$$

which says, "current is proportional to voltage drop."

For the long distance transmission of electric power, *alternating currents* are far more efficient than direct currents. An alternating current is produced by rotating a coiled wire in a fixed magnetic field. Charles Proteus Steinmetz, an engineer at General Electric when the United States was starting to bring electricity to the cities, found the algebra of alternating currents unwieldly. As he wrote in 1893,

> The current rises from zero to a maximum; then decreases again to nothing, reverses and rises to a maximum in the opposite direction; decreases to zero, again reverses and rises to a maximum in the first direction—and so on.
>
> Thus in all calculations with alternating current, instead of a simple mechanical value of direct current theory, the investigator had to use a complicated function of time to represent the alternating current. The theory of alternating current apparatus thereby became so complicated that the investigator never got very far ....
>
> The idea suggested itself at length of representing the alternating current by a single complex number .... This proved the solution of the alternating current calculation.
>
> It gave to the alternating current a single numerical value, just as to the direct current, instead of the complicated function of time of the previous theory; and thereby it made alternating current calculations as simple as direct current calculations.
>
> The introduction of the complex number has eliminated the function of time from the alternating current theory, and has made the alternating current theory the simple algebra of the complex number, just as the direct current theory leads to the simple algebra of the real number.

Let us see how Steinmetz, as has been said, "generated electricity out of the square root of -1."

To describe the motor, which produces the varying voltage, he used a single complex number **E**. The magnitude of **E** is the maximum voltage that the motor produces. The argument of **E** is determined by the initial position of the rotating coil.

Corresponding to the alternating current is the complex number **I**. The magnitude of **I** is the maximum current flowing through the circuit. The argument of **I** records the lag in the current. (It does not necessarily reach its maximum when the voltage does.)

Within the typical circuit are, in addition, a resistance, a capacitor (for storing electrical charge), and a fixed coil (which produces a magnetic field when current passes through it). The resistance is described by a real number R; the capacitor, by the real number X_C; the fixed coil, by the real number X_L. Steinmetz introduced the single complex number

$$j = \sqrt{-1}$$

$$\mathbf{R} = R + (X_L - X_C)j,$$

called the complex impedance, and then described the basic behavior of alternating currents in the single equation

$$\mathbf{E} = \mathbf{IR}.$$

(For instance, this equation records the fact that the maximum voltage is equal to the product of the maximum current and the number $\sqrt{R^2 + (X_L - X_C)^2}$. Why?)

Exercises

In Exercises 1 and 2 compute the given quantities:

1. (a) $(2 + 3i) + (\frac{1}{2} - 2i)$ (b) $(2 + 3i)(2 - 3i)$

   (c) $\dfrac{1}{2 - i}$ (d) $\dfrac{3 + 2i}{4 - i}$

2. (a) $(2 + 3i)^2$ (b) $\dfrac{4}{3 - i}$

   (c) $(1 + i)(3 - i)$ (d) $\dfrac{1 + 5i}{2 - 3i}$

3. Let z_1 be the complex number of magnitude 2 and argument $\pi/6$ and let z_2 be the complex number of magnitude 3 and argument $\pi/3$. (a) Plot z_1 and z_2. (b) Find $z_1 z_2$ using the polar form. (c) Write z_1 and z_2 in the form $x + iy$. (d) With the aid of (c) compute $z_1 z_2$.

4. (a) If $|z_1| = 1$ and $|z_2| = 1$, how large can $|z_1 + z_2|$ be? (Hint: Draw some pictures.)
   (b) If $|z_1| = 1$ and $|z_2| = 1$, what can be said about $|z_1 z_2|$?

5. The complex number z has argument $\pi/3$ and magnitude 1. Find and plot (a) z^2, (b) $z^3 = z^2 \times z$, (c) $z^4 = z^3 \times z$.

6. Find (a) i^3, (b) i^4, (c) i^5, (d) i^{73}.

7. If z has magnitude 2 and argument $\pi/6$, what are the magnitude and argument of (a) z^2? (b) z^3? (c) z^4? (d) z^n? (e) Sketch z, z^2, z^3, and z^4.

8. Let z have magnitude 0.9 and argument $\pi/4$. (a) Find and plot z^2, z^3, z^4, z^5, z^6. (b) What happens to z^n as $n \to \infty$.

9. Using the fact that $(\cos\theta + i\sin\theta)^n = \cos n\theta + i\sin n\theta$, find formulas for $\cos 3\theta$ and $\sin 3\theta$ in terms of $\cos\theta$ and $\sin\theta$.

10. Figure 10.30 shows three complex numbers on the unit circle, P, Q, and R. Using DeMoivre's law show that $P^3 = 1$, $Q^3 = 1$, and $R^3 = 1$. The equation $x^3 = 1$ has three complex roots (one of which is real).

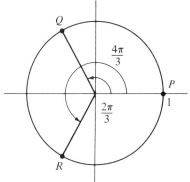

Figure 10.30

11. Let P have magnitude r and argument θ. Let Q have magnitude $1/r$ and argument $-\theta$. Show that $P \times Q = 1$. (Q is called the reciprocal of P, denoted P^{-1} or $1/P$.)

12. Find P^{-1} if $P = 4 + 4i$. (Use the definition in Exercise 11.)

13. (a) By substitution, verify that $2 + 3i$ is a solution of the equation $x^2 - 4x + 13 = 0$.
    (b) Use the quadratic formula to find all solutions of the equation $x^2 - 4x + 13 = 0$.

14. (a) Use the quadratic formula to find the solutions of the equation $x^2 + x + 1 = 0$.
    (b) Plot the solutions in (a).

15. Write in polar form: (a) $5 + 5i$, (b) $-\dfrac{1}{2} - \dfrac{\sqrt{3}}{2}i$, (c) $-\dfrac{\sqrt{2}}{2} + \dfrac{\sqrt{2}}{2}i$, (d) $3 + 4i$.

16 Write in rectangular form: (a) $3\left(\cos\dfrac{3\pi}{4}+i\sin\dfrac{3\pi}{4}\right)$,

(b) $2\left(\cos\dfrac{\pi}{6}+i\sin\dfrac{\pi}{6}\right)$, (c) $10(\cos\pi+i\sin\pi)$,

(d) $\dfrac{1}{5}(\cos 22°+i\sin 22°)$. (Express the answer to three places.)

17 Let z_1 have magnitude r_1 and argument θ_1 and let z_2 have magnitude r_2 and argument θ_2. (a) Explain why the magnitude of z_1/z_2 is r_1/r_2. (b) Explain why the argument of z_1/z_2 is $\theta_1-\theta_2$.

18 Compute

$$\dfrac{\cos\dfrac{5\pi}{4}+i\sin\dfrac{5\pi}{4}}{\cos\dfrac{3\pi}{4}+i\sin\dfrac{3\pi}{4}}$$

two ways: (a) by the result in Exercise 17, (b) by conjugating the denominator.

19 Compute (a) $(2+3i)(1+i)$, (b) $\dfrac{2+3i}{1+i}$,

(c) $(7-3i)(\overline{7-3i})$,

(d) $3(\cos 42°+i\sin 42°)\cdot 5(\cos 168°+i\sin 168°)$,

(e) $\dfrac{\sqrt{8}(\cos 147°+i\sin 147°)}{\sqrt{2}(\cos 57°+i\sin 57°)}$,

(f) $1/(3-i)$,

(g) $[3(\cos 52°+i\sin 52°)]^{-1}$,

(h) $\left(\cos\dfrac{\pi}{6}+i\sin\dfrac{\pi}{6}\right)^{12}$.

20 Compute (a) $(3+4i)(3-4i)$, (b) $\dfrac{3+5i}{-2+i}$,

(c) $\dfrac{1}{2+i}$, (d) $\left(\cos\dfrac{\pi}{12}+i\sin\dfrac{\pi}{12}\right)^{20}$,

(e) $[r(\cos\theta+i\sin\theta)]^{-1}$,

(f) $\operatorname{Re}((r(\cos\theta+i\sin\theta))^{10})$,

(g) $\dfrac{3\left(\cos\dfrac{\pi}{6}+i\sin\dfrac{\pi}{6}\right)}{5-12i}$.

21 Figure 10.31 shows five complex numbers of magnitude 2. Explain why each is a root of the equation $z^5=32$.

22 Sketch all complex numbers z such that (a) $z^6=1$, (b) $z^6=64$.

23 Show that (a) $\overline{z_1 z_2}=\bar{z}_1\bar{z}_2$, (b) $\overline{z_1+z_2}=\bar{z}_1+\bar{z}_2$.

24 If $\arg z$ is θ, what is the argument of (a) $\bar{z}$, (b) $1/z$?

25 For which complex numbers z is $\bar{z}=1/z$?

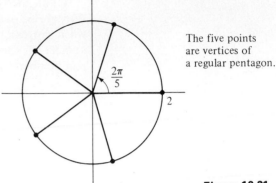

The five points are vertices of a regular pentagon.

Figure 10.31

26 Let $z=\dfrac{1}{\sqrt{2}}+\dfrac{i}{\sqrt{2}}$.

(a) Compute z^2 algebraically.

(b) Compute z^2 by putting z into polar form.

27 Let $z=\dfrac{1}{2}+\dfrac{i}{2}$. (a) Sketch the numbers z^n for $n=1,2,3,4$, and 5. (b) What happens to z^n as $n\to\infty$?

28 Let $z=1+i$. (a) Sketch the numbers $z^n/n!$ for $n=1,2,3,4$, and 5. (b) What happens to $z^n/n!$ as $n\to\infty$?

29 Let $z_1=1+2i$, and $z_2=2+3i$.

(a) Find $\arg z_1$. (Calculate to two decimals.)

(b) Find $\arg z_2$.

(c) Find $\arg z_1 z_2$.

(d) Check whether $\arg z_1 z_2=\arg z_1+\arg z_2$.

30 (a) If $z^3=-8$, find $|z|$.

(b) If $z^3=-8$, find $\arg z$ if $0\le\arg z<2\pi$.

(c) Plot the three complex numbers z such that $z^3=-8$.

31 Let a, b, and c be real numbers such that $b^2-4ac<0$.

(a) Show that the equation $az^2+bz+c=0$ has two complex roots, r_1 and r_2.

(b) Show that $az^2+bz+c=a(z-r_1)(z-r_2)$.

(c) Show that r_1 and r_2 are conjugates, $r_2=\bar{r}_1$.

32 Let a, b, and c be complex numbers such that $a\neq 0$ and $b^2-4ac\neq 0$. Show that the equation $az^2+bz+c=0$ has two roots.

∎ ∎

33 The real number system occupies a line and the complex number system occupies a plane. It might be conjectured that there is a number system containing the complex numbers that occupies space and obeys the usual rules of arithmetic. This exercise shows that no such system exists.

Assume that each point in space corresponds to a "number" $a+bi+cj$ where $i^2=-1$. Denote the product of i and j, ij, by $a+bi+cj$.

(a) Assuming the usual rules of arithmetic, show that $-j=ai-b+c(ij)$.

(b) From (a) deduce that $-j=ai-b+ca+cbi+c^2j$.

(c) From (b) obtain a contradiction.

34 Compute the roots of the following equations and plot them relative to the same axes.
(a) $x^2 - 3x + 2 = 0$, (b) $x^2 - 3x + 2.25 = 0$,
(c) $x^2 - 3x + 2.5 = 0$.

35 Let z_1 and z_2 be the roots of $ax^2 + bx + c = 0$, $a \neq 0$.
(a) Using the quadratic formula (or by other means), show that $z_1 + z_2 = -b/a$ and $z_1 z_2 = c/a$.
(b) From (a) deduce that $ax^2 + bx + c = a(x - z_1) \times (x - z_2)$.
(c) With the aid of (b) show that
$$\frac{1}{ax^2 + bx + c} = \frac{1}{a(z_1 - z_2)}\left(\frac{1}{x - z_1} - \frac{1}{x - z_2}\right).$$

Part (c) shows that the theory of partial fractions, described in Sec. 7.5, becomes much simpler when complex numbers are allowed as the coefficients of the polynomials. Only partial fractions of the type $k/(ax + b)^n$ are needed.

10.10 The relation between the exponential function and the trigonometric functions (optional)

With the aid of complex numbers, Euler in 1743 discovered that the trigonometric functions can be expressed in terms of the exponential function e^z, where z is complex. This section will retrace his discovery. In particular, it will show that

$$e^{i\theta} = \cos \theta + i \sin \theta, \quad \cos \theta = \frac{e^{i\theta} + e^{-i\theta}}{2}, \quad \text{and} \quad \sin \theta = \frac{e^{i\theta} - e^{-i\theta}}{2i}.$$

THE DEFINITION OF e^z

The Maclaurin series for e^x when x is real suggests the following definition.

DEFINITION e^z *for complex z.* Let z be a complex number. The sequence of complex numbers

$$1, \ 1 + z, \ 1 + z + \frac{z^2}{2!}, \ \ldots, \ 1 + z + \frac{z^2}{2!} + \cdots + \frac{z^n}{n!}, \ \ldots$$

approaches a complex number which will be called e^z. In short,

$$e^z = 1 + z + \frac{z^2}{2!} + \cdots + \frac{z^n}{n!} + \cdots = \sum_{n=0}^{\infty} \frac{z^n}{n!}.$$

We shall assume that the series converges, that is, that there is a complex number L such that

$$\lim_{k \to \infty} \left| \sum_{n=0}^{k} \frac{z^n}{n!} - L \right| = 0.$$

It can be shown that $e^{z_1 + z_2} = e^{z_1} e^{z_2}$ in accordance with the basic law of exponents.

The following theorem, due to Euler, provides the key link between the exponential function e^z and the trigonometric functions $\cos \theta$ and $\sin \theta$.

THEOREM 1 Let θ be a real number. Then

$$e^{i\theta} = \cos \theta + i \sin \theta.$$

660 SERIES AND RELATED TOPICS

PROOF By definition of e^z for any complex number,

$$e^{i\theta} = 1 + i\theta + \frac{(i\theta)^2}{2!} + \frac{(i\theta)^3}{3!} + \frac{(i\theta)^4}{4!} + \cdots.$$

Thus

$$e^{i\theta} = 1 + i\theta + \frac{i^2\theta^2}{2!} + \frac{i^3\theta^3}{3!} + \frac{i^4\theta^4}{4!} + \cdots$$

Recall that $i^2 = -1$, $i^3 = -i$, $i^4 = 1$, $i^5 = i, \ldots$.

$$= 1 + i\theta - \frac{\theta^2}{2!} - \frac{i\theta^3}{3!} + \frac{\theta^4}{4!} + \cdots$$

$$= \left(1 - \frac{\theta^2}{2!} + \frac{\theta^4}{4!} - \cdots\right) + i\left(\theta - \frac{\theta^3}{3!} + \cdots\right)$$

$$= \cos\theta + i\sin\theta.$$

Fig. 10.32 shows $e^{i\theta}$, which lies on the standard unit circle. ■

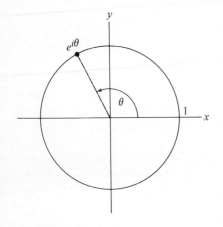

Figure 10.32

Theorem 1 asserts for instance that

$$e^{\pi i} = \cos\pi + i\sin\pi = -1 + i\cdot 0 = -1,$$

or

$$e^{\pi i} = -1,$$

an equation that links e (the fundamental number in calculus), π (the fundamental number in trigonometry), i (the fundamental complex number), and the negative number -1. The history of that short equation would recall the struggles of hundreds of mathematicians to create the number system that we now take for granted.

With the aid of Theorem 1, both $\cos\theta$ and $\sin\theta$ may be expressed in terms of the exponential function.

THEOREM 2 Let θ be a real number. Then

$$\cos\theta = \frac{e^{i\theta} + e^{-i\theta}}{2} \quad \text{and} \quad \sin\theta = \frac{e^{i\theta} - e^{-i\theta}}{2i}.$$

PROOF By Theorem 1, $e^{i\theta} = \cos\theta + i\sin\theta$. Replacing θ by $-\theta$, we obtain

$$e^{i(-\theta)} = \cos(-\theta) + i\sin(-\theta)$$

or

$$e^{-i\theta} = \cos\theta - i\sin\theta.$$

Addition of the equations

$$e^{i\theta} = \cos\theta + i\sin\theta \tag{1}$$

10.10 THE RELATION BETWEEN THE EXPONENTIAL AND TRIGONOMETRIC FUNCTIONS (OPTIONAL)

and

$$e^{-i\theta} = \cos\theta - i\sin\theta \qquad (2)$$

yields

$$e^{i\theta} + e^{-i\theta} = 2\cos\theta;$$

hence

$$\cos\theta = \frac{e^{i\theta} + e^{-i\theta}}{2}.$$

Subtraction of (2) from (1) yields

$$e^{i\theta} - e^{-i\theta} = 2i\sin\theta;$$

hence

$$\sin\theta = \frac{e^{i\theta} - e^{-i\theta}}{2i}.$$

This establishes the theorem. ∎

Recall Sec. 6.9, where cosh *x and* sinh *x were defined.*

The hyperbolic functions cosh x and sinh x were defined in terms of the exponential function by

$$\cosh x = \frac{e^x + e^{-x}}{2} \quad \text{and} \quad \sinh x = \frac{e^x - e^{-x}}{2}.$$

Theorem 2 shows that the trigonometric functions could be similarly defined in terms of the exponential function—if complex numbers were available.

COMPUTING e^z

The magnitude and argument of e^{x+iy}

If $z = x + iy$, the evaluation of e^z can be carried out as follows:

$$e^z = e^{x+iy} = e^x e^{iy} = e^x(\cos y + i\sin y).$$

Thus the magnitude of e^{x+iy} is e^x and the argument of e^{x+iy} is y.

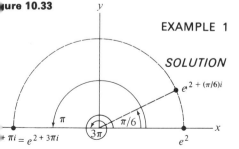

Figure 10.33

EXAMPLE 1 Compute and sketch (a) $e^{2+(\pi/6)i}$, (b) $e^{2+\pi i}$, and (c) $e^{2+3\pi i}$.

SOLUTION (a) $e^{2+(\pi/6)i}$ has magnitude e^2 and argument $\pi/6$. (b) $e^{2+\pi i}$ has magnitude e^2 and argument π; it equals $-e^2$. (c) $e^{2+3\pi i}$ has magnitude e^2 and argument 3π, so is the same number as the number in (b). The results are sketched in Fig. 10.33. ∎

The next example illustrates a typical computation in alternating currents.

662 SERIES AND RELATED TOPICS

EXAMPLE 2 Find the real part of $100\,e^{j(\pi/6)}e^{j\omega t}$. Here t refers to time, ω is a real constant related to frequency, and j is the electrical engineers' symbol for i.

SOLUTION
$$100\,e^{j(\pi/6)}e^{j\omega t} = 100\,e^{j(\pi/6)+j\omega t}$$
$$= 100\,e^{j[(\pi/6)+\omega t]}$$
$$= 100\left(\cos\left(\frac{\pi}{6}+\omega t\right) + i\sin\left(\frac{\pi}{6}+\omega t\right)\right)$$

Thus $\quad \text{Re}\,(100\,e^{j(\pi/6)}e^{j\omega t}) = 100\cos\left(\frac{\pi}{6}+\omega t\right).$ ∎

It is often convenient to think of $\cos\theta$ as $\text{Re}\,(e^{i\theta})$. The next example exploits this point of view.

EXAMPLE 3 Let A and B be real numbers. Let t be a real number, representing time. Show that there is a constant t_0 such that
$$A\cos t + B\sin t = \sqrt{A^2+B^2}\cos(t-t_0).$$

SOLUTION
$$A\cos t + B\sin t = A\cos t + B\cos\left(t-\frac{\pi}{2}\right)$$
$$= \text{Re}\,(Ae^{it}) + \text{Re}\,(Be^{i[t-(\pi/2)]})$$
$$= \text{Re}\,(Ae^{it} + Be^{i[t-(\pi/2)]})$$

The number $Ae^{it} + Be^{i[t-(\pi/2)]}$ is shown in Fig. 10.34.

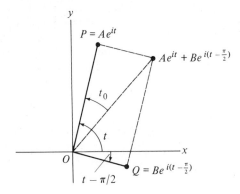

Figure 10.34

By the Pythagorean theorem,
$$\left|Ae^{it} + Be^{i[t-(\pi/2)]}\right| = \sqrt{A^2+B^2}.$$

Also,
$$\arg\left(Ae^{it} + Be^{i[t-(\pi/2)]}\right) = t - t_0,$$

where t_0 is shown in Fig. 10.34.

Consequently,

$$A \cos t + B \sin t = A \cos t + B \cos\left(t - \frac{\pi}{2}\right)$$
$$= \text{Re}\,(Ae^{it} + Be^{i(t-t_0)})$$
$$= \text{Re}\,(\sqrt{A^2 + B^2}\, e^{i(t-t_0)})$$
$$= \sqrt{A^2 + B^2} \cos(t - t_0),$$

which completes the solution. Note that t_0 does not depend on t. ■

Summary of notations for complex numbers

Notation	Remark
z	Symbol for a complex number
$x + iy$	Rectangular form (x and y real)
$r(\cos\theta + i\sin\theta)$	Polar form (magnitude r, argument θ)
$re^{i\theta}$	Exponential form (magnitude r, argument θ)
$r \angle \theta$ $x + jy$ $re^{j\theta}$	Electrical engineering forms

Exercises

In Exercises 1 to 6 sketch the numbers given and state their real and imaginary parts.

1. $e^{(5\pi i/4)}$
2. $5e^{(\pi i/4)}$
3. $2e^{(\pi i/4)} + 3e^{(\pi i/6)}$
4. e^{2+3i}
5. $e^{(\pi i/6)} e^{(3\pi i/4)}$
6. $2e^{\pi i} \cdot 3e^{-(\pi i/3)}$

In Exercises 7 to 10 express the given numbers in the form $re^{i\theta}$, for a positive real number r and argument θ, $-\pi < \theta \leq \pi$.

7. $\dfrac{e^2}{\sqrt{2}} - \dfrac{e^2}{\sqrt{2}}i$

8. $3\left(\cos\dfrac{\pi}{4} + i\sin\dfrac{\pi}{4}\right)$

9. $5\left(\cos\dfrac{\pi}{6} + i\sin\dfrac{\pi}{6}\right) \cdot 3\left(\cos\dfrac{\pi}{2} + i\sin\dfrac{\pi}{2}\right)$

10. $7\left(\cos\dfrac{7\pi}{3} + i\sin\dfrac{7\pi}{3}\right)$

In Exercises 11 to 14 plot the given numbers.

11. $e^{(\pi/y + 3\pi i)}$
12. $e^{1+(9\pi/4)i}$
13. $e^{2-(\pi/3)i}$
14. $e^{[-1+(17\pi/6)i]}$
15. Let $z = e^{a+bi}$. Find (a) $|z|$, (b) $\bar{z}$, (c) z^{-1}, (d) Re z, (e) Im z, (f) arg z.
16. For which values of a and b is $\lim_{n\to\infty} (e^{a+ib})^n = 0$?
17. Find all z such that $e^z = 1$.
18. Find all z such that $e^z = -1$.

■

19. Find all z such that $e^z = 3 + 4i$.
20. Assuming that $e^{z_1+z_2} = e^{z_1}e^{z_2}$ for complex numbers z_1 and z_2, obtain the trigonometric identities for $\cos(A+B)$ and $\sin(A+B)$.
21. Let $z = \dfrac{1+i}{\sqrt{2}}$. (a) Plot z, $z^2/2!$, $z^3/3!$, $z^4/4!$.

    (b) Plot $1 + z + z^2/2! + z^3/3! + z^4/4!$, which is an estimate of $e^{(1+i)/\sqrt{2}}$.

    (c) Plot $e^{(1+i)/\sqrt{2}}$ relative to the same axes.

■ ■

22. Use the fact that $1 + \cos \theta + \cos 2\theta + \cdots + \cos(n-1)\theta$ is the real part of $1 + e^{\theta i} + e^{2\theta i} + \cdots + e^{(n-1)\theta i}$, to find a short formula for the trigonometric sum. Hint: Look for a geometric series.

Exercises 23 and 24 treat the complex logarithms of a complex number. They show that $z = \ln w$ is not single valued.

23. Let w be a nonzero complex number. Show that there are an infinite number of z such that $e^z = w$.

24. (See Exercise 23.) If $e^z = w$, write $z = \ln w$, though $\ln w$ is not a uniquely defined number. If b is a nonzero complex number and q is a complex number, define b^q to be $e^{q \ln b}$. Since $\ln b$ is not unique, b^q is usually not unique. List all possible values of
    (a) $(-1)^i$, (b) $10^{1/2}$, (c) 10^3.

25. (a) How would you define $\sinh z$ and $\cosh z$ for a complex number z?
    (b) Show that $\sinh(ix) = i \sin x$ and $\cosh(ix) = \cos x$. (Note the close relation that (b) establishes between the hyperbolic and the circular functions.)

26. An integral table lists $\int xe^{ax} dx = e^{ax}(ax-1)/a^2$. At first glance, finding the integral of $xe^{ax} \cos bx$ may appear to be a much harder problem. However, by noticing that $\cos bx = \text{Re}(e^{ibx})$ we can reduce it to a simpler problem. Following this approach, find $\int xe^{ax} \cos bx \, dx$. Suggestion: The formula for $\int xe^{ax} \, dx$ holds when a is complex.

27. Evaluate $\sum_{n=0}^{\infty} \dfrac{\cos n\theta}{2^n}$.

10.11 Linear differential equations with constant coefficients (optional)

The differential equation $dy/dx = ky$, or equivalently,

$$\frac{dy}{dx} - ky = 0 \tag{1}$$

was solved in Sec. 6.7. Any solution has to be of the form $y = Ae^{kx}$ for some constant A. This section is concerned with generalizations of (1).

First, we consider the differential equation of the form

$$\frac{dy}{dx} + ay = f(x), \tag{2}$$

where a is real constant and $f(x)$ is some function of x. (Equation (1) is the special case where $f(x) = 0$.) Equation (2) is called a *first order linear differential equation with constant coefficients*.

Second, we consider the second-order equation

$$\frac{d^2y}{dx^2} + b\frac{dy}{dx} + cy = f(x), \tag{3}$$

where b and c are real constants. For some b and c, solving (3) may use complex numbers even though the solution will be a real function.

Many other types of differential equations are discussed in a differential equations course. Types (2) and (3), however, are used so often in physics and engineering that the student should be introduced to them early.

10.11 LINEAR DIFFERENTIAL EQUATIONS WITH CONSTANT COEFFICIENTS (OPTIONAL)

THE DIFFERENTIAL EQUATION $\dfrac{dy}{dx} + ay = f(x)$

The homogeneous case

Imagine for the moment that you have managed to find a particular solution y_p of (2) and a solution y_1 of the associated *homogeneous* equation, obtained by replacing $f(x)$ by 0,

$$\frac{dy}{dx} + ay = 0. \tag{4}$$

A straightforward computation then shows that $y_p + y_1$ is a solution of (2), as follows:

$$\begin{aligned}\frac{d}{dx}(y_p + y_1) + a(y_p + y_1) &= \frac{dy_p}{dx} + \frac{dy_1}{dx} + ay_p + ay_1 \\ &= \left(\frac{dy_p}{dx} + ay_p\right) + \left(\frac{dy_1}{dx} + ay_1\right) \\ &= f(x) + 0 \\ &= f(x).\end{aligned}$$

Now, the function $y_1 = Ce^{-ax}$, for any constant C, is a solution of (4). Thus, if y_p is a solution of (2), then so is $y_p + Ce^{-ax}$. In fact, each solution of (2) must be of the form $y_p + Ce^{-ax}$. To see why, assume that y_p and y both solve (2). Then

$$\begin{aligned}\frac{d}{dx}(y - y_p) + a(y - y_p) &= \left(\frac{dy}{dx} + ay\right) - \left(\frac{dy_p}{dx} + ay_p\right) \\ &= f(x) - f(x) \\ &= 0.\end{aligned}$$

Thus $y - y_p$, being a solution of (4), must be of the form Ce^{-ax} for some constant C. Thus $y = y_p + Ce^{-ax}$.

These observations are summarized in the following theorem.

THEOREM 1 Let y_p be a particular solution of the differential equation

$$\frac{dy}{dx} + ay = f(x).$$

Then the most general solution is

$$y_p + Ce^{-ax}. \quad\blacksquare$$

EXAMPLE 1 Solve the differential equation

$$\frac{dy}{dx} + 3y = 12.$$

SOLUTION One solution is the constant function $y_p = 4$. The most general solution is therefore $4 + Ce^{-3x}$ for any constant C. ∎

Once a particular solution y_p has been found, Theorem 1 provides the general solution. Example 2 illustrates one technique for finding y_p.

EXAMPLE 2 Find all solutions of the differential equation

$$\frac{dy}{dx} - y = \sin x. \tag{5}$$

SOLUTION First find one solution. Since $f(x) = \sin x$ let us see if there is a solution of the form $y_p = A \cos x + B \sin x$, for some constants A and B. Substitution in (5) yields

$$\frac{d}{dx}(A \cos x + B \sin x) - (A \cos x + B \sin x) = \sin x,$$

or

$$-A \sin x + B \cos x - A \cos x - B \sin x = \sin x,$$

or

$$(-A - B) \sin x + (B - A) \cos x = \sin x.$$

So choose A and B such that $-A - B = 1$ and $B - A = 0$. It follows that $B = A$ and that $-A - (A) = 1$ or $A = -1/2$. Consequently,

$$y_p = -\tfrac{1}{2} \cos x - \tfrac{1}{2} \sin x$$

is a solution of (5), as may be checked by substitution in (5).

The general solution of the homogeneous equation $dy/dx - y = 0$ is Ce^x, so the general solution of (5) is

$$y = -\tfrac{1}{2} \cos x - \tfrac{1}{2} \sin x + Ce^x. \quad \blacksquare$$

Example 2 uses the method of *undetermined coefficients*: Guess a general form of the solution and see if the unknown constants can be chosen properly to yield a solution of the differential equation.

Before turning to solutions of (3), consider the special case when $f(x)$ is identically 0, the so-called *homogeneous* case.

10.11 LINEAR DIFFERENTIAL EQUATIONS WITH CONSTANT COEFFICIENTS (OPTIONAL)

THE DIFFERENTIAL EQUATION $\dfrac{d^2y}{dx^2} + b\dfrac{dy}{dx} + cy = 0$

Let us find all solutions of the homogeneous equation

Homogeneous linear differential equation of second order

$$\frac{d^2y}{dx^2} + b\frac{dy}{dx} + cy = 0. \tag{6}$$

If y_1 and y_2 are both solutions of (6), a straightforward computation shows that $C_1 y_1 + C_2 y_2$ is also a solution of (6) for any choice of constants C_1 and C_2. (Since (6) involves the second derivative of y, we expect the general solution for y to contain two arbitrary constants.)

EXAMPLE 3 Solve

$$\frac{d^2y}{dx^2} - 3\frac{dy}{dx} + 2y = 0. \tag{7}$$

SOLUTION Recalling our experience with (1), we are tempted to look for a solution of the form e^{kx} for some constant k.

Substitution of e^{kx} into (7) yields

$$\frac{d^2}{dx^2}(e^{kx}) - 3\frac{d}{dx}(e^{kx}) + 2e^{kx} = 0$$

or

$$k^2 e^{kx} - 3k e^{kx} + 2e^{kx} = 0,$$

which is equivalent to

$$k^2 - 3k + 2 = 0. \tag{8}$$

By the quadratic formula, $k = 1$ or $k = 2$. Thus $y_1 = e^x$ and $y_2 = e^{2x}$ are solutions of (7). Consequently

$$C_1 e^x + C_2 e^{2x} \tag{9}$$

is a solution of (7) for any choice of constants C_1 and C_2. (It can be proved that there are no other solutions.) ∎

The most general solution of the differential equation

$$\frac{d^2y}{dx^2} + 6\frac{dy}{dx} + 9y = 0 \tag{10}$$

is of a different form. If we try $y = e^{kx}$ we obtain

$$k^2 e^{kx} + 6k e^{kx} + 9 e^{kx} = 0$$
$$e^{kx}(k^2 + 6k + 9) = 0$$
$$(k + 3)^2 = 0$$
$$k = -3.$$

This gives only the solutions of the form $y = Ce^{-3x}$. However, a second-order equation should possess a solution containing *two* arbitrary constants. Let us seek all solutions of the form

$$y = v(x)e^{-3x},$$

hoping to find some not of the form Ce^{-3x}.

Straightforward computations give

$$\frac{dy}{dx} = v(x)(-3e^{-3x}) + v'(x)e^{-3x} = -3v(x)e^{-3x} + v'(x)e^{-3x}$$

and

$$\frac{d^2y}{dx^2} = 9v(x)e^{-3x} - 6v'(x)e^{-3x} + v''(x)e^{-3x}.$$

Substituting into (10) yields

$$9v(x)e^{-3x} - 6v'(x)e^{-3x} + v''(x)e^{-3x} - 18v(x)e^{-3x} + 6v'(x)e^{-3x} + 9v(x)e^{-3x} = 0,$$

which simplifies to $\quad v''(x)e^{-3x} = 0;$

hence to $\quad v''(x) = 0.$

Therefore $v(x) = C_1 + C_2 x$ and our general solution is

$$y = C_1 e^{-3x} + C_2 x e^{-3x},$$

for arbitrary constants C_1 and C_2.

The associated quadratic The key to the nature of the solutions of (6) lies in the associated quadratic equation

$$t^2 + bt + c = 0. \tag{11}$$

The type of solution to (6) depends on the nature of the roots of (11). There are three cases: two distinct real roots; a repeated root (necessarily real); two distinct complex roots. Each case will be described by a corresponding theorem.

10.11 LINEAR DIFFERENTIAL EQUATIONS WITH CONSTANT COEFFICIENTS (OPTIONAL)

THEOREM 2 If $b^2 - 4c$ is positive, (11) has two distinct real roots, r_1 and r_2. In this case the general solution of (6) is

$$C_1 e^{r_1 x} + C_2 e^{r_2 x}. \quad \blacksquare$$

The proof that $C_1 e^{r_1 x} + C_2 e^{r_2 x}$ is a solution is left to the reader. Theorem 2 covers the differential equation (7).

EXAMPLE 4 Solve

$$\frac{d^2 y}{dx^2} + 5 \frac{dy}{dx} + y = 0.$$

SOLUTION In this case $b^2 - 4c = 21$, which is positive. The roots of the associated quadratic equation are

$$r_1 = \frac{-5 + \sqrt{21}}{2} \quad \text{and} \quad r_2 = \frac{-5 - \sqrt{21}}{2}.$$

The general solution of the differential equation is

$$C_1 e^{[(-5+\sqrt{21})/2]x} + C_2 e^{[(-5-\sqrt{21})/2]x}. \quad \blacksquare$$

The next theorem concerns the special case when the associated quadratic equation $t^2 + bt + c = 0$ has a repeated root, r.

THEOREM 3 If $b^2 - 4c = 0$, (11) has a repeated root r. In this case the general solution of (6) is

$$C_1 e^{rx} + C_2 x e^{rx} = (C_1 + C_2 x) e^{rx}. \quad \blacksquare$$

That $(C_1 + C_2 x) e^{rx}$ is a solution is left to the reader to check by substitution. Theorem 3 is illustrated by the solution of (10).

THEOREM 4 If $b^2 - 4c$ is negative, (11) has two distinct complex roots $r_1 = p + iq$ and $r_2 = p - iq$. In this case the general solution of (6) is

$$(C_1 \cos qx + C_2 \sin qx) e^{px}.$$

PROOF Just as in Theorem 2,

$$y = A_1 e^{r_1 x} + A_2 e^{r_2 x} \tag{12}$$

is a solution of (6) for any choice of constants A_1 and A_2, even complex.

Unfortunately, (12) will usually be complex. In order to find a *real* function that satisfies (6), expand (12):

$$A_1 e^{r_1 x} + A_2 e^{r_2 x} = A_1 e^{(p+iq)x} + A_2 e^{(p-iq)x}$$
$$= A_1 e^{px} e^{iqx} + A_2 e^{px} e^{-iqx}$$
$$= e^{px}[A_1 (\cos qx + i \sin qx) + A_2 (\cos(-qx) + i \sin(-qx))]$$
$$= e^{px}[(A_1 + A_2) \cos qx + i(A_1 - A_2) \sin qx]$$

Appropriate choices of A_1 and A_2 will generate the desired solution.

Choosing $A_1 = A_2 = 1/2$ produces the real solution $e^{px} \cos qx$. Next choose A_1 and A_2 so that $A_1 + A_2 = 0$ and $i(A_1 - A_2) = 1$. This will produce the real solution $e^{px} \sin qx$. (To find A_1 and A_2 solve the simultaneous equations $A_1 + A_2 = 0$ and $i(A_1 - A_2) = 1$. The solutions are $A_1 = -i/2$, and $A_2 = i/2$, which may be found by algebra.)

Thus

$$C_1 e^{px} \cos qx + C_2 e^{px} \sin qx \tag{13}$$

is a real valued solution of (6) for any choice of real constants C_1 and C_2. It can be proved that there are no other real solutions. ∎

EXAMPLE 5 Find the general solution of the differential equation of harmonic motion,

$$\frac{d^2 y}{dx^2} = -k^2 y, \tag{14}$$

where k is a constant.

SOLUTION Rewrite (14) in the form

$$\frac{d^2 y}{dx^2} + k^2 y = 0,$$

which has the associated quadratic equation $t^2 + k^2 = 0$. The roots of this equation are $0 + ki$ and $0 - ki$. By Theorem 4, the general solution of (14) is

$$C_1 e^{0x} \cos kx + C_2 e^{0x} \sin kx,$$

or simply

$$C_1 \cos kx + C_2 \sin kx. \quad \blacksquare$$

Equation (14) describes the motion of a mass bobbing at the end of a spring. The height of the mass at time x is y. Since the motion is oscillatory, it is plausible that it is described by a combination of $\cos kx$ and $\sin kx$.

10.11 LINEAR DIFFERENTIAL EQUATIONS WITH CONSTANT COEFFICIENTS (OPTIONAL)

THE DIFFERENTIAL EQUATION $\frac{d^2y}{dx^2} + b\frac{dy}{dx} + cy = f(x)$

If y_p is any particular solution of

$$\frac{d^2y}{dx^2} + b\frac{dy}{dx} + cy = f(x), \tag{15}$$

and y^* is a solution of the associated homogeneous equation (6), then $y_p + y^*$ is a solution of (15), as may be checked by a straightforward calculation. Since we know how to find the general solution of (6), all that remains is to find a particular solution of (15). This can often be accomplished by a shrewd guess and the use of undetermined coefficients, as illustrated by the following example.

EXAMPLE 6 Solve the differential equation $d^2y/dx^2 + dy/dx + 2y = 2x^2 + 5$. (16)

SOLUTION Let us seek a polynomial solution. If there is such a solution it cannot have degree greater than 2 since the right hand side of (16) has degree 2. So try $y = Ax^2 + Bx + C$, hence $y' = 2Ax + B$, and $y'' = 2A$. Substitution in (16) gives

$$2A + (2Ax + B) + 2(Ax^2 + Bx + C) = 2x^2 + 5,$$

or

$$2Ax^2 + (2A + 2B)x + (2A + B + 2C) = 2x^2 + 5.$$

Comparing coefficients gives $2A = 2$, $2A + 2B = 0$, and $2A + B + 2C = 5$. Thus $A = 1$, $B = -1$, and $C = 2$.

Consequently, $y_p = x^2 - x + 2$ is a particular solution of (16).

Next turn to solving the associated homogeneous equation

$$\frac{d^2y}{dx^2} + \frac{dy}{dx} + 2y = 0. \tag{17}$$

Here $b = 1$ and $c = 2$, so $b^2 - 4c = -7$. The roots of the associated quadratic equation $t^2 + t + 2 = 0$ are

$$\frac{-1 \pm \sqrt{-7}}{2} = \frac{-1}{2} \pm \frac{\sqrt{7}}{2}i.$$

By Theorem 4, the general solution of (17) is

$$y^* = C_1 e^{-x/2} \cos \frac{\sqrt{7}}{2}x + C_2 e^{-x/2} \sin \frac{\sqrt{7}}{2}x.$$

Putting everything together, we obtain the general solution of (16), namely

$$y = x^2 - x + 2 + C_1 e^{-x/2} \cos \frac{\sqrt{7}}{2} x + C_2 e^{-x/2} \sin \frac{\sqrt{7}}{2} x. \quad \blacksquare$$

Guessing a particular solution of (15) depends on the form of $f(x)$. This table describes the most common cases.

Form of $f(x)$	Guess for y_p
Polynomial	Arbitrary polynomial of same degree as f
e^{kx} (k not a root of associated quadratic equation)	Ae^{kx}
xe^{kx} (k not a root of the associated quadratic equation)	$(A + Bx)e^{kx}$
$e^{kx} \sin qx$ or $e^{kx} \cos qx$ ($k + qi$ not a root of the associated quadratic equation)	$Ae^{kx} \cos qx + Be^{kx} \sin qx$

A complete handbook of mathematical tables includes several pages of specific solutions for a much wider variety of functions $f(x)$ that appear on the right side of (15).

Exercises

In Exercises 1 to 18 find all solutions of the given differential equations.

1. $y' + 2y = 0$.
2. $y' + 2y = \cos x$
3. $3y' + 12y = x$
4. $y' - (1/3)y = e^{2x}$
5. $y' - y = x^2$
6. $y'/2 + y = xe^{2x}$
7. $y'' - 2y' - 3y = 0$
8. $y'' + 5y' + 6y = 0$
9. $y'' - 3y' + y = 0$
10. $3y'' - 2y' + 3y = 0$
11. $y'' - 6y' + 9y = 0$
12. $y'' + y' + y = 0$
13. $y'' - 2\sqrt{11}y' + 11y = 0$
14. $y'' - 3y' + 4y = 0$
15. $y'' - 2y' - 3y = e^{2x}$
16. $y'' + y' + y = x^2$
17. $y'' - 4y' + y = \cos 3x$
18. $y'' + 3y' + 2y = e^{-2x} \sin x$

19. (a) Show that $y = e^{-ax} \int e^{ax} f(x) \, dx$ is a solution of $y' + ay = f(x)$.

(b) Use (a) to find a solution of $y' + y = 1/(1 + e^x)$.
(c) Find all solutions of the equation in (b).

20. Check that $C_1 e^{-3x} + C_2 x e^{-3x}$ is a solution of (10).

21. (a) Show that if $b^2 - 4c = 0$, then $t^2 + bt + c = 0$ has only one root, r, and $(t - r)^2 = t^2 + bt + c$.

(b) Check that $C_1 e^{rx} + C_2 x e^{rx}$ is a solution of $y'' + by' + c = 0$.

22. Let k be a nonzero constant. Find all solutions of the equation $y'' = k^2 y$.

■ ■

In some tables of solutions to differential equations, y'' is written $D^2 y$ and y' is written Dy. The equation $y'' + by' + c = f(x)$ is then expressed as $(D^2 + bD + c)y = f(x)$. Similarly, $y'' - 2ay' + a^2 y = (D^2 - 2aD + a^2)y = (D - a)^2 y$.

23. Verify that $x^2 e^{ax}/2$ is a solution of $(D - a)^2 y = e^{ax}$.
24. Verify that $e^{rx}/(r - a)^2$ is a solution of $(D - a)^2 y = e^{rx}$, if $r \neq a$.
25. Verify that $(-x \cos bx)/(2b)$ is a solution of $(D^2 + b^2)y = \sin bx$, $b \neq 0$.
26. Verify that $(\sin sx)/(b^2 - s^2)$ is a solution of $(D^2 + b^2)y = \sin sx$, $s^2 \neq b^2$.

10.12 The binomial theorem for any exponent (optional)

If n is a positive integer and x is any number, then

$$(1 + x)^n = 1 + nx + \frac{n(n-1)}{1 \cdot 2} x^2 + \frac{n(n-1)(n-2)}{1 \cdot 2 \cdot 3} x^3 + \cdots + \frac{n(n-1) \cdots 1}{1 \cdot 2 \cdots n} x^n, \quad (1)$$

as shown in Exercises 49 to 51 in Sec. 4.3. In fact, the right side of (1) is the Maclaurin series for $(1 + x)^n$; all powers from x^{n+1} on have coefficient 0. In this section we examine the Maclaurin series for $(1 + x)^r$, where r is not a positive integer or 0. It turns out that for $|x| < 1$ the function $(1 + x)^r$ is indeed represented by its associated Maclaurin series. However, as shown in Exercise 13, the formula

$$R_n(x; 0) = \frac{f^{(n+1)}(c) x^{n+1}}{(n+1)!}$$

is inadequate for showing that $\lim_{n \to \infty} R_n(x; 0) = 0$. The integral form of the remainder is needed.

Consider the Maclaurin series for $f(x) = (1 + x)^r$, where r is *not* a positive integer or 0. The following table will help in computing $f^{(n)}(0)$.

n	$f^{(n)}(x)$	$f^{(n)}(0)$
0	$(1 + x)^r$	1
1	$r(1 + x)^{r-1}$	r
2	$r(r-1)(1 + x)^{r-2}$	$r(r-1)$
3	$r(r-1)(r-2)(1 + x)^{r-3}$	$r(r-1)(r-2)$
$\cdots$	$\cdots$	$\cdots$
n	$r(r-1) \cdots (r-n+1)(1 + x)^{r-n}$	$r(r-1)(r-2) \cdots (r-n+1)$

Consequently, the Maclaurin series associated with $(1 + x)^r$ is

$$1 + rx + \frac{r(r-1)}{1 \cdot 2} x^2 + \frac{r(r-1)(r-2)}{1 \cdot 2 \cdot 3} x^3 + \cdots. \quad (2)$$

The two key questions in this section

Note that the series does not stop, for r is not a positive integer or 0. For which x does series (2) converge? If it does converge, does it represent $(1 + x)^r$?

Just to get a feeling for (2), consider the case $r = -1$. When $r = -1$, (2) becomes

$$1 + (-1)x + \frac{(-1)(-2)}{1 \cdot 2} x^2 + \frac{(-1)(-2)(-3)}{1 \cdot 2 \cdot 3} x^3 + \cdots$$

or

$$1 - x + x^2 - x^3 + \cdots.$$

This series converges for $|x| < 1$. Moreover it does represent the function $(1 + x)^r = (1 + x)^{-1}$, for it is a geometric series with first term 1 and ratio $-x$.

EXAMPLE 1 Show that series (2) converges when $|x| < 1$.

SOLUTION For $x = 0$ the series clearly converges. So consider $0 < |x| < 1$. Let a_n be the term containing the power x^n. Then

$$a_n = \frac{r(r-1)(r-2)\cdots(r-n+1)}{1\cdot 2\cdot 3\cdots n} x^n,$$

and

$$a_{n+1} = \frac{r(r-1)(r-2)\cdots(r-n)}{1\cdot 2\cdot 3\cdots(n+1)} x^{n+1}.$$

Thus

$$\left|\frac{a_{n+1}}{a_n}\right| = \left|\frac{\dfrac{r(r-1)(r-2)\cdots(r-n)}{1\cdot 2\cdot 3\cdots(n+1)} x^{n+1}}{\dfrac{r(r-1)(r-2)\cdots(r-n+1)}{1\cdot 2\cdot 3\cdots n} x^n}\right|$$

$$= \left|\frac{r-n}{n+1} x\right|.$$

Since r is fixed, $\lim\limits_{n\to\infty}\left|\dfrac{a_{n+1}}{a_n}\right| = |x|$.

By the absolute ratio test, series (2) converges. ∎

The series (2) converges when $|x| < 1$, but does it converge to $f(x) = (1 + x)^r$? To show that (2) converges to $(1 + x)^r$, we must show that the remainder $R_n(x; 0)$ approaches 0 as $n \to \infty$. This will be done with the aid of the integral form of the remainder. Two lemmas will be needed.

LEMMA 1 Let x be a fixed number in $(-1, 1)$. Then if t is in the closed interval whose endpoints are 0 and x,

$$\left|\frac{x-t}{1+t}\right| \le |x|. \quad \blacksquare$$

The proof is outlined in Exercise 14.

LEMMA 2 Let f be continuous on an interval that includes a and b. Then

$$\left|\int_a^b f(x)\,dx\right| \le \left|\int_a^b |f(x)|\,dx\right|.$$

PROOF We prove the case $a < b$; the case $b < a$ can be obtained from it. For x in $[a, b]$,

$$-|f(x)| \le f(x) \le |f(x)|.$$

Thus
$$-\int_a^b |f(x)|\,dx \le \int_a^b f(x)\,dx \le \int_a^b |f(x)|\,dx.$$

Consequently, $\left|\int_a^b f(x)\,dx\right| \le \int_a^b |f(x)|\,dx = \left|\int_a^b |f(x)|\,dx\right|.$ ∎

To show that (2) converges to $(1+x)^r$, we must show that $R_n(x; 0)$ approaches 0 as $n \to \infty$.

By Sec. 10.8, $\quad R_n(x; 0) = \dfrac{1}{n!}\int_0^x f^{(n+1)}(t)(x-t)^n\,dt.$

Now, $f(x) = (1+x)^r$. Repeated differentiation shows that
$$f^{(n+1)}(t) = r(r-1)(r-2)\cdots(r-n)(1+t)^{r-n-1}.$$

Thus $\quad R_n(x; 0) = \dfrac{r(r-1)(r-2)\cdots(r-n)}{n!}\int_0^x \left(\dfrac{x-t}{1+t}\right)^n (1+t)^{r-1}\,dt. \quad (3)$

Consider x in $(-1, 1)$. By Lemmas 1 and 2,
$$\left|\int_0^x \left(\dfrac{x-t}{1+t}\right)^n (1+t)^{r-1}\,dt\right| \le \left|\int_0^x |x^n(1+t)^{r-1}|\,dt\right|$$
$$= |x|^n \left|\int_0^x (1+t)^{r-1}\,dt\right|.$$

Thus
$$|R_n(x; 0)| \le \left|\dfrac{r(r-1)(r-2)\cdots(r-n)}{n!}\right| |x|^n \left|\int_0^x (1+t)^{r-1}\,dt\right|. \quad (4)$$

The integral in (4) does not depend on n. The remaining part of (4) is simply the absolute value of a typical term in (2); hence it approaches 0 as $n \to \infty$. (Example 1 showed that (2) converges; therefore its nth term approaches 0 as $n \to \infty$.) Consequently $|R_n(x; 0)| \to 0$ as $n \to \infty$. This establishes the binomial theorem for arbitrary exponents and $|x| < 1$.

BINOMIAL THEOREM

If $|x| < 1$ and r is any real number, then
$$(1+x)^r = 1 + rx + \dfrac{r(r-1)}{2!}x^2 + \cdots + \dfrac{r(r-1)\cdots(r-n+1)}{n!}x^n + \cdots.$$

To put the binomial theorem in summation notation, introduce the generalized binomial coefficient

$$\binom{r}{n} = \frac{r(r-1)\cdots(r-n+1)}{n!}$$

for any real number r and nonnegative integer n.

Then
$$(1+x)^r = \sum_{n=0}^{\infty} \binom{r}{n} x^n \quad \text{for } |x| < 1.$$

EXAMPLE 2 Write out the first four terms of the binomial expansion of $1/\sqrt{1+x} = (1+x)^{-1/2}$, for $|x| < 1$.

SOLUTION In this case $r = -\frac{1}{2}$. Thus (2) becomes

$$(1+x)^{-1/2} = 1 + (-\tfrac{1}{2})x + \frac{(-\tfrac{1}{2})(-\tfrac{3}{2})}{2!} x^2 + \frac{(-\tfrac{1}{2})(-\tfrac{3}{2})(-\tfrac{5}{2})}{3!} x^3 + \cdots$$

$$= 1 - \frac{1}{2}x + \frac{1\cdot 3}{2^2 \cdot 2!} x^2 - \frac{1\cdot 3 \cdot 5}{2^3 \cdot 3!} x^3 + \cdots. \quad \blacksquare$$

EXAMPLE 3 Find the Maclaurin series for $1/\sqrt{1-x^2}$, if $|x| < 1$.

SOLUTION Replace x in $1/\sqrt{1+x}$ by $-x^2$. Then, by Example 2,

$$\frac{1}{\sqrt{1-x^2}} = (1-x^2)^{-1/2}$$

$$= 1 - \frac{1}{2}(-x^2) + \frac{1\cdot 3}{2^2 \cdot 2!}(-x^2)^2 - \frac{1\cdot 3 \cdot 5}{2^3 \cdot 3!}(-x^2)^3 + \cdots$$

$$= 1 + \frac{1}{2}x^2 + \frac{1\cdot 3}{2^2 \cdot 2!} x^4 + \frac{1\cdot 3 \cdot 5}{2^3 \cdot 3!} x^6 + \cdots. \quad \blacksquare$$

EXAMPLE 4 Write out the first four terms of the binomial series for $1/(1+x)^2$.

SOLUTION In this case $r = -2$. Thus

$$\frac{1}{(1+x)^2} = (1+x)^{-2}$$

$$= 1 + (-2)x + \frac{(-2)(-3)}{2!} x^2 + \frac{(-2)(-3)(-4)}{3!} x^3 + \cdots$$

$$= 1 - 2x + 3x^2 - 4x^3 + \cdots. \quad \blacksquare$$

Exercises

In Exercises 1 to 4 write out the first five terms of the binomial expansions of the given functions.

1 $(1+x)^{1/2}$ **2** $(1+x)^{1/3}$
3 $(1+x)^{-3}$ **4** $(1+x)^{-4}$

In Exercises 5 to 7 write the first four nonzero terms in the Maclaurin series for the given functions.

5 $(1+x^3)^{1/2}$ **6** $(1+x^2)^{1/3}$
7 $(1-x^2)^{1/3}$

8 Show that if $|x| > 1$ and r is not a nonnegative integer, the binomial series (2) does not converge.

9 Using the first four nonzero terms of the Maclaurin series for $\sqrt{1+x^3}$, estimate $\int_0^1 \sqrt{1+x^3}\,dx$, an integral that cannot be evaluated by the fundamental theorem of calculus.

Exercises 10 to 13 concern the Lagrange form of $R_n(x;0)$ for $f(x) = (1+x)^r$.

10 Let $f(x) = (1+x)^r$. Show that the Lagrange form of $R_n(x;0)$ is

$$R_n(x;0) = \frac{r(r-1)\cdots(r-n)}{(n+1)!}(1+c)^r\left(\frac{x}{1+c}\right)^{n+1}$$

where c is between 0 and x. Note that c depends both on x and n.

11 Let $0 < x < 1$. Using the formula in Exercise 10, show that $R_n(x;0) \to 0$ as $n \to \infty$.

12 Let $-\frac{1}{2} < x < 0$. Using the formula in Exercise 10, show that $R_n(x;0) \to 0$, following these steps:
(a) Show that $1+c \geq 1+x$.
(b) Show that $\left|\dfrac{x}{1+c}\right| \leq \dfrac{|x|}{1+x}$.
(c) Show that $|x|/(1+x) < \frac{1}{2}$.
(d) Show that $R_n(x;0) \to 0$ as $n \to \infty$.

13 Let $x = -\frac{1}{2}$. Explain why the Lagrange form of $R_n(x;0)$ is not of use in showing that $R_n(x;0)$ in Exercise 10 approaches 0 as $n \to \infty$. (The same trouble occurs for any x, $-1 < x \leq -1/2$.)

14 (Proof of Lemma 1) For x a fixed number in $(-1, 1)$ define the function g by $g(t) = (x-t)/(1+t)$.
(a) Show that if $0 \leq x < 1$ and $0 \leq t \leq x$, then $0 \leq g(t) \leq x$. (Show $g(t)$ decreases for $0 \leq t \leq x$.)
(b) Show that, if $-1 < x \leq 0$ and $x \leq t \leq 0$, then $x \leq g(t) \leq 0$.

(c) Combining (a) and (b), show that, if $|x| < 1$ and t is between 0 and x, then

$$\left|\frac{x-t}{1+t}\right| \leq |x|.$$

■ ■

Exercises 15 to 21 outline an argument due to Euler that

$$\frac{1}{1^2} + \frac{1}{2^2} + \frac{1}{3^2} + \cdots = \frac{\pi^2}{6}.$$

15 Show that, if

$$\frac{1}{1^2} + \frac{1}{3^2} + \frac{1}{5^2} + \frac{1}{7^2} + \cdots + \frac{1}{(2n-1)^2} + \cdots = \frac{\pi^2}{8},$$

then $\dfrac{1}{1^2} + \dfrac{1}{2^2} + \dfrac{1}{3^2} + \cdots + \dfrac{1}{n^2} + \cdots = \dfrac{\pi^2}{6}.$

16 Show that $\displaystyle\int_0^1 \frac{\sin^{-1} x}{\sqrt{1-x^2}}\,dx = \frac{\pi^2}{8}.$

17 Use Example 3 to show that, if $|t| < 1$, then

$$\frac{1}{\sqrt{1-t^2}} = 1 + \frac{1}{2}t^2 + \frac{1\cdot 3}{2\cdot 4}t^4 + \frac{1\cdot 3\cdot 5}{2\cdot 4\cdot 6}t^6 + \cdots.$$

18 (See Exercise 17.) Show that

$$\sin^{-1} x = x + \frac{1}{2}\frac{x^3}{3} + \frac{1\cdot 3}{2\cdot 4}\frac{x^5}{5} + \frac{1\cdot 3\cdot 5}{2\cdot 4\cdot 6}\frac{x^7}{7} + \cdots$$

for $|x| < 1$. This equation is also valid when $x = 1$ or -1.

19 Show that

$$\int_0^1 \frac{x^{2n+1}}{\sqrt{1-x^2}}\,dx = \int_0^{\pi/2} \sin^{2n+1}\theta\,d\theta.$$

20 Assuming that it is safe to integrate term by term, even in the case of an improper integral, show that

$$\int_0^1 \frac{\sin^{-1} x}{\sqrt{1-x^2}}\,dx = \frac{1}{1^2} + \frac{1}{3^2} + \cdots + \frac{1}{(2n-1)^2} + \cdots.$$

21 Deduce that $\displaystyle\sum_{n=1}^\infty n^{-2} = \frac{\pi^2}{6}.$

22 In *Introduction to Fluid Mechanics*, by Steven Whitaker, McGraw-Hill, the following argument appears in the discussion of flow through a nozzle:

The pressure p equals

$$\left(1 + \frac{\gamma - 1}{2} M^2\right)^{\frac{\gamma}{1-\gamma}}$$

By the binomial theorem and the fact that $v^2 = M^2 \gamma RT$,

$$p = 1 - \frac{1}{2} \frac{v^2}{RT} + \frac{\gamma(2\gamma - 1)}{8} M^4 + \cdots.$$

Fill in the steps. (γ is specific heat, which is about 1.4, and M is a Mach number, which is in the range 1 to 2.)

23 In R. P. Feynman, *Lectures on Physics*, Addison-Wesley, Reading, Mass., 1963, this statement appears in Sec. 15.8 of Vol. 1:

An approximate formula to express the increase of mass, for the case when the velocity is small, can be found by expanding $m_0/\sqrt{1 - v^2/c^2} = m_0(1 - v^2/c^2)^{-1/2}$ in a power series, using the binomial theorem. We get

$$m_0 \left(1 - \frac{v^2}{c^2}\right)^{-1/2} = m_0 \left(1 + \frac{1}{2} \frac{v^2}{c^2} + \frac{3}{8} \frac{v^4}{c^4} + \cdots\right).$$

We see clearly from the formula that the series converges rapidly when v is small and the terms after the first two or three are negligible.

Check the expansion and justify the equation.

10.13 Newton's method for solving an equation

Suppose that we wish to estimate a solution (or root) r of an equation $f(x) = 0$. If a first guess is, say, x_1, then Fig. 10.35 suggests that a better estimate of r may be x_2, the point at which the tangent line at $(x_1, f(x_1))$ crosses the x axis.

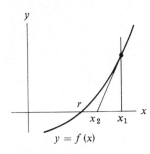

Figure 10.35

To find x_2 explicitly, observe that the slope of the tangent line at $(x_1, f(x_1))$ is $f(x_1)/(x_1 - x_2)$ and is also $f'(x_1)$. Equating these two forms of the slope and solving for x_2 yields

Newton's recursive formula for estimating a root of $f(x) = 0$

$$x_2 = x_1 - \frac{f(x_1)}{f'(x_1)}, \tag{1}$$

which is meaningful if $f'(x_1)$ is not 0.

Formula (1), suggested by the diagram above, is the basis of Newton's method for estimating a root of an equation. Generally, this formula is applied several times to increase accuracy.

EXAMPLE 1 Use Newton's method to estimate the square root of 3, that is, the positive root of the equation $x^2 - 3 = 0$.

SOLUTION Here $f(x) = x^2 - 3$ and $f'(x) = 2x$. According to (1), if the first guess is x_1, then the next estimate x_2 should be

$$x_2 = x_1 - \frac{f(x_1)}{f'(x_1)} = x_1 - \frac{x_1^2 - 3}{2x_1} = \frac{x_1 + 3/x_1}{2}.$$

If $x_1 = 2$, say, then

$$x_2 = \frac{2 + \frac{3}{2}}{2} = 1.75.$$

For a better estimate of $\sqrt{3}$ repeat the process, using 1.75 instead of 2. Thus

$$x_3 = \frac{x_2 + 3/x_2}{2} = \frac{1.75 + 3/1.75}{2} \approx 1.73214$$

is the third estimate, to five decimals. One more repetition of the process yields (to five decimals) $x_4 = 1.73205$, which is quite close to $\sqrt{3}$, whose decimal expansion begins 1.7320508. ∎

Since the recursive process represented by Newton's method is of practical use and easily programmed, it is important to know under what circumstances $|x_i - r|$ approaches 0 as $i \to \infty$. If the graph of f happens to be a nonhorizontal straight line, a quick sketch shows that no matter what choice of x_1 is made, the number x_2 is exactly the root r. In other words, if $f''(x)$ is identically 0, Newton's method is perfectly accurate. It is therefore reasonable to expect that the accuracy of Newton's method is influenced by $f''(x)$. (When $f''(x)$ is small, the method is probably more accurate.) On the other hand, if $f'(x_1)$ is near 0, the tangent line at $(x_1, f(x_1))$ is nearly horizontal, and may depart a great deal from the graph of f by the time it crosses the x axis. Hence $f'(x)$ also should influence the accuracy. (When $f'(x)$ is large, the method is probably more accurate).

The following theorem shows that, if $f''(x)$ is not too large nor $f'(x)$ too small, then $|x_i - r|$ does approach 0 as $i \to \infty$. Its proof is sketched in Exercise 17.

THEOREM Let r be a root of $f(x) = 0$ and x_i an estimate of r such that $f'(x_i)$ is not 0. Let

$$x_{i+1} = x_i - \frac{f(x_i)}{f'(x_i)}.$$

If f' and f'' are continuous and M is a number such that

$$\left|\frac{f''(x)}{f'(t)}\right| \le M$$

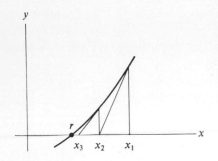

Figure 10.36

for all x and t in the interval from x_i to r, then

$$|x_{i+1} - r| \le \frac{M}{2}|x_i - r|^2. \quad \blacksquare \qquad (2)$$

If $f''(x), f'(x)$, and $f(x)$ are positive from $x = r$ to $x = x_1 > r$, then, as Fig. 10.36 shows, $x_1 > x_2 > x_3 > \cdots > r$. This means that the successive estimates $x_2, x_3, \ldots$ are situated between the initial estimate x_1 and the root r. Hence the hypotheses of the theorem apply to x_i for all i.

In this case how swiftly does the decreasing sequence $x_1, x_2, x_3, \ldots$ approach r? Notice that if x_1 is close to r, say $|x_1 - r| \le 0.1$, then

$$|x_2 - r| \le \frac{M}{2}(0.1)^2 = \frac{M}{2}(0.01).$$

Thus, if M is not too large, x_2 is a much better approximation to r than x_1 is. For instance, if $M = 2$, then

$$|x_2 - r| \le 0.01,$$

and

$$|x_3 - r| \le \frac{M}{2}(x_2 - r)^2 \le 0.0001,$$

and so on. Hence if x_1 is an estimate of r accurate to one decimal place, then x_2 is accurate to two decimal places, x_3 is accurate to four decimal places, and so on. The number of decimal places of accuracy tends to double at each step of the Newton recursion. For instance, the Newton recursion formula for $\sqrt{10}$ (≈ 3.162278) is

$$x_{i+1} = \frac{x_i + 10/x_i}{2}.$$

The following table shows the results of the recursive process when the initial estimate x_1 is 3.

Step	Estimate	Correct Digits	Number of Correct Decimal Digits
1	$x_1 = 3$	3	0
2	$x_2 = 3.166667$	3.16	2
3	$x_3 = 3.162281$	3.1622	4

EXAMPLE 2 The line $y = 2x/3$ crosses the curve $y = \sin x$ at a point P, whose x coordinate r is between 0 and π, as shown in Fig. 10.37. The number r is a solution of the equation $2x/3 = \sin x$, since the graphs have equal y coordinates at $x = r$. Use Newton's method to approximate r.

SOLUTION A glance at the graph in Fig. 10.37 suggests that r is approximately 1.5. To obtain a better estimate, note that r is a root of the equation

$$f(x) = \sin x - \frac{2x}{3} = 0.$$

Since $f'(x) = \cos x - \frac{2}{3}$, Newton's method provides this second estimate of r:

$$x_2 = 1.5 - \frac{f(1.5)}{f'(1.5)} = 1.5 - \frac{\sin 1.5 - 2(1.5)/3}{\cos 1.5 - \frac{2}{3}}.$$

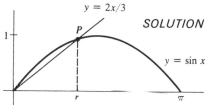

Figure 10.37

Angle in radians, not degrees

Computing $\sin 1.5$ and $\cos 1.5$, we find that

$$x_2 \approx 1.5 - \frac{0.9975 - 1}{0.0707 - 0.6667} = 1.5 - \frac{0.0025}{0.5960} \approx 1.496.$$

Incidentally, to three decimals, $r = 1.496$. ∎

Exercises

1 Let a be a positive number. Show that the Newton recursion formula for estimating $\sqrt{a}$ is given by

$$x_{i+1} = \frac{x_i + a/x_i}{2}.$$

2 Use the formula of Exercise 1 to estimate $\sqrt{15}$. Choose $x_1 = 4$ and compute x_2 and x_3 to three decimals.

3 Use the formula of Exercise 1 to estimate $\sqrt{19}$. Choose $x_1 = 4$ and compute x_2 and x_3 to three decimals.

4 In estimating $\sqrt{3}$, an electronic computer began with $x_1 = 50$. What does Newton's method give for x_2, x_3, and x_4?

5 (a) Show that Newton's method gives this recursion formula for estimating $\sqrt[3]{7}$:

$$x_{i+1} = \frac{2}{3}x_i + \frac{7}{3x_i^2}.$$

(b) Let $x_1 = 1$, and compute x_2 and x_3.
(c) Let $x_1 = 2$, and compute x_2 and x_3.

6 Let $f(x) = x^4 + x - 19$.
(a) Show that $f(2) < 0 < f(3)$ and that f must thus have a root r between 2 and 3.
(b) Apply Newton's method, starting with $x_1 = 2$. Compute x_2 and x_3.

7 Let $f(x) = x^5 + x - 1$.
(a) Show that there is exactly one root of the equation $f(x) = 0$ in the interval $[0, 1]$. (Examine f'.)
(b) Using $x_1 = \frac{1}{2}$ as a first estimate, apply Newton's method to find a second estimate x_2.

8 Let $f(x) = 2x^3 - x^2 - 2$.
(a) Show that there is exactly one root of the equation $f(x) = 0$ in the interval $[1, 2]$.
(b) Using $x_1 = \frac{3}{2}$ as a first estimate, apply Newton's method to find a second estimate x_2.

9 (a) Graph $y = e^x$ and $y = x + 2$ relative to the same axes.
(b) With the aid of (a), estimate a root of $e^x - x - 2 = 0$.
(c) Use Newton's method and a table of e^x or a calculator to estimate the root to two-decimal accuracy.

10 (a) Show that there is a number r in $[0, 1]$ such that $\ln(1 + r) = 1 - r$.
   (b) Show that there is only one such number r in $[0, 1]$.
   (c) Use Newton's method with $x_1 = 0.5$ to find x_2, a closer approximation to r.

■

11 (a) Graph $y = \ln x$ and $y = \sin x$ relative to the same axes.
   (b) With the aid of the graphs in (a), estimate the x coordinate of the point such that $\sin x - \ln x = 0$.
   (c) Using the estimate in (b) as x_1, find another estimate x_2 by Newton's method.
12 (a) Graph $y = x \sin x$ for x in $[0, \pi]$.
   (b) Using the first and second derivatives, show that it has a unique relative maximum in the interval $[0, \pi]$.
   (c) Show that the maximum value of $x \sin x$ occurs when $x \cos x + \sin x = 0$.
   (d) Use Newton's method, with $x_1 = \pi/2$, to find an estimate x_2 for a root of the equation $x \cos x + \sin x = 0$.
   (e) Use Newton's method again to find x_3.
13 (a) Graph $y = e^x$ and $y = \tan x$ relative to the same axes.
   (b) Show that the equation $e^x - \tan x = 0$ has a solution between 0 and $\pi/2$.
   (c) Choose x_1, an estimate of the solution in (b), on the basis of the graph in (a). Then determine x_2 by Newton's method.
14 (a) Show that the equation $3x + \sin x - e^x = 0$ has a root between 0 and 1.
   (b) Starting with $x_1 = 0.5$, compute x_2 and x_3, the estimates of the root in (a) by Newton's recursion.

■ ■

Exercises 15 and 16 show that care should be taken in applying Newton's method.

15 Let $f(x) = 2x^3 - 4x + 1$.
   (a) Show that $f(1) < 0 < f(0)$ and that there must be a root r of $f(x) = 0$ in $[0, 1]$.
   (b) Take $x_1 = 1$, and apply Newton's method to obtain x_2 and x_3, estimates of r.
   (c) Graph f, and show what is happening in the sequence of estimates.
16 Let $f(x) = x^2 + 1$.
   (a) Using Newton's method with $x_1 = 2$, compute x_2, x_3, x_4, and x_5 to two decimal places.
   (b) Using the graph of f, show geometrically what is happening in (a).
   (c) Using Newton's method with $x_1 = \sqrt{3}/3$, compute x_2 and x_3. What happens to x_n as $n \to \infty$?
   (d) Using Newton's method with $x_1 = (1 + \sqrt{2})/2$, examine x_2, x_3, and x_4.
17 This exercise justifies inequality (2). Recall that x_i is an approximation to a root r of the equation $f(x) = 0$.
   (a) Show that

$$0 = f(r) = f(x_i) + f'(x_i)(r - x_i) + f''(c)\frac{(r - x_i)^2}{2}$$

   for some c between x_i and r.
   (b) From (a) deduce that

$$x_{i+1} - r = \frac{(r - x_i)^2}{2} \frac{f''(c)}{f'(x_i)}.$$

   (c) From (b) deduce that

$$|x_{i+1} - r| \leq \frac{M}{2}|x_i - r|^2.$$

10.S Summary

Section 10.1 introduced the notion of a sequence. Sections 10.2 to 10.5 developed various tests for determining whether a series converges or diverges. These tests are collected in a table later in this section. Sections 10.6 to 10.8 concerned power series and their use in representing a function.

Sections 10.9 to 10.13, calling on ideas developed in Secs. 10.6 to 10.8, treated several independent topics. Section 10.9 introduced the complex numbers and the following section showed with the aid of power series that

$$\cos\theta = \frac{e^{i\theta} + e^{-i\theta}}{2} \quad \text{and} \quad \sin\theta = \frac{e^{i\theta} - e^{-i\theta}}{2i}.$$

Differential equations were treated in Sec. 10.11. Section 10.12, which depends on the integral form of the remainder, obtained the binomial theorem for exponents that are not positive integers. In Sec. 10.13 Newton's method for estimating a root of an equation was presented. That the method works in general was established by the use of the derivative form of the remainder. (See Exercise 17 in Sec. 10.13.) Sections 10.9 to 10.11 are not covered in this summary.

VOCABULARY AND SYMBOLS

sequence a_n or $\{a_n\}$
convergent, divergent sequence
limit of a sequence
series
partial sum S_n
convergent, divergent series
nth term of series
geometric series
nth-term test for divergence
integral test
p series
harmonic series
comparison test
limit-comparison test
ratio test
root test
alternating series
decreasing alternating series
alternating-series test
absolute convergence
conditional convergence
absolute-convergence test
absolute-ratio test
power series
Maclaurin series
radius of convergence
Taylor series $\sum_{n=0}^{\infty} f^{(n)}(a)(x-a)^n/n!$
Taylor polynomial $P_n(x; a)$
remainder (error) in Taylor series $R_n(x; a) = f(x) - P_n(x; a)$
complex number
magnitude and argument of a complex number
Newton's method $x_{i+1} = x_i - f(x_i)/f'(x_i)$

KEY FACTS

If $|r| < 1$, $r^n \to 0$ as $n \to \infty$.

$x^n/n! \to 0$ as $n \to \infty$.

$\sum_{n=0}^{\infty} ax^n$ converges to $a/(1-x)$ if $|x| < 1$.

A p series converges if $p > 1$ and diverges if $p \leq 1$.

Associated with a power series $\sum_{n=0}^{\infty} a_n(x-a)^n$ is its radius of convergence, R. (R may be 0, a positive number, or infinite.) The series converges for $|x - a| < R$ and diverges for $|x - a| > R$.

Within its radius of convergence a power series may be differentiated or integrated term by term.

The binomial series

$$1 + rx + \frac{r(r-1)}{2!}x^2 + \frac{r(r-1)(r-2)}{3!}x^3 + \cdots$$

converges to $(1+x)^r$ for $|x| < 1$ and any exponent r. The series has only a finite number of nonzero terms only if r is a nonnegative integer.

Tests for Convergence (or Divergence) of $\sum_{n=1}^{\infty} a_n$

Name	Brief Formulation	When Used
1. nth term test for divergence	If a_n does *not* approach 0 as $n \to \infty$, $\sum_{n=1}^{\infty} a_n$ diverges.	When you suspect a_n does not $\to 0$.
2. Integral test	If $f(x) > 0$ decreases and $\int_a^{\infty} f(x)\,dx$ converges, $\sum_{n=1}^{\infty} a_n$ converges. If $\int_a^{\infty} f(x)\,dx$ diverges, $\sum_{n=1}^{\infty} a_n$ diverges.	When you have a positive decreasing series $a_n = f(n)$ and $\int f(x)\,dx$ is easy to calculate.
3. Comparison test	If $0 \le a_n \le c_n$ and $\sum_{n=1}^{\infty} c_n$ converges, so must $\sum_{n=1}^{\infty} a_n$. If $a_n \ge c_n \ge 0$ and $\sum_{n=1}^{\infty} c_n$ diverges, so must $\sum_{n=1}^{\infty} a_n$.	When you have a positive series that can be compared to a series known to converge or diverge.
4. Limit-comparison test	If $a_n/c_n \to$ nonzero limit, then $\sum_{n=1}^{\infty} a_n$ converges if $\sum_{n=1}^{\infty} c_n$ does and diverges if $\sum_{n=1}^{\infty} c_n$ does.	When you have a positive series very much like a series known to converge or diverge.
5. Ratio test	If $a_{n+1}/a_n \to L < 1$, $\sum_{n=1}^{\infty} a_n$ converges. If $L > 1$, it diverges. If $L = 1$, no information.	When you have a positive series in which a_n involves powers or factorials.
6. Decreasing-alternating-series test	A decreasing alternating series whose nth term $\to 0$ converges.	When you have an alternating series whose terms diminish in absolute value from some point on and approach 0.
7. Absolute-convergence test	If $\sum_{n=1}^{\infty} |a_n|$ converges, so does $\sum_{n=1}^{\infty} a_n$.	When you feel that the series would converge even if its terms were all made positive.
8. Absolute-ratio test	If $|a_{n+1}/a_n| \to L < 1$, $\sum_{n=1}^{\infty} a_n$ converges (absolutely). If $|a_{n+1}/a_n| \to L > 1$, $\sum_{n=1}^{\infty} a_n$ diverges.	Especially suitable for power series.
9. Root test	If $\sqrt[n]{|a_n|} \to L < 1$, $\sum_{n=1}^{\infty} a_n$ converges.	Usually ratio test is preferable. However, if something like n^n appears, try root test.

Note that test 1 can only be used to show divergence; tests 6 and 7 can be used only to show convergence. Tests 2, 3, 4, and 5 are only for series with positive terms, but can be used to show absolute convergence.

ESTIMATING THE ERROR Assume that $\sum_{n=1}^{\infty} a_n = S$. Let $\sum_{i=1}^{n} a_i = S_n$. The error or remainder R_n is defined to be the difference $S - S_n$.

If $\sum_{n=1}^{\infty} a_n$ is a decreasing alternating series whose nth term approaches 0, then

$$|R_n| < |a_{n+1}|.$$

If $\sum_{n=1}^{\infty} a_n$ is a positive series to which the integral test applies, then

$$0 < R_n < \int_n^{\infty} f(x)\, dx.$$

If $\sum_{n=1}^{\infty} a_n$ is a convergent geometric series, $a_n = ar^{n-1}$, then

$$R_n = \frac{ar^n}{1-r}.$$

In the case of the Maclaurin series associated with the function f,

$$\sum_{n=0}^{\infty} \frac{f^{(n)}(0)x^n}{n!},$$

the error $R_n(x; 0)$ is the difference between $f(x)$ and the sum of the first $n+1$ terms,

$$R_n(x; 0) = f(x) - \sum_{k=0}^{n} \frac{f^{(k)}(0)x^k}{k!}.$$

There are two formulas for $R_n(x; 0)$:

$$R_n(x; 0) = \frac{1}{n!} \int_0^x f^{(n+1)}(t)(x-t)^n\, dt \qquad \text{(integral form)}$$

and $R_n(x; 0) = \dfrac{f^{(n+1)}(c)x^{n+1}}{(n+1)!},$ c between 0 and x (derivative or Lagrange form)

The second formula is easier to use and is strong enough to deal with e^x, $\sin x$, and $\cos x$. It is not strong enough for $(1+x)^r$, however. (See Sec. 10.12.) Similar formulas hold for Taylor series in powers of $x - a$. (See Sec. 10.8.)

Function	Maclaurin Series	Reference
e^x	$1 + x + \dfrac{x^2}{2!} + \dfrac{x^3}{3!} + \cdots = \sum_{n=0}^{\infty} \dfrac{x^n}{n!}$	Example 2, Sec. 10.8
$\sin x$	$x - \dfrac{x^3}{3!} + \dfrac{x^5}{5!} - \dfrac{x^7}{7!} + \cdots = \sum_{n=0}^{\infty} (-1)^n \dfrac{x^{2n+1}}{(2n+1)!}$	Secs. 10.7 and 10.8
$\cos x$	$1 - \dfrac{x^2}{2!} + \dfrac{x^4}{4!} - \dfrac{x^6}{6!} + \cdots = \sum_{n=0}^{\infty} (-1)^n \dfrac{x^{2n}}{(2n)!}$	Secs. 10.7 and 10.8
$\dfrac{1}{1-x}$	$1 + x + x^2 + \cdots = \sum_{n=0}^{\infty} x^n,\ |x| < 1$	Geometric series, Sec. 10.2
$\ln(1+x)$	$x - \dfrac{x^2}{2} + \dfrac{x^3}{3} - \cdots = \sum_{n=1}^{\infty} (-1)^{n+1} \dfrac{x^n}{n},\ -1 < x \leq 1$	Example 3, Sec. 10.7
$\tan^{-1} x$	$x - \dfrac{x^3}{3} + \dfrac{x^5}{5} - \cdots = \sum_{n=0}^{\infty} (-1)^n \dfrac{x^{2n+1}}{2n+1},\ |x| \leq 1$	Exercise 29, Sec. 10.7
$\sin^{-1} x$	$x + \dfrac{1}{2}\dfrac{x^3}{3} + \dfrac{1 \cdot 3}{2 \cdot 4}\dfrac{x^5}{5} + \dfrac{1 \cdot 3 \cdot 5}{2 \cdot 4 \cdot 6}\dfrac{x^7}{7} + \cdots,\ |x| \leq 1$	Exercise 18, Sec. 10.12
$(1+x)^r$	$1 + rx + \dfrac{r(r-1)}{2!}x^2 + \dfrac{r(r-1)(r-2)}{3!}x^3 + \cdots,\ |x| < 1$	General binomial theorem, Sec. 10.12

USES OF POWER SERIES

COMPUTATIONAL   A partial sum of a power series provides a numerical approximation of its sum.

THEORETICAL   The formula for the remainder is of use in analyzing numerical procedures. The formula for $R_1(x; 0)$ was used to show when Newton's method works. (Exercise 17 in Sec. 10.13.)   The formula for $R_3(x; 0)$ was used to estimate the size of

$$\int_{-1}^{1} f(x)\, dx - f\left(-\dfrac{1}{\sqrt{3}}\right) - f\left(\dfrac{1}{\sqrt{3}}\right). \quad \text{(Exercise 42 in Sec. 10.8.)}$$

The power series for e^x, $\cos x$, and $\sin x$ were used in Sec. 10.10 to obtain the fundamental equation $e^{i\theta} = \cos \theta + i \sin \theta$.

Incidentally, the series for e^x is not used in a calculator to estimate e^x since the series converges too slowly when $|x|$ is large. Instead, key values of e^x are stored (such as e^1, $e^{0.1}$, $e^{0.01}$, $e^{0.001}$, $e^{0.0001}$) and the basic law of exponents is exploited.   For instance, to compute $e^{2.3}$ a calculator may essentially write $e^{2.3} = e^1 e^1 e^{0.1} e^{0.1} e^{0.1}$ and multiply.

Guide quiz on chap. 10

1. A Maclaurin series converges for $x = 2$. For which other values of x must it converge?
2. The series $\sum_{n=0}^{\infty} a_n(x - 2)^n$ diverges for $x = 5$. For which other values of x must it diverge?
3. Give an example of a Maclaurin series that converges for $|x| < 1$ and no other values of x.
4. Estimate $\int_0^1 \sin x^2 \, dx$ to two decimal places.
5. State the integral and derivative forms of the remainder in a Taylor series, (a) $R_n(x; 0)$, (b) $R_n(x; a)$.
6. Give the Maclaurin series and their radii of convergence for the following functions.
   (a) $\dfrac{1}{1 + x^2}$ (b) e^{-x} (c) $\cos x$
   (d) $\ln(1 - x)$ (e) $1/(1 - 2x)$
7. Assume that $f(x)$ is represented by a Maclaurin series for some open interval $(-R, R)$, $f(x) = \sum_{n=0}^{\infty} a_n x^n$. Explain why $a_n = f^{(n)}(0)/n!$.
8. Determine for which x the following series converge or diverge.
   (a) $\sum_{n=1}^{\infty} \dfrac{n}{1 + n^2}(x - 2)^n$ (b) $\sum_{n=0}^{\infty} \dfrac{2^n}{n!}(x + 3)^n$
9. Using a picture, obtain Newton's recursive formula for finding a root of an equation.

Review exercises for chap. 10

1. Give the formula for the nth nonzero term of the Maclaurin series for
   (a) $\tan^{-1} 3x$ (b) $\ln(1 + x^2)$ (c) e^{-x}
   (d) $\sin x^2$ (e) $\cos x^2$.

   In some cases it may be simpler to start at $n = 0$.

2. (a) Show that
   $$\sum_{n=0}^{\infty} \frac{\cos(2n + 1)t}{(2n + 1)^2}$$
   converges for all t.
   (b) In the theory of Fourier series it is shown that, for $0 \leq t < \pi$, the sum of the series in (a) is $(\pi^2 - 2\pi t)/8$.
   Deduce that $\dfrac{\pi^2}{8} = \dfrac{1}{1^2} + \dfrac{1}{3^2} + \dfrac{1}{5^2} + \cdots$.

3. A text on the economics of equipment maintenance contains this argument:

   Adding the discounted net benefits over all the cycles yields
   $$B(N) = \sum_{n=1}^{\infty} e^{-(n-1)(N+K)\alpha}\left[\int_0^N e^{-\alpha t}b(t)\,dt - e^{-\alpha N}c(N)\right].$$

   Summing the geometric series gives
   $$B(N) = \frac{\int_0^N e^{-\alpha t}b(t)\,dt - e^{-\alpha N}c(N)}{1 - e^{-\alpha(N + K)}}.$$

   Sum the geometric series to check that the missing algebra is correct.

4. Obtain the first three nonzero terms of the Maclaurin series for $\sin 2x$:
   (a) by replacing x by $2x$ in the Maclaurin series for $\sin x$;
   (b) by using the formula $a_n = f^{(n)}(0)/n!$;
   (c) by using the identity $\sin 2x = 2 \sin x \cos x$ and the Maclaurin series for $\sin x$ and $\cos x$.

5. Obtain the first three nonzero terms of the Maclaurin series for $\sin^2 x$:
   (a) by using the formula $a_n = f^{(n)}(0)/n!$;
   (b) by using the identity $\sin^2 x = (1 - \cos 2x)/2$ and the series for $\cos 2x$.

6. An engineer wishes to use a partial sum of the Maclaurin series for e^x to approximate e^x. How many terms of the series should she use to be sure that the error is less than
   (a) 0.01 for $|x| \leq 1$? (b) 0.001 for $|x| \leq 1$?
   (c) 0.01 for $|x| \leq 2$? (d) 0.001 for $|x| \leq 2$?

In each of Exercises 7 to 31 determine whether the series converges or diverges. If you can give the sum of the series, do so.

7. $\sum_{n=0}^{\infty} e^{-n}$

8. $\sum_{n=1}^{\infty} \sin \dfrac{1}{n}$

9. $\sum_{n=1}^{\infty} \dfrac{1 + (-1)^n}{n^2}$

10. $\sum_{n=1}^{\infty} \dfrac{n^3}{2^n}$

11. $\sum_{n=1}^{\infty} \dfrac{\cos^3 n}{n^2}$

12. $\sum_{n=0}^{\infty} (-1)^n \dfrac{\pi^{2n+1}}{2^{2n+1}(2n + 1)!}$

13. $\sum_{n=1}^{\infty} \dfrac{1}{n\sqrt{n}}$

14. $\sum_{n=0}^{\infty} (-\tfrac{3}{4})^n$

15. $\sum_{n=1}^{\infty} (-1)^n \ln\left(\dfrac{n+1}{n}\right)$

16. $\sum_{n=1}^{\infty} \dfrac{(-2)^n}{n}$

17. $\sum_{n=1}^{\infty} n \sin \dfrac{1}{n}$

18. $\sum_{n=0}^{\infty} \dfrac{5n^2 - 3n + 1}{2n^3 + n^2 - 1}$

19. $\sum_{n=0}^{\infty} \dfrac{5n^3 + 6n + 1}{n^5 + n^3 + 2}$

20. $\sum_{n=1}^{\infty} \ln\left(\dfrac{n+1}{n}\right)$

21. $\sum_{n=1}^{\infty} \dfrac{2^{-n}}{n}$

22. $\sum_{n=0}^{\infty} (-1)^n \dfrac{\pi^{2n}}{(2n)!}$

23. $\sum_{n=0}^{\infty} \dfrac{10^n}{n!}$

24. $\sum_{n=1}^{\infty} \dfrac{\ln n}{n}$

25. $\sum_{n=0}^{\infty} \dfrac{(-1)^n (\tfrac{1}{2})^n}{n!}$

26. $\sum_{n=1}^{\infty} \dfrac{\sqrt{n+1} - \sqrt{n}}{n}$

27. $\sum_{n=0}^{\infty} \dfrac{n+2}{n+1} \left(\dfrac{2}{3}\right)^n$

28. $\sum_{n=1}^{\infty} \dfrac{n^2}{1+n^3}$

29. $\sum_{n=1}^{\infty} \dfrac{n \cos n}{1+n^4}$

30. $\sum_{n=1}^{\infty} \dfrac{(-1)^n n^2}{(2n)!}$

31. $\sum_{n=1}^{\infty} \dfrac{n-3}{n\sqrt{n}}$

In each of Exercises 32 to 35 use the first three nonzero terms of a Maclaurin series to estimate the quantity. Also put an upper bound on the error.

32. $\sqrt[10]{e}$
33. $\cos 1/3$
34. $\sin 28°$
35. $1/\sqrt{e}$

In each of Exercises 36 to 43 determine for which x the series diverges, converges absolutely, and converges conditionally. Give the radius of convergence in each case and the sum of the series if it is easily determined.

36. $\sum_{n=1}^{\infty} \dfrac{2^n x^n}{n}$

37. $\sum_{n=1}^{\infty} n x^{n-1}$

38. $\sum_{n=0}^{\infty} \dfrac{(x-3)^n}{n!}$

39. $\sum_{n=0}^{\infty} \dfrac{x^{2n}}{n!}$

40. $\sum_{n=1}^{\infty} (-n)^n x^n$

41. $\sum_{n=1}^{\infty} \dfrac{x^n}{n}$

42. $\sum_{n=0}^{\infty} \dfrac{3^n (x-\tfrac{2}{3})^n}{4^n}$

43. $\sum_{n=0}^{\infty} \dfrac{n^5 + 2}{n^3 + 1}(x+1)^n$

In each of Exercises 44 to 53 determine the radius of convergence and for which x the series converges or diverges.

44. $\sum_{n=1}^{\infty} \dfrac{(n!)^2 x^n}{n^{2n}}$

45. $\sum_{n=1}^{\infty} \dfrac{2^n (x-1)^n}{[1+(1/n)]^n}$

46. $\sum_{n=1}^{\infty} (n^2 x)^n$

47. $\sum_{n=0}^{\infty} \dfrac{(2n+1)x^n}{n!}$

48. $\sum_{n=1}^{\infty} \dfrac{(-1)^n n^2 (x-3)^n}{n^3 + 1}$

49. $\sum_{n=2}^{\infty} \dfrac{(x-1)^n}{\ln n}$

50. $\sum_{n=2}^{\infty} (\ln n)(x+2)^n$

51. $\sum_{n=1}^{\infty} \dfrac{(x+3)^n}{\sqrt[3]{n}}$

52. $\sum_{n=1}^{\infty} \dfrac{(-1)^n x^{2n}}{\sqrt{n}}$

53. $\sum_{n=1}^{\infty} \left(\dfrac{n+1}{n+3}\right)^{n^2} x^n$

54. If $\sum_{n=0}^{\infty} a_n (x-2)^n$ converges for $x = 7$ and diverges for $x = -3$, determine its radius of convergence.

In Exercises 55 to 57 determine the limits, using power series. It might be instructive to solve the problems by l'Hôpital's rule also.

55. $\lim_{x \to 0} \dfrac{\ln(1+x^2) - \sin^2 x}{\tan x^2}$

56. $\lim_{x \to 0} \dfrac{(e^{x^2} - 1)^2}{1 - x^2/2 - \cos x}$

57. $\lim_{x \to 0} \dfrac{(1 - \cos x^2)^5}{(x - \sin x)^{20}}$

58. Put an upper bound on the error in using the sum of the first 1000 terms to estimate the sum of the series:

    (a) $\sum_{n=1}^{\infty} n^{-2}$ (b) $\sum_{n=1}^{\infty} (-1)^n n^{-2}$ (c) $\sum_{n=1}^{\infty} (n!)^{-1}$.

59. Estimate or compute exactly

    $$\sum_{n=1}^{\infty} (-2)^{-n} - \sum_{n=1}^{10} (-2)^{-n}$$

    (a) by noticing that the series is alternating decreasing, (b) by noticing that the absolute value of the error is less than $\sum_{n=11}^{\infty} 2^{-n}$, (c) by noticing that $\sum_{n=1}^{\infty} (-2)^{-n}$ is a geometric series, (d) by considering $f(x) = (1+x)^{-1}$ for $x = \tfrac{1}{2}$ and using Lagrange's form of the remainder.

60. Estimate the positive root of the equation $e^x = 2x + 1$

    (a) using a Taylor polynomial of degree 3 to approximate e^x,
    (b) using Newton's method for the function $e^x - 2x - 1$.

61. Express $x^2 + x + 2$ as a polynomial in powers of $x - 5$.

62. In each of the following integrals, use the first three nonzero terms of the Maclaurin series for the integrand to estimate the integral. Put an upper bound on the size of the error.

    (a) $\int_0^{1/2} \cos x^3 \, dx$ (b) $\int_0^1 \sin x^2 \, dx$

    (c) $\int_0^{1/2} \sqrt[3]{1+x^3} \, dx$ (d) $\int_1^2 e^{-x^3} \, dx$

63 (a) Show that the equation $2x^5 - 10x + 5 = 0$ has precisely three real roots.
   (b) Show that one of the roots in (a) is between 0 and 1.
   (c) Use Newton's method to estimate the root in (b) to two-decimal accuracy.
   Galois, early in the nineteenth century, proved that the root in (c) cannot be expressed in terms of square roots, cube roots, fourth roots, fifth roots, etc. (His theorem applies to any polynomial with integer coefficients such that (i) its degree is a prime $p \geq 5$, (ii) it is not the product of two polynomials of lower degree with integer coefficients, and (iii) exactly $p - 2$ of its roots are real.)

64 Show that $x^3 - 3x - 3$ has exactly one real root and use Newton's method to estimate it to two decimal places.

65 Show that $x^3 + x - 6$ has exactly one real root and use Newton's method to estimate it to two decimal places.

∎

66 For $|x| \leq 1$, $\tan^{-1} x = \sum_{n=0}^{\infty} (-1)^n x^{2n+1}/(2n+1)$. Thus $\pi/4 = 1 - \frac{1}{3} + \frac{1}{5} - \frac{1}{7} + \cdots$, which converges slowly. The identity
$$\frac{\pi}{4} = 4\tan^{-1}\frac{1}{5} - \tan^{-1}\frac{1}{70} + \tan^{-1}\frac{1}{99} \quad (1)$$
provides a faster way to estimate π.
   (a) Using the first three nonzero terms of the Maclaurin series for $\tan^{-1} x$ and identity (1), estimate $\pi/4$ and hence π.
   (b) Discuss the size of the error in (a) in the estimate of $\pi/4$.
   (c) How would you show that (1) is true? (Describe a method, but don't carry out the computations.)

67 Let $f(x) = \sum_{n=0}^{\infty} 2^n x^n$. Find $f^{(33)}(0)$.

68 Give an example of a Maclaurin series whose radius of convergence is 1 and which
   (a) converges at 1 and -1;
   (b) diverges at 1 and -1;
   (c) converges at 1 and diverges at -1.

69 Show that, if $\sum_{n=0}^{\infty} a_n 3^n$ converges, then so does $\sum_{n=0}^{\infty} n a_n 2^n$.

70 Find $f^{(99)}(0)$, $f^{(100)}(0)$, and $f^{(101)}(0)$ if $f(x)$ is
   (a) $\tan^{-1} x$ (b) e^{x^2}.

71 A tennis ball, when dropped on a concrete surface, rebounds 80 percent of the distance that it falls. If a tennis ball is dropped from a height of 6 feet it bounces infinitely often. Is the total time for these bounces finite or infinite? (An object falling from rest drops $16t^2$ feet in t seconds. The time of each rise equals the time of the following descent.)

72 A certain function f has $f(0) = 3, f^{(1)}(0) = 2, f^{(2)}(0) = 5, f^{(3)}(0) = \frac{1}{2}$, and $f^{(j)}(x) = 0$ if $j > 3$. Give an explicit formula for $f(x)$.

73 (a) Show that, if $\{a_n\}$ is a sequence of positive terms and $\sum_{n=1}^{\infty} a_n$ converges, so does $\sum_{n=1}^{\infty} a_n^2$.
   (b) Give an example of a sequence $\{a_n\}$ such that $\sum_{n=1}^{\infty} a_n$ converges but $\sum_{n=1}^{\infty} a_n^2$ does not.

74 If $\sum_{n=0}^{\infty} a_n x^n = \sum_{n=0}^{\infty} b_n x^n$ for all x in $(-1, 1)$, must $a_n = b_n$?

75 Give the formula for the nth nonzero term of the Maclaurin series for
   (a) $\tan^{-1} x$ (b) $\ln(1 + x^2)$ (c) e^{-x^2}
   (d) $\sin x^2$ (e) $\cos x$.

∎ ∎

76 (a) Show that, if $\sum_{n=1}^{\infty} a_n^2$ and $\sum_{n=1}^{\infty} b_n^2$ converge, so does $\sum_{n=1}^{\infty} a_n b_n$.
   (b) Show that, if $\sum_{n=1}^{\infty} a_n^2$ converges, so does $\sum_{n=1}^{\infty} a_n/n$.

77 Prove that $\sin x$ cannot be written as a polynomial $P(x)$, even if we demand that $\sin x = P(x)$ only throughout a small interval $[a, b]$.

78 The zeta function $\zeta(p)$ is defined for $p > 1$ as $\zeta(p) = \sum_{n=1}^{\infty} n^{-p}$.
   (a) Examine $\lim_{p \to 1^+} \zeta(p)$.
   (b) Show that $(p-1)^{-1} < \zeta(p) < p(p-1)^{-1}$.
   (c) Show that $\lim_{p \to 1^+} \zeta(p)(p-1) = 1$.

79 Let f be a function having continuous $f^{(1)}, f^{(2)}$, and $f^{(3)}$ for all x. Assume that $\lim_{x \to \infty} f(x) = 1$ and $\lim_{x \to \infty} f^{(3)}(x) = 0$. Prove that $\lim_{x \to \infty} f^{(1)}(x) = 0 = \lim_{x \to \infty} f^{(2)}(x)$. Hint: Express $f(a+1)$ and $f(a-1)$ in terms of derivatives of f at a.

80 Let f be defined for all x and have a continuous $f^{(1)}$ and $f^{(2)}$. Prove that, if $|f(x)| \leq 1$ and $|f^{(2)}(x)| \leq 1$ for all x in $[0, 2]$, then $|f^{(1)}(x)| \leq 2$ for all x in $[0, 2]$. Hint: Express both $f(0)$ and $f(2)$ in terms of derivatives of f at x.

81 Assume that $f, f^{(1)}$, and $f^{(2)}$ are continuous and $f^{(2)}(a) \neq 0$. By the mean-value theorem, $f(a+h) = f(a) + hf'(a + \theta h)$ for some θ in $[0, 1]$.
   (a) When h is small, why is θ unique?
   (b) Prove that $\theta \to \frac{1}{2}$ as $h \to 0$.

82 Though $f'(a)$ is the limit of $[f(a + \Delta x) - f(a)]/\Delta x$, there is a better way to estimate $f'(a)$ than by that quotient. Assume that $f, f^{(1)}, f^{(2)}$, and $f^{(3)}$ are continuous. Show that
   (a) $\dfrac{f(a + \Delta x) - f(a)}{\Delta x} = f'(a) + \dfrac{f^{(2)}(c_1)}{2}\Delta x$
   for some c_1 between a and $a + \Delta x$, and
   (b) $\dfrac{f(a + \Delta x) - f(a - \Delta x)}{2\Delta x} = f'(a) + \dfrac{f^{(3)}(c_2)}{6}(\Delta x)^2$,

where c_2 is in $[a - \Delta x, a + \Delta x]$. [Since the error in using the quotient in (b) involves $(\Delta x)^2$, while the error in using the standard quotient involves Δx, the quotient in (b) is more accurate when Δx is small.] Test this observation on the function $y = x^3$ at $a = 2$.

83 (a) Using the inequality $e^x > 1 + x$ for $x > 0$, prove this theorem: If $u_1, u_2, u_3, \ldots$ is a sequence of positive numbers such that the sequence of sums $u_1, u_1 + u_2, u_1 + u_2 + u_3 + , \ldots$ has a limit, then the sequence $(1 + u_1), (1 + u_1)(1 + u_2), (1 + u_1)(1 + u_2)(1 + u_3), \ldots$ also has a limit.
(b) Prove the converse.

84 Consider $\int_0^b xe^{-x}\,dx$, when b is a small positive number. Since e^{-x} is then close to $1 - x$, the definite integral behaves like $\int_0^b (x - x^2)\,dx = b^2/2 - b^3/3$, hence approximately like $b^2/2$. On the other hand, $\int_0^b xe^{-x}\,dx = 1 - e^{-b}(1 + b)$, and, since e^{-b} is approximately $1 - b$, we have $1 - e^{-b}(1 + b)$ approximately equal to $1 - (1 - b)(1 + b) = b^2$. Hence $\int_0^b xe^{-x}\,dx$ behaves like b^2. Which is correct, $b^2/2$ or b^2? Find the error.

In Exercises 85 to 90 a short formula for estimating $n!$ is obtained.

85 Let f have the properties that, for $x \geq 1$, $f(x) \geq 0$, $f'(x) > 0$, and $f''(x) < 0$. Let a_n be the area of the region below the graph of $y = f(x)$ and above the line segment that joins $(n, f(n))$ with $(n + 1, f(n + 1))$.
(a) Draw a large-scale version of Fig. 10.38. The individual regions of areas a_1, a_2, a_3, and a_4 should be clear and not too narrow.

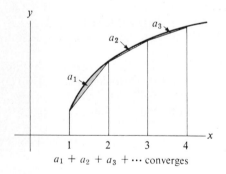

$a_1 + a_2 + a_3 + \cdots$ converges

Figure 10.38

(b) Using geometry, show that the series $a_1 + a_2 + a_3 + \cdots$ converges and has a sum no larger than the area of the triangle with vertices $(1, f(1))$, $(2, f(2))$, $(1, f(2))$.

86 Let $y = \ln x$.
(a) Using Exercise 85 show that, as $n \to \infty$,
$$\int_1^n \ln x\,dx - \left(\frac{\ln 1 + \ln 2}{2} + \frac{\ln 2 + \ln 3}{2} + \cdots + \frac{\ln(n-1) + \ln n}{2} \right)$$
has a limit; denote this limit by C.
(b) Show that (a) is equivalent to the assertion
$$\lim_{n \to \infty} (n \ln n - n + 1 - \ln n! + \ln \sqrt{n}) = C.$$

87 From Exercise 86(b) deduce that there is a constant k such that
$$\lim_{n \to \infty} \frac{n!}{k(n/e)^n \sqrt{n}} = 1.$$

Exercises 88 and 89 are related. Review Exercise 41 of Sec. 7.2 first.

88 Let $I_n = \int_0^{\pi/2} \sin^n x\,dx$.
(a) Evaluate I_0 and I_1.
(b) Show that
$$I_{2n} = \frac{2n - 1}{2n} \frac{2n - 3}{2n - 2} \cdots \frac{3}{4} \frac{1}{2} \frac{\pi}{2};$$
and
$$I_{2n+1} = \frac{2n}{2n + 1} \frac{2n - 2}{2n - 1} \cdots \frac{4}{5} \frac{2}{3}.$$
(c) Show that
$$\frac{I_7}{I_6} = \frac{6}{7} \frac{6}{5} \frac{4}{5} \frac{4}{3} \frac{2}{3} \frac{2}{1} \frac{2}{\pi}.$$
(d) Show that
$$\frac{I_{2n+1}}{I_{2n}} = \frac{2n}{2n+1} \frac{2n}{2n-1} \frac{2n-2}{2n-1} \cdots \frac{2}{3} \frac{2}{1} \frac{2}{\pi}.$$
(e) Show that
$$\frac{2n}{2n+1} I_{2n} < \frac{2n}{2n+1} I_{2n-1} = I_{2n+1} < I_{2n},$$
and thus
$$\lim_{n \to \infty} \frac{I_{2n+1}}{I_{2n}} = 1.$$
(f) From (d) and (e) deduce that
$$\lim_{n \to \infty} \frac{2 \cdot 2}{1 \cdot 3} \frac{4 \cdot 4}{3 \cdot 5} \frac{6 \cdot 6}{5 \cdot 7} \cdots \frac{(2n)(2n)}{(2n-1)(2n+1)} = \frac{\pi}{2}.$$
This is Wallis's formula, usually written in shorthand as
$$\frac{2 \cdot 2}{1 \cdot 3} \frac{4 \cdot 4}{3 \cdot 5} \frac{6 \cdot 6}{5 \cdot 7} \cdots = \frac{\pi}{2}.$$

89. (a) Show that $2 \cdot 4 \cdot 6 \cdot 8 \cdots 2n = 2^n n!$.
    (b) Show that $1 \cdot 3 \cdot 5 \cdots (2n-1) = (2n)!/(2^n n!)$.
    (c) From Exercise 88 deduce that
    $$\lim_{n \to \infty} \frac{(n!)^2 4^n}{(2n)! \sqrt{2n+1}} = \sqrt{\frac{\pi}{2}}.$$

90. (a) Using Exercise 89(c), show that k in Exercise 87 equals $\sqrt{2\pi}$. Thus a good estimate of $n!$ is provided by the formula
    $$n! \sim \sqrt{2\pi n}(n/e)^n$$
    This is known as Stirling's formula.
    (b) Using the factorial key on a calculator, compute $20!$. Then compute the ratio $\sqrt{2\pi n}(n/e)^n/n!$ for $n = 20$.

91. Determine for which x the series $\sum_{n=1}^{\infty} n^n x^n / n!$ converges or diverges. (Stirling's formula, which asserts that
    $$\lim_{n \to \infty} n!/\sqrt{2\pi n}(n/e)^n = 1,$$
    may be of use in part of the solution.)

This exercise is used in Exercise 93.

92. Let Δf denote $f(x+1) - f(x)$ and Df denote, as usual, the derivative of f. Just as we speak of $D^n f$, we can define $\Delta^n f$ by repeated applications of Δ to f. For instance,
    $$(\Delta^2 f)(x) = (\Delta f)(x+1) - (\Delta f)(x)$$
    $$= [f(x+2) - f(x+1)] - [f(x+1) - f(x)].$$
    (a) Show that $D\Delta f = \Delta Df$.
    (b) Show that $(\Delta f)(x) = (Df)(c_1)$ for some number c_1, $x < c_1 < x+1$.
    (c) Show that for any positive integer k there is a number c_k, $x < c_k < x+k$, such that
    $$(\Delta^k f)(x) = (D^k f)(c_k).$$
    (Assume that f has derivatives of all orders.)
    (d) Let r be a positive number that is not an integer. Show that
    $$\Delta^k(x^r) = r(r-1)\cdots(r-k+1)c_k^{r-k}$$
    for some c_k, $x < c_k < x+k$.

93. Let r be a positive number such that n^r is an integer for all positive integers n. Show that r is a positive integer. The preceding exercise may be of aid.

    Incidentally, it is known that, if p^r is an integer for three distinct primes p, then r is an integer. It is not known whether the assumption that p^r is an integer for two distinct primes forces r to be an integer.

94. (a) From the Maclaurin series for $\cos x$ in powers of x, obtain the Maclaurin series for $\cos 2x$.
    (b) Exploiting the identity $\sin^2 x = (1 - \cos 2x)/2$, obtain the Maclaurin series for $(\sin^2 x)/x^2$.
    (c) Estimate $\int_0^1 (\sin x/x)^2 \, dx$ using the first three nonzero terms of the series in (b).
    (d) Find a bound on the error in the estimate in (c).

95. In advanced mathematics a certain function $E(x)$ is defined as the sum $\sum_{n=0}^{\infty} x^n/n!$. Pretending that you have never heard of e or e^x, solve the following problems.
    (a) Show that $E(0) = 1$.
    (b) Show that $E'(x) = E(x)$.
    (c) Show that $E(x)E(-x) = 1$. Hint: Differentiate $E(x)E(-x)$ and use (a) and (b).
    (d) Deduce that $E(x+y)/E(x)$ is independent of x.
    (e) Deduce that $E(x+y) = E(x)E(y)$.

ALGEBRAIC OPERATIONS ON VECTORS

In this chapter we discuss the algebra of vectors in the plane and in space. Section 11.5 develops enough of the theory of determinants to permit a simple presentation of certain vector concepts. These concepts were developed primarily in response to James Clerk Maxwell's *Treatise on Electricity and Magnetism*, published in 1873. Josiah Gibbs, who in 1863 earned the first doctorate in engineering awarded in the United States and became a mathematical physicist, put vector analysis in its present form. His *Elements of Vector Analysis*, printed in 1881, introduced the notation used in this chapter.

11.1 The algebra of vectors

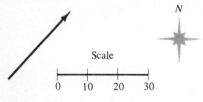

Figure 11.1

An adequate description of the wind indicates both its speed and its direction. One way to describe a wind of 30 miles per hour from the southwest is to draw an arrow aimed in the direction in which the wind blows, scaled so that its length represents a magnitude of 30, as in Fig. 11.1.

Relative to this same scale, some more wind arrows are shown in Fig. 11.2.

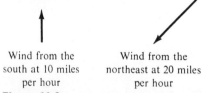

Figure 11.2

Similarly the flow of water on the surface of a stream is best indicated by a few sample arrows, as in Fig. 11.3. Of course, associated with *each* point on

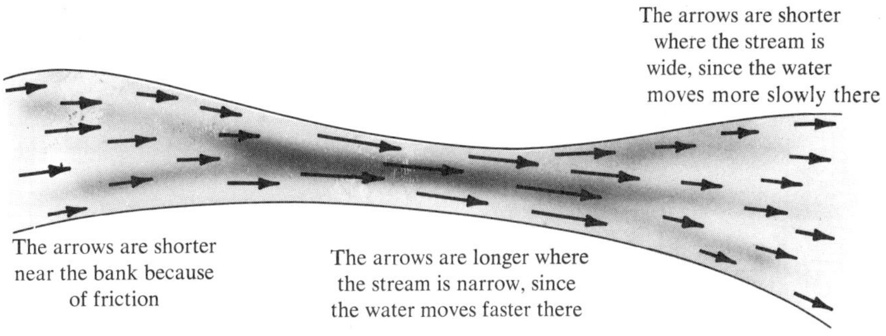

11.1 THE ALGEBRA OF VECTORS

The arrows are shorter where the stream is wide, since the water moves more slowly there

The arrows are shorter near the bank because of friction

The arrows are longer where the stream is narrow, since the water moves faster there

Figure 11.3

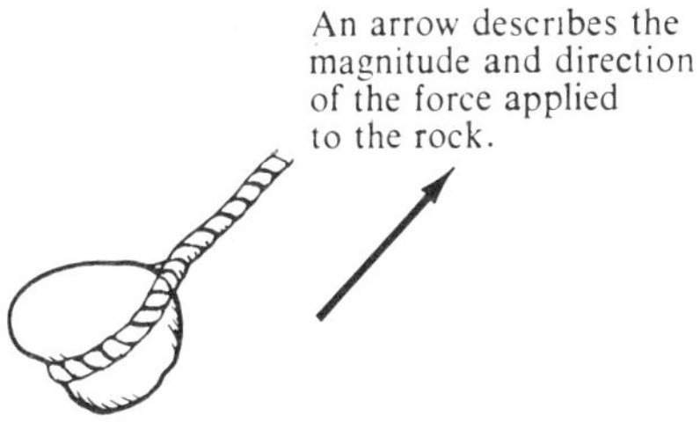

An arrow describes the magnitude and direction of the force applied to the rock.

Figure 11.4

the surface is an arrow representing the velocity of the water at that point.

Likewise when we pull a heavy rock by a rope, the force we exert has both a magnitude and a direction, as shown in Fig. 11.4. The magnitude describes how hard we pull; the direction of our pull is along the rope.

THE NOTION OF A VECTOR

Wind, fluid flow, and force introduce the mathematical notion of a *vector*, which may be represented by an arrow that records both a magnitude and a direction.

P is the "tail" and Q is the "head."

To be more precise, any two distinct points P and Q in the plane determine a *directed line segment* $\overrightarrow{PQ}$ from P to Q, as shown in Fig. 11.5. If $\overrightarrow{P_1Q_1}$ and $\overrightarrow{P_2Q_2}$ are two directed line segments that have both the same length and the same direction, then we say that they represent the same vector. (See Fig. 11.6.) A directed line segment has a particular location. A vector does not. The arrows in Fig. 11.6 all represent the *same* vector.

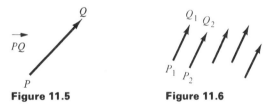

Figure 11.5   **Figure 11.6**

The zero vector

It is useful to introduce the *zero vector*. It has length 0 and no direction. In print, boldface letters, such as **A**, **B**, **F**, **r** and **v**, are used to denote vectors. In handwriting the symbols $\mathbb{A}$ and $\vec{A}$ are used. The length of **A** is denoted by $|\mathbf{A}|$. The length of **A** is also called the *magnitude* of **A**.

Vectors versus scalars

Lowercase letters, such as a, b, x, y, z will be used to name numbers. Numbers will also be called *scalars* to contrast them with vectors. Thus **A** is a vector, but $|\mathbf{A}|$ is a scalar. It is important to distinguish between the number 0 and the zero vector **0**.

In this section we consider only vectors **A** situated in the xy plane, which we will call *plane vectors*. The next section treats vectors in space.

If the origin of a rectangular coordinate system is at the tail of **A**, then the head of **A** has coordinates (x, y), as shown in Fig. 11.7. The numbers x and y are called the *scalar components* of **A** relative to the coordinate

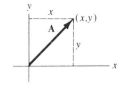

Figure 11.7

694 ALGEBRAIC OPERATIONS ON VECTORS

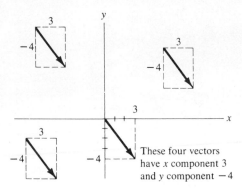

Figure 11.8

system. The two scalar components determine **A**, which lies along the diagonal of a rectangle whose sides have lengths $|x|$ and $|y|$. For instance, the vector represented by any of the four arrows in Fig. 11.8 has x component 3 and y component -4.

A vector whose head is at $(13, 18)$ and whose tail is at $(10, 22)$ also has scalar components 3 and -4, as a sketch will show. A vector with scalar components x and y will be denoted $\overrightarrow{(x, y)}$ to distinguish it from the point (x, y). The vector $\overrightarrow{(0, 0)}$ of length 0 is the zero vector **0**. Observe that $\overrightarrow{(x_1, y_1)} = \overrightarrow{(x_2, y_2)}$ if and only if $x_1 = x_2$ and $y_1 = y_2$.

By the Pythagorean theorem, the magnitude of $\mathbf{A} = \overrightarrow{(x, y)}$ is $\sqrt{|x|^2 + |y|^2}$, which is simply $\sqrt{x^2 + y^2}$, since x and y are squared.

EXAMPLE 1   Find the length of the vector represented by $\overrightarrow{PQ}$ if $P = (4, 1)$ and $Q = (7, -3)$.

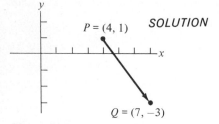

Figure 11.9

SOLUTION   Figure 11.9 shows the vector $\overrightarrow{PQ}$. The scalar components of $\overrightarrow{PQ}$ are $x = 7 - 4 = 3$ and $y = -3 - 1 = -4$. Thus

$$|\overrightarrow{PQ}| = \sqrt{3^2 + (-4)^2} = \sqrt{25} = 5. \quad \blacksquare$$

Any vector of length 1 is called a *unit vector*. For instance,

$$\overrightarrow{(1, 0)}, \quad \overrightarrow{(-1, 0)}, \quad \overrightarrow{(0, 1)}, \quad \overrightarrow{(0, -1)}, \quad \text{and} \quad \overrightarrow{\left(\frac{\sqrt{2}}{2}, \frac{\sqrt{2}}{2}\right)}$$

are unit vectors. More generally, for any angle θ, the vector $\overrightarrow{(\cos \theta, \sin \theta)}$ is a unit vector, since

$$\sqrt{(\cos \theta)^2 + (\sin \theta)^2} = \sqrt{1} = 1. \quad \blacksquare$$

ADDING AND SUBTRACTING VECTORS

Addition of vectors   The *sum of two vectors* **A** and **B** is defined as follows. Place **B** in such a way that its tail is at the head of **A**. Then the vector sum $\mathbf{A} + \mathbf{B}$ goes from the tail of **A** to the head of **B**. Observe that $\mathbf{B} + \mathbf{A} = \mathbf{A} + \mathbf{B}$, since both sums lie on the diagonal of a parallelogram, as shown in Fig. 11.10.

For example, if **W** is a wind vector (describing the motion of the air

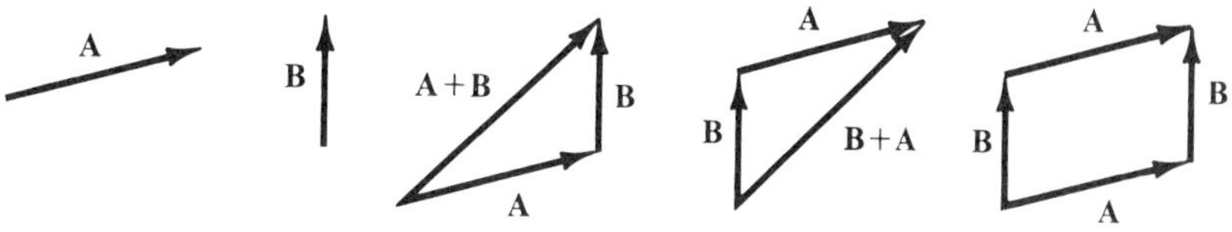

11.1 THE ALGEBRA OF VECTORS   695

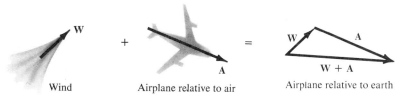

Figure 11.10

relative to the earth) and **A** is a vector describing the motion of an airplane relative to the air, then **W** + **A** is the vector describing the motion of the airplane relative to the earth. (See Fig. 11.11.)

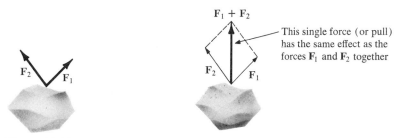

Figure 11.11

The concept of the sum of two vectors is also important in the study of force. If $\mathbf{F}_1$ and $\mathbf{F}_2$ describe the forces in two ropes lifting a heavy rock, as shown in Fig. 11.12, then a single rope with the force $\mathbf{F}_1 + \mathbf{F}_2$ pulling from the same point has the same effect on the rock.

Figure 11.12

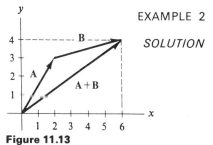

Figure 11.13

EXAMPLE 2   Find the sum of the vectors $\mathbf{A} = \overrightarrow{(2, 3)}$ and $\mathbf{B} = \overrightarrow{(4, 1)}$.

SOLUTION   Place the tail of **B** at the head of **A**. Inspection of Fig. 11.13 shows that   **A** + **B**   has the scalar components 6 and 4; that is,

$$\mathbf{A} + \mathbf{B} = \overrightarrow{(6, 4)}. \quad \blacksquare$$

Subtraction of vectors will be defined the way it is for numbers. If x and y are numbers, $x - y$ is that number that must be added to y to obtain x. (That is, $y + (x - y) = x$.) This suggests the following definition.

Let **A** and **B** be vectors. The vector **V** such that **B** + **V** equals **A** is called the *difference* of **A** and **B** and is denoted **A** − **B**.

Thus $$\mathbf{B} + (\mathbf{A} - \mathbf{B}) = \mathbf{A}.$$

To illustrate the definition, let **A** and **B** be given as in the first diagram in

696 ALGEBRAIC OPERATIONS ON VECTORS

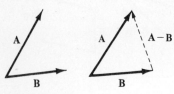

Figure 11.14

Fig. 11.14. $\mathbf{A} - \mathbf{B}$ is the dotted vector from the head of $\mathbf{B}$ to the head of $\mathbf{A}$.

The *negative* of the vector $\mathbf{A}$ is defined as the vector having the same magnitude as $\mathbf{A}$ but the opposite direction. It is denoted $-\mathbf{A}$. (See Fig. 11.15.) If $\mathbf{A} = \overrightarrow{PQ}$, then $-\mathbf{A} = \overrightarrow{QP}$. Observe that $\mathbf{A} + (-\mathbf{A}) = \mathbf{0}$, just as with scalars. More generally, subtracting a vector gives the same result as adding its negative; that is, $\mathbf{A} - \mathbf{B} = \mathbf{A} + (-\mathbf{B})$.

Figure 11.15

The definitions of addition and subtraction and the negative of a vector are geometric. The following theorem tells how to compute the sum and difference of vectors and the negative of a vector in terms of scalar components.

THEOREM

$$\overrightarrow{(x_1, y_1)} + \overrightarrow{(x_2, y_2)} = \overrightarrow{(x_1 + x_2, y_1 + y_2)} \quad \text{(add components)}$$

$$\overrightarrow{(x_1, y_1)} - \overrightarrow{(x_2, y_2)} = \overrightarrow{(x_1 - x_2, y_1 - y_2)} \quad \text{(subtract components)}$$

$$-\overrightarrow{(x, y)} = \overrightarrow{(-x, -y)} \quad \text{(change signs of components)}$$

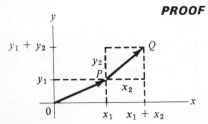

Figure 11.16

PROOF To justify the first assertion let $\overrightarrow{OP} = \overrightarrow{(x_1, y_1)}$ and $\overrightarrow{PQ} = \overrightarrow{(x_2, y_2)}$. What is the scalar component of $\overrightarrow{OP} + \overrightarrow{PQ}$ along, say, the x axis? Figure 11.16, in which x_1 and x_2 are positive, shows that $\overrightarrow{OQ} = \overrightarrow{OP} + \overrightarrow{PQ}$ has scalar components $x_1 + x_2$ and $y_1 + y_2$.

If some of the components are negative, the sketch needs to be modified slightly.

The second assertion follows from the first. For, what must x and y be if

$$\overrightarrow{(x_2, y_2)} + \overrightarrow{(x, y)} = \overrightarrow{(x_1, y_1)}?$$

We have, by the first assertion,

$$\overrightarrow{(x_2 + x, y_2 + y)} = \overrightarrow{(x_1, y_1)}.$$

Hence $\quad x_2 + x = x_1 \quad$ and $\quad y_2 + y = y_1$.

Thus $\quad x = x_1 - x_2 \quad$ and $\quad y = y_1 - y_2$.

Consequently, $\quad \overrightarrow{(x_1, y_1)} - \overrightarrow{(x_2, y_2)} = \overrightarrow{(x_1 - x_2, y_1 - y_2)}$.

To justify the third assertion, add the two vectors $\overrightarrow{(x, y)}$ and $\overrightarrow{(-x, -y)}$, obtaining $\overrightarrow{(0, 0)}$, the zero vector. Figure 11.17 shows a typical vector $\overrightarrow{(x, y)}$ and its negative $\overrightarrow{(-x, -y)}$ placed with their tails at the origin. ∎

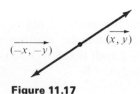

Figure 11.17

The following computations illustrate the theorem.

$$\overrightarrow{(1, 4)} + \overrightarrow{(2, 5)} = \overrightarrow{(3, 9)}$$

$$\overrightarrow{(2, 5)} - \overrightarrow{(3, 4)} = \overrightarrow{(-1, 1)}$$

$$-\overrightarrow{(4, -2)} = \overrightarrow{(-4, 2)}$$

Summary of terms   For convenience we list the terms defined in this section: vector, zero vector, length (magnitude) of a vector, scalar, scalar components, unit vector, sum of vectors, difference of vectors, negative of a vector.

Exercises

1. Draw the vector $\overrightarrow{(2, 3)}$, placing its tail at
   (a) $(0, 0)$, (b) $(-1, 2)$, (c) $(1, 1)$.
2. Draw the vector $\overrightarrow{(1, 2)}$, placing its tail at
   (a) $(0, 0)$, (b) $(-1, 2)$, (c) $(3, 3)$.
3. Find $|\mathbf{A}|$ if $\mathbf{A}$ is
   (a) $\overrightarrow{(4, 0)}$, (b) $\overrightarrow{(5, 12)}$,
   (c) $\overrightarrow{(-5, 12)}$ (d) $\overrightarrow{(-8, -6)}$.
4. Find the magnitude of $\mathbf{A} = \overrightarrow{PQ}$ if
   (a) $P = (2, 3)$ and $Q = (1, 4)$,
   (b) $P = (1, -3)$ and $Q = (-2, 4)$.
5. Find $\mathbf{A} + \mathbf{B}$ geometrically and express the result in the form $\overrightarrow{(x, y)}$ if
   (a) $\mathbf{A} = \overrightarrow{(2, 1)}$, $\mathbf{B} = \overrightarrow{(1, 4)}$;
   (b) $\mathbf{A} = \overrightarrow{(2, 2)}$, $\mathbf{B} = \overrightarrow{(1, -1)}$.
6. Give an example of plane vectors $\mathbf{A}$ and $\mathbf{B}$ such that
   (a) $|\mathbf{A} + \mathbf{B}|$ is not equal to $|\mathbf{A}| + |\mathbf{B}|$,
   (b) $|\mathbf{A} + \mathbf{B}| = |\mathbf{A}| + |\mathbf{B}|$.
7. Find $\mathbf{A} - \mathbf{B}$ geometrically and express the result in the form $\overrightarrow{(x, y)}$ if
   (a) $\mathbf{A} = \overrightarrow{(4, 3)}$, $\mathbf{B} = \overrightarrow{(2, 0)}$;
   (b) $\mathbf{A} = \overrightarrow{(3, 4)}$, $\mathbf{B} = \overrightarrow{(5, 1)}$;
   (c) $\mathbf{A} = \overrightarrow{(1, 1)}$, $\mathbf{B} = \overrightarrow{(-2, 4)}$.
8. Show pictorially that $\overrightarrow{(2, 3)} + \overrightarrow{(-1, 2)} = \overrightarrow{(1, 5)}$.
9. Write $\mathbf{A}$ in the form $\overrightarrow{(x, y)}$ if
   (a) its tail is at $(\tfrac{1}{2}, 5)$ and its head at $(\tfrac{7}{2}, \tfrac{13}{2})$;
   (b) its tail is at $(2, 7)$ and its head at $(2, 4)$;
   (c) its tail is at $(2, 4)$ and its head at $(2, 7)$;
   (d) its tail is at $(5, 3)$ and its head at $(-\tfrac{1}{2}, \tfrac{1}{3})$.
10. Find the scalar components of $\mathbf{A}$ if
    (a) $|\mathbf{A}| = 10$, and $\mathbf{A}$ points to the northwest;
    (b) $|\mathbf{A}| = 6$, and $\mathbf{A}$ points to the south;
    (c) $|\mathbf{A}| = 9$, and $\mathbf{A}$ points to the southeast;
    (d) $|\mathbf{A}| = 5$, and $\mathbf{A}$ points to the east
    (North is indicated by the positive y axis.)
11. Sketch a diagram to show that $\mathbf{A} + (\mathbf{B} + \mathbf{C}) = (\mathbf{A} + \mathbf{B}) + \mathbf{C}$.
12. Show with a diagram that $\mathbf{A} + (-\mathbf{B}) = \mathbf{A} - \mathbf{B}$.

■

13. The wind is 30 miles per hour to the northeast. An airplane is traveling 100 miles per hour relative to the wind, and the vector from the tail of the plane to its front tip points to the southeast, as shown in Fig. 11.18.

    (a) What is the speed of the plane relative to the ground?
    (b) What is the direction of the flight relative to the ground?

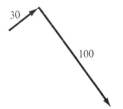

Figure 11.18

14. Let a and b be scalars, not both 0. Show that
$$\overrightarrow{\left(\frac{a}{\sqrt{a^2 + b^2}}, \frac{b}{\sqrt{a^2 + b^2}}\right)}$$
is a unit vector.

15. (a) Using an appropriate diagram, explain why $|\mathbf{A} + \mathbf{B}| \leq |\mathbf{A}| + |\mathbf{B}|$.
    (b) For which pairs of vectors $\mathbf{A}$ and $\mathbf{B}$ is $|\mathbf{A} + \mathbf{B}| = |\mathbf{A}| + |\mathbf{B}|$?

■ ■

16. From Exercise 15 deduce that, for any four real numbers $x_1, y_1, x_2,$ and y_2,
$$x_1 x_2 + y_1 y_2 \leq \sqrt{x_1^2 + y_1^2}\sqrt{x_2^2 + y_2^2}.$$
When does equality hold?

17. (a) What is the sum of the five vectors shown in Fig. 11.19?
    (b) Sketch the pentagon corresponding to the sum $\mathbf{A} + \mathbf{C} + \mathbf{D} + \mathbf{E} + \mathbf{B}$.

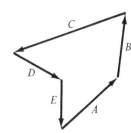

Figure 11.19

11.2 Vectors in space

All the notions developed in Sec. 11.1 extend with only slight changes to vectors in space. The main difference is that vectors in space have three components instead of two. In order to deal with the third component we introduce a coordinate system with which to describe points in space.

COORDINATES IN SPACE

By means of an *xy*-coordinate system every point in a plane is described by two numbers. To describe points in space, three numbers are required. The simplest way to do this is by an *xyz*-coordinate system, which is constructed as follows.

Introduce three mutually perpendicular lines, called the *x* axis, the *y* axis, and the *z* axis. They are usually chosen according to the "right-hand rule," shown in Fig. 11.20. If the fingers of the right hand curl in the direction from the positive *x* axis to the positive *y* axis, then the thumb points in the direction of the positive *z* axis.

The right-hand rule

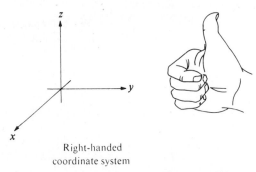

Right-handed coordinate system

Figure 11.20

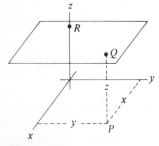

Figure 11.21

Any point Q in space is now described by three numbers: First, two numbers specify the x and y coordinates of the point P in the xy plane directly below (or above) Q; then the height of Q above (or below) the xy plane is recorded by the z coordinate of the point R where the plane through Q and parallel to the xy plane meets the z axis. The point Q is then denoted by (x, y, z). Note in Fig. 11.21 that points on the xy plane have $z = 0$.

Figure 11.22 displays the four points $(1, 0, 0)$, $(0, 1, 0)$, $(-1, 0, 0)$, and $(1, 2, 3)$. The rectangular box shown in the fourth of these diagrams helps to show the location of the point. It is also a convenient way of indicating spatial perspective.

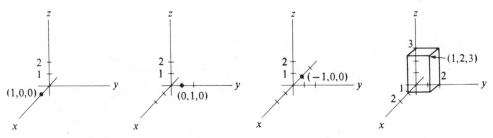

Figure 11.22

11.2 VECTORS IN SPACE

The distance between the points (x_1, y_1) and (x_2, y_2) in the xy plane is $\sqrt{(x_1 - x_2)^2 + (y_1 - y_2)^2}$, a formula based on the Pythagorean theorem. The next theorem generalizes this formula to space.

THEOREM 1 The distance between the points $Q_1 = (x_1, y_1, z_1)$ and $Q_2 = (x_2, y_2, z_2)$ is

$$\sqrt{(x_1 - x_2)^2 + (y_1 - y_2)^2 + (z_1 - z_2)^2}.$$

PROOF Figure 11.23 shows the two points Q_1 and Q_2 and the box that they determine with edges parallel to the three coordinate axes. Note that the dimensions of the box are $|x_1 - x_2|$, $|y_1 - y_2|$, and $|z_1 - z_2|$. Label two of the corners of the box B and C, as shown. Let d be the distance between Q_1 and Q_2. Let e be the distance between Q_2 and B.

From the right triangle Q_1BQ_2,

$$d^2 = e^2 + |z_1 - z_2|^2.$$

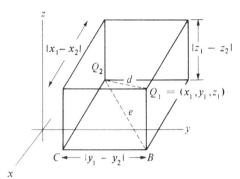

Figure 11.23

From the right triangle Q_2CB,

$$e^2 = |x_1 - x_2|^2 + |y_1 - y_2|^2.$$

Hence $\quad d^2 = |x_1 - x_2|^2 + |y_1 - y_2|^2 + |z_1 - z_2|^2,$

and $\quad d = \sqrt{|x_1 - x_2|^2 + |y_1 - y_2|^2 + |z_1 - z_2|^2}$

$\qquad = \sqrt{(x_1 - x_2)^2 + (y_1 - y_2)^2 + (z_1 - z_2)^2}.$

This completes the proof. ∎

EXAMPLE 1 Find the distance between the points $(5, 4, 3)$ and $(2, 1, 1)$.

SOLUTION The distance is

$$\sqrt{(5 - 2)^2 + (4 - 1)^2 + (3 - 1)^2} = \sqrt{3^2 + 3^2 + 2^2}$$
$$= \sqrt{22}. \quad \blacksquare$$

VECTORS IN SPACE

For vectors **A** and **B** in space, $\mathbf{A} + \mathbf{B}$, $\mathbf{A} - \mathbf{B}$, and $-\mathbf{A}$ are defined just as for plane vectors. A space vector **A**, however, has three scalar components, $\mathbf{A} = \overrightarrow{(x, y, z)}$. The magnitude of $\mathbf{A} = \overrightarrow{(x, y, z)}$ is $\sqrt{x^2 + y^2 + z^2}$. The formulas for $\mathbf{A} + \mathbf{B}$, $\mathbf{A} - \mathbf{B}$, and $-\mathbf{A}$ in terms of scalar components are similar to those for plane vectors:

Just throw in the third component.

$$\overrightarrow{(x_1, y_1, z_1)} + \overrightarrow{(x_2, y_2, z_2)} = \overrightarrow{(x_1 + x_2, y_1 + y_2, z_1 + z_2)}$$
$$\overrightarrow{(x_1, y_1, z_1)} - \overrightarrow{(x_2, y_2, z_2)} = \overrightarrow{(x_1 - x_2, y_1 - y_2, z_1 - z_2)}$$
$$-\overrightarrow{(x, y, z)} = \overrightarrow{(-x, -y, -z)}.$$

EXAMPLE 2 (a) Show that $\mathbf{A} = \overrightarrow{(\tfrac{4}{9}, -\tfrac{8}{9}, \tfrac{1}{9})}$ is a unit vector. (b) Draw **A**.

SOLUTION (a)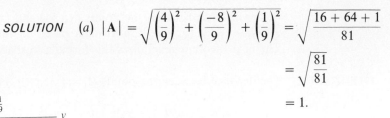

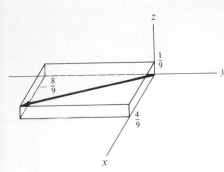

Figure 11.24

Thus **A** is a unit vector.

(b) To draw **A**, show the box that it determines. For convenience, place the tail of **A** at the origin, as shown in Fig. 11.24. ∎

Henceforth, vectors may be either in the plane or in space. The definitions will cover both cases together, but, for the sake of clarity, the diagrams will generally show only plane vectors.

THE PRODUCT OF A SCALAR AND A VECTOR

The algebra of vectors described in Sec. 11.1 closely resembles the algebra of scalars. The present section introduces a type of multiplication that does not resemble anything in ordinary algebra. It concerns the product of a scalar and a vector, and the result will be a vector. Another way to say this is, "Scalars will operate on vectors."

A scalar will operate on a vector by magnifying or shrinking it. Thus 3**A** shall mean a vector three times as long as **A** having the same direction as **A**. Also, $(-3)\mathbf{A}$ shall be $-(3\mathbf{A})$, a vector three times as long as **A** but in the opposite direction. (See Fig. 11.25.)

More generally, the product of a scalar c and a vector **A** is defined as follows. Note that c need not be an integer.

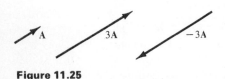

Figure 11.25

DEFINITION *The product of a scalar and a vector.* If c is a scalar and **A** a vector, the product $c\mathbf{A}$ is the vector whose length is $|c|$ times the length of **A** and whose direction is the same as that of **A** if c is positive and opposite that of **A** if c is negative.

Observe that $0\mathbf{A}$ has length 0 and thus is the zero vector **0** to which no direction is assigned. The vector $c\mathbf{A}$ is called a *scalar multiple* of the vector **A**.

EXAMPLE 3 If **A** is the vector shown in Fig. 11.26, sketch $(-1)\mathbf{A}$, $\tfrac{1}{2}\mathbf{A}$, $(-2)\mathbf{A}$.

Figure 11.26

SOLUTION $(-1)\mathbf{A}$ has the same length as **A** but the opposite direction. Thus $(-1)\mathbf{A} = -\mathbf{A}$. $\tfrac{1}{2}\mathbf{A}$ is half as long as **A** and has the same direction as **A**. $(-2)\mathbf{A}$ is twice as long as **A** but its direction is opposite that of **A**. These vectors are shown in Fig. 11.27.

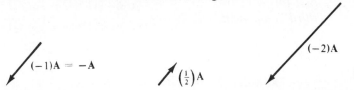

Figure 11.27

The next theorem shows how to express the scalar components of $c\mathbf{A}$ in terms of c and the scalar components of $\mathbf{A}$.

THEOREM 2
$$c\overrightarrow{(x, y)} = \overrightarrow{(cx, cy)},$$
and
$$c\overrightarrow{(x, y, z)} = \overrightarrow{(cx, cy, cz)}.$$

PROOF For the sake of simplicity, we will carry through the argument for plane vectors, showing that $c\overrightarrow{(x, y)} = \overrightarrow{(cx, cy)}$. The proof for vectors in space is similar, but the diagrams are messier.

First it will be shown that the length of $\overrightarrow{(cx, cy)}$ is $|c|$ times the length of $\overrightarrow{(x, y)}$. To do this, compute the length of $\overrightarrow{(cx, cy)}$:

$$|\overrightarrow{(cx, cy)}| = \sqrt{(cx)^2 + (cy)^2} = \sqrt{c^2(x^2 + y^2)} = \sqrt{c^2}\sqrt{x^2 + y^2}.$$

Now, $\sqrt{x^2 + y^2}$ is the length of $\overrightarrow{(x, y)}$.

Furthermore, $\sqrt{c^2} = |c|$,

whether c is positive or negative. (For instance, $\sqrt{(-3)^2} = \sqrt{9} = 3 = |-3|$.)

Thus the length of $\overrightarrow{(cx, cy)}$ is $|c|$ times the length of $\overrightarrow{(x, y)}$.

Next it will be shown that the direction of $\overrightarrow{(cx, cy)}$ is the same as that of $c\overrightarrow{(x, y)}$.

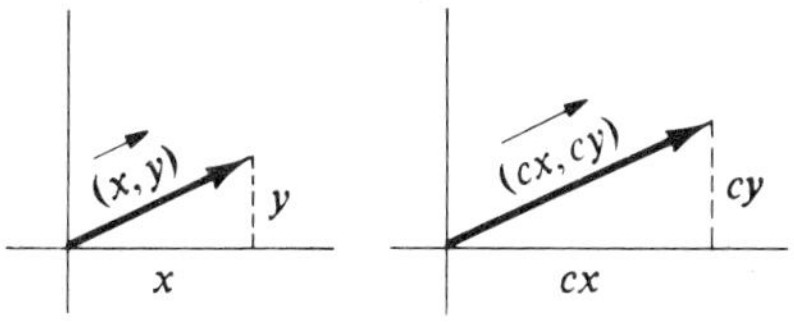

Consider the case $c > 0$. Then $\overrightarrow{(x, y)}$ and $\overrightarrow{(cx, cy)}$ lie on corresponding sides of similar triangles, as shown in Fig. 11.28. Hence they are parallel and point in the same direction. Thus if c is positive, $c\overrightarrow{(x, y)}$ and $\overrightarrow{(cx, cy)}$ have the same direction. Since $c\overrightarrow{(x, y)}$ and $\overrightarrow{(cx, cy)}$ have the same length and the same direction, they are the same vector.

If c is negative, similar reasoning shows that $c\overrightarrow{(x, y)} = \overrightarrow{(cx, cy)}$. ∎

Figure 11.28

By Theorem 2,
$$4\overrightarrow{(2, 5)} = \overrightarrow{(8, 20)};$$
$$\tfrac{1}{2}\overrightarrow{(4, 2, -5)} = \overrightarrow{(2, 1, -2.5)};$$
$$0.03\overrightarrow{(2, -1, 4)} = \overrightarrow{(0.06, -0.03, 0.12)};$$
$$(-1)\overrightarrow{(2, 3, 4)} = \overrightarrow{(-2, -3, -4)}.$$

DEFINITION *The division of a vector by a scalar.* Let $\mathbf{A}$ be a vector and let c be a scalar other than 0. Then

$$\frac{\mathbf{A}}{c}$$

is defined as $\left(\dfrac{1}{c}\right)\mathbf{A}.$

Figure 11.29

EXAMPLE 4 Sketch the vector $\mathbf{A}/2$ for a typical vector $\mathbf{A}$.

SOLUTION $\mathbf{A}/2$ is defined to be $\tfrac{1}{2}\mathbf{A}$, which is a vector half as long as $\mathbf{A}$ and in the same direction as $\mathbf{A}$. (See Fig. 11.29.) ∎

EXAMPLE 5 Compare the vectors $\mathbf{A}$ and $\mathbf{A}/0.1$.

SOLUTION $\mathbf{A}/0.1 = (1/0.1)\mathbf{A} = 10\mathbf{A}$, a vector 10 times as long as $\mathbf{A}$ and in the same direction as $\mathbf{A}$. ∎

EXAMPLE 6 Compute $\overrightarrow{(4, 7, 6)}/3$.

SOLUTION
$$\frac{\overrightarrow{(4, 7, 6)}}{3} = \frac{1}{3}\overrightarrow{(4, 7, 6)} = \overrightarrow{\left(\frac{4}{3}, \frac{7}{3}, 2\right)}. \quad \blacksquare$$

Example 6 generalizes to the formula

$$\frac{\overrightarrow{(x, y, z)}}{c} = \overrightarrow{\left(\frac{x}{c}, \frac{y}{c}, \frac{z}{c}\right)} \qquad \text{if } c \neq 0.$$

The next theorem will be used often to produce a unit vector with a prescribed direction.

THEOREM 3 For any vector $\mathbf{A}$ not equal to $\mathbf{0}$, the vector

$$\frac{\mathbf{A}}{|\mathbf{A}|}$$

is a unit vector (and has the same direction as $\mathbf{A}$).

PROOF Since $|\mathbf{A}|$ is a positive number, $\mathbf{A}/|\mathbf{A}|$ has the same direction as $\mathbf{A}$. What is the length of $\mathbf{A}/|\mathbf{A}|$? Since $\mathbf{A}/|\mathbf{A}|$ is defined as

$$\frac{1}{|\mathbf{A}|}\mathbf{A},$$

its length is $\dfrac{1}{|\mathbf{A}|}$ times the length of $\mathbf{A}$,

that is, $\dfrac{1}{|\mathbf{A}|}|\mathbf{A}|,$

which is 1.

Thus $\mathbf{A}/|\mathbf{A}|$ is a unit vector. This concludes the proof. ∎

EXAMPLE 7 Compute $\mathbf{A}/|\mathbf{A}|$ when $\mathbf{A} = \overrightarrow{(4, -5, 20)}$.

SOLUTION We have

$$|\mathbf{A}| = \sqrt{4^2 + (-5)^2 + (20)^2} = \sqrt{16 + 25 + 400} = \sqrt{441} = 21.$$

Thus $\dfrac{\mathbf{A}}{|\mathbf{A}|} = \overrightarrow{\left(\dfrac{4}{21}, \dfrac{-5}{21}, \dfrac{20}{21}\right)}. \quad \blacksquare$

The next definition introduces and names three special vectors which will be used often.

DEFINITION *The basic unit vectors.* The vectors

$$\mathbf{i} = \overrightarrow{(1, 0, 0)},$$
$$\mathbf{j} = \overrightarrow{(0, 1, 0)},$$
$$\text{and} \quad \mathbf{k} = \overrightarrow{(0, 0, 1)}$$

are called the *basic unit vectors* and are shown in Fig. 11.30.

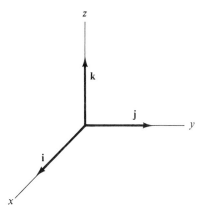

Figure 11.30

EXAMPLE 8 Express the vector $3\mathbf{i} + 4\mathbf{j} + 5\mathbf{k}$ in the form $\overrightarrow{(x, y, z)}$.

SOLUTION
$$3\mathbf{i} + 4\mathbf{j} + 5\mathbf{k} = 3\overrightarrow{(1, 0, 0)} + 4\overrightarrow{(0, 1, 0)} + 5\overrightarrow{(0, 0, 1)}$$
$$= \overrightarrow{(3, 0, 0)} + \overrightarrow{(0, 4, 0)} + \overrightarrow{(0, 0, 5)}$$
$$= \overrightarrow{(3, 4, 5)}.$$

Figure 11.31 shows the relation between $\mathbf{i}$, $\mathbf{j}$, and $\mathbf{k}$ and the vector $3\mathbf{i} + 4\mathbf{j} + 5\mathbf{k}$. ∎

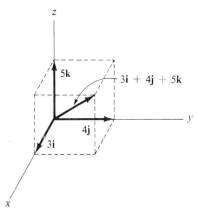

Figure 11.31

As Example 8 suggests,

$$x\mathbf{i} + y\mathbf{j} + z\mathbf{k} = \overrightarrow{(x, y, z)}.$$

Generally the notation $x\mathbf{i} + y\mathbf{j} + z\mathbf{k}$ is preferable to $\overrightarrow{(x, y, z)}$. First of all, it is more geometric. Second, when x, y, and z are messy expressions, the notation $x\mathbf{i} + y\mathbf{j} + z\mathbf{k}$ is easier to read.

When representing a plane vector $\mathbf{A} = \overrightarrow{(x, y)}$, we do not need the vector $\mathbf{k}$; we have simply

$$\mathbf{A} = x\mathbf{i} + y\mathbf{j},$$

since the coefficient of $\mathbf{k}$ is 0.

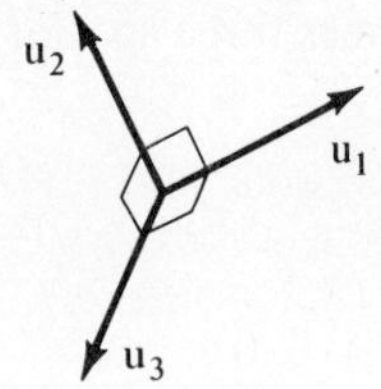

704 ALGEBRAIC OPERATIONS ON VECTORS

While **i**, **j**, and **k** will be the three mutually perpendicular unit vectors most often referred to, occasionally other such triplets will provide a convenient reference.

Let $\mathbf{u}_1$, $\mathbf{u}_2$, and $\mathbf{u}_3$ be three unit vectors that are perpendicular to each other as shown in Fig. 11.32. For any scalars x, y, and z, the vector $\mathbf{A} = x\mathbf{u}_1 + y\mathbf{u}_2 + z\mathbf{u}_3$ is the sum of three vectors, one parallel to $\mathbf{u}_1$, one parallel to $\mathbf{u}_2$, and one parallel to $\mathbf{u}_3$, as shown in Fig. 11.33. (If x, say, is negative, then $x\mathbf{u}_1$ points in the direction opposite that of $\mathbf{u}_1$.) Note that since $\mathbf{u}_1$ is a unit vector, $x\mathbf{u}_1$ has length $|x|$. The length of $x\mathbf{u}_1 + y\mathbf{u}_2 + z\mathbf{u}_3$ is $\sqrt{x^2 + y^2 + z^2}$.

Figure 11.32

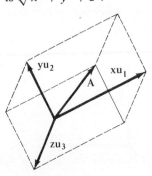
Figure 11.33

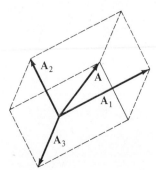
Figure 11.34

The procedure can be reversed. It is possible to start with a vector **A** and the three unit vectors $\mathbf{u}_1$, $\mathbf{u}_2$, and $\mathbf{u}_3$ (which are perpendicular to each other) and express **A** as the sum of a vector $\mathbf{A}_1$ parallel to $\mathbf{u}_1$, a vector $\mathbf{A}_2$ parallel to $\mathbf{u}_2$, and a vector $\mathbf{A}_3$ parallel to $\mathbf{u}_3$. Draw the box whose diagonal is **A** and whose edges are parallel to $\mathbf{u}_1, \mathbf{u}_2,$ and $\mathbf{u}_3$, as in Fig. 11.34. Thus, $\mathbf{A}_1 = x\mathbf{u}_1$, $\mathbf{A}_2 = y\mathbf{u}_2$, and $\mathbf{A}_3 = z\mathbf{u}_3$ for some scalars x, y, and z.

In the case of a plane vector **A** and a pair of perpendicular unit vectors $\mathbf{u}_1$ and $\mathbf{u}_2$, the diagram is much easier to sketch. (See Fig. 11.35.)

Figure 11.35

DEFINITION *Vector and scalar components.* Let $\mathbf{u}_1$, $\mathbf{u}_2$, and $\mathbf{u}_3$ be perpendicular unit vectors. Let **A** be an arbitrary vector. Then **A** is expressible in the form

$$\mathbf{A} = \mathbf{A}_1 + \mathbf{A}_2 + \mathbf{A}_3,$$

where $\mathbf{A}_1$ is parallel to $\mathbf{u}_1$,
$\mathbf{A}_2$ is parallel to $\mathbf{u}_2$,
$\mathbf{A}_3$ is parallel to $\mathbf{u}_3$.

Vector components $\mathbf{A}_1$ is the *vector component* of **A** parallel to $\mathbf{u}_1$. $\mathbf{A}_2$ is the *vector component* of **A** parallel to $\mathbf{u}_2$. $\mathbf{A}_3$ is the *vector component* of **A** parallel to $\mathbf{u}_3$.

Scalar components   Furthermore, if $A_1 = x\mathbf{u}_1$, $A_2 = y\mathbf{u}_2$, and $A_3 = z\mathbf{u}_3$, the number x is called the *scalar component* of $\mathbf{A}$ in the direction of $\mathbf{u}_1$; y, the *scalar component* of $\mathbf{A}$ in the direction of $\mathbf{u}_2$; and z, the *scalar component* of $\mathbf{A}$ in the direction of $\mathbf{u}_3$. A scalar component may be negative. By the scalar (or vector) component of a vector $\mathbf{A}$ along the vector $\mathbf{r}$ is meant the scalar (or vector) component of $\mathbf{A}$ in the direction of the unit vector $\mathbf{r}/|\mathbf{r}|$.

EXAMPLE 9   Find the vector components of

$$\mathbf{A} = \overrightarrow{(3, -4)} \quad \text{parallel to} \quad \mathbf{u}_1 = \overrightarrow{(1, 0)} \quad \text{and} \quad \mathbf{u}_2 = \overrightarrow{(0, 1)}.$$

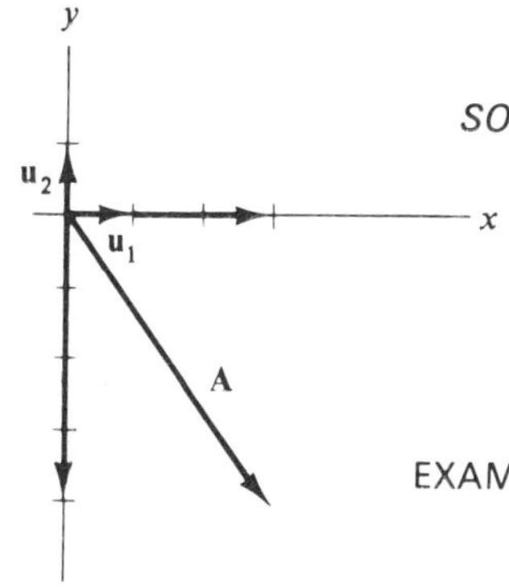

SOLUTION   Figure 11.36 displays $\mathbf{A}$, $\mathbf{u}_1$, and $\mathbf{u}_2$. Inspection of the diagram shows that

$$\mathbf{A} = 3\mathbf{u}_1 - 4\mathbf{u}_2.$$

Thus the vector component of $\mathbf{A}$ along $\mathbf{u}_1$ is $3\mathbf{u}_1$ and along $\mathbf{u}_2$ is $-4\mathbf{u}_2$. The scalar component of $\mathbf{A}$ along $\mathbf{u}_1$ is 3 and along $\mathbf{u}_2$ is -4. ∎

EXAMPLE 10   Find the vector components of $\mathbf{A} = \overrightarrow{(3, -4, 5)}$ along

$$\mathbf{u}_1 = \overrightarrow{(-1, 0, 0)}, \quad \mathbf{u}_2 = \overrightarrow{(0, 1, 0)}, \quad \text{and} \quad \mathbf{u}_3 = \overrightarrow{(0, 0, 1)}.$$

SOLUTION

$$\mathbf{A} = -3\mathbf{u}_1 - 4\mathbf{u}_2 + 5\mathbf{u}_3.$$

Figure 11.36

Thus the vector components are $-3\mathbf{u}_1$, $-4\mathbf{u}_2$, and $5\mathbf{u}_3$. The scalar components of $\mathbf{A}$ along $\mathbf{u}_1$, $\mathbf{u}_2$, and $\mathbf{u}_3$ are, respectively, -3, -4, and 5. ∎

The vectors $\mathbf{i}$, $\mathbf{j}$, $\mathbf{k}$ are sometimes denoted $\mathbf{e}_1$, $\mathbf{e}_2$, $\mathbf{e}_3$. The three scalar components of $\mathbf{A}$ are sometimes denoted A_x, A_y, A_z. Thus $\mathbf{A} = A_x\mathbf{i} + A_y\mathbf{j} + A_z\mathbf{k}$.

The scalar components of $\mathbf{A}$ are also denoted a_1, a_2, a_3; then we have

$$\mathbf{A} = a_1\mathbf{i} + a_2\mathbf{j} + a_3\mathbf{k} \quad \text{and} \quad \mathbf{A} = \overrightarrow{(a_1, a_2, a_3)}.$$

The next section shows how to find the scalar and vector components of any vector $\mathbf{A}$ along any unit vector $\mathbf{u}$.

Exercises

In Exercises 1 and 2 plot the given points P and indicate perspective by drawing the box they determine. (P and $O = (0, 0, 0)$ are opposite corners of the box. The edges of the box are parallel to the axes.)

1.   (a) $P = (2, 3, 1)$, (b) $P = (0, 4, -1)$, (c) $P = (1, -1, 2)$.
2.   (a) $P = (-1, -1, 2)$, (b) $P = (-1, -2, -3)$, (c) $P = (1, -2, 13)$.
3.   Find the distance between the points (a) (1, 4, 2) and (2, 1, 5), (b) (-3, 2, 1) and (4, 0, -2).
4.   Find the magnitude of the vector (a) $\overrightarrow{(1, 3, -2)}$, (b) $\overrightarrow{(2, 1, -5)}$.
5.   Find $\mathbf{A} - \mathbf{B}$ if
     (a) $\mathbf{A} = \overrightarrow{(2, 3, 4)}$ and $\mathbf{B} = \overrightarrow{(1, 5, 0)}$;
     (b) $\mathbf{A} = \overrightarrow{(3, 4, 2)}$ and $\mathbf{B} = \overrightarrow{(0, 0, 0)}$.
6.   Find $|\mathbf{A} + \mathbf{B}|$ if
     (a) $\mathbf{A} = \overrightarrow{(1, 2, 0)}$ and $\mathbf{B} = \overrightarrow{(2, 3, 5)}$;
     (b) $\mathbf{A} = \overrightarrow{(3, -2, 1)}$ and $\mathbf{B} = \overrightarrow{(-4, 3, 2)}$.
7.   Determine $|\overrightarrow{(\cos\theta, \sin\theta, 1)}|$, where θ is arbitrary.
8.   Find $|\mathbf{A}|$ if $\mathbf{A}$ is
     (a) $\overrightarrow{(-\tfrac{1}{2}, \tfrac{1}{2}, \sqrt{2}/2)}$,
     (b) $\overrightarrow{(-3, 4, 12)}$.

(c) $\left(\overrightarrow{\dfrac{2}{\sqrt{38}}, \dfrac{-3}{\sqrt{38}}, \dfrac{5}{\sqrt{38}}}\right)$.

9. Compute and sketch $c\mathbf{A}$ if $\mathbf{A} = 2\mathbf{i} + 3\mathbf{j} + \mathbf{k}$ and c is
   (a) 2 (b) -2 (c) $\tfrac{1}{2}$ (d) $-\tfrac{1}{2}$.

10. Express each of the following vectors in the form $c(2\mathbf{i} + 3\mathbf{j} + 4\mathbf{k})$ for suitable c:
    (a) $\overrightarrow{(4, 6, 8)}$
    (b) $-2\mathbf{i} - 3\mathbf{j} - 4\mathbf{k}$
    (c) $\mathbf{0}$
    (d) $\tfrac{2}{11}\mathbf{i} + \tfrac{3}{11}\mathbf{j} + \tfrac{4}{11}\mathbf{k}$.

11. If $\mathbf{u}$ is a unit vector, what is the length of $-3\mathbf{u}$?

12. Find $|5(2\mathbf{i} + 4\mathbf{j} - \mathbf{k})|$.

13. What is the vector component of $-4\mathbf{i} + 5\mathbf{j} + 0\mathbf{k}$ along
    (a) $\mathbf{i}$? (b) $\mathbf{j}$? (c) $\mathbf{k}$?

14. The vector component of $\mathbf{A}$ along $\mathbf{u}$ is $6\mathbf{u}$.
    (a) What is the scalar component of $\mathbf{A}$ along $\mathbf{u}$?
    (b) What is the vector component of $\mathbf{A}$ along $2\mathbf{u}$?
    (c) What is the scalar component of $\mathbf{A}$ along $2\mathbf{u}$?
    (d) What is the vector component of $\mathbf{A}$ along $-\mathbf{u}$?
    (e) What is the scalar component of $\mathbf{A}$ along $-\mathbf{u}$?

15. Sketch an appropriate diagram and find the vector component of $2\mathbf{i} + 2\mathbf{j} + 3\mathbf{k}$ along
    (a) $\mathbf{i}$ (b) $-\mathbf{i}$ (c) $3\mathbf{i}$ (d) $\dfrac{\mathbf{i}}{2} + \dfrac{\mathbf{j}}{2}$

16. (a) Find the vector component of $3\mathbf{i} - 4\mathbf{j} + 2\mathbf{k}$ along $-\mathbf{i}$.
    (b) Find the scalar component of $3\mathbf{i} - 4\mathbf{j} + 2\mathbf{k}$ along $-\mathbf{i}$.

17. (a) Show that the vectors
    $$\mathbf{u}_1 = \tfrac{1}{2}\mathbf{i} + \tfrac{\sqrt{3}}{2}\mathbf{j} \quad \text{and} \quad \mathbf{u}_2 = \tfrac{\sqrt{3}}{2}\mathbf{i} - \tfrac{1}{2}\mathbf{j}$$
    are perpendicular unit vectors. *Hint*: What angles do they make with the x axis?
    (b) Find scalars x and y such that $\mathbf{i} = x\mathbf{u}_1 + y\mathbf{u}_2$.

18. (a) Show that
    $$\mathbf{u}_1 = \tfrac{\sqrt{2}}{2}\mathbf{i} + \tfrac{\sqrt{2}}{2}\mathbf{j} \quad \text{and} \quad \mathbf{u}_2 = \tfrac{-\sqrt{2}}{2}\mathbf{i} + \tfrac{\sqrt{2}}{2}\mathbf{j}$$
    are perpendicular unit vectors. *Hint*: Draw them.
    (b) Express $\mathbf{i}$ in the form $x\mathbf{u}_1 + y\mathbf{u}_2$. *Hint*: Draw $\mathbf{i}$, $\mathbf{u}_1$, and $\mathbf{u}_2$.
    (c) Express $\mathbf{j}$ in the form $x\mathbf{u}_1 + y\mathbf{u}_2$.
    (d) Express $-2\mathbf{i} + 3\mathbf{j}$ in the form $x\mathbf{u}_1 + y\mathbf{u}_2$.

19. Draw a unit vector tangent to the curve $y = \sin x$ at $(0, 0)$. What are its vector components along $\mathbf{i}$ and $\mathbf{j}$?

20. Draw a unit vector tangent to the curve $y = x^3$ at $(1, 1)$. What are its scalar components along $\mathbf{i}$ and $\mathbf{j}$?

21. Find the unit vector that has the same direction as $\mathbf{i} + 2\mathbf{j} + 3\mathbf{k}$.

22. If the vector component of $\mathbf{A}$ along the unit vector $\mathbf{u}$ is $4\mathbf{u}$ and the vector component of $\mathbf{B}$ along $\mathbf{u}$ is $-3\mathbf{u}$, what is
    (a) the scalar component of $\mathbf{A}$ along $\mathbf{u}$?
    (b) the vector component of $\mathbf{A} + \mathbf{B}$ along $\mathbf{u}$?

23. Let $\mathbf{u}_1$ and $\mathbf{u}_2$ be the unit vectors in the xy plane shown in Fig. 11.37. Express in the form $a\mathbf{u}_1 + b\mathbf{u}_2$ the vectors
    (a) $\mathbf{i}$ (b) $\mathbf{j}$ (c) $3\mathbf{i} - 4\mathbf{j}$.

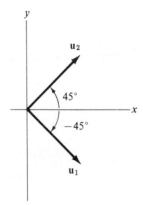

Figure 11.37

24. Find the vector and scalar components of $2\mathbf{i} + 2\mathbf{j} + \mathbf{k}$ along (a) $\mathbf{i} + \mathbf{j}$, (b) $-\mathbf{i} - \mathbf{j}$.

25. Let $\mathbf{u}$ be the unit vector that makes an angle $\pi/3$ with the x axis, as shown in Fig. 11.38.

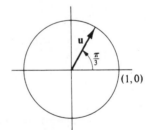

Figure 11.38

    (a) Express $\mathbf{u}$ in the form $x\mathbf{i} + y\mathbf{j}$.
    (b) Draw the vector components of $\mathbf{u}$ along $\mathbf{i}$ and $\mathbf{j}$.

26. Show that for any angles α and β, the vector
    $$\overrightarrow{(\cos\alpha \cos\beta, \cos\alpha \sin\beta, \sin\alpha)}$$
    has length 1.

27. Let a, b, and c be scalars, not all 0. Show that
    $$\overrightarrow{\left(\dfrac{a}{\sqrt{a^2 + b^2 + c^2}}, \dfrac{b}{\sqrt{a^2 + b^2 + c^2}}, \dfrac{c}{\sqrt{a^2 + b^2 + c^2}}\right)}$$
    is a unit vector.

28 (*Midpoint formula.*) Let A and B be two points in space. Let M be their midpoint. Let $\mathbf{A} = \overrightarrow{OA}$, $\mathbf{B} = \overrightarrow{OB}$, and $\mathbf{M} = \overrightarrow{OM}$.
  (a) Show that $\mathbf{M} = \mathbf{A} + \frac{1}{2}(\mathbf{B} - \mathbf{A})$.
  (b) Deduce that $\mathbf{M} = (\mathbf{A} + \mathbf{B})/2$.

29 Let $\mathbf{A}$ and $\mathbf{B}$ be two nonzero and nonparallel vectors in the plane. Let $\mathbf{C}$ be any vector in the plane. Show with the aid of a sketch that there are scalars x and y such that $\mathbf{C} = x\mathbf{A} + y\mathbf{B}$.

30 Let A and B be two distinct points in space. Let C be the point on the line segment AB that is twice as far from A as it is from B. Let $\mathbf{A} = \overrightarrow{OA}$, $\mathbf{B} = \overrightarrow{OB}$, and $\mathbf{C} = \overrightarrow{OC}$. Show that $\mathbf{C} = \frac{1}{3}\mathbf{A} + \frac{2}{3}\mathbf{B}$.

31 (See Exercises 28 and 30.) Let A, B, and C be the vertices of a triangle. Let $\mathbf{A} = \overrightarrow{OA}$, $\mathbf{B} = \overrightarrow{OB}$, and $\mathbf{C} = \overrightarrow{OC}$.
  (a) Let P be the point that is on the line segment joining A to the midpoint of the edge BC and twice as far from A as from the midpoint. Show that $\overrightarrow{OP} = (\mathbf{A} + \mathbf{B} + \mathbf{C})/3$.
  (b) Use (a) to show that the three medians of a triangle are concurrent.

32 (See Exercise 28.) Prove, using vectors, that the line segment joining the midpoints of two sides of a triangle is parallel to the third side and half as long.

33 (See Exercise 32.) The midpoints of a quadrilateral in space are joined to form another quadrilateral. Prove that this second quadrilateral is a parallelogram.

34 Draw two plane vectors $\mathbf{A}$ and $\mathbf{B}$ with their tails at the origin. Sketch the vectors $t\mathbf{A} + (1-t)\mathbf{B}$ with their tails at the origin for (a) $t = 2$, (b) $t = -2$, (c) $t = 0$, (d) $t = 1$, (e) $t = \frac{1}{2}$, (f) $t = \frac{1}{3}$. (g) On what familiar geometric object do the heads of the six vectors sketched seem to lie?

35 Consider two vectors $\mathbf{A}$ and $\mathbf{B}$, with their tails at the origin.
  (a) Why is the vector from the origin to any point on the line through their heads of the form $\mathbf{B} + t(\mathbf{A} - \mathbf{B})$ for some number t?
  (b) Use (a) to explain Exercise 34(g).

■ ■

There are two important properties of the set of vectors. First, you can add or subtract any two of them and get another vector. Moreover, this addition or subtraction satisfies the familiar rules of arithmetic. For instance, if $\mathbf{u}$ and $\mathbf{v}$ are vectors, then $\mathbf{u} + \mathbf{v} = \mathbf{v} + \mathbf{u}$. Second, each real number c "operates" on a vector $\mathbf{v}$ to produce a vector $c\mathbf{v}$. This operation satisfies such rules as $c(\mathbf{u} + \mathbf{v}) = c\mathbf{u} + c\mathbf{v}$, $(ab)\mathbf{u} = a(b\mathbf{u})$, and $(a + b)\mathbf{u} = a\mathbf{u} + b\mathbf{u}$ for any real numbers a, b, and c, and $1\mathbf{u} = \mathbf{u}$. Any set of objects that add like vectors and on which the real numbers can "operate" is called a "vector space."

36 (a) Show that the set of polynomials of degree at most three is a vector space.
  (b) Show that the set of polynomials of degree at least five is not a vector space.

37 (a) Show that the set of all differentiable functions with domain $(0, 1)$ is a vector space.
  (b) Show that the set of all differentiable functions f with domain $(0, 1)$ such that $f(\frac{1}{2}) = 0$ is a vector space.
  (c) Show that the set of all differentiable functions f with domain $(0, 1)$ such that $f(\frac{1}{2}) = 3$ is not a vector space.

38 (a) Show that, if f and g are solutions of the differential equation $d^2y/dx^2 = -9y$, then so are $f + g$ and cf for any real number c.
  (b) The set of solutions of the differential equation $d^2y/dx^2 = -9y$ forms a vector space. Show that $\sin 3x$ and $\cos 3x$ are in that space. It can be proved that every element in the space is of the form $a \sin 3x + b \cos 3x$ for some real numbers a and b.

39 Let $\mathbf{u}_1$ and $\mathbf{u}_2$ be unit vectors and $\mathbf{A} = a\mathbf{u}_1 + b\mathbf{u}_2$. Is $a\mathbf{u}_1$ the vector component of $\mathbf{A}$ parallel to $\mathbf{u}_1$?

11.3 The dot product of two vectors

Consider a rock being pulled along level ground by a rope inclined at a fixed angle to the ground. Let the force applied to the rock be represented by the vector $\mathbf{F}$. The force $\mathbf{F}$ can be expressed as the sum of a vertical force $\mathbf{F}_2$ and a horizontal force $\mathbf{F}_1$, as shown in Fig. 11.39.

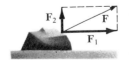

We may replace $\mathbf{F}$ with $\mathbf{F}_1$ and $\mathbf{F}_2$, which together have the same effect on the rock as $\mathbf{F}$.

Figure 11.39

How much work is done by the force $\mathbf{F}$ in moving the rock along the ground? The physicist defines the work accomplished by a constant force

F (whatever direction it may have) in moving a particle on a straight path from the tail to the head of the vector **R** as the following product:

$$\begin{pmatrix} \text{Scalar component of } \mathbf{F} \\ \text{in direction of } \mathbf{R} \end{pmatrix} (\text{Length of } \mathbf{R})$$

Thus the force $\mathbf{F}_2$ shown in Fig. 11.39 accomplishes no work in moving the rock; only the force $\mathbf{F}_1$ does so. The work that **F** accomplishes is the work done by $\mathbf{F}_1$ in overcoming friction and hence equals the product

$$|\mathbf{F}_1| \text{ (distance the rock moves)}.$$

More generally, let a force represented by the vector **F** move an object along a straight line from the tail to the head of a vector **R**, as in Fig. 11.40.

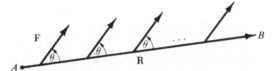

A constant force **F** moves an object from A to B, perhaps against gravity, air resistance, friction, etc.

Figure 11.40

The work accomplished is defined to be

$$\underbrace{|\mathbf{F}| \cos \theta}_{\substack{\text{Magnitude} \\ \text{of force in} \\ \text{direction} \\ \text{object is} \\ \text{moved}}} \cdot \underbrace{|\mathbf{R}|}_{\substack{\text{Distance} \\ \text{the object} \\ \text{is moved}}}$$

where θ is the angle between **R** and **F**. The angle θ can be anywhere in $[0, \pi]$.

This important physical concept illustrates the *dot product* of two vectors, which will be introduced after the following definition.

DEFINITION *Angle between two nonzero vectors.* Let **A** and **B** be two nonparallel and nonzero vectors. They determine a triangle and an angle θ, shown in Fig. 11.41. The *angle between* **A** *and* **B** is θ. Note that

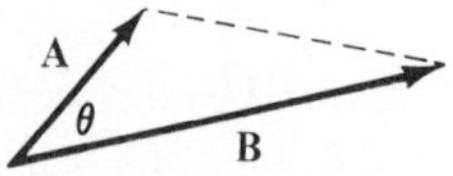

$$0 < \theta < \pi.$$

Figure 11.41

If **A** and **B** are parallel, the angle between them is 0 (if they have the same direction) or π (if they have opposite directions). The angle between **0** and another vector is not defined.

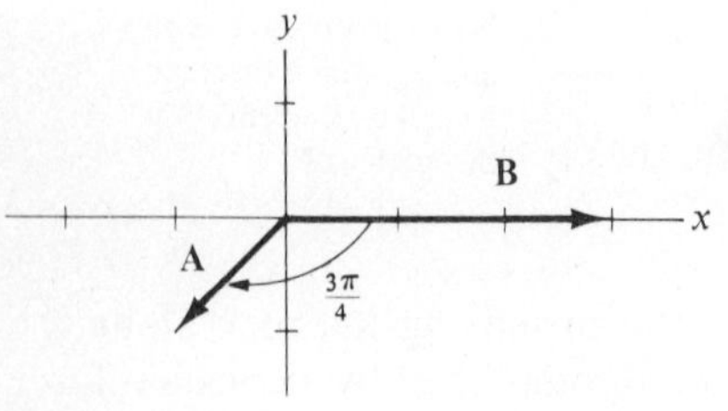

The angle between **i** and **j** is $\pi/2$. The angle between $\mathbf{A} = -\mathbf{i} - \mathbf{j}$ and $\mathbf{B} = 3\mathbf{i}$ is $3\pi/4$, as Fig. 11.42 shows. The angle between **k** and $-\mathbf{k}$ is π; the angle between $2\mathbf{i} + 2\mathbf{j} + 2\mathbf{k}$ and $5\mathbf{i} + 5\mathbf{j} + 5\mathbf{k}$ is 0.

The dot product of vectors can now be defined.

Figure 11.42

THE DOT PRODUCT

DEFINITION *Dot product.* Let **A** and **B** be two nonzero vectors. Their dot product is the number

$$|\mathbf{A}||\mathbf{B}|\cos\theta,$$

where θ is the angle between **A** and **B**. If **A** or **B** is **0**, their dot product is 0. The dot product is denoted $\mathbf{A}\cdot\mathbf{B}$. It is a scalar and is also called the *scalar* product of **A** and **B**.

EXAMPLE 1 Compute the dot product $\mathbf{A}\cdot\mathbf{B}$ if $\mathbf{A}=3\mathbf{i}+3\mathbf{j}$ and $\mathbf{B}=5\mathbf{i}$.

SOLUTION Inspection of Fig. 11.43 shows that θ, the angle between **A** and **B**, is $\pi/4$. Also,

$$|\mathbf{A}|=\sqrt{3^2+3^2}=\sqrt{18},$$

and

$$|\mathbf{B}|=\sqrt{5^2+0^2}=5.$$

Thus

$$\mathbf{A}\cdot\mathbf{B}=|\mathbf{A}||\mathbf{B}|\cos\theta$$

$$=\sqrt{18}\cdot 5\frac{\sqrt{2}}{2}=15.\quad\blacksquare$$

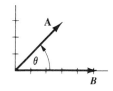

Figure 11.43

EXAMPLE 2 How much work is accomplished by the force $\mathbf{F}=6\mathbf{j}$, in moving an object from the tail to the head of $\mathbf{R}=4\mathbf{i}+4\mathbf{j}$?

SOLUTION The work is defined as the dot product

$$\mathbf{F}\cdot\mathbf{R}=|\mathbf{F}||\mathbf{R}|\cos\theta.$$

Figure 11.44 shows that $\theta=\pi/4$. Hence

$$\text{Work}=\sqrt{0^2+6^2}\sqrt{4^2+4^2}\cos\frac{\pi}{4}$$

$$=6\sqrt{32}\,\frac{\sqrt{2}}{2}$$

$$=\frac{6\sqrt{64}}{2}=24.$$

If force is measured in pounds and distance in feet, the work is 24 foot-pounds. ∎

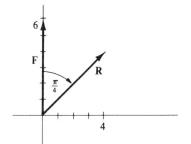

Figure 11.44

$\mathbf{A}\cdot\mathbf{B}=0$ *means that* **A** *is perpendicular to* **B**.

Observe that $\mathbf{A}\cdot\mathbf{A}=|\mathbf{A}|^2$, a fact that will be used often.

Note also that, if **A** is perpendicular to **B**, then $\mathbf{A}\cdot\mathbf{B}=0$. Thus the dot product of two perpendicular vectors is 0. This fact is an important property of the dot product. Moreover, its converse is valid: If $\mathbf{A}\cdot\mathbf{B}=0$ and neither **A** nor **B** is **0**, then the cosine of the angle between **A** and **B** is 0. Thus **A** and **B** are perpendicular. *Consequently, the vanishing of the dot product is a test for perpendicularity.*

The next example gives another physical interpretation of the dot product.

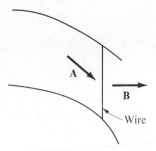

Figure 11.45

EXAMPLE 3   Let **A** represent the direction and the rate of flow of water in a stream as it passes an imaginary fixed wire as shown in Fig. 11.45. Let **B** represent a vector on the surface of the water and perpendicular to the wire such that $|\mathbf{B}|$ equals the length of the wire. Show that $\mathbf{A} \cdot \mathbf{B}$ is proportional to the rate at which water passes the wire.

SOLUTION   If **A** is parallel to the wire, no water passes over the wire. If **A** is perpendicular to the wire, hence parallel to **B**, water passes over the wire. If the angle θ between **A** and **B** is, say, $\pi/4$, water passes over the wire but not as quickly as when **A** is parallel to **B**. The rate at which water passes over the wire depends on the speed of the water, the direction in which the water moves, and the length of the wire. The simplest measure of this rate is the product

$$\begin{pmatrix} \text{Scalar component of } \mathbf{A} \\ \text{in direction of } \mathbf{B} \end{pmatrix} (\text{Length of wire}).$$

If θ denotes the angle between **A** and **B**, then this product equals

$$(|\mathbf{A}| \cos \theta)|\mathbf{B}|,$$

which is the dot product of **A** and **B**, $\mathbf{A} \cdot \mathbf{B}$. ∎

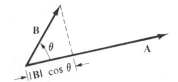

Figure 11.46

The dot product and scalar components

Note that the factor $|\mathbf{B}| \cos \theta$ in the definition of the dot product, $\mathbf{A} \cdot \mathbf{B} = |\mathbf{A}||\mathbf{B}| \cos \theta$, is the scalar component of **B** in the direction along **A**. If $0 \leq \theta < \pi/2$, this scalar component is positive, as in Fig. 11.46. If $\pi/2 < \theta \leq \pi$, this scalar component is negative. In any case,

$$\mathbf{A} \cdot \mathbf{B} = |\mathbf{A}|\,(\text{Scalar component of } \mathbf{B} \text{ along } \mathbf{A})$$

and, similarly,

$$\mathbf{A} \cdot \mathbf{B} = |\mathbf{B}|\,(\text{Scalar component of } \mathbf{A} \text{ along } \mathbf{B}).$$

The next theorem shows how to compute $\mathbf{A} \cdot \mathbf{B}$ in terms of the components of **A** and **B**.

COMPUTING THE DOT PRODUCT

THEOREM 1   *Formula for dot product in scalar components.* If $\mathbf{A} = x_1\mathbf{i} + y_1\mathbf{j}$ and $\mathbf{B} = x_2\mathbf{i} + y_2\mathbf{j}$, then

$$\mathbf{A} \cdot \mathbf{B} = x_1 x_2 + y_1 y_2.$$

If $\mathbf{A} = x_1 \mathbf{i} + y_1 \mathbf{j} + z_1 \mathbf{k}$ and $\mathbf{B} = x_2 \mathbf{i} + y_2 \mathbf{j} + z_2 \mathbf{k}$, then

$$\mathbf{A} \cdot \mathbf{B} = x_1 x_2 + y_1 y_2 + z_1 z_2.$$

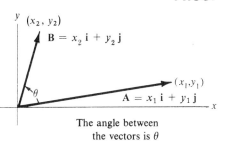

Figure 11.47

PROOF We will prove the theorem for plane vectors only, since the diagrams are easier and the reasoning is the same as for vectors in space.

If either $\mathbf{A}$ or $\mathbf{B}$ is $\mathbf{0}$, a simple computation verifies the theorem. In the remainder of the proof, assume that neither is $\mathbf{0}$. For convenience, put the tails of $\mathbf{A}$ and $\mathbf{B}$ at the origin, as in Fig. 11.47. Now, by the definition of the dot product,

$$\mathbf{A} \cdot \mathbf{B} = \sqrt{x_1^2 + y_1^2} \sqrt{x_2^2 + y_2^2} \cos \theta. \qquad (1)$$

To express $\cos \theta$ in terms of x_1, y_1, x_2, and y_2, apply the law of cosines to the triangle whose vertices are $(0, 0)$, (x_1, y_1), and (x_2, y_2), shown in Fig. 11.48. The law of cosines ($c^2 = a^2 + b^2 - 2ab \cos \theta$, where a, b, c are

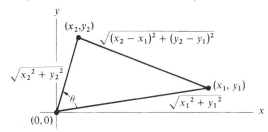

Figure 11.48

the lengths of the sides of a triangle and c is opposite angle θ) asserts in this case that

$$(\sqrt{(x_2 - x_1)^2 + (y_2 - y_1)^2})^2 =$$
$$(\sqrt{x_1^2 + y_1^2})^2 + (\sqrt{x_2^2 + y_2^2})^2 - 2\sqrt{x_1^2 + y_1^2} \sqrt{x_2^2 + y_2^2} \cos \theta.$$

Squaring and canceling yields

$$-2x_1 x_2 - 2y_1 y_2 = -2\sqrt{x_1^2 + y_1^2} \sqrt{x_2^2 + y_2^2} \cos \theta.$$

Dividing this equation by -2, we obtain

$$x_1 x_2 + y_1 y_2 = \sqrt{x_1^2 + y_1^2} \sqrt{x_2^2 + y_2^2} \cos \theta. \qquad (2)$$

Comparison of (1) and (2) establishes the theorem. ■

EXAMPLE 4 In Example 1 the dot product of $3\mathbf{i} + 3\mathbf{j}$ and $5\mathbf{i}$ was computed geometrically. Compute it now with the aid of Theorem 1.

SOLUTION $(3\mathbf{i} + 3\mathbf{j}) \cdot 5\mathbf{i} = 3 \cdot 5 + 3 \cdot 0 = 15.$ ■

EXAMPLE 5 Are the vectors $3\mathbf{i} + 7\mathbf{j}$ and $9\mathbf{i} - 4\mathbf{j}$ perpendicular? (Sketch them before reading the solution; do they seem perpendicular?)

SOLUTION   Recall that two nonzero vectors are perpendicular if their dot product is 0. So compute the dot product:

$$(3\mathbf{i} + 7\mathbf{j}) \cdot (9\mathbf{i} - 4\mathbf{j}) = 3 \cdot 9 + 7(-4)$$
$$= 27 - 28 = -1.$$

Since the dot product is *not* 0, the vectors are not perpendicular. ∎

EXAMPLE 6   Find the angle θ between the two vectors $\mathbf{A} = 2\mathbf{i} - \mathbf{j} + 3\mathbf{k}$ and $\mathbf{B} = \mathbf{i} + \mathbf{j} + 2\mathbf{k}$.

SOLUTION   Compute $\mathbf{A} \cdot \mathbf{B}$ in two ways: first by its geometric definition and then in terms of scalar components.

$$\mathbf{A} \cdot \mathbf{B} = |\mathbf{A}||\mathbf{B}| \cos \theta$$
$$= \sqrt{2^2 + (-1)^2 + 3^2} \sqrt{1^2 + 1^2 + 2^2} \cos \theta$$
$$= \sqrt{14} \sqrt{6} \cos \theta$$
$$= \sqrt{84} \cos \theta.$$

Also, by Theorem 1,

$$\mathbf{A} \cdot \mathbf{B} = 2 \cdot 1 + (-1) \cdot 1 + 3 \cdot 2$$
$$= 2 - 1 + 6$$
$$= 7.$$

Comparing the two versions of $\mathbf{A} \cdot \mathbf{B}$ shows that

$$\sqrt{84} \cos \theta = 7.$$

or
$$\cos \theta = \frac{7}{\sqrt{84}} = \frac{7\sqrt{84}}{84} = \frac{\sqrt{84}}{12} = \frac{\sqrt{21}}{6} \approx 0.764.$$

A calculator or trigonometric table shows that θ is about 40.2°, or 0.702 radian. ∎

How to find the angle between two vectors

The technique illustrated in Example 6 is a way to find the angle between two line segments if their endpoints are given or can be expressed simply in rectangular coordinates. Exercise 19, for instance, concerns the angle between diagonals in a cube. The angle between two lines can be determined by the formula

$$\cos \theta = \frac{\mathbf{A} \cdot \mathbf{B}}{|\mathbf{A}||\mathbf{B}|}.$$

An economic interpretation

In economics the dot product is used as an algebraic convenience, without any regard for its geometric significance. For instance, a shopper buys 20 pounds of potatoes at 5 cents a pound and 10 pounds of oranges at 12 cents a pound. The vector 20**i** + 10**j** records how much he bought of each item, while the vector 5**i** + 12**j** records the corresponding prices. The dot product

$$(20\mathbf{i} + 10\mathbf{j}) \cdot (5\mathbf{i} + 12\mathbf{j}) = 20 \cdot 5 + 10 \cdot 12 = 220 \text{ cents}$$

records the total cost of the purchase.

The dot product has several properties that can be demonstrated directly from the definition or with the aid of Theorem 1:

$$\mathbf{A} \cdot \mathbf{B} = \mathbf{B} \cdot \mathbf{A} \qquad \mathbf{0} \cdot \mathbf{A} = 0$$

$$\mathbf{A} \cdot x\mathbf{B} = x(\mathbf{A} \cdot \mathbf{B}) \qquad \mathbf{A} \cdot \frac{\mathbf{B}}{x} = \frac{\mathbf{A} \cdot \mathbf{B}}{x}$$

(x a scalar) \qquad (x a nonzero scalar)

$$\mathbf{A} \cdot (\mathbf{B} + \mathbf{C}) = \mathbf{A} \cdot \mathbf{B} + \mathbf{A} \cdot \mathbf{C}$$

$$\mathbf{A} \cdot (\mathbf{B} - \mathbf{C}) = \mathbf{A} \cdot \mathbf{B} - \mathbf{A} \cdot \mathbf{C}$$

COMPUTING SCALAR AND VECTOR COMPONENTS

The next theorem is useful in finding scalar components of a vector.

THEOREM 2 *Scalar component along a unit vector.* Let **A** be a vector and let **u** be a unit vector. Then the scalar component of **A** in the direction **u** is the dot product **A** · **u**.

PROOF By definition of the dot product of **A** and **u**,

$$\mathbf{A} \cdot \mathbf{u} = |\mathbf{A}| |\mathbf{u}| \cos \theta,$$

where θ is the angle between **u** and **A**. Since **u** is a unit vector, $|\mathbf{u}| = 1$. Thus

$$\mathbf{A} \cdot \mathbf{u} = |\mathbf{A}| \cos \theta,$$

which is the scalar component of **A** in the direction **u**. ∎

EXAMPLE 7 Find the scalar component of $\mathbf{A} = 2\mathbf{i} + 6\mathbf{j}$ in the direction of the unit vector

$$\mathbf{u} = \frac{\sqrt{2}}{2}\mathbf{i} - \frac{\sqrt{2}}{2}\mathbf{j}.$$

SOLUTION By Theorem 2, this scalar component is **A** · **u**. By Theorem 1,

$$\mathbf{A} \cdot \mathbf{u} = (2\mathbf{i} + 6\mathbf{j}) \cdot \left(\frac{\sqrt{2}}{2}\mathbf{i} - \frac{\sqrt{2}}{2}\mathbf{j}\right) = \frac{2\sqrt{2}}{2} - \frac{6\sqrt{2}}{2} = -2\sqrt{2}.$$

Hence the scalar component of **A** along **u** is $-2\sqrt{2}$. (The negative sign shows that the angle between **A** and **u** is greater than $\pi/2$.) The vector component of **A** along **u** is then $-2\sqrt{2}\,\mathbf{u}$.

Figure 11.49 shows that these results are plausible: $\overrightarrow{OP}$ is the vector component of **A** in the direction of **u**. Its length is $2\sqrt{2}$.

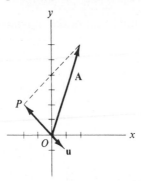

Figure 11.49 ■

Let **A** be a vector and **u** be a unit vector. Then **A** can be expressed as the sum of two vectors, $\mathbf{A} = \mathbf{A}_1 + \mathbf{A}_2$, where $\mathbf{A}_1$ is parallel to **u** and $\mathbf{A}_2$ is perpendicular to **u**. (See Fig. 11.50.)

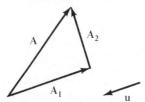

Figure 11.50

Vector components The vector component $\mathbf{A}_1$ is called the *projection* of **A** on **u**. The vector $\mathbf{A}_2$ is called the *vector component of* **A** *perpendicular to* **u**.

We have

$$\mathbf{A}_1 = (\text{scalar component of } \mathbf{A} \text{ on } \mathbf{u})\mathbf{u} = (\mathbf{A} \cdot \mathbf{u})\mathbf{u}.$$

Thus the vector $\mathbf{A}_2$ must be

$$\mathbf{A}_2 = \mathbf{A} - \mathbf{A}_1 = \mathbf{A} - (\mathbf{A} \cdot \mathbf{u})\mathbf{u}.$$

The formulas for $\mathbf{A}_1$ and $\mathbf{A}_2$ are of sufficient importance to be honored with a theorem.

THEOREM 3 Let **A** be a vector and **u** a unit vector. Then **A** is the sum of a vector parallel to **u** and a vector perpendicular to **u**. The first is $(\mathbf{A} \cdot \mathbf{u})\mathbf{u}$. The second is $\mathbf{A} - (\mathbf{A} \cdot \mathbf{u})\mathbf{u}$. ■

If **A** and **B** are vectors and **B** is not **0**, then **A** can be expressed as the sum of a vector $\mathbf{A}_1$ parallel to **B** and a vector $\mathbf{A}_2$ perpendicular to **B**. These two vectors can be found by applying Theorem 3 in the case $\mathbf{u} = \mathbf{B}/|\mathbf{B}|$:

$$\mathbf{A}_1 = \left(\mathbf{A} \cdot \frac{\mathbf{B}}{|\mathbf{B}|}\right)\frac{\mathbf{B}}{|\mathbf{B}|} \qquad \mathbf{A}_2 = \mathbf{A} - \left(\mathbf{A} \cdot \frac{\mathbf{B}}{|\mathbf{B}|}\right)\frac{\mathbf{B}}{|\mathbf{B}|}. \qquad (3)$$

Perpendicular vectors are frequently called "orthogonal."

A_1 is called the *vector component of* **A** *on* **B** or the *projection of* **A** *on* **B**. A_2 is the *vector component of* **A** *perpendicular to* **B**. Note that if **B** is a unit vector, these definitions agree with those just given. [It is more convenient to use the formula in Theorem 3 with $\mathbf{u} = \mathbf{B}/|\mathbf{B}|$ than memorizing formula (3).]

EXAMPLE 8 Let P, Q, and R be three points in space, such that $Q \neq R$. Find the distance from P to the line through Q and R. (See Fig. 11.51.)

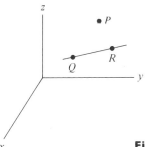

Figure 11.51

SOLUTION Draw the vectors $\overrightarrow{QP}$ and $\overrightarrow{QR}$, shown in Fig. 11.52. The distance from P to the line through Q and R is the length of the dashed line in Fig. 11.52. The length of that line is the magnitude of the vector component of $\overrightarrow{QP}$ perpendicular to the vector $\overrightarrow{QR}$. Thus

$$\text{Distance from } P \text{ to the line through } Q \text{ and } R = \left| \overrightarrow{QP} - \left(\overrightarrow{QP} \cdot \frac{\overrightarrow{QR}}{|\overrightarrow{QR}|} \right) \frac{\overrightarrow{QR}}{|\overrightarrow{QR}|} \right|.$$

If the coordinates of P, Q, and R were given, they could be substituted directly into that formula. ∎

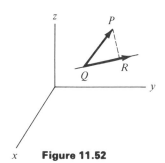

Figure 11.52

The final example illustrates the use of the dot product in establishing geometric properties.

EXAMPLE 9 Let $PQRS$ be a parallelogram. Show that if the two diagonals of the parallelogram are perpendicular, then the four sides have the same length.

SOLUTION Let $\mathbf{A} = \overrightarrow{PQ}$ and $\mathbf{B} = \overrightarrow{PS}$, as shown in Fig. 11.53. One diagonal is $\mathbf{A} + \mathbf{B}$ and the other is $\mathbf{A} - \mathbf{B}$. These diagonals are perpendicular if

$$(\mathbf{A} + \mathbf{B}) \cdot (\mathbf{A} - \mathbf{B}) = 0$$

or

$\mathbf{A} \cdot \mathbf{A} + \mathbf{B} \cdot \mathbf{A} - \mathbf{A} \cdot \mathbf{B} - \mathbf{B} \cdot \mathbf{B} = 0$ (expanding)

$\mathbf{A} \cdot \mathbf{A} - \mathbf{B} \cdot \mathbf{B} = 0$ $(\mathbf{A} \cdot \mathbf{B} = \mathbf{B} \cdot \mathbf{A})$

$\mathbf{A} \cdot \mathbf{A} = \mathbf{B} \cdot \mathbf{B}$

$|\mathbf{A}|^2 = |\mathbf{B}|^2$

$|\mathbf{A}| = |\mathbf{B}|.$

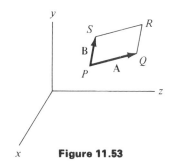

Figure 11.53

716 ALGEBRAIC OPERATIONS ON VECTORS

This shows that the sides PQ and PS are of equal length. Hence all the sides are of the same length. ∎

Exercises

In Exercises 1 to 6 compute $\mathbf{A} \cdot \mathbf{B}$.

1. $\mathbf{A}$ has length 3, $\mathbf{B}$ has length 4, and the angle between $\mathbf{A}$ and $\mathbf{B}$ is $\pi/4$.
2. $\mathbf{A}$ has length 2, $\mathbf{B}$ has length 3, and the angle between $\mathbf{A}$ and $\mathbf{B}$ is $3\pi/4$.
3. $\mathbf{A}$ has length 5, $\mathbf{B}$ has length $\frac{1}{2}$, and the angle between $\mathbf{A}$ and $\mathbf{B}$ is $\pi/2$.
4. $\mathbf{A}$ is the zero vector $\mathbf{0}$, and $\mathbf{B}$ has length 5.
5. $\mathbf{A} = 2\mathbf{i} - 3\mathbf{j} + 5\mathbf{k}$, and $\mathbf{B} = \mathbf{i} - \mathbf{j} - \mathbf{k}$.
6. $\mathbf{A} = \overrightarrow{PQ}$, and $\mathbf{B} = \overrightarrow{PR}$, where $P = (1, 0, 2)$, $Q = (1, 1, -1)$, $R = (2, 3, 5)$.
7. (a) Draw the vectors $7\mathbf{i} + 12\mathbf{j}$ and $9\mathbf{i} - 5\mathbf{j}$.
   (b) Do they seem to be perpendicular?
   (c) Determine whether they are perpendicular by examining their dot product.
8. (a) Draw the vectors $\mathbf{i} + 2\mathbf{j} + 3\mathbf{k}$ and $\mathbf{i} + \mathbf{j} - \mathbf{k}$.
   (b) Do they seem to be perpendicular?
   (c) Determine whether they are perpendicular by examining their dot product.
9. (a) Estimate the angle between $\mathbf{A} = 3\mathbf{i} + 4\mathbf{j}$ and $\mathbf{B} = 5\mathbf{i} + 12\mathbf{j}$ by drawing them.
   (b) Find the angle between $\mathbf{A}$ and $\mathbf{B}$.
10. Find the cosine of the angle between $2\mathbf{i} - 4\mathbf{j} + 6\mathbf{k}$ and $\mathbf{i} + 2\mathbf{j} + 3\mathbf{k}$.
11. (a) Show that the scalar component of $\mathbf{A}$ on $\mathbf{B}$ is given by the formula $\mathbf{A} \cdot \mathbf{B}/|\mathbf{B}|$.
    (b) Find the scalar component of $2\mathbf{i} - \mathbf{j}$ on $\mathbf{j} + 3\mathbf{k}$.
    (c) Find the scalar component of $2\mathbf{i} - \mathbf{j}$ on $(\mathbf{j} + 3\mathbf{k})/|\mathbf{j} + 3\mathbf{k}|$.
12. Find the scalar component of $\mathbf{i} + 4\mathbf{j} + 2\mathbf{k}$ on (a) $3\mathbf{j} + 2\mathbf{k}$, (b) $-3\mathbf{j} - 2\mathbf{k}$.
13. Find the cosine of the angle between $\overrightarrow{AB}$ and $\overrightarrow{CD}$ if $A = (1, 3)$, $B = (7, 4)$, $C = (2, 8)$, and $D = (1, -5)$.
14. Find the cosine of the angle between $\overrightarrow{AB}$ and $\overrightarrow{CD}$ if $A = (1, 2, -5)$, $B = (1, 0, 1)$, $C = (0, -1, 3)$, and $D = (2, 1, 4)$.

In Exercises 15 to 18 find the vector components of $\mathbf{A}$ parallel and perpendicular to $\mathbf{B}$.

15. $\mathbf{A} = 2\mathbf{i} + 3\mathbf{j} + 4\mathbf{k}$, $\mathbf{B} = \mathbf{i} + \mathbf{j} + \mathbf{k}$.
16. $\mathbf{A} = \mathbf{j} + \mathbf{k}$, $\mathbf{B} = \mathbf{i} - \mathbf{j}$.
17. $\mathbf{A} = \mathbf{i} + 2\mathbf{j} + 3\mathbf{k}$, $\mathbf{B} = 2\mathbf{j} + 3\mathbf{k}$.
18. $\mathbf{A} = \mathbf{j}$, $\mathbf{B} = 2\mathbf{i} + 3\mathbf{j} - \mathbf{k}$.

∎

Exercises 19 to 23 refer to the cube in Fig. 11.54.
19. Find the cosine of the angle between $\overrightarrow{AC}$ and $\overrightarrow{BD}$.

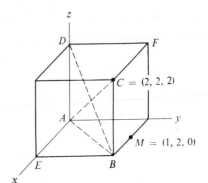

Figure 11.54

20. Find the cosine of the angle between $\overrightarrow{AF}$ and $\overrightarrow{BD}$.
21. Find the cosine of the angle between $\overrightarrow{AC}$ and $\overrightarrow{AM}$.
22. Find the cosine of the angle between $\overrightarrow{MD}$ and $\overrightarrow{MF}$.
23. Find the cosine of the angle between $\overrightarrow{EF}$ and $\overrightarrow{BD}$.
24. How far is the point $(1, 2, 3)$ from the line through the points $(1, 4, 2)$ and $(2, 1, -4)$?
25. If $\mathbf{A} \cdot \mathbf{B} = \mathbf{A} \cdot \mathbf{C}$ and $\mathbf{A}$ is not $\mathbf{0}$ must $\mathbf{B} = \mathbf{C}$?
26. The projection of the vector $\mathbf{A}$ on the vector $\mathbf{B}$ is the vector $\mathbf{i} - \mathbf{j} + 4\mathbf{k}$. What is the projection of the vector $\mathbf{A}$ on (a) $2\mathbf{B}$? (b) $-\mathbf{B}$?
27. The scalar component of the vector $\mathbf{A}$ on the vector $\mathbf{B}$ is 15. What is the scalar component of $\mathbf{A}$ on (a) $2\mathbf{B}$? (b) $-\mathbf{B}$?
28. Prove that for any vector $\mathbf{A}$, $\mathbf{A} \cdot \mathbf{A} = |\mathbf{A}|^2$, using
    (a) the definition of the dot product;
    (b) Theorem 1.
29. Use Theorem 1 to prove that $x\mathbf{A} \cdot y\mathbf{B} = xy(\mathbf{A} \cdot \mathbf{B})$.
30. Use Theorem 1 to prove that
    (a) $\mathbf{A} \cdot (\mathbf{B} + \mathbf{C}) = \mathbf{A} \cdot \mathbf{B} + \mathbf{A} \cdot \mathbf{C}$;
    (b) $(x\mathbf{A}) \cdot \mathbf{B} = \mathbf{A} \cdot (x\mathbf{B}) = x(\mathbf{A} \cdot \mathbf{B})$.
31. Draw two vectors $\mathbf{A}$ and $\mathbf{B}$ such that
    (a) $\mathbf{A} \cdot \mathbf{B} < 0$;
    (b) $\mathbf{A} \cdot \mathbf{B} > 0$;
    (c) $\mathbf{A} \cdot \mathbf{B} = 0$.
32. A firm sells x chairs at C dollars per chair and y desks at D dollars per desk. It costs the firm c dollars to make a chair and d dollars to make a desk. What is the economic interpretation of
    (a) Cx?
    (b) $(x\mathbf{i} + y\mathbf{j}) \cdot (C\mathbf{i} + D\mathbf{j})$?
    (c) $(x\mathbf{i} + y\mathbf{j}) \cdot (c\mathbf{i} + d\mathbf{j})$?
    (d) $(x\mathbf{i} + y\mathbf{j}) \cdot (C\mathbf{i} + D\mathbf{j}) > (x\mathbf{i} + y\mathbf{j}) \cdot (c\mathbf{i} + d\mathbf{j})$?

33 Let $\mathbf{u}_1$, $\mathbf{u}_2$, and $\mathbf{u}_3$ be perpendicular unit vectors, and let $\mathbf{A}$ be any vector. Show that

$$\mathbf{A} = (\mathbf{A} \cdot \mathbf{u}_1)\mathbf{u}_1 + (\mathbf{A} \cdot \mathbf{u}_2)\mathbf{u}_2 + (\mathbf{A} \cdot \mathbf{u}_3)\mathbf{u}_3$$

by finding the dot product of both sides with the vectors $\mathbf{u}_1$, $\mathbf{u}_2$, and $\mathbf{u}_3$.

34 (See Exercise 33.)

Let $\mathbf{u}_1 = \frac{\sqrt{2}}{2}\mathbf{i} - \frac{\sqrt{2}}{2}\mathbf{j}$, $\mathbf{u}_2 = \frac{\sqrt{2}}{2}\mathbf{i} + \frac{\sqrt{2}}{2}\mathbf{j}$ and $\mathbf{u}_3 = \mathbf{k}$.

Let $\mathbf{A} = 2\mathbf{i} + 3\mathbf{j} - 4\mathbf{k}$. Find x, y, and z such that $\mathbf{A} = x\mathbf{u}_1 + y\mathbf{u}_2 + z\mathbf{u}_3$.

35 (a) Using elementary geometry, show that $|\mathbf{A} + \mathbf{B}| \leq |\mathbf{A}| + |\mathbf{B}|$.
(b) From (a) deduce that for any real numbers x_1, y_1, z_1, x_2, y_2, z_2,

$$x_1 x_2 + y_1 y_2 + z_1 z_2 \leq \sqrt{x_1^2 + y_1^2 + z_1^2}\sqrt{x_2^2 + y_2^2 + z_2^2}.$$

Exercises 36 to 39 are related.

36 In a more abstract approach than the one followed in this book, a plane vector is defined as an ordered pair of real numbers (x, y). Taking this purely algebraic approach, define
(a) the vector $\mathbf{0}$,
(b) the product of a scalar and a vector,
(c) the sum of two vectors,
(d) the magnitude of a vector,
(e) the dot product.

Just as it is possible to define a plane vector as an ordered pair of real numbers (x, y), or a space vector as an ordered triple of real numbers (x, y, z), it is possible to define vectors algebraically in higher dimensions. A four-dimensional vector can be defined to be an ordered quadruple of numbers $\overrightarrow{(x_1, x_2, x_3, x_4)}$. The *dot product* of two vectors $\overrightarrow{(x_1, x_2, x_3, x_4)}$ and $\overrightarrow{(y_1, y_2, y_3, y_4)}$ is defined as the sum $x_1 y_1 + x_2 y_2 + x_3 y_3 + x_4 y_4$. The *magnitude* of a vector $\overrightarrow{(x_1, x_2, x_3, x_4)}$ is defined as $\sqrt{x_1^2 + x_2^2 + x_3^2 + x_4^2}$. This can be done in any number of dimensions.

37 (a) If the sequence $\overrightarrow{(x_1, x_2, x_3, \ldots, x_n, \ldots)}$ is considered an *infinite-dimensional vector*, how would its magnitude be defined? Would every vector have a magnitude?
(b) What is the magnitude of the vector

$$\overrightarrow{(\tfrac{1}{2}, \tfrac{1}{4}, \tfrac{1}{8}, \tfrac{1}{16}, \ldots, 1/2^n, \ldots)}?$$

38 The output of a firm that manufactures x_1 washing machines, x_2 refrigerators, x_3 dishwashers, x_4 stoves, and x_5 clothes dryers is recorded by the five-dimensional production vector $\mathbf{P} = \overrightarrow{(x_1, x_2, x_3, x_4, x_5)}$. Similarly, the cost vector $\mathbf{C} = \overrightarrow{(y_1, y_2, y_3, y_4, y_5)}$ records the cost of producing each item; for instance, each refrigerator costs the firm y_2 dollars.
(a) What is the economic significance of $\mathbf{P} \cdot \mathbf{C} = \overrightarrow{(20, 0, 7, 9, 15)} \cdot \overrightarrow{(50, 70, 30, 20, 10)}$?
(b) If the firm doubles its production of all items in (a), what is its new production vector?

39 Let P_1 be the profit from selling a washing machine and P_2, P_3, P_4, and P_5 be defined analogously for the firm of Exercise 38. (Some of the P's may be negative.) What does it mean to the firm to have $\overrightarrow{(P_1, P_2, P_3, P_4, P_5)}$ "perpendicular" to

$$\overrightarrow{(x_1, x_2, x_3, x_4, x_5)}?$$

40 By considering the dot product of the two unit vectors $\mathbf{u}_1 = \cos\theta_1\mathbf{i} + \sin\theta_1\mathbf{j}$ and $\mathbf{u}_2 = \cos\theta_2\mathbf{i} + \sin\theta_2\mathbf{j}$, prove that

$$\cos(\theta_1 - \theta_2) = \cos\theta_1 \cos\theta_2 + \sin\theta_1 \sin\theta_2.$$

41 Prove Theorem 1 for the two vectors in space, $\mathbf{A} = x_1\mathbf{i} + y_1\mathbf{j} + z_1\mathbf{k}$ and $\mathbf{B} = x_2\mathbf{i} + y_2\mathbf{j} + z_2\mathbf{k}$.

42 Consider a tetrahedron (not necessarily regular). It has six edges. Show that the line segment joining the midpoints of two opposite edges is perpendicular to the line segment joining another pair of opposite edges if and only if the remaining two edges are of the same length.

11.4 Lines and planes

The equations of lines and planes are easy to develop with the aid of vectors, especially with the dot-product test for perpendicularity. This section develops the basic geometry of lines and planes and, in particular, formulas for determining the distance between a point and a line or plane.

LINES IN THE PLANE

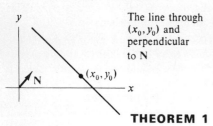

Figure 11.55

The line through (x_0, y_0) and perpendicular to **N**

Let $\mathbf{N} = A\mathbf{i} + B\mathbf{j}$ be a nonzero vector and (x_0, y_0) be a point in the xy plane. There is a unique line through (x_0, y_0) that is perpendicular to **N**, as shown in Fig. 11.55. **N** is called a *normal* to the line. The next theorem provides an algebraic criterion for determining whether the point (x, y) lies on this line.

THEOREM 1 An equation of the line (in the xy plane) passing through (x_0, y_0) and perpendicular to the nonzero vector $\mathbf{N} = A\mathbf{i} + B\mathbf{j}$ is given by

$$A(x - x_0) + B(y - y_0) = 0.$$

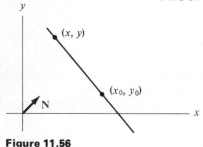

Figure 11.56

PROOF Let (x, y) be a point on the line perpendicular to **N**. (See Fig. 11.56.) Then the vector $(x - x_0)\mathbf{i} + (y - y_0)\mathbf{j}$ is perpendicular to **N**. Hence

$$0 = [(x - x_0)\mathbf{i} + (y - y_0)\mathbf{j}] \cdot \mathbf{N}$$
$$= A(x - x_0) + B(y - y_0).$$

Conversely, it must be shown that if $A(x - x_0) + B(y - y_0) = 0$, then (x, y) is on the line through (x_0, y_0) perpendicular to **N**. The number $A(x - x_0) + B(y - y_0)$ is the scalar product of the vectors **N** and $(x - x_0)\mathbf{i} + (y - y_0)\mathbf{j}$. If this scalar product is zero, the two vectors **N** and $(x - x_0)\mathbf{i} + (y - y_0)\mathbf{j}$ are perpendicular. Thus (x, y) lies on the line through (x_0, y_0) perpendicular to **N**. ∎

EXAMPLE 1 Find an equation of the line through $(2, -7)$ and perpendicular to the vector $4\mathbf{i} + \mathbf{j}$.

SOLUTION By Theorem 1 an equation is

$$4(x - 2) + 1(y + 7) = 0,$$

which, when multiplied out, is

$$4x + y = 1. \quad \blacksquare$$

As Theorem 1 and Example 1 show, to find a vector perpendicular to a given line $Ax + By + C = 0$, read off the coefficients of x and y in order, A and B, and form the vector $A\mathbf{i} + B\mathbf{j}$. It will be perpendicular to the line. The constant term C plays no role in determining the direction of the line or of a vector perpendicular to it.

THEOREM 2 The distance from the point $P_1 = (x_1, y_1)$ to the line L whose equation is $Ax + By + C = 0$ is

$$\frac{|Ax_1 + By_1 + C|}{\sqrt{A^2 + B^2}}.$$

PROOF Select a point $P_0 = (x_0, y_0)$ on the line L. Let $\mathbf{u}$ be a unit vector perpendicular to L. Inspection of the right triangle in Fig. 11.57 shows that the distance from P_1 to the line L is the absolute value of the scalar component of the vector $\overrightarrow{P_0 P_1}$ in the direction $\mathbf{u}$. Thus, by Theorem 2 of Sec. 11.3,

$$\text{Distance from } P_1 \text{ to } L = |\mathbf{u} \cdot \overrightarrow{P_0 P_1}|.$$

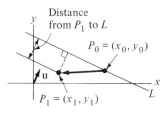

Figure 11.57

Since $A\mathbf{i} + B\mathbf{j}$ is perpendicular to L,

$$\mathbf{u} = \frac{A\mathbf{i} + B\mathbf{j}}{\sqrt{A^2 + B^2}}$$

is a unit normal to L. Thus

$$\begin{aligned}\text{Distance from } P_1 \text{ to } L &= \left| \frac{A\mathbf{i} + B\mathbf{j}}{\sqrt{A^2 + B^2}} \cdot \overrightarrow{P_0 P_1} \right| \\ &= \left| \frac{A\mathbf{i} + B\mathbf{j}}{\sqrt{A^2 + B^2}} \cdot [(x_1 - x_0)\mathbf{i} + (y_1 - y_0)\mathbf{j}] \right| \\ &= \frac{|A(x_1 - x_0) + B(y_1 - y_0)|}{\sqrt{A^2 + B^2}} \\ &= \frac{|Ax_1 + By_1 - (Ax_0 + By_0)|}{\sqrt{A^2 + B^2}}\end{aligned}$$

Since $P_0 = (x_0, y_0)$ is on L, $Ax_0 + By_0 + C = 0$, or $Ax_0 + By_0 = -C$. Thus

$$\text{Distance from } P_1 \text{ to } L = \frac{|Ax_1 + By_1 - (-C)|}{\sqrt{A^2 + B^2}} = \frac{|Ax_1 + By_1 + C|}{\sqrt{A^2 + B^2}},$$

and the theorem is established. ∎

EXAMPLE 2 Find the distance from the line $4x + y = 1$ to the origin.

SOLUTION The line has the equation

$$4x + y - 1 = 0.$$

In this case, $A = 4$, $B = 1$, $C = -1$, $x_1 = 0$, and $y_1 = 0$. By Theorem 2, the distance from the line to the origin is

$$\frac{|4 \cdot 0 + 1 \cdot 0 + (-1)|}{\sqrt{4^2 + 1^2}} = \frac{1}{\sqrt{17}}. \quad \blacksquare$$

EXAMPLE 3 Find the distance from the point $(3, 7)$ to the line $2x - 4y + 5 = 0$.

SOLUTION By Theorem 2, the distance is

$$\frac{|2 \cdot 3 - 4 \cdot 7 + 5|}{\sqrt{2^2 + 4^2}} = \frac{|6 - 28 + 5|}{\sqrt{20}}$$

$$= \frac{|-17|}{\sqrt{20}}$$

$$= \frac{17}{\sqrt{20}} \quad \blacksquare$$

PLANES

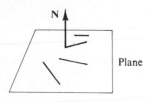

N is perpendicular to the plane.
(Four line segments in the plane are shown.)
Figure 11.58

Analogs of Theorems 1 and 2 hold for planes located in space. First of all, the notion of a vector being perpendicular to a plane must be defined. A vector **N** is said to be *perpendicular* to a plane if **N** is perpendicular to every line situated in the plane. (See Fig. 11.58.)

The next theorem is analogous to Theorem 1; its proof, which is practically the same as that of Theorem 1, is omitted. It gives an algebraic condition that a point (x, y, z) must satisfy to be in a particular plane.

THEOREM 3 An equation of the plane passing through (x_0, y_0, z_0) and perpendicular to the nonzero vector $A\mathbf{i} + B\mathbf{j} + C\mathbf{k}$ is given by

$$A(x - x_0) + B(y - y_0) + C(z - z_0) = 0. \quad \blacksquare$$

EXAMPLE 4 Find an equation of the plane through $(1, -2, 4)$ that is perpendicular to the vector $5\mathbf{i} + 3\mathbf{j} + 6\mathbf{k}$.

SOLUTION By Theorem 3, an equation of the plane is

$$5(x - 1) + 3(y - (-2)) + 6(z - 4) = 0,$$

which simplifies to

$$5x + 3y + 6z = 23. \quad \blacksquare$$

Example 4 shows that the equation $5x + 3y + 6z = 23$ describes a plane. (The plane passes through $(1, -2, 4)$ and is perpendicular to the vector $5\mathbf{i} + 3\mathbf{j} + 6\mathbf{k}$.) The next theorem generalizes this result.

THEOREM 4 Let A, B, C, and D be constants such that not all A, B, and C are 0. Then the equation $Ax + By + Cz + D = 0$ describes a plane. Moreover, the vector $A\mathbf{i} + B\mathbf{j} + C\mathbf{k}$ is perpendicular to this plane.

PROOF Choose a point $P_0 = (x_0, y_0, z_0)$ that satisfies the equation $Ax + By + Cz + D = 0$. Let $P = (x, y, z)$ be any point that satisfies the equation. Then compute the dot product of $\overrightarrow{P_0 P}$ with $A\mathbf{i} + B\mathbf{j} + C\mathbf{k}$, to show that it is 0. Now, $\overrightarrow{P_0 P}$ lies in the given plane and has the form

$$\overrightarrow{P_0 P} = (x - x_0)\mathbf{i} + (y - y_0)\mathbf{j} + (z - z_0)\mathbf{k}.$$

Thus $(A\mathbf{i} + B\mathbf{j} + C\mathbf{k}) \cdot \overrightarrow{P_0 P} = A(x - x_0) + B(y - y_0) + C(z - z_0)$

Recall that $Ax + By + Cz = -D$ and $Ax_0 + By_0 + Cz_0 = -D$.

$$= Ax + By + Cz - (Ax_0 + By_0 + Cz_0)$$
$$= -D - (-D)$$
$$= 0.$$

The steps are reversible. Thus the set of points that satisfy the equation $Ax + By + Cz + D = 0$ is the plane perpendicular to $A\mathbf{i} + B\mathbf{j} + C\mathbf{k}$ and passing through the point (x_0, y_0, z_0). ∎

The next theorem is a natural companion of Theorem 2. The proof, which is practically the same as that of Theorem 2, is omitted.

THEOREM 5 The distance from the point (x_1, y_1, z_1) to the plane

$$Ax + By + Cz + D = 0$$

is

$$\frac{|Ax_1 + By_1 + Cz_1 + D|}{\sqrt{A^2 + B^2 + C^2}}. \quad ∎$$

EXAMPLE 5 How far is the point $(2, 1, 5)$ from the plane $x - 3y + 4z + 8 = 0$?

SOLUTION By Theorem 5 the desired distance is

$$\frac{|1 \cdot 2 - 3 \cdot 1 + 4 \cdot 5 + 8|}{\sqrt{1^2 + (-3)^2 + 4^2}} = \frac{|2 - 3 + 20 + 8|}{\sqrt{26}}$$
$$= \frac{27}{\sqrt{26}}. \quad ∎$$

LINES IN SPACE

So far in this section we have been concerned with lines in the xy plane and planes in space. Vectors also provide a neat way to treat the geometry of lines in space, as will now be shown.

Consider the line L through the point $P_0 = (x_0, y_0, z_0)$ and parallel to the vector $\mathbf{A} = a_1\mathbf{i} + a_2\mathbf{j} + a_3\mathbf{k}$ shown in Fig. 11.59. A point $P = (x, y, z)$ is on this line if and only if the vector $\overrightarrow{P_0P}$ is parallel to $\mathbf{A}$. One way to express that $\overrightarrow{P_0P}$ is parallel to $\mathbf{A}$ is to assert that there is a scalar t such that

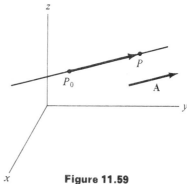

Figure 11.59

722 ALGEBRAIC OPERATIONS ON VECTORS

$$\overrightarrow{P_0P} = t\mathbf{A},$$

that is, $(x - x_0)\mathbf{i} + (y - y_0)\mathbf{j} + (z - z_0)\mathbf{k} = ta_1\mathbf{i} + ta_2\mathbf{j} + ta_3\mathbf{k}$.

In short, $x - x_0 = ta_1$, $y - y_0 = ta_2$, $z - z_0 = ta_3$.

Consequently, we have these *parametric equations* for the line through (x_0, y_0, z_0) parallel to $\mathbf{A} = a_1\mathbf{i} + a_2\mathbf{j} + a_3\mathbf{k}$:

Parametric equations of a line in space

$$\begin{cases} x = x_0 + a_1 t \\ y = y_0 + a_2 t \\ z = z_0 + a_3 t. \end{cases}$$

These equations can also be expressed vectorially. Let $\mathbf{P}$ be the vector $\overrightarrow{OP}$ and $\mathbf{P}_0$ be the vector $\overrightarrow{OP_0}$. The three scalar equations reduce to the single vector equation

Vector equation of a line

$$\mathbf{P} = \mathbf{P}_0 + t\mathbf{A}.$$

The equations of the line through (x_0, y_0, z_0) and parallel to $\mathbf{A} = a_1\mathbf{i} + a_2\mathbf{j} + a_3\mathbf{k}$ can be given in the form

$$\frac{x - x_0}{a_1} = \frac{y - y_0}{a_2} = \frac{z - z_0}{a_3},$$

if none of a_1, a_2, a_3 is 0. These nonparametric equations describe the line as the intersection of two planes, namely,

$$\frac{x - x_0}{a_1} = \frac{y - y_0}{a_2} \quad \text{and} \quad \frac{y - y_0}{a_2} = \frac{z - z_0}{a_3}.$$

EXAMPLE 6 Write parametric equations for the line through $(1, 2, 2)$ and parallel to the vector $3\mathbf{i} - \mathbf{j} + 5\mathbf{k}$. Does the point $(10, -1, 16)$ lie on the line?

SOLUTION Parametric equations of the line are given by

$$\begin{cases} x = 1 + 3t \\ y = 2 - t \\ z = 2 + 5t. \end{cases}$$

Does $(10, -1, 16)$ lie on this line? To find out, determine if there is a number t such that these three equations are *simultaneously* satisfied:

$$10 = 1 + 3t$$
$$-1 = 2 - t$$
$$16 = 2 + 5t.$$

The first equation, $10 = 1 + 3t$, has the solution $t = 3$. This value, 3, does satisfy the second equation, since $-1 = 2 - 3$. But it does *not* satisfy the third equation, since $16 \neq 2 + 5 \cdot 3$. Hence $(10, -1, 16)$ is *not* on the line. ∎

DEFINITION *Direction numbers of a line.* If the vector $\mathbf{A} = a_1 \mathbf{i} + a_2 \mathbf{j} + a_3 \mathbf{k}$ is parallel to the line L, then the numbers a_1, a_2, and a_3 are called *direction numbers* of L. (Note that direction numbers are not unique.)

The next definition is closely related to the preceding one.

The direction of a vector in the plane is described by a single angle, the angle it makes with the positive x axis. The direction of a vector in space involves three angles, two of which almost determine the third.

DEFINITION *Direction angles of a vector.* Let $\mathbf{A}$ be a nonzero vector in space. The angle between

$\mathbf{A}$ and $\mathbf{i}$ is denoted α,

$\mathbf{A}$ and $\mathbf{j}$ is denoted β,

$\mathbf{A}$ and $\mathbf{k}$ is denoted γ.

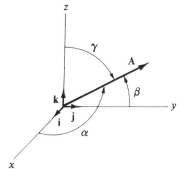

Figure 11.60

The angles α, β, γ are called the *direction angles* of $\mathbf{A}$. (See Fig. 11.60.)

The numbers $\cos \alpha$, $\cos \beta$, and $\cos \gamma$ are called the *direction cosines* of $\mathbf{A}$. If $\mathbf{A}$ is parallel to the line L, then the direction cosines of $\mathbf{A}$ are also called *direction cosines* for L. Note that the direction cosines of $-\mathbf{A}$ are the negatives of the direction cosines of $\mathbf{A}$.

The three direction angles are not independent of each other, as is shown by the next theorem, which concerns the *direction cosines*, $\cos \alpha$, $\cos \beta$, and $\cos \gamma$.

THEOREM 6 If α, β, and γ are the direction angles of the vector $\mathbf{A}$, then

$$\cos^2 \alpha + \cos^2 \beta + \cos^2 \gamma = 1.$$

PROOF It is no loss of generality to assume that $\mathbf{A}$ is a unit vector,

$$\mathbf{A} = x\mathbf{i} + y\mathbf{j} + z\mathbf{k},$$

where
$$x^2 + y^2 + z^2 = 1.$$

Now,
$$\mathbf{A} \cdot \mathbf{i} = |\mathbf{A}||\mathbf{i}| \cos \alpha$$
$$= 1 \cdot 1 \cdot \cos \alpha$$
$$= \cos \alpha.$$

The direction cosines of a unit vector are just its scalar components on $\mathbf{i}$, $\mathbf{j}$, and $\mathbf{k}$.

But $\mathbf{A} \cdot \mathbf{i} = (x\mathbf{i} + y\mathbf{j} + z\mathbf{k}) \cdot (\mathbf{i} + 0\mathbf{j} + 0\mathbf{k}) = x \cdot 1 + y \cdot 0 + z \cdot 0 = x.$

Thus
$$\cos \alpha = x.$$

Similarly,
$$\cos \beta = y,$$

and
$$\cos \gamma = z.$$

Hence
$$\cos^2 \alpha + \cos^2 \beta + \cos^2 \gamma = x^2 + y^2 + z^2 = 1. \blacksquare$$

EXAMPLE 7 The vector **A** makes angles of 60° with the x and y axes. What angle does it make with the z axis?

SOLUTION Here $\alpha = 60°$ and $\beta = 60°$; hence

$$\cos \alpha = \tfrac{1}{2}, \quad \text{and} \quad \cos \beta = \tfrac{1}{2}.$$

Since
$$\cos^2 \alpha + \cos^2 \beta + \cos^2 \gamma = 1,$$

it follows that
$$(\tfrac{1}{2})^2 + (\tfrac{1}{2})^2 + \cos^2 \gamma = 1,$$

so
$$\cos^2 \gamma = \tfrac{1}{2}.$$

Thus
$$\cos \gamma = \frac{\sqrt{2}}{2} \quad \text{or} \quad \cos \gamma = -\frac{\sqrt{2}}{2}.$$

Hence
$$\gamma = 45° \quad \text{or} \quad 135°.$$

Figures 11.61 and 11.62 show the two possibilities for the direction of **A**.

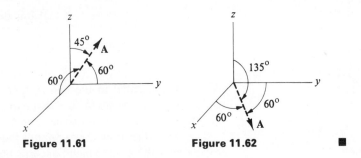

Figure 11.61   Figure 11.62

Exercises

In each of Exercises 1 to 4 find an equation of the line through the given point and perpendicular to the given vector.

1. $(2, 3)$, $4\mathbf{i} + 5\mathbf{j}$.
2. $(1, 0)$, $2\mathbf{i} - \mathbf{j}$.
3. $(4, 5)$, $2\mathbf{i} + 3\mathbf{j}$.
4. $(2, -1)$, $\mathbf{i} + 3\mathbf{j}$

In each of Exercises 5 to 8 find a vector perpendicular to the given line.

5. $2x - 3y + 8 = 0$.
6. $\pi x - \sqrt{2} y = 7$
7. $y = 3x + 7$.
8. $2(x - 1) + 5(y + 2) = 0$.

In each of Exercises 9 and 10 find the distance from the given point to the given line.

9. The point $(0, 0)$ to $3x + 4y - 10 = 0$.
10. The point $(\tfrac{3}{2}, \tfrac{2}{3})$ to $2x - y + 5 = 0$.

In each of Exercises 11 to 14 find the distance from the given point to the given plane.

11. The point $(0, 0, 0)$ to the plane $2x - 4y + 3z + 2 = 0$.
12. The point $(1, 2, 3)$ to the plane $x + 2y - 3z + 5 = 0$.
13. The point $(2, 2, -1)$ to the plane that passes through $(1, 4, 3)$ and has a normal $2\mathbf{i} - 7\mathbf{j} + 2\mathbf{k}$.
14. The point $(0, 0, 0)$ to the plane that passes through $(4, 1, 0)$ and is perpendicular to the vector $\mathbf{i} + \mathbf{j} + \mathbf{k}$.

15. Prove Theorem 3.
16. Prove Theorem 5.
17. Find the direction cosines of the vector $2\mathbf{i} + 3\mathbf{j} + 4\mathbf{k}$.
18. Find direction cosines of the line through the points $(1, 3, 2)$ and $(4, -1, 5)$.
19. A vector **A** has direction angles $\alpha = 70°$ and $\beta = 80°$. Find the third direction angle γ and show the possibilities on a diagram.
20. Suppose that the three direction angles of a vector are equal. What must they be?
21. Give parametric equations for the line through $(\tfrac{1}{2}, \tfrac{1}{3}, \tfrac{1}{2})$ and with direction numbers 2, -5, and 8 in (a) scalar form, (b) vector form.
22. Give parametric equations for the line through $(1, 2, 3)$ and $(4, 5, 7)$ in (a) scalar form, (b) vector form.

Parametric equations may be of use in Exercises 23 to 27.

23. Are the three points $(1, 2, -3)$, $(1, 6, 2)$, and $(7, 14, 11)$ on a single line?

24 Where does the line through $(1, 2, 4)$ and $(2, 1, -1)$ meet the plane $x + 2y + 5z = 0$?

25 What point on the line through $(1, 2, 5)$ and $(3, 1, 1)$ is closest to the point $(2, -1, 5)$?

26 Does the line through $(5, 7, 10)$ and $(3, 4, 5)$ meet the line through $(1, 4, 0)$ and $(3, 6, 4)$? If so, where?

27 Does the line through $(0, 0, 0)$ and $(2, 1, 1)$ meet the line through $(3, 5, 6)$ and $(13, 3, 4)$? If so, where?

28 Develop a general formula for determining the distance from the point $P_1 = (x_1, y_1, z_1)$ to the line through the point $P_0 = (x_0, y_0, z_0)$ and parallel to the vector $\mathbf{A} = a_1\mathbf{i} + a_2\mathbf{j} + a_3\mathbf{k}$. The formula should be expressed in terms of the vectors $\overrightarrow{P_0P_1}$ and $\mathbf{A}$.

29 How far is the point $(1, 2, -1)$ from the line through $(1, 3, 5)$ and $(2, 1, -3)$?
 (a) Solve by calculus, minimizing a certain function.
 (b) Solve by vectors.

■ ■

Exercises 30 to 32 are related.

30 Find the direction cosines of the vector $\mathbf{A}$ shown in Fig. 11.63.

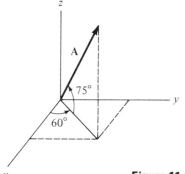

Figure 11.63

31 If $\mathbf{A}$ has direction angles α, β, and γ and $\mathbf{A}'$ has direction angles α', β', and γ', find the cosine of the angle between $\mathbf{A}$ and $\mathbf{A}'$.

32 (Contributed by Mark O'Donnell.) The following problem arose in the design of a solar collector. The front of a house is aligned 60° from south as shown in Fig. 11.64. The roof is inclined at 22° from the horizontal. At noon the sun has an elevation 75° above the horizon. (See Fig. 11.64.)

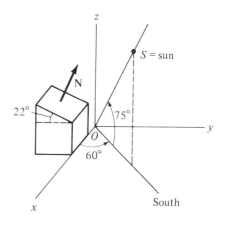

Figure 11.64

(a) Find the cosine of the angle between the vector $\overrightarrow{OS}$, which points to the sun, and the vector $\mathbf{N}$, a normal to the roof.
(b) Find the angle between $\mathbf{N}$ and $\overrightarrow{OS}$.

11.5 Determinants of orders 2 and 3

This section develops a small part of the extensive field of linear algebra. It should be kept in mind that the few definitions and theorems we present have broad generalizations and applications that only a full course could adequately explore.

DEFINITION *Matrix.* An array of four numbers a_1, a_2, b_1, b_2 forming a square is called a *matrix of order 2* and is denoted

$$\begin{pmatrix} a_1 & a_2 \\ b_1 & b_2 \end{pmatrix}.$$

Similarly, an array of nine numbers arranged in a square is called a *matrix of order 3*.

$$\begin{pmatrix} a_1 & a_2 & a_3 \\ b_1 & b_2 & b_3 \\ c_1 & c_2 & c_3 \end{pmatrix}.$$

For instance, $\begin{pmatrix} 2 & 3 \\ 4 & -1 \end{pmatrix}$ and $\begin{pmatrix} 1 & 5 & 0 \\ 0 & 2 & 3 \\ 8 & 4 & -1 \end{pmatrix}$

are matrices of orders 2 and 3, respectively.

The set of entries of a matrix in a line parallel to the bottom of the page is called a *row* of the matrix. The set of entries in a line parallel to the margin is called a *column*. A matrix of order 3 thus consists of three rows; it may also be thought of as being composed of three columns. For convenience, number the rows from top to bottom and the columns from left to right:

$$\begin{pmatrix} \text{row 1} \\ \text{row 2} \\ \text{row 3} \end{pmatrix} \quad \text{or} \quad \begin{pmatrix} c & c & c \\ o & o & o \\ l & l & l \\ u & u & u \\ m & m & m \\ n & n & n \\ 1 & 2 & 3 \end{pmatrix}.$$

In the matrix $\begin{pmatrix} 2 & 3 \\ 4 & -1 \end{pmatrix}$

the first row is (2, 3), and the second column is $\begin{pmatrix} 3 \\ -1 \end{pmatrix}$.

Associated with each matrix is an important number, called its *determinant*.

DEFINITION *Determinant of a second-order matrix.* The determinant of the matrix

$$\begin{pmatrix} a_1 & a_2 \\ b_1 & b_2 \end{pmatrix}$$

is the number $a_1 b_2 - a_2 b_1$.

It is denoted $\begin{vmatrix} a_1 & a_2 \\ b_1 & b_2 \end{vmatrix}.$

EXAMPLE 1 Compute the determinants:

(a) $\begin{vmatrix} 2 & 3 \\ 1 & 4 \end{vmatrix}$ (b) $\begin{vmatrix} 0 & -5 \\ 2 & 8 \end{vmatrix}.$

SOLUTION (a) $\begin{vmatrix} 2 & 3 \\ 1 & 4 \end{vmatrix} = 2 \cdot 4 - 3 \cdot 1 = 8 - 3 = 5.$

(b) $\begin{vmatrix} 0 & -5 \\ 2 & 8 \end{vmatrix} = 0 \cdot 8 - (-5) \cdot 2 = 0 + 10 = 10.$ ∎

The determinant of a third-order matrix may be defined with the aid of second-order matrices, as follows.

DEFINITION *Determinant of a third-order matrix.* The determinant of the matrix

$$\begin{pmatrix} a_1 & a_2 & a_3 \\ b_1 & b_2 & b_3 \\ c_1 & c_2 & c_3 \end{pmatrix}$$

is the number $a_1 \begin{vmatrix} b_2 & b_3 \\ c_2 & c_3 \end{vmatrix} - a_2 \begin{vmatrix} b_1 & b_3 \\ c_1 & c_3 \end{vmatrix} + a_3 \begin{vmatrix} b_1 & b_2 \\ c_1 & c_2 \end{vmatrix}.$

It is denoted $\begin{vmatrix} a_1 & a_2 & a_3 \\ b_1 & b_2 & b_3 \\ c_1 & c_2 & c_3 \end{vmatrix}.$

EXAMPLE 2 Compute $\begin{vmatrix} 3 & 2 & -4 \\ 1 & 5 & 0 \\ 2 & -7 & 3 \end{vmatrix}.$

SOLUTION By definition this determinant equals

$$3 \begin{vmatrix} 5 & 0 \\ -7 & 3 \end{vmatrix} - 2 \begin{vmatrix} 1 & 0 \\ 2 & 3 \end{vmatrix} + (-4) \begin{vmatrix} 1 & 5 \\ 2 & -7 \end{vmatrix}.$$

Now $\begin{vmatrix} 5 & 0 \\ -7 & 3 \end{vmatrix} = 15,$ $\begin{vmatrix} 1 & 0 \\ 2 & 3 \end{vmatrix} = 3,$ and $\begin{vmatrix} 1 & 5 \\ 2 & -7 \end{vmatrix} = -17.$

Hence the third-order determinant equals

$$3 \cdot 15 - 2 \cdot 3 + (-4)(-17) = 45 - 6 + 68 = 107. \quad \blacksquare$$

$\begin{pmatrix} \not{a}_1 & \not{a}_2 & \not{a}_3 \\ \not{b}_1 & b_2 & b_3 \\ \not{c}_1 & c_2 & c_3 \end{pmatrix}$ Note that in the definition of a third-order determinant a_2 has a negative sign. Also observe that the second-order matrix associated with a_1 is obtained by blotting out of the third-order matrix the row and column on which a_1 lies, as shown in the margin. A similar procedure works for a_2 and a_3 (but remember the minus sign that goes with a_2).

We now obtain some useful theorems about determinants that are true of matrices of either order 2 or 3. However, the proofs, which are straightforward computations, will be given in only one of the two cases.

THEOREM 1 If two rows (or two columns) of a matrix are identical, then the determinant of the matrix is 0.

PROOF Let us show that when two rows are identical, the matrix has determinant 0; for instance,

$$\begin{vmatrix} a_1 & a_2 & a_3 \\ a_1 & a_2 & a_3 \\ c_1 & c_2 & c_3 \end{vmatrix} = 0.$$

This determinant equals

$$a_1 \begin{vmatrix} a_2 & a_3 \\ c_2 & c_3 \end{vmatrix} - a_2 \begin{vmatrix} a_1 & a_3 \\ c_1 & c_3 \end{vmatrix} + a_3 \begin{vmatrix} a_1 & a_2 \\ c_1 & c_2 \end{vmatrix}$$

$$= a_1(a_2 c_3 - a_3 c_2) - a_2(a_1 c_3 - a_3 c_1) + a_3(a_1 c_2 - a_2 c_1)$$

$$= a_1 a_2 c_3 - a_1 a_3 c_2 - a_2 a_1 c_3 + a_2 a_3 c_1 + a_3 a_1 c_2 - a_3 a_2 c_1$$

$$= 0; \quad \text{the terms cancel in pairs.}$$

The other cases may be proved similarly. ∎

THEOREM 2 If two rows (or two columns) of a matrix are switched with each other, the determinant of the resulting matrix is the determinant of the original matrix with its sign changed.

PROOF Take the case, for instance, in which the second and third columns of a third-order matrix are switched. We will prove that

$$\begin{vmatrix} a_1 & a_3 & a_2 \\ b_1 & b_3 & b_2 \\ c_1 & c_3 & c_2 \end{vmatrix} = - \begin{vmatrix} a_1 & a_2 & a_3 \\ b_1 & b_2 & b_3 \\ c_1 & c_2 & c_3 \end{vmatrix}. \tag{1}$$

Simply calculate both determinants and compare the results:

$$\begin{vmatrix} a_1 & a_3 & a_2 \\ b_1 & b_3 & b_2 \\ c_1 & c_3 & c_2 \end{vmatrix} = a_1 \begin{vmatrix} b_3 & b_2 \\ c_3 & c_2 \end{vmatrix} - a_3 \begin{vmatrix} b_1 & b_2 \\ c_1 & c_2 \end{vmatrix} + a_2 \begin{vmatrix} b_1 & b_3 \\ c_1 & c_3 \end{vmatrix}, \tag{2}$$

while

$$\begin{vmatrix} a_1 & a_2 & a_3 \\ b_1 & b_2 & b_3 \\ c_1 & c_2 & c_3 \end{vmatrix} = a_1 \begin{vmatrix} b_2 & b_3 \\ c_2 & c_3 \end{vmatrix} - a_2 \begin{vmatrix} b_1 & b_3 \\ c_1 & c_3 \end{vmatrix} + a_3 \begin{vmatrix} b_1 & b_2 \\ c_1 & c_2 \end{vmatrix}. \tag{3}$$

Since $\begin{vmatrix} b_3 & b_2 \\ c_3 & c_2 \end{vmatrix} = b_3 c_2 - b_2 c_3$ and $\begin{vmatrix} b_2 & b_3 \\ c_2 & c_3 \end{vmatrix} = b_2 c_3 - b_3 c_2$,

direct comparison of the three summands in (2) and in (3) establishes (1). The other cases are proved similarly. ∎

Determinants can be used to solve linear equations. The next two theorems provide a formula, known as Cramer's rule, for solving two equations in two unknowns or three equations in three unknowns. The proofs, which are straightforward calculations, are left to the reader.

THEOREM 3 Assume that $\begin{vmatrix} a_1 & a_2 \\ b_1 & b_2 \end{vmatrix} \neq 0.$

Then the simultaneous equations

$$a_1 x + a_2 y = k_1$$
$$b_1 x + b_2 y = k_2$$

have a unique solution (x, y) given by

$$x = \frac{\begin{vmatrix} k_1 & a_2 \\ k_2 & b_2 \end{vmatrix}}{\begin{vmatrix} a_1 & a_2 \\ b_1 & b_2 \end{vmatrix}} \quad \text{and} \quad y = \frac{\begin{vmatrix} a_1 & k_1 \\ b_1 & k_2 \end{vmatrix}}{\begin{vmatrix} a_1 & a_2 \\ b_1 & b_2 \end{vmatrix}}.$$ ∎

Much faster ways of solving simultaneous equations with more variables are developed in linear algebra.

Theorem 3 tells us how to find where two nonparallel lines in the xy plane intersect. Notice that the two denominators (for x and for y) are the same, being the determinant of the matrix formed by the coefficients. The determinants in the numerators are obtained from that determinant by replacing an appropriate column by the column

$$\begin{pmatrix} k_1 \\ k_2 \end{pmatrix}.$$

EXAMPLE 3 Use determinants to find where the lines

$$2x - y = 5 \quad \text{and} \quad 3x + 4y = -1$$

intersect.

SOLUTION First check that the determinant of the coefficient matrix is not 0:

$$\begin{vmatrix} 2 & -1 \\ 3 & 4 \end{vmatrix} = 2 \cdot 4 - (-1) \cdot 3 = 8 + 3 = 11.$$

Thus $\quad x = \dfrac{\begin{vmatrix} 5 & -1 \\ -1 & 4 \end{vmatrix}}{11} \quad$ and $\quad y = \dfrac{\begin{vmatrix} 2 & 5 \\ 3 & -1 \end{vmatrix}}{11}.$

Straightforward calculations show that

$$x = \frac{19}{11} \quad \text{and} \quad y = \frac{-17}{11}.$$

730 ALGEBRAIC OPERATIONS ON VECTORS

The skeptical reader may check this solution by plugging it into both the original equations. $2x - y = 5$ and $3x + 4y = -1$. ∎

The next theorem is the companion for third-order determinants of Theorem 3.

THEOREM 4 Assume that $\begin{vmatrix} a_1 & a_2 & a_3 \\ b_1 & b_2 & b_3 \\ c_1 & c_2 & c_3 \end{vmatrix} \neq 0.$

Then the simultaneous equations

$$a_1 x + a_2 y + a_3 z = k_1$$
$$b_1 x + b_2 y + b_3 z = k_2$$
$$c_1 x + c_2 y + c_3 z = k_3$$

have a unique solution (x, y, z) given by

$$x = \frac{\begin{vmatrix} k_1 & a_2 & a_3 \\ k_2 & b_2 & b_3 \\ k_3 & c_2 & c_3 \end{vmatrix}}{\begin{vmatrix} a_1 & a_2 & a_3 \\ b_1 & b_2 & b_3 \\ c_1 & c_2 & c_3 \end{vmatrix}}, \quad y = \frac{\begin{vmatrix} a_1 & k_1 & a_3 \\ b_1 & k_2 & b_3 \\ c_1 & k_3 & c_3 \end{vmatrix}}{\begin{vmatrix} a_1 & a_2 & a_3 \\ b_1 & b_2 & b_3 \\ c_1 & c_2 & c_3 \end{vmatrix}}, \quad z = \frac{\begin{vmatrix} a_1 & a_2 & k_1 \\ b_1 & b_2 & k_2 \\ c_1 & c_2 & k_3 \end{vmatrix}}{\begin{vmatrix} a_1 & a_2 & a_3 \\ b_1 & b_2 & b_3 \\ c_1 & c_2 & c_3 \end{vmatrix}}.$$ ∎

Note that Theorem 4 tells how to find the point where three planes meet.

EXAMPLE 4 Find the point of intersection of the three planes

$$3x + 2y - 4z = 6$$
$$1x + 5y + 0z = -3$$
$$2x - 7y + 3z = 4.$$

SOLUTION The matrix of coefficients already appeared in Example 2, where its determinant was shown to be 107, which is not 0. Thus

$$x = \frac{\begin{vmatrix} 6 & 2 & -4 \\ -3 & 5 & 0 \\ 4 & -7 & 3 \end{vmatrix}}{107}, \quad y = \frac{\begin{vmatrix} 3 & 6 & -4 \\ 1 & -3 & 0 \\ 2 & 4 & 3 \end{vmatrix}}{107}, \quad z = \frac{\begin{vmatrix} 3 & 2 & 6 \\ 1 & 5 & -3 \\ 2 & -7 & 4 \end{vmatrix}}{107}.$$

Straightforward calculations show that

$$x = \frac{104}{107} \quad y = \frac{-85}{107} \quad z = -\frac{125}{107}.$$ ∎

The theory introduced in this brief section goes much further. For any positive integer n there are matrices and determinants of the nth order. The theorems of this section generalize to all orders. It is possible to define a determinant of order n without referring to matrices or determinants of lower orders. A linear algebra or vector space course carries the theory of matrices and determinants much further.

Exercises

In Exercises 1 to 6 evaluate the determinants. Use a shortcut if there is one.

1. $\begin{vmatrix} 3 & 4 \\ 7 & 2 \end{vmatrix}$

2. $\begin{vmatrix} 3 & 3 \\ 7 & 7 \end{vmatrix}$

3. $\begin{vmatrix} 4 & 2 & 0 \\ 5 & 6 & -1 \\ 1 & -1 & 2 \end{vmatrix}$

4. $\begin{vmatrix} 1 & 3 & 1 \\ 2 & 1 & 2 \\ 4 & 5 & 4 \end{vmatrix}$

5. $\begin{vmatrix} 2 & 1 & 4 \\ 2 & 1 & 4 \\ 3 & 5 & 7 \end{vmatrix}$

6. $\begin{vmatrix} 0 & 0 & 0 \\ 1 & 5 & 9 \\ 3 & -1 & 2 \end{vmatrix}$

7. Prove that if the first and third columns of a third-order matrix are identical, then the determinant of the matrix is 0.

8. Show that Theorem 2 implies Theorem 1.

9. Prove that if the first and third rows of a third-order matrix are switched, the determinant of the resulting matrix is the determinant of the original matrix with its sign changed.

10. Prove that

$$\begin{vmatrix} ka_1 & ka_2 \\ b_1 & b_2 \end{vmatrix} = k \begin{vmatrix} a_1 & a_2 \\ b_1 & b_2 \end{vmatrix}.$$

This illustrates a general theorem. If the entries in a single row (or single column) are all multiplied by the number k, the determinant of the resulting matrix is k times the determinant of the original matrix.

11. Prove that

$$\begin{vmatrix} a_1 & a_2 & a_3 \\ b_1 & b_2 & b_3 \\ c_1 & c_2 & c_3 \end{vmatrix} = \begin{vmatrix} a_1 & b_1 & c_1 \\ a_2 & b_2 & c_2 \\ a_3 & b_3 & c_3 \end{vmatrix}.$$

This says that if you spin a matrix around the diagonal that stretches from the top left corner to the bottom right, the determinant of the resulting matrix is the same as that of the original matrix. The second matrix is called the *transpose* of the first.

12. Prove that

$$\begin{vmatrix} a_1 + kb_1 & a_2 + kb_2 & a_3 + kb_3 \\ b_1 & b_2 & b_3 \\ c_1 & c_2 & c_3 \end{vmatrix} = \begin{vmatrix} a_1 & a_2 & a_3 \\ b_1 & b_2 & b_3 \\ c_1 & c_2 & c_3 \end{vmatrix}.$$

This illustrates the general theorem: If you multiply a row by a scalar and add the result to a different row, the resulting matrix has the same determinant as the original one. A corresponding theorem holds for columns.

In Exercises 13 to 16 use Theorems 3 and 4 to solve the equations.

13. $x - 2y = 5$
    $3x - y = 1$

14. $2x + y = 7$
    $4x - 5y = 7$

15. $x + 2y + 3z = 7$
    $3x - y + z = 6$
    $x + y + z = 4$

16. $x + 2y - z = 0$
    $3x + y + z = 3$
    $-x + y + 2z = 5$

17. (a) Use determinants to find x and y that satisfy both equations simultaneously:

$$2x - 3y = 6$$
$$3x + 5y = 15.$$

(b) Check your solution to (a) by substituting it in the two equations.

(c) Graph the two lines in (a) and show their intersection.

18. (a) Use determinants to find the point common to these three planes:

$$x - y + z = 2$$
$$2x + y - 3z = 0$$
$$x + 3y + 4z = 1.$$

(b) Check that your solution satisfies the three equations in (a).

■

19 (a) Let (x_1, y_1) and (x_2, y_2) be two distinct points in the xy plane. Show that the equation

$$\begin{vmatrix} x & y & 1 \\ x_1 & y_1 & 1 \\ x_2 & y_2 & 1 \end{vmatrix} = 0$$

is an equation for the line determined by the two given points.

(b) Use the formula in (a) to find an equation of the line through the points $(2, 3)$ and $(1, -4)$.

20 If the matrix of coefficients of the equations

$$a_1 x + a_2 y = k_1,$$
$$b_1 x + b_2 y = k_2 \qquad (4)$$

has determinant 0, Theorem 3 gives no information.

(a) Give an example where Eqs. (4) have no solution.
(b) Give an example where Eqs. (4) have an infinite number of solutions.

21 Sketch three planes whose common intersection is (a) a single point, (b) a line, (c) no points at all.

22 (a) Show that the planes

$$x + 2y + 3z = 1$$
$$2x + 4y + 6z = 3$$
$$x - y + z = 4$$

have no points in common. *Hint:* Look carefully at the first two equations.

(b) What must the determinant of the matrix of coefficients be?

■ ■

23 (a) Show that the planes

$$x + 2y + 3z = 1$$
$$x + 3y + 5z = 2$$
$$2x + 5y + 8z = 100$$

have no points in common.

(b) What must the determinant of the matrix of coefficients be?

24 Two plane vectors, $\mathbf{A} = a_1 \mathbf{i} + a_2 \mathbf{j}$ and $\mathbf{B} = b_1 \mathbf{i} + b_2 \mathbf{j}$, are located in the first quadrant in the relative positions shown in Fig. 11.65.

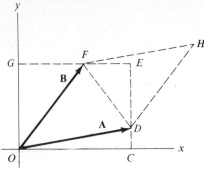

Figure 11.65

(a) Express in terms of $a_1, a_2, b_1,$ and b_2 the areas of the rectangle $OCEG$ and the triangles OCD, DEF, and OFG.

(b) Deduce that the area of triangle ODF is

$$\frac{1}{2} \begin{vmatrix} a_1 & a_2 \\ b_1 & b_2 \end{vmatrix}.$$

From (b) it follows that

$$\begin{vmatrix} a_1 & a_2 \\ b_1 & b_2 \end{vmatrix}$$

may be interpreted as the area of a parallelogram $ODHF$ determined by $\mathbf{A}$ and $\mathbf{B}$. This illustrates a general result valid for any plane vectors $\mathbf{A} = a_1 \mathbf{i} + a_2 \mathbf{j}$ and $\mathbf{B} = b_1 \mathbf{i} + b_2 \mathbf{j}$:

The determinant $\begin{vmatrix} a_1 & a_2 \\ b_1 & b_2 \end{vmatrix}$

is either $+$ or $-$ the area of the parallelogram determined by $\mathbf{A}$ and $\mathbf{B}$.

25 A driver wants to learn how many miles per gallon her car gets in the city and on the highway. On Monday she drove 30 miles in the city and 90 miles on the highway and used 6 gallons. During the 2-day period, Tuesday and Wednesday, she drove 75 miles in the city and 300 miles on the highway and used 17 gallons. On Thursday she drove 150 miles in the city and 210 miles on the highway and used 18 gallons.

(a) How much gasoline evaporates or leaks out of the tank per day?
(b) How many miles per gallon does her car get in the city and on the highway?

26 Show that

$$\begin{vmatrix} a_1 & a_2 & a_3 \\ b_1 & b_2 & b_3 \\ c_1 & c_2 & c_3 \end{vmatrix} = a_1 \begin{vmatrix} b_2 & b_3 \\ c_2 & c_3 \end{vmatrix} - b_1 \begin{vmatrix} a_2 & a_3 \\ c_2 & c_3 \end{vmatrix} + c_1 \begin{vmatrix} a_2 & a_3 \\ b_2 & b_3 \end{vmatrix}.$$

This is called "expansion along the first column."

27 According to Exercise 12, if you add a multiple of one row of a matrix to another row you do not change the value of the determinant. A similar statement holds for columns.

(a) Use this fact to justify the following computations.

$$\begin{vmatrix} 2 & 4 & 4 \\ 3 & 7 & 0 \\ 4 & 9 & 1 \end{vmatrix} = \begin{vmatrix} 2 & 0 & 4 \\ 3 & 1 & 0 \\ 4 & 1 & 1 \end{vmatrix} = \begin{vmatrix} 2 & 0 & 0 \\ 3 & 1 & -6 \\ 4 & 1 & -7 \end{vmatrix} = \begin{vmatrix} 2 & 0 & 0 \\ 0 & 1 & -6 \\ 4 & 1 & -7 \end{vmatrix} = \begin{vmatrix} 2 & 0 & 0 \\ 0 & 1 & -6 \\ 0 & 1 & -7 \end{vmatrix}.$$

(b) Evaluate the last determinant in (a).

28 Use the technique in Exercise 27 to evaluate

$$\begin{vmatrix} 1 & 3 & 4 \\ 1 & 2 & 5 \\ 1 & 1 & 7 \end{vmatrix}.$$

11.6 The cross product of two vectors

It is frequently necessary in applications of vectors in space to construct a nonzero vector perpendicular to two given vectors **A** and **B**. This section provides a formula for finding such a vector.

If **A** and **B** are not parallel and are drawn with their tails at a single point, they determine a plane, as in the left diagram in Fig. 11.66. Any vector **C**

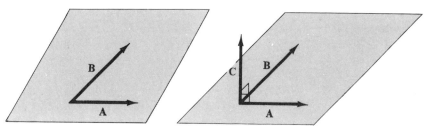

Figure 11.66

perpendicular to this plane is perpendicular to both **A** and **B**. There are many such vectors, all parallel to each other and having various lengths. For instance, any vector perpendicular to both **i** and **j** is of the form $c\mathbf{k}$, where c is an arbitrary scalar, positive or negative.

In the following definition vectors appear as entries in a matrix. Though such a matrix and its determinant do not fit into the general definition in the preceding section, the notation is so useful and unambiguous that we shall not hesitate to employ it.

ALGEBRAIC DEFINITION OF CROSS PRODUCT

DEFINITION *Cross product (vector product).* Let

$$\mathbf{A} = a_1\mathbf{i} + a_2\mathbf{j} + a_3\mathbf{k} \quad \text{and} \quad \mathbf{B} = b_1\mathbf{i} + b_2\mathbf{j} + b_3\mathbf{k}.$$

The vector

$$\begin{vmatrix} \mathbf{i} & \mathbf{j} & \mathbf{k} \\ a_1 & a_2 & a_3 \\ b_1 & b_2 & b_3 \end{vmatrix} = (a_2 b_3 - a_3 b_2)\mathbf{i} - (a_1 b_3 - a_3 b_1)\mathbf{j} + (a_1 b_2 - a_2 b_1)\mathbf{k}$$

is called the cross product (or vector product) of **A** and **B**. It is denoted

$$\mathbf{A} \times \mathbf{B}.$$

The determinant of **A** × **B** is expanded along its first row:

$$\begin{pmatrix} \mathbf{i} & \mathbf{j} & \mathbf{k} \\ a_1 & a_2 & a_3 \\ b_1 & b_2 & b_3 \end{pmatrix} \qquad \begin{pmatrix} \mathbf{i} & \mathbf{j} & \mathbf{k} \\ a_1 & a_2 & a_3 \\ b_1 & b_2 & b_3 \end{pmatrix} \qquad \begin{pmatrix} \mathbf{i} & \mathbf{j} & \mathbf{k} \\ a_1 & a_2 & a_3 \\ b_1 & b_2 & b_3 \end{pmatrix}$$

Delete the two lines through **i**. The determinant of the remaining square is the coefficient of **i** in **A** × **B**.

Delete the two lines through **j**. The negative of the determinant of the remaining square is the coefficient of **j** in **A** × **B**.

Delete the two lines through **k**. The determinant of the remaining square is the coefficient of **k** in **A** × **B**.

EXAMPLE 1 Compute **A** × **B** if **A** = $2\mathbf{i} - \mathbf{j} + 3\mathbf{k}$ and **B** = $3\mathbf{i} + 4\mathbf{j} + \mathbf{k}$.

SOLUTION By definition,

$$\mathbf{A} \times \mathbf{B} = \begin{vmatrix} \mathbf{i} & \mathbf{j} & \mathbf{k} \\ 2 & -1 & 3 \\ 3 & 4 & 1 \end{vmatrix} = \mathbf{i} \begin{vmatrix} -1 & 3 \\ 4 & 1 \end{vmatrix} - \mathbf{j} \begin{vmatrix} 2 & 3 \\ 3 & 1 \end{vmatrix} + \mathbf{k} \begin{vmatrix} 2 & -1 \\ 3 & 4 \end{vmatrix}$$

$$= -13\mathbf{i} + 7\mathbf{j} + 11\mathbf{k}. \blacksquare$$

Note that **A** × **B** is a vector, while **A** · **B** is a scalar. The most important property of the cross product is expressed in the following theorem.

THEOREM 1 **A** × **B** is a vector perpendicular to both **A** and **B**.

PROOF Let us show, for instance, that **A** · (**A** × **B**) is 0. Now, **A** = $a_1\mathbf{i} + a_2\mathbf{j} + a_3\mathbf{k}$, **B** = $b_1\mathbf{i} + b_2\mathbf{j} + b_3\mathbf{k}$

and

$$\mathbf{A} \times \mathbf{B} = \mathbf{i} \begin{vmatrix} a_2 & a_3 \\ b_2 & b_3 \end{vmatrix} - \mathbf{j} \begin{vmatrix} a_1 & a_3 \\ b_1 & b_3 \end{vmatrix} + \mathbf{k} \begin{vmatrix} a_1 & a_2 \\ b_1 & b_2 \end{vmatrix}.$$

Thus

$$\mathbf{A} \cdot (\mathbf{A} \times \mathbf{B}) = a_1 \begin{vmatrix} a_2 & a_3 \\ b_2 & b_3 \end{vmatrix} - a_2 \begin{vmatrix} a_1 & a_3 \\ b_1 & b_3 \end{vmatrix} + a_3 \begin{vmatrix} a_1 & a_2 \\ b_1 & b_2 \end{vmatrix},$$

which is precisely the definition of the determinant

$$\begin{vmatrix} a_1 & a_2 & a_3 \\ a_1 & a_2 & a_3 \\ b_1 & b_2 & b_3 \end{vmatrix}.$$

Since two rows are identical, the determinant equals 0. Thus **A** · (**A** × **B**) = 0.

A similar argument shows that **B** · (**A** × **B**) = 0. This completes the proof. ∎

EXAMPLE 2 Verify Theorem 1 in the case of the vectors **A**, **B**, and **A** × **B** of Example 1.

SOLUTION In the case of Example 1,

$$\mathbf{A} \cdot (\mathbf{A} \times \mathbf{B}) = (2\mathbf{i} - \mathbf{j} + 3\mathbf{k}) \cdot (-13\mathbf{i} + 7\mathbf{j} + 11\mathbf{k})$$
$$= -26 - 7 + 33$$
$$= 0.$$
$$\mathbf{B} \cdot (\mathbf{A} \times \mathbf{B}) = (3\mathbf{i} + 4\mathbf{j} + \mathbf{k}) \cdot (-13\mathbf{i} + 7\mathbf{j} + 11\mathbf{k})$$
$$= -39 + 28 + 11$$
$$= 0.$$

These results verify the theorem in this case. ∎

EXAMPLE 3 Compute **i** × **j** and **j** × **i**.

SOLUTION

$$\mathbf{i} \times \mathbf{j} = \begin{vmatrix} \mathbf{i} & \mathbf{j} & \mathbf{k} \\ 1 & 0 & 0 \\ 0 & 1 & 0 \end{vmatrix} = 0\mathbf{i} - 0\mathbf{j} + \mathbf{k} = \mathbf{k}.$$

$$\mathbf{j} \times \mathbf{i} = \begin{vmatrix} \mathbf{i} & \mathbf{j} & \mathbf{k} \\ 0 & 1 & 0 \\ 1 & 0 & 0 \end{vmatrix} = 0\mathbf{i} - 0\mathbf{j} - \mathbf{k} = -\mathbf{k}.$$

Thus **i** × **j** is the negative of **j** × **i**. ∎

The order of the factors in the vector product is critical.

Example 3 shows that **A** × **B** may be different from **B** × **A**. Indeed it is easy to show that for all vectors **A** and **B**

$$\mathbf{B} \times \mathbf{A} = -(\mathbf{A} \times \mathbf{B}).$$

This property corresponds to the fact that, when two rows of a matrix are interchanged, its determinant changes sign.

Another surprising property of the operation × is that, for all vectors **A**,

$$\mathbf{A} \times \mathbf{A} = \mathbf{0}.$$

This corresponds to the fact that, if two rows of a matrix are identical, then its determinant is 0. More generally, if **A** and **B** are parallel,

$$\mathbf{A} \times \mathbf{B} = \mathbf{0}.$$

After these shocks it may be comforting to know that

$$\mathbf{A} \times (\mathbf{B} + \mathbf{C}) = \mathbf{A} \times \mathbf{B} + \mathbf{A} \times \mathbf{C},$$

which is reminiscent of the arithmetic of numbers. This distributive law can be established by a straightforward computation.

GEOMETRIC DESCRIPTION OF THE CROSS PRODUCT

The definition of **A** × **B** is purely algebraic. For many applications it is important to have a *completely geometric* description of **A** × **B**, one that expresses the direction and magnitude of **A** × **B** in terms of those of **A** and **B**.

We know already that **A** × **B** is perpendicular to both **A** and **B**. However, there are two directions in which **A** × **B** may point. Which is the correct one? The clue is given by the case

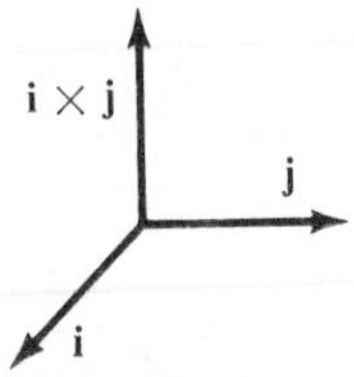

$$\mathbf{i} \times \mathbf{j} = \mathbf{k},$$

Figure 11.67

shown in Fig. 11.67.

Since the positive x, y, and z axes were chosen to match the right-hand rule, this rule describes in general the direction of **A** × **B**. If the fingers of the right hand curl from **A** to **B** through an angle less than 180°, the thumb points in the direction of **A** × **B**. (See Fig. 11.68.)

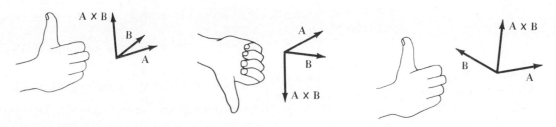

Figure 11.68

The direction of **A** × **B** is completely determined. But what is the length of **A** × **B**? It is given by the following theorem.

THEOREM 2 The magnitude of **A** × **B** is equal to the area of the parallelogram spanned by **A** and **B**.

PROOF Let θ be the angle between **A** and **B**. Then the area of the parallelogram spanned by **A** and **B** (shown in Fig. 11.69) is

$$|\mathbf{A}| |\mathbf{B}| \sin \theta.$$

To avoid square roots, consider the square of the area of the parallelogram. We wish to show that

$$|\mathbf{A}|^2 |\mathbf{B}|^2 \sin^2 \theta = |\mathbf{A} \times \mathbf{B}|^2;$$

that is, $\quad |\mathbf{A}|^2 |\mathbf{B}|^2 \sin^2 \theta = \begin{vmatrix} a_2 & a_3 \\ b_2 & b_3 \end{vmatrix}^2 + \begin{vmatrix} a_1 & a_3 \\ b_1 & b_3 \end{vmatrix}^2 + \begin{vmatrix} a_1 & a_2 \\ b_1 & b_2 \end{vmatrix}^2.$ (1)

Now, $\quad \sin^2 \theta = 1 - \cos^2 \theta$

$$= 1 - \frac{(\mathbf{A} \cdot \mathbf{B})^2}{|\mathbf{A}|^2 |\mathbf{B}|^2}$$

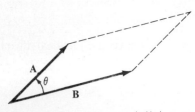

The parallelogram has area $|\mathbf{A}| |\mathbf{B}| \sin \theta$.

Figure 11.69

11.6 THE CROSS PRODUCT OF TWO VECTORS

$$= \frac{|\mathbf{A}|^2|\mathbf{B}|^2 - (\mathbf{A} \cdot \mathbf{B})^2}{|\mathbf{A}|^2|\mathbf{B}|^2}.$$

Thus,

$$\sin^2 \theta = \frac{(a_1^2 + a_2^2 + a_3^2)(b_1^2 + b_2^2 + b_3^2) - (a_1 b_1 + a_2 b_2 + a_3 b_3)^2}{|\mathbf{A}|^2 |\mathbf{B}|^2}. \tag{2}$$

Comparison of (1) and (2) with the aid of elementary algebra completes the proof. The details are left to the reader. ■

A × B described geometrically

In short, **A × B** is that vector perpendicular to both **A** and **B**, whose direction is obtained by the right-hand rule and whose length is the area of the parallelogram spanned by **A** and **B**. If **A** or **B** is **0**, or if **A** is parallel to **B**, **A × B** is the vector **0**.

SOME APPLICATIONS OF THE CROSS PRODUCT

EXAMPLE 4 A parallelogram in the plane has the vertices $(0, 0)$, (a_1, a_2), (b_1, b_2), and $(a_1 + b_1, a_2 + b_2)$. Find its area. (See Fig. 11.70.)

SOLUTION The parallelogram is spanned by the vectors

$$\mathbf{A} = a_1 \mathbf{i} + a_2 \mathbf{j} + 0\mathbf{k}$$

and

$$\mathbf{B} = b_1 \mathbf{i} + b_2 \mathbf{j} + 0\mathbf{k}.$$

Consequently, its area is the magnitude of the vector

$$\begin{vmatrix} \mathbf{i} & \mathbf{j} & \mathbf{k} \\ a_1 & a_2 & 0 \\ b_1 & b_2 & 0 \end{vmatrix} = 0\mathbf{i} - 0\mathbf{j} + \mathbf{k} \begin{vmatrix} a_1 & a_2 \\ b_1 & b_2 \end{vmatrix}.$$

Thus the area is
$$\pm \begin{vmatrix} a_1 & a_2 \\ b_1 & b_2 \end{vmatrix}. \quad ■$$

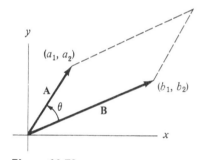

Figure 11.70

Example 4 provides a general result, namely that the absolute value of a second-order determinant may be interpreted as the area of a certain parallelogram. The next example is typical of the geometric applications of the cross product.

EXAMPLE 5 Find a vector normal to the plane determined by the three points

$$P = (1, 3, 2), \quad Q = (4, -1, 1), \quad R = (3, 0, 2).$$

Then find an equation of the plane.

SOLUTION The vectors $\overrightarrow{PQ}$ and $\overrightarrow{PR}$ lie in the plane (see Fig. 11.71). The vector $\overrightarrow{PQ} \times \overrightarrow{PR}$, being perpendicular to both $\overrightarrow{PQ}$ and $\overrightarrow{PR}$, is a normal to the plane. Now,

$$\overrightarrow{PQ} = 3\mathbf{i} - 4\mathbf{j} - \mathbf{k},$$

and

$$\overrightarrow{PR} = 2\mathbf{i} - 3\mathbf{j} + 0\mathbf{k}.$$

Thus

$$\mathbf{N} = \begin{vmatrix} \mathbf{i} & \mathbf{j} & \mathbf{k} \\ 3 & -4 & -1 \\ 2 & -3 & 0 \end{vmatrix} = -3\mathbf{i} - 2\mathbf{j} - \mathbf{k}$$

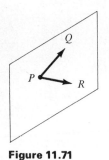

Figure 11.71

is a normal to the plane.

An equation of the plane can be obtained by using $\mathbf{N}$ and any of the three points P, Q, or R with the aid of Theorem 3 in Sec. 11.4. Using $P = (1, 3, 2)$, we obtain the equation

$$-3(x - 1) - 2(y - 3) - (z - 2) = 0,$$

which reduces to $\quad -3x + 3 - 2y + 6 - z + 2 = 0$

or $\quad 3x + 2y + z - 11 = 0.$

The reader may check that each of the three points lies in this plane. ■

The cross product is used in physics to express various concepts. We cite three examples.

Consider a particle of mass m moving in a straight line with constant speed. Let $\mathbf{v}$ be the vector whose length is the speed of the particle and which points in the direction of motion. The vector $m\mathbf{v}$ is called the *momentum* of the particle.

If O is a fixed point and P is the location of the particle at a certain instant, the vector

$$\overrightarrow{OP} \times m\mathbf{v}$$

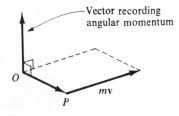

Figure 11.72

is called the *angular momentum* of the particle about O. (See Fig. 11.72.)

If $\mathbf{F}$ is the force applied at P, then

$$\overrightarrow{OP} \times \mathbf{F}$$

represents the *torque* or turning tendency of the force.

If the vector $\mathbf{I}$ represents the current in a wire (it has a direction and a magnitude) and $\mathbf{B}$ represents a magnetic field, then the force per unit length on the wire is the vector

$$\mathbf{I} \times \mathbf{B}.$$

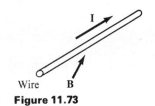

Figure 11.73

(See Fig. 11.73.)

Exercises

In Exercises 1 to 4 compute and sketch $\mathbf{A} \times \mathbf{B}$.
1. $\mathbf{A} = \mathbf{k}, \mathbf{B} = \mathbf{j}$.
2. $\mathbf{A} = \mathbf{i} + \mathbf{j}, \mathbf{B} = \mathbf{i} - \mathbf{j}$.
3. $\mathbf{A} = \mathbf{i} + \mathbf{j} + \mathbf{k}, \mathbf{B} = \mathbf{i} + \mathbf{j}$.
4. $\mathbf{A} = \mathbf{k}, \mathbf{B} = \mathbf{i} + \mathbf{j}$.

In Exercises 5 and 6 compute $\mathbf{A} \times \mathbf{B}$ and check that it is perpendicular to both $\mathbf{A}$ and to $\mathbf{B}$.

5. $\mathbf{A} = 2\mathbf{i} - 3\mathbf{j} + \mathbf{k}, \mathbf{B} = \mathbf{i} + \mathbf{j} + 2\mathbf{k}$.
6. $\mathbf{A} = \mathbf{i} - \mathbf{j}, \mathbf{B} = \mathbf{j} + 4\mathbf{k}$.

In Exercises 7 and 8 use the cross product.

7. Find the area of a parallelogram three of whose vertices are $(0, 0, 0)$, $(1, 5, 4)$, and $(2, -1, 3)$.
8. Find the area of a parallelogram three of whose vertices are $(1, 2, -1)$, $(2, 1, 4)$, and $(3, 5, 2)$.
9. Find a vector perpendicular to the plane determined by the points $(1, 2, 1)$, $(2, 1, -3)$, and $(0, 1, 5)$.
10. Find direction numbers of a line that is perpendicular to the line through $(1, 2, 1)$ and $(4, 1, 0)$ and also to the line through $(3, 5, 2)$ and $(2, 6, -3)$.
11. Find an equation of the plane determined by the points $(0, 0, 0)$, $(4, 1, 2)$, and $(2, 5, 0)$.
12. Find an equation of the plane determined by the points $(1, -1, 2)$, $(2, 1, 3)$, and $(3, 3, 5)$.
13. Prove that $\mathbf{B} \times \mathbf{A} = -(\mathbf{A} \times \mathbf{B})$ in two ways:
    (a) using the algebraic definition of the cross product;
    (b) using the geometric description of the cross product.
14. Show that, if $\mathbf{B} = c\mathbf{A}$, then $\mathbf{A} \times \mathbf{B} = \mathbf{0}$:
    (a) using the algebraic definition of the cross product;
    (b) using the geometric description of the cross product.
15. In the proof of Theorem 1 it was shown that $\mathbf{A}$ is perpendicular to $\mathbf{A} \times \mathbf{B}$. Show that $\mathbf{B}$ is perpendicular to $\mathbf{A} \times \mathbf{B}$.
16. Complete the algebra in the proof of Theorem 2 that shows that Eq. (1) holds.

■

17. We needed the fact that the area of a parallelogram of adjacent sides of length a and b and included angle θ is $ab \sin \theta$. Use elementary geometry to show that this formula is correct.
18. How far is the point $(3, 1, 5)$ from the plane determined by the points $(1, 3, 3)$, $(2, -1, 7)$, and $(1, 2, 4)$?
19. The planes $2x + 3y + 5z = 8$ and $x - 2y + 4z = 9$ meet in a line. Find a vector parallel to that line.
20. Show by an example that $\mathbf{A} \times (\mathbf{B} \times \mathbf{C})$ need not equal $(\mathbf{A} \times \mathbf{B}) \times \mathbf{C}$.
21. Prove that $\mathbf{A} \times (\mathbf{B} + \mathbf{C}) = \mathbf{A} \times \mathbf{B} + \mathbf{A} \times \mathbf{C}$.
22. Find the areas of the triangles whose vertices are
    (a) $(0, 0)$, $(3, 5)$, $(2, -1)$;
    (b) $(1, 4)$, $(3, 0)$, $(-1, 2)$;
    (c) $(1, 1, 1)$, $(2, 0, 1)$, $(3, -1, 4)$.

Hint: Use cross products of appropriate vectors.
Exercises 23 and 24 are related.

23. Let $\mathbf{A}$ and $\mathbf{B}$ be nonparallel vectors and let $\mathbf{C}$ be a vector.
    (a) Using a sketch, show that there are scalars x and y such that
    $$\mathbf{C} \times (\mathbf{A} \times \mathbf{B}) = x\mathbf{A} + y\mathbf{B}.$$
    (b) Why is $x(\mathbf{C} \cdot \mathbf{A}) + y(\mathbf{C} \cdot \mathbf{B}) = 0$?
    (c) From (a) and (b) deduce that there is a scalar z such that
    $$\mathbf{C} \times (\mathbf{A} \times \mathbf{B}) = z[(\mathbf{C} \cdot \mathbf{B})\mathbf{A} - (\mathbf{C} \cdot \mathbf{A})\mathbf{B}].$$

24. Using the formulas for dot products and cross products in terms of components, show that
    $$\mathbf{C} \times (\mathbf{A} \times \mathbf{B}) = (\mathbf{C} \cdot \mathbf{B})\mathbf{A} - (\mathbf{C} \cdot \mathbf{A})\mathbf{B}.$$

25. From Exercise 24 deduce that $\mathbf{A} \times (\mathbf{A} \times \mathbf{B}) = (\mathbf{A} \cdot \mathbf{B})\mathbf{A} - (\mathbf{A} \cdot \mathbf{A})\mathbf{B}$.
26. From Exercise 24 deduce that $(\mathbf{A} \times \mathbf{B}) \times (\mathbf{C} \times \mathbf{D}) = [(\mathbf{A} \times \mathbf{B}) \cdot \mathbf{D}]\mathbf{C} - [(\mathbf{A} \times \mathbf{B}) \cdot \mathbf{C}]\mathbf{D}$.
27. Prove that $(\mathbf{A} + \mathbf{B}) \times (\mathbf{A} - \mathbf{B}) = 2(\mathbf{B} \times \mathbf{A})$.
28. Prove that $\mathbf{A} \cdot (\mathbf{B} \times \mathbf{C}) = (\mathbf{A} \times \mathbf{B}) \cdot \mathbf{C}$.
29. Prove that $|\mathbf{A} \times \mathbf{B}|^2 = |\mathbf{A}|^2 |\mathbf{B}|^2 - (\mathbf{A} \cdot \mathbf{B})^2$.
    Suggestion: Take advantage of Exercises 25 and 28.

■ ■

Exercises 30 to 33 develop a geometric interpretation of the dot product, $\mathbf{A} \cdot (\mathbf{B} \times \mathbf{C})$.

30. Three vectors $\mathbf{A}$, $\mathbf{B}$, and $\mathbf{C}$ determine (span) a parallelepiped, as shown in Fig. 11.74.

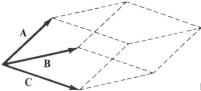

Figure 11.74

    (a) What is the area of the base spanned by $\mathbf{B}$ and $\mathbf{C}$?
    (b) What is the height of the parallelepiped?
    (c) Deduce that the volume of the parallelepiped is the absolute value of $\mathbf{A} \cdot (\mathbf{B} \times \mathbf{C})$.

31. (See Exercise 30.) Show that, if $\mathbf{A} = a_1\mathbf{i} + a_2\mathbf{j} + a_3\mathbf{k}$, $\mathbf{B} = b_1\mathbf{i} + b_2\mathbf{j} + b_3\mathbf{k}$, and $\mathbf{C} = c_1\mathbf{i} + c_2\mathbf{j} + c_3\mathbf{k}$, then the parallelepiped spanned by $\mathbf{A}$, $\mathbf{B}$, and $\mathbf{C}$ has a volume equal to the absolute value of the determinant

$$\begin{vmatrix} a_1 & a_2 & a_3 \\ b_1 & b_2 & b_3 \\ c_1 & c_2 & c_3 \end{vmatrix}.$$

Thus a third-order determinant can be thought of as $+$ or $-$ the volume of a certain parallelepiped.

32. (See Exercise 30.)
    (a) Show geometrically that these numbers all have the same absolute value: $(\mathbf{A} \times \mathbf{B}) \cdot \mathbf{C}$, $\mathbf{A} \cdot (\mathbf{B} \times \mathbf{C})$, $(\mathbf{A} \times \mathbf{C}) \cdot \mathbf{B}$.
    (b) Which of the three numbers in (a) are equal?

33. (See Exercise 31.) Show that the points $(0, 0, 0)$, (x_1, y_1, z_1), (x_2, y_2, z_2), and (x_3, y_3, z_3) lie on a plane if and only if
$$\begin{vmatrix} x_1 & y_1 & z_1 \\ x_2 & y_2 & z_2 \\ x_3 & y_3 & z_3 \end{vmatrix} = 0.$$

Exercises 34 to 37 illustrate the use of the cross product in the study of lines and planes.

34. (a) Assume that the points Q, R, and S determine a plane. Show that the vector $\overrightarrow{QR} \times \overrightarrow{QS}$ is normal to the plane.
    (b) Find a normal to the plane through the points $(1, 1, -1)$, $(0, 2, 3)$, and $(4, 1, 5)$.

35. (See Exercise 34.) How far is the point P from the plane through Q, R, and S?
    (a) Draw a picture that shows that the answer is given by the magnitude of the projection of the vector $\overrightarrow{PQ}$ on a vector normal to the plane.
    (b) Using (a), find a formula for the distance from P to the plane.

36. (See Exercise 35.) The points P and Q determine a line L, and the points R and S determine a line L'. Find the distance between the two lines. Disregard the case when L and L' are parallel. (Suggestion: Consider the plane through L' that is parallel to L. Then the distance from L to L' is the distance from any point on L to that plane, as a diagram will show.)

37. The angle between two planes is defined as the angle between their normals, so chosen that this angle is at most $\pi/2$. Find the cosine of the angle between the plane determined by the points P, Q, and R and the plane determined by the points S, T, and U.

38. (Contributed by Melvyn Stein) A symmetrically designed chute joins a 4-foot-square intake to a 3-foot-square opening, 10 feet lower, as shown in Fig. 11.75. Find the angle between two adjacent sides of the chute. (See Exercise 37.)

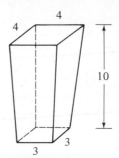

Figure 11.75

39. (From crystallography) Let $\mathbf{v}_1$, $\mathbf{v}_2$, and $\mathbf{v}_3$ be three vectors in space that are not parallel to any plane. Form the vectors
$$\mathbf{k}_1 = \frac{\mathbf{v}_2 \times \mathbf{v}_3}{\mathbf{v}_1 \cdot (\mathbf{v}_2 \times \mathbf{v}_3)}, \quad \mathbf{k}_2 = \frac{\mathbf{v}_3 \times \mathbf{v}_1}{\mathbf{v}_1 \cdot (\mathbf{v}_2 \times \mathbf{v}_3)},$$
$$\mathbf{k}_3 = \frac{\mathbf{v}_1 \times \mathbf{v}_2}{\mathbf{v}_1 \cdot (\mathbf{v}_2 \times \mathbf{v}_3)}.$$

Show that
(a) $\mathbf{k}_1 \cdot (\mathbf{k}_2 \times \mathbf{k}_3) = \dfrac{1}{\mathbf{v}_1 \cdot (\mathbf{v}_2 \times \mathbf{v}_3)}$;
(b) $\mathbf{k}_i$ is perpendicular to $\mathbf{v}_j$ if $i \neq j$;
(c) $\mathbf{k}_i \cdot \mathbf{v}_i = 1$ for each $i = 1, 2, 3$.

The vectors $x_1 \mathbf{v}_1 + x_2 \mathbf{v}_2 + x_3 \mathbf{v}_3$, for integer values of x_1, x_2, and x_3, form one lattice for the crystal. The vectors $\mathbf{k}_1, \mathbf{k}_2,$ and $\mathbf{k}_3$ generate what is called the "reciprocal lattice."

40. Let L be the line in which the planes $x + y + 3z = 5$ and $2x - y + z = 2$ intersect.
    (a) Find a vector parallel to L.
    (b) Find a point on L.
    (c) Find parametric equations for L.

11.S Summary

This chapter concerns the algebra of vectors, which will be applied in the following chapters.

A vector may be pictured as an arrow. Two arrows that point in the same direction and have the same length represent the same vector. The following table summarizes the basic concepts of vectors in space. (For plane vectors disregard the third component.)

11.S SUMMARY

Symbol	Name	Geometric Description	Algebraic formula if $\mathbf{A} = a_1\mathbf{i} + a_2\mathbf{j} + a_3\mathbf{k}$ and $\mathbf{B} = b_1\mathbf{i} + b_2\mathbf{j} + b_3\mathbf{k}$
$\mathbf{A}$	Vector	Direction and magnitude	$a_1\mathbf{i} + a_2\mathbf{j} + a_3\mathbf{k}$
$|\mathbf{A}|$	Magnitude of $\mathbf{A}$	Length or magnitude of $\mathbf{A}$	$\sqrt{a_1^2 + a_2^2 + a_3^2}$
$-\mathbf{A}$	Negative or opposite of $\mathbf{A}$		$-a_1\mathbf{i} - a_2\mathbf{j} - a_3\mathbf{k}$
$\mathbf{A} + \mathbf{B}$	Sum of $\mathbf{A}$ and $\mathbf{B}$		$(a_1 + b_1)\mathbf{i} + (a_2 + b_2)\mathbf{j} + (a_3 + b_3)\mathbf{k}$
$\mathbf{A} - \mathbf{B}$	Difference of $\mathbf{A}$ and $\mathbf{B}$		$(a_1 - b_1)\mathbf{i} + (a_2 - b_2)\mathbf{j} + (a_3 - b_3)\mathbf{k}$
$c\mathbf{A}$	Scalar multiple of $\mathbf{A}$		$ca_1\mathbf{i} + ca_2\mathbf{j} + ca_3\mathbf{k}$
$\mathbf{A} \cdot \mathbf{B}$	Dot or scalar product	$|\mathbf{A}||\mathbf{B}|$ (cosine of angle between $\mathbf{A}$ and $\mathbf{B}$)	$a_1b_1 + a_2b_2 + a_3b_3$
$\mathbf{A} \times \mathbf{B}$	Cross or vector product	Magnitude: area of parallelogram spanned by $\mathbf{A}$ and $\mathbf{B}$ Direction: perpendicular to $\mathbf{A}$ and $\mathbf{B}$, direction by right-hand rule	$\begin{vmatrix} \mathbf{i} & \mathbf{j} & \mathbf{k} \\ a_1 & a_2 & a_3 \\ b_1 & b_2 & b_3 \end{vmatrix}$ (This is the definition.)
	Vector components of $\mathbf{A}$ parallel and perpendicular to $\mathbf{B}$		$\mathbf{A}_1 = \left(\mathbf{A} \cdot \dfrac{\mathbf{B}}{|\mathbf{B}|}\right)\dfrac{\mathbf{B}}{|\mathbf{B}|}$ $\mathbf{A}_2 = \mathbf{A} - \mathbf{A}_1$

Matrices and determinants were introduced, primarily for the definition of $\mathbf{A} \times \mathbf{B}$.

VOCABULARY AND SYMBOLS

vector $\mathbf{A}$, $\mathbf{B}$, $\overrightarrow{(x, y)}$, $\overrightarrow{(x, y, z)}$, $\overrightarrow{PQ}$
sum of vectors
scalar
product of scalar and vector $c\mathbf{A}$
unit vector
basic unit vectors $\mathbf{i}, \mathbf{j}, \mathbf{k}$
dot product (scalar product) $\mathbf{A} \cdot \mathbf{B}$
scalar component
vector components parallel and perpendicular to a vector

projection
parametric equations of a line
direction angles α, β, γ
direction cosines
matrix
determinant
Cramer's rule
cross product (vector product) of vectors in space $\mathbf{A} \times \mathbf{B}$

KEY FACTS If **A** is not **0**, then $\mathbf{A}/|\mathbf{A}|$ is a unit vector.
If $\mathbf{A}\cdot\mathbf{B}=0$, then $\mathbf{A}=\mathbf{0}$, $\mathbf{B}=\mathbf{0}$, or **A** is perpendicular to **B**.
The cosine of the angle between **A** and **B** is

$$\frac{\mathbf{A}\cdot\mathbf{B}}{|\mathbf{A}||\mathbf{B}|}.$$

The scalar component of **A** on the unit vector **u** is $\mathbf{A}\cdot\mathbf{u}$. The scalar component of **A** on **B** is $\mathbf{A}\cdot(\mathbf{B}/|\mathbf{B}|)$.

An equation of the line through (x_0, y_0) and perpendicular to the vector $A\mathbf{i}+B\mathbf{j}$ is

$$A(x-x_0)+B(y-y_0)=0.$$

The vector $A\mathbf{i}+B\mathbf{j}$ is perpendicular to the line

$$Ax+By+C=0.$$

The distance from the point (x_1, y_1) to the line

$$Ax+By+C=0$$

is

$$\frac{|Ax_1+By_1+C|}{\sqrt{A^2+B^2}}.$$

An equation of the plane through (x_0, y_0, z_0) and perpendicular to the vector $A\mathbf{i}+B\mathbf{j}+C\mathbf{k}$ is

$$A(x-x_0)+B(y-y_0)+C(z-z_0)=0.$$

The vector $A\mathbf{i}+B\mathbf{j}+C\mathbf{k}$ is perpendicular to the plane

$$Ax+By+Cz+D=0.$$

The distance from the point (x_1, y_1, z_1) to the plane

$$Ax+By+Cz+D=0$$

is

$$\frac{|Ax_1+By_1+Cz_1+D|}{\sqrt{A^2+B^2+C^2}}.$$

The direction angles α, β, and γ of a vector are the angles it makes with **i**, **j**, and **k**; $\cos\alpha$, $\cos\beta$, and $\cos\gamma$ are the direction cosines of the vector (or of a line parallel to the vector). They are related by the equation

$$\cos^2\alpha+\cos^2\beta+\cos^2\gamma=1.$$

The line through $P_0=(x_0, y_0, z_0)$ parallel to $\mathbf{A}=a_1\mathbf{i}+a_2\mathbf{j}+a_3\mathbf{k}$ is given parametrically as

$$x = x_0 + a_1 t$$
$$y = y_0 + a_2 t$$
$$z = z_0 + a_3 t,$$

or vectorially as $\overrightarrow{OP} = \overrightarrow{OP_0} + t\mathbf{A}$.
Also, the line has the description

$$\frac{x - x_0}{a_1} = \frac{y - y_0}{a_2} = \frac{z - z_0}{a_3}.$$

An arrangement $\begin{pmatrix} a_1 & a_2 \\ b_1 & b_2 \end{pmatrix}$

is a matrix of order 2. A determinant of order 2 is defined by

$$\begin{vmatrix} a_1 & a_2 \\ b_1 & b_2 \end{vmatrix} = a_1 b_2 - a_2 b_1.$$

A determinant of order 3 is defined as

$$\begin{vmatrix} a_1 & a_2 & a_3 \\ b_1 & b_2 & b_3 \\ c_1 & c_2 & c_3 \end{vmatrix} = a_1 \begin{vmatrix} b_2 & b_3 \\ c_2 & c_3 \end{vmatrix} - a_2 \begin{vmatrix} b_1 & b_3 \\ c_1 & c_3 \end{vmatrix} + a_3 \begin{vmatrix} b_1 & b_2 \\ c_1 & c_2 \end{vmatrix}.$$

If two rows (or columns) of a matrix are the same, its determinant is 0. If two rows (or columns) are interchanged, the determinant changes sign.

Determinants provide formulas for the solution of simultaneous linear equations.

The cross product *disobeys* many of the familiar rules of arithmetic. For instance, $\mathbf{A} \times \mathbf{A} = \mathbf{0}$, $\mathbf{B} \times \mathbf{A} = -\mathbf{A} \times \mathbf{B}$, and $(\mathbf{A} \times \mathbf{B}) \times \mathbf{C}$ is usually not equal to $\mathbf{A} \times (\mathbf{B} \times \mathbf{C})$. In fact,

$$\mathbf{A} \times (\mathbf{B} \times \mathbf{C}) = (\mathbf{A} \cdot \mathbf{C})\mathbf{B} - (\mathbf{A} \cdot \mathbf{B})\mathbf{C}.$$

$\mathbf{A} \cdot (\mathbf{B} \times \mathbf{C}) = \pm$ volume of parallelepiped spanned by $\mathbf{A}$, $\mathbf{B}$, and $\mathbf{C}$

$$= \begin{vmatrix} a_1 & a_2 & a_3 \\ b_1 & b_2 & b_3 \\ c_1 & c_2 & c_3 \end{vmatrix}.$$

(See Exercises 30 and 31 in Sec. 11.6.)

Guide quiz on chap. 11

1 Given $\mathbf{A} = \mathbf{i} + 2\mathbf{j} - \mathbf{k}$ and $\mathbf{B} = 2\mathbf{i} - \mathbf{j} + 3\mathbf{k}$, find
   (a) $\mathbf{A} \cdot \mathbf{B}$,   (b) $|\mathbf{A}|$,
   (c) a unit vector in the direction of $\mathbf{A}$,
   (d) the scalar component of $\mathbf{B}$ on $\mathbf{A}$,

(e) the vector component of **B** on **A**,
(f) the scalar component of **A** on **B**,
(g) the vector component of **B** perpendicular to **A**.
(h) the cosine of the angle between **A** and **B**,
(i) the angle between **A** and **B**,
(j) **A** × **B**,
(k) **B** × **A**,
(l) a unit vector perpendicular to both **A** and **B**,
(m) the area of the parallelogram spanned by **A** and **B**.

2   Draw the necessary diagrams and explain why the distance from the point (x_1, y_1) to the line $Ax + By + C = 0$ is

$$\frac{|Ax_1 + By_1 + C|}{\sqrt{A^2 + B^2}}.$$

3   Find the direction cosines of a vector normal to the plane $x - 2y + 2z = 16$.

4   Prove that

$$\begin{vmatrix} a_1 & a_2 & a_3 \\ b_1 & b_2 & b_3 \\ c_1 + b_1 & c_2 + b_2 & c_3 + b_3 \end{vmatrix} = \begin{vmatrix} a_1 & a_2 & a_3 \\ b_1 & b_2 & b_3 \\ c_1 & c_2 & c_3 \end{vmatrix}.$$

5   (a) Find the volume of the parallelepiped shown in Fig. 11.76.

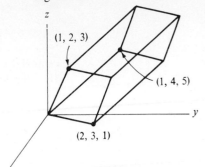

Figure 11.76

(b) Justify the general formula you used in (a).

6   Let **A** be a vector in space and P, Q, and R three points that determine a plane.
(a) Using a picture, show that there are vectors $\mathbf{A}_1$ and $\mathbf{A}_2$ such that $\mathbf{A} = \mathbf{A}_1 + \mathbf{A}_2$ and $\mathbf{A}_1$ is parallel to the plane and $\mathbf{A}_2$ is perpendicular to the plane.
(b) Find formulas for $\mathbf{A}_1$ and $\mathbf{A}_2$.

7   Where does the line through $(1, 2, 1)$ and $(3, 1, 1)$ meet the plane determined by the three points $(2, -1, 1)$, $(5, 2, 3)$, and $(4, 1, 3)$?

Review exercises for chap. 11

1   (a) Define **A** · **B** where **A** and **B** are vectors in space.
(b) What is the formula for **A** · **B** in terms of the components of **A** and **B**?
(c) Prove that if **A** · **B** = 0 and neither **A** nor **B** is the zero vector, then **A** is perpendicular to **B**.

2   What is the scalar component of $\mathbf{A} = -2\mathbf{i} + 3\mathbf{j}$ on the vector
(a) **i**?   (c) $0.6\mathbf{i} + 0.8\mathbf{j}$?
(b) **j**?   (d) $4\mathbf{i} - 5\mathbf{j}$?

3   What is the vector component of $\mathbf{A} = -2\mathbf{i} + 3\mathbf{j}$ on the vector
(a) **i**?   (c) $0.6\mathbf{i} + 0.8\mathbf{j}$?
(b) **j**?   (d) $4\mathbf{i} - 5\mathbf{j}$?

4   (a) Give an application of the dot product in physics and one in economics.
(b) Give an application of the cross product in physics.

5   Prove that $(a_1\mathbf{i} + a_2\mathbf{j} + a_3\mathbf{k}) \cdot (b_1\mathbf{i} + b_2\mathbf{j} + b_3\mathbf{k}) = a_1 b_1 + a_2 b_2 + a_3 b_3$.

6   Let **A** be a nonzero vector. (a) How would you produce a vector **B** perpendicular to **A**? (b) How would you produce a vector perpendicular to both **A** and **B**?

7   (a) Define **A** × **B**.
(b) Describe **A** × **B** geometrically.

8   Find the point on the plane $2x - y + 3z + 12 = 0$ that is nearest the origin. Use vectors, not calculus.

9   Let **A** and **B** be two nonparallel nonzero vectors in space.
(a) Explain why there are scalars x and y such that

$$\mathbf{A} \times (\mathbf{A} \times \mathbf{B}) = x\mathbf{A} + y\mathbf{B}.$$

(b) Show that $\mathbf{A} \times (\mathbf{A} \times \mathbf{B})$ is not **0**.
(c) Show that $(\mathbf{A} \times \mathbf{A}) \times \mathbf{B} = \mathbf{0}$.

10   Use determinants to find x and y such that

$$x(2\mathbf{i} + \mathbf{j}) + y(\mathbf{i} + 3\mathbf{j}) = 4\mathbf{i} - 2\mathbf{j}.$$

11   Use determinants to find x, y, and z such that

$$x(3\mathbf{i} + \mathbf{j} + 2\mathbf{k}) + y(\mathbf{i} - \mathbf{j} + 2\mathbf{k}) + z(2\mathbf{i} + 2\mathbf{j} + \mathbf{k}) = 3\mathbf{i} + 3\mathbf{j} + 3\mathbf{k}.$$

12   The determinant of a matrix in which some column has only 0's is 0. Prove this for the matrix

$$\begin{pmatrix} a_1 & 0 & a_3 \\ b_1 & 0 & b_3 \\ c_1 & 0 & c_3 \end{pmatrix}.$$

13 Show that
$$\begin{vmatrix} 0 & a & b \\ -a & 0 & c \\ -b & -c & 0 \end{vmatrix} = 0$$
for all numbers a, b, c.

14 Suppose that a vector from $(1, 3, 3)$ to the plane $x - 4y + 5z + 4 = 0$ makes an angle of $45°$ with that plane. Find the length of the vector.

15 Figure 11.77 shows a pyramid with a square base. Find the cosine of the angle between

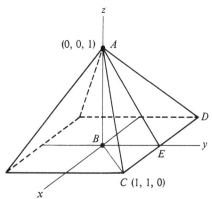

Figure 11.77

(a) $\overrightarrow{CA}$ and $\overrightarrow{CB}$,
(b) $\overrightarrow{EA}$ and $\overrightarrow{EB}$,
(c) $\overrightarrow{AD}$ and $\overrightarrow{AC}$.

16 The planes $2x + 5y + z = 10$ and $3x - y + 4z = 11$ meet in a line. For this line find
(a) direction numbers,
(b) direction cosines,
(c) direction angles,
(d) a point on the line.

17 How far apart are the planes parallel to the plane $Ax + By + Cz + D = 0$ that pass through the points (x_1, y_1, z_1) and (x_2, y_2, z_2)?

18 Prove that the numbers
$$x = \frac{\begin{vmatrix} k_1 & a_2 \\ k_2 & b_2 \end{vmatrix}}{\begin{vmatrix} a_1 & a_2 \\ b_1 & b_2 \end{vmatrix}} \quad \text{and} \quad y = \frac{\begin{vmatrix} a_1 & k_1 \\ b_1 & k_2 \end{vmatrix}}{\begin{vmatrix} a_1 & a_2 \\ b_1 & b_2 \end{vmatrix}}$$
satisfy both equations $a_1 x + a_2 y = k_1$ and $b_1 x + b_2 y = k_2$. Assume that $a_1 b_2 - a_2 b_1$ is not 0.

19 Find parametric equations of the line through $(1, 1, 2)$ that is parallel to the planes $x + 2y + 3z = 0$ and $2x - y + 3z + 4 = 0$.

20 Does the plane through $(1, 1, -1)$, perpendicular to $2\mathbf{i} + 4\mathbf{j} + 5\mathbf{k}$, pass through the point $(4, 5, -7)$?

21 Is the line through $(1, 4, 7)$ and $(5, 10, 15)$ perpendicular to the plane $2x + 3y + 4z = 17$?

22 Two planes that intersect in a line determine a *dihedral angle* θ, as shown in Fig. 11.78. Find the dihedral angle between the planes $x - 3y + 4z = 10$ and $2x + y + z = 11$.

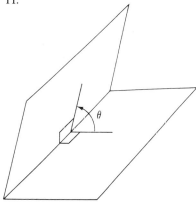

Figure 11.78

23 Find the point on the plane
$$\frac{x}{2} + \frac{y}{3} + \frac{z}{4} = 1$$
(a) nearest the origin,
(b) nearest the point $(1, 2, 3)$.

24 Let $\mathbf{A} = 2\mathbf{i} + 3\mathbf{j} + 4\mathbf{k}$. Find
(a) the scalar component of $\mathbf{A}$ in the direction $\mathbf{i}$,
(b) the vector component of $\mathbf{A}$ in the direction $\mathbf{i}$,
(c) the scalar component of $\mathbf{A}$ in the direction $-\mathbf{i}$,
(d) the vector component of $\mathbf{A}$ in the direction $-\mathbf{i}$,
(e) the projection of $\mathbf{A}$ on $\mathbf{i} + \mathbf{j}$.

25 How would you determine whether the points (x_1, y_1) and (x_2, y_2) are on the same side or on opposite sides of the line $Ax + By + C = 0$?

26 How would you determine whether the points (x_1, y_1, z_1) and (x_2, y_2, z_2) are on the same side or on opposite sides of the plane $Ax + By + Cz + D = 0$?

27 Express in terms of cross products and dot products the statement that the line through P_1 and P_2 is parallel to the plane through P_3, P_4, and P_5.

∎

28 A parallelepiped is spanned by the three vectors $\mathbf{A} = \mathbf{i} + 2\mathbf{j} + 3\mathbf{k}$, $\mathbf{B} = 2\mathbf{i} + \mathbf{j} + \mathbf{k}$, $\mathbf{C} = 3\mathbf{i} + 3\mathbf{j} + \mathbf{k}$.
(a) Find the volume of the parallelepiped.
(b) Find the area of the face spanned by $\mathbf{A}$ and $\mathbf{C}$.
(c) Find the angle between $\mathbf{A}$ and the face spanned by $\mathbf{B}$ and $\mathbf{C}$.

29 Find the angle between the line through $(0, 0, 0)$ and $(1, 1, 1)$ and the plane through $(1, 2, 3)$, $(4, 1, 5)$, and $(2, 0, 6)$.

30 Let $\mathbf{u}_1$ and $\mathbf{u}_2$ be perpendicular unit vectors in space. Find $(\mathbf{u}_1 \times \mathbf{u}_2) \times \mathbf{u}_1$ and $\mathbf{u}_1 \times (\mathbf{u}_1 \times \mathbf{u}_2)$.

31 Consider a company that produces x_1 washing machines, x_2 refrigerators, and x_3 dishwashers. The profit from each washing machine is p_1 dollars, from each refrigerator p_2 dollars, and from each dishwasher p_3 dollars. (The p's need not be positive.)

The three-dimensional vector $\mathbf{V} = \overrightarrow{(x_1, x_2, x_3)}$ is called the *production vector*, and $\mathbf{P} = \overrightarrow{(p_1, p_2, p_3)}$ the *profit vector*.

(a) Show that the total profit is $\mathbf{P} \cdot \mathbf{V}$.

(b) Assume that the company also manufactures x_4 freezers at a profit of p_4 dollars on each, and x_5 electric brooms at a profit of p_5 dollars on each. How would you define its production and profit vectors then?

32 Express in terms of cross products and dot products the assertion that the points P_4 and P_5 are situated on the same side of the plane through P_1, P_2, and P_3.

33 Let L_1 and L_2 be two lines in the xy plane through $(0, 0)$, one of slope m_1 and one of slope m_2.

(a) Show that the point $(1, m_1)$ is on L_1 and that $\mathbf{i} + m_1\mathbf{j}$ is a vector parallel to L_1.

(b) Show that the point $(1, m_2)$ is on L_2 and that $\mathbf{i} + m_2\mathbf{j}$ is a vector parallel to L_2.

(c) Using (a) and (b), prove that L_1 is perpendicular to L_2 if and only if $m_1 m_2 = -1$.

34 (a) Show that

$$(\mathbf{A} \times \mathbf{B}) \times (\mathbf{C} \times \mathbf{D}) = (\mathbf{A} \times \mathbf{B} \cdot \mathbf{D})\mathbf{C} - (\mathbf{A} \times \mathbf{B} \cdot \mathbf{C})\mathbf{D}.$$

Warning: There is only one meaningful way to interpret $\mathbf{A} \times \mathbf{B} \cdot \mathbf{D}$, and it is not $\mathbf{A} \times (\mathbf{B} \cdot \mathbf{D})$.

(b) Show that

$$(\mathbf{A} \times \mathbf{B}) \cdot (\mathbf{C} \times \mathbf{D}) = (\mathbf{B} \cdot \mathbf{D})(\mathbf{A} \cdot \mathbf{C}) - (\mathbf{B} \cdot \mathbf{C})(\mathbf{A} \cdot \mathbf{D}).$$

35 Show that

$$\mathbf{A} \times (\mathbf{B} \times \mathbf{C}) + \mathbf{B} \times (\mathbf{C} \times \mathbf{A}) + \mathbf{C} \times (\mathbf{A} \times \mathbf{B}) = 0.$$

36 Find the area of the triangle whose vertices are $(1, 1, 2)$, $(2, 1, 4)$, and $(3, 0, 5)$.

■ ■

37 Let $\mathbf{B}$ be a nonzero vector. Assume that $\mathbf{A}_1$ and $\mathbf{A}'_1$ are parallel to $\mathbf{B}$, $\mathbf{A}_2$ and $\mathbf{A}'_2$ are perpendicular to $\mathbf{B}$, and $\mathbf{A}_1 + \mathbf{A}_2 = \mathbf{A}'_1 + \mathbf{A}'_2$. Must $\mathbf{A}_1 = \mathbf{A}'_1$ and $\mathbf{A}_2 = \mathbf{A}'_2$? Explain both with and without pictures.

38 (a) Consider any triangle in the plane and a point P_0 inside the triangle. Let $f(P)$ denote the sum of the (positive) perpendicular distances from P to the three lines of the triangle. Show that there is a line segment whose ends lie on the border of the triangle such that $f(P) = f(P_0)$ for all points P on the segment.

(b) Does this generalize to all polygons in the plane?

39 Consider any tetrahedron and a point P_0 inside the tetrahedron. Let $f(P)$ denote the sum of the (positive) distances from P to the four planes of the tetrahedron. Show that there is a plane through P_0 such that, if P is on the plane and inside the tetrahedron, then $f(P) = f(P_0)$.

40 Let $\mathbf{A}$, $\mathbf{B}$, and $\mathbf{C}$ be three vectors in space which do not lie on one plane. That is, when located so that their tails are at the origin, they are not contained in a plane.

(a) Show with the aid of sketches that any vector $\mathbf{D}$ is of the form $x\mathbf{A} + y\mathbf{B} + z\mathbf{C}$ for suitable scalars x, y, z.

(b) For a given vector $\mathbf{D}$ are the scalars in (a) unique?

PARTIAL DERIVATIVES

The volume V of a cylindrical can of radius r and height h is given by the formula

$$V = \pi r^2 h.$$

The function $\pi r^2 h$ is an example of a function of two variables r and h. Almost all the work of the first 11 chapters concerns functions of one variable; Chap. 12 is devoted to functions of two variables, their graphs, their derivatives, and their maxima and minima.

12.1 Graphs of equations

The set of points (x, y, z) that satisfy some given equation in x, y, and z is called the *graph* of that equation. For instance, as shown in Theorem 4 of Sec. 11.4, the graph of the equation $Ax + By + Cz + D = 0$, where not all of A, B, and C are 0, is a plane. This section examines particular planes and then goes on to the equations of spheres and other common surfaces.

PLANES

EXAMPLE 1 Sketch the graph of the equation $z = 1$.

SOLUTION The graph consists of all points whose z coordinates are 1. This is a plane parallel to the plane $z = 0$ (the xy plane) and a distance 1 above it. The plane $z = 1$ is endless. The jagged border in Fig. 12.1 is one method of suggesting this.

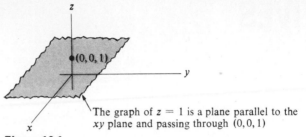

The graph of $z = 1$ is a plane parallel to the xy plane and passing through $(0, 0, 1)$

Figure 12.1

EXAMPLE 2   Sketch the graph of the equation $x = y$.

SOLUTION   The point (x, y, z) is on this graph if $x = y$. There is no restriction on z. For instance, the points $(1, 1, 0)$, $(1, 1, 3)$, $(1, 1, -7)$ are all on this graph. For convenience, first sketch the part of the graph for which $z = 0$. In other words, consider the part of the graph that lies in the xy plane.

In the xy plane, the equation

$$x = y$$

Warning: The graph of $x = y$ in space is a plane, not a line.

describes a line, as shown in Fig. 12.2. As observed above, there is no restriction on z. This means that if $(x, y, 0)$ is on the graph, so is (x, y, z) for any value of z. Thus the graph of the equation $x = y$ in space is the plane perpendicular to the xy plane, passing through the line $x = y$ in the xy plane. It is shown in Fig. 12.3.

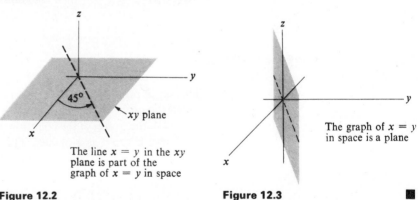

The line $x = y$ in the xy plane is part of the graph of $x = y$ in space

Figure 12.2

The graph of $x = y$ in space is a plane

Figure 12.3

EXAMPLE 3   Sketch the graph of $z = x + 2y$.

SOLUTION   Since the equation can be put in the form $x + 2y - z = 0$, the graph is a plane. This plane passes through the origin since $0 + 2 \cdot 0 - 0 = 0$.

One way to sketch the plane is to draw the vector $\mathbf{i} + 2\mathbf{j} - \mathbf{k}$, which is perpendicular to the plane, to suggest the tilt of the plane. This is done in Fig. 12.4.

12.1 GRAPHS OF EQUATIONS   749

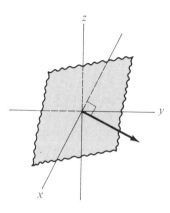

Figure 12.4 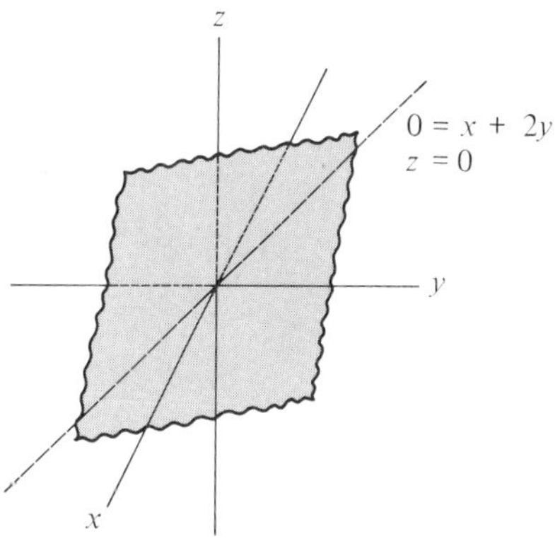   **Figure 12.5**

Also, it may help to show where the plane $z = 2x + y$ meets a coordinate plane. To determine the line of intersection with the xy plane, set the z-coordinate $z = 0$. The line therefore has the equation

$$0 = 2x + y$$

and is shown dashed in Fig. 12.5.

The intersection of the surface $z = x + 2y$ and the xz plane consists of all points (x, y, z) for which

$$z = x + 2y \quad \text{and} \quad y = 0.$$

It is therefore the line $z = x$

in the xz plane, shown in Fig. 12.6. The plane in Fig. 12.4 is determined by the dashed lines in Figs. 12.5 and 12.6. ∎

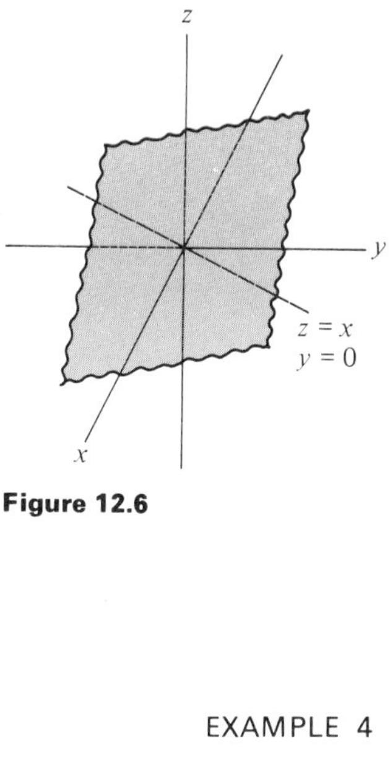

Figure 12.6

CYLINDERS

The next example illustrates a notion of "cylinder" which is far more general than the right-circular cylinder.

EXAMPLE 4   Sketch the graph of $y = x^2$.

SOLUTION   The graph of $y = x^2$ consists of all points (x, y, z) such that

$$y = x^2.$$

For instance $(3, 9, 0)$ is on the graph of $y = x^2$. So is $(3, 9, z)$ for any choice of z. The simplest way to graph the equation $y = x^2$ (in space) is to graph $y = x^2$ in the xy plane and then use the information that z is unrestricted. Since the graph of $y = x^2$ in the xy plane is a parabola, the graph of $y = x^2$ in space is a curved surface, as sketched in Fig. 12.7. ∎

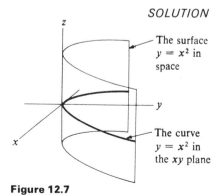

Figure 12.7

Example 4 is a special case of what is called a *cylinder*.

DEFINITION *Cylinder.* Let R be a set in a plane. The set formed by all lines that are perpendicular to the given plane and that meet R is called the *cylinder* determined by R.

The surface $y = x^2$ in space, of Example 4, is a cylinder. The set R in this case is the parabola $y = x^2$ in the xy plane. The plane $z = 0$ is also a cylinder, for it consists of all lines perpendicular to the yz plane that pass through the y axis. If R is a circle, then the cylinder it determines is like the cylinder of everyday life, except that it is infinite and therefore has no base or top.

How to spot an equation of a cylinder

If an equation involves at most two of the letters x, y, and z, its graph will be a cylinder.

SPHERES AND ELLIPSOIDS

The set of all points that are a fixed distance r from a given point (a, b, c) is a sphere of radius r and center (a, b, c). To sketch this sphere, dot in the horizontal equator in perspective. (See Fig. 12.8.)

A point (x, y, z) is on this sphere when the distance between it and (a, b, c) is r, that is, when

$$\sqrt{(x - a)^2 + (y - b)^2 + (z - c)^2} = r,$$

or, equivalently, when

$$(x - a)^2 + (y - b)^2 + (z - c)^2 = r^2.$$

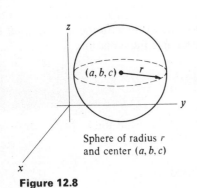

Figure 12.8 Sphere of radius r and center (a, b, c)

This last equation is an algebraic description of the sphere of radius r and center (a, b, c).

In practice, the origin $(0, 0, 0)$ of the xyz-coordinate system is usually placed at the center of the sphere. The equation of a sphere of radius r and center $(0, 0, 0)$ thus becomes

$$x^2 + y^2 + z^2 = r^2.$$

For instance, the equation

$$x^2 + y^2 + z^2 = 25$$

describes a sphere of radius 5 and center $(0, 0, 0)$. As another example, the equation

$$x^2 + y^2 + z^2 = 3$$

describes a sphere of radius $\sqrt{3}$ and center $(0, 0, 0)$.

ELLIPSOIDS

The graph of

$$\frac{x^2}{a^2} + \frac{y^2}{b^2} + \frac{z^2}{c^2} = 1 \tag{1}$$

where a, b, and c are positive constants, is called an *ellipsoid*. In the special case when $a = b = c$ the equation becomes

$$\frac{x^2}{a^2} + \frac{y^2}{a^2} + \frac{z^2}{a^2} = 1$$

or
$$x^2 + y^2 + z^2 = a^2,$$

the equation of a sphere of radius a. An ellipsoid meets the coordinate planes in ellipses (rather than circles). For instance, the ellipsoid (1) meets the xy plane ($z = 0$) in the ellipse

$$\frac{x^2}{a^2} + \frac{y^2}{b^2} = 1.$$

To find where the ellipsoid (1) meets a given axis, set the variables corresponding to the other two axes equal to 0. Points on the z axis, for example, are described by the simultaneous equations $x = 0$ and $y = 0$. Thus the ellipsoid (1) meets the z axis at the values of z for which

$$\frac{0^2}{a^2} + \frac{0^2}{b^2} + \frac{z^2}{c^2} = 1,$$

that is, for $z = c$ or $-c$. The graph of (1) is shown in Fig. 12.9.

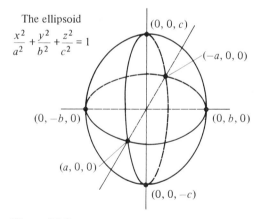

Figure 12.9

HYPERBOLOIDS

The graph of $x^2 + y^2 + z^2 = 1$ is the sphere of radius 1 and center at the origin. By changing some of the plus signs to minus signs, we get new equations and graphs that are quite different from spheres.

If we make all three coefficients negative, the equation becomes $-x^2 - y^2 - z^2 = 1$, or $x^2 + y^2 + z^2 = -1$. Since $x^2 + y^2 + z^2$ is the sum of squares of real numbers, it is never negative; thus there are no points on the graph of $-x^2 - y^2 - z^2 = 1$.

The graphs of $x^2 + y^2 - z^2 = 1$ and $x^2 - y^2 - z^2 = 1$ turn out to be of interest and will introduce the "hyperboloid of one sheet" and the "hyperboloid of two sheets."

EXAMPLE 5 Graph $x^2 + y^2 - z^2 = 1$.

SOLUTION To graph this surface, let us examine its intersections with various planes parallel to the xy plane.

The points (x, y, z) that lie on $x^2 + y^2 - z^2 = 1$ and in the plane $z = 0$ satisfy the equation $x^2 + y^2 - 0^2 = 1$, that is, $x^2 + y^2 = 1$. They form a circle of radius 1 in the xy plane. (See Fig. 12.10.)

The points (x, y, z) that lie on $x^2 + y^2 - z^2 = 1$ and the plane $z = 1$ satisfy the equation $x^2 + y^2 - 1^2 = 1$, or $x^2 + y^2 = 2$. They form a circle of radius $\sqrt{2}$. (See Fig. 12.10.)

In a similar way it can be shown that the intersection of the plane $z = k$, for any constant k, with the surface $x^2 + y^2 - z^2 = 1$ is a circle of radius $\sqrt{1 + k^2}$. Note that as $|k|$ increases, so does the radius of the circle.

To find the intersection of $x^2 + y^2 - z^2 = 1$ with the yz plane, set $x = 0$, obtaining the equation $y^2 - z^2 = 1$, which is the equation of a hyperbola. This hyperbola is a great help in graphing the surface, which is shown in Fig. 12.11.

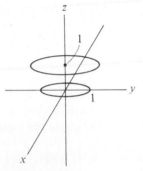

Figure 12.10

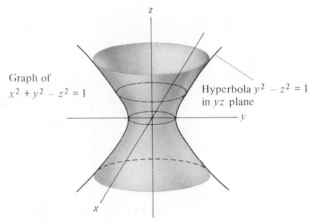

Graph of $x^2 + y^2 - z^2 = 1$

Hyperbola $y^2 - z^2 = 1$ in yz plane

Figure 12.11

For any positive numbers a, b, and c the graph of

$$-\frac{x^2}{a^2} + \frac{y^2}{b^2} + \frac{z^2}{c^2} = 1 \text{ and}$$

$$\frac{x^2}{a^2} - \frac{y^2}{b^2} + \frac{z^2}{c^2} = 1 \text{ are also}$$

$$\frac{x^2}{a^2} + \frac{y^2}{b^2} - \frac{z^2}{c^2} = 1 \quad (2)$$

hyperboloids of one sheet.

is called a *hyperboloid of one sheet*. (The case $a = 1$, $b = 1$, and $c = 1$ is shown in Fig. 12.11.) Cross sections by planes parallel to the xy plane are ellipses. For instance, the intersection of (2) with the xy plane itself is the ellipse

$$\frac{x^2}{a^2} + \frac{y^2}{b^2} = 1.$$

Consider next the case where two signs are negative, $x^2 - y^2 - z^2 = 1$.

EXAMPLE 6 Graph $x^2 - y^2 - z^2 = 1$.

SOLUTION Consider first the intersections of the surface with planes parallel to the yz plane.

When $x = 0$, the equation $x^2 - y^2 - z^2 = 1$ becomes

$$-y^2 - z^2 = 1,$$

which has no real solutions. Thus the yz plane does not meet the surface at all.

When $x = 1$, the equation $x^2 - y^2 - z^2 = 1$ becomes $1 - y^2 - z^2 = 1$, or $y^2 + z^2 = 0$. The equation $y^2 + z^2 = 0$ has only one solution, $y = 0$ and $z = 0$. Thus the plane $x = 1$ meets the surface in only a single point $(1, 0, 0)$.

More generally, consider the intersection of the surface with the plane $x = k$, where k is a constant. The intersection is described by the equation

$$k^2 - y^2 - z^2 = 1,$$

or
$$y^2 + z^2 = k^2 - 1. \tag{3}$$

If $|k| < 1$, $k^2 - 1$ is negative, so there are no points on the plane $x = k$ that lie on the surface $x^2 - y^2 - z^2 = 1$.

If $|k| \geq 1$, equation (3) describes a circle of radius $\sqrt{k^2 - 1}$. Thus the surface is composed of circles.

To see how these circles fit together, use the intersection of the surface with the xy plane, $z = 0$. On this plane the equation of the surface becomes $x^2 - y^2 - 0^2 = 1$, or $x^2 - y^2 = 1$, the equation of a hyperbola. The surface is sketched in Fig. 12.12.

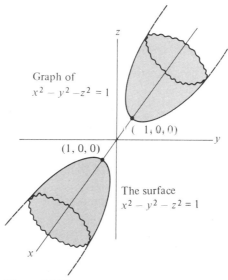

Figure 12.12

Two minuses and one plus in any arrangement give a hyperboloid of two sheets.

For any positive constants the graph of

$$\frac{x^2}{a^2} - \frac{y^2}{b^2} - \frac{z^2}{c^2} = 1$$

is called a *hyperboloid of two sheets*. (The case $a = 1$, $b = 1$, and $c = 1$ is shown in Fig. 12.12.) Cross sections by planes parallel to the yz plane are ellipses, single points, or else empty. The cross section by the xy plane is the hyperbola

$$\frac{x^2}{a^2} - \frac{y^2}{b^2} = 1.$$

THE SADDLE

The graph of $z = y^2 - x^2$ resembles a saddle or a pass between two hills.

EXAMPLE 7 Graph $z = y^2 - x^2$.

SOLUTION The intersection of the surface with the xz plane is obtained by setting $y = 0$. This intersection is the parabola

$$z = -x^2,$$

which lies below the xy plane.

The intersection of the surface with the yz plane is obtained by setting $x = 0$. This intersection is the parabola

$$z = y^2,$$

which lies above the xy plane.

Setting $z = 0$ determines the intersection of the surface with the xy plane. This intersection has the equation

$$0 = y^2 - x^2.$$

Since the right-hand side of this equation factors,

$$0 = (y - x)(y + x),$$

its graph in the xy plane consists of the two lines

$$0 = y - x \quad \text{and} \quad 0 = y + x,$$

that is, the lines $y = x$ and $y = -x$.

The surface is shown in Fig. 12.13. ∎

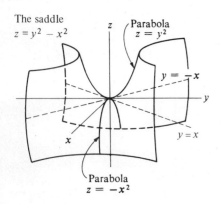

Figure 12.13

Incidentally, the graph of $z = xy$ is also a saddle. It is essentially the graph in Fig. 12.13 rotated 45° around the z axis. It meets the xy plane in the x and y axes. I have never seen a good perspective drawing of it relative

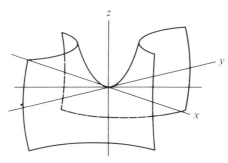

The saddle $z = xy$ contains the x and y axes.
Figure 12.14

to the standard position of the x and y axes whether made by hand or by computer. If the axes are rotated 45°, as in Fig. 12.14, then the saddle $z = xy$ can be sketched by making use of Fig. 12.13.

Another type of surface, the "paraboloid," is discussed in Exercise 17.

Exercises

In Exercises 1 to 4 graph the given planes.
1 $z = 3$ **2** $y = -1$ **3** $x + z = 2$ **4** $y = 2x$

In Exercises 5 and 6 graph the equations and show the intersections of the graphs with the three coordinate planes.

5 $x + y + z = 1$ **6** $\dfrac{x}{2} + \dfrac{y}{3} + \dfrac{z}{4} = 1$

7 Where does the sphere $x^2 + y^2 + z^2 = 49$ meet (a) the x axis? (b) the y axis? (c) the z axis?

8 What is the equation of the sphere (a) whose center is $(-3, 2, 1)$ and whose radius is 1? (b) whose center is $(1, 1, -1)$ and whose radius is $\sqrt{2}$?

In Exercises 9 to 16 graph the surfaces.

9 $x^2 - y^2 + z^2 = 1$ **10** $z^2 - x^2 - y^2 = 1$

11 $x^2 + \dfrac{y^2}{4} + \dfrac{z^2}{9} = 1$

12 $x^2 + 4y^2 + 9z^2 = 16$

$\left(\text{Hint. Rewrite in form } \dfrac{x^2}{a^2} + \dfrac{y^2}{b^2} + \dfrac{z^2}{c^2} = 1.\right)$

13 $x^2 + \dfrac{y^2}{4} - \dfrac{z^2}{25} = 1$ **14** $x^2 - \dfrac{y^2}{4} - \dfrac{z^2}{9} = 1$

15 $4x^2 + 9y^2 - z^2 = 25$

16 $(x - 1)^2 + (y - 2)^2 + z^2 = 16$

17 Graph $z = x^2 + y^2$ and show the intersection with the plane (a) $z = 0$, (b) $z = 1$, (c) $z = 4$, (d) $y = 0$.

This graph is called a *paraboloid*. More generally, for any positive constants a and b, the graph of

$$z = \dfrac{x^2}{a^2} + \dfrac{y^2}{b^2}$$

is called a paraboloid. If $a = b$, the cross sections by planes parallel to the xy plane are circles, and the paraboloid is called a *paraboloid of revolution*. In the general case the cross sections by planes parallel to the xy plane are ellipses. The paraboloid is then called an *elliptical paraboloid*.

18 Graph $z = x^2 + 2y^2$ and show the intersection with the planes (a) $z = 0$, (b) $z = 1$, (c) $x = 0$, (d) $y = 0$.

19 Graph the intersection of the saddle $z = xy$ with the planes (a) $y = 1$, (b) $y = 2$, (c) $x = y$, (d) $x = -y$. (Use the x and y axes in standard position, not as in Fig. 12.14.)

20 (See Exercise 19.) Show that the surface $z = xy$ is made up of straight lines. (This implies that a model of the surface can be made of sticks or stretched pieces of string or wire. Such a model is worth a thousand pictures.)

21 Graph $z = y^2 - x^2/4$ and show the intersection with the plane (a) $z = 0$, (b) $z = 1$, (c) $x = 0$, (d) $y = 0$.

22 Graph $z^2 = x^2 + y^2$ and show the intersection with the plane (a) $x = 0$, (b) $z = 0$, (c) $z = 1$, (d) $z = 2$, (e) $z = -1$.

■ ■

23 The line $z = 2x$ in the xz plane is revolved around the z axis to produce a surface of revolution. Find an equation for this surface.

24 The curve $z = e^{-y^2}$ in the yz plane is revolved around the z axis to produce a surface of revolution. (a)

Graph the surface. (b) Find an equation for the surface.

25 The curve $z = f(y)$ in the yz plane is revolved around the z axis to produce a surface of revolution. Assuming that $y \geq 0$, show that an equation for this surface is $z = f(\sqrt{x^2 + y^2})$.

26 (a) Sketch the part of the surface $x^2 + y^2 = 1$ that lies above the xy plane and below the plane $z = x$.
(b) Find its area.

27 (a) Sketch the triangle in the plane $y = x$ that lies above the xy plane, below the plane $z = y$, and to the left of the plane $y = 3$.
(b) Find the coordinates of its vertices.
(c) Find its area.

12.2 Functions and level curves

Consider a region R in the plane, occupied by a very thin piece of metal, which may be idealized to have zero thickness. Imagine that a match is put under it. Then at each point P in R the metal has a temperature which may be denoted $f(P)$. This is an example of a function whose domain is a set in the plane. The function f assigns to each point P in R a number $f(P)$.

Perhaps the metal is not homogeneous. Its density, in grams per square centimeter, may vary from point to point. The function that assigns to each point P in R the density at P is another example of a function whose domain is a set in the plane.

The next example shows how to graph a function whose domain is a set in the plane.

EXAMPLE 1 Let f be the function that assigns to the point $P = (x, y)$ in the xy plane the number $f(P) = x + 2y$. For instance, $f(1, 2) = 5$. Graph this function.

SOLUTION To record the information that $f(1, 2) = 5$, draw the point $Q = (1, 2, 5)$ in space. To do this, first plot the point $P = (1, 2)$ in the xy plane and then go up from it a distance 5. (See Fig. 12.15.) In brief, the z coordinate of Q is used to record the value of $f(P)$.

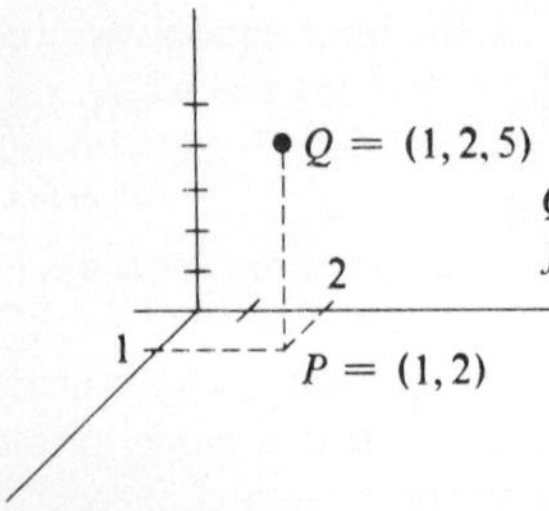

More generally, the information that $f(x, y) = x + 2y$ is recorded by plotting the point $Q = (x, y, x + 2y)$. The z coordinate of the point Q is equal to $x + 2y$. Thus Q lies on the plane

$$z = x + 2y.$$

$Q = (1, 2, 5)$
Q records that $f(P) = 5$.
$P = (1, 2)$

Figure 12.15

As P wanders through the xy plane, the point Q sweeps out the surface $z = x + 2y$. In this case the graph of f is a plane, sketched in Example 3 of the preceding section. ∎

Example 1 illustrates the following definition, which is similar to the definition of the graph of a function $y = f(x)$.

THE GRAPH OF A FUNCTION

DEFINITION *Graph.* The *graph* of a function f whose domain is a set R in the xy plane consists of all points in space of the form $(x, y, f(x, y))$, for (x, y) in R. In other words, the graph of f consists of all points (x, y, z) that satisfy the equation

$$z = f(x, y)$$

and for which (x, y) is in R. Such a function is frequently called a *function of two variables*. Its graph is usually a surface.

A precise definition of continuity of $f(x, y)$ is to be found in Appendix F.

It will be assumed that the function f is *continuous*, in the following sense. If (a, b) is in the domain of f and the point (x, y) in the domain of f approaches the point (a, b), then the number $f(x, y)$ approaches the number $f(a, b)$. Intuitively, nearby inputs produce nearby outputs.

The next two examples discuss the graphs of two functions that will be used later.

EXAMPLE 2   Let O be a fixed point in the plane and P an arbitrary point in the plane. Define $f(P)$ to be the square of the distance from P to O. Graph f.

SOLUTION   Introduce a rectangular coordinate system in the plane, with origin at O; then f has the formula

$$f(x, y) = x^2 + y^2.$$

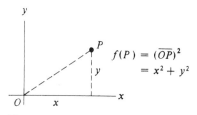

Figure 12.16

(See Fig. 12.16.)

The graph of f is the surface $z = x^2 + y^2$. Since $x^2 + y^2 \geq 0$, this surface lies above the xy plane and meets the xy plane only at $(0, 0, 0)$. The farther $P = (x, y)$ is from $(0, 0)$, the higher the point $Q = (x, y, x^2 + y^2)$ is on the surface.

In order to help visualize and sketch the graph of

$$z = x^2 + y^2,$$

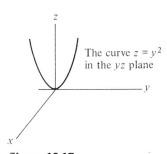

Figure 12.17

consider its intersections with the coordinate planes and planes parallel to these planes. As already mentioned, the plane $z = 0$ meets the surface $z = x^2 + y^2$ only in one point, $(0, 0, 0)$. The plane $x = 0$ meets the surface $z = x^2 + y^2$ in the parabola

$$z = y^2$$

in the yz plane. (See Fig. 12.17.)

Consider next the intersection of a plane parallel to the xy plane with the surface $z = x^2 + y^2$. For instance, consider the plane $z = 4$. If a point (x, y, z) is on the surface $z = x^2 + y^2$ and the plane $z = 4$, it satisfies the equation

$$4 = x^2 + y^2;$$

hence it lies on the cylinder determined by the circle $4 = x^2 + y^2$ in the xy plane. Thus the intersection of

the plane $z = 4$

and   the surface $z = x^2 + y^2$

is the same as the intersection of

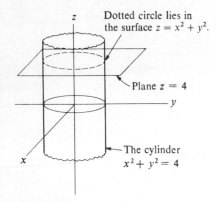

Figure 12.18

the plane $z = 4$ and the circular cylinder $x^2 + y^2 = 4$.

This intersection is dotted in Fig. 12.18. More generally, for any positive constant k, the intersection of the plane $z = k$ with the surface $z = x^2 + y^2$ is a circle of radius $\sqrt{k}$. As k increases, so does the radius of this circle. The surface $z = x^2 + y^2$ is built up of circles whose centers are on the z axis. The easiest way to think of it is as the surface obtained by revolving the parabola $z = y^2$ around the z axis. It has the shape of a headlight, as shown in Fig. 12.19, and is called a paraboloid of revolution, as mentioned in Exercise 17 of the preceding section.

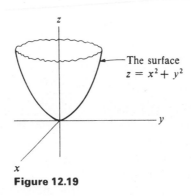

Figure 12.19 ∎

Though the function in the next example differs from the function graphed in Example 2 only in the replacement of x^2 by $-x^2$, the graphs of the two functions are quite different.

EXAMPLE 3   Let f be the function defined by the formula

$$f(x, y) = y^2 - x^2.$$

Graph f.

SOLUTION   The graph is the surface $z = y^2 - x^2$, which is the saddle shown in Fig. 12.13 in the preceding section. ∎

LEVEL CURVES

A U.S. Geodetic Survey map indicates height by contour lines. For instance, one curve may show all the points at the 300-foot altitude. This is the 300-foot contour. The map in Fig. 12.20 depicts a 530-foot hill by using four contour lines at intervals of 100 feet. A hiker walking along that part of the hill which corresponds to a contour line neither rises nor descends.

The hill along with its contours is shown in Fig. 12.21. Note that where the contour lines are close together the hill is steeper.

Let $f(x, y)$ be the altitude of the hill corresponding to the point (x, y) on the map. Then the 300-foot contour consists of those points where

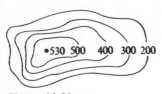

Figure 12.20

12.2 FUNCTIONS AND LEVEL CURVES

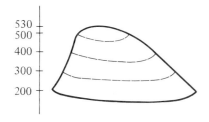

Figure 12.21

A Langley is a measure of energy: one gram-calorie per square centimeter.

$f(x, y) = 300$. On the 300-foot contour the altitude function f is constant: Its value is 300.

The same idea carries over to other functions of x and y. If f is a function of x and y, and c is a fixed number, the set of points where

$$f(x, y) = c$$

is called a *level curve* for f.

The behavior of a function $f(P)$ is often exhibited by showing a few of its level curves. For example, let $f(P)$ be the average daily solar radiation at the point P (measured in Langleys). The level curves corresponding to 350, 400, 450, and 500 Langleys are shown in Fig. 12.22. Inspection of Fig. 12.22 shows where $f(P)$ is large and where it is small, as well as the influence of the Sierra Nevada and Rocky Mountains.

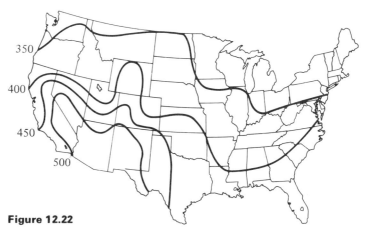

Figure 12.22

The distribution of radioactive fallout from the explosion of a 15-megaton H-bomb in 1954 can best be described in terms of level curves. In this case $f(P)$ is the total radiation dose at point P accumulated within 4 days after the explosion. (This dose is measured in rads; a dose of 700 rads in 4 days is lethal.) Figure 12.23 shows some of the level curves of this function. Note that they are not all at the same intervals.

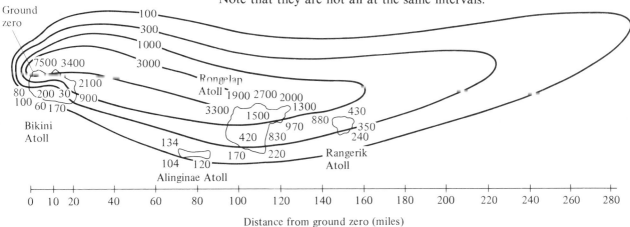

Figure 12.23 (*Source: The Effect of Nuclear Weapons, U. S. Atomic Energy Commission, 1962*)

760 PARTIAL DERIVATIVES

In Examples 4 and 5 the level curves of $f(x, y) = x^2 + y^2$ and $f(x, y) = y^2 - x^2$ are used to examine the behavior of the functions near $(0, 0)$.

EXAMPLE 4 Graph the level curves of $f(x, y) = x^2 + y^2$ corresponding to the values 0, 1, 2, 3, 4, 5, 6, 7, 8, and 9.

SOLUTION The level curve $x^2 + y^2 = 0$ is just a single point, the origin. For any constant $c \geq 0$ the level curve $x^2 + y^2 = c$ is the circle of radius $\sqrt{c}$ centered at the origin. Figure 12.24 shows the circles for the particular values of c listed above.

Level curves near a minimum

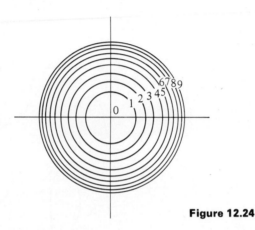

Figure 12.24

Figure 12.24 may be interpreted as follows. First of all, the function has a minimum at the origin. Second, the further P is from the origin, the larger is $f(P)$. Third, since the level curves get closer as one moves away from the origin, the function increases more rapidly further from the origin. ∎

EXAMPLE 5 Graph the level curves of the function $f(x, y) = y^2 - x^2$ corresponding to the values 0, 1, and -1.

SOLUTION The level curve $0 = y^2 - x^2$ or $0 = (y - x)(y + x)$ breaks into two lines, $y - x = 0$ and $y + x = 0$. The level curve $y^2 - x^2 = 1$ is a hyperbola, as is the level curve $y^2 - x^2 = -1$. They are shown in Fig. 12.25. ∎

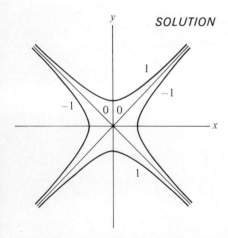

Figure 12.25

Level curves near a saddle point

As Fig. 12.25 indicates, the function $f(x, y) = y^2 - x^2$ has neither a maximum nor a minimum at $(0, 0)$. There are points P arbitrarily near the origin at which $f(P)$ is positive and also such points where $f(P)$ is negative. This conclusion could also be drawn by inspection of the graph of the function, the saddle shown in Fig. 12.13 of the preceding section.

Level curves go by many names, as this table shows:

Function f	Name of Level Curve
Altitude of land	Contour line
Air pressure	Isobar
Temperature	Isotherm
Utility (in economic theory)	Indifference curve
Gravitational potential	Equipotential curve

Exercises

In Exercises 1 to 6 graph the functions.
1. $f(x, y) = x + y$
2. $f(x, y) = x$
3. $f(x, y) = 3$ (a constant function)
4. $f(x, y) = x^2$
5. $f(x, y) = x^2 + 2y^2$
6. $f(x, y) = \dfrac{1}{x^2 + y^2}$

In Exercises 7 to 10 draw for the given functions the level curves corresponding to the values $-1, 0, 1,$ and 2 (if they are not empty).
7. $f(x, y) = x + y$
8. $f(x, y) = x + 2y$
9. $f(x, y) = x^2 + 2y^2$
10. $f(x, y) = x^2 - 2y^2$

In Exercises 11 to 14 draw the level curves for the given functions that pass through the given points.
11. $f(x, y) = x^2 + y^2$ through $(1, 1)$
    Hint: First compute $f(1, 1)$.
12. $f(x, y) = x^2 + 3y^2$ through $(1, 2)$
13. $f(x, y) = x^2 - y^2$ through $(3, 2)$
14. $f(x, y) = x^2 - y^2$ through $(2, 3)$

15. The graph of $f(x, y) = xy$ may be hard to draw, but its level curves are not. Draw the level curves corresponding to the values $0, 1, 2, 3, -1, -2,$ and -3.
16. Graph $f(x, y) = \sqrt{1 - x^2 - y^2}$.
17. Sketch the level curves of the function $f(x, y) = y/x$ corresponding to the values $-1, 0, 1,$ and 2.

18. Figure 12.26 shows some of the level curves (isobars) of the atmospheric pressure function throughout the United States at a given moment.
    (a) At which point is the pressure a maximum?
    (b) At which point is the pressure a minimum?
    (c) At which point would you expect the strongest wind?

19. For a function $g(x, y, z)$ and constant c, the set of points (x, y, z) where $g(x, y, z) = c$ is called a *level surface* for g. What type of surface is the level surface $g(x, y, z) = 1$ if $g(x, y, z)$ is
    (a) $x + y + z$,
    (b) $x^2 + y^2 + z^2$,
    (c) $x^2 + y^2 - z^2$,
    (d) $x^2 - y^2 - z^2$?

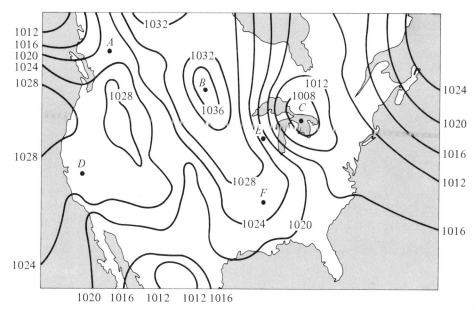

Figure 12.26

12.3 Partial derivatives

Let f be a function of x and y. The graph of $z = f(x, y)$ is a surface. Consider a point (a, b) in the xy plane. The graph of $z = f(x, y)$ for (x, y) near (a, b) may look like the surface in Fig. 12.27. Let P be the point on the surface directly above (a, b). The plane M through (a, b) perpendicular to the y axis meets the surface in a curve, labeled C.

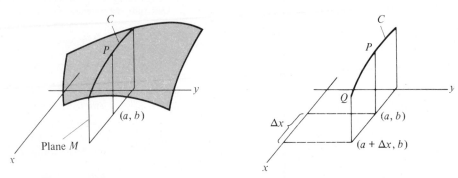

Figure 12.27

To find the slope of this curve in the plane M at the point P, consider, as in Sec. 3.1, a nearby point Q on the curve C directly above $(a + \Delta x, b)$. The limit of the slope of the line through P and Q, as Q approaches P, is

$$\lim_{\Delta x \to 0} \frac{f(a + \Delta x, b) - f(a, b)}{\Delta x}.$$

This is the definition of the slope of the curve C at P and suggests the following definition.

PARTIAL DERIVATIVES

DEFINITION *Partial derivatives.* If the domain of f includes the region within some circle around the point (a, b) and

$$\lim_{\Delta x \to 0} \frac{f(a + \Delta x, b) - f(a, b)}{\Delta x}$$

exists, this limit is called the *partial derivative of f with respect to x at (a, b)*. Similarly, if

$$\lim_{\Delta y \to 0} \frac{f(a, b + \Delta y) - f(a, b)}{\Delta y}$$

exists, it is called the *partial derivative of f with respect to y at (a, b)*.

Notation for partial derivatives   The following notations are used for the partial derivative of f with respect to x:

12.3 PARTIAL DERIVATIVES

$$f_x, f_1, z_x, \frac{\partial z}{\partial x} \text{ or } \frac{\partial f}{\partial x};$$

and the following for the partial derivative of f with respect to y:

$$f_y, f_2, z_y, \frac{\partial z}{\partial y} \text{ or } \frac{\partial f}{\partial y}.$$

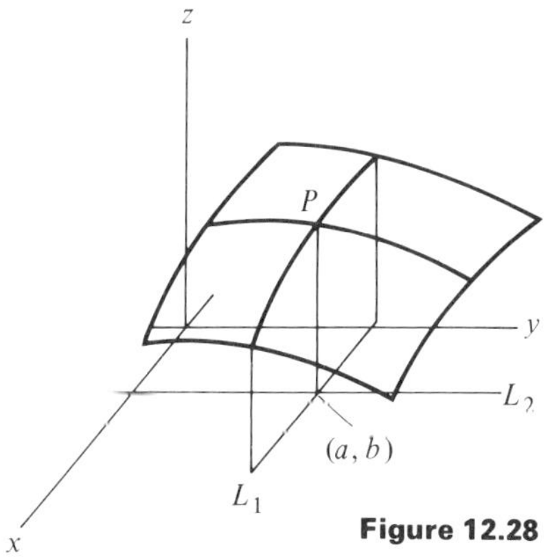

The curve above line L_1 in Fig. 12.28 has slope $f_x(a, b)$ at the point P. The curve above line L_2, which is parallel to the y axis, has slope $f_y(a, b)$ at P.

Note that on L_1 the y coordinate is constant while x is free to vary; on L_2 the x coordinate is constant while y is free to vary.

Figure 12.28

EXAMPLE 1 Find the partial derivative f_x at (a, b) if $f(x, y) = x^2 y$.

SOLUTION From the definition,

$$
\begin{aligned}
f_x(a, b) &= \lim_{\Delta x \to 0} \frac{f(a + \Delta x, b) - f(a, b)}{\Delta x} \\
&= \lim_{\Delta x \to 0} \frac{(a + \Delta x)^2 b - a^2 b}{\Delta x} \\
&= \lim_{\Delta x \to 0} b \left(\frac{(a + \Delta x)^2 - a^2}{\Delta x} \right) \\
&= \lim_{\Delta x \to 0} b \left(\frac{a^2 + 2a \Delta x + (\Delta x)^2 - a^2}{\Delta x} \right) \\
&= \lim_{\Delta x \to 0} b(2a + \Delta x) \\
&= 2ab. \quad \blacksquare
\end{aligned}
$$

As the computations in Example 1 suggest, to compute a partial derivative f_x, just treat y in the expression for f as a constant and differentiate with respect to x.

EXAMPLE 2 Compute f_x and f_y at (x, y), where f is given by the formula $x^2 y^3 + e^{x^2 y}$.

SOLUTION Treating y as a constant, differentiate with respect to x:

$$f_x(x, y) = 2xy^3 + 2xy e^{x^2 y}.$$

Similarly, treating x as a constant and differentiating with respect to y, we obtain

$$f_y(x, y) = 3x^2 y^2 + x^2 e^{x^2 y}. \quad \blacksquare$$

EXAMPLE 3 Find the two partial derivatives of f at $(1, 3)$ if f is given by the formula

$$f(x, y) = 2x + 4y + x^2 + \ln(x^2 + y^2).$$

SOLUTION To find $\partial f/\partial x$, treat y as a constant and compute the partial derivative,

$$\frac{\partial f}{\partial x} = 2 + 0 + 2x + \frac{2x}{x^2 + y^2}.$$

At $(x, y) = (1, 3)$, this yields

$$\frac{\partial f}{\partial x}(1, 3) = 2 + 2 \cdot 1 + \frac{2 \cdot 1}{1^2 + 3^2} = 4.2.$$

To find $\partial f/\partial y$, differentiate with respect to y, treating x as a constant:

$$\frac{\partial f}{\partial y} = 0 + 4 + 0 + \frac{2y}{x^2 + y^2}.$$

At $(x, y) = (1, 3)$, this yields

$$\frac{\partial f}{\partial y}(1, 3) = 4 + \frac{2 \cdot 3}{1^2 + 3^2} = 4.6. \blacksquare$$

Partial derivatives in economics

Slope is only one interpretation of the partial derivatives $\partial f/\partial x$ and $\partial f/\partial y$. If $f(x, y)$ represents the total production from x units of labor and y units of capital, then $\partial f/\partial x$ describes the rate of change of production f with respect to a change in labor x; $\partial f/\partial y$ records the rate of change of production f with respect to increases of capital y.

Partial derivatives and a vibrating string

Partial derivatives also are used in the study of a vibrating string. Figure 12.29 shows the position of a vibrating string at time t. The string is fixed at points A and B and vibrates parallel to the y axis. Let $f(x, t)$ be the height of the string at the point with abscissa x at time t, as shown in the figure. In this case the partial derivatives are denoted $\partial y/\partial x$ and $\partial y/\partial t$. The first one, $\partial y/\partial x$, is the slope of the curve in Fig. 12.29 at the point P. The second, $\partial y/\partial t$, is the velocity of the string along the vertical line with abscissa x. In other words, $\partial y/\partial t$ tells how fast the point P is moving.

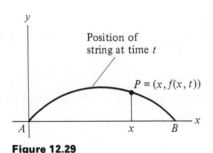

Figure 12.29

HIGHER PARTIALS   Just as there are derivatives of derivatives, so are there partial derivatives of partial derivatives. For instance, if

$$z = 2x + 5x^4 y^7,$$

then $\quad z_x = 2 + 20x^3 y^7 \quad$ and $\quad z_y = 35x^4 y^6.$

We may go on and compute partial derivatives of z_x and z_y:

$$(z_x)_x = 60x^2 y^7 \qquad (z_x)_y = 140x^3 y^6$$
$$(z_y)_x = 140x^3 y^6 \qquad (z_y)_y = 210x^4 y^5.$$

It is customary to denote $(z_x)_x$ simply by z_{xx}, $(z_x)_y$ by z_{xy}, and so on, omitting the parentheses. There are four possible partial derivatives of the second order.

$$z_{xx}, \qquad z_{xy}, \qquad z_{yx}, \qquad z_{yy}.$$

In the subscript notation, differentiate from left to right in the order of subscripts.

They are also denoted

$$f_{xx}, \qquad f_{xy}, \qquad f_{yx}, \qquad f_{yy}.$$

In the example just given, observe that z_{xy} equals z_{yx}. For most functions met in practice, the two "mixed partials" z_{xy} and z_{yx} are equal. In view of the importance of this remark we state it as a theorem. The proof, to be found in Appendix G, may be read at this point.

THEOREM   If $z = f(x, y)$ has continuous partial derivatives z_x, z_y, z_{xy}, and z_{yx}, then

$$z_{xy} = z_{yx}. \quad \blacksquare$$

The ∂ notation is also used for naming the four second-order partial derivatives. Thus

$$\frac{\partial}{\partial y}\left(\frac{\partial f}{\partial x}\right),$$

already denoted z_{xy}, is also denoted

$$\frac{\partial^2 f}{\partial y\, \partial x}.$$

In the ∂ notation, differentiate from right to left in the order that the variables occur.

Thus

$$z_{xx} = \frac{\partial^2 f}{\partial x^2}, \qquad z_{xy} = \frac{\partial^2 f}{\partial y\, \partial x}, \qquad z_{yx} = \frac{\partial^2 f}{\partial x\, \partial y}, \qquad z_{yy} = \frac{\partial^2 f}{\partial y^2}.$$

EXAMPLE 4   Compute $\partial^2 z/\partial x^2$ and $\partial^2 z/\partial y\, \partial x$ for $z = y \cos xy$.

SOLUTION First compute
$$\frac{\partial^2 z}{\partial x^2} = \frac{\partial}{\partial x}\left(\frac{\partial z}{\partial x}\right)$$
$$= \frac{\partial}{\partial x}(-y^2 \sin xy)$$
$$= -y^3 \cos xy;$$

and then
$$\frac{\partial^2 z}{\partial y\, \partial x} = \frac{\partial}{\partial y}\left(\frac{\partial z}{\partial x}\right)$$
$$= \frac{\partial}{\partial y}(-y^2 \sin xy)$$
$$= -2y \sin xy - xy^2 \cos xy. \quad \blacksquare$$

FUNCTIONS OF MORE THAN TWO VARIABLES

A quantity may depend on more than two variables. For instance, the volume of a box depends on three variables: the length l, width w, and height h, $V = lwh$. The "chill factor" depends on the temperature, humidity, and wind velocity. The temperature T at any point in the atmosphere is a function of the three space coordinates, x, y, and z: $T = f(x, y, z)$.

The notions and notations of partial derivatives carry over to functions of more than two variables. If $u = f(x, y, z, t)$, there are four first-order partial derivatives. For instance, the partial derivative of u with respect to x, holding y, z, and t fixed, is denoted

$$u_x, f_x, \text{ or } \frac{\partial u}{\partial x}, \text{ etc.}$$

Higher-order partial derivatives are defined and denoted similarly. For the functions commonly encountered, we also have the equality of the mixed partial derivatives,

$$\frac{\partial^2 u}{\partial x\, \partial y} = \frac{\partial^2 u}{\partial y\, \partial x}.$$

Exercises

In Exercises 1 to 10 compute f_x and f_y for the given functions f.

1. $x^3 y^4$
2. $[\ln(x+2y)]^2$
3. $\sqrt{x} \sec x^2 y$
4. x^y
5. $e^{3x} \tan^{-1} xy$
6. $\dfrac{x^2 + \cos 3y}{1+x}$
7. $\sin^{-1}(x+3y)$
8. $e^{xy}\sqrt{1+x^2} \cos(x/y)$
9. $\dfrac{2^x}{y^3}$
10. $\dfrac{x^3 y + 1}{\sin xy}$

In Exercises 11 to 16 compute the four partial derivatives of order two,

$$\frac{\partial^2 f}{\partial x^2}, \frac{\partial^2 f}{\partial y^2}, \frac{\partial^2 f}{\partial x\, \partial y}, \text{ and } \frac{\partial^2 f}{\partial y\, \partial x},$$

and check that the last two are equal.

11. $f(x,y) = 5x^2 - 3xy + 6y^2$
12. $f(x,y) = x^4 y^7$
13. $f(x,y) = 1/\sqrt{x^2+y^2}$
14. $f(x,y) = \sin(x^2 y)$
15. $f(x,y) = \tan(x+3y)$
16. $f(x,y) = x/y$

In each of Exercises 17 to 20 find the slope of the curve formed by the intersection of the given surface with a plane perpendicular to the y axis and passing through the given point.

17 $z = xy^2$ at $(1, 2)$. **18** $z = \cos(x + 2y)$ at $(\pi/4, \pi/2)$.
19 $z = x/y$ at $(1, 1)$. **20** $z = x^2 e^{xy}$ at $(1, 0)$.

In each of Exercises 21 and 22 find the slope of the curve formed by the intersection of the given surface with a plane perpendicular to the x axis and passing through the given point.

21 $z = ye^{xy}$ at $(1, 1)$. **22** $z = e^{x/y}$ at $(0, 1)$.

23 Let $f(x, y) = a + bx + cy + dx^2 + exy + ky^2$, where a, b, c, d, e, and k are constants. Show that

(a) $a = f(0, 0)$, (b) $b = \dfrac{\partial f}{\partial x}(0, 0)$,

(c) $c = \dfrac{\partial f}{\partial y}(0, 0)$, (d) $d = \dfrac{1}{2}\dfrac{\partial^2 f}{\partial x^2}(0, 0)$,

(e) $e = \dfrac{\partial^2 f}{\partial x \, \partial y}(0, 0)$, (f) $k = \dfrac{1}{2}\dfrac{\partial^2 f}{\partial y^2}(0, 0)$.

24 Let $z = f(x)g(y)$, where f and g are differentiable functions. Show that z satisfies the partial differential equation

$$z\frac{\partial^2 z}{\partial x \, \partial y} = \frac{\partial z}{\partial x} \cdot \frac{\partial z}{\partial y}.$$

25 A function $z = f(x, t)$, where z denotes temperature, x position, and t time, is said to satisfy the heat equation if

$$a^2 f_{xx} = f_t,$$

where a is a constant. Show that $f(x, t) = e^{-\pi^2 a^2 t} \sin \pi x$ satisfies the heat equation.

26 Let $T = f(x, y, z)$ be the temperature at the point (x, y, z) within a solid whose surface has a fixed temperature distribution. It can be shown that, if T does not vary with time, then $T_{xx} + T_{yy} + T_{zz} = 0$. Similarly, if $P(x, y, z)$ is the work done in moving a particle from a fixed base point in a gravitational field to the point (x, y, z), then $P_{xx} + P_{yy} + P_{zz} = 0$. The equation

$$f_{xx} + f_{yy} + f_{zz} = 0$$

is called Laplace's equation (in three dimensions). Verify that the functions $1/\sqrt{x^2 + y^2 + z^2}$, $x^2 - y^2 - z$, and $e^x \cos y + z$ satisfy Laplace's equation.

27 Let $u = x + x^3 yz$. Compute

(a) $\dfrac{\partial u}{\partial x}$ and (b) $\dfrac{\partial u}{\partial y}$.

28 Is there a function f such that

$$f_x = e^x \cos y \quad \text{and} \quad f_y = e^x \sin y?$$

29 Let $f(x, y) = \displaystyle\int_0^1 \cos(x + 2y + t)\, dt$. Find f_x and f_y.

30 Let $f(x, y) = \displaystyle\int_0^y \sqrt{x + t}\, dt$. Find f_x and f_y.

31 Let $f(x, y) = \int_x^y g(t)\, dt$, where g is continuous. Find f_x and f_y.

32 Let r and θ be polar coordinates of the point (x, y), in the first quadrant. Then $x = r \cos \theta$, $y = r \sin \theta$, $\theta = \tan^{-1}(y/x)$, and $r = \sqrt{x^2 + y^2}$.
(a) Show that $\partial x / \partial \theta = -r \sin \theta$.

(b) Show that $\partial \theta / \partial x = -\dfrac{\sin \theta}{r}$.

(c) If $y = f(x)$ and $x = f^{-1}(y)$, functions of one variable, dy/dx is the reciprocal of dx/dy. Why isn't $\partial x / \partial \theta$ the reciprocal of $\partial \theta / \partial x$?

33 Let x, y, z be functions of u, v, w. Assume that u, v, w can be found as functions of x, y, z. Thus $\partial x / \partial u$ can be expressed as a function of x, y, z. The following argument shows that $\partial x / \partial u$ is *independent* of x, that is,

$$\frac{\partial \left(\dfrac{\partial x}{\partial u}\right)}{\partial x} = 0. \quad \text{Argument:}$$

$$\frac{\partial \left(\dfrac{\partial x}{\partial u}\right)}{\partial x} = \frac{\partial^2 x}{\partial x \, \partial u} = \frac{\partial^2 x}{\partial u \, \partial x} = \frac{\partial}{\partial u}\left(\frac{\partial x}{\partial x}\right) = \frac{\partial}{\partial u}(1) = 0.$$

Apply the result to the case

$$x = (2u)^{1/2}, \quad y = (2v)^{1/2}, \quad z = (2w)^{1/2}.$$

We have $\dfrac{\partial x}{\partial u} = \dfrac{1}{(2u)^{1/2}} = \dfrac{1}{x}$.

Thus $1/x$ is independent of x. Where did the argument go wrong?

12.4 The change Δf and the differential df

In the case of a function of one variable it was shown in Sec. 4.7 that the differential is a good approximation to the change in the function: The change

$$\Delta f = f(x + \Delta x) - f(x)$$

is approximated by the differential

$$df = f'(x)\, \Delta x.$$

(See Fig. 12.30.)

This section considers the analogous problem for a function of two variables,

$$z = f(x, y).$$

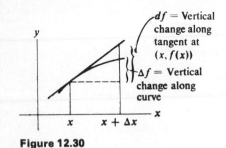

Figure 12.30

Throughout this chapter it will be assumed that all partial derivatives exist and are continuous.

The problem is to estimate the difference between $f(x + \Delta x, y + \Delta y)$ and $f(x, y)$, the lengths of the vertical segments shown in Fig. 12.31.

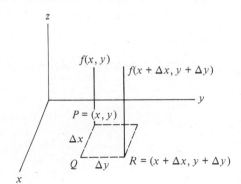

Figure 12.31

(For convenience, assume that both function values are positive.) Denote the difference by Δf or Δz:

$$\Delta f = \Delta z = f(x + \Delta x, y + \Delta y) - f(x, y).$$

Figure 12.31 suggests a way to estimate Δf. Consider the path in the xy plane from $P = (x, y)$ to $R = (x + \Delta x, y + \Delta y)$ that passes through the point $Q = (x + \Delta x, y)$ and consists of two line segments. From P to Q only the x coordinate changes; from Q to R only the y coordinate changes. The change in f on each of the two parts of the path can be estimated by a differential as discussed in Sec. 4.7 since from P to Q only one variable, namely x, changes, and from Q to R only one variable, namely y, changes. The sum of these two estimates provides an estimate of the total change Δf. (See Fig. 12.32). In other words, since

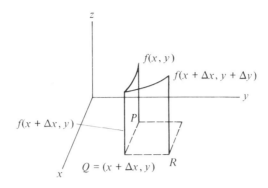

Figure 12.32

As a check, notice that $-f(x + \Delta x, y)$ cancels $f(x + \Delta x, y)$.

$$\Delta f = [f(x + \Delta x, y + \Delta y) - f(x + \Delta x, y)] + [f(x + \Delta x, y) - f(x, y)], \quad (1)$$

all that remains is to approximate each of the bracketed quantities in (1) in terms of partial derivatives of f.

Take the quantity $f(x + \Delta x, y) - f(x, y)$ first. This is the change in the function as we move from P to Q in Fig. 12.31. On this path y is constant. By the mean-value theorem there is a number c_1 between x and $x + \Delta x$ such that

The mean-value theorem is discussed in Sec. 4.1.

$$f(x + \Delta x, y) - f(x, y) = f_x(c_1, y) \Delta x. \quad (2)$$

Next treat the difference $f(x + \Delta x, y + \Delta y) - f(x + \Delta x, y)$, the other bracketed quantity in (1). This quantity is the change in the function as we go from Q to R, a path on which the x coordinate is constant, with value $x + \Delta x$. By the mean-value theorem, there is a number c_2 between y and $y + \Delta y$ such that

$$f(x + \Delta x, y + \Delta y) - f(x + \Delta x, y) = f_y(x + \Delta x, c_2) \Delta y. \quad (3)$$

Combining (1), (2), and (3) gives us the basic equation of this section, from which much will follow:

$$\Delta f = f_x(c_1, y) \Delta x + f_y(x + \Delta x, c_2) \Delta y. \quad (4)$$

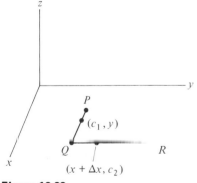

Figure 12.33

(See Fig. 12.33.)
When both Δx and Δy are small the points (c_1, y) and $(x + \Delta x, c_2)$ are near the point (x, y). If we assume that the partial derivatives f_x and f_y are continuous at $P = (x, y)$, then we may conclude that

For a formal definition of a limit of a function of two variables, see Appendix F.

$$\begin{aligned} f_x(c_1, y) &= f_x(x, y) + \varepsilon_1 \\ f_y(x + \Delta x, c_2) &= f_y(x, y) + \varepsilon_2, \end{aligned} \quad (5)$$

where both ε_1 and ε_2 approach 0 as Δx and Δy both approach 0.

Combining (4) and (5) gives the key result of this section. For emphasis we state it as a theorem.

THEOREM Let f have continuous partial derivatives f_x and f_y for all points within some disk with center at the point (x, y). Then Δf, which is the change $f(x + \Delta x, y + \Delta y) - f(x, y)$, can be written

$$\Delta f = f_x(x, y) \Delta x + f_y(x, y) \Delta y + \varepsilon_1 \Delta x + \varepsilon_2 \Delta y,$$

where ε_1 and ε_2 approach 0 as Δx and Δy approach 0. (Both ε_1 and ε_2 are functions of the four variables x, y, Δx, and Δy.) ∎

THE DIFFERENTIAL

When Δx and Δy are small, the quantities $\varepsilon_1 \Delta x$ and $\varepsilon_2 \Delta y$, being the products of small quantities, are usually negligible when compared with $f_x(x, y) \Delta x$ and $f_y(x, y) \Delta y$ [if $f_x(x, y)$ and $f_y(x, y)$ are not 0]. For this reason, $f_x(x, y) \Delta x + f_y(x, y) \Delta y$ is often a good estimate of Δf when Δx and Δy are small. The similarity with the case of a function of one variable suggests the following definition.

DEFINITION *Differential.* If f is a function of two variables, and x, y, Δx, and Δy are numbers, then the function of the four variables x, y, Δx, and Δy, given by

$$f_x(x, y) \Delta x + f_y(x, y) \Delta y,$$

is called the *differential* of f at x, y, Δx, and Δy. It is denoted df (or dz).

Though the differential df is a function of the four variables x, y, Δx, and Δy, it is usually applied only when Δx and Δy are small.

EXAMPLE 1 Let f be given by the formula $f(x, y) = x^2 y^3$. Compute Δf and df when $x = 2$, $y = 1$, $\Delta x = 0.1$, and $\Delta y = 0.05$.

SOLUTION In this case $x + \Delta x = 2.1$ and $y + \Delta y = 1.05$. Thus

$$\begin{aligned}\Delta f &= f(2.1, 1.05) - f(2, 1) \\ &= (2.1)^2(1.05)^3 - 2^2 \cdot 1^3 \\ &= 5.10512625 - 4 \\ &= 1.10512625.\end{aligned}$$

To compute $\qquad df = f_x(2, 1) \Delta x + f_y(2, 1) \Delta y,$

begin by finding the partial derivatives,

$$f_x(x, y) = 2xy^3 \quad \text{and} \quad f_y(x, y) = 3x^2 y^2.$$

Then
$$\begin{aligned}df &= 2 \cdot 2 \cdot 1^3 (0.1) + 3 \cdot 2^2 \cdot 1^2 (0.05) \\ &= 0.4 + 0.6 \\ &= 1. \quad \blacksquare\end{aligned}$$

The next example is similar to Example 1 but has a practical context.

EXAMPLE 2 The dimensions of a cylindrical tin can change from radius 3 inches and height 4 inches to radius 2.9 inches and height 4.2 inches. Estimate the change in the volume.

SOLUTION In this case the volume is a function of the two variables r and h: $V(r, h) = \pi r^2 h$. We wish to estimate

$$\Delta V = V(2.9, 4.2) - V(3, 4),$$

which can be thought of as

$$\Delta V = V(3 + (-0.1), 4 + 0.2) - V(3, 4).$$

Rather than compute ΔV, use $\quad dV = V_r(3, 4) \, \Delta r + V_h(3, 4) \, \Delta h \quad$ as an estimate of ΔV.

First of all, $\quad V_r = 2\pi r h, \quad V_h = \pi r^2, \quad \Delta r = -0.1, \quad$ and $\quad \Delta h = 0.2$. Thus

$$V_r(3, 4) = 24\pi \quad \text{and} \quad V_h(3, 4) = 9\pi.$$

Hence ΔV is approximately

$$(24\pi)(-0.1) + (9\pi)(0.2) = -0.6\pi.$$

A direct computation shows that $\quad \Delta V = -0.678\pi \quad$ (the minus sign indicates a decrease in volume). ∎

EXAMPLE 3 A box of height y has a square base whose sides have length x. If we measure x with a possible error of 2 percent and y with a possible error of 3 percent, approximately how large an error can these errors induce in an estimate of the volume $x^2 y$?

SOLUTION That x may be off by 2 percent and y by 3 percent is recorded in these inequalities:

$$\frac{|\Delta x|}{x} \leq 0.02 \quad \text{and} \quad \frac{|\Delta y|}{y} \leq 0.03.$$

The volume $x^2 y$ is a function of x and y:

$$V = f(x, y) = x^2 y.$$

Approximately how large can $\dfrac{|\Delta f|}{f}$ be?

Rather than use Δf, consider the more convenient quantity

$$df = f_x \, \Delta x + f_y \, \Delta y$$

Then
$$= 2xy\,\Delta x + x^2\,\Delta y.$$
$$\frac{df}{f} = \frac{2xy\,\Delta x + x^2\,\Delta y}{x^2 y}$$
$$= \frac{2xy\,\Delta x}{x^2 y} + \frac{x^2\,\Delta y}{x^2 y}$$
$$= 2\frac{\Delta x}{x} + \frac{\Delta y}{y}.$$

Note that exponents in the function $x^2 y$ show up as coefficients here.

Consequently, $\quad \dfrac{|df|}{f} \leq 2\dfrac{|\Delta x|}{x} + \dfrac{|\Delta y|}{y} \leq 2(0.02) + 0.03 = 0.07.$

Since $|df|/f$ is a good estimate of $|\Delta f|/f$, the error in estimating the volume is at most approximately 7 percent. Of course, if we underestimate x and overestimate y, the error in estimating the volume may be much less, perhaps even zero. (Incidentally, a little arithmetic shows that $|\Delta f|/f$ can be as large as 7.1612 percent.) ∎

This view of dy is discussed in Sec. 4.7.

In the case of a function of one variable the differential dy represents the change in y along a tangent line to the graph of the function. In the case of a function of two variables it turns out that the differential dz represents the change in z along a "tangent plane" to the graph of a function. The tangent plane is developed in Sec. 14.3 but a preview of it is presented in Exercise 21.

Exercises

In Exercises 1 to 4 compute Δf and df for the indicated f, x, y, Δx, and Δy.

1. $f(x, y) = xy$, $x = 2$, $y = 3$, $\Delta x = 0.1$, $\Delta y = 0.2$.
2. $f(x, y) = 2x + 3y$, $x = 3$, $y = 4$, $\Delta x = 0.1$, $\Delta y = -0.3$.
3. $f(x, y) = x/y$, $x = 2$, $y = 1$, $\Delta x = 0.1$, $\Delta y = -0.2$.
4. $f(x, y) = x\sqrt{y}$, $x = 1$, $y = 4$, $\Delta x = 0.1$, $\Delta y = 0.2$.

In Exercises 5 to 8 compute df for the indicated f, x, y, Δx, and Δy.

5. $f(x, y) = x^3 y^4$, $x = 1$, $y = 2$, $\Delta x = 0.1$, $\Delta y = -0.1$.
6. $f(x, y) = \sqrt{x^2 + y^2}$, $x = 3$, $y = 4$, $\Delta x = 0.02$, $\Delta y = 0.03$.
7. $f(x, y) = \ln(x + 2y)$, $x = 2$, $y = 3$, $\Delta x = -0.1$, $\Delta y = 0.2$.
8. $f(x, y) = 3x - 4y + 2$, $x = 7$, $y = 9$, $\Delta x = 0.2$, $\Delta y = -0.3$.

In Exercises 9 to 12 x is measured with a possible error of 3 percent and y with a possible error of 4 percent. Estimate the maximum possible percentage error in measuring the indicated quantity.

9. $x^3 y^2$
10. x^3/y^2
11. $x^5 y$
12. $x^m y^n$ (where m and n are constants)

In Exercises 13 to 16 use an appropriate differential to estimate the indicated difference.

13. $(3.1)^3(2.8)^2 - 3^3 3^2$ (*Suggestion*: Use $(x, y) = (3, 3)$, $\Delta x = 0.1$, $\Delta y = -0.2$.)
14. $\sqrt{(3.1)^2 + (4.2)^2} - \sqrt{3^2 + 4^2}$
15. $\ln[(1.1)^3 + (2.3)^3] - \ln 9$
16. $\dfrac{(1.1)^3}{(0.8)^2} - \dfrac{1^3}{1^2}$

∎

17. Let $f(x, y) = Ax + By + C$, where A, B, and C are constants. Show that $df = \Delta f$ for all values of Δx and Δy.
18. (This exercise continues Example 2 and shows how dV compares with ΔV.) Fill in the following table. [ΔV denotes $V(3 + \Delta r, 4 + \Delta h) - V(3, 4)$, while dV denotes $V_r(2, 4)\,\Delta r + V_h(3, 4)\,\Delta h$.]

Δr	Δh	dV	ΔV	$\Delta V/dV$
-0.1	0.2	-0.6π	-0.678π	1.13
0.1	0.2			
0.01	0.003			
0.001	-0.001			

19. The theorem in this section was proved by using a path from (x, y) to $(x + \Delta x, y)$, and then to the point $(x + \Delta x, y + \Delta y)$. Prove the theorem by using the path that passes through $(x, y + \Delta y)$ instead of $(x + \Delta x, y)$.

20 Let u be a function of x, y, and z. Assume that u_x, u_y, and u_z are continuous functions.
   (a) Obtain the formula for Δu analogous to that of the theorem for Δf.
   (b) From (a) show that $u_x \Delta x + u_y \Delta y + u_z \Delta z$ is a good approximation to Δu when Δx, Δy, and Δz are small.
   (c) Use the result in (a) to generalize the theorem to functions of three variables.

21 (The tangent plane) In order not to have too many x's and y's, we will consider the differential function df at (x_0, y_0) instead of at (x, y):

$$df = f_x(x_0, y_0) \Delta x + f_y(x_0, y_0) \Delta y.$$

Hold (x_0, y_0) fixed. Let

$$z_0 = f(x_0, y_0)$$

Define $g(x, y)$ to be

$$z_0 + f_x(x_0, y_0)(x - x_0) + f_y(x_0, y_0)(y - y_0).$$

Thus $g(x, y)$ is the approximation of $f(x, y)$ obtained by adding the differential dz to $z_0 = f(x_0, y_0)$.
   (a) Show that the point $(x, y, g(x, y))$ is on the surface

$$z = z_0 + f_x(x_0, y_0)(x - x_0) + f_y(x_0, y_0)(y - y_0).$$

   (b) Show that the plane in (a) contains the tangent line at the point (x_0, y_0, z_0) to the curve formed by the intersection of the surface $z = f(x, y)$ and the plane $x = x_0$.
   (c) Show that the plane in (a) contains the tangent line at the point (x_0, y_0, z_0) to the curve formed by the intersection of the surface $z = f(x, y)$ and the plane $y = y_0$.

The plane in (a) is called the *tangent plane* to the surface $z = f(x, y)$ at (x_0, y_0, z_0).

12.5 The chain rules

The theorem in Sec. 12.4 is the basis for the chain rules for differentiating composite functions of more than one variable. Theorem 1 of this section concerns the case in which z is a function of x and y, and x and y are functions of one variable. Theorem 2 concerns the case in which z is a function of x and y, and x and y are, in turn, functions of two variables.

THEOREM 1 *Chain rule.* Let $z = f(x, y)$ have continuous partial derivatives f_x and f_y, and let $x = g(t)$ and $y = h(t)$ be differentiable functions of t. Then z is a differentiable function of t and

$$\frac{dz}{dt} = \frac{\partial z}{\partial x} \frac{dx}{dt} + \frac{\partial z}{\partial y} \frac{dy}{dt}.$$

PROOF By definition,

$$\frac{dz}{dt} = \lim_{\Delta t \to 0} \frac{\Delta z}{\Delta t}.$$

Now, Δt induces changes Δx and Δy in x and y, respectively. According to the theorem in Sec. 12.4,

$$\Delta z = f_x(x, y) \Delta x + f_y(x, y) \Delta y + \varepsilon_1 \Delta x + \varepsilon_2 \Delta y,$$

where $\varepsilon_1 \to 0$ and $\varepsilon_2 \to 0$ as Δx and Δy approach 0. (Keep in mind that x and y are fixed.) Thus

$$\frac{\Delta z}{\Delta t} = f_x(x, y) \frac{\Delta x}{\Delta t} + f_y(x, y) \frac{\Delta y}{\Delta t} + \varepsilon_1 \frac{\Delta x}{\Delta t} + \varepsilon_2 \frac{\Delta y}{\Delta t},$$

and $\quad \dfrac{dz}{dt} = \lim\limits_{\Delta t \to 0} \dfrac{\Delta z}{\Delta t} = f_x(x, y)\dfrac{dx}{dt} + f_y(x, y)\dfrac{dy}{dt} + 0\dfrac{dx}{dt} + 0\dfrac{dy}{dt}.$

This proves the theorem. ■

We illustrate Theorem 1 by examples.

EXAMPLE 1 Let $z = x^2 y^3$, $x = 3t^2$, and $y = t/3$. Find dz/dt when $t = 1$.

SOLUTION In order to apply Theorem 1, compute z_x, z_y, dx/dt, and dy/dt:

$$z_x = 2xy^3, \qquad z_y = 3x^2 y^2,$$

$$\dfrac{dx}{dt} = 6t, \qquad \dfrac{dy}{dt} = \dfrac{1}{3}.$$

By Theorem 1, $\quad \dfrac{dz}{dt} = 2xy^3 \cdot 6t + 3x^2 y^2 \cdot \dfrac{1}{3}.$

In particular, when $t = 1$, x is 3 and y is $\tfrac{1}{3}$. Therefore, when $t = 1$,

$$\dfrac{dz}{dt} = 2 \cdot 3\left(\dfrac{1}{3}\right)^3 6 \cdot 1 + 3 \cdot 3^2 \left(\dfrac{1}{3}\right)^2 \dfrac{1}{3}.$$

$$= \tfrac{36}{27} + \tfrac{27}{27}$$

$$= \tfrac{7}{3}.$$

The derivative dz/dt can also be found without using Theorem 1. To do this, express z explicitly in terms of t:

$$z = x^2 y^3$$

$$= (3t^2)^2 \left(\dfrac{t}{3}\right)^3$$

$$= \dfrac{t^7}{3}.$$

Then $\quad \dfrac{dz}{dt} = \dfrac{7t^6}{3}.$

When $t = 1$, this gives $\dfrac{dz}{dt} = \dfrac{7}{3}$,

in agreement with the first computation. ■

EXAMPLE 2 At time t, measured in minutes, a bug walking on the xy plane is at the point $(g(t), f(t))$, where distances are measured in feet. (See Fig. 12.34.) The temperature at (x, y) is e^{-x-2y} degrees. When the bug is at the point $(0, 0)$, it is moving east at 2 feet per minute $(dx/dt = 2)$ and north at 3 feet per minute $(dy/dt = 3)$. From the bug's point of view, how quickly is the

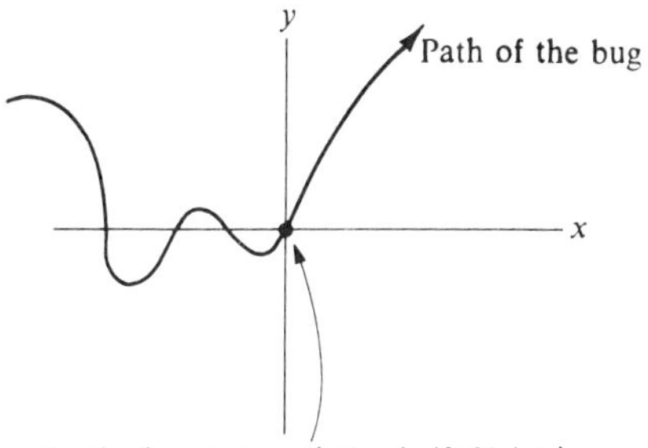

Path of the bug

As the bug passes through (0,0) he is traveling east at 2 feet per minute and north at 3 feet per minute.

Figure 12.34

temperature of the ground changing? That is, consider the temperature z to be a function of time, and find dz/dt.

SOLUTION In this case, $z = e^{-x-2y}$, and Theorem 1 says that

$$\frac{dz}{dt} = \frac{\partial z}{\partial x}\frac{dx}{dt} + \frac{\partial z}{\partial y}\frac{dy}{dt}.$$

Now, $\quad \dfrac{\partial z}{\partial x} = -e^{-x-2y} \quad$ and $\quad \dfrac{\partial z}{\partial y} = -2e^{-x-2y}$.

At $(x, y) = (0, 0)$, $\quad \dfrac{\partial z}{\partial x} = -1 \quad$ and $\quad \dfrac{\partial z}{\partial y} = -2$,

while $\quad \dfrac{dx}{dt} = 2 \quad$ and $\quad \dfrac{dy}{dt} = 3$.

Thus, as the bug passes through the origin it finds the temperature changing at the rate

$$\frac{dz}{dt} = (-1)(2) + (-2)(3)$$

$$= -8° \text{ per minute.}$$

The bug observes that the temperature is decreasing at the rate of 8° per minute. Note that this rate depends not only on the temperature function e^{-x-2y}, but also upon the speed and direction of the bug. ∎

The next theorem is a generalization of Theorem 1, in that x and y are assumed to be functions of *two* variables t and u.

THEOREM 2 *Chain rule.* Let $z = f(x, y)$ have continuous partial derivatives $\partial z/\partial x$ and $\partial z/\partial y$, and let x and y be differentiable functions of t and u. Then z is indirectly a function of t and u, and

$$\frac{\partial z}{\partial t} = \frac{\partial z}{\partial x}\frac{\partial x}{\partial t} + \frac{\partial z}{\partial y}\frac{\partial y}{\partial t} \quad \text{and} \quad \frac{\partial z}{\partial u} = \frac{\partial z}{\partial x}\frac{\partial x}{\partial u} + \frac{\partial z}{\partial y}\frac{\partial y}{\partial u}.$$

PROOF The proof, which is similar to that for Theorem 1, is only sketched. To examine

$$\frac{\partial z}{\partial t} = \lim_{\Delta t \to 0} \frac{\Delta z}{\Delta t}, \quad (u \text{ constant}),$$

hold u fixed and let t change by an amount Δt. Then x changes by Δx, and y changes by Δy. By the theorem in Sec. 12.4,

$$\Delta z = \frac{\partial z}{\partial x} \Delta x + \frac{\partial z}{\partial y} \Delta y + \varepsilon_1 \Delta x + \varepsilon_2 \Delta y, \tag{1}$$

$\dfrac{\partial z}{\partial x}$ and $\dfrac{\partial z}{\partial y}$ are evaluated at the fixed point (x, y).

where $\varepsilon_1 \to 0$ and $\varepsilon_2 \to 0$ as Δx and Δy approach 0. Divide (1) by Δt and, recalling that

$$\lim_{\Delta t \to 0} \frac{\Delta x}{\Delta t} = \frac{\partial x}{\partial t} \quad \text{and} \quad \lim_{\Delta t \to 0} \frac{\Delta y}{\Delta t} = \frac{\partial y}{\partial t},$$

finish the proof in the same manner as for Theorem 1. ■

In the statement of Theorem 2 no mention is made of where the various partial derivatives are to be evaluated. In case of any doubt, let us spell out the inputs. First write $x = g(t, u)$ and $y = h(t, u)$. Let (t_0, u_0) be the point at which $\partial z/\partial t$ and $\partial z/\partial u$ are evaluated. Let $x_0 = g(t_0, u_0)$ and $y_0 = h(t_0, u_0)$. Then the first equation in Theorem 2, when filled out completely, runs as follows:

$$\frac{\partial z}{\partial t}(t_0, u_0) = \frac{\partial z}{\partial x}(x_0, y_0) \frac{\partial x}{\partial t}(t_0, u_0) + \frac{\partial z}{\partial y}(x_0, y_0) \frac{\partial y}{\partial t}(t_0, u_0). \tag{2}$$

A similar equation holds for $\partial z/\partial u$. Since the host of symbols in (2) may obscure the basic form of the chain rule, the abbreviated form used in Theorem 2 has its advantages.

The chain rules generalize to any number of variables. For instance, let $w = f(x, y, z)$, $x = g(t, u)$, $y = h(t, u)$, and $z = j(t, u)$, where f, g, h, and j have continuous partial derivatives. Then w is a composite function of t and u, $w = F(t, u)$, and

$$\frac{\partial w}{\partial t} = \frac{\partial w}{\partial x} \frac{\partial x}{\partial t} + \frac{\partial w}{\partial y} \frac{\partial y}{\partial t} + \frac{\partial w}{\partial z} \frac{\partial z}{\partial t}.$$

The number of summands is determined by the number of variables of the function $w = f(x, y, z)$. Each summand is a product of two partial derivatives and is reminiscent of the chain rule for functions of one variable described in Sec. 3.6.

EXAMPLE 3 Given that $z = e^{xy}$, $x = 3t + 2u$, and $y = 4t - 2u$, find

$$\frac{\partial z}{\partial t} \quad \text{and} \quad \frac{\partial z}{\partial u}.$$

SOLUTION Begin by computing

$$z_x = ye^{xy}, \qquad z_y = xe^{xy},$$

$$\frac{\partial x}{\partial t} = 3, \qquad \frac{\partial y}{\partial t} = 4,$$

$$\frac{\partial x}{\partial u} = 2, \qquad \frac{\partial y}{\partial u} = -2.$$

By Theorem 2,
$$\frac{\partial z}{\partial t} = z_x \frac{\partial x}{\partial t} + z_y \frac{\partial y}{\partial t}$$
$$= ye^{xy} \cdot 3 + xe^{xy} \cdot 4$$
$$= (3y + 4x)e^{xy},$$

and
$$\frac{\partial z}{\partial u} = z_x \frac{\partial x}{\partial u} + z_y \frac{\partial y}{\partial u}$$
$$= ye^{xy} \cdot 2 + xe^{xy} \cdot (-2)$$
$$= (2y - 2x)e^{xy}.$$

It is also possible to compute $\partial z/\partial t$ and $\partial z/\partial u$ without using the chain rule. Write

$$z = e^{xy} = e^{(3t+2u)(4t-2u)}$$

and compute $\partial z/\partial t$ and $\partial z/\partial u$ directly. It will be instructive to do this and compare the solution with the one based on Theorem 2. ∎

APPLICATIONS OF THE CHAIN RULES Examples 1, 2, and 3 are intended to make the chain rules concrete. However, the main applications of the chain rules are of a more theoretical nature, whether in mathematics itself or in such applied fields as economics, chemistry, or physics. The remainder of this section illustrates some of these uses.

EXAMPLE 4 A function $y = g(x)$ is given implicitly as a solution of the equation $f(x, y) = 0$. Assuming that the derivatives and partial derivatives of g and f are continuous, show that

$$\frac{dy}{dx} = -\frac{f_x}{f_y}$$

SOLUTION Let $F(x) = f(x, g(x))$. By Theorem 1,

$$\frac{dF}{dx} = f_x \frac{dx}{dx} + f_y \frac{dy}{dx}. \tag{3}$$

By the definition of $g(x)$, $f(x, g(x)) = 0$. Thus $F(x) = 0$ and

$$\frac{dF}{dx} = 0.$$

778 PARTIAL DERIVATIVES

Moreover, $$\frac{dx}{dx} = 1.$$

Thus (3) reduces to

$$0 = f_x + f_y \frac{dy}{dx}.$$

Solving this last equation for the unknown dy/dx gives

$$\frac{dy}{dx} = -\frac{f_x}{f_y}. \quad \blacksquare$$

THE CHAIN RULE IN THERMODYNAMICS Partial derivatives play a central role in thermodynamics. The result in Example 5 is used there, as are the formulas developed in Exercises 34 to 36.

It is often important in thermodynamics to indicate which variables are fixed. If z is, say, a function of x and y, then $\partial z/\partial x$ is denoted

$$\left(\frac{\partial z}{\partial x}\right)_y.$$

The subscript y denotes the variable that is held fixed.

EXAMPLE 5 Let $z = f(x, y)$ and $y = h(x, u)$. Then z is a composite function of x and u, $z = F(x, u)$. Show that

$$\left(\frac{\partial z}{\partial x}\right)_u = \left(\frac{\partial z}{\partial x}\right)_y + \left(\frac{\partial z}{\partial y}\right)_x \left(\frac{\partial y}{\partial x}\right)_u. \tag{4}$$

SOLUTION Though it may not seem so at first glance, Theorem 2 will give this result immediately.

We have $z = f(x, y)$ and $y = h(x, u)$. But x may also be treated as a function of x and u, namely, $x = x + 0u$, which we could call $g(x, u)$.

So z is a function of x and y and each of these variables is a function of two variables, namely x and u. The chain rule of Theorem 2 then says that

$$\left(\frac{\partial z}{\partial x}\right)_u = \left(\frac{\partial z}{\partial x}\right)_y \left(\frac{\partial x}{\partial x}\right)_u + \left(\frac{\partial z}{\partial y}\right)_x \left(\frac{\partial y}{\partial x}\right)_u. \tag{5}$$

However $$\left(\frac{\partial x}{\partial x}\right)_u = 1. \tag{6}$$

Combining (5) and (6) gives (4). $\blacksquare$

If in (4) subscripts were not used, we would have gibberish:

$$\frac{\partial z}{\partial x} = \frac{\partial z}{\partial x} + \frac{\partial z}{\partial y}\frac{\partial y}{\partial x}.$$

Note that $\dfrac{\partial z}{\partial x}$ has two meanings.

In this and many other cases subscripts are obligatory.

The student of thermodynamics will see Example 5 expressed in the variables T (temperature), P (pressure), V (volume), and E (enthalpy). E may be considered a function of pressure and temperature, $E = f(P, T)$, but the pressure is a function of the temperature and volume, $P = g(T, V)$. Then

$$\left(\frac{\partial E}{\partial T}\right)_V = \left(\frac{\partial E}{\partial T}\right)_P + \left(\frac{\partial E}{\partial P}\right)_T \left(\frac{\partial P}{\partial T}\right)_V. \tag{7}$$

Equation (7) is obtained from (4) by replacing z with E, x with T, u with V, and y with P. It would be good practice to obtain (7) directly, from the chain rule.

THE CHAIN RULE IN ECONOMICS

Let $f(x, y)$ be the production (measured in dollars) of x units of labor and y units of capital. If the amounts of labor and capital are, say, doubled, we could expect that the output might double, that is, $f(2x, 2y) = 2f(x, y)$. More generally, we would expect that for any positive number k,

Homogeneous functions

$$f(kx, ky) = kf(x, y). \tag{8}$$

A function satisfying (8) is called *homogeneous*.

Marginal products
See Sec. 3.S.

The *marginal product* of labor is defined as $\partial f/\partial x$, and the *marginal product* of capital as $\partial f/\partial y$. These partials play a key role in economic theory. As Eugene Silberberg put it in *The Structure of Economics*,

> "The development of marginal productivity theory ... led to the conclusion that the factors of production would be paid the value of their marginal product. Roughly speaking, factors would be hired until their contributions to the output of the firm just equaled the cost of acquiring additional units of the factor ... [Would] the firm be capable of making these payments? ... Would enough output be produced (or perhaps would too much be produced) ...?
> A theorem developed by the great Swiss mathematician Euler ... came to the rescue of this analysis."

A more general form of Euler's theorem is given in Exercise 31.

Euler's theorem asserts that if f satisfies (8), then

$$f(x, y) = x f_x + y f_y.$$

In economic terms, it says that "production equals the cost of labor plus the cost of capital if each is paid at the rate of its marginal product."

As the next example shows, Euler's theorem depends on the chain rule. So it is the chain rule that rescued the economists.

EXAMPLE 6 Let $f(x, y)$ be homogeneous and possess continuous partial derivatives. Show that

$$f(x, y) = x f_x + y f_y.$$

SOLUTION Start with Eq. (8):

$$f(kx, ky) = kf(x, y).$$

Holding x and y fixed, we shall differentiate both sides of (8) with respect to k. By equating the results, the problem will be solved.

Differentiating the right side of (8) first gives

$$\left(\frac{\partial}{\partial k}(kf(x,y))\right)_{x,y} = f(x,y). \tag{9}$$

Differentiating the left side of (8) requires the chain rule. Let $z = f(u, v)$ and let $u = kx$ and $v = ky$. Then z is a composite function of k, x, and y, namely, $z = f(kx, ky)$. We wish to compute

$$\left(\frac{\partial z}{\partial k}\right)_{x,y},$$

a task for which the chain rule is handy. It says that

The notations f_1 and f_2 for the partial derivatives with respect to the first and second variables are useful here.

$$\left(\frac{\partial z}{\partial k}\right)_{x,y} = f_1(kx, ky)\frac{\partial(kx)}{\partial k} + f_2(kx, ky)\frac{\partial(ky)}{\partial k}. \tag{10}$$

But

$$\frac{\partial(kx)}{\partial k} = x \quad \text{and} \quad \frac{\partial(ky)}{\partial k} = y.$$

So

$$\left(\frac{\partial z}{\partial k}\right)_{x,y} = xf_1(kx, ky) + yf_2(kx, ky). \tag{11}$$

They're equal because they are the partial derivatives of the two sides of (8).

Since (9) must equal (11), we have

$$f(x, y) = xf_1(kx, ky) + yf_2(kx, ky). \tag{12}$$

Since (12) holds for all $k > 0$, it holds for $k = 1$. Substituting $k = 1$ into (12) gives

$$f(x, y) = xf_1(x, y) + yf_2(x, y),$$

which is Euler's theorem on homogeneous functions, and the economists are saved. ■

Exercises

In Exercises 1 and 2 check Theorem 1 by expressing z, dz/dt, z_x, z_y, dx/dt, and dy/dt explicitly in terms of t.

1 $z = x^2y^3$, $x = t^2$, $y = t^3$.
2 $z = xe^y$, $x = t$, $y = 1 + 3t$.
3 Find dz/dt if $z_x = 4$, $z_y = 3$, $\dot{x} = -2$, and $\dot{y} = 1$.
4 If z is a function of x and y, and x and y are functions of t and u, find $\partial z/\partial t$ and $\partial z/\partial u$ if $\partial z/\partial x = 3$, $\partial z/\partial y = 5$, $\partial x/\partial t = 2$, $\partial x/\partial u = -3$, $\partial y/\partial t = 5$, $\partial y/\partial u = 4$.

In Exercises 5 and 6 solve by Theorem 2 and also by expressing z explicitly in terms of r and θ.

5 (a) Find $\partial z/\partial r$ if $z = x^3 + y^2$, $x = r\cos\theta$, and $y = r\sin\theta$.

(b) In the partial derivative $\partial z/\partial r$ which variable is held constant?

6 Find $\partial z/\partial \theta$ and $\partial z/\partial r$ if $z = 1/\sqrt{x^2 + y^2}$, $x = r\cos\theta$, and $y = r\sin\theta$.

7 Let $z = f(x, y)$, $x = u + v$, and $y = u - v$. Show that $(\partial z/\partial x)^2 - (\partial z/\partial y)^2 = (\partial z/\partial u)(\partial z/\partial v)$.
(*Suggestion*: First use the chain rule to express $\partial z/\partial u$ and $\partial z/\partial v$ in terms of $\partial z/\partial x$ and $\partial z/\partial y$.)

8 Let $z = f(x, y)$, $x = u^2 - v^2$, and $y = v^2 - u^2$. Show that

$$u\frac{\partial z}{\partial v} + v\frac{\partial z}{\partial u} = 0.$$

9 Let $z = f(x, y)$, where $x = t - u$ and $y = -t + u$. Show that
$$\frac{\partial z}{\partial t} + \frac{\partial z}{\partial u} = 0.$$

10 Let $z = f(x, y)$, $x = r\cos\theta$, and $y = r\sin\theta$. Show that

(a) $\left(\dfrac{\partial z}{\partial r}\right)_\theta = f_x \cos\theta + f_y \sin\theta.$

(b) $\dfrac{\partial(f_x)}{\partial r} = f_{xx}\cos\theta + f_{xy}\sin\theta.$

(c) $\dfrac{\partial(f_y)}{\partial r} = f_{yx}\cos\theta + f_{yy}\sin\theta.$

(d) $\left(\dfrac{\partial^2 z}{\partial r^2}\right)_\theta = \dfrac{\partial(f_x)}{\partial r}\cos\theta + \dfrac{\partial(f_y)}{\partial r}\sin\theta.$

(e) From (b), (c), and (d) deduce that
$$\frac{\partial^2 z}{\partial r^2} = f_{xx}\cos^2\theta + 2f_{xy}\cos\theta\sin\theta + f_{yy}\sin^2\theta.$$

∎

11 Let $u = f(r)$ and $r = (x^2 + y^2 + z^2)^{1/2}$. Show that
$$u_{xx} + u_{yy} + u_{zz} = u_{rr} + \frac{2}{r} u_r.$$

12 Let z be a function of x and y and let $x = e^u$ and $y = e^v$. Show that
$$z_{uu} + z_{vv} = x^2 z_{xx} + y^2 z_{yy} + xz_x + yz_y.$$

13 Let $x = u\cos\theta - v\sin\theta$ and $y = u\sin\theta + v\cos\theta$. Let $f(x, y)$ be given and define
$$g(u, v) = f(u\cos\theta - v\sin\theta, u\sin\theta + v\cos\theta).$$
Show that
$$f_x^2 + f_y^2 = g_u^2 + g_v^2.$$

14 Let u and v be differentiable functions of t and let f be a continuous function.
(a) Using a chain rule, show that
$$\frac{d}{dt}\left(\int_u^v f(x)\,dx\right) = f(v)\frac{dv}{dt} - f(u)\frac{du}{dt}.$$

(b) Verify this for the special case $f(x) = \cos x$, $u = t$, and $v = t^2$.

Exercises 15 to 18 concern the differentiation of implicit functions.

15 Let $u = F(x, y, z)$ and $f(x, y)$ be functions such that $F(x, y, f(x, y)) = 0$. Show that (a) $f_x = -F_x/F_z$, (b) $f_y = -F_y/F_z$.

16 Let $u = F(x, y, z)$ and $f(x, z)$ be functions such that $F(x, f(x, z), z) = 0$. Obtain formulas for f_x and f_z in terms of F_x, F_y, and F_z.

17 Let $u = F(x, y, z)$ and $f(y, z)$ be functions such that $F(f(y, z), y, z) = 0$. Obtain formulas for f_y and f_z in terms of F_x, F_y, and F_z.

18 Let y be a function of x given implicitly by the equation $x^3 y + xy^5 - 2 = 0$.
(a) Find dy/dx by implicit differentiation, as in Sec. 4.6.
(b) Find dy/dx by the formula in Example 4.

19 Let $z = f(x, y)$, $x = r\cos\theta$, and $y = r\sin\theta$. Verify that
$$\frac{\partial^2 z}{\partial x^2} + \frac{\partial^2 z}{\partial y^2} = \frac{\partial^2 z}{\partial r^2} + \frac{1}{r^2}\frac{\partial^2 z}{\partial \theta^2} + \frac{1}{r}\frac{\partial z}{\partial r}.$$

Suggestion: Compute $\dfrac{\partial z}{\partial r}$, $\dfrac{\partial^2 z}{\partial r^2}$, and $\dfrac{\partial^2 z}{\partial \theta^2}$.

20 Prove Theorem 2, filling in the details.

21 Let $z = f(r)$ and $r = \sqrt{x^2 + y^2}$.
(a) Compute z_x and z_y in terms of f'.
(b) Show that $yz_x = xz_y$.

22 At what rate is the volume of a rectangular box changing when its width is 3 feet and increasing at the rate of 2 feet per second, its length is 8 feet and decreasing at the rate of 5 feet per second, and its height is 4 feet and increasing at the rate of 2 feet per second?

23 If T is the temperature at (x, y, z) in space, $T = f(x, y, z)$, and an astronaut is traveling in such a way that his x and y coordinates increase at the rate of 4 miles per second and his z coordinate decreases at the rate of 3 miles per second, compute dT/dt at a point where
$$\frac{\partial T}{\partial x} = 4, \quad \frac{\partial T}{\partial y} = 7, \quad \text{and} \quad \frac{\partial T}{\partial z} = 9.$$

24 Let $z = r^2 + s^2 + t^2$ and let $t = rsu$.
(a) The symbol $\partial z/\partial r$ has two interpretations. What are they?
(b) Evaluate $\partial z/\partial r$ in both cases in (a).

25 Let (r, θ) be polar coordinates for the point (x, y) given in rectangular coordinates.
(a) From the relation $r = \sqrt{x^2 + y^2}$, show that $\partial r/\partial x = \cos\theta$.
(b) From the relation $r = x/\cos\theta$, show that $\partial r/\partial x = 1/\cos\theta$.
(c) Explain why (a) and (b) are not contradictory.

26 Let $z = uv$, where u and v are differentiable functions of x. Use Theorem 1 to obtain the formula $(uv)' = uv' + vu'$.

27 Verify that each of the following functions is homoge-

neous and also satisfies the conclusion of Euler's theorem.
(a) $3x + 4y$, (b) $x^3 y^{-2}$, (c) $xe^{x/y}$.

28 Let $w = f(x, y, z)$, and let $h(t, u)$ be a function such that $f(t, u, h(t, u)) = 0$. Obtain a formula for h_t in terms of f_x, f_y, and f_z.

29 (This continues Exercise 28.) Obtain formulas for h_t and h_u if (a) $f(t, h(t, u), u) = 0$, (b) $f(h(t, u), t, u) = 0$.

■ ■

30 A function $f(x, y)$ is homogeneous of degree r if $f(kx, ky) = k^r f(x, y)$ for all $k > 0$. Show that each of the following functions is homogeneous of some degree r.
(a) $f(x, y) = x^2(\ln x - \ln y)$,
(b) $f(x, y) = 1/\sqrt{x^2 + y^2}$,
(c) $f(x, y) = \sin(y/x)$.

31 (See Exercise 30.) Show that if f is homogeneous of degree r then $xf_x + yf_y = rf$. This is the general form of Euler's theorem.

32 (See Exercise 31.) Verify Euler's theorem for the functions in Exercise 30.

33 (See Exercise 30.) Show that, if f is homogeneous of degree r, then f_x is homogeneous of degree $r - 1$.

Exercises 34 to 36 concern equations that appear in thermodynamics. They are related to Exercises 28 and 29.

34 The symbols P, T, V denote the pressure, temperature, and volume of a given amount of a gas. They are not independent. There is a function f of three variables such that the "equation of state"

$$f(P, T, V) = 0$$

determines any one of the three variables P, T, V in terms of the other two. Show that

$$\left(\frac{\partial P}{\partial T}\right)_V = \frac{-(\partial V/\partial T)_P}{(\partial V/\partial P)_T}.$$

35 Let E (internal energy) be a function of T and P. Recall that T is a function of V and P. Then E is indirectly a function of V and P. Show that

(a) $\left(\dfrac{\partial E}{\partial V}\right)_P = \left(\dfrac{\partial E}{\partial T}\right)_V \left(\dfrac{\partial T}{\partial V}\right)_P + \left(\dfrac{\partial E}{\partial V}\right)_T,$

(b) $\left(\dfrac{\partial E}{\partial P}\right)_V = \left(\dfrac{\partial E}{\partial T}\right)_P \left(\dfrac{\partial T}{\partial P}\right)_V + \left(\dfrac{\partial E}{\partial P}\right)_T.$

36 (a) Prove that $\left(\dfrac{\partial P}{\partial T}\right)_V \left(\dfrac{\partial T}{\partial P}\right)_V = 1.$

(b) Prove that $\left(\dfrac{\partial P}{\partial T}\right)_V \left(\dfrac{\partial T}{\partial V}\right)_P \left(\dfrac{\partial V}{\partial P}\right)_T = -1.$

12.6 Directional derivatives and the gradient

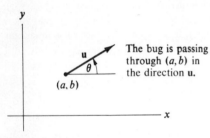

Figure 12.35

Imagine that $f(x, y)$ is the temperature at the point (x, y) of a thin layer of metal occupying the xy plane. A bug is passing through the point (a, b) in the direction given by the unit vector **u**, which makes an angle θ with the positive x axis. See Fig. 12.35. The bug observes that the temperature changes as it moves. The rate at which it changes per unit distance that the bug travels depends not only on the point (a, b) but on the direction **u** in which the bug is moving.

If, for example, $\mathbf{u} = \mathbf{i} = \overrightarrow{(1, 0)}$, the bug is walking east, and the rate of change in the temperature is given by the partial derivative $f_x(a, b)$. If the bug is going in the opposite direction (west) and $\mathbf{u} = -\mathbf{i}$, the rate of change is $-f_x(a, b)$. (For instance, if the temperature increases as the bug walks east, then it decreases as it walks west.) Similarly, when $\mathbf{u} = \mathbf{j} = \overrightarrow{(0, 1)}$, the rate of change is simply $f_y(a, b)$.

This section shows how to find the rate of change in the temperature in any direction **u** and introduces an important concept, the gradient.

DIRECTIONAL DERIVATIVES

To begin, let (a, b) be a point in the xy plane and consider a line in the plane through (a, b). See Fig. 12.36. On this line introduce a coordinate system with the same scale as the x or y axis: Call the line the t axis and place $t = 0$

12.6 DIRECTIONAL DERIVATIVES AND THE GRADIENT

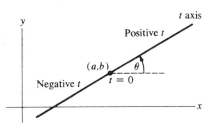

Figure 12.36

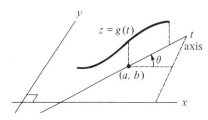

Figure 12.37

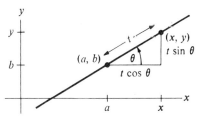

Figure 12.38

at (a, b). Let θ be the angle from the positive x axis to the positive t axis. Then, if we consider $z = f(x, y)$ only at points (x, y) on the t axis, z is a (composite) function of t; that is, $z = g(t)$. The graph of g is a curve situated in the graph of f, as shown in Fig. 12.37. When $t = 0$, $(x, y) = (a, b)$. More generally, the point on the t axis having coordinate t is

$$(x, y) = (a + t \cos \theta, b + t \sin \theta),$$

as inspection of the right triangle in Fig. 12.38 shows.

DEFINITION *Directional derivative of $f(x, y)$.* The derivative of f at (a, b) in the direction θ is $g'(t)$, where g is defined by $g(t) = f(a + t \cos \theta, b + t \sin \theta)$. A directional derivative at (a, b) is a number that depends on θ, as well as on a and b.

The directional derivative is the rate at which a function changes in a given direction.

There are two ways to think of the directional derivative: either as the slope of a certain curve or as the rate at which $f(x, y)$ changes as you move from the point (a, b) in a fixed direction in the xy plane. It is the latter interpretation which easily generalizes at the end of this section to the directional derivative of functions $f(x, y, z)$.

THEOREM 1 If $f(x, y)$ has continuous partial derivatives f_x and f_y, then the directional derivative of f at (a, b) in the direction θ is

$$f_x(a, b) \cos \theta + f_y(a, b) \sin \theta.$$

PROOF The directional derivative of f at (a, b) in the direction θ is the derivative of the function

$$g(t) = f(a + t \cos \theta, b + t \sin \theta),$$

when $t = 0$.

Now g is a composite function

$$g(t) = f(x, y) \quad \text{where} \quad \begin{cases} x = a + t \cos \theta \\ y = b + t \sin \theta. \end{cases}$$

By a chain rule, $\quad g'(t) = f_x \dfrac{dx}{dt} + f_y \dfrac{dy}{dt}.$

Now, $\quad \dfrac{dx}{dt} = \cos \theta \quad$ and $\quad \dfrac{dy}{dt} = \sin \theta.$

Thus $\quad g'(0) = f_x(a, b) \cos \theta + f_y(a, b) \sin \theta$,

and the theorem is proved. ∎

When $\theta = 0$, the formula given by Theorem 1 becomes

$$f_x(a, b) \cos 0 + f_y(a, b) \sin 0 = f_x(a, b)(1) + f_y(a, b)(0) = f_x(a, b).$$

Checking the formula for $\mathbf{u} = \mathbf{i}$ *and* $\mathbf{u} = -\mathbf{i}$

This agrees with the earlier observation that when the bug goes east the rate of change in the temperature is $f_x(a, b)$.

When $\theta = \pi$, Theorem 1 asserts that the directional derivative is

$$f_x(a, b) \cos \pi + f_y(a, b) \sin \pi = f_x(a, b)(-1) + f_y(a, b)(0) = -f_x(a, b).$$

This corresponds to the bug walking west and agrees with the observation that the rate of change in the temperature is $-f_x(a, b)$.

When $\theta = \pi/2$, Theorem 1 asserts that the directional derivative is

Checking when $\mathbf{u} = \mathbf{j}$
$$f_x(a, b) \cos \frac{\pi}{2} + f_y(a, b) \sin \frac{\pi}{2} = f_x(a, b)(0) + f_y(a, b)(1) = f_y(a, b),$$

which also agrees with an earlier observation.

EXAMPLE 1 Compute the derivative of $f(x, y) = x^2 y^3$ at $(1, 2)$ in the direction given by the angle $\pi/3$. Interpret the result if f describes a temperature distribution.

SOLUTION First of all, $\quad f_x = 2xy^3 \quad$ and $\quad f_y = 3x^2 y^2$.

Hence $\quad f_x(1, 2) = 16 \quad$ and $\quad f_y(1, 2) = 12$.

Second, $\quad \cos \dfrac{\pi}{3} = \dfrac{1}{2} \quad$ and $\quad \sin \dfrac{\pi}{3} = \dfrac{\sqrt{3}}{2}$.

Thus the derivative of f in the direction given by $\theta = \pi/3$ is

$$16\left(\frac{1}{2}\right) + 12\left(\frac{\sqrt{3}}{2}\right) = 8 + 6\sqrt{3}.$$

If $x^2 y^3$ is the temperature in degrees at the point (x, y), where x and y are measured in centimeters, the rate at which the temperature changes at $(1, 2)$, in the direction given by $\theta = \pi/3$, is $(8 + 6\sqrt{3})$ degrees per centimeter. ∎

Let $\mathbf{u} = \cos \theta \mathbf{i} + \sin \theta \mathbf{j}$ be the unit vector in the direction of angle θ, as pictured in Fig. 12.39. The derivative of $f(x, y)$ in the direction corresponding to θ is denoted

$$D_\mathbf{u} f.$$

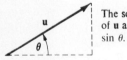

The scalar components of $\mathbf{u}$ are $\cos \theta$ and $\sin \theta$.

Figure 12.39

12.6 DIRECTIONAL DERIVATIVES AND THE GRADIENT

For example,
$$D_{\mathbf{i}}f = f_x,$$
$$D_{-\mathbf{i}}f = -f_x,$$
$$D_{\mathbf{j}}f = f_y.$$

In Example 1 $D_{\mathbf{u}}f$ was computed for $f(x, y) = x^2 y^3$ and

$$\mathbf{u} = \cos\frac{\pi}{3}\mathbf{i} + \sin\frac{\pi}{3}\mathbf{j} = \frac{1}{2}\mathbf{i} + \frac{\sqrt{3}}{2}\mathbf{j}.$$

THE GRADIENT Theorem 1 asserts that, if $\mathbf{u} = \cos\theta\,\mathbf{i} + \sin\theta\,\mathbf{j}$, then

$$D_{\mathbf{u}}f = f_x(a, b)\cos\theta + f_y(a, b)\sin\theta. \tag{1}$$

Formula (1) bears a close resemblance to the formula for the dot product. To exploit this similarity, it is useful to introduce the vector whose scalar components are $f_x(a, b)$ and $f_y(a, b)$.

DEFINITION *The gradient of $f(x, y)$.* The vector $f_x(a, b)\mathbf{i} + f_y(a, b)\mathbf{j}$ is the *gradient* of f at (a, b) and is denoted ∇f. (It is also called del f, because of the upside-down delta, ∇.)

The del symbol is in boldface to emphasize that the gradient of f is a vector.

For instance, let $f(x, y) = x^2 + y^2$. We compute and draw ∇f at a few points, listed in the following table.

(x, y)	$f_x = 2x$	$f_y = 2y$	∇f
$(1, 2)$	2	4	$2\mathbf{i} + 4\mathbf{j}$
$(3, 0)$	6	0	$6\mathbf{i}$
$(2, -1)$	4	-2	$4\mathbf{i} - 2\mathbf{j}$

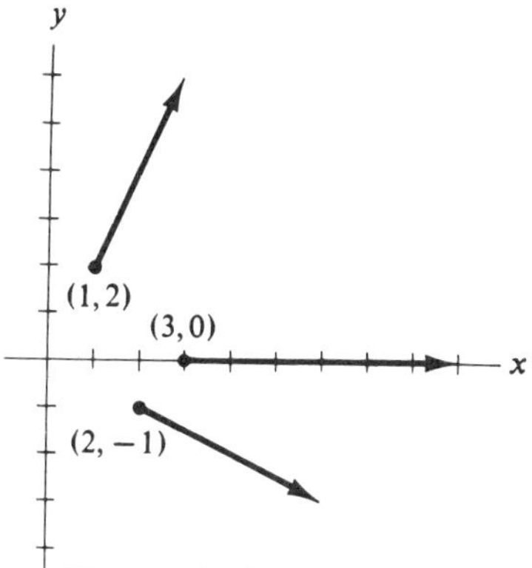

Figure 12.40 shows ∇f in each case, with the tail of ∇f placed at the point where ∇f is computed.

In vector notation, Theorem 1 reads as follows:

Figure 12.40

THEOREM 1 (*Rephrased*). If $z = f(x, y)$ has continuous partial derivatives f_x and f_y, then

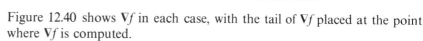
$$D_{\mathbf{u}}f = \nabla f \cdot \mathbf{u} \quad \blacksquare$$

The gradient is introduced not merely to simplify the statement of Theorem 1. Its importance is made clear in the next theorem.

THEOREM 2 (*Significance of ∇f*). Let $z = f(x, y)$ have continuous partial derivatives f_x and f_y. Let (a, b) be a point in the plane where ∇f is not $\mathbf{0}$. Then the magnitude of ∇f at (a, b) is the largest directional derivative of f at (a, b); the direction of ∇f is the direction in which the directional derivative at (a, b) has its largest value.

The meaning of $|\nabla f|$ and the direction of ∇f

PROOF By Theorem 1 (rephrased), if **u** is a unit vector, then at (a, b)

$$D_{\mathbf{u}}f = \nabla f \cdot \mathbf{u}.$$

By the definition of the dot product,

$$\nabla f \cdot \mathbf{u} = |\nabla f| \, |\mathbf{u}| \cos \theta,$$

where θ is the angle between ∇f and **u**, as shown in Fig. 12.41. Since $|\mathbf{u}| = 1$,

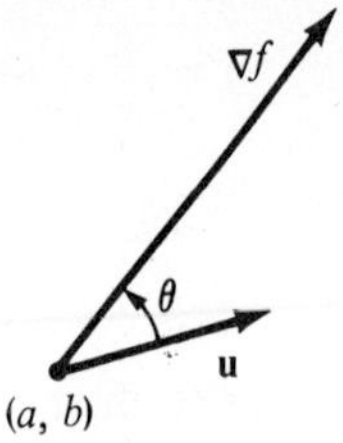

$$D_{\mathbf{u}}f = |\nabla f| \cos \theta. \tag{2}$$

Figure 12.41

The largest value of $\cos \theta$, for $0 \leq \theta \leq \pi$, occurs when $\theta = 0$; that is, $\cos \theta = 1$. Thus, by (2), the largest directional derivative of $f(x, y)$ at (a, b) occurs when the direction is that of ∇f at (a, b). For that choice of **u**, $D_{\mathbf{u}}f = |\nabla f|$. This proves the theorem. ∎

What does Theorem 2 tell the wandering bug about the flat piece of metal? If it is at the point (a, b) and wishes to get warmer as quickly as possible, it should compute the gradient of the temperature function, and then go in the direction indicated by the gradient. If, instead, it wishes to cool off as quickly as possible, it should go in the direction opposite the gradient.

EXAMPLE 2 What is the largest directional derivative of $f(x, y) = x^2 y^3$ at $(2, 3)$? In what direction does this maximum directional derivative occur?

SOLUTION At the point (x, y), $\quad \nabla f = 2xy^3 \mathbf{i} + 3x^2 y^2 \mathbf{j}.$

Thus, at $(2, 3)$, $\quad \nabla f = 108\mathbf{i} + 108\mathbf{j},$

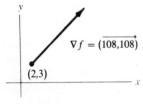

Figure 12.42

which is sketched in Fig. 12.42 (not to scale). Note that its angle θ is $\pi/4$. The maximal directional derivative of $x^2 y^3$ at $(2, 3)$ is $|\nabla f| = 108\sqrt{2}$. This is achieved at the angle $\theta = \pi/4$, relative to the x axis, that is, for

$$\mathbf{u} = \cos \frac{\pi}{4} \mathbf{i} + \sin \frac{\pi}{4} \mathbf{j} = \frac{\sqrt{2}}{2} \mathbf{i} + \frac{\sqrt{2}}{2} \mathbf{j}. \quad \blacksquare$$

Incidentally, if $f(x, y)$ denotes the temperature at (x, y), the gradient ∇f helps to indicate the direction in which heat flows. It tends to flow "toward the coldest," which boils down to the mathematical assertion, "Heat tends to flow in the direction of $-\nabla f$."

The gradient and directional derivative have been interpreted in terms of a temperature distribution in the plane and a wandering bug. It is also of interest to interpret these concepts in terms of a hiker on the surface of a mountain.

Consider a mountain above the xy plane. The altitude of the point on the surface above the point (x, y) will be denoted by $f(x, y)$. The directional derivative

$$D_{\mathbf{u}}f$$

indicates the rate at which altitude changes per unit change in *horizontal* distance. The gradient ∇f at (a, b) points in the compass direction the hiker should choose if she wishes to climb in the direction of steepest ascent. The magnitude of ∇f tells the hiker the steepest slope available to her. As Exercise 21 shows, the gradient evaluated at (a, b) is perpendicular to the contour that passes through (a, b).

THE DIRECTIONAL DERIVATIVES AND GRADIENT OF $f(x, y, z)$

The notions of directional derivative and gradient can be generalized with little effort to functions of three (or more) variables. However, the "slope of a curve" interpretation of the directional derivative no longer applies. It is easiest to interpret the directional derivative of $f(x, y, z)$ in a particular direction in space as indicating the rate of change of the function in that direction in space. A useful interpretation is how fast the temperature changes in a certain direction. The formal definition will now be given.

DEFINITION *Directional derivative of $f(x, y, z)$.* The derivative of f at (a, b, c) in the direction of the unit vector $\mathbf{u} = \cos \alpha \, \mathbf{i} + \cos \beta \, \mathbf{j} + \cos \gamma \, \mathbf{k}$ is $g'(t)$, where g is defined by

$$g(t) = f(a + t \cos \alpha, b + t \cos \beta, c + t \cos \gamma).$$

It is denoted $D_{\mathbf{u}}f$.

Note that t is simply the measure of length along the line through (a, b, c) with direction cosines $\cos \alpha$, $\cos \beta$, and $\cos \gamma$. So $D_{\mathbf{u}}f$ is just the derivative df/dt along the t axis.

The following theorem is proved just like Theorem 1.

THEOREM 3 If $f(x, y, z)$ has continuous partial derivatives f_x, f_y, and f_z, then the directional derivative of f at (a, b, c) in the direction of the unit vector $\cos \alpha \, \mathbf{i} + \cos \beta \, \mathbf{j} + \cos \gamma \, \mathbf{k}$ is

$$f_x(a, b, c) \cos \alpha + f_y(a, b, c) \cos \beta + f_z(a, b, c) \cos \gamma. \quad \blacksquare$$

DEFINITION *The gradient of $f(x, y, z)$.* The vector

$$f_x(a, b, c)\mathbf{i} + f_y(a, b, c)\mathbf{j} + f_z(a, b, c)\mathbf{k}$$

is the *gradient* of f at (a, b, c) and is denoted ∇f.

Theorem 3 thus asserts that the derivative of $f(x, y, z)$ in the direction of the unit vector **u** is simply

$$D_\mathbf{u} f = \nabla f \cdot \mathbf{u}.$$

Just as in the case of a function of two variables, ∇f, evaluated at (a, b, c), points in the direction **u** that produces the largest directional derivative at (a, b, c). Moreover $|\nabla f|$ is that largest directional derivative. The proof is practically identical with that of Theorem 2.

EXAMPLE 3 The temperature at the point (x, y, z) in a solid piece of metal is given by the formula $f(x, y, z) = e^{2x+y+3z}$ degrees. In what direction at the point $(0, 0, 0)$ does the temperature increase most rapidly?

SOLUTION First compute

$$\nabla f = 2e^{2x+y+3z}\mathbf{i} + e^{2x+y+3z}\mathbf{j} + 3e^{2x+y+3z}\mathbf{k}.$$

At $(0, 0, 0)$ $\qquad\qquad \nabla f = 2\mathbf{i} + \mathbf{j} + 3\mathbf{k}.$

Consequently, the direction of most rapid increase in temperature is that given by the vector $2\mathbf{i} + \mathbf{j} + 3\mathbf{k}$. The rate of increase is then

$$|2\mathbf{i} + \mathbf{j} + 3\mathbf{k}| = \sqrt{14} \text{ degrees per unit length.}$$

If the line through $(0, 0, 0)$ parallel to $2\mathbf{i} + \mathbf{j} + 3\mathbf{k}$ is given on a coordinate system so that it becomes the t axis, with $t = 0$ at the origin, then $df/dt = \sqrt{14}$ at 0. ∎

Exercises

In Exercises 1 and 2 compute the directional derivatives of $x^4 y^5$ at $(1, 1)$ in the indicated directions.

1 (a) $\theta = 0$, (b) $\theta = \pi$, (c) $\theta = \pi/4$.
2 (a) $\theta = \pi/2$, (b) $\theta = 3\pi/2$, (c) $\theta = 5\pi/4$.

In Exercises 3 and 4 compute and draw ∇f at the indicated points for the given functions.

3 $f(x, y) = x^2 y$ at (a) $(2, 5)$, (b) $(3, 1)$.
4 $f(x, y) = 1/\sqrt{x^2 + y^2}$ at (a) $(1, 2)$, (b) $(3, 0)$.
5 Find the directional derivative of e^{x+2y} at $(0, 0)$ in the direction of the vector $\mathbf{A} = 2\mathbf{i} + 3\mathbf{j}$.
6 Let $f(x, y, z) = 2x + 3y + z$.
(a) Compute ∇f at $(0, 0, 0)$ and at $(1, 1, 1)$.
(b) Draw ∇f for the two points in (a), in each case putting its tail at the point.
7 Let $f(x, y, z) = x^2 + y^2 + z^2$.
(a) Compute ∇f at $(2, 0, 0)$, $(0, 2, 0)$, and $(0, 0, 2)$.
(b) Draw ∇f for the three points in (a), in each case putting its tail at the point.

8 Find the directional derivative of $f(x, y, z) = x^3 y^2 z$ at $(1, 1, 1)$ in the direction of (a) **i**, (b) **j**, (c) **k**, (d) $-\mathbf{i}$, (e) $\mathbf{i} + \mathbf{j} + \mathbf{k}$.
9 If $f_x(a, b) = 2$ and $f_y(a, b) = 3$, in what direction should a directional derivative at (a, b) be computed in order that it is
(a) 0?
(b) as large as possible?
(c) as small as possible?
10 Assume that ∇f at (a, b) is not **0**. Show that there are two unit vectors $\mathbf{u}_1$ and $\mathbf{u}_2$ such that the directional derivatives of f at (a, b) in the directions of $\mathbf{u}_1$ and $\mathbf{u}_2$ are 0.

∎

11 (a) If $f_x(a, b, c) = 2$, $f_y(a, b, c) = 3$, and $f_z(a, b, c) = 1$, find three different unit vectors **u** such that $D_\mathbf{u} f$ at (a, b, c) is 0.

(b) How many unit vectors **u** are there such that $D_\mathbf{u} f$ at (a, b, c) is 0?

12. Let $f(x, y) = x^2 + y^2$. Prove that, if (a, b) is an arbitrary point on the curve $x^2 + y^2 = 9$, then ∇f computed at (a, b) is perpendicular to the tangent line to that curve at (a, b).

13. If, at (a, b), $D_\mathbf{u} f = 3$, find $D_{-\mathbf{u}} f$.

14. Let $f(x, y, z)$ equal temperature at (x, y, z). Let $P = (a, b, c)$ and Q be a point very near (a, b, c). Show that $\nabla f \cdot \vec{PQ}$ is a good estimate of the change in temperature from point P to point Q.

15. Let $f(x, y) = 1/\sqrt{x^2 + y^2}$; the function f is defined everywhere except at $(0, 0)$. (This function is the potential in a gravitational field due to a point-mass.) Let $\mathbf{R} = (x, y)$.
    (a) Show that $\nabla f = -\mathbf{R}/|\mathbf{R}|^3$.
    (b) Show that $|\nabla f| = 1/|\mathbf{R}|^2$.
    (The gradient is closely related to the gravitational force of attraction.)

16. What happens to ∇f when f has a local maximum? What happens to $D_\mathbf{u} f$ there? Explain. (Assume that f is defined in the entire plane and has continuous partial derivatives.)

17. If $f(P)$ is the electric potential at the point P, then the electric field $\mathbf{E}$ at P is given by $-\nabla f$. Calculate $\mathbf{E}$ if $f(x, y) = \sin \alpha x \cos \beta y$, where α and β are constants.

■ ■

18. Let f have continuous partial derivatives f_x, f_y, f_{xy}, and f_{yx} (hence $f_{xy} = f_{yx}$). Let $\mathbf{u}_1$ and $\mathbf{u}_2$ be two unit vectors. Prove that $D_{\mathbf{u}_2} D_{\mathbf{u}_1} f = D_{\mathbf{u}_1} D_{\mathbf{u}_2} f$.

19. Show that the maximum of $D_\mathbf{u} f$ at (a, b) is $\sqrt{f_x^2 + f_y^2}$, where f_x and f_y are evaluated at (a, b).

20. Prove the first part of Theorem 2 without the aid of vectors. That is, prove that the maximum value of $g(\theta) = f_x(a, b) \cos \theta + f_y(a, b) \sin \theta$ is $\sqrt{(f_x(a, b))^2 + (f_y(a, b))^2}$.

21. Show that the gradient of $f(x, y)$ evaluated at (a, b) is perpendicular to the level curve of f that passes through (a, b). *Suggestion:* First express dy/dx in terms of f_x and f_y.

12.7 Critical points

Just as in the case of a function of one variable, calculus provides tools for finding a maximum (or minimum) of a function of two variables. In Chap. 4 the derivative helped to find a highest (or lowest) point on a *curve*. Now, partial derivatives help find a highest (or lowest) point on a *surface*.

The number M is called the *maximum* (or *global maximum*) of f over a set R in the plane if it is the largest value of $f(x, y)$ for (x, y) in R. A *relative maximum* (or *local* maximum) of f occurs at a point (a, b) in R if there is a circle around (a, b) such that $f(a, b)$ is the maximum value of $f(x, y)$ for all points (x, y) within the circle. *Minimum* and *relative* (or *local*) *minimum* are defined similarly.

Let us look closely at the surface above a point (a, b) where a relative maximum of f occurs. Assume that f is defined for all points within some circle around (a, b) and possesses partial derivatives at (a, b). Let L_1 be the line $y = b$ in the xy plane; let L_2 be the line $x = a$ in the xy plane. (See Fig. 12.43. Assume, for convenience, that the values of f are positive.)

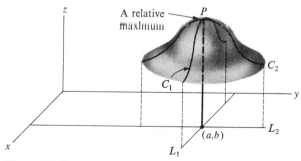

Figure 12.43

Let C_1 be the curve in the surface directly above the line L_1. Let C_2 be the curve in the surface directly above the line L_2. Let P be the point on the surface directly above (a, b).

Since f has a relative maximum at (a, b), no point on the surface near P is higher than P. Thus P is a highest point on the curve C_1 and on the curve C_2 (for points near P). The study of functions of one variable showed that both these curves have horizontal tangents at P. In other words, at (a, b) both partial derivatives of f must be 0:

$$f_x(a, b) = 0, \quad \text{and} \quad f_y(a, b) = 0.$$

This conclusion is summarized in the following theorem.

THEOREM Let f be defined on a domain that includes the point (a, b) and all points within some circle whose center is (a, b). If f has a relative maximum (or relative minimum) at (a, b) and f_x and f_y exist at (a, b), then both these partial derivatives are 0 at (a, b); that is,

$$f_x(a, b) = 0 = f_y(a, b). \quad \blacksquare$$

A function need not have a global maximum or a relative maximum. For instance, the function $f(x, y) = x$ has no maximum value. (Indeed, its graph is the plane $z = x$, which has no highest point.) Nor does this function have any relative maximum.

A point (a, b) where both partial derivatives, f_x and f_y, are 0, is clearly of importance. The following definition is analogous to that of a critical point of a function of one variable.

CRITICAL POINTS

DEFINITION *Critical point.* If $f_x(a, b) = 0$ and $f_y(a, b) = 0$, the point (a, b) is a *critical point* of the function $f(x, y)$.

EXAMPLE 1 Examine the critical points of

$$f(x, y) = 6x^2 + 2y^2 - 24x + 36y + 2.$$

SOLUTION Straightforward computations show that

$$\begin{cases} \dfrac{\partial f}{\partial x} = 12x - 24 \\ \dfrac{\partial f}{\partial y} = 4y + 36. \end{cases}$$

The only numbers x and y for which both partial derivatives are 0 are

$$x = 2 \quad \text{and} \quad y = -9.$$

There is only one critical point, namely $(2, -9)$.

To determine whether this point provides a relative maximum, a relative minimum, or neither requires further study of the function.

This particular function, a polynomial of degree 2 in x and y, can be analyzed by completing the square, as follows

$$\begin{aligned} f(x, y) &= 6x^2 + 2y^2 - 24x + 36y + 2 \\ &= 6x^2 - 24x + 2y^2 + 36y + 2 \\ &= 6(x^2 - 4x) + 2(y^2 + 18y) + 2 \\ &= 6[(x-2)^2 - 4] + 2[(y+9)^2 - 81] + 2 \\ &= 6(x-2)^2 + 2(y+9)^2 - 184. \end{aligned}$$

Since $(x-2)^2$ and $(y+9)^2$, being the squares of real numbers, are nonnegative, the minimum of $f(x, y)$ occurs when $x - 2$ and $y + 9$ are both 0. Thus the point $(2, -9)$ provides a global minimum of $f(x, y)$. The minimum value of $f(x, y)$ is $f(2, -9) = -184$. ∎

EXAMPLE 2 Examine the function f given by

$$f(x, y) = y^2 - x^2$$

for maximum and minimum values.

SOLUTION First of all, $\quad f(x, 0) = 0^2 - x^2 = -x^2.$

So when $|x|$ is large, $f(x, 0)$ is negative and of large absolute value; consequently f has no minimum value.

Second,

$$f(0, y) = y^2 - 0^2 = y^2.$$

When $|y|$ is large, so is $f(0, y)$, and f has no maximum value either.

Finally, let us see whether there are any relative maxima or minima. Since

$$f_x(x, y) = \frac{\partial}{\partial x}(y^2 - x^2) = -2x$$

and

$$f_y(x, y) = \frac{\partial}{\partial y}(y^2 - x^2) = 2y,$$

both partial derivatives are 0 only at

$$(x, y) = (0, 0).$$

Thus, if there is any relative maximum or minimum, it would have to be at (0, 0). However, inspection of the graph of

$$z = y^2 - x^2$$

shows that (0, 0) is neither a relative maximum nor a relative minimum. (Example 7 in Sec. 12.1 showed that in the vicinity of (0, 0) the surface is saddle-shaped.)

The second partial derivatives, $z_{yy} = 2$ and $z_{xx} = -2$ at (0, 0), also show what is happening. Considered only on the line $x = 0$, the function f has a relative minimum; on the line $y = 0$ it has a relative maximum. ∎

EXTREMA ON A POLYGON

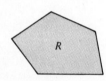

A continuous function on R (which includes the border) has a maximum value at some point in R.

Figure 12.44

If R consists of the region bounded by a polygon including its boundary, as shown in Fig. 12.44, and f is continuous throughout R, then f has a maximum value at some point in R. This situation is similar to the maximum-value theorem of Sec. 2.7, which concerns a continuous function defined on a closed interval $[a, b]$. Proofs of both results are to be found in any advanced calculus text. The theorem also holds if R is bounded by a curve. (R is also assumed to be "finite" in the sense that it lies within some circle.) A similar result holds for the minimum value of f on R.

To find a maximum in this case can be rather involved. The procedure is similar to that for maximizing a function on a closed interval:

How to maximize f on a region with a boundary

1 First find any points that are in R but not on the boundary of R where both f_x and f_y are 0. These are called *critical* points. (If there are no critical points, the maximum occurs on the boundary.)

2 If there are critical points, evaluate f at them. Also find the maximum of f on the boundary. (The next example shows how to do this.) The maximum of f on R is the largest value of f on the boundary and at critical points.

EXAMPLE 3 Let R be the triangle whose vertices are (0, 0), (2, 0), and (1, 1). Let $f(x, y) = 2x + 3y$. What are the maximum and minimum values of f for points in R?

SOLUTION The maximum or minimum may occur at a point in R that is not on the boundary. At such a point both f_x and f_y will be 0. Alternatively, the maximum or minimum may occur on the boundary. Let us see which cases occur for this particular function.

Begin by computing f_x and f_y. Since $f(x, y) = 2x + 3y$,

$$f_x = 2 \quad \text{and} \quad f_y = 3$$

at all points. The partial derivatives are never 0, and therefore neither the maximum nor minimum of f for points in R can occur at a point not on the

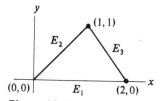

Figure 12.45

boundary. In other words, both the maximum and minimum must occur on the boundary.

It is necessary to consider the behavior of f on the boundary, which consists of three edges labeled E_1, E_2, and E_3 in Fig. 12.45.

On E_1, y is 0. Therefore, if (x, y) is a point on E_1,

$$f(x, y) = f(x, 0) = 2x + 3 \cdot 0 = 2x.$$

Since x goes from 0 to 2 on E_1, the minimum of f on E_1 is 0 and the maximum on E_1 is 4.

On E_2, $y = x$. Thus, if (x, y) is on E_2,

$$f(x, y) = f(x, x) = 2x + 3x = 5x.$$

Since x goes from 0 to 1 on E_2, the minimum of f on E_2 is 0 and the maximum on E_2 is 5. On E_3, $y = 2 - x$. Thus, if (x, y) is on E_3,

$$f(x, y) = f(x, 2 - x) = 2x + 3(2 - x) = 6 - x.$$

Since x goes from 1 to 2 on E_3, the minimum of f on E_3 is 4 and the maximum is 5. The table shows that the maximum of f on the border is 5 and the minimum is 0.

The minimum value of f on R is 0, occurring at $(0, 0)$. The maximum value is 5, occurring at $(1, 1)$. ∎

Edge	Minimum of f on edge	Maximum of f on edge
E_1	0	4
E_2	0	5
E_3	4	5

In the preceding example it was possible to determine the extrema of f on the edges of R by inspection. In more complicated situations, it is necessary to use the techniques developed in Chap. 4.

For instance, suppose the function in Example 3 had been $f(x, y) = 2x^2 - 2x + xy + 1$. Using the same triangular region R, we have $y = x$ on E_2 so $f(x, y) = f(x, x) = 3x^2 - 2x + 1$. For convenience, call this $g(x)$. Observe that we seek the extrema of $g(x) = 3x^2 - 2x + 1$ on the closed interval $[0, 1]$. Accordingly, compute $g'(x) = 6x - 2$, set it equal to 0, and find that $g'(\frac{1}{3}) = 0$. Since $g''(x) = 6 > 0$ for all x, $g(x)$ has a minimum at $x = \frac{1}{3}$, namely, $g(\frac{1}{3}) = \frac{2}{3}$.

Checking the endpoints yields $g(0) = 1$ and $g(1) = 2$, so the maximum value of $g(x)$ for x in $[0, 1]$ occurs at $x = 1$ and is equal to 2, while the minimum occurs at $x = \frac{1}{3}$ and is equal to $\frac{2}{3}$.

The other edges may be treated similarly. (The reader should show that the maximum value of f in the region R is $f(2, 0) = 5$ while the minimum value is $f(\frac{1}{2}, 0) = \frac{1}{2}$.) Incidentally, $f(0, 0)$ is a local maximum.

It would be natural to expect that, if at a critical point of f both second partial derivatives f_{xx} and f_{yy} are positive, then f would have a local minimum there. The next example destroys such a hope and is included now to prepare the reader for the message of the next section.

EXAMPLE 4 Let $z = f(x, y) = x^2 + 3xy + y^2$. Show that $(0, 0)$ is a critical point of f, $f_{xx}(0, 0) > 0$, and $f_{yy}(0, 0) > 0$, but $(0, 0)$ is *not* a local minimum of f.

SOLUTION Straightforward computations show that

$$z_x = 2x + 3y \qquad z_y = 3x + 2y$$
$$z_{xx} = 2 \qquad z_{yy} = 2.$$

At $(0, 0)$, then $z_x = 0$, $z_y = 0$, $z_{xx} = 2$, and $z_{yy} = 2$.

All that remains is to show that there are points (x, y) near $(0, 0)$ where $f(x, y)$ is negative. To see that such points exist, rewrite f:

$$f(x, y) = x^2 + 3xy + y^2$$
$$= (x + y)^2 + xy.$$

Thus, on the line $x + y = 0$ in the xy plane, $y = -x$ and

$$f(x, y) = 0^2 + xy$$
$$= -x^2.$$

For $(x, y) \neq (0, 0)$ on the line $x + y = 0$, f takes on negative values. Thus the critical point $(0, 0)$ is *not* a local minimum. It is neither a local maximum nor a local minimum, but a saddle point, like that illustrated in Example 7 of Sec. 12.1.

Note that, for (x, y) on the x axis $(y = 0)$, the function has the formula x^2; hence it has a local minimum at $(0, 0)$ when considered only on the x axis. Similarly, it has a local minimum when considered only on the y axis. The reader might pause to sketch the saddle in this case. ∎

The next section will show how to use the three second-order partial derivatives of f—f_{xx}, f_{xy}, and f_{yy}— to help determine whether a critical point provides a relative maximum, a relative minimum, or neither.

Exercises

For the functions given in Exercises 1 to 6 determine any minimum or maximum values and where they occur.

1 $2x^2 + 4x + y^2 + 6y$ 2 $-3x^2 - 5y^2 + 6x - 9y$
3 $xy - x - y$ 4 $x^2 - 3xy + y^2$
5 $-x^2 + 5xy - 2y^2$
6 $x^2 - 2xy + y^2$ (Note that this equals $(x - y)^2$.)
7 Find the maximum value of $f(x, y) = 3x^2 - 4y^2 + 2xy$ for points (x, y) in the square region whose vertices are $(0, 0)$, $(0, 1)$, $(1, 0)$, and $(1, 1)$.
8 Find the maximum value of $f(x, y) = xy$ for points in the triangular region whose vertices are $(0, 0)$, $(1, 0)$, and $(0, 1)$.
9 Maximize the function $-x + 3y + 6$ on the quadrilateral whose vertices are $(1, 1)$, $(4, 2)$, $(0, 3)$, and $(5, 6)$.

10 (a) Show that $z = x^2 - y^2 + 2xy$ has no maximum and no minimum.
  (b) Find the minimum and maximum of z if we consider only (x, y) on the circle of radius 1 and center $(0, 0)$, that is, all (x, y) such that $x^2 + y^2 = 1$. *Hint:* To deal with this, use $x = \cos \theta$ and $y = \sin \theta$.
  (c) Find the minimum and maximum of z if we consider all (x, y) in the disk of radius 1 and center $(0, 0)$, that is, all (x, y) such that $x^2 + y^2 \leq 1$.

∎

11 Find the dimensions of the rectangular box of volume 1 cubic meter which has the least surface area. *Sug-*

gestion: Call the dimensions x, y, and z; eliminate, say, z; and express the surface area as a function of x and y.

12 The maximum combined height and girth of a package that can be sent through the mails is 100 inches. Find the dimensions of a rectangular box of largest volume that can be sent through the mails. (The height is the largest dimension; the girth is the distance around a cross section perpendicular to the direction that gives the height.)

13 Let $(x_1, y_1), (x_2, y_2), \ldots, (x_n, y_n)$ be n points in the plane. Statisticians define the *line of regression* as the line that minimizes the sum of the squares of the distances between y_i and the ordinate of the line at x_i. (See Fig. 12.46.) Let the typical line in the plane have the equation $y = mx + b$.

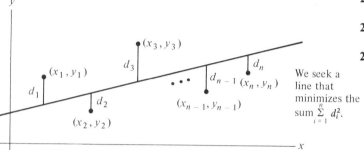

Figure 12.46

(a) Show that the line of regression minimizes the sum $\sum_{i=1}^{n} [y_i - (mx_i + b)]^2$ for all choices of m and b.
(b) Let $f(m, b) = \sum_{i=1}^{n} [y_i - (mx_i + b)]^2$. Compute f_m and f_b.
(c) Show that when $f_m = 0 = f_b$ we have

$$m \sum_{i=1}^{n} x_i^2 + b \sum_{i=1}^{n} x_i = \sum_{i=1}^{n} x_i y_i$$

and

$$m \sum_{i=1}^{n} x_i + nb = \sum_{i=1}^{n} y_i.$$

The simultaneous equations in (c) usually have a unique solution for m and b. (See Exercise 25.)

14 Let $P_1 = (x_1, y_1)$, $P_2 = (x_2, y_2)$, $P_3 = (x_3, y_3)$, and $P_4 = (x_4, y_4)$. Find the coordinates of the point P that minimizes the sum of the squares of the distances from P to the four points.

15 Find the shortest distance from the point $(1, 2, 3)$ to the plane $x + 3y + 5z = 6$ using (a) calculus, (b) vectors.

16 Find the shortest distance between the line through $(1, 0, 1)$ parallel to the vector $\mathbf{i} + 2\mathbf{j} + \mathbf{k}$ and the line through $(2, 1, 4)$ parallel to the vector $\mathbf{i} - \mathbf{j} + \mathbf{k}$ using (a) calculus, (b) vectors.

17 Find the dimensions of the rectangular box of largest volume if its total surface area is to be 12 square meters.

Exercises 18 to 21 are related.

18 Show that $x^2 + 3xy + 4y^2$ has a relative minimum at $(0, 0)$. *Hint:* Rewrite it as $(x + \tfrac{3}{2}y)^2 - 9y^2/4 + 4y^2$, that is, complete the square.

19 Show that $x^2 + 4xy + 4y^2$ has a local minimum at $(0, 0)$.

20 Show that $x^2 + 5xy + 4y^2$ does *not* have a local minimum at $(0, 0)$.

21 For which values of the constant k does $x^2 + kxy + 4y^2$ have a relative minimum at $(0, 0)$?

22 Find (a) the minimum value of xyz, and (b) the maximum value of xyz, for all triplets of nonnegative real numbers x, y, z such that $x + y + z = 1$.

23 (See Exercise 22.) (a) Deduce from Exercise 22 that for any three nonnegative numbers a, b, and c, $\sqrt[3]{abc} \leq (a + b + c)/3$. This asserts that the "geometric mean" of three numbers is not larger than the "arithmetic mean." (b) Obtain a corresponding result for four numbers.

24 Let $f(x, y) = (y - x^2)(y - 2x^2)$.
(a) Show that f has neither a local minimum nor a local maximum at $(0, 0)$.
(b) Show that f has a local minimum at $(0, 0)$ when considered only on any fixed line through $(0, 0)$.
Suggestion for (b): Graph $y = x^2$ and $y = 2x^2$ and show where $f(x, y)$ is positive and where it is negative.

25 When do the equations in Exercise 13 (c) have a unique solution for m and b? Express your answer as simply as possible.

12.8 Second-order partial derivatives and relative extrema

In the case of a function of one variable, $y = f(x)$, if $f'(a) = 0$ and $f''(a) > 0$, then $f(x)$ has a relative minimum at a. In this section the analogous test for a function of two variables is presented.

Recall Example 4 of the preceding section, $f(x, y) = x^2 + 3xy + y^2$.

Though at the critical point $(0, 0)$ both f_{xx} and f_{yy} are positive, $(0, 0)$ is not a local minimum for f. Contrast this with the following example, which sets the stage for the main result of this section.

EXAMPLE 1 Show that, if k is a constant, $k^2 < 4$, the function $f(x, y) = x^2 + kxy + y^2$ has a relative minimum at $(0, 0)$.

SOLUTION Completing the square shows that

$$x^2 + kxy + y^2 = x^2 + kxy + \left(\frac{ky}{2}\right)^2 + y^2 - \left(\frac{ky}{2}\right)^2$$

$$= \left(x + \frac{ky}{2}\right)^2 + \left(1 - \left(\frac{k}{2}\right)^2\right)y^2.$$

Since $k^2 < 4$, $\qquad 1 - \left(\frac{k}{2}\right)^2 > 0.$

Thus, recalling that the square of any real number is nonnegative, we obtain the inequality

$$\left(x + \frac{ky}{2}\right)^2 + \left(1 - \left(\frac{k}{2}\right)^2\right)y^2 \geq 0.$$

Since $f(0, 0) = 0$, the function has a relative minimum at $(0, 0)$; in fact, a global minimum. Incidentally, note that, if $f(x, y) = 0$, both y and $x + ky/2$ must be 0. So only at $(0, 0)$ does the function assume its minimum value. ∎

Example 1 suggests that the problem of determining for which (fixed) values of A, B, and C the function

The coefficient of xy is called 2B rather than B to simplify the proof of Theorem 1.

$$f(x, y) = Ax^2 + 2Bxy + Cy^2$$

has a relative minimum at $(0, 0)$ is worth studying. When $y = 0$, the function reduces to Ax^2. Thus A must be nonnegative. Similarly, C must be nonnegative. But as Example 4 of the preceding section illustrates, these conditions are not enough. The key is provided by the next theorem.

RELATIVE EXTREMA OF $Ax^2 + 2Bxy + Cy^2$

THEOREM 1 Let $f(x, y) = Ax^2 + 2Bxy + Cy^2$, where A, B, and C are constants. Let $\mathscr{D} = AC - B^2$.

1. If $\mathscr{D} > 0$ and $A > 0$, then f has a minimum at $(0, 0)$.

2. If $\mathscr{D} > 0$ and $A < 0$, then f has a maximum at $(0, 0)$.

For the case $\mathscr{D} = 0$ see Exercise 27.

3. If $\mathscr{D} < 0$, then f has neither a maximum nor a minimum at $(0, 0)$. [There is a saddle point at $(0, 0)$.]

PROOF We prove only case 1, leaving cases 2 and 3 as exercises. Since $f(0, 0) = 0$, it suffices to prove that for all $(x, y), f(x, y) \geq 0$. This amounts to showing that

$$A(Ax^2 + 2Bxy + Cy^2) \geq 0,$$

since A is positive. Completing the square establishes this inequality as follows:

$$A(Ax^2 + 2Bxy + Cy^2) = A^2x^2 + 2ABxy + ACy^2$$
$$= A^2x^2 + 2ABxy + B^2y^2 - B^2y^2 + ACy^2$$
$$= (Ax + By)^2 + (AC - B^2)y^2.$$

Recall that the square of a real number is never negative.

But

$$(Ax + By)^2 + (AC - B^2)y^2 \geq 0 \quad (1)$$

since $(Ax + By)^2$, $AC - B^2$, and y^2 are all nonnegative. ∎

Note that, in both the maximum and minimum cases, 1 and 2, $\mathscr{D}$ is positive; only the sign of A changes. Also, if $\mathscr{D}$ is positive, that is, $AC - B^2 > 0$, it follows that $AC > B^2$; hence AC is positive. Consequently A and C have the same sign. The assumptions $\mathscr{D} > 0$ and $A > 0$ together therefore imply that $C > 0$. Similarly, the assumptions $\mathscr{D} > 0$ and $A < 0$ imply that $C < 0$. It should also be pointed out that the minimum in case 1 and the maximum in case 2 are global, hence also relative.

EXAMPLE 2 Show that $f(x, y) = 2x^2 - xy + y^2$ has a minimum at $(0, 0)$.

SOLUTION Apply Theorem 1. In this case $A = 2$, $B = -\frac{1}{2}$, and $C = 1$. The hypotheses of Theorem 1 apply, for $A > 0$ and $AC - B^2 > 0$. Thus $f(x, y) = 2x^2 - xy + y^2$ has a minimum at $(0, 0)$. ∎

Before stating the test for a relative extremum of a more general function, we express Theorem 1 in terms of partial derivatives.

Observe that, if

$$f(x, y) = Ax^2 + 2Bxy + Cy^2,$$

then
$$f_x(x, y) = 2Ax + 2By, \qquad f_y(x, y) = 2Bx + 2Cy,$$
$$f_{xx}(x, y) = 2A, \qquad f_{xy}(x, y) = 2B, \qquad \text{and} \qquad f_{yy}(x, y) = 2C.$$

Thus
$$A = \frac{f_{xx}(0, 0)}{2}, \qquad B = \frac{f_{xy}(0, 0)}{2}, \qquad \text{and} \qquad C = \frac{f_{yy}(0, 0)}{2}.$$

Hence
$$AC - B^2 = \frac{f_{xx}(0, 0)f_{yy}(0, 0) - [f_{xy}(0, 0)]^2}{4}.$$

Thus Theorem 1 can be rephrased as follows.

THEOREM 2 *Theorem 1 rephrased.* Let $f(x, y) = Ax^2 + 2Bxy + Cy^2$, where A, B, and C are constants. Let

$$\mathscr{D} = f_{xx}(0, 0)f_{yy}(0, 0) - [f_{xy}(0, 0)]^2.$$

1. If $\mathscr{D} > 0$ and $f_{xx}(0, 0) > 0$, then f has a minimum at $(0, 0)$.
2. If $\mathscr{D} > 0$ and $f_{xx}(0, 0) < 0$, then f has a maximum at $(0, 0)$.
3. If $\mathscr{D} < 0$, then f has neither a maximum nor a minimum at $(0, 0)$. [There is a saddle point at $(0, 0)$.] ∎

$\mathscr{D} > 0$ means that the effects of f_{xx} and f_{yy} dominate.

Incidentally, the assumption that $\mathscr{D} > 0$ is equivalent to the assertion that $[f_{xy}(0, 0)]^2 < f_{xx}(0, 0)f_{yy}(0, 0)$, an inequality that can be interpreted to say "$f_{xy}(0, 0)$ must not be large in comparison to the product $f_{xx}(0, 0)f_{yy}(0, 0)$."

Theorem 2 generalizes to functions that are not necessarily of the form $Ax^2 + 2Bxy + Cy^2$.

RELATIVE EXTREMA OF $f(x, y)$

The proof appears in the next section.

The second-partial-derivative test for a relative maximum or minimum runs as follows.

THEOREM 3 *The second-partial-derivative test.* Let (a, b) be a critical point of the function $f(x, y)$. Assume that the partial derivatives f_x, f_y, f_{xx}, f_{xy}, and f_{yy} are continuous at and near (a, b). Let

This is the central result of the section.

$$\mathscr{D} = f_{xx}(a, b)f_{yy}(a, b) - [f_{xy}(a, b)]^2.$$

1. If $\mathscr{D} > 0$ and $f_{xx}(a, b) > 0$, then f has a relative minimum at (a, b).
2. If $\mathscr{D} > 0$ and $f_{xx}(a, b) < 0$, then f has a relative maximum at (a, b).
3. If $\mathscr{D} < 0$, then f has neither a relative minimum nor a relative maximum at (a, b). (There is a saddle point at (a, b).) ∎

If $\mathscr{D} = 0$, then anything can happen; there may be a relative minimum, a relative maximum, or a saddle. These possibilities are illustrated in Exercise 18.

EXAMPLE 3 Determine whether

$$f(x, y) = 3x^2 + 5y^2 + 6x - 20y$$

has any relative maxima or minima.

SOLUTION First compute f_x and f_y: $\quad f_x(x, y) = 6x + 6$

and

$$f_y(x, y) = 10y - 20.$$

Both of these vanish only when

$$6x + 6 = 0 \quad \text{and} \quad 10y - 20 = 0,$$

therefore only at the point $(-1, 2)$.

Now utilize the second-derivative test of Theorem 3 to help determine whether f does have a relative maximum or minimum at $(-1, 2)$.

The second-order partial derivatives are

$$f_{xx}(x, y) = 6, \quad f_{yy}(x, y) = 10, \quad f_{xy}(x, y) = 0$$

(independent of the point (x, y)). Since f_{xx} is positive and $f_{xx} f_{yy} - (f_{xy})^2 = 6 \cdot 10 - 0^2 > 0$, it follows from Theorem 3 that f has a relative minimum at $(-1, 2)$. ∎

The quantity $\mathscr{D} = f_{xx} f_{yy} - (f_{xy})^2$ is sometimes called the *discriminant* of f. Note that it also equals the determinant

$$\begin{vmatrix} f_{xx} & f_{xy} \\ f_{yx} & f_{yy} \end{vmatrix}$$

since $f_{xy} = f_{yx}$. This determinant is called the *Hessian* of f.

Exercises

Use Theorem 3 to determine any relative maxima or minima of the functions in Exercises 1 to 10.

1. $x^2 + 3xy + y^2$
2. $x^2 - y^2$
3. $x^2 - 2xy + 2y^2 + 4x$
4. $x^4 + 8x^2 + y^2 - 4y$
5. $x^2 - xy + y^2$
6. $x^2 + 2xy + 2y^2 + 4x$
7. $2x^2 + 2xy + 5y^2 + 4x$
8. $-4x^2 - xy - 3y^2$
9. $4/x + 2/y + xy$
10. $x^3 - y^3 + 3xy$

Let f be a function of x and y such that at (a, b) both f_x and f_y equal 0. In each of Exercises 11 to 16 values are specified for f_{xy}, f_{xx}, and f_{yy} at (a, b). Assume that all these partial derivatives are continuous. On the basis of the information decide whether (a) f has a relative maximum at (a, b), (b) f has a relative minimum at (a, b), (c) neither (a) nor (b) occurs, (d) there is inadequate information.

11. $f_{xy} = 4, f_{xx} = 2, f_{yy} = 8$.
12. $f_{xy} = 3, f_{xx} = 2, f_{yy} = 4$.
13. $f_{xy} = 3, f_{xx} = 2, f_{yy} = 4$.
14. $f_{xy} = 2, f_{xx} = 3, f_{yy} = 4$.
15. $f_{xy} = -2, f_{xx} = -3, f_{yy} = -4$.
16. $f_{xy} = -2, f_{xx} = 3, f_{yy} = -4$.

17. If, at (x_0, y_0), $z_x = 0 = z_y$, $z_{xx} = 3$, and $z_{yy} = 12$, for what values of z_{xy} is it certain that z has a relative minimum at (x_0, y_0)?

18. This exercise shows that, if $\mathscr{D} = f_{xx} f_{yy} - (f_{xy})^2 = 0$, then no conclusions of the type given in Theorem 3 can be drawn.
    (a) Let $f(x, y) = x^2 + 2xy + y^2$. Show that at $(0, 0)$ both f_x and f_y are 0, f_{xx} and f_{yy} are positive, $\mathscr{D} = 0$, and f has a relative minimum.
    (b) Let $f(x, y) = x^2 + 2xy + y^2 - x^4$. Show that at $(0, 0)$ both f_x and f_y are 0, f_{xx} and f_{yy} are positive, $\mathscr{D} = 0$, and f has neither a relative maximum nor a relative minimum at $(0, 0)$.
    (c) Give an example of a function $f(x, y)$ for which $(0, 0)$ is a critical point and $\mathscr{D} = 0$ there, but f has a relative maximum at $(0, 0)$.

19. The material for the top and bottom of a rectangular box costs 3 cents per square foot, and that for the sides 2 cents per square foot. What is the least expensive box that has a volume of 1 cubic foot? Use Theorem 3 as a check that the critical point provides a minimum.

20. Let $U(x, y, z) = x^{1/2} y^{1/3} z^{1/6}$ be the "utility" or "desirability" to a given consumer of the amounts x, y, and z of three different commodities. Their prices are, respectively, 2 dollars, 1 dollar, and 5 dollars, and the consumer has 60 dollars to spend. How much of each product should he buy to maximize the utility?

21. Find the dimensions of the open rectangular box of volume 1 of smallest surface area. Use Theorem 3 as a check that the critical point provides a minimum.

22 Find the dimensions of the rectangular box of largest volume that can be inscribed in a sphere of radius 1.

23 Establish case 2 in Theorem 1.

■ ■

24 Establish case 3 in Theorem 1 as follows. For convenience assume that none of A, B, and C are 0. Assume that $AC - B^2 < 0$, and that $A > 0$. Recall that

$$A(Ax^2 + 2Bxy + Cy^2) = (Ax + By)^2 + (AC - B^2)y^2.$$

(a) Show that there are points on the line $Ax + By = 0$ and arbitrarily near $(0, 0)$ at which $f(x, y) = Ax^2 + 2Bxy + Cy^2$ is negative.

(b) Show that there are points on the x axis arbitrarily near $(0, 0)$ where $f(x, y)$ is positive.

(c) Show that there are two lines through the origin on which $f(x, y)$ is 0. *Hint:* The quadratic formula might be of use.

25 A firm produces q_1 washers and q_2 driers at a cost of $2q_1^2 + q_1 q_2 + 2q_2^2$ monetary units. The revenue is $2q_1 + 3q_2$ monetary units.

(a) Find the profit $P(q_1, q_2)$, that is, the revenue less the cost.

(b) Find the combination of q_1 and q_2 that maximizes the profit. Use Theorem 3 as a check.

26 (The test for a relative maximum or relative minimum of $f(x, y, z)$) Let $f(x, y, z)$ have continuous partial derivatives of the first and second order within some sphere centered at (a, b, c). Let

$$H_1 = f_{xx}, \quad H_2 = \begin{vmatrix} f_{xx} & f_{xy} \\ f_{yx} & f_{yy} \end{vmatrix}, \quad \text{and} \quad H_3 = \begin{vmatrix} f_{xx} & f_{xy} & f_{xz} \\ f_{yx} & f_{yy} & f_{yz} \\ f_{zx} & f_{zy} & f_{zz} \end{vmatrix},$$

where the partial derivatives are evaluated at (a, b, c). Assume that f_x, f_y, and f_z are 0 at (a, b, c) and none of H_1, H_2, and H_3 is 0.

1 If $H_1 > 0$, $H_2 > 0$, and $H_3 > 0$, f has a relative minimum at (a, b, c).

2 If $H_1 < 0$, $H_2 > 0$, and $H_3 < 0$, f has a relative maximum at (a, b, c).

3 In the other of the six possible cases (e.g., $H_1 > 0$, $H_2 > 0$, $H_3 < 0$), f has neither a relative minimum nor a relative maximum at (a, b, c).

(Only when the signs of H_1, H_2, and H_3 are all the same or else alternate is there a relative extremum.)

Using this test, which depends on results in linear algebra, determine the behavior of the following functions at $(0, 0, 0)$.

(a) $\sin^2 x + y^2 + z^2 - y \sin x - yz + z \sin x$

(b) $-2 \tan^2 x - 10y^2 - z^2 + 6y \tan x - 2z \tan x + 2yz$

(c) $10x^2 + 3y^2 + 3z^2 + 12xy + 4xz + 2yz$

27 If $\mathscr{D} = 0$ in Theorem 1, show that f has either a maximum or minimum at $(0, 0)$. *Suggestion:* Use the technique of the proof of Theorem 1.

12.9 The Taylor series for $f(x, y)$ (optional)

The higher order partial derivatives The higher partial derivatives of $z = f(x, y)$ were introduced in Sec. 12.3. Recall that

$$\frac{\partial}{\partial y}\left(\frac{\partial z}{\partial x}\right) \quad \text{is denoted} \quad \frac{\partial^2 z}{\partial y\, \partial x}, \frac{\partial^2 f}{\partial y\, \partial x}, \quad \text{or } z_{xy}.$$

There are four partial derivatives of the second order:

$$z_{xx}, \quad z_{xy}, \quad z_{yx}, \quad \text{and} \quad z_{yy}.$$

If they are continuous, z_{xy} equals z_{yx}. Each of these four partial derivatives may be differentiated with respect to x or with respect to y. Thus there are eight possible partial derivatives of order 3. For instance, two of them are

$$\frac{\partial(z_{xx})}{\partial x} \quad \text{and} \quad \frac{\partial(z_{xx})}{\partial y}.$$

which will be denoted z_{xxx} and z_{xxy}.

The eight are

$$z_{xxx}, \quad z_{xxy}, \quad z_{xyx}, \quad z_{xyy}, \quad z_{yxx}, \quad z_{yxy}, \quad z_{yyx}, \quad \text{and} \quad z_{yyy}.$$

If they are continuous, however, many of them are equal. For instance,

$$z_{xxy} = z_{xyx}$$

for $\quad z_{xxy} = (z_x)_{xy} \quad$ and $\quad z_{xyx} = (z_x)_{yx}$.

In general, the order of differentiation does not affect the result; all differentiations with respect to x may be done first, and afterward the differentiations with respect to y. Thus

$$z_{xyy} = z_{yxy} = z_{yyx},$$

or, in the ∂ notation, $\quad \dfrac{\partial^3 z}{\partial y^2\, \partial x} = \dfrac{\partial^3 z}{\partial y\, \partial x\, \partial y} = \dfrac{\partial^3 z}{\partial x\, \partial y^2}.$

Similar statements and notations hold for partial derivatives of higher orders.

Thus $\quad z_{xyxyy} = z_{xxyyy} = \dfrac{\partial^5 z}{\partial y^3\, \partial x^2}.$

EXAMPLE 1 Compute the partial derivatives of $x^4 y^7$ up through order 3.

SOLUTION To begin: $\quad z_x = 4x^3 y^7 \quad$ and $\quad z_y = 7x^4 y^6$.

Then $z_{xx} = 12x^2 y^7$, $z_{xy} = z_{yx} = 28x^3 y^6$, $z_{yy} = 42x^4 y^5$. The third-order partial derivatives are

$$z_{xxx} = 24xy^7$$
$$z_{xxy} = z_{xyx} = z_{yxx} = 84x^2 y^6,$$
$$z_{xyy} = z_{yxy} = z_{yyx} = 168x^3 y^5,$$

and $\quad z_{yyy} = 210 x^4 y^4.$

Note that on account of duplication there are in practice only four partial derivatives of order 3. Similarly, there are in practice only five different partial derivatives of order 4, and $n+1$ different partial derivatives of order n. ■

TAYLOR SERIES FOR $f(x, y)$

Just as many functions $f(x)$ of a single variable can be expressed as power series in $x - a$, so can many functions $f(x, y)$ of two variables be expressed as a series in powers of $x - a$ and $y - b$. Such a series begins,

802 PARTIAL DERIVATIVES

$$c_1 + c_2(x-a) + c_3(y-b) + c_4(x-a)^2 + c_5(x-a)(y-b) + c_6(y-b)^2 + \cdots$$

where the c's are constants. Frequently in applications only this much of the series is used to approximate the function near the point (a, b). The typical term has the form $c(x-a)^r(y-b)^s$. The *degree* of the term is $r+s$. The terms are written from left to right in increasing degree; within terms of a given degree, the terms are written in increasing degree of $y-b$.

We shall obtain a formula for this series in terms of the partial derivatives of f evaluated at (a, b). This formula will then be used to establish the test for a relative maximum or minimum given in Theorem 3 of the preceding section. The argument begins by relating $f(x, y)$ to a function of one variable.

Let f be a function of x and y that possesses partial derivatives of all orders. Let $a, b, h,$ and k be fixed numbers. Define a function g as follows:

$$g(t) = f(a + th, b + tk).$$

The Taylor series for f can be obtained from that for g by expressing the derivatives $g'(0), g^{(2)}(0), g^{(3)}(0), \ldots$ in terms of partial derivatives of f.

To compute $g'(t)$, observe that g is a composite function:

$$g(t) = f(x, y) \qquad \text{where} \qquad x = a + th, \ y = b + tk. \tag{1}$$

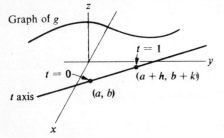

Figure 12.47

g' in terms of f_x and f_y

(See Fig. 12.47.) By a chain rule (Theorem 1 of Sec. 12.5),

$$g'(t) = \frac{\partial f}{\partial x}\frac{dx}{dt} + \frac{\partial f}{\partial y}\frac{dy}{dt}.$$

Now by (1) $\qquad \dfrac{dx}{dt} = h \qquad$ and $\qquad \dfrac{dy}{dt} = k.$

Thus $\qquad g'(t) = f_x \cdot h + f_y \cdot k, \tag{2}$

where f_x and f_y are evaluated at $(a + th, b + tk)$. In particular,

$$g'(0) = f_x(a, b)h + f_y(a, b)k. \tag{3}$$

Next, express $g^{(2)}(t)$ in terms of partial derivatives of the function f. To do so, differentiate (2) with respect to t:

$$g^{(2)}(t) = \frac{d}{dt}[g'(t)] = \frac{d}{dt}(f_x h + f_y k)$$

$$= \frac{\partial}{\partial x}(f_x h + f_y k)\frac{dx}{dt} + \frac{\partial}{\partial y}(f_x h + f_y k)\frac{dy}{dt}$$

$$= \frac{\partial}{\partial x}(f_x h + f_y k)h + \frac{\partial}{\partial y}(f_x h + f_y k)k$$

$$= (f_{xx}h + f_{yx}k)h + (f_{xy}h + f_{yy}k)k.$$

$g^{(2)}$ in terms of $f_{xx}, f_{xy},$ and f_{yy} Hence
$$g^{(2)}(t) = f_{xx}h^2 + 2f_{xy}hk + f_{yy}k^2, \qquad (4)$$

where all the partial derivatives are evaluated at $(a + th, b + tk)$. Thus

$$g^{(2)}(0) = f_{xx}(a, b)h^2 + 2f_{xy}(a, b)hk + f_{yy}(a, b)k^2. \qquad (5)$$

Notice the similarity of the right side of (5) to the binomial expansion,

$$(c + d)^2 = c^2 + 2cd + d^2,$$

in the coefficients, the powers of h and k, and the subscripts. To make use of this similarity, introduce the expression

$$(h\,\partial_x + k\,\partial_y)^2 f,$$

where $(h\,\partial_x + k\,\partial_y)^2$ is treated formally like an algebraic product: for instance, $(\partial_x \partial_x)f$ is interpreted as f_{xx}.

Thus (5) may be written in this shorthand as

The symbol $\big|_{(a,b)}$ means "evaluated at (a, b)."

$$g^{(2)}(0) = (h\,\partial_x + k\,\partial_y)^2 f \Big|_{(a,b)}. \qquad (6)$$

Differentiating (4) with respect to t sufficiently often and then setting $t = 0$, we can show similarly that

$$g^{(n)}(0) = (h\,\partial_x + k\,\partial_y)^n f \Big|_{(a,b)} \qquad (7)$$

for $n = 1, 2, 3, \ldots$.

THEOREM *Taylor series for a function of two variables.* Let f have continuous partial derivatives of all orders up to and including $n + 1$ at and near the point (a, b). If $(x, y) = (a + h, b + k)$ is sufficiently near (a, b), then

$$f(x, y) = f(a, b) + (h\,\partial_x + k\,\partial_y)f\Big|_{(a,b)} + \frac{(h\,\partial_x + k\,\partial_y)^2 f}{2}\Big|_{(a,b)} + \cdots$$
$$+ \frac{(h\,\partial_x + k\,\partial_y)^n f}{n!}\Big|_{(a,b)} + \frac{(h\,\partial_x + k\,\partial_y)^{n+1} f}{(n+1)!}\Big|_{(X,Y)}$$

where (X, Y) is some point on the line segment joining (a, b) and (x, y). In sigma notation this reads

$$f(x, y) = \sum_{i=0}^{n} \frac{(h\,\partial_x + k\,\partial_y)^i f}{i!}\Big|_{(a,b)} + \frac{(h\,\partial_x + k\,\partial_y)^{n+1} f}{(n+1)!}\Big|_{(X,Y)}.$$

PROOF This theorem follows from Theorem 2 of Sec. 10.8. Once again we make use of the function g, defined as follows:

$$g(t) = f(a + th, b + tk).$$

(See Fig. 12.47.) Observe that $g(0) = f(a, b)$ (8)

and $\qquad g(1) = f(a + h, b + k) = f(x, y).$ (9)

By Lagrange's formula for the remainder $R_n(1; 0)$,

$$g(1) = g(0) + g'(0) \cdot 1 + \frac{g^{(2)}(0)}{2!} 1^2 + \cdots + \frac{g^{(n)}(0)}{n!} 1^n + \frac{g^{(n+1)}(T)}{(n+1)!} 1^{n+1} \quad (10)$$

for a suitable number $T, 0 \leq T \leq 1$. Combining (7) and (10) completes the proof. ∎

As a consequence of the theorem, the coefficient of $h^r k^s$ in the Taylor series for f is

$$\frac{1}{(r+s)!} \binom{r+s}{r} \frac{\partial^{r+s} f}{\partial x^r \, \partial y^s},$$

where the partial derivative is evaluated at (a, b), and $\binom{r+s}{r}$ denotes the binomial coefficient

$$\frac{(r+s)!}{r! \, s!}.$$

Observe that $(r + s)!$ can be canceled in the numerator and denominator. Thus the coefficient of $h^r k^s$ is

$$\frac{1}{r! \, s!} \frac{\partial^{r+s} f}{\partial x^r \, \partial y^s}.$$

EXAMPLE 2 Use the theorem 1 to express $f(x, y) = x^2 y$ in powers of $x - 1$ and $y - 2$.

SOLUTION In this case $a = 1$, $b = 2$, $h = x - 1$, and $k = y - 2$. To begin, compute the partial derivatives of f at $(1, 2)$. We have

$$f_x = 2xy, \qquad f_{xx} = 2y, \qquad f_{xy} = 2x, \qquad f_{xxy} = 2, \qquad \text{and} \qquad f_y = x^2.$$

All higher partial derivatives of f are identically 0. Thus $f(1, 2) = 2$, $f_x(1, 2) = 2 \cdot 1 \cdot 2 = 4$, and so on. Therefore

$$f(x, y) = f(1 + h, 2 + k)$$
$$= f(1, 2) + [hf_x(1, 2) + kf_y(1, 2)] +$$
$$\left[\frac{h^2 f_{xx}(1, 2) + 2hk f_{xy}(1, 2) + k^2 f_{yy}(1, 2)}{2!} \right] +$$
$$\left[\frac{h^3 f_{xxx}(1, 2) + 3h^2 k f_{xxy}(1, 2) + 3hk^2 f_{xyy}(1, 2) + k^3 f_{yyy}(1, 2)}{3!} \right]$$

$$= 2 + 4h + k + \frac{4h^2 + 4hk + 0k^2}{2!} + \frac{6h^2k}{3!};$$

that is,

$$x^2y = 2 + 4(x-1) + (y-2) + 2(x-1)^2 + 2(x-1)(y-2) + (x-1)^2(y-2). \quad (11)$$

This can be checked by expanding the right side of the equation. ∎

According to the theorem the Taylor series associated with $f(x, y)$ in powers of $x - a$ and $y - b$ begins

$$f(a, b) + f_x(a, b)(x-a) + f_y(a, b)(y-b) + \frac{f_{xx}(a, b)}{2!}(x-a)^2 + \frac{2f_{xy}(a, b)}{2!}(x-a)(y-b) + \frac{f_{yy}(a, b)}{2!}(y-b)^2 + \cdots.$$

Whether the series converges and whether it converges to $f(x, y)$ depend on the behavior of the higher-order partial derivatives of f.

Taylor series for functions of three or more variables can be developed in a similar way.

Sometimes only the terms up through degree one are used to approximate the function. For instance, Harold A. Thomas in an essay entitled "Population Dynamics of Primitive Societies" (included in *Is there an Optimum Level of Population?*, edited by S. Fred Singer) wrote,

"For a small three-dimensional region about the equilibrium point $(\bar{P}, \bar{x}, \bar{w})$ the utility function may be approximated by the first terms of a Taylor's series:

$$U(P, x, w) = U(\bar{P}, \bar{x}, \bar{w}) + b_1(P - \bar{P}) + b_2(x - \bar{x}) + b_3(w - \bar{w}) + \cdots$$

where b_1, b_2, and b_3 represent the partial derivatives $\partial U/\partial P$, $\partial U/\partial x$, and $\partial U/\partial w$ evaluated at the equilibrium point."

The notation is different from that of this section, but the idea is the same: to approximate the behavior of a function near a point by a polynomial in several variables.

PROOF OF THE SECOND-PARTIAL-DERIVATIVE TEST

The theorem of this section, with $n = 1$, is the basis of the proof of Theorem 3 in Sec. 12.8. We shall prove only case 1, which is repeated here schematically:

Assumptions:
$\begin{cases} f_x(a, b) = 0, f_y(a, b) = 0 \\ f_{xx}(a, b) > 0 \\ f_{xx}(a, b)f_{yy}(a, b) - [f_{xy}(a, b)]^2 > 0 \\ f_x, f_y, f_{xx}, f_{xy}, \text{ and } f_{yy} \text{ are continuous} \\ \text{at and near } (a, b). \end{cases}$

Conclusion: f has a relative minimum at (a, b).

PROOF Using the theorem of this section, we conclude that

$$f(x, y) = f(a, b) + hf_x(a, b) + kf_y(a, b) + \tfrac{1}{2}[h^2 f_{xx}(X, Y) + 2hk f_{xy}(X, Y) + k^2 f_{yy}(X, Y)], \quad (12)$$

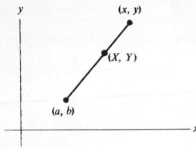

Figure 12.48

where (X, Y) is somewhere on the line segment joining (a, b) and $(x, y) = (a + h, b + k)$, shown in Fig. 12.48.

Now, $f_x(a, b) = 0$ and $f_y(a, b) = 0$. Thus (12) reduces to

$$f(x, y) = f(a, b) + \tfrac{1}{2}[h^2 f_{xx}(X, Y) + 2hk f_{xy}(X, Y) + k^2 f_{yy}(X, Y)]. \quad (13)$$

Since f_{xx}, f_{yy}, and f_{xy} are continuous, when (x, y) is sufficiently close to (a, b), $f_{xx}(X, Y) > 0$ and $f_{xx}(X, Y) f_{yy}(X, Y) - [f_{xy}(X, Y)]^2 > 0$ from the assumptions. Now let

$$A = f_{xx}(X, Y), \quad B = f_{xy}(X, Y), \quad \text{and} \quad C = f_{yy}(X, Y).$$

We have $A > 0$ and $AC - B^2 > 0$.

In view of (13), all that remains is to show that

$$Ah^2 + 2Bhk + Ck^2 \geq 0. \quad (14)$$

But inequality (14) was already established in the proof of Theorem 1 of the preceding section. This concludes the proof. ∎

Exercises

In each of Exercises 1 to 4 compute all eight partial derivatives of the third order of the given function.

1. $x^5 y^7$
2. x/y
3. e^{2x+3y}
4. $\sin(x^2 + y^3)$
5. Verify (11) by expanding the right side of the equation.
6. Using (11), compute the difference in the volumes of these two boxes: One has a square base of side 1 foot and height 2 feet; the other has a square base of side 1.1 feet and height 2.1 feet.
7. (a) Using partial derivatives, obtain the first four nonzero terms in the Taylor series for e^{x+y^2} in powers of x and y.
   (b) Noticing that $e^{x+y^2} = e^x e^{y^2}$ and using a few terms of the Maclaurin series for e^x and e^{y^2}, solve (a) again.
8. (a) Using partial derivatives, obtain the first four nonzero terms in the Taylor series for $\cos(x + y)$ in powers of x and y.
   (b) Using the Maclaurin series for $\cos t$ and replacing t by $x + y$, obtain the terms described in (a).
9. (a) Using partial derivatives, express $x^2 y^2$ as a polynomial in $x - 1$ and $y - 1$.
   (b) Verify your answer to (a) by expanding it.
10. (a) Using partial derivatives, express xy^3 as a polynomial in powers of $x + 1$ and $y - 2$.
    (b) Verify your answer in (a) by expanding it.
11. The binomial theorem for $n = 3$ asserts that
    $$(x + y)^3 = x^3 + 3x^2 y + 3xy^2 + y^3.$$
    Use a Taylor series for $f(x, y) = (x + y)^3$ to obtain this result.
12. The binomial theorem for $n = 4$ asserts that
    $$(x + y)^4 = x^4 + 4x^3 y + 6x^2 y^2 + 4xy^3 + y^4.$$
    Use a Taylor series for $f(x, y) = (x + y)^4$ to obtain this result.
13. Assume that f has continuous partial derivatives of all orders, $f(0, 0) = 2$, $f_x(0, 0) = 3$, $f_y(0, 0) = -5$, $f_{xx}(0, 0) = 6$, $f_{xy}(0, 0) = 7$, and $f_{yy}(0, 0) = 1$. Write out the Taylor series in powers of x and y associated with f up through terms of degree two.
14. The Taylor series for a certain function $f(x, y)$ begins $5 + 6x + 11y - 2x^2 - 3xy + 7y^2 + \cdots$. Use this information to determine $f(0, 0)$ and the first- and second-order partial derivatives of f at $(0, 0)$.

15 Verify that the expansion of $\sqrt{1+x+y}$ begins $1 + \frac{1}{2}x + \frac{1}{2}y - \frac{1}{8}x^2 - \frac{1}{4}xy - \frac{1}{8}y^2 + \cdots$, using the theorem of this section.

16 (a) Prove that, if all the partial derivatives of f are continuous, then

$$f_{xyxy} = f_{xyyx}.$$

(b) Prove that each of the 16 partial derivatives of f of the fourth order is equal to one of these five:

$$\frac{\partial^4 f}{\partial x^4}, \quad \frac{\partial^4 f}{\partial y \, \partial x^3}, \quad \frac{\partial^4 f}{\partial y^2 \, \partial x^2}, \quad \frac{\partial^4 f}{\partial y^3 \, \partial x}, \quad \frac{\partial^4 f}{\partial y^4}.$$

17 What is the coefficient of $x^5 y^7$ in the Taylor series expansion of the function f around the point $(a, b) = (0, 0)$?

18 Show that if the Taylor series for $f(x, y)$ in powers of $x - a$ and $y - b$ converges to $f(x, y)$, then

$$f(x, y) = \sum_{n=0}^{\infty} \left(\sum_{r=0}^{n} \left. \frac{\partial^n f(x, y)}{\partial x^r \, \partial y^{n-r}} \right|_{(a, b)} \frac{(x - a)^r (y - b)^{n-r}}{r!(n - r)!} \right).$$

12.S Summary

This chapter developed some analogs of properties of a function of one variable for a function of two variables. Most of the results are summarized in the table below. The basis of the chapter is the equation

$$\Delta f = f_x(x, y) \, \Delta x + f_y(x, y) \, \Delta y + \varepsilon_1 \, \Delta x + \varepsilon_2 \, \Delta y,$$

where $\varepsilon_1(x, y, \Delta x, \Delta y)$ and $\varepsilon_2(x, y, \Delta x, \Delta y)$ both approach 0 as Δx and Δy approach 0. The chain rules, which are based on this equation, are in turn the basis for the formulas for $D_{\mathbf{u}} f$ and the Taylor series associated with $f(x, y)$.

The Plane and $f(x)$	Space and $f(x, y)$
A point is determined by x and y. $x = k$ is a line perpendicular to x axis; $y = k$ is a line perpendicular to y axis.	A point is determined by x, y, and z. $x = k$ is a plane perpendicular to x axis; $y = k$ is a plane perpendicular to y axis; $z = k$ is a plane perpendicular to z axis.
$(x - a)^2 + (y - b)^2 = r^2$ describes a circle of radius r, and center (a, b).	$(x - a)^2 + (y - b)^2 + (z - c)^2 = r^2$ describes a sphere of radius r and center (a, b, c).
Ellipse: $\quad \dfrac{x^2}{a^2} + \dfrac{y^2}{b^2} = 1.$	Ellipsoid: $\quad \dfrac{x^2}{a^2} + \dfrac{y^2}{b^2} + \dfrac{z^2}{c^2} = 1.$
Hyperbola: $\quad \dfrac{x^2}{a^2} - \dfrac{y^2}{b^2} = 1.$	Hyperboloid of one sheet: $\dfrac{x^2}{a^2} + \dfrac{y^2}{b^2} - \dfrac{z^2}{c^2} = 1;$ Hyperboloid of two sheets: $\dfrac{x^2}{a^2} - \dfrac{y^2}{b^2} - \dfrac{z^2}{c^2} = 1.$
The graph of $y = f(x)$ is usually a curve. The graph of $y = Ax + B$ is a line. The derivative f' measures the rate of change of f.	The graph of $z = f(x, y)$ is usually a surface. The graph of $z = Ax + By + C$ is a plane. The partial derivatives f_x and f_y measure the rate of change of f in the x and y directions.

The Plane and $f(x)$	Space and $f(x, y)$
Notation: f', $\dfrac{dy}{dx}$, Df, y'	Notation: f_x, $\dfrac{\partial f}{\partial x}$, z_x, $\dfrac{\partial z}{\partial x}$, f_1, $\left(\dfrac{\partial f}{\partial x}\right)_y$ f_y, $\dfrac{\partial f}{\partial y}$, z_y, $\dfrac{\partial z}{\partial y}$, f_2, $\left(\dfrac{\partial f}{\partial y}\right)_x$ More generally, the directional derivative $D_{\mathbf{u}} f$ measures the rate of change of f in the direction of $\mathbf{u} = \cos\theta\,\mathbf{i} + \sin\theta\,\mathbf{j}$. In terms of the gradient ∇f, $$D_{\mathbf{u}} f = \mathbf{u} \cdot \nabla f.$$
A continuous function on $[a, b]$ has a maximum.	A continuous function on a region contained in some circle and bounded by a polygon or curve has a maximum.
The maximum of a differentiable f on $[a, b]$ is the largest of the values of f at a, b, and critical points.	The maximum of f is the largest of the values of f at points on the boundary and at critical points.
Higher derivatives: $d^2f/dx^2, \ldots$.	Higher partial derivatives: $\partial^2 f/\partial x^2$, $\partial^2 f/\partial y^2$, $\partial^2 f/\partial x\,\partial y$, $\partial^2 f/\partial y\,\partial x, \ldots$.
$\Delta y = f'(x)\,\Delta x + \varepsilon\,\Delta x$ ($\varepsilon \to 0$ as $\Delta x \to 0$) where ε is function of x and Δx. dy is defined as $f'(x)\,\Delta x$. dy is change along tangent line.	$\Delta z = f_x\,\Delta x + f_y\,\Delta y + \varepsilon_1\,\Delta x + \varepsilon_2\,\Delta y$ ($\varepsilon_1 \to 0$, $\varepsilon_2 \to 0$ as Δx and $\Delta y \to 0$) where ε_1 and ε_2 are functions of x, y, Δx, and Δy. dz is defined as $f_x\,\Delta x + f_y\,\Delta y$. dz is change along tangent plane.
Chain rule: $\dfrac{dy}{dx} = \dfrac{dy}{du}\dfrac{du}{dx}$.	Chain rule: $\dfrac{dz}{dt} = \dfrac{\partial z}{\partial x}\dfrac{dx}{dt} + \dfrac{\partial z}{\partial y}\dfrac{dy}{dt}$ (where x and y are functions only of t); Chain rule: $\dfrac{\partial z}{\partial t} = \dfrac{\partial z}{\partial x}\dfrac{\partial x}{\partial t} + \dfrac{\partial z}{\partial y}\dfrac{\partial y}{\partial t}$ and $\dfrac{\partial z}{\partial u} = \dfrac{\partial z}{\partial x}\dfrac{\partial x}{\partial u} + \dfrac{\partial z}{\partial y}\dfrac{\partial y}{\partial u}$ (where x and y are functions of t and u).
Critical point: $f'(a) = 0$.	Critical point: $f_x(a, b) = 0$ and $f_y(a, b) = 0$.
Second-derivative test for local maximum: $$f' = 0, \quad f'' < 0.$$	Second-partial-derivative test for local maximum: $$f_x = 0, \quad f_y = 0$$ $$f_{xx} < 0, \quad f_{xx} f_{yy} - (f_{xy})^2 > 0.$$
Second-derivative test for local minimum: $$f' = 0, \quad f'' > 0.$$	Second-partial derivative test for local minimum: $$f_x = 0, \quad f_y = 0$$ $$f_{xx} > 0, \quad f_{xx} f_{yy} - (f_{xy})^2 > 0.$$ If $f_x = 0$, $f_y = 0$, $f_{xx} f_{yy} - (f_{xy})^2 < 0$, the critical point is a saddle.
Taylor series in $x - a$: coefficient of $(x - a)^n$ is $$\dfrac{f^{(n)}(a)}{n!}.$$	Taylor series in $x - a$ and $y - b$: coefficient of $(x - a)^r (y - b)^s$, $r + s = n$ is $$\dfrac{1}{r!\,s!}\left.\dfrac{\partial^n f}{\partial x^r\,\partial y^s}\right|_{(a,\,b)}.$$

VOCABULARY

- function of two variables
- level curve
- partial derivative
- differential
- chain rule
- directional derivative
- gradient
- maximum, relative maximum
- minimum, relative minimum
- critical point
- Taylor series in $x - a$ and $y - b$

Guide quiz on chap. 12

1. Graph
   (a) $x^2 + y^2 + z^2 = 9$
   (b) $x^2 + \dfrac{y^2}{2} + \dfrac{z^2}{3} = 6$
   (c) $-x^2 - \dfrac{y^2}{4} + z^2 = 1$
   (d) $x^2 - 2y^2 + z^2 = 8$
   (e) $z = y^2 - 2x^2$

2. Explain why $f(x + \Delta x, y + \Delta y) - f(x, y) = f_x(x, y)\, \Delta x + f_y(x, y)\, \Delta y + \varepsilon_1\, \Delta x + \varepsilon_2\, \Delta y$, where ε_1 and ε_2 both approach 0 as Δx and Δy approach 0. (State the assumptions about f used.)

3. (a) Let $y(x, t) = y_0 \sin(kx - kvt)$, where y_0, k, and v are constants. Show that $y(x, t)$ satisfies the equation
$$\frac{\partial^2 y}{\partial x^2} = \frac{1}{v^2} \frac{\partial^2 y}{\partial t^2}.$$
   (b) Let $f(p)$ have first and second derivatives. Show that $f(x - vt)$ and $f(x + vt)$ both satisfy the equation in (a), if v is a constant.

4. Suppose $w = f(u, v)$ has continuous partial derivatives with respect to u and v, and $u = x + y$ and $v = x - y$. Show that:
   (a) $\dfrac{\partial w}{\partial x} \cdot \dfrac{\partial w}{\partial y} = \left(\dfrac{\partial f}{\partial u}\right)^2 - \left(\dfrac{\partial f}{\partial v}\right)^2$
   (b) $\dfrac{\partial^2 w}{\partial x\, \partial y} = \dfrac{\partial^2 f}{\partial u^2} - \dfrac{\partial^2 f}{\partial v^2}$

5. Give an example of a function f that has a critical point at $(0, 0)$, $f_{xx}(0, 0) > 0$, $f_{yy}(0, 0) > 0$, and
   (a) $(0, 0)$ is a local minimum,
   (b) $(0, 0)$ is not a local minimum.

6. A house in the form of a box is to hold 10,000 cubic feet. The glass walls admit heat at the rate of 5 units per minute per square foot, the roof at the rate of 3 units per minute per square foot, and the floor at a rate of 1 unit per minute per square foot. What should the shape of the house be in order to minimize the rate at which heat enters?

7. (a) Using a table of natural logarithms or a calculator, evaluate $(1.1)^2 \ln(1.2)$.
   (b) Using the differential of the function $f(x, y) = x^2 \ln y$ and making use of $f(1, 1)$, estimate $(1.1)^2 \ln(1.2)$.

8. The kinetic energy of a particle of mass m and velocity v is given by $K = \tfrac{1}{2} m v^2$. If the maximum error in measuring m is 1 percent and in measuring v is 3 percent, estimate the maximum error in measuring K.

9. Let $f(x, y) = x^3 y^2$.
   (a) What is the largest directional derivative of f at $(1, 1)$?
   (b) In what direction is the derivative in (a)?
   (c) Find the directional derivative of f at $(1, 1)$ in the direction given by $\theta = 5\pi/4$.
   (d) What is the equation of the level curve of f that passes through $(1, 1)$?
   (e) Find the slope of the curve in (d) at $(1, 1)$.
   (f) Show that the gradient of f evaluated at $(1, 1)$ is perpendicular to the tangent line to the level curve at $(1, 1)$.

Review exercises for chap. 12

1. Evaluate the following partial derivatives.
   (a) $\dfrac{\partial}{\partial y} \left(\sec^3(x + 2y) \ln(1 + 2xy) \right)$
   (b) $\dfrac{\partial}{\partial x} \left(x \tan^{-1} 3xy \right)$
   (c) $\dfrac{\partial (y e^{x^3 y})}{\partial x}$
   (d) $\dfrac{\partial^2}{\partial x\, \partial y} \left(\cos(2x + 3y) \right)$
   (e) $\dfrac{\partial^2}{\partial y^2} \left(\dfrac{3}{xy} \right)$
   (f) $\dfrac{\partial^2}{\partial x^2} \left(\sin^3 xy \right)$

2. On the graph of $z = f(x, y)$, sketch the curves whose slopes are given by f_x and f_y. In particular, draw the two tangent lines whose slopes are $f_x(a, b)$ and $f_y(a, b)$.

3. Compute:
   (a) $\dfrac{d}{dy} \left(\displaystyle\int_0^1 e^{xy}\, dx \right)$
   (b) $\displaystyle\int_0^1 \dfrac{\partial (e^{xy})}{\partial y}\, dx$

4. (a) Without using calculus, show that $f(x, y) = (x - y)^2$ has a minimum but no maximum. Where does the minimum occur?
   (b) Without using calculus, find the minimum value of $f(x, y) = (2x + 3y - 5)^2 + (x - y)^2$.

5. For what values of k does the function $x^2 + 3xy + ky^2$ have a minimum at $(0, 0)$?

6. For what values of the constant k does the function $x^2 + kxy + 9y^2$ have a relative minimum at $(0, 0)$?
7. Let f be a function of x and y. Assume that $f(2, 3) = 5$, $f_x(2, 3) = 4$, $f_y(2, 3) = -1$. Estimate $f(2.1, 2.8)$.
8. Using a differential, estimate the volume of a box of sides 3.02, 4.97, and 2.01 meters.
9. Using differentials, estimate:
   (a) $\sqrt{(3.01)^2 + (4.02)^2}$
   (b) $\sqrt{(3.04)^2 + (3.97)^2}$
10. Let $f_x(0, 0) = 0$ and $f_y(0, 0) = 0$. In each of these cases decide, if there is enough information, whether f at $(0, 0)$ has a relative maximum, relative minimum, or neither.
    (a) $f_{xx}(0, 0) = -1; f_{yy}(0, 0) = 2$.
    (b) $f_{xx}(0, 0) = 3; f_{yy}(0, 0) = 2; f_{xy}(0, 0) = -2.2$.
    (c) $f_{xx}(0, 0) = -3; f_{yy}(0, 0) = -5; f_{xy}(0, 0) = 4$.
    (d) $f_{xx}(0, 0) = 3; f_{yy}(0, 0) = 12; f_{xy}(0, 0) = -6$.
11. (a) Graph the level curve of $x^2 + 2y^2$ that passes through the point $(1, 2)$.
    (b) Compute the gradient of $x^2 + 2y^2$ at $(1, 2)$ and show that it is perpendicular to the level curve at $(1, 2)$.
12. Let $u = f(x, y)$ and $v = g(x, y)$. Assume that
$$u_x = v_y \quad \text{and} \quad v_x = -u_y.$$
Prove that $u_{xx} + u_{yy} = 0$
and $v_{xx} + v_{yy} = 0$.
13. Let $z = u^3 v^5$; where $u = x + y$ and $v = x - y$.
    (a) Express z explicitly as a function of x and y and use this explicit expression to find z_x and z_y.
    (b) Find z_x and z_y by the chain rule. Do your answers agree? Which method is easier to use?
14. Let $z = e^{uv}$, where $u = y \sin x$ and $v = x + \cos y$.
    (a) Compute z_x and z_y by the chain rule.
    (b) Express z explicitly in terms of x and y and use this expression to compute z_x and z_y.
    (c) Does your answer to (a) agree with your answer to (b)?
15. Graph the planes
    (a) $x + y = 1$ (b) $x + y + z = 1$ (c) $\dfrac{x}{2} + \dfrac{y}{3} + \dfrac{z}{4} = 1$
16. Sketch the surfaces
    (a) $x^2 + \dfrac{y^2}{4} + \dfrac{z^2}{9} = 1$ (b) $z = x^2 + 3y^2$
    (c) $z = y^2 - 3x^2$ (d) $\dfrac{x^2}{4} + \dfrac{y^2}{4} - z^2 = 1$
    (e) $x^2 - 4y^2 - 9z^2 = 36$
17. Sketch the cylinders
    (a) $y = x^2$ (b) $x^2 - y^2 = 1$
    (c) $z = x^3$
18. The pressure P, volume V, and temperature T of a gas are related by the equation $(P + a/V^2)(V - b) = cT$, where a, b, and c are constants. Thus any two of P, V, and T determine the third.
    (a) Compute $\partial V/\partial T$, $\partial T/\partial P$, and $\partial P/\partial V$.
    (b) Show that the product of the three partial derivatives in (a) is -1.
19. Let $f(x, y) = x^2 + 2xy - y^2$.
    (a) Show that f considered only on the x axis has a local minimum at $(0, 0)$.
    (b) Show that f considered only on the y axis has a local maximum at $(0, 0)$.
    (c) For which values of m does f have a local minimum at $(0, 0)$ when considered only on the line $y = mx$?
20. Consider Euler's partial differential equation
$$az_{xx} + 2bz_{xy} + cz_{yy} = 0,$$
where a, b, and c are constants and $b^2 \neq ac$. Show that
$$z = f(x + r_1 y) + g(x + r_2 y),$$
where r_1 and r_2 are the roots of $a + 2bx + cx^2 = 0$, is a solution of the differential equation. The functions f and g are differentiable.
21. Determine the minimum value of the function $f(x, y) = x^4 - x^2 y^2 + y^4$.
22. If $u = x^4 f(y/x, z/x)$ show that $xu_x + yu_y + zu_z = 4u$.
23. Let $z = \sin(x - 3y) + \cos(x - 3y)$. Show that
$$\dfrac{\partial^2 z}{\partial y^2} = 9 \dfrac{\partial^2 z}{\partial x^2}.$$
24. A wire of length 1 is to be cut into three pieces which will be bent into a square, a circle, and an equilateral triangle. How should this be done to (a) minimize their total area? (b) maximize their total area?
25. Consider $f(x, y) = 2x^2 + 4x + y^2 + 8y$.
    (a) Show that f has no maximum value.
    (b) Find the minimum value of f.
26. Let $z = -3x^2 - 5y^2 + 6x - 9y$.
    (a) Why does z have no minimum value?
    (b) Find the point (x, y) at which z is a maximum.
27. Find the maximum and minimum of $f(x, y) = 4x^2 - 3y^2 + 2xy$ on the square $0 \leq x \leq 1$, $0 \leq y \leq 1$.
28. Explain in detail why Δf is approximately equal to $f_x(x, y) \Delta x + f_y(x, y) \Delta y$ when Δx and Δy are small.

■

29. Assume that $z = f(x, y)$ satisfies the equation $xyz + x + y + z^5 + 3 = 0$.
    (a) Find $(\partial z/\partial x)_y$ by differentiating both sides of the equation.
    (b) Similarly, find $(\partial x/\partial y)_z$.

(c) Similarly, find $(\partial y/\partial z)_x$.

(d) Show that $(\partial z/\partial x)_y (\partial x/\partial y)_z (\partial y/\partial z)_x = -1$.

30 Sketch the intersection of the cylinder $x^2 + y^2 = 4$ and the plane $z = x$.

31 Sketch the intersection of the cylinders $x^2 + y^2 = 1$ and $x^2 + z^2 = 1$.

32 Let $z = x^2 y$, where $y = e^{3x} u$. Thus z may be considered a function of x and y or of x and u.

(a) Compute $\partial z/\partial x$, considering z to be a function of x and y.

(b) Compute $\partial z/\partial x$, considering z to be a function of x and u.

(c) Use subscript notation to distinguish the partial derivatives in (a) and (b).

33 Let $y = f(x, t)$ describe the vertical displacement of a particle in a wave corresponding to the horizontal coordinate x at time t. It can be shown on physical grounds that f satisfies the wave equation

$$a^2 f_{xx} = f_{tt},$$

where a is a constant. Show that any function $f(x, t)$ of the form $g(x + at)$ satisfies the wave equation, where g is a function of a single variable that possesses first and second derivatives.

34 The second-order directional derivative of $f(x, y)$ in the direction $\mathbf{u}$, $D_\mathbf{u}^2(f)$, is defined as $D_\mathbf{u}(D_\mathbf{u}(f))$. Show that $D_\mathbf{u}^2(f) = \cos^2 \theta f_{xx} + 2 \cos \theta \sin \theta f_{xy} + \sin^2 \theta f_{yy}$, where $\mathbf{u} = \cos \theta \mathbf{i} + \sin \theta \mathbf{j}$.

35 (See Exercise 34.) Let $z = f(x, y)$ describe a surface. Assume that $f_{xx} > 0$ and $f_{xx} f_{yy} - (f_{xy})^2 > 0$ for all points (x, y). Show that the curve formed by the intersection of the surface with a plane perpendicular to the xy plane is concave upward ("convex" in the language of economists and upper-division mathematics courses).

36 Using the ideas in Exercise 35, obtain the second-partial-derivative test for a local minimum of a function $f(x, y)$. (All partial derivatives are assumed to be continuous. For instance, if $f_{xx}(a, b) > 0$, there is a disk around (a, b) where f_{xx} remains positive.)

37 (Review Exercise 24 of Sec. 12.7.) What does the second-partial-derivative test say about the function $f(x, y) = (y - x^2)(y - 2x^2)$ at the critical point $(0, 0)$?

38 Let $z = f(u, v)$, where u and v are functions of x and y. Then, indirectly, $z = g(x, y)$. Show that, if $du = u_x \, dx + u_y \, dy$ and $dv = v_x \, dx + v_y \, dy$, then the two expressions for dz,

$$dz = z_u \, du + z_v \, dv \qquad \text{and} \qquad dz = z_x \, dx + z_y \, dy,$$

have equal values.

39 The partial differential equation

$$\frac{\partial^2 y}{\partial t^2} = a^2 \frac{\partial^2 y}{\partial x^2} \qquad (a \text{ constant})$$

describes a vibrating string. Assume that $y(x, t)$ has the form $f(x)g(t)$, that is, of a product of a function of x and a function of t.

(a) Show that $fg'' = a^2 f'' g$.

(b) From (a) it follows that

$$\frac{f''}{f} = \frac{g''}{a^2 g}.$$

The left side of this equation is a function of x and the right side is a function of t. Deduce that there is a constant k such that $f''/f = k$ and $g''/(a^2 g) = k$.

(c) Assume that k in (b) is positive. Show that any function of the form $f(x) = c_1 e^{\sqrt{k} x} + c_2 e^{-\sqrt{k} x}$ satisfies the equation $f''/f = k$.

(d) Show that any function $g(t)$ of the form $c_3 e^{a\sqrt{k} t} + c_4 e^{a\sqrt{k} t}$ satisfies the equation $g''/(a^2 g) = k$. Thus any function of the form

$$(c_1 e^{\sqrt{k} x} + c_2 e^{-\sqrt{k} x})(c_3 e^{a\sqrt{k} t} + c_4 e^{-a\sqrt{k} t})$$

satisfies the given partial differential equation.

40 The heat flow equation for $u(x, t)$ is

$$\frac{\partial u}{\partial t} = a^2 \frac{\partial^2 u}{\partial x^2}.$$

Following the approach in the preceding exercise, we try $u(x, t) = f(x)g(t)$.

(a) Show that there is a constant k such that $f''/f = k$ and $g'/(a^2 g) = k$. (Assume that k is negative.)

(b) What form must g have?

(c) Show that $c_1 \sin (\sqrt{-k} x) + c_2 \cos (\sqrt{-k} x)$ satisfies the equation $f''/f = k$.

Thus $(c_1 \sin (\sqrt{-k} x) + c_2 \cos (\sqrt{-k} x)) e^{a^2 k t}$ is a solution of the heat equation.

41 In the ideal gas $P = nRT/V$, where P is pressure, V is volume, T is temperature, R is a constant, and n is the number of moles, assumed constant.

(a) Compute $(\partial P/\partial V)_T$ and $(\partial P/\partial T)_V$.

(b) From the fact that there is a function $f(x, y, z)$ such that $f(P, V, T) = 0$, deduce that $(\partial T/\partial V)_P = -(\partial P/\partial V)_T/(\partial P/\partial T)_V$.

(c) Solve the equation $P = nRT/V$ for T and compute $(\partial T/\partial V)_P$ directly. The result should agree with that found in (b).

42 Let $u = f(x, y)$ and $v = g(x, y)$. Assume that u and v determine x and y, $x = h(u, v)$ and $y = j(u, v)$. Show that

$$\frac{\partial u}{\partial x} \frac{\partial x}{\partial u} + \frac{\partial u}{\partial y} \frac{\partial y}{\partial u} = 1$$

and

$$\frac{\partial u}{\partial x} \frac{\partial x}{\partial v} + \frac{\partial u}{\partial y} \frac{\partial y}{\partial v} = 0.$$

Exercises 43 to 50 concern partial differential equations.

43 Let $f(x, y)$ be a function defined throughout the xy plane such that $\partial f/\partial x = 0$.
  (a) Show that on each line parallel to the x axis, $y = b$, the function $g(x) = f(x, b)$ is constant.
  (b) Show that there is a function $h(y)$ such that $f(x, y) = h(y)$.

44 Let $f(x, y)$ be a function defined throughout the xy plane such that $\partial f/\partial x = x$. Show that $f(x, y) = x^2/2 + h(y)$ for some function $h(y)$. *Suggestion:* Make use of Exercise 43.

45 Let $f(x, y)$ be a function defined throughout the xy plane such that $\partial f/\partial x = a(x)$, a function of x. Deduce that there are functions $g(x)$ and $h(y)$ such that $f(x, y) = g(x) + h(y)$.

46 Let $f(x, y)$ be defined throughout the xy plane and assume that $\partial f/\partial y = 0$. Explain what form $f(x, y)$ must have.

47 Let $f(x, y)$ be defined throughout the xy plane and assume that $\partial f/\partial x = g(y)$, some function of y. Show that $f(x, y)$ is equal to $xg(y) + h(y)$ for some function $h(y)$.

48 Find the most general function $f(x, y)$ such that $\partial^2 f/\partial x^2 = 0$ throughout the xy plane.

49 Find the most general function $f(x, y)$ such that $\partial^2 f/\partial x \, \partial y = 0$ throughout the xy plane.

50 (a) Show that $\dfrac{\partial^2 (xy)}{\partial x \, \partial y} = 1$.

  (b) Find the most general function $f(x, y)$ such that
  $$\frac{\partial^2 f}{\partial x \, \partial y} = 1.$$

51 Show that, if $\nabla f = \mathbf{0}$ throughout the plane, then f is constant.

52 Let $z = f(x, y)$, $x = g(u, v)$, and $y = h(u, v)$; thus, $z = m(u, v)$. Assume that $g_u = h_v$ and $g_v = -h_u$. Show that
$$m_{uu} + m_{vv} = (f_{xx} + f_{yy})(g_u^2 + g_v^2).$$

53 A fence perpendicular to the xy plane has as its base the line segment whose ends are $(1, 0, 0)$ and $(0, 1, 0)$. It is bounded on the top by the parabolic cylinder $z = y^2$.
  (a) Draw the fence.
  (b) Find its area.

■ ■

54 Exercise 57 of Sec. 4.S shows that $\Delta y - dy$ is small when compared with Δx if Δx is small. Prove that $\Delta z - dz$ is small when compared with $\sqrt{(\Delta x)^2 + (\Delta y)^2}$ if Δx and Δy are small. (Assume f_x and f_y are not 0.)

55 Tell what is wrong with this "proof" that if $z = f(x, y)$ and x and y are functions of t, then dz/dt is just $(\partial z/\partial x)(dx/dt)$: "Since dz/dt is $\lim_{\Delta t \to 0} \Delta z/\Delta t$, $dz/dt = \lim_{\Delta t \to 0} (\Delta z/\Delta x)(\Delta x/\Delta t)$, where Δx is the change in x induced by the change in t. Thus $dz/dt = (\partial z/\partial x)(dx/dt)$.

56 Let (x, y) be rectangular coordinates in the plane, and (X, Y, Z) in space. Assume that F is a one-to-one correspondence between the plane and space, such that x and y depend continuously on X, Y, and Z and have continuous partial derivatives with respect to them. Similarly, assume that through the inverse function F^{-1}, X, Y, and Z are continuous functions of x and y and have continuous partial derivatives with respect to them. From this deduce that $2 = 3$. *Hint:* $2 = dx/dx + dy/dy$ and $3 = dX/dX + dY/dY + dZ/dZ$. Use the chain rule. Incidentally, there *is* a one-to-one correspondence between the plane and space, but it does not have the specified properties of continuity and differentiability.

57 (See Exercises 30 and 31 of Sec. 12.5.) Prove that, if f is homogeneous of degree n and has continuous second-order partial derivatives, then
$$x^2 f_{xx} + 2xy f_{xy} + y^2 f_{yy} = n(n-1)f.$$

58 You arrive early in the chemistry auditorium and see this left on the blackboard from the previous lecture:
$$\left(\frac{\partial H}{\partial T}\right)_P \left(\frac{\partial T}{\partial P}\right)_H + \left(\frac{\partial H}{\partial P}\right)_T = 0.$$

You have no idea what the letters refer to. Nevertheless, show that the formula is correct.

Introduction to Exercises 59 to 62. If f is a function of x and y, then by definition df is $(\partial f/\partial x) \Delta x + (\partial f/\partial y) \Delta y$. But x is a function of x and y, $x = x + 0y$. Thus $dx = 1 \, \Delta x + 0 \, \Delta y$, so $dx = \Delta x$. Similarly, $dy = \Delta y$. We may write $df = (\partial f/\partial x) \, dx + (\partial f/\partial y) \, dy$, which is a standard notation. Given any functions of x and y, $M(x, y)$ and $N(x, y)$, we may form the expression $M(x, y) \, dx + N(x, y) \, dy$, which is called a *differential form*. A differential form is called *exact* if there is a function f such that $df = M(x, y) \, dx + N(x, y) \, dy$, that is $\partial f/\partial x = M(x, y)$ and $\partial f/\partial y = N(x, y)$.

59 (a) Show that if $M \, dx + N \, dy$ is exact, then $\partial M/\partial y = \partial N/\partial x$.
  (b) Show that $x^2 y \, dx + xy^3 \, dy$ is not exact.
  (c) Show that $3x^2 y^2 \, dx + 2x^3 y \, dy$ is exact by exhibiting a function f such that $\partial f/\partial x = 3x^2 y^2$ and $\partial f/\partial y = 2x^3 y$.

Exercises 61 and 62 will show that, if $\partial M/\partial y = \partial N/\partial x$ and M and N are defined throughout the xy plane, then there is a function f such that $\partial f/\partial x = M$ and $\partial f/\partial y = N$. The argument needs a formula described in Exercise 60.

60 It is shown in Appendix G that

$$\frac{d}{dy}\int_a^b f(x, y)\,dx = \int_a^b \frac{\partial f}{\partial y}\,dx.$$

Check that this equation is valid if (a) $f(x, y) = x^3 y^2$, (b) $f(x, y) = e^{xy}$.

61 Let $M(x, y)$ and $N(x, y)$ be given. Define $f(a, b)$ to be $\int_0^b N(0, y)\,dy + \int_0^a M(x, b)\,dx$. This is the sum of an integral over the interval OP and an integral over the interval PQ, shown in Fig. 12.49. Show that $\partial f/\partial a = M(a, b)$.

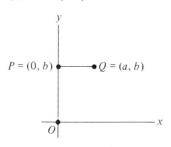

Figure 12.49

62 In Exercise 61 no relation between M and N was assumed. Now assume that $(\partial M/\partial y) = (\partial N/\partial x)$. Let f be defined as in Exercise 61.

(a) Show that

$$\frac{\partial f}{\partial b} = N(0, b) + \int_0^a \frac{\partial}{\partial b} M(x, b)\,dx.$$

(b) Show that

$$\int_0^a \frac{\partial}{\partial b} M(x, b)\,dx = \int_0^a \frac{\partial}{\partial x} N(x, b)\,dx.$$

(c) Why is $\int_0^a (\partial/\partial x) N(x, b)\,dx = N(a, b) - N(0, b)$?

(d) Deduce from (a), (b), and (c) that $\partial f/\partial b = N(a, b)$. Combining this exercise with the preceding one shows that, if $\partial M/\partial y = \partial N/\partial x$, then there is a function f such that $df = M\,dx + N\,dy$. It was assumed that M and N are defined throughout the xy plane. The proof would work if M and N were defined throughout some region accessible by paths of the form shown in Fig. 12.49, say a disk or rectangle.

DEFINITE INTEGRALS OVER PLANE AND SOLID REGIONS

In Chap. 5 the definite integral of a function f over an interval $[a, b]$ was discussed. We partitioned the interval into sections, chose a sampling point in each section, and then formed an approximating sum. The definite integral is the limit of these sums as the partitions are chosen finer and finer.

A similar procedure can be carried out for regions in the plane, such as triangles, disks, rectangles, etc. Such a region is partitioned into smaller regions, a sampling point selected in each of them, and an analogous sum formed. This leads to a definite integral over a plane region, a tool for computing volumes, centers of mass, moments of inertia, etc.

In Sec. 13.1 the definite integral over a plane region is defined and some of its applications described. Sections 13.2 to 13.4 show how to compute these integrals. Optional Secs. 13.5 and 13.6 develop the Jacobian, which plays a key role when working with coordinate systems other than the rectangular coordinate system. Sections 13.7 to 13.9 generalize the notion of a definite integral to integrals over sets in space, such as cubes and balls.

13.1 The definite integral of a function over a region in the plane

Two problems will introduce the definite integral of a function over a region in the plane.

A region in the plane shall be a set of points in the plane enclosed by a curve or polygon. When a region is partitioned into subsets, the smaller regions will also be bounded by curves or polygons.

PROBLEM 1   Estimate the volume of the solid S which we now describe. Above each point P in a 4-inch by 2-inch rectangle R erect a line segment whose length, in inches, is the square of the distance from P to the corner A. (R is shown

13.1 THE DEFINITE INTEGRAL OF A FUNCTION OVER A REGION IN THE PLANE

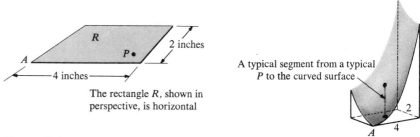

Figure 13.1

Figure 13.2

in Fig. 13.1.) These segments form a solid S which is shown in Fig. 13.2. Note that the highest point of S is above the corner of R opposite A; there its height, by the Pythagorean theorem, is $\sqrt{4^2 + 2^2} = \sqrt{20}$ inches.

APPROACH

To begin, observe that the volume of S is certainly less than $4 \cdot 2 \cdot \sqrt{20} = \sqrt{160}$ cubic inches, since S can be put into a box whose base has area $4 \cdot 2$ square inches and whose height is $\sqrt{20}$ inches.

In order to make more accurate estimates, cut the rectangular base into smaller pieces. For convenience, cut it into four congruent rectangles R_1, R_2, R_3, and R_4, as in Fig. 13.3. To estimate the volume of S, estimate the volume of that portion of S above each of the rectangles R_1, R_2, R_3, and R_4, and add these estimates. To do this, select a point in each of the four rectangles, say the center of each, and above each rectangle form a box whose height is the height of S above the center of the corresponding rectangle. (See Figs. 13.4 and 13.5.)

The set R is cut into smaller sets:

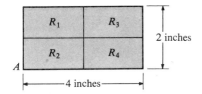

Figure 13.3

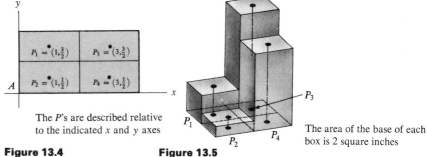

The P's are described relative to the indicated x and y axes

Figure 13.4

Figure 13.5

The area of the base of each box is 2 square inches

For instance, the height of the box above R_1 is $\sqrt{1^2 + (\frac{3}{2})^2} = \sqrt{\frac{13}{4}}$ inches, and the height of the box above R_2 is $\sqrt{1^2 + (\frac{1}{2})^2} = \sqrt{\frac{5}{4}}$ inches. Adding the volumes of the four boxes, we obtain the following estimate of the volume of S:

$$\sqrt{\frac{13}{4}} \cdot 2 + \sqrt{\frac{5}{4}} \cdot 2 + \sqrt{\frac{45}{4}} \cdot 2 + \sqrt{\frac{37}{4}} \cdot 2 = \sqrt{50} \text{ cubic inches.}$$

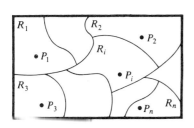

Figure 13.6

This is only an estimate. With the same partition of R we could make other estimates by choosing other P's to determine the heights of approximating boxes.

In general, to estimate the volume of S, begin by partitioning R into smaller subsets $R_1, R_2, \ldots, R_n$ and select points P_1 in R_1, P_2 in $R_2, \ldots, P_n$ in R_n. (See Fig. 13.6.)

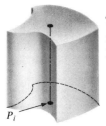

The cylinder has base R_i with area A_i, height $c(P_i)$, and volume $c(P_i) A_i$

Figure 13.7

A cylinder need not be a circular cylinder.

Denote the height of S above a typical point P_i in R_i by $c(P_i)$, and the area of R_i by A_i. Then

$$c(P_1)A_1 + c(P_2)A_2 + \cdots + c(P_n)A_n$$

is an estimate of the volume of S by a sum of the volumes of n solids; a typical one may look like the one in Fig. 13.7. Such a solid is called a cylinder. ∎

PROBLEM 2 Estimate the mass of the rectangular sheet R described as follows. Its dimensions are 4 centimeters by 2 centimeters. The material is sparse near the corner A and dense far from A. Indeed, assume that the density in the vicinity of any point P is numerically equal to the square of the distance from P to A (grams per square centimeter). See Fig. 13.8. Note that it is densest at the corner opposite A, where its density is $4^2 + 2^2 = 20$ grams per square centimeter.

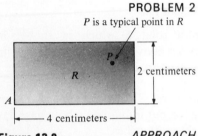

Figure 13.8

The region R is cut into smaller regions

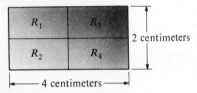

Figure 13.9

A point is chosen in each region of the partition

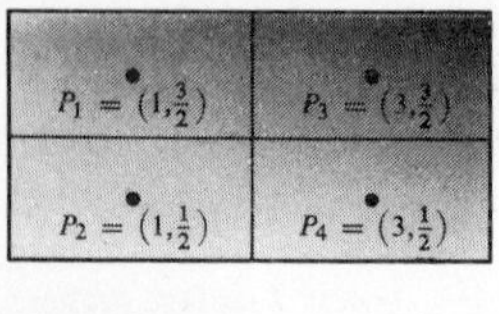

Figure 13.10

Though only estimates have been made, they show the similarity of the two problems.

APPROACH To begin, observe that the total mass is certainly less than $4 \cdot 2 \cdot 20 = 160$ grams, since the area of R is $4 \cdot 2$ square centimeters and the maximum density is 20 grams per square centimeter.

In order to make more accurate estimates, cut the rectangle into smaller pieces, to be specific, into the four congruent rectangles shown in Fig. 13.9. Estimate the total mass by estimating the mass in each of the rectangles R_1, R_2, R_3, and R_4. To do this, select a point, say the center, in each of the four rectangles and compute the density at each of these four points. (See Fig. 13.10.) The density at P_1 is $1^2 + (\frac{3}{2})^2 = \frac{13}{4}$ grams per square centimeter. As an estimate of the mass in R_1, we have $\frac{13}{4} \cdot 2$ grams, since the area of R_1 is 2 square centimeters. The sum of the estimates for each of the four rectangles,

$$\frac{13}{4}2 + \frac{5}{4}2 + \frac{45}{4}2 + \frac{37}{4}2 = 50 \text{ grams},$$

is an estimate of the total mass in R.

This is only an estimate. Just as for the volume in Prob. 1 other estimates can be made in the same way. To do so, partition R into small subsets $R_1, R_2, \ldots, R_n$ and select points P_1 in R_1, P_2 in R_2, ..., P_n in R_n. Denote the density at P_i by $f(P_i)$ and the area of R_i by A_i. Then

$$f(P_1)A_1 + f(P_2)A_2 + \cdots + f(P_n)A_n$$

is an estimate of the total mass. ∎

Even though we have found neither the volume in Problem 1 nor the mass in Problem 2, it is clear that, if we know the answer to one, we have the answer to the other: The arithmetic for calculating any estimate for the volume is the same as that for an estimate of the mass.

The similarity of the sums formed for both problems to the sums met in Chap. 5 suggests that the idea of the definite integral can be generalized from intervals $[a, b]$ to regions in the plane. First, in order to speak of "fine" partitions on the plane, two definitions are required.

DEFINITION *Diameter of a region.* Let S be a region in the plane bounded by a curve or polygon. The *diameter* of S is the largest distance between two points of S.

Note that the diameter of a square of side s is $s\sqrt{2}$, and the diameter of a circle whose radius is r is $2r$, its usual diameter.

DEFINITION *Mesh of a partition in the plane.* Let $R_1, R_2, \ldots, R_n$ be a partition of a region R in the plane into smaller regions. The mesh of this partition is the largest of the diameters of the regions $R_1, R_2, \ldots, R_n$.

The mesh of the partition used in both Problems 1 and 2 is $\sqrt{5}$.

THE DEFINITE INTEGRAL OVER A PLANE REGION

Now the definite integral over a plane region can be defined.

DEFINITION *Definite integral of a function f over a region R in the plane.* Let f be a function that assigns to each point P in a region R in the plane a number $f(P)$. Consider the typical sum

$$f(P_1)A_1 + f(P_2)A_2 + \cdots + f(P_n)A_n,$$

formed from a partition of R, where A_i is the area of R_i, and P_i is in R_i. If these sums approach a certain number as the mesh of the partitions shrinks toward 0 (no matter how P_i is chosen in R_i), that number is called *the definite integral of f over the set R* and is written

$$\int_R f(P)\, dA.$$

That is, in shorthand,

$$\lim_{\text{mesh} \to 0} \sum_{i=1}^{n} f(P_i)A_i = \int_R f(P)\, dA.$$

It is illuminating to compare this definition with the definition of the definite integral over an interval. Both are numbers that are approached by certain sums of products. The sums are formed in a similar manner, as the following table shows.

This table contrasts integration over an interval with integration over a planar region.

Given	For each subset in a partition compute	Select in each of the subsets	Take the limits of sums of the form
An interval and a function defined there	Its length $x_i - x_{i-1}$	A point (described by its coordinate c_i)	$\sum_{i=1}^{n} f(c_i)(x_i - x_{i-1})$
A set in the plane and a function defined there	Its area A_i	A point P_i	$\sum_{i=1}^{n} f(P_i)A_i$

The definite integral is not defined as a sum formed in this table, but rather as the number approached by these sums when the mesh approaches 0. The definite integral of f over R is sometimes called the *integral of f over R*, or the *integral of $f(P)$ over R*.

For example, the volume in Problem 1 and the mass in Problem 2 are both given by the definite integral

$$\int_R (x^2 + y^2) \, dA,$$

where R is the rectangle that has vertices, (0, 0), (4, 0), (0, 2), and (4, 2).

APPLICATIONS OF $\int_R f(P) \, dA$

The two illustrations given of the definite integral over a plane region are quite important, and we emphasize them by stating them in full generality.

VOLUME OF A SOLID EXPRESSED AS AN INTEGRAL OVER A PLANE REGION

Consider a solid set S and pick a line L in space. Assume that all lines parallel to L that meet S intersect S in a line segment or a point. Pick a plane perpendicular to L. Let R be the "shadow" or "projection" of S on that plane, that is, the set of all points where lines parallel to L that meet S intersect the plane. For each point P in R let $c(P)$ be the length of the intersection with S of the line through P parallel to L. (See Fig. 13.11.) Partition R into smaller regions $R_1, R_2, \ldots, R_n$ and pick a sampling point P_i in R_i for each $i = 1, 2, \ldots, n$. Let the area of R_i be A_i. Then approximate the volume of S above R_i by a cylinder of height $c(P_i)$ and base congruent to R_i, as shown in Fig. 13.12. The volume of the ith cylinder is

$$c(P_i) A_i.$$

Hence

$$\sum_{i=1}^{n} c(P_i) A_i$$

is an estimate of the volume of S. Thus

$$\boxed{\text{Volume of } S = \int_R c(P) \, dA.}$$

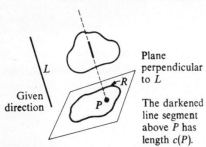

Plane perpendicular to L

Given direction

The darkened line segment above P has length $c(P)$.

Figure 13.11

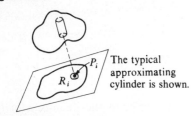

The typical approximating cylinder is shown.

Figure 13.12

MASS OF A FLAT REGION (LAMINA) EXPRESSED AS AN INTEGRAL OVER A PLANE REGION R

Consider a plane distribution of mass through a region R, as shown in Fig. 13.13. The density may vary throughout the region. Denote the density at P by $\delta(P)$ (in grams per square centimeter). To estimate the total

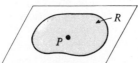

A plane distribution of matter. The density at P is $\delta(P)$ (grams per square centimeter).

Figure 13.13

mass in the region R, partition R into small regions $R_1, R_2, \ldots, R_n$ and pick a sampling point P_i in R_i for each $i = 1, 2, \ldots, n$. Then the mass in R_i is approximately $\delta(P_i)A_i$, since density times area gives mass if the density is constant. Thus

$$\sum_{i=1}^{n} \delta(P_i)A_i$$

is an estimate of the mass in R. Consequently,

δ is a standard symbol for density.

$$\text{Mass in } R = \int_R \delta(P)\, dA.$$

For engineers and physicists this is an important interpretation of the two-dimensional integral. The definite integral of density equals mass.

That the definite integral of density over an interval gives the total mass for matter distributed along a string was shown in Chap. 5. In Sec. 13.7 an analogous fact is shown to hold for matter distributed in space, once the definite integral over a region in space is defined.

To emphasize further the similarity between integrals over plane sets and integrals over intervals, we define the average of a function over a plane set.

DEFINITION *Average value.* The average value of f over the region R is

$$\frac{\int_R f(P)\, dA}{\text{Area of } R}.$$

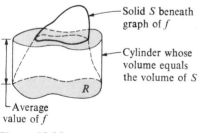

Figure 13.14

If $f(P)$ is positive for all P in R, a simple geometric interpretation of the average of f over R can be given. Let S be the solid situated below the graph of f (a surface) and above the region R. The average value of f over R is the height of the cylinder whose base is R and whose volume is the same as the volume of S. (See Fig. 13.14.) This is the analog of the observation in Sec. 8.8 concerning the average of $f(x)$ over $[a, b]$.

The integral $\int_R f(P)\, dA$ is called a *two-dimensional integral* to distinguish it from $\int_a^b f(x)\, dx$, which, for contrast, is called a *one-dimensional integral*.

Often $\int_R f(P)\, dA$ is denoted $\iint_R f(P)\, dA$, with the two integral signs emphasizing that the integral is over a plane set. However, the symbol dA, which calls to mind areas, is an adequate reminder.

The integral of the function $f(P) = 1$ over a region is of special interest. The typical approximating sum $\sum_{i=1}^{n} f(P_i)A_i$ then equals $\sum_{i=1}^{n} 1 A_i = A_1 + A_2 + \cdots + A_n$, which is the area of the region R that is being partitioned. Since *every* approximating sum has this same value, it follows that

$$\lim_{\text{mesh} \to 0} \sum_{i=1}^{n} f(P_i)A_i = \text{Area of } R.$$

The integral of the constant function, 1, gives area.

Consequently

$$\int_R 1 \, dA = \text{Area of } R.$$

This formula will come in handy on several occasions.

This table summarizes some of the main applications of the integral $\int_R f(P) \, dA$.

Integral	Interpretation
$\int_R 1 \, dA$	Area of R
$\int_R \delta(P) \, dA$, $\delta(P)$ = density	Mass of R
$\int_R c(P) \, dA$, $c(P)$ = cross section of a solid	Volume of the solid

PROPERTIES OF INTEGRALS   Integrals over plane regions have properties similar to those of integrals over intervals:

1. $\int_R cf(P) \, dA = c \int_R f(P) \, dA$ for any constant c.

2. $\int_R [f(P) + g(P)] \, dA = \int_R f(P) \, dA + \int_R g(P) \, dA$.

3. If $f(P) \le g(P)$ for all points P in R, then

$$\int_R f(P) \, dA \le \int_R g(P) \, dA.$$

4. If R is broken into two regions, R_1 and R_2, overlapping at most on their boundaries, then

$$\int_R f(P) \, dA = \int_{R_1} f(P) \, dA + \int_{R_2} f(P) \, dA.$$

Exercises

1. In the estimates for the volume in Problem 1, the centers of the subrectangles were used as the P_i's. Make an estimate for the volume in Problem 1 by using the same partition but taking as P_i (a) the lower left corner of each R_i, (b) the upper right corner of each R_i. (c) What do (a) and (b) tell about the volume of the solid?

2. Estimate the mass in Problem 2 using a partition of R into eight congruent squares and taking as the P_i's (a) centers, (b) upper right corners, (c) lower left corners.

3. Let R be a set in the plane whose area is A. Let f be the function such that $f(P) = 5$ for every point P in R.
   (a) What can be said about any approximating sum $\sum_{i=1}^{n} f(P_i) A_i$ formed for this R and this f?
   (b) What is the value of $\int_R f(P) \, dA$?

4. Let R be the square with vertices $(1, 1), (5, 1), (5, 5)$, and $(1, 5)$. Let $f(P)$ be the distance from P to the y axis.
   (a) Estimate $\int_R f(P) \, dA$ by partitioning R into four squares and using midpoints as sampling points.
   (b) Show that $16 \le \int_R f(P) \, dA \le 80$.

5. (a) Let f and R be as in Problem 2. Use the estimate

of $\int_R f(P)\,dA$ obtained in the text to estimate the average of f over R.
(b) Using the information from Exercise 2, show that the average is between 4 and 10.

6 Assume that, for all P in R, $m \leq f(P) \leq M$, where m and M are constants. Let A be the area of R. By examining approximating sums, show that

$$mA \leq \int_R f(P)\,dA \leq MA.$$

A calculator or tables would be of aid in Exercises 7 and 8.

7 (a) Let R be the rectangle with vertices $(0, 0)$, $(2, 0)$, $(2, 3)$, and $(0, 3)$. Let $f(x, y) = \sqrt{x + y}$. Estimate $\int_R \sqrt{x + y}\,dA$ by partitioning R into six squares and choosing the sampling points to be their centers.
(b) Use (a) to estimate the average value of f over R.

8 (a) Let R be the square with vertices $(0, 0)$, $(0.8, 0)$, $(0.8, 0.8)$, and $(0, 0.8)$. Let $f(P) = f(x, y) = e^{xy}$. Estimate $\int_R e^{xy}\,dA$ by partitioning R into 16 squares and choosing the sampling points to be their centers.
(b) Use (a) to estimate the average value of $f(P)$ over R.
(c) Show that $0.64 \leq \int_R f(P)\,dA \leq 0.64 e^{0.64}$.

9 (a) Let R be the triangle with vertices $(0, 0)$, $(4, 0)$, and $(0, 4)$, shown in Fig. 13.15.

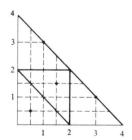

Figure 13.15

Let $f(x, y) = x^2 y$. Use the partition into four triangles and sampling points shown in the diagram to estimate $\int_R f(P)\,dA$.
(b) What is the maximum value of $f(x, y)$ in R?
(c) From (b) obtain an upper bound on $\int_R f(P)\,dA$.

10 This exercise involves estimating an integral by choosing points randomly.

A computing machine can be used to generate random numbers and thus random points in the plane which can be used to estimate definite integrals, as we now show. Say that a complicated region R lies in the square whose vertices are $(0, 0)$, $(2, 0)$, $(2, 2)$, and $(0, 2)$, and a complicated function f is defined in R. The machine generates 100 random points (x, y) in the square. Of these, 73 lie in R. The average value of f for these 73 points is 2.31.

(a) What is a reasonable estimate of the area of R?
(b) What is a reasonable estimate of the two-dimensional integral $\int_R f(P)\,dA$?

Techniques such as this one which utilize randomness are called *Monte Carlo methods*.

In Exercises 11 to 14 describe the general shape of the projection (shadow) of the given solid on the given plane.

11 A sphere; any plane.
12 A right circular cylinder; a plane parallel to the axis of the cylinder.
13 A right circular cone; a plane parallel to the axis of the cone.
14 A right circular cone; a plane parallel to the base.

15 The amount of rain that falls at point P during 1 year is $f(P)$ inches. Let R be some geographic region, and assume areas are measured in square inches.
(a) What is the meaning of $\int_R f(P)\,dA$?
(b) What is the meaning of

$$\frac{\int_R f(P)\,dA}{\text{Area of } R}?$$

16 *The Effects of Nuclear Weapons*, prepared by the U.S. Department of Defense, contains these remarks on the computation of the radioactive dose after a 1-kiloton atomic explosion.

> "If all the residues from 1-kiloton fission yield were deposited on a smooth surface in varying concentrations typical of an early fallout pattern, instead of uniformly, the product of the dose rate at 1 hour and the area would be replaced by the 'area integral' of the 1-hour dose rate defined by
>
> $$\text{Area Integral} = \int_A R_1\,dA,$$
>
> where R_1 is the 1-hour dose rate over an element of area dA and A square miles is the total area covered by the residues."

Explain why an integral is involved. (Note that in this case R_1 denotes the function and A the region. In the diverse applications of integrals many notations are employed.)

17 (a) Prove that, if the diameter of a set in the plane is d, then its area is less than d^2.
(b) Which has the largest area: a circle of diameter d, a square of diameter d, or an equilateral triangle of diameter d?
(c) How large an area do you think a set of diameter d can have?

(d) Can a circle of diameter d always be drawn to contain a given set of diameter d?

■ ■

18 If a square of side 1 is partitioned into n regions, what is the largest that the mesh can be? (It is not known in general how small the mesh can be. For $n = 9$, it can be as small as $\frac{5}{11}$.)

19 (a) Let R be a circle of radius 1. Let $f(P)$ be the distance from P to the center of the circle. By partitioning R with the aid of rays through the center and concentric circles, show that $\int_R f(P)\, dA = 2\pi/3$.

(b) Find the average value of f over R.

13.2 How to describe a plane region by coordinates

In Sec. 13.3 we shall evaluate definite integrals over plane regions by means of two integrals over intervals. The method requires a description of these regions in terms of a coordinate system. The present section is devoted to this aspect of analytic geometry. Examples illustrate the method first for rectangular coordinates and then for polar coordinates.

DESCRIBING R IN RECTANGULAR COORDINATES

EXAMPLE 1 Let R be the region bounded by $y = x^2$, the x axis, and the line $x = 2$. Describe R in terms of cross sections parallel to the y axis.

SOLUTION A glance at R in Fig. 13.16 shows that, for points (x, y) in R, x ranges from 0 to 2. To describe R completely, we shall describe the behavior of y for any x in the interval $[0, 2]$.

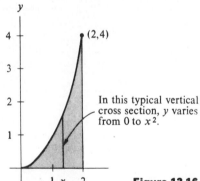

Figure 13.16

Hold x fixed and consider only the cross section above the point $(x, 0)$. It extends from the x axis to the curve $y = x^2$. For any x, the y coordinate varies from 0 to x^2. This is a complete description of R by vertical cross sections, written in compact notation:

$$0 \leq x \leq 2, \qquad 0 \leq y \leq x^2. \quad \blacksquare$$

EXAMPLE 2 Describe the region R of Example 1 by cross sections parallel to the x axis, that is, by horizontal cross sections.

SOLUTION A glance at R in Fig. 13.17 shows that y varies from 0 to 4. For any y in the interval $[0, 4]$, x varies from a smallest value $x_1(y)$ to a largest value

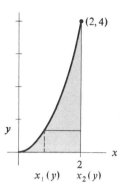

Figure 13.17

$x_2(y)$. Note that $x_2(y) = 2$ for each value of y in $[0, 4]$. To find $x_1(y)$, utilize the fact that the point $(x_1(y), y)$ is on the curve $y = x^2$,

$$x_1(y) = \sqrt{y}.$$

The description of R in terms of horizontal cross sections is

$$0 \leq y \leq 4, \qquad \sqrt{y} \leq x \leq 2. \quad \blacksquare$$

EXAMPLE 3 Describe the region R whose vertices are $(0, 0)$, $(6, 0)$, $(4, 2)$, and $(0, 2)$ by vertical cross sections and then by horizontal cross sections. (See Fig. 13.18.)

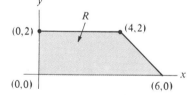

Figure 13.18

SOLUTION Clearly x varies between 0 and 6. For any x in the interval $[0, 4]$, y ranges from 0 to 2 (independently of x). For x in $[4, 6]$, y ranges from 0 to the value of y on the line through $(4, 2)$, and $(6, 0)$. This line has the equation

$$y = 6 - x.$$

The description of R by vertical cross sections therefore requires two separate statements:

$$0 \leq x \leq 4, \qquad 0 \leq y \leq 2;$$
and
$$4 \leq x \leq 6, \qquad 0 \leq y \leq 6 - x.$$

Use of horizontal cross sections provides a simpler description. First, y goes from 0 to 2. For each y in $[0, 2]$, x goes from 0 to the value of x on the line

$$y = 6 - x.$$

Solving this equation for x yields $x = 6 - y$.

The description is much shorter:

$$0 \leq y \leq 2, \qquad 0 \leq x \leq 6 - y. \quad \blacksquare$$

These three examples are typical. First determine the range of one coordinate and then how the other coordinate varies for any fixed value of the first coordinate.

DESCRIBING R IN POLAR COORDINATES

The method is the same when polar coordinates are used: First determine the range of θ, and then see how r varies for any fixed value of θ. In the descriptions keep r nonnegative. This will simplify the descriptions and the formulas for evaluating plane integrals with polar coordinates.

EXAMPLE 4   Let R be the circle of radius a and center at the pole of a polar coordinate system. (See Fig. 13.19.) Describe R in terms of cross sections by rays emanating from the pole.

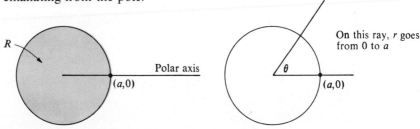

Figure 13.19   Figure 13.20

SOLUTION   To sweep out R, θ goes from 0 to 2π. Hold θ fixed and consider the behavior of r on the ray of angle θ. Clearly r goes from 0 to a, independently of θ. (See Fig. 13.20.) The complete description is

$$0 \leq \theta \leq 2\pi, \qquad 0 \leq r \leq a. \quad \blacksquare$$

EXAMPLE 5   Let R be the region between the circles $r = 2\cos\theta$ and $r = 4\cos\theta$. Describe R in terms of cross sections by rays from the pole. (See Fig. 13.21.)

SOLUTION   To sweep out this region, use the rays from $\theta = -\pi/2$ to $\theta = \pi/2$. For each such θ, r varies from $2\cos\theta$ to $4\cos\theta$. The complete description is

$$-\frac{\pi}{2} \leq \theta \leq \frac{\pi}{2}, \qquad 2\cos\theta \leq r \leq 4\cos\theta. \quad \blacksquare$$

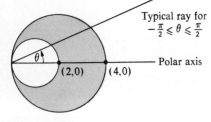

Figure 13.21

As Examples 4 and 5 suggest, polar coordinates provide simple descriptions for regions bounded by circles. The next example shows that polar coordinates may also provide simple descriptions of regions bounded by straight lines, especially if the lines pass through the origin.

EXAMPLE 6   Let R be the triangular region whose vertices, in rectangular coordinates, are $(0, 0)$, $(1, 1)$, and $(0, 1)$. Describe R in polar coordinates.

SOLUTION   Inspection of R in Fig. 13.22 shows that θ varies from $\pi/4$ to $\pi/2$. For each θ, r goes from 0 until the point (r, θ) is on the line $y = 1$, that is, on the

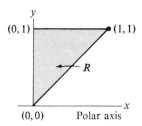

Figure 13.22

line $r \sin \theta = 1$. Thus the upper limit of r for each θ is $1/\sin \theta$. The description of R is:

$$\frac{\pi}{4} \leq \theta \leq \frac{\pi}{2}, \qquad 0 < r \leq \frac{1}{\sin \theta}. \quad \blacksquare$$

The three descriptions by cross sections

In practice, the region R often has the property that its intersection with a horizontal or vertical line or a ray from some pole is a line segment. This greatly simplifies the description of R if the appropriate coordinates are used.

If vertical cross sections are used, the description will be of the form

$$a \leq x \leq b, \qquad y_1(x) \leq y \leq y_2(x),$$

where $y_1(x)$ and $y_2(x)$ are functions of x (perhaps constant). In the case of horizontal cross sections, the description will have the form

$$c \leq y \leq d, \qquad x_1(y) \leq x \leq x_2(y).$$

Cross sections by rays will lead to this type of description:

$$\alpha \leq \theta \leq \beta, \qquad r_1(\theta) \leq r \leq r_2(\theta).$$

AREAS IN RECTANGULAR AND POLAR COORDINATES: A CONTRAST

Consider all points (x, y) in the plane that satisfy the inequalities

$$x_1 \leq x \leq x_2 \quad \text{and} \quad y_1 \leq y \leq y_2,$$

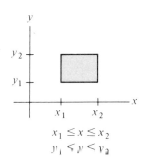

Figure 13.23

This formula will be contrasted with formula (2).

where $x_1, x_2, y_1,$ and y_2 are fixed numbers. This set is the rectangle of sides $x_2 - x_1$ and $y_2 - y_1$ shown in Fig. 13.23. The area of this rectangle is $(x_2 - x_1)(y_2 - y_1)$. Thus the area of the region consisting of all points whose x coordinates are between x and $x + \Delta x$ and whose y coordinates are between y and $y + \Delta y$ is the product

$$\Delta x \, \Delta y. \qquad (1)$$

Next consider the analogous figure for polar coordinates, namely the set consisting of all the points (r, θ) in the plane such that

$$r_1 \leq r \leq r_2 \quad \text{and} \quad \theta_1 \leq \theta \leq \theta_2,$$

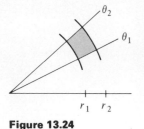

Figure 13.24

where r_1, r_2, θ_1, and θ_2 are fixed numbers, as shown in Fig. 13.24. The area of the region shaded in Fig. 13.24 may be computed as follows. The area of the ring between the circle of radius r_1 and the circle of radius r_2 is $\pi r_2^2 - \pi r_1^2$. The portion of this ring that lies between the rays $\theta = \theta_1$ and $\theta = \theta_2$ has area

$$\frac{\theta_2 - \theta_1}{2\pi}(\pi r_2^2 - \pi r_1^2)$$

which equals

$$\frac{\theta_2 - \theta_1}{2\pi}\pi(r_2 + r_1)(r_2 - r_1) = \frac{r_2 + r_1}{2}(r_2 - r_1)(\theta_2 - \theta_1).$$

Thus the region consisting of all points (r, θ) whose r coordinates are between r and $r + \Delta r$ and whose θ coordinates are between θ and $\theta + \Delta \theta$ is

$$\frac{(r + \Delta r) + r}{2} \Delta r \, \Delta \theta,$$

or

$$\left(r + \frac{\Delta r}{2}\right) \Delta r \, \Delta \theta.$$

When Δr and $\Delta \theta$ are both small, then $\Delta r/2$ is small in comparison to r; the area is approximately

The coefficient r is needed; contrast $r \, \Delta r \, \Delta \theta$ with $\Delta x \, \Delta y$.

$$r \, \Delta r \, \Delta \theta. \tag{2}$$

Recall that radians involve "length divided by length."

The contrast between (1) and (2) will be important in Sec. 13.4. In polar coordinates a coefficient r appears. The area is not just "$\Delta r \, \Delta \theta$." Indeed "$\Delta r \, \Delta \theta$" could not be right, since it does not even have the units of area. After all, $\Delta \theta$, the ratio of lengths, is a pure dimensionless number; Δr has the units of length. So $\Delta r \, \Delta \theta$ has the units of length. On the other hand, $r \, \Delta r \, \Delta \theta$ has the units of area since both r and Δr have the units of length.

Exercises

1 (a) Draw the triangular region whose vertices are $(0, 0)$, $(2, 1)$, and $(0, 1)$.
   (b) Describe it by vertical cross sections.
2 Describe the region in Exercise 1 by horizontal cross sections.
3 (a) Draw the region between $y = x$ and $y = x^2$, $0 \leq x \leq 1$.
   (b) Describe it by vertical cross sections.
4 Describe the region in Exercise 3 by horizontal cross sections.
5 Describe the circle in Example 4 by vertical cross sections (placing the xy-coordinate system in standard position).
6 Describe the region in Example 5 by (a) vertical cross sections, (b) horizontal cross sections.

In Exercises 7 to 12 draw the regions described.

7 $0 \leq x \leq 2$, $2x \leq y \leq 3x$.
8 $0 \leq x \leq 1$, $x^3 \leq y \leq x^2$.
9 $1 \leq y \leq e$, $\ln y \leq x \leq 1$.
10 $0 \leq \theta \leq \frac{\pi}{2}$, $0 \leq r \leq 2$.

11 $-\frac{\pi}{4} \leq \theta \leq \frac{\pi}{4}$, $1 \leq r \leq 2$.

12 $0 \leq \theta \leq 2\pi$, $1 \leq r \leq 2 + \cos \theta$.

13 Describe the region in Exercise 7 by horizontal cross sections.

14 Describe the region in Exercise 8 by horizontal cross sections.

15 Describe the region in Exercise 9 by vertical cross sections.

16 (a) Draw the region R whose description is $-2 \leq y \leq 2$, $-\sqrt{4 - y^2} \leq x \leq \sqrt{4 - y^2}$.
   (b) Describe R by vertical cross sections.
   (c) Describe R in polar coordinates.

17 (a) Draw the region R bounded by the four lines $y = 1$, $y = 2$, $y = x$, $y = x/3$.
   (b) Describe R in terms of cross sections. (Choose the direction that is most convenient.)

18 Let R be bounded by the ellipse $x^2/a^2 + y^2/b^2 = 1$. Describe the region R by (a) vertical cross sections, (b) horizontal cross sections.

13.3 Computing $\int_R f(P)\, dA$ by introducing rectangular coordinates in R

This section gives an intuitive development of formulas for the rapid computation of definite integrals over plane regions. Such questions as, "What properties of f and R ensure the existence of $\int_R f(P)\, dA$?" will not concern us. Just as the reasoning in Sec. 5.4 made the fundamental theorem of calculus plausible, the reasoning in this chapter will be more persuasive than rigorous. No proofs will be included. We were able to discuss the proof of the fundamental theorem of calculus because an interval $[a, b]$ offers less complication than a region R in the plane. It suffices to say that, if R is bounded by fairly simple curves and f is continuous, then the various formulas are valid.

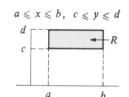

Figure 13.25

We first develop a method for computing a definite integral over a rectangle. After applying this formula in Example 1, we make the slight modification needed to evaluate integrals over more general regions.

Consider a rectangular region R whose description by vertical cross sections is

$$a \leq x \leq b, \qquad c \leq y \leq d,$$

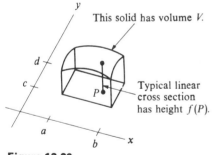

Figure 13.26

as shown in Fig. 13.25. If $f(P) \geq 0$ for all P in R, then $\int_R f(P)\, dA$ is the volume V of the solid whose base is R and which has, above P, a linear cross section of height $f(P)$. (See Fig. 13.26.) Let $A(x)$ be the area of the cross section made by a plane perpendicular to the x axis and having abscissa x, as in Fig. 13.27. As was shown in Sec. 5.3,

$$V = \int_a^b A(x)\, dx.$$

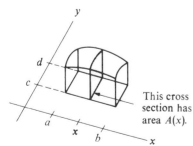

Figure 13.27

But the area $A(x)$ is itself expressible as a definite integral:

$$A(x) = \int_c^d f(x, y)\, dy.$$

Note that x is held fixed throughout this integration. This reasoning provides a repeated integral whose value is $V = \int_R f(P)\, dA$, namely,

828 DEFINITE INTEGRALS OVER PLANE AND SOLID REGIONS

$$\int_R f(P) \, dA = V$$
$$= \int_a^b A(x) \, dx$$
$$= \int_a^b \left[\int_c^d f(x, y) \, dy \right] dx.$$

An integral over a rectangle expressed as a repeated integral

Of course, cross sections by planes perpendicular to the y axis could be used. Then similar reasoning shows that

$$\int_R f(P) \, dA = \int_c^d \left[\int_a^b f(x, y) \, dx \right] dy.$$

The quantities $\int_a^b [\int_c^d f(x, y) \, dy] \, dx$ and $\int_c^d [\int_a^b f(x, y) \, dx] \, dy$ are called *double integrals*, *repeated integrals*, or *iterated integrals*. Usually the brackets are omitted and the integrals are written $\int_a^b \int_c^d f(x, y) \, dy \, dx$ and $\int_c^d \int_a^b f(x, y) \, dx \, dy$.

The order of dx and dy matters; the differential that is on the left tells which integration is performed first.

EXAMPLE 1 Compute the definite integral $\int_R f(P) \, dA$, where R is the rectangle shown in Fig. 13.28, and the function f is defined by $f(P) = \overline{AP}^2$. (This integral was estimated in Sec. 13.1.)

SOLUTION Introduce xy coordinates in the convenient manner depicted in Fig. 13.29. Then f has this description in rectangular coordinates:

$$f(x, y) = \overline{AP}^2 = x^2 + y^2.$$

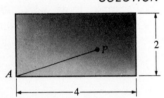

Figure 13.28

To describe R, observe that x takes all values from 0 to 4 and that for each x the number y takes all values between 0 and 2. Thus

$$\int_R f(P) \, dA = \int_0^4 \left[\int_0^2 (x^2 + y^2) \, dy \right] dx.$$

We must first compute $\int_0^2 (x^2 + y^2) \, dy$,

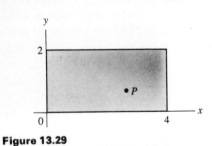

Figure 13.29

where x is fixed in [0, 4]. (This is the cross-sectional area $A(x)$.) To apply the fundamental theorem of calculus, first find a function $F(x, y)$ such that

$$\frac{\partial F}{\partial y} = x^2 + y^2.$$

Keep in mind that x is constant during this first integration.

$$F(x, y) = x^2 y + \frac{y^3}{3}$$

is such a function. The appearance of x in its formula should not disturb us, since x is fixed for the time being. By the fundamental theorem of calculus,

The notation $|_{y=0}^{y=2}$ tells which of the two variables is replaced by 0 and 2.

$$\int_0^2 (x^2 + y^2)\, dy = \left(x^2 y + \frac{y^3}{3}\right)\Bigg|_{y=0}^{y=2}$$

$$= \left[x^2 \cdot 2 + \frac{2^3}{3}\right] - \left[x^2 \cdot 0 + \frac{0^3}{3}\right]$$

$$= 2x^2 + \tfrac{8}{3}.$$

[The integral $2x^2 + \tfrac{8}{3}$ is the area $A(x)$ discussed earlier in this section.]

Now compute
$$\int_0^4 (2x^2 + \tfrac{8}{3})\, dx.$$

By the fundamental theorem of calculus,

$$\int_0^4 \left(2x^2 + \frac{8}{3}\right) dx = \left(\frac{2x^3}{3} + \frac{8x}{3}\right)\Bigg|_0^4 = \frac{160}{3}.$$

Hence the two-dimensional definite integral has the value $\tfrac{160}{3}$. The volume of the region in Problem 1 of Sec. 13.1 is $\tfrac{160}{3}$ cubic inches. The mass in Problem 2 is $\tfrac{160}{3}$ grams. ∎

The repeated integrals for $\int_R f(P)\, dA$ in rectangular coordinates

If R is not a rectangle, the repeated integral that equals $\int_R f(P)\, dA$ differs from that for the case where R is a rectangle only in the intervals of integration. If R has the description

$$a \le x \le b, \qquad y_1(x) \le y \le y_2(x),$$

by cross sections parallel to the y axis, then

$$\boxed{\int_R f(P)\, dA = \int_a^b \left[\int_{y_1(x)}^{y_2(x)} f(x, y)\, dy\right] dx.}$$

Similarly, if R has the description

$$c \le y \le d, \qquad x_1(y) \le x \le x_2(y),$$

by cross sections parallel to the x axis, then

$$\boxed{\int_R f(P)\, dA = \int_c^d \left[\int_{x_1(y)}^{x_2(y)} f(x, y)\, dx\right] dy.}$$

The intervals of integration are determined by R; the function f influences only the integrand.

The next example illustrates the method.

830 DEFINITE INTEGRALS OVER PLANE AND SOLID REGIONS

EXAMPLE 2 A triangular lamina is located as in Fig. 13.30. Its density at (x, y) is e^{y^2}. Find its mass, that is,

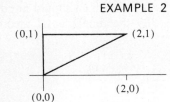

Figure 13.30

$$\int_R f(P) \, dA,$$

where $f(x, y) = e^{y^2}$.

SOLUTION The description of R by vertical cross sections is

$$0 \le x \le 2, \quad \frac{x}{2} \le y \le 1.$$

Hence
$$\int_R f(P) \, dA = \int_0^2 \left(\int_{x/2}^1 e^{y^2} \, dy \right) dx.$$

Unfortunately, the fundamental theorem of calculus is useless in computing

$$\int_{x/2}^1 e^{y^2} \, dy,$$

so we try horizontal cross sections instead.

The description of R is now

$$0 \le y \le 1, \quad 0 \le x \le 2y.$$

Thus
$$\int_R f(P) \, dA = \int_0^1 \left(\int_0^{2y} e^{y^2} \, dx \right) dy.$$

The first integration $\int_0^{2y} e^{y^2} \, dx$ is easy, since y is fixed; the integrand is constant. Thus

$$\int_0^{2y} e^{y^2} \, dx = e^{y^2} \int_0^{2y} 1 \, dx = e^{y^2} x \Big|_{x=0}^{x=2y}$$
$$= e^{y^2} 2y.$$

The second definite integral in the repeated integral is thus $\int_0^1 e^{y^2} 2y \, dy$, which luckily can be evaluated by the fundamental theorem of calculus, since $d(e^{y^2})/dy = e^{y^2} 2y$:

$$\int_0^1 e^{y^2} 2y \, dy = e^{y^2} \Big|_0^1 = e^{1^2} - e^{0^2} = e - 1.$$

The total mass is $e - 1$. ∎

Notice that computing a definite integral over R involves, first, a wise choice of an xy-coordinate system; second, a description of R and f relative to this coordinate system; and finally, the computation of two successive definite integrals over intervals. The order of these integrations should be

13.3 COMPUTING $\int_R f(P)\,dA$ BY INTRODUCING RECTANGULAR COORDINATES IN R

considered carefully since computation may be much simpler in one than in the other. This order is determined by the description of R by cross sections. For instance, if the description is in the form

$$a \le x \le b, \qquad y_1(x) \le y \le y_2(x),$$

then the repeated integral has the form

$$\int_a^b \left[\int_{y_1(x)}^{y_2(x)} f(x, y)\,dy \right] dx,$$

and the y integration is performed first.

EXAMPLE 3 Let R be the region in the plane bounded by $y = x^2$, $x = 2$, and $y = 0$. At each point $P = (x, y)$ erect a line segment of height $3xy$. What is the volume of the resulting solid? (See Fig. 13.31.)

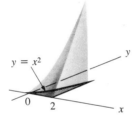

Figure 13.31

SOLUTION

R is discussed in Example 2 of Sec. 13.2.

The volume is $\int_R f(P)\,dA$, where $f(P)$ is the length of the line segment above P cut off by the solid. In this case, this integral is $\int_R 3xy\,dA$.

If cross sections parallel to the y axis are used, then R is described by

$$0 \le x \le 2, \qquad 0 \le y \le x^2.$$

Thus

$$\int_R 3xy\,dA = \int_0^2 \left(\int_0^{x^2} 3xy\,dy \right) dx,$$

which is easy to compute. First with x fixed,

$$\int_0^{x^2} 3xy\,dy = \left(3x \frac{y^2}{2} \right) \Big|_{y=0}^{y=x^2}$$

$$= 3x \frac{(x^2)^2}{2} - 3x \frac{0^2}{2}$$

$$= \frac{3x^5}{2}.$$

Then

$$\int_0^2 \frac{3x^5}{2}\,dx = \frac{3x^6}{12} \Big|_0^2 = 16.$$

R can also be described in terms of cross sections parallel to the x axis:

$$0 \le y \le 4, \quad \sqrt{y} \le x \le 2.$$

Then
$$\int_R 3xy\, dA = \int_0^4 \left(\int_{\sqrt{y}}^2 3xy\, dx \right) dy,$$

which, as the reader may verify, equals 16. ∎

MOMENTS AND CENTERS OF MASS

Review the definitions in Sec. 8.5.

In Sec. 8.5 moments, balancing lines, and the center of mass for a plane region furnished with a *constant density* (a *homogeneous* distribution of mass) were studied. Now, with the aid of integrals over plane sets, these concepts can be extended to cases where the density is not constant. Moreover, these integrals enable us to establish properties of moments and centers of mass. (See Exercise 33 for instance.)

Let L be a line in the plane of the region R, which is furnished with a distribution of matter that has density $\delta(P)$ at the point P. (Density may be measured in grams per square centimeter.) Introduce an x axis perpendicular to L with its origin on L, as in Fig. 13.32. Consider the moment of the mass in a small region of area ΔA. Let P be a point in the small region and let its x coordinate be x.

The mass in the small region is approximately

$$\underbrace{\delta(P)}_{\text{Density}} \underbrace{\Delta A}_{\text{Area}}.$$

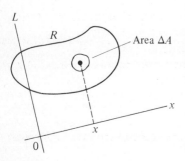

Figure 13.32

The lever arm of the mass around L is approximately x. So the moment about L of this small bit of mass is approximately

$$\underbrace{x}_{\text{Lever}} \underbrace{\delta(P)\, \Delta A}_{\text{Mass}}. \tag{1}$$

The local estimate (1) suggests the following definition.

DEFINITION *First moment.* The first moment of a plane distribution of matter around a line L in the plane is the integral

$$\int_R x\, \delta(P)\, dA.$$

Here $\delta(P)$ is the density of the matter at P and x is the coordinate of P measured on an axis perpendicular to L.

Frequently the term "first moment" is shortened to "moment."

Various (first) moments

The first moment is usually computed around either the x axis or the y axis. The moment around the x axis is denoted M_x:

$$M_x = \int_R y\, \delta(P)\, dA.$$

13.3 COMPUTING $\int_R f(P)\,dA$ BY INTRODUCING RECTANGULAR COORDINATES IN R

M_x has a y in the integrand; M_y has an x.

(In this case L is the x axis and the lever arms are measured along the y axis.) The moment around the y axis is denoted M_y:

$$M_y = \int_R x\,\delta(P)\,dA.$$

The center of mass for any continuous density function $\delta(P)$ is defined as in the case of a constant density, discussed in Sec. 8.5.

DEFINITION *Center of mass.* Let R be a region in the xy plane furnished with a density function $\delta(P)$. The center of mass $(\bar{x}, \bar{y})$ is defined by the formulas

$$\bar{x} = \frac{\int_R x\delta(P)dA}{\int_R \delta(P)dA} = \frac{\text{Moment around } y \text{ axis}}{\text{Total mass}}$$

and

$$\bar{y} = \frac{\int_R y\delta(P)dA}{\int_R \delta(P)dA} = \frac{\text{Moment around } x \text{ axis}}{\text{Total mass}}.$$

If $\delta = 1$, the center of mass is often called the *centroid of R*.

Moment of inertia

The *moment of inertia*, which is of use in studying the rotation of mass around a line, is defined as follows:

DEFINITION The *moment of inertia* of the mass in R around the x axis is

$$I_x = \int_R y^2\,\delta(P)\,dA.$$

The *moment of inertia* is of use in the study of an object spinning around an axis. Consider a point mass m located at a distance y centimeters from the x axis. Assume that it is spinning around the axis with angular velocity ω radians per second. Its speed is thus $y\omega$ centimeters per second and its kinetic energy is defined to be $\frac{1}{2}m(y\omega)^2$ ergs.

Consider instead of a point mass a planar distribution of mass spinning around the x axis at ω radians per second. The mass in a small region of area ΔA is approximately $\delta(P)\,\Delta A$ grams, where $\delta(P)$ is the density at some point P in the region. If the point P is a distance y from the x axis, the kinetic energy of the small portion of mass is approximately $\frac{1}{2}\delta(P)\,\Delta A(y\omega)^2$ ergs. This suggests that the kinetic energy of the mass in the region R is

$$\int_R \tfrac{1}{2}(y\omega)^2\,\delta(P)\,dA \text{ ergs}$$

or

$$\text{Kinetic energy} = \frac{\omega^2}{2}\int_R y^2\,\delta(P)\,dA \text{ ergs}.$$

Observe that the quantity $\int_R y^2\,\delta(P)\,dA$ is the moment of inertia of the mass around the x axis, denoted I_x. Thus

$$\text{Kinetic energy} = \frac{I_x \omega^2}{2} \text{ ergs}.$$

The larger the moment of inertia of a wheel is around its axle, the more work is required to bring it up to a certain angular velocity ω.

Similarly the *moment of inertia around the y axis* is

I_x has a y^2 in the integrand, I_y has an x^2, and I_z has both.

$$I_y = \int_R x^2 \, \delta(P) \, dA.$$

The moment of inertia around the z axis (the "polar moment of inertia") is

$$I_z = \int_R (x^2 + y^2) \, \delta(P) \, dA.$$

The integrand in each case is "(square of distance to axis) times (density)." Note that the moment of inertia is positive since the integrand is positive. A moment of inertia around any line in the xy plane can be defined similarly.

EXAMPLE 4 A lamina (a thin flat object), occupies the triangle whose vertices are $(0, 0)$, $(1, 0)$, and $(1, 2)$ and has density $\delta = x^2 y$ at the point (x, y). (See Fig. 13.33.) Find its center of mass.

SOLUTION In this case it will be necessary to compute three definite integrals:

$$\text{Mass} = \int_R x^2 y \, dA,$$

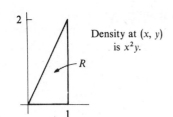

Figure 13.33

and the first moments,

$$M_y = \int_R x \delta \, dA = \int_R x \cdot x^2 y \, dA = \int_R x^3 y \, dA,$$

and

$$M_x = \int_R y \delta \, dA = \int_R y \cdot x^2 y \, dA = \int_R x^2 y^2 \, dA.$$

To evaluate these integrals, describe R by, say, vertical cross sections,

$$0 \le x \le 1, \qquad 0 \le y \le 2x.$$

Then

$$\text{Mass} = \int_R x^2 y \, dA = \int_0^1 \left(\int_0^{2x} x^2 y \, dy \right) dx.$$

The first integration goes as follows:

$$\int_0^{2x} x^2 y \, dy = \left. \frac{x^2 y^2}{2} \right|_{y=0}^{y=2x}$$

$$= \frac{x^2(2x)^2}{2} - \frac{x^2 0^2}{2}$$

$$= 2x^4.$$

The second integration gives

$$\int_0^1 2x^4 \, dx = \frac{2x^5}{5}\bigg|_0^1 = \frac{2}{5}.$$

Thus
$$\text{Mass} = \int_R x^2 y \, dA = \frac{2}{5}.$$

Next, compute $\int_R x^3 y \, dA = \int_0^1 (\int_0^{2x} x^3 y \, dy) \, dx$. The inner integral (the one in parentheses) is

$$\int_0^{2x} x^3 y \, dy = \frac{x^3 y^2}{2}\bigg|_{y=0}^{y=2x}$$

$$= \frac{x^3(2x)^2}{2} - \frac{x^3 0^2}{2}$$

$$= 2x^5.$$

The outer integral is then $\int_0^1 2x^5 \, dx = \frac{2x^6}{6}\bigg|_0^1 = \frac{1}{3}.$

Thus
$$\int_R x^3 y \, dA = \frac{1}{3}.$$

Consequently
$$\bar{x} = \frac{\frac{1}{3}}{\frac{2}{5}} = \frac{1}{3}\frac{5}{2} = \frac{5}{6}.$$

Next, compute $\int_R x^2 y^2 \, dA$. This equals $\int_0^1 (\int_0^{2x} x^2 y^2 \, dy) \, dx$. The inner integral is

$$\int_0^{2x} x^2 y^2 \, dy = \frac{x^2 y^3}{3}\bigg|_{y=0}^{y=2x}$$

$$= \frac{8x^5}{3}.$$

The outer integral is
$$\int_0^1 \frac{8x^5}{3} \, dx = \frac{8x^6}{18}\bigg|_0^1$$

$$= \frac{4}{9}.$$

Hence
$$\bar{y} = \frac{\frac{4}{9}}{\frac{2}{5}} = \frac{4}{9}\frac{5}{2} = \frac{10}{9}.$$

The center of mass is therefore

$$(\bar{x}, \bar{y}) = \left(\tfrac{5}{6}, \tfrac{10}{9}\right). \quad \blacksquare$$

EXAMPLE 5 A homogeneous rectangular plate has mass M and sides of lengths a and b. Find the moment of inertia of the plate around a line through its center and parallel to the side of length a.

SOLUTION The rectangle is shown in Fig. 13.34 along with x and y axes chosen parallel to the sides of the rectangle and with the origin at the center of the rectangle. The moment about the x axis is

$$I_x = \int_R y^2 \delta(P) \, dA.$$

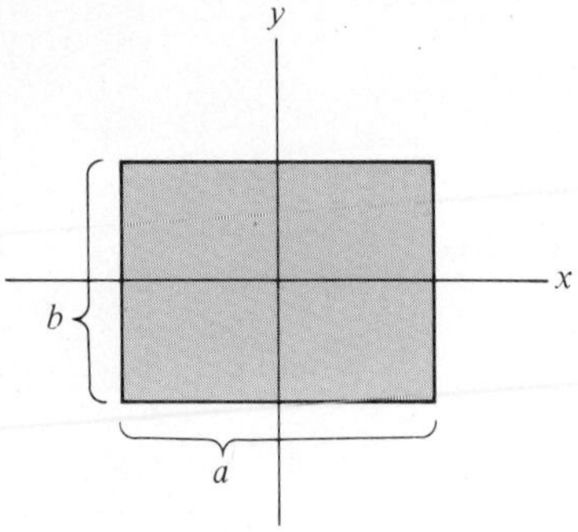

Since the density is constant, $\delta(P) = $ Total mass/Area $= M/ab$. Thus

$$I_x = \frac{M}{ab} \int_R y^2 \, dA.$$

To evaluate $\int_R y^2 \, dA$, first note by the symmetry of the rectangle and the fact that $(-y)^2 = y^2$ that it suffices to compute the integral of y^2 over the part of R in the first quadrant, R_I, and multiply by 4:

$$\int_R y^2 \, dA = 4 \int_{R_I} y^2 \, dA.$$

A repeated integration then gives

$$\int_{R_I} y^2 \, dA = \int_0^{a/2} \left(\int_0^{b/2} y^2 \, dy \right) dx$$

$$= \int_0^{a/2} \frac{y^3}{3} \bigg|_{y=0}^{y=b/2} dx$$

$$= \int_0^{a/2} \frac{b^3}{24} \, dx$$

$$= \frac{a}{2} \frac{b^3}{24}$$

$$= \frac{ab^3}{48}$$

Thus
$$I_x = \frac{M}{ab} \cdot 4 \cdot \frac{ab^3}{48} = \frac{Mb^2}{12}. \quad \blacksquare$$

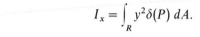

Figure 13.34

Recall that "homogeneous" means density is constant.

Exercises

In Exercises 1 to 6 evaluate the repeated integrals.

1 $\int_0^1 \left(\int_0^x (x + 2y) \, dy \right) dx$

2 $\int_1^2 \left(\int_x^{2x} dy \right) dx$

3 $\int_0^2 \left(\int_0^{x^2} xy^2 \, dy \right) dx$

4 $\int_1^2 \left(\int_0^y e^{x+y} \, dx \right) dy$

5 $\int_1^2 \left(\int_0^{\sqrt{y}} yx^2 \, dx \right) dy$

6 $\int_0^1 \left(\int_0^x y \sin \pi x \, dy \right) dx$

In each of Exercises 7 to 14 find the volume of the solid whose cross section by a line perpendicular to the xy plane has the given length, and whose projection on the xy plane is the given region R.

7 $3x$; R is between $x=0$ and $x=1$ and is bounded by $y=x^2$ and $y=x^3$.
8 $x+2y$; R is the triangle whose vertices are $(0,0)$, $(1,0)$, $(1,1)$.
9 xy; R is the finite region bounded by $y=x^2$, the x axis, and the line $x=2$.
10 $x+3y$; R is the finite region bounded by $y=x^2$, the x axis, and the line $x=2$.
11 x^2+2y^2; R is the parallelogram bounded by the lines $y=1$, $y=2$, $y=1+x$, and $y=x$.
12 $x+y$; R is the trapezoid bounded by the lines $x=0$, $y=1$, $y=2$, and $y=x$.
13 y; R is the portion of the circle of radius a, center at $(0,0)$ above the x axis.
14 xy; R is bounded by $y=\ln x$, the x axis, the y axis, and the line $y=1$.
15 Find the mass of a thin lamina occupying the finite region bounded by $y=2x^2$ and $y=5x-3$ and whose density at (x,y) is xy.
16 Find the mass of a thin lamina occupying the triangle whose vertices are $(0,0)$, $(1,0)$, $(1,1)$ and whose density at (x,y) is $1/(1+x^2)$.

In each of Exercises 17 to 22 find the center of mass of the lamina occupying the indicated region and having the indicated density.

17 The square whose vertices are $(0,0)$, $(1,0)$, $(1,1)$, $(0,1)$; density at (x,y) equal to $y \tan^{-1} x$.
18 The finite region bounded by $y=1+x$ and $y=2^x$; density at (x,y) is $x+y$.
19 The triangle whose vertices are $(0,0)$, $(1,2)$, $(1,3)$; density at (x,y) is xy.
20 The top half of the circle of radius 1, center $(0,0)$; density at (x,y) is $x+y+3$.
21 The finite region bounded by $y=x^2$, the x axis, and $x=2$; density at (x,y) is e^x.
22 The finite region bounded by $y=x^2$ and $y=x+6$, situated to the right of the y axis; density at (x,y) is $2x$.

■

23 Let $f(x,y) = y^2 e^{y^2}$ and let R be the triangle bounded by $y=a$, $y=x/2$, and $y=x$. Assume that a is positive.

(a) Set up two repeated integrals for $\int_R f(P) \, dA$.
(b) Evaluate the easier one.

24 Let R be the finite region bounded by the curve $y=\sqrt{x}$ and the line $y=x$. Let $f(x,y) = (\sin y)/y$ if $y \neq 0$ and $f(x,0)=1$. Compute $\int_R f(P) \, dA$.

In Exercises 25 and 26 replace the given repeated integral by another of the form

$$\int_a^b \left(\int_{x_1(y)}^{x_2(y)} f(x,y) \, dx \right) dy.$$

25 $\int_0^2 \left(\int_0^{x^2} x^3 y \, dy \right) dx.$

26 $\int_0^{\pi/2} \left(\int_0^{\cos x} x^2 \, dy \right) dx.$

27 In Sec. 8.5 it was shown that if the density is 1, then the moment about the x axis of the region under $y=f(x)$ and above $[a,b]$ is given by the integral $\int_a^b [(f(x))^2/2] \, dx$. Obtain that result by evaluating an appropriate plane integral by an iterated integral.

28 A homogeneous flat piece of sheet metal has mass M and forms a square of side a.
(a) Find its moment of inertia around the line perpendicular to the metal and passing through its center.
(b) Find its moment of inertia around a line perpendicular to the metal and passing through one of its four corners.

29 A flat piece of homogeneous sheet metal has mass M and forms a disk of radius a.
(a) Find its moment of inertia around the line perpendicular to the metal and passing through its center.
(b) Find its moment of inertia around a line perpendicular to the metal and passing through its border.
(c) Find its moment of inertia around a diameter (that is, a line in the plane of the metal and passing through its center).
(d) Find its moment of inertia around a tangent line.

30 A flat piece of metal has mass M and forms a semicircle of radius a. The density at a distance y from its diameter is $2y$.
(a) Find the total mass.
(b) Find the moment around its diameter.
(c) Describe its center of mass.
(d) Find the moment of inertia around its diameter.
(e) Find the moment of inertia around the line perpendicular to the metal and passing through the midpoint of its diameter.

■ ■

31 Show that the moment of inertia around the z axis of a plane distribution of matter is the sum of its moments of inertia around the x and y axes.

Exercises 32 and 33 are related.

32 Let $(\bar{x}, \bar{y})$ be the center of mass of a plane distribution of mass. Let L be a line in the plane parallel to the y axis and passing through the center of mass.
(a) Show that the moment of the mass around L is 0. (L is called a "balancing line.")
(b) Show that the moment of the mass around a line parallel to the x axis and passing through the center of mass is 0.

33 Consider a distribution of mass in a plane region R. Show that any line in the plane that passes through the center of mass is a balancing line, following these steps.
(a) For convenience place the origin of the xy-coordinate system at the center of mass. Show that $\int_R x\delta(P)\,dA = 0$ and $\int_R y\delta(P)\,dA = 0$.
(b) Let L be any line $ax + by = 0$ through the origin. Why is the lever arm of the point (x, y) about L equal to $(ax + by)/\sqrt{a^2 + b^2}$? (This number may be positive or negative.)
(c) Show that the moment of the mass about L is

$$\int_R \frac{ax + by}{\sqrt{a^2 + b^2}} \delta(P)\,dA.$$

(d) From (a) and (c) deduce that the moment of the mass about L is 0.

Thus all balancing lines for the mass pass through a single point. Any two of them therefore determine that point, which is called the center of mass. It is customary to use the two lines parallel to the x and y axes to determine that point.

34 Assume that the origin of the xy-coordinate system is placed at the center of mass. (Thus $\int_R x\delta(P)\,dA = 0$ and $\int_R y\delta(P)\,dA = 0$.) Let I_z be the moment of inertia of the mass around the z axis. Let L be a line parallel to the z axis and a distance k from it. Assume that it intersects the xy plane in the point (a, b).
(a) Show that the moment of inertia of the mass about L, I_L, is $\int_R [(x - a)^2 + (y - b)^2]\delta(P)\,dA$.
(b) Multiplying out the integrand in (a) and recalling that the center of mass is at the origin, show that

$$I_L = \int_R (x^2 + y^2 + a^2 + b^2)\delta(P)\,dA.$$

(c) From (b) deduce that

$$I_L = I_z + k^2 M,$$

where M is the total mass in R.

(d) In view of (c), explain why the center of mass is "that point around which the moment of inertia is least." The formula in (c) shows how to compute the moment of inertia about any line perpendicular to the xy plane if the moment of inertia around the line perpendicular to the xy plane and through the center of mass is known. The next exercise gives a similar result for lines in the xy plane.

35 (Parallel axis theorem) Consider a plane distribution of mass in a region R. Let the total mass be M. Assume that the moment of inertia about a line L in the xy plane that passes through the center of mass is known. Call it I. Let L' be a line in the plane parallel to L and a distance k from L. Let I' be the moment of inertia about L'. Show that $I' = I + k^2 M$.

36 A lamina (a plane distribution of mass) of mass M and center of mass $(\bar{x}, \bar{y})$ is cut into two objects, one of mass M_1 and center of mass $(\bar{x}_1, \bar{y}_1)$, the other of mass M_2 and center of mass $(\bar{x}_2, \bar{y}_2)$.
(a) Prove that the masses and centers of mass of the two smaller pieces completely determine the center of mass of the original lamina, i.e., express $(\bar{x}, \bar{y})$ in terms of $(\bar{x}_1, \bar{y}_1)$, $(\bar{x}_2, \bar{y}_2)$, M_1 and M_2.
(b) Why must the center of mass of the original lamina lie on the line segment joining $(\bar{x}_1, \bar{y}_1)$ to $(\bar{x}_2, \bar{y}_2)$? *Hint:* Do (b) by considering moments about a wisely chosen line.
(c) Find the center of mass of the region in Fig. 13.35. (Assume the density is constant.)

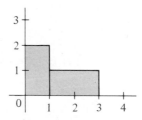

Figure 13.35

(d) How would (a) generalize from two masses, M_1 and M_2, to three masses, M_1, M_2, and M_3? (Justify your formula.)

13.4 Computing $\int_R f(P)\, dA$ by introducing polar coordinates in R

A planar integral $\int_R f(P)\, dA$ can also be computed by a repeated integral in polar coordinates. However, the area corresponding to small changes Δr and $\Delta\theta$ is not $\Delta r\, \Delta\theta$ but $r\, \Delta r\, \Delta\theta$. (This was shown in Sec. 13.2.) The coefficient r must be inserted in the integrand when setting up a repeated integral in polar coordinates. A more detailed explanation is presented at the end of this section.

Notice the factor r in the integrand.

> **HOW TO EVALUATE $\int_R f(P)\, dA$ IN POLAR COORDINATES**
>
> 1. Express $f(P)$ in terms of r and θ: $f(r, \theta)$.
>
> 2. Describe the region R in polar coordinates:
>
> $$\alpha \leq \theta \leq \beta, \quad r_1(\theta) \leq r \leq r_2(\theta).$$
>
> 3. Evaluate the repeated integral
>
> $$\int_\alpha^\beta \left[\int_{r_1(\theta)}^{r_2(\theta)} f(r, \theta) r\, dr \right] d\theta.$$

Notice that integration with respect to r is performed first. The other order is seldom used.

EXAMPLE 1 Let R be the semicircle of radius a shown in Fig. 13.36. Let $f(P)$ be the distance from a point P to the x axis. Evaluate $\int_R f(P)\, dA$ by a repeated integral in polar coordinates.

SOLUTION In polar coordinates, R has the description

$$0 \leq \theta \leq \pi, \quad 0 \leq r \leq a.$$

The distance from P to the x axis is, in rectangular coordinates, y. Since $y = r \sin \theta$, $f(P) = r \sin \theta$. Thus,

Notice this r, needed when using polar coordinates.

$$\int_R f(P)\, dA = \int_0^\pi \left[\int_0^a (r \sin \theta)\, r\, dr \right] d\theta.$$

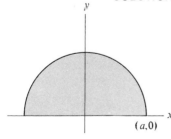

Figure 13.36

840 DEFINITE INTEGRALS OVER PLANE AND SOLID REGIONS

From here on the calculations are like those in the preceding section.

The calculation of the repeated integral is like that for a repeated integral in rectangular coordinates. First evaluate the inside integral:

$$\int_0^a r^2 \sin \theta \, dr = \sin \theta \int_0^a r^2 \, dr$$

$$= \sin \theta \left(\frac{r^3}{3}\right)\bigg|_0^a$$

$$= \frac{a^3 \sin \theta}{3}.$$

The outer integral is therefore

$$\int_0^\pi \frac{a^3 \sin \theta}{3} \, d\theta = \frac{a^3}{3} \int_0^\pi \sin \theta \, d\theta$$

$$= \frac{a^3}{3}(-\cos \theta)\bigg|_0^\pi$$

$$= \frac{a^3}{3}[(-\cos \pi) - (-\cos 0)]$$

$$= \frac{a^3}{3}(1 + 1) = \frac{2a^3}{3}.$$

Thus
$$\int_R y \, dA = \frac{2a^3}{3}. \quad \blacksquare$$

Distinction between "ball" and "sphere"

Example 2 refers to a ball of radius a. Generally, we will distinguish between a *ball*, which is a solid region, and a *sphere*, which is only the surface of a ball.

EXAMPLE 2 A ball of radius a has its center at the pole of a polar-coordinate system. Find the volume of the part of the ball that lies above the plane region bounded by the curve $r = a \cos \theta$. (See Fig. 13.37.)

SOLUTION It is necessary to describe R and f in polar coordinates, where $f(P)$ is the length of a cross section of the solid made by a vertical line through P. R is described as follows: r goes from 0 to $a \cos \theta$ for each θ in $[-\pi/2, \pi/2]$, that is,

$$-\frac{\pi}{2} \leq \theta \leq \frac{\pi}{2}, \qquad 0 \leq r \leq a \cos \theta.$$

To express $f(P)$ in polar coordinates, consider Fig. 13.38, which shows the top half of a ball of radius a. By the Pythagorean theorem

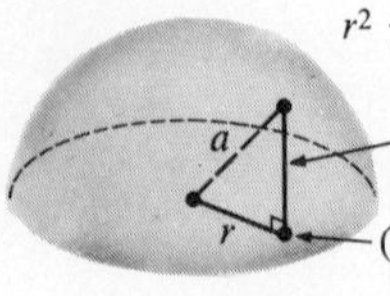

$$r^2 + [f(r, \theta)]^2 = a^2.$$

Thus
$$f(r, \theta) = \sqrt{a^2 - r^2}.$$

Consequently
$$\text{Volume} = \int_R f(P) \, dA = \int_{-\pi/2}^{\pi/2} \left(\int_0^{a \cos \theta} \sqrt{a^2 - r^2} \, r \, dr\right) d\theta.$$

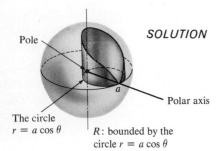

Figure 13.37
The circle $r = a \cos \theta$
R: bounded by the circle $r = a \cos \theta$

Figure 13.38
$r^2 + [f(r,\theta)]^2 = a^2$
$f(r,\theta)$
(r,θ)

Exploiting symmetry, compute half the volume, keeping θ in $[0, \pi/2]$, and then double the result:

$$\int_0^{a\cos\theta} \sqrt{a^2 - r^2}\, r\, dr = \frac{-(a^2 - r^2)^{3/2}}{3} \Big|_0^{a\cos\theta}$$

$$= -\left[\frac{(a^2 - a^2\cos^2\theta)^{3/2}}{3} - \frac{(a^2)^{3/2}}{3}\right]$$

$$= \frac{a^3}{3} - \frac{(a^2 - a^2\cos^2\theta)^{3/2}}{3}$$

$$= \frac{a^3}{3} - \frac{a^3(1 - \cos^2\theta)^{3/2}}{3} = \frac{a^3}{3}(1 - \sin^3\theta).$$

(The trigonometric formula used above, $\sin\theta = \sqrt{1 - \cos^2\theta}$, is true when $0 \le \theta \le \pi/2$ but not when $-\pi/2 \le \theta < 0$.)

The second integration is then carried out:

$$\int_0^{\pi/2} \frac{a^3}{3}(1 - \sin^3\theta)\, d\theta = \frac{a^3}{3} \int_0^{\pi/2} [1 - (1 - \cos^2\theta)\sin\theta]\, d\theta$$

$$= \frac{a^3}{3}\left(\theta + \cos\theta - \frac{\cos^3\theta}{3}\right)\Big|_0^{\pi/2}$$

$$= \frac{a^3}{3}\left[\frac{\pi}{2} - \left(1 - \frac{1}{3}\right)\right]$$

$$= \frac{a^3}{3}\left(\frac{3\pi - 6 + 2}{6}\right) = a^3\left(\frac{3\pi - 4}{18}\right).$$

The total volume is twice as large:

$$a^3\left(\frac{3\pi - 4}{9}\right). \quad \blacksquare$$

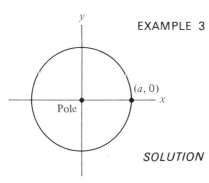

Figure 13.39

EXAMPLE 3 A circular disk of radius a is formed by a material which has a density at each point equal to the distance from that point to the center.
(a) Set up a repeated integral in rectangular coordinates for the total mass of the disk.
(b) Set up a repeated integral in polar coordinates for the total mass of the disk.
(c) Compute the easier one.

SOLUTION The disk is shown in Fig. 13.39.
(a) (Rectangular coordinates) The density $\delta(P)$ at $P = (x, y)$ is $\sqrt{x^2 + y^2}$. The disk has the description

$$-a \le x \le a, \qquad -\sqrt{a^2 - x^2} \le y \le \sqrt{a^2 - x^2}.$$

Thus Mass $= \int_R \delta(P)\, dA = \int_{-a}^{a} \left(\int_{-\sqrt{a^2-x^2}}^{\sqrt{a^2-x^2}} \sqrt{x^2+y^2}\, dy \right) dx.$

(b) (Polar coordinates) The density $\delta(P)$ at $P = (r, \theta)$ is r. The disk has the description

$$0 \leq \theta \leq 2\pi, \qquad 0 \leq r \leq a.$$

Thus $\quad$ Mass $= \int_R \delta(P)\, dA = \int_0^{2\pi} \left(\int_0^a r \cdot r\, dr \right) d\theta$

$$= \int_0^{2\pi} \left(\int_0^a r^2\, dr \right) d\theta.$$

The integrand is r^2, not just r.

(c) Even the first integration in the repeated integral in (a) would be tedious. However, the repeated integral in (b) is a delight:
The first integration gives

$$\int_0^a r^2\, dr = \left. \frac{r^3}{3} \right|_0^a$$
$$= \frac{a^3}{3}.$$

The second integration gives

$$\int_0^{2\pi} \frac{a^3}{3}\, d\theta = \left. \frac{a^3 \theta}{3} \right|_0^{2\pi}$$
$$= \frac{2\pi a^3}{3}.$$

The total mass is $2\pi a^3/3$. ∎

As Example 3 suggests, polar coordinates may be preferable to rectangular coordinates. If either $f(P)$ or R has a more convenient description in terms of polar coordinates, try polar coordinates. In Example 3 both $f(P)$ and R took simple forms in polar coordinates.

Rectangular coordinates are usually preferable when dealing with integrals over rectangles, as you might expect.

APPROXIMATING SUMS AND THE APPEARANCE OF r (OPTIONAL)

Let a and b be nonnegative numbers, $a < b$; let α and β satisfy $0 \leq \alpha < \beta \leq 2\pi$. Let $f(P)$ be defined throughout the region R consisting of those points (r, θ) such that $a \leq r \leq b$ and $\alpha \leq \theta \leq \beta$. (See Fig. 13.40.)

In order to devise an approximating sum for $\int_R f(P)\, dA$, select $r_0 = a < r_1 < \cdots < r_n = b$ and $\theta_0 = \alpha < \theta_1 < \cdots < \theta_n = \beta$. These numbers determine n^2 small regions of which the typical one, $R_{i,j}$, is shown in Fig. 13.41. By Sec. 13.2, the area of $R_{i,j}$ is exactly

13.4 COMPUTING $\int_R f(P)\, dA$ BY INTRODUCING POLAR COORDINATES IN R

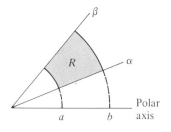

Figure 13.40

Figure 13.41

$$\Delta r_i = r_i - r_{i-1}$$
$$\Delta \theta_j = \theta_j - \theta_{j-1}$$

$$\frac{r_i + r_{i-1}}{2} \Delta r_i\, \Delta \theta_j.$$

Choose the sampling point $P_{i,j}$ in $R_{i,j}$ to be $r_i^* = (r_i + r_{i-1})/2$. Then

$$\sum_{i,j=1}^n f(r_i^*, \theta_j)\, r^*$$

is an approximation to $\int_R f($
each small region, R_i.

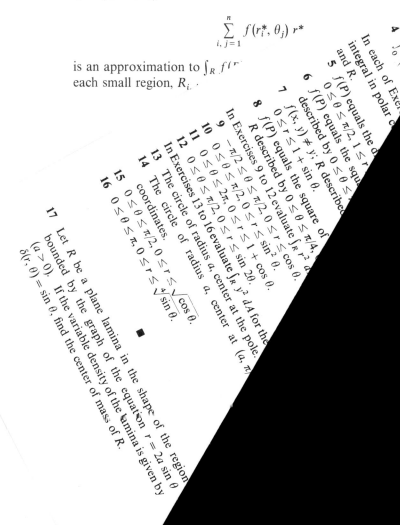

DEFINITE INTEGRALS OVER PLANE AND SOLID REGIONS

$$\int_\alpha^\beta \left(\int_a^b f(r, \theta) \, r \, dr \right) d\theta. \quad (4)$$

Thus it is plausible that as the mesh of the partition approaches 0, the sums (1), which approach $\int_R f(P) \, dA$, also approach (4).

In Sec. 13.6 the appearance of the factor r in the integrand is shown to be a special case of a theorem on the change from rectangular coordinates to another coordinate system.

Exercises

Evaluate the repeated integrals in Exercises 1 to 4.

1 $\int_0^\pi \left(\int_0^{\cos \theta} r \sin \theta \, dr \right) d\theta$

$\int_0^\pi \left(\int_0^3 r^2 \sin \theta \, dr \right) d\theta$

$\int \left(\int_0^{1+\cos\theta} r^3 \sin \theta \, dr \right) d\theta$

$\int \left(\int_0^\theta r \, dr \right) d\theta$

cises 5 to 8 set up and evaluate a repeated ordinates for $\int_R f(P) \, dA$, for the given f

istance from P to pole; R described by $\leq 2 + \cos \theta$.
re of distance from P to pole; R $/6$, $0 \leq r \leq \sin 3\theta$.
by $0 \leq \theta \leq \pi/2$,

distance from P to y axis; $0 \leq r \leq \sin 2\theta$.
A for the given regions R.

given regions R.

2) in polar

18 Find the volume of a ball of radius a using a repeated integral in polar coordinates.

19 Recall that $\int_R 1 \, dA$ equals the area of R. Using this fact, develop the formula for area in polar coordinates given in Sec. 9.2, namely,

$$\text{Area} = \int_\alpha^\beta \frac{[f(\theta)]^2}{2} \, d\theta.$$

20 A right circular cone has radius a and height h.
(a) Set up a repeated integral in polar coordinates for the volume of the cone.
(b) Evaluate the repeated integral in (a).

21 In Example 2 we computed half the volume and doubled the result. Evaluate the repeated integral

$$\int_{-\pi/2}^{\pi/2} \left(\int_0^{a \cos \theta} \sqrt{a^2 - r^2} \, r \, dr \right) d\theta$$

directly. The result should still be $a^3(3\pi - 4)/9$. *Caution:* Use trigonometric formulas with care.

22 The density of material at $P = (r, \theta)$ is r grams per square centimeter. The region is bounded by the cardioid $r = 1 + \cos \theta$. Find the total mass.

23 Find the average distance from points in the circle $r = 2 \sin \theta$ to the pole.

24 This exercise shows that $\int_0^\infty e^{-x^2} \, dx = \sqrt{\pi}/2$. Let R_1, R_2, R_3 be the three regions indicated in Fig. 13.42, and $f(P) = e^{-r^2}$, where r is the distance from P to the origin. Hence $f(r, \theta) = e^{-r^2}$ and $f(x, y) = e^{-x^2-y^2}$.

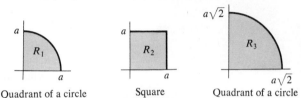

Figure 13.42

(Observe that R_1 is inside R_2, and R_2 is inside R_3.)
(a) Show that $\int_{R_1} f(P) \, dA = (\pi/4)(1 - e^{-a^2})$ and $\int_{R_3} f(P) \, dA = (\pi/4)(1 - e^{-2a^2})$.

(b) By considering $\int_{R_2} f(P)\, dA$ and the results in (a), show that

$$\frac{\pi}{4}(1 - e^{-a^2}) < \left(\int_0^a e^{-x^2}\, dx\right)^2 < \frac{\pi}{4}(1 - e^{-2a^2}).$$

(c) Show that $\int_0^\infty e^{-x^2}\, dx = \sqrt{\pi}/2$.

S. P. Thompson, in *Life of Lord Kelvin* (Macmillan, London, 1910), wrote,

"Once when lecturing to a class he [the physicist Lord Kelvin] used the word "mathematician" and then interrupting himself asked his class: "Do you know what a mathematician is?" Stepping to his blackboard he wrote upon it: $\int_{-\infty}^\infty e^{-x^2}\, dx = \sqrt{\pi}$. Then putting his finger on what he had written, he turned to his class and said, "A mathematician is one to whom that is as obvious as that twice two makes four is to you.""

On the other hand the mathematician Littlewood wrote,

"Many things are not accessible to intuition at all, the value of $\int_0^\infty e^{-x^2}\, dx$ for instance."

From J. E. Littlewood, "Newton and the Attraction of the Sphere," *Mathematical Gazette*, vol. 63, 1948.

25 Using the fact that $\int_0^\infty e^{-x^2}\, dx = \sqrt{\pi}/2$, show that

(a) $\displaystyle\int_0^\infty e^{-4x^2}\, dx = \frac{\sqrt{\pi}}{4}$; (b) $\displaystyle\int_0^\infty \frac{e^{-x}}{\sqrt{x}}\, dx = \sqrt{\pi}$;

(c) $\displaystyle\int_0^\infty x^2 e^{-x^2}\, dx = \frac{\sqrt{\pi}}{4}$; (d) $\displaystyle\int_0^\infty \sqrt{x}\, e^{-x}\, dx = \frac{\sqrt{\pi}}{2}$;

(e) $\displaystyle\int_0^1 \frac{dx}{\sqrt{\ln(1/x)}} = \sqrt{\pi}$; (f) $\displaystyle\int_0^1 \sqrt{\ln(1/x)}\, dx = \frac{\sqrt{\pi}}{2}$.

26 (The spread of epidemics.) In the theory of a spreading epidemic it is assumed that the probability that a contagious individual infects an individual D miles away depends only on D. Consider a population that is uniformly distributed in a circular city whose radius is 1 mile. Assume that the probability we mentioned is proportional to $2 - D$. For a fixed point Q let $f(P) = 2 - \overline{PQ}$. Let R be the region occupied by the city.

(a) Why is the exposure of a person residing at Q proportional to $\int_R f(P)\, dA$, assuming that contagious people are uniformly distributed throughout the city?

(b) Compute this definite integral when Q is the center of town and when Q is on the edge of town.

(c) In view of (b), which is the safer place?

27 Evaluate $\int_R \cos(x^2 + y^2)\, dA$ over the portion in the first quadrant of the disk of radius a centered at the origin.

28 Evaluate $\int_R \sqrt{x^2 + y^2}\, dA$ over the triangle bounded by the line $y = x$, the line $x = 2$, and the x axis.

■ ■

29 What is wrong with this reasoning?

We shall obtain a repeated integral in polar coordinates equal to $\int_R f(P)\, dA$, where R has the description $\alpha \leq \theta \leq \beta$, $r_1(\theta) \leq r \leq r_2(\theta)$. Assuming that $f(P) \geq 0$, construct the solid whose base is R and whose height at P is $f(P)$. The cross section of this solid by a plane perpendicular to R, and passing through the ray of angle θ is a plane section of area $A(\theta) = \int_{r_1(\theta)}^{r_2(\theta)} f(r, \theta)\, dr$. Since $V = \int_\alpha^\beta A(\theta)\, d\theta$,

$$\int_R f(P)\, dA = \int_\alpha^\beta \left(\int_{r_1(\theta)}^{r_2(\theta)} f(r, \theta)\, dr\right) d\theta.$$

This is a *wrong formula*, since r is not present in the integrand. Find the error.

30 (A gravitational paradox) Consider a mass distributed uniformly throughout the entire plane and a point mass a distance a from the plane. The gravitational attraction of the planar mass on the point mass is directed toward the plane and has magnitude

$$\int_R \frac{a}{(\sqrt{a^2 + r^2})^3}\, dA.$$

(The integral, which is taken over the entire plane, is improper in the sense that an integral over the x axis is improper. Treat it similarly by computing the integral over a circle of radius s centered at the origin, and letting s go to infinity.) In the integrand r refers to polar coordinates where the pole is the point in the plane closest to the point mass.

(a) Show that the integral has the value 2π.

(b) According to (a), the attractive force of the plane on the point mass is independent of the distance between the point mass and the plane. Does that make sense?

31 This exercise is based on "Sudden expansion in a pipeline," from *Introduction to Fluid Mechanics* by Stephen Whitaker, Prentice-Hall, 1968.

The velocity of a fluid at a distance r from the axis of a pipe of radius r_0 is given by a formula of the form $v(r) = a(1 - r/r_0)^{1/n}$, where a and n are constants, and r_0 is the radius of the pipe. Let R be a cross section of the pipe perpendicular to its axis. Find the average over R of (a) the velocity of the fluid, (b) the square of the velocity of the fluid.

13.5 Mappings from a plane to a plane (optional)

Review the magnification example in Sec. 3.1.

In the discussion of the linear slide and screen in Sec. 3.1 it was shown that the magnification at a point is represented by a derivative. If $x = f(u)$ is a function from the u axis (the slide) to the x axis (the screen), then the local magnification is given by the absolute value of the derivative. This means that a short interval on the u axis around the number u, of length Δu, appears on the screen as a section of length $|\Delta x|$ where Δx is approximately $f'(u)\Delta u$, as shown in Fig. 13.43. Keep in mind that a magnification greater than 1 in absolute value describes an expansion, while a magnification less than 1 in absolute value describes a shrinking.

In the so-called "real world," slides and screens are two dimensional surfaces, not lines. A projector may show the image of a square 35-millimeter slide on a large screen. The further the screen is from the slide, the greater is the magnification.

Instead of a slide and a screen, we may have a (flat) city and a map of that city. The map represents a "magnification" by a factor less than one, with perhaps 1 square mile of the city shown as 1 square inch on the map.

The slide and screen and the city and map are special cases of functions called "mappings," which will be examined in this section. The next section discusses their magnification, which may vary from point to point, and their use in evaluating planar integrals.

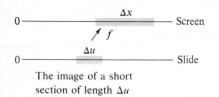

The image of a short section of length Δu has length Δx approximately $f'(u)\Delta u$.

Figure 13.43

MAPPINGS A *mapping* is a function F that assigns to points in some plane region points in another plane. The inputs are points in the uv plane; the outputs are points in the xy plane. (See Fig. 13.44.)

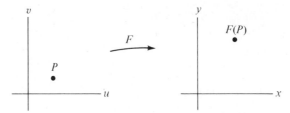

Figure 13.44

The uv plane plays the role of the slide or the flat city; the xy plane plays the role of the screen or map. The mapping F may be thought of as the projector or the means of assigning to each point in the city a point in the map. It may be defined for only part of the uv plane, perhaps on a rectangle, triangle, circle, or half plane.

It will be convenient to denote the "input" point (u, v) and the "output" point (x, y). Thus

$$F(u, v) = (x, y).$$

Notation for planar mapping F as a pair of real valued functions

Both x and y depend on u and v. So we may write

$$x = f(u, v) \quad \text{and} \quad y = g(u, v).$$

Thus
$$F(u, v) = (f(u, v), g(u, v)), \qquad (1)$$

and F is shorthand for a pair, f and g, of real-valued functions of two variables. Any mapping from the uv plane to the xy plane is described by a pair of such functions, $x = f(u, v)$ and $y = g(u, v)$.

It will be assumed that both f and g have continuous partial derivatives. However F may be quite general. The image of a line need not be another line; it may, for instance, be a parabola as in Example 3. In Example 1 the image of the given square is a rectangle and in Example 2 the image of a circle is an ellipse. Moreover, the magnification of the mapping F may vary from point to point.

EXAMPLE 1 Let F be the mapping that assigns to the point (u, v) the point $(2u, 3v)$.
(a) Describe this mapping geometrically.
(b) Find the image of the line $v = u$.
(c) Find the image of the square in the uv plane whose vertices are $(0, 0)$, $(1, 0)$, $(1, 1)$, and $(0, 1)$.

SOLUTION (a) In this case $x = 2u$ and $y = 3v$.

The table below records the effect of the mapping on the points listed in (c),

(u, v)	$(0, 0)$	$(1, 0)$	$(1, 1)$	$(0, 1)$
$(2u, 3v)$	$(0, 0)$	$(2, 0)$	$(2, 3)$	$(0, 3)$

In the notation $F(u, v) = (x, y)$, these data read:

$$F(0, 0) = (2 \cdot 0, 3 \cdot 0) = (0, 0);$$
$$F(1, 0) = (2 \cdot 1, 3 \cdot 0) = (2, 0);$$
$$F(1, 1) = (2 \cdot 1, 3 \cdot 1) = (2, 3);$$
$$F(0, 1) = (2 \cdot 0, 3 \cdot 1) = (0, 3).$$

Note that the first coordinate of $(x, y) = F(u, v)$ is $x = 2u$, twice the first coordinate of (u, v). Thus the mapping magnifies horizontally by a factor of 2. Similarly, it stretches vertically by a factor of 3. (This causes a sixfold magnification of areas.)

(b) Let $P = (u, v)$ be on the line $v = u$. Then $F(P) = F(u, v) = (x, y)$, with $x = 2u$ and $y = 3v$. Thus

$$u = \frac{x}{2} \quad \text{and} \quad v = \frac{y}{3}.$$

Since $v = u$, $\quad \dfrac{y}{3} = \dfrac{x}{2},$

or $\quad y = \tfrac{3}{2}x.$

The image of the line $v = u$ in the uv plane is the line $y = 3x/2$ in

848 DEFINITE INTEGRALS OVER PLANE AND SOLID REGIONS

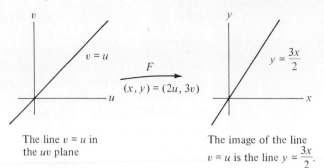

Figure 13.45

This mapping takes any line into a line

the xy plane. (See Fig. 13.45.)

A similar argument shows that, for this mapping, the image of any line $Au + Bv + C = 0$ in the uv plane is a line in the xy plane, namely the line $Ax/2 + By/3 + C = 0$.

(c) If P is a point in the square R whose vertices are

$$(0, 0), \quad (1, 0), \quad (1, 1), \quad (0, 1),$$

then the image of P is a point in the rectangle S whose vertices are

$$(0, 0), \quad (2, 0), \quad (2, 3), \quad (0, 3).$$

(See Fig. 13.46.)

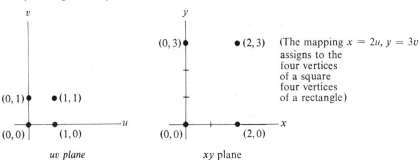

Figure 13.46

Think of (u, v) as a point on a slide and $(2u, 3v)$ as its image on the screen. Then the mapping F projects the square R on the slide onto a rectangle S on the screen. (See Fig. 13.47.)

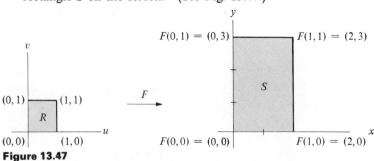

Figure 13.47

13.5 MAPPINGS FROM A PLANE TO A PLANE (OPTIONAL)

EXAMPLE 2 Let $F = (f, g)$ be the same as in Example 1. Let R be the set of points (u, v) in the uv plane (the slide) such that

$$u^2 + v^2 \leq 1.$$

In words, R is the disk of radius 1 centered at $(0, 0)$. What is the image of R on the xy plane (the screen)?

SOLUTION Since

$$u = \frac{x}{2} \quad \text{and} \quad v = \frac{y}{3},$$

Note once again how the image in the xy plane is calculated.

the inequality

$$u^2 + v^2 \leq 1$$

implies that

$$\left(\frac{x}{2}\right)^2 + \left(\frac{y}{3}\right)^2 \leq 1;$$

that is,

$$\frac{x^2}{4} + \frac{y^2}{9} \leq 1.$$

Consequently, the image of the disk of radius 1 consists of all points inside the ellipse

$$\frac{x^2}{4} + \frac{y^2}{9} = 1.$$

For this mapping the image of a circle is an ellipse.

The image of the disk R is the ellipse S in Fig. 13.48.

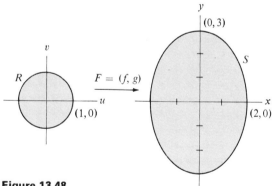

Figure 13.48 ∎

In the next mapping the image of a line is a parabola.

EXAMPLE 3 Let $(x, y) = (uv, v)$. If $P = (u, v)$ lies on the line $v = u$, where does $F(P) = (uv, v)$ lie?

SOLUTION In this case

$$x = uv \quad \text{and} \quad y = v.$$

To find what the mapping does to the line $u = v$, find an equation linking x and y if (x, y) is on the image of the line $u = v$.

850 DEFINITE INTEGRALS OVER PLANE AND SOLID REGIONS

P	Image of P
$P_1 = (-2, -2)$	$(4, -2)$
$P_2 = (-1, -1)$	$(1, -1)$
$P_3 = (0, 0)$	$(0, 0)$
$P_4 = (1, 1)$	$(1, 1)$
$P_5 = (2, 2)$	$(4, 2)$
$P_6 = (3, 3)$	$(9, 3)$

Observe that $x = uv = v^2 = y^2$.

Thus as $P = (u, v)$ wanders about the line $u = v$, the image point (uv, v) wanders about the parabola $x = y^2$.

In order to make this fact more concrete, let us compute $F(P)$ for a few points on the line $v = u$. For instance, if

$$P = (2, 2), \qquad F(P) = (2 \cdot 2, 2) = (4, 2).$$

The table in the margin records the results of six such computations.

Figure 13.49 shows these data (the dotted line joins the images).

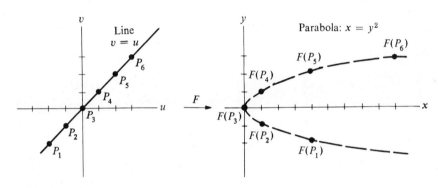

Figure 13.49

A MAPPING RELATED TO POLAR COORDINATES

The final example will play a critical role in the next section.

EXAMPLE 4

The restrictions on u and v make F one-to-one.

Let $F(u, v) = (u \cos v, u \sin v)$ for $u > 0$ and $0 \leq v < 2\pi$. Let R be the rectangle in the uv plane,

$$1 \leq u \leq 2, \qquad \frac{\pi}{6} \leq v \leq \frac{\pi}{4}.$$

Sketch the image S of R under the effect of the mapping F.

SOLUTION R is shown in Fig. 13.50. To find the image of R, examine the points $(x, y) = F(u, v)$ for (u, v) in the rectangle R.

By the definition of F,

$$x = u \cos v \qquad \text{and} \qquad y = u \sin v. \qquad (2)$$

Note the resemblance between (2) and the equations that link polar and rectangular coordinates,

$$x = r \cos \theta \qquad \text{and} \qquad y = r \sin \theta. \qquad (3)$$

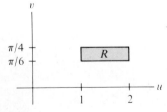

Figure 13.50

Assume $0 \leq \theta < 2\pi$.

In fact, u must equal r and v must equal θ if we assume that $0 \leq \theta < 2\pi$. To see this, note from (2) that

$$x^2 + y^2 = (u \cos v)^2 + (u \sin v)^2$$
$$= u^2(\cos^2 v + \sin^2 v)$$
$$= u^2.$$

So $u = \sqrt{x^2 + y^2}$. But $\sqrt{x^2 + y^2}$ equals the r of polar coordinates. So $u = r$. Thus $\cos v = \cos \theta$ and $\sin v = \sin \theta$; from this it follows that $v = \theta$, since both v and θ are between 0 and 2π.

So the image of R consists of those points in the xy plane whose polar coordinates satisfy the inequalities

$$1 \leq r \leq 2 \quad \text{and} \quad \frac{\pi}{6} \leq \theta \leq \frac{\pi}{4}.$$

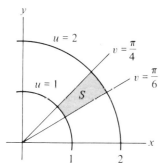

Figure 13.51

This region is shown in Fig. 13.51. ∎

The main point of Example 4

The point of Example 4 is that, while r and θ are polar coordinates in the xy plane, they are also rectangular coordinates in their own right in the $r\theta$ plane.

Exercises

It will be shown in Exercise 13 that the mapping

$$F(u, v) = (f(u, v), g(u, v)) = (au + bv, cu + dv),$$

for constants a, b, c, and d such that $ad - bc \neq 0$, takes any straight line into a straight line. This fact may be assumed when doing the other exercises.

Exercises 1 to 4 concern the mapping $F(u, v) = (2u, 3v)$.
1 (a) Plot the points $P_1 = (1, 0)$, $P_2 = (3, 0)$, and $P_3 = (6, 0)$ in the uv plane.
 (b) Plot $Q_1 = F(P_1)$, $Q_2 = F(P_2)$, and $Q_3 = F(P_3)$ in the xy plane.
 (c) What is the image in the xy plane of the line $v = 0$?
2 (a) Pick three points P_1, P_2, and P_3 on the line $u + v = 1$.
 (b) Plot their images in the xy plane.
 (c) Sketch the image in the xy plane of the line $u + v = 1$ and give its equation.
3 (a) Fill in this table.

(u, v)	$(x, y) = (2u, 3v)$
(0, 0)	
(2, 0)	
(1, 1)	

 (b) Plot the three points in the left column of the table in (a) in the uv plane.
 (c) Plot the three points entered in the right column of the table in (a) in the xy plane.
 (d) Draw the triangular region R whose vertices are listed in (a) in the uv plane.
 (e) Draw S, the image of R in the xy plane.
 (f) Compute the areas of R and S, and show that S is six times as large as R.
 (g) Show that R is a right triangle, but S is not.
4 Let R in the uv plane consist of all points (u, v) such that $u^2 + v^2 \leq 4$. (a) Draw R. (b) Draw the image in the xy plane of R.

In Exercises 5 and 6, let $F(u, v) = (u + v, u - v)$.
5 (a) Plot the points $P_1 = (1, 1)$, $P_2 = (3, 1)$, $P_3 = (1, 4)$, and $P_4 = (3, 4)$ in the uv plane.
 (b) Plot in the xy plane the respective images of the points in (a) and label these images Q_1, Q_2, Q_3, and Q_4.
 (c) Let R be the rectangle with vertices P_1, P_2, P_3, and P_4. Let S in the xy plane be the image of R. Sketch S. What kind of figure is S?
6 (a) Show that, if $F(u, v) = (x, y)$, then $u = (x + y)/2$ and $v = (x - y)/2$.
 (b) Use (a) to find the image of the line $u + 2v = 1$.
 (c) Use (a) to find the image of the circle $u^2 + v^2 = 9$.

7 Let $F(u, v) = (v, \sqrt{u})$ for positive u and v.
   (a) The points $(1, 2)$, $(2, 4)$, $(3, 6)$, and $(4, 8)$ lie on the line $v = 2u$. Plot their images in the xy plane.
   (b) Show that, if $(x, y) = (v, \sqrt{u})$, then $v = x$ and $u = y^2$.
   (c) Use (b) to find the equation of the image of the half line $v = 2u$. (Recall that $u, v > 0$.)
   (d) Sketch the image of the half line $v = 2u$.

8 Let $(f(u, v), g(u, v)) = (2u + 3v, u + v)$.
   (a) Fill in this table.

(u, v)	$(x, y) = (2u + 3v, u + v)$
$(1, 0)$	
$(0, 1)$	
$(-1, 0)$	
$(0, -1)$	

   (b) As (u, v) runs through the circle $u^2 + v^2 = 1$ counterclockwise, $(x, y) = (2u + 3v, u + v)$ sweeps out a curve C in the xy plane. Is C swept out clockwise or counterclockwise?

Exercises 9 to 12 concern the mapping given by $F(u, v) = (u \cos v, u \sin v)$ for $u > 0$ and $0 \leq v < 2\pi$.

9 (a) The points $(5, 0)$, $(5, \pi/6)$, $(5, \pi/2)$, and $(5, \pi)$ lie on the line $u = 5$ in the uv plane. Plot their images in the xy plane.
   (b) What is the image of the line segment $u = 5$?

10 (a) The points $(3, \pi/6)$, $(4, \pi/6)$, $(5, \pi/6)$, and $(6, \pi/6)$ lie on the line $v = \pi/6$. Plot their images in the xy plane.
   (b) What is the image of the half line $v = \pi/6$?

11 (a) Sketch the region R in the uv plane bounded by the lines $u = 2$, $u = 3$, $v = 1$, and $v = 4$.
   (b) Sketch the image of R.

12 Find (u, v) if $F(u, v)$ is
   (a) $(3, 3)$,   (b) $(4, 4\sqrt{3})$,
   (c) $(6, 0)$,   (d) $(0, -7)$.

∎

13 Let a, b, c, and d be constants such that $ad - bc$ is not 0. Let $F(u, v) = (f(u, v), g(u, v)) = (au + bv, cu + dv)$.
   (a) Show that, if $(x, y) = (au + bv, cu + dv)$, then

   $$u = \frac{dx - by}{ad - bc} \quad \text{and} \quad v = \frac{ay - cx}{ad - bc}.$$

   Suggestion: Solve the simultaneous equations $au + bv = x$, $cu + dv = y$ for the unknowns u and v.
   (b) Show that the image of a line $Au + Bv + C = 0$ in the uv plane is a line in the xy plane. (Because F takes lines to lines it is called a "linear" mapping.)
   (c) Why must the images of parallel lines be parallel?

14 Let $F(u, v) = (u \cos v, u \sin v)$ for $u > 0$ and $0 \leq v < 2\pi$. Let R be the rectangle $u_0 \leq u \leq u_0 + \Delta u$, $v_0 \leq v \leq v_0 + \Delta v$, where u_0, v_0, Δu, and Δv are constants.
   (a) What is the area of R?
   (b) What is the area of S, the image of R?
   (c) Find the limit of the quotient

   $$\frac{\text{Area of } S}{\text{Area of } R}$$

   as Δu and Δv approach 0.
   The limit in (c) depends on the point (u_0, v_0). It is called the "magnification" of the mapping at (u_0, v_0).

15 (a) The magnitude of the cross product $\mathbf{A} \times \mathbf{B}$ for two vectors $\mathbf{A}$ and $\mathbf{B}$ is the area of the parallelogram spanned by $\mathbf{A}$ and $\mathbf{B}$. (See Sec. 11.6.) Use this fact to show that the area of the parallelogram determined by the plane vectors $a_1 \mathbf{i} + a_2 \mathbf{j}$ and $b_1 \mathbf{i} + b_2 \mathbf{j}$ is the absolute value of

   $$\begin{vmatrix} a_1 & a_2 \\ b_1 & b_2 \end{vmatrix},$$

   that is, $|a_1 b_2 - a_2 b_1|$.
   (b) Let $F(u, v) = (2u + 3v, u + 2v)$. Let R be the rectangle $u_0 \leq u \leq u_0 + \Delta u$, $v_0 \leq v \leq v_0 + \Delta v$. Let S be its image. Find the ratio

   $$\frac{\text{Area of } S}{\text{Area of } R}.$$

   The ratio in (b) is constant, independent of Δu, Δv, u_0, and v_0. The magnification of the mapping is constant throughout the uv plane.

16 (See Exercise 15.) Let a, b, c, and d be constants such that $ad - bc \neq 0$. Let $F(u, v) = (au + bv, cu + dv)$.
   (a) Show that the image of a rectangle is a parallelogram. Suggestion: See Exercise 13.
   (b) Show that the image of the rectangle $u_0 \leq u \leq u_0 + \Delta u$, $v_0 \leq v \leq v_0 + \Delta v$ is a parallelogram whose area is $|ad - bc|$ times as large.

∎ ∎

17 Suppose that F, a mapping from the uv plane to the xy plane, "preserves lengths," that is, assume that for all pairs of points P and Q in the uv plane the distance between $F(P)$ and $F(Q)$ is the same as the distance between P and Q.
   (a) Prove that, if P_1, P_2, and P_3 lie on a line in the uv plane, then $F(P_1)$, $F(P_2)$, and $F(P_3)$ lie on a line in the xy plane.
   (b) Prove that, if P_1, P_2, and P_3 are points in the uv plane, then the angle $F(P_1)F(P_2)F(P_3)$ equals the angle $P_1 P_2 P_3$.

18 Sketch the image of the half line $v = 2u$, $u \geq 0$, under the mapping $F(u, v) = (\sqrt{u}, \sqrt[3]{v})$, $u \geq 0$.

19 Sketch the image of the half line $v = u$, $u \geq 0$, under the mapping $F(u, v) = (\ln u, v)$, $u > 0$.

13.6 Magnification, change of coordinates, and the Jacobian (optional)

As shown in Sec. 3.1, the magnification of the function $x = f(u)$ is described by its derivative. This section determines the magnification of the mapping $(x, y) = F(u, v) = (f(u, v), g(u, v))$. It turns out to be determined by the partial derivatives of f and g:

"abs" is short for "absolute value of."

$$\text{Local magnification} = \text{abs} \begin{vmatrix} f_u & f_v \\ g_u & g_v \end{vmatrix} = |f_u g_v - f_v g_u|. \tag{1}$$

The determinant in (1) is called the *Jacobian* of the mapping F in honor of Jacobi, who introduced it in 1841. After showing that the Jacobian gives the local magnification (which may vary from point to point), we apply it to integration over plane sets. It will then be shown that the extra r inserted in the integrand when polar coordinates are used is the Jacobian of a certain mapping.

MAGNIFICATION OF A SPECIAL MAPPING

We begin by computing the magnification of a mapping of the special form

$$F(u, v) = (au + bv, cu + dv), \tag{2}$$

where a, b, c, and d are constants and $ad - bc \neq 0$. This mapping takes lines into lines, hence is called a "linear" mapping. Note that the image of a rectangle is a parallelogram.

THEOREM 1 Let F be the mapping (2). Let R be the rectangle

$$u_1 \leq u \leq u_2, \qquad v_1 \leq v \leq v_2.$$

Let S be the image of R. Then

$$\frac{\text{Area of } S}{\text{Area of } R} = |ad - bc|.$$

PROOF Let P_1, P_2, P_3, and P_4 be the vertices of the rectangle R and let Q_1, Q_2, Q_3, and Q_4 be the corresponding vertices of the parallelogram S, as shown in Fig. 13.52.

Figure 13.52

The area of R is $(u_2 - u_1)(v_2 - v_1)$. All that remains is to compute the area of the parallelogram S.

The area of S is the magnitude of the cross product of the vectors $\overrightarrow{Q_1Q_2}$ and $\overrightarrow{Q_1Q_4}$. To find the cross product first find the coordinates of $Q_1, Q_2,$ and Q_4.

$$\left.\begin{array}{l}Q_1 = F(P_1) = F(u_1, v_1) = (au_1 + bv_1, cu_1 + dv_1) \\ Q_2 = F(P_2) = F(u_2, v_1) = (au_2 + bv_1, cu_2 + dv_1) \\ Q_4 = F(P_4) = F(u_1, v_2) = (au_1 + bv_2, cu_1 + dv_2)\end{array}\right\} \quad (3)$$

From (3) it follows that

$$\overrightarrow{Q_1Q_2} = (au_2 - au_1)\mathbf{i} + (cu_2 - cu_1)\mathbf{j} = a(u_2 - u_1)\mathbf{i} + c(u_2 - u_1)\mathbf{j}$$

and $\overrightarrow{Q_1Q_4} = (bv_2 - bv_1)\mathbf{i} + (dv_2 - dv_1)\mathbf{j} = b(v_2 - v_1)\mathbf{i} + d(v_2 - v_1)\mathbf{j}.$

Thus $\overrightarrow{Q_1Q_2} \times \overrightarrow{Q_1Q_4} = \begin{vmatrix} \mathbf{i} & \mathbf{j} & \mathbf{k} \\ a(u_2-u_1) & b(u_2-u_1) & 0 \\ c(v_2-v_1) & d(v_2-v_1) & 0 \end{vmatrix}$

$$= \begin{vmatrix} a(u_2-u_1) & b(u_2-u_1) \\ c(v_2-v_1) & d(v_2-v_1) \end{vmatrix}\mathbf{k}. \quad (4)$$

Since $\mathbf{k}$ is a unit vector, the magnitude of $\overrightarrow{Q_1Q_2} \times \overrightarrow{Q_1Q_4}$ is the absolute value of the coefficient of $\mathbf{k}$ in (4), that is, the absolute value of

$$a(u_2 - u_1)d(v_2 - v_1) - b(u_2 - u_1)c(v_2 - v_1),$$

or $\quad |ad - bc|(u_2 - u_1)(v_2 - v_1).$

Thus the area of the parallelogram S is $|ad - bc|$ times the area of the rectangle R. This proves the theorem. ■

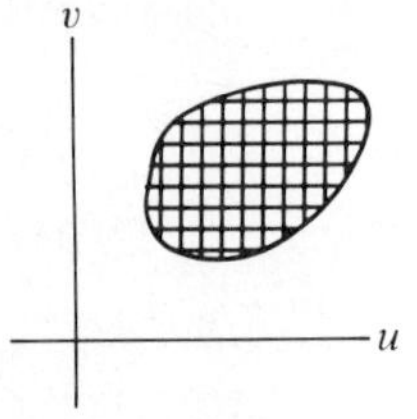

The mapping F in Theorem 1 magnifies the area of *any* region, not just the rectangle described in the theorem, by the same factor, $|ad - bc|$. To see why, consider a typical region R in the uv plane, as in Fig. 13.53. This region may be approximated by many small rectangles of the type considered in Theorem 1, as shown in Fig. 13.54. Since the area of each of the small rectangles in Fig. 13.54 is magnified by the factor $|ad - bc|$, it is plausible that the area of the image of R is $|ad - bc|$ times the area of R.

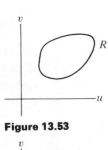

Figure 13.53

Figure 13.54

The magnification defined as a limit

MAGNIFICATION OF A GENERAL MAPPING

Let $F(u, v) = (f(u, v), g(u, v))$ be a mapping such that f and g have continuous partial derivatives. Let $P_0 = (u_0, v_0)$ be a point in the domain of F. The magnification of F at P_0 is defined as follows.

Let R be the rectangle $u_0 \leq u \leq u_0 + \Delta u$, $v_0 \leq v \leq v_0 + \Delta v$. Let S be its image. Then the *magnification* of F at P_0 is

$$\lim_{\Delta u, \Delta v \to 0} \frac{\text{Area of } S}{\text{Area of } R}; \quad (5)$$

13.6 MAGNIFICATION, CHANGE OF COORDINATES, AND THE JACOBIAN (OPTIONAL)

the limit, if it exists, is taken as both Δu and Δv approach 0.

To find the limit (5), we will approximate F near P_0 by a mapping of the special form to which Theorem 1 applies.

THEOREM 2 Let $F(u, v) = (f(u, v), g(u, v))$ be a mapping such that f and g have continuous partial derivatives. Let $P_0 = (u_0, v_0)$ be a point in the domain of F. Then the magnification of F at P_0 is

Recall that

$$f_u = \frac{\partial f}{\partial u}, \quad f_v = \frac{\partial f}{\partial v}.$$

$$\text{abs} \begin{vmatrix} f_u(P_0) & f_v(P_0) \\ g_u(P_0) & g_v(P_0) \end{vmatrix},$$

that is, $|f_u g_v - f_v g_u|$ evaluated at P_0.

PROOF (Informal sketch) In order to avoid a clutter of symbols, take P_0 to be $(0, 0)$, the origin of the uv plane, and assume that $F(P_0)$ is the origin of the xy plane. Thus $f(P_0) = 0$ and $g(P_0) = 0$.

Let R be the rectangle $0 \leq u \leq \Delta u$, $0 \leq v \leq \Delta v$, where Δu and Δv are fixed numbers. Let S be its image in the xy plane.

For (u, v) in R approximate $f(u, v)$ and $g(u, v)$ with the aid of differentials. This is possible since we shall be interested in the case when Δu and Δv are small. (See Fig. 13.55.)

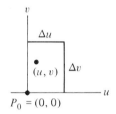

Figure 13.55

Note that $f_u(P_0), f_v(P_0),$ $g_u(P_0), g_v(P_0)$ are just constants.

For (u, v) near P_0, $f(u, v)$ is approximated by $f(P_0)$ plus the differential,

$$f(u, v) \approx f(P_0) + f_u(P_0)u + f_v(P_0)v.$$

Since $f(P_0) = 0$, $f(u, v) \approx f_u(P_0)u + f_v(P_0)v;$

similarly, $g(u, v) \approx g_u(P_0)u + g_v(P_0)v.$ (6)

Introduce the mapping F_* defined by

F_* is linear.

$$F_*(u, v) = (f_u(P_0)u + f_v(P_0)v, \, g_u(P_0)u + g_v(P_0)v), \quad (7)$$

a type of mapping to which Theorem 1 may be applied. By (7), F_* is a good approximation of F when (u, v) is near $(0, 0)$. Since the image of R under the mapping F_* is a parallelogram, the image S of R under the mapping F will closely resemble a parallelogram, as shown in Fig. 13.56.

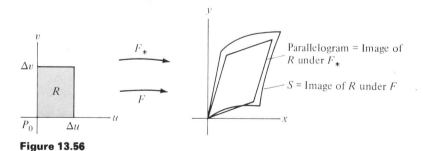

Figure 13.56

By Theorem 1, the area of the parallelogram is the absolute value of

$$\begin{vmatrix} f_u(P_0) & f_v(P_0) \\ g_u(P_0) & g_v(P_0) \end{vmatrix} \Delta u \, \Delta v. \tag{8}$$

Since F is closely approximated by F_* when Δu and Δv are small, it is reasonable to expect that the area of the image of R under F, though not a parallelogram, is close to (8). Consequently, we expect that

$$\lim_{\Delta u, \Delta v \to 0} \frac{\text{Area of } S}{\text{Area of } R} = \text{abs} \begin{vmatrix} f_u(P_0) & f_v(P_0) \\ g_u(P_0) & g_v(P_0) \end{vmatrix}.$$

This concludes the sketch of the argument. ∎

The Jacobian of $F = (f, g)$, that is, the determinant

$$\begin{vmatrix} f_u & f_v \\ g_u & g_v \end{vmatrix}$$

Notations for the Jacobian is denoted J, $J(P)$, $\dfrac{\partial(f, g)}{\partial(u, v)}$,   or $\dfrac{\partial(x, y)}{\partial(u, v)}$.

In the next example the Jacobian is computed for the mapping of Example 4 in the preceding section.

EXAMPLE 1 Let $F(u, v) = (u \cos v, u \sin v)$ for $u > 0$, $0 \leq v < 2\pi$. Find the magnification of F at (u, v).

SOLUTION The magnification is the absolute value of the Jacobian,

$$\left| \frac{\partial(x, y)}{\partial(u, v)} \right| = \text{abs} \begin{vmatrix} \dfrac{\partial}{\partial u}(u \cos v) & \dfrac{\partial}{\partial v}(u \cos v) \\ \dfrac{\partial}{\partial u}(u \sin v) & \dfrac{\partial}{\partial v}(u \sin v) \end{vmatrix}$$

$$= \text{abs} \begin{vmatrix} \cos v & -u \sin v \\ \sin v & u \cos v \end{vmatrix}$$

$$= |u \cos^2 v + u \sin^2 v|$$

$$= |u|.$$

Since $u > 0$, the local magnification at (u, v) is u. (Compare this with Exercise 14 in the preceding section.) ∎

THE JACOBIAN AND CHANGE OF VARIABLES

Let F be a one-to-one mapping from the uv plane to the xy plane. Let R be a region in the uv plane and S its image in the xy plane. Let $h(x, y)$ be a real-valued function defined on S. (See Fig. 13.57.)

We use the letter h since f and g are used for $F = (f, g)$.

The integral $\int_S h(x, y) \, dA$ can be expressed as an integral of a different function over the region R. If R is a simpler region than S, then the new

13.6 MAGNIFICATION, CHANGE OF COORDINATES, AND THE JACOBIAN (OPTIONAL)

integral may be easier to compute. This is analogous to the substitution technique, which expresses a definite integral of a given function over one interval as the definite integral of another function over another interval. To see how to express an integral over S as an integral over R, go back to the approximating sums used in the definition of an integral.

Use Q for points in S and P for points in R.

Consider a typical partition $S_1, S_2, \ldots, S_n$ of S and corresponding sampling points $Q_1, Q_2, \ldots, Q_n$. Let $R_1, R_2, \ldots, R_n$ be the corresponding partition of R such that S_i is the image of R_i. Let P_i be the point in R_i whose image is Q_i, $1 \leq i \leq n$. (See Fig. 13.58.)

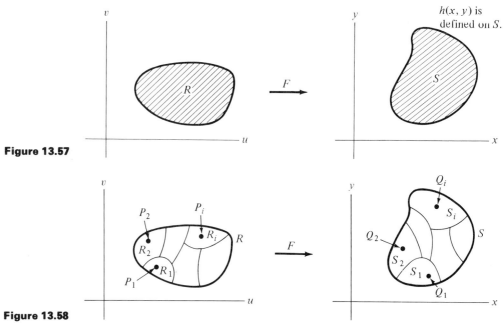

Figure 13.57

Figure 13.58

Then

$$\sum_{i=1}^{n} h(Q_i)(\text{Area of } S_i) \qquad (9)$$

is an approximating sum for $\int_S h(Q)\, dA$. But the area of S_i is approximately $|J(P_i)|$ times the area of R_i. So (9) is approximately

$$\sum_{i=1}^{n} h(Q_i)|J(P_i)|\, (\text{Area of } R_i). \qquad (10)$$

Since $Q_i = F(P_i)$, (10) equals

$$\sum_{i=1}^{n} h(F(P_i))|J(P_i)|\, (\text{Area of } R_i). \qquad (11)$$

The sum (11) is an approximating sum for the definite integral of the function $h(F(P))|J(P)|$ over the set R. Thus it is reasonable to expect that

$$\int_S h(Q)\, dA = \int_R h(F(P))|J(P)|\, dA. \qquad (12)$$

This argument suggests the following theorem, the proof of which is part of an advanced calculus course.

THEOREM 3 Let F be a one-to-one mapping from the uv plane to the xy plane, $x = f(u, v)$, $y = g(u, v)$. Let S be the image in the xy plane of the set R in the uv plane. Let h be a real-valued function defined on S. Then, if $J(P)$ is never 0 in R,

$$\int_S h(Q) \, dA = \int_R h(F(P)) |J(P)| \, dA. \quad \blacksquare$$

EXAMPLE 2 Let R be the triangle in the uv plane with vertices $(0, 0)$, $(1, 0)$, and $(0, 1)$. Let F be the mapping

$$x = 2u - 3v, \qquad y = 5u + 7v.$$

Let S be the image of R under F. Evaluate the integral $\int_S x \, dA$ over the set S in the xy plane.

SOLUTION Since F is a linear mapping, S is a triangle. Its vertices are

$$F(0, 0) = (0, 0), \qquad F(1, 0) = (2, 5), \qquad F(0, 1) = (-3, 7).$$

(See Fig. 13.59.)

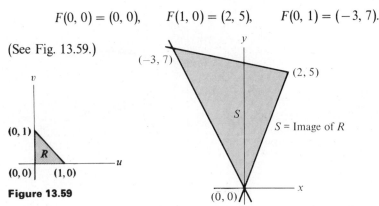

Figure 13.59

Since R is a simpler region over which to integrate, replace $\int_S x \, dA$ by an integral over R with the aid of Theorem 3:

$$\int_S x \, dA = \int_R (2u - 3v) \left| \frac{\partial(x, y)}{\partial(u, v)} \right| dA.$$

The Jacobian in this case is

$$\frac{\partial(x, y)}{\partial(u, v)} = \begin{vmatrix} \dfrac{\partial}{\partial u}(2u - 3v) & \dfrac{\partial}{\partial v}(2u - 3v) \\ \dfrac{\partial}{\partial u}(5u + 7v) & \dfrac{\partial}{\partial v}(5u + 7v) \end{vmatrix}$$

$$= \begin{vmatrix} 2 & -3 \\ 5 & 7 \end{vmatrix} = 29.$$

13.6 MAGNIFICATION, CHANGE OF COORDINATES, AND THE JACOBIAN (OPTIONAL)

The Jacobian is constant (which is consistent with Theorem 1). Thus

$$\int_S x \, dA = \int_R (2u - 3v)(29) \, dA$$

$$= 29 \int_0^1 \left(\int_0^{1-u} (2u - 3v) \, dv \right) du.$$

The first integration gives

$$\int_0^{1-u} (2u - 3v) \, dv = \left(2uv - \frac{3v^2}{2} \right) \Big|_{v=0}^{v=1-u}$$

$$= 2u(1 - u) - \frac{3(1 - u)^2}{2}.$$

The second integration gives, as can be checked,

If in doubt about the value of the Jacobian, compute $\int_S x \, dA$ directly.

$$\int_0^1 \left(2u(1 - u) - \frac{3(1 - u)^2}{2} \right) du = -\frac{1}{6}.$$

Thus

$$\int_S x \, dA = -\frac{29}{6}. \quad \blacksquare$$

Why the extra r in polar coordinates

Theorem 3 is what lies behind the appearance of the extra r when integration is carried out with polar coordinates, as will now be shown.

Let F be the mapping $F(u, v) = (u \cos v, u \sin v)$. Let R be the rectangle in the uv plane, $a \leq u \leq b$, $\alpha \leq v \leq \beta$. Let S be the image of R in the xy plane. As shown in Example 4 of the preceding section, S is bounded by parts of two rays and two circles. Let $h(x, y)$ be a real-valued function defined on S. Then, by Theorem 3 and Example 1,

$$\int_S h(x, y) \, dA = \int_R h(u \cos v, u \sin v) \left| \frac{\partial(x, y)}{\partial(u, v)} \right| dA$$

$$= \int_R h(u \cos v, u \sin v) \, u \, dA.$$

Here is where the r appears, well before considering repeated integrals.

As was shown in Example 4 in the preceding section, u and v are the polar coordinates r and θ of the point (x, y). Thus

$$\int_S h(x, y) \, dA = \int_R h(r \cos \theta, r \sin \theta) \, r \, dA. \tag{13}$$

The second integral in (13) is over a set R in the $r\theta$ (or uv) plane. But r and θ are the *rectangular* coordinates in the $r\theta$ plane. This second integral can be evaluated by repeated integration in rectangular coordinates:

$$\int_R h(r \cos \theta, r \sin \theta) \, r \, dA = \int_\alpha^\beta \left(\int_a^b h(r \cos \theta, r \sin \theta) \, r \, dr \right) d\theta. \tag{14}$$

So the extra r appears in (13) as a Jacobian. It appears in (14) simply as part of the integrand in a repeated integral in rectangular coordinates.

The Jacobian must be used whenever we shift to nonrectangular coordinates. This is the analog for the plane of the "substitution in the definite integral" theorem of Sec. 7.1. That theorem relates an integral over an interval $[a, b]$ to an integral over another interval $[A, B]$. We state that theorem in symbols similar to those used in Theorem 3:

Let $x = f(u)$ map the interval $[A, B]$ on the u axis to the interval $[a, b]$ on the x axis. Assume $f'(u) > 0$. Let $h(x)$ be defined on $[a, b]$. Then

$$\int_a^b h(x)\, dx = \int_A^B h(f(u)) f'(u)\, du.$$

See Fig. 13.60.

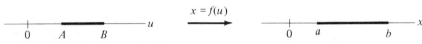

Figure 13.60

Exercises

In Exercises 1 to 6 compute the local magnification of area by the given mappings at the given points in the uv plane.

1. $F(u, v) = (u - v, u + 3v)$ at
  (a) $(1, 0)$ (b) $(2, 3)$.
2. $F(u, v) = (u - 3v, -u + 2v)$ at
  (a) $(0, 1)$ (b) $(2, 1)$.
3. $F(u, v) = (uv, v^2)$, $u, v > 0$, at
  (a) $(1, 2)$ (b) $(3, 1)$.
4. $F(u, v) = (1/u, 1/v)$, $u, v > 0$, at
  (a) $(2, 3)$ (b) $(\frac{1}{2}, 4)$.
5. $F(u, v) = (e^u \cos v, e^u \sin v)$, $0 \le v < 2\pi$, at
  (a) $(1, \pi/4)$ (b) $(2, \pi/6)$.
6. $F(u, v) = (u/(u^2 + v^2), v/(u^2 + v^2))$, $u^2 + v^2 \ne 0$, at
  (a) $(3, 1)$ (b) $(1, 0)$.

In each of Exercises 7 and 8 find the area of the image of the given region by the given mapping F.

7. R is the triangle with vertices $(0, 0)$, $(1, 0)$, and $(0, 1)$ and $F(u, v) = (u - v, u + 2v)$.
8. R is the rectangle bounded by the lines $u = 1$, $u = 2$, $v = 3$, and $v = 7$ and $F(u, v) = (3u + 2v, 2u - v)$.
9. Let a, b, c, and d be constants such that $ad - bc \ne 0$. Let

   $$x = au + bv, \qquad y = cu + dv.$$

   (a) Show that the Jacobian of the mapping is $ad - bc$ at all points.
   (b) Show that Theorem 1 is a special case of Theorem 2.
10. Let the Jacobian of a mapping be 3 at $(2, 4)$. Let R be a small region around $(2, 4)$ of area 0.05. Approximately how large is the image of R under the mapping?

11. Let R be the disk of radius 1 in the uv plane. Let S be the region within the ellipse $x^2/a^2 + y^2/b^2 = 1$. Let $F(u, v) = (au, bv)$.
    (a) Show that S is the image of R.
    (b) Express $\int_S y^2\, dA$ as an integral over R.
    (c) Evaluate the integral over R by polar coordinates. (If you have some spare time, evaluate $\int_S y^2\, dA$ directly by a repeated integral in rectangular coordinates. The experience will deepen your appreciation of the Jacobian.)
12. (See Exercise 11.) Evaluate $\int_S y\, dA$ where S is the region within the ellipse $x^2/a^2 + y^2/b^2 = 1$ that lies in the first quadrant.
13. Let S be the image of the region R under the one-to-one mapping F. Show that the area of S equals $\int_R |J|\, dA$.
14. Let R be the square with vertices $(1, 1)$, $(2, 1)$, $(2, 2)$, and $(1, 2)$ in the uv plane. Let $F(u, v) = (u^3 + v, v^3/3)$. Find the area of the image of R.

■

15. Let R be the triangle bounded by $u = v$, $v = 0$, and $u = 1$. Let S be the image of R under the mapping $x = u^2$, $y = 2v$.
    (a) Draw S, which has one curved side.
    (b) Compute the area of S with the aid of the formula in Exercise 13.
    (c) Compute the area of S directly.
16. Consider the mapping $x = u^2 - v^2$, $y = 2uv$. Let R be the square whose vertices are $(1, 0)$, $(2, 0)$, $(2, 1)$, and $(1, 1)$.

(a) Show that, when $u = 1$, the image of (u, v) lies on the curve $x = 1 - (y/2)^2$.
(b) Show that, when $u = 2$, the image of (u, v) lies on the curve $x = 4 - (y/4)^2$.
(c) Show that the image of the line $v = 0$ is the positive x axis.
(d) Show that the image of the line $v = 1$ is the curve $x = (y/2)^2 - 1$.
(e) Draw S, the image of R. (It has three curved sides and one straight side.)
(f) Find the area of S.

17 A mapping F is said to *preserve area* if the area of the image of R is equal to the area of R for all regions that have an area. Show that the mapping $F(u, v) = (4u + 3v, 3u + 2v)$ preserves area.

18 (See Exercise 17.)
(a) Show that the mapping defined by
$$x = u - v^2 - 2u^2v - u^4, \quad y = v + u^2$$
preserves area.
(b) Sketch the image of the square whose vertices are $(0, 0)$, $(1, 0)$, $(1, 1)$, and $(0, 1)$.
(c) What is the area of the region sketched in (b)?

19 Consider only positive u, v, x, y. Define a mapping by setting $x = u^{1/3}v^{2/3}$, $y = u^{2/3}v^{1/3}$.
(a) Show that $x^2 = vy$ and $y^2 = ux$.
(b) Let R in the uv plane be the rectangle bounded by the lines $u = 1, u = 2, v = 3$, and $v = 4$. Let S be the image in the xy plane of R. Show that S is bounded by the four parabolas $y^2 = x$, $y^2 = 2x$, $x^2 = 3y$, and $x^2 = 4y$.
(c) Draw S.

(d) Compute the area of S by integrating the Jacobian over R.

■ ■

20 Let $x = 2u + v$, $y = 3u + 2v$ be a mapping from the uv plane to the xy plane.
(a) Show that the mapping does not change areas (though, as (b) will show, it changes shapes).
(b) Sketch the image of the circle $u^2 + v^2 = 1$.
(c) Let R be the region in the uv plane bounded by the circle $u^2 + v^2 = 1$. Let S be its image in the xy plane. Evaluate $\int_S x^2 \, dA$.

21 Consider only positive u and $0 \le v \le \pi/2$.
(a) Show that, if $x = u \cos v$ and $y = u \sin v$, then $u = \sqrt{x^2 + y^2}$ and $v = \tan^{-1}(y/x)$.
(b) Show that $\partial(u, v)/\partial(x, y) = 1/\sqrt{x^2 + y^2}$.
(c) Show that $\partial(u, v)/\partial(x, y)$ is the reciprocal of $\partial(x, y)/\partial(u, v)$.
(d) Why is (c) to be expected? (Think optically.)

22 This exercise examines the significance of the sign of the Jacobian.
(a) Let $F(u, v) = (u - v, u + v)$. Let $P_1 = (0, 0)$, $P_2 = (1, 0)$, and $P_3 = (1, 1)$. Plot P_1, P_2, and P_3 in the uv plane and $F(P_1), F(P_2)$, and $F(P_3)$ in the xy plane. Note that the order that P_1, P_2, and P_3 is counterclockwise, the order of $F(P_1), F(P_2)$, and $F(P_3)$ is also counterclockwise, and J is positive.
(b) Let $F(u, v) = (u + 2v, u + v)$ and P_1, P_2, and P_3 be as in (a). Show that the order $F(P_1), F(P_2)$, and $F(P_3)$ is now clockwise and J is negative.
It can be shown that, in general, when the Jacobian is positive, the mapping F preserves "orientation" and, when the Jacobian is negative, it reverses "orientation."

13.7 The definite integral of a function over a region in space

The notion of a definite integral over an interval in the line or over a plane region generalizes to integrals over solids located in space. (These solids will be assumed to be bounded by smooth surfaces or planes.) Rather than plunge directly into the definition, let us first illustrate the idea with a problem.

INTRODUCTION

PROBLEM A cube of side 4 centimeters is made of a material of varying density. Near one corner A it is very light; at the opposite corner it is very dense. In fact, the density $f(P)$ (in grams per cubic centimeter) at any point P in the cube is the square of the distance from A to P (in centimeters). How do we estimate the mass of the cube? (See Fig. 13.61.)

862 DEFINITE INTEGRALS OVER PLANE AND SOLID REGIONS

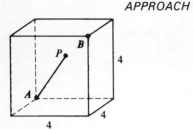

The density at P is the square of the distance AP.

Figure 13.61

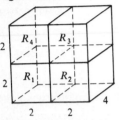

The cube is partitioned into 4 boxes, R_1, R_2, R_3, R_4.

Figure 13.62

APPROACH We can proceed exactly as in the case of the string of Sec. 5.1 and the rectangular plate of Sec. 13.1. First, partition the cube into regions R_1, $R_2, \ldots, R_d$; then compute the density at a selected point P_i in each R_i and form the sum

$$f(P_1)V_1 + f(P_2)V_2 + \cdots + f(P_n)V_n,$$

where V_i is the volume of R_i. As the R_i's become smaller, we obtain more reliable estimates of the total mass of the cube.

Observe first of all that the maximum density is the square of the length of the longest diagonal. This density is $\overline{AB}^2 = 4^2 + 4^2 + 4^2 = 48$ grams per cubic centimeter. Since the total volume is $4 \cdot 4 \cdot 4 = 64$ cubic centimeters, the total mass is less than $48 \cdot 64 = 3072$ grams.

The arithmetic in evaluating even the simplest approximating sum is tedious. It may be of value, though, to go through the drudgery of computing one such sum. The following is a sample.

Partition the cube into four 2- by 2- by 4-centimeter boxes, as shown and labeled in Fig. 13.62. This table displays the computation:

Region	Volume	Minimum density	Maximum density	Mass is between
R_1	16	0	$2^2 + 2^2 + 4^2 = 24$	0 and 384 grams
R_2	16	$2^2 = 4$	$4^2 + 2^2 + 4^2 = 36$	64 and 576 grams
R_3	16	$2^2 + 2^2 = 8$	$4^2 + 4^2 + 4^2 = 48$	128 and 768 grams
R_4	16	$2^2 = 4$	$2^2 + 4^2 + 4^2 = 36$	64 and 576 grams

Thus the mass of the cube is between

$$0 + 64 + 128 + 64 = 256 \text{ grams}$$

and

$$384 + 576 + 768 + 576 = 2304 \text{ grams}.$$

This is more precise information than the fact that the mass is less than 3072 grams. ∎

If the cube is cut into smaller regions, perhaps sixty-four 1- by 1- by 1-centimeter cubes, more accurate estimates can be made. The important idea is that the procedure for making an approximation is practically the same as the one that led to the sums

$$\sum_{i=1}^{n} f(c_i)(x_i - x_{i-1})$$

of Chap. 5 and to the sums $\quad \sum_{i=1}^{n} f(P_i)A_i$

of Sec. 13.1.

Two definitions are needed before defining the definite integral of a function over a region R in space.

DEFINITION *Diameter of a region in space.* Let S be a set of points in space bounded by some surface or polyhedron. The *diameter* of S is the largest distance between two points of S.

For instance, the diameter of a cube of side s is $s\sqrt{3}$, the length of its longest diagonal. The diameter of a ball is its customary diameter (twice the radius).

DEFINITION *Mesh of a partition in space.* Let $R_1, R_2, \ldots, R_n$ be a partition of a region R in space. The *mesh* of this partition is the largest of the diameters of the regions $R_1, R_2, \ldots, R_n$.

THE DEFINITE INTEGRAL $\int_R f(P) \, dV$

The typical function of interest will have some region R in space as its domain. A function f will assign to each point P in R a number, denoted $f(P)$. For the sake of concreteness, think of $f(P)$ as the density at P or temperature at P.

Don't try to graph a function $f(x, y, z)$.

The graph of a function of one variable, $y = f(x)$, is a curve in the xy plane, the set of points $(x, f(x))$. The graph of a function of two variables, $z = f(x, y)$, is a surface in space, the set of points $(x, y, f(x, y))$. The graph of a function of three variables (that is, a function defined on a region in space) is a set in four-dimensional space, the set of the points $(x, y, z, f(x, y, z))$. Since our eyes and intuition are accustomed to a three-dimensional world, this graph is of little use. For this reason, it is best to think of a function defined on a solid region as, say, describing the varying density of a distribution of matter.

DEFINITION *The definite integral of a function f over a set R in space.* Let f be a function that assigns to each point P of a region R in space a number $f(P)$. Consider the sum

$$f(P_1)V_1 + f(P_2)V_2 + \cdots + f(P_n)V_n.$$

formed from a partition $R_1, R_2, \ldots, R_n$ of R, where V_i is the volume of R_i, and P_i is in R_i. If these sums approach a certain number as the mesh of the partition shrinks toward 0 (no matter how P_i is chosen in R_i), we call that certain number the definite integral of f over the region R. The definite integral of f over R is denoted

$$\int_R f(P) \, dV.$$

If $f(P)$ is thought of as the density at P of some solid matter, the definite integral can be interpreted as the total mass of the solid.

EXAMPLE 1 If $f(P) = 1$ for each point P in a solid region R, compute $\int_R f(P)\, dV$.

SOLUTION Each approximating sum

$$\sum_{i=1}^{n} f(P_i) V_i$$

has the value

$$\sum_{i=1}^{n} 1 \cdot V_i = V_1 + V_2 + \cdots + V_n$$

$$= \text{Volume of } R.$$

Hence

$$\int_R f(P)\, dV = \text{Volume of } R,$$

a fact that will be useful for computing volumes. ∎

SOME APPLICATIONS OF $\int_R f(P)\, dV$

The average value of a function f defined on a region R in space is defined as

Average of a function

$$\frac{\int_R f(P)\, dV}{\text{Volume of } R}.$$

This is the analog of the definition of the average of a function over an interval (Sec. 8.8) or the average of a function over a plane region (Sec. 13.1). If f describes the density of matter in R, then the average value of f is the density of a *homogeneous* solid occupying R and having the same total mass as the given solid. (For if the number

$$\frac{\int_R f(P)\, dV}{\text{Volume of } R}$$

is multiplied by

$$\text{Volume of } R,$$

the result is

$$\int_R f(P)\, dV,$$

which is the total mass.)

The average value and the mass are important applications of the definite integral over a solid, as they are for definite integrals over an interval or planar region. The total gravitational attraction of the sun on the earth or of the earth on a satellite and the centers of mass of physical bodies will provide further applications.

Moments, centers of mass, and moments of inertia are defined for spatial regions very much as they are for planar regions. If $\delta(P)$ is the density of the material at P, then we have:

13.7 THE DEFINITE INTEGRAL OF A FUNCTION OVER A REGION IN SPACE

Name	Formula
M = Mass	$\int_R \delta(P)\,dV$
M_{xy} = Moment with respect to the xy plane	$\int_R z\delta(P)\,dV$
M_{xz} = Moment with respect to the xz plane	$\int_R y\delta(P)\,dV$
M_{yz} = Moment with respect to the yz plane	$\int_R x\delta(P)\,dV$
$(\bar{x}, \bar{y}, \bar{z})$ = Center of mass	$\bar{x} = \dfrac{\int_R x\delta(P)\,dV}{\text{Mass in } R}$,
	$\bar{y} = \dfrac{\int_R y\delta(P)\,dV}{\text{Mass in } R}$,
	$\bar{z} = \dfrac{\int_R z\delta(P)\,dV}{\text{Mass in } R}$
$(\bar{x}, \bar{y}, \bar{z})$ = Centroid	$\bar{x} = \dfrac{\int_R x\,dV}{\text{Volume of } R}$,
	$\bar{y} = \dfrac{\int_R y\,dV}{\text{Volume of } R}$,
	$\bar{z} = \dfrac{\int_R z\,dV}{\text{Volume of } R}$
I_x = Moment of inertia with respect to x axis	$\int_R (y^2 + z^2)\delta(P)\,dV$
I_y = Moment of inertia with respect to y axis	$\int_R (x^2 + z^2)\delta(P)\,dV$
I_z = Moment of inertia with respect to z axis	$\int_R (x^2 + y^2)\delta(P)\,dV$
Moment of inertia with respect to any line L	$\int_R f(P)\delta(P)\,dV$, where $f(P)$ = square of distance from P to L

In order to evaluate definite integrals over spatial regions, it is necessary to describe these regions in terms of a coordinate system.

DESCRIBING SOLID REGIONS WITH RECTANGULAR COORDINATES

Figure 13.63

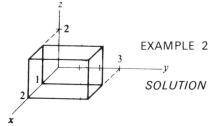

EXAMPLE 2 Describe in terms of x, y, and z the rectangular box shown in Fig. 13.63.

SOLUTION First, x may be any number between 1 and 2. For each x, the cross section of the box made by a plane through $(x, 0, 0)$ and parallel to the yz plane is a rectangle. On this typical rectangle y varies from 0 to 3, indepen-

866 DEFINITE INTEGRALS OVER PLANE AND SOLID REGIONS

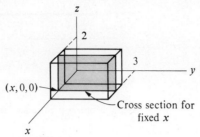

Figure 13.64

dently of x. (See Fig. 13.64.)

Each pair x and y that satisfies the above conditions, $1 \leq x \leq 2$ and $0 \leq y \leq 3$, determines a cross section of the box by the line parallel to the z axis passing through $(x, y, 0)$. On the cross section illustrated in Fig. 13.65, which is a line segment, z varies from 0 to 2, independently of x and y.

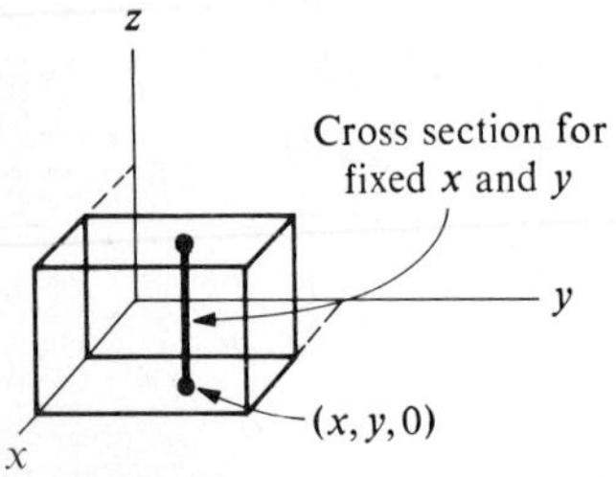

The description of the box is

$$1 \leq x \leq 2, \quad 0 \leq y \leq 3, \quad 0 \leq z \leq 2. \quad \blacksquare$$

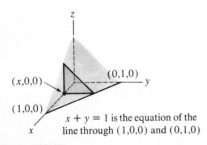

Figure 13.65

EXAMPLE 3 Describe the cross sections of the tetrahedron bounded by the planes $x = 0$, $y = 0$, $z = 0$, and $x + y + z = 1$, as shown in Fig. 13.66.

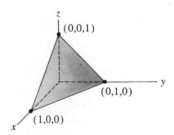

Figure 13.66

SOLUTION For any x in $[0, 1]$, the cross section of the tetrahedron made by the plane parallel to the yz plane and passing through $(x, 0, 0)$ is a triangle, such as the one shaded in Fig. 13.67.

On this typical triangle, y varies from 0 up to the value for y that satisfies the equation $x + y = 1$, that is, up to $y = 1 - x$.

Finally, for each x and y given above, z varies from 0 up to the value of z that satisfies the equation $x + y + z = 1$, that is, up to $z = 1 - x - y$. This is a description of the behavior of z on the line parallel to the z axis and passing through $(x, y, 0)$. The tetrahedron is described by the equations

$$0 \leq x \leq 1, \quad 0 \leq y \leq 1 - x, \quad 0 \leq z \leq 1 - x - y.$$

That is, x goes from 0 to 1; for each x, y goes from 0 to $1 - x$; for each x and y, z goes from 0 to $1 - x - y$. $\blacksquare$

$x + y = 1$ is the equation of the line through $(1,0,0)$ and $(0,1,0)$

Figure 13.67

In the next example the description of the solid is obtained by first holding x and y fixed and letting z vary.

EXAMPLE 4 Describe in rectangular coordinates the ball of radius 4 whose center is at the origin.

SOLUTION Hold (x, y) fixed in the xy plane and consider the way z varies on the line parallel to the z axis that passes through the point $(x, y, 0)$. Since the sphere that bounds the ball has the equation

$$x^2 + y^2 + z^2 = 16,$$

for each appropriate (x, y), z varies from

$$-\sqrt{16 - x^2 - y^2} \quad \text{to} \quad \sqrt{16 - x^2 - y^2}.$$

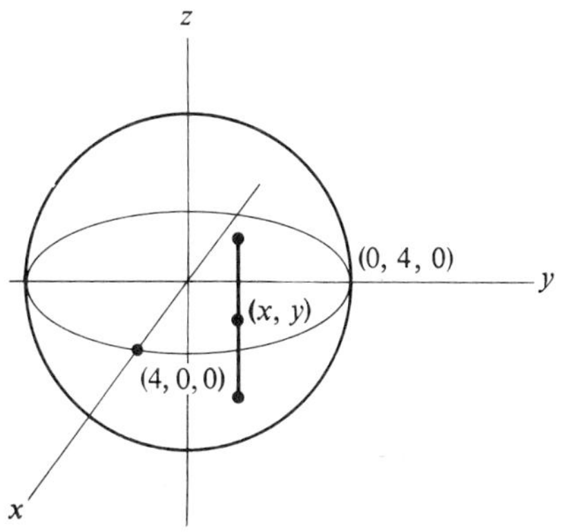

This describes the line segment shown in Fig. 13.68.

Next describe the possible values of x and y. Since (x, y) ranges over a disk of radius 4 and center $(0, 0)$ in the xy plane,

$$-4 \leq x \leq 4, \quad -\sqrt{16 - x^2} \leq y \leq \sqrt{16 - x^2}.$$

The ball, therefore, has the description

$$-4 \leq x \leq 4, \quad -\sqrt{16 - x^2} \leq y \leq \sqrt{16 - x^2}, \quad -\sqrt{16 - x^2 - y^2} \leq z \leq \sqrt{16 - x^2 - y^2}. \quad \blacksquare$$

Figure 13.68

COMPUTING $\int_R f(P) \, dV$ WITH RECTANGULAR COORDINATES

The repeated integral in rectangular coordinates for evaluating $\int_R f(P) \, dV$ is similar to that for evaluating integrals over plane sets. It involves three integrations instead of two. The limits of integration are determined by the description of R in rectangular coordinates. If R has the description

$$a \leq x \leq b, \quad y_1(x) \leq y \leq y_2(x), \quad z_1(x, y) \leq z \leq z_2(x, y),$$

then

$$\int_R f(P) \, dV = \int_a^b \left[\int_{y_1(x)}^{y_2(x)} \left(\int_{z_1(x, y)}^{z_2(x, y)} f(x, y, z) \, dz \right) dy \right] dx.$$

Some examples illustrate how this formula is applied. In Exercise 36 an argument for its plausibility is presented.

EXAMPLE 5 Compute $\int_R z \, dV$, where R is the tetrahedron in Example 3.

SOLUTION The description of the tetrahedron is

$$0 \leq x \leq 1, \quad 0 \leq y \leq 1 - x, \quad 0 \leq z \leq 1 - x - y.$$

Hence

$$\int_R z \, dV = \int_0^1 \int_0^{1-x} \int_0^{1-x-y} z \, dz \, dy \, dx.$$

Compute the inner integral first, treating x and y as constants. By the fundamental theorem,

$$\int_0^{1-x-y} z \, dz = \frac{z^2}{2} \bigg|_{z=0}^{z=1-x-y} = \frac{(1-x-y)^2}{2}.$$

The next integration, where x is fixed, is

$$\int_0^{1-x} \frac{(1-x-y)^2}{2} \, dy = -\frac{(1-x-y)^3}{6} \bigg|_{y=0}^{y=1-x}$$

$$= -\frac{0^3}{6} + \frac{(1-x)^3}{6}$$

$$= \frac{(1-x)^3}{6}.$$

The third integration is

$$\int_0^1 \frac{(1-x)^3}{6} \, dx = -\frac{(1-x)^4}{24} \bigg|_0^1$$

$$= -\frac{0^4}{24} + \frac{1^4}{24}$$

$$= \frac{1}{24}. \quad \blacksquare$$

Having computed $\int_R z \, dV$, we can find the z coordinate of the centroid of the tetrahedron in Example 3 with just a little more effort. By definition,

$$\bar{z} = \frac{\int_R z \, dV}{\text{Volume of } R}.$$

The volume of the tetrahedron is $\frac{1}{3} \cdot$ height $\cdot$ area of base $= \frac{1}{3} \cdot 1 \cdot \frac{1}{2} = \frac{1}{6}$. Thus

$$\bar{z} = \frac{\frac{1}{24}}{\frac{1}{6}} = \frac{1}{4},$$

which is one-fourth of the distance from the base to the opposite vertex.

EXAMPLE 6 Compute the moment of inertia of a homogeneous cube of side a and mass M about an edge.

SOLUTION For a convenient description of the cube place the origin of a rectangular coordinate system at a corner, as in Fig. 13.69. We shall compute the moment of inertia about the edge lying on the x axis.

The density at any point P in the cube is M/a^3, since the volume of the cube is a^3. The square of the distance from $P = (x, y, z)$ to the x axis is $y^2 + z^2$. The definition of moment of inertia in the table provides this integral:

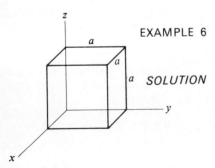

Figure 13.69

$$\int_R (y^2 + z^2)\frac{M}{a^3}\, dV.$$

To evaluate the integral, it is necessary to describe the cube R in rectangular coordinates:

$$0 \leq x \leq a, \quad 0 \leq y \leq a, \quad 0 \leq z \leq a.$$

Thus
$$\int_R (y^2 + z^2)\frac{M}{a^3}\, dV = \frac{M}{a^3}\int_0^a \int_0^a \int_0^a (y^2 + z^2)\, dz\, dy\, dx.$$

The three integrations can be carried out with the aid of the fundamental theorem of calculus.

The first integration is

$$\int_0^a (y^2 + z^2)\, dz = \left(y^2 z + \frac{z^3}{3}\right)\Big|_{z=0}^{z=a}$$
$$= y^2 a + \frac{a^3}{3}$$
$$= a\left(y^2 + \frac{a^2}{3}\right).$$

The second is:
$$\int_0^a a\left(y^2 + \frac{a^2}{3}\right) dy = a\left(\frac{y^3}{3} + \frac{a^2 y}{3}\right)\Big|_{y=0}^{y=a}$$
$$= a\left(\frac{a^3}{3} + \frac{a^3}{3}\right)$$
$$= \frac{2a^4}{3}.$$

Finally,
$$\int_0^a \frac{2a^4}{3}\, dx = \frac{2a^5}{3}.$$

The moment of inertia about an edge is thus $(M/a^3)(2a^5/3) = 2Ma^2/3$. (This is as though all the mass were concentrated at a point a distance $\sqrt{2/3}\,a$ from the edge.) ∎

Exercises

Exercises 1 to 8 concern the definition of $\int_R f(P)\, dV$.

1. Find upper and lower estimates for the mass of the cube in the opening problem of this section by partitioning it into eight cubes. (See Fig. 13.61.)
2. Using the same partition as in the text, estimate the mass of the cube, but select as the P_i's the centers of the four rectangular boxes.
3. If R is a ball of radius r and $f(P) = 5$ for each point in R, compute $\int_R f(P)\, dV$ by examining approximating sums. Recall that the ball has volume $\tfrac{4}{3}\pi r^3$.
4. How would you define the average distance from points of a certain set in space to a fixed point P_0?
5. Estimate the mass of the cube described in the opening problem by cutting it into eight congruent cubes and using their centers as the P_i's.
6. If R is a three-dimensional set, and $f(P)$ is never more than 8 for all P in R,
   (a) what can we say about the maximum possible value of $\int_R f(P)\, dV$?
   (b) what can we say about the average of f over R?

7 What is the mesh of the partition of the cube used in the text?
8 What is the mesh of the partition used in Exercise 1?

In Exercises 9 to 14 draw the solids described.

9 $1 \le x \le 3, 0 \le y \le 2, 0 \le z \le x$.
10 $0 \le x \le 1, 0 \le y \le 1, 1 \le z \le 1 + x + y$.
11 $0 \le x \le 1, 0 \le y \le x^2, 0 \le z \le 1$.
12 $0 \le x \le 1, x^2 \le y \le x, 0 \le z \le x + y$.
13 $-1 \le x \le 1, -1 \le y \le 1, 0 \le z \le \sqrt{4 - x^2 - y^2}$.
14 $0 \le x \le 1, 0 \le y \le \sqrt{9 - x^2}, 0 \le z \le \sqrt{9 - x^2 - y^2}$.

In Exercises 15 to 18 evaluate the repeated integrals.

15 $\int_0^1 \left[\int_0^2 \left(\int_0^x z \, dz \right) dy \right] dx.$

16 $\int_0^1 \left[\int_{x^3}^{x^2} \left(\int_0^{x+y} z \, dz \right) dy \right] dx.$

17 $\int_2^3 \left[\int_x^{2x} \left(\int_0^1 (x + z) \, dz \right) dy \right] dx.$

18 $\int_0^1 \left[\int_0^x \left(\int_0^3 (x^2 + y^2) \, dz \right) dy \right] dx.$

19 A homogeneous rectangular solid box has mass M and sides of lengths a, b, and c. Find its moment of inertia about an edge of length a.

20 A rectangular homogeneous box of mass M has dimensions a, b, and c. Show that the moment of inertia of the box around a line through its center and parallel to the side of length a is $M(b^2 + c^2)/12$.

21 Let R be the tetrahedron whose vertices are $(0, 0, 0)$, $(a, 0, 0)$, $(0, b, 0)$, and $(0, 0, c)$, where a, b, and c are positive.
(a) Sketch the tetrahedron.
(b) Find the equation of its top surface.
(c) Compute $\int_R z \, dV$.
(d) Find $\bar{z}$, the z coordinate of the centroid of R.

22 Assume that the tetrahedron in Exercise 21 is filled with a homogeneous material of total mass M. Find its moment of inertia about the z axis.

23 A right solid circular cone has altitude h, radius a, constant density, and mass M.
(a) Why is its moment of inertia about its axis less than Ma^2?
(b) Show that its moment of inertia about its axis is $3Ma^2/10$.

24 Compute $\int_R z \, dV$, where R is the region above the rectangle whose vertices are $(0, 0, 0)$, $(2, 0, 0)$, $(2, 3, 0)$, and $(0, 3, 0)$ and below the plane $z = x + 2y$.

25 Find the mass of the cube in the opening problem. (See Fig. 13.61.)

26 Find the average value of the square of the distance from a corner of a cube of side a to points in the cube.

27 Compute $\int_R xy \, dV$ for the tetrahedron of Example 3.

28 A solid consists of all points below the surface $z = xy$ that are above the triangle whose vertices are $(0, 0, 0)$, $(1, 0, 0)$, and $(0, 2, 0)$. If the density at (x, y, z) is $x + y$, find the total mass.

29 Without using a repeated integral, evaluate $\int_R x \, dV$, where R is a spherical ball whose center is $(0, 0, 0)$ and whose radius is a.

30 (a) Describe in rectangular coordinates the cylinder of radius a and height h shown in Fig. 13.70.

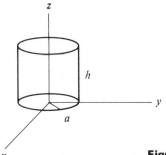

Figure 13.70

(b) Assume that the cylinder in (a) is made of a homogeneous material of total mass M. Find its moment of inertia about the z axis.

31 Find the moment of inertia of the cylinder in Exercise 30 about the x axis.

32 (a) Describe in rectangular coordinates the right circular cone of radius r and height h if its axis is on the positive z axis and its vertex is at the origin. Draw the cross sections for fixed x and for fixed x and y.
(b) Find the z coordinate of its centroid.

33 The center of mass of a solid of mass M is located at $(0, 0, 0)$. Let its moment of inertia about the x axis be I.
(a) Find the moment of inertia of the solid about a line parallel to the x axis and a distance k from it.
(b) About which line parallel to the x axis is the moment of inertia of the solid least?

■ ■

34 The work done in lifting a weight of w pounds a vertical distance of x feet is wx foot-pounds. Imagine that through geological activity a mountain is formed consisting of material originally at sea level. Let the density of the material near point P in the mountain be $g(P)$ pounds per cubic foot and the height of P be $h(P)$ feet. What definite integral represents the total work expended in forming the mountain? This type of problem is important in the geological theory of mountain formation.

35 Let P_0 be a fixed point in a solid of mass M. Show that, for all choices of three mutually perpendicular lines that meet at P_0, the sum of the moments of inertia of the solid about the lines is the same.

36 In section 13.3 an intuitive argument was presented for the equality $\int_R f P \, dA = \int_a^b \left(\int_{y_1(x)}^{y_2(x)} f(x, y) \, dy \right) dx$. Here is an intuitive argument for the equality

$$\int_R f(P) \, dV = \int_{x_1}^{x_2} \left[\int_{y_1(x)}^{y_2(x)} \left(\int_{z_1(x, y)}^{z_2(x, y)} f(x, y, z) \, dz \right) dy \right] dx.$$

To start, interpret $f(P)$ as "density."

(a) Let $R(x)$ be the plane cross section consisting of all points in R whose x coordinate is x. Show that the average density in $R(x)$ is

$$\frac{\int_{y_1(x)}^{y_2(x)} \left(\int_{z_1(x, y)}^{z_2(x, y)} f(x, y, z) \, dz \right) dy}{\text{Area of } R(x)}.$$

(b) Show that the mass of R between the plane sections $R(x)$ and $R(x + \Delta x)$ is approximately

$$\int_{y_1(x)}^{y_2(x)} \left(\int_{z_1(x, y)}^{z_2(x, y)} f(x, y, z) \, dz \right) dy \, \Delta x.$$

(c) From (b) obtain a repeated integral in rectangular coordinates for $\int_R f(P) \, dV$.

13.8 Describing solid regions with cylindrical or spherical coordinates

CYLINDRICAL COORDINATES

Cylindrical coordinates combine polar coordinates in the plane with the z of rectangular coordinates in space. Each point P in space receives the name (r, θ, z), as in Fig. 13.71. We are free to choose the direction of the polar axis; usually it will coincide with the x axis of an (x, y, z) system. Note that (r, θ, z) is directly above (or below) $P^* = (r, \theta)$ in the $r\theta$ plane. Since the set of all points $P = (r, \theta, z)$ for which $r = k$, some constant, is a circular cylinder, this coordinate system is especially convenient for describing such cylinders. Just as with polar coordinates, cylindrical coordinates of a point are not unique.

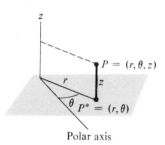

Figure 13.71

EXAMPLE 1 Describe a solid cylinder of radius a and height h in cylindrical coordinates. Assume that the axis of the cylinder is on the positive z axis and the lower base has its center at the pole, as in Fig. 13.72.

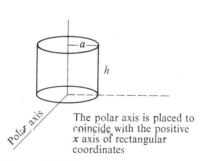

Figure 13.72

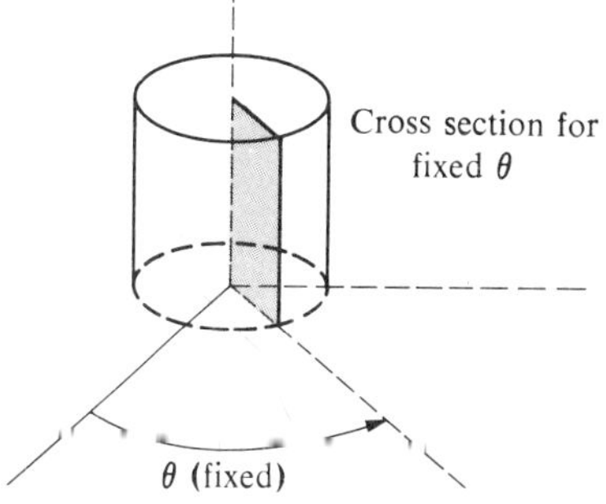

Figure 13.73

SOLUTION First of all, θ varies from 0 to 2π. If we hold θ fixed and consider only positive r, the cross section we obtain is a rectangle perpendicular to the $r\theta$ plane, as shown in Fig. 13.73.

When describing solid regions keep r nonnegative.

Next, examine the behavior of r and z on the cross section pictured. First of all, r goes from 0 to a independently of θ. Thus far it has been shown that

$$0 \leq \theta \leq 2\pi, \quad 0 \leq r \leq a,$$

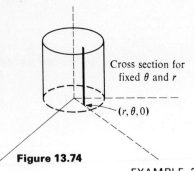

Figure 13.74

which is just a description of the shadow of the cylinder on the $r\theta$ plane cast by light parallel to the z axis.

Finally, hold r and θ fixed, and determine the behavior of z. The cross section for fixed r and θ is a line segment, as shown in Fig. 13.74. On this line segment z varies from 0 to h.

Hence the cylinder has this description:

$$0 \leq \theta \leq 2\pi, \quad 0 \leq r \leq a, \quad 0 \leq z \leq h. \quad \blacksquare$$

EXAMPLE 2 Describe in cylindrical coordinates the region in space formed by the intersection of a solid cylinder of radius 3 with a ball of radius 5 whose center is on the axis of the cylinder. Locate the cylindrical coordinate system as shown in Fig. 13.75.

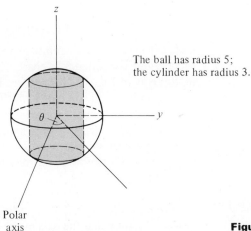

Figure 13.75

SOLUTION First, describe the surface of the ball in cylindrical coordinates. To do this, note that the point $P = (r, \theta, z)$ is a distance $\sqrt{r^2 + z^2}$ from the origin O. (See Fig. 13.76.) For, by the Pythagorean theorem,

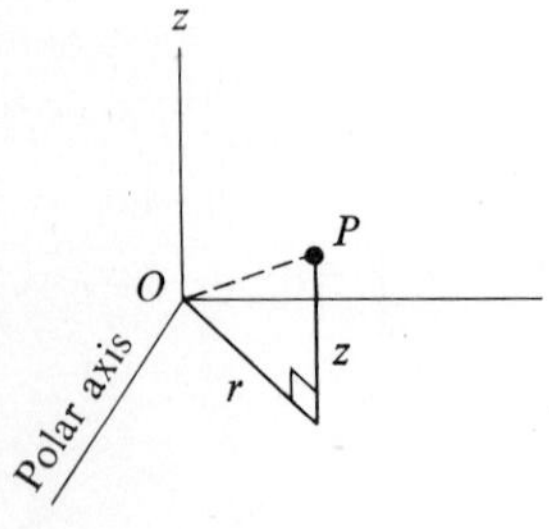

$$r^2 + z^2 = \overline{OP}^2.$$

Now consider the description of the solid. First of all, θ varies from 0 to 2π and r from 0 to 3, bounds determined by the cylinder. For fixed θ and r, the cross section of the solid is a line segment determined by the sphere that bounds the ball, as shown in Fig. 13.77. Now, since the sphere has radius 5, for any point (r, θ, z) on it,

$$r^2 + z^2 = 25,$$
$$z = \pm\sqrt{25 - r^2}.$$

Thus, on the line segment determined by fixed r and θ, z varies from $-\sqrt{25 - r^2}$ to $\sqrt{25 - r^2}$.

The solid has this description:

$$0 \leq \theta \leq 2\pi, \quad 0 \leq r \leq 3, \quad -\sqrt{25 - r^2} \leq z \leq \sqrt{25 - r^2}. \quad \blacksquare$$

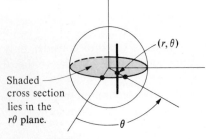

Figure 13.76

Figure 13.77

13.8 DESCRIBING SOLID REGIONS WITH CYLINDRICAL OR SPHERICAL COORDINATES

The set of all points (r, θ, z) whose r coordinates are between r and $r + \Delta r$, whose θ coordinates are between θ and $\theta + \Delta\theta$, and whose z coordinates are between z and $z + \Delta z$ is shown in Fig. 13.78. It is a solid with four flat surfaces and two curved surfaces.

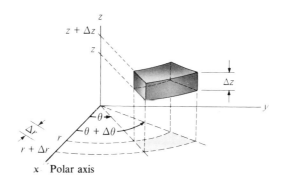

Figure 13.78

When Δr is small, the area of the base of the solid is approximately $r \, \Delta r \, \Delta\theta$, as shown in Sec. 13.2. Thus, when Δr, $\Delta\theta$, and Δz are small, the volume of the solid in Fig. 13.78 is approximately

$$r \, \Delta r \, \Delta\theta \, \Delta z.$$

Note the extra factor r. It will appear whenever cylindrical coordinates are used to evaluate integrals over three-dimensional sets.

SPHERICAL COORDINATES

The third standard coordinate system in space is *spherical coordinates*. A point P is described by three numbers:

ρ is pronounced "roe." ρ, the distance from P to the origin O,

θ is "longitude." θ, the same angle as in cylindrical coordinates,

ϕ is "colatitude." ϕ, the angle between the positive z axis and the ray from O to P.

Note that θ is not unique. The point P is denoted (ρ, θ, ϕ). Note the order: first ρ, then θ, and last ϕ. (See Fig. 13.79.)

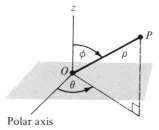

θ is the same as in cylindrical coordinates; ϕ is the indicated angle from the z axis $(0 \leq \phi \leq \pi)$; and ρ is the distance from P to the pole.

Figure 13.79

Observe that the sphere of radius a whose center is at the origin has the equation $\rho = a$. This equation is much shorter than the equation of the sphere in rectangular coordinates, $x^2 + y^2 + z^2 = a^2$ or, in cylindrical coor-

874 DEFINITE INTEGRALS OVER PLANE AND SOLID REGIONS

dinates, $r^2 + z^2 = a^2$. Spherical coordinates are, as their name suggests, ideal for integrating over spheres.

α is called the "half-vertex angle."

The set of points for which ϕ is a constant α is the surface of a cone, as shown in Fig. 13.80. For this reason, spherical coordinates may also be convenient when integrating over cones.

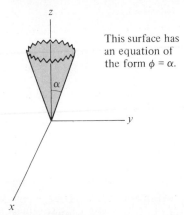

Figure 13.80

The set of points for which θ is prescribed is a half plane bounded by the z axis. Thus the set of points for which both ϕ and θ are fixed is a ray emanating from the origin, namely, the intersection of the surface of a cone and a half plane whose edge is the axis of the cone, as shown in Fig. 13.81.

Rectangular coordinates in terms of spherical

The rectangular coordinates of (ρ, θ, ϕ) can be found by inspection of Fig. 13.82.

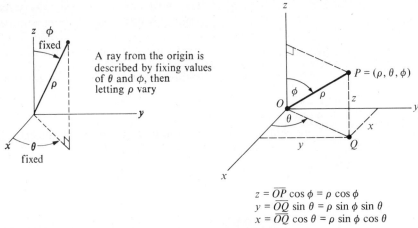

$z = \overline{OP} \cos \phi = \rho \cos \phi$
$y = \overline{OQ} \sin \theta = \rho \sin \phi \sin \theta$
$x = \overline{OQ} \cos \theta = \rho \sin \phi \cos \theta$

Figure 13.81 **Figure 13.82**

EXAMPLE 3 Find the equation of the plane $z = 4$ in spherical coordinates.

SOLUTION Since
$$z = \rho \cos \phi$$
describes z in terms of spherical coordinates, the equation of the plane $z = 4$

is
$$\rho \cos \phi = 4,$$
or
$$\rho = \frac{4}{\cos \phi},$$
or
$$\rho = 4 \sec \phi. \blacksquare$$

EXAMPLE 4 Describe in spherical coordinates the solid cone of height 4 shown in Fig. 13.83.

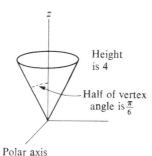

Figure 13.83

SOLUTION It is usually most convenient to examine first how θ varies. In this case θ goes from 0 to 2π. For each fixed θ the cross section is a triangle, as shown in Fig. 13.84.

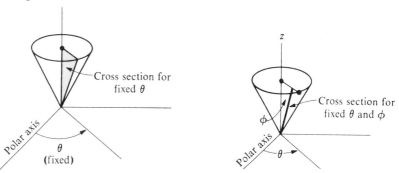

Figure 13.84 **Figure 13.85**

On this typical triangle ϕ varies from 0 to $\pi/6$. For each fixed θ and ϕ the cross section is a segment on a ray emanating from the origin, as noted in Fig. 13.85. On this cross section ρ varies from 0 to its value where the ray meets the plane $z = 4$. By Example 3, this plane has the equation

$$\rho = 4 \sec \phi.$$

Hence for fixed θ and ϕ, ρ varies from 0 to $4 \sec \phi$.

This is the description of the cone:

$$0 \leq \theta \leq 2\pi, \qquad 0 \leq \phi \leq \frac{\pi}{6}, \qquad 0 \leq \rho \leq 4 \sec \phi. \blacksquare$$

Notice the simplicity of the description of θ and ϕ for the cone in Example

4. Clearly, spherical coordinates are convenient for describing cones. They are also fine for a ball whose center is at the origin. To be specific,

$$0 \leq \theta \leq 2\pi, \qquad 0 \leq \phi \leq \pi, \qquad 0 \leq \rho \leq a$$

As a rule of thumb, describe regions by θ first, then ϕ, then ρ.

is the description of a ball of radius a whose center is at the origin.

The set of all points whose ρ coordinates are between ρ and $\rho + \Delta\rho$, whose θ coordinates are between θ and $\theta + \Delta\theta$, and whose ϕ coordinates are between ϕ and $\phi + \Delta\phi$ is a solid R with two flat surfaces and four curved surfaces. We shall estimate its volume when $\Delta\rho$, $\Delta\theta$, and $\Delta\phi$ are small. This estimate will be needed when setting up repeated integrals in spherical coordinates.

R is bounded by the six surfaces shown in Fig. 13.86: by spheres of radii ρ and $\rho + \Delta\rho$; by cones of half-vertex angles ϕ and $\phi + \Delta\phi$; by half planes of polar angles θ and $\theta + \Delta\theta$.

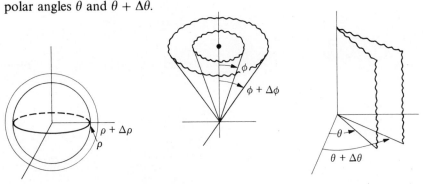

Figure 13.86

The little solid R appears as shown in Fig. 13.87. Label the eight corners as in Fig. 13.88. $ABCD$ and $EFGH$ are spherical. $BCGF$ and $ADHE$ are conical. $ABFE$ and $DCGH$ are flat.

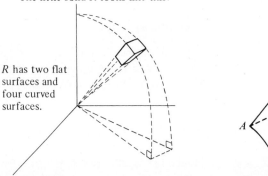

Figure 13.87   **Figure 13.88**

Since the small solid R resembles a rectangular box, its volume is approximated by the product

$\overset{\frown}{AB}$ denotes length of arc AB.

$$\overline{AE} \cdot \overset{\frown}{AB} \cdot \overset{\frown}{AD}.$$

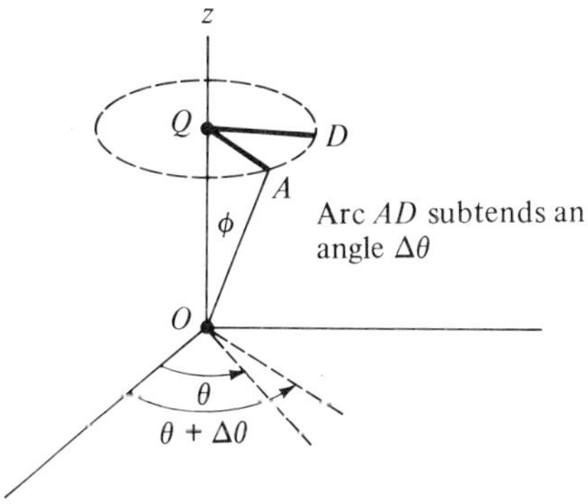

First of all, $\overline{AE}$ is just the difference in the radii of the two spheres:

$$\overline{AE} = \Delta\rho.$$

Next, AB is an arc of a circle of radius ρ and subtends an angle $\Delta\phi$. Thus

$$\widehat{AB} = \rho\,\Delta\phi.$$

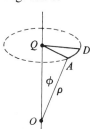

Finally, consider AD. It is an arc of a circle that is perpendicular to the z axis, as depicted in Fig. 13.89. AD subtends an angle $\Delta\theta$. The radius of the dashed circle is computed from right triangle AOQ, shown in Fig. 13.90.

Figure 13.89

Figure 13.90

Thus arc AD has length $\rho \sin\phi\,\Delta\theta$.

The volume of R is, therefore, approximately

$$\overline{AE} \cdot \widehat{AB} \cdot \widehat{AD} = \Delta\rho(\rho\,\Delta\phi)(\rho\sin\phi\,\Delta\theta)$$
$$= \rho^2 \sin\phi\,\Delta\rho\,\Delta\phi\,\Delta\theta.$$

Consequently, the solid consisting of all points (ρ, θ, ϕ) whose ρ coordinates are between ρ and $\rho + \Delta\rho$, whose θ coordinates are between θ and $\theta + \Delta\theta$, and whose ϕ coordinates are between ϕ and $\phi + \Delta\phi$, has volume approximately

The factor $\rho^2 \sin\phi$ will be needed in repeated integrals.

$$\rho^2 \sin\phi\,\Delta\rho\,\Delta\theta\,\Delta\phi. \tag{1}$$

Notice the factor $\rho^2 \sin\phi$. It will be needed in forming repeated integrals in spherical coordinates. Note that in (1) ρ and $\Delta\rho$ have units of length, whereas $\Delta\phi$ and $\Delta\theta$ are dimensionless (radian measure is defined as the quotient of two lengths). Thus the units of (1) are *length cubed*, as they should be if (1) is to measure volume.

Exercises

Exercises 1 to 18 concern cylindrical coordinates. In Exercises 1 to 6 sketch graphs of all points (r, θ, z) such that

1 $r = 1$.
2 $\theta = \pi/4$ (consider only nonnegative r).
3 $z = 1$.
4 $r = z$.
5 $r^2 + z^2 = 9$.
6 $r = 2\cos\theta$.

In Exercises 7 and 8 draw the cross sections corresponding to fixed θ of the regions described. (Restrict r to nonnegative values.)

7. The region R consists of all points within a distance a of the origin of the $r\theta z$-coordinate system.
8. The region R consists of all points of the solid of Exercise 7 that are within the cylinder whose equation is $r = a \cos \theta$.
9. Describe the solid in Exercise 7 in cylindrical coordinates.
10. Describe the solid in Exercise 8 in cylindrical coordinates.
11. (a) What are the cylindrical coordinates of the point $P = (x, y, z)$? Assume (x, y) is in the first quadrant.
    (b) What are the rectangular coordinates of the point $P = (r, \theta, z)$?
12. Describe in cylindrical coordinates the solid cone shown in Fig. 13.91.

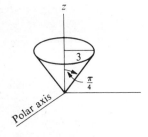

Figure 13.91

13. Give the equation in cylindrical coordinates of
    (a) the xy plane,
    (b) the plane $x = y$,
    (c) the z axis.

In Exercises 14 to 18 draw the solids described.

14. $0 \le \theta \le \pi/2$, $1 \le z \le 2$, $1 \le r \le z$.
15. $0 \le \theta \le 2\pi$, $0 \le r \le 1$, $r \le z \le 1$.
16. $0 \le \theta \le \pi/2$, $0 \le r \le \cos \theta$, $1 \le z \le 2$.
17. $0 \le \theta \le \pi/2$, $0 \le r \le 1$, $0 \le z \le r \cos \theta$.
18. $0 \le \theta \le \pi/2$, $0 \le r \le \cos \theta$, $r \le z \le \sqrt{4 - r^2}$.

Exercises 19 to 40 concern spherical coordinates. In each of Exercises 19 to 24 sketch the graph of all points (ρ, θ, ϕ) satisfying the given equation.

19. $\rho = 2$
20. $\phi = \pi/6$
21. $\theta = \pi/2$
22. $\phi = \pi/2$
23. $\phi = 0$
24. $\phi = \pi$
25. What are the cylindrical coordinates of the point (ρ, θ, ϕ)?

26. Sketch the set of all points (ρ, θ, ϕ) such that $\phi = \pi/2$ and $\theta = \pi/2$.

In Exercises 27 to 31 describe in spherical coordinates the regions R.

27. R is the ball of radius a centered at the origin.
28. R is the top half of the ball of radius a centered at the origin.
29. R is the ice cream cone-shaped intersection of a solid cone and a ball shown in Fig. 13.92.

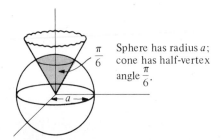

Sphere has radius a; cone has half-vertex angle $\frac{\pi}{6}$.

Figure 13.92

30. R is the region between two spheres, both with center at the origin, of radii a and b, $a < b$.
31. R is the region in the cone shown in Exercise 29 below the plane $z = 3a/5$.

In Exercises 32 to 35 sketch the regions described.

32. $0 \le \theta \le \pi/2$, $0 \le \phi \le \pi/2$, $0 \le \rho \le 1$.
33. $0 \le \theta \le 2\pi$, $\pi/2 \le \phi \le \pi$, $1 \le \rho \le 2$.
34. $0 \le \theta \le \pi$, $0 \le \phi \le \pi/4$, $0 \le \rho \le \sec \phi$.
35. $0 \le \theta \le \pi/2$, $0 \le \phi \le \pi/4$, $1 \le \rho \le 2 \sec \phi$.

36. Fill in the blanks and explain with the aid of a sketch: Rectangular coordinates describe a point by specifying three planes on which it lies. Spherical coordinates describe a point by specifying _____, _____, and _____ on which it lies.
Cylindrical coordinates describe a point by specifying _____, _____, and _____ on which it lies.

37. Find the spherical coordinates of the point whose cylindrical coordinates are (r, θ, z), $r > 0$.
38. Sketch the solid whose description is $0 \le \theta \le \pi/2$, $\pi/4 \le \phi \le \pi/2$, $1 \le \rho \le 2$.
39. Find the spherical coordinates of the point whose rectangular coordinates are (x, y, z).
40. Find the equation in spherical coordinates of the plane
    (a) $x = 2$,
    (b) $2x + 3y + 4z = 1$.

13.9 Computing $\int_R f(P)\, dV$ with cylindrical or spherical coordinates

Repeated integrals in polar coordinates in Sec. 13.4 involve the introduction of a factor r in the integrand. This must also be done when using cylindrical coordinates (r, θ, z), for the volume of the little solid obtained by letting each coordinate change a small amount is, as was shown in Sec. 13.8, approximately

$$r \, \Delta r \, \Delta\theta \, \Delta z,$$

where Δr, $\Delta\theta$, Δz are the changes in the coordinates.

When setting up repeated integrals in spherical coordinates, it is necessary to insert the extra factor

$$\rho^2 \sin \phi$$

in the integrand. The reason for this is that the little solid obtained by letting each coordinate change a small amount is approximately

$$\rho^2 \sin \phi \, \Delta\rho \, \Delta\theta \, \Delta\phi,$$

where $\Delta\rho$, $\Delta\theta$, and $\Delta\phi$ are the changes in the coordinates, as was shown in Sec. 13.8.

The limits of integration are determined by the description of the region R relative to the coordinate system. A few examples will show how to set up and compute these repeated integrals; we will not stop to justify the technique, which is studied in advanced calculus.

EXAMPLE 1 Find the volume of a ball R of radius a using cylindrical coordinates.

SOLUTION Place the origin of a cylindrical coordinate system at the center of the ball, as in Fig. 13.93.

The volume of the ball is $\int_R 1 \, dV$. The description of R in cylindrical coordinates is

$$0 \leq \theta \leq 2\pi, \qquad 0 \leq r \leq a, \qquad -\sqrt{a^2 - r^2} \leq z \leq \sqrt{a^2 - r^2}.$$

The repeated integral for the volume is thus

$$\int_R 1 \, dV = \int_0^{2\pi} \left[\int_0^a \left(\int_{-\sqrt{a^2-r^2}}^{\sqrt{a^2-r^2}} 1 \cdot r \, dz \right) dr \right] d\theta.$$

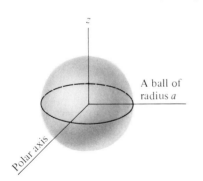

Figure 13.93

Note that the order of integration is determined by the order in describing R.

(Note the r in the integrand.)

Evaluation of the first integral, where r and θ are fixed, yields

$$\int_{-\sqrt{a^2-r^2}}^{\sqrt{a^2-r^2}} r \, dz = rz \Big|_{z=-\sqrt{a^2-r^2}}^{z=\sqrt{a^2-r^2}}$$

$$= 2r\sqrt{a^2 - r^2}.$$

Evaluation of the second integral, where θ is fixed, yields:

$$\int_0^a 2r\sqrt{a^2 - r^2} \, dr = \frac{-2(a^2 - r^2)^{3/2}}{3} \Big|_{r=0}^{r=a}$$

$$= \frac{2a^3}{3}.$$

Evaluation of the third integral yields

$$\int_0^{2\pi} \frac{2a^3}{3}\, d\theta = \frac{2a^3}{3}\, 2\pi = \frac{4\pi a^3}{3}. \quad \blacksquare$$

EXAMPLE 2 Find $\bar{z}$, the z coordinate of the centroid of a solid homogeneous hemisphere R of radius a whose base is on the xy plane.

SOLUTION Use spherical coordinates. By definition,

$$\bar{z} = \frac{\int_R z\, dV}{\text{Volume of hemisphere}}.$$

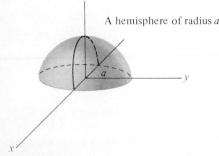

A hemisphere of radius a

Figure 13.94

The volume is $2\pi a^3/3$. All that remains is to compute $\int_R z\, dV$.

Placing the origin of the spherical coordinate system at the center of the hemisphere, as in Fig. 13.94, we have this description of R:

$$0 \leq \theta \leq 2\pi, \quad 0 \leq \phi \leq \pi/2, \quad 0 \leq \rho \leq a.$$

Before we set up a repeated integral for $\int_R z\, dV$, it is necessary to express z in spherical coordinates also,

$$z = \rho \cos \phi.$$

Then
$$\int_R z\, dV = \int_0^{2\pi} \left[\int_0^{\pi/2} \left(\int_0^a \rho \cos \phi\ \overbrace{\rho^2 \sin \phi}^{\text{Notice the extra } \rho^2 \sin \phi.}\, d\rho \right) d\phi \right] d\theta.$$

Evaluation of the first integral, where ϕ and θ are fixed, yields

$$\int_0^a \rho^3 \cos \phi \sin \phi\, d\rho = \frac{\rho^4 \cos \phi \sin \phi}{4} \Big|_{\rho=0}^{\rho=a}$$

$$= \frac{a^4 \cos \phi \sin \phi}{4}.$$

Evaluation of the second integral, where θ is fixed, yields

$$\int_0^{\pi/2} \frac{a^4 \cos \phi \sin \phi\, d\phi}{4} = \frac{a^4 \sin^2 \phi}{8} \Big|_{\phi=0}^{\phi=\pi/2}$$

$$= \frac{a^4}{8}.$$

Evaluation of the third integral yields

$$\int_0^{2\pi} \frac{a^4}{8}\, d\theta = \frac{\pi a^4}{4}.$$

Hence $\int_R z \, dV = \pi a^4/4$, and

$$\bar{z} = \frac{\pi a^4/4}{2\pi a^3/3} = \frac{3a}{8}. \quad \blacksquare$$

This was one of Newton's discoveries.

The next example is of importance in the theory of gravitational attraction. Students of the physical sciences will see later that it implies that a homogeneous ball attracts a particle (or satellite) as if all the mass of the ball were at its center.

EXAMPLE 3 Let R be a homogeneous ball of mass M and radius a. Let A be a point at a distance H from the center of the ball, $H > a$. Compute the potential

$$-\int_R \frac{\delta}{q} \, dV$$

where δ is density and q is the distance from a point P in R to A. (See Figs. 13.95 and 13.96.)

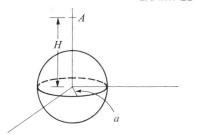

Figure 13.95

SOLUTION First, express q in terms of spherical coordinates. To do so, choose a spherical coordinate system whose origin is at the center of the sphere and such that the ϕ coordinate of A is 0. (See Fig. 13.96.)

Let $P = (\rho, \theta, \phi)$ be a typical point in the sphere. Applying the law of cosines to triangle AOP, we find that

$$q^2 = H^2 + \rho^2 - 2\rho H \cos \phi.$$

Hence $\quad q = \sqrt{H^2 + \rho^2 - 2\rho H \cos \phi}.$

Since the ball is homogeneous,

$$\delta = \frac{M}{\frac{4}{3}\pi a^3} = \frac{3M}{4\pi a^3}.$$

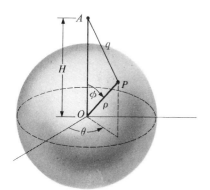

Figure 13.96

Hence $\quad \int_R \frac{\delta}{q} \, dV = \int_R \frac{3M}{4\pi a^3 q} \, dV,$

or $\quad \int_R \frac{\delta}{q} \, dV = \frac{3M}{4\pi a^3} \int_R \frac{1}{q} \, dV. \quad (1)$

Now evaluate $\quad \int_R \frac{1}{q} \, dV$

by a repeated integral in spherical coordinates:

A rare case, where integration with respect to ρ is not first

$$\int_R \frac{1}{q} \, dV = \int_0^{2\pi} \left[\int_0^a \left(\int_0^\pi \frac{\rho^2 \sin \phi}{\sqrt{H^2 + \rho^2 - 2\rho H \cos \phi}} \, d\phi \right) d\rho \right] d\theta.$$

(Integrate with respect to ϕ first, rather than ρ, because it is easier to do so in this case.)

Evaluation of the first integral, where ρ and θ are constants, is accomplished with the aid of the fundamental theorem:

$$\int_0^\pi \frac{\rho^2 \sin \phi}{\sqrt{H^2 + \rho^2 - 2\rho H \cos \phi}} \, d\phi = \frac{\rho\sqrt{H^2 + \rho^2 - 2\rho H \cos \phi}}{H} \bigg|_{\phi=0}^{\phi=\pi}$$

$$= \frac{\rho}{H}(\sqrt{H^2 + \rho^2 + 2\rho H} - \sqrt{H^2 + \rho^2 - 2\rho H}).$$

Now, $\sqrt{H^2 + \rho^2 + 2\rho H} = H + \rho$. Since $\rho \le a < H$, $H - \rho$ is positive and

$$\sqrt{H^2 + \rho^2 - 2\rho H} = H - \rho.$$

Thus the first integral equals

$$\frac{\rho}{H}[(H + \rho) - (H - \rho)] = \frac{2\rho^2}{H}.$$

Evaluation of the second integral yields

$$\int_0^a \frac{2\rho^2}{H} \, d\rho = \frac{2a^3}{3H}.$$

Evaluation of the third integral yields:

$$\int_0^{2\pi} \frac{2a^3}{3H} \, d\theta = \frac{4\pi a^3}{3H}.$$

Hence

$$\int_R \frac{1}{q} \, dV = \frac{4\pi a^3}{3H}.$$

By (1), $\quad$ Potential $= -\int_R \frac{\delta}{q} \, dV = \frac{-3M}{4\pi a^3} \frac{4\pi a^3}{3H} = \frac{-M}{H}.$

Though the computations are long, the answer is short. It says that the potential is just the negative of the mass of the ball times the reciprocal of the distance from A to the center. In other words, the potential is the same as if all the mass of the ball were concentrated at its center. ∎

JACOBIANS AND THE FACTOR $\rho^2 \sin \phi$ (OPTIONAL)

If you studied the Jacobian in Secs. 13.5 and 13.6, you might expect $\rho^2 \sin \phi$ to be the Jacobian of a certain mapping—not from a plane to a plane but from a space to a space. Indeed it is, and the explanation is quite similar to that for the presence of "r" in the integrand in polar coordinates.

Let ρ, θ, and ϕ be *rectangular* coordinates in a three dimensional space. Let x, y, and z be the usual rectangular coordinates in another copy of space. Define the mapping F from the first space to the second by the formulas

$$x = \rho \sin \phi \cos \theta, \qquad y = \rho \sin \phi \sin \theta, \qquad z = \rho \cos \phi.$$

If R is a set in the first space and S its image in the xyz space, then

$$\int_S f(Q)\, dV = \int_R f(F(P))\, |J(P)|\, dV$$

where $J(P)$ is the Jacobian of F at the point P. The Jacobian of F is defined as the third-order determinant:

$$\begin{vmatrix} \dfrac{\partial x}{\partial \rho} & \dfrac{\partial x}{\partial \phi} & \dfrac{\partial x}{\partial \theta} \\[4pt] \dfrac{\partial y}{\partial \rho} & \dfrac{\partial y}{\partial \phi} & \dfrac{\partial y}{\partial \theta} \\[4pt] \dfrac{\partial z}{\partial \rho} & \dfrac{\partial z}{\partial \phi} & \dfrac{\partial z}{\partial \theta} \end{vmatrix}.$$

After computing the partial derivatives we find that this determinant is

$$\begin{vmatrix} \sin \phi \cos \theta & \rho \cos \phi \cos \theta & -\rho \sin \phi \sin \theta \\ \sin \phi \sin \theta & \rho \cos \phi \sin \theta & \rho \sin \phi \cos \theta \\ \cos \phi & -\rho \sin \phi & 0 \end{vmatrix}. \tag{2}$$

A routine computation shows that (2) equals $\rho^2 \sin \phi$. So, if you used Jacobians, you would not need to go through the geometry of Sec. 13.8 which led up to the formula $\rho^2 \sin \phi\, \Delta\rho\, \Delta\phi\, \Delta\theta$.

The Jacobian also would provide the formula $r\, \Delta r\, \Delta\theta\, \Delta z$ of Sec. 13.8, needed for integration in cylindrical coordinates. The Jacobian is defined for mappings in space just as for mappings in the plane; the only difference is that the determinant is 3 by 3 instead of 2 by 2. It can be shown to describe the local magnification of volume (rather than area).

Exercises

In each of Exercises 1 to 4 evaluate the repeated integral.

1 $\int_0^{2\pi} \left(\int_0^1 \left(\int_r^1 zr^3 \cos^2 \theta\, dz \right) dr \right) d\theta$

2 $\int_0^{2\pi} \left(\int_0^1 \left(\int_{-\sqrt{a^2-r^2}}^{\sqrt{a^2-r^2}} z^2 r\, dz \right) dr \right) d\theta$

3 $\int_0^{\pi/2} \left(\int_0^{\pi/4} \left(\int_0^{\cos \phi} \rho^3 \sin^2 \theta \sin \phi\, d\rho \right) d\phi \right) d\theta$

4 $\int_0^{\pi/4} \left(\int_{\pi/6}^{\pi/2} \left(\int_0^{\sec \theta} \rho^3 \sin \phi \cos \phi\, d\rho \right) d\phi \right) d\theta$

5 Compute the volume of a ball of radius a using spherical coordinates.

6 Find the moment of inertia of a homogeneous ball of radius a and mass M about a diameter, using cylindrical coordinates.

7 Solve Exercise 6 using spherical coordinates.

8 A homogeneous right circular cone has altitude h, radius a, and mass M. Using cylindrical coordinates, show that the moment of inertia about its axis is $3Ma^2/10$. *Suggestion:* Place the z axis on the axis of the cone and the origin at the vertex of the cone.

9 Solve Exercise 8 using spherical coordinates.

10 A right circular cone of radius a and height h has a density at point P equal to the distance from P to the base of the cone. Find its mass, using
  (a) cylindrical coordinates,
  (b) spherical coordinates.

11 Find the moment of inertia of a homogeneous right circular cylinder of radius a, height h, and mass M about a line L on its surface and parallel to its axis.
(a) Use cylindrical coordinates, the z axis coinciding with L.
(b) Use cylindrical coordinates, the z axis coinciding with the axis of the cylinder.

12 Solve Example 2 using cylindrical coordinates.

13 A solid consists of that part of a ball of radius a that lies within a cone of half-vertex angle $\phi = \pi/6$, the vertex being at the center of the ball. Set up repeated integrals for $\int_R z\, dV$ in all three coordinate systems and evaluate the simplest.

14 Show by using a repeated integral that the volume of the little solid estimated in Sec. 13.8 is precisely

$$\frac{(\rho + \Delta\rho)^3 - \rho^3}{3} \Delta\theta(\cos\phi - \cos(\phi + \Delta\phi)).$$

15 Find the average distance from the center of a ball of radius a to other points of the ball by setting up appropriate repeated integrals in the three types of coordinate systems and evaluating the two easiest.

16 Find the average length of the shadows on the xy plane of all line segments whose lengths are at most a with one end at the origin and the other in or above the xy plane. Assume that the light is parallel to the z axis.

17 Find the moment of inertia of a homogeneous solid hemisphere of radius a and mass M about a diameter in its circular base,
(a) using cylindrical coordinates;
(b) using spherical coordinates.

18 A certain ball of radius a is *not* homogeneous. However, its density at P depends only on the distance from P to the center of the ball. That is, there is a function $g(\rho)$ such that the density at $P = (\rho, \theta, \phi)$ is $g(\rho)$. Using a repeated integral, show that the mass of the ball is

$$4\pi \int_0^a g(\rho)\rho^2\, d\rho.$$

■

19 Let R be the solid region inside both the sphere $x^2 + y^2 + z^2 = 1$ and the cone $z = \sqrt{x^2 + y^2}$. Let the density at (x, y, z) be $f(x, y, z) = z$. Set up repeated integrals for the mass in R using
(a) rectangular coordinates,
(b) cylindrical coordinates,
(c) spherical coordinates.
(d) Evaluate the repeated integral in (c).

20 A homogeneous right circular cone has radius a and height h. Find its centroid.

21 Show that the average of the reciprocal of the distance from a fixed point A outside a ball to points in the ball is equal to the reciprocal of the distance from A to the center of the ball.

22 Show that $\int_R (x^3 + y^3 + z^3)\, dV = 0$, where R is a ball whose center is the origin of a rectangular coordinate system. (Do not use a repeated integral.)

23 Let R be a solid ball of radius a with center at the origin of the coordinate system.
(a) Explain why $\int_R x^2\, dV = \frac{1}{3}\int_R (x^2 + y^2 + z^2)\, dV$.
(b) Evaluate the second integral by spherical coordinates.
(c) Use (b) to find $\int_R x^2\, dV$.

24 Combining the fact that the volume of R equals $\int_R 1\, dV$ with a repeated integral in rectangular coordinates,
(a) obtain the formula $V = \int_a^b A(x)\, dx$;
(b) obtain the formula for volume as a definite integral over a plane set.

25 Combining the fact that the volume of R equals $\int_R 1\, dV$ with a repeated integral in cylindrical coordinates, obtain the shell technique of Sec. 8.4.

■ ■

26 When the dimensions of the solid in Exercise 14 are small, does the formula in that exercise become, approximately, the formula $\rho^2 \sin\phi\, \Delta\rho\, \Delta\theta\, \Delta\phi$?

27 Using the method of Example 3, find the average value of q for all points P in the ball. Note that it is *not* the same as if the entire ball were placed at its center.

28 Show that the result of Example 3 holds if the density $\delta(P)$ depends only on ρ, the distance to the center. (This is approximately the case with the planet Earth, which is not homogeneous.) Call $\delta(\rho, \theta, \phi) = g(\rho)$. See Exercise 18 and Example 3.

29 Let R be the part of a ball of radius a removed by a cylindrical drill of diameter a whose edge passes through the center of the sphere.
(a) Sketch R.
(b) Notice that R consists of four congruent pieces. Find the volume of one of these pieces using cylindrical coordinates. Multiply by four to get the volume of R.

30 (See Exercise 29.) A person tried to get the volume of R directly, as follows:

$$V = \int_{-\pi/2}^{\pi/2} \int_0^{a\cos\theta} \int_{-\sqrt{a^2 - r^2}}^{\sqrt{a^2 - r^2}} r\, dz\, dr\, d\theta$$

$$= \int_{-\pi/2}^{\pi/2} \int_0^{a\cos\theta} 2r\sqrt{a^2 - r^2}\, dr\, d\theta$$

$$= \int_{-\pi/2}^{\pi/2} \left[-\tfrac{2}{3}(a^2 - r^2)^{3/2}\right]\Big|_0^{a\cos\theta} d\theta$$

$$= \int_{-\pi/2}^{\pi/2} \frac{2a^3}{3}(1 - \sin^3\theta)\, d\theta$$

$$= \frac{2\pi a^3}{3}.$$

The answer is wrong. What is the error?

31 Let R be the ball of radius a. For any point P in the ball other than the center of the ball, define $f(P)$ to be the reciprocal of the distance from P to the origin. The average value of f over R involves an improper integral, since the function blows up near the origin. Does this improper integral converge or diverge? What is the average value of f over R? *Suggestion:* Examine the integral over the region between concentric spheres of radii a and t, and let $t \to 0$.

Exercises 32 to 34 concern Jacobians.

32 Evaluate the determinant (2).
33 Compute the Jacobian for the transformation from cylindrical coordinates to rectangular, that is, for the mapping $x = r \cos \theta$, $y = r \sin \theta$, $z = z$.
34 It would be no fun (in any of the three coordinate systems) to integrate over the region S within the ellipsoid

$$\frac{x^2}{a^2} + \frac{y^2}{b^2} + \frac{z^2}{c^2} = 1.$$

However it may be fun to integrate over the ball R of radius 1 centered at the origin located in a space with rectangular coordinates u, v, and w. Let F be the mapping from R to S defined by $F(u, v, w) = (au, bv, cw)$. Assume that the integral of f over S equals

$$\int_R f(F(P)) |J(P)| \, dV,$$

where J is the Jacobian of F.
(a) Show that $J(P) = abc$.
(b) Evaluate $\int_S z^2 \, dA$. (Use spherical coordinates in the u, v, w space.)

13.S Summary

Let R be a region in the plane and f a function that assigns to each point P in R a number. Then the definite integral of f over R is defined with the aid of partitions of R and sampling points as

$$\lim_{\text{mesh} \to 0} \sum_{i=1}^{n} f(P_i) A_i.$$

This number is denoted $\int_R f(P) \, dA$.

The definite integral over a region R in space, $\int_R f(P) \, dV$, is defined similarly. Both definitions are analogous to the definition of $\int_a^b f(x) \, dx$, the integral over an interval.

These integrals are of use in physics for defining and/or computing mass, moments, the center of mass, the centroid, moment of inertia, and gravitational attraction. A few exercises illustrated some of the other interpretations and applications. Sec. 13.1, Exercise 15, total rainfall; Sec. 13.1, Exercise 16, total radiation; Sec. 13.4, Exercise 26, danger in an epidemic; and Sec. 13.7, Exercise 34, work.

Most of the chapter was concerned with the computation of these integrals by repeated integrals over intervals.

When polar coordinates or cylindrical coordinates are used, an extra r must be put in the integrand. In the case of spherical coordinates, $\rho^2 \sin \phi$ must be inserted.

The two optional sections, Secs. 13.5 and 13.6, developed the Jacobian to provide a general formula for the extra factor that must be put in the integrand when a coordinate system other than a rectangular one is used. The Jacobian measures the local magnification of a mapping.

DEFINITE INTEGRALS OVER PLANE AND SOLID REGIONS

VOCABULARY AND SYMBOLS

integrals over planar or spatial regions $\int_R f(P)\, dA$ or $\int_R f(P)\, dV$ (Also called multiple integrals, sometimes denoted $\iint_R f(P)\, dA$ or $\iiint_R f(P)\, dV$.)
partition, diameter, mesh
average value
moment (first moment)
moment of inertia (second moment)
center of mass
centroid
repeated integral (rectangular, polar, cylindrical, spherical) (Also called "iterated integral," "double integral," and "triple integral.")

KEY FACTS

Formula	Significance
$\int_R 1\, dA$	Area of R
$\int_R 1\, dV$	Volume of R
$\dfrac{\int_R f(P)\, dA}{\text{Area of } R}$ or $\dfrac{\int_R f(P)\, dV}{\text{Volume of } R}$	Average of f over R
$\int \delta(P)\, dA$ or $\int \delta(P)\, dV$ where $\delta(P)$ = density	M, total mass in R
$\int_R y\delta(P)\, dA$, $\int_R x\delta(P)\, dA$	Moments M_x and M_y about x and y axes respectively (A moment can be computed around any line in the plane.)
$\int_R f(P)\delta(P)\, dA$, $\int_R f(P)\delta(P)\, dV$ where $f(P)$ = square of distance from P to some fixed line L	Moment of inertia around L for planar and solid regions respectively
$\left(\dfrac{M_y}{M}, \dfrac{M_x}{M}\right)$ (a similar definition for space)	Center of mass $(\bar{x}, \bar{y})$
	If density is 1, the center of mass is called the centroid.

RELATIONS BETWEEN RECTANGULAR COORDINATES AND SPHERICAL OR CYLINDRICAL COORDINATES

$$x = \rho \sin \phi \cos \theta \qquad x = r \cos \theta$$
$$y = \rho \sin \phi \sin \theta \qquad y = r \sin \theta$$
$$z = \rho \cos \phi \qquad z = z$$

Repeated integrals can be used to evaluate $\int_R f(P)\, dA$. It is necessary to use the fundamental theorem of calculus twice in succession. Except in the case of the simplest integrals (which are fortunately the ones most often applied), one will meet a nonelementary integral. The repeated integrals are

$$\int_a^b \left[\int_{y_1(x)}^{y_2(x)} f(x, y)\, dy \right] dx \quad \text{and} \quad \int_c^d \left[\int_{x_1(y)}^{x_2(y)} f(x, y)\, dx \right] dy$$

in rectangular coordinates, and

$$\int_\alpha^\beta \left[\int_{r_1(\theta)}^{r_2(\theta)} f(r, \theta) \, r \, dr \right] d\theta$$

in polar coordinates (the other order is seldom convenient). Remember the extra r in the integrand of the repeated integral in polar coordinates. It is present because $r \, dr \, d\theta$ (not $dr \, d\theta$) is the approximate area of the little region corresponding to changes of dr and $d\theta$ in the coordinates. Figure 13.97 will serve as a reminder.

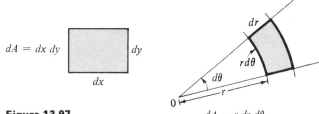

Figure 13.97

Similarly $\int_R f(P) \, dV$ may be evaluated by a repeated integral in one of the three coordinate systems: rectangular, cylindrical, or spherical. Rectangular is usually best for boxes or polyhedra, cylindrical for right circular cylinders or regions whose projections are disks, and spherical for cones or spheres. However, the formula for the integrand may also influence the choice.

Guide quiz on chap. 13

1 (a) Describe in rectangular coordinates the region whose description in polar coordinates is

$$0 \leq \theta \leq \pi, \quad 0 \leq r \leq a.$$

(b) Describe in polar coordinates the region whose description in rectangular coordinates is

$$1 \leq x \leq 2, \quad \frac{x}{\sqrt{3}} \leq y \leq x.$$

2 Find the moment about the x axis of the triangle whose vertices are $(0, 0)$, $(2, 0)$, and $(2, 2)$, using $\delta = 1$ and
(a) a repeated integral in rectangular coordinates,
(b) a repeated integral in polar coordinates.

3 (a) Find the average distance from points in a disc of radius a to the center.
(b) Why is the average larger than $a/2$? (Give an intuitive explanation.)

4 Transform this repeated integral to a repeated integral in polar coordinates, and evaluate the latter:

$$\int_0^a \left[\int_0^{\sqrt{a^2 - x^2}} (x^2 + y^2)^{3/2} \, dy \right] dx.$$

5 (a) Find the moment of inertia of one loop of the curve $r = \sin 2\theta$ centimeters about the pole if the density is 1 gram per square centimeter.
(b) Find the mass within the loop in (a) if the density at (r, θ) is r^2 grams per square centimeter.
(c) Find the volume of a solid whose base is the loop in (a) and whose cross section above (r, θ) has length r^2 centimeters.
(d) The temperature at the point (r, θ) inside the loop in (a) is r^2 degrees Celsius. What is the average temperature?

6 An agricultural sprinkler distributes water in a circle of radius 100 feet. By placing a few random cans in this circle, it is determined that the sprinkler supplies water at a depth of e^{-r} feet of water at a distance of r feet from the sprinkler in 1 hour. How much water does the sprinkler supply to the region within (a) 100 feet of the sprinkler? (b) 50 feet of the sprinkler?

7 A solid circular cylinder of radius a and height h is composed of a uniform material of mass M. Show that its moment of inertia about a line perpendicular to the axis and midway between the two ends of the cylinder is

$$\frac{Ma^2}{4} + \frac{Mh^2}{12}.$$

8 Find $\bar{z}$ of the homogeneous solid region bounded by two concentric spherical shells of radii a and b, $a < b$, centered at the origin and by the xy plane, as shown in Fig. 13.98.

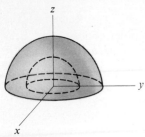

Figure 13.98

9 (a) What is meant by the symbol $\int_R f(P)\, dV$?
(b) Using the definition in (a), show that, if $2 \leq f(P) \leq 3$ for all P in R, then $2 \cdot \text{Volume of } R \leq \int_R f(P)\, dV \leq 3 \cdot \text{Volume of } R$.

10 (a) Find the cylindrical coordinates of the point whose rectangular coordinates are $(3, 4, -3)$.
(b) Find the rectangular coordinates of the point whose spherical coordinates are $(3, \pi/2, 2\pi/3)$.
(c) Find the spherical coordinates of the point whose cylindrical coordinates are $(2, \pi/4, 2)$.

11 Draw the set of points in a ball of radius 1, whose center is at the origin of the coordinate system, determined by
(a) $x = \frac{1}{2}$;
(b) $\phi = \pi/3$;
(c) $\rho = \frac{1}{2}$;
(d) $\theta = \pi/2$;
(e) $z = -\frac{1}{2}$.

12 (a) What extra factor must be introduced when setting up a repeated integral in cylindrical or in spherical coordinates?
(b) Why?

13 (a) Draw the little solid region corresponding to changes $\Delta\rho$, $\Delta\theta$, and $\Delta\phi$ in the spherical coordinates.
(b) Show why its volume is approximately
$$\rho^2 \sin \phi\, \Delta\rho\, \Delta\theta\, \Delta\phi.$$

14 A solid right circular cylinder has radius a and height h. Find the average over R of the function f, where $f(P)$ is the square of the distance from the axis of the cylinder to P:
(a) Set up repeated integrals in at least two of the three coordinate systems.
(b) Evaluate the easier repeated integral in (a).

15 (a) Evaluate the repeated integral
$$\int_0^1 \left[\int_0^1 \left(\int_0^x y e^{x^2}\, dz \right) dy \right] dx.$$
(b) Draw the region R described by the ranges of integration in (a).

16 A solid homogeneous right circular cone of radius a and height h has a mass M.
(a) What is meant by its "moment of inertia about a line through its vertex and parallel to its base"?
(b) Set up repeated integrals in all three coordinate systems for the moment of inertia in (a).
(c) Evaluate at least one of the repeated integrals in (b).

17 What is the equation of
(a) the plane $z = 3$ in spherical coordinates;
(b) the cylindrical surface $r = 2$ in rectangular coordinates;
(c) the cylindrical surface $r = 2$ in spherical coordinates;
(d) the spherical surface $\rho = 3$ in cylindrical coordinates?

18 (This concerns optional Sec. 13.6.) Explain in detail why the magnification of the mapping $F(u, v) = (au + bv, cu + dv)$, $ad - bc \neq 0$, is $|ad - bc|$.

Review exercises for chap. 13

1 Describe the finite region between $y = x^2$ and $y = 4$ by
(a) vertical cross sections,
(b) horizontal cross sections.

2 Compute:
(a) $\int_0^1 x^2 y\, dy$,
(b) $\int_0^1 x^2 y\, dx$.

3 Compute:
(a) $\int_1^{x^2} (x + y)\, dy$,
(b) $\int_y^{y^2} (x + y)\, dx$.

4 Compute the easier of
(a) $\int_0^1 \sin(x^2 y)\, dy$,
(b) $\int_0^1 \sin(x^2 y)\, dx$.

5 Describe the region in Fig. 13.99 in terms of

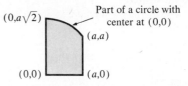

Figure 13.99

(a) rectangular coordinates and vertical cross sections,
(b) rectangular coordinates and horizontal cross sections,
(c) polar coordinates.

Translate the repeated integrals in Exercises 6 to 8 to repeated integrals in polar coordinates and in each case evaluate the latter.

6 $\int_0^1 \left(\int_0^x x^2 \, dy \right) dx$

7 $\int_0^{1/\sqrt{2}} \left(\int_x^{\sqrt{1-x^2}} \sqrt{x^2 + y^2} \, dy \right) dx$

8 $\int_0^2 \left(\int_0^{\sqrt{2x-x^2}} x \, dy \right) dx$

9 Consider the repeated integral $\int_{\pi/4}^{\pi/2} \left(\int_0^a r^2 \sin\theta \, dr \right) d\theta$.

(a) Draw R and describe f such that $\int_R f(P) \, dA$ is represented by the given repeated integral.
(b) Choose a convenient repeated integral in rectangular coordinates equal to $\int_R f(P) \, dA$.
(c) Evaluate $\int_R f(P) \, dA$ by the simplest method.

Translate the repeated integrals in Exercises 10 to 12 to repeated integrals in rectangular coordinates and evaluate. (Choose the more convenient direction.)

10 $\int_0^{\pi/4} \left(\int_0^a r^2 \cos\theta \, dr \right) d\theta$

11 $\int_0^{\pi/4} \left(\int_0^a r^3 \, dr \right) d\theta$

12 $\int_{\pi/4}^{3\pi/4} \left(\int_0^a r^3 \, dr \right) d\theta$

13 Find the moment of inertia of a homogeneous square of mass M and side a about
(a) a side,   (b) a diagonal.

14 Find the centroid of the region outside the circle $r = 1$ and inside the cardioid $r = 1 + \cos\theta$.

15 Find the moment of inertia of the finite region bounded by the curve $y = x^3$, $y = 8$, and the y axis about
(a) the x axis,   (b) the y axis,   (c) the z axis.
Assume that it is homogeneous and has mass M.

In Exercises 16 to 19 compute $\int_R f(P) \, dA$ if $f(x, y) = xy$ and R is described in coordinates as

16 $0 \le x \le 2, x^3 \le y \le 2x^3$.
17 $0 \le x \le \pi/2, 0 \le y \le \sin x$.
18 $0 \le \theta \le \pi/4, 0 \le r \le 2 \sin\theta$.
19 $0 \le \theta \le \pi/4, 0 \le r \le \cos 2\theta$.

20 Find the centroid of the finite region bounded by $y = x^2$ and $y = \sqrt{x}$.

21 The depth of water provided by a water sprinkler is approximately 2^{-r} feet at a distance of r feet from the sprinkler. Find the total amount of water within a distance of a feet of the sprinkler.

22 Find the centroid of the region bounded by the curve $y = \cos x$, the x axis, and the lines $x = \pi/2$ and $x = -\pi/2$.

23 Let R be a triangle. Place an xy-coordinate system in such a way that its origin is at one vertex of the triangle and the x axis is parallel to the opposite side. Call the coordinates of the two other vertices (a, b) and (c, b). Show that $\bar{y} = 2b/3$.

24 Let R be a circle of radius a and center $(0, 0)$.
(a) Without evaluating them, explain why the integrals $\int_R x^2 \, dA$ and $\int_R y^2 \, dA$ are equal.
(b) Without evaluating any of these integrals, show that $\int_R x^2 \, dA + \int_R y^2 \, dA = \int_R r^2 \, dA$.
(c) Evaluate $\int_R r^2 \, dA$ by using polar coordinates.
(d) Combining (a), (b), and (c), compute $\int_R x^2 \, dA$.

25 (a) Draw the region R whose description is
$$\frac{\sqrt{2}}{2} \le x \le 1, \quad \sqrt{1-x^2} \le y \le x.$$
(b) Describe R in polar coordinates.
(c) Transform the repeated integral
$$\int_{\sqrt{2}/2}^1 \left(\int_{\sqrt{1-x^2}}^x \frac{1}{\sqrt{x^2+y^2}} \, dy \right) dx$$
into polar coordinates.
(d) Evaluate the repeated integral in polar coordinates.

26 Evaluate $\int_R \ln(x^2 + y^2) \, dA$ over the region in Exercise 25.

27 A homogeneous right circular cylindrical shell has inner radius a, outer radius b, and height h. Its mass is M. Show that its moment of inertia
(a) about its axis is $M(a^2 + b^2)/2$;
(b) about a line through its center of mass and perpendicular to its axis is $M(a^2 + b^2 + h^2/3)/4$.

28 A homogeneous solid of mass M occupies the space between two concentric spheres of radii a and b, $a < b$. Show that its moment of inertia around a diameter is $2M(b^5 - a^5)/[5(b^3 - a^3)]$.

29 Let R be the region bounded by a circle of radius a. Let A be a point in the plane of the circle at a distance $H > a$ from the center of the circle. Define a function f by setting $f(P) = \overline{PA}^2$. Show that the average value of f over R is $H^2 + a^2/2$.

∎

30 The gravitational attraction between a homogeneous ball of radius s and a point mass, as shown in Fig. 13.100, involves evaluation of the integral
$$\int_S \frac{\cos\alpha}{x^2} \, dV.$$

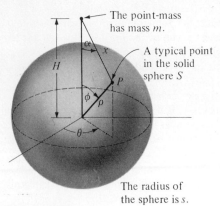

Figure 13.100

Show that its value is $\dfrac{4\pi s^3}{3H^2}$.

31 A right rectangular pyramid has a base of dimensions a by b and height h. Its mass is M. Show that the moment of inertia of the pyramid around the line that is perpendicular to the base and passes through its top vertex is $M(a^2 + b^2)/20$. (See Fig. 13.101.)

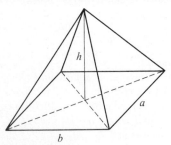

Figure 13.101

32 A doughnut (torus) is formed by revolving a circle of radius a in a plane around a line L in that plane that is a distance $b > a$ from the center of the circle. (See Fig. 13.102.) Its mass is M. Show that the moment of inertia of the doughnut around the line L is $M(b^2 + 3a^2/4)$. Assume that the density is constant.

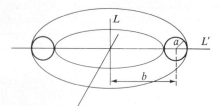

Figure 13.102

33 Show that the moment of inertia of the doughnut of Exercise 32 around line L' shown in Fig. 13.102 is $M(b^2/2 + 5a^2/8)$.

34 A homogeneous solid right circular cone of radius a and height h has mass M. Show that the moment of inertia of the cone around a line through its center of mass and parallel to its base is $3M(a^2 + h^2/4)/20$.

Transportation problems lead to integrals over plane sets, as Exercises 35 to 40 illustrate.

35 Show that the average travel distance from the center of a circle of area A to points in the circle is approximately $0.376\sqrt{A}$ (precisely $2\sqrt{A}/(3\sqrt{\pi})$).

36 Show that the average travel distance from the center of a regular hexagon of area A to points in the hexagon is approximately $0.377\sqrt{A}$ (precisely
$$\frac{\sqrt{2}\sqrt{A}}{\sqrt{3}\sqrt[4]{3}}\left(\frac{1}{3} + \frac{\ln 3}{4}\right)\text{).}$$

37 Show that the average travel distance from the center of a square of area A to points in the square is approximately $0.383\sqrt{A}$ (precisely $[(\sqrt{2} + \ln \tan 3\pi/8)\sqrt{A}]/6$).

38 Show that the average travel distance from the centroid of an equilateral triangle of area A to points in the triangle is approximately $0.404\sqrt{A}$ (precisely
$$\frac{\sqrt{A}}{9\sqrt[4]{3}}\left(2\sqrt{3} + \ln \tan \frac{5\pi}{12}\right)\text{).}$$

In Exercises 35 to 38 the distance is the ordinary straight-line distance. In cities the usual street pattern suggests that the "metropolitan" distance between the points (x_1, y_1) and (x_2, y_2) should be measured by $|x_1 - x_2| + |y_1 - y_2|$.

39 Show that, if in Exercise 35 metropolitan distance is used, then the average is $8\sqrt{A}/(3\pi^{3/2}) \approx 0.479\sqrt{A}$.

40 Show that, if in Exercise 37 metropolitan distance is used, then the average is $\sqrt{A}/2$. In most cities the metropolitan average tends to be about 25 percent larger than the direct-distance average.

41 In an ordinary map an area on the map is proportional to the area depicted. But some maps for airline pilots have a larger scale near airports. It has been suggested that maps should be distorted so that they represent accurately various quantities. Thus a map of population would show the cities as large and sparsely populated states as small. In a map of rainfall the Olympic peninsula of Washington would be large, while Arizona would be small.

Consider a square region R in the uv plane, furnished with a positive function $Q(u, v)$. (Perhaps $Q(u, v)$ is the number of inches of rain per year at (u, v).)

(a) Why might we seek a mapping $x = f(u, v)$, $y = g(u, v)$ whose Jacobian equals $Q(u, v)$?

(b) Let the vertices of the square region in the uv plane be $(0, 0)$ $(1, 0)$, $(1, 1)$, and $(0, 1)$. Define
$$x = f(u, v) = \int_0^u \left[\int_0^1 Q(u_1, v_1)\, dv_1\right] du_1$$

and
$$y = g(u, v) = \frac{\int_0^v Q(u, v_1)\, dv_1}{\int_0^1 Q(u, v_1)\, dv_1}.$$

Show that
$$f_u(u, v) = \int_0^1 Q(u, v_1)\, dv_1$$

and that $f_v(u, v) = 0$.

(c) Show that
$$g_v(u, v) = \frac{Q(u, v)}{\int_0^1 Q(u, v_1)\, dv_1}.$$

(There is no need to compute g_u.)

(d) Show that $\partial(x, y)/\partial(u, v) = Q(u, v)$, as desired.

(e) What is the image of R?

■ ■

42 Let $(\bar{x}, \bar{y}, \bar{z})$ be the center of mass of a solid region R with density function $\delta(P)$.

(a) Consider a plane parallel to a coordinate plane and passing through the center of mass. Let $f(P) = $ distance from P to the plane, positive if P is on one side of the plane, negative if P is on the other side. Show that $\int_R f(P)\, \delta(P)\, dV = 0$.

(b) Consider *any* plane through the center of mass. Let $f(P)$ be defined as in (a). Show that $\int_R f(P)\, \delta(P)\, dV = 0$.

43 Let $z = g(y)$ be a decreasing function of y such that $g(1) = 0$. Let R be the solid of revolution formed by revolving about the z axis the region in the yz plane bounded by $y = 0$, $z = 0$, and $z = g(y)$. Using appropriate repeated integrals in cylindrical coordinates, show that $\int_R z\, dV = \int_0^1 \pi y[g(y)]^2\, dy$ and $\int_R z\, dV = \int_0^{g(0)} \pi[g^{-1}(z)]^2 z\, dz$.

44 (See Exercise 43.)

(a) Show that the z coordinate of the centroid of the solid described in Exercise 43 is
$$\frac{\int_0^1 \{x[g(x)]^2/2\}\, dx}{\int_0^1 xg(x)\, dx},$$
while the z coordinate of the centroid of the plane region that was revolved is
$$\frac{\int_0^1 \{[g(x)]^2/2\}\, dx}{\int_0^1 g(x)\, dx}.$$

(b) By considering
$$\int_0^1 \left(\int_0^1 g(x)g(y)(x - y)[g(x) - g(y)]\, dx \right) dy,$$
show that the centroid of the solid of revolution is below that of the plane region. *Hint:* Why is the repeated integral less than or equal to 0?

45 We outline another proof of Schwarz's inequality given in Exercise 21 of Sec. 8.8. Let R be the square $a \le x \le b$, $a \le y \le b$.

(a) Why is $\int_R [f(x)g(y) - f(y)g(x)]^2\, dA \ge 0$?

(b) From (a) deduce Schwarz's inequality.

46 Let $f(x, y) = e^{y^3}$.

(a) Devise a region R in the plane such that $\int_R f(P)\, dA$ can be evaluated with the aid of a repeated integral.

(b) Devise a region R in the plane such that $\int_R f(P)\, dA$ *cannot* be evaluated with the aid of a repeated integral, and describe the difficulty.

47 Define $f(t)$ to be $\int_t^1 e^{x^2}\, dx$. Find the average value of f over the interval $[0, 1]$.

THE DERIVATIVE OF A VECTOR FUNCTION

The motion of a particle traveling along a straight line is most easily described with the aid of a coordinate system on the line. But the motion of a particle along a curved path, as well as the forces acting on that particle, are most easily described with vectors. For example, the attraction of the earth on an astronaut's spacecraft is represented by a vector directed toward the center of the earth, as shown in Fig. 14.1.

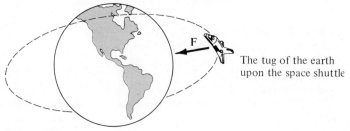

The tug of the earth upon the space shuttle

Figure 14.1

The astronaut influences his or her flight by firing small rockets. Their thrust is described by a vector, as shown in Fig. 14.2. When you spin an

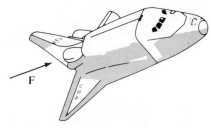

When the astronaut fires an engine, she experiences the shove F.

Figure 14.2

object in a circle at the end of a rope, it is subject to two forces: one is toward your hand, and the other is the pull of gravity, straight down.

This chapter uses vectors to investigate the motion of a particle whose direction and speed may both change with time, because of some external forces. In addition it develops an equation for the tangent plane to a surface and the method of Lagrange multipliers for finding extrema of functions of several variables. Throughout the chapter it is assumed that the functions have continuous partial derivatives of all orders.

14.1 The derivative of a vector function

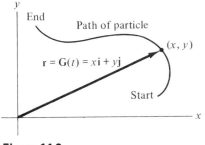

Figure 14.3

Consider an object moving in a plane. It might be a mass at the end of a rope, a ball, a satellite, a comet, a raindrop, or an astronaut's spacecraft. Call this object a "particle" and assume that all its mass is located at a single point. Denote the position of the particle at time t relative to an xy-coordinate system by (x, y). We shall describe its position with the position vector $\mathbf{r}$, whose tail is at $(0, 0)$ and whose head is at (x, y), as in Fig. 14.3. Thus $\mathbf{r} = x\mathbf{i} + y\mathbf{j}$, where x and y depend on time t. Therefore $\mathbf{r}$ depends on t and may be written as $\mathbf{r} = \mathbf{G}(t)$.

If the particle is moving on a curve in space, its position at time t is described by a position vector $\mathbf{r} = \mathbf{G}(t) = x\mathbf{i} + y\mathbf{j} + z\mathbf{k}$, where x, y, and z are functions of time.

This brings us to an important definition, which calls attention to a new type of function.

DEFINITION *Vector function.* A function whose inputs are scalars and whose outputs are vectors is called a *vector function.* It is denoted by a boldface letter, such as **F** or **G**.

Frequently $\mathbf{r}$ is a function of time t or arc length s.

Usually the scalar input may be thought of as time, and $\mathbf{G}(t)$ as the position vector at time t. The scalar input might sometimes be arc length along the curve, and $\mathbf{G}(s)$ would be the position vector when the particle has swept out a distance s along the curve.

EXAMPLE 1 Sketch the path of a particle that has the position vector

$$\mathbf{G}(t) = 2t\mathbf{i} + 4t^2\mathbf{j}$$

at time t.

SOLUTION At time t the particle is at the point (x, y) where

$$x = 2t \quad \text{and} \quad y = 4t^2.$$

The path is given parametrically by these equations. Elimination of t shows that $y = x^2$. The path is a parabola, sketched in Fig. 14.4, in which $\mathbf{G}(1)$ is also depicted. ∎

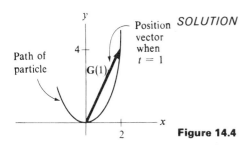

Figure 14.4

In the case of motion on a horizontal line the derivative of position with respect to time is sufficient to describe the motion of the particle. If the derivative is positive, the particle is moving to the right. If the derivative is negative, the particle is moving to the left. The speed is simply the absolute value of the derivative. But the study of motion in the plane depends on the concept of the derivative of a vector function. First, a definition.

LIMIT OF A VECTOR FUNCTION

DEFINITION *Limit of a vector function.* Let **H** be a vector function. Let a be a real number and **A** a vector. We say that

$$\lim_{t \to a} \mathbf{H}(t) = \mathbf{A}$$

if

$$\lim_{t \to a} |\mathbf{H}(t) - \mathbf{A}| = 0,$$

that is, the magnitude of $\mathbf{H}(t) - \mathbf{A}$ approaches 0 as $t \to a$.

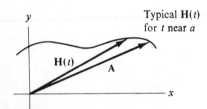

Typical **H**(t) for t near a

Figure 14.5

In other words, if $\mathbf{H}(t)$ and $\mathbf{A}$ are drawn with their tails at the origin, the head of $\mathbf{H}(t)$ gets close to the head of $\mathbf{A}$ as $t \to a$, as shown in Fig. 14.5.

EXAMPLE 2 Let $\mathbf{H}(t) = t^2 \mathbf{i} + 2t \mathbf{j}$. Use the above definition to show that

$$\lim_{t \to 1} \mathbf{H}(t) = \mathbf{i} + 2\mathbf{j}.$$

SOLUTION First compute $|\mathbf{H}(t) - (\mathbf{i} + 2\mathbf{j})|,$

which is

$$|(t^2 \mathbf{i} + 2t\mathbf{j}) - (\mathbf{i} + 2\mathbf{j})| = |(t^2 - 1)\mathbf{i} + (2t - 2)\mathbf{j}|$$
$$= \sqrt{(t^2 - 1)^2 + (2t - 2)^2}.$$

Since

$$\lim_{t \to 1} \sqrt{(t^2 - 1)^2 + (2t - 2)^2} = 0,$$

it follows that

$$\lim_{t \to 1} \mathbf{H}(t) = \mathbf{i} + 2\mathbf{j}. \blacksquare$$

Note in Example 2 that

$$\lim_{t \to 1} \mathbf{H}(t) = \left(\lim_{t \to 1} t^2\right)\mathbf{i} + \left(\lim_{t \to 1} 2t\right)\mathbf{j}.$$

This suggests a practical method for finding the limit of a vector function: If its scalar components are known, just find the limits of these components. This observation is justified by the following theorem.

THEOREM 1 Suppose a vector function **H** is described by two scalar functions g and h,

$$\mathbf{H}(t) = g(t)\mathbf{i} + h(t)\mathbf{j}.$$

To find a limit of $\mathbf{H}(t)$, find limits of its components.

Then
$$\lim_{t \to a} \mathbf{H}(t) = A\mathbf{i} + B\mathbf{j}$$

if and only if $\lim_{t \to a} g(t) = A$ and $\lim_{t \to a} h(t) = B$.

PROOF If $\lim_{t \to a} g(t) = A$ and $\lim_{t \to a} g(t) = B$,

then
$$\lim_{t \to a} |\mathbf{H}(t) - (A\mathbf{i} + B\mathbf{j})| = \lim_{t \to a} \sqrt{(g(t) - A)^2 + (h(t) - B)^2}$$
$$= 0.$$

Thus
$$\lim_{t \to a} \mathbf{H}(t) = A\mathbf{i} + B\mathbf{j}.$$

Conversely, if $\lim_{t \to a} \mathbf{H}(t) = A\mathbf{i} + B\mathbf{j}$, then

$$\lim_{t \to a} \sqrt{(g(t) - A)^2 + (h(t) - B)^2} = 0,$$

so that $\lim_{t \to a} g(t) = A$ and $\lim_{t \to a} h(t) = B$. ∎

Theorem 1 says: To find the limit of a vector function, find the limit of each of its component scalar functions. Though Theorem 1 concerns only functions whose values are *plane vectors*, it extends to functions whose values are vectors in space, that is, vectors that have three scalar components. The proof is essentially unchanged.

EXAMPLE 3 Find $\lim_{t \to 2} \mathbf{H}(t)$, where $\mathbf{H}(t) = t^2\mathbf{i} + t^3\mathbf{j} + t^4\mathbf{k}$.

SOLUTION
$$\lim_{t \to 2} \mathbf{H}(t) = \left(\lim_{t \to 2} t^2\right)\mathbf{i} + \left(\lim_{t \to 2} t^3\right)\mathbf{j} + \left(\lim_{t \to 2} t^4\right)\mathbf{k} = 4\mathbf{i} + 8\mathbf{j} + 16\mathbf{k}. \quad ∎$$

In Example 3, $\lim_{t \to 2} \mathbf{H}(t) = \mathbf{H}(2)$. This is similar to continuous scalar functions, where $\lim_{x \to a} f(x) = f(a)$, and suggests the following definition.

DEFINITION *Continuity of a vector function.* The vector function $\mathbf{H}$ is *continuous* at a if

$$\lim_{t \to a} \mathbf{H}(t) = \mathbf{H}(a).$$

By Theorem 1, the vector function $\mathbf{H}(t) = g(t)\mathbf{i} + h(t)\mathbf{j}$ is continuous at a if and only if the two scalar functions $g(t)$ and $h(t)$ are continuous at a.

DERIVATIVE OF A VECTOR FUNCTION

Now it is possible to define the derivative of a vector function. The definition will be modeled after that of the derivative of a scalar function,

$$f'(x) = \lim_{\Delta x \to 0} \frac{f(x + \Delta x) - f(x)}{\Delta x}.$$

After the definition is given, its geometric meaning will be examined.

DEFINITION *Derivative of a vector function.* Let **G** be a vector function. The limit

$$\lim_{\Delta t \to 0} \frac{\mathbf{G}(t + \Delta t) - \mathbf{G}(t)}{\Delta t} \qquad (1)$$

(if it exists) is called the derivative of **G** at t. It is denoted $\mathbf{G}'(t)$.

What does the quotient in (1) mean geometrically? The numerator, which may be called

$$\Delta \mathbf{G},$$

is a vector. The denominator Δt is a scalar. The quotient

$$\frac{\Delta \mathbf{G}}{\Delta t}$$

is a vector.

If the derivative exists, then, when Δt is small, $\Delta \mathbf{G}$ is a short vector. (See Fig. 14.6.) It is the vector from the head of $\mathbf{G}(t)$ to the head of $\mathbf{G}(t + \Delta t)$. That is, if $\mathbf{G}(t) = \overrightarrow{OP}$ and $\mathbf{G}(t + \Delta t) = \overrightarrow{OQ}$, then $\Delta \mathbf{G} = \overrightarrow{PQ}$. The vector $\Delta \mathbf{G}$ lies on a chord of the path; when Δt is small, the direction of $\Delta \mathbf{G}$ is close to that of the tangent line to the curve at P. Since Δt is a scalar (for simplicity, consider it to be positive), the vector $\Delta \mathbf{G}/\Delta t$ is parallel to the vector $\Delta \mathbf{G}$ and points in the same direction as $\Delta \mathbf{G}$. Therefore the vector $\Delta \mathbf{G}/\Delta t$, when Δt is small and positive, would presumably point almost along a tangent line at P.

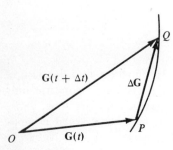

Figure 14.6

Thus $\mathbf{G}'(t)$ is presumably parallel to the tangent line at P. The direction of $\mathbf{G}'(t)$ is the direction in which the particle is moving as it passes through P. These observations suggest the following definition.

DEFINITION *Tangent vector to a curve.* Let $\mathbf{r} = \mathbf{G}(t)$ describe a curve in the plane or in space. The vector $\mathbf{G}'(t)$ is called a *tangent vector* to the curve.

For any parameterization of the curve, $\mathbf{G}'(t)$ is a tangent vector. Its length may depend on the particular parameterization.

To compute $\mathbf{G}'(t)$ just differentiate the scalar components of $\mathbf{G}(t)$ if they are known. For if $\mathbf{G}(t) = g(t)\mathbf{i} + h(t)\mathbf{j}$, then

$$\frac{\Delta \mathbf{G}}{\Delta t} = \frac{\Delta g \mathbf{i} + \Delta h \mathbf{j}}{\Delta t}$$

$$= \frac{\Delta g}{\Delta t} \mathbf{i} + \frac{\Delta h}{\Delta t} \mathbf{j}.$$

By Theorem 1, $\quad \mathbf{G}'(t) = \lim\limits_{\Delta t \to 0} \dfrac{\Delta \mathbf{G}}{\Delta t} = \lim\limits_{\Delta t \to 0} \dfrac{\Delta g}{\Delta t} \mathbf{i} + \lim\limits_{\Delta t \to 0} \dfrac{\Delta h}{\Delta t} \mathbf{j}$

$$= g'(t)\mathbf{i} + h'(t)\mathbf{j}.$$

The following theorem states this formally in the case in which $\mathbf{G}(t)$ is a plane vector. The principle applies just as well to the more general case when **G**(t) is a space vector.

THEOREM 2 Let $\mathbf{G}(t) = g(t)\mathbf{i} + h(t)\mathbf{j}$. If both g and h have derivatives at $t = a$, then **G** has a derivative at $t = a$, and

$$\mathbf{G}'(a) = g'(a)\mathbf{i} + h'(a)\mathbf{j}. \blacksquare$$

*To differentiate **G**(t),* If $\mathbf{G}(t) = g(t)\mathbf{i} + h(t)\mathbf{j}$, then $\mathbf{G}'(t) = g'(t)\mathbf{i} + h'(t)\mathbf{j}$ is a tangent vector to a
differentiate its scalar plane curve. Hence the slope of the curve at the point in question is
components. $h'(t)/g'(t)$. This formula agrees with the formula in Sec. 9.3 for finding the slope of a curve given parametrically:

$$\frac{dy}{dx} = \frac{dy/dt}{dx/dt}.$$

The vector approach to the definition of a tangent line is more general, for it also applies to curves in space. The next example illustrates this for a particular curve.

EXAMPLE 4 At time t a particle has the position vector

$$\mathbf{r} = \mathbf{G}(t) = 3 \cos 2\pi t \, \mathbf{i} + 3 \sin 2\pi t \, \mathbf{j} + 5t\mathbf{k}.$$

Describe its path and sketch a typical tangent vector $\mathbf{G}'(t)$.

SOLUTION At time t the particle is at the point

$$\begin{cases} x = 3 \cos 2\pi t \\ y = 3 \sin 2\pi t \\ z = 5t. \end{cases}$$

Notice that $x^2 + y^2 = (3 \cos 2\pi t)^2 + (3 \sin 2\pi t)^2 = 9$. Thus the point is always on the cylinder

$$x^2 + y^2 = 9.$$

Moreover, as t increases, $z = 5t$ increases.

The path is thus the spiral spring sketched in Fig. 14.7. When t increases by 1, the angle $2\pi t$ increases by 2π, and the particle goes once around the spiral. This type of corkscrew path is called a *helix*.

At time t

$$\mathbf{G}'(t) = -6\pi \sin 2\pi t \, \mathbf{i} + 6\pi \cos 2\pi t \, \mathbf{j} + 5\mathbf{k}. \blacksquare$$

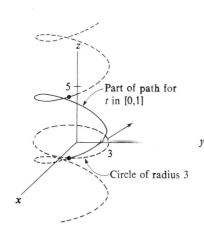

Figure 14.7

*The magnitude of **G**'(t)* The length of the vector $\mathbf{G}'(t) = g'(t)\mathbf{i} + h'(t)\mathbf{j}$ is
is the speed.

$$\sqrt{[g'(t)]^2 + [h'(t)]^2}$$

898 THE DERIVATIVE OF A VECTOR FUNCTION

or, equivalently,
$$\sqrt{\left(\frac{dx}{dt}\right)^2 + \left(\frac{dy}{dt}\right)^2}.$$

This is the formula obtained in Sec. 9.4 for the speed of a particle moving on a plane curve. Now, a similar assertion holds for curves in space. It can be shown that the arc length s swept out on a space curve during the time interval $[a, b]$ is

$$\int_a^b \sqrt{\left(\frac{dx}{dt}\right)^2 + \left(\frac{dy}{dt}\right)^2 + \left(\frac{dz}{dt}\right)^2}\, dt.$$

Consequently, the speed of the particle at time t is

$$\sqrt{\left(\frac{dx}{dt}\right)^2 + \left(\frac{dy}{dt}\right)^2 + \left(\frac{dz}{dt}\right)^2},$$

which a physicist prefers to write in the dot notation of Newton as

$$\sqrt{(\dot{x})^2 + (\dot{y})^2 + (\dot{z})^2}.$$

If $\quad \mathbf{r} = \mathbf{G}(t) = x(t)\mathbf{i} + y(t)\mathbf{j} + z(t)\mathbf{k},$
then $\quad |\mathbf{G}'(t)| = \sqrt{(\dot{x})^2 + (\dot{y})^2 + (\dot{z})^2} =$ Speed.

Thus the magnitude of the vector $\mathbf{G}'(t)$ may be interpreted as the *speed* of the particle:

$$|\mathbf{r}'| = |\mathbf{G}'(t)| = \text{Speed of the particle at time } t.$$

$\mathbf{G}'(t)$ is called the velocity vector. For this reason it is customary to call $\mathbf{r}'$ or $\mathbf{G}'(t)$ the *velocity vector*. It points in the direction the particle is moving at a given instant, and its magnitude is the speed of the particle. Frequently the velocity vector will be denoted $\mathbf{v}$.

EXAMPLE 5 Find the speed at time t of the particle described in Example 4.

SOLUTION
$$\text{Speed} = |\mathbf{G}'(t)| = \sqrt{(-6\pi \sin 2\pi t)^2 + (6\pi \cos 2\pi t)^2 + 5^2}$$
$$= \sqrt{36\pi^2(\sin^2 2\pi t + \cos^2 2\pi t) + 25}$$
$$= \sqrt{36\pi^2 + 25}.$$

The particle travels at a constant speed along its helical path. In t units of time it travels the distance $\sqrt{36\pi^2 + 25}\, t$.

Note that the velocity vector is not constant; its direction is always changing. However, its length in this example remains constant, for the speed is constant. ∎

EXAMPLE 6 Sketch the path of a particle whose position vector at time $t \geq 0$ is

$$\mathbf{G}(t) = \cos t^2\, \mathbf{i} + \sin t^2\, \mathbf{j}.$$

Find its speed at time t.

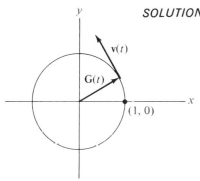

Figure 14.8

SOLUTION Note that $|G(t)| = \sqrt{\cos^2 t^2 + \sin^2 t^2} = 1$.

So the path of the particle is on the circle of radius 1 and center $(0, 0)$. (See Fig. 14.8.) The speed of the particle is

$$|v(t)| = |G'(t)| = |-2t \sin t^2 \, i + 2t \cos t^2 \, j|$$
$$= \sqrt{(-2t \sin t^2)^2 + (2t \cos t^2)^2}$$
$$= 2t\sqrt{\sin^2 t^2 + \cos^2 t^2}$$
$$= 2t.$$

The particle travels faster and faster around a circle of radius 1. ∎

Exercises

1 At time t a particle has the position vector $r = G(t) = t\,i + t^2\,j$.
  (a) Compute and draw $G(1)$, $G(2)$, and $G(3)$.
  (b) Show that the path is a parabola.

2 At time t a particle has the position vector $r = G(t) = (2t + 1)i + 4t\,j$.
  (a) Compute and draw $G(0)$, $G(1)$, and $G(2)$.
  (b) Show that the path is a straight line.

3 Let $G(t) = 2t\,i + t^2\,j$.
  (a) Compute and draw $G(1.1)$, $G(1)$, and $\Delta G = G(1.1) - G(1)$.
  (b) Compute and draw $\Delta G/0.1$, where ΔG is defined in part (a).
  (c) Compute $G'(1)$.

4 Let $G(t) = 3t\,i + 2t^2\,j$.
  (a) Compute and draw $\Delta G = G(2.01) - G(2)$.
  (b) Compute and draw $\Delta G/0.01$.
  (c) Compute $G'(2)$.

5 At time t the position vector of a thrown ball is $r(t) = 32t\,i - 16t^2\,j$.
  (a) Draw $r(1)$ and $r(2)$.
  (b) Sketch the path.
  (c) Compute and draw $v(0)$, $v(1)$, and $v(2)$. In each case place the tail of the vector at the head of the corresponding position vector.

6 At time $t \geq 0$ a particle is at the point $x = 2t$, $y = 4t^2$.
  (a) What is the position vector $r(t)$ at time t?
  (b) Sketch the path.
  (c) How fast is the particle moving when $t = 1$?
  (d) Draw $v(1)$ with its tail at the head of $r(1)$.

In Exercises 7 and 8 find the velocity vectors and the speeds for the given paths.

7 $r(t) = \cos 3t\,i + \sin 3t\,j + 6t\,k$.
8 $r(t) = 3 \cos 5t\,i + 2 \sin 5t\,j + t^2\,k$.
9 Let $G(t) = t^2\,i + t^3\,j$.
  (a) Sketch the vector $\Delta G = G(1.1) - G(1)$.
  (b) Sketch the vector $\Delta G/\Delta t$, where ΔG is given in (a) and $\Delta t = 0.1$.
  (c) Sketch $G'(1)$.
  (d) Find $|\Delta G/\Delta t - G'(1)|$, where ΔG is given in (a) and $\Delta t = 0.1$.

10 At time t the position vector of a particle is
$$G(t) = 2 \cos 4\pi t\,i + 2 \sin 4\pi t\,j + t\,k.$$
  (a) Sketch its path.
  (b) Find its speed.
  (c) Find a unit tangent vector to the path at time t.

11 At time t the position vector of a particle is
$$G(t) = t \cos 2\pi t\,i + t \sin 2\pi t\,j + t\,k.$$
  (a) Show that the particle lies on the surface of a cone whose equation in cylindrical coordinates is $z = r$.
  (b) Sketch the path of the particle.

12 At time t a particle is at $(4t, 16t^2)$.
  (a) Show that the particle moves on the curve $y = x^2$.
  (b) Draw $r(t)$ and $v(t)$ for $t = 0, \frac{1}{4}, \frac{1}{2}$.
  (c) What happens to $|v(t)|$ and the direction of $v(t)$ for large t?

13 At time $t \geq 1$ a particle is at the point (t, t^{-1}).
  (a) Draw the path of the particle.
  (b) Draw $r(1)$, $r(2)$, and $r(3)$.
  (c) Draw $v(1)$, $v(2)$, and $v(3)$.
  (d) As time goes on, what happens to dx/dt, dy/dt, $|v|$, and v?

14 At time t a particle is at $(\cos t^2, \sin t^2)$.
  (a) Show that it moves on the circle $x^2 + y^2 = 1$.
  (b) Compute $\mathbf{r}(t)$ and $\mathbf{v}(t)$.
  (c) How does $|\mathbf{v}(t)|$ behave for large t? What does this say about the particle?

15 An electron travels at constant speed clockwise in a circle of radius 100 feet 200 times a second. At time $t = 0$ it is at $(100, 0)$.
  (a) Compute $\mathbf{r}(t)$ and $\mathbf{v}(t)$.
  (b) Draw $\mathbf{r}(0)$, $\mathbf{r}(\tfrac{1}{800})$, $\mathbf{v}(0)$, $\mathbf{v}(\tfrac{1}{800})$.
  (c) How do $|\mathbf{r}(t)|$ and $|\mathbf{v}(t)|$ behave as time goes on?

16 A ball is thrown up at an initial speed of 200 feet per second and at an angle of 60° from the horizontal. If we disregard air resistance, then at time t it is at $(100t, 100\sqrt{3}\,t - 16t^2)$, as long as it is in flight. Compute and draw $\mathbf{r}(t)$ and $\mathbf{v}(t)$ (a) when $t = 0$; (b) when the ball reaches its maximum height; (c) when the ball strikes the ground.

17 Instead of time t, use arc length s along the path as a parameter, $\mathbf{r} = \mathbf{G}(s)$.
  (a) Show that $d\mathbf{r}/ds$ is a unit vector.
  (b) Sketch $\Delta \mathbf{r}$ and the arc of length Δs. Why is it reasonable that $|\Delta \mathbf{r}/\Delta s|$ is near 1 when Δs is small?

18 A particle at time $t = 0$ is at the point (x_0, y_0, z_0). It moves on a line through that point in the direction of the unit vector $\mathbf{u} = \cos \alpha\, \mathbf{i} + \cos \beta\, \mathbf{j} + \cos \gamma\, \mathbf{k}$. It travels at the constant speed of 3 feet per second.
  (a) Give a formula for its position vector $\mathbf{r} = \mathbf{G}(t)$.
  (b) Find its velocity vector $\mathbf{v} = \mathbf{G}'(t)$.

19 A particle moves in a circular orbit of radius a. At time t its position vector is

$$\mathbf{r}(t) = a \cos 2\pi t\, \mathbf{i} + a \sin 2\pi t\, \mathbf{j}.$$

  (a) Draw its position vector when $t = 0$ and when $t = \tfrac{1}{4}$.
  (b) Draw its velocity vector when $t = 0$ and when $t = \tfrac{1}{4}$.
  (c) Using the dot product, prove that its velocity vector is always perpendicular to its position vector.

20 (a) Show that the paths $\mathbf{G}(t) = t\mathbf{i} + t^2\mathbf{j} + t^3\mathbf{k}$ and $\mathbf{H}(t) = t^2\mathbf{i} + t^3\mathbf{j} + t^4\mathbf{k}$ intersect when $t = 1$.
  (b) At what angle do they intersect?

21 A rock is thrown up at an angle θ from the horizontal and at a speed v_0. Show that

$$\mathbf{r}(t) = (v_0 \cos \theta)t\mathbf{i} + ((v_0 \sin \theta)t - 16t^2)\mathbf{j}.$$

(At time $t = 0$, the rock is at $(0, 0)$; the x axis is horizontal. Time is in seconds and distance is in feet.)

22 (See Exercise 21.) The moment a ball is dropped straight down, you shoot an arrow directly at it. Assume that there is no air resistance. Show that, if you shoot the arrow fast enough, it will hit the ball.
  (a) Solve with the aid of the formulas in Exercise 21.
  (b) Solve with a maximum of intuition and a minimum of formula.
  (c) Must the acceleration be constant for the reasoning in (b) to be valid? Explain.

23 (See Exercise 21.)
  (a) Show that the horizontal distance that the rock travels by the time it reaches its initial height is the same whether the angle is θ or its complement $(\pi/2) - \theta$.
  (b) What value of θ yields the maximum range?

14.2 Properties of the derivative of a vector function

The theorems to follow concern some useful properties of the derivative of a vector function.

We now have two types of functions, vector functions and numerical functions. A vector function $\mathbf{G}$ assigns to a number t a vector $\mathbf{G}(t)$. A numerical function f assigns to a number t a number $f(t)$; for emphasis we call such a function a *scalar function*. The function that assigns to the number t the vector $f(t)\mathbf{G}(t)$ is a vector function. Thus the product of a scalar function and a vector function is a vector function. It is denoted $f\mathbf{G}$. The notation $\mathbf{G}f$, where the scalar appears second, is seldom used. The first theorem concerns the derivative of $f\mathbf{G}$. Note its similarity to the theorem about the derivative of the product of two scalar functions in Sec. 3.4.

14.2 PROPERTIES OF THE DERIVATIVE OF A VECTOR FUNCTION

THEOREM 1 If f is a scalar function and $\mathbf{G}$ is a vector function, the derivative of the vector function $f\mathbf{G}$ is

$$f\mathbf{G}' + f'\mathbf{G}.$$

PROOF The proof is almost the same as that which appears in Sec. 3.4 for the derivative of the product of two functions. First work with $\Delta(f\mathbf{G})$, as follows:

$$\begin{aligned}\Delta(f\mathbf{G}) &= f(t + \Delta t)\mathbf{G}(t + \Delta t) - f(t)\mathbf{G}(t) \\ &= [f(t) + \Delta f][\mathbf{G}(t) + \Delta \mathbf{G}] - f(t)\mathbf{G}(t) \\ &= f(t)\,\mathbf{G}(t) + f(t)\,\Delta \mathbf{G} + \Delta f\,\mathbf{G}(t) + \Delta f\,\Delta \mathbf{G} - f(t)\mathbf{G}(t) \\ &= f(t)\Delta \mathbf{G} + \Delta f\,\mathbf{G}(t) + \Delta f\,\Delta \mathbf{G}.\end{aligned}$$

Thus

$$\lim_{\Delta t \to 0} \frac{\Delta(f\mathbf{G})}{\Delta t} = \lim_{\Delta t \to 0} \left[f(t)\frac{\Delta \mathbf{G}}{\Delta t} + \frac{\Delta f}{\Delta t}\mathbf{G}(t) + \Delta f\,\frac{\Delta \mathbf{G}}{\Delta t} \right]$$

$$= f(t)\mathbf{G}'(t) + f'(t)\mathbf{G}(t) + 0\mathbf{G}'(t).$$

This proves the theorem. ∎

EXAMPLE 1 Find the derivative of the vector function

$$t^2(\cos t\,\mathbf{i} + \sin t\,\mathbf{j}).$$

SOLUTION By Theorem 1, the derivative is

$$t^2(\cos t\,\mathbf{i} + \sin t\,\mathbf{j})' + (t^2)'(\cos t\,\mathbf{i} + \sin t\,\mathbf{j})$$

$$= t^2(-\sin t\,\mathbf{i} + \cos t\,\mathbf{j}) + 2t(\cos t\,\mathbf{i} + \sin t\,\mathbf{j})$$

$$= (-t^2 \sin t + 2t \cos t)\mathbf{i} + (t^2 \cos t + 2t \sin t)\mathbf{j}.$$

Actually, Theorem 1 is not needed for this particular example. We could have written the function as

$$t^2 \cos t\,\mathbf{i} + t^2 \sin t\,\mathbf{j}$$

and differentiated each scalar component as in the preceding section. ∎

From two vector functions $\mathbf{G}$ and $\mathbf{H}$ we can obtain a scalar function $\mathbf{G} \cdot \mathbf{H}$ by defining the value of $\mathbf{G} \cdot \mathbf{H}$ at t to be the dot product

$$\mathbf{G}(t) \cdot \mathbf{H}(t).$$

THEOREM 2 If $\mathbf{G}$ and $\mathbf{H}$ are vector functions, then

$$(\mathbf{G} \cdot \mathbf{H})' = \mathbf{G} \cdot \mathbf{H}' + \mathbf{H} \cdot \mathbf{G}'. \quad \blacksquare$$

The proof of Theorem 2 is outlined in Exercise 4.

EXAMPLE 2 Use Theorem 2 to show that, if the magnitude of $\mathbf{G}(t)$ is independent of t, then $\mathbf{G}'(t)$ is perpendicular to $\mathbf{G}(t)$.

SOLUTION Theorem 2, with $\mathbf{H}(t) = \mathbf{G}(t)$, says that

$$(\mathbf{G} \cdot \mathbf{G})' = \mathbf{G} \cdot \mathbf{G}' + \mathbf{G} \cdot \mathbf{G}'$$
$$= 2\mathbf{G} \cdot \mathbf{G}'. \tag{1}$$

Now $\mathbf{G}(t) \cdot \mathbf{G}(t) = |\mathbf{G}(t)|^2$. Call the length $|\mathbf{G}(t)|$ simply r. Then (1) asserts that

$$\frac{d(r^2)}{dt} = 2\mathbf{G} \cdot \mathbf{G}';$$

hence

$$2r\frac{dr}{dt} = 2\mathbf{G} \cdot \mathbf{G}'.$$

Thus

$$r\frac{dr}{dt} = \mathbf{G} \cdot \mathbf{G}'. \tag{2}$$

If r is constant, (2) implies that $\mathbf{G}(t) \cdot \mathbf{G}'(t)$ is 0. Thus, if neither $\mathbf{G}(t)$ nor $\mathbf{G}'(t)$ is $\mathbf{0}$, they are perpendicular nonzero vectors. ∎

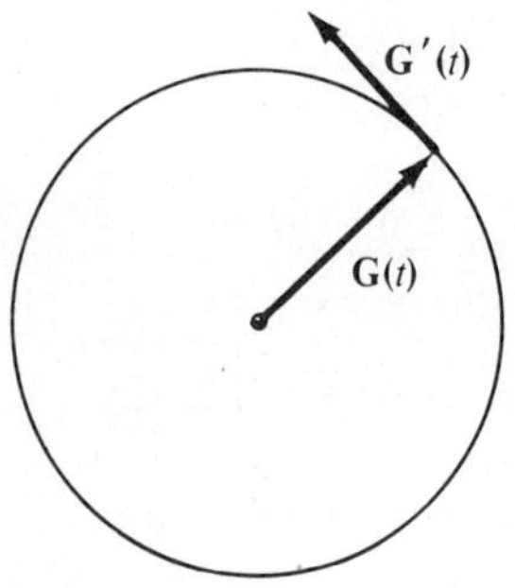

If $\mathbf{G}(t)$ has constant magnitude, $\mathbf{G}'(t)$ is perpendicular to $\mathbf{G}(t)$.

The result obtained in Example 2 also follows from high school geometry. If $|\mathbf{G}(t)|$ is constant, the particle moves on a sphere (or circle). $\mathbf{G}'(t)$ is tangent to the sphere and therefore perpendicular to $\mathbf{G}(t)$, since any ray drawn from the center of a sphere is perpendicular to the tangent at the point where the ray meets the sphere. (See Fig. 14.9.)

Let $\mathbf{G}$ be a vector function of the scalar s, $\mathbf{r} = \mathbf{G}(s)$, and let s in turn be a function of the scalar t, $s = f(t)$. Then we may consider $\mathbf{r}$ to be a function of t:

$$\mathbf{r} = \mathbf{G}(f(t)).$$

(Think of s as arc length along a curve and t as time.) Denote this (composite) function $\mathbf{G} \circ f$.

If $|\mathbf{G}(t)|$ is constant, then $\mathbf{G}'(t)$ is perpendicular to $\mathbf{G}(t)$.
Figure 14.9

THEOREM 3 *Chain rule.* If $\mathbf{G}$ is a vector function and f is a scalar function, then $\mathbf{G} \circ f$ is a vector function, and its derivative at t is

$$\mathbf{G}'(f(t))f'(t).$$

In other words, if $\mathbf{r} = \mathbf{G}(s)$ and $s = f(t)$,

then

$$\frac{d\mathbf{r}}{dt} = \frac{d\mathbf{r}}{ds}\frac{ds}{dt}. \quad ∎$$

The proof of Theorem 3 is outlined in Exercise 16.

EXAMPLE 3 Let $\mathbf{r} = s^2\mathbf{i} + s^3\mathbf{j}$ where $s = e^{2t}$. Compute $d\mathbf{r}/dt$.

SOLUTION
$$\frac{d\mathbf{r}}{ds} = 2s\mathbf{i} + 3s^2\mathbf{j} \quad \text{and} \quad \frac{ds}{dt} = 2e^{2t}.$$

Thus
$$\frac{d\mathbf{r}}{dt} = (2s\mathbf{i} + 3s^2\mathbf{j})2e^{2t},$$

which is usually written with the scalar coefficient in front:
$$\frac{d\mathbf{r}}{dt} = 2e^{2t}(2s\mathbf{i} + 3s^2\mathbf{j}) = 2e^{2t}(2e^{2t}\mathbf{i} + 3e^{4t}\mathbf{j}). \blacksquare$$

Exercises

1. At $t = 1$, $\mathbf{G}(t) = \mathbf{i} + \mathbf{j}$ and $\mathbf{G}'(t) = 2\mathbf{i} - 3\mathbf{j}$. Find $[t^2\mathbf{G}(t)]'$ at $t = 1$.

2. At $t = 0$, $\mathbf{G}(t) = 3\mathbf{i}$ and $\mathbf{G}'(t) = 2\mathbf{i} - 3\mathbf{j} + 4\mathbf{k}$. Find $[t^3\mathbf{G}(t)]'$ at $t = 0$.

3. Theorem 1 was proved without referring to scalar components. Prove Theorem 1 with the aid of scalar components as follows: First write $\mathbf{G}(t) = g(t)\mathbf{i} + h(t)\mathbf{j}$ and then differentiate $f(t)\mathbf{G}(t) = f(t)g(t)\mathbf{i} + f(t)h(t)\mathbf{j}$. (A similar argument would work for vectors in space.)

4. This outlines a proof of Theorem 2.
   (a) Express $\Delta(\mathbf{G} \cdot \mathbf{H})$ as simply as possible with the aid of the equations $\mathbf{G}(t + \Delta t) = \mathbf{G}(t) + \Delta\mathbf{G}$ and $\mathbf{H}(t + \Delta t) = \mathbf{H}(t) + \Delta\mathbf{H}$.
   (b) Using (a), find $\lim\limits_{\Delta t \to 0} \dfrac{\Delta(\mathbf{G} \cdot \mathbf{H})}{\Delta t}$.

5. (a) If $\mathbf{G}$ and $\mathbf{H}$ are vector functions, one obtains a new vector function $\mathbf{G} \times \mathbf{H}$ by defining $(\mathbf{G} \times \mathbf{H})(t)$ to be $\mathbf{G}(t) \times \mathbf{H}(t)$. Prove that
$$(\mathbf{G} \times \mathbf{H})' = \mathbf{G} \times \mathbf{H}' + \mathbf{G}' \times \mathbf{H}$$
   Suggestion: Proceed as in the proof of Theorem 1.
   (b) Would it be correct to write, as in the case of a derivative of the product of scalar functions,
$$(\mathbf{G} \times \mathbf{H})' = \mathbf{G} \times \mathbf{H}' + \mathbf{H} \times \mathbf{G}'?$$

6. Solve Exercise 5(a) by writing $\mathbf{G}(t) = g_1\mathbf{i} + g_2\mathbf{j} + g_3\mathbf{k}$ and $\mathbf{H}(t) = h_1\mathbf{i} + h_2\mathbf{j} + h_3\mathbf{k}$, expressing $\mathbf{G}(t) \times \mathbf{H}(t)$ in components, and then differentiating. Compare the result with $\mathbf{G}(t) \times \mathbf{H}'(t) + \mathbf{G}'(t) \times \mathbf{H}(t)$, also expressed in scalar components.

7. Assume that $\mathbf{G}'(t)$ is always perpendicular to $\mathbf{G}(t)$. Show that the path $\mathbf{r} = \mathbf{G}(t)$ lies on the surface of a sphere.

8. Let $\mathbf{G}(t) = \cos 2t\,\mathbf{i} + \sin 2t\,\mathbf{j}$.
   (a) Show that $\mathbf{G}(t)$ is a unit vector.
   (b) Is $\mathbf{G}'(t)$ a unit vector?

9. A particle at time $t \geq 0$ is at the point (t, t^2, t).
   (a) Plot its path.
   (b) Find its speed at time t.
   (c) Show that its path lies above a parabola in the xy plane.
   (d) Show that its path lies in a plane.

10. A particle moves in a path such that
$$\mathbf{r}(t) = e^t \cos t\,\mathbf{i} + e^t \sin t\,\mathbf{j}.$$
    Show that the angle between its position vector and velocity vector has the constant value $\pi/4$.

11. If the velocity vector $\mathbf{v}(t)$ is constant, show that the path lies on a straight line.

12. A particle moves on the path $\mathbf{r} = \mathbf{G}(t)$, which does not pass through the origin. Prove that, if P is a point on the path closest to the origin O, then the position vector $\overrightarrow{OP}$ is perpendicular to the velocity vector at P.

13. If $\mathbf{G}(t)$ has a constant magnitude, must $\mathbf{G}'(t)$ have constant magnitude?

14. Let $\mathbf{G}$ be a vector function such that $\mathbf{G}(t)$ is never $\mathbf{0}$. Show that $\mathbf{G}$ can be written in the form $f\mathbf{H}$, where $f(t) \geq 0$ and $|\mathbf{H}(t)| = 1$ for all t.

15. Let $\mathbf{G}$ be a vector function such that $\mathbf{G}(t)$ is never $\mathbf{0}$. Let $r(t) = |\mathbf{G}(t)|$.
    (a) Show that $\left|\dfrac{\mathbf{G}(t)}{r(t)}\right|$ is 1.
    (b) By Example 2, $\left(\dfrac{\mathbf{G}}{r}\right)' \cdot \dfrac{\mathbf{G}}{r} = 0$.
    Deduce that $r\mathbf{G} \cdot \mathbf{G}' = r'\mathbf{G} \cdot \mathbf{G}$.
    (c) Verify (b) by expressing $\mathbf{G}$ and $\mathbf{G}'$ in components.

16 To prove Theorem 3 write $\mathbf{G}(s) = g(s)\mathbf{i} + h(s)\mathbf{j}$ and work with components. Carry out the details.

17 Let $\mathbf{G}$ and $\mathbf{H}$ be two vector functions. Define $\mathbf{G} + \mathbf{H}$ by defining $(\mathbf{G} + \mathbf{H})(t)$ to be $\mathbf{G}(t) + \mathbf{H}(t)$. Show that $(\mathbf{G} + \mathbf{H})' = \mathbf{G}' + \mathbf{H}'$.

14.3 Vectors perpendicular to a surface; the tangent plane

If we can find a vector perpendicular to a surface, we can obtain the tangent plane.

In Sec. 12.4 it was stated that the differential dz of a function of two variables represents the change along a "tangent plane" to the graph of $z = f(x, y)$, which is a surface. The present section defines the tangent plane to a surface and shows how to compute it. The foundation of the section is a method for producing a vector that is perpendicular to a surface at a given point on the surface.

In Sec. 12.2 level curves of a function of two variables were discussed. In this section level surfaces of a function $u = F(x, y, z)$ of three variables will be needed.

LEVEL SURFACES

Let F be a function defined in some region in space. Let

$$u = F(x, y, z).$$

If c is a fixed number, the set of points where

$$F(x, y, z) = c$$

is called a *level surface* for F.

EXAMPLE 1 Consider the function F defined by $F(x, y, z) = x^2 + y^2 + z^2$. Sketch the level surface of f that passes through the point $(2, 6, 3)$.

SOLUTION Since $\quad F(2, 6, 3) = 2^2 + 6^2 + 3^2 = 4 + 36 + 9 = 49,$

the level surface through $(2, 6, 3)$ has the equation

$$x^2 + y^2 + z^2 = 49.$$

This equation describes a sphere of radius 7 and center $(0, 0, 0)$, shown in Fig. 14.10.

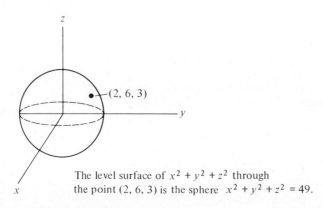

The level surface of $x^2 + y^2 + z^2$ through the point $(2, 6, 3)$ is the sphere $x^2 + y^2 + z^2 = 49.$ **Figure 14.10** ■

EXAMPLE 2   Consider the function $F(x, y, z) = x + 2y + 3z$. Sketch the level surface $F(x, y, z) = 6$.

SOLUTION   The level surface has the equation $x + 2y + 3z = 6$. It is a plane that meets the three axes at $(6, 0, 0)$, $(0, 3, 0)$, and $(0, 0, 2)$. It is shown in Fig. 14.11.

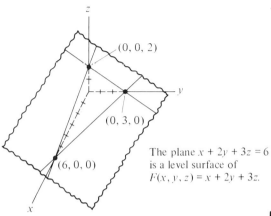

The plane $x + 2y + 3z = 6$ is a level surface of $F(x, y, z) = x + 2y + 3z$.

Figure 14.11

The vector ∇F, defined in Sec. 12.6, will be perpendicular to the surface $F(x, y, z) = k$.

How do we construct a vector perpendicular to a level surface $F(x, y, z) = k$ at a given point $P = (a, b, c)$? As will be shown, the gradient of F, $\nabla F = F_x \mathbf{i} + F_y \mathbf{j} + F_z \mathbf{k}$, evaluated at (a, b, c), is such a vector. Let us check that this is so in the special case $F(x, y, z) = x + 2y + 3z$ of Example 2. There $\nabla F = \mathbf{i} + 2\mathbf{j} + 3\mathbf{k}$. As shown in Sec. 11.4, the vector $\overrightarrow{(1, 2, 3)}$, formed of the coefficients in the expression $x + 2y + 3z$, is indeed perpendicular to the plane $x + 2y + 3z = k$ for any constant k.

NORMAL TO $F(x, y, z) = k$

Before proceeding, it is necessary to state what is meant by a vector being perpendicular to a surface.

DEFINITION   *Normal vector to a surface.* A vector is perpendicular to a surface at the point (a, b, c) on this surface if the vector is perpendicular to any curve on the surface through the point (a, b, c). (A vector is perpendicular to a curve at a point (a, b, c) on the curve if the vector is perpendicular to a tangent vector to the curve at (a, b, c).) Such a vector is called a *normal vector*.

The word "orthogonal" is frequently used instead of "perpendicular" or "normal."

THEOREM 1   The gradient ∇F at (a, b, c) is a normal to the level surface of F passing through (a, b, c).

PROOF   Let $\mathbf{G}(t) = g(t)\mathbf{i} + h(t)\mathbf{j} + j(t)\mathbf{k}$ be the parameterization of a curve in the level surface of F that passes through the point (a, b, c). Let $\mathbf{G}'(t_0)$ be the tangent vector to the curve at (a, b, c) and let $\nabla F = F_x(a, b, c)\mathbf{i} + F_y(a, b, c)\mathbf{j} + F_z(a, b, c)\mathbf{k}$ be the gradient at (a, b, c). We wish to show that

$$\nabla F \cdot \mathbf{G}'(t_0) = 0;$$

that is, $\quad F_x(a, b, c)g'(t_0) + F_y(a, b, c)h'(t_0) + F_z(a, b, c)j'(t_0) = 0.$ (1)

The left side of (1) is the kind of expression met in a chain rule for a function of three variables. This observation suggests introducing the function $u(t)$ defined as

$$u(t) = F(g(t), h(t), j(t)).$$

By a chain rule,

$$\left.\frac{du}{dt}\right|_{t_0} = F_x(a, b, c)g'(t_0) + F_y(a, b, c)h'(t_0) + F_z(a, b, c)j'(t_0). \quad (2)$$

However, since the curve $\mathbf{G}(t)$ lies on a level surface of F, $u(t)$ is constant. (In fact, $u(t) = F(a, b, c)$.) Thus $du/dt = 0$. Consequently, the right side of (2) is 0, as required. ∎

EXAMPLE 3 Find a normal vector to the ellipsoid $\quad x^2 + y^2/4 + z^2/9 = 3 \quad$ at the point $(1, 2, 3)$.

SOLUTION The ellipsoid is a level surface of the function

$$F(x, y, z) = x^2 + \frac{y^2}{4} + \frac{z^2}{9}.$$

The gradient of F is

$$\nabla F = 2x\mathbf{i} + \frac{y}{2}\mathbf{j} + \frac{2z}{9}\mathbf{k}.$$

At $(1, 2, 3)$, $\qquad \nabla F = 2\mathbf{i} + \mathbf{j} + \tfrac{2}{3}\mathbf{k}.$

This vector is normal to the ellipsoid at $(1, 2, 3)$. ∎

TANGENT PLANE TO $F(x, y, z) = k$

A tangent plane can now be defined.

DEFINITION *Tangent plane to a surface.* Consider a surface that is a level surface of a function $u = F(x, y, z)$. Let (a, b, c) be a point on this surface where ∇F is not $\mathbf{0}$. The tangent plane to the surface at the point (a, b, c) is that plane through (a, b, c) that is perpendicular to the vector ∇F evaluated at (a, b, c).

The tangent plane at (a, b, c) is the plane that best approximates the surface near (a, b, c). It consists of all the tangent lines at (a, b, c) to curves in the surface that pass through the point (a, b, c).

EXAMPLE 4 Find an equation of the tangent plane to the ellipsoid

$$x^2 + \frac{y^2}{4} + \frac{z^2}{9} = 3$$

at the point (1, 2, 3).

SOLUTION By Example 3, the vector $2\mathbf{i} + \mathbf{j} + \frac{2}{3}\mathbf{k}$ is normal to the surface at the point (1, 2, 3). The tangent plane consequently has an equation

$$2(x - 1) + 1(y - 2) + \tfrac{2}{3}(z - 3) = 0,$$

which can be written in the form

$$2x + y + \tfrac{2}{3}z = 6. \quad\blacksquare$$

NORMAL AND TANGENT PLANE TO $z = f(x, y)$

It may happen that a surface is described explicitly in the form $z = f(x, y)$ rather than implicitly in the form $F(x, y, z) = k$. It turns out that the techniques already developed enable us to find the normal and tangent plane in the case $z = f(x, y)$ as well.

THEOREM 2 Let f be a function of two variables, and let (a, b, c) be a point on the surface $z = f(x, y)$. Then the vector

$$f_x\mathbf{i} + f_y\mathbf{j} - \mathbf{k},$$

where f_x and f_y are evaluated at (a, b), is perpendicular to the surface at the point (a, b, c).

PROOF The proof consists of introducing a function F of three variables such that the given surface is a level surface of F. Application of Theorem 1 to this function F will prove the theorem.

Define $F(x, y, z)$ to be $f(x, y) - z$, and consider the particular level surface of F,

$$F(x, y, z) = 0.$$

This surface also has the description

$$f(x, y) - z = 0,$$

or, equivalently,

$$z = f(x, y).$$

This shows that the given surface $z = f(x, y)$ is the same as a level surface of F.

Now, Theorem 1 applied to F asserts that

$$F_x(a, b, c)\mathbf{i} + F_y(a, b, c)\mathbf{j} + F_z(a, b, c)\mathbf{k}$$

is a normal to the surface $F(x, y, z) = 0$. But, since

$$F(x, y, z) = f(x, y) - z,$$

it follows that

$$F_x(a, b, c) = f_x(a, b), \quad F_y(a, b, c) = f_y(a, b) \quad \text{and} \quad F_z(a, b, c) = -1.$$

Hence
$$f_x(a, b)\mathbf{i} + f_y(a, b)\mathbf{j} - \mathbf{k}$$

is perpendicular to the surface $z = f(x, y)$ at (a, b, c). ∎

EXAMPLE 5 Find a vector that is perpendicular to the surface $z = y^2 - x^2$ at the point $(1, 2, 3)$.

SOLUTION Apply Theorem 2 to $f(x, y) = y^2 - x^2$. Since $f_x(x, y) = -2x$ and $f_y(x, y) = 2y$, the vector

$$-2x\mathbf{i} + 2y\mathbf{j} - \mathbf{k}$$

is perpendicular to the surface at (x, y, z). In particular, at $(1, 2, 3)$, the vector

$$-2\mathbf{i} + 4\mathbf{j} - \mathbf{k}$$

is perpendicular to the surface.

This surface, which looks like a saddle near the origin, was graphed in Sec. 12.1. The surface and the normal vector $-2\mathbf{i} + 4\mathbf{j} - \mathbf{k}$ are shown in Fig. 14.12. ∎

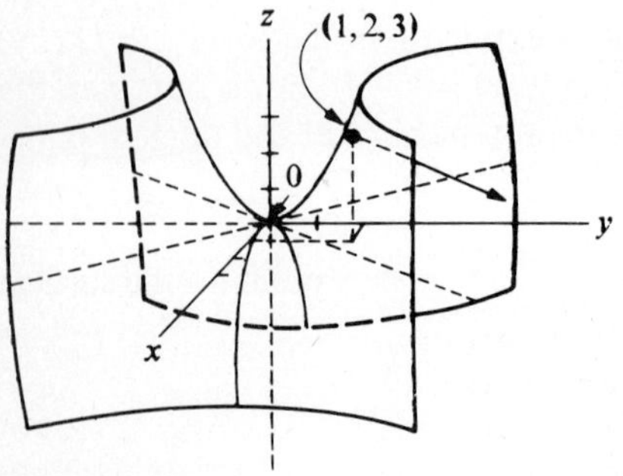

Figure 14.12

In conclusion we single out a result obtained in the course of proving Theorem 1. We state it as a theorem for later use.

THEOREM 3 Let $u = F(x, y, z)$. Let $x = g(t), y = h(t), z = j(t)$ describe the path $\mathbf{r} = \mathbf{G}(t)$. Then u is a composite function of t and

$$\frac{du}{dt} = \nabla F \cdot \mathbf{G}'(t). \quad ∎$$

For instance, let $u = F(x, y, z)$ be the temperature at the point (x, y, z) and let $\mathbf{G}$ describe the journey of a bug. Then the rate of change in the temperature as observed by the bug is the dot product of the temperature gradient ∇F and the velocity vector $\mathbf{v} = \mathbf{G}'$.

Exercises

In each of Exercises 1 to 6 find a normal vector to the given surface at the given point.
1 $x^2 + 2y^2 + 3z^2 = 6$ at $(1, 1, 1)$
2 $xy + xz + yz = 11$ at $(1, 2, 3)$
3 $z - 3xy = 1$ at $(1, 1, 4)$
4 $z - x^2 - y^2 = 0$ at $(1, 2, 5)$
5 $z = xy$ at $(1, 2, 2)$
6 $z = x^3 y$ at $(2, 1, 8)$

In each of Exercises 7 to 10 find an equation of the tangent plane to the given surface at the given point.

7. $xyz = 6$ at $(1, 2, 3)$
8. $x^2 + 3y^2 - z^2 = 8$ at $(3, 1, 2)$
9. $z = x^5 y + 7$ at $(1, 1, 8)$
10. $z = \dfrac{6}{xy}$ at $(2, 3, 1)$

∎

11. Let $T(x, y, z)$ be the temperature at the point (x, y, z). A level surface $T(x, y, z) = k$ is called an *isotherm*. Show that, if you are at the point (a, b, c) and wish to move in the direction in which the temperature changes most rapidly, you would move in a direction perpendicular to the isotherm that passes through the point (a, b, c).
12. The level surfaces of $F(x, y, z) = x^2 + y^2 + z^2$ are spheres whose centers are at the origin O.
    (a) Use the gradient ∇F to show that the vector $\overrightarrow{OP}$ is perpendicular to the sphere that passes through the point P.
    (b) What geometric theorem does (a) demonstrate?
13. In Chap. 11 it was shown that the vector $A\mathbf{i} + B\mathbf{j} + C\mathbf{k}$ is normal to the plane $Ax + By + Cz + D = 0$. Use Theorem 1 to establish this result.
14. Find the cosine of the angle between a normal to the surface $z = f(x, y)$ at (a, b, c) and the vector $\mathbf{k}$.
15. (a) Show that $(1, 1, 2)$ is on the surface whose equation is $z = x^2 + y^2$.
    (b) Sketch the surface.
    (c) Find a vector perpendicular to the surface at $(1, 1, 2)$.

The angle between two surfaces that pass through the point (a, b, c) is defined as the angle between the two lines through (a, b, c) that are perpendicular to the two surfaces at the point (a, b, c). This angle may be taken to be acute.

16. (a) Show that the point $(1, 1, 2)$ lies on the surfaces $xyz = 2$ and $x^3 y z^2 = 4$.
    (b) Find the cosine of the angle between the surfaces in (a) at the point $(1, 1, 2)$.
17. (a) Show that the point $(1, 2, 3)$ lies on the plane
$$2x + 3y - z = 5$$

and the sphere
$$x^2 + y^2 + z^2 = 14.$$

(b) Find the angle between them at the point $(1, 2, 3)$.

18. (a) Show that the surfaces $z = x^2 y^3$ and $z = 2xy$ pass through the point $(2, 1, 4)$.
    (b) At what angle do they cross at that point?
19. Two surfaces $f(x, y, z) = 0$ and $g(x, y, z) = 0$ both pass through the point (a, b, c). Their intersection is a curve. How would you find a tangent vector to that curve at (a, b, c)?

■ ■

What was done in this section for level surfaces $F(x, y, z) = k$ can also be done for level curves $f(x, y) = k$ in the plane. The following exercises concern normals to level curves in the plane.

20. Let $z = f(x, y)$. Show that ∇f evaluated at (a, b) is perpendicular to the level curve of $f(x, y)$ that passes through the point (a, b) by parameterizing the level curve as $\mathbf{r} = \mathbf{G}(t)$ and showing that $\nabla f \cdot \mathbf{G}'(t) = 0$.
21. Let $f(x, y) = xy$.
    (a) Draw ∇f at $(1, 1)$, $(1, 2)$, and $(2, 3)$, each time placing the tail of ∇f at the point where ∇f is evaluated.
    (b) Draw the level curves $xy = 1$, $xy = 2$, $xy = 6$, which pass through the respective points in (a).
22. (a) Draw three level curves of the function f defined by $f(x, y) = xy$. Include the curve through $(1, 1)$ as one of them.
    (b) Draw three level curves of the function g defined by $g(x, y) = x^2 - y^2$. Include the level curve through $(1, 1)$ as one of them.
    (c) Prove that each level curve of f intersects each level curve of g at a right angle.
    (d) If we think of f as air pressure, how may we interpret the level curves of g?
23. (a) Draw a level curve for the function $2x^2 + y^2$.
    (b) Draw a level curve for the function y^2/x.
    (c) Prove that any level curve of $2x^2 + y^2$ crosses any level curve of y^2/x at a right angle.

14.4 Lagrange multipliers (optional)

The goal of this section is to give an intuitive geometric explanation of the method of Lagrange multipliers. This method is of use in maximizing or minimizing functions subject to several constraints and is an important tool in economics and advanced mechanics.

In Sec. 4.5 we discussed the problem of finding the minimal surface area of a right circular can of volume 100. The problem was to minimize

$$2\pi r^2 + 2\pi r h, \tag{1}$$

subject to the constraint that

$$\pi r^2 h = 100. \tag{2}$$

The solution began by using (2) to eliminate h. Then, the expression $2\pi r^2 + 2\pi rh$ was written as a function of r alone.

In Sec. 4.6 the same problem was solved with the aid of implicit differentiation. Though (2) was not solved to give h explicitly as a function of r, it was clear that h could be considered a function of r. Thus the derivative dh/dr made sense.

In both solutions the variables r and h assumed quite different roles. We singled out one of them, r, to be the independent variable, and the other, h, to be the dependent variable. In the method of Lagrange multipliers *all the variables are treated the same*. None is distinguished from the others. Variables that play similar roles in the assumptions will play similar roles in the details of the solution. Furthermore, the method of Lagrange multipliers generalizes easily to several variables and several constraints.

An example of the method

First let us illustrate the method of Lagrange multipliers by using it to solve the can problem just cited. The explanation of why it works will follow.

The first step is to form a certain function L of the variables r, h, and λ.

λ ("lambda") is the Greek letter corresponding to l.

$$L(r, h, \lambda) = 2\pi r^2 + 2\pi rh - \lambda(\pi r^2 h - 100);$$

that is, $L(r, h, \lambda) = $ (function to be minimized) $- \lambda$(constraint).

(λ is called a Lagrange multiplier.) Then compute the partial derivatives

$$\frac{\partial L}{\partial r}, \quad \frac{\partial L}{\partial h}, \quad \text{and} \quad \frac{\partial L}{\partial \lambda}$$

and find where they are all simultaneously 0:

$$0 = \frac{\partial L}{\partial r} = 4\pi r + 2\pi h - 2\pi \lambda rh$$

$$0 = \frac{\partial L}{\partial h} = 2\pi r - \lambda \pi r^2$$

$$0 = \frac{\partial L}{\partial \lambda} = -(\pi r^2 h - 100).$$

This gives three equations in three unknowns, r, h, and λ:

$$0 = 4\pi r + 2\pi h - 2\pi \lambda rh \tag{3}$$

$$0 = 2\pi r - \lambda \pi r^2 \tag{4}$$

$$0 = \pi r^2 h - 100. \tag{5}$$

Note that (5) is just the given constraint (2).

Since $r = 0$ does not satisfy (5), we may divide (4) by πr, obtaining

$$0 = 2 - \lambda r;$$

hence
$$\lambda r = 2. \tag{6}$$

Combining (6) with (3) yields

$$0 = 4\pi r + 2\pi h - (2\pi)(2)h,$$

or
$$0 = 4\pi r - 2\pi h;$$

hence
$$2r = h. \tag{7}$$

Equation (7) already shows that the can of smallest surface area has its diameter equal to its height.

To find r and h explicitly, combine (5) and (7), obtaining

$$0 = \pi r^2 (2r) - 100.$$

Since there is no can of largest surface area, this must give the can of smallest surface area.

Hence
$$2\pi r^3 = 100,$$

or
$$r^3 = \frac{50}{\pi}.$$

Thus
$$r = \left(\frac{50}{\pi}\right)^{1/3}.$$

It is then possible to solve for h. This illustrates the method of Lagrange multipliers.

THE GEOMETRY BEHIND LAGRANGE MULTIPLIERS

Let us see why Lagrange's method works. Consider the following problem:

Maximize or minimize $u = f(x, y)$, given the constraint $g(x, y) = 0$.

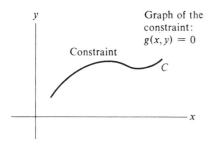

Figure 14.13

The graph of $g(x, y) = 0$ is in general a curve C, as shown in Fig. 14.13. Assume that f, *considered only on points of C*, takes a maximum (or minimum) value at the point P_0. Let C be parameterized by the vector function $\mathbf{G}$: $\mathbf{G}(t) = x(t)\mathbf{i} + y(t)\mathbf{j}$. Let $\mathbf{G}(t_0) = \overrightarrow{OP_0}$. Then u is a function of t:

$$u = f(x(t), y(t)),$$

and by Theorem 3 of Sec. 14.3,

$$\frac{du}{dt} = \nabla f \cdot \mathbf{G}'(t_0). \tag{8}$$

Since f, considered only on C, has a maximum at $\mathbf{G}(t_0)$,

$$\frac{du}{dt} = 0$$

at t_0. Thus, by (8), $\quad \nabla f \cdot \mathbf{G}'(t_0) = 0.$

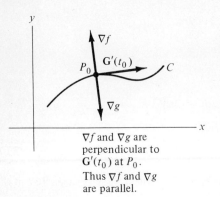

∇f and ∇g are perpendicular to $\mathbf{G}'(t_0)$ at P_0. Thus ∇f and ∇g are parallel.

Figure 14.14

This means that ∇f is perpendicular to $\mathbf{G}'(t_0)$ at P_0. But ∇g, evaluated at P_0, is also perpendicular to $\mathbf{G}'(t_0)$, since the gradient ∇g is perpendicular to the level curve $g(x, y) = 0$. (See Fig. 14.14.) Thus,

$$\nabla f \text{ is parallel to } \nabla g.$$

In other words, there is a scalar λ such that

$$\nabla f = \lambda \nabla g.$$

Thus at a maximum (or minimum) of f, subject to the constraint $g(x, y) = 0$, there is a scalar λ such that

$$f_x \mathbf{i} + f_y \mathbf{j} = \lambda(g_x \mathbf{i} + g_y \mathbf{j})$$

or, equivalently,
$$f_x = \lambda g_x$$
and
$$f_y = \lambda g_y.$$

Consequently, at such a maximum (or minimum), occurring at $P_0 = (x_0, y_0)$, there is a scalar λ such that these three conditions hold:

$$\begin{cases} f_x(x_0, y_0) = \lambda g_x(x_0, y_0), \\ f_y(x_0, y_0) = \lambda g_y(x_0, y_0), \end{cases} \quad (9)$$

and
$$g(x_0, y_0) = 0. \quad (10)$$

Conditions (9) and (10) provide three equations for three unknowns x_0, y_0, and λ.

Conditions (9) and (10) have a simple description in terms of the function L, defined as follows:

$$L(x, y, \lambda) = f(x, y) - \lambda g(x, y).$$

Conditions (9) are equivalent to

$$\frac{\partial L}{\partial x} = 0 \quad \text{and} \quad \frac{\partial L}{\partial y} = 0.$$

Condition (10) is equivalent to

$$\frac{\partial L}{\partial \lambda} = 0,$$

since this partial is $-g(x, y)$.

AN EXAMPLE WITH ONE CONSTRAINT

EXAMPLE 1 Maximize the function $x^2 y$ for points (x, y) on the circle

$$x^2 + y^2 = 1.$$

SOLUTION First, put the constraint in the form

$$x^2 + y^2 - 1 = 0.$$

The function L in this case is given by

$$L(x, y, \lambda) = x^2 y - \lambda(x^2 + y^2 - 1).$$

The three partial derivatives, which are set equal to 0, are

$$\frac{\partial L}{\partial x} = 2xy - 2\lambda x = 0$$

$$\frac{\partial L}{\partial y} = x^2 - 2\lambda y = 0$$

$$\frac{\partial L}{\partial \lambda} = -(x^2 + y^2 - 1) = 0.$$

Since at a maximum of $x^2 y$ the number x is not 0, $2x$ can be canceled from the first of the three equations. These equations then simplify to

$$y - \lambda = 0 \tag{11}$$

$$x^2 - 2\lambda y = 0 \tag{12}$$

$$x^2 + y^2 = 1. \tag{13}$$

By (11) and (12), $\qquad x^2 = 2y^2. \tag{14}$

By (13) and (14), $\qquad 2y^2 + y^2 = 1;$

hence $\qquad y^2 = \frac{1}{3}.$

Thus, $\qquad y = \frac{\sqrt{3}}{3} \quad \text{or} \quad y = -\frac{\sqrt{3}}{3}.$

By (14), $\qquad x = \sqrt{2}\, y \quad \text{or} \quad x = -\sqrt{2}\, y.$

There are only four points to be considered on the circle:

$$\left(\frac{\sqrt{6}}{3}, \frac{\sqrt{3}}{3}\right), \quad \left(\frac{-\sqrt{6}}{3}, \frac{\sqrt{3}}{3}\right), \quad \left(\frac{-\sqrt{6}}{3}, \frac{-\sqrt{3}}{3}\right), \quad \left(\frac{\sqrt{6}}{3}, \frac{-\sqrt{3}}{3}\right).$$

At the first and second points $x^2 y$ is positive, while at the third and fourth $x^2 y$ is negative. The first two points provide the maximum of $x^2 y$ on the circle $x^2 + y^2 = 1$, namely

$$\left(\frac{\sqrt{6}}{3}\right)^2 \frac{\sqrt{3}}{3} = \frac{2\sqrt{3}}{9}.$$

The third and fourth points provide the minimum value of $x^2 y$, namely

$$\frac{-2\sqrt{3}}{9}. \quad \blacksquare$$

914 THE DERIVATIVE OF A VECTOR FUNCTION

More variables

The same method applies to finding a maximum or minimum of $f(x, y, z)$ subject to the constraint $g(x, y, z) = 0$. In this case, form $L(x, y, z) = f(x, y, z) - \lambda g(x, y, z)$ and set the four partial derivatives L_x, L_y, L_z, and L_λ equal to 0.

More constraints

Lagrange multipliers can also be used to maximize $f(x, y, z)$ subject to more than one constraint; for instance, the constraints may be

$$g(x, y, z) = 0 \quad \text{and} \quad h(x, y, z) = 0. \tag{15}$$

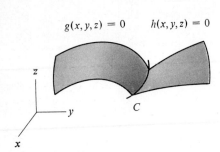

Figure 14.15

The two surfaces (15) in general meet in a curve C, as shown in Fig. 14.15. Assume that C is parameterized by the function $\mathbf{G}$. Then at a maximum (or minimum) of f at a point $P_0 = (x_0, y_0, z_0)$ on C,

$$\nabla f \cdot \mathbf{G}'(t) = 0.$$

Thus ∇f, evaluated at P_0, is perpendicular to $\mathbf{G}'(t_0)$. But ∇g and ∇h, being normal vectors at P_0 to the level surfaces $g(x, y, z) = 0$ and $h(x, y, z) = 0$, respectively, are both perpendicular to $\mathbf{G}'(t_0)$. Thus

$$\nabla f, \quad \nabla g, \quad \text{and} \quad \nabla h$$

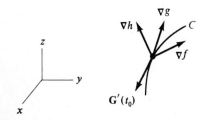

Figure 14.16

are all perpendicular to $\mathbf{G}'(t_0)$ at (x_0, y_0, z_0). (See Fig. 14.16.) Consequently, ∇f lies in the plane determined by the vectors ∇g and ∇h (which we assume are not parallel). Hence there are scalars λ and μ such that

$$\nabla f = \lambda \nabla g + \mu \nabla h.$$

This equation asserts that at P_0 there are scalars λ and μ such that

$$\begin{aligned} f_x &= \lambda g_x + \mu h_x \\ f_y &= \lambda g_y + \mu h_y \\ f_z &= \lambda g_z + \mu h_z. \end{aligned} \tag{16}$$

Now (15) and (16) together are equivalent to the brief assertion that the five partial derivatives of the function L defined by

$$L(x, y, z, \lambda, \mu) = f(x, y, z) - \lambda g(x, y, z) - \mu h(x, y, z) \tag{17}$$

are 0. Hence to maximize (or minimize) f subject to the two constraints $g(x, y, z) = 0$ and $h(x, y, z) = 0$, form the function L in (17) and proceed as in Example 1.

A rigorous development of the material in this section belongs in an advanced calculus course. If a maximum occurs at an endpoint of the curves in question or if the two surfaces do not meet in a curve, the method does not apply. We will content ourselves by illustrating the method with an example in which there are two constraints.

AN EXAMPLE WITH TWO CONSTRAINTS

EXAMPLE 2 Minimize $x^2 + y^2 + z^2$ subject to the constraints $x + 2y + 3z = 6$ and $x + 3y + 9z = 9$.

SOLUTION There are three variables and two constraints.

Each of the two constraints mentioned describes a plane. Thus the two constraints together describe a *line*. The function $x^2 + y^2 + z^2$ is the square of the distance from (x, y, z) to the origin. So the problem can be rephrased as, "How far is the origin from a certain line?" (It could be solved by vector algebra.) When viewed in this perspective, the problem certainly has a solution; that is, there is clearly a minimum.

Before forming the Lagrange function L, write the constraints with 0 on one side of the equation:

$$x + 2y + 3z - 6 = 0$$
$$x + 3y + 9z - 9 = 0.$$

The Lagrange function is

$$L(x, y, z, \lambda, \mu) = x^2 + y^2 + z^2 - \lambda(x + 2y + 3z - 6) - \mu(x + 3y + 9z - 9).$$

Setting the five partial derivatives of L equal to 0 gives

$$\left. \begin{array}{ll} L_x = 0 & 2x - \lambda - \mu = 0 \\ L_y = 0 & 2y - 2\lambda - 3\mu = 0 \\ L_z = 0 & 2z - 3\lambda - 9\mu = 0 \end{array} \right\} \quad (18)$$

$$\left. \begin{array}{ll} L_\lambda = 0 & x + 2y + 3z - 6 = 0 \\ L_\mu = 0 & x + 3y + 9z - 9 = 0. \end{array} \right\} \quad (19)$$

There are many ways to solve these five equations. Let us use the first three to express x, y, and z in terms of λ and μ. Then, after substituting the results in the last two equations, we will find λ and μ.

By (18)

$$x = \frac{\lambda + \mu}{2}, \quad y = \frac{2\lambda + 3\mu}{2}, \quad z = \frac{3\lambda + 9\mu}{2}.$$

Equations (19) then become

$$\frac{\lambda + \mu}{2} + \frac{2(2\lambda + 3\mu)}{2} + \frac{3(3\lambda + 9\mu)}{2} - 6 = 0$$

and

$$\frac{\lambda + \mu}{2} + \frac{3(2\lambda + 3\mu)}{2} + \frac{9(3\lambda + 9\mu)}{2} - 9 = 0,$$

which yield

$$14\lambda + 34\mu = 12 \atop 34\lambda + 91\mu = 18.\Bigg\}\qquad(20)$$

By Sec. 11.5,

$$\lambda = \frac{\begin{vmatrix}12 & 34\\18 & 91\end{vmatrix}}{\begin{vmatrix}14 & 34\\34 & 91\end{vmatrix}} \quad\text{and}\quad \mu = \frac{\begin{vmatrix}14 & 12\\34 & 18\end{vmatrix}}{\begin{vmatrix}14 & 34\\34 & 91\end{vmatrix}};$$

hence
$$\lambda = \frac{240}{59} \quad\text{and}\quad \mu = -\frac{78}{59}.$$

Thus
$$x = \frac{\lambda + \mu}{2} = \frac{81}{59}$$

$$y = \frac{2\lambda + 3\mu}{2} = \frac{123}{59}$$

$$z = \frac{3\lambda + 9\mu}{2} = \frac{9}{59}.$$

Since there is no maximum, this must be a minimum. The minimum of $x^2 + y^2 + z^2$ is thus

$$\left(\frac{81}{59}\right)^2 + \left(\frac{123}{59}\right)^2 + \left(\frac{9}{59}\right)^2 = \frac{21{,}771}{3{,}481} = \frac{369}{59}.\quad\blacksquare$$

In Example 2 there were three variables x, y, z and two constraints. There may in some cases be many variables, $x_1, x_2, \ldots, x_n$, and many constraints. Write each constraint in the form "something = 0" and introduce Lagrange multipliers $\lambda_1, \lambda_2, \ldots, \lambda_m$, one for each constraint.

Exercises

In all exercises use Lagrange multipliers.

1. Maximize xy for points on the circle $x^2 + y^2 = 4$.
2. Minimize $x^2 + y^2$ for points on the line $2x + 3y - 6 = 0$.
3. Minimize $2x + 3y$ on the portion of the hyperbola $xy = 1$ in the first quadrant.
4. Maximize $x + 2y$ on the ellipse $x^2 + 2y^2 = 8$.
5. Find the largest area of all rectangles whose perimeters are 12 centimeters.
6. A rectangular box is to have a volume of 1 cubic meter. Find its dimensions if its surface area is minimal.
7. Find the point on the plane $x + 2y + 3z = 6$ that is closest to the origin. *Suggestion:* Minimize the square of the distance in order to avoid square roots.
8. Maximize $x + y + 2z$ on the sphere $x^2 + y^2 + z^2 = 9$.
9. Minimize the distance from (x, y, z) to $(1, 3, 2)$ for points on the plane $2x + y + z = 5$.
10. Find the dimensions of the box of largest volume whose surface area is to be 6 square inches.
11. Maximize $x^2 y^2 z^2$ subject to the constraint

$$x^2 + y^2 + z^2 = 1.$$

12. Find the points on the surface $xyz = 1$ closest to the origin.

■

13. Minimize $x^2 + y^2 + z^2$ on the line common to the two planes $x + 2y + 3z = 0$ and $2x + 3y + z = 4$.
14. The plane $2y + 4z - 5 = 0$ meets the cone $z^2 = 4(x^2 + y^2)$ in a curve. Find the point on this curve nearest the origin.

15 Maximize $x^3 + y^3 + 2z^3$ on the intersection of the surfaces $x^2 + y^2 + z^2 = 4$ and $(x-3)^2 + y^2 + z^2 = 4$.

16 (a) Maximize $x_1 x_2 \cdots x_n$ subject to the constraint $\sum_{i=1}^n x_i = 1$ and all $x_i \geq 0$.
   (b) Deduce that, for nonnegative numbers $a_1, a_2, \ldots, a_n$, $\sqrt[n]{a_1 a_2 \cdots a_n} \leq (a_1 + a_2 + \cdots + a_n)/n$. (The *geometric mean* is less than or equal to the *arithmetic mean*.)

17 (a) Maximize $\sum_{i=1}^n x_i y_i$ subject to the constraints $\sum_{i=1}^n x_i^2 = 1$ and $\sum_{i=1}^n y_i^2 = 1$.
   (b) Deduce that, for any numbers $a_1, a_2, \ldots, a_n$ and $b_1, b_2, \ldots, b_n$, $\sum_{i=1}^n a_i b_i \leq (\sum_{i=1}^n a_i^2)^{1/2} (\sum_{i=1}^n b_i^2)^{1/2}$.
   Hint: Let
   $$x_i = \frac{a_i}{(\sum_{i=1}^n a_i^2)^{1/2}}$$
   and
   $$y_i = \frac{b_i}{(\sum_{i=1}^n b_i^2)^{1/2}}.$$
   (c) How would you justify the inequality in (b), for $n = 3$, by vectors?

   The inequality in (b) is a discrete version of the Schwarz inequality, discussed in Exercise 21 of Sec. 8.8.

18 Maximize $x + 2y + 3z$ subject to the constraints $x^2 + y^2 + z^2 = 1$ and $x + y + z = 0$.

19 Let $a_1, a_2, \ldots, a_n$ be fixed nonzero numbers. Maximize $\sum_{i=1}^n a_i x_i$ subject to $\sum_{i=1}^n x_i^2 = 1$.

■ ■

20 Let p and q be positive numbers that satisfy the equation $1/p + 1/q = 1$. Obtain Hölder's inequality for nonnegative numbers a_i and b_i,
$$\sum_{i=1}^n a_i b_i \leq \left(\sum_{i=1}^n a_i^p\right)^{1/p} \left(\sum_{i=1}^n b_i^q\right)^{1/q},$$
as follows.
   (a) Maximize $\sum_{i=1}^n x_i y_i$ subject to $\sum_{i=1}^n x_i^p = 1$ and $\sum_{i=1}^n y_i^q = 1$.
   (b) By letting
   $$x_i = \frac{a_i}{(\sum_{i=1}^n a_i^p)^{1/p}}$$
   and
   $$y_i = \frac{b_i}{(\sum_{i=1}^n b_i^q)^{1/q}},$$
   obtain Hölder's inequality.

   Note that Hölder's inequality, with $p = 2$ and $q = 2$, reduces to the Schwarz inequality in Exercise 17.

21 A consumer has a budget of B dollars and may purchase n different items. The price of the ith item is p_i dollars. When the consumer buys x_i units of the ith item, the total cost is $\sum_{i=1}^n p_i x_i$. Assume that $\sum_{i=1}^n p_i x_i = B$ and that the consumer wishes to maximize his utility $u(x_1, x_2, \ldots, x_n)$.
   (a) Show that, when $x_1, \ldots, x_n$ are chosen to maximize utility, then
   $$\frac{\partial u/\partial x_i}{p_i} = \frac{\partial u/\partial x_j}{p_j}.$$
   (b) Explain the result in (a) using just economic intuition. Hint: Consider a slight change in x_i and x_j, with the other x_k's held fixed.

22 The following is quoted from Colin W. Clark in *Mathematical Bioeconomics*, Wiley, 1976.

   [S]uppose there are N fishing grounds. Let $H^i = H^i(R^i, E^i)$ denote the production function for the total harvest H^i on the ith ground as a function of the recruited stock level R^i and effort E^i on the ith ground. The problem is to determine the least total cost $\sum_{i=1}^N c_i E^i$ at which a given total harvest $H = \sum_{i=1}^N H^i$ can be achieved. This problem can be easily solved by Lagrange multipliers. The result is simply
   $$\frac{1}{c_i} \frac{\partial H^i}{\partial E^i} = \text{constant},$$
   [independent of i].

   Verify his assertion. The c_i's are constants. The superscripts name the functions; they are not exponents.

14.5 The acceleration vector

If $\mathbf{r} = \mathbf{G}(t)$ is the position vector at time t, then $\mathbf{v} = \mathbf{G}'(t)$ is the velocity vector at time t. The definition of the acceleration vector is motivated by the definition of acceleration in the case of a particle moving on a line.

918 THE DERIVATIVE OF A VECTOR FUNCTION

DEFINITION *Acceleration vector.* The derivative of the velocity vector is called the *acceleration vector* and is denoted **a**:

$$\mathbf{a} = \frac{d\mathbf{v}}{dt}.$$

EXAMPLE Let $\mathbf{G}(t) = 32t\mathbf{i} - 16t^2\mathbf{j}$ be the position of a thrown ball at time t. Compute $\mathbf{v}(t)$ and $\mathbf{a}(t)$.

SOLUTION
$$\mathbf{v}(t) = 32\mathbf{i} - 32t\mathbf{j}$$
and
$$\mathbf{a}(t) = -32\mathbf{j}.$$

In this case the acceleration vector is constant in direction and length. It points directly downward, as does the vector that represents the force of gravity. Figure 14.17 displays **v** and **a** at two points on the path. ∎

As this example shows, it is a simple matter to compute the velocity vector and the acceleration vector **a** when the components of the position vector $\mathbf{r}(t)$ are given. If

$$\mathbf{r}(t) = x(t)\mathbf{i} + y(t)\mathbf{j} + z(t)\mathbf{k},$$

then
$$\mathbf{v}(t) = \frac{dx}{dt}\mathbf{i} + \frac{dy}{dt}\mathbf{j} + \frac{dz}{dt}\mathbf{k},$$

and
$$\mathbf{a}(t) = \frac{d^2x}{dt^2}\mathbf{i} + \frac{d^2y}{dt^2}\mathbf{j} + \frac{d^2z}{dt^2}\mathbf{k}.$$

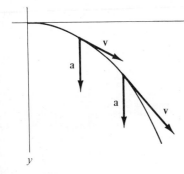

Figure 14.17

The dot notation for derivatives ($\dot{x} = dx/dt$, $\ddot{x} = d^2x/dt^2$) provides these simpler formulas: If

$$\mathbf{r}(t) = x(t)\mathbf{i} + y(t)\mathbf{j} + z(t)\mathbf{k},$$

then
$$\mathbf{v}(t) = \dot{x}\,\mathbf{i} + \dot{y}\,\mathbf{j} + \dot{z}\,\mathbf{k}$$
and
$$\mathbf{a}(t) = \ddot{x}\,\mathbf{i} + \ddot{y}\,\mathbf{j} + \ddot{z}\,\mathbf{k}.$$

If no forces act on a moving particle, **v** is constant in direction and magnitude; hence $\mathbf{a} = \mathbf{0}$. That is, if the vector **F**, representing the forces, is **0**, then $\mathbf{a} = \mathbf{0}$. Newton's second law asserts universally that **F**, **a**, and the mass m of the particle are related by the vector equation:

The relation between force and acceleration

$$\mathbf{F} = m\mathbf{a}.$$

This little equation says several things: (1) The direction of the acceleration vector **a** is the same as the direction of **F**. (2) A force **F** applied to a heavy mass produces a shorter acceleration vector **a** than the same force applied to a light mass. (3) For a given mass, the magnitude of **a** is proportional to the magnitude of **F**.

We may always think of the acceleration vector as representing the effect

14.5 THE ACCELERATION VECTOR

of a force on the particle, since the acceleration vector **a** and the force vector **F** point in the same direction. If the mass of the particle is 1, then **F** = **a**.

CIRCULAR MOTION AT CONSTANT SPEED

Consider now a particle moving in a circular orbit at constant speed v. It may be, perhaps, a heavy mass at the end of a rope or a satellite in a circular orbit around the earth. The following theorem describes the acceleration vector associated with this motion.

THEOREM If a particle moves in a circular path of radius r at a constant speed v, its acceleration vector is directed toward the center of the circle and has magnitude v^2/r.

PROOF Introduce an xy-coordinate system such that $(0, 0)$ is at the center of the orbit and the particle is at $(r, 0)$ at time 0. Let $\mathbf{r} = \mathbf{r}(t)$ be the position vector of the particle at time $t \geq 0$. Assume that the particle travels counterclockwise; then θ is positive, as shown in Fig. 14.18.

$r = |\mathbf{r}|$

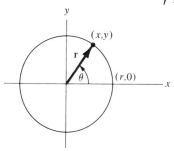

Figure 14.18

Since the particle moves at the constant speed v, it sweeps out an arc length vt up to time t. By definition of radian measure,

$$\theta = \frac{vt}{r},$$

where θ is the angle of **r** at time t. Thus

$$\mathbf{r} = x\mathbf{i} + y\mathbf{j}$$
$$= r \cos \theta \, \mathbf{i} + r \sin \theta \, \mathbf{j}$$
$$= r \cos \frac{vt}{r} \mathbf{i} + r \sin \frac{vt}{r} \mathbf{j}.$$

Now **v** and **a** can be computed explicitly. Remembering that r is constant, we have then

$$\mathbf{v} = \frac{d\mathbf{r}}{dt}$$

$v = |\mathbf{v}|$

$$= \frac{-rv}{r} \sin \frac{vt}{r} \mathbf{i} + \frac{rv}{r} \cos \frac{vt}{r} \mathbf{j}$$

$$= -v \sin \frac{vt}{r} \mathbf{i} + v \cos \frac{vt}{r} \mathbf{j}.$$

Hence

$$\mathbf{a} = \frac{d\mathbf{v}}{dt}$$

$$= \frac{-v^2}{r} \cos \frac{vt}{r} \mathbf{i} - \frac{v^2}{r} \sin \frac{vt}{r} \mathbf{j}$$

$$= \frac{v^2}{r}(-\cos \theta \, \mathbf{i} - \sin \theta \, \mathbf{j}).$$

From this last equation we can read off the direction and length of **a**. First of all,

$$-\cos\theta\,\mathbf{i} - \sin\theta\,\mathbf{j}$$

is a unit vector pointing in the direction opposite that of

$$\mathbf{r} = r\cos\theta\,\mathbf{i} + r\sin\theta\,\mathbf{j}.$$

Hence **a** points toward the center of the circle. (See Fig. 14.19.) Since **a** is v^2/r times a unit vector, its magnitude is v^2/r. This concludes the proof.

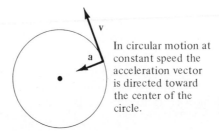

Figure 14.19

This theorem was discovered in 1657 by Huygens while developing a theory of clock mechanisms. Anyone who has spun a pail of water at the end of a rope should find the theorem plausible. First of all, to hold the pail in its orbit one must pull on the rope. Thus the force of the rope on the pail is directed toward the center of its circular orbit. Second, the faster one spins the pail (keeping the rope at fixed length), the harder one must pull on the rope. Hence the appearance of v^2 in the numerator is reasonable. Third, if the same speed is maintained but the radius of the circle is decreased, more force is required.

With the aid of the theorem we can determine the speed that a satellite requires to achieve a circular orbit around the earth. Disregard the resistance of the air and assume that the launch is made *horizontally* from a high tower, as shown in Fig. 14.20. The satellite would then be swung around the earth like a pail at the end of a rope. Instead of the tension on a rope, the force of gravity pulls the satellite in toward the earth from a linear path. Now, if a particle moves in a circle of radius 4000 miles with velocity v miles per second, it has an acceleration toward the center of the earth of $v^2/4000$ miles per second per second. This acceleration must coincide with the acceleration of 32 feet per second per second, approximately 0.006 miles per second per second, which gravity imparts to any object at the surface of the earth. Thus $0.006 \approx v^2/4000$, and

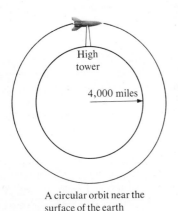

A circular orbit near the surface of the earth

Figure 14.20

$$\text{Orbital velocity} = v \approx \sqrt{(4000)(0.006)}$$
$$= \sqrt{24}$$
$$\approx 4.90 \text{ miles per second.}$$

Exercises

1. Let $r(t) = t^2 i + t^3 j$. Compute and sketch $r(1)$, $v(1)$, and $a(1)$.

2. Let
$$r(t) = \cos \frac{\pi e^t}{2} i + \sin \frac{\pi e^t}{2} j.$$
Compute and sketch $r(0)$, $v(0)$, and $a(0)$.

3. Let $r(t) = 4ti - 16t^2 j$ describe a falling ball. Compute and draw r, v, and a for $t = 0$, $t = 1$, and $t = 2$.

4. Let $r(t) = 10 \cos 2\pi t \, i + 10 \sin 2\pi t \, j$ denote the position vector of a particle at time t.
   (a) What is the shape of the path this particle follows?
   (b) Compute v and a.
   (c) Draw r, v, and a for $t = \frac{1}{4}$.
   (d) Show that a is always in the direction opposite that of r.

5. Let $r(t) = ti + (1/t)j$.
   (a) Sketch the path of the particle for $t > 0$.
   (b) Compute and sketch $r(1)$, $v(1)$, and $a(1)$.
   (c) What happens to $v(t)$ and $a(t)$ when t is large?
   (d) What happens to the speed when t is large?

6. At time t a particle has the position vector $r(t) = (t + \cos t)i + (t - \sin t)j$.
   (a) Show that a has constant magnitude.
   (b) Sketch the path corresponding to t in $[0, 4\pi]$.
   (c) Sketch r, v, and a for $t = 0, \pi/2, \pi, 3\pi/2,$ and 2π.

7. Let
$$r(t) = 5 \cos 3t \, i + 5 \sin 3t \, j + 6t \, k$$
describe the position at time t of a particle moving on a helical path.
   (a) Compute v and a at time t.
   (b) How would you interpret in terms of the motion the fact that a is parallel to the xy plane?

8. Let $r(t) = 2ti + 3tj + 13t^2 k$.
   (a) Show that the path lies on the paraboloid $z = x^2 + y^2$.
   (b) Compute a at time t.

9. (a) If v has constant direction, must a have constant direction? Explain.
   (b) If v has constant magnitude, must a have constant magnitude? Explain.

10. The momentum of a particle of mass m and velocity vector v is the vector mv. Newton stated his second law in the form $F = (mv)'$.
    (a) Prove that $F = ma + m'v$.
    (b) Deduce that, if m is constant, then $F = ma$. (In relativity theory m is not necessarily constant.)

11. Prove that, if a is always perpendicular to v on a certain path, the speed of the particle is constant.

12. (a) Prove that, if the speed of a particle is constant, a is perpendicular to v.
    (b) Does this make sense physically?

13. Let the position vector r be a function of arc length s on a curve. Prove that dr/ds is a *unit* vector tangent to the curve.

14. (a) Prove that, if a particle moves in a circular orbit (centered at $(0, 0)$), then $r \cdot a + v \cdot v = 0$.
    (b) From (a) deduce that $r \cdot a \leq 0$.
    (c) What does (b) say about the direction of the force vector F?

15. A particle moves in the circular orbit $r = \cos t^2 \, i + \sin t^2 \, j$.
    (a) Compute v and a.
    (b) Verify that $r \cdot a \leq 0$.

16. The acceleration that gravity imparts to an object decreases with the square of the distance of the object from the center of the earth.
    (a) Show that, if an object is r miles from the center of the earth, it has an acceleration of $(0.006)(4000/r)^2$ miles per second per second.
    (b) With the aid of (a), find the velocity of a satellite in orbit at an altitude of 1000 miles.

17. How long would it take a satellite to orbit the earth just above the earth's surface? One hundred miles above the earth's surface? One thousand miles above the surface?

18. What altitude must an orbiting satellite have in order to stay directly above a fixed spot on the equator?

19. A certain satellite in circular orbit goes around the earth once every 92 minutes. How high is it above the earth?

20. A boy is spinning a pail of water at the end of a rope.
    (a) If he doubles the speed of the pail, how many times as hard must he pull on the rope?
    (b) If instead he doubles the length of the rope, but keeps the speed of the pail the same, will he have to pull more or less? How much?

21. Two cars make the same circular turn around a corner. The first is traveling at 15 miles per hour; the

second, at 30 miles per hour. How many times as large is the force required to keep the second car from skidding, as compared to the force required to keep the first car from skidding? (Assume the cars have equal mass.)

22 (a) If a curve in the plane is given parametrically, x and y being functions of t, how are $\dot{x}$, $\dot{y}$, and dy/dx related? Express d^2y/dx^2 in terms of $\dot{x}$, $\dot{y}$, $\ddot{x}$, and $\ddot{y}$.
    (b) If at a certain instant $\mathbf{v} = 2\mathbf{i} + 3\mathbf{j}$ and $\mathbf{a} = \mathbf{i} + 4\mathbf{j}$, find dy/dx, d^2y/dx^2, and the radius of curvature at that instant.

23 Show that the orbital velocity at the surface of the earth is $1/\sqrt{2}$ times the escape velocity.

24 A particle in space moves under the influence of a force that is always directed toward the origin. (For instance, the particle may be moving under the influence of the gravitational field of the sun.) Prove that its path lies in a plane:
    (a) Let $\mathbf{r}$, $\mathbf{v}$, and $\mathbf{v}' = \mathbf{a}$ be the position, velocity, and acceleration vectors. Show that $\mathbf{a}(t) = f(t)\mathbf{r}(t)$, where f is a scalar function.
    (b) Assume that, if $\mathbf{G}$ and $\mathbf{H}$ are vector functions, then $(\mathbf{G} \times \mathbf{H})' = \mathbf{G}' \times \mathbf{H} + \mathbf{G} \times \mathbf{H}'$. Show that $(\mathbf{r} \times \mathbf{v})' = 0$.
    (c) From (b) it follows that $\mathbf{r} \times \mathbf{v}$ is a constant vector $\mathbf{C}$. Show that, if $\mathbf{C}$ is not $\mathbf{0}$, then the particle travels in a plane perpendicular to $\mathbf{C}$.
    (d) If $\mathbf{C}$ in part (c) is $\mathbf{0}$, what is the path of the particle?

14.6 The unit vectors T and N

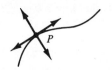

There are two unit vectors tangent to the curve at P and two unit vectors perpendicular to the curve at P.

Figure 14.21

In the study of motion along a curve in the plane, two unit vectors defined at a point P on the curve are of special importance. One is tangent to the curve at P. The other is perpendicular to the curve at P. There are two possible choices for each, as Fig. 14.21 shows.

This section is devoted to making a precise mathematical choice of one of the two unit tangent vectors and of one of the two unit vectors normal to the curve at P. Their physical interpretation will also be given in this section; in the next section they will be used to examine the acceleration vector.

Consider a particle whose position vector at time t is $\mathbf{r}(t)$. Then $\mathbf{r}'(t)$ is its velocity vector $\mathbf{v}(t)$. If $\mathbf{v}(t)$ is not the zero vector,

$$\frac{\mathbf{v}(t)}{|\mathbf{v}(t)|}$$

is a unit vector that points in the same direction as $\mathbf{v}(t)$. Denote $|\mathbf{v}(t)|$, the speed of the particle at time t, by v. The vector

$$\frac{\mathbf{v}(t)}{v}$$

is thus a unit vector tangent to the curve and pointing in the direction in which the particle is moving.

DEFINITION *Unit tangent vector* **T**. *The vector*

$$\frac{\mathbf{v}(t)}{v}$$

is denoted **T** *or* **T**(t) *and is called the unit tangent vector.* (It is defined whenever $\mathbf{v}(t)$ is not $\mathbf{0}$.)

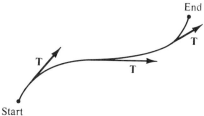

Figure 14.22

We have thus picked out one of the two tangent vectors, and the choice records the direction of motion. As the particle traverses the path, **T** remains of unit length, but changes in direction, as shown in Fig. 14.22.

EXAMPLE 1 Let $\mathbf{r}(t) = 3\cos 2t\,\mathbf{i} + 3\sin 2t\,\mathbf{j}$. Find $\mathbf{T}(t)$.

SOLUTION By differentiation, $\mathbf{v}(t) = -6\sin 2t\,\mathbf{i} + 6\cos 2t\,\mathbf{j}$. Thus

$$v = |\mathbf{v}(t)|$$
$$= \sqrt{(-6\sin 2t)^2 + (6\cos 2t)^2}$$
$$= 6.$$

Hence $\mathbf{T}(t) = \dfrac{\mathbf{v}(t)}{6} = -\sin 2t\,\mathbf{i} + \cos 2t\,\mathbf{j}.$

The curve being swept out is a circle of radius 3. The direction of $\mathbf{T}(t)$ shows that the particle moves counterclockwise. (See Fig. 14.23.) ∎

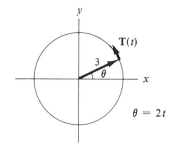

Figure 14.23

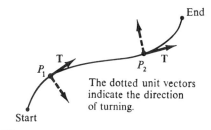

The dotted unit vectors indicate the direction of turning.

Figure 14.24

A unit vector perpendicular to **T** will also be needed. There are two choices. The one we shall take indicates the direction in which the particle is veering—to the right or to the left. For instance, at P_1 in Fig. 14.24 the particle is veering to the right, as if making a right turn; at P_2 it is veering to the left.

This choice of a vector perpendicular to **T** is intuitive and physical. The computations in the next section require that a more formal mathematical definition for this vector be given.

DEFINITION *Principal unit normal vector* **N**. If $\mathbf{T}'(t)$ is not **0**, the vector

$$\frac{\mathbf{T}'(t)}{|\mathbf{T}'(t)|}$$

is called the *principal unit normal* vector and is denoted $\mathbf{N}(t)$.

Let us show that

$$\mathbf{N}(t) = \frac{\mathbf{T}'(t)}{|\mathbf{T}'(t)|}$$

is a unit vector, is perpendicular to $\mathbf{T}(t)$, and points in the direction in which the particle is turning.

First of all, for any nonzero vector **B**,

$$\frac{\mathbf{B}}{|\mathbf{B}|}$$

is a unit vector. Thus **N**(*t*) is a unit vector.

Second, since **T**(*t*) is of constant length one, then, by Example 2 of Sec. 14.2, **T**′(*t*) is perpendicular to **T**(*t*). Since **N**(*t*) is just a scalar multiple of **T**′(*t*), **N**(*t*) is perpendicular to **T**(*t*).

*Why **N** points in the direction toward which **T** is turning*

Third, we show that **T**′ (hence **N**) points in the direction in which **T** is turning as *t* increases. To check that this is so, go back to the definition of **T**′(*t*):

$$\mathbf{T}'(t) = \lim_{\Delta t \to 0} \frac{\mathbf{T}(t + \Delta t) - \mathbf{T}(t)}{\Delta t}$$

$$= \lim_{\Delta t \to 0} \frac{\Delta \mathbf{T}}{\Delta t}.$$

When Δt is small, the direction of

$$\frac{\Delta \mathbf{T}}{\Delta t}$$

is approximately that of **T**′(*t*). Moreover, when Δt is positive,

$$\frac{\Delta \mathbf{T}}{\Delta t} \quad \text{and} \quad \Delta \mathbf{T}$$

have the same direction. Let us sketch $\Delta \mathbf{T}$ for $\Delta t > 0$. When both **T**(*t* + Δt) and **T**(*t*) are placed with their tails at the origin, $\Delta \mathbf{T}$ is the vector from the head of **T**(*t*) to the head of **T**(*t* + Δt). As Fig. 14.25 shows, $\Delta \mathbf{T}$, hence **T**′(*t*), points in the direction in which **T** is turning (to the right in this case).

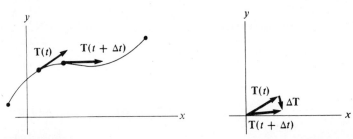

Figure 14.25

EXAMPLE 2 Compute **N**(*t*) for the curve described in Example 1, namely for

$$\mathbf{r}(t) = 3 \cos 2t \, \mathbf{i} + 3 \sin 2t \, \mathbf{j}.$$

SOLUTION As was shown in Example 1,

14.6 THE UNIT VECTORS T AND N

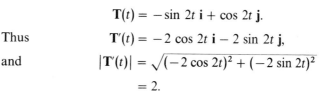

$$\mathbf{T}(t) = -\sin 2t\, \mathbf{i} + \cos 2t\, \mathbf{j}.$$

Thus
$$\mathbf{T}'(t) = -2\cos 2t\, \mathbf{i} - 2\sin 2t\, \mathbf{j},$$

and
$$|\mathbf{T}'(t)| = \sqrt{(-2\cos 2t)^2 + (-2\sin 2t)^2}$$
$$= 2.$$

Consequently,
$$\mathbf{N}(t) = \frac{\mathbf{T}'(t)}{|\mathbf{T}'(t)|}$$
$$= \frac{-2\cos 2t\, \mathbf{i} - 2\sin 2t\, \mathbf{j}}{2}$$
$$= -\cos 2t\, \mathbf{i} - \sin 2t\, \mathbf{j}.$$

Observe that in this example $\mathbf{N}(t) = -\frac{1}{3}\mathbf{r}(t)$. Thus $\mathbf{N}(t)$, pointing toward the center of the circular path, is, as was expected, perpendicular to the curve. It also indicates the direction in which $\mathbf{T}$ is turning: It is always veering toward the center of the circle. (See Fig. 14.26.) ∎

Figure 14.26

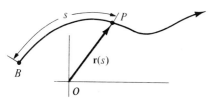

Figure 14.27

The symbol t denotes the parameter, which we have generally interpreted as time. But arc length s along the path can also be chosen as a parameter. Let s denote the arc length as measured from some base point B on the curve. If P is a point on the curve, the position vector $\overrightarrow{OP}$ depends on this arc length s and may be denoted $\overrightarrow{OP} = \mathbf{r}(s)$, as in Fig. 14.27.

With this special choice of parameterization, the formula for a unit tangent vector becomes very simple.

THEOREM 1 Let s denote arc length along a curve and let $\mathbf{r}(s)$ be the position vector corresponding to the arc length s. Then

$$\frac{d\mathbf{r}}{ds} = \mathbf{T},$$

a unit tangent vector.

PROOF By the definition of $\mathbf{T}$,
$$\mathbf{T} = \frac{d\mathbf{r}/ds}{|d\mathbf{r}/ds|}.$$

It suffices to show that
$$\left|\frac{d\mathbf{r}}{ds}\right| = 1.$$

Since $\mathbf{r} = x(s)\mathbf{i} + y(s)\mathbf{j}$,
$$\frac{d\mathbf{r}}{ds} = \frac{dx}{ds}\mathbf{i} + \frac{dy}{ds}\mathbf{j}.$$

Hence
$$\left|\frac{d\mathbf{r}}{ds}\right| = \sqrt{\left(\frac{dx}{ds}\right)^2 + \left(\frac{dy}{ds}\right)^2}$$

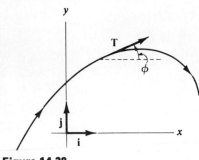

Figure 14.28

$$= \sqrt{\left(\frac{ds}{ds}\right)^2} \quad \text{by Sec. 9.4}$$

$$= 1. \blacksquare$$

There is another natural geometric parameter associated with a plane curve, namely the angle that **T** makes with the vector **i**. This is the angle ϕ in Fig. 14.28. (This angle ϕ is used in Sec. 9.7 for the study of curvature.) The next theorem is concerned with $d\mathbf{T}/d\phi$. It is assumed that the curve is not a straight line. (Otherwise ϕ is constant.)

THEOREM 2 Let $\mathbf{r}(t)$ be the position vector on a curve at time t. Let ϕ denote the angle that

$$\mathbf{T}(t) = \frac{\mathbf{v}(t)}{|\mathbf{v}(t)|}$$

makes with the vector **i**. Then **T** is a function of ϕ and

$$\left|\frac{d\mathbf{T}}{d\phi}\right| = 1.$$

PROOF Since **T** is a unit vector and makes an angle ϕ with **i**, the vector components of **T** along **i** and **j** are $\cos\phi\,\mathbf{i}$ and $\sin\phi\,\mathbf{j}$. (See Fig. 14.29.)

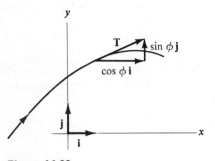

Figure 14.29

Thus $\quad\mathbf{T} = \cos\phi\,\mathbf{i} + \sin\phi\,\mathbf{j}.$

Hence $\quad\dfrac{d\mathbf{T}}{d\phi} = -\sin\phi\,\mathbf{i} + \cos\phi\,\mathbf{j}$

and $\quad\left|\dfrac{d\mathbf{T}}{d\phi}\right| = \sqrt{(-\sin\phi)^2 + (\cos\phi)^2}$

$$= \sqrt{1}$$

$$= 1. \blacksquare$$

The conclusion of Theorem 2 is similar to that of Theorem 1. In fact, Theorem 2 is a consequence of Theorem 1, as discussed in Exercise 22.

Exercises

In each of Exercises 1 to 6 compute and draw $\mathbf{T}(1)$ and $\mathbf{N}(1)$ for the given path at the point $\mathbf{r}(1)$. Also sketch the path for $t \geq 0$.

1. $\mathbf{r}(t) = t\mathbf{i} - t^2\mathbf{j}$
2. $\mathbf{r}(t) = 2t\mathbf{i} - 16t^2\mathbf{j}$
3. $\mathbf{r}(t) = \cos t\,\mathbf{i} + \sin t\,\mathbf{j}$
4. $\mathbf{r}(t) = t^2\mathbf{i} + t^3\mathbf{j}$
5. $\mathbf{r}(t) = e^t\mathbf{i} + e^{2t}\mathbf{j}$
6. $\mathbf{r}(t) = (t - \sin t)\mathbf{i} + (1 - \cos t)\mathbf{j}$ (a cycloid)
7. True or false: At a given point P on a curve, **T** depends on the direction in which the curve is being swept out.
8. True or false: At a given point P on a curve, **N** depends on the direction in which the curve is being swept out.
9. Let $\mathbf{r}(t) = t\mathbf{i} + t^2\mathbf{j}$ describe the path of a particle.
   (a) Graph the path.
   (b) Compute $\mathbf{r}(1)$, $\mathbf{v}(1)$, $\mathbf{T}(1)$, and $\mathbf{N}(1)$.
   (c) Draw the vectors in (b), placing the tails of the last three at the head of $\mathbf{r}(1)$.
10. Let $\mathbf{r}(t) = \cos 2\pi t\,\mathbf{i} - \sin 2\pi t\,\mathbf{j}$.
    (a) Sketch the path corresponding to $0 \leq t \leq 1$.

(b) Compute $\mathbf{r}(0)$, $\mathbf{v}(0)$, $\mathbf{T}(0)$, and $\mathbf{N}(0)$.
(c) Draw the four vectors in (b).
(d) Check that $\mathbf{T}(0)$ points in the direction of motion when the particle is at $\mathbf{r}(0)$.

11 A particle has the position vector $\mathbf{r}(t)$ at time t. Its speed when $t=1$ is 3 meters per second. Find $\mathbf{v}(1) \cdot \mathbf{T}(1)$.

12 If $\mathbf{v}$ is the velocity vector and $\mathbf{N}$ is the principal normal vector, what can be said about the magnitude of $\mathbf{v} \cdot \mathbf{N}$? $\mathbf{v} \cdot \mathbf{T}$?

In Exercises 13 and 14 use the fact that the scalar component of a vector $\mathbf{A}$ on a unit vector $\mathbf{u}$ is $\mathbf{A} \cdot \mathbf{u}$. Moreover, if $\mathbf{u}_1$ and $\mathbf{u}_2$ are perpendicular unit vectors in the plane, then $\mathbf{A} = (\mathbf{A} \cdot \mathbf{u}_1)\mathbf{u}_1 + (\mathbf{A} \cdot \mathbf{u}_2)\mathbf{u}_2$.

13 Let $\mathbf{r}(t) = 32t\mathbf{i} - 16t^2\mathbf{j}$ be the position vector of a thrown ball.
(a) Sketch the path.
(b) Compute and sketch $\mathbf{T}(1)$ and $\mathbf{N}(1)$.
(c) Find the scalar components of $\mathbf{v}(1)$ along $\mathbf{T}(1)$ and along $\mathbf{N}(1)$.
(d) Find the scalar components of $\mathbf{a}(1)$ along $\mathbf{T}(1)$ and $\mathbf{N}(1)$.

14 (This continues Exercise 13.)
(a) Express $\mathbf{v}(1)$ in the form $x\mathbf{i} + y\mathbf{j}$ for suitable scalars x and y.
(b) Express $\mathbf{v}(1)$ in the form $x\mathbf{T}(1) + y\mathbf{N}(1)$.
(c) Express $\mathbf{a}(1)$ in the form $x\mathbf{T}(1) + y\mathbf{N}(1)$. (See Exercise 13(b).)

15 A particle is moving in a given direction on a given path.
(a) Does $\mathbf{v}$ depend on the speed of the particle?
(b) Does $\mathbf{T}$ depend on the speed of the particle?

16 The interesting part of Theorem 1 is that $d\mathbf{r}/ds$ has length 1.
(a) Make a sketch showing $\mathbf{r}(s + \Delta s)$, $\mathbf{r}(s)$, $\Delta \mathbf{r}$, and Δs.
(b) Why is it plausible that

$$\lim_{\Delta s \to 0} \left|\frac{\Delta \mathbf{r}}{\Delta s}\right| = 1?$$

17 Let $\mathbf{r}(t) = \cos t^2 \mathbf{i} + \sin t^2 \mathbf{j}$. Show that $\mathbf{T}$ is not defined when $t = 0$.

18 Let $\mathbf{r}(t) = t\mathbf{i} + t^3\mathbf{j}$.
(a) Compute $\mathbf{T}(t)$.
(b) Show that $\mathbf{N}$ is not defined when $t = 0$.
(c) Sketch the path. What property of the path causes $\mathbf{N}$ not to be defined when $t = 0$?

19 Show that there is a scalar c, not necessarily constant, such that $\mathbf{N}' = c\mathbf{T}$.

20 Prove that

$$\mathbf{T}' = \frac{v\mathbf{a} - (dv/dt)\mathbf{v}}{v^2}.$$

21 Prove that if a particle travels at a constant speed, then

$$\mathbf{T}' = \frac{\mathbf{a}}{v}.$$

■ ■

22 A plane curve is given by the function $\mathbf{r}(t)$. Then $\mathbf{T}(t)$ is defined as $\mathbf{v}(t)/|\mathbf{v}(t)|$. Now parameterize the curve by the angle ϕ that was used in Theorem 2. Define $\mathbf{T}^*(\phi) = \mathbf{v}^*(\phi)/|\mathbf{v}^*(\phi)|$, where $\mathbf{v}^*(\phi)$ is now the derivative of the position vector with respect to ϕ.
(a) Show by a sketch that $\mathbf{T}(t)$ need not equal $\mathbf{T}^*(\phi)$ for corresponding values of t and ϕ. In fact, sketch an example in which $\mathbf{T}^*(\phi) = -\mathbf{T}(t)$.
(b) As ϕ varies, plot $\mathbf{T}^*(\phi)$ with its tail at the origin. In other words, consider $\mathbf{T}^*(\phi)$ to be the position vector of a new curve. What is that curve?
(c) Why may ϕ be interpreted as arc length on the new curve?
(d) Deduce from Theorem 1 that

$$\left|\frac{d\mathbf{T}^*(\phi)}{d\phi}\right| = 1,$$

as Theorem 2 asserts.

23 If $\mathbf{r}(t)$ describes a curve that is not restricted to a plane, the definition of $\mathbf{T}$ remains the same as in this section. Instead of two unit normals to choose $\mathbf{N}$ from, there are now an infinite number. The *principal* normal $\mathbf{N}$ is defined as in this section. The *binormal* $\mathbf{B}$ is defined as $\mathbf{T} \times \mathbf{N}$.
(a) Show that $\mathbf{B}$ is a unit vector.
(b) Compute and sketch $\mathbf{T}(\frac{\pi}{4})$, $\mathbf{N}(\frac{\pi}{4})$, and $\mathbf{B}(\frac{\pi}{4})$ for the helical curve $\mathbf{r}(t) = 3\cos 2t\, \mathbf{i} + 3\sin 2t\, \mathbf{j} + 5t\, \mathbf{k}$.

24 (This continues Exercise 23.)
(a) Show that $d\mathbf{B}/dt$ is perpendicular to $\mathbf{B}$. Thus there are scalars x and y such that $d\mathbf{B}/dt = x\mathbf{T} + y\mathbf{N}$.
(b) By differentiating $\mathbf{B} \cdot \mathbf{T} = 0$ show that $d\mathbf{B}/dt$ is perpendicular to $\mathbf{T}$.
(c) From (b) and (c) deduce that $d\mathbf{B}/dt = y\mathbf{N}$ for some scalar y.

25 (This continues Exercise 24.) Assume $\mathbf{B}(t)$ is constant, $\mathbf{B}(t) = \mathbf{k}$.
(a) Deduce that $\mathbf{k} \cdot \mathbf{v} = 0$, where $\mathbf{v}$ is the velocity vector.
(b) Deduce that the path lies in a plane.

(See Exercise 23 in Sec. 14.S for a further discussion of $\mathbf{T}$, $\mathbf{N}$, and $\mathbf{B}$.)

14.7 The scalar components of the acceleration vector along T and N

In this section all paths lie in the xy plane.

Let $\mathbf{r}(t) = x(t)\mathbf{i} + y(t)\mathbf{j}$ be the position vector at time t. The acceleration vector is then

$$\mathbf{a}(t) = \frac{d^2x}{dt^2}\mathbf{i} + \frac{d^2y}{dt^2}\mathbf{j}.$$

The scalar components of $\mathbf{a}(t)$ along $\mathbf{i}$ and $\mathbf{j}$ are simply

$$\frac{d^2x}{dt^2} \quad \text{and} \quad \frac{d^2y}{dt^2},$$

the second derivatives of each of the two scalar components.

In the study of a particle moving along a plane curve it is of use to find the scalar components of $\mathbf{a}(t)$ along $\mathbf{T}$ and $\mathbf{N}$. These scalars will be denoted a_T and a_N, respectively. In other words,

$$\mathbf{a}(t) = a_T\mathbf{T} + a_N\mathbf{N};$$

a_T is called the *tangential component* and a_N the *normal component* of the acceleration vector.

EXAMPLE 1 Find a_T and a_N if a particle moves at a constant speed v in a circular orbit of radius r.

SOLUTION As the theorem in Sec. 14.5 shows, $\mathbf{a}(t)$ is directed toward the center of the circle and has length

$$\frac{v^2}{r}.$$

Figure 14.30

The principal normal vector $\mathbf{N}$ points in the same direction, as shown in Fig. 14.30. Thus

$$\mathbf{a} = \frac{v^2}{r}\mathbf{N}.$$

The tangential component of $\mathbf{a}$ is 0, since $\mathbf{a}$ is perpendicular to $\mathbf{T}$. For emphasis we may write

$$\mathbf{a} = 0\mathbf{T} + \frac{v^2}{r}\mathbf{N}.$$

Thus $\quad a_T = 0 \quad \text{and} \quad a_N = \frac{v^2}{r}.$ ∎

Example 1 determines a_T and a_N for circular motion at constant

14.7 THE SCALAR COMPONENTS OF THE ACCELERATION VECTOR ALONG T AND N

speed. The goal of this section is to obtain a formula for a_T and a_N for *motion on any curve* and with *speed not necessarily constant*. The key will be the following lemma. Before reading it, review the definition of radius of curvature from Sec. 9.7.

LEMMA Let s represent arc length along the path of a moving particle. Let $v = ds/dt$ be the speed (which is always assumed to be positive). Then

Do not confuse this r with $|\mathbf{r}|$.

$$\frac{d\mathbf{T}}{dt} = \frac{v}{r}\mathbf{N},$$

where r is the radius of curvature.

PROOF By the definition of $\mathbf{N}$ given in Sec. 14.6,

$$\frac{d\mathbf{T}}{dt} = \left|\frac{d\mathbf{T}}{dt}\right|\mathbf{N}. \tag{1}$$

To compute $\left|\dfrac{d\mathbf{T}}{dt}\right|$,

use the chain rule (Theorem 3 in Sec. 14.2),

$$\frac{d\mathbf{T}}{dt} = \frac{d\mathbf{T}}{d\phi}\frac{d\phi}{dt}, \tag{2}$$

where ϕ is the angle between the tangent line and the positive x axis. Theorem 2 of Sec. 14.6 shows that

$$\left|\frac{d\mathbf{T}}{d\phi}\right| = 1.$$

Thus (2) implies that

$$\left|\frac{d\mathbf{T}}{dt}\right| = \left|\frac{d\phi}{dt}\right|. \tag{3}$$

By the chain rule for scalar functions,

$$\frac{d\phi}{dt} = \frac{d\phi}{ds}\frac{ds}{dt}.$$

Hence

$$\left|\frac{d\phi}{dt}\right| = \left|\frac{d\phi}{ds}\right|\left|\frac{ds}{dt}\right|. \tag{4}$$

By the definition of radius of curvature r in Sec. 9.7,

$$\left|\frac{d\phi}{ds}\right| = \frac{1}{r}. \tag{5}$$

Since

$$\frac{ds}{dt} = v,$$

it follows from Eqs. (3) to (5) that

$$\left|\frac{d\mathbf{T}}{dt}\right| = \left|\frac{d\phi}{dt}\right|$$

$$= \left|\frac{d\phi}{ds}\right|\left|\frac{ds}{dt}\right|$$

$$= \frac{1}{r}v$$

$$= \frac{v}{r}.$$

Thus (1) becomes $\quad\dfrac{d\mathbf{T}}{dt} = \dfrac{v}{r}\mathbf{N},$

and the lemma is proved. ∎

The lemma is the basis of the proof of the following theorem, which provides formulas for a_T and a_N for a parametrized plane curve. For simplicity the theorem is stated in terms of a moving particle.

THEOREM If a particle moves in a plane curve, and arc length s is measured in such a way that $v = ds/dt$ is positive, then

$$\mathbf{a} = \frac{d^2s}{dt^2}\mathbf{T} + \frac{v^2}{r}\mathbf{N},$$

where r is the radius of curvature. In other words,

$$a_T = \frac{d^2s}{dt^2} \quad\text{and}\quad a_N = \frac{v^2}{r}.$$

PROOF By definition of $\mathbf{T}$, $\quad\dfrac{\mathbf{v}}{v} = \mathbf{T};$

hence $\quad \mathbf{v} = v\mathbf{T}.$

By Theorem 1 of Sec. 14.2 $\quad \mathbf{v}' = v\mathbf{T}' + v'\mathbf{T}.$ \hfill (6)

The lemma of this section asserts that

$$\mathbf{T}' = \frac{v}{r}\mathbf{N}. \tag{7}$$

Consequently, $\quad \mathbf{a} = \mathbf{v}' = v\left(\dfrac{v}{r}\mathbf{N}\right) + v'\mathbf{T}$ \hfill (8)

$$= \frac{v^2}{r}\mathbf{N} + v'\mathbf{T}$$

14.7 THE SCALAR COMPONENTS OF THE ACCELERATION VECTOR ALONG T AND N

$$= \frac{d^2s}{dt^2}\mathbf{T} + \frac{v^2}{r}\mathbf{N}.$$

This proves the theorem. ∎

a_T is determined by change of speed, a_N by speed and radius of curvature.

Observe that a_T is determined by the "scalar" acceleration, the rate at which the speed is changing. On the other hand, a_N depends only on the speed and the radius of curvature; it is the normal component of the acceleration of a particle moving in a circle of radius r with constant speed v. Note that the scalar component of $\mathbf{a}$ along $\mathbf{N}$, being v^2/r, is positive. (See Example 1.)

The fact that a_N is positive is reasonable, since $\mathbf{a}$ points in the same direction as the external force $\mathbf{F}$; $\mathbf{F}$ causes the particle to veer, and $\mathbf{N}$ records the direction in which the particle veers. a_T can be positive or negative, depending on whether the particle is speeding up or slowing down, as shown in Fig. 14.31.

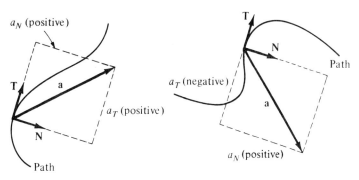

The particle veers to the right and is accelerating.

The particle veers to the right and is slowing down.

Figure 14.31

EXAMPLE 2 At time t a particle is at $(t^2, \tfrac{1}{2}t^4)$ on the curve $2y = x^2$. Determine $\mathbf{r}$, $\mathbf{v}$, $\mathbf{a}$, a_N, and a_T when $t = 1$.

SOLUTION First, since $\mathbf{r} = t^2\mathbf{i} + \tfrac{1}{2}t^4\mathbf{j}$, we have $\mathbf{v} = 2t\mathbf{i} + 2t^3\mathbf{j}$. Thus,

$$\mathbf{a} = 2\mathbf{i} + 6t^2\mathbf{j}.$$

From this it follows that $|\mathbf{a}| = \sqrt{4 + 36t^4}$.

Let us sketch $\mathbf{a}$, $\mathbf{v}$, and $\mathbf{r}$ at $t = 1$. At that time,

$$\mathbf{a} = 2\mathbf{i} + 6\mathbf{j}, \quad \mathbf{v} = 2\mathbf{i} + 2\mathbf{j}, \quad \text{and} \quad \mathbf{r} = \mathbf{i} + \tfrac{1}{2}\mathbf{j}.$$

They are shown in Fig. 14.32. a_T is simply dv/dt, where

$$v = |\mathbf{v}| = \sqrt{4t^2 + 4t^6}.$$

Thus

$$a_T = \frac{4t + 12t^5}{\sqrt{4t^2 + 4t^6}}.$$

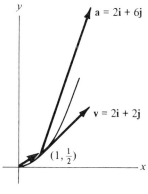

Figure 14.32

Finding a_N without using the formula v^2/r

Rather than compute a_N by the formula $a_N = v^2/r$, which involves computation of the radius of curvature, use the Pythagorean relation

$$a_N^2 + a_T^2 = |\mathbf{a}|^2.$$

At $t = 1$, for instance, this becomes

$$a_N^2 + \left(\frac{16}{\sqrt{8}}\right)^2 = (\sqrt{4 + 36 \cdot 1^4})^2 = 40.$$

Hence $a_N^2 = 40 - 32 = 8$, and $a_N = 2\sqrt{2}$. Figure 14.33 shows $\mathbf{a}$, $a_T \mathbf{T}$, and $a_N \mathbf{N}$ at $t = 1$.

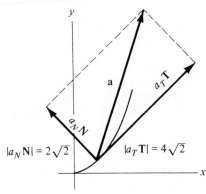

Figure 14.33   ∎

Exercises

In Exercises 1 to 4 find a_N and a_T, given that at a certain value of t:

1. $\dfrac{dv}{dt} = 3$, $v = 4$, $r = 5$.

2. $\dfrac{d^2s}{dt^2} = -2$, $v = 3$, curvature $= -\frac{1}{3}$.

3. $\dfrac{d^2s}{dt^2} = 0$, $v = 1$, $r = 3$.

4. $\dfrac{d^2s}{dt^2} = 3$, $v = 2$, and $|\mathbf{a}| = 5$.

In Exercises 5 and 6 note that the path is a circle. Find a_T and a_N in each case when $t = 1$, if

5. $\mathbf{r}(t) = 5 \cos 3\pi t\, \mathbf{i} + 5 \sin 3\pi t\, \mathbf{j}$.
6. $\mathbf{r}(t) = 5 \cos \pi t^2\, \mathbf{i} + 5 \sin \pi t^2\, \mathbf{j}$.
7. Show that
   (a) $\mathbf{a} \cdot \mathbf{T} = d^2s/dt^2$;
   (b) $\mathbf{a} \cdot \mathbf{N} = v^2/r$.

8. What can be said about $\mathbf{a} \cdot \mathbf{v}$ if the particle is
   (a) slowing down?
   (b) speeding up?
9. At a certain moment $\mathbf{r} = \mathbf{i} + \mathbf{j}$, $\mathbf{v} = 3\mathbf{i} + 4\mathbf{j}$, $\mathbf{a} = 3\mathbf{i} - 3\mathbf{j}$.
   (a) Draw $\mathbf{r}$, $\mathbf{v}$, and $\mathbf{a}$.
   (b) Is the particle speeding up or slowing down? Explain.
   (c) Estimate a_T and a_N graphically.
   (d) Compute a_T and a_N.
10. Repeat Exercise 9 for $\mathbf{r} = 0$, $\mathbf{v} = -5\mathbf{i} + 12\mathbf{j}$, $\mathbf{a} = 3\mathbf{i} + \mathbf{j}$.
11. Let $\mathbf{r}$ describe the journey in Example 2 of this section.
    (a) Compute and draw $\mathbf{a}$ for $t = 1/\sqrt{6}$.
    (b) From your drawing estimate a_T and a_N at $t = 1/\sqrt{6}$.
    (c) Compute a_T and a_N at $t = 1/\sqrt{6}$.
    (d) Compare (b) and (c) in decimal form.
12. At time $t \geq 0$ a particle is at the point $(\cos t + t \sin t,\ \sin t - t \cos t)$.
    (a) Show that $|\mathbf{v}| = t$ and $|\mathbf{a}| = \sqrt{1 + t^2}$.
    (b) Show that $a_T = 1$ and $a_N = t$.
    (c) Show that the radius of curvature equals t.

13 Let $\mathbf{r}(t) = e^t\mathbf{i} + e^{2t}\mathbf{j}$.
   (a) Compute $\mathbf{v}(t)$ and $\mathbf{a}(t)$ and express them in terms of $\mathbf{i}$ and $\mathbf{j}$.
   (b) Compute $v(t)$ and dv/dt.
   (c) Find a_T.
   (d) Find a_N.

14 At a certain instant $\mathbf{v} = 5\mathbf{i} + 12\mathbf{j}$ and $\mathbf{a} = \mathbf{i} + 2\mathbf{j}$.
   (a) Draw $\mathbf{v}$ and $\mathbf{a}$.
   (b) On the basis of the sketch for (a) estimate a_T and a_N.
   (c) Compute a_T and a_N.

15 A moving particle has the position vector $\mathbf{r}(t) = t^2\mathbf{i} + t^3\mathbf{j}$ at time t. Find a_T and a_N.

16 A moving particle has the position vector $\mathbf{r}(t) = 3\cos t\,\mathbf{i} + 4\sin t\,\mathbf{j}$ at time t. Find a_T and a_N.

17 At a certain moment, a particle headed northeast is speeding up and veering to the left. Taking north as the positive y axis and east as the positive x axis, draw (a) $\mathbf{T}$, (b) $\mathbf{N}$, (c) enough typical $\mathbf{a}$'s to indicate the directions possible for $\mathbf{a}$.

18 Repeat Exercise 17 for a particle headed north, slowing down, and veering to the left.

19 At a certain instant $\mathbf{v} = 2\mathbf{i} + 3\mathbf{j}$ and $\mathbf{a} = 3\mathbf{i} - 4\mathbf{j}$.
   (a) Draw $\mathbf{v}$, $\mathbf{a}$, $\mathbf{T}$, and $\mathbf{N}$.
   (b) Find v and d^2s/dt^2.
   (c) Find the radius of curvature.

20 Prove that if a particle travels with constant speed, then $\mathbf{a} \cdot \mathbf{v} = 0$, using
   (a) Example 2 of Sec. 14.2,
   (b) the theorem in this section.

21 From the theorem in this section deduce this formula for obtaining r, the radius of curvature:

$$\frac{v^4}{r^2} = \left(\frac{d^2x}{dt^2}\right)^2 + \left(\frac{d^2y}{dt^2}\right)^2 - \left(\frac{d^2s}{dt^2}\right)^2.$$

22 According to Sec. 9.7, the radius of curvature of the curve $y = f(x)$ is given by the formula

$$r = \frac{[1 + (dy/dx)^2]^{3/2}}{|d^2y/dx^2|}.$$

Obtain this formula from the result in Exercise 21.

14.8 Newton's law implies Kepler's three laws (optional)

After hundreds of pages of computation based on the observations made by the astronomer Tycho Brahe in the last three decades of the sixteenth century, plus lengthy detours and lucky guesses, Kepler arrived at these three laws of planetary motion:

Kepler's three laws

1 Every planet travels around the sun in an elliptical orbit such that the sun is situated at one focus (discovered 1605, published 1609).

2 The velocity of a planet varies in such a way that the line joining the planet to the sun sweeps out equal areas in equal times (discovered 1602, published 1609).

3 The square of the time required by a planet for one revolution around the sun is proportional to the cube of its mean distance from the sun (discovered 1618, published 1619).

The work of Kepler shattered the crystal spheres which for 2000 years had carried the planets. Before him astronomers admitted only circular motion and motion compounded of circular motion. Copernicus, for instance, used five circles to describe the motion of Mars.

The ellipse got a cold reception.

The ellipse was not welcomed. In 1605 Kepler complained to a skeptical astronomer:

> You have disparaged my oval orbit.... If you are enraged because I cannot take away oval flight how much more you should be enraged by the motions assigned by the ancients, which I did take away.... You disdain my oval, a single cart of dung, while you endure a whole stable. (If indeed my oval is a cart of dung.)

But the astronomical tables that Kepler based on his theories, and published in 1627, proved to be more accurate than any other and the ellipse gradually gained acceptance.

The three laws stood as mysteries alongside a related question: If there are no crystal spheres, what propels the planets? Bullialdus, a French mathematician, suggested in 1645:

The inverse square law was conjectured.

> That force with which the sun seizes or pulls the planets, a physical force which serves as hands for it, is sent out in straight lines into all the world's space...; since it is physical it is decreased in greater space;... the ratio of this decrease is the same as that for light, namely as the reciprocal of the square of the distance.

In 1666 Hooke, more of an experimental scientist than a mathematician, wondered,

> ...why the planets should move about the sun... being not included in any solid orbs... nor tied to it... by any visible strings.... I cannot imagine any other likely cause besides these two: The first may be from an unequal density of the medium...; if we suppose that part of the medium, which is farthest from the centre, or sun, to be more dense outward, than that which is more near, it will follow, that the direct motion will be always deflected inwards, by the easier yielding of the inwards....
>
> But the second cause of inflecting a direct motion into a curve may be from an attractive property of the body placed in the centre; whereby it continually endeavours to attract or draw it to itself. For if such a principle be supposed all the phenomena of the planets seem possible to be explained by the common principle of mechanic motions.... By this hypothesis, the phenomena of the comets as well as of the planets may be solved.

In 1674 Hooke, in an announcement to the Royal Society, went further:

> All celestial bodies have an attraction towards their own centres, whereby they attract not only their own parts but also other celestial bodies that are within the sphere of their activity.... All bodies that are put into direct simple motion will so continue to move forward in a straight line till they are, by some other effectual powers, deflected and bent into a motion describing a circle, ellipse, or some other more compound curve.... These attractive powers are much more powerful in operating by how much the nearer the body wrought upon is to their own centres.... It is a notion which if fully prosecuted as it ought to be, will mightily assist the astronomer to reduce all the celestial motions to a certain rule....

Hooke pressed Newton to work on the problem.

Trying to interest Newton in the question, Hooke wrote on November 24, 1679: "I shall take it as a great favor if ... you will let me know your thoughts of that of compounding the celestial motion of planets of a direct motion by the tangent and an attractive motion toward the central body." But four days later, Newton replied:

> ... my affection to philosophy [science] being worn out, so that I am almost as little concerned about it as one tradesman uses to be about another man's trade or a countryman about learning, I must acknowledge myself averse from spending that time in writing about it which I think I can spend otherwise more to my own content and the good of others....

In a letter to Newton, January 17, 1680, Hooke returned to the problem of planetary motion:

> ... It now remains to know the properties of a curved line (not circular...) made by a central

attractive power which makes the velocities of descent from the tangent line or equal straight motion at all distances in a duplicate proportion to the distances reciprocally taken. I doubt not that by your excellent method you will easily find out what that curve must be, and its properties, and suggest a physical reason of this proportion.

Hooke succeeded in drawing Newton back to science, as Newton himself admitted in his *Principia*, published in 1687: "I am beholden to him only for the diversion he gave me from my other studies to think on these things and for his dogmaticalness in writing as if he had found the motion in the ellipse, which inclined me to try it."

It seems that Newton then obtained a proof—perhaps containing a mistake (the history is not clear)—that the motion would be elliptical. In 1684, at the request of the astronomer Halley, Newton provided a correct proof. With Halley's encouragement, Newton spent the next year and a half writing the *Principia*.

In the *Principia*, which develops the science of mechanics and applies it to celestial motions, Newton begins with two laws:

1. Every body continues in its state of rest, or of uniform motion in a straight line, unless it is compelled to change this state by forces impressed upon it.

2. The change of momentum is proportional to the motive force impressed; and is made in the direction of the straight line in which that force is impressed.

To state these in the language of vectors, let **v** be the velocity of the body, **F** the impressed force, and m the mass of the body. The first law asserts that **v** is constant if **F** is **0**. *Momentum* is defined as $m\mathbf{v}$; the second law asserts that

$$\mathbf{F} = \frac{d}{dt}(m\mathbf{v}).$$

If m is constant, this reduces to

$$\mathbf{F} = m\mathbf{a}$$

where **a** is the acceleration vector.

From Kepler's laws Newton deduced his universal *law of gravitation*: Any particle P exerts an attractive force on any other particle Q, and the magnitude of the force is proportional to the product of the masses of the particles and inversely proportional to the square of the distance between them. The direction of the force is from Q toward P. In addition, Newton deduced Kepler's three laws from his single law. This section will do the latter.

We will assume that the sun is fixed at O.

Newton's universal law of gravitation asserts that any particle, of mass M, exerts a force on any other particle, of mass m. The magnitude of this force is proportional to the product of the two masses, Mm, and is directed toward the particle of mass M.

Assume that the sun has mass M and is located at point O and that the

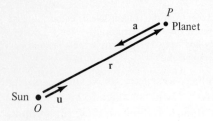

Figure 14.34

planet has mass m and is located at point P. (See Fig. 14.34.) Let $\mathbf{r} = \overrightarrow{OP}$ and $r = |\mathbf{r}|$. Then the sun exerts a force $\mathbf{F}$ on the planet given by the formula

$$\mathbf{F} = -\frac{GMm}{r^3}\mathbf{r}, \tag{1}$$

where G is a universal constant. It is convenient to introduce the unit vector $\mathbf{u} = \mathbf{r}/|\mathbf{r}|$, which points in the direction of $\mathbf{r}$. Then (1) reads

$$\mathbf{F} = -\frac{GMm}{r^2}\mathbf{u}.$$

Now, $\mathbf{F} = m\mathbf{a}$, where $\mathbf{a}$ is the acceleration vector of the planet. Thus

$$m\mathbf{a} = -\frac{GMm}{r^2}\mathbf{u},$$

from which it follows that

$$\mathbf{a} = -\frac{q\mathbf{u}}{r^2}, \tag{2}$$

where $q = GM$ is independent of the planet.

The vectors $\mathbf{u}$, $\mathbf{r}$, and $\mathbf{a}$ are indicated in Fig. 14.34.

The following exercises show how to obtain Kepler's three laws from the single law of Newton, $\mathbf{a} = -q\mathbf{u}/r^2$.

PRELIMINARIES

1. Let $\mathbf{r}(t)$ be the position vector of a given planet at time t. Let $\Delta\mathbf{r} = \mathbf{r}(t + \Delta t) - \mathbf{r}(t)$. Show that for small Δt,

$$\tfrac{1}{2}|\mathbf{r} \times \Delta\mathbf{r}|$$

approximates the area swept out by the position vector during the small interval of time Δt. *Hint:* Draw a picture.

2. From Exercise 1 deduce that

$$\frac{1}{2}\left|\mathbf{r} \times \frac{d\mathbf{r}}{dt}\right|$$

is the rate at which the position vector $\mathbf{r}$ sweeps out area. (See Fig. 14.35.)

Let $\mathbf{v} = d\mathbf{r}/dt$. The vector $\mathbf{r} \times \mathbf{v}$, introduced in Exercise 2, will play a central role in the argument.

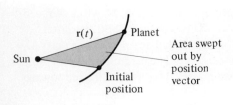

NEWTON'S LAW IMPLIES THE "EQUAL AREA" LAW

3. With the aid of (2), show that the vector **r** × **v** is constant, independent of time.

Since **r** × **v** is constant, $\frac{1}{2}|\mathbf{r} \times \mathbf{v}|$ is constant. In view of Exercise 2, it follows that the radius vector of a given planet sweeps out area at a constant rate. To put it another way, the radius vector sweeps out equal areas in equal times. This is Kepler's second law.

NEWTON'S LAW IMPLIES ELLIPTICAL ORBITS

The vector **r** × **v** is constant. Introduce an xyz coordinate system such that the unit vector **k**, which points in the direction of the positive z axis, has the same direction as **r** × **v**. Thus there is a positive constant h such that

$$\mathbf{r} \times \mathbf{v} = h\mathbf{k}. \qquad (3)$$

4. Show that h in (3) is twice the rate at which the position vector of the planet sweeps out area.

5. Show that the planet remains in the plane perpendicular to **k** that passes through the sun.

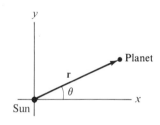

Figure 14.36

By Exercise 5, the orbit of the planet is planar. We may assume that the orbit lies in the xy plane; for convenience, locate the origin of the xy coordinates at the sun. Also introduce polar coordinates in this plane, with the pole at the sun and the polar axis along the positive x axis, as in Fig. 14.36.

6. (a) Show that during the time interval $[t_0, t]$ the position vector of the planet sweeps out the area

$$\frac{1}{2}\int_{t_0}^{t} r^2 \frac{d\theta}{dt} dt.$$

(b) From (a) deduce that the radius vector sweeps out area at the rate $\frac{1}{2}r^2 \, d\theta/dt$.

Henceforth use the dot notation for differentiation with respect to time. Thus $\dot{\mathbf{r}} = \mathbf{v}$, $\dot{\mathbf{v}} = \mathbf{a}$, and $\dot{\theta} = d\theta/dt$.

7. Show that $\mathbf{r} \times \mathbf{v} = r^2 \dot{\theta} \mathbf{k}$.

8. Show that

$$\dot{\mathbf{u}} = \frac{d\mathbf{u}}{d\theta}\dot{\theta}$$

and is perpendicular to **u**.

9. Recalling that $\mathbf{r} = r\mathbf{u}$, show that $h\mathbf{k} = r^2(\mathbf{u} \times \dot{\mathbf{u}})$.

10. Using (2) and Exercise 9, show that $\mathbf{a} \times h\mathbf{k} = q\dot{\mathbf{u}}$. (*Hint:* What is the vector identity for **A** × (**B** × **C**)?)

11. Deduce from Exercise 10 that $\mathbf{v} \times h\mathbf{k}$ and $q\mathbf{u}$ differ by a constant vector.

By Exercise 11, there is a constant vector $\mathbf{C}$ such that

$$\mathbf{v} \times h\mathbf{k} = q\mathbf{u} + \mathbf{C}. \tag{4}$$

Rotate the coordinate systems shown in Fig. 14.36 in such a way that the polar axis points in the direction of $\mathbf{C}$. Then the angle between $\mathbf{r}$ and $\mathbf{C}$ is the angle θ of polar coordinates.

The next exercise depends on the identity $(\mathbf{A} \times \mathbf{B}) \cdot \mathbf{C} = \mathbf{A} \cdot (\mathbf{B} \times \mathbf{C})$, which is valid for any three vectors $\mathbf{A}$, $\mathbf{B}$, and $\mathbf{C}$.

12. (a) Show that $(\mathbf{r} \times \mathbf{v}) \cdot h\mathbf{k} = h^2$.
    (b) Show that $\mathbf{r} \cdot (\mathbf{v} \times h\mathbf{k}) = rq + \mathbf{r} \cdot \mathbf{C}$.
    (c) Combining (a) and (b), deduce that $h^2 = rq + rc \cos \theta$, where $c = |\mathbf{C}|$.

It follows from Exercise 12 that the polar equation for the orbit of the planet is given by

$$r(\theta) = \frac{h^2}{q + c \cos \theta}. \tag{5}$$

13. By expressing (5) in rectangular coordinates, show that it describes a conic section.

Since the orbit of a planet is bounded and is also a conic section, it must be an ellipse. This establishes Kepler's first law.

NEWTON'S LAW IMPLIES KEPLER'S THIRD LAW

Kepler's third law asserts that the square of the time required for a planet to complete one orbit is proportional to the cube of its mean distance from the sun.

First the term "mean distance" must be defined. For Kepler this meant the average of the shortest distance and the longest distance from the planet to the sun in its orbit. Let us compute this average for the ellipse of semi-major axis a and semiminor axis b, shown in Fig. 14.37. The sun is at the focus F, which is also the pole of the polar coordinate system we are using. The line through the two foci contains the polar axis.

Recall that an ellipse is the set of points P such that the sum of the distances from P to the two foci F and F' is constant, $2a$. The shortest distance from the planet to the sun is $\overline{FQ} = a - d$ and the longest distance is $\overline{EF} = a + d$. Thus Kepler's mean distance is

$$\frac{(a - d) + (a + d)}{2} = a.$$

Now let T be the time required by the given planet to complete one

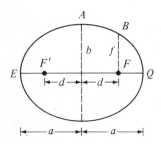

Figure 14.37

By (5), r increases as θ goes from 0 to π.

orbit. Kepler's third law asserts that T^2 is proportional to a^3. The following exercises establish this law by showing that T^2/a^3 is the same for all planets.

14. Using the fact that the area of the ellipse in Fig. 14.37 is πab, show that $Th/2 = \pi ab$, hence that

$$T = \frac{2\pi ab}{h} \qquad (6)$$

The rest of the argument depends only on (5), (6), and the "fixed sum of two distances" property of an ellipse.

15. Using (5), show that f in Fig. 14.37 equals h^2/q.
16. Show that $b^2 = af$, as follows:
    (a) From the fact that $\overline{F'A} + \overline{FA} = 2a$, deduce that $a^2 = b^2 + d^2$.
    (b) From the fact that $\overline{F'B} + \overline{FB} = 2a$, deduce that $d^2 = a^2 - af$.
    (c) From (a) and (b) deduce that $b^2 = af$.
17. From Exercises 15 and 16 deduce that $b^2 = ah^2/q$.
18. Combining (6) and Exercise 17, show that

$$\frac{T^2}{a^3} = \frac{4\pi^2}{q}.$$

Since $4\pi^2/q$ is a constant, the same for all planets, Kepler's third law is established.

14.S Summary

In this chapter we defined and applied the derivative of a vector function $\mathbf{G}$, which assigns to a scalar t a vector $\mathbf{G}(t)$. $\mathbf{G}(t)$ may be thought of as the position vector in the plane or in space of a moving particle at time t.

The derivative of $\mathbf{G}$ is defined as

$$\lim_{\Delta t \to 0} \frac{\Delta \mathbf{G}}{\Delta t}.$$

The mathematician denotes it $\mathbf{G}'$, but a physicist denotes it $\mathbf{v}$, and calls it the velocity vector, for its length is the speed of the moving particle and its direction is that in which the particle is moving at a given instant.

The derivative of a vector function has many useful properties that resemble those of the ordinary derivative. For instance

$$(\mathbf{G} \cdot \mathbf{H})' = \mathbf{G} \cdot \mathbf{H}' + \mathbf{H} \cdot \mathbf{G}'.$$

In particular, if $\mathbf{G} = \mathbf{H}$, then

$$(\mathbf{G} \cdot \mathbf{G})' = 2\mathbf{G} \cdot \mathbf{G}'.$$

From this it follows that if $|\mathbf{G}|$ is constant, then $\mathbf{G}'(t)$ is perpendicular to $\mathbf{G}(t)$ for all t.

The velocity vector is defined as the first derivative of $\mathbf{G}$, $\mathbf{v} = \mathbf{G}'$.

The acceleration vector is the second derivative of $\mathbf{G}$,

$$\mathbf{a} = (\mathbf{G}')' = \mathbf{v}'.$$

If $\mathbf{G}$ is given in components relative to the basic unit vectors $\mathbf{i}$ and $\mathbf{j}$,

$$\mathbf{G}(t) = g(t)\mathbf{i} + h(t)\mathbf{j};$$

then

$$\mathbf{a}(t) = \frac{d^2g}{dt^2}\mathbf{i} + \frac{d^2h}{dt^2}\mathbf{j}.$$

In the study of a particle moving in the xy plane two other perpendicular unit vectors are of interest: $\mathbf{T}$, pointing straight ahead, and $\mathbf{N}$, pointing in the direction in which the particle is turning. $\mathbf{T}$ and $\mathbf{N}$ depend on time. It turns out that

$$\mathbf{a} = \frac{d^2s}{dt^2}\mathbf{T} + \frac{v^2}{r}\mathbf{N}.$$

where s is arc length and r, in this case, is radius of curvature.

The gradient, defined in Sec. 12.6, appears again in the present chapter. When evaluated at (a, b, c), it provides a vector perpendicular to the level surface of F that passes through (a, b, c). The tangent plane to a level surface is defined in terms of ∇F.

A normal to a surface $z = f(x, y)$ can be found by rewriting the equation as $f(x, y) - z = 0$, a level surface of the function of three variables $F(x, y, z) = f(x, y) - z$. Thus $f_x\mathbf{i} + f_y\mathbf{j} - \mathbf{k}$ is a normal to the surface $z = f(x, y)$.

The gradient is of use in rephrasing the chain rule:

$$\frac{du}{dt} = \frac{\partial F}{\partial x}\frac{dx}{dt} + \frac{\partial F}{\partial y}\frac{dy}{dt} + \frac{\partial F}{\partial z}\frac{dz}{dt},$$

where $u = F(x, y, z)$ and x, y, and z are functions of t. In vector notation this rule reads

$$\frac{du}{dt} = \nabla F \cdot \mathbf{G}'(t).$$

This rephrasing in terms of the dot product is useful in showing that ∇F is perpendicular to $\mathbf{G}'$ if $\mathbf{G}$ parameterizes a curve in a surface on which F is constant.

Lagrange multipliers provide a method of finding a maximum or minimum of a function of several variables subject to one or more constraints.

The concluding section on Newton and Kepler is a special and optional treat. It is not reviewed in this summary.

Vectors are used for two purposes in this chapter. First, they are a bookkeeping device. For instance, the two equations

$$\frac{dx}{dt} = 3 \quad \text{and} \quad \frac{dy}{dt} = 4$$

are summarized in a single vector equation

$$\mathbf{v} = 3\mathbf{i} + 4\mathbf{j}.$$

Second, they provide a language in which the symbols closely correspond to intuitive concepts. For instance, instead of saying. "The x coordinate changes at the rate of 3 feet per second and the y coordinate changes at the rate of 4 feet per second," one may say simply, "The velocity vector $\mathbf{v}$ is $3\mathbf{i} + 4\mathbf{j}$." The longer $\mathbf{v}$ is, the faster the particle is moving; the direction of $\mathbf{v}$ is the direction in which the particle moves.

Throughout the chapter we have switched back and forth between mathematical and physical terminology. This table records some of the concepts in both languages.

Mathematical Formulation	Physical Interpretation
Vector function $\mathbf{G}$	Parameterized path of moving particle
Derivative of $\mathbf{G}$	Velocity vector $\mathbf{v}$
Second derivative of $\mathbf{G}$	Acceleration vector $\mathbf{a}$
$\mathbf{T} = \dfrac{\mathbf{v}}{|\mathbf{v}|}$	Unit vector pointing straight ahead
$\mathbf{N} = \dfrac{d\mathbf{T}/dt}{|d\mathbf{T}/dt|}$	Unit vector perpendicular to path and pointing in the direction the path is veering
$\mathbf{a} = \dfrac{d^2 s}{dt^2}\mathbf{T} + \dfrac{v^2}{r}\mathbf{N}$	Acceleration vector, sum of two vectors: one, the acceleration if the particle were moving in a straight line; the other, the acceleration if the particle were moving at a constant speed on a circle whose radius is the radius of curvature

VOCABULARY AND SYMBOLS

vector function $\mathbf{G}$, $\mathbf{r}$
limit of a vector function
continuity of a vector function
position vector $\mathbf{r}$
derivative of $\mathbf{G}$, $\mathbf{G}'$
velocity vector $\mathbf{G}'$, $\mathbf{v}$
level surface $F(x, y, z) = k$
tangent plane
Lagrange multipliers
acceleration vector $\mathbf{G}''$, $\mathbf{a}$
unit tangent vector $\mathbf{T}$
principal unit normal vector $\mathbf{N}$
scalar components of $\mathbf{a}$, a_T and a_N, along $\mathbf{T}$ and $\mathbf{N}$

Guide quiz on chap. 14

1. Let $\mathbf{r} = \mathbf{G}(t) = t\mathbf{i} + t^2\mathbf{j} + \mathbf{k}$ parametrize a curve.
   (a) Sketch the curve.
   (b) Calculate and sketch $\Delta\mathbf{r} = \mathbf{G}(1.1) - \mathbf{G}(1)$.
   (c) Calculate and sketch $\Delta\mathbf{r}/0.1$.
2. Explain why the gradient of $F(x, y, z)$ is perpendicular to the level surface $F(x, y, z) = k$.
3. Find a unit vector perpendicular to
   (a) the surface $z = x^2 + y^2$ at $(1, 1, 2)$;
   (b) the surface $x + y^2 + z^3 = 3$ at $(1, 1, 1)$;
   (c) the sphere $x^2 + y^2 + z^2 = 9$ at (x, y, z).
4. The temperature T at the point (x, y, z) is $x^2 + y^2 + z^2$. As a bird flies on the path $t\mathbf{i} + t^2\mathbf{j} + t^3\mathbf{k}$, it notices that the air is getting hotter. Find the rate of change in the temperature along the bird's flight at the point $(1, 1, 1)$ (a) with respect to time; (b) with respect to distance.
5. The surfaces $2x^2 + 3y^2 + z^2 = 6$ and $x^3 + y^3 + z^3 = 3$ pass through the point $(1, 1, 1)$. At what angle do they cross there?
6. Find an equation of the
   (a) tangent line to $x^5y + x^3y^2 = 2$ at $(1, 1)$;
   (b) tangent plane to $xy + x^3z + xy^3 = 3$ at $(1, 1, 1)$.
7. (a) If a plane curve is given in the form $\mathbf{r} = \mathbf{G}(t)$, how would you find a vector parallel to the curve? Perpendicular to the curve?
   (b) If a plane curve is given in the form $f(x, y) = 0$, how would you find a vector perpendicular to the curve?
   (c) If a curve is given as the intersection of two surfaces, how could you find a vector perpendicular to the curve? Parallel to the curve?
   (d) If a space curve is given in the form $\mathbf{r} = \mathbf{G}(t)$, how would you find a vector parallel to the curve? Perpendicular to the curve?
8. At a certain time a particle moving on the curve $y = f(x)$ has

   $$\mathbf{v} = 2\mathbf{i} + 3\mathbf{j} \quad \text{and} \quad \mathbf{a} = \mathbf{i} - \mathbf{j}.$$

   (a) Is the particle speeding up or slowing down?
   (b) Estimate a_T and a_N graphically.
   (c) Compute a_T and a_N.
   (d) Find the radius of curvature.
   (e) Find dy/dx.
   (f) Find d^2y/dx^2.
9. Consider a curve given as the intersection of the surfaces $g(x, y, z) = 0$ and $h(x, y, z) = 0$. The function $u = f(x, y, z)$, considered only on this curve, has a maximum at (a, b, c). Why might we expect there to be, in general, scalars λ and μ such that

   $$\nabla f = \lambda \nabla g + \mu \nabla h,$$

   where the gradients are evaluated at (a, b, c)? (Include a sketch of the curve and the gradients.)
10. Find the point on the surface $z = x^2 + 2y^2$ closest to the plane $x + y - z = 1$, as follows:
    (a) Sketch the surface and the plane.
    (b) Prove that they do not meet.
    (c) Use the method of Lagrange multipliers to find the point on the surface closest to the plane.

Review exercises for chap. 14

1. (a) Find a vector normal to the surface $z = x^2 + y^3$ at $(1, 1, 2)$.
   (b) Find a vector parallel to the surface $z = x^2 + y^3$ at $(1, 1, 2)$.
2. An astronaut traveling in the path $\mathbf{G}(t) = t\mathbf{i} + t^2\mathbf{j} + t^3\mathbf{k}$ shuts off her engine when at the point $(1, 1, 1)$.
   (a) Show that she passes through the point $(3, 5, 7)$.
   (b) How near does she get to the point $(5, 8, 9)$?
3. Let

   $$\mathbf{G}(t) = \frac{e^t + e^{-t}}{2}\mathbf{i} + \frac{e^t - e^{-t}}{2}\mathbf{j}.$$

   (a) Show that the particle moves on the hyperbola $x^2 - y^2 = 1$.
   (b) Find a_T and a_N when $t = 1$.
4. (a) Draw the level surface of $F(x, y, z) = 2x + 3y + 4z$ that passes through $(1, 0, 0)$.
   (b) Compute ∇F at $(1, 0, 0)$.
   (c) Find two points on the level surface mentioned in (a).
   (d) Using these two points, construct a vector $\mathbf{B}$ lying on the surface.
   (e) Compute $\mathbf{B} \cdot \nabla F$.
   (f) Is the answer to (e) reasonable? Why?
5. At time $t = 0$ the velocity vector of a certain particle is $0\mathbf{i} + 1\mathbf{j}$. If the acceleration vector $\mathbf{a}(t)$ is $6t\mathbf{i} + e^t\mathbf{j}$, find
   (a) $\mathbf{v}$ and $\mathbf{a}$ when $t = 1$;
   (b) v, a_T, a_N, and r (the radius of curvature) when $t = 1$.
6. (a) Show that, when $t = 1$, the particles on the paths $\mathbf{G}(t) = t^2\mathbf{i} + 6t\mathbf{j} + \mathbf{k}$ and $\mathbf{H}(t) = t^3\mathbf{i} + (t + 5)\mathbf{j} + \mathbf{k}$ collide.
   (b) At what angle do they collide?
7. Let $\mathbf{r}$ be the position vector of a moving particle and let $r = |\mathbf{r}|$.
   (a) Show that $r\, dr/dt = \mathbf{r} \cdot d\mathbf{r}/dt$.

(b) Using the definition of $\mathbf{r}'$ as $\lim_{\Delta t \to 0} \Delta \mathbf{r}/\Delta t$, show why $\mathbf{r}'$ is tangent to the curve whose position vector at time t is $\mathbf{r}(t)$.

8 If $\mathbf{r} = \mathbf{G}(t)$ is a curve in space, how would you find a unit vector tangent to it?

9 Find a unit normal to the surface $x^2 y + 2xz = 4$ at the point $(2, -2, 3)$.

10 At time t the position vector of a particle is

$$\mathbf{G}(t) = t\mathbf{i} + 3t\mathbf{j} + 4t\mathbf{k}.$$

(a) Show that any point in the path lies in the plane $3x = y$ and also in the plane $4y = 3z$.
(b) Sketch the planes in (a) and indicate their intersection, which is the path of the particle.
(c) Find the velocity vector at time t.
(d) Find the speed at time t.

11 Let $\mathbf{G}(t) = e^t \cos t\, \mathbf{i} + e^t \sin t\, \mathbf{j} + e^t\, \mathbf{k}$.
(a) Find the speed at time t.
(b) Find the distance traveled during the time interval $[0, 1]$.

12 A particle moves on a straight line at a constant speed of 4 feet per second in the direction of the unit vector

$$\frac{\mathbf{i}}{3} + \frac{2\mathbf{j}}{3} + \frac{2\mathbf{k}}{3}.$$

At time $t = 0$ the particle is at the point $(2, 5, 7)$. Find its position vector at time t.

13 A particle moves along a straight line at a constant speed. At time $t = 0$ it is at the point $(1, 2, 1)$, and at time $t = 2$ it is at $(4, 1, 5)$.
(a) Find its speed.
(b) Find the direction cosines of the line.
(c) Find the position vector of the particle at time t.

■

14 Let $P_0 = (a, b, c)$ be a point not on the smooth curve given as the intersection of the surfaces $F(x, y, z) = 0$ and $G(x, y, z) = 0$. Let P_1 be a nearest point to P_0 on the curve. Prove that $\overrightarrow{P_0 P_1}$ is perpendicular to the curve at P_1. (Use Lagrange multipliers.)

15 At a certain moment a particle moving on a curve has $\mathbf{v} = 2\mathbf{i} + 3\mathbf{j}$ and $\mathbf{a} = -\mathbf{i} + 2\mathbf{j}$.
(a) What is the speed of the particle?
(b) Find $\mathbf{T}$ and $\mathbf{N}$, expressing each in the form $x\mathbf{i} + y\mathbf{j}$, for suitable scalars x and y.
(c) Draw $\mathbf{v}$, $\mathbf{a}$, $\mathbf{T}$, and $\mathbf{N}$.
(d) Compute $\mathbf{a} \cdot \mathbf{N}$.
(e) From (a) and (d) obtain the radius of curvature.
(f) Compute $\mathbf{a} \cdot \mathbf{T}$.
(g) Is the particle speeding up or slowing down? Compute $d^2 s/dt^2$.

16 (a) Explain why ∇F is perpendicular to the level $F(x, y, z) = 0$.
(b) Explain why ∇f is perpendicular to the level $f(x, y) = 0$.

17 How would you find a vector perpendicular to the face $z = f(x, y)$? Explain.

18 (a) Explain why $\dfrac{d\mathbf{r}}{ds}$

is a unit vector. (s denotes arc length.)
(b) From (a) deduce *as a special case* that

$$\frac{d\mathbf{T}}{d\phi}$$

is a unit vector. (ϕ denotes the angle between the unit tangent vector $\mathbf{T}$ and the vector $\mathbf{i}$.)

19 Assume that the surfaces $F(x, y, z) = 0$ and $G(x, y, z) = 0$ meet in a curve and that the point (a, b, c) is on this curve. Explain why

$$\nabla F \times \nabla G$$

is a tangent vector to this curve at (a, b, c). (∇F and ∇G are evaluated at (a, b, c).)

20 Find a tangent vector to the curve of intersection of the surfaces $x^2 + 2y^2 + 3z^2 = 36$ and $2x^2 - y^2 + z^2 = 7$ at the point $(1, 2, 3)$.

21 (a) Sketch the level curve of the function $5x^2 + 3y^2$ that passes through $(1, 1)$.
(b) Do the same for the function y^5/x^3.
(c) Show that the two curves cross at a right angle at the point $(1, 1)$.
(d) Prove that each level curve of $5x^2 + 3y^2$ crosses each level curve of y^5/x^3 at a right angle.

22 Let $\mathbf{G}(s)$, where s denotes arc length, parametrize a curve in space. Let $u = F(x, y, z)$ be a function of x, y, and z. Then u may be considered a composite function of s. Show that
(a) $du/ds = \nabla F \cdot \mathbf{G}'(s)$, where ∇F is evaluated at $\mathbf{G}(s)$.
(b) $du/ds = D_{\mathbf{T}}(u)$, where $\mathbf{T}$ is a unit tangent vector to the curve at $\mathbf{G}(s)$.

■ ■

23 (This continues Exercises 23 to 25 of Sec. 14.6.) Let a curve in space be parametrized by arc length s, $\mathbf{r} = \mathbf{G}(s)$. Define $\mathbf{T}$ as the unit vector $d\mathbf{r}/ds$. Define $\mathbf{N}$ as the unit vector such that $d\mathbf{T}/ds = k\mathbf{N}$ for some positive scalar k, which turns out to be the curvature of the curve. (Note that $\mathbf{T}$ and $\mathbf{N}$ are perpendicular.) Define $\mathbf{B}$ as $\mathbf{T} \times \mathbf{N}$. By Exercise 24 of Sec. 14.6, $d\mathbf{B}/ds = -\tau \mathbf{N}$ for some scalar τ. By Exercise 25 of

is constant, then the curve remains in ...dicular to **B**. Thus τ serves to measure ... of the curve to twist out of the plane ...by **T** and **N**. For this reason τ is called the ...he curve.

...that $d\mathbf{N}/ds = p\mathbf{T} + q\mathbf{B}$ for some scalars p and ...

... differentiating the equation $\mathbf{B} \cdot \mathbf{N} = 0$, show that ... $= \tau$.

...y differentiating the equation $\mathbf{N} \cdot \mathbf{T} = 0$, show that $p = -k$.

...us $d\mathbf{N}/ds = -k\mathbf{T} + \tau\mathbf{B}$. The formulas for $d\mathbf{T}/ds$, $d\mathbf{B}/ds$, and $d\mathbf{N}/ds$ are known as the *Frenet formulas*: $d\mathbf{T}/ds = k\mathbf{N}$, $d\mathbf{N}/ds = -k\mathbf{T} + \tau\mathbf{B}$, $d\mathbf{B}/ds = \tau\mathbf{N}$.

Though defined at first algebraically, the gradient turned out to be an intrinsic geometric property of the function, for its direction and magnitude are expressible in terms of the function, without reference to any particular coordinate system.

Consider, then, a function $f(P)$, where P runs over a portion of the plane. If f is given in rectangular coordinates, then

$$\nabla f = f_x \mathbf{i} + f_y \mathbf{j}.$$

What if f is described in *polar coordinates*, $f(r, \theta)$? How would ∇f be expressed? The natural unit vectors to use, instead of $\mathbf{i}$ and $\mathbf{j}$, are $\mathbf{u}_r$ and $\mathbf{u}_\theta$ shown in Fig. 14.38.

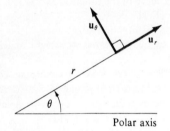

Figure 14.38

(a) If $\nabla f = A\mathbf{u}_r + B\mathbf{u}_\theta$ for scalars A and B, show that $A = D_{\mathbf{u}_r}(f)$ and $B = D_{\mathbf{u}_\theta}(f)$.

(b) Show that $D_{\mathbf{u}_r}(f) = \dfrac{\partial f}{\partial r}$.

(c) Why would $D_{\mathbf{u}_\theta}(f) = \dfrac{1}{r}\dfrac{\partial f}{\partial \theta}$? $\left(\text{Not } \dfrac{\partial f}{\partial \theta}\,!!\right)$

Give a persuasive argument, not necessarily a rigorous proof.

(d) From (a), (b), and (c) deduce that

$$\nabla f = \frac{\partial f}{\partial r}\mathbf{u}_r + \frac{1}{r}\frac{\partial f}{\partial \theta}\mathbf{u}_\theta.$$

25 (See Exercise 24.) Let $f(P)$ be the reciprocal of the distance from P in the plane to the origin.

(a) In rectangular coordinates f has the formula $f(x, y) = 1/\sqrt{x^2 + y^2}$. Calculate ∇f using rectangular coordinates.

(b) In polar coordinates f has the formula $f(r, \theta) = 1/r$. Calculate ∇f using polar coordinates.

(c) Sketch ∇f as calculated in (a) and in (b). Show that the results are the same.

26 (See Exercise 24.) Consider a spherical coordinate system ρ, θ, ϕ. At a given point let $\mathbf{u}_\rho, \mathbf{u}_\phi$, and $\mathbf{u}_\theta$ be unit vectors pointing in the directions of increasing ρ, ϕ, and θ respectively. (See Fig. 14.39.) Let f be a function defined on space. Show that

$$\nabla f = \frac{\partial f}{\partial \rho}\mathbf{u}_\rho + \frac{1}{\rho}\frac{\partial f}{\partial \phi}\mathbf{u}_\phi + \frac{1}{\rho \sin \phi}\frac{\partial f}{\partial \theta}\mathbf{u}_\theta.$$

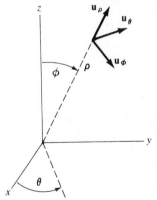

Figure 14.39

GREEN'S THEOREM, THE DIVERGENCE THEOREM, AND STOKES' THEOREM

This chapter develops two main theorems, the divergence theorem and Stokes' theorem. The first relates an integral over a spatial region to an integral over its bounding surface or surfaces. The second relates an integral over a surface to an integral over its bounding curve or curves. Both are generalizations of Green's theorem, which relates an integral over a planar region to an integral over its bounding curve or curves.

The functions in this chapter will be assumed to have partial derivatives of all orders. The curves and surfaces will be assumed to be smooth (in the sense that the curves locally resemble straight lines and the surfaces locally resemble planes) or to be made up of a finite number of such curves or surfaces. The spatial regions will be assumed to be bounded by smooth surfaces.

The main objectives of this chapter are to convey what the theorems of Green and Stokes and the divergence theorem say, show why they are plausible, and illustrate their practical importance through a variety of examples.

Engineering students, especially, should keep these words in mind, quoted from the preface of Stephen Whitaker's upper-division text, *Introduction to Fluid Mechanics*, "Vector notation is used freely throughout the text, not because it leads to elegance or rigor but simply because fundamental concepts are best expressed in a form which attempts to connect them with reality."

15.1 Vector and scalar fields

Imagine a loop of wire C held firmly in place on the surface of a stream. At some points of C water is entering the region bounded by C, and at some points of C water is leaving the region. In Fig. 15.1 the arrows indicate the velocity at various points on the stream, and the curve is the wire C. How can the net amount entering or leaving be computed? Presumably some kind of integration is required. This chapter will develop methods for answering questions such as this.

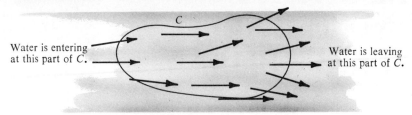

Figure 15.1

The flow of water in a stream provides an example of a *vector field*, which will now be defined.

VECTOR AND SCALAR FIELDS

DEFINITION *Vector field.* A function that assigns a vector to each point in some region in the plane (or space) is called a *vector field*. It will usually be denoted **F**.

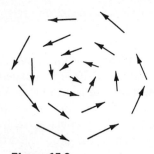

For example, the function **F** that assigns to each point on the surface of the stream the velocity vector of the water at that point is a vector field.

The daily weather map displays a few of the (vector) values of the vector field that assigns to each point on the surface of a portion of the earth (considered to be flat) the wind vector at that point. Figure 15.2 shows a few of the wind vectors of the vector field associated with a hurricane. Near the eye of the hurricane the wind vectors are shorter: The air is relatively calm.

Figure 15.2

DEFINITION *Scalar field.* A function that assigns a number to each point in some region in the plane (or in space) is called a *scalar field*. It will usually be denoted f.

The function that assigns the temperature at a point is a scalar function; so is the function that describes the density at a point.

A vector field **F** in the plane is described by two scalar fields, the scalar components of **F**:

$$\mathbf{F}(x, y) = P(x, y)\mathbf{i} + Q(x, y)\mathbf{j}.$$

Both P and Q are scalar fields.

OBTAINING ONE FIELD FROM ANOTHER Several important vector or scalar fields can be defined in terms of other vector or scalar fields. This section presents these definitions; their physical interpretation will be discussed later in the chapter.

DEFINITION *The gradient field.* Let f be a scalar field. The vector field that assigns to each point (x, y) the gradient of f at (x, y), ∇f, is called the *gradient field* associated with f. A similar construction can be carried out on $f(x, y, z)$, again producing a vector field ∇f.

15.1 VECTOR AND SCALAR FIELDS

EXAMPLE 1 Compute and sketch the gradient field associated with the scalar field $f(x, y, z) = 1/\sqrt{x^2 + y^2 + z^2}$. (These two fields are of importance in gravitational and electromagnetic theory.)

SOLUTION
$$\nabla f = f_x \mathbf{i} + f_y \mathbf{j} + f_z \mathbf{k}$$
$$= \frac{-x}{(\sqrt{x^2 + y^2 + z^2})^3} \mathbf{i} + \frac{-y}{(\sqrt{x^2 + y^2 + z^2})^3} \mathbf{j} + \frac{-z}{(\sqrt{x^2 + y^2 + z^2})^3} \mathbf{k}.$$

Let
$$\mathbf{r} = x\mathbf{i} + y\mathbf{j} + z\mathbf{k}$$

be the position vector of the point (x, y, z).

Then
$$|\mathbf{r}| = \sqrt{x^2 + y^2 + z^2}.$$

Thus
$$\nabla f = -\frac{\mathbf{r}}{|\mathbf{r}|^3}$$

The magnitude of $\mathbf{r}/|\mathbf{r}|^3$ equals "the inverse square," $|\mathbf{r}|^{-2}$.

and is therefore pointed toward the origin. Moreover,

$$|\nabla f| = \left| -\frac{\mathbf{r}}{|\mathbf{r}|^3} \right| = \frac{|\mathbf{r}|}{|\mathbf{r}|^3} = \frac{1}{|\mathbf{r}|^2}.$$

When $\mathbf{r}$ is short, ∇f is long and, when $\mathbf{r}$ is long, ∇f is short. In physics, the gradient field

$$\frac{-\mathbf{r}}{|\mathbf{r}|^3}$$

corresponds to a force of attraction. Figure 15.3 shows a few of the values of this vector field. ∎

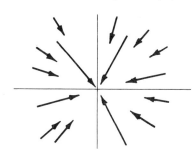

Figure 15.3

In Example 1 a vector field was obtained from a scalar field by "taking the gradient." (However not every vector field is the gradient of some scalar field, as Exercise 27 shows.) Also of importance in mathematics and physics is the following procedure, which obtains a scalar field from a vector field.

DIVERGENCE

DEFINITION *Divergence of a vector field.* Let

$$\mathbf{F}(x, y) = P(x, y)\mathbf{i} + Q(x, y)\mathbf{j}$$

be a vector field in the plane. The scalar field

$$\frac{\partial P}{\partial x} + \frac{\partial Q}{\partial y}$$

The divergence of a vector field is a scalar field.

is called the *divergence* of $\mathbf{F}$. Similarly, if

$$\mathbf{F}(x, y, z) = P(x, y, z)\mathbf{i} + Q(x, y, z)\mathbf{j} + R(x, y, z)\mathbf{k},$$

a vector field in space, then the scalar field

$$\frac{\partial P}{\partial x} + \frac{\partial Q}{\partial y} + \frac{\partial R}{\partial z}$$

is called the *divergence* of **F**.

Section 15.4 will show why this particular function is called the *divergence*. It turns out that, if **F** describes the velocity of a fluid, then the divergence of **F** at a point describes the tendency of that fluid to accumulate or disperse near that point.

EXAMPLE 2 Compute the divergence of the vector field

$$5x^2 y\mathbf{i} + xy\mathbf{j} + x^2 z\mathbf{k}.$$

SOLUTION By definition, the divergence of the vector field $5x^2 y\mathbf{i} + xy\mathbf{j} + x^2 z\mathbf{k}$ is

$$\frac{\partial}{\partial x}(5x^2 y) + \frac{\partial}{\partial y}(xy) + \frac{\partial}{\partial z}(x^2 z) = 10xy + x + x^2. \blacksquare$$

NOTATION For convenience introduce the formal "vector"

$$\nabla = \frac{\partial}{\partial x}\mathbf{i} + \frac{\partial}{\partial y}\mathbf{j}.$$

If $\mathbf{F} = P\mathbf{i} + Q\mathbf{j},$

compute $\nabla \cdot \mathbf{F}$ as a dot product of two ordinary vectors:

$$\nabla \cdot \mathbf{F} = \frac{\partial}{\partial x} P + \frac{\partial}{\partial y} Q.$$

Interpret this to mean $\quad\dfrac{\partial P}{\partial x} + \dfrac{\partial Q}{\partial y}.$

This explains the customary notation for the divergence of **F**,

$$\nabla \cdot \mathbf{F}.$$

$\nabla \cdot \mathbf{F}$ *is the usual notation for divergence.*

Similarly, to provide a short notation for the divergence of a vector field in space, introduce the formal "vector"

$$\nabla = \frac{\partial}{\partial x}\mathbf{i} + \frac{\partial}{\partial y}\mathbf{j} + \frac{\partial}{\partial z}\mathbf{k}.$$

Then if $\quad \mathbf{F} = P\mathbf{i} + Q\mathbf{j} + R\mathbf{k},$

$\nabla \cdot \mathbf{F}$ is a shorthand for the divergence

15.1 VECTOR AND SCALAR FIELDS

$$\frac{\partial P}{\partial x} + \frac{\partial Q}{\partial y} + \frac{\partial R}{\partial z}.$$

The divergence of **F** is also written "div **F**."

The next example, which is important in the study of gravitational and electrostatic forces, will be referred to later in the chapter.

EXAMPLE 3 Let $\mathbf{F}(P)$, for any point P in space other than the origin O, be defined as $\mathbf{r}/|\mathbf{r}|^3$, where $\mathbf{r}$ is the position vector $\overrightarrow{OP}$. Show that the divergence of **F** is 0, that is, $\nabla \cdot \mathbf{F} = 0$.

SOLUTION
$$\mathbf{F}(x, y, z) = \frac{x\mathbf{i} + y\mathbf{j} + z\mathbf{k}}{r^3},$$

where $r = |\mathbf{r}| = \sqrt{x^2 + y^2 + z^2}$. By the definition of the divergence,

$$\nabla \cdot \mathbf{F} = \frac{\partial}{\partial x}\left(\frac{x}{r^3}\right) + \frac{\partial}{\partial y}\left(\frac{y}{r^3}\right) + \frac{\partial}{\partial z}\left(\frac{z}{r^3}\right).$$

It is necessary to compute three partial derivatives. The first one is

$$\frac{\partial}{\partial x}\left(\frac{x}{r^3}\right) = \frac{r^3 \frac{\partial x}{\partial x} - x\frac{\partial}{\partial x}(r^3)}{r^6} = \frac{r^3 - 3xr^2\frac{\partial r}{\partial x}}{r^6}$$

$$= \frac{r - 3x\frac{\partial r}{\partial x}}{r^4}.$$

Now $\quad \dfrac{\partial r}{\partial x} = \dfrac{\partial}{\partial x}(\sqrt{x^2 + y^2 + z^2}) = \dfrac{x}{\sqrt{x^2 + y^2 + z^2}} = \dfrac{x}{r}.$

Thus, $\quad \dfrac{\partial}{\partial x}\left(\dfrac{x}{r^3}\right) = \dfrac{r - 3x\dfrac{x}{r}}{r^4} = \dfrac{1}{r^3} - \dfrac{3x^2}{r^5}.$

Similarly, $\quad \dfrac{\partial}{\partial y}\left(\dfrac{y}{r^3}\right) = \dfrac{1}{r^3} - \dfrac{3y^2}{r^5} \quad$ and $\quad \dfrac{\partial}{\partial z}\left(\dfrac{z}{r^3}\right) = \dfrac{1}{r^3} - \dfrac{3z^2}{r^5}.$

Consequently,

$$\nabla \cdot \mathbf{F} = \left(\frac{1}{r^3} - \frac{3x^2}{r^5}\right) + \left(\frac{1}{r^3} - \frac{3y^2}{r^5}\right) + \left(\frac{1}{r^3} - \frac{3z^2}{r^5}\right)$$

$$= \frac{3}{r^3} - \frac{3(x^2 + y^2 + z^2)}{r^5}$$

$$= \frac{3}{r^3} - \frac{3r^2}{r^5} = 0. \quad \blacksquare$$

How nicely everything canceled in Example 3 and reduced to 0. If a vector field in space has the form $\mathbf{r}/|\mathbf{r}|^k$ for some constant k, the divergence of that field is generally not 0. As Exercise 23 asks the reader to check, only for $k = 3$ (the "inverse square law") is the divergence 0.

Only for $k = 3$ does $\mathbf{r}/|\mathbf{r}|^k$ have divergence 0. ("The inverse square law")

Note how much information is packed into the compact notation $\nabla \cdot \mathbf{F}$, read as "del dot $\mathbf{F}$." When you read $\nabla \cdot \mathbf{F}$, think "$\mathbf{F}$ is a vector field in the plane or space. It is described by scalar fields P, Q, and, if in space, R:

$$\mathbf{F} = P\mathbf{i} + Q\mathbf{j} + R\mathbf{k}.$$

Then $\nabla \cdot \mathbf{F}$ is the sum of three partial derivatives,

$$\frac{\partial P}{\partial x} + \frac{\partial Q}{\partial y} + \frac{\partial R}{\partial z}."$$

CURL

It is possible to derive from a vector field $\mathbf{F}$ another vector field, called the *curl* of $\mathbf{F}$, which is important in physics. Section 15.7 will show why it is called the *curl*.

DEFINITION *Curl of a vector field.* Let $\mathbf{F} = P\mathbf{i} + Q\mathbf{j} + R\mathbf{k}$ be a vector field. The function that assigns to each point the vector

$$\left(\frac{\partial R}{\partial y} - \frac{\partial Q}{\partial z}\right)\mathbf{i} + \left(\frac{\partial P}{\partial z} - \frac{\partial R}{\partial x}\right)\mathbf{j} + \left(\frac{\partial Q}{\partial x} - \frac{\partial P}{\partial y}\right)\mathbf{k}$$

is called the *curl* of $\mathbf{F}$ and is denoted **curl F**.

If you expand the formal determinant

$$\begin{vmatrix} \mathbf{i} & \mathbf{j} & \mathbf{k} \\ \frac{\partial}{\partial x} & \frac{\partial}{\partial y} & \frac{\partial}{\partial z} \\ P & Q & R \end{vmatrix},$$

The curl of a vector field is again a vector field.

you obtain **curl F**. The similarity to the cross product of vectors suggests the notation $\nabla \times \mathbf{F}$ for **curl F**. This second notation is read, "del cross $\mathbf{F}$."

The definition also applies to a vector field $\mathbf{F} = P(x, y)\mathbf{i} + Q(x, y)\mathbf{j}$ in the plane. Writing $\mathbf{F}$ as $P(x, y)\mathbf{i} + Q(x, y)\mathbf{j} + 0\mathbf{k}$, we find that

$$\nabla \times \mathbf{F} = \left(\frac{\partial Q}{\partial x} - \frac{\partial P}{\partial y}\right)\mathbf{k}.$$

EXAMPLE 4 Compute the curl of $\mathbf{F} = xyz\mathbf{i} + x^2\mathbf{j} - xy\mathbf{k}$.

SOLUTION The curl of $\mathbf{F}$ is given by

$$\begin{vmatrix} \mathbf{i} & \mathbf{j} & \mathbf{k} \\ \dfrac{\partial}{\partial x} & \dfrac{\partial}{\partial y} & \dfrac{\partial}{\partial z} \\ xyz & x^2 & -xy \end{vmatrix},$$

which is short for

$$\left[\frac{\partial}{\partial y}(-xy) - \frac{\partial}{\partial z}(x^2)\right]\mathbf{i} - \left[\frac{\partial}{\partial x}(-xy) - \frac{\partial}{\partial z}(xyz)\right]\mathbf{j} + \left[\frac{\partial}{\partial x}(x^2) - \frac{\partial}{\partial y}(xyz)\right]\mathbf{k}$$
$$= (-x - 0)\mathbf{i} - (-y - xy)\mathbf{j} + (2x - xz)\mathbf{k}$$
$$= -x\mathbf{i} + (y + xy)\mathbf{j} + (2x - xz)\mathbf{k}. \blacksquare$$

THE LAPLACIAN

The divergence of a gradient field: the Laplacian

There is another scalar field of general importance in engineering, physics, and mathematics. Say that you start with a scalar field f and form the gradient field ∇f. Then you may take the divergence of this vector field, obtaining a scalar field. This scalar field is called the *Laplacian* of f. If f is a function of two variables, then $\nabla f = f_x \mathbf{i} + f_y \mathbf{j}$ and $\nabla \cdot \nabla f$ equals

$$\left(\frac{\partial}{\partial x}\mathbf{i} + \frac{\partial}{\partial y}\mathbf{j}\right) \cdot (f_x \mathbf{i} + f_y \mathbf{j}) = \frac{\partial f_x}{\partial x} + \frac{\partial f_y}{\partial y} = f_{xx} + f_{yy}.$$

The Laplacian

Thus the Laplacian of f is simply $f_{xx} + f_{yy}$. In the case of a function $f(x, y, z)$, the Laplacian equals

$$f_{xx} + f_{yy} + f_{zz}.$$

The symbols $\nabla \cdot \nabla f$ and $\nabla^2 f$ are standard notations for the Laplacian.

EXAMPLE 5 Compute the Laplacian of $f(x, y) = x^3 - 3xy^2$.

SOLUTION First compute f_x and f_y:

$$f_x = 3x^2 - 3y^2 \quad \text{and} \quad f_y = -6xy.$$

Then $f_{xx} = 6x$ and $f_{yy} = -6x$.

Consequently, $\nabla^2 f = f_{xx} + f_{yy} = 6x + (-6x) = 0.$

The Laplacian in this special case has the constant value 0. $\blacksquare$

A function f whose Laplacian is identically 0 is called *harmonic*. Harmonic functions are important in the study of electricity and temperature distributions.

SUMMARY OF SECTION

This table summarizes the key operations described in this section.

Type of field	Operation	Notation for new field	Type
Scalar, f	Gradient	∇f	Vector
Vector, $\mathbf{F}$	Divergence	$\nabla \cdot \mathbf{F} = \operatorname{div} \mathbf{F}$	Scalar
Vector, $\mathbf{F}$	Curl	$\nabla \times \mathbf{F} = \operatorname{curl} \mathbf{F}$	Vector
Scalar, f	Divergence of gradient	$\nabla \cdot \nabla f = \nabla^2 f$ (Laplacian of f)	Scalar

Recall that the magnitude of the vector field $\mathbf{r}/|\mathbf{r}|^3$ varies inversely as the *square* of the magnitude of $\mathbf{r}$. As a field defined in space ($\mathbf{r} = x\mathbf{i} + y\mathbf{j} + z\mathbf{k}$), its divergence is 0.

Exercises

In Exercises 1 to 4 compute $\nabla \cdot \mathbf{F} = \operatorname{div} \mathbf{F}$.

1. $\mathbf{F} = 2x\mathbf{i} + 3y\mathbf{j} + 4z\mathbf{k}$.

2. $\mathbf{F} = y^2\mathbf{i} - z^2\mathbf{j} + x^2\mathbf{k}$.

3. $\mathbf{F} = \sin xy\, \mathbf{i} + \cos xy^2\, \mathbf{j} + \tan^{-1} 3xz\, \mathbf{k}$.

4. $\mathbf{F} = \dfrac{x\mathbf{i}}{x^2 + y^2} + \dfrac{y\mathbf{j}}{x^2 + y^2} + \dfrac{z\mathbf{k}}{x^2 + y^2}$.

In Exercises 5 to 8 compute $\operatorname{curl} \mathbf{F} = \nabla \times \mathbf{F}$.

5. $\mathbf{F} = x^3\mathbf{i} - y^3\mathbf{j} + z^2\mathbf{k}$.

6. $\mathbf{F} = yz\mathbf{i} + z^2 x\mathbf{j} + xyz\mathbf{k}$.

7. $\mathbf{F} = \sec yz\, \mathbf{i} + \tan(x+z)\, \mathbf{j} + \ln x\, \mathbf{k}$.

8. $\mathbf{F} = \dfrac{x\mathbf{i}}{r} + \dfrac{y\mathbf{j}}{r} + \dfrac{z\mathbf{k}}{r}$, where $r = \sqrt{x^2 + y^2 + z^2}$.

9. If f is a scalar field and $\mathbf{F}$ is a vector field, which type of field is $f\mathbf{F}$?

10. If $\mathbf{F}$ and $\mathbf{G}$ are vector fields, which type of field is the function $\mathbf{F} \cdot \mathbf{G}$?

11. By a straightforward computation verify that the curl of the gradient of f is $\mathbf{0}$, that is,
$$\nabla \times (\nabla f) = \mathbf{0},$$
    (a) in case f is defined on the plane, $f = f(x, y)$;
    (b) in case f is defined in space, $f = f(x, y, z)$.

12. By a straightforward computation, verify that the divergence of the curl of $\mathbf{F}$ is 0, that is, $\nabla \cdot (\nabla \times \mathbf{F}) = 0$ where $\mathbf{F} = P\mathbf{i} + Q\mathbf{j} + R\mathbf{k}$. Thus div ($\operatorname{curl} \mathbf{F}$) = 0.

13. Which of the following are scalar fields? vector fields?
    (a) $\operatorname{curl} \mathbf{F}$ (b) $|\mathbf{F}|$ (c) $\mathbf{F} \cdot \mathbf{F}$ (d) $\operatorname{div} \mathbf{F}$
    (e) $\nabla \cdot \mathbf{F}$ (f) $\mathbf{F} \times \mathbf{i}$ (g) $\nabla \times \mathbf{F}$.

14. Which of these expressions make sense? Which do not?
    (a) the curl of the curl of $\mathbf{F}$,
    (b) the curl of the gradient of f,
    (c) the divergence of the curl of $\mathbf{F}$,
    (d) the gradient of the curl of $\mathbf{F}$,
    (e) the divergence of the divergence of $\mathbf{F}$.

15. The vector field $\mathbf{F}(x, y, z) = x\mathbf{i} + y\mathbf{j} + z\mathbf{k}$ is usually denoted $\mathbf{r}$. It represents the position vector of the point (x, y, z). The same notation is used for $\mathbf{F}(x, y) = x\mathbf{i} + y\mathbf{j}$. Show that (a) in space $\nabla \cdot \mathbf{r} = 3$; (b) in the plane $\nabla \cdot \mathbf{r} = 2$.

16. (a) Show that $f(x, y) = \ln(x^2 + y^2)$ is harmonic.
    (b) Show that $f(x, y, z) = \ln(x^2 + y^2 + z^2)$ is not harmonic.

17. (a) Prove that the divergence of a constant vector field is 0.
    (b) Give an example of a nonconstant vector field $\mathbf{F}$ for which $\nabla \cdot \mathbf{F} = 0$.

18. (a) If f is a scalar field and $\mathbf{F}$ is a vector field, prove that
$$\nabla \cdot (f\mathbf{F}) = f \nabla \cdot \mathbf{F} + \nabla f \cdot \mathbf{F}.$$
    (b) Express the equation in (a) in a sentence, using such terms as "divergence," "gradient," and "dot product."

19. If f and g are scalar fields, so is fg, their product. Prove

that the gradient of fg equals f times the gradient of g plus g times the gradient of f, that is,

$$\nabla(fg) = f\,\nabla g + g\,\nabla f.$$

20 Letting $\mathbf{F} = P\mathbf{i} + Q\mathbf{j} + R\mathbf{k}$, prove that

$$\nabla \times f\mathbf{F} = f\nabla \times \mathbf{F} + \nabla f \times \mathbf{F}.$$

21 Letting $\mathbf{F} = P\mathbf{i} + Q\mathbf{j} + R\mathbf{k}$ and $\mathbf{G} = L\mathbf{i} + M\mathbf{j} + N\mathbf{k}$, prove that

$$\operatorname{div}(\mathbf{F} \times \mathbf{G}) = \mathbf{G} \cdot \operatorname{curl} \mathbf{F} - \mathbf{F} \cdot \operatorname{curl} \mathbf{G}.$$

22 For scalar fields f and g show that

$$\operatorname{div}(\nabla f \times \nabla g) = 0.$$

Exercises 23 to 25 are related. They continue Example 3.

23 Let k be a fixed real number other than 3. Show that the vector field $\mathbf{r}/|\mathbf{r}|^k$, defined in space except at the origin, has a nonzero divergence.

24 Let $\mathbf{F}$ be the vector field $\mathbf{r}/|\mathbf{r}|^k$, defined in the plane except at the origin. Show that (a) if $k = 2$, $\nabla \cdot \mathbf{F} = 0$, (b) if k is not 2, $\nabla \cdot \mathbf{F}$ is not 0. (This should be contrasted with Exercise 23 and Example 3.)

25 A vector field $\mathbf{F}$ in the plane or in space is *central* if $\mathbf{F}(P)$ is parallel to $\overrightarrow{OP}$ for each point where $\mathbf{F}(P)$ is defined. A central vector field $\mathbf{F}$ is called *radially symmetric* if the magnitude of $\mathbf{F}(P)$ is the same for all points P at the same distance from O.

(a) Show that a radially symmetric vector field $\mathbf{F}$ has the form $\mathbf{F}(P) = f(r)\mathbf{r}$, where $\mathbf{r} = \overrightarrow{OP}$, $r = |\mathbf{r}|$, and f is a scalar function.

(b) Assume that $\mathbf{F}$ is defined everywhere in space except at O and that f is differentiable. Show that, if the divergence of $\mathbf{F}$ is 0, then $f(r) = kr^{-3}$ for some constant k.

(Thus the gravitational field can be described up to a constant factor as "the radially symmetric vector field with divergence equal to 0.")

26 Show that the curl of the vector field $f(r)\mathbf{r}$ is $\mathbf{0}$ if the field is defined in (a) the plane, (b) space.

■ ■

27 (a) Show that, if $\mathbf{F} = P\mathbf{i} + Q\mathbf{j}$ is a vector field and equals ∇f for some scalar field, then

$$\frac{\partial P}{\partial y} = \frac{\partial Q}{\partial x}.$$

(b) Show that the vector field $x^2 y\mathbf{i} + x^2 y^3\mathbf{j}$ is not of the form ∇f for any scalar field f.

15.2 Line integrals

Consider a curve C in the plane (or in space). Assume that C is parametrized by $\mathbf{r} = \mathbf{G}(t)$, defined for all t in the interval $[a, b]$; think of t as time. Let f be a scalar function in the plane (or in space) defined at least on every point of the curve C. This section defines the definite integral of f over the curve C, and illustrates its use.

LINE INTEGRAL OF A SCALAR FUNCTION

To begin, let $s(t)$ denote arc length along the curve, measured from some base point B on the curve. Then consider a typical partition of the interval $[a, b]$,

$$t_0 = a, t_1, \ldots, t_n = b.$$

Let $\Delta s_i = s(t_i) - s(t_{i-1})$. This is positive since we assume that s is increasing. Let P_i be a sample point on the curve corresponding to some instant chosen in the ith time interval. (See Fig. 15.4.) Form the sum

$$\sum_{i=1}^{n} f(P_i)\,\Delta s_i.$$

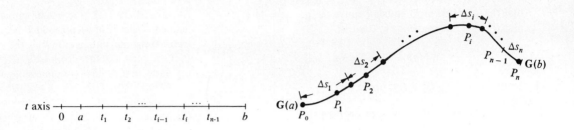

Figure 15.4

DEFINITION *Line integral of a scalar function.* Assume that the scalar function f is defined at least on the curve C. The limit of sums of the form

$$\sum_{i=1}^{n} f(P_i) \, \Delta s_i,$$

as the mesh of the partition of the interval $[a, b]$ approaches 0, is the *line integral* of f over the curve given parametrically as $\mathbf{r} = \mathbf{G}(t)$. If the curve is called C, the line integral is denoted

$$\int_C f(P) \, ds.$$

"Curve integral" might be more apt, but "line integral" is the traditional term.

If the curve is thought of as a wire and $f(P)$ as the density of the wire at P, then $\int_C f(P) \, ds$ is the mass of the wire.

The integral $\int_C f(P) \, ds$ does not depend on the orientation of the curve C, that is, on the direction in which it is swept out. Nor does it depend on the particular parametrization.

The average of a function over a curve is defined just like the average of a function over an interval: The average of $f(P)$ over the curve C is the quotient,

$$\frac{\int_C f(P) \, ds}{\text{Length of } C}.$$

EXAMPLE 1 A wire occupies the portion of the parabola $y = x^2$ that lies between $(1, 1)$ and $(2, 4)$. The density of the wire at the point (x, y) is x grams per centimeter. Find the mass of the wire. (See Fig. 15.5.)

SOLUTION We have

$$\text{Mass of wire} = \int_C x \, ds \text{ grams}.$$

To evaluate the line integral use x, rather than s, as the independent variable. The computations then run like this:

$$\int_C x \, ds = \int_1^2 x \, \frac{ds}{dx} \, dx$$

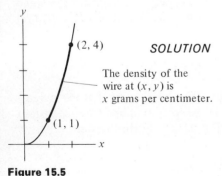

Figure 15.5

$$= \int_1^2 x\sqrt{1 + \left(\frac{dy}{dx}\right)^2}\, dx$$

$$= \int_1^2 x\sqrt{1 + (2x)^2}\, dx$$

$$= \int_1^2 x\sqrt{1 + 4x^2}\, dx$$

$$= \tfrac{1}{12}(1 + 4x^2)^{3/2}\Big|_1^2$$

$$= \tfrac{1}{12}(17^{3/2} - 5^{3/2}). \blacksquare$$

As Example 1 illustrates, for computational purposes a line integral is usually transformed to an integral over an interval. If the curve is parametrized by $x = g(t)$ and $y = h(t)$, t in $[a, b]$, then

$$\int_C f(P)\, ds = \int_a^b f(g(t), h(t))\, \frac{ds}{dt}\, dt.$$

In Example 1 the parameter was x; the curve was parametrized as $y = x^2$, $x = x$. For a curve C in space one would have

$$\int_C f(P)\, ds = \int_a^b f(x, y, z)\, \frac{ds}{dt}\, dt.$$

The remainder of this section describes two line integrals of importance in physics, thermodynamics, and in the theory of this chapter. The first concerns the work accomplished by a varying force in pushing a particle along a curve C. The second concerns fluid flow.

THE LINE INTEGRAL AND WORK

Consider a particle pushed along the curve C (in the plane or in space) by a varying force **F** (which may be a mixture of gravity, wind, friction, ropes, etc.). How much work is done? (See Fig. 15.6.)

Figure 15.6

An object moves from A to B on curve C.

In case the force **F** remains constant in magnitude and direction and pushes a particle in a straight path from the tail to the head of a vector **r**, the work is defined (in Sec. 11.3) as the dot product

$$\mathbf{F} \cdot \mathbf{r}.$$

This is the product of the scalar component of **F** in the direction of **r** and the distance the particle moves.

956 GREEN'S THEOREM, THE DIVERGENCE THEOREM, AND STOKES' THEOREM

To find the work accomplished by the varying force **F** pushing a particle along a curve, a line integral is required. First, break up the path into short sections of length Δs_i. Select a point P_i in the ith section, evaluate the force **F** at P_i, and find the unit tangent vector, $\mathbf{T}(P_i)$, at P_i. (See Fig. 15.7.)

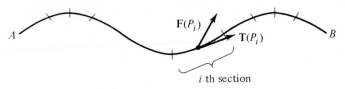

Figure 15.7

T *points in the direction the curve is traversed.*

Then a reasonable estimate of the work required to push the particle over the ith section is the product

$$\underbrace{\mathbf{F}(P_i) \cdot \mathbf{T}(P_i)}_{\substack{\text{Scalar component} \\ \text{of } \mathbf{F} \text{ in direction} \\ \text{of motion}}} \underbrace{\Delta s_i}_{\substack{\text{Distance the} \\ \text{particle is} \\ \text{pushed}}}.$$

Thus, when all Δs_i are small, the sum

$$\sum_{i=1}^{n} \mathbf{F}(P_i) \cdot \mathbf{T}(P_i) \, \Delta s_i$$

is an estimate of the total work. Taking limits, as all Δs_i are chosen smaller, shows that the line integral

$$\int_C \mathbf{F}(P) \cdot \mathbf{T}(P) \, ds$$

represents the total work. (Note that $\mathbf{F}(P) \cdot \mathbf{T}(P)$ is a scalar function $f(P)$, defined on the curve C.)

In short,
$$\int_C \mathbf{F} \cdot \mathbf{T} \, ds = \text{Work}.$$

The effect of sweeping out the curve in the opposite direction

If the path is parametrized in the opposite direction, then the tangent vector **T** is replaced by $-\mathbf{T}$, for this vector records the direction of motion along the curve. The integrand therefore is the negative of the original integrand since $\mathbf{F} \cdot (-\mathbf{T}) = -\mathbf{F} \cdot \mathbf{T}$. This means that $\int_C \mathbf{F} \cdot \mathbf{T} \, ds$ depends on the direction in which the curve is parametrized. Reversing the direction changes the sign of the line integral. (See Exercises 7 and 8 for an example.) However, two parametrizations of the same curve with the same orientation will lead to the same value for $\int_C \mathbf{F} \cdot \mathbf{T} \, ds$.

If $\mathbf{F} = P\mathbf{i} + Q\mathbf{j} + R\mathbf{k}$, then the integral $\int_C \mathbf{F} \cdot \mathbf{T} \, ds$ can be written

Recall that $\dfrac{dx}{ds}\mathbf{i} + \dfrac{dy}{ds}\mathbf{j} + \dfrac{dz}{ds}\mathbf{k}$ *is a unit tangent vector.*

$$\int_C (P\mathbf{i} + Q\mathbf{j} + R\mathbf{k}) \cdot \left(\frac{dx}{ds}\mathbf{i} + \frac{dy}{ds}\mathbf{j} + \frac{dz}{ds}\mathbf{k}\right) ds = \int_C \left(P \frac{dx}{ds} + Q \frac{dy}{ds} + R \frac{dz}{ds}\right) ds.$$

Hence we write $\int_C \mathbf{F} \cdot \mathbf{T}\, ds = \int_C (P\, dx + Q\, dy + R\, dz)$,

which is a useful form for computation. A similar formula,

$$\int_C (P\, dx + Q\, dy),$$

holds in the case of a vector field $\mathbf{F} = P\mathbf{i} + Q\mathbf{j}$ and curve C in the plane.

There is another common notation for the integral $\int_C \mathbf{F} \cdot \mathbf{T}\, ds$ or $\int_C (P\, dx + Q\, dy + R\, dz)$. Let $\mathbf{r}(t)$ be the position vector on the curve C. Then $d\mathbf{r} = dx\,\mathbf{i} + dy\,\mathbf{j} + dz\,\mathbf{k}$. Thus $\int_C (P\, dx + Q\, dy + R\, dz)$ may be denoted $\int_C \mathbf{F} \cdot d\mathbf{r}$.

The sign of the line integral $\int_C (P\, dx + Q\, dy + R\, dz)$ depends on the direction in which the path is traversed. To see this, recall that $dx = (dx/dt)\, dt$. If in one parametrization x is increasing (and dx/dt is positive), then in the reverse direction x is a decreasing function (and dx/dt is negative).

EXAMPLE 2 Let $\mathbf{F} = xy\mathbf{i} + x\mathbf{j}$. Let $A = (1, 1)$ and $B = (2, 4)$. Compute $\int_C \mathbf{F} \cdot \mathbf{T}\, ds$ on each of these two curves that start at A and end at B:
(a) the straight line from A to B,
(b) the parabola $y = x^2$.
(See Fig. 15.8.)

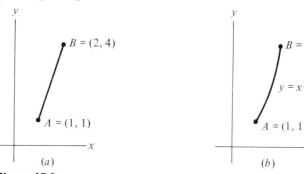

Figure 15.8

SOLUTION
$$\mathbf{F} \cdot \mathbf{T} = (xy\mathbf{i} + x\mathbf{j}) \cdot \left(\frac{dx}{ds}\mathbf{i} + \frac{dy}{ds}\mathbf{j}\right)$$

$$= xy\,\frac{dx}{ds} + x\,\frac{dy}{ds}.$$

Thus $\int_C \mathbf{F} \cdot \mathbf{T}\, ds = \int_C \left(xy\,\frac{dx}{ds} + x\,\frac{dy}{ds}\right) ds,$

which we abbreviate to

$$\int_C \mathbf{F} \cdot \mathbf{T}\, ds = \int_C (xy\, dx + x\, dy). \tag{1}$$

(a) To evaluate (1) over the straight path, parametrize the path using x as a parameter: $x = x$, $y = 3x - 2$, with x going from 1 to 2. Then (1) equals

$$\int_C [x(3x-2)\, dx + x\, d(3x-2)] = \int_C [(3x^2 - 2x)\, dx + 3x\, dx]$$

$$= \int_1^2 (3x^2 + x)\, dx \quad \text{(collecting)}$$

$$= \tfrac{17}{2}.$$

(b) To evaluate (1) over the path $y = x^2$, again use x as parameter. Then (1) equals

$$\int_C [xx^2\, dx + x\, d(x^2)] = \int_C (x^3 + 2x^2)\, dx$$

$$= \int_1^2 (x^3 + 2x^2)\, dx$$

$$= \tfrac{101}{12}. \quad \blacksquare$$

THE LINE INTEGRAL AND FLUID FLOW

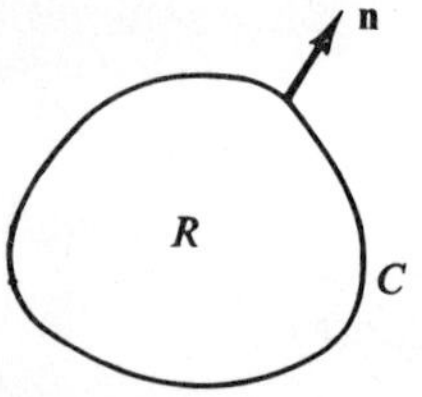

The curve C bounds the region R.

Figure 15.9

Let C be a curve enclosing some region R in the plane. At each point P on C let $\mathbf{n}$ be the unit vector perpendicular to the curve and pointing away from R. The vector $\mathbf{n}$ is called the *exterior normal*. This vector $\mathbf{n}$ should not be confused with the principal normal $\mathbf{N}$ defined in Sec. 14.6. (See Fig. 15.9.)

The curve C is given parametrically by the vector function $\mathbf{G}(t)$ for t in $[a, b]$. Since the curve forms a closed loop, its start or initial point $\mathbf{G}(a)$ coincides with its end or terminal point $\mathbf{G}(b)$,

$$\mathbf{G}(a) = \mathbf{G}(b).$$

Closed curves

C is called a *closed curve*. An integral over a closed curve, $\int_C f(P)\, ds$, is often denoted $\oint_C f(P)\, ds$.

Imagine that a closed curve C is placed on top of a stream and kept fixed. Let $\mathbf{v}$ be the velocity vector of the fluid at each point on the surface. Let f be the density of the liquid. Let $\mathbf{F} = f\mathbf{v}$. This vector function records the product of the density of the fluid and the velocity vector and its direction. Note that $\mathbf{F}$ and $\mathbf{v}$ point in the same direction. Both the density f and the velocity $\mathbf{v}$ may vary with the point in the stream and with time.

We raise the question: At what rate is fluid escaping or entering the region R whose boundary is C?

If the fluid tends to escape, then it is thinning out in R, becoming less dense at some points. If the fluid tends to accumulate, it is becoming denser at some points. Think of this ideal fluid as resembling a gas rather than water; it can vary in density.

Since the fluid is escaping or entering R only along its boundary, it suffices to consider the total loss or gain past C. Where $\mathbf{v}$ is tangent to C, fluid neither enters nor leaves. Where $\mathbf{v}$ is not tangent to C, fluid is either entering or leaving across C, as indicated in Fig. 15.10.

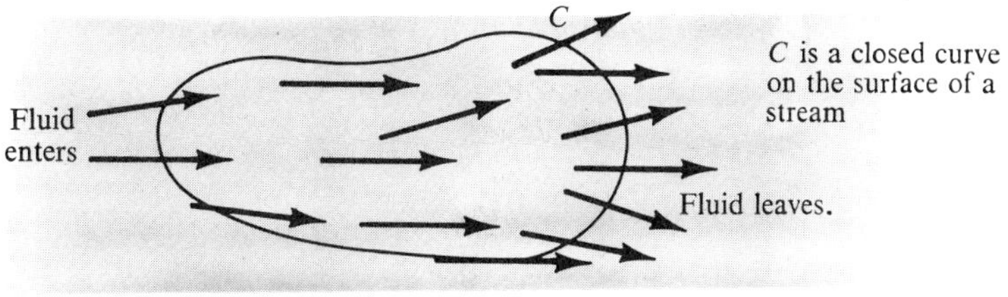

15.2 LINE INTEGRALS

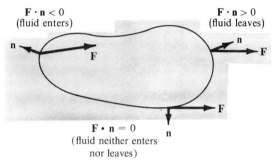

Fluid enters

C is a closed curve on the surface of a stream

Fluid leaves.

Fluid neither enters nor leaves.

Figure 15.10

The critical component of **F** is that along **n**. Indeed, **F** · **n** is a measure of how much fluid is leaving or entering near a certain point on the boundary, as shown in Fig. 15.11.

F · **n** < 0 (fluid enters)

F · **n** > 0 (fluid leaves)

F · **n** = 0 (fluid neither enters nor leaves)

Figure 15.11

Let us now compute the total net loss of fluid across C. The amount of fluid lost crossing a short (nearly straight) section of the curve, of length Δs, is presumably proportional to Δs and is approximately

$$(\mathbf{F} \cdot \mathbf{n}) \, \Delta s$$

where **F** · **n** is evaluated at some point in the section.

Hence the line integral

$$\oint_C \mathbf{F} \cdot \mathbf{n} \, ds$$

represents the rate at which fluid is leaving the region R. If this integral is positive, the amount of fluid in R tends to decrease; if it is negative, the amount of fluid in R tends to increase. We shall be interested in $\int_C \mathbf{F} \cdot \mathbf{n} \, ds$ only for closed curves C.

The line integral of **F** · **n**

For any vector field **F** the integral $\oint_C \mathbf{F} \cdot \mathbf{n} \, ds$ is called the *line integral of the normal component* of **F** around C.

Do not confuse P with "point P."

In order to compute $\oint_C \mathbf{F} \cdot \mathbf{n} \, ds$ where $\mathbf{F} = P\mathbf{i} + Q\mathbf{j}$ (P and Q being functions of x and y), it will be useful to have a formula for **n**, the exterior normal. *Assume that the curve C is swept out counterclockwise.*

If arc length s is used as a parameter, then the vector

$$\mathbf{T} = \frac{dx}{ds}\mathbf{i} + \frac{dy}{ds}\mathbf{j}$$

GREEN'S THEOREM, THE DIVERGENCE THEOREM, AND STOKES' THEOREM

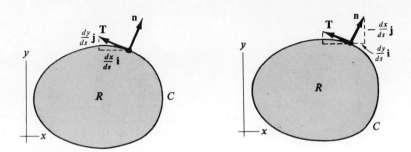

Figure 15.12   **Figure 15.13**

is tangent to the curve, has length 1, and points in the direction in which the curve is swept out. A typical **T** and **n** are shown in Fig. 15.12. As Fig. 15.13 shows, the exterior normal **n** has its x component equal to the y component of **T** and its y component equal to the negative of the x component of **T**.

$\mathbf{n} = \dfrac{dy}{ds}\mathbf{i} - \dfrac{dx}{ds}\mathbf{j}$ *for counterclockwise curves.*

Thus
$$\mathbf{n} = \frac{dy}{ds}\mathbf{i} - \frac{dx}{ds}\mathbf{j}.$$

Consequently, if $\mathbf{F} = P\mathbf{i} + Q\mathbf{j}$, then

$$\oint_C \mathbf{F} \cdot \mathbf{n}\, ds = \oint_C (P\mathbf{i} + Q\mathbf{j}) \cdot \left(\frac{dy}{ds}\mathbf{i} - \frac{dx}{ds}\mathbf{j}\right) ds$$

$$= \oint_C \left[P\left(\frac{dy}{ds}\right) + Q\left(-\frac{dx}{ds}\right)\right] ds$$

$$= \oint_C (P\, dy - Q\, dx)$$

$$= \oint_C (-Q\, dx + P\, dy).$$

To compute $\oint_C (-Q\, dx + P\, dy)$, parametrize the curve in a counterclockwise direction, $\mathbf{r} = \mathbf{G}(t)$, and then express P, Q, dx, and dy in terms of t and dt. The parameter need not be arc length. The next example illustrates how $\oint_C \mathbf{F} \cdot \mathbf{n}\, ds$ can be computed.

EXAMPLE 3 Compute
$$\oint_C \mathbf{F} \cdot \mathbf{n}\, ds,$$

where $\mathbf{F} = P\mathbf{i} + Q\mathbf{j} = x\mathbf{i} + 2y\mathbf{j}$ and the curve C encloses the ellipse

$$\frac{x^2}{2^2} + \frac{y^2}{3^2} = 1.$$

SOLUTION First parametrize the ellipse counterclockwise, as in Exercise 4 of Sec. 9.3:

$$x = 2\cos t, \qquad y = 3\sin t;$$

The ellipse $\dfrac{x^2}{a^2} + \dfrac{y^2}{b^2} = 1$ can be parametrized $x = a \cos t$, $y = b \sin t$, $0 \le t \le 2\pi$.

that is, $\qquad \mathbf{G}(t) = 2 \cos t\, \mathbf{i} + 3 \sin t\, \mathbf{j}, \qquad t \text{ in } [0, 2\pi].$

In this case $P = x$ and $Q = 2y$, and

$$\oint_C \mathbf{F} \cdot \mathbf{n}\, ds = \oint_C (P\, dy - Q\, dx)$$

$$= \oint_C (x\, dy - 2y\, dx)$$

$$= \int_0^{2\pi} (2 \cos t)(3 \cos t\, dt) - 2(3 \sin t)(-2 \sin t\, dt)$$

Check that as t goes from 0 to 2π the ellipse is swept out counterclockwise.

$$= \int_0^{2\pi} (6 \cos^2 t + 12 \sin^2 t)\, dt$$

$$= 18\pi. \quad \blacksquare$$

GENERAL REMARKS

The line integral $\int_C \mathbf{F} \cdot \mathbf{T}\, ds$ is called the "integral of the tangential component of $\mathbf{F}$ over C" or, more tersely, "the integral of $\mathbf{F}$ over C." If $\mathbf{F}$ is defined in the plane, $\mathbf{F} = P\mathbf{i} + Q\mathbf{j}$, and C is a curve in the plane, then

$$\int_C \mathbf{F} \cdot \mathbf{T}\, ds = \int_C (P\, dx + Q\, dy). \tag{2}$$

If C is parametrized by $x = g(t)$, $y = h(t)$, $a \le t \le b$, then

$$\int_C (P\, dx + Q\, dy) = \int_a^b \left(P\, \frac{dx}{dt} + Q\, \frac{dy}{dt} \right) dt,$$

the usual definite integral over an interval.

Similarly, in space, if $\mathbf{F} = P\mathbf{i} + Q\mathbf{j} + R\mathbf{k}$, then

$$\int_C \mathbf{F} \cdot \mathbf{T}\, ds = \int_C (P\, dx + Q\, dy + R\, dz),$$

which, like (2), is evaluated by expressing the line integral as a definite integral over an interval determined by the parameter, which may be x, t, θ, etc.

EXAMPLE 4 Evaluate $\int_C (x^2 z\, dx + xy\, dy + x\, dz)$ over each of these two curves:
(a) The straight path, C_1, from $(1, 0, 0)$ to $(1, 0, 3)$;
(b) The helical path, C_2, from $(1, 0, 0)$ to $(1, 0, 3)$,

$$x = \cos 2\pi t, \qquad y = \sin 2\pi t, \qquad z = 3t.$$

The paths are shown in Fig. 15.14.

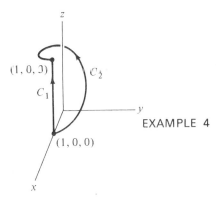

Figure 15.14

SOLUTION (a) Parametrize the straight path C_1 by $x = 1$, $y = 0$, $z = 3t$, $0 \le t \le 1$. Thus

$$\int_{C_1} (x^2 z \, dx + xy \, dy + x \, dz) = \int_0^1 (1^2 \cdot 3t \cdot d(1) + 1 \cdot 0 \cdot d(0) + 1 \cdot d(3t))$$

$$= \int_0^1 3 \, dt = 3.$$

(b) $$\int_{C_2} (x^2 z \, dx + xy \, dy + x \, dz)$$

$$= \int_0^1 [(\cos 2\pi t)^2 \cdot 3t \cdot d(\cos 2\pi t) + \cos 2\pi t \sin 2\pi t \, d(\sin 2\pi t) + \cos 2\pi t \, d(3t)]$$

$$= \int_0^1 [-6\pi t \cos^2 2\pi t \sin 2\pi t + 2\pi \cos^2 2\pi t \sin 2\pi t + 3 \cos 2\pi t] \, dt$$

$$= 1.$$

(To integrate $t \cos^2 2\pi t \sin 2\pi t$, use integration by parts with $u = t$ and $dv = \cos^2 2\pi t \sin 2\pi t$.)

Even though C_1 and C_2 have the same initial point and terminal point, the line integrals in (a) and (b) are different. ∎

SUMMARY OF SECTION

The section began with the notion of the line integral of a scalar function, $\int_C f(P) \, ds$. It then examined two particular scalar functions, one involving the work accomplished by a force, the other involving the net loss of fluid past a closed curve. This table summarizes the main ideas of the section.

Integrand	Curve	Line integral	Line integral expressed in components
$\mathbf{F} \cdot \mathbf{T}$	Any curve C in the plane or in space. C is assumed to be composed of a finite number of curves over each of which $\mathbf{T}$ varies continuously.	$\int_C \mathbf{F} \cdot \mathbf{T} \, ds$ is the work accomplished by the force $\mathbf{F}$ along the path C.	If $\mathbf{F} = P\mathbf{i} + Q\mathbf{j} + R\mathbf{k}$, $\int_C \mathbf{F} \cdot \mathbf{T} \, ds = \int_C (P \, dx + Q \, dy + R \, dz)$
$\mathbf{F} \cdot \mathbf{n}$	Any *closed* curve in the plane. The normal $\mathbf{n}$ is the *exterior* normal.	$\oint_C \mathbf{F} \cdot \mathbf{n} \, ds$ is the net loss of fluid across C, where $\mathbf{F}$ denotes fluid flow.	If $\mathbf{F} = P\mathbf{i} + Q\mathbf{j}$, and the curve C is oriented counterclockwise, then $\mathbf{n} = \dfrac{dy}{ds}\mathbf{i} - \dfrac{dx}{ds}\mathbf{j}$ and $\oint_C \mathbf{F} \cdot \mathbf{n} \, ds = \oint_C (-Q \, dx + P \, dy)$.

Exercises

In each of Exercises 1 and 2 find the mass of the wire.

1. Density at (x, y) is y; the wire occupies the portion of the cubic $y = x^3$ between $(1, 1)$ and $(2, 8)$.
2. Density at (x, y) is x; the wire occupies the portion of the parabola $y = x^2/2$ from $(0, 0)$ to $(1, \tfrac{1}{2})$.
3. A fence perpendicular to the xy plane has as its base the portion of the cardioid $r = 1 + \cos\theta$ in the first two quadrants. The height of the fence at the point (r, θ) is $\sin\theta$. Find the area of one side of the fence.
4. A wire occupies the part of the spiral $r = e^\theta$ corresponding to θ in $[0, 2\pi]$. At the point (r, θ) the temperature is r. Find the average temperature in the wire.

In Exercises 5 and 6 compute the work accomplished by the force $\mathbf{F} = x^2 y \mathbf{i} + y \mathbf{j}$ along the given curve.

5. From $(0, 0)$ to $(2, 4)$ along the parabola $y = x^2$.
6. From $(0, 0)$ to $(2, 4)$ along the line $y = 2x$.
7. Let C_1 be the curve parameterized by $\mathbf{G}_1(t) = (t - 1)\mathbf{i} + (t + 1)\mathbf{j}$ for $0 \le t \le 2$; let C_2 be parameterized by $\mathbf{G}_2(t) = (1 - t)\mathbf{i} + (3 - t)\mathbf{j}$ for $0 \le t \le 2$; let C_3 be parameterized by $\mathbf{G}_3(t) = -\cos t \mathbf{i} + (2 - \cos t)\mathbf{j}$, for $0 \le t \le \pi/2$. Evaluate $\int_{C_1} xy\, ds$, $\int_{C_2} xy\, ds$, and $\int_{C_3} xy\, ds$. Are the answers the same? Why or why not?
8. Let C_1, C_2, and C_3 be the curves given in Exercise 7 and let $\mathbf{F}(x, y) = x\mathbf{i} + y\mathbf{j}$. Evaluate $\int_{C_1} \mathbf{F} \cdot \mathbf{T}\, ds$, $\int_{C_2} \mathbf{F} \cdot \mathbf{T}\, ds$, and $\int_{C_3} \mathbf{F} \cdot \mathbf{T}\, ds$. Are the answers the same? Why or why not?

In each of Exercises 9 to 12 evaluate $\int_C \mathbf{F} \cdot \mathbf{T}\, ds$ for the given vector field $\mathbf{F}$ and curve C with the given parametrization $\mathbf{G}$.

9. $\mathbf{F}(x, y) = 2x\mathbf{i}$ and the curve is the circle, $\mathbf{G}(\theta) = 3\cos\theta\, \mathbf{i} + 3\sin\theta\, \mathbf{j}$, θ in $[0, 2\pi]$.
10. $\mathbf{F}(x, y) = x\mathbf{i} + y\mathbf{j}$ and the curve is the ellipse, $\mathbf{G}(t) = a\cos t\, \mathbf{i} + b\sin t\, \mathbf{j}$, t in $[0, 2\pi]$.
11. $\mathbf{F}(x, y, z) = y\mathbf{i} + z y\mathbf{k}$ and the curve is the helix, $\mathbf{G}(t) = \cos t\, \mathbf{i} + \sin t\, \mathbf{j} + 3t\mathbf{k}$, $0 \le t \le 2\pi$.
12. $\mathbf{F}(x, y, z) = y\mathbf{i} + z y\mathbf{k}$ and the curve is the straight line, $\mathbf{G}(t) = 6\pi t\mathbf{k}$, $0 \le t \le 1$.

In each of Exercises 13 to 15 compute $\oint_C \mathbf{F} \cdot \mathbf{n}\, ds$ for the given vector field $\mathbf{F}$ and closed curve C. (The vector $\mathbf{n}$ is the *exterior* normal.)

13. $\mathbf{F}(x, y) = x^2\mathbf{i} + y^2\mathbf{j}$; C is the circle $x^2 + y^2 = 49$.
14. $\mathbf{F}(x, y) = x\mathbf{i} + y\mathbf{j}$; C is the circle $x^2 + y^2 = a^2$.
15. $\mathbf{F}(x, y) = x\mathbf{i} + y^2\mathbf{j}$; C consists of three parts, the parabola $y = x^2$ from $(0, 0)$ to $(1, 1)$, the line $y = 1$ from $(1, 1)$ to $(0, 1)$, the line $x = 0$ from $(0, 1)$ back to $(0, 0)$.

■

16. Compute:
$$\int_C \left(\frac{-y\, dx}{x^2 + y^2} + \frac{x\, dy}{x^2 + y^2} \right),$$

where C goes from $(1, 0)$ to $(1, 1)$ along (a) the straight line $x = 1$, parametrized as $x = 1$, $y = t$; (b) the circular path parametrized as $x = \cos 2\pi t$, $y = \sin 2\pi t$, t in $[0, \tfrac{1}{4}]$ and then followed by the path $x = t$, $y = 1$, t in $[0, 1]$.

17. Evaluate $\int_C (e^x y\, dx + x \sin \pi y\, dy + xy \tan^{-1} z\, dz)$, where C is composed of the line segment $(0, 0, 0)$ to $(1, 1, 0)$ and the line segment from $(1, 1, 0)$ to $(1, 1, 1)$.
18. Evaluate
$$\int_C \left(\frac{x\, dx}{yz} + \frac{y\, dy}{xz} + \frac{8z\, dz}{xy} \right)$$

where C is parameterized by $\mathbf{G}(t) = \cos t\, \mathbf{i} + \sin t\, \mathbf{j} + \cos t\, \mathbf{k}$ for $\pi/6 \le t \le \pi/3$.

19. (a) Let $x = f(t)$, $y = g(t)$ be any parametrization of a curve from $(0, 0)$ to $(1, 2)$. Show that $\oint_C (y\, dx + x\, dy)$ equals 2.
    (b) Show that $\int_C (y\, dx + x\, dy)$ depends only on the endpoints of any curve C.
20. Compute $\oint_C (-y\, dx + x\, dy)$, where C is the ellipse
$$\begin{cases} x = a\cos t, \\ y = b\sin t \end{cases} \quad 0 \le t \le 2\pi.$$

21. Let the vector field describing a fluid flow have at the point (x, y) the value $(x + 1)^2\mathbf{i} + y\mathbf{j}$. Let C be the unit circle described parametrically as $x = \cos t$, $y = \sin t$, for t in $[0, 2\pi]$.
    (a) Draw $\mathbf{F}$ at eight convenient equally spaced points on the circle.
    (b) Is fluid tending to leave or enter the region bounded by C; that is, is the net outward flow positive or negative? (Answer on the basis of your diagram in (a).)
    (c) Compute the net outward flow with the aid of a line integral.
22. Compute $\int_C (xy\, dx + x^2\, dy)$ if C goes from $(0, 0)$ to $(1, 1)$ on
    (a) the line $y = x$, parametrized as $x = t$, $y = t$;
    (b) the line $y = x$, parametrized as $x = t^2$, $y = t^2$;
    (c) the parabola $y = x^2$, parametrized as $y = t^2$, $x = t$;
    (d) the polygonal path from $(0, 0)$ to $(0, 1)$ to $(1, 1)$, parametrized conveniently.

■ ■

23. Verify that the integral $\oint_C (-y\, dx + x\, dy)$, where C is swept out counterclockwise, is twice the area of the region enclosed by C, when C is
    (a) the square path from $(a, 0)$ to $(0, a)$ to $(-a, 0)$ to $(0, -a)$ and back to $(a, 0)$;
    (b) the triangular path from $(0, 0)$ to $(a, 0)$ to $(0, b)$ and back to $(0, 0)$. Assume that a and b are positive.

24. Compute $\oint_C (-y\, dx + x\, dy)$, where C sweeps out the boundary of the rectangle whose vertices are $(0, 0)$, $(a, 0)$, (a, b), $(0, b)$. Assume that a and b are positive.

25. Let $\mathbf{r} = x\mathbf{i} + y\mathbf{j}$.
    (a) Compute $\oint \mathbf{r} \cdot \mathbf{n}\, ds$ around the triangular curve going from $(0, 0)$ to $(a, 0)$, then to $(0, b)$, and then back to $(0, 0)$. (Here a and b are positive numbers.)
    (b) Drawing a picture of the "local estimate" $\mathbf{r} \cdot \mathbf{n}\, ds$, suggest why $\frac{1}{2} \oint \mathbf{r} \cdot \mathbf{n}\, ds$ should be the area of the region enclosed by the curve.
    (c) Use (b) to explain why $\frac{1}{2} \oint_C (-y\, dx + x\, dy)$, where C is swept out counterclockwise, equals the area within C.

26. Let $\mathbf{r} = \mathbf{G}(t)$ describe a curve C in the plane or in space. What is the geometric interpretation of $\frac{1}{2} \int_C |\mathbf{r} \times \mathbf{T}| ds$?

27. Let $\mathbf{F}$ be a vector field describing fluid flow. Why do you think $\oint_C \mathbf{F} \cdot \mathbf{T}\, ds$ is called the *circulation* around C?

28. The gravitational force $\mathbf{F}$ of the earth, located at the origin $(0, 0)$ of a rectangular coordinate system, on a certain particle at the point (x, y) is

$$\frac{-x\mathbf{i}}{(\sqrt{x^2 + y^2})^3} + \frac{-y\mathbf{j}}{(\sqrt{x^2 + y^2})^3}.$$

Compute the total work done by $\mathbf{F}$ if the particle goes from $(2, 0)$ to $(0, 1)$ along (a) the portion of the ellipse $x = 2 \cos t$, $y = \sin t$ in the first quadrant; (b) the line $x = 2 - 2t$, $y = t$.

15.3 Conservative vector fields

For some vector fields $\mathbf{F}$ the line integral $\int_C \mathbf{F} \cdot \mathbf{T}\, ds$ depends only on the endpoints of C. Such fields, which are important in physics and thermodynamics, are the subjects of this section.

CONSERVATIVE VECTOR FIELDS

DEFINITION
Conservative vector field

A vector field $\mathbf{F}$ defined in some region is called *conservative* if, whenever C_1 and C_2 are two curves in the region with the same initial points and terminal points, then

$$\int_{C_1} \mathbf{F} \cdot \mathbf{T}\, ds = \int_{C_2} \mathbf{F} \cdot \mathbf{T}\, ds.$$

Another way to view "conservative"

An equivalent definition of a conservative vector field $\mathbf{F}$ is that for any closed curve C in the region, $\oint_C \mathbf{F} \cdot \mathbf{T}\, ds = 0$. This alternative viewpoint is justified by the following theorem.

THEOREM 1 A vector field $\mathbf{F}$ is conservative if and only if $\oint_C \mathbf{F} \cdot \mathbf{T}\, ds = 0$ for every closed curve in the region where $\mathbf{F}$ is defined.

PROOF Assume that $\mathbf{F}$ is conservative and let C be a closed curve that starts and ends at the point A. Pick a point B on the curve and break C into two curves: C_1 from A to B and C_2^* from B to A, as indicated in Fig. 15.15.

Let C_2 be the curve C_2^* traversed in the opposite direction, from A to B. Then

Note the sign change.

$$\oint_C \mathbf{F} \cdot \mathbf{T}\, ds = \int_{C_1} \mathbf{F} \cdot \mathbf{T}\, ds + \int_{C_2^*} \mathbf{F} \cdot \mathbf{T}\, ds$$

$$= \int_{C_1} \mathbf{F} \cdot \mathbf{T}\, ds - \int_{C_2} \mathbf{F} \cdot \mathbf{T}\, ds$$

$$= 0,$$

since $\mathbf{F}$ is conservative.

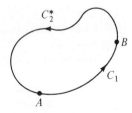

Figure 15.15

On the other hand, assume that $\mathbf{F}$ has the property that $\oint_C \mathbf{F} \cdot \mathbf{T}\, ds = 0$ for any closed curve C in the region. Let C_1 and C_2 be two curves in the

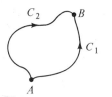

Figure 15.16 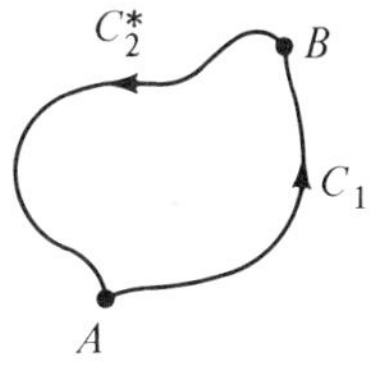   Figure 15.17

region, starting at A and ending at B. Let C_2^* be C_2 taken in the reverse direction. (See Figs. 15.16 and 15.17.) Then C_1 followed by C_2^* is a closed curve C from A back to A. Thus

$$0 = \oint_C \mathbf{F} \cdot \mathbf{T}\, ds = \int_{C_1} \mathbf{F} \cdot \mathbf{T}\, ds + \int_{C_2^*} \mathbf{F} \cdot \mathbf{T}\, ds$$

$$= \int_{C_1} \mathbf{F} \cdot \mathbf{T}\, ds - \int_{C_2} \mathbf{F} \cdot \mathbf{T}\, ds.$$

Consequently, $\quad\displaystyle\int_{C_1} \mathbf{F} \cdot \mathbf{T}\, ds = \int_{C_2} \mathbf{F} \cdot \mathbf{T}\, ds.$

This concludes both directions of the argument. ∎

GRADIENT FIELDS ARE CONSERVATIVE

The criteria mentioned so far for determining whether a vector field $\mathbf{F}$ is conservative are hardly practical. They require the evaluation of an infinite number of integrals. The next theorem provides a way of constructing conservative vector fields. Before discussing it we mention two conditions that will be assumed about the domains of vector fields of interest.

First, let A and B be any two points in the region where the vector field $\mathbf{F}$ is defined. It will be assumed that there is a curve C lying completely in the region that joins A to B. Such a region is called *arcwise connected*. A disk is arcwise connected, but the region shaded in Fig. 15.18 is not.

An arcwise connected region comes in a single piece.

The shaded region is not arcwise connected.

Figure 15.18

Second, it is assumed that for any point P in the region there is a disk with nonzero radius and center at P that lies entirely within the region (if the region is planar). Such a region is called *open*. If the region is spatial, it is assumed that for each point P in the region there is a ball with nonzero radius and center at P that lies entirely within the region. This condition assures us that partial derivatives can be defined at each point in the region. If f is a scalar function defined in the region and $P = (x, y)$ is in the region, then for Δx suitably small $(x + \Delta x, y)$ is also in the region and the difference quotient

$$\frac{f(x + \Delta x, y) - f(x, y)}{\Delta x}$$

used in defining the partial derivative $\partial f/\partial x$ at P makes sense.

THEOREM 2 Let f be a scalar function defined in some region in the plane or in space. Then the gradient field, $\mathbf{F} = \nabla f$ is conservative. In fact, for any points A and B in the region and any curve C from A to B in the region

$$\int_C \nabla f \cdot \mathbf{T} \, ds = f(B) - f(A).$$

PROOF For simplicity take the planar case. Let C be given by the parametrization $\mathbf{r} = \mathbf{G}(t)$ for t in $[a, b]$.

Let $\mathbf{G}(t) = x(t)\mathbf{i} + y(t)\mathbf{j}$. Then

$$\int_C \nabla f \cdot \mathbf{T} \, ds = \int_C (f_x \, dx + f_y \, dy)$$

$$= \int_a^b \left(f_x \frac{dx}{dt} + f_y \frac{dy}{dt} \right) dt.$$

The integrand

$$f_x \frac{dx}{dt} + f_y \frac{dy}{dt}$$

is reminiscent of the first chain rule in Sec. 12.5. To be specific, introduce the function H defined by the formula

$$H(t) = f(x(t), y(t)).$$

Then, by Theorem 1 of Sec. 12.5,

$$\frac{dH}{dt} = f_x(x(t), y(t)) \frac{dx}{dt} + f_y(x(t), y(t)) \frac{dy}{dt}.$$

Thus

$$\int_a^b \left(f_x \frac{dx}{dt} + f_y \frac{dy}{dt} \right) dt = \int_a^b \frac{dH}{dt} dt = H(b) - H(a)$$

by the fundamental theorem of calculus.

But $\qquad H(b) = f(x(b), y(b)) = f(B)$

and similarly $\qquad H(a) = f(A).$

Consequently $\qquad \int_C \nabla f \cdot \mathbf{T} \, ds = f(B) - f(A),$

and the theorem is proved. ∎

Theorem 2 resembles the fundamental theorem of calculus,

$$\int_a^b \frac{df}{dx} dx = f(b) - f(a),$$

and extends it from closed intervals on a straight line to curves.

When the curve C is closed, Theorem 2 takes the following form.

THEOREM 3 If f is a scalar function defined throughout some region, then the integral of the tangential component of the vector field ∇f around any closed curve in that region is 0,

$$\oint \nabla f \cdot \mathbf{T} \, ds = 0.$$

PROOF This is a consequence of Theorem 2. For in the case of a closed path, the initial point A coincides with the terminal point B. Hence $f(B) - f(A) = 0$. This proves the theorem. ∎

In the theory of gravitational attraction, the scalar function f defined by

$$f(x, y, z) = \frac{1}{\sqrt{x^2 + y^2 + z^2}}$$

is of great importance. It is defined everywhere except at the origin and is called a *potential function*. The gradient ∇f is called the *force field*. A direct calculation shows that

$$\nabla f = -\frac{x\mathbf{i} + y\mathbf{j} + z\mathbf{k}}{(\sqrt{x^2 + y^2 + z^2})^3}$$

$$= -\frac{\mathbf{r}}{|\mathbf{r}|^3},$$

where $\mathbf{r}$ is the position vector $x\mathbf{i} + y\mathbf{j} + z\mathbf{k}$. Since

$$\nabla f = -\frac{\mathbf{r}}{|\mathbf{r}|^3},$$

∇f is pointed toward the origin, and its magnitude is inversely proportional to $|\mathbf{r}|^2$.

Theorem 2 tells the physicist that the work accomplished by the gravitational field when moving a particle from one point to another is independent of the path. Theorem 3 says that the work accomplished by gravity on a satellite during one orbit is 0. (It is this fact that permits the satellite to remain in orbit indefinitely.)

A CONSERVATIVE FIELD IS A GRADIENT

The question may come to mind, "If $\mathbf{F}$ is conservative, is it necessarily the gradient of some scalar function?" The answer is "yes." This is the substance of the next theorem.

THEOREM 4 Let $\mathbf{F}$ be a conservative vector field defined in some arcwise-connected region in the plane (or in space). Then there is a scalar function f defined in that region such that $\mathbf{F} = \nabla f$.

PROOF Consider the case when **F** is planar, $\mathbf{F} = P(x, y)\mathbf{i} + Q(x, y)\mathbf{j}$. (The case where **F** is defined in space is similar.) For convenience, assume that the region contains the origin. Define a scalar function f as follows.

Let (x, y) be a point in the region. Select a curve C in the region that starts at $(0, 0)$ and ends at (x, y). Define $f(x, y)$ to be $\int_C \mathbf{F} \cdot \mathbf{T}\, ds$. Since **F** is conservative, the number $f(x, y)$ depends only on the point (x, y) and not on the choice of C. (See Fig. 15.19.)

All that remains is to show that $\nabla f = \mathbf{F}$, that is, $f_x = P$ and $f_y = Q$. We will go through the details for the first case, $f_x = P$. The reasoning for the other is similar.

Let (x_0, y_0) be a fixed point in the region and consider the difference quotient whose limit is $f_x(x_0, y_0)$, namely,

$$\frac{f(x_0 + h, y_0) - f(x_0, y_0)}{h},$$

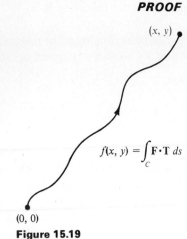

Figure 15.19

for h small enough so that $(x_0 + h, y_0)$ is also in the region.

Let C_1 be a curve from $(0, 0)$ to (x_0, y_0) and let C_2 be the straight path from (x_0, y_0) to $(x_0 + h, y_0)$. (See Fig. 15.20.) Let C be the curve from $(0, 0)$ to $(x_0 + h, y_0)$ formed by taking C_1 first and then continuing on C_2. Then

$$f(x_0, y_0) = \int_{C_1} \mathbf{F} \cdot \mathbf{T}\, ds$$

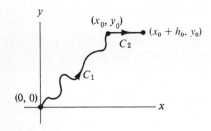

Figure 15.20

and $\quad f(x_0 + h, y_0) = \int_C \mathbf{F} \cdot \mathbf{T}\, ds = \int_{C_1} \mathbf{F} \cdot \mathbf{T}\, ds + \int_{C_2} \mathbf{F} \cdot \mathbf{T}\, ds.$

Thus

$$\frac{f(x_0 + h, y_0) - f(x_0, y_0)}{h} = \frac{\int_{C_2} \mathbf{F} \cdot \mathbf{T}\, ds}{h} = \frac{\int_{C_2} (P(x, y)\, dx + Q(x, y)\, dy)}{h}.$$

But on C_2, y is constant, $y = y_0$; hence $dy = 0$. Thus $\int_{C_2} Q(x, y)\, dy = 0$. Also,

$$\int_{C_2} P(x, y)\, dx = \int_{x_0}^{x_0 + h} P(x, y_0)\, dx.$$

Now, $\quad \lim_{h \to 0} \dfrac{\int_{x_0}^{x_0 + h} P(x, y_0)\, dx}{h}$

is the derivative of $\int_{x_0}^t P(x, y_0)\, dx$ with respect to t at the value x_0. By the second fundamental theorem of calculus, this derivative is the value of the integrand when $x = x_0$.

Consequently, $\quad f_x(x_0, y_0) = P(x_0, y_0),$

as was to be shown.

In a similar manner it can be shown that

$$f_y(x_0, y_0) = Q(x_0, y_0). \blacksquare$$

THE CURL OF A CONSERVATIVE FIELD

For a vector field **F** defined throughout some region in the plane (or space) the following three properties are therefore equivalent;

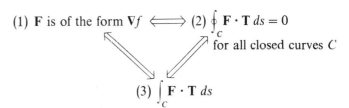

depends only on the endpoints of C

An arrow $\Rightarrow$ means "implies." Any one of the three properties, (1), (2), or (3), describes a conservative field. In our approach, (3) was used as the defining property.

None of the three criteria offers a practical test to determine whether an arbitrary vector field is conservative. However, the next theorem does half the job.

THEOREM 5   If **F** is a conservative vector field, then **curl F** is **0**.

PROOF   If **F** is conservative, then it is of the form ∇f for some scalar function f. By Exercise 11 of Sec. 15.1, the curl of a gradient field is **0**. In other words, the curl of **F** is **0**. $\blacksquare$

The curl of a conservative field is **0**.

According to Theorem 5, if $\mathbf{F}(x, y) = P(x, y)\mathbf{i} + Q(x, y)\mathbf{j}$ is a conservative vector field in the plane, then

$$\begin{vmatrix} \mathbf{i} & \mathbf{j} & \mathbf{k} \\ \dfrac{\partial}{\partial x} & \dfrac{\partial}{\partial y} & \dfrac{\partial}{\partial z} \\ P(x, y) & Q(x, y) & 0 \end{vmatrix} = \mathbf{0},$$

that is,
$$-\mathbf{i}\,\frac{\partial Q}{\partial z} + \mathbf{j}\,\frac{\partial P}{\partial z} + \mathbf{k}\left(\frac{\partial Q}{\partial x} - \frac{\partial P}{\partial y}\right) = \mathbf{0}.$$

Since Q and P are functions of x and y only, $\partial Q/\partial z$ and $\partial P/\partial z$ are both 0. Consequently, if $P(x, y)\mathbf{i} + Q(x, y)\mathbf{j}$ is conservative, then

$$\frac{\partial P}{\partial y} = \frac{\partial Q}{\partial x}. \tag{1}$$

A similar calculation shows that, if $P(x, y, z)\mathbf{i} + Q(x, y, z)\mathbf{j} + R(x, y, z)\mathbf{k}$ is conservative, then

$$\frac{\partial R}{\partial y} = \frac{\partial Q}{\partial z}, \qquad \frac{\partial R}{\partial x} = \frac{\partial P}{\partial z}, \qquad \text{and} \qquad \frac{\partial Q}{\partial x} = \frac{\partial P}{\partial y}. \qquad (2)$$

Warning: The converse of Theorem 5 is false. All would be delightful if the converse of Theorem 5 were true. Unfortunately, it is not. There are vector fields $\mathbf{F}$, whose curls are $\mathbf{0}$, which are not conservative. The next example provides one such $\mathbf{F}$.

EXAMPLE 1 Let
$$\mathbf{F} = \frac{-y\mathbf{i}}{x^2 + y^2} + \frac{x\mathbf{j}}{x^2 + y^2}.$$

Show that (a) $\nabla \times \mathbf{F} = \mathbf{0}$, but (b) $\mathbf{F}$ is not conservative.

SOLUTION (a) To solve (a) it is necessary to show that $\mathbf{F}$ satisfies (1), that is,

$$\frac{\partial}{\partial y}\left(\frac{-y}{x^2 + y^2}\right) = \frac{\partial}{\partial x}\left(\frac{x}{x^2 + y^2}\right).$$

Straightforward computations show that both sides equal the quotient $(y^2 - x^2)/(x^2 + y^2)^2$, hence are equal.

(b) To show that $\mathbf{F}$ is *not* conservative, it suffices to exhibit a closed curve C such that $\oint_C \mathbf{F} \cdot \mathbf{T}\, ds$ is not 0. As C we use the unit circle parametrized by

$$x = \cos \theta, \qquad y = \sin \theta, \qquad 0 \leq \theta \leq 2\pi.$$

Then
$$\oint_C \mathbf{F} \cdot \mathbf{T}\, ds = \oint_C \left(\frac{-y\, dx}{x^2 + y^2} + \frac{x\, dy}{x^2 + y^2}\right)$$

$$= \int_0^{2\pi} \left[\frac{-\sin\theta\, d(\cos\theta)}{\cos^2\theta + \sin^2\theta} + \frac{\cos\theta\, d(\sin\theta)}{\cos^2\theta + \sin^2\theta}\right]$$

$$= \int_0^{2\pi} \frac{(\sin^2\theta + \cos^2\theta)\, d\theta}{\sin^2\theta + \cos^2\theta}$$

$$= \int_0^{2\pi} d\theta$$

$$= 2\pi.$$

This establishes (b). ■

Example 1 shows that even though **curl F** can be identically **0**, **F** need not be conservative. However, under suitable circumstances, if **curl F** = **0**, it will follow that **F** is conservative. The extra condition involves the domain of **F**.

15.3 CONSERVATIVE VECTOR FIELDS

WHEN curl F = 0 DOES IMPLY THAT F IS CONSERVATIVE

The next theorem asserts that if **curl F = 0** and the domain of **F** is the entire plane or all of space, then **F** is indeed conservative. (Note that the vector field in Example 1 is not defined at (0, 0).) In Sec. 15.7 this result will be considerably generalized by stating a specific condition that the domain of **F** must satisfy if **curl F = 0** is to imply that **F** is conservative.

THEOREM 6 Let $\mathbf{F} = P\mathbf{i} + Q\mathbf{j}$ be defined throughout the plane or let $\mathbf{F} = P\mathbf{i} + Q\mathbf{j} + R\mathbf{k}$ be defined throughout space. If **curl F = 0**, then **F** is conservative.

PROOF Take the planar case, $\mathbf{F} = P\mathbf{i} + Q\mathbf{j}$, and assume that $\nabla \times \mathbf{F} = \mathbf{0}$, that is, by (1),

$$\frac{\partial P}{\partial y} = \frac{\partial Q}{\partial x}.$$

We wish to construct a scalar function f such that $\mathbf{F} = \nabla f$, that is,

$$\frac{\partial f}{\partial x} = P \quad \text{and} \quad \frac{\partial f}{\partial y} = Q.$$

$\int P(x, y)\,dx$ is meaningful since P is defined throughout the plane.

Begin by constructing a function whose partial derivative with respect to x is P. To do this, let $\int P(x, y)\,dx$ be any antiderivative of $P(x, y)$ with respect to x, holding y constant. Note that

$$\frac{\partial}{\partial x}\left(\int P(x, y)\,dx\right) = P(x, y).$$

In fact, for any function of y, $g(y)$, we have

$$\frac{\partial}{\partial x}\left(\int P(x, y)\,dx + g(y)\right) = P(x, y).$$

All that remains is to show that it is possible to choose $g(y)$ in such a way that

$$\frac{\partial}{\partial y}\left(\int P(x, y)\,dx + g(y)\right) = Q(x, y).$$

that is,

$$\frac{\partial}{\partial y}\left(\int P(x, y)\,dx\right) + \frac{\partial g(y)}{\partial y} = Q(x, y).$$

Noticing that $\partial g/\partial y = dg/dy$, an ordinary derivative, we rewrite this last equation in the form

$$\frac{dg}{dy} = Q(x, y) - \frac{\partial}{\partial y}\left(\int P(x, y)\, dx\right).$$

The left side of this equation is to be a function of y. In order for such a function g to exist, the right side of the equation must be a function of y alone. (Then g would just be an antiderivative of the right side.) All that remains is to show that the right side is actually a function of y alone. To check this, we show that its partial derivative with respect to x is 0:

$$\frac{\partial}{\partial x}\left[Q(x, y) - \frac{\partial}{\partial y}\left(\int P(x, y)\, dx\right)\right]$$

$$= \frac{\partial Q}{\partial x} - \frac{\partial}{\partial x}\left[\frac{\partial}{\partial y}\left(\int P(x, y)\, dx\right)\right]$$

$$= \frac{\partial Q}{\partial x} - \frac{\partial}{\partial y}\left[\frac{\partial}{\partial x}\left(\int P(x, y)\, dx\right)\right] \quad \text{since} \quad \frac{\partial^2}{\partial x\, \partial y} = \frac{\partial^2}{\partial y\, \partial x}$$

$$= \frac{\partial Q}{\partial x} - \frac{\partial P}{\partial y} \quad\quad\quad\quad\quad\quad\quad\quad\quad \text{by definition of } \int P(x, y)\, dx$$

$$= 0 \quad\quad\quad\quad\quad\quad\quad\quad\quad\quad\quad\quad\quad\quad \text{by assumption.}$$

Since the right side of the equation

$$\frac{dg}{dy} = Q(x, y) - \frac{\partial}{\partial y}\left(\int P(x, y)\, dx\right)$$

is just a function of y, $g(y)$ can be found by integration with respect to y. For such a choice of $g(y)$ the function

$$f(x, y) = \int P(x, y)\, dx + g(y)$$

has a gradient equal to $P\mathbf{i} + Q\mathbf{j}$, as desired.

A similar proof works if $\mathbf{F} = P\mathbf{i} + Q\mathbf{j} + R\mathbf{k}$. ∎

EXAMPLE 2 Is there a function $f(x, y)$ such that $\nabla f = e^x y\mathbf{i} + (e^x + 2y)\mathbf{j}$? If so, construct such an f.

SOLUTION In this case $P = e^x y$ and $Q = e^x + 2y$. We have

$$\frac{\partial P}{\partial y} = e^x \quad \text{and} \quad \frac{\partial Q}{\partial x} = e^x.$$

Since $\mathbf{F} = e^x y\mathbf{i} + (e^x + 2y)\mathbf{j}$ is defined throughout the plane and $\nabla \times \mathbf{F} = \mathbf{0}$, $\mathbf{F}$ is the gradient of some function $f(x, y)$.

To construct such an f, use the technique employed in the proof of Theorem 6. First compute

$$\int e^x y\, dx = e^x y + g(y),$$

15.3 CONSERVATIVE VECTOR FIELDS

where $g(y)$ will be determined so that

$$\frac{\partial}{\partial y}(e^x y + g(y)) = e^x + 2y. \quad (3)$$

Differentiation of (3) gives

$$e^x + g'(y) = e^x + 2y,$$

hence

$$g(y) = \int 2y \, dy = y^2 + C.$$

For any choice of the constant C,

$$f(x, y) = e^x y + y^2 + C$$

is a scalar function whose gradient is $e^x y \mathbf{i} + (e^x + 2y)\mathbf{j}$. ∎

Once we know that $\mathbf{F}(x, y)$ is conservative, we can define a scalar function $f(x, y)$ as $\int_C \mathbf{F} \cdot \mathbf{T} \, ds$, where C is any curve that starts at $(0, 0)$ and ends at (x, y). By the proof of Theorem 4, the gradient of f is $\mathbf{F}$. The next example uses this idea to provide another solution for Example 2.

EXAMPLE 3 Let $\mathbf{F} = e^x y \mathbf{i} + (e^x + 2y)\mathbf{j}$. Since **curl F = 0** and **F** is defined throughout the plane, there is a scalar function f such that $\nabla f = \mathbf{F}$. Use line integrals to construct such an f.

SOLUTION Define $f(x_1, y_1)$ to be $\int_C \mathbf{F} \cdot \mathbf{T} \, ds$, where C is any curve from $(0, 0)$ to (x_1, y_1). For ease of integration choose C to be the curve that goes from $(0, 0)$ to $(x_1, 0)$ in a straight line and then from $(x_1, 0)$ to (x_1, y_1) in a straight line. (See Fig. 15.21.)

We use x_1 and y_1 here in order to have x and y available as variables.

Call the first part of the path C_1 and the second part C_2. Then

$$f(x_1, y_1) = \int_{C_1} [e^x y \, dx + (e^x + 2y) \, dy] + \int_{C_2} [e^x y \, dx + (e^x + 2y) \, dy].$$

On C_1, $dy = 0$; on C_2, $dx = 0$. Thus

$$\int_{C_1} [e^x y \, dx + (e^x + 2y) \, dy] = \int_{C_1} e^x y \, dx$$

$$= \int_0^{x_1} e^x y \, dx$$

$$= \int_0^{x_1} e^x \cdot 0 \, dx \quad (y = 0 \text{ on } C_1)$$

$$= 0$$

and

$$\int_{C_2} [e^x y \, dx + (e^x + 2y) \, dy] = \int_{C_2} (e^x + 2y) \, dy$$

$$= \int_0^{y_1} (e^{x_1} + 2y) \, dy \quad (x = x_1 \text{ on } C_2)$$

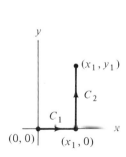

Figure 15.21

$$= (e^{x_1}y + y^2)\Big|_0^{y_1}$$

$$= e^{x_1}y_1 + y_1^2.$$

Thus $\int_C \mathbf{F} \cdot \mathbf{T}\, ds = 0 + e^{x_1}y_1 + y_1^2 = e^{x_1}y_1 + y_1^2.$ ∎

The answers to Examples 2 and 3 differ at most by a constant. Actually, any two functions f and g defined throughout the plane and having the same gradient must differ by a constant. To show this, note that $\nabla(f - g) = \mathbf{0}$. Thus $\dfrac{\partial}{\partial x}(f - g) = 0$ and $\dfrac{\partial}{\partial y}(f - g) = 0$. Hence $f - g$ must be constant.

EXACT DIFFERENTIALS

A different view of a conservative vector field

The vector field $\mathbf{F}(x, y) = P(x, y)\mathbf{i} + Q(x, y)\mathbf{j}$ is conservative if there is a scalar function $f(x, y)$ such that $f_x = P$ and $f_y = Q$. Thus the differential of f,

$$df = f_x\, dx + f_y\, dy,$$

coincides with the expression

$$P\, dx + Q\, dy.$$

For a random choice of P and Q, $P\, dx + Q\, dy$ is seldom exact.

The expression $P\, dx + Q\, dy$ is then said to be an *exact differential*. Thus "exact differentials" and "conservative vector fields" are different views of the same concept. The contrast between an exact differential and an inexact one is of fundamental importance in thermodynamics. See Exercise 24.

OTHER NOTATIONS FOR $\int_C \mathbf{F} \cdot \mathbf{T}\, ds$

In various applications of $\int_C \mathbf{F} \cdot \mathbf{T}\, ds$ several notations are commonly used. They are listed in this table.

Notation	Remark
$\int_C \mathbf{F} \cdot \mathbf{T}\, ds$	Integral of scalar component of $\mathbf{F}$ along $\mathbf{T}$
$\int_C \mathbf{F} \cdot d\mathbf{r}$	Think of $d\mathbf{r}$ as $\mathbf{T}\, ds$
$\int_C \mathbf{F} \cdot d\mathbf{s}$	$d\mathbf{s}$ is another notation for $d\mathbf{r}$
$\int_C (P\, dx + Q\, dy + R\, dz)$	Differential form

Exercises

In Exercises 1 and 2 determine which vector fields are conservative and which are not.

1. (a) ∇f, where $f(x, y, z) = x^2 y e^z$
   (b) $\dfrac{-y\mathbf{i} + x\mathbf{j}}{(x^2 + y^2)^2}$

2. (a) $y e^z \cos xy\, \mathbf{i} + x e^z \cos xy\, \mathbf{j} + e^z \sin xy\, \mathbf{k}$
   (b) $x^3 \mathbf{i} + y^2 \mathbf{j}$

In each of Exercises 3 to 6 show that the vector field is conservative and then construct a scalar function of which it is the gradient. Use the method of Example 2.

3. $2xy\mathbf{i} + x^2 \mathbf{j}$
4. $e^{-x}(\cos 2xy + 2y \sin 2xy)\mathbf{i} + (2x e^{-x} \sin 2xy + 3y^2)\mathbf{j}$
5. $(y + 1)\mathbf{i} + (x + 1)\mathbf{j}$
6. $3y \sin^2 xy \cos xy\, \mathbf{i} + (1 + 3x \sin^2 xy \cos xy)\mathbf{j}$
7. Solve Exercise 3 by the method of Example 3.
8. Solve Exercise 4 by the method of Example 3.
9. Show that $2xy^2\mathbf{i} + x^2y\mathbf{j}$ is not conservative by
   (a) computing its curl,
   (b) exhibiting a closed path over which its integral is not 0.
10. Show that $2xy^2\mathbf{i} + 2x^2y\mathbf{j} + xyz\mathbf{k}$ is not conservative.

In Exercises 11 to 12 compute $\int_C \nabla f \cdot \mathbf{T}\, ds$ for the given f and curve C.

11. $f(x, y, z) = x^2 + y^2 + z^2$; C is parametrized by $\mathbf{G}(t) = 2 \cos t\, \mathbf{i} + 3 \sin t\, \mathbf{j} + t^2 \mathbf{k}$, $\pi/2 \le t \le \pi$.
12. $f(x, y, z) = xyz^2$; C is parametrized by $\mathbf{G}(t) = t\mathbf{i} + t^2\mathbf{j} + t^3\mathbf{k}$, $1 \le t \le 2$.

■

13. Let $\mathbf{F}(x, y, z) = 2xyz\mathbf{i} + x^2z\mathbf{j} + x^2y\mathbf{k}$. Evaluate $\int_C \mathbf{F} \cdot \mathbf{T}\, ds$ on each of these three paths from $(0, 0, 0)$ to $(1, 1, 1)$.
    (a) C_1, the straight path from $(0, 0, 0)$ to $(1, 1, 1)$;
    (b) C_2, the polygonal path from $(0, 0, 0)$ to $(1, 0, 0)$, then to $(1, 1, 0)$, and then to $(1, 1, 1)$;
    (c) C_3, the polygonal path from $(0, 0, 0)$ to $(0, 1, 0)$ and then directly to $(1, 1, 1)$.

14. Let $\mathbf{F}(x, y) = y\mathbf{i} + x\mathbf{j}$. Compute $\int_C \mathbf{F} \cdot \mathbf{T}\, ds$ on each of the following two curves from $(0, 0)$ to $(1, 2)$:
    (a) the path along the parabola $y = 2x^2$.
    (b) the path which goes in a straight line from $(0, 0)$ to $(1, 0)$ and then in a straight line from $(1, 0)$ to $(1, 2)$.
    (c) Show that $\mathbf{F}$ is conservative by exhibiting a scalar field f such that $\mathbf{F} = \nabla f$.

15. Show that
    (a) $3x^2 y\, dx + x^3\, dy$ is exact,
    (b) $3x^2 y\, dx + x^3\, dy + xyz\, dz$ is not exact.

16. Show that $x\, dx/(x^2 + y^2) + y\, dy/(x^2 + y^2)$ is exact by exhibiting a function f such that df equals the given expression.

17. Let $\mathbf{F} = xz\mathbf{i} + x^2\mathbf{j} + xy\mathbf{k}$. Let C be the path around the square whose vertices are $(0, 0, 1)$, $(1, 0, 1)$, $(1, 1, 1)$, and $(0, 1, 1)$. The path starts at $(0, 0, 1)$, sweeps out the vertices in the indicated order, and returns to $(0, 0, 1)$. Evaluate $\oint_C \mathbf{F} \cdot \mathbf{T}\, ds$. (See Fig. 15.22.)

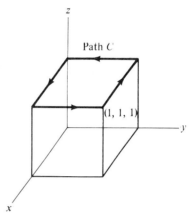

Figure 15.22

In each of Exercises 18 to 20 (a) show that the given vector field $\mathbf{F}$ is conservative, and (b) use the technique of Example 3 to construct a scalar field f such that $\mathbf{F} = \nabla f$.

18. $\mathbf{F} = (3x^2 + \sin z + yz)\mathbf{i} + xz\,\mathbf{j} + (x \cos z + xy)\mathbf{k}$
19. $\mathbf{F} = \dfrac{x\mathbf{i} + 2y\mathbf{j} + 3z\mathbf{k}}{1 + x^2 + 2y^2 + 3z^2}$
20. $\mathbf{F} = yze^{xy}\mathbf{i} + xze^{xy}\mathbf{j} + e^{xy}\mathbf{k}$

■ ■

21. Complete the proof of Theorem 4, showing that
    $$f_y(x_0, y_0) = Q(x_0, y_0).$$

Exercises 22 and 23 are related.

22. By a *special* curve in the plane we shall mean a polygonal path whose segments are parallel to either the x axis or the y axis. Let $\mathbf{F}(x, y)$ have the property that $\int_C \mathbf{F} \cdot \mathbf{T}\, ds$ depends only on the endpoints of C for all *special* curves. Deduce that $\mathbf{F}$ is conservative.

23. By a *rectangular* curve in the plane we shall mean a closed curve that surrounds a rectangle, as in Fig. 15.23.
    Let $\mathbf{F}$ be a vector field defined throughout the plane with the property that $\oint \mathbf{F} \cdot \mathbf{T}\, ds = 0$ whenever the curve is rectangular and has diameter at most 1. Deduce that

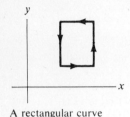

A rectangular curve

Figure 15.23

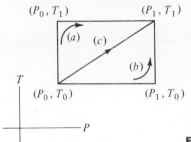

Figure 15.24

$\oint \mathbf{F} \cdot \mathbf{T}\, ds = 0$ for any rectangular path. *Hint:* Partition a large rectangle into small rectangles and draw some paths.

24 A gas at temperature T_0 and pressure P_0 is brought to the temperature $T_1 > T_0$ and pressure $P_1 > P_0$. The work done in this process is given by the line integral in the TP plane

$$\int_C \left(\frac{RT\, dP}{P} - R\, dT \right),$$

where R is a constant and C is the curve that records the various combinations of T and P during the process. Evaluate this integral over the following paths, shown in Fig. 15.24:

(a) The pressure is kept constant at P_0, while the temperature is raised from T_0 to T_1; then the temperature is kept constant at T_1 while the pressure is raised from P_0 to P_1.

(b) The temperature is kept constant at T_0 while the pressure is raised from P_0 to P_1; then the temperature is raised from T_0 to T_1 while the pressure is kept constant at P_1.

(c) Both pressure and temperature are raised simultaneously in such a way that the path from (P_0, T_0) to (P_1, T_1) is straight.

Because the integrals are path dependent, the differential expression $RT\, dP/P - R\, dT$ defines a thermodynamic quantity that depends on the process, not just on the state. Mathematically speaking, the differential expression $RT\, dP/P - R\, dT$ is not exact, that is, the vector field $(RT/P)\mathbf{i} - R\mathbf{j}$ is not conservative.

15.4 Green's theorem

Before stating Green's theorem, we treat a matter of notation. The remainder of this chapter concerns integrals over curves, regions in the plane, curved surfaces, and regions in space. In order to distinguish them clearly we use the following notation.

Curve C	Arc length s	Locally, ds
Plane region $\mathscr{A}$	Measure of area A	Locally, dA
Surface $\mathscr{S}$	Measure of surface area S	Locally, dS
Spatial region $\mathscr{V}$	Measure of volume V	Locally, dV

Green's theorem asserts that the integral of a certain function over a plane region $\mathscr{A}$ is equal to another integral of another function over its boundary C. We state it in two forms, the second being more general than the first.

GREEN'S THEOREM

The boundary consists of just one curve.

Let $\mathscr{A}$ be a region in the plane bounded by the single curve C. Let $\mathbf{n}$ denote the exterior normal along the boundary. Then

$$\oint_C \mathbf{F} \cdot \mathbf{n}\, ds = \int_{\mathscr{A}} \nabla \cdot \mathbf{F}\, dA$$

for any vector field $\mathbf{F}$ defined on $\mathscr{A}$. ∎

(The integral of the normal component of **F** over a closed curve equals the integral of the divergence of **F** over the plane region that the curve bounds.)
If $\mathbf{F} = P\mathbf{i} + Q\mathbf{j}$, then Green's theorem reads

Recall that
$$\oint_C (P\mathbf{i} + Q\mathbf{j}) \cdot \left(\frac{dy}{ds}\mathbf{i} - \frac{dx}{ds}\mathbf{j}\right) ds = \int_{\mathscr{A}} (P_x + Q_y) \, dA$$

$P_x = \dfrac{\partial P}{\partial x}$, $Q_y = \dfrac{\partial Q}{\partial y}$.

or, more briefly,
$$\oint_C (P \, dy - Q \, dx) = \int_{\mathscr{A}} (P_x + Q_y) \, dA,$$

the curve being traversed counterclockwise.

Before examining why Green's theorem is plausible, let us first check it in an example.

EXAMPLE 1 Let $\mathbf{F} = 2x\mathbf{i} + 3y\mathbf{j}$ and let $\mathscr{A}$ be the region bounded by the unit circle $x^2 + y^2 = 1$. Verify Green's theorem in this case.

SOLUTION The region $\mathscr{A}$ and its bounding curve C are shown in Fig. 15.25.
First evaluate the integral of the normal component of **F** around the boundary by the formula

$$\int_C (2x\mathbf{i} + 3y\mathbf{j}) \cdot \mathbf{n} \, ds = \int_C (-3y \, dx + 2x \, dy),$$

where C is traversed counterclockwise. To compute this integral use the parametrization

$$x = \cos \theta, \quad y = \sin \theta, \quad 0 \le \theta \le 2\pi.$$

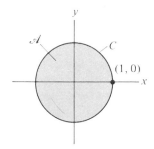

Figure 15.25

Hence,

$$\int_C (-3y \, dx + 2x \, dy) = \int_0^{2\pi} [-3 \sin \theta \, d(\cos \theta) + 2 \cos \theta \, d(\sin \theta)]$$

$$= \int_0^{2\pi} (3 \sin^2 \theta + 2 \cos^2 \theta) \, d\theta$$

$$= 5\pi.$$

On the other hand,

$$\int_{\mathscr{A}} \nabla \cdot \mathbf{F} \, dA = \int_{\mathscr{A}} \nabla \cdot (2x\mathbf{i} + 3y\mathbf{j}) \, dA$$

$$= \int_{\mathscr{A}} (2 + 3) \, dA = 5\pi.$$

Thus
$$\int_C \mathbf{F} \cdot \mathbf{n} \, ds = 5\pi = \int_{\mathscr{A}} \nabla \cdot \mathbf{F} \, dA,$$

verifying Green's theorem. ∎

THE PLAUSIBILITY OF GREEN'S THEOREM

If **F** is interpreted as representing a fluid flow, Green's theorem is plausible. Letting $\mathbf{F} = P\mathbf{i} + Q\mathbf{j}$, we have

$$\oint_C \mathbf{F} \cdot \mathbf{n} \, ds = \oint_C (P \, dy - Q \, dx),$$

which represents the net rate of loss of fluid across the curve C.

Think of the fluid as a gas moving between two parallel panes of glass.

Now, the net loss of fluid past C is the same as the net loss from the region $\mathcal{A}$ bounded by C, since the only way that fluid enters or leaves $\mathcal{A}$ is across C. Let us calculate the net loss from the viewpoint of $\mathcal{A}$ rather than of C. Consider the net loss from a typical small rectangle of dimensions Δx and Δy in $\mathcal{A}$. Let its lower-left-hand corner D have the coordinates (a, b), and let the other vertices be E, F, G, as in Fig. 15.26.

What is the net loss of fluid from this small rectangle?

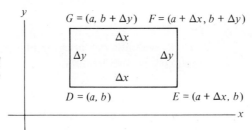

Figure 15.26

The loss from this small rectangle is precisely

$$\int_D^E (P \, dy - Q \, dx) + \int_E^F (P \, dy - Q \, dx) + \int_F^G (P \, dy - Q \, dx) + \int_G^D (P \, dy - Q \, dx),$$

where the integrals are over straight paths. Let us estimate this sum when Δx and Δy are small.

Consider first the integrals over the parallel edges DE and FG. They are

$$\int_D^E (P \, dy - Q \, dx) = \int_D^E -Q \, dx$$

and

$$\int_F^G (P \, dy - Q \, dx) = \int_F^G -Q \, dx,$$

since in both cases y is constant and $dy = 0$. Now

$$\int_D^E -Q \, dx = \int_a^{a+\Delta x} -Q(x, b) \, dx,$$

and

$$\int_F^G -Q \, dx = \int_G^F Q \, dx = \int_a^{a+\Delta x} Q(x, b + \Delta y) \, dx.$$

Hence $\int_D^E -Q\,dx + \int_F^G -Q\,dx = \int_a^{a+\Delta x} [Q(x, b + \Delta y) - Q(x, b)]\,dx.$ (1)

By the mean-value theorem,

$$Q(x, b + \Delta y) - Q(x, b) = Q_y(x, Y)\,\Delta y$$

for some Y, with $b < Y < b + \Delta y$.

From the continuity of Q_y, it follows that $Q_y(x, Y)$ is approximately equal to $Q_y(a, b)$. Hence the integral on the right side of (1) is approximately

$$\int_a^{a+\Delta x} Q_y(a, b)\,\Delta y\,dx = Q_y(a, b)\,\Delta y\,\Delta x.$$

Thus $\int_D^E (P\,dy - Q\,dx) + \int_F^G (P\,dy - Q\,dx)$

is approximately $Q_y(a, b)\,\Delta y\,\Delta x.$

Similarly $\int_E^F (P\,dy - Q\,dx) + \int_G^D (P\,dy - Q\,dx)$

is approximately $P_x(a, b)\,\Delta x\,\Delta y.$

Thus the flow out of a typical small rectangle, whose area is $\Delta A = \Delta x\,\Delta y$, is approximately

$$[P_x(a, b) + Q_y(a, b)]\,\Delta A.$$

Since the net flow out of $\mathscr{A}$ is the sum of the net flows out of all such small rectangles, we may expect that

$$\text{Net flow out of } \mathscr{A} = \int_{\mathscr{A}} (P_x + Q_y)\,dA.$$

Since   Net flow out of $\mathscr{A}$ = Net flow past C,

it seems reasonable that

$$\int_{\mathscr{A}} (P_x + Q_y)\,dA = \oint_C (P\,dy - Q\,dx).$$

A physical interpretation of $\nabla \cdot \mathbf{F}$ This fluid-flow interpretation of Green's theorem also provides a physical interpretation of $\nabla \cdot \mathbf{F} = P_x + Q_y$. As we saw in the fluid-flow reasoning,

$$\nabla \cdot \mathbf{F} = P_x + Q_y$$

is a local measure of net loss—or "divergence"—of the fluid. This explains why $\nabla \cdot \mathbf{F}$ is called the divergence of $\mathbf{F}$.

GREEN'S THEOREM: GENERAL FORM

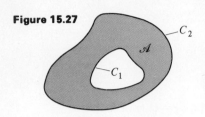

Figure 15.27

Green's theorem holds for regions in the plane whose boundaries consist of any finite number of curves. However, we state it in the special case in which the boundary consists of two nonoverlapping curves, since this case is especially useful. A region $\mathscr{A}$ is shown in Fig. 15.27. The region has a hole bounded by the curve C_1 and an outer boundary C_2.

GREEN'S THEOREM *Two-curve case.* Let $\mathscr{A}$ be a region in the plane bounded by the curves C_1 and C_2. Let $\mathbf{n}^*$ denote the exterior normal along the boundary. Then

$$\oint_{C_1} \mathbf{F} \cdot \mathbf{n}^* \, ds + \oint_{C_2} \mathbf{F} \cdot \mathbf{n}^* \, ds = \int_{\mathscr{A}} \nabla \cdot \mathbf{F} \, dA$$

for any vector field defined on $\mathscr{A}$. ∎

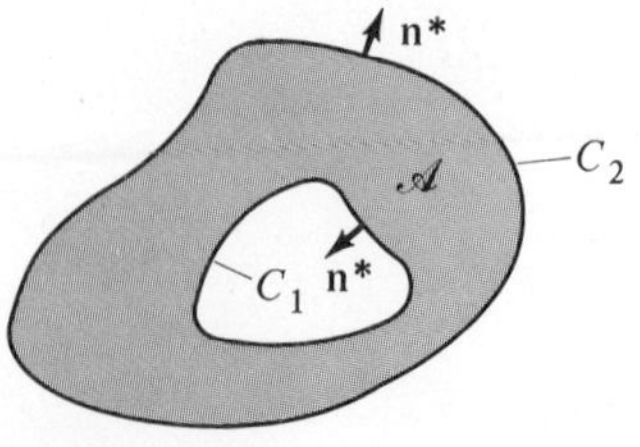

Figure 15.28

$\mathbf{n}^*$ *points away from* $\mathscr{A}$

As shown in Fig. 15.28, the exterior normal $\mathbf{n}^*$ points into the hole for the inner curve. We call the normal $\mathbf{n}^*$ rather than $\mathbf{n}$ to simplify the terminology in the proof of Corollary 1.

On the inner curve C_1, $\mathbf{n}^*$ is not the exterior normal $\mathbf{n}$ relative to the region that C_1 surrounds. Relative to the hole bounded by C_1, the exterior normal $\mathbf{n}$ points away from the hole, as shown in Fig. 15.29. Inspection of Fig. 15.29 shows that, on C_1, $\mathbf{n}^* = -\mathbf{n}$.

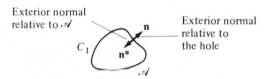

Figure 15.29

Simply connected

A region $\mathscr{A}$ in the plane is called *simply connected* if each loop in $\mathscr{A}$ can be continuously shrunk to a point while staying within $\mathscr{A}$. (Thus $\mathscr{A}$ has no holes.) A region in the plane bounded by a single curve is simply connected; a region bounded by at least two nonintersecting closed curves is not simply connected.

The two-curve case of Green's theorem is the basis of the following corollary.

COROLLARY 1 Let C_1 and C_2 be two closed curves that form the boundary of the region $\mathscr{A}$. Let $\mathbf{F}$ be a vector field defined on $\mathscr{A}$ such that the divergence of $\mathbf{F}$, $\nabla \cdot \mathbf{F}$, is 0 throughout $\mathscr{A}$. Then

$$\oint_{C_1} \mathbf{F} \cdot \mathbf{n} \, ds = \oint_{C_2} \mathbf{F} \cdot \mathbf{n} \, ds.$$

15.4 GREEN'S THEOREM

PROOF Assume that C_1 is the inner curve and C_2 is the outer curve. By Green's theorem in the two-curve case,

$$\oint_{C_1} \mathbf{F} \cdot \mathbf{n}^* \, ds + \oint_{C_2} \mathbf{F} \cdot \mathbf{n}^* \, ds = \int_{\mathcal{A}} \nabla \cdot \mathbf{F} \, dA$$

$$= \int_{\mathcal{A}} 0 \, dA = 0 \quad (\text{since } \nabla \cdot \mathbf{F} = 0).$$

Thus
$$\oint_{C_1} \mathbf{F} \cdot (-\mathbf{n}) \, ds + \oint_{C_2} \mathbf{F} \cdot \mathbf{n} \, ds = 0$$

or
$$-\oint_{C_1} \mathbf{F} \cdot \mathbf{n} \, ds + \oint_{C_2} \mathbf{F} \cdot \mathbf{n} \, ds = 0,$$

from which the corollary follows. ■

The power of the corollary is illustrated in the following example.

EXAMPLE 2 Let $\mathbf{F}(x, y) = \mathbf{r}/|\mathbf{r}|^2$, where $\mathbf{r} = x\mathbf{i} + y\mathbf{j}$. Let C be the curve shown in Fig. 15.30. Evaluate $\oint_C \mathbf{F} \cdot \mathbf{n} \, ds$.

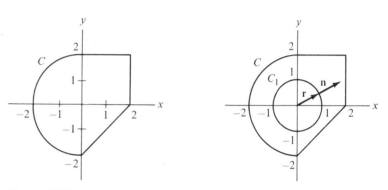

Figure 15.30   **Figure 15.31**

SOLUTION By Exercise 24 of Sec. 15.1 the divergence of $\mathbf{F}$ is 0 wherever $\mathbf{F}$ is defined. (It is not defined at the origin.) Motivated by Corollary 1, consider $\oint_{C_1} \mathbf{F} \cdot \mathbf{n} \, ds$, where C_1 is the unit circle with center at the origin. The curve C_1 is easier to deal with than the original curve C. (See Fig. 15.31.)

Since div $\mathbf{F} = 0$, *By Corollary 1,*
$\oint_C \mathbf{F} \cdot \mathbf{n} \, ds$ *can be replaced*
by $\oint_{C_1} \mathbf{F} \cdot \mathbf{n} \, ds.$

$$\oint_C \mathbf{F} \cdot \mathbf{n} \, ds = \oint_{C_1} \mathbf{F} \cdot \mathbf{n} \, ds$$

$$= \oint_{C_1} \frac{\mathbf{r} \cdot \mathbf{n}}{|\mathbf{r}|^2} \, ds$$

On the unit circle $|\mathbf{r}| = 1$. Also, $\mathbf{n}$ points in the same direction as $\mathbf{r}$. Hence, $\mathbf{n} = \mathbf{r}$, and $\mathbf{r} \cdot \mathbf{n} = \mathbf{r} \cdot \mathbf{r} = 1$. Thus

$$\oint_{C_1} \frac{\mathbf{r} \cdot \mathbf{n}}{|\mathbf{r}|^2} \, ds = \oint_{C_1} 1 \, ds = 2\pi.$$

Consequently
$$\oint_C \frac{\mathbf{r} \cdot \mathbf{n}}{|\mathbf{r}|^2} \, ds = 2\pi. \quad \blacksquare$$

The integral required in Example 2 could have been computed directly, but only with much greater difficulty.

COROLLARY 2 Let $\mathbf{F}$ be a vector field defined everywhere in the plane except perhaps at the point P_0. Assume that the divergence of $\mathbf{F}$ is 0. Let C_1 and C_2 be two closed curves each of which encloses the point P_0. Then

$$\oint_{C_1} \mathbf{F} \cdot \mathbf{n} \, ds = \oint_{C_2} \mathbf{F} \cdot \mathbf{n} \, ds.$$

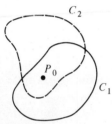

Figure 15.32

PROOF The two typical curves, which may overlap, are shown in Fig. 15.32.
Choose a very small circle C around P_0, so small that it meets neither C_1 nor C_2. (See Fig. 15.33.) By Corollary 1,

$$\oint_{C_1} \mathbf{F} \cdot \mathbf{n} \, ds = \oint_C \mathbf{F} \cdot \mathbf{n} \, ds \quad \text{and} \quad \oint_{C_2} \mathbf{F} \cdot \mathbf{n} \, ds = \oint_C \mathbf{F} \cdot \mathbf{n} \, ds.$$

Thus
$$\oint_{C_1} \mathbf{F} \cdot \mathbf{n} \, ds = \oint_{C_2} \mathbf{F} \cdot \mathbf{n} \, ds. \quad \blacksquare$$

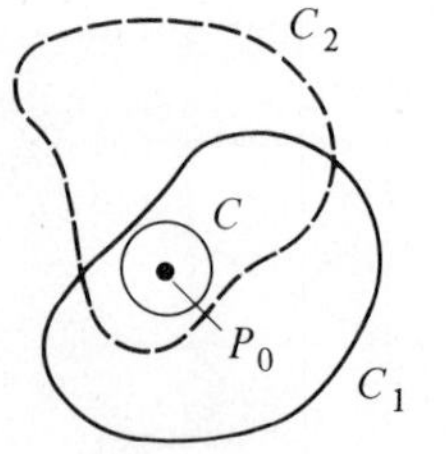

Figure 15.33

According to Corollary 2, as you gradually move and distort a curve C the integral $\oint_C \mathbf{F} \cdot \mathbf{n} \, ds$ remains constant, as long as you remain in a region where the divergence of $\mathbf{F}$ is 0.

ANOTHER VIEW OF GREEN'S THEOREM

Green's theorem asserts that $\oint_C \mathbf{E} \cdot \mathbf{n} \, ds = \int_{\mathcal{A}} \nabla \cdot \mathbf{E} \, dA$, where $\mathbf{E}$ is a vector field in the plane and $\mathbf{n}$ is the external normal. (We call the vector field $\mathbf{E}$ now for in a moment $\mathbf{F}$ will be needed to denote another vector field.) Let $\mathbf{E} = Q\mathbf{i} - P\mathbf{j}$. If C is swept out counterclockwise, Green's theorem says that

$$\oint_C (Q\mathbf{i} - P\mathbf{j}) \cdot \left(\frac{dy}{ds} \mathbf{i} - \frac{dx}{ds} \mathbf{j} \right) ds = \int_{\mathcal{A}} \nabla \cdot \mathbf{E} \, dA$$

or
$$\oint_C \left(Q \frac{dy}{ds} + P \frac{dx}{ds} \right) ds = \int_{\mathcal{A}} \left(\frac{\partial Q}{\partial x} - \frac{\partial P}{\partial y} \right) dA,$$

(2) is a very common way of expressing Green's Theorem.

or
$$\oint_C (P \, dx + Q \, dy) = \int_{\mathcal{A}} \left(\frac{\partial Q}{\partial x} - \frac{\partial P}{\partial y} \right) dA. \quad (2)$$

Now introduce a second vector field $\mathbf{F} = P\mathbf{i} + Q\mathbf{j}$. The left side of (2) is just

$$\oint_C \mathbf{F} \cdot \mathbf{T} \, ds.$$

15.4 GREEN'S THEOREM

The right side of (2) can be expressed in terms of the curl of **F**. By Sec. 15.1,

$$\nabla \times \mathbf{F} = \left(\frac{\partial Q}{\partial x} - \frac{\partial P}{\partial y}\right)\mathbf{k}.$$

Thus
$$(\nabla \times \mathbf{F}) \cdot \mathbf{k} = \frac{\partial Q}{\partial x} - \frac{\partial P}{\partial y}.$$

Equation (2) therefore can be rephrased as

Green's theorem expressed in terms of curl

$$\oint_C \mathbf{F} \cdot \mathbf{T} \, ds = \int_{\mathcal{A}} (\nabla \times \mathbf{F}) \cdot \mathbf{k} \, dA.$$

This formulation of Green's theorem will be generalized in Sec. 15.7 to curves in space.

Stated this way, Green's theorem relates the integral of $\mathbf{F} \cdot \mathbf{T}$ around the boundary of a region to an integral involving **curl F** over the region. It shows that if **curl F** = **0** *throughout* a planar region $\mathcal{A}$ whose boundary is C, then $\oint_C \mathbf{F} \cdot \mathbf{T} \, ds = 0$ (since $\oint_C \mathbf{F} \cdot \mathbf{T} \, ds = \int_{\mathcal{A}} (\nabla \times \mathbf{F}) \cdot \mathbf{k} \, dA = 0$). In this argument it is essential that **F** be defined throughout the region $\mathcal{A}$. (Compare this with Example 1 in the preceding section, where the function is not defined at $(0, 0)$.) In Sec. 15.7 we will return to the question, "When does **curl F** = **0** imply that **F** is conservative?", and treat it both in the plane and in space.

PROOF OF GREEN'S THEOREM (OPTIONAL)

We prove Green's theorem in the special case where $\mathcal{A}$ is convex and each of the four tangent lines to the boundary C that are parallel to the axes meets C at only one point.

Letting $\mathbf{F} = P\mathbf{i} + Q\mathbf{j}$, we wish to show that

$$\int_{\mathcal{A}} (P_x + Q_y) \, dA = \oint_C (P \, dy - Q \, dx).$$

PROOF It will be proved that

$$\int_{\mathcal{A}} Q_y \, dA = \oint_C - Q \, dx. \tag{3}$$

A similar proof will show that

$$\int_{\mathcal{A}} P_x \, dA = \oint_C P \, dy.$$

Green's theorem follows immediately from these two equations.

Let the region $\mathcal{A}$ have the following description:

$$a \le x \le b, \qquad y_1(x) \le y \le y_2(x),$$

as shown in Fig. 15.34.

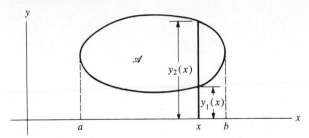

Figure 15.34

(Note by the assumption on C that $y_1(a) = y_2(a)$ and $y_1(b) = y_2(b)$.) Then

$$\int_{\mathscr{A}} Q_y \, dA = \int_a^b \left[\int_{y_1(x)}^{y_2(x)} \frac{\partial Q}{\partial y} \, dy \right] dx.$$

By the fundamental theorem of calculus,

$$\int_{y_1(x)}^{y_2(x)} \frac{\partial Q}{\partial y} \, dy = Q(x, y_2(x)) - Q(x, y_1(x)).$$

Hence
$$\int_{\mathscr{A}} Q_y \, dA = \int_a^b [Q(x, y_2(x)) - Q(x, y_1(x))] \, dx. \tag{4}$$

Now consider the right side of (3),

$$\oint_C - Q \, dx.$$

Break the closed path C into two successive paths, one along the bottom part of $\mathscr{A}$, described by $y = y_1(x)$, the other along the top part of $\mathscr{A}$, described by $y = y_2(x)$. Denote the bottom path C_1 and the top path C_2. (See Fig. 15.35.)

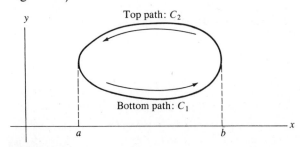

Figure 15.35

Then
$$\oint_C - Q \, dx = \int_{C_1} - Q \, dx + \int_{C_2} - Q \, dx. \tag{5}$$

But
$$\int_{C_1} -Q\,dx = \int_{C_1} -Q(x, y_1(x))\,dx$$
$$= \int_a^b -Q(x, y_1(x))\,dx,$$

and
$$\int_{C_2} -Q\,dx = \int_{C_2} -Q(x, y_2(x))\,dx$$
$$= \int_b^a -Q(x, y_2(x))\,dx$$
$$= \int_a^b Q(x, y_2(x))\,dx.$$

Thus by (5),
$$\oint_C -Q\,dx = \int_a^b -Q(x, y_1(x))\,dx + \int_a^b Q(x, y_2(x))\,dx$$
$$= \int_a^b [Q(x, y_2(x)) - Q(x, y_1(x))]\,dx.$$

This is precisely (4) and concludes the proof. ∎

Exercises

In Exercises 1 and 2 compute $\int_{\mathcal{A}} \nabla \cdot \mathbf{F}\,dA$ and $\oint_C \mathbf{F} \cdot \mathbf{n}\,ds$ and verify Green's theorem.

1. $\mathbf{F} = 3x\mathbf{i} + 2y\mathbf{j}$, and $\mathcal{A}$ is the disk of radius 1 with center $(0, 0)$.
2. $\mathbf{F} = 5y^3\mathbf{i} - 6x^2\mathbf{j}$, and $\mathcal{A}$ is the disk of radius 2 with center $(0, 0)$.

In Exercises 3 to 6 use Green's theorem to evaluate $\oint_C \mathbf{F} \cdot \mathbf{n}\,ds$ for the given $\mathbf{F}$, where C is the boundary of the given region $\mathcal{A}$.

3. $\mathbf{F} = e^x \sin y\,\mathbf{i} + e^{2x} \cos y\,\mathbf{j}$ and $\mathcal{A}$ is the rectangle with vertices $(0, 0)$, $(1, 0)$, $(0, \pi/2)$, $(1, \pi/2)$.
4. $\mathbf{F} = y \tan x\,\mathbf{i} + y^2\,\mathbf{j}$ and $\mathcal{A}$ is the square with vertices $(0, 0)$, $(1, 0)$, $(1, 1)$, and $(0, 1)$.
5. $\mathbf{F} = 2x^3 y\mathbf{i} - 3x^2 y^2\mathbf{j}$ and $\mathcal{A}$ is the triangle with vertices $(0, 1)$, $(3, 4)$, and $(2, 7)$.
6. $\mathbf{F} = -\mathbf{i}/(xy^2) + \mathbf{j}/(x^2 y)$ and $\mathcal{A}$ is the triangle with vertices $(1, 1)$, $(2, 2)$, and $(1, 2)$.

In Exercises 7 and 8 verify Green's theorem in its two-curve form.

7. $\mathbf{F} = x\mathbf{i} + y^2\mathbf{j}$ and $\mathcal{A}$ is the region between the circles of radii 1 and 2 with centers at $(0, 0)$.
8. $\mathbf{F} = 3x^2\mathbf{i} + xy\mathbf{j}$ and $\mathcal{A}$ is the region between the circle $x^2 + y^2 = 1$ and the square with vertices $(2, 2)$, $(-2, 2)$, $(-2, -2)$, and $(2, -2)$.

9. Let $\mathcal{A}$ be a convex region and let C be its boundary. Let $\mathbf{F}$ be a vector field defined on $\mathcal{A}$ such that at any point P on C, $\mathbf{F}(P)$ is tangent to the curve C. Find $\int_{\mathcal{A}} \nabla \cdot \mathbf{F}\,dA$.

10. Let $\mathbf{F}$ and $\mathbf{G}$ be two vector fields defined throughout the plane with the property that $\mathbf{F}(P) = \mathbf{G}(P)$ for points P on a certain curve C that bounds a region $\mathcal{A}$. Show that $\int_{\mathcal{A}} \nabla \cdot \mathbf{F}\,dA = \int_{\mathcal{A}} \nabla \cdot \mathbf{G}\,dA$.

11. Use Green's theorem to evaluate $\oint_C (xy\,dx + e^x\,dy)$, where C is the curve that goes from $(0, 0)$ to $(2, 0)$ on the x axis and returns from $(2, 0)$ to $(0, 0)$ on the parabola $y = 2x - x^2$.

12. Let C be the circle of radius 1 with center $(0, 0)$.
    (a) What does Green's theorem say about the line integral $\oint_C [(x^2 - y^3)\,dx + (y^2 + x^3)\,dy]$?
    (b) Use Green's theorem to evaluate the integral in (a).
    (c) Evaluate the integral in (a) directly.

13. The vector field $\mathbf{F} = \mathbf{r}/|\mathbf{r}|^2$, defined for every point in the plane except the origin, has divergence 0.
    (a) Evaluate $\oint_C \mathbf{F} \cdot \mathbf{n}\,ds$ for a circle of radius a and center $(0, 0)$.
    (b) The answer in (a) is not 0. Does that violate Green's theorem?
    (c) The answer in (a) does not depend on the radius a. How could you have predicted this without any computations of integrals?

14. Let $\mathbf{F} = y\mathbf{i}/(x^2 + y^2)^3 - x\mathbf{j}/(x^2 + y^2)^3$.
    (a) Compute $\nabla \cdot \mathbf{F}$.
    (b) Evaluate $\oint_C \mathbf{F} \cdot \mathbf{n} \, ds$ over each of the curves C_1, C_2, and C_3 in Fig. 15.36 as easily as possible.

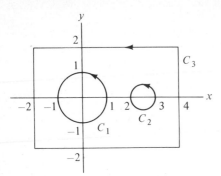

Figure 15.36

■

15. Verify Green's theorem if $\mathbf{F} = 2y \ln(1 + x + x^2)\mathbf{i} + (1/(1 + y))\mathbf{j}$ and $\mathscr{A}$ is the region bounded by the curves $y = x^2$ and $y = \sqrt{x}$.

16. Let f be a scalar function defined throughout the region $\mathscr{A}$. Using Green's theorem, show that
$$\int_{\mathscr{A}} \nabla \cdot (\nabla f) \, dA = \oint_C (f_x \, dy - f_y \, dx),$$
where C is the boundary of $\mathscr{A}$.

17. Let f be a scalar function. Using Green's theorem, show that
$$\int_{\mathscr{A}} \left(\frac{\partial^2 f}{\partial x^2} + \frac{\partial^2 f}{\partial y^2}\right) dA = \oint_C \left(\frac{\partial f}{\partial x} dy - \frac{\partial f}{\partial y} dx\right),$$
where $\mathscr{A}$ is a convex region and C its boundary taken counterclockwise.

18. Let $\mathbf{F}$ be a vector field defined throughout the region bounded by the curve C. Show that
$$\oint_C (\text{curl } \mathbf{F}) \cdot \mathbf{n} \, ds = 0.$$

19. Evaluate $\oint_C \mathbf{F} \cdot \mathbf{n} \, ds$ where $\mathbf{F} = \mathbf{r}/|\mathbf{r}|^2$ and C is
    (a) the circle of radius 1 and center $(2, 0)$,
    (b) the circle of radius 3 and center $(2, 0)$.

20. Evaluate $\oint_C \mathbf{F} \cdot \mathbf{n} \, ds$ where $\mathbf{F} = \mathbf{r}/|\mathbf{r}|^3$ and C is
    (a) the circle of radius 1 and center $(0, 0)$,
    (b) the circle of radius 3 and center $(0, 0)$.

21. Assume that Green's theorem is known to hold for triangular regions.
    (a) Deduce that it then holds for regions bounded by quadrilaterals $ABCD$, as in Fig. 15.37. *Hint:* Draw the line AC.

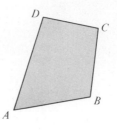

Figure 15.37

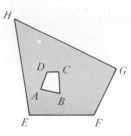

Figure 15.38

   (b) Deduce that Green's theorem holds for the region shown in Fig. 15.38.

Exercises 22 to 27 are related.

22. Let $\mathscr{A}$ be a simply connected plane region with boundary C swept out counterclockwise. Use Green's theorem to show that the area of $\mathscr{A}$ equals
$$\tfrac{1}{2} \oint_C (-y \, dx + x \, dy).$$

23. Use Exercise 22 to find the area of the region bounded by the line $y = x$ and the curve
$$\begin{cases} x = t^6 + t^4, \\ y = t^3 + t \end{cases}$$
for t in $[0, 1]$.

24. A curve is given parametrically by $x = t(1 - t^2)$, $y = t^2(1 - t^3)$, for t in $[0, 1]$.
    (a) Sketch the points corresponding to $t = 0$, 0.2, 0.4, 0.6, 0.8, and 1.0, and use them to sketch the curve.
    (b) Let $\mathscr{A}$ be the region enclosed by the curve. What difficulty arises when you try to compute the area of $\mathscr{A}$ by a definite integral involving vertical or horizontal cross sections?
    (c) Use Exercise 22 to find the area of $\mathscr{A}$.

25. Repeat Exercise 24 for $x = \sin \pi t$ and $y = t - t^2$. In (a), let $t = 0, \tfrac{1}{4}, \tfrac{1}{2}, \tfrac{3}{4}, 1$.

26. (a) Using Green's theorem, prove that $\oint_C x \, dy = A = -\oint_C y \, dx$, where A is the area of the region bounded by a simple closed curve C.
    (b) Why is (a) a generalization of Exercise 22?

27. Use Exercise 22 to obtain the formula for area in polar coordinates,
$$\text{Area} = \frac{1}{2} \int_\alpha^\beta r^2 \, d\theta.$$

■ ■

28. Assuming Green's theorem for regions bounded by a single curve, deduce Green's theorem for regions bounded by two curves, as in Fig. 15.39.
    Hint: Draw the lines AB and CD.

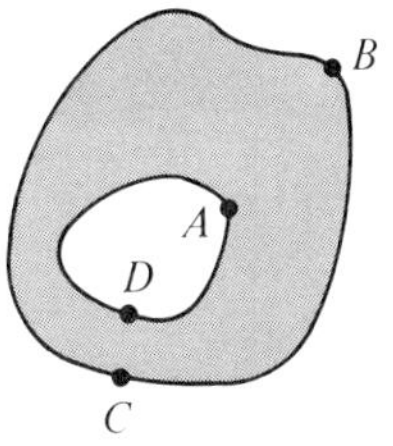

Figure 15.39

29 The moment of inertia about the z axis of a region $\mathscr{A}$ in the xy plane is defined as the integral $\int_{\mathscr{A}} (x^2 + y^2)\, dA$. Show that this equals

$$\tfrac{1}{3} \oint_C (-y^3\, dx + x^3\, dy),$$

where C bounds $\mathscr{A}$.

30 Show that, as stated in the physical motivation of Green's theorem,

$$\int_E^F (P\, dy - Q\, dx) + \int_G^D (P\, dy - Q\, dx)$$

is approximately $P_x(a, b)\, \Delta x\, \Delta y$.

31 Complete the proof of Green's theorem by showing that

$$\int_{\mathscr{A}} P_x\, dA = \oint_C P\, dy.$$

32 Let f be a continuous scalar field defined throughout the plane. Assume that for any convex region $\mathscr{A}$,

$$\int_{\mathscr{A}} f(P)\, dA = 0.$$

Explain why $f(P)$ is 0 for all points P. *Hint:* Consider arbitrarily small regions around a fixed point P_0 and assume that $f(P_0)$ is, say, positive.

33 Let $\mathbf{F}$ be a vector field with the property that, for any closed curve C,

$$\oint_C \mathbf{F} \cdot \mathbf{n}\, ds = 0.$$

Show that the divergence of $\mathbf{F}$ is 0 everywhere. *Hint:* See Exercise 32.

34 Let g be a scalar field with the property that

$$\oint_C \nabla g \cdot \mathbf{n}\, ds = 0$$

for all closed curves. Prove that $g_{xx} + g_{yy} = 0$. *Hint:* See Exercise 33.

35 The convex plane region $\mathscr{A}$ is occupied by a sheet of metal. By various heating and cooling devices the temperature along the border is kept fixed (independent of time). Assume that the temperature in $\mathscr{A}$ eventually stabilizes also. The steady-state temperature at point P is denoted $T(P)$.
  (a) Why is it plausible that heat in the metal at P tends to flow in the direction of $-\nabla T$?
  (b) Why is it plausible that the rate at which heat moves through P is proportional to $|-\nabla T|$?
  (c) If C is any curve in $\mathscr{A}$ that surrounds some region, why is $\oint_C (-\nabla T \cdot \mathbf{n})\, ds$ a plausible measure of the heat lost across C?
  (d) Why is $\oint_C \nabla T \cdot \mathbf{n}\, ds = 0$ for any such curve C?
  (e) Use Exercise 34 to show that the Laplacian of T, $T_{xx} + T_{yy}$, is 0.

Part (e) describes an important property of the temperature function.

36 (a) Show that the curve parametrized by

$$\mathbf{G}(t) = \frac{e^t + e^{-t}}{2}\mathbf{i} + \frac{e^t - e^{-t}}{2}\mathbf{j} = \cosh t\, \mathbf{i} + \sinh t\, \mathbf{j}$$

for t in $[0, a]$, $a > 0$, lies on the hyperbola $x^2 - y^2 = 1$ and joins the point $A = (1, 0)$ to the point $B = (\cosh a, \sinh a)$.
  (b) Let $\mathscr{A}$ be the region bounded by the line segment OA, the line segment OB, and the curve in (a) joining A and B. Using Exercise 22, show that the area of $\mathscr{A}$ is $a/2$.

Part (b) brings out another analogy between the circular and the hyperbolic functions. An angle θ corresponds to a sector of area $\theta/2$ in the unit circle. So $\cos\theta$ and $\sin\theta$ could be defined as the coordinates of the point on $x^2 + y^2 = 1$ corresponding to the sector of area $\theta/2$ one arm of which lies on the positive x axis. Similarly $\cosh a$ and $\sinh a$ could be defined as the coordinates of the point on $x^2 - y^2 = 1$ corresponding to a "sector" of area $a/2$ as described in part (b). However, this analogy is not as revealing as that in Sec. 10.10.

37 Let C_1 and C_2 be two closed curves, similarly oriented, that together form the boundary of the region $\mathscr{A}$ in the plane. Assume that $\nabla \times \mathbf{F} = \mathbf{0}$ in $\mathscr{A}$. Deduce that $\oint_{C_1} \mathbf{F} \cdot \mathbf{T}\, ds = \oint_{C_2} \mathbf{F} \cdot \mathbf{T}\, ds$.

38 (a) Prove that Green's theorem holds for right triangles. (*Hint:* introduce a coordinate system in such a way that the vertices of the right triangle are $(0, 0)$, $(a, 0)$, and (a, b).)
  (b) Deduce from (a) that Green's theorem holds for any triangle.

15.5 Surface integrals

The main object of this section is to introduce the notion of an integral over a surface. This definition is very similar to that given in Sec. 13.1 for the integral of a function over a plane region, that is, over a flat surface.

We shall not delve into fine points. The subject of surface area is fraught with far more difficulties than the area of a plane region. We shall assume that the surfaces we deal with are smooth, or composed of a finite number of smooth pieces. A sphere or the surface of a cube would be typical.

INTEGRATION OVER A SURFACE

DEFINITION

We use δ instead of f, because f will appear in the equations of some surfaces. Think of δ as "density."

Definite integral of a function δ over a surface $\mathscr{S}$. Let δ be a function that assigns to each point P in a surface $\mathscr{S}$ a number $\delta(P)$. Consider the typical sum

$$\delta(P_1)S_1 + \delta(P_2)S_2 + \cdots + \delta(P_n)S_n,$$

formed from a partition of $\mathscr{S}$, where S_i is the area of the ith region in the partition, and P_i is a point in the ith region. If these sums approach a certain number as the mesh of the partitions shrinks toward 0 (no matter how P_i is chosen in the ith region), the number is called the definite integral of δ over $\mathscr{S}$ and is written

$$\int_{\mathscr{S}} \delta(P) \, dS.$$

The definitions of partition and mesh, used in the preceding definition, being like those in Sec. 13.1, will be omitted.

If $\delta(P)$ is the density of matter at P, then $\int_{\mathscr{S}} \delta(P) \, dS$ is the total mass of the surface. If $\delta(P) = 1$ for all P, then $\int_{\mathscr{S}} \delta(P) \, dS$ is the area of $\mathscr{S}$. Though the primary interest in the surface integral is conceptual, we will show how to compute it.

One method of computing a surface integral is to express it as an integral over a region in the plane. This method is based on the following geometric lemma, which, in turn, depends on the relation between the hypotenuse and leg of a right triangle, as shown in Fig. 15.40.

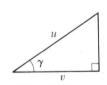

Figure 15.40

LEMMA

The angle between two planes is chosen to be acute.

Let U be a region in a plane that is inclined at the angle $\gamma < \pi/2$ to the xy plane. Let V be the set in the xy plane directly below (or above) U. Then

$$\text{Area of } U = (\sec \gamma) \text{ Area of } V.$$

(See Fig. 15.41.)

PROOF Introduce a t axis parallel both to the xy plane and to the plane of U. Let $c(t)$ be the cross-sectional length of U for a given t. Then the corresponding cross-sectional length for V is

$$\cos \gamma \, c(t).$$

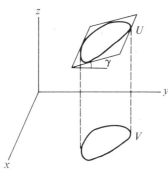

Figure 15.41   **Figure 15.42**

(See Fig. 15.42.) Thus

$$\text{Area of } V = \int_a^b \cos \gamma \; c(t) \, dt$$

$$= \cos \gamma \int_a^b c(t) \, dt$$

$$= (\cos \gamma) \text{ Area of } U.$$

The lemma follows immediately. ∎

EXAMPLE 1 Find the area of that portion of the plane $x + 2y + 3z = 12$ that lies inside the cylinder $x^2 + y^2 = 9$.

SOLUTION Let U be the planar region in question, and V its projection on the xy plane. The region V, being a circle of radius 3, has area 9π. To find the area of U, the portion of the plane above V, it is necessary to find $\sec \gamma$, where γ is the angle between the xy plane and the plane $x + 2y + 3z = 12$. Now, the angle between two planes is the same as the angle between their normals. (See Fig. 15.43.) The vector

$$\mathbf{i} + 2\mathbf{j} + 3\mathbf{k}$$

is a normal to the tilted plane, and the vector $\mathbf{k}$ is a normal to the xy plane. Therefore,

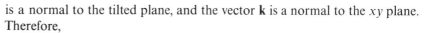

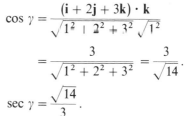

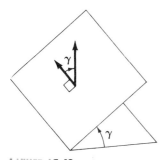

Figure 15.43

Hence $\sec \gamma = \dfrac{\sqrt{14}}{3}.$

By the lemma, the area of the tilted flat region U is thus

$$(\sec \gamma) \text{ Area of } V = \frac{\sqrt{14}}{3} \cdot 9\pi = 3\sqrt{14}\,\pi. \quad \blacksquare$$

The lemma is the key to the following theorem.

THEOREM 1 Let $\mathscr{S}$ be a surface and let $\mathscr{A}$ be a region in the xy plane with the property that for each point Q in $\mathscr{A}$ the line through Q parallel to the z axis meets $\mathscr{S}$ in exactly one point P. Let δ be a function defined on $\mathscr{S}$. Define a function g on $\mathscr{A}$ by

$$g(Q) = \delta(P).$$

Think of this as saying that $dS = (\sec \gamma)\, dA$.

Then
$$\int_{\mathscr{S}} \delta(P)\, dS = \int_{\mathscr{A}} g(Q) \sec \gamma \, dA.$$

In this equation γ denotes the angle between the tangent plane to the surface at P and the xy plane. (See Fig. 15.44.)

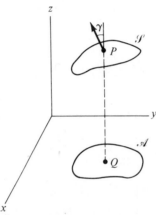

Figure 15.44

PROOF Imagine a typical set U_i in a partition of $\mathscr{S}$. Let its area be S_i. Choose a point P_i in this set. Let V_i be the "shadow," or the corresponding set in $\mathscr{A}$. Let the area of V_i be A_i. Let Q_i be the point located in $\mathscr{A}$ that corresponds to P_i. (See Fig. 15.45.)

Now, if U_i is small, it looks very much like part of a plane (just as a small portion of a curve looks much like a line). Hence it is reasonable to assume that, if S is smooth, then

$$\text{Area of } U_i \approx (\sec \gamma_i) \text{ Area of } V_i,$$

where γ_i is the angle between the tangent plane at P_i and the xy plane.

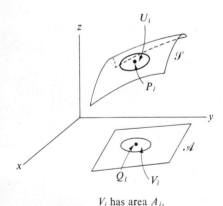

V_i has area A_i.

Figure 15.45

Thus
$$S_i \approx (\sec \gamma_i) A_i,$$
and
$$\delta(P_i) S_i \approx \delta(P_i)(\sec \gamma_i) A_i$$
$$= g(Q_i)(\sec \gamma_i) A_i.$$

Adding and taking limits as the mesh approaches 0, shows that

$$\int_{\mathscr{S}} \delta(P)\, dS = \int_{\mathscr{A}} g(Q) \sec \gamma \, dA. \quad \blacksquare$$

15.5 SURFACE INTEGRALS

Just as we replace a line integral by an integral over an interval, we have replaced an integral over a surface by an integral over a plane region.

The following example will show how Theorem 1 is applied. Note that the factor $\sec \gamma$ permits the evaluation of a surface integral by a plane integral. In particular the case $\delta(P) = 1$ gives

$$\int_{\mathscr{S}} 1 \, dS = \int_{\mathscr{A}} 1 \sec \gamma \, dA$$

How to find surface area   or   Area of $\mathscr{S} = \int_{\mathscr{A}} \sec \gamma \, dA.$

EXAMPLE 2   Let $\mathscr{S}$ be the top half of the sphere $x^2 + y^2 + z^2 = a^2$, as shown in Fig. 15.46. Let $\delta(P)$ be the distance from point P on the sphere to the xy plane. Evaluate $\int_{\mathscr{S}} \delta(P) \, dS$.

SOLUTION   In this case $\delta(P) = f(x, y, z) = z$. We wish to evaluate

$$\int_{\mathscr{S}} z \, dS.$$

By Theorem 1, $\int_{\mathscr{S}} z \, dS = \int_{\mathscr{A}} z \sec \gamma \, dA,$

where $\mathscr{A}$ is the shaded disk in Fig. 15.46.

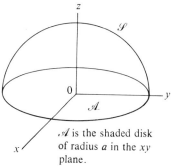

$\mathscr{A}$ is the shaded disk of radius a in the xy plane.

Figure 15.46

To find γ at $P = (x, y, z)$, recall that a radius vector $\overrightarrow{OP}$ is normal to the sphere. Thus γ is the angle between the vectors $x\mathbf{i} + y\mathbf{j} + z\mathbf{k}$ and $\mathbf{k}$.

Consequently, $\cos \gamma = \dfrac{(x\mathbf{i} + y\mathbf{j} + z\mathbf{k}) \cdot \mathbf{k}}{\sqrt{x^2 + y^2 + z^2} \sqrt{1^2}}$

$$= \dfrac{z}{a}.$$

From this it follows that $\sec \gamma = \dfrac{a}{z}.$

Thus $\int_{\mathscr{A}} z \sec \gamma \, dA = \int_{\mathscr{A}} z \cdot \dfrac{a}{z} \, dA = \int_{\mathscr{A}} a \, dA = a \int_{\mathscr{A}} dA = a \cdot \pi a^2.$

In short, $\int_{\mathscr{S}} z \, dS = \pi a^3.$ ∎

Incidentally, the z coordinate $\bar{z}$ of the center of mass of the hemispherical surface in Example 2 is defined as $\int_{\mathscr{S}} z \, dS / \text{Area of } \mathscr{S}$. Thus it is

$$\frac{\pi a^3}{2\pi a^2} = \frac{a}{2}.$$

HOW TO FIND sec γ

In Examples 1 and 2, sec γ was found by using the fact that the surface was a plane or sphere. The next theorem shows how to find sec γ in general.

THEOREM 2 (a) If the surface $\mathscr{S}$ is given as the level surface $g(x, y, z) = c$, then

$$\sec \gamma = \frac{\sqrt{(g_x)^2 + (g_y)^2 + (g_z)^2}}{|g_z|}.$$

(b) If the surface $\mathscr{S}$ is given in the form $z = f(x, y)$, then

$$\sec \gamma = \sqrt{(f_x)^2 + (f_y)^2 + 1}.$$

PROOF (a) A normal vector to $\mathscr{S}$ at a given point is provided by the gradient $\nabla g = g_x \mathbf{i} + g_y \mathbf{j} + g_z \mathbf{k}$. The cosine of the angle between $\mathbf{k}$ and ∇g is

$$\frac{\mathbf{k} \cdot \nabla g}{|\mathbf{k}||\nabla g|} = \frac{\mathbf{k} \cdot (g_x \mathbf{i} + g_y \mathbf{j} + g_z \mathbf{k})}{|\mathbf{k}| \cdot |g_x \mathbf{i} + g_y \mathbf{j} + g_z \mathbf{k}|}$$

$$= \frac{g_z}{\sqrt{(g_x)^2 + (g_y)^2 + (g_z)^2}}.$$

sec γ is positive. Thus

$$\sec \gamma = \frac{\sqrt{(g_x)^2 + (g_y)^2 + (g_z)^2}}{|g_z|}$$

(b) Rewrite $z = f(x, y)$ as $z - f(x, y) = 0$. The surface $z = f(x, y)$ is thus the level surface $g(x, y, z) = 0$ of the function $g(x, y, z) = z - f(x, y)$. Note that

$$g_x = -f_x, \qquad g_y = -f_y, \qquad \text{and} \qquad g_z = 1.$$

By the formula in (a),

$$\sec \gamma = \frac{\sqrt{(-f_x)^2 + (-f_y)^2 + 1^2}}{1}$$

$$= \sqrt{(f_x)^2 + (f_y)^2 + 1}.$$

This establishes the second formula for sec γ. ∎

The next two examples illustrate the use of these two formulas.

EXAMPLE 3 Find the area of that portion of the surface $z - xy = 0$ that is inside the cylinder $x^2 + y^2 = a^2$.

SOLUTION Let $\mathscr{S}$ be the surface described. The area of $\mathscr{S}$ equals the integral $\int_{\mathscr{A}} \sec \gamma \, dA$, where $\mathscr{A}$ is the projection of $\mathscr{S}$ on the xy plane, that is, the disk of radius a and center $(0, 0)$.

The reader may sketch the saddle $z - xy = 0$ and $\mathscr{S}$.

The surface is given in the form $g(x, y, z) = 0$, where $g(x, y, z)$ is $z - xy$. By part (a) of Theorem 2,

$$\sec \gamma = \frac{\sqrt{g_x^2 + g_y^2 + g_z^2}}{|g_z|} = \frac{\sqrt{(-y)^2 + (-x)^2 + 1^2}}{1}.$$

Hence
$$\text{Area of } \mathscr{S} = \int_{\mathscr{A}} \sqrt{y^2 + x^2 + 1} \, dA.$$

Switching to polar coordinates in $\mathscr{A}$, we obtain

$$\int_{\mathscr{A}} \sqrt{y^2 + x^2 + 1} \, dA = \int_0^{2\pi} \left(\int_0^a \sqrt{r^2 + 1} \, r \, dr \right) d\theta.$$

First of all,
$$\int_0^a \sqrt{r^2 + 1} \, r \, dr = \frac{(r^2 + 1)^{3/2}}{3} \Big|_0^a$$
$$= \frac{(a^2 + 1)^{3/2} - 1^{3/2}}{3}$$

Then
$$\int_0^{2\pi} \frac{(a^2 + 1)^{3/2} - 1}{3} \, d\theta = \frac{2\pi}{3} [(a^2 + 1)^{3/2} - 1]. \quad \blacksquare$$

In the next example part (b) of Theorem 2 will be used to compute $\sec \gamma$.

EXAMPLE 4 Let $\mathscr{S}$ be the portion of the surface $z = 9 - x^2 - y^2$ above the xy plane. Let

$$\delta(x, y, z) = \frac{2x^2 + 2y^2 + z}{\sqrt{4x^2 + 4y^2 + 1}}.$$

Compute
$$\int_{\mathscr{S}} \delta(P) \, dS.$$

SOLUTION The region $\mathscr{S}$, as well as its shadow $\mathscr{A}$ on the xy plane, is shown in Fig. 15.47. In order to transform $\int_{\mathscr{S}} \delta(P) \, dS$ to an integral over $\mathscr{A}$ we must compute $\sec \gamma$. Use the formula

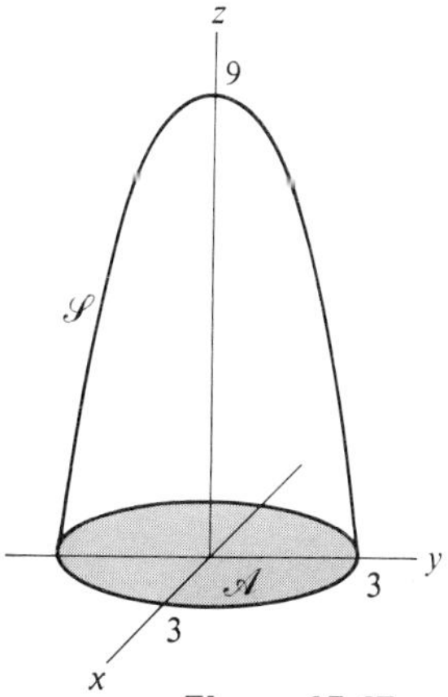

$$\sec \gamma = \sqrt{f_x^2 + f_y^2 + 1}$$
$$= \sqrt{(-2x)^2 + (-2y)^2 + 1}$$
$$= \sqrt{4x^2 + 4y^2 + 1}.$$

Thus
$$\int_{\mathscr{S}} \frac{2x^2 + 2y^2 + z}{\sqrt{4x^2 + 4y^2 + 1}} \, dS = \int_{\mathscr{A}} \frac{2x^2 + 2y^2 + z}{\sqrt{4x^2 + 4y^2 + 1}} \cdot \sqrt{4x^2 + 4y^2 + 1} \, dA$$
$$= \int_{\mathscr{A}} (2x^2 + 2y^2 + z) \, dA.$$

Figure 15.47

Remember that z depends on x and y.
The integrand, $2x^2 + 2y^2 + z$, which appears to be a function of x, y, and z, is actually a function only of x and y because $z = 9 - x^2 - y^2$. Consequently, the integral to be evaluated is

$$\int_{\mathscr{A}} [2x^2 + 2y^2 + (9 - x^2 - y^2)] \, dA = \int_{\mathscr{A}} (9 + x^2 + y^2) \, dA.$$

This final integral is typical of the integrals met in Chap. 13. Since $\mathscr{A}$ is a disk and $x^2 + y^2$ appears in the integrand, it is natural to evaluate the integral using polar coordinates in $\mathscr{A}$. We have then

$$\int_{\mathscr{A}} (9 + x^2 + y^2) \, dA = \int_0^{2\pi} \left[\int_0^3 (9 + r^2) r \, dr \right] d\theta$$

$$= \int_0^{2\pi} \left[\int_0^3 (9r + r^3) \, dr \right] d\theta.$$

A direct computation shows

$$\int_0^3 (9r + r^3) \, dr = \frac{243}{4},$$

and then that

$$\int_0^{2\pi} \frac{243}{4} \, d\theta = \frac{243\pi}{2}.$$

Thus
$$\int_{\mathscr{S}} \frac{2x^2 + 2y^2 + z}{\sqrt{4x^2 + 4y^2 + 1}} \, dS = \frac{243\pi}{2}. \blacksquare$$

INTEGRATING OVER A SPHERE

If $\mathscr{S}$ is a sphere or part of a sphere, it is sometimes possible to evaluate an integral over it with the aid of spherical coordinates.

If the center of a spherical coordinate system (ρ, θ, ϕ) is at the center of a sphere, then ρ is constant on the sphere. As Fig. 15.48 suggests, the area of the small region on the sphere corresponding to slight changes $d\phi$ and $d\theta$ is approximately

$$(\rho \, d\phi)(\rho \sin \phi \, d\theta) = \rho^2 \sin \phi \, d\theta \, d\phi.$$

See Sec. 13.8 for a similar argument, where ρ was not constant. Thus we may write

$$dS = \rho^2 \sin \phi \, d\theta \, d\phi$$

and evaluate
$$\int_{\mathscr{S}} \delta(P) \, dS$$

in terms of a repeated integral in ϕ and θ. The next example illustrates this technique.

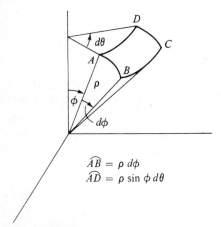

$\widehat{AB} = \rho \, d\phi$
$\widehat{AD} = \rho \sin \phi \, d\theta$

Figure 15.48

EXAMPLE 5 Let $\mathscr{S}$ be the top half of the sphere $x^2 + y^2 + z^2 = a^2$. Evaluate

$$\int_{\mathscr{S}} z^3 \, dS.$$

SOLUTION In spherical coordinates $z = \rho \cos \phi$. Since the sphere has radius a, $\rho = a$. Thus

$$\int_{\mathscr{S}} z^3 \, dS = \int_{\mathscr{S}} (a \cos \phi)^3 \, dS$$

$$= \int_0^{2\pi} \left[\int_0^{\pi/2} (a \cos \phi)^3 a^2 \sin \phi \, d\phi \right] d\theta.$$

Now,

$$\int_0^{\pi/2} (a \cos \phi)^3 a^2 \sin \phi \, d\phi = a^5 \int_0^{\pi/2} \cos^3 \phi \sin \phi \, d\phi$$

$$= a^5 \left. \frac{(-\cos^4 \phi)}{4} \right|_0^{\pi/2}$$

$$= \frac{a^5}{4} [-0 - (-1)]$$

$$= \frac{a^5}{4},$$

so that

$$\int_{\mathscr{S}} z^3 \, dS = \int_0^{2\pi} \frac{a^5}{4} \, d\theta$$

$$= \frac{\pi a^5}{2}. \quad \blacksquare$$

Exercises

1. A triangle in the plane $z = x + y$ is directly above the triangle in the xy plane whose vertices are $(1, 2)$, $(3, 4)$, and $(2, 5)$. Find the area of (a) the triangle in the xy plane, (b) the triangle in the plane $z = x + y$.

2. Let $\mathscr{S}$ be the triangle with vertices $(1, 1, 1)$, $(2, 3, 4)$, and $(3, 4, 5)$.
   (a) Using vectors, find the area of $\mathscr{S}$.
   (b) Using the formula, Area of $\mathscr{S} = \int_{\mathscr{A}} \sec \gamma \, dA$, find the area of $\mathscr{S}$.

3. Find the area of the portion of the cone $z^2 = x^2 + y^2$ that lies above one loop of the curve $r = \sqrt{\cos 2\theta}$.

4. Let $\mathscr{S}$ be the triangle whose vertices are $(1, 0, 0)$, $(0, 2, 0)$, and $(0, 0, 3)$. Let $\delta(x, y, z) = 3x + 2y + 2z$. Evaluate $\int_{\mathscr{S}} \delta(P) \, dS$.

5. Let $\mathscr{S}$ be the triangle whose vertices are $(1, 0, 0)$, $(0, 1, 0)$, and $(0, 0, 1)$. Let
   $$\delta(x, y, z) = (1/\sqrt{3})(-3x + z - 1).$$
   Evaluate $\int_{\mathscr{S}} \delta(P) \, dS$.

6. Let $\mathscr{S}$ be the portion of the surface $z = 9 - x^2 - 2y^2$ that lies above the xy plane. Let $\delta(x, y, z) = (2x^2 + 4y^2 + 2z)/\sqrt{1 + 4x^2 + 16y^2}$. Evaluate
   $$\int_{\mathscr{S}} \delta(P) \, dS.$$

The z coordinate of the centroid of a surface $\mathscr{S}$ is defined as $\bar{z} = \int_{\mathscr{S}} z \, dS/\text{Area of } \mathscr{S}$. In Exercises 7 to 10 find $\bar{z}$ for the given surfaces.

7. The portion of the paraboloid $2z = x^2 + y^2$ below the plane $z = 9$.

8. The portion of the plane $x + 2y + 3z = 6$ above the triangle in the xy plane whose vertices are $(0, 0)$, $(4, 0)$, and $(0, 1)$.

9. The portion of the sphere $x^2 + y^2 + z^2 = 9$ that lies within the cone $z = \sqrt{x^2 + y^2}$.

10. The portion of the cylindrical surface $z = 2\sqrt{x}$ above the square in the xy plane whose vertices are $(0, 0)$, $(1, 0)$, $(1, 1)$, and $(0, 1)$.

In Exercises 11 and 12 evaluate $\int_{\mathscr{S}} \mathbf{F} \cdot \mathbf{n}\, dS$ for the given spheres and vector fields. (**n** is an exterior normal.)

11 The sphere $x^2 + y^2 + z^2 = 9$ and $\mathbf{F} = x^2\mathbf{i} + y^2\mathbf{j} + z^2\mathbf{k}$.
12 The sphere $x^2 + y^2 + z^2 = 1$ and $\mathbf{F} = x^3\mathbf{i} + y^2\mathbf{j}$.
13 Find the area of that portion of the parabolic cylinder $z = \tfrac{1}{2}x^2$ between the three planes $y = 0$, $y = x$, and $x = 2$.
14 Find the area of the part of the spherical surface $x^2 + y^2 + z^2 = 1$ that lies within the vertical cylinder erected on the circle $r = \cos\theta$ and above the xy plane.

■

15 Let $\mathscr{S}$ be the portion of the surface $z = x^2 + 3y^2$ below the surface $z = 1 - x^2 - y^2$.
(a) Set up a plane integral for the area of $\mathscr{S}$.
(b) Set up repeated integrals in rectangular coordinates and in polar coordinates for the integral in (a), but do not evaluate them.
16 A sphere of radius $2a$ has its center at the origin of a rectangular coordinate system. A circular cylinder of radius a has its axis parallel to the z axis and passes through the z axis. Find the area of that part of the sphere that lies within the cylinder and is above the xy plane.

Consider a distribution of mass on the surface $\mathscr{S}$. Let its density at P be $\delta(P)$. The moment of inertia of the mass around the z axis is defined as $\int_{\mathscr{S}} (x^2 + y^2)\, \delta(P)\, dS$. Exercises 17 and 18 concern this integral.

17 Find the moment of inertia of a homogeneous distribution of mass on the surface of a ball of radius a around a diameter. Let the total mass be M.
18 Find the moment of inertia about the z axis of a homogeneous distribution of mass on the triangle whose vertices are $(a, 0, 0)$, $(0, b, 0)$, and $(0, 0, c)$. Take a, b, and c to be positive. Let the total mass be M.

In Exercises 19 and 20 let $\mathscr{S}$ be a sphere of radius a with center at the origin of a rectangular coordinate system.

19 Evaluate each of these integrals with a minimum amount of labor.

(a) $\int_{\mathscr{S}} x\, dS$ (b) $\int_{\mathscr{S}} x^3\, dS$ (c) $\int_{\mathscr{S}} \dfrac{2x + 4y^5}{\sqrt{2 + x^2 + 3y^2}}\, dS$.

20 (a) Why is $\int_{\mathscr{S}} x^2\, dS = \int_{\mathscr{S}} y^2\, dS$?
(b) Evaluate $\int_{\mathscr{S}} (x^2 + y^2 + z^2)\, dS$ with a minimum amount of labor.
(c) In view of (a) and (b), evaluate $\int_{\mathscr{S}} x^2\, dS$.
(d) Evaluate $\int_{\mathscr{S}} (2x^2 + 3y^2)\, dS$.
21 An electric field radiates power at the rate of $k (\sin^2 \phi)/\rho^2$ units per square meter to the point $P = (\rho, \theta, \phi)$. Find the total power radiated to the sphere $\rho = a$.

The average of a function $\delta(P)$ over a surface $\mathscr{S}$ is defined as

$$\dfrac{\int_{\mathscr{S}} \delta(P)\, dS}{\text{Area of } \mathscr{S}}.$$

This concept is illustrated in Exercises 22 and 23.

22 Let $\mathscr{S}$ be a sphere of radius a. Let A be a point at distance $b > a$ from the center of $\mathscr{S}$. For P in $\mathscr{S}$ let $\delta(P)$ be $1/q$, where q is the distance from P to A. Show that the average of $\delta(P)$ over $\mathscr{S}$ is $1/b$.
23 The data are the same as in Exercise 22 but $b < a$. Show that in this case the average of $1/q$ is $1/a$. (The average does *not* depend on b in this case.)

■ ■

24 Let C be a curve in the xy plane whose equation is $y = f(x)$ for x in $[a, b]$. Let γ be the acute angle between a line perpendicular to the curve at (x, y) and the y axis. Show that the length of C is $\int_a^b \sec \gamma\, dx$.
(This brings out the similarity of the formulas for arc length and surface area. Compare also the formula $\int_a^b \sqrt{1 + (df/dx)^2}\, dx$ for arc length with the formula $\int_{\mathscr{A}} \sqrt{1 + f_x^2 + f_y^2}\, dA$ for surface area.)
25 Let $z = g(y)$ be positive for $a \leq y \leq b$. The area of the surface formed by revolving this curve around the y axis is $2\pi \int_a^b g(y)\sqrt{1 + (dg/dy)^2}\, dy$. Derive this formula from the general formula, Area of $\mathscr{S} = \int_{\mathscr{A}} \sec \gamma\, dA$, as follows:
(a) Show that the surface of revolution has the equation $z^2 + x^2 = [g(y)]^2$.
(b) Show that $x\mathbf{i} - g(y)\, dg/dy\, \mathbf{j} + z\mathbf{k}$ is a normal to the surface at (x, y, z).
(c) Let $\mathscr{A}$ be the region in the xy plane bounded by the interval $a \leq y \leq b$ on the y axis, the curve $x = g(y)$, and the lines $y = a$ and $y = b$. Using (b), evaluate $\int_{\mathscr{A}} \sec \gamma\, dA$.
26 Let $\mathscr{S}$ be the portion of the surface $g(x, y, z) = c$ above the region $\mathscr{A}$ in the xy plane. Let $\mathbf{F} = P\mathbf{i} + Q\mathbf{j} + R\mathbf{k}$. Show that

$$\int_{\mathscr{S}} \mathbf{F} \cdot \mathbf{n}\, dS = \int_{\mathscr{A}} \left(P\dfrac{g_x}{g_z} + Q\dfrac{g_y}{g_z} + R \right) dA,$$

where **n** has a positive z component.
27 Let $\mathscr{S}$ be the portion of the surface $z = f(x, y)$ above the region $\mathscr{A}$ in the xy plane. Let $\mathbf{F} = P\mathbf{i} + Q\mathbf{j} + R\mathbf{k}$. Show that $\int_{\mathscr{S}} \mathbf{F} \cdot \mathbf{n}\, dS = \int_{\mathscr{A}} (-Pf_x - Qf_y + R)\, dA$, where **n** has a positive z component.

15.6 The divergence theorem

A set is connected if any two points in it can be joined by a curve lying in the set.

The divergence theorem is the analog in space of Green's theorem in the plane. Instead of a region in the plane and its bounding curve or curves, the divergence theorem concerns a finite region $\mathscr{V}$ in space and its bounding surface or surfaces. The region $\mathscr{V}$ may, for instance, be a ball, cube, or doughnut; in these cases the bounding surface consists of a single connected piece. Or perhaps the region $\mathscr{V}$ is bounded by two surfaces $\mathscr{S}_1$ and $\mathscr{S}_2$, one inside the other like two concentric spheres. Figure 15.49 illustrates this type of region.

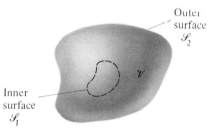

Figure 15.49

It's not easy to make a convincing drawing of one surface inside another, so let us suggest the idea by using curves to represent surfaces, as in Fig. 15.50, and not indicating perspective.

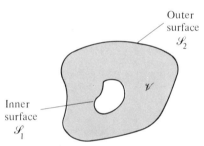

Figure 15.50

Just as Green's theorem was stated separately for a plane region $\mathscr{A}$ bounded by a single curve C and for a plane region bounded by two curves C_1 and C_2, the divergence theorem will be stated in two forms.

DIVERGENCE THEOREM   *One-surface case*   Let $\mathscr{V}$ be a region in space bounded by the single connected surface $\mathscr{S}$. Let $\mathbf{n}$ denote the exterior normal of $\mathscr{V}$ along its boundary $\mathscr{S}$. Then

$$\int_{\mathscr{S}} \mathbf{F} \cdot \mathbf{n} \, dS = \int_{\mathscr{V}} \nabla \cdot \mathbf{F} \, dV$$

for any vector field $\mathbf{F}$ defined on $\mathscr{V}$. ∎

Direction cosines were defined in Sec. 11.4.

In words: The integral of the normal component of $\mathbf{F}$ over a surface equals the integral of the divergence of $\mathbf{F}$ over the solid region that the surface bounds. If $\mathbf{F} = P\mathbf{i} + Q\mathbf{j} + R\mathbf{k}$, then the divergence theorem reads

998 GREEN'S THEOREM, THE DIVERGENCE THEOREM, AND STOKES' THEOREM

$$\int_{\mathscr{S}} (P\mathbf{i} + Q\mathbf{j} + R\mathbf{k}) \cdot (\cos \alpha \, \mathbf{i} + \cos \beta \, \mathbf{j} + \cos \gamma \, \mathbf{k}) \, dS = \int_{\mathscr{V}} (P_x + Q_y + R_z) \, dV,$$

where $\cos \alpha$, $\cos \beta$, and $\cos \gamma$ are the direction cosines of the exterior normal.

EXAMPLE 1 Let $\mathbf{F} = (z^2 + 2)\mathbf{k}$ and let $\mathscr{S}$ consist of $\mathscr{H}$, the top half of the sphere $x^2 + y^2 + z^2 = a^2$, together with its base $\mathscr{B}$, the disk of radius a centered at $(0, 0)$ in the xy plane, as shown in Fig. 15.51. Verify the divergence theorem in this case.

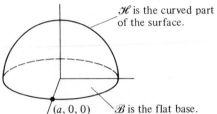

Figure 15.51

SOLUTION Let $\mathscr{V}$ be the solid region below $\mathscr{H}$ and above $\mathscr{B}$. Then

$\mathscr{V}$ is a hemisphere.

$$\int_{\mathscr{V}} \nabla \cdot \mathbf{F} \, dV = \int_{\mathscr{V}} \left[\frac{\partial}{\partial x}(0) + \frac{\partial}{\partial y}(0) + \frac{\partial}{\partial z}(z^2 + 2) \right] dV$$

$$= \int_{\mathscr{V}} 2z \, dV$$

In spherical coordinates
$z = \rho \cos \phi$.

$$= \int_0^{2\pi} \int_0^{\pi/2} \int_0^a 2(\rho \cos \phi) \rho^2 \sin \phi \, d\rho \, d\phi \, d\theta$$

$$= \int_0^{2\pi} \int_0^{\pi/2} \int_0^a 2\rho^3 \cos \phi \sin \phi \, d\rho \, d\phi \, d\theta.$$

Straightforward evaluation of the repeated integral gives the value $\pi a^4/2$.

Next evaluate $\int_{\mathscr{S}} \mathbf{F} \cdot \mathbf{n} \, dS$. It is necessary to compute $\int_{\mathscr{B}} \mathbf{F} \cdot \mathbf{n} \, dS$ and $\int_{\mathscr{H}} \mathbf{F} \cdot \mathbf{n} \, dS$. On $\mathscr{B}$ the exterior normal is $-\mathbf{k}$. Thus

$$\int_{\mathscr{B}} \mathbf{F} \cdot \mathbf{n} \, dS = \int_{\mathscr{B}} (z^2 + 2)\mathbf{k} \cdot (-\mathbf{k}) \, dS.$$

On $\mathscr{B}$, however, $z = 0$. So

$$\int_{\mathscr{B}} \mathbf{F} \cdot \mathbf{n} \, dS = \int_{\mathscr{B}} -2 \, dS = -2 \, (\text{Area of } \mathscr{B})$$

$$= -2\pi a^2.$$

To evaluate $\int_{\mathscr{H}} \mathbf{F} \cdot \mathbf{n} \, dS$, recall that, on a sphere of radius a centered at the origin,

$$\frac{x\mathbf{i} + y\mathbf{j} + z\mathbf{k}}{|x\mathbf{i} + y\mathbf{j} + z\mathbf{k}|} = \frac{x\mathbf{i} + y\mathbf{j} + z\mathbf{k}}{a}$$

is a unit exterior normal. Thus

$$\int_{\mathcal{H}} \mathbf{F} \cdot \mathbf{n}\, dS = \int_{\mathcal{H}} (z^2 + 2)\mathbf{k} \cdot \frac{(x\mathbf{i} + y\mathbf{j} + z\mathbf{k})}{a}\, dS$$

$$= \int_{\mathcal{H}} \frac{z^3 + 2z}{a}\, dS = \frac{1}{a} \int_{\mathcal{H}} (z^3 + 2z)\, dS.$$

Using spherical coordinates on the sphere of radius a, we have

$$\int_{\mathcal{H}} (z^3 + 2z)\, dS = \int_0^{2\pi} \int_0^{\pi/2} [(a \cos \phi)^3 + 2a \cos \phi] a^2 \sin \phi\, d\phi\, d\theta$$

$$= a^3 \int_0^{2\pi} \int_0^{\pi/2} (a^2 \cos^3 \phi \sin \phi + 2 \cos \phi \sin \phi)\, d\phi\, d\theta$$

$$= a^3 \left(\frac{a^2}{4} + 1 \right)(2\pi).$$

Thus
$$\int_{\mathcal{H}} \mathbf{F} \cdot \mathbf{n}\, dS = \frac{1}{a} \left(a^3 \left(\frac{a^2}{4} + 1 \right)(2\pi) \right)$$

$$= \frac{\pi a^4}{2} + 2\pi a^2.$$

All told,

$$\int_{\mathcal{S}} \mathbf{F} \cdot \mathbf{n}\, dS = \int_{\mathcal{B}} \mathbf{F} \cdot \mathbf{n}\, dS + \int_{\mathcal{H}} \mathbf{F} \cdot \mathbf{n}\, dS = -2\pi a^2 + \left(\frac{\pi a^4}{2} + 2\pi a^2 \right)$$

$$= \frac{\pi a^4}{2},$$

which agrees with the result for $\int_{\mathcal{V}} \nabla \cdot \mathbf{F}\, dV$. ∎

An argument for the divergence theorem in terms of net loss of fluid (similar to that for Green's theorem) is presented at the end of the section. A mathematical proof is outlined in Exercise 25. In practice the divergence theorem is used to replace an integral of $\mathbf{F} \cdot \mathbf{n}$ over a surface by an integral of $\nabla \cdot \mathbf{F}$ over a solid region, or vice versa.

The divergence theorem also holds if the solid region has several holes like a piece of Swiss cheese. In this case the boundary consists of several separate connected surfaces. The most important case is when there is just one hole and hence an inner surface $\mathcal{S}_1$ and an outer surface $\mathcal{S}_2$, as shown in Fig. 15.49.

DIVERGENCE THEOREM *Two-surface case.* Let $\mathcal{V}$ be a region in space bounded by the surfaces $\mathcal{S}_1$ and $\mathcal{S}_2$. Let $\mathbf{n}^*$ denote the exterior normal along the boundary. Then

$$\int_{\mathcal{S}_1} \mathbf{F} \cdot \mathbf{n}^*\, dS + \int_{\mathcal{S}_2} \mathbf{F} \cdot \mathbf{n}^*\, dS = \int_{\mathcal{V}} \nabla \cdot \mathbf{F}\, dV$$

for any vector field defined on $\mathcal{V}$. ∎

The importance of this form of the divergence theorem is that it provides us with Corollary 1 in the same way that the two-curve form of Green's theorem provided Corollary 1 of Sec. 15.4.

COROLLARY 1 Let $\mathscr{S}_1$ and $\mathscr{S}_2$ be two connected surfaces that form the boundary of the region $\mathscr{V}$. Let $\mathbf{F}$ be a vector field defined on $\mathscr{V}$ such that the divergence of $\mathbf{F}$, $\nabla \cdot \mathbf{F}$, is 0 throughout $\mathscr{V}$. Then

$$\int_{\mathscr{S}_1} \mathbf{F} \cdot \mathbf{n} \, dS = \int_{\mathscr{S}_2} \mathbf{F} \cdot \mathbf{n} \, dS. \quad \blacksquare$$

The proof, similar to that for Corollary 1 in Sec. 15.4, is omitted. The next corollary and its proof are similar to Corollary 2 of Sec. 15.4 and its proof.

COROLLARY 2 Let $\mathbf{F}$ be a vector field defined everywhere in space except perhaps at the point P_0. Assume that the divergence of $\mathbf{F}$ is 0. Let $\mathscr{S}_1$ and $\mathscr{S}_2$ be two surfaces each of which encloses the point P_0. Then

$$\int_{\mathscr{S}_1} \mathbf{F} \cdot \mathbf{n} \, dS = \int_{\mathscr{S}_2} \mathbf{F} \cdot \mathbf{n} \, dS. \quad \blacksquare$$

The next example plays an important role in electrostatics and gravitational theory.

EXAMPLE 2 Let $\mathbf{F} = \mathbf{r}/|\mathbf{r}|^3$, where $\mathbf{r} = x\mathbf{i} + y\mathbf{j} + z\mathbf{k}$ and let $\mathscr{S}$ be a surface that surrounds the origin, as in Fig. 15.52. Show that $\int_{\mathscr{S}} \mathbf{F} \cdot \mathbf{n} \, dS = 4\pi$.

SOLUTION By Example 3 of Sec. 15.1, the divergence of $\mathbf{F}$ is 0. According to Corollary 2, then, we may learn the value of $\int_{\mathscr{S}} \mathbf{F} \cdot \mathbf{n} \, dS$ by computing $\int_{\mathscr{S}_1} \mathbf{F} \cdot \mathbf{n} \, dS$ for any surface $\mathscr{S}_1$ that encloses the origin. The simplest surface to use is a sphere $\mathscr{S}_1$ of radius a and center at the origin. On $\mathscr{S}_1$, $|\mathbf{r}| = a$ and $\mathbf{r} \cdot \mathbf{n} = |\mathbf{r}||\mathbf{n}| \cos 0 = a \cdot 1 \cdot 1 = a$. Thus

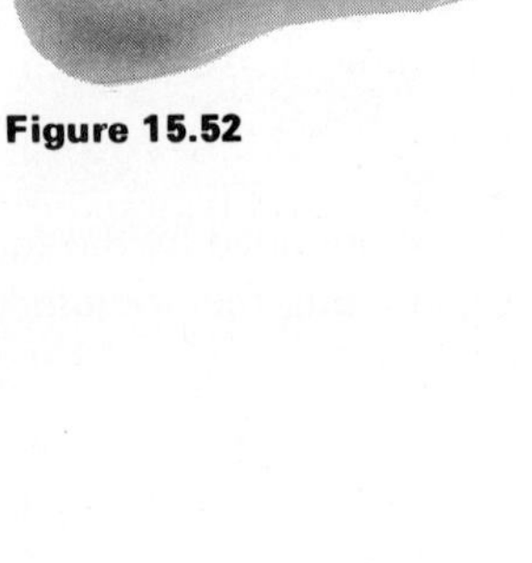

$$\int_{\mathscr{S}_1} \mathbf{F} \cdot \mathbf{n} \, dS = \int_{\mathscr{S}_1} \frac{a \, dS}{a^3}$$

$$= \frac{1}{a^2} \int_{\mathscr{S}_1} 1 \, dS$$

$$= \frac{\text{Area of } \mathscr{S}_1}{a^2}$$

$$= \frac{4\pi a^2}{a^2} \quad \text{(The area of a sphere is 4 times the area of its equatorial cross section.)}$$

$$= 4\pi. \quad \blacksquare$$

$\mathscr{S}$ encloses the origin.

Figure 15.52

GAUSS'S LAW

Example 2 is the basis of Gauss's law, which concerns the electric field $\mathbf{E}$ induced by a distribution of charge in space. A charge q at the point P_0 creates an electric field $\mathbf{E}$ whose magnitude at the point P is proportional to q and inversely proportional to the square of the distance between P and P_0; for positive q the field at P points in the direction from P_0 to P. Thus, if $\overrightarrow{P_0 P} = \mathbf{r}$, then

Observe that $\nabla \cdot \mathbf{E} = 0$.
$$\mathbf{E}(P) = \frac{kq\mathbf{r}}{|\mathbf{r}|^3},$$

where k is a constant. Moreover, the electric field due to several charges at different points is just the vector sum of the fields due to the individual charges. Gauss's law relates the total charge within a surface $\mathscr{S}$ to the integral $\int_{\mathscr{S}} \mathbf{E} \cdot \mathbf{n} \, dS$, which is called the "flux" of the field through the surface.

GAUSS'S LAW Let a distribution of charge occupy the region $\mathscr{V}$ whose surface is $\mathscr{S}$. Let Q be the total charge in $\mathscr{V}$. Let $\mathbf{E}$ be the electric field due to charges in $\mathscr{V}$ and elsewhere. Then

$$\int_{\mathscr{S}} \mathbf{E} \cdot \mathbf{n} \, dS = 4\pi k Q.$$

ARGUMENT Consider first the contribution to $\mathbf{E}$ of any charge outside the region bounded by $\mathscr{S}$. The electric field of such a charge is defined at every point of $\mathscr{V}$ and has divergence 0. By the divergence theorem, the integral of the normal component of such a field over $\mathscr{S}$ is therefore 0. Thus we need only consider charge within $\mathscr{V}$.

For a single point charge q in $\mathscr{V}$, with associated field $\mathbf{E}^*$, we have, by Example 2,

Note that $\mathbf{r}$ depends on the location of the charge q as well as the point on $\mathscr{S}$.
$$\int_{\mathscr{S}} \mathbf{E}^* \cdot \mathbf{n} \, dS = \int_{\mathscr{S}} \frac{kq\mathbf{r} \cdot \mathbf{n} \, dS}{|\mathbf{r}|^3}$$
$$= 4\pi k q.$$

Assume that the charge distribution in $\mathscr{V}$ is continuous, with charge density $\delta(P)$. Partition $\mathscr{V}$ into n small regions $\mathscr{V}_1, \mathscr{V}_2, \ldots, \mathscr{V}_n$ of volumes $V_1, V_2, \ldots, V_n$. Pick a point P_i in $\mathscr{V}_i$, $1 \leq i \leq n$. The charge in $\mathscr{V}_i$ is approximately $\delta(P_i)V_i$. This typical charge, which resembles a point charge located at P_i, contributes approximately $4\pi k \, \delta(P_i)V_i$ to the surface integral $\int_{\mathscr{S}} \mathbf{E} \cdot \mathbf{n} \, dS$. Hence $\sum_{i=1}^{n} 4\pi k \, \delta(P_i)V_i$ is an estimate of $\int_{\mathscr{S}} \mathbf{E} \cdot \mathbf{n} \, dS$. Taking limits as the mesh approaches 0, we conclude that

$$4\pi k \int_{\mathscr{V}} \delta(P) \, dV = \int_{\mathscr{S}} \mathbf{E} \cdot \mathbf{n} \, dS.$$

Since $\int_{\mathscr{V}} \delta(P) \, dV$ is the total charge Q in $\mathscr{V}$, Gauss's law is established. ∎

TEMPERATURE DISTRIBUTION

Gauss's law concerns an electric (or gravitational) field. The next application of the divergence theorem concerns the temperature distribution in a solid object. The following lemma will be needed. Its proof makes use of Appendix F.2.

LEMMA Let $f(P)$ be a continuous function defined throughout some region $\mathscr{V}$ bounded by some surface. Assume that $\int_{\mathscr{V}^*} f(P)\, dV = 0$ for all balls $\mathscr{V}^*$ located in the region. Then $f(P) = 0$ throughout $\mathscr{V}$.

PROOF Let P_0 be a point in $\mathscr{V}$. It will be shown that $f(P_0) = 0$. Assume that $f(P_0)$, on the contrary, is *not* 0. To be specific, assume that $f(P_0) = a > 0$. Since f is continuous at P_0, there is some ball $\mathscr{V}^*$ with center P_0 so small that $f(P)$ remains larger than $a/2$ for all points in that ball. Consequently $\int_{\mathscr{V}^*} f(P)\, dV \geq \int_{\mathscr{V}^*} (a/2)\, dV$, which is positive, not 0. This contradicts the assumption that $\int_{\mathscr{V}^*} f(P)\, dV$ is 0 for all balls $\mathscr{V}^*$ in $\mathscr{V}$. ∎

$\int_{\mathscr{V}^*} (a/2)\, dV = (a/2)\, Volume\ of\ \mathscr{V}^*$

On the surface of a solid piece of metal a fixed temperature distribution is maintained (by a mixture of ice cubes, heating coils, etc.). The temperature on the surface does not vary with time, but it may vary from point to point. Assume that the temperature distribution inside the metal reaches a steady state—independent of time. Let $T(x, y, z)$ be the temperature at (x, y, z). We will show that T satisfies the partial differential equation

$$T_{xx} + T_{yy} + T_{zz} = 0.$$

At each point in the metal ∇T points in the direction of most rapid increase in temperature. Thus heat flows in the direction of $-\nabla T$. Moreover, let us assume that the rate of flow (as a function of time) is proportional to the magnitude of the vector $-\nabla T$.

Heat flows from warm regions to cool.

Consider any solid region $\mathscr{V}$ of the metal. Let $\mathscr{S}$ be the surface of $\mathscr{V}$. Since the temperature distribution in the metal is in a steady state, heat neither accumulates in $\mathscr{V}$ nor disappears from $\mathscr{V}$. Thus

$$\int_{\mathscr{S}} (\nabla T) \cdot \mathbf{n}\, dS = 0.$$

By the divergence theorem,

$$\int_{\mathscr{V}} \nabla \cdot \nabla T\, dV = 0,$$

so

$$\int_{\mathscr{V}} \nabla \cdot \left(\frac{\partial T}{\partial x} \mathbf{i} + \frac{\partial T}{\partial y} \mathbf{j} + \frac{\partial T}{\partial z} \mathbf{k} \right) dV = 0,$$

and finally

$$\int_{\mathscr{V}} \left(\frac{\partial^2 T}{\partial x^2} + \frac{\partial^2 T}{\partial y^2} + \frac{\partial^2 T}{\partial z^2} \right) dV = 0.$$

By the lemma, the integrand must be 0 throughout the metal. Thus the Laplacian of T, $T_{xx} + T_{yy} + T_{zz}$, is identically 0.

15.6 THE DIVERGENCE THEOREM

Recall: A function whose Laplacian is 0 is called "harmonic."

The problem of finding the temperature distribution in a metal object is therefore reduced to the problem of finding a solution to a partial differential equation $(T_{xx} + T_{yy} + T_{zz} = 0)$ that satisfies certain conditions on the boundary.

PHYSICAL ARGUMENT FOR THE DIVERGENCE THEOREM

As with Green's theorem, consideration of fluid flow makes the divergence theorem plausible. First of all, think of

Review the argument for Green's theorem in Sec. 15.4 first.

$$\int_{\mathscr{S}} \mathbf{F} \cdot \mathbf{n} \, dS$$

as the net flow out of $\mathscr{V}$ across $\mathscr{S}$. Now consider a typical small box in $\mathscr{V}$, as shown in Fig. 15.53.

Let $\mathbf{F}(x, y, z) = P\mathbf{i} + Q\mathbf{j} + R\mathbf{k}$, and estimate the loss of fluid from the little box shown in Fig. 15.53. One corner is (a, b, c); the opposite corner is $(a + \Delta x, b + \Delta y, c + \Delta z)$. The volume of the box is $\Delta x \, \Delta y \, \Delta z$.

Let us estimate the net flow out across the two planes parallel to the yz plane. These planes have the equations $x = a$ and $x = a + \Delta x$. For the plane $x = a + \Delta x$ the outer unit normal is $\mathbf{i}$. The net loss across this surface is thus approximately

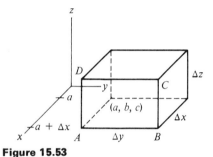

Figure 15.53

$[\mathbf{F}(a + \Delta x, b, c) \cdot \mathbf{i}] \cdot$ Area of face $ABCD$
$= [P(a + \Delta x, b, c)\mathbf{i} + Q(a + \Delta x, b, c)\mathbf{j} + R(a + \Delta x, b, c)\mathbf{k}] \cdot \mathbf{i} \, \Delta y \, \Delta z$
$= P(a + \Delta x, b, c) \, \Delta y \, \Delta z.$

For the rectangular face corresponding to $x = a$ the outer unit normal is $-\mathbf{i}$. By similar reasoning the net flow past that rectangle is approximately

$$-P(a, b, c) \, \Delta y \, \Delta z.$$

The net flow across the two faces is thus approximately

$$[P(a + \Delta x, b, c) - P(a, b, c)] \, \Delta y \, \Delta z,$$

which is approximately

$$[P_x(a, b, c) \, \Delta x] \, \Delta y \, \Delta z = P_x(a, b, c) \, \Delta x \, \Delta y \, \Delta z.$$

Similar reasoning for the other four faces suggests that the loss of fluid out of the little box is approximately

$$[P_x(a, b, c) + Q_y(a, b, c) + R_z(a, b, c)] \, \Delta x \, \Delta y \, \Delta z.$$

Since the volume of the box is $\Delta x \, \Delta y \, \Delta z$, this suggests that the net loss of fluid out of $\mathscr{V}$ is the integral

$$\int_{\mathscr{V}} (P_x + Q_y + R_z)\, dV,$$

which is

$$\int_{\mathscr{V}} \nabla \cdot \mathbf{F}\, dV.$$

Hence

$$\int_{\mathscr{S}} \mathbf{F} \cdot \mathbf{n}\, dS = \text{Net loss} = \int_{\mathscr{V}} \nabla \cdot \mathbf{F}\, dV,$$

as the divergence theorem asserts.

Exercises

1. State Green's theorem in words.
2. State the divergence theorem in words.

In Exercises 3 and 4 verify the divergence theorem for the given vector fields $\mathbf{F}$ and solid regions $\mathscr{V}$.

3. $\mathbf{F} = x\mathbf{i} + y\mathbf{j} + z\mathbf{k}$; $\mathscr{V}$ is the ball of radius a centered at the origin.
4. $\mathbf{F} = z\mathbf{k}$; $\mathscr{V}$ is the top half of a ball of radius a centered at the origin.

In Exercises 5 to 10 use the divergence theorem to evaluate $\int_{\mathscr{S}} \mathbf{F} \cdot \mathbf{n}\, dS$ for the given $\mathbf{F}$ where $\mathscr{S}$ is the boundary of the given region $\mathscr{V}$.

5. $\mathbf{F} = x^2\mathbf{i}$; $\mathscr{V}$ is the rectangular box bounded by the planes $x=0$, $x=2$, $y=0$, $y=3$, $z=0$, and $z=4$.
6. $\mathbf{F} = x^3\mathbf{i}$; $\mathscr{V}$ is the solid region between the spheres of radii a and b ($a < b$) with centers at the origin.
7. $\mathbf{F} = 3x\mathbf{i} + 2y\mathbf{j} + 6z\mathbf{k}$; $\mathscr{V}$ is the tetrahedron with the four vertices $(0, 0, 0)$, $(1, 0, 0)$, $(0, 2, 0)$, and $(0, 0, 3)$.
8. $\mathbf{F} = x\mathbf{i} + y\mathbf{j} + z\mathbf{k}$; $\mathscr{V}$ is bounded by the surface $z = 9 - x^2 - 2y^2$ and the xy plane.
9. $\mathbf{F} = x^3\mathbf{i} + y^3\mathbf{j} + 3z\mathbf{k}$; $\mathscr{V}$ is the region above the xy plane and below the surface $z = x^2$ which lies within the cylinder $r = \sin\theta$.
10. $\mathbf{F} = x^3\mathbf{i}$; $\mathscr{V}$ is bounded by the plane $z = 1$ and the cone $\phi = \pi/6$.
11. Show that, if $\mathbf{F}$ is a constant vector field and $\mathscr{S}$ is a surface bounding a region in space, then $\int_{\mathscr{S}} \mathbf{F} \cdot \mathbf{n}\, dS = 0$.
12. Let $\mathbf{F} = 2x\mathbf{i} + 3y\mathbf{j} + (5z + 6x)\mathbf{k}$, and let

    $$\mathbf{G} = (3x + 4z^2)\mathbf{i} + (2y + 5x)\mathbf{j} + 5z\mathbf{k}.$$

    Show that

    $$\int_{\mathscr{S}} \mathbf{F} \cdot \mathbf{n}\, dS = \int_{\mathscr{S}} \mathbf{G} \cdot \mathbf{n}\, dS,$$

    where $\mathscr{S}$ is any surface bounding a region in space.
13. If the length of $\mathbf{F}(P)$ is at most 5 for all points P on the surface $\mathscr{S}$ of the region $\mathscr{V}$, what can be said about

    $$\int_{\mathscr{V}} \nabla \cdot \mathbf{F}\, dV?$$

14. Let $\mathbf{F}(x, y, z) = x^3\mathbf{i} + y^3\mathbf{j} + z^3\mathbf{k}$. Let $\mathscr{S}$ be the sphere of radius a and center $(0, 0, 0)$. Use the divergence theorem to help evaluate $\int_{\mathscr{S}} \mathbf{F} \cdot \mathbf{n}\, dS$.
15. Evaluate $\int_{\mathscr{S}} \mathbf{F} \cdot \mathbf{n}\, dS$ where $\mathbf{F} = 4xz\mathbf{i} - y^2\mathbf{j} + yz\mathbf{k}$ and $\mathscr{S}$ is the surface of the cube bounded by the planes $x = 0$, $x = 1$, $y = 0$, $y = 1$, $z = 0$, and $z = 1$, with the face corresponding to $x = 1$ removed.
16. Compute $\int_{\mathscr{S}} \mathbf{F} \cdot \mathbf{n}\, dS$ where $\mathbf{F} = x\mathbf{i} + y\mathbf{j} + z\mathbf{k}$ and $\mathscr{S}$ is the surface of the cube bounded by the planes $x = 0$, $x = 2$, $y = 0$, $y = 2$, $z = 0$, and $z = 2$
    (a) without using the divergence theorem,
    (b) using the divergence theorem.

In Exercises 17 to 20 evaluate $\int_{\mathscr{S}} \mathbf{F} \cdot \mathbf{n}\, dS$ for $\mathbf{F} = \mathbf{r}/|\mathbf{r}|^3$ and the given surfaces, doing as little calculation as possible.

17. $\mathscr{S}$ is the sphere of radius 2 and center $(5, 3, 1)$.
18. $\mathscr{S}$ is the sphere of radius 3 and center $(1, 0, 1)$.
19. $\mathscr{S}$ is the surface of the box bounded by the planes $x = -1$, $x = 2$, $y = 2$, $y = 3$, $z = -1$, and $z = 6$.
20. $\mathscr{S}$ is the surface of the box bounded by the planes $x = -1$, $x = 2$, $y = -1$, $y = 3$, $z = -1$, and $z = 4$.

■

21. Explain these two steps, paraphrased from a fluid-mechanics text,

    (a) Since

    $$\int_{\mathscr{V}} \frac{\partial f}{\partial t}\, dV + \int_{\mathscr{S}} (f\mathbf{v} \cdot \mathbf{n}) = dS = 0,$$

    it follows that

    $$\int_{\mathscr{V}} \left[\frac{\partial f}{\partial t} + \nabla \cdot f\mathbf{v} \right] dV = 0.$$

    (b) Since $\mathscr{V}$ can be chosen arbitrarily small and the integrand

    $$\frac{\partial f}{\partial t} + \nabla \cdot f\mathbf{v}$$

is continuous, it follows that the integrand must be 0 everywhere. This provides the *fundamental continuity equation*

$$\frac{\partial f}{\partial t} + \mathbf{V} \cdot f\mathbf{v} = 0.$$

22 Explain why the fluid flow past the two surfaces of the box described in the text parallel to the xz plane is approximately $Q_y \, \Delta x \, \Delta y \, \Delta z$.

23 Show that $\mathbf{V} \cdot \mathbf{F}$, evaluated at P_0, equals

$$\lim_{a \to 0} \frac{\int_{\mathscr{S}} \mathbf{F} \cdot \mathbf{n} \, dS}{\text{Volume of } \mathscr{V}}$$

where $\mathscr{V}$ is the ball of radius a and center P_0 and $\mathscr{S}$ is its surface. This gives a definition of the divergence of a vector field without referring to its components.

24 Let $\mathscr{S}$ be the surface of a region $\mathscr{V}$ in which the vector field $\mathbf{F}$ is defined. Evaluate $\int_{\mathscr{S}} (\text{curl } \mathbf{F}) \cdot \mathbf{n} \, dS$.

■ ■

25 This exercise outlines a proof of the divergence theorem.
 (a) Review the proof of Green's theorem in Sec. 15.4.
 (b) Let $\mathbf{F} = P\mathbf{i} + Q\mathbf{j} + R\mathbf{k}$, where P, Q, and R are three functions of x, y, and z. Let the surface $\mathscr{S}$ bound the solid region $\mathscr{V}$. Show that the divergence theorem for $\mathscr{V}$ and $\mathscr{S}$ is equivalent to the assertion that these three equations all hold:

$$\int_{\mathscr{V}} P_x \, dV = \int_{\mathscr{S}} P(\mathbf{i} \cdot \mathbf{n}) \, dS, \quad \int_{\mathscr{V}} Q_y \, dV = \int_{\mathscr{S}} Q(\mathbf{j} \cdot \mathbf{n}) \, dS,$$

and

$$\int_{\mathscr{V}} R_z \, dV = \int_{\mathscr{S}} R(\mathbf{k} \cdot \mathbf{n}) \, dS.$$

 (c) Since the diagram corresponding to the third equation in (b) is easiest to sketch, establish that equation. Divide $\mathscr{S}$ into an upper surface $\mathscr{S}_2$ and a lower surface $\mathscr{S}_1$. Then reason as in Sec. 15.4, expressing the various integrals as integrals over a region $\mathscr{A}$ in the xy plane.
 Hint: Think of $\mathbf{k} \cdot \mathbf{n} \, dS$ as the area dA of the projection on the xy plane of the portion of $\mathscr{S}_1$ of area dS. Similarly $(-\mathbf{k}) \cdot \mathbf{n} \, dS$ is the projection of dS in $\mathscr{S}_2$ on the xy plane.

26 Let f and g be scalar functions defined throughout space. Let $\mathscr{S}$ be the boundary of the convex solid $\mathscr{V}$. Prove that

$$\int_{\mathscr{S}} f(\nabla g \cdot \mathbf{n}) \, dS = \int_{\mathscr{V}} (f \nabla^2 g + \nabla f \cdot \nabla g) \, dV,$$

where $\nabla^2 f$ is short for $\mathbf{V} \cdot (\nabla f) = f_{xx} + f_{yy} + f_{zz}$. ($\nabla^2$ is read "del squared.")

27 (See Exercise 26.) Prove that, if $\nabla^2 f$ and $\nabla^2 g$ are both 0 everywhere, then

$$\int_{\mathscr{S}} f(\nabla g \cdot \mathbf{n}) \, dS = \int_{\mathscr{S}} g(\nabla f \cdot \mathbf{n}) \, dS.$$

28 Verify the divergence theorem for $\mathbf{F} = 3x\mathbf{i} + 4y\mathbf{j} + 2\mathbf{k}$ where $\mathscr{V}$ is the cylinder bounded by the surfaces $z = 3$, $z = 0$, and $x^2 + y^2 = 9$. *Hint:* To integrate on the curved part of the surface, use $dS = 3 \, dz \, d\theta$.

29 In the text it was assumed that the temperature distribution had reached a steady state. If that assumption is removed, what is the physical significance of $T_{xx} + T_{yy} + T_{zz}$? That is, what does it measure?

30 A continuous vector field $\mathbf{F}$ has the property that at each point P at a distance 1 from the origin O, $\mathbf{F}(P)$ is perpendicular to the vector $\overrightarrow{OP}$. Show that the divergence of $\mathbf{F}$ must be 0 at at least one point.

31 We discussed the motivation for the part of the divergence theorem corresponding to the component $P\mathbf{i}$ (because the diagram for this case is easiest to draw and visualize). Making the necessary sketches, carry out the proof for the component $R\mathbf{k}$.

15.7 Stokes' theorem

Green's theorem relates the integral of $\mathbf{F} \cdot \mathbf{n}$ around a closed curve in the plane to the integral of the divergence of $\mathbf{F}$ over the plane region that the curve bounds. Stokes' theorem relates the integral of $\mathbf{F} \cdot \mathbf{T}$ around a closed curve, not necessarily in the plane, to the integral of $(\text{curl } \mathbf{F}) \cdot \mathbf{n}$ over any surface of which the curve is the boundary. It therefore is a generalization of Green's theorem in its second formulation in Sec. 15.4, namely,

$$\int_{\mathscr{A}} (\nabla \times \mathbf{F}) \cdot \mathbf{k} \, dA = \oint_C \mathbf{F} \cdot \mathbf{T} \, ds,$$

1006 GREEN'S THEOREM, THE DIVERGENCE THEOREM, AND STOKES' THEOREM

where the line integral is counterclockwise and C bounds the region $\mathscr{A}$ in the plane. Here $\mathbf{F} = P(x, y)\mathbf{i} + Q(x, y)\mathbf{j}$.

Stokes announced his theorem in 1854 (without proof, for it appeared as a question on a Cambridge University examination). By 1870 it was in common use. It is the most recent of the three major theorems in this chapter, for Green published his theorem in 1828 and Gauss published the divergence theorem in 1839.

STATEMENT OF STOKES' THEOREM

Figure 15.54

Let $\mathbf{F}$ be a vector field, let C be a curve in space, and let $\mathscr{S}$ be a surface of which C is the boundary, as shown in Fig. 15.54.

Stokes' theorem asserts that

$$\int_{\mathscr{S}} (\operatorname{curl} \mathbf{F}) \cdot \mathbf{n} \, dS = \oint_{C} \mathbf{F} \cdot \mathbf{T} \, ds, \tag{1}$$

The choice of $\mathbf{n}$ will be determined by the direction in which C is traversed.

where $\mathbf{n}$ is a unit normal to the surface. In order to state this theorem precisely, we must describe what kind of surface $\mathscr{S}$ is permitted and which of the two possible normals $\mathbf{n}$ is chosen.

In the case of a typical surface $\mathscr{S}$ that comes to mind, it is possible to assign at each point P a unit normal $\mathbf{n}$ in a continuous manner. On the surface shown in Fig. 15.55 there are two ways to do this. They are shown in Fig. 15.56.

Figure 15.55

Figure 15.56

But, for the surface shown in Fig. 15.57 (a Möbius band), it is impossible to make such a choice.

Follow the choice through nine stages—there's trouble.

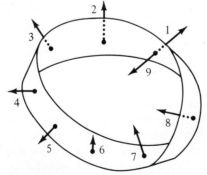

Figure 15.57

If you start with choice (1) and move the normal continuously along the surface, by the time you return to the initial point on the surface at stage (9),

Orientable surfaces

you have the opposite normal. A surface for which a continuous choice can be made is called *orientable* or *two-sided*. Stokes' theorem holds for orientable surfaces, which include, for instance, any part of the surface of a convex body, such as a ball or cube.

Consider an orientable surface $\mathscr{S}$, bounded by a parametrized curve C so that the curve is swept out in a definite direction. Choose the unit normal **n** to $\mathscr{S}$ in accordance with the right-hand rule. The direction of **n** should match the thumb of the right hand if the fingers curl in the direction of C. (See Fig. 15.58.) Figure 15.59 illustrates the choice of **n**.

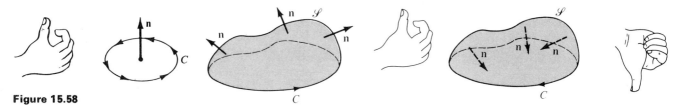

Figure 15.58

Figure 15.59

In the case of an orientable surface that is not sufficiently "flat," the proper choice of the normal **n** may be made according to the following rule. Imagine walking along the curve C in the direction of its orientation but standing on the side of the surface such that *nearby points* on the surface are on your *left*. Then choose the normal **n** to be the one on the same side of the surface as you are. Fig. 15.60 illustrates this choice for a surface that consists of five faces of a cube and a bounding curve oriented as shown.

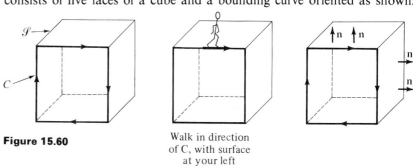

Figure 15.60

Walk in direction of C, with surface at your left

Now Stokes' theorem can be stated precisely.

THEOREM 1 *Stokes' theorem.* Let $\mathscr{S}$ be an orientable surface bounded by the parametrized curve C. At each point of $\mathscr{S}$ let **n** be the unit normal chosen by the right-hand rule. Let **F** be a vector field defined on some region in space including $\mathscr{S}$. Then

$$\int_{\mathscr{S}} (\operatorname{curl} \mathbf{F}) \cdot \mathbf{n} \, dS = \oint_C \mathbf{F} \cdot \mathbf{T} \, ds. \quad \blacksquare$$

Before applying Stokes' theorem, let us prove it for two special types of surfaces.

PROOFS OF STOKES' THEOREM IN TWO SPECIAL CASES

SPECIAL CASE 1 $\mathscr{S}$ *lies in the xy plane.* Since $\mathscr{S}$ lies in the xy plane, we may speak of "clockwise" and "counterclockwise." Assume that C is swept out counterclockwise. The right-hand rule then gives as the unit normal in Stokes' theorem the vector **k**. (See Fig. 15.61.)

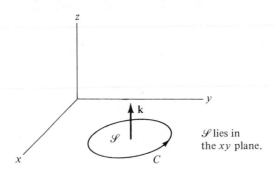

Figure 15.61

Let $\mathbf{F}(x, y, z) = P(x, y, z)\mathbf{i} + Q(x, y, z)\mathbf{j} + R(x, y, z)\mathbf{k}$. We shall show that

$$\int_{\mathscr{S}} (\nabla \times \mathbf{F}) \cdot \mathbf{k} \, dS = \oint_C \mathbf{F} \cdot \mathbf{T} \, ds.$$

To do this, introduce the vector field in the plane

$$\mathbf{G}(x, y) = \mathbf{F}(x, y, 0).$$

We already know, by Green's theorem, that Stokes' theorem is true for **G**. All that remains is to show that for points in the xy plane

$$(\nabla \times \mathbf{F}) \cdot \mathbf{k} = (\nabla \times \mathbf{G}) \cdot \mathbf{k} \quad \text{and} \quad \mathbf{F} \cdot \mathbf{T} = \mathbf{G} \cdot \mathbf{T}.$$

Straightforward computations establish both equations, as follows:

$$(\nabla \times \mathbf{F}) \cdot \mathbf{k} = [(R_y - Q_z)\mathbf{i} - (R_x - P_z)\mathbf{j} + (Q_x - P_y)\mathbf{k}] \cdot \mathbf{k}$$
$$= Q_x - P_y$$
$$= (\nabla \times \mathbf{G}) \cdot \mathbf{k}.$$

Also, in the xy plane $z = 0$. Thus

$$\mathbf{F} \cdot \mathbf{T} = \mathbf{F}(x, y, 0) \cdot \mathbf{T} = \mathbf{G}(x, y) \cdot \mathbf{T}.$$

Thus the validity of Stokes' theorem for **G** implies that it holds for **F**. This establishes special case 1. ∎

SPECIAL CASE 2 $\mathscr{S}$ *is a surface whose boundary C lies in the xy plane.* Let $\mathscr{S}$ be the surface and let $\mathscr{A}$ be the region in the xy plane bounded by C, as shown in Fig. 15.62.

Orient C counterclockwise. $\mathscr{S}$ and $\mathscr{A}$ together bound a spatial region $\mathscr{V}$. Assume that **F** is defined throughout some region that includes $\mathscr{S}$, $\mathscr{A}$, and $\mathscr{V}$. In accord with the right-hand rule, the exterior normal is **n** on $\mathscr{S}$, and $-\mathbf{k}$ on $\mathscr{A}$.

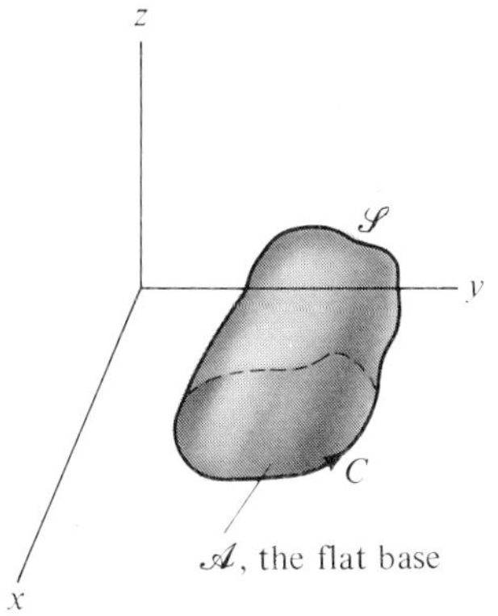

Since the divergence of the curl of **F** is 0, the divergence theorem implies that

$$\int_{\mathscr{S}} (\text{curl } \mathbf{F}) \cdot \mathbf{n} \, dS + \int_{\mathscr{A}} (\text{curl } \mathbf{F}) \cdot (-\mathbf{k}) \, dA = 0.$$

Thus
$$\int_{\mathscr{S}} (\text{curl } \mathbf{F}) \cdot \mathbf{n} \, dS = \int_{\mathscr{A}} (\text{curl } \mathbf{F}) \cdot \mathbf{k} \, dA. \qquad (2)$$

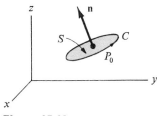

$\mathscr{A}$, the flat base

Figure 15.62

But by special case 1,

$$\int_{\mathscr{A}} (\text{curl } \mathbf{F}) \cdot \mathbf{k} \, dA = \oint_C \mathbf{F} \cdot \mathbf{T} \, ds. \qquad (3)$$

Combining (2) and (3) shows that

$$\int_{\mathscr{S}} (\text{curl } \mathbf{F}) \cdot \mathbf{n} \, dS = \oint_C \mathbf{F} \cdot \mathbf{T} \, ds$$

and again verifies Stokes' theorem. ∎

Stokes' theorem will now be applied, first to fluid flow and then to conservative vector fields.

WHY CURL IS CALLED CURL

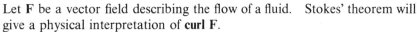

Figure 15.63

Let **F** be a vector field describing the flow of a fluid. Stokes' theorem will give a physical interpretation of **curl F**.

Consider a fixed point P_0 in space. Imagine a *small* circular disk S with center P_0. Let C be the boundary of S oriented in such a way that C and **n** fit the right-hand rule. (See Fig. 15.63.)

Now examine the two sides of the equation

$$\int_S (\text{curl } \mathbf{F}) \cdot \mathbf{n} \, dS = \oint_C \mathbf{F} \cdot \mathbf{T} \, ds. \qquad (4)$$

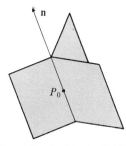

A paddle wheel in the fluid.

Figure 15.64

The right side of (4) measures the tendency of the fluid to move along C (rather than, say, perpendicular to it). Thus $\oint_C \mathbf{F} \cdot \mathbf{T} \, ds$ might be thought of as the "circulation" or "whirling tendency" of the fluid along C. For each placement of the small disk S at P_0—or, equivalently, each choice of unit normal vector **n**—$\oint_C \mathbf{F} \cdot \mathbf{T} \, ds$ measures a corresponding circulation. It records the tendency of a paddle wheel at P_0 with axis along **n** to rotate. (See Fig. 15.64.)

Now consider the left side of (4). If S is small, the integrand is almost constant and the integral is approximately

$$(\text{curl } \mathbf{F})_{P_0} \cdot \mathbf{n} \cdot \text{Area of } S, \tag{5}$$

where $(\text{curl } \mathbf{F})_{P_0}$ denotes the curl at P_0.

Keeping the center of S at P_0, vary the vector $\mathbf{n}$. For which choice of $\mathbf{n}$ will (5) be largest? Answer: for that $\mathbf{n}$ which has the same direction as the fixed vector $(\text{curl } \mathbf{F})_{P_0}$. With that choice of $\mathbf{n}$, (5) becomes

$$|(\text{curl } \mathbf{F})_{P_0}| \text{Area of } S.$$

The physical interpretation of curl

Thus a paddle wheel placed in the fluid at P_0 *rotates most quickly when its axis is in the direction of* $\text{curl } \mathbf{F}$ at P_0. The *magnitude* of $\text{curl } \mathbf{F}$ is a measure of *how fast* the paddle wheel can rotate when placed at P_0. Thus $\text{curl } \mathbf{F}$ records the direction and magnitude of maximum circulation at a given point.

In a letter to the mathematician Tait written on November 7, 1870, Maxwell offered some names for the function $\nabla \times \mathbf{F}$,

"Here are some rough hewn names. Will you like a good Divinity shape their ends properly so as to make them stick? . . .

The vector part $[\nabla \times \mathbf{F}]$ I would call the twist of the vector function. Here the word twist has nothing to do with a screw or helix. [T]he word *turn* ... would be better than twist, for twist suggests a screw. Twirl is free from the screw motion and is sufficiently racy. Perhaps it is too dynamical for pure mathematicians, so for Cayley's sake I might say Curl (after the fashion of Scroll)."

His proposal, "curl," has stuck.

A TEST FOR A CONSERVATIVE FIELD

Review Example 1 in Sec. 15.3. Note where $\mathbf{F}$ *is defined.*

Section 15.3 presented several tests for a conservative field $\mathbf{F}$. Since the curl of a conservative field is $\mathbf{0}$, it was hoped that, if $\text{curl } \mathbf{F} = \mathbf{0}$, then the field $\mathbf{F}$ would be conservative. Example 1 in that section blasted such a hope. However, it was shown that if the domain of $\mathbf{F}$ is all the plane or all of space, then the curl of $\mathbf{F}$ being $\mathbf{0}$ does imply that $\mathbf{F}$ is conservative.

With the aid of Stokes' theorem, this result can be vastly strengthened. What is essential is that the domain of $\mathbf{F}$ be "simply connected," a concept that will now be defined.

Simply connected

An arcwise-connected region in the plane or in space is *simply connected* if each closed curve in the region can be gradually shrunk to a point while staying within the region. The shaded planar region in Fig. 15.65 is *not* simply connected since the curve C cannot be shrunk to a point while remaining within the region.

Similarly, the region consisting of the entire xy plane except the origin is not simply connected. However, a disk is simply connected. So is the region in space consisting of all of space except the origin. However, the region in space consisting of all of space except the z axis is not simply connected; the curve C shown in Fig. 15.66 cannot be shrunk to a point without crossing the z axis.

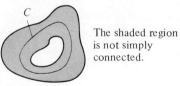

The shaded region is not simply connected.

Figure 15.65

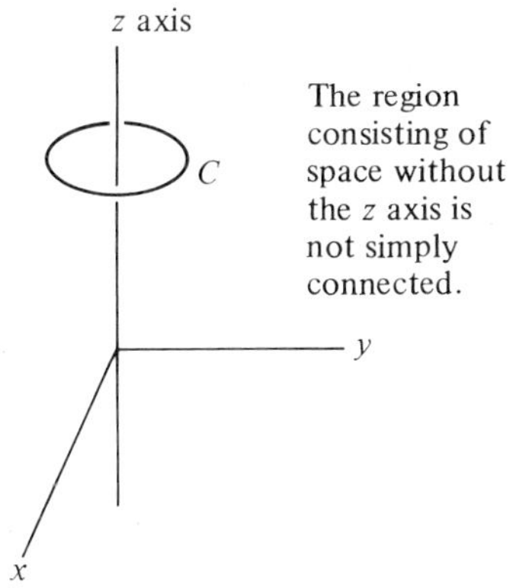

The region consisting of space without the z axis is not simply connected.

Figure 15.66

Now we are ready to give a simple test for determining whether a vector field is conservative.

THEOREM 2 Let **F** be defined on a simply connected region. If **curl F** = **0**, then **F** is conservative.

SKETCH OF PROOF Let C be a closed curve in the region. We wish to show that $\oint_C \mathbf{F} \cdot \mathbf{T} \, ds = 0$. Since the region where **F** is defined is simply connected, C can be shrunk to a point in that region. It is proved in advanced topology that C bounds an orientable surface $\mathscr{S}$ (which may intersect itself). By Stokes' theorem,

$$\oint_C \mathbf{F} \cdot \mathbf{T} \, ds = \int_{\mathscr{S}} (\mathbf{curl\ F}) \cdot \mathbf{n} \, dS$$

$$= \int_{\mathscr{S}} \mathbf{0} \cdot \mathbf{n} \, dS = 0.$$

This concludes the sketch of the proof. ∎

EXAMPLE Show that

$$\mathbf{F} = \frac{x\mathbf{i}}{x^2 + y^2 + z^2} + \frac{y\mathbf{j}}{x^2 + y^2 + z^2} + \frac{z\mathbf{k}}{x^2 + y^2 + z^2}$$

is conservative.

SOLUTION The domain of **F** consists of all points in space except the origin. Thus the domain of **F** is simply connected. All that remains is to show that **curl F** = **0**:

$$\nabla \times \mathbf{F} = \begin{vmatrix} \mathbf{i} & \mathbf{j} & \mathbf{k} \\ \dfrac{\partial}{\partial x} & \dfrac{\partial}{\partial y} & \dfrac{\partial}{\partial z} \\ \dfrac{x}{x^2 + y^2 + z^2} & \dfrac{y}{x^2 + y^2 + z^2} & \dfrac{z}{x^2 + y^2 + z^2} \end{vmatrix}$$

$$= \left[\frac{\partial}{\partial y}\left(\frac{z}{x^2+y^2+z^2}\right) - \frac{\partial}{\partial z}\left(\frac{y}{x^2+y^2+z^2}\right)\right]\mathbf{i}$$
$$- \left[\frac{\partial}{\partial x}\left(\frac{z}{x^2+y^2+z^2}\right) - \frac{\partial}{\partial z}\left(\frac{x}{x^2+y^2+z^2}\right)\right]\mathbf{j}$$
$$+ \left[\frac{\partial}{\partial x}\left(\frac{y}{x^2+y^2+z^2}\right) - \frac{\partial}{\partial y}\left(\frac{x}{x^2+y^2+z^2}\right)\right]\mathbf{k}$$
$$= \mathbf{0},$$

as straightforward differentiation will show. ∎

Exercises

1 State Stokes' theorem in words, not in mathematical symbols.

2 Show that Stokes' theorem for the vector field $Q\mathbf{i} - P\mathbf{j}$ implies Green's theorem for the vector field $P\mathbf{i} + Q\mathbf{j}$.

3 Let $\mathscr{S}$ be the top half of the sphere of radius 1 and with center $(0, 0, 0)$. Let C be the unit circle in the xy plane with center $(0, 0, 0)$, swept out counterclockwise. Let $\mathbf{F}(x, y, z) = y^2 x\mathbf{i} + y^3\mathbf{j} + y^2 z\mathbf{k}$. Verify Stokes' theorem in this case. (See Fig. 15.67.)

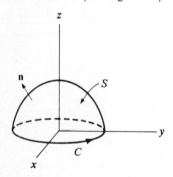

Figure 15.67

4 Let $\mathbf{F}(x, y, z) = y\mathbf{i} + xz\mathbf{j} + x^2\mathbf{k}$. Let $\mathscr{S}$ be the triangle with vertices $(1, 0, 0)$, $(0, 1, 0)$, and $(0, 0, 1)$. Verify Stokes' theorem in this case.

5 Verify Stokes' theorem for the special case when $\mathbf{F}$ has the form ∇f, that is, is a gradient field.

6 Assume that Stokes' theorem is true for triangles. Deduce that it then holds for the surface $\mathscr{S}$ in Fig. 15.68, consisting of the three triangles DAB, DBC, DCA, and the curve $ABCA$.

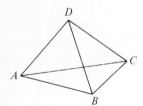

Figure 15.68

7 Let $\mathbf{F} = P\mathbf{i} + Q\mathbf{j} + R\mathbf{k}$. A convex region has a flat base situated on the xy plane. Let $\mathscr{S}$ be the part of its surface other than the base $\mathscr{B}$. (See Fig. 15.69.) Show that

$$\int_{\mathscr{S}} (\text{curl } \mathbf{F}) \cdot \mathbf{n} \, dS = \int_{\mathscr{B}} (Q_x - P_y) \, dA.$$

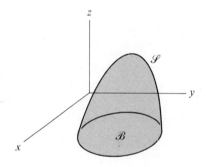

Figure 15.69

8 Using Stokes' theorem, evaluate $\int_{\mathscr{S}} (\nabla \times \mathbf{F}) \cdot \mathbf{n} \, dS$, where $\mathbf{F} = (x^2 + y - 4)\mathbf{i} + 3xy\mathbf{j} + (2xz + z^2)\mathbf{k}$, and $\mathscr{S}$ is the portion of the surface $z = 4 - (x^2 + y^2)$ above the xy plane. (Let $\mathbf{n}$ be the upward normal.)

9 Evaluate as simply as possible $\int_{\mathscr{S}} \mathbf{F} \cdot \mathbf{n} \, dS$, where $\mathscr{S}$ and $\mathbf{F}$ are the following: $\mathbf{F}(x, y, z) = x\mathbf{i} - y\mathbf{j}$, and $\mathscr{S}$ is the surface of the cube bounded by the three coordinate planes and the planes $x = 1$, $y = 1$, $z = 1$, exclusive of the surface in the plane $x = 1$. (Let $\mathbf{n}$ be outward from the cube.)

10 Let $\mathbf{F}$ be a vector field throughout space such that $\mathbf{F}(P)$ is perpendicular to the curve C at each point P on C, the boundary of a surface $\mathscr{S}$. What can one conclude about

$$\int_{\mathscr{S}} (\text{curl } \mathbf{F}) \cdot \mathbf{n} \, dS?$$

11 Let $\mathbf{F}(x, y, z) = \mathbf{r}/|\mathbf{r}|^a$, where $\mathbf{r} = x\mathbf{i} + y\mathbf{j} + z\mathbf{k}$ and a is a fixed real number.
   (a) Show that **curl** $\mathbf{F} = \mathbf{0}$.
   (b) Show that $\mathbf{F}$ is conservative.
   (c) Exhibit a scalar function f such that $\mathbf{F} = \nabla f$.

12 Let $\mathbf{F}(x, y)$ be a vector field defined everywhere in the plane except at the origin. Assume that **curl** $\mathbf{F} = \mathbf{0}$. Let C_1 be the circle $x^2 + y^2 = 1$ counterclockwise; let C_2 be the circle $x^2 + y^2 = 4$ clockwise; let C_3 be the circle $(x - 2)^2 + y^2 = 1$ counterclockwise; let C_4 be the circle $(x - 1)^2 + y^2 = 9$ clockwise. Assuming that $\oint_{C_1} \mathbf{F} \cdot \mathbf{T}\, ds$ is 5, evaluate

   (a) $\oint_{C_2} \mathbf{F} \cdot \mathbf{T}\, ds$, (b) $\oint_{C_3} \mathbf{F} \cdot \mathbf{T}\, ds$, (c) $\oint_{C_4} \mathbf{F} \cdot \mathbf{T}\, ds$.

In each of Exercises 13 to 16 state whether or not the region is simply connected.

13 All points (x, y) in the plane such that $x^2 + y^2 > 1$.
14 All points (x, y, z) in space such that $x^2 + y^2 + z^2 > 1$.
15 All points (x, y) in the plane such that $1 < x^2 + y^2 < 4$.
16 A doughnut (obtained by rotating around the z axis a disk in the yz plane that does not meet the z axis.)

In each of Exercises 17 to 20 determine whether the given vector field is conservative. Use Theorem 2, if it applies.

17 $\dfrac{x\mathbf{i}}{\sqrt{x^2 + y^2}} + \dfrac{y\mathbf{j}}{\sqrt{x^2 + y^2}}$

18 $\dfrac{x\mathbf{i}}{\sqrt{x^2 + y^2 + z^2}} + \dfrac{y\mathbf{j}}{\sqrt{x^2 + y^2 + z^2}} + \dfrac{z\mathbf{k}}{\sqrt{x^2 + y^2 + z^2}}$

19 $e^{x^2 + y^2 + z^2}(x\mathbf{i} + y\mathbf{j} + z\mathbf{k})$

20 $\dfrac{y}{z}\mathbf{i} + \dfrac{x}{z}\mathbf{j} - \dfrac{xy}{z^2}\mathbf{k}$

In each of Exercises 21 to 24 use Stokes' theorem to evaluate $\oint_C \mathbf{F} \cdot \mathbf{T}\, ds$ for the given $\mathbf{F}$ and C. In each case assume that C is oriented counterclockwise when viewed from above.

21 $\mathbf{F} = \sin xy\,\mathbf{i}$; C is the intersection of the plane $x + y + z = 1$ and the cylinder $x^2 + y^2 = 1$.
22 $\mathbf{F} = e^x\mathbf{j}$; C is the triangle with vertices $(2, 0, 0)$, $(0, 3, 0)$, and $(0, 0, 4)$.
23 $\mathbf{F} = xy\,\mathbf{k}$; C is the intersection of the plane $z = y$ with the cylinder $x^2 - 2x + y^2 = 0$.
24 $\mathbf{F} = \cos(x + z)\mathbf{j}$; C is the boundary of the rectangle with vertices $(1, 0, 0)$, $(1, 1, 1)$, $(0, 1, 1)$, and $(0, 0, 0)$.

25 (a) Make a Möbius strip. One way is to cut a long narrow strip of paper, pull its ends around together, as in forming a cylinder, but give one end a 180° twist at the last moment. Then glue the two ends together.

(b) Verify that you can move a normal vector continuously on the surface and, upon returning to the initial position, have the vector point in the opposite direction.

26 Let $\mathbf{F}$ be a vector field in space with the property that, for any points A and B and any curve C from A to B, the line integral

$$\int_C \mathbf{F} \cdot \mathbf{T}\, ds$$

depends only on the endpoints A and B. Show that

$$\nabla \times \mathbf{F} = \mathbf{0}$$

at all points.

27 Let $\mathbf{F}(x, y, z) = P(x, y)\mathbf{i} + Q(x, y)\mathbf{j} + 0\mathbf{k}$ for some scalar functions P and Q.
   (a) Show that each vector $\mathbf{F}(x, y, z)$ is parallel to the xy plane.
   (b) Show that **curl** $\mathbf{F}$ is parallel to the z axis.
   (c) Interpret (b) in terms of a paddle wheel.

28 Let $\mathbf{F}(x, y, z) = x\mathbf{j}$. Fig. 15.70 shows a few values of $\mathbf{F}$ near $(0, 0, 0)$.

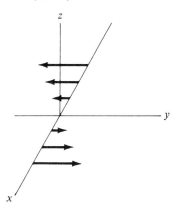

Figure 15.70

(a) The diagram shows $\mathbf{F}(x, y, z)$ for a few cases when $z = 0$. Sketch $\mathbf{F}(x, y, z)$ for a few cases when $z = 0.1$.
(b) Using your "fluid" intuition, decide in what direction the axis of a paddle wheel at $(0, 0, 0)$ should be placed in order that the paddle wheel rotates most quickly.
(c) Is your opinion in (b) compatible with **curl** $\mathbf{F}$ at the point $(0, 0, 0)$?

29 Let $\mathscr{S}_1$ be the top half and $\mathscr{S}_2$ the bottom half of a sphere of radius a in space. Let $\mathbf{F}$ be a vector field defined on the sphere and let $\mathbf{n}$ denote an exterior normal to the sphere. What relation, if any, is there between $\int_{\mathscr{S}_1} (\textbf{curl }\mathbf{F}) \cdot \mathbf{n}\, dS$ and $\int_{\mathscr{S}_2} (\textbf{curl }\mathbf{F}) \cdot \mathbf{n}\, dS$.

30 Let C be the curve formed by the intersection of the plane $z = x$ and the paraboloid $z = x^2 + y^2$. Orient C to be counterclockwise when viewed from above. Evaluate $\oint_C (xyz\, dx + x^2\, dy + xz\, dz)$.

31 In Example 1 of Sec. 15.3 there is a vector field whose curl is $\mathbf{0}$, yet the field is not conservative.
   (a) What does this imply about the domain of the field?
   (b) Check your answer to (a) by examining the domain of the field.

Exercises 32 to 34 outline a proof of Stokes' theorem in the case that $\mathscr{S}$ is part of a surface $z = f(x, y)$.

32 Let $\mathbf{F} = P(x, y, z)\mathbf{i} + Q(x, y, z)\mathbf{j} + R(x, y, z)\mathbf{k}$ be a vector field in space. Let $\mathscr{S}$ be a surface on $z = f(x, y)$ and C its bounding curve. Let $\mathbf{n}$ be the unit normal to $\mathscr{S}$ that has *positive z* component. Orient C in accord with the right-hand rule. (See Fig. 15.71.)

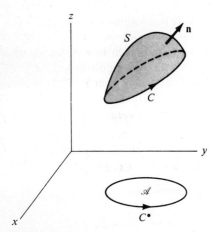

Figure 15.71

Let $\mathscr{A}$ be the projection (shadow) of $\mathscr{S}$ in the xy plane and let C^* be the projection of C, with a counterclockwise orientation. Show that

$$\int_{\mathscr{S}} (\nabla \times \mathbf{F})\, dS = \oint_C \mathbf{F} \cdot \mathbf{n}\, ds,$$

by expressing the first integral as an integral over $\mathscr{A}$ and the second integral as an integral over C^*. Then use Green's theorem. The details are sketched in the rest of this exercise and in Exercises 33 and 34.
(a) Show that

$$\mathbf{n} = \frac{-f_x \mathbf{i} - f_y \mathbf{j} + \mathbf{k}}{\sqrt{1 + f_x^2 + f_y^2}}.$$

(b) Show that

$$\int_{\mathscr{S}} (\nabla \times \mathbf{F}) \cdot \mathbf{n}\, dS$$
$$= \int_{\mathscr{A}} [(Q_z - R_y)f_x + (R_x - P_z)f_y + (Q_x - P_y)]\, dA.$$

33 This continues Exercise 32. Show that

$$\oint_C \mathbf{F} \cdot \mathbf{T}\, ds = \int_C (P\, dx + Q\, dy + R\, dz)$$
$$= \int_{C^*} [P(x, y, f(x, y))\, dx + Q(x, y, f(x, y))\, dy$$
$$+ R(x, y, f(x, y))(f_x\, dx + f_y\, dy)]$$
$$= \int_{C^*} \{[P(x, y, f(x, y)) + f_x R(x, y, f(x, y))]\, dx$$
$$+ [Q(x, y, f(x, y)) + f_y R(x, y, f(x, y))]\, dy\}.$$

34 Exercises 32 and 33 express the integrals in Stokes' theorem as integrals over a plane region $\mathscr{A}$ and a plane curve C^*. Use Green's theorem to complete the proof of Stokes' theorem. Be careful in computing partial derivatives. For instance,

$$\frac{\partial}{\partial x} P(x, y, f(x, y)) = P_x + f_x P_z$$

by a chain rule.

15.8 Maxwell's equations (optional)

In addition to the gravitational field, which is associated with matter, there are two other vector fields of fundamental physical significance. These are the electric and magnetic fields, which are associated with fixed or moving charges.

15.8 MAXWELL'S EQUATIONS (OPTIONAL)

The electric field **E** at a point P_0 can be detected and measured as follows. Hold a charge q at P_0 and measure the force **F** on the charge (after subtracting the gravitational or other forces). This force **F** is ascribed to the electric field **E** at P_0, which is related to the force **F** by the equation $\mathbf{E} = \mathbf{F}/q$.

The magnetic field **B** at a point P_0 can be calculated by observing its effect on a *moving* test charge at P_0. The force **F** on a charge q moving with velocity **v** is $\mathbf{F} = q\mathbf{E} + \mathbf{B} \times q\mathbf{v}$. Since **E** is known, **B** can be found by experimenting with various velocities **v**. (Note that the force exerted by **B** is perpendicular to the path of the charge.)

$\mathbf{B} \times q\mathbf{v}$ is perpendicular to $\mathbf{B}$ and $\mathbf{v}$.

The magnetic field **B** is created by moving charges, and its effect is registered only on moving charges. The electric field **E** is associated with charges, moving or not, and it operates on charges, moving or not.

At the time that Newton published his *Principia* on the gravitational field (1687), electricity and magnetism were the subjects of little scientific study. But the experiments of Franklin, Oersted, Henry, Ampère, Faraday, and others in the eighteenth and early nineteenth centuries gradually built up a mass of information subject to mathematical analysis. All the phenomena could be summarized in four equations, which in their final form appeared in Maxwell's *Treatise on Electricity and Magnetism*, published in 1873. We state these four equations in the special case when neither **E** nor **B** varies with time.

For a fuller treatment, see *The Feynman Lectures on Physics*, vol. 2, Addison-Wesley, 1964.

MAXWELL'S EQUATIONS IN INTEGRAL FORM

c is the velocity of light.

In the following equations ε_0 and c are constants.

(1) $\int_{\mathscr{S}} \mathbf{E} \cdot \mathbf{n} \, dS = Q/\varepsilon_0$, where $\mathscr{S}$ is a surface bounding a spatial region and Q is the total charge in that region.

(This implies that **E** originates in charges. It also implies that the magnitude of the electric field associated with a point charge varies inversely as the square of the distance from the charge. See Exercise 8.)

If there is no charge within a surface, then the "net flow" of **E** across the surface is 0. See Fig. 15.72.

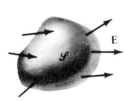

Figure 15.72

(2) $\int_C \mathbf{E} \cdot \mathbf{T} \, ds = 0$ for any closed curve C.

(3) $\int_{\mathscr{S}} \mathbf{B} \cdot \mathbf{n} \, dS = 0$ where $\mathscr{S}$ is any surface bounding a spatial region. (Unlike the electric field, the magnetic field has no point source.)

(4) $c^2 \int_C \mathbf{B} \cdot \mathbf{T} \, ds = (1/\varepsilon_0) \int_{\mathscr{S}} \mathbf{i} \cdot \mathbf{n} \, dS$, where $\mathscr{S}$ is a surface whose boundary is the closed curve C. The vector **i** records the current density at points of $\mathscr{S}$. (It is not the unit vector $(1, 0, 0)$.)

This implies that current flowing within a loop C generates a magnetic field in the vicinity of the loop. (See Fig. 15.73.)

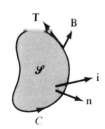

Figure 15.73

MAXWELL'S EQUATIONS IN DIFFERENTIAL FORM

Maxwell's equations are frequently expressed in terms of the behavior of **E** and **B** locally, at each point in space, rather than globally with integrals. The analogs of the four integral equations are the following four differential equations, which can be obtained from the integral equations with the aid of the divergence theorem and Stokes' theorem.

(1') $\quad \nabla \cdot \mathbf{E} = q/\varepsilon_0$, where q is the charge density per unit volume.

(2') $\quad \nabla \times \mathbf{E} = \mathbf{0}$.

(3') $\quad \nabla \cdot \mathbf{B} = 0$.

(4') $\quad c^2 \nabla \times \mathbf{B} = \mathbf{i}/\varepsilon_0$, where $\mathbf{i}$ is the current density.

For instance, let us derive (1') from (1).

According to (1), $\int_{\mathscr{S}} \mathbf{E} \cdot \mathbf{n} \, dS = Q/\varepsilon_0$, where Q is the charge contained in the solid region $\mathscr{V}$ surrounded by $\mathscr{S}$. This total charge is the integral of the charge density q over the region $\mathscr{V}$,

$$Q = \int_{\mathscr{V}} q \, dV.$$

Consequently,
$$\int_{\mathscr{S}} \mathbf{E} \cdot \mathbf{n} \, dS = \frac{1}{\varepsilon_0} \int_{\mathscr{V}} q \, dV. \tag{1}$$

The divergence theorem transforms (1) into an equation relating two integrals over $\mathscr{V}$.

$$\int_{\mathscr{V}} \nabla \cdot \mathbf{E} \, dV = \frac{1}{\varepsilon_0} \int_{\mathscr{V}} q \, dV \tag{2}$$

or
$$\int_{\mathscr{V}} \left(\nabla \cdot \mathbf{E} - \frac{q}{\varepsilon_0} \right) dV = 0. \tag{3}$$

Since (3) holds for all spatial regions $\mathscr{V}$, however small, the integrand must be identically 0. Consequently

$$\nabla \cdot \mathbf{E} - \frac{q}{\varepsilon_0} = 0, \tag{4}$$

from which (1') follows.

The derivations of (2'), (3') and (4') are similar and are left as exercises. It is also possible to derive the integral form of Maxwell's equations from their differential form by reversing the steps.

THE MAGNETIC FIELD OF AN INFINITE STRAIGHT WIRE

Most electromagnetic phenomena can be explained on the basis of Maxwell's equations. As an illustration, we shall use (3) and (4) to compute the magnetic field around an infinitely long straight wire carrying a current $\mathbf{j}$, as shown in Fig. 15.74.

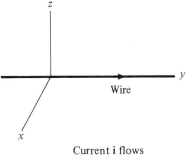

Current **i** flows
along the y axis.

Figure 15.74

Some observations of symmetry will simplify the argument. For convenience, we will assume that the magnetic field approaches **0** at points arbitrarily far from the wire, that is, $\mathbf{B}(P) \to \mathbf{0}$ as the distance from P to the wire approaches infinity.

B *has rotational and translational symmetry.*

Observe that the wire and current remain unchanged under rotation around the y axis or by a shift parallel to the y axis. Thus $\mathbf{B}(x, y, z) = \mathbf{B}(x, y', z)$, that is, the magnetic field depends only on x and z. (In Fig. 15.75 $\mathbf{B}(P) = \mathbf{B}(P'')$.) Also, if the point P' is obtained from the point P by a rotation of angle θ around the y axis, then $\mathbf{B}(P')$ is obtained from $\mathbf{B}(P)$ by the same rotation. (See Fig. 15.75.)

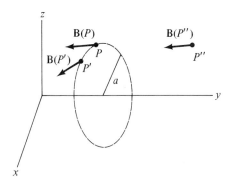

Figure 15.75

According to (3) $\int_{\mathscr{S}} \mathbf{B} \cdot \mathbf{n}\, dS = 0$ over any surface surrounding a solid region. Choose $\mathscr{S}$ to be a right circular cylinder with the y axis as axis. (See Fig. 15.76.)

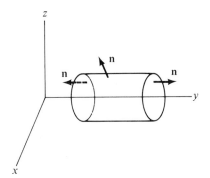

Figure 15.76

n *is the exterior normal.*

The integrals of **B** · **n** over the two circular bases of the cylinder cancel out since the normals **n** point in opposite directions. On the curved surface of the cylinder, **B** · **n** is constant (by the symmetry arguments). Since $\int_{\mathscr{S}} \mathbf{B} \cdot \mathbf{n}\, dS = 0$, it follows that **B** · **n** = 0. So we have already found that the magnetic field **B** is always perpendicular to **n**.

In a moment the edge FG will be moved far from the wire.

We still must determine the exact direction of **B** and its magnitude. To do so, consider the integral $\oint_C \mathbf{B} \cdot \mathbf{T}\, ds$ over a closed rectangular loop of the type shown in Fig. 15.77.

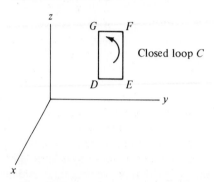

Figure 15.77

Since there is no current within this rectangle, (4) implies that

$$\oint_C \mathbf{B} \cdot \mathbf{T}\, ds = 0. \tag{5}$$

Keep in mind that $\oint_C \mathbf{B} \cdot \mathbf{T}\, ds$ is the sum of integrals over the four edges of C.

Since **B** is perpendicular to **n**, a normal to the cylinder in Fig. 15.76, the integrals of **B** · **T** over the portions of the loop parallel to the z axis (that is, the edges EF and GD) are 0. The part of the integral (5) corresponding to the edge FG approaches 0 as the edge FG is moved arbitrarily far from the wire. In view of (5), therefore, the integral of **B** · **T** over the fixed edge DE must be 0. From this and the symmetry conditions, **B** · **T** = 0.

At any point P off the wire $\mathbf{B}(P)$ is perpendicular to the vectors **A** and **C** shown in Fig. 15.78.

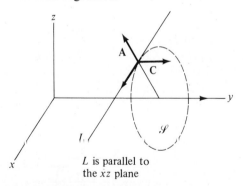

Figure 15.78

Consequently $\mathbf{B}(P)$ is parallel to the line L through P that is parallel to the xz plane and tangent to the dotted circle shown in Fig. 15.78. (The circle is parallel to the xz plane and centered on the y axis.)

So **B**(P) points in one of two directions parallel to L. With the aid of **(4)** the direction can be determined, as well as the magnitude of **B**(P).

Let the circle in Fig. 15.78 have radius a. The circle bounds a disk $\mathscr{S}$. Sweep out the circle C in the direction compatible with the right-hand rule if **j** gives the direction of the unit normal to $\mathscr{S}$. According to **(4)**,

$$c^2 \oint_C \mathbf{B} \cdot \mathbf{T} \, ds = \frac{1}{\varepsilon_0} \int_{\mathscr{S}} \mathbf{i} \cdot \mathbf{n} \, dS. \tag{6}$$

The integral $\int_{\mathscr{S}} \mathbf{i} \cdot \mathbf{n} \, dS$ is the total current in the wire, whose magnitude is $|\mathbf{j}|$. By **(6)**,

$$c^2 \, \mathbf{B} \cdot \mathbf{T} \, 2\pi a = \frac{|\mathbf{j}|}{\varepsilon_0}. \tag{7}$$

From **(7)** it follows that **B** points in the *same* direction as **T** in Fig. 15.78 and that the magnitude of **B** is inversely proportional to a, the distance from the wire. The magnetic field is shown in Fig. 15.79. It matches the fingers of the right hand if the thumb points in the direction of the current.

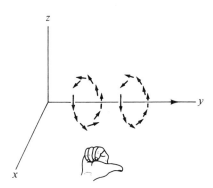

Figure 15.79

Exercises

1. Obtain **(1)** from **(1′)** as follows:

   (a) Show that
   $$\int_{\mathscr{V}} \frac{q}{\varepsilon_0} \, dV = \int_{\mathscr{S}} \mathbf{E} \cdot \mathbf{n} \, dS.$$

   (b) Deduce that
   $$\frac{Q}{\varepsilon_0} = \int_{\mathscr{S}} \mathbf{E} \cdot \mathbf{n} \, dS.$$

2. Obtain **(3′)** from **(3)**.
3. Obtain **(3)** from **(3′)**.
4. Obtain **(2′)** from **(2)**.
5. Obtain **(2)** from **(2′)**.
6. Obtain **(4′)** from **(4)**.
7. Obtain **(4)** from **(4′)**.
8. This exercise outlines the computation of the electric field **E** due to a point charge Q placed at the origin O in space.

   (a) Using symmetry, show that **E**(P) is parallel to $\overrightarrow{OP}$.

   (b) Using symmetry, show that the magnitude of **E**(P) depends only on the magnitude of $\overrightarrow{OP}$.

   (c) Using **(1)**, show that **E**(P) equals $k\mathbf{r}/|\mathbf{r}|^3$ for some constant k, where $\mathbf{r} = \overrightarrow{OP}$.

9. Show that, if the fields **E** and **B** satisfy Maxwell's equations, so do the vector fields **E** + **K** and **B** + **C**, where **K** and **C** are constant vector fields (independent of position and time).

15.S Summary

This chapter concerns line integrals, surface integrals, Green's theorem, the divergence theorem, Stokes' theorem, and conservative vector fields (which are equivalent to exact differential expressions).

Theorem	Statement	Remarks
Green's theorem	$\oint_C \mathbf{F} \cdot \mathbf{n}\, ds = \int_{\mathscr{A}} \nabla \cdot \mathbf{F}\, dA$	$\mathscr{A}$ planar, C its boundary, $\mathbf{n}$ the exterior normal
	$\oint_C (P\,dx + Q\,dy) = \int_{\mathscr{A}} (Q_x - P_y)\, dA$	C counterclockwise
	$\oint_C \mathbf{F} \cdot \mathbf{T}\, ds = \int_{\mathscr{A}} (\nabla \times \mathbf{F}) \cdot \mathbf{k}\, dA$	restated to be special case of Stokes' theorem
Divergence theorem	$\int_{\mathscr{S}} \mathbf{F} \cdot \mathbf{n}\, dS = \int_{\mathscr{V}} \nabla \cdot \mathbf{F}\, dV$	$\mathscr{V}$ in space, $\mathscr{S}$ its boundary, $\mathbf{n}$ the exterior normal
Stokes' theorem	$\int_{\mathscr{S}} (\nabla \times \mathbf{F}) \cdot \mathbf{n}\, dS = \oint_C \mathbf{F} \cdot \mathbf{T}\, ds$	$\mathscr{S}$ in plane or space, C its boundary, choice of $\mathbf{n}$ determined by right-hand rule and orientation of C

The integral $\int_{\mathscr{S}} \mathbf{F} \cdot \mathbf{n}\, dS$ can be interpreted in terms of fluid flow. In electromagnetic theory it is called the flux of the field through the surface $\mathscr{S}$. If $\mathbf{F}$ is interpreted as a force, then $\int_C \mathbf{F} \cdot \mathbf{T}\, ds$ is a measure of work.

Conservative vector fields were defined and various criteria given for identifying them.

In many texts $\mathbf{F} \cdot \mathbf{n}\, dS$ is written as $\mathbf{F} \cdot d\mathbf{S}$.

VOCABULARY AND SYMBOLS

scalar field f
vector field $\mathbf{F}$
gradient of scalar field ∇f
divergence of a vector field $\nabla \cdot \mathbf{F}$, div $\mathbf{F}$
curl of a vector field $\nabla \times \mathbf{F}$, **curl F**
line integral of a scalar field $\int_C f\, ds$
closed curve
integral over a closed curve $\oint_C$
exterior normal $\mathbf{n}$
conservative vector field
exact differential
Green's theorem
surface integral
divergence theorem
arcwise-connected
orientable (two-sided)
simply connected

KEY FACTS To integrate $\delta(P)$ over a surface $\mathscr{S}$, compute $\int_{\mathscr{A}} \delta(P) \sec \gamma \, dA$, where $\mathscr{A}$ is the shadow or projection of $\mathscr{S}$ on the xy plane. If $\mathscr{S}$ is part of the level surface $g(x, y, z) = 0$, then

$$\sec \gamma = \frac{\sqrt{g_x^2 + g_y^2 + g_z^2}}{|g_z|}.$$

If $\mathscr{S}$ is part of the surface $z = f(x, y)$, then

$$\sec \gamma = \sqrt{f_x^2 + f_y^2 + 1}.$$

These three conditions on **F** are equivalent: (1) $\int_C \mathbf{F} \cdot \mathbf{T} \, ds$ depends only on the endpoints of C, (2) $\oint_C \mathbf{F} \cdot \mathbf{T} \, ds = 0$ for closed curves C, and (3) **F** is of the form ∇f. If the domain of **F** is simply connected, then condition (4), **curl F** = **0**, is equivalent to any of the three conditions given.

Guide quiz on chap. 15

1. Why is $\nabla \cdot \mathbf{F}$ called the divergence of **F**?
2. Why is $\nabla \times \mathbf{F}$ called the curl of **F**?
3. Show that Green's theorem is a special case of Stokes' theorem.
4. (a) State Green's theorem.
   (b) What is its physical interpretation?
5. (a) State the divergence theorem.
   (b) What is its physical interpretation?
6. The field $\mathbf{F} = \mathbf{r}/|\mathbf{r}|^3$ is met often in physics; it is defined at every point in space except at the origin. Show that
   (a) div **F** = 0   (b) **curl F** = **0**.
7. (a) How could you produce lots of conservative vector fields?
   (b) How would you go about determining whether a given vector field is conservative?
8. Compute $\oint_C \mathbf{F} \cdot \mathbf{T} \, ds$, where $\mathbf{F}(x, y, z) = xy\mathbf{i} + 3z\mathbf{j} + y\mathbf{k}$, and C is the polygonal curve that starts at $(0, 0, 0)$, goes to $(1, 1, 1)$, then to $(0, 1, 1)$, and then back to $(0, 0, 0)$.
9. Verify that

$$\int_{\mathscr{V}} \nabla \cdot \mathbf{F} \, dV = \int_{\mathscr{S}} \mathbf{F} \cdot \mathbf{n} \, dS.$$

where $\mathscr{V}$ is the box bounded by the three coordinate planes and the three planes $x = a$, $y = b$, and $z = c$ (a, b, and c are positive), $\mathscr{S}$ is the surface of $\mathscr{V}$, and $\mathbf{F}(x, y, z) = x^2\mathbf{i} + y^2\mathbf{j} + z^2\mathbf{k}$.

10. Let $\mathscr{S}$ be the top half of the ellipsoid whose equation is $x^2/4 + y^2/4 + z^2/25 = 1$. Let

$$\mathbf{F}(x, y, z) = (x^2 + y)\mathbf{i} + (y^2 + x)\mathbf{j} + (y^3 + x^3)\mathbf{k}.$$

Find the integral of the normal component of **curl F** over $\mathscr{S}$. (Use the exterior normal to the ellipsoid.)

11. Let **F** be a vector field defined everywhere in the xy plane except at the two points $(1, 0)$ and $(2, 0)$. Assume that the curl of **F** is **0**. In Fig. 15.80 six curves are shown, swept out in the indicated directions. If $\int_{C_1} \mathbf{F} \cdot \mathbf{T} \, dS = 1$ and $\int_{C_2} \mathbf{F} \cdot \mathbf{T} \, ds = 3$, find

(a) $\int_{C_3} \mathbf{F} \cdot \mathbf{T} \, ds$ (b) $\int_{C_4} \mathbf{F} \cdot \mathbf{T} \, ds$

(c) $\int_{C_5} \mathbf{F} \cdot \mathbf{T} \, ds$ (d) $\int_{C_6} \mathbf{F} \cdot \mathbf{T} \, ds$.

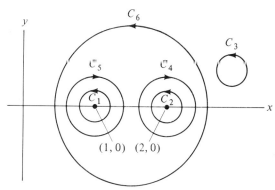

Figure 15.80

Review exercises for chap. 15

1. Check Green's theorem for $\mathbf{F} = (x - 5y)\mathbf{i} + xy\mathbf{j}$ and $\mathscr{A}$ the region bounded by $y = x^2$ and $y = \sqrt{x}$.

2. $\mathscr{A}$ is a convex set in the plane and C its boundary (taken counterclockwise). $\mathbf{F}$ is a vector field in the plane. Complete these equations.

   (a) $\oint_C \mathbf{F} \cdot \mathbf{n}\, ds = \int_{\mathscr{A}} \underline{}$;

   (b) $\oint_C \mathbf{F} \cdot \mathbf{T}\, ds = \int_{\mathscr{A}} \underline{}$.

3. Let $\mathbf{F}(x, y, z)$ be conservative. Show that $\text{curl } \mathbf{F} = \mathbf{0}$.

4. Let C be the curve $\mathbf{G}(t) = (t^4 + t)\mathbf{i} + t^5\mathbf{j}$ for t in $[0, 1]$. Evaluate $\int_C \nabla f \cdot d\mathbf{r}$ if
   (a) $f(x, y) = 3x^2 - 4y^2$;
   (b) $f(x, y) = \cos 2\pi xy$;
   (c) $f(x, y) = e^{x^3 y}$.

5. Let C be the curve $\mathbf{G}(t) = \cos t\, \mathbf{i} + \sin t\, \mathbf{j} + (t - \pi)^2\, \mathbf{k}$ for t in $[0, \pi]$. Evaluate $\int_C \mathbf{F} \cdot \mathbf{T}\, ds$, where $\mathbf{F}$ is the gradient of
   (a) $x^2 yz$,
   (b) $x^5 + y^5 + z^5$,
   (c) $e^{xz} + \tan xyz$.

6. (a) Show that if $\mathbf{F} = P\mathbf{i} + Q\mathbf{j}$ is a gradient field, then
   $$P_y = Q_x.$$
   (b) Show that $\mathbf{F}(x, y) = x^2 y\mathbf{i} - xy^2\mathbf{j}$ is *not* a gradient field.

7. The electric field $\mathbf{E}$ at any point (x, y) due to a point charge q at $(0, 0)$ is equal to
   $$\mathbf{E} = \frac{q\mathbf{u}}{4\pi\varepsilon_0 r^2},$$
   where r is the distance from the charge to the point (x, y), and $\mathbf{u}$ is the unit vector directed from the charge to the point (x, y). Evaluate the work done by the field when a unit charge is moved from $(1, 0)$ to $(2, 0)$ along (a) the x axis; (b) the rectangular path from $(1, 0)$ to $(1, \tfrac{1}{2})$ to $(2, \tfrac{1}{2})$ to $(2, 0)$. Is there a difference between the work done in (a) and that done in (b)?

8. Show that
   (a) $\mathbf{F}(x, y) = ye^{xy}\mathbf{i} + xe^{xy}\mathbf{j}$ is conservative.
   (b) $\mathbf{F}(x, y, z) = e^y\mathbf{i} + e^x\mathbf{j} + z^5\mathbf{k}$ is not conservative by exhibiting a circular closed path over which $\oint \mathbf{F} \cdot \mathbf{T}\, ds$ is not 0.

9. Let $f(x, y, z) = e^{xy} + \ln(1 + z^2)$ be a scalar function defined for points in space.
   (a) Compute ∇f.
   (b) Compute the divergence of ∇f.
   (c) Compute the curl of ∇f.

10. Let $\mathbf{F} = P\mathbf{i} + Q\mathbf{j}$ be a vector field.
    (a) Explain why $\int_C (P\, dx + Q\, dy)$ represents the work accomplished by the force $\mathbf{F}$ in moving a particle along C.
    (b) Explain why $\int_C (-Q\, dx + P\, dy)$ represents the net loss of fluid past the closed curve C bounding a convex region if the flux is $\mathbf{F}$.

11. Using a diagram, show that, if a closed planar curve C is swept out counterclockwise, then $(dy/ds)\mathbf{i} - (dx/ds)\mathbf{j}$ is a unit exterior normal.

12. Which of these vector fields in the plane are conservative? Explain.
    (a) $\mathbf{F}(x, y) = 3\mathbf{i} + 4\mathbf{j}$;
    (b) $\mathbf{F}(x, y) = y \sin xy\, \mathbf{i} + x \cos xy\, \mathbf{j}$;
    (c) $\mathbf{F}(x, y) = x\mathbf{i} + y\mathbf{j}$.

13. Let $\mathbf{F}(x, y, z) = x^3\mathbf{i} + y^3\mathbf{j} + z^3\mathbf{k}$. Let $\mathscr{S}$ be the sphere $x^2 + y^2 + z^2 = 1$, and $\mathscr{V}$ be the ball $x^2 + y^2 + z^2 \leq 1$. By computing both sides verify that
    $$\int_{\mathscr{V}} \text{div } \mathbf{F}\, dV = \int_{\mathscr{S}} \mathbf{F} \cdot \mathbf{n}\, dS.$$

14. Which of these vector fields in space are conservative? Not conservative? Explain.
    (a) $\mathbf{F}(x, y, z) = 3\mathbf{i} + 4\mathbf{j} + 5\mathbf{k}$
    (b) $\mathbf{F}(x, y, z) = (2xy^2 z + xy^3)\mathbf{i} + (2x^2 yz + 3x^2 y^2/2)\mathbf{j} + (x^2 y^2 + yz^2)\mathbf{k}$
    (c) $\mathbf{F}(x, y, z) = \sin^2 5x \cos^2 5x\, \mathbf{i} + y^2 \sin y\, \mathbf{j} + (z + 1)^2 e^z \mathbf{k}$

15. Let $\mathbf{F}(x, y, z) = ze^{xz}\mathbf{i} + xe^{xz}\mathbf{k}$.
    (a) Exhibit a function f such that $\mathbf{F} = \nabla f$.
    (b) Evaluate $\int_C \mathbf{F} \cdot \mathbf{T}\, ds$ for the curve given parametrically as
    $$\mathbf{r}(t) = t^2 \mathbf{i} + \cos t\, \mathbf{j} + e^t \mathbf{k} \qquad 0 \leq t \leq 2\pi.$$

16. Let $\mathbf{F}$ be a vector function defined at all points of a curve C and suppose $|\mathbf{F}| \leq M$ on C, where M is some positive number. Show that $\left| \int_C \mathbf{F} \cdot \mathbf{T}\, ds \right| \leq Ml$, where l is the length of C.

17. Let $\mathbf{F}(x, y, z) = 4xz\mathbf{i} - y^2\mathbf{j} + yz\mathbf{k}$. Let $\mathscr{V}$ be the cube bounded by the three coordinate planes and the planes $x = 1$, $y = 1$, $z = 1$. Let $\mathscr{S}$ be the surface of $\mathscr{V}$. Verify that
    $$\int_{\mathscr{V}} \text{div } \mathbf{F}\, dV = \int_{\mathscr{S}} \mathbf{F} \cdot \mathbf{n}\, dS.$$

18. Let $\mathbf{F}(x, y, z) = 3xy\mathbf{i} + 4\mathbf{j} + z\mathbf{k}$. Let C be the closed polygonal curve that starts at $(0, 0, 0)$, goes to $(1, 0, 1)$, then to $(1, 1, 1)$, then to $(0, 1, 0)$, and back to $(0, 0, 0)$.
    (a) Evaluate $\int_C \mathbf{F} \cdot \mathbf{T}\, ds$.
    (b) Is $\mathbf{F}$ conservative?

19 Let C be the triangle with vertices $(0, 0)$, $(1, 0)$, and $(0, 1)$ oriented counterclockwise. Evaluate

$$\oint_C [(3x^2 + y)\, dx + 4y^2\, dy]$$

by (a) direct calculation, (b) Green's theorem.

20 Let $\mathscr{S}$ be the sphere of radius 3 centered at $(4, 1, 2)$. Evaluate $\int_{\mathscr{S}} \mathbf{F} \cdot \mathbf{n}\, dS$ where

$$\mathbf{F} = 7x\mathbf{i} + x^2 z\mathbf{j} + 5z\mathbf{k}.$$

21 Let $\mathscr{S}$ be a surface that encloses a solid region $\mathscr{V}$. Assume that the origin O is in $\mathscr{V}$. Let $\mathbf{F}(P)$ be the position vector $\overrightarrow{OP}$. Using approximating sums and a sketch, show that $\int_{\mathscr{S}} \mathbf{F}(P) \cdot \mathbf{n}\, dS = 3 \cdot$ Volume of $\mathscr{V}$.

22 For the function $\mathbf{F}$ of Exercise 21 verify that

$$\int_{\mathscr{S}} \mathbf{F} \cdot \mathbf{n}\, dS = \int_{\mathscr{V}} \nabla \cdot \mathbf{F}\, dV.$$

23 State all the ways you can think of to show that the expression $P\, dx + Q\, dy + R\, dz$ is (a) exact, (b) not exact.

24 If the length of $\mathbf{F}(P)$ is at most 5 for all points P on the surface $\mathscr{S}$, which bounds the region $\mathscr{V}$, what can be said about

$$\int_{\mathscr{V}} \nabla \cdot \mathbf{F}\, dV?$$

25 Let C be the square with vertices $(0, 0, 0)$, $(0, 2, 2)$, $(0, 0, 4)$, and $(0, -2, 2)$ swept out in the given order. Evaluate $\oint_C \mathbf{F} \cdot d\mathbf{r}$ where

$$\mathbf{F} = x^2 y e^z \mathbf{i} + (x + y + z)\mathbf{j} + x^2 z\mathbf{k}.$$

26 Let $\mathbf{F}(x, y, z) = xz\mathbf{i} + x^2\mathbf{j} + xy\mathbf{k}$. Let $\mathscr{S}$ be all of the surface of the cube shown in Fig. 15.81 *other than the top*. Verify Stokes' theorem in this case.

27 Let $\mathbf{F}(x, y, z) = \mathbf{r}/|\mathbf{r}|^n$ for a fixed integer n and $\mathbf{r} = x\mathbf{i} + y\mathbf{j} + z\mathbf{k}$.
(a) For which n is **curl F** = **0**?
(b) For which n is div $\mathbf{F} = 0$?

28 Using Stokes' theorem, evaluate $\int_{\mathscr{S}} (\mathbf{curl\ F}) \cdot \mathbf{n}\, dS$, where $\mathbf{F} = (x^2 + y)\mathbf{i} + 3xy\mathbf{j} + (2xz + z^2)\mathbf{k}$, $\mathscr{S}$ is the surface of the paraboloid $z = 4 - x^2 - y^2$ above the xy plane, and $\mathbf{n}$ has a positive z component.

29 Let C be the closed curve $x = 2 \cos t$, $y = 3 \sin t$, for t in $[0, 2\pi]$.
(a) Graph C.
(b) Compute $\oint_C [x^2\, dx + (y + 1)\, dy]$.
(c) Devise a work problem whose answer is the integral in (b).
(d) Devise a fluid-flow problem whose answer is the integral in (b).

30 Let $\mathbf{F}(x, y)$ be a vector field defined everywhere except at the origin. Assume that the value of $\oint_C \mathbf{F} \cdot \mathbf{n}\, ds$ for any curve that encloses the origin and bounds a region is the same (not necessarily 0, however). Must the divergence of $\mathbf{F}$ be 0?

31 The following argument appears in *Introduction to Fluid Mechanics*, by Stephen Whitaker, Prentice Hall, New Jersey: We have

$$\int_{\mathscr{V}} \frac{\partial \rho}{\partial t}\, dV + \int_{\mathscr{S}} \rho \mathbf{v} \cdot \mathbf{n}\, dS = 0$$

for any region $\mathscr{V}$ and its bounding surface $\mathscr{S}$. With the aid of the divergence theorem it follows that $\partial \rho / \partial t + \nabla \cdot (\rho \mathbf{v}) = 0$ identically. Fill in the details.

32 Let $\mathscr{S}$ be the surface of a convex region $\mathscr{V}$. Consider a very small loop C on $\mathscr{S}$ and let $\mathscr{S}_1$ be the large part of $\mathscr{S}$ not inside C. Let $\mathbf{F}$ be a vector field in space. (See Fig. 15.82.)

Figure 15.82

(a) Show that

$$\int_{\mathscr{S}_1} (\nabla \times \mathbf{F}) \cdot \mathbf{n}\, dS = \oint_C \mathbf{F} \cdot \mathbf{n}\, ds.$$

(b) Letting C get arbitrarily small, deduce that

$$\int_{\mathscr{S}} (\nabla \times \mathbf{F}) \cdot \mathbf{n}\, dS = 0.$$

(c) Deduce that $\int_{\mathscr{V}} \nabla \cdot (\nabla \times \mathbf{F})\, dV = 0$.
(d) Deduce that the divergence of the curl is 0.

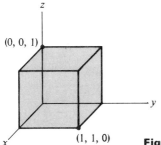

Figure 15.81

33 Let $\mathscr{V}$ be the solid cylinder consisting of all points (x, y, z) such that $x^2 + y^2 \leq a^2$ $(a > 0)$ and $0 \leq z \leq h$ $(h > 0)$. Let $\mathscr{S}$ be the surface of $\mathscr{V}$. Let $\mathbf{F}(x, y, z) = x\mathbf{i} + y\mathbf{j} + z\mathbf{k}$. Verify that

$$\int_{\mathscr{V}} \nabla \cdot \mathbf{F} \, dV = \int_{\mathscr{S}} \mathbf{F} \cdot \mathbf{n} \, dS.$$

Comparison of Exercises 34 and 35 presents an instructive contrast.

34 Let $\mathscr{D}$ consist of those points (x, y, z) not on the z axis. Let

$$\mathbf{F}(x, y, z) = \frac{x\mathbf{i}}{x^2 + y^2} + \frac{y\mathbf{j}}{x^2 + y^2}$$

for (x, y, z) in $\mathscr{D}$. Let $f(x, y, z) = \ln(x^2 + y^2)$.
(a) Show that div $\mathbf{F} = 0$.
(b) Show that curl $\mathbf{F} = \mathbf{0}$.
(c) Show that $\mathbf{F} = \nabla f$.
(d) Is $\mathbf{F}$ conservative?

35 Let $\mathscr{D}$ be as in Exercise 34. Let

$$\mathbf{F}(x, y, z) = \frac{-y\mathbf{i}}{x^2 + y^2} + \frac{x\mathbf{j}}{x^2 + y^2}.$$

(a) Show that div $\mathbf{F} = 0$.
(b) Show that curl $\mathbf{F} = \mathbf{0}$.
(c) Let $f(x, y, z) = \tan^{-1}(y/x)$. Is $\mathbf{F} = \nabla f$ where both are defined?
(d) Evaluate $\oint_C \mathbf{F} \cdot \mathbf{T} \, ds$, where C is the circle $x^2 + y^2 = 1$ in the xy plane.
(e) Show that $\mathbf{F}$ is not conservative.
(f) Does (e) contradict (c)? Explain.

36 Let $f(x, y, z, t)$ be the density of a fluid or gas at the point (x, y, z) at time t. Let $\mathscr{V}(t)$ be a solid region which depends on time, and let $\mathscr{S}(t)$ be the surface that bounds it. Let $\mathbf{W} = \mathbf{F}(x, y, z, t)$ be the velocity vector of the point (x, y, z) on $\mathscr{S}(t)$ at time t. The equation

$$\frac{d}{dt} \int_{\mathscr{V}(t)} f(P) \, dV = \int_{\mathscr{V}(t)} \frac{\partial f}{\partial t} \, dV - \int_{\mathscr{S}(t)} f(P)(\mathbf{W} \cdot \mathbf{n}) \, dS$$

is known as the *general transport theorem* in fluid mechanics.

(a) Interpret $\int_{\mathscr{V}(t)} f(P) \, dV$ physically.

(b) Interpret $\dfrac{d}{dt} \int_{\mathscr{V}(t)} f(P) \, dV$ physically.

(c) Interpret $\dfrac{\partial f}{\partial t}$ physically.

(d) Interpret $\int_{\mathscr{V}(t)} \dfrac{\partial f}{\partial t} \, dV$ physically.

(e) Interpret $\int_{\mathscr{S}(t)} f(P)(\mathbf{W} \cdot \mathbf{n}) \, dS$ physically.

(f) Is the general transport theorem plausible?

■ ■

37 Consider a function that assigns to each point in the uv plane a point (x, y, z) in space. There are consequently three scalar functions, $x = f(u, v)$, $y = g(u, v)$, $z = h(u, v)$. Let $\mathbf{r}(u, v)$ be the position vector $x\mathbf{i} + y\mathbf{j} + z\mathbf{k}$ corresponding to the point (u, v). Assume that all the functions f, g, h are "nice" (in particular, have continuous partial derivatives). Then as (u, v) varies in a region $\mathscr{A}$ in the plane, (x, y, z) sweeps out a smooth surface $\mathscr{S}$.

If v is held fixed, $\mathbf{r}(u, v)$ parametrizes a curve on $\mathscr{S}$. Then $\partial \mathbf{r}/\partial u$ is a tangent vector to that curve. Similarly, $\partial \mathbf{r}/\partial v$ is a tangent vector to another curve on $\mathscr{S}$.
(a) How would you obtain a normal vector to $\mathscr{S}$?
(b) Consider a very small rectangle in the uv plane with vertices (a, b), $(a + \Delta u, b)$, $(a, b + \Delta v)$, and $(a + \Delta u, b + \Delta v)$. Show that the area of the image in $\mathscr{S}$ of this rectangle is approximately

$$\left| \frac{\partial \mathbf{r}}{\partial u} \times \frac{\partial \mathbf{r}}{\partial v} \right|.$$

(c) Express the area of $\mathscr{S}$ as an integral over $\mathscr{A}$.
(d) Taking the case $x = a \sin u \cos v$, $y = a \sin u \sin v$, $z = a \cos u$, obtain the formula for integrating over a spherical surface of radius a.
(e) Taking the case $x = u$, $y = v$, $z = f(x, y)$, obtain the formula for integrating over a surface $z = f(x, y)$ by integrating over a suitable region in the xy plane.

38 Let $\mathbf{F} = P\mathbf{i} + Q\mathbf{j} + R\mathbf{k}$ be a vector field whose divergence is 0 throughout space. This exercise will construct a vector field $\mathbf{G}$ such that $\mathbf{F} = \text{curl } \mathbf{G}$. In fact, $\mathbf{G}$ will be constructed to be of the form $u\mathbf{i} + v\mathbf{j} + 0\mathbf{k}$.
(a) Show that we are seeking functions u and v such that

$$P = -v_z, \qquad Q = u_z, \qquad R = v_x - u_y.$$

(b) Define $v(x, y, z) = -\int_0^z P(x, y, z_1) \, dz_1$. Show that $P = -v_z$.
(c) Let $A(x, y)$ be a function of x and y to be determined later. Define

$$u(x, y, z) = \int_0^z Q(x, y, z_1) \, dz_1 + A(x, y).$$

Show that $Q = u_z$.
(d) Show that $v_x - u_y =$

$$-\int_0^z [P_x(x, y, z_1) + Q_y(x, y, z_1)] \, dz_1 - A_y(x, y).$$

You will need the theorem in Appendix G.2.
(e) Recalling that $\mathbf{V} \cdot \mathbf{F} = 0$, show that

$$v_x - u_y = \int_0^z R_z(x, y, z_1)\, dz_1 - A_y(x, y)$$
$$= R(x, y, z) - R(0, y, z) - A_y(x, y).$$

(f) Define $A(x, y)$ to be $-\int_0^y R(x, y_1, 0)\, dy_1$. Show that $R = v_x - u_y$.

39 If $\int_{\mathscr{V}} (\mathbf{V}f \cdot \mathbf{V}f)\, dV = 0$, where $\mathscr{V}$ is a ball, what can we conclude about f? Explain.

40 This exercise is an introduction to Exercise 41. A motorboat goes back and forth on a straight measured mile at a constant speed v relative to the water. Show that the time required for a round trip is always less when there is no current than when there is a constant current $\mathbf{W}$. (Assume that the component of $\mathbf{W}$ along the route is less than v and greater than 0. If the component along the route were larger than v, the boat could not make a round trip.)

41 (See Exercise 40.) An aircraft traveling at constant air speed v traverses a closed horizontal curve marked on the ground. Show that the time required for one complete trip is always less when there is no wind than when there is a constant wind $\mathbf{W}$. (Assume that $W = |\mathbf{W}|$ is less than v, so that the plane never meets an insuperable head wind.)

42 This argument appears in Colin W. Clark, *Mathematical Bioeconomics*, Wiley, New York. Let $G(x, y)$ and $H(x, y)$ be two functions such that $H_x \geq G_y$. Let C_1 and C_2 be the two curves directed as shown in Fig. 15.83. (Each represents a way of harvesting a crop; x is time and y is the rate of harvest.)

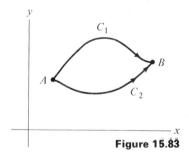

Figure 15.83

Show that

$$\int_{C_1} (G\, dx + H\, dy) \leq \int_{C_2} (G\, dx + H\, dy).$$

Exercises 43 and 44 concern harmonic functions. A function f is harmonic if $\nabla^2 f = 0$, that is, $f_{xx} + f_{yy} + f_{zz} = 0$.

43 Let $\mathscr{V}$ be a solid and let $\mathscr{S}$ be its surface. Let f and g be two scalar functions defined on $\mathscr{V}$.
(a) Show that

$$\int_{\mathscr{V}} \mathbf{V} \cdot (f\, \mathbf{V}g)\, dV = \int_{\mathscr{V}} (f\, \mathbf{V}^2 g + \mathbf{V}f \cdot \mathbf{V}g)\, dV$$
$$= \int_{\mathscr{S}} (f\, \mathbf{V}g) \cdot \mathbf{n}\, dS.$$

(b) Show that

$$\int_{\mathscr{V}} \mathbf{V} \cdot (g\, \mathbf{V}f)\, dV = \int_{\mathscr{V}} (g\, \mathbf{V}^2 f + \mathbf{V}g \cdot \mathbf{V}f)\, dV$$
$$= \int_{\mathscr{S}} (g\, \mathbf{V}f) \cdot \mathbf{n}\, dS.$$

(c) Deduce that

$$\int_{\mathscr{V}} (f\, \mathbf{V}^2 g - g\, \mathbf{V}^2 f)\, dV = \int_{\mathscr{S}} [(f\, \mathbf{V}g - g\, \mathbf{V}f) \cdot \mathbf{n}]\, dS.$$

This result, also, is called Green's theorem.

44 Assume that f and g are harmonic on $\mathscr{V}$ and equal on $\mathscr{S}$. This exercise will show that they must be equal on $\mathscr{V}$. Let $h = f - g$.
(a) Use Exercise 43(a) to show that

$$\int_{\mathscr{V}} h\, \mathbf{V}^2 h\, dV + \int_{\mathscr{V}} \mathbf{V}h \cdot \mathbf{V}h\, dV = \int_{\mathscr{S}} (h\, \mathbf{V}h) \cdot \mathbf{n}\, dS.$$

(b) Deduce that $\mathbf{V}h = \mathbf{0}$ throughout $\mathscr{V}$.
(c) Deduce that h is constant throughout $\mathscr{V}$.
(d) Show that $f = g$ throughout $\mathscr{V}$.

45 Consider an object submerged in water. It occupies a region $\mathscr{V}$ with surface $\mathscr{S}$. The force $\mathbf{F}$ of the water against the object at each point is perpendicular to $\mathscr{S}$. Assume that the surface of the water is the xy plane. Thus the z coordinate of a point in the water is negative. The force against a small patch of $\mathscr{S}$ of area ΔS and depth $|z|$ is approximately $z\mathbf{n}\, \Delta S$, where $\mathbf{n}$ is the unit exterior normal. The z component of this force is $cz\mathbf{k} \cdot \mathbf{n}\, \Delta S$, where c is the density of water.
(a) What integral over $\mathscr{S}$ represents the vertical component of the total force of the water against the submerged object?
(b) Show that the vertical component of the force against the object is equal to the weight of the water displaced by the object.
(c) Show that the x and y components of the total force against the object are 0.

Throughout this chapter only integrals of scalar functions were considered. But integrals of vector functions are sometimes of use and are easily defined. If $\mathbf{F} = P\mathbf{i} + Q\mathbf{j} + R\mathbf{k}$ is defined throughout the solid $\mathscr{V}$, define $\int_{\mathscr{V}} \mathbf{F}(x, y, z)\, dV$ to be

$$\left(\int_{\mathscr{V}} P\, dV\right)\mathbf{i} + \left(\int_{\mathscr{V}} Q\, dV\right)\mathbf{j} + \left(\int_{\mathscr{V}} R\, dV\right)\mathbf{k}.$$

In short, to integrate a vector function, integrate each of its scalar components. This concept is the subject of Exercises 46 to 51.

46 Let $\mathbf{F}(x, y, z) = f(x, y, z)(x\mathbf{i} + y\mathbf{j} + z\mathbf{k})$, where $f(x, y, z)$ denotes the density of matter at point (x, y, z) in the solid that occupies the region $\mathscr{V}$. What is the physical significance of the equation

$$\int_{\mathscr{V}} \mathbf{F}(x, y, z) \, dV = \mathbf{0}?$$

47 Let $\mathbf{F} = P\mathbf{i} + Q\mathbf{j} + R\mathbf{k}$ be a vector field and $\mathscr{V}$ a solid region. Let $\mathbf{c}$ be a fixed vector. Show that $\int_{\mathscr{V}} \mathbf{c} \cdot \mathbf{F} \, dV = \mathbf{c} \cdot \int_{\mathscr{V}} \mathbf{F} \, dV$. A similar theorem holds for surface integrals.

48 Let $\mathbf{A}$ and $\mathbf{B}$ be vectors such that for all vector $\mathbf{c}$, $\mathbf{c} \cdot \mathbf{A} = \mathbf{c} \cdot \mathbf{B}$. Show that $\mathbf{A} = \mathbf{B}$.

49 Let $\mathbf{c}$ be any fixed vector. Let $\mathscr{V}$ be a solid region and $\mathscr{S}$ the surface that bounds it. Let f be a scalar function on $\mathscr{V}$.
   (a) Show that $\int_{\mathscr{V}} \nabla \cdot f\mathbf{c} \, dV = \int_{\mathscr{S}} f\mathbf{c} \cdot \mathbf{n} \, dS$.
   (b) Show that $\int_{\mathscr{V}} \nabla \cdot f\mathbf{c} \, dV = \int_{\mathscr{V}} \mathbf{c} \cdot \nabla f \, dV$.
   (c) Show that $\mathbf{c} \cdot \int_{\mathscr{V}} \nabla f \, dV = \mathbf{c} \cdot \int_{\mathscr{S}} f\mathbf{n} \, dS$.
   (d) Deduce that $\int_{\mathscr{V}} \nabla f \, dV = \int_{\mathscr{S}} f\mathbf{n} \, dS$.

50 Let $\mathscr{S}$ be a surface bounding a solid region $\mathscr{V}$. Let $\mathbf{c}$ be a fixed vector.
   (a) Show that $\int_{\mathscr{S}} \mathbf{c} \cdot \mathbf{n} \, dS = 0$.
   (b) Deduce that $\int_{\mathscr{S}} \mathbf{n} \, dS = \mathbf{0}$.

51 (See Exercise 50.) On each of the four faces of a tetrahedron a vector is constructed that is perpendicular to the face, points outward, and has magnitude equal to the area of the face. Show that the sum of the four vectors is $\mathbf{0}$.

52 A parallelepiped $\mathscr{V}$ is spanned by the vectors $\mathbf{i} + \mathbf{j} + \mathbf{k}$, $2\mathbf{i} + \mathbf{j} + \mathbf{k}$, and $2\mathbf{i} + 3\mathbf{j} + 4\mathbf{k}$. Let $\mathbf{F} = 2x\mathbf{i} + 3y\mathbf{j} + 4z\mathbf{k}$. Evaluate $\int_{\mathscr{S}} \mathbf{F} \cdot \mathbf{n} \, dS$, where $\mathscr{S}$ is the surface of $\mathscr{V}$.

53 At the end of Sec. 15.5 we obtained the formula $dS = \rho^2 \sin \phi \, d\phi \, d\theta$ for integrating over a portion of a sphere with center at the origin.
   (a) Show that the corresponding formula for integrating over the surface of a cylinder of radius r with its axis along the z axis is $dS = r \, dz \, d\theta$.
   (b) Use (a) to find the area of the portion of the cylinder $x^2 + y^2 = r^2$ that lies between the xy plane and the plane $z = y$.

54 (See Exercise 53.)
   (a) Show that the corresponding formula for integration over the surface of the cone with vertex at the origin and axis along the z axis is $dS = \rho \sin \phi \, d\rho \, d\theta$.
   (b) Use (a) to set up an integral for the portion of the cone $z = \sqrt{x^2 + y^2}$ that lies below the plane $z = 1 + y$. (Do not evaluate the integral, but describe how it could be evaluated.)

55 (a) Let $\mathscr{S}_1$ and $\mathscr{S}_2$ be bounded by the same curve C. Choose an orientation for C and then choose the unit normals $\mathbf{n}_1$ for $\mathscr{S}_1$ and $\mathbf{n}_2$ for $\mathscr{S}_2$ by the right-hand rule. Let $\mathbf{F}$ be a vector field defined on $\mathscr{S}_1$ and $\mathscr{S}_2$. Show that $\int_{\mathscr{S}_1} \text{curl } \mathbf{F} \cdot \mathbf{n}_1 \, dS = \int_{\mathscr{S}_2} \text{curl } \mathbf{F} \cdot \mathbf{n}_2 \, dS$.
   (b) The result in (a) permits $\int_{\mathscr{S}} \text{curl } \mathbf{F} \cdot \mathbf{n} \, dS$ to be replaced by a similar integral over the same boundary as $\mathscr{S}$ has. Use this fact to evaluate $\int_{\mathscr{S}} \text{curl } \mathbf{F} \cdot \mathbf{n} \, dS$, where $\mathbf{F} = x(x^2 + y^2 + z^2)^{3/2}\mathbf{i} - y(x^2 + y^2 + z^2)^{3/2}\mathbf{j} + x^2 \sin \pi y \, e^{\tan^{-1} z}\mathbf{k}$ and $\mathscr{S}$ is the top half of the ellipsoid $x^2/16 + y^2/9 + z^2/4 = 1$.

APPENDIX A
REAL NUMBERS

This appendix discusses rational and irrational numbers, division by 0, inequalities, intervals, and absolute value.

The *positive integers* are the counting numbers 1, 2, 3, ...; the *negative integers* are $-1, -2, -3, \ldots$. The set of positive and negative integers, together with 0, is called the set of *integers*. The integers appear on the number line as regularly spaced points, as shown in Fig. A.1.

The integers

$\cdots, -3, -2, -1, 0, 1, 2, 3, \cdots$

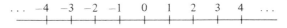

Figure A.1

Every point on the number line corresponds to a *real number*. For instance, $\frac{11}{3}$, $-\sqrt{2}$, π, and 1.13 are real numbers and their positions are shown in Fig. A.2.

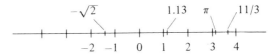

Figure A.2

Rational and irrational

A real number that can be written as a fraction or ratio p/q, where p is an integer and q is a positive integer, is called a *rational number*. For instance, $\frac{11}{3}$ is rational, as are $8 = \frac{8}{1}$ and $7\sqrt{2}/(3\sqrt{2}) = \frac{7}{3}$. A real number that is not rational is called *irrational*. Greek mathematicians some 2400 years ago showed that $\sqrt{2}$ is irrational; Johann Lambert in 1761 showed that π is irrational. This means that neither $\sqrt{2}$ nor π can be written as the quotient of two integers, p/q. (See Exercise 37.)

If a and b are real numbers, we may always form their sum $a + b$, their difference $a - b$, and their product ab. If b is not 0, then there is a unique

Why division by zero is meaningless

number x such that $bx = a$. This number x, the *quotient* of a by b, is denoted a/b. For instance, since the equation $2x = 6$ has the unique solution 3, we write $\frac{6}{2} = 3$.

However, when $b = 0$, solving the equation $bx = a$ runs into some problems. For instance, the equation

$$0x = 6$$

has no solution whatsoever, since the product of 0 and any real number is 0. Thus the symbol 6/0 is totally meaningless. When both a and b are 0, the equation $bx = a$ runs into a different trouble: there are too many solutions. The equation is now

$$0x = 0.$$

Every real number is a solution; for instance, $0 \cdot 5 = 0$, $0 \cdot \pi = 0$, and $0(-3) = 0$. The symbol 0/0 is meaningless because it does not describe a single number.

In short, "division by zero" makes no sense. If you find yourself dividing by zero while working a problem, turn back, as though you had met a sign warning "WRONG WAY." In particular, resist the temptation to say that 0/0 is equal to 1. This temptation will be placed before you often in the study of limits and derivatives. Though $3/3 = 1$, the expression 0/0 is utterly devoid of meaning.

```
WRONG
 WAY
  0/0
```

INEQUALITIES

If the point that represents the number a lies to the left of the point that represents the number b on the number line, we write $a < b$ ("a is less than b") or $b > a$ ("b is greater than a"). For instance, $5 < 7$, $7 > 5$, $-8 < 2$, and $-8 < -3$.

The expression $a \leq b$ means that a is either less than b or equal to b. Thus $3 \leq 4$ and $4 \leq 4$. Read "$a \leq b$" as "a is less than or equal to b." If $a \leq b$, we also write $b \geq a$.

Inequalities behave nicely with respect to addition but demand great care in the case of multiplication and division. Inspection of the following list of properties of inequalities shows the difference.

The properties of inequalities

1 If $a < b$ and $c < d$, then $a + c < b + d$. (You can add two inequalities that are in the same direction.)

2 If $a < b$ and c is any number, then $a + c < b + c$. (You can add the same number to both sides of an inequality.)

3 If $a < b$ and c is any number, then $a - c < b - c$. (You can subtract the same number from both sides of an inequality.)

Multiplication or division by a positive number preserves an inequality but multiplication or division by a negative number reverses an inequality.

4 If $a < b$ and c is a *positive* number, then $ac < bc$. (Multiplication by a positive number preserves an inequality.)

5 If $a < b$ and c is a *negative* number, then $ac > bc$. (Multiplication by a negative number reverses an inequality.)

6 If $a < b$ and $c < d$ and $a, b, c,$ and d are *positive*, then $ac < bd$. (You can multiply two inequalities, if they are in the same direction and involve only positive numbers.)

7 If $a < b$ and c is a *positive* number, then $a/c < b/c$. (Division by a positive number preserves an inequality.)

8 If $a < b$ and c is *negative*, then $a/c > b/c$. (Division by a negative number reverses the inequality.)

$2 < 3$ *but* $\frac{1}{2} > \frac{1}{3}$.

9 If a and b are positive numbers and $a < b$, then

$$\frac{1}{a} > \frac{1}{b}.$$

(Taking reciprocals of an inequality between positive numbers reverses the inequality.)

INTERVALS The notation $a < x < b$ in the following definition is short for the two inequalities $a < x$ and $x < b$. Thus, $a < x < b$ is short for "x is larger than a and less than b."

DEFINITION *Open interval.* Let a and b be real numbers, with $a < b$. The *open interval* (a, b) consists of all numbers x such that $a < x < b$.

DEFINITION *Closed interval.* Let a and b be real numbers, with $a < b$. The *closed interval* $[a, b]$ consists of all numbers x such that $a \leq x \leq b$.

The open interval (a, b) is obtained from the closed interval $[a, b]$ by the deletion of the endpoints a and b. Some intervals are shown in Fig. A.3.

Closed interval $[a, b]$ Open interval (a, b) Half-open interval $(a, b]$

Figure A.3

The meaning of ● and ○ in diagrams

The solid dot (●) indicates the presence of a point; the hollow dot (○) indicates the absence of a point. This notation will be used in diagrams throughout the text.

Infinite intervals

Infinite intervals will also be of use. For convenience they are listed in the following table:

Symbol	Description	Picture	In Words
$[a, \infty)$	$x \geq a$		Closed interval to the right of a
(a, ∞)	$x > a$		Open interval to the right of a
$(-\infty, a]$	$x \leq a$		Closed interval to the left of a
$(-\infty, a)$	$x < a$		Open interval to the left of a
$(-\infty, \infty)$	all x		Entire x axis

The symbols ∞ and $-\infty$ do not refer to numbers but provide a convenient shorthand. Thus the symbol $[a, \infty]$ is meaningless.

MANIPULATING INEQUALITIES

The next three examples apply the properties of inequalities and illustrate the interval notation.

EXAMPLE 1 Let x and y be positive numbers with $x < y$. Show that

$$3x^2 < x^2 + xy + y^2 < 3y^2.$$

SOLUTION These steps establish the first inequality.

$x^2 < xy$ ("multiplying $x < y$ by the positive number x," rule 4)

$x^2 < y^2$ ("multiplying $x < y$ by $x < y$," rule 6)

$x^2 + x^2 < xy + y^2$ (adding two inequalities, rule 1)

$x^2 + x^2 + x^2 < x^2 + xy + y^2$ (adding x^2 to both sides of preceding inequality, rule 2)

Thus $3x^2 < x^2 + xy + y^2.$

Similar reasoning establishes the second inequality. ∎

EXAMPLE 2 Find all numbers x such that

$$3x + 1 < 5x + 2.$$

SOLUTION Gradually transform the given inequality until one side of the inequality consists of just x and the other side does not involve x.

$3x + 1 < 5x + 2$ given

$3x < 5x + 1$ (subtracting 1 from both sides, rule 3)

$-2x < 1$ (subtracting $5x$ from both sides, rule 3)

$x > -\frac{1}{2}$ (dividing by the negative number -2, rule 8)

Each of these steps can be reversed. So the given inequality is equivalent to the inequality $x > -\frac{1}{2}$ and so has as solutions all numbers in the interval $(-\frac{1}{2}, \infty)$. ∎

EXAMPLE 3 Describe the set of numbers x such that

$$(x + 2)(x - 1)(x - 3) > 0.$$

SOLUTION First check where each of the three factors is positive or negative. For example, $x + 2$ is 0 at $x = -2$, positive if $x > -2$, and negative if $x < -2$. This information is recorded in Fig. A.4.

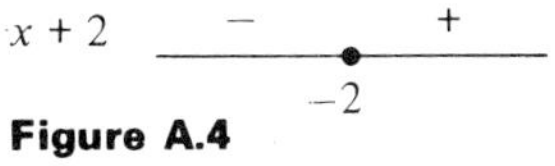

$x + 2$ $\quad\underset{-2}{\underline{\quad - \quad\bullet\quad + \quad}}$

Figure A.4

Similarly, $x - 1$ changes sign at $x = 1$ and $x - 3$ changes sign at $x = 3$. This information is collected in the following table.

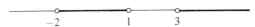

	-2		1		3	
$x + 2$	$-$		$+$		$+$	$+$
$x - 1$	$-$		$-$		$+$	$+$
$x - 3$	$-$		$-$		$-$	$+$
$(x + 2)(x - 1)(x - 3)$	$-$		$+$		$-$	$+$

The product of the three factors is positive when x is in $(-2, 1)$ or $(3, \infty)$. The solutions of the inequality $(x + 2)(x - 1)(x - 3) > 0$ thus fill out the two intervals shown in Fig. A.5.

Figure A.5

ABSOLUTE VALUE The absolute value of a real number tells "how far it is from 0." The following definition makes this precise.

DEFINITION *Absolute value.* The absolute value of a positive number x is x itself. The absolute value of a negative number x is its negative $-x$. The absolute value of 0 is 0.

If $x \geq 0$, $|x| = x$. If $x < 0$, $|x| = -x$.

The absolute value of x is denoted $|x|$. By definition, $|3| = 3$ and $|-3| = -(-3) = 3$. For all x, $|-x| = |x|$; x and $-x$ have the same absolute value. On some programmable calculators, the "absolute value" key is labeled "abs." On occasion, we will use abs x to denote $|x|$.

The absolute value behaves better with respect to multiplication than with respect to addition. For any real numbers x and y,

$$|xy| = |x| \, |y|$$ (The absolute value of the product is the product of the absolute values.)

The triangle inequality and $$|x + y| \leq |x| + |y|$$ (The absolute value of the sum is less than or equal to the sum of the absolute values.)

The second assertion is known as the *triangle inequality*. For instance, $|(-3) + 7| \leq |-3| + |7|$, as a little arithmetic verifies.

EXAMPLE 4 Sketch on the number line the set of numbers x such that $|x| < 2$.

SOLUTION A positive number has an absolute value less than 2 if the number itself is less than 2. A negative number has an absolute value less than 2 if the number itself is greater than -2. Also, the absolute value of 0 is less than 2. All told, the set of numbers x such that $|x| < 2$ is the open interval $(-2, 2)$, sketched in Fig. A.6.

Figure A.6

The interval $(-2, 2)$ consists of all numbers within a distance 2 of the origin. ∎

Observe that for any number $a > 0$, $|x| < a$ describes the open interval $(-a, a)$.

EXAMPLE 5 Sketch on the number line the set of numbers x such that $|5(x-1)| < 3$.

SOLUTION

$|5(x-1)| < 3$	given
$5|x-1| < 3$	absolute value of product
$|x-1| < \frac{3}{5}$	dividing by 5
$-\frac{3}{5} < x - 1 < \frac{3}{5}$	$x - 1$ has absolute value less than $\frac{3}{5}$
$1 - \frac{3}{5} < x < 1 + \frac{3}{5}$	adding 1 to an inequality
$\frac{2}{5} < x < \frac{8}{5}$	arithmetic

In short, x is in the open interval $(\frac{2}{5}, \frac{8}{5})$, sketched in Fig. A.7.

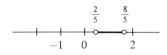

Figure A.7 ∎

Exercises

In Exercises 1 to 6 assume that a and b are positive numbers and $a < b$. Place the correct symbol, $<$ or $>$, in each blank.

1. $a + 3$ ___ $b + 3$
2. $a - 2$ ___ $b - 2$
3. $5a$ ___ $5b$
4. $(-2)a$ ___ $(-2)b$
5. $1/a$ ___ $1/b$
6. $3 - a$ ___ $3 - b$

In Exercises 7 to 20 describe where the inequalities hold. (Use interval notation.)

7. $2x + 7 < 4x + 9$
8. $3x - 5 < 7x + 11$
9. $-3x + 2 > 5x + 18$
10. $2x > 3x + 7$
11. $(x - 1)(x - 3) > 0$
12. $(x + 1)(x - 3) < 0$
13. $(x + 2)(x + 3) < 0$
14. $(2x + 6)(x + 3) < 0$
15. $x(x - 1)(x + 1) > 0$
16. $(x - 2)(x - 3)(x - 4) > 0$
17. $x(x + 3)(x + 5) > 0$

18 $(x-1)^2(x-2) > 0$
19 $(3x-1)(2x-1) > 0$
20 $x(2x+1)(3x-1) > 0$

In each of Exercises 21 to 24 sketch the intervals for which the inequality holds.

21 $|x-3| < 2$
22 $|2x-4| < 1$
23 $|3(x-1)| < 6$
24 $|4(x+2)| < 2$

■

25 For which x is $x^2(x-3) > 0$?
26 For which x is $(x-1)^2(x-2)^2 < 0$?
27 For which positive numbers x is $x < x^2$?
28 For which positive numbers x is $x^2 < x^3$?

In Exercises 29 and 30 give an example of numbers a and b, $a < b$, neither 0, for which the stated inequality is *false*.

29 $a^2 < b^2$ 30 $1/a > 1/b$
31 Give an example of numbers a, b, c, and d such that $a < b$, $c < d$, and $ac > bd$.

■ ■

32 Express 3.1416 in the form p/q, where p and q are integers. This shows that 3.1416 is a rational number. It is commonly used as a rational approximation of the irrational number π.
33 Let $a = 2.3474747...$, an endless decimal in which the block "47" continues to repeat without end. This exercise shows that a is rational.
    (a) Compute $100a$.
    (b) Compute $100a - a$.
    (c) From (b) deduce that $a = 232.4/99$.
    (d) From (c) deduce that a can be written as the quotient of two integers and is therefore rational. It can be shown that any number whose decimal representation from some point on consists of a block repeated endlessly is rational.
34 Here is a proposed proof that $0 = 1$. Let $x = 0$. Then $x = x^2$. Cancellation of x from both sides of the equation shows that $1 = x$. Thus we have $x = 0$ and $x = 1$ and conclude that $0 = 1$. Where is the error?
35 Here is another proof that $0 = 1$. Let $x = 1$. Then $x^2 = x^3$ or $x^2(x-1) = 0$. Thus $x^2 = 0$, from which it follows that $x = 0$. Since $x = 1$ and $x = 0$, we conclude that $0 = 1$. Where is the error?
36 (a) Show that, if x and y are rational numbers, so is their product. (*Hint*: $x = m/n$ and $y = p/q$, where m, n, p, and q are integers.)

(b) Give an example of irrational numbers x and y whose product is rational.

37 This is how the Greeks proved that $\sqrt{2}$ is irrational. Assume that $\sqrt{2}$ is rational. Then $\sqrt{2} = m/n$ where m and n are positive integers. Moreover we may assume that m and n have no common factor other than 1. (For if there were a common factor, we could cancel it in numerator and denominator.) They then showed that both m and n must be even, which is a contradiction, since two even numbers have the common factor 2. Their argument depends on the fact that, if the square of an integer is even, then the integer itself must be even.
    (a) Show that $2n^2 = m^2$.
    (b) From (a) deduce that m is even; hence $m = 2p$, for some integer p.
    (c) Deduce that $n^2 = 2p^2$.
    Thus m and n are both even, a contradiction.

Exercises 38 and 39 concern an inequality that arose in the study of bias in admissions at a university.

38 Let a_1, b_1, a_2, b_2, A_1, B_1, A_2, and B_2 be positive numbers such that

$$\frac{a_1}{b_1} < \frac{A_1}{B_1} \quad \text{and} \quad \frac{a_2}{b_2} < \frac{A_2}{B_2}.$$

Does it follow that

$$\frac{a_1 + a_2}{b_1 + b_2} < \frac{A_1 + A_2}{B_1 + B_2}?$$

One department admits a_1 out of b_1 women applicants and A_1 out of B_1 men applicants. Since $a_1/b_1 < A_2/B_2$, we may say that the department seems biased against women. Another department admits a_2 out of b_2 women and A_2 out of B_2 men; it also appears to be biased against women. The two departments, considered together, admit $a_1 + a_2$ out of $b_1 + b_2$ women and $A_1 + A_2$ out of $B_1 + B_2$ men. Can we be sure that the two, taken as a whole, are biased against women, that is, that

$$\frac{a_1 + a_2}{b_1 + b_2} < \frac{A_1 + A_2}{B_1 + B_2}?$$

39 (See Exercise 38.) Jane has a higher batting average than Jim in May. In June she again has a higher batting average. Does this imply that for the 2-month period taken as a whole, Jane has a higher batting average than Jim? (Batting average = hits/at bats.)

APPENDIX B
SOME TOPICS IN ALGEBRA

This appendix presents several topics from algebra: rationalizing an expression, completing the square, the quadratic formula, the binomial theorem, geometric series, and a factorization of $d^n - c^n$.

RATIONALIZING AN EXPRESSION The identities $\sqrt{a}\sqrt{a} = a$ and $(a + b)(a - b) = a^2 - b^2$ enable us to remove a square root from certain algebraic expressions. Three examples illustrate the technique.

EXAMPLE 1 Remove the square root from the denominator of the expression $4/\sqrt{2}$. ("Rationalize" the denominator.)

SOLUTION
$$\frac{4}{\sqrt{2}} = \frac{4}{\sqrt{2}} \cdot \frac{\sqrt{2}}{\sqrt{2}} = \frac{4\sqrt{2}}{2} = 2\sqrt{2} \quad \blacksquare$$

EXAMPLE 2 Remove the square root from the denominator of $2/(1 + \sqrt{5})$. (Rationalize the denominator.)

SOLUTION
$$\frac{2}{1+\sqrt{5}} = \frac{2}{1+\sqrt{5}} \cdot \frac{1-\sqrt{5}}{1-\sqrt{5}} = \frac{2-2\sqrt{5}}{1^2-(\sqrt{5})^2} = \frac{2-2\sqrt{5}}{1-5} = \frac{2-2\sqrt{5}}{-4}$$
$$= \frac{\sqrt{5}-1}{2}. \quad \blacksquare$$

In the next example the numerator is rationalized.

EXAMPLE 3 Remove the square roots from the numerator of $(\sqrt{x} - \sqrt{2})/(x - 2)$, $x \neq 2$. (Rationalize the numerator.)

SOLUTION

$$\frac{\sqrt{x}-\sqrt{2}}{x-2} = \frac{\sqrt{x}-\sqrt{2}}{x-2} \cdot \frac{\sqrt{x}+\sqrt{2}}{\sqrt{x}+\sqrt{2}} = \frac{(\sqrt{x})^2 - \sqrt{2}^2}{(x-2)(\sqrt{x}+\sqrt{2})}$$

$$= \frac{x-2}{(x-2)(\sqrt{x}+\sqrt{2})} = \frac{1}{\sqrt{x}+\sqrt{2}}. \blacksquare$$

COMPLETING THE SQUARE The expression $x^2 + bx + c$ can be written in the form $(x+k)^2 + d$ for a suitable choice of constants k and d. Finding the numbers k and d is called "completing the square."

Recall that $\qquad (x+k)^2 = x^2 + 2kx + k^2.$

Therefore, if $\qquad (x+k)^2 + d = x^2 + bx + c,$

then we have $\quad x^2 + 2kx + k^2 + d = x^2 + bx + c.$

Consequently $2k$ must be b. In other words, k must be $b/2$. The next two examples show why this observation is the key to completing the square.

EXAMPLE 4 Complete the square in $x^2 + 6x + 11$.

SOLUTION In this case $b = 6$ and $b/2 = 3$. Thus $k = 3$. To find d proceed as follows:

$$x^2 + 6x + 11 = (x^2 + 6x \quad) + 11$$
$$= (x^2 + 6x + 3^2) + 11 - 3^2$$
$$= (x+3)^2 + 11 - 9$$
$$= (x+3)^2 + 2. \blacksquare$$

EXAMPLE 5 Complete the square in $x^2 - 5x + 2$.

SOLUTION First of all, $k = -5/2$. Then

$$x^2 - 5x + 2 = (x^2 - 5x + (-\tfrac{5}{2})^2) + 2 - (-\tfrac{5}{2})^2$$
$$= (x - \tfrac{5}{2})^2 + 2 - \tfrac{25}{4}$$
$$= (x - \tfrac{5}{2})^2 - \tfrac{17}{4}. \blacksquare$$

A slight variation in the technique illustrated in Examples 4 and 5 can be used to complete the square in the expression $ax^2 + bx + c$, that is, to write it in the form $e(x+k)^2 + d$, for suitable constants k, d, and e. The next example shows how to do this.

EXAMPLE 6 Complete the square in $2x^2 + 5x + 3$.

SOLUTION First factor the coefficient of x^2 out of the first two terms, obtaining

$$2x^2 + 5x + 3 = 2\left(x^2 + \frac{5x}{2}\right) + 3.$$

Then complete the square in the expression $x^2 + 5x/2$, as follows:

$$x^2 + \tfrac{5}{2}x = x^2 + \tfrac{5}{2}x + (\tfrac{5}{4})^2 - (\tfrac{5}{4})^2$$
$$= (x + \tfrac{5}{4})^2 - (\tfrac{5}{4})^2.$$

Thus
$$2x^2 + 5x + 3 = 2[(x + \tfrac{5}{4})^2 - (\tfrac{5}{4})^2] + 3$$
$$= 2[(x + \tfrac{5}{4})^2 - \tfrac{25}{16}] + 3$$
$$= 2(x + \tfrac{5}{4})^2 - \tfrac{25}{8} + 3$$
$$= 2(x + \tfrac{5}{4})^2 - \tfrac{1}{8}. \blacksquare$$

THE QUADRATIC FORMULA Completing the square permits us to analyze the real solutions, if there are any, of the quadratic equation

$$ax^2 + bx + c = 0.$$

The quadratic formula,

$$x = \frac{-b \pm \sqrt{b^2 - 4ac}}{2a}, \tag{1}$$

which describes the solutions, is obtained as follows.

$$ax^2 + bx + c = 0 \quad \text{given}$$

$$a\left(x^2 + \frac{b}{a}x \quad\right) + c = 0$$

$$a\left[\left(x + \frac{b}{2a}\right)^2 - \left(\frac{b}{2a}\right)^2\right] + c = 0 \quad \text{completing the square}$$

$$a\left(x + \frac{b}{2a}\right)^2 - a\left(\frac{b}{2a}\right)^2 + c = 0$$

$$a\left(x + \frac{b}{2a}\right)^2 = a\left(\frac{b}{2a}\right)^2 - c$$

$$\left(x + \frac{b}{2a}\right)^2 = \left(\frac{b}{2a}\right)^2 - \frac{c}{a}$$

$$= \frac{b^2}{4a^2} - \frac{c}{a}$$

$$= \frac{b^2 - 4ac}{4a^2}$$

$$x + \frac{b}{2a} = \pm\sqrt{\frac{b^2 - 4ac}{4a^2}} \quad \text{taking square roots}$$

The fact that $\sqrt{4a^2} = 2|a|$ is covered by the "$\pm$" sign.

$$= \pm\frac{\sqrt{b^2 - 4ac}}{2a}$$

and finally
$$x = \frac{-b}{2a} \pm \frac{\sqrt{b^2 - 4ac}}{2a}$$
$$x = \frac{-b \pm \sqrt{b^2 - 4ac}}{2a}.$$

The discriminant   The number $b^2 - 4ac$ which appears under the square root in (1) is called the *discriminant* of the quadratic expression $ax^2 + bx + c$. Note that, if the discriminant is negative, the equation $ax^2 + bx + c = 0$ has no real solutions; complex solutions are treated in Sec. 10.9.

INFORMATION PROVIDED BY THE DISCRIMINANT
$b^2 - 4ac$

If $b^2 - 4ac < 0$, $ax^2 + bx + c = 0$ has no real solutions.

If $b^2 - 4ac > 0$, $ax^2 + bx + c = 0$ has two distinct real solutions.

If $b^2 - 4ac = 0$, $ax^2 + bx + c = 0$ has one real solution.

EXAMPLE 7   Discuss the solutions of the equations

(a) $2x^2 + x + 3 = 0$,

(b) $2x^2 - 7x + 4 = 0$,

and (c) $x^2 - 6x + 9 = 0$.

SOLUTION   (a) Here $a = 2$, $b = 1$, and $c = 3$. Thus the discriminant $b^2 - 4ac$ equals $1^2 - 4 \cdot 2 \cdot 3 = -23$. Since the discriminant is negative there are no real solutions.

(b) Here $a = 2$, $b = -7$, and $c = 4$. The discriminant is therefore $(-7)^2 - 4 \cdot 2 \cdot 4 = 17$, which is positive. Thus there are two real solutions, $(7 + \sqrt{17})/4$ and $(7 - \sqrt{17})/4$.

(c) Here $a = 1$, $b = -6$, and $c = 9$. Thus

$$x = \frac{-(-6) \pm \sqrt{(-6)^2 - 4 \cdot 1 \cdot 9}}{2 \cdot 1} = \frac{6 \pm \sqrt{36 - 36}}{2} = \frac{6}{2}.$$

There is only one solution, namely 3. This reflects the fact that, when $x^2 - 6x + 9$ is factored, the factor $x - 3$ is repeated, that is, $x^2 - 6x + 9 = (x - 3)^2$. ∎

THE BINOMIAL THEOREM   Let n be a positive integer and let x be a real number. When $(1 + x)^n$ is multiplied out, it leads to a sum of terms, each of which is of the form "a coefficient times a power of x." For instance,

$$(1 + x)^2 = 1 + 2x + x^2,$$

and
$$(1 + x)^3 = 1 + 3x + 3x^2 + x^3,$$
$$(1 + x)^4 = 1 + 4x + 6x^2 + 4x^3 + x^4.$$

The *binomial theorem* provides a formula for the expansion of $(1 + x)^n$ for any positive integer n. It asserts that

$$(1 + x)^n = 1 + nx + \frac{n(n-1)}{1 \cdot 2}x^2 + \frac{n(n-1)(n-2)}{1 \cdot 2 \cdot 3}x^3 + \cdots + x^n. \quad (2)$$

The coefficient of x^k in the expansion of $(1 + x)^n$ is

Formula for the binomial coefficients
$$\frac{n(n-1)(n-2) \cdots (n-k+1)}{1 \cdot 2 \cdot 3 \cdots k}. \quad (3)$$

This coefficient is denoted $\binom{n}{k}$ or C_k^n and called "a binomial coefficient." To help remember formula (3) for $\binom{n}{k}$ note that the denominator is the product of the integers from 1 through k. The numerator has the same number of factors as the denominator, starting with n and going down to its smallest factor, which turns out to be $n - k + 1$. Simply write each factor in the numerator directly above a factor in the denominator, and you will stop at the right number. For instance,

$$\binom{7}{3} = C_3^7 = \frac{7 \cdot 6 \cdot 5}{1 \cdot 2 \cdot 3}.$$

In the $\binom{n}{k}$ notation the binomial theorem reads

$$(1 + x)^n = 1 + \binom{n}{1}x + \binom{n}{2}x^2 + \binom{n}{3}x^3 + \cdots + x^n,$$

where n is a positive integer.

EXAMPLE 8 Find the coefficient of x^4 in the expansion of $(1 + x)^9$.

SOLUTION In this case $n = 9$ and $k = 4$. Thus the coefficient is

$$\frac{9 \cdot 8 \cdot 7 \cdot 6}{1 \cdot 2 \cdot 3 \cdot 4},$$

which equals 126. ∎

The factorial $k!$ The product of all the integers from 1 through k, $1 \cdot 2 \cdots k$, is called "k factorial" and denoted "$k!$". Thus $5! = 1 \cdot 2 \cdot 3 \cdot 4 \cdot 5 = 120$. Formula (3) for the binomial coefficient can also be written

$$\frac{n(n-1)(n-2) \cdots (n-k+1)}{k!}. \quad (4)$$

In fact, (3) can be replaced by a formula that uses only factorials. To see this, note that

$$n(n-1)(n-2)\cdots(n-k+1) = \frac{n(n-1)(n-2)\cdots(n-k+1)(n-k)\cdots\cdot 3\cdot 2\cdot 1}{(n-k)\cdots\cdot 3\cdot 2\cdot 1}$$

$$= \frac{n!}{(n-k)!}.$$

Combining this information with (3) shows that the binomial coefficient $\binom{n}{k}$ can be written completely in terms of factorials,

$$\binom{n}{k} = \frac{n!}{k!(n-k)!}. \tag{5}$$

When using a hand-held calculator you may find this the most useful of the three formulas, since many calculators have a factorial key that gives $k!$ for k up to around 69.

$0! = 1$ It is necessary to define $0!$ to be 1 in order that formula (5) remains valid even in the cases $k = 0$ and $k = n$. For instance, if $k = n$, (5) gives

$$\frac{n!}{n!0!} = 1,$$

which is what we want: the coefficient of x^n in the expansion of $(1 + x)^n$ is 1.

EXAMPLE 9 Find the coefficient of x^5 in the expansion of $(1 + x)^{11}$.

SOLUTION By formula (3)

$$\binom{11}{5} = \frac{11\cdot 10\cdot 9\cdot 8\cdot 7}{1\cdot 2\cdot 3\cdot 4\cdot 5} = 462.$$

If instead you use formula (5) with a calculator, the computations would be

$$\binom{11}{5} = \frac{11!}{5!6!} = \frac{39{,}916{,}800}{120\cdot 720} = 462. \quad\blacksquare$$

The binomial theorem is frequently stated in terms of the expansion of $(a + b)^n$ instead of that of $(1 + x)^n$, which is the special case in which $a = 1$ and $b = x$. It asserts that

The binomial theorem for $(a+b)^n$

$$(a+b)^n = a^n + na^{n-1}b + \frac{n(n-1)}{1\cdot 2}a^{n-2}b^2 + \cdots + b^n. \tag{6}$$

The coefficient of $a^{n-k}b^k$ is

$$\binom{n}{k} = \frac{n(n-1)(n-2)\cdots(n-k+1)}{1\cdot 2\cdot 3\cdot\ \cdots\ \cdot k} = \frac{n!}{k!(n-k)!}.$$

For instance, $(a+b)^4 = a^4 + 4a^3b + 6a^2b^2 + 4ab^3 + b^4$.

FINITE GEOMETRIC SERIES

Another algebraic identity provides a short formula for the sum

$$a + ax + ax^2 + \cdots + ax^{n-1} \tag{7}$$

So named by the Greeks because corresponding lengths in similar geometric figures are in the same ratio.

where a and x are real numbers and n is a positive integer. Note that there are n summands in (7). Such an expression is called a *finite geometric series* with first term a and *ratio* x. Each term in (7) is obtained from the one before by multiplying by the ratio x. We present an algebraic identity that gives a shortcut for evaluating the sum $a + ax + ax^2 + \cdots + ax^{n-1}$.

A short formula for the sum of a finite geometric series

> Let a and x be real numbers, with $x \neq 1$. Let n be a positive integer. Then
>
> $$a + ax + ax^2 + \cdots + ax^{n-1} = \frac{a(1-x^n)}{1-x}. \tag{8}$$

To show that (8) is true, multiply both sides by $1 - x$. The multiplication of $a + ax + ax^2 + \cdots + ax^{n-1}$ by $1 - x$ leads, happily, to many cancellations:

$$(1-x)(a + ax + ax^2 + \cdots + ax^{n-1})$$
$$= (a + ax + ax^2 + \cdots + ax^{n-1}) - (ax + ax^2 + \cdots + ax^n)$$
$$= a - ax^n$$
$$= a(1 - x^n).$$

Division by $1 - x$ then establishes (8).

EXAMPLE 10 Use (8) to evaluate

$$1 + \tfrac{1}{2} + (\tfrac{1}{2})^2 + (\tfrac{1}{2})^3 + (\tfrac{1}{2})^4.$$

SOLUTION In this case $a = 1$, $x = \tfrac{1}{2}$, and $n = 5$. Thus

$$1 + \tfrac{1}{2} + (\tfrac{1}{2})^2 + (\tfrac{1}{2})^3 + (\tfrac{1}{2})^4 = \frac{1[1 - (\tfrac{1}{2})^5]}{1 - \tfrac{1}{2}}$$

$$= \frac{1 - (\tfrac{1}{2})^5}{\tfrac{1}{2}}$$

$$= 2(1 - \tfrac{1}{32}) = 2 - \tfrac{1}{16} = \tfrac{31}{16}. \quad\blacksquare$$

A FACTORING OF $d^n - c^n$

The well known factorization $d^2 - c^2 = (d + c)(d - c)$ is just a special case of a factorization of $d^n - c^n$ (where n is a positive integer) that will come in handy on several occasions.

It can be shown that

$$d^3 - c^3 = (d^2 + dc + c^2)(d - c)$$
$$d^4 - c^4 = (d^3 + d^2c + dc^2 + c^3)(d - c)$$

and, more generally, for any positive integer n,

A factorization of $d^n - c^n$

$$d^n - c^n = (d^{n-1} + d^{n-2}c + d^{n-3}c^2 + \cdots + dc^{n-2} + c^{n-1})(d - c). \quad (9)$$

For instance, to check the factorization of $d^3 - c^3$, multiply out $(d^2 + dc + c^2)(d - c)$:

$$(d^2 + dc + c^2)(d - c) = (d^2 + dc + c^2)d - (d^2 + dc + c^2)c$$
$$= (d^3 + d^2c + dc^2) - (d^2c + dc^2 + c^3)$$
$$= d^3 - c^3.$$

The next example, based on the factorization of $d^3 - c^3$, develops an inequality that is used in Chap. 5.

EXAMPLE 11 Let d and c be nonnegative numbers, with $d > c$. Show that

$$c^2(d - c) < \frac{d^3}{3} - \frac{c^3}{3} < d^2(d - c).$$

SOLUTION To establish the first inequality,

$$c^2(d - c) < \frac{d^3}{3} - \frac{c^3}{3},$$

start with the identity

$$(d^2 + dc + c^2)(d - c) = d^3 - c^3.$$

Recall Example 1 in Appendix A. Since $c < d$, $c^2 + c^2 + c^2 < d^2 + dc + c^2$;
hence $3c^2 < d^2 + dc + c^2.$ (10)

Multiplying both sides of (10) by the *positive* number $d - c$ yields

$$3c^2(d - c) < (d^2 + dc + c^2)(d - c);$$

hence $3c^2(d - c) < d^3 - c^3.$

Thus $c^2(d - c) < \dfrac{d^3 - c^3}{3},$

from which it follows that $\quad c^2(d-c) < \dfrac{d^3}{3} - \dfrac{c^3}{3}.$

This establishes the first inequality.

To establish the second inequality,

$$\frac{d^3}{3} - \frac{c^3}{3} < d^2(d-c),$$

observe that $\quad d^2 + dc + c^2 < d^2 + d^2 + d^2 = 3d^2.$

Thus $\quad (d^2 + dc + c^2)(d-c) < 3d^2(d-c).$

Hence $\quad d^3 - c^3 < 3d^2(d-c).$

Division by 3 establishes the second inequality. ∎

A similar argument shows that, if $0 \le c < d$, then

$$c^{n-1}(d-c) < \frac{d^n}{n} - \frac{c^n}{n} < d^{n-1}(d-c)$$

for any positive integer n.

Exercises

In Exercises 1 to 8 rationalize the denominators.

1. $10/\sqrt{5}$
2. $3/\sqrt{x}$
3. $(2 + 4\sqrt{2})/\sqrt{2}$
4. $x^3/\sqrt{x}$
5. $4/(3 - \sqrt{3})$
6. $2/(3 + \sqrt{2})$
7. $x/(1 - \sqrt{x})$
8. $1/(\sqrt{2} - \sqrt{3})$

In Exercises 9 to 12 rationalize the numerators.

9. $(3 + \sqrt{2})/5$
10. $(\sqrt{x+1} - \sqrt{x})/x$
11. $(\sqrt{x} - \sqrt{5})/(x - 5)$
12. $(\sqrt{u} - \sqrt{v})/5$

In Exercises 13 to 18 complete the squares.

13. (a) $x^2 + 8x + 13$
    (b) $x^2 - 8x + 23$
    (c) $x^2 - x + 2$
14. (a) $x^2 + 6x + 3$
    (b) $x^2 - 6x + 3$
    (c) $x^2 + 3x + 5$
15. (a) $x^2 + 3x - 2$
    (b) $x^2 + 3x + 7$
    (c) $x^2 + 5x/2 + 4$
16. (a) $x^2 + x/3 + 1$
    (b) $x^2 + 2x/3 - 7$
    (c) $x^2 + 10x + 25$
17. (a) $2x^2 - 5x + 3$
    (b) $2x^2 + 6x + 7$
    (c) $3x^2 + 5x + 1$
18. (a) $4x^2 - 8x + 5$
    (b) $3x^2 + 2x + 1$
    (c) $3x^2 - 2x + 1$

In Exercises 19 and 20 use the quadratic formula to find all real solutions of the equations.

19. (a) $x^2 + x + 1 = 0$
    (b) $x^2 + x - 1 = 0$
    (c) $x^2 + 2x + 1 = 0$
20. (a) $2x^2 + 5x - 7 = 0$
    (b) $3x^2 + x + 5 = 0$
    (c) $4x^2 - 4x + 1 = 0$

Using the discriminant, determine how many distinct real solutions there are for each of the equations in Exercises 21 and 22.

21. (a) $2x^2 + 5x + 6 = 0$
    (b) $2x^2 + 5x + 2 = 0$
    (c) $4x^2 - 12x + 9 = 0$
22. (a) $9x^2 - 30x + 25 = 0$
    (b) $x^2 + x + 1 = 0$
    (c) $x^2 - x - 1 = 0$
23. Find the coefficient of x^2 in the expansion of
    (a) $(1 + x)^5$, (b) $(1 + x)^6$, (c) $(1 + x)^{10}$.
24. Find the coefficient of x^3 in the expansion of
    (a) $(1 + x)^5$, (b) $(1 + x)^6$, (c) $(1 + x)^{10}$.
25. Use the binomial theorem to expand $(1 + x)^7$.
26. Use the binomial theorem to find the first four terms (starting with 1) in the expansion of $(1 + x)^{12}$.
27. What is the coefficient of $a^8 b^2$ in the expansion of $(a + b)^{10}$?
28. What is the coefficient of $a^8 b^3$ in the expansion of $(a + b)^{11}$?

In Exercises 29 and 30 use the formula for the sum of a geometric series to obtain short expressions for the given sums.

29. (a) $1 + \dfrac{1}{3} + \dfrac{1}{3^2} + \dfrac{1}{3^3} + \dfrac{1}{3^4} + \dfrac{1}{3^5} + \dfrac{1}{3^6}$
    (b) $5 + 5 \cdot 2 + 5 \cdot 2^2 + 5 \cdot 2^3 + 5 \cdot 2^4 + 5 \cdot 2^5 + 5 \cdot 2^6 + 5 \cdot 2^7$

(c) $6 - 6 \cdot \frac{1}{2} + 6 \cdot (\frac{1}{2})^2 - 6 \cdot (\frac{1}{2})^3 + 6 \cdot (\frac{1}{2})^4$

30 (a) $1 - \frac{1}{2} + (\frac{1}{2})^2 - (\frac{1}{2})^3 + (\frac{1}{2})^4 - (\frac{1}{2})^5$

(b) $8 + \frac{8}{10} + \frac{8}{10^2} + \frac{8}{10^3} + \frac{8}{10^4} + \frac{8}{10^5} + \frac{8}{10^6} + \frac{8}{10^7} + \frac{8}{10^8} + \frac{8}{10^9}$

(c) $\frac{2}{3} + (\frac{2}{3})^2 + (\frac{2}{3})^3 + (\frac{2}{3})^4 + (\frac{2}{3})^5 + (\frac{2}{3})^6 + (\frac{2}{3})^7 + (\frac{2}{3})^8$

■

31 The decimal 0.333333333 is actually a finite geometric progression in disguise, namely

$$\frac{3}{10} + \frac{3}{10^2} + \frac{3}{10^3} + \frac{3}{10^4} + \frac{3}{10^5} + \frac{3}{10^6} + \frac{3}{10^7} + \frac{3}{10^8} + \frac{3}{10^9}.$$

Using this fact, find a shorter expression for 0.333333333.

32 A flea, to amuse himself, jumps $\frac{1}{2}$ meter to the right, then $\frac{1}{4}$ meter to the left, then $\frac{1}{8}$ meter to the right, then $\frac{1}{16}$ meter to the left, and so on, as shown in Fig. B.1. On his nth jump he travels $1/2^n$ meter, but continues alternating right and left. The flea starts at the number 0 on the number line.

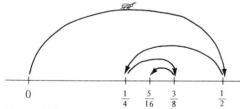

Figure B.1

(a) Show that after n jumps, where n is odd, he is at the number $\frac{1}{3}(1 + 1/2^n)$.

(b) Show that after n jumps, where n is even, he is at the number $\frac{1}{3}(1 - 1/2^n)$.

(c) As he keeps on jumping, what single number does he keep jumping over?

33 From the fact that

$$(1 + x)^4 = 1 + 4x + 6x^2 + 4x^3 + x^4,$$

deduce that

$$(a + b)^4 = a^4 + 4a^3b + 6a^2b^2 + 4ab^3 + b^4.$$

Hint: First show that

$$(a + b)^4 = a^4[1 + (b/a)]^4.$$

(Assume $a \ne 0$.)

34 Use the binomial theorem to show that, when $|x|$ is very small, $(1 + x)^{10}$ is approximately $1 + 10x$. If you have a calculator, compare the two quantities when $x = 0.01$.

■ ■

35 Show that, for all values of x, $x^2 + x + 1 \ge \frac{3}{4}$. *Hint:* Complete the square.

36 Show that if $0 \le c < d$, then

$$c^4(d - c) < \frac{d^5}{5} - \frac{c^5}{5} < d^4(d - c).$$

37 Let a be a positive number. Show that the smallest value of $y = ax^2 + bx + c$ occurs when $x = -b/(2a)$. *Hint:* Complete the square.

38 Using a formula for the binomial coefficients, show that in the expansion of $(1 + x)^n$ the powers x^k and x^{n-k} have equal coefficients.

39 The formula for the sum of a geometric series is closely connected to the factorization of $d^n - c^n$. Replacing d by x and c by 1 in the factorization of $d^n - c^n$, deduce that

$$1 + x + x^2 + \cdots + x^{n-1}$$

is equal to $(x^n - 1)/(x - 1)$ if $x \ne 1$.

APPENDIX C
EXPONENTS

For each fixed positive number b there is an exponential function $f(x) = b^x$. In both the theory and applications of calculus these are among the most important functions. It is no coincidence that every scientific calculator has a y^x key and, before the advent of the calculator, mathematical tables listing the values of exponential functions to many decimal places were common. This appendix first reviews the definition of b^x for positive b and then examines the situation when b is replaced by a negative number.

THE DEFINITION OF b^x FOR $b > 0$

Base and exponent

In the expression b^x, where b is positive and x is any real number, b is called the *base* and x is called the *exponent*. We will review the definition of b^x in a sequence of steps:

1   x is a positive integer.

2   x is 0.

3   x is a negative integer.

"n" is the usual symbol for an integer.

4   x is the reciprocal of a positive integer, $x = 1/n$.

5   x is rational, $x = m/n$, where m is an integer and n is a positive integer.

6   x is irrational.

For the sake of simplicity, let us take the special case $b = 2$. All the ideas show up here.

2^x when $x = 1, 2, 3, \cdots$

Definition of 2^x when $x = 1, 2, 3, \ldots$. If x is a positive integer, define 2^x to be the product of x 2's.

For instance $\qquad 2^5 = 2 \cdot 2 \cdot 2 \cdot 2 \cdot 2 = 32,$

$$2^3 = 2 \cdot 2 \cdot 2 = 8, \qquad 2^2 = 2 \cdot 2 = 4, \quad \text{and} \quad 2^1 = 2.$$

Note that $\qquad 2^2 \cdot 2^3 = (2 \cdot 2)(2 \cdot 2 \cdot 2) = 2^{2+3}.$

More generally, for any positive integers x and y

The basic law of exponents
$$2^{x+y} = 2^x \cdot 2^y.$$

This *basic law of exponents* (which holds for any positive base) will serve as a guide to the definition of 2^x when x is not a positive integer. The demand that $2^{x+y} = 2^x \cdot 2^y$ for *all real numbers* x and y will tell us how the exponential function 2^x must be defined. Note that $2^{x+y+z} = 2^{(x+y)+z} = 2^{x+y} \cdot 2^z = (2^x \cdot 2^y)2^z = 2^x \cdot 2^y \cdot 2^z$. Thus the basic law of exponents extends to the case when the exponent is the sum of three numbers: $2^{x+y+z} = 2^x \cdot 2^y \cdot 2^z$. In a similar manner, it can be extended to the case where the exponent is the sum of any finite number of numbers.

Why 2^0 should be 1 For instance, what should the numerical value of 2^0 be? If the basic law of exponents is to be true for all exponents, then, in particular, this equation must hold:
$$2^{0+1} = 2^0 \cdot 2^1;$$
hence $\qquad 2^1 = 2^0 \cdot 2^1.$

But $2^1 = 2$, so 2^0 must satisfy the equation
$$2 = 2^0 \cdot 2.$$

There is no choice; 2^0 must be 1.

What should $2^{-1}, 2^{-2}, 2^{-3}, \ldots$ be? To preserve the basic law of exponents, we must have, for instance,
$$2^3 \cdot 2^{-3} = 2^{3+(-3)}.$$
But $2^{3+(-3)} = 2^0 = 1$. Thus $\quad 2^3 \cdot 2^{-3} = 1.$

Why $2^{-3} = \tfrac{1}{8}$ This shows that, if 2^{-3} is to have meaning, it must be the reciprocal of 2^3, that is,
$$2^{-3} = \frac{1}{2^3} = \frac{1}{8}.$$

For this reason we define
$$2^{-n} \quad \text{to be} \quad \frac{1}{2^n}$$

for any positive integer n.

The exponential 2^x has now been defined for any *integer* x. Let us graph $y = 2^x$ for integer values of x, as in Fig. C.1.

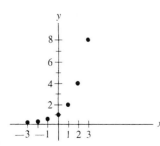

Figure C.1

It is tempting to draw a smooth curve through these points, but 2^x has not yet been defined if x is not an integer.

How should $2^{1/2}$ be defined? To preserve the basic law of exponents, it is necessary that

$$2^{1/2} \cdot 2^{1/2} = 2^{1/2 + 1/2};$$

thus
$$2^{1/2} \cdot 2^{1/2} = 2^1 = 2.$$

Hence $2^{1/2}$ is a solution of the equation

$$x^2 = 2.$$

Why $2^{1/2}$ should be the square root of 2

Should it be the positive or the negative solution? The graph just sketched suggests that $2^{1/2}$ should be positive. Thus define $2^{1/2}$ to be $\sqrt{2}$, the square root of 2. Note that $2^{1/2} \approx 1.4$, which fits nicely into the preceding graph, as is shown in Fig. C.2.

Similarly, $2^{1/3}$ can be determined by the basic law of exponents:

$$2^{1/3} \cdot 2^{1/3} \cdot 2^{1/3} = 2^{1/3 + 1/3 + 1/3}$$
$$= 2^1$$
$$= 2.$$

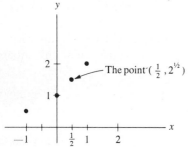

Figure C.2

Thus $2^{1/3}$ must be a solution of the equation

$$x^3 = 2.$$

Why $2^{1/3}$ should be the cube root of 2

There is only one solution, and it is called the *cube root of 2*, denoted also $\sqrt[3]{2}$. Thus define

$$2^{1/3} = \sqrt[3]{2},$$

which is about 1.26.

Similarly, define $2^{1/n}$ for any positive integer n to be the positive solution of the equation $x^n = 2$, that is,

$$2^{1/n} = \sqrt[n]{2}.$$

What should $2^{3/4}$ be?

How should $2^{3/4}$ be defined? To preserve the basic law of exponents, we must have

$$2^{1/4} \cdot 2^{1/4} \cdot 2^{1/4} = 2^{3/4}.$$

In short we must have
$$2^{3/4} = (2^{1/4})^3.$$

Defining $2^{m/n}$

This suggests that for any integer m and positive integer n we should define

$$2^{m/n} \quad \text{to be} \quad (2^{1/n})^m.$$

With this step we have defined 2^x for every rational number x.

How to define $2^{\sqrt{2}}$

How should 2^x be defined when x is not rational? For instance, how should $2^{\sqrt{2}}$ be defined? To begin, consider the decimal representation of $\sqrt{2}$, namely

$$\sqrt{2} = 1.41421356\ldots.$$

The successive decimals 1.4, 1.41, 1.414, ... are rational (for instance, 1.41 = 141/100). Thus $2^{1.4}$, $2^{1.41}$, $2^{1.414}$, ... are already defined. This table indicates some of their values.

x	1.4	1.41	1.414	1.4142	1.41421
2^x	2.63901...	2.65737...	2.66474...	2.66511...	2.66513...

By going further in this table we can get better approximations of $2^{\sqrt{2}}$. This procedure, though cumbersome, would enable us to find $2^{\sqrt{2}}$ to any number of decimal places. To five decimal places,

The symbol $\approx$ stands for "is approximately."

$$2^{\sqrt{2}} \approx 2.66514.$$

The graph of $y = 2^x$ is shown in Fig. C.3.

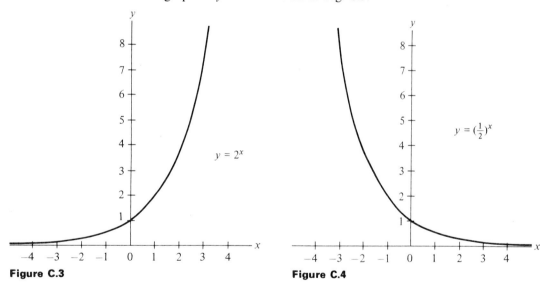

Figure C.3

Figure C.4

The definition of b^x for $b > 0$ is similar to that for 2^x.

For any positive number b the definition of b^x follows the same outline as the definition of 2^x. For $b > 1$, the graph of $f(x) = b^x$ resembles that of 2^x. Far to the left it gets near the x axis; far to the right it gets arbitrarily high. Moreoever, as x increases, so does b^x, if $b > 1$.

For $0 < b < 1$ the graph of $f(x) = b^x$ resembles the graph of $(\frac{1}{2})^x$, which is shown in Fig. C.4.

EXAMPLE 1 Evaluate $64^{1/3}$, $64^{2/3}$, $64^{1/2}$, and $64^{-2/3}$.

SOLUTION $64^{1/3}$ is the cube root of 64, $\sqrt[3]{64}$, which is 4.

APPENDIX C EXPONENTS

$$64^{2/3} = (64^{1/3})^2 = 4^2 = 16;$$
$$64^{1/2} = \sqrt{64} = 8;$$
$$64^{-2/3} = (64^{1/3})^{-2} = 4^{-2} = \tfrac{1}{16} = 0.0625. \blacksquare$$

EXAMPLE 2 Find a decimal approximation of $2^{1/3}$, the cube root of 2.

SOLUTION The quickest way is to use a calculator. Either the y^x key, with $y = 2$ and $x = \tfrac{1}{3}$, or the $\sqrt[x]{y}$ key, with $y = 2$ and $x = 3$, will do the trick, and show that $2^{1/3} \approx 1.25992105$.

If a calculator is not available, $2^{1/3}$ may be looked up in a cube-root table, in a mathematical handbook.

If worst comes to worst, brute-force arithmetic could be used. For instance, by straightforward multiplication, $1.2^3 = 1.728$ and $1.3^3 = 2.197$. Thus

$$1.2 < 2^{1/3} < 1.3.$$

To find the next decimal place, compute $1.21^3, 1.22^3, \ldots, 1.29^3$. Since $1.25^3 = 1.953125$ and $1.26^3 = 2.000376$,

$$1.25 < 2^{1/3} < 1.26.$$

With enough time and pencils, the curious could find as many decimal places of $2^{1/3}$ as needed. $\blacksquare$

In addition to the basic law of exponents, there are other laws that hold for the exponential functions for any positive base. They are displayed here.

Properties of exponential functions

LAWS OF EXPONENTS

The bases are positive, the exponents any real numbers.

$b^{x+y} = b^x b^y$	basic law of exponents
$b^{x-y} = b^x/b^y$	difference of exponents
$(b^x)^y = b^{xy}$	power of a power
$(ab)^x = a^x b^x$	power of a product
$\left(\dfrac{a}{b}\right)^x = \dfrac{a^x}{b^x}$	power of a quotient
$b^0 = 1$	definition
$b^1 = b$	definition

The laws of exponents help in simplifying certain algebraic expressions, as is illustrated by Example 3.

EXAMPLE 3 Let b be a positive number. Write each of the following in the form b^x for a suitable exponent x:

$$(a)\ (\sqrt{b})^3 \qquad (b)\ b^7/b^3 \qquad (c)\ \sqrt{b}\sqrt[3]{b} \qquad (d)\ 1/\sqrt{b}.$$

SOLUTION
(a) $(\sqrt{b})^3 = (b^{1/2})^3 = b^{(1/2)(3)} = b^{3/2}$

(b) $\dfrac{b^7}{b^3} = b^{7-3} = b^4$

(c) $\sqrt{b}\sqrt[3]{b} = b^{1/2}b^{1/3} = b^{1/2 + 1/3} = b^{5/6}$

(d) $\dfrac{1}{\sqrt{b}} = (\sqrt{b})^{-1} = (b^{1/2})^{-1} = b^{-1/2}$

In parts (a) and (d) the "power of a power" law was used. ∎

THE DEFINITION OF b^x WHEN b IS NOT POSITIVE

For certain values of the exponent x it is possible to define b^x even if b is not positive.

First of all, 0^x is defined for *positive* x to be 0. However, 0^0 is not defined, nor is 0^x, when x is negative.

0^x is defined for $x > 0$.

If the base b is negative and n is an *odd* integer, $b^{1/n}$ is defined to be $\sqrt[n]{b}$, just as in the case when b is positive. For instance,

$$(-8)^{1/3} = -2.$$

$b^{m/n}$ can also be defined when b is negative and n is odd.

More generally, b^x can be defined for any negative base and any exponent x of the form m/n where m is an integer and n is a positive *odd* integer, just as in the case b positive. First define b^m just as when the base was positive.

DEFINITION Let b be negative, m an integer, and n a positive odd integer. Then $b^{m/n}$ is

$$(\sqrt[n]{b})^m.$$

Warning! The equation $(b^x)^y = b^{xy}$ does not always hold when b is negative. For instance, $[(-1)^2]^{1/2} = 1^{1/2} = 1$, but $(-1)^{2 \cdot 1/2} = (-1)^1 = -1$.

EXAMPLE 4 Graph $f(x) = x^{1/3}$, the cube-root function.

SOLUTION This table records a few values of the function.

x	-8	-1	0	1	8
$f(x) = x^{1/3}$	-2	-1	0	1	2

The graph of $y = x^{1/3}$

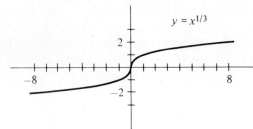

Figure C.5

The graph is shown in Fig. C.5. The curve passes through the origin vertically, not at an angle. ∎

EXAMPLE 5 Graph $f(x) = x^{2/3}$.

SOLUTION Again begin with a few convenient inputs x. For instance,

$$(-8)^{2/3} = (\sqrt[3]{-8})^2 = (-2)^2 = 4.$$

x	-8	-1	0	1	8
$x^{2/3}$	4	1	0	1	4

The graph is symmetric with respect to the y axis and comes to a point or "cusp" at the origin, as shown in Fig. C.6.

The graph of $y = x^{2/3}$

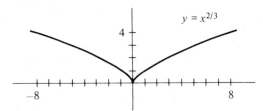

Figure C.6 ∎

Note that when b is negative and n is an *even* positive integer, b has no real nth root. For instance $(-9)^{1/2}$ makes no sense since -9 does not have a real square root.

Exercises

1. Evaluate 32^x for x equal to
   (a) 0 (b) 1 (c) $\frac{1}{5}$ (d) $-\frac{1}{5}$ (e) $\frac{3}{5}$.
2. Evaluate 64^x for x equal to
   (a) $\frac{1}{2}$ (b) $-\frac{1}{2}$ (c) $\frac{2}{3}$ (d) 0 (e) $\frac{5}{6}$.

In Exercises 3 to 6 graph the given functions.

3. $f(x) = 4^x$
4. $f(x) = 5^x$
5. $f(x) = (\frac{1}{4})^x$
6. $f(x) = (\frac{1}{5})^x$
7. Express in the form 2^x for a suitable exponent x:
   (a) 16 (b) $\frac{1}{8}$ (c) $\sqrt{2}$ (d) 1 (e) 0.25.
8. Express in the form 4^x for a suitable exponent x:
   (a) 16 (b) 8 (c) $1/\sqrt{2}$ (d) $4\sqrt{2}$ (e) 16^{35}.
9. Let b be a positive number. Express in the form b^x:
   (a) $b\sqrt{b}$ (b) $(\sqrt{b})^3$ (c) $1/\sqrt[3]{b}$
   (d) $\sqrt[3]{b^2}$ (e) $\sqrt{\sqrt{b}}$.
10. Let a be a positive number. Express in the form a^x:
    (a) $(\sqrt{a})^2$ (b) $1/a^2$ (c) 1
    (d) $(a^3/a^5)^{10}$ (e) $a/\sqrt[3]{a}$.

In Exercises 11 and 12 give the domains and ranges of the functions. (See Sec. 1.3 for definitions of "domain" and "range.")

11 (a) 2^x (b) $x^{1/3}$ (c) $x^{1/4}$ (d) $x^{2/3}$

12 (a) $(\tfrac{1}{2})^x$ (b) x^3 (c) $x^{4/5}$ (d) $x^{3/4}$

13 (a) Compute $x^{3/5}$ for $x = -32, -1, 0, 1, 32$.
   (b) Using (a), graph $y = x^{3/5}$.

14 (a) Compute $x^{4/5}$ for $x = -32, -1, 0, 1, 32$.
   (b) Using (a), graph $y = x^{4/5}$. (Use a large graph.)

15 A square-root table lists 2.236 for $\sqrt{5}$. Which is larger, 2.236 or $\sqrt{5}$?

16 A square-root table lists 2.646 for $\sqrt{7}$. Which is larger, 2.646 or $\sqrt{7}$?

■

17 For which positive numbers x is
   (a) $x < x^2$?   (b) $x = x^2$?   (c) $x > x^2$?

18 There is only one pair of positive integers x and y, $x \neq y$, such that $x^y = y^x$. What is that pair? (It is not necessary to prove that it is unique.)

19 Graph $y = 2^{-x}$.

20 Graph $y = 2^x + 2^{-x}$.

21 Express each of these numbers in the form 10^x:
   (a) 1000   (b) 0.0001   (c) 1,000,000
   (d) 0.0000001.

22 Each positive number can be written in the form $A \cdot 10^x$, where $1 \leq A < 10$ and x is an integer. (This is the scientific notation for real numbers.) Express each of the following in scientific notation.
   (a) 900   (b) 957   (c) 0.095
   (d) 15,000   (e) 0.0015

23 There are two ways to calculate $64^{2/3}$ by hand, either as $(64^2)^{1/3}$ or as $(64^{1/3})^2$. Which is easier?

■ ■

24 If a calculator has a square root key and a reciprocal key, how could you use them to calculate
   (a) $5^{1/4}$,   (b) $5^{1/8}$,   (c) $5^{3/8}$,   (d) $5^{-1/4}$?

25 (Calculator)
   (a) Using the y^x key, evaluate $(2^h - 1)/h$ for $h = 0.1$, 0.01, and 0.001.
   (b) Continue part (a) with smaller h. The calculator may give erroneous values when h is chosen too small. Experiment with your calculator with h as small as 0.0000001 or 0.00000001 and smaller. (It is shown in Chap. 6 that as h is chosen smaller and smaller, the quotient in (a) approaches a fixed number whose decimal expansion begins 0.693.)

26 This exercise shows why b^x is not defined when b is negative and x is irrational. For instance, consider $(-1)^{\sqrt{2}}$.
   (a) Show that $15/11 < \sqrt{2} < 16/11$.
   (b) Evaluate $(-1)^{15/11}$ and $(-1)^{16/11}$.
   (c) What trouble arises when you use rational numbers, m/n, near $\sqrt{2}$ and the values of $(-1)^{m/n}$ as a way to find out what $(-1)^{\sqrt{2}}$ "should be"?

27 The ability of a nuclear weapon to destroy hard missile targets is called its lethality, denoted K. K is a function of the explosive power of the warhead, denoted Y, measured in megatons, and the accuracy of the missile. A megaton is equivalent to 1 million tons of TNT. The accuracy is measured by the radius of the circle around the target such that the warhead has a 50 percent chance of landing within that circle. This radius, measured in nautical miles, is denoted CEP (circle of error probable). The lethality is given by the formula

$$K = Y^{2/3}(CEP)^{-2}.$$

A nautical mile is about 6076 feet. In the following computations take it to be 6000 feet.
   (a) Find the total lethality of one Trident 2 missile with fourteen 150,000-ton-yield warheads and CEP = 200 feet. (Multiply the lethality of one warhead by 14.)
   (b) Find the total lethality of one SS-NX-18 missile with three 500,000-ton-yield warheads and CEP = 0.25 nautical miles.
   (c) Which increases K more: to make the warhead 8 times as large (in megatonnage) or to make CEP 3 times as small?

Incidentally, estimates of CEP may be quite unreliable. They are based on test firings from California to the South Pacific, while wartime launches would presumably be over the north pole. Gravitational variations and weather conditions (e.g., jet stream and storms) can alter the path by hundreds of feet.

APPENDIX D
TRIGONOMETRY

The part of trigonometry most useful in calculus concerns primarily three functions: sine, cosine, and tangent. This appendix emphasizes the relation of angles to a circle in order to obtain the trigonometric functions quickly.

THE RADIAN MEASURE OF AN ANGLE

In daily life angles are measured in degrees, with 360° measuring the full circular angle. The number 360 was chosen by the Babylonian astronomers, perhaps because there are close to 360 days in a year and 360 has many divisors. This somewhat arbitrary measure would complicate the calculus of trigonometric functions. In calculus a much more natural and geometric measure is used, called "radian measure," which is defined in terms of how much arc an angle cuts off on a circle.

Radian measure

To measure the size of an angle, such as the angle ABC shown in Fig. D.1, draw a circle with center at B.

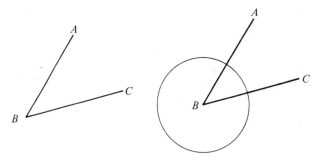

Figure D.1

If the circle has radius r, and the angle intercepts an arc of length s, then the quotient

$$\frac{s}{r}$$

APPENDIX D TRIGONOMETRY S-27

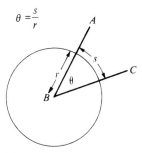

Figure D.2

shall be the measure of the angle, and we say that the angle has a measure of *s/r radians*. It is frequently convenient to denote the measure of an angle by θ, and then write

$$\theta = \frac{s}{r}.$$

(See Fig. D.2.) Since both s and r measure lengths, their ratio is dimensionless. Moreover, the radian measure of an angle does not depend on the size of the circle.

EXAMPLE 1   Find the radian measure of the right angle ABC in Fig. D.3.

SOLUTION   Draw a circle of radius r centered at B and compute the quotient s/r. The circumference of a circle of radius r is $2\pi r$. The right angle intercepts a quarter of the circumference.

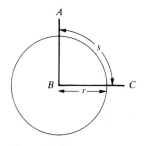

Figure D.3

Thus
$$s = \frac{1}{4} 2\pi r = \frac{\pi r}{2},$$

and
$$\theta = \frac{s}{r} = \frac{\pi r/2}{r} = \frac{\pi}{2}.$$

So a right angle has the measure $\pi/2$ radians (about 1.57 radians). ∎

How to translate degrees to radians and vice versa

The straight angle, which has 180°, consists of two right angles and therefore has a measure of π radians. This fact is helpful in translating from degrees to radians and from radians to degrees: Simply use the proportion

$$\frac{\text{Degrees}}{180} = \frac{\text{Radians}}{\pi}.$$

EXAMPLE 2   What is the measure in radians of the 30° angle?

SOLUTION   The proportion in this case becomes

$$\frac{30}{180} = \frac{\text{Radians}}{\pi},$$

from which it follows that

$$\text{Radians} = \pi \frac{30}{180} = \frac{\pi}{6}. \quad \blacksquare$$

EXAMPLE 3   How many degrees are there in an angle of 1 radian?

SOLUTION   In this case the proportion is

$$\frac{\text{Degrees}}{180} = \frac{1}{\pi};$$

hence
$$\text{Degrees} = \frac{180}{\pi} \approx \frac{180}{3.14} \approx 57.3°.$$

So an angle of 1 radian is about 57.3°, a little less than 60°. ∎

The unit circle

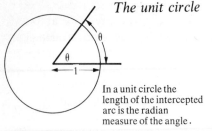

In a unit circle the length of the intercepted arc is the radian measure of the angle.

Figure D.4

In the case of the *unit circle*, the circle whose radius is 1, the formula
$$\theta = \frac{s}{r}$$
becomes
$$\theta = \frac{s}{1} = s.$$

In that case the length of arc intercepted equals the measure of the angle in radians, as shown in Fig. D.4.

Angles θ larger than 2π

So far, an angle has been associated with each number θ in the interval $[0, 2\pi]$. Next, associate an angle with any *positive* number θ. For convenience, a unit circle will be used, and one arm of the angle will be placed along the positive x axis. To draw the second arm of the angle of θ radians, go around the unit circle in a counterclockwise direction a distance θ. The point P reached determines the second arm. For instance, if $\theta = 5\pi/2$ radians, it is necessary to travel clear around the circle once and then reach the point P above the center of the circle. In this case we obtain the right angle, also described by $\pi/2$ radians and shown in Fig. D.5.

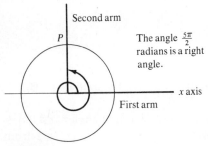

The angle $\frac{5\pi}{2}$ radians is a right angle.

Figure D.5

Every time we travel around the unit circle, we increase the measure of an angle by 2π radians. Thus the right angle of $\pi/2$ radians has an endless supply of descriptions:

$$\frac{\pi}{2}, \quad \frac{\pi}{2} + 2\pi = \frac{5\pi}{2}, \quad \frac{\pi}{2} + 4\pi = \frac{9\pi}{2}, \quad \ldots.$$

Negative θ

To associate angles with the negative number θ, go *clockwise* around the unit circle through an angle $|\theta|$. For instance, to draw the angle $-\pi/2$, start at the point $(1, 0)$ and move along the unit circle clockwise through a right angle until reaching the point P directly below the center of the circle. Note in Fig. D.6 that an angle of $-\pi/2$ radians coincides with an angle of $3\pi/2$ radians.

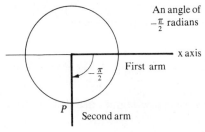

Figure D.6

THE SINE AND COSINE FUNCTIONS

The two fundamental functions of trigonometry, sine and cosine, can now be defined.

DEFINITION *The sine and cosine functions.* For each number θ, the sine of θ, denoted by $\sin \theta$, and the cosine of θ, denoted by $\cos \theta$, are defined as follows. Draw the angle of θ radians, whose first arm is the positive x axis and whose vertex is at $(0, 0)$. The second arm meets the unit circle whose center is at $(0, 0)$ in a point P. The x coordinate of P is called $\cos \theta$. The y coordinate of P is called $\sin \theta$. (See Fig. D.7.)

This diagram is the most important in this section. It is the basis of trigonometry.

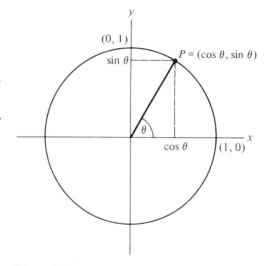

Figure D.7

EXAMPLE 4 Find $\cos(\pi/2)$ and $\sin(\pi/2)$.

SOLUTION If $\theta = \pi/2$, then the angle is a right angle, and $P = (0, 1)$. Hence

$$\cos \frac{\pi}{2} = 0 \quad \text{and} \quad \sin \frac{\pi}{2} = 1. \quad \blacksquare$$

EXAMPLE 5 Find $\cos(-\pi)$ and $\sin(-\pi)$.

SOLUTION If $\theta = -\pi$, P is the point $(-1, 0)$.

Hence $\cos(-\pi) = -1$ and $\sin(-\pi) = 0$. $\blacksquare$

TRIGONOMETRIC IDENTITIES

The trigonometric functions satisfy various identities. First of all, since a change of 2π in θ leads to the same point P on the circle,

$$\cos(\theta + 2\pi) = \cos \theta,$$

and

$$\sin(\theta + 2\pi) = \sin \theta.$$

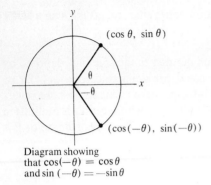

Figure D.8

Diagram showing that $\cos(-\theta) = \cos\theta$ and $\sin(-\theta) = -\sin\theta$

One says that the cosine and sine functions have *period* 2π. Second, inspection of the unit circle in Fig. D.8 shows that

$$\cos(-\theta) = \cos\theta,$$

and

$$\sin(-\theta) = -\sin\theta.$$

The numbers $\cos\theta$ and $\sin\theta$ are related by the equation

$$\cos^2\theta + \sin^2\theta = 1.$$

($\cos^2\theta$ is short for $(\cos\theta)^2$.) To establish this, apply the Pythagorean theorem to the right triangle OAP, shown in Fig. D.9.

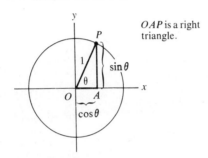

Figure D.9

OAP is a right triangle.

With the aid of this relation between $\cos\theta$ and $\sin\theta$ the next two examples determine $\cos(\pi/4)$, $\sin(\pi/4)$, $\cos(\pi/3)$, and $\sin(\pi/3)$.

EXAMPLE 6 Find $\cos(\pi/4)$ and $\sin(\pi/4)$.

SOLUTION When the angle is $\pi/4$ (45°), a quick sketch shows that the cosine equals the sine:

$$\cos\frac{\pi}{4} = \sin\frac{\pi}{4}.$$

Thus

$$\cos^2\frac{\pi}{4} + \cos^2\frac{\pi}{4} = 1,$$

from which it follows that

$$\cos^2\frac{\pi}{4} = \frac{1}{2}.$$

Since $\cos(\pi/4)$ is positive,

$$\cos\frac{\pi}{4} = \sqrt{\frac{1}{2}}$$

$$= \frac{\sqrt{2}}{2} \approx 0.707$$

and

$$\sin\frac{\pi}{4} = \frac{\sqrt{2}}{2}. \blacksquare$$

EXAMPLE 7 Find cos $(\pi/3)$ and sin $(\pi/3)$.

SOLUTION The angle $\pi/3$ (60°) is the angle in an equilateral triangle. Place such a triangle in the unit circle as shown in Fig. D.10. Inspection of the figure shows that

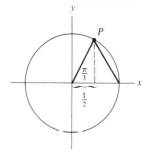

Figure D.10

$$\cos \frac{\pi}{3} = \frac{1}{2}.$$

Then
$$\left(\frac{1}{2}\right)^2 + \sin^2 \frac{\pi}{3} = 1$$

$$\sin^2 \frac{\pi}{3} = \frac{3}{4}$$

$$\sin \frac{\pi}{3} = \frac{\sqrt{3}}{2} \approx 0.866. \blacksquare$$

Once cos $(\pi/4)$ is known, the cosines of multiples of $\pi/4$ can be found by sketching the unit circle. For instance, to find cos $(3\pi/4)$ draw an angle of $3\pi/4$ radians, as in Fig. D.11. It is clear that cos $(3\pi/4)$ is negative and that $|\cos (3\pi/4)| = \sqrt{2}/2$. Hence

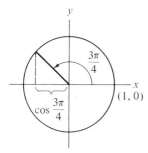

Figure D.11

$$\cos \frac{3\pi}{4} = -\frac{\sqrt{2}}{2}.$$

A similar method can be used to compute the cosines of multiples of $\pi/6$. With the aid of such computations, this table for the cosine function can be obtained.

θ	0	$\frac{\pi}{6}$	$\frac{\pi}{4}$	$\frac{\pi}{3}$	$\frac{\pi}{2}$	$\frac{2\pi}{3}$	$\frac{3\pi}{4}$	$\frac{5\pi}{6}$	π	$\frac{7\pi}{6}$	$\frac{4\pi}{3}$	$\frac{3\pi}{2}$	2π
$\cos \theta$	1	$\frac{\sqrt{3}}{2}$	$\frac{\sqrt{2}}{2}$	$\frac{1}{2}$	0	$\frac{-1}{2}$	$\frac{-\sqrt{2}}{2}$	$\frac{-\sqrt{3}}{2}$	-1	$\frac{-\sqrt{3}}{2}$	$\frac{-1}{2}$	0	1

(It is easier to draw the unit circle and the angle each time you need cos θ than to memorize this table.) This table provides enough information to graph the cosine function. Since cos $(\theta + 2\pi) = \cos \theta$, the graph consists of the portion from 0 to 2π endlessly repeated. It is sketched in Fig. D.12. The graph of the sine function can be sketched in a similar manner. It is shown in Fig. D.13.

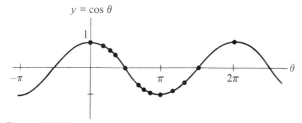

Figure D.12

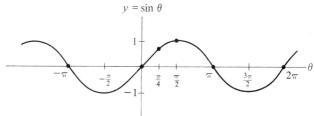

Figure D.13

Four important identities relate the cosine and sine of the sum and difference of two angles to the values of the cosine and sine of the angles:

These identities are obtained in Exercises 43 to 48.

$$\cos(A + B) = \cos A \cos B - \sin A \sin B;$$
$$\sin(A + B) = \sin A \cos B + \cos A \sin B;$$
$$\cos(A - B) = \cos A \cos B + \sin A \sin B;$$
$$\sin(A - B) = \sin A \cos B - \cos A \sin B.$$

From these follow the "double angle" and "half angle" identities:

$$\cos 2\theta = \cos^2 \theta - \sin^2 \theta$$
$$\cos 2\theta = 2\cos^2 \theta - 1$$
$$\cos 2\theta = 1 - 2\sin^2 \theta$$
$$\sin 2\theta = 2 \sin \theta \cos \theta$$
$$\sin^2 \theta = \frac{1 - \cos 2\theta}{2}$$
$$\cos^2 \theta = \frac{1 + \cos 2\theta}{2}.$$

THE TANGENT FUNCTION   The trigonometric function next in importance to the cosine and sine is the tangent function.

DEFINITION   *The tangent function.*   For each number θ that does not differ from $\pi/2$ by a multiple of π, the tangent of θ, denoted $\tan \theta$, is defined as follows: Draw the angle of θ radians whose first arm is the positive x axis and whose vertex is $(0, 0)$. Let L be the line through $(1, 0)$ parallel to the y axis. The line on the second arm of the angle meets the line L at a point Q. The y coordinate of Q is called $\tan \theta$.   (See Fig. D.14.)

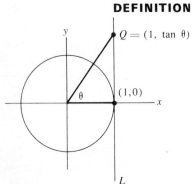

Figure D.14

Note from Fig. D.14 that for θ near $\pi/2$ but less than $\pi/2$, $\tan \theta$ becomes very large. While $\cos \theta$ and $\sin \theta$ never exceed 1, $\tan \theta$ takes arbitrarily large values. Note that for

$$\frac{\pi}{2} < \theta < \pi$$

(a second-quadrant angle) $\tan \theta$ is negative, as shown in Fig. D.15. The behavior of $\tan \theta$ for θ near $\pi/2$ is described by these two limits:

$$\lim_{\theta \to (\pi/2)^-} \tan \theta = \infty \quad \text{and} \quad \lim_{\theta \to (\pi/2)^+} \tan \theta = -\infty.$$

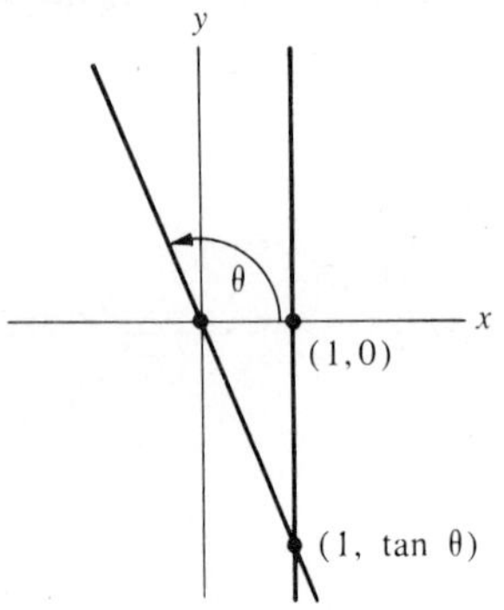

It follows from the definition of the tangent function that

$$\tan(\theta + \pi) = \tan \theta.$$

Figure D.15

While cosine and sine have period 2π, the tangent function has period π.

Tables for the functions cos θ, sin θ, and tan θ are included in Appendix H; these functions are also available on many calculators.

The functions tan θ, sin θ, and cos θ are related by the equation

$$\tan\theta = \frac{\sin\theta}{\cos\theta}.$$

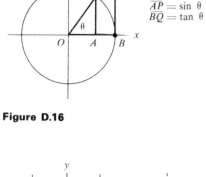

Figure D.16

This can easily be deduced from inspection of Fig. D.16. By the similarity of triangles OBQ and OAP,

$$\frac{\overline{BQ}}{\overline{OB}} = \frac{\overline{AP}}{\overline{OA}}$$

or

$$\frac{\tan\theta}{1} = \frac{\sin\theta}{\cos\theta};$$

thus

$$\tan\theta = \frac{\sin\theta}{\cos\theta}.$$

The following identities involving the tangent functions are of use:

$$\tan(A+B) = \frac{\tan A + \tan B}{1 - \tan A \tan B},$$

$$\tan(A-B) = \frac{\tan A - \tan B}{1 + \tan A \tan B},$$

$$\tan 2\theta = \frac{2\tan\theta}{1 - \tan^2\theta}.$$

It is easy to see that $\tan 0 = 0$

and $\tan\dfrac{\pi}{4} = 1.$

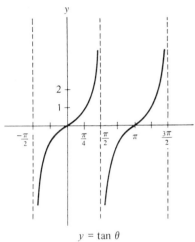

Figure D.17

Slope and the angle of inclination

Figure D.17 presents the graph of the tangent function.

The tangent function can be used to describe the slope of a line. Let a line L make an angle θ with the positive x axis. (This angle is taken counterclockwise from the x axis upward to the line; thus $0 \le \theta < \pi$.) The angle θ is called the *angle of inclination* of the line L. Then the slope of L is equal to $\tan\theta$. If the slope is known, then the angle of inclination can be estimated with the aid of a table of $\tan\theta$ or a calculator. For instance, when the slope is $\frac{1}{2}$, as in the case of the line L in Fig. D.18, the angle is about 0.46 radians or 26.6°.

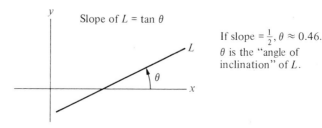

Figure D.18

THE SECANT, COSECANT, AND COTANGENT FUNCTIONS

Three other trigonometric functions will be needed in calculus. They are the reciprocals of the cosine, sine, and tangent functions, and are called, respectively, secant, cosecant, and cotangent:

$$\sec \theta = \frac{1}{\cos \theta} \qquad \csc \theta = \frac{1}{\sin \theta} \qquad \cot \theta = \frac{\cos \theta}{\sin \theta}.$$

Properties of $\sec \theta$

Consider the function $\sec \theta$. First of all, it is not defined when $\cos \theta = 0$. For instance, $\pi/2$ is not in the domain of $\sec \theta$. However, for θ near $\pi/2$, $|\sec \theta|$ is large. Second, since $|\cos \theta| \leq 1$ for all θ, $|\sec \theta| \geq 1$ for all θ for which $\sec \theta$ is defined. Third, $\sec \theta$ has period 2π since $\cos \theta$ does. Fourth, $\sec \theta$ is positive when $\cos \theta$ is positive, and negative when $\cos \theta$ is negative. Figure D.19 shows the graph of $y = \sec \theta$, and, for comparison, the graph of $\cos \theta$, which is shown dashed.

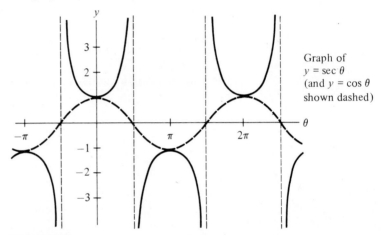

Graph of $y = \sec \theta$ (and $y = \cos \theta$ shown dashed)

Figure D.19

The graph of $y = \csc \theta$ bears a similar relation to the graph of $\sin \theta$. It is simply the graph of $\sec \theta$ moved $\pi/2$ units to the right. (See Fig. D.20.)

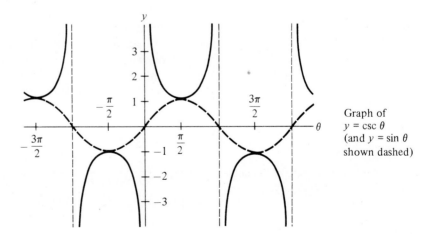

Graph of $y = \csc \theta$ (and $y = \sin \theta$ shown dashed)

Figure D.20

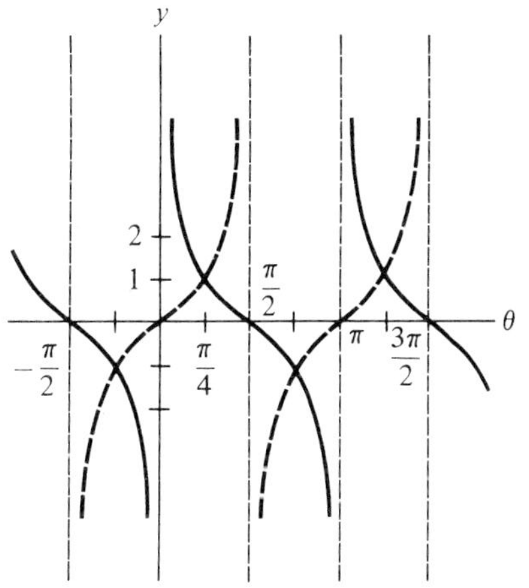

Figure D.21

The graph of $y = \cot \theta$, along with that of $\tan \theta$, is shown in Fig. D.21. Note that, when $\tan \theta$ is large, $\cot \theta$ is near 0; when $\tan \theta$ is near 0, $|\cot \theta|$ is large.

THE LAW OF COSINES

When $\theta = \dfrac{\pi}{2}$ the law of cosines is the Pythagorean theorem.

If the lengths of two sides of a triangle, a and b, and the angle θ between these sides are known, the length of the third side c is determined. The formula for finding c, the *law of cosines*, is

$$c^2 = a^2 + b^2 - 2ab \cos \theta.$$

This formula will be needed several times in the text. A proof is outlined in Exercise 53. Further trigonometric identities are developed in the exercises.

Exercises

1. What are the radian measures of the following angles?
   (a) 90°,   (b) 30°,   (c) 120°,   (d) 360°.
2. What are the radian measures of the following angles?
   (a) 180°,   (b) 60°,   (c) 45°,   (d) 0°.
3. How many degrees are there in an angle whose radian measure is
   (a) $3\pi/4$,   (b) $\pi/3$,   (c) $2\pi/3$,   (d) 4π?
4. How many degrees are there in an angle whose radian measure is
   (a) $\pi/4$,   (b) $\pi/6$,   (c) 2π,   (d) π?
5. An angle intercepts an arc of 5 inches in a circle of radius 3 inches.
   (a) What is the measure of the angle in radians?
   (b) What is the measure of the angle in degrees?
6. How long an arc of a circle of radius 3 inches is intercepted by an angle of 0.5 radian?
7. (a) Express an angle of 50° in radians.
   (b) Express an angle of 2 radians in degrees.
8. (a) Express an angle of 3 radians in degrees.
   (b) Express an angle of 1 degree in radians.
9. How long an arc does an angle of 1.5 radians intercept in a circle of radius
   (a) 3 inches,   (b) 4 inches,   (c) 5 inches?
10. How would you draw an angle of 2 radians
    (a) with a protractor that shows angles in degrees?
    (b) with a string?
11. Draw an angle of
    (a) $13\pi/6$ radians,   (b) -3π radians.
12. Draw an angle of
    (a) 6π radians,   (b) $-\pi/3$ radians.
13. In Example 7 it was shown that $\cos(\pi/3) = \frac{1}{2}$. Using the identity $\sin \theta = \cos(\pi/2 - \theta)$, deduce that $\sin(\pi/6) = \frac{1}{2}$.
14. In Example 7 it was shown that $\sin(\pi/3) = \sqrt{3}/2$. Using the identity $\cos \theta = \sin(\pi/2 - \theta)$, deduce that $\cos(\pi/6) = \sqrt{3}/2$.

15 Fill in this table by making a quick sketch of the unit circle:

θ	0	$\frac{\pi}{6}$	$\frac{\pi}{4}$	$\frac{\pi}{3}$	$\frac{\pi}{2}$	π	$\frac{3\pi}{2}$	2π
$\sin \theta$								

16 By sketching the angles in a unit circle and using the information that $\cos (\pi/6) = \sqrt{3}/2$ and $\sin (\pi/6) = \frac{1}{2}$, find
(a) $\cos (-\pi/6)$, (b) $\sin (-\pi/6)$,
(c) $\cos (5\pi/6)$, (d) $\sin (5\pi/6)$,
(e) $\cos (13\pi/6)$, (f) $\sin (13\pi/6)$.

17 Check the identity for $\cos (A + B)$ when $A = \pi/6$ and $B = \pi/3$.

18 Check the identity for $\sin (A + B)$ when $A = \pi/4$ and $B = \pi/4$.

19 Find $\sin \frac{5\pi}{12}$. Hint: $\frac{5\pi}{12} = \frac{\pi}{4} + \frac{\pi}{6}$.

20 Find $\sin \frac{\pi}{12}$. Hint: $\frac{\pi}{12} = \frac{\pi}{4} - \frac{\pi}{6}$.

21 Deduce the identity $\cos 2\theta = \cos^2 \theta - \sin^2 \theta$ from the identity for $\cos (A + B)$.

22 Deduce the identity $\sin 2\theta = 2 \sin \theta \cos \theta$ from the identity for $\sin (A + B)$.

23 Deduce the identity $\cos 2\theta = 2 \cos^2 \theta - 1$ from the identity for $\cos (A + B)$.

24 Deduce the identity $\cos 2\theta = 1 - 2 \sin^2 \theta$ from the identity for $\cos (A + B)$.

25 Deduce from Exercise 23 that $\cos^2 \theta = (1 + \cos 2\theta)/2$.

26 Deduce from Exercise 24 that $\sin^2 \theta = (1 - \cos 2\theta)/2$.

27 Use a sketch of the angle in a unit circle to determine the sign ($+$ or $-$) of the function in each case:
(a) $\sin \theta, \pi < \theta < 2\pi$; (b) $\tan \theta, \pi < \theta < 3\pi/2$;
(c) $\cos \theta, -\pi/2 < \theta < \pi/2$; (d) $\tan \theta, \pi/2 < \theta < \pi$.

28 Give the sign ($+$ or $-$) of $\cos \theta$, $\sin \theta$, and $\tan \theta$ for
(a) $0 < \theta < \pi/2$, (b) $-\pi/2 < \theta < 0$.

29 (a) Using the definition of $\tan \theta$ and a sketch, show that $\tan (\pi/4) = 1$.
(b) Using the identity, $\tan \theta = \sin \theta/\cos \theta$, show that $\tan (\pi/4) = 1$.

30 Find (a) $\tan (\pi/6)$, (b) $\tan (\pi/3)$.

31 Using a calculator, find the slope of a line whose angle of inclination is
(a) $10°$, (b) $70°$, (c) $110°$,
(d) $135°$, (e) $0°$.

32 Using a calculator, find the angle of inclination, both in degrees and in radians, of a line whose slope is
(a) -1, (b) 1, (c) 2, (d) 3, (e) -3.

33 Find the angle of inclination in degrees and in radians of a line whose slope is
(a) $\sqrt{3}$, (b) $1/\sqrt{3}$.

34 (Calculator) Find the angle of inclination in degrees and in radians of a line whose slope is
(a) $1/10$, (b) 10.

35 Consider an acute angle θ in any right triangle, as in Fig. D.22. With the aid of the similar triangles in Fig. D.23, show that
(a) $\cos \theta = a/c$, (b) $\sin \theta = b/c$, (c) $\tan \theta = b/a$.

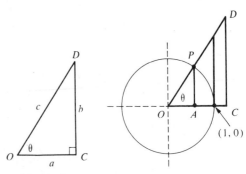

Figure D.22   **Figure D.23**

These formulas for cosine, sine, and tangent are sometimes taken as their definitions for $0 < \theta < \pi/2$.

36 (See Exercise 35, Fig. D.22.)
(a) Express b in terms of θ and c.
(b) Express a in terms of θ and c.
(c) Express b in terms of θ and a.

37 In Fig. D.24 the two acute angles in a right triangle are labeled α and β.

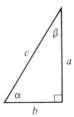

Figure D.24

(a) Express $\cos \alpha$ in terms of the lengths of the sides.
(b) Express $\sin \beta$ in terms of the lengths of the sides.
(c) Express $\tan \alpha$ in terms of the lengths of the sides.

38 (See Exercise 35.) Use the triangle OCD in Fig. D.25 to obtain $\cos (\pi/3)$, $\sin (\pi/3)$, and $\tan (\pi/3)$. (ODE is equilateral.) A quick sketch of OCD will remind you of the values of $\cos (\pi/3)$, $\sin (\pi/3)$, and $\tan (\pi/3)$ when you need them.

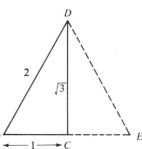

Figure D.25

39 Use OCD in Exercise 38 to compute $\cos(\pi/6)$, $\sin(\pi/6)$, and $\tan(\pi/6)$. *Hint:* See Exercise 35.

40 (See Exercise 35.) Solve for the length x in each triangle in Fig. D.26.

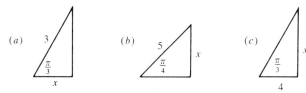

Figure D.26

41 (a) Evaluate $\sec \theta$ for $\theta = 0, \pi/6, \pi/4, \pi/3$.
(b) Plot the four points given by (a) and graph $\sec \theta$.

42 (a) Evaluate $\csc \theta$ for $\theta = \pi/6, \pi/4, \pi/3, \pi/2$.
(b) Plot the four points given by (a) and graph $\csc \theta$.

■

43 This exercise outlines a proof that $\cos(A + B) = \cos A \cos B - \sin A \sin B$. Figure D.27 shows two diagrams involving the unit circle.

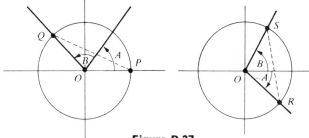

Figure D.27

(a) Show that the line segments PQ and RS in Fig. D.27 have the same length.
(b) Show that $Q = (\cos(A + B), \sin(A + B))$
$S = (\cos B, \sin B)$
$R = (\cos A, -\sin A)$.
(c) Using the coordinates in (b), show that

$$\overline{PQ}^2 = 2 - 2\cos(A + B).$$

(d) Using the coordinates in (b), show that

$$\overline{RS}^2 = 2 - 2\cos A \cos B + 2\sin A \sin B.$$

(e) Deduce the identity for $\cos(A + B)$.

44 Replacing B by $-B$ in the identity for $\cos(A + B)$, obtain the identity

$$\cos(A - B) = \cos A \cos B + \sin A \sin B.$$

45 Using the identity for $\cos(A - B)$, show that

$$\sin \theta = \cos\left(\frac{\pi}{2} - \theta\right).$$

46 Using the identity $\sin \theta = \cos(\pi/2 - \theta)$ from Exercise 45, show that $\cos \theta = \sin(\pi/2 - \theta)$. *Hint:* Replace θ by $\pi/2 - \theta$ in the identity, $\sin \theta = \cos(\pi/2 - \theta)$.

47 With the aid of the identities in Exercises 44 to 46, show that $\sin(A + B) = \sin A \cos B + \cos A \sin B$, as follows:
(a) Show that $\sin(A + B) = \cos[\pi/2 - (A + B)] = \cos((\pi/2 - A) - B)$.
(b) Apply the identity in Exercise 44 to $\cos(x - y)$ with $x = \pi/2 - A$ and $y = B$.
(c) Use the identities in Exercises 45 and 46 to remove $\pi/2$ from the identity you obtain in (b).

48 Use the identity for $\sin(A + B)$ obtained in Exercise 47 to show that $\sin(A - B) = \sin A \cos B - \cos A \sin B$. *Hint:* $\sin(A - B) = \sin[A + (-B)]$.

49 (a) From the identity $\cos 2\theta = 2\cos^2 \theta - 1$ deduce that $\cos \theta = \pm\sqrt{(1 + \cos 2\theta)/2}$.
(b) Use the identity in (a) to find $\cos(\pi/4)$.
(c) Use the identity in (a) to find $\cos(3\pi/4)$.

50 (a) From the identity $\cos 2\theta = 1 - 2\sin^2 \theta$ deduce that $\sin \theta = \pm\sqrt{(1 - \cos 2\theta)/2}$.
(b) Use the identity in (a) to find $\sin(\pi/4)$.
(c) Use the identity in (a) to find $\sin(-\pi/4)$.

51 Using the identities for $\cos(A - B)$ and $\sin(A - B)$, prove that

$$\tan(A - B) = \frac{\tan A - \tan B}{1 + \tan A \tan B}.$$

52 Show that
(a) $\sec^2 \theta = 1 + \tan^2 \theta$,
(b) $\csc^2 \theta = 1 + \cot^2 \theta$.

■ ■

53 This exercise outlines a proof of the law of cosines in the case $0 < \theta < \pi/2$. Consider Fig. D.28.
(a) Show that $\overline{CD} = a\cos\theta$ and $\overline{AD} = b - a\cos\theta$.
(b) Show that

$$a^2 - a^2\cos^2\theta = h^2 = c^2 - (b - a\cos\theta)^2.$$

(c) From (b) deduce the law of cosines.

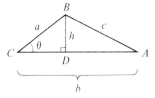

Figure D.28

54 (Calculator)
(a) Calculate $(\sin x)/x$ for $x = 1.5, 1, 0.1, 0.001$.
(b) What do you think happens to $(\sin x)/x$ as $x \to 0$?

APPENDIX E
CONIC SECTIONS

This appendix defines the ellipse, hyperbola, and parabola and develops their equations in rectangular coordinates (E.1, E.2, and E.3) and in polar coordinates (E.4).

E.1 Conic sections

The intersection of a plane and the surface of a double cone is called a *conic section*. If the plane cuts off a bounded curve, that curve is called an *ellipse*. (In particular a circle is an ellipse.) See Fig. E.1.

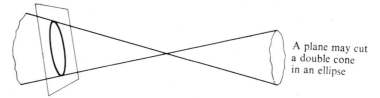

A plane may cut a double cone in an ellipse

Figure E.1

If the plane is parallel to the edge of the double cone, as in Fig. E.2, the intersection is called a *parabola*. In the cases of the ellipse and the parabola, the plane generally meets just one of the two cones.

If the plane meets both parts of the cone and is not parallel to an edge, the intersection is called a *hyperbola*. The hyperbola consists of two separate pieces. It can be proved that these two pieces are congruent and that they are *not* congruent to parabolas. (See Fig. E.3.)

For the sake of simplicity we shall use a definition of the conic sections that depends only on the geometry of the plane. It is shown in geometry courses that the two approaches yield the same curves.

Figure E.2

Figure E.3

THE ELLIPSE

DEFINITION

"*Foci*" (*pronounced "foe-sigh"*) *is the plural of "focus."*

Ellipse. Let F and F' be points in the plane and let a be a fixed positive number such that $2a$ is greater than the distance between F and F'. A point P in the plane is on the *ellipse* determined by F, F', and $2a$ if and only if the sum of the distances from P to F and from P to F' equals $2a$. Points F and F' are the *foci* of the ellipse.

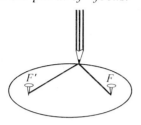

The string has length $2a$, greater than the distance between the tacks.

Figure E.4

To construct an ellipse, place two tacks in a piece of paper, tie a string of length $2a$ to them, and trace out a curve with a pencil held against the string, keeping the string taut by means of the pencil point. (See Fig. E.4.) The foci are at the tacks. (Note that when $F = F'$ the ellipse is a circle of radius a.)

The equation of a circle whose center is at $(0, 0)$ and whose radius is a is

$$x^2 + y^2 = a^2,$$

or

$$\frac{x^2}{a^2} + \frac{y^2}{a^2} = 1. \qquad (1)$$

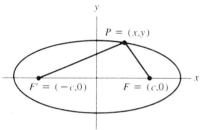

Figure E.5

Let us generalize this result by determining the equation of an ellipse. To make the equation as simple as possible, introduce the x and y axes in such a way that the x axis contains the foci and the origin is midway between them, as in Fig. E.5. Thus $F = (c, 0)$ and $F' = (-c, 0)$, where $c \geq 0$ and $2c < 2a$; hence $c < a$.

Now translate into algebra this assertion: The sum of the distances from $P = (x, y)$ to $F = (c, 0)$ and from P to $F' = (-c, 0)$ equals $2a$. By the distance formula, the distance from P to F is

$$\sqrt{(x - c)^2 + (y - 0)^2}.$$

Similarly, the distance from P to F' is

$$\sqrt{(x + c)^2 + (y - 0)^2}.$$

Thus the point (x, y) is on the ellipse if and only if

$$\sqrt{(x - c)^2 + y^2} + \sqrt{(x + c)^2 + y^2} = 2a.$$

The rest of the argument is getting rid of square roots.

A few algebraic steps will transform this equation into an equation without square roots.

First, write the equation as

$$\sqrt{(x + c)^2 + y^2} = 2a - \sqrt{(x - c)^2 + y^2}.$$

Then square both sides, obtaining

$$(x + c)^2 + y^2 = 4a^2 - 4a\sqrt{(x - c)^2 + y^2} + (x - c)^2 + y^2.$$

Expanding yields

$$x^2 + 2cx + c^2 + y^2 = 4a^2 - 4a\sqrt{(x - c)^2 + y^2} + x^2 - 2cx + c^2 + y^2,$$

which a few cancellations reduce to

$$2cx = 4a^2 - 4a\sqrt{(x-c)^2 + y^2} - 2cx$$

or

$$4cx - 4a^2 = -4a\sqrt{(x-c)^2 + y^2}.$$

Dividing by -4 yields

$$a^2 - cx = a\sqrt{(x-c)^2 + y^2}.$$

Squaring gets rid of the square root:

$$a^4 - 2a^2cx + c^2x^2 = a^2(x^2 - 2cx + c^2 + y^2),$$

or

$$a^4 - 2a^2cx + c^2x^2 = a^2x^2 - 2a^2cx + a^2c^2 + a^2y^2,$$

or

$$a^4 + c^2x^2 = a^2x^2 + a^2c^2 + a^2y^2.$$

This equation can be transformed to

$$(a^2 - c^2)x^2 + a^2y^2 = a^2(a^2 - c^2).$$

Dividing both sides by $a^2(a^2 - c^2)$ results in the equation

$$\frac{x^2}{a^2} + \frac{y^2}{a^2 - c^2} = 1. \tag{2}$$

Since $a^2 - c^2 > 0$, there is a number b such that

$$b^2 = a^2 - c^2 \qquad b > 0,$$

and thus (2) takes the shorter form

Equation of ellipse in standard position

$$\boxed{\frac{x^2}{a^2} + \frac{y^2}{b^2} = 1.} \tag{3}$$

(Note that (3) generalizes (1), the equation for a circle.)

Setting $y = 0$ in (3), we obtain $x = a$ or $-a$; if we set $x = 0$ in (3), we obtain $y = b$ or $-b$. Thus the four "extreme" points of the ellipse have coordinates $(a, 0)$, $(-a, 0)$, $(0, b)$, and $(0, -b)$, as shown in Fig. E.6. Observe that the distance from F or F' to $(0, b)$ is a, which is half the length of string. The right triangle in the diagram is a reminder of the fact that $b^2 = a^2 - c^2$. Keep in mind that in the above ellipse a is larger than b. The *semimajor axis* is said to have length a; the *semiminor axis* has length b.

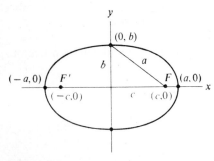

Figure E.6

EXAMPLE 1 Discuss the foci and "length of string" of the ellipse whose equation is

$$\frac{x^2}{25} + \frac{y^2}{9} = 1.$$

SOLUTION Since the larger denominator is with the x^2, the foci lie on the x axis. In this case $a = 5$ and $b = 3$. The length of string is $2a$ or 10. The foci are at a distance

$$\sqrt{a^2 - b^2} = \sqrt{25 - 9} = 4$$

from the origin, as shown in Fig. E.7.

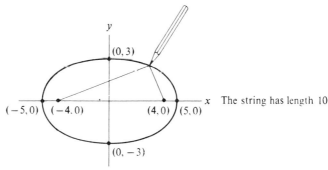

Figure E.7

EXAMPLE 2 Discuss the foci and "length of string" of the ellipse whose equation is

$$\frac{x^2}{9} + \frac{y^2}{25} = 1.$$

SOLUTION This is similar to Example 1. The only difference is that the roles of x and y are interchanged. The foci are at $(0, 4)$ and $(0, -4)$, and the ellipse is longer in the y direction than in the x direction. ∎

THE HYPERBOLA

The definition of the hyperbola is similar to that of the ellipse.

DEFINITION *Hyperbola.* Let F and F' be points in the plane and let a be a fixed positive number such that $2a$ is less than the distance between F and F'. A point P in the plane is on the *hyperbola* determined by F, F', and $2a$ if and only if the difference between the distances from P to F and from P to F' equals $2a$ (or $-2a$). Points F and F' are the *foci* of the hyperbola. (See Fig. E.8.)

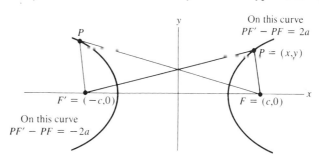

Figure E.8

A hyperbola consists of two separate curves. On one curve $\overline{PF'} - \overline{PF} = 2a$; on the other, $\overline{PF'} - \overline{PF} = -2a$. If the distance $\overline{FF'}$ is $2c$, then $2a < 2c$; hence $a < c$. Again place the axes in such a way that $F = (c, 0)$ and $F' = (-c, 0)$. Let $P = (x, y)$ be a typical point on the hyperbola. Then x and y satisfy the equation

$$\sqrt{(x-c)^2 + y^2} - \sqrt{(x+c)^2 + y^2} = \pm 2a. \qquad (4)$$

Get rid of square roots. Some algebra similar to that used in simplifying the equation of the ellipse transforms (4) into

$$\frac{x^2}{a^2} + \frac{y^2}{a^2 - c^2} = 1. \qquad (5)$$

But now $a^2 - c^2$ is *negative* and can be expressed as $-b^2$ for some number $b > 0$. Hence the hyperbola has the equation

Equation of hyperbola in standard position

$$\boxed{\frac{x^2}{a^2} - \frac{y^2}{b^2} = 1.} \qquad (6)$$

If the foci are on the y axis, the equation is

$$\boxed{\frac{y^2}{a^2} - \frac{x^2}{b^2} = 1.}$$

In both cases, $c^2 = a^2 + b^2$.

EXAMPLE 3 Sketch the hyperbola $\dfrac{x^2}{9} - \dfrac{y^2}{16} = 1.$

SOLUTION Since the minus sign is with the y^2, the foci are on the x axis. In this case $a^2 = 9$ and $b^2 = 16$. Observe that the hyperbola meets the x axis at $(3, 0)$ and $(-3, 0)$. The hyperbola does not meet the y axis. The distance c from the origin to a focus is determined by the equation

$$c^2 = 9 + 16 = 25;$$

hence

$$c = 5.$$

The hyperbola $\dfrac{x^2}{9} - \dfrac{y^2}{16} = 1$

is shown in Fig. E.9. ∎

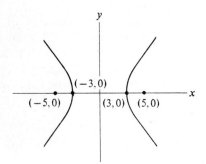

Figure E.9

THE PARABOLA

The definition of a parabola involves the distance to a point and the distance to a line.

DEFINITION *Parabola.* Let L be a line in the plane and let F be a point in the plane which is not on the line. A point P in the plane is on the *parabola* determined by F and L if and only if the distance from P to F equals the distance from P to the line L. Point F is the *focus* of the parabola; line L is its *directrix*. (See Fig. E.10.)

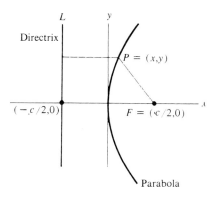

Figure E.10

To obtain an algebraic equation for a parabola, denote the distance from point F to line L by c, and introduce axes in such a way that $F = (c/2, 0)$ and L has the equation $x = -c/2$ with $c > 0$. The distance from P to F is $\sqrt{(x - c/2)^2 + (y - 0)^2}$. Now, if $P = (x, y)$ is on the parabola, x is clearly not negative. The distance from P to the line L is therefore $x + c/2$. Thus the equation of the parabola is

$$\sqrt{\left(x - \frac{c}{2}\right)^2 + y^2} = x + \frac{c}{2}. \tag{7}$$

Squaring and simplifying reduces (7) to

$$\boxed{y^2 = 2cx \qquad c > 0,} \tag{8}$$

which is the equation of a parabola in "standard position."

If the focus is at $(0, c/2)$ and the directrix is the line $y = -c/2$, the parabola has the equation

$$\boxed{x^2 = 2cy.}$$

EXAMPLE 4 Sketch the parabola $y = x^2$, showing its focus and directrix.

SOLUTION The equation $y = x^2$ is equivalent to $x^2 = 2cy$, where $c = \frac{1}{2}$.

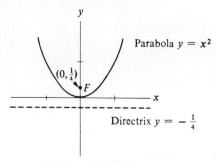

Figure E.11

The focus is on the y axis at $(0, \frac{1}{4})$. The directrix is the line $y = -\frac{1}{4}$. The parabola is sketched in Fig. E.11. ∎

SUMMARY

Conic section	Equation in standard position	Location of foci
Ellipse	$\dfrac{x^2}{a^2} + \dfrac{y^2}{b^2} = 1$	$a > b$, foci on x axis $a < b$, foci on y axis $a = b$, circle, both foci at $(0, 0)$
Hyperbola	$\dfrac{x^2}{a^2} - \dfrac{y^2}{b^2} = 1$	Foci on x axis
	$\dfrac{y^2}{a^2} - \dfrac{x^2}{b^2} = 1$	Foci on y axis
Parabola	$y^2 = 2cx$ $x^2 = 2cy$	Focus at $(c/2, 0)$ Focus at $(0, c/2)$

Exercises

1. Sketch the ellipse $\dfrac{x^2}{49} + \dfrac{y^2}{25} = 1$ and its foci.

2. Sketch the ellipse $\dfrac{x^2}{4} + \dfrac{y^2}{36} = 1$ and its foci.

3. What is the equation of the ellipse whose foci are at $(2, 0)$ and $(-2, 0)$ and such that the sum of the distances from a point on the ellipse to the two foci is 10?

4. What is the equation of the ellipse whose foci are at $(0, 3)$ and $(0, -3)$ and such that the sum of the distances from a point on the ellipse to the two foci is 14?

5. Sketch the hyperbola $x^2 - y^2 = 1$ and its foci.

6. Sketch the hyperbola $y^2 - x^2 = 1$ and its foci.

7. Sketch the parabola $y = 6x^2$, its focus, and its directrix.

8. Sketch the parabola $x = -6y^2$, its focus, and its directrix.

9. What is the equation of the parabola whose focus is at $(3, 0)$ and whose directrix is the line $x = -3$?

10. What is the equation of the parabola whose focus is at $(0, -5)$ and whose directrix is the line $y = 5$?

11. Obtain (5) from (4).

12. Obtain (8) from (7). ∎

13. (a) Using the definition of the hyperbola, show that the hyperbola that has its foci at $(\sqrt{2}, \sqrt{2})$ and $(-\sqrt{2}, -\sqrt{2})$ and for which $2a = 2\sqrt{2}$ has the equation $xy = 1$.
    (b) Graph $xy = 1$ and show the foci.

14. How would you inscribe an elliptical garden in a rectangle whose dimensions are 8 by 10 feet? (The line through the foci is to be parallel to an edge of the rectangle.)

15. In the definition of the hyperbola it was assumed that $2a$ is less than the distance between the foci. Show that, if $2a$ were greater than the distance between the foci, the hyperbola would have no points.

16. Find the equation of the parabola whose focus is $(2, 4)$ and whose directrix is the line $y = -3$.

Exercises 17 and 18 concern the determination of the position of an object with the aid of conic sections.

17 The location of a submarine can be found as follows. A small explosion is set off at point D. The sound from this explosion bounces off the submarine and is picked up by hydrophones at points A, B, and C. The time of the explosion and the times of reception of the sound at A, B, and C are known. Show how to locate the submarine as the point where three ellipses meet.

18 The sound of the shooting of a cannon arrives 1 second later at point B than at point A.
   (a) Show that the cannon is somewhere on a certain hyperbola whose foci are A and B.
   (b) On which of the two pieces (branches) of the hyperbola is the cannon located?

With the aid of a third listening post the location of the cannon can be more precisely determined. LORAN, a system for long-range navigation, is based on a similar use of hyperbolas.

19 A plane intersects the surface of a right circular cylinder in a curve. Prove that this curve is an ellipse, as defined in terms of foci and sum of distances. *Hint:* Consider the two spheres inscribed in the cylinder and tangent to the plane, the spheres being on opposite sides of the plane. Let $2a$ denote the distance between the equators of the spheres perpendicular to the axis of the cylinder, and let F and F' be the points at which they touch the plane.

E.2 Translation of axes and the graph of $Ax^2 + Cy^2 + Dx + Ey + F = 0$

Section E.3 treats the more general equation $Ax^2 + Bxy + Cy^2 + Dx + Ey + F = 0$.

The equation of a particular geometric object such as a line, a circle, or a conic section depends on where we choose to place the axes. Consider, for instance, the line L in Fig. E.12. Relative to the axes in Fig. E.13 it has the equation $y = 3$. Relative to the axes in Fig. E.14 it has the equation $y = 0$.

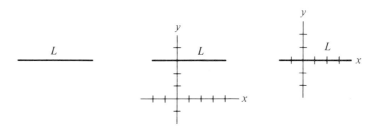

Figure E.12   **Figure E.13**   **Figure E.14**

Clearly, a wise choice of axes may yield a simpler equation for a given line or curve. This section shows one way to choose convenient axes and uses the method to analyze equations of the form $Ax^2 + Cy^2 + Dx + Ey + F = 0$.

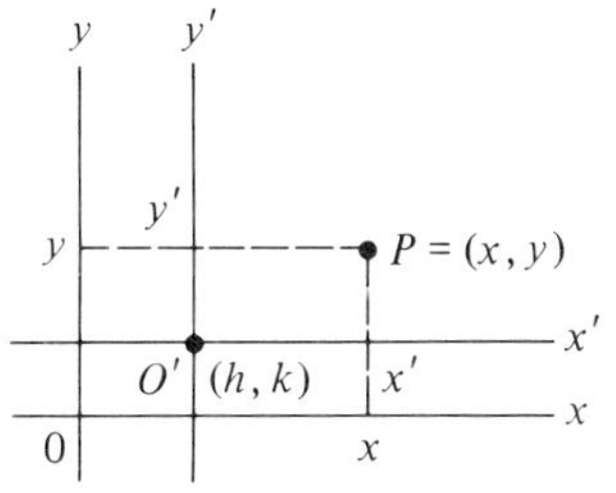

THE EFFECT OF TRANSLATION OF AXES

A point P has coordinates (x, y) relative to a given choice of axes. Another pair of axes is chosen parallel to the first pair with its origin at the point (h, k). Call the second pair of axes the $x'y'$ system. (See Fig. E.15.) Inspection of Fig. E.15 shows that

Figure E.15

$x' = x - h$ and $y' = y - k$ are the key to this whole section.

$$x' = x - h \quad \text{and} \quad y' = y - k \qquad (1)$$

or, equivalently, $\quad x = x' + h \quad$ and $\quad y = y' + k. \qquad (2)$

The coordinates change by fixed amounts, h and k. This observation is applied in the following three examples.

EXAMPLE 1 Find an equation of the parabola whose focus is at (3, 7) and whose directrix is the line $y = 1$, as shown in Fig. E.16.

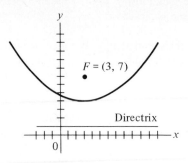

Figure E.16

SOLUTION Introduce an $x'y'$-coordinate system whose origin is at the point (3, 4), which is midway between the focus and the directrix, as shown in Fig. E.17.

Relative to the $x'y'$ axes the parabola is in standard position. Since the distance from the focus to the directrix is $c = 6$, the parabola has an equation $(x')^2 = 2 \cdot 6y'$, that is, $(x')^2 = 12y'$ relative to the $x'y'$ axes. By (1),

$$(x - 3)^2 = 12(y - 4). \tag{3}$$

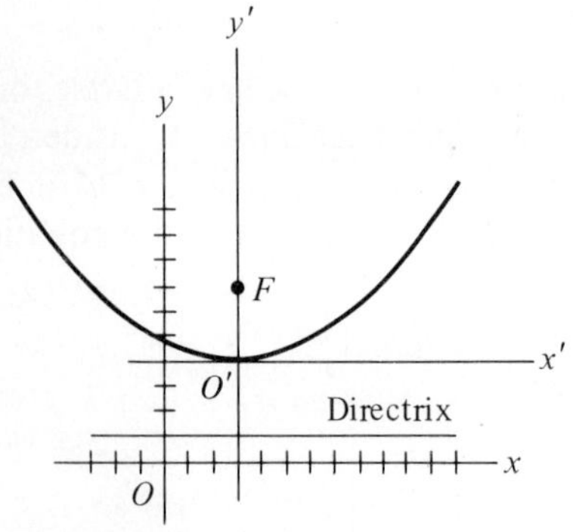

Equation (3) describes the parabola relative to the xy axes. It could be rewritten as

$$y = \frac{x^2 - 6x + 57}{12}. \blacksquare$$

Figure E.17

EXAMPLE 2 Graph the equation $x^2 - 2x + y^2 - 6y - 15 = 0$.

SOLUTION Complete the square in order to find axes relative to which the equation of the curve is simpler. The details run like this:

We are looking for h and k, the coordinates of the "new" origin.

$$x^2 - 2x + y^2 - 6y - 15 = 0 \tag{4}$$
$$(x^2 - 2x \quad) + (y^2 - 6y \quad) = 15$$
$$(x^2 - 2x + 1) + (y^2 - 6y + 9) = 15 + 1 + 9$$
$$(x - 1)^2 + (y - 3)^2 = 25. \tag{5}$$

Equation (5) suggests that we introduce $x'y'$ axes with origin at the point (1, 3). In this case $h = 1$, $k = 3$ and then $x' = x - 1$ and $y' = y - 3$. Thus on the graph x' and y' satisfy the equation

$$(x')^2 + (y')^2 = 25. \tag{6}$$

E.2 TRANSLATION OF AXES AND THE GRAPH OF $Ax^2 + Cy^2 + Dx + Ey + F = 0$

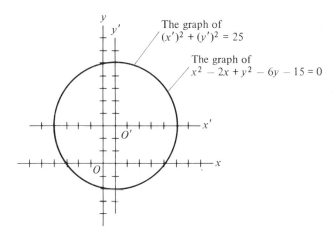

Figure E.18

The graph of (6) is a circle of radius 5 centered at the origin of the $x'y'$-coordinate system. This is also the graph of (4). It is shown in Fig. E.18. ∎

THE GRAPH OF $Ax^2 + Cy^2 + Dx + Ey + F = 0$

The method of completing the square illustrated by Example 2 works equally well for any equation which has the general form $Ax^2 + Cy^2 + Dx + Ey + F = 0$. It can be shown that the graph is a conic section (perhaps empty or one or two intersecting lines) or a pair of parallel lines.

EXAMPLE 3 Show that the graph of $x^2 + 4y^2 - 6x + 8y + 9 = 0$ is a conic section and graph it.

SOLUTION Begin by completing the square, much as in Example 2:

$$x^2 + 4y^2 - 6x + 8y + 9 = 0 \tag{7}$$

$$(x^2 - 6x \quad) + (4y^2 + 8y \quad) = -9$$

$$(x^2 - 6x \quad) + 4(y^2 + 2y \quad) = -9$$

$$(x^2 - 6x + 9) + 4(y^2 + 2y + 1) = -9 + 9 + 4$$

$$(x - 3)^2 + 4(y + 1)^2 = 4$$

$$\frac{(x - 3)^2}{4} + \frac{(y + 1)^2}{1} = 1. \tag{8}$$

Equation (8) suggests that we choose $x'y'$ axes with origin at $(3, -1)$. The graph of (7) or (8) then has the equation

$$\frac{(x')^2}{4} + \frac{(y')^2}{1} = 1 \tag{9}$$

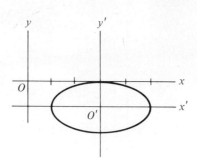

Ellipse $\dfrac{(x')^2}{4} + \dfrac{(y')^2}{1} = 1$

or $\dfrac{(x-3)^2}{4} + \dfrac{(y+1)^2}{1} = 1$

Figure E.19

The foci are on the x' axis since $4 > 1$. relative to the $x'y'$ axes. Equation (9) describes an ellipse in standard position with $a^2 = 4$ and $b^2 = 1$. It is graphed in Fig. E.19. ∎

Exercises

Using a suitable translation of axes, graph the equations in Exercises 1 to 14 relative to the xy axes.

1. $y = (x - 2)^2$
2. $y - 1 = 4(x - 1)^2$
3. $y + 1 = 2(x - 3)^2$
4. $y = x^2 + 6x + 9$
5. $y = 3x^2 + 12x + 13$
6. $y = 2x^2 - 12x + 20$
7. $x^2 + y^2 - 2x - 4y + 4 = 0$
8. $x^2 + y^2 + 6x - 7 = 0$
9. $x^2 - y^2 - 4x + 4y - 1 = 0$
10. $9x^2 - 4y^2 - 18x - 27 = 0$
11. $-4x^2 + 9y^2 + 24x + 36y - 36 = 0$
12. $4x^2 + y^2 - 16x + 12 = 0$
13. $25x^2 + 4y^2 + 100x + 24y + 36 = 0$
14. $x^2 + 4y^2 - 2x - 16y + 21 = 0$

∎

15. This exercise concerns the graph of $Ax^2 + Cy^2 + F = 0$ for nonzero constants A, C, and F.

    (a) Show that, if A and C are positive and F is negative, then the graph is an ellipse.
    (b) Show that, if A, C, and F are positive, then the graph is empty.
    (c) Show that, if A and C have opposite signs, then the graph is a hyperbola.
    (d) When is the graph a circle?

16. This exercise concerns the graph of the equation $Ax^2 + Cy^2 + Dx + Ey + F = 0$, where A, C, and F are nonzero constants and D and E are constants which may be 0.

    (a) Show that if A and C have the same signs, then the graph is an ellipse or else is empty.
    (b) Show that, if A and C have opposite signs, then the graph is a hyperbola.
    (c) When is the graph a circle?

17. Show that, if A is 0 and C and D are not 0, then the graph of $Ax^2 + Cy^2 + Dx + Ey + F = 0$ is a parabola.

18. Show that the graph of $y = ax^2 + bx + c$, $a \neq 0$, is a parabola by finding the equation of the graph relative to $x'y'$ axes whose origin is at the point $(-b/2a, c - b^2/4a)$.

E.3 Rotation of axes and the graph of $Ax^2 + Bxy + Cy^2 + Dx + Ey + F = 0$

This section examines the graph of $Ax^2 + Bxy + Cy^2 + Dx + Ey + F = 0$, which is a conic (including degenerate cases when it is a line, a pair of lines, a point, or empty). The key idea is to remove the term Bxy by choosing $x'y'$ axes that are tilted with respect to the xy axes.

THE EFFECT OF AXIS ROTATION

A point P has coordinates (x, y) relative to a given choice of axes. Another pair of axes is chosen with the same origin but rotated by an angle θ. Call

E.3 ROTATION OF AXES AND THE GRAPH OF $Ax^2 + Bxy + Cy^2 + Dx + Ey + F = 0$

these tilted axes the $x'y'$ system. How are x' and y', the coordinates of P in the $x'y'$ system, related to x and y? (See Fig. E.20.)

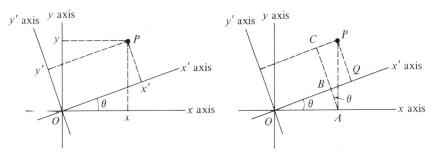

Figure E.20 **Figure E.21**

To determine the relation, introduce the line shown in Fig. E.21. By Fig. E.21,

$$x' = \overline{OB} + \overline{BQ} = \overline{OB} + \overline{CP} = \overline{OA}\cos\theta + \overline{AP}\sin\theta = x\cos\theta + y\sin\theta$$

and $\quad y' = \overline{AC} - \overline{AB} = \overline{AP}\cos\theta - \overline{OA}\sin\theta = y\cos\theta - x\sin\theta.$

Thus
$$\begin{cases} x' = x\cos\theta + y\sin\theta \\ y' = -x\sin\theta + y\cos\theta. \end{cases} \quad (1)$$

Just as the $x'y'$ axes are obtained from the xy axes through a rotation by the angle θ, the xy axes are obtained from the $x'y'$ axes by the rotation by the angle $-\theta$. In view of (1), then,

$$\begin{cases} x = x'\cos(-\theta) + y'\sin(-\theta) \\ y = -x'\sin(-\theta) + y'\cos(-\theta), \end{cases}$$

which reduces to

These are the key equations of this section.
$$\begin{cases} x = x'\cos\theta - y'\sin\theta \\ y = x'\sin\theta + y'\cos\theta. \end{cases} \quad (2)$$

EXAMPLE 1 Find an equation of the graph of $xy = 1$ relative to the $x'y'$ axes obtained by rotating the xy axes $45°$ ($= \pi/4$ radians).

The equation $xy = 1$ was graphed in Sec. 1.1.

SOLUTION Use (2) with $\theta = \pi/4$ radians. We have

$$\begin{cases} x = x'\cos\dfrac{\pi}{4} - y'\sin\dfrac{\pi}{4} \\[6pt] y = x'\sin\dfrac{\pi}{4} + y'\cos\dfrac{\pi}{4} \end{cases}$$

or
$$\begin{cases} x = \dfrac{\sqrt{2}}{2}(x' - y') \\ y = \dfrac{\sqrt{2}}{2}(x' + y'). \end{cases} \quad (3)$$

Substitution of (3) into the equation $xy = 1$ yields

$$\frac{\sqrt{2}}{2}(x' - y')\frac{\sqrt{2}}{2}(x' + y') = 1,$$

$$\frac{1}{2}[(x')^2 - (y')^2] = 1, \quad \text{multiplying out}$$

$$(x')^2 - (y')^2 = 2, \quad \text{clearing denominator} \quad (4)$$

which is the equation of a hyperbola with foci on the x' axis.

To graph (4), rewrite it as

$$\frac{(x')^2}{(\sqrt{2})^2} - \frac{(y')^2}{(\sqrt{2})^2} = 1. \quad (5)$$

This is the hyperbola $(x')^2/a^2 - (y')^2/b^2 = 1$ with $a = \sqrt{2}$ and $b = \sqrt{2}$. Its foci are at $(x', y') = (c, 0)$ and $(-c, 0)$ where $c = \sqrt{a^2 + b^2} = \sqrt{2 + 2} = 2$. The hyperbola is sketched in Fig. E.22. In the xy system, the foci are at $(\sqrt{2}, \sqrt{2})$ and $(-\sqrt{2}, -\sqrt{2})$. ∎

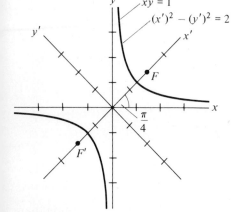

Figure E.22

WE CAN ALWAYS GET RID OF THE xy TERM

In Example 1 rotation of the axes by $\pi/4$ radians resulted in an equation with no $x'y'$ term. (In other words, the coefficient of the product $x'y'$ is 0 in (4).) Less precisely, we "got rid of the xy term." It turns out that, for any equation of the form $Ax^2 + Bxy + Cy^2 + Dx + Ey + F = 0$, it is always possible to rotate the axes in such a way that the $x'y'$ term disappears.

We sketch the algebra, leaving the details to be filled in by the reader. Start with the equation

We will worry later about which θ to use.

$$Ax^2 + Bxy + Cy^2 + Dx + Ey + F = 0 \quad (6)$$

and make the substitution (2). After some multiplications and collections, we obtain an equation in x' and y' of the form

$$A'(x')^2 + B'x'y' + C'(y')^2 + D'x' + E'y' + F' = 0, \quad (7)$$

for certain constants A', B', C', D', E', and F'.

At this point only the coefficient of $x'y'$, namely B', draws our interest since we want to find θ that makes B' equal to 0. The formula for B' is

$$B' = 2(C - A)\sin\theta\cos\theta + B(\cos^2\theta - \sin^2\theta), \quad (8)$$

which simplifies to

$$B' = (C - A)\sin 2\theta + B\cos 2\theta$$

E.3 ROTATION OF AXES AND THE GRAPH OF $Ax^2 + Bxy + Cy^2 + Dx + Ey + F = 0$

To make $B' = 0$, find θ such that

$$(C - A)\sin 2\theta + B \cos 2\theta = 0;$$

thus

$$\frac{\sin 2\theta}{\cos 2\theta} = \frac{B}{A - C} \quad \text{or} \quad \tan 2\theta = \frac{B}{A - C}. \tag{9}$$

For θ in $(0, \pi/4)$, $\tan 2\theta$ sweeps through all positive numbers. For θ in $(\pi/4, \pi/2)$, $\tan 2\theta$ sweeps through all negative numbers. If $C = A$, use use $\theta = \pi/4$; by (8), B' then is 0. Thus it is always possible to find a first quadrant angle θ such that (9) holds. For that θ, B' is 0 and (7) takes the form

$$A'(x')^2 + C'(y')^2 + D'x' + E'y' + F' = 0. \tag{10}$$

This type of equation was discussed in the preceding section. Its graph is a conic section except in certain degenerate cases. Thus the same holds for the graph of (6).

GETTING RID OF THE xy TERM IN A SPECIFIC EXAMPLE

We have seen that there is an angle θ such that, relative to $x'y'$ axes inclined at an angle θ to the xy axes, Eq. (7) takes the form of (10). First find θ such that $\tan 2\theta = B/(A - C)$. Then find $\cos \theta$ and $\sin \theta$ and use the substitution (2). The steps are:

1 Find $\tan 2\theta$.
2 Find $\cos 2\theta$.

Since θ is in the first quadrant, the positive square roots are used.

3 Find $\sin \theta = \sqrt{\dfrac{1 - \cos 2\theta}{2}}$ and $\cos \theta = \sqrt{\dfrac{1 + \cos 2\theta}{2}}$.

Example 2 illustrates the technique.

EXAMPLE 2 Choose $x'y'$ axes so that the equation for the graph of

$$-7x^2 + 48xy + 7y^2 - 25 = 0 \tag{11}$$

has no $x'y'$ term. Then graph the curve.

SOLUTION In this case $A = -7$, $B = 48$, $C = 7$. Hence θ must be chosen so that

$$\tan 2\theta = \frac{48}{(-7) - 7} = \frac{48}{-14} = \frac{24}{-7}. \tag{12}$$

Figure E.23 shows 2θ and θ. By the Pythagorean theorem, $\overline{OB}$ in Fig. E.23 is $\sqrt{(24)^2 + (-7)^2} = 25$. Thus $\cos 2\theta = (-7)/25$. Consequently

$$\sin \theta = \sqrt{\frac{1 - \left(-\dfrac{7}{25}\right)}{2}} = \frac{4}{5}$$

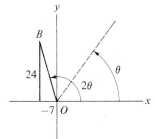

Figure E.23

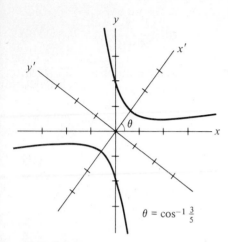

Figure E.24

A shortcut for telling what type of conic

and
$$\cos\theta = \sqrt{\frac{1 + \left(\frac{-7}{25}\right)}{2}} = \frac{3}{5}.$$

Thus (2) becomes
$$\begin{cases} x = \tfrac{3}{5}x' - \tfrac{4}{5}y' \\ y = \tfrac{4}{5}x' + \tfrac{3}{5}y'. \end{cases} \tag{13}$$

Substitution of (13) into (11) gives
$$-7(\tfrac{3}{5}x' - \tfrac{4}{5}y')^2 + 48(\tfrac{3}{5}x' - \tfrac{4}{5}y')(\tfrac{4}{5}x' + \tfrac{3}{5}y') + 7(\tfrac{4}{5}x' + \tfrac{3}{5}y')^2 - 25 = 0,$$
which simplifies to
$$625(x')^2 - 625(y')^2 = 625$$
or
$$(x')^2 - (y')^2 = 1,$$
which describes a hyperbola in standard position relative to the $x'y'$ axes. Its foci are on the x' axis. It is sketched in Fig. E.24. ∎

The reasoning in this section and the one preceding shows that if the graph of (6) is not degenerate, it is a conic. To determine what type of conic it is, compute the *discriminant*, $\mathscr{D} = B^2 - 4AC$. If it is negative, the conic is an ellipse; if it is positive, the conic is a hyperbola; if it is zero, the conic is a parabola. Remember this with the nonsense word *neph* (*n*egative, *e*llipse, *p*ositive, *h*yperbola). Exercises 14 to 16 explain why this test works. For emphasis, we state this test formally: To determine what type of conic the equation

$$Ax^2 + Bxy + Cy^2 + Dx + Ey + F = 0$$

describes, compute $\mathscr{D} = B^2 - 4AC$. When $\mathscr{D} < 0$, the conic is an ellipse; when $\mathscr{D} > 0$, the conic is a hyperbola. If $\mathscr{D} = 0$, the conic is a parabola.

Exercises

Graph each of the curves in Exercises 1 to 8, using the technique described in this section.

1. $x^2 - 4xy - 2y^2 - 6 = 0$
2. $41x^2 + 24xy + 34y^2 - 25 = 0$
3. $x^2 + xy + y^2 - 12 = 0$
4. $5x^2 + 6xy + 5y^2 - 8 = 0$
5. $23x^2 + 26\sqrt{3}\,xy - 3y^2 - 144 = 0$
6. $3x^2 + 2\sqrt{3}xy + y^2 + 2x + 2\sqrt{3}y = 0$
7. $6x^2 - 12xy + 6y^2 - \sqrt{2}x + \sqrt{2}y = 0$
8. $7x^2 - 48xy - 7y^2 - 25 = 0$

In each of Exercises 9 to 12 use the discriminant $B^2 - 4AC$ to determine what type of conic the equation describes.

9. $-x^2 + 24xy + 6y^2 - 10x + 13y + 5 = 0$
10. $x^2 + xy + y^2 + 3x + 2y = 0$
11. $x^2 - 2xy + y^2 + x + 3y + 5 = 0$
12. $3x^2 - xy - y^2 = 1$
13. For each of the following equations the graph is degenerate; it is either one or two lines, a point, or empty. Graph each of them.
    (a) $x^2 - y^2 = 0$ (b) $x^2 + 2xy + y^2 = 0$
    (c) $3x^2 + 4y^2 = 0$ (d) $3x^2 + 2xy + 3y^2 + 1 = 0$.

Exercises 14 to 16 are related.

14. Show that A', B', and C' in (7) are given by
    $A' = A\cos^2\theta + B\cos\theta\sin\theta + C\sin^2\theta$,
    $B' = 2(C - A)\sin\theta\cos\theta + B(\cos^2\theta - \sin^2\theta)$, and
    $C' = A\sin^2\theta - B\cos\theta\sin\theta + C\cos^2\theta$.
15. Use Exercise 14 to show that $(B')^2 - 4A'C' = B^2 - 4AC$.
16. Assume that $x'y'$ axes have been chosen to make $B' = 0$. Thus the graph of (6) is given by $A'(x')^2 + C'(y')^2 + D'x' + E'y' + F' = 0$. Assume this graph is a conic (that is, is not degenerate).

(a) Show that, if $(B')^2 - 4A'C'$ is negative, the graph is an ellipse. (Keep in mind that $B' = 0$.)
(b) Show that, if $(B')^2 - 4A'C'$ is positive, the graph is a hyperbola.
(c) Show that, if $(B')^2 - 4A'C'$ is 0, the graph is a parabola.
(d) Combining these facts with Exercise 15, explain the *neph* memory device.

17 Show that, if the graph of $Ax^2 + Bxy + Cy^2 + Dx + Ey + F = 0$ is a circle, then $B = 0$ and $A = C$.

E.4 Conic sections in polar coordinates

This section depends on Sec. 9.1. It is used in Sec. 14.8.

For the study of the conic sections in terms of polar coordinates, it is convenient to use definitions that depend on the ratios of distances, rather than on their sums or differences. (Note that the definition of the parabola involves essentially the ratio of two distances being equal to 1.)

A DIFFERENT APPROACH TO CONICS

Consider the ellipse whose foci are at $F = (c, 0)$ and $F' = (-c, 0)$ and whose "length of string" is $2a$. In the algebraic treatment of the ellipse in Sec. E.1, the equation

$$a^2 - cx = a\sqrt{(x-c)^2 + y^2}$$

appears.

Some algebraic manipulations of this equation will show that this ellipse can be defined in terms of the focus F, a line, and a fixed ratio of the distances from P to F and from P to the line.

First observe that this equation asserts that, for $P = (x, y)$,

$$a^2 - cx = a\overline{PF}$$

or, equivalently,

$$\overline{PF} = a - \frac{c}{a}x.$$

Eccentricity of an ellipse

e is not to be confused with the base of natural logarithms.

Denote the quotient c/a, which is less than 1, by e. The number e is called the *eccentricity* of the ellipse. (When $e = 0$, the ellipse is a circle.) Thus

$$\overline{PF} = a - ex = e\left(\frac{a}{e} - x\right),$$

an equation which is meaningful if the ellipse is not a circle. Now, $(a/e) - x$ is the distance from P to the vertical line through $(a/e, 0)$, a line which will be denoted L. Letting $Q = (a/e, y)$, the point on L and on the horizontal line through P, we have

$$\overline{PF} = e\overline{PQ}.$$

(See Fig. E.25.)

APPENDIX E CONIC SECTIONS

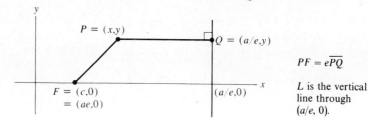

Figure E.25

In other words, *the ratio $\overline{PF}\ \overline{PQ}$ has a constant value*, less than 1. Thus the ellipse, like the parabola, can be defined in terms of a point F and a line L.

EXAMPLE 1 Find the eccentricity and draw the line L for the ellipse

$$\frac{x^2}{25} + \frac{y^2}{9} = 1.$$

SOLUTION In this ellipse $a = 5$ and $b = 3$. Thus $c = \sqrt{a^2 - b^2} = \sqrt{25 - 9} = 4$. Consequently

$$e = \frac{c}{a} = \frac{4}{5}.$$

The line L has the equation $\quad x = \dfrac{a}{e}$

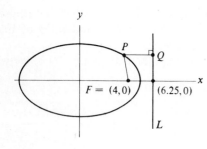

or $\quad x = \dfrac{5}{\frac{4}{5}};$

hence $\quad x = \dfrac{25}{4} = 6.25.$

The ellipse is sketched in Fig. E.26. Note that for each point P on the ellipse

$$\frac{\overline{PF}}{\overline{PQ}} = \frac{4}{5}. \quad \blacksquare$$

Figure E.26

The hyperbola in polar coordinates

The hyperbola can be treated in a similar manner. The main difference is that the eccentricity of a hyperbola, again defined as c/a, is greater than 1. With this background, we now describe the approach to the conic sections in terms of the ratios of certain distances.

DEFINITION *Conic section.* Let L be a line in the plane, and let F be a point in the plane but not on the line. Let e be a positive number. A point P in the plane is on the conic section determined by F, L, and e if and only if

$$\frac{\text{Distance from } P \text{ to } F}{\text{Distance from } P \text{ to } L} = e.$$

E.4 CONIC SECTIONS IN POLAR COORDINATES

Note that each of the three types of conics has a directrix.

When $e = 1$, the conic section is a parabola (this is the definition of the parabola used in the preceding section). When $e < 1$, it is an ellipse. When $e > 1$, it is a hyperbola. The point F is called a *focus*; the line L is called the *directrix*.

CONICS IN POLAR COORDINATES

To obtain the simplest description of the conic sections in polar coordinates, place the pole at the focus F. Let the polar axis make an angle B with a line perpendicular to the directrix. Figure E.27 shows a typical point $P = (r, \theta)$ on the conic section, as well as the point Q on the directrix nearest P.

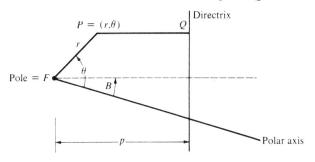

Figure E.27

Let the distance from F to the directrix be p. Then $\overline{PF}/\overline{PQ} = e$. But $\overline{PF} = r$ and $\overline{PQ} = p - r \cos(\theta - B)$. Thus

$$\frac{r}{p - r \cos(\theta - B)} = e. \tag{1}$$

Solving (1) for r yields *the equation of a conic section in polar coordinates*,

$$r = \frac{ep}{1 + e \cos(\theta - B)}. \tag{2}$$

Usually B is chosen to be 0 or π.

EXAMPLE 2 Show that the graph of the equation

$$r = \frac{8}{5 + 6 \cos \theta}$$

is a conic section.

SOLUTION This can be put in the form (2) by dividing both numerator and denominator by 5:

$$r = \frac{\frac{8}{5}}{1 + \frac{6}{5} \cos \theta} = \frac{(\frac{6}{5})(\frac{8}{6})}{1 + \frac{6}{5} \cos \theta}.$$

Hence the graph is a conic section for which $p = \frac{8}{6}$ and $e = \frac{6}{5}$. It is a hyperbola, since $e > 1$. ∎

Exercises

1. (a) Sketch the four points on the graph of $r = 10/(3 + 2 \cos \theta)$ corresponding to $\theta = 0, \pi/2, \pi, 3\pi/2$.
   (b) Using (2), show that the curve in (a) is an ellipse.
2. Obtain (2) from (1).
3. Find the eccentricities of these conics:
   (a) $r = 5/(3 + 4 \cos \theta)$;   (b) $r = 5/(4 + 3 \cos \theta)$;
   (c) $r = 5/(3 + 3 \cos \theta)$;   (d) $r = 5/(3 - 4 \cos \theta)$.
4. (a) Show that $r = 8/(1 - \frac{1}{2} \cos \theta)$ is the equation of an ellipse. *Hint:* Set $B = \pi$ in (2).
   (b) Graph the ellipse and its foci.
   (c) Find a, where $2a$ is the fixed sum of the distances from points on the ellipse to the foci.
5. In rectangular coordinates the focus of a certain parabola is at $(-1, 0)$ and its directrix is the line $x = 1$.
   (a) Show that its equation is $y^2 = -4x$.
   (b) Find the equation of the parabola relative to a polar coordinate system whose pole is at F and whose polar axis contains the positive x axis.
   (c) Find the equation of the parabola relative to a polar coordinate system whose pole is at the origin of the rectangular coordinate system and whose polar axis coincides with the positive x axis.
6. Assume that an ellipse has the equation $x^2/a^2 + y^2/b^2 = 1$ relative to a rectangular coordinate system. Place a polar coordinate system so that the polar axis coincides with the positive x axis (the pole thus being at the center of the ellipse, *not at a focus*). Show that the polar equation of the ellipse is (the relatively complicated)

$$r^2 = \frac{a^2 b^2}{b^2 \cos^2 \theta + a^2 \sin^2 \theta}.$$

APPENDIX F
THEORY OF LIMITS

In Sec. 2.4 a precise definition of the limit of a function $f(x)$ was given. This definition is the basis of this appendix, where the fundamental properties of limits will be obtained. In particular it will be proved that any polynomial is continuous.

F.1 Proofs of some theorems about limits

With the aid of the ε-δ definition of limit in Sec. 2.4 it can be established that, if

$$\lim_{x \to a} f(x) = A \quad \text{and} \quad \lim_{x \to a} g(x) = B,$$

then

$$\lim_{x \to a} [f(x) + g(x)] = A + B, \quad \lim_{x \to a} [f(x)g(x)] = AB,$$

and

$$\lim_{x \to a} \frac{f(x)}{g(x)} = \frac{A}{B} \quad \text{(if } B \text{ is not 0)}.$$

The first two assertions will be proved in this section.

THE LIMIT OF $f(x) + g(x)$

THEOREM 1 If $\lim_{x \to a} f(x) = A$ and $\lim_{x \to a} g(x) = B$, then $\lim_{x \to a} [f(x) + g(x)] = A + B$.

PROOF It must be shown that given any $\varepsilon > 0$, no matter how small, there exists a number $\delta > 0$ depending on ε, such that

$$|[f(x) + g(x)] - (A + B)| < \varepsilon \qquad (1)$$

whenever $|x - a| < \delta$ and $x \neq a$.

Rewrite $[f(x) + g(x)] - (A + B)$ as $[f(x) - A] + [g(x) - B]$, which is a sum of two quantities known to be small when x is near a. Since the absolute value of the sum of two numbers is not larger than the sum of their absolute values,

$$|[f(x) - A] + [g(x) - B]| \leq |f(x) - A| + |g(x) - B|. \qquad (2)$$

Since $\lim_{x \to a} f(x) = A$, there is a positive number δ_1 such that

$$|f(x) - A| < \frac{\varepsilon}{2}$$

when $|x - a| < \delta_1$ and $x \neq a$. (Why we pick $\varepsilon/2$ rather than ε will be clear in a moment.) Similarly there is a positive number δ_2 such that

$$|g(x) - B| < \frac{\varepsilon}{2}$$

when $|x - a| < \delta_2$ and $x \neq a$.

Now let δ be the smaller of δ_1 and δ_2. For any x (not equal to a) such that $|x - a| < \delta$ we have both

$$|x - a| < \delta_1 \quad \text{and} \quad |x - a| < \delta_2$$

and therefore, simultaneously,

$$|f(x) - A| < \frac{\varepsilon}{2} \quad \text{and} \quad |g(x) - B| < \frac{\varepsilon}{2}. \qquad (3)$$

Combining (2) and (3), we conclude that, when $|x - a| < \delta$ but $x \neq a$,

$$|[f(x) + g(x)] - (A + B)| < \frac{\varepsilon}{2} + \frac{\varepsilon}{2} = \varepsilon.$$

Thus for each $\varepsilon > 0$ there exists a suitable $\delta > 0$, depending, of course, on ε, f, and g. This ends the proof. ∎

Theorem 1 is the basis of the proof of the next theorem.

THEOREM 2 The sum of two functions that are continuous at a is itself continuous at a.

PROOF Let f and g be continuous at a. Let h be their sum; that is, let $h(x) = f(x) + g(x)$. We wish to show that h is continuous at a.

In view of the definition of continuity given in Sec. 2.6, it must be shown that $h(a)$ is defined and that $\lim_{x \to a} h(x) = h(a)$.

Since f and g are defined at a, so is h, and $h(a) = f(a) + g(a)$. All that remains is to show that $\lim_{x \to a} [f(x) + g(x)] = f(a) + g(a)$. But Theorem 1 assures us that $\lim_{x \to a} [f(x) + g(x)] = \lim_{x \to a} f(x) + \lim_{x \to a} g(x)$. Since f and g are continuous at a, $\lim_{x \to a} f(x) = f(a)$ and $\lim_{x \to a} g(x) = g(a)$. This concludes the proof. ∎

THE LIMIT OF $f(x)g(x)$

THEOREM 3 If $\lim_{x \to a} f(x) = A$ and $\lim_{x \to a} g(x) = B$, then $\lim_{x \to a} f(x)g(x) = AB$.

Plan of proof: We know that $|f(x) - A|$ and $|g(x) - B|$ are small when x is near a. We wish to conclude that $|f(x)g(x) - AB|$ is small when x is near a. The algebraic identity

$$f(x)g(x) - AB = f(x)[g(x) - B] + B[f(x) - A] \tag{4}$$

will be of use. From (4) and properties of the absolute value, it follows that

$$|f(x)g(x) - AB| \le |f(x)||g(x) - B| + |B||f(x) - A|. \tag{5}$$

Now $|B|$ is fixed, and $|f(x) - A|$ and $|g(x) - B|$ are small when x is near a. The real problem is to control $|f(x)|$. Watch carefully the way in which $|f(x)|$ is treated in the proof.

PROOF Consider the case for $B \ne 0$. Let $\varepsilon > 0$ be given. We wish to show that there is a number $\delta > 0$ such that $|f(x)g(x) - AB| < \varepsilon$ when $|x - a| < \delta$ but $x \ne a$. Observe that

$$\begin{aligned} |f(x)g(x) - AB| &= |f(x)[g(x) - B] + B[f(x) - A]| \\ &\le |f(x)||g(x) - B| + |B||f(x) - A|. \end{aligned} \tag{6}$$

Since $\lim_{x \to a} f(x) = A$, there is a number $\delta_1 > 0$ such that

It will be clear in a moment why we want $|f(x) - A|$ to be less than $\varepsilon/(2|B|)$.

$$|f(x) - A| < \frac{\varepsilon}{2|B|},$$

when $|x - a| < \delta_1$ but $x \ne a$. Thus the second summand in (6) is less than

$$\frac{|B|\varepsilon}{2|B|} = \frac{\varepsilon}{2}.$$

For x such that $|x - a| < \delta_1$, $|f(x)|$ does not become arbitrarily large, since $|f(x) - A| < \varepsilon/(2|B|)$. Indeed for such x,

$$|f(x)| < |A| + \frac{\varepsilon}{2|B|}.$$

Letting $C = |A| + \varepsilon/(2|B|)$, we have $|f(x)| < C$ when $|x - a| < \delta_1$ but $x \ne a$. (This controls the size of $|f(x)|$.)

Since $\lim_{x \to a} g(x) = B$, there is a $\delta_2 > 0$ such that $|g(x) - B| < \varepsilon/(2C)$ when $|x - a| < \delta_2$ but $x \neq a$.

Now let δ be the smaller of δ_1 and δ_2. When $|x - a| < \delta$ and $x \neq a$, both $|x - a| < \delta_1$ and $|x - a| < \delta_2$ hold; hence $|f(x) - A| < \varepsilon/(2|B|)$, $|f(x)| < C$, and $|g(x) - B| < \varepsilon/(2C)$. Inspection of (6) then shows that for such x,

$$|f(x)g(x) - AB| < C\frac{\varepsilon}{2C} + |B|\frac{\varepsilon}{2|B|} = \frac{\varepsilon}{2} + \frac{\varepsilon}{2} = \varepsilon.$$

After reviewing the proof, do it yourself without looking here.

The proof is completed. (The case $B = 0$ is left to the reader as Exercise 7.) ∎

THEOREM 4 The product of two functions that are continuous at a is itself continuous at a. ∎

The proof is similar to that of Theorem 2, but depends on Theorem 3 instead of Theorem 1.

ANY POLYNOMIAL IS CONTINUOUS

The fact that the sum and product of continuous functions are continuous is the basis of the proof that any polynomial is continuous. It is necessary to prove first that the function $f(x) = x$ is continuous and then that any constant function is continuous.

THEOREM 5 The function f, such that $f(x) = x$, is continuous everywhere. So are the functions $x^2, x^3, x^4, \ldots$.

PROOF Since $f(a) = a$, it must be shown that $|f(x) - a|$ is small whenever $|x - a|$ is sufficiently small. More precisely, for $\varepsilon > 0$ we wish to exhibit $\delta > 0$ such that $|f(x) - a| < \varepsilon$ whenever $|x - a| < \delta$. But $f(x) = x$; hence $|f(x) - a|$ is simply $|x - a|$. Let $\delta = \varepsilon$. Thus when $|x - a| < \delta$, it follows that $|f(x) - a| < \varepsilon$. This shows that the function x is continuous.

Since the function x^2 is the product of the function x and the function x, Theorem 4 implies that x^2 is continuous. Similarly, x^3 is continuous, for $x^3 = x^2 x$. Similarly, x^4 and x^5 can be shown to be continuous. Mathematical induction establishes the continuity of x^n for all positive integers n. ∎

THEOREM 6 Any constant function is continuous everywhere.

PROOF Let $f(x) = c$ for all x. For any $\varepsilon > 0$, choose $\delta = 1776$, a perfectly fine positive number. Now $|f(x) - f(a)| = |c - c| = 0 < \varepsilon$ for any x, and hence for x such that $|x - a| < 1776$. Thus f is continuous at any number a. ∎

Any $\delta > 0$ will do.

THEOREM 7 Any polynomial is continuous everywhere.

PROOF We illustrate the idea of the proof by showing that $6x^2 - 5x + 1$ is continuous everywhere.

By Theorem 6, the constant functions $f(x) = 6$, $g(x) = -5$, and $h(x) = 1$

are continuous everywhere. By Theorem 5, the functions x and x^2 are continuous everywhere. By Theorem 4, the functions $6x^2$ and $(-5)x$ are continuous. By Theorem 2, the function $[6x^2 + (-5)x]$ is continuous. Again, by Theorem 2, the function $[6x^2 + (-5)x] + 1$ is continuous. Thus $6x^2 - 5x + 1$ is continuous. The same argument applies to any polynomial. ∎

Exercises

1. Prove that $f(x) = 1/x$ is continuous at any number $a \neq 0$.
2. Prove that, if g is continuous at a and f is continuous at $g(a)$, then the composite function $h = f \circ g$ is continuous at a.
3. Using Exercises 1 and 2, show that the function $1/(x^2 + 1)$ is continuous.
4. Using Exercises 1 and 2, show that, if $g(x)$ is continuous at a and $g(a) \neq 0$, then $1/g(x)$ is continuous at a.
5. Let $f(x)$ and $g(x)$ be continuous at a and $g(a) \neq 0$. Combining Exercise 4 with a result in this section, show that $f(x)/g(x)$ is continuous at a.
6. Let $\lim_{x \to a} g(x) = B \neq 0$. Show that $\lim_{x \to a} [1/g(x)] = 1/B$.
7. Prove Theorem 3 in the case $B = 0$.
8. What theorems assure us that the function $(x^3 - 1)/(x^2 - 5)$ is continuous throughout its domain?
9. Assume that f is continuous at a and $f(a)$ is positive. Prove that there is an open interval (b, c) including a such that $f(x)$ is positive for all x in that interval.

In the remaining exercises use the definitions from Sec. 2.4.

10. Prove that if $\lim_{x \to \infty} f(x) = \infty$ and $\lim_{x \to \infty} g(x) = \infty$, then $\lim_{x \to \infty} (f(x) + g(x)) = \infty$.
11. Prove that if $\lim_{x \to \infty} f(x) = \infty$ and $\lim g(x) = A$, then $\lim_{x \to \infty} (f(x) + g(x)) = \infty$.
12. Prove that if $\lim_{x \to \infty} f(x) = A$ and $\lim_{x \to \infty} g(x) = B$, then $\lim_{x \to \infty} (f(x) + g(x)) = A + B$.
13. Prove that if $\lim_{x \to \infty} f(x) = \infty$ and $\lim_{x \to \infty} g(x) = \infty$, then $\lim_{x \to \infty} f(x)g(x) = \infty$.
14. Prove that if $\lim_{x \to \infty} f(x) = \infty$ and $\lim_{x \to \infty} g(x) = A > 0$, then $\lim_{x \to \infty} f(x)g(x) = \infty$.
15. Prove that if $\lim_{x \to \infty} f(x) = A$ and $\lim_{x \to \infty} g(x) = B$, then $\lim_{x \to \infty} f(x)g(x) = AB$.

F.2 Definitions of other limits

Other formal definitions in the spirit of the ε-δ definition of Sec. 2.4 are given in the text. The formal definition of $\int_a^b f(x)\,dx$ is given at the end of Sec. 5.3 and a formal definition of the limit of a sequence at the end of Sec. 10.1. However, the formal definition of a limit of $f(x, y)$, a function defined in the plane, was not given. The definition, similar to that for the limit of $f(x)$, will be given now.

LIMIT OF $f(x, y)$ Let $P_0 = (a, b)$ be a point in the xy plane. Let $f(P) = f(x, y)$ be defined for all points $P = (x, y)$ in the plane sufficiently near P_0, except perhaps at P_0 itself. To be precise, let $d(P_0, P)$ denote the distance between P_0 and P. Assume that there is a positive number r such that f is defined for all points P satisfying the inequality

$$0 < d(P_0, P) < r.$$

See Fig. F.1.

Figure F.1

f is defined for all points in this disk except perhaps at P_0.

This assumption on f assures us that the domain of f includes points arbitrarily near P_0.

The definition of "$\lim_{P \to P_0} f(P) = L$" can now be given. It will say that, if P is close enough to P_0, then $f(P)$ is as close as we please to $f(P_0)$.

APPENDIX F THEORY OF LIMITS

DEFINITION

ε is the challenge, δ is the response.

($\lim_{P \to P_0} f(P) = L$.) Let $P_0 = (a, b)$ be a point in the xy plane and let L be a real number. (Assume that $f(P)$ is defined for all P such that $0 < d(P_0, P) < r$ for some positive real number r.) Assume that for each positive number ε there is a positive number δ such that, for all points P that satisfy the inequality

$$0 < d(P_0, P) < \delta,$$

it is true that
$$|f(P) - L| < \varepsilon.$$

Then we say that "the limit of $f(P)$ as P approaches P_0 is L" and write

$$\lim_{P \to P_0} f(P) = L$$

or
$$\lim_{(x, y) \to (a, b)} f(x, y) = L.$$

Figure F.2 shows the meaning of the challenge ε and the response δ.

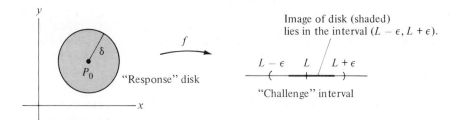

Figure F.2

No matter how small the challenge interval around L is, a disk around P_0 can be found such that, for any point P in the disk (except perhaps P_0), $f(P)$ lies within that interval.

Continuity of $f(x, y)$

Continuity of $f(x, y)$ at $P_0 = (a, b)$ is then defined in the same way as continuity of $f(x)$ at a. First of all, f must be defined at P_0 and throughout some disk centered at P_0. Then,

1 $\lim_{P \to P_0} f(P)$ exists, and
2 that limit is $f(P_0)$.

CONTINUITY OF A MAPPING

In Sec. 13.5 mappings F from the uv plane to the xy plane were discussed. Limits and continuity for such functions are defined very much like the corresponding concepts for $f(x)$ (a function from a line to a line) and $f(x, y)$ (a function from a plane to a line).

Rather than define the limit of such a function and then define continuity, let us go directly to continuity.

Let F be a mapping from the uv plane to the xy plane. Assume that F is defined at P_0 and throughout some disk centered at P_0. We shall say that F is *continuous* at P_0 if, for every $\varepsilon > 0$, there is a $\delta > 0$ such that if

$$d(P, P_0) < \delta,$$

then $\qquad d(F(P), F(P_0)) < \varepsilon.$

Geometrically, the definition asserts that no matter how small a disk is drawn around $F(P_0)$, it is possible to find a (response) disk around P_0 such that every point in the second disk is mapped by F into the first (challenge) disk. See Fig. F.3.

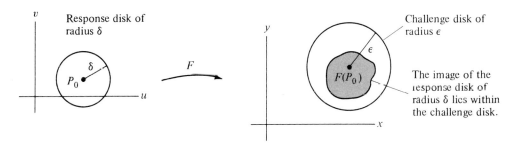

Figure F.3

It can be shown that, if $F(u, v) = (f(u, v), g(u, v))$, then the continuity of F is equivalent to the continuity of f and g.

Exercises

1 Prove that, if $f(x, y)$ is positive and continuous at $P_0 = (a, b)$, then there is a disk with center P_0 where f remains positive.

2 Prove that the mapping $F = (f, g)$ is continuous at P_0 if and only if both f and g are continuous.

APPENDIX G
THE INTERCHANGE OF LIMITS

This appendix provides proofs of two theorems whose validity was assumed in the text. In Sec. G.1 the equality of the mixed partial derivatives is examined. It can be read after Sec. 12.3. Section G.2 concerns the differentiation of $\int_a^b f(x, y)\, dx$ with respect to y. Section G.3, besides presenting an example of a function $f(x, y)$ whose two mixed partials are not equal, gives a gentle introduction to advanced calculus.

G.1 The equality of f_{xy} and f_{yx}

For most common functions $f(x, y)$,

$$\frac{\partial}{\partial y}\left(\frac{\partial f}{\partial x}\right) \quad \text{equals} \quad \frac{\partial}{\partial x}\left(\frac{\partial f}{\partial y}\right);$$

that is, the order in which we compute partial derivatives does not affect the result. This assertion is justified in the following theorem.

THEOREM Let f be a function defined on the xy plane. If f_{xy} and f_{yx} exist and are continuous at all points, then they are equal.

PROOF To keep the proof uncluttered, it will be shown that $f_{xy}(0, 0)$ equals $f_{yx}(0, 0)$. The identical argument holds for any point (a, b).

To begin, consider the definition of $f_{xy}(0, 0)$:

$$f_{xy}(0, 0) = \left.\frac{\partial(f_x)}{\partial y}\right|_{(0,0)} = \lim_{k \to 0} \frac{f_x(0, k) - f_x(0, 0)}{k}.$$

But, by definition of the partial derivative f_x,

$$f_x(0, k) = \lim_{h \to 0} \frac{f(h, k) - f(0, k)}{h} \quad \text{and} \quad f_x(0, 0) = \lim_{h \to 0} \frac{f(h, 0) - f(0, 0)}{h}.$$

Thus

$$f_{xy}(0, 0) = \lim_{k \to 0} \frac{f_x(0, k) - f_x(0, 0)}{k}$$

$$= \lim_{k \to 0} \frac{\lim_{h \to 0} \frac{f(h, k) - f(0, k)}{h} - \lim_{h \to 0} \frac{f(h, 0) - f(0, 0)}{h}}{k}$$

$$= \lim_{k \to 0} \left\{ \lim_{h \to 0} \frac{[f(h, k) - f(0, k)] - [f(h, 0) - f(0, 0)]}{hk} \right\}. \quad (1)$$

Let us focus our attention on the numerator in (1):

$$\text{Numerator} = [f(h, k) - f(0, k)] - [f(h, 0) - f(0, 0)]. \quad (2)$$

Note that the second bracketed expression is obtained from the first bracketed expression by replacing k by 0. Define, for fixed h, a function

$$u(y) = f(h, y) - f(0, y). \quad (3)$$

Then (2) takes the simple form

$$u(k) - u(0). \quad (4)$$

By the mean-value theorem (see Sec. 4.1),

$$u(k) - u(0) = u'(K)k \quad (5)$$

for some K between 0 and k. But by the definition of the function u, given in (3),

$$u'(K) = f_y(h, K) - f_y(0, K). \quad (6)$$

Thus, by the mean-value theorem, applied to the function $f_y(x, K)$, for fixed K,

$$u'(K) = f_{yx}(H, K)h \quad (7)$$

for some H between 0 and h.
Thus (2) becomes

$$\text{Numerator} = f_{yx}(H, K)hk \quad (8)$$

S-66 APPENDIX G THE INTERCHANGE OF LIMITS

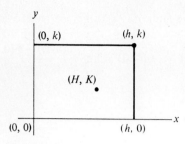

Figure G.1

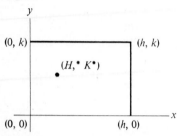

Figure G.2

for some point (H, K) in the rectangle with vertices $(0, 0)$, $(h, 0)$, (h, k), and $(0, k)$. (See Fig. G.1.) Substituting (8) in (1), we obtain

$$f_{xy}(0, 0) = \lim_{k \to 0} (\lim_{h \to 0} f_{yx}(H, K)).$$

Since H is between 0 and h and K is between 0 and k, it follows that H and K both approach 0 as $h \to 0$ and $k \to 0$. By the continuity of f_{yx}, $f_{yx}(H, K)$ approaches $f_{yx}(0, 0)$. Hence,

$$f_{xy}(0, 0) = f_{yx}(0, 0), \text{ as asserted.} \blacksquare$$

In fact, we have obtained a stronger result: the assumption that f_{xy} is continuous is not needed.

Example 3 in Sec. G.3 presents a function whose two mixed partial derivatives f_{xy} and f_{yx} are *not* equal at $(0, 0)$.

Exercises

1. Prove the theorem without looking at the text.
2. Prove the theorem at an arbitrary point (a, b).
3. Read Example 3 in Sec. G.3.

G.2 The derivative of $\int_a^b f(x, y)\, dx$ with respect to y

The integral

$$\int_a^b f(x, y)\, dx$$

depends on y. Let

$$F(y) = \int_a^b f(x, y)\, dx.$$

It makes sense to speak of the derivative of F with respect to y. It turns out that for most common functions f,

$$\frac{dF}{dy} = \int_a^b \frac{\partial f}{\partial y}\, dx.$$

That is,

$$\frac{d}{dy}\left[\int_a^b f(x, y)\, dx\right] = \int_a^b \frac{\partial f}{\partial y}\, dx.$$

Generally it is safe to differentiate the integral by differentiating the integrand.

The reader may check this assertion for the function $f(x, y) = x^3 + xy^2$ before going through the proof of the theorem.

THEOREM Let f be defined on the xy plane, and assume that f and f_y are continuous. Assume also that f_{yy} is defined on the xy plane and that it is bounded on each rectangle. (That is, if R is a rectangle, then there is a number M, depending on R, such that $|f_{yy}(x, y)| \leq M$ for all (x, y) in R.) Let F be defined by

$$F(y) = \int_a^b f(x, y)\, dx.$$

Then F is differentiable, and

$$\frac{dF}{dy} = \int_a^b \frac{\partial f}{\partial y}\, dx.$$

PROOF To show that
$$\lim_{h \to 0} \frac{F(y+h) - F(y)}{h} = \int_a^b \frac{\partial f}{\partial y}\, dx,$$

consider, for a fixed y, the difference

$$\frac{F(y+h) - F(y)}{h} - \int_a^b f_y(x, y)\, dx, \tag{1}$$

which, by the definition of F, equals

$$\frac{\int_a^b f(x, y+h)\, dx - \int_a^b f(x, y)\, dx}{h} - \int_a^b f_y(x, y)\, dx. \tag{2}$$

Now, (2) equals

$$\int_a^b \left[\frac{f(x, y+h) - f(x, y)}{h} - f_y(x, y) \right] dx. \tag{3}$$

To show that (1) approaches 0 as $h \to 0$, it suffices to show that the integrand in (3) is small when h is small. It may be assumed now that $|h| \leq 1$.

First of all, the expression

$$\frac{f(x, y+h) - f(x, y)}{h}$$

in the integrand in (3) equals

$$\frac{h f_y(x, y+H)}{h} = f_y(x, y+H)$$

for some number H between 0 and h, by the mean-value theorem. (H depends on x, y, and h.)

Thus the integrand in (3) equals

$$f_y(x, y + H) - f_y(x, y). \tag{4}$$

By the mean-value theorem, (4) equals

$$Hf_{yy}(x, y + H^*) \tag{5}$$

for some number H^* between 0 and H.

Since $|H^*| \leq |H| \leq |h| \leq 1$, the point $(x, y + H^*)$ lies somewhere in the rectangle whose vertices are

$$(a, y - 1), \quad (a, y + 1), \quad (b, y - 1), \quad (b, y + 1).$$

By assumption, $|f_{yy}| \leq M$ on this rectangle. By (4) and (5), the integrand in (3) has absolute value at most

$$|H|M \leq |h|M.$$

Thus the absolute value of (3) is at most

$$|h|M(b - a),$$

which approaches 0 as $h \to 0$, since M and $b - a$ are fixed numbers. This proves the theorem. ∎

The assumption made in the theorem that f_{yy} is bounded in each rectangle is satisfied if f_{yy} is continuous. So the theorem does cover the cases commonly encountered. In advanced calculus the theorem is proved without any assumption on f_{yy}. (See for instance, R. C. Buck, *Advanced Calculus*, 3d ed., p. 118, McGraw-Hill, New York, 1978.)

Exercises

1. Verify the theorem for $f(x, y) = x^3 y^4$.
2. Verify the theorem for $f(x, y) = \cos xy$.
3. For what value of y does the function

$$F(y) = \int_0^{\pi/2} (y - \cos x)^2 \, dx$$

   have a minimum? (Use the theorem of this section.)
4. Let $G(u, v, w) = \int_u^v f(w, x) \, dx$. Find
   (a) $\partial G/\partial v$ (b) $\partial G/\partial u$ (c) $\partial G/\partial w$.

5. Let $G(u) = \int_0^u f(u, x) \, dx$. Find dG/du.
6. Let $G(u, v) = \int_0^u e^{-vx^2} \, dx$. Find
   (a) $\partial G/\partial u$ (b) $\partial G/\partial v$.
7. Let $F(y) = \int_0^1 [(x^y - 1)/\ln x] \, dx$ for $y \geq 0$.
   (a) Assuming that one may differentiate F by differentiating under the integral sign, show that $dF/dy = 1/(1 + y)$.
   (b) From (a) deduce that $F(y) = \ln(1 + y) + C$.
   (c) Show that the constant C in part (b) is 0 by examining the case $y = 0$.

G.3 The interchange of limits

Though the two theorems in Secs. G.1 and G.2 are independent, they both illustrate a certain type of problem which students who go on to advanced calculus will study, namely, the interchange of limits. To see what this means, let us take a new look at the theorem in Sec. G.1, which concerns the equality of the mixed partials. By (1) in Sec. G.1,

$$f_{xy}(0, 0) = \lim_{k \to 0} \left[\lim_{h \to 0} \frac{f(h, k) - f(0, k) - f(h, 0) + f(0, 0)}{hk} \right].$$

Similarly, from the definition of f_{yx}, it can be shown that

$$f_{yx}(0, 0) = \lim_{h \to 0} \left[\lim_{k \to 0} \frac{f(h, k) - f(0, k) - f(h, 0) + f(0, 0)}{hk} \right].$$

Note that the two quotients are identical, but that f_{xy} involves

$$\lim_{k \to 0} \left(\lim_{h \to 0} \right),$$

while f_{yx} involves

$$\lim_{h \to 0} \left(\lim_{k \to 0} \right).$$

It is tempting to claim that the order of taking limits should not matter. But it does. (In Example 3 it is shown that f_{xy} does not always equal f_{yx}.) This instance raises the general question, "When can one interchange limits?" Example 1 presents a simple case in which the order of taking the limits *does* matter.

EXAMPLE 1 Let $f(x, y) = x^y$ for $x > 0$ and $y > 0$. Evaluate the two "repeated limits"

$$\lim_{y \to 0^+} \left(\lim_{x \to 0^+} x^y \right)$$

and

$$\lim_{x \to 0^+} \left(\lim_{y \to 0^+} x^y \right).$$

SOLUTION

$$\lim_{y \to 0^+} \left(\lim_{x \to 0^+} x^y \right) = \lim_{y \to 0^+} 0 = 0.$$

On the other hand,

$$\lim_{x \to 0^+} \left(\lim_{y \to 0^+} x^y \right) = \lim_{x \to 0^+} 1 = 1.$$

This shows that *the order of taking limits may affect the result.* Moreover, it suggests why the symbol 0^0 is not given any meaning ■

The next example also illustrates the effect of switching the order of taking limits. In Example 3 it becomes the basis of an illustration that shows the mixed partial derivatives f_{xy} and f_{yx} are not always equal.

EXAMPLE 2 Let

$$g(x, y) = \begin{cases} \dfrac{x^2 - y^2}{x^2 + y^2} & \text{if } (x, y) \neq (0, 0); \\ 0 & \text{if } (x, y) = (0, 0). \end{cases}$$

Show that $$\lim_{x \to 0} \left[\lim_{y \to 0} g(x, y) \right] \neq \lim_{y \to 0} \left[\lim_{x \to 0} g(x, y) \right].$$

SOLUTION
$$\lim_{y \to 0} g(x, y) = \lim_{y \to 0} \frac{x^2 - y^2}{x^2 + y^2}$$
$$= \frac{x^2}{x^2}$$
$$= 1.$$

Thus $$\lim_{x \to 0} \left[\lim_{y \to 0} g(x, y) \right] = \lim_{x \to 0} 1 = 1.$$

On the other hand, $$\lim_{x \to 0} g(x, y) = \lim_{x \to 0} \frac{x^2 - y^2}{x^2 + y^2}$$
$$= \frac{-y^2}{y^2}$$
$$= -1.$$

Thus $$\lim_{y \to 0} \left[\lim_{x \to 0} g(x, y) \right] = \lim_{y \to 0} (-1) = -1. \blacksquare$$

EXAMPLE 3 Let $f(x, y) = xy g(x, y)$, where g is given in Example 2. Show that

$$f_{xy}(0, 0) \neq f_{yx}(0, 0).$$

SOLUTION That f_x, f_y, f_{xy}, f_{yx} exist at all points is left for the reader to demonstrate. (See Exercise 8.) Note that $f(x, y) = 0$ whenever x or y is 0. By (1) in Sec. G.1,

$$f_{xy}(0, 0) = \lim_{k \to 0} \left\{ \lim_{h \to 0} \frac{[f(h, k) - f(0, k)] - [f(h, 0) - f(0, 0)]}{hk} \right\}$$
$$= \lim_{k \to 0} \left[\lim_{h \to 0} \frac{f(h, k)}{hk} \right]$$
$$= \lim_{k \to 0} \left[\lim_{h \to 0} g(h, k) \right] = -1$$

from Example 2. Similarly,

$$f_{yx}(0, 0) = \lim_{h \to 0} \left[\lim_{k \to 0} g(h, k) \right] = 1.$$

Therefore, $f_{xy}(0, 0) \neq f_{yx}(0, 0)$. $\blacksquare$

The theorem in Sec. G.2, which asserts that in general

$$\frac{d}{dy} \left[\int_a^b f(x, y) \, dx \right] = \int_a^b f_y(x, y) \, dx,$$

Exercises

1. Let $f(x, y) = 1$ if $y \geq x$ and let $f(x, y) = 0$ if $y < x$.
   (a) Shade in the part of the plane where $f(x, y) = 1$.
   (b) Show that $\lim_{x \to \infty} [\lim_{y \to \infty} f(x, y)] = 1$.
   (c) Show that $\lim_{y \to \infty} [\lim_{x \to \infty} f(x, y)] = 0$.

2. Show that $\lim_{x \to 0} [\lim_{n \to \infty} nx/(1 + nx)] = 1$, while $\lim_{n \to \infty} [\lim_{x \to 0} nx/(1 + nx)] = 0$. (Assume $x > 0$.)

3. Let $f_n(x) = nx/(1 + n^2 x^4)$. Show that

$$\int_0^\infty \lim_{n \to \infty} f_n(x)\, dx = 0,$$

but

$$\lim_{n \to \infty} \int_0^\infty f_n(x)\, dx = \frac{\pi}{4}.$$

4. Show that $\lim_{x \to 0} [\lim_{y \to 0} x^2/(x^2 + y^2)]$ is not equal to $\lim_{y \to 0} [\lim_{x \to 0} x^2/(x^2 + y^2)]$.

5. Let $f_n(x) = n\pi \sin(n\pi x)$ if $0 \leq x \leq 1/n$, and 0 otherwise.
   (a) Graph $f_1, f_2,$ and f_3.
   (b) Show that $\lim_{n \to \infty} \int_0^1 f_n(x)\, dx = 2$, but that $\int_0^1 \lim_{n \to \infty} f_n(x)\, dx = 0$.

6. Compare $\lim_{x \to \infty} \left(\lim_{y \to \infty} \frac{x^2}{x^2 + y^2 + 1} \right)$

   and $\lim_{y \to \infty} \left(\lim_{x \to \infty} \frac{x^2}{x^2 + y^2 + 1} \right)$.

7. Let $f_n(x) = (1/n) \sin nx$ for all x and all positive integers n.
   Show that $\lim_{n \to \infty} \left[\lim_{h \to 0} \frac{f_n(h) - f_n(0)}{h} \right] = 1$,

   while $\lim_{h \to 0} \left[\lim_{n \to \infty} \frac{f_n(h) - f_n(0)}{h} \right] = 0$.

8. Show that f_{xy} and f_{yx} exist at all points in the plane, where f is the pathological function in Example 3.

9. Show that

$$\int_0^\infty \left[\int_0^1 (2xy - x^2 y^2) e^{-xy}\, dx \right] dy = 1,$$

but

$$\int_0^1 \left[\int_0^\infty (2xy - x^2 y^2) e^{-xy}\, dy \right] dx = 0.$$

10. Show that l'Hôpital's rule in the zero-over-zero case concerns the equality of these two limits:

$$\lim_{\Delta t \to 0} \left[\lim_{t \to a} \frac{f(t + \Delta t) - f(t)}{g(t + \Delta t) - g(t)} \right]$$

and

$$\lim_{t \to a} \left[\lim_{\Delta t \to 0} \frac{f(t + \Delta t) - f(t)}{g(t + \Delta t) - g(t)} \right].$$

11. Let $f_n(x)$ be defined for each x in $[0, 1]$ and each positive integer n. Assume that f_n is continuous for each n and that $\lim_{n \to \infty} f_n(x)$ exists for each x in $[0, 1]$. Call this limit $f(x)$,

$$\lim_{n \to \infty} f_n(x) = f(x).$$

(a) Show that the statement, "f is continuous at a," is equivalent to the equation

$$\lim_{n \to \infty} \left[\lim_{x \to a} f_n(x) \right] = \lim_{x \to a} \left[\lim_{n \to \infty} f_n(x) \right].$$

(b) In particular, let $f_n(x) = x^n$. Show that in this case the two repeated limits in (a) are not necessarily equal.

APPENDIX H
TABLES

H.1 Exponential function

x	e^x	e^{-x}	x	e^x	e^{-x}	x	e^x	e^{-x}	x	e^x	e^{-x}
0.00	1.0000	1.00000	0.35	1.4191	.70469	0.70	2.0138	.49659	1.50	4.4817	.22313
0.01	1.0101	0.99005	0.36	1.4333	.69768	0.71	2.0340	.49164	1.60	4.9530	.20190
0.02	1.0202	.98020	0.37	1.4477	.69073	0.72	2.0544	.48675	1.70	5.4739	.18268
0.03	1.0305	.97045	0.38	1.4623	.68386	0.73	2.0751	.48191	1.80	6.0496	.16530
0.04	1.0408	.96079	0.39	1.4770	.67706	0.74	2.0959	.47711	1.90	6.6859	.14957
0.05	1.0513	.95123	0.40	1.4918	.67032	0.75	2.1170	.47237	2.00	7.3891	.13534
0.06	1.0618	.94176	0.41	1.5068	.66365	0.76	2.1383	.46767	3.00	20.086	.04979
0.07	1.0725	.93239	0.42	1.5220	.65705	0.77	2.1598	.46301	4.00	54.598	.01832
0.08	1.0833	.92312	0.43	1.5373	.65051	0.78	2.1815	.45841	5.00	148.41	.00674
0.09	1.0942	.91393	0.44	1.5527	.64404	0.79	2.2034	.45384	6.00	403.43	.00248
0.10	1.1052	.90484	0.45	1.5683	.63763	0.80	2.2255	.44933	7.00	1096.6	.00091
0.11	1.1163	.89583	0.46	1.5841	.63128	0.81	2.2479	.44486	8.00	2981.0	.00034
0.12	1.1275	.88692	0.47	1.6000	.62500	0.82	2.2705	.44043	9.00	8103.1	.00012
0.13	1.1388	.87810	0.48	1.6161	.61878	0.83	2.2933	.43605	10.00	22026.5	.00005
0.14	1.1503	.86936	0.49	1.6323	.61263	0.84	2.3164	.43171			
0.15	1.1618	.86071	0.50	1.6487	.60653	0.85	2.3396	.42741			
0.16	1.1735	.85214	0.51	1.6653	.60050	0.86	2.3632	.42316			
0.17	1.1853	.84366	0.52	1.6820	.59452	0.87	2.3869	.41895			
0.18	1.1972	.83527	0.53	1.6989	.58860	0.88	2.4109	.41478			
0.19	1.2092	.82696	0.54	1.7160	.58275	0.89	2.4351	.41066			
0.20	1.2214	.81873	0.55	1.7333	.57695	0.90	2.4596	.40657			
0.21	1.2337	.81058	0.56	1.7507	.57121	0.91	2.4843	.40252			
0.22	1.2461	.80252	0.57	1.7683	.56553	0.92	2.5093	.39852			
0.23	1.2586	.79453	0.58	1.7860	.55990	0.93	2.5345	.39455			
0.24	1.2712	.78663	0.59	1.8040	.55433	0.94	2.5600	.39063			
0.25	1.2840	.77880	0.60	1.8221	.54881	0.95	2.5857	.38674			
0.26	1.2969	.77105	0.61	1.8404	.54335	0.96	2.6117	.38289			
0.27	1.3100	.76338	0.62	1.8589	.53794	0.97	2.6379	.37908			
0.28	1.3231	.75578	0.63	1.8776	.53259	0.98	2.6645	.37531			
0.29	1.3364	.74826	0.64	1.8965	.52729	0.99	2.6912	.37158			
0.30	1.3499	.74082	0.65	1.9155	.52205	1.00	2.7183	.36788			
0.31	1.3634	.73345	0.66	1.9348	.51685	1.10	3.0042	.33287			
0.32	1.3771	.72615	0.67	1.9542	.51171	1.20	3.3201	.30119			
0.33	1.3910	.71892	0.68	1.9739	.50662	1.30	3.6693	.27253			
0.34	1.4049	.71177	0.69	1.9937	.50158	1.40	4.0552	.24660			

Excerpted from Burington's *Handbook of Mathematical Tables and Formulas*. 5th edition. McGraw-Hill, New York, 1973.

H.2 Natural logarithms (base e)

x	$\ln x$	x	$\ln x$	x	$\ln x$	x	$\ln x$
0.01	−4.60517	0.50	−0.69315	1.00	0.00000	1.5	0.40547
.02	−3.91202	.51	.67334	1.01	.00995	1.6	.47000
.03	.50656	.52	.65393	1.02	.01980	1.7	.53063
.04	.21888	.53	.63488	1.03	.02956	1.8	.58779
		.54	.61619	1.04	.03922	1.9	.64185
.05	−2.99573	.55	.59784	1.05	.04879	2.0	.69315
.06	.81341	.56	.57982	1.06	.05827	2.1	.74194
.07	.65926	.57	.56212	1.07	.06766	2.2	.78846
.08	.52573	.58	.54473	1.08	.07696	2.3	.83291
.09	.40795	.59	.52763	1.09	.08618	2.4	.87547
0.10	−2.30259	0.60	−0.51083	1.10	.09531	2.5	.91629
.11	.20727	.61	.49430	1.11	.10436	2.6	.95551
.12	.12026	.62	.47804	1.12	.11333	2.7	.99325
.13	.04022	.63	.46204	1.13	.12222	2.8	1.02962
.14	−1.96611	.64	.44629	1.14	.13103	2.9	1.06471
.15	.89712	.65	.43078	1.15	.13976	3.0	1.09861
.16	.83258	.66	.41552	1.16	.14842		
.17	.77196	.67	.40048	1.17	.15700	4.0	1.3863
.18	.71480	.68	.38566	1.18	.16551		
.19	.66073	.69	.37106	1.19	.17395	5.0	1.6094
0.20	−1.60944	0.70	−0.35667	1.20	.18232	6.0	1.7918
.21	.56065	.71	.34249	1.21	.19062		
.22	.51413	.72	.32850	1.22	.19885	7.0	1.9459
.23	.46968	.73	.31471	1.23	.20701		
.24	.42712	.74	.30111	1.24	.21511	8.0	2.0794
.25	.38629	.75	.28768	1.25	.22314	9.0	2.1972
.26	.34707	.76	.27444	1.26	.23111		
.27	.30933	.77	.26136	1.27	.23902	10.0	2.3026
.28	.27297	.78	.24846	1.28	.24686		
.29	.23787	.79	.23572	1.29	.25464	20.0	2.9957
0.30	−1.20397	0.80	−0.22314	1.30	.26236	30.0	3.4012
.31	.17118	.81	.21072	1.31	.27003		
.32	.13943	.82	.19845	1.32	.27763	40.0	3.6889
.33	.10866	.83	.18633	1.33	.28518		
.34	.07881	.84	.17435	1.34	.29267	50.0	3.9120
.35	−1.04982	.85	−0.16252	1.35	.30010	60.0	4.0943
.36	.02165	.86	.15082	1.36	.30748		
.37	−0.99425	.87	.13926	1.37	.31481	70.0	4.2485
.38	.96758	.88	.12783	1.38	.32208		
.39	.94161	.89	.11653	1.39	.32930	80.0	4.3820
0.40	−0.91629	0.90	−0.10536	1.40	.33647	90.0	4.4998
.41	.89160	.91	.09431	1.41	.34359		
.42	.86750	.92	.08338	1.42	.35066	100.0	4.6052
.43	.84397	.93	.07257	1.43	.35767		
.44	.82098	.94	.06188	1.44	.36464		
.45	.79851	.95	.05129	1.45	.37156		
.46	.77653	.96	.04082	1.46	.37844		
.47	.75502	.97	.03046	1.47	.38526		
.48	.73397	.98	.02020	1.48	.39204		
.49	.71335	.99	.01005	1.49	.39878		

Warning: The integer to the left of the decimal point is not shown in each line.

H.3 Common logarithms (base 10)[1]

N	0	1	2	3	4	5	6	7	8	9	N	0	1	2	3	4	5	6	7	8	9
10	0000	0043	0086	0128	0170	0212	0253	0294	0334	0374	55	7404	7412	7419	7427	7435	7443	7451	7459	7466	7474
11	0414	0453	0492	0531	0569	0607	0645	0682	0719	0755	56	7482	7490	7497	7505	7513	7520	7528	7536	7543	7551
12	0792	0828	0864	0899	0934	0969	1004	1038	1072	1106	57	7559	7566	7574	7582	7589	7597	7604	7612	7619	7627
13	1139	1173	1206	1239	1271	1303	1335	1367	1399	1430	58	7634	7642	7649	7657	7664	7672	7679	7686	7694	7701
14	1461	1492	1523	1553	1584	1614	1644	1673	1703	1732	59	7709	7716	7723	7731	7738	7745	7752	7760	7767	7774
15	1761	1790	1818	1847	1875	1903	1931	1959	1987	2014	60	7782	7789	7796	7803	7810	7818	7825	7832	7839	7846
16	2041	2068	2095	2122	2148	2175	2201	2227	2253	2279	61	7853	7860	7868	7875	7882	7889	7896	7903	7910	7917
17	2304	2330	2355	2380	2405	2430	2455	2480	2504	2529	62	7924	7931	7938	7945	7952	7959	7966	7973	7980	7987
18	2553	2577	2601	2625	2648	2672	2695	2718	2742	2765	63	7993	8000	8007	8014	8021	8028	8035	8041	8048	8055
19	2788	2810	2833	2856	2878	2900	2923	2945	2967	2989	64	8062	8069	8075	8082	8089	8096	8102	8109	8116	8122
20	3010	3032	3054	3075	3096	3118	3139	3160	3181	3201	65	8129	8136	8142	8149	8156	8162	8169	8176	8182	8189
21	3222	3243	3263	3284	3304	3324	3345	3365	3385	3404	66	8195	8202	8209	8215	8222	8228	8235	8241	8248	8254
22	3424	3444	3464	3483	3502	3522	3541	3560	3579	3598	67	8261	8267	8274	8280	8287	8293	8299	8306	8312	8319
23	3617	3636	3655	3674	3692	3711	3729	3747	3766	3784	68	8325	8331	8338	8344	8351	8357	8363	8370	8376	8382
24	3802	3820	3838	3856	3874	3892	3909	3927	3945	3962	69	8388	8395	8401	8407	8414	8420	8426	8432	8439	8445
25	3979	3997	4014	4031	4048	4065	4082	4099	4116	4133	70	8451	8457	8463	8470	8476	8482	8488	8494	8500	8506
26	4150	4166	4183	4200	4216	4232	4249	4265	4281	4298	71	8513	8519	8525	8531	8537	8543	8549	8555	8561	8567
27	4314	4330	4346	4362	4378	4393	4409	4425	4440	4456	72	8573	8579	8585	8591	8597	8603	8609	8615	8621	8627
28	4472	4487	4502	4518	4533	4548	4564	4579	4594	4609	73	8633	8639	8645	8651	8657	8663	8669	8675	8681	8686
29	4624	4639	4654	4669	4683	4698	4713	4728	4742	4757	74	8692	8698	8704	8710	8716	8722	8727	8733	8739	8745
30	4771	4786	4800	4814	4829	4843	4857	4871	4886	4900	75	8751	8756	8762	8768	8774	8779	8785	8791	8797	8802
31	4914	4928	4942	4955	4969	4983	4997	5011	5024	5038	76	8808	8814	8820	8825	8831	8837	8842	8848	8854	8859
32	5051	5065	5079	5092	5105	5119	5132	5145	5159	5172	77	8865	8871	8876	8882	8887	8893	8899	8904	8910	8915
33	5185	5198	5211	5224	5237	5250	5263	5276	5289	5302	78	8921	8927	8932	8938	8943	8949	8954	8960	8965	8971
34	5315	5328	5340	5353	5366	5378	5391	5403	5416	5428	79	8976	8982	8987	8993	8998	9004	9009	9015	9020	9025
35	5441	5453	5465	5478	5490	5502	5514	5527	5539	5551	80	9031	9036	9042	9047	9053	9058	9063	9069	9074	9079
36	5563	5575	5587	5599	5611	5623	5635	5647	5658	5670	81	9085	9090	9096	9101	9106	9112	9117	9122	9128	9133
37	5682	5694	5705	5717	5729	5740	5752	5763	5775	5786	82	9138	9143	9149	9154	9159	9165	9170	9175	9180	9186
38	5798	5809	5821	5832	5843	5855	5866	5877	5888	5899	83	9191	9196	9201	9206	9212	9217	9222	9227	9232	9238
39	5911	5922	5933	5944	5955	5966	5977	5988	5999	6010	84	9243	9248	9253	9258	9263	9269	9274	9279	9284	9289
40	6021	6031	6042	6053	6064	6075	6085	6096	6107	6117	85	9294	9299	9304	9309	9315	9320	9325	9330	9335	9340
41	6128	6138	6149	6160	6170	6180	6191	6201	6212	6222	86	9345	9350	9355	9360	9365	9370	9375	9380	9385	9390
42	6232	6243	6253	6263	6274	6284	6294	6304	6314	6325	87	9395	9400	9405	9410	9415	9420	9425	9430	9435	9440
43	6335	6345	6355	6365	6375	6385	6395	6405	6415	6425	88	9445	9450	9455	9460	9465	9469	9474	9479	9484	9489
44	6435	6444	6454	6464	6474	6484	6493	6503	6513	6522	89	9494	9499	9504	9509	9513	9518	9523	9528	9533	9538
45	6532	6542	6551	6561	6571	6580	6590	6599	6609	6618	90	9542	9547	9552	9557	9562	9566	9571	9576	9581	9586
46	6628	6637	6646	6656	6665	6675	6684	6693	6702	6712	91	9590	9595	9600	9605	9609	9614	9619	9624	9628	9633
47	6721	6730	6739	6749	6758	6767	6776	6785	6794	6803	92	9638	9643	9647	9652	9657	9661	9666	9671	9675	9680
48	6812	6821	6830	6839	6848	6857	6866	6875	6884	6893	93	9685	9689	9694	9699	9703	9708	9713	9717	9722	9727
49	6902	6911	6920	6928	6937	6946	6955	6964	6972	6981	94	9731	9736	9741	9745	9750	9754	9759	9763	9768	9773
50	6990	6998	7007	7016	7024	7033	7042	7050	7059	7067	95	9777	9782	9786	9791	9795	9800	9805	9809	9814	9818
51	7076	7084	7093	7101	7110	7118	7126	7135	7143	7152	96	9823	9827	9832	9836	9841	9845	9850	9854	9859	9863
52	7160	7168	7177	7185	7193	7202	7210	7218	7226	7235	97	9868	9872	9877	9881	9886	9890	9894	9899	9903	9908
53	7243	7251	7259	7267	7275	7284	7292	7300	7308	7316	98	9912	9917	9921	9926	9930	9934	9939	9943	9948	9952
54	7324	7332	7340	7348	7356	7364	7372	7380	7388	7396	99	9956	9961	9965	9969	9974	9978	9983	9987	9991	9996
N	0	1	2	3	4	5	6	7	8	9	N	0	1	2	3	4	5	6	7	8	9

[1] This table lists the logarithms to the base 10 to four decimal places of numbers in the range from 1 to 9.99. For instance, log 3.57 = 0.5527. (First find 35 under N.) Logarithms of numbers outside this range can be calculated as shown in these two examples:

$$\log_{10} 357 = \log_{10} (10^2)(3.57) = \log_{10} 10^2 + \log_{10} 3.57 \approx 2 + 0.5527 = 2.5527$$

$$\log_{10} 0.357 = \log_{10} (10^{-1})(3.57) = \log_{10} 10^{-1} + \log_{10} 3.57 \approx (-1) + 0.5527 = -0.4473$$

H.4 Trigonometric functions (degrees)

Deg	Rad	Sin	Cos	Tan	Cot	Sec	Csc		
0	0.0000	0.0000	1.0000	0.0000	⋯⋯	1.0000	⋯⋯	1.5708	90
1	0.0175	0.0175	0.9998	0.0175	57.290	1.0002	57.299	1.5533	89
2	0.0349	0.0349	0.9994	0.0349	28.636	1.0006	28.654	1.5359	88
3	0.0524	0.0523	0.9986	0.0524	19.081	1.0014	19.107	1.5184	87
4	0.0698	0.0698	0.9976	0.0699	14.301	1.0024	14.336	1.5010	86
5	0.0873	0.0872	0.9962	0.0875	11.430	1.0038	11.474	1.4835	85
6	0.1047	0.1045	0.9945	0.1051	9.5144	1.0055	9.5668	1.4661	84
7	0.1222	0.1219	0.9925	0.1228	8.1443	1.0075	8.2055	1.4486	83
8	0.1396	0.1392	0.9903	0.1405	7.1154	1.0098	7.1853	1.4312	82
9	0.1571	0.1564	0.9877	0.1584	6.3138	1.0125	6.3925	1.4137	81
10	0.1745	0.1736	0.9848	0.1763	5.6713	1.0154	5.7588	1.3963	80
11	0.1920	0.1908	0.9816	0.1944	5.1446	1.0187	5.2408	1.3788	79
12	0.2094	0.2079	0.9781	0.2126	4.7046	1.0223	4.8097	1.3614	78
13	0.2269	0.2250	0.9744	0.2309	4.3315	1.0263	4.4454	1.3439	77
14	0.2443	0.2419	0.9703	0.2493	4.0108	1.0306	4.1336	1.3265	76
15	0.2618	0.2588	0.9659	0.2679	3.7321	1.0353	3.8637	1.3090	75
16	0.2793	0.2756	0.9613	0.2867	3.4874	1.0403	3.6280	1.2915	74
17	0.2967	0.2924	0.9563	0.3057	3.2709	1.0457	3.4203	1.2741	73
18	0.3142	0.3090	0.9511	0.3249	3.0777	1.0515	3.2361	1.2566	72
19	0.3316	0.3256	0.9455	0.3443	2.9042	1.0576	3.0716	1.2392	71
20	0.3491	0.3420	0.9397	0.3640	2.7475	1.0642	2.9238	1.2217	70
21	0.3665	0.3584	0.9336	0.3839	2.6051	1.0711	2.7904	1.2043	69
22	0.3840	0.3746	0.9272	0.4040	2.4751	1.0785	2.6695	1.1868	68
23	0.4014	0.3907	0.9205	0.4245	2.3559	1.0864	2.5593	1.1694	67
24	0.4189	0.4067	0.9135	0.4452	2.2460	1.0946	2.4586	1.1519	66
25	0.4363	0.4226	0.9063	0.4663	2.1445	1.1034	2.3662	1.1345	65
26	0.4538	0.4384	0.8988	0.4877	2.0503	1.1126	2.2812	1.1170	64
27	0.4712	0.4540	0.8910	0.5095	1.9626	1.1223	2.2027	1.0996	63
28	0.4887	0.4695	0.8829	0.5317	1.8807	1.1326	2.1301	1.0821	62
29	0.5061	0.4848	0.8746	0.5543	1.8040	1.1434	2.0627	1.0647	61
30	0.5236	0.5000	0.8660	0.5774	1.7321	1.1547	2.0000	1.0472	60
31	0.5411	0.5150	0.8572	0.6009	1.6643	1.1666	1.9416	1.0297	59
32	0.5585	0.5299	0.8480	0.6249	1.6003	1.1792	1.8871	1.0123	58
33	0.5760	0.5446	0.8387	0.6494	1.5399	1.1924	1.8361	0.9948	57
34	0.5934	0.5592	0.8290	0.6745	1.4826	1.2062	1.7883	0.9774	56
35	0.6109	0.5736	0.8192	0.7002	1.4281	1.2208	1.7434	0.9599	55
36	0.6283	0.5878	0.8090	0.7265	1.3764	1.2361	1.7013	0.9425	54
37	0.6458	0.6018	0.7986	0.7536	1.3270	1.2521	1.6616	0.9250	53
38	0.6632	0.6157	0.7880	0.7813	1.2799	1.2690	1.6243	0.9076	52
39	0.6807	0.6293	0.7771	0.8098	1.2349	1.2868	1.5890	0.8901	51
40	0.6981	0.6428	0.7660	0.8391	1.1918	1.3054	1.5557	0.8727	50
41	0.7156	0.6561	0.7547	0.8693	1.1504	1.3250	1.5243	0.8552	49
42	0.7330	0.6691	0.7431	0.9004	1.1106	1.3456	1.4945	0.8378	48
43	0.7505	0.6820	0.7314	0.9325	1.0724	1.3673	1.4663	0.8203	47
44	0.7679	0.6947	0.7193	0.9657	1.0355	1.3902	1.4396	0.8029	46
45	0.7854	0.7071	0.7071	1.0000	1.0000	1.4142	1.4142	0.7854	45
		Cos	Sin	Cot	Tan	Csc	Sec	Rad	Deg

For degrees indicated in the left-hand column use the column headings at the top. For degrees indicated in the right-hand column use the column headings at the bottom.

H.5 Trigonometric functions (radians)

Rad	Sin	Tan	Cot	Cos	Rad	Sin	Tan	Cot	Cos
.00	.0000	.0000	......	1.0000	.50	.4794	.5463	1.830	.8776
.01	.0100	.0100	99.997	1.0000	.51	.4882	.5594	1.788	.8727
.02	.0200	.0200	49.993	.9998	.52	.4969	.5726	1.747	.8678
.03	.0300	.0300	33.323	.9996	.53	.5055	.5859	1.707	.8628
.04	.0400	.0400	24.987	.9992	.54	.5141	.5994	1.668	.8577
.05	.0500	.0500	19.983	.9988	.55	.5227	.6131	1.631	.8525
.06	.0600	.0601	16.647	.9982	.56	.5312	.6269	1.595	.8473
.07	.0699	.0701	14.262	.9976	.57	.5396	.6410	1.560	.8419
.08	.0799	.0802	12.473	.9968	.58	.5480	.6552	1.526	.8365
.09	.0899	.0902	11.081	.9960	.59	.5564	.6696	1.494	.8309
.10	.0998	.1003	9.967	.9950	.60	.5646	.6841	1.462	.8253
.11	.1098	.1104	9.054	.9940	.61	.5729	.6989	1.431	.8196
.12	.1197	.1206	8.293	.9928	.62	.5810	.7139	1.401	.8139
.13	.1296	.1307	7.649	.9916	.63	.5891	.7291	1.372	.8080
.14	.1395	.1409	7.096	.9902	.64	.5972	.7445	1.343	.8021
.15	.1494	.1511	6.617	.9888	.65	.6052	.7602	1.315	.7961
.16	.1593	.1614	6.197	.9872	.66	.6131	.7761	1.288	.7900
.17	.1692	.1717	5.826	.9856	.67	.6210	.7923	1.262	.7838
.18	.1790	.1820	5.495	.9838	.68	.6288	.8087	1.237	.7776
.19	.1889	.1923	5.200	.9820	.69	.6365	.8253	1.212	.7712
.20	.1987	.2027	4.933	.9801	.70	.6442	.8423	1.187	.7648
.21	.2085	.2131	4.692	.9780	.71	.6518	.8595	1.163	.7584
.22	.2182	.2236	4.472	.9759	.72	.6594	.8771	1.140	.7518
.23	.2280	.2341	4.271	.9737	.73	.6669	.8949	1.117	.7452
.24	.2377	.2447	4.086	.9713	.74	.6743	.9131	1.095	.7385
.25	.2474	.2553	3.916	.9689	.75	.6816	.9316	1.073	.7317
.26	.2571	.2660	3.759	.9664	.76	.6889	.9505	1.052	.7248
.27	.2667	.2768	3.613	.9638	.77	.6961	.9697	1.031	.7179
.28	.2764	.2876	3.478	.9611	.78	.7033	.9893	1.011	.7109
.29	.2860	.2984	3.351	.9582	.79	.7104	1.009	.9908	.7038
.30	.2955	.3093	3.233	.9553	.80	.7174	1.030	.9712	.6967
.31	.3051	.3203	3.122	.9523	.81	.7243	1.050	.9520	.6895
.32	.3146	.3314	3.018	.9492	.82	.7311	1.072	.9331	.6822
.33	.3240	.3425	2.919	.9460	.83	.7379	1.093	.9146	.6749
.34	.3335	.3537	2.827	.9428	.84	.7446	1.116	.8964	.6675
.35	.3429	.3650	2.740	.9394	.85	.7513	1.138	.8785	.6600
.36	.3523	.3764	2.657	.9359	.86	.7578	1.162	.8609	.6524
.37	.3616	.3879	2.578	.9323	.87	.7643	1.185	.8437	.6448
.38	.3709	.3994	2.504	.9287	.88	.7707	1.210	.8267	.6372
.39	.3802	.4111	2.433	.9249	.89	.7771	1.235	.8100	.6294
.40	.3894	.4228	2.365	.9211	.90	.7833	1.260	.7936	.6216
.41	.3986	.4346	2.301	.9171	.91	.7895	1.286	.7774	.6137
.42	.4078	.4466	2.239	.9131	.92	.7956	1.313	.7615	.6058
.43	.4169	.4586	2.180	.9090	.93	.8016	1.341	.7458	.5978
.44	.4259	.4708	2.124	.9048	.94	.8076	1.369	.7303	.5898
.45	.4350	.4831	2.070	.9004	.95	.8134	1.398	.7151	.5817
.46	.4439	.4954	2.018	.8961	.96	.8192	1.428	.7001	.5735
.47	.4529	.5080	1.969	.8916	.97	.8249	1.459	.6853	.5653
.48	.4618	.5206	1.921	.8870	.98	.8305	1.491	.6707	.5570
.49	.4706	.5334	1.875	.8823	.99	.8360	1.524	.6563	.5487
.50	.4794	.5463	1.830	.8776	1.00	.8415	1.557	.6421	.5403

Rad	Sin	Tan	Cot	Cos	Rad	Sin	Tan	Cot	Cos
1.00	.8415	1.557	.6421	.5403	1.30	.9636	3.602	.2776	.2675
1.01	.8468	1.592	.6281	.5319	1.31	.9662	3.747	.2669	.2579
1.02	.8521	1.628	.6142	.5234	1.32	.9687	3.903	.2562	.2482
1.03	.8573	1.665	.6005	.5148	1.33	.9711	4.072	.2456	.2385
1.04	.8624	1.704	.5870	.5062	1.34	.9735	4.256	.2350	.2288
1.05	.8674	1.743	.5736	.4976	1.35	.9757	4.455	.2245	.2190
1.06	.8724	1.784	.5604	.4889	1.36	.9779	4.673	.2140	.2092
1.07	.8772	1.827	.5473	.4801	1.37	.9799	4.913	.2035	.1994
1.08	.8820	1.871	.5344	.4713	1.38	.9819	5.177	.1931	.1896
1.09	.8866	1.917	.5216	.4625	1.39	.9837	5.471	.1828	.1798
1.10	.8912	1.965	.5090	.4536	1.40	.9854	5.798	.1725	.1700
1.11	.8957	2.014	.4964	.4447	1.41	.9871	6.165	.1622	.1601
1.12	.9001	2.066	.4840	.4357	1.42	.9887	6.581	.1519	.1502
1.13	.9044	2.120	.4718	.4267	1.43	.9901	7.055	.1417	.1403
1.14	.9086	2.176	.4596	.4176	1.44	.9915	7.602	.1315	.1304
1.15	.9128	2.234	.4475	.4085	1.45	.9927	8.238	.1214	.1205
1.16	.9168	2.296	.4356	.3993	1.46	.9939	8.989	.1113	.1106
1.17	.9208	2.360	.4237	.3902	1.47	.9949	9.887	.1011	.1006
1.18	.9246	2.427	.4120	.3809	1.48	.9959	10.983	.0910	.0907
1.19	.9284	2.498	.4003	.3717	1.49	.9967	12.350	.0810	.0807
1.20	.9320	2.572	.3888	.3624	1.50	.9975	14.101	.0709	.0707
1.21	.9356	2.650	.3773	.3530	1.51	.9982	16.428	.0609	.0608
1.22	.9391	2.733	.3659	.3436	1.52	.9987	19.670	.0508	.0508
1.23	.9425	2.820	.3546	.3342	1.53	.9992	24.498	.0408	.0408
1.24	.9458	2.912	.3434	.3248	1.54	.9995	32.461	.0308	.0308
1.25	.9490	3.010	.3323	.3153	1.55	.9998	48.078	.0208	.0208
1.26	.9521	3.113	.3212	.3058	1.56	.9999	92.620	.0108	.0108
1.27	.9551	3.224	.3102	.2963	1.57	1.0000	1255.8	.0008	.0008
1.28	.9580	3.341	.2993	.2867	1.58	1.0000	−108.65	−.0092	−.0092
1.29	.9608	3.467	.2884	.2771	1.59	.9998	−52.067	−.0192	−.0192
1.30	.9636	3.602	.2776	.2675	1.60	.9996	−34.233	−.0292	−.0292

ANSWERS TO SELECTED ODD-NUMBERED PROBLEMS AND TO GUIDE QUIZZES

CHAPTER 0. AN OVERVIEW OF CALCULUS AND THIS BOOK

SEC. 0.1. THE DERIVATIVE

1 (a)

x	$\sqrt{x^2 + 2x}$	$\sqrt{x^2 + 2x} - x$
1	1.7321	0.7321
5	5.9161	0.9161
10	10.9545	0.9545
100	100.9950	0.9950
1000	1000.9995	0.9995

(b) It appears that $\sqrt{x^2 + 2x} - x$ approaches 1 as x grows large.

3 (a)

x	$x^3 - 1$	$x - 1$	$\dfrac{x^3 - 1}{x - 1}$
0.5	-0.8750	-0.5000	1.7500
0.9	-0.2710	-0.1000	2.7100
0.99	-0.0297	-0.0100	2.9701
0.999	-0.0030	-0.0010	2.9970

(b) The ratio $(x^3 - 1)/(x - 1)$ appears to approach 3 as x approaches 1.

5 (a) 0.030301 ft (b) 3.0301 ft/min (c) 3.003001 ft/min (d) 2.997001 ft/min (e) 3 ft/min

7 (a)

x	2^x	$2^x - 1$	$\dfrac{2^x - 1}{x}$
1	2	1	1
0.5	1.4142	0.4142	0.8284
0.1	1.0718	0.0718	0.7177
0.01	1.0070	0.0070	0.6956
0.001	1.0007	0.0007	0.6934
-0.001	0.9993	-0.0007	0.6929

(b) The ratio $(2^x - 1)/x$ approaches a number near 0.693 as x approaches 0.

SEC. 0.2. THE INTEGRAL

1 (a)

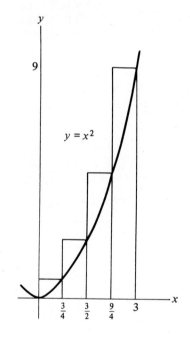

(b)

Rectangle	Height	Width	Area
First	$(3/4)^2$	3/4	27/64
Second	$(3/2)^2$	3/4	108/64
Third	$(9/4)^2$	3/4	243/64
Fourth	3^2	3/4	432/64

(c) $405/32 \approx 12.6563$

CHAPTER 1. PRELIMINARIES

SEC. 1.1. COORDINATE SYSTEMS AND GRAPHS

1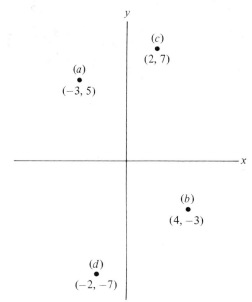

(a) Second (b) Fourth (c) First (d) Third

3 (a) 10 (b) 13 (c) 3

5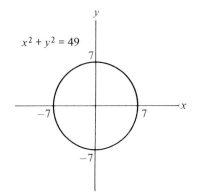

7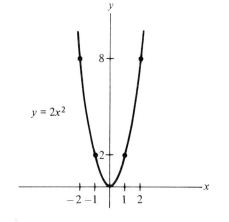

3 (a)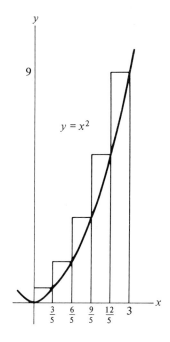

(b)

Rectangle	Height	Width	Area
First	$(3/5)^2$	3/5	27/125
Second	$(6/5)^2$	3/5	108/125
Third	$(9/5)^2$	3/5	243/125
Fourth	$(12/5)^2$	3/5	432/125
Fifth	3^2	3/5	675/125

(c) $297/25 = 11.88$

7 (a) $100/256 = 0.390625$ (b) $1{,}296/4{,}096 \approx 0.3164$
(c) $784/4{,}096 \approx 0.1914$

9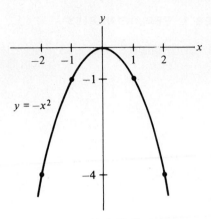

23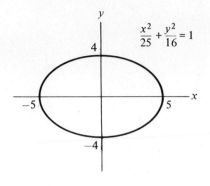

11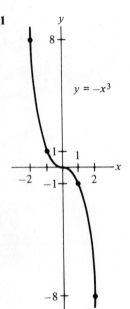

13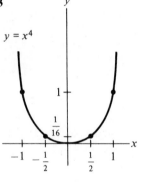

25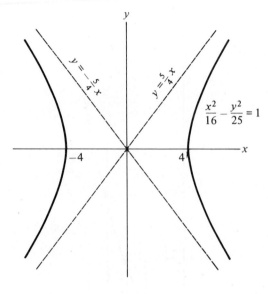

15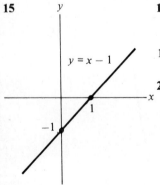

17 x intercept, -3
y intercept, 6

19 No intercepts

21 x intercepts, -3 and $1/2$
y intercept, -3

27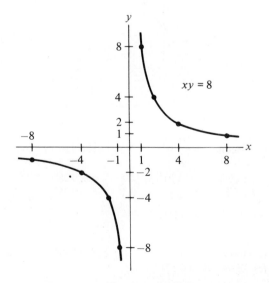

29 (*a*) Symmetric with respect to the x axis

(*b*)

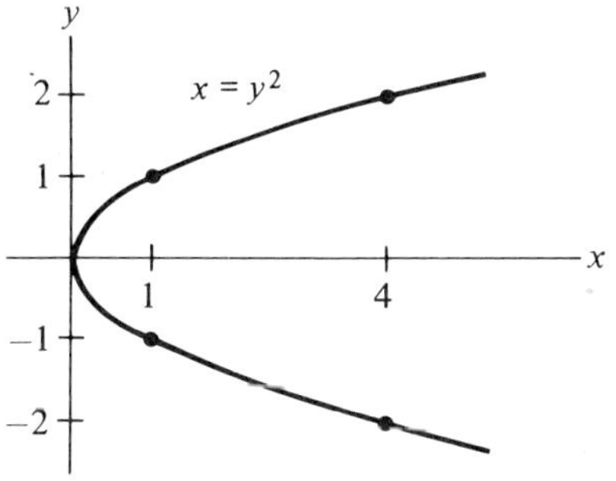

31

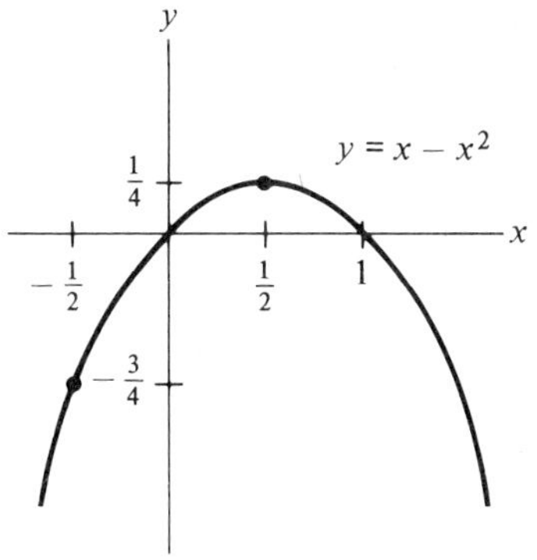

33

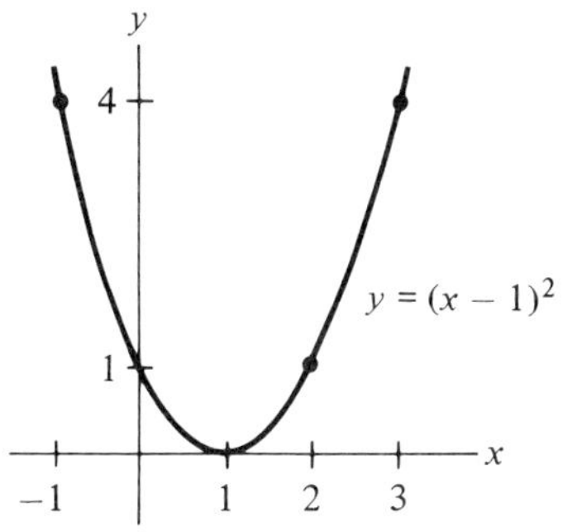

35

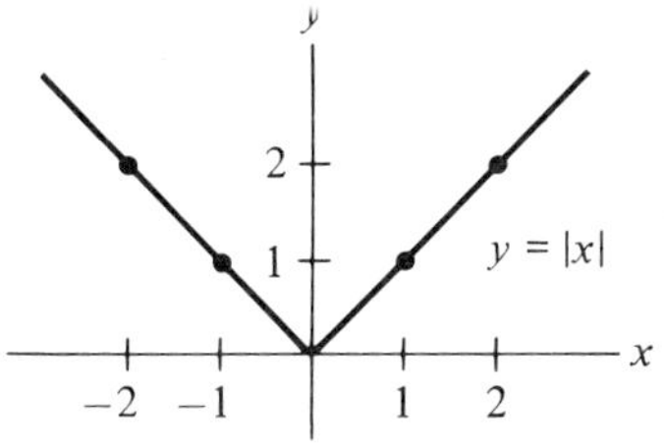

37 $(x-2)^2 + (y-1)^2 = 49$

39 (*a*) $x = -b/2a$

SEC. 1.2. LINES AND THEIR SLOPES

1 $m = 1/5$

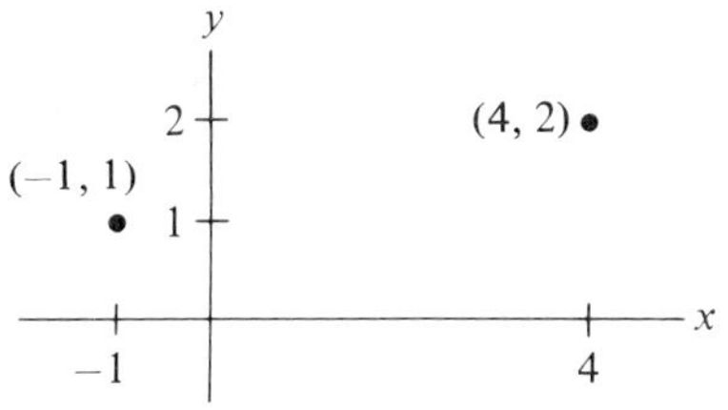

3 $m = -5/4$

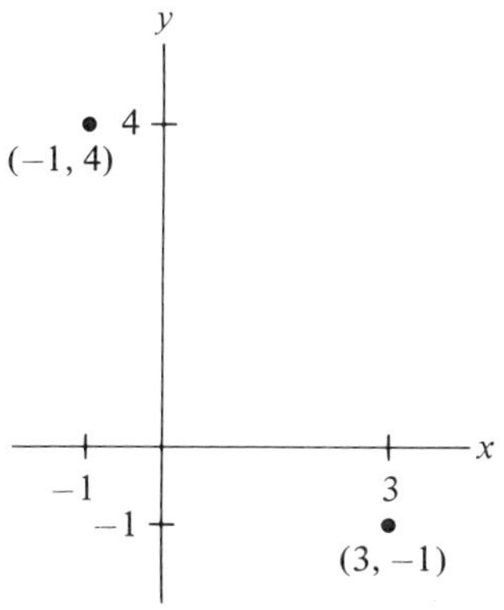

5 $m = 0$

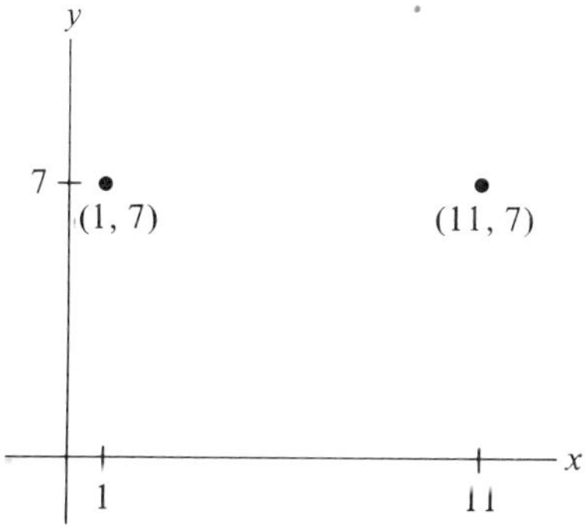

7 (*a*) Negative (*b*) Negative (*c*) Positive
(*d*) Positive (*e*) Zero

9 (*a*) 4 (*b*) $-1/4$

11 Yes

13 No

15 (*a*) $y = 3x + 2$ (*b*) $y = \frac{2}{3}x - 2$ (*c*) $y = -3x$

17 (a) $m = 3, b = -1$

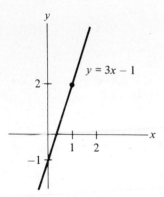

(b) $m = -2, b = 1$

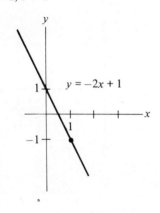

(c) $m = \frac{3}{5}, b = 0$

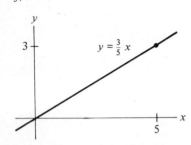

19 (a) $y - 2 = 3(x - 1)$ (b) $y + 1 = -2(x - 3)$

21 (a) $y - 2 = \frac{1}{4}(x - 1)$ (b) $y - 2 = -\frac{1}{4}(x + 1)$
(c) $y - 5 = x - 4$

23 (a)

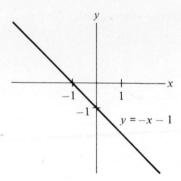

(b)

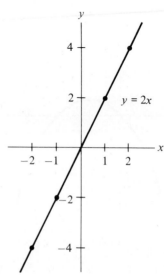

(c)

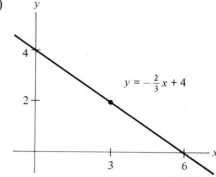

25 (a) Yes   (b) No

27 $y = \frac{x}{2} + \frac{5}{2}$

29 $y = 3x + 5$

31 $m = -b/a$

33 (a) $\frac{7}{4}$ (b) $\frac{11}{2}$ (c) 3   (d) -6

35 (a) $m = \dfrac{y_2 - y_1}{x_2 - x_1}$ (b) $m' = \dfrac{y_2 - y_1}{x_2 - x_1}$

SEC. 1.3. FUNCTIONS

1 $f(x) = 3x$

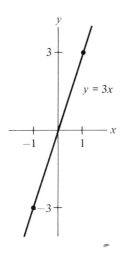

3 $f(x) = 3x^2$

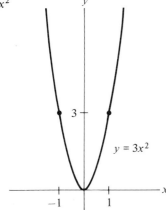

5 $f(x) = -x^2 + 1$

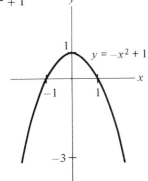

7 $f(x) = x^2 - x$

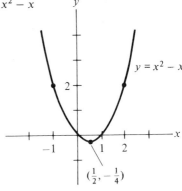

9 $f(x) = \dfrac{2}{1 + x^2}$

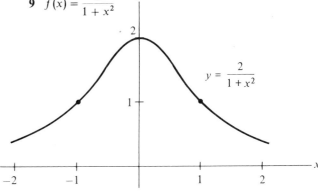

11 Domain: $[0, \infty)$; range: $[0, \infty)$

13 Domain: $[-2, 2]$; range: $[0, 2]$

15 Domain: All nonzero x; range: $(0, \infty)$

17 Domain: All nonzero x; range: All nonzero y

19 Domain: $(0, \infty)$; range: $(0, \infty)$

21 (b) and (c) are the graphs of functions; (a) is not.

23 (a) 6   (b) -3 (c) 0   (d) 1

25 (a) 0   (b) 4   (c) 2.25   (d) 1

27 (a) 27   (b) 27

29 $3a^2 + 3a + 1$

31 $-(c + d)/c^2 d^2$

33 3

35 $1 - \dfrac{1}{uv}$

37

x	-2	-1	0	$\frac{1}{2}$	1	2	3
$x^2 + x$	2	0	0	$\frac{3}{4}$	2	6	12

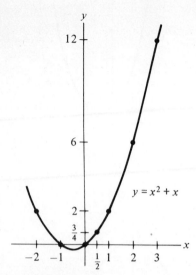

SEC. 1.S. SUMMARY: GUIDE QUIZ ON CHAPTER 1

1 $\sqrt{12}$

2 (a)

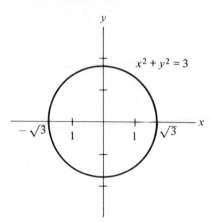

(b)

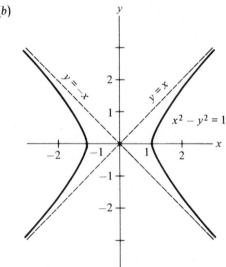

39 (a) 0, 3, 6, 9, respectively

(b)
x	0	1	2	3
$f(x)$	0	3	6	9

(c)

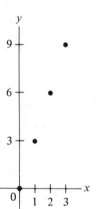

(d) Neither

41 (a) 7.0000   (b) 6.0100   (c) 5.9900   (d) 6.0001

43 (b) and (c)

(c)

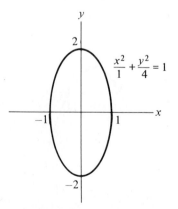

SEC. 1.4. COMPOSITE FUNCTIONS

1 $y = x^6$

3 $y = \sqrt{1 + 2x^3}$

5 $y = u^2, u = 1 + x^3$

7 $y = \sqrt[3]{u}, u = 1/v, v = 1 + x^2$

9 $w = 27t^3$

11 $w = \sqrt[5]{1 + \dfrac{1}{x^2}}$

13 $b = a + 1$

17 $f(0) = 0$

(d)
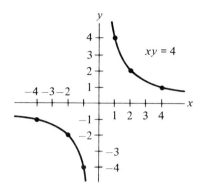

(e) (f)
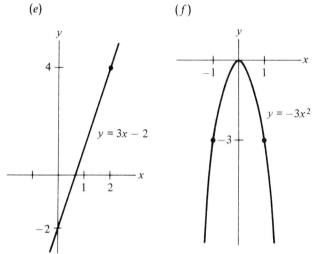

3 (a)
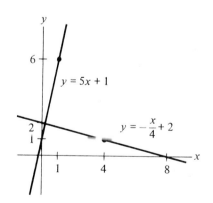

(b) No

4 $4x + 2h + \dfrac{1}{x(x+h)}$

5 $y = u^3 + u,\ u = \sqrt{v},\ v = 1 + 2x$

REVIEW EXERCISES FOR CHAPTER 1

1 Domain: All real numbers; range: All real numbers

3 Domain: All real numbers; range: All nonnegative real numbers

5 Domain: All real numbers greater than -1; range: All positive real numbers

7 0.41

9 39

11 $\dfrac{-1}{(a+h+1)(a+1)}$

13 $3 + 3h + h^2$

15 $u^2 + uv + v^2 - 3$

17
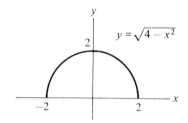

19
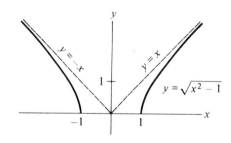

21
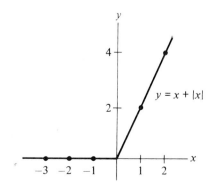

23

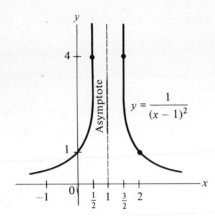

29

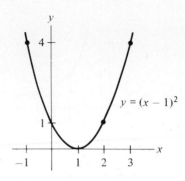

31 (b) and (c)

33 $-2 < x < 0$ or $x > 1$

35 (a)

x	0.001	0.01	0.1	1	2	10	100
$f(x)$	2.717	2.705	2.594	2	1.732	1.271	1.047

25

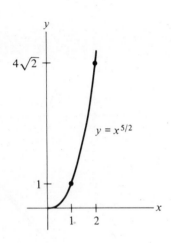

(b)

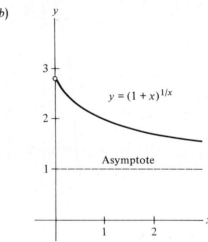

27

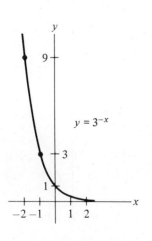

37

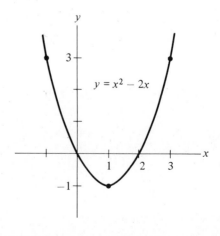

ANSWERS TO SELECTED ODD-NUMBERED PROBLEMS AND TO GUIDE QUIZZES   **S-87**

39 (a)

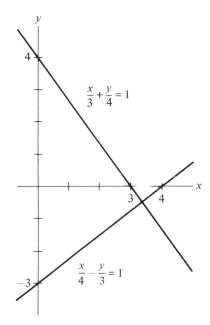

(b) $-\frac{4}{3}, \frac{3}{4}$
(c) Yes
41 (a) $|m|$ is large.   (b) m is close to zero.
(c) $m < 0$ (d) $m > 0$
43 $y = -3x + 5$
45 $P = (-1, 0)$

CHAPTER 2. LIMITS AND CONTINUOUS FUNCTIONS

SEC. 2.1. THE LIMIT OF A FUNCTION

1 12
3 4
5 $\frac{4}{3}$
7 $\frac{1}{5}$
9 25
11 0
13 1
15 3
17 $-\frac{1}{4}$
19 $\frac{1}{4}$
21 2
23 1
25 (a) 2   (b) 1   (c) 1   (d) 2
27 (a) 0   (b) 0   (c) 4
29 (a)

x	1	0.1	0.01	0.001	-1	-0.01	-0.001
$f(x)$	2	1.618	1.588	1.585	1.333	1.582	1.585

31 1

33 (a)

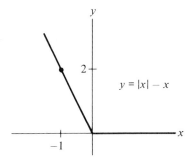

(b) $\lim_{x \to a} f(x)$ exists for all values of a.

35 (a)

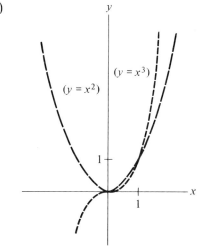

$$f(x) = \begin{cases} x^2, & x \text{ rational} \\ x^3, & x \text{ irrational} \end{cases}$$

(b) $\lim_{x \to 2} f(x)$ does not exist.

(c) $\lim_{x \to 1} f(x) = 1$

(d) $\lim_{x \to 0} f(x) = 0$

(e) $\lim_{x \to a} f(x)$ exists only for $a = 0, 1$.

SEC. 2.2. COMPUTATIONS OF LIMITS

1 ∞ **11** 0   **21** $\frac{2}{3}$
3 $-\infty$ **13** ∞ **23** $\frac{2}{3}$
5 ∞ **15** ∞ **25** ∞
7 0   **17** 0   **27** 0
9 ∞ **19** 50   **29** 0

31 (a) ∞ (b) $-\infty$ (c) Does not exist.
33 (a) 0 (b) 1 (c) ∞
35 $\frac{1}{2}$
37 1
39 (a) 1 (b) -1 (c) Does not exist.
41 (a) ∞ (b) Indeterminate (c) ∞
(d) Indeterminate
43 (a) a/b (b) ∞ (c) 0
45 No limit; oscillates

SEC. 2.3. ASYMPTOTES AND THEIR USE IN GRAPHING

1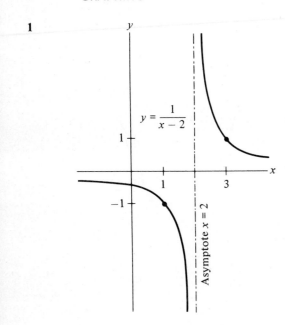

3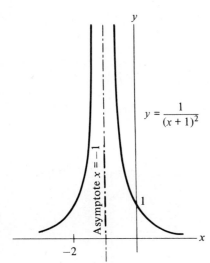

5 $y = \dfrac{1}{x^2 - x} = \dfrac{1}{x(x-1)}$

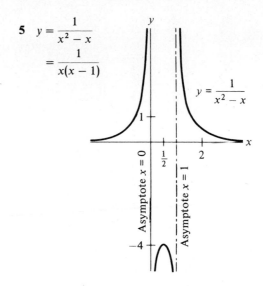

7 $y = \dfrac{1}{x^4 - x^2} = \dfrac{1}{x^2(x^2 - 1)}$

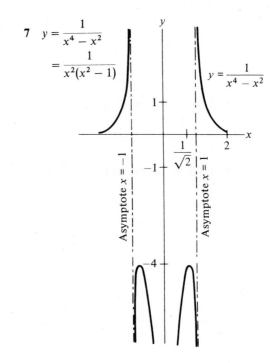

9 $y = \dfrac{x(x-1)}{x^2 + 1}$

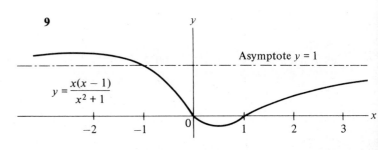

11

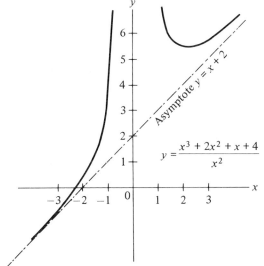

13

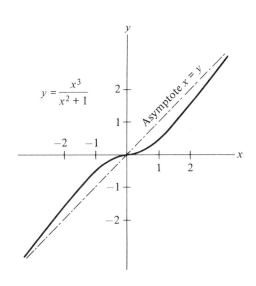

15

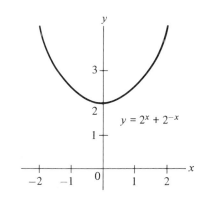

17

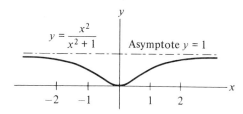

19 $y = ax + (b - a)$
21 (b) $-\frac{1}{4}$ (c) -4 (d) $(\frac{1}{2}, -4)$

SEC. 2.4. PRECISE DEFINITIONS OF LIMITS (OPTIONAL)

1 (a) 200   (b) Any number greater than 200
   (c) 200
3 (a) 400   (b) 2,000
5 Let $D = E/3$.
7 Let $D = E - 5$.
9 Let $D = \frac{1}{2}(E - 4)$.
11 Let $D = \frac{1}{4}(E + 100)$.
13 (a) 10   (b) $\sqrt{E}$ (c) Any value will work.
   (d) Let $D = \sqrt{|E|}$.
15 (a) 10   (b) 11   (c) 10   (d) Let $D = 1/\varepsilon$.
17 Let $D = 1/\varepsilon$.
19 Let $D = 2/\sqrt{\varepsilon}$.
21 Let $D = 100 + 1/\varepsilon$
23 (a) $\frac{1}{30}$ (b) $\sqrt{\varepsilon/3}$
25 (a) 15   (b) $\frac{1}{2}$ (c) $\frac{1}{20}$
27 For each number E there is a number D such that $f(x) < E$ for all $x > D$.

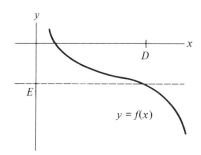

29 For each positive number ε there is a number D such that $|f(x) - L| < \varepsilon$ for all $x < D$.

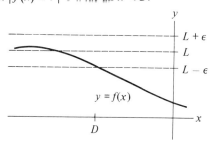

31 For each positive number ε there is a positive number δ such that $|f(x) - L| < \varepsilon$ for all x between a and $a + \delta$.

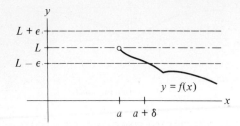

33 For any number E there is a positive number δ such that $f(x) < E$ for all x between a and $a + \delta$.

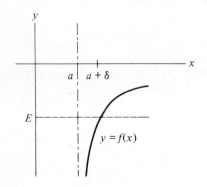

35 For any number E there is a positive number δ such that $f(x) < E$ for all x satisfying $0 < |x - a| < \delta$.

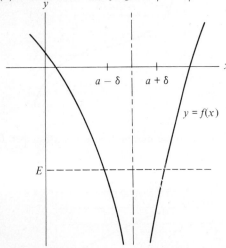

37 (b) Let $\delta = \varepsilon/7$.
39 (b) Let $\delta = \varepsilon/12$.
43 Let $\varepsilon = \frac{1}{2}$.

SEC. 2.5. THE LIMIT OF $(\sin \theta)/\theta$ AS θ APPROACHES 0

1 (a) $9\pi/4$ (b) $\theta/2$ (c) 2θ
3 $\frac{1}{2}$
5 $\frac{3}{5}$
7 0
9 0
11 $\frac{1}{2}$
13 ∞
15 ∞
17 (a) All nonzero values of x
(c) 0
(d) $x = n\pi$, n a nonzero integer
(e)

x	0.1	$\pi/2$	$3\pi/2$	2π	$5\pi/2$	3π	$7\pi/2$
$\sin x$	0.10	1.00	-1.00	0	1.00	0	-1.00
$\dfrac{\sin x}{x}$	1.00	0.64	-0.21	0	0.13	0	-0.09

(f), (g)

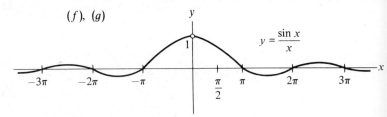

(h) 1
19 0
21 (b)

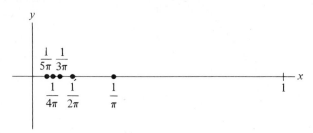

23 (b)

$(2/5\pi, 1)$ $(2/\pi, 1)$
$(2/9\pi, 1)$

25 (a)

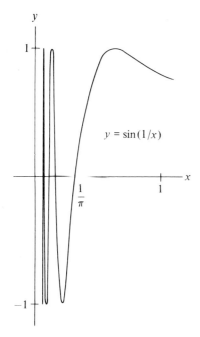

(b) No
27 Near 0.1667.
29 (b) $\cos 0.1 \approx 0.99500$, $\cos 0.01 \approx 0.99995$
(c) Calculator agrees to all 5 places.

SEC. 2.6. CONTINUOUS FUNCTIONS

1 (a) Yes; 1   (b) Yes
3 (a) No   (b) No
5 (a) No   (b) No
7 (a) Yes   (b) No   (c) No   (d) No
  (e) Yes   (f) No
9 (a)

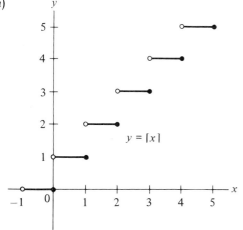

(b) Yes; 4
(c) Yes; 5
(d) No
(e) No
(f) Nonintegers
(g) Integers
11 Yes
13 (a)

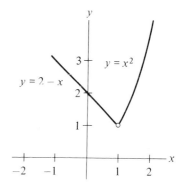

(b) Yes
15 (a)

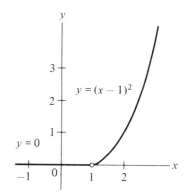

(b) Yes
17 (a) 1   (b) 1   (c) No   (d) Yes
(e)

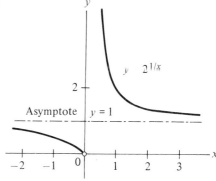

(f) No

19 (a)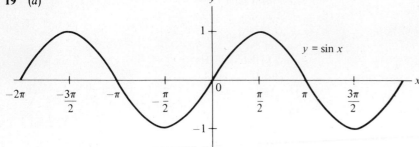

(b) Yes

21 (a)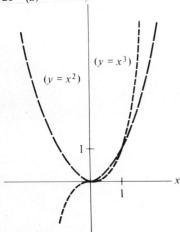

(b) 0, 1

21 (a) $f(x) = \sin x$

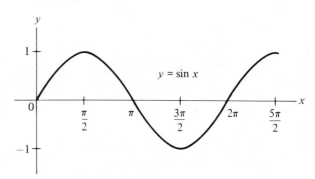

(b) $c = 0, \pi, 2\pi$
(c) Three choices for $0 \le m < 1$; two choices for $m = 1$

SEC. 2.S. SUMMARY: GUIDE QUIZ ON LIMITS AND CONTINUOUS FUNCTIONS

3 (a) Yes (b) No (c) No (d) No
  (e) No (f) Yes
4 (a) 6 (b) $\tfrac{3}{5}$ (c) -3 (d) 0 (e) $\tfrac{1}{2}$
  (f) ∞ (g) 0 (h) 4 (i) 12
  (j) Does not exist. (k) Does not exist.
  (l) 0 (m) 1 (n) $\tfrac{1}{8}$
5 (a) 12 (b) $\tfrac{3}{4}$ (c) 7 (d) Indeterminate
  (e) Indeterminate
6 (a) ∞ (b) 0 (c) Indeterminate (d) ∞
  (e) Indeterminate (f)

SEC. 2.7. THE MAXIMUM-VALUE THEOREM AND THE INTERMEDIATE-VALUE THEOREM

1 (a) Yes (b) Yes
3 (a) Yes (b) Yes (c) No
5 (a) Yes (b) No
9 $c = \tfrac{5}{3}$
11 $c = \pi, 2\pi, 3\pi, 4\pi, 5\pi$
13 $c = \dfrac{\pi}{3}, \dfrac{5\pi}{3}, \dfrac{7\pi}{3}, \dfrac{11\pi}{3}, \dfrac{13\pi}{3}$
15 $c = -1, 0, 1$

REVIEW EXERCISES FOR CHAPTER 2

1 1   **9** $\tfrac{1}{4}$ **17** ∞ **25** $\tfrac{1}{3}$
3 $\tfrac{8}{3}$ **11** 2   **19** 1   **27** 0
5 $\tfrac{1}{2}$ **13** ∞ **21** 1   **29** 0
7 $-\infty$ **15** 5   **23** 0   **31** 0

33 $\dfrac{\pi^2}{16\sqrt{2}}$

35 1

37 $f(x) = 5x, g(x) = x$

39 $f(x) = 1/x, g(x) = x^2$

41 $f(x) = x^2, g(x) = x$

43 (a) Yes   (b) No

45 (a) Yes   (b) No

49 They approach 1 and 0, respectively.

51 $c = 2$

53 $c = \pi/4$

55 0

57 Yes; let $f(0) = 0$.

59 Both are mistaken; the limit is -13.

CHAPTER 3. THE DERIVATIVE

SEC. 3.1. FOUR PROBLEMS WITH ONE THEME

1 6

3 -4

5 12

7 (a) 0

(b)
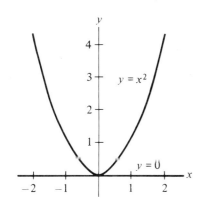

9 96 ft/sec

11 32 ft/sec

13 (a) 1.261 ft   (b) 12.61 ft/sec   (c) $t^2 + 2t + 4$
   (d) 12 ft/sec

15 (a) 3.9   (b) 3.99   (c) $2 + x$ (d) 4

17 (a),(b)
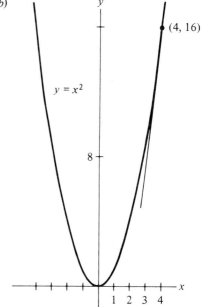

(d) 8.01
(e) 7.99

19 (a) 2.1   (b) 2.01   (c) 2.001   (d) 2

21 (a) 0.99   (b) 0.999   (c) 1

23 (a) 0.0601   (b) 6.01   (c) 5.99   (d) 6   (e) 6

25 (a) 0.0401   (b) 4.01   (c) 4

27 (a), (b), (c) $y = 2x^2 + x$
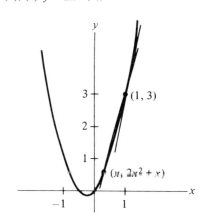

(d) $2x + 3$
(e) 5

29 $2x$

31 $2x$

33 No

35 (a)

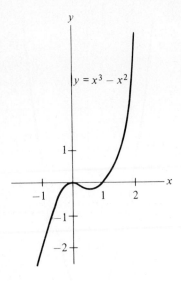

(b) $3x^2 - 2x$
(c) $(0, 0), (\frac{2}{3}, -\frac{4}{27})$
(d) $(-\frac{1}{3}, -\frac{4}{27}), (1, 0)$

37 $(1, 0)$ and $(1 - \sqrt{3}, 9 - 5\sqrt{3})$

SEC. 3.2. THE DERIVATIVE

1 $4x^3$ **9** $7/(2\sqrt{x})$ **17** -4
3 2 **11** $-1/x^2$ **19** $5a^4$
5 $2x$ **13** $-2/x^3$ **21** $\frac{1}{12}$
7 $-10x + 4$ **15** $6/x^2$ **23** $\frac{1}{32}$

25 (a) 4.641 (b) 4
27 (a) 4.060401 (b) 4

29 (b)

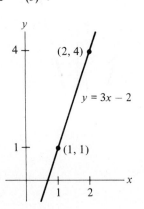

31 (a) $a = 2$ (b) $a = 1, 2$ (c) $a = 1, 2, 3$

SEC. 3.3. THE DERIVATIVE AND CONTINUITY; ANTIDERIVATIVES

1 $3x^2$ **5** $10x$
3 $1/(2\sqrt{x})$ **7** $-3/x^2$

9 $-\dfrac{3}{x^2} - 4$

11 (a) 0.21 (b) -0.59
13 (a) 6 (b) 6 (c) 6 (d) 6
(e) $6x + C$; choose different values of C.
15 (a) $4x^3$ (b) $68x^3$ (c) $4kx^3$ (d) $x^4/4$
17 (a) $6kx^5$ (b) $\dfrac{x^6}{6} + C$; choose different values of C.

19 (a), (b) $f(x) = x^3$

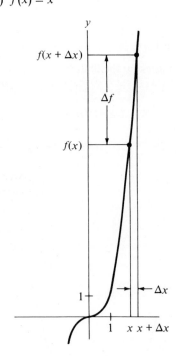

21 $\dfrac{-3}{(3x + 5)^2}$

23 $-\sin x$

SEC. 3.4. THE DERIVATIVES OF THE SUM, DIFFERENCE, PRODUCT, AND QUOTIENT

1 $5x^4 - 4x$ **7** $\frac{1}{2}(12x^3 - 2x + 5)$
3 $8x^3 - 12x + 5$ **9** $5/(2\sqrt{x})$
5 $5x^4 + 12x^3 - x^2 - 12x - 2$ **11** $-12/x^2$

13 $\dfrac{3 - 6x - x^2}{(3 + x^2)^2}$

15 $\dfrac{-t^4 + 6t^3 - 3t^2 + 2t - 3}{(t^3 + 1)^2}$

17 $3x^2 + \tfrac{7}{2}x^{5/2}$ 19 $24x^2$ 21 $\dfrac{1}{x^2} - \dfrac{2}{x^3}$

23 $\dfrac{-3x^2 - 2}{(x^3 + 2x + 1)^2}$

25 $\dfrac{15x^6 + 15x^4 + 5x^2 + 16x - 3}{(5x^2 + 3)^2}$

27 $8x + 4$ 29 $\dfrac{-2}{(x - 1)^2}$ 31 $\dfrac{-1}{2x^{3/2}}$

33 $\dfrac{-x^5 - 9x^4 - 20x^3 - 5x^2 + 23x + 2}{2\sqrt{x}(x^3 - 5x + 2)^2}$

35 3   41 $\tfrac{1}{12}$
37 $\tfrac{41}{2}$ 43 $\tfrac{23}{4}$
39 $\tfrac{5}{6}, \tfrac{5}{6}$ 45 $\tfrac{1}{9}$

SEC. 3.5. THE DERIVATIVES OF THE TRIGONOMETRIC FUNCTIONS

1 $x \sin x$ 5 $\tan^2 x$ 9 $\dfrac{1 + \sin x}{\cos^2 x}$

3 $x^2 \sin x$ 7 $x^2 \cos x$ 11 $\dfrac{-1 - 3 \cos x}{\sin^2 x}$

13 $\dfrac{-\csc x (3x \cot x + 1)}{3x^{4/3}}$

15 $\sin x (\sec^2 x + 1)$

17 $\dfrac{-(1 + x^2) \csc^2 x - 2x \cot x}{(1 + x^2)^2}$

21 (a) 0   (b) $-\tfrac{1}{2}$ (c) $-1/\sqrt{2}$ (d) $-\sqrt{3}/2$
   (e) -1

23 (a) 3   (b) -3 (c) 3, -3 (d) 3, 3

25 $\pi/4$

27 (a) $\sin x, \sin x + 1$
   (b) $5 \sin x, 5 \sin x + 1$
   (c) $-\cos x, -\cos x + 1$
   (d) $3 \cos x, 3 \cos x + 1$

SEC. 3.6. COMPOSITE FUNCTIONS AND THE CHAIN RULE

1 $y = u^{50}, u = x^3 + x^2 - 2$
3 $y = \sqrt{u}, u = x + 3$

5 $y = \sin u, u = 2x$
7 $y = u^3, u = \cos v, v = 2x$
9 $40(2x^3 - x)^{39}(6x^2 - 1)$

11 $-\dfrac{12}{x^2}\left(1 + \dfrac{3}{x}\right)^3$

13 $\dfrac{3x^2 + 1}{2\sqrt{x^3 + x + 2}}$

15 $3 \cos 3x$

17 $\dfrac{-(x + x^3) \sec^2 (1/x) + (3x^2 + x^4) \tan (1/x)}{(1 + x^2)^2}$

19 $(2x + 1)^4(3x + 1)^6(72x + 31)$
21 $x \sin^4 3x (15x \cos 3x + 2 \sin 3x)$

23 $\dfrac{-10}{(2x + 3)^6}$

25 $\dfrac{-6(2x + 1)^2(x + 1)}{(3x + 1)^5}$

27 $\dfrac{(x^3 - 1)^5 \cot^3 5x}{x}\left[\dfrac{15x^2}{x^3 - 1} - \dfrac{15 \csc^2 5x}{\cot 5x} - \dfrac{1}{x}\right]$

29 $\dfrac{-4(1 + 2x)^3}{(1 + 3x)^5}$

31 $\dfrac{1}{\sqrt{x}} \tan \sqrt{x} \sec^2 \sqrt{x}$

33 $\dfrac{x}{(1 - x^2)^{3/2}}$

35 $\dfrac{30(3x - 2)^3}{\sqrt{5(3x - 2)^4 + 1}}$

37 $4 \cos 3x \cos 4x - 3 \sin 3x \sin 4x$

39 $\tfrac{7}{5}x^{2/5}$

41 $\dfrac{7x^2}{3}(x^3 + 2)^{-2/9}$

43 $\tfrac{10}{3}x^{7/3}$

45 $5y^4 \dfrac{dy}{dx}$

47 $(\cos y)\dfrac{dy}{dx}$

49 (a) 15   (b) 15
51 $\tfrac{1}{3} \sin 3x$
53 $-\tfrac{1}{2} \cos 2x$
55 $\sin^3 3x$
57 (b) 35

SEC. 3.S. SUMMARY: GUIDE QUIZ ON CHAPTER 3

2 (a) $15x^2 - 2$ (b) $\dfrac{-15}{(3x + 2)^2} + 6$ (c) $6 \cos 2x$
   (d) $-2x^{-3}$

3 (a) $5/(2\sqrt{x})$ (b) $\dfrac{6x(1-x^2)}{\sqrt{3-2x^2}}$ (c) $-5\sin 5x$

(d) $\dfrac{3x}{2}(1+x^2)^{-1/4}$ (e) $\dfrac{2\sec^2 6x}{(\tan 6x)^{2/3}}$

(f) $5x^3\cos 5x + 3x^2 \sin 5x$ (g) $\dfrac{-1}{(2x+1)^{3/2}}$

(h) $\dfrac{-4(10x^4 - 3x^2)}{(2x^5 - x^3)^5}$ (i) $\dfrac{x^2}{(x^3-3)^{2/3}}$

(j) $\dfrac{6(2x^3 + x^2 - 1)}{(3x+1)^2}$ (k) $\dfrac{-10x}{(5x^2+1)^2}$ (l) $\dfrac{-30}{(3x+2)^{11}}$

(m) $x^2(1+2x)^4 \sec 3x\,[16x + 3 + 3x(1+2x)\tan 3x]$

(n) $\dfrac{-\csc\sqrt{x}\cot\sqrt{x}}{2\sqrt{x}}$ (o) $\dfrac{24\csc^2 4x}{(1+3\cot 4x)^3}$

4 (a), (b)

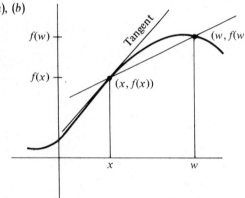

5 (a)

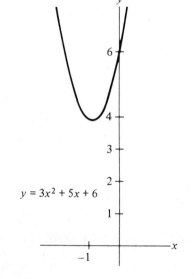

(c) $x = -\tfrac{5}{6}$

6

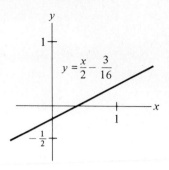

7 (a)

x	-2	-1	0	1	2	3
$x^3 - 12x$	16	11	0	-11	-16	-9

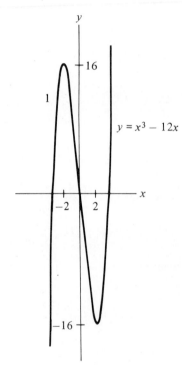

(b) $3x^2 - 12$ (c) $x = \pm 2$ (d) $(-2, 16), (2, -16)$

8 (a) $f(w) - f(x)$ is the change in height during the time interval of duration $w - x$. The quotient $[f(w) - f(x)]/(w - x)$ is the average velocity.

(b) $f(w) - f(x)$ is the growth of the culture during the time interval of duration $w - x$. The quotient is the average rate of growth.

(c) See Problem 4 in the text of Sec. 3.1.

(d) See Problem 3 in the text of Sec. 3.1.

9 (a) $2t - 2$ (b) $-\tfrac{3}{2}$ m/sec (c) $\tfrac{3}{2}$ m/sec

(d) Left

10 (a) $x^3/3$ (b) $-1/x$ (c) $-\tfrac{1}{3}\cos 3x$

(d) $x^3 + 2x^2 + 5x$ (e) $\tan x$ (f) $\tfrac{1}{2}\sin^2 x$

REVIEW EXERCISES FOR CHAPTER 3

1. $15x^2$
3. $\dfrac{-1}{(x+3)^2}$
5. $-3 \sin 3x$
7. $10x^4 + 3x^2 - 1$
9. $\dfrac{4x^2 + 2x}{(4x+1)^2}$
11. $\dfrac{3x+1}{\sqrt{3x^2 + 2x + 4}}$
13. $\dfrac{4}{3\sqrt[3]{2t-1}}$
15. $10 \sin 5x \cos 5x$
17. $\tfrac{20}{7}(5x+1)^3$
19. $3x \cos 3x + \sin 3x$
21. $\dfrac{4 \tan \sqrt[3]{1+2x} \sec^2 \sqrt[3]{1+2x}}{3(1+2x)^{2/3}}$
23. $\dfrac{(x^4 + 3x^2) \cos 2x - (2x^3 + 2x^5) \sin 2x}{(1+x^2)^2}$
25. $\dfrac{1}{6\sqrt{x}\,(1+\sqrt{x})^{2/3}}$
27. $-\tfrac{35}{3} \csc^2 5x (\cot 5x)^{4/3}$
29. $\dfrac{4}{\sqrt{8x+3}}$
31. $\dfrac{2x - x^4}{(x^3+1)^2}$
33. $\dfrac{-5[4(x^2+3x)^3(2x+3)+1]}{7[(x^2+3x)^4+x]^{12/7}}$
35. $\dfrac{3x+1}{5\sqrt{2x+1}} \cos^2 6x - \dfrac{12x}{5}\sqrt{2x+1}\cos 6x \sin 6x$
37. $x \sin ax$
39. $\sin^2 ax$
41. (b) 64, 32, 0, -32 (c) 64, 32, 0, 32
    (d) Rising: $0 < t < 2$; falling: $t > 2$
43. $y = -x$
45. (a) 12 g/cm
47. (a) $\tfrac{3}{2}$ (b) $\tfrac{1}{2}$ (c) $\tfrac{1}{4}$ (d) $\tfrac{1}{6}$
49. (a) 1,000   (b) $5 + x/100$ (c) 5.1   (d) 5.105
51. (a) 12.61   (b) 11.11   (c) 12
53. Rate of weight change
55. (a) 5   (b) 2.25   (c) 4.5   (d) 4
57. (a) $3y^2 \dfrac{dy}{dx}$ (b) $(-\sin y)\dfrac{dy}{dx}$ (c) $\dfrac{-1}{y^2}\dfrac{dy}{dx}$
59. (a) $x^4 + C$, $C = 0, 1$
    (b) $\dfrac{x^4}{4} + C$, $C = 0, 1$
    (c) $\dfrac{x^5}{5} + \dfrac{x^4}{4} + \sin x + C$, $C = 0, 1$
    (d) $\dfrac{x^4}{4} - \cos x + C$, $C = 0, 1$
    (e) $\dfrac{x^5}{5} + \dfrac{2x^3}{3} + x + C$, $C = 0, 1$
61. (a) $\lim_{x \to a} f(x)$ exists for all values of a.   (b) $a = 1$
    (c) $a = 0, 2, 3, 4$
67. $\dfrac{\cos \sqrt{3}}{2\sqrt{3}}$
69. (a) 0.6932   (b) 1.0987   (c) 1.00005
71. (a), (b)

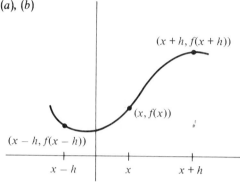

CHAPTER 4. APPLICATIONS OF THE DERIVATIVE

SEC. 4.1. ROLLE'S THEOREM AND THE MEAN-VALUE THEOREM

1. (a)

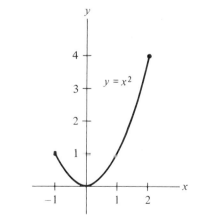

(b) 4
(c) Yes
(d) No
(e) Yes

3 (a)

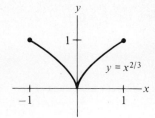

(c) No

5 $\pi/2, 3\pi/2$

7 $-1, 0, 1$

9 $\dfrac{1-\sqrt{7}}{3}, \dfrac{1+\sqrt{7}}{3}$

11 All values in $(1, 3)$

13 $\pi, 2\pi, 3\pi, 4\pi, 5\pi$

15 $(-3+\sqrt{57})/3$

17 (a) $2\sec^2 x \tan x$ (b) 1

19 (a) $2x^4 + C$ (b) $-\tfrac{1}{2}\cos 2x + C$

(c) $-\dfrac{1}{x} + C$ (d) $-\tfrac{1}{2}(x+1)^{-2} + C$

25 (a) $3\sqrt{1-x^2}\cos 3x - \dfrac{x}{\sqrt{1-x^2}}\sin 3x$

(b) $\dfrac{1-5x^2}{3x^{2/3}(x^2+1)^2}$ (c) $\dfrac{-4}{(2x+1)^3}\sec^2\dfrac{1}{(2x+1)^2}$

27 (a) Corollary 2 (b) Corollary 1

29 (a) $-1, 1$ (c) 110

11 Critical point and global minimum at $(1, 0)$

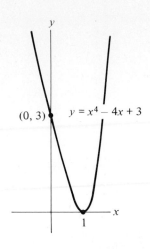

SEC. 4.2. USING THE DERIVATIVE AND LIMITS WHEN GRAPHING A FUNCTION

1 0, neither **3** 1, neither

5 $-\tfrac{1}{4}$, minimum; 0, neither

7 0, minimum; $(4k+1)\dfrac{\pi}{2}$ is a maximum and $(4k+3)\dfrac{\pi}{2}$ is a minimum, $k = 0, \pm 1, \pm 2, \pm 3, \ldots$

9 Critical point at $(1, 1)$

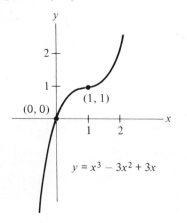

13 Critical point and global minimum at $(3, -4)$

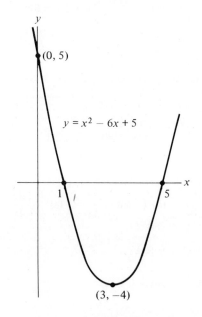

15 Critical point and local maximum at $(0, 0)$; critical number for $x_1 = (-3 + \sqrt{33})/4$ and local minimum at $(x_1, f(x_1))$; critical number for $x_2 = (-3 - \sqrt{33})/4$ and global minimum at $(x_2, f(x_2))$

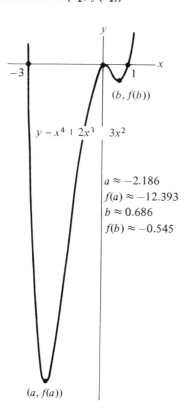

17 Asymptotes $x = \frac{1}{3}$ and $y = 1$

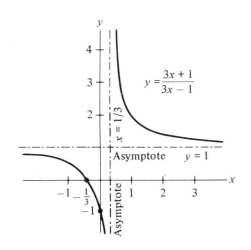

19 Critical point and global maximum at $(1, \frac{1}{2})$; critical point and global minimum at $(-1, -\frac{1}{2})$; asymptote $y = 0$

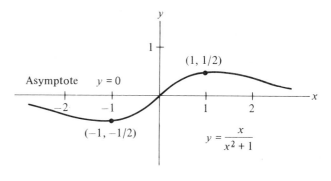

21 Critical point and local maximum at $(\frac{1}{4}, -8)$; asymptotes $x = 0$, $x = \frac{1}{2}$, $y = 0$

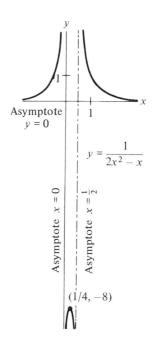

23 Critical point and local maximum at $(0, -\frac{3}{4})$; asymptotes $x = -2$, $x = 2$, $y = 1$

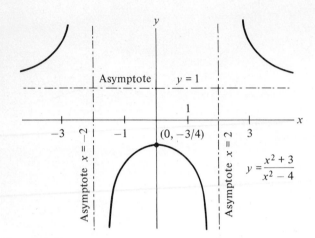

25 $\frac{1}{4}$, 0
27 3, 0
29 24, -8
31 2, 0

33

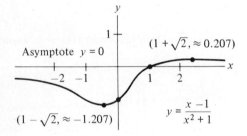

35

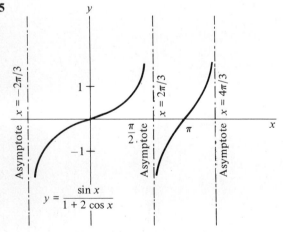

37

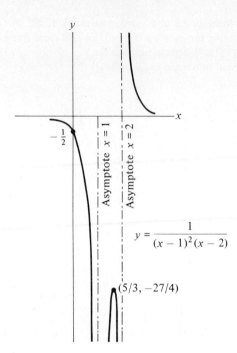

39

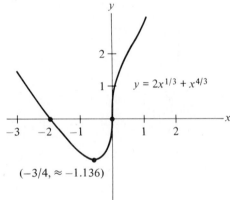

41

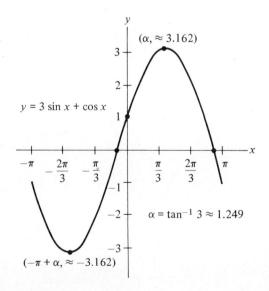

SEC. 4.3. CONCAVITY AND THE SECOND DERIVATIVE

1 $2x - 1, 2, 0, 0$

3 $3x^2, 6x, 6, 0$

5 $2\cos 2x, -4\sin 2x, -8\cos 2x, 16\sin 2x$

7 $-2/x^3, 6/x^4, -24/x^5, 120/x^6$

9 (a) $8\sec^2 2x \tan 2x$
(b) $-(1 + 2x)^{-3/2}$
(c) $\dfrac{\csc^2 \sqrt{x}}{4x^{3/2}} + \dfrac{\csc^2 \sqrt{x} \cot \sqrt{x}}{2x}$

11 Maximum at $x = 0$; minimum at $x = \frac{2}{3}$; inflection point at $x = \frac{1}{3}$

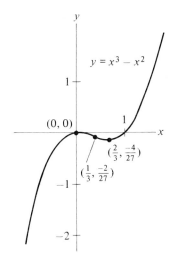

13 Minimum at $x = -\frac{3}{2}$; inflection points at $x = 0$ and $x = -1$

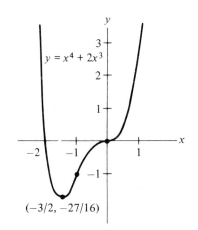

15 Minimum at $x = 3$; inflection points at $x = 0$ and $x = 2$

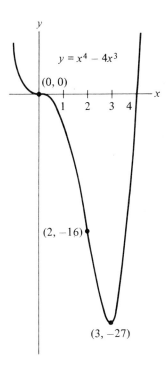

17 Maximum at $x = 0$; inflection points at $x = \pm 1/\sqrt{3}$

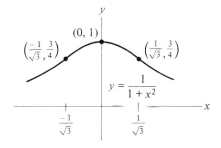

19 Maximum at $x = -1$; minimum at $x = 5$; inflection point at $x = 2$

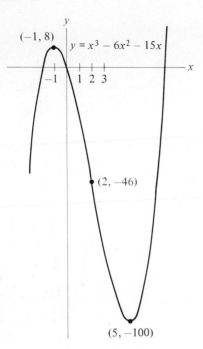

23 Inflection points at $x = n\pi$, $n = 0, \pm 1, \pm 2, \pm 3, \ldots$

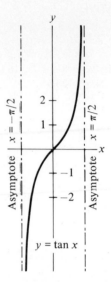

25 Maximum at $x = 0$; inflection points at $x = \pm\frac{1}{3}$

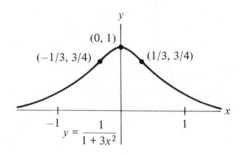

21 Inflection points at $x = 0$ and $x = 1/\sqrt[3]{2}$

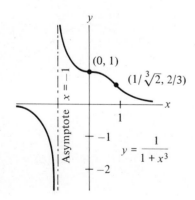

27 Minimum at $x = 1$

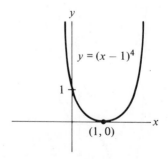

29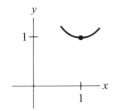

31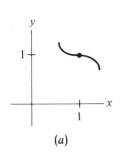
(a) (b)

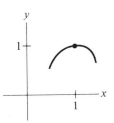 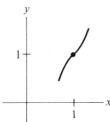
(c) (d)

33

35

41 (a) $x = 1; x = 2$ (b) 1

43 (a)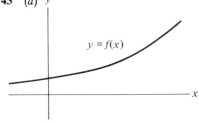

(b)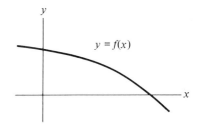

(c) No

49 (a) 1, 2, 2 (b) $a_0, a_1, 2a_2$

SEC. 4.4. MOTION AND THE SECOND DERIVATIVE

1 $f'' > 0$
3 (a) 4 sec (b) Velocity $= -64$ ft/sec; speed $= 64$ ft/sec
9 $3t^2 - 3t + 3$ ft
11 $-16t^2$
13 (a) $f(x) = c_1 x + c_2$ (b) $f(x) = 5x + 4$
15 $-\frac{1}{9}\cos 3x + c_1 x + c_2$
19 (a) 294.8 ft (b) 86.9 ft (c) 44.5 ft
21 (a)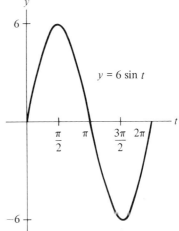

(b) 6 in
(d) $t = 0, \pm\pi, \pm 2\pi, \pm 3\pi, \ldots$
(e) $t = (2n+1)\dfrac{\pi}{2}; n = 0, \pm 1, \pm 2, \pm 3, \ldots$

23 16,000 mi

25 $\lim\limits_{r\to\infty} v = 0$

27 No

SEC. 4.5. APPLIED MAXIMUM AND MINIMUM PROBLEMS

1. $\frac{5}{6}$
3. $(6 - 2\sqrt{3})/3$
5. $x = 25$, $y = 50$
7. $h = \sqrt[3]{\dfrac{400}{\pi}}$
9. $10 \times 10 \times 10$
11. $\theta = \pi/4$; $w = h = \sqrt{2}\, a$
15. 40×60
17. (a) $x = 1$, $y = 0$ (b) $x = \frac{1}{2}$, $y = \frac{1}{2}$
19. $x = 8$
21. $r = 100/3\pi$, $h = 100/3$
23. Length = $50/3$, height = $100/3$
25. (c) $\sqrt[3]{\dfrac{25}{\pi}}$ (d) $4\sqrt[3]{\dfrac{25}{\pi}}$
27. Length = $2\sqrt[3]{\dfrac{75}{7}}$, height = $\dfrac{7}{3}\sqrt[3]{\dfrac{75}{7}}$
29. 414 on the first road and 586 on the second
31. $r = \sqrt{\dfrac{2}{3}}\, a$, $h = \dfrac{2}{\sqrt{3}}\, a$
33. Length = $\sqrt[3]{\dfrac{2V}{3}}$, height = $\dfrac{3}{2}\sqrt[3]{\dfrac{2V}{3}}$
35. (a) SD (b) 63
39. $13\sqrt{13}$
41. $\sqrt{a(a+b)}$
43. $130
45. (b) $\dfrac{RLv}{2A} x - K + \dfrac{ALc}{2v} x^3 = 0$

SEC. 4.6. IMPLICIT DIFFERENTIATION

1. -4
3. $-\frac{7}{15}$
5. $-\pi/2$
7. $-\frac{8}{9}$
9. $r = \sqrt[3]{\dfrac{50}{\pi}}$, $h = 2\sqrt[3]{\dfrac{50}{\pi}}$
11. $x = 25$, $y = 50$
15. $\dfrac{50}{3} \times \dfrac{50}{3} \times \dfrac{100}{3}$
17. (b) $-\frac{3}{5}$ (d) $-\frac{168}{125}$
19. $-2, -6$

SEC. 4.7. THE DIFFERENTIAL

1. $dy = 0.6$, $\Delta y = 0.69$

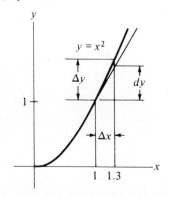

3. $dy = -0.3$, $\Delta y = -0.271$

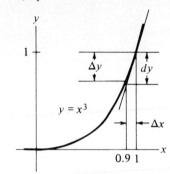

5. $dy = -\frac{1}{9}$, $\Delta y \approx -0.115501$

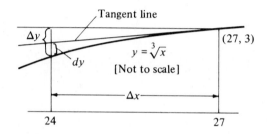

7. (a) -0.02
9. (a) $h/2$ (c) 0.005

11.

x	dx	dy	Δy	$\Delta y/dy$
3	1	27	37	1.3704
3	-0.5	-13.5	-11.3750	0.8426
1	0.1	0.3	0.3310	1.1033
2	-0.1	-1.2	-1.1410	0.9508

13. $-\dfrac{3}{x^4}\, dx$
15. $2 \cos 2x \, dx$
17. $-\csc x \cot x \, dx$
19. $-\dfrac{1}{x^2}(5x \csc^2 5x + \cot 5x)\, dx$
21. 10 percent
23. 7.8125
25. 9.9
27. 2.9259
29. 0.98
31. 0.8560
33. 0.13
35. 0.248125
37. $\dfrac{1}{2} + \dfrac{\sqrt{3}\pi}{180} \approx 0.5302$
39. $\dfrac{\sqrt{3}}{2} + \dfrac{\pi}{180} \approx 0.8835$

SEC. 4.S. SUMMARY: GUIDE QUIZ ON CHAPTER 4

3 (a) If f is continuous, $f(a) = 0$.
   (b) There is a maximum at $x = a$.
   (c) The concavity changes from upward to downward; $x = a$ is an inflection number.

6 (a) $-\frac{1}{3}\cos 3x + C$ (b) Corollary 2 in Sec. 4.1

7 $(4 - \sqrt{7})/3$ is squared, $(-1 + \sqrt{7})/3$ is cubed.

8 Maximum at $x = 0$; minimum at $x = L/2$ $(r = 0)$

9 (c) 8.1

10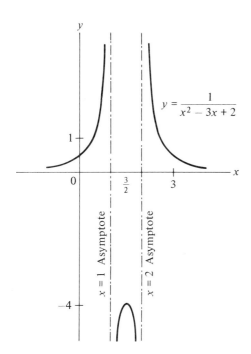

11 $0, -\frac{1}{3}$

12 (a) $\dfrac{x^4}{4} + \dfrac{2x^3}{3} + C$

   (b) $\dfrac{-1}{2x^2} + C$

   (c) $-\frac{5}{2}\cos 2x + C$

   (d) $\sqrt{1 + x^2} + C$

14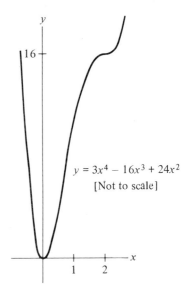

15 (a) $240x - \dfrac{24}{x^5}$ (b) $16\cos 2x$ (c) 0
   (d) $-\frac{15}{16}x^{-7/2}$

REVIEW EXERCISES FOR CHAPTER 4

3 (b) No

5 $y = \dfrac{1}{x^2} + \dfrac{1}{x - 1}$

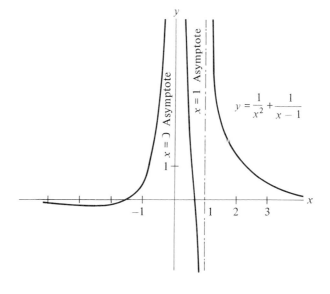

7 (a), (d)

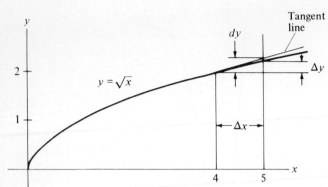

(e) $\dfrac{3}{x^4} \sin \dfrac{1}{x^3}$

(f) $\dfrac{-x}{\sqrt{1-x^2}} \sec^2 \sqrt{1-x^2}$

23 In all cases, travel $\tfrac{5}{4}$ mi on the grass.

25 (a) 0.0873   (b) 0.988

27 (c) There is only one intersection in $(0, \pi)$.

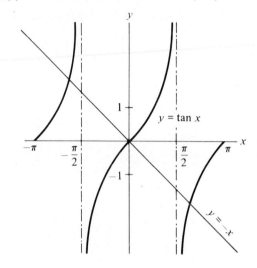

(b) $\tfrac{1}{4}$

(c) $\sqrt{5} - 2 \approx 0.236$

11 (a) $\dfrac{9L}{4\sqrt{3}+9}$ for the triangle, $\dfrac{4\sqrt{3}L}{4\sqrt{3}+9}$ for the square

(b) All for the square

13 $f(3) \geq 3$

15 -1

17 $y = \sqrt{x}/(1+x)$

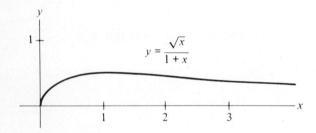

21 (a) $\dfrac{4x^3 + 12x^2 - 2}{(x+2)^2}$

(b) $\dfrac{3x^5}{2\sqrt{1+3x}} + 5x^4\sqrt{1+3x}$

(c) $\tfrac{10}{7}(2x-1)^4$

(d) $\dfrac{2}{\sqrt{x}} \sin^3 \sqrt{x} \cos \sqrt{x}$

35 $\sqrt{2a} \times \sqrt{2b}$

37 $a/\sqrt{2}$

43 (a) 2, $\tfrac{11}{2}$ (b) $-1, 4, 6$ (c) $-1, 1, 4, \tfrac{11}{2}, 6$
(d) None   (e) -1

45 (a) $36x - 3x^2$ (b) midpoint

47 $h = \sqrt{8}\, r$

49 (a) 5   (b) $\sqrt{A^2 + B^2}$

55 $f(x) = x^3$

61 No

63 (c)

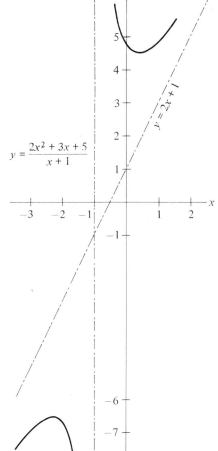

65 (a) $y = x + 1$, $y = -x - 1$

(b)

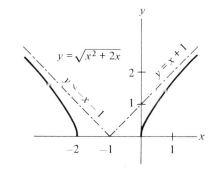

67 (a) $m > n$ (b) $m = n$ (c) $n = m + 1$
71 Yes; no; it has precisely one of the two, never both.
75 $r = \left(\dfrac{1 + \sqrt{5}}{4}\right)h$
77 (b) f is constant.

CHAPTER 5. THE DEFINITE INTEGRAL

SEC. 5.1. ESTIMATES IN FOUR PROBLEMS

1 (a) 8.75
(b) 14

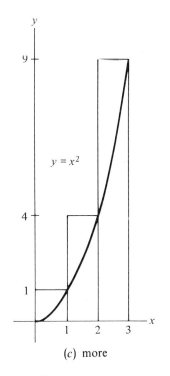
(c) more

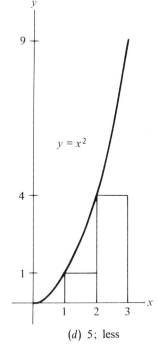
(d) 5; less

(e) 14, 5

3 (a) 44.55 g (b) 59.4 g (c) 32.4 g
(d) 59.4 g, 32.4 g

5 Problem 1: $\dfrac{2{,}413}{300} \approx 8.043$

Problem 2: $\dfrac{2{,}413}{60} \approx 40.217$ g

Problem 3: $\dfrac{4{,}826}{75} \approx 64.347$ mi

Problem 4: $\dfrac{2{,}413}{300} \approx 8.043$ ft^3

7 (a) 7.695 ft^3 (b) 10.395 ft^3

9 $\dfrac{143}{16} = 8.9375$ (millions of dollars)

11 (a)

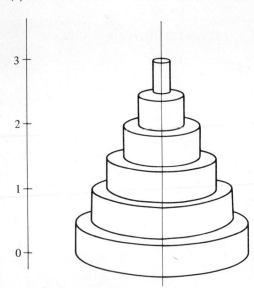

(b) $143\pi/16 \approx 28.078$ ft^3

13 (a) $\dfrac{1,879}{2,520} \approx 0.7456$

(b) $\dfrac{1,627}{2,520} \approx 0.6456$

15 (a) $\tfrac{1}{4}(x+2)^4$
(b) $\tfrac{1}{5}x^5 + \tfrac{2}{3}x^3 + x$
(c) $-\tfrac{1}{2}\cos x^2$
(d) $\dfrac{x^4}{4} - \dfrac{1}{2x^2}$
(e) $2\sqrt{x}$

17 Underestimates (for a perfect graph)

SEC. 5.2. SUMMATION NOTATION AND APPROXIMATING SUMS

1 (a) 6 (b) 20 (c) 14
3 (a) 4 (b) 1 (c) 450
5 (a) $\displaystyle\sum_{i=0}^{100} 2^i$ (b) $\displaystyle\sum_{j=3}^{7} x^j$ (c) $\displaystyle\sum_{k=3}^{102} \dfrac{1}{k}$
7 (a) $\displaystyle\sum_{i=1}^{3} x_{i-1}^2(x_i - x_{i-1})$ (b) $\displaystyle\sum_{i=1}^{3} x_i^2(x_i - x_{i-1})$
9 (a) $2^{100} - 1$ (b) $-\tfrac{99}{100}$ (c) $-\tfrac{100}{101}$
13 13.50
15 10.00
17 1.07
19 $11(b-a)$
21 6
23 $\tfrac{117}{160} = 0.73125$

25 $\displaystyle\sum_{i=1}^{n} c^{n-i}d^{i-1}$ and $\displaystyle\sum_{i=0}^{n-1} c^{n-i-1}d^i$ are two acceptable answers; there are others.
27 (b) n^2
29 (c) n (d) $n(n+1)/2$
31 (a) 1.5, 1.083, 0.95, and 0.885, respectively

SEC. 5.3. THE DEFINITE INTEGRAL

1 (a) 1 (b) 2 (c) 3
3 (a) Length of the ith section of string
(b) Density at a point of the ith section
(c) Approximate mass of the ith section
(d) Approximate mass of the string
(e) Actual mass of the string
5 (a) Length of the ith time interval
(b) Rate of water flow at an instant in the ith time interval
(c) Approximate amount of water flowing into the lake during the ith time interval
(d) Estimated total volume of water flow into the lake
(e) Actual volume of water flow
7 (a) 30 (b) 54
11 (a) $\tfrac{77}{60} \approx 1.2833$ (b) $\tfrac{19}{20} = 0.9500$
13 (a) $\tfrac{4}{25} = 0.16$ (b) $\tfrac{9}{25} = 0.36$
15 (a) 0.77 (b) 1.20
17 $\tfrac{1669}{1800} \approx 0.9272$
19 39 21 $\tfrac{26}{3}$ g 27 (b) $p \leq \dfrac{0.001}{8} = 0.000125$
29 (a) 20 (b) 20 ft (c) 20π
31 9 ft^3
33 (c) Logarithmic functions
35 (a)

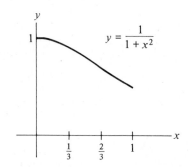

37 (a)

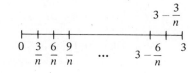

(b) 0, $3/n$, $6/n$, $3i/n$, respectively
(e) $\tfrac{9}{2}$

SEC. 5.4. THE FUNDAMENTAL THEOREMS OF CALCULUS

3   63
5   $\frac{12}{5}$
7   10.5 ft
9   18.75 g
11 936 cm^3
13 3.75
15 27
17 $\frac{190}{3}$
19 18.6
21 x^4
23 $20x^2$
25 x^5
27 $(\sqrt{1 + \sin^3 x})(\cos x)$
29 $\sin x^2$
31 (a) $\pi(f(x))^2$ (b) $\int_a^b \pi(f(x))^2 \, dx$
33 $\pi/4$
35 $37\pi/3$
37 (b) 1
39 (a) Cannot apply FTC   (b) $\frac{3}{4}(2^{4/3} - 1)$
    (c) $\frac{3}{2}(2^{2/3} - 1)$
41 (c) $\frac{4}{3}\pi r^3$

SEC. 5.5. PROPERTIES OF THE ANTIDERIVATIVE AND THE DEFINITE INTEGRAL

1   $x^2 - \frac{1}{4}x^4 + \frac{1}{6}x^6 + C$
3   $\frac{3}{2}x^2 - 2\cos 2x + \frac{1}{4x} + C$
5   $\frac{1}{3}(2x + 5)^{3/2} + C$
7   $\frac{1}{2}\sec 2x + C$
9   $x - \frac{1}{2}\sin 2x + C$
11 (a) x^5 (b) $\sin x^5$
13 (a) $\frac{7}{3}$ (b) $\frac{1}{6}$ (c) $-\frac{1}{3}(\cos 6 - \cos 3)$
15 (a) 3   (b) 0   (c) 0   (d) -2
17 (a) Function   (b) Number   (c) Number
19 (a) True   (b) False
23 0
25 -42
27 $\sin^3 x$
29 $9x \tan 3x - 4x \tan 2x$
31 $2x \cos x^2$
33 $-\frac{x^2}{2}(1 + 3x)^{-3/2}(6 \sin 2x + 15x \sin 2x + 4x \cos 2x + 12x^2 \cos 2x)$
35 (a) $12 \leq \int_2^6 f(x) \, dx \leq 20$
    (b) The mass of a 4-cm string whose density is between 3 and 5 grams per centimeter is between 12 and 20 g.
37 $c = 0, \pi, 2\pi, 3\pi,$ or 4π
39 $\sqrt[3]{\frac{9}{4}}$

SEC. 5.S. SUMMARY: GUIDE QUIZ ON CHAPTER 5

2   (a) $\frac{29}{18} \approx 1.6111$ (b) $\frac{163}{252} \approx 0.6468$
3   (a) $-2/(2x + 3)^2$ (b) $\frac{1}{15}$
4   $3\pi/4$
5   (a) $\sin 25$
    (b) $\frac{8x^2}{1 - 5x^6} + \frac{40x^6 + 4}{(1 - 5x^6)^{3/2}} \int_0^{x^2} \frac{dt}{\sqrt{1 - 5t^3}}$
6   (b) $f(x) = x \tan x$ (c) $g(x) = \int_0^x t \tan t \, dt$

REVIEW EXERCISES FOR CHAPTER 5

1   $\frac{15}{2}$
3   $\frac{1}{3}$
5   $\frac{5}{72}$
7   $\tan x + C$
9   $\sec x + C$
11 $-4 \csc x + C$
13 $\frac{1}{7}x^7 + \frac{1}{2}x^4 + x + C$
15 $5x^{20} + C$
17 (a) $18x^2(x^3 + 1)^5$ (b) $\frac{1}{18}(x^3 + 1)^6 + C$
19 (a) $d + d^2 + d^3$
    (b) $x + x^2 + x^3 + x^4$
    (c) $\frac{11}{8} = 0 + \frac{1}{2} + \frac{1}{2} + \frac{3}{8}$
    (d) $\frac{317}{60} = \frac{3}{2} + \frac{4}{3} + \frac{5}{4} + \frac{6}{5}$
    (e) $\frac{3}{10} = (\frac{1}{2} - \frac{1}{3}) + (\frac{1}{3} - \frac{1}{4}) + (\frac{1}{4} - \frac{1}{5})$
    (f) $1 + \sqrt{2} = \sin\frac{\pi}{4} + \sin\frac{\pi}{2} + \sin\frac{3\pi}{4} + \sin\pi$
21 (b) $1/n$ (c) $1/n$ (d) right endpoint
23 (a) $\int_0^2 x^3 \, dx$ (b) $\int_0^1 x^4 \, dx$ (c) $\int_1^3 x^5 \, dx$
25 (a) $3x^2 \sin 2x^3$
    (b) $(\sin 2x^3)(3x^2)$
27 $\int_4^x \sqrt[3]{1 + t^2} \, dt$
29 Distance traveled
33 It makes no difference.

CHAPTER 6. TOPICS IN DIFFERENTIAL CALCULUS

SEC. 6.1. REVIEW OF LOGARITHMS; THE NUMBER e

1. (a) $5 = \log_2 32$ (b) $4 = \log_3 81$
   (c) $-3 = \log_{10}(0.001)$ (d) $0 = \log_5 1$
   (e) $\frac{1}{3} = \log_{1000}(10)$ (f) $\frac{1}{2} = \log_{49} 7$

3. (a)

x	$\frac{1}{9}$	$\frac{1}{3}$	1	3	9
$\log_3 x$	-2	-1	0	1	2

(b)

[Graph of $y = \log_3 x$]

5. (a) $2^x = 7$ (b) $5^s = 2$
   (c) $3^{-1} = \frac{1}{3}$ (d) $7^2 = 49$
7. (a) 16   (b) $\frac{1}{2}$ (c) 7   (d) g
9. (a) $\frac{1}{2}$ (b) 5   (c) -3
11. $x = \log_3 \frac{7}{2} \approx 1.1403$
13. $x = 0$
15. (a) $\log_2 8 = 3$, $\log_8 2 = \frac{1}{3}$
19. (a) -0.6309 (b) 1.1398 (c) -2.5850
21. e^8
23. (a) $\log_b c$ (b) $c^{1/3}$

SEC. 6.2. THE DERIVATIVES OF THE LOGARITHMIC FUNCTIONS

1. $\dfrac{2x}{1 + x^2}$
3. $\dfrac{1 - \ln x}{x^2}$
5. $\dfrac{1}{x} \sin 4x + 4 \cos 4x \ln 3x$
7. $2 \cot x \ln(\sin x)$
9. $\dfrac{2}{2x + 3}$
11. $\dfrac{x}{(5x + 2)^2}$
13. $\dfrac{1}{\sqrt{x^2 - 5}}$
15. $\dfrac{1}{x(3x + 5)}$
17. $\dfrac{1}{25 - x^2}$
19. $\dfrac{6x}{x^2 + 1} + \dfrac{20x^4}{x^5 + 1}$
21. $(1 + 3x)^5 (\sin 3x)^6 \left(\dfrac{15}{1 + 3x} + 18 \cot 3x \right)$
23. $x^{-1/2} (\sec 4x)^{5/3} \sin^3 2x \left[\dfrac{20}{3} \tan 4x + 6 \cot 2x - \dfrac{1}{2x} \right]$
25. $\dfrac{1}{3x} \log_{10} e$
27. (a) n
29. (a) $\ln|5x + 1| + C$ (b) $\frac{1}{2} \ln(x^2 + 5) + C$
    (c) $\ln|\sin x| + C$ (d) $\ln|\ln x| + C$
31. (a) 1   (b) $\ln 10$ (c) 13
33. (a) $\ln b$
    (c) $\pi \left(1 - \dfrac{1}{b} \right)$
    (e) See the solutions manual.

SEC. 6.3. THE NATURAL LOGARITHM DEFINED AS A DEFINITE INTEGRAL

3. (a) $\frac{28271}{27720} \approx 1.0199$

SEC. 6.4. INVERSE FUNCTIONS AND THE DERIVATIVE OF b^x

1.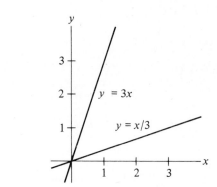

3.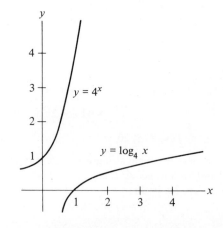

5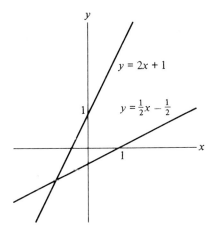

7 (a) No (b) Yes (c) Yes (d) Yes
9 (a) No (b) Yes (c) No (d) Yes
11 $\dfrac{1}{\sqrt{x}}$
13 $\sqrt{\log_{10} x}$
15 $\dfrac{x}{3} - \dfrac{5}{3}$
17 $\sqrt[5]{x^3 - 1}$
19 $2xe^{x^2}$
21 $2(x^2 + x)e^{2x}$
23 $-x(\ln 2)2^{-x^2+1}$
25 $x^{(x^2)}(x + 2x \ln x)$
27 3
29 $x^{\tan 3x}\left(3 \sec^2 3x \ln x + \dfrac{1}{x} \tan 3x\right)$
31 $\dfrac{-(1+e^x)(5\sin 5x) - (\cos 5x)(4 + 5e^x)}{e^{4x}(1+e^x)^2}$
33 $x^{\sqrt{3}-1}e^{x^2}(2x^2 \sin 3x + 3x \cos 3x + \sqrt{3} \sin 3x)$
35 xe^{ax}
37 $e^{ax} \sin bx$
39 (a) $\dfrac{1}{3}(e^{15} - e^3)$ (b) $\dfrac{\pi}{6}(e^{30} - e^6)$
41 (a) $\dfrac{999}{\ln 10}$ (b) $\dfrac{999999\pi}{2 \ln 10}$
43 (a) (0, 1) (b) $(\pm 1/\sqrt{2}, 1/\sqrt{e})$ (c) $y = 0$

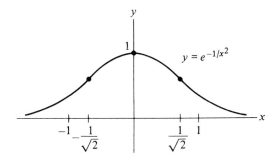

45 (a) (0, 0) (b) (1, 1/e)
(c) (1, 1/e), global maximum (d) $(2, 2/e^2)$
(e) $y = 0$
(f)

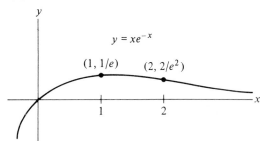

47 (a) (0, 0)
(b) (0, 0), $(3, 27/e^3)$
(c) $(3, 27/e^3)$, global maximum
(d) (0, 0), $(3 - \sqrt{3}, \approx 0.574), (3 + \sqrt{3}, \approx 0.933)$
(e) $y = 0$
(f)

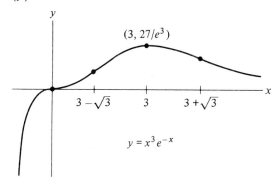

49 (a) (0, 0), (1, 0)
(b) $\left(\dfrac{3-\sqrt{5}}{2}, \approx 0.161\right), \left(\dfrac{3+\sqrt{5}}{2}, \approx -0.309\right)$,
(c) $\left(\dfrac{3-\sqrt{5}}{2}, \approx 0.161\right)$, global maximum;
$\left(\dfrac{3+\sqrt{5}}{2}, \approx -0.309\right)$, local minimum
(d) (1, 0), $(4, -12/e^4)$
(e) $y = 0$
(f)

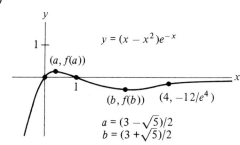

51 (a) 1   (b) 2 ln 2   (c) ln 10
55 (a)

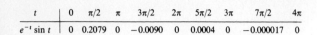

(b)

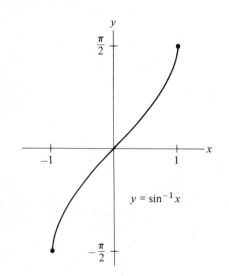

[Vertical scale exaggerated]

(c) $(\pi/4, \approx 0.3224)$, $(5\pi/4, \approx -0.0139)$
57 (b) Yes   (c) No

SEC. 6.5. THE DERIVATIVES OF THE INVERSE TRIGONOMETRIC FUNCTIONS

3 (a) 1.2490   (b) -1.1071 (c) 0.4115   (d) 1.2310
5 (a) and (d)
7

x	-1	-0.8	-0.6	-0.4	-0.2	0	0.2	0.4	0.6	0.8	1
$\sin^{-1} x$	-1.57	-0.93	-0.64	-0.41	-0.20	0	0.20	0.41	0.64	0.93	1.57

11 $1/\sqrt{2}$
13 -1
15 0.3
17 $\dfrac{5}{\sqrt{1 - 25x^2}}$
19 $\dfrac{1}{|x|\sqrt{9x^2 - 1}}$
21 $\dfrac{1}{3x^{2/3}(1 + x^{2/3})}$
23 $\dfrac{x}{2\sqrt{x-1}} + 2x \sec^{-1}\sqrt{x}$
25 $\dfrac{3 \sin 3x}{\sqrt{1 - 9x^2}} + 3 \cos 3x \sin^{-1} 3x$
27 $e^{-2x}\left[\dfrac{x}{|x|\sqrt{9x^2 - 1}} + (1 - 2x)\sec^{-1} 3x\right]$
29 $\dfrac{1}{2\sqrt{x}(1 + x)}$
31 $\dfrac{1}{2x\sqrt{x-1} \sec^{-1}\sqrt{x}}$
33 $\dfrac{1}{\tan^{-1} 10^x} - \dfrac{x \, 10^x \ln 10}{(\tan^{-1} 10^x)^2(1 + 10^{2x})}$
35 $\sqrt{\dfrac{1 + x}{1 - x}}$
37 $\dfrac{6(\tan^{-1} 2x)^2}{1 + 4x^2}$
39 $\sqrt{2 - x^2}$
41 $\dfrac{5}{3x\sqrt{9x^{10} - 1}}$
43 $\sqrt{\dfrac{2 - x}{1 + x}}$
45 $(\sin^{-1} 2x)^2$
47 (a) $\dfrac{1}{\sqrt{x^2 - 9}}$ (b) $\dfrac{1}{\sqrt{9 - x^2}}$
49 (a) $\dfrac{1}{x\sqrt{2x^2 + 1}}$ (b) $\dfrac{1}{|x|\sqrt{2x^2 - 1}}$
51 (b) $\sin^{-1}\dfrac{x}{5} + C$ (c) $\sin^{-1}\dfrac{x}{\sqrt{5}} + C$
53 (a)

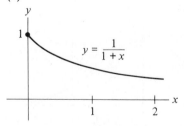

(b) $\ln(1 + b)$ (c) Infinite

55 (a) $\dfrac{dx}{1+x^2}$ (b) 0.835

57 (a) $\dfrac{dx}{|x|\sqrt{x^2-1}}$ (b) 1.07

59 (a) $\dfrac{a}{1+a^2x^2}$ (b) $\dfrac{\pi}{3\sqrt{3}}$

SEC. 6.6. RELATED RATES

1 $2\sqrt{901}$ ft/sec ≈ 60.03 ft/sec
3 (a) $\tfrac{3}{4}$ ft/sec   (b) $\tfrac{4}{3}$ ft/sec   (c) $9/\sqrt{19}$ ft/sec
5 (b) $x = 3\tan\theta$
7 (a) -1 ft²/sec   (b) 8 ft/sec   (c) $\tfrac{69}{13}$ ft/sec
9 (a) $\dfrac{5}{2\pi}$ yd/hr   (b) $\dfrac{1}{10\pi}$ yd/hr
11 $\dfrac{27}{2\sqrt{2}}$ ft²/sec, increasing
13 (a) -0.00014 ft/sec²   (b) -7.56 ft/sec²
15 (a) $6x$ (b) 18
17 (a) $\dfrac{-25}{\sqrt{29}}$ ft/sec   (b) Decreasing
19 (a) $\dfrac{5\pi}{2}$ ft/sec   (b) $\dfrac{5\sqrt{3}\pi}{2}$ ft/sec
21 (a) -562.47 mb/sec   (b) -1.07 mb/sec

SEC. 6.7. SEPARABLE DIFFERENTIAL EQUATIONS AND GROWTH

7 $\tfrac{1}{5}y^5 = \tfrac{1}{4}x^4 + C$
9 $y = kx$
11 $\sin^{-1}y = \tan^{-1}x + C$, or
   $y = \dfrac{kx}{\sqrt{1+x^2}} \pm \dfrac{\sqrt{1-k^2}}{\sqrt{1+x^2}}$, where $k = \cos C$
13 $\tfrac{1}{3}y^3 = -\tfrac{1}{2}x^2 + C$
15 $\tfrac{1}{2}\ln(1+y^2) = \tfrac{1}{3}x^3 + C$
17 $\sec^{-1}y = \tfrac{1}{3}x^3 + C$
19 $\sin^{-1}y = -2(3-x)^{1/2} + C$
23 (a) 0.10   (b) 1.10   (c) 0.0953   (d) 7.2725 hours
25 (a) 10 grams   (b) 1.0986   (c) 200%
27 (a) 38.85 years   (b) 77.71 years   (c) 173.83 years
29 77 years
31 (a) 6:18 p.m.   (b) 0.2618   (c) 29.93%
33 (a) $0.25A$ (b) $0.785A$
35 282.8 grams
37 (a) $Ae^{-0.005t}$ (b) 138.6 days
41 (a) $\dfrac{di}{dt} = \dfrac{E - Ri}{L}$ (b) $i = \dfrac{1}{R}(E - (E - Ri_0)e^{-Rt/L})$
43 (a) 1001.65 million   (b) 0.0226   (c) 0.0228
47 (a) $Y = \tfrac{3}{2}x + 1$ (b) $y = 2^{(3/2)x+1}$

49 (a)

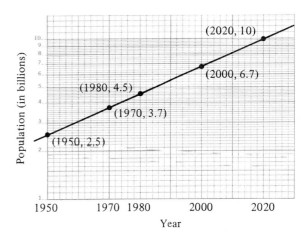

(b) 6.7 billion   (c) 2020
53 (b) $f(t) = f(0)e^{kt}$

SEC. 6.8. L'HÔPITAL'S RULE

1 3
3 $\tfrac{3}{2}$
5 0
7 $\tfrac{1}{2}$
9 3
11 0
13 e^{-2}
15 1
17 1
19 0
21 $(\log_2 e)/(\log_3 e)$
23 Does not exist
25 1
27 -1
29 Does not exist
31 0
33 1
35 Does not exist
37 1
39 0
41 $\ln(\tfrac{5}{3})$
43 $\tfrac{16}{9}$
45 $\tfrac{1}{6}$
47 Does not exist
49 0

51

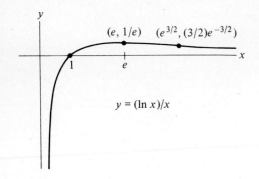

53

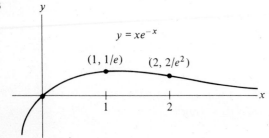

55

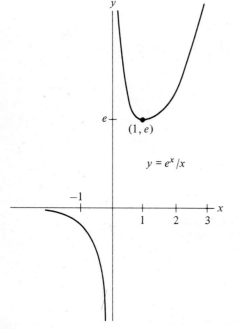

59 $\sqrt{2}$
61 (b) 0

SEC. 6.9. THE HYPERBOLIC FUNCTIONS AND THEIR INVERSES

3 $\cosh t$

5 $-\operatorname{sech} t \tanh t$
7 $3 \sinh 3x$
9 $\dfrac{\operatorname{sech}^2 \sqrt{x}}{2\sqrt{x}}$
11 $e^{3x}(\cosh x + 3 \sinh x)$
13 $(4 \sinh 4x)(\coth 5x)(\operatorname{csch} x^2)$
$+ (\cosh 4x)(-5 \operatorname{csch}^2 5x)(\operatorname{csch} x^2)$
$+ (\cosh 4x)(\coth 5x)(-2x \operatorname{csch} x^2 \coth x^2)$
31 (a)

t	-3	-2	-1	0	1	2	3
$\cosh t$	10.068	3.762	1.543	1	1.543	3.762	10.068
$\sinh t$	-10.018	-3.627	-1.175	0	1.175	3.627	10.018

(b)

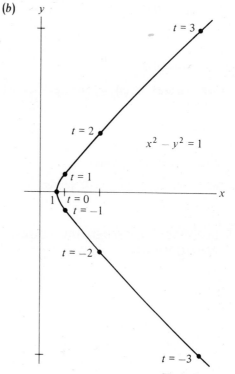

33 (a)

x	0	1	2	3
$\tanh x$	0	0.762	0.964	0.995

(b)

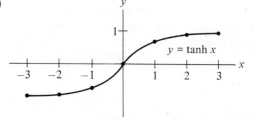

35 (a) $v(t) = V \tanh(gt/V)$
(c) $dv/dt = g \operatorname{sech}^2(gt/V)$
(e) 0

SEC. 6.10. EXPONENTIAL FUNCTIONS DEFINED IN TERMS OF LOGARITHMS (OPTIONAL)

SEC. 6.S. SUMMARY: GUIDE QUIZ ON CHAPTER 6

1 (a) $\dfrac{3e^{\sin^{-1} 3x}}{\sqrt{1-9x^2}}$ (b) $\dfrac{8\cos((\tan^{-1} 4x)^2)\tan^{-1} 4x}{1+16x^2}$

(c) $\left(\dfrac{3x^2+5}{x^3+5x} + \dfrac{2+\ln x}{2\sqrt{x}}\right)(x^3+5x)x^{\sqrt{x}}$

(d) $(\sec^{-1} 2x)^{-2}\left[(\ln 5)5^x \sec^{-1} 2x \sec^2 5^x\right.$

$\left. - \dfrac{\tan 5^x}{|x|\sqrt{4x^2-1}}\right]$

(e) $5\left(\dfrac{16x}{(x^2+1)\ln(x^2+1)} - \dfrac{3}{x} + \dfrac{4}{3}\tan 2x\right) \times$

$\left(x^{-3} \dfrac{(\ln(x^2+1))^8}{\sqrt[3]{\cos^2 2x}}\right)^5$

2 (a) $e^{ax} \cos bx$ (b) $\sin^{-1} ax$
(c) $\tan^{-1} ax$ (d) $\sec^{-1} ax$

3 $\dfrac{-2}{e+1}$, $\dfrac{-(3e^2+2e-5)}{(e+1)^3}$

4 (a) $\dfrac{5\sqrt{5}}{3}$ ft/sec (b) $\dfrac{3\sqrt{11}}{2}$ ft/sec

5 (a) $\frac{1}{2}e^{2y} = \frac{1}{4}x^4 + C$ (b) $\frac{1}{8}\ln(4y^2+1) = x + C$

6 12.706 days

7 (a) $\frac{2}{3}$ (b) Does not exist (c) 0 (d) 1 (e) 1

8 $\log_{10} x$, $\ln x$, x^3, $(1.001)^x$, 2^x

9 (a) $x > 0$
(b) $x = 1$
(c) $x = e$
(d) $(e, 1/e)$ is a maximum.
(e) $0 < x < e, x > e$
(f) No; yes
(g) $-\infty$
(h) 0

(i)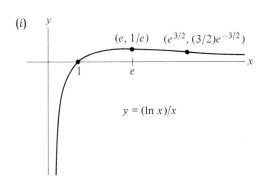
$y = (\ln x)/x$

10 (a) $e = \lim\limits_{x \to 0}(1+x)^{1/x}$ (b) $e \approx 2.718$

11 Divide $\log_{10} 7$ by $\log_{10} e$.

REVIEW EXERCISES FOR CHAPTER 6

1 $\dfrac{3x^2}{2(1+x^3)^{1/2}}$

3 $1/(2\sqrt{x})$

5 $-6\cos 3x \sin 3x$

7 $3\sqrt{x}/2$

9 $\dfrac{\cos x}{2\sqrt{\sin x}}$

11 $-2x \csc^2 x^2$

13 $e^{x^2}\left(\dfrac{2x^2-1}{2x^2}\right)$

15 $\dfrac{x^{5/6}}{\sqrt{1-x^2}} + \frac{5}{6}x^{-1/6} \sin^{-1} x$

17 $e^{3x}(3x^2 + 2x)$

19 $3 \sec 3x$

21 $\dfrac{-\sin\sqrt{x}}{2\sqrt{x}}$

23 $\sec x$

25 $\dfrac{-3x}{(6+3x^2)^{3/2}}$

27 $\sqrt{\frac{15}{2}(5x+7)}$

29 $\dfrac{x}{3x+4}$

31 $(1+x^2)^4[(1+x^2)3\cos 3x + 10x \sin 3x]$

33 $(24x^2 - 8)(2x^3 - 2x + 5)^3$

35 $(1+2x)^4[10\cos 3x - 3(1+2x)\sin 3x]$

37 $4\cos 3x \cos 4x - 3 \sin 3x \sin 4x$

39 $-6x \csc 3x^2 \cot 3x^2$

41 $\left(\dfrac{x}{1+x}\right)^4 (7x^2 + 4x^3)$

43 $(\sec 3x)(3x \tan 3x + 1)$

45 $\dfrac{(x^3-1)^3(x^{10}+1)^4}{(2x+1)^5}\left[\dfrac{9x^2}{x^3-1} + \dfrac{40x^9}{x^{10}+1} - \dfrac{10}{2x+1}\right]$

47 $1/\sqrt{x^2+1}$

49 $e^{-x}\left[\dfrac{2x}{1+x^4} - \tan^{-1} x^2\right]$

51 $e^{\sqrt{x}}$

53 $\dfrac{1-2\ln x}{x^3}$

55 $\dfrac{(\sin x)(\cos^2 x + 1)}{\cos^2 x}$

57 $\dfrac{x^2 \ln(1+x^2)}{|x|\sqrt{9x^2-1}} + 2x \sec^{-1} 3x \ln(1+x^2)$

$+ \dfrac{2x^3}{1+x^2} \sec^{-1} 3x$

59 $1 - \dfrac{2}{x-1} - \dfrac{1}{(x+1)^2}$

61 $\dfrac{-12x + 3}{6x^2 + 3x + 1}$

63 $\dfrac{15}{5x+1} + \dfrac{12}{6x+1} - \dfrac{8}{2x+1}$

65 $\sqrt{9 - x^2}$

67 $3 \tan^2 3x$

69 $1/\sqrt{x^2 + 25}$

71 $\sin^{-1} x$

73 $\sec 3x$

75 $\sin^3 3x$

77 $e^{3x} \sin 2x [3 \sin 2x \tan x + 4 \cos 2x \tan x + \sin 2x \sec^2 x]$

79 $\sqrt{4x^2 + 3}$

81 $\dfrac{x^2}{(x+1)^3} [5x^5 + 7x^4 - 2x^2 - x + 9]$

83 $(1 + 3x)^{x^2} \left[\dfrac{3x^2}{1 + 3x} + 2x \ln (1 + 3x) \right]$

85 $\dfrac{1 + x^2 + \ln x - 9x^2 \ln x}{(1 + x^2)^6}$

87 (a)

x	$\tfrac{1}{8}$	$\tfrac{1}{4}$	$\tfrac{1}{2}$	1	2	4	8
$\log_2 x$	-3	-2	-1	0	1	2	3

(b)

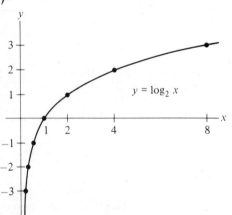

(c) ∞ (d) $-\infty$

89 (b) 0.22

101 (a) $\dfrac{2}{1 - x^2}$ (b) $\tfrac{1}{2} \ln 3$

103 $\dfrac{ab}{\sqrt{ax + b}(ax + b - b^2)}$

105 (a) $\ln |x^3 + x - 6| + C$
    (b) $\tfrac{1}{2} \ln |\sin 2x| + C$
    (c) $\tfrac{1}{3} \ln |5x + 3| + C$
    (d) $\dfrac{-1}{5(5x + 3)} + C$

107 (a) $\dfrac{4}{3} \sec 4x \csc 4x - \dfrac{10}{1 - 2x} - \dfrac{2}{1 + 3x}$

(b) $\dfrac{(1 + x^2)^3 \sqrt{1 + x}}{\sin 3x} \left[\dfrac{6x}{1 + x^2} + \dfrac{1}{2(1 + x)} - 3 \cot 3x \right]$

109 (a) Does not exist   (b) 0

113 $\tfrac{1}{2}$

115

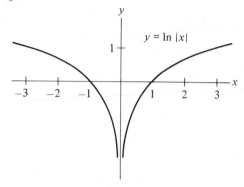

117 $(e, 1)$ 119 $y'(0) = -2$

121 2

123 e^3

125 $-\tfrac{1}{2}$

127 e^2

129 -1

131 0

133 $\tfrac{2}{3}$

135 3

137 3

139 1

141 $\tfrac{8}{3}$

143 $-\infty$

145 0

147 $\ln \left(\tfrac{5}{3}\right)$

149 e^6

151 1

153 0

155 2

157 1

159 0

163 (a) 0   (b) 1   (c) Can't tell
    (d) Can't tell   (e) Can't tell   (f) 0

167 (a) e^x (b) $\ln x$ (c) $\sqrt[3]{x}$
    (d) $x/3$ (e) x^3 (f) $\sin x$

169 1

177 (b) $k < 0$

179 29 days

181 $\tfrac{3}{4}$

183 1

185 $\tfrac{2}{3}$

187 (b) $-e/2$

191 (a) Yes   (b) No   (c) Yes
    (d) No   (e) Yes   (f) Yes

CHAPTER 7. COMPUTING ANTIDERIVATIVES

SEC. 7.1. SUBSTITUTION IN AN ANTIDERIVATIVE AND IN A DEFINITE INTEGRAL

1   $\frac{5}{4}x^4 + C$
3   $\frac{3}{4}x^{4/3} + C$
5   $6\sqrt{x} + C$
7   $-\frac{5}{2}e^{-2x} + C$
9   $6 \sec^{-1} x + C$
11   $\frac{1}{4}\ln(1 + x^4) + C$
13   $-\ln(1 + \cos x) + C$
15   $\ln|x + x^2| + C$
17   $\frac{x^5}{5} + 2x^3 + 9x + C$
19   $\frac{x^3}{3} + \frac{3}{4}x^4 + C$
21   $\frac{2}{7}x^{7/2} + C$
23   $\ln|x| + 2\sqrt{x} + C$
25   $\frac{(1 + 3x)^6}{6} + C$
27   $\sqrt{1 + x^2} + C$
29   $-\frac{1}{2}\cos 2x + C$
31   $\frac{1}{3}e^{3x} + C$
33   $\frac{1}{3}\sin^{-1} 3x + C$
35   $\frac{\tan^2 \theta}{2} + C$
37   $\frac{(\ln x)^5}{5} + C$
39   $-\frac{1}{12}(1 - x^2)^6 + C$
41   $\frac{3}{8}(1 + x^2)^{4/3} + C$
43   $2e^{\sqrt{x}} + C$
45   $-\frac{1}{3}\cos 3\theta + C$
47   $\frac{2}{7}(x - 3)^{7/2} + C$
49   $\ln|x^2 + 3x + 2| + C$
51   $\frac{1}{2}e^{2x} + C$
53   $-\frac{1}{5}\cos x^5 + C$
55   $\frac{1}{2}\tan^{-1}(x^2) + C$
57   $x - \ln|1 + x| + C$
59   $\frac{(\ln(3x))^2}{2} + C$
61   $\frac{1}{3}\int_1^8 e^u\, du$
63   $\int_1^2 (u^{-2} - 2u^{-3} - 2u^{-4})\, du$
65   $\int_0^1 u^3\, du$
67   $\frac{1}{a^3}\left(\frac{a^2 x^2}{2} - abx + b^2 \ln|ax + b|\right) + C$
69   $\frac{1}{a^3}\left(ax - 2b \ln|ax + b| - \frac{b^2}{ax + b}\right) + C$
71   They are both right.
73   Jill is right.

SEC. 7.2. INTEGRATION BY PARTS

1   $\frac{1}{4}e^{2x}(2x - 1) + C$
3   $-\frac{x}{2}\cos 2x + \frac{1}{4}\sin 2x + C$
5   $\frac{x^2}{2}\ln 3x - \frac{x^2}{4} + C$
7   $\frac{5e - 10}{e^2}$
9   $\sin^{-1} 1 - 1$
11   $\frac{x^3}{3}\ln x - \frac{1}{9}x^3 + C$
13   $\frac{(3x + 5)^{11}}{33}\left(\frac{33x - 5}{36}\right) + C$
15   $3(\ln 3)^2 - 6 \ln 3 - 2(\ln 2)^2 + 4 \ln 2 + 2$
17   $\frac{e - 2}{2}$
19   $\frac{1}{2}e^x(\sin x - \cos x) + C$
21   $\frac{-\ln(1 + x^2)}{x} + 2\tan^{-1} x + C$
25   (a), (b) $(x + 1)\ln(x + 1) - x + C$
    (c)   (b) is easier.
27   $\int (\ln x)^n\, dx = x(\ln x)^n - n \int (\ln x)^{n-1}\, dx$
29   $2(\sin \sqrt{x} - \sqrt{x}\cos \sqrt{x}) + C$
31   $2(\exp \sqrt{x})(\sqrt{x} - 1) + C$
33   (a) $\frac{1}{2}(x - \sin x \cos x) + C$
    (b) $\frac{-\sin^3 x \cos x}{4} + \frac{3}{8}(x - \sin x \cos x) + C$
    (c) $\frac{-\sin^5 x \cos x}{6} - \frac{5 \sin^3 x \cos x}{24}$
        $+ \frac{5}{16}(x - \sin x \cos x) + C$
37   $\frac{2}{9}(3x + 7)^{3/2}\left(\frac{3x}{5} - \frac{14}{15}\right) + C$
39   $\frac{1}{20a^2}(ax + b)^4(4ax - b) + C$
41   (a) $I_0 = \frac{\pi}{2}, I_1 = 1$
    (c) $I_2 = \frac{\pi}{4}, I_3 = \frac{2}{3}$
    (d) $I_4 = \frac{3\pi}{16}, I_5 = \frac{8}{15}$
    (e) $I_n = \frac{2 \cdot 4 \cdots (n - 1)}{3 \cdot 5 \cdots n}$ when n is odd, $n \geq 3$
    (f) $I_n = \left(\frac{1 \cdot 3 \cdots (n - 1)}{2 \cdot 4 \cdots n}\right)\left(\frac{\pi}{2}\right)$ when n is even, $n \geq 2$
45   (e) $f(b) = f(0) + f^{(1)}(0)b + \frac{f^{(2)}(0)b^2}{2} + \frac{f^{(3)}(0)b^3}{6}$
        $+ \frac{1}{6}\int_0^b f^{(4)}(x)(b - x)^3\, dx$

SEC. 7.3. USING A TABLE OF INTEGRALS

1. $\dfrac{(3x+5)^6}{18} + C$

3. $\dfrac{-5}{4(2x-5)} + \dfrac{1}{4}\ln|2x-5| + C$

5. $\dfrac{2}{15}(5x-7)^{3/2} + C$

7. $\dfrac{2}{3}\sqrt{5x+4} + C$

9. $\sqrt{2}\tan^{-1}\sqrt{\dfrac{3x-2}{2}} + C$

11. $\dfrac{1}{3}\tan^{-1}\dfrac{x}{3} + C$

13. $\dfrac{1}{4}\ln\left|\dfrac{2+x}{2-x}\right| + C$

15. $\dfrac{1}{\sqrt{15}}\tan^{-1}x\sqrt{\dfrac{5}{3}} + C$

17. $\dfrac{1}{6}\ln(3x^2+1) + C$

19. $\dfrac{1}{2}[x\sqrt{x^2-1} - \ln|x+\sqrt{x^2-1}|] + C$

21. $\dfrac{1}{2}[x\sqrt{x^2-3} - 3\ln|x+\sqrt{x^2-3}|] + C$

23. $\dfrac{1}{4}\ln|2x^2+x+3| - \dfrac{1}{2\sqrt{23}}\tan^{-1}\dfrac{4x+1}{\sqrt{23}} + C$

25. $\sin^{-1}\dfrac{2x-3}{\sqrt{29}} + C$

27. $\sqrt{x^2+x-6} + \dfrac{5}{2}\ln|2x+1+2\sqrt{x^2+x-6}| + C$

29. $\dfrac{-1}{14}\sin^6 2x\cos 2x - \dfrac{3}{35}\sin^4 2x\cos 2x - \dfrac{4}{35}\sin^2 2x\cos 2x - \dfrac{8}{35}\cos 2x + C$

31. $\dfrac{1}{5}\sin 5x - \dfrac{1}{15}\sin^3 5x + C$

33. $\dfrac{1}{5}\ln|\sec 5x + \tan 5x| + C$

35. $\dfrac{x^2}{3}\sin 3x - \dfrac{2}{27}\sin 3x + \dfrac{2x}{9}\cos 3x + C$

37. $-e^{-x}(x^2+2x+2) + C$

39. $\dfrac{e^{-3x}}{34}(-3\sin 5x - 5\cos 5x) + C$

41. $x\sin^{-1}3x + \dfrac{1}{3}\sqrt{1-9x^2} + C$

43. $x\sec^{-1}4x - \dfrac{1}{4}\ln|4x+\sqrt{16x^2-1}| + C$

45. $-\dfrac{1}{3}\tan\left(\dfrac{\pi}{4} - \dfrac{3x}{2}\right) + C$

47. $\dfrac{2}{45}(5x^3+2)^{3/2} + C$

49. $\dfrac{1}{2}\tan^{-1}\left(\dfrac{\sin x}{2}\right) + C$

51. $\dfrac{3x+1}{6}\sqrt{(1+3x)^2-5} - \dfrac{5}{6}\ln|1+3x+\sqrt{(1+3x)^2-5}| + C$

53. $\dfrac{x^2}{4}\sqrt{x^4+9} + \dfrac{9}{4}\ln|x^2+\sqrt{x^4+9}| + C$

SEC. 7.4. HOW TO INTEGRATE CERTAIN RATIONAL FUNCTIONS

1. $\dfrac{1}{3}\ln|3x-4| + C$

3. $-\dfrac{5}{2(2x+7)} + C$

5. $\dfrac{1}{3}\tan^{-1}\left(\dfrac{x}{3}\right) + C$

7. $\dfrac{1}{2}\ln(x^2+9) + C$

9. $\ln(x^2+9) + \tan^{-1}\left(\dfrac{x}{3}\right) + C$

11. $\dfrac{1}{20}\tan^{-1}\left(\dfrac{4x}{5}\right) + C$

13. $\dfrac{1}{32}\ln(16x^2+25) + C$

15. $\dfrac{1}{18}\ln(9x^2+4) + \dfrac{1}{3}\tan^{-1}\left(\dfrac{3x}{2}\right) + C$

17. $\dfrac{1}{\sqrt{6}}\tan^{-1}(\sqrt{\tfrac{2}{3}}x) + C$

19. $\dfrac{1}{\sqrt{2}}\tan^{-1}\left(\dfrac{x+1}{\sqrt{2}}\right) + C$

21. $\dfrac{1}{\sqrt{2}}\tan^{-1}\left(\dfrac{x-1}{\sqrt{2}}\right) + C$

23. $\dfrac{2}{\sqrt{23}}\tan^{-1}\left(\dfrac{4x+1}{\sqrt{23}}\right) + C$

25. $\dfrac{1}{\sqrt{3}}\tan^{-1}\left(\dfrac{x+2}{\sqrt{3}}\right) + C$

27. $\dfrac{1}{\sqrt{10}}\tan^{-1}\left(\dfrac{2x+2}{\sqrt{10}}\right) + C$

29. $\ln(x^2+2x+3) - \sqrt{2}\tan^{-1}\left(\dfrac{x+1}{\sqrt{2}}\right) + C$

31. $\dfrac{3}{10}\ln(5x^2+3x+2) - \dfrac{9}{5\sqrt{31}}\tan^{-1}\left(\dfrac{10x+3}{\sqrt{31}}\right) + C$

33. $\dfrac{1}{2}\ln(x^2+x+1) + \dfrac{1}{\sqrt{3}}\tan^{-1}\left(\dfrac{2x+1}{\sqrt{3}}\right) + C$

35. $\dfrac{1}{2}\ln(3x^2+2x+1) + 2\sqrt{2}\tan^{-1}\left(\dfrac{3x+1}{\sqrt{2}}\right) + C$

37. $\dfrac{1}{\sqrt{ac}}\tan^{-1}\left(\sqrt{\dfrac{a}{c}}x\right) + K$

45. $\dfrac{x}{2(4x^2+1)} + \dfrac{1}{4}\tan^{-1}2x + C$

SEC. 7.5. THE INTEGRATION OF RATIONAL FUNCTIONS BY PARTIAL FRACTIONS

1. $\dfrac{k_1}{x+1} + \dfrac{k_2}{x+2}$

3. $\dfrac{k_1}{x-2} + \dfrac{k_2}{x+2}$

5. $\dfrac{k_1}{x+1} + \dfrac{k_2}{(x+1)^2} + \dfrac{k_3}{(x+1)^3}$

7 $1 + \dfrac{k_1}{x-1} + \dfrac{k_2}{x+1}$

9 $x - 1 + \dfrac{k_1}{x-1} + \dfrac{k_2}{x+2}$

11 $\dfrac{c_1 x + d_1}{x^2 + x + 1} + \dfrac{c_2 x + d_2}{(x^2 + x + 1)^2} + \dfrac{c_3 x + d_3}{(x^2 + x + 1)^3}$

13 $x + \dfrac{3}{x+1}$

15 $x + \dfrac{1}{2} + \dfrac{5/2}{2x+1}$

17 $\dfrac{1}{x+1} - \dfrac{2}{x+2}$

19 $\dfrac{-1}{x} + \dfrac{2}{x+1} - \dfrac{3}{(x+1)^2}$

21 $\dfrac{2}{x+1} - \dfrac{1}{(x+1)^2} + \dfrac{2}{(x+1)^3}$

23 $\dfrac{1}{x+1} + \dfrac{3}{(x+1)^2}$

25 $x - 3 + \dfrac{4x+15}{x^2 + 3x + 5}$

27 $x + \dfrac{1}{x} + \dfrac{2x-1}{x^2 + x + 1}$

29 $\ln\left|\dfrac{x+1}{(x+2)^2}\right| + C$

31 $\ln\left|\dfrac{(x+1)^2}{x}\right| + \dfrac{3}{x+1} + C$

33 $\dfrac{x^2}{2} - 3x + 2\ln(x^2 + 3x + 5) + \dfrac{18}{\sqrt{11}}\tan^{-1}\left(\dfrac{2x+3}{\sqrt{11}}\right) + C$

35 $\dfrac{x^2}{2} + \ln|x(x^2 + x + 1)| - \dfrac{4}{\sqrt{3}}\tan^{-1}\left(\dfrac{2x+1}{\sqrt{3}}\right) + C$

37 $\tfrac{3}{2}x^2 + \ln|x^2 - 1| + C$

39 $\tfrac{1}{2}\ln\tfrac{32}{27}$

41 $\dfrac{1}{2}\ln\left|\dfrac{y-1}{y+1}\right| = \ln\left|\dfrac{x+1}{x}\right| - \dfrac{1}{x} + C$

43 (a) $\tfrac{1}{4}\ln(x^4 + 1) + C$
    (b) $\tfrac{1}{2}\tan^{-1} x^2 + C$
    (c) $\dfrac{\sqrt{2}}{8}\ln\left(\dfrac{x^2 + \sqrt{2}x + 1}{x^2 - \sqrt{2}x + 1}\right) + \dfrac{\sqrt{2}}{4}\tan^{-1}\left(\dfrac{\sqrt{2}x}{1 - x^2}\right) + C$

45 (a) One choice is $\tfrac{5}{6} = \tfrac{1}{2} + \tfrac{1}{3}$
    (b) One choice is $\tfrac{4}{15} = \tfrac{2}{3} - \tfrac{2}{5}$
    (c) One choice is $\tfrac{19}{15} = 1 + \tfrac{2}{3} - \tfrac{2}{5}$
    (d) One choice is $\tfrac{7}{27} = \tfrac{2}{9} + \tfrac{1}{27}$

47 $k_1 = k_3 = 2,\ k_2 = -1$

SEC. 7.6. HOW TO INTEGRATE POWERS OF TRIGONOMETRIC FUNCTIONS

1 $\dfrac{\theta}{2} - \dfrac{\sin 2\theta \cos 2\theta}{4} + C$

3 $\dfrac{\pi}{8}$

5 $-\tfrac{1}{2}(\cos 2\theta - \tfrac{2}{3}\cos^3 2\theta + \tfrac{1}{5}\cos^5 2\theta) + C$

7 $\tfrac{2}{3}$

9 $\dfrac{\cos^7 \theta}{7} - \dfrac{\cos^5 \theta}{5} + C$

11 $\dfrac{\sin^4 \theta}{4} - \dfrac{\sin^6 \theta}{6} + C$

13 $\dfrac{\pi}{32} + \dfrac{1}{4} - \dfrac{9\sqrt{3}}{64} \approx 0.1046$

15 $\dfrac{\tan^5 \theta}{5} + C$

17 $\dfrac{\sec^3 \theta}{3} - \sec \theta + C$

19 $\dfrac{\tan^2 \theta}{2} + \ln|\cos \theta| + C$

21 $4/3$

23 $\dfrac{\sin^3 \theta}{3} - \dfrac{\sin^5 \theta}{5} + C$

25 $\ln|\sin \theta| + C$

27 $\tfrac{1}{3}\ln|\sec 3\theta + \tan 3\theta| + C$

29 $\dfrac{\cos^8 \theta}{8} - \dfrac{\cos^6 \theta}{6} + C$

31 $\dfrac{\csc^4 \theta}{4} - \dfrac{\csc^6 \theta}{6} + C$ or $-\left(\dfrac{\cot^6 \theta}{6} + \dfrac{\cot^4 \theta}{4}\right) + C$

33 $-\left(\dfrac{\cot^5 \theta}{5} + \dfrac{\cot^7 \theta}{7}\right) + C$

35 $\sec \theta - \dfrac{\cos^3 \theta}{3} + 2\cos \theta + C$

37 $\tfrac{8}{15}$

39 $\dfrac{5\pi}{8}$

41 $\dfrac{1}{8}\left(\theta - \dfrac{\sin 4\theta}{4}\right) + C$

45 $-\ln|\csc \theta + \cot \theta| + C = \ln|\csc \theta - \cot \theta| + C$

47 (b) $\dfrac{\sin(m-n)x}{2(m-n)} + \dfrac{\sin(m+n)x}{2(m+n)} + C$

SEC. 7.7. HOW TO INTEGRATE ANY RATIONAL FUNCTION OF $\sin \theta$ AND $\cos \theta$

1 $2\displaystyle\int \dfrac{u^2 - 2u - 1}{(u^2 - 3)(u^2 + 1)}\,du$

3 $\displaystyle\int \dfrac{3u^2 - 2u - 3}{(u^2 - u - 1)(1 + u^2)}\,du$

5 $\tfrac{1}{5}\ln\left|\dfrac{1 + 2\tan\theta/2}{2 - \tan\theta/2}\right| + C$

7 $\dfrac{1}{2}\ln\left(\dfrac{3 + 2\sqrt{2}}{3}\right)$

9 $\dfrac{1}{7}\ln\left|\dfrac{2\tan(\theta/2) + 1}{\tan(\theta/2) - 3}\right| + C$

11 $\frac{1}{5} \ln \left| \tan \frac{\theta}{2} \right| - \frac{8}{15} \tan^{-1} \left[\frac{5 \tan (\theta/2) + 4}{3} \right] + C$

13 $2\theta - 2 \tan \frac{\theta}{2} + 2 \ln \left| \sec \frac{\theta}{2} \right| + C$

15 $\frac{2}{125} \left\{ 6 \ln \left| \frac{\tan (\theta/2) + 2}{2 \tan (\theta/2) - 1} \right| - \left[\frac{20 + 15 \tan (\theta/2)}{\sec^2 (\theta/2)} \right] \right\} + C$

17 $\frac{1}{3} \ln \left| \frac{\tan (\theta/2) + 3}{\tan (\theta/2) - 3} \right| + C$

19 $\frac{1}{6} \tan^{-1} \left(\frac{2}{3} \tan \frac{\theta}{2} \right) + C$

21 $\frac{1}{50} \ln (25 \tan^2 (\theta/2) + 4 \tan (\theta/2) + 13)$
  $- \frac{2}{75\sqrt{39}} \tan^{-1} \left[\frac{25 \tan (\theta/2) + 2}{3\sqrt{39}} \right] + C$

SEC. 7.8. HOW TO INTEGRATE RATIONAL FUNCTIONS INVOLVING ROOTS

1 $\sqrt{1 - x^2} - \ln \left(\frac{1 + \sqrt{1 - x^2}}{x} \right) + C$

3 $\frac{2 + \sqrt{2}}{3}$

5 $\ln \left(\frac{\sqrt{2} + 1}{\sqrt{3}} \right)$

7 $\frac{(1 - x^2)^{5/2}}{5} - \frac{(1 - x^2)^{3/2}}{3} + C$

9 $\frac{2\pi}{3} - \frac{\sqrt{3}}{2}$

11 $10 - 2\sqrt{7} + \frac{9}{2} \ln \left(\frac{9}{4 + \sqrt{7}} \right)$

13 $\ln |\sqrt{9 + x^2} + x| + C$

15 $\sqrt{27} - \pi$

17 $2(\sqrt{x} - 3 \ln (\sqrt{x} + 3)) + C$

19 $2 \ln \left(\frac{\sqrt{x}}{\sqrt{x + 1}} \right) + C$

21 $2 + 4 \ln (3/2)$

23 $2\sqrt{2x + 1} + \ln \left| \frac{\sqrt{2x + 1} - 1}{\sqrt{2x + 1} + 1} \right| + C$

25 $\frac{(3x + 2)^{7/3}}{21} - \frac{(3x + 2)^{4/3}}{6} + C$

27 $x + 10\sqrt{x} + 20 \ln |\sqrt{x} - 2| + C$

29 $\frac{1}{2} \ln \left| \frac{\sqrt{x + 2} - 2}{\sqrt{x + 2} + 2} \right| + C$

31 $\frac{1}{2}(a^2 \sin^{-1} \frac{x}{a} + x\sqrt{a^2 - x^2}) + C$

33 $\frac{1}{2}(x\sqrt{a^2 + x^2} + a^2 \ln (\sqrt{a^2 + x^2} + x)) + C$

35 $\frac{1}{5} \ln |5x + \sqrt{25x^2 - 16}| + C$

37 $\frac{64 - 33\sqrt{3}}{480}$

39 $2\sqrt{x^2 - 4} - 4 \sec^{-1} \frac{x}{2} + C$

SEC. 7.9. WHAT TO DO IN THE FACE OF AN INTEGRAL

1 Division, power rule
3 Partial fractions
5 Integration by parts
7 Repeated integrations by parts
9 Substitution
11 Substitution, power rule
13 Substitution, division, power rule
15 Trigonometric identity
17 Substitution, power rule
19 Substitution, integration by parts
21 Recursion formula, complete square, substitution
23 (Recursion formula), partial fractions
25 Substitution, power rule
27 Substitution, division, methods of section 7.4
29 Trigonometric substitution
31 Substitution, partial fractions
33 Log rules, power rule
35 Repeated integration by parts, trig identity
37 Division, substitution, power rule
39 Substitution, power rule
41 Substitution, power rule
43 Substitution, partial fractions
45 Substitution, division, power rule
47 Partial fractions
49 Substitution, power rule
51 Rules of logs and exponents
53 Break into two pieces, substitution on both
55 Substitution, power rule
57 Substitution, power rule
59 Substitution, power rule
61 Recursion formula of Example 7 in Section 7.2, trig identity
63 Recursion formula of Example 7 in Section 7.2
65 Trigonometric substitution, substitution
67 Substitution, power rule
69 Substitution
71 Substitution, power rule
73 Complete the square, substitution
75 Substitution, power rule
77 Substitution, power rule
79 $n = 1: \frac{2}{3}(1 + x)^{3/2} + C$
  $n = 2: \frac{x\sqrt{1 + x^2}}{2} + \frac{1}{2} \ln (\sqrt{1 + x^2} + x) + C$
81 $n = 1: -\frac{4}{3}(1 - x)^{3/2} + \frac{2}{5}(1 - x)^{5/2} + C$
  $n = 2: \frac{x}{2}\sqrt{1 - x^2} + \frac{1}{2} \sin^{-1} x + C$

83 $-2\sqrt{2} \cos \dfrac{x}{2} + C$

85 $\dfrac{1}{20} \ln \left(\dfrac{x^{20}}{x^{20}+1} \right) + C$

SEC. 7.S. SUMMARY: GUIDE QUIZ ON CHAPTER 7

1 $\tfrac{1}{4} \ln (17/2)$

2 $\tfrac{2}{3} \sin^{-1} \left(\dfrac{3x}{2} \right) + \dfrac{x\sqrt{4-9x^2}}{2} + C$

3 $\tfrac{1}{4} \ln \left| \dfrac{x-1}{x+1} \right| - \tfrac{1}{2} \tan^{-1} x + C$

4 $\dfrac{\tan^6 2x}{12} + C$

5 $x + \tfrac{1}{4} \ln \left| \dfrac{x-1}{x+1} \right| - \tfrac{1}{2} \tan^{-1} x + C$

6 $\tfrac{1}{3} \ln (1 + \sqrt{2})$

7 $\sqrt{x} + \sqrt[4]{x} + \tfrac{1}{2} \ln |2\sqrt[4]{x} - 1| + C$

8 $\dfrac{1}{\sqrt{6}} \ln \left| \dfrac{\sqrt{x+3} - \sqrt{6}}{\sqrt{x+3} + \sqrt{6}} \right| + C$

9 $\dfrac{1}{2\sqrt{3}} \ln \left| \dfrac{x + \sqrt{3}}{x - \sqrt{3}} \right| + C$

10 $-\tfrac{1}{2}(\cos 2x - \tfrac{2}{3} \cos^3 2x + \tfrac{1}{5} \cos^5 2x) + C$

11 $\tfrac{1}{5} e^x(\cos 2x + 2 \sin 2x) + C$

12 $-\tfrac{1}{12} \csc^3 3x \cot 3x - \tfrac{1}{8} \csc 3x \cot 3x$
$+ \tfrac{1}{8} \ln |\csc 3x - \cot 3x| + C$

13 $\tfrac{1}{4} \tan^{-1} x^4 + C$

14 $\tfrac{1}{2} \ln \left| \dfrac{\sqrt{4+x^2} - 2}{x} \right| + C$

15 $\dfrac{x\sqrt{x^2-9}}{2} + \tfrac{9}{2} \ln |x + \sqrt{x^2-9}| + C$

16 $\dfrac{x}{8}(-2x^2 + 45)\sqrt{9-x^2} + \tfrac{243}{8} \sin^{-1} \dfrac{x}{3} + C$

17 $\dfrac{1}{\sqrt{2}} \tan^{-1} \left[\dfrac{\tan (x/2)}{\sqrt{2}} \right] + C$

18 $-\dfrac{\sqrt{x^2+25}}{25x} + C$

19 $\dfrac{x^2}{2} + \tfrac{2}{3} x^{-3/2} + C$

REVIEW EXERCISES FOR CHAPTER 7

1 (a) $\displaystyle\int_{1}^{2} u^{3/2} \, du$
   (b) $\tfrac{2}{5}(4\sqrt{2} - 1)$

3 (a) $\dfrac{373}{14}$
   (b) $\dfrac{721}{9}$

5 (a) $\dfrac{-1}{2x^2} + C$ (b) $2\sqrt{x+1} + C$ (c) $\tfrac{1}{5} \ln (1 + 5e^x) + C$

7 $-\tfrac{1}{4}(1+x^2)^{-2} + \tfrac{1}{6}(1+x^2)^{-3} + C$

9 $\dfrac{x^3+1}{3} \ln (1+x) - \dfrac{x^3}{9} + \dfrac{x^2}{6} - \dfrac{x}{3} + C$

11 (a) $\dfrac{1}{2\sqrt{2}} \ln \left| \dfrac{(2-\sqrt{2}) + (1-\sqrt{2}) \cos \theta}{(2+\sqrt{2}) + (1+\sqrt{2}) \cos \theta} \right| + C$

    (b) $\dfrac{1}{2\sqrt{2}} \ln \left| \dfrac{\cos \theta - \sqrt{2}}{\cos \theta + \sqrt{2}} \right| + C$

13 (a) Not elementary
    (b) $2 \sin \theta + C$
    (c) $2\sqrt{1 - \cos \theta} + C$

15 (a) $\tfrac{6}{11}$

17 $\dfrac{2}{x-1} + \dfrac{1}{x^2+x+1}$

19 $\dfrac{-1}{x+1} + \dfrac{x}{x^2-x+1}$

21 $\dfrac{x}{x^2 + \sqrt{2}x + 1} - \dfrac{1}{x^2 - \sqrt{2}x + 1}$

23 $5x + 6 + \dfrac{1}{x} - \dfrac{1}{x+1}$

25 (a) m odd
    (b) n even
    (c) Either m or n odd
    (d) m even or n odd
    (e) m odd or n even

27 $\displaystyle\int \dfrac{8 \cos^3 \theta + 1}{64 \cos^6 \theta + 5} \, d\theta$

29 $5^7 \sqrt{5} \displaystyle\int \dfrac{\sec \theta \tan^{15} \theta \, d\theta}{5 \sec^2 \theta + 3 + \sqrt{5} \tan \theta}$

31 $\displaystyle\int \dfrac{3\left(\dfrac{2u}{1-u^2}\right)^2 + \left(\dfrac{1+u^2}{1-u^2}\right) + 1}{2 + \left(\dfrac{2u}{1-u^2}\right) + \left(\dfrac{1-u^2}{1+u^2}\right)} \cdot \dfrac{2du}{1+u^2}$

33 $\tfrac{1}{12} \ln |\sin x - 2| - \tfrac{1}{24} \ln (\sin^2 x + 2 \sin x + 4)$
$- \dfrac{1}{4\sqrt{3}} \tan^{-1} \left(\dfrac{\sin x + 1}{\sqrt{3}} \right) + C$

35 $\dfrac{-(x^2+1)^{3/2}}{3x^3} + C$

37 $-\ln (3 + \cos x) + C$

39 $\tfrac{2}{9}(x^3 - 1)^{3/2} + C$

41 $\tfrac{1}{16} \left(\tan^{-1} \dfrac{x}{2} + \dfrac{2x}{4+x^2} \right) + C$

43 $\dfrac{x}{8} - \dfrac{\sin 12x}{96} + C$

45 $\tfrac{1}{9} \tan^3 3\theta - \tfrac{1}{3} \tan 3\theta + \theta + C$

47 $\dfrac{x^2}{2} - \dfrac{1}{2x^2} + \ln |x| + C$

49 $\dfrac{10^x}{\ln 10} + C$

51 $\dfrac{x^2}{4(x^4+1)} + \tfrac{1}{4} \tan^{-1} x^2 + C$

53 $\frac{1}{4}(2x + \sin 2x) + C$
55 $\frac{1}{3}(x^2 + 4)^{3/2} + C$
57 $\frac{1}{3} \tan^{-1}(x^3) + C$
59 $-\frac{1}{3} \cos(x^3) + C$
61 $\dfrac{x^5}{5} \ln x - \frac{1}{25}x^5 + C$
63 $2e^{\sqrt{x}} + C$
65 $(x - 1) \ln(x^3 - 1) - 3x + \frac{3}{2} \ln(x^2 + x + 1)$
$\qquad + \sqrt{3} \tan^{-1}\left(\dfrac{2x + 1}{\sqrt{3}}\right) + C$

67 $-\dfrac{1}{\sqrt{x^2 + 1}} + C$
69 $-\dfrac{2}{\sqrt{x + 1}} + C$
71 $3 \tan^{-1}(x + 2) + C$

73 $\frac{3}{5}x^{5/3} - \frac{3}{4}x^{4/3} + x - \frac{3}{2}x^{2/3} + 3x^{1/3}$
$\qquad - 3 \ln|1 + x^{1/3}| + C$
75 $\frac{1}{7}(x^2 + 1)^{7/2} - \frac{3}{5}(x^2 + 1)^{5/2}$
$\qquad + (x^2 + 1)^{3/2} - (x^2 + 1)^{1/2} + C$
77 $-\dfrac{\tan^{-1} x}{x} + \ln|x| - \frac{1}{2} \ln(x^2 + 1) + C$
79 $\frac{1}{10}e^x(\sin 3x - 3 \cos 3x) + C$
81 $\dfrac{x}{4\sqrt{4 - x^2}} + C$

83 $\frac{1}{8} \ln\left|\dfrac{x^2 - 3}{x^2 + 1}\right| + C$
85 $\frac{3}{8}(x - 1)^{8/3} + \frac{6}{5}(x - 1)^{5/3} + \frac{3}{2}(x - 1)^{2/3} + C$
87 $\sqrt{x^2 + 4} + 2 \ln\left|\dfrac{\sqrt{x^2 + 4} - 2}{x}\right| + C$
89 $\dfrac{\sec^5 \theta}{5} + C$
91 $-\frac{1}{3} \ln\left|\dfrac{3 + \sqrt{x^2 + 9}}{x}\right| + C$
93 $\frac{3}{2}x^{2/3} - \frac{6}{5}x^{5/3} + \frac{3}{8}x^{8/3} + C$
95 $\frac{4}{3} \sin^3 x - \frac{4}{5} \sin^5 x + C$
97 $\frac{1}{2}e^{2x} - 2x - \frac{1}{2}e^{-2x} + C$
99 $\frac{1}{2}(x^2 \sin^{-1}(x^2) + \sqrt{1 - x^4}) + C$
101 $-\dfrac{x}{25} - \frac{1}{5}e^{-x} + \frac{1}{25} \ln(e^x + 5) + C$
103 $\frac{1}{135}(36x + 14)(3x + 2)^{3/2} + C$
105 $\ln|x - 1| - \dfrac{2}{x - 1} - \dfrac{1}{2(x - 1)^2} + C$
107 $-x + 2 \ln|e^x - 1| + C$
109 $x + 2x^3 + \frac{9}{5}x^5 + C$
111 $x - \frac{1}{3} \ln|x + 1| + \frac{1}{6} \ln(x^2 - x + 1)$
$\qquad - \dfrac{1}{\sqrt{3}} \tan^{-1}\left(\dfrac{2x - 1}{\sqrt{3}}\right) + C$

113 $\sqrt{2x + 1} + C$

115 $\dfrac{1}{2\sqrt{17}} \ln\left|\dfrac{2x^2 - 3 - \sqrt{17}}{2x^2 - 3 + \sqrt{17}}\right| + C$
117 $\tan^{-1} e^x + C$
119 $\ln\left|\dfrac{x + 2}{x + 3}\right| + C$
121 $2 \ln|x^2 + 5x + 6| + C$
123 $\dfrac{2}{\sqrt{23}} \tan^{-1}\left(\dfrac{4x + 5}{\sqrt{23}}\right) + C$
125 $\dfrac{1}{\sqrt{73}} \ln\left|\dfrac{4x + 5 - \sqrt{73}}{4x + 5 + \sqrt{73}}\right| + C$
127 $-\cot x + C$
129 $-\dfrac{\cot^3 x}{3} - \cot x + C$
131 $\sin^{-1}(2x - 5) + C$
133 $\frac{1}{4}(\sin^{-1} 2x + 2x\sqrt{1 - 4x^2}) + C$
135 $2\sqrt{x^2 + 1} + C$
137 $x + 3 \ln|x - 1| + \ln|x + 1| + 4 \tan^{-1} x + C$
139 $-\dfrac{1}{x + 2} - 3 \ln|x + 1| + C$
141 $3 \ln|x| + \tan^{-1} 2x + C$
143 $x^2 + 4 \ln|\sqrt{3}x - 1| + \ln|\sqrt{3}x + 1| + C$
145 $\frac{1}{3} \sin^{-1} 3x + C$
147 $\dfrac{x}{2\sqrt{3x^2 + 2}} + C$
149 $\frac{1}{4} \ln|\sec 4x + \tan 4x| + C$
151 $\frac{1}{2}e^x - \frac{1}{10}e^x \cos 2x - \frac{1}{5}e^x \sin 2x + C$
153 $\frac{8}{315}(1 + \sqrt{1 + \sqrt{x}})^{5/2}$
$\qquad \times (61 + 35\sqrt{x} - 65\sqrt{1 + \sqrt{x}}) + C$
155 $(4x^{3/4} - 12x^{1/2} + 24x^{1/4} - 24)e^{\sqrt[4]{x}} + C$
157 $\frac{1}{2}(\sec x \tan x + \ln|\sec x + \tan x|) + C$
159 $\frac{1}{3}((x^3 + 1) \ln(x^3 + 1) - x^3) + C$
161 $9 \ln|x| + 3x^2 + \dfrac{x^4}{4} + C$
163 $10 \ln|\csc x - \cot x| + 6 \ln|\sin x| + 9 \cos x + C$
165 $e^x - e^{-x} + C$
167 $\ln|x| + \frac{1}{2} \ln(x^2 + 2x + 3) + \dfrac{1}{\sqrt{2}} \tan^{-1}\left(\dfrac{x + 1}{\sqrt{2}}\right) + C$
169 $\ln|\tan \theta| + C$
171 $\frac{1}{4}(x^2 - 2x \sin 2x - \cos 2x) + C$
173 $x \tan x - \dfrac{x^2}{2} + \ln|\cos x| + C$

175 (a) $\frac{1}{2} \ln\left|\dfrac{x + 1}{x + 3}\right| + C$

(b) $-\dfrac{1}{x + 2} + C$

(c) $\tan^{-1}(x + 2) + C$

(d) $\dfrac{1}{2\sqrt{6}} \ln\left|\dfrac{x + 2 - \sqrt{6}}{x + 2 + \sqrt{6}}\right| + C$

177 (a) $\frac{1}{8}(2 \sec^3 x \tan x + 3 \sec x \tan x + 3 \ln |\sec x + \tan x|) + C$
(b) $\frac{\sec^5 x}{5} + C$
(c) $\frac{1}{2} \sec^2 x + C$

179 $\frac{\sqrt{1+x^2}}{3}(x^2 - 2) + C$

181 (a) $2 \int (u^2 - 1)^2 \, du$
(b) $\int \frac{(u-1)^2}{u^{1/2}} \, du$
(c) $2 \int \tan^5 \theta \sec \theta \, d\theta$
(d) $\frac{2}{5}(x+1)^{5/2} - \frac{4}{3}(x+1)^{3/2} + 2(x+1)^{1/2} + C$

183 $\frac{21}{8}$

185 $2(\sqrt{2} - 1)$

187 (a) $x^2\sqrt{1+x^2} - \int 2x\sqrt{1+x^2} \, dx$
(b) $\int \tan^3 \theta \sec \theta \, d\theta$
(c) $\int (u^2 - 1) \, du$

189 $\frac{2}{5}(1+x)^{5/2} - \frac{2}{3}(1+x)^{3/2} + C$

191 (a) e^{x^2} does not have an elementary antiderivative.
(b) 1.4537

193 0

195 (a) $\frac{1}{3}\left(\frac{1}{x-1} - \frac{x-1}{x^2+x+1}\right)$
(b) $\frac{1}{9}\left(-\frac{1}{x-1} + \frac{3}{(x-1)^2} + \frac{1}{x+2}\right)$
(c) $x^2 - \frac{1}{3}\left(\frac{1}{x+1} + \frac{2x-1}{x^2-x+1}\right)$
(d) $\frac{1}{12}\left(\frac{1}{x+2} - \frac{x-4}{x^2-2x+4}\right)$

197 No, there is no error.

203 $\frac{2}{15}[8\sqrt{2} - 4\sqrt{3} - \sqrt{6} + 1] + \frac{16}{15\sqrt{30}}\left[\tan^{-1}\frac{3\sqrt{5}}{4} - \tan^{-1}\frac{\sqrt{30}}{8} - \tan^{-1} 2\sqrt{15} + \tan^{-1} 3\sqrt{\frac{5}{2}}\right]$

211 (b) $n = 1: \frac{2}{3}(1+x)^{3/2} - 2(1+x)^{1/2} + C$
$n = 2: \sqrt{1+x^2} + C$
$n = 4: \frac{1}{2} \ln(x^2 + \sqrt{1+x^4}) + C$

213 (a) $n = 3 + 4k$ or $5 + 4k$, k any positve integer or zero.
(b) $n = 3: \frac{1}{2}\sqrt{1+x^4} + C$
$n = 5: \frac{x^2}{4}\sqrt{1+x^4} - \frac{1}{4}\ln(x^2 + \sqrt{1+x^4}) + C$

215 (b) Constant

CHAPTER 8. APPLICATIONS OF THE DEFINITE INTEGRAL

SEC. 8.1. COMPUTING AREA BY PARALLEL CROSS SECTIONS

1 $\frac{1}{12}$
3 $\frac{1}{3}$
5 $\frac{21}{4}$
7 $\sqrt{2} - 1$
9 $\frac{1}{6}$
11 $\frac{8}{3}$
13 $\frac{2}{3}$
15 1

17 $c(x) = \begin{cases} \sqrt{9-x^2} + x & \frac{-3\sqrt{2}}{2} \le x \le \frac{3\sqrt{2}}{2} \\ 2\sqrt{9-x^2} & \frac{3\sqrt{2}}{2} \le x \le 3 \end{cases}$

19 (b) $c(x) = \sin x, 0 \le x \le \pi$
(c) $c(y) = \pi - 2 \sin^{-1} y, 0 \le y \le 1$

21 (b) $c(x) = \begin{cases} \dfrac{x}{e} & 0 \le x \le 1 \\ \dfrac{x}{e} - \ln x & 1 \le x \le e \end{cases}$
(c) $c(y) = e^y - ey, 0 \le y \le 1$

23 $\frac{\pi}{4} + \frac{1}{2}$ 25 $\frac{2\pi}{3} - \frac{\sqrt{3}}{2}$

27 $c(y) = 3y + 8 \quad 0 \le y \le 2$

29 (a) $c(x) = \begin{cases} 6 & -4 \le x \le 4 \\ 2\sqrt{25-x^2} & -5 \le x \le -4, 4 \le x \le 5 \end{cases}$
(b) $c(x) = 2\sqrt{25-x^2} \quad -3 \le x \le 3$

SEC. 8.2. COMPUTING VOLUME BY PARALLEL CROSS SECTIONS

1 (b) $\int_0^1 x^4 \, dx$
(c) $1/5$

3 (b) $\int_0^1 (\sqrt{y} - y)^2 \, dy$
(c) $1/30$

5 $\int_0^2 \frac{3}{2} x^2 \, dx = 4$

7 $\int_0^4 \frac{3}{16} x^2 \, dx = 4$

13 $A(x) = \dfrac{h(a-x)\sqrt{a^2-x^2}}{a}, \quad -a \le x \le a$

15 (b) $A(x) = \dfrac{\pi a^2 x^2}{h^2}, \quad 0 \le x \le h$

(c) $\int_0^h \frac{\pi a^2 x^2}{h^2} dx$ (d) $\frac{\pi a^2 h}{3}$

17 (b) $A(x) = \frac{(a^2 - x^2)\tan\theta}{2}, \quad -a \le x \le a$

(c) $\frac{2a^3 \tan\theta}{3}$

19 (b) $A(x) = \begin{cases} 9\pi & -4 \le x \le 4 \\ (25 - x^2)\pi & -5 \le x \le -4, \; 4 \le x \le 5 \end{cases}$

(c) $\frac{244\pi}{3}$

21 $\frac{2}{3}$

SEC. 8.3. HOW TO SET UP A DEFINITE INTEGRAL

3 (a) $(g(r))(2\pi r \, dr)$ cubic feet

(b) $\int_{1000}^{2000} 2\pi r g(r) \, dr$

5 (a) $2000\pi^2 x^2 \, dx$ ergs

(b) $\int_{-3}^{3} 2000\pi^2 x^2 \, dx$ ergs

(c) $36000\pi^2$ ergs

7 $6(62.4)$ pounds

9 $(144)(62.4)$ pounds

13 $(64\pi)(62.4)$ pounds

15 $(\frac{15}{2})(62.4)$ pounds

17 (a) $2\pi r g(r) \, dr$ cm³/sec

(b) $\int_0^b 2\pi r g(r) \, dr$

21 (a) $\$1.00$

(b) Effects of inflation

(c) $qg(t)$ dollars

(d) $g(t) f(t) \, dt$ dollars

(e) $\int_0^b f(t) g(t) \, dt$ dollars

23 $(62.4)\left(\frac{175\sqrt{3}}{4}\right)$ pounds

SEC. 8.4. COMPUTING THE VOLUME OF A SOLID OF REVOLUTION

1 (d) $\int_1^2 \pi x \, dx$

(e) $\frac{3\pi}{2}$

3 (d) $\int_0^1 \pi(x^4 - x^6) \, dx$

(e) $\frac{2\pi}{35}$

5 (d) $\pi \int_0^{\pi/4} (\tan^2 x - \sin^2 x) \, dx$

(e) $\frac{\pi}{8}(10 - 3\pi)$

7 (d) $\int_0^1 2\pi x^3 \, dx$

(e) $\frac{\pi}{2}$

9 (d) $\int_0^1 2\pi(y^3 - y^4) \, dy$

(e) $\frac{\pi}{10}$

11 (d) $\int_0^{\pi/2} 2\pi x \sin x \, dx$

(e) 2π

13 $\frac{4}{3}\pi a^3$

15 $\frac{4\pi(22\sqrt{2} - 8)}{15}$

17 (a) $2\pi(99\pi^2 + 2)$

(b) $\frac{81\pi^2}{2}$

19 $\frac{\pi^2}{4}$

21 $\frac{\pi}{3}(7 - 3\sqrt{2} - 3\ln(1 + \sqrt{2}))$

25 Let R be bounded by $y = e^{x^2}$, $y = 0$, $x = 0$, and $x = 1$.

SEC. 8.5. THE CENTROID OF A PLANE REGION

1 $\frac{\pi}{4}$

3 π

5 $\frac{ab^2}{2}$

7 $(\frac{2}{3}, \frac{1}{3})$

9 $(\frac{4}{3}, 2)$

11 $(0, \frac{12}{5})$

13 $\left(\frac{\pi}{4}, \frac{\pi}{8}\right)$

15 $\left(\frac{e^2 + 1}{4}, \frac{e - 2}{2}\right)$

17 $\left(\frac{a}{3}, \frac{b}{3}\right)$

19 (a) $\int_a^b x(f(x) - g(x)) \, dx$

(b) $\int_a^b \frac{(f(x))^2 - (g(x))^2}{2} \, dx$

21 (a) $\frac{15}{\ln 3} - \frac{6}{(\ln 3)^2} - \frac{6}{\ln 2} + \frac{2}{(\ln 2)^2}$

(b) $3\left(\dfrac{6}{\ln 3} - \dfrac{1}{\ln 2}\right)$

(c) $2\left(\dfrac{3}{\ln 3} - \dfrac{1}{\ln 2}\right)$

(d) $\dfrac{[(15/\ln 3) - (6/(\ln 3)^2) - (6/\ln 2) + (2/(\ln 2)^2)]}{2[(3/\ln 3) - (1/\ln 2)]}$

(e) $\dfrac{3(6 \ln 2 - \ln 3)}{2(3 \ln 2 - \ln 3)}$

23 Let R be the triangle bounded by the positive axes and the line $x + y = 1$.

25 (b) $\bar{y}_R = \tfrac{1}{3}$
$\bar{y}_{R^*} = \tfrac{29}{90}$

(c) $\dfrac{2a^3 + 1}{3(a^2 + 1)}$

(f) ≈ 0.322185

SEC. 8.6. IMPROPER INTEGRALS

1 convergent, $\tfrac{1}{2}$
3 divergent
5 convergent, $\dfrac{\pi}{4}$
7 divergent
9 convergent, 100
11 convergent, $\tfrac{1}{8}$
13 divergent
15 divergent
17 convergent, 2
19 convergent, -1
21 convergent, $\dfrac{\pi}{3}$
23 convergent, $\tfrac{3}{10}$
25 divergent
27 divergent
29 divergent
31 divergent
33 convergent, π
35 infinite
45 (a) $\Gamma(1) = 1$
(c) $\Gamma(2) = 1, \Gamma(3) = 2! = 2, \Gamma(4) = 3! = 6,$
$\Gamma(5) = 4! = 24$
(d) $\Gamma(n) = (n - 1)!$

47 $u = \dfrac{1}{x}$ is not defined at $x = 0$.

51 $P(r) = \dfrac{1}{r - 1}$

53 $P(r) = \dfrac{r}{r^2 + 1}$

SEC. 8.7. WORK

1 $(280)(62.4)$ foot-pounds
3 $(180)(62.4)$ foot-pounds
5 $(18000)(62.4)$ foot-pounds
7 2000 mile-pounds
9 2000 mile-pounds
11 $(\tfrac{880}{3})(62.4)$ foot-pounds
13 $\left(\dfrac{81\pi}{4}\right)(62.4)$ foot-pounds
15 (a) $\dfrac{k}{2}$ foot-pounds

(b) $\dfrac{3k}{2}$ foot-pounds

SEC. 8.8. THE AVERAGE OF A FUNCTION OVER AN INTERVAL

1 $\tfrac{7}{3}$
3 $\dfrac{1}{e - 1}$
5 $\ln(\tfrac{3}{2})$
7 $2 \ln 3 - 3 \ln 2$
9 $\dfrac{\pi}{4}$
11 $\dfrac{\pi}{4} - \dfrac{\ln 2}{2}$
13 (a) 30 mph
(b) $\tfrac{280}{9}$ mph
15 (a) $v(t) = 32t$ feet per second
(b) $v(y) = 8\sqrt{y}$ feet per second
(c) $16t$ feet per second
(d) $\dfrac{64t}{3}$ feet per second
17 0
19 It approaches 3.
21 (c) $a = \displaystyle\int_A^B (f(x))^2 \, dx$
$b = -2 \displaystyle\int_A^B f(x)g(x) \, dx$
$c = \displaystyle\int_A^B (g(x))^2 \, dx$
(e) $f = 0$ or $g = Kf$ for some constant K

SEC. 8.9. ESTIMATES OF DEFINITE INTEGRALS

1 (a) $\tfrac{19}{54}$
(b) $\tfrac{1}{3}$; error $= \tfrac{1}{54}$
3 (a) $\tfrac{28}{15}$
(b) $\ln 5$; error ≈ 0.25723

5 $\dfrac{54{,}493}{65{,}520} \approx 0.83170$

7 $\dfrac{76}{45} \approx 1.68889$

9 $\dfrac{15}{7} \approx 2.14286$

11 $\dfrac{82{,}141}{98{,}280} \approx 0.83579$

13 (a) 3.06339
    (b) 3.05914

19 (a) $\tfrac{1}{3}\ln 2 + \dfrac{\pi}{3\sqrt{3}} \approx 0.835649$

    (b) $\dfrac{3{,}231{,}532}{3{,}866{,}940} \approx 0.835682$

    (c) ≈ 0.000033

27 $\tfrac{3}{2}$

29 (a) $\dfrac{101}{60} \approx 1.6833$

    (b) $\dfrac{481}{300} \approx 1.6033$ (c) $\dfrac{73}{45} \approx 1.6222$

    (d) Error ≈ 0.0739, Bound $= \tfrac{2}{3} \approx 0.6667$;
    Error ≈ 0.0061, Bound $= \tfrac{2}{15} \approx 0.1333$;
    Error ≈ 0.0128, Bound $= \tfrac{8}{15} \approx 0.5333$

SEC. 8.S. SUMMARY: GUIDE QUIZ ON CHAPTER 8

1 $e(e-1)$

2 (a) $\dfrac{e^2(e-1)(e+1)}{4}$

    (b) e^2

    (c) $\left(\dfrac{e}{e-1},\ \dfrac{e(e+1)}{4}\right)$

3 $\dfrac{\pi e}{2}(e^3 + 3e - 4)$

4 $2\pi e^2$

6 $\tfrac{1}{6}\ln(\tfrac{5}{2})$

7 (a) $a < -1$
    (b) $a > -1$
    (c) Never convergent

9 (a) $\tfrac{1}{6}[f(1) + 2f(\tfrac{4}{3}) + 2f(\tfrac{5}{3}) + 2f(\tfrac{6}{3}) + 2f(\tfrac{7}{3})$
    $+ 2f(\tfrac{8}{3}) + f(3)]$

    (b) $\tfrac{1}{9}[f(1) + 4f(\tfrac{4}{3}) + 2f(\tfrac{5}{3}) + 4f(\tfrac{6}{3}) + 2f(\tfrac{7}{3})$
    $+ 4f(\tfrac{8}{3}) + f(3)]$

REVIEW EXERCISES FOR CHAPTER 8

1 (a) $\displaystyle\int_0^2 (2x - x^2)\, dx$

   (b) $\displaystyle\int_0^4 \left(\sqrt{y} - \dfrac{y}{2}\right) dy$

3 $\displaystyle\int_0^1 \pi y^2\, dy + \int_1^2 \pi y(2-y)\, dy$

5 (a) $1 + \ln(\tfrac{2}{3})$
   (b) $\pi(\tfrac{7}{6} + 2\ln(\tfrac{2}{3}))$
   (c) $2\pi(\tfrac{1}{2} + \ln(\tfrac{3}{2}))$
   (d) $\pi(\tfrac{19}{6} + 4\ln(\tfrac{2}{3}))$

7 (a) 1
   (b) $\dfrac{\pi^2}{4}$
   (c) $\dfrac{\pi^2}{2}$
   (d) $\dfrac{\pi^2}{4} + 2\pi$

9 (a) $\tfrac{1}{2}\ln 3$
   (b) $\dfrac{\pi}{3}$
   (c) $\dfrac{\pi}{2}(2 - \ln 3)$
   (d) $\dfrac{\pi}{3}(1 + 3\ln 3)$

11 1
13 $\sqrt{2} - 1$
15 $\tfrac{1}{4}(2 + 2\ln 2 - \pi)$
17 $\dfrac{5\sqrt{2}}{6} - \dfrac{2}{3}$

19 (a) $\tfrac{1}{2}\ln\tfrac{3}{2}$
    (b) $\dfrac{\ln 2}{2}$

21 (a) $\dfrac{3\pi - 8}{12}$
    (b) $\dfrac{\pi}{64}$
    (c) $\dfrac{3\pi}{16(3\pi - 8)}$

23 $\dfrac{4\pi}{3}$

25 (a) 1
    (b) $\pi/2$
    (c) 2π

27 (b)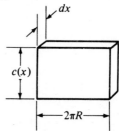

35 $\tfrac{1}{3}(\sin 3 + 2\cos 3)$
37 (a) $\dfrac{1369}{3}$
    (b) $\dfrac{1369}{18}$
41 The integral is improper; the integrand is undefined at 0.
43 $\dfrac{-1}{25}$
47 $(\tfrac{3}{2}, \tfrac{12}{5})$
49 (a) $\dfrac{192{,}199}{179{,}129} \approx 1.072964$

(b) $\dfrac{567,896}{537,387} \approx 1.056773$

(c) $\dfrac{1}{4\sqrt{2}} \left[\ln\left(\dfrac{17 + 4\sqrt{2}}{17 - 4\sqrt{2}}\right) + 2 \tan^{-1}(4\sqrt{2} + 1) \right.$
$\left. + 2 \tan^{-1}(4\sqrt{2} - 1) \right] \approx 1.105521$

55 (a) $\pi r^2 \sqrt{h}(2 - \sqrt{2})$ seconds
(b) $2\pi r^2 \sqrt{h}$ seconds

57 (a) $G(0) - G(1) - G(2) = \dfrac{\pi}{4}$

59 $a = 0, -1$

61 $b = -1, 0, 1$

63 $a = 0$

65 (b) $\mu = 0$
(c) $\mu_2 = k^2$
(d) $\sigma = k$

CHAPTER 9. PLANE CURVES AND POLAR COORDINATES

SEC. 9.1. POLAR COORDINATES

1
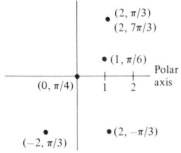

3 (a) $\left(3, \dfrac{\pi}{4} + 2\pi k\right)$ for any integer k

(b) $\left(-3, \dfrac{5\pi}{4} + 2\pi k\right)$ for any integer k

5 $x^2 + y^2 = y$

7 $4x + 5y = 3$

9 $(x^2 + y^2)^3 = 4x^2 y^2$

11 $r = \dfrac{3}{\cos\theta + 2\sin\theta}$

13 $r^2 = \dfrac{2}{\sin 2\theta}$

15 $r = -2\sec\theta$

17
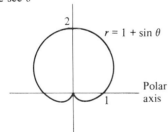

19
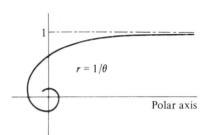

21
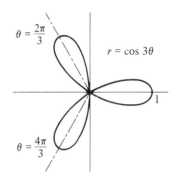

23 (a) The cardioid was graphed in Example 5. The circle was graphed in Example 2.
(b) $(0, 0), (2, 0)$

25
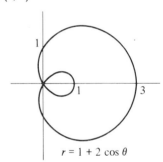

27 $(x^2 + y^2)^2 = x^2 - y^2$

29 $y^2 = 1 - 2x$

31 (a)
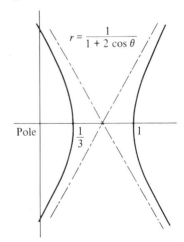

(b) $\pm \dfrac{2\pi}{3}$

(c) $y^2 = 3x^2 - 4x + 1$

SEC. 9.2. AREA IN POLAR COORDINATES

1. $\dfrac{\pi^3}{12}$

3. $\dfrac{\pi}{(4+\pi)(2+\pi)}$

5. $\dfrac{4-\pi}{8}$

7. (a)

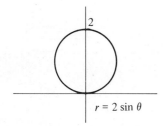

$r = 2\sin\theta$

(b) π

9. $18 + \dfrac{9\pi}{4}$

11. $\dfrac{\pi}{12}$

13. $\dfrac{\pi}{16}$

15. $\dfrac{\pi}{6} + \dfrac{\sqrt{3}}{4}$

17. $\dfrac{\pi+2}{8}$

21. (a)

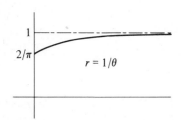

$r = 1/\theta$

(b) Infinite

23. Yes

SEC. 9.3. PARAMETRIC EQUATIONS

1. (a)

t	-2	-1	0	1	2
x	-3	-1	1	3	5
y	-3	-2	-1	0	1

(b), (c)

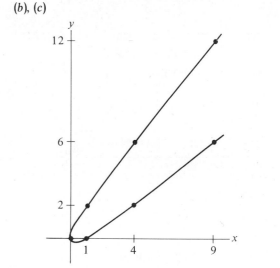

(d) $x - 2y = 3$

3. (a)

t	-3	-2	-1	0	1	2	3
x	9	4	1	0	1	4	9
y	6	2	0	0	2	6	12

(b), (c)

(d) $x^2 + y^2 - 2xy - x = 0$

5. $x = t$
$y = \sqrt{1 + t^3}$

7. $\theta = t$
$r = \cos 2t$

9. $\dfrac{dy}{dx} = \dfrac{7t^6 + 1}{3t^2 + 1}$

$\dfrac{d^2y}{dx^2} = \dfrac{84t^7 + 42t^5 - 6t}{(3t^2 + 1)^3}$

11. $\dfrac{dy}{dx} = \dfrac{3\sin 3\theta \sin\theta - \cos 3\theta \cos\theta}{3\sin 3\theta \cos\theta + \cos 3\theta \sin\theta}$

$\dfrac{d^2y}{dx^2} = \dfrac{10 + 8\sin^2 3\theta}{(3\sin 3\theta \cos\theta + \cos 3\theta \sin\theta)^3}$

13. $\dfrac{1+e}{5+2\pi}$

15. $y = x\tan\alpha - \dfrac{16x^2}{v_0^2 \cos^2\alpha}$

17 (b) πab
19 (a) π
   (b) 2π
   (c) 3π
   (d) 4π
21 (a) $-\dfrac{\cos\theta + \cos 2\theta}{\sin\theta + \sin 2\theta}$
   (b) It approaches 0.
   (c) 0
23 $5\pi^2 a^3$
33 $x = (a+b)\cos\theta - b\cos\left(\dfrac{a+b}{b}\theta\right)$

   $y = (a+b)\sin\theta - b\sin\left(\dfrac{a+b}{b}\theta\right)$

   (θ is the polar angle of the center of the circle of radius b.)

SEC. 9.4. ARC LENGTH AND SPEED ON A CURVE

1 $\dfrac{22\sqrt{22} - 13\sqrt{13}}{27}$ 3 $\frac{59}{24}$

5 $\frac{1}{64}[\frac{205}{2} - \frac{81}{8}\ln 3]$ 7 $\dfrac{e^b - e^{-b}}{2}$ 9 $\frac{3}{2}$

11 $\sqrt{2}(e^{2\pi} - 1)$
13 2
15 $2\sqrt{625 + 256t^2}$
17 $\sqrt{6 - 2\sin t - 4\cos t}$
19 (a) $\frac{1}{27}(40^{3/2} - 13^{3/2})$
   (b) $t\sqrt{4 + 9t^2}$
   (c)

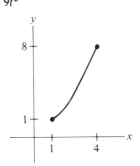

   (d) $y = x^{3/2}$
23 (a) $\sqrt{(g'(t))^2 + (g(t)h'(t))^2}$
   (b) $\sqrt{26}\, e^t$
27 (a) 1
   (b) $\frac{4}{3}$

SEC. 9.5. AREA OF A SURFACE OF REVOLUTION

1 $\displaystyle\int_1^2 2\pi x^3 \sqrt{1 + 9x^4}\, dx$

3 $\displaystyle\int_1^8 2\pi y^{1/3}\sqrt{1 + \tfrac{1}{9}y^{-4/3}}\, dy$

5 $\pi[e\sqrt{1+e^2} + \ln(e + \sqrt{1+e^2}) - \sqrt{2} - \ln(1 + \sqrt{2})]$

7 $\dfrac{64\pi}{3}$

11 (a) $\dfrac{(32 - 4\sqrt{2})\pi}{5}$
   (b) $\dfrac{24\sqrt{2}\,\pi}{5}$

13 (a) The volume of the sphere is $\frac{2}{3}$ times the volume of the cylinder.
   (b) Both surface areas are $4\pi a^2$.

15 $\dfrac{6\pi}{5}$

17 $\displaystyle 2\pi \int_1^2 \dfrac{\sqrt{1 + x^4}\, dx}{x^3}$; use trig substitution

19 $\displaystyle \int_1^8 \dfrac{2\pi}{3} x\sqrt{9 + 16x^{2/3}}\, dx$; use trig substitution

21 $\displaystyle \pi \int_1^3 \left(x^3 + \dfrac{1}{x}\right) dx$; use power rule

23 $\displaystyle \int_{-1}^2 2\pi(x^2 + 1)\sqrt{1 + 4x^2}\, dx$

27 It will be a section between two concentric circles and two rays from their center. The area is $\pi RL - \pi rl$.

29 $\bar{x} = 0,\ \bar{y} = 2a/\pi$
31 $4\pi^2 ab$

SEC. 9.6. THE ANGLE BETWEEN A LINE AND A TANGENT LINE

1 $\dfrac{\pi}{2}$

3 $\tan^{-1}\left(\tfrac{1}{7}\right) \approx 0.142$ radians

5 $\tan^{-1}\left(\tfrac{1}{7}\right) \approx 0.142$ radians

7 $-2\sqrt{2}$

9 The tangent is undefined; the curves are perpendicular.

11 $\gamma = \dfrac{\pi}{6},\ \varphi = \dfrac{\pi}{3}$.

13 $\gamma = \tan^{-1}\left(\dfrac{\sqrt{3}}{2}\right) \approx 0.714$ radians;

   $\varphi = \dfrac{\pi}{6} + \tan^{-1}\left(\dfrac{\sqrt{3}}{2}\right) \approx 1.24$ radians

19 $r = a \sin \theta$ for some constant a

23 (a) $r = \dfrac{a}{\sqrt{2}} e^{-\theta}$, $s = a$

SEC. 9.7. THE SECOND DERIVATIVE AND THE CURVATURE OF A CURVE

1 Curvature $= \dfrac{2\sqrt{5}}{25}$

Radius of curvature $= \dfrac{5\sqrt{5}}{2}$

3 Curvature $= \dfrac{e^{-1}}{(1 + e^{-2})^{3/2}}$

Radius of curvature $= \dfrac{(1 + e^{-2})^{3/2}}{e^{-1}}$

5 Curvature $= \dfrac{4}{5\sqrt{5}}$

Radius of curvature $= \dfrac{5\sqrt{5}}{4}$

7 $\dfrac{1}{2}$

9 $\dfrac{e^{\pi/6}}{\sqrt{2}}$

11 (a) Curvature $= \dfrac{4}{(e^x + e^{-x})^2}$

Radius of curvature $= \dfrac{(e^x + e^{-x})^2}{4}$

13 $x = -\dfrac{\ln 2}{2}$

17 $\dfrac{(a^4 y^2 + b^4 x^2)^{3/2}}{a^4 b^4}$

19

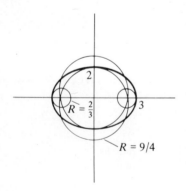

23 $\dfrac{5 + 3 \sin^2 2\theta}{(1 + 3 \sin^2 2\theta)^{3/2}}$

SEC. 9.S. SUMMARY: GUIDE QUIZ ON CHAPTER 9

1 $\displaystyle\int_{-\pi/10}^{\pi/10} \dfrac{\cos^2 5\theta}{2} d\theta$

2 $\displaystyle\int_1^2 \sqrt{1 + 16x^6}\, dx$

3 $\displaystyle\int_1^2 2\pi x^4 \sqrt{1 + 16x^6}\, dx$

4 $\displaystyle\int_1^2 2\pi(x - 1)\sqrt{1 + 16x^6}\, dx$

5 $\displaystyle\int_0^{2\pi} \sqrt{5 + 4 \sin \theta}\, d\theta$

6 $\displaystyle\int_0^{\pi} 2\pi(2 + \sin \theta) \sin \theta \sqrt{5 + 4 \sin \theta}\, d\theta$

7 $\displaystyle\int_0^1 \sqrt{100t^2 + \dfrac{t^{-1}}{4}}\, dt$

8 $\sqrt{873}$

9 $y^2 = x^2 + 1$

10 (a) $\varphi = \dfrac{\pi}{8} + \gamma \approx 3.07$ radians ($\approx 176°$)

(b)

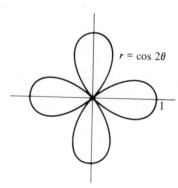

11 $\sqrt{2}$

REVIEW EXERCISES FOR CHAPTER 9

1 (a) See Sec. 9.2
(b) Figure 9.13

3 (a) See Sec. 9.5.
(b) Figure 9.43

5 (a) Curvature $= \dfrac{d\varphi}{ds}$

(b) See Theorem 2 of Sec. 9.7.

7 Arc length $= \displaystyle\int_0^{\pi} \sqrt{1 + \cos^2 x}\, dx$

Surface area (about x axis)

$= \displaystyle\int_0^{\pi} 2\pi \sin x \sqrt{1 + \cos^2 x}\, dx$

$= 2\pi(\sqrt{2} + \ln(1 + \sqrt{2}))$

Surface area (about y axis)

$= \displaystyle\int_0^{\pi} 2\pi x \sqrt{1 + \cos^2 x}\, dx$

9 $\dfrac{3\sqrt{3}}{4}$
11 Infinite
13 $\sqrt{2}$
15 $\dfrac{dy}{dx} = -\dfrac{3\cos 3t}{2\sin 2t}$
   $\dfrac{d^2 y}{dx^2} = \dfrac{18\sin 2t \sin 3t + 12\cos 2t \cos 3t}{(-2\sin 2t)^3}$
17 (a)

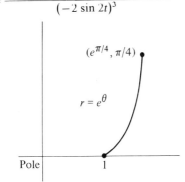

(b) $\displaystyle\int_0^{\pi/4} \dfrac{e^{2\theta}}{2}\, d\theta$

(c) $\displaystyle\int_0^{\pi/4} 2\pi e^{3\theta} \sin\theta\,(\cos^2\theta - \sin^2\theta)\, d\theta$

(d) $\displaystyle\int_0^{\pi/4} \pi e^{3\theta}(\cos^2\theta - \sin^2\theta)(\sin\theta + \cos\theta)\, d\theta$

(e) $\displaystyle\int_0^{\pi/4} 2\sqrt{2}\,\pi e^{2\theta} \sin\theta\, d\theta$

(f) $\displaystyle\int_0^{\pi/4} 2\sqrt{2}\,\pi e^{2\theta} \cos\theta\, d\theta$

23 At (1000, 0), $A = \tan^{-1}(0.968) \approx 0.77$ radians ($\approx 44°$)
   At (0, 500), $A = \tan^{-1}(0.121) \approx 0.12$ radians ($\approx 7°$)

CHAPTER 10. SERIES AND RELATED TOPICS

SEC. 10.1. SEQUENCES

1 0
3 1
5 0
7 0
9 ∞
11 3
13 Diverges; oscillates
15 e^2
17 0
21 (a) 1 (b) 0 (c) 0
23 (a) 0.6561 (b) 0

25 (a) 0, 1, $\tfrac{1}{2}$
   (b) All x

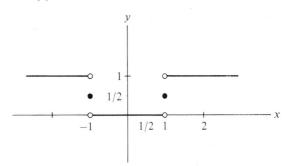

(c) $x \neq \pm 1$
27 ln 2
29 (a) $\tfrac{1}{2}, \tfrac{2}{3}, \tfrac{3}{4}, \tfrac{4}{5}$
   (b) $\dfrac{n}{n+1}$
33 The degree of $Q(x)$ is greater than or equal to the degree of $P(x)$.

SEC. 10.2. SERIES

1 2
3 $\tfrac{1}{9}$
5 495
7 8
9 Diverges
11 Diverges
13 24
15 $\tfrac{3}{2}$
19 (a) 30.8940 ft (b) 114 ft
23 (b) 251/128
25 (b) 2
27 2848/825
29 $5,000

SEC. 10.3. THE INTEGRAL TEST

1 Converges
3 Diverges
5 Diverges
7 Diverges
9 Diverges
11 Converges
13 (a) 1.1777
   (b) $1/50 \leq R_4 \leq 1/32$
   (c) $1.1977 \leq \displaystyle\sum_{n=1}^{\infty} \dfrac{1}{n^3} \leq 1.2089$
15 (a) 0.8588
   (b) $0.1974 \leq R_4 \leq 0.2450$
   (c) $1.0562 \leq \displaystyle\sum_{n=1}^{\infty} \dfrac{1}{n^2 + 1} \leq 1.1038$

SEC. 10.4. THE COMPARISON TEST AND THE RATIO TEST

1. Converges
3. Converges
5. Converges
7. Diverges
9. Converges
11. Converges
13. Converges
15. Converges
17. Converges
19. $0 < x < 1$
21. $0 < x < \infty$
31. Converges

SEC. 10.5. THE ALTERNATING-SERIES AND ABSOLUTE-CONVERGENCE TESTS

1. Diverges
3. Converges
5. Diverges
7. Diverges
9. (a) 0.78333 (b) Larger
   (c) $0.58333 < S < 0.78334$
13. Converges conditionally
15. Converges absolutely
17. Converges absolutely
19. Converges absolutely
21. Converges conditionally
23. Converges absolutely
25. Converges absolutely
27. (a) 0.328
    (b) $R_6 < 2^{-7} \approx 0.0078$
    (c) $1/192 \approx 0.0052$

SEC. 10.6. POWER SERIES

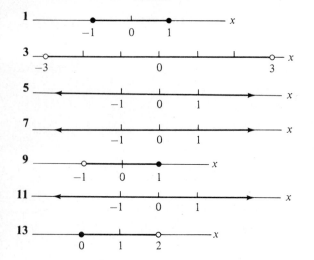

15.

17. (graph from -8 to 0, 1, 2)

19. (a) Converges absolutely
    (b) Converges absolutely
    (c) Cannot be determined
    (d) Cannot be determined
    (e) Cannot be determined
    (f) Diverges
    (g) Diverges
23. (a) $x \geq 3$ and $x < -3$
    (b) $x < -8$ and $x \geq -2$

SEC. 10.7. MANIPULATING POWER SERIES

1. (a)

n	$f^{(n)}(x)$	$f^{(n)}(0)$	$f^{(n)}(0)/n!$
0	$\sin x$	0	0
1	$\cos x$	1	1
2	$-\sin x$	0	0
3	$-\cos x$	-1	$-1/6$
4	$\sin x$	0	0
5	$\cos x$	1	$1/120$

   (b) $x - \dfrac{x^3}{6} + \dfrac{x^5}{120}$

5. $x + \dfrac{x^3}{3} + \dfrac{2x^5}{15}$

7. $x + \dfrac{x^3}{6} + \dfrac{3x^5}{40}$

9. (c) $a_n = \dfrac{(-1)^n x^{2n+1}}{2n+1}$, $n = 0, 1, 2, \ldots$

11. $x + x^2 + \dfrac{x^3}{3}$

13. ∞

15. 1

17. $\dfrac{1}{2} + \dfrac{\sqrt{3}}{2}\left(x - \dfrac{\pi}{6}\right) - \dfrac{1}{4}\left(x - \dfrac{\pi}{6}\right)^2$

19. $e + e(x-1) + \dfrac{e}{2}(x-1)^2$

23. (a) 1.6484   (b) Error $< 1/3520$

25. 2.7182815

27. (a) $1 - x^2 + \dfrac{x^4}{2!} - \dfrac{x^6}{3!} + \cdots$
    (c) 0.747

SEC. 10.8. TAYLOR'S FORMULA

1 $\int_0^x \dfrac{4(x-t)^3}{(1+t)^5}\,dt$

3 $\int_0^x \dfrac{-(x-t)^3}{(1+t)^4}\,dt$

5 $\dfrac{1}{6}\int_0^x (x-t)^3 \cos t\,dt$

7 $x - \dfrac{x^3}{6}$

9 $x + \dfrac{x^3}{3}$

11 $P_0(x;0) = 1$
$P_1(x;0) = 1 - x$
$P_2(x;0) = 1 - x + x^2$
$P_3(x;0) = 1 - x + x^2 - x^3$

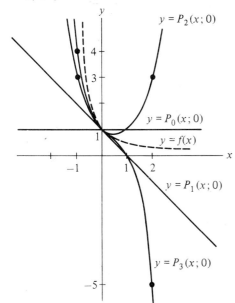

13 $P_0(x;0) = 0$
$P_1(x;0) = x$
$P_2(x;0) = x$
$P_3(x;0) = x - \dfrac{x^3}{6}$

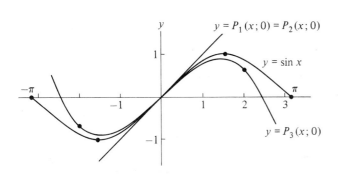

23 $\sqrt[3]{e} \approx 1.3889$, Error $< \tfrac{1}{81} \approx 0.0123$

25 $\sin 1 \approx 0.841667$, $|\text{Error}| < \dfrac{1}{7!} \approx 0.0002$

27 $\cos 10° \approx 0.98480779$, Error $< 3.9 \times 10^{-8}$

33

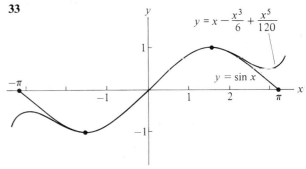

35 (a) $1 - \dfrac{x^4}{2} + \dfrac{x^8}{4!} - \dfrac{x^{12}}{6!} + \cdots$

(b) 0.470

37 0.36

SEC. 10.9. THE COMPLEX NUMBERS

1 (a) $\tfrac{5}{2} + i$ (b) 13

(c) $\dfrac{2}{5} + \dfrac{i}{5}$ (d) $\tfrac{10}{17} + \tfrac{11}{17} i$

3 (a)

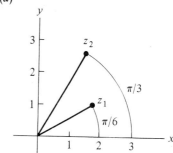

(b) $6i$

(c) $z_1 = \sqrt{3} + i,\ z_2 = \dfrac{3}{2} + \dfrac{3\sqrt{3}}{2} i$

(d) $6i$

5

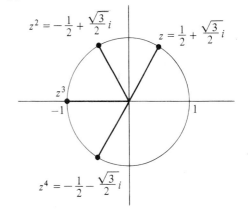

7 (a) $|z^2| = 4$, arg $z^2 = \pi/3$
(b) $|z^3| = 8$, arg $z^3 = \pi/2$
(c) $|z^4| = 16$, arg $z^4 = 2\pi/3$
(d) $|z^n| = 2^n$, arg $z^n = n\pi/6$
(e)

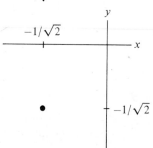

9 $\cos 3\theta = 4\cos^3\theta - 3\cos\theta$
$\sin 3\theta = 3\sin\theta - 4\sin^3\theta$

13 (b) $2 + 3i$, $2 - 3i$

15 (a) $(5\sqrt{2}, \pi/4)$ (b) $(1, 4\pi/3)$
(c) $(1, 3\pi/4)$ (d) $(5, \tan^{-1}(4/3))$
(g) $\frac{1}{3}(\cos 308° + i \sin 308°)$ (h) 1

19 (a) $-1 + 5i$ (b) $\frac{5}{2} + \frac{i}{2}$
(c) 58   (d) $15\left(-\frac{\sqrt{3}}{2} - \frac{i}{2}\right)$
(e) $2i$ (f) $\frac{3}{10} + \frac{i}{10}$

25 $\pm 1, \pm i$

27 (a)

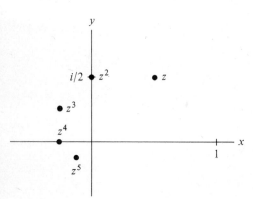

(b) $\lim_{n \to \infty} z^n = 0$

29 (a) 1.11   (b) 0.98   (c) $-4 + 7i$

SEC. 10.10. THE RELATION BETWEEN THE EXPONENTIAL FUNCTION AND THE TRIGONOMETRIC FUNCTIONS

1 Re $z = -1/\sqrt{2}$

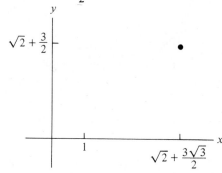

Im $z = -1/\sqrt{2}$

3 Re $z = \sqrt{2} + \frac{3\sqrt{3}}{2}$

Im $z = \sqrt{2} + \frac{3}{2}$

5 Re $z = \cos\frac{11\pi}{12}$

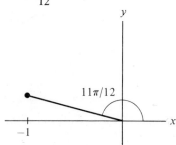

Im $z = \sin\frac{11\pi}{12}$

7 $e^2 e^{-\pi i/4}$

9 $15e^{2\pi i/3}$

11

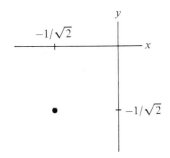

13

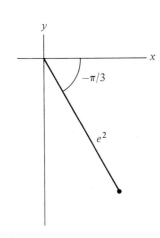

15 (a) e^a (b) e^{a-bi} (c) e^{-a-bi}
   (d) $e^a \cos b$ (e) $e^a \sin b$ (f) b

17 $z = 2\pi ni$, $n = 0, 1, 2, \ldots$

19 $z = \ln 5 + (\alpha + 2\pi n)i$, $\alpha = \tan^{-1}(\frac{4}{3})$, $n = 0, 1, 2, \ldots$

21 (a)

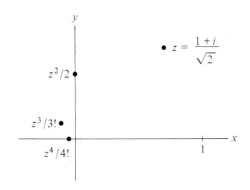

(b), (c)

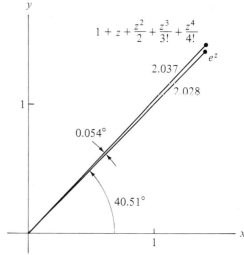

[Difference exaggerated; numbers rounded]

25 (a) $\sinh z = \dfrac{e^z - e^{-z}}{2}$

$\cosh z = \dfrac{e^z + e^{-z}}{2}$

27 $\dfrac{4 - 2\cos\theta}{5 - 4\cos\theta}$

SEC. 10.11. LINEAR DIFFERENTIAL EQUATIONS WITH CONSTANT COEFFICIENTS

1 $y = Ce^{-2x}$

3 $y = \dfrac{x}{12} - \dfrac{1}{48} + Ce^{-4x}$

5 $y = -x^2 - 2x - 2 + Ce^x$

7 $y = C_1 e^{-x} + C_2 e^{3x}$

9 $y = e^{3x/2}[C_1 e^{\sqrt{5}x/2} + C_2 e^{-\sqrt{5}x/2}]$

11 $y = C_1 e^{3x} + C_2 x e^{3x}$

13 $y = C_1 e^{\sqrt{11}x} + C_2 x e^{\sqrt{11}x}$

15 $y = -\frac{1}{3}e^{2x} + C_1 e^{-x} + C_2 e^{3x}$

17 $y = -\frac{1}{26}\cos 3x - \frac{3}{52}\sin 3x + e^{2x}[C_1 e^{\sqrt{3}x} + C_2 e^{-\sqrt{3}x}]$

19 (b) $y_p = e^{-x} \ln(1 + e^x)$
   (c) $y = e^{-x} \ln(1 + e^x) + Ce^{-x}$

SEC. 10.12. THE BINOMIAL THEOREM FOR ANY EXPONENT

1 $1 + \frac{1}{2}x - \frac{1}{8}x^2 + \frac{1}{16}x^3 - \frac{5}{128}x^4$

3 $1 - 3x + 6x^2 - 10x^3 + 15x^4$

5 $1 + \frac{1}{2}x^3 - \frac{1}{8}x^6 + \frac{1}{16}x^9$

7 $1 - \frac{1}{3}x^2 - \frac{1}{9}x^4 - \frac{5}{81}x^6$

9 $\frac{1247}{1120} \approx 1.1134$

SEC. 10.13. NEWTON'S METHOD FOR SOLVING AN EQUATION

3 $x_2 = 4.375$, $\quad x_3 \approx 4.359$
5 (b) $x_2 = 3$, $\quad x_3 \approx 2.259$
  (c) $x_2 \approx 1.917$, $\quad x_3 \approx 1.913$
7 (b) $x_2 \approx 0.857$
9 (a)

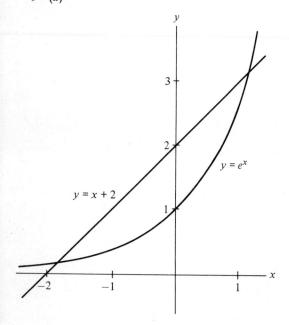

(c) ≈ 1.15

11 (a)

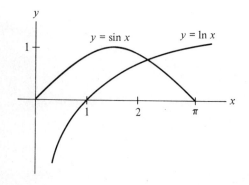

(c) If $x_1 = 2$, then $x_2 \approx 2.236$.

13 (a)

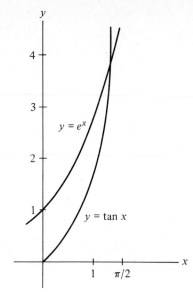

(c) If $x_1 = 1.3$, then $x_2 \approx 1.307$.

15 (b) $x_2 = 1.5$, $x_3 \approx 1.316$ (c)

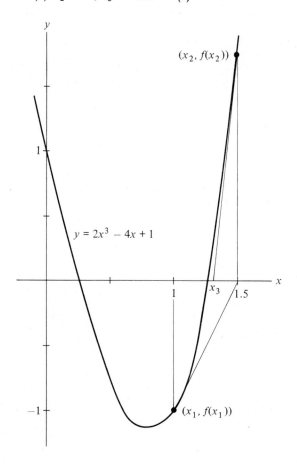

SEC. 10.S. SUMMARY: GUIDE QUIZ ON CHAP. 10

1. $-2 < x < 2$
2. $x > 5$ and $x < -1$
3. $\dfrac{1}{1-x} = 1 + x + x^2 + \cdots$
4. 0.31
6. (a) $1 - x^2 + x^4 - x^6 + \cdots, R = 1$
   (b) $1 - x + \dfrac{x^2}{2} - \dfrac{x^3}{3!} + \cdots, R = \infty$
   (c) $1 - \dfrac{x^2}{2} + \dfrac{x^4}{4!} - \dfrac{x^6}{6!} + \cdots, R = \infty$
   (d) $-x - \dfrac{x^2}{2} - \dfrac{x^3}{3} - \dfrac{x^4}{4} - \cdots, R = 1$
   (e) $1 + 2x + 4x^2 + 8x^3 + \cdots, R = \tfrac{1}{2}$
8. (a) $1 \le x < 3$ (b) All x

REVIEW EXERCISES FOR CHAP. 10

1. (a) $\dfrac{(-1)^{n+1}(3x)^{2n-1}}{2n-1}$
   (b) $(-1)^{n+1}\dfrac{x^{2n}}{n}$
   (c) $\dfrac{(-1)^n x^n}{n!}$ (Sum starts at $n = 0$.)
   (d) $\dfrac{(-1)^{n+1} x^{2(2n-1)}}{(2n-1)!}$
   (e) $\dfrac{(-1)^n x^{4n}}{(2n)!}$ (Sum starts at $n = 0$.)
5. (a), (b) $\dfrac{(2x)^2}{2 \cdot 2} - \dfrac{(2x)^4}{2 \cdot 4!} + \dfrac{(2x)^6}{2 \cdot 6!} - \cdots$
7. Converges to $\dfrac{e}{e-1}$
9. Converges to $\pi^2/12$
11. Converges
13. Converges
15. Converges
17. Diverges
19. Converges
21. Converges to $\ln 2$
23. Converges to e^{10}
25. Converges to $e^{-1/2}$
27. Converges to $3 + \tfrac{3}{2}\ln 3$
29. Converges
31. Diverges
33. $\cos \tfrac{1}{3} \approx 0.94496$, Error $< 3.4 \times 10^{-5}$
35. $1/\sqrt{e} \approx \tfrac{5}{8}$, Error $< \tfrac{1}{48} \approx 0.0208$
37. Converges absolutely for $-1 < x < 1$ and diverges elsewhere. $R = 1$ and the sum is $1/(1-x)^2$.
39. Converges absolutely for all x; $R = \infty$ and the sum is e^{x^2}.
41. Converges absolutely for $-1 < x < 1$ and conditionally at $x = -1$. $R = 1$ and the sum is $-\ln(1-x)$.
43. Converges absolutely for $-2 < x < 0$ and diverges elsewhere; $R = 1$.
45. Converges absolutely for $\tfrac{1}{2} < x < \tfrac{3}{2}$ and diverges elsewhere; $R = \tfrac{1}{2}$.
47. Converges absolutely for all x.
49. Converges absolutely for $0 < x < 2$ and conditionally at $x = 0$; $R = 1$.
51. Converges absolutely for $-4 < x < -2$ and conditionally at $x = -4$; $R = 1$.
53. Converges absolutely for $-e^2 < x < e^2$; $R = e^2$
55. 0
57. ∞
59. $\dfrac{-1}{3 \cdot 2^{10}}$
61. $32 + 11(x - 5) + (x - 5)^2$
63. (c) 0.51
65. 1.63
67. $(33!)2^{33}$
71. Finite $\left(\dfrac{\sqrt{6}}{4} + \dfrac{\sqrt{3}}{2} \right) \left(\dfrac{2}{\sqrt{5} - 2} \right)$ seconds
73. (b) $a_n = \dfrac{(-1)^n}{\sqrt{n}}$
75. (a) $(-1)^{n+1}\dfrac{x^n}{2n-1}$
    (b) $(-1)^{n+1}\dfrac{x^{2n}}{n}$
    (c) $(-1)^n\dfrac{x^{2n}}{n!}$ (Sum starts at $n = 0$.)
    (d) $(-1)^{n+1}\dfrac{x^{2(2n-1)}}{(2n-1)!}$
    (e) $(-1)^n\dfrac{x^{2n}}{(2n)!}$ (Sum starts at $n = 0$.)
91. Converges for $-\dfrac{1}{e} < x < \dfrac{1}{e}$ and diverges elsewhere.

CHAPTER 11. ALGEBRAIC OPERATIONS ON VECTORS

SEC. 11.1. THE ALGEBRA OF VECTORS

1.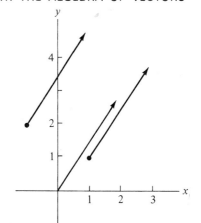

S-138 ANSWERS TO SELECTED ODD-NUMBERED PROBLEMS AND TO GUIDE QUIZZES

3 (a) 4   (b) 13   (c) 13   (d) 10

5 (a) $\overrightarrow{(3, 5)}$ (b) $\overrightarrow{(3, 1)}$

7 (a) $\overrightarrow{(2, 3)}$ (b) $\overrightarrow{(-2, 3)}$ (c) $\overrightarrow{(3, -3)}$

9 (a) $\overrightarrow{(3, 3\tfrac{1}{2})}$ (b) $\overrightarrow{(0, -3)}$ (c) $\overrightarrow{(0, 3)}$
   (d) $\overrightarrow{(\tfrac{11}{2}, -\tfrac{8}{3})}$

11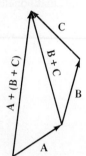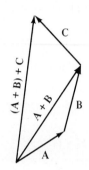

13 (a) $\approx$ 104.40 mi/hr   (b) $\approx$ 28.30° south of east

15 (b) Parallel vectors with the same direction

17 (a) 0   (b)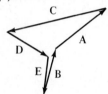

SEC. 11.2. VECTORS IN SPACE

1 (a)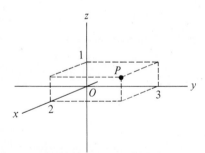

(b)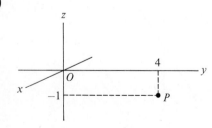

(c)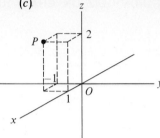

3 (a) $\sqrt{19}$ (b) $\sqrt{62}$

5 (a) $\overrightarrow{(1, -2, 4)}$ (b) $\overrightarrow{(3, 4, 2)}$

7 $\sqrt{2}$

9 (a), (b)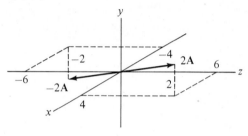

(c), (d)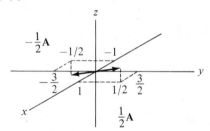

11 3

13 (a) $-4\mathbf{i}$ (b) $5\mathbf{j}$ (c) $0\mathbf{k}$

15 (a) $2\mathbf{i}$ (b) $2\mathbf{i}$ (c) $2\mathbf{i}$ (d) $2\mathbf{i} + 2\mathbf{j}$

17 (b) $x = \tfrac{1}{2}, y = \dfrac{\sqrt{3}}{2}$

19 $\dfrac{1}{\sqrt{2}}\mathbf{i}, \dfrac{1}{\sqrt{2}}\mathbf{j}$

21 $\dfrac{1}{\sqrt{14}}\mathbf{i} + \dfrac{2}{\sqrt{14}}\mathbf{j} + \dfrac{3}{\sqrt{14}}\mathbf{k}$

23 (a) $a = b = 1/\sqrt{2}$
   (b) $a = -1/\sqrt{2}, b = 1/\sqrt{2}$
   (c) $a = 7/\sqrt{2}, b = -1/\sqrt{2}$

25 (a) $\mathbf{u} = \dfrac{1}{2}\mathbf{i} + \dfrac{\sqrt{3}}{2}\mathbf{j}$

(b)
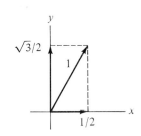

39 Only if $\mathbf{u}_1$ is perpendicular to $\mathbf{u}_2$.

SEC. 11.3. THE DOT PRODUCT OF TWO VECTORS

1 $6\sqrt{2}$
3 0
5 0
7 (a)
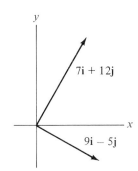
(c) Not perpendicular
9 (b) 0.2487 radians $\approx 14.25°$
11 (b) $-1/\sqrt{10}$ (c) $-1/\sqrt{10}$
13 $\dfrac{-19}{\sqrt{37}\sqrt{170}} \approx -0.2396$
15 $3\mathbf{i} + 3\mathbf{j} + 3\mathbf{k}$ and $-\mathbf{i} + \mathbf{k}$, respectively
17 $2\mathbf{j} + 3\mathbf{k}$ and $\mathbf{i}$, respectively
19 $-\frac{1}{3}$
21 $3/\sqrt{15}$
23 $\frac{1}{3}$
25 No
27 (a) 15   (b) -15
31 (a)
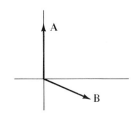

(b)
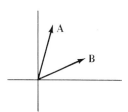

(c)
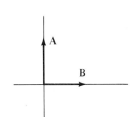

37 (a) $\overline{|(x_1, x_2, x_3, \ldots, x_n, \ldots)|} = \left(\displaystyle\sum_{n=1}^{\infty} x_n^2\right)^{1/2}$,
if it is convergent. No.
(b) $1/\sqrt{3}$
39 There is zero profit.

SEC. 11.4. LINES AND PLANES

1 $4(x - 2) + 5(y - 3) = 0$
3 $2(x - 4) + 3(y - 5) = 0$
5 $2\mathbf{i} - 3\mathbf{j}$
7 $3\mathbf{i} - \mathbf{j}$
9 2
11 $2/\sqrt{29}$
13 $8/\sqrt{57}$
17 $2/\sqrt{29}, 3/\sqrt{29}, -4/\sqrt{29}$
19
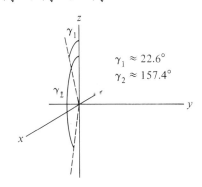
$\gamma_1 \approx 22.6°$
$\gamma_2 \approx 157.4°$

21 (a) $x = 2t + \frac{1}{2}, y = -5t + \frac{1}{3}, z = 8t + \frac{1}{2}$
(b) $(x - \frac{1}{2})\mathbf{i} + (y - \frac{1}{3})\mathbf{j} + (z - \frac{1}{2})\mathbf{k} = t(2\mathbf{i} - 5\mathbf{j} + 8\mathbf{k})$
23 No
25 $(\frac{31}{21}, \frac{37}{21}, \frac{85}{21})$
27 No
29 $\sqrt{3657}/69 \approx 0.8764$
31 $\cos \alpha \cos \alpha' + \cos \beta \cos \beta' + \cos \gamma \cos \gamma'$

SEC. 11.5. DETERMINANTS OF ORDERS 2 AND 3

1. -22
3. 22
5. 0
13. $x = -\frac{3}{5}, y = -\frac{14}{5}$
15. $x = 2, y = 1, z = 1$
17. (a) $x = 75/19, y = 12/19$
    (c)

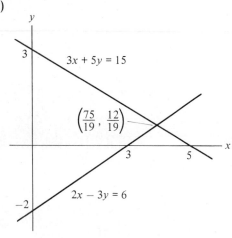

19. (b) $7x - y = 11$
21. (a)

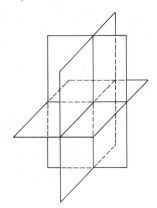

    (b)

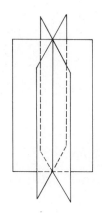

    (c)

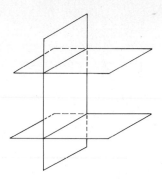

23. (b) 0
25. (a) 1 gallon/day
    (b) 15 mi/gallon, city; 30 mi/gallon, highway
27. (b) -2

SEC. 11.6. THE CROSS PRODUCT OF TWO VECTORS

1. $-\mathbf{i}$

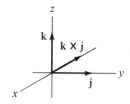

3. $-\mathbf{i} + \mathbf{j}$

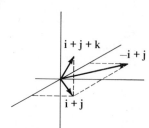

5. $-7\mathbf{i} - 3\mathbf{j} + 5\mathbf{k}$
7. $\sqrt{507} \approx 22.52$
9. $-8\mathbf{i} - 2\mathbf{k}$
11. $-10x + 4y + 18z = 0$
19. $22\mathbf{i} - 3\mathbf{j} - 7\mathbf{k}$
35. (a)

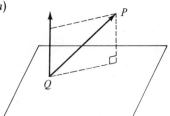

    (b) $d = \left| \dfrac{\vec{QP} \cdot (\vec{QR} \times \vec{QS})}{|\vec{QR} \times \vec{QS}|} \right|$

37 $\cos\theta = \dfrac{(\overrightarrow{PQ}\times\overrightarrow{PR})\cdot(\overrightarrow{ST}\times\overrightarrow{SU})}{|\overrightarrow{PQ}\times\overrightarrow{PR}||\overrightarrow{ST}\times\overrightarrow{SU}|}$

CHAPTER 12. PARTIAL DERIVATIVES

SEC. 12.1. GRAPHS OF EQUATIONS

1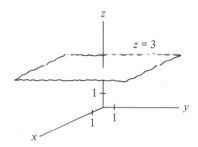

SEC. 11.5. SUMMARY: GUIDE QUIZ

1 (a) -3
  (b) $\sqrt{6}$
  (c) $\dfrac{1}{\sqrt{6}}\mathbf{i} + \dfrac{2}{\sqrt{6}}\mathbf{j} - \dfrac{1}{\sqrt{6}}\mathbf{k}$
  (d) $-3/\sqrt{6}$
  (e) $-\tfrac{1}{2}\mathbf{i} - \mathbf{j} + \tfrac{1}{2}\mathbf{k}$
  (f) $-3/\sqrt{14}$
  (g) $\tfrac{5}{2}\mathbf{i} + \tfrac{5}{2}\mathbf{j}$
  (h) $\dfrac{-3}{2\sqrt{21}}$
  (i) 1.9043 radians $\approx 109.11°$
  (j) $5\mathbf{i} - 5\mathbf{j} - 5\mathbf{k}$
  (k) $-5\mathbf{i} + 5\mathbf{j} + 5\mathbf{k}$
  (l) $\dfrac{\mathbf{i} - \mathbf{j} - \mathbf{k}}{\sqrt{3}}$
  (m) $5\sqrt{3}$

3 $\tfrac{1}{3}, -\tfrac{2}{3}, \tfrac{2}{3}$

5 (a) 8

6 (b) $\mathbf{A}_2 = \dfrac{\mathbf{A}\cdot(\overrightarrow{PQ}\times\overrightarrow{PR})}{|\overrightarrow{PQ}\times\overrightarrow{PR}|^2}(\overrightarrow{PQ}\times\overrightarrow{PR})$
  $\mathbf{A}_1 = \mathbf{A} - \mathbf{A}_2$

7 $(\tfrac{11}{3}, \tfrac{2}{3}, 1)$

3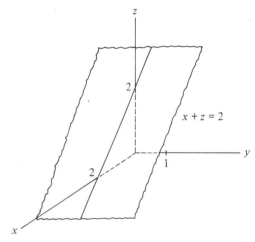

REVIEW EXERCISES

3 (a) $-2\mathbf{i}$ (b) $3\mathbf{j}$ (c) $0.72\mathbf{i} + 0.96\mathbf{j}$
  (d) $-\tfrac{23}{41}(4\mathbf{i} - 5\mathbf{j})$
11 $x = -\tfrac{3}{2}, y = \tfrac{3}{2}, z = 3$
15 (a) $2/\sqrt{6}$ (b) $1/\sqrt{2}$ (c) $\tfrac{1}{9}$
17 $\dfrac{|A(x_2 - x_1) + B(y_2 - y_1) + C(z_2 - z_1)|}{\sqrt{A^2 + B^2 + C^2}}$
19 $x = 1 + 9t, y = 1 + 3t, z = 2 - 5t$
21 Yes
23 (a) $(\tfrac{72}{61}, \tfrac{48}{61}, \tfrac{36}{61})$
  (b) $\left(\dfrac{-5}{61}, \dfrac{78}{61}, \dfrac{150}{61}\right)$
27 $\overrightarrow{P_1P_2}\cdot(\overrightarrow{P_3P_4}\times\overrightarrow{P_3P_5}) = 0$
29 The line makes an angle of approximately $42.83°$ (0.7476 radians) with the normal to the plane.
31 (b) $\overrightarrow{(x_1, x_2, x_3, x_4, x_5)}$ and $\overrightarrow{(P_1, P_2, P_3, P_4, P_5)}$, respectively.
37 Yes

5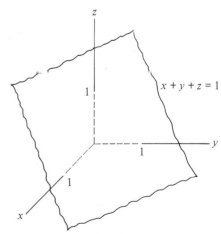

7 (a) ± 7 (b) ± 7 (c) ± 7

9

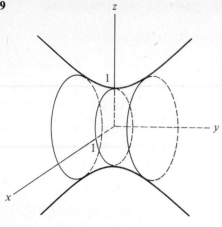

11

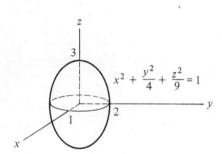

13

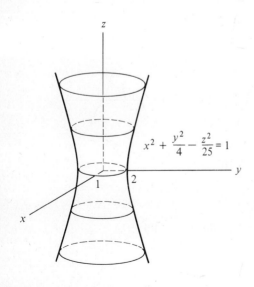

15

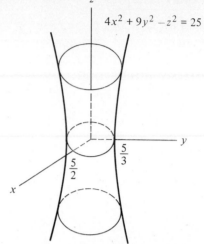

17

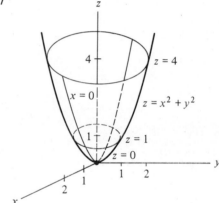

19

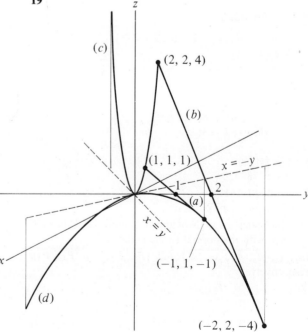

21

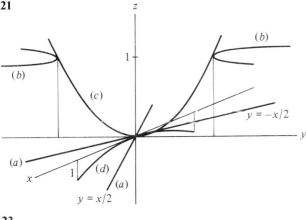

23

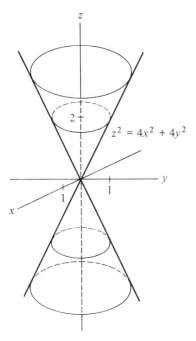

27 (a)

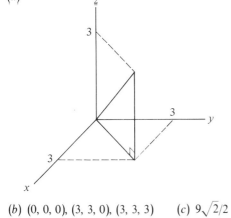

(b) (0, 0, 0), (3, 3, 0), (3, 3, 3)   (c) $9\sqrt{2}/2$

SEC. 12.2. FUNCTIONS AND LEVEL CURVES

1

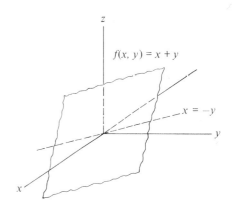

3

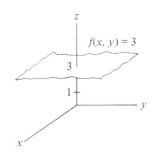

5

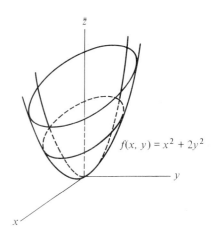

7
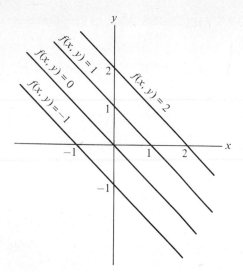

9
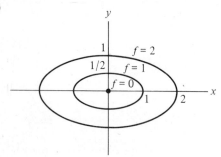

11
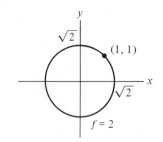

13
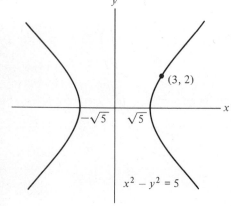

15
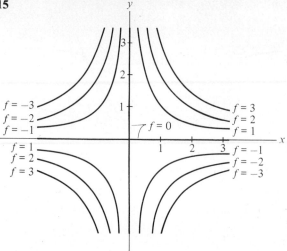

17
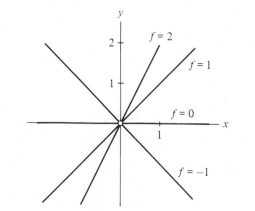

SEC. 12.3. PARTIAL DERIVATIVES

1 $f_x = 3x^2y^4, f_y = 4x^3y^3$

3 $f_x = \frac{1}{2}x^{-1/2} \sec x^2y + 2x^{3/2}y \sec x^2y \tan x^2y$
$f_y = x^{5/2} \sec x^2y \cdot \tan x^2y$

5 $f_x = e^{3x}\left(3 \tan^{-1} xy + \dfrac{y}{1+(xy)^2}\right)$
$f_y = e^{3x} \dfrac{x}{1+(xy)^2}$

7 $f_x = \dfrac{1}{\sqrt{1-(x+3y)^2}}, f_y = \dfrac{3}{\sqrt{1-(x+3y)^2}}$

9 $f_x = \dfrac{(\ln 2)2^x}{y^3}, f_y = \dfrac{-3(2^x)}{y^4}$

11 $f_x = 10x - 3y, f_{xx} = 10, f_{xy} = -3$
$f_y = -3x + 12y, f_{yy} = 12, f_{yx} = -3$

13 $f_x = -x(x^2+y^2)^{-3/2}$
$f_{xx} = -(x^2+y^2)^{-3/2} + 3x^2(x^2+y^2)^{-5/2}$
$f_{xy} = 3xy(x^2+y^2)^{-5/2}$
$f_y = -y(x^2+y^2)^{-3/2}$
$f_{yy} = -(x^2+y^2)^{-3/2} + 3y^2(x^2+y^2)^{-5/2}$
$f_{xy} = 3xy(x^2+y^2)^{-5/2}$

15 $f_x = \sec^2(x + 3y)$
    $f_{xx} = 2\sec^2(x + 3y)\tan(x + 3y)$
    $f_{xy} = 6\sec^2(x + 3y)\tan(x + 3y)$
    $f_y = 3\sec^2(x + 3y)$
    $f_{yy} = 18\sec^2(x + 3y)\tan(x + 3y)$
    $f_{xy} = 6\sec^2(x + 3y)\tan(x + 3y)$
17 4
19 1
21 $2e$
27 (a) $1 + 3x^2yz$
    (b) x^3z
29 $\dfrac{\partial f}{\partial x} = \cos(x + 2y + 1) - \cos(x + 2y)$
    $\dfrac{\partial f}{\partial y} = 2\cos(x + 2y + 1) - 2\cos(x + 2y)$
31 $fx = -g(x), fy = g(y)$

SEC. 12.4. THE CHANGE Δf AND THE DIFFERENTIAL df

1 $\Delta f = 0.72$,
   $df = 0.70$
3 $\Delta f = 0.625$, $df = 0.5$
5 1.6
7 0.0375
9 17%
11 19%
13 -8.1
15 0.433

SEC. 12.5. THE CHAIN RULES

3 -5
5 (a) $3r^2\cos^3\theta + 2r\sin^2\theta$
   (b) θ
17 $f_y = \dfrac{-F_y}{F_x}, f_z = \dfrac{-F_z}{F_x}$
21 (a) $\dfrac{\partial z}{\partial x} = f' \cdot \dfrac{x}{\sqrt{x^2 + y^2}}$
   $\dfrac{\partial z}{\partial y} = f' \cdot \dfrac{y}{\sqrt{x^2 + y^2}}$
23 17 degrees/sec
29 (a) $h_t = \dfrac{-f_x}{f_y}, h_u = \dfrac{-f_z}{f_y}$
   (b) $h_t = \dfrac{-f_y}{f_x}, h_u = \dfrac{-f_z}{f_x}$

SEC. 12.6. DIRECTIONAL DERIVATIVES AND THE GRADIENT

1 (a) 4   (b) -4 (c) $\dfrac{9\sqrt{2}}{2}$

3 (a) $20\mathbf{i} + 4\mathbf{j}$ (b) $6\mathbf{i} + 9\mathbf{j}$
5 $\dfrac{8}{\sqrt{13}}$
7 (a) $4\mathbf{i}$ at (2, 0, 0)
       $4\mathbf{j}$ at (0, 2, 0)
       $4\mathbf{k}$ at (0, 0, 2)
   (b)

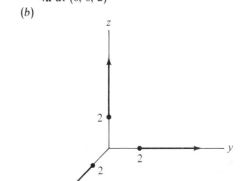

9 (a) $\pm\dfrac{(3\mathbf{i} - 2\mathbf{j})}{\sqrt{13}}$ (b) $\dfrac{(2\mathbf{i} + 3\mathbf{j})}{\sqrt{13}}$ (c) $\dfrac{-(2\mathbf{i} + 3\mathbf{j})}{\sqrt{13}}$
11 (a) (i) $\dfrac{-3\mathbf{i} + 2\mathbf{j}}{\sqrt{13}}$ (ii) $\dfrac{3\mathbf{i} - 2\mathbf{j}}{\sqrt{13}}$ (iii) $\dfrac{-\mathbf{j} + 3\mathbf{k}}{\sqrt{10}}$
   (b) Infinitely many
13 -3
17 $\mathbf{E} = -\nabla f = -\alpha\cos\alpha x\cos\beta y\,\mathbf{i} + \beta\sin\alpha x\sin\beta y\,\mathbf{j}$

SEC. 12.7. CRITICAL POINTS

1 No maximum
   Minimum is -11 at $(-1, -3)$
3 No maximum
   No minimum
5 No maximum, no minimum
7 Maximum is $3\frac{1}{4}$ at $(1, \frac{1}{4})$.
9 Maximum is 19 at (5, 6).
11 $x = y = z = 1$
13 (b) $f_m = -2\sum\limits_{i=1}^{n} x_i[y_i - (mx_i + b)]$
       $f_b = -2\sum\limits_{i=1}^{n} [y_i - (mx_i + b)]$
15 (a) $16/\sqrt{5}$ (b) $16/\sqrt{35}$
17 $x = y = z = \sqrt{2}$
21 $|k| \leq 4$
23 (b) $\sqrt[4]{abcd} \leq (a + b + c + d)/4$
25 x_i's are not all equal.

SEC. 12.8. RELATIVE EXTREMA OF $f(x, y)$ AND SECOND-ORDER PARTIAL DERIVATIVES

1. None
3. Relative minimum at $(-4, -2)$
5. Relative minimum at $(0, 0)$
7. Relative minimum at $(-\frac{10}{9}, \frac{2}{9})$
9. Relative minimum at $(2, 1)$
11. Can't determine
13. Neither
15. Maximum
17. $-6 < z_{xy} < 6$
19. $x = y = (\frac{2}{3})^{1/3}, z = (\frac{3}{2})^{2/3}$
21. $l = w = 2^{1/3}, h = 2^{-2/3}$
25. (a) Profit $= 2g_1 + 3g_2 - 2g_1^2 - g_1 g_2 - 2g_2^2$
    (b) $\dfrac{q_1}{q_2} = \dfrac{1}{2}$

SEC. 12.9. THE TAYLOR SERIES FOR $f(x, y)$ (OPTIONAL)

1. $f_{xxx} = 60x^2 y^7$
   $f_{yyy} = 210x^5 y^4$
   $f_{xxy} = f_{xyx} = f_{yxx} = 140x^3 y^6$
   $f_{yyx} = f_{yxy} = f_{xyy} = 210x^4 y^5$
3. $f_{xxx} = 8e^{2x+3y}, f_{yyy} = 27e^{2x+3y}$
   $f_{xxy} = f_{xyx} = f_{yxx} = 12e^{2x+3y}$
   $f_{yyx} = f_{xyy} = f_{yxy} = 18e^{2x+3y}$
7. (a), (b) $e^{x+y^2} = 1 + x + \dfrac{x^2}{2} + y^2 + \cdots$
9. (a) $x^2 y^2 = 1 + 2(x-1) + 2(y-1) + (x-1)^2$
       $+ 4(y-1)(x-1) + (y-1)^2$
       $+ 2(x-1)^2(y-1) + 2(x-1)(y-1)^2$
       $+ (x-1)^2(y-1)^2$
13. $2 + 3x - 5y + 3x^2 + 7xy + \frac{1}{2}y^2$
17. $\dfrac{1}{5!7!} \dfrac{\partial f^{12}}{\partial x^5 \partial y^7}(0, 0)$

SEC. 12.S. SUMMARY: GUIDE QUIZ

1. (a)
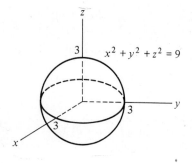

(b)
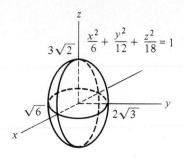

(c)
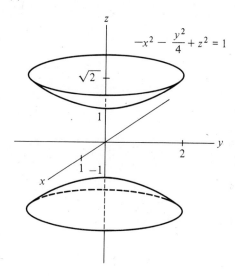

(d)
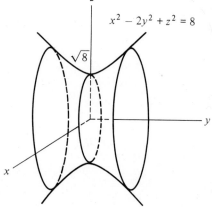

(e)

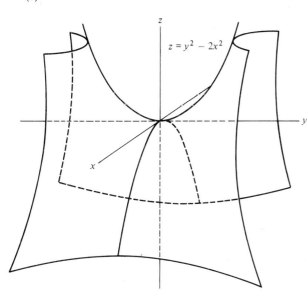

5 (a) $f(x, y) = x^2 + y^2$ (b) $f(x, y) = x^2 - 5xy + y^2$
6 $(10)(25)^{1/3} \times (10)(25)^{1/3} \times (100)/(25)^{2/3}$
7 (a) 0.2206 (b) 0.20
8 7%
9 (a) $\sqrt{13}$ (b) $\dfrac{3\mathbf{i} + 2\mathbf{j}}{\sqrt{13}}$ (c) $-5/\sqrt{2}$
   (d) $x^3 y^2 = 1$ (e) $-\tfrac{3}{2}$

REVIEW EXERCISES FOR CHAP. 12

1 (a) $2 \sec^3 (x + 2y) \left[3 \tan (x + 2y) \ln (1 + 2xy) + \dfrac{x}{1 + 2xy} \right]$
   (b) $\dfrac{3xy}{1 + 9x^2 y^2} + \tan^{-1} 3xy$
   (c) $3x^2 y^2 e^{x^3 y}$
   (d) $-6 \cos (2x + 3y)$
   (e) $6/xy^3$
   (f) $-3y^2 (\sin^3 xy - 2 \sin xy \cos^2 xy)$
3 (a) $\dfrac{1}{y^2}(ye^y - e^y + 1)$ (b) $\dfrac{1}{y^2}(ye^y - e^y + 1)$
5 $k > \tfrac{9}{4}$
7 5.6
9 (a) 5.022 (b) 5

11 (a)
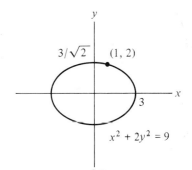

(b) $2\mathbf{i} + 8\mathbf{j}$
13 $z_x = (x + y)^2 (x - y)^4 (8x + 2y)$
    $z_y = (x + y)^2 (x - y)^4 (-2x - 8y)$
15 (a)
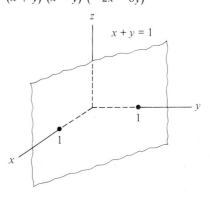

(b)
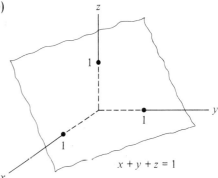

(c)
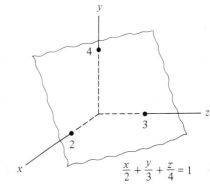

17 (a)

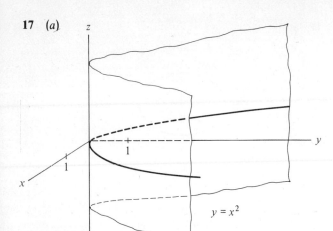

(b)

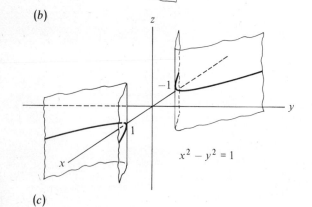

(c)

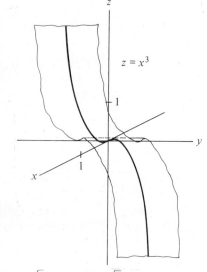

19 (c) $1 - \sqrt{2} < m < 1 + \sqrt{2}$
21 0
25 -18 at $(-1, -4)$
27 Minimum: -3 at $(0, 1)$
Maximum: $\frac{13}{3}$ at $(1, \frac{1}{3})$

29 (a) $\dfrac{-1 - yz}{xy + 5z^4}$ (b) $\dfrac{-1 - xz}{1 + yz}$ (c) $\dfrac{-xy - 5z^4}{xz + 1}$

31

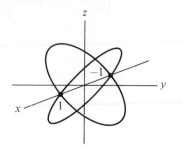

37 Nothing
41 (a) $\left(\dfrac{\partial P}{\partial V}\right)_T = \dfrac{-nRT}{V^2}$
(b) $\left(\dfrac{\partial P}{\partial V}\right)_V = \dfrac{nR}{V}$
(c) $\left(\dfrac{\partial T}{\partial V}\right)_P = \dfrac{T}{V}\left(= \dfrac{P}{nR}\right)$
49 $f(x, y) = g(x) + h(y)$
53 (a)

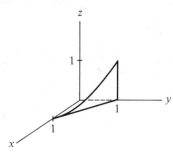

(b) $\sqrt{2}/3$
55 A change in t also induces a change in y, but $\lim_{\Delta t \to 0} (\Delta z/\Delta x) = \partial z/\partial x$ only if y is held constant.
59 (c) $f(x, y) = x^3 y^2$

CHAPTER 13. DEFINITE INTEGRALS OVER PLANE AND SOLID REGIONS

SEC. 13.1. THE DEFINITE INTEGRAL OF A FUNCTION OVER A REGION IN THE PLANE

1 (a) 20 cubic inches
(b) 100 cubic inches
(c) The volume is more than 20 but less than 100 cubic inches.
3 (a) It equals $5A$. (b) $5A$
5 (a) 6.25
7 (a) $3 + 2(\sqrt{2} + \sqrt{3}) \approx 9.29$ (b) ≈ 1.55
9 (a) 31 (b) $\frac{256}{27}$ (c) $\frac{2048}{27}$
11 A disk
13 An isosceles triangle

15 (a) Total rainfall on R
  (b) Average rainfall over R
17 (b) A circle
  (c) $\pi d^2/4$
  (d) No; consider the equilateral triangle of diameter d.
19 (b) $\frac{2}{3}$

SEC. 13.2. HOW TO DESCRIBE A PLANE REGION BY COORDINATES

1 (a)

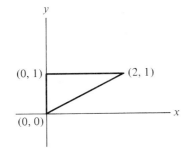

(b) $0 \le x \le 2$, $x/2 \le y \le 1$

3 (a)

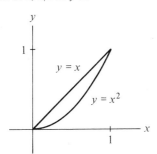

(b) $0 \le x \le 1$, $x^2 \le y \le x$

5 $-a \le x \le a$, $-\sqrt{a^2 - x^2} \le y \le \sqrt{a^2 - x^2}$

7

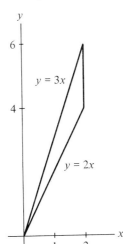

9

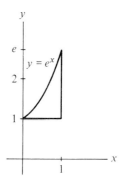

11

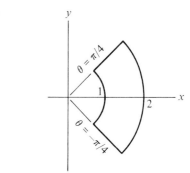

13 $0 \le y \le 4$, $y/3 \le x \le y/2$
  $4 \le y \le 6$, $y/3 \le x \le 2$

15 $0 \le x \le 1$, $1 \le y \le e^x$

17 (a)

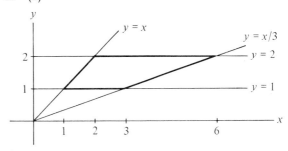

(b) $1 \le y \le 2$, $y \le x \le 3y$

SEC. 13.3. COMPUTING $\int_R f(P)\, dA$ BY INTRODUCING RECTANGULAR COORDINATES IN R

1 $\frac{2}{3}$
3 $\frac{32}{3}$
5 $\frac{2}{21}(2^{7/2} - 1)$
7 $\frac{3}{20}$
9 $\frac{16}{3}$
11 $\frac{35}{6}$

13 $2a^3/3$

15 $\frac{65}{384}$

17 $\bar{x} = \dfrac{\pi - 2}{\pi - 2\ln 2}$
$\bar{y} = \frac{2}{3}$

19 $\bar{x} = \frac{4}{5}, \bar{y} = \frac{152}{75}$

21 $\bar{x} = \dfrac{e^2 + 3}{e^2 - 1}$ $\bar{y} = \dfrac{2(e^2 - 3)}{e^2 - 1}$

23 (a) $\int_0^a \int_{x/2}^x y^2 e^{y^2}\, dy\, dx + \int_a^{2a} \int_{x/2}^a y^2 e^{y^2}\, dy\, dx$,
$\int_0^a \int_y^{2y} y^2 e^{y^2}\, dx\, dy$
(b) $\frac{1}{2}e^{a^2}(a^2 - 1) + \frac{1}{2}$

25 $\int_0^4 \int_{\sqrt{y}}^2 x^3 y\, dx\, dy$

29 (a) $Ma^2/2$ (b) $3Ma^2/2$ (c) $Ma^2/4$
(d) $5Ma^2/4$

SEC. 13.4. COMPUTING $\int_R f(P)\, dA$ BY INTRODUCING POLAR COORDINATES IN R

1 $\frac{1}{3}$

3 $\frac{8}{5}$

5 $\dfrac{5\pi}{3} + \dfrac{38}{9}$

7 $1 + \dfrac{5\pi}{16}$

9 $3\pi/32$

11 $35\pi/16$

13 $\pi a^4/4$

15 $\pi/64$

17 $\bar{x} = 0, \bar{y} = 16a/15$

23 $32/9\pi$

27 $\dfrac{\pi}{4} \sin a^2$

31 (a) $\dfrac{2an^2}{(1+n)(1+2n)}$

(b) $\dfrac{a^2 n^2}{(1+n)(2+n)}$

SEC. 13.5. MAPPINGS FROM A PLANE TO A PLANE

1 (a)

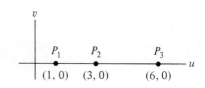

(b)

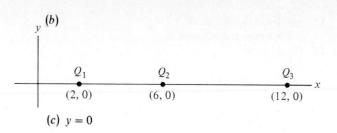

(c) $y = 0$

3 (a)

(u, v)	$(x, y) = (2u, 3v)$
$(0, 0)$	$(0, 0)$
$(2, 0)$	$(4, 0)$
$(1, 1)$	$(2, 3)$

(b), (d)

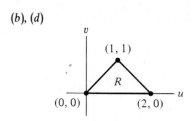

(c), (e)

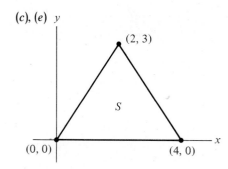

(f) Area of $R = 1$, Area of $S = 6$

5 (a)

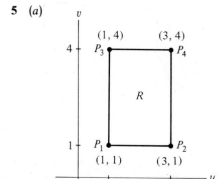

(b), (c) S is a rectangle.

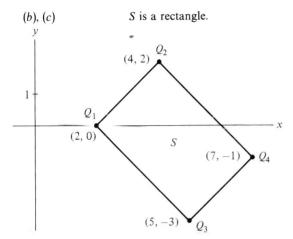

7 (a)

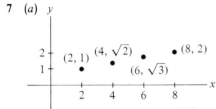

(c) $x = 2y^2$, $y > 0$

9 (a)

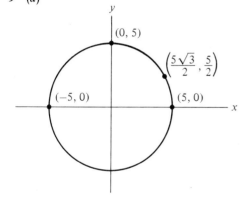

(b) $x^2 + y^2 = 25$

11 (a)

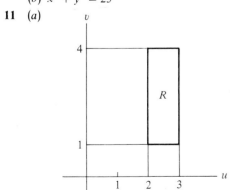

(b)

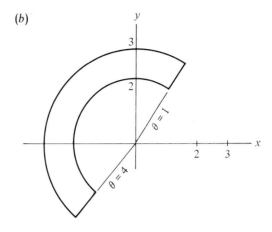

13 (c) Lines with equal slopes are mapped onto lines with equal slopes.

15 (b) 1

19

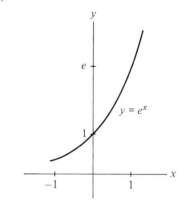

SEC. 13.6. MAGNIFICATION, CHANGE OF COORDINATES, AND THE JACOBIAN

1 (a) 4 (b) 4
3 (a) 8 (b) 2
5 (a) e^2 (b) e^4
7 $\frac{3}{2}$
11 (b) $\int_{-1}^{1} \int_{-\sqrt{1-u^2}}^{\sqrt{1-u^2}} ab^3 v^2 \, du \, dv$
   (c) $\pi ab^3/4$
15 (a) (b), (c) $\frac{4}{3}$

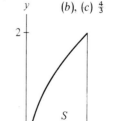

19 (c)

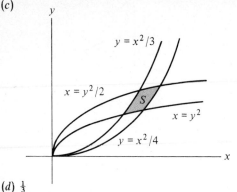

(d) $\frac{1}{3}$

21 (d) The inverse of a magnification by a factor k is a magnification by a factor $1/k$.

13

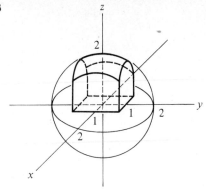

15 $\frac{1}{3}$

17 $\frac{91}{12}$

19 $\frac{M}{3}(b^2 + c^2)$

21 (a)

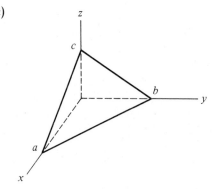

SEC. 13.7. THE DEFINITE INTEGRAL OF A FUNCTION OVER A REGION IN SPACE

1 384, 1920

3 $\frac{20\pi r^3}{3}$

5 960

7 $2\sqrt{6}$

9

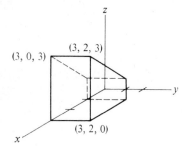

11

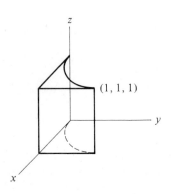

(b) $\frac{x}{a} + \frac{y}{b} + \frac{z}{c} = 1$

(c) $abc^2/24$

(d) $c/4$

25 1024 g

27 $\frac{1}{120}$

29 0

31 $M\left(\frac{h^2}{3} - \frac{a^2}{4}\right)$

33 (a) $I + k^2 M$

(b) The x axis itself, where $k = 0$.

SEC. 13.8. DESCRIBING SOLID REGIONS WITH CYLINDRICAL OR SPHERICAL COORDINATES

1 $r = 1$

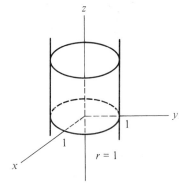

3 $z = 1$

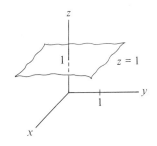

5 $r^2 + z^2 = 9$

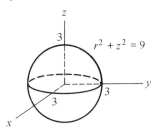

7

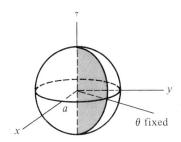

9 $0 \leq \theta \leq 2\pi, 0 \leq r \leq a, -\sqrt{a^2 - r^2} \leq z \leq \sqrt{a^2 - r^2}$
11 (a) $r = \sqrt{x^2 + y^2}, \theta = \tan^{-1}(y/x), z = z$
(b) $x = r \cos \theta, y = r \sin \theta, z = z$
13 (a) $z = 0$ (b) $\theta = \pi/4$ and $\theta = 5\pi/4$ (c) $r = 0$

15

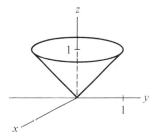

17

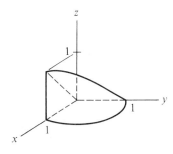

19 $\rho = 2$

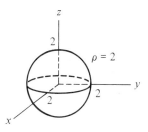

21 $\theta = \pi/2$

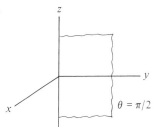

23 $\varphi = 0$

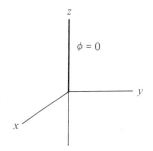

25 $r = \rho \sin \varphi, \theta = \theta, z = \rho \cos \varphi$
27 $0 \leq \theta \leq 2\pi, 0 \leq \varphi \leq \pi, 0 \leq \rho \leq a$
29 $0 \leq \theta \leq 2\pi, 0 \leq \varphi \leq \pi/6, 0 \leq \rho \leq a$

31 $0 \le \theta \le 2\pi, 0 \le \varphi \le \pi/6, 0 \le \rho \le \dfrac{3a}{5} \sec \varphi$

33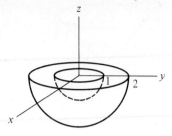

35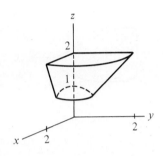

37 $\rho = \sqrt{r^2 + z^2}, \theta = \theta, \varphi = \cos^{-1} \dfrac{z}{\sqrt{r^2 + z^2}}$

39 $\rho = \sqrt{x^2 + y^2 + z^2},$
$\theta = \tan^{-1}(y/x)$ (1st and 4th quadrants),
$\theta = \pi + \tan^{-1}(y/x)$ (2nd and 3rd quadrants)
$\varphi = \cos^{-1} \dfrac{z}{\sqrt{x^2 + y^2 + z^2}}$

SEC. 13.9. COMPUTING $\int_R f(P)\, dV$ WITH CYLINDRICAL OR SPHERICAL COORDINATES

1 $\pi/24$

3 $\dfrac{\pi(4\sqrt{2} - 1)}{320\sqrt{2}}$

5 $\dfrac{4\pi a^3}{3}$

7 $2Ma^2/5$

9 $3Ma^2/10$

11 (a), (b) $3Ma^2/2$

13 $\pi a^4/16$

15 $3a/4$

17 $2Ma^2/5$

19 (a) $\displaystyle\int_{-1/\sqrt{2}}^{1/\sqrt{2}} \int_{-\sqrt{1/2-x^2}}^{\sqrt{1/2-x^2}} \int_{\sqrt{x^2+y^2}}^{\sqrt{1-x^2-y^2}} z\, dz\, dy\, dx$

(b) $\displaystyle\int_0^{2\pi} \int_0^{1/\sqrt{2}} \int_r^{\sqrt{1-r^2}} zr\, dz\, dr\, d\theta$

(c) $\displaystyle\int_0^{2\pi} \int_0^{\pi/4} \int_0^1 \rho^3 \cos\varphi \sin\varphi\, d\rho\, d\varphi\, d\theta$

(d) $\pi/8$

23 (b) $4\pi a^5/5$ (c) $4\pi a^5/15$

27 $H + \dfrac{a^2}{5H}$

29 (a)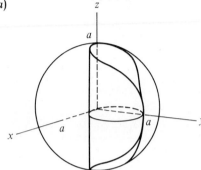

(b) $\dfrac{2a^3}{9}(3\pi - 4)$

31 Converges; the average value of f over R is $3/2a$.

33 r

SEC. 13.S. SUMMARY: GUIDE QUIZ

1 (a) $-a \le x \le a, 0 \le y \le \sqrt{a^2 - x^2}$
   (b) $\pi/6 \le \theta \le \pi/4, \sec\theta \le r \le 2\sec\theta$

2 (a), (b) $\tfrac{4}{3}$

3 (a) $\dfrac{2a}{3}$

4 $\displaystyle\int_0^{\pi/2} \int_0^a r^4\, dr\, d\theta = \dfrac{\pi a^5}{10}$

5 (a) $3\pi/64$ (b) $3\pi/64$ (c) $3\pi/64$ (d) $\tfrac{3}{512}$

6 (a) $2\pi(1 - 101e^{-100})$ (b) $2\pi(1 - 51e^{-50})$

8 $\dfrac{3b^4 - a^4}{8b^3 - a^3}$

10 (a) $(5, \tan^{-1}(\tfrac{4}{3}), -3)$
    (b) $(-\tfrac{3}{2}, 3\sqrt{3}/2, 0)$
    (c) $(2\sqrt{2}, \pi/4, \pi/4)$

11 (a) $x = \tfrac{1}{2}$

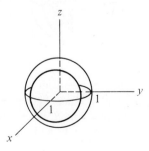

(b) $\varphi = \pi/3$

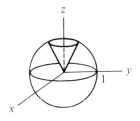

(b)

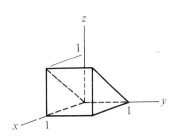

(c) $\rho = \tfrac{1}{2}$

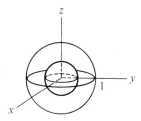

16 (b) $\dfrac{3M}{\pi a^2 h}\displaystyle\int_{-a}^{a}\int_{-\sqrt{a^2-x^2}}^{\sqrt{a^2-x^2}}\int_{(h/a)\sqrt{x^2+y^2}}^{h}(x^2+y^2)\,dz\,dy\,dx$

$\dfrac{3M}{\pi a^2 h}\displaystyle\int_{0}^{2\pi}\int_{0}^{a}\int_{(h/a)r}^{h} r^3\,dz\,dr\,d\theta$

$\dfrac{3M}{\pi a^2 h}\displaystyle\int_{0}^{2\pi}\int_{0}^{\alpha}\int_{0}^{h\sec\varphi}\rho^4\sin^3\varphi\,d\rho\,d\varphi\,d\theta,\ \alpha=\tan^{-1}(a/h)$

(c) $3Ma^2/10$

17 (a) $\rho = 3\sec\varphi$
(b) $x^2 + y^2 = 4$
(c) $\rho = 2\csc\varphi$
(d) $r^2 + z^2 = 9$

(d) $\theta = \pi/2$

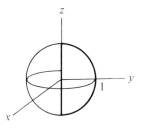

REVIEW EXERCISES

1 (a) $-2 \le x \le 2,\ x^2 \le y \le 4$
 (b) $0 \le y \le 4,\ -\sqrt{y}\le x \le \sqrt{y}$

3 (a) $\dfrac{x^4}{2} + x^3 - x - \dfrac{1}{2}$
 (b) $\tfrac{1}{2}y^4 + y^3 - \tfrac{3}{2}y^2$

5 (a) $0 \le x \le a,\ 0 \le y \le \sqrt{2a^2 - x^2}$
 (b) $0 \le y \le a,\ 0 \le x \le a$
  $a \le y \le a\sqrt{2},\ 0 \le x \le \sqrt{2a^2 - y^2}$
 (c) $0 \le \theta \le \pi/4,\ 0 \le r \le a\sec\varphi$
  $\pi/4 \le \theta \le \pi/2,\ 0 \le r \le a\sqrt{2}$

7 $\displaystyle\int_{\pi/4}^{\pi/2}\int_{0}^{1} r^2\,dr\,d\theta = \pi/12$

9 (a) $f(P) = y = r\sin\theta$

(e) $z = -\tfrac{1}{2}$

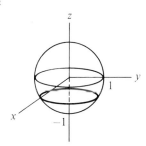

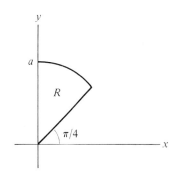

12 (a) Cylindrical: r
  Spherical: $\rho^2\sin\varphi$
13 See Sec. 13.8.
14 (a) $\dfrac{1}{\pi a^2 h}\displaystyle\int_{0}^{2\pi}\int_{0}^{a}\int_{0}^{h} r^3\,dz\,dr\,d\theta$

  $\dfrac{1}{\pi a^2 h}\displaystyle\int_{-a}^{a}\int_{-\sqrt{a^2-x^2}}^{\sqrt{a^2-x^2}}\int_{0}^{h}(x^2+y^2)\,dz\,dy\,dx$

 (b) $a^2/2$
15 (a) $\tfrac{1}{4}(e-1)$

(b) $\int_0^{a/\sqrt{2}} \int_x^{\sqrt{a^2-x^2}} y \, dy \, dx$

(c) $\dfrac{a^3}{3\sqrt{2}}$

11 $\int_0^{a/\sqrt{2}} \int_y^{\sqrt{a^2-y^2}} (x^2 + y^2) \, dx \, dy = \pi a^4/16$

13 (a) $Ma^2/3$ (b) $Ma^2/12$

15 (a) $\dfrac{128M}{5}$ (b) $\dfrac{8M}{9}$ (c) $\dfrac{1192M}{45}$

17 $\dfrac{\pi^2 + 8}{64}$

19 $\frac{1}{80}$

21 $\dfrac{2\pi}{(\ln 2)^2} [1 - 2^{-a}(1 + a \ln 2)]$

25 (a)

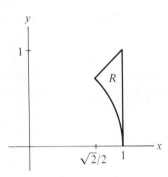

(b) $0 \leq \theta \leq \pi/4$, $1 \leq r \leq \sec \theta$

(c) $\int_0^{\pi/4} \int_1^{\sec \theta} dr \, d\theta$ (d) $\ln (\sqrt{2} + 1)$

45 (a) The integrand is nonnegative.

47 $\frac{1}{2}(e - 1)$

CHAPTER 14. THE DERIVATIVE OF A VECTOR FUNCTION

SEC. 14.1. THE DERIVATIVE OF A VECTOR FUNCTION

1 (a) $\mathbf{G}(1) = \mathbf{i} + \mathbf{j}$
$\mathbf{G}(2) = 2\mathbf{i} + 4\mathbf{j}$
$\mathbf{G}(3) = 3\mathbf{i} + 9\mathbf{j}$

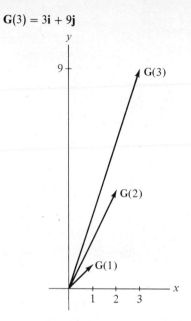

(b) The parabola is $y = x^2$.

3 (a) $\mathbf{G}(1.1) = 2.2\mathbf{i} + 1.21\mathbf{j}$
$\mathbf{G}(1) \;\; = 2\mathbf{i} + \mathbf{j}$
$\Delta \mathbf{G} \;\; = 0.2\mathbf{i} + 0.21\mathbf{j}$

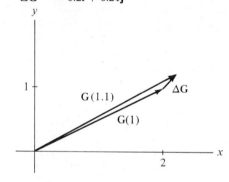

(b) $\dfrac{\Delta \mathbf{G}}{0.1} = 2\mathbf{i} + 2.1\mathbf{j}$

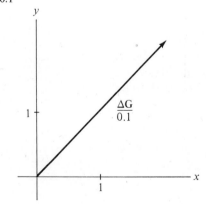

(c) $\mathbf{G}'(1) = 2\mathbf{i} + 2\mathbf{j}$
5 (a) $\mathbf{r}(1) = 32\mathbf{i} - 16\mathbf{j}$
$\mathbf{r}(2) = 64\mathbf{i} - 64\mathbf{j}$

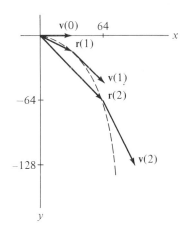

(b), (c) The parabola is $y = \dfrac{-x^2}{64}$.

5 (c) $\mathbf{v}(0) = 32\mathbf{i}$
$\mathbf{v}(1) = 32\mathbf{i} - 32\mathbf{j}$
$\mathbf{v}(2) = 32\mathbf{i} - 64\mathbf{j}$
7 $\mathbf{v}(t) = (-3\sin 3t)\mathbf{i} + (3\cos 3t)\mathbf{j} + 6\mathbf{k}$
$v = 3\sqrt{5}$
9 (a) $\Delta\mathbf{G} = 0.21\mathbf{i} + 0.331\mathbf{j}$

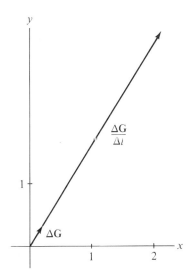

(b) $\dfrac{\Delta\mathbf{G}}{\Delta t} = 2.1\mathbf{i} + 3.31\mathbf{j}$

(c) $\mathbf{G}'(1) = 2\mathbf{i} + 3\mathbf{j}$

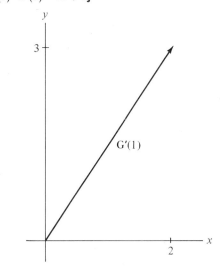

(d) $\left|\dfrac{\Delta\mathbf{G}}{\Delta t} - \mathbf{G}'(1)\right| = \sqrt{0.1061}$

11 (b)

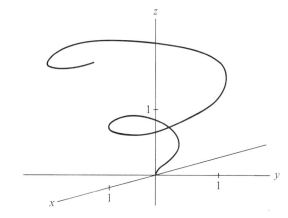

13 (a), (b) $\mathbf{r}(1) = \mathbf{i} + \mathbf{j}$
$\mathbf{r}(2) = 2\mathbf{i} + \tfrac{1}{2}\mathbf{j}$
$\mathbf{r}(3) = 3\mathbf{i} + \tfrac{1}{3}\mathbf{j}$

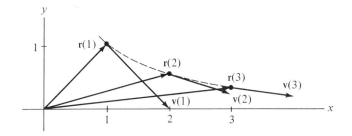

(c) $\mathbf{v}(1) = \mathbf{i} - \mathbf{j}$
$\mathbf{v}(2) = \mathbf{i} - \frac{1}{4}\mathbf{j}$
$\mathbf{v}(3) = \mathbf{i} - \frac{1}{9}\mathbf{j}$
(d) $\frac{dx}{dt} \to 1, \frac{dy}{dt} \to 0, \mathbf{v}(t) \to \mathbf{i}, |\mathbf{v}(t)| \to 1$

15 (a) $\mathbf{r}(t) = 100 \cos(400\pi t)\mathbf{i} - 100 \sin(400\pi t)\mathbf{j}$
$\mathbf{v}(t) = -40000\pi(\sin(400\pi t)\mathbf{i} + \cos(400\pi t)\mathbf{j})$
(b) $\mathbf{r}(0) = 100\mathbf{i}$
$\mathbf{r}(\frac{1}{800}) = -100\mathbf{j}$
$\mathbf{v}(0) = -40000\pi\mathbf{j}$
$\mathbf{v}(\frac{1}{800}) = -40000\pi\mathbf{i}$

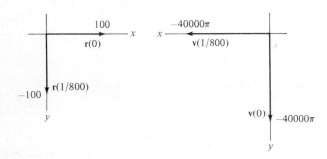

(c) They are constant.

17 (b)

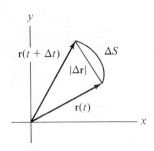

19 (a) $\mathbf{r}(0) = a\mathbf{i}, \mathbf{r}(\frac{1}{4}) = a\mathbf{j}$
(b) $\mathbf{v}(0) = 2\pi a\mathbf{j}, \mathbf{v}(\frac{1}{4}) = -2\pi a\mathbf{i}$

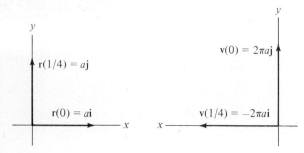

23 (b) $\theta = \frac{\pi}{4}$

SEC. 14.2. PROPERTIES OF THE DERIVATIVE OF A VECTOR FUNCTION

1 $4\mathbf{i} - \mathbf{j}$
5 (b) No, the cross product is not commutative.
9 (a)

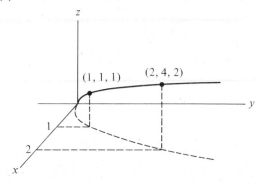

(b) $\sqrt{2 + 4t^2}$

13 No. The particle may travel in a circular path with non-constant speed.

SEC. 14.3. VECTORS PERPENDICULAR TO A SURFACE; THE TANGENT PLANE

1 $2\mathbf{i} + 4\mathbf{j} + 6\mathbf{k}$
3 $-3\mathbf{i} - 3\mathbf{j} + \mathbf{k}$
5 $2\mathbf{i} + \mathbf{j} - \mathbf{k}$
7 $6x + 3y + 2z = 18$
9 $5x + y - z = -2$
15 (b)

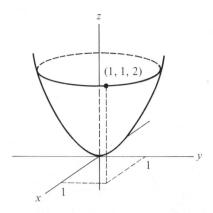

(c) $2\mathbf{i} + 2\mathbf{j} - \mathbf{k}$
17 (b) $\cos^{-1}(\frac{5}{14}) \approx 1.2056$ radians $\approx 69.08°$
19 $\nabla f \times \nabla g$
21 (a) $\nabla \mathbf{f}(1, 1) = \mathbf{i} + \mathbf{j}$
$\nabla \mathbf{f}(1, 2) = 2\mathbf{i} + \mathbf{j}$
$\nabla f(2, 3) = 3\mathbf{i} + 2\mathbf{j}$

ANSWERS TO SELECTED ODD-NUMBERED PROBLEMS AND TO GUIDE QUIZZES   S-159

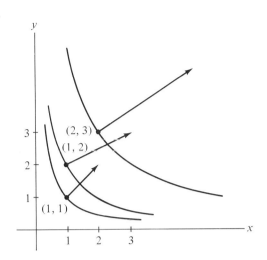

(b) The level curves are hyperbolas.

23 (a), (b)

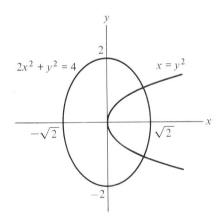

SEC. 14.4. LAGRANGE MULTIPLIERS

1 The maximum value is 2; it occurs at $(\sqrt{2}, \sqrt{2})$ and $(-\sqrt{2}, -\sqrt{2})$.

3 The minimum value is $2\sqrt{6}$; it occurs at $\left(\frac{\sqrt{6}}{2}, \frac{\sqrt{6}}{3}\right)$.

5 The maximal area is 9; it occurs for a square of side length 3.

7 $\left(\frac{3}{7}, \frac{6}{7}, \frac{9}{7}\right)$.

9 The minimal distance $\sqrt{\frac{2}{3}}$ occurs at the point $\left(\frac{1}{3}, \frac{8}{3}, \frac{5}{3}\right)$.

11 The maximal value $\frac{1}{27}$ occurs when $x = \pm \frac{\sqrt{3}}{3}$, $y = \pm \frac{\sqrt{3}}{3}$, $z = \pm \frac{\sqrt{3}}{3}$ (eight different points).

13 The minimum value $\frac{224}{75}$ occurs at $\left(\frac{68}{75}, \frac{16}{15}, \frac{-76}{75}\right)$.

15 The maximum value $\frac{27 + 14\sqrt{7}}{8}$ occurs at $\left(\frac{3}{2}, 0, \frac{\sqrt{7}}{2}\right)$.

17 (a) 1
(c) $\mathbf{A} \cdot \mathbf{B} = |\mathbf{A}| |\mathbf{B}| \cos \theta \leq |\mathbf{A}| |\mathbf{B}|$

19 The maximum value $\left(\sum_{i=1}^{n} a_i^2\right)^{1/2}$ occurs for $x_i = \dfrac{a_i}{\left(\sum_{j=1}^{n} a_j^2\right)^{1/2}}$.

SEC. 14.5. THE ACCELERATION VECTOR

1 $\mathbf{r}(1) = \mathbf{i} + \mathbf{j}$, $\mathbf{v}(1) = 2\mathbf{i} + 3\mathbf{j}$, $\mathbf{a}(1) = 2\mathbf{i} + 6\mathbf{j}$

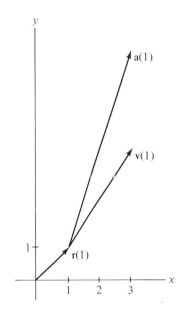

3 $\mathbf{r}(0) = 0$, $\mathbf{v}(0) = 4\mathbf{i}$, $\mathbf{a}(0) = -32\mathbf{j}$
$\mathbf{r}(1) = 4\mathbf{i} - 16\mathbf{j}$, $\mathbf{v}(1) = 4\mathbf{i} - 32\mathbf{j}$, $\mathbf{a}(1) = -32\mathbf{j}$

$r(2) = 8i - 64j$, $v(2) = 4i - 64j$, $a(2) = -32j$

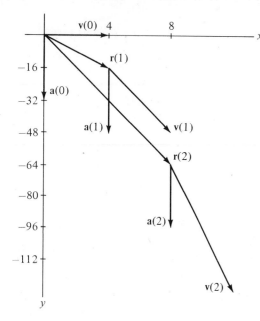

5 (a)

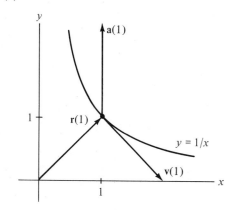

(b) $r(1) = i + j$, $v(1) = i - j$, $a(1) = 2j$

(c) $v(t) \to i$, $a(t) \to 0$
(d) $v \to 1$

7 (a) $v(t) = -15 \sin 3t\, i + 15 \cos 3t\, j + 6k$
$a(t) = -45 \cos 3t\, i - 45 \sin 3t\, j$
(b) The velocity in the z direction is constant.

9 (a) Yes
(b) No, consider circular motion of varying speed.

15 (a) $v(t) = -2t \sin t^2\, i + 2t \cos t^2\, j$
$a(t) = [-2 \sin t^2 - 4t^2 \cos t^2]i$
$+ [2 \cos t^2 - 4t^2 \sin t^2]j$

17 $\dfrac{200\pi}{3\sqrt{6}}$ minutes ≈ 88.7 minutes

$\dfrac{8200\pi}{290.334}$ minutes ≈ 85.5 minutes

$\dfrac{5000\pi}{30\sqrt{19.2}}$ minutes ≈ 119.5 minutes

19 About 200 miles

21 Four times as much

SEC. 14.6. THE UNIT VECTORS **T** AND **N**

1 $T(1) = \dfrac{i - 2j}{\sqrt{5}}$, $N(1) = \dfrac{-2i - j}{\sqrt{5}}$ graph z.321

3 $T(1) = -\sin 1\, i + \cos 1\, j$, $N(1) = -\cos 1\, i - \sin 1\, j$

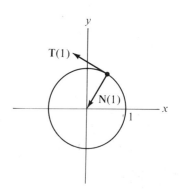

5 $T(1) = \dfrac{i + 2ej}{\sqrt{1 + 4e^2}}$, $N(1) = \dfrac{-2ei + j}{\sqrt{1 + 4e^2}}$

7 True

9 (a) The parabola $y = x^2$; see (c).
(b) $\mathbf{r}(1) = \mathbf{i} + \mathbf{j}$, $\mathbf{v}(1) = \mathbf{i} + 2\mathbf{j}$,
$\mathbf{T}(1) = \dfrac{1}{\sqrt{5}}(\mathbf{i} + 2\mathbf{j})$, $\mathbf{N}(1) = \dfrac{1}{\sqrt{5}}(-2\mathbf{i} + \mathbf{j})$.
(c)

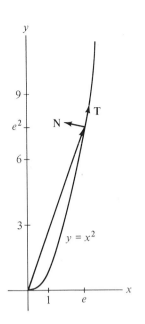
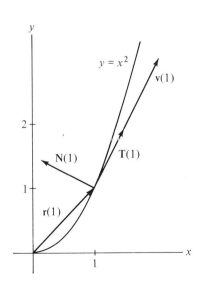

11 3

13 (a) $y = \dfrac{-x^2}{64}$

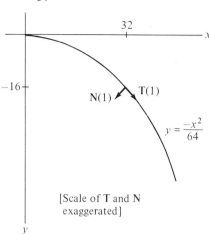

[Scale of **T** and **N** exaggerated]

(b) $\mathbf{T}(1) = \dfrac{1}{\sqrt{2}}(\mathbf{i} - \mathbf{j})$, $\mathbf{N}(1) = \dfrac{-1}{\sqrt{2}}(\mathbf{i} + \mathbf{j})$
(c) $32\sqrt{2}$, 0
(d) $16\sqrt{2}$, $16\sqrt{2}$

15 (a) Yes (b) No

23 (b) $\mathbf{T}\left(\dfrac{\pi}{4}\right) = \dfrac{1}{\sqrt{61}}(-6\mathbf{i} + 5\mathbf{k})$
$\mathbf{N}\left(\dfrac{\pi}{4}\right) = -\mathbf{j}$
$\mathbf{B}\left(\dfrac{\pi}{4}\right) = \dfrac{1}{\sqrt{61}}(5\mathbf{i} + 6\mathbf{k})$

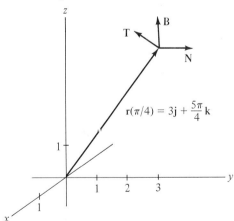

SEC. 14.7. THE SCALAR COMPONENTS OF THE ACCELERATION VECTOR ALONG T AND N

1 $a_N = \frac{16}{5}$, $a_T = 3$
3 $a_N = \frac{1}{3}$, $a_T = 0$
5 $a_T = 0$, $a_N = 45\pi^2$

9 (a)

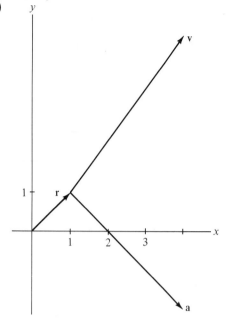

(b) Slowing down

(d) $a_N = \dfrac{21}{5}, a_T = \dfrac{-3}{5}$

11 (a) $\mathbf{a}\left(\dfrac{1}{\sqrt{6}}\right) = 2\mathbf{i} + \mathbf{j}$

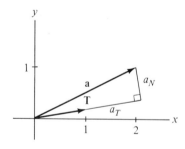

(c) $a_N = \dfrac{4}{\sqrt{37}}, a_T = \dfrac{13}{\sqrt{37}}$

13 (a) $\mathbf{v}(t) = e^t\mathbf{i} + 2e^{2t}\mathbf{j},\ \mathbf{a}(t) = e^t\mathbf{i} + 4e^{2t}\mathbf{j}$

(b) $v(t) = e^t\sqrt{1+4e^{2t}},\ \dfrac{dv}{dt} = \dfrac{e^t(1+8e^{2t})}{\sqrt{1+4e^{2t}}}$

(c) $a_T = \dfrac{e^t(1+8e^{2t})}{\sqrt{1+4e^{2t}}}$

(d) $a_N = \dfrac{2e^{2t}}{\sqrt{1+4e^{2t}}}$

15 $a_T = \dfrac{2t(2+9t^2)}{\sqrt{4t^2+9t^4}},\ a_N = \dfrac{6|t|}{\sqrt{4+9t^2}}$

19 (a)

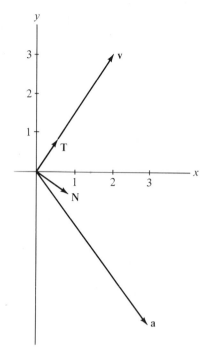

(b) $v = \sqrt{13},\ \dfrac{d^2s}{dt^2} = \dfrac{-6}{\sqrt{13}}$

(c) $r = \dfrac{13\sqrt{13}}{17}$

SEC. 14.8. NEWTON'S LAW IMPLIES KEPLER'S THREE LAWS

See the solutions manual.

SEC. 14.S. SUMMARY: GUIDE QUIZ ON CHAP. 14

1 (a), (b) $\Delta \mathbf{r} = 0.1\mathbf{i} + 0.2\mathbf{j}$

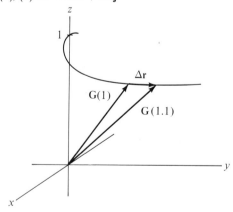

(c) $\mathbf{i} + 2.1\mathbf{j}$

3 (a) $\frac{2}{3}\mathbf{i} + \frac{2}{3}\mathbf{j} - \frac{1}{3}\mathbf{k}$ (b) $\dfrac{\mathbf{i} + 2\mathbf{j} + 3\mathbf{k}}{\sqrt{14}}$

   (c) $\dfrac{x\mathbf{i} + y\mathbf{j} + z\mathbf{k}}{\sqrt{x^2 + y^2 + z^2}}$

4 (a) 12   (b) $\dfrac{12}{\sqrt{14}}$

5 $\cos^{-1}\left(\dfrac{6}{\sqrt{42}}\right) \approx 0.3876$ radians $\approx 22.21°$

6 (a) $8x + 3y = 11$ (b) $5x + 4y + z = 10$

7 (a) $\mathbf{T}(t) = \dfrac{\mathbf{G}'(t)}{|\mathbf{G}'(t)|}$, $\mathbf{N}(t) = \dfrac{\mathbf{T}'(t)}{|\mathbf{T}'(t)|}$

   (b) ∇f

   (c) Suppose the surfaces are $f(x, y, z) = 0$ and $g(x, y, z) = 0$. Both ∇f and ∇g are perpendicular to the curve of intersection. $\nabla f \times \nabla g$ is parallel to it.

   (d) $\mathbf{G}'(t)$ and $\mathbf{M} \times \mathbf{G}'(t)$, respectively, where $\mathbf{M}$ is any vector not parallel to $\mathbf{G}'(t)$.

8 (a) Slowing down

   (c) $a_T = \dfrac{-1}{\sqrt{13}}$, $a_N = \dfrac{5}{\sqrt{13}}$ (d) $\dfrac{13\sqrt{13}}{5}$

   (e) $\frac{3}{2}$ (f) $\dfrac{-5}{8}$

9 ∇f lies in the plane determined by ∇g and ∇h.

10 (a)

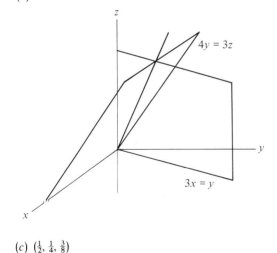

   (c) $(\frac{1}{2}, \frac{1}{4}, \frac{3}{8})$

(b) $v = \sqrt{9 + e^2}$, $a_T = \dfrac{18 + e^2}{\sqrt{9 + e^2}}$,

$a_N = \dfrac{3e}{\sqrt{9 + e^2}}$, $r = \dfrac{(9 + e^2)^{3/2}}{3e}$

9 $\dfrac{-\mathbf{i} + 2\mathbf{j} + 2\mathbf{k}}{3}$

11 (a) $\sqrt{3}e^t$ (b) $\sqrt{3}(e - 1)$

13 (a) $\dfrac{\sqrt{26}}{2}$

   (b) $\cos \alpha = \dfrac{3}{\sqrt{26}}$, $\cos \beta = \dfrac{-1}{\sqrt{26}}$, $\cos \gamma = \dfrac{4}{\sqrt{26}}$

   (c) $(1 + \frac{3}{2}t)\mathbf{i} + (2 - \frac{1}{2}t)\mathbf{j} + (1 + 2t)\mathbf{k}$

15 (a) $\sqrt{13}$

   (b) $\mathbf{T} = \dfrac{2}{\sqrt{13}}\mathbf{i} + \dfrac{3}{\sqrt{13}}\mathbf{j}$ $\mathbf{N} = \dfrac{-3}{\sqrt{13}}\mathbf{i} + \dfrac{2}{\sqrt{13}}\mathbf{j}$

   (c)

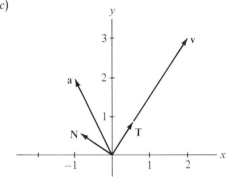

   (d) $\dfrac{7}{\sqrt{13}}$ (e) $\dfrac{13\sqrt{13}}{7}$ (f) $\dfrac{4}{\sqrt{13}}$

   (g) $\dfrac{d^2s}{dt^2} = \dfrac{4}{\sqrt{13}}$; speeding up

17 $f_x\mathbf{i} + f_y\mathbf{j} - \mathbf{k}$

21 (a), (b)

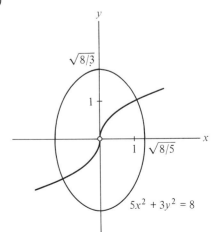

REVIEW EXERCISES FOR CHAP. 14

1 (a) $2\mathbf{i} + 3\mathbf{j} - \mathbf{k}$ (b) $-3\mathbf{i} + 2\mathbf{j}$

3 (b) $a_T = \dfrac{e^2 - e^{-2}}{\sqrt{2(e^2 + e^{-2})}}$, $a_N = \sqrt{\dfrac{2}{e^2 + e^{-2}}}$

5 (a) $\mathbf{v}(1) = 3\mathbf{i} + e\mathbf{j}$, $\mathbf{a}(1) = 6\mathbf{i} + e\mathbf{j}$

25 (a) $\dfrac{-1}{(x^2+y^2)^{3/2}}(x\mathbf{i}+y\mathbf{j})$ (b) $\dfrac{-1}{r^2}\mathbf{u}_r$

CHAPTER 15. GREEN'S THEOREM, THE DIVERGENCE THEOREM, AND STOKES' THEOREM

SEC. 15.1. VECTOR AND SCALAR FIELDS

1 9
3 $y\cos xy - 2xy \sin xy^2 + \dfrac{3x}{1+9x^2z^2}$
5 0
7 $-\sec^2(x+z)\mathbf{i} + \left(y\sec yz \tan yz - \dfrac{1}{x}\right)\mathbf{j} + (\sec^2(x+z) - z\sec yz \tan yz)\mathbf{k}$
9 Vector
13 (a) Vector (b) Scalar (c) Scalar (d) Scalar
(e) Scalar (f) Vector (g) Vector
17 (b) $\mathbf{F} = x\mathbf{i} + y\mathbf{j} - 2z\mathbf{k}$ is one example.

SEC. 15.2. LINE INTEGRALS

1 $\tfrac{1}{54}(145^{3/2} - 10^{3/2})$
3 $\tfrac{8}{3}$
5 $\tfrac{72}{5}$
7 All equal $\dfrac{2\sqrt{2}}{3}$ because the endpoints are the same.
9 0
11 -19π
13 0
15 22/15
17 $1 + \dfrac{1}{\pi} + \dfrac{\pi}{4} - \dfrac{1}{2}\ln 2$
21 (b) Fluid is leaving; net outward flow is positive.
(c) 3π
25 (a) ab

SEC. 15.3. CONSERVATIVE VECTOR FIELDS

1 (a) Conservative (b) Nonconservative
3 $x^2y + C$
5 $xy + x + y + C$
7 x^2y
11 $\tfrac{15}{16}\pi^4 - 5$
13 (a) 1 (b) 1 (c) 1
17 1
19 (b) $\tfrac{1}{2}\ln(1 + x^2 + 2y^2 + 3z^2)$

SEC. 15.4. GREEN'S THEOREM

1 5π
3 $e - \tfrac{1}{2}e^2 - \tfrac{1}{2}$
5 0 **7** 3π **9** 0
11 8/3
13 (a) 2π (b) No **15** $\dfrac{7}{3} - 2\ln 2 - \dfrac{\pi}{4}$
19 (a) 0 (b) 2π
23 $\tfrac{94}{105}$
25 (a)   (c) 0

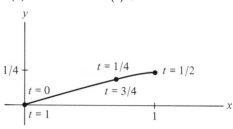

SEC. 15.5. SURFACE INTEGRALS

1 (a) 2 (b) $2\sqrt{3}$
3 $1\|\sqrt{2}$
5 $-\tfrac{5}{6}$
7 $\dfrac{3 \cdot 19^{5/2} - 5 \cdot 19^{3/2} + 2}{10(19^{3/2} - 1)}$
9 $4 - \sqrt{2}$
11 0
13 $\tfrac{1}{3}[5^{3/2} - 1]$
15 (a) $\displaystyle\int_{\mathscr{A}} \sqrt{4x^2 + 36y^2 + 1}\,dA$, where $\mathscr{A}$ is the ellipse $2x^2 + 4y^2 = 1$.
(b) $4\displaystyle\int_0^{1/\sqrt{2}} \int_0^{(1/2)\sqrt{1-2x^2}} \sqrt{4x^2 + 36y^2 + 1}\,dy\,dx$
$4\displaystyle\int_0^{\pi/2} \int_0^{1/\sqrt{2+2\sin^2\theta}} \sqrt{4r^2(1+8\sin^2\theta)+1}\,r\,dr\,d\theta$
17 $\tfrac{2}{3}Ma^2$
19 (a) 0 (b) 0 (c) 0
21 $8\pi k/3$

SEC. 15.6. THE DIVERGENCE THEOREM

5 48
7 11
9 $\dfrac{51\pi}{768}$
13 It is no greater than five times the surface area of $\mathscr{S}$.
15 $\tfrac{3}{2}$
17 0
19 0

SEC. 15.7. STOKE'S THEOREM

9 -1
11 (c) $f(x,y,z) = \dfrac{1}{2-a}(x^2 + y^2 + z^2)^{(2-a)/2}$

13 Not simply connected
15 Not simply connected
17 Conservative
19 Conservative
21 0
23 0
29 $\int_{\mathscr{S}_1} (\nabla \times \mathbf{F}) \cdot \mathbf{n}\, dS = -\int_{\mathscr{S}_2} (\nabla \times \mathbf{F}) \cdot \mathbf{n}\, dS$
31 (a) It is not simply connected.

SEC. 15.8. MAXWELL'S EQUATIONS

See the solutions manual.

SEC. 15.S. SUMMARY: GUIDE QUIZ ON CHAP. 15

8 $-\frac{1}{6}$
10 0
11 (a) 0   (b) 3   (c) -1 (d) 4

REVIEW EXERCISES FOR CHAP. 15

1 29/60
5 (a) 0   (b) $-(2 + \pi^{10})$ (c) $1 - e^{\pi^2}$
7 (a) $\dfrac{q}{8\pi\varepsilon_0}$ (b) $\dfrac{q}{8\pi\varepsilon_0}$
9 (a) $ye^{xy}\mathbf{i} + xe^{xy}\mathbf{j} + \dfrac{2z}{1+z^2}\mathbf{k}$
  (b) $y^2 e^{xy} + x^2 e^{xy} + \dfrac{2(1-z^2)}{(1+z^2)^2}$
  (c) **0**
13 $\dfrac{12\pi}{5}$
15 (a) e^{xz} (b) $e^{4\pi^2 e^{2\pi}} - 1$
17 3/2
19 (a) $-\frac{1}{2}$ (b) $-\frac{1}{2}$
25 -8
27 (a) All integers n (b) $n = 3$
29 (b) $0\Delta t$
33 $3\pi a^2 h$
35 (c) Yes   (d) 2π
   (f) No; the domain is not simply connected.
37 (a) $\dfrac{\partial \mathbf{r}}{\partial u} \times \dfrac{\partial \mathbf{r}}{\partial v}$ (c) $\int_{\mathscr{A}} \left| \dfrac{\partial \mathbf{r}}{\partial u} \times \dfrac{\partial \mathbf{r}}{\partial v} \right| dA$
39 f is constant.
45 (a) $\int_{\mathscr{S}} cz\mathbf{k} \cdot \mathbf{n}\, dS$

APPENDIX A. REAL NUMBERS

1 $a + 3 < b + 3$
3 $5a < 5b$
5 $1/a > 1/b$
7 $(-1, \infty)$
9 $(-\infty, -2)$
11 $(-\infty, 1), (3, \infty)$
13 $(-3, -2)$
15 $(-1, 0), (1, \infty)$
17 $(-5, -3), (0, \infty)$
19 $(-\infty, \frac{1}{3}), (\frac{1}{2}, \infty)$
21
23
25 $x > 3$
27 $x > 1$
29 $a = -2, b = 1$
31 $a = -1, b = 1, c = -2, d = 1$
33 (a) 234.74747...
   (b) 232.4
   (d) $a = \dfrac{2324}{990}$
35 Cancellation of $x - 1$ is division by zero.
39 No

APPENDIX B. SOME TOPICS IN ALGEBRA

1 $2\sqrt{5}$
3 $\sqrt{2} + 4$
5 $(6 + 2\sqrt{3})/3$
7 $\dfrac{x(1 + \sqrt{x})}{1 - x}$
9 $\dfrac{7}{5(3 - \sqrt{2})}$
11 $\dfrac{1}{\sqrt{x} + \sqrt{5}}$
13 (a) $(x + 4)^2 - 3$ (b) $(x - 4)^2 + 7$
   (c) $(x - \frac{1}{2})^2 + \frac{7}{4}$
15 (a) $(x + \frac{3}{2})^2 - \frac{17}{4}$ (b) $(x + \frac{3}{2})^2 + \frac{19}{4}$
   (c) $(x + \frac{5}{4})^2 + \frac{39}{16}$
17 (a) $2(x - \frac{5}{4})^2 - \frac{1}{8}$ (b) $2(x + \frac{3}{2})^2 + \frac{5}{2}$
   (c) $3(x + \frac{5}{6})^2 - \frac{13}{12}$
19 (a) No real solutions   (b) $x = \frac{1}{2}(-1 \pm \sqrt{5})$
   (c) $x = 1$
21 (a) No real solutions
   (b) Two real solutions
   (c) One real solution
23 (a) 10   (b) 15   (c) 45
25 $1 + 7x + 21x^2 + 35x^3 + 35x^4 + 21x^5 + 7x^6 + x^7$
27 45
29 (a) $\frac{1093}{729}$ (b) 1275   (c) $\frac{33}{8}$
31 $\dfrac{1}{3}\left(1 - \dfrac{1}{10^9}\right)$

APPENDIX C. EXPONENTS

1 (a) 1   (b) 32   (c) 2   (d) $\frac{1}{2}$ (e) 8

3

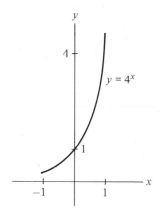

5

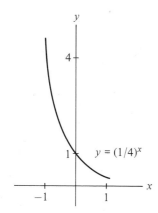

7 (a) 2^4 (b) 2^{-3} (c) $2^{1/2}$ (d) 2^0 (e) 2^{-2}
9 (a) $b^{3/2}$ (b) $b^{3/2}$ (c) $b^{-1/3}$
 (d) $b^{2/3}$ (e) $b^{1/4}$
11 (a) Domain: all x; range: $y > 0$
 (b) Domain: all x; range: all y
 (c) Domain: $x \geq 0$; range: $y \geq 0$
 (d) Domain: all x; range: $y \geq 0$
13 (a)

x	-32	-1	0	1	32
$x^{3/5}$	-8	-1	0	1	8

(b)

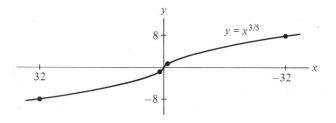

15 $\sqrt{5}$
17 (a) $x > 1$ (b) $x = 1$ (c) $0 < x < 1$

19

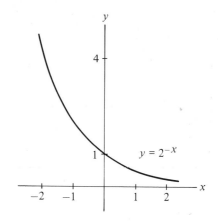

21 (a) 10^3 (b) 10^{-4} (c) 10^6 (d) 10^{-7}
23 The second way
25 (a)

h	0.1	0.01	0.001
$\dfrac{2^h - 1}{h}$	0.7177	0.6956	0.6934

27 (a) 3557 (b) 30.238 (c) Reduce CEP

APPENDIX D. TRIGONOMETRY

1 (a) $\pi/2$ (b) $\pi/6$ (c) $2\pi/3$ (d) 2π
3 (a) $135°$ (b) $60°$ (c) $120°$ (d) $720°$
5 (a) $\frac{5}{3}$ (b) $\approx 95.49°$
7 (a) $\dfrac{5\pi}{18}$ (b) $\approx 114.59°$
9 (a) 4.5 in (b) 6 in (c) 7.5 in
11 (a)

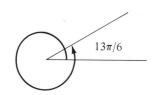

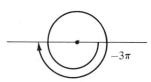

15

θ	0	$\pi/6$	$\pi/4$	$\pi/3$	$\pi/2$	π	$3\pi/2$	2π
$\sin \theta$	0	1/2	$1/\sqrt{2}$	$\sqrt{3}/2$	1	0	-1	0

19 $\dfrac{1 + \sqrt{3}}{2\sqrt{2}}$

27 (a) −   (b) +   (c) +   (d) −
31 (a) 0.1763   (b) 2.7475   (c) −2.7475
   (d) −1.0000   (e) 0
33 (a) $60° = \pi/3$ radians   (b) $30° = \pi/6$ radians
37 (a) $\cos \alpha = b/c$ (b) $\sin \beta = b/c$ (c) $\tan \alpha = a/b$

41 (a)

θ	0	$\pi/6$	$\pi/4$	$\pi/3$
$\sec \theta$	1	$2/\sqrt{3}$	$\sqrt{2}$	2

(b)

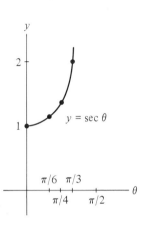

49 (b) $1/\sqrt{2}$ (c) $-1/\sqrt{2}$

APPENDIX E. CONIC SECTIONS

SEC. E.1. CONIC SECTIONS

1

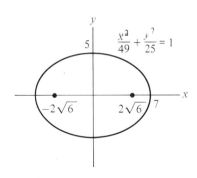

3 $\dfrac{x^2}{25} + \dfrac{y^2}{21} = 1$

5

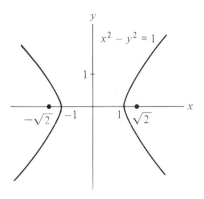

7

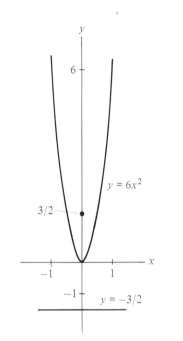

9 $y^2 = 12x$

13 (b)

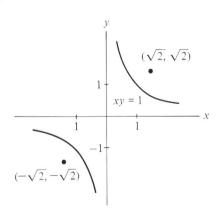

SEC. E.2. TRANSLATION OF AXES AND THE GRAPH OF $Ax^2 + Cy^2 + Dx + Ey + F = 0$

1

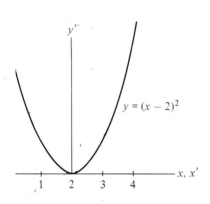

7

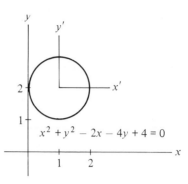

3

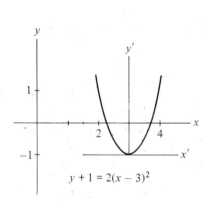

9

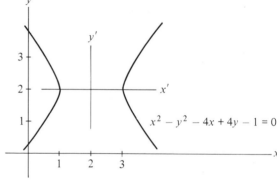

5

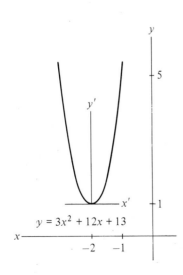

11

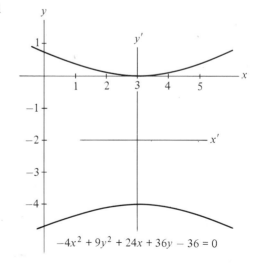

13

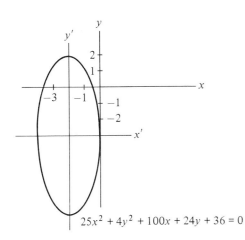

$25x^2 + 4y^2 + 100x + 24y + 36 = 0$

15 (d) When $A = C$ and their sign is opposite that of F.

SEC. E.2. ROTATION OF AXES AND THE GRAPH OF $Ax^2 + Bxy + Cy^2 + Dx + Ey + F = 0$

1

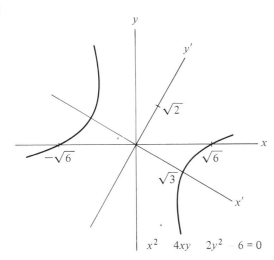

$x^2 - 4xy + 2y^2 - 6 = 0$

3

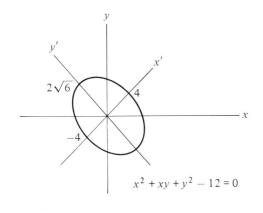

$x^2 + xy + y^2 - 12 = 0$

5

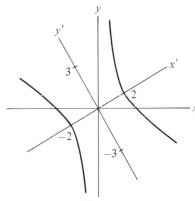

$23x^2 + 26\sqrt{3}xy - 3y^2 - 144 = 0$

7

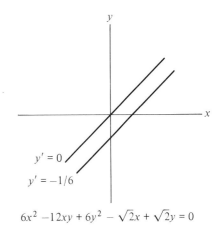

$6x^2 - 12xy + 6y^2 - \sqrt{2}x + \sqrt{2}y = 0$

9 Hyperbola ($\mathscr{D} = 600$)
11 Parabola ($\mathscr{D} = 0$)
13 (a)

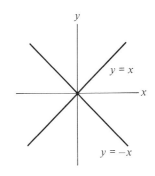

(b)

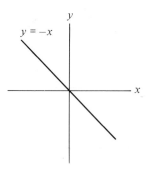

(c)

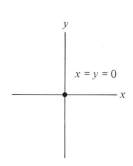

SEC. E.4. CONIC SECTIONS IN POLAR COORDINATES

1 (a)

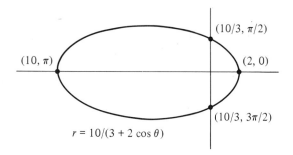

3 (a) $e = \frac{4}{3}$ (b) $e = \frac{3}{4}$ (c) $e = 1$ (d) $e = -\frac{4}{3}$
5 (b) $r^2 \sin^2 \theta + 4r \cos \theta - 4 = 0$
  (c) $r = -4 \csc \theta \cot \theta$

APPENDIX F. THEORY OF LIMITS

See the solutions manual.

APPENDIX G. THE INTERCHANGE OF LIMITS

SEC. G.2. THE DERIVATIVE OF $\int_a^b f(x, y) \, dx$ WITH RESPECT TO y

3 $y = -2/\pi$
5 $f(u, u) + \int_0^u f_u(u, x) \, dx$

SEC. G.3. THE INTERCHANGE OF LIMITS

1 (a)

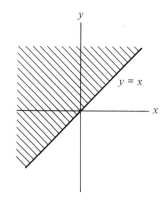

5 (a)

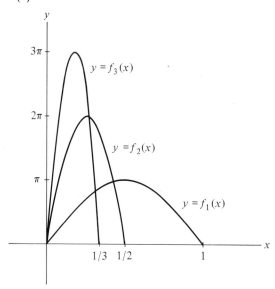

LIST OF SYMBOLS

Note: Numbers in parentheses refer to exercises on the indicated pages. Numbers prefixed by "S" refer to the pages at the end of the book.

Symbol	Description	Page
$\lvert x \rvert$, abs x	absolute value of x	853, S5
a_N	normal component of acceleration	928
a_T	tangential component of acceleration	928
$\overline{AB}$	length of AB	87
$\widehat{AB}$	length of arc AB	876
$\to$	approaches	50
$x \to a$	x approaches a	51
$x \to a^+$	x approaches a from the right	52
$x \to a^-$	x approaches a from the left	52
$\mathscr{D}$	discriminant	798, S52
$\bullet$	included point	53
$\circ$	excluded point	50, 53
$\approx$	approximate equality	2, S21
$\stackrel{?}{=}$	conjectured equality	648
ϵ	epsilon (challenge)	77
δ	delta (response)	79
$F(x)\Big\vert_a^b$	$F(b) - F(a)$	272
$\Big\vert_{x=a}^{x=b}$	evaluation limits	390
$\Big\vert_{(a,b)}$	evaluation at (a, b)	803
$[F(x)]_a^b$	$F(b) - F(a)$	387
FTC	Fundamental Theorem of Calculus	263
$\Gamma(x)$	gamma function	502(44)
∞	infinity	S3
(a, b)	open interval	S3
$[a, b]$	closed interval	S3
$[a, b), (a, b]$	"half-open" intervals	S3
lim	limit	49
$J, \dfrac{\partial(x, y)}{\partial(u, v)}$	Jacobian	856

LIST OF SYMBOLS

Symbol	Description	Page
$n!$	n factorial	S12
(x, y)	ordered pair	11
$R(x, y)$	rational function	427
$R(\cos \theta, \sin \theta)$	rational function of $\cos \theta$ and $\sin \theta$	427
Σ	sigma notation	241
ω	angular velocity	833
ω	frequency	648
$f(x)$	function of x	31
Δf	change in f	127
df, dy	differentials	221
f^{-1}	inverse of f	312
$f \circ g$	composite function	40
$A(x)$	cross-sectional area	256
$c(x), c(y)$	cross-sectional length	454
e^x	exponential function	311
$E(x)$	exponential function	366
$\exp x$	exponential function	398
$\ln x$	natural logarithm	299
$L(x)$	natural logarithm	305
$\log_b x$	logarithm base b	288
$\tan^{-1} x$	inverse tangent	320
$\arctan x$	inverse tangent	320
$\sinh x, \cosh x, \tanh x$	hyperbolic functions	359–361
$\sinh^{-1} x$	inverse hyperbolic sine	362
$[x], \lfloor x \rfloor$	greatest integer (floor) function	91
$\lceil x \rceil$	ceiling function	96(9)
$\dfrac{dy}{dx}, f', Df$	first derivative	128
$\dfrac{d^2 y}{dx^2}, f'', D^2 f$	second derivative	185
$\dfrac{d^n y}{dx^n}, f^{(n)}$	nth derivative	186
$\dfrac{\partial y}{\partial x}, f_x, f_1$	partial derivative	763
$dy \div dx$	quotient of differentials	222
$\dot{y}, \ddot{y}$	Newton's dot notation	185
$\int f(x) dx$	antiderivative	270
$\int_a^b f(x) dx$	definite integral	247
$\int_R f(P) dA$	definite integral	817
$\int_R f(P) dV$	definite integral	863
$\int_C f(P) ds$	line integral	954
$\int_C (P\, dx + Q\, dy + R\, dz)$	line integral	957
$\oint_C f(P) ds$	line integral	958
$\int_C \mathbf{F} \cdot d\mathbf{r}$	line integral	957
$\int_{\mathscr{S}} \delta(P) dS$	surface integral	988

Symbol	Description	Page
$\int_{\mathcal{S}} \mathbf{F} \cdot \mathbf{n}\, dS$	surface integral	997
i	$\sqrt{-1}$	646
z	complex number	646
$|z|$	magnitude of z	649
Im z	imaginary part of z	647
Re z	real part of z	647
z^{-1}	reciprocal of z	657(11)
$\bar{z}$	conjugate of z	647
$x + iy$	complex number	646
$e^{i\theta}$	complex exponential	659
$re^{i\theta}$	exponential form	663
$\mathbf{A}$	vector	693
$|\mathbf{A}|$	magnitude of $\mathbf{A}$	709
$\overrightarrow{PQ}$	directed line segment (vector)	693
(x, y)	vector	694
$\mathbf{A} \cdot \mathbf{B}$	dot product	709
$\mathbf{A} \times \mathbf{B}$	cross product	734
$\mathbf{i}, \mathbf{j}, \mathbf{k}$	basic unit vectors	703
$\mathbf{T}$	unit tangent vector	922
$\mathbf{N}$	principal unit normal vector	923
$\mathbf{B}$	binormal vector	927(23)
∇	del	948
∇f	gradient of f	785, 787
$\nabla \cdot \mathbf{F}$	divergence of $\mathbf{F}$	948
div $\mathbf{F}$	divergence of $\mathbf{F}$	949
$\nabla \times \mathbf{F}$	curl of $\mathbf{F}$	950
curl $\mathbf{F}$	curl of $\mathbf{F}$	950
$\nabla^2 f$	Laplacian of f	951
$D_{\mathbf{u}} f$	directional derivative	784
(x, y)	rectangular coordinates	11
(x, y, z)	rectangular coordinates	698
(r, θ)	polar coordinates	529
(r, θ, z)	cylindrical coordinates	871
(r, θ, φ)	spherical coordinates	873
$\bar{x}, \bar{y}, \bar{z}$	centroid or center of mass	488, 833, 865
M_x, M_y, M_z	moment	832, 833, 865
I_x, I_y, I_z	moment of inertia	834, 865
$t_{1/2}$	half-life	343
t_2	doubling time	342
$\begin{pmatrix} a_1 & a_2 \\ b_1 & b_2 \end{pmatrix}$	matrix	725
$\begin{vmatrix} a_1 & a_2 \\ b_1 & b_2 \end{vmatrix}$	determinant	726
$\binom{n}{k}, C_k^n$	binomial coefficient	S12

INDEX

Numbers in parentheses refer to exercises on the indicated pages. Numbers prefixed by "S" refer to the pages at the end of the book.

Abell, Steve, 483(26)
Abscissa, 11
Absolute convergence, 613–615
Absolute value, 94, S5–S6
  of a complex number (*see* Magnitude, of a complex number)
  derivative of, 129–130
  of a vector (*see* Magnitude, of a vector)
Acceleration, 196–201
Acceleration vector, 917–920, 928–932
  normal component, 928–932
  tangential component, 928–932
Agriculture, 195(39)
Alternating current, 656–657
Alternating series, 608
  decreasing, 610
  test for convergence, 608–612
Ampère, André Marie (1775–1836), 1015
Analytic geometry, 11–30, 747–755, S38–S56
Angle(s), 529
  between curves, 564
  direction, 723
  half-vertex, 874
  of inclination, 563, S33
  between lines, 563–564
  between planes, 740(37), 745(22)
  between surfaces, 909(16–18)
  between vectors, 708, 712
Angular momentum, 738
Angular velocity, 833
Antiderivatives, 130–131
  of $cf(x)$, 382
  of $\cos x$, 381
  of $\cos^m\theta \sin^n\theta$, 418–421
  of $\csc x$, 427(45)

Antiderivatives (*Cont.*):
  of e^x, 381
  of f'/f, 301, 381
  of $f(x) + g(x)$, 382
  integrand of, 272
  notation for, 270, 384
  of $1/x$, 288, 300–301
  by partial fractions, 410
  by parts, 392–398
  of a polynomial, 382
  of a power, 271–272
  of a power series, 628–629
  properties of, 270–272
  of rational functions, 404–408, 431–437
  of $\cos\theta$ and $\sin\theta$, 427–430
  of $\sec x$, 424–425, 430(23)
  of $\sin x$, 381
  and substitution, 384–390, 428, 431–437
  of $\tan^{-1} x$, 396
  of $\tan x$, 423
  of $\tan^m\theta \sec^n\theta$, 422–424
  of x^n, 381
Applications:
  agriculture, 195(39)
  biology, 158, 163(56)
  cartography, 426(44)
  crystallography, 740(39)
  ecology, 377(177), 378(186), 917(22)
  economics, 142(52), 158–159, 163(55), 214(43), 277(46), 358(63), 474(21), 500–501, 594(28–29), 712–713, 716(32), 717(38–39), 779–780, 917(21)
  electricity, 347(41), 655–657, 1001–1002, 1015–1016

Applications (*Cont.*):
  energy, 159, 195(39), 451(186)
  entropy, 375(110)
  epidemics, 845(26)
  flexure formula, 578(22)
  fluid flow, 958–959, 978–979, 1009–1010
  forestry, 213(36)
  gasoline prices, 39(40)
  Gauss's law, 1001–1002
  general transport theorem, 1024(36)
  geology, 509(16), 870(34)
  gravitation, 845(30), 881–882, 921(16–19), 933–939, 964(28), 967
  information content, 375(110)
  LORAN, S45(18)
  Monte Carlo methods, 821(10)
  nuclear weapons, 163(58), 759, 821(16), S25(27)
  physics, 200, 203(7), 234(78), 240(16), 317(12), 365(35), 470–172, 174(18), 738, 767(25), 778–779, 782(34–36), 809(8), 811(39–41), 833–834, 976(22), 987(35), 1002–1003
  physiology, 158, 474(17)
  probability, 593(23)
  rainfall, 473(3)
  reflector, 565–566
  replacement policy, 643
  resources, 377(171–175)
  rocket motion, 200–203, 507–508
  rotary engine, 544–545, 547(28)
  seismology, 296(26)
  solar collector, 725(32)
  statistics, 795(13)

Applications (*Cont.*):
  submarine detection, S45(17)
  thermodynamics, 778–779, 782(34–36)
  traffic flow, 108(58), 213(35)
  transportation, 890(35–40)
  volcanic ash settling, 470, 473(1–2)
  (*See also specific entries*)
Approximation:
  of area, 6, 236
  of a definite integral, 514–518
  by a differential, 223–225
  of functions, 580
  local, 469–472
  of root of an equation, 678–681
Arc length, 548–555, 996(24)
Arccosine, 323–324
Archimedes (ca. 287–212 B.C.), 9, 461(31), 561(13)
Arcsecant, 324–325
Arcsine, 321–323
Arctangent, 319–321
Arcwise connected, 965, 1010
Area, 236
  and cross product, 736
  as a definite integral (*see* Definite integral, and area)
  as a line integral, 963(23), 986(22–27)
  in polar coordinates, 534–539, 986(27)
  of a sector, 85, 534
  of a sphere, 560–561
  of a surface of revolution, 556–561
  of a surface in space, 991
  between two curves, 455, 537–539
Argument:
  of a complex number, 649
  of a function, 33
Arithmetic mean, 795(23), 917(16)
Asymptote, 19, 67–72, 177
  horizontal, 67
  of a hyperbola, 19
  tilted, 70–71
  vertical, 67
Average of a function, 509–513, 819, 864, 954, 996(22–23)
  (*See also* Mean)
Average speed, 3
Average velocity, 170

Bacteria growth, 163(56)
Baier, Othmar, 545
Balancing line, 486–490, 838(32–33)
Ball, 840, 867, 872
Barcellos, Anthony, xvii
Base, 288, 298–299, S18
Behrens, W. W., III, 377
Bell, E. T. (1883–1961), 10

Berkeley, Bishop George (1685–1753), 222
Bernoulli brothers, 9
Bernoulli's inequality, 232(52)
Binomial coefficients, 676, S12
Binomial series, 625(25)
Binomial theorem, 196(49–51), 673–676, S11–S14
Binormal vector, 927(23–25), 943(23)
Biology, 158, 163(56)
Bolzano, Bernhard (1781–1848), 160
Bond, Henry (ca. 1600–1678), 424–425, 430(23)
Boyer, Carl B., 10
Brahe, Tycho (1546–1601), 933
Bullialdus, Ismael (1605–1694), 934

Carbon 14 dating, 343, 347(38)
Cardioid, 532
  area of, 538
Cartesian coordinates, 11–12, 698
Cartography, 426(44)
Catenary, 359
Cavalieri, F. B. (1598–1647), 6
Cayley, Arthur (1821–1895), 110
Ceiling function, 96(9)
Center of gravity (*see* Center of mass)
Center of mass, 485–486, 833, 865
Central force field, 934–936, 953(25)
Centroid, 488–493, 833, 865
  of a curve, 562(29)
  and Pappus's theorem, 491–493
  of a surface, 995(7–10)
  and work, 506–507
Chain rule, 149–157, 773–780, 902
  in D notation, 151
  in Leibniz notation, 150–151
  and magnification, 150
  for partial derivatives, 773–780
  proof of, 155–157
  and substitution in an integral, 389–390
  for vector functions, 902
Change of variables (*see* Substitution)
Chebyshev, Pafnuti L. (1821–1894), 452(204), 556(26), 578(26)
Chord, 109(66), 167
Cicero, Marcus Tullius (106–43 B.C.), 561(13)
Circle, 13, S38
  equations of, 13–14, 530–531
  osculating, 573–574
  unit, S28–S33
Circular motion, 919–920
Circulation, 964(27)
Clark, Colin W., 917(22), 1025(42)
Closed interval, S3
Coefficients, 60

Colatitude, 873
Comparison test, 600–601
Completeness, 586
Completing the square, S9–S10
  in integration problems, 406–407
Complex numbers, 646–657
  absolute value of (*see* magnitude of, *below*)
  and alternating current, 656–657
  argument of, 649
  conjugate of, 647
  DeMoivre's law, 654
  imaginary part of, 647
  logarithm of, 664(23–24)
  magnitude of, 649
  notations for, 663
  polar form of, 649
  powers of, 654–655
  products of: algebraic, 646–647
    geometric, 650–651
  quotients of, 647
  real part of, 647
  roots of, 648
  sums of: algebraic, 646
    geometric, 649
Complex series, 659
Composite function, 39–43
  derivative of, 149–157
Composition, 40
Concave downward, 187–191
Concave upward, 187–191
Cone, 468(20), 556–557
Conic sections, S38–S44, S53–S55
Conjugate, 647
Connected, 997
  arcwise, 965, 1010
  simply, 980, 1010
Conservative vector fields (*see* Vector field, conservative)
Constant function, 54, 132
Constant of integration, 270–271, 382
Constraint, 218
Continuity:
  derivatives and, 129–130
  of $f(x)$, 93
  of $f(x, y)$, S62–S63
  from left, 91
  of a mapping, S62–S63
  at a point, 90–93
  from right, 90
  of a vector function, 895
Continuous function, 90–95
  definition of, 93
  nowhere differentiable, 160
Contour lines, 758–759, 761
Convergence:
  absolute, 613–615
  of an improper integral, 495, 497–498

Convergence (*Cont.*):
   of a power series, 619–620, 622–623
   and remainder, 636
   of a sequence, 582
   of a series, 589
Convex, 102, 811(35)
Convex sets, 102(25–30)
Coordinates:
   Cartesian, 11–12, 698
   cylindrical, 871–873
   polar, 529–533
   rectangular, 11–12, 698
   spherical, 873–877
Copernicus, Nicolaus (1473–1543), 933
Cosecant, S34
   antiderivative of, 427(45)
   derivative of, 145
   graph of, S34
Cosine(s), S29–S32
   antiderivative of, 381
   derivative of, 144
   direction, 723
   graph of, S31
   hyperbolic, 359–361
   law of, S35
Cotangent, S34–S35
   derivative of, 146
   graph of, S35
Courant, Richard (1888–1972), 10
Cramer's rule, 728–729
Critical number, 177
Critical point, 177, 790–794
Cross product, 733–738
   and area, 736
   geometric description of, 736–737
Cross sections, 253–256, 454–459, 461–467
   horizontal, 459
   vertical, 459
   volume by, 256, 461–467, 475–477
Crowe, M. J., 10
Cube root, 309, S23–S24
Cubic $y = x^3$, 14–15
Cubing function, 309
Curl, 950–951
   and conservative vector fields, 969–974
   physical interpretation of, 1009–1010
Curvature, 569–574
   formulas for, 571–572, 575(20)
   radius of, 573
Curve(s):
   centroid of, 562(29)
   closed, 958
   integral, 954
   length of, 548–555, 996(24)
   level, 758–761
   logistic, 344, 347(39)
   normal vector to, 923–925, 958

Curve(s) (*Cont.*):
   parameterized (*see* Parameterized curve)
   rectangular, 975(23)
   slope of, 125
   special, 975(22)
   tangent vector to, 896, 922–926
   work along, 955–956
Cusp, S24
Cycloid, 542
   length of, 550–551
Cylinder, 749–750
Cylindrical coordinates, 871–873

Decreasing function, 173
Definite integral, 235, 247–253, 454, 468–472, 514–518, 814–820, 861–869, 988–995
   and area, 5–8, 249, 250, 253–254, 454–459, 535, 819–820, 991
   and distance, 237–238, 250, 255
   existence of, 248
   formal definition of, 257, 468–469, 817, 863
   informal approach to, 469–472
   integrand of, 272
   as limit of a sequence, 585, 588(26–28)
   and mass, 249, 250, 254–255, 818–819
   mean-value theorem for, 275
   notation for, 247, 817, 863
   by parts, 396
   over a planar region, 814–820, 827–844
   properties of, 273–275, 820
   over a region in space, 861–869
   substitution in, 388, 856–860, 990–991
   over a surface, 988–995
   and volume, 238–239, 250, 255–256, 461–467, 818, 864
   (*See also* Integrals)
Degree, S26
Degree of a polynomial, 60
Del, 785
Delta and epsilon, 77, 79
Delta notation, 126–128
DeMoivre's law, 654
Density, 115–116, 125, 236–237
   linear, 115, 236
Dependent variable, 33
Derivative(s), 1–3, 110, 118–125, 782–788, 895–899
   of b^x, 315
   of cf, 135–136
   of a composite function, 149–157
   of a constant function, 132
   and continuity, 129–130
   of $\cos^{-1} x$, 323–324
   of $\cos x$, 144

Derivative(s) (*Cont.*):
   of $\cosh x$, 361
   of $\cot^{-1} x$, 326
   of $\cot x$, 146
   of $\coth x$, 361
   of $\csc^{-1} x$, 326
   of $\csc x$, 145
   of $\csch x$, 361
   definition of, 119
   in delta notation, 127
   directional, 782–787, 811(34)
   of e^x, 313–314
   at an endpoint, 119
   at an extremum, 166–167
   of $f + g$, 133–134
   of $f - g$, 133–134
   of fg, 134–135
   of f/g, 137–139
   of $f(x)^{g(x)}$, 316–317
   of a function, 119
   and graphing, 177–184
   higher, 185–187
   of inverse functions, 313, 319–327
   of $\ln x$, 299
   of $\log_b x$, 297–298
   notations for, 128, 222
   of $1/g$, 140
   of $1/x$, 140
   parametric equations, 543, 897
   partial (*see* Partial derivatives)
   of a polynomial, 137
   of a power series, 626
   as rate of change, 158–159, 329–333
   of $\sec^{-1} x$, 324–325
   of $\sec x$, 145
   of $\sech x$, 361
   second (*see* Second derivative)
   sign of, 119
   of $\sin^{-1} x$, 321–323
   of $\sin x$, 143–144
   of $\sinh x$, 361
   of sums, 133–134
   of $\tan^{-1} x$, 320
   of $\tan x$, 146
   of $\tanh x$, 361
   of vector function, 893–899, 901–903
   of x^a, 316
   of $x^{m/n}$, 154–155
   of $x^{1/n}$, 122–124
   of x^n, 122–124
   of $\sqrt{x}$, 121–122
Determinants, 725–731
   and simultaneous equations, 728–730
Diameter of a region, 817, 863
Difference of a function, 93
Difference quotient, 118
Differentiable function, 129–130

Differential(s), 219–225, 770–772
　applied to estimates, 223–225, 771–772
　exact, 812, 974
　of $f(x)$, 221
　of $f(x, y)$, 770–772
Differential equations, 335–346, 664–672
　associated quadratic, 668–670
　characteristic equation (see associated quadratic, above)
　exact (see Differential form, exact)
　first-order, 664
　general solution, 665–672
　harmonic motion, 203(7)
　homogeneous, 665
　inhibited growth, 338–339, 344
　natural growth, 337–338, 340
　order, 335
　particular solution, 665–666, 671, 672
　second-order, 667–672
　separable, 336–337
　solution of, 335
　undetermined coefficients, 666
Differential form, 812
　exact, 812, 974
Differentiation, 132–141, 143–147, 149–157
　implicit (see Implicit differentiation)
　logarithmic, 302–303
Dihedral angle, 745(22)
Direct current, 656
Directed line segment, 693
Direction angles, 723
Direction cosines, 723
Direction numbers, 723
Directional derivative, 782–788
　and gradient, 785, 788
　in polar coordinates, 944(24)
　second-order, 811(34)
Directrix, S43–S44, S55
Discriminant, 22(39), 799, S11, S52
Distance, 237–238
　as an integral, 5, 235, 255
　to a line, 718–719
　to a plane, 721
　between two lines, 740(36)
　between two points, 12–13, 699
Distance formula, 13, 699
Divergence, 947–951
　and gradient, 951
　physical interpretation of, 979
Divergence theorem (Gauss's theorem), 997–1004
　proof of, 1005(25)
Divergent improper integral, 495–498
Divergent series, 589
　nth term test for, 590
Division by zero, S2
Domain of a function, 31

Doomsday equation, 348(46)
Dot notation, S3
　Newton's, 185
Dot product, 707–716, 717(36), 732
Doubling time, 342
Dummy index, 241
Dyadic number, 286(30)

e, 292–295, 296(30), 299, 308
　as an integral, 308
　as a limit, 292, 296(30)
Eccentricity, S53–S55
Economics:
　compound interest, 293–295
　dot product, 712–713, 716(32), 717(38–39)
　elasticity of demand, 142(52)
　and Euler's theorem, 779–780
　marginal cost, 158–159
　marginal product, 779
　money supply, 594(29)
　multiplier effect, 594(28)
　present value, 474(21), 500–501
　semilog, 378(188)
　utility, 917(21)
Edwards, C. H., Jr., 10, 215(44), 494(23)
Einstein, Albert (1879–1955), 358(61)
Elasticity of demand, 142(52)
Electricity, 347(41), 655–657, 1001–1002, 1015–1016
Elementary functions, 264
Ellipse, 17–18, 568(21), S38–S41, S56(6)
　directrices of, S55
　eccentricity of, S53–S54
　foci of, S39–S41
　"length of string," S39–S41
　in standard position, 18, S40
Ellipsoid, 751
Elliptic integral, 447
Elliptic paraboloid, 755(17)
Enthalpy, 778
Epicycloid, 548(33)
Epitrochoid, 545, 547(28), 578(25)
Epsilon and delta, 77, 79
Equations:
　roots of, 678
　　approximation of, 678–681
Equipotential curves, 761
Equipotential surfaces (see Level surfaces)
Error, relative (see Relative error)
Escape velocity, 200–203, 921, 922(23)
Estimating an integral, 514–518
　Gaussian quadrature, 519–520(26)
　left-point method, 519
　midpoint method, 527(51)
　prismoidal formula, 519(22)
　right-point method, 519

Estimating an integral (Cont.):
　Simpson's method, 516–518
　trapezoidal method, 515–516
　error in, 518
Euler, Leonhard (1707–1783), 9, 292, 659
Euler's constant, 618(34–35)
Euler's theorem, 779–780
Even function, 42
　symmetry of, 43
Exact differentials, 812, 974
Exponents, S18–S24
　basic law of, S19
　laws of, S22
Exponential functions [see Function(s), exponential]
Exponential index, 377(172)
Extrema, 99
　on a polygon, 792–793
Extreme values, 99
Extremum, 99
　absolute, 177
　derivative at, 166–167
　edge, 792–793
　of $f(x)$, 99
　　derivative tests for, 178–179, 192–193
　of $f(x, y)$, 789–799
　　partial derivative test for, 796, 798
　　proof of, 805–806
　of $f(x, y, z)$, partial derivative test for, 800(24)
　global, 177, 789
　interior, 167
　local, 177
　relative (see Relative extremum)

Factor theorem, 102(23), 451(194)
Factorial, 502(45), S12–S13
Faraday, Michael (1791–1867), 1015
Fermat, Pierre de (1601–1665), 2, 6
Feynman, R. P., 358(58), 502(44), 646(43), 678(23)
Field:
　gradient, 946–947, 965–970
　line integral of, 966–967
　scalar (see Scalar field)
　vector (see Vector field)
Firebaugh, Morris W., 195(39)
First moment, 832
Floor function, 91
Flux, 1001
Focus (foci), 566
　of ellipse, S39–S41
　of hyperbola, S41–S42
　of parabola, 566, S43–S44

Force:
  and acceleration, 918
  of water, 471, 1025(45)
  and work, 707–708, 955–956
Force field, central, 934–936, 953(25)
Franklin, Benjamin (1706–1790), 1015
Frenet formulas, 943(23)
Frustum, 557
Function(s), 31–38, 756
  absolute value, 94
  approximation of, 580
  average of, 510
  average value, 510
  ceiling, 96(9)
  chord of, 109(66)
  composite, 39–43
  constant, 54, 132
  continuous, 93
  cubing, 309
  decreasing, 173
  derivative of, 119
    and continuity, 129–130
  difference of, 93
  differentiable, 129
    is continuous, 129–130
  domain of, 31
  elementary, 264
  even, 42
  exponential, 65, 310, 313–315, 365–369, 659–663, S18–S24
    relation to trigonometric functions, 659–661
  extreme values, 99
  floor, 91
  gamma, 502(44)
  graph of, 33, 747–761
    rising or falling, 119
  greatest integer, 91–92
  harmonic, 951, 1003
  homogeneous, 779
  hyperbolic (see Hyperbolic functions)
  implicit, 215
  increasing, 173
  input, 33
  inverse, 42, 309–313
    on a calculator, 311
    derivative of, 313
    graph of, 312
    hyperbolic, 362–364
    trigonometric (see Inverse trigonometric functions)
  Laplacian of, 951
  limit of, 48–54
  logarithmic, 289–290, 297–302, 368
  as a machine, 36
  numerical, 33, 35–36
  odd, 42

Function(s) (*Cont.*):
  one-to-one, 310
  output, 33
  positive, 595
  product of, 93
  quotient of, 93
  range, 31
  rational (*see* Rational function)
  real, 45
  scalar, 900
  sum of, 93
  as a table, 35
  trigonometric, S29–S35
    relation to exponential functions, 659–661
  value, 31
  vector [*see* Vector function(s)]
Fundamental theorem of calculus, 6, 260–268
  first, 260–264
    proof of, 279–280, 280(1)
  second, 264–268
    proof of, 277–279

Galois, Évariste (1811-1832), 689(63)
Gamma function, 502(44)
Gasoline prices, 39(40)
Gauss, C. F. (1777–1855), 1006
Gauss's law, 1001–1002
Gauss's theorem (*see* Divergence theorem)
Geology, 509(16), 870(34)
Geometric mean, 795(23), 917(18)
Geometric series (*see* Series, geometric)
Gibbs, Josiah W. (1839–1903), 692
Gradient, 785–788
  and curl, 952(11), 969
  and directional derivative, 785, 788
  and divergence, 951
  and level curves, 789(21)
  and level surfaces, 905
  in polar coordinates, 944(24)
  of scalar field, 946–947
  significance of, 785
  in spherical coordinates, 944(26)
Gradient field, 946–947, 965–970
  divergence of, 951
  line integral of, 966–967
Graph, 14, 747–761
  of $f(x)$, 33
  of $f(x, y)$, 756–758
  of a line, 15, 28–29
  of a natural logarithm, 300
  of a numerical function, 33–34
  of a sine, S31
  of a tangent, S35
Gravitation, law of, 935
Greatest integer function, 91–92

Green, George (1793–1841), 1006
Green's theorem, 976–985, 1025(43)
  and area, 986(22–27)
  and fluid flow, 978–979
  proof of, 983–985
  special case of Stokes' theorem, 1005–1006
Gregory of St. Vincent (1584–1667), 305
Gregory, James (1638–1675), 6, 424
Growth:
  inhibited, 232(41), 344–345
    differential equation of, 338–339, 344
  natural (*see* Natural growth)

Half-life, 343
Half-vertex angle, 874
Halley, Edmund (1656–1742), 935
Harmonic function, 951, 1003
Harmonic motion, 203(7)
Harmonic series, 592
  alternating, 614
Hayes, David, 195(48)
Heat equation, 767(25), 811(40)
Helix, 897
Henry, Joseph (1797–1878), 1015
Hessian, 799
Hickerson, Dean, xvii
Hölder's inequality, 917(20)
Homogeneous differential equation, 665
Homogeneous functions, 779
  and Euler's theorem, 779–780
Honsberger, Ross, 10
Hooke, Robert (1635–1703), 934–935
Huygens, Christiaan (1629–1695), 494(23), 920
Hydrostatic force, 471–472, 494(26), 1025(45)
Hyperbola, 19–21, S38, S41–S42
  asymptotes of, 19
  directrix of, S55
  eccentricity of, S54–S55
  foci of, S41–S42
  in standard position, 19
Hyperbolic cosine, 359–361
Hyperbolic functions, 359–361, 987(36)
  inverses of, 362–364
Hyperbolic sine, 360–361
Hyperbolic tangent, 361
Hyperboloid, 751–754
Hypocycloid, 547(31)

i, 646
Ideal gas, 811(41)
Imaginary number, 647
Imaginary part, 647
Impedance, 657

Implicit differentiation, 215–216
  and extrema, 217–218
Implicit function, 215
Improper integral, 494–501, 885(31)
  convergent, 495, 497–501
  divergent, 495–498
Increasing function, 173
Increment (see Delta notation)
Indefinite integral, 270
Independent variable, 33
Index:
  exponential, 377(172)
  summation, 241
Inequalities, S2–S6
  of integrals, 917(20)
Infinite series (see Series)
Infinity, S3
Inflection number, 189
Inflection point, 189
Inhibited growth (see Growth, inhibited)
Input, 33
Integers, S1
Integral(s):
  and area, 5–8, 249, 250, 253–254
  approximation of (see Estimating an integral)
  curve, 954
  definite (see Definite integral)
  of density, mass as, 254–255, 954–955
  distance as, 5, 235, 255
  double, 828
  e as, 308
  elliptic, 447
  estimating (see Estimating an integral)
  and fluid loss, 978–979, 958–959
  improper, 494–501, 885(31)
  indefinite, 270
  inequalities of, 917(20)
  iterated, 828
  line (see Line integrals)
  multiple (see repeated, below)
  nonelementary, 284
  over a planar region, 814–820, 827–844
    in polar coordinates, 839–844, 859–860
    in rectangular coordinates, 827–836, 867–869
  repeated, 828, 867
  over a solid region, 861–869, 878–883
    in cylindrical coordinates, 878–880
    in rectangular coordinates, 867–869
    in sperical coordinates, 879–883
  substitution in, chain rule and, 389–390
  over a surface, 988–995
  triple (see repeated, above)
  of vector functions, 1025, 1026(46–51)
  volume as, 255–256, 461–467, 476, 477
  (See also Antiderivatives)

Integral table, 380, 399–403
Integral test, 594–599
Integrand, 272
Integration, 272
  constant of, 270–271, 382
  limit of, 267
  numerical (see Estimating an integral)
  by partial fractions, 410
  by parts, 392–398
  in polar coordinates, 839–844, 859–860
  of rational functions, 404–408, 431–437
    of $\cos\theta$ and $\sin\theta$, 427–430
  of series, 628–629
  strategy of, 438–444
  by substitution, 384–390
  by trigonometric substitution, 431–437
Intercept, 16–17
Interest, compound, 293–295
Interior extremum, theorem of, 167
  proof of, 173–174
Intermediate value theorem, 100–101
Interval, S3
  closed, S3
  of convergence (see Radius, of convergence)
  open, S3
Inverse function [see Function(s), inverse]
Inverse hyperbolic functions, 362–364
Inverse square law, 934–936, 950
Inverse trigonometric functions, 319–327
  notation for, 312–313
Irrational numbers, S1
Irrationality of $\sqrt{2}$, S7(37)
Irreducible polynomial, 404, 409(42), 411
Isobar, 761
Isotherm, 761, 909(11)
Iterated integral, 828

Jacobians, 853, 856–860, 882–883
  and change of variables, 856–860
  and magnification, 853, 855
  and substitution in an integral, 856–860
Jorgenson, D. W., 643

Kelvin, Lord (1824–1907), 845(24)
Kepler, Johann (1571–1630), 933–934
Kepler's laws, 933–939
Kinetic energy, 240(16), 470–471, 474(18), 809(8), 833–834
King, J. E., 562(13)
Kline, Morris, 10
Kochen, Manfred, 215(45)

Lagrange multipliers, 909–916

Lagrange's formula for the remainder, 640–642
Lambert, Johann (1728–1777), S1
Lamina, 834–835, 838(36)
Langley, 759
Laplace transform, 501, 503(50–55)
Laplace's equation, 767(26)
Laplacian, 951
Law of cosines, S35
Least-squares method, 795(13)
Leibniz, Gottfried (1646–1716), 6, 9, 135, 214(44), 222, 305, 494(23)
Length of curve, 548–555, 996(24)
Level curves, 758–761
  and gradient, 789(21)
Level surfaces, 904–908
l'Hôpital's rule, 349–357, 547(25)
  proof of, 379(190)
Lichtman, Jeff, 494(25)
Limaçon, 533
Limit, 3, 8, S57–S63
  computation of, 56–65
  ϵ, δ definition of, 79–83, S57–S61
  formal definition of, 73–82
  of a function, 48–54
  of $f(x, y)$, S61–S63
  of integration, 267
  left-hand, 52
  of $(1 + h)^{1/h}$, 292, 296(30)
  notation, 49
  of a polynomial, 60
  precise definition of, 73–82
  of a product, 56
  properties of, 56, S57–S61
  of a quotient (see Quotient, limit of)
  right-hand, 52
  of a sequence, 581, 587
  of $(\sin\theta)/\theta$, 84–86
  of a sum, 56
  of vector functions, 894–895
Limits, interchange of, S68–S71
Line, 15, 22–29, 717–719, 721–723
  direction angles of, 723
  direction cosines of, 723
  direction numbers of, 723
  distance to, 718–719
  equation of, 15, 28–29, 718, 722
  graph of, 15, 28–29
  intercept equation, 30(30)
  parametric equation of, 722
  point-slope equation, 27
  of regression, 795(13)
  slope of, 22–23
  slope-intercept equation, 26
  in space, 721–723
  two-point formula, 28
  vector equation of, 722

Line integrals, 953–962
　and area, 963(23), 986(22–27)
　and fluid flow, 958–959
　of normal component, 959
　path independent, 966
　of tangential component, 961
　and work, 955–956
Linear algebra, 725–731
Linear density, 115, 236
Linear mapping, 852(13)
Liouville, Joseph (1809–1882), 451(192), 452(201)
Littlewood, J. E. (1885–1977), 845(24)
Local approximation, 469–472
Local estimate, 469, 535
Logarithms, 288–292
　base, 288, 298–299
　common, 289–290
　of complex numbers, 664(23–24)
　derivative of, 297–302
　natural (see Natural logarithm)
　properties of, 290–291
　and semilog paper, 343–344, 348(47–49), 378(188)
　and slide rule, 296(31)
Logarithmic differentiation, 302–303
Logarithmic functions [see Function(s), logarithmic]
Logistic curve, 344, 347(39)
Logistic growth, 344–345
Longitude, 873
Loomis, Lynn H., 520(28)
LORAN, S45(18)

McCall, J. J., 643
Maclaurin series, 618–620, 686
Magnetic field, 1015–1019
Magnification, 113–114, 125, 853–860
　and chain rule, 150
　and Jacobian, 853, 855
　and Mercator's map, 426(44)
Magnitude:
　of a complex number, 649
　of a vector, 693, 717(36)
Mappings, 846–851, S62–S63
Marginal cost, 158–159
Mass, 236–237
　center of, 485–486, 833, 865
　definite integral and, 249, 250, 254–255, 818–819
　as integral of density, 254–255, 954–955
Matrix, 725–726
　column of, 726
　determinant of, 726–727
　row of, 726
　transpose of, 731(11)

Maximum, 166, 204–211, 789–794
　absolute, 177
　over closed interval, 183–184
　edge, 792–793
　endpoint, 183–184
　of $f(x)$, 99, 177
　　first-derivative test for, 178–179
　　second derivative test for, 192–193
　of $f(x, y)$, 789–799
　　partial derivative test for, 796, 798
　　proof of, 805–806
　global, 177, 789
　local, 177
　relative, 177, 789
Maximum-value theorem, 98–100
Maxwell, James Clerk (1831–1879), 692, 1010
Maxwell's equations, 1014–1019
Mead, David G., 165(72)
Meadows, D. H., 377
Meadows, D. L., 377
Mean, 527, 795(23), 917(16)
Mean-value theorem, 169–175
　for definite integrals, 275
　generalized, 379(189)
　proof of, 174–175
Median, 494(28)
Mercator, Gerhardus (1512–1594), 424
Mercator, Nicolaus (1602–1687), 424
Mercator's map, 424, 426(44)
Mesh, 247, 817, 863
Minimum (see Maximum)
Minimum-value theorem, 98
Mixed partial derivatives, 765, 800–801, S64–S66, S70
Möbius strip, 1006–1007, 1013(25)
Moment, 484, 487, 832–836, 864–865
　and center of mass, 486, 833, 865
　and centroid, 488, 833, 865
　of a curve, 562(29)
　of inertia, 833–834, 865, 987(29), 996(17–18)
Momentum, 738, 935
Monte Carlo methods, 821(10)
Motion:
　under constant acceleration, 200
　harmonic, 203(7)
　against resistance, 365(35)
　rocket, 200–203, 507–508
Mountain building, 509(16), 870(10)
Multiple integral (see Repeated integrals)
Multiplier effect, 594(28)

Natural growth, 339–344
　differential equation of, 337–338, 340
　doubling time, 342

Natural growth (Cont.):
　growth constant, 340
　half-life, 343
　relative growth rate, 341
Natural logarithm, 298–302
　as a definite integral, 304–308
　definition of, 299, 305
　derivative of, 299–301
　graph of, 300
　properties of, 306–307
Negative, S1
Neil, William (1637–1670), 549
Newton, Isaac (1642–1727), 6, 9, 305, 934–935, 1015
Newton's dot notation, 185
Newton's law of cooling, 347(42)
Newton's method, 678–681
Newton's second law, 918, 921(10)
Nonelementary integrals, 284
Nonuniform string, 236–237
Normal distribution, 528(65)
Normal vector, 718, 923–925
　to a closed curve, 958
　exterior, 958–960
　to a line, 718
　to a plane, 720, 733, 737–738
　principal unit, 923–925
　to a surface, 905–908
nth term, 581, 589
nth-term test for divergence, 590
Nuclear weapons, 163(58), 759, 821(16), S25(27)
Number(s):
　complex (see Complex numbers)
　critical, 177
　direction, 723
　dyadic, 286(30)
　imaginary, 647
　inflection, 189
　irrational, S1
　prime, 600(23)
　rational, S1
　real, 646, 647, S1–S6
　sampling, 211
Numerical function, 33, 35–36
Numerical integration (see Estimating an integral)

Octant, 468(21)
Odd function, 42
　symmetry of, 43
O'Donnell, Mark A., 725(32)
Oersted, Hans Christian (1777–1851), 1015
One-to-one function, 310
Open interval, S3
Open region, 965

Optics:
    refraction, 214(44)
    Snell's law, 214(44)
Orbital velocity, 203, 920–921, 922(23)
Orbits of planets, 933–939
Ordinate, 11
Oresme, Nicolas of (ca. 1323–1382), 592, 594(25)
Orientable surfaces, 1007
Origin, 11
Orthogonal vector, 715, 905
Osculating circle, 573–574

p series, 596–597
Pappus's theorem, 491–493, 557, 562(30)
    volume by, 491–492
Parabola, 14, 461(31), S38, S42–S44
    directrix of, S43–S44
    eccentricity of, S55
    equations of, 14, 534, 541
    focus of, 566, S43–S44
    in polar form, 534
    reflecting property, 565–566
    in standard position, 14, S43
Parabolic cylinder, 749
Paraboloid, 755(17)
Parallel axis theorem, 838(35)
Parallel lines, 24–25
Parallelepiped, volume of, 739(30–31)
Parallelogram, area of, 732(24), 736–737
Parameter, 540
Parameterized curve, 540–545
    curvature of, 572
    length of, 548–555
    normal to, 922–925
    second derivative of, 543, 575(25)
    slope of, 543, 897
    tangent to, 896–897, 922–926
Parametric equations, 540–545, 722, 897
Partial derivatives, 164(62), 762–766, 800–801
    chain rules for, 773–780
    higher-order, 765, 800–801
    mixed, 765, 800–801, S64–S66, S70
    notations for, 762–763
    test for extrema, 796, 798, 800(24), 805–806
Partial differential equations, 767(24), 812(43–50)
Partial fractions, 410–416
    integration technique, 410
    for rational functions, 411–416
    for rational numbers, 417(45), 418(46)
Partition, 243–244
Period, S30
Perpendicular lines, 24–25
Perpendicular vector, 709

Petroleum imports, 35
Physics:
    heat equation, 767(25), 811(40)
    ideal gas, 811(41)
    kinetic energy (see Kinetic energy)
    momentum, 738, 935
    motion (see Motion)
    Newton's law of cooling, 347(42)
    potential energy, 234(78)
    steady-state temperature, 987(35), 1002–1003
    thermodynamics, 778–779, 782(34–36), 976(22)
    vibrating string, 764, 811(39)
    water pressure, 471–472, 1025(45)
Plane, 720–721
    distance to, 721
    equation of, 720–721
        normal form, 720
    tangent, 773(21), 906–908
Plane vector, 693
Plankton, 502(38)
Point-slope equation, 27
Points of inflection, 189
Poiseuille's law of blood flow, 474(17)
Polar axis, 529
Polar coordinates, 529–533, 850–851
    area in, 534–539, 986(27)
    describing regions in, 824–826
    directional derivative in, 944(24)
    gradient in, 944(24)
    graphing in, 530–533
    integration in, 839–844, 859–860
    and rectangular coordinates, 530
Polar equations:
    of circles, 530–531
    of conics, S53–S56
    of ellipses, S53–S54, S56(6)
    of lines, 531–532
    of roses, 532–533
Pole, 529
Polynomial, 60
    antiderivative of, 382
    coefficients of, 60
    continuity of, 93, S60–S61
    degree of, 60
    derivative of, 137
    irreducible, 404, 409(42), 411
    limit of, 60
    root of, 101, 176(36), 451(194), 648
Population growth, 339–342, 474(22)
Positive, S1
Potential, 967
    of a sphere, 881–882
Power series (see Series, power)
Present value, 474(21), 500–501
Prime numbers, 600(23)
Principia, 200, 935, 1015

Probability distribution, 527
Product(s):
    of complex numbers (see Complex numbers, products of)
    derivative of, 134–135
    dot, 707–716, 717(36)
        and direction cosines, 732
        and perpendicularity, 709
        and scalar components, 713
    of functions, 93
    limit of, 56
    triple, 739(30–33)
    vector, 733
Projectile, 201
Pythagorean theorem, 457, S35

Quadrants, 12
Quadratic formula, S10–S11
Quotient:
    of complex numbers, 647
    derivative of, 137–139
    difference, 118
    of a function, 93
    limit of, 56
        l'Hôpital's rule, 349–355
        of polynomials, 61–62
        using series, 631

Rabinowitz, Philip, 519(26)
Radian, S26–S28
    use in calculus, 148–149
Radioactive decay, 343
Radius:
    of convergence, 620, 622–624
    of curvature, 573
Radner, R., 643
Radon, 373(6)
Rainfall, 473(3)
Ralston, Anthony, 519(26)
Randers, J., 377
Range, 31
Rate of change, 158–159, 329–333
Rates, related, 329–333
Ratio, 589
Ratio test, 603–606
Rational function, 61, 403, 404–409, 427
    of $\cos\theta \sin\theta$, 427–430
    integration of (see Integration, of rational functions)
    limit of, 61–62
    partial fractions for, 411–416
    of x and y, 427
Rational number(s), S1
    partial fractions for, 417(45), 418(46)
Rationalizing, S8–S9
Real function, 45

Real number, 646, 647, S1–S6
Real part, 647
Reciprocal, 657(11)
Rectangular coordinates:
　integration in, 827–836, 867–869
　in plane, 11–12
　and polar coordinates, 530
　regions in, 822–824, 865–867
　in space, 698
Recursion formulas, 397
Reduction formulas, 397
Reflector, 565–566
Refraction, 214(44)
Regions:
　in cylindrical coordinates, 871–873
　in polar coordinates, 824–826
　in rectangular coordinates, 822–824, 865–867
　in spherical coordinates, 873–877
Regression, line of, 795(123)
Related rates, 329–333
Relative error, 224
　and differentials, 224–225
Relative extremum:
　first derivative test for, 178–179
　partial derivative test for, 796, 798, 800(24)
　second derivative test for, 192–193
Relative growth rate, 341
Relative maximum, 177
　(See also Relative extremum)
Relative minimum, 177
　(See also Relative extremum)
Remainder (of a Taylor series), 636–642
　and convergence, 636
　derivative form, 640–642
　integral form, 636–640
　Lagrange's formula, 640–642
Repeated integrals, 828, 867
Resources, 377(171–175)
Richter scale, 296(26)
Riemann, Georg F. B. (1826–1866), 245
Riemann sum, 245
Right-hand rule, 698, 736
Rise, 22
Robbins, Herbert, 10
Rocket motion, 200–203, 507–508
Rolle, Michel (1652–1719), 167
Rolle's theorem, 167–169
　proof of, 174
Root(s), 101, 678
　of complex numbers, 648
　of polynomials, 101, 176(36), 451(194), 648
Root test, 606
Rose, 533
Rotary engine, 544–545, 547(28)
Rotation of axes, S48–S52

Ruedisili, Lon C., 195(39)
Run, 22

Saddle, 754–755
Saddle point, 760
St. Vincent, Gregory (1584–1667), 305
Sampling number, 244
Sarasa, A. A. de, 305
Satellites, 920–921, 921(16–19)
Scalar field, 946–952
　gradient of, 946–947
　Laplacian of, 951
Scalar function, 900
Scalars, 693
Schoenfeld, Alan H., 431(24), 439
Schwarz inequality, 513(21), 891(45), 917(17)
Scientific notation, S25(22)
Secant, S34
　derivative of, 145
　graph of, S34
Second derivative, 185–186
　and concavity, 187–190
　and motion, 196–200
　sign of, 188–190
　test for relative extrema, 192–193
Section, 244
Semilog, 343–344, 348(47–49), 378(188)
Separable differential equations, 335–336
Sequence, 581–587
　convergent, 582
　divergent, 582
　limit of, 581, 587
Series, 580, 588–644
　absolute convergence, 613–615
　absolute-convergence test, 613
　absolute-ratio test, 615
　alternating (see Alternating series)
　binomial, 625(25)
　comparison test, 600–601
　complex, 659
　conditional convergence, 614, 616
　convergent, 589
　differentiation of, 626
　divergent, 589
　error, 597–598, 685
　geometric, 589, S14
　　initial term of, 589
　　ratio of, 589, S14
　　sum of, 589, S14
　harmonic (see Harmonic series)
　integral test, 594–599
　integration of, 628–629
　limit-comparison test, 602–603
　Maclaurin, 618–620
　nth term of, 589
　nth-term test for divergence, 590

Series (Cont.):
　p, 596–597
　partial sum, 589
　power, 618–624, 626–632, 635–644
　　algebra of, 629–630
　　applied to limits, 631
　　convergence of, 619–620, 622–623
　　differentiation of, 626
　　formula for a_n, 627, 632
　　integration of, 628–629
　　and l'Hôpital's rule, 631
　　manipulation of, 626–632
　　radius of convergence, 620, 622–624
　ratio test, 603–606
　remainder, 597–598
　root test, 606
　sum of, 589
　tail end of, 592
　Taylor (see Taylor series)
Shell technique, 477–482
Sigma Beta Beta (ΣBB), S183
Sigma notation, 241–243
　for approximating sums, 243–245
　properties of, 242
Sigmoidal growth, 344
Silberberg, Eugene, 779
Similar triangles, 457
Simply connected, 980
Simpson's method, 516–518
　error in, 518
Sine, S29–S32
　antiderivative of, 381
　derivative of, 143–144
　graph of, S31
　hyperbolic, 360–361
Singer, S. Fred, 473(4)
Slide rule, 296(31)
Slope, 110–112
　and angle of inclination, S33
　of a curve, 125
　of a line, 22–23
　negative, 24
　no, 23
　of a parameterized curve, 543, 897
　positive, 24
　zero, 24
Slope-intercept equation, 26
Snell's law, 214(44)
Solid of revolution, 269(31)
　volume of, 475–482
Space coordinates, 698, 871–877
Space curve, 893
Space shuttle, 892
Speed, 2, 112–113, 125, 898–899
　average, 3
　on a curve, 551–552
Sphere, 560–561, 750, 840
　potential of, 881–882

Spherical coordinates, 873–877
   and cylindrical coordinates, 873–874
   gradient in, 944(26)
   and rectangular coordinates, 874
Spiral, 532, 553
Spring, 509(15)
Squaring the circle, 540(20)
Standard deviation, 527
Static index, 377(172)
Statistics, 795(13)
Steen, Lynn Arthur, 10
Stein, Melvyn Kopald, 740(38)
Stein, Sherman K., 10
Steinhart, John S. and Carol E., 195(39)
Steinmetz, Charles Proteus (1865–1923), 656–657
Stirling's formula, 691(90)
Stokes, George G. (1819–1903), 1006
Stokes' theorem, 1005–1012
   generalization of Green's theorem, 1005–1006
   physical interpretation of, 1009–1010
   proof of, 1008–1009
Substitution, 384–390, 428, 431–437, 856–860, 990–991
Summation index, 241
Summation notation, 241–243
   for approximating sums, 243–245
   properties of, 242
Surface(s):
   centroid of, 995(7–10)
   integrals over, 988–995
   level, 904–908
   moment of inertia of, 996(17–18)
   normal vector to, 905–908
   orientable, 1007
   of revolution, 556
   tangent plane to, 907–908
   two-sided, 1007
Surface area, 556–561, 991
Surface integrals, 988–995
   over a cone, 1026(54)
   over a cylinder, 1026(53)
   over a sphere, 994–995
Symmetry, 16
   of even functions, 43
   of odd functions, 43
   radial, 953(25)

Tail end of a series, 592
Tait, Peter G. (1831–1901), 1010
Tangent, S32–S35
   antiderivative of, 423
   to a curve, 110, 125
   derivative of, 146
   graph of, S35
   hyperbolic, 361

Tangent line, 1, 110
Tangent plane, 773(21), 906–908
Tangent vector to a curve, 896, 922–926
Taylor, Brook (1685–1731), 9
Taylor polynomial, 635–636
Taylor series, 635–644, 800–805
   to approximate functions, 636, 800–805
   remainder in, 636–642
      derivative (Lagrange) form, 640–642
      integral form, 636–640
Telescoping sum, 243
Test(s):
   for convergence of series, 684
   for extrema (see Extremum; Relative extremum)
Thermodynamics, 778–779, 782(34–36), 976(22)
Thomas, Harold A., 805
Thompson, S. P., 845(24)
Thomson, William (see Kelvin, Lord)
Torque, 738
Torsion, 943(23)
Torus, 483(18), 890(32–33)
Toynbee, Arnold (1889–1975), v
Traffic flow, 108(58), 213(35)
Translation of axes, S45–S48
Transportation, 890(35–40)
Trapezoidal method, 515–516
   error in, 518
Triangle inequality, S5
Trigonometric functions [see Function(s), trigonometric]
Trigonometric identities, S29–S30, S32–S33
Trigonometric substitution, 431–437
Trigonometry, S26–S35
Triple integrals (see Repeated integrals)
Triple product, 739(30–33)
Trochoid, 547(30)
Tuchinsky, Philip M., 427(44)
Two-sided surfaces, 1007

UMAP, 427(44), 439
Undetermined coefficients, 666
Unit circle, S28–S33
Unit vector, 694
   binormal, 927(23–25), 943(23)
   normal, 923–925
   polar form, 694
   tangent, 922–926

Variables:
   dependent, 33
   independent, 33

Variables (*Cont.*):
   separation of, 336
Variance, 527
Vector(s), 692–746
   acceleration (see Acceleration vector)
   addition of, 694–695, 699
   angle between, 708, 712
   basic unit, 703–705
   binormal, 927(23–25), 943(23)
   components, 693–694, 704–705, 713–715
   cross product of, 733–738
   differentiation of, 897, 901–903
   direction of, 693
   direction angles of, 723
   direction cosines of, 723
   dot product of, 707–716, 717(36)
   equality of, 693, 694
   infinite-dimensional, 717(37)
   length of, 693
   magnitude of, 693, 717(36)
   negative of, 696, 699
   normal (see Normal vector)
   orthogonal, 715, 905
   perpendicular, 709
   plane, 693
   principal unit normal, 923–925
   projection of, 714
   scalar components of, 693–694, 705, 713–714
   scalar multiple of, 700
   scalar product of, 709
   in space, 698–705
   subtraction of, 695–696, 699
   tangent, 896, 922–926
   unit (see Unit vector)
   vector components, 704, 714–715
   velocity, 898
   zero, 693
Vector field, 945–952
   conservative, 964–974, 1010–1012
   and curl, 969–974
   curl of, 950–951
   divergence of, 947–951
   line integral of, 959, 961
Vector function(s), 892–941
   continuity of, 895
   definition of, 893
   derivative of, 893–899, 901–903
   integral of, 1025, 1026(46–51)
   limit of, 894–895
Vector product, 733
Vector space, 707(36–38)
Velocity, 125, 898
   angular, 833
   average, 170
   escape, 200–203, 921, 922(23)
   orbital, 203, 920–921, 922(23)

Velocity vector, 898
Vibrations, 764, 811(39)
Volcanic ash settling, 470, 473(1–2)
Volume, 238–239, 461–467, 475–482
    by cross section, 256, 461–467, 475–477
    as an integral, 255–256, 461–467, 476, 477 (*see also* Definite integral and volume)
    by Pappus's theorem, 491–492
    of a parallelepiped, 739(30–31)
    by shell technique, 477–482
    of a solid of revolution, 475–482
    by washer method, 476–477

Wallis's formula, 690(88)
Wankel, Felix, 544
Washer method, 476–477
Water pressure, 471–472, 1025(45)
Wave equation, 811(33)
Weierstrass, Karl (1815–1897), 73
Whispering rooms, 568(21)
Whitaker, Steven, 678(22), 845(31), 945, 1023(31)
Work, 503–508, 707–708
    and centroid, 506–507
    along a curve, 955–956
    with gravity varying, 507–508
    in pumping water, 503–507

x axis, 11
x intercept, 17, 177

y axis, 11
y intercept, 17, 26, 177

Zero, division by, S2
Zero of a polynomial (*see* Polynomial, root of)
Zero slope, 24
Zero vector, 693